AF662523

# Higher Engineering Mathematics

45th Edn.

# Preface to the 45th Edition

This book having a rich legacy, has run through its 44th edition, with twenty eight subsequent reprints, which itself is a testament to its enduring popularity.

Despite maintaining the familiar format and features that have garnered such widespread acclaim, we have meticulously revised the text, introducing additions and alterations to enhance its utility. We have expanded the number of solved examples, integrating problems sourced from the latest university question papers. It is our sincere hope that these revisions will further elevate the book's standing among its readers.

The authors extend their heartfelt thanks to the esteemed readers for their letters of appreciation and to the renowned professors for their invaluable suggestions and insightful opinions which helped in shaping this edition. In particular, they express gratitude to Dr. K. L. Sai Prasad, of G.V.P. College of Engineering for Women, Visakhapatnam., Dr. Rashmi Bhardwaj of G.G.S. Indraprastha University, Delhi; Dr. B.B. Singh of Vidyavihar, Ionere (Raigarh); Dr. V. Vasudeva Murthy of S.R.K.R. Engineering College, Bhimavaram (A.P.); Prof. P. Veeraiah of Sri Venkateswara College of Engineering, Sriperumbudur (T.N.); Dr. Jeavargi Phakirappa of R.Y.M. Engineering College, Bellary (Ka.); Dr. T. Subedi of the National Institute of Technology, Shillong; Dr. S. Mayilaganan of S.R.M. University, Kuttankulathur (T.N.); Dr. Garima Aggarwal of Manipal University, Jaipur; Dr. Barun Kumar Yadav, Head of the Department of Mathematics at Chouksey Engineering College, Bilaspur (C.G.); Dr. Arvind Kumar Sinha of the National Institute of Technology, Raipur; Dr. T. Yogalakshmi of Vellore Institute of Technology, Vellore; Dr. C.N.B. Rao of S.R.K.R. Engineering College, Bhimavaram (A.P.); Dr. V. Dhanapalan of Government College of Technology, Pudar (T.N.); and Prof. Shivani Mehra of Chandigarh University, Mohali.

Suggestions for improvement of the text and intimation of misprints will be thankfully acknowledged.

*New Delhi*

*B.S. GREWAL*
*J.S. GREWAL*
*J.K. DHANOA*

# Contents

**Note :** *The references given alongside the problems pertain to the* **Degree Engineering Examinations** *of the various universities and professional bodies. The abbreviations used for some of these are given below :*

| | | |
|---|---|---|
| *Andhra* | *stands for* | Andhra University, Waltair |
| *Anna* | " | Anna University, Chennai |
| *Bhopal* | " | Rajiv Gandhi Technical University, Bhopal |
| *B.P.T.U.* | " | Biju Patnaik Technical University, Rourkela |
| *C.S.V.T.U.* | " | Chhattisgarh Swami Vivekanand Technical University, Raipur |
| *Coimbatore* | " | Bharathiyar University, Coimbatore |
| *Delhi* | " | Guru Gobind Singh Indraprastha University, Delhi |
| *D.T.U.* | " | Delhi Technological University (Formerly Delhi College of Engineering) |
| *Hazaribag* | " | Vinoba Bhave University, Hazaribag |
| *J.N.T.U.* | " | Jawahar Lal Nehru Technological University, Hydrabad |
| *K.T.U.* | " | APJ Abdul Kalam Kerala Technological University, Thiruvananthapuram |
| *Kurukshetra* | " | National Institute of Technology, Kurukshetra |
| *Marathwada* | " | B.A.M. University, Aurangabad |
| *P.T.U.* | " | Punjab Technical University, Jalandhar |
| *R.T.U.* | " | Rajasthan Technical University, Kota |
| *Rohtak* | " | Maharishi Dayanand University, Rohtak |
| *U.P.T.U.* | " | UP Technical University, Lucknow |
| *U.K.T.U.* | " | Uttarakhand Technical University, Dehradun |
| *V.T.U.* | " | Visveswaraiah Technological University, Belgaum |
| *Warangal* | " | Warangal University of Technology, Warangal |
| *W.B.T.U.* | " | Maulana Abdul Kalam Azad, West Bengal University of Technology, Kolkata |

CHAPTER 1

# Solution of Equations

**1.** Introduction. **2.** General properties. **3.** Transformation of equations. **4.** Reciprocal equations. **5.** Solution of cubic equations–Cardan's method. **6.** Solution of biquadratic equations—Ferrari's method ; Descarte's method. **7.** Graphical solution of equations. **8.** Objective Type of Questions.

## 1.1 INTRODUCTION

The expression $f(x) = a_0x^n + a_1x^{n-1} + ... + a_{n-1}x + a_n$

where $a$'s are constants ($a_0 \neq 0$) and $n$ is a positive integer, is called a *polynomial in x* of degree $n$. The polynomial $f(x) = 0$ is called an *algebraic equation of degree n*. If $f(x)$ contains some other functions such as trigonometric, logarithmic, exponential etc. ; then $f(x) = 0$ is called a *transcendental equation*.

The value of $x$ which satisfies $f(x) = 0$, ...(1)

is called its root. Geometrically, a root of (1) is that value of $x$ where the graph of $y = f(x)$ crosses the $x$-axis. The process of finding the roots of an equation is known as *solution* of that equation. This is a problem of basic importance in applied mathematics. We often come across problems in deflection of beams, electrical circuits and mechanical vibrations which depend upon the solution of equations. As such, a brief account of solution of equations is given in this chapter.

## 1.2 GENERAL PROPERTIES

**I.** *If α is a root of the equation f(x) = 0, then the polynomial f(x) is exactly divisible by x – α and conversely.*

For instance, 3 is a root of the equation $x^4 - 6x^2 - 8x - 3 = 0$, because $x = 3$ satisfies this equation.

$\therefore$ $x - 3$ divides $x^4 - 6x^2 - 8x - 3$ completely, *i.e.*, $x - 3$ is its factor.

**II.** *Every equation of the nth degree has n roots (real or imaginary).*

Conversely if $\alpha_1, \alpha_2, ..., \alpha_n$ be the roots of the $n$th degree equation $f(x) = 0$, then

$$f(x) = A(x - \alpha_1)(x - \alpha_2) ... (x - \alpha_n) \text{ where } A \text{ is a constant.}$$

**Obs.** If a polynomial of degree $n$ vanishes for more than $n$ values of $x$, it must be identically zero.

**Example 1.1.** *Solve the equation* $2x^3 + x^2 - 13x + 6 = 0$.

**Solution:** By inspection, we find $x = 2$ satisfies the given equation.

$\therefore$ 2 is its root, *i.e.* $x - 2$ is a factor of $2x^3 + x^2 - 13x + 6$. Dividing this polynomial by $x - 2$, we get the quotient $2x^2 + 5x - 3$ and remainder 0.

Equating the quotient to zero, we get $2x^2 + 5x - 3 = 0$.

Solving this quadratic, we get $x = \dfrac{-5 \pm \sqrt{[5^2 - 4.(2).(-3)]}}{2 \times 2} = \dfrac{-5 \pm 7}{4} = -3, \dfrac{1}{2}$.

Hence, the roots of the given equation are 2, –3, 1/2.

**Note.** *The labour of dividing the polynomial by x – 2 can be saved considerably by the following simple device called* **synthetic division.**

| 2 | 1 | –13 | 6 | 2 |
|---|---|---|---|---|
| | 4 | 10 | – 6 | |
| 2 | 5 | – 3 | 0 | |

[*Explanation* : (*i*) Write down the coefficient of the powers of $x$ in order (supplying the missing powers of $x$ by zero co-efficients and write 2 on extreme right.

(*ii*) Put 2 as the first term of 3rd row and multiply it by 2, write 4 under 1 and add, giving 5.

(*iii*) Multiply 5 by 2, write 10 under –13 and add, giving –3.

(*iv*) Multiply –3 by 2, write –6 under 6 and add given zero].

Thus the quotient is $2x^2 + 5x - 3$ and remainder is zero.

**Obs.** To divide a polynomial by $x + h$, we write $-h$ on the extreme right.

**III. Intermediate value property.** If $f(a)$ and $f(b)$ *have different signs, then the equation* $f(x) = 0$ *has atleast one root between* $x = a$ *and* $x = b$.

The polynomial $f(x)$ is a continuous function of $x$ (Fig. 1.1). So while $x$ changes from $a$ to $b$, $f(x)$ must pass through all the values from $f(a)$ to $f(b)$. But since one of these quantities $f(a)$ or $f(b)$ is positive and the other negative, it follows that at least for one value of $x$(say $\alpha$) lying between $a$ and $b$, $f(x)$ must be zero. Then $\alpha$ is the required root.

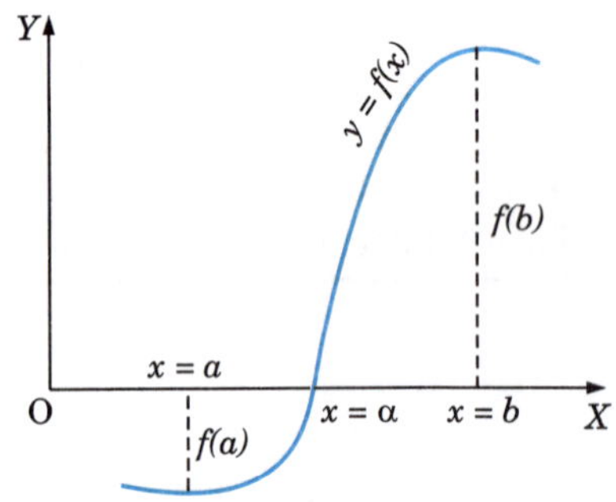


Fig. 1.1

**IV.** *In an equation with real coefficients, imaginary roots occur in conjugate pairs, i.e.*, if $\alpha + i\beta$ is a root of the equation $f(x) = 0$, then $\alpha - i\beta$ must also be its root.

Similarly if $a + \sqrt{b}$ is an irrational root of an equation, then $a - \sqrt{b}$ must also be its root.

**Obs.** *Every equation of the odd degree has at least one real root.*

This follows from the fact that imaginary roots occur in conjugate pairs.

**Example 1.2.** *Solve the equation* $3x^3 - 4x^2 + x + 88 = 0$, *one root being* $2 + \sqrt{7}i$.

**Solution:** Since one root is $2 + \sqrt{7}i$, the other root must be $2 - \sqrt{7}i$.

∴ The factors corresponding to these roots are

$$(x - 2 - \sqrt{7}i) \text{ and } (x - 2 + \sqrt{7}i)$$

or $$(x - 2 - \sqrt{7}i)(x - 2 + \sqrt{7}i) = (x - 2)^2 + 7 = x^2 - 4x + 11,$$

which is a divisor of $3x^3 - 4x^2 + x + 88$ ...(*i*)

∴ Division of (*i*) by $x^2 - 4x + 11$ gives $3x + 8$ as the quotient.

Thus the depressed equation is $3x + 8 = 0$. Its root is $-8/3$. Hence the roots of the given equation are $2 \pm \sqrt{7}i, -8/3$.

**V. Descarte's rule of signs.*** *The equation* $f(x) = 0$ *cannot have more positive roots than the changes of signs in* $f(x)$; *and more negative roots than the changes of signs in* $f(-x)$.

For instance, consider the equation $f(x) = 2x^7 - x^5 + 4x^3 - 5 = 0$ ...(1)

Sign of $f(x)$ are + – + –

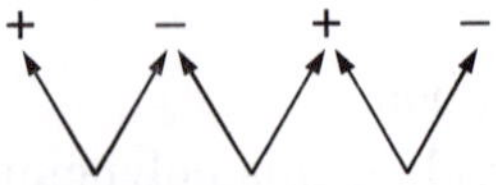

Clearly, $f(x)$ has 3 changes of signs (from + to – or – to +).

Thus (*i*) cannot have more than 3 positive roots.

*After the French mathematician and philosopher *Rene Descartes* (1596–1650), who invented Analytic geometry in 1637.

Also, $\quad f(-x) = 2(-x)^7 - (-x)^5 + 4(-x)^3 - 5$
$\qquad = -2x^7 + x^5 - 4x^3 - 5$

This shows that $f(x)$ has 2 changes of signs. Thus $(i)$ cannot have more than 2 negative roots.

**Obs. Existence of imaginary roots.** *If an equation of the nth degree has at the most p positive roots and at the most q negative roots, then it follows that the equation has at least n – (p + q) imaginary roots.*

Evidently (1) above is an equation of the 7th degree and has at the most 3 positive roots and 2 negative roots. Thus (1) has at least 2 imaginary roots.

**VI. Relations between roots and coefficients.** *If* $\alpha_1, \alpha_2, \alpha_3, ..., \alpha_n$ *be the roots of the equation*

$$a_0x^n + a_1x^{n-1} + a_2x^{n-2} + ... + a_{n-1}x + a_n = 0 \qquad ...(1)$$

*then*
$$\Sigma\alpha_1 = -\frac{a_1}{a_0}, \quad \Sigma\alpha_1\alpha_2 = \frac{a_2}{a_0}, \quad \Sigma\alpha_1\alpha_2\alpha_3 = -\frac{a_3}{a_0}$$

..............................................................................

$$\alpha_1\alpha_2\alpha_3 \ldots\ldots \alpha_n = (-1)^n \frac{a_n}{a_0}.$$

**Example 1.3.** *Solve the equation $x^3 - 7x^2 + 36 = 0$, given that one root is double of another.*

**Solution:** Let the roots be $\alpha, \beta, \gamma$ such that $\beta = 2\alpha$.

Also $\qquad \alpha + \beta + \gamma = 7, \alpha\beta + \beta\gamma + \gamma\alpha = 0, \alpha\beta\gamma = -36$

$\therefore \qquad 3\alpha + \gamma = 7 \qquad ...(i)$

$\qquad 2\alpha^2 + 3\alpha\gamma = 0 \qquad ...(ii)$

$\qquad 2\alpha^2\gamma = -36 \qquad ...(iii)$

Solving $(i)$ and $(ii)$, we get $\alpha = 3, \gamma = -2$.

[The values $\alpha = 0, \gamma = 7$ are inadmissible, as they do not satisfy $(iii)$].

Hence the required roots are 3, 6 and – 2.

**Example 1.4.** *Solve the equation $x^4 - 2x^3 + 4x^2 + 6x - 21 = 0$, given that the sum of two of its roots is zero.*

**Solution:** Let the roots be $\alpha, \beta, \gamma, \delta$ such that $\alpha + \beta = 0$.

Also $\qquad \alpha + \beta + \gamma + \delta = 2 \quad \therefore \quad \gamma + \delta = 2$

Thus the quadratic factor corresponding to $\alpha, \beta$ is of the form $x^2 - 0x + p$, and that corresponding to $\gamma, \delta$ is of the form of $x^2 - 2x + q$.

$\therefore \quad x^4 - 2x^3 + 4x^2 + 6x - 21 = (x^2 + p)(x^2 - 2x + q) \qquad ...(i)$

Equating the coefficients of $x^2$ and $x$ from both sides of $(i)$, we get

$$4 = p + q, \quad 6 = -2p.$$

$\therefore \qquad p = -3, \quad q = 7.$

Hence the given equation is equivalent to $(x^2 - 3)(x^2 - 2x + 7) = 0$

$\therefore$ The roots are $x = \pm\sqrt{3}, 1 \pm i\sqrt{6}$.

**Example 1.5.** *Find the condition that the cubic $x^3 - lx^2 + mx - n = 0$ should have its roots in*

*(a) arithmetical progression.*

*(b) geometrical progression.*

**Solution:** $(a)$ Let the roots be $a - d, a, a + d$ so that the sum of the roots $= 3a = l$ *i.e.*, $a = l/3$.

Since $a$ is the root of the given equation

$\therefore \qquad a^3 - la^2 + ma - n = 0 \qquad ...(i)$

Substituting $a = l/3$, we get $(l/3)^3 - l(l/3)^2 + m(l/3) - n = 0$.

or $\qquad 2l^3 - 9lm + 27n = 0, \qquad$ which is the required condition.

(*b*) Let the roots be $a/r$, $a$, $ar$, so that the product of the roots $= a^3 = n$.

Putting $a = (n)^{1/3}$, in (*i*), we get $n - ln^{2/3} + mn^{1/3} - n = 0$ or $m = ln^{1/3}$

Cubing both sides, we get $m^3 = l^3 n$, which is the required condition.

**Example 1.6.** *Solve the equation $x^4 - 2x^3 - 21x^2 + 22x + 40 = 0$ whose roots are in A.P.*

**Solution:** Let the roots be $a - 3d$, $a - d$, $a + d$, $a + 3d$, so that the sum of the roots $= 4a = 2$.

$$\therefore \quad a = 1/2$$

Also product of the roots $= (a^2 - 9d^2)(a^2 - d^2) = 40$

or $$\left(\frac{1}{4} - 9d^2\right)\left(\frac{1}{4} - d^2\right) = 40 \quad \text{or} \quad 144d^4 - 40d^2 - 639 = 0$$

$$\therefore \quad d^2 = 9/4 \quad \text{or} \quad -7/36$$

Thus, $d = \pm 3/2$, the other value is not admissible.

Hence the required roots are $-4, -1, 2, 5$.

**Example 1.7.** *Solve the equation $2x^4 - 15x^3 + 35x^2 - 30x + 8 = 0$, whose roots are in G.P.*

**Solution:** Let the roots be $a/r^3$, $a/r$, $ar$, $ar^3$, so that product of the roots $= a^4 = 4$.

Also the product of $a/r^3$, $ar^3$ and $a/r$, $ar$ are each $= a^2 = 2$.

$\therefore$ The factors corresponding to $a/r^3$, $ar^3$ and $a/r$, $ar$ are $x^2 + px + 2$, $x^2 + qx + 2$.

Thus, $2x^4 - 15x^3 + 35x^2 - 30x + 8 = 2(x^2 + px + 2)(x^2 + qx + 2)$

Equating the coefficients of $x^3$ and $x^2$

$$-15 = 2p + 2q \quad \text{and} \quad 35 = 8 + 2pq$$

$$\therefore \quad p = -9/2, \; q = -3.$$

Thus the given equation is $2\left(x^2 - \frac{9}{2}x + 2\right)(x^2 - 3x + 2) = 0$

Hence the required roots are 1/2, 4 and 1, 2 *i.e.*, $\frac{1}{2}$, 1, 2, 4.

**Example 1.8.** *If α, β, γ be the roots of the equation $x^3 + px + q = 0$, find the value of*
*(a)* $\Sigma\alpha^2\beta$ *(b)* $\Sigma\alpha^4$ *(c)* $\Sigma\alpha^2\beta$.

**Solution:** We have
$$\alpha + \beta + \gamma = 0 \qquad ...(i)$$
$$\alpha\beta + \beta\gamma + \gamma\alpha = p \qquad ...(ii)$$
$$\alpha\beta\gamma = -q \qquad ...(iii)$$

(*a*) Multiplying (*i*) and (*ii*), we get

$\alpha^2\beta + \alpha^2\gamma + \beta^2\gamma + \beta^2\alpha + \gamma^2\alpha + \gamma^2\beta + 3\alpha\beta\gamma = 0$

or $$\Sigma\alpha^2\beta = -3\alpha\beta\gamma = 3q \qquad [\text{By } (iii)]$$

(*b*) Multiplying the given equation by $x$, we get $x^4 + px^2 + qx = 0$

Putting $x = \alpha, \beta, \gamma$ successively and adding, we get $\Sigma\alpha^4 + p\Sigma\alpha^2 + q\Sigma\alpha = 0$

or $$\Sigma\alpha^4 = -p\Sigma\alpha^2 - q(0) \qquad ...(iv)$$

Now squaring (*i*), we get $\alpha^2 + \beta^2 + \gamma^2 + 2(\alpha\beta + \beta\gamma + \gamma\alpha) = 0$

or $$\Sigma\alpha^2 = -2p \qquad [\text{By } (ii)]$$

Hence, substituting the value of $\Sigma\alpha^2$ in (*iv*), we obtain

$$\Sigma\alpha^4 = -p(-2p) = 2p^2.$$

(*c*) $\Sigma\alpha^3\beta = \Sigma\alpha^2\,\Sigma\alpha\beta - \alpha\beta\gamma\Sigma\alpha = -2p(p) - (-q)(0) = -2p^2$.

## PROBLEMS 1.1

1. Form the equation of the fourth degree whose roots are $3 + i$ and $\sqrt{7}$.
2. Solve the equation (*i*) $x^3 + 6x + 20 = 0$, one root being $1 + 3i$. (*C.S.V.T.U., 2012*)
   (*ii*) $x^4 - 2x^3 - 22x^2 + 62x - 15 = 0$, given that $2 + \sqrt{3}$ is a root.
3. Show that $x^7 - 3x^4 + 2x^3 - 1 = 0$ has at least four imaginary roots.
4. Show that the equation $x^4 + 15x^2 + 7x - 11 = 0$ has one positive, one negative and two imaginary roots.
5. Find the number and position of real roots of $x^4 + 4x^3 - 4x - 13 = 0$.
6. Solve the equation $3x^3 - 11x^2 + 8x + 4 = 0$, given that two of its roots are equal.
7. If $r_1, r_2, r_3$ are the roots of the equation $2x^3 - 3x^2 + kx - 1 = 0$, find constant $k$ if sum of two roots is 1.
8. The equation $x^4 - 4x^3 + ax^2 + 4x + b = 0$ has two pairs of equal roots. Find the values of $a$ and $b$.

*Solve the following equations 9–14 :*

9. $x^3 - 9x^2 + 14x + 24 = 0$, given that two of its roots are in the ratio 3 : 2.
10. $x^3 - 4x^2 - 20x + 48 = 0$ given that the roots $\alpha$ and $\beta$ are connected by the relation $\alpha + 2\beta = 0$. (*C.S.V.T.U., 2012*)
11. $x^4 - 6x^3 + 13x^2 - 12x + 4 = 0$, given that it has two parts of equal roots.
12. $x^4 - 8x^3 + 21x^2 - 20x + 5 = 0$ given that the sum of two of the roots is equal to the sum of the other two.
13. $x^3 - 12x^2 + 39x - 28 = 0$, roots being in arithmetical progression.
14. $8x^3 - 14x^2 + 7x - 1 = 0$, roots being in geometrical progression.
15. *O, A, B, C* are the four points on a straight line such that the distances of *A, B, C* from *O* are the roots of equation $ax^3 + 3bx^2 + 3cx + d = 0$. If *B* is the middle point of *AC*, show that $a^2d - 3abc + 2b^3 = 0$. (*C.S.V.T.U., 2011*)
16. Solve the equations (*i*) $x^4 + 2x^3 - 21x^2 - 22x + 40 = 0$ whose roots are in A.P.
   (*ii*) $x^4 + 5x^3 - 30x^2 + 40x + 64 = 0$ whose roots are in G.P.
17. If $\alpha, \beta, \gamma$ be the roots of the equation $x^3 - lx^2 + mx - n = 0$, find the value of
   (*i*) $\Sigma\alpha^2\beta^2$, (*ii*) $(\beta + \gamma)(\gamma + \alpha)(\alpha + \beta)$
18. Find the sum of the cubes of the roots of the equation $x^3 - 6x^2 + 11x - 6 = 0$.
19. If $\alpha, \beta, \gamma$ are the roots of $x^3 + 4x - 3 = 0$, find the value of (*i*) $\alpha^{-1} + \beta^{-1} + \gamma^{-1}$ (*ii*) $\alpha^{-2} + \beta^{-2} + \gamma^{-2}$.
20. If $\alpha, \beta, \gamma$ be the roots of $x^3 + px + q = 0$, show that
   (*i*) $\alpha^5 + \beta^5 + \gamma^5 = 5\alpha\beta\gamma(\beta\gamma + \gamma\alpha + \alpha\beta)$, (*ii*) $3\Sigma\alpha^2\, \Sigma\alpha^5 = 5\Sigma\alpha^3\, \Sigma\alpha^4$.

## 1.3 TRANSFORMATION OF EQUATIONS

**(1) To find an equation whose roots are *m* times the roots of the given equation,** *multiply the second term by m, third term by $m^2$ and so on* (all missing terms supplied with zero coefficients).

For instance, let the given equation be

$$3x^4 + 6x^3 + 4x^2 - 8x + 11 = 0 \quad ...(i)$$

To multiply its roots by $m$, put $y = mx$ (or $x = y/m$) in (*i*).

Then $$3(y/m)^4 + 6(y/m)^3 + 4(y/m)^2 + 8(y/m) + 11 = 0$$

Multiplying by $m^4$, we get $$3y^4 + m(6y^3) + m^2(4y^2) - m^3(8y) + m^4(11) = 0$$

This is same as multiplying the second term by $m$, third term by $m^2$ and so on in Eqn. (*i*).

**Cor. To find an equation whose roots are with opposite signs to those of the given equation,** *change the signs of the every alternative term of the given equation beginning with the second.*

Changing the signs of the roots of Eqn. (*i*) is same as multiplying its roots by –1.

$\therefore$ The required equation will be

$$3x^4 + (-1)6x^3 + (-1)^2\, 4x^2 - (-1)^3\, 8x + (-1)^4\, 11 = 0$$

or $$3x^4 - 6x^3 + 4x^3 + 8x + 11 = 0$$

which is Eqn. (*i*) with signs of every alternate term changed beginning with the second.

**(2) To find an equation whose roots are reciprocal of the root of the given equation,** *change x to 1/x.*

**Example 1.9.** *Solve $6x^3 - 11x^2 - 3x + 2 = 0$, given that its roots are in harmonic progression.* (C.S.V.T.U., 2014 S)

**Solution:** Since the roots of the given equation are in H.P., the roots of the equation having reciprocal roots will be in A.P.

The equation with reciprocal roots is $6(1/x)^3 - 11(1/x)^2 - 3(1/x) + 2 = 0$

or $$2x^3 - 3x^2 - 11x + 6 = 0 \quad ...(i)$$

Since the roots of the given equation are in H.P., therefore, the roots of Eqn. (*i*) are in A.P. Let the root be $a - d, a, a + d$. Then

$$3a = 3/2 \text{ and } a(a^2 - d^2) = -3.$$

Solving these equations, we get $a = 1/2$, $d = 5/2$.

Thus, the roots of Eqn. (*i*) are $-2, 1/2, 3$.

Hence the required roots of the given equation are $-1/2, 2, 1/3$.

**Example 1.10.** *If α, β, γ be the roots of the cubic equation $x^3 - px^2 + qx - r = 0$, form the equation whose roots are $\beta\gamma + 1/\alpha$, $\gamma\alpha + 1/\beta$, $\alpha\beta + 1/\gamma$.*
*Hence evaluate $\Sigma(\alpha\beta + 1/\gamma)(\beta\gamma + 1/\alpha)$.*

**Solution:** If $x$ is a root of the given equation and $y$ a root of the required equation, then

$$y = \beta\gamma + 1/\alpha = \frac{\alpha\beta\gamma + 1}{\alpha} = \frac{r+1}{\alpha} \qquad [\because \alpha\beta\gamma = r]$$

or $$y = \frac{r+1}{x} \Rightarrow x = \frac{r+1}{y}$$

Thus substituting $x = (r + 1)/y$ in the given equation, we get

$$\left(\frac{r+1}{y}\right)^3 - p\left(\frac{r+1}{y}\right)^2 + q\left(\frac{r+1}{y}\right) - r = 0$$

or $ry^3 - q(r + 1)y^2 + p(r + 1)^2 y - (r + 1)^3 = 0$, which is the required equation.

Hence $\Sigma(\alpha\beta + 1/\gamma)(\beta\gamma + 1/\alpha) = p(r + 1)^2/r$.

**Example 1.11.** *Form an equation whose roots are cubes of the roots of $x^3 - 3x^2 + 1 = 0$.* ...(i)

**Solution:** If $y$ be a root of the required equation, then $y = x^3$ ...(ii)

Now we have to eliminate $x$ from Eqns. (*i*) and (*ii*)

∴ Rewriting Eqn. (*i*) as $x^3 + 1 = 3x^2$

Cubing both sides, $x^9 + 3x^6 + 3x^3 + 1 = 27x^6$

Substituting $x^3 = y$, we get $y^3 - 24y^2 + 3y + 1 = 0$, which is the required equation.

**(3) To diminish the roots of an equation $f(x) = 0$ by $h$,** *divide $f(x)$ by $x - h$ successively. Then the successive remainders determine the coefficients of the required equation.*

Let the given equation be

$$a_0x^n + a_1x^{n-1} + ... + a_{n-1}x + a_n = 0 \quad ...(i)$$

To diminish its roots by $h$, put $y = x - h$ (or $x = y + h$) in Eqn. (*i*) so that

$$a_0(y + h)^n + a_1(y + h)^{n-1} + ... + a_n = 0 \quad ...(ii)$$

On simplification, it takes the form

$$A_0 y^n + A_1 y^{n-1} + ... + A_n = 0 \quad ...(iii)$$

Its coefficient $A_0, A_1, ... A_n$ can easily be found with the help of *synthetic division* (p. 2). For this, we put $y = x - h$ in Eqn. (*iii*) so that

$$A_0 (x - h)^n + A_1(x - h)^{n-1} + ... + A_n = 0 \quad ...(iv)$$

Clearly, Eqns. (*i*) and (*iv*) are identical. If we divide L.H.S. of Eqn. (*iv*) by $x - h$, the remainder is $A_n$ and the quotient $Q = A_0(x - h)^{n-1} + A_1(x - h)^{n-2} + ... + A_{n-1}$. Similarly, if we divide $Q$ by $x - h$, the remainder is $A_{n-1}$ and the quotient is $Q_1$(say). Again dividing $Q_1$ by $x - h$, $A_{n-2}$ will be obtained as remainder and so on.

**Obs.** To increase the roots by $h$, we take $h$ negative.

**Example 1.12.** *Transform the equation $x^3 - 6x^2 + 5x + 8 = 0$ into another in which the second term is missing.* (C.S.V.T.U., 2014)

*Hence find the equation of its squared differences.*

**Solution:** Sum of the roots of the given equation = 6.

In order that the second term in the transformed equation is missing, the sum of the roots is to be zero.

Since the equation has 3 roots, if we decrease each root by 2, the sum of the roots of the new equation will become zero.

$\therefore$ Dividing $x^3 - 6x^2 + 5x + 8$ by $x - 2$ successively, we have

| | | | | |
|---|---|---|---|---|
| 1 | –6 | 5 | 8 | (2 |
| | 2 | –8 | –6 | |
| | –4 | –3 | **2** | |
| | 2 | –4 | | |
| | –2 | **–7** | | |
| | 2 | | | |
| **1** | **0** | | | |

Thus the transformed equation is $x^3 - 7x + 2 = 0$. ...(*i*)

If $\alpha, \beta, \gamma$ be the roots of the given equation, then the roots of Eqn. (*i*) are $\alpha - 2, \beta - 2, \gamma - 2$.

Let these roots be denoted by $a, b, c$.

Then $b - c = \beta - \gamma$. Also $a + b + c = 0$, $abc = -2$.

Now $(b - c)^2 = (b + c)^2 - 2bc = (a + b + c - a)^2 - \dfrac{2abc}{a} = a^2 + 4/a$

$\therefore$ The equation of squared differences of Eqn. (*i*) is given by the transformation $y = x^2 + 4/x$

or $$x^3 - xy + 4 = 0 \qquad ...(ii)$$

Subtracting Eqns. (*ii*) from (*i*), we get $-7x + xy - 2 = 0$ or $x = 2/(y - 7)$

Substituting for $x$ in Eqn. (*i*), the equation becomes

$$[2/(y - 7)]^3 - 7[2/(y - 7)] + 2 = 0 \quad \text{or} \quad y^3 - 28y^2 + 245y - 682 = 0 \qquad ...(iii)$$

Roots of this equation are $(b - c)^2, (c - a)^2, (a - b)^2$ *i.e.*, $(\beta - \gamma)^2, (\gamma - \alpha)^2, (\alpha - \beta)^2$.

Hence Eqn. (*iii*) is the required equation.

## 1.4 RECIPROCAL EQUATIONS

*If an equation remains unaltered on changing $x$ to $1/x$, it is called a* **reciprocal equation.**

*Such equations are of the following standard types :*

I. *A reciprocal equation of an odd degree having coefficients of terms equidistant from the beginning and end equal.* It has a root = –1.

II. *A reciprocal equation of an odd degree having coefficients of terms equidistant from the beginning and end equal but opposite in sign.* It has root = 1.

III. *A reciprocal equation of an even degree having coefficients of terms equidistant from the beginning and end equal but opposite in sign.* Such an equation has two roots = 1 and –1.

The substitution $x + 1/x = y$ reduces the degree of the equation of half its former degree.

**Example 1.13.** *Solve $6x^5 - 41x^4 + 97x^3 - 97x^2 + 41x - 6 = 0$.* (C.S.V.T.U., 2011)

**Solution:** This is a reciprocal equation of odd degree with opposite signs. $\therefore$ $x = 1$ is a root.

Dividing L.H.S. by $x - 1$, the given equation reduces to

$$6x^4 - 35x^3 + 62x^2 - 35x + 6 = 0$$

Dividing by $x^2$, we have

$$6(x^2 + 1/x^2) - 35(x + 1/x) + 62 = 0$$

Putting $x + 1/x = y$ and $x^2 + 1/x^2 = y^2 - 2$, we get

$$6(y^2 - 2) - 35y + 62 = 0 \quad \text{or} \quad 6y^2 - 35y + 50 = 0 \quad \text{or} \quad (3y - 1)(2y - 5) = 0$$

$$\therefore \qquad x + 1/x = y = 1/3 \quad \text{or} \quad 5/2$$

*i.e.,* $3x^2 - 10x + 3 = 0$ or $2x^2 - 5x + 2 = 0$

*i.e.,* $(3x - 1)(x - 3) = 0$ or $(2x - 1)(x - 2) = 0$

$\therefore$ $x = 1/3, 3$ or $1/2, 2$

Hence the required roots are 1, 1/3, 3, 1/2, 2.

**Example 1.14.** *Solve* $6x^6 - 25x^5 + 31x^4 - 31x^2 + 25x - 6 = 0$.

**Solution:** This is a reciprocal equation of even degree with opposite signs. $\therefore$ $x = 1, -1$ are its roots. Dividing L.H.S. by $x - 1$ and $x + 1$, the given equation reduces to

$$6x^4 - 25x^3 + 37x^2 - 25x + 6 = 0$$

Dividing by $x^2$, we get

$$6(x^2 + 1/x^2) - 25(x + 1/x) + 37 = 0.$$

Putting $x + 1/x = y$ and $x^2 + 1/x^2 = y^2 - 2$, it becomes

$$6(y^2 - 2) - 25y + 37 = 0 \quad \text{or} \quad 6y^2 - 25y + 25 = 0$$

or $(2y - 5)(3y - 5) = 0$

$\therefore$ $x + 1/x = y = 5/2$ or $5/3$.

*i.e.,* $2x^2 - 5x + 2 = 0$ or $3x^2 - 5x + 3 = 0$

$\therefore$ $x = 2, 1/2$ or $x = \dfrac{5 \pm i\sqrt{11}}{6}$

Hence the required roots of the given equation are 1, –1, 2, 1/2, $\dfrac{5 \pm i\sqrt{11}}{6}$.

## PROBLEMS 1.2

1. Find the equation whose roots are 3 times the roots of $x^3 + 2x^2 - 4x + 1 = 0$.
2. Form the equation whose roots are the reciprocals of the roots of $2x^5 + 4x^3 - 13x^2 + 7x - 6 = 0$.
3. Find the equation whose roots are the negative reciprocals of the roots of
$$x^4 + 7x^3 + 8x^2 - 9x + 10 = 0.$$
4. By suitable transformation, reduce the equation $x^4 + 16x^3 + 83x^2 + 152x + 84 = 0$ to an equation in which term in $x^3$ is absent and hence solve it.
5. Find the equation whose roots are the roots of
   (*i*) $x^3 - 6x^2 + 11x - 6 = 0$ each increased by 1.
   (*ii*) $x^4 + x^3 - 3x^2 - x + 2 = 0$ each diminished by 3.
   (*iii*) $x^5 - 5x^4 + 10x^3 - 10x^2 + 5x + 6 = 0$ each diminished by 1.
6. Find the equation whose roots are the squares of the roots of $x^3 - x^2 + 8x - 6 = 0$.
7. Find the equation whose roots are the cubes of the roots of $x^3 + px^2 + q = 0$.
8. If $\alpha, \beta, \gamma$ are the roots of the equation $2x^3 + 3x^2 - x - 1 = 0$, form the equation whose roots are $(1 - \alpha)^{-1}$, $(1 - \beta)^{-1}$ and $(1 - \gamma)^{-1}$.
9. If $a, b, c$ are the roots of the equation $x^3 + px^2 + qx + r = 0$, find the equation whose roots are $ab$, $bc$ and $ca$.
10. If $\alpha, \beta, \gamma$ be the roots of $x^3 + mx + n = 0$, form the equation whose roots are
    (*a*) $\alpha + \beta - \gamma, \beta + \gamma - \alpha, \gamma + \alpha - \beta$, (*b*) $\beta\gamma/\alpha, \gamma\alpha/\beta, \alpha\beta/\gamma$ (*c*) $\dfrac{\beta+\gamma}{\alpha^2}, \dfrac{\gamma+\alpha}{\beta^2}, \dfrac{\alpha+\beta}{\gamma^2}$ (*C.S.V.T.U., 2015*)
11. Find the equation of squared differences of the roots of the cubic $x^3 + 6x^2 + 7x + 2 = 0$. (*C.S.V.T.U., 2011*)
12. Solve the equations :
    (*i*) $6x^4 + 5x^3 - 38x^2 + 5x + 6 = 0$ (*ii*) $4x^4 - 20x^3 + 33x^2 - 20x + 4 = 0$.
    (*iii*) $8x^5 - 22x^4 - 55x^3 + 55x^2 + 22x - 8 = 0$. (*iv*) $6x^5 + x^4 - 43x^3 - 43x^2 + x + 6 = 0$
    (*v*) $3x^6 + x^5 - 27x^4 + 27x^2 - x - 3 = 0$.
13. Show that the equation $x^4 - 10x^3 + 23x^2 - 6x - 15 = 0$ can be transformed into reciprocal equation by diminishing the roots by 2. Hence solve the equation.

## 1.5 SOLUTION OF CUBIC EQUATIONS–CARDAN'S METHOD*

Consider the equation $ax^3 + bx^2 + cx + d = 0$ ...(1)

Dividing by $a$, we get an equation of the form $x^3 + lx^2 + mx + n = 0$.

*To remove the* $x^2$ term, put $y = x - (-l/3)$ or $x = y - l/3$ so that the resulting equation is of the form

$$y^3 + py + q = 0 \quad ...(2)$$

*To solve* Eqn. (2), put $y = u + v$

so that $y^3 = u^3 + v^3 + 3uv(u + v) = u^3 + v^3 + 3uvy$

or $y^3 - 3uvy - (u^3 + v^3) = 0$ ...(3)

Comparing Eqns. (2) and (3), we get

$$uv = -p/3,\ u^3 + v^3 = -q \text{ or } u^3 + v^3 = -q \text{ and } u^3v^3 = -p^3/27$$

$\therefore$ $u^3, v^3$ are the roots of the equation $t^2 + qt - p^3/27 = 0$

which gives $u^3 = \frac{1}{2}(-q + \sqrt{q^2 + 4p^3/27}) = \lambda^3$ (say)

and $v^3 = \frac{1}{2}(-q - \sqrt{q^2 + 4p^3/27})$

$\therefore$ The three values of $u$ are $\lambda, \lambda\omega, \lambda\omega^2$, where $\omega$ is one of the imaginary cube roots of unity.

From $uv = -p/3$, we have $v = -p/3u$

$\therefore$ When $u = \lambda, \lambda\omega$ and $\lambda\omega^2$,

$$v = -\frac{p}{3\lambda}, -\frac{p\omega^2}{3\lambda} \text{ and } -\frac{p\omega}{3\lambda}. \quad [\because \omega^3 = 1]$$

Hence the three roots of (2) are $\lambda - \frac{p}{3\lambda}, \lambda\omega - \frac{p\omega^2}{3\lambda}, \lambda\omega^2 - \frac{p\omega}{3\lambda}$ $(y = u + v)$

Having known $y$, the corresponding values of $x$ can be found from the relation $x = y - l/3$.

**Obs. 1.** If one value of $u$ is found to be a rational number, find the corresponding value of $v$ giving one root $y = u + v$. Then find the corresponding root $x = \alpha$ (say). Finally, divide the left hand side of Eqn. (1) by $x - \alpha$, giving the remaining quadratic equation from which the other two roots can be found readily.

**Obs. 2.** If $u^3$ and $v^3$ turn out to be conjugate complex numbers, the roots of the given cubic can be obtained in neat forms by employing De Moivre's theorem. (§ 19.5)

**Example 1.15.** *Solve by Cardan's method* $x^3 - 3x^2 + 12x + 16 = 0$.

**Solution:** Given equation is $x^3 - 3x^2 + 12x + 16 = 0$ ...(*i*)

*To remove the second term from Eqn.* (*i*), diminish each root of Eqn. (*i*) by $3/3 = 1$, *i.e.*,

put $y = x - 1$ or $x = y + 1$ [$\therefore$ Sum of roots = 3].

Then Eqn. (*i*) becomes

$$(y + 1)^3 - 3(y + 1) + 12(y + 1) + 16 = 0 \text{ or } y^3 + 9y^2 + 26 = 0 \quad ...(ii)$$

To solve Eqn. (*ii*), put $y = u + v$ so that $y^3 - 3uvy - (u^3 + v^3) = 0$ ...(*iii*)

Comparing Eqns. (*ii*) and (*iii*), we get $uv = -3$ and $u^3 + v^3 = -26$

$\therefore$ $u^3, v^3$ are the roots of the equation $t^2 + 26t - 27 = 0$

or $(t + 27)(t - 1) = 0$ whence $t = -27, t = 1$.

or $u^3 = -27$ *i.e.*, $u = -3$ and $v^3 = 1$ *i.e.*, $v = 1$

$\therefore$ $y = u + v = -3 + 1 = -2$ and $x = y + 1 = -1$

Dividing L.H.S. of (*i*) by $x + 1$, we obtain $x^2 - 4x + 16 = 0$

or $x = \frac{4 \pm \sqrt{(16 - 64)}}{2} = 2 \pm i\ 2\sqrt{3}$

Hence the required roots of the given equation are $-1, 2 \pm i\ 2\sqrt{3}$.

*Named after an Italian mathematician *Girolamo Cardan* (1501–1576), who was the first to use complex numbers as roots of an equation.

**Example 1.16.** *Solve the cubic equation $28x^3 - 9x^2 + 1 = 0$ by Cardan's method.*

**Solution:** Since the term in $x$ is missing, let us put $x = 1/y$ in the given equation so that the transformed equation is $y^3 - 9y + 28 = 0$ ...(i)

*To solve* Eqn. (i), put $y = u + v$ so that $y^3 - 3uvy - (u^3 + v^3) = 0$ ...(ii)

Comparing Eqns. (ii) and (iii), we get $uv = 3$ and $u^3 + v^3 = -28$.

$\therefore$ $u^3, v^3$ are the roots of $t^2 + 28t + 27 = 0$

or $(t + 1)(t + 27) = 0$ or $t = -1, -27$ or $u = -1, v = -3$

$\therefore$ $y = u + v = -4$. Dividing L.H.S. of Eqn. (i) by $y + 4$, we obtain $y^2 - 4y + 7 = 0$ whence $y = 2 \pm i\sqrt{3}$.

$\therefore$ Roots of Eqn. (i) are $-4, 2 \pm i\sqrt{3}$.

Hence the roots of the given cubic equation are $-\frac{1}{4}, \frac{1}{2 \pm i\sqrt{3}}$ or $-\frac{1}{4}, (2 - i\sqrt{3})/7, (2 + i\sqrt{3})/7$.

**Example 1.17.** *Solve the equation $x^3 + x^2 - 16x + 20 = 0$.*

**Solution:** Instead of diminishing the roots of the given equation by $-1/3$, we first multiply its roots by 3, so that the equation becomes

$$x^3 + 3x^2 - 144x + 540 = 0 \quad \text{...(i)}$$

*To remove the $x^2$ term,* put $y = x - (-3/3)$ or $x = y - 1$ in Eqn.(i)

so that $(y - 1)^3 + 3(y - 1)^2 - 144(y - 1) + 540 = 0$

or $$y^3 - 147y + 686 = 0 \quad \text{...(ii)}$$

*To solve Eqn. (iii),* let $y = u + v$, so that

$$y^3 - 3uvy - (u^3 + v^3) = 0 \quad \text{...(iii)}$$

Comparing Eqns. (ii) and (iii), we get

$$uv = 49, u^3 + v^3 = -686, \text{ so that } u^3 v^3 = (343)^2.$$

$\therefore$ $u^3, v^3$ are the roots of the quadratic

$$t^2 + 686t + (343)^2 = 0 \quad \text{or} \quad (t + 343)^2 = 0$$

$\therefore$ $$t = -343 \quad i.e., u^3 = v^3 = -343 \quad \text{or} \quad u = v = -7.$$

Thus $$y = u + v = -14 \text{ and } x = y - 1 = -15.$$

Dividing L.H.S. of (i) by $x + 15$, we get

$$(x - 6)^2 = 0 \quad \text{or} \quad x = 6, 6.$$

$\therefore$ The root of Eqn. (i) are $-15, 6, 6$.

Hence the roots of the given equation are $-5, 2, 2$.

**Example 1.18.** *Solve $x^3 - 3x^2 + 3 = 0$.* *(C.S.V.T.U., 2011 S)*

**Solution:** Given equation is $x^3 - 3x^2 + 3 = 0$ ...(i)

*To remove the $x^2$ term, put* $y = x - 3/3$ or $x = y + 1$,

so that Eqn. (i) becomes $(y + 1)^3 - 3(y + 1)^2 + 3 = 0$

or $$y^3 - 3y + 1 = 0 \quad \text{...(ii)}$$

*To solve it,* put $y = u + v$

so that $$y^3 - 3uvy - (u^3 + v^3) = 0 \quad \text{...(iii)}$$

Comparing Eqns. (ii) and (iii), we get $uv = 1, u^3 + v^3 = -1$

$\therefore$ $u^3, v^3$ are the roots of the equation $t^2 + t + 1 = 0$

Hence $$u^3 = \frac{-1 + i\sqrt{3}}{2} \text{ and } v^3 = \frac{-1 - i\sqrt{3}}{2}$$

$\therefore$ $$u = \left(\frac{-1 + i\sqrt{3}}{2}\right)^{1/3} \quad \text{put } -\frac{1}{2} = r\cos\theta \text{ and } \sqrt{3}/2 = r\sin\theta$$

$$= [r(\cos\theta + i\sin\theta)]^{1/3} \quad \text{so that} \quad r = 1, \theta = 2\pi/3$$

$$= [\cos(\theta + 2n\pi) + i\sin(\theta + 2n\pi)]^{1/3},$$

where $n$ is any integer or zero. Using De Moivre's theorem (p. 647).

$$u = \cos\left(\frac{\theta + 2n\pi}{3}\right) + i \sin\left(\frac{\theta + 2n\pi}{3}\right)$$

Giving $n$ the value 0, 1, 2 successively we get the three values of $u$ to be

$$\cos\frac{\theta}{3} + i\sin\frac{\theta}{3}, \cos\frac{\theta + 2\pi}{3} + i\sin\frac{\theta + 2\pi}{3}, \cos\frac{\theta + 4\pi}{3} + i\sin\frac{\theta + 4\pi}{3}$$

*i.e.,* $$\cos\frac{2\pi}{9} + i\sin\frac{2\pi}{9}, \cos\frac{8\pi}{9} + i\sin\frac{8\pi}{9}, \cos\frac{14\pi}{9} + i\sin\frac{14\pi}{9}.$$

The corresponding values of $v$ are

$$\cos\frac{2\pi}{9} - i\sin\frac{2\pi}{9}, \cos\frac{8\pi}{9} - i\sin\frac{8\pi}{9}, \cos\frac{14\pi}{9} - i\sin\frac{14\pi}{9}.$$

$\therefore$ The three values of $y = u + v$ are $2\cos 2\pi/9, 2\cos 8\pi/9, 2\cos 14\pi/9$.

Hence the roots of Eqn. ($i$) are found from $x = 1 + y$ to be

$$1 + 2\cos 2\pi/9, 1 + 2\cos 8\pi/9, 1 + 2\cos 14\pi/9.$$

**PROBLEMS 1.3**

Solve the following equations by Cardan's method :

1. $x^3 - 27x + 54 = 0$.
2. $x^3 - 18x + 35 = 0$ (C.S.V.T.U., 2012)
3. $x^3 - 15x = 126$ (C.S.V.T.U., 2014 S)
4. $2x^3 + 5x^2 + x - 2 = 0$
5. $9x^3 + 6x^2 - 1 = 0$ (C.S.V.T.U., 2008)
6. $x^3 - 6x^2 + 6x - 5 = 0$
7. $x^3 - 3x + 1 = 0$ (C.S.V.T.U., 2014)
8. $27x^3 + 54x^2 + 198x - 73 = 0$ (C.S.V.T.U., 2015)

## 1.6 SOLUTION OF BIQUADRATIC EQUATIONS

**(1) Ferrari's method***

This method of solving a biquadratic equation is illustrated by the following examples :

**Example 1.19.** *Solve the equation $x^4 - 12x^3 + 41x^2 - 18x - 72 = 0$ by Ferrari's method.* (C.S.V.T.U., 2011 S)

**Solution:** Combining $x^4$ and $x^3$ terms into a perfect square, the given equation can be written as

$$(x^2 - 6x + \lambda)^2 + (5 - 2\lambda)x^2 + (12\lambda - 18)x - (\lambda^2 + 72) = 0$$

or $$(x^2 - 6x + \lambda)^2 = \{(2\lambda - 5)x^2 + (18 - 12\lambda)x + (\lambda^2 + 72)\} \quad ...(i)$$

This equation can be factorised if R.H.S. is a perfect square

*i.e.,* if $$(18 - 12\lambda)^2 = 4(2\lambda - 5)(\lambda^2 + 72) \quad [b^2 = 4ac]$$

*i.e.,* if $2\lambda^3 - 41\lambda^2 + 252\lambda - 441 = 0$ which gives $\lambda = 3$.

$\therefore$ Eqn. ($i$) reduces to $(x^2 - 6x + 3)^2 = (x - 9)^2$

*i.e.,* $(x^2 - 5x - 6)(x^2 - 7x + 12) = 0$.

Hence the roots of the given equation are $-1, 3, 4$ and $6$.

**Example 1.20.** *Solve the equation $x^4 - 2x^3 - 5x^2 + 10x - 3 = 0$ by Ferrari's method.*

**Solution:** Combining $x^4$ and $x^3$ terms into a perfect square, the given equation can be written as $(x^2 - x + \lambda)^2 = (2\lambda + 6)x^2 - (2\lambda + 10)x + (\lambda^2 + 3)$. This equation can be factorised, if R.H.S. is a perfect square *i.e.,* if $(2\lambda + 10)^2 = 4(2\lambda + 6)(\lambda^2 + 3)$ $[b^2 = 4ac]$

or $2\lambda^3 + 5\lambda^2 - 4\lambda - 7 = 0$, which gives $\lambda = -1$.

$\therefore$ Eqn. ($i$) reduces to $(x^2 - x - 1)^2 = 4x^2 - 8x + 4$

or $(x^2 - x - 1)^2 - (2x - 2)^2 = 0$ or $(x^2 + x - 3)(x^2 - 3x + 1) = 0$

$$\therefore \quad x = \frac{-1 \pm \sqrt{1 + 12}}{2} \text{ or } \frac{3 \pm \sqrt{9 - 4}}{2}$$

Hence the roots are $\dfrac{-1 \pm \sqrt{13}}{2}, \dfrac{3 \pm \sqrt{5}}{2}$.

*Named after an Italian mathematician *Scipio Ferrari,* a student of Girolamo Cardan.

**(2) Descarte's method**

This method of solving a biquadratic equations consists in removing the term in $x^3$ and then expressing the new equation as product of two quadratics. It has been best illustrated by the following examples :

**Example 1.21.** *Solve the equation $x^4 - 8x^2 - 24x + 7 = 0$ by Descarte's method.*

**Solution:** In the given equation, the term in $x^3$ is already absent so we assume that

$$x^4 - 8x^2 - 24x + 7 = (x^2 + px + q)(x^2 - px + q') \qquad ...(i)$$

Equating coefficients of the like powers of $x$ in Eqn. $(i)$, we get

$$-8 = q + q' - p^2, - 24 = p(q' - q) \ ; 7 = qq'$$

$$\therefore \quad q + q' = p^2 - 8, q - q' = 24/p$$

$$\therefore \quad (p^2 - 8)^2 - (24/p)^2 = 4 \times 7$$

$$p^2 - 16p^4 + 36p^2 - 576 = 0 \quad \text{or} \quad t^3 - 16t^2 + 36t - 576 = 0 \text{ where } t = p^2$$

Now $t = 16$ satisfies this cubic so that $p = 4$.

$$\therefore \quad q + q' = 8, q - q' = 6 \quad \therefore \quad q = 7, q' = 1$$

Thus Eqn. $(i)$ takes the form

$$(x^2 + 4x + 7)(x^2 - 4x + 1) = 0$$

whence

$$x = \frac{-4 \pm \sqrt{(16-28)}}{2} \text{ or } x = \frac{4 \pm \sqrt{(16-4)}}{2}$$

Hence $x = -2 \pm \sqrt{3}i, 2 \pm \sqrt{3}$.

**Example 1.22.** *Solve the equation $x^4 - 6x^3 - 3x^2 + 22x - 6 = 0$ by Descarte's method.*

**Solution:** Here sum of roots = 6 and number of roots = 4

Therefore, to remove the second term, we have to diminish the roots by 6/4 (= 3/2) which will be a problem. Therefore, we first multiply the roots by 2. Therefore, $y^4 - 12y^3 + 12y^2 + 176y - 96 = 0$, where $y = 2x$. Now diminishing the roots by 3, we obtain $z^4 - 42z^2 + 32z + 297 = 0$, where $z = y - 3$.

Assuming that $z^4 - 42z^2 + 32z + 297 = (z^2 + pz + q)(z^2 - pz + q')$ ...$(i)$

and comparing coefficients, we get

$$-42 = q + q' - p^2 \ ; 32 = p(q' - q) \ ; 297 = q\,q'$$

$$\therefore \quad q + q' = p^2 - 42 \ ; q - q' = -32/p, q\,q' = 297$$

$$\therefore \quad (p^2 - 42)^2 - (-32/p)^2 = 4 \times 297$$

or

$$t^3 - 84\,t^2 + 576t - 1024 = 0 \quad \text{where } t = p^2$$

Now $t = 4$ satisfies this cubic so that $p = 2$.

$$\therefore \quad q + q' = -38, q - q' = -16, \quad \therefore \quad q = -27, q' = -11.$$

Thus $(i)$ takes the form $(z^2 + 2z - 27)(z^2 - 2z - 11) = 0$

Whence

$$z = \frac{-2 \pm \sqrt{(4+108)}}{2} \quad \text{or} \quad z = \frac{2 \pm \sqrt{(4+44)}}{2}$$

or

$$x = \frac{1}{2}y = \frac{1}{2}(z + 3) = \frac{1}{2}(2 \pm \sqrt{28}) = \frac{1}{2}(4 \pm \sqrt{12})$$

Hence

$$x = 1 \pm \sqrt{7}, 2 \pm \sqrt{3}.$$

**PROBLEMS 1.4**

Solve by Ferrari's method, the equations :

1. $x^4 - 10x^3 + 35x^2 - 50x + 24 = 0$ *(C.S.V.T.U., 2015)*
2. $x^4 + 2x^3 - 7x^2 - 8x + 12 = 0$
3. $x^4 - 10x^2 - 20x - 16 = 0$
4. $x^4 - 8x^3 - 12x^2 + 60x + 63 = 0$

Solve the following equations by Descartes method :

5. $x^4 - 6x^3 + 3x^2 + 22x - 6 = 0$
6. $x^4 + 12x - 5 = 0$
7. $x^4 - 8x^3 - 24x + 7 = 0$
8. $x^4 - 10x^3 + 44x^2 - 104x + 96 = 0$

**Obs.** We have obtained algebraic solutions of cubic and biquadratic equations. But the need often arises to solve higher degree or transcendental equations for which no algebraic methods are available in general. Such equations can be best solved by *graphical method* (explained below) or by *numerical methods* (§28.2).

## 1.7 GRAPHICAL SOLUTION OF EQUATIONS

Let the equation be $f(x) = 0$.

(i) *Find the interval (a, b) in which a root of $f(x) = 0$ lies.*

[At least one root of $f(x) = 0$ lies in $(a, b)$ if $f(a)$ and $f(b)$ are of opposite signs—§1.2(III) p. 2].

(ii) *Write the equation $f(x) = 0$ as $\phi(x) = \psi(x)$ where $\psi(x)$ contains only terms in $x$* and the constants.

(iii) *Draw the graphs of $y = \phi(x)$ and $y = \psi(x)$ on the same scale and with respect to the same axes.*

(iv) *Read the abscissae of the points of intersection of the curves $y = \phi(x)$ and $y = \psi(x)$. These are required real roots of $f(x) = 0$.*

Sometimes it may not be convenient to write the given equation $f(x) = 0$ in the form $\phi(x) = \psi(x)$. In such cases, we proceed as follows :

(i) Form a table for the value of $x$ and $y = f(x)$ directly.

(ii) Plot these points and pass a smooth curve through them.

(iii) Read the abscissae of the points where this curve cuts the $x$-axis. These are the required roots of $f(x) = 0$.

**Obs.** The roots, thus located graphically are approximate and to improve their accuracy, the curves are replotted on the larger scale in the immediate vicinity of each point of intersection. This gives a better approximation to the value of desired root. The above graphical operation may be repeated until the root is obtained correct upto required number of decimal places. But this method of repeatedly drawing graphs is very tedious. It is, therefore, advisable to improve upon the accuracy of an approximate root by numerical method of §28.2.

**Example 1.23.** *Find graphically an approximate value of the root of the equation.*

$$3 - x = e^{x-1}.$$

**Solution:** Let $f(x) = e^{x-1} + x - 3 = 0$ ...(i)

$f(1) = 1 + 1 - 3 = -\text{ve}$

and $f(2) = e + 2 - 3 = 2.718 - 1 = +\text{ve}$

∴ A root of Eqn. (i) lies between $x = 1$ and $x = 2$.

Let us rewrite Eqn. (i) as $e^{x-1} = 3 - x$.

The abscissa of the point of intersection of the curves

$$y = e^{x-1} \quad ...(ii)$$

and

$$y = 3 - x \quad ...(iii)$$

will give the required root.

*To plot* Eqn. (ii), we form the following table of values :

| $x =$ | $y = e^{x-1}$ |
|---|---|
| 1.1 | 1.11 |
| 1.2 | 1.22 |
| 1.3 | 1.35 |
| 1.4 | 1.49 |
| 1.5 | 1.65 |
| 1.6 | 1.82 |
| 1.7 | 2.01 |
| 1.8 | 2.23 |
| 1.9 | 2.46 |
| 2.0 | 2.72 |

Taking the origin at (1, 1) and 1 small unit along either axis = 0.02, we plot these points and pass a smooth curve through them as shown in Fig. 1.2.

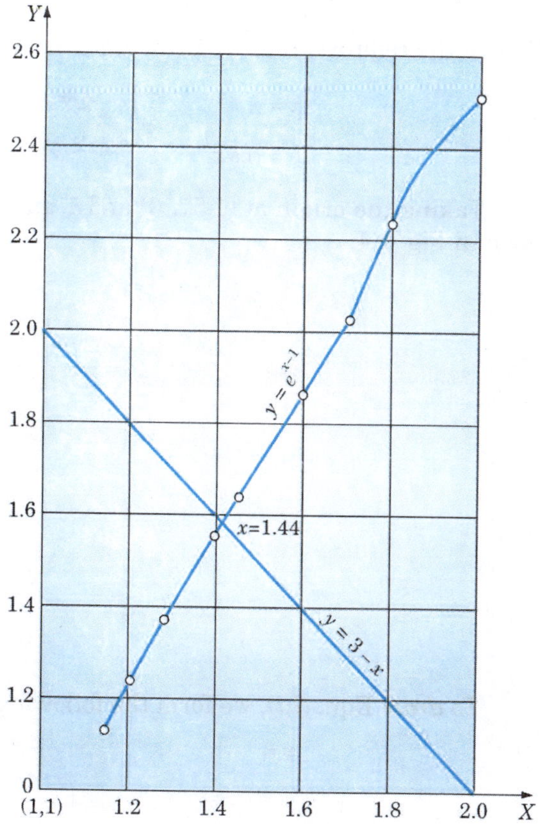


Fig. 1.2

*To draw the line Eqn. (iii), we join the points (1, 2) and (2, 1) on the same scale and with the same axes.* From the figure, we get the required root to be $x = 1.44$ nearly.

**Example 1.24.** *Obtain graphically an approximate value of the root of $x = \sin x + \pi/2$.*

**Solution:** Let us write the given equation as $\sin x = x - \pi/2$

The abscissa of the point of intersection of the curve $y = \sin x$ and the line $y = x - \pi/2$ will give a rough estimate of the root.

*To draw a curve $y = \sin x$, we form the following table :*

| $x$ | 0 | $\pi/4$ | $\pi/2$ | $3\pi/4$ | $\pi$ |
|---|---|---|---|---|---|
| $y$ | 0 | 0.71 | 1 | 0.71 | 0 |

Taking 1 unit along either axis $= \pi/4 = 0.8$ nearly, we plot the curve as shown in Fig. 1.3.

Also we draw the line $y = x - \pi/2$ to the same scale and with the same axis.

From the graph, we get $x = 2.3$ radians approximately.

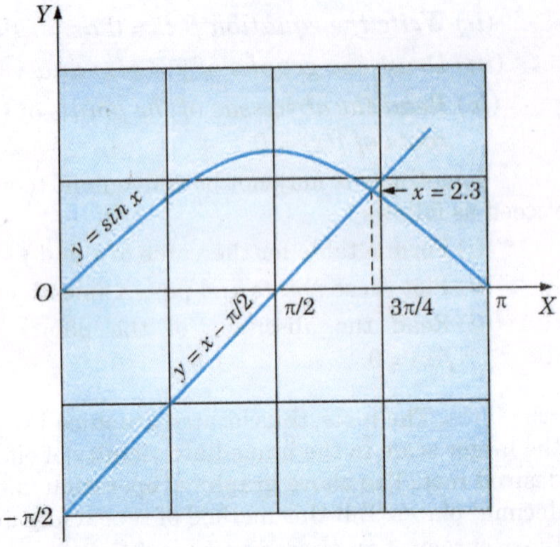

Fig. 1.3

**Example 1.25.** *Obtain graphically the lowest root of $\cos x \cosh x = -1$.*

**Solution:** Let $f(x) = \cos x \cosh x + 1 = 0$ ...(i)

$\because$ $f(0) = +\text{ ve}, f(\pi/2) = +\text{ve}$ and $f(\pi) = -\text{ve}$.

$\therefore$ The lowest root of Eqn. (*i*) lies between $x = \pi/2$ and $x = \pi$.

Let us write Eqn. (*i*) as $\cos x = -\text{sech}\, x$.

The abscissa of the point of intersection of the curves

$y = \cos x$ ...(ii) and $y = -\text{sech}\, x$ ...(iii)

will give the required root. To draw Eqn. (*ii*), we form the following table of values :

| $x =$ | $\pi/2 = 1.57$ | $3\pi/4 = 2.36$ | $\pi = 3.14$ |
|---|---|---|---|
| $y = \cos x$ | 0 | −0.71 | −1 |

Taking the origin at (1.57, 0) and 1 unit along either axes $= \pi/8 = 0.4$ nearly, we plot the cosine curve as shown in Fig. 1.4.

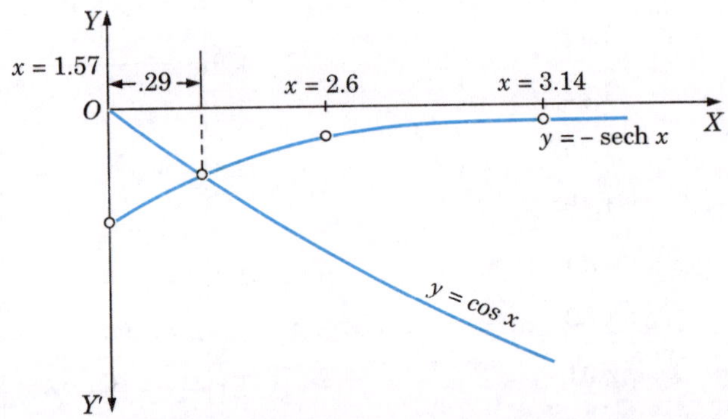

Fig. 1.4

*To draw* Eqn. (*iii*), we form the following table :

| $x =$ | 1.57 | 2.36 | 3.14 |
|---|---|---|---|
| $\cosh x =$ | 2.58 | 5.56 | 11.12 |
| $y = -\text{sech}\, x$ | −0.39 | −0.18 | −0.09 |

Then we plot the curve Eqn. (*iii*) to the same scale with the same axes.

From the figure we get the lowest root to be approximately $x = 1.57 + 0.29 = 1.86$.

**PROBLEMS 1.5**

Solve the following equations graphically :

1. $x^3 - x - 1 = 0$
2. $x^3 - 3x - 5 = 0$
3. $x^3 - 6x^2 + 9x - 3 = 0$.
4. $\tan x = 1.2\, x$
5. $x = 3 \cos (x - \pi/4)$
6. $e^x = 5x$ which is near $x = 0.2$.

## 1.8 OBJECTIVE TYPE OF QUESTIONS

**PROBLEMS 1.6**

*Choose the correct answer or fill up the blanks in the following problems :*

1. If for the equation $x^3 - 3x^2 + kx + 3 = 0$, one root is the negative of another, then the value of $k$ is
   (*a*) 3 (*b*) –3 (*c*) 1 (*d*) –1.
2. If the roots of the equation $x^n - 1 = 0$ are $1, \alpha_1, \alpha_2, ......, \alpha_{n-1}$, then $(1 - \alpha_1)(1 - \alpha_2) ...... (1 - \alpha_{n-1})$ is equal to
   (*a*) 0 (*b*) 1 (*c*) $n$ (*d*) $n + 1$.
3. If $\alpha, \beta, \gamma$ are the roots of $2x^3 - 3x^2 + 6x + 1 = 0$, then $\alpha^2 + \beta^2 + \gamma^2$ is
   (*a*) 15/4 (*b*) –3 (*c*) –15/4 (*d*) 33/4.
4. $x + 2$ is a factor of
   (*a*) $x^4 + 2$ (*b*) $x^4 - x^2 + 12$
   (*c*) $x^4 - 2x^3 - x + 2$ (*d*) $x^4 + 2x^3 - x - 2$
5. If $\alpha + \beta + \gamma = 5$ ; $\alpha\beta + \beta\gamma + \gamma\alpha = 7$ ; $\alpha\beta\gamma = 3$, then the equation whose roots are $\alpha$, $\beta$ and $\gamma$ is
   (*a*) $x^3 - 7 = 0$ (*b*) $x^3 - 7x^2 + 3 = 0$
   (*c*) $x^3 - 5x^2 + 7x - 3 = 0$ (*d*) $x^3 + 7x^2 - 3 = 0$.
6. If one of the roots of the equation $x^3 - 6x^2 + 11x - 6 = 0$ is 2, then the other two roots are
   (*a*) 1 and 3 (*b*) 0 and 4
   (*c*) – 1 and 5 (*d*) – 2 and 6.
7. The equation whose roots are the reciprocals of the roots of $x^3 + px^2 + r = 0$ is
   (*a*) $x^3 + 1/p.x^2 + 1/r = 0$ (*b*) $1/r \,.\, x^3 + 1/p.x + 1 = 0$
   (*c*) $rx^3 + px^2 + 1 = 0$ (*d*) $rx^3 + px + 1 = 0$.
8. If 1 and 2 are two roots of the equation $x^4 - x^3 - 19x^2 + 49x - 30 = 0$, then the remaining two roots are
   (*a*) –3 and 5 (*b*) 3 and –5
   (*c*) –6 and 5 (*d*) 6 and –5.
9. If the roots of $x^3 - 3x^2 + px + 1 = 0$, are in arithmetic progression, then the sum of squares of the largest and the smallest roots is
   (*a*) 3 (*b*) 5 (*c*) 6 (*d*) 10.
10. A root of $x^3 - 8x^2 + px + q = 0$ where $p$ and $q$ are real numbers is $3 + i\sqrt{3}$. The real root is
    (*a*) 2 (*b*) 6 (*c*) 9 (*d*) 12.
11. One of the roots of the equation $f(x) = x^n + a_{n-1}x^{n-1} + ... + a_1x + a_0 = 0$ where $a_0, a_1, ... a_{n-1}$ are real, is given to be $2 - 3i$. Of the remaining, the next $n - 2$ roots are given to be 1, 2, 3, ..., $n - 2$. The $n$th root is
    (*a*) $n$ (*b*) $n - 1$ (*c*) $2 + 3i$ (*d*) $- 2 + 3i$.
12. If a real root of $f(x) = 0$ lies in $[a, b]$, then the sign of $f(a).f(b)$ is ......
13. Descartes rule of signs states that ......
14. If $\alpha, \beta, \gamma$ are the roots of the equation $x^3 - px + q = 0$, then $\Sigma\, 1/\alpha$ = ......
15. If $\alpha, \beta, \gamma$ are the roots of $x^3 = 7$, then $\Sigma\alpha^3$ is ......
16. One real root of the equation $x^3 + 2x^2 + 5 = 0$ lies between .......

17. In an equation with real coefficients, imaginary roots must occur in ......
18. If $f(\alpha)$ and $f(\beta)$ are of opposite signs, then $f(x) = 0$ has at least one root between $\alpha$ and $\beta$ provided ......
19. If $\alpha, \beta, \gamma$ are the roots of the equation $x^3 + 2x + 3 = 0$, then $\alpha + 3, \beta + 3, \gamma + 3$ are the roots of the equation ......
20. If one root is double of another in $x^3 - 7x^2 + 36 = 0$, then its roots are ......
21. The equation whose roots are 10 times those $x^3 - 2x - 7 = 0$, is ......
22. If $\alpha, \beta, \gamma$ are the roots of $x^3 + px^2 + qx + r = 0$, then $\Sigma\ (1/\alpha\beta) =$ ......
23. $\sqrt{3}$ and $-1 + i$ are the roots of the biquadratic equation ......
24. If $\alpha, \beta, \gamma$ are the roots of $x^3 - 3x + 2 = 0$, then the value of $\alpha^2 + \beta^2 + \gamma^2$ is ......
25. If there is a root of $f(x) = 0$ in the interval $[a, b]$, then sign of $f(a)/f(b)$ is ......
26. If $\alpha, \beta, \gamma$ are the roots of $x^3 + px^2 + qx + r = 0$, then the condition for $\alpha + \beta = 0$ is ......
27. The three roots of $x^3 = 1$ are ......
28. One real root of the equation $x^3 + x - 5 = 0$ lies in the interval
    (*i*) (2, 3), (*ii*) (3, 4), (*iii*) (1, 2), (*iv*) (–3, –2)
29. If two roots of $x^3 - 3x^2 + 2 = 0$ are equal, then its roots are ......
30. The cubic equation whose two roots are 5 and $1 - i$ is ......
31. The sum and product of the roots of the equation $x^5 = 2$ are ...... and .....
32. If the roots of the equation $x^4 + 2x^3 - ax^2 - 22x + 40 = 0$ are – 5, – 2, 1 and 4, then $a =$ ......
33. A root of $x^3 - 3x^2 + 2.5 = 0$ lies between 1.1 and 1.2. (True or False)
34. The equation $x^6 - x^5 - 10x + 7 = 0$ has four imaginary roots. (True or False)

CHAPTER 2

# Linear Algebra : Determinants, Matrices

**1.** Introduction. **2.** Determinants, Cofactors, Laplace's expansion. **3.** Properties of determinants. **4.** Matrices, Special matrices. **5.** Matrices operations. **6.** Related matrices. **7.** Rank of a matrix, Elementary transformations, Elementary matrices, Inverse from elementary matrices, Normal form of a matrix. **8.** Partition method of finding the inverse. **9.** Solution of linear system of equations. **10.** Consistency of linear system of equations. **11.** Linear and orthogonal transformations. **12.** Vectors ; Linear dependence. **13.** Eigen values and eigen vectors. **14.** Properties of eigen values. **15.** Cayley-Hamilton theorem. **16.** Reduction to diagonal form. **17.** Reduction of quadratic form to canonical form. **18.** Nature of quadratic form. **19.** Complex matrices. **20.** Objective Types of Questions.

## 2.1 INTRODUCTION

*Linear algebra* comprises of the theory and applications of linear system of equation, linear transformations and eigen value problems. In linear algebra, we make a systematic use of matrices and to a lesser extent determinants and their properties.

Determinants were first introduced for solving linear systems and have important engineering applications in systems of differential equations, electrical networks, eigen-value problems and so on. Many complicated expressions occurring in electrical and mechanical systems can be elegantly simplified by expressing them in the form of determinants.

Cayley* discovered matrices in the year 1860. But it was not until the twentieth century was well-advanced that engineers heard of them. These days, however, matrices have been found to be of great utility in many branches of applied mathematics such as algebraic and differential equations, mechanics theory of electrical circuits, nuclear physics, aerodynamics and astronomy. With the advent of computers, the usage of matrix methods has been greatly facilitated.

## 2.2 DETERMINANTS

(1) **Definition.** The expression $\begin{vmatrix} a_1 & b_1 \\ a_2 & b_2 \end{vmatrix}$ is called a *determinant of the second order* and stands for '$a_1b_2 - a_2b_1$'. It contains 4 numbers $a_1, b_1, a_2, b_2$ (called *elements*) which are arranged along two horizontal lines (called *rows*) and two vertical lines (called *columns*).

Similarly, $\begin{vmatrix} a_1 & b_1 & c_1 \\ a_2 & b_2 & c_2 \\ a_3 & b_3 & c_3 \end{vmatrix}$ is called a *determinant of the third order*. It consists of 9 *elements* which are arranged in 3 *rows* and 3 *columns*.

---

**Arthur Cayley* (1821–1895) was a professor at Cambridge and is known for his important contributions to algebra, matrices and differential equations.

In general, *a determinant of the nth order* is denoted by $\begin{vmatrix} a_1 & b_1 & c_1 & d_1 \ldots l_1 \\ a_2 & b_2 & c_2 & d_2 \ldots l_2 \\ \ldots & \ldots & \ldots & \ldots \\ \ldots & \ldots & \ldots & \ldots \\ a_n & b_n & c_n & d_n \ldots l_n \end{vmatrix}$

which is a block of $n^2$ *elements* arranged in the form of a square along *n-rows* and *n-columns.* The diagonal through the left hand top corner which contains the elements $a_1, b_2, c_3, \ldots l_n$ is called the *leading or principal diagonal.*

**(2) Cofactors**

**The cofactor** *of any element in a determinant is obtained by deleting the row and column which intersect in that element with the proper sign. The sign of an element in the ith row and jth column is* $(-1)^{i+j}$. The cofactor of an element is usually denoted by the corresponding capital letter.

For instance, in $\Delta = \begin{vmatrix} a_1 & b_1 & c_1 \\ a_2 & b_2 & c_2 \\ a_3 & b_3 & c_3 \end{vmatrix}$, the cofactor of $b_3$ *i.e.*, $B_3 = (-1)^{3+2} \begin{vmatrix} a_1 & c_1 \\ a_2 & c_2 \end{vmatrix}$ and $C_2 = -\begin{vmatrix} a_1 & b_1 \\ a_3 & b_3 \end{vmatrix}$.

**(3) Laplace's expansion.*** *A determinant can be expanded in terms of any row (or column) as follows :*

*Multiply each element of the row (or column) in terms of which we intend expanding the determinant, by its cofactor and then add up all these terms.*

∴ Expanding by $R_1$ (*i.e.*, 1st row),

$$\Delta = a_1A_1 + b_1B_1 + c_1C_1 = a_1 \begin{vmatrix} b_2 & c_2 \\ b_3 & c_3 \end{vmatrix} - b_1 \begin{vmatrix} a_2 & c_2 \\ a_3 & c_3 \end{vmatrix} + c_1 \begin{vmatrix} a_2 & b_2 \\ a_3 & b_3 \end{vmatrix}$$

$$= a_1(b_2c_3 - b_3c_2) - b_1(a_2c_3 - a_3c_2) + c_1(a_2b_3 - a_3b_2)$$

Similarly, expanding by $C_2$ (*i.e.*, 2nd column)

$$\Delta = b_1B_1 + b_2B_2 + b_3B_3 = -b_1 \begin{vmatrix} a_2 & c_2 \\ a_3 & c_3 \end{vmatrix} + b_2 \begin{vmatrix} a_1 & c_1 \\ a_3 & c_3 \end{vmatrix} - b_3 \begin{vmatrix} a_1 & c_1 \\ a_2 & c_2 \end{vmatrix}$$

$$= -b_1(a_2c_3 - a_3c_2) + b_2(a_1c_3 - a_3c_1) - b_3(a_1c_2 - a_2c_1)$$

and expanding by $R_3$ (*i.e.*, 3rd row), $\Delta = a_3A_3 + b_3B_3 + c_3C_3$.

*Thus* Δ *is the sum of the products of the elements of any row (or column) by the corresponding cofactors.*

*If, however, the sum of the products of the elements of any row (or column) by the cofactors of another row (or column) be taken, the result is zero.*

*e.g.*, in Δ, $$a_3A_2 + b_3B_2 + c_3C_2 = -a_3 \begin{vmatrix} b_1 & c_1 \\ b_3 & c_3 \end{vmatrix} + b_3 \begin{vmatrix} a_1 & c_1 \\ a_3 & c_3 \end{vmatrix} - c_3 \begin{vmatrix} a_1 & b_1 \\ a_3 & b_3 \end{vmatrix}$$

$$= -a_3(b_1c_3 - b_3c_1) + b_3(a_1c_3 - a_3c_1) - c_3(a_1b_3 - a_3b_1) = 0$$

In general, $a_iA_j + b_iB_j + c_iC_j = \Delta$ when $i = j = 0$ when $i \neq j$

**Example 2.1.** *Expand* $\Delta = \begin{vmatrix} a & h & g \\ h & b & f \\ g & f & c \end{vmatrix}$.

**Solution:** Expanding by $R_1$, $\Delta = a \begin{vmatrix} b & f \\ f & c \end{vmatrix} - h \begin{vmatrix} h & f \\ g & c \end{vmatrix} + g \begin{vmatrix} h & b \\ g & f \end{vmatrix}$

$$= a(bc - f^2) - h(hc - gf) + g(hf - gb) = abc + 2fgh - af^2 - bg^2 - ch^2.$$

*Named after a great French mathematician *Pierre Simon Marquis De Laplace* (1749–1827). He made important contributions to probability theory, special functions, potential theory and astronomy. While a professor in Paris, he taught Napolean Bonapart for a year.

**Example 2.2.** *Find the value of* $\Delta = \begin{vmatrix} 0 & 1 & 2 & 3 \\ 1 & 0 & 3 & 0 \\ 2 & 3 & 0 & 1 \\ 3 & 0 & 1 & 2 \end{vmatrix}$.

**Solution:** Since there are two zeros in the second row, therefore, expanding by $R_2$, we get

$$\Delta = -\begin{vmatrix} 1 & 2 & 3 \\ 3 & 0 & 1 \\ 0 & 1 & 2 \end{vmatrix} + 0 - 3\begin{vmatrix} 0 & 1 & 3 \\ 2 & 3 & 1 \\ 3 & 0 & 2 \end{vmatrix} + 0$$

(Expand by $C_1$) (Expand by $R_1$)

$$= -[1(0 \times 2 - 1 \times 1) - 3(2 \times 2 - 1 \times 3) + 0] - 3[0 - (2 \times 2 - 3 \times 1) + 3(2 \times 0 - 3 \times 3)]$$

$$= -(-1 - 3) - 3(-1 - 27) = 4 + 84 = 88.$$

## 2.3 PROPERTIES OF DETERMINANTS

The following properties, are proved for determinants of the third order, but these hold good for determinants of any order. These properties enable us to simplify a given determinant and evaluate it without expanding the given determinant.

**I.** *A determinant remains unaltered by changing its rows into columns and columns into rows.*

Let $$\Delta = \begin{vmatrix} a_1 & b_1 & c_1 \\ a_2 & b_2 & c_2 \\ a_3 & b_3 & c_3 \end{vmatrix} \quad \text{[Expand by } R_1\text{]}$$

$$= a_1(b_2c_3 - b_3c_2) - b_1(a_2c_3 - a_3c_2) + c_1(a_2b_3 - a_3b_2)$$

Then $$\Delta' = \begin{vmatrix} a_1 & a_2 & a_3 \\ b_1 & b_2 & b_3 \\ c_1 & c_2 & c_3 \end{vmatrix} \quad \text{[Expand by } R_1\text{]}$$

$$= a_1(b_2c_3 - b_3c_2) - a_2(b_1c_3 - b_3c_1) + a_3(b_1c_2 - b_2c_1)$$

$$= a_1(b_2c_3 - b_3c_2) - b_1(a_2c_3 - a_3c_2) + c_1(a_2b_3 - a_3b_2) = \Delta.$$

**Obs. 1.** Any theorem concerning the rows of a determinant, therefore, applies equally to its columns and *vice-versa*.

**2.** When a *row* or a *column* is referred to in a general manner, it is called a *line*.

**II.** *If two parallel lines of a determinant are interchanged, the determinant retains its numerical value but changes in sign.*

Let $$\Delta = \begin{vmatrix} a_1 & b_1 & c_1 \\ a_2 & b_2 & c_2 \\ a_3 & b_3 & c_3 \end{vmatrix} \quad \text{[Expand by } R_1\text{]}$$

$$= a_1(b_2c_3 - b_3c_2) - b_1(a_2c_3 - a_3c_2) + c_1(a_2b_3 - a_3b_2)$$

Interchanging $C_2$ and $C_3$, we have

$$\Delta' = \begin{vmatrix} a_1 & c_1 & b_1 \\ a_2 & c_2 & b_2 \\ a_3 & c_3 & b_3 \end{vmatrix} \quad \text{[Expand by } R_1\text{]}$$

$$= a_1(c_2b_3 - c_3b_2) - c_1(a_2b_3 - a_3b_2) + b_1(a_2c_3 - a_3c_2)$$

$$= -[a_1(b_2c_3 - b_3c_2) - b_1(a_2c_3 - a_3c_2) + c_1(a_2b_3 - a_3b_2)] = -\Delta.$$

**Cor.** *If a line of* $\Delta$ *be passed over two parallel lines, i.e., if the resulting determinant is like*

$$\Delta' = \begin{vmatrix} b_1 & c_1 & a_1 \\ b_2 & c_2 & a_2 \\ b_3 & c_3 & a_3 \end{vmatrix}, \quad \text{then} \quad \Delta' = (-1)^2 \Delta.$$

*In general, if any line of a determinant be passed over m parallel lines, the resulting determinant*

$$\Delta' = (-1)^m \Delta.$$

**III.** *A determinant vanishes if two parallel lines are identical.*

Consider a determinant $\Delta$ in which two parallel lines are identical.

Interchange of the identical lines leaves the determinant unaltered yet by the previous property, the interchanges of two parallel lines changes the sign of the determinant.

Hence $\Delta = \Delta' = -\Delta$ or $2\Delta = 0$, or $\Delta = 0$.

**IV.** *If each element of a line be multiplied by the same factor, the whole determinant is multiplied by that factor.*

*i.e.,*
$$\begin{vmatrix} a_1 & pb_1 & c_1 \\ a_2 & pb_2 & c_2 \\ a_3 & pb_3 & c_3 \end{vmatrix} = p \begin{vmatrix} a_1 & b_1 & c_1 \\ a_2 & b_2 & c_2 \\ a_3 & b_3 & c_3 \end{vmatrix}$$

For on expanding by $C_2$,

$$\text{L.H.S.} = -pb_1(a_2c_3 - a_3c_2) + pb_2(a_1c_3 - a_3c_1) - pb_3(a_1c_2 - a_2c_1)$$
$$= p\{-b_1B_1 + b_2B_2 - b_3B_3\} = \text{R.H.S.}$$

Similarly,
$$\begin{vmatrix} a_1 & b_1 & c_1 \\ ka_1 & kb_2 & kc_2 \\ a_3 & b_3 & c_3 \end{vmatrix} = k \begin{vmatrix} a_1 & b_1 & c_1 \\ a_2 & b_2 & c_2 \\ a_3 & b_3 & c_3 \end{vmatrix}$$

**Cor.** *If two parallel lines be such that the elements of one are equi-multiples of the elements of the other, the determinant vanishes.*

*i.e.,*
$$\begin{vmatrix} a_1 & b_1 & pb_1 \\ a_2 & b_2 & pb_2 \\ a_3 & b_3 & pb_3 \end{vmatrix} = p \begin{vmatrix} a_1 & b_1 & b_1 \\ a_2 & b_2 & b_2 \\ a_3 & b_3 & b_3 \end{vmatrix} = p(0) = 0$$

**V.** *If each element of a line consists of m terms, the determinant can be expressed as the sum of m determinants.*

Consider the determinant
$$\Delta = \begin{vmatrix} a_1 & b_1 & c_1 + d_1 - e_1 \\ a_2 & b_2 & c_2 + d_2 - e_2 \\ a_3 & b_3 & c_3 + d_3 - e_3 \end{vmatrix}$$

end of whose third column elements consists of three terms.

Expanding $\Delta$ by $C_3$, we have

$$\Delta = (c_1 + d_1 - e_1)(a_2b_3 - a_3b_2) - (c_2 + d_2 - e_2)(a_1b_3 - a_3b_1) + (c_3 + d_3 - e_3)(a_1b_2 - a_2b_1)$$
$$= [c_1(a_2b_3 - a_3b_2) - c_2(a_1b_3 - a_3b_1) + c_3(a_1b_2 - a_2b_1)] + [d_1(a_2b_3 - a_3b_2) - d_2(a_1b_3 - a_3b_1)$$
$$+ d_3(a_1b_2 - a_2b_1)] - [e_1(a_2b_3 - a_3b_2) - e_2(a_1b_3 - a_3b_1) + e_3(a_1b_2 - a_2b_1)]$$
$$= \begin{vmatrix} a_1 & b_1 & c_1 \\ a_2 & b_2 & c_2 \\ a_3 & b_3 & c_3 \end{vmatrix} + \begin{vmatrix} a_1 & b_1 & d_1 \\ a_2 & b_2 & d_2 \\ a_3 & b_3 & d_3 \end{vmatrix} - \begin{vmatrix} a_1 & b_1 & e_1 \\ a_2 & b_2 & e_2 \\ a_3 & b_3 & e_3 \end{vmatrix}$$

Further, *if the elements of three parallel lines consist of m, n and p terms respectively, the determinants can be expressed as the sum of m × n × p determinants.*

**Example 2.3.** *If* $\begin{vmatrix} a & a^2 & a^3 - 1 \\ b & b^2 & b^3 - 1 \\ c & c^2 & c^3 - 1 \end{vmatrix} = 0$ *in which a, b, c are different, show that abc = 1.*

**Solution:** As each term of $C_3$ in the given determinant consists of two terms, we express it as a sum of two determinants.

$$\begin{vmatrix} a & a^2 & a^3 - 1 \\ b & b^2 & b^3 - 1 \\ c & c^2 & c^3 - 1 \end{vmatrix} = \begin{vmatrix} a & a^2 & a^3 \\ b & b^2 & b^3 \\ c & c^2 & c^3 \end{vmatrix} + \begin{vmatrix} a & a^2 & -1 \\ b & b^2 & -1 \\ c & c^2 & -1 \end{vmatrix} = abc \begin{vmatrix} 1 & a & a^2 \\ 1 & b & b^2 \\ 1 & c & c^2 \end{vmatrix} - \begin{vmatrix} a & a^2 & 1 \\ b & b^2 & 1 \\ c & c^2 & 1 \end{vmatrix}$$

[Taking common $a, b, c$ from $R_1, R_2, R_3$ respectively of the first determinant and –1 from $C_3$ of the second determinant.]

$$= abc \begin{vmatrix} 1 & a & a^2 \\ 1 & b & b^2 \\ 1 & c & c^2 \end{vmatrix} - \begin{vmatrix} 1 & a & a^2 \\ 1 & b & b^2 \\ 1 & c & c^2 \end{vmatrix}$$

[Passing $C_3$ over $C_2$ and $C_1$ in the second determinant]

$\therefore \quad \begin{vmatrix} 1 & a & a^2 \\ 1 & b & b^2 \\ 1 & c & c^2 \end{vmatrix} (abc - 1) = 0$. Hence $abc = 1$, since $\begin{vmatrix} 1 & a & a^2 \\ 1 & b & b^2 \\ 1 & c & c^2 \end{vmatrix} \neq 0$ as $a, b, c$ are all different.

**VI.** *If to each elements of a line be added equi-multiples of the corresponding elements of one or more parallel lines, the determinants remains unaltered.*

Let $$\Delta = \begin{vmatrix} a_1 & b_1 & c_1 \\ a_2 & b_2 & c_2 \\ a_3 & b_3 & c_3 \end{vmatrix}$$

Then $$\Delta' = \begin{vmatrix} a_1 + pb_1 - qc_1 & b_1 & c_1 \\ a_2 + pb_2 - qc_2 & b_2 & c_2 \\ a_3 + pb_3 - qc_3 & b_3 & c_3 \end{vmatrix}$$

$$= \begin{vmatrix} a_1 & b_1 & c_1 \\ a_2 & b_2 & c_2 \\ a_3 & b_3 & c_3 \end{vmatrix} + \begin{vmatrix} pb_1 & b_1 & c_1 \\ pb_2 & b_2 & c_2 \\ pb_3 & b_3 & c_3 \end{vmatrix} + \begin{vmatrix} -qc_1 & b_1 & c_1 \\ -qc_2 & b_2 & c_2 \\ -qc_3 & b_3 & c_3 \end{vmatrix}$$

$= \Delta + 0 + 0 = \Delta$. [by *IV-Cor.*]

**Obs.** *This property is very useful for simplifying determinants.* To add equi-multiples of parallel lines, we shall employ the following **notation :**

Suppose to the elements of the second row, we add $p$ times the elements of the first row and $q$ times the element of the third row ; then we say :

Operate $R_2 + pR_1 + qR_3$.

Similarly Operate '$C_3 + mC_1 - nC_2$'

means that to the elements of the third column add $m$ times the elements of the first column and $-n$ times the elements of the second column.

**Example 2.4.** *Evaluate* $\begin{vmatrix} 21 & 17 & 7 & 10 \\ 24 & 22 & 6 & 10 \\ 6 & 8 & 2 & 3 \\ 6 & 7 & 1 & 2 \end{vmatrix}$.

**Solution:** Operating $R_1 - R_2 - R_4$, $R_2 - 3R_3$, $R_3 - 2R_4$, the given determinant becomes

$$\Delta = \begin{vmatrix} -8 & -12 & 0 & -2 \\ 6 & -2 & 0 & 1 \\ -4 & -6 & 0 & -1 \\ 5 & 7 & 1 & 2 \end{vmatrix} \qquad \text{[Expand by } C_1\text{]}$$

$$= -\begin{vmatrix} -8 & -12 & -2 \\ 6 & -2 & 1 \\ -4 & -6 & -1 \end{vmatrix} = 0 \qquad [\because \; R_1 = 2R_2]$$

**Example 2.5.** *Solve the equation* $\begin{vmatrix} x+2 & 2x+3 & 3x+4 \\ 2x+3 & 3x+4 & 4x+5 \\ 3x+5 & 5x+8 & 10x+17 \end{vmatrix} = 0.$

**Solution:** Operating $R_3 - (R_1 + R_2)$, we get

$$\begin{vmatrix} x+2 & 2x+3 & 3x+4 \\ 2x+3 & 3x+4 & 4x+5 \\ 0 & 1 & 3x+8 \end{vmatrix} = 0 \qquad \text{(Operate } R_2 - R_1 \text{ and } R_1 + R_3\text{)}$$

or $$\begin{vmatrix} x+2 & 2x+4 & 6x+12 \\ x+1 & x+1 & x+1 \\ 0 & 1 & 3x+8 \end{vmatrix} = 0 \quad \text{or} \quad (x+1)(x+2)\begin{vmatrix} 1 & 2 & 6 \\ 1 & 1 & 1 \\ 0 & 1 & 3x+8 \end{vmatrix} = 0$$

To bring one more zero in $C_1$, operate $R_1 - R_2$.

$$\therefore \quad (x+1)(x+2)\begin{vmatrix} 0 & 1 & 5 \\ 1 & 1 & 1 \\ 0 & 1 & 3x+8 \end{vmatrix} = 0$$

Now expand by $C_1$. $\therefore$ $-(x+1)(x+2)(3x+8-5) = 0$ or $-3(x+1)(x+2)(x+1) = 0$

Thus, $x = -1, -1, -2$.

**Example 2.6.** *Prove that* $\begin{vmatrix} 1+a & 1 & 1 & 1 \\ 1 & 1+b & 1 & 1 \\ 1 & 1 & 1+c & 1 \\ 1 & 1 & 1 & 1+d \end{vmatrix} = abcd\left(1+\frac{1}{a}+\frac{1}{b}+\frac{1}{c}+\frac{1}{d}\right)$.

**Solution:** Let $\Delta$ be the given determinant. Taking $a, b, c, d$ common from $R_1, R_2, R_3, R_4$ respectively, we get

$$\Delta = abcd\begin{vmatrix} a^{-1}+1 & a^{-1} & a^{-1} & a^{-1} \\ b^{-1} & b^{-1}+1 & b^{-1} & b^{-1} \\ c^{-1} & c^{-1} & c^{-1}+1 & c^{-1} \\ d^{-1} & d^{-1} & d^{-1} & d^{-1}+1 \end{vmatrix}$$

[Operate $R_1 + (R_2 + R_3 + R_4)$ and take out the common factor from $R_1$]

$$= abcd\,(1 + a^{-1} + b^{-1} + c^{-1} + d^{-1})\begin{vmatrix} 1 & 1 & 1 & 1 \\ b^{-1} & b^{-1}+1 & b^{-1} & b^{-1} \\ c^{-1} & c^{-1} & c^{-1}+1 & c^{-1} \\ d^{-1} & d^{-1} & d^{-1} & d^{-1}+1 \end{vmatrix}$$

[Operate $C_2 - C_1, C_3 - C_1, C_4 - C_1$]

$$= abcd\left(1+\frac{1}{a}+\frac{1}{b}+\frac{1}{c}+\frac{1}{d}\right)\begin{vmatrix} 1 & 0 & 0 & 0 \\ b^{-1} & 1 & 0 & 0 \\ c^{-1} & 0 & 1 & 0 \\ d^{-1} & 0 & 0 & 1 \end{vmatrix} = abcd\left(1+\frac{1}{a}+\frac{1}{b}+\frac{1}{c}+\frac{1}{d}\right)$$

**Obs.** If all elements on one side of the leading diagonal are zero, then the determinant is equal to the product of leading diagonal elements and such a determinants is called a *triangular determinant.*

**VII. Factor Theorem.** *If the elements of a determinant $\Delta$ are functions of $x$ and two parallel lines become identical when $x = a$, then $x - a$ is a factor of $\Delta$.*

Let $\Delta = f(x)$

Since $\Delta = 0$ when $x = a$, $\therefore$ $f(a) = 0$.

*i.e.*, $(x - a)$ is a factor of $f(x)$.

Hence $x - a$ is a factor of $\Delta$.

**Obs.** *If k parallel lines of a determinant $\Delta$ become identical when $x = a$, then $(x - a)^{k-1}$ is a factor of $\Delta$.*

**Example 2.7.** *Factorize* $\Delta = \begin{vmatrix} a^3 & a^2 & a & 1 \\ b^3 & b^2 & b & 1 \\ c^3 & c^2 & c & 1 \\ d^3 & d^2 & d & 1 \end{vmatrix}$.

**Solution:** Putting $a = b$, $R_1 \equiv R_2$ and hence $\Delta = 0$. $\therefore$ $a - b$ is a factor of $\Delta$.

Similarly, $a - c$ and $a - d$ are also factors of $\Delta$.

Again putting $b = c$, $R_2 \equiv R_3$ and hence $\Delta = 0$. $\therefore$ $b - c$ is a factor of $\Delta$.

Similarly $b - d$ and $c - d$ are also factors of $\Delta$.

Also $\Delta$ is of the sixth degree in $a, b, c, d$ and therefore, there cannot be any other algebraic factor of $\Delta$.

$\therefore$ Suppose $\Delta = k(a-b)(a-c)(a-d)(b-c)(b-d)(c-d)$, where $k$ is a numerical constant.

The leading term in $\Delta = a^3b^2c$. The corresponding term on R.H.S. $= ka^3b^2c$.

$\therefore$ $k = 1$.

Hence, $\Delta = (a-b)(a-c)(a-d)(b-c)(b-d)(c-d)$.

**Example 2.8.** *Prove that* $\begin{vmatrix} (b+c)^2 & a^2 & a^2 \\ b^2 & (c+a)^2 & b^2 \\ c^2 & c^2 & (a+b)^2 \end{vmatrix} = 2abc\,(a+b+c)^3.$

**Solution:** Let the given determinant be $\Delta$. If we put $a = 0$,

$$\Delta = \begin{vmatrix} (b+c)^2 & 0 & 0 \\ 0 & c^2 & b^2 \\ c^2 & c^2 & b^2 \end{vmatrix} = 0$$

$\therefore$ $a$ is a factor of $\Delta$. Similarly $b$ and $c$ are its factors.

Again if we put $a + b + c = 0$.

$$\Delta = \begin{vmatrix} (-a)^2 & a^2 & a^2 \\ b^2 & (-b)^2 & b^2 \\ c^2 & c^2 & (-c)^2 \end{vmatrix} = 0$$

In this, three columns being identical, $(a + b + c)^2$ is a factor of $\Delta$.

As $\Delta$ is of the sixth degree and is symmetrical in $a, b, c$ the remaining factor must therefore, be of the first degree and of the form $k\,(a + b + c)$.

Thus $\Delta = kabc\,(a + b + c)^3$

To determine $k$, put $a = b = c = 1$, then

$$\begin{vmatrix} 4 & 1 & 1 \\ 1 & 4 & 1 \\ 1 & 1 & 4 \end{vmatrix} = 27k \quad \text{or} \quad 54 = 27k \quad i.e., k = 2$$

Hence $\Delta = 2abc\,(a + b + c)^3$.

**Otherwise :** Operating $C_1 - C_3$ and $C_2 - C_3$, we have

$$\Delta = \begin{vmatrix} (b+c)^2 - a^2 & 0 & a^2 \\ 0 & (c+a)^2 - b^2 & b^2 \\ c^2 - (a+b)^2 & c^2 - (a+b)^2 & (a+b)^2 \end{vmatrix} \qquad [\text{Take } (a+b+c) \text{ common from } C_1 \text{ and } C_2]$$

$$= (a+b+c)^2 \begin{vmatrix} b+c-a & 0 & a^2 \\ 0 & c+a-b & b^2 \\ c-a-b & c-a-b & (a+b)^2 \end{vmatrix} \qquad [\text{Operate } R_3 - R_1 - R_2]$$

$$= (a+b+c)^2 \begin{vmatrix} b+c-a & 0 & a^2 \\ 0 & c+a-b & b^2 \\ -2b & -2a & 2ab \end{vmatrix} \qquad \left[\text{Operate } C_1 + \frac{1}{a}C_3, C_2 + \frac{1}{b}C_3\right]$$

$$= (a+b+c)^2 \begin{vmatrix} b+c & a^2/b & a^2 \\ b^2/a & c+a & b^2 \\ 0 & 0 & 2ab \end{vmatrix} \qquad [\text{Expand by } R_3]$$

$$= 2ab(a+b+c)^2\,[(b+c)(c+a) - ab] = 2abc\,(a+b+c)^3.$$

**VIII. Multiplication of Determinants.** *The product of two determinants of the same order is itself a determinant of that order.*

Let $\Delta_1 = \begin{vmatrix} a_1 & b_1 & c_1 \\ a_2 & b_2 & c_2 \\ a_3 & b_3 & c_3 \end{vmatrix}$ and $\Delta_2 = \begin{vmatrix} l_1 & m_1 & n_1 \\ l_2 & m_2 & n_2 \\ l_3 & m_3 & n_3 \end{vmatrix}$

then their product is defined as

$$\Delta_1\Delta_2 = \begin{vmatrix} a_1l_1 + b_1m_1 + c_1n_1, & a_1l_2 + b_1m_2 + c_1n_2, & a_1l_3 + b_1m_3 + c_1n_3 \\ a_2l_1 + b_2m_1 + c_2n_1, & a_2l_2 + b_2m_2 + c_2n_2, & a_2l_3 + b_2m_3 + c_2n_3 \\ a_3l_1 + b_3m_1 + c_3n_1, & a_3l_2 + b_3m_2 + c_3n_2, & a_3l_3 + b_3m_3 + c_3n_3 \end{vmatrix}$$

Similarly, the product of two determinants of the $n$th order is a determinant of the $n$th order.

**Example 2.9.** *Evaluate* $\begin{vmatrix} a^2+\lambda^2 & ab+c\lambda & ca-b\lambda \\ ab-c\lambda & b^2+\lambda^2 & bc+a\lambda \\ ca+b\lambda & bc-a\lambda & c^2+\lambda^2 \end{vmatrix} \times \begin{vmatrix} \lambda & c & -b \\ -c & \lambda & a \\ b & -a & \lambda \end{vmatrix}$

**Solution:** By the rule of multiplication of determinants, the resulting determinant

$$\Delta = \begin{vmatrix} d_{11} & d_{12} & d_{13} \\ d_{21} & d_{22} & d_{23} \\ d_{31} & d_{32} & d_{33} \end{vmatrix}$$

where $d_{11} = (a^2+\lambda^2)\,\lambda + (ab+c\lambda)c + (ca-b\lambda)(-b) = \lambda(a^2+b^2+c^2+\lambda^2)$

$d_{12} = (a^2+\lambda^2)(-c) + (ab+c\lambda)\lambda + (ca-b\lambda)a = 0$

$d_{13} = 0,$

$d_{21} = 0,\ d_{22} = \lambda(a^2+b^2+c^2+\lambda^2),\ d_{23} = 0.$

$d_{31} = 0,\ d_{32} = 0,\ d_{33} = \lambda(a^2+b^2+c^2+\lambda^2).$

Hence
$$\Delta = \begin{vmatrix} \lambda(a^2+b^2+c^2+\lambda^2) & 0 & 0 \\ 0 & \lambda(a^2+b^2+c^2+\lambda^2) & 0 \\ 0 & 0 & \lambda(a^2+b^2+c^2+\lambda^2) \end{vmatrix}$$

$$= \lambda^3(a^2+b^2+c^2+\lambda^2)^3.$$

**Example 2.10.** *Show that* $\begin{vmatrix} A_1 & B_1 & C_1 \\ A_2 & B_2 & C_2 \\ A_3 & B_3 & C_3 \end{vmatrix} = \begin{vmatrix} a_1 & b_1 & c_1 \\ a_2 & b_2 & c_2 \\ a_3 & b_3 & c_3 \end{vmatrix}^2$ *where A, B etc. are the co-factors of a, b, etc. respectively in the determinant* $(a_1b_2c_3)$.

**Solution:** Let $\Delta = \begin{vmatrix} a_1 & b_1 & c_1 \\ a_2 & b_2 & c_2 \\ a_3 & b_3 & c_3 \end{vmatrix}$ and $\Delta' = \begin{vmatrix} A_1 & B_1 & C_1 \\ A_2 & B_2 & C_2 \\ A_3 & B_3 & C_3 \end{vmatrix}$

Then
$$\Delta\Delta' = \begin{vmatrix} a_1A_1+b_1B_1+c_1C_1, & a_1A_2+b_1B_2+c_1C_2, & a_1A_3+b_1B_3+c_1C_3 \\ a_2A_1+b_2B_1+c_2C_1, & a_2A_2+b_2B_2+c_2C_2, & a_2A_3+b_2B_3+c_2C_3 \\ a_3A_1+b_3B_1+c_3C_1, & a_3A_2+b_3B_2+c_3C_2, & a_3A_3+b_3B_3+c_3C_3, \end{vmatrix} = \begin{vmatrix} \Delta & 0 & 0 \\ 0 & \Delta & 0 \\ 0 & 0 & \Delta \end{vmatrix} = \Delta^3$$

Hence $\Delta' = \Delta^2$.

**Obs.** $\Delta'$ is called the **reciprocal** *or* **adjugate determinant** *of* $\Delta$.

**Example 2.11.** *Express* $\begin{vmatrix} 2bc-a^2 & c^2 & b^2 \\ c^2 & 2ca-b^2 & a^2 \\ b^2 & a^2 & 2ab-c^2 \end{vmatrix}$ *as the square of a determinant, and hence find its value.*

**Solution:** Given determinant

$$= \begin{vmatrix} a.(-a)+b.c+c.b, & a.(-b)+b.a+c.c, & a.(-c)+b.b+c.a \\ b.(-a)+c.c+a.b, & b.(-b)+c.a+a.c, & b.(-c)+c.b+a.a \\ c.(-a)+a.c+b.b, & c.(-b)+a.a+b.c, & c.(-c)+a.b+b.a \end{vmatrix} = \begin{vmatrix} a & b & c \\ b & c & a \\ c & a & b \end{vmatrix} \times \begin{vmatrix} -a & c & b \\ -b & a & c \\ -c & b & a \end{vmatrix}$$

[Taking out (–1) common from $C_1$ and interchange $C_2$, $C_3$.]

$$= \begin{vmatrix} a & b & c \\ b & c & a \\ c & a & b \end{vmatrix} \times (-1)^2 \begin{vmatrix} a & b & c \\ b & c & a \\ c & a & b \end{vmatrix} = \Delta^2$$

where $\Delta = \begin{vmatrix} a & b & c \\ b & c & a \\ c & a & b \end{vmatrix} = -(a^3+b^3+c^3-3abc)$

Hence the given determinant $= \Delta^2 = (a^3+b^3+c^3-3abc)^2$.

## PROBLEMS 2.1

1. Prove, without expanding, that $\begin{vmatrix} 1 & \alpha & \alpha^2 & -\beta\gamma \\ 1 & \beta & \beta^2 & -\gamma\alpha \\ 1 & \gamma & \gamma^2 & -\alpha\beta \end{vmatrix}$ vanishes. *(W.B.T.U., 2016)*

2. If $\begin{vmatrix} x & x^2 & 1+x^3 \\ y & y^2 & 1+y^3 \\ z & z^2 & 1+z^3 \end{vmatrix} = 0$, then prove, without expansion, that $xyz = -1$ where $x, y, z$ are unequal.

3. Show that (*i*) $\begin{vmatrix} x & l & m & 1 \\ \alpha & x & n & 1 \\ \alpha & \beta & x & 1 \\ \alpha & \beta & \gamma & 1 \end{vmatrix} = (x-\alpha)(x-\beta)(x-\gamma)$.

(*ii*) $\begin{vmatrix} a & b & c \\ b+c & c+a & a+b \\ a^2 & b^2 & c^2 \end{vmatrix} = -(a-b)(b-c)(c-a)(a+b+c)$.

4. If $a, b, c$ are all different and $\begin{vmatrix} a & a^3 & a^4-1 \\ b & b^3 & b^4-1 \\ c & c^3 & c^4-1 \end{vmatrix} = 0$, then show that $abc(bc+ca+ab) = a+b+c$.

5. Evaluate (*i*) $\begin{vmatrix} 1 & 2 & 3 & 4 \\ 1 & 3 & 3 & 4 \\ 1 & 2 & 4 & 4 \\ 1 & 2 & 3 & 5 \end{vmatrix}$ (*ii*) $\begin{vmatrix} 1^2 & 2^2 & 3^2 & 4^2 \\ 2^2 & 3^2 & 4^2 & 5^2 \\ 3^2 & 4^2 & 5^2 & 6^2 \\ 4^2 & 5^2 & 6^2 & 7^2 \end{vmatrix}$

Prove the following results : (6 to 13)

6. $\begin{vmatrix} a+b & b+c & c+a \\ l+m & m+n & n+l \\ p+q & q+r & r+p \end{vmatrix} \div \begin{vmatrix} a & b & c \\ l & m & n \\ p & q & r \end{vmatrix} = 2$

7. $\begin{vmatrix} a-b-c & 2b & 2c \\ 2a & b-c-a & 2c \\ 2a & 2b & c-a-b \end{vmatrix} = (a+b+c)^3$

8. $\begin{vmatrix} 1+a^2-b^2 & 2b & -2b \\ 2ab & 1-a^2+b^2 & 2a \\ 2b & -2a^2 & 1-a^2-b^2 \end{vmatrix}$ is a perfect cube.

9. $\begin{vmatrix} 1 & \cos A & \sin A \\ 1 & \cos B & \sin B \\ 1 & \cos C & \sin C \end{vmatrix} = 4 \sin \dfrac{B-C}{2} \sin \dfrac{C-A}{2} \sin \dfrac{A-B}{2}$.

10. $\begin{vmatrix} 4 & 5 & 6 & x \\ 5 & 6 & 7 & y \\ 6 & 7 & 8 & z \\ x & y & z & 0 \end{vmatrix}$ is a perfect square.

11. $\begin{vmatrix} 0 & a & b & c \\ -a & 0 & d & e \\ -b & -d & 0 & f \\ -c & -e & -f & 0 \end{vmatrix} = (af - be + cd)^2$ *(W.B.T.U., 2016)*

12. $\begin{vmatrix} 1 & a & a^2 & a^3+bcd \\ 1 & b & b^2 & b^3+cda \\ 1 & c & c^2 & c^3+dab \\ 1 & d & d^2 & d^3+abc \end{vmatrix}$ vanishes.

13. $\begin{vmatrix} a^2+\lambda & ab & ac & ad \\ ab & b^2+\lambda & bc & bd \\ ac & bc & c^2+\lambda & cd \\ ad & bd & cd & d^2+\lambda \end{vmatrix} = \lambda^3(a^2+b^2+c^2+d^2+\lambda)$

Factorize each of the following determinants : (14 to 17)

14. $\begin{vmatrix} 1 & 1 & 1 \\ a & b & c \\ a^2 & b^2 & c^2 \end{vmatrix}$

15. $\begin{vmatrix} 1 & 1 & 1 \\ a^2 & b^2 & c^2 \\ a^3 & b^3 & c^3 \end{vmatrix}$.

**16.** $\begin{vmatrix} b^2c^2+a^2 & bc+a & 1 \\ c^2a^2+b^2 & ca+b & 1 \\ a^2b^2+c^2 & ab+c & 1 \end{vmatrix}$

**17.** $\begin{vmatrix} a^2 & b^2 & c^2 & d^2 \\ a & b & c & d \\ 1 & 1 & 1 & 1 \\ bcd & cda & dab & abc \end{vmatrix}$

**18.** If $a+b+c=0$, solve $\begin{vmatrix} a-x & c & b \\ c & b-x & a \\ b & a & c-x \end{vmatrix} = 0$

**19.** Solve the equation $\begin{vmatrix} x+1 & 2x+1 & 3x+1 \\ 2x & 4x+3 & 6x+3 \\ 4x+1 & 6x+4 & 8x+4 \end{vmatrix} = 0.$

**20.** Show that $\begin{vmatrix} b^2+c^2 & a^2 & a^2 \\ b^2 & c^2+a^2 & b^2 \\ c^2 & c^2 & a^2+b^2 \end{vmatrix} = 4a^2b^2c^2.$ (*W.B.T.U., 2012*)

## 2.4 MATRICES

**(1) Definition.** *A system of mn numbers arranged in a rectangular formation along m rows and n columns and bounded by the brackets [ ] is called an m by n* **matrix** ; *which is written as m × n matrix.* A matrix is also denoted by a single capital letter.

Thus
$$A = \begin{bmatrix} a_{11} & a_{12} & ...a_{1j} & ...a_{1n} \\ a_{21} & a_{22} & ...a_{2j} & ...a_{2n} \\ ... & ... & ... & ... \\ a_{i1} & a_{i2} & ...a_{ij} & ...a_{in} \\ ... & ... & ... & ... \\ a_{m1} & a_{m2} & a_{mj} & ...a_{mn} \end{bmatrix}$$

is a matrix of *order mn*. It has *m rows* and *n columns*. Each of the *mn* numbers is called an element of the matrix.

*To locate any particular element of a matrix, the elements are denoted by a letter followed by two suffixes* which respectively specify the rows and columns. Thus $a_{ij}$ is the element in the *i*-th row and *j*-th column of *A*. In this notation, the matrix *A* is denoted by $[a_{ij}]$.

*A matrix should be treated as a single entity with a number of components,* rather than a collection of numbers. For example, the coordinates of a point in solid geometry, are given by a set of three numbers which can be represented by the matrix $[x, y, z]$. Unlike a determinant, a matrix cannot reduce to a single number and the question of finding the value of a matrix never arises. The difference between a determinant and a matrix is brought out by the fact that an interchange of rows and columns does not alter the determinant but gives an entirely different matrix.

**(2) Special matrices**

**Row and column matrices.** *A matrix having a single row is called a row matrix, e.g.,* $[1 \;\; 3 \;\; 5 \;\; 7]$.

*A matrix having a single column is called a column matrix, e.g.,* $\begin{bmatrix} 2 \\ 3 \\ 5 \end{bmatrix}$

Row and column matrices are sometimes called *row vectors* and *column vectors*.

**Square matrix.** *A matrix having n rows and n columns is called a square matrix of order n.*

The determinant having the same elements as the square matrix *A* is called the *determinant of the matrix* and is denoted by the symbol $|A|$. For example, if

$$A = \begin{bmatrix} 1 & 2 & 3 \\ 2 & 3 & 4 \\ 3 & 4 & 5 \end{bmatrix}, \quad \text{then } |A| = \begin{vmatrix} 1 & 2 & 3 \\ 2 & 3 & 4 \\ 3 & 4 & 5 \end{vmatrix}$$

The diagonal of this matrix containing the elements 1, 3, 5 is called the *leading or principal diagonal. The sum of the diagonal elements of a square matrix A is called the* **trace** *of A.*

*A square matrix is said to be* **singular** *if its determinant is zero otherwise* **non-singular.**

**Diagonal matrix.** *A square matrix all of whose elements except those in the leading diagonal, are zero is called a diagonal matrix.*

A diagonal matrix whose all the leading diagonal elements are equal is called a *scalar matrix.* For example,

$$\begin{bmatrix} 3 & 0 & 0 \\ 0 & -2 & 0 \\ 0 & 0 & 6 \end{bmatrix} \text{ and } \begin{bmatrix} 3 & 0 & 0 \\ 0 & 3 & 0 \\ 0 & 0 & 3 \end{bmatrix}$$

are the diagonal and scalar matrices respectively.

**Unit matrix.** *A diagonal matrix of order n which has unity for all its diagonal elements, is called a unit matrix or an identity matrix of order n* and is denoted by $I_n$. For example, unit matrix of order 3 is

$$\begin{bmatrix} 1 & 0 & 0 \\ 0 & 1 & 0 \\ 0 & 0 & 1 \end{bmatrix}$$

**Null matrix.** *If all the elements of a matrix are zero, it is called a null or zero matrix* and is denoted by **'0'** ; *e.g.*,

$$\begin{bmatrix} 0 & 0 & 0 & 0 \\ 0 & 0 & 0 & 0 \end{bmatrix} \text{ is a null matrix.}$$

**Symmetric and skew-symmetric matrices.** *A square matrix $A = [a_{ij}]$ is said to be symmetric when $a_{ij} = a_{ji}$ for all i and j.*

If $a_{ij} = -a_{ji}$ *for all i and j so that all the leading diagonal elements are zero, then the matrix is called a skew-symmetric matrix.* Examples of symmetric and skew-symmetric matrices are

$$\begin{bmatrix} a & h & g \\ h & b & f \\ g & f & c \end{bmatrix} \text{ and } \begin{bmatrix} 0 & h & -g \\ -h & 0 & f \\ g & -f & 0 \end{bmatrix} \text{ respectively.}$$

**Triangular matrix.** A square matrix all of whose elements below the leading diagonal are zero, is called an *upper triangular matrix.* A square matrix all of whose elements above the leading diagonal are zero, is called a *lower triangular matrix.* Thus

$$\begin{bmatrix} a & h & g \\ 0 & b & f \\ 0 & 0 & c \end{bmatrix} \text{ and } \begin{bmatrix} 1 & 0 & 0 \\ 2 & 3 & 0 \\ 1 & -5 & 4 \end{bmatrix}$$

are upper and lower triangular matrices respectively.

**Transposed matrix:** This matrix obtained from any given matrix $A$ be changing rows into columns and columns into rows is called transposed matrix or transpose of $A$ and it is denoted by $A'$ or $A^t$.

Thus if

$$A = \begin{bmatrix} 1 & 2 & 3 \\ 4 & 5 & 6 \\ 7 & 8 & 9 \\ 10 & 11 & 12 \end{bmatrix}; \text{ then } A^t = \begin{bmatrix} 1 & 4 & 7 & 10 \\ 2 & 5 & 8 & 11 \\ 3 & 6 & 9 & 12 \end{bmatrix}$$

It follows from the definition that $(A^t)^t = A$

**Adjoint matrix:** Let $A$ be the square matrix

$$\begin{bmatrix} a_{11} & a_{12} & a_{13} \\ a_{21} & a_{22} & a_{23} \\ a_{31} & a_{33} & a_{33} \end{bmatrix}$$

Now consider the determinant $|A|$. Let $A_{11}, A_{12}, A_{13}, A_{21}, \ldots$ be the cofactors of the elements $a_{11}, a_{12}, a_{13}, a_{21} \ldots$

Then the matrix formed by these cofactors is given by

$$\begin{bmatrix} A_{11} & A_{12} & A_{13} \\ A_{21} & A_{22} & A_{23} \\ A_{31} & A_{32} & A_{33} \end{bmatrix}$$

The transpose of this co-factor matrix is called adjoint of the matrix $A$. It is denoted by *adj. A*.

Thus, $$adj.\ A = \begin{bmatrix} A_{11} & A_{21} & A_{31} \\ A_{12} & A_{22} & A_{32} \\ A_{13} & A_{23} & A_{33} \end{bmatrix}$$

**Equal matrixes:** Two matrices $A$ and $B$ are said to be equal when they are of the same order and the elements in the corresponding positions are equal. In other words the two matrices are identical. We write then $A = B$.

Thus if $$\begin{bmatrix} a & b & c \\ d & e & f \end{bmatrix} = \begin{bmatrix} 4 & 0 & 7 \\ 3 & 1 & 6 \end{bmatrix}$$

then $a = 4, b = 0, c = 7, d = 3, e = 1$ and $f = 6$.

Also note that the matrices are of the same order *i.e.*, $2 \times 3$.

## 2.5 MATRICES OPERATIONS

**(1) Equality of matrices**

*Two matrices A and B are said to equal if and only if*

*(i) they are of the same order*

*and (ii) each element of A is equal to the corresponding element of B.*

**(2) Addition and subtraction of matrices.** *If A, B be two matrices of the same order, then their sum A + B is defined as the matrix each element of which is the sum of the corresponding elements of A and B.*

Thus, $$\begin{bmatrix} a_1 & b_1 \\ a_2 & b_2 \\ a_3 & b_3 \end{bmatrix} + \begin{bmatrix} c_1 & d_1 \\ c_2 & d_2 \\ c_3 & d_3 \end{bmatrix} = \begin{bmatrix} a_1 + c_1 & b_1 + d_1 \\ a_2 + c_2 & b_2 + d_2 \\ a_3 + c_3 & b_3 + d_3 \end{bmatrix}$$

Similarly, $A - B$ *is defined as a matrix whose elements are obtained by subtracting the elements of B from the corresponding elements of A.*

Thus, $$\begin{bmatrix} a_1 & b_1 \\ a_2 & b_2 \end{bmatrix} - \begin{bmatrix} c_1 & d_1 \\ c_2 & d_2 \end{bmatrix} = \begin{bmatrix} a_1 - c_1 & b_1 - d_1 \\ a_2 - c_2 & b_2 - d_2 \end{bmatrix}$$

**Obs.** 1. Only matrices of the same order can be added or subtracted.

2. Addition of matrices is *commutative*,

*i.e.*, $A + B = B + A$.

3. Addition and subtraction of matrices is *associative*.

*i.e.* $(A + B) - C = A + (B - C) = B + (A - C)$.

**(3) Multiplication of matrix by a scalar.** *The product of a matrix A by a scalar k is a matrix whose each element is k times the corresponding elements of A.*

Thus, $$k \begin{bmatrix} a_1 & b_1 & c_1 \\ a_2 & b_2 & c_2 \end{bmatrix} = \begin{bmatrix} ka_1 & kb_1 & kc_1 \\ ka_2 & kb_2 & kc_2 \end{bmatrix}$$

*The distributive law* holds for such products, *i.e.*, $k(A + B) = kA + kB$.

**Obs.** All the laws of ordinary algebra hold for the addition or subtraction of matrices and their multiplication by scalars.

**Example 2.12.** *Find x, y, z and w given that*

$$3\begin{bmatrix} x & y \\ z & w \end{bmatrix} = \begin{bmatrix} x & 5 \\ -1 & 2w \end{bmatrix} + \begin{bmatrix} 6 & x + y \\ z + w & 5 \end{bmatrix}$$

**Solution:** We have $$\begin{bmatrix} 3x & 3y \\ 3z & 3w \end{bmatrix} = \begin{bmatrix} x + 6 & 5 + x + y \\ -1 + z + w & 2w + 5 \end{bmatrix}$$

Equating the corresponding elements, we get

$$3x = x + 6,\ 3y = 5 + x + y,\ 3z = -1 + z + w,\ 3w = 2w + 5.$$

or $$2x = 6,\ 2y = 5 + x,\ 2z = w - 1,\ w = 5$$

Hence $x = 3, y = 4, z = 2, w = 5$.

**Example 2.13.** *Express* $\begin{bmatrix} 3 & 5 & -7 \\ -8 & 11 & 4 \\ 13 & -14 & 6 \end{bmatrix}$ *as the sum of a lower triangular matrix with zero leading diagonal and an upper triangular matrix.*

**Solution:** Let $L = \begin{bmatrix} 0 & 0 & 0 \\ a & 0 & 0 \\ b & c & 0 \end{bmatrix}$ be the lower triangular matrix with zero leading diagonal.

and $U = \begin{bmatrix} l & m & n \\ 0 & p & q \\ 0 & 0 & r \end{bmatrix}$ be the upper triangular matrix.

Then $$\begin{bmatrix} 3 & 5 & -7 \\ -8 & 11 & 4 \\ 13 & -14 & 6 \end{bmatrix} = \begin{bmatrix} 0 & 0 & 0 \\ a & 0 & 0 \\ b & c & 0 \end{bmatrix} + \begin{bmatrix} l & m & n \\ 0 & p & q \\ 0 & 0 & r \end{bmatrix}$$

Equating corresponding elements from both sides, we obtain $3 = l$, $5 = m$, $-7 = n$, $-8 = a$, $11 = p$, $4 = q$, $13 = b$, $-14 = c$, $6 = r$.

Hence $$L = \begin{bmatrix} 0 & 0 & 0 \\ -8 & 0 & 0 \\ 13 & -14 & 0 \end{bmatrix} \text{ and } U = \begin{bmatrix} 3 & 5 & -7 \\ 0 & 11 & 4 \\ 0 & 0 & 6 \end{bmatrix}$$

**(4) Multiplication of matrices.** *Two matrices can be multiplied only when the number of columns in the first is equal to the number of rows in the second. Such matrices are said to be* **conformable.**

For instance, the product $\begin{bmatrix} a_1 & b_1 & c_1 \\ a_2 & b_2 & c_2 \\ a_3 & b_3 & c_3 \\ a_4 & b_4 & c_4 \end{bmatrix} \times \begin{bmatrix} l_1 & l_2 \\ m_1 & m_2 \\ n_1 & n_2 \end{bmatrix}$

*is defined as the matrix* $\begin{bmatrix} a_1l_1 + b_1m_1 + c_1n_1 & a_1l_2 + b_1m_2 + c_1n_2 \\ a_2l_1 + b_2m_1 + c_2n_1 & a_2l_2 + b_2m_2 + c_2n_2 \\ a_3l_1 + b_3m_1 + c_3n_1 & a_3l_2 + b_3m_2 + c_3n_2 \\ a_4l_1 + b_4m_1 + c_4n_1 & a_4l_2 + b_4m_2 + c_4n_2 \end{bmatrix}$

In general, if $A = \begin{bmatrix} a_{11} & a_{12} & \cdots & a_{1n} \\ a_{21} & a_{22} & \cdots & a_{2n} \\ \cdots & \cdots & \cdots & \cdots \\ a_{m1} & a_{m2} & \cdots & a_{mn} \end{bmatrix}$ and $B = \begin{bmatrix} b_{11} & b_{12} & \cdots & b_{1p} \\ b_{21} & b_{22} & \cdots & b_{2p} \\ \cdots & \cdots & \cdots & \cdots \\ b_{n1} & b_{n2} & \cdots & b_{np} \end{bmatrix}$

be two $m \times n$ and $n \times p$ conformable matrices, then their product is defined as the $m \times p$ matrix

$$AB = \begin{bmatrix} c_{11} & c_{12} & \cdots & c_{1p} \\ c_{21} & c_{22} & \cdots & c_{2p} \\ \cdots & \cdots & \cdots & \cdots \\ c_{m1} & c_{m2} & \cdots & c_{mp} \end{bmatrix}$$

where $c_{ij} = a_{i1} b_{1j} + a_{i2}b_{2j} + a_{i3}b_{3j} + \ldots + a_{in}b_{nj}$, *i.e.*, the element in the $i$th row and the $j$th column of the matrix $AB$ is obtained by weaving the $i$th row of $A$ with $j$th column of $B$. The expression for $c_{ij}$ is known as the *inner product* of the $i$th row with the $j$th column.

*Post-multiplication and Pre-multiplication*. In the product $AB$, the matrix $A$ is said to be *post-multiplied* by the matrix $B$. Whereas in $BA$, the matrix $A$ is said to be *pre-multiplied* by $B$. In one case the product may exist and in the other case it may not. Also the product in both cases may exist yet may or may not be equal.

**Obs. 1.** *Multiplication of matrices is associative. i.e.,* $(AB)C = A(BC)$
provided $A$, $B$ are conformable for the product $AB$ and $B$, $C$ are conformable for the product $BC$. (Ex. 2.16).

**Obs. 2.** *Multiplication of matrices is distributive. i.e.,* $A(B + C) = AB + AC$.
provided $A$, $B$ are conformable for the product $AB$ and $A$, $C$ are conformable for the product $AC$.

**Obs. 3. Power of a matrix.** If $A$ be a square matrix, then the product $AA$ is defined as $A^2$. Similarly, we define higher powers of $A$. *i.e.*, $A \cdot A^2 = A^3$, $A^2 \cdot A^2 = A^4$ etc.

*If* $A^2 = A$, *then the matrix A is called idempotent.*

**Example 2.14.** *If* $A = \begin{bmatrix} 0 & 1 & 2 \\ 1 & 2 & 3 \\ 2 & 3 & 4 \end{bmatrix}$ *and* $B = \begin{bmatrix} 1 & -2 \\ -1 & 0 \\ 2 & -1 \end{bmatrix}$, *form the product of AB. Is BA defined?*

**Solution:** Since the number of columns of $A$ = the number of rows of $B$ (each being = 3).

$\therefore$ The product $AB$ is defined and

$$= \begin{bmatrix} 0.1+1.-1+2.2, & 0.-2+1.0+2.-1 \\ 1.1+2.-1+3.2, & 1.-2+2.0+3.-1 \\ 2.1+3.-1+4.2, & 2.-2+3.0+4.-1 \end{bmatrix} = \begin{bmatrix} 3 & -2 \\ 5 & -5 \\ 7 & -8 \end{bmatrix}$$

Again since the number of columns of $B \neq$ the number of rows of $A$.

$\therefore$ The product $BA$ is not possible.

**Example 2.15.** *If* $A = \begin{bmatrix} 3 & 2 & 2 \\ 1 & 3 & 1 \\ 5 & 3 & 4 \end{bmatrix}$, *find the matrix B such that* $AB = \begin{bmatrix} 3 & 4 & 2 \\ 1 & 6 & 1 \\ 5 & 6 & 4 \end{bmatrix}$. (*Mumbai, 2005*)

**Solution:** Let $AB = \begin{bmatrix} 3 & 2 & 2 \\ 1 & 3 & 1 \\ 5 & 3 & 4 \end{bmatrix} \begin{bmatrix} l & m & n \\ p & q & r \\ u & v & w \end{bmatrix}$

$$= \begin{bmatrix} 3l+2p+2u & 3m+2q+2v & 3n+2r+2w \\ l+3p+u & m+3q+v & n+3r+w \\ 5l+3p+4u & 5m+3q+4v & 5n+3r+4w \end{bmatrix} = \begin{bmatrix} 3 & 4 & 2 \\ 1 & 6 & 1 \\ 5 & 6 & 4 \end{bmatrix} \quad \text{(given)}$$

Equating corresponding elements, we get

$$3l + 2p + 2u = 3, \quad l + 3p + u = 1, \quad 5l + 3p + 4u = 5 \qquad ...(i)$$

$$3m + 2q + 2v = 4, \quad m + 3q + v = 6, \quad 5m + 3q + 4v = 6 \qquad ...(ii)$$

$$3n + 2r + 2w = 2, \quad n + 3r + w = 1, \quad 5n + 3r + 4w = 4 \qquad ...(iii)$$

Solving Eqn. (*i*), we get $l = 1, p = 0, u = 0$

Similarly Eqn. (*ii*) give $m = 0, q = 2, v = 0$

and Eqn. (*iii*) give $n = 0, r = 0, w = 1$

Thus, $B = \begin{bmatrix} 1 & 0 & 0 \\ 0 & 2 & 0 \\ 0 & 0 & 1 \end{bmatrix}$.

**Example 2.16.** *Prove that* $A^3 - 4A^2 - 3A + 11I = 0$, *where* $A = \begin{bmatrix} 1 & 3 & 2 \\ 2 & 0 & -1 \\ 1 & 2 & 3 \end{bmatrix}$.

**Solution:** $A^2 = A \times A = \begin{bmatrix} 1 & 3 & 2 \\ 2 & 0 & -1 \\ 1 & 2 & 3 \end{bmatrix} \begin{bmatrix} 1 & 3 & 2 \\ 2 & 0 & -1 \\ 1 & 2 & 3 \end{bmatrix} = \begin{bmatrix} 1+6+2 & 3+0+4 & 2-3+6 \\ 2+0-1 & 6+0-2 & 4+0-3 \\ 1+4+3 & 3+0+6 & 2-2+9 \end{bmatrix} = \begin{bmatrix} 9 & 7 & 5 \\ 1 & 4 & 1 \\ 8 & 9 & 9 \end{bmatrix}$

$$A^3 = A^2 \times A = \begin{bmatrix} 9 & 7 & 5 \\ 1 & 4 & 1 \\ 8 & 9 & 9 \end{bmatrix} \times \begin{bmatrix} 1 & 3 & 2 \\ 2 & 0 & -1 \\ 1 & 2 & 3 \end{bmatrix} = \begin{bmatrix} 9+14+5 & 27+0+10 & 18-7+15 \\ 1+8+1 & 3+0+2 & 2-4+3 \\ 8+18+9 & 24+0+18 & 16-9+27 \end{bmatrix} = \begin{bmatrix} 28 & 37 & 26 \\ 10 & 5 & 1 \\ 35 & 42 & 34 \end{bmatrix}$$

$A^3 - 4A^2 - 3A + 11I$

$$= \begin{bmatrix} 28 & 37 & 26 \\ 10 & 5 & 1 \\ 35 & 42 & 34 \end{bmatrix} - 4\begin{bmatrix} 9 & 7 & 5 \\ 1 & 4 & 1 \\ 8 & 9 & 9 \end{bmatrix} - 3\begin{bmatrix} 1 & 3 & 2 \\ 2 & 0 & -1 \\ 1 & 2 & 3 \end{bmatrix} + 11\begin{bmatrix} 1 & 0 & 0 \\ 0 & 1 & 0 \\ 0 & 0 & 1 \end{bmatrix}$$

$$= \begin{bmatrix} 28-36-3+11 & 37-28-9+0 & 26-20-6+0 \\ 10-4-6-0 & 5-16+0+11 & 1-4+3+0 \\ 35-32-3+0 & 42-36-6+0 & 34-36-9+11 \end{bmatrix} = \begin{bmatrix} 0 & 0 & 0 \\ 0 & 0 & 0 \\ 0 & 0 & 0 \end{bmatrix} = 0.$$

**Example 2.17.** *By mathematical induction, prove that if*

$$A = \begin{bmatrix} 11 & -25 \\ 4 & -9 \end{bmatrix}, \text{ then } A^n = \begin{bmatrix} 1+10n & -25n \\ 4n & 1-10n \end{bmatrix}$$

**Solution.** When $n = 1$, $A^n$ gives $A^1 = \begin{bmatrix} 11 & -25 \\ 4 & -9 \end{bmatrix}$ ...(*i*)

Let us assume that the result is true for any positive integer $k$, so that

$$A^k = \begin{bmatrix} 1+10k & -25k \\ 4k & 1-10k \end{bmatrix}$$

$$\therefore \quad A^{k+1} = A^k \cdot A^1 = \begin{bmatrix} 1+10k & -25k \\ 4k & 1-10k \end{bmatrix}\begin{bmatrix} 11 & -25 \\ 4 & -9 \end{bmatrix}$$

$$= \begin{bmatrix} 11(1+10k)-100k & -25(1+10k)+225k \\ 44k+4(1-10k) & -100k-9(1-10k) \end{bmatrix} = \begin{bmatrix} 1+10(k+1) & -25(k+1) \\ 4(k+1) & 1-10(k+1) \end{bmatrix}$$

This is true for $n = k + 1$ ...(*ii*)

We have seen in Eqn. (*i*) that the result is true for $n = 1$.

$\therefore$ It is true for $n = 1 + 1 = 2$ [by (*ii*)]

Similarly, it is true for $n = 2 + 1 = 3$ and so on.

Hence by mathematical induction, the result is true for all positive integers $n$.

**Example 2.18.** *Prove that (AB)C = A(BC), where A, B, C are matrices conformable for the products.*

**Solution.** Let $A = [a_{ij}]$ be of order $m \times n$, $B = [b_{ij}]$ be of order $n \times p$ and $C = [c_{ij}]$ be of order of $p \times q$.

Then $\quad AB = [a_{ik}]\,[b_{kj}] = \sum_{k=1}^{n} a_{ik}\, b_{kj}$

$$\therefore \quad (AB)C = \left[\sum_{k=1}^{n} a_{ik}\, b_{kl}\right] \cdot [c_{lj}] = \left[\sum_{l=1}^{p}\left(\sum_{k=1}^{n} a_{ik}\, b_{kl}\right) c_{lj}\right] = \left[\sum_{k=1}^{n}\sum_{l=1}^{p} a_{ik}\, b_{kl}\, c_{lj}\right]$$

Similarly, $\quad BC = [b_{kl}] \cdot [c_{lj}] = \sum_{l=1}^{p} b_{kl}\, c_{lj}$

$$\therefore \quad A(BC) = [a_{ik}]\left[\sum_{l=1}^{p} b_{kl}\, c_{lj}\right] = \left[\sum_{k=1}^{n} a_{ik}\left(\sum_{l=1}^{p} b_{kl}\, c_{lj}\right)\right] = \left[\sum_{k=1}^{n}\left(\sum_{l=1}^{p} a_{ik}\, b_{kl}\, c_{lj}\right)\right]$$

Hence $\quad (AB)C = A(BC)$.

## PROBLEMS 2.2

1. For what values of $x$, the matrix $\begin{bmatrix} 3-x & 2 & 2 \\ 2 & 4-x & 1 \\ -2 & -4 & -1-x \end{bmatrix}$ is singular ?
2. Find the values of $x, y, z$ and $a$ which satisfy the matrix equation $\begin{bmatrix} x+3 & 2y+x \\ z-1 & 4a-6 \end{bmatrix} = \begin{bmatrix} 0 & -7 \\ 3 & 2a \end{bmatrix}$.
3. Matrix $A$ has $x$ rows and $x + 5$ columns. Matrix $B$ has $y$ rows and $11 - y$ columns. Both $AB$ and $BA$ exist. Find $x$ and $y$.
4. If $A + B = \begin{bmatrix} 1 & -1 \\ 3 & 0 \end{bmatrix}$ and $A - B = \begin{bmatrix} 3 & 1 \\ 1 & 4 \end{bmatrix}$, calculate the product $AB$.
5. If $A = \begin{bmatrix} 1 & 2 & 3 & 4 \\ 2 & 0 & 1 & 2 \\ 3 & 1 & 0 & 5 \end{bmatrix}$ and $B = \begin{bmatrix} 2 & 1 & 0 \\ 3 & 2 & 1 \\ 1 & 0 & 1 \end{bmatrix}$, find $AB$ or $BA$, whichever exists.
6. If $A = \begin{bmatrix} 1 & 2 \\ -2 & 3 \end{bmatrix}$, $B = \begin{bmatrix} 2 & 1 \\ 2 & 3 \end{bmatrix}$ and $C = \begin{bmatrix} -3 & 1 \\ 2 & 0 \end{bmatrix}$, verify that $(AB)C = A(BC)$ and $A(B + C) = AB + AC$.
7. Evaluate (*i*) $[x, y, z]\begin{bmatrix} a & h & g \\ h & b & f \\ g & f & c \end{bmatrix}\begin{bmatrix} x \\ y \\ z \end{bmatrix}$ ; (*ii*) $\begin{bmatrix} 2 & 1 & -1 \\ 4 & -5 & 6 \\ -3 & 7 & 3 \end{bmatrix} \times \begin{bmatrix} 3 & 1 \\ -6 & 4 \\ -2 & 5 \end{bmatrix} \times \begin{bmatrix} 5 & 3 \\ -2 & 1 \end{bmatrix}$ ; (*iii*) $\begin{bmatrix} 1 \\ -2 \\ 3 \end{bmatrix} \times [4 \;\; 5 \;\; 2] \times \begin{bmatrix} 2 \\ -3 \\ 5 \end{bmatrix} \times [3 \;\; 2]$

8. Prove that the product of two matrices

$$\begin{bmatrix} \cos^2 \theta & \cos \theta \sin \theta \\ \cos \theta \sin \theta & \sin^2 \theta \end{bmatrix} \text{ and } \begin{bmatrix} \cos^2 \phi & \cos \phi \sin \phi \\ \cos \phi \sin \phi & \sin^2 \phi \end{bmatrix}$$

is a null matrix when $\theta$ and $\phi$ differ by an odd multiple of $\pi/2$.

9. If $A = \begin{bmatrix} 0 & -\tan \alpha/2 \\ \tan \alpha/2 & 0 \end{bmatrix}$, show that $I + A = (I - A) \begin{bmatrix} \cos \alpha & -\sin \alpha \\ \sin \alpha & \cos \alpha \end{bmatrix}$.

10. If $A = \begin{bmatrix} 3 & 2 \\ 2 & 3 \end{bmatrix}$, find the value of $A^2 - 6A + 8I$, where $I$ is a unit matrix of second order.

11. If $A = \begin{bmatrix} 1 & -2 & 3 \\ 2 & 3 & -1 \\ -3 & 1 & 2 \end{bmatrix}$, and $I$ is the unit matrix of order 3, evaluate $A^2 - 3A + 9I$.

12. If $A = \begin{bmatrix} 1 & 2 & -1 \\ 2 & 0 & 3 \\ 0 & 1 & 2 \end{bmatrix}, B = \begin{bmatrix} 3 & -1 & 1 \\ 0 & 0 & 2 \\ 4 & -3 & 2 \end{bmatrix}$, verify the result $(A + B)^2 = A^2 + BA + AB + B^2$.

13. If $E = \begin{bmatrix} 0 & 1 & 0 \\ 0 & 0 & 1 \\ 0 & 0 & 0 \end{bmatrix}$ and $F = \begin{bmatrix} 0 & 0 & 0 \\ 1 & 0 & 0 \\ 0 & 1 & 0 \end{bmatrix}$,

calculate the products $EF$ and $FE$ and show that $E^2F + FE^2 = E$.

14. If $A = \begin{bmatrix} \cos \alpha & \sin \alpha \\ -\sin \alpha & \cos \alpha \end{bmatrix}$, show that $A^n = \begin{bmatrix} \cos n\alpha & \sin n\alpha \\ -\sin n\alpha & \cos n\alpha \end{bmatrix}$, when $n$ is a positive integer.

15. Factorize the matrix $A = \begin{bmatrix} 5 & -2 & 1 \\ 7 & 1 & -5 \\ 3 & 7 & 4 \end{bmatrix}$ into the form $LU$, where $L$ is lower triangular and $U$ is upper triangular matrix.

## 2.6 RELATED MATRICES

**(1) Transpose of a matrix.** *The matrix obtained from any given matrix A, by interchanging rows and columns is called the* **transpose** *of A and is denoted by A′.*

Thus the transposed matrix of $A = \begin{bmatrix} 1 & 2 \\ 4 & 5 \\ 7 & 8 \end{bmatrix}$ is $A' = \begin{bmatrix} 1 & 4 & 7 \\ 2 & 5 & 8 \end{bmatrix}$

Clearly, the transpose of an $m \times n$ matrix is an $n \times m$ matrix. Also the transpose of the transpose of a matrix coincides with itself, *i.e.*, $(A')' = A$.

*For a symmetric matrix, $A' = A$ and for a skew-symmetric matrix, $A' = -A$.*

**Obs. 1.** *The transpose of the product of the two matrices is the product of their transposes taken in the reverse order i.e.,* $(\mathbf{AB})' = \mathbf{B'A'}$.

For, the element in the $i$th row and $j$th col. of $(AB)'$

= element in the $j$th row and $i$th col. of $AB$ = inner product of $j$th row of $A$ with $i$th col. of $B$

= inner product of $j$th col. of $A'$ with $i$th row of $B'$ = element in the $i$th row and $j$th col. of $B'A'$

Hence $(AB)' = B'A'$.

**Obs. 2.** *Every square matrix can be uniquely expressed as a sum of a symmetric and a skew-symmetric matrix.*

Let $A$ be the given square matrix, then $A = \frac{1}{2}(A + A') + \frac{1}{2}(A - A')$.

Let $B = \frac{1}{2}(A + A')$ and $C = \frac{1}{2}(A - A')$

$\therefore$ $B' = \left[\frac{1}{2}(A + A')\right]' = \frac{1}{2}[A' + (A')'] = \frac{1}{2}(A' + A) = B$, *i.e.*, $B = \frac{1}{2}(A + A')$ is a symmetric matrix.

Again, $C' = \left[\frac{1}{2}(A - A')\right]' = \frac{1}{2}[A' - (A')'] = \frac{1}{2}(A' - A) = -C$, *i.e.*, $C = \frac{1}{2}(A - A')$ is a skew-symmetric matrix.

Hence $A$ can be expressed as the sum of a symmetric and a skew-symmetric matrix.

To prove the uniqueness, assume that $P$ is a symmetric matrix and $Q$ is a skew-symmetric matrix such that $A = P + Q$.

Then $A' = (P + Q)' = P' + Q' = P - Q$

Thus, $P = \frac{1}{2}(A + A')$ and $Q = \frac{1}{2}(A - A')$

which shows that there is one and only one way of expressing $A$ as the sum of a symmetric and skew-symmetric matrix.

**Example 2.19.** *Express the matrix A as the sum of a symmetric and a skew-symmetric matrix where*

$$A = \begin{bmatrix} 4 & 2 & -3 \\ 1 & 3 & -6 \\ -5 & 0 & -7 \end{bmatrix}$$

**Solution:** We have $A' = \begin{bmatrix} 4 & 1 & -5 \\ 2 & 3 & 0 \\ -3 & -6 & -7 \end{bmatrix}$

Then $A + A' = \begin{bmatrix} 8 & 3 & -8 \\ 3 & 6 & -6 \\ -8 & -6 & -14 \end{bmatrix}$ and $A - A' = \begin{bmatrix} 0 & 1 & 2 \\ -1 & 0 & -6 \\ -2 & 6 & 0 \end{bmatrix}$

$$\therefore \quad A = \frac{1}{2}(A + A') + \frac{1}{2}(A - A') = \begin{bmatrix} 4 & 1.5 & -4 \\ 1.5 & 3 & -3 \\ -4 & -3 & -7 \end{bmatrix} + \begin{bmatrix} 0 & 0.5 & 1 \\ -0.5 & 0 & -3 \\ -1 & 3 & 0 \end{bmatrix}.$$

(2) **Adjoint of a square matrix.** The determinant of the square matrix

$$A = \begin{bmatrix} a_1 & b_1 & c_1 \\ a_2 & b_2 & c_2 \\ a_3 & b_3 & c_3 \end{bmatrix} \text{ is } \Delta = \begin{vmatrix} a_1 & b_1 & c_1 \\ a_2 & b_2 & c_2 \\ a_3 & b_3 & c_3 \end{vmatrix}$$

The matrix formed by the cofactors of the elements in $\Delta$ is

$\begin{bmatrix} A_1 & B_1 & C_1 \\ A_2 & B_2 & C_2 \\ A_3 & B_3 & C_3 \end{bmatrix}$. Then the transpose of this matrix, *i.e.*, $\begin{bmatrix} A_1 & A_2 & A_3 \\ B_1 & B_2 & B_3 \\ C_1 & C_2 & C_3 \end{bmatrix}$

is called the *adjoint of the matrix A* and is written as *Adj. A*.

*Thus the* **adjoint** *of A is the transposed matrix of cofactors of A.*

(3) **Inverse of a matrix.** *If A be any matrix, then a matrix B if it exists, such that AB = BA = I, is called the* **Inverse** *of A* which is denoted by $A^{-1}$ so that $AA^{-1} = I$.

Also $$A^{-1} = \frac{Adj.A}{|A|}$$

For $$A\,(Adj\,A) = \begin{bmatrix} a_1 & b_1 & c_1 \\ a_2 & b_2 & c_2 \\ a_3 & b_3 & c_3 \end{bmatrix}\begin{bmatrix} A_1 & A_2 & A_3 \\ B_1 & B_2 & B_3 \\ C_1 & C_2 & C_3 \end{bmatrix} = \begin{bmatrix} |A| & 0 & 0 \\ 0 & |A| & 0 \\ 0 & 0 & |A| \end{bmatrix} = |A| \begin{bmatrix} 1 & 0 & 0 \\ 0 & 1 & 0 \\ 0 & 0 & 1 \end{bmatrix}$$

or $A \cdot \frac{Adj.A}{|A|} = I \quad [\because \; |A| \neq 0]$ or $\frac{Adj.A}{|A|}$ is the inverse of $A$.

**Obs. 1.** *Inverse of a matrix, is unique.*

If possible, let the two inverses of the matrix $A$ be $B$ and $C$,

then $AB = BA = I$ and $AC = CA = I$

$\therefore$ $CAB = (CA)B = IB = B$ and $CAB = C(AB) = CI = C$

Thus, $B = C$.

**Obs. 2.** *The reciprocal of the product of two matrices is the product of their reciprocals taken in the reverse order i.e.,* $(\mathbf{AB})^{-1} = \mathbf{B}^{-1}\mathbf{A}^{-1}$

If $A$, $B$ be two matrices, then the reciprocal of their product is $(AB)^{-1}$.

Clearly, $(AB) . (B^{-1} A^{-1}) = A(BB^{-1}) A^{-1}$ [by Associative law]

$= AIA^{-1} = AA^{-1} = I.$

Similarly, $(B^{-1}A^{-1}) . (AB) = I$

Hence $B^{-1} A^{-1}$ is the reciprocal of $AB$.

**Obs. 3.** Multiplication by an inverse matrix plays the same role in matrix algebra that division plays in ordinary algebra.

*i.e.*, if $[A][B] = [C][D]$, then $[A]^{-1} [A][B] = [A^{-1}][C][D]$

or $B = A^{-1}[C][D]$, *i.e.*, $\dfrac{[C][D]}{[A]} = A^{-1} [C][D]$.

**Note.** For other methods see Examples 2.25 ; 2.28 and 2.46.

**Example 2.20.** *Find the matrix A if* $\begin{bmatrix} 2 & 1 \\ 3 & 2 \end{bmatrix} A \begin{bmatrix} -3 & 2 \\ 5 & -3 \end{bmatrix} = \begin{bmatrix} -2 & 4 \\ 3 & -1 \end{bmatrix}$

**Solution:** If $\begin{bmatrix} 2 & 1 \\ 3 & 2 \end{bmatrix} = B, \begin{bmatrix} -3 & 2 \\ 5 & -3 \end{bmatrix} = C$ and $\begin{bmatrix} -2 & 4 \\ 3 & -1 \end{bmatrix} = D$, then

$$BAC = D \quad \text{or} \quad AC = B^{-1}D$$

$\therefore$
$$A = B^{-1} DC^{-1}$$

Now,
$$B^{-1} = \frac{adj\ B}{|B|} = \begin{bmatrix} 2 & -1 \\ -3 & 2 \end{bmatrix}$$

Similarly,
$$C^{-1} = \begin{bmatrix} 3 & 2 \\ 5 & 3 \end{bmatrix}$$

Hence,
$$A = \begin{bmatrix} 2 & -1 \\ -3 & 2 \end{bmatrix}\begin{bmatrix} -2 & 4 \\ 3 & -1 \end{bmatrix}\begin{bmatrix} 3 & 2 \\ 5 & 3 \end{bmatrix} = \begin{bmatrix} 2 & -1 \\ -3 & 2 \end{bmatrix}\begin{bmatrix} 14 & 8 \\ 4 & 3 \end{bmatrix} = \begin{bmatrix} 24 & 13 \\ -34 & -18 \end{bmatrix}.$$

**Example 2.21 :** *Show that* $\begin{vmatrix} (b+c)^2 & a^2 & a^2 \\ b^2 & (c+a)^2 & b^2 \\ c^2 & c^2 & (a+b)^2 \end{vmatrix} = 2abc(a+b+c)^3$ (GTU, 2019)

**Solution:** Applying $C_1 - C_3$ and $C_2 - C_3$ and denoting the given determinant by $\Delta$. we have

$$\Delta = \begin{vmatrix} (b+c)^2 - a^2 & 0 & a^2 \\ 0 & c+a-b & b^2 \\ c-a-b & c-a-b & (a+b)^2 \end{vmatrix}$$

on taking $(a + b + c)$ common from $C_1$ and $C_2$

Next applying $R_3 - R_2 - R_1$, we have

$$\Delta = (a+b+c)^2 \begin{vmatrix} b+c-a & 0 & a^2 \\ 0 & c+a-b & b^2 \\ -2b & -2a & 2ab \end{vmatrix}$$

finally, applying $\left(c + \dfrac{1}{a} c_3 \text{ and } c_2 + \dfrac{1}{b} c_3\right)$. We have

$$\Delta = (a+b+c)^2 \begin{vmatrix} b+c & a^2/b & a^2 \\ \dfrac{b^2}{a} & c+a & b^2 \\ 0 & 0 & 2ab \end{vmatrix}$$

$$\Delta = (a+b+c)^2 \cdot 2ab \begin{vmatrix} b+c & a^2/b \\ b^2/a & c+a \end{vmatrix}$$

$$= 2ab\,(a+b+c)^2\,[(b+c)(c+a) - ab]$$

$$= 2abc\,(a+b+c)^3$$

**Example 2.22.** *Show that*

$$\begin{vmatrix} 1+a & 1 & 1 & 1 \\ 1 & 1+b & 1 & 1 \\ 1 & 1 & 1+c & 1 \\ 1 & 1 & 1 & 1+d \end{vmatrix} = abcd\left(1+\frac{1}{a}+\frac{1}{b}+\frac{1}{c}+\frac{1}{d}\right). \qquad \text{(B.P.T.U. 2017)}$$

**Solution:** Denoting the given determinant by $\Delta$ and taking $a, b, c, d$ common from $R_1, R_2, R_3, R_4$ respectively. We have.

$$\Delta = abcd \begin{vmatrix} a^{-1}+1 & a^{-1} & a^{-1} & a^{-1} \\ b^{-1} & b^{-1}+1 & +b^{-1} & b^{-1} \\ c^{-1} & c^{-1} & c^{-1}+1 & c^{-1} \\ d^{-1} & d^{-1} & d^{-1} & d^{-1}+1 \end{vmatrix}$$

Now applying $R_1 + (R_2 + R_3 + R_4)$ and taking out the common factor from $R_1$, we have

$$\Delta = abcd\,(1 + a^{-1} + b^{-1} + c^{-1} + d^{-1}) \times \begin{vmatrix} 1 & 1 & 1 & 1 \\ b^{-1} & b^{-1}+1 & b^{-1} & b^{-1} \\ c^{-1} & c^{-1} & c^{-1}+1 & c^{-1} \\ d^{-1} & d^{-1} & d^{-1} & d^{-1}+1 \end{vmatrix}$$

Next, applying $C_2 - C_1$, $C_3 - C_1$ and $C_4 - C_1$, we have

$$\Delta = abcd\,(1 + a^{-1} + b^{-1} + c^{-1} + d^{-1}) \begin{vmatrix} 1 & 0 & 0 & 0 \\ b^{-1} & 1 & 0 & 0 \\ c^{-1} & 0 & 1 & 0 \\ d^{-1} & 0 & 0 & 1 \end{vmatrix}$$

$$= abcd\left(1+\frac{1}{a}+\frac{1}{b}+\frac{1}{c}+\frac{1}{d}\right)$$

**Example 2.23.** $\Delta = \begin{vmatrix} (a-x)^2 & (a-y)^2 & (a-z)^2 \\ (b-x)^2 & (b-y)^2 & (b-z)^2 \\ (c-x)^2 & (c-y)^2 & (c-z)^2 \end{vmatrix}$ *as product of two determinants, and find its value.*

**Solution:** Since $(x-y)$, $(y-z)$, $(y-x)$ and also $(a-b)$, $(b-c)$, $(c-a)$ are factors of $\Delta$ by factor theorem, we have by trial and adjustment.

$$\Delta = \begin{vmatrix} x^2 & -2x & 1 \\ y^2 & -2y & 1 \\ z^2 & -2z & 1 \end{vmatrix} \times \begin{vmatrix} 1 & a & a^2 \\ 1 & b & b^2 \\ 1 & c & c^2 \end{vmatrix}$$

$$= 2\begin{vmatrix} 1 & 1 & 1 \\ x & y & z \\ x^2 & y^2 & z^2 \end{vmatrix} \times \begin{vmatrix} 1 & 1 & 1 \\ a & b & c \\ a^2 & b^2 & c^2 \end{vmatrix}$$

$$= 2(x-y)(y-z)(z-x) \times (a-b)(b-c)(c-a)$$

## PROBLEMS 2.3

1. If $A = \begin{bmatrix} \cos\alpha & \sin\alpha \\ -\sin\alpha & \cos\alpha \end{bmatrix}$, verify that $AA' = I = A'A$, where $I$ is the unit matrix.

2. Express each of the following matrices as the sum of a symmetric and a skew-symmetric matrix :

(i) $\begin{bmatrix} 3 & -2 & 6 \\ 2 & 7 & -1 \\ 5 & 4 & 0 \end{bmatrix}$ (ii) $\begin{bmatrix} a & a & b \\ c & b & b \\ c & a & c \end{bmatrix}$

3. Show that

$$\begin{vmatrix} 3 & a+b+c & a^2+b^2+c^2 \\ a+b+c & a^2+b^2+c^2 & a^3+b^3+c^3 \\ a^2+b^2+c^2 & a^3+b^3+c^3 & a^4+b^4+c^4 \end{vmatrix} = \{(a-b)(b-c)(c-a)\}^2$$ (U.P.T.U. 2018)

4. Show that

$$\begin{vmatrix} a^2+\lambda & ab & ac & ad \\ ab & b^2+\lambda & bc & bd \\ ac & bc & c^2+\lambda & cd \\ ad & bd & cd & d^2+\lambda \end{vmatrix} = \lambda^3(a^2+b^2+c^2+d^2+\lambda).$$ (Mumbai, 2018)

5. If $A = \begin{bmatrix} 3 & 2 & 2 \\ 1 & 3 & 1 \\ 5 & 3 & 4 \end{bmatrix}$, compute *adj* $A$ and $A^{-1}$. Also find $B$ such that $AB = \begin{bmatrix} 3 & 4 & 2 \\ 1 & 6 & 1 \\ 5 & 6 & 4 \end{bmatrix}$.

6. If $A = \begin{bmatrix} 3 & -3 & 4 \\ 2 & -3 & 4 \\ 0 & -1 & 1 \end{bmatrix}$, (i) find $A^{-1}$ ; (ii) show that $A^3 = A^{-1}$.

7. Find the inverse of the matrix

$$S = \begin{bmatrix} 0 & 1 & 1 \\ 1 & 0 & 1 \\ 1 & 1 & 0 \end{bmatrix} \text{ and if } A = \frac{1}{2}\begin{bmatrix} 4 & -1 & 1 \\ -2 & 3 & -1 \\ 2 & 1 & 5 \end{bmatrix},$$

show that $SAS^{-1}$ is a diagonal matrix dig (2, 3, 1).

8. If $A = \frac{1}{3}\begin{bmatrix} 1 & 2 & 2 \\ 2 & 1 & -2 \\ -2 & 2 & -1 \end{bmatrix}$, prove that $A^{-1} = A'$.

9. Show that $\begin{bmatrix} \cos\theta & -\sin\theta \\ \sin\theta & \cos\theta \end{bmatrix} = \begin{bmatrix} 1 & -\tan\theta/2 \\ \tan\theta/2 & 1 \end{bmatrix}\begin{bmatrix} 1 & \tan\theta/2 \\ -\tan\theta/2 & 1 \end{bmatrix}^{-1}$.

10. If $A = \begin{bmatrix} 1 & 2 & -1 \\ 3 & 0 & 2 \\ 4 & 5 & 0 \end{bmatrix}$ and $B = \begin{bmatrix} 1 & 0 & 0 \\ 2 & 1 & 0 \\ 0 & 1 & 3 \end{bmatrix}$, verify that $(AB)' = B'A'$, where $A'$ is the transpose of $A$.

11. Show that

$$\begin{vmatrix} a_1 & b_1 & c_1 \\ a_2 & b_2 & c_2 \\ a_3 & b_3 & c_3 \end{vmatrix}^2 = \begin{vmatrix} A_1 & B_1 & C_1 \\ A_2 & B_2 & C_2 \\ A_3 & B_3 & C_3 \end{vmatrix}$$

where $A$, $B$ etc are the co-factors of $a$, $b$ etc. in the determinant. $(a_1\, b_2\, c_3)$. (RGTU, 2018)

12. If $A$ is a square matrix, show that (i) $A + A'$ is symmetric, and (ii) $A - A'$ is skew-symmetric.

13. If $A$ and $B$ are square matrices of the same order and $A$ is symmetrical, show that $B'AB$ is also symmetrical.

[**Hint.** Show that $(B'AB)' = B'AB$]

14. If a non-singular matrix $A$ is symmetric, show that $A^{-1}$ is also symmetric.

15. Let $\Delta = \begin{vmatrix} (1+ax)^2 & (1+ay)^2 & (1+az)^2 \\ (1+bx)^2 & (1+by)^2 & (1+bz)^2 \\ (1+cx)^2 & (1+cy)^2 & (1+cz)^2 \end{vmatrix}$

Express $\Delta$ as the product of two determinats. Hence show that $\Delta = 2(a-b)(b-c)(c-a)(x-y)(y-z)(z-x)$.

16. Show that

(*i*) $\begin{vmatrix} 1 & \cos(\beta-\alpha) & \cos(\gamma-\alpha) \\ \cos(\alpha-\beta) & 1 & \cos(\gamma-\beta) \\ \cos(\alpha-\gamma) & \cos(\beta-\gamma) & 1 \end{vmatrix} = 0$

(*ii*) $\begin{vmatrix} 3 & a+b+c & a^2+b^2+c^2 \\ a+b+c & a^2+b^2+c^2 & a^3+b^3+c^3 \\ a^2+b^2+c^2 & a^3+b^3+c^3 & a^4+b^4+c^4 \end{vmatrix} = \{(a-b)(b-c)(c-a)\}^2$

## 2.7 (1) RANK OF A MATRIX

If we select any $r$ rows and $r$ columns from any matrix $A$, deleting all the other rows and columns, then the determinant formed by these $r \times r$ elements is called the *minor of A of order r*. Clearly, there will be a number of different minors of the same order, got by deleting different rows and columns from the same matrix.

**Def.** *A matrix is said to be of rank r when*

(*i*) *it has at least one non-zero minor of order r, and*

(*ii*) *every minor of order higher than r vanishes.*

*Briefly, the rank of a matrix is the largest order of any non-vanishing minor of the matrix.*

*The rank of a matrix A shall be denoted by* $\rho(A)$.

**Obs. 1.** *Rank of A and its transpose is the same i.e.* $\rho(A) = \rho(A')$.

**Obs. 2.** *Rank of a null matrix is zero.*

**Obs. 3.** *Rank of a non-singular square matrix of order r is r.*

**Obs. 4.** If a matrix has a non-zero minor of order $r$, its rank is $\geq r$.

If all minors of a matrix of order $r + 1$ are zero, its rank is $\leq r$.

**(2) Elementary transformation of a matrix.** The following operations, three of which refer to rows and three to columns are known as *elementary transformations :*

*I. The interchange of any two rows (columns).*

*II. The multiplication of any row (column) by a non-zero number.*

*III. The addition of a constant multiple of the elements of any row (column) to the corresponding elements of any other row (column).*

**Notation.** The *elementary row transformations* will be denoted by the following symbols :

(*i*) $R_{ij}$ for the interchange of the $i$th and $j$th rows.

(*ii*) $kR_i$ for multiplication of the $i$th row by $k$.

(*iii*) $R_i + pR_j$ for addition to the $i$th row, $p$ times the $j$th row.

The corresponding column transformation will be denoted by writing $C$ in place of $R$.

*Elementary transformations do not change either the order or rank of a matrix.* While the value of the minors may get changed by the transformation I and II, their zero or non-zero character remains unaffected.

**(3) Equivalent matrix.** *Two matrices A and B are said to be equivalent if one can be obtained from the other by a sequence of elementary transformations.* Two equivalent matrices have the same order and *the same rank*. The symbol ~ is used for equivalence.

**Example 2.24.** *Determine the rank of the following matrices :*

$$(i)\ \begin{bmatrix} 1 & 2 & 3 \\ 1 & 4 & 2 \\ 2 & 6 & 5 \end{bmatrix} \qquad (ii)\ \begin{bmatrix} 0 & 1 & -3 & -1 \\ 1 & 0 & 1 & 1 \\ 3 & 1 & 0 & 2 \\ 1 & 1 & -2 & 0 \end{bmatrix}$$ (V.T.U., 2011)

**Solution:** (*i*) Operate $R_2 - R_1$ and $R_3 - 2R_1$ so that the given matrix

$$\sim \begin{bmatrix} 1 & 2 & 3 \\ 0 & 2 & -1 \\ 0 & 2 & -1 \end{bmatrix} = A \text{ (say)}$$

Obviously, the 3rd order minor of $A$ vanishes. Also its 2nd order minors formed by its 2nd and 3rd rows are all zero. But another 2nd order minor is $\begin{bmatrix} 1 & 3 \\ 0 & -1 \end{bmatrix} = -1 \neq 0.$

$\therefore \quad \rho(A) = 2$. Hence the rank of the given matrix is 2.

(*ii*) Given matrix

$$\sim \begin{bmatrix} 0 & 1 & -3 & -1 \\ 1 & 0 & 0 & 0 \\ 3 & 1 & -3 & -1 \\ 1 & 1 & -3 & -1 \end{bmatrix} \qquad \sim \begin{bmatrix} 0 & 1 & -3 & -1 \\ 1 & 0 & 0 & 0 \\ 3 & 0 & 0 & 0 \\ 1 & 0 & 0 & 0 \end{bmatrix}$$

[Operating $C_3 - C_1, C_4 - C_1$] [Operating $R_3 - R_1, R_4 - R_1$]

$$\sim \begin{bmatrix} 0 & 1 & -3 & -1 \\ 1 & 0 & 0 & 0 \\ 0 & 0 & 0 & 0 \\ 0 & 0 & 0 & 0 \end{bmatrix} \qquad \sim \begin{bmatrix} 0 & 1 & 0 & 0 \\ 1 & 0 & 0 & 0 \\ 0 & 0 & 0 & 0 \\ 0 & 0 & 0 & 0 \end{bmatrix} = A \text{ (say)}$$

[Operating $R_3 - 3R_2, R_4 - R_2$] [Operating $C_3 + 3C_2, C_4 + C_2$]

Obviously, the 4th order minor of $A$ is zero. Also every 3rd order minor of $A$ is zero. But, of all the 2nd order minors, only $\begin{bmatrix} 0 & 1 \\ 1 & 0 \end{bmatrix} = -1 \neq 0. \quad \therefore \quad \rho(A) = 2.$

Hence the rank of the given matrix is 2.

**(4) Elementary matrices.** *An elementary matrix is that, which is obtained from a unit matrix, by subjecting it to any of the elementary transformations.*

Examples of elementary matrices obtained from

$$I_3 = \begin{bmatrix} 1 & 0 & 0 \\ 0 & 1 & 0 \\ 0 & 0 & 1 \end{bmatrix} \text{ are } R_{23} = \begin{bmatrix} 1 & 0 & 0 \\ 0 & 0 & 1 \\ 0 & 1 & 0 \end{bmatrix} = C_{23}\ ;\ kR_2 = \begin{bmatrix} 1 & 0 & 0 \\ 0 & k & 0 \\ 0 & 0 & 1 \end{bmatrix}\ ;\ R_1 + pR_2 = \begin{bmatrix} 1 & p & 0 \\ 0 & 1 & 0 \\ 0 & 0 & 1 \end{bmatrix}$$

**(5) Theorem.** *Elementary row (column) transformations of a matrix A can be obtained by pre-multiplying (post-multiplying) A by the corresponding elementary matrices.*

Consider the matrix $A = \begin{bmatrix} a_1 & b_1 & c_1 \\ a_2 & b_2 & c_2 \\ a_3 & b_3 & c_3 \end{bmatrix}$

Then $$R_{23} \times A = \begin{bmatrix} 1 & 0 & 0 \\ 0 & 0 & 1 \\ 0 & 1 & 0 \end{bmatrix} \times \begin{bmatrix} a_1 & b_1 & c_1 \\ a_2 & b_2 & c_2 \\ a_3 & b_3 & c_3 \end{bmatrix} = \begin{bmatrix} a_1 & b_1 & c_1 \\ a_3 & b_3 & c_3 \\ a_2 & b_2 & c_2 \end{bmatrix}$$

So a pre-multiplication by $R_{23}$ has interchanged the 2nd and 3rd rows of $A$. Similarly, pre-multiplication by $kR_2$ will multiply the 2nd row of $A$ by $k$ and pre-multiplication by $R_1 + pR_2$ will result in the addition of $p$ times the 2nd row of $A$ to its 1st row.

Thus the pre-multiplication of $A$ by elementary matrices results in the corresponding elementary row transformation of $A$. It can easily be seen that post mutiplication will perform the elementary column transformations.

**(6) Gauss-Jordan method of finding the inverse*.** *Those elementary row transformations which reduce a given square matrix A to the unit matrix, when applied to unit matrix I give the inverse of A.*

Let the successive row transformations which reduce $A$ to $I$ result from pre-multiplication by the elementary matrices $R_1, R_2, \ldots R_i$ so that

$$R_i R_{i-1} \ldots R_2 R_1 A = I$$

$$\therefore \quad R_i R_{i-1} \ldots R_2 R_1 A A^{-1} = I A^{-1}$$

or $$R_i R_{i-1} \ldots R_2 R_1 I = A^{-1} \qquad [\because \; AA^{-1} = I]$$

Hence the result.

**Working rule to evaluate $A^{-1}$.** *Write the two matrices A and I side by side. Then perform the same row transformations on both. As soon as A is reduced to I, the other matrix represents $A^{-1}$.*

**Example 2.25.** *Using the Gauss-Jordan method, find the inverse of the matrix*

$$\begin{bmatrix} 1 & 1 & 3 \\ 1 & 3 & -3 \\ -2 & -4 & -4 \end{bmatrix}. \qquad \text{(C.S.V.T.U., 2015 ; Anna, 2012)}$$

**Solution:** Writing the same matrix side by side with the unit matrix of order 3, we have

$$\left[\begin{array}{ccc:ccc} 1 & 1 & 3 & 1 & 0 & 0 \\ 1 & 3 & -3 & 0 & 1 & 0 \\ -2 & -4 & -4 & 0 & 0 & 1 \end{array}\right] \qquad \text{(Operate } R_2 - R_1 \text{ and } R_3 + 2R_1\text{)}$$

$$\sim \left[\begin{array}{ccc:ccc} 1 & 1 & 3 & 1 & 0 & 0 \\ 0 & 2 & -6 & -1 & 1 & 0 \\ 0 & -2 & 2 & 2 & 0 & 1 \end{array}\right] \qquad \left(\text{Operate } \frac{1}{2}R_2 \text{ and } \frac{1}{2}R_3\right)$$

$$\sim \left[\begin{array}{ccc:ccc} 1 & 1 & 3 & 1 & 0 & 0 \\ 0 & 1 & -3 & -\frac{1}{2} & \frac{1}{2} & 0 \\ 0 & -1 & 1 & 1 & 0 & \frac{1}{2} \end{array}\right] \qquad \text{(Operate } R_1 - R_2 \text{ and } R_3 + R_2\text{)}$$

$$\sim \left[\begin{array}{ccc:ccc} 1 & 0 & 6 & \frac{3}{2} & -\frac{1}{2} & 0 \\ 0 & 1 & -3 & -\frac{1}{2} & \frac{1}{2} & 0 \\ 0 & 0 & -2 & \frac{1}{2} & \frac{1}{2} & \frac{1}{2} \end{array}\right] \qquad \left[\text{Operate } R_1 + 3R_3,\ R_2 - \frac{3}{2}R_3 \text{ and } \left(-\frac{1}{2}\right)R_2\right]$$

$$\sim \left[\begin{array}{ccc:ccc} 1 & 0 & 0 & 3 & 1 & \frac{3}{2} \\ 0 & 1 & 0 & -\frac{5}{4} & -\frac{1}{4} & -\frac{3}{4} \\ 0 & 0 & 1 & -\frac{1}{4} & -\frac{1}{4} & -\frac{1}{4} \end{array}\right]$$

*Named after the great German mathematician *Carl Friedrich Gauss* (1777–1855) who made his first great discovery as a student at Gottingen. His important contributions are to algebra, number theory, mechanics, complex analysis, differential equations, differential geometry, non-Euclidean geometry, numerical analysis, astronomy and electromagnetism. He became director of the observatory at Gottingen in 1807.

Name after another German mathematician and geodesist *Wilhelm Jordan* (1842–1899).

Hence the inverse of the given matrix is $\begin{bmatrix} 3 & 1 & \frac{3}{2} \\ -\frac{5}{4} & -\frac{1}{4} & -\frac{3}{4} \\ -\frac{1}{4} & -\frac{1}{4} & -\frac{1}{4} \end{bmatrix}$ [cf. Example 2.21]

**(7) Normal form of a matrix.** *Every non-zero matrix A of rank r, can be reduced by a sequence of elementary transformations, to the form*

$$\begin{bmatrix} I_r & 0 \\ 0 & 0 \end{bmatrix}$$ *called the* **normal form** *of A.* ...(*i*)

**Cor. 1.** *The rank of a matrix A is r if and only if it can be reduced to the normal form (i).*

**Cor. 2.** *Since each elementary transformation can be affected by pre-multiplication or post-multiplication with a suitable elementary matrix and each elementary matrix is non-singular, therefore, we have the following result :*

*Corresponding to every matrix A of rank r, there exist non-singular matrices P and Q such that PAQ equals (i).*

If $A$ be a $m \times n$ matrix, then $P$ and $Q$ are square matrices of orders $m$ and $n$ respectively.

**Example 2.26.** *Reduce the following matrix into its normal form and hence find its rank.*

$$A = \begin{bmatrix} 2 & 3 & -1 & -1 \\ 1 & -1 & -2 & -4 \\ 3 & 1 & 3 & -2 \\ 6 & 3 & 0 & -7 \end{bmatrix}.$$ *(J.N.T.U., 2015 ; V.T.U., 2012 ; U.P.T.U., 2011)*

**Solution:**

$$A \sim \begin{bmatrix} 1 & -1 & -2 & -4 \\ 2 & 3 & -1 & -1 \\ 3 & 1 & 3 & -2 \\ 6 & 3 & 0 & -7 \end{bmatrix} \quad [\text{By } R_{12}]$$

$$\sim \begin{bmatrix} 1 & -1 & -2 & -4 \\ 0 & 5 & 3 & 7 \\ 0 & 4 & 9 & 10 \\ 0 & 9 & 12 & 17 \end{bmatrix} \quad [\text{By } R_2 - 2R_1, R_3 - 3R_1, R_4 - 6R_1]$$

$$\sim \begin{bmatrix} 1 & 0 & 0 & 0 \\ 0 & 5 & 3 & 7 \\ 0 & 4 & 9 & 10 \\ 0 & 9 & 12 & 17 \end{bmatrix} \quad [\text{By } C_2 + C_1, C_3 + 2C_1, C_4 + 4C_1]$$

$$\sim \begin{bmatrix} 1 & 0 & 0 & 0 \\ 0 & 5 & 3 & 7 \\ 0 & 4 & 9 & 10 \\ 0 & 0 & 0 & 0 \end{bmatrix} \quad [\text{By } R_4 - R_2 - R_3]$$

$$\sim \begin{bmatrix} 1 & 0 & 0 & 0 \\ 0 & 1 & -6 & -3 \\ 0 & 4 & 9 & 10 \\ 0 & 0 & 0 & 0 \end{bmatrix} \quad [\text{By } R_2 - R_3]$$

$$\sim \begin{bmatrix} 1 & 0 & 0 & 0 \\ 0 & 1 & -6 & -3 \\ 0 & 0 & 33 & 22 \\ 0 & 0 & 0 & 0 \end{bmatrix} \quad [\text{By } R_3 - 4R_2]$$

$$\sim \begin{bmatrix} 1 & 0 & 0 & 0 \\ 0 & 1 & 0 & 0 \\ 0 & 0 & 33 & 22 \\ 0 & 0 & 0 & 0 \end{bmatrix} \quad [\text{By } C_3 + 6C_2, C_4 + 3C_2]$$

$$\sim \begin{bmatrix} 1 & 0 & 0 & 0 \\ 0 & 1 & 0 & 0 \\ 0 & 0 & 1 & 22 \\ 0 & 0 & 0 & 0 \end{bmatrix} \qquad \left[\text{By } \frac{1}{33} C_3\right]$$

$$\sim \begin{bmatrix} 1 & 0 & 0 & 0 \\ 0 & 1 & 0 & 0 \\ 0 & 0 & 1 & 0 \\ 0 & 0 & 0 & 0 \end{bmatrix} \qquad [\text{By } C_4 - 22C_3]$$

$$\sim \begin{bmatrix} I_3 & 0 \\ 0 & 0 \end{bmatrix}$$

Hence $\rho(A) = 3$.

**Example 2.27.** *Find non-singular matrices P and Q such that* $A = \begin{bmatrix} 1 & 2 & 3 & 4 \\ 2 & 1 & 4 & 3 \\ 3 & 0 & 5 & -10 \end{bmatrix}$ *is reduced to normal form. Also find the rank of A.*

**Solution:** Total number of rows = 3 $\therefore$ Consider unit matrix $I_3$

Total number of columns = 4 $\therefore$ Consider unit matrix $I_4$

$$A_{3 \times 4} = I_3 A I_4$$
$$A = I_3 A I_4$$

$$\begin{bmatrix} 1 & 2 & 3 & 4 \\ 2 & 1 & 4 & 3 \\ 3 & 0 & 5 & -10 \end{bmatrix} = \begin{bmatrix} 1 & 0 & 0 \\ 0 & 1 & 0 \\ 0 & 0 & 1 \end{bmatrix} A \begin{bmatrix} 1 & 0 & 0 & 0 \\ 0 & 1 & 0 & 0 \\ 0 & 0 & 1 & 0 \\ 0 & 0 & 0 & 1 \end{bmatrix}$$

Our aim is to reduce $A$ to normal form by using elementary transformations.

$R_2 - 2R_1, R_3 - 3R_1$

$$\begin{bmatrix} 1 & 2 & 3 & 4 \\ 2 & -3 & -2 & -5 \\ 0 & -6 & -4 & -22 \end{bmatrix} = \begin{bmatrix} 1 & 0 & 0 \\ -2 & 1 & 0 \\ -3 & 0 & 1 \end{bmatrix} A \begin{bmatrix} 1 & 0 & 0 & 0 \\ 0 & 1 & 0 & 0 \\ 0 & 0 & 1 & 0 \\ 0 & 0 & 0 & 1 \end{bmatrix}$$

$C_2 - 2C_1, C_3 - 3C_1, C_4 - 4C_1$

$$\begin{bmatrix} 1 & 0 & 0 & 0 \\ 0 & -3 & -2 & -5 \\ 0 & -6 & -4 & -22 \end{bmatrix} = \begin{bmatrix} 1 & 0 & 0 \\ -2 & 1 & 0 \\ -3 & 0 & 1 \end{bmatrix} A \begin{bmatrix} 1 & -2 & -3 & -4 \\ 0 & 1 & 0 & 0 \\ 0 & 0 & 1 & 0 \\ 0 & 0 & 0 & 1 \end{bmatrix}$$

By $R_3 - 2R_2$

$$\begin{bmatrix} 1 & 0 & 0 & 0 \\ 0 & -3 & -2 & -5 \\ 0 & 0 & 0 & -12 \end{bmatrix} = \begin{bmatrix} 1 & 0 & 0 \\ -2 & 1 & 0 \\ 1 & -2 & 1 \end{bmatrix} A \begin{bmatrix} 1 & -2 & -3 & -4 \\ 0 & 1 & 0 & 0 \\ 0 & 0 & 1 & 0 \\ 0 & 0 & 0 & 1 \end{bmatrix}$$

By $-R_2, -R_3$

$$\begin{bmatrix} 1 & 0 & 0 & 0 \\ 0 & 3 & 2 & 5 \\ 0 & 0 & 0 & 12 \end{bmatrix} = \begin{bmatrix} 1 & 0 & 0 \\ 2 & -1 & 0 \\ -1 & 2 & -1 \end{bmatrix} A \begin{bmatrix} 0 & 1 & 0 & 0 \\ 0 & 0 & 1 & 0 \\ 0 & 0 & 0 & 1 \end{bmatrix}$$

By $C_2 - C_3$

$$\begin{bmatrix} 1 & 0 & 0 & 0 \\ 0 & 1 & 2 & 5 \\ 0 & 0 & 0 & 12 \end{bmatrix} = \begin{bmatrix} 1 & 0 & 0 \\ 2 & -1 & 0 \\ -1 & 2 & -1 \end{bmatrix} A \begin{bmatrix} 1 & 1 & -3 & -4 \\ 0 & 1 & 0 & 0 \\ 0 & -1 & 1 & 0 \\ 0 & 0 & 0 & 1 \end{bmatrix}$$

By $C_3 - 2\,C_2, C_4 - 5\,C_2$
$$\begin{bmatrix} 1 & 0 & 0 & 0 \\ 0 & 1 & 0 & 0 \\ 0 & 0 & 0 & 12 \end{bmatrix} = \begin{bmatrix} 1 & 0 & 0 \\ 2 & -1 & 0 \\ -1 & 2 & -1 \end{bmatrix} A \begin{bmatrix} 1 & 1 & -3 & -4 \\ 0 & 1 & 0 & 0 \\ 0 & -1 & 1 & 0 \\ 0 & 0 & 0 & 1 \end{bmatrix}$$

By $C_3 - 2C_2, C_4 - 5C_2$
$$\begin{bmatrix} 1 & 0 & 0 & 0 \\ 0 & 1 & 0 & 0 \\ 0 & 0 & 0 & 12 \end{bmatrix} = \begin{bmatrix} 1 & 0 & 0 \\ 2 & -1 & 0 \\ -1 & 2 & -1 \end{bmatrix} A \begin{bmatrix} 1 & 1 & -5 & -5 \\ 0 & 1 & -2 & -5 \\ 0 & -1 & 3 & 5 \\ 0 & 0 & 0 & 1 \end{bmatrix}$$

By $\frac{1}{12} R_3$
$$\begin{bmatrix} 1 & 0 & 0 & 0 \\ 0 & 1 & 0 & 0 \\ 0 & 0 & 0 & 1 \end{bmatrix} = \begin{bmatrix} 1 & 0 & 0 \\ 2 & -1 & 0 \\ \frac{-1}{12} & \frac{1}{6} & \frac{-1}{12} \end{bmatrix} A \begin{bmatrix} 1 & 1 & -5 & -9 \\ 0 & 1 & -2 & -5 \\ 0 & -1 & 3 & 5 \\ 0 & 0 & 0 & 1 \end{bmatrix}$$

By $C_{34}$
$$\begin{bmatrix} 1 & 0 & 0 & : & 0 \\ 0 & 1 & 0 & : & 0 \\ 0 & 0 & 1 & : & 0 \end{bmatrix} = \begin{bmatrix} 1 & 0 & 0 \\ 2 & -1 & 0 \\ -\frac{1}{12} & \frac{1}{6} & -\frac{1}{12} \end{bmatrix} A \begin{bmatrix} 1 & 1 & -9 & -5 \\ 0 & 1 & -5 & -2 \\ 0 & -1 & 5 & 3 \\ 0 & 0 & 1 & 0 \end{bmatrix}$$

$$[I_3 : O] = PAQ$$

Here $PAQ$ is in normal form $P(A) = 3$

$$P = \begin{bmatrix} 1 & 0 & 0 \\ 2 & -1 & 0 \\ -\frac{1}{12} & \frac{1}{6} & -\frac{1}{12} \end{bmatrix}, Q = \begin{bmatrix} 1 & 1 & -9 & -5 \\ 0 & 1 & -5 & -2 \\ 0 & -1 & 5 & 3 \\ 0 & 0 & 1 & 0 \end{bmatrix}$$

Here $P$ and $Q$ are non-singular matrices

**Example 2.28.** *For the matrix A, find non-singular matrices P and Q in the normal form.*

$$A = \begin{bmatrix} 2 & 1 & -3 & -6 \\ 3 & -3 & 1 & 2 \\ -1 & 1 & 1 & 2 \end{bmatrix}$$

**Solution:** Order of $A = 3 \times 4$

Total number of rows in $A = 3$ $\quad\therefore$ Consider Unit matrix $I_3$

Total number of columns in $A = 4$ $\quad\therefore$ Consider unit matrix $I_4$

$A_{3\times4} = I_3 A I_4$

$$\begin{bmatrix} 2 & 1 & -3 & -6 \\ 3 & -3 & 1 & 2 \\ -1 & 1 & 1 & 2 \end{bmatrix} = \begin{bmatrix} 1 & 0 & 0 \\ 0 & 1 & 0 \\ 0 & 0 & 1 \end{bmatrix} A \begin{bmatrix} 1 & 0 & 0 & 0 \\ 0 & 1 & 0 & 0 \\ 0 & 0 & 1 & 0 \\ 0 & 0 & 0 & 1 \end{bmatrix}$$

By $R_{13}$
$$\begin{bmatrix} 1 & 1 & 1 & 2 \\ 3 & -3 & 1 & 2 \\ 2 & 1 & -3 & -6 \end{bmatrix} = \begin{bmatrix} 0 & 0 & 1 \\ 0 & 1 & 0 \\ 1 & 0 & 0 \end{bmatrix} A \begin{bmatrix} 1 & 0 & 0 & 0 \\ 0 & 1 & 0 & 0 \\ 0 & 0 & 1 & 0 \\ 0 & 0 & 0 & 1 \end{bmatrix}$$

By $R_2 - 3R_1, R_3 - 2R_1$

$$\begin{bmatrix} 1 & 1 & 1 & 2 \\ 0 & -6 & -2 & -4 \\ 0 & -1 & -5 & -10 \end{bmatrix} = \begin{bmatrix} 0 & 0 & 1 \\ 0 & 1 & -3 \\ 1 & 0 & -2 \end{bmatrix} A \begin{bmatrix} 1 & 0 & 0 & 0 \\ 0 & 1 & 0 & 0 \\ 0 & 0 & 1 & 0 \\ 0 & 0 & 0 & 1 \end{bmatrix}$$

By $C_2 - C_1, C_3 - C_1, C_4 - 2C_1$

$$\begin{bmatrix} 1 & 0 & 0 & 0 \\ 0 & -6 & -2 & -4 \\ 0 & -1 & -5 & -10 \end{bmatrix} = \begin{bmatrix} 0 & 0 & 1 \\ 0 & 1 & -3 \\ 1 & 0 & -2 \end{bmatrix} A \begin{bmatrix} 1 & -1 & -1 & -2 \\ 0 & 1 & 0 & 0 \\ 0 & 0 & 1 & 0 \\ 0 & 6 & 0 & 1 \end{bmatrix}$$

By $-R_2$, $-R_3$ and then followed by $R_{23}$

$$\begin{bmatrix} 1 & 0 & 0 & 0 \\ 0 & 1 & 5 & 10 \\ 0 & 6 & 2 & 4 \end{bmatrix} = \begin{bmatrix} 0 & 0 & 1 \\ -1 & 0 & 2 \\ 0 & -1 & 3 \end{bmatrix} A \begin{bmatrix} 1 & -1 & -1 & -2 \\ 0 & 1 & 0 & 0 \\ 0 & 0 & 1 & 0 \\ 0 & 0 & 0 & 1 \end{bmatrix}$$

By $R_3 - 6R_2$
$$\begin{bmatrix} 1 & 0 & 0 & 0 \\ 0 & 1 & 5 & 10 \\ 0 & 0 & -28 & -56 \end{bmatrix} = \begin{bmatrix} 0 & 0 & 1 \\ -1 & 0 & 2 \\ 6 & -1 & -9 \end{bmatrix} A \begin{bmatrix} 1 & -1 & -1 & -2 \\ 0 & 1 & 0 & 0 \\ 0 & 0 & 1 & 0 \\ 0 & 0 & 0 & 0 \end{bmatrix}$$

By $C_3 - 5C_2, C_4 - 10\,C_2$

$$\begin{bmatrix} 1 & 0 & 0 & 0 \\ 0 & 1 & 0 & 0 \\ 0 & 0 & -28 & -56 \end{bmatrix} = \begin{bmatrix} 0 & 0 & 1 \\ -1 & 0 & 2 \\ 6 & -1 & -9 \end{bmatrix} A \begin{bmatrix} 1 & -1 & 4 & -8 \\ 0 & 1 & -5 & -10 \\ 0 & 0 & 1 & 0 \\ 0 & 0 & 0 & 1 \end{bmatrix}$$

By $-\frac{1}{28}R_3$
$$\begin{bmatrix} 1 & 0 & 0 & 0 \\ 0 & 1 & 0 & 0 \\ 0 & 0 & 1 & 2 \end{bmatrix} = \begin{bmatrix} 0 & 0 & 1 \\ -1 & 0 & 2 \\ -\frac{6}{8} & \frac{1}{28} & \frac{9}{28} \end{bmatrix} A \begin{bmatrix} 1 & -1 & 4 & 8 \\ 0 & 1 & -5 & -10 \\ 0 & 0 & 1 & 0 \\ 0 & 0 & 0 & 1 \end{bmatrix}$$

By $C_4 - 2C_3$
$$\begin{bmatrix} 1 & 0 & 0 & 0 \\ 0 & 1 & 0 & 0 \\ 0 & 0 & 1 & 0 \end{bmatrix} = \begin{bmatrix} 0 & 0 & 1 \\ -1 & 0 & 2 \\ -\frac{6}{28} & \frac{1}{28} & \frac{9}{28} \end{bmatrix} A \begin{bmatrix} 1 & -1 & 4 & 0 \\ 0 & 1 & -5 & 0 \\ 0 & 0 & 1 & -2 \\ 0 & 0 & 0 & 1 \end{bmatrix}$$

$$[I_3 : 0] = PAQ$$

$PAQ$ is in normal from $P$, $Q$ are non-singular matrices.

$$P(A) = 3.$$

## PROBLEMS 2.4

*Determine the rank of the following matrix :*

1. $\begin{bmatrix} 1 & 3 \\ 2 & 4 \end{bmatrix}$ (*P.T.U., 2014*)

2. Reduce the following matrix to its normal form and hence find its rank.

(*i*) $\begin{vmatrix} 3 & 2 & -1 & 5 \\ 5 & 1 & 4 & -2 \\ 1 & -4 & 11 & -19 \end{vmatrix}$ (*ii*) $\begin{bmatrix} 1 & 3 & 6 & -1 \\ 1 & 4 & 5 & 1 \\ 1 & 5 & 4 & 3 \end{bmatrix}$ (*iii*) $\begin{bmatrix} 1 & -2 & 3 & 4 \\ -2 & 4 & -1 & -3 \\ -1 & 2 & 7 & 6 \end{bmatrix}$ (Andhra, 2022)

3. $\begin{bmatrix} 1 & 3 & 4 & 3 \\ 3 & 9 & 12 & 3 \\ 1 & 3 & 4 & 1 \end{bmatrix}$ 4. $\begin{bmatrix} 2 & 3 & 4 & 5 \\ 3 & 4 & 5 & 6 \\ 4 & 5 & 6 & 7 \\ 9 & 10 & 11 & 12 \end{bmatrix}$ (*JNTU, 2022, C.S.V.T.U., 2012*)

5. Determine the values of $p$ such that the rank of $A = \begin{bmatrix} 1 & 1 & -1 & 0 \\ 4 & 4 & -3 & 1 \\ p & 2 & 2 & 2 \\ 9 & 9 & p & 3 \end{bmatrix}$ is 3.

6. Use Gauss-Jordan method to find the inverse of the following matrices :

(*i*) $\begin{bmatrix} 2 & 3 & 4 \\ 4 & 3 & 1 \\ 1 & 2 & 4 \end{bmatrix}$ (*U.P.T.U., 2012*) (*ii*) $\begin{bmatrix} 2 & 3 & 1 \\ 1 & 3 & 3 \\ 0 & 1 & 2 \end{bmatrix}$ (*P.T.U., 2015*)

7. Find non singular matrices $P$ and $Q$ such that $PAQ$ in normal form. Hence find the rank of $A$ where,

(*i*) $A = \begin{bmatrix} 2 & 3 & 1 & 4 \\ 1 & 2 & 2 & 3 \\ 0 & -1 & -3 & -2 \end{bmatrix}$ (*Rohtak, 2019*) (*ii*) $A = \begin{bmatrix} 3 & 2 & -1 & 5 \\ 5 & 1 & 4 & -2 \\ 1 & -4 & 11 & -19 \end{bmatrix}$ (*Bhopal, 2019*)

(*iii*) $A = \begin{bmatrix} 0 & 1 & 2 & -2 \\ 4 & 0 & 2 & 6 \\ 2 & 1 & 3 & 1 \end{bmatrix}$ (*Kurukhetra, 2018*) (*iv*) $A = \begin{bmatrix} 1 & 2 & -1 & 3 \\ 3 & 4 & 0 & -1 \\ -1 & 0 & -2 & 7 \end{bmatrix}$ (*UTU, 2018*)

8. If $A = \begin{bmatrix} 3 & -3 & 4 \\ 2 & -3 & 4 \\ 0 & -1 & 1 \end{bmatrix}$, find $A^{-1}$. Also find two non-singular matrices $P$ and $Q$ such that $PAQ = I$, where $I$ is the unit matrix and verify that $A^{-1} = QP$.

9. Reduce each of the following matrices to normal form and hence find their ranks :

(*i*) $\begin{bmatrix} 8 & 1 & 3 & 6 \\ 0 & 3 & 2 & 2 \\ -8 & -1 & -3 & 4 \end{bmatrix}$ (*Delhi., 2015 ; Kurukshetra, 2005*) (*ii*) $\begin{bmatrix} 1 & 2 & 3 & 2 \\ 2 & 3 & 5 & 1 \\ 1 & 3 & 4 & 5 \end{bmatrix}$ (*V.T.U., 2014 ; Delhi, 2012*)

(*iii*) $\begin{bmatrix} 1 & 2 & -1 & 4 \\ 2 & 4 & 3 & 4 \\ 1 & 2 & 3 & 4 \\ -1 & -2 & 6 & -7 \end{bmatrix}$ (*U.K.T.U., 2010*)

## 2.8 PARTITION METHOD OF FINDING THE INVERSE

According to this method of finding the inverse, if the inverse of a matrix $A_n$ of order $n$ is known, then the inverse of the matrix $A_{n+1}$ can easily be obtained by adding $(n + 1)$th row and $(n + 1)$th column to $A_n$.

Let $$A = \begin{bmatrix} A_1 & : & A_2 \\ \cdots & : & \cdots \\ A_3' & : & \alpha \end{bmatrix} \text{ and } A^{-1} = \begin{bmatrix} X_1 & : & X_2 \\ \cdots & : & \cdots \\ X_3' & : & x \end{bmatrix}$$

where $A_2$, $X_2$ are column vectors and $A_3'$, $X_3'$ are row vectors (being transposes of column vectors $A_3$, $X_3$) and $\alpha$, $x$ are ordinary numbers. We also assume that $A_1^{-1}$ is known.

Then, $AA^{-1} = I_{n+1}$, *i.e.*, $\begin{bmatrix} A_1 & : & A_2 \\ \cdots & : & \cdots \\ A_3' & : & \alpha \end{bmatrix}\begin{bmatrix} X_1 & : & X_2 \\ \cdots & : & \cdots \\ X_3' & : & x \end{bmatrix} = \begin{bmatrix} I_n & 0 \\ 0 & 1 \end{bmatrix}$

gives

$$A_1X_1 + A_2X_3' = I_n \quad \ldots(i)$$
$$A_1X_2 + A_2x = 0 \quad \ldots(ii)$$
$$A_3'X_1 + \alpha X_3' = 0 \quad \ldots(iii)$$
$$A_3'X_2 + \alpha x = 1 \quad \ldots(iv)$$

From Eqn. (*ii*), $X_2 = -A_1^{-1} A_2 x$ and using this, Eqn. (*iv*) gives $x = (\alpha - A_3' A_1^{-1} A_2)^{-1}$

Hence $x$ and then $X_2$ are given.

Also from Eqn. (*i*), $X_1 = A_1^{-1}(I_n - A_2X_3')$

and using this, Eqn. (*iii*) gives $X_3' = -A_3' A_1^{-1}(\alpha - A_3'A_1^{-1}A_2)^{-1} = -A_3'A_1^{-1}x$

Then $X_1$ is determined and hence $A^{-1}$ is computed.

**Obs.** This is also known as the '*Escalator method*'. For evaluation of $A^{-1}$ we only need to determine two inverse matrices $A_1^{-1}$ and $(\alpha - A_3'A_1^{-1}A_2)^{-1}$.

**Example 2.29.** *Using the partition method, find the inverse of* $\begin{bmatrix} 1 & 1 & 1 \\ 4 & 3 & -1 \\ 3 & 5 & 3 \end{bmatrix}$.

**Solution:** Let $A = \begin{bmatrix} 1 & 1 & : & 1 \\ 4 & 3 & : & -1 \\ \cdots & \cdots & : & \cdots \\ 3 & 5 & : & 3 \end{bmatrix} = \begin{bmatrix} A_1 & : & A_2 \\ \cdots & : & \cdots \\ A_3' & : & \alpha \end{bmatrix}$

so that $A_1^{-1} = \begin{bmatrix} 1 & 1 \\ 4 & 3 \end{bmatrix}^{-1} = -\begin{bmatrix} 3 & -1 \\ -4 & 1 \end{bmatrix}$

Let $A^{-1} = \begin{bmatrix} X_1 & : & X_2 \\ \cdots & : & \cdots \\ X_3' & : & x \end{bmatrix}$ so that $AA^{-1} = I$.

$$\alpha - A_3' A_1^{-1} A_2 = 3 + [3 \quad 5] = \begin{bmatrix} 3 & -1 \\ -4 & 1 \end{bmatrix}\begin{bmatrix} 1 \\ -1 \end{bmatrix} = -10$$

$\therefore$ $$x = (\alpha - A_3' A_1^{-1} A_2)^{-1} = -\frac{1}{10}$$

Also, $$X_2 = -A_1^{-1} A_2 x = \begin{bmatrix} 3 & -1 \\ -4 & 1 \end{bmatrix}\begin{bmatrix} 1 \\ -1 \end{bmatrix}\left(-\frac{1}{10}\right) = -\frac{1}{10}\begin{bmatrix} 4 \\ -5 \end{bmatrix}$$

Then $$X_3' = -A_3' A_1^{-1} x = [3 \quad 5]\begin{bmatrix} 3 & -1 \\ -4 & 1 \end{bmatrix}\left(-\frac{1}{10}\right) = -\frac{1}{10}[-11 \quad 2]$$

Finally, $$X_1 = A_1^{-1}(I - A_2 X_3') = -\begin{bmatrix} 3 & -1 \\ -4 & 1 \end{bmatrix} - \frac{1}{10}\begin{bmatrix} 3 & -1 \\ -4 & 1 \end{bmatrix}\begin{bmatrix} 1 \\ -1 \end{bmatrix}[-11 \quad 2]$$

$$= \begin{bmatrix} -3 & 1 \\ 4 & -1 \end{bmatrix} - \frac{1}{10}\begin{bmatrix} -44 & 8 \\ 55 & -10 \end{bmatrix} = \begin{bmatrix} 1.4 & 0.2 \\ -1.5 & 0 \end{bmatrix}$$

Hence $$A^{-1} = \begin{bmatrix} 1.4 & 0.2 & -0.4 \\ -1.5 & 0 & 0.5 \\ 1.1 & -0.2 & -0.1 \end{bmatrix}.$$

**Example 2.30.** *If A and C are non-singular matrices, then show that* $\begin{bmatrix} A & 0 \\ B & C \end{bmatrix}^{-1} = \begin{bmatrix} A^{-1} & 0 \\ -C^{-1}BA^{-1} & C^{-1} \end{bmatrix}$

*Hence find inverse of* $\begin{bmatrix} 1 & 0 & 0 & 0 \\ 0 & 2 & 0 & 0 \\ 3 & 0 & 4 & 0 \\ 0 & 1 & 0 & 3 \end{bmatrix}$.

**Solution:** Let the given matrix be $M = \begin{bmatrix} A & O \\ B & C \end{bmatrix}$ and its inverse be $M^{-1} = \begin{bmatrix} P & Q \\ R & S \end{bmatrix}$ both in the partioned form where $A, B, C, P, Q, R, S$ are all matrices.

$$\therefore \quad MM^{-1} = \begin{bmatrix} A & O \\ B & C \end{bmatrix}\begin{bmatrix} P & Q \\ R & S \end{bmatrix} = I$$

or $$\begin{bmatrix} AP + OR & AQ + OS \\ BP + CR & BQ + CS \end{bmatrix} = \begin{bmatrix} I & O \\ O & I \end{bmatrix}$$

$\therefore$ *Equating corresponding elements, we have*

$$AP + OR = I, AQ + OS = 0, BP + CR = 0, BQ + CS = I. \qquad \text{[By § 2.8]}$$

Second relation gives $AQ = 0$, *i.e.*, $Q = 0$ as $A$ is non-singular.

First relation gives $AP = I$, *i.e.*, $P = A^{-1}$.

From third equation, $BP + CR = 0$, *i.e.*, $CR = -BP = -BA^{-1}$

$\therefore$ $C^{-1}CR = -C^{-1}BA^{-1}$ or $IR = -C^{-1}BA^{-1}$ or $R = -C^{-1}BA^{-1}$

From fourth equation, $BQ + CS = I$, or $CS = I$ or $S = C^{-1}$

Hence $$M^{-1} = \begin{bmatrix} A^{-1} & 0 \\ -C^{-1}BA^{-1} & C^{-1} \end{bmatrix}.$$

Let $$M = \begin{bmatrix} 1 & 0 & 0 & 0 \\ 0 & 2 & 0 & 0 \\ 3 & 0 & 4 & 0 \\ 0 & 1 & 0 & 3 \end{bmatrix} = \begin{bmatrix} A & O \\ B & C \end{bmatrix}$$

Whence $$A = \begin{bmatrix} 1 & 0 \\ 0 & 2 \end{bmatrix}, B = \begin{bmatrix} 3 & 0 \\ 0 & 1 \end{bmatrix}, C = \begin{bmatrix} 4 & 0 \\ 0 & 3 \end{bmatrix}$$

$$\therefore \quad A^{-1} = \frac{1}{2}\begin{bmatrix} 2 & 0 \\ 0 & 1 \end{bmatrix}, C^{-1} = \frac{1}{12}\begin{bmatrix} 3 & 0 \\ 0 & 4 \end{bmatrix}$$

$$\therefore \quad -C^{-1}(BA^{-1}) = -\frac{1}{12}\begin{bmatrix} 3 & 0 \\ 0 & 4 \end{bmatrix}\left\{\frac{1}{2}\begin{bmatrix} 3 & 0 \\ 0 & 1 \end{bmatrix}\begin{bmatrix} 2 & 0 \\ 0 & 1 \end{bmatrix}\right\}$$

$$= -\frac{1}{24}\begin{bmatrix} 3 & 0 \\ 0 & 4 \end{bmatrix}\begin{bmatrix} 6 & 0 \\ 0 & 1 \end{bmatrix} = -\frac{1}{24}\begin{bmatrix} 18 & 0 \\ 0 & 4 \end{bmatrix}$$

Hence, $$M^{-1} = \begin{bmatrix} 1 & 0 & 0 & 0 \\ 0 & 1/2 & 0 & 0 \\ -3/4 & 0 & 1/4 & 0 \\ 0 & -1/6 & 0 & 1/3 \end{bmatrix}.$$

## PROBLEMS 2.5

Find the inverse of each of the following matrices using the partition method :

1. $\begin{bmatrix} 3 & 2 & 4 \\ 2 & 1 & 1 \\ 1 & 3 & 5 \end{bmatrix}$

2. $\begin{bmatrix} 1 & 2 & 3 & 1 \\ 1 & 3 & 3 & 2 \\ 2 & 4 & 3 & 3 \\ 1 & 1 & 1 & 1 \end{bmatrix}$

## 2.9 SOLUTION OF LINEAR SYSTEM OF EQUATIONS

**(1) Method of determinants—Cramer's* rule**

Consider the equations $\left\{\begin{matrix} a_1x + b_1y + c_1z = d_1 \\ a_2x + b_2y + c_2z = d_2 \\ a_3x + b_3y + c_3z = d_3 \end{matrix}\right\}$ ...(*i*)

If the determinant of coefficient be $\Delta = \begin{vmatrix} a_1 & b_1 & c_1 \\ a_2 & b_2 & c_2 \\ a_3 & b_3 & c_3 \end{vmatrix}$

then $x\Delta = \begin{vmatrix} xa_1 & b_1 & c_1 \\ xa_2 & b_2 & c_2 \\ xa_3 & b_3 & c_3 \end{vmatrix}$ [Operate $C_1 + yC_2 + zC_3$]

$$= \begin{vmatrix} a_1x + b_1y + c_1z & b_1 & c_1 \\ a_2x + b_2y + c_2z & b_2 & c_2 \\ a_3x + b_3y + c_3z & b_3 & c_3 \end{vmatrix} = \begin{vmatrix} d_1 & b_1 & c_1 \\ d_2 & b_2 & c_2 \\ d_3 & b_3 & c_3 \end{vmatrix}$$ [By (*i*)]

Thus $x = \begin{vmatrix} d_1 & b_1 & c_1 \\ d_2 & b_2 & c_2 \\ d_3 & b_3 & c_3 \end{vmatrix} \div \begin{vmatrix} a_1 & b_1 & c_1 \\ a_2 & b_2 & c_2 \\ a_3 & b_3 & c_3 \end{vmatrix}$ provided $\Delta \neq 0$. ...(*ii*)

Similarly, $y = \begin{vmatrix} a_1 & d_1 & c_1 \\ a_2 & d_2 & c_2 \\ a_3 & d_3 & c_3 \end{vmatrix} \div \begin{vmatrix} a_1 & b_1 & c_1 \\ a_2 & b_2 & c_2 \\ a_3 & b_3 & c_3 \end{vmatrix}$ ...(*iii*)

and $z = \begin{vmatrix} a_1 & b_1 & d_1 \\ a_2 & b_2 & d_2 \\ a_3 & b_3 & d_3 \end{vmatrix} \div \begin{vmatrix} a_1 & b_1 & c_1 \\ a_2 & b_2 & c_2 \\ a_3 & b_3 & c_3 \end{vmatrix}$ ...(*iv*)

*Equations (ii), (iii) and (iv) giving the values of x, y, z constitute the* **Cramer's rule,** which reduces the solution of the linear Eqn. (*i*) to a problem in evaluation of determinants.

**(2) Matrix inversion method**

If $A = \begin{bmatrix} a_1 & b_1 & c_1 \\ a_2 & b_2 & c_2 \\ a_3 & b_3 & c_3 \end{bmatrix}, X = \begin{bmatrix} x \\ y \\ z \end{bmatrix}$ and $D = \begin{bmatrix} d_1 \\ d_2 \\ d_3 \end{bmatrix}$

then the Eqn. (*i*) are equivalent to the matrix equation $AX = D$ ...(*v*)

where $A$ is the *coefficient matrix*.

Multiplying both sides of Eqn. (*v*) by the reciprocal matrix $A^{-1}$, we get

$A^{-1}AX = A^{-1}D$ or $IX = A^{-1}D$ [$\because A^{-1}A = I$]

or $X = A^{-1}D$ *i.e.*, $\begin{bmatrix} x \\ y \\ z \end{bmatrix} = \frac{1}{\Delta}\begin{bmatrix} A_1 & A_2 & A_3 \\ B_1 & B_2 & B_3 \\ C_1 & C_2 & C_3 \end{bmatrix} \times \begin{bmatrix} d_1 \\ d_2 \\ d_3 \end{bmatrix}$ ...(*vi*)

where, $A_1$, $B_1$ etc. are the cofactors of $a_1$, $b_1$ etc. in the determinant $\Delta = \begin{vmatrix} a_1 & b_1 & c_1 \\ a_2 & b_2 & c_2 \\ a_3 & b_3 & c_3 \end{vmatrix}$ $(\Delta \neq 0)$

Hence equating the values of $x, y, z$ to the corresponding elements in the product on the right side of (*vi*), we get the desired solutions.

**Obs.** *When A is a singular matrix, i.e., $\Delta = 0$, the above methods fail. These also fail when the number of equations and the number of unknowns are unequal. Matrices can, however, be usefully applied to deal with such equations as will be seen in § 2.10(2).*

**Gabriel Cramer* (1704–1752), a Swiss mathematician.

**Example 2.31.** *Solve the equations $3x + y + 2z = 3$, $2x - 3y - z = -3$, $x + 2y + z = 4$ by (i) determinants (ii) matrices.*

**Solution:** (*i*) *By determinants :*

Here $\Delta = \begin{vmatrix} 3 & 1 & 2 \\ 2 & -3 & -1 \\ 1 & 2 & 1 \end{vmatrix} = 3(-3+2) - 2(1-4) + (-1+6) = 8$ [Expanding by $C_1$]

$\therefore$ $x = \dfrac{1}{\Delta}\begin{vmatrix} 3 & 1 & 2 \\ -3 & -3 & -1 \\ 4 & 2 & 1 \end{vmatrix}$ [Expand by $C_1$]

$= \dfrac{1}{8}[3(-3+2) + 3(1-4) + 4(-1+6)] = 1$

Similarly, $y = \dfrac{1}{\Delta}\begin{vmatrix} 3 & 3 & 2 \\ 2 & -3 & -1 \\ 1 & 4 & 1 \end{vmatrix} = 2$ and $z = \dfrac{1}{\Delta}\begin{vmatrix} 3 & 1 & 3 \\ 2 & -3 & -3 \\ 1 & 2 & 4 \end{vmatrix} = -1$

Hence $x = 1, y = 2, z = -1$.

**Note.** The use of Cramer's rule involves a lot of labour when the number of equations exceeds four. In such and other cases, the numerical methods given in § 28.4 to 28.6 are perferable.

(*ii*) *By matrices :*

Here $\Delta = \begin{vmatrix} 3 & 1 & 2 \\ 2 & -3 & -1 \\ 1 & 2 & 1 \end{vmatrix} = \begin{vmatrix} a_1 & b_1 & c_1 \\ a_2 & b_2 & c_2 \\ a_3 & b_3 & c_3 \end{vmatrix}$ (say).

Then $A_1 = -1, A_2 = 3, A_3 = 5 ; B_1 = -3, B_2 = 1, B_3 = 7 ; C_1 = 7, C_2 = -5, C_3 = -11$.

Also $\Delta = a_1A_1 + a_2A_2 + a_3A_3 = 8$.

$$\therefore \quad \begin{bmatrix} x \\ y \\ z \end{bmatrix} = \frac{1}{\Delta}\begin{vmatrix} A_1 & A_2 & A_3 \\ B_1 & B_2 & B_3 \\ C_1 & C_2 & C_3 \end{vmatrix} \times \begin{bmatrix} d_1 \\ d_2 \\ d_3 \end{bmatrix}$$

$$= \frac{1}{8}\begin{bmatrix} -1 & 3 & 5 \\ -3 & 1 & 7 \\ 7 & -5 & -11 \end{bmatrix} \times \begin{bmatrix} 3 \\ -3 \\ 4 \end{bmatrix} = \frac{1}{8}\begin{bmatrix} -3-9+20 \\ -9-3+28 \\ 21+15-44 \end{bmatrix} = \begin{bmatrix} 1 \\ 2 \\ -1 \end{bmatrix}$$

Hence $x = 1, y = 2, z = -1$.

**Example 2.32.** *Solve the equations $x_1 - x_2 + x_3 + x_4 = 2$ ; $x_1 + x_2 - x_3 + x_4 = -4$ ; $x_1 + x_2 + x_3 - x_4 = 4$ ; $x_1 + x_2 + x_3 + x_4 = 0$, by finding the inverse by elementary row operations.*

**Solution:** Given system can be written as $AX = B$, where

$$A = \begin{bmatrix} 1 & -1 & 1 & 1 \\ 1 & 1 & 1 & 1 \\ 1 & 1 & -1 & -1 \\ 1 & 1 & 1 & 1 \end{bmatrix}, X = \begin{bmatrix} x_1 \\ x_2 \\ x_3 \\ x_4 \end{bmatrix}, B = \begin{bmatrix} 2 \\ -4 \\ 4 \\ 0 \end{bmatrix}$$

To find $A^{-1}$, we write

$$[A : I] = \begin{bmatrix} 1 & -1 & 1 & 1 : 1 & 0 & 0 & 0 \\ 1 & 1 & -1 & 1 : 0 & 1 & 0 & 0 \\ 1 & 1 & 1 & -1 : 0 & 0 & 1 & 0 \\ 1 & 1 & 1 & 1 : 0 & 0 & 0 & 1 \end{bmatrix} \begin{bmatrix} R_2 - R_1 \\ R_3 + R_1 \\ R_4 + R_1 \end{bmatrix}$$

$$= \begin{bmatrix} 1 & -1 & 1 & 1 : & 1 & 0 & 0 & 0 \\ 0 & 2 & -2 & 0 : & -1 & 1 & 0 & 0 \\ 2 & 0 & 2 & 0 : & 1 & 0 & 1 & 0 \\ 2 & 0 & 2 & 2 : & 1 & 0 & 0 & 1 \end{bmatrix} \begin{bmatrix} \frac{1}{2}R_2 \\ \frac{1}{2}R_3 \\ \frac{1}{2}R_4 \end{bmatrix}$$

$$= \begin{bmatrix} 1 & -1 & 1 & 1 : & 1 & 0 & 0 & 0 \\ 0 & 1 & -1 & 0 : & -1/2 & 1/2 & 0 & 0 \\ 1 & 0 & 1 & 0 : & 1/2 & 0 & 1/2 & 0 \\ 1 & 0 & 1 & 1 : & 1/2 & 0 & 0 & 1/2 \end{bmatrix} \begin{bmatrix} R_3 - R_2 \\ R_4 - R_3 \end{bmatrix}$$

$$= \begin{bmatrix} 1 & 0 & 0 & 1 : & 1/2 & 1/2 & 0 & 0 \\ 0 & 1 & -1 & 0 : & -1/2 & 1/2 & 0 & 0 \\ 1 & 0 & 1 & 0 : & 1/2 & 0 & 1/2 & 0 \\ 0 & 0 & 0 & 1 : & 0 & 0 & -1/2 & 1/2 \end{bmatrix} \begin{bmatrix} R_1 - R_4 \\ R_2 + R_3 \end{bmatrix}$$

$$= \begin{bmatrix} 1 & 0 & 0 & 0 : & 1/2 & 1/2 & +1/2 & -1/2 \\ 1 & 1 & 0 & 0 : & 0 & 1/2 & 1/2 & 0 \\ 1 & 0 & 1 & 0 : & 1/2 & 0 & 1/2 & 0 \\ 0 & 0 & 0 & 1 : & 0 & 0 & -1/2 & 1/2 \end{bmatrix} \begin{bmatrix} R_2 - R_1 \\ R_3 - R_1 \end{bmatrix}$$

$$= \begin{bmatrix} 1 & 0 & 0 & 0 : & 1/2 & 1/2 & 1/2 & -1/2 \\ 0 & 1 & 0 & 0 : & -1/2 & 0 & 0 & 1/2 \\ 0 & 0 & 1 & 0 : & 0 & -1/2 & 0 & 1/2 \\ 0 & 0 & 0 & 1 : & 0 & 0 & -1/2 & 1/2 \end{bmatrix}$$

Thus, $$A^{-1} = \begin{bmatrix} 1/2 & 1/2 & 1/2 & -1/2 \\ -1/2 & 0 & 0 & 1/2 \\ 0 & -1/2 & 0 & 1/2 \\ 0 & 0 & -1/2 & 1/2 \end{bmatrix}$$

Hence, $$X = A^{-1}B = \begin{bmatrix} 1/2 & 1/2 & 1/2 & -1/2 \\ -1/2 & 0 & 0 & 1/2 \\ 0 & -1/2 & 0 & 1/2 \\ 0 & 0 & -1/2 & 1/2 \end{bmatrix} \begin{bmatrix} 2 \\ -4 \\ 4 \\ 0 \end{bmatrix} = \begin{bmatrix} 1 \\ -1 \\ 2 \\ -2 \end{bmatrix}$$

*i.e.,* $x_1 = 1, x_2 = -1, x_3 = 2, x_4 = -2.$

## PROBLEMS 2.6

Solve the following equations with the help of determinants (1 to 5) :

1. $x + y + z = 1,\ 3x + y - 3z = 5,\ x - 2y - 5z = 0$ *(Andhra, 2016)*
2. $x + 3y + 6z = 2\ ;\ 3x - y + 4z = 9\ ;\ x - 4y + 2z = 7.$
3. $x + y + z = 6.6\ ;\ x - y + z = 2.2\ ;\ x + 2y + 3z = 15.2.$
4. $x^2 z^3/y = e^8\ ;\ y^2 z/x = e^4\ ;\ x^3 y/z^4 = 1.$
5. $2vw - wu + uv = 3uvw\ ;\ 3vw + 2wu + 4uv = 19uvw\ ;\ 6vw + 7wu - uv = 17uvw.$

Solve the following system of equations by matrix method (6 to 8) :

6. $x_1 + x_2 + x_3 = 1,\ x_1 + 2x_2 + 3x_3 = 6,\ x_1 + 3x_2 + 4x_3 = 6.$
7. $x + y + z = 3\ ;\ x + 2y + 3z = 4\ ;\ x + 4y + 9z = 6.$
8. $x + y + z = 6$
   $2x + y + 3z = 13$
   $5x + 2y + z = 12$
   $2x - 3y - 2z = -0$ *(UPTU. 2019)*
9. $2x - y + 3z = 8\ ;\ x - 2y - z = -4\ ;\ 3x + y - 4z = 0.$
10. $2x_1 + x_2 + 2x_3 + x_4 = 6,\ 4x_1 + 3x_2 + 3x_3 - 3x_4 = -1,\ 6x_1 - 6x_2 + 6x_3 + 12x_4 = 36,\ 2x_1 + 2x_2 - x_3 + x_4 = 10.$
11. By finding $A^{-1}$, solve the linear equation $AX = B$, where $A = \begin{bmatrix} 3 & 4 & 5 \\ 1 & 2 & 0 \\ 5 & 1 & 1 \end{bmatrix}, X = \begin{bmatrix} x \\ y \\ z \end{bmatrix}$ and $B = \begin{bmatrix} 4 \\ -1 \\ 5 \end{bmatrix}$.
12. In a given electrical network, the equations for the currents $i_1, i_2, i_3$ are
    $3i_1 + i_2 + i_3 = 8\ ;\ 2i_1 - 3i_2 - 2i_3 = -5\ ;\ 7i_1 + 2i_2 - 5i_3 = 0.$
    Calculate $i_1$ and $i_3$ by Cramer's rule.
13. Using the loop current method on a circuit, the follwoing equations are obtained :
    $7i_1 - 4i_2 = 12,\ -4i_1 + 12i_2 - 6i_3 = 0,\ -6i_2 + 14i_3 = 0.$
    By matrix method, solve for $i_1, i_2$ and $i_3$.
14. Solve the following equations by calculating the inverse by elementary row operations :
    $2x_1 + 2x_2 + 2x_3 - 3x_4 = 2\ ;\ 3x_1 + 6x_2 - 2x_3 + x_4 = 8\ ;\ x_1 + x_2 - 3x_3 - 4x_4 = -1\ ;\ 2x_1 + x_2 + 5x_3 + x_4 = 5.$

## 2.10 (1) CONSISTENCY OF LINEAR SYSTEM OF EQUATIONS

Consider the system of $m$ linear equations

$$\left.\begin{aligned} a_{11}x_1 + a_{12}x_2 + \ldots + a_{1n}x_n &= k_1 \\ a_{21}x_1 + a_{22}x_2 + \ldots + a_{2n}x_n &= k_2 \\ \ldots\ldots\ldots\ldots\ldots\ldots\ldots\ldots\ldots\ldots\ldots\ldots\ldots\ldots\ldots \\ a_{m1}x_1 + a_{m2}x_2 + \ldots + a_{mn}x_n &= k_m \end{aligned}\right\} \qquad \ldots(i)$$

containing the $n$ unknowns $x_1, x_2, \ldots, x_n$. To determine whether the Eqn. (*i*) are consistent (*i.e.*, possess a solution) or not, we consider the ranks of the matrices

$$A = \begin{bmatrix} a_{11} & a_{12} & \ldots & a_{1n} \\ a_{21} & a_{22} & \ldots & a_{2n} \\ \ldots & \ldots & \ldots & \ldots \\ a_{m1} & a_{m2} & \ldots & a_{mn} \end{bmatrix} \quad \text{and} \quad K = \begin{bmatrix} a_{11} & a_{12} & \ldots & a_{1n} & k_1 \\ a_{21} & a_{22} & \ldots & a_{2n} & k_2 \\ \ldots & \ldots & \ldots & \ldots & \ldots \\ a_{m1} & a_{m2} & \ldots & a_{mn} & k_m \end{bmatrix}$$

$A$ is the *co-efficient matrix* and $K$ is called the *augmented matrix* of the Eqn. (*i*).

(2) **Rouche's theorem.** *The system of Eqn. (i) is consistent if and only if the coefficient matrix A and the augmented matrix K are of the same rank otherwise the system is inconsistent.*

*Proof.* We consider the following two possible cases :

*I. Rank of* $A$ = *rank of* $K = r$ ($r \leq$ the smaller of the numbers $m$ and $n$). The Eqn. (*i*) can, be suitable row operations, be reduced to

$$\left.\begin{aligned} b_{11}x_1 + b_{12}x_2 + \ldots + b_{1n}x_n &= l_1 \\ 0.x_1 + b_{22}x_2 + \ldots + b_{2n}x_n &= l_2 \\ \ldots\ldots\ldots\ldots\ldots\ldots\ldots\ldots\ldots\ldots\ldots\ldots\ldots\ldots\ldots \\ 0.x_1 + 0.x_2 + \ldots + b_{rn}x_n &= l_r \end{aligned}\right\} \qquad \ldots(ii)$$

and the remaining $m - r$ equations being all of the form $0.x_1 + 0.x_2 + \ldots\ldots + 0.x_n = 0$.

The Eqn. (*ii*) will have a solution, though $n - r$ of the unknowns may be chosen arbitrarily. The solution, will be unique only when $r = n$. Hence the Eqn. (*i*) are consistent.

*II. Rank of A (i.e., r) < rank of K.* In particular, let the rank of $K$ be $r + 1$. In this case, the Eqn. (*i*) will reduce, by suitable row operations, to

$$\begin{aligned} b_{11}x_1 + b_{12}x_2 + \ldots + b_{1n}x_n &= l_1, \\ 0.x_1 + b_{22}x_2 + \ldots + b_{2n}x_n &= l_2, \\ \ldots\ldots\ldots\ldots\ldots\ldots\ldots\ldots\ldots\ldots\ldots\ldots\ldots\ldots\ldots&, \\ 0.x_1 + 0.x_2 + \ldots + b_{rn}x_n &= l_r, \\ 0.x_1 + 0.x_2 + \ldots + 0.x_n &= l_{r+1}, \end{aligned}$$

and the remaining $m - (r + 1)$ equations are of the form $0.x_1 + 0.x_2 + \ldots + 0.x_n = 0$.

Clearly, the $(r + 1)$th equation cannot be satisfied by any set of values for the unknowns. Hence the Eqn. (*i*) are inconsistent.

[**Procedure to test the consistency of a system of equations in *n* unknowns :**

*Find the ranks of the coefficient matrix A and the augmented matrix K, by reducing A to the triangular form by elementary row operations.* Let the rank of $A$ be $r$ and that of $K$ be $r'$.

(*i*) *If* $r \neq r'$, *the equations are inconsistent, i.e., there is no solution.*

(*ii*) *If* $r = r' = n$, *the equations are consitent and there is a unique solution.*

(*iii*) *If* $r = r' < n$, *the equations are consistent and there are infinite number of solutions*. Giving arbitrary values to $n - r$ of the unknowns, we may express the other $r$ unknowns in terms of these.]

**Example 2.33.** *Test for consistency and solve*

$$5x + 3y + 7z = 4,\ 3x + 26y + 2z = 9,\ 7x + 2y + 10z = 5.$$

(*Andhra, 2021; V.T.U., 2016 ; P.T.U., 2014*)

**Solution:** We have $\begin{bmatrix} 5 & 3 & 7 \\ 3 & 26 & 2 \\ 7 & 2 & 10 \end{bmatrix}\begin{bmatrix} x \\ y \\ z \end{bmatrix} = \begin{bmatrix} 4 \\ 9 \\ 5 \end{bmatrix}$

Operate $3R_1, 5R_2$, $\begin{bmatrix} 15 & 9 & 21 \\ 15 & 130 & 10 \\ 7 & 2 & 10 \end{bmatrix}\begin{bmatrix} x \\ y \\ z \end{bmatrix} = \begin{bmatrix} 12 \\ 45 \\ 5 \end{bmatrix}$

Operate $R_2 - R_1$, $\begin{bmatrix} 15 & 9 & 21 \\ 0 & 121 & -11 \\ 7 & 2 & 10 \end{bmatrix}\begin{bmatrix} x \\ y \\ z \end{bmatrix} = \begin{bmatrix} 12 \\ 33 \\ 5 \end{bmatrix}$

Operate $\frac{7}{3}R_1, 5R_3, \frac{1}{11}R_2$, $\begin{bmatrix} 35 & 21 & 49 \\ 0 & 11 & -1 \\ 35 & 10 & 50 \end{bmatrix}\begin{bmatrix} x \\ y \\ z \end{bmatrix} = \begin{bmatrix} 28 \\ 3 \\ 25 \end{bmatrix}$

Operate $R_3 - R_1 + R_2, \frac{1}{7}R_1$, $\begin{bmatrix} 5 & 3 & 7 \\ 0 & 11 & -1 \\ 0 & 0 & 0 \end{bmatrix}\begin{bmatrix} x \\ y \\ z \end{bmatrix} = \begin{bmatrix} 4 \\ 3 \\ 0 \end{bmatrix}$

The ranks of coefficient matrix and augmented matrix for the last set of equations, are both 2. Hence the equations are consistent. Also the given system is equivalent to

$$5x + 3y + 7z = 4,\ 11y - z = 3, \quad \therefore \quad y = \frac{3}{11} + \frac{z}{11} \quad \text{and} \quad x = \frac{7}{11} - \frac{16}{11}z$$

where $z$ is a parameter.

Hence $x = \frac{7}{11}, y = \frac{3}{11}$ and $z = 0$, is a particular solution.

**Obs.** In the above solution, the coefficient matrix is reduced to an upper triangular matrix by row-transformations.

**Example 2.34.** *Investigate the values of* $\lambda$ *and* $\mu$ *so that the equations*

$$2x + 3y + 5z = 9,\ 7x + 3y - 2z = 8,\ 2x + 3y + \lambda z = \mu,$$

*have (i) no solution, (ii) a unique solution and (iii) an infinite number of solutions.*

*(Andhra, 2022; Hazaribag, 2015 ; C.S.V.T.U., 2014 ; V.T.U., 2013 ; D.T.U., 2011)*

**Solution:** We have $\begin{bmatrix} 2 & 3 & 5 \\ 7 & 3 & -2 \\ 2 & 3 & \lambda \end{bmatrix}\begin{bmatrix} x \\ y \\ z \end{bmatrix} = \begin{bmatrix} 9 \\ 8 \\ \mu \end{bmatrix}$

The system admits of unique solution if, and only if, the coefficient matrix is of rank 3. This requires that

$$\begin{vmatrix} 2 & 3 & 5 \\ 7 & 3 & -2 \\ 2 & 3 & \lambda \end{vmatrix} = 15(5 - \lambda) \neq 0$$

Thus for a unique solution $\lambda \neq 5$ and $\mu$ may have any value. If $\lambda = 5$, the system will have no solution for those values of $\mu$ for which the matrices

$$A = \begin{bmatrix} 2 & 3 & 5 \\ 7 & 3 & -2 \\ 2 & 3 & 5 \end{bmatrix}$$

and $$K = \begin{bmatrix} 2 & 3 & 5 & 9 \\ 7 & 3 & -2 & 8 \\ 2 & 3 & 5 & \mu \end{bmatrix}$$

are not of the same rank. But $A$ is of rank 2 and $K$ is not of rank 2 unless $\mu = 9$. Thus if $\lambda = 5$ and $\mu \neq 9$, the system will have no solution.

If $\lambda = 5$ and $\mu = 9$, the system will have an infinite number of solutions.

**Example 2.35.** *Test for consistency the following equations and solve them if consistent :* $x - 2y + 3t = 2$, $2x + y + z + t = -4$ ; $4x - 3y + z + 7t = 8$.

**Solution:** Given equation can be written as

$$\begin{bmatrix} 1 & -2 & 0 & 3 \\ 2 & 1 & 1 & 1 \\ 4 & -3 & 1 & 7 \end{bmatrix} \begin{bmatrix} x \\ y \\ z \\ t \end{bmatrix} = \begin{bmatrix} 2 \\ -4 \\ 8 \end{bmatrix}$$

Operate $R_2 - 2R_1$, $R_3 - 4R_1$,

$$\begin{bmatrix} 1 & -2 & 0 & 3 \\ 0 & 5 & 1 & -5 \\ 0 & 5 & 1 & -5 \end{bmatrix} \begin{bmatrix} x \\ y \\ z \\ t \end{bmatrix} = \begin{bmatrix} 2 \\ 0 \\ 0 \end{bmatrix}$$

Operate $R_3 - R_2$,
$$\begin{bmatrix} 1 & -2 & 0 & 3 \\ 0 & 5 & 1 & -5 \\ 0 & 0 & 0 & 0 \end{bmatrix} \begin{bmatrix} x \\ y \\ z \\ t \end{bmatrix} = \begin{bmatrix} 2 \\ 0 \\ 0 \end{bmatrix}$$

Clearly, rank of the coefficient matrix is 2 and the rank of augmented matrix is also 2. Hence the given equations are consistent. But the rank 2 < 4, the number of unknowns.

$\therefore$ The number of parameters is $4 - 2 = 2$

Thus the equations have doubly infinite solutions. Now putting $t = k_1$ and $z = k_2$ in

$$x - 2y + 3t = 2 \quad \text{and} \quad 5y + z - 5t = 0,$$

we get
$$x - 2y + 3k_1 = 2 \quad \text{and} \quad 5y + k_2 - 5k_1 = 0$$

Hence
$$y = k_1 - k_2/5$$

and
$$\begin{aligned} x &= 2 + 2y - 3k_1 \\ &= 2 + 2(k_1 - k_2/5) - 3k_1 \\ &= 2 - k_1 - \frac{2}{5}k_2 \end{aligned}$$

**(3) System of linear homogeneous equations.** Consider the homogeneous linear equations

$$\left.\begin{aligned} a_{11}x_1 + a_{12}x_2 + \ldots + a_{1n}x_n &= 0 \\ a_{21}x_1 + a_{22}x_2 + \ldots + a_{2n}x_n &= 0 \\ \ldots\ldots\ldots\ldots\ldots\ldots\ldots\ldots\ldots & \\ a_{m1}x_1 + a_{m2}x_2 + \ldots + a_{mn}x_n &= 0 \end{aligned}\right\} \quad \ldots(iii)$$

Find the rank $r$ of the coefficient matrix $A$ by reducing it to the triangular form by elementary row operations.

I. *If $r = n$, the equations (iii) have only a trivial zero solution*

$$x_1 = x_2 = \ldots = x_n = 0$$

*If $r < n$, the equation (iii) have $(n - r)$ linearly independent solutions.*

The number of linearly independent solutions is $(n - r)$ means, if arbitrary values are assigned to $(n - r)$ of the variables, the values of the remaining variables can be uniquely found.

Thus the equations (*iii*) will have an infinite number of solutions.

II. *When $m < n$* (*i.e.*, the number of equations is less than the number of variables), *the solution is always other than $x_1 = x_2 = \ldots = x_n = 0$.* The number of solutions is infinite.

III. *When $m = n$* (*i.e.*, the number of equations = the number of variables), *the necessary and sufficient condition for solutions other than $x_1 = x_2 = \ldots = x_n = 0$, is that the determinant of the coefficient matrix is zero.* In this case the equations are said to be *consistent* and such a solution is called *non-trivial solution. The determinant is called the* **eliminant** *of the equations.*

**Example 2.36.** *Solve the equations*

*(i)* $x + 2y + 3z = 0$, $3x + 4y + 4z = 0$, $7x + 10y + 12z = 0$

*(ii)* $4x + 2y + z + 3w = 0$, $6x + 3y + 4z + 7w = 0$, $2x + y + w = 0$.

**Solution:** *(i)* Rank of the coefficient matrix

$$\begin{bmatrix} 1 & 2 & 3 \\ 3 & 4 & 4 \\ 7 & 10 & 12 \end{bmatrix} \sim \begin{bmatrix} 1 & 2 & 3 \\ 0 & -2 & -5 \\ 7 & 10 & 12 \end{bmatrix} \quad [\text{Operating } R_3 - 3R_1]$$

$$\sim \begin{bmatrix} 1 & 2 & 3 \\ 0 & -2 & -5 \\ 0 & 0 & 1 \end{bmatrix} \quad [\text{Operating } R_3 - 7R_1 - 2R_2]$$

is 3 which = the number of variables (*i.e.*, $r = n$)

$\therefore$ The equations have only a trivial solution : $x = y = z = 0$.

*(ii)* Rank of the coefficient matrix

$$\begin{bmatrix} 4 & 2 & 1 & 3 \\ 6 & 3 & 4 & 7 \\ 2 & 1 & 0 & 1 \end{bmatrix} \sim \begin{bmatrix} 4 & 2 & 1 & 3 \\ 0 & 0 & 5/2 & 5/2 \\ 0 & 0 & -1/2 & -1/2 \end{bmatrix} \quad \left[\text{Operating } R_2 - \frac{3}{2}R_1, R_3 - \frac{1}{2}R_1\right.$$

$$\sim \begin{bmatrix} 4 & 2 & 1 & 3 \\ 0 & 0 & 5/2 & 5/2 \\ 0 & 0 & 0 & 0 \end{bmatrix} \quad \left[\text{Operating } R_3 + \frac{1}{5}R_2\right.$$

is 2 which < the number of variable (*i.e.*, $r < n$)

$\therefore$ Number of independent solutions = 4 – 2 = 2. Given system is equivalent to

$$4x + 2y + z + 3w = 0, z + w = 0.$$

$\therefore$ We have $z = -w$ and $y = -2x - w$

which give an infinite number of non-trivial solutions, $x$ and $w$ being the parameters.

**Example 2.37.** *Show that if $\mu \neq 0$, the system of equations.*

$$2x + y = a$$
$$x + \mu y - z = b$$
$$y + 2z = c$$

*has unique solution for every choice of a, b, c. If $\mu = 0$, determine the relation satisfy by a, b, c such that the system is inconsistent. Find the general solution by taking $\mu = 0$, $a = 1$, $b = 1$, $c = -1$.* (JBTU, 2019)

**Solution: Part 1.** The above system of equations in matrix from can be written as

$$\begin{bmatrix} 2 & 1 & 0 \\ 1 & \mu & -1 \\ 0 & 1 & 2 \end{bmatrix}\begin{bmatrix} x \\ y \\ z \end{bmatrix} = \begin{bmatrix} a \\ b \\ c \end{bmatrix}$$

Here 3 equations in 3 unknowns are there, hence we will apply another method.

$$|A| = \begin{vmatrix} 2 & 1 & 0 \\ 1 & \mu & -1 \\ 0 & 1 & 2 \end{vmatrix} = 2(2\mu + 1) - 1(2)$$

$$|A| = 4\mu$$

If $\mu = 0$, $\therefore$ $|A| \neq 0$ the system $AX = B$ always possesses a unique solution for every choice of $a, b, c$.

**Part 2:** If $\mu = 0$ then.

$$\begin{bmatrix} 2 & 1 & 0 \\ 1 & 0 & -1 \\ 0 & 1 & 2 \end{bmatrix} = \begin{bmatrix} x \\ y \\ z \end{bmatrix} = \begin{bmatrix} a \\ b \\ c \end{bmatrix}$$

$$\text{Perform } R_{12}, (A, B) \sim \begin{bmatrix} 1 & 0 & -1 & | & b \\ 2 & 1 & 0 & | & a \\ 0 & 1 & 2 & | & c \end{bmatrix}$$

$$\text{Perform } R_2 - 2R_1 \sim \begin{bmatrix} 1 & 0 & -1 & | & b \\ 0 & 1 & 2 & | & a - 2b \\ 0 & 1 & 2 & | & c \end{bmatrix}$$

$$\text{Perform } R_3 - R_2 \sim \begin{bmatrix} 1 & 0 & -1 & | & b \\ 0 & 1 & 2 & | & a - 2b \\ 0 & 0 & 0 & | & c - a + 2b \end{bmatrix}$$

If $\mu = 0$ and $c - a + 2b \neq 0$ then

$\rho(A) = 3 - 1 = 2, \rho(A, B) = 3 - 0 = 3$

$\rho(A) \neq \rho(A, B)$ system is inconsistent.

Hence if $\mu = 0$ and $c - a + 2b \neq 0$ then system posesses no solution.

**Part 3:** If $\mu = 0, c - a + 2b = 0$ then $\rho(A) = \rho(A, B) = 2 < 3$. System possesses an infinite number of solutions. Hence, system is consistent if $\mu = 0$ and

$$c - a + 2b = 0.$$

**Part 4:** Take $\mu = 0, a = 1, b = 1, c = -1$, substituting in equation (*i*), we get

$$(A, B) \sim \begin{bmatrix} 1 & 0 & -1 & | & 1 \\ 0 & 1 & 2 & | & -1 \\ 0 & 0 & 0 & | & 0 \end{bmatrix}$$

$\rho(A) = \rho(A, B) = 2 < 3$, system possesses an infinite no. of solutions.

By $R_2$ $\qquad y + 2z = -1$ let $z = t, y = -1 - 2t$

By $R_1$ $\qquad x - z = 1 \quad x = 1 + t$

Hence, if $\mu = 0, a = 1, b = 1, c = -1$,

then solution, set is $x = 1 + t$

$$y = -1 - 2t,$$
$$z = t.$$

**Example 2.38.** *By considering the rank of relevant matrices, exanine for consistency the system of equations.*

$$2x - y - z = 2$$
$$x + 2y + z = 2$$
$$4x - 7y - 5z = 2 \text{ and solve if found consistant.}$$

*(MDU, 2018)*

**Solution:** Given system of equations in matrix form can be written as

$$\begin{bmatrix} 2 & -1 & -1 \\ 1 & 2 & 1 \\ 4 & -7 & -5 \end{bmatrix} \begin{bmatrix} x \\ y \\ z \end{bmatrix} = \begin{bmatrix} 2 \\ 2 \\ 2 \end{bmatrix}$$

*i.e.,* $\qquad AX = B$

Consiter an augmented matrix

$$(A, B) = \begin{bmatrix} 2 & -1 & -1 & | & 2 \\ 1 & 2 & 1 & | & 2 \\ 4 & -7 & -5 & | & 2 \end{bmatrix}$$

$$\text{Perform } R_{12} \sim \begin{bmatrix} 1 & 2 & 1 & | & 2 \\ 2 & -1 & -1 & | & 2 \\ 4 & -7 & -5 & | & 2 \end{bmatrix}$$

$$\text{Perform } R_2 - 2R_1, R_3 - 4R_1 \sim \begin{bmatrix} 1 & 2 & 1 & | & 2 \\ 0 & -5 & -3 & | & -2 \\ 0 & 15 & -5 & | & -6 \end{bmatrix}$$

$$\text{Perform } R_3 - 3R_2 \sim \begin{bmatrix} 1 & 2 & 1 & | & 2 \\ 0 & -5 & -3 & | & -2 \\ 0 & 0 & 0 & | & 0 \end{bmatrix}$$

Here, $\rho(A) = 3 - 1 = 2, \quad \rho(A, B) = 3 - 1 = 2$

$\rho(A) = \rho(A, B) \quad \therefore$ System is Consistent

$\rho(A) = \rho(A, B) = 2 < 3$, number of Variables

System possesses an infinite number of solutions given as follows.

By $R_2$, $\quad -5y - 3z = -2 \quad$ Let $z = t$, $y = \dfrac{2 - 3t}{5}$

By $R_1$, $\quad x + 2y + z = 2x = 2 - \dfrac{4 - 6t}{t} - t = \dfrac{10 - 4 + 6t - 5t}{5} = \dfrac{6 + t}{5}$

Solution set is $x = \dfrac{6 + t}{5}, y = \dfrac{2 - 3t}{5}, z = t.$

**Example 2.39.** *Examine for consistency the following set of equations and obtains the solution if consistent.*

$$2x_1 + x_2 - x_3 + 3x_4 = 11$$
$$x_1 - 2x_2 + x_3 + x_4 = 8$$
$$4x_1 + 7x_2 + 2x_3 - x_4 = 0$$
$$3x_1 + 5x_2 + 4x_3 + 4x_4 = 17$$

**Solution:** Given system of equation in matrix form can be written as

$$\begin{bmatrix} 2 & 1 & -1 & 3 \\ 1 & 2 & 1 & 1 \\ 4 & 7 & 2 & -1 \\ 3 & 5 & 4 & 4 \end{bmatrix} \begin{bmatrix} x_1 \\ x_2 \\ x_3 \\ x_4 \end{bmatrix} = \begin{bmatrix} 11 \\ 8 \\ 0 \\ 17 \end{bmatrix}$$

$$AX = B$$

Consider an augmented matrix

$$(A, B) = \begin{bmatrix} 2 & 1 & -1 & 3 & | & 11 \\ 1 & -2 & 1 & 1 & | & 18 \\ 4 & 7 & 2 & -1 & | & 0 \\ 3 & 5 & 4 & 4 & | & 17 \end{bmatrix}$$

$$\text{By } R_{12} \quad \sim \begin{bmatrix} 1 & -2 & 1 & 1 & | & 11 \\ 2 & 1 & -1 & 3 & | & 0 \\ 4 & 7 & 2 & -1 & | & 17 \\ 3 & 5 & 4 & 4 & | & 8 \end{bmatrix}$$

$$\text{By } R_2 - 2R_1, R_3 - 4R_1, R_4 - 3R_1 \quad \sim \begin{bmatrix} 1 & -2 & 1 & 1 & | & 8 \\ 0 & 5 & -3 & 1 & | & -5 \\ 0 & 15 & -2 & -5 & | & -32 \\ 0 & 11 & 1 & 1 & | & -7 \end{bmatrix}$$

$$\text{By } R_4 - \frac{6}{5}R_2 \quad \sim \begin{bmatrix} 1 & -2 & 1 & 1 & | & 8 \\ 0 & 5 & -3 & 1 & | & -5 \\ 0 & 0 & 7 & -8 & | & -17 \\ 0 & 0 & \frac{38}{5} & \frac{-6}{5} & | & 4 \end{bmatrix}$$

By $R_3 - 3R_2, R_4 - R_2$ $\sim \left[\begin{array}{cccc|c} 1 & -2 & 1 & 1 & 8 \\ 0 & 5 & -3 & 1 & -5 \\ 0 & 0 & 7 & -8 & -17 \\ 0 & 6 & 4 & 0 & -2 \end{array}\right]$

By $R_4 - \frac{38}{35} R_3$ $\sim \left[\begin{array}{cccc|c} 1 & -2 & 1 & 1 & 8 \\ 0 & 5 & -3 & 1 & -5 \\ 0 & 0 & 7 & -8 & -17 \\ 0 & 0 & 0 & \frac{262}{35} & \frac{786}{35} \end{array}\right]$

Here $\rho(A) = 4 - 0 = 4, \quad \rho(A, B) = 4 - 0 = 4$

$\rho(A) = \rho(A, B) = 4 = 4,$

Number of variable system is consistant and posses a unique solution.

By $R_4$ $\frac{262}{35} x_4 = \frac{786}{35}, x_4 = 3$

By $R_3$ $7x_3 - 8x_4 = -17, \quad 7x_3 = -17 + 24 = 7, \quad x_3 = 1$

By $R_2$ $5x_2 - 3x_3 + x_4 = -5, \quad 5x_2 = -5 + 3 - 3 = -5, \quad x_2 = -1,$

$x_1 = -2 - 1 - 3 + 8 = 2, x_1 = 2$

Required solution set is $x_1 = 2, x_2 = -1, x_3 = 1, x_4 = 3$.

**Example 2.40.** *Find the values of $\lambda$ for which the equations*

$$(\lambda - 1)x + (3\lambda + 1)y + 2\lambda z = 0$$
$$(\lambda - 1)x + (4\lambda - 2)y + (\lambda + 3)z = 0$$
$$2x + (3\lambda + 1)y + 3(\lambda - 1)z = 0$$

*are consistent, and find the ratios of $x : y : z$ when $\lambda$ has the smallest of these values. What happens when $\lambda$ has the greatest of these values.* (J.N.T.U., 2015)

**Solution:** The given equations will be consistent, if

$$\begin{vmatrix} \lambda - 1 & 3\lambda + 1 & 2\lambda \\ \lambda - 1 & 4\lambda - 2 & \lambda + 3 \\ 2 & 3\lambda + 1 & 3(\lambda - 1) \end{vmatrix} = 0 \qquad [\text{Operate } R_2 - R_1]$$

or if,
$$\begin{vmatrix} \lambda - 1 & 3\lambda + 1 & 2\lambda \\ 0 & \lambda - 3 & 3 - \lambda \\ 2 & 3\lambda + 1 & 3(\lambda - 1) \end{vmatrix} = 0 \qquad [\text{Operate } C_3 + C_2]$$

or if,
$$\begin{vmatrix} \lambda - 1 & 3\lambda + 1 & 5\lambda + 1 \\ 0 & \lambda - 3 & 0 \\ 2 & 3\lambda + 1 & 6\lambda - 2 \end{vmatrix} = 0 \qquad [\text{Expand by } R_2]$$

or if,
$$(\lambda - 3) \begin{vmatrix} \lambda - 1 & 5\lambda + 1 \\ 2 & 2(3\lambda - 1) \end{vmatrix} = 0 \quad \text{or} \quad \text{if, } 2(\lambda - 3)\,[(\lambda - 1)(3\lambda - 1) - (5\lambda + 1)] = 0$$

or if,
$$6\lambda(\lambda - 3)^2 = 0 \quad \text{or} \quad \text{if, } \lambda = 0 \text{ or } 3.$$

(*a*) *When* $\lambda = 0$, the equations become $-x + y = 0$ ...(*i*)

$-x - 2y + 3z = 0$ ...(*ii*)

$2x + y - 3z = 0$ ...(*iii*)

Solving (*ii*) and (*iii*), we get $\frac{x}{6 - 3} = \frac{y}{6 - 3} = \frac{z}{-1 + 4}$. Hence $x = y = z$.

(*b*) *When* $\lambda = 3$, equations becomes identical.

## PROBLEMS 2.7

1. Investigate for consistency of the following equations and if possible find the solutions :
$$4x - 2y + 6z = 8,\ x + y - 3z = -1,\ 15x - 3y + 9z = 21.$$ *(D.T.U., 2012)*
2. Find the values of $a, b$ for which the equations $x + ay + z = 3, x + 2y + 27 = b, x + 5y + 3z = 9$ are consistent. When will these equations have a unique solution. *(Andhra, 2021)*
3. Investigate for what values of $\lambda$ and $\mu$ the simultaneous equations *(Andhra, 2022, 2018)*
$$x + y + z = 6,\ x + 2y + 3z = 10,\ x + 2y + \lambda z = \mu,$$
have (*i*) no solution, (*ii*) a unique solution, (*iii*) an infinite number of solutions. *(C.S.V.T.U., 2015 ; V.T.U., 2014 ; P.T.U., 2013)*
4. Test for consistency and solve,
(*i*) $x + 2y + z = 3, 2x + 3y + 2z = 5, 3x - 5y + 5z = 2, 3x + 9y - z = 4.$ *(C.S.V.T.U., 2011S)*
(*ii*) $2x + 6y + 11 = 0, 6x + 20y - 6z + 3 = 0, 6y - 18z + 1 = 0.$ *(Rohtak, 2012)*
(*iii*) $3x + 3y + 2z = 1, x + 2y = 4, 10y + 3z = -2, 2x - 3y - z = 5.$ *(U.K.T.U., 2010)*
5. Investigate for what values of $a$ and $b$, the system of simultaneous equations. *(DTU, 2018)*
$$2x - y + 3z = 2$$
$$x + y + 2z = 2$$
$$5x - y + az = b$$
have (*i*) No solution (*ii*). A unique solution (*iii*) An infinite number of solutions.
6. Show that if $\lambda \neq -5$, the system of equations
$$3x - y + 4z = 3,\ x + 2y - 3z = -2,\ 6x + 5y + \lambda z = -3,$$
have a unique solution. If $\lambda = -5$, show that the equations are consistent. Determine the solutions in each case.
7. Show that the equations
$$3x + 4y + 5z = a,\ 4x + 5y + 6z = b,\ 5x + 6y + 7z = c$$
do not have a solution unless $a + c = 2b$ *i.e.* $a, b, c$ are in A.P. *(Andhra., 2015 ; J.N.T.U., 2015 ; D.T.U., 2013)*
8. Prove that the equations $5x + 3y + 2z = 12, 2x + 4y + 5z = 2, 39x + 43y + 45z = c$ are incompatible unless $c = 74$ ; and in that case the equations are satisfied by $x = 2 + t, y = 2 - 3t, z = -2 + 2t$, where $t$ is any arbitrary quantity.
9. Find the values of $\lambda$ for which the equations $(2 - \lambda)x + 2y + 3 = 0, 2x + (4 - \lambda)y + 7 = 0, 2x + 5y + (6 - \lambda) = 0$ are consistent and find the values of $x$ and $y$ corresponding to each of these values of $\lambda$. *(Andhra, 2014)*
10. Show that there are three real values of $\lambda$ for which the equations $(a - \lambda) x + by + cz = 0, bx + (c - \lambda)y + az = 0, cx + ay + (b - \lambda)z = 0$ are simultaneously true and that the product of these values of $\lambda$ is
$$D = \begin{vmatrix} a & b & c \\ b & c & a \\ c & a & b \end{vmatrix}.$$
11. Determine the values of $k$ for which the following system of equations has non-trivial solutions and find them :
$(k - 1) x + (4k - 2) y + (k + 3) z = 0, (k - 1) x + (3k + 1) y + 2kz = 0, 2x + (3k + 1) y + 3(k - 1) z = 0.$ *(Mumbai, 2005)*
12. Show that the following system of equations.
$$3x + 4y + 5z = a$$
$$4x + 5y + 6z = b$$
$$5x + 6y + 7z = c$$
will be consistent only when $a + c = 2b$. *(Mumbai, 2019)*
13. Investigate for what values of $k$ the equations. *(Andhra, 2022; Kurukshetra, 2019)*
$$x + y + z = 1$$
$$2x + y + 4z = k$$
$$4x + y + 10z = k^2$$
have infinite number of solution? Hence find solutions
14. Solve completely the system of equations
(*i*) $x + y - 2z + 3w = 0 ; x - 2y + z - w = 0 ; 4x + y - 5z + 8w = 0 ; 5x - 7y + 2z - w = 0.$
(*ii*) $3x + 4y - z - 6w = 0 ; 2x + 3y + 2z - 3w = 0 ; 2x + y - 14z - 9w = 0 ; x + 3y + 13z + 3w = 0.$
15. Show that the system
$$3x + 4y + 5z = \alpha$$
$$4x + 5y + 6z = \beta$$
$$5x + 6y + 7z = \gamma$$
is consistent only when $\alpha, \beta, \gamma$ are in arithmetic progression. *(UPTU, 2018)*
16. Examine for consistency and if consistent then solve it.
(*i*)
$$x + 2y + z = -1$$
$$6x + y + z = -4$$
$$2x - 3y - z = 0$$
$$-x - 7y - 2z = 7$$
$$x - y = 1$$

(ii)
$$\begin{aligned} x + y + z &= 6 \\ 2x + y + 3z &= 13 \\ 5x + 2y + z &= 12 \\ 2x - 3y - 2z &= -10 \end{aligned}$$

(iii)
$$\begin{aligned} 2x_1 + x_2 + 2x_3 + x_4 &= 6 \\ 6x_1 - 6x_2 + 6x_3 + 12x_4 &= 36 \\ 4x_1 + 3x_2 + 3x_3 - 3x_4 &= -1 \\ 2x_1 + 2x_2 - x_3 + x_4 &= 10 \end{aligned}$$

(*Calicut-2017*)

## 2.11 (1) LINEAR AND ORTHOGONAL TRANSFORMATIONS

Let $(x, y)$ be the co-ordinates of a point $P$ referred to set of rectangular axes $OX$, $OY$. Then its co-ordinates $(x', y')$ referred to $OX'$, $OY'$, obtained by rotating the former axes through an angle $\theta$ given by

$$\left.\begin{aligned} x' &= x\cos\theta + y\sin\theta, \\ y' &= -x\sin\theta + y\cos\theta \end{aligned}\right\} \quad ...(i)$$

A more general transformation than (*i*) is

$$\left.\begin{aligned} x' &= a_1 x + b_1 y \\ y' &= a_2 x + b_2 y \end{aligned}\right\} \quad ...(ii)$$

which in matrix notation is $\begin{bmatrix} x' \\ y' \end{bmatrix} = \begin{bmatrix} a_1 & b_1 \\ a_2 & b_2 \end{bmatrix} \begin{bmatrix} x \\ y \end{bmatrix}$

Such transformations as Eqns. (*i*) and (*ii*), are called *linear transformations* in two dimensions.

Similarly, the relations of the type
$$\left.\begin{aligned} x' &= l_1 x + m_1 y + n_1 z \\ y' &= l_2 x + m_2 y + n_2 z \\ z' &= l_3 x + m_3 y + n_3 z \end{aligned}\right\} \quad ...(iii)$$

give a *linear transformation* from $(x, y, z)$ to $(x', y', z')$ in three dimensional problems.

In general, the relation $Y = AX$ where $Y = \begin{bmatrix} y_1 \\ y_2 \\ \cdots \\ y_n \end{bmatrix}$, $A = \begin{bmatrix} a_1 & b_1 & c_1 & \cdots & k_1 \\ a_2 & b_2 & c_2 & \cdots & k_2 \\ \cdots & \cdots & \cdots & \cdots & \cdots \\ a_n & b_n & c_n & \cdots & k_n \end{bmatrix}$, $X = \begin{bmatrix} x_1 \\ x_2 \\ \cdots \\ x_n \end{bmatrix}$ ...(iv)

give *linear transformation* from $n$ variables $x_1, x_2, ..., x_n$ to the variables $y_1, y_2, ..., y_n$ *i.e.*, the transformation of the vector $X$ to the vector $Y$.

This transformation is called linear because the linear relations $A(X_1 + X_2) = AX_1 + AX_2$ and $A(bX) = bAX$, hold for this transformation.

If the transformation matrix $A$ is singular, the transformation also is said to be singular otherwise non-singular. For a non-singular transformation $Y = AX$, we can also write the inverse transformation $X = A^{-1}Y$. A non-singular transformation is also called a *regular* transformation.

**Cor.** If a transformation from $(x_1, x_2, x_3)$ to $(y_1, y_2, y_3)$ is given by $Y = AX$ and another transformation of $(y_1, y_2, y_3)$ to $(z_1, z_2, z_3)$ is given by $Z = BY$, then the transformation from $(x_1, x_2, x_3)$ to $(z_1, z_2, z_3)$ is given by

$$Z = BY = B(AX) = (BA)X.$$

(2) **Orthogonal transformation.** *The linear transformation* Eqn. *(iv), i.e.,* $Y = AX$, *is said to be* ***orthogonal*** *if, it transforms*

$$y_1^2 + y_2^2 + \ldots + y_n^2 \text{ into } x_1^2 + x_2^2 + \ldots + x_n^2$$

*The matrix of an orthogonal transformation is called an* **orthogonal matrix.**

We have, $$X'X = [x_1 x_2 \ldots\ldots x_n] \times \begin{bmatrix} x_1 \\ x_2 \\ \cdots \\ x_n \end{bmatrix} = x_1^2 + x_2^2 + \ldots + x_n^2$$

and similarly, $$Y'Y = y_1^2 + y_2^2 + \ldots + y_n^2.$$

$\therefore$ If $Y = AX$ is an orthogonal transformation, then

$X'X = Y'Y = (AX)'(AX) = X'A'AX$ which is possible only if $A'A = I$.

But $A^{-1}A = I$, therefore, $A' = A^{-1}$ for an orthogonal transformation.

Hence a *square matrix A is said to be orthogonal if* $AA' = A'A = I$.

**Obs. 1.** *If A is orthogonal, A' and $A^{-1}$ are also orthogonal.*

Since $A$ is orthogonal, $A' = A^{-1}$.

$\therefore$ $(A')' = (A^{-1})' = (A')^{-1}$, *i.e.*, $B' = B^{-1}$ where $B = A'$

Hence $B$ (*i.e.*, $A'$) is orthogonal. As $A' = A^{-1}$, $A^{-1}$ is also orthogonal.

**Obs. 2.** *If A is orthogonal, then* $|A| = \pm 1$.

Since $AA' = A'A = I \quad \therefore \quad |A||A'| = |I|$

But $|A'| = |A|, \quad \therefore \quad |A||A| = |1|$

or $|A|^2 = 1 \quad$ *i.e.,* $\quad |A| = \pm 1$.

**Example 2.41.** *Show that the transformation*

$$y_1 = 2x_1 + x_2 + x_3,\ y_2 = x_1 + x_2 + 2x_3,\ y_3 = x_1 - 2x_3$$

*is regular. Write down the inverse transformation.* (*V.T.U., 2016*)

**Solution:** The given transformation may be written as

$$Y = AX$$

where, $X = \begin{bmatrix} x_1 \\ x_2 \\ x_3 \end{bmatrix}, Y = \begin{bmatrix} y_1 \\ y_2 \\ y_3 \end{bmatrix}, A = \begin{bmatrix} 2 & 1 & 1 \\ 1 & 1 & 2 \\ 1 & 0 & -2 \end{bmatrix}$

Now $$|A| = \begin{vmatrix} 2 & 1 & 1 \\ 1 & 1 & 2 \\ 1 & 0 & -2 \end{vmatrix} = -1$$

Thus the matrix $A$ is non-singular and hence the transformatiuon is regular.

$\therefore$ The inverse transformation is given by

$$X = A^{-1} Y$$

where, $A^{-1} = \begin{bmatrix} 2 & -2 & -1 \\ -4 & 5 & 3 \\ 1 & -1 & -1 \end{bmatrix}$

Thus $x_1 = 2y_1 - 2y_2 - y_3$ ; $x_2 = -4y_1 + 5y_2 + 3y_3$ ; $x_3 = y_1 - y_2 - y_3$

is the inverse transformation.

**Example 2.42.** *Prove that the following matrix is orthogonal :*

$$\begin{bmatrix} -2/3 & 1/3 & 2/3 \\ 2/3 & 2/3 & 1/3 \\ 1/3 & -2/3 & 2/3 \end{bmatrix}$$ (*Delhi, 2012*)

**Solution:** We have $AA' = \begin{bmatrix} -2/3 & 1/3 & 2/3 \\ 2/3 & 2/3 & 1/3 \\ 1/3 & -2/3 & 2/3 \end{bmatrix} \times \begin{bmatrix} -2/3 & 2/3 & 1/3 \\ 1/3 & 2/3 & -2/3 \\ 2/3 & 1/3 & 2/3 \end{bmatrix}$

$$= \begin{bmatrix} 4/9 + 1/9 + 4/9 & -4/9 + 2/9 + 2/9 & -2/9 - 2/9 + 4/9 \\ -4/9 + 2/9 + 2/9 & 4/9 + 4/9 + 1/9 & 2/9 - 4/9 + 2/9 \\ -2/9 - 2/9 + 4/9 & 2/9 - 4/9 + 2/9 & 1/9 + 4/9 + 4/9 \end{bmatrix} = I.$$

Hence the matrix is orthogonal.

**Example 2.43.** *If* $A = \frac{1}{3}\begin{bmatrix} 1 & 2 & a \\ 2 & 1 & b \\ 2 & -2 & c \end{bmatrix}$ *is orthogonal, find a, b, c and* $A^{-1}$.

**Solution:** As $A$ is orthogonal, $AA' = I$

$$\therefore \quad \frac{1}{3}\begin{bmatrix} 1 & 2 & a \\ 2 & 1 & b \\ 2 & -2 & c \end{bmatrix} \frac{1}{3}\begin{bmatrix} 1 & 2 & 2 \\ 2 & 1 & -2 \\ a & b & c \end{bmatrix} = \begin{bmatrix} 1 & 0 & 0 \\ 0 & 1 & 0 \\ 0 & 0 & 1 \end{bmatrix}$$

or $$\begin{bmatrix} 1+4+a^2 & 2+2+ab & 2-4+ac \\ 2+2+ab & 4+1+b^2 & 4-2+bc \\ 2-4+ac & 4-2+bc & 4+4+c^2 \end{bmatrix} = \begin{bmatrix} 9 & 0 & 0 \\ 0 & 9 & 0 \\ 0 & 0 & 9 \end{bmatrix}$$

$\therefore$ $5 + a^2 = 9, 5 + b^2 = 9, 8 + c^2 = 9$, *i.e.,* $a^2 = 4, b^2 = 4, c^2 = 1$. Thus $a = 2, b = 2, c = 1$.

Since $A$ is orthogonal, $A^{-1} = A' = \frac{1}{3}\begin{bmatrix} 1 & 2 & 2 \\ 2 & 1 & -2 \\ 2 & 2 & 1 \end{bmatrix}$.

## 2.12 (1) VECTORS

Any quantity having $n$-components is called a *vector of order n*. Therefore, the coefficients in a linear equation or the elements in a row or column matrix will form a vector. Thus any $n$ numbers $x_1, x_2, ..., x_n$ written in a particular order, constitute a vector $\mathbf{x}$.

**(2) Linear dependence.** *The vectors* $\mathbf{x}_1, \mathbf{x}_2, ..., \mathbf{x}_n$ *are said to be* **linearly dependent,** *if there exist numbers* $\lambda_1, \lambda_2, ..., \lambda_r$ *not all zero, such that*

$$\lambda_1\mathbf{x}_1 + \lambda_2\mathbf{x}_2 + ... + \lambda_r\mathbf{x}_r = \mathbf{0}. \qquad ...(i)$$

*If no such numbers, other than zero, exist, the vectors are said to be* **linearly independent.** If $\lambda_1 \neq 0$, transposing $\lambda_1\mathbf{x}_1$ to the other side and dividing by $-\lambda_1$, we write Eqn. $(i)$ in the form

$$\mathbf{x}_1 = \mu_2\mathbf{x}_2 + \mu_3\mathbf{x}_3 + ... + \mu_r\mathbf{x}_r.$$

Then the vector $\mathbf{x}_1$ is said to be a linear combination of the vectors $\mathbf{x}_2, \mathbf{x}_3, ..., \mathbf{x}_r$.

**Example 2.44.** *Are the vectors* $\mathbf{x}_1 = (1, 3, 4, 2)$, $\mathbf{x}_2 = (3, -5, 2, 2)$ *and* $\mathbf{x}_3 = (2, -1, 3, 2)$ *linearly dependent ? If so express one of these as a linear combination of the others.*

**Solution:** The relation $\lambda_1\mathbf{x}_1 + \lambda_2\mathbf{x}_2 + \lambda_3\mathbf{x}_3 = \mathbf{0}$.

*i.e.,* $\lambda_1(1, 3, 4, 2) + \lambda_2(3, -5, 2, 2) + \lambda_3(2, -1, 3, 2) = \mathbf{0}$

is equivalent to $\lambda_1 + 3\lambda_2 + 2\lambda_3 = 0,\ 3\lambda_1 - 5\lambda_2 - \lambda_3 = 0,$

$4\lambda_1 + 2\lambda_2 + 3\lambda_3 = 0,\ 2\lambda_1 + 2\lambda_2 + 2\lambda_3 = 0$

As these are satisfied by the values $\lambda_1 = 1, \lambda_2 = 1, \lambda_3 = -2$ which are not zero, the given vectors are linearly dependent. Also we have the relation,

$$\mathbf{x}_1 + \mathbf{x}_2 - 2\mathbf{x}_3 = \mathbf{0}$$

by means of which any of the given vectors can be expressed as a linear combination of the others.

**Obs.** Applying elementary row operations to the vectors $\mathbf{x}_1, \mathbf{x}_2, \mathbf{x}_3$, we see that the matrices

$$A = \begin{bmatrix} \mathbf{x}_1 \\ \mathbf{x}_2 \\ \mathbf{x}_3 \end{bmatrix} \quad \text{and} \quad B = \begin{bmatrix} \mathbf{x}_1 \\ \mathbf{x}_2 \\ \mathbf{x}_1 + \mathbf{x}_2 - 2\mathbf{x}_3 \end{bmatrix}$$

have the same rank. The rank of $B$ being 2, the rank of $A$ is also 2. Moreover $\mathbf{x}_1, \mathbf{x}_2$ are linearly independent and $\mathbf{x}_3$ can be expressed as a linear combination of $\mathbf{x}_1$ and $\mathbf{x}_2$ $\left[\because\ \mathbf{x}_3 = \frac{1}{2}(\mathbf{x}_1 + \mathbf{x}_2)\right]$. Similar results will hold for column operations and for any matrix. In general, we have the following results :

*If a given matrix has r linearly independent vectors (rows or columns) and the remaining vectors are linear combinations of these r vectors, then rank of the matrix is r. Conversely, if a matrix is of rank r, it contains r linearly independent vectors are remaining vectors (if any) can be expressed as a linear combination of these vectors.*

### PROBLEMS 2.8

1. Represent each of the transformations
$$x_1 = 3y_1 + 2y_2, y_1 = z_1 + 2z_2 \text{ and } x_2 = -y_1 + 4y_2, y_2 = 3z_1$$
by the use of matrices and find the composite transformation which express $x_1, x_2$ in terms of $z_1, z_2$.
2. If $\xi = x \cos \alpha - y \sin \alpha$, $\eta = x \sin \alpha + y \cos \alpha$, write the matrix $A$ of transformation and prove that $A^{-1} = A'$. Hence write the inverse transformation. *(V.T.U., 2015)*
3. A transformation from the variables $x_1, x_2, x_3$ to $y_1, y_2, y_3$ is given by $Y = AX$, and another transformation from $y_1, y_2, y_3$ to $z_1, z_2, z_3$ is given by $Z = BY$, where
$$A = \begin{bmatrix} 2 & 1 & 0 \\ 0 & 1 & -2 \\ -1 & 2 & 1 \end{bmatrix}, B = \begin{bmatrix} 1 & 1 & 1 \\ 1 & 2 & 3 \\ 1 & 3 & 5 \end{bmatrix}.$$ Obtain the transformation from $x_1, x_2, x_3$ to $z_1, z_2, z_3$.
4. Find the inverse transformation of $y_1 = x_1 + 2x_2 + 5x_3$ ; $y_2 = 2x_1 + 4x_2 + 11x_3$ ; $y_3 = -x_2 + 2x_3$. *(V.T.U., 2015)*
5. Show that the transformation
$$y_1 = 2x_1 - 2x_2 - x_3, y_2 = -4x_1 + 5x_2 + 3x_3, y_3 = x_1 - x_2 - x_3$$
is regular and find the inverse transformation. *(V.T.U., 2013)*

6. Verify that the following matrix is orthogonal :

(i) $\begin{bmatrix} 1/3 & 2/3 & 2/3 \\ 2/3 & 1/3 & -2/3 \\ 2/3 & -2/3 & 1/3 \end{bmatrix}$  (ii) $\begin{bmatrix} \cos\theta & 0 & \sin\theta \\ 0 & 1 & 0 \\ \sin\theta & 0 & \cos\theta \end{bmatrix}$

7. Find the values of $a, b, c$ if $A = \begin{bmatrix} 0 & 2b & c \\ a & b & -c \\ a & -b & c \end{bmatrix}$ is orthogonal ?

8. Prove that $\begin{bmatrix} l & m & n & 0 \\ 0 & 0 & 0 & -1 \\ n & l & -m & 0 \\ -m & n & -l & 0 \end{bmatrix}$ is orthogonal, when $l = 2/7$, $m = 3/7$, $n = 6/7$.

9. If $A$ and $B$ are orthogonal matrices, prove that $AB$ is also orthogonal.

10. Are the following vectors linearly dependent. If so, find the relation between them :
(i) (2, 1, 1), (2, 0, –1), (4, 2, 1).
(ii) (1, –1, 0), (0, 1, –1), (0, 0, 1). (P.T.U., 2015 S)
(iii) $\mathbf{x}_1 = (1, 1, -1, 1)$, $\mathbf{x}_2 = (1, -1, 2, -1)$, $\mathbf{x}_3 = (3, 1, 0, 1)$ (U.P.T.U., 2013)
(iv) $\mathbf{x}_1 = (1, 2, 4)$, $\mathbf{x}_2 = (2, -1, 3)$, $\mathbf{x}_3 = (0, 1, 2)$, $\mathbf{x}_4 = (-3, 7, 2)$.

## 2.13 (1) EIGEN VALUES

If $A$ is any square matrix of order $n$, we can form the matrix $A - \lambda I$, where $I$ is the $n$th order unit matrix. The determinant of this matrix equated to zero,

*i.e.,*
$$|A - \lambda I| = \begin{vmatrix} a_{11}-\lambda & a_{12} & \dots & a_{1n} \\ a_{21} & a_{22}-\lambda & \dots & a_{2n} \\ \dots & \dots & \dots & \dots \\ a_{n1} & a_{n2} & \dots & a_{nn}-\lambda \end{vmatrix} = 0$$

is called the *characteristic equation of A*. On expanding the determinant, the characteristic equation takes the form

$$(-1)^n \lambda^n + k_1\lambda^{n-1} + k_2\lambda^{n-2} + \dots + k_n = 0,$$

where $k$'s are expressible in terms of the elements $a_{ij}$. The roots of this equation are called the *eigenvalues* or *latent roots* or *characteristic roots* of the matrix $A$.

**(2) Eigen vectors**

If $X = \begin{bmatrix} x_1 \\ x_2 \\ \dots \\ x_n \end{bmatrix}$ and $A = \begin{bmatrix} a_{11} & a_{12} & \dots & a_{1n} \\ a_{21} & a_{22} & \dots & a_{2n} \\ \dots & \dots & \dots & \dots \\ a_{n1} & a_{n2} & \dots & a_{nn} \end{bmatrix}$, then the linear transformation $Y = AX$ ...(i)

carries the column vector $X$ into the column vector $Y$ by means of the square-matrix $A$. In practice, it is often required to find such vectors which transform into themselves or to a scalar multiple of themselves.

Let $X$ be such a vector which transforms into $\lambda X$ by means of the transformation (i).

Then $\lambda X = AX$ or $AX - \lambda IX = 0$ or $[A - \lambda I]X = 0$ ...(ii)

This matrix equation represents $n$ homogeneous linear equations

$$\left.\begin{aligned} (a_{11}-\lambda)x_1 + a_{12}x_2 + \dots + a_{1n}x_n &= 0 \\ a_{21}x_1 + (a_{22}-\lambda)x_2 + \dots + a_{2n}x_n &= 0 \\ \dots\dots\dots\dots\dots\dots\dots\dots\dots\dots \\ a_{n1}x_1 + a_{n2}x_2 + \dots + (a_{nn}-\lambda)x_n &= 0 \end{aligned}\right\} \quad \text{...(iii)}$$

which will have a non-trivial solution only if the coefficient matrix is singular, *i.e.*, if $|A - \lambda I| = 0$.

This is called the characteristic equation of the transformation and is same as the characteristic equation of the matrix $A$. It has $n$ roots and corresponding to each root, the equation (ii) [or (iii)] will have a non-zero solution.

$X = [x_1, x_2, \dots\dots, x_n]'$, which is known as the *eigen vector* or *latent vector*.

**Obs. 1.** Corresponding to $n$ distinct eigen values, we get $n$ independent eigen vectors. But when two or more eigen values are equal, it may or may not be possible to get linearly independent eigen vectors corresponding to the repeated roots.

**Obs. 2.** If $X_i$ is a solution for a eigen value $\lambda_i$ then it follows from (*ii*) that $cX_i$ is also a solution, where $c$ is arbitrary constant. Thus *the eigen vector corresponding to a eigen value is not unique* but may be any one of the vectors $cX_i$.

**Example 2.45.** $A = \begin{bmatrix} 4 & 6 & 6 \\ 1 & 3 & 2 \\ -1 & -4 & -3 \end{bmatrix}$ (*DTU, 2018*)

*Determine the Eigen values and Eigen vectors of the following matrix.*

**Solution: Step 1:** The characteristic equation of $A$ is $[A - \lambda I] = 0$

*i.e.*, $$\begin{bmatrix} 4-\lambda & 6 & 6 \\ 1 & 3-\lambda & 2 \\ -1 & -4 & -3-\lambda \end{bmatrix} = 0 \quad \text{...(1)}$$

$\lambda^3 - S_1\lambda^2 + S_2\lambda - |A| = 0$, where $S_1 = 4 + 3 - 3 = 4$

$$S_2 = \begin{vmatrix} 3 & 2 \\ -4 & -3 \end{vmatrix} + \begin{vmatrix} 4 & 6 \\ -1 & -3 \end{vmatrix} + \begin{vmatrix} 4 & 6 \\ 1 & 3 \end{vmatrix}$$

$$S_2 = -1 - 6 + 6 = -1$$

$$|A| = 4(-1) - 6(-1) + 6(-1) = -4$$

The characteristic equation of $A$ is $\lambda^3 - 4\lambda^2 - \lambda + 4 = 0$

**Step 2:** To find the roots of (1) we note that coefficients = 0

$\therefore$ $(\lambda - 1) = 0$ must be the factor of (1)

Now use synetheic division

| 1 | 1 | –4 | –1 | 4 |
|---|---|---|---|---|
| | | +1 | –3 | –4 |
| | 1 | –3 | –4 | 0 |

The factors are $(\lambda - 1)(\lambda^2 - 3\lambda - 4) = 0$

$$(\lambda - 1)(\lambda + 1)(\lambda - 4) = 0$$

Hence $\lambda = -1, \lambda = 1, \lambda = 4$ are the required Eigen values of $A$.

Here $\lambda_1 + \lambda_2 + \lambda_3 = a_{11} + a_{22} + a_{33}$, *i.e.*, $-1 + 1 + 4 = 4 + 3 - 3$.

**Step 3:** Let $\lambda_1 = -1, \lambda_2 = 1, \lambda_3 = 4$, be the Eigen values of $A$.

To find the Eigen vectors for the correspacting Eigen values we will consider the matrix equation $(A - \lambda I) X = 0$, where,

$$X = \begin{bmatrix} x \\ y \\ z \end{bmatrix} \text{ i.e., } \begin{bmatrix} 4-\lambda & 6 & 6 \\ 1 & 3-\lambda & 2 \\ -1 & -4 & -3-\lambda \end{bmatrix} \begin{bmatrix} x \\ y \\ z \end{bmatrix} = \begin{bmatrix} 0 \\ 0 \\ 0 \end{bmatrix} \quad \text{...(2)}$$

**Step 4:** For $\lambda = \lambda_1 = -1$, Let the corresponding eigen vector be $X_1 = \begin{bmatrix} x \\ y \\ z \end{bmatrix}$.

put $\lambda = \lambda_1 = -1$ in (2). The matrix equation (2) will become $\begin{bmatrix} 5 & 6 & 6 \\ 1 & 4 & 2 \\ -1 & -4 & -2 \end{bmatrix} \begin{bmatrix} x \\ y \\ z \end{bmatrix} = \begin{bmatrix} 0 \\ 0 \\ 0 \end{bmatrix}$ which is homogeneous of equations. Consider first two equations.

$$5x + 6y + 6z = 0$$

$$x + 4y + 2z = 0$$

Solving by Cramer's rule

$$\frac{x}{\begin{vmatrix} 6 & 6 \\ 4 & 2 \end{vmatrix}} = \frac{-y}{\begin{vmatrix} 5 & 6 \\ 1 & 2 \end{vmatrix}} = \frac{z}{\begin{vmatrix} 5 & 6 \\ 1 & 4 \end{vmatrix}}$$

$$\frac{x}{-12} = \frac{-y}{4} = \frac{z}{14} \text{ or } \frac{x}{-6} = \frac{y}{-2} = \frac{z}{7} = t$$

$\therefore$ $x = -6t, y = -2t, z = 7t$

For $\lambda = -1$

$$X_1 = \begin{bmatrix} -6t \\ -2t \\ 7t \end{bmatrix} \text{ or } t\begin{bmatrix} -6 \\ -2 \\ 7 \end{bmatrix}$$

Hence for $\lambda_1 = -1$, corresponding Eigen vector $X_1 = \begin{bmatrix} -6 \\ -2 \\ 7 \end{bmatrix}$.

**Step 5:** For $\lambda = \lambda_2 = 1$ let the corresponding Eigen vector be $X_2 = \begin{bmatrix} x \\ y \\ z \end{bmatrix}$.

put $\lambda = \lambda_2 = 1$ in (2)

$\therefore$ The matrix equation (2) will become

$$\begin{bmatrix} 3 & 6 & 6 \\ 1 & 2 & 2 \\ -1 & -4 & -4 \end{bmatrix}\begin{bmatrix} x \\ y \\ z \end{bmatrix} = \begin{bmatrix} 0 \\ 0 \\ 0 \end{bmatrix}$$

$$AX = Z$$

which is homogeneous system of equations. Here first two equaions are identical.

$\therefore$ Consider last two equations

$$x + 2y + 2z = 0$$
$$-x - 4y - 4z = 0$$

Solving by Cramer's rule, $\dfrac{x}{\begin{vmatrix} 2 & 2 \\ -4 & -4 \end{vmatrix}} = \dfrac{y}{\begin{vmatrix} 1 & 2 \\ -1 & -4 \end{vmatrix}} = \dfrac{z}{\begin{vmatrix} 1 & 2 \\ -1 & -4 \end{vmatrix}}$

$$\frac{x}{0} = \frac{-y}{-2} = \frac{z}{-z} \text{ or, } \frac{x}{0} = \frac{y}{1} = \frac{z}{-1} = t$$

$x = 0, y = t, z = -t$

For $\lambda_2 = 1, X_2 = \begin{bmatrix} 0 \\ t \\ -t \end{bmatrix}$ or $t\begin{bmatrix} 0 \\ 1 \\ -1 \end{bmatrix}$

Hence for $\lambda_1 = -1$, corresponding eigen vectors is $X_2 = \begin{bmatrix} 0 \\ 1 \\ -1 \end{bmatrix}$.

**Step 6:** For $\lambda = \lambda_3 = 4$, let the corresponding Eigen vector be $X_3 = \begin{bmatrix} x \\ y \\ z \end{bmatrix}$.

$$\lambda = \lambda_3 = 4 \text{ in } (2)$$

∴ The matrix equation (2) will become

$$\begin{bmatrix} 0 & 6 & 6 \\ 1 & -1 & 2 \\ -1 & -4 & -7 \end{bmatrix}\begin{bmatrix} x \\ y \\ z \end{bmatrix} = \begin{bmatrix} 0 \\ 0 \\ 0 \end{bmatrix} \quad i.e., AX = -Z$$

which is Homogeneous system of equations.

By $R_1$ $\quad 6y + 6z = 0, \quad y + z = 0 \quad$ Let $z = t, y = -t$

$$x - y + 2z = 0, \quad x = -3t$$

For $\lambda_3 = 4$, $X_3 = \begin{bmatrix} -3t \\ -t \\ t \end{bmatrix} = t\begin{bmatrix} -3 \\ -1 \\ 1 \end{bmatrix}$

Hence for $\lambda_3 = 4$, the Eigen vector is $X_3 = \begin{bmatrix} -3 \\ -1 \\ 1 \end{bmatrix}$.

**Step 7.** $\lambda_1 = -1 \to X_1 = \begin{bmatrix} -6 \\ -2 \\ 7 \end{bmatrix}$; $\lambda_2 = 1 \to X_2 = \begin{bmatrix} 0 \\ 1 \\ -1 \end{bmatrix}$

$$\lambda_3 = 4 \to X_3 = \begin{bmatrix} -3 \\ -1 \\ 1 \end{bmatrix}$$

**Example 2.46.** *Find the eigen values and eigen vectors of the matrix* $\begin{bmatrix} 1 & 1 & 3 \\ 1 & 5 & 1 \\ 3 & 1 & 1 \end{bmatrix}$.

*(Andhra, 2018; Anna, 2016 ; V.T.U., 2013 ; Calicut, 2012 ; C.S.V.T.U., 2012)*

**Solution.** The characteristic equation is $|A - \lambda I| = \begin{vmatrix} 1-\lambda & 1 & 3 \\ 1 & 5-\lambda & 1 \\ 3 & 1 & 1-\lambda \end{vmatrix}$, *i.e.*, $\lambda^3 - 7\lambda^2 + 36 = 0$

Since, $\lambda = -2$ satifies it, we can write this equation as

$$(\lambda + 2)(\lambda^2 - 9\lambda + 18) = 0 \quad \text{or} \quad (\lambda + 2)(\lambda - 3)(\lambda - 6) = 0.$$

Thus the eigen values of $A$ are $\lambda = -2, 3, 6$.

If $x$, $y$ $z$ be the components of an eigen vector corresponding to the eigen value $\lambda$, we have

$$[A - \lambda I]X = \begin{bmatrix} 1-\lambda & 1 & 3 \\ 1 & 5-\lambda & 1 \\ 3 & 1 & 1-\lambda \end{bmatrix}\begin{bmatrix} x \\ y \\ z \end{bmatrix} = 0 \quad \text{...}(i)$$

Putting $\lambda = -2$, we have $3x + y + 3z = 0$, $x + 7y + z = 0$, $3x + y + 3z = 0$.

The first and third equations being the same, we have from the first two

$$\frac{x}{-20} = \frac{y}{0} = \frac{z}{20} \quad \text{or} \quad \frac{x}{-1} = \frac{y}{0} = \frac{z}{1}$$

Hence the eigen vector is (–1, 0, 1). Also every non-zero multiple of this vector is an eigen vector corresponding to $\lambda = -2$.

Similarly, the eigen vectors corresponding to $\lambda = 3$ and $\lambda = 6$ are the arbitrary non-zero multiples of the vectors (1, –1, 1) and (1, 2, 1) which are obtained from (*i*).

Hence the three eigen vectors may be taken as (–1, 0, 1), (1, –1, 1), (1, 2, 1).

**Example 2.47.** *Find the eigen values and eigen vectors of the matrix* $A = \begin{bmatrix} 3 & 1 & 4 \\ 0 & 2 & 6 \\ 0 & 0 & 5 \end{bmatrix}$ (AU, 2023)

**Solution.** The characteristic equation is

$$|A - \lambda I| = 0, \quad i.e., \quad \begin{vmatrix} 3-\lambda & 1 & 4 \\ 0 & 2-\lambda & 6 \\ 0 & 0 & 5-\lambda \end{vmatrix} = 0$$

or $$(3 - \lambda)(2 - \lambda)(5 - \lambda) = 0$$

Thus the eigen values of $A$ are 2, 3, 5.

If $x, y, z$ be the components of an eigen vector corresponding to the eigen value $\lambda$, we have

$$[A - \lambda I]\, X = \begin{bmatrix} 3-\lambda & 1 & 4 \\ 0 & 2-\lambda & 6 \\ 0 & 0 & 5-\lambda \end{bmatrix} \begin{bmatrix} x \\ y \\ z \end{bmatrix} = 0$$

Putting $\lambda = 2$, we have $x + y + 4z = 0, 6z = 0, 3z = 0$, *i.e.*, $x + y = 0$ and $z = 0$.

$$\therefore \quad \frac{x}{1} = \frac{y}{-1} = \frac{z}{0} = k_1 \text{ (say)}$$

Hence the eigen vector corresponding to $\lambda = 2$ is $k_1$ (1, –1, 0).

Putting $\lambda = 3$, we have $y + 4z = 0, -y + 6z = 0, 2z = 0$, *i.e.*, $y = 0$, $z = 0$.

$$\therefore \quad \frac{x}{1} = \frac{y}{0} = \frac{z}{0} = k_2$$

Hence the eigen vector corresponding to $\lambda = 3$ is $k_2$ (1, 0, 0).

Similarly, the eigen vector corresponding to $\lambda = 5$ is $k_3$ (3, 2, 1).

## 2.14 PROPERTIES OF EIGEN VALUES

*I. Any square matrix A and its transpose A′ have the same eigen values.*

We have $$(A - \lambda I)' = A' - \lambda I' = A' - \lambda I$$

$$|(A - \lambda I)'| = |A' - \lambda I|$$

$$|A - \lambda I| = |A' - \lambda I| \qquad [\because |B'| = |B|$$

$\therefore \quad |A - \lambda I| = 0$ if and only if $|A' - \lambda I| = 0$

*i.e.*, $\lambda$ is an eigen value of $A$ if and only if it is an eigen value of $A'$.

*II. The eigen values of a triangular matrix are just the diagonal elements of the matrix.*

Let $$A = \begin{vmatrix} a_{11} & a_{12} & \dots & a_{1n} \\ 0 & a_{22} & \dots & a_{2n} \\ \dots & \dots & \dots & \dots \\ \dots & \dots & \dots & \dots \\ 0 & 0 & \dots & a_{nn} \end{vmatrix}$$ be a triangular matrix of order $n$.

Then $|A - \lambda I| = (a_{11} - \lambda)(a_{22} - \lambda) \,......(a_{nn} - \lambda)$.

$\therefore$ Roots of $|A - \lambda I| = 0$ are $\lambda = a_{11}, a_{22}, ..., a_{nn}$.

Hence the eigen values of $A$ are the diagonal elements of $A$, *i.e.*, $a_{11}, a_{22}, ..., a_{nn}$.

**Cor.** *The eigen values of a diagonal matrix are just the diagonal elements of the matrix.*

*III. The eigen values of an idempotent matrix are either zero or unity.*

Let $A$ be an idempotent matrix so that $A^2 = A$. If $\lambda$ be an eigen value of $A$, then there exists a non-zero vector $X$ such that

$$AX = \lambda X \qquad ...(1)$$

$$\therefore \quad A(AX) = A(\lambda X), \quad i.e., \quad A^2X = \lambda(AX)$$

*i.e.* $$AX = \lambda(\lambda X) \qquad [\because A^2 = A \text{ and } AX = \lambda X$$

$$\therefore \quad AX = \lambda^2 X \qquad ...(2)$$

From (1) and (2), we get $\lambda^2 X = \lambda X$ or $(\lambda^2 - \lambda) X = 0$

or $$\lambda^2 - \lambda = 0 \text{ whence } \lambda = 0 \text{ or } 1.$$

Hence the result.

*IV. The sum of the eigen values of a matrix is its trace i.e. the sum of the elements of the principal diagonal.*

[This property will be proved for a matrix of order 3, but the method will be capable of easy extension to matrices of any order.]

Consider the square matrix

$$A = \begin{bmatrix} a_{11} & a_{12} & a_{13} \\ a_{21} & a_{22} & a_{23} \\ a_{31} & a_{32} & a_{33} \end{bmatrix} \quad \text{...}(i)$$

so that $$|A - \lambda I| = \begin{vmatrix} a_{11} - \lambda & a_{12} & a_{13} \\ a_{21} & a_{22} - \lambda & a_{23} \\ a_{31} & a_{32} & a_{33} - \lambda \end{vmatrix} \quad \text{(On expanding)}$$

$$= -\lambda^3 + \lambda^2(a_{11} + a_{22} + a_{33}) - \ldots \quad \text{...}(ii)$$

If $\lambda_1, \lambda_2, \lambda_3$ be the eigen values of $A$, then $|A - \lambda I| = (-1)^3 (\lambda - \lambda_1)(\lambda - \lambda_2)(\lambda - \lambda_3)$

$$= -\lambda^3 + \lambda^2 (\lambda_1 + \lambda_2 + \lambda_3) - \ldots \quad \text{...}(iii)$$

Equating the right hand sides of Eqns. (*ii*) and (*iii*) and comparing coefficients of $\lambda^2$, we get

$\lambda_1 + \lambda_2 + \lambda_3 = a_{11} + a_{22} + a_{33}$. Hence the result. (*Rohtak, 2011*)

*V. The product of the eigen values of a matrix A is equal to its determinant.*

Putting $\lambda = 0$ in Eqn. (*iii*), we get the result. (*Anna, 2013*)

*VI. If $\lambda$ is an eigen value of a matrix A, then $1/\lambda$ is the eigen value of $A^{-1}$.*

If $X$ be the eigen vector corresponding to $\lambda$, then $AX = \lambda X$ ...(*i*)

Premultiplying both sides by $A^{-1}$, we get $A^{-1} AX = A^{-1}\lambda X$

*i.e.,* $$IX = \lambda A^{-1} X \quad \text{or} \quad X = \lambda(A^{-1} X), \quad i.e., \quad A^{-1}X = (1/\lambda) X$$

This being of the same form as Eqn. (*i*), shows that $1/\lambda$ is an eigen value of the inverse matrix $A^{-1}$.

*VII. If $\lambda$ is an eigen value of an orthogonal matrix, then $1/\lambda$ is also its eigen value.*

We know that if $\lambda$ is an eigen value of a matrix $A$, then $1/\lambda$ is an eigen value of $A^{-1}$. [Property V]. Since $A$ is an orthogonal matrix, $A^{-1}$ is same as its transpose $A'$.

$\therefore$ $1/\lambda$ is an eigen value of $A'$.

But the matrices $A$ and $A'$ have the same eigen values, since the determinants $|A - \lambda I|$ and $|A' - \lambda I|$ are the same.

Hence $1/\lambda$ is also an eigen value of $A$

*VIII. If $\lambda_1, \lambda_2 \ldots, \lambda_n$ are the eigen values of a matrix A, then $A^m$ has the eigen values $\lambda_1^m, \lambda_2^m, \ldots, \lambda_n^m$ (m being a positive integer).* (*J.N.T.U., 2015 ; Mumbai, 2006*)

Let $A_i$ be the eigen value of $A$ and $X_i$ the corresponding eigen vector. Then

$$AX_i = \lambda_i X_i \quad \text{...}(i)$$

We have $$A^2X_i = A(AX_i) = A(\lambda_i X_i) = \lambda_i(AX_i) = \lambda(\lambda_i X_i) = \lambda_i^2 X_i$$

Similarly, $A^3X_i = \lambda_i^3 X_i$. In general, $A^mX_i = \lambda_i^m X_i$ which is of the same form as Eqn. (*i*).

Hence $\lambda_i^m$ is an eigen value of $A^m$.

The corresponding eigen vector is the same $X_i$.

## 2.15 CAYLEY-HAMILTON THEOREM*

*Every square matrix satisfies its own characteristic equation ; i.e.,* if the characteristic equation for the $n$th order square matrix $A$ is

$$|A - \lambda I| = (-1)^n \lambda^n + k_1\lambda^{n-1} + \ldots + k_n = 0 \quad \text{...}(i)$$

then $$(-1)^n A^n + k_1A^{n-1} + \ldots + k_n = 0.$$

Let the adjoint of the matrix $A - \lambda I$ be $P$. Clearly, the elements of $P$ will be polynomials of the $(n - 1)$th degree in $\lambda$, for the cofactors of the elements in $|A - \lambda I|$ will be such polynomials.

$\therefore$ $P$ can be split up into a number of matrices, containing terms with the same powers of $\lambda$, such that

$$P = P_1\lambda^{n-1} + P_2\lambda^{n-2} + \ldots + P_{n-1}\lambda + P_n \quad \text{...}(ii)$$

where $P_1, P_2, \ldots, P_n$ are all the square matrices of order $n$ whose elements are functions of the elements of $A$.

---

**William Rowan Hamilton* (1805–1865) an Irish mathematician who is known for his work in dynamics.

Since the product of a matrix by its adjoint = determinant of the matrix × unit matrix.

$\therefore \qquad [A - \lambda I]P = |A - \lambda I| \times I$

$\therefore$ by Eqns. (*i*) and (*ii*),
$$[A - \lambda I]\,[P_1\lambda^{n-1} + P_2\lambda^{n-2} + \ldots + P_{n-1}\lambda + P_n] = [(-1)^n\lambda^n + k_1\lambda^{n-1} + \ldots + k_{n-1}\lambda + k_n]\,I.$$

Equating the coefficients of various powers of $\lambda$, we get
$$-P_1 = (-1)^n I \qquad [\because\ IP_1 = P_1]$$
$$AP_1 - P_2 = k_1 I,$$
$$AP_2 - P_3 = k_2 I,$$
$$\ldots\ldots\ldots\ldots\ldots\ldots$$
$$AP_{n-1} - P_n = k_{n-1}I,$$
$$AP_n = k_n I.$$

Now pre-multiplying the equations by $A^n, A^{n-1}, \ldots, A, I$ respectively and adding, we get
$$(-1)^n A^n + k_1 A^{n-1} + \ldots + k_{n-1}A + k_n I = 0, \qquad \ldots(iii)$$
for the terms on the left cancel in pairs. This proves the theorem.

**Cor. Another method of finding the inverse.**

Multiplying (*iii*) by $A^{-1}$, we get
$$(-1)^n A^{n-1} + k_1 A^{n-2} + \ldots + k_{n-1}I + k_n A^{-1} = 0$$
whence
$$A^{-1} = -\frac{1}{k_n}[(-1)^n A^{n-1} + k_1 A^{n-2} + \ldots + k_{n-1}I].$$

*This result gives the inverse of A in terms of* $\overline{n-1}$ *powers of A and is considered as a practical method for the computation of the inverse of the large matrices.* As a by-product of the computation, the characteristic equation and the determinant of the matrix are also obtained.

**Example 2.48.** *Verify Cayley-Hamilton theorem for the matrix* $A = \begin{bmatrix} 1 & 4 \\ 2 & 3 \end{bmatrix}$ *and find its inverse. Also express* $A^5 - 4A^4 - 7A^3 + 11A^2 - A - 10I$ *as a linear polynomial in A*

(*Andhra, 2022; P.T.U., 2015 ; Delhi, 2012*)

**Solution.** The characteristic equation of $A$ is
$$\begin{vmatrix} 1-\lambda & 1 \\ 2 & 3-\lambda \end{vmatrix} = 0 \quad \text{or} \quad \lambda^2 - 4\lambda - 5 = 0 \qquad \ldots(i)$$

By Cayley-Hamilton theorem, $A$ must satisfy its characteristic equation (*i*), so that
$$A^2 - 4A - 5I = 0 \qquad \ldots(ii)$$

Now
$$A^2 - 4A - 5I = \begin{bmatrix} 1 & 4 \\ 2 & 3 \end{bmatrix}\begin{bmatrix} 1 & 4 \\ 2 & 3 \end{bmatrix} - 4\begin{bmatrix} 1 & 4 \\ 2 & 3 \end{bmatrix} - 5\begin{bmatrix} 1 & 0 \\ 0 & 1 \end{bmatrix}$$
$$= \begin{bmatrix} 9 & 16 \\ 8 & 17 \end{bmatrix} - \begin{bmatrix} 4 & 16 \\ 8 & 12 \end{bmatrix} - \begin{bmatrix} 5 & 0 \\ 0 & 5 \end{bmatrix} = \begin{bmatrix} 0 & 0 \\ 0 & 0 \end{bmatrix} = 0$$

This verifies the theorem.

Multiplying (*ii*) by $A^{-1}$, we get $A - 4I - 5A^{-1} = 0$

or
$$A^{-1} = \frac{1}{5}(A - 4I) = \frac{1}{5}\left\{\begin{bmatrix} 1 & 4 \\ 2 & 3 \end{bmatrix} - 4\begin{bmatrix} 1 & 0 \\ 0 & 1 \end{bmatrix}\right\} = \frac{1}{5}\begin{bmatrix} -3 & 4 \\ 2 & -1 \end{bmatrix}$$

Now dividing the polynomial $\lambda^5 - 4\lambda^4 - 7\lambda^3 + 11\lambda^2 - \lambda - 10I$ by the polynomial $\lambda^2 - 4\lambda - 5$, we obtain
$$\lambda^5 - 4\lambda^4 - 7\lambda^3 - \lambda - 10I = (\lambda^2 - 4\lambda - 5)(\lambda^3 - 2\lambda + 3) + \lambda + 5$$
$$= \lambda + 5 \qquad [\text{By } (i)]$$

Hence $A^5 - 4A^4 - 7A^3 + 11A^2 - A - 10I = A + 5I$, which is a linear polynomial in $A$.

**Example 2.49.** *Find the characteristic equation of the matrix* $A = \begin{bmatrix} 1 & 1 & 3 \\ 1 & 3 & -3 \\ -2 & -4 & -4 \end{bmatrix}$ *and hence find its inverse.*

(*C.S.V.T.U., 2015 ; P.T.U., 2014*)

**Solution.** The characteristic equation is $\begin{vmatrix} 1-\lambda & 1 & 3 \\ 1 & 3-\lambda & -3 \\ -2 & -4 & -4-\lambda \end{vmatrix} = 0,$ *i.e.*, $\lambda^3 - 20\lambda + 8 = 0.$

By Cayley-Hamilton theorem, $A^3 - 20A + 8I = 0$, whence $A^{-1} = \frac{5}{2}I - \frac{1}{8}A^2,$

$$= \frac{5}{2}\begin{bmatrix} 1 & 0 & 0 \\ 0 & 1 & 0 \\ 0 & 0 & 1 \end{bmatrix} - \frac{1}{8}\begin{bmatrix} -4 & -8 & -12 \\ 10 & 22 & 6 \\ 2 & 2 & 22 \end{bmatrix} = \begin{bmatrix} 3 & 1 & 3/2 \\ -5/4 & -1/4 & -3/4 \\ -1/4 & -1/4 & -1/4 \end{bmatrix}$$ [cf. Ex. 2.21

**Example 2.50.** *Find the characteristic equation of the matrix,* $A = \begin{bmatrix} 2 & 1 & 1 \\ 0 & 1 & 0 \\ 1 & 1 & 2 \end{bmatrix}$ *and hence compute* $A^{-1}$.

*(U.T.U., 2010)*

*Also find the matrix represented by*
$A^8 - 5A^7 + 7A^6 - 3A^5 + A^4 - 5A^3 + 8A^2 - 2A + I.$ *(Andhra, 2021, 2018; C.S.V.T.U., 2010S)*

**Solution.** The characteristic equation of the matrix $A$ is

$$\begin{vmatrix} 2-\lambda & 1 & 1 \\ 0 & 1-\lambda & 0 \\ 1 & 1 & 2-\lambda \end{vmatrix} = 0 \quad \text{or} \quad [\lambda^3 - 5\lambda^2 + 7\lambda - 3 = 0]$$

According to Cayley-Hamilton theorem, we have $A^3 - 5A^2 + 7A - 3I = 0$ ...(*i*)
Multiplying Eqn. (*i*) by $A^{-1}$, we get

$$A^2 - 5A + 7I - 3A^{-1} = 0 \quad \text{or} \quad A^{-1} = \frac{1}{3}[A^2 - 5A + 7I] \qquad ...(ii)$$

But $$A^2 = \begin{bmatrix} 2 & 1 & 1 \\ 0 & 1 & 0 \\ 1 & 1 & 2 \end{bmatrix}\begin{bmatrix} 2 & 1 & 1 \\ 0 & 1 & 0 \\ 1 & 1 & 2 \end{bmatrix} = \begin{bmatrix} 4+0+1 & 2+1+1 & 2+0+2 \\ 0+0+0 & 0+1+0 & 0+0+0 \\ 2+0+2 & 1+1+2 & 1+0+4 \end{bmatrix} = \begin{bmatrix} 5 & 4 & 4 \\ 0 & 1 & 0 \\ 4 & 4 & 5 \end{bmatrix}$$

$$\therefore \quad A^2 - 5A + 7I = \begin{bmatrix} 5 & 4 & 4 \\ 0 & 1 & 0 \\ 4 & 4 & 5 \end{bmatrix} - 5\begin{bmatrix} 2 & 1 & 1 \\ 0 & 1 & 0 \\ 1 & 1 & 2 \end{bmatrix} + 7\begin{bmatrix} 1 & 0 & 0 \\ 0 & 1 & 0 \\ 0 & 0 & 1 \end{bmatrix} = \begin{bmatrix} 2 & -1 & -1 \\ 0 & 3 & 0 \\ -1 & -1 & 2 \end{bmatrix}$$

Hence from Eqn. (*ii*), $A^{-1} = \frac{1}{3}\begin{bmatrix} 2 & -1 & -1 \\ 0 & 3 & 0 \\ -1 & -1 & 2 \end{bmatrix}$

$$A^8 - 5A^7 + 7A^6 - 3A^5 + A^4 - 5A^3 + 8A^2 - 2A + I$$
$$= A^5(A^3 - 5A^2 + 7A - 3I) + A(A^3 - 5A^2 + 7A - 3I) + A^2 + A + I$$
$$= A^2 + A + I \qquad [\because \; A^3 - 5A^2 + 7A - 3I = 0]$$
$$= \begin{bmatrix} 5 & 4 & 4 \\ 0 & 1 & 0 \\ 4 & 4 & 5 \end{bmatrix} + \begin{bmatrix} 2 & 1 & 1 \\ 0 & 1 & 0 \\ 1 & 1 & 2 \end{bmatrix} + \begin{bmatrix} 1 & 0 & 0 \\ 0 & 1 & 0 \\ 0 & 0 & 1 \end{bmatrix} = \begin{bmatrix} 8 & 5 & 5 \\ 0 & 3 & 0 \\ 5 & 5 & 8 \end{bmatrix}.$$

**Example 2.51.** *Find the Eigen values and Eigen vectors of the following matrix* $A = \begin{bmatrix} 14 & -10 \\ 5 & -1 \end{bmatrix}$.

**Solution. Step 1:** The characteristic equation of $A$ is $|A - \lambda I| = 0$.

$$\begin{vmatrix} 14-\lambda & -10 \\ 5 & -1-\lambda \end{vmatrix} = 0$$

$(14 - \lambda)(-1 - \lambda) + 50 = 0$ or $\lambda^2 - 13\lambda + 36 = 0$ or $(\lambda - 9)(\lambda - 4) = 0$
or, $\lambda^2 - S_1\lambda + |A| = 0$, where $S_1 = 14 - 1 = 13$ and $|A| = -14 + 50 = 36$

$$\lambda^2 - 13\lambda + 36 = 0$$

Hence $\lambda = 9, \lambda = 4$ are the required Eigen values of $A$.

$\lambda_1 + \lambda_2 = a_{11} + a_{22}$ *i.e.*, $\quad 9 + 4 = 14 - 1.$

**Step 2:** Let $\lambda_1 = 9$, $\lambda_2 = 4$ be the Eigen values of $A$. To find the Eigen vectors for the corresponding Eigen values, we will consider the matrix equation.

$$(A - \lambda I)X = 0$$

Where $$X = \begin{bmatrix} x \\ y \end{bmatrix} \text{ i.e., } \begin{bmatrix} 14-\lambda & -10 \\ 5 & -1-\lambda \end{bmatrix}\begin{bmatrix} x \\ y \end{bmatrix} = \begin{bmatrix} 0 \\ 0 \end{bmatrix} \quad ...(1)$$

**Step 3:** For $\lambda = \lambda_1 = 9$, Let the corresponding Eigen vector be $X_1 = \begin{bmatrix} x \\ y \end{bmatrix}$

Put $\lambda = \lambda_1 = 9$ in Eqns. (1).

$\therefore$ The matrix Eqns. (1) will be become $\begin{bmatrix} 5 & -10 \\ 5 & -10 \end{bmatrix}\begin{bmatrix} x \\ y \end{bmatrix} = \begin{bmatrix} 0 \\ 0 \end{bmatrix}$

By $R_1$; $5x - 10y = 0$, $x - 2y = 0$, we set $y = t$, $x = 2t$

For $\lambda_1 = 9$, $X_1 = \begin{bmatrix} 2t \\ t \end{bmatrix}$ or $t\begin{bmatrix} 2 \\ 1 \end{bmatrix}$. Hence for $\lambda_1 = 9$ the eigen vector is $X_1 \begin{bmatrix} 2 \\ 1 \end{bmatrix}$.

**Step 4:** For $\lambda = \lambda_2 = 4$, Let the corresponding Eigen vector be $X_2 = \begin{bmatrix} x \\ y \end{bmatrix}$

put $\lambda = \lambda_2 = 4$ in Eqns. (1)

$\therefore$ The matrix Eqns. (1) will become $\begin{bmatrix} 10 & -10 \\ 5 & -5 \end{bmatrix}\begin{bmatrix} x \\ y \end{bmatrix} = \begin{bmatrix} 0 \\ 0 \end{bmatrix}$

By $R_1$ $\quad 10x - 10y = 0, x - y = 0, x = y.$

We set $\quad y = t, x = t.$

For $\quad \lambda_2 = 4, X_2 = \begin{bmatrix} t \\ t \end{bmatrix}$ or $t\begin{bmatrix} 1 \\ 1 \end{bmatrix}$

Hence for $\lambda_2 = 4$, the Eigen vector is $X_2 = \begin{bmatrix} 1 \\ 1 \end{bmatrix}$

**Step 5.** $\lambda_1 = 9 \to X_1 = \begin{bmatrix} 2 \\ 1 \end{bmatrix}$,

$$\lambda_2 = 4 \to X_2 = \begin{bmatrix} 1 \\ 1 \end{bmatrix}$$

**Example 2.52.** *Find the Eigen values and Eigen vectors of the following matrix* $A = \begin{bmatrix} 1 & 0 & -1 \\ 1 & 2 & 1 \\ 2 & 2 & 3 \end{bmatrix}$.

**Solution. Step-1 :** The characteristic equation of $A$ is $|A - \lambda I| = 0$.

i.e., $$\begin{vmatrix} 1-\lambda & 0 & -1 \\ 1 & 2-\lambda & 1 \\ 2 & 2 & 3-\lambda \end{vmatrix} = 0$$

$$\lambda^3 - S_1\lambda^2 + S_2\lambda - |A| = 0$$

Where, $\quad S_1 = 1 + 2 + 3, S_1 = 6$

$$S_2 = \begin{vmatrix} 2 & 1 \\ 2 & 3 \end{vmatrix} + \begin{vmatrix} 1 & -1 \\ 2 & 3 \end{vmatrix} + \begin{vmatrix} 1 & 0 \\ 1 & 2 \end{vmatrix}$$

$$= 4 + 5 + 2$$
$$S_2 = 11$$
$$|A| = |(6-2) - 1(2-4)| = 6$$

$\therefore$ The characteristic equation of $A$ is $\lambda^3 - 6\lambda^2 + 11\lambda - 6 = 0$. ...(1)

**Step 2:** To find the roots of (1), we note that sum of coefficients = 0.

$\lambda - 1 = 0$ must be the factor of (1).

$$\begin{array}{c|cccc} 1 & 1 & -6 & 11 & -6 \\ & & 1 & -5 & 6 \\ \hline & 1 & -5 & 6 & 0 \end{array}$$

The factors are $(\lambda - 1)(\lambda^2 - 5\lambda + 6) = 0$

$$(\lambda - 1)(\lambda - 2)(\lambda - 3) = 0$$

$\lambda = 1, \lambda = 2, \lambda = 3$ are the required Eigen value of $A$,

$$\lambda_1 + \lambda_2 + \lambda_3 = a_{11} + a_{22} + a_{33}$$
$$1 + 2 + 3 = 1 + 2 + 3$$

**Step 3:** Let $\lambda_1 = 1, \lambda_2 = 2, \lambda_3 = 3$ be the Eigen values of $A$.

To find the Eigen vectors for the corresponding Eigen values. We will consider the matrix equation.

$$(A - \lambda I)X = 0, \qquad \text{where } X = \begin{bmatrix} x \\ y \\ z \end{bmatrix}$$

$$\begin{bmatrix} 1-\lambda & 0 & -1 \\ 1 & 2-\lambda & 1 \\ 2 & 2 & 3-\lambda \end{bmatrix} \begin{bmatrix} x \\ y \\ z \end{bmatrix} = \begin{bmatrix} 0 \\ 0 \\ 0 \end{bmatrix} \quad ...(2)$$

**Step 4:** For Eigen values $\lambda = \lambda_1 = 1$, Let the corresponding Eigen vector be $X_1 = \begin{bmatrix} x \\ y \\ z \end{bmatrix}$.

put $\lambda = \lambda_1 = 1$ $\therefore$ The matrix Eqn. (2) will become

$$\begin{bmatrix} 0 & 0 & -1 \\ 1 & 1 & 1 \\ 2 & 2 & 2 \end{bmatrix} \begin{bmatrix} x \\ y \\ z \end{bmatrix} = \begin{bmatrix} 0 \\ 0 \\ 0 \end{bmatrix} \text{ i.e., } AX = Z$$

which is Homogeneous system of equations.

By $R_1$ $\quad 0(x) + 0(y) - 1(z) = 0 \quad \therefore Z = 0$

$$x + y + z = 0 \quad x + y = 0$$

Let $y = t, x = -t$

Hence for $\lambda_1 = 1$, the corresponding Eigen Vector is

$$X_1 = \begin{bmatrix} -t \\ t \\ 0 \end{bmatrix} = t \begin{bmatrix} -1 \\ 1 \\ 0 \end{bmatrix}$$

By considering the particular solution (vector), we have of $\lambda_1 = 1$,

$$X_1 = \begin{bmatrix} -1 \\ 1 \\ 0 \end{bmatrix}$$

**Step 5:** For Eigen value $\lambda = \lambda_2 = 2$. Let the corresping Eigen vector be $X_2 = \begin{bmatrix} x \\ y \\ z \end{bmatrix}$.

put $\lambda = \lambda_2 = 2$, $\therefore$ The matrix equation (2) will become $\begin{bmatrix} -1 & 0 & -1 \\ 1 & 0 & 1 \\ 2 & 2 & 1 \end{bmatrix} \begin{bmatrix} x \\ y \\ z \end{bmatrix} = \begin{bmatrix} 0 \\ 0 \\ 0 \end{bmatrix}$.

*i.e.*, $AX = 7$, which is homogeneous system of equation by $R_1 - x - y = 0$

Let $Z = t$, $\therefore x = -t$

By $R_3$, $2x + 2y + z = 0$, $y = t/2]$

$$\therefore \qquad x_2 = \begin{bmatrix} -t \\ t/2 \\ t \end{bmatrix} \text{or} \begin{bmatrix} -2t \\ t \\ 2t \end{bmatrix} \text{or } t \begin{bmatrix} -2 \\ 1 \\ 2 \end{bmatrix}$$

Hence for $\lambda_2 = 2$, the corresponding Eigen vector is $X_2 = \begin{bmatrix} -2 \\ 1 \\ 2 \end{bmatrix}$.

**Step 6:** For Eigen value $\lambda = \lambda_3 = 3$. Let the corresponding.

Eigen vector be $X_3 = \begin{bmatrix} X \\ Y \\ Z \end{bmatrix}$. Put $\lambda = \lambda_3 = 3$.

$\therefore$ The matrix equation (2) will become

$$\begin{bmatrix} -2 & 0 & -1 \\ 1 & -1 & 1 \\ 2 & 2 & 0 \end{bmatrix} \begin{bmatrix} X \\ Y \\ Z \end{bmatrix} = \begin{bmatrix} 0 \\ 0 \\ 0 \end{bmatrix}$$

$AX = Z$, which is homogeneous system of equations.

By $R_1$, $-2x - z = 0$ Let $x = t$, $z = -2t$

By $R_3$, $2x + 2y = 0$, $x + y = 0, y = -t$

$$X_3 = \begin{bmatrix} t \\ -t \\ 2t \end{bmatrix} \quad \text{or} \quad t = \begin{bmatrix} 1 \\ -1 \\ -2 \end{bmatrix}$$

Hence for $\lambda_3 = 3$ the correspoing Eigen vector is $X_3 = \begin{bmatrix} 1 \\ -1 \\ -2 \end{bmatrix}$

**Step 7:** $\lambda_1 = 1 \rightarrow X_1 = \begin{bmatrix} -1 \\ 1 \\ 0 \end{bmatrix}$, $\lambda_2 = 2 \rightarrow X_2 = \begin{bmatrix} -2 \\ 1 \\ 2 \end{bmatrix}$, $\lambda_3 = 3 \rightarrow X_3 = \begin{bmatrix} 1 \\ -1 \\ -2 \end{bmatrix}$

## PROBLEMS 2.9

1. Find the sum and product of the eigen values of $\begin{bmatrix} 3 & 1 & 4 \\ 0 & 2 & 6 \\ 0 & 0 & 5 \end{bmatrix}$. *(Andhra, 2015 ; C.S.V.T.U., 2014)*

2. Find the eigen values and eigen vectors of the matrices :

   (*a*) $\begin{bmatrix} 4 & 3 \\ 2 & 9 \end{bmatrix}$

   (*b*) Find the Eigen values and the corresponding Eigen vectors for the following matrices. $\begin{bmatrix} 15 & 0 & -15 \\ -3 & 6 & 9 \\ 5 & 0 & -5 \end{bmatrix}$

   *(UPTU, 2019)*

3. Find the latent roots and the latent vectors of the matrices :

   (*a*) $\begin{bmatrix} 1 & 1 & 0 \\ 0 & 1 & 1 \\ 0 & 0 & 1 \end{bmatrix}$ *(P.T.U., 2015)*

   (*b*) $\begin{bmatrix} 8 & -6 & 2 \\ -6 & 7 & -4 \\ 2 & -4 & 3 \end{bmatrix}$ *(Delhi, 2015 ; V.T.U., 2014 ; Calicut, 2013 ; U.P.T.U., 2011)*

   (*c*) $\begin{bmatrix} 2 & 0 & 1 \\ 0 & 2 & 0 \\ 1 & 0 & 2 \end{bmatrix}$ *(Andhra, 2022; C.S.V.T.U., 2014 ; D.T.U., 2011)*

   (*d*) $\begin{bmatrix} -2 & 2 & -3 \\ 2 & 1 & -6 \\ -1 & -2 & 0 \end{bmatrix}$ *(Andhra, 2021; D.T.U., 2015 ; P.T.U., 2013 S ; Rohtak, 2012)*

   (*e*) $\begin{bmatrix} 6 & -2 & 2 \\ -2 & 3 & -1 \\ 2 & -1 & 3 \end{bmatrix}$ *(Andhra, 2022; JNTUK, 2022, D.T.U., 2016 ; C.S.V.T.U., 2015 ; V.T.U., 2013)*

4. If $\lambda$ be an eigen value of a non-singular matrix $A$, show that $|A|/\lambda$ is an eigen value of the matrix *adj A*.

5. Find the eigen values of adj $A$ and of $A^2 - 2A + I$, where $A = \begin{bmatrix} 2 & 3 & 4 \\ 0 & 4 & 2 \\ 0 & 0 & 3 \end{bmatrix}$

6. Two eigen values of the matrix $A = \begin{bmatrix} 2 & 2 & 1 \\ 1 & 3 & 1 \\ 1 & 2 & 2 \end{bmatrix}$ are = 1 each. Find the eigen values of $A^{-1}$.

7. Show that if $\lambda_1, \lambda_2, ..., \lambda_n$ are the latent roots of a matrix $A$, then (*i*) $A^2$ has the latent roots $\lambda_1^2, \lambda_2^2, ... \lambda_n^2$. *(Anna., 2016)*

   (*ii*) $1k_a\lambda_i$ are the eigen values of $k\lambda_{-i}$.

   (*iii*) $1/\lambda_i$ are the eigen values of $A^{-i}$. *(Anna, 2012)*

8. The eigen vectors of a 3 × 3 matrix A corresponding to the eigen values 1, 1, 3 are $[1, 0, -1]^T$, $[0, 1, -1]^T$ and $[1, 1, 0]^T$ respectively, find the matrix . *(U.K.T.U., 2013)*

9. Using Cayley-Hamilton theorem, find the inverse of

   (*i*) $\begin{bmatrix} 5 & 3 \\ 3 & 2 \end{bmatrix}$ (*ii*) $\begin{bmatrix} 1 & 1 & 2 \\ 0 & -2 & 0 \\ 0 & 0 & 3 \end{bmatrix}$

   (*iii*) $\begin{bmatrix} 1 & 2 & 3 \\ 2 & 4 & 5 \\ 3 & 5 & 6 \end{bmatrix}$ *(Andhra, 2022)* (*iv*) $\begin{bmatrix} 1 & 1 & 0 \\ 1 & 0 & 1 \\ 0 & 1 & 1 \end{bmatrix}$ *(Andhra, 2023)*

10. (*i*) Matrix $A$ is Non-symmetric, Eigen values are different. Find the Eigen values and the corresponding Eigen vectors for the following matrices.

$$\begin{bmatrix} 9 & -1 & 9 \\ 3 & -1 & 3 \\ -7 & 1 & -7 \end{bmatrix}$$ *(Bhopal, 2018)*

(*ii*) Find the Eigen values and the corresponding Eigen vectors of the following matrices.

$$\begin{bmatrix} 3 & 1 & 1 \\ 1 & 3 & -1 \\ 1 & -1 & 3 \end{bmatrix}$$

11. Find the characteristic equation of the matrix $A = \begin{bmatrix} 1 & 3 & 7 \\ 4 & 2 & 3 \\ 1 & 2 & 1 \end{bmatrix}$. Show that the equation is satisfied by $A$ and hence obtain the inverse of the given matrix. *(Calicut, 2013 ; Bhopal, 2008 ; Anna, 2005)*

12. Verify Cayley-Hamilton theorem for the matrix $A$ and find its inverse.

(*a*) $\begin{bmatrix} 2 & -1 & 1 \\ -1 & 2 & -1 \\ 1 & -1 & 2 \end{bmatrix}$ *(Andhra, 2021; Anna, 2013 ; C.S.V.T.U., 2012 ; U.P.T.U., 2012)*

(*b*) $\begin{bmatrix} 1 & 2 & 3 \\ 2 & -1 & 4 \\ 3 & 1 & 1 \end{bmatrix}$ *(Anna, 2016 ; Andhra., 2014)*

(*c*) $\begin{bmatrix} 7 & -1 & 3 \\ 6 & 1 & 4 \\ 2 & 4 & 8 \end{bmatrix}$ *(J.N.T.U., 2015)* (*d*) $\begin{bmatrix} 7 & 2 & -2 \\ -6 & -1 & 2 \\ 6 & 2 & -1 \end{bmatrix}$ *(D.T.U., 2016)*

13. Using Cayley-Hamilton theorem, find $A^8$, if $A = \begin{bmatrix} 1 & 2 \\ 2 & -1 \end{bmatrix}$.

14. If $A = \begin{bmatrix} 2 & -1 & 2 \\ -1 & 2 & -1 \\ 1 & -1 & 2 \end{bmatrix}$, find $A^4$.

15. Using Cayley-Hamilton theorem, find $A^{-2}$, where $A = \begin{bmatrix} 1 & 2 & 0 \\ 2 & -1 & 0 \\ 0 & 0 & -1 \end{bmatrix}$. *(Andhra, 2022)*

16. If $A = \begin{bmatrix} 4 & 6 & 6 \\ 1 & 3 & 2 \\ -1 & -4 & -3 \end{bmatrix}$, evaluate $A^{-1}, A^{-2}$ and $A^{-3}$.

17. If $A = \begin{bmatrix} 1 & 0 & 0 \\ 1 & 0 & 1 \\ 0 & 1 & 0 \end{bmatrix}$, show that $A^n = A^{n-2} + A^2 - 1$. Hence find $A^{50}$.

18. Matrix $A$ is Non-Symmetric, Eigen values are different. Find the Eigen values and the correspond Eigen vectors for the following matries. *(CSVTU, 2017)*

$$\begin{bmatrix} 1 & 20 & 0 \\ -1 & 7 & 1 \\ 3 & 6 & -2 \end{bmatrix}$$

## 2.16 (1) REDUCTION TO DIAGONAL FORM

*If a square matrix A of order n has n linearly independent eigen vectors, then a matrix P can be found such that $P^{-1}AP$ is a diagonal matrix.*

[This result will be proved for a square matrix of order 3 but the method will be capable of easy extension to matrices of any order.]

Let $A$ be a square matrix of order 3. Let $\lambda_1, \lambda_2, \lambda_3$ be its eigen values and

$$X_1 = \begin{bmatrix} x_1 \\ y_1 \\ z_1 \end{bmatrix}, X_2 = \begin{bmatrix} x_2 \\ y_2 \\ z_2 \end{bmatrix} \text{ and } X_3 = \begin{bmatrix} x_3 \\ y_3 \\ z_3 \end{bmatrix} \text{ be the corresponding eigen vectors.}$$

Denoting the square matrix $[X_1 X_2 X_3] = \begin{bmatrix} x_1 & x_2 & x_3 \\ y_1 & y_2 & y_3 \\ z_1 & z_2 & z_3 \end{bmatrix}$ by $P$, we have

$$AP = A[X_1 X_2 X_3] = [AX_1, AX_2, AX_3] = [\lambda_1 X_1, \lambda_2 X_2, \lambda_3 X_3]$$

$$= \begin{bmatrix} \lambda_1 x_1 & \lambda_2 x_2 & \lambda_3 x_3 \\ \lambda_1 y_1 & \lambda_2 y_2 & \lambda_3 y_3 \\ \lambda_1 z_1 & \lambda_2 z_2 & \lambda_3 z_3 \end{bmatrix} = \begin{bmatrix} x_1 & x_2 & x_3 \\ y_1 & y_2 & y_3 \\ z_1 & z_2 & z_3 \end{bmatrix} \times \begin{bmatrix} \lambda_1 & 0 & 0 \\ 0 & \lambda_2 & 0 \\ 0 & 0 & \lambda_3 \end{bmatrix} = PD, \text{ where } D \text{ is the diagonal matrix.}$$

$\therefore \quad P^{-1}AP = P^{-1}PD = D$, which proves the theorem.

**Obs. 1.** *The matrix P which diagonalises A is called the* ***modal matrix*** *of A and the resulting diagonal matrix D is known as the* ***spectral matrix*** *of A.*

**2.** *The diagonal matrix has the eigen values of A as its diagonal elements.*

**3.** *The matrix P, which diagonalise A, constitutes the eigen vectors of A.*

**(2) Similarity of matrices.** *A square matrix $\hat{A}$ of order n is called* **similar** *to a square matrix A of order n if $\hat{A} = P^{-1}AP$ for some non-singular $n \times n$ matrix P.*

This transformation of a matrix $A$ by a non-singular matrix $P$ to $\hat{A}$ is called a **similarity transformation.**

**Obs.** *If the matrix $\hat{A}$ is similar to the matrix A, then $\hat{A}$ has the same eigen values as A.*

*If* **x** *is an eigen vector of A, then $y = P^{-1}\mathbf{x}$ is an eigen vector of $\hat{A}$ corresponding to the same eigen value.*

**(3) Powers of a matrix.** *Diagonalisation of a matrix is quite useful for obtaining powers of a matrix.* Let $A$ be the square matrix. Then a non-singular matrix $P$ can be found such that

$$D = P^{-1}AP$$

$\therefore \quad D^2 = (P^{-1}AP)(P^{-1}AP) = P^{-1}A^2P \qquad [\because PP^{-1} = I]$

Similarly, $\quad D^3 = P^{-1}A^3P$ and in general, $D^n = P^{-1}A^nP \qquad ...(i)$

To obtain $A^n$, premultiply Eqn. $(i)$ by $P$ and post-multiply by $P^{-1}$.

Then $\quad PD^nP^{-1} = PP^{-1}A^nPP^{-1} = A^n$ which gives $A^n$.

Thus, $\quad A^n = PD^nP^{-1}$ where, $D^n = \begin{bmatrix} \lambda_1^n & 0 & 0 \\ 0 & \lambda_2^n & 0 \\ 0 & 0 & \lambda_3^n \end{bmatrix}$

**Working procedure :**

1. Find the eigen values of the square matrix $A$.
2. Find the corresponding eigen vectors and write the modal matrix $P$.
3. Find the diagonal matrix $D$ from $D = P^{-1}AP$
4. Obtain $A^n$ from $A^n = PD^nP^{-1}$.

**Example 2.53.** *Reduce the matrix* $A = \begin{bmatrix} -1 & 2 & -2 \\ 1 & 2 & 1 \\ -1 & -1 & 0 \end{bmatrix}$ *to the diagonal form.*

*(Andhra, 2023, 2021; V.T.U., 2011 ; U.T.U., 2010)*

**Solution:** The characteristic equation of $A$ is

$$\begin{vmatrix} -1-\lambda & 2 & -2 \\ 1 & 2-\lambda & 1 \\ -1 & -1 & -\lambda \end{vmatrix} = 0 \quad \text{or} \quad \lambda^3 - \lambda^2 - 5\lambda + 5 = 0.$$

Solving, we get $\lambda_1 = 1, \lambda_2 = \sqrt{5}, \lambda_3 = -\sqrt{5}$ as the eigen values of $A$.

When $\lambda = 1$, the corresponding eigen vector is given by

$$-2x + 2y - 2z = 0, x + y + z = 0, -x - y - z = 0$$

Solving the first two equations, we get $\frac{x}{2} = \frac{y}{0} = \frac{z}{-2}$ giving the eigen vector (1, 0, –1)

When $\lambda = \sqrt{5}$, the corresponding eigen vector is given by

$$(-1 - \sqrt{5})x + 2y - 2z = 0, x + (2 - \sqrt{5})y + z = 0, -x - y - \sqrt{5}z = 0.$$

Solving 2nd and 3rd equations, we get

$$\frac{x}{6 - 2\sqrt{5}} = \frac{y}{-1 + \sqrt{5}} = \frac{z}{1 - \sqrt{5}} \quad \text{or} \quad \frac{x}{\sqrt{5} - 1} = \frac{y}{1} = \frac{z}{-1}$$

giving the eigen vector $(\sqrt{5} - 1, 1, -1)$.

Similarly the eigen vector corresponding to $\lambda = -\sqrt{5}$, is $(\sqrt{5} + 1, -1, 1)$.

Writing the three eigen vectors as the three columns, we get the transformation (*modal*) matrix as

$$P = \begin{bmatrix} 1 & \sqrt{5}-1 & \sqrt{5}+1 \\ 0 & 1 & -1 \\ 1 & -1 & 1 \end{bmatrix}.$$

Hence the diagonal matrix is

$$D = \begin{bmatrix} \lambda_1 & 0 & 0 \\ 0 & \lambda_2 & 0 \\ 0 & 0 & \lambda_3 \end{bmatrix} = \begin{bmatrix} 1 & 0 & 0 \\ 0 & \sqrt{5} & 0 \\ 0 & 0 & -\sqrt{5} \end{bmatrix}.$$

**Example 2.54.** *Find the matrix P which transforms the matrix* $A = \begin{bmatrix} 1 & 1 & 3 \\ 1 & 5 & 1 \\ 3 & 1 & 1 \end{bmatrix}$ *to the diagonal form. Hence calculate* $A^4$.

**Solution:** The eigen values of $A$ (found in Ex. 2.43) are –2, 3, 6 and the eigen vectors are (–1, 0, 1), (1, –1, 1), (1, 2, 1). Writing these eigen vectors as the three columns, the required transformation matrix (*modal matrix*) is

$$P = \begin{bmatrix} -1 & 1 & 1 \\ 0 & -1 & 2 \\ 1 & 1 & 1 \end{bmatrix}$$

To find $P^{-1}$,
$$|P| = \begin{vmatrix} -1 & 1 & 1 \\ 0 & -1 & 2 \\ 1 & 1 & 1 \end{vmatrix} = \begin{vmatrix} a_1 & b_1 & c_1 \\ a_2 & b_2 & c_2 \\ a_3 & b_3 & c_3 \end{vmatrix} \quad \text{(say)}$$

$$A_1 = -3, B_1 = 2, C_1 = 1, A_2 = 0, B_2 = -2, C_2 = 2, A_3 = 3, B_3 = 2, C_3 = 1$$

Also
$$|P| = a_1A_1 + b_1B_1 + c_1C_1 = 6$$

$\therefore$
$$P^{-1} = \frac{1}{|P|}\begin{bmatrix} A_1 & A_2 & A_3 \\ B_1 & B_2 & B_3 \\ C_1 & C_2 & C_3 \end{bmatrix} = \frac{1}{6}\begin{bmatrix} -3 & 0 & 3 \\ 2 & -2 & 2 \\ 1 & 2 & 1 \end{bmatrix}$$

Thus $$D = P^{-1} AP = \begin{bmatrix} -2 & 0 & 0 \\ 0 & 3 & 0 \\ 0 & 0 & 6 \end{bmatrix}$$

$\therefore$ $$D^4 = \begin{bmatrix} (-2)^4 & 0 & 0 \\ 0 & 3^4 & 0 \\ 0 & 0 & 6^4 \end{bmatrix} = \begin{bmatrix} 16 & 0 & 0 \\ 0 & 81 & 0 \\ 0 & 0 & 1296 \end{bmatrix}$$

Hence $$A^4 = PD^4P^{-1} = \frac{1}{6}\begin{bmatrix} -1 & 1 & 1 \\ 0 & -1 & 2 \\ 1 & 1 & 1 \end{bmatrix}\begin{bmatrix} 16 & 0 & 0 \\ 0 & 81 & 0 \\ 0 & 0 & 1296 \end{bmatrix}\begin{bmatrix} -3 & 0 & 3 \\ 2 & -2 & 2 \\ 1 & 2 & 1 \end{bmatrix}$$

$$= \begin{bmatrix} -1 & 1 & 1 \\ 0 & -1 & 2 \\ 1 & 1 & 1 \end{bmatrix}\begin{bmatrix} -8 & 0 & 8 \\ 27 & -27 & 27 \\ 216 & 512 & 216 \end{bmatrix} = \begin{bmatrix} 251 & 485 & 235 \\ 485 & 1051 & 485 \\ 235 & 485 & 251 \end{bmatrix}$$

**Example 2.55.** *Find $e^A$ and $4^A$ if* $A = \begin{bmatrix} 3/2 & 1/2 \\ 1/2 & 3/2 \end{bmatrix}$.

**Solution:** The characteristic equation of $A$ is

$$\begin{vmatrix} 3/2-\lambda & 1/2 \\ 1/2 & 3/2-\lambda \end{vmatrix} = 0, \quad i.e., (3/2-\lambda)^2 - 1/4 = 0.$$

$\therefore$ $$\lambda^2 - 3\lambda + 2 = 0 \quad \text{whence } \lambda = 1, 2.$$

*When* $\lambda = 1$, $[A - \lambda I]\, X = 0$, gives

$$\begin{bmatrix} 1/2 & 1/2 \\ 1/2 & 1/2 \end{bmatrix}\begin{bmatrix} x_1 \\ x_2 \end{bmatrix} = \begin{bmatrix} 0 \\ 0 \end{bmatrix} \quad \text{or} \quad \begin{bmatrix} 1 & 1 \\ 1 & 1 \end{bmatrix}\begin{bmatrix} x_1 \\ x_2 \end{bmatrix} = \begin{bmatrix} 0 \\ 0 \end{bmatrix} \qquad [\text{By } 2R_1, 2R_2]$$

or $$\begin{bmatrix} 1 & 1 \\ 0 & 0 \end{bmatrix}\begin{bmatrix} x_1 \\ x_2 \end{bmatrix} = \begin{bmatrix} 0 \\ 0 \end{bmatrix} \qquad [\text{By } R_2 - R_1]$$

$\therefore$ $x_1 + x_2 = 0$. If $x_2 = -1$, $x_1 = 1$, *i.e.*, the eigen vector is $[1, -1]'$.

*When* $\lambda = 2$, $[A - \lambda I]\, X = 0$, gives $\begin{bmatrix} -1/2 & 1/2 \\ 1/2 & -1/2 \end{bmatrix}\begin{bmatrix} x_1 \\ x_2 \end{bmatrix} = \begin{bmatrix} 0 \\ 0 \end{bmatrix}$

or $$\begin{bmatrix} -1 & 1 \\ 1 & -1 \end{bmatrix}\begin{bmatrix} x_1 \\ x_2 \end{bmatrix} = \begin{bmatrix} 0 \\ 0 \end{bmatrix} \qquad \begin{bmatrix} \text{By } 2R_1 \\ 2R_2 \end{bmatrix}$$

or $$\begin{bmatrix} -1 & 1 \\ 0 & 0 \end{bmatrix}\begin{bmatrix} x_1 \\ x_2 \end{bmatrix} = \begin{bmatrix} 0 \\ 0 \end{bmatrix} \qquad [\text{By } R_2 - R_1]$$

$\therefore$ $$-x_1 + x_2 = 0, \quad i.e., \quad x_1 = x_2$$

If $x_2 = 1$, $x_1 = 1$, *i.e.*, the eigen vector is $[1, 1]'$

Now $D = \begin{bmatrix} 1 & 0 \\ 0 & 2 \end{bmatrix}$ and $P = \begin{bmatrix} 1 & 1 \\ -1 & 1 \end{bmatrix}$

$\therefore$ $$P^{-1} = \frac{adj\, P}{|P|} = \frac{1}{2}\begin{bmatrix} 1 & -1 \\ 1 & 1 \end{bmatrix}$$

If $$f(A) = e^A, f(D) = e^D = \begin{bmatrix} e^1 & 0 \\ 0 & e^2 \end{bmatrix}$$

$\therefore$ $$e^A = P\, f(D)\, P^{-1} = \begin{bmatrix} 1 & 1 \\ -1 & 1 \end{bmatrix}\begin{bmatrix} e^1 & 0 \\ 0 & e^2 \end{bmatrix}\frac{1}{2}\begin{bmatrix} 1 & -1 \\ 1 & 1 \end{bmatrix}$$

$$= \frac{1}{2}\begin{bmatrix} e & e^2 \\ -e & e^2 \end{bmatrix}\begin{bmatrix} 1 & -1 \\ 1 & 1 \end{bmatrix} = \frac{1}{2}\begin{bmatrix} e+e^2 & -e+e^2 \\ -e+e^2 & e+e^2 \end{bmatrix}$$

Replacing $e$ by 4, we get

$$4^A = \frac{1}{2}\begin{bmatrix} 20 & 12 \\ 12 & 20 \end{bmatrix} = \begin{bmatrix} 10 & 6 \\ 6 & 10 \end{bmatrix}.$$

## 2.17 REDUCTION OF QUADRATIC FORM TO CANONICAL FORM

A homogeneous expression of the second degree in any number of variables is called a *quadratic form.*

For instance, if $A = \begin{bmatrix} a & h & g \\ h & b & f \\ g & f & c \end{bmatrix}, X = \begin{bmatrix} x \\ y \\ z \end{bmatrix}$ and $X' = [x \;\; y \;\; z]$, then

$$X'AX = ax^2 + by^2 + cz^2 + 2fyz + 2gzx + 2hxy \qquad ...(i)$$

which is a *quadratic form.*

Let $\lambda_1, \lambda_2, \lambda_3$ be the eigen values of the matrix $A$ and

$$X_1 = \begin{bmatrix} x_1 \\ y_1 \\ z_1 \end{bmatrix}, X_2 = \begin{bmatrix} x_2 \\ y_2 \\ z_2 \end{bmatrix}, X_3 = \begin{bmatrix} x_3 \\ y_3 \\ z_3 \end{bmatrix}$$

be its corresponding eigen vectors in the normalized form (*i.e.*, each element is divided by square root of sum of the squares of all the three elements in the eigen vector).

Then by § 2.16(**1**), $P^{-1}AP = \begin{bmatrix} \lambda_1 & 0 & 0 \\ 0 & \lambda_2 & 0 \\ 0 & 0 & \lambda_3 \end{bmatrix}$ where, $P = \begin{bmatrix} x_1 & x_2 & x_3 \\ y_1 & y_2 & y_3 \\ z_1 & z_2 & z_3 \end{bmatrix}$

Hence the *quadratic form (i) is reduced to a* **canonical form (or sum of squares form** or **Principal axes form).**

$$\lambda_1 x^2 + \lambda_2 y^2 + \lambda_3 z^2$$

*and P is the* **matrix of transformation** which is an orthogonal matrix.

**Note. Congruent (or orthogonal) transformation.** The diagonal matrix $D$ and the matrix $A$ are called *congruent matrices* and *the above method of reduction is called* **congruent (or orthogonal) transformation.**

**Remember that** the matrix $A$ corresponding to the quadratic form

$$ax^2 + by^2 + cz^2 + 2fyz + 2gzx + 2hxy$$

is

$$\begin{bmatrix} \text{coeff. of } x^2 & \frac{1}{2}\text{coeff. of } xy & \frac{1}{2}\text{coeff. of } zx \\ \frac{1}{2}\text{coeff. of } xy & \text{coeff. of } y^2 & \frac{1}{2}\text{coeff. of } yz \\ \frac{1}{2}\text{coeff. of } zx & \frac{1}{2}\text{coeff. of } yz & \text{coeff. of } z^2 \end{bmatrix}, \quad i.e., \begin{bmatrix} a & h & g \\ h & b & f \\ g & f & c \end{bmatrix}.$$

**Example 2.56.** *Reduce the quadratic form $3x^2 + 5y^2 + 3z^2 - 2yz + 2zx - 2xy$ to the canonical form and specify the matrix of transformation.*

*(JNTUK, 2022; V.T.U., 2016 ; Andhra, 2014 ; Anna, 2013 ; D.T.U., 2013)*

**Solution:** The matrix of the given quadratic form is $A = \begin{bmatrix} 3 & -1 & 1 \\ -1 & 5 & -1 \\ 1 & -1 & 3 \end{bmatrix}$

Its characteristic equation is $|A - \lambda I| = 0$, *i.e.*, $\begin{vmatrix} 3-\lambda & -1 & 1 \\ -1 & 5-\lambda & -1 \\ 1 & -1 & 3-\lambda \end{vmatrix} = 0$

which gives $\lambda = 2, 3, 6$ as its eigen values. Hence the given quadratic form reduces to the canonical form

$$\lambda_1 x^2 + \lambda_2 y^2 + \lambda_3 z^2, \qquad i.e., \qquad 2x^2 + 3y^2 + 6z^2.$$

*To find the matrix of transformation*

From $[A - \lambda I]\, X = 0$, we obtain the equations

$$(3 - \lambda)\, x - y + z = 0\ ;\ -x + (5 - \lambda)\, y - z = 0\ ;\ x - y + (3 - \lambda)\, z = 0.$$

Now corresponding to $\lambda = 2$, we get $x - y + z = 0$, $-x + 3y - z = 0$, and $x - y + z = 0$,

whence
$$\frac{x}{1} = \frac{y}{0} = \frac{z}{-1}$$

$\therefore$ The eigen vector is $X_1$ (1, 0, –1) and its normalised form is $(1/\sqrt{2}, 0, -1/\sqrt{2})$.

Similarly, corresponding to $\lambda = 3$, the eigen vector is $X_2$ (1, 1, 1) and its normalised form is $(1/\sqrt{3}, 1/\sqrt{3}, 1/\sqrt{3})$.

Finally, corresponding to $\lambda = 6$, the eigen vector is $X_3$ (1, –2, 1) and its normalised form is $(1/\sqrt{6}, -2/\sqrt{6}, 1/\sqrt{6})$.

Hence the matrix of transformation is $P = \begin{bmatrix} 1/\sqrt{2} & 1/\sqrt{3} & 1/\sqrt{6} \\ 0 & 1/\sqrt{3} & -2/\sqrt{6} \\ -1/\sqrt{2} & 1/\sqrt{3} & 1/\sqrt{6} \end{bmatrix}$.

## 2.18 NATURE OF A QUADRATIC FORM

Let $Q = X'AX$ be a quadratic form in $n$ variables $x_1, x_2, \ldots x_n$.

**Index.** *The number of positive terms in its canonical form is called the index of the quadratic form.*

**Signature** (*s*) *of the quadratic form is the difference of positive and negative terms in the canonical form. If the rank of the matrix A is r and the index of the quadratic form Q is s, then the quadratic form is said to be*

(*i*) *positive definite* if $r = n$ and $s = n$

(*ii*) *negative definite* if $r = n$ and $s = 0$

(*iii*) *positive semidefinite* if $r < n$ and $s = r$

(*iv*) *negative semidefinite* if $r < n$ and $s = 0$

(*v*) *indefinite* in all other cases.

**In other words** *a real quadratic form X'AX in a variable is said to be*

(*i*) **positive definite** *if all the eigen values of A > 0.*

(*ii*) **negative definite** *if all the eigen values of A < 0.*

(*iii*) **positive semidefinite** *if all the eigen values of A ≥ 0 and at least one eigen value = 0.*

(*iv*) **negative semidefinite** *if all the eigen values of A ≤ 0 and at least one eigen value = 0.*

(*v*) **indefinite** *if some of the eigen values of A are positive and others negative.*

**Example 2.57.** *Reduce the quadratic form $2x_1x_2 + 2x_1x_3 - 2x_2x_3$ to a canonical form by an orthogonal reduction and discuss its nature.*

*Also find the modal matrix.* (*Andhra, 2022*)

**Solution:** (*i*) The matrix of the given quadratic form is $A = \begin{bmatrix} 0 & 1 & 1 \\ 1 & 0 & -1 \\ 1 & -1 & 0 \end{bmatrix}$

Its characteristic equation is $[A - \lambda I] = 0$, *i.e.*, $\begin{bmatrix} -\lambda & 1 & 1 \\ 1 & -\lambda & -1 \\ 1 & -1 & -\lambda \end{bmatrix} = 0$

which gives $\lambda^3 - 3\lambda + 2 = 0$

Solving, we get $\lambda = 1, 1, -2$ as the eigen values. Hence the given quadratic form reduces to the canonical form

$$\lambda_1 x^2 + \lambda_2 y^2 + \lambda_3 z^2 = 0, \quad i.e., \quad x^2 + y^2 - 2z^2 = 0$$

(*ii*) Since some of the eigen values of $A$ are positive and others are negative, the given quadratic form is *Indefinite*.

(*iii*) *To find the matrix of transformation*

From $[A - \lambda I]\, X = 0$, we get the equations

$$-\lambda x + y + z = 0, x - \lambda y + z = 0, x - y - \lambda z = 0$$

*When* $\lambda = -2$, we get $\quad 2x + y + z = 0, x + 2y - z = 0, x - y + 2z = 0.$

Solving first and second equations, we get

$$\frac{x}{-1} = \frac{y}{1} = \frac{z}{1}$$

$\therefore$ The corresponding eigen vector $X_1 = (-1, 1, 1)$ and its normalised form is $(-1/\sqrt{3}, 1/\sqrt{3}, 1/\sqrt{3})$

*When* $\lambda = 1$, we get $\quad -x + y + z = 0, x - y - z = 0, x - y - z = 0.$

These equations are same. Take $y = 0$ so that $x = z$.

$\therefore$ The corresponding eigen vector $X_2 = (1, 0, 1)$ and its normalised form is $(1/\sqrt{2}, 0, 1/\sqrt{2})$

*To find the eigen vector* $X_3 = (l, m, n)$ (say)

Since $X_3$ is orthogonal to $X_1$, $\therefore \quad -l + m + n = 0$

Since $X_3$ is orthogonal to $X_2$, $\therefore \quad l + n = 0$

These equations give $\dfrac{l}{1} = \dfrac{m}{2} = \dfrac{n}{-1}$.

$\therefore$ The eigen vector $X_3 = (1, 2, -1)$ and normalised form is $(1/\sqrt{6}, 2/\sqrt{6}, -1/\sqrt{6})$.

Hence the modal matrix is

$$P = \begin{bmatrix} -1/\sqrt{3} & 1/\sqrt{2} & 1/\sqrt{6} \\ 1/\sqrt{3} & 0 & 2/\sqrt{6} \\ 1/\sqrt{3} & 1/\sqrt{2} & -1/\sqrt{6} \end{bmatrix}.$$

**Example 2.58.** *Find the characteristic equation for* $A = \begin{bmatrix} 1 & 4 \\ 2 & 3 \end{bmatrix}$ *and use it to find the simplified expression for* $A^5 + 5A^4 - 6A^3 + 2A^2 - 4A + 7I$. *(V.T.U. 2019)*

**Solution.** The characteristic equation of $A$ is $|A - \lambda I| = 0$

**Step 1:** $\begin{vmatrix} 1-\lambda & 4 \\ 2 & 3-\lambda \end{vmatrix} = 0$

$$\lambda^2 - S_1\lambda + |A| = 0$$

$$|A| = 3 - 8 = -5$$

$\lambda^2 - 4\lambda - 5 = 0$ which is the required characteristic equation of $A$. ...(*i*)

**Step 2:** By Cayley - Hamilton theorem $A^2 - 4A - 5I = 0$

**Step 3:** To verify it, we require $A^2$

$$A^2 = \begin{bmatrix} 1 & 4 \\ 2 & 3 \end{bmatrix}\begin{bmatrix} 1 & 4 \\ 2 & 3 \end{bmatrix} = \begin{bmatrix} 9 & 16 \\ 8 & 17 \end{bmatrix}$$

$$A^2 - 4A - 5I = \begin{bmatrix} 9 & 16 \\ 8 & 17 \end{bmatrix} - 4\begin{bmatrix} 1 & 4 \\ 2 & 3 \end{bmatrix} - \begin{bmatrix} 5 & 0 \\ 0 & 5 \end{bmatrix} = \begin{bmatrix} 0 & 0 \\ 0 & 0 \end{bmatrix}$$

**Step 4:** From Eqn. (*i*) $A^2 = 4A + 5I$

$$A^3 = 4A^2 + 5A = 4(4A + 5I) + 5A = 21A + 20I$$

$$A^4 = 21A^2 + 20A = 21(4A + 5I) + 20A = 104A + 105I$$

$$A^5 = 104A^2 + 105A = 104(4A + 5I) + 105A = 521A + 520I$$

$\therefore \quad A^5 + 5A^4 - 6A^3 + 2A^3 - 4A + 7I = 521A + 520I + 5(104A + 105I) - 6(21A + 20I) + 2(4A + 5I) - 4A + 7I$

$$= 919A + 942I = 919\begin{bmatrix} 1 & 4 \\ 2 & 3 \end{bmatrix} + \begin{bmatrix} 942 & 0 \\ 0 & 942 \end{bmatrix}$$

$$= \begin{bmatrix} 1861 & 3676 \\ 1838 & 3699 \end{bmatrix}.$$

## PROBLEMS 2.10

1. If $A = \begin{bmatrix} -1/2 & -\sqrt{3}/2 & 0 \\ -\sqrt{3}/2 & 1/2 & 0 \\ 0 & 0 & 0 \end{bmatrix}$ and $P = \begin{bmatrix} 1/2 & \sqrt{3}/2 & 0 \\ -\sqrt{3}/2 & 1/2 & 0 \\ 0 & 0 & 1 \end{bmatrix}$, show that $P^{-1}AP$ is a diagonal matrix.

2. Show that the linear transformation

$H = \begin{bmatrix} \cos\theta & \sin\theta \\ -\sin\theta & \cos\theta \end{bmatrix}$, where $\theta = \frac{1}{2}\tan^{-1}\frac{2h}{a-b}$, changes the matrix

$C = \begin{bmatrix} a & h \\ h & b \end{bmatrix}$ to the diagonal form $D = HCH'$.

3. Reduce the following matrices to the diagonal form :

(a) $A = \begin{bmatrix} 19 & 7 \\ -42 & 16 \end{bmatrix}$ *(Andhra, 2022)*  (b) $\begin{bmatrix} -1 & 2 & -2 \\ 1 & 2 & 1 \\ -1 & -1 & 0 \end{bmatrix}$ *(Andhra, 2023; V.T.U., 2013)*

4. If $A = \begin{bmatrix} 1 & 2 & 3 \\ 0 & 3 & 1 \\ 0 & 0 & 1 \end{bmatrix}$, then obtain the matrix $P$ such that $P^{-1}AP$ is a diagonal matrix. *(Andhra, 2022, Delhi, 2012)*

5. If $A = \begin{bmatrix} 5 & 3 \\ 1 & 3 \end{bmatrix}$, find $A^n$ and $A^4$.

6. If $A = \begin{bmatrix} -1 & 4 \\ 2 & 1 \end{bmatrix}$, then prove that $3 \tan A = A \tan 3$.

7. Find the eigen vectors of the matrix $\begin{bmatrix} 6 & -2 & 2 \\ -2 & 3 & -1 \\ 2 & -1 & 3 \end{bmatrix}$ and hence reduce $6x^2 + 3y^2 + 3z^2 - 2yz + 4zx - 4xy$ to a 'sum of squares'. Also write the nature of the matrix. *(JNTUK, 2022; Andhra, 2018; Anna, 2016)*

8. Reduce the quadratic form $2xy + 2yz + 2zx$ into canonical form. *(Anna, 2013)*

9. (a) Find the eigen values, eigen vectors and the modal of matrix $\begin{bmatrix} 1 & 0 & 0 \\ 0 & 3 & -1 \\ 0 & -1 & 3 \end{bmatrix}$.

(b) Reduce the quadratic form $x_1^2 + 3x_2^2 + 3x_3^2 - 2x_2x_3$ to a canonical form. *(J.N.T.U., 2015)*

10. Reduce the following quadratic forms into a 'sum of squares' by an orthogonal transformation and give the matrix of transformation.

(i) $3x_1^2 + 3x_2^2 + 3x_3^2 + 2x_1x_2 + 2x_1x_3 - 2x_2x_3$. *(Andhra, 2021)*

(ii) $x^2 + 5y^2 + z^2 + 2yz + 6zx + 2xy$. *(V.T.U., 2013)*

11. Find the index and signature of the quadratic form $x_1^2 + 2x_2^2 - 3x_3^2$.

12. Find the nature of the quadratic form $x^2 + 5y^2 + z^2 + 2xy + 2yz + 6zx$.

13. Show that the form $5x_1^2 + 26x_2^2 + 10x_3^2 + 4x_2x_3 + 14x_3x_1 + 6x_1x_2$ is a positive semi-definite and find a non-zero set of values of $x_1, x_2, x_3$ which make the form zero.

## 2.19 COMPLEX MATRICES

So far, we have considered matrices whose elements were real numbers. The elements of a matrix can, however, be complex numbers also.

**(1) Conjugate of a matrix.** *If the elements of a matrix* $A = [a_{rs}]$ *are complex numbers* $\alpha_{rs} + i\beta_{rs}$, $\alpha_{rs}$ *and* $\beta_{rs}$ *being real, then the matrix*

$\overline{A} = [\overline{a}_{rs}] = [\alpha_{rs} - i\beta_{rs}]$ *is called the* **conjugate matrix** *of A.*

The transpose of a conjugate of a matrix $A$ is denoted by $A^*$ or $A^\theta$, *i.e.*, $(\overline{A})' = A^*$.

**(2) Hermitian matrix.** *A square matrix A such that* $A' = \overline{A}$ *is said to be a* **Hermitian matrix*.** The elements of the leading diagonal of a Hermitian matrix are evidently real, while every other element is the complex conjugate of the element in the transposed position. For instance $A = \begin{bmatrix} 2 & 3+4i \\ 3-4i & -5 \end{bmatrix}$ is a Hermitian matrix, since $A' = \begin{bmatrix} 2 & 3-4i \\ 3+4i & -5 \end{bmatrix} = \overline{A}$

**(3) Skew-Hermitian matrix.** *A square matrix A such that* $A' = -\overline{A}$ *is said to be a* **skew-Hermitian matrix.** This implies that the leading diagonal elements of a skew-Hermitian matrix are either all zeros or all purely imaginary.

**Obs.** *A Hermitian matrix is a generalisation of a real symmetric matrix* as every real symmetric matrix is Hermitian. Similarly, *a skew-Hermitian matrix is a generalisation of a real skew-symmetric matrix.*

**Properties**

*I. Any square matrix A can be written as the sum of a Hermitian and skew-Hermitian matrices.*

Take $B = \frac{1}{2}(A + \overline{A}')$ and $C = \frac{1}{2}(A - \overline{A}')$

Then $B' = \frac{1}{2}(A + \overline{A}')' = \frac{1}{2}(A' + \overline{A})$

and $\overline{B} = \frac{1}{2}\overline{(A + \overline{A}')} = \frac{1}{2}(\overline{A} + A') = B'$

*i.e.*, $B$ is a Hermitian matrix.

Again, $C' = \frac{1}{2}(A - \overline{A}')' = \frac{1}{2}(A' - \overline{A})$

and $\overline{C} = \frac{1}{2}\overline{(A - \overline{A}')} = \frac{1}{2}(\overline{A} - A') = -C'$

$\therefore$ $C' = -C$, *i.e.*, $C$ is a skew-Hermitian matrix.

Thus, $A = \frac{1}{2}(A + \overline{A}') + \frac{1}{2}(A - \overline{A}') = B + C$

Hence the result.

*II. If A is a Hermitian matrix, then (iA) is a skew-Hermitian matrix.*

We have $(\overline{iA})' = (\overline{i}\,\overline{A})' = (-i\overline{A})' = -i\overline{A}'$

$= -iA$ $\quad [\because \overline{A}' = A]$

Thus $(iA)$ is a skew-Hermitian matrix.

Similarly *if A is a skew-Hermitian matrix then (iA) is a Hermitian matrix.*

*III. The eigen values of a Hermitian matrix are real. (see Fig. 2.1)*

Let $\lambda$ be the eigen value and $X$ the corresponding eigen vector of a Hermitian matrix $A$, so that

$$AX = \lambda X$$

$$\overline{X}'AX = \overline{X}'\lambda X = \lambda\,\overline{X}'X \quad \text{or} \quad \lambda = \overline{X}'AX/\overline{X}'X$$

Since $\overline{X}'X = \overline{x}_1x_1 + \overline{x}_2x_2 + \ldots + \overline{x}_nx_n = |x_1|^2 + |x_2|^2 + \ldots + |x_n|^2$ is real and non-zero. Also $\overline{X}'AX$ is a Hermitian form which is always real.

$\therefore$ $\lambda$, the eigen value of a Hermitian matrix is real.

*IV. The eigen values of a skew-Hermitian matrix are purely imaginary or zero.*

Let $\lambda$ be the eigen value and $X$ the corresponding eigen vector of a skew-Hermitian matrix $B$ so that $BX = \lambda X$.

$$\therefore \quad \overline{X}'BX = \overline{X}'\lambda X = \lambda\overline{X}'X \quad \text{or} \quad \lambda = \overline{X}'BX/\overline{X}'X$$

Since $\overline{X}'X$ is real and non-zero. Also $\overline{X}BX$ is a skew-Hermitian form which is purely imaginary or zero.

$\therefore$ $\lambda$, the eigen value of a skew-Hermitian matrix is purely imaginary or zero.

* Named after the French mathematician *Charles Hermite* (1822–1901), known for his contributions to algebra and number theory.

4. **Unitary matrix.** A square matrix $U$ such that $\overline{U'} = U^{-1}$ is called a **unitary matrix.** For a unitary matrix, $U$, $U \cdot U^* = U^* \cdot U = I$.

This is a generalisation of the orthogonal matrix in the complex field.

**Properties**

*I. Inverse of a unitary matrix is unitary*

If $U$ is a unitary matrix, then

$$\overline{U'} = U^{-1}$$

or $$U' = \overline{U^{-1}}$$

$\therefore$ $$[(U^{-1})^{-1}]' = \overline{U^{-1}}$$

Writing, $U^{-1} = V$, we have

$$[V^{-1}]' = \overline{V} \quad \text{or} \quad V^{-1} = \overline{V'}$$

Thus $V (= U^{-1})$ is also unitary.

**Cor.** *Inverse of an orthogonal matrix is orthogonal.*

*II. Transpose of a unitary matrix is unitary*

If $U$ is a unitary matrix, $\overline{U'} = U^{-1}$

or $$(\overline{U'}) = U^{-1}$$

or $$[(\overline{U'})]' = [U^{-1}]' = [U']^{-1}$$

Writing $U' = V$, we have $\overline{V'} = V^{-1}$

Thus $V$ (*i.e.*, $U'$) is also unitary.

**Cor.** *Transpose of an orthogonal matrix is orthogonal.*

*III. Product of two unitary matrices is a unitary matrix.* (*Delhi., 2015*)

If $U$ and $V$ are unitary matrices then

$$U' = \overline{U}^{-1}, V' = \overline{V}^{-1}$$

Now, $$(\overline{UV})^{-1} = (\overline{U}\overline{V})^{-1} = \overline{V}^{-1}\overline{U}^{-1}$$

$$= V'U'$$

$$= (UV)' \qquad [\because \ U, V \text{ are unitary.}]$$

Thus, $UV$ is a unitary matrix.

**Cor.** *Product of two orthogonal matrices is an orthogonal matrix.*

*IV. The eigen value of a unitary matrix has absolute value 1.* (*P.T.U., 2014 S ;U.K.T.U., 2010*)

If $U$ is a unitary matrix then $UX = \lambda X$ ...(1)

Taking conjugate transpose of Eqn. (1),

$$(\overline{UX})' = (\overline{U}\overline{X})' = \overline{X}'\overline{U}' = \overline{X}'\overline{U}^{-1}$$

Also $$(\overline{UX})' = (\overline{\lambda X})' = \overline{\lambda}\overline{X}'$$

*i.e.,* $$\overline{X}' U^{-1} = \overline{\lambda}\overline{X}' \qquad \text{...(2)}$$

Post-multiplying Eqn. (2) by Eqn. (1), we get

$$(\overline{X}' U^{-1})(UX) = (\overline{\lambda}\overline{X}') = (\lambda X)$$

$$\overline{X}'(U^{-1}U)X = (\overline{\lambda}\lambda)(\overline{X}'X) \qquad [\because \ U^{-1}U = I]$$

$$\overline{X}'X = (\lambda\lambda')\overline{X}'X$$

Thus $$\lambda\lambda' = |\lambda|^2 = 1. \qquad [\because \ \overline{X}X \neq 0]$$

Hence the result.

**Cor.** *The eigen value of an orthogonal matrix has absolute value 1.*

**Example 2.59.** *If* $A = \begin{bmatrix} 2+i & 3 & -1+3i \\ -5 & i & 4-2i \end{bmatrix}$, *show that* $AA^*$ *is a Hermitian matrix, where* $A^*$ *is the conjugate transpose of* $A$..

**Solution.** We have $A' = \begin{bmatrix} 2+i & -5 \\ 3 & i \\ -1+3i & 4-2i \end{bmatrix}$

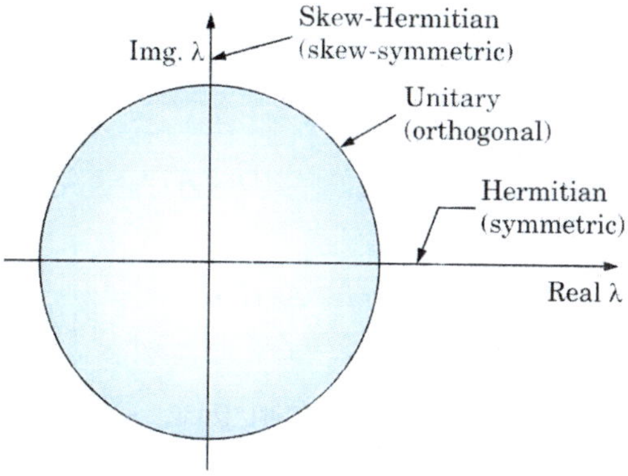


Fig. 2.1. Eigen values of various matrices.

and $$A^* = \begin{bmatrix} 2-i & -5 \\ 3 & -i \\ -1-3i & 4+2i \end{bmatrix}$$

$$\therefore \quad AA^* = \begin{bmatrix} 2-i & 3 & -1+3i \\ -5 & i & 4-2i \end{bmatrix} \begin{bmatrix} 2-i & -5 \\ 3 & -i \\ -1-3i & 4+2i \end{bmatrix}$$

$$= \begin{bmatrix} 4-i^2+9+1-9i^2 & -10-5i-3i-10+10i \\ -10+5i+3i-10-10i & 25-i^2+16-4i^2 \end{bmatrix}$$

$$= \begin{bmatrix} 24 & -20+2i \\ -20-2i & 46 \end{bmatrix}$$, which is a Hermitian matrix.

**Example 2.60.** *Prove that the matrix* $A = \begin{bmatrix} \frac{1}{2}(1+i) & \frac{1}{2}(-1+i) \\ \frac{1}{2}(1+i) & \frac{1}{2}(1-i) \end{bmatrix}$ *is unitary and find* $A^{-1}$.

*(Andhra, 2021; D.T.U., 2013)*

**Solution.** Conjugate of $A$, *i.e.*, $\overline{A} = \begin{bmatrix} \frac{1}{2}(1-i) & \frac{1}{2}(-1-i) \\ \frac{1}{2}(1-i) & \frac{1}{2}(1+i) \end{bmatrix}$

$\therefore$ Transpose of $\overline{A}$, *i.e.*, $$A^\theta = \begin{bmatrix} \frac{1}{2}(1-i) & \frac{1}{2}(1-i) \\ \frac{1}{2}(-1-i) & \frac{1}{2}(1+i) \end{bmatrix}$$

Now $$A^\theta.A = \begin{bmatrix} \frac{1}{2}(1-i) & \frac{1}{2}(1-i) \\ \frac{1}{2}(-1-i) & \frac{1}{2}(1+i) \end{bmatrix} \begin{bmatrix} \frac{1}{2}(1+i) & \frac{1}{2}(-1+i) \\ \frac{1}{2}(1+i) & \frac{1}{2}(1-i) \end{bmatrix}$$

$$= \begin{bmatrix} \frac{1}{4}(1+1)+\frac{1}{4}(1+1) & -\frac{1}{4}(1-i)^2+\frac{1}{4}(1-i)^2 \\ -\frac{1}{4}(1+i)^2+\frac{1}{4}(1+i)^2 & \frac{1}{4}(1+1)+\frac{1}{4}(1+1) \end{bmatrix} = \begin{bmatrix} 1 & 0 \\ 0 & 1 \end{bmatrix} = I$$

Similarly, $AA^\theta = I$.

Hence $A$, is a unitary matrix.

Also $$A^{-1} = A^\theta = \begin{bmatrix} \frac{1}{2}(1-i) & \frac{1}{2}(1-i) \\ -\frac{1}{2}(1+i) & \frac{1}{2}(1+i) \end{bmatrix}$$

**Example 2.61.** *Given that* $A = \begin{bmatrix} 0 & 1+2i \\ -1+2i & 0 \end{bmatrix}$, *show that* $(1 - A)(1 + A)^{-1}$ *is a unitary matrix.*

(*Andhra, 2021; U.P.T.U., 2011 ; Mumbai, 2007*)

**Solution.** $I + A = \begin{bmatrix} 1 & 1+2i \\ -1+2i & 1 \end{bmatrix}$, $|I + A| = 1 - (-1 - 4) = 6$

$$(I + A)^{-1} = \begin{bmatrix} 1 & -1-2i \\ 1-2i & 1 \end{bmatrix} \div 6. \text{ Also } I - A = \begin{bmatrix} 1 & -1-2i \\ 1-2i & 1 \end{bmatrix}$$

$$\therefore \quad (I - A)(I + A)^{-1} = \begin{bmatrix} 1 & -1-2i \\ 1-2i & 1 \end{bmatrix}\begin{bmatrix} 1 & -1-2i \\ 1-2i & 1 \end{bmatrix} \div 6 = \frac{1}{6}\begin{bmatrix} -4 & -2-4i \\ 2-4i & -4 \end{bmatrix} \quad ...(i)$$

Its conjugate-transpose $= \dfrac{1}{6}\begin{bmatrix} -4 & 2+4i \\ -2+4i & -4 \end{bmatrix}$ ...(*ii*)

$\therefore$ Product of Eqns. (*i*) and (*ii*) $= \dfrac{1}{36}\begin{bmatrix} -4 & -2-4i \\ 2-4i & -4 \end{bmatrix}\begin{bmatrix} -4 & 2+4i \\ -2+4i & -4 \end{bmatrix} = \dfrac{1}{36}\begin{bmatrix} 36 & 0 \\ 0 & 36 \end{bmatrix} = I.$

Hence the result.

## PROBLEMS 2.11

1. Prove that every Hermitian matrix can be written as $A + iB$, where $A$ is real and symmetric and $B$ is real and skew-symmetric.
2. Show that every square matrix can be uniquely expressed as $P + iQ$, where $P$ and $Q$ are Hermitian matrices.
3. Show that a Hermitian matrix remains Hermitian when transformed by an orthogonal matrix.
4. Show that the matrix $\begin{bmatrix} \alpha + i\gamma & -\beta + i\delta \\ \beta + i\delta & \alpha - i\gamma \end{bmatrix}$ is a unitary matrix, if $\alpha^2 + \beta^2 + \gamma^2 + \delta^2 = 1$. (*U.P.T.U., 2011*)
5. Show that $\begin{bmatrix} 3 & 7-4i & -2+5i \\ 7+4i & -2 & 3+i \\ -2-5i & 3-i & 4 \end{bmatrix}$ is a Hermitian matrix.
6. Prove that $\begin{bmatrix} i & 0 & 0 \\ 0 & 0 & i \\ 0 & i & 0 \end{bmatrix}$ is a skew Hermitian matrix and also unitary. (*Andhra, 2016*)
7. If $A = \begin{bmatrix} -1 & 2+i & 5-3i \\ 2-i & 7 & 5i \\ 5+3i & -5i & 2 \end{bmatrix}$, show that $A$ is a Hermitian matrix and $iA$ is a skew-Hermitian matrix.
8. Show that the following matrix is unitary

   (*a*) $\dfrac{1}{\sqrt{3}}\begin{bmatrix} 1 & 1+i \\ 1-i & -1 \end{bmatrix}$ (*U.P.T.U., 2013*) (*b*) $\begin{bmatrix} \frac{2+i}{3} & \frac{2i}{3} \\ \frac{2i}{3} & \frac{2-i}{3} \end{bmatrix}$
9. Express $A = \begin{bmatrix} 2i & 2+i & 1-i \\ -2+i & -i & 3i \\ -1-i & 3i & 0 \end{bmatrix}$ as $P + iQ$ where $P$ is real and skew-symmetric and $Q$ is real and symmetric.
10. If $S = \begin{bmatrix} 1 & 1 & 1 \\ 1 & a^2 & a \\ 1 & a & a^2 \end{bmatrix}$, where $a = e^{2i\pi/3}$, prove that $S^{-1} = \dfrac{1}{3}\bar{S}$.

## 2.20 OBJECTIVE TYPE OF QUESTIONS

### PROBLEMS 2.12

*Choose the correct answer or fill up the blanks in the following problems :*

1. To multiply a matrix by scalar $k$, multiply
   (*a*) any row by $k$ (*b*) every element by $k$ (*c*) any column by $k$.

2. If $A = \begin{bmatrix} 3 & -4 \\ 1 & -1 \end{bmatrix}$, then $A^n$ is

   (*a*) $\begin{bmatrix} 1+2n & -4n \\ n & 1-2n \end{bmatrix}$ (*b*) $\begin{bmatrix} 3^n & (-4)^n \\ 1 & (-1)^n \end{bmatrix}$ (*c*) $\begin{bmatrix} 1+3n & 1-4n \\ 1+n & 1-n \end{bmatrix}$ (*d*) $\begin{bmatrix} 1+2n & -4n \\ 1+n & 1-2n \end{bmatrix}$

3. The inverse of the matrix $\begin{bmatrix} -0.5 & 0 & 0 \\ 0 & 4 & 0 \\ 0 & 0 & 1 \end{bmatrix}$ is

   (*a*) $\begin{bmatrix} 0.5 & 0 & 0 \\ 0 & -4 & 0 \\ 0 & 0 & 1 \end{bmatrix}$ (*b*) $\begin{bmatrix} 0.5 & 0 & 0 \\ 0 & -4 & 0 \\ 0 & 0 & 1 \end{bmatrix}$ (*c*) $\begin{bmatrix} -2 & 0 & 0 \\ 0 & 0.25 & 0 \\ 0 & 0 & 1 \end{bmatrix}$ (*d*) $\begin{bmatrix} 2 & 0 & 0 \\ 0 & -0.25 & 0 \\ 0 & 0 & -1 \end{bmatrix}$

4. If $A = \begin{bmatrix} 2 & 0 & 0 \\ 0 & 2 & 0 \\ 0 & 0 & 2 \end{bmatrix}$ and $B = \begin{bmatrix} 1 & 2 & 3 \\ 0 & 1 & 3 \\ 0 & 0 & 2 \end{bmatrix}$, then the determinant $AB$ has the value

   (*a*) 4 (*b*) 8 (*c*) 16 (*d*) 32

5. The system of equations $x + 2y + z = 9$, $2x + y + 3z = 7$ can be expressed as

   (*a*) $\begin{bmatrix} 1 & 2 & 1 \\ 2 & 1 & 3 \end{bmatrix} = \begin{bmatrix} 9 \\ 7 \end{bmatrix} \begin{bmatrix} x \\ y \\ z \end{bmatrix}$ (*b*) $\begin{bmatrix} 1 & 2 & 1 \\ 2 & 1 & 3 \end{bmatrix} = \begin{bmatrix} 9 \\ 7 \end{bmatrix} = \begin{bmatrix} x \\ y \\ z \end{bmatrix}$

   (*c*) $\begin{bmatrix} 1 & 2 & 1 \\ 2 & 1 & 3 \end{bmatrix} \begin{bmatrix} x \\ y \\ z \end{bmatrix} = \begin{bmatrix} 9 \\ 7 \end{bmatrix}$ (*d*) none of the above.

6. If $\begin{bmatrix} 5 & 4 \\ 1 & 1 \end{bmatrix} X = \begin{bmatrix} 1 & -2 \\ 1 & 3 \end{bmatrix}$, then $X$ equals

   (*a*) $\begin{bmatrix} -3 & -14 \\ 4 & 17 \end{bmatrix}$ (*b*) $\begin{bmatrix} 1 & -2 \\ 3 & 1 \end{bmatrix}$ (*c*) $\begin{bmatrix} 1 & 3 \\ -2 & 1 \end{bmatrix}$ (*d*) $\begin{bmatrix} 3 & -14 \\ 4 & -17 \end{bmatrix}$

7. If $A = \begin{bmatrix} 3 & 2 \\ 1 & 4 \end{bmatrix}$, then A (adj $A$) equals

   (*a*) $\begin{bmatrix} 10 & 0 \\ 0 & 10 \end{bmatrix}$ (*b*) $\begin{bmatrix} 0 & 10 \\ 10 & 0 \end{bmatrix}$ (*c*) $\begin{bmatrix} 10 & 1 \\ 1 & 10 \end{bmatrix}$ (*d*) none of the above.

8. If $3x + 2y + z = 0$, $x + 4y + z = 0$, $2x + y + 4z = 0$, be a system of equations, then
   (*a*) it is inconsistent
   (*b*) it has only the trivial solution $x = 0, y = 0, z = 0$.
   (*c*) it can be reduced to a single equation and so a solution does not exist.
   (*d*) determinant of the matrix of coefficients is zero.

9. If $A = \begin{bmatrix} 1 & 0 \\ 0 & 1 \end{bmatrix}$, $B = \begin{bmatrix} 0 & 1 \\ -1 & 0 \end{bmatrix}$ and $C = \begin{bmatrix} \cos\theta & \sin\theta \\ -\sin\theta & \cos\theta \end{bmatrix}$, then

   (*a*) $C = A\cos\theta - B\sin\theta$ (*b*) $C = A\sin\theta + B\cos\theta$
   (*c*) $C = A\sin\theta - B\cos\theta$ (*d*) $C = A\cos\theta + B\sin\theta$.

10. Let $A = \begin{bmatrix} 1 & 0 & 0 \\ \alpha & 1 & 0 \\ \beta & \gamma & 1 \end{bmatrix}$ and $B = \begin{bmatrix} 1 & 0 & 0 \\ 2 & 1 & 0 \\ 3 & 4 & 1 \end{bmatrix}$, then

(*a*) $A$ is row equivalent to $B$ only when $\alpha = 2$, $\beta = 3$, and $\gamma = 4$
(*b*) $A$ is row equivalent to $B$ only when $\alpha \neq 0$, $\beta \neq 0$, and $\gamma = 0$
(*c*) A is not row equivalent to $B$
(*d*) $A$ is row equivalent to $B$ for all value of $\alpha, \beta, \gamma$.

11. If $A \begin{bmatrix} 0 & 1 \\ 2 & -1 \end{bmatrix} = \begin{bmatrix} 2 & 1 \\ -1 & 0 \end{bmatrix}$ where $A = \begin{bmatrix} a & b \\ c & d \end{bmatrix}$, then $A$ is

(*a*) $\begin{bmatrix} 2 & 1 \\ 0 & 0 \end{bmatrix}$ (*b*) $\begin{bmatrix} 0 & 1 \\ 2 & -1 \end{bmatrix}$ (*c*) $\begin{bmatrix} 2 & 1 \\ -1 & 0 \end{bmatrix}$ (*d*) $\begin{bmatrix} 2 & 1 \\ -1/2 & -1/2 \end{bmatrix}$

12. Matrix has a value. This statement

(*a*) is always true (*b*) depends upon the matrices
(*c*) is false

13. If $A$ is a square matrix such that $AA' = I$, then value of $A'A$ is

(*a*) $A^2$ (*b*) $I$ (*c*) $A^{-1}$

14. If every minor of order $r$ of a matrix $A$ is zero, then rank of $A$ is

(*a*) greater than $r$ (*b*) equal to $r$ (*c*) less than or equal to $r$ (*d*) less than $r$.

15. A square matrix $A$ is called orthogonal if

(*a*) $A = A^2$ (*b*) $A' = A^{-1}$ (*c*) $AA^{-1} = I$

16. The rank of matrix $\begin{bmatrix} 2 & -1 & 3 & 1 \\ 1 & 4 & -2 & 1 \\ 5 & 2 & 4 & 3 \end{bmatrix}$ is .........

17. The sum of the eigen values of a matrix is the ......... of the elements of the principal diagonal.

18. The sum and product of the eigen values of the matrix $\begin{bmatrix} 2 & -3 \\ 4 & -2 \end{bmatrix}$ are ......... and ......... respectively.

19. Inverse of $\begin{bmatrix} 1 & 2 & -2 \\ -1 & 3 & 0 \\ 0 & -2 & 1 \end{bmatrix}$ is $\begin{bmatrix} 3 & 2 & 6 \\ 1 & 1 & k \\ 2 & 2 & 5 \end{bmatrix}$ then $k$ is .........

20. Using Cayley-Hamilton theorem, the value of $A^4 - 4A^3 - 5A^2 - A + 2I$ when $A = \begin{bmatrix} 1 & 2 \\ 4 & 2 \end{bmatrix}$ is .........

21. If two eigen values of $\begin{bmatrix} 8 & -6 & 2 \\ -6 & 7 & -4 \\ 2 & -4 & 3 \end{bmatrix}$ are 3 and 15, then the third eigen value is .........

22. A quadratic form is positive semi-definite when .........

23. $A_{m \times n}$ and $B_{p \times q}$ are two matrices. When will

(*a*) $A \cdot B$ exist (*b*) $A + B$ exist ?

24. The product of the eigen values of (*a*) $\begin{bmatrix} 1 & 0 & 0 \\ 0 & 3 & -1 \\ 0 & -1 & 3 \end{bmatrix}$ is ......... (*b*) $\begin{bmatrix} 1 & 1 & 3 \\ 1 & 5 & 1 \\ 3 & 1 & 1 \end{bmatrix}$ = ......... (*C.S.V.T.U., 2011 S*)

25. The quadratic form corresponding to the diagonal matrix diag $(\lambda_1, \lambda_2, ..., \lambda_n)$ is

(*a*) $x_1^2 + x_2^2 + ... + x_n^2$ (*b*) $\lambda_1 x_1^2 + \lambda_2 x_2^2 + ... + \lambda_n x_n^2$
(*c*) $\lambda_1^2 x_1^2 + \lambda_2^2 x_2^2 + ... + \lambda_n^2 x_n^2$

26. An example of a 3 × 3 matrix of rank one is .........

27. The quadratic form corresponding to the symmetric matrix $\begin{bmatrix} 1 & 2 \\ 2 & -4 \end{bmatrix}$ is .........

28. Solving the equations $x + 2y + 3z = 0$, $3x + 4y + 4z = 0$, $7x + 10y + 12z = 0$, $x =$ ........., $y =$ ........., $z =$ .........

29. If 3 and 6 are two eigen values of $A = \begin{bmatrix} 1 & 1 & 3 \\ 1 & 5 & 1 \\ 3 & 7 & 7 \end{bmatrix}$, then the eigen values of $A^{-1}$ are ......... *(Anna, 2012)*

30. A matrix $A$ is *idempotent* if .........

31. The rank of the matrix $\begin{bmatrix} 1 & 1 & -1 \\ 2 & -3 & 4 \\ 3 & -2 & 3 \end{bmatrix}$ is .........

32. If $A = \begin{bmatrix} -1 & 2 & 3 \\ 0 & 3 & 5 \\ 0 & 0 & -2 \end{bmatrix}$, then the eigen values of $A^2$ are .........

33. The sum of the eigen values of $\begin{bmatrix} 1 & 1 & 3 \\ 1 & 5 & 1 \\ 3 & -1 & -1 \end{bmatrix}$ is

(*i*) –2 (*ii*) 3 (*iii*) 6 (*iv*) 7

34. The maximum value of the rank of a 4 × 5 matrix is .........

35. The sum of two eigen values and trace of a 3 × 3 matrix are equal, then the value of $|A|$ is .........

36. If the sum of the eigen values of the matrix of the quadratic form is zero, then the nature of the quadratic form is .........

37. The eigen values of matrix $\begin{bmatrix} \cos\theta & -\sin\theta \\ -\sin\theta & -\cos\theta \end{bmatrix}$ are .........

38. The eigen values of a triangular matrix are .........

39. If the product of two eigen values of the matrix $\begin{bmatrix} 6 & -2 & 2 \\ -2 & 3 & -1 \\ 2 & -1 & 3 \end{bmatrix}$ is 16, then the third eigen value is .........

40. If $\lambda_i$, $i = 1, 2, \ldots\ldots\ldots n$ are the eigen values of a square matrix $A$, then the eigen values of $A^T$ are .........

41. By applying elementary transformations to a matrix, its rank

(*a*) increases (*b*) decreases (*c*) does not change

42. If $\lambda$ is an eigen value of $A$, then it is an eigen value of $B$, only if $B$ = .........

43. If $A = \begin{bmatrix} 3 & 1 & 4 \\ 0 & 2 & 6 \\ 0 & 0 & 5 \end{bmatrix}$, then eigen values of $A^{-1}$ are ......... *(Anna, 2013)*

44. The characteristic equation of $\begin{bmatrix} 3 & 10 & 5 \\ -2 & -3 & -4 \\ 3 & 5 & 7 \end{bmatrix}$ is .........

45. If $A = \begin{bmatrix} 2 & 1 \\ 1 & 2 \end{bmatrix}$, then eigen values of $A^{-1}$ are .........

46. Matrix $\begin{bmatrix} x & 2 \\ 1 & x-1 \end{bmatrix}$ is singular for $x$ = .........

47. Every Hermitian matrix can be written as $A + iB$, where $A$ is real and ......... and $B$ is real and .........

48. The sum and product of the eigen values of $\begin{bmatrix} 2 & 2 & 1 \\ 1 & 3 & 1 \\ 1 & 2 & 2 \end{bmatrix}$ are ......... and .........

49. If $A = \begin{bmatrix} \cos\theta & \sin\theta \\ -\sin\theta & \cos\theta \end{bmatrix}$, then $A^3$ = .........

50. Pick the correct answer :

(i) The eigen values of $\begin{bmatrix} 1 & 1 & 1 \\ 1 & 1 & 1 \\ 1 & 1 & 1 \end{bmatrix}$ are

(a) 0, 0, 0 (b) 0, 0, 1 (c) 0, 0, 3 (d) 1, 1, 1. (*U.P.T.U. 2011*)

(ii) Two square matrices $A$ and $B$ are similar if

(a) $A = B$ (b) $B = P^{-1} AP$ (c) $A' = B'$ (d) $A^{-1} = B^{-1}$. (*V.T.U., 2013*)

51. The eigen values of $A = \begin{bmatrix} 1 & 1 \\ 2 & 5 \end{bmatrix}$ are the roots of the equation .........

52. A system of linear non-homogeneous equations is consistent, if and only if the rank of coefficient matrix is equal to rank of .........

53. The matrix of the quadratic form $q = 4x^2 - 2y^2 + z^2 - 2xy + 6zx$ is .........

54. If $\lambda_1, \lambda_2, \lambda_3$ are the eigen values of a matrix $A$, then $A^3$ has the eigen values ........

55. If $\lambda$ is an eigen value of a non-singular matrix $A$, then the eigen value of $A^{-1}$ is ........

56. Value of $P$ for which $A = \begin{bmatrix} 3 & P & P \\ P & 3 & P \\ P & P & 3 \end{bmatrix}$ is of rank 1 is ......... (*U.P.T.U., 2012*)

57. The sum of the squares of the eigen values of $\begin{bmatrix} 3 & 1 & 4 \\ 0 & 2 & 6 \\ 0 & 0 & 5 \end{bmatrix}$ is .........

58. If the rank of a matrix $A$ is 2, then the rank of $A'$ is .........

59. The index and signature of the quadratic form $x_1^2 + 2x_2^2 - 3x_3^2$ are respectively ......... and ..........

60. The equations $x + 2y = 1$, $7x + 14y = 12$ are consistent. (True or False)

61. If rank $(A) = 2$, rank $(B) = 3$, then rank $(AB) = 6$. (True or False)

62. Any set of vectors which includes the zero vector is linearly independent. (True or False)

63. If $\lambda$ is an eigen value of a symmetric matrix, then $\lambda$ is real. (True or False)

64. Every square matrix does not satisfy its own characteristic equation. (True or False)

65. If $\lambda$ is an eigen value of an orthogonal matrix, then $1/\lambda$ is also its eigen value. (*J.N.T.U., 2015*) (True or False)

66. If the rank of a matrix $A$ is 3, then the rank of $3A^T$ is 1. (True or False)

67. The vectors [1, 1, –1, 1], [1, –1, 2, –1], [3, 1, 0, 1] are linearly dependent. (True or False)

68. The eigen values of a triangular matrix are just the diagonal elements of the matrix.(*D.T.U., 2016*) (True or False)

69. Inverse of a unitary matrix is a unitary matrix. (True or False)

70. $A$ is a non-zero column matrix and $B$ is a non-zero row matrix, then rank of $AB$ is one. (True or False)

71. The sum of the eigen values of $A$ equals to the trace of $A = \begin{bmatrix} 1 & 0 & 0 \\ 0 & 3 & -1 \\ 0 & -1 & 3 \end{bmatrix}$. (True or False)

72. The matrix, $A = \begin{bmatrix} i & 0 & 0 \\ 0 & 0 & i \\ 0 & i & 0 \end{bmatrix}$ is Skew-Hermitian and unitary. (True or False) (*U.P.T.U., 2012*)

73. The quadratic form corresponding to the matrix $\begin{bmatrix} 2 & 0 & -2 \\ 0 & 2 & 1 \\ -2 & 1 & -2 \end{bmatrix}$ is ......... (*Anna, 2013*)

CHAPTER 3

# Vector Algebra & Solid Geometry



## VECTOR ALGEBRA

### 3.1 (1) VECTORS

A quantity which is completely specified by its magnitude only is called a *scalar*. Length, time, mass, volume, temperature, work, electric charge and numerical data in Statistics are all examples of scalar quantities.

*A quantity which is completely specified by its magnitude and direction is called a* **vector***. Weight, force, displacement, velocity, acceleration, momentum and electric current density are all vector quantities for each involves magnitude and direction.

A vector is represented by a directed line segment. Thus $\vec{PQ}$ represents a vector whose magnitude is the length $PQ$ and direction is from $P$ (starting point) to $Q$ (end point). We denote a vector by a single letter in capital bold type (or with an arrow on it) and its magnitude (length) by the corresponding small letter in italics type. Thus if **V** is the velocity vector, its magnitude is $v$, the speed.

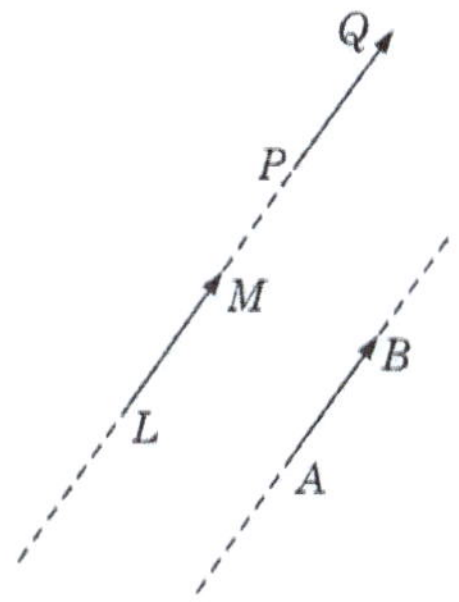

Fig. 3.1

A vector of unit magnitude is called a *unit vector*. The idea of unit vector is often used to represent concisely the direction of any vector. Unit vector corresponding to the vector **A** is written as $\hat{\mathbf{A}}$.

A vector of zero magnitude (which can have no direction associated with it) is called a *zero (or null) vector* and is denoted by **0**–a thick zero.

The vector $\vec{QP}$ represents the negative of $\vec{PQ}$, *i.e.*, – **A.**

*Two vectors* **A** *and* **B** *having the same magnitude and the same* (or parallel) directions are said to be equal and we write **A** = **B.** Clearly the vectors $\vec{AB}$, $\vec{LM}$ and $\vec{PQ}$ are all equal (Fig. 3.1).

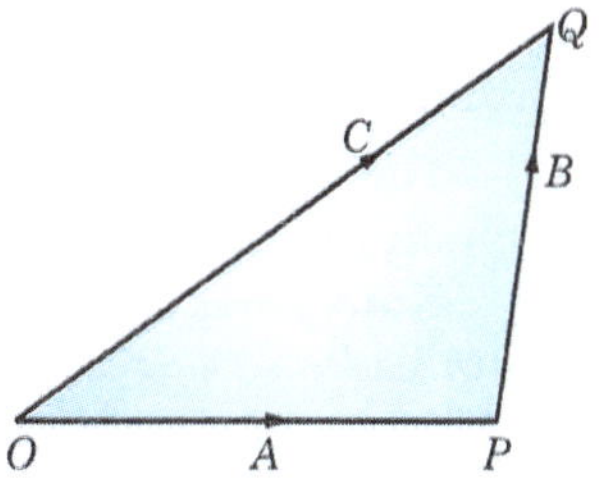

Fig. 3.2

**(2) Addition of vectors.** Vectors are added according to the *triangle law*

*Vector analysis was introduced by an American mathematician Josiah Willard Gibbs (1839-1903).

*of addition*, which is a matter of common knowledge. Let **A** and **B** be represented by two vectors $\vec{OP}$ and $\vec{PQ}$ respectively then $\vec{OQ}$ = **C** is called the sum or resultant of **A** and **B**. Symbolically, we write,

$$\mathbf{C} = \mathbf{A} + \mathbf{B}$$

**(3) Subtraction of vectors.** *The subtraction of a vector* **B** *from* **A** *is taken to be the addition of* –**B** *to* **A** *and we write*

$$\mathbf{A} + (-\mathbf{B}) = \mathbf{A} - \mathbf{B}$$

**(4) Multiplication of vectors by scalars.**

We have just seen that $\mathbf{A} + \mathbf{A} = 2\mathbf{A}$

and $-\mathbf{A} + (-\mathbf{A}) = -2\,\mathbf{A}$

where both 2**A** and – 2**A** denote vectors of magnitude twice that of **A** ; the former having the same direction as **A** and the latter the opposite direction.

In general, *the product m***A** *of a vector* **A** *and a scalar m is a vector whose magnitude is m times that of* **A** *and direction is the same or opposite to* **A** *according as m is positive or negative.*

Thus $\mathbf{A} = a\,\hat{\mathbf{A}}$.

**Example 3.1.** *If* **A** *and* **B** *are the vectors determined by two adjacent sides of a regular hexagon. What are the vectors represented by the other sides taken in order ?*

**Solution.** Let *ABCDEF* be the given hexagon, such that

$$\vec{AB} = \mathbf{A} \text{ and } \vec{BC} = \mathbf{B}$$

$\therefore \quad \vec{AC} = \vec{AB} + \vec{BC} = \mathbf{A} + \mathbf{B}$

Also $\quad \vec{AD} = 2\vec{BC} = 2\mathbf{B}$

$\therefore \quad \vec{CD} = \vec{AD} - \vec{AC} = 2\mathbf{B} - (\mathbf{A} + \mathbf{B}) = \mathbf{B} - \mathbf{A}$

Now $\quad \vec{DE} = -\vec{AB} = -\mathbf{A} \qquad [\because \; AB = \text{and} \parallel ED]$

$\quad \vec{EF} = -\vec{BC} = -\mathbf{B} \qquad [\because \; BC = \text{and} \parallel FE]$

and $\quad \vec{FA} = -\vec{CD} = -(\mathbf{B} - \mathbf{A}) = \mathbf{A} - \mathbf{B} \qquad [\because \; CD = \text{and} \parallel AF]$

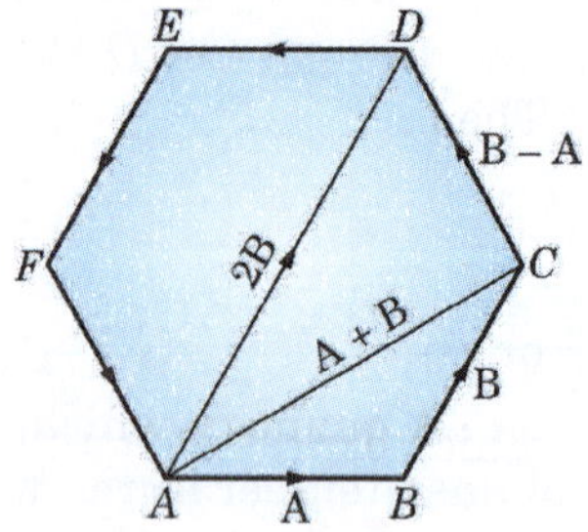

Fig. 3.3

## 3.2 SPACE COORDINATES, RESOLUTION OF VECTOR DIRECTION COSINE

**(1) Space Coordinates.** Let *X′OX* and *Y′OY*, *Z′OZ* be three mutually perpendicular lines which intersect at *O*. Then *O* is called the origin.

*X′OX is called the* **x-axis,** *Y′OY the* **y-axis**, *Z′OZ the* **z-axis** *and taken together these are called the* **coordinate axes.**

*The plane YOZ is called the* **yz-plane,** *the plane ZOX the* **zx-plane,** *the plane XOY the* **xy-plane** *and taken together these are called the* **coordinate planes.**

Let *P* be any point in space, Draw *PL*, *PM*, *PN* ⊥s to the *yz*, *zx* and *xy*-planes. Then *LP*, *MP*, *NP* are respectively called the coordinates of *P* (Fig. 3.4). In practice, if $OA = x$, $AN = y$, $NP = z$, then $(x, y, z)$ are the coordinates of *P* which are positive along *OX*, *OY*, *OZ* respectively and negative along *OX′*, *OY′*, *OZ′*.

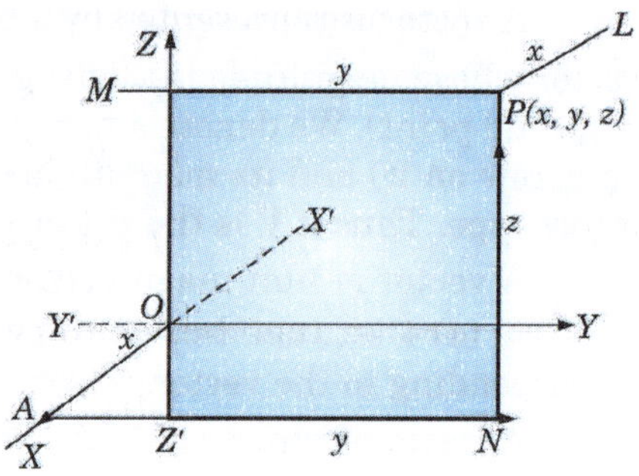

Fig. 3.4

The three coordinate planes divide the space into eight compartments called **octants.** The octant *OXYZ* in which all the coordinates are positive is called the **positive** or **first octant.**

**Note.** *Three non-coplanar vectors* **A, B, C** *are said to form a* **right-handed** *(or a* **left-handed***) system according as a right threaded screw rotated through an angle less than* 180° *from* **A** *to* **B** *will advance along (or opposite to)* **C** as shown in Fig. 3.5.

*An area of a closed curve described in a given manner is represented by a vector whose magnitude is the given area and direction normal to the plane of the area.* Thus the vector **A** representing the area is taken to be positive or negative according as the direction of description of the boundary of the curve and the sense of **A** correspond to a right-handed or a left-handed system.

We have explained the most commonly used system of coordinates namely the *Rectangular Cartesian Coordinates.* The other two systems of coordinates often used to locate a point in space are the *Polar spherical coordinates and Cylindrical coordinates,* which are explained in § 8.21(1) and 8.20.

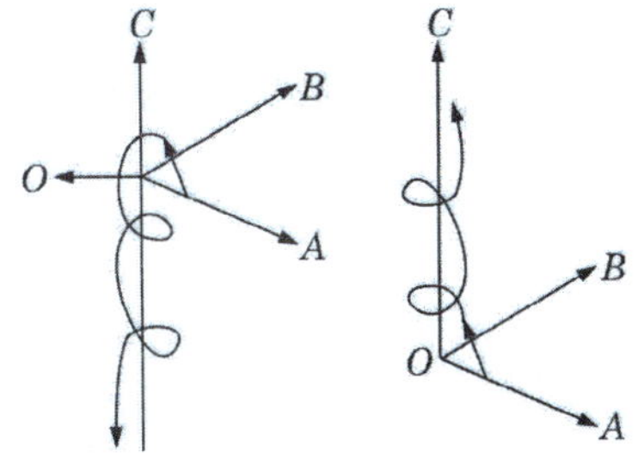


Fig. 3.5

**(2) Resolution of vectors.** Let **I, J, K** denote unit vectors along $OX$, $OY$, $OZ$ respectively. Let $P(x, y, z)$ be a point in space. On $OP$ as diagonal, construct a rectangular parallelopiped with edges $OA$, $OB$, $OC$ along the axes so that

$$\vec{OA} = x\mathbf{I},\ \vec{OB} = y\mathbf{J},\ \vec{OC} = z\mathbf{K}$$

Then $$\mathbf{R} = \vec{OP} = \vec{OC'} + \vec{C'P}$$

$$= \vec{OA} = \vec{AC'} + \vec{OC} = \vec{OA} + \vec{OB} + \vec{OC}$$

Hence $\mathbf{R} = x\mathbf{I} + y\mathbf{J} + z\mathbf{K}$ is called the *position vector* of $P$ relative to origin $O$ and.

$$r = |\mathbf{R}| = \sqrt{(x^2 + y^2 + z^2)}$$

$$[\because\ r^2 = OP^2 = OC'^2 + C'P^2 = OA^2 + AC'^2 + C'P^2]$$

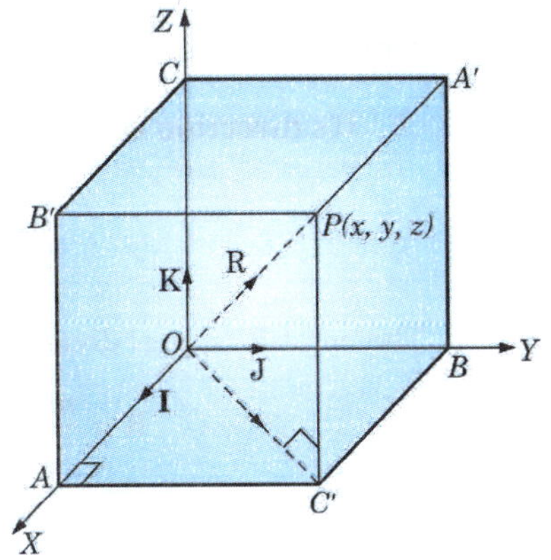


Fig. 3.6

**(3) Direction cosines.** Let any line $L$ or its parallel $OP$, make angles $\alpha$, $\beta$, $\gamma$ with $OX$, $OY$, $OZ$ respectively, then $\cos \alpha$, $\cos \beta$, $\cos \gamma$ *are called the direction cosines of this line* which are usually denoted by $l$, $m$, $n$.

*If* $l$, $m$, $n$ *are direction cosines of a vector* **R**, then

$(i)\ \hat{\mathbf{R}} = l\mathbf{I} + m\mathbf{J} + n\mathbf{K}$, $(ii)\ \mathbf{l^2 + m^2 + n^2 = 1}$

*Proof.* Let $D$ be the foot of the perpendicular from $P(x, y, z)$ on $OY$. Then

$$y = OD = r \cos \beta = mr. \text{ Similarly, } z = nr \text{ and } x = lr.$$

$$\therefore \quad \mathbf{R} = x\mathbf{I} + y\mathbf{J} + z\mathbf{K} = r(l\mathbf{I} + m\mathbf{J} + n\mathbf{K})$$

or $$\hat{\mathbf{R}} = \frac{\mathbf{R}}{r} = l\mathbf{I} + m\mathbf{J} + n\mathbf{K}$$

which expresses a unit vector in terms of its direction cosines.

Also $$1 = |\hat{\mathbf{R}}| = \sqrt{(l^2 + m^2 + n^2)} \text{ thus } \mathbf{l^2 + m^2 + n^2 = 1}$$

*i.e.,* $$\cos^2 \alpha + \cos^2 \beta + \cos^2 \gamma = 1 \qquad (V.T.U.,\ 2010)$$

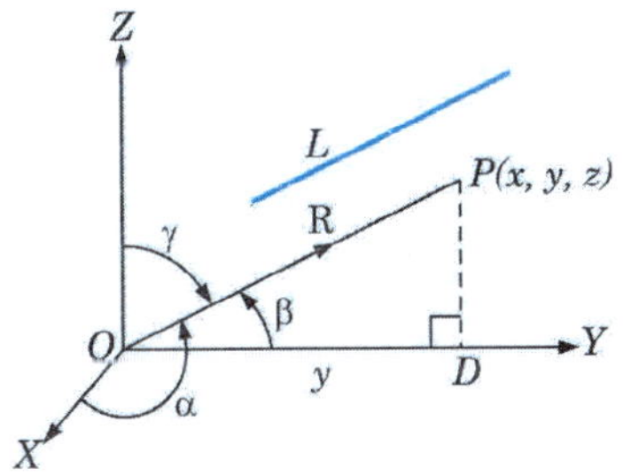


Fig. 3.7

**Obs. Directions ratios.** If the direction cosines of a line be proportional to $a$, $b$, $c$, then these are called proportional direction cosines or direction ratios of the line.

If the direction cosines be $l$, $m$, $n$, then

$$\frac{l}{a} = \frac{m}{b} = \frac{n}{c} = \frac{\sqrt{(l^2 + m^2 + n^2)}}{\sqrt{(a^2 + b^2 + c^2)}} = \frac{1}{\sqrt{(\Sigma a^2)}}$$

$$\therefore \quad l = \frac{a}{\sqrt{(\Sigma a^2)}},\ m = \frac{b}{\sqrt{(\Sigma a^2)}},\ n = \frac{c}{\sqrt{(\Sigma a^2)}}$$

**(4) Distance between two points** $P(x_1, y_1, z_1)$ and $Q(x_2, y_2, z_2)$ is

$$\sqrt{[(\mathbf{x_2 - x_1})^2 + (\mathbf{y_2 - y_1})^2 + (\mathbf{z_2 - z_1})^2]}$$

*and* **direction ratios** *of* $\vec{PQ}$ *are* $\mathbf{x_2 - x_1, y_2 - y_1, z_2 - z_1}$

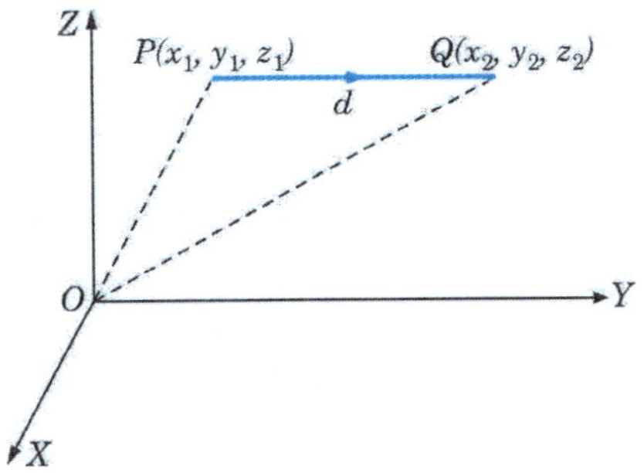


Fig. 3.8

We have

$$\vec{OP} = x_1\mathbf{I} + y_1\mathbf{J} + z_1\mathbf{K}$$

and $$\vec{OQ} = x_2\mathbf{I} + y_2\mathbf{J} + z_2\mathbf{K}$$

$$\therefore \quad \vec{PQ} = \vec{OQ} - \vec{OP} = (x_2 - x_1)\mathbf{I} + (y_2 - y_1)\mathbf{J} + (z_2 - z_1)\mathbf{K}$$

Thus, $$d = |\vec{PQ}| = \sqrt{[(x_2 - x_1)^2 + (y_2 - y_1)^2 + (z_2 - z_1)^2]}$$

and direction cosines of $\vec{PQ}$ are proportional to $x_2 - x_1, y_2 - y_1, z_2 - z_1$.

**Example 3.2.** *Show that the points A(–4, 9, 6), B(–1, 6, 6) and C(0, 7, 10) form a right angled isosceles triangle. Also find the direction cosines of AB.*

**Solution.** We have
$$AB = \sqrt{[(-1+4)^2 + (6-9)^2 + (6-6)^2]} = 3\sqrt{2}$$
$$BC = \sqrt{[(0+1)^2 + (7-6)^2 + (10-6)^2]} = 3\sqrt{2}$$
and
$$CA = \sqrt{[(-4-0)^2 + (9-7)^2 + (6-10)^2]} = 6$$

Since $AB^2 + BC^2 = CA^2$ and $AB = BC$, it follows that $\Delta ABC$ is a right-angled isosceles triangle. The direction ratios of $\vec{AB}$ are $-1 + 4, 6 - 9, 6 - 6$.

∴ Its direction cosines are $\frac{1}{\sqrt{2}}, -\frac{1}{\sqrt{2}}, 0$.

## 3.3 SECTION FORMULAE

The point **R** $(x, y, z)$ dividing the join of the points $\mathbf{A}(x_1, y_1, z_1)$ *and* $\mathbf{B}(x_2, y_2, z_2)$ *in the ratio* $m_1 : m_2$ *is*
$$\mathbf{R} = \frac{m_1\mathbf{B} + m_2\mathbf{A}}{m_1 + m_2}, \quad i.e., \quad \left(\frac{m_1x_2 + m_2x_1}{m_1 + m_2}, \frac{m_1y_2 + m_2y_1}{m_1 + m_2}, \frac{m_1z_2 + m_2z_1}{m_1 + m_2}\right) \quad ...(i)$$

Let $P(\mathbf{A})$ and $Q(\mathbf{B})$ be the given points referred to origin $O$. Let $R(\mathbf{R})$ be the point dividing the line joining $P$ and $Q$ in the ratio $m_1 : m_2$ so that
$$\frac{PR}{RQ} = \frac{m_1}{m_2}, \quad i.e., \quad m_2 . PR = m_1 . RQ$$

∴ We have
$$m_2\vec{PR} = m_1\vec{RQ}$$
or
$$m_2(\vec{OR} - \vec{OP}) = m_1(\vec{OQ} - \vec{OR})$$
or
$$m_2(\mathbf{R} - \mathbf{A}) = m_1(\mathbf{B} - \mathbf{R})$$
whence
$$\mathbf{R} = \frac{m_1\mathbf{B} + m_2\mathbf{A}}{m_1 + m_2}$$
Since
$$\mathbf{A} = x_1\mathbf{I} + y_1\mathbf{J} + z_1\mathbf{K}, \mathbf{B} = x_2\mathbf{I} + y_2\mathbf{J} + z_2\mathbf{K}$$
and
$$\mathbf{R} = x\mathbf{I} + y\mathbf{J} + z\mathbf{K}$$
∴
$$x\mathbf{I} + y\mathbf{J} + z\mathbf{K} = \frac{m_1(x_2\mathbf{I} + y_2\mathbf{J} + z_2\mathbf{K}) + m_2(x_1\mathbf{I} + y_1\mathbf{J} + z_1\mathbf{K})}{m_1 + m_2}$$

Equating coefficient of **I, J, K,** we get the desired Eqn. (*i*).

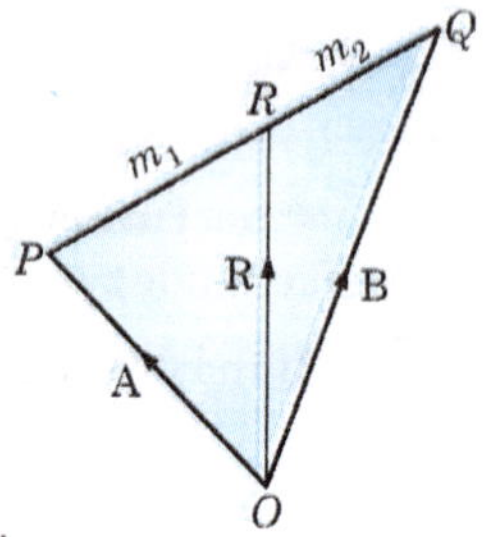

Fig. 3.9

**Cor. 1.** *Mid-point of* $P(\mathbf{A})$ *and* $Q(\mathbf{B})$ *is* $\frac{1}{2}(\mathbf{A} + \mathbf{B})$.

**2.** Point $R$ dividing the join of $P(\mathbf{A})$ and $Q(\mathbf{B})$ in the ratio $m_1 : m_2$ externally is $\mathbf{R} = \frac{m_1\mathbf{B} - m_2\mathbf{A}}{m_1 - m_2}$.

**Obs.** Rewriting (*i*) as $m_2\mathbf{A} + m_1\mathbf{B} - (m_1 + m_2)\mathbf{R} = 0$, we note that the sum of the coefficients of **A, B** and **R** is zero. Hence it follows that any *three points with position vectors* **A, B** *and* **C** *are collinear if*
$$\lambda\mathbf{A} + \mu\mathbf{B} + \gamma\mathbf{C} = 0, \text{ where } \lambda + \mu + \gamma = 0.$$

**Example 3.3.** *In a trapezium, prove that the straight line joining the mid-points of the diagonals is parallel to the parallel sides and half their difference.*

**Solution.** Consider a trapezium $OABC$ with parallel sides $OA$ and $BC$. Take $O$ as the origin and let the other vertices be $A(\mathbf{A})$, $B(\mathbf{B})$, $C(\mathbf{C})$.

Since $CB$ is parallel to $OA$, therefore,
$$\mathbf{B} - \mathbf{C} = \vec{CB} = \lambda\vec{OA} = \lambda\mathbf{A}.$$

The mid-points of the diagonals $OB$ and $AC$ are $D(\mathbf{B}/2)$ and $E(\mathbf{A}+\mathbf{C})/2$.

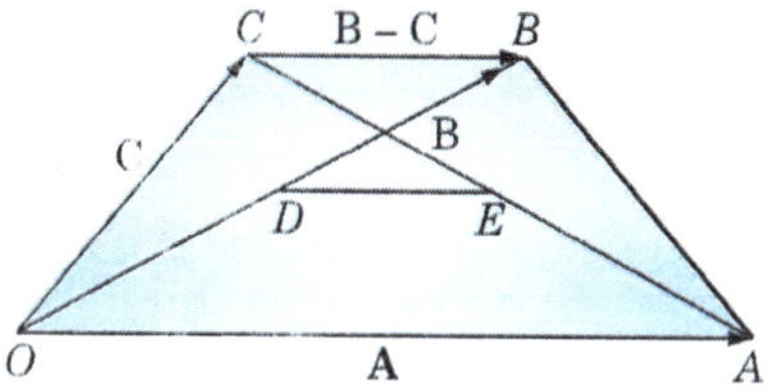

Fig. 3.10

$$\therefore \quad \overrightarrow{DE} = \overrightarrow{OE} - \overrightarrow{OD} = \frac{1}{2}(\mathbf{A}+\mathbf{C}) - \frac{1}{2}\mathbf{B} = \frac{1}{2}[\mathbf{A} - (\mathbf{B}-\mathbf{C})] \quad ...(i)$$

$$= \frac{1}{2}(1-\lambda)\mathbf{A} \quad ...(ii)$$

From (*ii*), it is clear that $\overrightarrow{DE}$ is parallel to $\overrightarrow{OA}$ : from (*i*), it follows that $DE = \frac{1}{2}(OA - CB)$.

Hence the result.

**Example 3.4.** *Show that the line joining one vertex of a parallelogram to the mid-point of an opposite side trisects the diagonal and is itself trisected there at.*

**Solution.** Consider a parallelogram $OABC$. Take $O$ as the origin and let the other vertices be $A(\mathbf{A})$, $B(\mathbf{B})$ and $C(\mathbf{C})$.

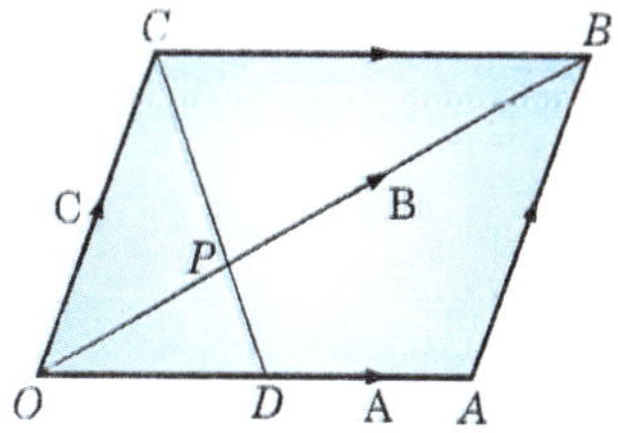

Fig. 3.11

The mid-point $D$ of $OA$ is $\mathbf{A}/2$.

Now since $OA$ is equal to and parallel to $CB$.

$$\therefore \quad \overrightarrow{OA} = \overrightarrow{CB}, \text{ i.e., } \mathbf{A} = \mathbf{B} - \mathbf{C}$$

which may be written as $\dfrac{2(\mathbf{A}/2) + 1 \cdot \mathbf{C}}{2+1} = \dfrac{\mathbf{B}}{3} = \mathbf{P}$ so that $P$ trisects $DC$ and $OB$.

## PROBLEMS 3.1

1. Given $\mathbf{R}_1 = 5\mathbf{I} - 2\mathbf{J} + 4\mathbf{K}$ and $\mathbf{R}_2 = \mathbf{I} + 3\mathbf{J} + 7\mathbf{K}$, find the magnitude and direction cosines of the vectors $\mathbf{R}_1 + \mathbf{R}_2$ and $2\mathbf{R}_1 - \mathbf{R}_2$.
2. Show that (*i*) the position vectors of the vertices of a triangle $A = 3(\sqrt{3}\mathbf{I} - \mathbf{J})$, $B = 6\mathbf{J}$, $C = 3(\sqrt{3}\mathbf{I} + \mathbf{J})$ form as isosales triangle. (*V.T.U., 2013*)
   (*ii*) The points (0, 4, 1) ; (2, 3, –1) ; (4, 5, 0) and (2, 6, 2) are the vertices of a square.
3. A straight line is inclined to the axes of $x$ and $y$ at angles of 30° and 60°. Find the inclination of the line to the $z$-axis.
4. If a line makes angles $\alpha, \beta, \gamma$ with the axes, prove that
   (*i*) $\sin^2\alpha + \sin^2\beta + \sin^2\gamma = 2$.
   (*ii*) $\cos 2\alpha + \cos 2\beta + \cos 2\gamma = -1$. (*V.T.U., 2016*)
5. If **A** and **B** are non-collinear vectors and $\mathbf{P} = (2x + 3y - 2)\mathbf{A} + (3x + 2y + 5)\mathbf{B}$ and $\mathbf{Q} = (-x + 4y - 2)\mathbf{A} + (3x - 4y + 7)\mathbf{B}$, find $x, y$ such that $7\mathbf{P} = 3\mathbf{Q}$.
6. Prove that the line joining the mid-points of the two sides of a triangle is parallel to the third side and half of it.
7. Prove that (*i*) the diagonals of a parallelogram bisect each other ;
   (*ii*) a quadrilateral whose diagonals bisect each other is a parallelogram.
8. In a skew quadrilateral, prove that :
   (*i*) the figure formed by joining the mid-points of the adjacent sides is a parallelogram.
   (*ii*) the joins of the mid-points of opposite sides bisect each other.
9. In a trapezium, prove that the straight line joining the mid-points of the non-parallel sides is parallel to the parallel sides and half their sum.
10. Prove that the vectors $\mathbf{A} = 3\mathbf{I} + \mathbf{J} - 2\mathbf{K}$, $\mathbf{B} = -\mathbf{I} + 3\mathbf{J} + 4\mathbf{K}$, $\mathbf{C} = 4\mathbf{I} - 2\mathbf{J} - 6\mathbf{K}$ can form the sides of a triangle. Also find the length of the median bisecting the vector **C**.
11. Find the ratio in which the line joining (2, 4, 16) and (3, 5, – 4) is divided by the plane $2x - 3y + z + 6 = 0$.
12. Show that the three points $\mathbf{I} - 2\mathbf{J} + 3\mathbf{K}$, $2\mathbf{I} + 3\mathbf{J} - 4\mathbf{K}$, $-7\mathbf{J} + 10\mathbf{K}$ are collinear.
13. If **A, B, C** be the position vectors of the vertices $A, B, C$ of the triangle $ABC$, show that the three
    (*i*) medians concur at the point $\frac{1}{3}(\mathbf{A} + \mathbf{B} + \mathbf{C})$, called the *centroid*.
    (*ii*) internal bisectors of the angles concur at the point $\dfrac{a\mathbf{A} + b\mathbf{B} + c\mathbf{C}}{a+b+c}$, called the *incentre*.
14. Show that the coordinates of the *centroid of the triangle* whose vertices are $(x_1, y_1, z_1)$, $(x_2, y_2, z_2)$, $(x_3, y_3, z_3)$ are

$$\left[\frac{x_1 + x_2 + x_3}{3}, \frac{y_1 + y_2 + y_3}{3}, \frac{z_1 + z_2 + z_3}{3}\right].$$

**15.** Show that the coordinates of the *centroid of the tetrahedron* whose vertices are $(x_r, y_r, z_r) : r = 1, 2, 3, 4$ are

$$\left[\frac{1}{4}(x_1 + x_2 + x_3 + x_4), \frac{1}{4}(y_1 + y_2 + y_3 + y_4); \frac{1}{4}(z_1 + z_2 + z_3 + z_4)\right].$$

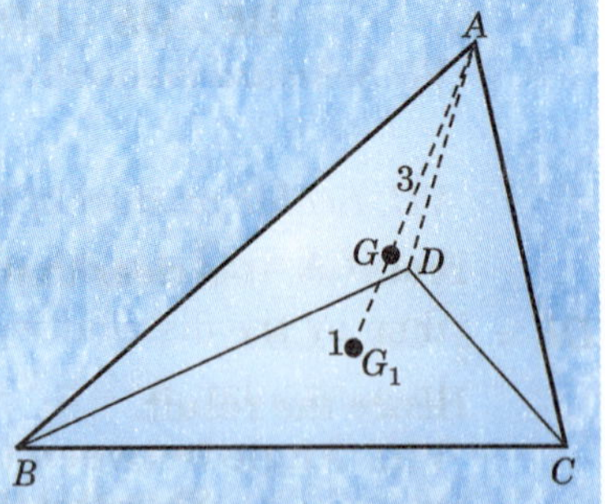

Fig. 3.12

[**Def.** *A* **tetrahedron** *is a solid bounded by four triangular faces.* Thus the tetrahedron *ABCD* has four faces—the Δs *ABC, ACD, ADB, BCD*. (Fig. 3.12.)

It has four **vertices** *A, B, C, D* and three pairs of **opposite edges** *AB, CD* ; *BC, AD* ; *CA, BD*.

*The centroid of the tetrahedron divides the join of each vertex to the centroid of the opposite triangular face in the ratio 3 : 1*].

**16.** *M* and *N* are the mid-points of the diagonals *AC* and *BD* respectively of a quadrilateral *ABCD*. Show that the resultant of the vectors $\vec{AB}, \vec{AD}, \vec{CB}, \vec{CD}$ is $4\vec{MN}$.

## 3.4 PRODUCTS OF TWO VECTORS

Unlike the product of two scalars or that of a vector by a scalar, the product of two vectors is sometimes seen to result in a scalar quantity and sometimes in a vector. As such, we are led to define two types of such products, called the *scalar product* and the *vector product* respectively.

The scalar and vector products of two vectors **A** and **B** are usually written as **A . B** and **A × B** respectively and are read as **A** dot **B** and **A** cross **B**. In view of this notation, the former is sometimes called the *dot product* and the latter the *cross product*.

In vector algebra, the division of a vector by another vector is not defined.

## 3.5 SCALAR OR DOT PRODUCT

**(1) Definition.** *The scalar or dot product of two vectors* **A** *and* **B** *is defined as the scalar ab* cos θ, *where* θ *is the angle between* **A** *and* **B**.

Thus $\quad\quad \mathbf{A.B} = ab \cos \theta.$

**(2) Geometrical interpretation. A.B** *is the product of the length of one vector and the length of the projection of the other in the direction of the former.*

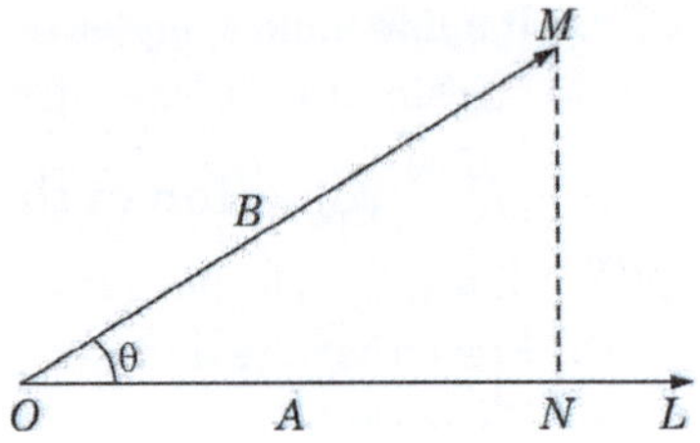

Fig. 3.13

Let $\quad \vec{OL} = \mathbf{A}, \vec{OM} = \mathbf{B} \quad$ then

$\mathbf{A} \,.\, \mathbf{B} = ab \cos \theta = a(OM \cos \theta) = a(ON) = |\mathbf{A}|$ Proj. of $|\mathbf{B}|$ in the direction of **A**.

Similarly, $\mathbf{A} \,.\, \mathbf{B} = |\mathbf{B}|$ Proj. of $|\mathbf{A}|$ in the direction of **B**.

**(3) Properties and other results.**

*I. Scalar product of two vectors is commutative.*

*i.e.*, $\mathbf{A} \,.\, \mathbf{B} = \mathbf{B} \,.\, \mathbf{A}$ for $\mathbf{A} \,.\, \mathbf{B} = ab \cos \theta = ba \cos (-\theta) = \mathbf{B} \,.\, \mathbf{A}$

*II. The necessary and sufficient condition for two vectors to be perpendicular is that their scalar product should be zero.*

When the vectors **A** and **B** are perpendicular, $\mathbf{A} \,.\, \mathbf{B} = ab \cos 90° = 0$.

Conversely, when $\mathbf{A} \,.\, \mathbf{B} = 0$, $ab \cos \theta = 0$, *i.e.*, $\cos \theta = 0$. $\quad\quad (\because \; a \neq 0, b \neq 0)$, or $\theta = 90°$.)

*III.* $\mathbf{A} \,.\, \mathbf{A} = a^2$ which is written as $\mathbf{A}^2$. Thus *the square of a vector is a scalar which stands for the square of its magnitude.*

*IV. For the mutually perpendicular unit vectors,* **I, J, K,** we have the relations.

$$\mathbf{I} \,.\, \mathbf{J} = \mathbf{J} \,.\, \mathbf{K} = \mathbf{K} \,.\, \mathbf{I} = 0$$

and
$$\mathbf{I}^2 = \mathbf{J}^2 = \mathbf{K}^2 = 1$$

which are of great utility.

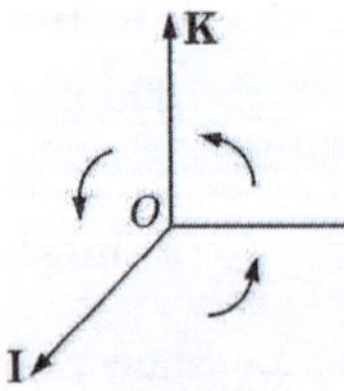

Fig. 3.14

*V. Scalar product of two vectors is distributive i.e.,*

$$(\mathbf{A} + \mathbf{B}) \cdot \mathbf{C} = \mathbf{A} \cdot \mathbf{C} + \mathbf{B} \cdot \mathbf{C}$$

*VI. Schwarz inequality* :* $|\mathbf{A} \cdot \mathbf{B}| \le |\mathbf{A}|\,|\mathbf{B}|$

$$|\mathbf{A} \cdot \mathbf{B}| = |\mathbf{A}|\,|\mathbf{B}|\,|\cos\theta| \le |\mathbf{A}|\,|\mathbf{B}| \qquad [\because \ |\cos\theta| \le 1]$$

*VII. Scalar product of two vectors is equal to the sum of the products of their corresponding components.*

For if $\mathbf{A} = a_1\mathbf{I} + a_2\mathbf{J} + a_3\mathbf{K}$, $\mathbf{B} = b_1\mathbf{I} + b_2\mathbf{J} + b_3\mathbf{K}$

then by the distributive law, $\mathbf{A} \cdot \mathbf{B} = a_1b_1 + a_2b_2 + a_3b_3$

In particular, $\mathbf{A}^2 = a_1^2 + a_2^2 + a_3^2$.

*VIII.* **Angle between two lines** *whose direction cosines are l, m, n and l′, m′, n′ is* $\mathbf{\cos^{-1} (ll' + mm' + nn')}$.

The unit vectors in the direction of the given lines are $\mathbf{U} = l\mathbf{I} + m\mathbf{J} + n\mathbf{K}$ and $\mathbf{U}' = l'\mathbf{I} + m'\mathbf{J} + n'\mathbf{K}$.

If θ be the angle between the lines, then

$$\mathbf{U} \cdot \mathbf{U}' = (l\mathbf{I} + m\mathbf{J} + n\mathbf{K}) \cdot (l'\mathbf{I} + m'\mathbf{J} + n'\mathbf{K})$$

or
$$1 \cdot 1 \cdot \cos\theta = ll' + mm' + nn'$$

Hence
$$\cos\theta = ll' + mm' + nn' \qquad \textit{(V.T.U., 2013)} \quad ...(i)$$

**Cor. 1.**
$$\sin^2\theta = 1 - \cos^2\theta = 1 - (ll' + mm' + nn')^2$$
$$= (l^2 + m^2 + n^2)(l'^2 + m'^2 + n'^2) - (ll' + mm' + nn')^2$$
$$= (mn' - nm')^2 + (nl' - ln')^2 + (lm' - ml')^2$$

$$\therefore \quad \sin\theta = \pm\sqrt{\Sigma (mn' - nm')^2}. \qquad ...(ii)$$

**Cor. 2.** *The condition that the lines whose direction cosines are l, m, n and l′, m′, n′ should be* **perpendicular** *is*

$$\mathbf{ll' + mm' + nn' = 0} \qquad ...(iii)$$

*and* **parallel** *is*
$$\mathbf{l = l',\ m = m',\ n = n'} \qquad ...(iv)$$

These conditions easily follow from Eqns. (*i*) and (*ii*).

**Cor. 3.** *The angle θ between two lines whose direction ratios are a, b, c, and a′, b′, c′ is given by*

$$\cos\theta = \frac{aa' + bb' + cc'}{\sqrt{(\Sigma a^2)}\sqrt{(\Sigma a'^2)}}$$

or
$$\sin\theta = \frac{\sqrt{(bc' - cb')^2 + (ca' - ac')^2 + (ab' - ba')^2}}{\sqrt{(\Sigma a^2)}\sqrt{(\Sigma a'^2)}}$$

*These lines are Eqn. (i) perpendicular if aa′ + bb′ + cc′ = 0, Eqn. (ii) parallel if a/a′ = b/b′ = c/c′.*

*IX.* **Projection of the line joining two points** $(x_1, y_1, z_1)$ *and* $(x_2, y_2, z_2)$ *on a line whose direction cosines are l, m, n is*
$$l(x_2 - x_1) + m(y_2 - y_1) + n(z_2 - z_1).$$

Let
$$\vec{OP} = x_1\mathbf{I} + y_1\mathbf{J} + z_1\mathbf{K}, \quad \vec{OQ} = x_2\mathbf{I} + y_2\mathbf{J} + z_2\mathbf{K}$$

$$\therefore \quad \vec{PQ} = (x_2 - x_1)\mathbf{I} + (y_2 - y_1)\mathbf{J} + (z_2 - z_1)\mathbf{K}$$

Also unit vector **U** along the given lines is $l\mathbf{I} + m\mathbf{J} + n\mathbf{K}$.

$\therefore$ Projection of PQ on the given line $= \vec{PQ} \cdot \mathbf{U}$.

$$= l(x_2 - x_1) + m(y_2 - y_1) + n(z_2 - z_1).$$

**Example 3.5.** *Find the sides and angles of the triangle whose vertices are* $\mathbf{I} - 2\mathbf{J} + 2\mathbf{K}$, $2\mathbf{I} + \mathbf{J} - \mathbf{K}$, *and* $3\mathbf{I} - \mathbf{J} + 2\mathbf{K}$.

**Solution.** Let $\vec{OA} = \mathbf{I} - 2\mathbf{J} + 2\mathbf{K}$, $\vec{OB} = 2\mathbf{I} + \mathbf{J} - \mathbf{K}$, $\vec{OC} = 3\mathbf{I} - \mathbf{J} + 2\mathbf{K}$

Then
$$\vec{BC} = \mathbf{I} - 2\mathbf{J} + 3\mathbf{K}$$
$$\vec{CA} = -2\mathbf{I} - \mathbf{J}$$

* Named after the German mathematician *Hermann Amandus Schwarz* (1843—1921) who is known for his work in conformal mapping, calculus of variations and differential geometry. He succeeded *Weierstrass* in Berlin University.

and $\vec{AB} = \mathbf{I} + 3\mathbf{J} - 3\mathbf{K}$

$\therefore$ $BC = \sqrt{14}, CA = \sqrt{5}, AB = \sqrt{19}.$

Now d.c.'s of $AB$ and $AC$ being

$$1/\sqrt{19}, 3/\sqrt{19}, -3/\sqrt{19} \text{ and } 2/\sqrt{5}, 1/\sqrt{5}, 0,$$

We have $\cos A = \frac{1}{\sqrt{19}} \cdot \frac{2}{\sqrt{5}} + \frac{3}{\sqrt{19}} \cdot \frac{1}{\sqrt{5}} + \frac{-3}{\sqrt{19}} \cdot 0 = \sqrt{(5/19)}$

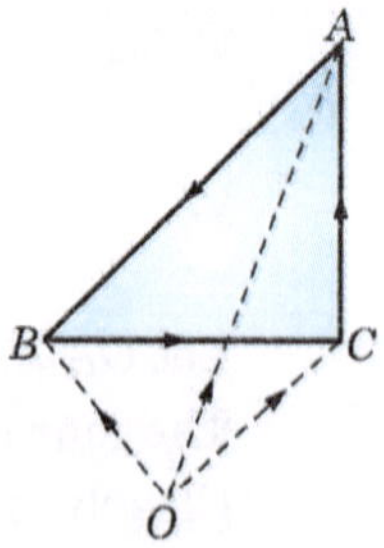

Fig. 3.15

*i.e.*, $\angle A = \cos^{-1} \sqrt{(5/19)}$. Again d.c.'s of $BC$ and $BA$ being

$$1/\sqrt{14}, -2/\sqrt{14}, 3/\sqrt{14} \text{ and } -1/\sqrt{19}, -3/\sqrt{19}, 3/\sqrt{19};$$

we have $\cos B = \frac{1}{\sqrt{14}} \cdot \frac{-1}{\sqrt{19}} + \frac{-2}{\sqrt{14}} \cdot \frac{-3}{\sqrt{19}} + \frac{3}{\sqrt{14}} \cdot \frac{3}{\sqrt{19}} = \sqrt{(14/19)}$, *i.e.*, $\angle B = \cos^{-1} \sqrt{(14/19)}$

Finally, d.c.'s of $CA$ and $CB$ being $-2/\sqrt{5}, -1/\sqrt{5}, 0$ and $-1/\sqrt{14}, 2/\sqrt{14}, -3/\sqrt{14}$;

we have $\cos C = \frac{-2}{\sqrt{5}} \cdot \frac{-1}{\sqrt{14}} + \frac{-1}{\sqrt{5}} \cdot \frac{2}{\sqrt{14}} + 0 \cdot \frac{-3}{\sqrt{14}} = 0$, *i.e.*, $\angle C = 90°$

**Example 3.6.** *Prove that the right -bisectors of the sides of a triangle concur at its circumcentre.*

**Solution.** Let $A(\mathbf{A})$, $B(\mathbf{B})$, $C(\mathbf{C})$ be the vertices of any triangle $ABC$. The mid-points of the sides $BC$, $CA$ and $AB$ are

$$D\left(\frac{\mathbf{B}+\mathbf{C}}{2}\right), E\left(\frac{\mathbf{C}+\mathbf{A}}{2}\right), F\left(\frac{\mathbf{A}+\mathbf{B}}{2}\right)$$

Let the perpendicular at $D$ and $E$ to $BC$ and $CA$ respectively intersect at the point $P(\mathbf{R})$. Then $\vec{DP} \cdot \vec{BC} = 0$

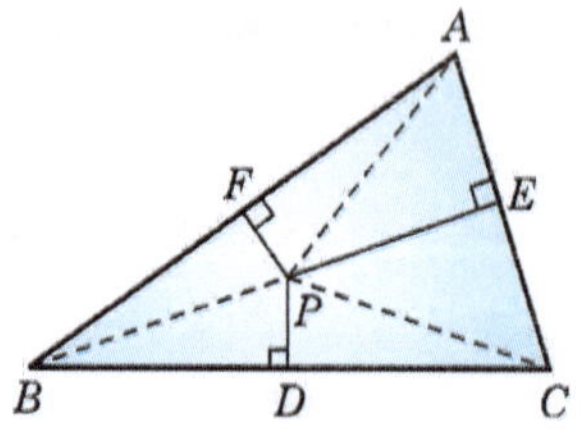

Fig. 3.16

*i.e.*, $\left(\mathbf{R} - \frac{\mathbf{B}+\mathbf{C}}{2}\right) \cdot (\mathbf{C} - \mathbf{B}) = 0$ ...(*i*)

and $\vec{EP} \cdot \vec{CA} = 0$, *i.e.*, $\left(\mathbf{R} - \frac{\mathbf{C}+\mathbf{A}}{2}\right) \cdot (\mathbf{A} - \mathbf{C}) = 0$ ...(*ii*)

Adding Eqns. (*i*) and (*ii*), we get $\left(\mathbf{R} - \frac{\mathbf{A}+\mathbf{B}}{2}\right) \cdot (\mathbf{A} - \mathbf{B}) = 0$

which shows that $FP$ is perpendicular to $AB$. Hence the result.

Further $PA = PB$ if $|\mathbf{A} - \mathbf{R}| = |\mathbf{B} - \mathbf{R}|$

or if, $(\mathbf{A} - \mathbf{R})^2 = (\mathbf{B} - \mathbf{R})^2$ or if, $\mathbf{A}^2 - 2\mathbf{A} \cdot \mathbf{R} = \mathbf{B}^2 - 2\mathbf{B} \cdot \mathbf{R}$

of if, $\left(\mathbf{R} - \frac{\mathbf{A}+\mathbf{B}}{2}\right) \cdot (\mathbf{A} - \mathbf{B}) = 0$, which is true.

**Example 3.7.** *If the distance between two points P and Q is d and the lengths of the projections of PQ on the coordinate planes $d_1, d_2, d_3$, show that $2d^2 = d_1^2 + d_2^2 + d_3^2$.*

**Solution.** Let $P$ be $(x_1, y_1, z_1)$ and $Q$ be $(x_2, y_2, z_2)$, then

$$d^2 = (x_1 - x_2)^2 + (y_1 - y_2)^2 + (z_1 - z_2)^2.$$

The feet of the perpendiculars drawn from $P$ and $Q$ on the $XY$-plane are the projections of $P$ and $Q$ on this plane. If these are $L$ and $M$, then $L$ is $(x_1, y_1, 0)$ and $M$ is $(x_2, y_2, 0)$.

$\therefore$ $d_1$ = projection of $PQ$ on $XY$-plane, *i.e.*, $LM$

or $d_1^2 = (x_1 - x_2)^2 + (y_1 - y_2)^2$

Similarly, $d_2^2 = (y_1 - y_2)^2 + (z_1 - z_2)^2$ and $d_3^2 = (z_1 - z_2)^2 + (x_1 - x_2)^2$

$\therefore$ $d_1^2 + d_2^2 + d_3^2 = 2[(x_1 - x_2)^2 + (y_1 - y_2)^2 + (z_1 - z_2)^2] = 2d^2.$

**Example 3.8.** *A line makes angles α, β, γ, δ with diagonals of a cube, prove that*

$$\cos^2 \alpha + \cos^2 \beta + \cos^2 \gamma + \cos^2 \delta = 4/3.$$

**Solution.** Take $O$, a corner of the cube as origin and $OA$, $OB$, $OC$ the three edges through it, as the axes. Let $OA = OB = OC = a$. Then the coordinates of the corners are as shown in Fig. 3.17.

The four diagonals are $OP$, $AA'$, $BB'$ and $CC'$.

Clearly, direction cosines of $OP$ are

$$\frac{a-0}{\sqrt{(\Sigma a^2)}}, \frac{a-0}{\sqrt{(\Sigma a^2)}}, \frac{a-0}{\sqrt{(\Sigma a^2)}} \text{ i.e., } \frac{1}{\sqrt{3}}, \frac{1}{\sqrt{3}}, \frac{1}{\sqrt{3}}.$$

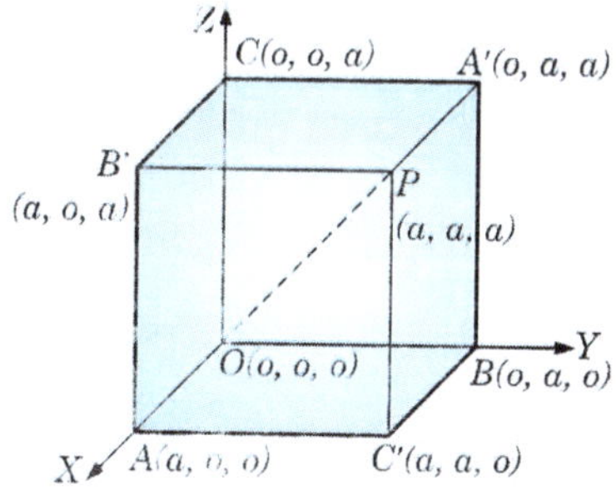

Fig. 3.17

Similarly, direction cosines of $AA'$ are $-\frac{1}{\sqrt{3}}, \frac{1}{\sqrt{3}}, \frac{1}{\sqrt{3}}$;

Similarly, direction cosines of $BB'$ are $\frac{1}{\sqrt{3}}, -\frac{1}{\sqrt{3}}, \frac{1}{\sqrt{3}}$;

and Similarly direction cosines of $CC'$ are $\frac{1}{\sqrt{3}}, \frac{1}{\sqrt{3}}, -\frac{1}{\sqrt{3}}$.

Let $l, m, n$ be the direction cosines of the given line which makes angles α, β, γ, δ with $OP$, $AA'$, $BB'$, $CC'$ respectively. Then

$$\cos \alpha = \frac{1}{\sqrt{3}}(l+m+n)\ ;\ \cos \beta = \frac{1}{\sqrt{3}}(-l+m+n)$$

$$\cos \gamma = \frac{1}{\sqrt{3}}(l-m+n);\ \cos \delta = \frac{1}{\sqrt{3}}(l+m-n)$$

Squaring and adding, we get

$$\cos^2 \alpha + \cos^2 \beta + \cos^2 \gamma + \cos^2 \delta = \frac{1}{3}[(l+m+n)^2 + (-l+m+n)^2 + (l-m+n)^2 + (l+m-n)^2]$$

$$= \frac{1}{3}[4(l^2+m^2+n^2)] = \frac{4}{3}. \qquad [\because l^2+m^2+n^2 = 1]$$

**Example 3.9.** *If the edges of a rectangular parallelopiped are a, b, c, show that the angle between the four diagonals are* $\cos^{-1}\left(\frac{\pm a^2 \pm b^2 \pm c^2}{a^2+b^2+c^2}\right)$.

**Solution.** Let $OA = a$, $OB = b$, $OC = c$ be the edges of the rectangular parallelopiped. Then the coordinates of the corners are as shown in Fig. 3.18. The four diagonals taken in pairs are (*i*) ($OP$, $AA'$), (*ii*) ($OP$, $BB'$), (*iii*) ($OP$, $CC'$), (*iv*) ($AA'$, $BB'$), (*v*) ($AA'$, $CC'$) and (*vi*) ($BB'$, $CC'$).

Let the angles between these pairs of diagonals be $\theta_1, \theta_2, \ldots \theta_6$ respectively. Clearly d.r.'s $OP$ are $a, b, c$ ; d.r.'s, of $AA'$ are $-a, b, c$, d.r.'s of $BB'$ are $a, -b, c$ and d.r.'s of $CC'$ are $a, b, -c$.

∴ For the pair (*i*) *i.e.*, ($OP$, $AA'$) ;

$$\cos \theta_1 = \frac{-a^2+b^2+c^2}{\sqrt{(a^2+b^2+c^2)}\sqrt{(a^2+b^2+c^2)}} = \frac{-a^2+b^2+c^2}{a^2+b^2+c^2}$$

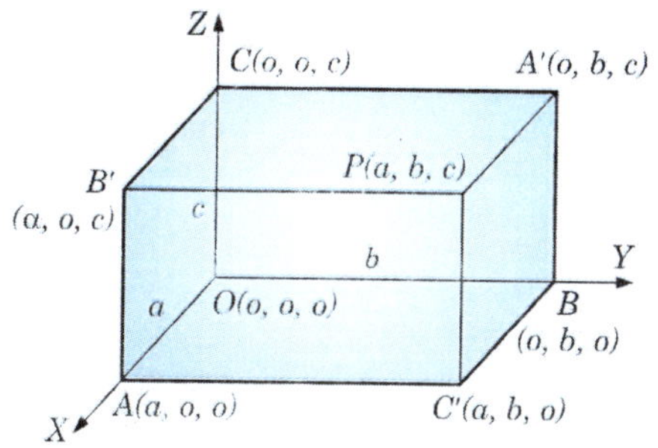

Fig. 3.18

Similarly, $\cos \theta_2 = \frac{a^2-b^2+c^2}{a^2+b^2+c^2}$ ; $\quad \cos \theta_3 = \frac{a^2+b^2-c^2}{a^2+b^2+c^2}$ ;

$\cos \theta_4 = \frac{-a^2-b^2+c^2}{a^2+b^2+c^2}$; $\quad \cos \theta_5 = \frac{-a^2+b^2-c^2}{a^2+b^2+c^2}$;

$$\cos \theta_6 = \frac{a^2-b^2-c^2}{a^2+b^2+c^2}$$

Thus, noting that at least one term in the numerator is negative, we have in general

$$\cos \theta = \frac{\pm a^2 \pm b^2 \pm c^2}{a^2+b^2+c^2}.$$

**Example 3.10.** *Prove that the lines whose direction cosines are given by the relations $al + bm + cn = 0$ and $mn + nl + lm = 0$ are*

*(i) Perpendicular if $a^{-1} + b^{-1} + c^{-1} = 0$*

*(ii) parallel if $\sqrt{a} + \sqrt{b} + \sqrt{c} = 0$.*

**Solution.** Eliminating $n$ from the given relations, we have

$$(m + l)\left(-\frac{al + bm}{c}\right) + lm = 0 \quad \text{or} \quad al^2 + (c - a - b)lm + bm^2 = 0$$

or $$a(l/m)^2 + (c - a - b)(l/m) + b = 0 \qquad ...(1)$$

If $l_1, m_1, n_1; l_2, m_2, n_2$, are the direction cosines of these lines then $l_1/m_1, l_2/m_2$ are the roots of the quadratic Eqn. (1).

$$\therefore \quad \frac{l_1}{m_1}\cdot\frac{l_2}{m_2} = \frac{b}{a} \quad \text{or} \quad \frac{l_1 l_2}{1/a} = \frac{m_1 m_2}{1/b} = \frac{n_1 n_2}{1/c} \quad \text{(by symmetry)} = k \text{ (say).}$$

*The lines will be perpendicular if* $l_1l_2 + m_1m_2 + n_1n_2 = k\left(\frac{1}{a} + \frac{1}{b} + \frac{1}{c}\right) = 0$

or if, $$\frac{1}{a} + \frac{1}{b} + \frac{1}{c} = 0.$$

*The lines will be parallel if* $l_1 = l_2, m_1 = m_2, n_1 = n_2$.

*i.e.,* if, $l_1/m_1 = l_2/m_2$; *i.e.* if, $(c - a - b)^2 = 4ab$

or if, $c - a - b = \pm 2\sqrt{(ab)}$ or if, $c = a + b \pm 2\sqrt{(ab)} = (\sqrt{a} \pm \sqrt{b})^2$

or if, $\pm\sqrt{c} = \sqrt{a} \pm \sqrt{b}$ or if, $\sqrt{a} + \sqrt{b} + \sqrt{c} = 0$ [Taking necessary signs]

**Example 3.11.** *Find the angle between the lines whose direction cosines are given by the equation $l + 3m + 5n = 0$ and $5lm - 2mn + 6nl = 0$.*

**Solution.** Let us eliminate $l$ from the given relations, by substituting $l = -3m - 5n$ in the second relation

$$5m(-3m - 5n) - 2mn + 6n(-3m - 5n) = 0$$

*i.e.,* $$15m^2 + 45mn + 30n^2 = 0 \quad \text{or} \quad m^2 + 3mn + 2n^2 = 0$$

or $$(m + n)(m + 2n) = 0, \quad i.e., \quad m + n = 0 \text{ or } m + 2n = 0$$

Now let us first solve the equations $l + 3m + 5n = 0$ and $m + n = 0$

These give $m = -n$ and $l = -2n$, *i.e.*, $= \frac{l}{-2} = \frac{m}{-1} = \frac{n}{1}$ ...(*i*)

Similarly, solving the equations $l + 3m + 5n = 0$ and $m + 2n = 0$,

We get $$\frac{l}{1} = \frac{m}{-2} = \frac{n}{1} \qquad ...(ii)$$

(*i*) and (*ii*) give the direction ratios of the two lines.

If $\theta$ be the angle between these two lines, then

$$\cos\theta = \frac{(-2)\times 1 + (-1)\times(-2) + 1\times 1}{\sqrt{(2^2 + 1^2 + 1^2)}\sqrt{(1^2 + 2^2 + 1^2)}} = \frac{1}{6}, \quad i.e., \quad \theta = \cos^{-1}\left(\frac{1}{6}\right).$$

## PROBLEMS 3.2

1. (*i*) If $\mathbf{A} = \mathbf{I} + 2\mathbf{J} + 3\mathbf{K}$, $\mathbf{B} = -\mathbf{I} + 2\mathbf{J} + \mathbf{K}$ and $\mathbf{C} = 3\mathbf{I} + \mathbf{J}$, find $t$ such that $\mathbf{A} + t\mathbf{B}$ is perpendicular to $\mathbf{C}$.

   (*ii*) If $|\mathbf{A} + \mathbf{B}| = |\mathbf{A} - \mathbf{B}|$, show that $\mathbf{A}$ and $\mathbf{B}$ are mutually perpendicular.

2. (*i*) Show that $\left(\frac{\mathbf{A}}{a^2} - \frac{\mathbf{B}}{b^2}\right)^2 = \left(\frac{\mathbf{A} - \mathbf{B}}{ab}\right)^2$.

   (*ii*) Interpret geometrically $(\mathbf{C} - \mathbf{A})\cdot(\mathbf{B} - \mathbf{C}) = 0$.

3. Find the direction cosines of the line perpendicular to the lines with direction cosines (3, –1, 1) and (–3, 2, 4). (*V.T.U., 2016*)

4. If $\mathbf{A} = \mathbf{I} + 2\mathbf{J} - 3\mathbf{K}$ and $\mathbf{B} = 3\mathbf{I} - \mathbf{J} + 2\mathbf{K}$, show that $\mathbf{A} + \mathbf{B}$ is perpendicular to $\mathbf{A} - \mathbf{B}$. Also calculate the angle between $2\mathbf{A} + \mathbf{B}$ and $\mathbf{A} + 2\mathbf{B}$.
5. Show that the three concurrent lines with direction cosines $(l_1, m_1, n_1)$, $(l_2, m_2, n_2)$, $(l_3, m_3, n_3)$ are coplanar if
$$\begin{vmatrix} l_1 & m_1 & n_1 \\ l_2 & m_2 & n_2 \\ l_3 & m_3 & n_3 \end{vmatrix} = 0.$$
6. Find the projection of the vector $\mathbf{I} - 2\mathbf{J} + \mathbf{K}$ on $4\mathbf{I} - 4\mathbf{J} + 7\mathbf{K}$.
7. The projection of a line on the coordinate axes are 12, 4, 3. Find the length and direction cosines of the line.
8. Show (by vector methods) that the mid-point of the hypotenuse of a right-angled triangle is equidistant from its vertices.
9. Prove (by vector methods) that the angle in a semi-circle is a right angle.
10. Show (by vector methods) that the diagonals of a rhombus intersect at right angles.
11. Show that the altitudes of a triangle meet in a point (called the *orthocentre*).
12. $ABCD$ is a tetrahedron having the edges $BC$ and $AC$ at right angles to opposite edges $AD$ and $BD$ respectively. Show that the third pair of opposite edges $AB$ and $CD$ are also at right angles.
13. Find the angle between the lines whose direction cosines are given by the equations $l + m + n = 0$, $l^2 + m^2 + n^2 = 0$.
14. Show that the lines whose direction cosines are given by the equations $4/n - 3/l - 1/m = 0$, and $3l + m + 2n = 0$ are perpendicular. *(Anna, 2013)*
15. Show that the lines whose direction cosines are given by the equations $l + m + n = 0$, $al^2 + bm^2 + cn^2 = 0$ are
    (*i*) perpendicular, if $a + b + c = 0$, (*ii*) parallel, if $a^{-1} + b^{-1} + c^{-1} = 0$.
16. Show that the straight lines whose direction cosines are given by the equations
$$al + bm + cn = 0,\ fmn + gnl + hlm = 0 \text{ are } (i) \text{ perpendicular if } \frac{f}{a} + \frac{g}{b} + \frac{h}{c} = 0$$
    (*ii*) parallel if $\sqrt{af} \pm \sqrt{bg} \pm \sqrt{ch} = 0$.
17. Show that the angle between any two diagonals of a cube is $\cos^{-1} 1/3$.
18. $(l_1, m_1, n_1)$, $(l_2, m_2, n_2)$ and $(l_3, m_3, n_3)$ are the direction cosines of three mutually perpendicular lines. Prove that the line whose d.c.'s are proportional to $l_1 + l_2 + l_3$, $m_1 + m_2 + m_3$, $n_1 + n_2 + n_3$ makes equal angles with the axes.
19. $AB$, $BC$ are the diagonals of adjacent faces of a rectangular box with its centre at the origin $O$, its edges are parallel to the axes. If the angles $BOC$, $COA$ and $AOB$ are equal to $\theta$, $\phi$, $\psi$ respectively, prove that
$$\cos\theta + \cos\phi + \cos\psi = -1.$$

## 3.6 VECTOR, OR CROSS PRODUCT

**(1) Definition.** *The vector, or cross product of two vectors* **A** *and* **B** *is defined as a vector such that*
(*i*) *its magnitude is ab sin θ, θ being the angle between* **A** *and* **B**,
(*ii*) *its direction is perpendicular to the plane of* **A** *and* **B**,
and (*iii*) *it forms with* **A** *and* **B** *a right-handed system.*

If **N** be a unit vector normal to the plane of **A** and **B** (**A, B, N** forming a right-handed system), then
$$\mathbf{A} \times \mathbf{B} = ab \sin\theta\, \mathbf{N}.$$

**(2) Geometrical interpretation.** **A × B** *represents twice the vector area of the triangle having the vectors* **A** *and* **B** *as its adjacent sides.*

If **N** be a unit vector normal to the plane of the triangle $OAB$, then
$$\mathbf{A} \times \mathbf{B} = ab \sin\theta\, \mathbf{N}$$
$$= 2\left(\frac{1}{2} ab \sin\theta\right) \mathbf{N} = 2\Delta\, OAB\, \mathbf{N} = 2\Delta\, \overrightarrow{OA}\, B.$$

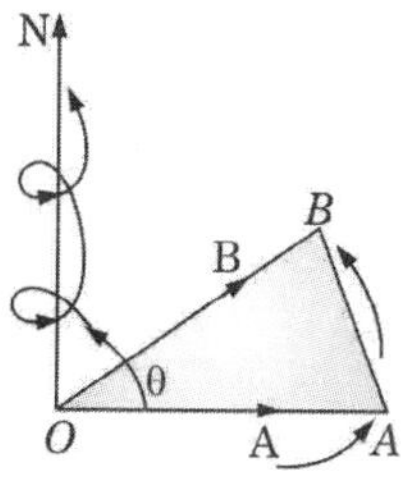

Fig. 3.19

**(3) Properties and other results**

*I. Vector product of two vectors is not commutative,*
$$\mathbf{A} \times \mathbf{B} \neq \mathbf{B} \times \mathbf{A}. \text{ In fact, } \mathbf{A} \times \mathbf{B} = -\mathbf{B} \times \mathbf{A}.$$

for $\mathbf{A} \times \mathbf{B} = ab \sin\theta\, \mathbf{N}$ or $2\Delta\, \overrightarrow{OAB}$.

and $\mathbf{B} \times \mathbf{A} = ab \sin(-\theta)\, \mathbf{N} = -ab \sin\theta\, \mathbf{N}$ or $2\Delta\overrightarrow{OBA}$.

*II. The necessary and sufficient condition for two non-zero vectors to be parallel is that their vector product should be zero.*

When the vectors **A** and **B** are parallel, the angle θ between them is 0 and 180° so that sin θ = 0, and as such **A × B = 0.**

Conversely, when $\mathbf{A} \times \mathbf{B} = \mathbf{0}$ ; $ab \sin\theta = 0$

*i.e.,* $\sin\theta = 0$ $\quad (\because\ a \neq 0, b \neq 0)$

or $\theta = 0$ or 180°. In particular. $\mathbf{A} \times \mathbf{A} = 0$.

*III. For the orthonormal vector trial* **I, J, K,** we have the relations :

$$\mathbf{I} \times \mathbf{I} = \mathbf{J} \times \mathbf{J} = \mathbf{K} \times \mathbf{K} = \mathbf{0}$$
$$\mathbf{I} \times \mathbf{J} = \mathbf{K}, \quad \mathbf{J} \times \mathbf{I} = -\mathbf{K}$$
$$\mathbf{J} \times \mathbf{K} = \mathbf{I}, \quad \mathbf{K} \times \mathbf{J} = -\mathbf{I}$$
$$\mathbf{K} \times \mathbf{I} = \mathbf{J}, \quad \mathbf{I} \times \mathbf{K} = -\mathbf{J}.$$

*IV. Relation between scalar and vector products.*

We have $(\mathbf{A} \cdot \mathbf{B})^2 = a^2b^2 \cos\theta = a^2b^2 - a^2b^2 \sin^2\theta = a^2b^2 - (\mathbf{A} \times \mathbf{B}) \cdot (\mathbf{A} \times \mathbf{B})$

$$(\mathbf{A} \times \mathbf{B})^2 = \mathbf{A}^2\mathbf{B}^2 - (\mathbf{A} \cdot \mathbf{B})^2.$$

*V. Vector product of two vectors is distributive*

*i.e.,* $(\mathbf{A} + \mathbf{B}) \times \mathbf{C} = \mathbf{A} \times \mathbf{C} + \mathbf{B} \times \mathbf{C}$.

*VI. Analytical expression for the vector product.*

If $\mathbf{A} = a_1\mathbf{I} + a_2\mathbf{J} + a_3\mathbf{K}$, $\mathbf{B} = b_1\mathbf{I} + b_2\mathbf{J} + b_3\mathbf{K}$ *then* $\mathbf{A} \times \mathbf{B} = \begin{vmatrix} \mathbf{I} & \mathbf{J} & \mathbf{K} \\ a_1 & a_2 & a_3 \\ b_1 & b_2 & b_3 \end{vmatrix}$

For we get

$$\mathbf{A} \times \mathbf{B} = (a_2b_3 - a_3b_2)\mathbf{I} + (a_3b_1 - a_1b_3)\mathbf{J} + (a_1b_2 - a_2b_1)\mathbf{K}$$

whence follows the required result.

**Example 3.12.** *If* $\mathbf{A} = 4\mathbf{I} + 3\mathbf{J} + \mathbf{K}$, $\mathbf{B} = 2\mathbf{I} - \mathbf{J} + 2\mathbf{K}$, *find a unit vector* **N** *perpendicular to vectors* **A** *and* **B** *such that* **A, B, N** *form a right handed system. Also find the angle between the vectors A and B.*

**Solution.** Since $\mathbf{A} \times \mathbf{B} = \begin{vmatrix} \mathbf{I} & \mathbf{J} & \mathbf{K} \\ 4 & 3 & 1 \\ 2 & -1 & 2 \end{vmatrix} = 7\mathbf{I} - 6\mathbf{J} - 10\mathbf{K}$

and $|\mathbf{A} \times \mathbf{B}| = \sqrt{[(7)^2 + (-6)^2 + (-10)^2]} = \sqrt{185}$

$\therefore$ Unit vector $\mathbf{N} \perp$ to **A** and **B** $= \dfrac{\mathbf{A} \times \mathbf{B}}{|\mathbf{A} \times \mathbf{B}|} = (7\mathbf{I} - 6\mathbf{J} - 10\mathbf{K})/\sqrt{185}$

Also $a = \sqrt{4^2 + 3^2 + 1^2} = \sqrt{26}$ and $b = 3$.

If θ be the angle between **A** and **B**, then $|\mathbf{A} \times \mathbf{B}| = ab \sin\theta$, *i.e.,* $\sin\theta = |\mathbf{A} \times \mathbf{B}|/ab$

Thus $\sin\theta = \sqrt{185}/3\sqrt{26}$ whence $\theta = 62°\ 40'$.

**Example 3.13.** (*i*) *Prove that the area of the triangle whose vertices are* **A, B, C** *is*

$$\frac{1}{2}\,|\mathbf{B} \times \mathbf{C} + \mathbf{C} \times \mathbf{A} + \mathbf{A} \times \mathbf{B}|$$

(*ii*) *Calculate the area of the triangle whose vertices are A(1, 0 – 1), B(2, 1, 5) and C(0, 1, 2).*

**Solution.** (*i*) Let $A(\mathbf{A})$, $B(\mathbf{B})$, $C(\mathbf{C})$ be the vertices of the triangle *ABC* (Fig. 3.20) and *O*, the origin so that

$$\overrightarrow{BC} = \overrightarrow{OC} - \overrightarrow{OB} = \mathbf{C} - \mathbf{B}$$

and $\vec{BA} = \vec{OA} - \vec{OB} = \mathbf{A} - \mathbf{B}$

$\therefore$ Vector area of $\Delta ABC$

$$= \frac{1}{2}[\vec{BC} \times \vec{BA}] = \frac{1}{2}[(\mathbf{C} - \mathbf{B}) \times (\mathbf{A} - \mathbf{B})]$$

$$= \frac{1}{2}[\mathbf{C} \times \mathbf{A} - \mathbf{C} \times \mathbf{B} - \mathbf{B} \times \mathbf{A} + \mathbf{B} \times \mathbf{B}]$$

$$= \frac{1}{2}[\mathbf{B} \times \mathbf{C} + \mathbf{C} \times \mathbf{A} + \mathbf{A} \times \mathbf{B}] \quad [\because \ \mathbf{B} \times \mathbf{B} = 0]$$

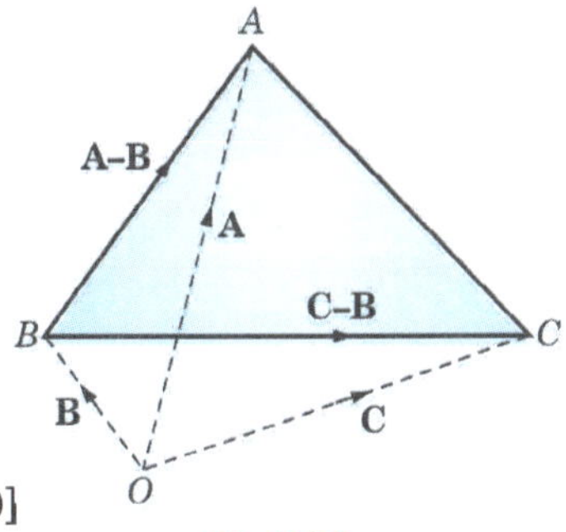


Fig. 3.20

Thus area of $\Delta ABC = \frac{1}{2} \, |\mathbf{B} \times \mathbf{C} + \mathbf{C} \times \mathbf{A} + \mathbf{A} \times \mathbf{B}|$.

(*ii*) Let $O$ be the origin so that

$$\vec{OA} = \mathbf{I} - \mathbf{K}, \ \vec{OB} = 2\mathbf{I} + \mathbf{J} + 5\mathbf{K} \text{ and } \vec{OC} = \mathbf{J} + 2\mathbf{K}$$

Then $\vec{BC} = \vec{OC} - \vec{OB} = -2\mathbf{I} - 3\mathbf{K}$

and $\vec{BA} = \vec{OA} - \vec{OB} = -\mathbf{I} - \mathbf{J} - 6\mathbf{K}$

$\therefore$ Vector area of $\Delta ABC = \frac{1}{2}(\vec{BC} \times \vec{BA}) = \frac{1}{2}\begin{vmatrix} \mathbf{I} & \mathbf{J} & \mathbf{K} \\ -2 & 0 & -3 \\ -1 & -1 & -6 \end{vmatrix}$

Thus area of $\Delta ABC = \frac{1}{2} \, |-3\mathbf{I} - 9\mathbf{J} + 2\mathbf{K}| = \frac{1}{2}\sqrt{94}$.

**Example 3.14.** *In a triangle ABC; D, E, F are the mid-points of the sides BC, CA, AB; prove that*

$$\Delta DEF = \Delta FCE = \frac{1}{4}\Delta ABC.$$

**Solution.** Take $B$ as the origin and let the position vectors of $C$ and $A$ be $\mathbf{C}$ and $\mathbf{A}$ (Fig. 3.21); so that the position vectors of $D, E, F$ are

$$\mathbf{C}/2, (\mathbf{C} + \mathbf{A})/2, \mathbf{A}/2.$$

$$\therefore \quad \Delta \vec{DEF} = \frac{1}{2}(\vec{DE} \times \vec{DF}) = \frac{1}{2}\left(\frac{\mathbf{C} + \mathbf{A}}{2} - \frac{\mathbf{C}}{2}\right)\left(\frac{\mathbf{A}}{2} - \frac{\mathbf{C}}{2}\right)$$

$$= \frac{1}{8}[\mathbf{A} \times (\mathbf{A} - \mathbf{C})] = \frac{1}{8} \ \mathbf{C} \times \mathbf{A} = \frac{1}{4}\Delta \vec{ABC}$$

$$\Delta \vec{FCE} = \frac{1}{2}(\vec{FC} \times \vec{FE}) = \frac{1}{2}[\mathbf{C} - \mathbf{A}/2) \times \mathbf{C}/2]$$

$$= \frac{1}{8} \ \mathbf{C} \times \mathbf{A} = \frac{1}{4}\Delta \vec{ABC}. \text{ Hence the result.}$$

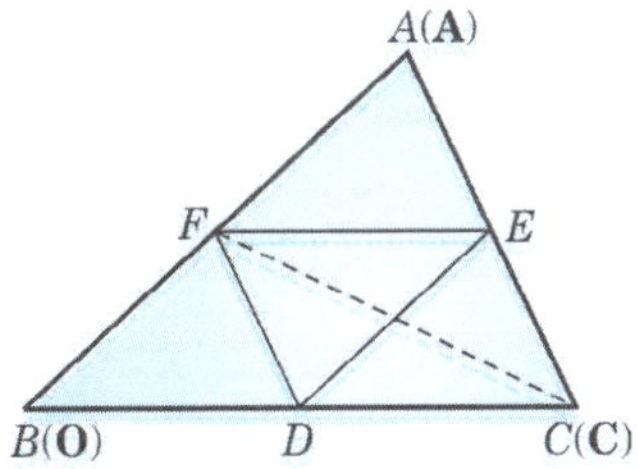


Fig. 3.21

**Example 3.15.** *Prove that*

(*i*) *sin* $(A + B) =$ *sin* $A$ *cos* $B +$ *cos* $A$ *sin* $B$.

(*ii*) *cos* $(A + B) =$ *cos* $A$ *cos* $B -$ *sin* $A$ *sin* $B$.

**Solution.** Let $\mathbf{I}$, $\mathbf{J}$ denote unit vectors along two perpendicular lines $OX$, $OY$ so that

$$\mathbf{I}^2 = \mathbf{J}^2 = 1, \mathbf{I}.\mathbf{J} = 0$$

and $\mathbf{I} \times \mathbf{I} = \mathbf{J} \times \mathbf{J} = \mathbf{0}$

Let $\angle POX = A$ and $\angle XOQ = B$,

so that $\angle POQ = A + B$.

If $OP = p$ and $OQ = q$, then the coordinates of $P$ are $(p \cos A, -p \sin A)$ and those of $Q$ are $(q \cos B, q \sin B)$ so that

$$\vec{OP} = (p \cos A)\mathbf{I} - (p \sin A)\mathbf{J}$$

$$\vec{OQ} = (q \cos B)\mathbf{I} + (q \sin B)\mathbf{J}$$

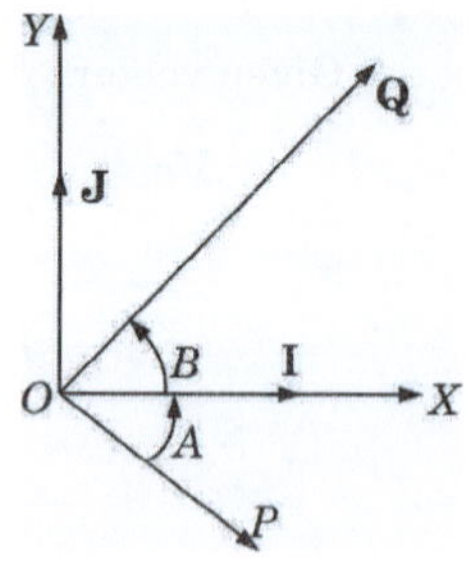

Fig. 3.22

Then $|\vec{OP} \times \vec{OQ}| = |[(p \cos A)\mathbf{I} - (p \sin A)\mathbf{J}] \times [(q \cos B)\mathbf{I} + (q \sin B)]\mathbf{J}|$

$= pq\ |\cos A \sin B\ (\mathbf{I} \times \mathbf{J}) - \sin A \cos B\ (\mathbf{J} \times \mathbf{I})|$

$= pq\ (\cos A \sin B + \sin A \cos B)$ for $|\mathbf{I} \times \mathbf{J}| = 1$

Also $|\vec{OP} \times \vec{OQ}| = pq \sin (A + B)$. Equating the two expressions, we get (*i*).

Similarly, (*ii*) follows from $\vec{OP} . \vec{OQ} = pq \cos (A + B)$.

**Example 3.16.** *In any triangle ABC, prove that*

(*i*) $a/\sin A = b/\sin B = c/\sin C$. *(Sine formula)*

(*ii*) $a = b \cos C + c \cos B$. *(Projection formula)*

(*iii*) $a^2 = b^2 + c^2 - 2bc \cos A$. *(Cosine formula)*

**Solution.** From $\Delta ABC$, we have $\vec{BC} + \vec{CA} + \vec{AB} = 0$

or $$\vec{CA} + \vec{AB} = -\vec{BC} \quad ...(\lambda)$$

(*i*) Multiplying ($\lambda$) vectorially by $\vec{AB}$, we get

$$\vec{CA} \times \vec{AB} = -\vec{BC} \times \vec{AB}$$

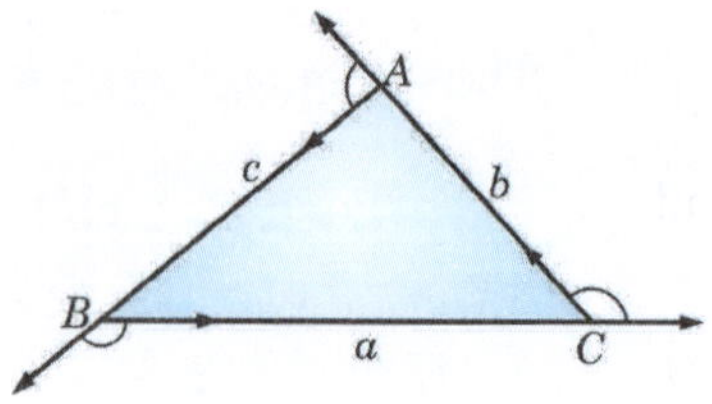

Fig. 3.23

or $$|\vec{CA} \times \vec{AB}| = |\vec{BC} \times \vec{AB}|$$

$\therefore$ $$bc \sin (\pi - A) = ac \sin (\pi - B)$$

or $$a/\sin A = b/\sin B.$$

Similarly, multiplying ($\lambda$) vectorially by $\vec{CA}$, we get

$a/\sin A = c/\sin C$, whence follows the result.

(*ii*) Multiplying ($\lambda$) scalarly by $\vec{BC}$, we get $\vec{CA} . \vec{BC} + \vec{AB} . \vec{BC} = -(\vec{BC})^2$

$\therefore$ $ba \cos (\pi - C) + ca \cos (\pi - B) = -a^2$ or $a = b \cos C + c \cos B$.

(*iii*) Squaring ($\lambda$), we get

$$(\vec{CA})^2 + (\vec{AB})^2 + 2\vec{CA} . \vec{AB} = (\vec{BC})^2$$

*i.e.,* $$b^2 + c^2\ 2bc \cos (\pi - A) = a^2 \text{ or } a^2 = b^2 + c^2 - 2bc \cos A.$$

## PROBLEMS 3.3

1. Find the unit normal vector to both the vectors $4\mathbf{I} - \mathbf{J} + 3\mathbf{K}$ and $-2\mathbf{I} + \mathbf{J} - 2\mathbf{K}$. Find also the sine of the angle between them. *(V.T.U., 2016)*
2. If $\mathbf{A}$ and $\mathbf{B}$ are unit vectors and $\theta$ is the angle between them, show that $\sin \frac{\theta}{2} = \frac{1}{2} |\mathbf{A} - \mathbf{B}|$.
3. Find a unit vector normal to the plane of $\mathbf{A} = 3\mathbf{I} - 2\mathbf{J} + 4\mathbf{K}$ and $\mathbf{B} = \mathbf{I} + \mathbf{j} - 2\mathbf{K}$.
4. For any vector $\mathbf{A}$, show that $|\mathbf{A} \times \mathbf{I}|^2 + |\mathbf{A} \times \mathbf{J}|^2 + |\mathbf{A} \times \mathbf{K}|^2 = 2\ |\mathbf{A}|^2$.
5. By vector method, find the area of the triangle whose vertices are (3, –1, 2), (1, –1, –3) and (4, –3, 1).
6. (*a*) Prove that the vector area of the quadrilateral *ABCD* is $\frac{1}{2}\ \vec{AC} \times \vec{BD}$.

   (*b*) If $3\mathbf{I} + \mathbf{J} - 2\mathbf{K}$ and $\mathbf{I} - 3\mathbf{J} - 4\mathbf{K}$ are the diagonals of a parallelogram. Find its area.

7. Given vectors $\mathbf{A} = \mathbf{I} - 3\mathbf{J} + 2\mathbf{K}$ and $\mathbf{B} = 2\mathbf{I} + \mathbf{J} - \mathbf{K}$. Find the projection of $\mathbf{A} \times \mathbf{B}$ parallel to $5\mathbf{I} - \mathbf{K}$.
8. If $\mathbf{A} + \mathbf{B} + \mathbf{C} = \mathbf{0}$, prove that $\mathbf{A} \times \mathbf{B} = \mathbf{B} \times \mathbf{C} = \mathbf{C} \times \mathbf{A}$, and interpret it geometrically.
9. Show that the perpendicular distance of the point **C** from the line joining **A** and **B** is $|\mathbf{B} \times \mathbf{C} + \mathbf{C} \times \mathbf{A} + \mathbf{A} \times \mathbf{B}| + |\mathbf{B} - \mathbf{A}|$
10. In $AC$, diagonal of the parallelogram $ABCD$, a point **P** is taken. Prove that $\Delta BAP = \Delta DAP$.
11. Prove by vector methods, that
(*i*) $\sin (A - B) = \sin A \cos B - \cos A \sin B$; (*ii*) $\cos (A - B) = \cos A \cos B + \sin A \sin B$.
12. In any triangle $ABC$, prove by vector methods, that
(*i*) $b = c \cos A + a \cos C$; (*ii*) $c^2 = a^2 + b^2 - 2ab \cos C$.

## 3.7 PHYSICAL APPLICATIONS

**(1) Work done as a scalar product.** If constant force **F** acting on a particle displaces it from the position $A$ to position $B$, then

Work done = (resolved part of $F$ in the direction of $AB$). $AB$

$$= F \cos \theta \,.\, AB = \mathbf{F} . \overrightarrow{AB}$$

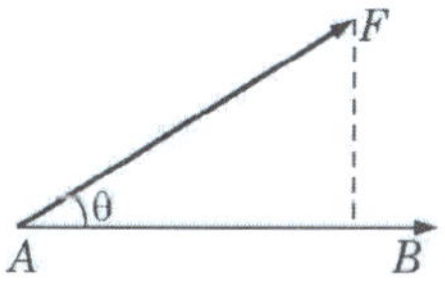

Fig. 3.24

*Thus, the work done by a constant force is the scalar (or dot) product of the vectors representing the force and the displacement.*

**Example 3.17.** *Constant forces* $\mathbf{P} = 2\mathbf{I} - 5\mathbf{J} + 6\mathbf{K}$ *and* $\mathbf{Q} = -\mathbf{I} + 2\mathbf{J} - \mathbf{K}$ *act on a particle. Determine the work done when the particle is displaced from A to B the position vectors of A and B being* $4\mathbf{I} - 3\mathbf{J} - 2\mathbf{K}$ *and* $6\mathbf{I} + \mathbf{J} - 3\mathbf{K}$ *respectively.*

**Solution.** Resultant force $\qquad \mathbf{F} = \mathbf{P} + \mathbf{Q} = \mathbf{I} - 3\mathbf{J} + 5\mathbf{K}$

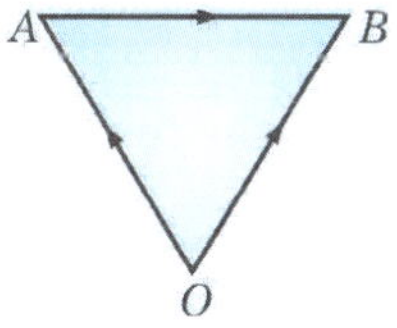

Fig. 3.25

and $\qquad \overrightarrow{AB} = \overrightarrow{OB} - \overrightarrow{OA} = (6\mathbf{I} + \mathbf{J} - 3\mathbf{K}) - (4\mathbf{I} - 3\mathbf{J} - 2\mathbf{K}) = 2\mathbf{I} + 4\mathbf{J} - \mathbf{K}$

$\therefore$ *Work done* $\qquad = \mathbf{F} . \overrightarrow{AB} = (\mathbf{I} - 3\mathbf{J} + 5\mathbf{K}) . (2\mathbf{I} + 4\mathbf{J} - \mathbf{K})$

$\qquad = 1 . 2 - 3 . 4 + 5 . (-1) = -15$ units.

**(2) Normal flux.** Consider the flow of a liquid through an element of area $\delta s$ with a velocity **V** inclined at an angle $\theta$ to the out ward unit normal **N** to the surface $\delta s$ (Fig. 3.26).

$\therefore$ Normal flux of the liquid through $\delta s$ in unit time

$$\mathbf{V} \cos \theta \,.\, \delta s = \mathbf{V} . \mathbf{N} \delta s.$$

Thus, *the rate of normal flux per unit area* = **V** . **N**

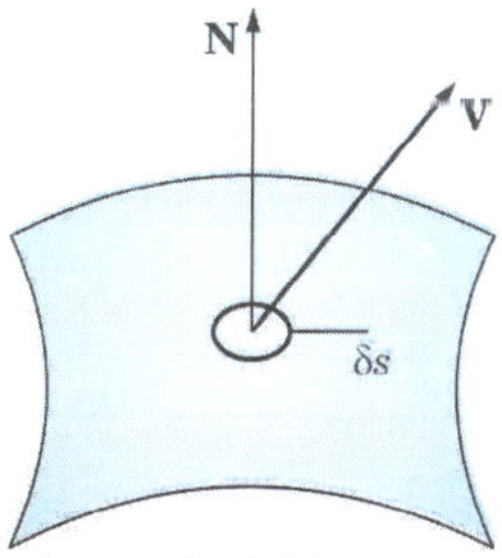

Fig. 3.26

**Obs.** We can also apply this result to the case of electric or magnetic flux.

**(3) Moment of a force about a point.** Suppose the moment of the force **F** acting at the point $P$ about the point $A$ is required.

Draw $AM \perp$ the line of action of **F** (Fig. 3.27). If $\theta$ be the angle between $\overrightarrow{AP}$ and **F** and **N** be a unit vector $\perp$ to their plane, then $\overrightarrow{AP} \times \mathbf{F} = (AP . F \sin \theta)\, \mathbf{N} = F(AP \sin \theta)\, \mathbf{N} = (F . AM)\, \mathbf{N}$

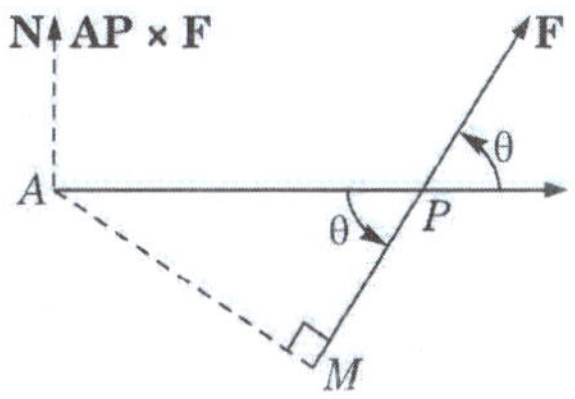

Fig. 3.27

Clearly, (*i*) the magnitude of $\overrightarrow{AP} \times \mathbf{F} = F . AM$ which is the numerical measure of the moment of **F** about $A$.

and (*ii*) the direction of $\overrightarrow{AP} \times \mathbf{F}$ is the direction of the moment of **F** about $A$.

Hence the moment (or torque) of **F** about $A$ is $\overrightarrow{AP} \times \mathbf{F}$.

**Example 3.18.** *Find the torque about the point* $2\mathbf{I} + \mathbf{J} - \mathbf{K}$ *of a force represented by* $4\mathbf{I} + \mathbf{K}$ *acting through the point* $\mathbf{I} - \mathbf{J} + 2\mathbf{K}$.

**Solution.** Let $O$ be the origin and $P$ be the point, moment about which of the force $\overrightarrow{AB}$ through $A$, is required (Fig. 3.28).

$\therefore \quad \vec{OP} = 2\mathbf{I} + \mathbf{J} - \mathbf{K},$

$\vec{OA} = \mathbf{I} - \mathbf{J} + 2\mathbf{K}$, and $\vec{AB} = 4\mathbf{I} + \mathbf{K}$.

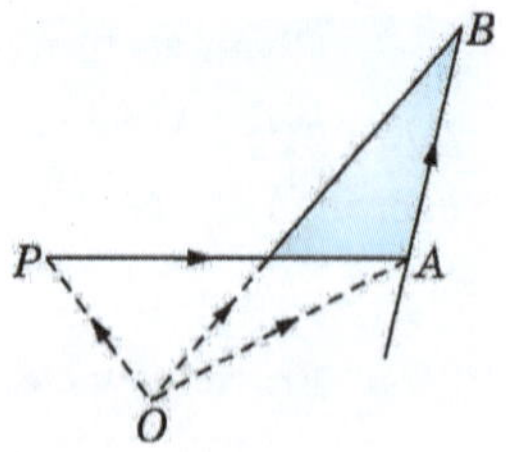

Fig. 3.28

Then, $\vec{PA} = \vec{OA} - \vec{OP} = -\mathbf{I} - 2\mathbf{J} + 3\mathbf{K}$

$\therefore$ Moment of the force $\vec{AB}$ about $P$

$$= \vec{PA} \times \vec{AB} = (-\mathbf{I} - 2\mathbf{J} + 3\mathbf{K}) \times (4\mathbf{I} + \mathbf{K})$$

$$= \begin{vmatrix} \mathbf{I} & \mathbf{J} & \mathbf{K} \\ -1 & -2 & 3 \\ 4 & 0 & 1 \end{vmatrix} = -2\mathbf{I} + 13\mathbf{J} + 8\mathbf{K}$$

$\therefore$ Magnitude of the moment $= \sqrt{(4 + 169 + 64)} = 15.4$

**(4) Moment of a force about a line.**

**Def.** *The moment of a force* **F** *about a line* **D** *is the resolved part along* **D** *of the moment of* **F** *about any point on* **D**.

**Example 3.19.** *Find the moment about a line through the origin having direction of* $2\mathbf{I} + 2\mathbf{J} + \mathbf{K}$, *due to a 30 kg force acting at a point (– 4, 2, 5) in the direction of* $12\mathbf{I} - 4\mathbf{J} - 3\mathbf{K}$.

**Solution.** Let **D** be the given line through the origin $O$ and **F** the force through **A**(– 4, 2, 5).

Clearly, $\vec{OA} = -4\mathbf{I} + 2\mathbf{J} + 5\mathbf{K}$

and the force $\mathbf{F} = 30\left(\dfrac{12\mathbf{I} - 4\mathbf{J} - 3\mathbf{K}}{13}\right)$

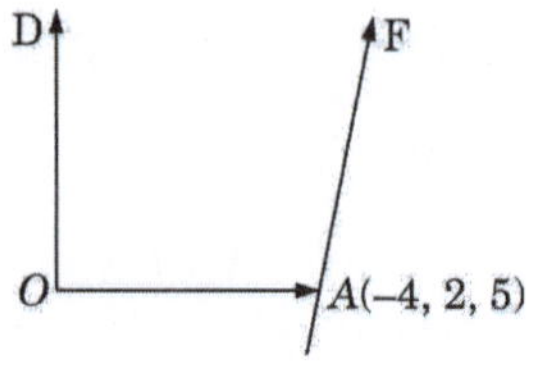

Fig. 3.29

$\therefore$ Moment of **F** about $O = \vec{OA} \times \mathbf{F}$

$$= \begin{vmatrix} \mathbf{I} & \mathbf{J} & \mathbf{K} \\ -4 & 2 & 5 \\ \frac{360}{13} & \frac{-120}{13} & \frac{-90}{13} \end{vmatrix} = \frac{60}{13}(7\mathbf{I} + 24\mathbf{J} - 4\mathbf{K})$$

*Thus the moment of* **F** *about the line* **D**

= resolved part of the moment of **F** about $O$ along **D**,

*i.e.*, $\dfrac{60}{13}(7\mathbf{I} + 24\mathbf{J} - 4\mathbf{K}) \cdot \hat{\mathbf{D}}$

$$= \frac{60}{13}(7\mathbf{I} + 24\mathbf{J} - 4\mathbf{K}) \cdot \frac{2\mathbf{I} + 2\mathbf{J} + \mathbf{K}}{\sqrt{(4 + 4 + 1)}} = \frac{20}{13}(7 \times 2 + 24 \times 2 - 4 \times 1) = 89.23.$$

**(5) Angular velocity of a rigid body**

Let a rigid body be rotating about the axis $OM$ with angular velocity $\omega$ radians per second (Fig. 3.30). Let P be a point of the body such that $\vec{OP} = \mathbf{R}$ and $\angle MOP = \theta$. Draw $PM \perp OM$.

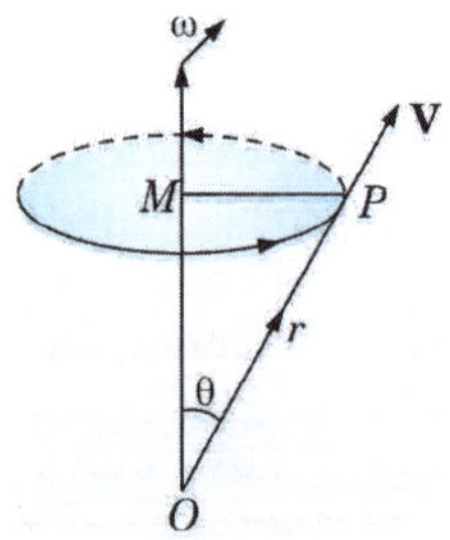

Fig. 3.30

Now if **N** be a unit vector $\perp \vec{\omega}$ **R** then

$$\vec{\omega} \times \mathbf{R} = \omega r \sin\theta \,.\, \mathbf{N} = \omega MP \,.\, \mathbf{N}$$

= (speed of **P**) **N**

= velocity **V** of **P** in a direction $\perp$ to the plane $MOP$.

Hence $\mathbf{V} = \vec{\omega} \times \mathbf{R}$.

**Example 3.20.** *A rigid body is spinning with angular velocity 27 radians per second about an axis parallel to* $2\mathbf{I}+\mathbf{J}-2\mathbf{K}$ *passing through the point* $\mathbf{I}+3\mathbf{J}-\mathbf{K}$. *Find the velocity of the point of the body whose position vector is* $4\mathbf{I}+8\mathbf{J}+\mathbf{K}$.

**Solution.** Unit vector along the direction of $\vec{\omega} = \frac{2\mathbf{I}+\mathbf{J}-2\mathbf{K}}{\sqrt{(4+1+4)}} = \frac{1}{3}(2\mathbf{I}+\mathbf{J}-2\mathbf{K})$

$\therefore$ Angular velocity $\vec{\omega} = \frac{27}{3}(2\mathbf{I}+\mathbf{J}-2\mathbf{K}) = 9(2\mathbf{I}+\mathbf{J}-2\mathbf{K})$

Let $A$ be the point $\mathbf{I}+3\mathbf{J}-\mathbf{K}$ and the point $P$ of the body be $(4\mathbf{I}+8\mathbf{J}-\mathbf{K})$ so that

$$\vec{AP} = (4\mathbf{I}+8\mathbf{J}+\mathbf{K})-(\mathbf{I}+3\mathbf{J}-\mathbf{K}) = 3\mathbf{I}+5\mathbf{J}+2\mathbf{K}$$

$\therefore$ Velocity vector of $P = \mathbf{V} = \vec{\omega}\times\vec{AP} = 9(2\mathbf{I}+\mathbf{J}-2\mathbf{K})\times(3\mathbf{I}+5\mathbf{J}+2\mathbf{K})$

$$= 9\begin{vmatrix} \mathbf{I} & \mathbf{J} & \mathbf{K} \\ 2 & 1 & -2 \\ 3 & 5 & 2 \end{vmatrix} = 9(12\mathbf{I}-10\mathbf{J}+7\mathbf{K})$$

and its magnitude $9\sqrt{(144+100+49)} = 9\sqrt{293}$.

**Example 3.21.** *Show that the four points whose position vectors are* $3i-2j+4k$, $6i+3j+2k$, $5i+7j+3k$, $2i+2j+5k$ *are coplaner.* *(PTU, 2017)*

**Solution.** Let the given points be $A, B, C, D$ respectively with respect to some origin $O$. Then

$$\overline{AB} = \overline{OB}-\overline{OA} = (6i+3j+2k)-(3i-2j+4k) = 3i+5j-2k$$

$$\overline{BC} = \overline{OC}-\overline{OB} = (5i+7i+3k)-(6i+3j+2k) = -i+4j+k$$

$$\overline{CD} = \overline{OD}-\overline{OC} = (2i+2j+5k)-(5i+7j+3k) = -3i-5j+2k$$

If $A, B, C, D$ are coplanar, then $\overline{AB}, \overline{BC}, \overline{CD}$ are also coplaner and in that case there exist scalars $x$ and $y$ such that

$$\overline{AB} = x\overline{BC}+y\overline{CD}$$

$$(3i+5j-2k) = x(-i+4j+k)+y(-3i-5j+2k)$$

$$= (-x-3y)i+(4x-5y)j+(x+2y)k$$

$$\Rightarrow \quad -x-3y = 3,\ 4x-5y = 5,\ x+2y = -2$$

Solving first two equations, we get $x = 0$, $y = -1$ which also satisfy the third equation. So, the given points are coplaner.

**Example 3.22.** *Forces* $\vec{P}, \vec{Q}$ *act at O and have a resultant* $\vec{R}$. *if any transversal cuts their lines of action at A, B, C respectively, show that* $\frac{P}{OA}+\frac{Q}{OB} = \frac{R}{OC}$ *(KTU, 2018)*

**Solution.** Let the force $\vec{P}, \vec{Q}$ and $\vec{R}$ act along $OA$, $OB$ and $OC$ respectively.

Let $\vec{a}, \vec{b}$ and $\vec{c}$ be the position vectors of $A$, $B$ and $C$ respectively referred to $O$ as origin.

Then the unit vectors along $OA$, $OB$, $OC$ are given by

$\frac{\vec{a}}{OA}, \frac{\vec{b}}{OB}$ and $\frac{\vec{c}}{OC}$ and therefore the forces $P, Q, R$ along $OA, OB, OC$ are given by

$$\vec{P} = \frac{P\vec{a}}{OA},\ \vec{Q} = \frac{Q\vec{b}}{OB},\ \vec{R} = \frac{R\vec{c}}{OC}$$

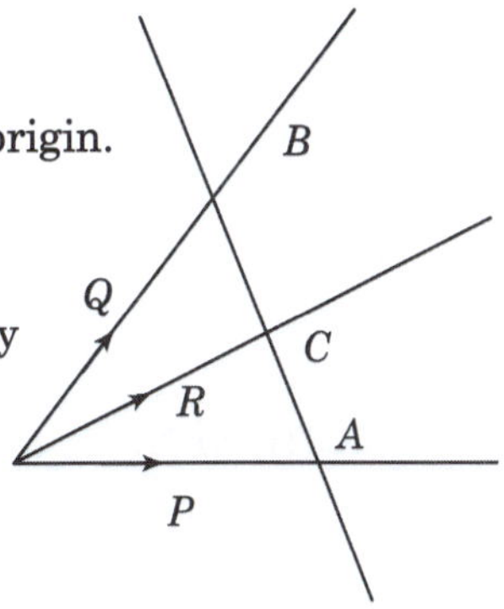

Fig. 3.31

Since $\vec{R}$ is resultant of $\vec{P}$ and $\vec{Q}$ *i.e.*, $\vec{R} = \vec{P}+\vec{Q}$, we have

$$\frac{R\vec{c}}{OC} = \frac{P\vec{a}}{OA} + \frac{Q\vec{b}}{OB} \text{ or } \frac{P\vec{a}}{OA} + \frac{Q\vec{b}}{OB} - \frac{R\vec{c}}{OC} = 0$$

But the points $A$, $B$, $C$ are collinear, hence the sum of the coefficients is zero, when *i.e.*,

$$\frac{P}{OA} + \frac{Q}{OB} - \frac{R}{OC} \text{ or } \frac{P}{OA} + \frac{Q}{OB} + \frac{R}{OC}.$$

## PROBLEMS 3.4

1. A particle acted on by constant forces $4\mathbf{I} + \mathbf{J} - 3\mathbf{K}$ and $3\mathbf{I} + \mathbf{J} - \mathbf{K}$ is displace from the point $\mathbf{I} + 2\mathbf{J} + 3\mathbf{K}$ to the point $5\mathbf{I} + 4\mathbf{J} + \mathbf{K}$. Find the total work done by the forces.
2. Forces $2\mathbf{I} - 5\mathbf{J} + 6\mathbf{K}$, $-\mathbf{I} + 2\mathbf{J} - \mathbf{K}$ and $2\mathbf{I} + 7\mathbf{J}$ act on a particle $P$ whose position vector is $4\mathbf{I} - 3\mathbf{J} - 2\mathbf{K}$. Determine the work done by the forces in a displacement of the particle to the point $Q$ (6, 1, –3).

   Also find the vector moment of the resultant of three forces acting at $P$ about the point $Q$.
3. Prove that the necessary and sufficient condition that the three points whose position vectors are $\vec{a}, \vec{b}, \vec{c}$ should be collinear is $x\vec{a} + y\vec{b} + z\vec{c} = 0$.
4. The point of application of the force (–2, 4, 7) is displaced from the point (3, –5, 1) to the point (5, 9, 7). But the force is suddenly halved when the point of application moves half the distance. Find the work done.
5. A force $\mathbf{F} = 3\mathbf{I} + 2\mathbf{J} - 4\mathbf{K}$ is applied at the point (1, –1, 2). Find the moment of the force about the point (2, –1, 3).
6. A force with components (5, –4, 2) acts at a point $P$ which is at a distance 3 units from the origin. If the moment of the force about origin has components (12, 8, –14), find the co-ordinates of $P$.
7. Find the moment of the force $\mathbf{F} = 2\mathbf{I} + 2\mathbf{J} - \mathbf{K}$ acting at the point (1, –2, 1) about z-axis.
8. A force of 10 kg acts in a direction equally inclined to the co-ordinate axes through the point (3, –2, 5). Find the magnitude of the moment of the force about a line through the origin and whose direction ratios are (2, –3, 6).
9. A rigid body is rotating at 2.5 radian per second about an axis $AB$, where $A$ and $B$ are the points, (1, –2, 1) and (3, –4, 2). Find the velocity of the particle $P$ of the body at the point (5, –1, –1). *(Delhi, 2016)* (**Ans.** 9.2 Units)
10. The velocity of a boat relative to the water is represented by $3i + 4j$ and that of the water relative to the earth is $i - 3j$. If $i$ and $j$ represented velocities of 1 km/h in east and north directions respectively show that the velocity of the boat relative to the earth is $\sqrt{17}$ km/h at $\tan^{-1}\frac{1}{4}$ north to east.

    *(Garakhpur, 2017)*

## 3.8 PRODUCTS OF THREE OR MORE VECTORS

With any three vectors **A**, **B**, **C**, we can form the products (**A** . **B**) **C**, (**A** × **B**) . **C** and (**A** × **B**) × **C**. The first being the product of a scalar **A** . **B** and a vector **C**, represents a vector in the direction of **C**. The second being the scalar product of vectors **A** × **B** and **C**, represents a scalar and is usually called the *scalar product of three vectors*. The third being the vector product of the vectors **A** × **B** and **C**, represents a vector and is usually known as the *vector product of three vectors*.

The reader must, however, note that the products of the form **A** . (**B** . **C**), **A** × (**B** . **C**) and **A**(**B** × **C**) are meaningless.

In practical applications, we seldom come across products of more than three vectors. Such products if they occur can, in general, be reduced by using successively the expansion formula for vector triple products. As an illustration, we shall consider two products (**A** × **B**). (**C** × **D**) and (**A** × **B**) × (**C** × **D**) of any four vectors, the former being a scalar and latter a vector.

## 3.9 SCALAR PRODUCT OF THREE VECTORS

**(1) Definition.** *If* **A**, **B**, **C** *be any three vectors then the scalar or dot product of* **A** × **B** *with* **C** *is called the scalar product of the three vectors* **A**, **B**, **C** *and is written as* (**A** × **B**) . **C** *or* [**ABC**].

No ambiguity can arise by omitting the brackets in $(A \times B).\,C$ as $A \times (B\,.\,C)$ would be meaningless .

**(2) Geometrical interpretation.** *The Product* **A** × **B** . **C** *represents numerically the volume of a parallelopiped having* **A**, **B**, **C** *as coterminous edges.*

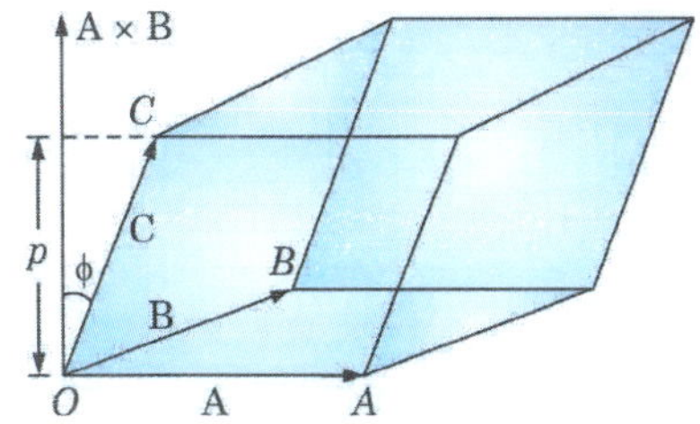


Fig. 3.32.

Consider a parallelopiped with $\vec{OA} = \mathbf{A}$, $\vec{OB} = \mathbf{B}$, $\vec{OC} = \mathbf{C}$ as coterminous edges (Fig. 3.32).

Let $V$ be its volume, $\alpha$ the area of each of the two faces parallel to the vectors **A** and **B** and $p$ the perpendicular distance between these faces.

Then $|\mathbf{A} \times \mathbf{B}| = \alpha$ and $|\mathbf{C}| \cos \phi = p$ or $-p$ according as **A, B, C** form a right-handed or left-handed triad.

$\therefore \quad \mathbf{A} \times \mathbf{B} . \mathbf{C} = |\mathbf{A} \times \mathbf{B}| . |\mathbf{C}| \cos \phi = \pm \alpha p = \pm V.$

*Thus* [**ABC**] = *V or –V according as* **A, B, C** *form a right-handed or left-handed triad.*

*(Kerala, 1990; J.N.T.U., 1988)*

In particular, for an orthonormal right-handed vector triad **I, J, K,**

$$[\mathbf{IJK}] = \mathbf{I} \times \mathbf{J} . \mathbf{K} = \mathbf{K} . \mathbf{K} = I.$$

**(3) Properties and other results.**

*I. The condition for three vectors to be coplanar is that their scalar triple product should vanish.*

If three vectors **A, B, C** are coplanar, then the volume of the parallelopiped so formed is zero, *i.e.,* [**ABC**] = 0.

*II. If any two vectors of a scalar triple product are equal, the product vanishes, i.e.,* [**ABC**] = 0 when either **A** = **B** or **B** = **C**, or **C** = **A**, for in this case the parallelopiped has zero volume.

*III.* **Two important rules** *(for evaluating a scalar triple product). Every scalar triple product*

*(i) is independent of the position of the dot or cross.*

*and (ii) depends upon the cyclic order of the vectors.*

It is easy to note that if **A, B, C** is a right-handed triad so are **B, C, A** and **C, A, B.**

Moreover a parallelopiped having **A, B, C** as coterminous edges is the same as that having **B, C, A** or **C, A, B** as coterminous edges.

Thus, if $V$ be the volume of this parallelopiped,

$$\mathbf{A} \times \mathbf{B} . \mathbf{C} = V, \mathbf{B} \times \mathbf{C} . \mathbf{A} = V, \mathbf{C} \times \mathbf{A} . \mathbf{B} = V$$

Also, since **A . B = B . A**, we have

$$\mathbf{C} . \mathbf{A} \times \mathbf{B} = \mathbf{A} \times \mathbf{B} . \mathbf{C} = V$$
$$\mathbf{A} . \mathbf{B} \times \mathbf{C} = \mathbf{D} \times \mathbf{C} . \mathbf{A} - V$$
$$\mathbf{B} . \mathbf{C} \times \mathbf{A} = \mathbf{C} \times \mathbf{A} . \mathbf{B} = V$$

Thus
$$\mathbf{AA} \times \mathbf{B} . \mathbf{C} = \mathbf{A} . \mathbf{B} \times \mathbf{C}$$
$$\mathbf{B} \times \mathbf{C} . \mathbf{A} = \mathbf{B} . \mathbf{C} \times \mathbf{A} = V \qquad ...(\alpha)$$
$$\mathbf{C} \times \mathbf{A} . \mathbf{B} = \mathbf{C} . \mathbf{A} \times \mathbf{B}$$

Further a right-handed triad becomes left-handed when the cyclic order of the vectors is changed. Therefore **A, C, B; B, A, C; C, B, A** being left-handed triads, it follows that

$$\mathbf{A} \times \mathbf{C} . \mathbf{B} = -V, \mathbf{B} \times \mathbf{A} . \mathbf{C} = -V, \mathbf{C} \times \mathbf{B} . \mathbf{A} = V.$$

Thus
$$\mathbf{AA} \times \mathbf{C} . \mathbf{B} = \mathbf{A} . \mathbf{C} \times \mathbf{B}$$
$$\mathbf{B} \times \mathbf{A} . \mathbf{C} = \mathbf{B} . \mathbf{A} \times \mathbf{C} = V \qquad ...(\beta)$$
$$\mathbf{C} \times \mathbf{B} . \mathbf{A} = \mathbf{C} . \mathbf{B} \times \mathbf{A}$$

**Obs.** In support of the above rules, our notation [**ABC**] indicates the cyclic order of the factors and has nothing to do with position of the dot or the cross.

$\therefore$ The relations ($\alpha$) and ($\beta$) can be compactly written as

$$[\mathbf{ABC}] = [\mathbf{BCA}] = [\mathbf{CAB}] = V \quad \text{and} \quad [\mathbf{ACB}] = [\mathbf{BAC}] = [\mathbf{CBA}] = -V.$$

*IV. Scalar triple product is distributive*

*i.e.,*
$$[\mathbf{A}, \mathbf{B} + \mathbf{C}, \mathbf{D} - \mathbf{E}] = [\mathbf{ABD}] - [\mathbf{ABE}] + [\mathbf{ACD}] - [\mathbf{ACE}]$$

*V. If* $\mathbf{A} = a_1\mathbf{I} + a_2\mathbf{J} + a_3\mathbf{K}$, $\mathbf{B} = b_1\mathbf{I} + b_2\mathbf{J} + b_3\mathbf{K}$, $\mathbf{C} = c_1\mathbf{I} + c_2\mathbf{J} + c_3\mathbf{K}$

*then*
$$[\mathbf{ABC}] = \begin{vmatrix} a_1 & a_2 & a_3 \\ b_1 & b_2 & b_3 \\ c_1 & c_2 & c_3 \end{vmatrix}$$

As $\quad \mathbf{A} \times \mathbf{B} = (a_2b_3 - a_3b_2)\,\mathbf{I} + (a_3b_1 - a_1b_3)\,\mathbf{J} + (a_1b_2 - a_2b_1)\mathbf{K}$

$\therefore \quad [\mathbf{ABC}] = [(a_2b_3 - a_3b_2)\,\mathbf{I} + (a_3b_1 - a_1b_3)\,\mathbf{J} + (a_1b_2 - a_2b_1)\,\mathbf{K}]\,.\,(c_1\mathbf{I} + c_2\mathbf{J} + c_3\mathbf{K})$

$= c_1(a_2b_3 - a_3b_2) + c_2(a_3b_1 - a_1b_3) + c_3(a_1b_2 - a_2b_1)$ which is the required result.

**Obs. Linear dependence of vectors.** Any three vectors **A, B, C** are said to be *linearly dependent* if one of these can be expressed as a linear combination of other two *i.e.,*

$$\mathbf{A} = m\mathbf{B} + n\mathbf{C}$$

where $m$, $n$ are constants. This means that **A** lies in the plane of **B, C** *i.e.,* [**ABC**] = 0. Thus *three vectors are linearly dependent if their scalar triple product is zero. Otherwise these vectors are linearly independent.*

## 3.10 VECTOR PRODUCT OF THREE VECTORS

**(1) Definition.** *If* **A, B, C** *be any three vectors, then the vector or cross product of* **A × B** *with* **C** *is called the vector product of three vectors* **A, B, C** *and is written as* (**A × B**) × **C**.

Here the brackets are essential as **A** × (**B** × **C**) ≠ (**A** × **B**) × **C**, expressing the fact that vector triple product is not associative.

**(2) Expansion formula.** *If* **A, B, C** *be any three vectors,* (**A × B**) × **C** = (**C . A**)**B** – (**C . B**) **A**

*In words (extreme × adjacent) × outer = (outer . extreme) adjacent – (outer . adjacent) extreme.*

The vector (**A × B**) × **C** is perpendicular to the vector **A × B** and the latter is perpendicular to the plane containing **A** and **B**. Hence (**A × B**) × **C** lies in the plane of **A** and **B**. As such we can write

$$(\mathbf{A} \times \mathbf{B}) \times \mathbf{C} = l\mathbf{A} + m\mathbf{B} \qquad ...(1)$$

where $l$ and $m$ are some scalars.

Multiply both sides scalarly by **C**, then $\mathbf{C} \,.\, (\mathbf{A} \times \mathbf{B}) \times \mathbf{C} = l\mathbf{C} \,.\, A + m\mathbf{C} \,.\, \mathbf{B}$

The scalar triple product on the left-hand side is zero, since two of its vectors are equal.

$\therefore \quad l(\mathbf{C}\,.\,\mathbf{A}) + m(\mathbf{C}\,.\,\mathbf{B}) = 0$

or $$\frac{l}{\mathbf{C}\,.\,\mathbf{B}} = \frac{m}{-\mathbf{C}\,.\,\mathbf{A}} = n, \text{ say.}$$

Substituting the values of $l$ and $m$ in Eqn. (1), we get

$$(\mathbf{A} \times \mathbf{B}) \times \mathbf{C} = n(\mathbf{C}\,.\,\mathbf{B})\mathbf{A} - n(\mathbf{C}\,.\,\mathbf{A})\mathbf{B} \qquad ...(2)$$

Evidently $n$ is some numerical constant. To find it, take the special case **A** = **I**, **B** = **C** = **J**. Then Eqn. (2) gives

$$(\mathbf{I} \times \mathbf{J}) \times \mathbf{J} = n(\mathbf{J}\,.\,\mathbf{J})\mathbf{I} - n(\mathbf{J}\,.\,\mathbf{I})\mathbf{J}$$

*i.e.,* $$\mathbf{K} \times \mathbf{J} = n\mathbf{I} \text{ or } -\mathbf{I} = n\mathbf{I}.$$

This gives $n = -1$. Hence (2) reduces to the required result.

Similarly, it can be shown that **A** × (**B** × **C**) = (**A . C**)**B** – (**A . B**)**C** *(V.T.U., 2013)*

**Cor.** **A** × (**B** × **C**) + **B** × (**C** × **A**) + **C** × (**A** × **B**) = **0.**

For L.H.S. = (**A . C**)**B** – (**A . B**)**C** + (**B . A**)**C** – (**B . C**)**A** + (**C . B**)**A** – (**C . A**)**B** which vanishes identically.

**Example 3.23.** *If* **A, B, C, D** *be any four vectors, prove that*

(*i*) $(\boldsymbol{A} \times \boldsymbol{B})\,.\,(\boldsymbol{C} \times \boldsymbol{D}) = \begin{vmatrix} \boldsymbol{A}\,.\,\boldsymbol{C} & \boldsymbol{B}\,.\,\boldsymbol{C} \\ \boldsymbol{A}\,.\,\boldsymbol{D} & \boldsymbol{B}\,.\,\boldsymbol{D} \end{vmatrix}$ (*ii*) $(\boldsymbol{A} \times \boldsymbol{B}) \times (\boldsymbol{C} \times \boldsymbol{D}) = [\boldsymbol{ACD}]\,\boldsymbol{B} - [\boldsymbol{BCD}]\boldsymbol{A}$

**Solution.** (*i*) $(\mathbf{A} \times \mathbf{B})\,.\,(\mathbf{C} \times \mathbf{D}) = [\mathbf{A} \times \mathbf{B}) \times \mathbf{C}]\,.\,\mathbf{D}$ (interchanging the dot and cross)

$= [(\mathbf{A}\,.\,\mathbf{C})\mathbf{B} - (\mathbf{B}\,.\,\mathbf{C})\mathbf{A}]\,.\,\mathbf{D}$

$= (\mathbf{A}\,.\,\mathbf{C})(\mathbf{B}\,.\,\mathbf{D}) - (\mathbf{A}\,.\,\mathbf{D})(\mathbf{B}\,.\,\mathbf{C})$ whence follows the result.

In particular, we have $(\mathbf{A} \times \mathbf{B})^2 = \mathbf{A}^2\mathbf{B}^2 - (\mathbf{A}\,.\,\mathbf{B})^2$ which has already been proved in § 3.6 (3) – IV.

(*ii*) $(\mathbf{A} \times \mathbf{B}) \times (\mathbf{C} \times \mathbf{D}) = (\mathbf{A} \times \mathbf{B}) \times \mathbf{P}$, where $\mathbf{P} = \mathbf{C} \times \mathbf{D}$

$= (\mathbf{A}\,.\,\mathbf{P})\mathbf{B} - (\mathbf{B}\,.\,\mathbf{P})\mathbf{A} = (\mathbf{A}\,.\,\mathbf{C} \times \mathbf{D})\mathbf{B} - (\mathbf{B}\,.\,\mathbf{C} \times \mathbf{D})\mathbf{A}$

$= [\mathbf{ACD}]\mathbf{B} - [\mathbf{BCD}]\mathbf{A}.$

**Obs.** In particular if **A, B, C, D** are coplanar, than (**A × B**) × (**C × D**) = **0.**

**Example 3.24.** *Show that the components of a vector* **B** *along and perpendicular to a vector* **A**, *in the plane of* **A** *and* **B**, *are*

$$\frac{\mathbf{A}.\mathbf{B}}{\mathbf{A}^2} \text{ and } \frac{(\mathbf{A}\times\mathbf{B})\times\mathbf{A}}{\mathbf{A}^2}.$$

**Solution.** Let $\vec{\mathbf{OA}} = \mathbf{A}$, $\vec{\mathbf{OB}} = \mathbf{B}$ and **OM** be the projection of **B** on **A** (Fig. 3.33)

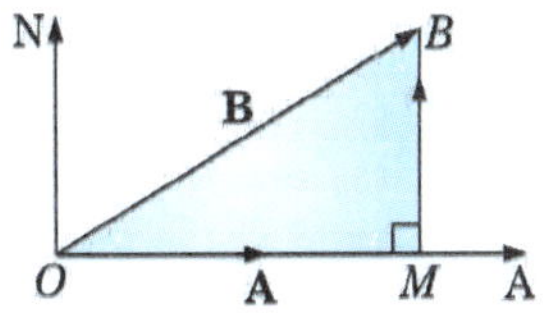

Fig. 3.33

$\therefore$ Component of **B** along **A** = $OM$ (unit vector along **A**)

$$= (\mathbf{B}.\hat{\mathbf{A}})\hat{\mathbf{A}} = \left(\frac{\mathbf{B}.\mathbf{A}}{a}\right)\frac{\mathbf{A}}{a} \qquad [\because \mathbf{A} = a\,\hat{\mathbf{A}}]$$

$$= \frac{\mathbf{B}.\mathbf{A}}{\mathbf{A}^2}\mathbf{A} \qquad [\because a^2 = \mathbf{A}^2]$$

Also component of $\mathbf{B} \perp \mathbf{A} = \vec{\mathbf{MB}}$

$$= \vec{OB} - \vec{OM} = \mathbf{B} - \frac{\mathbf{B}.\mathbf{A}}{\mathbf{A}^2}\mathbf{A} = \frac{(\mathbf{A}.\mathbf{A})\mathbf{B} - (\mathbf{A}.\mathbf{B})\mathbf{A}}{\mathbf{A}^2} = \frac{(\mathbf{A}\times\mathbf{B})\times\mathbf{A}}{\mathbf{A}^2}.$$

**Example 3.25.** *Prove the formula*

$$(\mathbf{B}\times\mathbf{C}).(\mathbf{A}\times\mathbf{D}) + (\mathbf{C}\times\mathbf{A}).(\mathbf{B}\times\mathbf{D}) + (\mathbf{A}\times\mathbf{B}).(\mathbf{C}\times\mathbf{D}) = 0.$$

*and hence show that* $\sin(\theta+\phi)\sin(\theta-\phi) = \sin^2\theta - \sin^2\phi$.

**Solution.** We know that

$$(\mathbf{B}\times\mathbf{C}).(\mathbf{A}\times\mathbf{D}) = (\mathbf{B}.\mathbf{A})(\mathbf{C}.\mathbf{D}) - (\mathbf{B}.\mathbf{D})(\mathbf{C}.\mathbf{A})$$
$$(\mathbf{C}\times\mathbf{A}).(\mathbf{B}\times\mathbf{D}) = (\mathbf{C}.\mathbf{B})(\mathbf{A}.\mathbf{D}) - (\mathbf{C}.\mathbf{D})(\mathbf{A}.\mathbf{B})$$
$$(\mathbf{A}\times\mathbf{B}).(\mathbf{C}\times\mathbf{D}) = (\mathbf{A}.\mathbf{C})(\mathbf{B}.\mathbf{D}) - (\mathbf{A}.\mathbf{D})(\mathbf{B}.\mathbf{C})$$

Adding, we get

$$(\mathbf{B}\times\mathbf{C}).(\mathbf{A}\times\mathbf{D}) + (\mathbf{C}\times\mathbf{A}).(\mathbf{B}\times\mathbf{D}) + (\mathbf{A}\times\mathbf{B}).(\mathbf{C}\times\mathbf{D}) = 0 \qquad ...(i)$$

Let the vectors **A, B, C, D** be acting along coplanar lines $OA$, $OB$, $OC$, $OD$ respectively (Fig. 3.34).

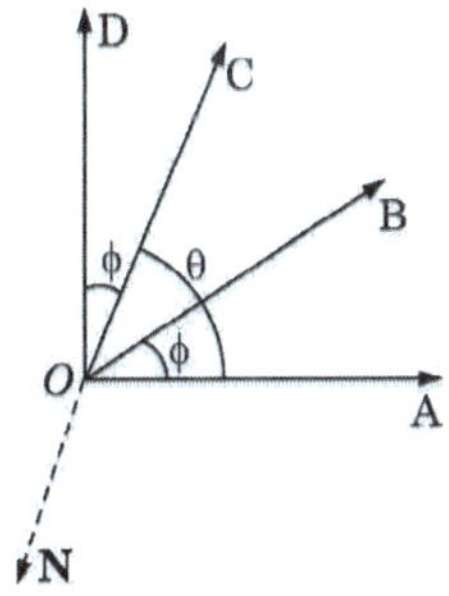

Fig. 3.34

Take $\angle AOC = \theta$ and $\angle AOB = \angle COD = \phi$,

so that $\angle AOD = \theta + \phi$ and $\angle BOC = \theta - \phi$

If **N** be a unit vector normal to the plane of these lines, then

$$(\mathbf{B}\times\mathbf{C}).(\mathbf{A}\times\mathbf{D}) = [bc\sin(\theta-\phi)\mathbf{N}].[ad\sin(\theta+\phi)\mathbf{N}]$$
$$= abcd\sin(\theta+\phi)\sin(\theta-\phi) \qquad ...(ii)$$
$$(\mathbf{C}\times\mathbf{A}).(\mathbf{B}\times\mathbf{D}) = [ca\sin(-\theta)\mathbf{N}].[bd\sin\theta\mathbf{N}]$$
$$= -abcd\sin^2\theta \qquad ...(iii)$$

and

$$(\mathbf{A}\times\mathbf{B}).(\mathbf{C}\times\mathbf{D}) = [ab\sin\phi\mathbf{N}].[cd\sin\phi\mathbf{N}]$$
$$= abcd\sin^2\phi \qquad ...(iv)$$

Substituting the values from (*ii*), (*iii*), (*iv*) in (*i*), we get

$abcd\sin(\theta+\phi)\sin(\theta-\phi) - abcd\sin^2\theta + abcd\sin^2\phi = 0$ whence follows the required result.

**Example 3.26.** *Prove that*

(*i*) $[\mathbf{B}\times\mathbf{C}, \mathbf{C}\times\mathbf{A}, \mathbf{A}\times\mathbf{B}] = [\mathbf{ABC}]^2$.

(*ii*) $\mathbf{A}\times\{\mathbf{B}\times(\mathbf{C}\times\mathbf{D})\} = \mathbf{B}.\mathbf{D}(\mathbf{A}\times\mathbf{C}) - \mathbf{B}.\mathbf{C}(\mathbf{A}\times\mathbf{D})$.

**Solution.** (*i*) $[\mathbf{B}\times\mathbf{C}, \mathbf{C}\times\mathbf{A}, \mathbf{A}\times\mathbf{B}] = (\mathbf{B}\times\mathbf{C}).(\mathbf{C}\times\mathbf{A})\times(\mathbf{A}\times\mathbf{B})$

$$= (\mathbf{B}\times\mathbf{C}).\{[(\mathbf{C}\times\mathbf{A}).\mathbf{B}]\mathbf{A} - [(\mathbf{C}\times\mathbf{A}).\mathbf{A}]\mathbf{B}\}$$
$$= (\mathbf{B}\times\mathbf{C}).\{[\mathbf{B}.(\mathbf{C}\times\mathbf{A})]\mathbf{A}\} \qquad [\because [(\mathbf{C}\times\mathbf{A}).\mathbf{A}] = 0$$
$$= [\mathbf{B}\times\mathbf{C}).\mathbf{A}]\,[(\mathbf{B}\times\mathbf{C}).\mathbf{A}] = [\mathbf{BCA}]^2 = [\mathbf{ABC}]^2 \qquad [\because [\mathbf{BCA}] = \{\mathbf{ABC}\}]$$

(*ii*)

$$\mathbf{A}\times\{\mathbf{B}\times(\mathbf{C}\times\mathbf{D})\} = \mathbf{A}\times\{(\mathbf{B}.\mathbf{D})\mathbf{C} - (\mathbf{B}.\mathbf{C})\mathbf{D}\}$$
$$= (\mathbf{A}\times\mathbf{C})(\mathbf{B}.\mathbf{D}) - (\mathbf{A}\times\mathbf{D})(\mathbf{B}.\mathbf{C}) = (\mathbf{B}.\mathbf{D})(\mathbf{A}\times\mathbf{C}) - \mathbf{B}.\mathbf{C}(\mathbf{A}\times\mathbf{D}).$$

## PROBLEMS 3.5

1. Find the volume of the parallelopied whose edges are represented by the vectors $\mathbf{A} = 2\mathbf{I} - 3\mathbf{J} + 4\mathbf{K}$, $\mathbf{B} = \mathbf{I} + 2\mathbf{J} - \mathbf{K}$, $\mathbf{C} = 3\mathbf{I} - \mathbf{J} + 2\mathbf{K}$.
2. Find $a$ such that the vectors $2\mathbf{I} - \mathbf{J} + \mathbf{K}$, $\mathbf{I} + 2\mathbf{J} - 3\mathbf{K}$ and $3\mathbf{I} + a\mathbf{J} + 5\mathbf{K}$ are coplanar. (*V.T.U., 2016*)
3. (*i*) Prove that the vectors $\mathbf{I} - 2\mathbf{J} + 3\mathbf{K}$, $-2\mathbf{I} + 3\mathbf{J} - 4\mathbf{K}$ and $\mathbf{I} - 3\mathbf{J} + 5\mathbf{K}$ are coplanar.
   (*ii*) Do the points (4, –2, 1), (5, 1, 6), (2, 2 – 5) and (3, 5, 0) lie in a plane.
4. (*a*) Test the linear dependency of the vectors (1, 2, 1), (2, 1, 4), (4, 5, 6) and (1, 8, –5).
   (*b*) Verify whether the following set of vectors are linearly independent (4, 2, 9), (3, 2, 1), (–4, 6, 9).
5. Find the volume of the tetrahedron, three of whose coterminus edges are $3\mathbf{I} - \mathbf{J} + 2\mathbf{K}$, $2\mathbf{I} + \mathbf{J} - \mathbf{K}$ and $\mathbf{I} - 2\mathbf{J} + 2\mathbf{K}$.
6. Find the volume of the tetrahedron formed by the points
   (*i*) (1, 3, 6), (3, 7, 12), (8, 8, 9) and (2, 2, 8).
   (*ii*) (2, 1, 1), (1, –1, 2), (0, 1, –1) and (1, –2, 1).
7. If $\mathbf{A} \cdot \mathbf{N} = 0$, $\mathbf{B} \cdot \mathbf{N} = 0$, $\mathbf{C} \cdot \mathbf{N} = 0$, prove that $[\mathbf{ABC}] = 0$. Interpret this result geometrically.
8. (*a*) Prove that $[\mathbf{A} + \mathbf{B}, \mathbf{B} + \mathbf{C}, \mathbf{C} + \mathbf{A}] = 2[\mathbf{ABC}]$. (*V.T.U., 2016 ; W.B.T.U., 2016*)
   (*b*) Show that volume of the tetrahedron having $\mathbf{A} + \mathbf{B}$, $\mathbf{B} + \mathbf{C}$ and $\mathbf{C} + \mathbf{A}$ as concurrent edges is twice the volume of the tetrahedron having $\mathbf{A}$, $\mathbf{B}$, $\mathbf{C}$ as concurrent edges.
9. If $\mathbf{A} \times (\mathbf{B} \times \mathbf{C}) = (\mathbf{A} \times \mathbf{B}) \times \mathbf{C}$, show that $(\mathbf{A} \times \mathbf{C}) \times \mathbf{B} = 0$.
10. Show that $\mathbf{I} \times (\mathbf{R} \times \mathbf{I}) + \mathbf{J} \times (\mathbf{R} \times \mathbf{J}) + \mathbf{K} \times (\mathbf{R} \times \mathbf{K}) = 2\mathbf{R}$.
11. If $\mathbf{A} = \mathbf{I} - 2\mathbf{J} - 3\mathbf{K}$, $\mathbf{B} = 2\mathbf{I} + \mathbf{J} - \mathbf{K}$, $\mathbf{C} = \mathbf{I} + 3\mathbf{J} - \mathbf{K}$, find
   (*i*) $\mathbf{A} \times (\mathbf{B} \times \mathbf{C})$ (*ii*) $(\mathbf{A} \times \mathbf{B}) \times (\mathbf{B} \times \mathbf{C})$.
12. (*a*) Given $\mathbf{A} = 2\mathbf{I} - \mathbf{J} + 3\mathbf{K}$, $\mathbf{B} = -\mathbf{I} + 3\mathbf{J} + 3\mathbf{K}$, $\mathbf{C} = \mathbf{I} + \mathbf{J} - 2\mathbf{K}$, find the reciprocal triad $(\mathbf{A}', \mathbf{B}', \mathbf{C}')$ and verify that $[\mathbf{ABC}]\,[\mathbf{A}'\mathbf{B}'\mathbf{C}'] = 1$.
   (*b*) Prove that $\mathbf{A} \times \mathbf{A}' + \mathbf{B} \times \mathbf{B}' + \mathbf{C} \times \mathbf{C}' = 0$
13. Prove that (*i*) $[\mathbf{A} \times \mathbf{B}, \mathbf{C} \times \mathbf{D}, \mathbf{E} \times \mathbf{F}] = [\mathbf{ABD}][\mathbf{CEF}] - [\mathbf{ABC}][\mathbf{DEF}]$
   (*ii*) $\{(\mathbf{A} + \mathbf{B} + \mathbf{C}) \times (\mathbf{B} + \mathbf{C})\} \cdot \mathbf{C} = [\mathbf{ABC}]$.
14. Show that
   (*i*) $(\mathbf{B} \times \mathbf{C}) \times (\mathbf{A} \times \mathbf{D}) + (\mathbf{C} \times \mathbf{A}) \times (\mathbf{B} \times \mathbf{D}) + (\mathbf{A} \times \mathbf{B}) \times (\mathbf{C} \times \mathbf{D}) = -2[\mathbf{ABC}]\mathbf{D}$.
   (*ii*) $\mathbf{A} \times [\mathbf{F} \times \mathbf{B}) \times (\mathbf{G} \times \mathbf{C})] + \mathbf{B} \times [\mathbf{F} \times \mathbf{C}) \times (\mathbf{G} \times \mathbf{A}) + \mathbf{C} \times [(\mathbf{F} \times \mathbf{A}) \times (\mathbf{G} \times \mathbf{B})] = 0$,
15. (*a*) Prove that $[\mathbf{LMN}][\mathbf{ABC}] = \begin{vmatrix} \mathbf{L}\cdot\mathbf{A} & \mathbf{L}\cdot\mathbf{B} & \mathbf{L}\cdot\mathbf{C} \\ \mathbf{M}\cdot\mathbf{A} & \mathbf{M}\cdot\mathbf{B} & \mathbf{M}\cdot\mathbf{C} \\ \mathbf{N}\cdot\mathbf{A} & \mathbf{N}\cdot\mathbf{B} & \mathbf{N}\cdot\mathbf{C} \end{vmatrix}$
   (*b*) The length of the edges $OA$, $OB$, $OC$ of the tetrahedron $OABC$ are $a$, $b$, $c$ and the angles $BOC$, $COA$, $AOB$ are $\theta$, $\phi$, $\psi$, find its volume.

# SOLID GEOMETRY

## 3.11 (1) EQUATION OF A PLANE

Let $P(x, y, z)$ be any point on the plane through $A(x_1, y_1\, z_1)$ which is normal to the vector $\mathbf{N} = a\mathbf{I} + b\mathbf{J} + c\mathbf{K}$.

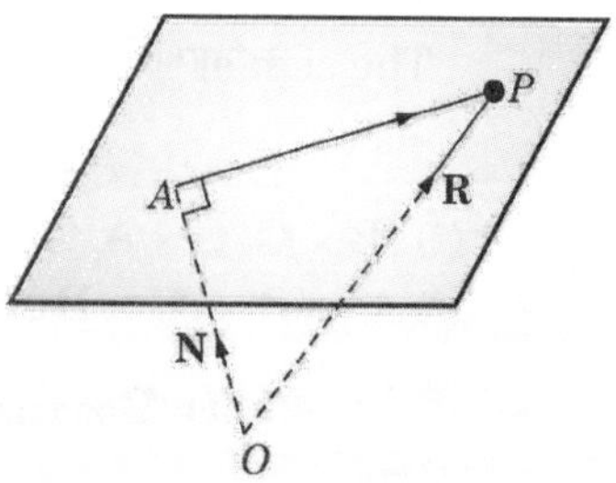

Fig. 3.35

Then $\quad \vec{OP} = x\mathbf{I} + y\mathbf{J} + z\mathbf{K}$ and $\vec{OA} = x_1\mathbf{I} + y_1\mathbf{J} + z_1\mathbf{K}$

Clearly the vectors $\vec{AP} = (x - x_1)\,\mathbf{I} + (y - y_1)\,\mathbf{J} + (z - z_1)\,\mathbf{K}$ and $\mathbf{N}$ are perpendicular to each other.

$$\therefore \quad \vec{AP} \cdot \mathbf{N} = 0 \quad \quad ...(i)$$

or $$[x - x_1)\,\mathbf{I} + (y - y_1)\,\mathbf{J} + (z - z_1)\,\mathbf{K}] \cdot (a\mathbf{I} + b\mathbf{J} + c\mathbf{K}) = 0$$

or $$\mathbf{a(x - x_1) + b(y - y_1) + c(z - z_1) = 0} \quad \quad ...(ii)$$

which is *the equation of any plane through the point* $(x_1, y_1, z_1)$.

**Obs.** Equation (*ii*) written as $ax + by + cz + d = 0$ *is the general equation of a plane.*

Conversely, *every linear equation in x, y, z represents a plane and the coefficients of x, y, z are the direction ratios of the normal to the plane.*

**Cor.** If $l, m, n$ be the direction cosines of the normal to the plane, then

$$\mathbf{lx + my + nz = p} \quad ...(iii)$$

which is called the **normal form** *of the equation of the plane* where $p$ is the perpendicular distance from the origin.

(2) **Angle between two planes. Def.** *The angle between two planes is equal to the angle between their normals.*

Let the two planes be

$$ax + by + cz + d = 0 \quad \text{and} \quad a'x + b'y + c'z + d' = 0.$$

Now the direction ratio of their normals are $a, b, c$ and $a', b', c'$.

Hence the angle $\theta$ between the planes is given by $\cos\theta = \dfrac{aa' + bb' + cc'}{\sqrt{(a^2 + b^2 + c^2)}\sqrt{a'^2 + b'^2 + c'^2}}$

*The planes will be perpendicular* (if their normal are parallel), *i.e., if* $aa' + bb' + cc' = 0$

*The planes will be parallel* (if their normals are parallel), *i.e., if* $\dfrac{a}{a'} = \dfrac{b}{b'} = \dfrac{c}{c'}$.

**Cor.** *Any plane parallel to the plane* $ax + by + cz + d = 0$

is $$\mathbf{ax + by + cz + k = 0} \quad (k \text{ being any constant})$$

for the direction-ratios of their normals are the same.

(3) **Perpendicular distance of the point** $(x_1, y_1, z_1)$ *from the plane*

$$ax + by + cz + d = 0 \quad ...(i)$$

is $$\frac{\mathbf{ax_1 + by_1 + cz_1 + d}}{\sqrt{(\mathbf{a^2 + b^2 + c^2})}}$$

Let $PL$ be the perpendicular distance of $P(x_1, y_1, z_1)$ from the plane Eqn. (*i*) so that the direction cosines of $\vec{LP}$ are

$$\frac{a}{\sqrt{(\Sigma a^2)}}, \frac{b}{\sqrt{(\Sigma a^2)}}, \frac{c}{\sqrt{(\Sigma a^2)}}.$$

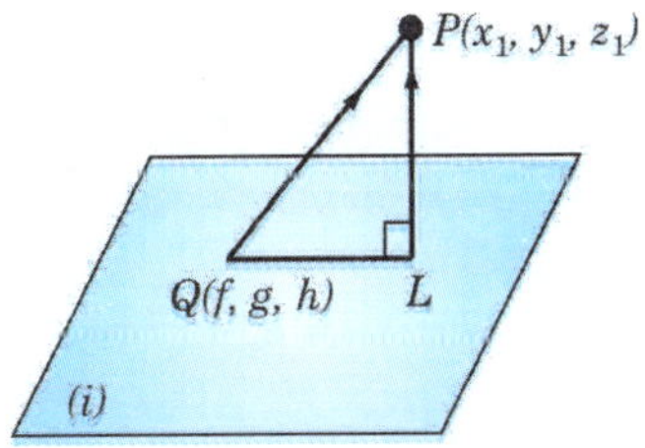

Fig. 3.36

If $Q$ $(f, g, h)$ be a point on (*i*) then

$$af + bg + ch + d = 0 \quad ...(ii)$$

$$\therefore \quad PL = \text{projection of } \vec{QP} \text{ on } \vec{LP} = \vec{QP} \cdot \vec{LP}$$

$$= \frac{(x_1 - f)a + (y_1 - g)b + (z_1 - h)c}{\sqrt{(a^2 + b^2 + c^2)}}$$

$$= \frac{ax_1 + by_1 + cz_1 + d}{\sqrt{(a^2 + b^2 + c^2)}} \text{ by virtue of Eqn. } (ii) \quad ...(iii)$$

The sign of the radical in Eqn. (*iii*) is taken to be positive or negative according as $d$ is positive or negative.

**Obs.** The perpendicular to a plane from two points are taken to be of the same sign if the points lie on the same side and of different signs if they lie on the opposite sides of the plane.

$\therefore$ *The two points* $(x_1, y_1, z_1)$ *and* $(x_2, y_2, z_2)$ *lie on the same side or on opposite sides of the plane* $ax + by + cz + d = 0$, according as $ax_1 + by_1 + cz_1 + d$ and $ax_2 + by_2 + cz_2 + d$ are of the same sign or of opposite signs.

**Cor. Planes bisecting the angles between two planes.**

Let $$ax + by + cz + d = 0 \quad ...(i)$$

and $$a'x + b'y + c'z + d' = 0 \quad ...(ii)$$

be the given planes.

Let $P(x, y, z)$ be any point on either of the planes bisecting the angles between the planes Eqns. (*i*) and (*ii*).

Then $\perp$ distance of $P$ from Eqn. (*i*) = $\perp$ distance of $P$ from Eqn. (*ii*),

$$\therefore \qquad \frac{ax + by + cz + d}{\sqrt{(a^2 + b^2 + c^2)}} = \pm \frac{a'x + b'y + c'z + d'}{\sqrt{(a'^2 + b'^2 + c'^2)}}$$

which are the required equations of the bisecting planes.

**Example 3.27.** *Find the equation of the plane which*
*(i) cuts off intercepts a, b, c from the axes.*
*(ii) passes through the points A(0, 1, 1), B(1, 1, 2) and C(–1, 2, –2).*

**Solution.** (*i*) **Intercept form** *of the equation of the plane.* Let the required equation of the plane be

$$\alpha x + \beta y + \gamma z + \delta = 0 \qquad \text{...(1)}$$

The plane cuts the axes at $A, B, C$ such that $OA = a$, $OB = b$, $OC = c$, *i.e.*, it passes through the points $A(a, 0, 0)$, $B(0, b, 0)$, $C(0, 0, c)$.

$$\therefore \qquad \alpha a + \delta = 0,\ \beta b + \delta = 0,\ \gamma c + \delta = 0$$

when $\qquad \alpha = -\delta/a,\ \beta = -\delta/b,\ \gamma = -\delta/c$

Substituting these values of $\alpha, \beta, \gamma$ in Eqn. (1), $-\frac{\delta}{a}x - \frac{\delta}{b}y - \frac{\delta}{c}z + \delta = 0$ or $\frac{\mathbf{x}}{\mathbf{a}} + \frac{\mathbf{y}}{\mathbf{b}} + \frac{\mathbf{z}}{\mathbf{c}} = 1.$

(*ii*) **Three points form** *of the equation of the plane.*

Any plane through (0, 1, 1) is $a(x - 0) + b(y - 1) + c(z - 1) = 0$ ...(2)

It will pass through (1, 1, 2) and (–1, 2, –2), if $a + c = 0$ and $-a + b - 3c = 0$.

By cross-multiplication, $\frac{a}{-1} = \frac{b}{2} = \frac{c}{1}$.

Substituting these values in (2), we get $-1 . x + 2(y - 1) + 1(z - 1) = 0$

or $\qquad x - 2y - z + 3 = 0$, which is the required equation of the plane.

**Example 3.28.** *Find the equation of the plane which passes through the point (3, –3, 1) and is*
*(i) parallel to the plane 2x + 3y + 5z + 6 = 0.*
*(ii) normal to the line joining the points (3, 2, –1) and (2, –1, 5).* (V.T.U., 2013)
*(iii) Perpendicular to the planes 7x + y + 2z = 6 and 3x + 5y – 6z = 8.*

**Solution.** (*i*) Any plane parallel to the given plane is

$2x + 3y + 5z + k = 0$ which goes through (3, –3, 1), if $k = -2$

Thus the required plane is $2x + 3y + 5z - 2 = 0$

(*ii*) Any plane through (3, –3, 1) is $a(x - 3) + b(y + 3) + c(z - 1) = 0$

The direction cosines of the line joining the points (3, 2, –1) and (2, –1, 5) are proportional to 1, 3, –6.

This line is normal to the plane (1). $\therefore$ $a, b, c$ are proportional to 1, 3, –6.

Substituting these values in (1), the required equation is

$$1(x - 3) + 3(y + 3) - 6(z - 1) = 0 \quad \text{or} \quad x + 3y - 6z + 12 = 0.$$

(*iii*) Any plane through (3, –3, 1) is

$a(x - 3) + b(y + 3) + c(z - 1) = 0$ which will be $\perp$ to the planes

$$7x + y + 2z = 6 \text{ and } 3x + 5y - 6z = 8$$

if $\qquad 7a + b + 2c = 0$ and $3a + 5b - 6c = 0$.

Solving these by cross-multiplication, we get $\frac{a}{1} = \frac{b}{-3} = \frac{c}{-2}$.

Hence the required equation is $1(x - 3) - 3(y + 3) - 2(z - 1) = 0$ or $x - 3y - 2z - 10 = 0$.

**Example 3.29.** *The plane 4x + 5y – z = 7 is rotated through a right angle about its line of intersection with the plane 2x + 3y – 3z = 5. Find the equation of this plane in its new position.*

**Solution.** Any plane through the line of intersection of

$$4x + 5y - z = 7 \qquad \text{...}(i)$$

and

$$2x + 3y - 3z = 5 \qquad \text{...}(ii)$$

is $$4x + 5y - z - 7 + k(2x + 3y - 3z - 5) = 0$$

*i.e.,* $$(4 + 2k)x + (5 + 3k)y - (1 + 3k)z - (7 + 5k) = 0 \qquad ...(iii)$$

Then new position of $(i)$ when rotated through a right angle, is such that Eqns. $(i)$ and $(iii)$ are perpendicular. This requires that

$$4(4 + 2k) + 5(5 + 3k) + (1 + 3k) = 0$$

*i.e.,* $$26k + 42 = 0 \quad \text{or} \quad k = -21/13$$

Substituting $k = -21/13$ in Eqn. $(iii)$, we get $10x + 2y + 50z + 14 = 0$.

or $5x + y + 25x + 7 = 0$, which is the required plane.

**Example 3.30.** *Find the distance between the parallel planes $2x - 2y + z + 3 = 0$ and $4x - 4y + 2z + 9 = 0$. Find also the equation of the parallel plane that lies mid-way between the given planes.*

**Solution.** The distance between the given planes is the perpendicular distance of any point on one of the planes from the other.

A point on the first plane is $(0, 0, -3)$.

$\therefore$ Required distance = $\perp$ distance of $(0, 0, -3)$ from $4x - 4y + 2z + 9 = 0$

$$= \frac{-6+9}{\sqrt{(16+16+4)}} = \frac{3}{6} = \frac{1}{2}$$

Let the equation of the parallel plane that lies mid-way between the given planes be

$$2x - 2y + z + k = 0 \qquad ...(i)$$

Now distance of any point $(0, 0, -3)$ on the first plane from Eqn. $(i)$ should be 1/4.

$$\therefore \quad \pm\frac{-3+k}{\sqrt{(4+4+1)}} = 1/4 \quad i.e., \quad k = 15/4 \text{ or } 9/4.$$

Thus the required plane is $2x - 2y + z + 15/4 = 0$.

Assume that $k = 15/4$ and verify that the distance of a point on this plane $4x - 4y + 2z + 9 = 0$ is also 1/4.

A point on this plane is $(0, 0, -9/4)$. Its distance from the plane Eqn. $(i)$ = $\frac{-9/2+15/4}{3} = \frac{1}{4}$ (in magnitude)

Thus $k = 9/4$ is not admissible.

$\therefore$ The required plane is $2x - 2y + z + 15/4 = 0$.

**Example 3.31.** *A variable plane is at a constant distance p from the origin and meets the axes at A, B, C. Find the locus of the centroid of the tetrahedron OABC.*

**Solution.** As the given plane is at a $\perp$ distance $p$ from the origin, therefore its equation is of the form

$$lx + my + nz = p \qquad ...(i)$$ where $l, m, n$ are the d.c's of the $\perp$ from the origin.

$(i)$ may be rewritten as $\frac{x}{(p/l)} + \frac{y}{(p/m)} + \frac{z}{(p/n)} = 1$

so that $OA = p/l$, $OB = p/m$, $OC = p/n$.

$$\therefore \quad A = (p/l, 0, 0), B = (0, p/m, 0), C = (0, 0, p/n).$$

Thus the coordinates of the centroid $G$ of the tetrahedron $OABC$ are

$$(x_1, y_1, z_1) = (p/4l, p/4m, p/4n)$$

$$\therefore \quad \frac{1}{x_1^2} + \frac{1}{y_1^2} + \frac{1}{z_1^2} = \frac{16}{p^2}(l^2 + m^2 + n^2) = \frac{16}{p^2}$$

Thus the locus of $G$ is $x^{-2} + y^{-2} + z^{-2} = 16p^{-2}$.

**Example 3.32.** *A variable plane at a constant distance p from the origin meets the axes in A, B, C. Planes are drawn through A, B, C parallel to the coordinate planes, Show that the locus of their point of intersection is given by $x^{-2} + y^{-2} + z^{-2} = p^{-2}$.*

**Solution.** Let the variable plane be $\frac{x}{a} + \frac{y}{b} + \frac{z}{c} = 1$.

Its distance from origin = $\frac{1}{\sqrt{a^{-2} + b^{-2} + c^{-2})}} = p$ (given)

*i.e.,* $$a^{-2} + b^{-2} + c^{-2} = p^{-2} \qquad ...(i)$$

Since $OA = a$, $OB = b$ and $OC = c$, therefore equations of the planes through $A, B, C$ parallel to $yz$, $zx$ and $xy$-planes are $x = a, y = b, z = c$

Let the point of intersection of these three planes be $(x_1, y_1, z_1)$.

Then $$x_1 = a, y_1 = b, z_1 = c \quad ...(ii)$$

Substituting Eqns. (*ii*) in (*i*), we get $x_1^{-2} + y_1^{-2} + z_1^{-2} = p^{-2}$

Thus the locus of $(x_1, y_1, z_1)$ is $x^{-2} + y^{-2} + z^{-2} = p^{-2}$.

**Example 3.33.** *A variable plane passes through the fixed point (a, b, c) and meets the coordinate axes in A, B, C. Show that the locus of the point common to the planes through A, B, C parallel to the coordinate planes is*

$$\frac{a}{x} + \frac{b}{y} + \frac{c}{z} = 1.$$

**Solution.** Let $ABC$ be any plane through the fixed point $H(a, b, c)$ such that $OA = x_1$, $OB = y_1$, $OC = z_1$.

Then its equation is

$$\frac{x}{x_1} + \frac{y}{y_1} + \frac{z}{z_1} = 1$$

Since $H$ lies on it.

$$\therefore \quad \frac{a}{x_1} + \frac{b}{y_1} + \frac{c}{z_1} = 1. \quad ...(1)$$

The planes through $A, B, C$ parallel to the coordinate planes are $x = x_1$, $y = y_1$, $z = z_1$, which meet in $P(x_1, y_1, z_1)$.

Thus changing $x_1$ to $x$, $y_1$, to $y$ and $z$, to $z$ in the locus of the $P$ is

$$\frac{a}{x} + \frac{b}{y} + \frac{c}{z} = 1.$$

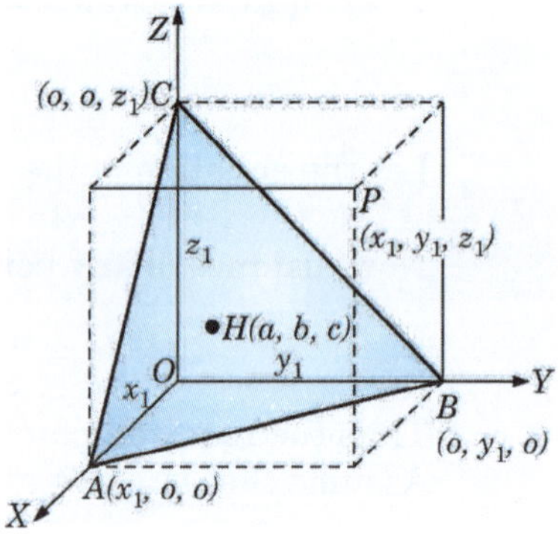


Fig. 3.37

**Example 3.34.** *Find the equations to the two planes which bisect the angles between the planes 3x – 4y + 5z = 3, 5x + 3y – 4z = 9.*

*Also point out which of the planes bisects the acute angle.*

**Solution.** The equations of the planes bisecting the angles between the given planes are

$$\frac{3x - 4y + 5z - 3}{\sqrt{[3^2 + (-4)^2 + 5^2]}} = \pm \frac{5x + 3y - 4z - 9}{\sqrt{[5^2 + 3^2 + (-4)^2]}}$$

or $$2x + 7y - 9z - 6 = 0 \quad ...(i)$$

and $$8x - y + z - 12 = 0 \quad ...(ii)$$

which are the required planes.

Let θ be the angle between Eqn. (*i*) and either of the given planes, say:

$$5x + 3y - 4z = 9.$$

Then, $$\cos\theta = \frac{2 \times 5 + 7 \times 3\,(-9) \times (-4)}{\sqrt{[2^2 + 7^2 + (-9)^2]}\sqrt{[5^2 + 3^2 + (-4)^2]}} = \frac{67}{5\sqrt{(268)}}$$

$$\therefore \quad \tan\theta = \frac{\sqrt{2211}}{67} \text{ which is less than 1.}$$

*i.e.,* $$\theta < 45°.$$

Now θ is half the angle between the given planes, so that Eqn. (*i*) bisects that angle between the planes which is $2\theta < 90°$.

Hence the plane $2x + 7y - 9z = 6$, bisects the acute angle.

## PROBLEMS 3.6

1. Find the equation of the plane passing through the point (1, 2, 3) and having the vector **N** = 2**I** – **J** + 3**K** normal to it.
2. Find the equation of the plane through the points
   (*i*) $(x_1, y_1, z_1), (x_2, y_2, z_2)$ and $(x_3, y_3, z_3)$ (*V.T.U., 2013*) (*ii*) (3, –1, 1), (1, 2, –1) and (1, 1, 1).
3. Find a unit vector normal to the plane through the points (– 1, 2, 3), (1, 1, 1) and (2, –1, 3).
4. Find the distance of the point (1, 4, 5) from the plane passing through the points (2, –1, 5), (0, –4, 1) and (2, –6, 0).
5. Show that the four points (0, –1, 0), (2, 1, –1), (1, 1, 1) and (3, 3, 0) are coplanar. Find the equation of the plane through them.
6. Show that the point (–1/2, 2, 0) is the circumcentre of the triangle formed by the points (1, 1, 0), (1, 2, 1), (–2, 2, –1).
   [**Hint.** Show that the point (–1/2, 2, 0) lies in the plane of the triangle and is equidistant from its vertices.]
7. Find the equation of the plane through the point (2, 1, 0) and perpendicular to the planes $2x - y - z = 5$ and $x + 2y - 3z = 5$.
8. Find the equations of the plane through (0, 0, 0) parallel to the plane $x + 2y = 3z + 4$.
9. Find the equation of the plane which bisects the join of the points $(x_1, y_1, z_1)$ and $(x_2, y_2, z_2)$ at right angles.
10. Find the equation of the plane through the points (–1, 2, 1), (–3, 2, –3) and parallel to *y*-axis (*V.T.U., 2010*)
11. Find the equation of the plane through the points (1, –2, 2) and (–3, 1, –2) and perpendicular to the plane, $2x - y - z + 6 = 0$. (*V.T.U., 2016*)
12. A plane contains the points *A*(–4, 9, –9) and *B*(5, –9, 6) and is perpendicular to the line which joins *B* and *C*(4, –6, *k*). Evaluate *k* and find the equation of the plane.
13. Find the distance between the parallel planes
    $$2x - 3y + 6z + 12 = 0 \text{ and } 6x - 9y + 18z - 6 = 0.$$
    Also find the equation of the parallel plane that lies mid-way between the given planes.
14. Find the angle between the plane $x + y + z = 8$ and $2x + y - z = 3$.
15. Two planes are given by $x + 2y - 3z - 2 = 0$ and $2x + y + z + 3 = 0$, find
    (*i*) direction cosines of their line of intersection,
    (*ii*) acute angle between the planes, and
    (*iii*) equation of the plane perpendicular to both of them through the point (2, 2, 1).
16. The plane $lx + my = 0$ is rotated about its line of intersection with the plane $z = 0$, through an angle α.
    Prove that the equation of the plane is $lx + my + z\sqrt{(l^2 + m^2)}\tan\alpha = 0$. (*Anna, 2013 S*)
17. Find the equations of the two planes through the points (0, 4, –3), (6, –4, 3) other than the plane through the origin which cut off from the axes intercepts whose sum is zero.
18. A plane meets the coordinates axes at *A, B, C,* such that the centroid of the triangle *ABC* is the point (*a, b, c*), show that the equation of the plane is $\frac{x}{a} + \frac{y}{b} + \frac{z}{c} = 3$.
19. A plane passes through a fixed point (*a, b, c*), show that the locus of the foot of the perpendicular from the origin on the plane is a sphere.
20. A variable plane is at a constant distance *p* from the origin and meets the axes at *A, B, C*. Find the locus of the centroid of the triangle *ABC*.
21. A variable plane makes with the coordinate axes a tetrahedron of constant volume 64 $k^3$. Find the locus of the centroid of the tetrahedron.
22. Find equations of the planes bisecting the angle between the planes
    $$x + 2y + 2z = 9,\ 4x - 3y + 12z + 12 = 0$$
    and specify the one which bisects the acute angle.

## 3.12 EQUATIONS OF A STRAIGHT LINE

**(1) General form.** Two linear equations in *x, y, z*

*i.e.*, $$ax + by + cz + d = 0 \quad ...(i)$$

and $$a'x + b'y + c'z + d' = 0 \quad ...(ii)$$

taken together represent a straight line which is the line of intersection of the planes Eqns. (*i*) and (*ii*). (Fig. 3.38).

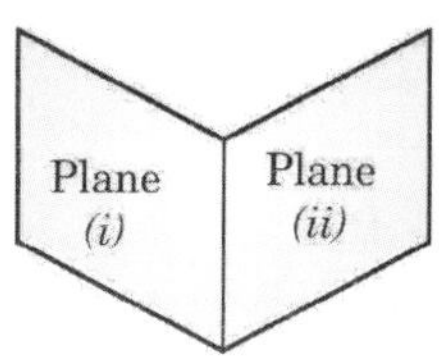


Fig. 3.38

**(2) Symmetrical form.** *Equations of the line through the point* $A(x_1, y_1, z_1)$ *and having direction cosines* $l$, $m$, $n$ *are*

$$\frac{\mathbf{x} - \mathbf{x_1}}{1} = \frac{\mathbf{y} - \mathbf{y_1}}{\mathbf{m}} = \frac{\mathbf{z} - \mathbf{z_1}}{\mathbf{n}}$$

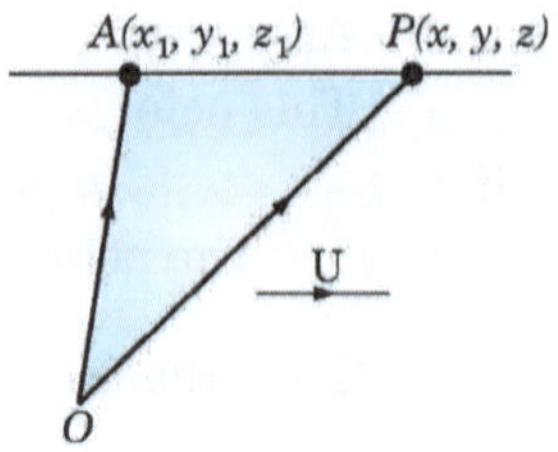

Fig. 3.39

Let $P(x, y, z)$ be any point on the given line through $A(x_1, y_1, z_1)$ and parallel to the unit vector $\mathbf{U} = l\mathbf{I} + m\mathbf{J} + n\mathbf{K}$.

Since $\overrightarrow{AP}$ is parallel to $\mathbf{U}$, we can write $\overrightarrow{AP} = t\mathbf{U}$, where $t$ is a parameter. ...(*i*)

or $$(x - x_1)\,\mathbf{I} + (y - y_1)\,\mathbf{J} + (z - z_1)\,\mathbf{K} = t\,(l\mathbf{I} + m\mathbf{J} + n\mathbf{k})$$

$\therefore$ $$x - x_1 = tl,\ y - y_1 = tm,\ z - z_1 = tn \quad \text{...(ii)}$$

Every point $P$ on the line is given by Eqn. (*ii*) for some value of $t$. Thus these are the parametric equations of the given line. Eliminating $t$, we get

$$\frac{x - x_1}{l} = \frac{y - y_1}{m} = \frac{z - z_1}{n} \quad \text{...(iii)}$$

which are the *symmetrical form of the equations of the line.*

**Obs.** Any point on the line Eqn. (*iii*) is $(x_1 + lt, y_1 + mt, z_1 + nt)$.

**Cor.** *The equations of the line joining the points* $(x_1, y_1, z_1)$ *and* $(x_2, y_2, z_2)$ *are*

$$\frac{\mathbf{x} - \mathbf{x_1}}{\mathbf{x_2} - \mathbf{x_1}} = \frac{\mathbf{y} - \mathbf{y_1}}{\mathbf{y_2} - \mathbf{y_1}} = \frac{\mathbf{z} - \mathbf{z_1}}{\mathbf{z_2} - \mathbf{z_1}}$$

for the direction-ratios of the line joining the given points are

$$x_2 - x_1, y_2 - y_1, z_2 - z_1.$$

[**To reduce the general equation of a line of the symmetrical form:**

(*i*) *find a point on the line,* by putting z = 0 in the given equations and solving the resulting equations for $x$ and $y$.

(*ii*) *find the direction cosines of the line,* from the fact that it is perpendicular to the normals to the given planes.

(*iii*) *write the equations of the line in the symmetrical form.*]

**Example 3.35.** *Find in symmetrical form, the equations of the line*

$$x + y + z + 1 = 0,\ 4x + y - 2z + 2 = 0.$$

**Solution.** (*i*) To find a point on the line.

Putting $z = 0$ in the given equations, we have

$$x + y + 1 = 0;\ 4x + y + 2 = 0$$

Solving, $\dfrac{x}{1} = \dfrac{y}{2} = \dfrac{1}{-3}$ $\therefore$ A point on the line is (–1/3, –2/3, 0).

(*ii*) *To find the direction cosines l, m, n of the line.*

Since the line lies on both the given planes.

$\therefore$ It is perpendicular to their normals whose direction cosines are proportional to 1, 1, 1 and 4, 1, –2.

*i.e.,* $$l + m + n = 0;\ 4l + m - 2n = 0.$$

Solving, $$\frac{l}{-1} = \frac{m}{2} = \frac{n}{-1}$$

$\therefore$ The direction cosines of the given line are proportional to –1, 2, –1.

(*iii*) Thus the equations of the line in the symmetrical form are

$$\frac{x + 1/3}{-1} = \frac{y + 2/3}{2} = \frac{z}{-1}.$$

**Example 3.36.** *Find the distance of the point (1, –2, 3) from the plane x – y + z = 5 measured parallel to the line*

$$\frac{x}{2} = \frac{y}{3} = \frac{z}{-6}$$

**Solution.** The line through $P(1, -2, 3)$ having direction ratios $(2, 3, -6)$ is

$$\frac{x-1}{2} = \frac{y+2}{3} = \frac{z-3}{-6} = r.$$

Any point on this line is $(2r + 1, 3r - 2, 3 - 6r)$.

This point will lie on the plane $x - y + z = 5$

if $\quad 2r + 1 - (3r - 2) + 3 - 6r = 5 \quad$ or $\quad r = 1/7$.

$\therefore$ The point of intersection is $Q(9/7, -11/7, 15/7)$

Thus the required distance $= PQ = \sqrt{\left(\frac{4}{49} + \frac{9}{49} + \frac{36}{49}\right)} = 1$

$x + 2y + 2z = 9$, $4x - 3y + 12z + 12 = 0$ and specify the one which bisects the acute angle.

**Example 3.37.** (a) *Find the image (reflection) of the point $(p, q, r)$ in the plane $2x + y + z = 6$.*

(b) *Find the image (reflection) of the line $\frac{x-1}{2} = \frac{y-2}{1} = \frac{z-3}{4}$ in the same plane.*

*[If two points P, P′ be such that the line PP′ is bisected perpendicularly by a plane then either of the points is the image (or reflection) of the other in the plane.]*

**Solution.** (a) Let $P'(p', q', r')$ be the image of $P(p, q, r)$. Then the mid-point of $PP'$ must lie on the given plane.

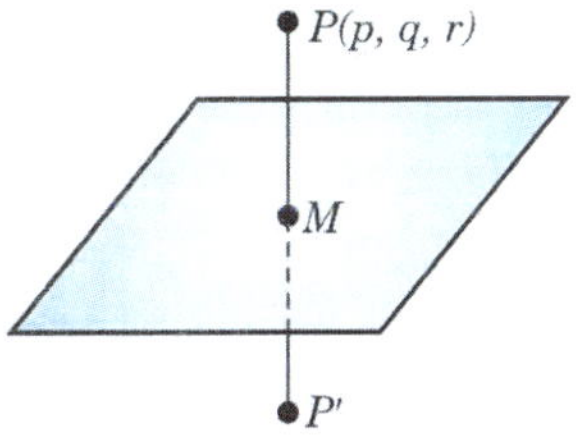

Fig. 3.40(a)

$$\therefore \quad \frac{p+p'}{2} + \frac{q+q'}{2} + \frac{r+r'}{2} = 6 \qquad ...(i)$$

Also the line $PP'$ must be perpendicular to the plane. The direction ratios of $PP'$ being $p - p', q - q', r - r'$, we therefore, have

$$\frac{p-p'}{2} = \frac{q-q'}{1} = \frac{r-r'}{1} = k \text{ (say)}$$

whence $p' = p - 2k$, $q' = q - k$, $r' = r - k$.

Substituting these in Eqn. $(i)$ and solving, we get

$$k = \frac{1}{3}(2p + q + r - 6).$$

Hence $P'$ is

$$\left[\frac{1}{3}(12 - p - 2q - 2r), \frac{1}{3}(6 - 2p + 2q - r), \frac{1}{3}(6 - 2p - q + 2r)\right] \qquad ...(ii)$$

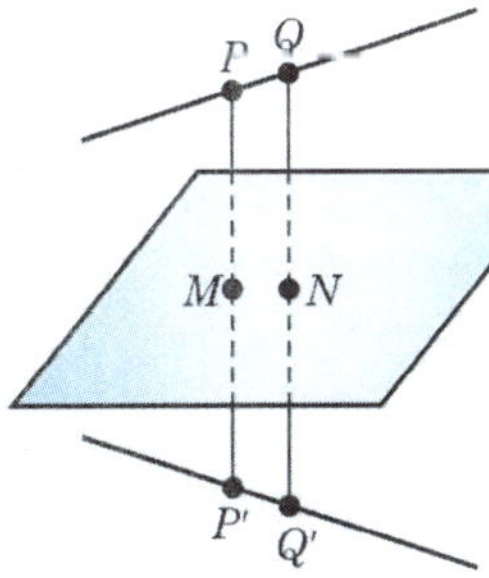

Fig. 3.40(b)

(b) Any two points on the given line are evidently $P(1, 2, 3)$ and (on putting $z = 7$) $Q(3, 3, 7)$. Their images are [by using Eqn. $(ii)$] $P'\left(\frac{1}{3}, \frac{5}{3}, \frac{8}{3}\right)$ and $Q'\left(-\frac{11}{3}, -\frac{1}{3}, \frac{11}{3}\right)$. The line joining $P'$ and $Q'$ is, therefore

$$\frac{x - \frac{1}{3}}{-\frac{11}{3} - \frac{1}{3}} = \frac{y - \frac{5}{3}}{-\frac{1}{3} - \frac{5}{3}} = \frac{z - \frac{8}{3}}{\frac{11}{3} - \frac{8}{3}}, \text{ i.e., } \frac{3x-1}{4} = \frac{3y-5}{2} = \frac{3z-8}{-1}$$

which is the required image of the given line $PQ$ [Fig. 3.40$(b)$].

**Example 3.38.** *Find the angle between the line*

$$\frac{x - x_1}{l} = \frac{y - y_1}{m} = \frac{z - z_1}{n}$$

*and the plane $ax + by + cz + d = 0$.* *(Andhra., 2016)*

**Solution.** If $\theta$ be the angle between the line and the plane, then $90° - \theta$ is the angle between the line and the normal to the plane (Fig. 3.41).

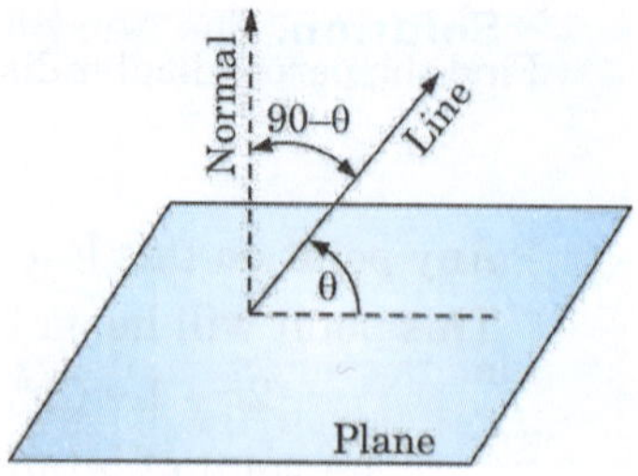

Fig. 3.41

Now the direction ratios of the line are $l, m, n$ and the direction ratios of the normal to the plane are $a, b, c$.

$$\therefore \quad \cos(90° - \theta) = \frac{la + mb + nc}{\sqrt{(l^2 + m^2 + n^2)}\sqrt{(a^2 + b^2 + c^2)}}$$

or
$$\sin\theta = \frac{la + mb + nc}{\sqrt{(\Sigma l^2)}\,\sqrt{(\Sigma a^2)}}$$

Hence the required angle $\theta = \sin^{-1}\left(\dfrac{al + bm + cn}{\sqrt{(\Sigma l^2)}\,\sqrt{(\Sigma a^2)}}\right)$

**Cor.** *If the line is parallel to the plane,* $\sin\theta = 0$

$$\therefore \quad \mathbf{al + bm + cn = 0}$$

*If the line is perpendicular to the plane, it will be parallel to its normal.*

$$\therefore \quad l/a = m/b = n/c.$$

**Example 3.39.** *Find the equations of the two straight lines through the origin, each of which intersects the straight line* $\frac{1}{2}(x - 3) = y - 3 = z$ *and is inclined at an angle of 60° to it.*

**Solution.** Let $AB$ be the given line so that any point $A$ on it is $(2r + 3, r + 3, r)$. (Fig. 3.42)

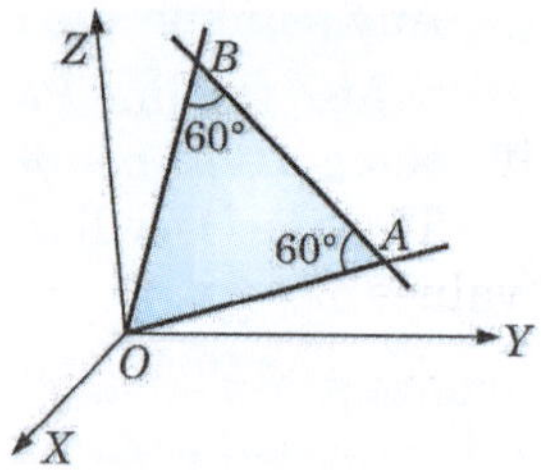

Fig. 3.42

$\therefore$ Direction ratios of $OA$ are $2r + 3 - 0, r + 3 - 0, r - 0$.

Angle between $AO$ and $AB$ has to be 60°,

$$\therefore \quad \cos 60° = \frac{2(2r + 3) + 1(r + 3) + 1.(r)}{\sqrt{2^2 + 1^2 + 1^2)}\,\sqrt{[2r + 3)^2 + (r + 3)^2 + r^2]}}$$

or
$$\frac{1}{2} = \frac{6r + 9}{\sqrt{[6(6r^2 + 18r + 18)]}} \text{ or } r^2 + 3r + 2 = 0 \text{ i.e., } r = -1, -2$$

$\therefore$ Coordinates of $A$ and $B$ are $(1, 2, -1)$ and $(-1, 1, -2)$.

Hence the equations of the required lines $OA$ and $OB$ are $\dfrac{x}{1} = \dfrac{y}{2} = \dfrac{z}{-1}$ and $\dfrac{x}{-1} = \dfrac{y}{1} = \dfrac{z}{-2}$

## PROBLEMS 3.7

1. Prove that the points (3, 2, 4), (4, 5, 2) and (5, 8, 0) are collinear. Find the equations of the line through them.
2. Find the angle between the line of intersection of the planes
$$2x + 2y - z + 15 = 0,\ 4y + z + 29 = 0 \text{ and the line } \frac{x+4}{4} = \frac{y-3}{-3} = \frac{z+2}{1}.$$
3. Find the angle between the line of intersection of the planes $3x + 2y + z = 5$ and $x + y - 2z = 3$ and the line of intersection of the plane $2x = y + z$ and $7x + 10y = 8z$.
4. Find the equation of the line through the point $(-2, 3, 4)$ and parallel to the planes $2x + 3y + 4z = 5$ and $4x + 3y + 5z = 6$.
5. Show that the line $\dfrac{x-1}{3} = \dfrac{y+2}{-2} = \dfrac{z-1}{2}$ is parallel to the plane $2x + 2y - z = 6$, and find the distance between them.
6. Find the equation of the line through $(1, 2, -1)$ perpendicular to each of the lines
$$\frac{x}{1} = \frac{y}{0} = \frac{z}{-1} \text{ and } \frac{x}{3} = \frac{y}{4} = \frac{z}{5}.$$
7. Find the equation of the lines bisecting the angle between the lines
$$\frac{x-1}{2} = \frac{y+2}{-2} = \frac{z-3}{1};\ \frac{x-1}{12} = \frac{y+2}{4} = \frac{z-3}{-3}.$$
8. Find the foot of the perpendicular from (1, 1, 1) to the line joining the points (1, 4, 6) and (5, 4, 4). (*V.T.U., 2010*)

9. Find the perpendicular distance of the point (1, 1, 1) from the line

$$\frac{x-2}{2}=\frac{y+3}{2}=\frac{z}{-1}.$$

10. Find the distance of the point (3, –4, 5) from the plane $2x + 5y - 6z = 16$, measured parallel to the line $x/2 = y/1 = z/- 2$. *(Andhra, 2015)*

11. Find the reflection (image) of the point
   (*i*) (1, 2, 3) in the plane $x + y + z = 9$. *(V. T.U., 2010)*
   (*ii*) (2, –1, 3) in the plane $3x - 2y - z - 9 = 0$. *(Andhra, 2016)*

12. Find the angle between the line $\frac{x+1}{2}=\frac{y}{3}=\frac{z-3}{6}$ and the plane $3x + y + z = 7$. *(Andhra, 2016)*

13. Find the equation of the plane through the points (1, 0, –1), (3, 2, 2) and parallel to the line $x-1=\frac{1}{2}(1-y)=\frac{1}{3}(z-2)$.

14. Find the equations of the straight line which passes through the point (2, –1, –1), is parallel to the plane $4x + y + z + 2 = 0$ and is perpendicular to the line $2x + y = 0 = x - z + 5$.

## 3.13 CONDITIONS FOR A LINE TO LIE IN A PLANE

*To find the conditions that the line* $\frac{x-x_1}{l}=\frac{y-y_1}{m}=\frac{z-z_1}{n}$ ...(1)

*may lie in the plane* $ax + by + cz + d = 0$ ...(2)

Any point on the line Eqn. (1) is $(lr + x_1, mr + y_1, nr + z_1)$ which will lie on the plane Eqn. (2), if

$$a(lr + x_1) + b(mr + y_1) + c(nr + z_1) + d = 0.$$

or if $$(al + bm + cn)\, r + (ax_1 + by_1 + cz_1 + d) = 0 \qquad ...(3)$$

The line (1) will lie in the plane (2), if every point of the line lies in the plane so that Eqn. (3) is satisfied by all values of $r$.

$\therefore$ The coefficient of $r = 0$ and the constant term = 0.

*i.e.,* $$\mathbf{al + bm + cn = 0} \qquad ...(4)$$

and $$\mathbf{ax_1 + by_1 + cz_1 + d = 0} \qquad ...(5)$$

These are the required conditions which state that

(*i*) *the line should be parallel to the plane,* (*ii*) *a point of line should lie in the plane.*

**Thus the equation of any plane through the line** $\frac{x-x_1}{l}=\frac{y-y_1}{m}=\frac{z-z_1}{n}$

*is* $$\mathbf{a(x - x_1) + b(y - y_1) + c(z - z_1) = 0} \textit{ where } \mathbf{al + bm + cn = 0.}$$

**Obs. The equation of any plane through the line of intersection of the planes**

$$ax + by + cz + d = 0 \qquad ...(i)$$

and $$a'x + b'y + c'z + d' = 0. \qquad ...(ii)$$

**is** $$\mathbf{ax + by + cz + d + k(a'x + b'y + c'z + d') = 0.}$$

For Eqn. (*i*) is an equation of the first degree in $x, y, z$ representing a plane and Eqn. (*ii*) it is satisfied by the coordinates of the points which satisfy both the given planes, *i.e.*, it contains all the points common to these planes.

**Example 3.40.** *Obtain the equation of a plane passing through the line of intersection of the planes $7x - 4y + 7z + 16 = 0$ and $4x + 3y - 2z + 13 = 0$ and perpendicular to the plane $x - y - 2z + 5 = 0$.*

**Solution.** The equation of any plane through the line of intersection of the two given planes is

$$7x - 4y + 7z + 16 + k(4x + 3y - 2z + 13) = 0$$

or $$(7 + 4k)x + (-4 + 3k)y + (7 - 2k)z + (16 + 13k) = 0 \qquad ...(i)$$

The plane (*i*) will be perpendicular to the plane

$x - y - 2z + 5 = 0$ if their normals are perpendicular,

*i.e.*, if $(7 + 4k) . 1 + (-4 + 3k) . (-1) + (7 - 2k) . (-2) = 0$ or if, $k = 3/5$.

Substituting this value of $k$ in Eqn. (*i*), we get

$$(7 + 12/5)x + (-4 + 9/5)y + (7 - 6/5)x + (16 + 39/5) = 0$$

or $47x - 11y + 29z + 119 = 0$ which is the required equation.

**Example 3.41.** *Find the equation in the symmetrical form of the projection of the line* $\frac{x-1}{2} = -(y+1) = \frac{z-3}{4}$ *on the plane* $x + 2y + z = 12$.

**Solution.** Any plane through the given line is

$$A(x-1) + B(y+1) + C(z-3) = 1 \quad ...(i)$$

where $$2A - B + 4C = 0 \quad ...(ii)$$

The plane (*i*) will be perpendicular to the given plane, if

$$A + 2B + C = 0 \quad ...(iii)$$

Solving Eqns. (*ii*) and (*iii*), we get $\frac{A}{-9} = \frac{B}{2} = \frac{C}{5}$.

Substituting these values in Eqn. (*i*), we get $9x - 2y - 5z + 4 = 0$ ...(*iv*)

which cuts the given plane $x + 2y + z = 12$ ...(*v*)

along the required line of projection.

One point on this line is got by putting $z = 0$ in Eqns. (*iv*) and (*v*) and solving, it is (4/5, 28/5, 0).

The direction ratios of the line are found, by solving

$l + 2m + n = 0$ and $9l - 2m - 5n = 0$ to be 4, –7, 10.

Hence the required equations of the line of projection are

$$\frac{x-4/5}{4} = \frac{y-28/5}{-7} = \frac{z}{10}$$

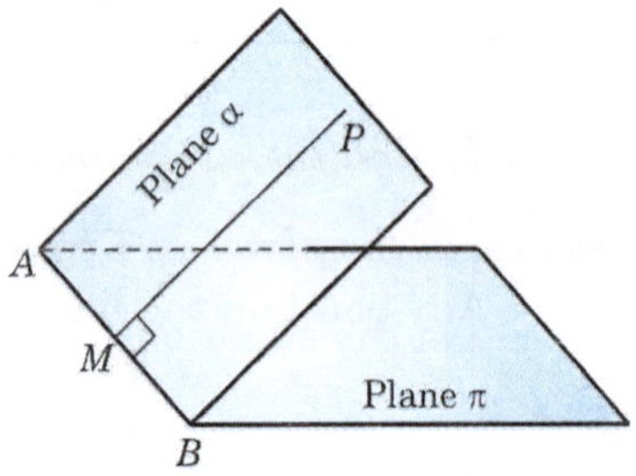

Fig. 3.43

[**The line of greatest slope** in a plane is a line which lies in the plane and is perpendicular to the line of intersection of the plane with the horizontal plane.

In Fig. 3.43, *AB* is the line of intersection of the given plane α with the horizontal plane π. Then *PM* drawn perpendicular to *AB*, is the line of greatest slope on the plane α through the point *P*.]

**Example 3.42.** *Assuming the line x/4 = y/–3 = z/7 as vertical, find the equations of the line of greatest slope in the plane 2x + y – 5z = 12 and passing through the point (2, 3, –1).*

**Solution.** The equation of the horizontal plane through the origin is $4x - 3y + 7z = 0$ ...(*i*)

[The direction ratios of the normal are those of the given vertical line.]

If $l, m, n$ be the direction ratios of the line of intersection of the plane (*i*) and

$$2x + y - 5z = 12 \quad ...(ii)$$

then solving, $4l - 3m + 7n = 0$ and $2l + m - 5n = 0$, we have $l/4 = m/17 = n/5$ ...(*iii*)

Let $l', m', n'$ be the direction ratios of the line of greatest slope which lies in the plane (*ii*).

$$\therefore \quad 2l' + m' - 5n' = 0 \quad ...(iv)$$

Also the line of greatest slope is perpendicular to the line of intersection of the planes (*i*) and (*ii*).

$$\therefore \quad 4l' + 17m' + 5n' = 0 \quad ...(v)$$

Solving (*iv*) and (*v*), $\frac{l'}{3} = \frac{m'}{-1} = \frac{n'}{1}$.

Hence the equations of the line of greatest slope through (2, 3, –1) and having direction ratios 3, –1, 1 are

$$\frac{x-2}{3} = \frac{y-3}{-1} = \frac{z+1}{1}.$$

## PROBLEMS 3.8

1. Find the equation of the plane which contains the line $\frac{x-1}{2} = y+1 = \frac{z-3}{4}$ and is perpendicular to the plane $x + 2y + z = 12$. (*Andhra, 2019*)

2. Find the equation of the plane through the line $\frac{x-1}{3} = \frac{y-4}{2} = \frac{z-4}{-2}$ and parallel to the line $\frac{x+1}{3} = \frac{y-1}{-4} = \frac{z+2}{1}$.
3. Find the equation of the plane passing through the line of intersection of the planes $x + y + z = 1$ and $2x + 3y - z + 4 = 0$ and perpendicular to the plane $2y - 3z = 4$. *(Andhra, 2021)*
4. Find the equation of the plane which contains the line of intersection of the planes $x + y + z = 3$ and $2x - y + 3z = 4$ and is parallel to the line joining the points (2, 1, 1) and (3, 2, 4).
5. Find in symmetric form the equations of the line which lies in the plane $2x - y - 3z = 4$ and is perpendicular to the line
$$\frac{x+1}{3} = \frac{y-1}{3} = \frac{z+4}{2}$$
at the point where the line pierces the plane.
6. A plane is drawn through the line $x + y = 1, z = 0$ to make an angle $\sin^{-1}(1/3)$ with the plane $x + y + z = 0$. Prove that two such planes can be drawn and find their equations. Prove also that the angle between the planes is $\cos^{-1}(7/9)$.
7. Find the equations of the projection of the line $3x - y + 2z - 1 = x + 2y - z - 2 = 0$ on the plane $3x + 2y + z = 0$ in the symmetrical form.
8. Assuming the plane $4x - 3y + 7z = 0$ to be horizontal, find the equations of the line of greatest slope through the point (2, 1, 1) in the plane $2x + y - 5z = 0$.

## 3.14 CONDITION FOR THE TWO LINES TO INTERSECT (OR TO BE COPLANAR)

Let the equations of the lines be $\frac{x-x_1}{l_1} = \frac{y-y_1}{m_1} = \frac{z-z_1}{n_1}$ ...(1)

$$\frac{x-x_2}{l_2} = \frac{y-y_2}{m_2} = \frac{z-z_2}{n_2} \quad ...(2)$$

The equation of any plane through the line (1) is $a(x - x_1) + b(y - y_1) + c(z - z_1) = 0$ ...(3)

where $al_1 + bm_1 + cn_1 = 0$ ...(4)

The line (2) will lie in the plane (3), if it is parallel to the plane and its point $(x_2, y_2, z_2)$ lies on this plane.

$\therefore$ $al_2 + bm_2 + cn_2 = 0$ ...(5)

and $a(x_2 - x_1) + b(y_2 - y_1) + c(z_2 - z_1) = 0$ ...(6)

Eliminating $a, b, c$ from (6), (4) and (5), we get

$$\begin{vmatrix} \mathbf{x_2 - x_1} & \mathbf{y_2 - y_1} & \mathbf{z_2 - z_1} \\ \mathbf{l_1} & \mathbf{m_1} & \mathbf{n_1} \\ \mathbf{l_2} & \mathbf{m_2} & \mathbf{n_2} \end{vmatrix} = 0 \text{ which is the required condition.}$$

Also eliminating $a, b, c$ from Eqns. (3), (4) and (5), we get $\begin{vmatrix} \mathbf{x - x_1} & \mathbf{y - y_1} & \mathbf{z - z_1} \\ \mathbf{l_1} & \mathbf{m_1} & \mathbf{n_1} \\ \mathbf{l_2} & \mathbf{m_2} & \mathbf{n_2} \end{vmatrix} = 0$

*which is the equation of the plane containing the lines (1) and (2).*

**Example 3.43.** *Show that the lines* $\frac{x-5}{4} = \frac{y-7}{4} = \frac{z+3}{-5}; \frac{x-8}{7} = \frac{y-4}{1} = \frac{z-5}{3}$ *are coplanar; find their common point and the equation of the plane in which they lie.*

**Solution.** Any point on the first line is $(5 + 4r, 7 + 4r, -3 - 5r)$ ...(*i*)

which lies on the second line if $\frac{-3+4r}{7} = \frac{3+4r}{7} = \frac{-8-5r}{3}$ ...(*ii*)

$\therefore$ From $\frac{-3+4r}{7} = 3 + 4r$, we have $r = -1$.

This value clearly satisfies the equation $\frac{3+4r}{7} = \frac{-8-5r}{3}$

Hence the lines intersect, (*i.e.*, are coplanar) and from Eqn. (*i*) their point of intersection is (1, 3, 2).

The equation of the plane in which they lie, is $\begin{vmatrix} x-5 & y-7 & z+3 \\ 4 & 4 & -5 \\ 7 & 1 & 3 \end{vmatrix} = 0$

*i.e.,* $17x - 47y - 24z + 172 = 0.$

**Example 3.44.** *Show that the lines*

$$\frac{x+4}{3} = \frac{y+6}{5} = \frac{z-1}{-2} \text{ and } 3x - 2y + z + 5 = 0 = 2x + 3y + 4z - 4$$

*are coplanar. Find their point of intersection and the plane in which they lie.* (Andhra, 2020)

**Solution.** Any point on the first line is $P(3r-4, 5r-6, -2r+1)$, which lie in the plane

$$3x - 2y + z + 5 = 0$$

if $3(3r-4) - 2(5r-6) + (-2r+1) + 5 = 0$ or $r = 2,$

The point $P$ will also lie in the plane $2x + 3y + 4z - 4 = 0$

if $2(3r-4) + 3(5r-6) + 4(-2r+1) - 4 = 0$ or $r = 2.$

Since the two values of $r$ are equal, the given lines intersect, *i.e.*, are coplanar.

Putting $r = 2$ in the coordinates of $P$, we get $(2, 4, -3)$ as their point of intersection.

The equation of a plane containing the second line is

$$3x - 2y + z + 5 + k(2x + 3y + 4z - 4) = 0$$

which will contain the first line if its point $(-4, -6, 1)$ lies on it.

$\therefore$ $-12 + 12 + 1 + 5 + k(-8 - 18 + 4 - 4) = 0$

*i.e.,* $k = 3/13$

Substituting this value of $k$, $(i)$ becomes $45x - 17y + 25z + 53 = 0$, which is the required plane.

**Example 3.45.** *Find the equations of the line drawn through the point (1, 0, –1) and intersecting the lines*

$$x = 2y = 2z \quad \text{and} \quad 3x + 4y = 1,\ 4x + 5z = 2.$$

**Solution.** The required line will comprise of

($a$) the plane containing the first line and the point (1, 0, –1).

($b$) the plane containing the second line and the point (1, 0, –1).

The equation of any plane containing the first line

*i.e.,* $\frac{x}{2} = \frac{y}{1} = \frac{z}{1}$

is $a(x-0) + b(y-0) + c(z-0) = 0$ ...($i$)

where $2a + b + c = 0$ ...($ii$)

Also (1, 0, –1) lies on Eqn. ($i$) $\therefore a - c = 0$ ...($iii$)

Solving Eqns. ($ii$) and ($iii$), we have $\frac{a}{1} = \frac{b}{-3} = \frac{c}{1}$.

Substituting these values in Eqn. ($i$), we get $x - 3y + z = 0$ ...($iv$)

Again, the equation of any plane containing the second line is

$3x + 4y - 1 + k(4x + 5z - 2) = 0$. Also (1, 0, –1) lies on it. ...($v$)

$\therefore$ $3 + 0 - 1 + k(4 - 5 - 2) = 0,$ *i.e.,* $k = \frac{2}{3}.$

Substituting $k = 2/3$ in Eqn. ($v$), we get $17x + 12y + 10z - 7 = 0$ ...($iv$)

Hence Eqns. ($iv$) and ($vi$) constitute the required line.

## PROBLEMS 3.9

1. Prove that the lines
$$\frac{x-1}{2}=\frac{y-2}{3}=\frac{z-3}{4} \text{ and } \frac{x-2}{3}=\frac{y-3}{4}=\frac{z-4}{5}$$
are coplanar and find the equation of the plane containing them. *(Andhra., 2013)*
2. Prove that the lines $\frac{x-4}{1}=\frac{y+3}{-4}=\frac{1+z}{7}$ and $\frac{x-1}{2}=\frac{y+1}{-3}=\frac{z+10}{8}$ intersect and find the coordinates of their point of intersection.
3. Find the condition that the lines $\frac{x-\alpha}{l}=\frac{y-\beta}{m}=\frac{z-\gamma}{n}$ and $ax+by+cz+d=0=a'x+b'y+c'z+d'$ are coplanar.
4. Show that the lines $\frac{x+1}{1}=\frac{y+1}{2}=\frac{z+1}{3}$ and $x+2y+3z-8=0=2x+3y+4z-11$ intersect. Find their point of intersection and the equation of the plane containing them.
5. Show that the lines $x-3y+2z-4=0=2x+y+4z+1$ and
$$3x+2y+5z-1=0=2y-z, \text{ are coplanar.}$$
6. Prove that the lines $x=ay+b=cz+d$ and $x=\alpha y+\beta=\gamma z+\delta$ are coplanar if
$$(\gamma-c)(a\beta-b\alpha)-(\alpha-a)(c\delta-d\gamma)=0$$
7. Obtain the equations of the straight line lying in the plane $x-2y+4z-51=0$ and interesecting the line $\frac{x-2}{3}=\frac{y+1}{4}=\frac{z-6}{7}$ at right angles.
8. Find the equation of the straight line perpendicular to both the lines
$$\frac{x-1}{1}=\frac{y-1}{2}=\frac{z+2}{3} \quad \text{and} \quad \frac{x+2}{2}=\frac{y-5}{-1}=\frac{z+3}{2}$$
and passing through their point of intersection.
9. A line with direction cosines proportional to 2, 7, – 5 is drawn to intersect the lines
$$\frac{x-8}{3}=\frac{y-6}{-1}=\frac{z+1}{1} \quad \text{and} \quad \frac{x+3}{-3}=\frac{y-3}{2}=\frac{z-6}{4}.$$
Find the coordinates of the point of intersection and the length intercepted.

## 3.15 SHORTEST DISTANCE BETWEEN TWO LINES

[Two straight lines which do not lie in one plane are called *skew lines*. Such lines possess a common perpendicular which is the *shortest distance* between them.

Let the given skew lines $AB$ and $CD$ be

$$\frac{x-x_1}{l_1}=\frac{y-y_1}{m_1}=\frac{z-z_1}{n_1} \quad \text{and} \quad \frac{x-x_2}{l_2}=\frac{y-y_2}{m_2}=\frac{z-z_2}{n_2}$$

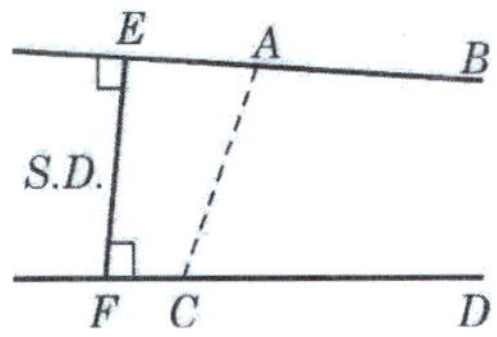

Fig. 3.44

so that $A\equiv(x_1, y_1, z_1)$ and $C\equiv(x_2, y_2, z_2)$.

Let $l, m, n$ be the direction cosines of the shortest distance $EF$.

Since $EF \perp$ to both $AB$ and $CD$.

$\therefore \quad ll_1+mm_1+nn_1=0$ and $ll_2+mm_2+nn_2=0.$

Solving,
$$\frac{l}{m_1n_2-m_2n_1}=\frac{m}{n_1l_2-n_2l_1}=\frac{n}{l_1m_2-l_2m_1}$$
$$=\frac{\sqrt{(l^2+m^2+n^2)}}{\sqrt{[\Sigma(m_1n_2-m_2n_1)^2]}}=\frac{1}{\sin\theta} \quad ...(1)$$

where $\theta$ is the angle between the lines $AB$ and $CD$.

$\therefore$ *Length of S.D.* $(EF)$ = projection of $AC$ on $EF$

$=l(x_2-x_1)+m(y_2-y_1)+n(z_2-z_1)$ where $l, m, n$ have the values as given by Eqn. (1).

To find the equations of the line of shortest distance, we observe that it is coplanar with both $AB$ and $CD$.

Plane containing the lines $AB$ and $EF$ is, $\begin{vmatrix} x - x_1 & y - y_1 & z - z_1 \\ l_1 & m_1 & n_1 \\ l & m & n \end{vmatrix} = 0$ ...(2)

Plane containing the lines $CD$ and $EF$ is $\begin{vmatrix} x - x_2 & y - y_2 & z - z_2 \\ l_2 & m_2 & n_2 \\ l & m & n \end{vmatrix} = 0$ ...(3)

Hence (2) and (3) are the equations of the line of shortest distance.

**Obs.** The condition for the given lines to be coplanar is also obtained by equating the shortest distance ($EF$) to zero.

**Example 3.46.** *Find the magnitude and the equations of the shortest distance between the lines*

$$\frac{x}{2} = \frac{y}{-3} = \frac{z}{1} \quad \textit{and} \quad \frac{x-2}{3} = \frac{y-1}{-5} = \frac{z+2}{2}.$$

(*Andhra.* 2022, 2016)

**Solution.** Let $l, m, n$ be the direction cosines of the shortest distance $EF$.

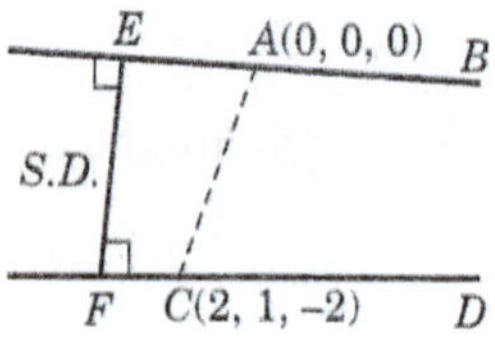

Fig. 3.45

$\because$ $EF \perp$ to both $AB$ and $CD$,

$\therefore$ $2l - 3m + n = 0,\ 3l - 5m + 2n = 0.$

Solving $$\frac{l}{1} = \frac{m}{1} = \frac{n}{1} = \frac{\sqrt{(l^2 + m^2 + n^2)}}{\sqrt{(1+1+1)}} = \frac{1}{\sqrt{3}}.$$

$\therefore$ *Length of S.D.* ($EF$) = projection of $AC$ on $EF$

$$= (2-0)\frac{1}{\sqrt{3}} + (1-0)\frac{1}{\sqrt{3}} + (-2-0)\frac{1}{\sqrt{3}} = \frac{1}{\sqrt{3}}.$$

The equations of the line of shortest distance ($EF$) are

$$\begin{vmatrix} x & y & z \\ 2 & -3 & 1 \\ 1 & 1 & 1 \end{vmatrix} = 0 \text{ and } \begin{vmatrix} x-2 & y-1 & z+2 \\ 3 & -5 & 2 \\ 1 & 1 & 1 \end{vmatrix} = 0$$

*i.e.,* $4x + y - 5z = 0$ and $7x + y - 8z = 31.$

**Example 3.47.** *Find the points on the lines*

$$\frac{x-6}{3} = \frac{y-7}{-1} = \frac{z-4}{1} \quad ...(i)$$

$$\frac{x}{-3} = \frac{y+9}{2} = \frac{z-2}{4} \quad ...(ii)$$

*which are nearest to each other. Hence find the shortest distance between the lines and its equations.*

(*Andhra,* 2019)

**Solution.** Any point on the line (*i*) is $E(6 + 3r, 7 - r, 4 + r)$ ...(*iii*)

and any point on the line (*ii*) is $F(-3r', -9 + 2r', 2 + 4r')$ ...(*iv*)

Then the direction cosines of $EF$ are proportional to $6 + 3r + 3r',\ 16 - r - 2r',\ 2 + r - 4r'$

Since $EF \perp$ both the lines (*i*) and (*ii*), $\therefore$ $3(6 + 3r + 3r') - (16 - r - 2r') + (2 + r - 4r') = 0$

and $-3(6 + 3r + 3r') + 2(16 - r - 2r') + 4(2 + r - 4r') = 0$

or $11r + 7r' + 4 = 0,\ 7r + 29r' - 22 = 0$, whence $r = -1,\ r' = 1.$

Substituting $r = -1$ in Eqn. (*iii*) and $r' = 1$ in Eqn. (*iv*), we get $E = (3, 8, 3)$ and $F = (-3, -7, 6)$ which are the points on Eqns. (*i*) and (*ii*) nearest to each other.

$\therefore$ Length of the shortest distance ($EF$) = $\sqrt{[(3+3)^2 + (8+7)^2 + (3-6)^2]} = 3\sqrt{30}$

The equations of the shortest distance ($EF$) is $\dfrac{x-3}{2} = \dfrac{y-8}{5} = \dfrac{z-3}{-1}.$

**Obs.** This method is sometimes very convenient and is especially useful when the points of intersection of the line of shortest distance with the given lines are required.

**Example 3.48.** *Two control cables in the form of straight lines AB and CD are laid such that the coordinates of A, B, C and D are respectively (1, 2, 3), (2, 1, 1), (–1, 1, 2) and (2, –1, –3). Determine the amount of clearance between the cables.*

**Solution.** The direction ratios of $AB$ are 1, –1, –2 and those of $CD$ are 3, –2, –5.

The amount of clearance between $AB$ and $CD$ is nothing but the shortest distance $PQ$ between the cables. If the direction cosines of $PQ$ be $l, m, n$ then

$$l - m - 2n = 0 \text{ and } 3l - 2m - 5n = 0$$

$$\therefore \quad \frac{l}{1} = \frac{m}{-1} = \frac{n}{1} \qquad [\because \; PQ \perp \text{ to both } AB + CD]$$

Thus the clearance between the cables

= shortest distance between $AB$ and $CD$

= projection of $AC$ (or $BD$) on $PQ$

$$= \frac{1(-1-1) - 1(1-2) + 1(2-3)}{\sqrt{(1+1+1)}} = \frac{2}{\sqrt{3}} \text{ (in magnitude)}$$

**Example 3.49.** *Find the equation of the plane through the line*

$$\frac{x-1}{3} = \frac{y-4}{2} = \frac{z-4}{-2} \qquad \ldots(i)$$

*and parallel to the line* $\frac{x+1}{2} = \frac{y-1}{-4} = \frac{z+2}{1}$ ...(*ii*)

*Hence find the shortest distance between them.*

**Solution.** The equation of the plane containing the line (*i*) and parallel to (*ii*) is

$$\begin{vmatrix} x-1 & y-4 & z-4 \\ 3 & 2 & -2 \\ 2 & -4 & 1 \end{vmatrix} = 0$$

or

$$6x + 7y + 16z = 98 \qquad \ldots(iii)$$

Now the shortest distance between the lines Eqns. (*i*) and (*ii*)

= Length of the perpendicular drawn from the point (–1, 1, –2) of Eqn. (*ii*) on the plane Eqn. (*iii*)

$$= \frac{-6 + 7 - 32 - 98}{\sqrt{(6^2 + 7^2 + 16^2)}} = \frac{120}{\sqrt{(341)}}, \text{ numerically.}$$

**Example 3.50.** *Show that the shortest distance between z-axis and the lines* $ax + by + cz + d = 0 = a'x + b'y + c'z + d'$ *is* $\frac{dc' - d'c}{\sqrt{[(ac' - a'c)^2 + (bc' - b'c)^2]}}$.

**Solution.** The plane containing the given line is

$$(ax + by + cz + d) + k(a'x + b'y + c'z + d') = 0 \qquad \ldots(i)$$

or

$$(a + ka')x + (b + kb')y + (c + kc')z + (d + kd') = 0$$

This plane is parallel to the $z$-axis ($d.\ c's$, 0, 0, 1) if $c + kc' = 0$ or $k = -c/c'$. Then Eqn. (*i*) becomes

$$(ac' - a'c)x + (bc' - b'c)y + (dc' - d'c) = 0 \qquad \ldots(ii)$$

A point on the $z$-axis is the origin.

$\therefore$ $\perp$ distance of the origin from the plane (*ii*)

$$= \frac{dc' - d'c}{\sqrt{[(ac' - a'c)^2 + (bc' - b'c)^2]}} \text{ which is the required S.D.}$$

**Example 3.51.** *A square ABCD of diagonal 2a is folded along the diagonal AC, so that the planes DAC and BAC are at right angles. Find the shortest distance between DC and AB.*

**Solution.** Let the diagonals $AC$ and $BD$ intersect at $O$ the folded position of the square. Let $OB$, $OC$ and $OD$ be the axes. Then equations of $DC$ are

$$\frac{x-0}{0-0}=\frac{y-a}{a-0}=\frac{z-0}{0-a} \quad \text{or} \quad \frac{x}{0}=\frac{y-a}{a}=\frac{z}{-a}$$

and those of $AB$ are $\frac{x-a}{a}=\frac{y}{a}=\frac{z}{0}$

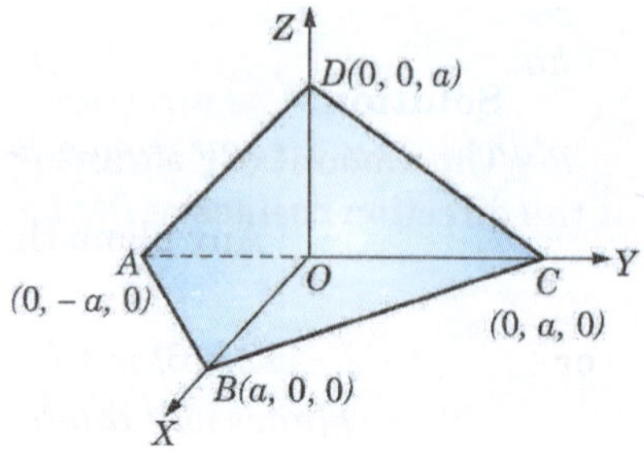

Fig. 3.46

The equation of the plane through $DC$ and parallel to $AB$ is

$$\begin{vmatrix} x & y-a & z \\ 0 & a & -a \\ a & a & 0 \end{vmatrix}=0 \quad \text{or} \quad x-y-z+a=0 \qquad ...(i)$$

A point on the line $AB$ is $(a, 0, 0)$.

Hence required S.D. = $\perp$ distance of $(a, 0, 0)$ from the plane Eqn. $(i)$

$$=\frac{a+a}{\sqrt{1+1+1}}=\frac{2a}{\sqrt{3}}.$$

## PROBLEMS 3.10

1. Find the shortest distance between the lines $\frac{x-8}{3}=\frac{y+9}{-16}=\frac{z-10}{7}$ and $\frac{x-15}{3}=\frac{y-29}{8}=\frac{z-5}{5}$. *(V.T.U., 2016)*
2. Find the magnitude and the equations of the shortest distance between the lines.
$$\frac{x-1}{2}=\frac{y-2}{3}=\frac{z-3}{4} \text{ and } \frac{x-2}{3}=\frac{y-4}{4}=\frac{z-5}{5}.$$
*(Andhra., 2016)*
3. Find the magnitude and equations of the shortest distance between the lines
$$\frac{x-3}{1}=\frac{y-5}{-2}=\frac{z-7}{1} \text{ and } \frac{x+1}{7}=\frac{y+1}{-6}=\frac{z+1}{1}$$ *(Andhra., 2014)*
Find also the points where it intersects the lines.
4. Find the shortest distance and the equation of the line of shortest distance between the line $\frac{x-1}{2}=\frac{y-2}{3}=\frac{z-3}{4}$ and the $y$-axis. *(Andhra, 2021; V.T.U., 2010)*
5. Show that the shortest distance between the lines $y-mx=0=z-c$ and $y+mx=0=z+c$ is $c$ units.
6. If the shortest distance between the lines $\frac{y}{b}+\frac{z}{c}=1, x=0$ and $\frac{x}{a}-\frac{z}{c}=1, y=0$ be $2d$, then show that $d^{-2}=a^{-2}+b^{-2}+c^{-2}$.
7. Show that the shortest distance between $x$-axis and the lines $ax+by+cz+d=0=a'x+b'y+c'z+d'$ is
$$\frac{da'-d'a}{\sqrt{[(ba'-b'a)^2+(ca'-c'a)^2]}}.$$
8. Show that the shortest distance between a diagonal of a rectangular parallelopiped whose edges are $a, b, c$ and the edges not meeting it, are
$bc/(b^2+c^2)^{1/2}$, $ca/(c^2+a^2)^{1/2}$, $ab/(a^2+b^2)^{1/2}$.
9. Show that the shortest distance between two opposite edges of the tetrahedron formed by the planes $x+y=0$, $y+z=0$, $z+x=0$ and $x+y+z=a$ is $2a/\sqrt{6}$.

## 3.16 INTERSECTION OF THREE PLANES

Any three planes (no two of which are parallel) intersect in one of the following ways :

(1) *The planes may meet in a point,* if the line of section of two of them is not parallel to the third.

(2) *The planes may have a common line of section,* if the line of section of two of them lies on the third (Fig. 3.47).

(3) *The planes may form a triangular prism,* if the line of section of two of them is parallel to the third but does not lie on it. (See Fig. 3.48)

**Example 3.52.** *Prove that the planes*
*(i)* $12x - 15y + 16z - 28 = 0$, *(ii)* $6x + 6y - 7z - 8 = 0$. *and (iii)* $2x + 35y - 39z + 92 = 0$,
*have a common line of intersection. Prove that the point in which the line* $\frac{x-1}{3} = \frac{y}{-2} = \frac{z-3}{1}$ *meets the third plane is equidistant from other two planes.*

**Solution.** Any plane through the line of intersection of the planes (*i*) and (*ii*) is

$$12x - 15y + 16z - 28 + \lambda(6x + 6y - 7z - 8) = 0$$

or
$$(12 + 6\lambda)x + (-15 + 6\lambda)y + (16 - 7\lambda)z - (28 + 8\lambda) = 0 \quad ...(iv)$$

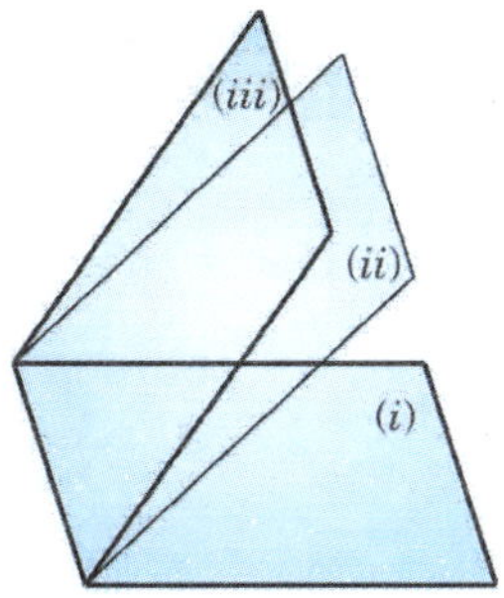

Fig. 3.47

*Three planes will intersect in a common line if the planes (iii) and (iv) represent the same plane.*

$$\therefore \quad \frac{12+6\lambda}{2} = \frac{-15+6\lambda}{35} = \frac{16-7\lambda}{-39} = \frac{-28-8\lambda}{12} \quad ...(v)$$

From $\frac{12+6\lambda}{2} = \frac{-15+6\lambda}{35}$, we have $\lambda = \frac{-25}{11}$ which satisfies all the equations (*v*).

Hence the given planes intersect in a line.

Any point on the line $\frac{x-1}{3} = \frac{y}{-2} = \frac{z-3}{1} = r$ (say) ...(*vi*)

is $(3r + 1, -2r, r + 3)$ which lies in the plane (*iii*)

if $2(3r + 1) + 35(-2r) - 39(r + 3) + 12 = 0$, *i.e.* if $r = -1$.

$\therefore$ The coordinates of the point $P$ in which (*vi*) meets (*iii*) are $(-2, 2, 2)$.

*Distance of P from plane (i)* $= \frac{12(-2) - 15(2) + 16(2) - 28}{\sqrt{144 + 225 + 256}} = \frac{-50}{\sqrt{(625)}} = 2$ (in magnitude)

*Distance of P from plane (ii)* $= \frac{6(-2) + 6(2) - 7(2) - 8}{\sqrt{36 + 36 + 49}} = 2$ (in magnitude)

Hence the point $P$ is equidistant from the planes (*i*) and (*ii*).

**Example 3.53.** *Prove that the three planes*
*(i)* $2x + y + z = 3$, *(ii)* $x - y + 2z = 4$, *(iii)* $x + z = 2$,
*form a triangular prism and find the area of the normal section of the prism.*

**Solution.** Let $l, m, n$ be the direction cosines of the line of intersection of the planes (*ii*) and (*iii*) so that $l - m + 2n = 0, l + n = 0$,

whence
$$\frac{l}{1} = \frac{m}{-1} = \frac{n}{-1}.$$

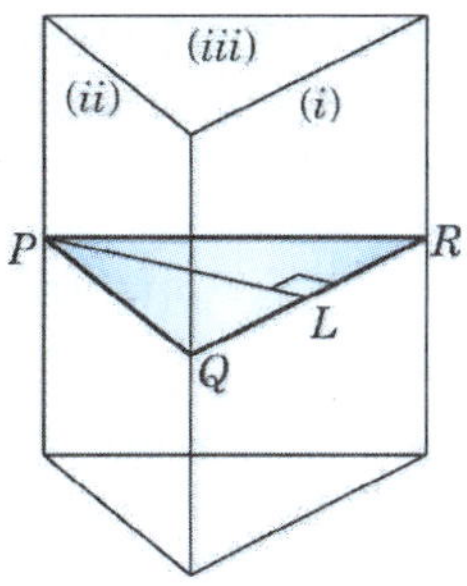

Fig. 3.48

To find a point $P$ on this line, put $x = 0$ in planes (*ii*) and (*iii*), $-y + 2z = 4$ and $z = 2$. Thus the point $P$ is $(0, 0, 2)$.

Now the line of intersection of planes (*ii*) and (*iii*) is parallel to the plane (*i*).

$[\because \quad 2 \times 1 + 1 \times (-1) + 1 \times (-1) = 0]$

Also the point $P$ does not lie on the plane (*i*).

Hence the given planes form a triangular prism.

Let $\Delta PQR$ be its normal section through $P$.

The equation of the plane through $P$ perpendicular to the line of intersection of the planes (*i*) and (*iii*) is,

$$1(x - 0) - 1(y - 0) - 1(z - 2) = 0$$

or
$$x - y - z + 2 = 0 \quad ...(iv)$$

Solving the Eqns. (*i*), (*ii*) and (*iv*), we get

$$Q \equiv \left(\frac{1}{3}, \frac{1}{3}, 2\right).$$

Solving the Eqns. (*i*), (*iii*) and (*iv*), we get

$$R \equiv \left(\frac{1}{3}, \frac{2}{3}, \frac{5}{3}\right).$$

$$\therefore \qquad QR = \sqrt{\left\{\left(\frac{1}{3}-\frac{1}{3}\right)^2 + \left(\frac{1}{3}-\frac{2}{3}\right)^2 + \left(2-\frac{5}{3}\right)^2\right\}} = \sqrt{\left(\frac{2}{9}\right)}$$

Also $PL \perp$ from $P$ on the plane (*i*) $= \dfrac{3-2}{\sqrt{(4+1+1)}} = \dfrac{1}{\sqrt{6}}$.

Hence the area of $\Delta PQR = \dfrac{1}{2} QR \times PL = \dfrac{1}{2} \cdot \dfrac{\sqrt{2}}{3} \cdot \dfrac{1}{\sqrt{6}} = \dfrac{1}{6\sqrt{3}}$.

## PROBLEMS 3.11

1. Prove that the three planes $2x - 3y - 7z = 0$, $3x - 14y - 13z = 0$, $8x - 31y - 33z = 0$ pass through one line.
2. Prove that the planes $x = cy + bz$, $y = az + cx$, $z = bx + ay$ intersect in a line if $a^2 + b^2 + c^2 + 2abc = 1$ and show that the equations of this line are
$$\frac{x}{\sqrt{1-a^2}} = \frac{y}{\sqrt{1-b^2}} = \frac{z}{\sqrt{1-c^2}}$$
3. Show that the planes $x + 2y - 3 = 0$, $3x - 4y + z - 4 = 0$ and $4x + 3y - 2z - 24 = 0$ from a triangular prism.
4. Prove that the planes $2x + 3y + 4z = 6$, $3x + 4y + 5z = 20$, $x + 2y + 3z = 0$ form a prism : obtain the equation of one of its edges in the symmetrical form.

## 3.17 SPHERE

**(1) Def.** *A **sphere** is the locus of a point which remains at a constant distance from a fixed point.*
The fixed point is called the *centre* and the constant distance the *radius* of the sphere

**(2) The equation of the sphere whose centre is (a, b, c) and radius r, is**

$$\mathbf{(x - a)^2 + (y - b)^2 + (z - c)^2 = r^2.}$$

For the distance of any point $P(x, y, z)$ on the sphere from the centre $C(a, b, c)$ = the radius $r$.

In particular the *equation of the sphere whose centre is the origin and radius a, is*

$$\mathbf{x^2 + y^2 + z^2 = a^2}$$

**(3) The equation $x^2 + y^2 + z^2 + 2ux + 2vy + 2wz + d = 0$ represents a sphere whose centre is $(-u, -v, -w)$ and radius**

$$= \sqrt{\mathbf{u^2 + v^2 + w^2 - d}}.$$

For on writing it as $(x^2 + 2ux) + (y^2 + 2vy) + (z^2 + 2wz) = -d$

or as $\qquad (x + u)^2 + (y + v)^2 + (z + w)^2 = u^2 + v^2 + w^2 - d$

and comparing with $\qquad (x - a)^2 + (y - b)^2 + (z - c)^2 = r^2,$

it clearly represents a sphere whose centre is

$$(a, b, c) = (-u, -v, -w) \text{ and radius } r = \sqrt{(u^2 + v^2 + w^2 - d)}$$

Thus *the general equation of a sphere is such that*

(*i*) *it is the second degree in x, y, z,*

(*ii*) *the coefficient of $x^2, y^2, z^2$ are equal,*

and (*iii*) *there are no terms containing yz, zx or xy*

**(4) Section of a sphere by a plane** *is a circle and the section of a sphere by a plane through its centre is called a* **great circle.**

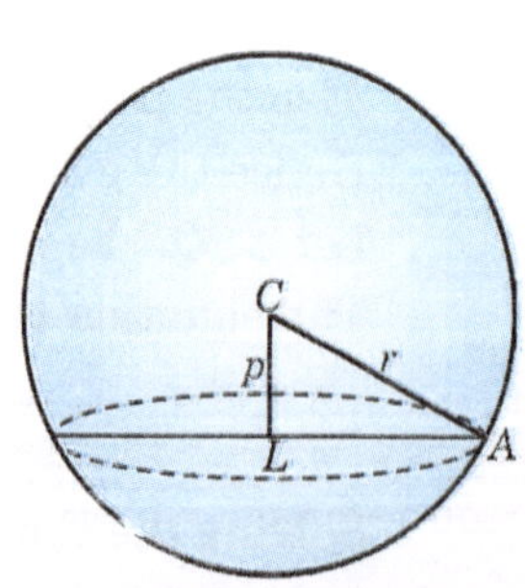

Fig. 3.49

Thus the equations $\qquad x^2 + y^2 + z^2 + 2ux\ 2vy + 2wz + d = 0 \quad$ [Sphere]

and $\qquad Ax + By + Cz + D = 0 \quad$ [Plane]

taken together represent a circle (Fig. 3.49) having centre $L$ and radius $LA = \sqrt{(r^2 - p^2)}$.

**(5) The equation of any sphere through the circle** *of intersection of*

*the sphere* $S = 0$

*and the plane* $U = 0$

*is* $S + kU = 0$

For the equation $S + kU = 0$

represents a sphere and the points of intersection of the sphere $S = 0$ and the plane $U = 0$ satisfy it.

**Example 3.54.** *Find the equation of the sphere through the points (0, 0, 0), (0, 1, –1), (–1, 2, 0) and (1, 2, 3). Locate its centre and find the radius.*

**Solution.** Let the required equation of the sphere be

$$x^2 + y^2 + z^2 + 2ux + 2vy + 2wz + d = 0 \qquad ...(i)$$

It passes through (0, 0, 0), (0, 1, –1), (–1, 2, 0) and (1, 2, 3).

$\therefore$ $d = 0,$

$$1 + 1 + 2v - 2w + d = 0 \quad \text{or} \quad v - w + 1 = 0 \qquad ...(ii)$$

$$1 + 4 - 2u + 4v + d = 0 \quad \text{or} \quad -2u + 4v + 5 = 0 \qquad ...(iii)$$

$$1 + 4 + 9 + 2u + 4v + 6w + d = 0 \quad \text{or} \quad u + 2v + 3w + 7 = 0 \qquad ...(iv)$$

Multiplying Eqns. (*ii*) by (*iii*) and adding to Eqn. (*iv*), we get

$$u + 5v + 10 = 0 \qquad ...(v)$$

Solving Eqns. (*iii*) and (*v*), we get $u = -\dfrac{15}{14},\ v = -\dfrac{25}{14}$

From Eqn. (*ii*), $w = v + 1 = \dfrac{-25}{14} + 1 = \dfrac{-11}{14}$

Substituting these values of $u, v, w, d$ in Eqn. (*i*), we get

$$x^2 + y^2 + z^2 - \frac{15}{7}x - \frac{25}{7}y - \frac{11}{7}z = 0 \qquad ...(vi)$$

which is the required equation of the sphere.

Its centre is (15/14, 25/14, 11/14) $[(-u, -v, -w)]$

and the radius = $[(-15/14)^2] + (-25/14)^2 + (-11/14)^2 - 0] = \sqrt{971/14}$.

**Example 3.55.** (*a*) *Find the equation of the sphere which has $(x_1, y_1, z_1)$ and $(x_2, y_2, z_2)$ as the extremities of a diameter.*

(*b*) *Deduce the equation of the sphere described on the line joining the points (2, –1, 4) and (–2, 2, –2) as diameter. Find the area of the circle in which the sphere is intersected by the plane $2x + y - z = 3$.*

**Solution.** (*a*) Let $P(x, y, z)$ be any point on the sphere having $A(x_1, y_1, z_1)$ and $B(x_2, y_2, z_2)$ as ends of diameter (Fig. 3.50. (*a*)). Then $AP$ and $BP$ are at right angles.

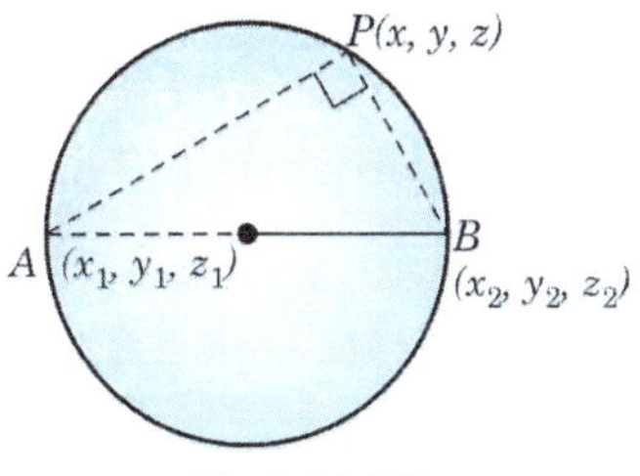

Fig. 3.50 (*a*)

Now direction ratio of $AP$ are $x - x_1, y - y_1, z - z_1$ and those of $BP$ are $x - x_2, y - y_2, z - z_2$.

Hence

$$(\mathbf{x} - \mathbf{x_1})\,(\mathbf{x} - \mathbf{x_2}) + (\mathbf{y} - \mathbf{y_1})\,(\mathbf{y} - \mathbf{y_2}) + (\mathbf{z} - \mathbf{z_1})\,(\mathbf{z} - \mathbf{z_2}) = 0$$

which is the required equation.

(*b*) The equation of the required sphere is

$$(x - 2)(x + 2) + (y + 1)(y - 2) + (z - 4)(z + 2) = 0$$

or $$x^2 + y^2 + z^2 - y - 2z - 14 = 0 \qquad ...(i)$$

Its centre is $C(0, 1/2, 1)$

and radius $(r) = \sqrt{(0, 1/4 + 1 + 14)} = \sqrt{(61/4)}$.

Let the given plane $2x + y - z - 3 = 0$ $\qquad ...(ii)$

cut the sphere (1) in the circle $PP'$ having centre $L$.

$\therefore$ $p$ = perpendicular $CL$ from $C$ on the plane (2)

$$= \frac{1/2 - 1 - 3}{\sqrt{4+1+1}} = \frac{7}{2\sqrt{6}} \text{ (in magnitude)}$$

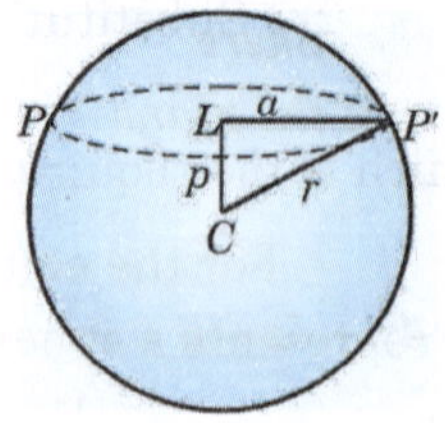


Fig. 3.50 (*b*)

If $a$ be the radius of the circle $PP'$, then

$$a^2 = r^2 - p^2 = \frac{61}{4} - \frac{49}{24} = \frac{317}{24}$$

Hence the area of circle $PP' = \pi a^2 = \frac{317}{24}\pi$.

**Example 3.56.** *A plane passes through a fixed point (a, b, c) and cuts the axes in A, B, C. Show that the locus of the centre of the sphere OABC is*

$$\frac{a}{x} + \frac{b}{y} + \frac{c}{z} = 2.$$

**Solution.** Let the centre of the sphere $OABC$ be $P(f, g, h)$ so that its radius $OP = \sqrt{(f^2 + g^2 + h^2)}$.

$\therefore$ The equation of the sphere is

$$(x - f)^2 + (y - g)^2 + (z - h)^2 = f^2 + g^2 + h^2$$

or $$x^2 + y^2 + z^2 - 2fx - 2gy - 2hz = 0 \quad ...(i)$$

To find $OA$, putting $y = 0, z = 0$ in Eqn. (*i*), we have

$$x^2 - 2fx = 0, \text{ i.e., } OA = x = 2f. \text{ Similarly, } OB = 2g, OC = 2h.$$

Thus the equation of the plane $ABC$ is $\frac{x}{2f} + \frac{y}{2g} + \frac{z}{2h} = 1$

Since the plane passes through $(a, b, c)$ $\therefore$ $\frac{a}{2f} + \frac{b}{2g} + \frac{c}{2h} = 1.$

Hence the locus of the centre $(f, g, h)$ of the sphere is,

$$\frac{a}{2x} + \frac{b}{2y} + \frac{c}{2z} = 1 \quad \text{or} \quad \frac{a}{x} + \frac{b}{y} + \frac{c}{z} = 2.$$

**Example 3.57.** *Find the equation of the sphere having the circle*

$$x^2 + y^2 + z^2 + 10y - 4z - 8 = 0, x + y + z = 3$$

*as a great circle.* (*Andhra., 2016 ; Anna, 2012*)

**Solution.** The equation of any sphere through the given circle is

$$x^2 + y^2 + z^2 + 10y - 4z - 8 + k(x + y + z - 3) = 0$$

*i.e.,* $$x^2 + y^2 + z^2 + kx + (10 + k)y - (4 - k)z - (8 + 3k) = 0 \quad ...(i)$$

In order that (*i*) may have the given circle as its great circle, its centre $[-k/2, -(10 + k)/2, (4 - k)/2]$ must lie on the plane $x + y + z = 3$

$$\therefore \quad -\frac{k}{2} - \frac{10+k}{2} + \frac{4-k}{2} = 3, \text{ i.e., } k = -4$$

whence (*i*) becomes, $x^2 + y^2 + z^2 - 4x + 6y - 8z + 4 = 0$ which is the required equation.

**Example 3.58.** *Find the equation of the smallest sphere which contains the circle $x^2 + y^2 + z^2 + 2x + 6y + 4z - 11 = 0$ and $2x + 2y + z + 1 = 0$.*

**Solution.** Equation of any sphere containing the given circle is

$$x^2 + y^2 + z^2 + 2x + 6y + 4z - 11 + \lambda (2x + 2y + z + 1) = 0$$

or $$x^2 + y^2 + z^2 + (2 + 2\lambda) x + (6 + 2\lambda) y + (4 + \lambda) z - 11 + \lambda = 0 \quad ...(i)$$

Its radius $r$ is given by

$$r^2 = (1 + \lambda)^2 + (3 + \lambda)^2 + (2 + \frac{1}{2}\lambda)^2 - (\lambda - 11) = \frac{9}{4}\left[\lambda^2 + 4\lambda + \frac{100}{9}\right] = \frac{9}{4}\left[(\lambda + 2)^2 + \frac{64}{9}\right]$$

Now $r^2$ has the least value when $\lambda = -2$.

$\therefore$ Substituting $\lambda = -2$ in Eqn. (*i*), we get

$$x^2 + y^2 + z^2 - 2x + 2y + 2z - 13 = 0$$

which is the required smallest sphere.

**Example 3.59.** *Prove that the circles $x^2 + y^2 + z^2 - 2x + 3y + 4z - 5 = 0$, $5y + 6z + 1 = 0$ and $x^2 + y^2 + z^2 - 3x - 4y + 5z - 6 = 0$, $x + 2y - 7z = 0$ lie on the same sphere and find its equation.* *(Andhra, 2022)*

**Solution.** Equation of any sphere containing the first circle is

$$x^2 + y^2 + z^2 - 2x + 3y + 4z - 5 + \lambda (5y + 6z + 1) = 0$$

or

$$x^2 + y^2 + z^2 - 2x + (3 + 5\lambda)y + (4 + 6\lambda)z - 5 + \lambda = 0 \quad ...(i)$$

Similarly equation of any sphere containing the second given circle is

$$x^2 + y^2 + z^2 - 3x - 4y + 5z - 6 + \lambda' (x + 2y - 7z) = 0$$

or

$$x^2 + y^2 + z^2 + (-3 + \lambda')x + (-4 + 2\lambda')y + (5 - 7\lambda') z - 6 = 0 \quad ...(ii)$$

Eqns. (*i*) and (*ii*) will represent the same sphere when

$$-2 = -3 + \lambda' \quad ...(iii); \qquad 3 + 5\lambda = -4 + 2\lambda' \quad ...(iv)$$

$$4 + 6\lambda = 5 - 7\lambda' \quad ...(v); \qquad -5 + \lambda = -6 \quad ...(vi)$$

Now Eqn. (*iii*) gives $\lambda' = 1$ and Eqn. (*vi*) gives $\lambda = -1$.

Clearly $\lambda = -1$ and $\lambda' = 1$ also satisfy Eqns. (*iv*) and (*v*). This shows that the given circles lie on the same sphere.

Substituting $\lambda = -1$ in Eqn. (*i*) or $\lambda' = 1$ in Eqn. (*ii*), we get

$$x^2 + y^2 + z^2 - 2x - 2y - 2z - 6 = 0$$

which is the desired sphere.

## PROBLEMS 3.12

1. Find the equation of the sphere through the points (2, 0, 1), (1, –5, –1), (0, –2, 3) and (4, –1, 2). Also find its centre and radius. *(Anna, 2013)*
2. Find the equation of the sphere whose diameter is the line joining the origin to the point (2, –2, 4). Also find its centre and radius. *(Andhra., 2016)*
3. Obtain the equation of the sphere which passes through the points (1, 0, 0), (0, 1, 0), (0, 0, 1) and
   (*a*) has its centre on the plane $x + y + z = 6$.
   (*b*) has its radius as small as possible.
4. A sphere of constant radius $k$ passes through the origin and meets the axes in $A, B, C$. Prove that the centroid of the
   (*i*) triangle $ABC$ lies on the sphere $9(x^2 + y^2 + z^2) = 4k^2$.
   (*ii*) tetrahedron $OABC$ lies on the sphere $x^2 + y^2 + z^2 = k^2/4$.
5. A plane passes through a fixed points $(a, b, c)$, show that the locus of the foot of the perpendicular from the origin on the plane is the sphere $x^2 + y^2 + z^2 - ax - by - cz = 0$.
6. A sphere of constant radius $r$ passes through the origin $O$ and cuts the axes in $A, B, C$. Prove that the locus of the foot of the perpendicular from $O$ on the plane $ABC$ is given by
   $$(x^2 + y^2 + z^2)^2 (x^{-2} + y^{-2} + z^{-2}) = 4r^2.$$
7. A plane cuts the coordinate axes at $A, B, C$. If $OA = a$, $OB = b$, $OC = c$, find the equation of the
   (*i*) circumsphere of the tetrahedron $OABC$,
   (*ii*) circum-circle of the triangle $ABC$. Also obtain the coordinates of its centre.
8. Find the centre and radius of the circle $x^2 + y^2 + z^2 - 2y - 4z = 11$, $x + 2y + 2z = 15$.
9. Show that the points (2, –6, 0), (4, –9, 6), (5, 0, 2), (7, –3, 8) are concyclic.
10. Find the equation of the sphere for which the circle $x^2 + y^2 + z^2 - 3x + 4y - 2z - 5 = 0$, $5x - 2y + 4z + 7 = 0$ is a great circle.
11. Find the equation of the sphere having its centre on the plane $4x - 5y - z = 3$ and passing through the circle $x^2 + y^2 + z^2 - 2x - 3y + 4z + 8 = 0$, $x - 2y + z = 8$. *(Andhra., 2016)*
12. Prove that the plane $x + 2y - z = 4$ cuts the sphere $x^2 + y^2 + z^2 - x - z - 2 = 0$ in a circle of radius unity. Find also the equation of the sphere which has this circle as one of its great circles.
13. Find the equation of the sphere passing through the circle $x^2 + y^2 + z^2 + 2x + 3y + 6 = 0$, $x - 2y + 4z = 9$ and the centre of the sphere $x^2 + y^2 + z^2 - 2x + 4y - 6z + 5 = 0$.

## 3.18 EQUATION OF THE TANGENT PLANE

*The equation of the tangent plane at any point* $(x_1, y_1, z_1)$ *of the sphere*

$$x^2 + y^2 + z^2 = a^2 \text{ is } \mathbf{xx_1 + yy_1 + zz_1 = a^2}.$$

If $P(x, y, z)$ be any point on the tangent plane at $P_1(x_1, y_1, z_1)$ to the given sphere, the direction ratios of $P_1P$ are $x - x_1, y - y_1, z - z_1$. Also the direction ratios of radius $OP_1$ are $x_1 - 0, y_1 - 0, z_1 - 0$.

Since $OP_1$ is normal to the tangent plane at $P_1$, $OP_1 \perp P_1P$.

$\therefore$
$$x_1(x - x_1) + y_1(y - y_1) + z_1(z - z_1) = 0$$

or
$$xx_1 + yy_1 + zz_1 = x_1^2 + y_1^2 + z_1^2 = a^2 \qquad [\because P_1(x_1, y_1, z_1) \text{ lies on the sphere.}$$

This is the desired equation of the tangent plane.

Similarly, *the tangent plane at* $(x_1, y_1, z_1)$ *to the sphere*

$$x^2 + y^2 + z^2 + 2ux + 2vy + 2wz + d = 0$$

is
$$\mathbf{xx_1 + yy_1 + zz_1 + u(x + x_1) + v(y + y_1) + w(z + z_1) + d = 0}$$

*Thus to write the equation of the tangent plane at* $(x_1, y_1, z_1)$ *to a sphere, change* $x^2$ *to* $xx_1$, $y^2$ to $yy_1$, $z^2$ *to* $zz_1$, $2x$ *to* $x + x_1$, $2y$ *to* $y + y_1$, $2z$ *to* $z + z_1$.

**Obs.** *The condition for a plane (or a line) to touch a sphere is that the perpendicular distance of the centre from the plane (or the line) = the radius.*

**Example 3.60.** *Find the equations of the spheres passing through the circle* $x^2 + y^2 + z^2 - 6x - 2z + 5 = 0$, $y = 0$ *and touching the plane* $3y + 4z + 5 = 0$.

**Solution.** The equation of any sphere through the given circle is

$$x^2 + y^2 + z^2 - 6x - 2z + 5 + ky = 0$$

or
$$x^2 + y^2 + z^2 - 6x + ky - 2z + 5 = 0 \qquad ...(i)$$

$\therefore$ Its centre = $(3, -k/2, 1)$ and radius = $\sqrt{[9 + (k^2/4) + 1 - 5]} = \sqrt{(5 + k^2/4)}$.

The sphere $(i)$ will touch the plane $3y + 4z + 5 = 0$, if $\perp$ distance of the centre $(3, -k/2, 1)$ from the plane = radius.

*i.e.,*
$$\frac{3(-k/2) + 4 + 5}{\sqrt{(9 + 16)}} = \sqrt{\left(5 + \frac{k^2}{4}\right)} \quad \text{or if,} \quad 4k^2 + 27k + 44 = 0$$

$\therefore$
$$k = \frac{-27 \pm \sqrt{[(27)^2 - 704]}}{8} = -\frac{11}{2} \text{ or } -4$$

Substituting the value of $k$ in (1), we get

$$x^2 + y^2 + z^2 - 6x - \frac{11}{4}y + 2z + 5 = 0 \text{ and } x^2 + y^2 + z^2 - 6x - 4y - 2z + 5 = 0$$

as the two required spheres.

**Example 3.61.** *Find the equation of the sphere which touches the plane* $x - 2y - 2z = 7$ *at the point* $L(3, -1, -1)$ *and passes through the point* $M(1, 1, -3)$.

**Solution.** If $C$ is the centre of the sphere, then $CL$ is perpendicular to the given plane $x - 2y - 2z = 7$.

$\therefore$ The direction ratios of $CL$ being $1, -2, -2$, the equation of $CL$ is (Fig. 3.51)

$$\frac{x - 3}{1} = \frac{y + 1}{-2} = \frac{z + 1}{-2} = k \text{ (say)}$$

Any point on $CL$ is $(k + 3, -2k - 1, -2k - 1)$ which will represent $C$ for some value of $k$.

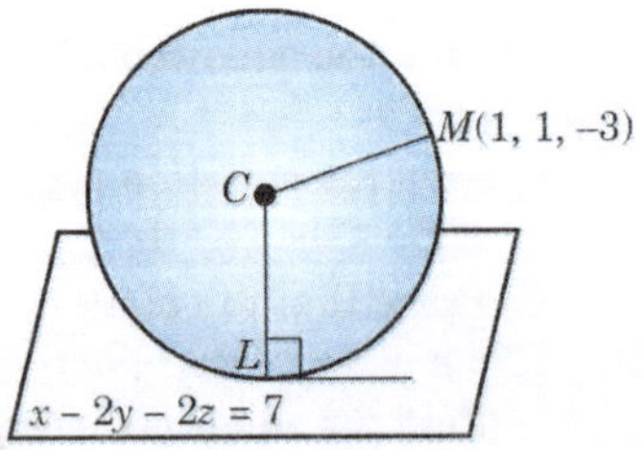

Fig. 3.51

Since $M$ lies on the sphere, therefore its radius $CL = CM$ or $(CL)^2 = (CM)^2$

*i.e.*
$$(k + 3 - 3)^2 + (-2k - 1 + 1)^2 + (-2k - 1 + 1)^2$$
$$= (k + 3 - 1)^2 + (-2k - 1 - 1)^2 + (-2k - 1 + 3)^2$$

or
$$4k = -12 \quad \text{or} \quad k = -3.$$

$\therefore$ The centre $C$ is $(0, 5, 5)$ and radius $CL = \sqrt{(9 + 36 + 36)} = 9$.

Hence the required equation of the sphere is

$$(x - 0)^2 + (y - 5)^2 + (2 - 5)^2 = (9)^2$$

or $$x^2 + y^2 + z^2 - 10y - 10z - 31 = 0$$

[**Orthogonal spheres.** Two spheres are said to cut orthogonally if the tangent planes at a point of intersection are at right angles (Fig. 3.52).

The radii of such spheres through their point of intersection $P$, being $\perp$ to the tangent planes at $P$ are also at right angles. *Thus two spheres cut orthogonally, if the square of the distance between their centre = sum of the squares of their radii*].

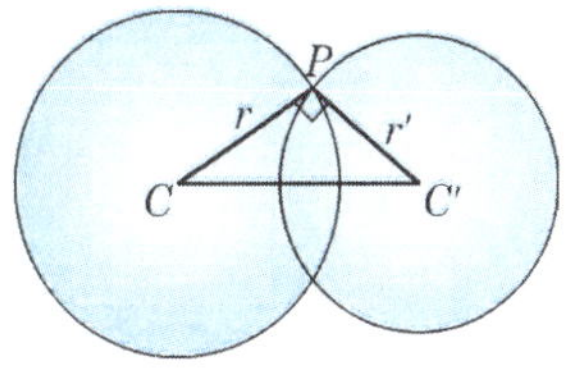

Fig. 3.52

**Example 3.62.** *Show that the condition for spheres*

$$x^2 + y^2 + z^2 + 2ux + 2vy + 2wz + d = 0$$

*and* $$x^2 + y^2 + z^2 + 2u'x + 2v'y + 2w'z + d' = 0$$

*to cut orthogonally is* $\mathbf{2uu' + 2vv' + 2ww' = d + d'}$

**Solution.** The centres of the spheres are

$C(-u, -v, -w)$, $C'(-u', -v', -w')$ and their radii are

$$r = \sqrt{(u^2 + v^2 + w^2 - d)},$$

$$r' = \sqrt{(u'^2 + v'^2 + w'^2 - d')}.$$

Now these spheres will cut orthogonally, if $(CC')^2 = r^2 + r'^2$

*i.e.,* $$(u - u')^2 + (v - v')^2 + (w - w')^2$$
$$= u^2 + v^2 + w^2 - d + u'^2 + v'^2 + w'^2 - d'$$

or $2uu' + 2vv' + 2ww' = d + d'$ which is the required condition.

**Example 3.63.** *Find the equation of the sphere which touches the plane $3x + 2y - z + 2 = 0$ at the point $(1, -2, 1)$ and cuts the sphere $R^2 - 2(2i - 3j) . R + 4 = 0$ orthogonally.* (*Andhra, 2019*)

**Solution.** The given plane $3x + 2y - z + 2 = 0$ ...(*i*)

will touch the required sphere at $A$ (1, – 2, 1) if its centre lies on the normal to (*i*) at $A$ (Fig. 3.53). The equations of the normal to (*i*) at $A$ are $\dfrac{x-1}{3} = \dfrac{y+2}{2} = \dfrac{z-1}{-1}$

Any point on this line is $C\,(3r + 1, 2r - 2, \pi r + 1)$

Also radius $(AC)$ of the required sphere.

$$= \sqrt{[(3r)^2 + (2r)^2 + (-r)^2]} = r\sqrt{14}.$$

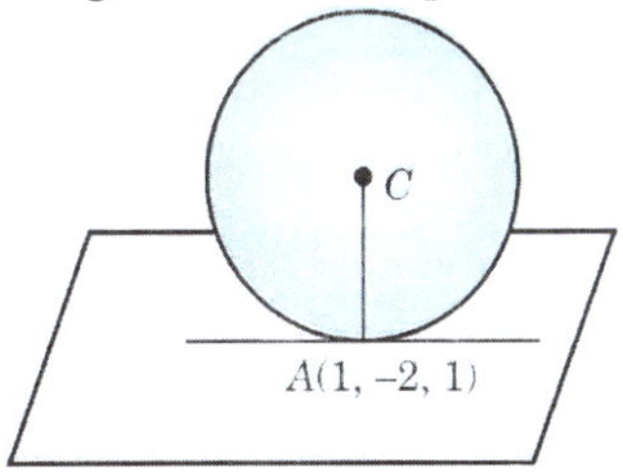

Fig. 3.53

Since the required sphere cuts the given sphere

$x^2 + y^2 + z^2 - 4x + 6y + 4 = 0$ [Centre = (2, –3, 0) and + radius = 3

orthogonally, therefore (distance between their centres)$^2$ = $\Sigma$ of squares of their radii

*i.e.,* $$(3r + 1 - 2)^2 + (2r - 2 + 3)^2 + (-r + 1)^2 = 14r^2 + 9 \text{ or } r = -3/2.$$

Thus centre $C$ is $(-7/2, -5, 5/2)$ and radius $= \dfrac{3\sqrt{14}}{2}$.

Hence the required sphere is

$$(x + 7/2)^2 + (y + 5)^2 + (z - 5/2)^2 = (3\sqrt{14/2})^2$$

or $$x^2 + y^2 + z^2 + 7x + 10y - 5z + 12 = 0.$$

P.S.: $R^2 - 2(2i - 3j) . R + 4 = 0$ is same as the sphere $x^2 + y^2 + z^2 - 4x + 6y + 4 = 0$

## PROBLEMS 3.13

1. Find the equations of the tangent planes to the sphere
   (*i*) $x^2 + y^2 + z^2 - 4x - 2y + 6z + 5 = 0$ which are parallel to the plane $x + 4y + 8z = 0$. Also find their points of contact. (*Anna, 2013*)
   (*ii*) $x^2 + y^2 + z^2 = 9$ which pass through the line $x + y = 6$, $x - 2z = 3$.

2. Find the equations of the spheres which pass through the circle
$$x^2 + y^2 + z^2 = 5x + 2y + 3z = 3, \text{ and touch the plane } 4x + 3y = 15.$$
3. Prove that the equation of the sphere which lies in the first octant and touches the coordinates planes is of the form $(x^2 + y^2 + z^2) - 2\lambda(x + y + z) + 2\lambda^2 = 0$.
4. Find the equation of the sphere passing through (1, 4, 9) and touching the coordinate planes.
5. Tangent plane at any point of the sphere $x^2 + y^2 + z^2 = r^2$ meets the coordinates axes at $A, B, C$. Show that the locus of the point of intersection of the planes drawn parallel to the coordinate planes through $A, B, C$ is the surface $x^{-2} + y^{-2} + z^{-2} = r^{-2}$.
6. Find the equation of the tangent line to the circle $x^2 + y^2 + z^2 = 3$, $3x - 2y + 4z + 3 = 0$ at the point (1, 1, –1).
7. Show that the sphere $x^2 + y^2 + z^2 - 2x + 6y + 14z + 3 = 0$ divides the line joining the points (2, –1, –4) and (5, 5, 5) internally and externally in the ratio 1 : 2.
8. Find the shortest and the longest distance from the point (1, 2, –1) to the sphere $x^2 + y^2 + z^2 = 24$.
9. Show that the spheres $x^2 + y^2 + z^2 + 6y + 14z + 8 = 0$ and $x^2 + y^2 + z^2 + 6x + 8y + 4z + 20 = 0$, intersect at right angles. Find their plane of intersection. *(Andhra., 2014)*
10. Find the equation to the sphere passing through the circle $x^2 + y^2 + z^2 = 9$, $x + y + z = 1$ and cuts orthogenally the sphere $x^2 + y^2 + z^2 + 2x - 4y - 10z + 17 = 0$. *(Anna, 2013)*
11. Show that the spheres $x^2 + y^2 + z^2 = 25$ and $x^2 + y^2 + z^2 - 24x - 40y - 18z + 225 = 0$ touch externally and find their point of contact.

## 3.19 (1) CONE

**Def.** *A cone is a surface generated by a straight line which passes through a fixed point and satisfies one more condition e.g.,* it may intersect a given curve (called the *guiding curve*).

*The fixed point is called the* **vertex** *and the straight line in any position is called a* **generator.**

The degree of the equation of a cone depends upon the nature of its guiding curve. In case the guiding curve is a conic, the equation of the cone shall be of the second degree. Such cones are called *Quadric cones.* In what follows, we shall be concerned only with quadric cones.

**Example 3.64.** *Find the equation of the cone whose vertex is (3, 1, 2) and base the circle*
$$2x^2 + 3y^2 = 1, z = 1.$$ *(Andhra, 2022)*

**Solution.** Any line through (3, 1, 2) is
$$\frac{x-3}{l} = \frac{y-1}{m} = \frac{z-2}{n} \quad ...(i)$$
It meets $z = 1$, where $\dfrac{x-3}{l} = \dfrac{y-1}{m} = \dfrac{-1}{n}$

whence $x = 3 - l/n$, $y = 1 - m/n$.

Substituting these values of $x$ and $y$ in $2x^2 + 3y^2 = 1$,
$$2(3 - l/n)^2 + 3(1 - m/n)^2 = 1 \quad ...(ii)$$
Eliminating $l, m, n$ from $(i)$ and $(ii)$, the locus of the line $(i)$ is
$$2\left(3 - \frac{x-3}{z-2}\right)^2 + 3\left(1 - \frac{y-1}{z-2}\right)^2 = 1$$
or $\quad 2x^2 + 3y^2 + 20z^2 - 6yz - 12xz + 12x + 6y - 38z + 17 = 0$ which is the required equation.

**Example 3.65.** *Find the equation of the cone whose vertex is at the origin and guiding curve is* $\dfrac{x^2}{4} + \dfrac{y^2}{9} + \dfrac{z^2}{1} = 1, x + y + z = 1.$

**Solution.** Any line through (0, 0, 0) is $x/l = y/m = z/n \quad ...(i)$

Any point on it is $P(lr, mr, nr)$.

If $(i)$ intersects the given curve, the coordinates of $P$ should satisfy its equations.

$\therefore \qquad \dfrac{l^2r^2}{4} + \dfrac{m^2r^2}{9} + \dfrac{n^2r^2}{1} = 1$ and $lr + mr + nr = 1$.

Eliminating $r$, $\qquad \left(\dfrac{l^2}{4} + \dfrac{m^2}{9} + n^2\right) \Big/ (l + m + n)^2 = 1.$

Simplifying, $27l^2 + 32m^2 + 72(lm + mn + nl) = 0$ ...(ii)

Eliminating $l, m, n$ from Eqns. (i) and (ii), the locus of the line (i) is

$$27x^2 + 32y^2 + 72(xy + yz + zx) = 0 \text{ which is the required equation.}$$

**Obs.** *The equation of a cone with vertex at the origin is a homogeneous equation of the second degree in x, y, z (i.e., all* terms are of the same degree). The reason is that every generator will have the equation of the form (i) above. So the point $(lr, mr, nr)$ will satisfy the equation of the cone for every value of $r$. This is possible only if the equation is homogeneous.

**Example 3.66.** *A variable plane parallel to the plane $\dfrac{x}{a} + \dfrac{y}{b} + \dfrac{z}{c} = 0$ meets the coordinate axes in A, B, C. Find the equation of the cone whose vertex is the origin and guiding curve the circle ABC.*

**Solution.** Let the plane $\qquad \dfrac{x}{a} + \dfrac{y}{b} + \dfrac{z}{c} = k$ ...(i)

meet the axes at $A, B, C$, so that $A = (ka, 0, 0)$, $B = (0, kb, 0)$ and $C = (0, 0, kc)$.

$\therefore$ The equation of the sphere through $O(0, 0, 0)$ and $A, B, C$ is

$$x^2 + y^2 + z^2 - k(ax + by + cz) = 0 \qquad \text{...(ii)}$$

Since the equation of the cone with vertex at $O$ is a homogeneous equation of the second degree, therefore, it must be satisfied by points lying on the circle $ABC$, *i.e.*, on Eqns. (i) and (ii) both.

$\therefore$ Making Eqn. (ii) homogeneous with the help of Eqn. (i), we have

$$x^2 + y^2 + z^2 - (ax + by + cz)\left(\frac{x}{a} + \frac{y}{b} + \frac{z}{c}\right) = 0$$

or $\qquad yz\left(\dfrac{b}{c} + \dfrac{c}{b}\right) + zx\left(\dfrac{c}{a} + \dfrac{a}{c}\right) + xy\left(\dfrac{a}{b} + \dfrac{b}{a}\right) = 0$ which is the required equation.

**Example 3.67.** *Show that the general equation of the cone of the second degree which passes through the axes is of the form $fyz + gzx + hxy = 0$.*

**Solution.** Any cone which passes through the axes will have origin $V$ as its vertex. The general equation of a cone of the second degree having vertex at the origin is of the form

$$ax^2 + by^2 + cz^2 + 2fyz + 2gzx + 2hxy = 0 \qquad \text{...(i)}$$

Since it passes through $x$-axis

$\therefore$ The direction cosines of $x$-axis (*i.e.*, 1, 0, 0) must satisfy Eqn. (i). This gives $a = 0$.

As the cone passes through $y$-axis, $b = 0$.

Similarly, as the cone passes through $z$-axis, $c = 0$.

Hence Eqn. (i) reduces to $fyz + gzx + hxy = 0$.

(2) **Right circular cone. Def.** *A right circular cone is a surface generated by a straight line which passes through a fixed point (vertex) and makes a constant angle with a fixed line* (Fig. 3.54).

*The constant angle ($\angle AVC$) is called its* **semi-vertical angle** *and the fixed line (VC) is called the* **axis.** The section of a right circular cone by a plane perpendicular to its axis is a circle.

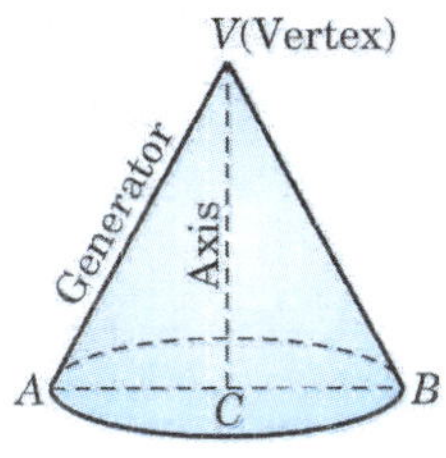

Fig. 3.54

**Example 3.68.** *Find the equation of the right circular cone whose vertex is the origin, whose axis is the line $x/1 = y/2 = z/3$ and which has semi-vertical angle of 30°.*

**Solution.** Let $P(x, y, z)$ be any point on the cone with vertex $O$ and axis $(OC)$

$$\frac{x}{1} = \frac{y}{2} = \frac{z}{3}, \text{ so that } \angle POC = 30°. \qquad \text{(Fig. 3.55)}$$

Now the direction ratios of $OP$ are $x, y, z$ and those of $OC$ are 1, 2, 3.

$$\therefore \quad \cos 30° = \frac{x(1) + y(2) + z(3)}{\sqrt{(x^2 + y^2 + z^2)} \cdot \sqrt{(1 + 4 + 9)}}$$

or
$$\frac{\sqrt{3}}{2} = \frac{x + 2y + 3z}{\sqrt{[14 (x^2 + y^2 + z^2)]}}$$

Squaring, $3 \times 14 (x^2 + y^2 + z^2) = 4(x + 2y + 3z)^2$

or
$$19x^2 + 13y^2 + 3z^2 - 8xy - 24yz - 12zx = 0$$

which is the required equation of the cone.

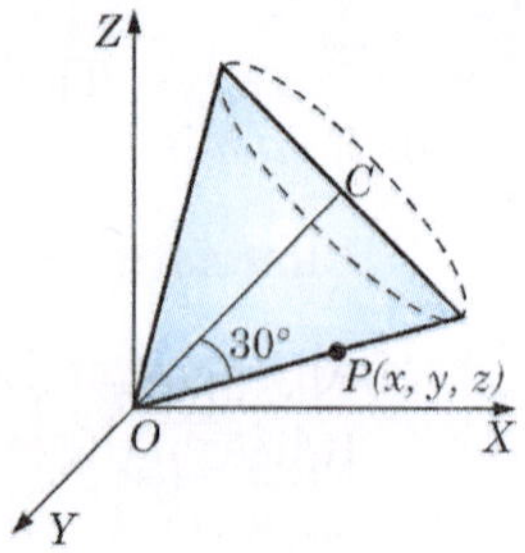

Fig. 3.55

**Example 3.69.** *Find the equation of the right circular cone generated when the straight line 2y + 3z = 6, x = 0 revolves about z-axis.* (Andhra., 2016)

**Solution.** The vertex is the point of intersection of the line $2y + 3z = 6$, $x = 0$ and the $z$-axis, *i.e.*, $x = 0$, $y = 0$ (Fig. 3.56).

$\therefore$ Vertex is $A(0, 0, 2)$. A generator of the cone is

$$\frac{x}{0} = \frac{y}{3} = \frac{z-2}{-2}$$

$\therefore$ Direction ratios of the generator are 0, 3, –2 and the axis ($z$-axis) are 0, 0, 1. The semi-vertical angle $\alpha$ is, therefore, given by

$$\cos \alpha = \frac{0.0 + 3.0 + (-2).1}{\sqrt{13}} = \frac{-2}{\sqrt{13}}$$

Let $P(x, y, z)$ be any point on the cone so that the direction ratios of $AP$ are $x, y, z - 2$. Since $AP$ makes an angle $\alpha$ with $AZ$, we have

$$\cos \alpha = \frac{x.0 + y.0 + (z-2).1}{\sqrt{[x^2 + y^2 + (z-2)^2]}}$$

Thus
$$\frac{(z-2)^2}{x^2 + y^2 + (z-2)^2} = \cos^2 \alpha = \frac{4}{13}$$

or
$$4x^2 + 4y^2 - 9z^2 + 36z - 36 = 0$$

which is the required equation of the cone.

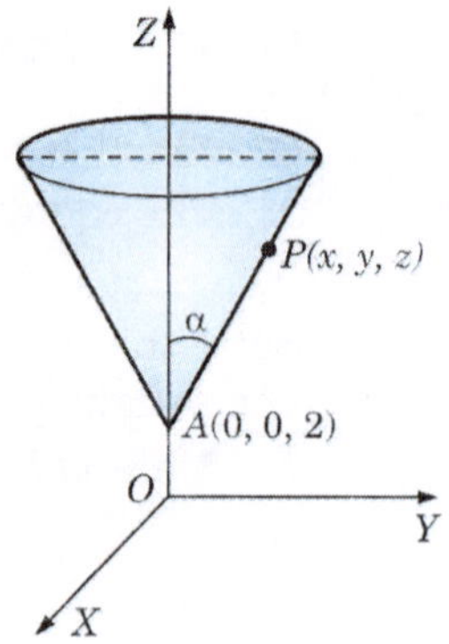

Fig. 3.56

**Example 3.70.** *Find the equations to the lines in which the plane 2x + y – z = 0 cuts the cone* $4x^2 - y^2 + 3z^2 = 0.$ (Andhra, 2020)

**Solution.** Let $x/l = y/m = z/n$ be one of the two lines in which the given plane $2x + y - z = 0$ ...(*i*)

cuts the given cone $4x^2 - y^2 + 3z^2 = 0$ ...(*ii*)

$\because$ This line lies on Eqn. (*i*), $\therefore$ $2l + m - n = 0$ ...(*iii*)

and it lies on Eqn. (*ii*), $\therefore$ $4l^2 - m^2 + 3n^2 = 0$ ...(*iv*)

To eliminate $n$ from Eqns. (*iii*) and (*iv*), put $n = 2l + m$ in Eqn. (*iv*).

$$4l^2 - m^2 + 3(2l + m)^2 = 0 \quad \text{or} \quad (4l + m)(2l + m) = 0$$

| | | | |
|---|---|---|---|
| $\therefore$ | Either $4l + m = 0$ | or | $2l + m = 0$ |
| From Eqn. (*iii*) | $2l + m - n = 0$ | and | $2l + m - n = 0$ |
| $\therefore$ | $\frac{l}{-1} = \frac{m}{4} = \frac{n}{2}$ | $\therefore$ | $\frac{l}{-1} = \frac{m}{2} = \frac{n}{0}$ |

Hence the required lines are

$$\frac{x}{-1} = \frac{y}{4} = \frac{z}{2} \text{ and } \frac{x}{-1} = \frac{y}{2} = \frac{z}{0}.$$

**Example 3.71.** *Find the equation of the enveloping cone of the sphere $x^2 + y^2 + z^2 = a^2$ with vertex at the point $(x_1, y_1, z_1)$.*

**Solution.** The equation of any generator through $V(x_1, y_1, z_1)$ having direction ratios $l, m, n$ is

$$\frac{x - x_1}{l} = \frac{y - y_1}{m} = \frac{z - z_1}{n} = r \text{ (say)} \quad ...(i)$$

Any point on Eqn. (*i*) is $P(x_1 + lr, y_1 + mr, z_1 + nr)$.

It lies on the given sphere if

$(x_1 + lr)^2 + (y_1 + mr)^2 + (z_1 + nr)^2 = a^2$

or $$(l^2 + m^2 + n^2) r^2 + 2(lx_1 + my_1 + nz_1) r + x_1^2 + y_1^2 + z_1^2 - a^2 = 0 \quad ...(ii)$$

The line (*i*) will touch the given sphere if (*ii*) has equal roots.

$$\therefore \quad (lx_1 + my_1 + nz_1)^2 = (l^2 + m^2 + n^2)(x_1^2 + y_1^2 + z_1^2 - a^2) \quad ...(iii)$$

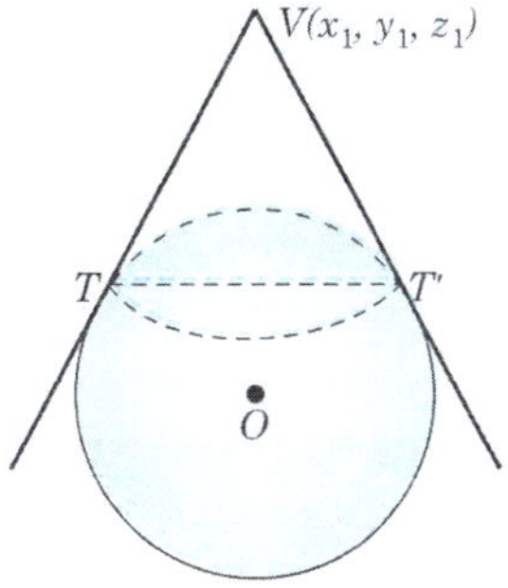

Fig. 3.57

The locus of all such lines is the enveloping cone of the given sphere which is obtained by eliminating $l, m, n$ from Eqns. (*i*) and (*iii*).

Thus $[(x - x_1) x_1 + (y - y_1)y_1 + (z - z_1)z_1]^2 = [(x - x_1)^2 + (y - y_1)^2 + (z - z_1)^2] (x_1^2 + y_1^2 + z_1^2 - a^2)$

which is the equation of the enveloping cone. (Fig. 3.57)

**Obs.** It can be reduced to the form $SS_1 = T^2$

where $S = x^2 + y^2 + z^2 - a^2, S_1 = x_1^2 + y_1^2 + z_1^2 - a^2, T = xx_1 + yy_1 + zz_1 - a^2$.

*Thus the enveloping cone of the surface $S = 0$ with vertex $(x_1, y_1, z_1)$ is $SS_1 = T^2$*

## PROBLEMS 3.14

1. Find the equation of the cone with vertex $(\alpha, \beta, \gamma)$ and base $y^2 - 4ax = 0, z = 0$.
2. Find the equation of the cone whose vertex is (3, 4, 5) and base is the conic $3y^2 + 4z^2 = 16, z + 2x = 0$.
3. Find the equation of the cone whose vertex is (1, 2, 3) and whose guiding curve is the circle $x^2 + y^2 + z^2 = 4$, $x + y + z = 1$. *(Andhra, 2019; P.T.U., 2010)*
4. The generators of a cone pass through the point (1, 1, 1) and their direction cosines $l, m, n$ satisfy the relation $l^2 + m^2 = 3n^2$. Obtain the equation of the cone.
5. Find the equation of the right circular cone whose vertex is at the origin and semi-vertical angle is $\alpha$ and having axis of $z$ as its axis. *(Anna, 2012)*
6. Find the equation of the cone whose vertical angle is $\pi/2$, which has its vertex at the origin and its axis along the line $x = -2y = z$.
   Also show that the plane $z = 0$ cuts the cone in two straight lines inclined at an angle $\cos^{-1} 4/5$.
7. Find the equation of the circular cone which passes through the point (1, 1, 2) and has its vertex at the origin and axis the line $x/2 = -y/4 = z/3$.
8. Find the equation of the right circular cone generated by revolving the line $2x + 3y = 5, z = 0$ about the $y$-axis. *(Anna, 2013)*
9. Find the equation of the right circular cone generated by rotating the line $\frac{x}{1} = \frac{y}{2} = \frac{z}{3}$ about the line $\frac{x}{-1} = \frac{y}{1} = \frac{z}{2}$. *(Andhra., 2016)*
10. Find the equation of the right circular cone passing through the coordinate axes having vertex at the origin. Obtain the semi-vertical angle and the equation of the axis.
11. Find the semi-vertical angle and the equation of the right circular cone having its vertex at the origin and passing through the circle $y^2 + z^2 = 25, x = 4$.
12. Find the equation of the right circular cone which has its vertex at (0, 0, 10) whose intersection with the $XY$-plane is a circle of radius 5.
13. Find the equations to the lines in which the plane $3x + y + 5z = 0$ cuts the cone $6yz - 2zx + 5xy = 0$.
14. Show that the plane $z = a$, meets any enveloping cone of the sphere $x^2 + y^2 + z^2 = a^2$, in a conic which has focus at the point $(0, 0, a)$. *(V T U 2017)*
15. Find the equation of the enveloping cone of the sphere $x^2 + y^2 + z^2 + 2x - 4y + 2z - 1 = 0$ with vertex at (1, 1, 1).

## 3.20 (1) CYLINDER

**Def.** *A cylinder is a surface generated by a straight line which is parallel to a fixed line and satisfies one more condition e.g.,* it may intersect a given curve (called the *guiding curve*).

The straight line in any position is called the *generator* and the fixed line the *axis* of the cylinder.

**Example 3.72.** *Find the equation of the right circular cylinder of radius 2, with axis.* (*Delhi, 2017*)

$$\frac{x-1}{2} = \frac{y}{3} = \frac{z-3}{1}$$

**Solution:** Let $P\,(x, y, z)$ be any point on the cylinder. A point on the axis of the cylinder is $A(1, 0, 3)$ and direction ratios of the axis are 2, 3, 1.

So, the direction cosines of the axis are

$$\frac{2}{\sqrt{14}}, \frac{3}{\sqrt{14}}, \frac{1}{\sqrt{14}}$$

Draw $PM$ perpendicular to the axis. Then

$AM$ = Projection of $AP$ on $AM$

$$= (x-1)\frac{2}{\sqrt{14}} + (y-0)\frac{3}{\sqrt{14}} + (z-3)\frac{1}{14}$$

$$= \frac{1}{\sqrt{14}}(2x + 3y + z - 5)$$

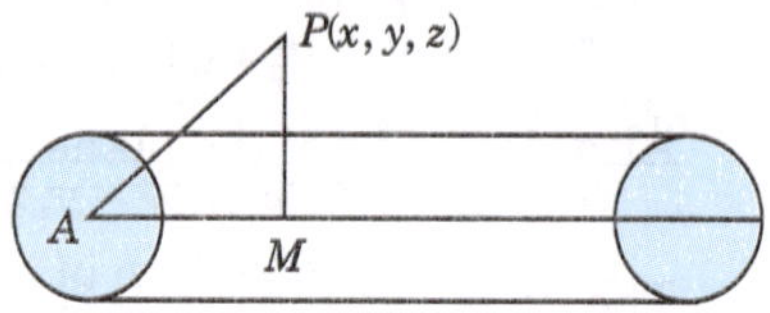

Fig. 3.58

Also, $AP = \sqrt{(x-1)^2 + (y-0)^2 - (z-3)^2}$

and $PM = 2$

Now, $AP^2 - AM^2 = PM^2$

$$(x-1)^2 + y^2 + (z-3)^2 - \frac{1}{14}(2x + 3y + z - 5)^2 = 2^2$$

$$10x^2 + 5y^2 + 13z^2 - 12xy - 6yz - 4zx - 8x + 30y - 74z + 59 = 0$$

which is the equation of cylinder.

**Example 3.73.** *The radius of a normal section of a right circular cylinder is 2 units ; the axis lies along the straight line*

$$\frac{x-1}{2} = \frac{y+3}{-1} = \frac{z-2}{5}, \text{ find its equation.}$$

**Solution.** A point on the axis of the cylinder is $A(1, -3, 2)$ and its direction ratios are 2, –1, 5.

$\therefore$ Its actual direction cosines are $\frac{2}{\sqrt{30}}, \frac{-1}{\sqrt{30}}, \frac{5}{\sqrt{30}}$.

Let $P(x, y, z)$ be any point on the cylinder. Draw $PM \perp$ to the axis $AM$. Then $MP = 2$. Now $AM$ = Projection of $AP$ on $AM$ (axis)

$$= (x-1)\frac{2}{\sqrt{30}} + (y+3)\frac{-1}{\sqrt{30}} + (z-2)\frac{5}{\sqrt{30}}$$

$$= \frac{2x - y + 5z - 15}{\sqrt{30}}$$

Also $AP = \sqrt{(x-1)^2 + (y+3)^2 + (z-2)^2}$

$\therefore$ From the rt. $\angle d\ \Delta\, AMP$, $(AM)^2 + (MP)^2 = (AP)^2$

or $$\frac{1}{30}(2x - y + 5z - 15)^2 + 4 = (x-1)^2 + (y+3)^2 + (z-2)^2$$

or $$26x^2 + 29y^2 + 5z^2 + 4xy + 10yz - 20zx + 150y + 30z + 75 = 0.$$

This is the required equation of the right circular cylinder. (Fig. 3.59)

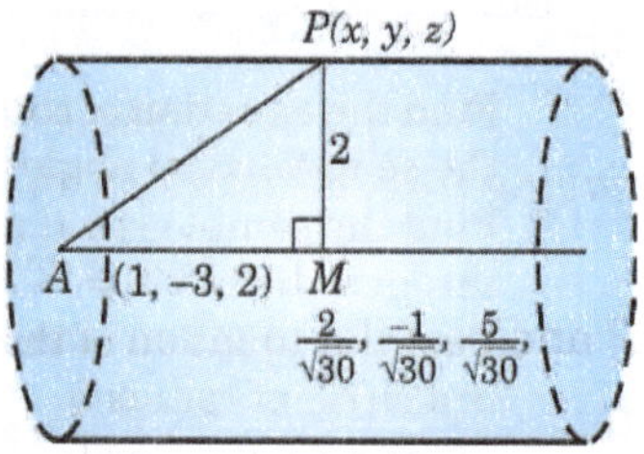

Fig. 3.59

**Example 3.74.** *Find the equation of the right circular cylinder described on the circle through the points* (1, 0, 0)(0, 1, 0) *and* (0, 0, 1) *as the guiding curve.*

(*Cuttak, 2018*)

**Solution.** Let the three given points (1, 0, 0) (0, 1, 0) and (0, 0, 1) be denoted by $A$, $B$ and $C$ respectively. Let us take $O$, the origin, as the fourth point and find the equation of the sphere $OABC$.

Let the equation of the sphere be

$$x^2 + y^2 + z^2 + 2ux + 2vy + 2wz + d = 0 \quad ...(i)$$

Since the sphere ($i$) passes through the point $O, A, B$ and $C$, we have,

$d = 0, 1 + 2u = 0, 1 + 2v = 0, 1 + 2w = 0$

The last three equations given

$$u = v = w = \frac{-1}{2}$$

Substituting these in Eqn. ($i$), the equation of the sphere $OABC$ is given by

$$x^2 + y^2 + z^2 - x - y - z = 0 \quad ...(ii)$$

Also, equation of the plane $ABC$ is

$$x + y + z = 1 \quad ...(iii)$$

The Eqns. ($ii$) and ($iii$) taken together represent the circle, which is the guiding curve of the required cylinder.

Since the axis of the cylinder is perpendicular to the plane ($iii$), its direction ratios are proportional to 1, 1, 1. Let $(\alpha, \beta, \gamma)$ be a point on the cylinder.

The equation of the generater through this points are

$$\frac{x-\alpha}{1} = \frac{y-\beta}{1} - \frac{z-\gamma}{1} = (r)$$

Any point on the generator is $(\alpha + r), (\beta + r)(\gamma + r)$.

If this point lies on the circle, then from Eqns. ($ii$) and ($iii$), we get

$$(\alpha + r)^2 + (\beta + r)^2 + (\gamma + r)^2 - (\alpha + r) - (\beta + r) - (\gamma + r) = 0 \quad ...(iv)$$

and $\quad (\alpha + r) + (\beta + r) + (\gamma + r) = 1 \quad ...(v)$

From Eqn. ($v$), we have

$$r = \frac{1-\alpha-\beta-\gamma}{3} \quad ...(vi)$$

Equation ($iv$) can be written as

$$(\alpha^2 + \beta^2 + \gamma^2 - \alpha - \beta - \gamma) + r(2\alpha + 2\beta + 2\gamma - 3) + 3r^2 = 0$$

Eliminating $r$ between this equation and Eqn. ($vi$), we get.

$$(\alpha^2 + \beta^2 + \gamma^2 - \alpha - \beta - \gamma) + \frac{1}{3}(1 - \alpha - \beta - \gamma)(2\alpha + 2\beta + 2\gamma - 3) + \frac{1}{3}(1 - \alpha - \beta - \gamma)^2 = 0$$

$$(\alpha^2 + \beta^2 + \gamma^2) - \alpha\beta - \beta\gamma - \gamma\alpha - 1 = 0$$

So, the locus of the point $P(\alpha, \beta, \gamma)$ is $x^2 + y^2 + z^2 - xy - yz - zx - 1 = 0$

So, the required equation of the cylinder.

**Example 3.75.** *Show that the equation of the enveloping cylinder of the sphere $x^2 + y^2 + z^2 - 2x + 4y = 1$, having its generators parallel to the line $x = y = 7$ is $x^2 + y^2 + z^2 - yz - zx - xy - 4x + 5y - z = 2$.* (*Mumbai, 2019*)

**Solution:** Let $P(\alpha, \beta, \gamma)$ be any point on the enveloping cylinder, equation of the generator through the point $P$ and parallel to given line are

$$\frac{x-\alpha}{1} = \frac{y-\beta}{1} = \frac{z-\gamma}{1} (r) \quad ...(i)$$

Let the line ($i$) interest the sphere

$$x^2 + y^2 + z^2 - 2x + 4y - 1 = 0$$

in the points $(r + \alpha, r + \beta, r + \gamma)$. them

$$(r + \alpha)^2 + (r + \beta)^2 + (r + \gamma)^2 - (2r + \alpha) + 4(r + \beta) - 1 = 0$$

$$3r^2 + 2r(\alpha + \beta + \gamma + 1) + (\alpha^2 + \beta^2 + \gamma^2 - 2\alpha + 4\beta - 1) = 0 \quad ...(ii)$$

If the line ($i$) is a tangent to the given sphere, it must meet the sphere in two coincident points and the Eqn. ($ii$) must have equal roots.

So, $4(\alpha + \beta + \gamma + 1)^2 = 12(\alpha^2 + \beta^2 + \gamma^2 - 2\alpha + 2\beta - 1)$

$\alpha^2 + \beta^2 + \gamma^2 - \beta\gamma - \alpha\gamma - \alpha\beta - 4\alpha + 5\beta - \gamma - 2 = 0$

The locus of the point $p(\alpha, \beta, \gamma)$ is

$x^2 + y^2 + z^2 - yz - zx - xy - 4x + 5y - z = 0$

the required equation of the enveloping cylinder.

## PROBLEMS 3.15

1. Find the equation of the right circular cylinder whose axis is the line $x = 2y = -z$ and radius 4.
2. Find the equation of the cylinder whose generators are parallel to the line
   (*i*) $x = -y/2 = z/3$ and whose guiding curve is the ellipse $x^2 + 2y^2 = 1$, $z = 3$.
   (*ii*) $x/2 = y/2 = z/-3$ and whose guiding curve is the ellipse $3x^2 + y^2 = 3$, $z = 2$. (*Anna, 2013*)
3. Find the equation of the right circular cylinder of radius 2 whose axis passes through (1, 2, 3) and has direction ratios (2, –3, 6).
4. Obtain the equation of a right circular cylinder whose guiding curve in the circle through the points $(a, 0, 0)$, $(0, b, 0)$, (0, 0, C). (*DTU 2019*)
5. Find the equation of the cylinder whose directing curve is $x^2 + z^2 - 4x - 2z + 4 = 0$, $y = 0$ and whose axis contains the point (0, 3, 0). Find also the area of the section of the cylinder by a plane parallel to $xz$-plane.
6. Find the equation of the enveloping cylinder of the sphere $x^2 + y^2 + z^2 - 2x - 4y - 6z - 2 = 0$ whose generators are perpendicular to the lines $\frac{x}{3} = \frac{y}{-1} = \frac{z}{0}$ and $\frac{x}{1} = \frac{y}{2} = \frac{z}{0}$. (*Andhra, 2019*)
7. Find the equation of the right circular cylinder whose axis is $\frac{x-2}{2} = \frac{y-1}{2} = \frac{z}{1}$, and which passes through the point (0, 0, 3). (*KTU, 2018*)
8. Find the equation of the enveloping cylinder of the ellipsoid $\frac{x^2}{y^2} + \frac{y^2}{b^2} + \frac{z^2}{c^2} = 1$. Whose generators are parallel to the line $\frac{x}{l} = \frac{y}{m} = \frac{z}{h}$. (*Karnataka, 2018*)

## 3.21 QUADRIC SURFACES

*The surface represented by general equation of the second degree in x, y, z is called a* **quadric surface** *or a* **conicoid.**

Thus the general equation of a *quadric surface* is of the form

$$ax^2 + by^2 + cz^2 + 2fyz + 2gzx + 2hxy + 2ux + 2vy + 2wz + d = 0$$

which can be reduced to any of the following standard forms so useful in engineering problems. We now, proceed to study their shapes.

**(1) Ellipsoid :** $\frac{x^2}{a^2} + \frac{y^2}{b^2} + \frac{z^2}{c^2} = 1$.

(*i*) It is symmetrical about each of the coordinate planes for only even powers of $x, y, z$ occur in its equation.

(*ii*) It meets the $x$-axis at $A$ $(a, 0, 0)$, $A'$ $(-a, 0, 0)$ ;
the $y$-axis at $B$ $(0, b, 0)$, $B'$ $(0, -b, 0)$ ;
and the $z$-axis at $C(0, 0, c)$, $C'$ $(0, 0, -c)$.

(*iii*) Its sections by the coordinate planes are ellipses. For the section by the $yz$-plane $(x = 0)$ is the ellipse.

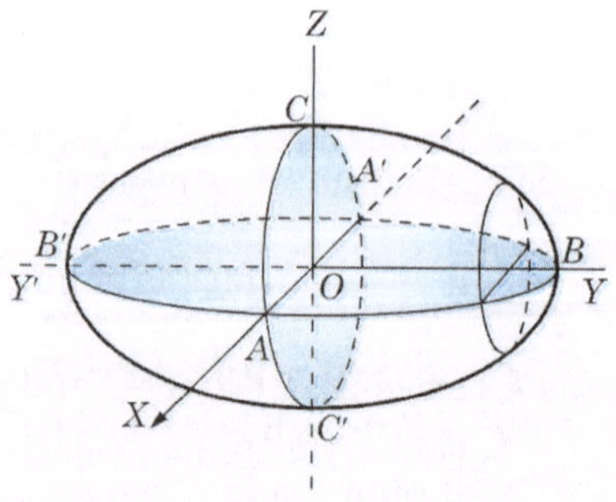


Fig. 3.60

$$\frac{y^2}{b^2} + \frac{z^2}{c^2} = 1, \text{ etc.}$$

(*iv*) The surface is generated by a variable ellipse

$$\frac{x^2}{a^2} + \frac{y^2}{b^2} = 1 - \frac{k^2}{c^2}, z = k;$$

(as $k$ varies from $-c$ to $c$) and is limited in every direction.

Hence its shape is as shown in Fig. 3.61 which is like that of an *egg*.

**(2) Hyperboloid of one sheet :** $\frac{x^2}{a^2} + \frac{y^2}{b^2} - \frac{z^2}{c^2} = 1.$

(*i*) It is symmetrical about each of the coordinate planes for only even powers of $x$, $y$, $z$ occur in its equation.

(*ii*) It meets the $x$-axis at $A(a, 0, 0)$, $A'(-a, 0, 0)$ ; the $y$-axis at $B(0, b, 0)$, $B'(0, -b, 0)$ ; and the $z$-axis in imaginary points.

(*iii*) Its section by the $yz$-plane ($x = 0$) is the hyperbola $\frac{y^2}{b^2} - \frac{z^2}{c^2} = 1$, (*i.e.*, $DE$, $D'E'$)

Its section by the $zx$-plane ($y = 0$) is the hyperbola $\frac{x^2}{a^2} - \frac{z^2}{c^2} = 1$. (*i.e.*, $FG$, $F'G'$)

Its section by the $xy$-plane ($z = 0$) is the ellipse $\frac{x^2}{a^2} + \frac{y^2}{b^2} = 1$.

(*iv*) The surface is generated by a variable ellipse

$\frac{x^2}{a^2} + \frac{y^2}{b^2} = 1 + \frac{k^2}{c^2}$, $z = k$ (as $k$ varies from $-\infty$ to $\infty$) and extends to infinity on both sides of the $xy$-plane.

Hence its shape is as shown in Fig. 3.61 which is like that of *juggler's dabru*.

**(3) Hyperboloid of two sheets :** $\frac{x^2}{a^2} + \frac{y^2}{b^2} - \frac{z^2}{c^2} = -1.$

(*i*) It is symmetrical about each of the coordinate planes for only even powers of $x$, $y$, $z$ occur in its equation.

(*ii*) It meets the $z$-axis at $C(0, 0, c)$, $C'(0, 0, -c)$ and the $x$ and $y$-axes in imaginary points.

(*iii*) Its section by the $yz$-plane ($x = 0$) is the hyperbola $\frac{z^2}{c^2} - \frac{y^2}{b^2} = 1$. (*i.e.*, $ACB$, $A'C'B'$)

Its section by the $zx$-plane ($y = 0$) is the hyperbola $\frac{z^2}{c^2} - \frac{x^2}{a^2} = 1$. (*i.e.*, $DCE$, $D'C'E'$)

Its section by the $xy$-plane ($z = 0$), is the imaginary ellipse $\frac{x^2}{a^2} - \frac{y^2}{b^2} = -1$.

Its section by the $xy$-plane ($z = 0$) is the ellipse $\frac{x^2}{a^2} + \frac{y^2}{b^2} = 1$.

(*iv*) The surface is generated by a variable ellipse $\frac{x^2}{a^2} + \frac{y^2}{b^2} = \frac{k^2}{c^2} - 1$, $z = k$,

(as $k$ varies from $-\infty$ to $-c$ and $c$ to $+\infty$) and extends to infinity on both sides of the $xy$-plane.

Hence its shape is as shown in Fig. 3.62.

**(4) Cone :** $\dfrac{x^2}{a^2}+\dfrac{y^2}{b^2}-\dfrac{z^2}{c^2}=0.$

(*i*) It is symmetrical about each of the coordinate planes.

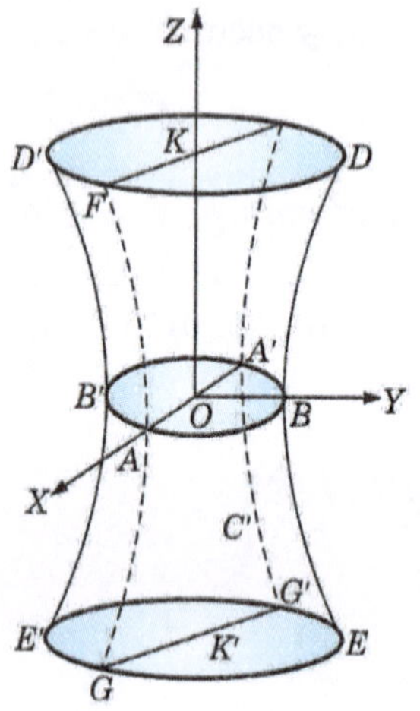


Fig. 3.61

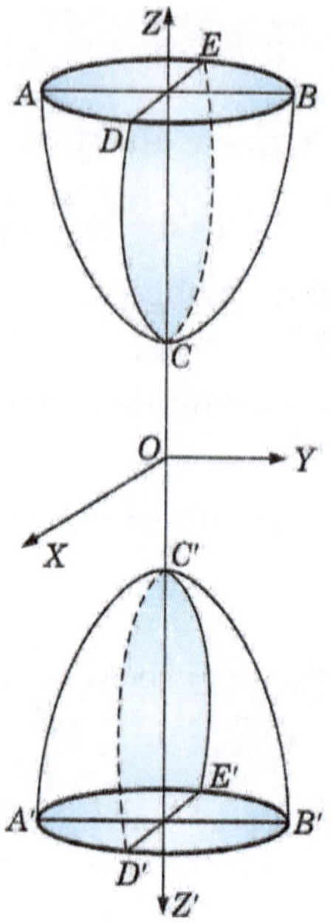


Fig. 3.62

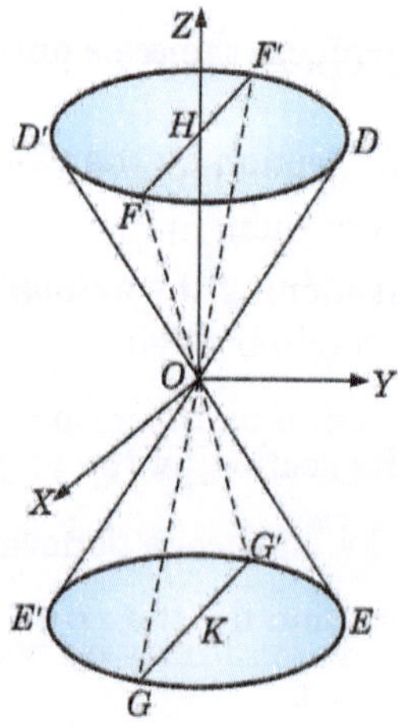


Fig. 3.63

(*ii*) It meets the axes only at the origin.

(*iii*) Its section by the *yz*-plane ($x = 0$) is the pair of straight lines

$$y = \pm \frac{b}{c} z \qquad (i.e., DOE' \text{ and } D'OE).$$

Its section by the *zx*-plane ($y = 0$) is the pair of straight lines

$$x = \pm \frac{a}{c} z \qquad (i.e., FOG' \text{ and } F'OG).$$

Its section by the *zx*-plane ($z = 0$) is the point ellipse $\dfrac{x^2}{a^2}+\dfrac{y^2}{b^2}=0.$

(*iv*) The surface is generated by a variable ellipse $\dfrac{x^2}{a^2}+\dfrac{y^2}{b^2}=\dfrac{k^2}{c^2}$, $z = k$ ($k$ varies) and extends to infinity on both sides of the *xy*-plane. Hence its shape is as shown in Fig. 3.63.

**(5) Elliptic paraboloid :** $\dfrac{x^2}{a^2}+\dfrac{y^2}{b^2}=\dfrac{2z}{c}$

(*i*) It is symmetrical about *yz*- and *zx*-planes for only even powers of $x$ and $y$ occur in its equation

(*ii*) It meets the axes at the origin only and touches the *xy*-plane threat.

(*iii*) Its section by the *yz*-plane ($x = 0$) is the parabola $y^2 = \dfrac{2b^2}{c} z$, (*i.e.*, $DOD'$).

Its section by the *zx*-plane ($y = 0$) is the parabola $x^2 = \dfrac{2a^2}{c} z$ (*i.e.*, $EOE'$).

Its section by the *xy*-plane ($z = 0$) is the point ellipse $\dfrac{x^2}{a^2}+\dfrac{y^2}{b^2}=0$

(*iv*) The surface is generated by a variable ellipse $\frac{x^2}{a^2}+\frac{y^2}{b^2}=\frac{2k}{c}$, $z = k$ (as $k$ varies from 0 to $\infty$) and it extends to infinity above the $xy$-plane.

Hence its shape is as shown in Fig. 3.64 and is like that of *tabla*.

**(6) Hyperbolic paraboloid : $\frac{x^2}{a^2}-\frac{y^2}{b^2}=\frac{2z}{c}$.**

(*i*) It is symmetrical about the $yz$ and $zx$-planes for only even powers of $x$ and $y$ occur in its equation.

(*ii*) It meets the axes only at the origin and touches the $xy$-plane threat.

(*iii*) Its section by the $yz$-plane ($x = 0$) is the parabola $y^2 = -\frac{2b^2}{c}z$. (*i.e., DOD′*)

Its section by the $zx$-plane ($y = 0$) is the parabola $x^2 = \frac{2a^2}{c}z$ (*i.e., EOE′*).

Its section by the $xy$-plane ($z = 0$) is the part of lines $y = \pm\frac{b}{a}x$ (not shown in Fig. 3.65.)

(*iv*) The surface is generated by a variable hyperbola $\frac{x^2}{a^2}-\frac{y^2}{b^2}=\frac{2k}{c}$, $z = k$ and it extends to infinity on both sides of $xy$-plane. Hence its shape is as shown in Fig. 3.65.

**(7) Cylinder.** An equation of the form $f(x, y) = 0$ represents a cylinder generated by a straight line which is parallel to the $z$-axis and its section by the $xy$-plane is the curve $f(x, y) = 0$ (Fig. 3.66).

In particular (*i*) $y^2 = 4ax$ represents a *parabolic cylinder*,

(*ii*) $\frac{x^2}{a^2}+\frac{y^2}{b^2}=1$ represents an *elliptic cylinder*, (*iii*) $\frac{x^2}{a^2}-\frac{y^2}{b^2}=1$ represents a *hyperbolic cylinder*.

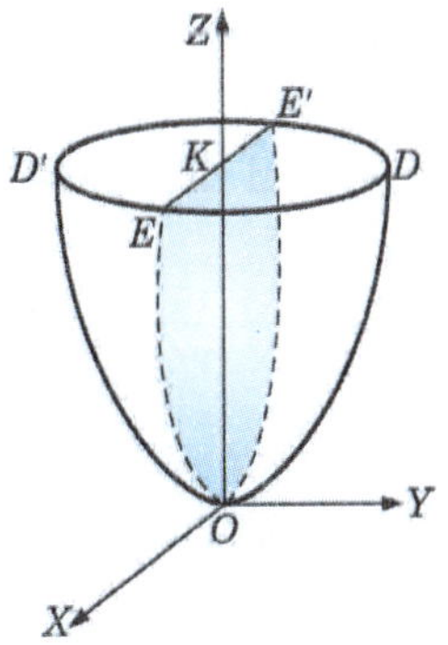

Fig. 3.64

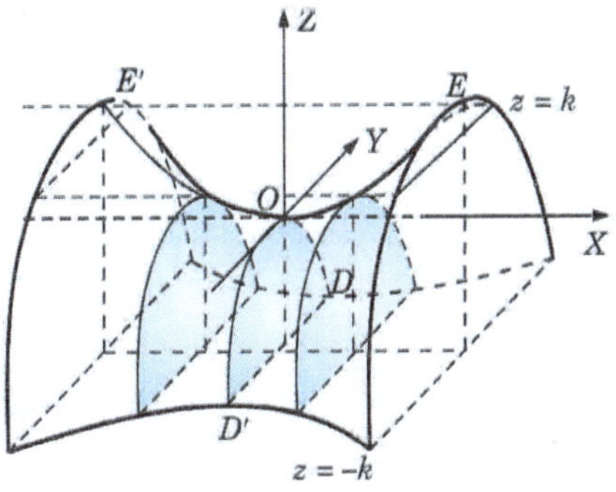

Fig. 3.65

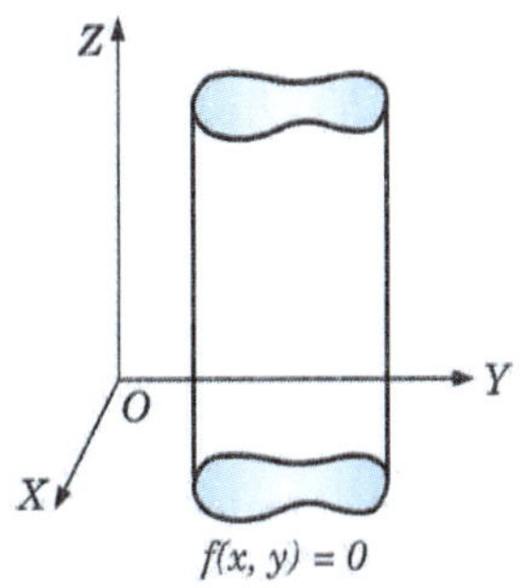

Fig. 3.66

## 3.22 SURFACES OF REVOLUTION

Let $P(x, y)$ be any point on the curve $y = f(x)$ in the $xy$-plane. Draw $PM \perp$ to $x$-axis so that $OM = x$ and $MP = y$. Thus the equation of this curve can be written as

$$MP = f(OM) \quad ...(1)$$

As this curve revolves about the $x$-axis, the point $P$ describe a circle with centre $M$ and radius $MP$. Let $Q(x, y, z)$ be any other position of $P$. Draw $QN \perp$ to $zx$-plane and join $MN$ so that $OM = x$, $MN = z$, $NQ = y$

and $\angle MNQ = 90°$. $\therefore$ $MP^2 = MQ^2 = MN^2 + NQ^2 = z^2 + y^2$.

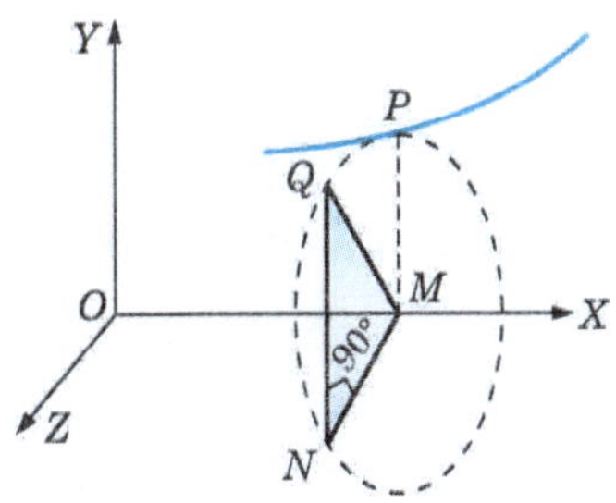

Fig. 3.67

Now substituting the values of $MP$ and $MO$ in (1), we have

$$\sqrt{(y^2+z^2)} = f(x) \quad \text{or} \quad \mathbf{y^2 + z^2 = [f(x)]^2}$$

which is *the equation of the surface generated by the revolution of the curve y = f(x) about the x-axis* (Fig. 3.67)

Similarly, *the surface generated by the revolution of the curve*

(*i*) $x = f(y)$ *about y-axis is* $\mathbf{z^2 + x^2 = [f(y)]^2}$, (*ii*) $x = f(z)$ about $z$-axis is $\mathbf{x^2 + y^2 = [f(z)]^2}$

The given revolving curve is called the *generating curve*.

**Some standard surfaces of revolution :**

Let the generating curve be $y = f(x)$ in the $xy$-plane and the axis of rotation be the $x$-axis ; then the surface generated is $y^2 + z^2 = [f(x)]^2$.

**(1) Right-circular cylinder.** *When* $f(x) = a$, *the generating curve is a straight line* ($y = a$) *parallel to the x-axis.*

∴ The surface generated is $\mathbf{y^2 + z^2 = a^2}$

which represents a *right-circular cylinder of radius a and axis as x-axis* (Fig. 3.68).

**(2) Right-circular cone.** *When* $f(x) = mx$, the generating curve is a straight line ($y = mx$) passing through the origin.

∴ The surface generated is $y^2 + z^2 = m^2x^2$ or $\mathbf{y^2 + z^2 = x^2 \tan^2 \alpha}$

which represents *a right-circular cone of semi-vertical angle α and axis as the x-axis* (Fig. 3.69).

**(3) Sphere.** *When* $f(x) = \sqrt{(a^2 - x^2)}$, the generating curve is a circle ($x^2 + y^2 = a^2$).

∴ The surface generated is

$$y^2 + z^2 = a^2 - x^2 \quad i.e., \quad \mathbf{x^2 + y^2 + z^2 = a^2},$$

which is *a sphere of radius a and centre (0, 0, 0).*

**(4) Ellipsoid of revolution.** *When* $f(x) = b\sqrt{(1 - x^2/a^2)}$, the generating curve is an ellipse

$$\left(\frac{x^2}{a^2} + \frac{y^2}{b^2} = 1\right). \qquad \therefore \text{ The surface generated is } y^2 + z^2 = b^2(1 - x^2/a^2)$$

or $$\mathbf{\frac{x^2}{a^2} + \frac{y^2}{b^2} + \frac{z^2}{b^2} = 1,}$$ which is called an *ellipsoid of revolution.*

If $a^2 > b^2$, the major axis of the generating ellipse is along the $x$-axis—the axis of revolution and the surface generated, in this case, is called a **prolate spheroid** (Fig. 3.70).

If $a^2 < b^2$, the minor axis of the ellipse lies along the $x$-axis—the axis of revolution and the surface thus generated is called an **oblate spheroid** (Fig. 3.71).

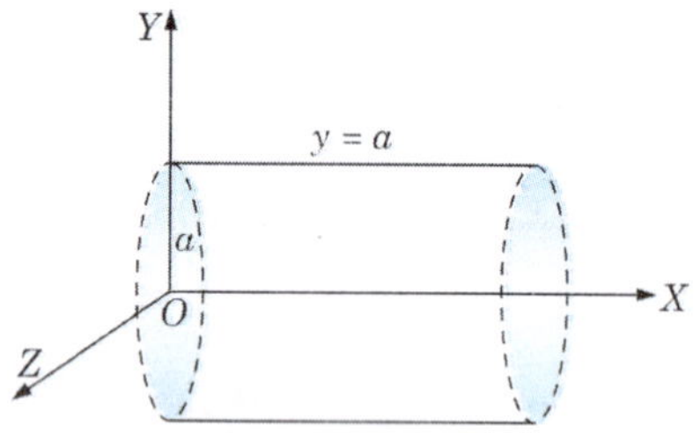

Fig. 3.68

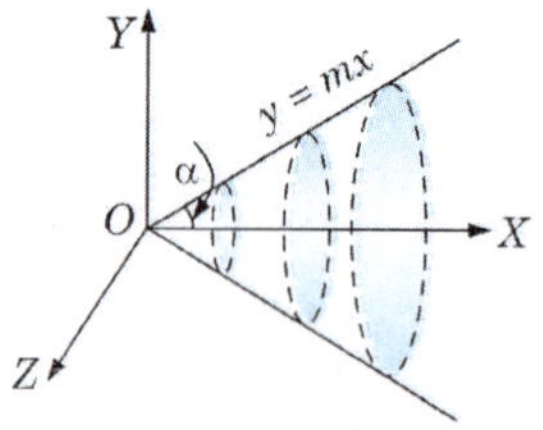

Fig. 3.69

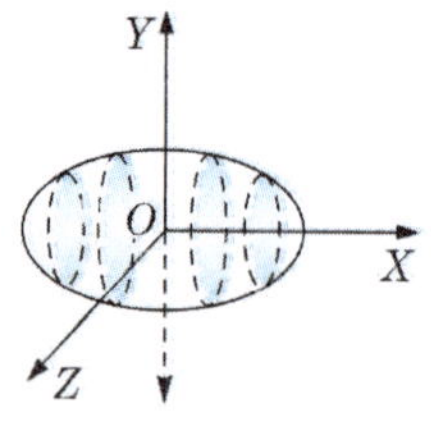

Prolate spheroid
Fig. 3.70

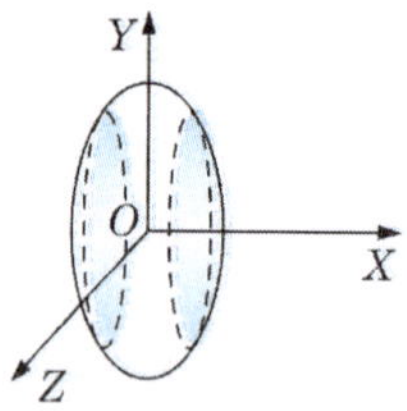

Oblate spheroid
Fig. 3.71

**(5) Hyperboloids of revolution**

(*i*) *When* $f(x) = b\sqrt{(1 + x^2/a^2)}$, the generating curve is $\dfrac{y^2}{b^2} - \dfrac{z^2}{a^2} = 1$, which represents a hyperbola having conjugate axis along the $x$-axis. (Fig. 3.72)

∴ The surface generated is $y^2 + z^2 = b^2(1 + x^2/a^2)$

or $$\mathbf{\frac{y^2}{b^2} + \frac{z^2}{b^2} - \frac{x^2}{a^2} = 1}$$ which is called a *hyperboloid of revolution of one sheet* (Fig. 3.72).

(*ii*) When $f(x) = b\sqrt{(x^2/a^2 - 1)}$, the generating curve is $\frac{x^2}{a^2} - \frac{y^2}{b^2} = 1$ which represents a hyperbola having transverse axis along the $x$-axis.

$\therefore$ The surface generated is $y^2 + z^2 = b^2(x^2/a^2 - 1)$

or $$\mathbf{\frac{x^2}{a^2} - \frac{y^2}{b^2} - \frac{z^2}{b^2} = 1},$$ which is called a *hyperboloid of revolution of two sheets* (Fig. 3.73).

**(6) Paraboloid of revolution.** When $f(x) = \sqrt{(ax)}$, the generating curve is a parabola ($y^2 = ax$). The surface generated is $\mathbf{y^2 + z^2 = ax}$. which is called a *paraboloid of revolution* (Fig. 3.74).

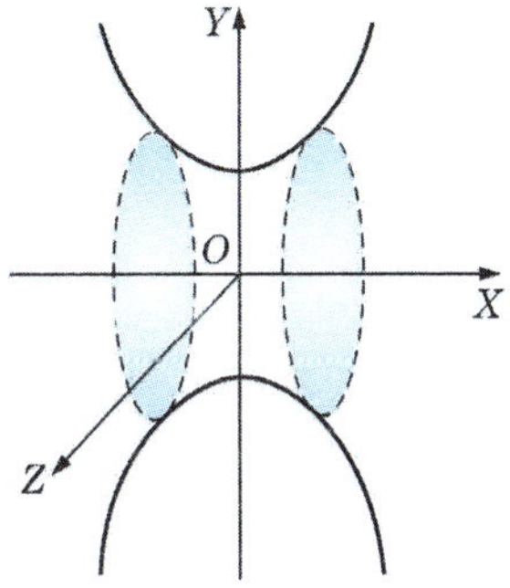


Fig. 3.72

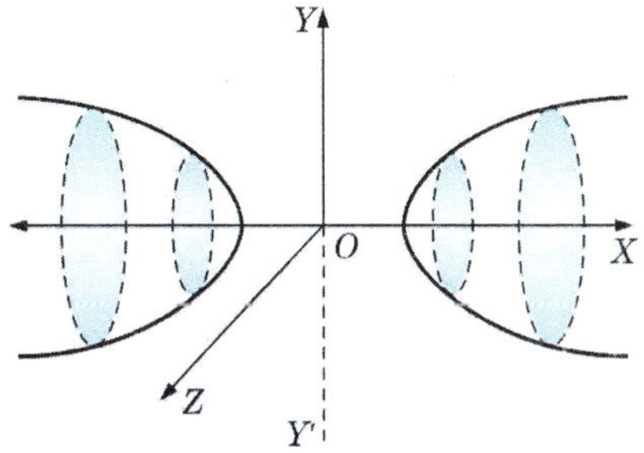


Fig. 3.73

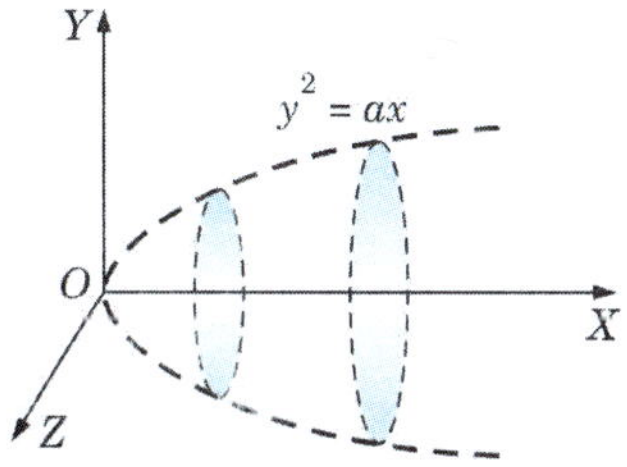


Fig. 3.74

### PROBLEMS 3.16

1. Identify the surfaces represented by $4x^2 + 9y^2 + 16z^2 = 144$ ? Find the area of the plane curve in which $y = 2$ cuts it.
2. Find the surface represented by $5(x^2 + z^2) - y^2 = 6$.

   In what curve does the plane $z = 2$ intersect it ? Find the area of the curve of intersection ?

   Identify surfaces represented by the following equations ?
3. $x^2 + y^2 = 16$.
4. $x^2/2 - y^2/3 = z$.
5. $z^2 = 4(1 + x^2 + y^2)$.
6. $y^2 = 4z - 8$
7. $x^2 + y^2 = 5 - 2y$.
8. $x^2 + y^2 = 9z^2$.
9. $\frac{x^2}{a^2} - \frac{y^2}{b^2} = \frac{2z}{c}$. (*P.T.U., 2009*)
10. $5x^2 - 4y^2 + 20z^2 = 0$. (*K.T.U., 2016*)

**Note.** For the equations of the tangent plane and the normal line to *a* surface refer to § 5.8 (2).

## 3.23 OBJECTIVE TYPE OF QUESTIONS

### PROBLEMS 3.17

*Select the correct answer or fill up the blanks in each of the following questions :*

1. The line $x = ay + b, z = cy + d$ and $x = a'y + b', z = c'y + d'$ are perpendicular if

   (*a*) $aa' + cc' = 1$ (*b*) $aa' + cc' = -1$ (*c*) $bb' + dd' = 1$ (*d*) $bb' + dd' = -1$.
2. The coordinates of the point of intersection of the line $\frac{x+1}{1} = \frac{y+3}{3} = \frac{z+2}{-2}$ with the plane $3x + 4y + 5z = 5$ is

   (*a*) (5, 15, –14) (*b*) (3, 4, 5) (*c*) (1, 3, –2) (*d*) (3, 12, –10).
3. The equation of a right circular cylinder, whose axis is the $z$-axis and radius $a$ is

   (*a*) $x^2 + y^2 + z^2 = a^2$ (*b*) $z^2 + y^2 = a^2$ (*c*) $x^2 + y^2 = a^2$ (*d*) $z^2 + x^2 = a^2$. (*Anna, 2013*)
4. The equation $\sqrt{fx} + \sqrt{gy} + \sqrt{hz} = 0$ represent a

   (*a*) sphere (*b*) cylinder (*c*) cone (*d*) pair of planes.

5. The plane $ax + by + cz = 0$ cuts the cone $xy + yz + zx = 0$ in perpendicular lines if
   (a) $a + b + c = 0$ (b) $1/a + 1/b + 1/c = 0$
   (c) $a^2 + b^2 + c^2 = 0$ (d) $abc = 0$.
6. The equation of the cylinder which intersects the curve $x^2 + y^2 + z^2 = 1$, $x + y + z = 1$ and whose generators are parallel to the axis of $z$, is
   (a) $x^2 + y^2 + xy - x - y = 0$ (b) $x^2 + y^2 + xy + x + y = 0$
   (c) $x^2 + y^2 - xy - x - y = 0$ (d) $x^2 + y^2 - xy + x + y = 0$.
7. The equation $x^2 + y^2 + z^2 + xy + yz - zx = 9$ represents
   (a) a sphere with $x^2 + y^2 + z^2 = 9$, $x - y + z = 3$ as a great circle
   (b) a cone with $x^2 + y^2 + z^2 = 9$, $x - y + z = 3$ as a guiding circle
   (c) a cylinder with $x^2 + y^2 + z^2 = 9$, $x - y + z = 3$ as a guiding circle
   (d) none of the above.
8. The sum of the direction cosines of a straight line is
   (a) zero (b) one (c) constant (d) $n$ one of the above.
9. The angle between the planes $x - y + 2z - 9 = 0$ and $2x + y + z = 7$ is
   (a) 30° (b) 90° (c) 60° (d) 120° (*V.T.U., 2010 S*)
10. The equation of the right circular cone whose axis is $x = y = z$, vertex is the origin and the semi-vertical angle is 45° is given as
    (a) $x^2 + y^2 + z^2 = 0$ (b) $2(x^2 + y^2 + z^2) = 3(x + y + z)^2$
    (c) $3(x^2 + y^2 + z^2) = 2(x + y + z)^2$ (d) $x^2 + y^2 + z^2 + xy + yz + zx = 0$.
11. The equation of a straight line parallel to the $x$-axis is given by
    (a) $\frac{x-a}{1} = \frac{y-b}{1} = \frac{z-c}{1}$ (b) $\frac{x-a}{0} = \frac{y-b}{1} = \frac{z-c}{1}$
    (c) $\frac{x-a}{0} = \frac{y-b}{0} = \frac{z-c}{1}$ (d) $\frac{x-a}{1} = \frac{y-b}{0} = \frac{z-c}{0}$.
12. The equation of the plane passing through (4, –2, 1) and perpendicular to the line with direction ratios 7, 2, – 3 is
    (a) $x + 3y - 4z - 8 = 0$ (b) $2x + 7y - 3z - 24 = 0$
    (c) $7x + 2y - 3z - 21 = 0$. (d) $7x + 3y - 2z + 21 = 0$.
13. The equation to the axis of the right circular cylinder whose guiding circle is $x^2 + y^2 + z^2 = 9$, $x - y + z = 3$ is given by
    (a) $x = y = z$ (b) $x = -y = z$ (c) $x = y = -z$ (d) $x = -y = -z$.
14. In three dimensions, the equation $x^2 - y^2 = a^2$ represents
    (a) a pair of straight lines (b) a hyperbola
    (c) a cylinder (d) a cone.
15. Section of a sphere by a plane is
    (a) parabola (b) ellipse (c) circle.
16. A line makes angles $\alpha$, $\beta$, $\gamma$ with the coordinate axes, then $\cos 2\alpha + \cos 2\beta + \cos 2\gamma$ is equal to
    (a) 1 (b) 2 (c) –1 (d) –2. (*V.T.U., 2010 S*)
17. Three lines are coplanar if
    (a) they are concurrent
    (b) a line is perpendicular to each of them
    (c) they are concurrent and a line is perpendicular to each of them.
18. The distance between the planes $2x + 2y + z - 6 = 0$ and $4x + 4y + 2z - 7 = 0$ is
    (a) 1/3 (b) 5/6 (c) 13/3.
19. The line $\frac{x-1}{1} = \frac{y-2}{2} = \frac{z-3}{3}$ and the plane $x + y - z = 0$ are
    (a) parallel (b) perpendicular (c) such that the line lies in the plane.

20. The radius of a great circle of a sphere is
(*a*) greater than the radius of the sphere (*b*) less than the radius of the sphere
(*c*) equal to the radius of the sphere.

21. Which of the following lines are generators of the cone $yz + 4zx + 3xy = 0$ ?
(*a*) $x = y = z$ (*b*) $x = -y = z$ (*c*) $x = 2y = -z$.

22. The semi-vertical angle of the cone generated by revolving the line $x + y = 0, z = 0$ about the $x$-axis is
(*a*) 90° (*b*) 45° (*c*) 30°.

23. All cones passing through the coordinate axes are given by the equation
(*a*) $x^2 + y^2 + z^2 - yz - zx - xy = 0$ (*b*) $ax^2 + by^2 + cz^2 - yz - zx - xy = 0$
(*c*) $ayz + bzx + cxy = 0$.

24. The line $\frac{x+1}{3} = \frac{y-2}{6} = \frac{z-3}{9}$ is perpendicular to the plane $ax + by + cz + d = 0$, if
(*a*) $a = 2b, b = 3c$ (*b*) $2a = b, b = 3c$ (*c*) $2a = b, 3b = 2c$ (*d*) $a = 3b, 2b = c$.

25. The equation $2(x^2 + y^2 + z^2) - 2xy + 2yz + 2zx = 3a^2$ represents a
(*a*) cone (*b*) right-circular cylinder
(*c*) sphere (*d*) pair of planes.

26. The equation of the plane through the point (2, –3, 1) and parallel to the plane $3x - 4y + 2z = 5$ is
(*a*) $3x - 4y + 2z - 20 = 0$ (*b*) $3x + 4y - 2z + 20 = 0$
(*c*) $3x - 4y - 2z + 20 = 0$ (*d*) $3x + 4y + 2z - 20 = 0$.

27. The direction cosines of a line which is equally inclined to the coordinate axes are ........

28. The direction cosines of the line $x = 0 = y$ are .......

29. The equation of the axis of the cylinder $x^2 + y^2 = 25$ is ......

30. The image of the point (3, 2, –1) in the YOZ plane is ......

31. The plane $x - 2y - 2z = k$ touches the sphere $x^2 + y^2 + z^2 - 2x + 4y - 6z + 5 = 0$ for $k =$ ...... (*P.T.U., 2010*)

32. The condition for the three concurrent lines to be coplanar is ......

33. The equation of the cone whose vertex is at the origin and base the circle $x = a, y^2 + z^2 = b^2$ is given by .......

34. The plane through points (2, 2, 1), (9, 3, 6) and perpendicular to the plane $2x + 6y + 6z = 9$ is ......

35. Volume of the sphere $x^2 + y^2 + z^2 + 2x - 4y + 8z - 2 = 0$ is ......

36. Angle between the planes $x - y + z = 1$ and $2x - 3y + z = 7$ is ......

37. The equation of the cone whose vertex is the origin and grinding curve is $\frac{x^2}{4} + \frac{y^2}{9} + \frac{z^2}{1} = 1, x + y + z = 1$, is ......

38. Any two points on the line $\frac{x-1}{2} = \frac{y-2}{3} = \frac{z-3}{4}$ other than (1, 2, 3) are ......

39. The equation of the line joining the points (1, 2, 3) and (2, 1, – 3) is ......

40. The equation of the sphere whose centre is (1, 2, – 1) and which touches the plane $2x - y + z + 3 = 0$ is ......
(*Anna, 2013*)

41. The conditions for the line $\frac{x - x_1}{l} = \frac{y - y_1}{m} = \frac{z - z_1}{n}$ to lie on the plane $ax + by + cz + d = 0$ are ......

42. The distance between the planes $4x + 3y + z + 4 = 0$ and $8x + 6y + 2z + 12 = 0$ is .......

43. The centre and radius of the sphere $2x^2 + 2y^2 + 2z^2 - 6x + 8y - 8z - 1 = 0$ are ......

44. The radius of the circle $x^2 + y^2 + z^2 - 2x - 4y - 11 = 0, x + 2y + 2z = 15$ is ......

45. The symmetric form of the line $x + y + z + 1 = 0 = 4x + y - 2z + 2$ is ......

46. The equation $y^2 = 4z - 8$ represents a .......

47. The equation $x^2 + y^2 = \frac{1}{4}z^2 - 1$ represents a ......

48. Angle between the lines whose d.r.s. are 1, 2, 3 and –1, 1, 2 is ......

49. The intercepts of the plane $2x - 3y + z = 12$ on the coordinate axes are .......

50. The radius of the sphere whose centre is (4, 4, –2) and which passes through the origin is ......

51. The points (0, 4, 1), (2, 3, –1), (4, 5, 0) and (2, 6, 2) are the vertices of a square. (True or False)

52. The points (3, –1, 1), (5, –4, 2) and (11, –13, 5) are collinear. (True or False)

53. The plane $5x + 6y + 7z = 110$, $2x + 3y - 4z = 29$ are perpendicular to each other. (True or False)

54. In three dimensional space, $9x^2 + 16y^2 = 144$ represents .........

55. Equation of the right circular cone with vertex at origin and passing through the curve $x^2 + y^2 + z^2 = 9$, $x + y + z = 1$ is .......... .

56. A unit vector perpendicular to the vectors $-2\mathbf{I} + 3\mathbf{J} + \mathbf{K}$ and $4\mathbf{I} + 2\mathbf{J}$ is ........... .

57. The equation of the plane which is parallel to $x + 3y + 5z + 1 = 0$ and 5 units from the origin is ......... . *(Anna, 2013)*

CHAPTER 4

# Differential Calculus and its Applications



## 4.1 (1) SUCCESSIVE DIFFERENTIATION

The reader is already familiar with the process of differentiating a function $y = f(x)$. For ready reference, a list of derivatives of some standard functions is given in the beginning.

The derivative $dy/dx$ is, in general, another function of $x$ which can be differentiated. The derivative of $dy/dx$ is called the *second derivative* of $y$ and is denoted by $d^2y/dx^2$. Similarly, the derivative of $d^2y/dx^2$ is called the *third derivative* of $y$ and is denoted by $d^3y/dx^3$. In general, the $n$th derivative of $y$ is denoted by $d^ny/dx^n$.

Alternative notations for the successive derivatives of $y = f(x)$ are

$$Dy, D^2y, D^3y, \ldots, D^ny\ ;$$

or $$y_1, y_2, y_3, \ldots, y_n\ ;$$

or $$f'(x), f''(x), f'''(x) \ldots, f^n(x).$$

The $n$th derivative of $y = f(x)$ at $x = a$ is denoted by $(d^ny/dx^n)_a$, $(y_n)_a$ or $f^n(a)$.

**Example 4.1.** *If $y = e^{ax} \sin bx$, prove that $y_2 - 2ay_1 + (a^2 + b^2)y = 0$.*

**Solution.** We have $$y = e^{ax} \sin bx \quad \ldots(i)$$

$\therefore$ $$y_1 = e^{ax} (\cos bx \,.\, b) + \sin bx\, (e^{ax} \,.\, a) = be^{ax} \cos bx + ay \quad [\text{By } (i)]$$

or $$y_1 - ay = be^{ax} \cos bx \quad \ldots(ii)$$

Again differentiating both sides,

$$y_2 - ay_1 = be^{ax} (-\sin bx \,.\, b) + b \cos bx\, (e^{ax} \,.\, a) = -b^2y + a(y_1 - ay)$$

or $$y_2 - 2ay_1 + (a^2 + b^2)y = 0.$$

**Example 4.2.** *If $x = a (\cos t + t \sin t)$, $y = a (\sin t - t \cos t)$, find $d^2y / dx^2$.*

**Solution.** We have $$\frac{dx}{dt} = a (-\sin t + t \cos t + \sin t) = at \cos t$$

and $$\frac{dy}{dt} = a (\cos t + t \sin t - \cos t) = at \sin t$$

$$\therefore \quad \frac{dy}{dx} = \frac{dy}{dt}\Big/\frac{dx}{dt} = \frac{at \sin t}{at \cos t} = \tan t$$

$$\therefore \quad \frac{d^2y}{dx^2} = \frac{d}{dt}(\tan t) \cdot \frac{dt}{dx} = \sec^2 t \cdot \frac{1}{at \cos t} = 1/at \cos^3 t.$$

**Example 4.3.** *Given $y^2 = f(x)$, a polynomial of third degree, then evaluate $\frac{d}{dx}\left(y^3 \frac{d^2y}{dx^2}\right)$.*

**Solution.** Differentiating $y^2 = f(x)$ w.r.t. $x$, we get

$$2y \frac{dy}{dx} = f(x) \quad \ldots(i)$$

Differentiating Eqn. (*i*) w.r.t. $x$ again, we obtain

$$2\left(\frac{dy}{dx} \cdot \frac{dy}{dx} + y \frac{d^2y}{dx^2}\right) = f''(x) \quad \text{or} \quad 2\left(\frac{dy}{dx}\right)^2 + 2y \frac{d^2y}{dx^2} = f''(x)$$

Again differentiating, we get

$$4 \cdot \frac{dy}{dx} \frac{d^2y}{dx^2} + 2 \frac{dy}{dx} \frac{d^2y}{dx^2} + 2y \frac{d^3y}{dx^3} = f'''(x)$$

or
$$3y^2 \frac{dy}{dx} \frac{d^2y}{dx^2} + y^3 \frac{d^3y}{dx^3} = \frac{1}{2} y^2 f'''(x) \quad \text{[Multiplying by } y^2\text{]}$$

Hence
$$\frac{d}{dx}\left(y^3 \frac{d^2y}{dx^2}\right) = \frac{1}{2} f(x) f'''(x). \quad [\because \; y^2 = f(x)]$$

**Example 4.4.** *If $ax^2 + 2hxy + by^2 = 1$, prove that $\frac{d^2y}{dx^2} = \frac{h^2 - ab}{(hx + by)^3}$.*

**Solution.** Differentiating the given equation w.r.t. $x$,

$$2ax + 2h\left(x \frac{dy}{dx} + y\right) + 2by \frac{dy}{dx} = 0 \quad \text{or} \quad \frac{dy}{dx} = -\frac{ax + hy}{hx + by} \quad \ldots(i)$$

Differentiating both sides of Eqn. (*i*) w.r.t. $x$,

$$\frac{d^2y}{dx^2} = -\frac{(hx + by)(a + h\,dy/dx) - (ax + hy)(h + b\,dy/dx)}{(hx + by)^2}$$

[Substituting the value of $dy/dx$ from Eqn. (*i*)]

$$= -\frac{(hx + by)\left(a - h \cdot \frac{ax + hy}{hx + by}\right) - (ax + hy)\left(h - b \cdot \frac{ax + hy}{hx + by}\right)}{(hx + by)^2}$$

$$= \frac{(h^2 - ab)(ax^2 + 2hxy + by^2)}{(hx + by)^3}$$

$$= (h^2 - ab)/(hx + by)^3 \quad [\because \; ax^2 + 2hxy + by^2 = 1]$$

## PROBLEMS 4.1

1. If $y = (ax + b)/(cx + d)$, show that $2y_1 y_3 = 3y_2^2$.
2. If $y = \sin(\sin x)$, prove that $\frac{d^2y}{dx^2} + \tan x \frac{dy}{dx} + y \cos^2 x = 0$.
3. If $y = e^{-kt} \cos(lt + c)$, show that $\frac{d^2y}{dx^2} + 2k \frac{dy}{dx} + n^2 y = 0$, where $n^2 = k^2 + l^2$.

4. If $y = \sinh [m \log \{x + \sqrt{(x^2+1)}\}]$, show that $(x^2+1)\dfrac{d^2y}{dx^2} + x\dfrac{dy}{dx} = m^2y$.
5. If $y = \sin^{-1} x$, show that $(1 - x^2)\, y_5 - 7xy_4 - 9y_3 = 0$.
6. If $x = \dfrac{1}{2}\left(t + \dfrac{1}{t}\right), y = \dfrac{1}{2}\left(t - \dfrac{1}{t}\right)$, find $\dfrac{d^2y}{dx^2}$.
7. If $x = 2\cos t - \cos 2t$, $y = 2 \sin t - \sin 2t$, find the value of $d^2y/dx^2$ when $t = \pi/2$.
8. If $x = a(\cos t + \log \tan t/2)$, $y = a \sin t$, find $d^2y/dx^2$.
9. If $x = \sin t$, $y = \sin pt$, prove that $(1 - x^2)\dfrac{d^2y}{dx^2} - x\dfrac{dy}{dx} + p^2y = 0$.
10. If $x^3 + y^3 = 3axy$, prove that $\dfrac{d^2y}{dx^2} = -\dfrac{2a^2xy}{(y^2-ax)^3}$.

**(2) Standard Results**

*We have* (1) $\mathbf{D^n (ax + b)^m = m(m-1)(m-2) \ldots (m-n+1)\, a^n (ax+b)^{m-n}}$

(2) $\mathbf{D^n \left(\dfrac{1}{ax+b}\right) = \dfrac{(-1)^n (n!)\, a^n}{(ax+b)^{n+1}}}$

(3) $\mathbf{D^n \log (ax+b) = \dfrac{(-1)^{n-1}(n-1)!\, a^n}{(ax+b)^n}}$

(4) $\mathbf{D^n (a^{mx}) = m^n (\log a)^n \,.\, a^{mx}}$

(5) $\mathbf{D^n (e^{mx}) = m^n e^{mx}}$

(6) $\mathbf{D^n \sin (ax+b) = a^n \sin (ax + b + n\pi/2)}$

(7) $\mathbf{D^n \cos (ax+b) = a^n \cos (ax + b + n\pi/2)}$

(8) $\mathbf{D^n [e^{ax} \sin (bx + c)] = (a^2+b^2)^{n/2} e^{ax} \sin (bx + c + n \tan^{-1} b/a)}$

(9) $\mathbf{D^n [e^{ax} \cos (bx + c)] = (a^2+b^2)^{n/2} e^{ax} \cos (bx + c + n \tan^{-1} b/a)}$

*To prove* (1), let $y = (ax+b)^m$

$y_1 = m \,.\, a(ax+b)^{m-1}$

$y_2 = m(m-1)a^2 (ax+b)^{m-2}$

$y_3 = m(m-1)(m-2)\, a^3 (ax+b)^{m-3}$

..................................................

Hence $y_n = m(m-1)(m-2) \ldots (m-n+1)\, a^n (ax+b)^{m-n}$

In particular, $\mathbf{D^n (x^n) = n!}$.

(2) follows from (1) by taking $m = -1$. The proof of (3) is left as an exercise for the student.

*To prove* (4), let $y = a^{mx}$

$y_1 = m \log a \,.\, a^{mx}$, $y_2 = (m \log a)^2 a^{mx}$, etc.

In general $y_n = (m \log a)^n a^{mx}$.

(5) follows from (4) by taking $a = e$.

*To prove* (6), let $y = \sin (ax+b)$

$\therefore$ $y_1 = a \cos (ax+b) = a \sin (ax + b + \pi/2)$

$y_2 = a^2 \cos (ax + b + \pi/2) = a^2 \sin (ax + b + 2\pi/2)$

$y_3 = a^3 \cos (ax + b + 2\pi/2) = a^3 \sin (ax + b + 3\pi/2)$

..................................................

In general, $y_n = a^n \sin (ax + b + n\pi/2)$.

The proof of (7), is left as an exercise for the reader.

*To prove* (8), let $y = e^{ax} \sin (bx + c)$

$\therefore$ $y_1 = e^{ax} \cos (bx + c) \,.\, b + ae^{ax} \sin (bx + c)$

$= e^{ax} [a \sin (bx + c) + b \cos (bx + c)]$

Put $a = r\cos\alpha$, $b = r \sin \alpha$ so that $r = \sqrt{(a^2+b^2)}$, $\alpha = \tan^{-1} b/a$

$\therefore$ $y_1 = re^{ax} [\sin (bx + c) \cos \alpha + \cos (bx + c) \sin \alpha]$

$= re^{ax} \sin (bx + c + \alpha)$

Similarly, $y_2 = r^2 e^{ax} \sin (bx + c + 2\alpha)$

$y_3 = r^3 e^{ax} \sin (bx + c + 3\alpha)$

..................................................

In general, $y_n = r^n e^{ax} \sin(bx + c + n\alpha)$

where, $r = \sqrt{(a^2 + b^2)}$ and $\alpha = \tan^{-1} b/a$.

Proceeding as in (8), the student should prove (9) himself.

(3) **Preliminary transformations.** Quite often preliminary simplification reduces the given function to one of the above standard forms and then the $n$th derivative can be written easily.

*To find the nth derivative of the powers of sines or cosines or their products, we first express each of these as a series of sines or cosines of multiple angles and then use the above formulae (6) and (7).*

**Example 4.5.** *If* $y = x \log \frac{x-1}{x+1}$, *show that* $y_n = (-1)^{n-2}(n-2)!\left[\frac{x-n}{(x-1)^n} - \frac{x+n}{(x+1)^n}\right]$.

**Solution.** Differentiating $y$ w.r.t. $x$, we have

$$y_1 = \log\frac{x-1}{x+1} + x\left[\frac{1}{x-1} - \frac{1}{x+1}\right]$$

$$= \log(x-1) - \log(x+1) + \frac{1}{x-1} + \frac{1}{x+1} \quad \text{...}(i)$$

Now, differentiating Eqn. $(i)$ $(n-1)$ times w.r.t. $x$,

$$y_n = \frac{(-1)^{n-2}(n-2)!}{(x-1)^{n-1}} - \frac{(-1)^{n-2}(n-2)!}{(x+1)^{n-1}} + \frac{(-1)^{n-1}(n-1)!}{(x-1)^n} + \frac{(-1)^{n-1}(n-1)!}{(x+1)^n}$$

$$= (-1)^{n-2}(n-2)!\left\{\frac{x-1}{(x-1)^n} - \frac{x+1}{(x+1)^n} + \frac{-(n-1)}{(x-1)^n} + \frac{-(n-1)}{(x+1)^n}\right\}$$

$$= (-1)^{n-2}(n-2)!\left[\frac{x-n}{(x-1)^n} - \frac{x+n}{(x+1)^n}\right].$$

**Example 4.6.** *Find the nth derivative of* $(i)$ $\cos x \cos 2x \cos 3x$
$(ii)$ $e^{2x} \cos^2 x \sin x$.

**Solution.** $(i)$ $y = \cos x \cos 2x \cos 3x = \frac{1}{2}\cos x(\cos 5x + \cos x)$

$$= \frac{1}{4}(2\cos x \cos 5x + 2\cos^2 x) = \frac{1}{4}[(\cos 6x + \cos 4x) + (1 + \cos 2x)]$$

$$= \frac{1}{4}(1 + \cos 2x + \cos 4x + \cos 6x)$$

$$\therefore \quad y_n = \frac{1}{4}[2^n \cos(2x + n\pi/2) + 4^n \cos(4x + n\pi/2 + 6^n \cos(6x + n\pi/2)]$$

$(ii)$ $\cos^2 x \sin x = \cos x(\sin x \cos x) = \cos x \,.\, \frac{1}{2}\sin 2x$

$$= \frac{1}{4}(2\sin 2x \cos x) = \frac{1}{4}(\sin 3x + \sin x)$$

$$\therefore \quad D^n(e^{2x}\cos^2 x \sin x) = \frac{1}{4}[D^n(e^{2x}\sin 3x) + D^n(e^{2x}\sin x)]$$

$$= \frac{1}{4}[(2^2 + 3^2)^{n/2}\sin(3x + n\tan^{-1} 3/2) + (2^2 + 1^2)^{n/2}\sin(x + n\tan^{-1}\tfrac{1}{2})]$$

$$= \frac{1}{4}[(13)^{n/2}\sin(3x + n\tan^{-1} 3/2) + (5)^{n/2}\sin(x + n\tan^{-1}\tfrac{1}{2})].$$

**(4) Use of partial fractions.** To find the $n$th derivative of any rational algebraic fraction, we first split it up into partial fractions. Even when the denominator cannot be resolved into real factors, the method of partial fractions can still be used after breaking the denominator into complex linear factors. Then to put the result back in a real form, we apply De Moivre's theorem (p. 756).

**Example 4.7.** *Find the nth derivative of* $\frac{x}{(x-1)(2x+3)}$. *(V.T.U., 2016)*

**Solution.** $$\frac{x}{(x-1)(2x+3)} = \frac{1}{(x-1)(2.1+3)} + \frac{-3/2}{(-3/2-1)(2x+3)}$$
$$= \frac{1}{5}\cdot\frac{1}{x-1} + \frac{3}{5}\cdot\frac{1}{2x+3}$$

Hence $$D^n\left[\frac{x}{(x-1)(2x+3)}\right] = \frac{1}{5}\cdot\frac{(-1)^n\, n!}{(x-1)^{n+1}} + \frac{3}{5}\cdot\frac{(-1)^n\,(n!)2^n}{(2x+3)^{n+1}}$$
$$= \frac{(-1)^n\, n!}{5}\left\{\frac{1}{(x-1)^{n+1}} + \frac{3.2^n}{(2x+3)^{n+1}}\right\}.$$

**Example 4.8.** *Find the nth derivative of* $\frac{1}{x^2+a^2}$.

**Solution.** We have $$y = \frac{1}{x^2+a^2} = \frac{1}{(x+ia)(x-ia)} = \frac{1}{2ia}\left(\frac{1}{x-ia} - \frac{1}{x+ia}\right)$$

$\therefore$ $$y_n = \frac{1}{2ia}\left\{\frac{(-1)^n\, n!}{(x-ia)^{n+1}} - \frac{(-1)^n\, n!}{(x+ia)^{n+1}}\right\}$$

[Put $x = r\cos\theta$, $a = r\sin\theta$ so that $r = \sqrt{(x^2+a^2)}$, $\theta = \tan^{-1}(a/x)$]

$$= \frac{(-1)^n\, n!}{2ia}\left\{\frac{1}{r^{n+1}(\cos\theta - i\sin\theta)^{n+1}} - \frac{1}{r^{n+1}(\cos\theta + i\sin\theta)^{n+1}}\right\}$$
$$= \frac{(-1)^n\, n!}{2iar^{n+1}}\{(\cos\theta - i\sin\theta)^{-(n+1)} - (\cos\theta + i\sin\theta)^{-(n+1)}\}$$
$$= \frac{(-1)^n\, n!}{2iar^{n+1}}[\cos(n+1)\theta + i\sin(n+1)\theta - [\cos(n+1)\theta - i\sin(n+1)\theta]\}$$

[By De Moivre's theorem]

$$= \frac{(-1)^n\, n!}{2iar^{n+1}}\,.\,2i\sin(n+1)\theta \qquad \left[\text{Put } \frac{1}{r} = \frac{\sin 0}{a}\right]$$
$$= \frac{(-1)^n\, n!}{a^{n+2}}\sin(n+1)\theta\,\sin^{n+1}\theta.$$

## PROBLEMS 4.2

Find the $n$th derivative of (1 to 11) :

1. $\log(4x^2-1)$ *(V.T.U., 2010)*
2. $\frac{x+2}{x+1} + \log\frac{x+2}{x+1}$
3. $\cos 2x\cos 3x$ *(V.T.U., 2016)*
4. $x^3\cos x$
5. $e^x(2x+3)^3$
6. $\sinh 2x\sin 4x$ *(V.T.U., 2010 S)*
7. $e^{5x}\cos x\cos 3x$
8. $\frac{x^4}{(x-1)(x-2)}$ *(Delhi, 2012, VTU, 2019)*
9. $\frac{x^2}{2x^2+7x+6}$ *(V.T.U., 2013)*
10. $\frac{1}{1+x+x^2+x^3}$
11. $e^x[a^2x^2 - 2nax + n(n+1)]$
12. Find the $n$th derivative of $\tan^{-1}\frac{2x}{1-x^2}$ in terms of $r$ and $\theta$.
13. If $y(x) = \sin px + \cos px$, prove that $y_n(x) = p^n\{1+(-1)^n\sin 2px\}^{1/2}$.
Hence, show that $y_8(\pi) = (1/2)^{31/2}$ when $p = 1/4$. *(U.P.T.U., 2012)*

## 4.2 LEIBNITZ'S THEOREM for the *n*th Derivative of the product of two functions*

*If u, v be two function of x possessing derivatives of the nth order, then*

$$(\mathbf{uv})_n = \mathbf{u}_n \mathbf{v} + {}^n\mathbf{C}_1 \mathbf{u}_{n-1} \mathbf{v}_1 + {}^n\mathbf{C}_2 \mathbf{u}_{n-2} \mathbf{v}_2 + \ldots + {}^n\mathbf{C}_r \mathbf{u}_{n-r} \mathbf{v}_r + \ldots + {}^n\mathbf{C}_n \mathbf{uv}_n$$

We shall prove this theorem by mathematical induction.

*Step I.* By actual differentiation,

$$(uv)_1 = u_1 v + uv_1$$

$$(uv)_2 = (u_2 v + u_1 v_1) + (u_1 v_1 + uv_2)$$

$$= u_2 v + {}^2C_1 u_1 v_1 + {}^2C_2 uv_2 \qquad [\because \; 2 = {}^2C_1, 1 = {}^2C_2]$$

Thus we see that the theorem is true for $n = 1, 2$.

*Step II.* Assume the theorem to be true for $n = m$ (say) so that

$$(uv)_m = u_m v + {}^mC_1 u_{m-1} v_1 + {}^mC_2 u_{m-2} v_2 + \ldots + {}^mC_{r-1} u_{m-r+1} v_{r-1} + {}^mC_r u_{m-r} v_r + \ldots + {}^mC_m uv_m$$

Differentiating both sides,

$$(uv)_{m+1} = (u_{m+1} v + u_m v_1) + {}^mC_1 (u_m v_1 + u_{m-1} v_2) + {}^mC_2 (u_{m-1} v_2 + u_{m-2} v_3) + \ldots$$

$$+ {}^mC_{r-1} (u_{m-r+2} v_{r-1} + u_{m-r+1} v_r) + {}^mC_r (u_{m-r+1} v_r + u_{m-r} v_{r+1}) + \ldots$$

$$+ {}^mC_m (u_1 v_m + uv_{m+1})$$

$$= u_{m+1} v + (1 + {}^mC_1) u_m v_1 + ({}^mC_1 + {}^mC_2) u_{m-1} v_2 + \ldots$$

$$+ ({}^mC_{r-1} + {}^mC_r) u_{m-r+1} v_r + \ldots + {}^mC_m uv_{m+1}$$

But $\quad 1 + {}^mC_1 = {}^mC_0 + {}^mC_1 = {}^{m+1}C_1, {}^mC_1 + {}^mC_2 = {}^{m+1}C_2 \ldots$

$${}^mC_{r-1} + {}^mC_r = {}^{m+1}C_r, \ldots \text{ and } {}^mC_m = 1 = {}^{m+1}C_{m+1}$$

$\therefore$

$$(uv)_{m+1} = u_{m+1} v + {}^{m+1}C_1 u_m v_1 + {}^{m+1}C_2 u_{m-1} v_2 + \ldots + {}^{m+1}C_r u_{m+1-r} v_r + \ldots + {}^{m+1}C_{m+1} uv_{m+1}$$

which is of exactly the same form as the given formula with $n$ replaced by $m + 1$. Hence if the theorem is true for $n = m$, it is also true for $n = m + 1$.

*Step III.* In Step I, the theorem has been seen to be true for $n = 2$, and by Step II, it must be true for $n = 2 + 1$ *i.e.*, 3 and so for $n = 3 + 1$ *i.e.*, 4 and so on.

Hence the theorem is true for all positive integral values of $n$.

**Obs.** Leibnitz's theorem is obtained by expanding $(u + v)^n$ by Binomial theorem and replacing powers of $u$ and $v$ by subscripts which are order of derivatives.

**Example 4.9.** *Find the nth derivative of $e^x (2x + 3)^3$.*

**Solution.** Take $u = e^x$ and $v = (2x + 3)^3$, so that $u_n = e^x$ for all integral values of $n$, and $v_1 = 6(2x + 3)^2$, $v_2 = 24(2x + 3)$, $v_3 = 48$, $v_4$, $v_5$ etc. are all zero.

$\therefore$ By Leibnitz's theorem,

$$(uv)_n = u_n v + {}^nC_1 u_{n-1} v_1 + {}^nC_2 u_{n-2} v_2 + {}^nC_3 u_{n-3} v_3$$

*i.e.*,

$$[e^x (2x + 3)^3]_n = e^x (2x + 3)^3 + ne^x [6 (2x + 3)^2]$$

$$+ \frac{n(n-1)}{1.2} e^x [24(2x + 3)] + \frac{n(n-1)(n-2)}{1.2.3} e^x [48]$$

$$= e^x \{(2x + 3)^3 + 6n(2x + 3)^2 + 12n(n-1)(2x + 3) + 8n(n-1)(n-2)\}.$$

**Example 4.10.** *If $y = (\sin^{-1} x)^2$, show that $(1 - x^2) y_{n+2} - (2n + 1) xy_{n+1} - n^2 y_n = 0$. Hence find $(y_n)_0$.*
*(Delhi., 2015 ; C.S.V.T.U, 2012)*

**Solution.** We have $\qquad y = (\sin^{-1} x)^2$

---

*Named after the German mathematician and philosopher *Gottfried Wilhelm Leibnitz* (1646–1716) who invented the differential and integral calculus independent of Sir Issac Newton.

Differentiating, $y_1 = \frac{2\sin^{-1}x}{\sqrt{(1-x^2)}}$ or $(1-x^2)\,y_1^2 = 4\,(\sin^{-1}x)^2 = 4y$ ...(*i*)

Again differentiating, $(1-x^2)\,2y_1y_2 + (-2x)\,y_1^2 = 4y_1$ ...(*ii*)

Dividing by $2y_1$, $(1-x^2)\,y_2 - xy_1 - 2 = 0$

Differentiating it $n$ times by Leibnitz's theorem,

$$(1-x^2)\,y_{n+2} + n(-2x)\,y_{n+1} + \frac{n(n-1)}{2}(-2)y_n - [xy_{n+1} + n(1)\,y_n] = 0$$

or $$(1-x^2)\,y_{n+2} - (2n+1)\,xy_{n+1} - n^2\,y_n = 0$$

which is the required result.

Putting $x = 0$, $(y_{n+2})_0 = n^2(y_n)_0$ ...(*iii*)

From Eqn. (*i*), $(y_1)_0 = 0$. From Eqn. (*ii*), $(y_2)_0 = 2$.

Putting $n = 1, 3, 5, 7, \ldots$ in Eqn. (*iii*), $0 = y_1 = y_3 = y_5 = y_7 = \ldots$

*i.e.*, if $n$ is odd, $(y_n)_0 = 0$

Again putting $n = 2, 4, 6, \ldots$ in Eqn. (*iii*)

$$(y_4)_0 = 2^2\,(y_2)_0 = 2\,.\,2^2$$
$$(y_6)_0 = 4^2\,(y_6)_0 = 2.\,2^2\,.\,4^2$$
$$(y_8)_0 = 6^2\,(y_6)_0 = 2.\,2^2\,.\,4^2\,.\,6^2$$

In general, if $n$ is even, $(y_n)_0 = 2\,.\,2^2\,.\,4^2\,.\,6^2 \ldots (n-2)^2,\ (n \neq 2).$

**Example 4.11.** *If $y = e^{a\sin^{-1}x}$, prove that $(1-x^2)\,y_{n+2} - (2n+1)\,x\,y_{n+1} - (n^2+a^2)y_n = 0$. Hence find the value of $y_n$ when $x = 0$.* (*W.B.T.U., 2016 ; C.S.V.T.U., 2013 ; Kerala, 2010*)

**Solution.** We have $y = e^{a\sin^{-1}x}$ ...(*i*)

Differentiating, $y_1 = e^{a\sin^{-1}x}\frac{a}{\sqrt{(1-x^2)}} = \frac{ay}{\sqrt{(1-x^2)}}$ ...(*ii*)

or $(1-x^2)\,y_1^2 = a^2\,y^2.$

Again differentiating, $(1-x^2)\,2y_1y_2 + (-2x)y_1^2 = 2a^2\,yy_1.$

Dividing by $2y_1$, $(1-x^2)y_2 - xy_1 - a^2y = 0$ ...(*iii*)

Differentiating it $n$ times by Leibnitz's theorem,

$$(1-x^2)\,y_{n+2} + n\,.\,(-2x)\,y_{n+1} + \frac{n(n-1)}{2}\,.\,(-2)\,y_n - [xy_{n+1} + n\,.\,1\,.\,y_n] - a^2\,y_n = 0$$

or $$(1-x^2)y_{n+2} - (2n+1)\,xy_{n+1} - (n^2+a^2)\,y_n = 0$$

which is the required result.

Putting $x = 0$, $(y_{n+2})_0 = (n^2+a^2)\,(y_n)_0$ ...(*iv*)

From Eqns. (*i*), (*ii*), (*iii*) : $(y)_0 = 1,\ (y_1)_0 = a,\ (y_2)_0 = a^2$

Putting $n = 1, 2, 3, 4 \ldots$ in Eqn. (*iv*),

$$(y_3)_0 = (1^2+a^2)\,(y_1)_0 = a(1^2+a^2)$$
$$(y_4)_0 = (2^2+a^2)\,(y_2)_0 = a^2(2^2+a^2)$$
$$(y_5)_0 = (3^2+a^2)\,(y_3)_0 = a(1^2+a^2)\,(3^2+a^2)$$
$$(y_6)_0 = (4^2+a^2)\,(y_4)_0 = a^2(2^2+a^2)\,(4^2+a^2).$$

Hence in general, $(y_n)_0 = a(1^2+a^2)\,(3^2+a^2) \ldots [(n-2)^2+a^2]$, when $n$ is odd.

$= a^2(2^2+a^2)\,(4^2+a^2) \ldots\ldots [(n-2)^2+a^2]$, when $n$ is even.

**Example 4.12.** *If $y^{1/m} + y^{-1/m} = 2x$, prove that*

$$(x^2-1)y_{n+2} + (2n+1)\,xy_{n+1} + (n^2-m^2)\,y_n = 0.$$

(*Gorakhpur, 2019 ; V.T.U., 2013 ; W.B.T.U., 2012 ; C.S.V.T.U., 2011*)

**Solution.** We have, $y^{1/m} + \frac{1}{y^{1/m}} = 2x$

or $(y^{1/m})^2 - 2x(y^{1/m}) + 1 = 0$

$\therefore$ $$y^{1/m} = \frac{2x \pm \sqrt{(4x^2-4)}}{2} = x \pm \sqrt{(x^2-1)}$$

Hence $$y = [x \pm \sqrt{(x^2 - 1)}\,]^m$$

Taking logarithm, $$\log y = m \log [x \pm \sqrt{(x^2 - 1)}\,]$$

Differentiating both sides w.r.t. $x$,

$$\frac{1}{y} y_1 = m \cdot \frac{1}{x \pm \sqrt{(x^2 - 1)}} \cdot \left\{1 \pm \frac{x}{\sqrt{(x^2 - 1)}}\right\} = \pm \frac{m}{\sqrt{(x^2 - 1)}}$$

Squaring, $y_1{}^2 (x^2 - 1) = m^2 y^2$

Again differentiating, $(x^2 - 1)\, 2y_1 y_2 + y_1{}^2 (2x) = m^2 \cdot 2y \cdot y_1$

Dividing by $2y_1$, $(x^2 - 1)\, y_2 + xy_1 - m^2 y = 0$

Differentiating it $n$ times by Leibnitz's theorem,

$$(x^2 - 1)\, y_{n+2} + ny_{n+1} (2x) + \frac{n(n-1)}{2} y_n (2) + xy_{n+1} + n \cdot y_n (1) - m^2 y_n = 0$$

or $$(x^2 - 1)\, y_{n+2} + (2n + 1)\, xy_{n+1} + (n^2 - m^2)\, y_n = 0.$$

**Example 4.13.** *If $x = \tan (\log y)$, Prove that*

$$(1 + x)^2 y_{n+1} + (2nx - 1)y_n + n(n-1)y_{n-1} = 0$$

*(KTU 2018)*

**Solution.** We have $x = \tan (\log y)$

$$\tan^{-1} x = \log y$$

$$y = e^{\tan^{-1} x} \qquad ...(i)$$

Differentiating Eqn. (*i*) *w.r.t.* $x$, we get $y_1 = e^{\tan^{-1}} \times \left(\frac{1}{1 + x^2}\right)$

$$(1 + x^2)y_1 = y \qquad ...(ii)$$

Differentiating Eqn. (*ii*) $n$ times by Leibnitz's theorem,

We get,

$$(1 + x^2)\, y_{n+1} + ny_n(2x) + \frac{n(n-1)}{2} y_{n-1}(2) = y_n$$

$$(1 + x^2)y_{n+1} + (2nx - 1)y_n + n(n-1)y_{n-1} = 0.$$

## PROBLEMS 4.3

1. Find the $n$th derivative of (*i*) $x^2 \log 3x$. *(C.S.V.T.U., 2014)*

   (*ii*) $x^2 e^{3x} \sin 4x$

2. If $y = a \cos (\log x) + b \sin (\log x)$, show that $x^2 y_2 + xy_1 + y = 0$ and $x^2 y_{n+2} + (2n + 1)\, xy_{n+1} + (n^2 + 1)\, y_n = 0$. Find $(y_n)_0$. *(U.P.T.U., 2012)*

3. If $y = \tan^{-1} x$, prove that $(1 + x^2)\, y_{n+1} + 2nxy_n + n(n-1)\, y_{n-1} = 0$. Find $y_{n(0)}$.

4. If $y = \cos (m \sin^{-1} x)$, prove that $(1 - x^2)\, y_{n+2} - (2n + 1)\, xy_{n+1} + (m^2 - n^2)y_n = 0$.

5. If $y = \sin (m \sin^{-1} x)$, prove that $(1 - x^2)\, y_2 - xy_1 + m^2 y = 0$

   and $(1 - x^2)\, y_{n+2} - 2(n + 1)\, xy_{n+1} + (m^2 - n^2)\, y_n = 0.$ *(D.T.U., 2013 ; U.K.T.U., 2013)*

   Also find $(y_n)_0$. *(U.P.T.U., 2011)*

6. If $y = e^{m \cos^{-1} x}$, prove that (*i*) $(1 - x^2)\, y_2 - xy_1 = m^2 y$

   (*ii*) $(1 - x^2)\, y_{n+2} - (2n + 1)\, xy_{n+1} - (n^2 + m^2)\, y_n = 0.$ *(V.T.U., 2013)*

   Also find $(y_n)_0$.

7. If $y = (x^2 - 1)^n$, prove that $(x^2 - 1) y_{n+2} + 2xy_{n+1} - n(n+1) y_n = 0$. (W.B.T.U., 2016)

Hence if $P_n = \frac{d^n}{dx^n}\{(x^2-1)^n\}$, show that $\frac{d}{dx}\left\{(1-x^2)\frac{dP_n}{dx}\right\} + n(n+1)P_n = 0$. (U.K.T.U., 2010)

8. If $\sin^{-1} y = 2\log(x+1)$, prove that $(x+1)^2 y_{n+2} + (2n+1)(x+1) y_{n+1} + (x^2+4) y_n = 0$. (V.T.U., 2015)

9. If $V_n = \frac{d^n}{dx^n}(x^n \log x)$, show that $V_n = nV_{n-1} + (n-1)!$

Hence, show that $V_n = n!\left(\log x + 1 + \frac{1}{2} + \frac{1}{3} + \ldots + \frac{1}{n}\right)$. (C.S.V.T.U., 2015)

10. If $x = \sin t$, $y = \cos pt$, show that $(1-x^2) y_2 - xy_1 + p^2 y = 0$. Hence prove that

$$(1-x^2) y_{n+2} - (2n+1) xy_{n+1} - (n^2 - p^2) y_n = 0.$$ (V.T.U., 2016; C.S.V.T.U., 2015)

11. If $y = [x + \sqrt{(x^2+1)}\,]^m$, prove that (i) $(x^2+1) y_2 + xy_1 - m^2 y = 0$, (ii) $y_{n+2} + (n^2 - m^2) y_n = 0$ at $x = 0$. (V.T.U., 2016)
Hence find $y_n(0)$. (U.P.T.U., 2013)

12. If $y = \cos(m \log x)$, prove that $x^2 y_{n+2} + (2n+1) xy_{n+1} + (m^2 + n^2) y_n = 0$. (Rohtak, 2011S)

13. If $y = \sin\log(x^2 + 2x + 1)$, prove that (i) $(x+1)^2 y_2 + (x+1) y_1 + 4y = 0$ (V.T.U., 2014)
(ii) $(x+1)^2 y_{n+2} + (2n+1)(x+1) y_{n+1} + (n^2+4) y_n = 0$. (U.P.T.U., 2006)

14. Find the 4th differential coefficient of $x^2 \sin 3x$. (Rohtak, 2019)

15. Find the 5th differential coefficieient of $\frac{\log x}{x}$ (Kurukshetra 2017)

16. Prove that $\frac{d^n}{dx^n}\left(\frac{\sin x}{x}\right) = \frac{1}{x^{n+1}}\left[P\sin\left(x + \frac{n\pi}{2}\right) + Q\cos\left(x + \frac{n\pi}{2}\right)\right]$

where $P = x^n - n(n-1)x^{n-2} + \ldots$. and $Q = nx^{n-1} - n(n-1)(n-2)x^{n-3} + \ldots$. (Cochin, 2019)

17. If $y = \frac{\sinh^{-1} x}{\sqrt{(1+x^2)}}$, show that $(1+x^2) y_{n+2} + (2n+3) xy_{n+1} + (n+1)^2 y_n = 0$. (V.T.U., 2010)

18. If $y = \sinh[m\log(x + \sqrt{x^2+1})]$, prove that $(x^2+1) y_{n+2} + (2n+1) xy_{n+1} + (n^2 - m^2) y_n = 0$. (V.T.U., 2010 S)

## 4.3 FUNDAMENTAL THEOREMS

### (1) Rolle's Theorem

If (i) *$f(x)$ is continuous in the closed interval $[a, b]$, (ii) $f'(x)$ exists for every value of $x$ in the open interval $(a, b)$ and (iii) $f(a) = f(b)$, then there is at least one value $c$ of $x$ in $(a, b)$ such that $f'(c) = 0$.*

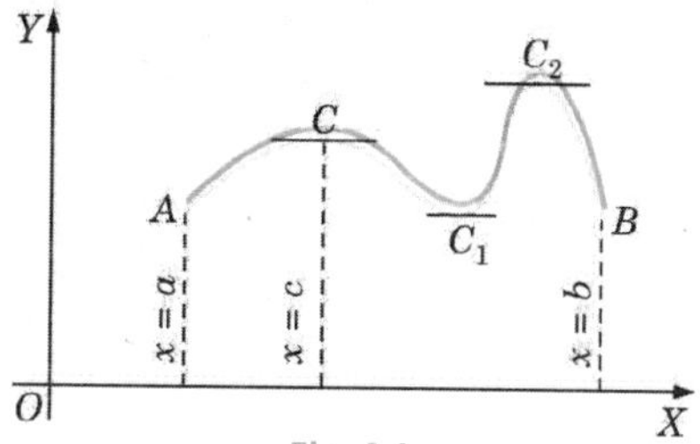

Fig. 4.1

Consider the portion $AB$ of the curve $y = f(x)$, lying between $x = a$ and $x = b$, such that

(i) it goes continuously from $A$ to $B$,
(ii) it has a tangent at every point between $A$ and $B$, and
(iii) ordinate of $A$ = ordinate of $B$.

From the Fig. 4.1, it is self-evident that there is at least one point $C$ (may be more) of the curve at which the tangent is parallel, to the $x$-axis.

*i.e.,* slope of the tangent at $C$ $(x = c) = 0$

But the slope of the tangent at $C$ is the value of the differential coefficient of $f(x)$ w.r.t. $x$ thereat, therefore $f'(c) = 0$.

Hence the theorem is proved.

**Example 4.14.** *Verify Rolle's theorem for (i) $\sin x / e^x$ in $(0, \pi)$.*
*(ii) $(x-a)^m (x-b)^n$ where $m, n$ are positive integers in $[a, b]$.* (V.T.U., 2010)

**Solution.** (i) Let $f(x) = \sin x / e^x$.

$f(x)$ is derivable in $(0, \pi)$.

Also $f(0) = f(\pi) = 0$.

Hence the conditions of Rolle's theorem are satisfied.

$\therefore$ $$f'(x) = \frac{e^x \cos x - e^x \sin x}{e^{2x}}$$ vanishes, where $e^x (\cos x - \sin x) = 0$

or $$\tan x = 1, \text{ i.e., } x = \pi/4.$$

The value $x = \pi/4$ lies in $(0, \pi)$, so that Rolle's theorem is verified.

(*ii*) Let $f(x) = (x - a)^m (x - b)^n$.

Since every polynomial is continuous for all values, $f(x)$ is also continuous in $[a, b]$.

$$f'(x) = m(x - a)^{m-1} (x - b)^n + (x - a)^m . n(x - b)^{n-1}$$
$$= (x - a)^{m-1} (x - b)^{n-1} [(m + n) x - (mb + na)]$$

which, exists, *i.e.*, $f(x)$ is derivable in $(a, b)$.

Also $$f(a) = 0 = f(b).$$

Thus all the conditions of Rolle's theorem are satisfied and there exists $c$ in $(a, b)$ such that $f'(c) = 0$.

$\therefore$ $(c - a)^{m-1} (c - b)^{n-1} [(m + n) c - (mb + na)] = 0$ or $c = (mb + na)/(m + n)$.

Hence, Rolle's theorem is verified.

**(2) Lagrange's Mean-Value Theorem***

**First form.** *If (i) f(x) is continuous in the closed interval [a, b], and*

*(ii) f'(x) exists in the open interval (a, b), then there is at least one value c of x in (a, b), such that*

$$\frac{\mathbf{f(b) - f(a)}}{\mathbf{b - a}} = \mathbf{f'(c)}.$$

Consider the function $$\phi(x) = f(x) - \frac{f(b) - f(a)}{b - a} x$$

Since $f(x)$ is continuous in $[a, b]$ ; $\therefore$ $\phi(x)$ is also continuous in $[a, b]$.

Since $f'(x)$ exists in $(a, b)$ ;

$\therefore$ $\phi'(x)$ also exists in $(a, b)$ and $= f'(x) - \dfrac{f(b) - f(a)}{b - a}$ ...(*i*)

Clearly, $$\phi(a) = \frac{b f(a) - a f(b)}{b - a} = \phi(b).$$

Thus $\phi(x)$ satisfies all the conditions of Rolle's theorem.

$\therefore$ There is at least one value $c$ of $x$ between $a$ and $b$ such that $\phi'(c) = 0$. Substituting $x = c$ in Eqn. (1), we get

$$f'(c) = \frac{f(b) - f(a)}{b - a} \quad ...(2)$$

which proves the theorem.

**Second form.** If we write $b = a + h$, then since $a < c < b$,

$$c = a + \theta h \text{ where } 0 < \theta < 1.$$

Thus the mean value theorem may be stated as follows :

*If (i) f(x) is continuous in the closed interval [a, a + h] and (ii) f'(x) exists in the open interval (a, a + h), then there is at least one number θ(0 < θ < 1) such that*

$$\mathbf{f(a + h) = f(a) + hf'(a + \theta h)}$$

**Geometrical Interpretation.** Let $A$, $B$ be the points on the curve $y = f(x)$ corresponding to $x = a$ and $x = b$ so that $A = [a, f(a)]$ and $B = [b, f(b)]$. (Fig. 4.2)

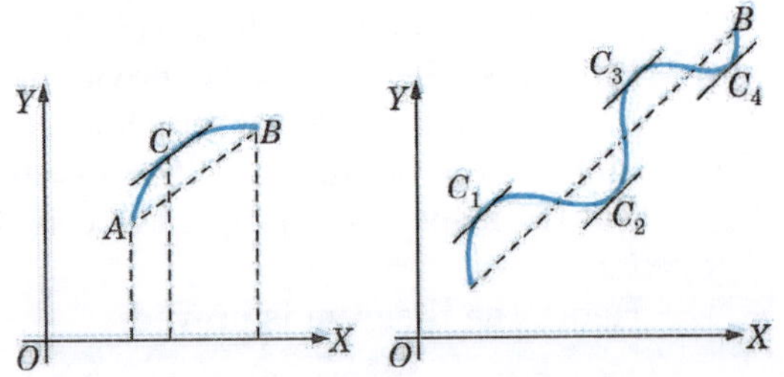

Fig. 4.2

$\therefore$ Slope of chord $AB = \dfrac{f(b) - f(a)}{b - a}$

By Eqn. (2), the slope of the chord $AB = f'(c)$, the slope of the tangent of the curve at $C(x = c)$.

---

*Named after the great French mathematician *Joseph Louis Lagrange* (1736–1813) who became professor at Military Academy, Turin when he was just 19 and director of Berlin Academy in 1766. His important contribution are to algebra, number theory, differential equations, mechanics, approximation theory and calculus of variations.

Hence the Lagrange's mean value theorem asserts that *if a curve AB has a tangent at each of its points, then there exists at least one point C on this curve, the tangent at which is parallel to the chord AB.*

**Cor.** *If $f'(x) = 0$ in the interval $(a, b)$ then $f(x)$ is constant in $[a, b]$.* For, if $x_1, x_2$ be any two values of $x$ in $(a, b)$, then by (2),

$$f(x_2) - f(x_1) = (x_2 - x_1) f'(c) = 0 \ (x_1 < c < x_2)$$

Thus, $f(x_1) = f(x_2)$ *i.e.*, $f(x)$ has the same value for every value of $x$ in $(a, b)$.

**Example 4.15.** *In the Mean value theorem $f(b) - f(a) = (b - a) f'(c)$,* ...(*i*)
*determine c lying between a and b, if $f(x) = x(x - 1)(x - 2)$, $a = 0$ and $b = 1/2$.*

**Solution.** $f(a) = 0,\qquad f(b) = \frac{1}{2}\left(-\frac{1}{2}\right)\left(-\frac{3}{2}\right) = \frac{3}{8}$

$f'(x) = 3x^2 - 6x + 2,\qquad f'(c) = 3c^2 - 6c + 2$

Substituting in Eqn. (*i*), $\frac{3}{8} - 0 = \left(\frac{1}{2} - 0\right)(3c^2 - 6c + 2)$

or $12c^2 - 24c + 5 = 0$

whence $c = \dfrac{24 \pm \sqrt{(24)^2 - 12 \times 5 \times 4}}{24} = 1 \pm 0.764 = 1.764 ; 0.236.$

Hence $c = 0.236$, since it only lies between 0 and 1/2.

**Example 4.16.** *If $f(x) = \tan^{-1}x$, prove that $(0 < a < b < 1)$, $\dfrac{b-a}{1+b^2} < \tan^{-1} b - \tan^{-1} a < \dfrac{b-a}{1+a^2}$.*

*Hence show that* $\frac{\pi}{4} + \frac{3}{25} < \tan^{-1}\frac{4}{3} < \frac{\pi}{4} + \frac{1}{6}$. (*Andhra, 2022; V.T.U., 2013*)

**Solution.** Let $f(x) = \tan^{-1} x$, so that $f'(x) = \dfrac{1}{1+x^2}$.

By Mean value theorem, $\dfrac{\tan^{-1} b - \tan^{-1} a}{b - a} = \dfrac{1}{1+c^2}$, $a < c < b$ ...(*i*)

Now, $a < c < b$, $\therefore$ $1 + a^2 < 1 + c^2 < 1 + b^2$.

$\therefore$ $\dfrac{1}{1+a^2} > \dfrac{1}{1+c^2} > \dfrac{1}{1+b^2}$ *i.e.*, $\dfrac{1}{1+b^2} < \dfrac{1}{1+c^2} < \dfrac{1}{1+a^2}$

*i.e.*, $\dfrac{1}{1+b^2} < \dfrac{\tan^{-1} b - \tan^{-1} a}{b-a} < \dfrac{1}{1+a^2}$ [By (*i*)]

Hence $\dfrac{b-a}{1+b^2} < \tan^{-1} b - \tan^{-1} a < \dfrac{b-a}{1+a^2}$

Now, let $a = 1$, $b = 4/3$.

Then $\dfrac{1/3}{1+16/9} < \tan^{-1}\dfrac{4}{3} - \dfrac{\pi}{4} < \dfrac{1/3}{1+1}$

*i.e.*, $\dfrac{\pi}{4} + \dfrac{3}{25} < \tan^{-1}\dfrac{4}{3} < \dfrac{\pi}{4} + \dfrac{1}{6}$.

**Example 4.17.** *Prove that $\log(1 + x) = x/(1 + \theta x)$, where $0 < \theta < 1$ and hence deduce that*

*$\dfrac{x}{1+x} < \log(1 + x) < x$, $x > a$* (*W.B.T.U., 2016*)

**Solution.** Let $f(x) = \log(1 + x)$, then by second form of Lagrange's mean value theorem

$f(a + h) = f(a) + h f'(a + \theta h)$, $(0 < \theta < 1)$

we have $f(x) = f(0) + x f'(\theta x)$ [Taking $a = 0$, $h = x$]

or $\log(1 + x) = \log(1) + x \cdot 1/(1 + \theta x)$ [$\because$ $f'(x) = 1/1 + x)$]

Hence $\log(1 + x) = x/(1 + \theta x)$ ...(*i*) [$\because$ $\log(1) = 0$]

Since $0 < \theta < 1$, $\therefore$ $0 < \theta x < x$ for $x > 0$.

or $$1 < 1 + \theta x < 1 + x \quad \text{or} \quad 1 > \frac{1}{1+\theta x} > \frac{1}{1+x}$$

or $$x > \frac{x}{1+\theta x} > \frac{x}{1+x}$$

or $$\frac{x}{1+x} < \log(1+x) < x, \; x > 0. \qquad \text{[By } (i)]$$

**(3) Cauchy's Mean-value Theorem***

*If (i) $f(x)$ and $g(x)$ be continuous in $[a, b]$*

*(ii) $f'(x)$ and $g'(x)$ exist in $(a, b)$*

*and (iii) $g'(x) \neq 0$ for any value of $x$ in $(a, b)$,*

*then there is at least one value c of x in (a, b), such that* $\dfrac{f(b)-f(a)}{g(b)-g(a)} = \dfrac{f'(c)}{f'(c)}$

Consider the function $\phi(x) = f(x) - \dfrac{f(b)-f(a)}{g(b)-g(a)} g(x)$

Since $f(x)$ and $g(x)$ are continuous in $[a, b]$

$\therefore$ $\phi(x)$ is also continuous in $[a, b]$.

Again since $f'(x)$ and $g'(x)$ exist in $(a, b)$.

$\therefore$ $\phi'(x)$ also exists in $(a, b)$ and $= f'(x) - \dfrac{f(b)-f(a)}{g(b)-g(a)} g'(x)$

Clearly, $\phi(a) = \phi(b)$.

Thus, $\phi(x)$ satisfies all the conditions of Rolle's theorem. Therefore, at least one value $c$ of $x$ between $a$ and $b$, such that $\phi'(c) = 0$

*i.e.,* $$0 = f'(c) - \frac{f(b)-f(a)}{g(b)-g(a)} g'(c) \text{ whence follows the result.}$$

(*V.T.U., 2015*)

**Obs.** Cauchy's mean value theorem is a generalisation of Lagrange's mean value theorem, where $g(x) = x$.

**Example 4.18.** *Verify Cauchy's Mean-value theorem for the functions $e^x$ and $e^{-x}$ in the interval (a, b).*

**Solution.** $f(x) = e^x$ and $g(x) = e^{-x}$ are both continuous in $[a, b]$ and both functions are differentiable in $(a, b)$.

$\therefore$ $$f'(x) = e^x, \; g'(x) = -e^{-x}$$

By Cauchy's mean value theorem.

$$\frac{f(b)-f(a)}{g(b)-g(a)} = \frac{f'(c)}{g'(c)}$$

$\therefore$ $$\frac{e^b - e^a}{e^{-b} - e^{-a}} = \frac{e^c}{-e^{-c}} \quad i.e., \; c = \frac{1}{2}(a+b)$$

Thus, $c$ lies in $(a, b)$ which verifies the Cauchy's Mean value theorem.

**(4) Taylor's Theorem** (Generalised mean value theorem)**

*If (i) $f(x)$ and its first $(n-1)$ derivatives be continuous in $[a, a+h]$, and (ii) $f^n(x)$ exists for every value of x in $(a, a+h)$, then there is at least one number $\theta$ $(0 < \theta < 1)$, such that*

$$\mathbf{f(a+h) = f(a) + hf'(a) + \frac{h^2}{2!} f''(a) + \ldots + \frac{h^n}{n!} f^n(a + \theta h)} \qquad \ldots(1)$$

*Named after the great French mathematician *Augustin-Louis Cauchy* (1789–1857) who is considered as the father of modern analysis and creator of complex analysis. He published nearly 800 research papers of basic importance. Cauchy is also well known for his contributions to differential equations, infinite series, optics and elasticity.

**Named after an English mathematician, *Brooke Taylor* (1685–1731).

*which is called Taylor's theorem with Lagrange's form remainder, the remainder $R_n$ being* $\frac{h^n}{n!} f^n(a + \theta h)$.

**Proof.** Consider the function

$$\phi(x) = f(x) + (a + h - x) f'(x) + \frac{(a + h - x)^2}{2!} f''(x) + \dots + \frac{(a + h - x)^n}{n!} K$$

where $K$ is defined by

$$f(a + h) = f(a) + hf'(a) + \frac{h^2}{2!} f''(a) + \dots + \frac{h^n}{n!} K \qquad \dots(2)$$

(*i*) Since $f(x), f'(x), \dots, f^{n-1}(x)$ are continuous in $[a, a + h]$, therefore $\phi(x)$ is also continuous in $[a, a + h]$,

(*ii*) $\phi'(x)$ exists and $= \frac{(a + h - x)^{n-1}}{(n-1)!} [f^n(x) - K]$

(*iii*) Also $\phi(a) = \phi(a + h)$. [By (2)]

Hence $\phi(x)$ satisfies all the conditions of Rolle's theorem, and therefore, there exists at least one number $\theta (0 < \theta < 1)$, such that $\phi'(a + \theta h) = 0$ *i.e.*, $K = f^n(a + \theta h)$ $(0 < \theta < 1)$

Substituting this value of $K$ in Eqn.(2), we get Eqn. (1).

**Cor. 1.** Taking $n = 1$ in Eqn. (1), *Taylor's theorem reduces to Lagrange's Mean-value theorem.* (*V.T.U., 2015*)

**Cor. 2.** Putting $a = 0$ and $h = x$ in Eqn. (1), we get

$$f(x) = f(0) + x f'(0) + \frac{x^2}{2!} f''(0) + \dots + \frac{x^n}{n!} f^n(\theta x). \qquad \dots(3)$$

which is known as *Maclaurin's theorem with Lagrange's form of remainder.*

**Example 4.19.** *Find the Maclaurin's theorem with Lagrange's form of remainder for $f(x) = \cos x$.*

**Solution.** $f^n(x) = \frac{d^n}{dx^n} (\cos x) = \cos\left(\frac{n\pi}{2} + x\right)$ so that $f^n_{(0)} = \cos(n\pi/2)$

Thus $f(0) = 1$,

$$f^{2n}(0) = \cos(2n\pi/2) = (-1)^n$$

$$f^{2n+1}(0) = \cos[(2n + 1)\pi/2] = 0$$

Substituting these values in the Maclaurin's theorem with Lagrange's form of remainder *i.e.*,

$$f(x) = f(0) + xf'(0) + \frac{x^2}{2!} f''(0) + \dots + \frac{x^{2n}}{(2n)!} f^{2n}(0) + \frac{x^{2n+1}}{(2n+1)!} f^{2n+1}(\theta x)$$

We get $$\cos x = 1 + 0 + \frac{x^2}{2!}(-1) + 0 + \dots + \frac{x^{2n}}{(2n)!}(-1)^n + \frac{x^{2n+1}}{(2n+1)!}(-1)^n(-1)\cos(\theta x)$$

*i.e.*, $$\cos x = 1 - \frac{x^2}{2!} + \frac{x^4}{4!} + \dots + (-1)^n \frac{x^{2n}}{(2n)!} + \frac{(-1)^{n+1} x^{2n+1}}{(2n+1)!} \cos(\theta x)$$

**Example 4.20.** *If $f(x) = \log(1 + x)$, $x > 0$, using Maclaurin's theorem, show that for $0 < \theta < 1$,*

$$\log(1 + x) = x - \frac{x^2}{2} + \frac{x^3}{3(1 + \theta x)^3}.$$

*Deduce that* $\log(1 + x) < x - \frac{x^2}{2} + \frac{x^3}{3}$ *for* $x > 0$.

**Solution.** By Maclaurin's theorem with remainder $R_3$, we have

$$f(x) = f(0) + x f'(0) + \frac{x^2}{2!} f''(0) + \frac{x^3}{3!} f'''(\theta x) \qquad \dots(i)$$

Here $f(x) = \log(1 + x)$, $\quad f(0) = 0$

$\therefore$ $f'(x) = \frac{1}{1 + x}$, $\quad f'(0) = 1$

$$f''(x) = \frac{-1}{(1+x)^2}, \qquad f''(0) = -1$$

and
$$f'''(x) = \frac{2}{(1+x)^3}, \qquad f'''(\theta x) = \frac{2}{(1+\theta x)^3}$$

Substituting in Eqn. (*i*), we get $\log(1+x) = x - \frac{x^2}{2} + \frac{x^3}{3(1+\theta x)^3}$ ...(*ii*)

Since $x > 0$ and $\theta > 0$, $\theta x > 0$

or
$$(1+\theta x)^3 > 1 \quad i.e., \quad \frac{1}{(1+\theta x)^3} < 1$$

$$\therefore \quad x - \frac{x^2}{2} + \frac{x^3}{3(1+\theta x)^3} < x - \frac{x^2}{2} + \frac{x^3}{3}$$

Hence $\log(1+x) < x - \frac{x^2}{2} + \frac{x^3}{3}$ [By (*ii*)]

## PROBLEMS 4.4

1. Verify Rolle's theorem for (*i*) $f(x) = (x+2)^3 (x-3)^4$ in $(-2, 3)$.
   (*ii*) $y = e^x (\sin x - \cos x)$ in $(\pi/4, 5\pi/4)$. (*iii*) $f(x) = x(x+3)\, e^{-1/2x}$ in $(-3, 0)$.
   (*iv*) $f(x) = \log\left\{\frac{x^2 + ab}{x(a+b)}\right\}$ in $(a, b)$.
2. Using Rolle's theorem for $f(x) = x^{2m-1}(a-x)^{2n}$, find the value of $x$ between $o$ and $a$ where $f'(x) = 0$.
3. Verify Lagrange's Mean value theorem for the following functions and find the appropriate value of $c$ in each case :
   (*i*) $f(x) = (x-1)(x-2)(x-3)$ in $(0, 4)$
   (*ii*) $f(x) = \sin x$ in $[0, \pi]$
   (*iii*) $f(x) = \log_e x$ in $[1, e]$.
   (*iv*) $f(x) = e^x$ in $[0, 1]$.
4. By applying Mean value theorem to
   $f(x) = \log 2 \,.\, \sin\frac{\pi x}{2} + \log x$, prove that $\frac{\pi}{2}\log 2 \,.\, \cos\frac{\pi x}{2} + \frac{1}{x} = 0$ for some $x$ between 1 and 2.
5. In the Mean value theorem : $f(x+h) = f(x) + h f'(x + \theta h)$,
   show that $\theta = 1/2$ for $f(x) = ax^2 + bx + c$ in $(0, 1)$.
6. If $f(h) = f(0) + h f'(0) + \frac{h^2}{2!} f''(\theta h)$, $0 < \theta < 1$, find $\theta$ when $h = 1$ and $f(x) = (1-x)^{5/2}$.
7. (*a*) If $x$ is positive, show that $x > \log(1+x) > x - \frac{1}{2}x^2$. (*J.N.T.U., 2015*)
   (*b*) Prove that, $x > \tan^{-1}x > \frac{x}{1+x^2}$, $0 < x < \pi/2$. (*W.B.T.U., 2016*)
8. If $f(x) = \sin^{-1} x$, $0 < a < b < 1$, use Mean value theorem to prove that
   $$\frac{b-a}{\sqrt{(1-a^2)}} < \sin^{-1} b - \sin^{-1} a < \frac{b-a}{\sqrt{(1-b^2)}}$$
9. Prove that $\frac{b-a}{b} < \log\left(\frac{b}{a}\right) < \frac{b-a}{a}$ for $0 < a < b$.
   Hence show that $\frac{1}{4} < \log\frac{4}{3} < \frac{1}{3}$.
10. Verify the result of Cauchy's mean value theorem for the functions
    (*i*) $\sin x$ and $\cos x$ in the interval $[a, b]$.
    (*ii*) $\log_e x$ and $1/x$ in the interval $[1, e]$.

11. If $f(x)$ and $g(x)$ are respectively $e^x$ and $e^{-x}$, prove that 'c' of Cauchy's mean value theorem is the arithmetic mean between $a$ and $b$.
12. Verify Maclaurin's theorem $f(x) = (1 - x)^{5/2}$ with Lagrange's form of remainder upto 3 terms where $x = 1$.
13. Using Taylor's theorem, prove that

$$x - \frac{x^3}{6} < \sin x < x - \frac{x^3}{6} + \frac{x^5}{120}, \quad \text{for } x > 0.$$

## 4.4 EXPANSIONS OF FUNCTIONS

**(1) Maclaurin's series.** *If $f(x)$ can be expanded as an infinite series, then*

$$\mathbf{f(x) = f(0) + x\, f'(0) + \frac{x^2}{2!} f''(0) + \frac{x^3}{3!} f'''(0) + \ldots \infty} \quad \ldots(1)$$

If $f(x)$ possess derivatives of all orders and the remainder $R_n$ in (3) on page 161 tends to zero as $n \to \infty$, then the Maclaurin's theorem becomes the Maclaurin's series (1).

**Example 4.21.** *Using Maclaurin's series, expand tan x upto the term containing $x^5$.* (V.T.U., 2006)

**Solution.** Let $f(x) = \tan x$ $\qquad f(0) = 0$

$\therefore$ $f'(x) = \sec^2 x = 1 + \tan^2 x$ $\qquad f'(0) = 1$

$$f''(x) = 2 \tan x \sec^2 x = 2 \tan x\,(1 + \tan^2 x)$$
$$= 2 \tan x + 2 \tan^3 x \qquad f''(0) = 0$$
$$f'''(0) = 2 \sec^2 x + 6 \tan^2 x \sec^2 x$$
$$= 2\,(1 + \tan^2 x) + 6 \tan^2 x\,(1 + \tan^2 x)$$
$$= 2 + 8 \tan^2 x + 6 \tan^4 x \qquad f'''(0) = 2$$
$$f^{iv}(0) = 16 \tan x \sec^2 x + 24 \tan^3 x \sec^2 x$$
$$= 16 \tan x\,(1 + \tan^2 x) + 24 \tan^3 x\,(1 + \tan^2 x)$$
$$= 16 \tan x + 40 \tan^3 x + 24 \tan^5 x \qquad f^{iv}(0) = 0$$
$$f^{v}(0) = 16 \sec^2 x + 120 \tan^2 x \sec^2 x + 120 \tan^4 x \sec^2 x. \qquad f^{v}(0) = 16$$

and so on.

Substituting the values of $f(0)$, $f'(0)$, etc. in the Maclaurin's series, we get

$$\tan x = 0 + x\,.\,1 + \frac{x^2}{2!} + \frac{x^3}{3!}\,.\,2 + \frac{x^4}{4!}\,.\,0 + \frac{x^5}{5!}\,.\,16 + \ldots = x + \frac{x^3}{3} + \frac{2x^5}{15} + \ldots$$

**(2) Expansion by use of known series.** When the expansion of a function is required only upto first few terms, it is often convenient to employ the following well-known series :

1. $\sin \theta = \theta - \frac{\theta^3}{3!} + \frac{\theta^5}{5!} - \frac{\theta^7}{7!} + \ldots$

2. $\sinh \theta = \theta + \frac{\theta^3}{3!} + \frac{\theta^5}{5!} + \frac{\theta^7}{7!} + \ldots$

3. $\cos \theta = 1 - \frac{\theta^2}{2!} + \frac{\theta^4}{4!} - \frac{\theta^6}{6!} + \ldots$

4. $\cosh \theta = 1 + \frac{\theta^2}{2!} + \frac{\theta^4}{4!} + \frac{\theta^6}{6!} + \ldots$

5. $\tan \theta = \theta + \frac{\theta^3}{3} + \frac{2}{15}\theta^5 + \ldots$

6. $\tan^{-1} x = x - \frac{x^3}{3} + \frac{x^5}{5} - \ldots$

7. $e^x = 1 + x + \frac{x^2}{2!} + \frac{x^3}{3!} + \frac{x^4}{4!} + \ldots$

8. $\log(1 + x) = x - \frac{x^2}{2} + \frac{x^3}{3} - \frac{x^4}{4} + \ldots$

9. $\log(1 - x) = -\left(x + \frac{x^2}{2} + \frac{x^3}{3} + \frac{x^4}{4} + \ldots\right)$

10. $(1 + x)^n = 1 + nx + \frac{n(n-1)}{2!}x^2 + \frac{n(n-1)(n-2)}{3!}x^3 + \ldots$

**Example 4.22.** *Expand $e^{\sin x}$ by Maclaurin's series or otherwise upto the term containing $x^4$.*

*(C.S.V.T.U., 2015 ; Rohtak, 2011)*

**Solution.** We have $e^{\sin x} = 1 + \sin x + \frac{(\sin x)^2}{2!} + \frac{(\sin x)^3}{3!} + \frac{(\sin x)^4}{4!} + \ldots$

$$= 1 + \left(x - \frac{x^3}{3!} + \ldots\right) + \frac{1}{2!}\left(x - \frac{x^3}{3!} + \ldots\right)^2 + \frac{1}{3!}\left(x - \frac{x^3}{3!} + \ldots\right)^3 + \frac{1}{4!}(x - \ldots)^4 + \ldots$$

$$= 1 + \left(x - \frac{x^3}{6} + \ldots\right) + \frac{1}{2}\left(x^2 - \frac{x^4}{3} + \ldots\right) + \frac{1}{6}\left(x^3 - \ldots\right) + \frac{1}{24}(x^4 + \ldots) + \ldots$$

$$= 1 + x + \frac{x^2}{2} - \frac{x^4}{8} + \ldots$$

**Otherwise,** let $f(x) = e^{\sin x}$ $\qquad f(0) = 1$

$\therefore$ $f'(x) = e^{\sin x} \cos x = f(x) \,.\, \cos x$ $\qquad f'(0) = 1$

$f''(x) = f'(x) \cos x - f(x) \sin x,$ $\qquad f''(0) = 1$

$f'''(x) = f''(x) \cos x - 2f'(x) \sin x - f(x) \cos x,$ $\qquad f'''(0) = 0$

$f^{iv}(x) = f'''(x) \cos x - 3f''(x) \sin x - 3f'(x) \cos x + f(x) \sin x,$ $\qquad f^{iv}(0) = -3$

and so on.

Substituting the values of $f(0)$, $f'(0)$ etc., in the Maclaurin's series, we obtain

$$e^{\sin x} = 1 + x \,.\, 1 + \frac{x^2}{2!} \,.\, 1 + \frac{x^3}{3!} \,.\, 0 + \frac{x^4}{4!} \,.\, (-3) + \ldots$$

$$= 1 + x + \frac{x^2}{2} - \frac{x^4}{8} + \ldots$$

**Example 4.23.** *Expand $\log (1 + \sin^2 x)$ in powers of $x$ as far as the term in $x^6$.*

**Solution.** We have $\sin^2 x = \left(x - \frac{x^3}{3!} + \frac{x^5}{5!} - \ldots\right)^2 = \left[x - \left(\frac{x^3}{6} - \frac{x^5}{120} + \ldots\right)\right]^2$

$$= x^2 - 2x\left(\frac{x^3}{6} - \frac{x^5}{120} + \ldots\right) + \left(\frac{x^3}{6} - \frac{x^5}{120} + \ldots\right)^2$$

$$= x^2 - \frac{x^4}{3} + \frac{x^6}{60} + \frac{x^6}{36} + \ldots = x^2 - \frac{x^4}{3} + \frac{2x^6}{45} + \ldots = t, \text{ say.}$$

Now $\qquad \log (1 + \sin^2 x) = \log (1 + t) = t - \frac{t^2}{2} + \frac{t^3}{3} - \frac{t^4}{4} + \ldots$

Substituting the value of $t$, we get

$$\log (1 + \sin^2 x) = x^2 - \frac{x^4}{3} + \frac{2x^6}{45} + \ldots - \frac{1}{2}\left(x^2 - \frac{x^4}{3} + \ldots\right)^2 - \frac{1}{3}(x^2 - \ldots)^3 - \ldots$$

$$= x^2 - \frac{x^4}{3} + \frac{2x^6}{45} - \frac{1}{2}\left(x^4 - \frac{2x^6}{3} + \ldots\right) + \frac{1}{3}(x^6 + \ldots) + \ldots$$

$$= x^2 - \frac{5}{6}x^4 + \frac{32}{45}x^6 + \ldots$$

**Obs.** As it is very cumbersome to find the successive derivatives of $\log(1 + \sin^2 x)$, therefore, the above method is preferable to Maclaurin's series method.

**Example 4.24.** *Expand $e^{a\sin^{-1} x}$ by Maclaurin's theorem and find the general term. Hence prove that*

$$e^{\theta} = 1 + \sin \theta + \frac{1}{2!} \sin^2 \theta + \frac{2}{3!} \sin^3 \theta + \frac{5}{4!} \sin^4 \theta + \ldots..$$

*(Cochine-2019)*

**Solution:** Let $y = e^{a\sin^{-1} x}$ ...(i)

$$y_1 = \frac{ae^{a\sin^{-1} x}}{\sqrt{1-x^2}}$$

$$y_1^2(1-x^2) = a^2y^2 \quad ...(ii)$$

Differentiating and Cancelling $2y_1$, we have

$$(1-x^2)y_2 - xy_1 - a^2y = 0 \quad ...(iii)$$

Differentiating $n$ times by Leibnitz's theorem

$$(1-x)^2 y_{n+2} - 2(n+1)xy_{n+1} - (n^2+a^2)y_n = 0 \quad ...(iv)$$

If we put $x = 0$, in Eqns. (*i*), (*ii*), (*iii*) and (*iv*)

$$(y)_0 = 1, (y_1)_0 = a, (y_2)_0 = a^2$$

$$(y_{n+2})_0 = (n^2+a^2)(y_n)_0 \quad ...(v)$$

when $n$ is odd, put $n = 1, 3, 5, .....$ in Eqn. (*v*) for general form

$$(y_n)_0 = [(n-2)^2 + a^2 .... (3^2+a^2)(1^2+a^2). a$$

when $n$ is even, put $n = 2, 4, 6, .....$ in Eqn. (*v*)

$$(y_n)_0 = [(n-2)^2 + a^2 ..... (4^2+a^2)(2^2+a^2)]\, a^2$$

$$(y_3)_0 = (1^2+a^2)\, a \text{ and } (y_4)_0 = (2^2+a^2)\, a^2$$

Now by Maclaurin's theorem, we have

$$y = y_0 + x(y_1)_0 + \frac{x^2}{2!}(y_2)_0 + \frac{x^3}{3!}(y_3)_0 + \frac{x^4}{4!}(y_4)_0 + \frac{x^n}{n!}(y_n)_0$$

$$\therefore \quad e^{a\sin^{-1} x} = 1 + ax + \frac{a^2}{2!}x^2 + a\frac{(1^2+a^2)}{3!}x^3 + \frac{a^2(2^2+a^2)}{4!}x^4 + .... \quad ...(vi)$$

general term $\frac{(y_n)_0}{n!}x^n$ can be found depending upon $n$, being odd or even

Next, put $x = \sin\theta$ and $a = 1$ in Eqn. (*vi*)

$$e^\theta = 1 + \sin\theta + \frac{1}{2!}\sin^2\theta + \frac{2}{3!}\sin^3\theta + \frac{5}{4!}\sin^4\theta + .....$$

**Example 4.25** *Expand $\log_e x$ in powers of $(x-1)$ and hence evaluate $\log_e 1{\cdot}1$ correct to 4 decimal places.* *(C.S.V.T.U., 2012 ; Rohtak, 2011)*

**Solution.** Let

| | |
|---|---|
| $f(x) = \log_e x$ | $f(1) = 0$ |
| $\therefore \quad f'(x) = \frac{1}{x},$ | $f'(1) = 1$ |
| $f''(x) = -\frac{1}{x^2},$ | $f''(1) = -1$ |
| $f'''(x) = \frac{2}{x^3},$ | $f'''(1) = 2$ |
| $f^{iv}(x) = -\frac{6}{x^4},$ | $f^{iv}(0) = -6$ |
| etc. | etc. |

Substituting these values in the Taylor's series

$$f(x) = f(1) + (x-1)f'(1) + \frac{(x-1)^2}{2!}f''(1) + \frac{(x-1)^3}{3!}f'''(1) + ...,$$

we get

$$\log x = (x-1) - \frac{(x-1)^2}{2} + \frac{(x-1)^3}{3} - \frac{(x-1)^4}{4} + ...$$

Now putting $x = 1.1$, so that $x - 1 = 0.1$, we have

$$\log(1.1) = 1.1 - \tfrac{1}{2}(0.1)^2 + \tfrac{1}{3}(0.1)^3 - \tfrac{1}{4}(0.1)^4 + ...$$

$$= 0.1 - 0.005 + 0.0003 - 0.00002 + ... = 0.0953.$$

**Example 4.26.** *Use Taylor's series, to prove that*

$$\tan^{-1}(x+h) = \tan^{-1}x + (h \sin z) \cdot \frac{\sin z}{1} - (h \sin z)^2 \cdot \frac{\sin 2z}{2} + (h \sin z)^3 \cdot \frac{\sin 3z}{3} - \ldots$$

*where* $z = \cot^{-1}x$. *(C.S.V.T.U., 2014 ; Rohtak, 2011S)*

**Solution.** We have $\cot z = x$ ...(*i*)

$\therefore$ $-\operatorname{cosec}^2 z.\, dz/dx = 1$ or $dz/dx = -\sin^2 z$ ...(*ii*)

Now let $f(x+h) = \tan^{-1}(x+h)$, so that $f(x) = \tan^{-1}x$

$$\therefore \quad f'(x) = \frac{1}{1+x^2} = \frac{1}{1+\cot^2 z} = \sin^2 z \qquad \text{[By (i)]}$$

$$f''(x) = 2 \sin z \cos z \frac{dz}{dx} = \sin 2z \cdot (-\sin^2 z) \qquad \text{[By (ii)]}$$

$$f'''(x) = -[2 \cos 2z \cdot \sin^2 z + \sin 2z \cdot 2 \sin z \cos z]\frac{dz}{dx}$$
$$= -2 \sin z\, [\sin z \cos 2z + \sin 2z \cos z]\,(-\sin^2 z) = 2 \sin^3 z \sin 3z$$

and so on.

Substituting these values in the Taylor's series

$$f(x+h) = f(x) + hf(x) + \frac{h^2}{2!}f''(x) + \frac{h^3}{3!}f'''(x) + \ldots,$$

we get the required result.

**Example 4.27.** *Prove that* $e^x = 1 + \tan x + \frac{1}{2!}\tan^2 x - \frac{1}{3!}\tan^3 x \frac{-7}{4!}\tan^4 x \ldots$ *(Poona, 2019)*

**Solution:** $\tan x = \theta; \quad \therefore \quad x = \tan^{-1}\theta$

and $y = e^x = e^{\tan^{-1}\theta}$ ...(*i*)

Differentiating Eqn. (*i*) *w.r.t.* θ we get

$$y_1 = \frac{e^{\tan^{-1}\theta}}{1+\theta^2} \qquad \ldots(ii)$$

Differentiating Eqn. (*ii*) *n* times *w.r.t.* θ using Leibnitz's theorem, we get

$$y_{n+1}(\theta^2+1)\, ny_n(2\theta) + \frac{n(n-1)}{2!}y_{n-1}(2) = y_n$$

$$\therefore \quad (\theta^2+1)y_{n+1} + (2n-1)y_{n+n}(n-1)y_{n-1} = 0 \qquad \ldots(iii)$$

If we put θ = 0 in Eqns. (*i*), (*ii*), (*iii*) we get

$$(y)_0 = 1, \quad (y_1)_0 = 1$$
$$(y_{n+1})_0 = (y_n) - n(n-1)(y_{n-1})_0 \qquad \ldots(iv)$$

If we put $n = 1, 2, 3, \ldots$ in Eqn. (*iv*)

$$(y_2)_0 = (y_1)_0 = 1,\ (y_3)_0 = [(y_2)_0 - 2(2-1)(y_1)_0] = -1$$
$$(y_4)_0 = [(y_3)_0 - 3(3-1)(y_2)_0] = -7$$

By Maclaurin's theorem, we have

$$y = (y_0) + \theta(y_1)_0 + \frac{\theta^2}{2!}(y_2)_0 + \frac{\theta^3}{3!}(y_3)_0 + \frac{\theta^4}{4!}(y_4)_0 + \ldots.$$

Substituting these results in Eqn. (*v*), we get

$$y = e^{\tan^{-1}}\theta = 1 + \theta + \frac{\theta^2}{2!}(1) + \frac{\theta^3}{3!}(-1) + \frac{\theta^4}{4!}(-7) + \ldots.$$

$$e^x = 1 + \tan x + \frac{\tan^2 x}{2!} - \frac{\tan^3 x}{3!} - \frac{7}{4!}\tan^4 x$$

## PROBLEMS 4.5

Using Maclaurin's series, expand the following functions :

1. $\log(1+x)$. Hence deduce that $\log\sqrt{\dfrac{1+x}{1-x}} = x + \dfrac{x^3}{3} + \dfrac{x^5}{5} + \ldots$ (*Rohtak, 2011S*)
2. $\sin^{-1} x$ (*Rohtak, 2010 S*)
3. $\tan^{-1} x$
4. $\sqrt{(1+\sin 2x)}$ (*V.T.U., 2016*)
5. $\log \sec x$ (*V.T.U., 2013*)
6. $e^x \sin x$ (*V.T.U., 2016*)

Prove that :

7. $\sec x = 1 + \dfrac{x^2}{2!} + \dfrac{5x^4}{4!} + \ldots$ (*V.T.U., 2016*)
8. $x \operatorname{cosec} x = 1 + \dfrac{x^2}{6} + \dfrac{7x^4}{360} + \ldots$
9. $\sin^{-1}\dfrac{2x}{1+x^2} = 2\left\{x - \dfrac{x^3}{3} + \dfrac{x^5}{5} - \dfrac{x^7}{7} + \ldots\right\}$
10. $\tan^{-1}\dfrac{\sqrt{(1+x^2)}-1}{x} = \dfrac{1}{2}\left(x - \dfrac{x^3}{3} + \dfrac{x^5}{5} - \dfrac{x^7}{7} + \ldots\right)$
11. $\sin^{-1}(3x - 4x^3) = 3\left(x + \dfrac{x^3}{3} + \dfrac{3x^5}{40} + \ldots\right)$
12. $e^x \cos x = 1 + x - \dfrac{2x^3}{3!} - \dfrac{2^2 x^4}{4!} \ldots$ (*Delhi., 2015 ; C.S.V.T.U., 2011 S*)
13. $e^{x \sin x} = 1 + x^2 + \dfrac{x^4}{3} + \dfrac{x^6}{120} + \ldots$
14. $e^{\cos^{-1} x} = e^{\pi/2}\left(1 - x + \dfrac{x^2}{3} - \dfrac{x^3}{3} + \ldots\right)$
15. $\log\dfrac{\sin x}{x} = -\left(\dfrac{x^2}{6} + \dfrac{x^4}{180} + \dfrac{x^6}{2835} + \ldots\right)$
16. $\log(1 + \sin x) = x - \dfrac{x^2}{2} + \dfrac{x^3}{6} - \dfrac{x^4}{12} + \ldots$ (*Delhi, 2012*)
17. $\log(1 + e^x) = \log 2 + \dfrac{x}{2} + \dfrac{x^2}{8} - \dfrac{x^4}{192} + \ldots$ (*C.S.V.T.U., 2015 ; J.N.T.U., 2015 ; V.T.U., 2015*)
18. $\dfrac{x}{e^{x-1}} = e\left(x - x^2 + \dfrac{x^3}{2} - \dfrac{x^4}{6} + \ldots\right)$ (*V.T.U., 2015*)
19. $\dfrac{e^x}{e^x + 1} = \dfrac{1}{2} + \dfrac{x}{4} - \dfrac{x^3}{48} + \ldots$ (*J.N.T.U., 2015*)
20. $\dfrac{x}{2}\left(\dfrac{e^x+1}{e^x-1}\right) = 1 + \dfrac{1}{6}\cdot\dfrac{x^2}{2!} - \dfrac{1}{30}\cdot\dfrac{x^4}{4!} + \ldots$
21. $\sin x \cosh x = x + \dfrac{x^3}{3} - \dfrac{x^5}{30} + \ldots$

By forming a differential equation, show that

22. $(\sin^{-1} x)^2 = 2\dfrac{x^2}{2!} + 2\cdot 2^2\,\dfrac{x^4}{4!} + 2\cdot 2^2\cdot 4^2\cdot\dfrac{x^6}{6!} + \ldots$
23. $\log[1 + \sqrt{(1+x^2)}] = x - \dfrac{1}{2}\dfrac{x^3}{3} + \dfrac{1\cdot 3}{2\cdot 4}\dfrac{x^5}{5} - \dfrac{1\cdot 3\cdot 5}{2\cdot 4\cdot 6}\dfrac{x^7}{7} + \ldots$
24. If $y = \sin(m \sin^{-1} x)$, show that $(1 - x^2)\,y_2 - xy_1 + m^2 y = 0$
Hence expand $\sin m\theta$ in powers of $\sin\theta$. (*D.T.U., 2012*)
25. Using Taylor's theorem, express the polynomial $2x^3 + 7x^2 + x - 6$ in powers of $(x - 1)$
26. Expand (*i*) $e^x$ (*ii*) $\tan^{-1} x$, in powers of $(x - 1)$ upto four terms. (*V.T.U., 2016 ; Delhi, 2012*)
27. Expand $\sin x$ in powers of $(x - \pi/2)$. (*V.T.U., 2016*)
Hence find the value of sin 91° correct to 4 decimal places. (*D.T.U., 2013*)
28. Prove that $\log \sin x = \log \sin a + (x - a)\cot a - \frac{1}{2}(x - a)^2 \operatorname{cosec}^2 a + \ldots$
29. Find the Taylor's series expansion for (*i*) $1/x$ about 1. (*K.T.U., 2016*) (*ii*) $\log \cos x$ about the point $\pi/3$.
30. Compute to four decimal places, the value of cos 32°, by the use of Taylor's series. (*D.T.U., 2015*)
31. Calculate approximately (*i*) $\log_{10} 404$, given $\log 4 = 0.6021$.
(*ii*) $(1.04)^{3.01}$

## 4.5 INDETERMINATE FORMS

In general $\underset{x \to a}{\text{Lt}} [f(x)/\phi(x)] = \underset{x \to a}{\text{Lt}} f(x) / \underset{x \to a}{\text{Lt}} \phi(x)$. But when $\underset{x \to a}{\text{Lt}} f(x)$ and $\underset{x \to a}{\text{Lt}} \phi(x)$ are both zero, then the quotient reduces to the indeterminate form 0/0. This does not imply that $\underset{x \to a}{\text{Lt}} [f(x)/\phi(x)]$ is meaningless or it does not exist. In fact, in many cases, it has a finite value. We shall now, study the methods of evaluating the limits in such and similar other cases :

**(1) Form 0/0.** *If $f(a) = \phi(a) = 0$, then*

$$\underset{x \to a}{\text{Lt}} \frac{f(x)}{\phi(x)} = \underset{x \to a}{\text{Lt}} \frac{f'(x)}{\phi'(x)}$$

By Taylor's series,

$$\underset{x \to a}{\text{Lt}} \frac{f(x)}{\phi(x)} = \underset{x \to a}{\text{Lt}} \frac{f(a) + (x-a)f'(a) + \frac{1}{2!}(x-a)^2 f''(a) + \ldots}{\phi(a) + (x-a)\phi'(a) + \frac{1}{2!}(x-a)^2 \phi''(a) + \ldots}$$

$$= \underset{x \to a}{\text{Lt}} \frac{f'(a) + \frac{1}{2}(x-a) f'(a) + \ldots}{\phi'(a) + \frac{1}{2}(x-a)\phi'(a) + \ldots}$$

$$= \frac{f'(a)}{\phi'(a)} = \underset{x \to a}{\text{Lt}} \frac{f'(x)}{\phi'(x)} \qquad \ldots(1)$$

This is known as *L'Hospital's rule.*

In general, if

$$f(a) = f'(a) = f''(a) = \ldots = f^{n-1}(a) = 0, \text{ but } f^n(a) \neq 0,$$

and

$$\phi(a) = \phi'(a) = \phi''(a) = \ldots = \phi^{n-1}(a) = 0, \text{ but } \phi^n(a) \neq 0,$$

then from (1),

$$\underset{x \to a}{\textbf{Lt}} \frac{\mathbf{f(x)}}{\boldsymbol{\phi}\mathbf{(x)}} = \frac{\mathbf{f^n(a)}}{\boldsymbol{\phi}^\mathbf{n}\mathbf{(a)}} = \underset{x \to a}{\textbf{Lt}} \frac{\mathbf{f^n(x)}}{\boldsymbol{\phi}^\mathbf{n}\mathbf{(x)}}$$

[**Rule to evaluate Lt [$f(x)/\phi(x)$] in 0/0 form :**

*Differentiating the numerator and denominator separately as many times as would be necessary to arrive a determinate form*].

**Example 4.28.** *Evaluate (i)* $\underset{x \to 0}{Lt} \dfrac{xe^x - \log(1+x)}{x^2}$. *(V.T.U., 2016)*

*(ii)* $\underset{x \to 1}{Lt} \dfrac{x^x - x}{x - 1 - \log x}$

**Solution.** *(i)*

$$\underset{x \to 0}{\text{Lt}} \frac{xe^x - \log(1+x)}{x^2} \qquad \left(\text{form } \frac{0}{0}\right)$$

$$= \underset{x \to 0}{\text{Lt}} \frac{(xe^x + e^x \cdot 1) - 1/(1+x)}{2x} \qquad \left(\text{form } \frac{0}{0}\right)$$

$$= \underset{x \to 0}{\text{Lt}} \frac{xe^x + e^x + e^x + 1/(1+x)^2}{2} = \frac{0+1+1+1}{2} = 1\frac{1}{2}.$$

*(ii)*

$$\underset{x \to 1}{\text{Lt}} \frac{x^x - x}{x - 1 - \log x} \qquad \left(\text{form } \frac{0}{0}\right)$$

$$= \underset{x \to 1}{\text{Lt}} \frac{d(x^x)/dx - 1}{1 - 0 - 1/x}$$

$$= \underset{x \to 1}{\text{Lt}} \frac{x^x(1 + \log x) - 1}{1 - 1/x} \qquad \left(\text{form } \frac{0}{0}\right)$$

Let $y = x^x$ so that

$$\log y = x \log x$$

$$\therefore \quad \frac{1}{y}\frac{dy}{dx} = x \cdot \frac{1}{x} + 1 \,.\, \log x$$

or $$\frac{d}{dx}(x^x) = x^x(1 + \log x) \qquad \ldots(i)$$

$$= \underset{x \to 1}{\text{Lt}} \frac{d(x^x)/dx\,.(1+\log x) + x^x(1/x) - 0}{1/x^2}$$

$$= \underset{x \to 1}{\text{Lt}} \frac{x^x(1+\log x)^2 + x^x(1/x)}{x^{-2}} \qquad \text{[By } (i)\text{]}$$

$$= \frac{1(1+0)^2 + 1 \cdot 1}{1} = 2.$$

**Example 4.29.** *Find the values of a and b such that* $\underset{x \to 0}{Lt} \dfrac{x(a + b\cos x) - c\sin x}{x^5} = 1.$

**Solution.** $\underset{x \to 0}{\text{Lt}} \dfrac{x(a+b\cos x) - c\sin x}{x^5} \qquad \left(\text{form } \dfrac{0}{0}\right)$

$$= \underset{x \to 0}{\text{Lt}} \frac{a + b\cos x - bx\sin x - c\cos x}{5x^4} \qquad ...(i)$$

As the denominator is 0 for $x = 0$, Eqn. $(i)$ will tend to a finite limit if and only if the numerator also becomes 0 for $x = 0$. This requires $a + b - c = 0$ ...$(ii)$

With this condition, Eqn. $(i)$ assumes the form 0/0.

$\therefore$ Eqn. $(i)$ $= \underset{x \to 0}{\text{Lt}} \dfrac{-b\sin x - b(\sin x + x\cos x) + c\sin x}{20x^3}$

$$= \underset{x \to 0}{\text{Lt}} \frac{(c-2b)\sin x - bx\cos x}{20x^3} \qquad \left(\text{form } \frac{0}{0}\right)$$

$$= \underset{x \to 0}{\text{Lt}} \frac{(c-2b)\cos x - b(\cos x - x\sin x)}{60x^2} \qquad ...(iii)$$

$$= \frac{c - 2b - b}{0} = \frac{c-3b}{0} = 1 \qquad \text{(Given)}$$

$\therefore$ $c - 3b = 0$ *i.e.,* $c = 3b$.

Now Eqn. $(iii)$ $= \underset{x \to 0}{\text{Lt}} \dfrac{b\cos x - b\cos x + bx\sin x}{60x^2}$

$$= \underset{x \to 0}{\text{Lt}} \frac{b\sin x}{60x} = \frac{b}{60} \underset{x \to 0}{\text{Lt}} \left(\frac{\sin x}{x}\right) = \frac{b}{60} = 1.$$

*i.e.,* $b = 60$, and $\therefore$ $c = 180$.

Form Eqn. $(ii)$, $a = 120$.

**(2) Form** $\infty/\infty$**.** It can be shown that L'Hospital's rule can also be applied to this case by differentiating the numerator and denominator separately as many times as would be necessary.

**Example 4.30.** *Evaluate* $\underset{x \to 0}{Lt} \dfrac{\log x}{\cot x}.$

**Solution.** $\underset{x \to 0}{\text{Lt}} \dfrac{\log x}{\cot x} = \underset{x \to 0}{\text{Lt}} \dfrac{1/x}{-\text{cosec}^2 x} = -\underset{x \to 0}{\text{Lt}} \dfrac{\sin^2 x}{x} \qquad \left(\text{form } \dfrac{0}{0}\right)$

$$= -\underset{x \to 0}{\text{Lt}} \frac{2\sin x\cos x}{1} = 0$$

**Obs. Use of known series and standard limits.** In many cases, it would be found more convenient to use expansions of known functions and standard limits for evaluating the indeterminate forms. For this purpose, remember the series of § 4.4 (2) and the following limits :

$$\underset{x \to 0}{\text{Lt}} \frac{\sin x}{x} = 1, \qquad \underset{x \to 0}{\text{Lt}} (1+x)^{1/x} = e$$

**Example 4.31.** *Evaluate* $\underset{x \to 0}{Lt} \dfrac{e^x \sin x - x - x^2}{x^2 + x \log(1 - x)}$.

**Solution.** Using the expansions of $e^x$, $\sin x$ and $\log(1 - x)$, we get

$$\underset{x \to 0}{\text{Lt}} \frac{e^x \sin x - x - x^2}{x^2 + x \log(1 - x)}$$

$$= \underset{x \to 0}{\text{Lt}} \frac{\left(1 + x + \frac{1}{2!}x^2 + \frac{1}{3!}x^3 + \ldots\right)\left(x - \frac{1}{3!}x^3 + \ldots\right) - x - x^2}{x^2 + x\left(-x - \frac{1}{2}x^2 - \frac{1}{3}x^3 - \ldots\right)}$$

$$= \underset{x \to 0}{\text{Lt}} \frac{\left(x + x^2 + \frac{1}{3}x^3 - 0 \cdot x^4 + \ldots\right) - x - x^2}{x^2 - \left(x^2 + \frac{1}{2}x^3 + \frac{1}{3}x^4 + \ldots\right)} = \underset{x \to 0}{\text{Lt}} \frac{\frac{1}{3}x^3 - 0 \cdot x^4 + \ldots}{-\frac{1}{2}x^3 - \frac{1}{3}x^4 - \ldots} = \underset{x \to 0}{\text{Lt}} \frac{\frac{1}{3} + \ldots}{-\frac{1}{2} - \frac{1}{3}x - \ldots} = -\frac{2}{3}.$$

**Example 4.32.** *Evaluate* $\underset{x \to 0}{Lt} \dfrac{(1 + x)^{1/x} - e}{x}$.

**Solution.** Let $y = (1 + x)^{1/x}$

$\therefore$
$$\log y = \frac{1}{x} \log(1 + x) = \frac{1}{x}\left(x - \frac{x^2}{2} + \frac{x^3}{3} - \ldots\right) = 1 - \frac{x}{2} + \frac{x^2}{3} - \ldots$$

or
$$y = e^{1 - \frac{1}{2}x + \frac{1}{3}x^2 - \ldots} = e \cdot e^{-\frac{1}{2}x + \frac{1}{3}x^2 - \ldots}$$

$$= e\left[1 + \left(-\frac{1}{2}x + \frac{1}{3}x^2 - \ldots\right) + \frac{1}{2!}\left(-\frac{1}{2}x + \frac{1}{3}x^2 - \ldots\right)^2 + \ldots\right] = e\left(1 - \frac{x}{2} + \frac{11}{24}x^2 + \ldots\right)$$

$\therefore$
$$\underset{x \to 0}{\text{Lt}} \frac{(1 + x)^{1/x} - e}{x} = \underset{x \to 0}{\text{Lt}} \frac{e\left(1 - \frac{x}{2} + \frac{11}{24}x^2 + \ldots\right) - e}{x}$$

$$= \underset{x \to 0}{\text{Lt}} \frac{e\left(-\frac{1}{2}x + \frac{11}{24}x^2 + \ldots\right)}{x} = \underset{x \to 0}{\text{Lt}} \left(\frac{-e}{2} + \frac{11}{24}ex + \ldots\right) = -\frac{e}{2}.$$

## PROBLEMS 4.6

Evaluate the following limits :

1. $\underset{x \to 0}{\text{Lt}} \dfrac{a^x - b^x}{x}$

2. $\underset{x \to 0}{\text{Lt}} \dfrac{x \cos x - \sin x}{x^2 \sin x}$

3. $\underset{\theta \to 0}{\text{Lt}} \dfrac{\theta - \sin\theta}{\sin\theta(1 - \cos\theta)}$

4. $\underset{x \to \pi/2}{\text{Lt}} \dfrac{a^{\sin x} - a}{\log_e \sin x}$

5. $\underset{x \to 0}{\text{Lt}} \dfrac{\sin x \sin^{-1} x - x^2}{x^6}$

6. $\underset{x \to 0}{\text{Lt}} \dfrac{x \cos x - \log(1 + x)}{x^2}$

7. $\underset{x \to 0}{\text{Lt}} \dfrac{\log \sec x - \frac{1}{2}x^2}{x^4}$

8. $\underset{x \to 0}{\text{Lt}} \dfrac{e^x + \sin x - 1}{\log(1 + x)}$

9. $\underset{x \to 0}{\text{Lt}} \dfrac{x e^x - \log(1 + x)}{\cosh x - \cos x}$

10. $\underset{x \to 0}{\text{Lt}} \dfrac{\cos x - \log(1 + x) - 1 + x}{\sin^2 x}$

11. $\underset{x \to 0}{\text{Lt}} \dfrac{e^x + 2\sin x - e^{-x} - 4x}{x^5}$

12. $\underset{x \to 0}{\text{Lt}} \dfrac{\log(x - a)}{\log(e^x - e^a)}$

13. $\underset{x \to 0}{\text{Lt}} \dfrac{\tan x - x}{x^2 \tan x}$

14. $\underset{x \to 0}{\text{Lt}} \dfrac{e^x - e^{-x} - 2x}{x - \sin x}$

15. $\underset{x \to 0}{\text{Lt}} \dfrac{e^x - e^{\sin x}}{x - \sin x}$

16. $\underset{x \to 0}{\text{Lt}} \dfrac{\sin\{(\log(1 + x)\}}{\log(1 + \sin x)}$

17. $\underset{x \to 0}{\text{Lt}} \dfrac{(1+x)^{1/x} - e + \frac{1}{2}e^x}{x^2}$

18. If $\underset{x \to 0}{\text{Lt}} \dfrac{\sin 2x + a \sin x}{x^3}$ is finite, find the value of $a$ and the limit.

19. Find $a, b$ if $\underset{x \to 0}{\text{Lt}} \dfrac{a \sinh x + b \sin x}{x^3} = \dfrac{5}{3}$.

20. Find $a, b, c$ so that $\underset{x \to 0}{\text{Lt}} \dfrac{ae^x - b \cos x + ce^{-x}}{x \sin x} = 2$.

**(3) Forms reducible to 0/0 form.** Each of the following indeterminate forms can be easily reduced to the form 0/0 (or ∞/∞) by suitable transformation and then the limits can be found as usual.

**I. Form 0 × ∞. If** $\underset{x \to 0}{Lt} f(x) = 0$ and $\underset{x \to \infty}{Lt} \phi(x) = \infty$, then

$$\underset{x \to a}{\text{Lt}} [f(x) . \phi(x)] \text{ assumes the form } 0 \times \infty.$$

To evaluate this limit, we write

$$f(x) \cdot \phi(x) = f(x)/[1/\phi(x)] \text{ to take the form } 0/0.$$
$$= \phi(x)/[1/f(x)] \text{ to take the form } \infty/\infty.$$

**Example 4.33.** *Evaluate* $\underset{x \to 0}{Lt} (\tan x \log x)$

**Solution.**
$$\underset{x \to 0}{\text{Lt}} (\tan x \log x) = \underset{x \to 0}{\text{Lt}} \left(\frac{\log x}{\cot x}\right) \qquad \left(\text{form } \frac{\infty}{\infty}\right)$$
$$= \underset{x \to 0}{\text{Lt}} \left(\frac{1/x}{-\text{cosec}^2 x}\right) = -\underset{x \to 0}{\text{Lt}} \left(\frac{\sin^2 x}{x}\right) \qquad \left(\text{form } \frac{0}{0}\right)$$
$$= \underset{x \to 0}{\text{Lt}} \frac{2 \sin x \cos x}{1} = 0.$$

**II. Form** ∞ to ∞. If $\underset{x \to a}{Lt} f(x) = \infty = \underset{x \to a}{Lt} \phi(x)$, then $\underset{x \to a}{\text{Lt}} [f(x) - \phi(x)]$ assumes the form ∞ to ∞.

It can be reduced to the form 0/0 by writing

$$f(x) - \phi(x) = \left[\frac{1}{\phi(x)} - \frac{1}{f(x)}\right] \Big/ \frac{1}{f(x)\,\phi(x)}$$

**Example 4.34.** *Evaluate* $\underset{x \to 0}{Lt} \left(\dfrac{1}{\sin x} - \dfrac{1}{x}\right)$.

**Solution.**
$$\underset{x \to 0}{\text{Lt}} \left(\frac{1}{\sin x} - \frac{1}{x}\right) = \underset{x \to 0}{\text{Lt}} \frac{x - \sin x}{x \sin x} = \underset{x \to 0}{\text{Lt}} \frac{1 - \cos x}{x \cos x + \sin x} \qquad \left(\text{form } \frac{0}{0}\right)$$
$$= \underset{x \to 0}{\text{Lt}} \frac{\sin x}{x(-\sin x) + \cos x + \cos x} = \frac{0}{0+1+1} = 0.$$

**III. Forms** $0^0, 1^\infty, \infty^0$. If $y = \underset{x \to a}{\text{Lt}} [f(x)]^{\phi(x)}$ assumes one of these forms, then $\log y = \underset{x \to a}{\text{Lt}} \phi(x) \log f(x)$ takes the form 0 × ∞, which can be evaluated by the method given in **I** above. If $\log y = l$, then $y = e^l$.

**Example 4.35.** *Evaluate (i)* $\underset{x \to \pi/2}{Lt} (\sin x)^{\tan x}$ *(ii)* $\underset{x \to 0}{Lt} \left(\dfrac{a^x + b^x + c^x}{3}\right)^{1/x}$ *(V.T.U., 2011)*

*(iii)* $\underset{x \to 0}{Lt} \left(\dfrac{\tan x}{3}\right)^{1/x^2}$

**Solution.** (*i*) Let $$y = \underset{x \to \pi/2}{\text{Lt}} (\sin x)^{\tan x}.$$

$$\therefore \quad \log y = \underset{x \to \pi/2}{\text{Lt}} \tan x \log \sin x = \underset{x \to \pi/2}{\text{Lt}} \frac{\log \sin x}{\cot x} \qquad \left(\text{form } \frac{0}{0}\right)$$

$$= \underset{x \to \pi/2}{\text{Lt}} \frac{(1/\sin x)\cos x}{-\text{cosec}^2 x} = -\underset{x \to \pi/2}{\text{Lt}} (\sin x \cos x) = 0$$

Hence $$y = e^0 = 1.$$

(*ii*) Let $$y = \underset{x \to 0}{\text{Lt}} \left(\frac{a^x + b^x + c^x}{3}\right)^{1/x}$$

so that $$\log y = \underset{x \to 0}{\text{Lt}} \frac{\log(a^x + b^x + c^x) - \log 3}{x} \qquad \text{form } \left(\frac{0}{0}\right)$$

$$= \underset{x \to 0}{\text{Lt}} \frac{(a^x + b^x + c^x)^{-1}(a^x \log a + b^x \log b + c^x \log c)}{1}$$

$$= (1 + 1 + 1)^{-1}(\log a + \log b + \log c) = \frac{1}{3}\log(abc) = \log(abc)^{1/3}.$$

$$\therefore \quad y = (abc)^{1/3}$$

(*iii*) $$\underset{x \to 0}{\text{Lt}} \left(\frac{\tan x}{x}\right)^{1/x^2} = \underset{x \to 0}{\text{Lt}} \left(\frac{x + \frac{x^3}{3} + \frac{2}{15}x^5 + \ldots}{x}\right)^{1/x^2}$$

$$= \underset{x \to 0}{\text{Lt}} \left(1 + \frac{x^2}{3} + \frac{2}{15}x^4 + \ldots\right)^{1/x^2}$$

$$= \underset{x \to 0}{\text{Lt}} (1 + tx^2)^{1/x^2} \qquad \text{where } t = \frac{1}{3} + \frac{2}{15}x^2 + \ldots$$

$$= \underset{x \to 0}{\text{Lt}} [(1 + tx^2)^{1/tx^2}]^t = \underset{x \to 0}{\text{Lt}} e^t = e^{1/3}. \qquad \left[\because \underset{z \to 0}{\text{Lt}} (1 + z)^{1/z} = e\right]$$

## PROBLEMS 4.7

Evaluate the following limits :

1. $\underset{x \to 0}{\text{Lt}} \left(\frac{1}{x^2} - \frac{1}{\sin^2 x}\right)$
2. $\underset{x \to 0}{\text{Lt}} \left(\frac{1}{x} - \frac{1}{e^x - 1}\right)$
3. $\underset{x \to 1}{\text{Lt}} (2x \tan x - \pi \sec x)$
4. $\underset{x \to 0}{\text{Lt}} \left(\frac{\cot x - 1/x}{x}\right)$
5. $\underset{x \to 0}{\text{Lt}} \left(\frac{1}{x^2} - \cot^2 x\right)$
6. $\underset{x \to 1}{\text{Lt}} (x)^{1/(1-x)}$
7. $\underset{x \to 0}{\text{Lt}} (a^x + x)^{1/x}$
8. $\underset{x \to \pi/2}{\text{Lt}} (\sec x)^{\cot x}$
9. $\underset{x \to 0}{\text{Lt}} (1 + \sin x)^{\cot x}$
10. $\underset{x \to 0}{\text{Lt}} (\cos x)^{1/x^2}$
11. $\underset{x \to \pi/2}{\text{Lt}} (\tan x)^{\tan 2x}$
12. $\underset{x \to 0}{\text{Lt}} (\cot x)^{1/\log x}$
13. $\underset{x \to \pi/2}{\text{Lt}} (\cos x)^{\frac{\pi}{2} - x}$
14. $\underset{x \to 0}{\text{Lt}} \left(\frac{\sin x}{x}\right)^{1/x}$ (V.T.U., 2014)
15. $\underset{x \to 0}{\text{Lt}} \left(\frac{\sin x}{x}\right)^{1/x^2}$ (V.T.U., 2013)
16. $\underset{x \to 1}{\text{Lt}} (1 - x^2)^{1/\log(1-x)}$

**17.** $\underset{x \to a}{\text{Lt}} \left(2 - \frac{x}{a}\right)^{\tan(\pi x/2a)}$ (*V.T.U., 2015*)

**18.** $\underset{x \to 0}{\text{Lt}} \left\{\frac{2(\cosh x - 1)^{1/x^2}}{x^2}\right\}$

**19.** $\underset{x \to 2}{\text{Lt}} \left\{\frac{1}{x-2} - \frac{1}{\log(x-1)}\right\}$

**20.** $\underset{x \to 0}{\text{Lt}} \left(\frac{1^x + 2^x + 3^x}{3}\right)^{1/x}$.

**21.** $\underset{x \to 0}{\text{Lt}} \left(\frac{a^x + b^x + c^x + d^x}{4}\right)^{1/x}$ (*V.T.U. 2016*)

## 4.6 TANGENTS AND NORMALS – CARTESIAN CURVES

**(1) Equation of the tangent** *at the point (x, y) of the curve y = f(x) is*

$$\mathbf{Y - y = \frac{dy}{dx}(X - x).}$$

The equation of any line through $P(x, y)$ is

$$Y - y = m(X - x)$$

where $X, Y$ are the current coordinates of any point on the line (Fig. 4.3).

If this line is the tangent $PT$, then

$$m = \tan \psi = dy/dx$$

Hence the equation of the tangent at $(x, y)$ is

$$Y - y = \frac{dy}{dx}(X - x) \qquad ...(2)$$

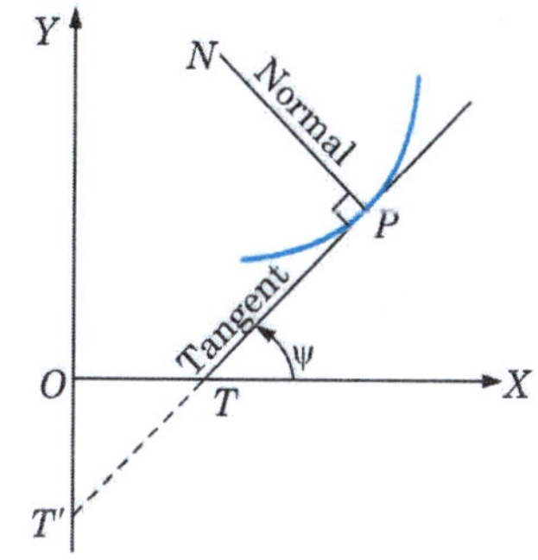

Fig. 4.3

**Cor. Intercepts.** Putting $Y = 0$ in (2)

$$-y = \frac{dy}{dx}(X - x) \quad \text{or} \quad X = x - y/\frac{dy}{dx}$$

∴ *Intercept which the tangent cuts off from x-axis* $(= OT) = \mathbf{x - y\, dx/dy}$

*Similarly putting X = 0 in (2), we see, that*

*the intercept which the tangent cuts off from the y-axis*

$$(= OT') = \mathbf{y - x\frac{dy}{dx}}$$

**(2) Equation of the normal** *at the point (x, y) of the curve y = f(x)* is

$$\mathbf{Y - y = -\frac{dx}{dy}(X - x)}$$

**A normal** *to the curve y = f(x) at P (x, y) is a line through P perpendicular to the tangent there at.*

∴ Its equation is $Y - y = m'(X - x)$

where $$m' \cdot dy/dx = -1 \quad \text{or} \quad m' = -1/\frac{dy}{dx} = -dx/dy$$

Hence the equation of the normal at $(x, y)$ is $Y - y = -\frac{dx}{dy}(X - x)$.

**Example 4.36.** *Find the equation of the tangent at any point (x, y) to the curve $x^{2/3} + y^{2/3} = a^{2/3}$. Show that the portion of the tangent intercepted between the axes is of constant length.*

**Solution.** Equation of the curve is $x^{2/3} + y^{2/3} = a^{2/3}$. ...(*i*)

Differentiating (*i*) w.r.t. $x$,

$$\frac{2}{3}x^{-1/3} + \frac{2}{3}y^{-1/3}\frac{dy}{dx} = 0$$

∴ Slope of the tangent at $(x, y) = \frac{dy}{dx} = -\left(\frac{y}{x}\right)^{1/3}$

∴ Equation of the tangent at $(x, y)$ is

$$Y - y = -(y/x)^{1/3}(X - x) \qquad ...(ii)$$

Put $Y = 0$ in Eqn. (*ii*). Then $X = x + x^{1/3} . y^{2/3}$

*i.e.*, Intercept on $x$-axis $= (x^{2/3} + y^{2/3})x^{1/3} = a^{2/3} . x^{1/3}$ [By (*i*)]

Put $X = 0$ in Eqn. (*ii*). Then $Y = y + y^{1/3} . x^{2/3}$

*i.e.*, Intercept on $y$-axis $= (x^{2/3} + y^{2/3})\, y^{1/3} = a^{2/3} . y^{1/3}$ [By (*i*)]

Thus the portion of the tangent intercepted between the axes

$$= \sqrt{[(\text{Intercept on } x\text{-axis})^2 + (\text{Intercept on } y\text{-axis})^2]}$$

$$= \sqrt{[(a^{2/3}.x^{1/3})^2 + (a^{2/3}.y^{1/3})^2]}$$

$$= \sqrt{[a^{4/3}(x^{2/3} + y^{2/3})]} = a^{2/3}\sqrt{(a)^{2/3}} \quad \text{[By (i)]}$$

$= a$, which is a constant length.

**Example 4.37.** *Show that the conditions for the line $x \cos \alpha + y \sin \alpha = p$ to touch the curve $(x/a)^m + (y/b)^m = 1$ is $(a \cos \alpha)^{m/(m-1)} + (b \sin \alpha)^{m/(m-1)} = p^{m/(m-1)}$.*

**Solution.** Equation of the curve is $\dfrac{x^m}{a^m} + \dfrac{y^m}{b^m} = 1$ ...(*i*)

Differentiating (*i*) w.r.t. $x$, $\dfrac{mx^{m-1}}{a^m} + \dfrac{my^{m-1}}{b^m}\dfrac{dy}{dx} = 0$

$\therefore$ Slope of the tangent at $(x, y) = \dfrac{dy}{dx} = -\left(\dfrac{b}{a}\right)^m \left(\dfrac{x}{y}\right)^{m-1}$

$\therefore$ Equation of the tangent at $(x, y)$ is

$$Y - y = -\left(\frac{b}{a}\right)^m \left(\frac{x}{y}\right)^{m-1} (X - x)$$

or
$$\frac{x^{m-1}X}{a^m} + \frac{y^{m-1}Y}{b^m} = \frac{x^m}{a^m} + \frac{y^m}{b^m} = 1 \quad \text{...(ii) [By (i)]}$$

If the given line touches Eqn. (*i*) at $(x, y)$ then Eqn. (*ii*) must be same as $X \cos \alpha + Y \sin \alpha = p$ ...(*iii*)

Comparing coefficients in Eqns. (*ii*) and (*iii*),

$$\frac{x^{m-1}}{a^m}\Big/\cos\alpha = \frac{y^{m-1}}{b^m}\Big/\sin\alpha = \frac{1}{p}$$

or
$$\left(\frac{x}{a}\right)^{m-1} = \frac{a\cos\alpha}{p}, \left(\frac{y}{b}\right)^{m-1} = \frac{b\sin\alpha}{p}$$

or
$$\left(\frac{a\cos\alpha}{p}\right)^{\frac{m}{m-1}} + \left(\frac{b\sin\alpha}{p}\right)^{\frac{m}{m-1}} = \left(\frac{x}{a}\right)^m + \left(\frac{y}{b}\right)^m = 1 \quad \text{[By (i)]}$$

whence follows the required condition.

**Example 4.38.** *Find the equation of the normal at any point $\theta$ to the curve $x = a (\cos \theta + \theta \sin \theta)$, $y = a (\sin \theta - \theta \cos \theta)$. Verify that these normals touch a circle with its centre at the origin and whose radius is constant.*

**Solution.** We have $\dfrac{dx}{d\theta} = a(-\sin\theta + \sin\theta + \theta\cos\theta) = a\theta\cos\theta$

$$\frac{dy}{d\theta} = a(\cos\theta - \cos\theta + \theta\sin\theta) = a\theta\sin\theta$$

$$\therefore \quad \frac{dy}{dx} = \frac{dy}{d\theta}\Big/\frac{dx}{d\theta} = \frac{\sin\theta}{\cos\theta}$$

$\therefore$ Slope of the normal at $\theta = -\dfrac{\cos\theta}{\sin\theta}$

Hence the equation of the normal at $\theta$

$$y - a(\sin\theta - \theta\cos\theta) = -\frac{\cos\theta}{\sin\theta}[x - a(\cos\theta + \theta\sin\theta)]$$

*i.e.,* $y\sin\theta - a\sin^2\theta + a\theta\sin\theta\cos\theta = -x\cos\theta + a\cos^2\theta + a\theta\sin\theta\cos\theta$

*i.e.,* $x\cos\theta + y\sin\theta = a(\cos^2\theta + \sin^2\theta) = a.$

Now the perpendicular distance of this normal from (0, 0) = $a$, which is a constant. Hence it touches a circle of radius $a$ having its centre at (0, 0).

**(3) Angle of intersection of two curves** *is the angle between the tangents to the curves at their point of intersection.*

*To find this angle θ, proceed as follows :*

(*i*) Find $P$, the point of intersection of the curves by solving their equations simultaneously.

(*ii*) Find the values of $dy/dx$ at $P$ for the two curves (say : $m_1, m_2$).

(*iii*) Find $\angle\theta$, using the $\tan\theta = \dfrac{m_1 - m_2}{1 + m_1 m_2}$.

When $m_1 m_2 = -1$, $\theta = 90°$ *i.e.,* the curves cut orthogonally.

**Example 4.39.** *Find the angle of intersection of the curves* $x^2 = 4y$ ...(*i*)

*and* $y^2 = 4x.$ ...(*ii*)

**Solution.** We have $x^4 = 16y^2 = 16.4\,x = 64x$

or $x(x^3 - 64) = 0$ whence $x = 0$ and 4.

Substituting these values in Eqn. (*i*), $y = 0$ and 4.

∴ The curves intersect at (0, 0) and (4, 4).

For the curve Eqn. (*i*), $dy/dx = x/2$. For the curve Eqn. (*ii*), $dy/dx = 2/y$

At (0, 0), slope of tangent to Eqn. (*i*) (= $m_1$) = 0/2 = 0 and slope of tangent to Eqn. (*ii*) (= $m_2$) = 2/0 = ∞.

Evidently the curves intersect at right angles.

At (4, 4), slope of tangent to Eqn. (*i*) (= $m_1$) = 4/2 = 2 and slope of tangent to Eqn. (*ii*) (= $m_2$) = 2/4 = $\frac{1}{2}$

∴ Angle of intersection of the curves

$$= \tan^{-1}\frac{m_1 - m_2}{1 + m_1 m_2} = \tan^{-1}\frac{2 - \frac{1}{2}}{1 + 2\cdot\frac{1}{2}} = \tan^{-1}\frac{3}{4}.$$

**Example 4.40.** *Show that the condition that the curves* $ax^2 + by^2 = 1$ *and* $a'x^2 + b'y^2 = 1$ *should intersect orthogonally is that*

$$\frac{1}{a} - \frac{1}{b} = \frac{1}{a'} - \frac{1}{b'}.$$

**Solution.** Given curves are $ax^2 + by^2 = 1$ ...(*i*) and $a'x^2 + b'y^2 = 1$ ...(*ii*)

Let $P(h, k)$ be a point of intersection of Eqns. (*i*) and (*ii*) so that

$$ah^2 + bk^2 = 1 \quad\text{and}\quad a'h^2 + b'k^2 = 1$$

$$\therefore \quad \frac{h^2}{-b + b'} = \frac{k^2}{-a' + a} = \frac{1}{ab' - a'b}$$

or $h^2 = (b' - b)/(ab' - a'b),\ k^2 = (a - a')/(ab' - a'b)$ ...(*iii*)

Differentiating (*i*) w.r.t. $x$,

$$2ax + 2by\,dy/dx = 0 \quad\text{or}\quad dy/dx = -ax/by.$$

Similarly for (*ii*), $dy/dx = -a'x/b'y$

∴ $m_1$ = slope of tangent to (*i*) at $P = -ah/bk$ ; $m_2$ = slope of tangent to (*ii*) at $P = -a'h/b'k$

For orthogonal intersection, we should have $m_1 m_2 = -1$.

*i.e.,* $\dfrac{-ah}{bk} \times \dfrac{-a'h}{b'k} = 1$ *i.e.,* $aa'h^2 + bb'k^2 = 0$

Substituting the values of $h^2$ and $k^2$ from (*iii*),

$$\frac{aa'(b'-b)}{ab'-a'b}+\frac{bb'(a-a')}{ab'-a'b}=0 \quad \text{or} \quad \frac{b'-b}{bb'}+\frac{a-a'}{aa'}=0$$

*i.e.,* $\frac{1}{b}-\frac{1}{b'}=\frac{1}{a}-\frac{1}{a'}$ which leads to the required condition.

**(4) Lengths of tangent, normal, subtangent and subnormal.** Let the tangent and the normal at any point $P(x, y)$ of the curve meet the $x$-axis at $T$ and $N$ respectively. (Fig. 4.4). Draw the ordinate $PM$. Then *PT and PN are called the lengths of the tangent and the normal respectively. Also TM and MN are called the subtangent and subnormal respectively.*

Let $\angle MTP = \psi$ so that $\tan\psi = dy/dx$.

Clearly, $\angle MPN = \psi$.

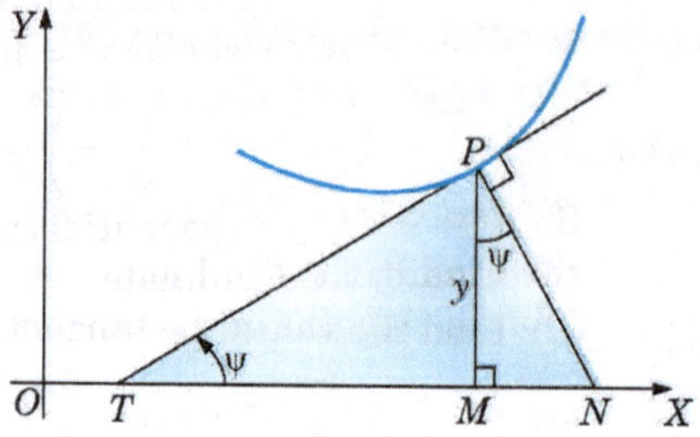


Fig. 4.4

(1) **Tangent** $= TP = MP \operatorname{cosec}\psi = y\sqrt{(1+\cot^2\psi)} = \mathbf{y\sqrt{[1+(dx/dy)^2]}}$

(2) **Normal** $= NP = MP \sec\psi = y\sqrt{(1+\tan^2\psi)} = \mathbf{y\sqrt{[1+(dy/dx)^2]}}$

(3) **Subtangent** $= TM = y\cot\psi = \mathbf{y\ dx/dy}$

(4) **Subnormal** $= MN = y\tan\psi = \mathbf{y\ dy/dx}$.

**Example 4.41.** *For the curve $x = a(\cos t + \log\tan t/2)$, $y = a\sin t$, prove that the portion of the tangent between the curve and x-axis is constant.*

*Also find its subtangent.*

**Solution.** Differentiating with respect to $t$,

$$\frac{dx}{dt}=a\left(-\sin t+\frac{1}{\tan t/2}\cdot\sec^2\frac{t}{2}\cdot\frac{1}{2}\right)=a\left(-\sin t+\frac{\cos t/2}{2\sin t/2}\cdot\frac{1}{\cos^2 t/2}\right)$$

$$=a\left(-\sin t+\frac{1}{\sin t}\right)=\frac{a(1-\sin^2 t)}{\sin t}=a\cos^2 t/\sin t\ ; \quad \frac{dy}{dt}=a\cos t$$

$$\therefore \quad \frac{dy}{dx}=\frac{dy}{dt}\Big/\frac{dx}{dt}=a\cos t\cdot\frac{\sin t}{a\cos^2 t}=\tan t.$$

Thus *length of the tangent between the curve and x-axis*

$$= y\sqrt{[1+(dx/dy)^2]} = a\sin t\cdot\sqrt{(1+\cot^2 t)} = a\sin t\ .\ \operatorname{cosec} t = a \text{ which is a constant.}$$

Also *subtangent* $= y\frac{dx}{dy} = a\sin t\cdot\cot t = a\cos t$.

## PROBLEMS 4.8

1. (a) Find the tangent to the curve $\mathbf{R}(t) = (t^2-1)\mathbf{I} + t\mathbf{J}$ at $t = 1$. *(K.T.U., 2016)*

   (b) Find the equation of the tangent and the normal to the curve $y(x-2)(x-3)-x+7=0$ at the point where it cuts the $x$-axis.

2. The straight line $x/a + y/b = 2$ touches the curve $(x/a)^n + (y/b)^n = 2$ for all values of $n$. Find the point of contact.

3. Prove that $\frac{x}{a}+\frac{y}{b}=1$ touches the curve $y = be^{-x/a}$ at the point where the curve crosses the axis of $y$.

4. If $p = x\cos\alpha + y\sin\alpha$, touches the curve $(x/a)^{n/(n-1)} + (y/b)^{n/(n-1)} = 1$, prove that

   $$p^n = (a\cos\alpha)^n + (b\sin\alpha)^n.$$

5. Prove that the condition for the line $x\cos\alpha + y\sin\alpha = p$ to touch the curve $x^m y^n = a^{m+n}$, is

   $$p^{m+n}\cdot m^m\cdot n^n = (m+n)^{m+n}\, a^{m+n}\cos^m\alpha\sin^n\alpha.$$

6. Show that the sum of the intercepts on the axes of any tangent to the curve $\sqrt{x}+\sqrt{y}=a$ is a constant.

7. If $x, y$ be the parts of the axes of $x$ and $y$ intercepted by the tangent at any point $(x, y)$ on the curve $(x/a)^{2/3} + (x/b)^{2/3} = 1$, then show that $(x_1/a)^2 + (y_1/b)^2 = 1$.

8. If the tangent at $(x_1, y_1)$ to the curve $x^3 + y^3 = a^3$ meets the curve again in $(x_2, y_2)$, show that
$$\frac{x_2}{x_1} + \frac{y_2}{y_1} = -1.$$
9. If the normal to the curve $x^{2/3} + y^{2/3} = a^{2/3}$ makes an angle $\phi$ with the axis of $x$, show that its equation is $y \cos \phi - x \sin \phi = a \cos 2\phi$.
10. Find the angle of intersection of the curves $x^2 - y^2 = a^2$ are $x^2 + y^2 = a^2\sqrt{2}$.
11. Show that the parabolas $y^2 = 4ax$ and $2x^2 = ay$ intersect at an angle $\tan^{-1}(3/5)$.
12. Prove that the curves $\frac{x^2}{a} + \frac{y^2}{b} = 1$ and $\frac{x^2}{a'} + \frac{y^2}{b'} = 1$ will cut orthogonally if $a - b = a' - b'$.
13. Show that in the exponential curve $y = be^{x/a}$, the subtangent is of constant length and that the subnormal varies as the square of the ordinate.
14. Find the lengths of the tangent, normal, subtangent and subnormal for the cycloid:
$$x = a(t + \sin t),\ y = a(1 - \cos t),$$
15. For the curve $x = a \cos^3 \theta, y = a \sin^3 \theta$, show that the portion of the tangent intercepted between the point of contact and the $x$-axis is $y \operatorname{cosec} \theta$. Also find the length of the subnormal.

## 4.7 POLAR CURVES

**(1) Angle between radius vector and tangent.** *If $\phi$ be the angle between the radius vector and the tangent at any point of the curve $r = f(\theta)$,* $\mathbf{\tan \theta = r\frac{d\theta}{dr}}$.

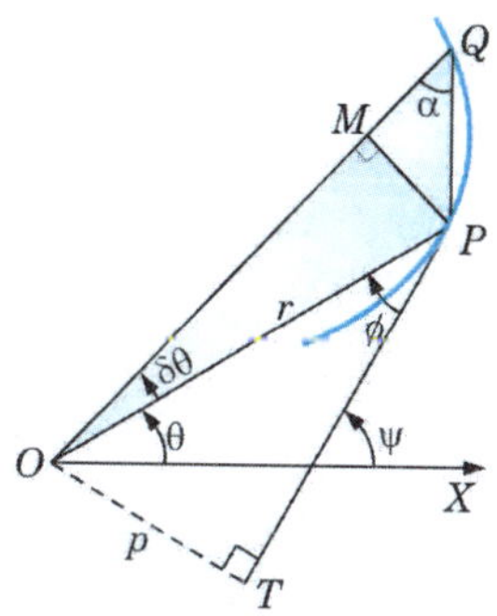
Fig. 4.5

Let $P(r, \theta)$ and $Q(r + \delta r, \theta + \delta\theta)$ be two neighbouring points on the curve (Fig. 4.5). Join $PQ$ and draw $PM \perp OQ$. Then from the rt. angled $\Delta OMP$, $MP = r \sin \delta\theta$, $OM = r \cos \delta\theta$.

$$\therefore \quad MQ = OQ - OM = r + \delta r - r \cos \delta\theta$$
$$= \delta r + r(1 - \cos \delta\theta) = \delta r + 2r \sin^2 \delta\theta/2.$$

If $\angle MQP = \alpha$, then
$$\tan \alpha = \frac{MP}{MQ} = \frac{r \sin \delta\theta}{\delta r + 2r \sin^2 \delta\theta/2}$$

In the limit as $Q \to P$ (*i.e.*, $\delta\theta \to 0$), the chord $PQ$ turns about $P$ and becomes the tangent at $P$ and $\alpha \to \phi$.

$$\therefore \quad \tan \phi = \underset{Q \to P}{\text{Lt}} (\tan \alpha) = \underset{\delta\theta \to 0}{\text{Lt}} \frac{r \sin \delta\theta}{\delta r + 2r \sin^2 \delta\theta/2}$$
$$= \underset{\delta\theta \to 0}{\text{Lt}} \frac{r(\sin \delta\theta/\delta\theta)}{(\delta r/\delta\theta) + r \sin \delta\theta/2 \cdot (\sin \delta\theta/2 \div \delta\theta/2)}$$
$$= \frac{r \cdot 1}{(dr/d\theta) + r \cdot 0 \cdot 1} = r\frac{d\theta}{dr} \qquad (V.T.U., 2013)$$

**Cor. Angle of intersection of two curves.** *If $\phi_1$, $\phi_2$ be the angles between the common radius vector and the tangents to the two curves at their point of intersection, then the angle of intersection of these curves is $\phi_1 \sim \phi_2$.*

**(2) Length of the perpendicular from pole on the tangent.** *If p be the perpendicular from the pole on the tangent, then*

(*i*) $\mathbf{p = r \sin \phi}$ (*ii*) $\mathbf{\frac{1}{p^2} = \frac{1}{r^2} + \frac{1}{r^4}\left(\frac{dr}{d\theta}\right)^2}$

From the rt. $\angle$ed $\Delta OTP$, $p = r \sin \phi$

$$\therefore \quad \frac{1}{p^2} = \frac{1}{r^2} \operatorname{cosec}^2 \phi = \frac{1}{r^2}(1 + \cot^2 \phi)$$
$$= \frac{1}{r^2}\left[1 + \frac{1}{r^2}\left(\frac{dr}{d\theta}\right)^2\right] = \frac{1}{r^2} + \frac{1}{r^4}\left(\frac{dr}{d\theta}\right)^2 \qquad \text{[By (1)]}$$

**(3) Polar subtangent and subnormal.** Let the tangent and the normal at any point $P(r, \theta)$ of a curve meet the line through the pole perpendicular to the radius vector $OP$ in $T$ and $N$ respectively (Fig. 4.6). Then $OT$ is called the *polar subtangent* and $ON$ the *polar subnormal.*

Let $\angle OTP = \phi$ so that $\tan \phi = r d\theta/dr$

Clearly, $\angle PNO = \phi$.

$\therefore$ (*i*) **Polar subtangent**

$$= OT = r \tan \phi = r \cdot r d\theta/dr = \mathbf{r^2 \frac{d\theta}{dr}}$$

(*ii*) **Polar subnormal**

$$= ON = r \cot \phi = r \cdot \frac{1}{r}\frac{dr}{d\theta} = \mathbf{\frac{dr}{d\theta}}$$

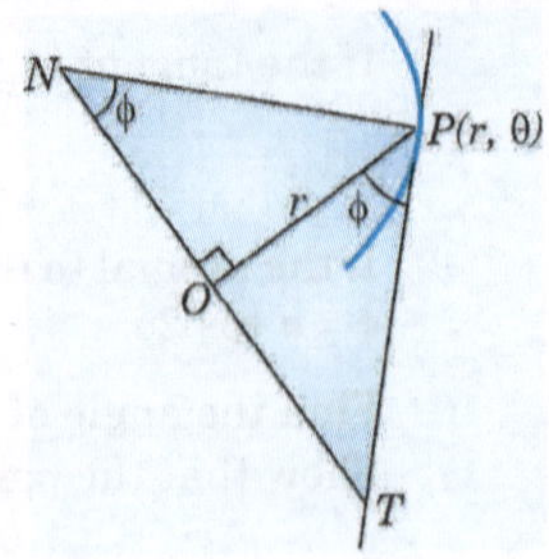


Fig. 4.6

**Example 4.42.** *For the cardioid $r = a(1 - \cos \theta)$, prove that*

(*i*) $\phi = \theta/2$ (*ii*) $p = 2a \sin^3 \theta/2$

(*iii*) *polar subtangent* $= 2a \sin^2 \frac{\theta}{2} \tan \frac{\theta}{2}$.

**Solution.** We have $$\frac{dr}{d\theta} = a \sin \theta$$

$$\therefore \quad \tan \phi = r\frac{d\theta}{dr} = a(1 - \cos \theta) \cdot \frac{1}{a \sin \theta}$$

$$= 2 \sin^2 \theta/2 \div 2 \sin \theta/2 \cos \theta/2 = \tan \theta/2. \text{ Thus } \phi = \theta/2 \quad ...(i)$$

Also $$p = r \sin \phi = a(1 - \cos \theta) \cdot \sin \theta/2 = a \cdot 2 \sin^2 \theta/2 \cdot \sin \theta/2$$

$$= 2a \sin^3 \theta/2 \quad ...(ii)$$

Polar subtangent $$= r^2 \, d\theta/dr = [a(1 - \cos \theta)]^2 \div a \sin \theta$$

$$= 4a \sin^4 \theta/2 \div 2 \sin \theta/2 \cos \theta/2 = 2a \sin^2 \theta/2 \tan \theta/2. \quad ...(iii)$$

**Example 4.43.** *Find the angle of intersection of the curves $r = \sin \theta + \cos \theta$, $r = 2 \sin \theta$.*

**Solution.** To find the point of intersection of the curves $r = \sin \theta + \cos \theta$

and $r = 2 \sin \theta$, ...(*ii*), we eliminate $r$.

Then $2 \sin \theta = \sin \theta + \cos \theta$ or $\tan \theta = 1$ *i.e.*, $\theta = \pi/4$.

For Eqn. (*i*), $$\frac{dr}{d\theta} = \cos \theta - \sin \theta$$

$$\therefore \quad \tan \phi = r\frac{d\theta}{dr} = \frac{\sin \theta + \cos \theta}{\cos \theta - \sin \theta} \text{ which } \to \infty \text{ at } \theta = \pi/4. \text{ Thus } \phi = \pi/2.$$

For Eqn. (*ii*), $dr/d\theta = 2 \cos \theta$ $\therefore$ $\tan \phi' = r\dfrac{d\theta}{dr} = \dfrac{2 \sin \theta}{2 \cos \theta} = 1$ at $\theta = \pi/4$. Thus $\phi' = \pi/4$

Hence the angle of intersection of Eqns. (*i*) and (*ii*) $= \phi - \phi' = \pi/4$.

## PROBLEMS 4.9

1. For a curve in Cartesian form, show that $\tan \phi = \dfrac{xy' - y}{x + yy'}$.
2. Show that in the equiangular spiral $r = ae^{\theta \cot \alpha}$, the tangent is inclined at a constant angle to the radius vector.
3. Show that the tangent to the cardioid $r = a\,(1 + \cos \theta)$ at the points $\theta = \pi/3$ and $\theta = 2\pi/3$ are respectively parallel and perpendicular to the initial line.
4. Prove that, in the parabola $2a/r = 1 - \cos \theta$,

   (*i*) $\phi = \pi - \theta/2$ (*ii*) $\pi = a \operatorname{cosec} \theta/2$, and (*iii*) polar subtangent $= 2\,a \operatorname{cosec} \theta$.
5. Show that the angle between the tangent at any point $P$ and the line joining $P$ to the origin is the same at all points of the curve

$$\log(x^2 + y^2) = k \tan^{-1}(y/x).$$

6. Show that in the curve $r = a\theta$, the polar subnormal is constant and in the curve $r\,\theta = a$ the polar subtangent is constant.
7. Find the angle of intersection of the curves
   (i) $r = 2\sin\theta$, and $r = 2\cos\theta$
   (ii) $r = a/(1 + \cos\theta)$ and $r = b/(1 - \cos\theta)$.
   (iii) $r^2 \sin 2\theta = 4$ and $r^2 = 16 \sin 2\theta$. (V.T.U., 2016)
8. Prove that the curves $r = a(1 + \cos\theta)$ and $r = b(1 - \cos\theta)$ intersect at right angles. (V.T.U., 2016)
9. Show that the curves $r^n = a^n \cos n\theta$ and $r^n = b^n \sin n\theta$ cut each other orthogonally.
10. Show that the angle of intersection of the curves $r = a \log\theta$ and $r = a/\log\theta$ is $\tan^{-1} [2e/(1 - e^2)]$. (V.T.U., 2015)

## 4.8 PEDAL EQUATION

If $r$ be the radius vector of any point on the curve and $p$ the length of the perpendicular from the pole on the tangent at that point, then *the relation between p and r is called pedal equation of the curve.*

Given the cartesian or polar equation of a curve, we can derive its pedal equation. The method is explained through the following examples.

**Example 4.44.** *Find the pedal equation of the ellipse* $\dfrac{x^2}{a^2} + \dfrac{y^2}{b^2} = 1.$ ...(i)

**Solution.** Equation of the tangent at $(x, y)$ is $\dfrac{Xx}{a^2} + \dfrac{Yy}{b^2} = 1$ ...(ii)

$p$, length of $\perp$ from (0, 0) on Eqn. (ii) $= \dfrac{-1}{\sqrt{[(x/a^2)^2 + (y/b^2)^2]}}$

or
$$\frac{1}{p^2} = \frac{x^2}{a^4} + \frac{y^2}{b^4} \quad ...(iii)$$

Also
$$r^2 = x^2 + y^2 \quad ...(iv)$$

Substituting the value of $y^2$ from Eqn. (iv) in (i),
$$\frac{x^2}{a^2} = \frac{r^2 - b^2}{a^2 - b^2}.$$

Then from Eqn. (i),
$$\frac{y^2}{b^2} = \frac{a^2 - r^2}{a^2 - b^2}$$

Now substituting these values of $x^2/a^2$ and $y^2/b^2$ in Eqn. (iii),
$$\frac{1}{p^2} = \frac{1}{a^2}\left(\frac{r^2 - b^2}{a^2 - b^2}\right) + \frac{1}{b^2}\left(\frac{a^2 - r^2}{a^2 - b^2}\right)$$

or
$$\frac{a^2 b^2}{p^2} = \frac{r^2 b^2 - b^4 + a^4 - a^2 r^2}{a^2 - b^2} = a^2 + b^2 - r^2$$

Hence $\dfrac{a^2 b^2}{p^2} = a^2 + b^2 - r^2$ which is the required pedal equation.

**Example 4.45.** *Find the pedal equation of the curves*
*(i)* $2a/r = 1 - \cos\theta.$ *(ii)* $r^n = a^n \cos n\,\theta$ (V.T.U., 2016)

**Solution.** (i) We have $2a/r = 1 - \cos\theta$

Taking logs of both sides,
$$\log 2a - \log r = \log(1 - \cos\theta)$$

Differentiating both sides with respect to θ, we get

$$-\frac{1}{r}\frac{dr}{d\theta} = \frac{1}{1-\cos\theta}\cdot\sin\theta = \cot\frac{\theta}{2}$$

∴ $$\tan\phi = r\frac{d\theta}{dr} = -\tan\theta/2 = \tan(\pi-\theta/2) \text{ i.e., } \phi = \pi - \theta/2$$

Also $$p = r\sin\phi = r\sin(\pi-\theta/2) \text{ i.e., } p = r\sin\theta/2$$

or $$p^2 = r^2\sin^2\theta/2 = r^2\left(\frac{1-\cos\theta}{2}\right) = r^2 . a/r$$ [By (*i*)]

Hence $p^2 = ar$, which is the required pedal equation.

(*ii*) From the given equation, $nr^{n-1}\dfrac{dr}{d\theta} = -na^n \sin n\theta$

so that $$\tan\phi = r\,d\theta/dr = r\frac{nr^{n-1}}{-na^n\sin n\theta} = -\cot n\theta = \tan\left(\frac{\pi}{2}+n\theta\right)$$

*i.e.,* $$\phi = \pi/2 + n\theta$$

∴ $$p = r\sin\phi = r\sin\left(\frac{\pi}{2}+n\theta\right) = r\cos n\theta = r\,.\,(r^n/a^n) = r^{n+1}/a^n.$$

Hence $p\,a^n = r^{n+1}$, which is the required pedal equation.

## 4.9 DERIVATIVE OF ARC

**(1)** *For the curve y = f (x), we have*

$$\frac{ds}{dx} = \sqrt{\left[1+\left(\frac{dy}{dx}\right)^2\right]}$$

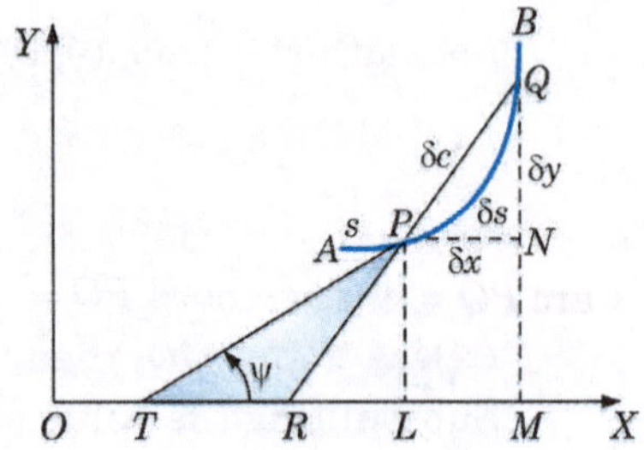

Fig. 4.7

Let $P(x, y)$, $Q(x + \delta x, y + \delta y)$ be two neighbouring points on the curve $AB$ (Fig. 4.7). Let arc $AP = s$, arc $PQ = \delta s$ and chord $PQ = \delta c$.

Draw $PL$, $QM$ ⊥s on the $x$-axis and $PN \perp QM$.

∴ From the rt. ∠ed Δ$PNQ$,

$$PQ^2 = PN^2 + NQ^2$$

*i.e.,* $$\delta c^2 = \delta x^2 + \delta y^2$$

or $$\left(\frac{\delta c}{\delta x}\right)^2 = 1 + \left(\frac{\delta y}{\delta x}\right)^2$$

∴ $$\left(\frac{\delta s}{\delta x}\right)^2 = \left(\frac{\delta s}{\delta c}\cdot\frac{\delta c}{\delta x}\right)^2$$

$$= \left(\frac{\delta s}{\delta c}\right)^2 = \left[1+\left(\frac{\delta y}{\delta x}\right)^2\right]$$

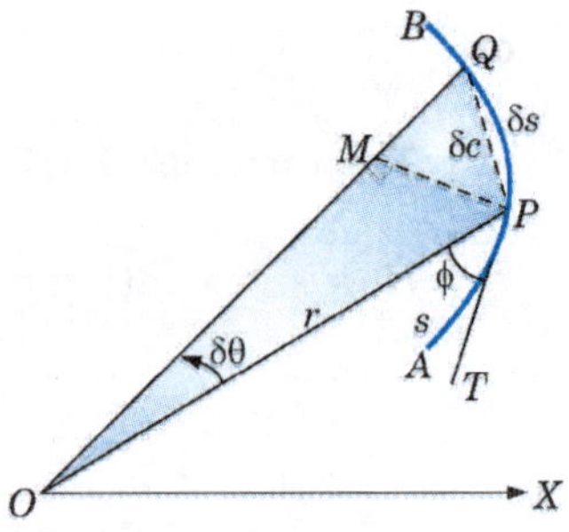

Fig. 4.8

Taking limits as $Q \to P$ (*i.e.,* $\delta c \to 0$),

$$\left(\frac{ds}{dx}\right)^2 = 1\cdot\left[1+\left(\frac{dy}{dx}\right)^2\right] \qquad \left[\because \underset{\delta c\to 0}{\text{Lt}}\frac{\delta s}{\delta c} = 1\right]$$

If $s$ increases with $x$ as in Fig. 4.7, $dy/dx$ is positive.

Thus $$\frac{ds}{dx} = \sqrt{\left[1+\left(\frac{dy}{dx}\right)^2\right]}$$, taking positive sign before the radical. ...(1)

**Cor. 1.** *If the equation of the curve is x = f(y), then*

$$\frac{ds}{dy} = \frac{ds}{dx}\cdot\frac{dx}{dy} = \sqrt{\left[1+\left(\frac{dy}{dx}\right)^2\right]}\cdot\frac{dx}{dy}$$

$$\therefore \quad \frac{\mathbf{ds}}{\mathbf{dx}} = \sqrt{\left[1+\left(\frac{\mathbf{dy}}{\mathbf{dx}}\right)^2\right]} \qquad ...(2)$$

**Cor. 2.** *If the equation of the curve is in parametric form $x = f(t)$, $y = \phi(t)$, then*

$$\frac{ds}{dt} = \frac{ds}{dx}\cdot\frac{dx}{dt} = \sqrt{\left[1+\left(\frac{dy}{dx}\right)^2\right]}\cdot\frac{dx}{dt}$$

$$= \sqrt{\left[\left(\frac{dx}{dt}\right)^2 + \left(\frac{dy}{dx}\cdot\frac{dx}{dt}\right)^2\right]}$$

$$\therefore \quad \frac{\mathbf{ds}}{\mathbf{dt}} = \sqrt{\left[\left(\frac{\mathbf{dx}}{\mathbf{dt}}\right)^2 + \left(\frac{\mathbf{dy}}{\mathbf{dt}}\right)^2\right]} \qquad ...(3)$$

**Cor. 3.** We have $\qquad \dfrac{ds}{dx} = \sqrt{\left[1+\left(\dfrac{dy}{dx}\right)^2\right]} = \sqrt{(1+\tan^2\psi)} = \sec\psi$

$$\therefore \quad \cos\psi = \frac{\mathbf{dx}}{\mathbf{ds}}. \qquad ...(4)$$

Also $\qquad \sin\psi = \tan\psi\cos\psi = \dfrac{dy}{dx}\cdot\dfrac{dx}{ds}$

$$\therefore \quad \sin\psi = \frac{\mathbf{dy}}{\mathbf{ds}} \qquad ...(5)$$

**(2)** *For the curve $r = f(\theta)$, we have* $\dfrac{\mathbf{ds}}{\mathbf{d\theta}} = \sqrt{\left[\mathbf{r}^2 + \left(\dfrac{\mathbf{dr}}{\mathbf{d\theta}}\right)^2\right]}$.

Let $P(r, \theta)$, $Q(r + \delta r, \theta + \delta\theta)$ be two neighbouring points on the curve $AB$ (Fig. 4.8). Let arc $AP = s$, arc $PQ = \delta s$ and chord $PQ = \delta c$.

Draw $PM \perp OQ$, then

$$PM = r\sin\delta\theta \text{ and } MQ = OQ - OM = r + \delta r - r\cos\delta\theta = \delta r + 2r\sin^2\delta\theta/2$$

From the rt. $\angle$ed $\Delta PMQ$,

$$PQ^2 = PM^2 + MQ^2$$

or
$$\delta c^2 = (r\sin\delta\theta)^2 + (\delta r + 2r\sin^2\delta\theta/2)^2$$

$$\therefore \quad \left(\frac{\delta s}{\delta\theta}\right)^2 = \left(\frac{\delta s}{\delta c}\cdot\frac{\delta c}{\delta\theta}\right)^2 = \left(\frac{\delta s}{\delta c}\right)^2\left[\left(\frac{r\sin\delta\theta}{\delta\theta}\right)^2 + \left(\frac{\delta r}{\delta\theta} + \frac{2r\sin^2\delta\theta/2}{\delta\theta}\right)^2\right]$$

$$= \left(\frac{\delta s}{\delta c}\right)^2\left[r^2\left(\frac{\sin\delta\theta}{\delta\theta}\right)^2 + \left(\frac{\delta r}{\delta\theta} + r\sin\frac{\delta\theta}{2}\cdot\frac{\sin\delta\theta/2}{\delta\theta/2}\right)^2\right]$$

Taking limits as $Q \to P$

$$\left(\frac{ds}{d\theta}\right)^2 = 1^2 \,.\left[r^2\cdot 1^2 + \left(\frac{dr}{d\theta} + r\cdot 0\cdot 1\right)^2\right] = r^2 + \left(\frac{dr}{d\theta}\right)^2$$

As $s$ increases with the increase of $\theta$, $ds/d\theta$ is positive. Thus

$$\frac{ds}{d\theta} = \sqrt{\left[r^2 + \left(\frac{dr}{d\theta}\right)^2\right]} \qquad ...(1)$$

**Cor. 1.** If the equation of the curve is $\theta = f(r)$, then

$$\frac{ds}{dr} = \frac{ds}{d\theta}\cdot\frac{d\theta}{dr} = \sqrt{\left[r^2 + \left(\frac{dr}{d\theta}\right)^2\right]}\cdot\frac{d\theta}{dr}$$

$$\therefore \qquad \mathbf{\frac{ds}{dr} = \sqrt{\left[1 + r^2 + \left(\frac{dy}{dt}\right)^2\right]}} \qquad ...(2)$$

**Cor. 2.** We have $\quad \dfrac{ds}{dr} = \sqrt{\left[1+\left(r\dfrac{d\theta}{dr}\right)^2\right]} = \sqrt{[1+\tan^2\phi]}\dfrac{ds}{dr} = \sqrt{\left[1+\left(r\dfrac{d\theta}{dr}\right)^2\right]} = \sqrt{[1+\tan^2\phi]} = \sec\phi$

$$\therefore \qquad \mathbf{\cos\phi = \frac{dr}{ds}} \qquad ...(3)$$

Also $\quad \sin\phi = \tan\phi \,.\, \cos\phi = r\dfrac{d\theta}{dr}\cdot\dfrac{dr}{ds}$

$$\therefore \qquad \mathbf{\sin\phi = r\frac{d\theta}{ds}} \qquad ...(4)$$

## PROBLEMS 4.10

Prove that the pedal equation of :

1. the parabola $y^2 = 4a(x + a)$ is $p^2 = ar$.
2. the hyperbola $\dfrac{x^2}{a^2} - \dfrac{y^2}{b^2} = 1$ is $a^2b^2/p^2 = r^2 - a^2 + b^2$.
3. the astroid $x = a\cos^3 t, y = a\sin^3 t$ is $r^2 = a^2 - 3p^2$.

Find the pedal equations of the following curves :

4. $r = a(1 + \cos\theta)$ *(V.T.U., 2016)*
5. $r^2 = a^2\sin^2\theta$
6. $r^m \cos m\theta = a^m$.
7. $r^m = a^m(\cos m\theta + \sin m\theta)$ *(V.T.U., 2010)*
8. $r^n = \text{sech}\, n\theta$ *(V.T.U., 2015)*
9. $r = ae^{m\theta}$.
10. Calculate $ds/dx$ for the following curves :
    (*i*) $ay^2 = x^3$. (*ii*) $y = c\cosh x/c$.
11. Find $ds/d\theta$ for the curve $x = a(\cos\theta + \theta\sin\theta), y = a(\sin\theta - \theta\cos\theta)$
12. Find $ds/d\theta$ for the following curves :
    (*i*) $r = a(1 - \cos\theta)$ (*ii*) $r^2 = a^2\cos^2 2\theta$ *(V.T.U., 2014S)*
    (*iii*) $r = \frac{1}{2}\sec^2\theta$
13. For the curves $\theta = \cos^{-1}(r/k) - \sqrt{(k^2 - r^2)}/r$, prove that $r\dfrac{ds}{dr}$ = constant.
14. With the usual meanings for $r, s, \theta$ and $\phi$ for the polar curve $r = f(\theta)$, show that $\dfrac{d\phi}{d\theta} + r\,\text{cosec}^2\theta\,\dfrac{d^2r}{ds^2} = 0$.

## 4.10 CURVATURE

Let $P$ be any point on a given curve and $Q$ a neighbouring point. Let arc $AP = s$ and arc $PQ = \delta s$. Let the tangents at $P$ and $Q$ make angle $\psi$ and $\psi + \delta\psi$ with the $x$-axis, so that the angle between the tangents at $P$ and $Q = \delta\psi$ (Fig. 4.9).

In moving from $P$ to $Q$ through a distance $\delta s$, the tangent has turned through the angle $\delta\psi$. This is called the *total bending or total curvature* of the arc $PQ$.

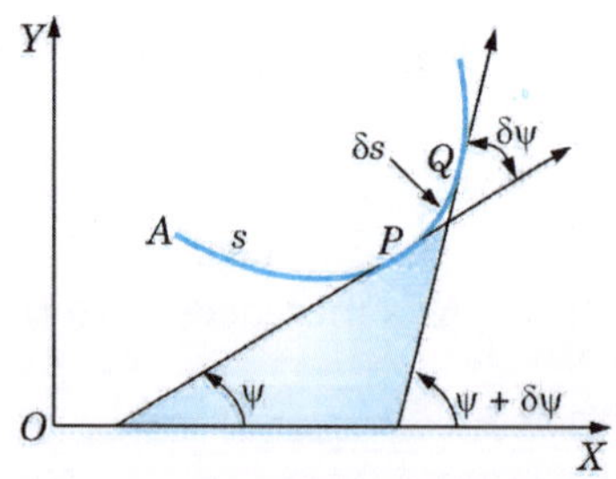

Fig. 4.9

$\therefore$ The *average curvature* of arc $PQ = \dfrac{\delta\psi}{\delta s}$

*The limiting value of average curvature when Q approaches P (i.e., $\delta s \to 0$) is defined as the curvature of the curve at P.*

Thus **curvature** $K$ (at $P$) $= \mathbf{\dfrac{d\psi}{ds}}$

**Obs.** Since $\delta\psi$ is measured in radians, the *unit of curvature is radians per unit length e.g.*, radians per centimetre.

**(1) Radius of curvature.** *The reciprocal of the curvature of a curve at any point P is called the* **radius of curvature** *at P and is denoted by* $\rho$, so the $\rho = \mathbf{ds/d\psi}$.

**(2) Centre of curvature.** *A point C on the normal at any point P of a curve distant* $\rho$ *from it, is called the* **centre of curvature** *at P.*

**(3) Circle of curvature.** *A circle with centre C (centre of curvature at P) and radius* $\rho$ *is called the* **circle of curvature** *or* **osculating circle** *at P.*

## 4.11 (1) RADIUS OF CURVATURE FOR CARTESIAN CURVE y = f (x), is given by

$$\rho = \frac{(1 + \mathbf{y_1}^2)^{3/2}}{\mathbf{y_2}}$$

We know that $\tan\psi = dy/dx = y_1$ or $\psi = \tan^{-1}(y_1)$

Differentiating both sides w.r.t. $x$,

$$\frac{d\psi}{dx} = \frac{1}{1+y_1^2} \cdot \frac{d(y_1)}{dx} = \frac{y_2}{1+y_1^2}$$

$$\therefore \qquad \rho = \frac{ds}{d\psi} = \frac{ds}{dx} \cdot \frac{dx}{d\psi} = \sqrt{(1+y_1^2)} \cdot \frac{1+y_1^2}{y_2} = \frac{(1+y_1^2)^{3/2}}{y_2} \qquad ...(1)$$

**(2) Radius of curvature for parametric equations**

$$x = f(t), \qquad y = \phi(t).$$

Denoting differentiations with respect to $t$ by dashes,

$$y_1 = \frac{dy}{dx} = \frac{dy}{dt}\bigg/\frac{dx}{dt} = y'/x'.$$

$$y_2 = \frac{d}{dx}(y_1) = \frac{d}{dt}\left(\frac{y'}{x'}\right) \cdot \frac{dt}{dx} = \frac{x'y'' - y'x''x}{(x')^2} \cdot \frac{1}{x'}$$

Substituting the values of $y_1$ and $y_2$ in (1)

$$\rho = \left[1 + \left(\frac{y'}{x'}\right)^2\right]^{3/2} \bigg/ \left[\frac{x'y'' - y'x''}{(x')^3}\right] = \frac{(x'^2 + y'^2)^{3/2}}{x'y'' - y'x''}$$

**(3) Radius of curvature at the origin. Newton's formulae***

(*i*) *If x-axis is tangent to a curve at the origin, then*

$$\rho \text{ at } (0, 0) = \underset{\mathbf{x \to 0}}{\mathbf{Lt}} \left(\frac{\mathbf{x^2}}{\mathbf{2y}}\right)$$

Since $x$-axis is a tangent at (0, 0), $(dy/dx)_0$ or $(y_1)_0 = 0$

Also $$\underset{x \to 0}{\text{Lt}} \left(\frac{x^2}{2y}\right) = \underset{x \to 0}{\text{Lt}} \left(\frac{2x}{2dy/dx}\right) = \underset{x \to 0}{\text{Lt}} \frac{1}{d^2y/dx^2} = \frac{1}{(y_2)_0} \qquad \left(\frac{0}{0} \text{ form}\right)$$

$$\therefore \qquad \rho \text{ at } (0,0) = \frac{\{1 + (y_1^2)_0\}^{3/2}}{(y_2)_0} = \frac{1}{(y_2)_0} = \underset{x \to 0}{\text{Lt}} \frac{x^2}{2y} \qquad \text{[From Eqn. (1)]}$$

(*ii*) Similarly, *if y-axis is tangent to a curve at the origin, then*

$$\rho \text{ at } (0, 0) = \underset{x \to 0}{\text{Lt}} \left(\frac{\mathbf{y^2}}{\mathbf{2x}}\right)$$

---

* Named after the great English mathematician and physicist *Sir Issac Newton* (1642–1727) whose contributions are of utmost importance. He discovered many physical laws, invented Calculus alongwith *Leibnitz* (see footnote p. 154) and created analytical methods of investigating physical problems. He became professor at Cambridge in 1699, but his '*Mathematical Principles of Natural Philosophy*' containing development of classical mechanics had been completed in 1687.

(*iii*) In case the curve passes through the origin but neither $x$-axis nor $y$-axis is tangent at the origin, we write the equation of the curve as

$$y = f(x) = f(0) + xf'(0) + \frac{x^2}{2!} f''(0) + \ldots \qquad \text{[By Maclaurin's series]}$$

$$= px + qx^2/2 + \ldots \qquad [\because \; f(0) = 0]$$

where, $p = f'(0)$ and $q = f''(0)$

Substituting this in the equation $y = f(x)$, we find the values of $p$ and $q$ by equating coefficients of like powers of $x$. Then $\rho\,(0, 0) = (1 + p^2)^{3/2}/q$.

**Obs.** *Tangents at the origin to a curve are found by equating to zero the lowest degree terms in its equation.*

**Example 4.46.** *Find the radius of curvature at the point (i) (3a/2, 3a/2) of the Folium $x^3 + y^3 = 3axy$.* (*V.T.U., 2016 ; Anna, 2012; Kurukshetra, 2009 S*)

*(ii) (a, 0) on the curve $xy^2 = a^3 - x^3$.* (*Anna, 2009 ; Kerala, 2005*)

**Solution.** (*i*) Differentiating with respect to $x$, we get

$$3x^2 + 3y^2 \frac{dy}{dx} = 3a\left(y + x\frac{dy}{dx}\right)$$

or $$(y^2 - ax)\frac{dy}{dx} = ay - x^2 \qquad \ldots(i) \qquad \therefore \quad \frac{dy}{dx} \text{ at } (3a/2, 3a/2) = -1$$

Differentiating (*i*),

$$\left(2y\frac{dy}{dx} - a\right)\frac{dy}{dx} + (y^2 - ax)\frac{d^2y}{dx^2} = a\frac{dy}{dx} - 2x \quad \therefore \quad \frac{d^2y}{dx^2} \text{ at } (3a/2, 3a/2) = -32/3a$$

Hence $\rho$ at $(3a/2, 3a/2)$ $$= \frac{[1 + (dy/dx)^2]^{3/2}}{d^2y/dx^2} = \frac{[1 + (-1)^2]^{3/2}}{-32/3a} = \frac{3a}{8\sqrt{2}} \quad \text{(in magnitude).}$$

**Note.** *Curvature at $(3a/2, 3a/2) = 8\sqrt{2}/3a$.* (*Delhi, 2012*)

(*ii*) We have $$y^2 = a^3 x^{-1} - x^2$$

$$\therefore \quad 2y\, dy/dx = -a^3x^{-2} - 2x \quad \text{or} \quad dy/dx = -a^3/(2x^2y) - x/y$$

At $(a, 0)$, $dy/dx \to \infty$, so we find $dx/dy$ from $xy^2 = a^3 - x^3$

$$\therefore \quad x - 2y + y^2\frac{dx}{dy} = -3x^2\frac{dx}{dy}$$

or $$\frac{dx}{dy} = \frac{-2xy}{3x^2 + y^2} \quad \text{or} \quad \frac{dx}{dy} \text{ at } (a, 0) = 0.$$

$$\therefore \quad \frac{d^2x}{dy^2} = \frac{(3x^2 + y^2)\left(-2y\frac{dx}{dy} - 2x\right) - (-2xy)\left(6x\frac{dx}{dy} + 2y\right)}{(3x^2 + y^2)^2}$$

or $$\frac{d^2x}{dy^2} \text{ at } (a, 0) = \frac{(3a^2 + 0)(0 - 2a) - 0}{(3a^2 + 0)^2} = \frac{-2}{3a}$$

Hence $$\rho \text{ at } (a, 0) = \frac{\left[1 + \left(\frac{dx}{dy}\right)_{(a,\,0)}\right]^{3/2}}{\left(\frac{d^2x}{dy^2}\right)_{(a,\,0)}} = \frac{(1 + 0)^{3/2}}{(-2/3a)} = -\frac{3a}{2}.$$

**Example 4.47.** *Show that the radius of curvature at any point of the cycloid $x = a\,(\theta + \sin\theta)$, $y = a(1 - \cos\theta)$ is $4a\cos\theta/2$.* (*V.T.U., 2016 ; Anna, 2013 ; D.T.U., 2013*)

**Solution.** We have $$\frac{dx}{d\theta} = a\,(1 + \cos\theta), \quad \frac{dy}{d\theta} = a\sin\theta.$$

$$\therefore \quad \frac{dy}{dx} = \frac{dy}{d\theta} \div \frac{dx}{d\theta} = \frac{a \sin \theta}{a(1+\cos\theta)} = \frac{2 \sin \theta/2 \cos \theta/2}{2 \cos^2 \theta/2} = \tan \theta/2$$

$$\frac{d^2y}{dx^2} = \frac{d}{d\theta}\left(\frac{dy}{dx}\right) \cdot \frac{d\theta}{dx} = \frac{1}{2} \sec^2 \frac{\theta}{2} \cdot \frac{1}{a(1+\cos\theta)}$$

$$= \frac{1}{2} \sec^2 \frac{\theta}{2} \cdot \frac{1}{2a \cos^2 \theta/2} = \frac{1}{4a} \sec^4 \frac{\theta}{2}.$$

$$\therefore \quad \rho = \frac{[1+(dy/dx)^2]^{3/2}}{d^2y/dx^2} = \frac{4a(1+\tan^2 \theta/2)^{3/2}}{\sec^4 \theta/2}$$

$$= 4a \cdot (\sec^2 \theta/2)^{3/2} \cdot \cos^4 \theta/2 = 4a \cos \theta/2.$$

**Example 4.48.** *Prove that the radius of curvature at any point of the astroid $x^{2/3} + y^{2/3} = a^{2/3}$, is three times the length of the perpendicular from the origin to the tangent at that point.*

(*D.T.U., 2012 ; P.T.U., 2011*)

**Solution.** The parametric equation of the curve is

$$x = a \cos^3 t, y = a \sin^3 t.$$

$$\therefore \quad x' \ (= dx/dt) = -3a \cos^2 t \sin t, \ y' = 3a \sin^2 t \cos t.$$

$$x'' = -3a (\cos^3 t - 2 \cos t \sin^2 t) = 3a \cos t (2 \sin^2 t - \cos^2 t)$$

$$y'' = 3a (2 \sin t \cos^2 t - \sin^3 t) = 3a \sin t (2 \cos^2 t - \sin^2 t)$$

$$x'^2 + y'^2 = 9a^2 (\cos^4 t \sin^2 t + \sin^4 t \cos^2 t) = 9a^2 \sin^2 t \cos^2 t$$

$$x' y'' - y' x'' = -9a^2 \cos^2 t \sin^2 t (2 \cos^2 t - \sin^2 t)$$

$$-9a^2 \cos^2 t \sin^2 t (2 \sin^2 t - \cos^2 t) = -9a^2 \sin^2 t \cos^2 t$$

$$\therefore \quad \rho = \frac{(x'^2 + y'^2)^{3/2}}{x'y'' - y'x''} = \frac{27a^3 \sin^3 t \cos^3 t}{-9a^2 \sin^2 t \cos^2 t} = -3a \sin t \cos t.$$

Since $\quad dy/dx = y'/x' = -\tan t,$

$\therefore$ Equation of the tangent at $(a \cos^3 t, a \sin^3 t)$ is $y - a \sin^3 t = -\tan t (x - a \cos^3 t)$

*i.e.*, $\quad x \tan t + y - a \sin t = 0 \quad$ ...(*i*)

$p$, length of $\perp$ from (0, 0) on (*i*) $= \dfrac{0 + 0 - a \sin t}{\sqrt{(\tan^2 t + 1)}} = -a \sin t \cos t$. Thus $\rho = 3p$.

**Example 4.49.** *If $\rho_1$ and $\rho_2$ be the radii of curvature at the ends of a focal chord of the parabola $y^2 = 4ax$, then show that $\rho_1^{-2/3} + \rho_2^{-2/3} = (2a)^{-2/3}$.* (*Rohtak, 2011 S*)

**Solution.** Given parabola is $y^2 = 4ax$ or $x = at^2$, $y = 2at$. If dashes denote differentiation w.r.t. $t$, then

$$x' = 2at, y' = 2a \ ; x'' = 2a, y'' = 0.$$

$$\therefore \quad \rho \text{ at } (at^2, 2at) = \frac{(x'^2 + y'^2)^{3/2}}{x'y'' - x''y'} = \frac{(4a^2 t^2 + 4a^2)^{3/2}}{0 - 4a^2} = 2a(1+t^2)^{3/2} \quad \text{(Numerically)}$$

If $P(t_1)$ and $Q(t_2)$ be the extremities of the focal chord of the parabola, then

$$t_1 t_2 = -1 \quad i.e., \quad t_2 = -1/t_1 \quad ...(i)$$

$$\therefore \quad \rho_1 \text{ at } P(t_1) = 2a (1 + t_1^2)^{3/2} ; \rho_2 \text{ at } Q(t_2) = 2a (1 + t_2^2)^{3/2}$$

Thus, $\quad \rho_1^{-2/3} + \rho_2^{-2/3} = (2a)^{-2/3} = [(1 + t_1^2)^{-1} + (1 + t_2^2)^{-1}]$

$$= (2a)^{-2/3} \left[\frac{1}{1+t_1^2} + \frac{t_1^2}{1+t_1^2}\right] \quad \text{[By (i)]}$$

$$= (2a)^{-2/3}$$

**Example 4.50.** *Show that the radius of curvature of P on an ellipse $x^2/a^2 + y^2/b^2 = 1$ is $CD^3/ab$ where CD is the semi-diameter conjugate to CP.* (*Rohtak, 2010 S*)

**Solution.** Two diameters of an ellipse are said to be conjugate if each bisects chords parallel to the other. If $CP$ and $CD$ are two semi-conjugate diameters and $P$ is $(a \cos \theta, b \sin \theta)$ then $D$ is $a \cos \left(\theta + \frac{\pi}{2}\right), b \sin \left(\theta + \frac{\pi}{2}\right)$

*i.e.,* $(-a \sin \theta, b \cos \theta)$.

Also $C(0, 0)$ is the centre of the ellipse.

$$\therefore \quad CD = \sqrt{(a^2 \sin^2 \theta + b^2 \cos^2 \theta)}$$

At $P$, we have $x = a \cos \theta$, $y = b \sin \theta$.

$$\therefore \quad \frac{dy}{dx} = \frac{dy/d\theta}{dx/d\theta} = \frac{b \cos \theta}{-a \sin \theta} = \frac{-b}{a} \cot\theta \,; \frac{d^2y}{dx^2} = \frac{b}{a} \operatorname{cosec}^2 \theta \,. \frac{d\theta}{dx} = \frac{-b}{a^2} \operatorname{cosec}^3\theta \,.$$

$$\therefore \quad \rho = \frac{[1 + (dy/dx)^2]^{3/2}}{d^2y/dx^2} = \frac{\left(1 + \frac{b^2}{a^2} \cot^2 \theta\right)^{3/2}}{-\frac{b}{a^2} \operatorname{cosec}^3 \theta}$$

$$= \frac{a^2}{b \operatorname{cosec}^3 \theta} \frac{(a^2 \sin^2 \theta + b^2 \cos^2 \theta)^{3/2}}{a^3 \sin^3 \theta} \qquad \text{(Numerically)}$$

$$= \frac{(a^2 \sin^2 \theta + b^2 \cos^2 \theta)^{3/2}}{ab} = \frac{CD^3}{ab} \,.$$

**Example 4.51.** *Find $\rho$ at the origin for the curves*
*(i)* $y^4 + x^3 + a(x^2 + y^2) - a^2 y = 0$ *(ii)* $y - x = x^2 + 2xy + y^2$ *(Rohtak, 2011)*

**Solution.** *(i)* Equating to zero the lowest degree terms, we get $y = 0$.

$\therefore$ $x$-axis is the tangent at the origin. Dividing throughout by $y$, we have

$$y^3 + x \,. \frac{x^2}{y} + a\left(\frac{x^2}{y} + y\right) - a^2 = 0$$

Let $x \to 0$, so that $\qquad \underset{x\to\infty}{\text{Lt}} (x^2/2y) = \rho.$

$$\therefore \quad 0 + 0.2\rho + a(2\rho + 0) - a^2 = 0 \quad \text{or} \quad \rho = a/2.$$

*(ii)* Equating to zero the lowest degree terms, we get $y = x$, as the tangent at the origin, which is neither of the coordinates axes.

$\therefore$ Putting $y = px + qx^2/2 + \ldots$ in the given equation, we get

$$px + qx^2/2 + \ldots - x = x^2 + 2x\,(px + qx^2/2 + \ldots) + (px + qx^2/2 + \ldots)^2$$

Equating coefficients of $x$ and $x^2$,

$$p - 1 = 0, q/2 = 1 + 2p + p^2 \quad i.e., \quad p = 1 \text{ and } q = 2 + 4 \cdot 1 + 2 \cdot 1^2 = 8.$$

$$\therefore \quad \rho\,(0, 0) = (1 + p^2)^{3/2}/q = (1 + 1)^{3/2}/8 = 1/2\sqrt{2}\,.$$

**(4) Radius of curvature for polar curve** $r = f(\theta)$ is given by

$$\rho = \frac{(\mathbf{r}^2 + \mathbf{r}_1^2)^{3/2}}{\mathbf{r}^2 + 2\mathbf{r}_1^2 - \mathbf{r}\mathbf{r}_2}$$

With the usual notations, we have from Fig. 4.10.

$$\psi = \theta + \phi$$

Differentiating w.r.t. $s$,

$$\frac{1}{\rho} = \frac{d\psi}{ds} = \frac{d\theta}{ds} + \frac{d\phi}{ds} = \frac{d\theta}{ds} + \frac{d\phi}{d\theta} \cdot \frac{d\theta}{ds}$$

$$= \frac{d\theta}{ds}\left(1 + \frac{d\phi}{d\theta}\right) \qquad \ldots(1)$$

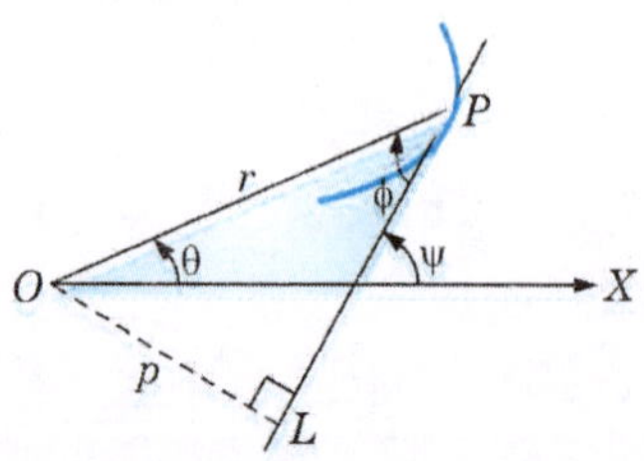

Fig. 4.10

Also we know that

$$\tan\phi = r\frac{d\theta}{dr} = \frac{r}{r_1} \quad \text{or} \quad \phi = \tan^{-1}\left(\frac{r}{r_1}\right) \qquad \text{where, } r_1 = \frac{dr}{d\theta}$$

Differentiating w.r.t. θ,

$$\frac{d\phi}{d\theta} = \frac{1}{1+(r/r_1)^2}\cdot\frac{r_1\cdot r_1 - rr_2}{r_1^2} = \frac{r_1^2 - rr_2}{r^2 + r_1^2} \qquad ...(2)$$

Also, $$\frac{ds}{d\theta} = \sqrt{(r^2 + r_1^2)} \qquad ...(3)$$

Substituting the value from (2) and (3) in (1),

$$\frac{1}{\rho} = \frac{1}{\sqrt{(r^2+r_1^2)}}\cdot\left(1 + \frac{r_1^2 - rr_2}{r^2 + r_1^2}\right)$$

Hence $$\rho = \frac{(r^2 + r_1^2)^{3/2}}{r^2 + 2r_1^2 - rr_2}$$ (*V.T.U., 2016*)

**(5) Radius of curvature for pedal curve** *$p = f(r)$ is given by*

$$\rho = \mathbf{r}\frac{\mathbf{dr}}{\mathbf{dp}}$$

With the usual notation (Fig. 4.10), we have $\psi = \theta + \phi$

Differentiating w.r.t. $s$,

$$\frac{1}{\rho} = \frac{d\psi}{ds} = \frac{d\theta}{ds} + \frac{d\phi}{ds} \qquad ...(1)$$

Also we know that $p = r\sin\phi$

$$\therefore \quad \frac{dp}{dr} = \sin\phi + r\cos\phi\,\frac{d\phi}{ds}$$

$$= r\frac{d\theta}{ds} + r\frac{dr}{ds}\cdot\frac{d\phi}{dr} \qquad \text{[By (3) and (4) of § 4.9 (2)]}$$

$$= r\left(\frac{d\theta}{ds} + \frac{d\phi}{ds}\right) = \frac{r}{\rho} \qquad \text{[By (1)]}$$

Hence $$\rho = r\frac{dr}{dp}.$$

**Example 4.52.** *Show that the radius of curvature at any point of the cardioid $r = a(1 - \cos\theta)$ varies as $\sqrt{r}$.*

**Solution.** Differentiating w.r.t. θ, we get

$$r_1 = a\sin\theta,\ r_2 = a\cos\theta$$

$$\therefore \quad (r^2 + r_1^2)^{3/2} = [a^2(1-\cos\theta)^2 + a^2\sin^2\theta]^{3/2} = a^3[2(1-\cos\theta)]^{3/2}$$

$$r^2 - rr_2 + 2r_1^2 = a^2(1-\cos\theta)^2 - a^2(1-\cos\theta)\cos\theta + 2a^2\sin^2\theta = 3a^2(1-\cos\theta)$$

Thus $$\rho = \frac{(r^2 + r_1^2)^{3/2}}{r^2 - rr_2 + 2r_1^2} = \frac{a^3\, 2\sqrt{2}(1-\cos\theta)^{3/2}}{3a^2(1-\cos\theta)}$$

$$= \frac{2\sqrt{2}}{3}a(1-\cos\theta)^{1/2} = \frac{2\sqrt{2}a}{3}\left(\frac{r}{a}\right)^{1/2} \propto \sqrt{r}.$$

**Otherwise.** The pedal equation of this cardioid is $2ap^2 = r^3$ ...(*i*)

Differentiating w.r.t. $p$, we get

that $$4ap = 3r^2\frac{dr}{dp} \text{ whence } \rho = r\frac{dr}{dp} = \frac{4ap}{3r} = \frac{4ar^{3/2}}{3r\cdot\sqrt{(2a)}} \propto \sqrt{r}.$$

[∵ $p = r^{3/2}/\sqrt{(2a)}$ from (*i*)]

## PROBLEMS 4.11

1. Find the radius of curvature at any point
   (i) $(at^2, 2at)$ of the parabola $y^2 = 4ax$. (D.T.U., 2012)
   (ii) $(o, c)$ of the catenary $y = c \cosh x/c$.
   (iii) $(a, 0)$ of the curve $y = x^3 (x - a)$. (V.T.U., 2010)
2. Show that for (i) the rectangular hyperbola $xy = c^2$, $\rho = \dfrac{(x^2 + y^2)^{3/2}}{2c^2}$.
   (ii) the curve $y = ae^{x/a}$, $\rho = a \sec^2 \theta \operatorname{cosec} \theta$ where $\theta = \tan^{-1} (y/a)$.
3. Show that the radius of curvature at
   (i) $(a, 0)$ on the curve $y^2 = a^2 (a - x)/x$ is $a/2$.
   (ii) $(a/4, a/4)$ on the curve $\sqrt{x} + \sqrt{y} = \sqrt{a}$ is $a/\sqrt{2}$.
   (iii) $x = \pi/2$ of the curve $y = 4 \sin x - \sin 2x$ is $5\sqrt{5}/4$.
4. For the curve $y = \dfrac{ax}{a + x}$, show that $\left(\dfrac{2\rho}{a}\right)^{2/3} = \left(\dfrac{x}{y}\right)^2 + \left(\dfrac{y}{x}\right)^2$. (V.T.U., 2015 ; Calicut, 2012 ; D.T.U., 2011)
5. Find the radius of curvature at any point on the
   (i) *ellipse* : $x = a \cos \theta, y = b \sin \theta$.
   (ii) *cycloid* : $x = a (\theta - \sin \theta), y = a (1 - \cos \theta)$.
   (iii) *curve* : $x = a (\cos t + t \sin t), y = a (\sin t - t \cos t)$. (Anna, 2013)
6. Show that the radius of curvature (i) at the point $(a \cos^3 \theta, a \sin^3 \theta)$ on the curve $x^{2/3} + y^{2/3} = a^{2/3}$ is $3a \sin \theta \cos \theta$. (Calicut, 2013)
   (ii) at the point $t$ on the curve $x = e^t \cos t, y = e^t \sin t$ is $\sqrt{2}e^t$. (Anna., 2016)
7. If $\rho$ be the radius of curvature at any point $P$ on the parabola, $y^2 = 4ax$ and $S$ be its focus, then show that $\rho^2$ varies as $(SP)^3$.
8. Prove that for the ellipse $\dfrac{x^2}{a^2} + \dfrac{y^2}{b^2} = 1$ or in pedal form $\dfrac{1}{p^2} = \dfrac{1}{a^2} + \dfrac{1}{b^2} - \dfrac{1}{a^2b^2}$, the radius of curvature at the point $(p, r)$ is $\rho = a^2 b^2/p^3$. (Delhi, 2012 ; V.T.U., 2010 S)
9. Show that the radius of curvature at an end of the major axis of the ellipse $x^2/a^2 + y^2/b^2 = 1$ is equal to the semi-latus rectum.
10. If $P_1$ and $P_2$ be the radii of curvature at the extremities of two conjugate semi-diameters of the ellipse $\dfrac{x^2}{a^2} + \dfrac{y^2}{b^2} = 1$, prove that $(P_1^{2/3} + P_2^{2/3}) (ab)^{2/3} = a^2 + b^2$. (P.T.U., 2013)
11. Show that the radius of curvature at each point of the curve $x = a (\cos t + \log \tan t/2)$, $y = a \sin t$, is inversely proportional to the length of the normal intercepted between the point on the curve and the $x$-axis. (Delhi., 2015 ; P.T.U., 2010)
12. Find the radius of curvature at the origin for
    (i) $x^3 + y^3 - 2x^2 + 6y = 0$
    (ii) $2x^4 + 3y^4 + 4x^2y + xy - y^2 + 2x = 0$
    (iii) $y^2 = x^2 (a + x)/(a - x)$.
13. Find the radius of the curvature at the point $(r, \theta)$ on each of the curves :
    (i) $r = a (1 - \cos \theta)$
    (ii) $r^n = a^n \cos n \theta$. (P.T.U., 2010)
    (iii) $r^n = a^n \sin n\theta$. (V.T.U., 2016)
14. For the cardioid $r = a(1 + \cos \theta)$, show that $\rho^2/r$ is constant.
15. Find the radius of curvature for the parabola $2a/r = 1 + \cos \theta$.
16. If $\rho_1$, $\rho_2$ be the radii of curvature at the extremities of any chord of the cardioid $r = a (1 + \cos \theta)$ which passes through the pole, show that $\rho_1^2 + \rho_2^2 = 16 a^2/9$. (D.T.U., 2013 ; Rohtak, 2012)
17. For any curve $r = f(\theta)$, prove that $\dfrac{r}{\rho} = \sin \phi \left(1 + \dfrac{d\phi}{d\theta}\right)$.

## 4.12 (1) CENTRE OF CURVATURE at any point P (x, y) on the curve y = f (x) is given by

$$\bar{x} = x - \frac{y_1(1 + y_1^2)}{y_2}, \quad \bar{y} = y + \frac{1 + y_1^2}{y_2}.$$

Let $C(\bar{x}, \bar{y})$ be the centre of curvature and $\rho$ the radius of curvature of the curve at $P(x, y)$ (Fig. 4.11). Draw $PL$ and $CM$ ⊥s to $OX$ and $PN \perp CM$. Let the tangent at $P$ make an $\angle\psi$ with the $x$-axis. Then $\angle NCP = 90° - \angle NPC = \angle NPT = \psi$

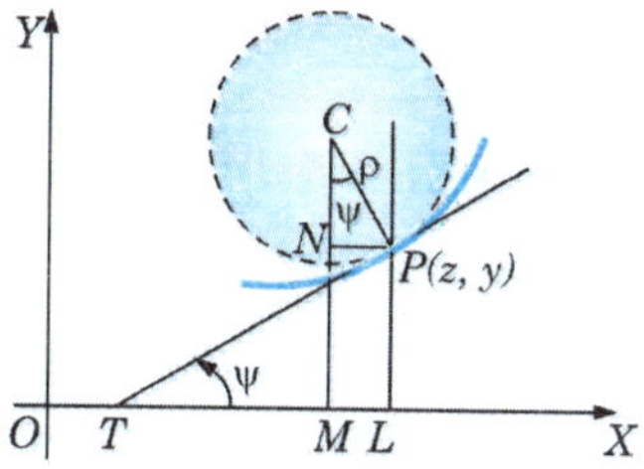

Fig. 4.11

$$\therefore \quad \bar{x} = OM = OL - ML = OL - NP$$

$$= x - \rho \sin\psi = x - \frac{(1+y_1^2)^{3/2}}{y_2} \cdot \frac{y_1}{\sqrt{(1+y_1^2)}}$$

$$\left[\because \tan\psi = y_1, \quad \therefore \sin\psi = \frac{y_1}{\sqrt{(1+y_1^2)}}\right]$$

$$= x - \frac{y_1(1+y_1^2)}{y_2}$$

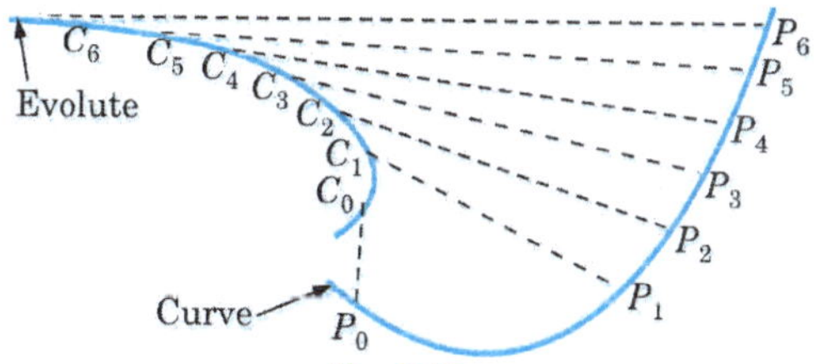

Fig. 4.12

and $$\bar{y} = MC = MN + NC = LP + \rho\cos\psi$$

$$\left[\because \sec\psi = \sqrt{(1+\tan^2\psi)} = \sqrt{(1+y_1^2)}\right]$$

$$= y + \frac{(1+y_1^2)^{3/2}}{y_2} \cdot \frac{1}{\sqrt{(1+y_1^2)}} = y + \frac{1+y_1^2}{y_2}$$

**Cor. Equation of the circle of curvature at** $P$ is $(x - \bar{x})^2 + (y - \bar{y})^2 = \rho^2$.

**(2) Evolute.** *The locus of the centre of curvature for a curve is called its* **evolute** *and the curve is called an* **involute** *of its evolute.* (*Fig. 4.12*)

**Obs. 1.** *The normal to the curve (i.e., involute) is the tangent to the evolute.*

**Obs. 2.** The length of the arc of the evolute between two points $C_1$ and $C_2$ is equal to the difference between the radii of curvature at $P_1$ and $P_2$ on the curve corresponding to $C_1$ and $C_2$.

**Example 4.53.** *Find the coordinates of the centre of curvature at any point of the parabola $y^2 = 4ax$. Hence show that its evolute is*

$$27ay^2 = 4(x - 2a)^3. \qquad \text{(Anna, 2013)}$$

**Solution.** We have $\quad 2yy_1 = 4a \quad i.e., \quad y_1 = 2a/y$

and $$y_2 = -\frac{2a}{y^2} \cdot y_1 = -\frac{4a^2}{y^3}$$

If $(\bar{x}, \bar{y})$ be the centre of curvature, then

$$\bar{x} = x - \frac{y_1(1+y_1^2)}{y_2} = x - \frac{2a/y(1+4a^2/y^2)}{-4a^2/y^3}$$

$$= x + \frac{y^2+4a^2}{2a} = x + \frac{4ax+4a^2}{2a} = 3x + 2a \qquad [\because y^2 = 4ax] \qquad ...(i)$$

and $$\bar{y} = y + \frac{1+y_1^2}{y_2} = y + \frac{1+4a^2/y^2}{-4a^2/y^2}$$

$$= y - \frac{y(y^2+4a^2)}{4a^2} = \frac{-y^3}{4a^2} = -\frac{2x^{3/2}}{\sqrt{a}} \qquad ...(ii)$$

To find the evolute, we have to eliminate $x$ from (*i*) and (*ii*)

$$\therefore \quad (\bar{y})^2 = \frac{4x^3}{a} = \frac{4}{a}\left(\frac{\bar{x}-2a}{3}\right)^3 \quad \text{or} \quad 27a(\bar{y})^2 = 4(\bar{x} - 2a)^3.$$

Thus the locus of $(\bar{x}, \bar{y})$ *i.e.*, evolute, is $27ay^2 = 4(x - 2a)^3$.

**Example 4.54.** *Show that the evolute of the cycloid $x = a(\theta - \sin\theta)$, $y = a(1 - \cos\theta)$ is another equal cycloid.*

**Solution.** We have $y_1 = \dfrac{dy}{d\theta} \div \dfrac{dx}{d\theta} = \dfrac{a\sin\theta}{a(1-\cos\theta)} = \cot\dfrac{\theta}{2}$.

$$y^2 = \frac{d}{dx}(y_1) = \frac{d}{d\theta}\left(\cot\frac{\theta}{2}\right)\cdot\frac{d\theta}{dx}$$

$$= -\operatorname{cosec}^2\frac{\theta}{2}\cdot\frac{1}{2}\cdot\frac{1}{a(1-\cos\theta)} = -\frac{1}{4a\sin^4\theta/2}$$

If $(\bar{x}, \bar{y})$ be the centre of curvature, then

$$\bar{x} = x - \frac{y_1(1+y_1^2)}{y_2} = a(\theta - \sin\theta) + \cot\frac{\theta}{2}\left(-4a\sin^4\frac{\theta}{2}\right)\left(1+\cot^2\frac{\theta}{2}\right)$$

$$= a(\theta - \sin\theta) + \frac{\cos\theta/2}{\sin\theta/2}\cdot 4a\sin^4\frac{\theta}{2}\cdot\operatorname{cosec}^2\frac{\theta}{2}$$

$$= a(\theta - \sin\theta) + 4a\sin\theta/2\cos\theta/2 = a(\theta - \sin\theta) + 2a\sin\theta = a(\theta + \sin\theta)$$

$$\bar{y} = y + \frac{1+y_1^2}{y_2} = a(1-\cos\theta) + \left(1+\cot^2\frac{\theta}{2}\right)\left(-4a\sin^4\frac{\theta}{2}\right)$$

$$= a(1-\cos\theta) - 4a\sin^4\theta/2\,.\operatorname{cosec}^2\theta/2$$

$$= a(1-\cos\theta) - 4a\sin^2\theta/2$$

$$= a(1-\cos\theta) - 2a(1-\cos\theta) = -a(1-\cos\theta)$$

Hence the locus of $(\bar{x}, \bar{y})$ *i.e.*, the evolute, is given by

$x = a(\theta + \sin\theta)$, $y = -a(1-\cos\theta)$ which is another equal cycloid.

**(3) Chord or curvature** *at a given point of a curve*

(*i*) parallel to $x$-axis $= 2\rho\sin\psi$

(*ii*) parallel to $y$-axis $= 2\rho\cos\psi$

Consider the circle of curvature at a given point $P$ on a curve. Let $C$ be the centre and $\rho$ the radius of curvature at $P$ so that $PQ = 2\rho$. (Fig. 4.13)

Let $PL$, $PM$ be the chords of curvature parallel to the axes of $x$ and $y$ respectively. Let the tangent $PT$ make an $\angle\psi$ with the $x$-axis so that $\angle LQP = \angle QPM = \psi$.

Then from the rt. $\angle$ed $\Delta PLQ$,

$$PL = 2\rho\sin\psi\sigma$$

and
$$PM = 2\rho\cos\psi.$$

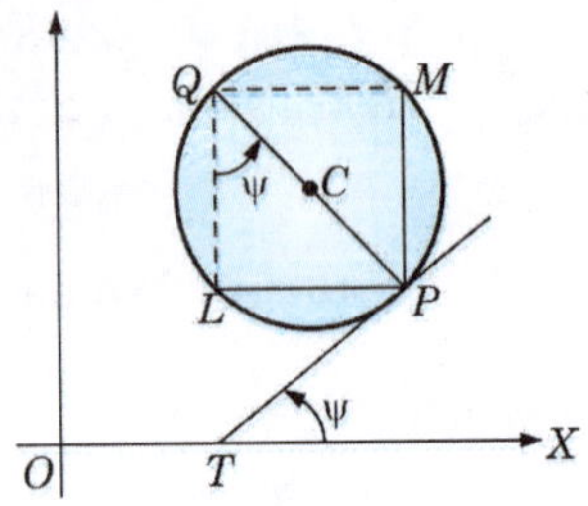

Fig. 4.13

## 4.13 (1) ENVELOPE

The equation $x\cos\alpha + y\sin\alpha = 1$ ...(1)

represents a straight line for a given value of $\alpha$. If different values are given to $\alpha$, we get different straight lines. All these straight lines thus obtained are said to constitute a family of straight lines.

In general, the *curves corresponding to the equation* $f(x, y, \alpha) = 0$ *for different values of* $\alpha$*, constitute a* **family of curves** *and* $\alpha$ *is called the* **parameter** *of the family.*

The envelope of a family of curves is the curve which touches each member of the family. For example, we know that all the straight lines of the family (1) touch the circle

$$x^2 + y^2 = 1 \qquad ...(2)$$

*i.e.*, the envelope of the family of lines (1) is the circle (2)—Fig. 4.14, which may also be seen as the locus of the ultimate points of intersection of the consecutive members of the family of lines (1). This leads to the following :

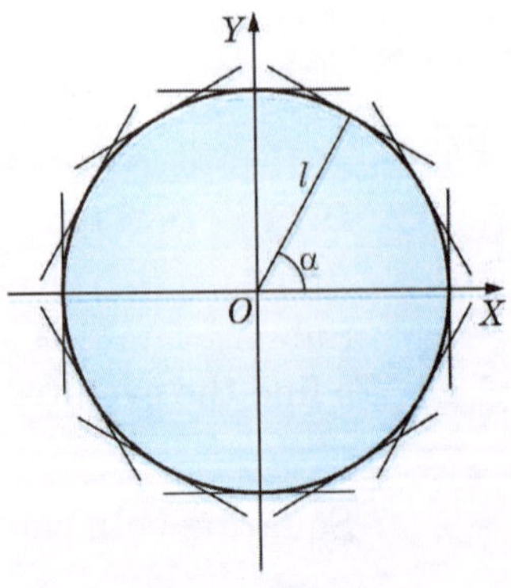

Fig. 4.14

**Def.** *If $f(x, y, \alpha) = 0$, and $f(x, y, \alpha + \delta\alpha) = 0$ be two consecutive members of a family of curves, then the locus of their ultimate points of intersection is called the* **envelope** *of that family.*

**(2) Rule to find the envelope of the family of curves $f(x, y, \alpha) = 0$ :**

*Eliminate $\alpha$ from $f(x, y, \alpha) = 0$ and* $\dfrac{\partial f(x, y, \alpha)}{\partial \alpha} = 0$.

**Example 4.55.** *Find the envelope of the family of lines $y = mx + \sqrt{(1+m^2)}$, m being the parameter.*

**Solution.** We have $\quad (y - mx)^2 = 1 + m^2 \qquad \ldots(i)$

Differentiating (*i*) partially with respect to $m$,

$$2(y - mx)(-x) = 2m \quad \text{or} \quad m = xy/(x^2 - 1) \qquad \ldots(ii)$$

Now eliminating $m$ from (*i*) and (*ii*)

Substituting the value of $m$ in (*i*), we get

$$\left(y - \frac{x^2 y}{x^2 - 1}\right)^2 = 1 + \left(\frac{xy}{x^2 - 1}\right)^2 \quad \text{or} \quad y^2 = (x^2 - 1)^2 + x^2y^2$$

or $\quad x^2 + y^2 = 1 \quad$ which is the required equation of the envelope.

**Obs.** Sometimes the equation to the family of curves contains two parameters which are connected by a relation. In such cases, we eliminate one of the parameters by means of the given relation, then proceed to find the envelope.

**Example 4.56.** *Find the envelope of a system of concentric and coaxial ellipses of constant area.*

**Solution.** Taking the common axes of the system of ellipses as the coordinate axes, the equation to an ellipse of the family is

$$\frac{x^2}{a^2} + \frac{y^2}{b^2} = 1 \text{ where } a \text{ and } b \text{ are the parameters.} \qquad \ldots(i)$$

The area of the ellipse $= \pi ab$ which is given to be constant, say $= \pi c^2$.

$$\therefore \qquad ab = c^2 \quad \text{or} \quad b = c^2/a. \qquad \ldots(ii)$$

Substituting in (*i*), $\dfrac{x^2}{a^2} + \dfrac{y^2}{(c^2/a^2)} = 1 \quad$ or $\quad x^2 a^{2} + (y^2/c^4)\, a^2 = 0 \qquad \ldots(iii)$

which is the given family of ellipses with $a$ as the only parameter.

Differentiating partially (*iii*) with respect to $a$,

$$-2x^2 a^{-3} + 2(y^2/c^4)\, a = 0 \quad \text{or} \quad a^2 = c^2 x/y \qquad \ldots(iv)$$

Eliminate $a$ from (*iii*) and (*iv*).

Substituting the value of $a^2$ in (*iii*), we get

$$x^2(y/c^2x) + (y^2/c^4)(c^2x/y) = 1 \text{ or } 2xy = c^2$$

which is the required equation of the envelope.

**(3) Evolute** *of a curve is the envelope of the normals to that curve* (Fig. 4.12).

**Example 4.57.** *Find the evolute of the parabola $y^2 = 4ax$.* *(Anna, 2016)*

**Solution.** Any normal to the parabola is $y = mx - 2am - am^3 \qquad \ldots(i)$

Differentiating it with respect to $m$ partially,

$$0 = x - 2a - 3am^2 \text{ or } m = [(x - 2a)/3a]^{1/2}$$

Substituting this value of $m$ in (*i*),

$$y = \left(\frac{x - 2a}{3a}\right)^{1/2} \left[x - 2a - a \cdot \frac{x - 2a}{3a}\right]$$

Squaring both sides, we have

$$27ay^2 = 4(x - 2a)^3$$

which is the evolute of the parabola. (*cf.* Example 4.51).

## PROBLEMS 4.12

1. Find the coordinates of the centre of curvature at $(at^2, 2at)$ on the parabola $y^2 = 4ax$.
2. If the centre of curvature of the ellipse $x^2/a^2 + y^2/b^2 = 1$ at one end of the minor axis lies at the other end, then show that the eccentricity of the ellipse is $1/\sqrt{2}$.
3. Show that the equation of the evolute of the
   (*i*) *parabola* $x^2 = 4ay$ is $4(y - 2a)^3 = 27ax^2$. (*Anna, 2013*)
   (*ii*) *ellipse* $x = a \cos \theta, y = b \sin \theta$ (*i.e.*, $x^2/a^2 + y^2/b^2 = 1$) is
   $$(ax)^{2/3} + (by)^{2/3} = (a^2 - b^2)^{2/3}.$$
   (*iii*) *rectangular hyperbola* $xy = c^2$, (*i.e.*, $x = ct, y = c/t$) is
   $$(x + y)^{2/3} - (x - y)^{2/3} = (4c)^{2/3}.$$
4. Find the evolute of (*i*) cycloid $x = a(t + \sin t), y = a(1 - \cos t)$
   (*ii*) the curve $x = a(\cos \theta + \theta \sin \theta), y = a(\sin \theta - \theta \cos \theta)$.
5. Find the evolute of the curve $x = a \cos^3 \theta, y = a \sin^3 \theta$ *i.e.*, $x^{2/3} + y^{2/3} = a^{2/3}$. (*Anna, 2013*)
6. Show that the evolute of the curve $x = a(\cos t + \log \tan t/2), y = a \sin t$ is $y = a \cosh x/a$. (*Anna, 2016*)
7. Find the circle of curvature at the point (*i*) $(a/4, a/4)$ of the curve $\sqrt{x} + \sqrt{y} = \sqrt{a}$. (*Anna, 2012*)
   (*ii*) $(3/2, 3/2)$ of the curve $x^3 + y^3 = 3xy$
8. Show that the circle of curvature at the origin for the curve $x + y = ax^2 + by^2 + ex^3$ is
   $$(a + b)(x^2 + y^2) = 2(x + y).$$
9. If $C_x$, $C_y$ be the chords of curvature parallel to the axes at any point on the curve $y = ae^{x/a}$, prove that
   $$\frac{1}{C_x^2} + \frac{1}{C_y^2} = \frac{1}{2aC_x}.$$
10. In the curve $y = a \cosh x/a$, prove that the chord of curvature parallel to $y$-axis is the double the ordinate.

Find the envelope of the following family of lines :

11. $y = mx + a/m$, $m$ being the parameter. (*Anna., 2016*)
12. $\frac{x}{a} \cos \alpha + \frac{y}{b} \sin \alpha = 1$, $\alpha$ being the parameter.
13. $y = mx - 2am - am^3$. (*Anna., 2016*)
14. $y = mx + \sqrt{(a^2m^2 + b^2)}$, $m$ being the parameter.
15. Find the envelope of the family of parabolas $y = x \tan \alpha - \frac{gx^2}{2u^2 \cos \alpha}$, $\alpha$ being the parameter.
16. Find the envelope of the straight line $x/a + y/b = 1$, where the parameters $a$ and $b$ are connected by the relation :
   (*i*) $a + b = c$. (*ii*) $ab = c^2$
   (*iii*) $a^2 + b^2 = c^2$. (*Anna, 2013*)
17. Find the envelope of the family of ellipses $x^2/a^2 + y^2/b^2 = 1$ for which $a + b = c$.

Prove that the evolute of the

18. *ellipse* $\frac{x^2}{a^2} + \frac{y^2}{b^2} = 1$ is $(ax)^{2/3} + (by)^{2/3} = (a^2 - b^2)^{2/3}$.
19. *hyperbola* $\frac{x^2}{a^2} - \frac{y^2}{b^2} = 1$ is $(ax)^{2/3} - (by)^{2/3} = (a^2 + b^2)^{2/3}$. (*Anna, 2012*)
20. *parabola* $x^2 = 4by$ is $27bx^2 = 4(y - 2b)^3$. (*Anna, 2013*)

## 4.14 (1) INCREASING AND DECREASING FUNCTIONS

In the function $y = f(x)$, *if $y$ increases as $x$ increases* (as at $A$), *it is called an* **increasing function of** $x$. On the contrary, *if $y$ decreases as $x$ increases* (as at $C$), *it is called a* **decreasing function** *of $x$*.

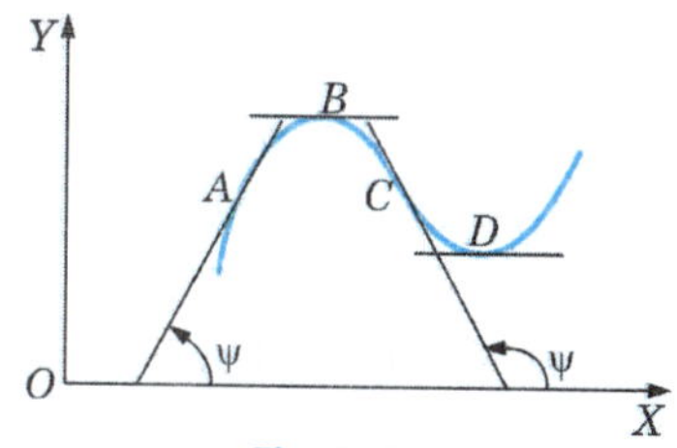

Fig. 4.15

Let the tangent at any point on the graph of the function make an $\angle\psi$ with the $x$-axis (Fig. 4.15) so that

$$dy/dx = \tan\psi$$

At any point such as $A$, where the function is increasing $\angle\psi$ is acute *i.e.*, $dy/dx$ is positive. At a point such as $C$, where the function is decreasing $\angle\psi$ is obtuse *i.e.*, $dy/dx$ is negative.

Hence the *derivative of an increasing function is + ve,* and *the derivative of a decreasing function is –ve.*

**Obs.** *If the derivative is zero* (as at $B$ or $D$), *then y is neither increasing nor decreasing.* In such cases, we say *that the function is* **stationary.**

**(2) Concavity, Convexity and Point of Inflexion**

(*i*) *If a portion of the curve on both sides of a point, however small it may be, lies above the tangent* (as at $D$), *then the curve is said to be* **concave upwards** at $D$ where $d^2y/dx^2$ is positive.

(*ii*) *If a portion of the curve on both sides of a point lies below the tangent* (as at $B$), *then the curve is said to be* **Convex upwards** at $B$ where $d^2y/dx^2$ is negative.

(*iii*) *If the two portions of the curve lie on different sides of the tangent thereat* (*i.e.*, the curve crosses the tangent (as at $C$), *then the point C* is said to be a **point of inflexion** of the curve.

At a point of inflexion $\dfrac{d^2y}{dx^2} = 0$ and $\dfrac{d^3y}{dx^3} \neq 0$.

## 4.15 (1) MAXIMA AND MINIMA

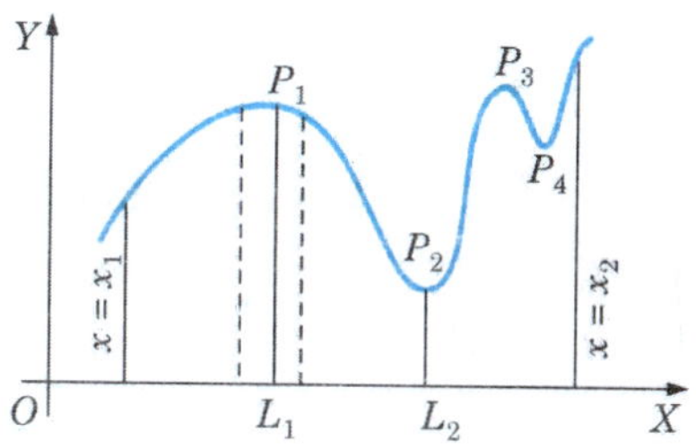

Fig. 4.16

Consider the graph of the continuous function $y = f(x)$ in the interval $(x_1, x_2)$ (Fig. 4.16). Clearly the point $P_1$ is the highest in its own immediate neighbourhood. So also is $P_3$. At each of these points $P_1, P_3$ the function is said to have a *maximum* value.

On the other hand, the point $P_2$ is the lowest in its own immediate neighbourhood. So also is $P_4$. At each of these points $P_2, P_4$ the function is said to have a *minimum* value.

Thus, we have

**Def.** *A function f(x) is said to have a* **maximum** *value at x = a, if there exists a small number h, however small, such that f(a) > both f(a – h) and f(a + h).*

*A function f(x) is said to have a* **minimum** *value at x = a, if there exists a small number h, however small, such that f(a) < both f(a – h) and f(a + h).*

**Obs. 1.** *The maximum and minimum values of a function taken together are called its* **extreme** *values and the points at which the function attains the extreme values are called the* **turning points** *of the function.*

**Obs. 2.** *A maximum or minimum value of a function is not necessarily the greatest or least value of the function in any finite interval.* The maximum value is simply the greatest value in the immediate neighbourhood of the maxima point or the minimum value is the least value in the immediate neighbourhood of the minima point. In fact, *there may be several maximum and minimum values of a function in an interval and a minimum value may be even greater than a maximum value.*

**Obs. 3.** It is seen from the Fig. 4.16 that *maxima and minima values occur alternately.*

**(2) Conditions for maxima and minima.** At each point of extreme value, it is seen from Fig. 4.16 that the tangent to the curve is parallel to the $x$-axis, *i.e.*, its slope (= $dy/dx$) is zero. Thus if *the function is maximum or minimum at* $x = a$, *then* $(dy/dx)_a = 0$.

Around a maximum point say, $P_1$ $(x = a)$, the curve is increasing in a small interval $(a - h, a)$ before $L_1$ and decreasing in $(a, a + h)$ after $L_1$ where $h$ is positive and small.

*i.e.*, in $(a - h, a)$, $dy/dx \geq 0$ ; at $x = a$, $dy/dx = 0$ and in $(a, a + h)$, $dy/dx \leq 0$.

Thus $dy/dx$ (which is a function of $x$) changes sign from positive to negative in passing through $P_1$, *i.e.*, it is a decreasing function in the interval $(a - h, a + h)$ and therefore, its derivative $d^2y/dx^2$ is negative at $P_1(x = a)$.

Similarly, around a minimum point say $P_2$, $dy/dx$ changes sign from negative to positive in passing through $P_2$, *i.e.*, it is an increasing function in the small interval around $L_2$ and therefore its derivative $d^2y/dx^2$ is positive at $P_2$.

Hence (*i*) *f(x) is maximum at x = a if f'(a) = 0 and f''(a) is –ve [i.e., f'(a) changes sign from +ve to –ve]*

(*ii*) *f(x) is minimum at x = a, if f'(a) = 0 and f''(a) is + ve [i.e., f'(a) changes sign from –ve to +ve]*

**Obs.** A maximum or a minimum value is a stationary value but a stationary value may neither be a maximum nor a minimum value.

**(3) Procedure for finding maxima and minima**

(*i*) *Put the given function = f(x)*

(*ii*) *Find f'(x) and equate it to zero. Solve this equation and let its roots be a, b, c, ...*

(*iii*) *Find f''(x) and substitute in it by turns x = a, b, c, ...*

*If f''(a) is –ve, f(x) is maximum at x = a.*

*If f''(a) is +ve, f''(x) is minima at x = a.*

(*iv*) Sometimes $f''(x)$ may be difficult to find out or $f''(x)$ may be zero at $x = a$. In such cases, see if $f'(x)$ *changes sign from +ve to –ve as x passes through a, then f(x) is maximum at x = a.*

*If f'(x) changes sign from –ve to +ve as x passes through a, f(x) is minimum at x = a.*

*If f'(x) does not change sign while passing through x = a, f(x) is neither maximum nor minimum at x = a.*

**Example 4.58.** *Find the maximum and minimum values of* $3x^4 - 2x^3 - 6x^2 + 6x + 1$ *in the interval (0, 2).*

**Solution.** Let $f(x) = 3x^4 - 2x^3 - 6x^2 + 6x + 1$

Then $f'(x) = 12x^3 - 6x^2 - 12x + 6 = 6(x^2 - 1)(2x - 1)$

$\therefore$ $f'(x) = 0$ when $x = \pm 1, \frac{1}{2}$.

So in the interval (0, 2) $f(x)$ can have maximum or minimum at $x = \frac{1}{2}$ or 1.

Now $f''(x) = 36x^2 - 12x - 12 = 12(3x^2 - x - 1)$ so that $f''\left(\frac{1}{2}\right) = -9$ and $f''(1) = 12$.

$\therefore$ $f(x)$ has a maximum at $x = \frac{1}{2}$ and a minimum at $x = 1$.

Thus the maximum value $= f\left(\frac{1}{2}\right) = 3\left(\frac{1}{2}\right)^4 - 2\left(\frac{1}{2}\right)^3 - 6\left(\frac{1}{2}\right)^2 + 6\left(\frac{1}{2}\right) + 1 = 2\frac{7}{16}$

and the minimum value $= f(1) = 3(1)^4 - 2(1)^3 - 6(1)^2 + 6(1) + 1 = 2$.

**Example 4.59.** *Show that sin x (1 + cos x) is a maximum when x = π/3.*

**Solution.** Let $f(x) = \sin x\,(1 + \cos x)$

Then $f'(x) = \cos x\,(1 + \cos x) + \sin x\,(-\sin x)$

$= \cos x\,(1 + \cos x) - (1 - \cos^2 x) = (1 + \cos x)(2\cos x - 1)$

$\therefore$ $f'(x) = 0$ when $\cos x = \frac{1}{2}$ or $-1$ *i.e.*, when $x = \pi/3$ or $\pi$.

Now $f''(x) = -\sin x\,(2\cos x - 1) + (1 + \cos x)(-2\sin x) = -\sin x\,(4\cos x + 1)$

so that $f''(\pi/3) = -3\sqrt{2}/2$ and $f''(\pi) = 0$.

Thus $f(x)$ has a maximum at $x = \pi/3$.

Since $f''(\pi)$ is 0, let us see whether $f'(x)$ changes sign or not.

When $x$ is slightly $< \pi$, $f'(x)$ is –ve, then when $x$ is slightly $> \pi$, $f'(x)$ is again –ve *i.e.*, $f'(x)$ does not change sign as $x$ passes through $\pi$. So $f(x)$ is neither maximum nor minimum at $x = \pi$.

**(4) Practical Problems**

In many problems, the function (whose maximum or minimum value is required) is not directly given. It has to be formed from the given data. If the function contains two variables, one of them has to be eliminated with the help of the other conditions of the problem. A number of problems deal with triangles, rectangles, circles, spheres, cones, cylinders etc. The student is therefore, advised to remember the formulae for areas, volumes, surfaces etc. of such figures.

**Example 4.60.** *A window has the form of a rectangle surmounted by a semi-circle. If the perimeter is 40 ft., find its dimensions so that the greatest amount of light may be admitted.*

**Solution.** The greatest amount of light may be admitted means that the area of the window may be maximum.

Let $x$ ft. be the radius of the semi-circle so that one side of the rectangle is $2x$ ft. (Fig. 4.17). Let the other side of the rectangle $y$ ft. Then the perimeter of the whole figure

$$= \pi x + 2x + 2y = 40 \text{ (given) and the area } A = \tfrac{1}{2}\pi x^2 + 2xy. \quad ...(i)$$

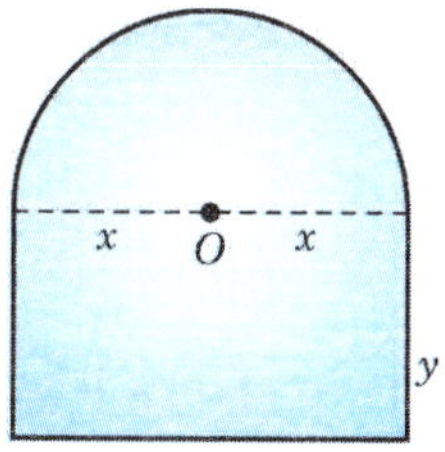

Fig. 4.17

Here $A$ is a function of two variables $x$ and $y$. To express $A$ in terms of one variable $x$ (say), we substitute the value of $y$ from $(i)$ in it.

$$\therefore \quad A = \tfrac{1}{2}\pi x^2 + x[40 - (\pi + 2)x] = 40\,x - \left(\frac{\pi}{2} + 2\right)x^2$$

Then $$\frac{dA}{dx} = 40 - (\pi + 4)x$$

For $A$ to be maximum or minimum, we must have $dA/dx = 0$ *i.e.*, $40 - (\pi + 4)x = 0$

or $$x = 40/(\pi + 4)$$

$\therefore$ From $(i)$, $$y = \tfrac{1}{2}[40 - (\pi + 2)x] = \tfrac{1}{2}[40 - (\pi + 2)\,40/(\pi + 4)] = 40/(\pi + 4) \text{ i.e., } x = y$$

Also $$\frac{d^2A}{dx^2} = -(\pi + 4), \text{ which is negative.}$$

Thus the area of the window is maximum when the radius of the semi-circle is equal to the height of the rectangle.

**Example 4.61.** *A rectangular sheet of metal of length 6 metres and width 2 metres is given. Four equal squares are removed from the corners. The sides of this sheet are now turned up to form an open rectangular box. Find approximately, the height of the box, such that the volume of the box is maximum.*

**Solution.** Let the side of each of the squares cut off be $x$ m so that the height of the box is $x$ m and the sides of the base are $6 - 2x$, $2 - 2x$ m (Fig. 4.18).

$\therefore$ Volume $V$ of the box

$$= x(6 - 2x)(2 - 2x) = 4(x^3 - 4x^2 + 3x)$$

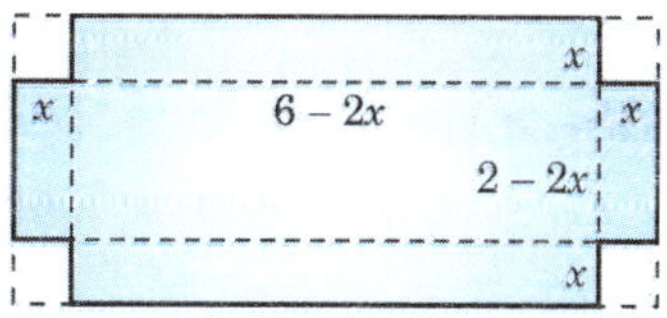

Fig. 4.18

Then $$\frac{dV}{dx} = 4(3x^2 - 8x + 3)$$

For $V$ to be maximum or minimum, we must have

$$dV/dx = 0 \text{ i.e., } 3x^2 - 8x + 3 = 0$$

$$\therefore \quad x = \frac{8 \pm \sqrt{[64 - 4 \times 3 \times 3]}}{6} = 2.2 \text{ or } 0.45 \text{ m.}$$

The value $x = 2.2$ m is inadmissible, as no box is possible for this value.

Also $\dfrac{d^2V}{dx^2} = 4(6x - 8)$, which is –ve for $x = 0.45$ m.

Hence the volume of the box is maximum when its height is 45 cm.

**Example 4.62.** *Show that the right circular cylinder of given surface (including the ends) and maximum volume is such that its height is equal to the diameter of the base.*

**Solution.** Let $r$ be the radius of the base and $h$, the height of the cylinder.

Then given surface $S = 2\pi rh + 2\pi r^2 \quad ...(i)$ and the volume $V = \pi r^2 h \quad ...(ii)$

Hence $V$ is a function of two variables $r$ and $h$. To express $V$ in terms of one variable only (say $r$), we substitute the value of $h$ from $(i)$ in $(ii)$.

Then $$V = \pi r^2\left(\frac{S - 2\pi r^2}{2\pi r}\right) = \frac{1}{2}Sr - \pi r^3 \quad \therefore \quad \frac{dV}{dr} = \frac{1}{2}S - 3\pi r^2.$$

For $V$ to be maximum or minimum, we must have $dV/dr = 0$,

*i.e.*, $$\tfrac{1}{2}S - 3\pi r^2 = 0 \quad \text{or} \quad r = \sqrt{(S/6\pi)}.$$

Also $\dfrac{d^2V}{dr^2} = -6\pi r$, which is negative for $r = \sqrt{(S/6\pi)}$.

Hence $V$ is maximum for $r = \sqrt{(S/6\pi)}$.

*i.e.*, for $6\pi r^2 = S = 2\pi rh + 2\pi r^2$ *i.e.*, for $h = 2r$, which proves the required result. [By (*i*)]

**Example 4.63.** *Show that the diameter of the right circular cylinder of greatest curved surface which can be inscribed in a given cone is equal to the radius of the cone.*

**Solution.** Let $r$ be the radius $OA$ of the base and $\alpha$ the semi-vertical angle of the given cone (Fig. 4.19). Inscribe a cylinder in it with base-radius $OL = x$.

Then the height of the cylinder $LP$
$$= LA \cot\alpha = (r - x)\cot\alpha$$

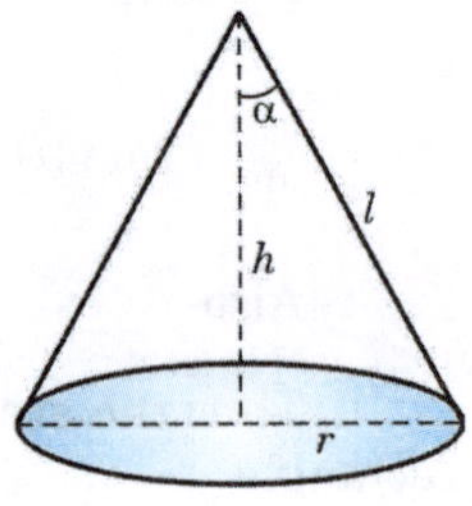

Fig. 4.19

$\therefore$ The curved surface $S$ of the cylinder
$$= 2\pi x.\ LP = 2\pi x(r - x)\cot\alpha$$
$$= 2\pi\cot\alpha\,(rx - x^2)$$

$\therefore$ $$\frac{dS}{dx} = 2\pi\cot\alpha\,(r - 2x) = 0 \text{ for } x = r/2.$$

and $$\frac{d^2S}{dx^2} = -4\pi\cot\alpha$$

Hence $S$ is maximum when $x = r/2$.

**Example 4.64.** *Find the altitude and the semi-vertical angle of a cone of least volume which can be circumscribed to a sphere of radius a.*

**Solution.** Let $h$ be the height and $\alpha$ the semi-vertical angle of the cone so that its radius $BD = h\tan\alpha$ (Fig. 4.20).

$\therefore$ The volume $V$ of the cone is given by
$$V = \frac{1}{3}\pi(h\tan\alpha)^2 h = \frac{1}{3}\pi h^3\tan^2\alpha.$$

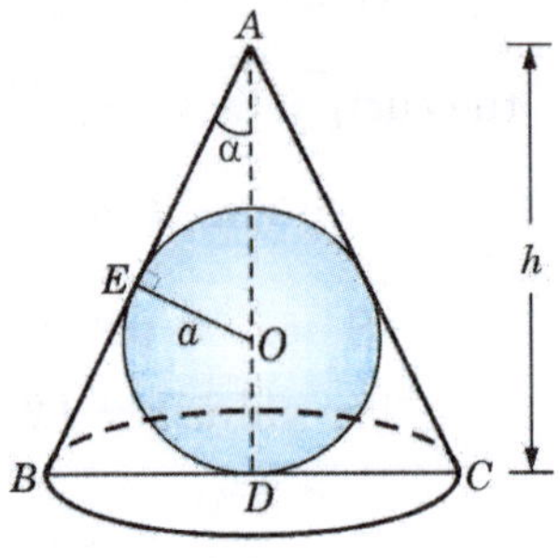

Fig. 4.20

Now we must express $\tan\alpha$ in terms of $h$.

In the rt. $\angle d\ \Delta AEO$,
$$EA = \sqrt{(OA - a^2)} = \sqrt{[(h - a)^2 - a^2]} = \sqrt{(h^2 - 2ha)}$$

$\therefore$ $$\tan\alpha = \frac{EO}{EA} = \frac{a}{\sqrt{(h^2 - 2ha)}}$$

Thus $$V = \frac{1}{3}\pi h^3 \cdot \frac{a^2}{h^2 - 2ha} = \frac{1}{3}\pi a^3 \cdot \frac{h^2}{h - 2a}$$

$\therefore$ $$\frac{dV}{dh} = \frac{1}{3}\pi a^2 \cdot \frac{(h - 2a)2h - h^2 \cdot 1}{(h - 2a)^2} = \frac{1}{3}\pi a^2 \cdot \frac{h(h - 4a)}{(h - 2a)^2}$$

Thus $\dfrac{dV}{dh} = 0$ for $h = 4a$, the other value ($h = 0$) being not possible.

Also $dV/dh$ is –ve when $h$ is slightly $< 4a$, and it is + ve when $h$ is slightly $> 4a$.

Hence $V$ is minimum (*i.e.* least) when $h = 4a$

and $$\alpha = \sin^{-1}\left(\frac{a}{OA}\right) = \sin^{-1}\left(\frac{a}{3a}\right) = \sin^{-1}\frac{1}{3}.$$

**Example 4.65.** *Find the volume of the largest possible right-circular cylinder that can be inscribed in a sphere of radius a.*

**Solution.** Let $O$ be the centre of the sphere of radius $a$. Construct a cylinder as shown in Fig. 4.21. Let $OA = r$.

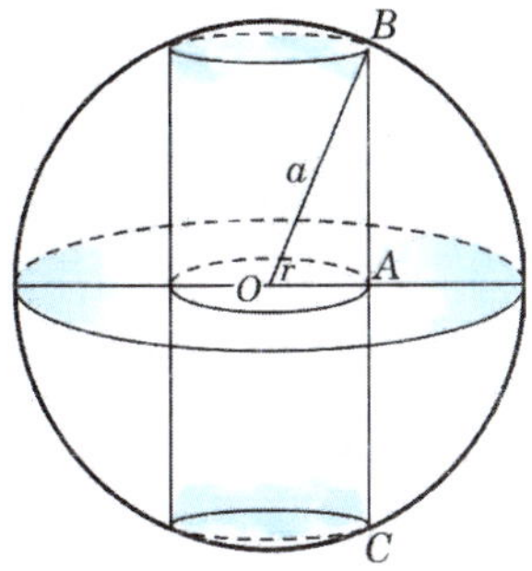

Fig. 4.21

Then $\quad AB = \sqrt{(OB^2 - OA^2)} = \sqrt{(a^2 - r^2)}$

$\therefore$ Height $h$ of the cylinder $= 2 \cdot AB = 2\sqrt{(a^2 - r^2)}$.

Thus volume $V$ of the cylinder

$$= \pi r^2 h = 2\pi r^2 \sqrt{(a^2 - r^2)}$$

$$\therefore \quad \frac{dV}{dr} = 2\pi \{2r\sqrt{(a^2 - r^2)} + r^2 \cdot \frac{1}{2}(a^2 - r^2)^{-1/2}(-2r)\}$$

$$= \frac{2\pi r(2a^2 - 3r^2)}{\sqrt{(a^2 - r^2)}}$$

The $dV/dr = 0$, when $r^2 = 2a^2/3$, the other value $(r = 0)$ being not admissible.

Now
$$\frac{d^2V}{dr^2} = 2\pi \frac{\sqrt{(a^2 - r^2)}(2a^2 - 9r^2) - r(2a^2 - 3r^2) \times \frac{1}{2}(a^2 - r^2)^{-1/2} \cdot (-2r)}{(a^2 - r^2)}$$

$$= 2\pi \frac{(a^2 - r^2)(2a^2 - 9r^2) + r^2(2a^2 - 3r^2)}{(a^2 - r^2)^{3/2}}$$ which is – ve for $r^2 = 2a^2/3$.

Hence, $V$ is maximum for $r^2 = 2a^2/3$ and maximum volume

$$= 2\pi r^2 \sqrt{(a^2 - r^2)} = 4\pi a^3/3\sqrt{3}.$$

**Example 4.66.** *Assuming that the petrol burnt (per hour) in driving a motor boat varies as the cube of its velocity, show that the most economical speed when going against a current of c miles per hour is $\frac{3}{2}c$ miles per hour.*

**Solution.** Let $v$ m.p.h. be the velocity of the boat so that its velocity relative to water (when going against the current) is $(v - c)$ m.p.h.

$\therefore$ Time required to cover a distance of $s$ miles $= \dfrac{s}{v - c}$ hours.

Since the petrol burnt per hour $= kv^3$, $k$ being a constant.

$\therefore$ The total petrol burnt, $y$, is given by

$$y = k\frac{v^3 s}{v - c} = ks\frac{v^3}{v - c} \quad \therefore \quad \frac{dy}{dv} = ks \cdot \frac{(v - c)3v^2 - v^3 \cdot 1}{(v - c)^2}$$

$$= ks \cdot \frac{v^2(2v - 3c)}{(v - c)^2}$$

Thus $dy/dv = 0$ for $v = 3c/2$, the other value $(v = 0)$ is inadmissible.

Also $dy/dv$ is $-ve$, when $v$ is slightly $< 3c/2$ and it is $+ve$, when $v$ is slightly $> 3c/2$.

Hence $y$ is minimum for $v = 3c/2$.

## PROBLEMS 4.13

1. (*i*) Test the curve $y = x^4$ for points of inflexion ?

   (*ii*) Show that the points of inflexion of the curve $y^2 = (x - a)^2 (x - b)$ lie on the straight line

   $$3x + a = 4b.$$

2. The function $f(x)$ defined by $f(x) = a/x + bx$, $f(2) = 1$, has an extremum at $x = 2$. Determine $a$ and $b$. Is this point (2, 1), a point of maximum or minimum on the graph of $f(x)$ ?
3. Show that $\sin^p\theta \cos^q\theta$ attains a maximum when $\theta = \tan^{-1}(p/q)$.
4. If a beam of weight $w$ per unit length is built-in horizontally at one end $A$ and rests on a support $O$ at the other end the deflection $y$ at a distance $x$ from $O$ is given by
$$EIy = \frac{w}{48}(2x^4 - 3lx^3 + l^3x),$$
where, $l$ is the distance between the ends. Find $x$ for $y$ to be maximum.
5. The horse-power developed by an aircraft travelling horizontally with velocity $v$ feet per second is given by
$$H = \frac{aw^2}{v} + bv,$$
where $a$, $b$ and $w$ are constants. Find for what value of $v$ the horse-power is maximum.
6. The velocity of waves of wave-length $\lambda$ on deep water is proportional to $\sqrt{(\lambda / a + a / \lambda)}$, where $a$ is a certain constant, prove that the velocity is minimum when $\lambda = a$.
7. In a submarine telegraph cable, the speed of signalling varies as $x^2 \log_e (1/x)$, where $x$ is the ratio of the radius of the core to that of the covering. Show that the greatest speed is attained when this ratio is $1/\sqrt{e}$.
8. The efficiency $e$ of a screw-jack is given by $e = \tan\theta/\tan(\theta + \alpha)$, where $\alpha$ is a constant. Find $\theta$ if this efficiency is to be maximum. Also find the maximum efficiency.
9. Show that of all rectangles of given area, the square has the least parameter.
10. Find the rectangle of greatest perimeter that can be inscribed in a circle of radius $a$.
11. A gutter of rectangular section (open at the top) is to be made by bending into shape of a rectangular strip of metal. Show that the capacity of the gutter will be greatest if its width is twice its depth.
12. Show that the triangle of maximum area that can be inscribed in a given circle is an equilateral triangle.
13. An open box is to be made from a rectangular piece of sheet metal 12 cms × 18 cms, by cutting out equal squares from each corner and folding up the sides. Find the dimensions of the box of largest volume that can be made in this manner.
14. An open tank is to be constructed with a square base and vertical sides to hold a given quantity of water. Find the ratio of its depth to the width so that the cost of lining the tank with lead is least.
15. A corridor of width $b$ runs perpendicular to a passageway of width $a$. Find the longest beam which can be moved in a horizontal plane along the passageway into the corridor ?
16. One corner of a rectangular sheet of paper of width $a$ is folded so as to reach the opposite edge of the sheet. Find the minimum length of the crease.
17. Show that the height of closed cylinder of given volume and least surface is equal to its diameter.
18. Prove that a conical vessel of a given storage capacity requires the least material when its height is $\sqrt{2}$ times the radius of the base.
19. Show that the semi-vertical angle of a cone of maximum volume and given slant height is $\tan^{-1}\sqrt{2}$.
20. The shape of a hole bored by a drill is cone surmounting a cylinder. If the cylinder be of height $h$ and radius $r$ and the semi-vertical angle of the cone be $\alpha$ where $\tan\alpha = h/r$, show that for a total fixed depth $H$ of the hole, the volume removed is maximum if $h = \frac{H}{6}(1 + \sqrt{7})$.
21. A cylinder is inscribed in a cone of height $h$. If the volume of the cylinder is maximum, show that its height is $h/3$.
22. Show that the volume of the biggest right circular cone that can be inscribed in a sphere of given radius is 8/27 times that of the sphere.
23. A given quantity of metal is to be cast into a half-cylinder with a rectangular base and semi-circular ends. Show that in order that the total surface area may be a minimum, the ratio of the length of the cylinder to the diameter of its semi-circular ends is $\pi/(\pi + 2)$.
24. A person being in a boat $a$ miles from the nearest point of the beach, wishes to reach as quickly as possible a point $b$ miles from that point along the shore. The ratio of his rate of walking to his rate of rowing is $\sec\alpha$. Prove that he should land at a distance $b - a\cot\alpha$ from the place to be reached.
25. The cost per hour of propelling a steamer is proportional to the cube of her speed through water. Find the relative speed at which the steamer should be run against a current of 5 km per hour to make a given trip at the least cost.

## 4.16 ASYMPTOTES

**(1) Def.** *An asymptote of a curve is a straight line at a finite distance from the origin, to which a tangent to the curve tends as the point of contact recedes to infinity.*

In other words, an asymptote is a straight line which cuts a curve on two points, at an infinite distance from the origin and yet is not itself wholly at infinity.

**(2) Asymptotes parallel to axes.** Let the equation of the curve arranged according to powers of $x$ be

$$a_0x^n + (a_1y + b_1)x^{n-1} + (a_2y^2 + b_2y + c_2)x^{n-2} + \dots 0 \quad \dots(1)$$

If $a_0 = 0$ and $y$ be so chosen that $a_1y + b_1 = 0$, then the coefficients of two highest powers of $x$ in Eqn.(1) vanish and therefore, two of its roots are infinite. Hence $a_1y + b_1 = 0$ is an asymptote of Eqn. (1) which is parallel to $x$-axis.

Again if $a_0$, $a_1$, $b_1$ are all zero and if $y$ be so chosen that $a_2y^2 + b_2y + c_2 = 0$, then three roots of Eqn.(1) become infinite. Therefore, the two lines represented by $a_2y^2 + b_2y + c_2 = 0$ are the asymptotes of Eqn.(1) which are parallel to $x$-axis, and so on.

Similarly, for asymptotes parallel to $y$-axis.

*Thus we have the following* **rules :**

**I.** *To find the asymptotes parallel to x-axis, equate to zero the coefficient of the highest power of x in the equation,* provided this is not merely a constant.

**II.** *To find the asymptotes parallel to y-axis, equate to zero the coefficient of the highest power of y in the equation,* provided this is not merely a constant.

**Example 4.67.** *Find the asymptotes of the curve*

$$x^2y^2 - x^2y - xy^2 + x + y + 1 = 0.$$

**Solution.** The highest power of $x$ is $x^2$ and its coefficient is $y^2 - y$.

$\therefore$ The asymptotes parallel to the $x$-axis are given by

$$y(y - 1) = 0 \text{ i.e., by } y = 0 \text{ and } y = 1.$$

The highest power of $y$ is $y^2$ and its coefficient is $x^2 - x$.

$\therefore$ The asymptotes parallel to the $y$-axis are given by

$$x(x - 1) = 0 \text{ i.e., by } x = 0 \text{ and } x = 1.$$

Hence the asymptotes are $x = 0$, $x = 1$, $y = 0$ and $y = 1$.

**(3) Inclined asymptotes.** Let the equation of the curve be of the form

$$x^n\phi_n(y/x) + x^{n-1}\phi_{n-1}(y/x) + x^{n-2}\phi_{n-2}(y/x) + \dots = 0 \quad \dots(1)$$

where $\phi_r(y/x)$ is an expression of degree $r$ is $y/x$.

To find where this curve is cut by the line $y = mx + c$, ...(2)

put $y/x = m + c/x$ in Eqn. (1). The resulting equation is

$$x^n\phi_n(m + c/x) + x^{n-1}\phi_{n-1}(m + c/x) + x^{n-2}\phi_{n-2}(m + c/x) + \dots = 0$$

which gives the abscissae of the points of intersection.

Expanding each of the $\phi$-functions by Taylor's series,

$$x^n\left\{\phi_n(m) + \frac{c}{x}\phi_n'(m) + \frac{c^2}{2!\,x^2}\phi_n''(m) + \dots\right\} + x^{n-1}\left\{\phi_{n-1}(m) + \frac{c}{x}\phi'_{n-1}(m) + \dots\right\}$$
$$+ x^{n-2}\{\phi_{n-2}(m) + \dots\} = 0$$

or

$$x^n\phi_n(m) + x^{n-1}\{c\phi'_n(m) + \phi_{n-1}(m)\}$$
$$+ x^{n-2}\left\{\frac{c^2}{2!}\phi_n''(m) + c\phi'_{n-1}(m) + \phi_{n-2}(m)\right\} + \dots = 0 \quad \dots(3)$$

If the line (2) is an asymptote to the curve, it cuts the curve in two points at infinity *i.e.*, the Eqn. (3) has two infinite roots for which the coefficients of two highest terms should be zero.

*i.e.*, $\phi_n(m) = 0$ ...(4) and $c\,\phi'_n(m) + \phi_{n-1}(m) = 0$ ...(5)

If the roots of Eqn. (4) be $m_1, m_2, \dots m_n$, then the corresponding values of $c$ (*i.e.* $c_1, c_2, \dots\dots c_n$) are given by Eqn. (5). Hence the asymptotes are

$$y = m_1x + c_1, y = m_2x + c_2, \dots\dots, y = m_nx + c_n.$$

**Obs.** If (4) gives two equal values of $m$, then the corresponding values of $c$ cannot be found from Eqn. (5). Then $c$ is determined by equating to zero the coefficient of $x^{n-2}$ *i.e.*, from

$$\frac{c^2}{2!}\phi''_n(m) + c\,\phi'_{n-1}(m) + \phi_{n-2}(m) = 0 \qquad \text{...(6)}$$

In this case, *there will be two parallel asymptotes.*

**Working rule :**

1. Put $x = 1, y = m$ in the highest degree terms, thus getting $\phi_n(m)$. Equate it to zero and solve for $m$. Let its roots be $m_1, m_2, \ldots\ldots$
2. Form $\phi_{n-1}(m)$ by putting $x = 1$ and $y = m$ in the $(n-1)$th degree terms.
3. Find the values of $c$ (*i.e.* $c_1, c_2, \ldots\ldots$) by substituting $m = m_1, m_2, \ldots\ldots$ in turn in the formula
$$c = -\phi_{n-1}(m)/\phi'_n(m)$$
[Sometimes it take (0/0) form, then find c from (6).]
4. Substitute the values of $m$ and $c$ in $y = mx + c$ in turn.

**Example 4.68.** *Find the asymptotes of the curve*

(*i*) $y^3 - 2xy^2 - x^2y + 2x^3 + 3y^2 - 7xy + 2y^2 + 2y + 2x + 1 = 0.$ (*Delhi, 2012 ; U.P.T.U., 2012*)

(*ii*) $x^3 + 3x^2y - 4y^3 - x + y + 3 = 0.$ (*D.T.U., 2013 ; Rohtak, 2011S*)

(*iii*) $(x + y)^2 (x + y + 2) = x + 9y - 2.$

**Solution.** (*i*) Putting $x = 1$ and $y = m$ in the third degree terms,

$$\phi_3(m) = m^3 - 2m^2 - m + 2, \quad \therefore \quad \phi_3(m) = 0 \text{ gives } m^3 - 2m^2 - m + 2 = 0$$

or $$(m^2 - 1)(m - 2) = 0 \text{ whence } m = 1, -1, 2.$$

Also putting $x = 1$ and $y = m$ in the 2nd degree terms, $\phi_2(m) = 3m^2 - 7m + 2$

$$\therefore \quad c = -\frac{\phi_2(m)}{\phi_3'(m)} = -\frac{3m^2 - 7m + 2}{3m^2 - 4m - 1}$$

$$= -1 \text{ when } m = 1, = -2 \text{ when } m = -1, = 0 \text{ when } m = 2.$$

Hence the asymptotes are $y = x - 1$, $y = -x - 2$ and $y = 2x$.

(*ii*) Putting $x = 1$ and $y = m$ in the third degree terms,

$$\phi_3(m) = 1 + 3m - 4m^3$$

$$\therefore \quad \phi_3(m) = 0 \text{ gives } 4m^3 - 3m - 1 = 0, \quad \text{or} \quad (m-1)(2m+1)^2 = 0$$

whence $$m = 1, -1/2, -1/2.$$

Similarly, $$\phi_2(m) = 0$$

$$\therefore \quad c = -\frac{\phi_2(m)}{\phi_3'(m)} = -\frac{0}{3 - 12m^2}$$

$$= 0 \text{ when } m = 1, = \frac{0}{0} \text{ form when } m = -\frac{1}{2}.$$

Thus $\left(\text{when } m = -\frac{1}{2}\right)$ $c$ is to be obtained from

$$\frac{c^2}{2!}\phi''_3(m) + c\,\phi'_2(m) + \phi_1(m) = 0$$

or $$\frac{c^2}{2}(-24m) + c\,.\,0 + (-1 + m) = 0$$

Putting $m = -1/2$, $6c^2 - 3/2 = 0$ whence $c = \pm 1/2$.

Hence the asymptotes are $y = x$, $y = -\frac{1}{2}x + \frac{1}{2}$, $y = -\frac{1}{2}x - \frac{1}{2}$.

(*iii*) Putting $x = 1$ and $y = m$ in the third degree terms, $\phi_3(m) = (1 + m)^3$.

$$\therefore \quad \phi_3(m) = 0 \text{ gives } (m+1)^3 = 0 \text{ whence } m = -1, -1, -1.$$

Similarly, $$\phi_2(m) = 2(1 + m)^2, \ \phi_1(m) = -1 - 9m, \ \phi_0(m) = 2.$$

For these three equal values of $m = -1$, values of $c$ are obtained from

$$\frac{c^3}{3!}\phi_3'''(m) + \frac{c^2}{2!}\phi_2''(m) + c\,\phi_1'(m) + \phi_0(m) = 0$$

or $$\frac{c^3}{6}(6) + \frac{c^2}{2}(4) + c(-9) + 2 = 0 \quad \text{or} \quad c^3 + 2c^2 - 9c + 2 = 0.$$

Solving for $c$, we have $c = 2, -2 \pm \sqrt{5}$.

Hence the three asymptotes are

$$y = -x + 2,\ y = -x - 2 + \sqrt{5},\ y = -x - 2 - \sqrt{5}.$$

**4. Asymptotes of polar curves.** It can be shown that an *asymptote of the curve* $1/r = f(\theta)$ *is*

$$r \sin(\theta - \alpha) = 1/f'(\alpha),$$

*where* $\alpha$ *is a root of the equation* $f(\theta) = 0$ *and* $f'(\alpha)$ *is the derivative of* $f(\theta)$ *w.r.t.* $\theta$ *at* $\theta = \alpha$.

**Example 4.69.** *Find the asymptote of the spiral* $r = a/\theta$.

**Solution.** Equation of the curve can be written as $1/r = \theta/a = f(\theta)$, say.

$$f(\theta) = 0, \text{ if } \theta = 0\ (= \alpha). \text{ Also } f'(\theta) = 1/a \quad \therefore \quad f'(\alpha) = 1/a.$$

$\therefore$ The asymptote is $r \sin(\theta - 0) = 1/f'(0)$ or $r \sin\theta = a$.

## PROBLEMS 4.14

Find the asymptotes of (1–6) :

1. $x^3 + y^3 = 3axy$
2. $(x^2 - a^2)(y^2 - b^2) = a^2 b^2$
3. $(a/x)^2 + (b/y)^2 = 1$ (*D.T.U., 2011*)
4. $x^2y + xy^2 + xy + y^2 + 3x = 0$.
5. $4x^3 + 2x^2 - 3xy^2 - y^3 - 1 - xy - y^2 = 0$.
6. $x^2(x - y)^2 - a^2(x^2 + y^2) = 0$
7. $(x + y)^2(x + 2y + 2) = (x + 9y - 2)$ (*Rohtak, 2012*)
8. Show that the asymptotes of the curve $x^2 y^2 = a^2(x^2 + y^2)$ form a square of side $2a$.
9. Find the asymptotes of the curve $x^2y - xy^2 + xy + y^2 + x - y = 0$ and show that they cut the curve again in three points which lie on the line $x + y = 0$.

Find the asymptotes of the following curves :

10. $r = a \tan\theta$.
11. $r = a(\sec\theta + \tan\theta)$
12. $r \sin\theta = 2\cos 2\theta$.
13. $r \sin n\theta = a$.

## 4.17 (1) CURVE TRACING

In many practical applications, a knowledge about the shapes of given equations is desirable. On drawing a sketch of the given equation, we can easily study the behaviour of the curve as regards its symmetry asymptotes, the number of branches passing through a point etc.

*A point through which two branches of a curve pass is called a* **double point.** At such a point $P$, the curve has two tangents, one for each branch.

*If the tangents are real and distinct, the double point is called a* **node** [Fig. 4.22 (*a*)].

*If the tangents are real and coincident, the double point is called a* **cusp** [Fig. 4.22 (*b*)].

If the tangents are imaginary, the double point is called a *conjugate point* (or an *isolated point*). Such a point cannot be shown in the figure.

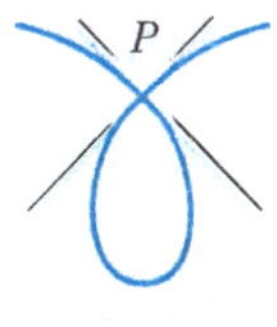

Fig. 4.22 (*a*)

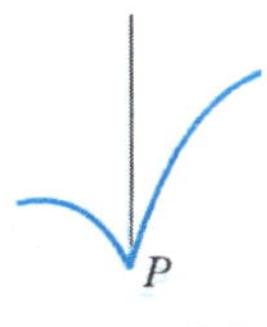

Fig. 4.22 (*b*)

**(2) Procedure for tracing cartesian curves.**

**1. Symmetry.** *See if the curve is symmetrical about any line.*

(*i*) *A curve is symmetrical about the x-axis, if only even powers of y occur in its equation.* (*e.g.*, $y^2 = 4ax$ is symmetrical about $x$-axis).

(*ii*) *A curve is symmetrical about the y-axis, if only even powers of x occur in its equation.*
(*e.g.*, $x^2 = 4ay$ is symmetrical about $y$-axis).

(*iii*) *A curve is symmetrical about the line $y = x$, if on interchanging $x$ and $y$ its equation remains unchanged,* (*e.g.*, $x^3 + y^3 = 3axy$ *is symmetrical about the line* $y = x$).

**2. Origin.** (*i*) *See if the curve passes through the origin.*

(A curve passes through the origin if there is no constant term in its equation).

(*ii*) *If it does, find the equation of the tangents thereat,* by equating to zero the lowest degree terms.

(*iii*) *If the origin is a double point, find whether the origin is a node, cusp or conjugate point.*

**3. Asymptotes.** (*i*) *See if the curve has any asymptote parallel to the axes* (p. 199).

(*ii*) *Then find the inclined asymptotes,* if need be. (p. 199).

**4. Points.** (*i*) *Find the points where the curve crosses the axes and the asymptotes.*

(*ii*) *Find the points where the tangent is parallel or perpendicular to the x-axis,*
(*i.e.* the points where $dy/dx = 0$ or $\infty$).

(*iii*) *Find the region (or regions) in which no portion of the curve exists.*

**Example 4.70.** *Trace the curve* $y^2 (2a - x) = x^3$.
(*C.S.V.T.U., 2015 ; V.T.U., 2015 ; D.T.U., 2013 ; U.P.T.U., 2011*)

**Solution.** (*i*) *Symmetry*: The curve is symmetrical about the $x$-axis.

[$\because$ only even powers of $y$ occur in the equation.]

(*ii*) *Origin* : The curve passes through the origin

[$\because$ there is no constant term in its equation.]

The tangents at the origin are $y = 0, y = 0$ [Equating to zero the lowest degree terms.]

$\therefore$ Origin is a *cusp*

(*iii*) *Asymptotes* : The curve has an asymptote $x = 2a$.

[$\because$ co-eff. of $y^3$ is absent, co-eff. of $y^2$ is an asymptote.

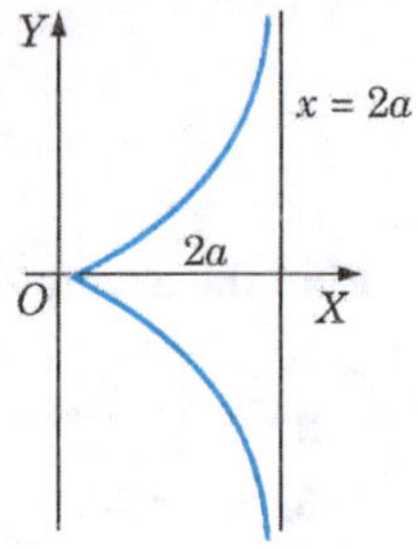

Fig. 4.23

(*iv*) *Points* : (*a*) curve meets the axes at (0, 0) only. (*b*) $y^2 = x^3/(2a - x)$]

When $x$ is –ve, $y^2$ is –ve (*i.e.* $y$ is imaginary) so that no portion of the cuve lies to the left of the $y$-axis. Also when, $x > 2a$, $y^2$ is again –ve, so that no portion of the curve lies to the right of the line $3x = 2a$.

Hence, the shape of the curve is as shown in Fig. 4.23. This curve is known as *Cissoid*.

**Example 4.71.** *Trace the curve* $y^2 (a - x) = x^2(a + x)$. (*V.T.U., 2016 ; C.S.V.T.U., 2014 S ; D.T.U., 2012*)

**Solution.** (*i*) *Symmetry* : The curve is symmetrical about the $x$-axis.

(*ii*) *Origin* : The curve passes through the origin and the tangents at the origin are $y^2 = x^2$,

*i.e.,* $y = x$ and $y = -x$. $\therefore$ Origin is a *node*.

(*iii*) *Asymptotes* : The curve has an asymptote $x = a$

(*iv*) *Points* : (*a*) When $x = 0, y = 0$ ; when $y = 0, x = 0$ or $-a$.

$\therefore$ The curve crosses the axes at (0, 0) and ( $-a$, 0).

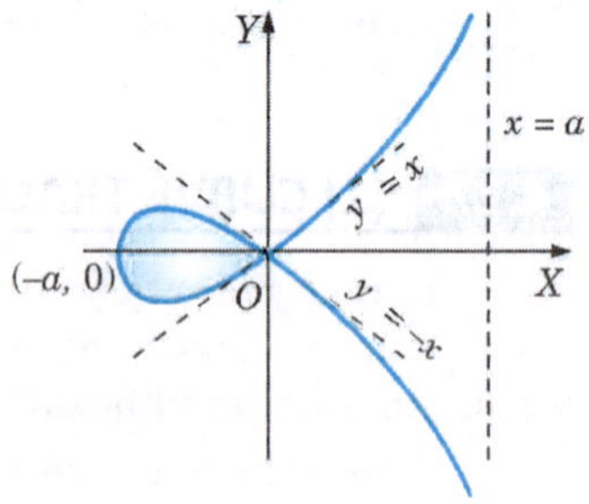

Fig. 4.24

We have $y = \pm x \sqrt{\left(\dfrac{a + x}{a - x}\right)}$

When, $x > a$ or $< -a$, $y$ is imaginary.

$\therefore$ No portion of the curve lies to the right of the line $x = a$ or to the left of the line $x = -a$.

Hence the shape of the curve is as shown in Fig. 4.24. This curve is known as *Strophoid*.

**Example 4.72.** *Trace the curve* $y = x^2/(1 - x^2)$.

**Solution.** (*i*) *Symmetry* : The curve is symmetrical about $y$-axis.

(*ii*) *Origin* : It passes through the origin and the tangent at the origin is $y = 0$ (*i.e.*, $x$-axis).

(*iii*) *Asymptotes* : The asymptotes are given by $1 - x^2 = 0$ or $x = \pm 1$ and $y = -1$.

(*iv*) *Points* : (*a*) The curve crosses the axes at the origin only. (*b*) When $x \to 1$ from left, $y \to \infty$

When $x \to 1$ from right $y \to -\infty$

When $x > 1$, $y$ is –ve

Hence the curve is as shown in Fig. 4.25.

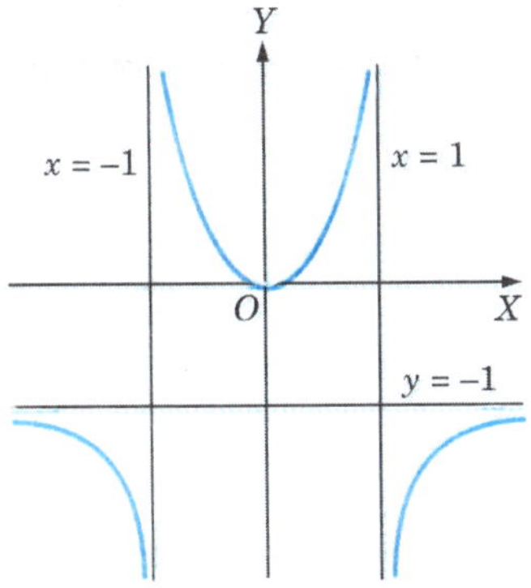

Fig. 4.25

**Example 4.73.** *Trace the curve* $a^2y^2 = x^2(a^2 - x^2)$. *(D.T.U., 2011)*

**Solution.** (*i*) *Symmetry*. The curve is symmetrical about $x$-axis, $y$-axis and origin.

(*ii*) *Origin*. The curve passes through the origin and the tangents at the origin are $a^2y^2 = a^2x^2$ *i.e.*, $y = \pm x$.

(*iii*) *Asymptotes*. The curve has no asymptote.

(*iv*) *Points*. (*a*) The curve cuts $x$-axis ($y = 0$) at $x = 0, \pm a$. and cuts $y$-axis ($x = 0$) at $y = 0$ *i.e.*, (0, 0) only.

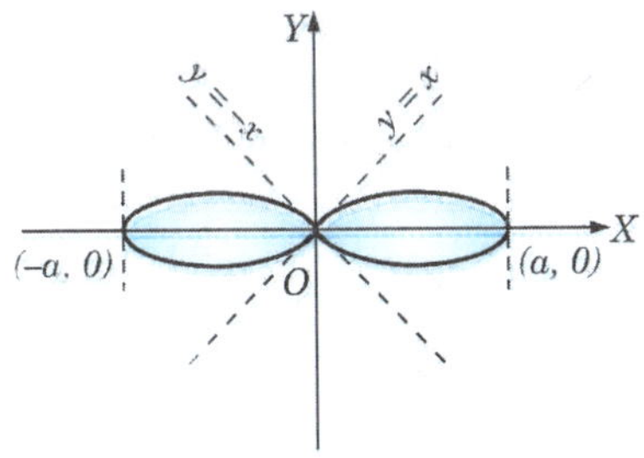

Fig. 4.26

(*b*) $\dfrac{dy}{dx} = \dfrac{x(a^2 - 2x^2)}{a^2 y} \to \infty$ at $(a, 0)$

*i.e.*, tangent to the curve at $(a, 0)$ is parallel to $y$-axis. Similarly the tangent at $(-a, 0)$ is parallel to $y$-axis.

(*c*) We have $y = \dfrac{x}{a}\sqrt{a^2 - x^2}$ which is real for $x^2 < a^2$ *i.e.*, $-a < x < a$

$\therefore$ The curve lies between $x = a$ and $x = -a$

Hence the shape of the curve is as shown is Fig. 4.26.

**Example 4.74.** *Trace the curve* $9ay^2 = x(x - 3a)^2$. *(Kochine, 2018)*

**Solution.** **1. Symmetry:** The curve is symmetrical about $x$-axis, as there are even and only even powers of $y$ in the equation of the curve.

**2. Origin** (*a*) The curve passes through the origin.

(*b*) The tangents at the origin are given by $9a^2 x = 0$ or $x = 0$.

**3. Intersection with axes:** The curve meets the $x$-axis at (0, 0) and at $(3a, 0)$ and the $y$-axis at the origin only. We shall find tangents at the points $(3a, 0)$.

To shift the origin to the point $(3a, 0)$ let

$$X = x - 3a, Y = y$$

The equation of the curve becomes

$$9aY^2 = (X + 3a)X^2$$

Now tangents at the new origin are obtained by equating the lowest degress terms to zero,

Thus tangents are

$$9aY^2 - 3aX^2 = 0$$

$$Y = \pm \frac{1}{\sqrt{3}} X$$

$$y = \pm \frac{1}{\sqrt{3}}(x - 3a)$$

The tangents at $(3a, 0)$ are inclined at an angle of $\tan^{-1}\left(\dfrac{1}{\sqrt{3}}\right)$ *i.e.*, 30° to $x$-axis.

The ordinate of the curve in neighbourhood of the point $(3\,a, 0)$ is greater than the ordinate of tangent at this point.

**4. Special point:** From the equation of the curve.

$$y = \frac{1}{3\sqrt{a}}(x - 3a)\cdot\sqrt{x} \qquad \text{(taking positive value only)}$$

$$= \frac{1}{3\sqrt{a}}(x^{3/2} - 3ax^{1/2})$$

$$\frac{dy}{dx} = \frac{1}{3\sqrt{a}}\left(\frac{3}{2}x^{1/2} - \frac{3a}{2}x^{-1/2}\right) = \frac{1}{2\sqrt{ax}}(x - a)$$

Now $\frac{dy}{dx} = 0$ at $x = a$, thus the tangent at $x = a$ is parallel to $x$-axis.

Also when $x = 0$, $\frac{dy}{dx}$ becomes infinite and hence the tangent at the origin is parallel to $y$-axis. *i.e.*, the $y$ axis it self is the tangent to the curve at the origin.

**5. Imaginary Values:** If $x < 0$, $y^2$ becomes negative and $y$ is imaginary. Hence no part of the curve lies in second and third quadrants.

**6. Asymptotes:** The curve has no asymptotes parallel to the axis.

**7. Region.** As $x$ increases from 0 to $a$, $y$ also increases and when $x$ increases from $a$ to $3a$, $y$ decreases to zero. When $x$ increases beyond $3a$, $y$ also goes on increasing and as $x \to \infty, y \to \infty$.

Also $x = 0, \quad y = 0$

Hence the shape of curve is shown in figure.

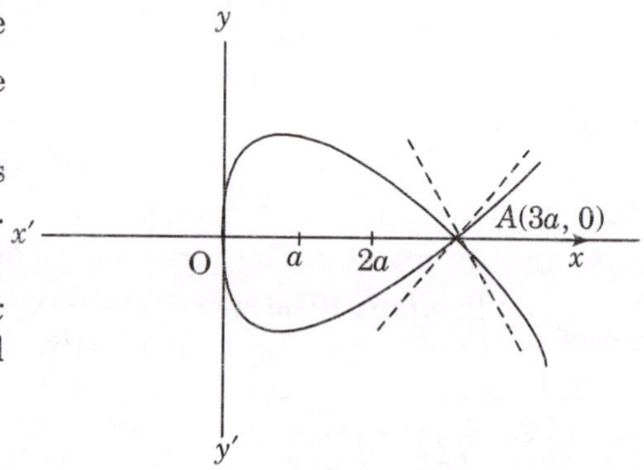


Fig. 4.27.

**Example 4.75.** *Trace the curve* $y = \frac{8a^3}{x^3 + 4a^2}$ *(Pondicherry, 2019)*

**Solution. 1. Symmetry:** The curve is symmetrical about $y$-axis as there are even and only even powers of $x$ in the equation of the curve.

**2. Origin:** The curve does not pass through the origin.

**3. Intersection with axis:** The curve meets the $y$-axis at the point $(0, 2a)$ which is obtained by putting $x = 0$, in the equation of curve and it does not meet the $x$-axis. We shall find the tangents to the curve at the point $(0, 2a)$, let

To shift the origin to the point $(0, 2a)$, Let

$$X = x,\ Y = y - 2a$$

The equation of the curve becomes

$$(Y + 2a)\,(X^2 + 4a)^2 = 8a^2$$

The tangents at the new origin are obtained by equating the lowest degree terms to zero. Thus the tangent is $4a^2\,Y = 0$ or $Y = 0$

$$\therefore \qquad y - 2a = 0$$

The tangent at the point $(0, 2a)$ is a line parallel to $x$-axis at a distance $2a$ from it.

**4. Special points:** From the equation of the curve.

$$y = \frac{8a^3}{(x^2 + 4a^2)}$$

$$\therefore \qquad \frac{dy}{dx} = -16a^3\,x(x^2 + 4a^2)^{-2} = \frac{-16a^3x}{(x^2 + 4a^2)^2}$$

w $\frac{dy}{dx} = 0$. When $x = 0$ and therefore, $y = 2a$ (from the equation of the curve).

nce, the tangent is parallel to $x$ axis at the point $(0, 2a)$ a fact already established in (3).

**5. Imaginary Values:** Solving the given equation of the curve for $x$.

$$x = 2a\sqrt{\frac{2a - y}{y}}$$

If $y < 0$ or if $y > 2a$, $x$ is imaginary. Thus no part of the curve lies below $x$-axis or above the line $y = 2a$.

**6. Asymtotes:** Equating to zero, the co-efficient of $x^2$, the highest degree term in $x$, we get $y = 0$. Hence $x$-axis is an asymptote of the curve.

The curve has no real asymtotes parallel to $y$-axis.

**7. Region:** As $x$-increases from 0 to $\infty$, $y$ decreases from $2a$ to 0.

Also when $x = 0$, $y = 2a$.

Hence the shape of the curve is shown in the Fig. 4.28.

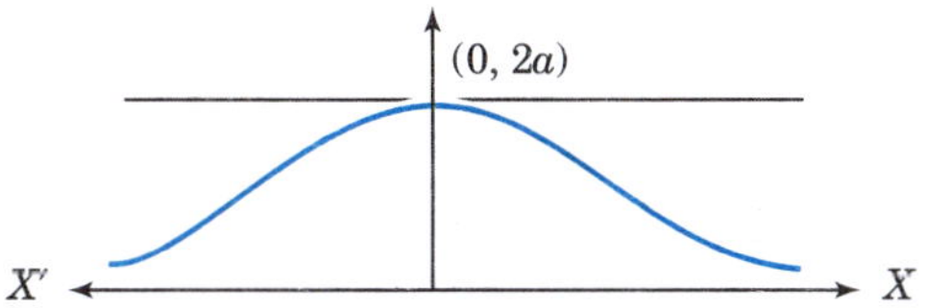


Fig. 4.28.

**Example 4.76.** *Trace the curve* $y^2(a^2 + x^2) = x^2(a^2 - x^2)$

**Solution: 1. Symmetry:** The curve is symmetrical about both the axes, as there are even and only even power of both $x$ and $y$, in the equation of the curve.

**2. Origin:** ($a$) The curve passes through the origin.

($b$) Tangent at the origin are

$$y^2 = x^2 \text{ or } y = \pm x$$

Since the tangents are real and different, the origin is passing through same axes.

**3. Intersection with the axes:** The curve meets the $x$-axis at $(a, 0)$ and $(-a, 0)$. To find tangents at $(0, 0)$ shift origin to this point, let $X = x - a, y = y$.

The equation of the curve becomes

$$Y[a^2 + (X + a)^2] = (X + a)^2[a^2 - (X + a)^2]$$

$$Y^2(2a^2 + 2aX + X^2) = -(X + a)^2(2aX + x)^2$$

The tangent at the new origin are obtained by equating the lowest degree terms to zero. Thus the tangent at $(a, 0)$

$$-2a^3X = 0 \text{ or } X = 0$$

$$\therefore \quad x - a = 0$$

Thus the tangents at $(a, 0)$ is a line parallel to $y$-axis at a distance a from it. By symmetry tangent at $(-a, 0)$ is also parallel to $y$-axis.

**4. Special point:**

$$y = x\sqrt{\frac{a^2 - x^2}{a^2 + x^2}}$$

$$\frac{dy}{dx} = \frac{a^4 - 2a^2x^2 - x^4}{(a^4 - x^4)^{1/2}(a^2 + x^2)}$$

$$\therefore \quad \frac{dy}{dx} = 0, \text{ when } a^4 - 2a^2x^2 - x^4 = 0$$

$$x^2 = \frac{2a^2 \pm \sqrt{4a^4 + 4a^4}}{-2} = a^2(1 - 1 \pm \sqrt{2})$$

$$\therefore \quad x = \pm\sqrt{(\sqrt{2} - 1)}$$

Thus the tangents are parallel to $x$-axis at

$$x = \pm\, a\sqrt{(\sqrt{2}-1)}$$

Also $$\frac{dy}{dx} = \infty, \text{ at } x = \pm\, a$$

tangents are parallel to $y$-axis at $(\pm a, 0)$

**5. Imaginary Value:** From the equation of the curve we have

$$y = \pm x\sqrt{\frac{a^2 - x^2}{a^2 + x^2}}$$

If $x^2 > a^2$ $|x| > a$, then $y$ is imaginary.

Hence no part of the curve lies beyond the lines $x = \pm\, a$.

**6. Asymptotes:** The curve has no asymptotes.

**7. Region:** From the equation of the curve.

$$y = x\sqrt{\frac{a^2 - x^2}{a^2 + x^2}} \qquad \text{(Taking +}ve\text{ value)}$$

When $x = 0, y = 0$

As $x$ increases from 0 to $a\sqrt{(\sqrt{2}-1)}$, $y$ goes on increasing At $\bar{x} = a\sqrt{(\sqrt{2}-1)}$; $\dfrac{dy}{dx} = 0$

Further as $x$ increases from to $a\sqrt{(\sqrt{2}-1)}$ to $a$, $y$ decreases ultimately to zero.

The shape of the curve is as shown in the Fig. 4.29.

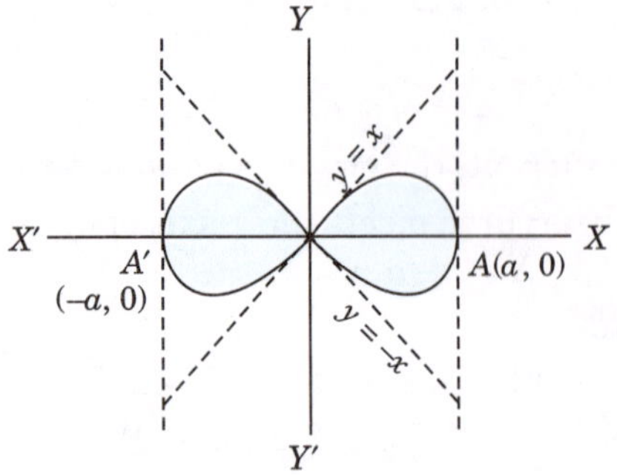


Fig. 4.29.

**Example 4.77.** *Trace the curve* $y = \dfrac{x^2+1}{x^2-1}$ (V.T.U, 2019)

**Solution. 1. Symmetry:** The curve is symmetrical about $y$-axis as there are even and only even powers of $x$.

**2. Origin:** The curve does not pass through origin.

**3. Intersection with the axes:** The curve does not meet the $x$-axis but meet $y$-axis at the point(0, –1).

To find the tangents at new origin, we shift the origin to (0, –1)

Let
$$X = x$$
$$Y = y + 1$$

The equation of the curve becomes

$$Y - 1 = \frac{X^2+1}{X^2-1}$$

$$(Y-1)(X^2-1) = X^2 + 1$$

$$X - Y - 2X^2 - Y = 0$$

Tangent the new origin is obtained by equating the lowest degree terms to zero. So the tangent at (0 – 1).

$$Y = 0$$

$$Y + 1 = 0$$

Hence the tangent at (0, –1) is a line parallel to $x$-axis at a distance –1 from it.

**4. Special points:** From the equation of the curve we have

$$y = \frac{x^2+1}{x^2-1} \qquad ...(i)$$

$$\frac{dy}{dx} = \frac{(x^2-1)2x - 2x(x^2+1)}{(x^2-1)^2} = -\frac{4x}{(x^2-1)^2}$$

Now $\frac{dy}{dx} = 0$, when $x = 0$

From $(i)$ when $x = 0, y = -1$

Hence tangent is parallel to $x$-axis at $(0, -1)$

$$\frac{dy}{dx} \to \infty, \text{ when } x = \pm 1$$

$$x = \pm 1 \text{ from Eqn. } (i), \text{ we have } y \to \infty.$$

Thus the tangents are parallel to $y$-axis at $\infty$, hence $x = \pm 1$ will be asymptotes.

**5. Imaginary values:** From Eqn. $(i)$

$$x^2 = \frac{y+1}{y-1}, \text{ when } -1 > y > 1, x^2 \text{ is } -ve \text{ and } x \text{ is imaginary.}$$

**6. Asymptotes:** $(i)$ Equating to zero the coefficient of the highest degree term in $x$, $x^2$, we have.

$$y - 1 = 0$$

the asymptote is parallel to $x$-axis.

$(ii)$ Equating to zero the coefficient of the highest degree term in $y$,

$$x^2 - 1 = 0$$

$$x = \pm 1$$

as the asymptotes parallel to $y$-axis.

**7. Region:** Solving the equation of the curve for $x$, we have

$$x = \sqrt{\frac{y+1}{y-1}}$$

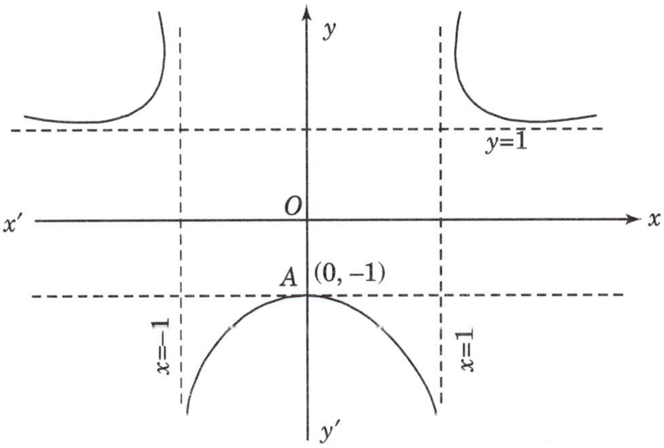

Fig. 4.30.

In first quadrant as $y$ decreases from $\infty$ to 1, $x$ increases from 1 to $\infty$.

Also for the portion of the curve in fourth quadrant as $y$ decreases from $-1$ to $-\infty$, $x$ increases from 0 to 1.

**Example 4.78.** *Trace the curve $x^3 + y^3 = 3axy$* *(Kurukshetra, 2013 ; C.S.V.T.U., 2012)*

*or* $$r = \frac{3a \sin\theta \cos\theta}{\sin^3\theta + \cos^3\theta}.$$

**Solution.** $(i)$ *Symmetry* : The curve is symmetrical about the line $y = x$.

$(ii)$ *Origin* : It passes through the origin and tangents at the origin are $xy = 0$, *i.e.*, $x = 0, y = 0$.

$\therefore$ Origin is a *node*.

$(iii)$ *Asymptotes* : $(a)$ It has no asymptote parallel to the axes.

$(b)$ Putting $y = m$ and $x = 1$ in the third degree terms,

$$\phi_3(m) = 1 + m^3, \phi_3(m) = 0 \text{ gives } m = -1.$$

$$\therefore \quad c = -\frac{\phi_2(m)}{\phi_3'(m)} = -\left(\frac{-3am}{3m^2}\right) = \frac{a}{m}.$$

$$= -a, \text{ when } m = -1.$$

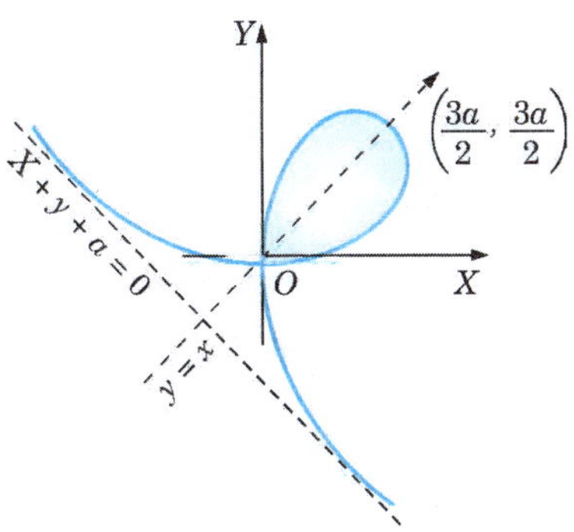

Fig. 4.31.

Hence $y = -x - a$ $\left(i.e., \dfrac{x}{-a} + \dfrac{y}{-a} = 1\right)$ is an asymoptote.

(*iv*) *Points* : (*a*) It meets the axes at the origin only.

(*b*) When $y = x$, $2x^3 = 3ax^2$, *i.e.* $x = 0$ or $3a/2$. *i.e.*, the curve crosses the line $y = x$ at $(3a/2, 3a/2)$.

Hence the shape of the curve is as shown in Fig. 4.29. This curve is known as *Folium of Descartes*.

**Example 4.79.** *Trace the curve* $x^3 + y^3 = 3ax^2$.

**Solution.** (*i*) *Symmetry* : The curve has no symmetry.

(*ii*) *Origin* : The curve passes through the origin and the tangents at the origin are $x = 0$ and $x = 0$.

$\therefore$ The origin is a *cusp*.

(*iii*) *Asymptotes* : (*a*) The curve has no asymptote parallel to the axes.

(*b*) Putting $x = 1$, $y = m$ in the third degree terms, we get

$$\phi_3(m) = m^3 + 1\ ; \quad \therefore \quad \phi_3(m) = 0, \text{ gives } m = -1.$$

$$\therefore \qquad c = -\frac{\phi_2(m)}{\phi_3'(m)} = -\frac{-3a}{3m^2} = a \text{ for } m = 1.$$

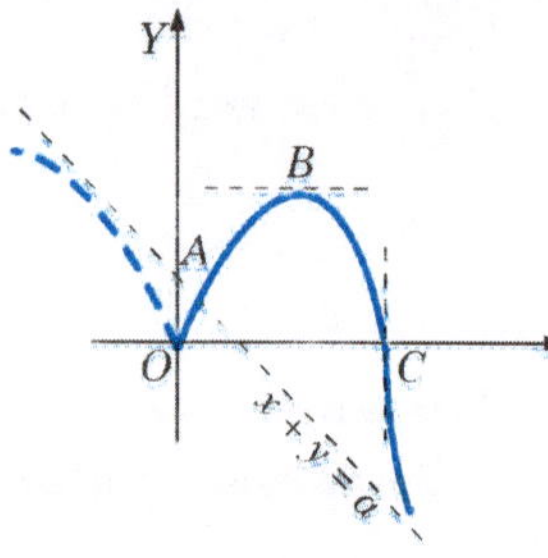

Fig. 4.32.

Thus $x + y = a$ is the only asymptote.

The curve lies above the asymptote when $x$ is positive and large and it lies below the asymptote when $x$ is negative.

(*iv*) *Points.* (*a*) The curve crosses the axes at $O$ (0, 0) and $C(3a, 0)$. It crosses the asymptote at $A(a/3, 2a/3)$.

(*b*) Since $y^2\, dy/dx = x(2a - x)$. $\therefore$ $dy/dx = 0$ for $x = 2a$.

(*c*) Now $y = [x^2 (3a - x)]^{1/3}$.

When $0 < x < 3a$, $y$ is positive. As $x$ increases from 0, $y$ also increases till $x = 2a$ where the tangent is parallel to the $x$-axis. As $x$ increases from $2a$ to $3a$, $y$ constantly decreases to zero.

When $x > 3a$, $y$ is negative.

When $x < 0$, $y$ is positive and constantly increases as $x$ varies from 0 to $-\infty$.

Combining all these facts we see that the shape of the curve is as shown in Fig. 4.32.

**Example 4.80.** *Trace the curve* $y^2 (x - a) = x^2(x + a)$.

**Solution.** (*i*) *Symmetry* : The curve is symmetrical about the $x$-axis.

(*ii*) *Origin* : The curve passes through the origin and the tangents at the origin are $y^2 = -x^2$ *i.e.*, $y = \pm ix$, which are imaginary lines. $\therefore$ The origin is an *isolated point.*

(*iii*) *Asymptotes* : (*a*) $x = a$ is the only asymptote parallel to the $y$-axis.

(*b*) Putting $x = 1$ and $y = m$ in the third degree terms, we get

$$\phi_3(m) = m^2 - 1.$$

$$\therefore \qquad \phi_3(m) = 0 \text{ gives } m = \pm 1$$

$$\therefore \qquad c = \frac{\phi_2(m)}{\phi_3'(m)}$$

$$= -\frac{-a(m^2 + 1)}{2m}$$

$$= \pm a \text{ for } m = \pm 1.$$

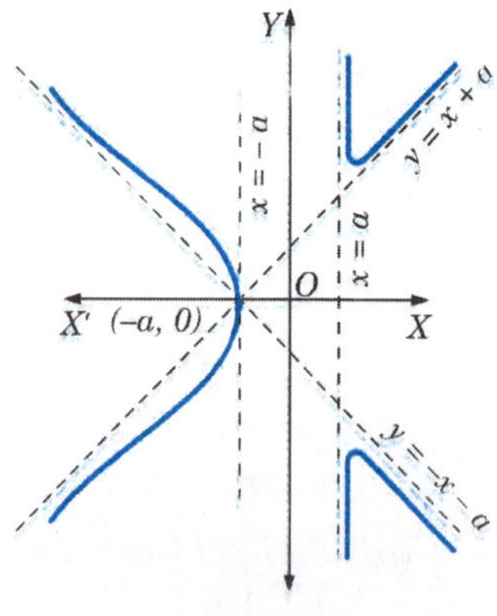

Fig. 4.33.

Thus the other two asymptotes are $y = x + a$ ; $y = -x - a$.

(*iv*) *Points* : (*a*) The curve crosses the axes at $(-a, 0)$ and $(0, 0)$.

It crosses the asymptotes $y = x + a$ and $y = -x - a$ at $(-a, 0)$.

(*b*) $y = \pm x \sqrt{\left(\dfrac{x + a}{x - a}\right)}$

When $x < a$ and $x > -a$, $y$ is imaginary.

$\therefore$ no portion of the curve lies between the lines $x = a$ and $x = -a$. Thus the vertical asymptote must be approached from the right.

$$(c)\ \frac{dy}{dx} = \pm \frac{x^2 - ax + a^2}{(x-a)^{3/2}(x+a)^{1/2}}$$

$\therefore$ $dy/dx = 0$, when $x = \frac{1}{2}(1 + \sqrt{5})\,a = 1.6a$ approx.

[rejecting the value $\frac{1}{2}(1 - \sqrt{5})a$ which lies between $-a$ and $a$]

and $dy/dx \to \infty$, when $x = \pm a$.

Thus the tangent is parallel to the $x$-axis at $x = 1.6a$ and perpendicular to the $x$-axis at $x = \pm a$.

Hence the shape of the curve is as shown in Fig. 4.31.

## 4.17 (3) PROCEDURE FOR TRACING CURVES IN PARAMETRIC FORM : $x = f(t)$ and $y = \phi(t)$

1. **Symmetry.** *See if the curve has any symmetry.*

(*i*) *A curve is symmetrical about the x-axis, if on replacing t by –t, f(t) remains unchanged and φ(t) changes to –φ(t).*

(*ii*) *A curve is symmetrical about the y-axis if on replacing t by –t, f(t) changes to –f(t) and φ(t) remains unchanged.*

(*iii*) *A curve is symmetrical in the opposite quadrants, if on replacing t by –t, both f(t) and φ(t) remains unchanged.*

2. **Limits.** *Find the greatest and least values of x and y so as to determine the strips, parallel to the axes, within or outside which the curve lies.*

3. **Points.** (*a*) *Determine the points where the curve crosses the axes.*

The points of intersection of the curve with the $x$-axis given by the roots of $\phi(t) = 0$, while those with the $y$-axis are given by the roots of $f(t) = 0$.

(*b*) *Giving t a series of value, plot the corresponding values of x and y, noting whether x and y increase or decrease for the intermediates values of t.* For this purpose, we consider the sign of $dx/dt$ and $dy/dt$ for the different values of $t$.

(*c*) *Determine the points where the tangent is parallel or perpendicular to the x-axis,* (*i.e.*, where $dy/dx = 0$ or $\to \infty$).

(*d*) *When x and y are periodic functions of t with a common period, we need to study the curve only for one period,* because the other values of $t$ will repeat the same curve over and over again.

**Obs.** Sometimes it is convenient to eliminate $t$ between the given equations and use the resulting cartesian equation to trace the curve.

**Example 4.81.** *Trace the curve* $x = a\cos^3 t$, $y = a\sin^3 t$ *or* $x^{2/3} + y^{2/3} = a^{2/3}$.

(*C.S.V.T.U., 2015*)

**Solution.** (*i*) *Symmetry.* The curve is symmetrical about the $x$-axis.

[$\because$ On changing $t$ to $-t$, $x$ remains unchanged but $y$ changed to $-y$]

(*ii*) *Limits :* $\because$ $|x| \le a$ and $|y| \le a$.

$\therefore$ The curve lies entirely within the square bounded by the lines $x = \pm a$, $y = \pm a$.

(*iii*) *Points* : We have $\frac{dx}{dt} = -3a\cos^2 t \sin t$,

$$\frac{dy}{dt} = 3a\sin^2 t\cos t,\ \frac{dy}{dx} = -\tan t.$$

$\therefore$ $dy/dx = 0$ when $t = 0$ or $\pi$

and $dy/dx \to \infty$, when $t = \pi/2$.

The following table gives the corresponding values of $t$, $x$, $y$ and $dy/dx$.

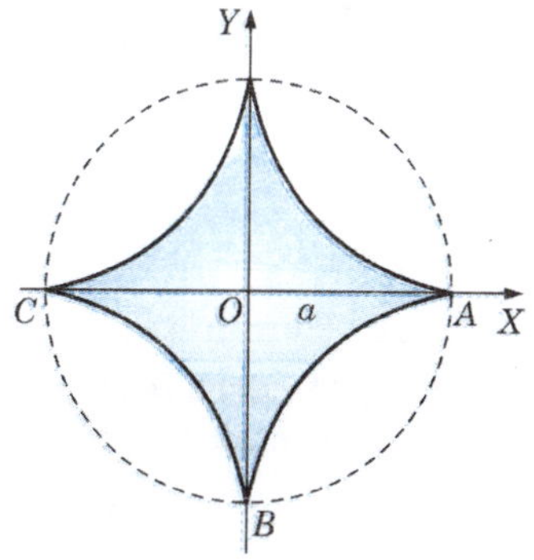


Fig. 4.34

| As $t$ increases | $x$ | $y$ | $dy/dx$ varies | Portion traced |
|---|---|---|---|---|
| from 0 to $\pi/2$ | +ve and decreases from $a$ to 0 | +ve and increases from 0 to $a$ | from 0 to $\infty$ | $A$ to $B$ |
| from $\pi/2$ to $\pi$ | + ve and increases numerically from 0 to $-a$ | +ve and decreases from $a$ to 0 | from $\infty$ to 0 | $B$ to $C$ |

As $t$ increases from $\pi$ to $2\pi$, we get the reflection of the curve $ABC$ in the $x$-axis. The values of $t > 2\pi$ give no new points.

Hence the shape of the curve is as shown in Fig. 4.34 and is known as **Astroid.**

**Example 4.82.** *Trace the curve $x = a(\theta + \sin\theta)$, $y = a(1 + \cos\theta)$.*

**Solution.** (*i*) *Symmetry*. The curve is symmetrical about the $y$-axis.

[$\because$ On changing $\theta$ to $-\theta$, $x$ changes to $-x$ and $y$ remains unchanged]

Thus we may consider the curve only for positive value of $x$, *i.e.*, for $\theta > 0$.

(*ii*) *Limits*. The greatest value of $y$ is $2a$ and the least value is zero.

Hence the curve lies entirely between the lines $y = 2a$ and $y = 0$.

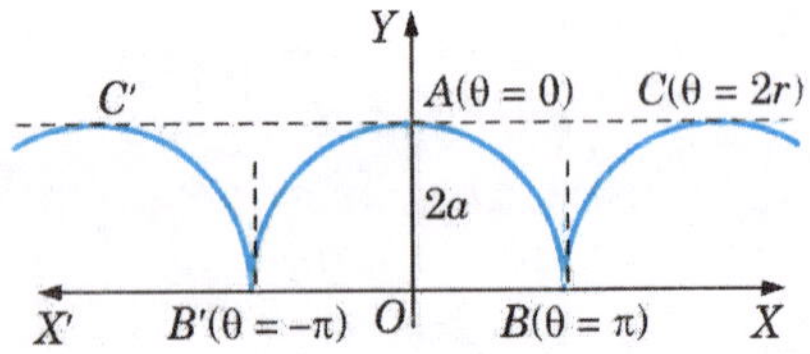


Fig. 4.35.

(*iii*) *Points*. We have

$$\frac{dx}{d\theta} = a(1 + \cos\theta),\ \frac{dy}{d\theta} = a\sin\theta \text{ and } \frac{dy}{dx} = -\tan\theta/2.$$

$\therefore$ $dy/dx = 0$ when $\theta = 0$ or $2\pi$ and $dy/dx \to \infty$ when $\theta = \pi$.

The following table gives the corresponding values of $\theta$, $x$, $y$ and $\dfrac{dy}{dx}$ :

| As $\theta$ increases | $x$ | $y$ | $dy/dx$ varies | Portion traced |
|---|---|---|---|---|
| from 0 to $\pi$ | increases from 0 to $a\pi$ | decreases from $2a$ to 0 | from 0 to $\infty$ | $A$ to $B$ |
| from $\pi$ to $2\pi$ | increases from $a\pi$ to $2a\pi$ | increases from 0 to $2a$ | from $\infty$ to 0 | $B$ to $C$ |

As $\theta$ decreases from 0 to $-2\pi$, we get the reflection of the curve $ABC$ in the $y$-axis.

The curve consists of congruent arches extending to infinity in both the directions of the $x$-axis in the intervals ... $(-3\pi, -\pi)$ $(-\pi, \pi)$ $(\pi, 3\pi)$, ...

Hence the shape of the curve is as shown in Fig. 4.35 and is known as **Cycloid.**

**Obs. 1. Cycloid** is the curve described by a point on the circumference of a circle which rolls without sliding on a fixed straight line. This fixed line ($x$-axis) is called the *base* and the farthest point ($A$) from it the *vertex* of the cycloid.

The complete cycloid consists of the arch $B'AB$ and its endless repetitions on both sides.

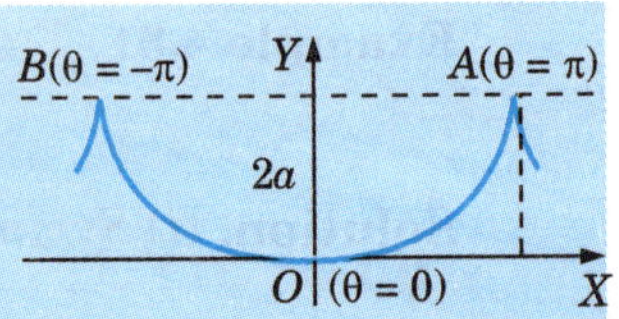


Fig. 4.36

**2. Inverted cycloid:** $x = a(\theta + \sin\theta)$, $y = a(1 - \cos\theta)$.

The complete inverted cycloid consists of the arch $BOA$ and an endless repetitions of the same on both sides. Here $AB$ is the base and $O$ the *vertex* of this cycloid. (Fig. 4.36).

## 4.17 (4) PROCEDURE FOR TRACING POLAR CURVES

1. **Symmetry.** *See if the curve is symmetrical about any line.*

(*i*) *A curve is symmetrical about the initial line OX, if only* $\cos\theta$ *(or* $\sec\theta$*) occur in its equation.* (*i.e.*, it remains unchanged when $\theta$ is changed to $-\theta$) *e.g.*, $r = a(1 + \cos\theta)$ is symmetrical about the initial line.

(*ii*) *A curve is symmetrical about the line through the pole $\perp$ to the initial line (i.e., OY), if only* $\sin\theta$ *(or* $\text{cosec}\,\theta$*) occur in its equation.* (*i.e.*, it remains unchanged when $\theta$ is changed to $\pi - \theta$) *e.g.*, $r = a\sin 3\theta$ is symmetrical about $OY$.

(iii) *A curve is symmetrical about the pole, if only even powers of r occur in the equation (i.e.,* it remains unchanged when $r$ is changed to $-r$) *e.g.,* $r^2 = a^2 \cos 2\theta$ is symmetrical about the pole.

**2. Limits.** See if $r$ and $\theta$ are confined between certain limits.

(i) *Determine the numerically greatest value of r, so as to notice whether the curve lies within a circle or not e.g.,* $r = a \sin 3\theta$ lies wholly within the circle $r = a$.

(ii) *Determine the region in which no portion of the curve lies by finding those values of θ for which r is imaginary e.g.,* $r^2 = a^2 \cos 2\theta$ does not lie between the lines $\theta = \pi/4$ and $\theta = 3\pi/4$.

**3. Asymptotes.** *If the curve possesses an infinite branch, find the asymptotes (p. 183).*

**4. Points.** (i) *Giving successive values to θ, find the corresponding values of r.*

(ii) *Determine the points where the tangent coincides with the radius vector or is perpendicular to it (i.e.,* the points where $\tan \phi = r\, d\theta/dr = 0$ or $\infty$).

**Example 4.83.** *Trace the curve $r = a \sin 3\theta$.* (U.P.T.U., 2011)

**Solution.** (i) *Symmetry.* The curve is symmetrical about the line through the pole ⊥ to the initial line.

(ii) *Limits.* The curve wholly lies within the curve $r = a$. ($\because$ $r$ is never $> a$)

(iii) *Asymptotes.* It has no asymptotes.

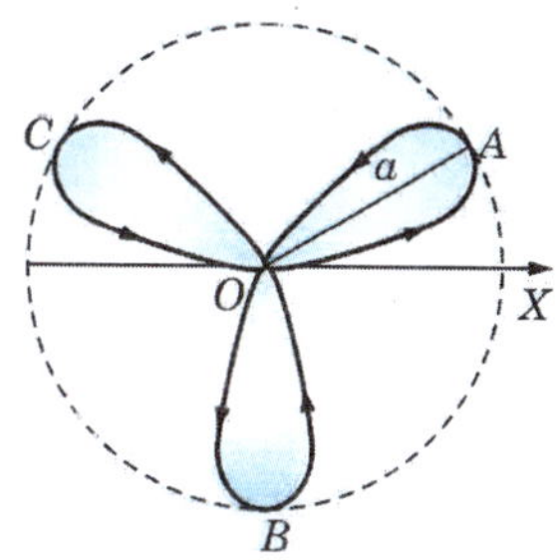

Fig. 4.37

(iv) *Points.* (a) $\tan \phi = r \dfrac{d\theta}{dr} = \dfrac{a \sin 3\theta}{3a \cos 3\theta} = \dfrac{1}{3} \tan 3\theta$

$$\phi = 0, \text{ when } \theta = 0, \pi/3, \ldots\ldots$$
$$\phi = \pi/2, \text{ when } \theta = \pi/6, \pi/2, \ldots\ldots$$

(b) The following table gives the variations of $r$, $\theta$ and $\phi$ :

| *As θ varies from* | *r varies from* | *ϕ varies from* | *Portion traced from* |
|---|---|---|---|
| 0 to $\pi/6$ | 0 to $a$ | 0 to $\pi/2$ | $O$ to $A$ |
| $\pi/6$ to $\pi/3$ | $a$ to 0 | $\pi/2$ to 0 | $A$ to $O$ |
| $\pi/3$ to $\pi/2$ | 0 to $-a$ | 0 to $\pi/2$ | $O$ to $B$ |

As $\theta$ increases from $\pi/2$ to $\pi$, portions of the curve from $B$ to $O$, $O$ to $C$ and $C$ to $O$ are traced by symmetry about the line $\theta = \pi/2$.

Hence the curve consists of three loops as shown in Fig. 4.35 and is known as *three-leaved rose.*

**Obs.** *The curves of the form $r = a \sin n\theta$ or $r = a \cos n\theta$ are called **Roses** having*

(i) $n$ leaves (loops) when $n$ is odd,

(ii) $2n$ leaves (loops) when $n$ is even.

**Example 4.84.** *Trace the curve $r = a \sin 2\theta$.(Four Leaved Rose)*

**Solution.** (i) *Symmetry.* The curve is symmetrical about the line through the pole, ⊥ to the initial line.

(ii) *Limits* : The curve lies wholly within the circle $r = a$ ($\because$ $r$ is never $> a$)

(iii) *Points* : (a)

| *As θ increases from* | *r varies from* | *Loop* |
|---|---|---|
| 0 to $\dfrac{\pi}{4}$ | 0 to $a$ | |
| | | no : 1, |
| $\dfrac{\pi}{4}$ to $\dfrac{\pi}{2}$ | $a$ to 0 | |
| $\dfrac{\pi}{2}$ to $\dfrac{3\pi}{4}$ | 0 to $-a$ | |
| | | no : 2, |
| $\dfrac{3\pi}{4}$ to $\dfrac{\pi}{2}$ | $-a$ to 0 | |

etc. etc.

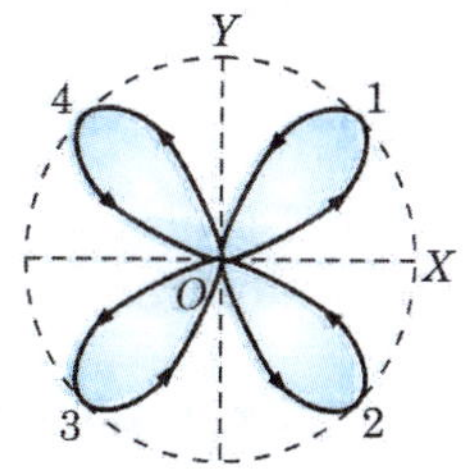

Fig. 4.38.

(b) $\tan\phi = r\dfrac{d\theta}{dr} = \dfrac{1}{2}\tan 2\theta$ ;

$\therefore$ $\phi = 0$, when $\theta = 0, \dfrac{\pi}{2}, \pi, 3\dfrac{\pi}{2}, 2\pi$ ...

$\phi = \dfrac{\pi}{2}$, when $\theta = \dfrac{\pi}{4}, \dfrac{3\pi}{4}, \dfrac{5\pi}{4}, \dfrac{7\pi}{4}$ ...

Hence, the shape of the curve is as shown in Fig. 4.36.

**Example 4.85.** *Trace the curve $r^2 = a^2 \cos 2\theta$.* *(Delhi, 2012 ; U.P.T.U., 2012)*

**Solution.** (*i*) *Symmetry*. The curve is symmetrical about the pole.

(*ii*) *Limits* : (*a*) The curve lies wholly within the circle $r = a$.

(*b*) No portion of the curve lies between the lines $\theta = \pi/4$ and $\theta = 3\pi/4$.

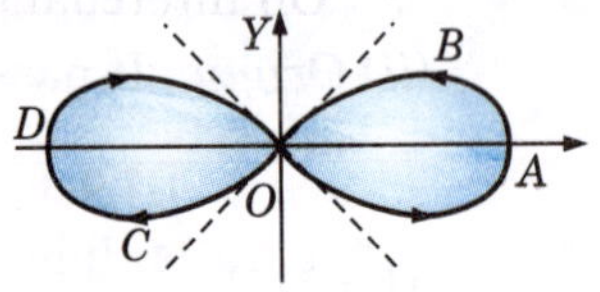

Fig. 4.39

(*iii*) *Points* : (*a*) $\tan\phi = r\dfrac{d\theta}{dr} = -\cot 2\theta = \tan\left(\dfrac{\pi}{2} + 2\theta\right)$

*i.e.*, $\phi = \dfrac{\pi}{2} + 2\theta$ $\therefore$ $\phi = 0$, when $\theta = -\pi/4$ ; $\phi = \pi/2$, when $\theta = 0$.

Thus, the tangent at $O$ is $\theta = -\pi/4$ and the tangent at $A$ is $\perp$ to the initial line.

(*b*) The variations of $r$ and $\theta$ are given below :

| *As θ varies from* | *r varies from* | *Portion traced* |
|---|---|---|
| 0 to $\pi/4$ | $a$ to 0 | *ABO* |
| $3\pi/4$ to $\pi$ | 0 to $a$ | *OCD* |

As $\theta$ increase from $\pi$ to $2\pi$, we get the reflection of the arc *ABOCD* in the initial line. Hence the shape of the curve is as shown in Fig. 4.37. This curve is known as *Lemniscate of Bernoulli.*

**Example 4.86.** *Trace the curve $r = a + b\cos\theta$ (Limacon)*

**Solution.** (*i*) *Symmetry*. It is symmetrical about the initial line.

(*ii*) *Limits* : The curve wholly lies within the circle $r = a + b$

($\because$ $r$ is never $> a + b$)

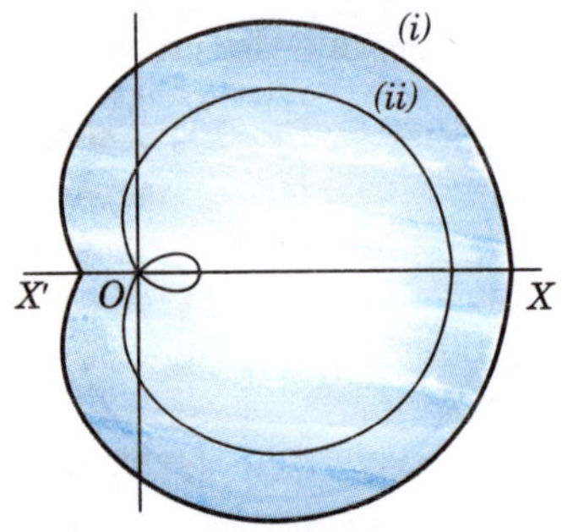

Fig. 4.40

(*iii*) *Points* : ($\alpha$) when $a > b$.

As $\theta$ increases from 0 to $\pi/2$ ; $r$ decreases from $a + b$ to $a$

As $\theta$ increases from $\pi/2$ to $\pi$ ; $r$ decreases from $a$ to $a - b$

The shape of the curve is as shown in Fig. 4.38 (*i*).

($\beta$) when $a < b$.

As $\theta$ increases from 0 to $\pi/2$ ; $r$ decreases from $a + b$ to $a$

As $\theta$ increases from $\pi/2$ to $\alpha$ ; $r$ decreases from $a$ to 0

As $\theta$ increases from $\alpha$ to $\pi$ ; $r$ decreases from 0 to $a - b$

when $\alpha = \cos^{-1}\left(-\dfrac{a}{b}\right)$

In this case, the curve consists of two parts, one of which forms a loop within the other and the shape is as shown in Fig. 4.38 (*ii*).

**Example 4.87.** *Trace the curve $r\theta = a$. (Spiral)* *(Kurukshetra, 2013)*

**Solution.** (*i*) *Symmetry*. There is no symmetry.

(*ii*) *Limits* : There are no limits to the values of $r$.

The curve does not pass through the pole for $r$ does not become zero for any real value of $\theta$.

(iii) *Asymptotes* : $\frac{1}{r} = \frac{\theta}{a} = f(\theta)$

$f(\theta) = 0$ for $\theta = 0$ ; $f'(\theta) = 1/a, f'(0) = 1/a$.

$\therefore$ Asymptote is $r \sin(\theta - 0) = 1/f'(0)$

*i.e.*, $y = r \sin\theta = a$ is an asymptote.

(iv) *Points* : As θ increases from 0 to ∞, r to positive and decreases from ∞ to 0. Hence the space of the curve is as shown in Fig. 4.41.

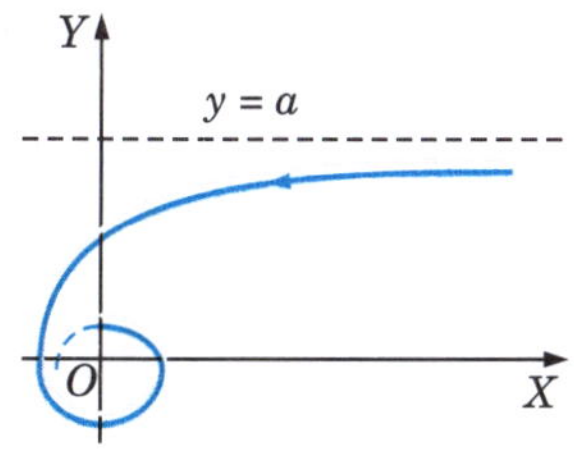

Fig. 4.41

**Example 4.88.** *Trace the curve $x^5 + y^5 = 5ax^2y^2$.*

**Solution.** (i) *Symmetry*. The curve is symmetrical about the line $y = x$.

$\because$ On interchanging x and y, it remains unchanged.]

(ii) *Origin* : It passes through the origin and the tangents at the origin are given by

$$x^2 y^2 = 0, \text{ i.e., } x = 0, x = 0 ; y = 0, y = 0.$$

Hence the curve has both *node* and the *cusp* at the origin.

(iii) *Asymptotes* : (a) It has no asymptotes parallel to the axes.

(b) Putting $x = 1, y = m$ in the fifth degree terms, we get

$$\phi_5(m) = 1 + m^5. \quad \therefore \quad \phi_5(m) = 0 \text{ gives } m = -1.$$

$$\therefore \quad c = -\frac{\phi_4(m)}{\phi_5'(m)} = -\frac{-5am^2}{5m^4} = a \text{ for } m = -1.$$

Hence, $y = -x + a$ or $x + y = a$ is an asymptote.

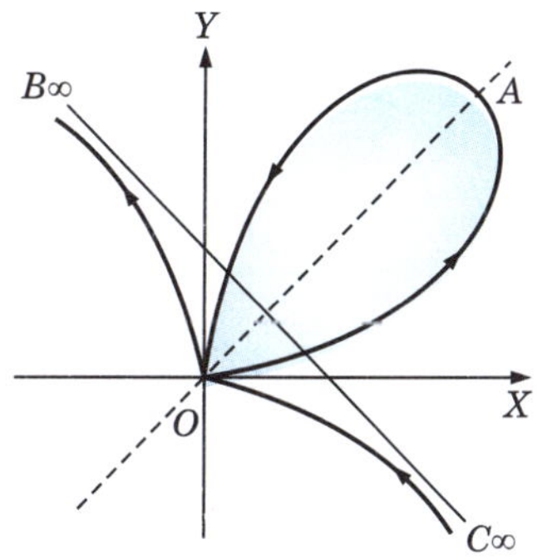

Fig. 4.42

(iv) *Points* : Since it is not convenient to express y as a function of x or *vice versa*, hence we change the equation into polar coordinates by putting, $x = r\cos\theta$ and $y = r\sin\theta$. The equation of the curve becomes :

$$r = \frac{5a\sin^2\theta\cos^2\theta}{\cos^5\theta + \sin^5\theta} = \frac{5a}{4} \cdot \frac{\sin^5 2\theta}{\cos^5\theta + \sin^5\theta}$$

| *As θ increases from* | *r* | *Portion traced from* |
|---|---|---|
| 0 to π/4 | is +ve and increases from 0 to $\frac{5\sqrt{2}}{2}a$ | 0 to A |
| π/4 to π/2 | is +ve and decreases from $\frac{5\sqrt{2}}{2}a$ to 0 | A to 0 |
| π/2 to 3π/4 | is +ve and increases from 0 to ∞ | 0 to $B_\infty$ |
| 3π/4 to π | is –ve and decreases from ∞ to 0 | $C_\infty$ to 0 |

As θ increases from π to 2π, the curve will retraced.

Hence the shape of the curve is as shown in Fig. 4.42.

## PROBLEMS 4.15

Trace the following curves :

1. $y^2(a + x) = x^2(a - x)$. (*P.T.U., 2013 ; C.S.V.T.U., 2011*)
2. $y^2(a^2 + x^2) = x^2(a^2 - x^2)$ (*D.T.U., 2013 ; V.T.U., 2010*)
3. $ay^2 = x^2(\alpha - x)$ (*C.S.V.T.U., 2011*)
4. $x^2y^2 = a^2(y^2 - x^2)$ (*Delhi., 2015*)
5. $x = a\cos^3\theta, y = b\sin^3\theta$
6. $x = a(\theta - \sin\theta), y = a(1 - \cos\theta)$ $(0 < \theta < 2\pi)$
7. $x = (a\cos t + \log\tan t/2), y = a\sin t$.
8. $r = a\cos 2\theta$
9. $r = 2 + 3\cos\theta$
10. $r^2\cos 2\theta = a^2$. (*P.T.U., 2010*)
11. $r = a(1 + \cos\theta)$ (*V.T.U., 2016 ; P.T.U., 2011*)

12. $a^2 x^2 = y^3 (2a - y)$

13. $r^2 \theta = a^2$

14. $(x^2 - a^2)(y^2 - b^2) = a^2 b^2$

15. $y^2(x + 3a) = x(x - a)(x - 2a)$ (*D.T.U. 2019*)

16. $r = \dfrac{a \sin^2 \theta}{\cos \theta}$ (*V.T.U. 2016*)

17. $r^2 = \dfrac{a\theta^2}{1 + \theta^2}$ (*Kurukshetra, 2018*)

18. $r = \dfrac{a \sin^2 \theta}{\cos \theta}$

[**Hint.** Changing to Cartesian form $x^2 - y^2 = a^2$. This is a rectangular hyperbola with asymptotes $x + y = 0$ and $x - y = 0$]

## 4.18 OBJECTIVE TYPES OF QUESTIONS

### PROBLEMS 4.16

*Select the correct answer or fill up the blanks in each of the following questions :*

1. The radius of curvature of the catenary $y = c \cosh x/c$ at the point where it crosses the $y$-axis is ......
2. The envelope of the family of straight lines $y = mx + am^2$, ($m$ being the parameter) is ......
3. The curvature of the circle $x^2 + y^2 = 25$ at the point (3, 4) is ......
4. The value of $\underset{x \to \pi/2}{\text{Lt}} \dfrac{\log \sin x}{(\pi/2 - x)^2}$ is

   (*a*) zero (*b*) 1/2 (*c*) –1/2 (*d*) –2. (*V.T.U., 2010*)

5. Taylor's expansion of the function $f(x) = \dfrac{1}{1 + x^2}$ is

   (*a*) $\sum_{n=0}^{\infty} (-1)^n x^{2n}$ for $-1 < x < 1$ (*b*) $\sum_{n=0}^{\infty} x^{2n}$ for $-1 < x < 1$

   (*c*) $\sum_{n=0}^{\infty} (-1)^n x^{2n}$ for any real $x$ (*d*) $\sum_{n=0}^{\infty} (-1)^n x^n$ for $-1 < x \le 1$.

6. A triangle of maximum area inscribed in a circle of radius $r$

   (*a*) is a right angled triangle with hypotenuse measuring $2r$

   (*b*) is an equilateral triangle

   (*c*) is an isosceles triangle of height $r$

   (*d*) does *not* exist.

7. The extreme value of $(x)^{1/x}$ is

   (*a*) $e$ (*b*) $(1/e)^e$ (*c*) $(e)^{1/e}$ (*d*) 1.

8. The percentage error in computing the area of an ellipse when an error of 1 per cent is made in measuring the major and minor axes is

   (*a*) 0.2% (*b*) 2% (*c*) 0.02%.

9. The length of subtangent of the rectangular hyperbola $x^2 - y^2 = a^2$ at the point $(a, \sqrt{2}a)$ is

(a) $\sqrt{2}\,a$ (b) $2a$ (c) $\frac{1}{2a}$ (d) $\frac{a^{3/2}}{\sqrt{2}}$.

10. The length of subnormal to the curve $y = x^2$ at (2, 8) is

(a) 2/3 (b) 32 (c) 96 (d) 64.

11. If the normal to the curve $y^2 = 5x - 1$ at the point $(1, -2)$ is of the form $ax - 5y + b = 0$, then $a$ and $b$ are

(a) 4, 14 (b) 4, –14 (c) –4, 14 (d) –4, –14.

12. The radius of curvature of the curve $y = e^x$ at the point where it crosses the $y$-axis is

(a) 2 (b) $\sqrt{2}$ (c) $2\sqrt{2}$ (d) $\frac{1}{2}\sqrt{2}$.

13. The equation of the asymptotes of $x^3 + y^3 = 3axy$, is

(a) $x + y - a = 0$ (b) $x - y + a = 0$ (c) $x + y + a = 0$ (d) $x - y - a = 0$.

14. If $\phi$ be the angle between the tangent and radius vector at any point on the curve $r = f(\theta)$, then $\sin\phi$ equals to

(a) $\frac{dr}{ds}$ (b) $r\frac{d\theta}{ds}$ (c) $r\frac{d\theta}{dr}$.

15. Envelope of the family of lines $x = my + 1/m$ is ...

16. The chord of curvature parallel to $y$-axis for the curve $y = a \log \sec x/a$ is ....

17. $\sinh x = \ldots x + \ldots x^3 + \ldots x^5 + \ldots$

18. The $n$th derivative of $(\cos 2x \cos 6x)$ = ......... *(C.S.V.T.U., 2011 S)*

19. If $x^3 + y^3 - 3axy = 0$, then $d^2y/dx^2$ at $(3a/2, 3a/2)$ = .........

20. When the tangent at a point on a curve is parallel to $x$-axis, then the curvature at that point is same as the second derivative at that point. *(True or False)*

21. If $x = at^2$, $y = 2at$, $t$ being the parameter, then $xy\, d^2y/dx^2$ = .........

22. The radius of curvature for the curve $x^2 + y^2 - 4x + 2y - 8 = 0$ at any point is ......... *(Anna, 2013)*

23. If $(a, b)$ are the coordinates of the centre of curvature whose curvature is $k$, then the equation of the circle of curvature is

24. Evolute is defined as the ......... of the normals for a given curve.

25. Envelope of the family of lines $\frac{x}{t} + yt = 2c$ (where $t$ is the parameter) is ..........

26. The angle between the radius vector and tangent for the curve $r = ae^{\theta \cot \alpha}$ is .........

27. The subnormal of the parabola $y^2 = 4ax$ is .........

28. The fourth derivative of $(e^{-x} x^3)$ is .........

29. If $y^2 = P(x)$, a polynomial of degree 3, then $\frac{2d}{dx}\left(y^3 \frac{d^2y}{dx^2}\right)$ equals .........

(a) $P'''(x) + P'(x)$ (b) $P''(x) + P'''(x)$ (c) $P(x)\,P'''(x)$.

30. The envelope of the family of straight line $\frac{x}{a}\cos\theta + \frac{y}{b}\sin\theta = 1$.

31. Curvature of a straight line is

(A) $\infty$ (B) zero (C) Both (A) and (B) (D) None of these.

32. The value of '$c$' of the Cauchy's Mean value theorem for $f(x) = e^x$ and $g(x) = e^{-x}$ in [2, 3] is .........

33. If the equation of a curve remains unchanged when $x$ and $y$ are interchanged, then the curve is symmetrical about ......... .

34. For the curve $y^2 (1 + x) = x^2 (1 - x)$, the origin is a ......... (node/cusp/conjugate point).

35. The number of loops of $r = a \sin 2\theta$ are ......... and these of $r = a \cos 3\theta$ are .......... .

36. Tangents at the origin for the curve $y^2 (x^2 + y^2) + a^2 (x^2 - y^2) = 0$ are .......... .

37. The asymptote to the curve $y^2 (4 - x) = x^3$ is ......... .

38. The points of inflexion of the curve $y^2 = (x - a)^2 (x - b)$ lie on the line $3x + a =$ ........ .

39. The curve $r = a/(1 + \cos \theta)$ intersects orthogonally with the curve

(A) $r = b/(1 - \cos \theta)$ (B) $r = b/(1 + \sin \theta)$ (C) $r = b/(1 + \sin^2 \theta)$ (D) $r = b/(1 + \cos^2 \theta)$. (V.T.U., 2010)

40. The region where the curve $r = a \sin \theta$ does not lie is ......... .

41. If $f(x)$ is continuous in the closed interval $[a, b]$, differentiable in $(a, b)$ and $f(a) = f(b)$, then there exists at least one value $c$ of $x$ in $(a, b)$ such that $f'(c)$ is equal to

(A) 1 (B) –1 (C) 2 (D) 0.

42. If two curves intersect orthogonally in cartesian form, then the angle between the same two curves in polar form is

(A) $\pi/4$ (B) Zero (C) 1 radian (D) None of these.

43. If the angle between the radius vector and the tangent is constant, then the curve is,

(A) $r = a \cos \theta$ (B) $r^2 = a^2 \cos^2 \theta$ (C) $r = ae^{b\theta}$.

44. The assumptotes parallel to the axes of the curve $x^2y^3 + x^3y^2 = x^3 + y^3$ are .......... . (Delhi, 2012)

CHAPTER 5

# Partial Differentiation and Its Applications

**1.** Functions of two or more variables. **2.** Partial derivatives. **3.** Which variable is to be treated as constant. **4.** Homogeneous functions—Euler's theorem. **5.** Total derivative—Differentiation of implicit functions. **6.** Change of variables. **7.** Jacobians. **8.** Geometrical interpretation—Tangent plane and normal to a surface. **9.** Taylor's theorem for functions of two variables. **10.** Errors and approximations; Total differential. **11.** Maxima and minima of functions of two variables. **12.** Lagrange's method of undetermined multipliers. **13.** Differentiation under the integral sign—Leibnitz Rule. **14.** Objective Type of Questions.

## 5.1 (1) FUNCTIONS OF TWO OR MORE VARIABLES

We often come across quantities which depend on two or more variables. For example, the area of a rectangle of length $x$ and breadth $y$ is given by $A = xy$. For a given pair of values of $x$ and $y$, $A$ has a definite value. Similarly, the volume of a parallelopiped $(= xyh)$ depends on the three variables $x$(= length), $y$(= breadth) and $h$(=height).

**Def.** *A symbol z which has a definite value for every pair of values of x and y is called a function of two independent variables x and y and we write $z = f(x, y)$ or $\phi(x, y)$.*

We may interpret $(x, y)$ as the coordinates of a point in the $XY$-plane and $z$ as the height of the surface $z = f(x, y)$. We have come across several examples of such surfaces in Chapter 4.

The set $R$ of points $(x, y)$ such that any two points $P_1$ and $P_2$ of $R$ can be so joined that any arc $P_1P_2$ wholly lies in $R$, is called as *region* in the $XY$-plane. A region is said to be a *closed region* if it includes all the points of its boundary, otherwise it is called an *open region*.

A set of points lying within a circle having centre at $(a, b)$ and radius $\delta > 0$, is said to be *neighbourhood* of $(a, b)$ in the circular region $R : (x - a)^2 + (y - b)^2 < \delta^2$.

When $z$ is a function of three or more variables $x, y, t, \ldots$, we represent the relation by writing $z = f(x, y, t, \ldots)$. For such functions, no geometrical representation is possible. However, the concepts of a region and neighbourhood can easily be extended to functions of three or more variables.

**(2) Limits.** *The function f(x, y) is said to tend to the limit l as $x \to a$ and $y \to b$ if and only if the limit l is independent of the path followed by the point (x, y) as $x \to a$ and $y \to b$ and we write*

$$\underset{\substack{x \to a \\ y \to b}}{Lt}\ f(x, y) = l$$

In terms of a circular neighbourhood, we have the following *definition of the limit :*

*The function f(x, y) defined in a region R, is said to tend to the limit l as $x \to a$ and $y \to b$ if and only if corresponding to a positive number ε, there exists another positive number δ such that $|f(x, y) - l| < \varepsilon$ for $0 < (x - a)^2 + (y - b)^2 < \delta^2$ for every point (x, y) in R.*

**(3) Continuity.** *A function f(x, y) is said to be continuous at the point (a, b) if*

$$\underset{\substack{x \to a \\ y \to b}}{Lt}\ f(x, y) \text{ exists and } = f(a, b)$$

If a function is continuous at all points of a region, then it is said to be *continuous in that region.* A function which is not continuous at a point is said to be *discontinuous* at that point.

**Obs.** *Usually, the limit is the same irrespective of the path along which the point (x, y) approaches (a, b) and*

$$\underset{x\to a}{Lt}\left\{\underset{y\to b}{Lt} f(x, y)\right\} = \underset{y\to b}{Lt}\left\{\underset{x\to a}{Lt} f(x, y)\right\}$$

*But it is not always so, as the following examples show :*

$$\text{Lt}\left(\frac{x-y}{x+y}\right) \text{ as } (x, y) \to (0, 0) \text{ along the line } y = mx$$

$$= \underset{x\to 0}{\text{Lt}} \frac{x-mx}{x+mx} = \frac{1-m}{1+m} \text{ which is different for lines with different slopes.}$$

Also $\underset{x\to 0}{\text{Lt}}\left[\underset{y\to 0}{\text{Lt}}\left(\frac{x-y}{x+y}\right)\right] = \underset{x\to 0}{\text{Lt}}\left(\frac{x}{x}\right) = 1$, whereas $\underset{y\to 0}{\text{Lt}}\left[\underset{x\to 0}{\text{Lt}}\left(\frac{x-y}{x+y}\right)\right] = \underset{y\to 0}{\text{Lt}}\left(\frac{-y}{y}\right) = 1.$

$\therefore$ As $(x, y)$ is made to approach $(0, 0)$ along different paths, $f(x, y)$ approaches different limits. Hence the two repeated limits are not equal and $f(x, y)$ is discontinuous at the origin.

Also the function is not defined at $(0, 0)$ since $f(x, y) = 0/0$ for $x = 0, y = 0$.

**(4)** *As in the case of functions of one variable, the following results hold :*

I. $\underset{\substack{x\to a\\ y\to b}}{Lt} f(x, y) = l$ and $\underset{\substack{x\to a\\ y\to b}}{Lt} g(x, y) = m,$

*then* (*i*) $\underset{\substack{x\to a\\ y\to b}}{Lt} [f(x, y) \pm g(x, y)] = l \pm m \quad (m \neq 0)$ (*ii*) $\underset{\substack{x\to a\\ y\to b}}{Lt} [f(x,y) \, . \, g(x, y)] = l \, . \, m \quad (m \neq 0)$

(*iii*) $\underset{\substack{x\to a\\ y\to b}}{Lt} [f(x,y)/g(x, y)] = l/m \qquad (m \neq 0)$

II. *If $f(x, y), g(x, y)$ are continuous at $(a, b)$ then these are the functions*

$f(x, y) \pm g(x, y), f(x, y) \, . \, g(x, y)$ and $f(x, y)/g(x, y)$

provided $g(x, y) \neq 0$ in the last case.

## PROBLEMS 5.1

Evaluate the following limits :

1. $\underset{\substack{x\to 1\\ y\to 2}}{\text{Lt}} \frac{2x^2y}{x^2+y^2+1}$ 2. $\underset{\substack{x\to 0\\ y\to 0}}{\text{Lt}} \frac{xy}{x^2+y^2}$ 3. $\underset{\substack{x\to \infty\\ y\to 2}}{\text{Lt}} \frac{xy+1}{x^2+2y^2}$ 4. $\underset{\substack{x\to 1\\ y\to 1}}{\text{Lt}} \frac{x(y-1)}{y(x-1)}$

5. If $f(x, y) = \dfrac{x-y}{2x+y}$, show that $\underset{x\to 0}{\text{Lt}}\left[\underset{y\to 0}{\text{Lt}} f(x, y)\right] \neq \underset{y\to 0}{\text{Lt}}\left[\underset{x\to 0}{\text{Lt}} f(x, y)\right]$

Also show that the function is discontinuous at the origin.

6. Show that the function $f(x, y) = x^2 + 2y, \quad (x, y) \neq (1, 2)$

$3(x, y) = (1, 2) \qquad = 0$

is discontinuous at (1, 2).

7. Investigate the continuity of the function

$$f(x, y) = xy/(x^2 + y^2), \qquad (x, y) \neq (0, 0)$$
$$= 0 \qquad (x, y) = (0, 0)$$

at the origin.

**Note.** In whatever follows, all the functions considered are continuous and their partial derivatives (as defined below) exist.

## 5.2 PARTIAL DERIVATIVES

Let $z = f(x, y)$ be a function of two variables $x$ and $y$.

If we keep $y$ as constant and vary $x$ alone, then $z$ is a function of $x$ only. The derivative of $z$ with respect to $x$, treating $y$ as constant, is called the *partial derivative of z with respect to x* and is denoted by one of the symbols

$$\frac{\partial z}{\partial x}, \frac{\partial f}{\partial x}, f_x(x, y), D_x f. \qquad \text{Thus } \frac{\partial z}{\partial x} = \underset{\delta x \to 0}{\text{Lt}} \frac{f(x + \delta x, y) - f(x, y)}{\delta x}$$

Similarly, the derivative of $z$ with respect to $y$, keeping $x$ as constant, is called the *partial derivative of $z$ with respect to $y$* and is denoted by one of the symbols.

$$\frac{\partial z}{\partial y}, \frac{\partial f}{\partial y}, f_y(x, y), D_y f. \qquad \text{Thus } \frac{\partial z}{\partial y} = \underset{\delta y \to 0}{\text{Lt}} \frac{f(x, y + \delta y) - f(x, y)}{\delta y}$$

Similarly, if $z$ is a function of three or more variables $x_1, x_2, x_3, \ldots$ *the partial derivative of $z$ with respect to $x_1$, is obtained by differentiating $z$ with respect to $x_1$, keeping all other variables constant* and is written as $\partial z/\partial x_1$.

In general $f_x$ and $f_y$ are also functions of $x$ and $y$ and so these can be differentiated further partially with respect to $x$ and $y$.

Thus,
$$\frac{\partial}{\partial x}\left(\frac{\partial z}{\partial x}\right) = \frac{\partial^2 z}{\partial x^2} \text{ or } \frac{\partial^2 z}{\partial x^2} \text{ or } f_{xx}, \frac{\partial}{\partial x}\left(\frac{\partial z}{\partial y}\right) = \frac{\partial^2 z}{\partial x \partial y} \text{ or } \frac{\partial^2 f}{\partial x \partial y} \text{ or } f_{yx}{}^*$$

$$\frac{\partial}{\partial y}\left(\frac{\partial z}{\partial x}\right) = \frac{\partial^2 z}{\partial y \partial x} \text{ or } \frac{\partial^2 f}{\partial y \partial x} \text{ or } f_{xy}, \text{ and } \frac{\partial}{\partial y}\left(\frac{\partial z}{\partial y}\right) = \frac{\partial^2 z}{\partial y^2} \text{ or } \frac{\partial^2 f}{\partial y^2} \text{ or } f_{yy}.$$

It can easily be verified that, in all ordinary cases,

$$\frac{\partial^2 z}{\partial x \partial y} = \frac{\partial^2 z}{\partial y \partial x}.$$

*Sometimes we use the following notation*

$$\frac{\partial z}{\partial x} = p, \ \frac{\partial z}{\partial y} = q, \ \frac{\partial^2 z}{\partial x^2} = r, \ \frac{\partial^2 z}{\partial x \partial y} = s, \ \frac{\partial^2 z}{\partial y^2} = t.$$

**Obs.** A function of two variables has 2 first order derivatives, $(2^2)$ second order derivatives, $(2^3)$ third order derivatives and so on.

**Example 5.1.** *Find the first and second partial derivatives of $z = x^3 + y^3 - 3axy$.*

**Solution.** We have $z = x^3 + y^3 - 3axy$.

$$\therefore \quad \frac{\partial z}{\partial x} = 3x^2 + 0 - 3ay(1) = 3x^2 - 3ay, \text{ and } \frac{\partial z}{\partial y} = 0 + 3y^2 - 3ax(1) = 3y^2 - 3ax$$

Also
$$\frac{\partial^2 z}{\partial x^2} = \frac{\partial}{\partial x}(3x^2 - 3ay) = 6x, \ \frac{\partial^2 z}{\partial y \partial x} = \frac{\partial}{\partial y}(3x^2 - 3ay) = -3a$$

$$\frac{\partial^2 z}{\partial y^2} = \frac{\partial}{\partial y}(3y^2 - 3ax) = 6y, \text{ and } \frac{\partial^2 z}{\partial x \partial y} = \frac{\partial}{\partial x}(3y^2 - 3ax) = -3a.$$

We observe that, $\dfrac{\partial^2 z}{\partial y \partial x} = \dfrac{\partial^2 z}{\partial x \partial y}$.

**Example 5.2.** *If $u = x^2 \tan^{-1} \dfrac{y}{x} - y^2 \tan^{-1} \dfrac{x}{y}$,*

*show that* $\dfrac{\partial^2 u}{\partial x \partial y} = \dfrac{x^2 - y^2}{x^2 + y^2}$. (U.T.U., 2012)

*and* $\dfrac{\partial^2 u}{\partial x \partial y} = \dfrac{\partial^2 u}{\partial y \partial x}$ (W.B.T.U., 2012)

**Solution.** We have
$$\frac{\partial u}{\partial y} = x^2 \cdot \frac{1}{1 + (y/x)^2} \cdot \frac{1}{x} - \left\{2y \cdot \tan^{-1} \frac{x}{y} + y^2 \cdot \frac{1}{1 + (x/y)^2} \cdot \left(-\frac{x}{y^2}\right)\right\}$$

$$= \frac{x^3}{x^2 + y^2} - 2y \tan^{-1} \frac{x}{y} + \frac{xy^2}{x^2 + y^2} = x - 2y \tan^{-1} \frac{x}{y}.$$

---

*It is important to note that in the subscript notation the subscripts are written in the same order in which we differentiate whereas in the '$\partial$' notation the order is opposite.

$\therefore \quad \dfrac{\partial^2 u}{\partial x \partial y} = \dfrac{\partial}{\partial x}\left\{x - 2y \tan^{-1}\dfrac{x}{y}\right\} = 1 - 2y \cdot \dfrac{1}{1+(x/y)^2} \cdot \dfrac{1}{y} = 1 - \dfrac{2y^2}{x^2+y^2} = \dfrac{x^2-y^2}{x^2+y^2}.$

Similarly, $\dfrac{\partial u}{\partial x} = 2x \tan^{-1} y/x - y$

and $\dfrac{\partial^2 u}{\partial y \partial x} = \dfrac{\partial}{\partial y}\left\{2x \tan^{-1}\dfrac{y}{x} - y\right\} = \dfrac{x^2-y^2}{x^2+y^2}$. Hence the result.

**Example 5.3.** *If $z = f(x + ct) + \phi(x - ct)$, prove that*

$$\frac{\partial^2 z}{\partial t^2} = c^2 \frac{\partial^2 z}{\partial x^2}$$

**Solution.** We have $\dfrac{\partial z}{\partial x} = f'(x + ct) \cdot \dfrac{\partial}{\partial x}(x + ct) + \phi'(x - ct)\dfrac{\partial}{\partial x}(x - ct) = f'(x + ct) + \phi'(x - ct)$

and $\dfrac{\partial^2 z}{\partial x^2} = f''(x + ct) + \phi''(x - ct)$ ...(*i*)

Again $\dfrac{\partial z}{\partial t} = f'(x + ct)\dfrac{\partial}{\partial t}(x + ct) + \phi'(x - ct)\dfrac{\partial}{\partial t}(x - ct) = cf'(x - ct) - c\phi'(x - ct)$

and $\dfrac{\partial^2 z}{\partial t^2} = c^2 f''(x + ct) + c^2\phi''(x - ct) = c^2[f''(x + ct) + \phi''(x - ct)]$ ...(*ii*)

From Eqns. (*i*) and (*ii*), it follows that $\dfrac{\partial^2 z}{\partial t^2} = c^2\dfrac{\partial^2 z}{\partial x^2}$.

**Obs.** This is an important partial differential equation, known as *wave equation* (§ 18.4).

**Example 5.4.** *If $\theta = t^n e^{-r^2/4t}$, what value of n will make $\dfrac{1}{r^2}\dfrac{\partial}{\partial r}\left(r^2\dfrac{\partial\theta}{\partial r}\right) = \dfrac{\partial\theta}{\partial t}$?*

*(C.S.V.T.U., 2011)*

**Solution.** We have $\dfrac{\partial\theta}{\partial r} = t^n \cdot e^{-r^2/4t} \cdot \left(\dfrac{-2r}{4t}\right) = -\dfrac{r}{2}t^{n-1}e^{-r^2/4t}$

$\therefore \quad r^2\dfrac{\partial\theta}{\partial r} = -\dfrac{r^3}{2}t^{n-1} \cdot e^{-r^2/4t}$

and $\dfrac{\partial}{\partial r}\left(r^2\dfrac{\partial\theta}{\partial r}\right) = -\dfrac{3r^2}{2}t^{n-1}e^{-r^2/4t} - \dfrac{r^3}{2}t^{n-1} \cdot e^{-r^2/4t}\left(-\dfrac{2r}{4t}\right)$

$\therefore \quad \dfrac{1}{r^2}\dfrac{\partial}{\partial r}\left(r^2\dfrac{\partial\theta}{\partial r}\right) = \left(-\dfrac{3}{2}t^{n-1} + \dfrac{r^2}{4}t^{n-2}\right)e^{-r^2/4t}$

Also $\dfrac{\partial\theta}{\partial t} = nt^{n-1} \cdot e^{-r^2/4t} + t^n \cdot e^{-r^2/4t} \cdot \dfrac{r^2}{4t^2} = \left(nt^{n-1} + \dfrac{1}{4}r^2t^{n-2}\right)e^{-r^2/4t}$

Since $\dfrac{1}{r^2}\dfrac{\partial}{\partial r}\left(r^2\dfrac{\partial\theta}{\partial r}\right) = \dfrac{\partial\theta}{\partial t}$,

$\therefore \quad \left(-\dfrac{3}{2}t^{n-1} + \dfrac{1}{4}r^2t^{n-2}\right)e^{-r^2/4t} = \left(nt^{n-1} + \dfrac{1}{4}r^2t^{n-2}\right)e^{-r^2/4t}$ or $\left(n + \dfrac{3}{2}\right)t^{n-1}e^{-r^2/4t} = 0.$

Hence $n = -3/2$.

**Example 5.5.** *If $v = (x^2 + y^2 + z^2)^{-1/2}$, prove that*

$\dfrac{\partial^2 v}{\partial x^2} + \dfrac{\partial^2 v}{\partial y^2} + \dfrac{\partial^2 v}{\partial z^2} = 0$. *(Laplace equation)** *(Andhra, 2022; U.P.T.U., 2011)*

---

*See footnote p. 219.

**Solution.** We have $\frac{\partial v}{\partial x} = -\frac{1}{2}(x^2+y^2+z^2)^{-3/2} . 2x = -x(x^2+y^2+z^2)^{-3/2}$

and $$\frac{\partial^2 v}{\partial x^2} = -1[1 . (x^2+y^2+z^2)^{-3/2} + x(-3/2)(x^2+y^2+z^2)^{-5/2} . 2x]$$
$$= -(x^2+y^2+z^2)^{-5/2}[x^2+y^2+z^2-3x^2] = (x^2+y^2+z^2)^{-5/2}(2x^2-y^2-z^2)$$

Similarly, $\frac{\partial^2 v}{\partial y^2} = (x^2+y^2+z^2)^{-5/2}(-x^2+2y^2-z^2)$ and $\frac{\partial^2 v}{\partial z^2} = (x^2+y^2+z^2)^{-5/2}(-x^2-y^2+2z^2)$

Hence $$\frac{\partial^2 v}{\partial x^2} + \frac{\partial^2 v}{\partial y^2} + \frac{\partial^2 v}{\partial z^2} = (x^2+y^2+z^2)^{-5/2} . (0) = 0.$$

**Obs.** *A function v satisfying the Laplace equation is said to be a* **harmonic function.**

**Example 5.6.** *If* $u = \log(x^3+y^3+z^3-3xyz)$, *show that* $\left(\frac{\partial}{\partial x}+\frac{\partial}{\partial y}+\frac{\partial}{\partial z}\right)^2 u = \frac{-9}{(x+y+z)^2}$.

*(Andhra, 2017; C.S.V.T.U., 2015 ; V.T.U., 2015 ; Delhi, 2013 ; U.K.T.U., 2011 ; P.T.U., 2010)*

**Solution.** We have, $\frac{\partial u}{\partial x} = \frac{3x^2-3yz}{x^3+y^3+z^3-3xyz}$, $\frac{\partial u}{\partial y} = \frac{3y^2-3zx}{x^3+y^3+z^3-3xyz}$, $\frac{\partial u}{\partial z} = \frac{3z^2-3xy}{x^3+y^3+z^3-3xyz}$

$$\therefore \quad \frac{\partial u}{\partial x}+\frac{\partial u}{\partial y}+\frac{\partial u}{\partial z} = \frac{3(x^2+y^2+z^2-xy-yz-zx)}{x^3+y^3+z^3-3xyz}$$
$$= \frac{3(x^2+y^2+z^2-xy-yz-zx)}{(x+y+z)(x^2+y^2+z^2-xy-yz-zx)} = \frac{3}{x+y+z}$$

Now, $$\left(\frac{\partial}{\partial x}+\frac{\partial}{\partial y}+\frac{\partial}{\partial z}\right)^2 u = \left(\frac{\partial}{\partial x}+\frac{\partial}{\partial y}+\frac{\partial}{\partial z}\right)\left(\frac{\partial u}{\partial x}+\frac{\partial u}{\partial y}+\frac{\partial u}{\partial z}\right)$$
$$= \left(\frac{\partial}{\partial x}+\frac{\partial}{\partial y}+\frac{\partial}{\partial z}\right)\left(\frac{3}{x+y+z}\right)$$
$$= -\frac{3}{(x+y+z)^2} - \frac{3}{(x+y+z)^2} - \frac{3}{(x+y+z)^2} = -\frac{9}{(x+y+z)^2}.$$

**Example 5.7.** *If* $\frac{x^2}{a^2+u}+\frac{y^2}{b^2+u}+\frac{z^2}{c^2+u} = 1$, *prove that*
$$\left(\frac{\partial u}{\partial x}\right)^2+\left(\frac{\partial u}{\partial y}\right)^2+\left(\frac{\partial u}{\partial z}\right)^2 = 2\left(x\frac{\partial u}{\partial x}+y\frac{\partial u}{\partial y}+z\frac{\partial u}{\partial z}\right)$$

**Solution.** We have $x^2(a^2+u)^{-1}+y^2(b^2+u)^{-1}+z^2(c^2+u)^{-1} = 1$ ...(*i*)

Differentiating Eqn. (*i*) partially w.r.t. $x$, we get
$$2x(a^2+u)^{-1} - x^2(a^2+u)^{-2}\frac{\partial u}{\partial x} - y^2(b^2+u)^{-2}\frac{\partial u}{\partial y} - z^2(c^2+u)^{-2}\frac{\partial u}{\partial z} = 0$$

or $$\frac{2x}{a^2+u} = \left\{\frac{x^2}{(a^2+u)^2}+\frac{y^2}{(b^2+u)^2}+\frac{y^2}{(c^2+u)^2}\right\}\frac{\partial u}{\partial x}$$

or $$\frac{\partial u}{\partial x} = \frac{2x}{(a^2+u)v} \text{ where } v = \frac{x^2}{(a^2+u)^2}+\frac{y^2}{(b^2+u)^2}+\frac{y^2}{(c^2+u)^2}$$

Similarly differentiating Eqn. (*i*) partially w.r.t. $y$, we get
$$\frac{2y}{b^2+u} = \left\{\frac{x^2}{(a^2+u)^2}+\frac{y^2}{(b^2+u)^2}+\frac{z^2}{(c^2+u)^2}\right\}\frac{\partial u}{\partial y} \text{ or } \frac{\partial u}{\partial y} = \frac{2y}{(b^2+u)v}$$

Similarly, differentiating (*i*) partially w.r.t. $z$, we get

$$\frac{2z}{(b^2+u)} = \left\{\frac{x^2}{(a^2+u)^2} + \frac{y^2}{(b^2+u)^2} + \frac{z^2}{(c^2+u)^2}\right\}\frac{\partial u}{\partial z} \text{ or } \frac{\partial u}{\partial z} = \frac{2z}{(c^2+u)v}$$

$$\therefore \quad \left(\frac{\partial u}{\partial x}\right)^2 + \left(\frac{\partial u}{\partial y}\right)^2 + \left(\frac{\partial u}{\partial z}\right)^2 = \frac{4}{v^2}\left\{\frac{x^2}{(a^2+u)^2} + \frac{y^2}{(b^2+u)^2} + \frac{y^2}{(c^2+u)^2}\right\} = \frac{4}{v} \qquad ...(ii)$$

Also,
$$2\left(x\frac{\partial u}{\partial x} + y\frac{\partial u}{\partial y} + z\frac{\partial u}{\partial z}\right) = 2\left\{\frac{2x^2}{(a^2+u)v} + \frac{2y^2}{(b^2+u)v} + \frac{2z^2}{(c^2+u)v}\right\}$$

$$= \frac{4}{v}\left\{\frac{x^2}{(a^2+u)} + \frac{y^2}{(b^2+u)} + \frac{z^2}{(c^2+u)}\right\} = \frac{4}{v} \qquad \text{[By } (i)] \; ...(iii)$$

Hence the equality of Eqns. (*ii*) and (*iii*) proves the result.

**Example 5.8.** *If* $u = x^y$, *show that* $\dfrac{\partial^3 u}{\partial x^2 \partial y} = \dfrac{\partial^3 u}{\partial x \partial y \partial x}$.

**Solution.** We have $\dfrac{\partial u}{\partial y} = x^y \log_e x$ and $\dfrac{\partial^2 u}{\partial x \partial y} = yx^{y-1} . \log x + x^y . \dfrac{1}{x} = x^{y-1}(y \log x + 1)$

$$\therefore \quad \frac{\partial^2 u}{\partial x^2 \partial y} = \frac{\partial}{\partial x}[x^{y-1}(y \log x + 1)] \qquad ...(i)$$

Again
$$\frac{\partial u}{\partial x} = yx^{y-1} \text{ and } \frac{\partial^2 u}{\partial y \partial x} = 1 . x^{y-1} + y\left(\frac{1}{x}x^y \log x\right) = x^{y-1}(1 + y \log x)$$

$$\therefore \quad \frac{\partial^3 u}{\partial x \partial y \partial x} = \frac{\partial}{\partial x}[x^{y-1}(y \log x + 1)] \qquad ...(ii)$$

From Eqns. (*i*) and (*ii*) follows the required result.

## PROBLEMS 5.2

1. Evaluate $\partial z/\partial x$ and $\partial z/\partial y$, if
   (*i*) $z = x^2 y - x \sin xy$ ; (*ii*) $z = \log(x^2 + y^2)$ ;
   (*iii*) $z = \tan^{-1}\{(x^2+y^2)/(x+y)\}$ ; (*iv*) $x + y + z = \log z$.
2. If $z(x+y) = x^2 + y^2$, show that $\left(\dfrac{\partial z}{\partial x} - \dfrac{\partial z}{\partial y}\right)^2 = 4\left(1 - \dfrac{\partial z}{\partial x} - \dfrac{\partial z}{\partial y}\right)$. (*AU, 2023; V.T.U., 2012 ; D.T.U., 2011*)
3. If $z = e^{ax+by} f(ax - by)$, prove that $b\dfrac{\partial z}{\partial x} + a\dfrac{\partial z}{\partial y} = 2abz$. (*C.S.V.T.U., 2015 ; V.T.U., 2010*)
4. Given $u = e^{r\cos\theta}\cos(r\sin\theta)$, $v = e^{r\cos\theta}\sin(r\sin\theta)$ ; prove that $\dfrac{\partial u}{\partial r} = \dfrac{1}{r}\dfrac{\partial v}{\partial \theta}$ and $\dfrac{\partial v}{\partial r} = -\dfrac{1}{r}\dfrac{\partial u}{\partial \theta}$.
5. If $z = \tan(y + ax) - (y - ax)^{3/2}$, show that $\dfrac{\partial^2 z}{\partial x^2} = a^2\dfrac{\partial^2 z}{\partial y^2}$.
6. Verify that $f_{xy} = f_{yx}$, when $f$ is equal to (*i*) $\sin^{-1}(y/x)$ ; (*ii*) $\log x \tan^{-1}(x^2 + y^2)$.
7. If $f(x, y) = (1 - 2xy + y^2)^{-1/2}$, show that $\dfrac{\partial}{\partial x}\left[(1-x^2)\dfrac{\partial f}{\partial x}\right] + \dfrac{\partial}{\partial y}\left[y^2\dfrac{\partial f}{\partial y}\right] = 0$.
8. Prove that $\dfrac{\partial^2 u}{\partial x^2} + \dfrac{\partial^2 u}{\partial y^2} = 0$ if (*i*) $u = \tan^{-1}\left[\dfrac{2xy}{x^2 - y^2}\right]$. (*ii*) $u = e^x \sin y$ (*K.T.U., 2016*)
   (*iii*) $u = \log(x^2 + y^2) + \tan^{-1}(y/x)$.
9. If $v = \dfrac{1}{\sqrt{t}}e^{-x^2/4a^2t}$, prove that $\dfrac{\partial v}{\partial t} = a^2\dfrac{\partial^2 v}{\partial x^2}$.

10. The equation $\frac{\partial u}{\partial t} = \mu \frac{\partial^2 u}{\partial x^2}$ refers to the conduction of heat along a bar without radiation, show that, if $u = Ae^{-gx} \sin (nt - gx)$, where $A, g, n$ are positive constants then $g = \sqrt{(n/2\mu)}$.
11. Find the value of $n$ so that the equation $V = r^n (3 \cos^2 \theta - 1)$ satisfies the relation
$$\frac{\partial}{\partial r}\left(r^2 \frac{\partial V}{\partial r}\right) + \frac{1}{\sin \theta} \frac{\partial}{\partial \theta}\left(\sin \theta \frac{\partial V}{\partial \theta}\right) = 0.$$
12. If $z = \log (e^x + e^y)$, show that $rt - s^2 = 0$ where $r = \partial^2 z/\partial x^2$, $s = \partial^2 z/\partial x \partial y$, $t = \partial^2 z/\partial y^2$. (J.N.T.U., 2015)
13. If $u = \frac{y}{z} + \frac{z}{x}$, show that $x\frac{\partial u}{\partial x} + y\frac{\partial u}{\partial y} + z\frac{\partial u}{\partial z} = 0$.
14. Let $r^2 = x^2 + y^2 + z^2$ and $V = r^m$, prove that $V_{xx} + V_{yy} + V_{zz} = m(m + 1)r^{m-2}$. (P.T.U., 2010)
15. If $v = \log (x^2 + y^2 + z^2)$, prove that $(x^2 + y^2 + z^2)\left(\frac{\partial^2 v}{\partial x^2} + \frac{\partial^2 v}{\partial y^2} + \frac{\partial^2 v}{\partial z^2}\right) = 2$.
16. If $v = x^y \cdot y^x$, prove that $x\frac{\partial v}{\partial x} + y\frac{\partial v}{\partial y} = v(x + y + \log v)$.
17. If $x^x y^y z^z = c$, show that at $x = y = z$, $\frac{\partial^2 z}{\partial x \partial y} = -(x \log ex)^{-1}$. (C.S.V.T.U., 2014 ; Kurukshetra, 2013 ; V.T.U., 2013)
18. If $u = e^{xyz}$, find the value of $\frac{\partial^3 u}{\partial x \partial y \partial z}$. (Delhi, 2013 ; Rohtak, 2012)

## 5.3 WHICH VARIABLE IS TO BE TREATED AS CONSTANT

(1) Consider the equation $x = r \cos \theta, y = r \sin \theta$ ...(1)

To find $\partial r/\partial x$, we need a relation between $r$ and $x$. Such a relation will contain one more variable $\theta$ or $y$, for we can eliminate only one variable out of four from the relations (1). Thus the two possible relations are

$$r = x \sec \theta \quad ...(2)$$

and

$$r^2 = x^2 + y^2 \quad ...(3)$$

Now we can find $\partial r/\partial x$ either from (2) by treating $\theta$ as constant or from (3) by regarding $y$ as constant. And there is no reason to suppose that the two values of $\partial r/\partial x$ so found, are equal. *To avoid confusion as to which variable is regarded constant, we introduce the following :*

**Notation :** $(\partial r/\partial x)_\theta$ *means the partial derivative of r with respect to x keeping* $\theta$ *constant in* a relation expressing $r$ as a function of $x$ and $\theta$.

Thus from (2), $(\partial r/\partial x)_\theta = \sec \theta$.

When no indication is given regarding the variable to be kept constant, then according to convention $(\partial/\partial x)$ always means $(\partial/\partial x)_y$ and $\partial/\partial y$ means $(\partial/\partial y)_x$. Similarly, $\partial/\partial r$ means $(\partial/\partial r)_\theta$ and $\partial/\partial\theta$ means $(\partial/\partial\theta)_r$.

(2) In thermodynamics, we come across ten variables such as $p$ (pressure), $v$ (volume), $T$ (temperature), $W$ (work), $\phi$ (entropy) etc. Any one of these can be expressed as a function of other two variables *e.g.*, $T = f(p, v)$, $T = g(p, \phi)$

As we shall see, these respectively give rise to the following results :

$$dT = \frac{\partial T}{\partial p} dp + \frac{\partial T}{\partial v} dv \quad ...(i)$$

$$dT = \frac{\partial T}{\partial p} dp + \frac{\partial T}{\partial \phi} d\phi \quad ...(ii)$$

Now, $\partial T/\partial p$ appearing in (*i*), has been obtained from $T$ as function of $p$ and $v$, treating $v$ as constant, we write it as $(\partial T/\partial p)_v$.

Similarly, $\partial T/\partial p$ occurring in (*ii*), is written as $(\partial T/\partial p)_\phi$.

**Example 5.9.** *If $u = f(r)$ and $x = r \cos \theta$, $y = r \sin \theta$, prove that*

$$\frac{\partial^2 u}{\partial x^2} + \frac{\partial^2 u}{\partial y^2} = f''(r) + \frac{1}{r} f'(r).$$ (J.N.T.U., 2015)

**Solution.** We have $\frac{\partial u}{\partial x} = f'(r) \cdot \frac{\partial r}{\partial x}$ and $\frac{\partial^2 u}{\partial x^2} = f''(r) \cdot \left(\frac{\partial r}{\partial x}\right)^2 + f'(r) \cdot \frac{\partial^2 r}{\partial x^2}$

Similarly, $$\frac{\partial^2 u}{\partial y^2} = f''(r) \cdot \left(\frac{\partial r}{\partial y}\right)^2 + f'(r) \cdot \frac{\partial^2 r}{\partial y^2}$$

$$\therefore \quad \frac{\partial^2 u}{\partial x^2} + \frac{\partial^2 u}{\partial y^2} = f''(r) \cdot \left[\left(\frac{\partial r}{\partial x}\right)^2 + \left(\frac{\partial r}{\partial y}\right)^2\right] + f'(r)\left[\frac{\partial^2 r}{\partial x^2} + \frac{\partial^2 r}{\partial y^2}\right]$$

Now to find $\frac{\partial r}{\partial x}, \frac{\partial r}{\partial y}$ etc., we write $r = (x^2 + y^2)^{1/2}$

$$\therefore \quad \frac{\partial r}{\partial x} = \frac{1}{2}(x^2 + y^2)^{-1/2} \cdot 2x = \frac{x}{r} \quad \text{and} \quad \frac{\partial^2 r}{\partial x^2} = \frac{r \cdot 1 - x \cdot \partial r/\partial x}{r^2} = \frac{r - x^2/r}{r^2} = \frac{y^2}{r^3}.$$

Similarly, $$\frac{\partial r}{\partial y} = \frac{y}{x} \text{ and } \frac{\partial^2 r}{\partial y^2} = \frac{x^2}{r^3}$$

Substituting the values of $\partial r/\partial x$ etc. in (*i*), we get

$$\frac{\partial^2 u}{\partial x^2} + \frac{\partial^2 u}{\partial y^2} = f''(r)\left[\frac{x^2}{r^2} + \frac{y^2}{r^2}\right] + f'(r)\left[\frac{y^2}{r^3} + \frac{x^2}{r^3}\right] = f''(r) + \frac{1}{r}f'(r).$$

**Example 5.10.** *If $x^x y^y z^z = C$. Show that at $x = y = z$*

$$\frac{\delta^2 z}{\delta x\, dy} = -(x \log_e x)^{-1}$$ (*VIT, 2019*)

**Solution:** We have $$x^x y^y z^z = C$$

Taking log both side

$$x\log x + y \log y + z \log z = \log c$$
$$z \log z = \log c - x \log x - y \log y$$

Differentiating partially *w.r. t. x.* we get

$$\left(z \cdot \frac{1}{z} + 1.\log z\right)\frac{dz}{dz} = -\left(x \cdot \frac{1}{x} + 1.\log x\right)$$

$$(1 + \log z)\left(\frac{\delta z}{\delta z}\right) = -(1 + \log x)$$

$$\frac{\delta z}{\delta x} = \frac{-(1 + \log x)}{(1 + \log z)}$$

Similarly, $$\frac{\delta z}{\delta y} = -\frac{(1 + \log y)}{(1 + \log z)} \quad \ldots(i)$$

Thus, when $x = y = z$

$$\frac{\delta z}{\delta x} = -1 \text{ and } \frac{\delta z}{\delta y} = -1$$

Now, differentiating Eqn. (*i*) partially *w.r.t x.* we get

$$\frac{\delta^2 z}{\delta x\, \delta y} = -(1 + \log y)\left\{-(1 + \log z)^{-2} \cdot \frac{1}{z} \cdot \frac{\delta z}{\delta x}\right\}$$

$$= \frac{1 + \log y}{z(1 + \log z)^2}\frac{\delta z}{\delta x}$$

When $x = y = z$, we have

$$\frac{\delta^2 z}{\delta x\, \delta y} = \frac{1 + \log x}{x(1 + \log x)^2}(-1) \quad \left[\because \frac{\delta z}{\delta x} = -1\right]$$

$$= -\frac{1}{x(1+\log x)} = -\frac{1}{x(\log_e + \log x)}$$

$$= -\frac{1}{x(\log_e x)} = -(x \log_e x)^{-1}$$

## PROBLEMS 5.3

1. If $x = r\cos\theta$, $y = r\sin\theta$, show that (i) $\frac{\partial r}{\partial x} = \frac{\partial x}{\partial r}$ (*D.T.U., 2013*) (ii) $\frac{1}{r}\frac{\partial x}{\partial \theta} = r\frac{\partial \theta}{\partial x}$, (*J.N.T.U., 2015*)

   (iii) $\left(\frac{\partial r}{\partial x}\right)^2 + \left(\frac{\partial r}{\partial y}\right)^2 = 1.$

2. If $x^2 = au + bv$, $y^2 = au - bv$, prove that $\left(\frac{\partial u}{\partial x}\right)_y \cdot \left(\frac{\partial x}{\partial u}\right)_v = \frac{1}{2} = \left(\frac{\partial v}{\partial y}\right)_x \cdot \left(\frac{\partial y}{\partial v}\right)_u$. (*U.P.T.U., 2013*)

3. If $u = lx + my$, $v = mx - ly$, show that $\left(\frac{\partial u}{\partial x}\right)_y \cdot \left(\frac{\partial x}{\partial u}\right)_v = \frac{l^2}{l^2+m^2}, \left(\frac{\partial y}{\partial v}\right)_x \left(\frac{\partial v}{\partial y}\right)_u = \frac{l^2+m^2}{l^2}$.

4. If $x = r\cos\theta$, $y = r\sin\theta$, prove that

   (i) $\frac{\partial^2 r}{\partial x^2} + \frac{\partial^2 r}{\partial y^2} = \frac{1}{r}\left[\left(\frac{\partial r}{\partial x}\right)^2 + \left(\frac{\partial r}{\partial y}\right)^2\right]$ (ii) $\frac{\partial^2 \theta}{\partial x^2} + \frac{\partial^2 \theta}{\partial y^2} = 0$ $(x \neq 0, y \neq 0)$.

5. If $u = e^{xy}$, show that $\frac{\partial^2 u}{\partial x^2} + \frac{\partial^2 u}{\partial y^2} = \frac{1}{u}\left\{\left(\frac{\partial u}{\partial x}\right)^2 + \left(\frac{\partial u}{\partial y}\right)^2\right\}$. (*Anna, 2013*)

6. Show that $\frac{\delta^2 u}{\delta x\, \delta y} = \frac{\delta^2 u}{\delta x \delta y}$, where (*PTU 2017*)

   (a) $u = c\tan^{-1}\left(\frac{x}{y}\right)$ (b) $u = \log(x^2 + y^2)$ (c) $u = x\cos y + y\cos x$

7. Prove that if $z = \frac{x^2y^2}{x+y}$, then show that $x\frac{\delta^2 z}{\delta x^2} + y\frac{\delta^2 z}{\delta x \delta y} = 2\frac{\delta z}{\delta x}$ (*KTU, 2019*)

8. If $u = e^{xyz}$, show that $\frac{\delta^3 u}{\delta x \delta y \delta z} = (1 + 3xyz + x^2y^2z^2)\, e^{xyz}$

9. If $z = f(x + ay) + \phi(x - ay)$, prove that $\frac{\delta^2 z}{\delta y^2} = a^2\frac{\delta^2 z}{\delta x^2}$

10. If $r^2 = x^2 + y^2 + z^2$ and $v = r^m$, prove that $\frac{\delta^2 v}{\delta x^2} + \frac{\delta^2 v}{\delta y^2} + \frac{\delta^2 v}{\delta z^2} = r(m+1)\, r^{m+2}$ (*CSVTU. Bhilai, 2019*)

11. The equation $\frac{\delta u}{\delta t} = \mu\frac{\delta^2 u}{\delta x^2}$, refers to the conduction of heat along a bar, without radiation. If $u = Ae^{-gx}$ $\sin(nt - gx)$, where $A, g, n$ are positive constants, show that $g = \sqrt{\frac{n}{2\mu}}$ (*VIT, A.P., 2020*)

12. If $u = f(r)$ where, $r = \sqrt{x^2 + y^2 + z^2}$, show that, $\frac{\delta^2 u}{\delta x^2} + \frac{\delta^2 u}{\delta y^2} + \frac{\delta^2 u}{\delta z^2} = f''(r) + \frac{2}{r}f'(r)$ (*Aligarh, 2019*)

## 5.4 (1) HOMOGENEOUS FUNCTIONS

An expression of the form $a_0x^n + a_1x^{n-1}y + a_2x^{n-2}y^2 + ... + a_ny^n$ in which every term is of the $n$th degree, is called a homogeneous function of degree $n$. This can be rewritten as

$$x^n[a_0 + a_1(y/x) + a_2(y/x)^2 + ... + a_n (y/x)^n].$$

Thus *any function f(x, y) which can be expressed in the form* $x^n \phi(y/x)$, *is called a* **homogeneous function** of degree $n$ in $x$ and $y$.

For instance, $x^3 \cos(y/x)$ is a homogeneous function of degree 3, in $x$ and $y$.

In general, a function $f(x, y, z, t, ...)$ is said to be a homogeneous function of degree $n$ in $x, y, z, t, ...$, if it can be expressed in the form $x^n\phi(y/x, z/x, t/x, ...)$.

**(2) Euler's theorem on homogeneous functions***. *If u be a homogeneous function of degree n in x and y, then*

$$\mathbf{x\frac{\partial u}{\partial x} + y\frac{\partial u}{\partial y} = nu.}$$

Since $u$ is a homogeneous function of degree $n$ in $x$ and $y$, therefore,

$$u = x^n f(y/x)$$

$$\therefore \quad \frac{\partial u}{\partial x} = nx^{n-1}f\left(\frac{y}{x}\right) + x^n f'\left(\frac{y}{x}\right)\cdot y\left(-\frac{1}{x^2}\right) = nx^{n-1}f\left(\frac{y}{x}\right) - yx^{n-2}f'\left(\frac{y}{x}\right)$$

and
$$\frac{\partial u}{\partial y} = x^n f'\left(\frac{y}{x}\right)\cdot\frac{1}{x} = x^{n-1} f'\left(\frac{y}{x}\right). \text{ Hence } x\frac{\partial u}{\partial x} + y\frac{\partial u}{\partial y} = nx^n f\left(\frac{y}{x}\right) = nu.$$

In general, if $u$ be a homogeneous function of degree $n$ in $x, y, z, t, ...$, then,

$$x\frac{\partial u}{\partial x} + y\frac{\partial u}{\partial y} + z\frac{\partial u}{\partial z} + t\frac{\partial u}{\partial t} \; ... = nu.$$

**Example 5.11.** *Show that* $x\frac{\partial u}{\partial x} + y\frac{\partial u}{\partial y} = 2u \log u$ *where* $\log u = (x^3 + y^3)/(3x + 4y)$.

**Solution.** Since
$$z = \log u = \frac{x^3 + y^3}{3x + 4y} = x^2 \cdot \frac{1 + (y/x)^3}{3 + 4(y/x)},$$

$\therefore$ $z$ is a homogeneous function of degree 2 in $x$ and $y$.

By Euler's theorem, we get

$$x\frac{\partial z}{\partial x} + y\frac{\partial z}{\partial y} = 2z \qquad ...(i)$$

But
$$\frac{\partial z}{\partial x} = \frac{1}{u}\frac{\partial u}{\partial x} \text{ and } \frac{\partial z}{\partial y} = \frac{1}{u}\frac{\partial u}{\partial y}$$

Hence $(i)$ becomes

$$x\cdot\frac{1}{u}\frac{\partial u}{\partial x} + y\cdot\frac{1}{u}\frac{\partial u}{\partial y} = 2\log u \quad \text{or} \quad x\frac{\partial u}{\partial x} + y\frac{\partial u}{\partial y} = 2u\log u.$$

**Example 5.12.** *If* $u = \sin^{-1}\frac{x + 2y + 3z}{x^8 + y^8 + z^8}$, *find the value of* $x\frac{\partial u}{\partial x} + y\frac{\partial u}{\partial y} + z\frac{\partial u}{\partial z}$.

*(Andhra, 2022; V.T.U., 2016 ; U.K.T.U., 2013)*

**Solution.** Here $u$ is not a homogeneous function. We therefore, write

$$\omega = \sin u = \frac{x + 2y + 3z}{x^8 + y^8 + z^8} = x^{-7}\cdot\frac{1 + 2(y/x) + 3(z/x)}{1 + (y/x)^8 + (z/x)^8}$$

Thus $\omega$ is a homogeneous function of degree $-7$ in $x, y, z$. Hence by Euler's theorem

$$x\frac{\partial \omega}{\partial x} + y\frac{\partial \omega}{\partial y} + z\frac{\partial \omega}{\partial z} = (-7)\,\omega \qquad ...(i)$$

But
$$\frac{\partial \omega}{\partial x} = \cos u\frac{\partial u}{\partial x}, \frac{\partial \omega}{\partial y} = \cos u\frac{\partial u}{\partial y}, \frac{\partial \omega}{\partial z} = \cos u\frac{\partial u}{\partial z}$$

* After an enormously creative Swiss mathematician *Leonhard Euler* (1707–1783). He studied under John *Bernoulli* and became a professor of mathematics in St. Petersburg, Russia. Even after becoming totally blind in 1771, he contributed to almost all branches of mathematics.

$\therefore$ (*i*) becomes $\quad x \cos u \frac{\partial u}{\partial x} + y \cos u \frac{\partial u}{\partial y} + z \cos u \frac{\partial u}{\partial z} = -7 \sin u \quad$ or $\quad x\frac{\partial u}{\partial x} + y\frac{\partial u}{\partial y} + z\frac{\partial u}{\partial z} = -7 \tan u.$

**Example 5.13.** *If u(x, y) is homogeneous function of degree n in x and y with all first and second derivatives continuous, then prove that*

$$x^2 \frac{\partial^2 u}{\partial x^2} + 2xy \frac{\partial^2 u}{\partial x \partial y} + y^2 \frac{\partial^2 u}{\partial y^2} = n(n-1)u. \qquad \textit{(Andhra, 2022)}$$

**Solution :** Given $u(x, y)$ is a homogeneous function of $x$ and $y$ of degree $n$. So, by Eulers theore, we have

$$x\frac{\partial u}{\partial x} + y\frac{\partial u}{\partial y} = nu. \qquad ...(i)$$

Differentiating Eqn. (*i*) partially w.r.t. $x$, we get

$$x\frac{\partial^2 u}{\partial x^2} + \frac{\partial u}{\partial x} + y\frac{\partial^2 u}{\partial x \partial y} = n\frac{\partial u}{\partial x}$$

$$\Rightarrow \qquad x\frac{\partial^2 u}{\partial x^2} + y\frac{\partial^2 u}{\partial x \partial y} = (n-1)\frac{\partial u}{\partial x}$$

On multiplying with $x$, we have

$$x^2 \frac{\partial^2 u}{\partial x \partial y} + y^2 \frac{\partial^2 u}{\partial y^2} = (n-1)x\frac{\partial u}{\partial x} \qquad ...(ii)$$

Similarly differentiating Eqn. (*i*) partially w.r.t. $y$ and on multiplying with $y$, we get

$$xy \frac{\partial^2 u}{\partial x \partial y} + y^2 \frac{\partial^2 u}{\partial y^2} = (n-1)y\frac{\partial u}{\partial x} \qquad ...(iii)$$

Then Eqn. (*ii*) + (*iii*)

$$\Rightarrow \qquad x^2 \frac{\partial^2 u}{\partial x^2} + 2xy \frac{\partial^2 u}{\partial x \partial y} + y^2 \frac{\partial^2 u}{\partial y^2} = (n-1)\left[x\frac{\partial u}{\partial x} + y\frac{\partial u}{\partial y}\right]$$

$$= (n-1)\, nu, \qquad \text{[using Eqn. } (i)]$$

which proves the result.

**Example 5.14.** *Given* $\theta = t^n . e^{-r^2/4t}$. *Find the value of n, for which* $\frac{1}{r^2}\frac{\delta}{\delta x}\left(r^2 - \frac{\delta\theta}{\delta r}\right) = \frac{\delta\theta}{dt}$. *(Roorkee, 2019)*

**Solution :** We have
$$\theta = t^n . e^{-r^2/4t}$$

$$\therefore \qquad \frac{d\theta}{dr} = t^n . e^{-r^2/4t}\left(\frac{-2r}{4t}\right)$$

$$= \frac{-1}{2} t^{n-1} . r e^{-r^2/4t}$$

Now,
$$r^2 - \frac{d\theta}{dr} = -\frac{1}{2} t^{n-1} . r^3 . e^{-r^2/4t}$$

$$\therefore \qquad \frac{\delta}{\delta r}\left(r^2 \frac{d\theta}{dr}\right) = -\frac{1}{2} t^{n-1}\left[3r^2 . e^{-r^2/4t} + r^3 e^{-r^2/4t}\left(\frac{-2r}{4t}\right)\right]$$

$$\frac{1}{r^2}\frac{\delta}{\delta r}\left(r^2\frac{\delta\theta}{\delta\pi}\right) = t^{n-1}\left(\frac{-3}{2}+\frac{r^2}{4t}\right)e^{-r^2/4t} \quad ...(i)$$

$$\frac{\delta\theta}{\delta t} = t^{n-1}.e^{-r^2/4t} + t^n e^{-r^2/4t}\left(\frac{r^2}{4t^2}\right),$$

$$= t^{n-1}\left(n+\frac{r^2}{4t}\right)e^{-r^2}/4t \quad ...(ii)$$

Now, $\dfrac{1}{r^2}\dfrac{d}{dr}\left(r^2\dfrac{d\theta}{4t}\right) = \dfrac{d\theta}{dt}$

So, using Eqn. (*i*) and (*ii*)

$$t^{n-1}\left(-\frac{3}{2}+\frac{r^2}{4t}\right)e^{-r^2/4t} = t^{n-1}\left(n+\frac{r^2}{4t}\right)e^{-r^2/4t}$$

Thus $n = -3/2.$

**Example 5.15.** *If* $u = \sin^{-1}\dfrac{x+y}{\sqrt{x}+\sqrt{y}}$, *prove that* $x\dfrac{\partial u}{\partial x}+y\dfrac{\partial u}{\partial y}=\dfrac{1}{2}\tan u.$

*(P.T.U., 2013)*

*and* $x^2\dfrac{\partial^2 u}{\partial x^2}+2xy\dfrac{\partial^2 u}{\partial x\partial y}+y^2\dfrac{\partial^2 u}{\partial y^2}=-\dfrac{\sin u\cos 2u}{4\cos^3 u}.$ *(D.T.U., 2012 ; Rohtak, 2012)*

**Solution :** Here $u$ is not a homogeneous function but $z = \sin u = \dfrac{x+y}{\sqrt{x}+\sqrt{y}}$ is a homogeneous function of degree 1/2 in $x$ and $y$.

$\therefore$ By Euler's theorem, $x\dfrac{\partial z}{\partial x}+y\dfrac{\partial z}{\partial y} = \dfrac{1}{2}z$

or $$x\cos u\frac{\partial u}{\partial x}+y\cos u\frac{\partial u}{\partial y} = \frac{1}{2}\sin u$$

Thus $$x\frac{\partial u}{\partial x}+y\frac{\partial u}{\partial y} = \frac{1}{2}\tan u \quad ...(i)$$

Differentiating Eqn. (*i*) w.r.t. $x$ partially, we get

$$x\frac{\partial^2 u}{\partial x^2}+\frac{\partial u}{\partial x}+y\frac{\partial^2 u}{\partial x\partial y} = \frac{1}{2}\sec^2 u\frac{\partial u}{\partial x} \quad \text{or} \quad x\frac{\partial^2 u}{\partial x^2}+y\frac{\partial^2 u}{\partial x\partial y} = \left(\frac{1}{2}\sec^2 u-1\right)\frac{\partial u}{\partial x} \quad ...(ii)$$

Again differentiating Eqn. (*i*) w.r.t. $y$ partially, we get

$$x\frac{\partial^2 u}{\partial y\partial x}+y\frac{\partial^2 u}{\partial y^2}+\frac{\partial u}{\partial y}=\frac{1}{2}\sec^2 u\frac{\partial u}{\partial y} \quad \text{or} \quad x\frac{\partial^2 u}{\partial x\partial y}+y\frac{\partial^2 u}{\partial y^2}=\left(\frac{1}{2}\sec^2 u-1\right)\frac{\partial u}{\partial x} \quad ...(iii)$$

Multiplying Eqn. (*ii*) by $x$ and Eqn. (*iii*) by $y$ and adding, we obtain

$$x^2\frac{\partial^2 u}{\partial x^2}+2xy\frac{\partial^2 u}{\partial x\partial y}+y^2\frac{\partial^2 u}{\partial y^2}=\left(\frac{1}{2}\sec^2 u-1\right)\left(x\frac{\partial u}{\partial x}+y\frac{\partial u}{\partial y}\right)$$

or $$x^2\frac{\partial^2 u}{\partial x^2}+2xy\frac{\partial^2 u}{\partial x\partial y}+y^2\frac{\partial^2 u}{\partial y^2}=\left(\frac{1}{2}\sec^2 u-1\right)\left(\frac{1}{2}\tan u\right) \quad \text{[By (i)]}$$

$$=\frac{1}{4}\frac{\sin u}{\cos^3 u}-\frac{1}{2}\frac{\sin u}{\cos u} = -\frac{\sin u\,(2\cos^2 u-1)}{4\cos^3 u}$$

Hence, $$x^2\frac{\partial^2 u}{\partial x^2}+2xy\frac{\partial^2 u}{\partial x\partial y}+y^2\frac{\partial^2 u}{\partial y^2}=-\frac{\sin u\cos 2u}{4\cos^3 u}.$$

**Example 5.16 :** *If* $u = \tan^{-1}\left(\frac{x^3 + y^3}{x - y}\right)$.

*Prove that* $x\frac{\delta u}{\delta x} + y\frac{\delta u}{\delta y} = \sin 2u$ *and* $x^2\frac{\delta^2 u}{\delta x^2} + 2xy\frac{\delta^2 u}{\delta x \delta y} + y^2\frac{\delta^2 u}{\delta y^2} = \sin 4u - \sin 2u.$

*(JNTUK, 2020; Lucknow, 2017)*

**Solution :** Here $\tan u = \frac{x^3 + y^3}{x - y} = z(\text{say})$

$z$ is a homogeneous function of degree 2.

Therefore $x\frac{\delta z}{\delta x} + y\frac{\delta z}{\delta y} = 2z$ ...(*i*)

Now $z = \tan u$

$$\therefore \quad \frac{\delta z}{\delta x} = \sec^2 u \frac{\delta u}{\delta x}$$

and $$\frac{\delta z}{\delta y} = \sec^2 u \frac{\delta u}{\delta x}$$ ...(*ii*)

from equation Eqns. (*i*) and (*ii*) we get

$$x\left(\sec^2 u\frac{\delta u}{\delta x}\right) + y\left(\sec^2 u\frac{\delta u}{\delta y}\right) = 2\tan u$$

$$x\frac{\delta u}{\delta x} + y\frac{\delta u}{\delta y} = \frac{2\tan u}{\sec^2 u} = 2\sin u \cos u = \sin 2u$$ ...(*iii*)

Differentiating Eqn. (*iii*) partially *w.r.t* $x$ and $y$ respectively. We have

$$x\frac{\delta^2 u}{\delta x^2} + \frac{\delta u}{\delta x} + y\frac{\delta^2 u}{\delta x \delta y} = 2\cos 2u\frac{\delta u}{\delta x}$$ ...(*iv*)

$$x\frac{\delta^2 u}{\delta y\,\delta x} + y\frac{\delta^2 u}{\delta y^2} + \frac{\delta u}{\delta y} = 2\cos 2u\frac{\delta u}{\delta y}$$ ...(*v*)

Multiplying Eqn. (*iv*) by $x$ and Eqn. (*v*) by $y$ and adding

$$x^2\frac{\delta^2 u}{\delta x^2} + 2xy\frac{\delta^2 u}{\delta x \delta y} + y^2\frac{\delta^2 y}{\delta y^2} + \left(x\frac{\delta u}{\delta x} + y\frac{\delta u}{\delta y}\right)$$

$$= 2\cos 2u\left(x\frac{\delta u}{\delta x} + y\frac{\delta u}{\delta y}\right) \qquad \left[\because \frac{\delta^2 u}{\delta x\,\delta y} = \frac{\delta^2 u}{\delta y\,\delta x}\right]$$

Substituting $x\frac{\delta u}{\delta x} + y\frac{\delta u}{\delta y} = \sin 2u$ from Eqn. (*iii*), we get

$$x^2\frac{\delta^2 u}{\delta x^2} + 2xy\frac{\delta^2 u}{\delta x\,\delta y} + y^2\frac{\delta^2 u}{\delta y^2} + \sin 2u = 2\cos 2u \cdot \sin 2u$$

$$x^2\frac{\delta^2 u}{\delta x^2} + 2xy\frac{\delta^2 u}{\delta x \delta y} + y^2\frac{\delta^2 y}{\delta y^2} = \sin 4u - \sin 2u$$

which is the required result.

## PROBLEMS 5.4

1. Verify Euler's theorem, when
   (*i*) $f(x, y) = ax^2 + 2hxy + by^2$ (*ii*) $f(x, y) = x^2 (x^2 - y^2)^3/(x^2 + y^2)^3$.
   (*iii*) $f(x, y) = 3x^2 yz + 5xy^2z + 4z^4$
2. If $u = \cos^{-1} \dfrac{x+y}{\sqrt{x}+\sqrt{y}}$, prove that $x\dfrac{\partial u}{\partial x} + y\dfrac{\partial u}{\partial y} = -\dfrac{1}{2}\cot u$. (*JNTUK, 2022; V.T.U., 2015*)
3. If $u = \tan^{-1}\dfrac{x^2+y^2}{x-y}$, show that $x\dfrac{\partial u}{\partial x} + y\dfrac{\partial u}{\partial y} = \dfrac{1}{2}\sin 2u$. (*W.B.T.U., 2016*)
4. Show that $x\dfrac{\partial u}{\partial x} + y\dfrac{\partial u}{\partial y} = 2u \log u$, where, $u = e^{x^2+y^2}$ (*P.T.U., 2010*)
5. If $u = e^{x^3+y^3}$, then prove that $x\dfrac{\partial u}{\partial x} + y\dfrac{\partial u}{\partial y} = 3u\log u$. (*V.T.U., 2016*)
6. Show that the function (*C. S. V. T. V, 2019*)

$$u(x, y, z, t) = \frac{1}{(2a\sqrt{\pi t})^3}\, e\, \frac{(x-x_o)^2 + (y-y_o)^2 + (z-z_o)^2}{4a^2 t}$$

   where $x_o, y_o, z_o$ and $a$ are constants, satisfy the equation of conduction of heat.

$$\frac{\delta u}{\delta t} = a^2\left(\frac{\delta^2 u}{\delta x^2} + \frac{\delta^2 u}{\delta y^2} + \frac{\delta^2 u}{\delta z^2}\right)$$ (*Bhillai, 2018*)

7. If $u = \log\dfrac{x^4+y^4}{x+y}$, show that $x\dfrac{\partial u}{\partial x} + y\dfrac{\partial u}{\partial y} = 3$. (*D.T.U., 2012 ; U.K.T.U., 2012*)
8. If $u = \dfrac{x}{y+z} + \dfrac{y}{z+x} + \dfrac{z}{x+y}$, show that $x\dfrac{\partial u}{\partial x} + y\dfrac{\partial u}{\partial y} + z\dfrac{\partial u}{\partial z} = 0$.
9. If $\sin u = \dfrac{x+2y+3z}{\sqrt{(x^8+y^8+z^8)}}$, show that $xu_x + yu_y + zu_z + 3\tan u = 0$.
10. If $u = \tan^{-1}\dfrac{x^3+y^3}{x+y}$, prove that $x\dfrac{\partial u}{\partial x} + y\dfrac{\partial u}{\partial y} = \sin 2u$. (*V.T.U., 2013 ; P.T.U., 2011*)
    and $x^2\dfrac{\partial^2 u}{\partial x^2} + 2xy\dfrac{\partial^2 u}{\partial x\partial y} + y^2\dfrac{\partial^2 u}{\partial y^2} = 2\cos 3u \sin u$. (*C.S.V.T.U., 2012 ; D.T.U., 2011*)
11. Given $z = x^n f_1(y/x) + y^{-n} f_2(x/y)$, prove that $x^2\dfrac{\partial^2 z}{\partial x^2} + 2xy\dfrac{\partial^2 z}{\partial x\partial y} + y^2\dfrac{\partial^2 z}{\partial y^2} + x\dfrac{\partial z}{\partial x} + y\dfrac{\partial z}{\partial y} = n^2 z$. (*Kurukshetra, 2013*)
12. If $u$ is a homogeneous function of $x, y$ of degree $n$, prove by using Euler's theorem, that

$$x^2\frac{\delta^2 u}{\delta x^2} + 2xy\frac{\delta^2 u}{\delta x\delta y} + y^2\frac{\delta^2 u}{\delta y^2} = \frac{\tan u}{144}(13 + \tan^2 u)$$ (*UTU, 2019*)

13. Given, $u = \sin^{-1}\left[\dfrac{x^{1/3}+y^{1/3}}{x^{1/2}+y^{1/2}}\right]^{1/2}$ (*D.T.U, 2018*)

    Show that $x\dfrac{\delta u}{\delta x} + y\dfrac{\delta u}{\delta y} = -\dfrac{1}{12}\tan u$.

    and $\dfrac{x^2\,\delta^2 u}{\delta x^2} + 2xy\dfrac{\delta^2 u}{\delta x\,\delta y} + y^2\dfrac{\delta^2 u}{\delta y^2} = \dfrac{\tan u}{144}(13 + \tan^2 u)$
14. If $u = \operatorname{cosec}^{-1}\left(\dfrac{x^{1/2}+y^{1/2}}{x^{1/3}+y^{1/3}}\right)^{1/2}$, prove that $x^2\dfrac{\partial^2 u}{\partial x^2} + 2xy\dfrac{\partial^2 u}{\partial x\partial y} + y^2\dfrac{\partial^2 u}{\partial y^2} = \dfrac{\tan u}{12}\left(\dfrac{13}{12} + \dfrac{\tan^2 u}{12}\right)$.
    (*D.T.U., 2015 ; Rohtak, 2012*)

**15.** Show that the function $u(x, t) = k \sin (P\lambda_t + \phi) \sin \lambda x$ satisfies the equation $\frac{\delta^2 u}{\delta t^2} = a^2 \frac{\delta^2 u}{\delta x^2}$ *(Bhopal, 2019)*

**16.** If $u = \sin^{-1}\left(\frac{x^2 + y^2}{x + y}\right)$, prove that $x\frac{\delta u}{\delta x} + y\frac{\delta u}{\delta y} = \tan u,$ *(JNTUK, 2022; Hissar, 2019)*

**17.** If $u = \tan^{-1}\left(\frac{x^2 + y^2}{x + y}\right)$, show that $x\frac{\delta u}{\delta x} + y\frac{\delta u}{\delta y} = \frac{1}{2}\sin 2u$ *(Kottayam, 2018)*

## 5.5 (1) TOTAL DERIVATIVE

If $u = f(x, y)$, where $x = \phi(t)$ and $y = \psi(t)$, then we can express $u$ as a function of $t$ alone by substituting the values of $x$ and $y$ in $f(x, y)$. Thus we can find the ordinary derivative $du/dt$ which is called the *total derivative* of $u$ to distinguish it from the partial derivatives $\partial u/\partial x$ and $\partial u/\partial y$.

Now to find $du/dt$ without actually substituting the values of $x$ and $y$ in $f(x, y)$, we establish the following **Chain rule :**

$$\frac{du}{dt} = \frac{\partial u}{\partial x}\cdot\frac{dx}{dt} + \frac{\partial u}{\partial y}\cdot\frac{dy}{dt} \qquad ...(i)$$

***Proof.*** We have $u = f(x, y)$

Giving increment $\delta t$ to $t$, let the corresponding increments of $x, y$ and $u$ be $\delta x$, $\delta y$ and $\delta u$ respectively.

Then $u + \delta u = f(x + \delta x, y + \delta y)$

Subtracting, $\delta u = f(x + \delta x, y + \delta y) - f(x, y)$

$$= \{f(x + \delta x, y + \delta y) - f(x, y + \delta y)\} + \{f(x, y + \delta y) - f(x, y)\}$$

$$\therefore \quad \frac{\delta u}{\delta t} = \frac{f(x + \delta x, y + \delta y) - f(x, y + \delta y)}{\delta x}\cdot\frac{\delta x}{\delta t} + \frac{f(x, y + \delta y) - f(x, y)}{\delta y}\cdot\frac{\delta y}{\delta t}$$

Taking limits as $\delta t \to 0$, $\delta x$ and $\delta y$ also $\to 0$, we have

$$\frac{du}{dt} = \underset{\delta y \to 0}{\text{Lt}}\left[\underset{\delta y \to 0}{\text{Lt}}\left\{\frac{f(x + \delta x, y + \delta y) - f(x, y + \delta y)}{\delta x}\right\}\right]\frac{dx}{dt} + \underset{\delta y \to 0}{\text{Lt}}\left\{\frac{f(x, y + \delta y) - f(x, y)}{\delta y}\right\}\frac{dy}{dt}$$

$$= \underset{\delta y \to 0}{\text{Lt}}\left\{\frac{\partial f(x, y + \delta y)}{\delta y}\right\}\cdot\frac{dx}{dt} + \frac{\partial f(x, y)}{\partial y}\cdot\frac{dy}{dt}$$

[Supposing $\partial f(x, y)/\partial x$ to be a continuous function of $y$]

$$= \frac{\partial f(x, y)}{\partial x}\cdot\frac{dx}{dt} + \frac{\partial f(x, y)}{\partial y}\cdot\frac{dy}{dt}$$ which is the desired formula.

**Cor.** Taking $t = x$, Eqn. $(i)$ becomes, $\frac{du}{dx} = \frac{\partial u}{\partial x} + \frac{\partial u}{\partial y}\cdot\frac{dy}{dx}$ ...(ii)

**Obs.** If $u = f(x, y, z)$, where $x, y, z$ are all functions of a variable $t$, then **Chain rule** is

$$\frac{du}{dt} = \frac{\partial u}{\partial x}\cdot\frac{dx}{dt} + \frac{\partial u}{\partial y}\cdot\frac{dy}{dt} + \frac{\partial u}{\partial z}\cdot\frac{dz}{dt} \qquad (iii)$$

**(2) Differentiation of implicit functions.** If $f(x, y) = c$ be an implicit relation between $x$ and $y$ which defines as a differentiable function of $x$, then Eqn. $(ii)$ becomes

$$0 = \frac{df}{dx} = \frac{\partial f}{\partial x} + \frac{\partial f}{\partial y}\cdot\frac{dy}{dx}$$

This gives the *important formula* $\frac{dy}{dx} = -\frac{\partial f}{\partial x}\Big/\frac{\partial f}{\partial y}$ $\left[\frac{\partial f}{\partial y} \neq 0\right]$

*for the first differential coefficient of an implicit function.*

**Example 5.17.** *Given $u = \sin (x/y)$, $x = e^t$ and $y = t^2$, find $du/dt$ as a function of $t$. Verify your result by direct substitution.*

**Solution.** We have $\dfrac{du}{dt} = \dfrac{\partial u}{\partial x}\cdot\dfrac{dx}{dt} + \dfrac{\partial u}{\partial y}\cdot\dfrac{dy}{dt} = \left(\cos\dfrac{x}{y}\right)\dfrac{1}{y}\cdot e^t + \left(\cos\dfrac{x}{y}\right)\left(-\dfrac{x}{y^2}\right)2t$

$$= \cos(e^t/t^2)\cdot e^t/t^2 - 2\cos(e^t/t^2)\cdot e^t/t^3 = \{(t-2)/t^3\}e^t\cos(e^t/t^2)$$

Also $u = \sin(x/y) = \sin(e^t/t^2)$

$$\therefore \quad \frac{du}{dt} = \cos\left(\frac{e^t}{t^2}\right)\cdot\frac{t^2e^t - e^t\cdot 2t}{t^4} = \frac{t-2}{t^3}e^t\cos\left(\frac{e^t}{t^2}\right) \text{ as before.}$$

**Example 5.18.** *If x increases at the rate of 2 cm/sec at the instant when x = 3 cm. and y = 1 cm., at what rate must y be changing in order that the function $2xy - 3x^2y$ shall be neither increasing nor decreasing ?*

**Solution.** Let $u = 2xy - 3x^2y$, so that

$$\frac{du}{dt} = \frac{\partial u}{\partial x}\cdot\frac{dx}{dt} + \frac{\partial u}{\partial y}\cdot\frac{dy}{dt} = (2y - 6xy)\frac{dx}{dt} + (2x - 3x^2)\frac{dy}{dt} \quad ...(i)$$

when $x = 3$ and $y = 1$, $dx/dt = 2$, and $u$ is neither increasing nor decreasing, *i.e.*, $du/dt = 0$.

$\therefore$ (*i*) becomes $\quad 0 = (2 - 6\times 3)\,2 + (2\times 3 - 3\times 9)\dfrac{dy}{dt}$

or $\quad \dfrac{dy}{dt} = -\dfrac{32}{21}$ cm/sec. Thus $y$ is decreasing at the rate of 32/21 cm/sec.

**Example 5.19.** *If $u = x\log xy$ where $x^3 + y^3 + 3xy = 1$, find du/dx.*

**Solution.** From $f(x, y) = x^3 + y^3 + 3xy - 1$, we have

$$\frac{dy}{dx} = -\frac{\partial f}{\partial x}\bigg/\frac{\partial f}{\partial y} = -\frac{3x^2 + 3y}{3y^2 + 3x} = -\frac{x^2 + y}{y^2 + x} \quad ...(i)$$

Also $\quad \dfrac{du}{dx} = \dfrac{\partial u}{\partial x} + \dfrac{\partial u}{\partial y}\cdot\dfrac{dy}{dx} = (1\cdot\log xy + x\cdot 1/x) + (x/y)\cdot dy/dx.$

Hence $\quad du/dx = 1 + \log xy - x(x^2 + y)/y(y^2 + x)$ [By (*i*)]

**Example 5.20.** *If $u = \sin^{-1}(x - y)$, $x = 3t$, $y = 4t^3$. Show that $\dfrac{du}{dt} = \dfrac{3}{\sqrt{1 - t^2}}$* (*Rohtak, 2019*)

**Solution:** $\quad \dfrac{\delta u}{\delta x} = \dfrac{1}{\sqrt{1-(x-y)^2}}$

and $\quad \dfrac{\delta u}{\delta y} = -\dfrac{1}{\sqrt{1-(x-y)^2}}$

Also, $\quad \dfrac{\delta x}{dt} = 3, \dfrac{dy}{dt} = 12t^2$

So, using the relation $\quad \dfrac{\delta u}{\delta x} = \dfrac{\delta u}{\delta x}\cdot\dfrac{\delta u}{\delta t} + \dfrac{\delta u}{\delta y}\cdot\dfrac{\delta y}{\delta t},$

We get

$$\frac{\delta u}{\delta t} = \frac{1}{\sqrt{1-(x-y)^2}}\,3 - \frac{1}{\sqrt{1-(x-y)^2}}\,12t^2$$

$$= \frac{3(1-4t^2)}{\sqrt{1-(3t-4t^3)^2}} = \frac{3(1-4t)^2}{\sqrt{(1-t^2)(1-4t^2)^2}}$$

$$\frac{\delta u}{\delta t} = \frac{3(1-4t^2)}{(1-4t^2)\sqrt{1-t^2}} = \frac{3}{\sqrt{1-}}$$

**Example 5.21.** *If* $(x, y) = 0$, $\phi(y, z) = 0$, *show that*

$$\frac{\delta f}{\delta y}\cdot\frac{\delta\phi}{\phi z}\cdot\frac{\delta z}{\delta x} = \frac{\delta f}{\delta x}\cdot\frac{\delta\phi}{\delta y}$$ *(Bhillai, 2019)*

**Solution:** Since $f(x, y) = 0$

$$\frac{dy}{dx} = \frac{-\left(\frac{\delta f}{\delta x}\right)}{\left(\frac{\delta f}{\delta y}\right)} \quad ...(i)$$

Also, $\phi(y, z) = 0$

$$\frac{\delta z}{\delta y} = \frac{-\frac{\delta\phi}{\delta y}}{\frac{\delta\phi}{\delta z}} \quad ...(ii)$$

From Eqns. (*i*) and (*ii*), we get

$$\frac{\delta z}{\delta x} = \frac{\delta y}{\delta x}\cdot\frac{\delta z}{\delta y} = \frac{\left(\frac{\delta f}{\delta x}\right)\cdot\left(\frac{\delta\phi}{\delta y}\right)}{\left(\frac{\delta f}{\delta y}\right)\left(\frac{\delta\phi}{\delta z}\right)}$$

$$\therefore \quad \frac{\delta f}{\delta y}\cdot\frac{\delta\phi}{\delta z}\cdot\frac{\delta z}{\delta x} = \frac{\delta f}{\delta x}\cdot\frac{\delta\phi}{\delta y}.$$

## PROBLEMS 5.5

1. If $x^2 + y^2 + 2gx + 2fy + c = 0$, then show that $\frac{d^2y}{dx^2} = \frac{c^2 - g^2 - f^2}{(y + f)^3}$ *(Kurukshetra, 2019)*
2. If $u = \tan^{-1}(y/x)$ where $x = e^t - e^{-t}$, and $y = e^t + e^{-t}$, find $du/dt$.
3. If $u = e^{3x + 3y}$, where $x = \cos t$, $y = t^2$. Find $\frac{du}{dt}$. *(PTU, 2018)*
4. At a given instant the sides of a rectangle are 4 ft. and 3 ft. respectively and they are increasing at the rate of 1.5 ft./sec. and 0.5 ft./sec. respectively, find the rate at which the area is increasing at that instant.
5. If $z = 2xy^2 - 3x^2y$ and if $x$ increases at the rate of 2 cm. per second and it passes through the value $x = 3$ cm., show that if $y$ is passing through the value $y = 1$ cm., $y$ must be decreasing at the rate of $2\frac{2}{15}$ cm. per second, in order that $z$ shall remain constant.
6. Let $w = 4x^2 + 4y^2 + z^2$ whose $x = \rho\sin\phi\cos\theta$, $y = \rho\sin\phi\sin\theta$, $z = \rho\cos\phi$. Find $\frac{\partial w}{\partial\rho}$ using chain rule. *(K.T.U., 2016)*
7. If $u = x^2 + y^2 + z^2$ and $x = e^{2t}$, $y = e^{2t}\cos 3t$, $z = e^{2t}\sin 3t$. Find $\frac{du}{dt}$ as a total derivative and verify the result by direct substitution. *(V.T.U., 2016)*
8. If $\phi(cx - az, cy - bz) = 0$, show that $\frac{a\partial z}{\partial x} = \frac{b\partial z}{\partial y} = c$.
9. If $f(x, y) = 0$, $\phi(y, z) = 0$, show that $\frac{\partial f}{\partial y}\cdot\frac{\partial\phi}{\partial z}\cdot\frac{dz}{dx} = \frac{\partial f}{\partial x}\cdot\frac{\partial\phi}{\partial y}$.
10. If the curves $f(x, y) = 0$ and $\phi(y, z) = 0$ touch, show that at the point of contact, $\frac{\partial f}{\partial x}\cdot\frac{\partial\phi}{\partial y} = \frac{\partial f}{\partial y}\cdot\frac{\partial\varphi}{\partial\gamma}$.
11. If $f(x, y) = 0$, show that $\left(\frac{\partial f}{\partial y}\right)^3\frac{d^2y}{dx^2} = 2\left(\frac{\partial f}{\partial x}\right)\left(\frac{\partial f}{\partial y}\right)\left(\frac{\partial^2 f}{\partial x\partial y}\right) - \left(\frac{\partial f}{dy}\right)^2\frac{\partial^2 f}{\partial x^2} - \left(\frac{\partial f}{\partial x}\right)^2\left(\frac{\partial^2 f}{\partial y^2}\right)$.
12. Find $\frac{du}{dt}$, if $u = \frac{x}{y}$, where $x = e^t$, $y = \log t$.

13. If $z = x^3 + y^3$, where $x = a \cos t, y = b \sin t$, find $\dfrac{dz}{dt}$ and verify the result by actual substitution.

14. Using partial derivatives, find $\dfrac{dy}{dx}$ for the following implicit functions. *(JNTU, 2014)*

(a) $\dfrac{x^2}{a^2} + \dfrac{y^2}{b^2} = 1$ (b) $\sin(xy) - e^{xy} - x^2y = 0$ (c) $e^x + e^y = 2xy$

15. Find $\dfrac{d^2y}{dx^2}$ (using partial derivatives) for the following functions.

(a) $ax^2 + 2hxy + by^2 = 1$ (b) $x^3 - 3ax^2 + y^3 = 0$ (c) $x^4 + y^4 = 4a^2 xy$

## 5.6 CHANGE OF VARIABLES

If $$u = f(x, y) \quad \text{...(1)}$$

where $$x = \phi(s, t) \text{ and } y = \Psi(s, t) \quad \text{...(2)}$$

it is often necessary to change expressions involving $u, x, y, \partial u/\partial x, \partial u/\partial y$ etc. to expressions involving $u, s, t, \partial u/\partial s, \partial u/\partial t$ etc.

The necessary formulae for the change of variables are easily obtained. If $t$ is regarded as a constant, then $x, y, u$ will be functions of $s$ alone. Therefore, by Eqn. (1) of page 223, we have

$$\frac{\partial u}{\partial s} = \frac{\partial u}{\partial x} \cdot \frac{\partial x}{\partial s} + \frac{\partial u}{\partial y} \cdot \frac{\partial y}{\partial s} \quad \text{...(3)}$$

where the ordinary derivatives have been replaced by the partial derivatives because $x, y$ are functions of two variables $s$ and $t$.

Similarly, regarding $s$ as constant, we obtain $$\frac{\partial u}{\partial t} = \frac{\partial u}{\partial x} \cdot \frac{\partial x}{\partial t} + \frac{\partial u}{\partial y} \cdot \frac{\partial y}{\partial t} \quad \text{...(4)}$$

On solving Eqns. (3) and (4) as simultaneous equations in $\partial u/\partial x$ and $\partial u/\partial y$, we get their values in terms of $\partial u/\partial s, \partial u/\partial t, u, s, t$.

If instead of the Eqn. (2), $s$ and $t$ are given in terms of $x$ and $y$, say: $s = \xi(x, y)$ and $t = \eta(x, y)$, ...(5)

then it is easier to use the formulae $$\frac{\partial u}{\partial x} = \frac{\partial u}{\partial s} \cdot \frac{\partial s}{\partial x} + \frac{\partial u}{\partial t} \cdot \frac{\partial t}{\partial x} \quad \text{...(6)}$$

$$\frac{\partial u}{\partial y} = \frac{\partial u}{\partial s} \cdot \frac{\partial s}{\partial y} + \frac{\partial u}{\partial t} \cdot \frac{\partial t}{\partial y} \quad \text{...(7)}$$

The higher derivatives of $u$ can be found by repeated application of formulae Eqns. (3) and (4) or of Eqns. (6) and (7).

**Example 5.22.** *If $u = F(x - y, y - z, z - x)$, prove that*

$$\frac{\partial u}{\partial x} + \frac{\partial u}{\partial y} + \frac{\partial u}{\partial z} = 0$$ *(Andhra, 2023; Anna, 2016 ; K.T.U., 2016 ; P.T.U., 2013)*

**Solution.** Put $x - y = r, y - z = s$ and $z - x = t$, so that $u = f(r, s, t)$.

$$\therefore \quad \frac{\partial u}{\partial x} = \frac{\partial u}{\partial r} \cdot \frac{\partial r}{\partial x} + \frac{\partial u}{\partial s} \cdot \frac{\partial s}{\partial x} + \frac{\partial u}{\partial t} \cdot \frac{\partial t}{\partial x}$$

$$= \frac{\partial u}{\partial r} \cdot (1) + \frac{\partial x}{\partial s} \cdot (0) + \frac{\partial u}{\partial t} \cdot (-1) = \frac{\partial u}{\partial r} - \frac{\partial u}{\partial t} \quad \text{...}(i)$$

Similarly, $$\frac{\partial u}{\partial y}=\frac{\partial u}{\partial r}\cdot\frac{\partial r}{\partial y}+\frac{\partial u}{\partial s}\cdot\frac{\partial s}{\partial y}+\frac{\partial u}{\partial t}\cdot\frac{\partial t}{\partial y}=-\frac{\partial u}{\partial r}+\frac{\partial u}{\partial s} \quad ...(ii)$$

and $$\frac{\partial u}{\partial z}=\frac{\partial u}{\partial r}\cdot\frac{\partial r}{\partial z}+\frac{\partial u}{\partial s}\cdot\frac{\partial s}{\partial z}+\frac{\partial u}{\partial t}\cdot\frac{\partial t}{\partial z}=-\frac{\partial u}{\partial s}+\frac{\partial u}{\partial t} \quad ...(iii)$$

Adding Eqns. (*i*), (*ii*) and (*iii*), we get the required result.

**Example 5.23.** *If $z = f(x, y)$ and $x = e^u \cos v$, $y = e^u \sin v$,*

*prove that* $$x\frac{\partial z}{\partial v}+y\frac{\partial z}{\partial u}=e^{2u}\frac{\partial z}{\partial y}$$

*and* $$\left(\frac{\partial z}{\partial x}\right)^2+\left(\frac{\partial z}{\partial y}\right)^2=e^{-2u}\left[\left(\frac{\partial z}{\partial u}\right)^2+\left(\frac{\partial u}{\partial v}\right)^2\right] \quad \textit{(U.P.T.U., 2013)}$$

**Solution.** We have $$\frac{\partial z}{\partial u}=\frac{\partial z}{\partial x}\cdot\frac{\partial x}{\partial u}+\frac{\partial z}{\partial y}\cdot\frac{\partial y}{\partial u}$$

$$=\frac{\partial z}{\partial x}(e^u\cos v)+\frac{\partial z}{\partial y}(e^u\sin v) \quad ...(i)$$

$$\frac{\partial z}{\partial v}=\frac{\partial z}{\partial x}\cdot\frac{\partial x}{\partial v}+\frac{\partial z}{\partial y}\cdot\frac{\partial y}{\partial v}$$

$$=\frac{\partial z}{\partial x}(-e^u\sin v)+\frac{\partial x}{\partial y}(e^u\cos v) \quad ...(ii)$$

$$x\frac{\partial x}{\partial v}+y\frac{\partial z}{\partial u}=(e^u\cos v)\left[-e^u\sin v\frac{\partial z}{\partial x}+e^u\cos v\frac{\partial z}{\partial y}\right]+(e^u\sin v)\left[e^u\cos v\frac{\partial z}{\partial x}+e^u\sin v\frac{\partial z}{\partial y}\right]$$

$$=(e^{2u}\cos^2 v+e^{2u}\sin^2 v)\frac{\partial z}{\partial y}=e^{2u}\frac{\partial z}{\partial y}$$

Now squaring Eqns. (*i*) and (*ii*) and adding, we get

$$\left(\frac{\partial z}{\partial u}\right)^2+\left(\frac{\partial z}{\partial v}\right)^2=e^{2u}\left(\cos v\frac{\partial z}{\partial x}+\sin v\frac{\partial z}{\partial y}\right)^2+e^{2u}\left(-\sin v\frac{\partial z}{\partial x}+\cos v\frac{\partial z}{\partial y}\right)^2$$

or $$e^{-2u}\left[\left(\frac{\partial z}{\partial u}\right)^2+\left(\frac{\partial z}{\partial v}\right)^2\right]=\cos^2 v\left(\frac{\partial z}{\partial x}\right)^2+\sin^2 v\left(\frac{\partial z}{\partial y}\right)^2+2\sin v\cos v\frac{\delta z}{\delta z}\frac{\delta z}{\delta y}$$

$$+\sin^2 v\left(\frac{\partial z}{\partial x}\right)^2+\cos^2 v\left(\frac{\partial z}{\partial v}\right)^2-2\sin v\cos v\frac{\partial z}{\partial x}\frac{\partial z}{\partial y}$$

$$=(\cos^2 v+\sin^2 v)\left(\frac{\partial z}{\partial x}\right)^2+\left(\frac{\partial z}{\partial y}\right)^2$$

Hence $$\left(\frac{\partial z}{\partial x}\right)^2+\left(\frac{\partial z}{\partial y}\right)^2=e^{-2u}\left[\left(\frac{\partial z}{\partial u}\right)^2+\left(\frac{\partial z}{\partial v}\right)^2\right].$$

**Example 5.24.** *If f(x, y) = c, obtain* $\frac{\delta^2 y}{\delta x^2}$ *(UTU, 2019)*

**Solution.** We have $f(x, y) = c$

Differentiating, we get

$$\frac{\delta f}{\delta x} + \frac{\delta f}{\delta y} \cdot \frac{\delta y}{\delta x} = 0$$

$$\frac{\delta y}{\delta x} = -\frac{fx}{fy} \quad ...(i)$$

$$\therefore \quad \frac{\delta^2 y}{\delta x^2} = \frac{\delta}{\delta x}\left(\frac{dy}{dx}\right) = \frac{d}{dx}\left(-\frac{fx}{fy}\right)$$

$$= \frac{-\left[fy \frac{d}{dx}(fx) - fx \cdot \frac{d}{dy}(fy)\right]}{(fy)^2} \quad ...(ii)$$

Now,

$$\frac{\delta}{\delta x}(fx) = \frac{\delta}{\delta x}(fx) + \frac{\delta}{\delta y}(fx)\frac{dy}{dx} = fxx + fyx\left(-\frac{fx}{fy}\right)$$

$$= \frac{fy\ fx\ x - fx\ fy\ y}{fy} \quad (\because f_{yx} = f_{xy})$$

Also,

$$\frac{\delta}{dx}(fy) = \frac{\delta}{\delta x}(fy) + \frac{\delta y}{\delta y}(fu) \cdot \frac{\delta y}{\delta x} = fxy + fyy\left(-\frac{fx}{fy}\right)$$

$$= \frac{fy\ fxy - fx\ fyy}{fy}$$

Substituting these values in equation (*ii*), we get

$$\frac{\delta^2 y}{dx^2} = \frac{(fxx\ fy^2 - 2fx\ fy\ fxy + fyy\ fx^2)}{(fy)^3}$$

**Example 5.25.** *Transform the equation* $\frac{\partial^2 u}{\partial x^2} + \frac{\partial^2 u}{\partial y^2} = 0$ *into polar coordinates.*

*(W.B.T.U., 2016 ; Delhi, 2013 ; P.T.U., 2013)*

**Solution.** We have $x = r\cos\theta,\ y = r\sin\theta$ and $r = \sqrt{(x^2 + y^2)},\ \theta = \tan^{-1}(y/x)$

$$\frac{\partial r}{\partial x} = \frac{x}{\sqrt{(x^2 + y^2)}} = \cos\theta \text{ and } \frac{\partial \theta}{\partial x} = -\frac{y}{x^2 + y^2} = -\frac{\sin\theta}{r}$$

Thus,

$$\frac{\partial u}{\partial x} = \frac{\partial u}{\partial r} \cdot \frac{\partial r}{\partial x} + \frac{\partial u}{\partial \theta} \cdot \frac{\partial \theta}{\partial x} = \cos\theta \frac{\partial u}{\partial r} - \frac{\sin\theta}{r}\frac{\partial u}{\partial \theta}$$

*i.e.,* $\frac{\partial}{\partial x} = \cos\theta \frac{\partial}{\partial r} - \frac{\sin\theta}{r}\frac{\partial}{\partial \theta}$. Similarly, $\frac{\partial}{\partial y} = \sin\theta \frac{\partial}{\partial r} + \frac{\cos\theta}{r}\frac{\partial}{\partial \theta}$.

$$\therefore \quad \frac{\partial^2 u}{\partial x^2} = \frac{\partial}{\partial x}\left(\frac{\partial u}{\partial x}\right) = \left(\cos\theta \frac{\partial}{\partial r} - \frac{\sin\theta}{r}\frac{\partial}{\partial \theta}\right)\left(\cos\theta \frac{\partial u}{\partial r} - \frac{\sin\theta}{r}\frac{\partial u}{\partial \theta}\right)$$

$$= \cos^2\theta \frac{\partial^2 u}{\partial r^2} - \frac{2\sin\theta\cos\theta}{r}\frac{\partial^2 u}{\partial r\partial\theta} + \frac{\sin^2\theta}{r^2}\frac{\partial^2 u}{\partial \theta^2} + \frac{\sin^2\theta}{r}\frac{\partial u}{\partial r} + \frac{2\sin\theta\cos\theta}{r^2}\frac{\partial u}{\partial \theta} \quad ...(i)$$

and

$$\frac{\partial^2 u}{\partial y^2} = \frac{\partial}{\partial y}\left(\frac{\partial u}{\partial y}\right) = \left(\sin\theta \frac{\partial}{\partial r} + \frac{\cos\theta}{r}\frac{\partial}{\partial \theta}\right)\left(\sin\theta \frac{\partial u}{\partial r} + \frac{\cos\theta}{r}\frac{\partial u}{\partial \theta}\right)$$

$$= \sin^2\theta\frac{\partial^2 u}{\partial r^2} + \frac{2\sin\theta\cos\theta}{r}\frac{\partial^2 u}{\partial r\partial\theta} + \frac{\cos^2\theta}{r^2}\frac{\partial^2 u}{\partial\theta^2} + \frac{\cos^2\theta}{r}\frac{\partial u}{\partial r} - \frac{2\sin\theta\cos\theta}{r^2}\frac{\partial u}{\partial r} \quad ...(ii)$$

Adding Eqns. (*i*) and (*ii*), we get $\frac{\partial^2 u}{\partial x^2} + \frac{\partial^2 u}{\partial y^2} = \frac{\partial^2 u}{\partial r^2} + \frac{1}{r^2}\frac{\partial^2 u}{\partial\theta^2} + \frac{1}{r}\frac{\partial u}{\partial r}$.

Hence the transformed equation is $\frac{\partial^2 u}{\partial r^2} + \frac{1}{r}\frac{\partial u}{\partial r} + \frac{1}{r^2}\frac{\partial^2 u}{\partial\theta^2} = 0$.

**Example 5.26.** *If $v = f(x, y)$ and $x = r\cos\theta$, $y = r\sin\theta$ then show that* $\frac{\delta^2 y}{\delta r^2} + \frac{1}{r}\frac{\delta v}{\delta r} + \frac{\delta^2 v}{\delta\theta^2} = \left(\frac{\delta^2 v}{\delta x^2} = \frac{\delta^2 v}{\delta y^2}\right)$

**Solution.** We have $x = r\cos\theta$, $y = r\sin\theta$

$$\therefore \quad \frac{\delta x}{\delta r} = \cos\theta\ \frac{x}{r},\ \frac{\delta y}{\delta r} = \sin\theta = \frac{y}{r} \quad ...(i)$$

$$\frac{\delta x}{\delta\theta} = -\sin\theta = -y,\ \frac{\delta y}{d\theta} = r\cos\theta = x$$

Now
$$\frac{\delta v}{dr} = \frac{\delta v}{dx}\cdot\frac{\delta x}{\delta r} + \frac{\delta v}{\delta y}\cdot\frac{\delta y}{\delta r} = \frac{x}{\delta r}\cdot\frac{\delta v}{\delta x} + \frac{y}{r}\cdot\frac{\delta v}{\delta y}$$

$$r - \frac{\delta v}{\delta r} = x\frac{\delta v}{\delta x} + y\frac{\delta v}{\delta y} \quad ...(ii)$$

Similarly
$$\frac{\delta v}{\delta\theta} = \frac{\delta v}{\delta x}\cdot\frac{\delta x}{\delta\theta} + \frac{\delta v}{\delta y}\cdot\frac{\delta y}{\delta\theta}$$

$$\frac{\delta v}{\delta\theta} = -y\frac{\delta y}{\delta x} + x\cdot\frac{\delta v}{\delta y} \quad ...(iii)$$

Differentiating Eqn. (*ii*) *w.r.t.* and Eqn. (*iii*) *w.r.t.* $\theta$

We get
$$r\frac{\delta^2 v}{\delta r^2} + \frac{\delta v}{\delta r} = x\frac{\delta^2 v}{\delta x^2}\cdot\frac{\delta x}{\delta r} + \frac{\delta x}{\delta r}\cdot\frac{\delta v}{\delta x} + y\frac{\delta^2 v}{\delta y^2}\cdot\frac{\delta y}{\delta r} + \frac{\delta y}{\delta r}\cdot\frac{\delta v}{\delta y} \quad ...(iv)$$

$$\frac{\delta^2 v}{\delta\theta^2} = -y\frac{\delta^2 v}{\delta x^2}\cdot\frac{\delta x}{\delta\theta} - \frac{\delta y}{\delta\theta},\frac{\delta v}{\delta x} + x\frac{\delta^2 v}{\delta y^2}\cdot\frac{\delta y}{\delta\theta} + \frac{\delta x}{\delta\theta}\cdot\frac{\delta v}{\delta y} \quad ...(v)$$

Substituting the values from Eqn. (*i*), we have from Eqn. (*iv*)

$$r\frac{\delta^2 v}{\delta r^2} + \frac{\delta v}{\delta r} = \frac{x^2}{r}\cdot\frac{\delta^2 v}{\delta x^2} + \frac{x}{r}\frac{\delta\theta}{\delta x} + \frac{y^2}{r}\frac{\delta^2 v}{\delta^2 y} + \frac{y}{r}\frac{\delta v}{\delta y}$$

or
$$r^2\frac{\delta^2 v}{\delta r^2} + r\frac{\delta v}{\delta r} = \left(x^2\frac{\delta^2 v}{\delta x^2} + y^2\frac{\delta^2 v}{\delta y^2}\right) + \left(x\frac{\delta v}{\delta x} + y\frac{\delta v}{\delta y}\right) \quad ...(vi)$$

Similarly from Eqn. (*v*), we have

$$\frac{\delta^2 y}{\delta\theta^2} = y^2\frac{\delta^2 v}{\delta x^2} + x^2\frac{\delta^2 y}{\delta y^2} - \left(x\frac{\delta v}{\delta x} + y\frac{\delta v}{\delta y}\right) \quad ...(vii)$$

Adding Eqns. (*vi*) and (*vii*), we get

$$r^2\frac{\delta^2 v}{\delta r^2} + r\frac{\delta v}{\delta r} + \frac{\delta^2 v}{\delta\theta^2} = (x^2 + y^2)\frac{\delta^2 v}{\delta x^2} + (x^2 + y^2)\frac{\delta^2 v}{\delta y^2}$$

$$\Rightarrow \quad r^2\frac{\delta^2 v}{\delta r^2} + r\frac{\delta v}{\delta r} + \frac{\delta^2 v}{\delta\theta^2} = r^2\left(\frac{\delta^2 v}{\delta x^2} + \frac{\delta^2 v}{\delta y^2}\right) \quad (\because x^2 + y^2 = r^2)$$

$$\Rightarrow \qquad \frac{\delta^2 y}{\delta r^2} + \frac{1}{r}\frac{\delta v}{\delta r} + \frac{\delta^2 v}{\delta \theta^2} = \left(\frac{\delta^2 v}{\delta x^2} = \frac{\delta^2 v}{\delta y^2}\right).$$

**Example 5.27.** *If $z = f(x, y)$ where $x = e^v \cos v$, $y = e^u \sin u$ show that $x\frac{\delta f}{\delta v} + y\frac{\delta f}{\delta u} = e^{2u}\frac{\delta f}{\delta y}$*

**Solution.** We know,

$$\frac{\delta f}{\delta u} = \frac{\delta f}{\delta x}\cdot\frac{\delta x}{\delta v} + \frac{\delta f}{\delta y}\cdot\frac{\delta y}{\delta u} \qquad ...(i)$$

$$\frac{\delta f}{\delta v} = \frac{\delta f}{\delta x}\cdot\frac{\delta x}{\delta v} + \frac{\delta f}{\delta y}\cdot\frac{\delta y}{\delta v} \qquad ...(ii)$$

Since $x = e^u \cos v$ and $y = e^u \sin v$

$$\frac{\delta x}{\delta u} = e^u \cos v = x \quad \frac{\delta y}{\delta v} = e^u \sin v = y$$

$$\frac{\delta x}{\delta u} = -e^u \sin v = -y \text{ and } \frac{\delta y}{\delta v} = e^u \cos v = x$$

Substituting these values in Eqns. (*i*) and (*ii*) we have,

$$\frac{\delta f}{\delta v} = x\frac{\delta f}{\delta x} + y\frac{\delta f}{\delta y} \qquad ...(iii)$$

$$\frac{\delta f}{\delta v} = -y\frac{\delta f}{\delta x} + x\frac{\delta f}{\delta y} \qquad ...(iv)$$

Multiplying Eqn. (*iii*) by $y$ and Eqn. (*iv*) by $x$ and adding we get

$$\frac{y\delta f}{\delta u} + x\frac{\delta f}{\delta v} = (x^2 + y^2)\frac{\delta f}{\delta y}$$

$$= (e^{2u}\cos^2 u + e^{2u}\sin^2 v)\frac{\delta f}{\delta y}$$

$$= x\frac{\delta f}{\delta u} + y\frac{\delta f}{\delta v} = e^{2u}\frac{\delta f}{\delta y}.$$

**Example. 5.28.** *If $u = f(r, s, t)$ and $r = \frac{x}{y}, s = \frac{y}{z}, t = \frac{z}{x}$. Show that $x\frac{\delta u}{\delta x} + y\frac{\delta u}{\delta y} + z\frac{\delta u}{\delta z} = 0$.*

**Solution.** We have

$$\frac{\delta u}{\delta x} = \frac{\delta f}{\delta r}\cdot\frac{\delta r}{\delta x} + \frac{\delta f}{\delta s}\cdot\frac{\delta s}{\delta x} + \frac{\delta f}{\delta t}\cdot\frac{\delta t}{\delta x} \qquad ...(i)$$

Since $\qquad r = \frac{x}{y}, s = \frac{y}{z}, t = \frac{z}{x}$

$$\frac{\delta r}{\delta x} = \frac{1}{y}, \frac{\delta s}{\delta x} = 0 \quad \frac{\delta t}{\delta x} = \frac{-z}{x^2}$$

Substituting these values in Eqn. (*i*), we have

$$\frac{\delta u}{\delta x} = \frac{1}{y}\frac{\delta f}{\delta r} - \frac{z}{x^2}\frac{\delta f}{\delta t} \qquad ...(ii)$$

Similarly,

$$\frac{\delta u}{\delta y} = \left(-\frac{x}{y^2}\right)\frac{\delta f}{\delta r} + \frac{1}{z}\frac{\delta f}{\delta s} \qquad ...(iii)$$

and

$$\frac{\delta u}{\delta z} = \left(\frac{-y}{z^2}\right)\left(\frac{\delta f}{\delta s}\right) + \frac{1}{x}\frac{\delta f}{\delta t} \qquad ...(iv)$$

Multiplying Eqn. (*ii*) by $x$, Eqn. (*iii*) by $y$, Eqn. (*iv*) by $z$ and adding we get

$$x\frac{\delta u}{\delta x}+y\frac{\delta u}{\delta y}+z\frac{\delta u}{\delta z}=0.$$

## PROBLEMS 5.6

1. If $z = f(x, y)$ and $x = e^u + e^{-v}$, $y = e^{-u} - e^v$, prove that $\frac{\partial z}{\partial u}-\frac{\partial z}{\partial v}=x\frac{\partial z}{\partial x}-y\frac{\partial z}{\partial y}$. (*V.T.U., 2016 ; D.T.U., 2015*)
2. If $u = f(r, s)$, $r = x + at$, $s = y + bt$ and $x, y, t$ are independent variables, show that $\frac{\partial u}{\partial t}=a\frac{\partial u}{\partial x}+b\frac{\partial u}{\partial y}$.
3. If $u = f(x, y)$ and $x = r\cos\theta$, $y = r\sin\theta$, prove that $\left(\frac{\partial u}{\partial x}\right)^2+\left(\frac{\partial u}{\partial y}\right)^2=\left(\frac{\partial u}{\partial r}\right)^2+\frac{1}{r^2}\left(\frac{\partial u}{\partial \theta}\right)^2$.
   (*W.B.T.U., 2016 ; V.T.U., 2015 ; Anna, 2013*)
4. If $u = f(2x - 3y, 3y - 4z, 4z - 2x)$, prove that $\frac{1}{2}\frac{\partial u}{\partial x}+\frac{1}{3}\frac{\partial u}{\partial y}+\frac{1}{4}\frac{\partial u}{\partial z}=0$. (*V.T.U., 2014 ; P.T.U., 2011*)
5. If $z = f(x, y)$ and $x = e^{-u} + e^{-v}$, $y = e^{-u} - e^v$ show that $\frac{\delta z}{\delta u}-\frac{\delta z}{\delta v}=x\frac{\delta z}{\delta x}-y\frac{\delta z}{\delta y}$. (*AKTU, 2019*)
6. If $u = f(r, s, t)$ and $r = x/y$, $s = y/z$, $t = z/x$, prove that $x\frac{\partial u}{\partial x}+y\frac{\partial u}{\partial y}+z\frac{\partial u}{\partial z}=0$.
   (*V.T.U., 2016 ; U.K.T.U., 2012*)
7. If $x = u + v + w$, $y = vw + wu + uv$, $z = uvw$ and $F$ is a function of $x, y, z$, show that
   $$u\frac{\partial F}{\partial u}+v\frac{\partial F}{\partial v}+w\frac{\partial F}{\partial w}=x\frac{\partial F}{\partial x}+2y\frac{\partial F}{\partial y}+3z\frac{\partial F}{\partial z}.$$
8. Given that $u(x, y, z) = f(x^2 + y^2 + z^2)$ where $x = r\cos\theta\cos\phi$, $y = r\cos\theta\sin\phi$ and $z = r\sin\theta$, find $\frac{\partial u}{\partial \theta}$ and $\frac{\partial u}{\partial \phi}$.
9. If the three thermodynamic variables $P, V, T$ are connected by a relation $f(P, V, T) = 0$, show that
   $$\left(\frac{\partial P}{\partial T}\right)_V\left(\frac{\partial T}{\partial V}\right)_P\left(\frac{\partial V}{\partial P}\right)_T=-1.$$ (*P.T.U., 2011 S*)
10. If by the substitution $u = x^2 - y^2$, $v = 2xy$, $f(x, y) = \theta(u, v)$, show that
    $$\frac{\partial^2 f}{\partial x^2}+\frac{\partial^2 f}{\partial y^2}=4(x^2+y^2)\left(\frac{\partial^2\theta}{\partial u^2}+\frac{\partial^2\theta}{\partial v^2}\right).$$
11. Transform $\frac{\partial^2 z}{\partial x^2}+2xy^2\frac{\partial z}{\partial x}+2(y-y^3)\frac{\partial z}{\partial y}+x^2y^2z=0$ by the substitution $x = uv$, $y = 1/v$. Hence show that $z$ is the same function of $u$ and $v$ as of $x$ and $y$.
12. Given, $V = f(x, y, z)$, where $x = r\sin\theta\cos\phi$, $y = r\sin\theta\sin\phi$ and $z = r\cos\theta$, show that
    $$\frac{\delta^2 V}{\delta x^2}+\frac{\delta^2 V}{\delta y^2}+\frac{\delta^2 V}{\delta z^2}=\frac{\delta^2 V}{\delta r^2}+\frac{2}{r}\frac{\delta V}{\delta r}+\frac{1}{r^2}\frac{\delta^2 V}{\delta\theta^2}+\frac{\cot\theta}{r^2}\frac{\delta V}{\delta\theta}+\frac{1}{r^2\sin^2\phi}\frac{\delta^2 V}{\delta\phi^2}.$$ (*S.P.U., Pune, 2017*)
13. If $u = f(r, s)$; where $r = x + y$, $s = x - y$, Prove that $\frac{\delta u}{\delta x}+\frac{\delta u}{\delta y}=2\frac{\delta u}{\delta r}$.

## 5.7 (1) JACOBIANS

If $u$ and $v$ are functions of two independent variables $x$ and $y$, then the determinant

$$\begin{vmatrix}\partial u/\partial x & \partial u/\partial y\\ \partial v/\partial x & \partial v/\partial y\end{vmatrix}$$ is called the *Jacobian** of $u, v$ with respect to $x, y$

* Called after the German mathematician *Carl Gustav Jacob Jacobi* (1804–1851), who made significant contributions to mechanics, partial differential equations, astronomy, elliptic functions and the calculus of variations.

and is written as $\dfrac{\partial(u, v)}{\partial(x, y)}$ or $J\left(\dfrac{u, v}{x, y}\right)$.

Similarly the Jacobian of $u, v, w$ with respect to $x, y, z$ is

$$\frac{\partial(u, v, w)}{\partial(x, y, z)} = \begin{vmatrix} \partial u/\partial x & \partial u/\partial y & \partial u/\partial z \\ \partial v/\partial x & \partial v/\partial y & \partial v/\partial z \\ \partial w/\partial x & \partial w/\partial y & \partial w/\partial z \end{vmatrix}$$

Likewise, we can define Jacobians of four or more variables. An important application of Jacobians is in connection with the change of variables in multiple integrals (§ 7.7).

**(2) Properties of Jacobians.** We give below two of the important properties of Jacobians. For simplicity, the properties are stated in terms of two variables only, but these are evidently true in general.

*I.* If $J = \partial(u, v)/\partial(x, y)$ and $J' = \partial(x, y)/\partial(u, v)$ then $JJ' = 1$.

Let $u = f(x, y)$ and $v = g(x, y)$.

Suppose, on solving for $x$ and $y$, we get $x = \phi(u, v)$ and $y = \psi(u, v)$.

Then
$$\frac{\partial u}{\partial u} = 1 = \frac{\partial u}{\partial x}\cdot\frac{\partial x}{\partial u} + \frac{\partial u}{\partial y}\cdot\frac{\partial y}{\partial u},$$
$$\frac{\partial u}{\partial v} = 0 = \frac{\partial u}{\partial x}\cdot\frac{\partial x}{\partial v} + \frac{\partial u}{\partial y}\cdot\frac{\partial y}{\partial v}, \qquad ...(i)$$
$$\frac{\partial v}{\partial u} = 0 = \frac{\partial v}{\partial x}\cdot\frac{\partial x}{\partial u} + \frac{\partial v}{\partial y}\cdot\frac{\partial y}{\partial u},$$
and
$$\frac{\partial v}{\partial v} = 1 = \frac{\partial v}{\partial x}\cdot\frac{\partial x}{\partial v} + \frac{\partial v}{\partial y}\cdot\frac{\partial y}{\partial v},$$

$\therefore$
$$JJ' = \begin{vmatrix} \partial u/\partial x & \partial u/\partial y \\ \partial v/\partial x & \partial v/\partial y \end{vmatrix} \times \begin{vmatrix} \partial x/\partial u & \partial x/\partial v \\ \partial y/\partial u & \partial y/\partial v \end{vmatrix} = \begin{vmatrix} \partial u/\partial x & \partial u/\partial y \\ \partial y/\partial x & \partial v/\partial y \end{vmatrix} \times \begin{vmatrix} \partial x/\partial u & \partial y/\partial u \\ \partial x/\partial v & \partial y/\partial v \end{vmatrix}$$

(Interchanging rows and columns of the 2nd determinant).

$$= \begin{vmatrix} 1 & 0 \\ 0 & 1 \end{vmatrix} = 1. \qquad \text{[By virtue of Eqn. } (i)]$$

*II.* **Chain rule for Jacobians.** *If $u, v$ are functions of $r, s$ and $r, s$ are functions of $x, y$, then*

$$\frac{\partial(u, v)}{\partial(x, y)} = \frac{\partial(u, v)}{\partial(r, s)}\cdot\frac{\partial(r, s)}{\partial(x, y)}.$$

$$\frac{\partial(u, v)}{\partial(r, s)}\cdot\frac{\partial(r, s)}{\partial(x, y)} = \begin{vmatrix} \partial u/\partial r & \partial u/\partial s \\ \partial v/\partial r & \partial v/\partial s \end{vmatrix} \times \begin{vmatrix} \partial r/\partial x & \partial s/\partial x \\ \partial r/\partial y & \partial s/\partial y \end{vmatrix}$$

[Interchanging rows and columns of the 2nd det.]

$$= \begin{vmatrix} \dfrac{\partial u}{\partial r}\cdot\dfrac{\partial r}{\partial x} + \dfrac{\partial u}{\partial s}\cdot\dfrac{\partial s}{\partial x} & \dfrac{\partial u}{\partial r}\cdot\dfrac{\partial r}{\partial y} + \dfrac{\partial u}{\partial s}\cdot\dfrac{\partial s}{\partial y} \\ \dfrac{\partial v}{\partial r}\cdot\dfrac{\partial r}{\partial x} + \dfrac{\partial v}{\partial s}\cdot\dfrac{\partial s}{\partial x} & \dfrac{\partial v}{\partial r}\cdot\dfrac{\partial r}{\partial y} + \dfrac{\partial v}{\partial s}\cdot\dfrac{\partial s}{\partial y} \end{vmatrix} = \begin{vmatrix} \partial u/\partial x & \partial u/\partial y \\ \partial v/\partial x & \partial v/\partial y \end{vmatrix} = \frac{\partial(u, v)}{\partial(x, y)}.$$

**Example 5.29.** *(i) In polar coordinates, $x = r\cos\theta$, $y = r\sin\theta$, show that*

$$\frac{\partial(x, y)}{\partial(r, \theta)} = r. \qquad \textit{(Anna., 2016 ; Andhra., 2015)}$$

*(ii) In cylindrical coordinates (Fig. 8.18), $x = \rho\cos\phi$, $y = \rho\sin\phi$, $z = z$, show that*

$$\frac{\partial(x, y, z)}{\partial(\rho, \phi, z)} = \rho.$$

*(iii) In spherical polar coordidnates (Fig. 8.19), $x = r\sin\theta\cos\phi$, $y = r\sin\theta\sin\phi$, $z = r\cos\theta$, show that*

$$\frac{\partial(x, y, z)}{\partial(r, \theta, \phi)} = r^2\sin\theta. \qquad \textit{(Anna., 2016 ; V.T.U., 2013)}$$

**Solution.** (*i*) We have $\frac{\partial x}{\partial r} = \cos\theta, \frac{\partial x}{\partial \theta} = -r\sin\theta$ and $\frac{\partial y}{\partial r} = \sin\theta, \frac{\partial y}{\partial \theta} = -r\cos\theta$

$$\therefore \quad \frac{\partial(x, y)}{\partial(r, \theta)} = \begin{vmatrix} \cos\theta & -r\sin\theta \\ \sin\theta & r\cos\theta \end{vmatrix} = r$$

(*ii*) We have $\frac{\partial x}{\partial \rho} = \cos\phi, \frac{\partial x}{\partial \phi} = -\rho\sin\phi, \frac{\partial x}{\partial z} = 0,$

$$\frac{\partial y}{\partial \rho} = \sin\phi, \frac{\partial y}{\partial \phi} = \rho\cos\phi, \frac{\partial y}{\partial z} = 0 \text{ and } \frac{\partial z}{\partial \rho} = 0, \frac{\partial z}{\partial \phi} = 0, \frac{\partial z}{\partial z} = 1$$

$$\therefore \quad \frac{\partial(x, y, z)}{\partial(\rho, \phi, z)} = \begin{vmatrix} \cos\phi & -\rho\sin\theta & 0 \\ \sin\phi & \rho\cos\phi & 0 \\ 0 & 0 & 1 \end{vmatrix} = \rho.$$

(*iii*) We have $\frac{\partial x}{\partial r} = \sin\theta\cos\theta, \frac{\partial x}{\partial \theta} = r\cos\theta\cos\phi, \frac{\partial x}{\partial \phi} = -r\sin\theta\sin\phi,$

$$\frac{\partial y}{\partial r} = \sin\theta\sin\phi, \frac{\partial y}{\partial \theta} = r\cos\theta\sin\phi, \frac{\partial y}{\partial \phi} = r\sin\theta\cos\phi,$$

and $$\frac{\partial z}{\partial r} = \cos\theta, \frac{\partial z}{\partial \theta} = -r\sin\theta, \frac{\partial z}{\partial \phi} = 0.$$

$$\therefore \quad \frac{\partial(x, y, z)}{\partial(r, \theta, \phi)} = \begin{vmatrix} \sin\theta\cos\phi & r\cos\theta\cos\phi & -r\sin\theta\sin\phi \\ \sin\theta\sin\phi & r\cos\theta\sin\phi & r\sin\theta\cos\phi \\ \cos\theta & -r\sin\theta & 0 \end{vmatrix} = r^2\sin\theta.$$

**Obs.** $\frac{\partial(\gamma, \theta, \phi)}{\partial(x, y, z)} + \frac{1}{r^2\sin\theta}$ *(Rohtak, 2011 S)*

**Example 5.30.** *If* $y_1 = \frac{x_2x_3}{x_1}, y_2 = \frac{x_3x_1}{x_2}, y_3 = \frac{x_1x_2}{x_3}$, *show that the Jacobian of* $y_1, y_2, y_3$ *with respect to* $x_1, x_2, x_3$ *is 4.* *(Patna, 2019; Anna, 2013 ; Calicut, 2013 , U.P.T.U., 2012)*

**Solution.** We have $\frac{\partial y_1}{\partial x_1} = -\frac{x_2x_3}{x_1^2}, \frac{\partial y_1}{\partial x_2} = \frac{x_3}{x_1}, \frac{\partial y_1}{\partial x_3} = \frac{x_2}{x_1}$

$$\frac{\partial y_2}{\partial x_1} = \frac{x_3}{x_2}, \frac{\partial y_2}{\partial x_2} = -\frac{x_3x_1}{x_2^2}, \frac{\partial y_2}{\partial x_3} = \frac{x_1}{x_2}$$

and $$\frac{\partial y_3}{\partial x_1} = \frac{x_2}{x_3}, \frac{\partial y_3}{\partial x_2} = \frac{x_1}{x_3}, \frac{\partial y_3}{\partial x_3} = -\frac{x_1x_2}{x_3^2}$$

$$\therefore \quad \frac{\partial(y_1y_2y_3)}{\partial(x_1x_2x_3)} = \begin{vmatrix} \frac{\partial y_1}{\partial x_1} & \frac{\partial y_1}{\partial x_2} & \frac{\partial y_1}{\partial x_3} \\ \frac{\partial y_2}{\partial x_1} & \frac{\partial y_2}{\partial x_2} & \frac{\partial y_2}{\partial x_3} \\ \frac{\partial y_3}{\partial x_1} & \frac{\partial y_3}{\partial x_2} & \frac{\partial y_3}{\partial x_3} \end{vmatrix} \begin{vmatrix} \frac{-x_2x_3}{x_1^2} & \frac{x_3}{x_1} & \frac{x_2}{x_1} \\ \frac{x_3}{x_2} & \frac{-x_3x_1}{x_2^2} & \frac{x_1}{x_1} \\ \frac{x_2}{x_3} & \frac{x_1}{x_3} & \frac{-x_1x_2}{x_3^2} \end{vmatrix}$$

$$= \frac{1}{x_1^2x_2^2x_3^2}\begin{vmatrix} -x_2x_3 & x_3x_1 & x_1x_2 \\ x_2x_3 & -x_3x_1 & x_1x_2 \\ x_2x_3 & x_3x_1 & -x_1x_2 \end{vmatrix} = -\frac{x_1^2x_2^2x_3^2}{x_1^2x_2^2x_3^2}\begin{vmatrix} -1 & 1 & 1 \\ 1 & -1 & 1 \\ 1 & 1 & -1 \end{vmatrix}$$

$$= \quad 1 - 1) - 1(-1 - 1) + 1(1 + 1) = 0 + 2 + 2 = 4.$$

**Example 5.31.** *If* $x = r\sin\theta\cos\theta, y = r\sin\theta\sin\phi, z = r\cos\theta$, *Show that* $\frac{\delta(x, y, z)}{\delta(r, \theta, \phi)} = r^2\sin\theta.$

*(Bengaluru, 2018)*

**Solution.** We know that

$$\frac{\delta(x, y, z)}{\delta(r, \theta, \phi)} = \begin{vmatrix} \frac{\delta x}{\delta r} & \frac{\delta x}{\delta \theta} & \frac{\delta x}{\delta \phi} \\ \frac{\delta y}{\delta r} & \frac{\delta y}{\delta \theta} & \frac{\delta y}{\delta \phi} \\ \frac{\delta z}{\delta r} & \frac{\delta z}{\delta \theta} & \frac{\delta z}{\delta \phi} \end{vmatrix}$$

$$= \begin{vmatrix} \sin\theta\cos\phi & r\cos\theta\cos\phi & -r\sin\theta\sin\phi \\ \sin\theta & r\cos\theta\sin\phi & -r\sin\theta\cos\phi \\ \cos\theta & -r\sin\theta & 0 \end{vmatrix}$$

$$= \frac{r^2}{\cos\phi}\begin{vmatrix} \sin\theta\cos^2\phi & \cos\theta\cos^2\phi & -\sin\theta\sin\phi\cos\phi \\ \sin\theta\sin\phi & \cos\theta\sin\phi & \sin\theta\cos\phi \\ \cos\theta & -\sin\theta & 0 \end{vmatrix}$$

$R_1 \to R_1 \times \cos\phi$
$$= \frac{r^2}{\cos\theta}\begin{bmatrix} \sin\theta(\cos^2\phi + \sin^2\phi) & \cos\theta(\cos^2\phi + \sin^2\phi) & 0 \\ \sin\theta\sin\phi & \cos\theta\sin\phi & \sin\theta\cos\phi \\ \cos\theta & -\sin\theta & 0 \end{bmatrix}$$

$R_1 \to R_1 + R_2 \sin\phi$
$$= \frac{r^2}{\cos\phi}\begin{vmatrix} \sin\theta & \cos\theta & 0 \\ \sin\theta\sin\phi & -\sin\phi & \sin\theta\cos\phi \\ \cos\theta & -\sin\theta & 0 \end{vmatrix}$$

$$= \frac{r^2}{\cos\phi}(-\sin\theta\cos\phi)(-\sin^2\theta - \cos^2\theta) = r^2\sin\theta$$

**Example 5.32.** *If $x = r\cos\theta$, $y = r\sin\theta$,*

*Show that* $\frac{\delta(x,y)}{\delta(r,\theta)} \cdot \frac{\delta(r,\theta)}{\delta(x,y)} = 1$ *(JNTUK, 2020; A.M.I.E., 2019)*

**Solution:** We have $x = r\cos\theta, y = r\sin\theta$ ...(*i*)

$$r^2 = x^2 + y^2, \theta = \tan^{-1}(y/x) \quad ...(ii)$$

From equation (*i*) $\frac{\delta x}{\delta \theta} = \cos\theta$, $\frac{\delta y}{\delta r} = \sin\theta$

$$\frac{\delta x}{\delta \theta} = -r\sin\theta,$$

$$\frac{\delta y}{\delta \theta} = r\cos\theta$$

From eqn (*ii*),
$$\frac{\delta \theta}{\delta x} = \frac{1}{1+\frac{y^2}{x^2}}\left(\frac{-y}{x^2}\right) = -\frac{y}{x^2+y^2} = \frac{-\sin\theta}{r}$$

$$\frac{\delta \theta}{\delta y} = \frac{1}{1+\frac{y^2}{x^2}}, \frac{1}{x} = \frac{x}{x^2+y^2} = \frac{\cos\theta}{r}$$

$$\frac{\delta r}{\delta x} = \frac{x}{r} = \cos\theta, \frac{\delta r}{\delta y} = \frac{y}{r} = \sin\theta$$

$$\therefore \quad \frac{\delta(x, y)}{\delta(r, \theta)} = \begin{vmatrix} \frac{\delta x}{\delta r} & \frac{\delta x}{\delta \theta} \\ \frac{\delta y}{\delta r} & \frac{\delta y}{\delta \theta} \end{vmatrix} = \begin{vmatrix} \cos\theta & -r\sin\theta \\ -\sin\frac{\theta}{r} & r\cos\frac{\theta}{r} \end{vmatrix}$$

Also
$$\frac{\delta(r, \theta)}{\delta(x, y)} = \begin{vmatrix} \frac{\delta r}{\delta x} & \frac{\delta r}{\delta y} \\ \frac{\delta \theta}{\delta x} & \frac{\delta \theta}{\delta y} \end{vmatrix}$$

$$= \begin{vmatrix} \cos\theta & \sin\theta \\ -\sin\frac{\theta}{r} & \cos\frac{\theta}{r} \end{vmatrix}$$

$$= \frac{1}{r}(\cos^2\theta + \sin^2\theta) = \frac{1}{r}$$

$$\therefore \quad \frac{\delta(x, y)}{\delta(r, \theta)} \cdot \frac{\delta(r, \theta)}{\delta(x, y)} = r : \frac{1}{r} = 1.$$

**Example 5.33.** *If $u = xy + yz + zx$, $v = x^2 + y^2 + z^2$ and $w = x + y + z$, determine whether there is a functional relationship shop between u, v, w and if so find it.* *(Bhilai, 2019)*

**Solution:** Since $u = xy + yz + zx$, $v = x^2 + y^2 + z^2$ and $w = x + y + z$

$$\frac{\delta u}{\delta x} = y + z, \quad \frac{\delta u}{\delta y} = z + x, \quad \frac{\delta u}{\delta z} = x + y$$

$$\frac{\delta v}{\delta x} = 2x, \quad \frac{\delta v}{\delta y} = 2y, \quad \frac{\delta v}{\delta z} = 2z$$

and
$$\frac{\delta w}{\delta x} = 1, \quad \frac{\delta w}{\delta y} = 1, \quad \frac{\delta w}{\delta z} = 1$$

Now,
$$\frac{\delta(u, v, w)}{\delta(x, y, z)} = \begin{vmatrix} \frac{\delta u}{\delta x} & \frac{\delta u}{\delta y} & \frac{\delta u}{\delta z} \\ \frac{\delta v}{\delta x} & \frac{\delta v}{\delta y} & \frac{\delta v}{\delta z} \\ \frac{\delta w}{\delta x} & \frac{\delta w}{\delta y} & \frac{\delta w}{\delta z} \end{vmatrix}$$

$$= \begin{vmatrix} y+z & z+x & x+y \\ 2x & 2y & 2z \\ 1 & 1 & 1 \end{vmatrix}$$

$$= 2\begin{vmatrix} y+z & z+x & x+y \\ x & y & z \\ 1 & 1 & 1 \end{vmatrix}$$

$$= 2\begin{vmatrix} x+y+z & x+y+z & x+y+z \\ x & y & z \\ 1 & 1 & 1 \end{vmatrix}$$

$$R_1 \to R_1 + R_2 = 2(x+y+z)\begin{vmatrix} 1 & 1 & 1 \\ x & y & z \\ 1 & 1 & 1 \end{vmatrix} = 0 \qquad (\because R_1 = R_2)$$

Hence the functional relationship exists between $u, v, w$

Now $$w^2 = (x + y + z)^2$$
$$= (x^2 + y^2 + z^2) + 2(xy + yz + zx)$$
$$= v + 2u$$

or, $$w^2 - v - 24 = 0 \text{ is required relationship.}$$

**Example 5.34.** *If $u = x\sqrt{(1-y^2)} + y\sqrt{(1-x^2)}$, $v = \sin^{-1} x + \sin^{-1} y$, show that $u, v$ are functionally related and find the relationship.* (J.N.T.U., 2015)

**Solution.** We have $$\frac{\partial u}{\partial x} = \sqrt{(1-y^2)} - \frac{xy}{\sqrt{(1-x^2)}}, \frac{\partial u}{\partial y} = \frac{-xy}{\sqrt{(1-y^2)}} + \sqrt{(1-x^2)}$$

and $$\frac{\partial v}{\partial x} = \frac{1}{\sqrt{(1-x^2)}}, \frac{\partial v}{\partial y} = \frac{1}{\sqrt{(1-y^2)}}$$

$$\therefore \quad \frac{\partial(u, v)}{\partial(x, y)} = \begin{vmatrix} \frac{\partial u}{\partial x} & \frac{\partial u}{\partial y} \\ \frac{\partial v}{\partial x} & \frac{\partial v}{\partial y} \end{vmatrix} = \begin{vmatrix} \sqrt{(1-y^2)} - \frac{xy}{\sqrt{(1-x^2)}}, & \sqrt{(1-x^2)} - \frac{xy}{\sqrt{(1-y^2)}} \\ \frac{1}{\sqrt{(1-x^2)}}, & \frac{1}{\sqrt{(1-y^2)}} \end{vmatrix}$$

$$= 1 - \frac{xy}{\sqrt{[(1-x^2)(1-y^2)]}} - 1 + \frac{xy}{\sqrt{[(1-x^2)(1-y^2)]}} = 0$$

Hence $u$ and $v$ are functionally related *i.e.*, they are not independent.

We have $$v = \sin^{-1} x + \sin^{-1} y = \sin^{-1}[x\sqrt{(1-y^2)} + y\sqrt{(1-x^2)}]$$

*i.e.*, $$u = \sin v$$

which is the required relationship between $u$ and $v$.

## PROBLEMS 5.7

1. If $x = r\cos\theta, y = r\sin\theta$, evaluate $\partial(r, \theta)/\partial(x, y)$, $\partial(x, y)/\partial(r, \theta)$ and prove that $\{\partial(r, \theta)\, \partial(x, y)\}.\{\partial(x, y)/\partial(r, \theta)\} = 1$. (V.T.U., 2010)
2. If $x = u(1 + v), y = v(1 + u)$, evaluate $\partial(u, v)/\partial(x, y)$. (Andhra, 2018; U.P.T.U., 2013)
3. If $x = u(1 - v), y = uv$, prove that $JJ' = 1$. (Andhra, 2017; J.N.T.U., 2015 ; V.T.U. 2014)
4. If $u = x + 3y^2 - z^3, v = 4x^2yz, w = 2z^2 - xy$, evaluate $\dfrac{\partial(u, v, w)}{\partial(x, y, z)}$ at $(1, -1, 0)$
5. If $u^2 + v^3 = x + y, u^2 + v^2 = x^3 + y^2$. Show that $\dfrac{\delta(u, v)}{\delta(x, y)} = \dfrac{1}{2}\dfrac{(y^3 - x)^2}{uv(u - v)}$ (MCA, 2019)
6. If $u = x^2 - 2y^2, v = 2x^2 - y^2$, where $x = r\cos\theta, y = r\sin\theta$, show that $\dfrac{\partial(u, v)}{\partial(r, \theta)} = 6r^3 \sin 2\theta$.
7. If $u = x^2 - 2y, v = x + y + z, w = x - 2y + 3z$, find $\partial(u, v, w)/\partial(x, y, z)$ (W.B.T.U., 2016)
8. If $u = x^2 + y^2 + z^2, v = xy + yz + zx, w = x + y + z$, find $\dfrac{\partial(u, v, w)}{\partial(x, y, z)}$. (V.T.U., 2016)
9. If $F = xu + v - y, G = u^2 + vy + w, H = zu - v + vw$, compute $\partial(F, G, H)/\partial(u, v, w)$.
10. If $u = x + y + z, uv = y + z, uvw = z$, show that $\partial(x, y, z)/\partial(u, v, w). = u^2v$. (JNTUK, 2020; Andhra, 2018; U.P.T.U., 2013 ; P.T.U., 2011S)
11. If $(x, y, z) = 0$ and $g(x, y, z) = 0$, show that $\dfrac{dx}{\frac{\delta(f.g)}{\delta(y.z)}} = \dfrac{\delta y}{\frac{\delta(f.g)}{\delta(z.x)}} = \dfrac{\delta z}{\frac{\delta(f, g)}{\delta(x, y)}}$ (Nagpur. 2019)
12. If $u = \dfrac{x + y}{1 - xy}$ and $v = \tan^{-1} x + \tan^{-1} y$, find $\dfrac{\partial(u, v)}{\partial(x, y)}$. (V.T.U., 2013)

    Are $u$ and $v$ functionally related. If so, find this relationship.

13. Given that $u = x + y + z$, $u^2 v = y + z$, $u^2 w = z$, show that $\dfrac{\delta(u, v, w)}{\delta(x, y, z)} = u^5$. (*Mumbai, 2018*)

14. Show that $u = y + z$, $v = x + 2z^2$, $w = x - 4yz - 2y^2$ are not independent. Find the relation between them. (*U.P.T.U. 2013*)

15. If $u, v, w$ are the roots of the cubic $(\lambda - x)^3 + (\lambda - y)^3 + (\lambda - z)^3 = 0$ in $\lambda$, then find $\dfrac{\partial(u, v, w)}{\partial(x, y, z)}$. (*U.T.U., 2011*)

16. If $x = a \cosh\theta \cos\phi$, $y = a \sinh\theta \sin\phi$ show that $\dfrac{\delta(x, y)}{\delta(\theta, \phi)} = \dfrac{1}{2}a^2(\cosh 2\theta - \cos 2\phi)$.

17. In cylinderical co-ordinates $x = r\cos\theta$, $y = r\sin\theta$ and $z = z$. Find the value of $\dfrac{\delta(x, y, z)}{\delta(r, \theta, z)}$.

18. If $x = u(1 - v)$, $y = uv$ verify that $\dfrac{\delta(u, v)}{\delta(r, y)} \times \dfrac{\delta(x, y)}{\delta(u, v)} = 1$. (*SVTU, 2019*)

19. If $u = xyz$, $v = x^2 + y^2 + z^2$, $w = x + y + z$, find $J\left(\dfrac{x, y, z}{u, v, w}\right)$

## 5.8 (1) GEOMETRICAL INTERPRETATION

If $P(x, y, z)$ be the coordinates of a point referred to axes $OX$, $OY$, $OZ$ then the equation $z = f(x, y)$ represents a surface. (Fig. 5.1)

Let a plane $y = b$ parallel to the $XZ$-plane pass through $P$ cutting the surface along the curve $APB$ given by

$$z = f(x, b).$$

As $y$ remains equal to $b$ and $x$ varies then $P$ moves along the curve $APB$ and $\partial z/\partial x$ is the ordinary derivative of $f(x, b)$ w.r.t. $x$.

Hence *$\partial z/\partial x$ at $P$ is the tangent of the angle which the tangent at $P$ to the section of the surface $z = f(x, y)$ by a plane through $P$ parallel to the plane $XOZ$, makes with a line parallel to the $x$-axis.*

Similarly, *$\partial z/\partial y$ at $P$ is the tangent of the angle which the tangent at $P$ to the curve of intersection of the surface $z = f(x, y)$ and the plane $x = a$, makes with a line parallel to the $y$-axis.*

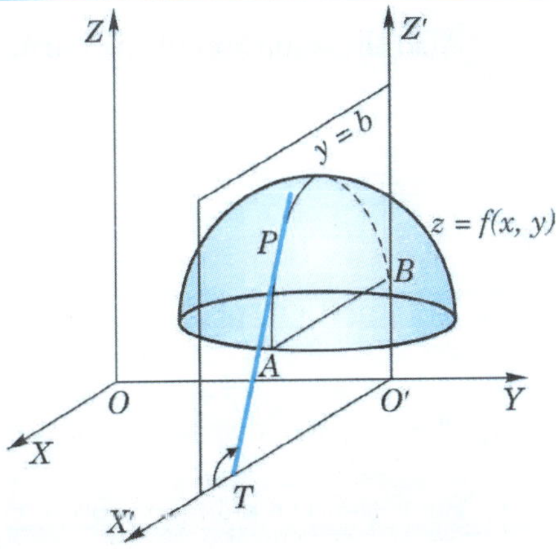

Fig. 5.1

**(2) Tangent plane and Normal to a surface.** Let $P(x, y, z)$ and $Q(x + \delta x, y + \delta y, z + \delta z)$ be two neighbouring points on the surface $F(x, y, z) = 0$. (Fig. 5.2) ...(*i*)

Let the arc $PQ$ be $\delta s$ and the chord $PQ$ be $\delta c$, so that (as for plane curves)

$$\underset{Q \to P}{\text{Lt}} \ (\delta s/\delta c) = 1.$$

The direction cosines of $PQ$ are $\dfrac{\delta x}{\delta c}, \dfrac{\delta y}{\delta c}, \dfrac{\delta z}{\delta c}$ *i.e.,* $\dfrac{\delta x}{\delta s} \cdot \dfrac{\delta s}{\delta c}, \dfrac{\delta y}{\delta s} \cdot \dfrac{\delta s}{\delta c}, \dfrac{\delta z}{\delta s} \cdot \dfrac{\delta s}{\delta c}$

When $\delta s \to 0$, $Q \to P$ and $PQ$ tends to tangent line $PT$. Then noting that the coordinates of any point on arc $PQ$ are functions of $s$ only, the direction cosines of $PT$ are

$$\frac{dx}{ds}, \frac{dy}{ds}, \frac{dz}{ds} \qquad ...(ii)$$

Differentiating Eqn. (*i*) with respect to $s$, we obtain

$$\frac{\partial F}{\partial x} \cdot \frac{dx}{ds} + \frac{\partial F}{\partial y} \cdot \frac{dy}{ds} + \frac{\partial F}{\partial z} \cdot \frac{dz}{ds} = 0.$$

This shows that the tangent line whose direction cosines are given by Eqn. (*ii*), is perpendicular to the line having direction ratios

$$\frac{\partial F}{\partial x}, \frac{\partial F}{\partial y}, \frac{\partial F}{\partial z} \qquad ...(iii)$$

Since we can take different curves joining $Q$ to $P$, we get a number of tangent lines at $P$ and the line having direction ratios Eqn. (*iii*) will be perpendicular to *all* these tangent lines at $P$. Thus all the tangent lines at $P$ lie in a plane through $P$ perpendicular to line (*iii*).

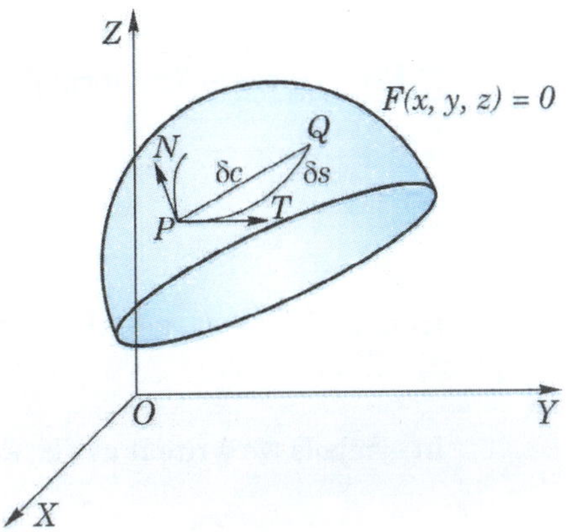

Fig. 5.2

Hence *the equation of the tangent plane to (i) at the point P is*

$$\frac{\partial \mathbf{F}}{\partial \mathbf{x}}(\mathbf{X}-\mathbf{x})+\frac{\partial \mathbf{F}}{\partial \mathbf{y}}(\mathbf{Y}-\mathbf{y})+\frac{\partial \mathbf{F}}{\partial \mathbf{z}}(\mathbf{Z}-\mathbf{z})=0$$

where $(X, Y, Z)$ are the current coordinates of any point on this tangent plane.

Also *the equation of the normal to the surface at P (i.e., the line through P,* perpendicular to the tangent plane at *P*) is

$$\frac{\mathbf{X}-\mathbf{x}}{\partial \mathbf{F}/\partial \mathbf{x}}=\frac{\mathbf{Y}-\mathbf{y}}{\partial \mathbf{F}/\partial \mathbf{y}}=\frac{\mathbf{Z}-\mathbf{z}}{\partial \mathbf{F}/\partial \mathbf{z}}.$$

**Example 5.35.** *Find the equations of the tangent plane and the normal to the surface $z^2 = 4(1 + x^2 + y^2)$ at (2, 2, 6).*

**Solution.** We have $F(x, y, z) = 4x^2 + 4y^2 - z^2 + 4.$

$\therefore$ $\partial F/\partial x = 8x$, $\partial F/\partial y = 8y$, $\partial F/\partial z = -2z$, and at the point (2, 2, 6)

$\partial F/\partial x = 16$, $\partial F/\partial y = 16$, $\partial F/\partial z = -12$

Hence the equation of the tangent plane at (2, 2, 6) is $16(X-2) + 16(Y-2) -12(Z-6) = 0$

*i.e.*, $4X + 4Y - 3Z + 2 = 0$ ...(*i*)

Also the equation of the normal at (2, 2, 6) [being perpendicular to (*i*)] is

$$\frac{X-2}{4}=\frac{Y-2}{4}=\frac{Z-6}{-3}.$$

## PROBLEMS 5.8

Find the equations of the tangent plane and normal to each of the following surfaces at the given points :

1. $2x^2 + y^2 = 3 - 2z$ at (2, 1, –3)
2. $x^3 + y^3 + 3xyz = 3$ at (1, 2, –1)
3. $xyz = a^2$ at $(x_1, y_1, z_1)$.
4. $2xz^2 - 3xy - 4x = 7$ at (1, –1, 2).
5. Show the plane $3x + 12y - 6z - 17 = 0$ touches the conicoid $3x^2 - 6y^2 + 9z^2 + 17 = 0$. Find also the point of contact.
6. Show that the plane $ax + by + cz + d = 0$ touches the surface $px^2 + qy^2 + 2z = 0$, if $\dfrac{a^2}{p}+\dfrac{b^2}{q}+2cd=0$.
7. Find the equation of the normal to the surface $x^2 + y^2 + z^2 = a^2$.

## 5.9 TAYLOR'S THEOREM FOR FUNCTIONS OF TWO VARIABLES

Considering $f(x + h, y + k)$ as a function of a single variable $x$, we have by Taylor's theorem*

$$f(x+h, y+k)=f(x, y+k)+h\frac{\partial f(x, y+k)}{\partial x}+\frac{h^2}{2!}\frac{\partial^2 f(x, y+k)}{\partial x^2}+\ldots \quad \ldots(i)$$

Now expanding $f(x, y + k)$ as a function of $y$ only,

$$f(x, y+k)=f(x, y)+k\frac{\partial f(x, y)}{\partial y}+\frac{k^2}{2!}\frac{\partial^2 f(x, y)}{\partial y^2}+\ldots$$

$\therefore$ Eqn. (*i*) takes the form $f(x+h, y+k)=f(x, y)+k\dfrac{\partial f(x, y)}{\partial y}+\dfrac{k^2}{2!}\dfrac{\partial^2 f(x, y)}{\partial y^2}+\ldots$

$$+h\frac{\partial}{\partial x}\left\{f(x, y)+k\frac{\partial f(x, y)}{\partial y}+\frac{k^2}{2!}\frac{\partial^2 f(x, y)}{\partial y^2}+\ldots\right\}+\frac{h^2}{2!}\frac{\partial^2}{\partial x^2}\left\{f(x, y)+k\frac{\partial(x, y)}{\partial y}+\ldots\right\}$$

Hence, $$f(x+h, y+k)=f(x, y)+h\frac{\partial f}{\partial x}+k\frac{\partial f}{\partial y}+\frac{1}{2!}\left(h^2\frac{\partial^2 f}{\partial x^2}+2hk\frac{\partial^2 f}{\partial x\partial y}+k^2\frac{\partial^2 f}{\partial y^2}\right)+\ldots \quad \ldots(1)$$

In symbols we write it as $f(x+h, y+k)=f(x, y)+\left(h\dfrac{\partial}{\partial x}+k\dfrac{\partial}{\partial y}\right)f+\dfrac{1}{2!}\left(h\dfrac{\partial}{\partial x}+k\dfrac{\partial}{\partial y}\right)^2 f+\ldots$

---

*See footnote on page 160.

Taking $x = a$ and $y = b$, Eqn. (1) becomes

$$f(a+h, b+k) = f(a, b) + [hf_x(a, b) + kf_y(a, b)] + \frac{1}{2!}[h^2 f_{xx}(a, b) + 2hkf_{xy}(a, b) + k^2 f_{yy}(a, b)] + \ldots$$

Putting $a + h = x$ and $b + k = y$ so that $h = x - a$, $k = y - b$, we get

$$\mathbf{f(x, y) = f(a, b) + [(x - a) f_x (a, b) + (y - b) f_y (a, b)]}$$

$$\mathbf{+ \frac{1}{2!}[(x - a)^2 f_{xx} (a, b) + 2(x - a)(y - b) f_{xy} (a, b) + (y - b)^2 f_{yy}(a, b)] + \ldots} \quad \ldots(2)$$

*This is Taylor's expansion of f(x, y) in powers of (x – a) and (y – b). It is used to expand f(x, y) in the neighbourhood of (a, b).*

**Cor.** Putting $a = 0$, $b = 0$, in (2), we get

$$\mathbf{f(x, y) = f(0, 0) + [xf_x(0, 0) + yf_y (0, 0)] + \frac{1}{2!}[x^2 f_{xx} (0, 0) + 2xy f_{xy} (0, 0) + y^2 f_{yy} (0, 0)] + \ldots} \quad \ldots(3)$$

This is Maclaurin's expansion of $f(x, y)$.

**Example 5.36.** *Expand $e^x \log(1 + y)$ in powers of x and y upto terms of third degree.* *(Anna., 2016 ; D.T.U., 2012)*

**Solution.** Here

$$\begin{aligned}
f(x, y) &= e^x \log(1 + y) & \therefore \quad f(0, 0) &= 0\\
f_x(x, y) &= e^x \log(1 + y) & f_x(0, 0) &= 0\\
f_y(x, y) &= e^x \frac{1}{1+y} & f_y(0, 0) &= 1\\
f_{xx}(x, y) &= e^x \log(1 + y) & f_{xx}(0, 0) &= 0\\
f_{xy}(x, y) &= e^x \frac{1}{1+y} & f_{xy}(0, 0) &= 1\\
f_{yy}(x, y) &= -e^x (1 + y)^{-2} & f_{yy}(0, 0) &= -1\\
f_{xxx}(x, y) &= e^x \log(1 + y) & f_{xxx}(0, 0) &= 0\\
f_{xxy}(x, y) &= e^x \frac{1}{1+y} & f_{xxy}(0, 0) &= 1\\
f_{xyy}(x, y) &= -e^x (1 + y)^{-2} & f_{xyy}(0, 0) &= -1\\
f_{yyy}(x, y) &= 2e^x (1 + y)^{-3} & f_{yyy}(0, 0) &= 2
\end{aligned}$$

Now Maclaurin's expansion of $f(x, y)$ gives

$$f(x, y) = f(0, 0) + xf_x(0, 0) + yf_y(0, 0) + \frac{1}{2!}\{x^2 f_{xx}(0, 0) + 2xy f_{xy}(0, 0) + y^2 f_{yy}(0, 0)\}$$

$$+ \frac{1}{3!}\{x^3 f_{xxx}(0, 0) + 3x^2 y f_{xxy}(0, 0) + 3xy^2 f_{xyy}(0, 0) + y^3 f_{yyy}(0, 0)\} + \ldots$$

$$\therefore \quad e^x \log(1 + y) = 0 + x(0) + y(1) + \frac{1}{2!}\{x^2 (0) + 2xy(1) + y^2 (-1)\}$$

$$+ \frac{1}{3!}\{x^3(0) + 3x^2y(1) + 3xy^2(-1) + y^3(2)\} + \ldots$$

$$= y + xy - \frac{1}{2}y^2 + \frac{1}{2}(x^2y - xy^2) + \frac{1}{3}y^3 + \ldots$$

**Example 5.37.** *Expand $x^2y + 3y - 2$ in powers of (x – 1) and ( y + 2) using Taylor's theorem.* *(Andhra., 2023, 2016 ; Anna, 2012 ; P.T.U., 2011)*

**Solution.** Taylor's expansion of $f(x, y)$ in powers of $(x - a)$ and $(x - b)$ is given by

$$f(x, y) = f(a, b) + [(x - a) f_x(a, b) + (y - b) f_y(a, b)] + \frac{1}{2!}[(x - a)^2 f_{xx}(a, b)$$

$$+ 2(x - a)(y - b) f_{xy}(a, b) + (y - b)^2 f_{yy}(a, b)] + \frac{1}{3!}[(x - a)^3 f_{xxx}(a, b)$$

$$+ 3(x - a)^2 (y - b) f_{xxy}(a, b) + 3(x - a)(y - b)^2 f_{xyy}(a, b)$$

$$+ (y - b)^3 f_{yyy}(a, b)] + \ldots \quad \ldots(i)$$

Hence $a = 1$, $b = -2$ and $f(x, y) = x^2y + 3y - 2$

$\therefore$ $f(1, -2) = -10, f_x = 2xy, f_x(1, -2) = -4; f_y = x^2 + 3, f_y(1, -2) = 4; f_{xx} = 2y,$
$f_{xx}(1, -2) = -4; f_{xy} = 2x, f_{xy}(1, -2) = 2; f_{yy} = 0, f_{yy}(1, -2) = 0; f_{xxx} = 0, f_{xxx} = (1, -2) = 0;$
$f_{xxy}(1, -2) = 2, f_{xyy}(1, -2) = 0, f_{yyy}(1, -2) = 0$

All partial derivatives of higher order vanish.

Substituting these in Eqn.($i$), we get

$$x^2y + 3y - 2 = -10 + [(x-1)(-4) + (y+2)4] + \frac{1}{2}[(x-1)^2(-4) + 2(x-1)(y+2)(2)$$
$$+ (y+2)^2(0)] + \frac{1}{6}[(x-1)^3(0) + 3(x-1)^2(y+2)(2) + 3(x-1)(y+2)^2(0) + (y+2)^3(0)]$$
$$= -10 - 4(x-1) + 4(y+2) - 2(x-1)^2 + 2(x-1)(y+2) + (x-1)^2(y+2).$$

**Example 5.38.** *Expand $f(x, y) = \tan^{-1}(y/x)$ in powers of $(x - 1)$ and $(y - 1)$ upto third-degree terms. Hence compute $f(1.1, 0.9)$ approximately.* *(JNTUK, 2020; U.K.T.U., 2012 ; V.T.U., 2010)*

**Solution.** Here $a = 1$, $b = 1$ and $f(1, 1) = \tan^{-1}(1) = \pi/4$.

$$f_x = \frac{-y}{x^2 + y^2} \qquad f_x(1, 1) = -\frac{1}{2}; \quad f_y = \frac{x}{x^2 + y^2}, \qquad f_y(1, 1) = \frac{1}{2}$$

$$f_{xx} = \frac{2xy}{(x^2 + y^2)^2}, \qquad f_{xx}(1, 1) = \frac{1}{2}; \qquad f_{xy} = \frac{y^2 - x^2}{(x^2 + y^2)^2}, \qquad f_{xy}(1, 1) = 0$$

$$f_{yy} = \frac{-2xy}{(x^2 + y^2)^2}, \qquad f_{yy}(1, 1) = -\frac{1}{2};$$

$$f_{xxx} = \frac{2y^3 - 6x^2y}{(x^2 + y^2)^3}, \qquad f_{xxx}(1, 1) = -\frac{1}{2}; f_{xxy} = \frac{2x^3 - 6xy^2}{(x^3 + y^2)^3}, \qquad f_{xxy}(1, 1) = -\frac{1}{2}$$

$$f_{xyy} = \frac{6x^2y - 2y^3}{(x^2 + y^2)^3}, \qquad f_{xyy}(1, 1) = \frac{1}{2}; \qquad f_{yyy} = \frac{6xy^2 - 2x^3}{(x^2 + y^2)^3}, \qquad f_{yyy}(1, 1) = \frac{1}{2}$$

Taylor's expansion of $f(x, y)$ in powers of $(x - 1)$ and $(y - 1)$ is given by

$$f(x, y) = f(1, 1) + \frac{1}{1!}\{(x-1) f_x(1, 1) + (y-1) f_y(1, 1)\} + \frac{1}{2!}\{(x-1)^2 f_{xx}(1, 1) + 2(x-1)(y-1)$$
$$f_{xy}(1, 1) + (y-1)^2 f_{yy}(1, 1) + \frac{1}{3!}\{(x-1)^3 f_{xxx}(1, 1) + 3(x-1)^2(y-1) f_{xxy}(1, 1)$$
$$+ 3(x-1)(y-1)^2 f_{xyy}(1, 1) + (y-1)^3 f_{yyy}(1, 1)\} + \ldots$$

$$\therefore \quad \tan^{-1}\left(\frac{y}{x}\right) = \frac{\pi}{4} + \left\{(x-1)\left(-\frac{1}{2}\right) + (y-1)\frac{1}{2}\right\} + \frac{1}{2!}\left\{(x-1)^2\frac{1}{2} + 2(x-1)(y-1)(0) + (y-1)^2\left(-\frac{1}{2}\right)\right\}$$
$$+ \frac{1}{3!}\left\{(x-1)^3\left(-\frac{1}{2}\right) + 3(x-1)^2(y-1)\left(-\frac{1}{2}\right) + 3(x-1)(y-1)^2\frac{1}{2} + (y-1)^3\frac{1}{2}\right\} + \ldots$$
$$= \frac{\pi}{4} - \frac{1}{2}\{(x-1) - (y-1)\} + \frac{1}{4}\{(x-1)^2 - (y-1)^2\} - \frac{1}{12}\{(x-1)^3 + 3(x-1)^2(y-1)$$
$$-3(x-1)(y-1)^2 - (y-1)^3\} + \ldots$$

Putting $x = 1.1$ and $y = 0.9$, we get

$$f(1.1, 0.9) = \frac{\pi}{4} - \frac{1}{2}(0.2) + \frac{1}{4}(0) - \frac{1}{12}\{(0.1)^3 - 3(0.1)^3 - 3(0.1)^3 - (-0.1)^3\}$$
$$= 0.7854 - 0.1000 + 0.0003 = 0.6857.$$

## 5.10 (1) ERRORS AND APPROXIMATIONS

Let $f(x, y)$ be a continuous function of $x$ and $y$. If $\delta x$ and $\delta y$ be the increments of $x$ and $y$, then the new value of $f(x, y)$ will be $f(x + \delta x, y + \delta y)$. Hence

$$\delta f = f(x + \delta x, y + \delta y) - f(x, y).$$

Expanding $f(x + \delta x, y + \delta y)$ by Taylor's theorem and supposing $\delta x$, $\delta y$ to be so small that their products, squares and higher powers can be neglected, we get

$$\boldsymbol{\delta f = \frac{\partial f}{\partial x}\,\delta x + \frac{\partial f}{\partial y}\,\delta y}, \text{ approximately.}$$

Similarly, if $f$ be a function of several variables $x, y, z, t, \ldots$, then

$$\delta f = \frac{\partial f}{\partial x}\,\delta x + \frac{\partial f}{\partial y}\,\delta y + \frac{\partial f}{\partial z}\,\delta z + \frac{\partial f}{\partial t}\,\delta t + \ldots \text{ approximately.}$$

These formulae are very useful in correcting the effect of small errors in measured quantities.

**(2) Total Differential**

If $u$ is a function of two variables $x$ and $y$, the *total differential* of $u$ is defined as

$$\mathbf{du = \frac{\partial u}{\partial x}\,dx + \frac{\partial u}{\partial y}\,dy} \qquad \ldots(1)$$

The differentials $dx$ and $dy$ are respectively the increments $\delta x$ and $\delta y$ in $x$ and $y$. If $x$ and $y$ are not independent variables but functions of another variable $t$ even then the formula (1) holds and we write $dx = \frac{dx}{dt}\,dt$ and $dy = \frac{dy}{dt}\,dt$. Similar definition can be given for a function of three or more variables.

**Example 5.39.** *The diameter and altitude of a can in the shape of a right circular cylinder are measured as 4 cm and 6 cm respectively. The possible error in each measurement is 0.1 cm. Find approximately the maximum possible error in the values computed for the volume and the lateral surface.* (V.T.U., 2016)

**Solution.** Let $x$ be the diameter and $y$ the height of the can. Then its volume $V = \frac{\pi}{4}\,x^2 y$

$$\therefore \qquad \delta V = \frac{\partial V}{\partial x}\,\delta x + \frac{\partial V}{\partial y}\,\delta y = \frac{\pi}{4}\,(2xy\delta x + x^2\,\delta y)$$

When $x = 4$ cm., $y = 6$ cm. and $\delta x = \delta y = 0.1$ cm.

$$\therefore \qquad \delta V = \frac{\pi}{4}(2 \times 4 \times 6 \times 0.1 + 4^2 \times 0.1) = 1.6\pi \text{ cm}^3$$

Also its lateral surface $S = \pi xy$

$$\therefore \qquad \delta S = \pi(y\,\delta x + x\,\delta y)$$

When $x = 4$ cm., $y = 6$ cm. and $\delta x = \delta y = 0.1$ cm., we have $\delta S = \pi(6 \times 0.1 + 4 \times 0.1) = \pi \text{ cm}^2$.

**Example 5.40.** *The period of a simple pendulum is $T = 2\pi\sqrt{(l/g)}$, find the maximum error in T due to the possible error upto 1% in l and 2.5% in g.* (P.T.U., 2013)

**Solution.** We have $T = 2\pi\sqrt{(l/g)}$

$$\text{or} \qquad \log T = \log 2\pi + \frac{1}{2}\log l - \frac{1}{2}\log g$$

$$\therefore \qquad \frac{1}{T}\,\delta T = 0 + \frac{1}{2}\frac{1}{l}\,\delta l - \frac{1}{2}\frac{1}{g}\,\delta g$$

$$\frac{\delta T}{T}\,100 = \frac{1}{2}\left(\frac{\delta l}{l}\,100 - \frac{\delta g}{g}\,100\right) = \frac{1}{2}(1 \pm 2.5) = 1.75 \text{ or } -0.75$$

Thus the maximum error in $T = 1.75\%$.

**Example 5.41.** *A balloon is in the form of right circular cylinder of radius 1.5 m and length 4 m and is surmounted by hemispherical ends. If the radius is increased by 0.01 m and length by 0.05 m, find the percentage change in the volume of balloon.*

**Solution.** Let the volume of the balloon (Fig. 5.3) be $V$, so that

$$V = \pi r^2 h + \frac{2}{3}\pi r^3 + \frac{2}{3}\pi r^3 = \pi r^2 h + \frac{4}{3}\pi r^3$$

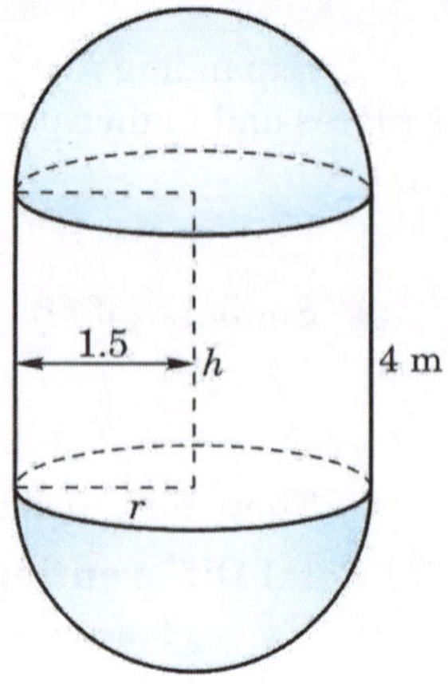

Fig. 5.3

$$\therefore \quad \delta V = 2\pi r \delta r h + \pi r^2 \delta h + \frac{4}{3}\pi 3r^2 \delta r$$

or
$$\frac{\delta V}{V} = \frac{\pi r[2h\delta r + r\delta h + 4r\delta r]}{\pi r^2 h + \frac{4}{3}\pi r^3}$$

$$= \frac{2(h + 2r)\,\delta r + r\delta h}{rh + \frac{4}{3}r^2} = \frac{2(4 + 3)\,(.01) + 1.5(.05)}{1.5 \times 4 + \frac{4}{3}(1.5)^2}$$

$$= \frac{0.14 + 0.075}{6 + 3} = \frac{0.215}{9}$$

Hence, the percentage change in $V = 100\,\dfrac{\delta V}{V} = \dfrac{21.5}{9} = 2.39\%$.

**Example 5.42.** *In estimating the cost of a pile of bricks measured as 2 m × 15 m × 1.2 m, the tape is stretched 1% beyond the standard length. If the count is 450 bricks to 1 cu. m. and bricks cost ₹ 530 per 1000, find the approximate error in the cost.*

**Solution.** Let $x$, $y$ and $z$ $m$ be the length, breadth and height of the pile so that its volume $V = xyz$

or
$$\log V = \log x + \log y + \log z \quad \therefore \quad \frac{\delta V}{V} = \frac{\delta x}{x} + \frac{\delta y}{y} + \frac{\delta z}{z}$$

Since
$$V = 2 \times 15 \times 1.2 = 36 \text{ m}^3, \text{ and } \frac{\delta x}{x} = \frac{\delta y}{y} = \frac{\delta z}{z} = \frac{1}{100}$$

$$\therefore \quad \delta V = 36\left(\frac{3}{100}\right) = 1.08 \text{ m}^3.$$

Number of bricks in $\delta V = 1.08 \times 450 = 486$

Thus error in the cost $= 486 \times \dfrac{530}{1000} =$ ₹ 257.58 which is a loss to the brick seller.

**Example 5.43.** *The height $h$ and semi-vertical angle $\alpha$ of a cone are measured and from them $A$, the total area of the surface of the cone including the base is calculated. If $h$ and $\alpha$ are in error by small quantities $\delta h$ and $\delta\alpha$ respectively, find the corresponding error in the area. Show further that if $\alpha = \pi/6$, an error of + 1% in $h$ will be approximately compensated by an error of – 0.33 degrees in $\alpha$.*

**Solution.** If $r$ be the base radius and $l$ the slant height of the cone, (Fig. 5.4), then total area

$$A = \text{area of base} + \text{area of curved surface}$$
$$= \pi r^2 + \pi rl = \pi r(r + l)$$
$$= \pi h \tan\alpha\,(h\tan\alpha + h\sec\alpha)$$
$$= \pi h^2\,(\tan^2\alpha + \tan\alpha\sec\alpha)$$

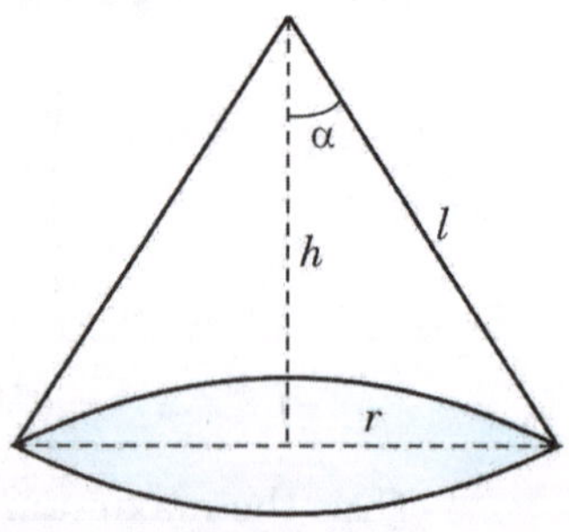

Fig. 5.4

$$\therefore \quad \delta A = \frac{\delta A}{\delta h}\delta h + \frac{\delta A}{\delta \alpha}\delta\alpha$$
$$= 2\pi h(\tan^2\alpha + \tan\alpha\sec\alpha)\delta h$$
$$+ \pi h^2\,(2\tan\alpha\sec^2\alpha + \sec^3\alpha + \tan\alpha\sec\alpha\tan\alpha)\delta\alpha$$

which gives the error in the area $A$.

Putting $\delta h = h/100$ and $\alpha = \pi/6$, we get

$$\delta A = 2\pi h\left[\left(\frac{1}{\sqrt{3}}\right)^2 + \frac{1}{\sqrt{3}}\cdot\frac{2}{\sqrt{3}}\right]\frac{h}{100} + \pi\,h^2\left[2.\frac{1}{\sqrt{3}}\cdot\frac{4}{3} + \frac{8}{3\sqrt{3}} + \frac{1}{\sqrt{3}}\cdot\frac{2}{\sqrt{3}}\cdot\frac{1}{\sqrt{3}}\right]\delta\alpha$$

$$= \frac{2\pi h^2}{100} + 2\sqrt{3}\pi h^2\ \delta\alpha$$

The error in $h$ will be compensated by the error in $\alpha$, when

$$\delta A = 0 \text{ i.e., } \frac{2\pi h^2}{100} + 2\sqrt{3}\pi h^2\,\delta\alpha = 0$$

or $$\delta\alpha = -\frac{1}{100\sqrt{3}} \text{ radians} = -\frac{.01}{1.732} \times 57.3° = -0.33°.$$

**Example 5.44.** *Show that the approximate change in the angle A of a triangle ABC due to small changes $\delta a$, $\delta b$, $\delta c$ in the sides a, b, c respectively, is given by*

$$\delta A = \frac{a}{2\Delta}(\delta a - \delta b \cos C - \delta c \cos B)$$

*where $\Delta$ is the area of the triangle. Verify that $\delta A + \delta B + \delta C = 0$.* (U.P.T.U., 2012)

**Solution.** We know that $a^2 = b^2 + c^2 - 2bc \cos A$

so that $$2a\delta a = 2b\delta b + 2c\delta c - 2(c\delta b \cos A - b\delta c \cos A + bc \sin A\,\delta A)$$

$\therefore$ $$bc \sin A\,\delta A = a\delta a - (b - c \cos A)\,\delta b - (c - b \cos A)\,\delta c$$

or $$2\,\Delta\delta A = a\delta a - (c \cos A + a \cos C - c \cos A)\,\delta b - (a \cos B + b \cos A - b \cos A)\,\delta c$$

[$\because$ $b = c \cos A + a \cos C$ etc. ...(*i*)]

$$= a\delta a - a \cos C\,\delta b - a \cos B\,\delta c$$

or $$\delta A = \frac{a}{2\Delta}(\delta a - \delta b \cos C - \delta c \cos B)$$

By symmetry, we have

$$\delta B = \frac{b}{2\Delta}(\delta b - \delta c \cos A - \delta a \cos C)$$

$$\delta C = \frac{c}{2\Delta}(\delta c - \delta a \cos B - \delta b \cos A)$$

$\therefore$ $$\delta A + \delta B + \delta C = \frac{1}{2\Delta}(a - b \cos C - c \cos B)\,\delta a + (b - c \cos A - a \cos C)\,\delta b + (c - a \cos B - b \cos A)]$$

$$= \frac{1}{2\Delta}[(a - a)\,\delta a + (b - b)\,\delta b + (c - c)\,\delta c] = 0$$ [By (*i*)]

**Example 5.45.** *If the sides of a plane triangle ABC vary in such a way that its circumradius remains constant, prove that* $\frac{da}{\cos A} + \frac{db}{\cos B} + \frac{dc}{\cos C} = 0$.

**Solution.** The circumradius $R$ of $\Delta ABC$ is given by

$$R = \frac{a}{2 \sin A} = \frac{b}{2 \sin B} = \frac{c}{2 \sin C}$$

$\therefore$ $$a = 2R \sin A$$ [$\because$ $R$ is constant

Taking differentials, $da = 2R \cos A\, dA$ or $\frac{da}{\cos A} = 2RdA$

Similarly, $$\frac{db}{\cos B} = 2RdB,\ \frac{dc}{\cos C} = 2RdC$$

$\therefore$ $$\frac{da}{\cos A} + \frac{db}{\cos B} + \frac{dc}{\cos C} = 2R\,(dA + dB + dC)$$

Now $A + B + C = \pi$, gives $dA + dB + dC = 0$ .. (*i*)

Thus $$\frac{da}{\cos A} + \frac{db}{\cos B} + \frac{dc}{\cos C} = 0$$ [By (*i*)]

## PROBLEMS 5.9

1. Expand the following functions as far as terms of third degree :
   (*i*) $\sin x \cos y$ (*ii*) $e^x \sin y$ at $(-1, \pi/4)$
2. Expand $f(x, y) = x^y$ in powers of $(x - 1)$ and $(y - 1)$.
3. Find the possible percentage error in computing the resistance $r$ from the formula $1/r = 1/r_1 + 1/r_2$, if $r_1, r_2$ are both in error by 2%. (*V.T.U., 2012*)
4. (*a*) Find percentage error made in calculating volume and surface of a sphere due to an error of $x$ % in radius. (*V.T.U., 2015*)
   (*b*) Find the percentage error in the area of an ellipse if one per cent error is made in measuring the major and minor axes. (*U.P.T.U., 2011 ; V.T.U., 2011*)
5. If $pv^2 = k$ and the relative errors in $p$ and $v$ are respectively 0.05 and 0.025, show that the error in $k$ is 10%. (*Kurukshetra, 2012*)
6. If the H.P. required to propel a steamer varies as the cube of the velocity and square of the length. Prove that a 3% increase in velocity and 4% increase in length will require an increase of about 17% in H.P.
7. The range $R$ of a projectile which starts with a velocity $v$ at an elevation $\alpha$ is given by $R = (v^2 \sin 2\alpha)/g$. Find the percentage error in $R$ due to an error of 1% in $v$ and an error of $\frac{1}{2}$% in $\alpha$.
8. The work that must be done to propel a ship of displacement $D$ for a distance $s$ in time $t$ is proportional to $(s^2D^{2/3}/t^2)$. Find approximately the increase of work necessary when the displacement is increased by 1%, the time is diminished by 1% and the distance diminished by 2%.
9. The torsional rigidity of a length of wire is obtained from the formula $N = 8\pi\, Il/t^2r^4$. If $l$ is decreased by 2%, $r$ is increased by 2%, $t$ is increased by 1.5%, show that the value of $N$ is diminished by 13% approximately.
10. The two sides of a triangle are measured as 50 cm and 70 cm and the angle between them is 30°. If there are possible errors of 0.5 % in the measurement of sides and 0.5° in that of the angle, find the maximum approximate percentage error in measuring the area of the triangle. (*U.P.T.U., 2013*)
11. If $x^2 + y^2 + z^2 - 2xyz = 1$, show that $\dfrac{dx}{\sqrt{(1-x^2)}} + \dfrac{dy}{\sqrt{(1-y^2)}} + \dfrac{dz}{\sqrt{(1-z^2)}} = 0$.
    [**Hint.** $2(x - yz)\, dx + 2(y - zx)\, dy + 2(z - xy)\, dz = 0$. Also $(x - yz)^2 = (1 - y^2)(1 - z^2)$, ...]
12. The sides of an acute angled triangle are measured. Prove that the increment in $A$ due to small increments in $a, b, c$ is given by the equation.
    $$bc \sin A\delta A = -(\cos c\, \delta b + \cos b\, \delta c - \delta a)$$
    Supposing that the limits of error in the length of any side are $\pm\, \mu$ percent, where $\mu$ is small,
    Prove that the limit of error in $A$ are approximately $\pm 1.15\left(\dfrac{\mu a^2}{bc \sin A}\right)$ (*Calicut, 2019*)
13. The height $h$ and semivertical angle $\alpha$ of a cone are measured and from them $A$, the total area of the cone including the base is calculated. If $h$ and $\alpha$ are in error by small quantities $h$ and $\alpha$ respectively, find the corresponding error in the area, show further that if $\alpha = \dfrac{\pi}{6}$ , an error of $\pm 1$% in the $h$ will be approximately compensated by an error of $-$ 0.33 degreee in $\alpha$. (*Chennai, 2019*)
14. In a plane triangle, if the sides $a, b, c$ be constant, Prove that the variations of its angles are given by the relations
    $$\frac{dA}{\sqrt{a^2 - b^2 \sin^2 A}} = \frac{dB}{\sqrt{b^2 - a^2 \sin^2 B}} = \frac{-dC}{c}$$ (*Magadh. 2017*)
15. The diameter and depth of a cylinderical well are measured as 4 m and 16 m respectively. If the measuring tape is streched 1% beyond its standard length. Find the percentage error in the volume of the well. (*Hissar, 2019*)
16. The indicated horse power $I$ of an engine is calculated by the formula $I = \dfrac{\text{PLAN}}{33,000}$ where $A = \dfrac{\pi d^2}{4}$. Assuming that errors of $r$ percent may have been made in measuring $P, L, N, d$ find the greatest possible error in $I$. (*Cochin, 2018*)

17. The period of simple pendulum with small oscillations is given by $T = 2\pi\sqrt{\dfrac{l}{g}}$

Find the maximum percentage error in $T$ due to possible errors if 1% and 2% in $l$ and $g$ respectively.

(*Anna, 2018*)

18. A pile of bricks measures 2 m × 20 m × 6 m and the count is 430 bricks to 1 $m^3$. Find the error in the cost when the tape is stretched by 1% beyond its standard length, the cost of bricks being ₹400 per thousand.

(*UTU, 2019*)

19. If $ax^2 + by^2 + cz^2 = 1$ and $lx + my + nz = 0$,

Prove that $$\frac{dx}{bny - cmz} = \frac{dy}{clz - anx} = \frac{dz}{amx - bly}$$

## 5.11 (1) MAXIMA AND MINIMA OF FUNCTIONS OF TWO VARIABLES

**Def.** *A function $f(x, y)$ is said to have a* **maximum** *or* **minimum** *at $x = a, y = b$, according as*

$$f(a + h, b + k) < \text{or} > f(a, b),$$

*for all positive or negative small values of h and k.*

In other words, if $\Delta = f(a + h, b + k) - f(a, b)$, *is of the same sign for all small values of h, k, and if this sign is negative, then $f(a, b)$ is a maximum. If this sign is positive, $f(a, b)$ is a minimum.*

Considering $z = f(x, y)$ as a surface, maximum value of $z$ occurs at the top of an elevation (*e.g.*, a dome) from which the surface descends in every direction and a minimum value occurs at the bottom of a depression (*e.g.*, a bowl) from which the surface ascends in every direction. Sometimes the maximum or minimum value may form a *ridge* such that the surface descends or ascends in all directions except that of the ridge. Besides these, we have such a point of the surface, where the tangent plane is horizontal and the surface looks like leather seat on the horse's back [Fig. 5.5 (c)] which falls for displacement in certain directions and rises for displacements in other directions. Such a point is called a **saddle point.**

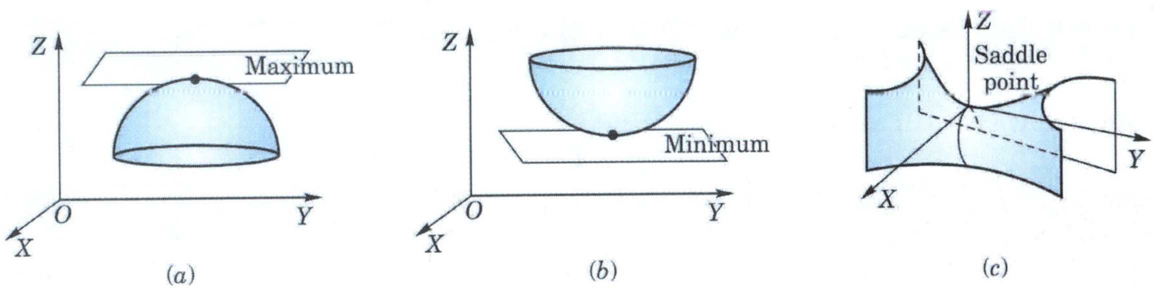


Fig. 5.5

**Note.** *A maximum or minimum value of a function is called its* **extreme value.**

### (2) Conditions for f(x, y) to be maximum or minimum

Using Taylor's theorem page 246, we have $\Delta = f(a + h, b + k) - f(a, b)$

$$= \left(h\frac{\partial f}{\partial x} + k\frac{\partial f}{\partial y}\right)_{a,b} + \frac{1}{2!}\left(h^2\frac{\partial^2 f}{\partial x^2} + 2hk\frac{\partial^2 f}{\partial x \partial y} + k^2\frac{\partial^2 f}{\partial y^2}\right) + \ldots \quad \ldots(i)$$

For small values of $h$ and $k$, the second and higher order terms are still smaller and hence may be neglected. Thus

$$\text{sign of } \Delta = \text{sign of } [hf_x(a, b) + kf_y(a, b)].$$

Taking $h = 0$ we see that the right hand side changes sign when $k$ changes sign. Hence $f(x, y)$ cannot have a maximum or a minimum at $(a, b)$ unless $f_y(a, b) = 0$.

Similarly taking $k = 0$, we find that $f(x, y)$ cannot have a maximum or minimum at $(a, b)$ unless $f_x(a, b) = 0$.

Hence *the necessary conditions for $f(x, y)$ to have a maximum or minimum at $(a, b)$ are that*

$$\mathbf{f_x(a, b) = 0, \; f_y(a, b) = 0.}$$

If these conditions are satisfied, then for small value of $h$ and $k$, Eqn. $(i)$ gives sign of $\Delta$ = sign of $\left[\frac{1}{2!}(h^2r + 2hks + k^2t)\right]$ where $r = f_{xx}(a, b)$, $s = f_{xy}(a, b)$ and $t = f_{yy}(a, b)$.

Now $$h^2r + 2hks + k^2t = \frac{1}{r}\left[(h^2r^2 + 2hkrs + k^2rt)\right] = \frac{1}{r}\left[(hr + ks)^2 + k^2(rt - s^2)\right]$$

Thus sign of $\Delta$ = sign of $\frac{1}{2r}\left\{(hr + ks)^2 + k^2(rt - s^2)\right\}$ ...(*ii*)

In Eqn. (*ii*), $(hr + ks)^2$ is always positive and $k^2(rt - s^2)$ will be positive if $rt - s^2 > 0$. In this case, $\Delta$ will have the same sign as that of $r$ for all values of $h$ and $k$.

Hence if $rt - s^2 > 0$, then *$f(x, y)$ has a maximum or a minimum at $(a, b)$ according as $r <$ or $> 0$.*

If $rt - s^2 < 0$, then $\Delta$ will change with $h$ and $k$ and hence there is no maximum or minimum at $(a, b)$ *i.e.,* it is a *saddle point.*

If $rt - s^2 = 0$, further investigation is required to find whether there is a maximum or minimum at $(a, b)$ or not.

**Note. Stationary value.** *$f(a, b)$ is said to be a stationary value of $f(x, y)$, if $f_x(a, b) = 0$* and $f_y(a, b) = 0$ *i.e.* the function is stationary at $(a, b)$.

Thus every extreme value is a stationary value but the converse may not be true.

### (3) Working rule to find the maximum and minimum values of f(x, y)

1. Find $\partial f/\partial x$ and $\partial f/\partial y$ *and equate each to zero. Solve these as simultaneous equations in x and y.* Let $(a, b)$, $(c, d)$, ... *be the pairs of values.*
2. *Calculate the value of* $r = \partial^2 f/\partial x^2$, $s = \partial^2 f/\partial x\partial y$, $t = \partial^2 f/\partial y^2$ *for each pair of values.*
3. (*i*) *If* $rt - s^2 > 0$ *and* $r < 0$ *at* $(a, b)$, $f(a, b)$ *is a max. value.*

   (*ii*) *If* $rt - s^2 > 0$ *and* $r > 0$ *at* $(a, b)$, $f(a, b)$ *is a min. value.*

   (*iii*) *If* $rt - s^2 < 0$ *at* $(a, b)$, $f(a, b)$ *is not an extreme value, i.e.,* $(a, b)$ *is a saddle point.*

   (*iv*) *If* $rt - s^2 = 0$ *at* $(a, b)$, *the case is doubtful and needs further investigation.*

Similarly examine the other pairs of values one by one.

**Example 5.46.** *Examine the following function for extreme values :*

$$f(x, y) = x^4 + y^4 - 2x^2 + 4xy - 2y^2.$$ (*Delhi, 2013 ; P.T.U., 2010 S*)

**Solution.** We have $f_x = 4x^3 - 4x + 4y$ ; $f_y = 4y^3 + 4x - 4y$

and $r = f_{xx} = 12x^2 - 4$, $s = f_{xy} = 4$, $t = f_{yy} = 12y^2 - 4$

Now $f_x = 0, f_y = 0$ give $x^3 - x + y = 0$, ...(*i*) $y^3 + x - y = 0$ ...(*ii*)

Adding these, we get $4(x^3 + y^3) = 0$ or $y = -x$.

Putting $y = -x$ in Eqn. (*i*), we obtain $x^3 - 2x = 0$, *i.e.* $x = \sqrt{2}, -\sqrt{2}, 0$.

$\therefore$ Corresponding values of $y$ are $-\sqrt{2}, \sqrt{2}, 0$.

*At* $(\sqrt{2}, -\sqrt{2})$, $rt - s^2 = 20 \times 20 - 4^2$ = + ve and $r$ is also +ve. Hence $f(\sqrt{2}, -\sqrt{2})$ is a minimum value.

*At* $(-\sqrt{2}, \sqrt{2})$ also both $rt - s^2$ and $r$ are + ve.

Hence $f(-\sqrt{2}, \sqrt{2})$, is also a minimum value.

*At (0, 0)*, $rt - s^2 = 0$ and, therefore, further investigation is needed.

Now $f(0, 0) = 0$ and for points along the $x$-axis, where $y = 0$, $f(x, y) = x^4 - 2x^2 = x^2(x^2 - 2)$, which is negative for points in the neighbourhood of the origin.

Again for points along the line $y = x$, $f(x, y) = 2x^4$ which is positive.

Thus in the neighbourhood of (0, 0) there are points where $f(x, y) < f(0, 0)$ and there are points where $f(x, y) > f(0, 0)$.

Hence $f(0, 0)$ is not an extreme value *i.e.,* it is a *saddle point.*

**Example 5.47.** *Discuss the maxima and minima of* $f(x, y) = x^3y^2(1 - x - y)$.

(*JNTUK, 2022; V.T.U., 2013 ; D.T.U., 2011*)

**Solution.** We have $f_x = 3x^2y^2 - 4x^3y^2 - 3x^2y^3$ ; $f_y = 2x^3y - 2x^4y - 3x^3y^2$

and $r = f_{xx} = 6xy^2 - 12x^2y^2 - 6xy^3$ ; $s = f_{xy} = 6x^2y - 8x^3y - 9x^2y^2$ ; $t = f_{yy} = 2x^3 - 2x^4 - 6x^3y$.

When $f_x = 0, f_y = 0$, we have $x^2y^2(3 - 4x - 3y) = 0, x^3y(2 - 2x - 3y) = 0$

Solving these, the stationary points are (1/2, 1/3), (0, 0).

Now $rt - s^2 = x^4y^2[12(1 - 2x - y)(1 - x - 3y) - (6 - 8x - 9y)^2]$

*At (1/2, 1/3),* $rt - s^2 = \frac{1}{16} \cdot \frac{1}{9}\left[12\left(1-1-\frac{1}{3}\right)\left(1-\frac{1}{2}-1\right)-(6-4-3)^2\right] = \frac{1}{14} > 0$

Also $r = 6\left(\frac{1}{2}\cdot\frac{1}{9} - \frac{2}{4}\cdot\frac{1}{9} - \frac{1}{2}\cdot\frac{1}{27}\right) = -\frac{1}{9} < 0$

Hence $f(x, y)$ has a maximum at (1/2, 1/3) and maximum value $= \frac{1}{8}\cdot\frac{1}{9}\left(1-\frac{1}{2}-\frac{1}{3}\right) = \frac{1}{432}$.

*At (0, 0)*, $rt - s^2 = 0$ and therefore further investigation is needed.

For points along the line $y = x$, $f(x, y) = x^5(1 - 2x)$ which is positive for $x = 0.1$ and negative for $x = -0.1$ *i.e.*, in the neighbourhood of (0, 0) there are points where $f(x, y) > f(0, 0)$ and there are points where $f(x, y) < f(0, 0)$. Hence $f(0, 0)$ is not an extreme value.

**Example 5.48.** *Locate the stationary points of the function $2(x - y)^2 - x^4 - y^4$ and determine their nature.* *(JNTUK, 2022; Patna, 2019)*

**Solution :** Let $f(x, y) = 2(x - y)^2 - x^4 - y^4$

For the stationary points, we have

$$\frac{\delta f}{\delta x} = 4(x - y) - 4x^3 = 0 \quad ...(i)$$

$$\frac{\delta f}{\delta y} = -4(x - y) - 4y^3 = 0 \quad ...(ii)$$

Adding Eqns. (*i*) and (*ii*), we get

$$-4(x^3 + y^3) = 0$$

or, $(x + y)(x^2 - xy + y^2) = 0$

$\therefore$ $x + y = 0$ or $x^2 - xy + y^2 = 0$

$$y = -x \quad ...(iii)$$

Now (Rejecting the imaginary values given by $x^2 - xy + y^2 = 0$)

Substituting the value $y = -x$ in eq (*i*)

$$8x - 4x^3 = 0 \quad \text{or} \quad 4x(2 - x^2) = 0$$

$$x = 0 \qquad x = \pm\sqrt{2}$$

From Eqn. (*iii*), $y = 0, y = \pm\sqrt{2}$.

So, the stationary points are (0, 0) $(\sqrt{2}, -\sqrt{2})(-\sqrt{2}, \sqrt{2})$ the other combinations satisfy Eqn. (*iii*).

To discuss the nature

$$r = \frac{\delta^2 f}{\delta x^2} = 4 - 12x^2,$$

$$s = \frac{\delta^2 f}{\delta x\, \delta y} = -4$$

$$t = \frac{\delta^2 f}{\delta y^2} = 4 - 12y^2$$

At the point (0, 0)

$$rt - s^2 = (4)(4) - (-4)^2$$
$$= 16 - 16 = 0$$

Since $rt - s^2 = 0$, at this point, so its needs further investigation.

(*ii*) At the point $(\sqrt{2},-\sqrt{2})$

$$r = -20, s = -4, t = -20$$

$$\therefore \quad rt - s^2 = (-20)(-20) - 16 = 384 \ (+ve)$$

Since $rt - s^2 \to 0$, and $r < 0$, there is a maximum at the point $(\sqrt{2},-\sqrt{2})$.

(*iii*) At the part $(-\sqrt{2},\sqrt{2})$

$$r = -20, s = -4, t = -20$$

$$\therefore \quad rt - 5^2 = (-20)(-20) - 16 = 384 \ (+ve)$$

Since $rt - s^2 > 70$, $r < 0$, there is a maximum at the point $(-\sqrt{2},\sqrt{2})$

Hence at the point $(\mp\sqrt{2}, \mp\sqrt{2})$ $rt - s^2 > 0$ and $r < 0$. So maximum at the points $(\mp\sqrt{2}, \mp\sqrt{2})$.

Now the point (0, 0), where $rt - s^2 = 0$. We investigating the $f(x, y)$ in the origin.

Now, $\quad f(x, y) = 2(x - y^2) > 0$

Also when $\quad y = x,$

We have $\quad y = x$

$$f(x, y) = -2x^4 < 0$$

So the point (0, 0), $f(x, y)$ is $+ve$ in one direction, So there is neither a maximum nor a minima at (0, 0).

Hence the point (0, 0) is a saddle point.

**Example 5.49.** *Prove that the rectangular solid of maximum volume which can be inscribed in a sphere is cube, Find the maximum volume also.* (*UTV. 2019*)

**Solution :** Let $d$ be the diameter of sphere and $x, y, z$ the three edges of the rectangular solid of volume V (say)

$$d^2 = x^2 + y^2 + z^2 \quad \text{...(i)}$$

$$u = xyz \quad \text{...(ii)}$$

From Eqns. (*i*) and (*ii*), we have

$$u = xy\sqrt{d^2 - x^2 - y^2}$$

$$u^2 = x^2y^2(d^2 - x^2 - y^2)$$

For maximum or minimum of $v^2$ and hence that of $v$.

$$\frac{\delta}{\delta x}(u^2) = 2xy^2(d^2 - 2x^2 - y^2) = 0 \quad \text{...(iii)}$$

$$\frac{\delta}{\delta x}(u^2) = 2x^2y(d^2 - x^2 - 2y^2) = 0 \quad \text{...(iv)}$$

From Eqns. (*iii*) and (*iv*), we have

$$d^2 - 2x^2 - y^2 = 0 \quad \text{...(v)}$$

and $\quad d^2 - x^2 - 2y^2 = 0 \quad$ ...(*vi*)

From Eqns. (*v*) and (*vi*), we get

$$x^2 = y^2 = \frac{d^2}{3}$$

Putting $x = y^2 = \dfrac{d^2}{3}$ in Eqn. (*i*), we get

$$z^2 = \frac{d^2}{3}$$

$$x^2 = y^2 = z^2 = \frac{d^2}{3}$$

Now at $x = y = z = \dfrac{d}{\sqrt{3}}$

$$r = \frac{\delta^2}{\delta x^2}(u^2) = 2y^2(d^2 - 6x^2 - y^2) = \frac{-8d^4}{9}$$

$$s = \frac{\delta^2}{\delta x\, \delta y}(u^2) = 4xy\,(d^2 - 2x^2 - 2y^2) = \frac{-4d^4}{9}$$

$$t = \frac{\delta^2}{\delta y^2}(u^2) = 2x^2(d^2 - x^2 - 6y^2) = \frac{-8d^4}{9}$$

Hence
$$rt - s^2 = \frac{64d^8}{81} - \frac{16d^8}{8}$$
$$= \frac{48\,d^8}{81} \text{ (+ ve)}$$

Since $rt - s^2 > 0$ and $r < 0$

$u^2$ and hence $u$ is maximum at $x = y = z = \dfrac{d}{\sqrt{3}}$

Thus the solid is a cube and maximum volume $= \dfrac{d^3}{3\sqrt{3}}$ cubic unit.

## 5.12 LAGRANGE'S METHOD OF UNDERTERMINED MULTIPLIERS

Sometimes it is required to find the stationary values of a function of several variables which are not all independent but are connected by some given relations. Ordinarily, we try to convert the given function to the one, having least number of independent variables with the help of given relations. Then solve it by the above method. When such a procedure becomes impracticable, Lagrange's method proves very convenient. Now we explain this method.

Let $\quad u = f(x, y, z) \quad$ ...(1)

be a function of three variables $x, y, z$ which are connected by the relation.

$\phi(x, y, z) = 0 \quad$ ...(2)

For $u$ to have stationary values, it is necessary that

$$\partial u/\partial x = 0,\ \partial u/\partial y = 0,\ \partial u/\partial z = 0.$$

$$\therefore \quad \frac{\partial u}{\partial x}dx + \frac{\partial u}{\partial y}dy + \frac{\partial u}{\partial z}dz = du = 0 \quad \text{...(3)}$$

Also differentiating (2), we get $\dfrac{\partial \phi}{\partial x}dx + \dfrac{\partial \phi}{\partial y}dy + \dfrac{\partial \phi}{\partial z}dz = d\phi = 0 \quad$ ...(4)

Multiply Eqn. (4) by a parameter $\lambda$ and add to Eqn. (3). Then

$$\left(\frac{\partial u}{\partial x} + \lambda\frac{\partial \phi}{\partial x}\right)dx + \left(\frac{\partial u}{\partial y} + \lambda\frac{\partial \phi}{\partial y}\right)dy + \left(\frac{\partial u}{\partial z} + \lambda\frac{\partial \phi}{\partial z}\right)dz = 0$$

This equation will be satisfied if $\dfrac{\partial u}{\partial x} + \lambda\dfrac{\partial \phi}{\partial x} = 0,\ \dfrac{\partial u}{\partial y} + \lambda\dfrac{\partial \phi}{\partial y} = 0,\ \dfrac{\partial u}{\partial z} + \lambda\dfrac{\partial \phi}{\partial z} = 0.$

These three equations together with Eqn. (2) will determine the values of $x, y, z$ and $\lambda$ for which $u$ is stationary.

**Working rule : 1.** *Write* $F = f(x, y, z) + \lambda\phi(x, y, z)$

**2.** *Obtain the equations* $\dfrac{\partial F}{\partial x} = 0, \dfrac{\partial F}{\partial y} = 0, \dfrac{\partial F}{\partial z} = 0.$

**3.** *Solve the above equations together with* $\phi(x, y, z) = 0$.

*The values of x, y, z so obtained will give the stationary value of f (x, y, z).*

**Obs.** Although the Lagrange's method is often very useful in application yet *the drawback is that we cannot determine the nature of the stationary point.* This can sometimes, be determined from physical considerations of the problem.

**Example 5.50.** *A rectangular box open at the top is to have volume of 32 cubic ft. Find the dimensions of the box requiring least material for its construction.*

(*Andhra, 2022 ; U.P.T.U., 2016 ; V.T.U., 2015 ; P.T.U., 2013 ; Anna, 2012*)

**Solution :** Let $x, y$ and $z$ ft. be the edges of the box and $S$ be its surface.

Then $$S = xy + 2yz + 2zx \quad ...(i)$$

and $$xyz = 32 \quad ...(ii)$$

Eliminating $z$ from Eqn. (*i*) with the help of Eqn. (*ii*), we get $S = xy + 2(y + x)\dfrac{32}{xy} = xy + 64\left(\dfrac{1}{x} + \dfrac{1}{y}\right)$

$$\therefore \quad \partial S/\partial x = y - 64/x^2 = 0 \quad \text{and} \quad \partial S/\partial y = x - 64/y^2 = 0.$$

Solving these, we get $x = y = 4$.

Now $\quad r = \partial^2 S/\partial x^2 = 128/x^3,\ s = \partial^2 S/\partial x \partial y = 1,\ t = \partial^2 S/\partial y^2 = 128/y^3$.

At $\quad x = y = 4,\ rt - s^2 = 2 \times 2 - 1 = +$ ve and $r$ is also + ve.

Hence $S$ is minimum for $x = y = 4$. Then from (*ii*), $z = 2$.

*Otherwise* (*by Lagrange's method*) :

Write $$F = xy + 2yz + 2zx + \lambda(xyz - 32)$$

Then $$\frac{\partial F}{\partial x} = y + 2z + \lambda yz = 0 \quad ...(iii)$$

$$\frac{\partial F}{\partial y} = x + 2z + \lambda zx = 0 \quad ...(iv)$$

$$\frac{\partial F}{\partial z} = 2y + 2x + \lambda xy = 0 \quad ...(v)$$

Multiplying Eqn. (*iii*) by $x$ and Eqn. (*iv*) by $y$ and subtracting, we get $2zx - 2zy = 0$ or $x = y$.

[The value $z = 0$ is neglected, as it will not satisfy Eqn. (*ii*)]

Again multiplying Eqn. (*iv*) by $y$ and Eqn. (*v*) by $z$ and subtracting, we get $y = 2z$.

Hence the dimensions of the box are $x = y = 2z = 4$ ...(*vi*)

Now let us see what happens as $z$ increases from a small value to a large one. When $z$ is small, the box is flat with a large base showing that $S$ is large. As $z$ increases, the base of the box decreases rapidly and $S$ also decreases. After a certain stage, $S$ again starts increasing as $z$ increases. Thus $S$ must be a minimum at some intermediate stage which is given by Eqn. (*vi*). Hence $S$ is minimum when $x = y = 4$ ft and $z = 2$ ft.

**Example 5.51.** *(i) Given $x + y + z = a$, find the maximum value of $x^m y^n z^p$.* (*JNTUK, 2020*)
*(ii) Find the minimum value of $x^2 + y^2 + z^2$ having given $ax + by + cz = p$*

(*Anna 2018 ; Andhra., 2015 ; J.N.T.U., 2015 S ; Anna, 2009*)

**Solution :** (*i*) Let $f(x, y, z) = x^m y^n z^p$ and $\phi(x, y, z) = x + y + z - a$.

Then $$F(x, y, z) = f(x, y, z) + \lambda\, \phi(x, y, z)$$
$$= x^m y^n z^p + \lambda(x + y + z - a).$$

For stationary values of $F$, $\dfrac{\partial F}{\partial x} = 0, \dfrac{\partial F}{\partial y} = 0, \dfrac{\partial F}{\partial z} = 0$

$$\therefore \quad mx^{m-1}y^n z^p + \lambda = 0,\ nx^m y^{n-1} z^p + \lambda = 0,\ px^m y^n z^{p-1} + \lambda = 0$$

or $$-\lambda = mx^{n-1}y^n z^p = nx^m y^{n-1} z^p = px^m y^n z^{p-1}$$

*i.e.* $$\frac{m}{x} = \frac{n}{y} = \frac{p}{z} = \frac{m+n+p}{x+y+z} = \frac{m+n+p}{a} \qquad [\because \ x + y + z = a]$$

$\therefore$ The maximum value of $f$ occurs when

$$x = am/(m + n + p),\ y = an/(m + n + p),\ z = ap/(m + n + p)$$

Hence the maximum value of $f(x, y, z) = \dfrac{a^{m+n+p} \cdot m^m \cdot n^n \cdot p^p}{(m + n + p)^{m+n+p}}$.

(*ii*) Let $$u = x^2 + y^2 + z^2 \quad ...(i)$$

and $$ax + by + cz - p = 0 \quad ...(ii)$$

Let us consider the Lagrange's function

$$F(x, y, z) = x^2 + y^2 + z^2 + \lambda(ax + by + cz - p)$$

For stationary values of $u$,

$$dF = (2x\,dx + 2y\,dy + 2z\,dz) + \lambda(adx + bdy + cdz) = 0$$
$$= (2x + \lambda a)\,dx + (2y + \lambda b)\,dy + (2z + \lambda c)dz = 0$$

$$2x + \lambda a = 0 \quad \text{...(iii)}$$
$$2y + \lambda b = 0 \quad \text{...(iv)}$$
$$2z + \lambda c = 0 \quad \text{...(v)}$$

Multiply Eqns. (*iii*), (*iv*), (*v*) by $x, y, z$ respectively and adding, we get $2(x^2 + y^2 + z^2) + \lambda(ax + by + cz) = 0$

$$2u + \lambda p = 0$$

From Eqns. (*i*) and (*ii*)

$$\lambda = \frac{-2u}{p} \quad \text{...(vi)}$$

From Eqns. (*iii*), (*iv*), (*v*), (*vi*),

$$x = \frac{au}{p},\ y = \frac{bu}{p},\ z = \frac{cu}{p} \quad \text{...(vii)}$$

From Eqns. (*i*) and (*vii*), we get

$$u = \frac{a^2u^2}{p^2} + \frac{b^2u^2}{p^2} + \frac{c^2u^2}{p^2}$$

or $$u = \frac{p^2}{a^2 + b^2 + c^2}$$, the stationary value of $u$.

Since $x, y, z$ are related by only one relation, let $z$ be a function of two independent variable $x$ only from Eqn. (*ii*)

$$cz = p - ax - by$$

$$\therefore \quad \frac{\delta z}{\delta x} = -\frac{a}{c},\ \frac{\delta z}{\delta y} = -\frac{b}{c}$$

Also, $$\frac{\delta u}{\delta x} = 2x + 2z\frac{\delta z}{\delta x} - 2x - \frac{2a}{c}z$$

$$r = \frac{\delta^2 u}{\delta x^2} = 2 - \frac{2a}{c}\frac{\delta z}{\delta x} = 2 + \frac{2a^2}{c^2} = 2\left(\frac{a^2 + c^2}{c^2}\right)$$

$$\frac{\delta u}{\delta y} = 2y + 2z\cdot\frac{\delta z}{\delta y} = 2y - \frac{2b}{c}z$$

$$\therefore \quad s = \frac{\delta^2 u}{\delta x \delta y} = \frac{-2b}{c}\cdot\frac{\delta z}{\delta x} = \frac{2ab}{c^2}$$

$$t = \frac{\delta^2 u}{\delta y^2} = 2 - \frac{bc}{c}\frac{\delta z}{\delta y} = 2 + \frac{2b^2}{c^2} = 2\left(\frac{b^2 + c^2}{c^2}\right)$$

$$\therefore \quad rt - s^2 = \frac{4}{c^4}(a^2 + c^2)(b^2 + c^2) - \frac{4a^2b^2}{c^4}$$

$$= \frac{4}{c^2}(a^2 + b^2 + c^2) > 0$$

Also, $r$ is positive. Since $r$ and $rt - s^2$ are both positive.

$$\therefore \quad u = \frac{p^2}{a^2 + b^2 + c^2} \text{ is minimum.}$$

**Example 5.52.** *Find the maximum and minimum distances of the point (3, 4, 12) from the sphere $x^2 + y^2 + z^2 = 4$.* *(Andhra, 2023; U.K.T.U., 2010)*

**Solution :** Let $P(x, y, z)$ be any point on the sphere and $A(3, 4, 12)$ the given point so that

$$AP^2 = (x - 3)^2 + (y - 4)^2 + (z - 12)^2 = f(x, y, z), \text{ say} \quad \text{...(i)}$$

We have to find the maximum and minimum values of $f(x, y, z)$ subject to the condition

$$\phi(x, y, z) = x^2 + y^2 + z^2 - 4 = 0 \quad ...(ii)$$

Let $F(x, y, z) = f(x, y, z) + \lambda\phi(x, y, z)$

$$= (x-3)^2 + (y-4)^2 + (z-12)^2 + \lambda(x^2 + y^2 + z^2 - 4)$$

Then $\frac{\partial F}{\partial x} = 2(x-3) + 2\lambda x, \frac{\partial F}{\partial y} = 2(y-4) + 2\lambda y, \frac{\partial F}{\partial z} = 2(z-12) + 2\lambda z$

$\therefore$ $\frac{\partial F}{\partial x} = 0, \frac{\partial F}{\partial y} = 0$ and $\frac{\partial F}{\partial z} = 0$ give

$$x - 3 + \lambda x = 0, y - 4 + \lambda y = 0, z - 12 + \lambda z = 0 \quad ...(iii)$$

which give $\lambda = -\frac{x-3}{x} = -\frac{y-4}{y} = -\frac{z-12}{z}$

$$= \pm\frac{\sqrt{[(x-3)^2 + (y-4)^2 + (z-12)^2]}}{\sqrt{(x^2 + y^2 + z^2)}} = \pm\frac{\sqrt{f}}{1}$$

Substituting for $\lambda$ in Eqn. $(iii)$, we get

$$x = \frac{3}{1+\lambda} = \frac{3}{1 \pm \sqrt{f}}, y = \frac{4}{1 \pm \sqrt{f}}, z = \frac{12}{1 \pm \sqrt{f}}$$

$\therefore$ $x^2 + y^2 + z^2 = \frac{9 + 16 + 144}{(1 \pm \sqrt{f})^2} = \frac{169}{(1 \pm \sqrt{f})^2}$

Using Eqn. $(ii)$, $1 = \frac{169}{(1 \pm \sqrt{f})^2}$ or $1 \pm \sqrt{f} = \pm 13, \sqrt{f} = 12, 14.$

[We have left out the negative values of $\sqrt{f}$, because $\sqrt{f} = AP$ is +ve by $(i)$]

Hence maximum $AP = 14$ and minimum $AP = 12$.

**Example 5.53.** *Show that the rectangular solid of maximum volume that can be inscribed in a sphere is a cube.* (Andhra, 2017)

**Solution.** Let $2x$, $2y$, $2z$ be the length, breadth and height of the rectangular solid so that its volume

$$V = 8xyz \quad ...(i)$$

Let $R$ be the radius of the sphere so that

$$x^2 + y^2 + z^2 = R^2 \quad ...(ii)$$

Then $F(x, y, z) = 8xyz + \lambda(x^2 + y^2 + z^2 - R^2)$

and $\partial F/\partial x = 0,$

$\partial F/\partial y = 0$

and $\partial F/\partial z = 0$ give

$$8yz + 2x\lambda = 0,$$
$$8zx + 2y\lambda = 0,$$
$$8xy + 2z\lambda = 0$$

or $2x^2\lambda = -8xyz$

$$= 2y^2\lambda = 2z^2\lambda$$

Thus for a maximum volume $x = y = z$.

*i.e.*, the rectangular solid is a cube.

**Example 5.54.** *A tent on a square base of side x, has its sides vertical of height y and the top is a regular pyramid of height h. Find x and y in terms of h, if the canvas required for its construction is to be minimum for the tent to have a given capacity.*

**Solution :** Let $V$ be the volume enclosed by the tent and $S$ be its surface area (Fig. 5.6).

Then $V$ = cuboid $(ABCD, A'B'C'D')$ + pyramid $(K, A'B'C'D')$

$$= x^2y + \frac{1}{3}x^2h = x^2(y + h/3)$$

$$S = 4(ABGF) + 4\Delta\, KGH = 4xy + 4\frac{1}{2}(x \cdot KM)$$

$$= 4xy + x\sqrt{(x^2 + 4h^2)} \qquad [\because KM = \sqrt{(KL^2 + LM^2)} = \sqrt{(h^2 + (x/2))}]$$

For constant $V$, we have

$$\delta V = 2x(y + h/3)\,\delta x + x^2(\delta y) + \frac{x^2}{3}\delta h = 0$$

For minimum $S$, we have

$$\delta S = [4y + \sqrt{(x^2 + 4h^2)} + x\cdot\frac{1}{2}(x^2 + 4h^2)^{-1/2}\cdot 2x]\delta x$$

$$+ 4x\delta y + x\cdot\frac{1}{2}(x^2 + 4h^2)^{-1/2}\cdot 8h\delta h = 0$$

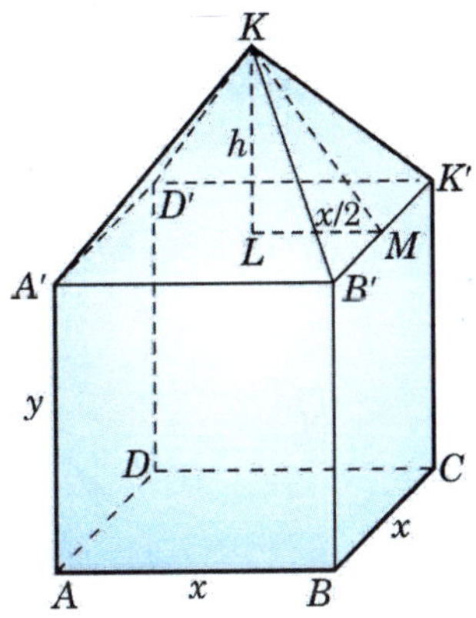

Fig. 5.6

By Lagrange's method,

$$[4y + \sqrt{(x^2 + 4h^2)} + x^2(x^2 + 4h^2)^{-1/2}] + \lambda\cdot 2x(y + h/3) = 0 \qquad ...(i)$$

$$4x + \lambda\cdot x^2 = 0 \qquad ...(ii)$$

$$4hx(x^2 + 4h^2)^{-1/2} + \lambda\cdot x^2/3 = 0 \qquad ...(iii)$$

$(ii)$ gives $\lambda = -4/x$. Then Eqn. $(iii)$ becomes

$$4hx(x^2 + 4h^2)^{-1/2} - 4x/3 = 0 \quad \text{or} \quad x = \sqrt{5}\,h$$

Now putting $x = \sqrt{5}\,h$, $\lambda = -4/x$ in $(i)$, we get

$$4y + 3h + \frac{5}{3}h - \frac{4}{x}\cdot 2x(y + h/3) = 0$$

or $$4y + \frac{14}{3}h - 8y - \frac{8h}{3} = 0, \quad i.e., \quad y = h/2.$$

**Example 5.55.** *If $u = a^3x^2 + b^3y^2 + c^3z^2$ where $x^{-1} + y^{-1} + z^{-1} = 1$, show that the stationary value of $u$ is given by $x = \Sigma a/a$, $y = \Sigma a/b$, $z = \Sigma a/c$.* (Anna, 2013)

**Solution :** Let $$u = f(x, y, z) = a^3x^2 + b^3y^2 + c^3z^2$$

and $$\phi(x, y, z) = x^{-1} + y^{-1} + z^{-1} - 1$$

Let $$F(x, y, z) = f(x, y, z) + \lambda\,\phi(x, y, z)$$

$$= a^3x^2 + b^3y^2 + c^3z^2 + \lambda(x^{-1} + y^{-1} + z^{-1} - 1)$$

Then $\frac{\partial F}{\partial x} = 0, \frac{\partial F}{\partial y} = 0$ and $\frac{\partial F}{\partial z} = 0$ give

$$2a^3x^2 - \lambda/x^2 = 0, \quad 2b^3y - \lambda/y^2 = 0, \quad 2c^3z - \lambda/z^2 = 0$$

or $$2a^3x^3 = \lambda,\ 2b^3y^3 = \lambda,\ 2c^3z^3 = \lambda$$

which give $ax = by = cz = k$ (say) *i.e.*, $x = k/a$, $y = k/b$, $z = k/c$.

Substituting these in $x^{-1} + y^{-1} + z^{-1} = 1$, we get $k = a + b + c$

Hence the stationary value of $u$ is given by

$$x = \Sigma a/a, \quad y = \Sigma a/b \quad \text{and} \quad z = \Sigma a/c.$$

**Example 5.56.** *Find the volume of the greatest rectangular parallelopiped that can be inscribed in the ellipsoid*

$$\frac{x^2}{a^2} + \frac{y^2}{b^2} + \frac{z^2}{c^2} = 1.$$

*(Andhra, 2022 ; C.S.V.T.U., 2016 ; D.T.U., 2015 ; Andhra., 2013 ; U.K.T.U., 2010)*

**Solution :** Let the edges of the parallelopiped be $2x$, $2y$ and $2z$ which are parallel to the axes. Then its volume $V = 8xyz$.

Now we have to find the maximum value of $V$ subject to the condition that

$$\frac{x^2}{a^2} + \frac{y^2}{b^2} + \frac{z^2}{c^2} - 1 = 0 \qquad ...(i)$$

Write $$F = 8xyz + \lambda\left(\frac{x^2}{a^2} + \frac{y^2}{b^2} + \frac{z^2}{c^2} - 1\right)$$

Then $$\frac{\partial F}{\partial x} = 8yz + \lambda\left(\frac{2x}{a^2}\right) = 0 \qquad ...(ii)$$

$$\frac{\partial F}{\partial y} = 8zx + \lambda\left(\frac{2y}{b^2}\right) = 0 \qquad ...(iii)$$

$$\frac{\partial F}{\partial z} = 8xy + \lambda\left(\frac{2z}{c^2}\right) = 0 \qquad ...(iv)$$

Equating the values of $\lambda$ from Eqns. (*ii*) and (*iii*), we get $x^2/a^2 = y^2/b^2$

Similarly from Eqns. (*iii*) and (*iv*), we obtain $y^2/b^2 = z^2/c^2 \quad \therefore \quad x^2/a^2 = y^2/b^2 = z^2/c^2$

Substituting these in Eqn. (*i*), we get $x^2/a^2 = \frac{1}{3}$ *i.e.* $\frac{x^2}{a^2} = \frac{y^2}{b^2} = \frac{z^2}{c^2} = \frac{1}{3}$

These give $x = a/\sqrt{3}$, $y = b/\sqrt{3}$, $z = c/\sqrt{3}$ ...(*v*)

When $x = 0$, the parallelopiped is just a rectangular sheet and as such its volume $V = 0$.

As $x$ increases, $V$ also increases continuously.

Thus $V$ must be greatest at the stage given by Eqn. (*v*).

Hence the greatest volume $= \dfrac{8abc}{3\sqrt{3}}$.

**Example 5.57 :** *Find the stationary values of*

$$\frac{x^2}{a^4} + \frac{y^2}{b^4} + \frac{z^2}{c^4}$$

*when* $lx + my + nz = 0$ *and* $\frac{x^2}{a^2} + \frac{y^2}{b^2} + \frac{z^2}{c^2} = 1$

*Interpret the result geometrically.*

**Solution :** Let $$u = \frac{x^2}{a^4} + \frac{y^2}{b^4} + \frac{z^2}{c^4} \qquad ...(i)$$

and $$lx + my + nz = 0 \qquad ...(ii)$$

$$\frac{x^2}{a^2} + \frac{y^2}{b^2} + \frac{z^2}{c^2} - 1 = 0 \qquad ...(iii)$$

Let us consider the Lagrange's function

$$F(x, y, z) = \left(\frac{x^2}{a^4}+\frac{y^2}{b^4}+\frac{z^2}{c^4}\right)+\lambda_1(lx+my+nz)+\lambda_2\left(\frac{x^2}{a^2}+\frac{y^2}{b^2}+\frac{z^2}{c^2}-1\right)$$

For stationary values, $dF = 0$

Now
$$dF = \left(\frac{2x}{a^4}dx+\frac{2y}{b^4}dy+\frac{2z}{c^4}dz\right)+\lambda_1(ldx+mdy+ndz)$$
$$+\lambda_2\left(\frac{2x}{a^2}dx+\frac{2y}{b^2}dy+\frac{2z}{c^2}dz\right) = 0$$

or
$$\left(\frac{2x}{a^4}+\lambda_1 l+\lambda_2\frac{2x}{a^2}\right)dx+\left(\frac{2y}{b^4}+\lambda_1 m+\lambda_2\frac{2y}{b^2}\right)dy+\left(\frac{2z}{c^4}+\lambda_1 n+\lambda_2\frac{2z}{c^2}\right) = 0$$

$$\therefore \quad \frac{2x}{a^4}+\lambda_1 l+\lambda_2\cdot\frac{2x}{a^2} = 0 \qquad \text{...(iv)}$$

$$\frac{2y}{b^4}+\lambda_1 m+\lambda_2\frac{2y}{b^2} = 0 \qquad \text{...(v)}$$

$$\frac{2z}{c^4}+\lambda_1 n+\lambda_2\cdot\frac{2z}{c^2} = 0 \qquad \text{...(vi)}$$

multiplying Eqns. (*iv*), (*v*), (*vi*) by $x, y, z$ respectively and adding, we get

$$2\left(\frac{x^2}{a^4}+\frac{y^2}{b^4}+\frac{z^2}{c^4}\right)+\lambda_1(lx+my+nz)+2\lambda_2\left(\frac{x^2}{a^2}+\frac{y^2}{b^2}+\frac{z^2}{c^2}\right) = 0 \qquad \text{...(vii)}$$

From Eqns. (*ii*), (*iii*) and (*vii*), we get

$$2u+\lambda_1\cdot 0+2\lambda_2(1) = 0$$
$$\lambda_2 = -u$$

From Eqns. (*iv*), (*v*), (*vi*) we have

$$\frac{2x}{a^4}+\lambda_1 l-u\frac{2x}{a^2} = 0$$

$$\therefore \quad x = \frac{-\lambda_1 l\,a^4}{2(1-a^2u)},\; y = \frac{-\lambda_1 mb^4}{2(1-b^2u)},\; z = \frac{\lambda_1 nc^4}{2(1-c^2u)}$$

Putting these values of $x, y, z$ in the relation Eqn. (*ii*), we get

$$\frac{l^2a^u}{1-a^2u}+\frac{m^2b^u}{1-b^2u}+\frac{n^2c^u}{1-c^2u} = 0 \qquad \text{...(viii)}$$

The relation Eqn. (*viii*) gives the stationary value of $u$.

**Geometrical Interpretation :** The equation $\frac{x^2}{a^2}+\frac{y^2}{b^2}+\frac{z^2}{c^2} = 1$ represented an ellipsoid and equation of the tangent plane at any point $(x, y, z)$ on it is

$$\frac{xX}{a^2}+\frac{yY}{b^2}+\frac{zZ}{c^2} = 1$$

The perpendicular distance $p$ from (0, 0, 0) to this tangent plane is

$$p = \frac{1}{\sqrt{\frac{x^2}{a^4} + \frac{y^2}{b^4} + \frac{z^2}{c^4}}}$$

$$\therefore \qquad \frac{1}{p^2} = \frac{x^2}{a^4} + \frac{y^2}{b^4} + \frac{z^2}{c^4}$$

If the point $(x, y, z)$ also lies on $lx + my + nz = 0$, it satisfies both the conditions (*ii*) and (*iii*)

So, the ellipsoid $\frac{x^2}{a^2} + \frac{y^2}{b^2} + \frac{z^2}{c^2} = 1$ at point that lie on the plane $lx + my + nz = 0$.

**Example 5.58 :** *Show that the volume of the greatest rectangular parallelopiped that can be inscribed in the ellipsoid*

$$\frac{x^2}{a^2} + \frac{y^2}{b^2} + \frac{z^2}{c^2} = 1 \text{ is } \frac{8\,abc}{3\sqrt{3}}$$ *(Rohtak, 2019)*

**Solution :** Let $(x, y, z)$ be the coordinate of an angular point of the rectangular parallelopiped. Then the lengths of the edges are $2x$, $2y$, $2z$ and volume $V$ is given by

$$V = 2x, \quad 2y, \quad 2z = 8\,xyz \qquad \text{...}(i)$$

where $$\frac{x^2}{a^2} + \frac{y^2}{b^2} + \frac{z^2}{c^2} = 1 \qquad \text{...}(ii)$$

Consider the Lagrange's function

$$f(x, y, z) = 8xyz + \lambda\left(\frac{x^2}{y^2} + \frac{y^2}{b^2} + \frac{z^2}{c^2} - 1\right)$$

For maximum or minimum values of $V$

$$dF = 0$$

Now, $dF = 8(yz\,dx + z\,xdy + xy\,dz) + 2\lambda\left(\frac{x}{a^2}dx + \frac{y}{b^2}dy + \frac{z}{c^2}dz\right) = 0$

$$\left(8yz + \frac{2\lambda x}{a^2}\right)dx + \left(8zx + \frac{2\lambda y}{b^2}\right) + \left(8xy + \frac{2\lambda z}{c^2}\right)dz = 0$$

$$\therefore \qquad 8yz + \frac{2\lambda x}{a^2} = 0 \qquad \text{...}(iii)$$

$$8zx + \frac{2\lambda y}{b^2} = 0 \qquad \text{...}(iv)$$

$$8xy + \frac{2\lambda z}{c^2} = 0 \qquad \text{...}(v)$$

multiplying Eqn. (*iii*) by $x$, $y$, $z$ respectively and adding, we get $3V + 2\lambda\left(\frac{x^2}{a^2} + \frac{y^2}{b^2} + \frac{z^2}{c^2}\right) = 0$ (using $8xyz = V$)

$$3V + 2\lambda - 1 = 0$$

$$\therefore \qquad \lambda = -\frac{3V}{2}$$

So, from Eqns. (*iii*), (*iv*), (*v*) by substituting $\lambda = -\frac{3V}{2}$

So, $$x = \frac{a}{\sqrt{3}},\ y = \frac{b}{\sqrt{3}},\ z = \frac{c}{\sqrt{3}}$$

$\therefore$ $$V = \frac{8abc}{3\sqrt{3}} \qquad ...(vi)$$

**Example 5.59 :** *Find the shortest and the longest distance from the point (1, 2, –1) to the sphere $x^2 + y^2 + z^2 = 24$* *(Bhilai, 2018)*

**Solution :** Let $(x, y, z)$ be any point on the given sphere and $d$ its distance from the point $(1, 2, -1)$, then

$$d^2 = (x-1)^2 + (y-2)^2 + (z+1)^2 = u \qquad ...(i)$$

$$x^2 + y^2 + z^2 - 24 = 0 \qquad ...(ii)$$

Lagrange's function

$$F(x, y, z) = u + \lambda(x^2 + y^2 + z^2 - 24)$$
$$= (x-1)^2 + (y-2)^2 + (z+1)^2 + \lambda(x^2 + y^2 + z^2 - 24)$$

For maximum or minimum value of $d^2$

$$dF = 0$$

Now, $$dF = 2(x-1)dx + 2(y-2)dy + 2(z+1)dz + 2\lambda\,(x\,dx + y\,dy + z\,dz) = 0$$
$$= [2x(1+\lambda) - 2]dx + [2y(1+\lambda) - 4]dy + [2z(1+\lambda) + 2]\,dz = 0$$

$\therefore$ $$2x(1+\lambda) - 2 = 0 \qquad ...(iii)$$
$$2y(1+\lambda) - 4 = 0 \qquad ...(iv)$$
$$2z(1+\lambda) + 2 = 0 \qquad ...(v)$$

From Eqns. $(iii)$, $(iv)$ and $(v)$ on equating the value of $(1 + \lambda)$ we get,

$$2x = y - 2z \qquad ...(vi)$$

Substituting in Eqn. $(ii)$, we have

$$x^2 + 4x^2 + x^2 - 24 = 0$$
$$x = \pm 2 \qquad ...(vii)$$

From Eqns. $(vi)$ and $(vii)$, we get

$$x = \pm 2, \quad y = \pm 4, \quad z = \pm 2$$

$\therefore$ $$d^2 = 6, \text{ or } 54$$
$$d = \sqrt{6} \text{ or } 3\sqrt{6} \text{ from Eqn. } (i)$$

which are the required shortest and the longest distance respectively.

## PROBLEMS 5.10

1. Find the maximum and minimum values of
   $(i)$ $x^3 + y^3 - 3axy$ *(Anna, 2013 ; U.P.T.U., 2012)* $(ii)$ $2(x^2 - y^2) - x^4 + y^4$
   $(iii)$ $x^3 + 3xy^2 - 15x^2 - 15y^2 + 72x$ *(JNTUK, 2022; Andhra, 2023)*
2. If $xyz = 8$, find the values of $x, y$ for which $u = 5xyz/(x + 2y + 4z)$ is a maximum.
3. Find the minimum value of $x^2 + y^2 + z^2$, given that
   $(i)$ $xyz = a^3$ $(ii)$ $ax + by + cz = p$. *(D.T.U., 2012)*
   $(iii)$ $xy + yz + zx = 3a^2$
4. If $\frac{3}{x} + \frac{4}{y} + \frac{5}{z} = 6$, find the values of $x, y, z$ which make $x + y + z$ minimum. *(Delhi, 2013)*
5. Find the dimensions of the rectangular box, open at the top, of maximum capacity whose surface is 432 sq. cm. *(C.S.V.T.U., 2015 ; Anna, 2013)*
6. The sum of three numbers is constant. Prove that their product is maximum when they are equal.
7. Find the points on the surface $z^2 = xy + 1$ nearest to the origin. *(J.N.T.U., 2015 ; Rohtak, 2012)*
8. Show that, if the perimeter of a triangle is constant, the triangle has maximum area when it is equilateral. *(Rohtak, 2011 S)*
9. Find the maximum and minimum distances from the origin to the curve $5x^2 + 6xy + 5y^2 - 8 = 0$.

10. The temperature $T$ at any point $(x, y, z)$ in space is $T = 400\, xyz^2$. Find the highest temperature on the surface of the unit sphere $x^2 + y^2 + z^2 = 1$.
11. Divide 24 into three parts such that the continued product of the first, square of the second and the cube of the third may be maximum. *(C.S.V.T.U., 2012 ; Kurukshetra, 2010 S)*
12. Find the stationary values of $u = x^2 + y^2 + z^2$ subject to $ax^2 + by^2 + cz^2 = 1$ and $lx + my + nz = 0$. *(C.S.V.T.U., 2015)*
13. The beds of two rivers have the shapes represented by a parabola $y = x^2$ and straight line $x - y - 2 = 0$. These two rivers are to be connected by a straight canal of least length. Through what points will it pass? What is the least length? *(Rohtak, 2019)*
14. A tent on a square of side $2a$ has its four sides vertical of height $h$ is surmounted by a regular pyramid of height $b$. Find the values of $a$ and $h$ in terms of $b$, if the cost of canvas required is to be least for its construction. *(Burdwan, 2019)*
15. Prove that the shortest distance from the origin both curve of intersection of the surface $xyz = a$ and $y = bx$, where $a > 0, b > 0$ is $3\sqrt{a(b^2+1)/2b}$. *(CSVTU, 2019)*
16. The temperature $T$ at any point $(x, y, z)$ in space is $T = 400\, xyz^2$. Find the highest temperature on the surface of the unit sphere $x^2 + y^2 + z^2 = 1$ *(Bhilai, 2018)*
17. Find the dimensions of closed rectangular bath tub of a given capacity $V$ having the smallest surface.
18. The heat generated in unit time, in an electric circuit having a resistance $R$ and the current flowing in it being $I$ is given by $I^2R$. Determine how to distribute the current $I$, into the currents $I_1, I_2, I_3$ by means of three wires, whose resistances are $R_1, R_2, R_3$ respectively. So that the heat generated is minimum.
19. Decompose a given positive number $a$ in to three parts, so that their product is maximum.
20. Show that the maximum and minimum values of $r^2$, where $r^2 = a^2x^2 + b^2y^2 + c^2z^2$ and $x^2 + y^2 + z^2 = 1$

    $lx + my + nz = 0$ are given by the equation $\dfrac{l^2}{a^2 - x^2} + \dfrac{m^2}{b^2 - x^2} + \dfrac{n^2}{c^2 - x^2}$.

## 5.13 DIFFERENTIATION UNDER THE INTEGRAL SIGN

If a function $f(x, \alpha)$ of two variables $x$ and $\alpha$ (called a parameter), be integrated with respect to $x$ between the limits $a$ and $b$, then $\int_a^b f(x, \alpha)\, dx$ is a function of $\alpha : F(\alpha)$, say. To find the derivative of $F(\alpha)$, when it exists, it is not always possible to first evaluate this integral and then to find the derivative. Such problems are solved by the following rules :

**(1) Leibnitz's rule***

*If $f(x, \alpha)$ and $\dfrac{\partial f(x, \alpha)}{\partial \alpha}$ be continuous functions of $x$ and $\alpha$, then*

$$\frac{d}{d\alpha}\left[\int_a^b f(x, \alpha)\, dx\right] = \int_a^b \frac{\partial f(x, \alpha)}{\partial \alpha}\, dx \quad \text{where, } a, b \text{ are constants independent of } \alpha.$$

*(i.e., the order of differentiation and integration can be interchanged.)*

Let $\quad \int_a^b f(x, \alpha)\, dx = F(\alpha),$

then $\quad F(\alpha + \delta\alpha) - F(\alpha) = \int_a^b f(x, \alpha + \delta\alpha)\, dx - \int_a^b f(x, \alpha)\, dx = \int_a^b [f(x, \alpha + \delta\alpha) - f(x, \alpha)]\, dx$

$$= \delta\alpha \int_a^b \frac{\partial f(x, \alpha + \theta\delta\alpha)}{\partial \alpha}\, dx, \; (0 < \theta < 1) \qquad \left\{\begin{array}{l}\because\; f(x, \alpha + h) - f(x, \alpha) = hf'(x, \alpha + \theta h) \\ \text{where } 0 < \theta < 1, \text{ by Mean Value Theorem}\end{array}\right.$$

Proceeding to limits as $\delta\alpha \to 0$, $\displaystyle \mathop{\text{Lt}}_{\delta\alpha \to 0} \frac{F(\alpha + \delta\alpha) - F(\alpha)}{\delta\alpha} = \int_a^b \frac{\partial f(x, \alpha + \theta \cdot 0)}{\partial \alpha}\, dx$

or $\quad \dfrac{dF}{d\alpha} = \displaystyle\int_a^b \frac{\partial f(x, \alpha)}{\partial \alpha}\, dx$ which is the desired result.

**Obs. 1.** Leibnitz's rule enables us to derive from the value of a simple definite integral, the value of another definite integral which it may otherwise be difficult, or even impossible, to evaluate.

**Obs. 2.** The rule for differentiation under the integral sign of an infinite integral is the same as for a definite integral.

*See footnote page 154.

**Example 5.60.** *Evaluate* $\int_0^1 \frac{x^\alpha - 1}{\log x} dx$, $\alpha \geq 0$. *(V.T.U., 2016 ; C.S.V.T.U., 2011)*

**Solution.** Let $F(\alpha) = \int_0^1 \frac{x^\alpha - 1}{\log x} dx$ ...(*i*)

then $$F'(\alpha) = \int_0^1 \frac{\partial}{\partial \alpha}\left(\frac{x^\alpha - 1}{\log x}\right) dx = \int_0^1 \frac{x^\alpha \log x}{\log x} dx$$

$$= \int_0^1 x^\alpha \, dx = \left| \frac{x^{\alpha+1}}{\alpha+1} \right|_0^1 = \frac{1}{1+\alpha} \qquad \left[\because \frac{d}{dt}(n^t) = n^t \log n\right]$$

Now integrating both sides w.r.t. $\alpha$, $F(\alpha) = \log(1+\alpha) + c$ ...(*ii*)

From Eqn. (*i*), when $\alpha = 0$, $F(0) = 0$

$\therefore$ From Eqn. (*ii*), $F(0) = \log(1) + c$, *i.e.*, $c = 0$. Hence Eqn. (*ii*) gives, $F(\alpha) = \log(1+\alpha)$.

**Example 5.61.** *Given* $\int_0^\pi \frac{dx}{a + b\cos x} = \frac{\pi}{\sqrt{(a^2 - b^2)}}$ $(a > b)$,

*evaluate* $\int_0^\pi \frac{dx}{(a + b\cos x)^2}$ *and* $\int_0^\pi \frac{\cos x}{(a + b\cos x)^2} dx$

**Solution.** We have $$\int_0^\pi \frac{dx}{a + b\cos x} = \frac{\pi}{\sqrt{(a^2 - b^2)}} \qquad ...(i)$$

Differentiating both sides of Eqn. (*i*) w.r.t. $a$,

$$\int_0^\pi \frac{\partial}{\partial a}\left(\frac{1}{a + b\cos x}\right) dx = \frac{\partial}{\partial a}\left\{\frac{\pi}{\sqrt{(a^2 - b^2)}}\right\}$$

*i.e.* $$\int_0^\pi \frac{-dx}{(a + b\cos x)^2} = \pi \cdot \left(-\frac{1}{2}\right)(a^2 - b^2)^{-3/2} \, 2a$$

$$\therefore \quad \int_0^\pi \frac{dx}{(a + b\cos x)^2} = \frac{\pi a}{(a^2 - b^2)^{3/2}}$$

Now differentiating both sides of Eqn. (*i*) w.r.t. $b$,

$$\int_0^\pi -(a + b\cos x)^{-2} \cdot \cos x \, dx = \pi\left(-\frac{1}{2}\right)(a^2 - b^2)^{-3/2} \cdot (-2b)$$

$$\therefore \quad \int_0^\pi \frac{\cos x}{(a + b\cos x)^2} dx = \frac{\pi b}{(a^2 - b^2)^{3/2}}.$$

**(2) Leibnitz's rule** *for variable limits of integration*

*If* $f(x, \alpha)$, $\frac{\partial f(x, \alpha)}{\partial \alpha}$ *be continuous functions of x and* $\alpha$, *then*

$$\frac{d}{d\alpha}\left\{\int_{\phi(\alpha)}^{\psi(\alpha)} f(x, \alpha) dx\right\} = \int_{\phi(\alpha)}^{\psi(\alpha)} \frac{\partial f(x, \alpha)}{\partial \alpha} dx + \frac{d\psi}{d\alpha} f[\psi(\alpha), \alpha] - \frac{d\phi}{d\alpha} f[\phi(\alpha), \alpha]$$

*provided* $\phi(\alpha)$ *and* $\psi(\alpha)$ *possesses continuous first order derivatives w.r.t.* $\alpha$.

Its proof is beyond the scope of this book.

**Example 5.62.** *Evaluate* $\int_0^\alpha \frac{\log(1 + \alpha x)}{1 + x^2} dx$ *and hence show that*

$$\int_0^1 \frac{\log(1 + x)}{1 + x^2} dx = \frac{\pi}{8} \log_e 2$$

**Solution.** Let $$F(\alpha) = \int_0^{\alpha} \frac{\log(1+\alpha x)}{1+x^2} dx \quad ...(i)$$

Then by the above rule, $$F'(\alpha) = \int_0^{\alpha} \frac{\partial}{\partial \alpha}\left(\frac{\log(1+\alpha x)}{1+x^2}\right) dx + \frac{d(\alpha)}{d\alpha} \cdot \frac{\log(1+\alpha^2)}{1+\alpha^2} - 0$$

$$= \int_0^{\alpha} \frac{x}{(1+\alpha x)(1+x^2)} dx + \frac{\log(1+\alpha^2)}{1+\alpha^2} \quad ...(ii)$$

Breaking the integrand into partial fractions,

$$\int_0^{\alpha} \frac{x\,dx}{(1+\alpha x)(1+x^2)} = -\frac{\alpha}{1+\alpha^2}\int_0^{\alpha} \frac{dx}{1+\alpha x} + \frac{1}{2(1+\alpha^2)}\int_0^{\alpha} \frac{2x}{1+x^2} dx + \frac{\alpha}{1+\alpha^2}\int_0^{\alpha} \frac{dx}{1+x^2}$$

$$= -\frac{1}{1+\alpha^2}\Big|\log(1+\alpha x)\Big|_0^{\alpha} + \frac{1}{2(1+\alpha^2)} \times \Big|\log(1+x^2)\Big|_0^{\alpha} + \frac{\alpha}{1+\alpha^2}\Big|\tan^{-1} x\Big|_0^{\alpha}$$

$$= -\frac{\log(1+\alpha^2)}{1+\alpha^2} + \frac{\log(1+\alpha^2)}{2(1+\alpha^2)} + \frac{\alpha \tan^{-1}\alpha}{1+\alpha^2}$$

Substituting this value in Eqn. (*ii*), $F'(\alpha) = \dfrac{\log(1+\alpha^2)}{2(1+\alpha^2)} + \dfrac{\alpha \tan^{-1}\alpha}{1+\alpha^2}$

Now integrating both sides w.r.t. $\alpha$,

$$F(\alpha) = \frac{1}{2}\int \log(1+\alpha^2) \cdot \frac{1}{1+\alpha^2} d\alpha + \int \frac{\alpha \tan^{-1}\alpha}{1+\alpha^2} d\alpha \quad \text{[Integrating by parts]}$$

$$= \frac{1}{2}\left[\log(1+\alpha^2)\cdot \tan^{-1}\alpha - \int \frac{2\alpha}{1+\alpha^2}\cdot \tan^{-1}\alpha\, d\alpha\right] + \int \frac{\alpha \tan^{-1}\alpha}{1+\alpha^2} d\alpha + c$$

$$= \frac{1}{2}\log(1+\alpha^2)\cdot \tan^{-1}\alpha + c \quad ...(iii)$$

But from Eqn. (*i*), when $\alpha = 0$, $F(0) = 0$.

$\therefore$ From Eqn. (*iii*), $F(0) = 0 + c$, *i.e.*, $c = 0$. Hence (*iii*) gives, $F(\alpha) = \dfrac{1}{2}\log(1+\alpha^2)\tan^{-1}\alpha$

Putting $\alpha = 1$, we get $\displaystyle\int_0^1 \frac{\log(1+x)}{1+x^2} dx = F(1) = \frac{\pi}{8}\log_e 2.$

## PROBLEMS 5.11

1. Differentiating $\displaystyle\int_0^x \frac{dx}{x^2+a^2} = \frac{1}{a}\tan^{-1}\frac{x}{a}$ under the integral sign, find the value of $\displaystyle\int_0^x \frac{dx}{(x^2+a^2)^2}$.
2. By successive differentiation of $\displaystyle\int_0^1 x^m\,dx = \frac{1}{m+1}$ w.r.t. $m$, evaluate $\displaystyle\int_0^1 x^m (\log x)^n\,dx$. (*C.S.V.T.U., 2014*)
3. Evaluate $\displaystyle\int_0^{\pi} \log(1 + a\cos x)\,dx$, using the method of differentiation under the sign of integration. (*C.S.V.T.U., 2015 ; V.T.U., 2014 ; Rohtak, 2010 S*)
4. Given that $\displaystyle\int_0^{\pi} \frac{dx}{a-\cos x} = \frac{\pi}{\sqrt{(a^2-1)}}$, evaluate $\displaystyle\int_0^{\pi} \frac{dx}{(a-\cos x)^2}$.

Prove that :

5. $\int_0^\infty e^{-x} \cdot \frac{\sin ax}{x} dx = \tan^{-1} a.$ $\left[\textbf{Hint.} \text{ Use } \int e^{ax} \cos bx\, dx = \frac{e^{ax}}{a^2+b^2}(a \cos bx + b \sin bx)\right]$

6. $\int_0^\infty e^{-ax} \frac{\sin x}{x} dx = \tan^{-1} \frac{1}{a}.$ *(V.T.U., 2015)*

Hence show that $\int_0^\infty \frac{\sin x}{x} = \frac{\pi}{2}.$ *(C.S.V.T.U., 2015)*

7. $\int_0^\infty \frac{\tan^{-1} ax}{x(1+x^2)} dx = \frac{\pi}{2} \log(1+a)$ where $a \geq 0$. *(Andhra., 2016 ; V.T.U., 2016 ; C.S.V.T.U., 2014)*

8. $\int_0^\infty \frac{e^{-x}}{x} (1 - e^{-ax})\, dx = \log(1+a), (a > -1).$

9. $\int_0^{\pi/2} \log(\alpha^2 \cos^2 \theta + \beta^2 \sin^2 \theta)\, d\theta = \pi \log \frac{\alpha+\beta}{2}$

10. $\int_0^{\pi/2} \frac{\log(1 + y \sin^2 x)}{\sin^2 x} dx = \pi [\sqrt{(1+y)} - 1]$ *(Andhra., 2014 ; V.T.U., 2013)*

11. $\int_0^{\pi/2} \frac{\log(1+\sin\alpha\cos x)}{\cos x} = \pi\, \alpha.$ *(V.T.U., 2014)*

12. $\int_0^\infty e^{-x^2} \cos \alpha x\, dx = \frac{\sqrt{\pi}}{2} e^{-\alpha^2/4}$ *(Andhra., 2015 ; C.S.V.T.U., 2011S)*

13. $\frac{d}{da}\int_0^{a^2} \tan^{-1} \frac{x}{a} dx = 2a \tan^{-1} a - \frac{1}{2} \log(a^2+1)$. Verify your result by direct integration.

14. $\int_{\pi/2-\alpha}^{\pi/2} \sin\theta \cos^{-1}(\cos\alpha \cos\theta)\, d\theta = \frac{\pi}{2}(1 - \cos\alpha).$

15. Evaluate the following integral by differentiating under the integral sign $\int_0^\infty e^{-(x^2+a^2/x^2)} dx.$ *(Kurukshetra, 2012)*

## 5.14 OBJECTIVE TYPE OF QUESTIONS

### PROBLEMS 5.12

*Select the correct answer or fill up the blanks in each of the following problems :*

1. If $u = e^x(x \cos y - y \sin y)$, then $\frac{\partial^2 u}{\partial x^2} + \frac{\partial^2 u}{\partial y^2} = \ldots$

2. If $x = uv, y = u/v$, then $\frac{\partial(x, y)}{\partial(u, v)}$ is

(a) $-2u/v$ (b) $-2v/u$ (c) 0 (d) 1. *(V.T.U., 2010)*

3. If $J_1 = \frac{\partial(u, v)}{\partial(x, y)}$ and $J_2 = \frac{\partial(x, y)}{\partial(u, v)}$, then $J_1 J_2 = \ldots$ *(P.T.U., 2013)*

4. If $u = \log(x/y)$, then

(a) $x\frac{\partial u}{\partial x} - y\frac{\partial u}{\partial y} = 0$ (b) $x\frac{\partial u}{\partial x} + y\frac{\partial u}{\partial y} = 0$ (c) $x\frac{\partial u}{\partial x} + y\frac{\partial u}{\partial y} = 2u$ (d) $x\frac{\partial u}{\partial x} + y\frac{\partial u}{\partial y} = 1.$

*(C.S.V.T.U., 2015)*

5. If $u = x^y$, then $\partial u/\partial x$ is
(a) 0 (b) $yx^{y-1}$ (c) $x^y \log x$.

6. If $x = r\cos\theta$, $y = r\sin\theta$, then
(a) $\frac{\partial x}{\partial r} = 1\Big/\frac{\partial r}{\partial x}$ (b) $\frac{\partial x}{\partial r} = \frac{\partial r}{\partial x}$ (c) $\frac{\partial x}{\partial r} = 0$.

7. If $u = y^x$, then $\frac{\partial u}{\partial x}$ = ...... and $\frac{\partial u}{\partial y}$ = ...... (Anna, 2013)

8. If $u = x^3 + y^3$, then $\frac{\partial^2 u}{\partial x \partial y}$ is equal to
(a) –3 (b) 3 (c) 0 (d) $3x + 3y$ (V.T.U., 2010 S)

9. If $u = x^2 + 2xy + y^2 + x + y$, then $x\frac{\partial u}{\partial x} + y\frac{\partial u}{\partial y}$ is equal to
(a) $2u$ (b) $u$ (c) 0 (d) none of these.

10. If $u = \log \frac{x^2}{y}$, then $x\frac{\partial u}{\partial x} + y\frac{\partial u}{\partial y}$ is equal to
(a) $2u$ (b) $3u$ (c) $u$ (d) 1. (V.T.U., 2010 S)

11. If $x = r\cos\theta$, $y = r\sin\theta$, then $\frac{\partial(x, y)}{\partial(r, \theta)}$ is equal to
(a) 1 (b) $r$ (c) $1/r$ (d) 0. (V.T.U., 2010 S)

12. If $A = f_{xx}(a, b)$, $B = f_{xy}(a, b)$, $C = f_{yy}(a, b)$, then $f(x, y)$ will have a maximum at $(a, b)$ if
(a) $f_x = 0, f_y = 0,\ AC < B^2$ and $A < 0$ (b) $f_x = 0, f_y = 0,\ AC = B^2$ and $A > 0$
(c) $f_x = 0, f_y = 0,\ AC > B^2$ and $A > 0$ (d) $f_x = 0, f_y = 0,\ AC > B^2$ and $A < 0$.

13. If $z = \sin^{-1}\frac{\sqrt{x^2 + y^2}}{x + y}$, then $x\frac{\partial z}{\partial x} + y\frac{\partial z}{\partial y}$ is
(a) 0 (b) 1/2 (c) 1 (d) 2.

14. If $u = \sin^{-1}(x/y) + \tan^{-1}(y/x)$, then $x\,\partial u/\partial x + y\,\partial u/\partial y$ equals
(a) $\sin^{-1}(x/y) + \tan^{-1}(y/x)$ (b) $2[\sin^{-1}(x/y) + \tan^{-1}(y/x)]$
(c) $3[\sin^{-1}(x/y) + \tan^{-1}(y/x)]$ (d) zero.

15. If an error of 1% is made in measuring its length and breadth, the percentage error in the area of a rectangle is
(a) 0.2% (b) 0.02% (c) 2% (d) 1%. (V.T.U., 2010)

16. $\frac{\partial(u, v)}{\partial(x, y)} \times \frac{\partial(x, y)}{\partial(u, v)}$ equals
(a) –1 (b) 1 (c) zero (d) none of these.

17. $\frac{\sqrt{x} - \sqrt{y}}{\sqrt{x} + \sqrt{y}}$ is a homogeneous function of degree ......

18. If $z = \log(x^3 + y^3 - x^2y - xy^2)$, then $\frac{\partial z}{\partial x} + \frac{\partial z}{\partial y}$ is equal to ......

19. If $r = \partial^2 f/\partial x^2$, $s = \partial^2 f/\partial x \partial y$ and $t = \partial^2 f/\partial y^2$, then the condition for the saddle point is ......

20. If $f(x, y) = \frac{1}{x^3} + \frac{1}{y^3} + \frac{1}{x^3 + y^3}$, then $x\frac{\partial f}{\partial x} + y\frac{\partial f}{\partial y}$ is
(a) 0 (b) $3f$ (c) 9 (d) $-3f$.

21. If $u = x^4 + y^4 + 3x^2y^2$, then $x\frac{\partial u}{\partial x} + y\frac{\partial u}{\partial y}$ = ......

22. If $u$ and $v$ are functions of $r, s$ where $r, s$ are functions of $x, y$, then $\frac{\partial(u, v)}{\partial(r, s)} \cdot \frac{\partial(r, s)}{\partial(x, y)}$ = ......

23. Given $f_{\lambda x} = 6x$, $f_{xy} = 0$, $f_{xy} = 6y$, find the nature of the stationary point at (–1, 2). (Anna, 2012)

24. If $u = (x - y)^4 + (y - z)^4 + (z - x)^4$, then $\frac{\partial u}{\partial x} + \frac{\partial u}{\partial y} + \frac{\partial u}{\partial z}$ is

(a) 1 (b) $u$ (c) $4u$ (d) 0. (V.T.U., 2010)

25. If $u$ is a composite function of $t$, defined by the relations $u = f(x, y); x = \phi(t), y = \psi(t)$, then total derivative $\frac{du}{dt}$ = ......

26. If $u = \cos^{-1}(x/y) + \tan^{-1}(y/x)$, then $x^2u_{xx} + 2xyu_{xy} + y^2u_{yy}$ is

(a) $u$ (b) $2u$ (c) 0 (d) 1. (V.T.U., 2010)

27. If $f(x, y, z) = 0$, then $\frac{\partial x}{\partial y} \cdot \frac{\partial y}{\partial z} \cdot \frac{\partial z}{\partial x}$ = ......

28. If $u = f(x + ay) + g(x - ay)$, then $\frac{\partial^2 u}{\partial y^2}$ equals

(a) $\frac{\partial^2 u}{\partial x^2}$ (b) $a\frac{\partial^2 u}{\partial x^2}$ (c) $a^2\frac{\partial^2 u}{\partial x^2}$ (d) $\frac{\partial^2 u}{\partial x \partial y}$. (V.T.U., 2010)

29. If sum of three numbers is constant, then their product is a maximum when the numbers are ......

30. $y = \cosh(\lambda x)\cosh(-\lambda at)$ is a solution of $\frac{\partial^2 y}{\partial t^2} = a^2\frac{\partial^2 y}{\partial x^2}$. (True or False)

CHAPTER 6

# Integral Calculus and its Curve Tracing

1. Reduction formulae. 2. Reduction formulae for $\int \sin^n x\, dx, \int \cos^n x\, dx$ and evaluation of $\int_0^{\pi/2} \sin^n x\, dx$, $\int_0^{\pi/2} \cos^n x\, dx$. 3. Reduction formula for $\int \sin^m x \cos^n x\, dx$ and evaluation of $\int_0^{\pi/2} \sin^m x \cos^n x\, dx$. 4. Reduction formulae for $\int \tan^n x\, dx, \int \cot^n x\, dx$. 5. Reduction formulae for $\int \sec^n x\, dx, \int \text{cosec}^n x\, dx$. 6. Reduction formulae for $\int x^n e^{ax}\, dx, \int x^m (\log x)^n\, dx$. 7. Reduction formulae for $\int x^n \sin mx\, dx$, $\int x^n \cos nx\, dx$ and $\int \cos^m x \sin nx\, dx$. 8. Definite integrals. 9. Integral as the limit of a sum. 10. Areas of curves. 11. Lengths of curves. 12. Volumes of revolution. 13. Surface areas of revolution. 14. Objective Type of Questions.

## 6.1 REDUCTION FORMULAE

The reader is already familiar with some standard methods of integrating functions of a single variable. However, there are some integrals which cannot be evaluated by the afore-said methods. In such cases, the method of reduction formulae proves useful. A reduction formula connects an integral with another of the same type but of lower order. The successive application of the reduction formula enables us to evaluate the given integral. Now we shall derive some standard reduction formulae.

## 6.2 (1) REDUCTION FORMULAE for

**(a)** $\int \sin^n x\, dx$ **(b)** $\int \cos^n x\, dx.$

(a) $\int \sin^n x\, dx = \int \sin^{n-1} x \cdot \sin x\, dx$ [Integrated by parts]

$$= \sin^{n-1} x \cdot (-\cos x) - \int (n-1) \sin^{n-2} x \cos x\, (-\cos x)\, dx$$

$$= -\sin^{n-1} x \cos x + (n-1) \int \sin^{n-2} x\, (1 - \sin^2 x)\, dx$$

$$= -\sin^{n-1} x \cos x + (n-1) \int \sin^{n-2} x\, dx - (n-1) \int \sin^n x\, dx$$

Transposing

$$n \int \sin^n x\, dx = -\sin^{n-1} x \cos x + (n-1) \int \sin^{n-2} x\, dx$$

or

$$\int \sin^n x\, dx = -\frac{\sin^{n-1} x \cos x}{n} + \frac{n-1}{n} \int \sin^{n-2} x\, dx \qquad ...(i)$$

(*b*) Similarly, $\int \cos^n x\, dx = \frac{\sin x \cos^{n-1} x}{n} + \frac{n-1}{n} \int \cos^{n-2} x\, dx$

Thus we have the required reduction formulae.

**Obs. To integrate $\int \sin^n x\, dx$ or $\int \cos^n x\, dx$,**

(*a*) *when the index of sin x is odd put cos x = t*

*when the index of cos x is odd, put sin x = t*

(*b*) *when the index is an even positive integer,* express the integrand as a series of cosines of multiple angles and integrate term by term if *n* is small, *otherwise use the method of reduction formulae.*

**(2) Wallis formula*:** $\int_0^{\pi/2} \sin^n x\, dx = \int_0^{\pi/2} \cos^n x\, dx$

$$= \frac{(n-1)(n-3)(n-5)\ldots}{n(n-2)(n-4)\ldots} \times \left(\frac{\pi}{2}, \textit{ only if } n \textit{ is even}\right)$$

From (*i*), we have

$$I_n = \int_0^{\pi/2} \sin^n x\, dx = -\left|\frac{\sin^{n-1} x \cos x}{n}\right|_0^{\pi/2} + \frac{n-1}{n}\int_0^{\pi/2} \sin^{n-2} x\, dx$$

*i.e.* $$I_n = \frac{n-1}{n} I_{n-2}$$

**Case I.** *When n is odd*

Similarly $$I_{n-2} = \frac{n-3}{n-2} I_{n-4}, \quad I_{n-4} = \frac{n-5}{n-4} I_{n-6}$$

..................................

$$I_5 = \frac{4}{5} I_3, I_3 = \frac{2}{3} I_1 = \frac{2}{3}\int_0^{\pi/2} \sin x\, dx = \frac{2}{3}\left| -\cos x \right|_0^{\pi/2} = \frac{2}{3}.$$

Form these, we get $I_n = \frac{(n-1)(n-3)(n-5)\ldots 2}{n(n-2)(n-4)\ldots 3}$ ...(*ii*)

**Case II.** *When n is even*

We have $$I_{n-2} = \frac{n-3}{n-2} I_{n-4}, I_{n-4} = \frac{n-5}{n-4} I_{n-6}$$

..........................................................

$$I_4 = \frac{3}{4} I_2, I_2 = \frac{1}{2} I_0 = \int_0^{\pi/2} \sin^0 x\, dx = \frac{1}{2}\int_0^{\pi/2} dx = \frac{1}{2}\cdot\frac{\pi}{2}.$$

Form these, we obtain $I_n = \frac{(n-1)(n-3)(n-5)\ldots 3\cdot 1}{n(n-2)(n-4)\ldots 4\cdot 2}\cdot\frac{\pi}{2}$ ...(*iii*)

Combining Eqns. (*ii*) and (*iii*), we get the required result for $\int_0^{\pi/2} \sin^n x\, dx$.

Proceeding exactly as above, we get the result for $\int_0^{\pi/2} \cos^n x\, dx$.

**Example 6.1.** *Integrate* (*i*) $\int \sin^4 x\, dx$ (*ii*) $\int_0^{\pi/2} \cos^6 x\, dx$.

**Solution.** (*i*) We have the reduction formula

$$\int \sin^n x\, dx = \frac{\sin^{n-1} x \cos x}{n} + \frac{n-1}{n}\int \sin^{n-2} x\, dx$$

Putting $n = 4, 2$ successively,

$$\int \sin^4 x\, dx = -\frac{\sin^3 x \cos x}{4} + \frac{3}{4}\int \sin^2 x\, dx \quad \ldots(\alpha)$$

*After an English mathematician *John Wallis* (1616–1703).

$$\int \sin^2 x\,dx = -\frac{\sin x \cos x}{2} + \frac{1}{2}\int (\sin x)^0\,dx$$

But $\int (\sin x)^0\,dx = \int dx = x. \quad \therefore \quad \int \sin^2 x\,dx = -\frac{\sin x \cos x}{2} + \frac{x}{2}$

Substituting this in (α), we get

$$\int \sin^4 x\,dx = -\frac{\sin^3 x \cos x}{4} + \frac{3}{4}\left(-\frac{\sin x \cos x}{2} + \frac{x}{2}\right)$$

(*ii*) We know that $\int_0^{\pi/2} \cos^n x\,dx = \frac{(n-1)(n-3)(n-5)\ldots}{n(n-2)(n-4)\ldots}\left(\frac{\pi}{2} \text{ if } n \text{ is even}\right)$

Putting $n = 6$, we get

$$\int_0^{\pi/2} \cos^6 x\,dx = \frac{5.3.1}{6.4.2}\frac{\pi}{2} = \frac{5\pi}{16}.$$

**Example 6.2.** *Evaluate*

(*a*) $\int_0^\infty \frac{dx}{(1+x^2)^5}$ (*b*) $\int_0^1 x^2(1-x^2)^{3/2}\,dx$ (*c*) $\int_0^{2a} x^3(2a-x)^{1/2}\,dx$ (*d*) $\int_0^4 x^3\sqrt{(4x-x^3)}\,dx$

**Solution.** (*a*) $\int_0^\infty \frac{dx}{(1+x^2)^5}$

Let $x = \tan\theta$

$dx = \sec^2\theta \cdot d\theta$

So, $\int_0^\infty \frac{dx}{(1+x^2)^5} = \int_0^{\pi/2} \frac{\sec^2\theta}{(1+\tan^2\theta)^5} = \int_0^{\pi/2} \cos^2\theta \cdot d\theta = \frac{7}{8}\cdot\frac{5}{6}\cdot\frac{3}{4}\cdot\frac{1}{2}\cdot\frac{\pi}{2} = \frac{35\pi}{256}$

(*b*) Let $x = \sin\theta$, $dx = \cos\theta\,d\theta$

The limit for θ are from 0 to $\frac{\pi}{2}$

So, $\int_0^1 x^2(1-x^2)^{3/2}\,dx = \int_0^{\pi/2} \sin^2\theta \cdot \cos^2\theta \cdot d\theta = \frac{\left|\overline{3/2}\right.\left|\overline{5/2}\right.}{2\left|\overline{(4)}\right.} = \frac{\pi}{32}$

(*c*) Let $x = 2a\sin^2\theta$, $dx = 4a\sin\theta\cos\theta.\,d\theta$

Also, $\theta = 0$, when $x = 0$, and $\theta = \frac{\pi}{2}$ when $x = 2a$

So, $\int_0^{2a} x^{3/2}(2a-x)^{1/2}\cdot d\theta = 16a^3\int_0^{\pi/2} \sin^4\theta \cdot \cos^2\theta \cdot d\theta = 16a^3.\frac{\pi}{32} = \frac{\pi a^3}{2}$

(*d*) Let $x = 4\sin^2\theta$, $dx = 8\sin\theta\cdot\cos\theta\cdot d\theta$

Also $\theta = 0$ when $x = 0$ and $\theta = \frac{\pi}{2}$, when $x = 4$

$$\int_0^4 x^3\sqrt{(4x-x^2)}\,dx = \int_0^4 x^{7/2}\sqrt{(4-x)}\,dx = \int_0^{\pi/2} (4\sin^2\theta)^{7/2}\sqrt{4-4\sin^2\theta}\cdot 8\sin\theta\cdot\cos\theta\,d\theta$$

$$= (4)^{7/2}\sqrt{4}\cdot 8\int_0^{\pi/2} \sin^8\theta\cos^2\theta\cdot d\theta = 2^{11}\left[\left(\frac{7}{10}\cdot\frac{5}{8}\cdot\frac{3}{6}\cdot\frac{1}{4}\right)\left(\frac{1}{2}\cdot\frac{\pi}{2}\right)\right]$$

$$= 28\,\pi$$

**Example 6.3.** *Evaluate* $\int_0^a \frac{x^n}{\sqrt{(a^2-x^2)}}dx$. *Hence find the value of* $\int_0^1 x^n \sin^{-1} x\,dx$.

**Solution.** Putting $x = a\sin\theta$, we get

$$\int_0^a \frac{x^n}{\sqrt{(a^2 - x^2)}} dx = \int_0^{\pi/2} \frac{(a \sin \theta)^n}{a \cos \theta} (a \cos \theta)\, d\theta = a^n \int_0^{\pi/2} \sin^n \theta \, d\theta$$

$$= \frac{(n-1)(n-3) \ldots. 2}{n(n-2) \ldots. 3} a^n, \text{ if } n \text{ is odd}$$

$$= \frac{(n-1)(n-3) \ldots 1}{n(n-2) \ldots 2} \cdot \frac{\pi}{2} a^n, \text{ if } n \text{ is even} \qquad \ldots(i)$$

Now integrating by parts, we have

$$\int_0^1 x^n \sin^{-1} x \, dx = \left|(\sin^{-1} x) \cdot \frac{x^{n+1}}{n+1}\right|_0^1 - \int_0^1 \frac{x^{n+1}}{n+1} \frac{1}{\sqrt{(1-x^2)}} dx$$

$$= \frac{1}{(n+1)} \left[\frac{\pi}{2} - \int_0^1 \frac{x^{n+1}}{(1-x^2)}\right]$$

$$= \frac{1}{n+1} \left\{\frac{\pi}{2} - \frac{n(n-2)(n-4) \ldots 1}{(n+1)(n-1)(n-3) \ldots 2} \frac{\pi}{2}\right\} \qquad \text{when } n \text{ is odd}$$

$$= \frac{1}{n+1} \left\{\frac{\pi}{2} - \frac{n(n-2)(n-4) \ldots 2}{(n+1)(n-1)(n-3) \ldots 3}\right\} \qquad \text{when } n \text{ is even}$$

**Evaluate 6.4.** *Evaluate* $I_n = \int_0^a (a^2 - x^2)^n \, dx$ *where n is a positive integer. Hence show that*

$$I_n = \frac{2n}{2n+1} a^2 I_{n-1}.$$

**Solution.** Putting $n = a \sin \theta$, we get

$$I_n = \int_0^a (a^2 - x^2)^n \, dx = \int_0^{\pi/2} (a^2 - a^2 \sin^2 \theta)^n \, a \cos \theta \, d\theta - a^{2n+1} \int_0^{\pi/2} \cos^{2n+1} \theta \, d\theta$$

$$= a^{2n+1} \cdot \frac{(2n)(2n-2)(2n-4) \ldots 4.2}{(2n+1)(2n-1)(2n-3) \ldots 5.3} \qquad [\because \ (2n+1) \text{ is always odd}]$$

Now replacing $n$ by $n - 1$, we get

$$I_{n-1} = a^{2n-1} \frac{(2n-2)(2n-4) \ldots 4.2}{(2n-1)(2n-3) \ldots 5.3} \quad \therefore \quad \frac{I_n}{I_{n-1}} = a^2 \cdot \frac{2n}{2n+1} \quad \text{or} \quad I_n = \frac{2n}{2n+1} a^2 I_{n-1}.$$

which is the second desired result.

## 6.3 (1) REDUCTION FORMULAE for $\int \sin^m x \cos^n x \, dx$

$$\int \sin^m x \cos^n x \, dx = \int \sin^{m-1} x \cdot \cos^n x \cdot \sin x \, dx \qquad \text{[Integrate by parts]}$$

$$= \sin^{m-1} x \cdot \left(\frac{-\cos^{n+1} x}{n+1}\right) - \int (m-1) \sin^{m-2} x \cos x \cdot \left(-\frac{\cos^{n+1} x}{n+1}\right) dx.$$

$$= -\frac{\sin^{m-1} x \cos^{n+1} x}{n+1} + \frac{m-1}{m+1} \int \sin^{m-2} x (1 - \sin^2 x) \cos^n x \, dx$$

$$= -\frac{\sin^{m-1} x \cos^{n+1} x}{n+1} + \frac{m-1}{n+1} \int \sin^{m-2} x \cos^n x \, dx - \frac{m-1}{n+1} \int \sin^m x \cos^n x \, dx$$

Transposing the last term to the left and dividing by $1 + (m-1)/(n+1)$, *i.e.*, $(m+n)/(n+1)$, we obtain the reduction formula

$$\int \sin^m x \cos^n x \, dx = -\frac{\sin^{m-1} x \cos^{n+1} x}{m+n} + \frac{m-1}{m+n} \int \sin^{m-2} x \cos^n x \, dx \qquad \ldots(1) \quad (\textit{V.T.U., 2012})$$

**Obs. To integrate $\int \sin^m x \cos^n x\,dx$,**

(*a*) *when m is odd,* put $\cos x = t$

*when n is odd,* put $\sin x = t$

(*b*) *when m and n both are even integers,* express the integrand as a series of cosines of multiple angles and integrate term by term if $m$ and $n$ are small, *otherwise* use the method of reduction formulae.

**(2) To show that**

$$\int_0^{\pi/2} \sin^m x \cos^n x\,dx = \frac{(m-1)(m-3)\ldots\times(n-1)(n-3)\ldots}{(m+n)(m+n-2)(m+n-4)\ldots} \times \left(\frac{\pi}{2},\ \textit{only if both m and n are even}\right)$$

From (*i*), we have

$$I_{m,n} = \int_0^{\pi/2} \sin^m x \cos^n x\,dx = \left| -\frac{\sin^{m-1} x \cos^{n+1} x}{m+n} \right|_0^{\pi/2} + \frac{m-1}{m+n}\int_0^{\pi/2} \sin^{m-2} x \cos^n x\,dx$$

*i.e.,* $$I_{m,n} = \frac{m-1}{m+n} I_{m-2,n}.$$

**Case I.** *When m is odd*

Similarly, $$I_{m-2,n} = \frac{m-3}{m+n-2} I_{m-4,n},\ I_{m-4,n} = \frac{m-5}{m+n-4} I_{m-6,n}$$

.........................................................................................

$$I_{5,n} = \frac{4}{n+5} I_{3,n}$$

Finally $$I_{3,n} = \frac{2}{n+3} I_{1,n} = \frac{2}{n+3}\int_0^{\pi/2} \sin x \cos^n x\,dx$$

$$= \frac{2}{n+3}\left| -\frac{\cos^{n+1} x}{n+1} \right|_0^{\pi/2} = \frac{2}{(n+3)(n+1)} \qquad \ldots(ii)$$

From these, we obtain

$$I_{m,n} = \frac{(m-1)(m-3)(m-5)\ldots 4.2}{(m+n)(m+n-2)(m+n-4)\ldots(n+3)(n+1)}$$

**Case II.** *When m is even*

We have, $$I_{m-2,n} = \frac{m-3}{m+n-2} I_{m-4,n},\quad I_{m-4,n} = \frac{m-5}{m+n-4} I_{m-6,n}$$

.........................................................................................

$$I_{4,n} = \frac{3}{n+4} I_{2,n},\ I_{2,n} = \frac{1}{n+2} I_{0,n} = \frac{1}{n+2}\int_0^{\pi/2} \cos^n x\,dx$$

From these, we have $$I_{m,n} = \frac{(m-1)(m-3)(m-5)\ldots 1}{(m+n)(m+n-2)(m+n-4)\ldots(n+2)}\int_0^{\pi/2} \cos^n x\,dx$$

$$= \frac{(m-1)(m-3)\ldots 1}{(m+n)(m+n-2)\ldots(n+2)} \cdot \frac{(n-1)(n-3)\ldots}{n(n-2)\ldots} \times (\pi/2 \textit{ only if n is even}) \qquad \ldots(iii)$$

Combining (*ii*) and (*iii*), we get the desired result.

**Example 6.5.** *Integrate* (*i*) $\int \sin^4 x \cos^2 x\,dx$

(*ii*) $\int_0^{\infty} \frac{t^6}{(1+t^2)^7}\,dt$ (*iii*) $\int_0^{\infty} \frac{x^2}{(1+x^2)^{7/2}}\,dx$ (*V.T.U., 2010 S*)

**Solution.** (*i*) Taking $n = 2$, in (*i*), we have the reduction formula :

$$\int \sin^m x \cos^2 x\,dx = \frac{\sin^{m-1} x \cos^3 x}{m+2} + \frac{m-1}{m+2}\int \sin^{m-2} x \cos^2 x\,dx$$

Putting $m = 4, 2$ successively,

$$\int \sin^4 x \cos^2 x\, dx = -\frac{\sin^3 x \cos^3 x}{6} + \frac{3}{6}\int \sin^2 x \cos^2 x\, dx \quad ...(1)$$

$$\int \sin^2 x \cos^2 x\, dx = -\frac{\sin x \cos^3 x}{4} + \frac{1}{4}\int \cos^2 x\, dx$$

But $$\int \cos^2 x\, dx = \frac{1}{2}\int (1 + \cos 2x)\, dx = \frac{1}{2}\left(x + \frac{1}{2}\sin 2x\right)$$

$$\therefore \quad \int \sin^2 x \cos^2 x\, dx = -\frac{\sin x \cos^3 x}{4} + \frac{1}{16}(2x + \sin 2x)$$

Substituting this in Eqn. (1), we get

$$\int \sin^4 x \cos^2 x\, dx = -\frac{\sin^3 x \cos^3 x}{6} + \frac{1}{2}\left\{-\frac{\sin x \cos^3 x}{4} + \frac{1}{16}(2x + \sin 2x)\right\}$$

(*ii*) Putting $t = \tan\theta$, so that

$$\int_0^\infty \frac{t^6}{(1+t^2)^7}\, dt = \int_0^{\pi/2} \frac{\tan^6\theta}{\sec^{14}\theta}\sec^2\theta\, d\theta = \int_0^{\pi/2} \sin^6\theta \cos^6\theta\, d\theta = \frac{5\cdot 3\cdot 1\times 5\cdot 3\cdot 1}{12\cdot 10\cdot 8\cdot 6\cdot 4\cdot 2}\times\frac{\pi}{2} = \frac{5\pi}{2048}.$$

(*iii*) Putting $x = \tan\theta$, so that

$$\int_0^\infty \frac{x^2}{(1+x^2)^{7/2}}\, dx = \int_0^{\pi/2} \frac{\tan^2\theta}{\sec^7\theta}\sec^2\theta\, d\theta = \int_0^{\pi/2} \sin^2\theta \cos^3\theta\, d\theta = \frac{1.2}{53.1} = \frac{2}{15}.$$

**Example 6.6.** *Evaluate :* (*i*) $\int_0^{\pi/6} \cos^4 3\theta \sin^3 6\theta\, d\theta$

(*ii*) $\int_0^1 x^4(1-x^2)^{3/2}\, dx$ (*iii*) $\int_0^{2a} x^2\sqrt{(2ax - x^2)}\, dx.$ (*V.T.U., 2014*)

**Solution.** (*i*) $\int_0^{\pi/6} \cos^4 3\theta \sin^3 6\theta\, d\theta = \int_0^{\pi/6} \cos^4 3\theta\,(2\sin 3\theta\cos 3\theta)^3\, d\theta$

$$= 8\int_0^{\pi/6} \sin^3 3\theta \cos^7 3\theta\, d\theta$$

Put $3\theta = x$ so that $3d\theta = dx$

Also when $\theta = 0, x = 0$; when $\theta = \pi/6, x = \pi/2$.

$$= \frac{8}{3}\int_0^{\pi/2} \sin^3 x \cos^7 x\, dx$$

$$= \frac{8}{3}\cdot\frac{2\times 6\cdot 4\cdot 2}{10\cdot 8\cdot 6\cdot 4\cdot 2} = \frac{1}{15}.$$

(*ii*) $\int_0^1 x^4(1-x^2)^{3/2}\, dx$

Put $x = \sin t$ so that $dx = \cos t\, dt$
When $x = 0, t = 0$; when $x = 1, t = \pi/2$

$$= \int_0^{\pi/2} \sin^4 t\,(\cos^2 t)^{3/2}\cdot\cos t\, dt = \int_0^{\pi/2} \sin^4 t \cos^4 t\, dt$$

$$= \frac{3\cdot 1\times 3\cdot 1}{8\cdot 6\cdot 4\cdot 2}\cdot\frac{\pi}{2} = \frac{3\pi}{256}.$$

(*iii*) $\int_0^{2a} x^2\sqrt{(2ax - x^2)}\, dx$

$$= \int_0^{\pi/2} x^{5/2}\sqrt{(2a - x)}\, dx$$

Put $x = 2a\sin^2\theta$
$\therefore \quad dx = 4a\sin\theta\cos\theta\, d\theta$

$$= \int_0^{\pi/2} (2a\sin^2\theta)^{5/2}\sqrt{(2a)}\cos\theta\cdot 4a\sin\theta\cos\theta\, d\theta$$

$$= 2^5 a^4\int_0^{\pi/2} \sin^6\theta\cos^2\theta\, d\theta = 32\, a^4\cdot\frac{5\cdot 3\cdot 1\times 1}{8\cdot 6\cdot 4\cdot 2}\cdot\frac{\pi}{2} = \frac{5\pi a^4}{8}.$$

## PROBLEMS 6.1

Evaluate :

1. (i) $\int_0^{\frac{\pi}{2}} \sin^9 x\, dx$ (ii) $\int_0^{\pi/6} \cos^5 3\theta\, d\theta$

2. (i) $\int_0^1 \frac{x^9}{\sqrt{(1-x^2)}} dx$ (ii) $\int_0^1 x^5 \sin^{-1} x\, dx$

3. (i) $\int_0^\infty \frac{dx}{(1+x^2)^n} (n > 1)$ (ii) $\int_0^{\pi/4} \cos^6 x \sin^6 x\, dx.$ (V.T.U., 2016)

4. If $I_{m,n} = \int_0^{\pi/2} \sin^m x \cos^n x\, dx\ (m > 0, n > 0)$, show that $I_{m,n} = \frac{m-1}{m+n} I_{m-2,n}$.

Hence evaluate (i) $\int_0^{\pi/2} \sin^4 x \cos^8 x\, dx$ (ii) $\int_0^{\pi/2} \sin^7 x \cos^4 x\, dx$ (Delhi, 2012)

Evaluate :

5. (i) $\int_0^{\pi/2} \sin^4 x \cos^6 x\, dx$ (ii) $\int_0^{\pi/2} \sin^{15} x \cos^3 x\, dx$

6. (i) $\int_0^1 x^6 \sqrt{(1-x^2)}\, dx$ (ii) $\int_0^{\pi/2} \cos^4 3\theta \sin^3 6\theta\, d\theta$

7. (i) $\int_0^{2a} x^{7/2} (2a-x)^{-1/2}\, dx$ (ii) $\int_0^{2a} \frac{x^3\, dx}{\sqrt{(2ax - x^2)}}$

8. (i) $\int_0^2 x^{5/2} \sqrt{(2-x)}\, dx$ (V.T.U., 2013) (ii) $\int_0^4 x^3 \sqrt{(4x - x^2)}\, dx$

(iii) $\int_0^1 x^{3/2} (1-x)^{3/2} dx$ (Delhi., 2015) (iv) $\int_0^4 x^{3/2} (4-x)^{5/2} dx$ (V.T.U., 2016)

9. If $I_n = \int x^n \sqrt{(a-x)}\, dx$, prove that $(2n+3) I_n = 2an\, I_{n-1} - 2x^n (a-x)^{3/2}$

10. If $n$ is a positive integer, show that $\int_0^{2a} x^n \sqrt{(2ax - x^2)}\, dx = \frac{2n+1}{(n+2)\, n!} \cdot \frac{a^{n+2}}{2n} \pi$

## 6.4 REDUCTION FORMULAE for (a) $\int \tan^n x\, dx$ (b) $\int \cot^n x\, dx$

(a) Let
$$I_n = \int \tan^n x\, dx = \int \tan^{n-2} x \cdot \tan^2 x\, dx = \int \tan^{n-2} x \cdot (\sec^2 x - 1)\, dx$$
$$= \int \tan^{n-2} x \sec^2 x\, dx - \int \tan^{n-2} x\, dx$$

Thus, $I_n = \frac{\tan^{n-1} x}{n-1} - I_{n-2}$, which is the required reduction formula.

(b) Let
$$I_n = \int \cot^n x\, dx = \int \cot^{n-2} x \cot^2 x\, dx = \int \cot^{n-2} x\, (\text{cosec}^2 x - 1)\, dx$$
$$= \int \cot^{n-2} x\, \text{cosec}^2 x\, dx - \int \cot^{n-2} x\, dx$$

Thus
$$I_n = -\frac{\cot^{n-1} x}{n-1} - I_{n-2}$$
which is the required reduction formula.

**Example 6.7.** *Evaluate (i) $\tan^5 x\, dx$, (ii) $\int \cot^6 x\, dx$.*

**Solution.** (i) Putting $n = 5, 3$ successively in the reduction formula for $\int \tan^n x\, dx$, we get

$$I_5 = \frac{1}{4} \tan^4 x - I_3;\quad I_3 = \frac{1}{2} \tan^2 x - I_1$$

Thus
$$I_5 = \frac{1}{4} \tan^4 x - \frac{1}{2} \tan^2 x + I_1$$

*i.e.,* $$\int \tan^5 x\, dx = \frac{1}{4}\tan^4 x - \frac{1}{2}\tan^2 x + \int \tan x\, dx = \frac{1}{4}\tan^4 x - \frac{1}{2}\tan^2 x - \log\cos x.$$

(*ii*) Putting $n = 6, 4, 2$ successively in the reduction formula for $\int \cot^n x\, dx$, we get

$$I_6 = -\frac{1}{5}\cot^5 x - I_4\,;\ I_4 = -\frac{1}{3}\cot^3 x - I_2\,;\ I_2 = -\cot x - I_0$$

Thus $$I_6 = \frac{-1}{5}\cot^5 x + \frac{1}{3}\cot^3 x - \cot x - \int dx$$

*i.e.,* $$\int \cot^6 x\, dx = -\frac{1}{5}\cot^5 x + \frac{1}{3}\cot^3 x - \cot x - x.$$

**Example 6.8.** *If* $I_n = \int_0^{\pi/4} \tan^n \theta\, d\theta$, *prove that* $n(I_{n-1} + I_{n+1}) = 1.$

**Solution.** The reduction formula for $\int_0^{\pi/4} \tan^n \theta\, d\theta$ is

$$I_n = \frac{1}{n-1}\Big| \tan^n x \Big|_0^{\pi/4} - I_{n-2} = \frac{1}{n-1} - I_{n-2} \quad \text{or} \quad I_n + I_{n-2} = \frac{1}{n-1}$$

Changing $n$ to $n + 1$, we obtain

$$I_{n+1} + I_{n-1} = \frac{1}{(n+1)} \quad \text{or} \quad (n+1)(I_{n+1} + I_{n-1}) = 1.$$

## 6.5 REDUCTION FORMULAE for (a) $\int \sec^n x\, dx$ (b) $\int \text{cosec}^n x\, dx$

(*a*) Let $$I_n = \int \sec^n x\, dx = \int \sec^{n-2} x \,.\, \sec^2 x\, dx$$

Integrating by parts, we have

$$I_n = \sec^{n-2} x.\tan x - \int \{(n-2)\sec^{n-3} x \cdot \sec x \tan x\} \tan x\, dx$$

$$= \sec^{n-2} x \tan x - (n-2)\int \sec^{n-2} x \cdot \tan^2 x\, dx$$

$$= \sec^{n-2} x \tan x - (n-2)\int \sec^{n-2} x \cdot (\sec^2 x - 1)\, dx$$

$$= \sec^{n-2} x \tan x - (n-2) I_n + (n-2) I_{n-2}$$

Transposing, we have

$$(n-1)I_n = \sec^{n-2} x \tan x + (n-2)I_{n-2}$$

Thus $$I_n = \frac{\sec^{n-2} x \tan x}{n-1} + \frac{n-2}{n-1} I_{n-2},$$ which is the desired reduction formula.

(*b*) Let $$I_n = \int \text{cosec}^n x\, dx = \int \text{cosec}^{n-2} x \cdot \text{cosec}^2 x\, dx$$

Integrating by parts, we have

$$I_n = \text{cosec}^{n-2} x.(-\cot x) - \int [(n-2)\,\text{cosec}^{n-3} x \cdot (-\text{cosec}\, x \cot x) \cdot (-\cot x)\, dx$$

$$= -\cot x\, \text{cosec}^{n-2} x - (n-2)\int \text{cosec}^{n-2} x (\text{cosec}^2 x - 1)\, dx$$

$$= -\cot x\, \text{cosec}^{n-2} x - (n-2)I_n + (n-2)I_{n-2}$$

or $$[1 + (n-2)]I_n = -\cot x\, \text{cosec}^{n-2} x + (n-2)I_{n-2}$$

Thus $$I_n = -\frac{\cot x\, \text{cosec}^{n-2} x}{n-1} + \frac{n-2}{n-1} I_{n-2}$$

which is the required reduction formula.

**Example 6.9.** *Evaluate* (*i*) $\int_0^{\pi/4} \sec^4 x\, dx$ (*ii*) $\int_{\pi/3}^{\pi/2} \text{cosec}^3 \theta\, d\theta.$

**Solution.** (*i*) Putting $n = 4$ in the reduction formula for $\int \sec^n x\, dx$, we get $I_4 = \dfrac{\sec^2 x \tan x}{3} + \dfrac{2}{3} I_2$

$$\therefore \quad \int_0^{\pi/4} \sec^4 x\, dx = \left| \frac{\sec^2 x \tan x}{3} \right|_0^{\pi/4} + \frac{2}{3} \int_0^{\pi/4} \sec^2 x\, dx$$

$$= \frac{2}{3} + \frac{2}{3} \left| \tan x \right|_0^{\pi/4} = \frac{2}{3}(1+1) = 4/3.$$

(*ii*) Putting $n = 3$ in the reduction formula for $\int \text{cosec}^n x\, dx$, we get

$$I_3 = -\frac{1}{2} \cot x \,\text{cosec}\, x + \frac{1}{2} I_1$$

$$\therefore \quad \int_{\pi/3}^{\pi/2} \text{cosec}^3 x\, dx = -\frac{1}{2} \left| \cot x \,\text{cosec}\, x \right|_{\pi/3}^{\pi/2} + \frac{1}{2} \int_{\pi/3}^{\pi/2} \text{cosec}\, x\, dx$$

$$= -\frac{1}{2}\left(0 - \frac{2}{3}\right) + \frac{1}{2} \left| \log(\text{cosec}\, x - \cot x) \right|_{\pi/3}^{\pi/2}$$

$$= \frac{1}{3} + \frac{1}{2}\left[\log 1 - \log\left(\frac{2}{\sqrt{3}} - \frac{1}{\sqrt{3}}\right)\right] = \frac{1}{3} + \frac{1}{4} \log 3.$$

## PROBLEMS 6.2

1. Evaluate (*i*) $\int \tan^6 x\, dx$ (*ii*) $\int \cot^5 x\, dx$.
2. Show that $\int_0^{\pi/4} \tan^7 x\, dx = \frac{1}{12}(5 - 6 \log 2)$
3. If $I_n = \int_0^{\pi/4} \tan^n x\, dx$, prove that $(n-1)(I_n + I_{n-2}) = 1$. Hence evaluate $I_5$.
4. If $I_n = \int_{\pi/4}^{\pi/2} \cot^n \theta\, d\theta$ $(n > 2)$, prove that $I_n = \dfrac{1}{n-1} - I_{n-1}$. Hence evaluate $I_4$.
5. Obtain the reduction formula for $\int_0^{\pi/4} \sec^n \theta\, d\theta$. (*V.T.U., 2010 S*)
6. Evaluate (*i*) $\int \sec^6 \theta\, d\theta$ (*ii*) $\int_{\pi/6}^{\pi/2} \text{cosec}^5\, d\theta$.
7. Evaluate $\int_0^a (a^2 + x^2)^{5/2}\, dx$.
8. If $I_n = \int \dfrac{t^n}{1+t^2}\, dt$, show that $I_{n+2} = \dfrac{t^{n+1}}{n+1} - I_n$. Hence evaluate $I_6$.

## 6.6 REDUCTION FORMULAE for

(*a*) $\int \mathbf{x^n\, e^{ax}\, dx}$ (*b*) $\int \mathbf{x^m (\log x)^n\, dx}$.

(*a*) Let $\quad I_n = \int x^n e^{ax}\, dx$

Integrating by parts, we have

$$I_n = x^n \cdot \frac{e^{ax}}{a} - \int n x^{n-1} \cdot \frac{e^{ax}}{a}\, dx$$

or $\quad I_n = \dfrac{x^n e^{ax}}{a} - \dfrac{n}{a} I_{n-1}$ which is the required reduction formula. (*Madras, 2006*) (*b*) Let $$I_{m,n} = \int x^m (\log x)^n \, dx = \int (\log x)^n \, . \, x^m \, dx$$

Integrating by parts, we have

$$I_{m,n} = (\log x)^n \cdot \frac{x^{m+1}}{m+1} - \int n(\log x)^{n-1} \cdot \frac{1}{x} \cdot \frac{x^{m+1}}{m+1} dx$$

$$= \frac{x^{m+1}}{m+1}(\log x)^n - \frac{n}{m+1}\int x^m (\log x)^{n-1} dx \quad \text{or} \quad I_{m,n} = \frac{x^{m+1}}{m+1}(\log x)^n - \frac{n}{m+1} I_{m,n-1}$$

which is the desired reduction formula.

## 6.7 REDUCTION FORMULAE for

(*a*) $\int x^n \sin mx \, dx$ (*b*) $\int x^n \cos mx \, dx$ (*c*) $\int \cos^m x \sin nx \, dx$

(*a*) Let $$I_n = \int x^n \sin mx \, dx$$

Integrating by parts, we get

$$I_n = x^n \left(\frac{-\cos mx}{m}\right) - \int n x^{n-1} \left(\frac{-\cos mx}{m}\right) dx$$

$$= -\frac{x^n \cos mx}{m} + \frac{n}{m}\int x^{n-1} \cos mx \, dx \qquad \text{[Again integrate by parts]}$$

$$= -\frac{x^n \cos mx}{m} + \frac{n}{m}\left\{x^{n-1} \cdot \frac{\sin mx}{m} - \left\{\int (n-1)x^{n-2} \cdot \frac{\sin mx}{m} dx\right\}\right\}$$

or $$I_n = -\frac{x^n \cos mx}{m} + \frac{n}{m^2} x^{n-1} \sin mx - \frac{n(n-1)}{m^2} I_{n-2}$$

which is the desired reduction formula.

(*b*) Let $$I_n = \int x^n \cos mx \, dx$$

Integrating twice by parts as above, we get

$$I_n = \frac{x^n \sin mx}{m} + \frac{n}{m^2} x^{n-1} \cos mx - \frac{n(n-1)}{m^2} I_{n-2}$$

(*c*) Let $$I_{m,n} = \int \cos^m x \sin nx \, dx$$

Integrating by parts,

$$I_{m,n} = -\cos^m x \, . \, \frac{\cos nx}{n} - \int m \cos^{m-1} x \, (-\sin x) \, . \left(\frac{-\cos nx}{n}\right) dx$$

$$= -\frac{1}{n}\cos^m x \cos nx - \frac{m}{n}\int \cos^{m-1} x \, . \, \cos nx \sin x \, dx$$

$\because$ $\sin(n-1)x = \sin nx \cos x - \cos nx \sin x$
or $\cos nx \sin x = \sin nx \cos x - \sin(n-1)x$

$$= -\frac{1}{n}\cos^m x \cos nx - \frac{m}{n}\int \cos^{m-1} x \, \{\sin nx \cos x - \sin(n-1)x\} \, dx$$

$$= -\frac{1}{n}\cos^m x \cos nx - \frac{m}{n}(I_{m,n} - I_{m-1,n-1})$$

Transposing, we get

$$\left(1 + \frac{m}{n}\right) I_{m,n} = -\frac{1}{n}\cos^m x \cos nx + \frac{m}{n} I_{m-1,n-1}$$

or $$I_{m,n} = -\frac{\cos^m x \cos nx}{m+n} + \frac{m}{m+n} I_{m-1,n-1}$$

which is the desired reduction formula.

**Example 6.10.** *Show that* $\int_0^{\pi/2} \cos^m x \cos nx\, dx = \frac{m}{m+n}\int_0^{\pi/2} \cos^{m-1} x \cos (n-1) x\, dx$

*Hence deduce that* $\int_0^{\pi/2} \cos^n x \cos nx\, dx = \frac{\pi}{2^{n+1}}$. *(C.S.V.T.U., 2015)*

**Solution.** Let $I_{m,n} = \int_0^{\pi/2} \cos^m x \cos nx\, dx$

Integrating by parts

$$I_{m,n} = \left| \cos^m x \cdot \frac{\sin nx}{n} \right|_0^{\pi/2} - \int_0^{\pi/2} m \cos^{m-1} x (-\sin x) \times \frac{\sin nx}{n}\, dx$$

$$= \frac{m}{n}\int_0^{\pi/2} \cos^{m-1} x \sin nx \sin x\, dx \qquad \left[\begin{array}{l}\because \cos (n-1)x = \cos nx \cos x + \sin nx \sin x \\ \text{or } \sin nx \sin x = \cos (n-1)x - \cos nx \cos x\end{array}\right]$$

$$= \frac{m}{n}\int_0^{\pi/2} \cos^{m-1} x [\cos (n-1)x - \cos nx \cos x]\, dx = \frac{m}{n}(I_{m-1,\, n-1} - I_{m,\, n})$$

Transposing and dividing by $(1 + m/n)$, we get

$$I_{m,n} = \frac{m}{m+n} I_{m-1,\, n-1}$$

which is the required result.

Putting $\quad m = n,\ I_n \left( = \int_0^{\pi/2} \cos^n x \cos nx\, dx \right) = \frac{1}{2} I_{n-1}$

Changing $n$ to $n-1$,

$$I_{n-1} = \frac{1}{2} I_{n-2}$$

$$\therefore \quad I_n = \frac{1}{2}\left(\frac{1}{2} I_{n-2}\right) = \frac{1}{2^2} I_{n-2} = \frac{1}{2^3} I_{n-3} \ldots\ldots = \frac{1}{2^n} I_{n-n} = \frac{1}{2^n} \cdot \int_0^{\pi/2} (\cos x)^0\, dx$$

Hence $\quad I_n = \frac{1}{2^n} \cdot \frac{\pi}{2} = \frac{\pi}{2^{n+1}}$.

**Example 6.11.** *Find a reduction formula for* $\int e^{ax} \sin x\, dx$. *Hence evaluate* $\int e^x \sin^3 x\, dx$.

**Solution.** Let $I_n = \int e^{ax} \sin^n x\, dx = \int \underset{I}{\sin^n x} \cdot \underset{II}{e^{ax}}\, dx$

Integrating by parts,

$$I_n = \sin^n x \cdot \frac{e^{ax}}{a} - \int (n \sin^{n-1} x \cos x) \cdot \frac{e^{ax}}{a}\, dx$$

$$= \frac{e^{ax} \sin^n x}{a} - \frac{n}{a}\int (\sin^{n-1} x \cos x) \cdot e^{ax}\, dx \qquad \text{[Again integrating by parts]}$$

$$= \frac{e^{ax} \sin^n x}{a} - \frac{n}{a}\left[\sin^{n-1} x \cos x \cdot \frac{e^{ax}}{a} - \int \{(n-1) \sin^{n-2} x \times \cos x \cdot \cos x + \sin^{n-1} x (-\sin x)\} \frac{e^{ax}}{a}\, dx\right]$$

$$= \frac{e^{ax} \sin^{n-1} x}{a^2} (a \sin x - n \cos x) + \frac{n}{a^2}\int [(n-1) \sin^{n-2} x \times (1 - \sin^2 x) - \sin^n x]\, e^{ax}\, dx$$

$$= \frac{e^{ax} \sin^{n-1} x}{a} (a \sin x - n \cos x) + \frac{n(n-1)}{a^2} I_{n-2} - \frac{n^2}{a^2} I_n$$

Transposing and dividing by $(1 + n^2/a^2)$, we get

$$I_n = \frac{e^{ax} \sin^{n-1} x\, (a \sin x - n \cos x)}{a^2 + n^2} + \frac{n(n-1)}{a^2 + n^2} I_{n-2}$$

which is the required reduction formula.

Putting $a = 1$ and $n = 3$, we get

$$I_3 = \frac{e^x \sin^2 x (\sin x - 3\cos x)}{1^2 + 9} + \frac{3 \cdot 2}{1^2 + 9} I_1$$

But $$I_1 = \int e^x \sin x \, dx = \frac{e^{ax}}{\sqrt{2}} \sin (x - \tan^{-1} 1).$$

$\therefore$ $$I_3 = \frac{e^x \sin^2 x (\sin x - 3\cos x)}{10} + \frac{3}{5} \cdot \frac{e^{ax}}{\sqrt{2}} \sin (x - \pi/4).$$

## PROBLEMS 6.3

1. If $I_n = \int x^n e^x \, dx$, show that $I_n + n I_{n-1} = x^n e^x$. Hence find $I_4$. *(C.S.V.T.U., 2015)*
2. If $u_n = \int_0^a x^n e^{-x} \, dx$, prove that $u_n - (n + a) u_{n-1} + a(n-1) u_{n-2} = 0$.
3. Obtain a reduction formula for $\int x^m (\log x)^n \, dx$. Hence evaluate $\int_0^1 x^5 (\log x)^3 \, dx$.
4. If $n$ is a positive integer, show that $\int_0^1 x^m (\log x)^n \, dx = \frac{(-1)^n n!}{(m+1)^{n+1}}$, $m > -1$. *(C.S.V.T.U., 2014)*
5. If $I_n = \int_0^{\pi/2} x \sin^n x \, dx$ $(n > 1)$, prove that $n^2 I_n = n(n-1) I_{n-2} + 1$. Hence evaluate $I_5$. *(C.S.V.T.U., 2015)*
6. If $I_n = \int_0^{\pi/2} x \cos^n x \, dx$ $(n > 1)$, prove that $I_n = \frac{n-1}{n} I_{n-2} - \frac{1}{n^2}$. Hence evaluate $I_4$.
7. If $u_n = \int_0^{\pi/2} x^n \sin x \, dx$, $(n > 1)$, prove that $u_n + n(n-1) u_{n-2} = n(\pi/2)^{n-1}$. Hence evaluate $u_2$. *(Delhi, 2012 ; C.S.V.T.U., 2011)*
8. If $I_n = \int x^n \sin ax \, dx$, show that $a^2 I_n = -ax^n \cos ax + nx^{n-1} \sin ax - n(n-1) I_{n-2}$.
9. Prove that $\int_0^{\pi/2} \cos^{n-2} x \sin nx \, dx = \frac{1}{n-1}$, $n > 1$. *(Hazaribag., 2015)*
10. If $I_{m,n} = \int_0^{\pi/2} \cos^m x \cos nx \, dx$, prove that (a) $I_{m,n} = \frac{m(m-1)}{m^2 - n^2} I_{m-2,n}$ (b) $I_{m,n} = \frac{m}{m+n} I_{m-1,n-1}$. *(C.S.V.T.U., 2011 S)*
11. Find a reduction formula for $\int e^{ax} \cos^n x \, dx$. Hence evaluate $\int_0^{\pi/2} e^{2x} \cos^3 x \, dx$.
12. Obtain a reduction formula for $I_m = \int_0^\infty e^{-x} \sin^m x \, dx$ where $m \geq 2$ in the form

    $(1 + m^2) I_m = m(m-1) I_{m-2}$. Hence evaluate $I_4$.

## 6.8 DEFINITE INTEGRALS

**Property I.** $\int_a^b f(x)\,dx = \int_a^b f(t)\,dt$

*(i.e., the value of a definite integral depends on the limits and not on the variable of integration).*

Let $\int f(x)\,dx = \phi(x)$ ; $\therefore \int_a^b f(x)\,dx = \phi(b) - \phi(a)$.

Then $\int f(t)\,dt = \phi(t)$ ; $\therefore \int_a^b f(t)\,dt = \phi(b) - \phi(a)$.

Hence the result.

**Property II.** $\int_a^b f(x)\,dx = -\int_b^a f(x)\,dx$

*(i.e., the interchange of limits changes the sign of the integral).*

Let $\int f(x)\,dx = \phi(x)$; $\therefore \int_a^b f(x)\,dx = \phi(b) - \phi(a)$

and $-\int_b^a f(x)\,dx = -\mid \phi(x)\mid_b^a = -[\phi(a) - \phi(b)] = \phi(b) - \phi(a).$

Hence the result.

**Property III.** $\int_a^b f(x)\,dx = \int_a^c f(x)\,dx + \int_c^b f(x)\,dx$

Let $\int f(x)\,dx = \phi(x)$, so that $\int_a^b f(x)\,dx = \phi(b) - \phi(a)$ ...(1)

Also $\int_a^c f(x)\,dx + \int_c^b f(x)\,dx = \mid \phi(x)\mid_a^c + \mid \phi(x)\mid_c^b$

$$= [\phi(c) - \phi(a)] + [\phi(b) - \phi(c)] = \phi(b) - \phi(a) \quad ...(2)$$

Hence the result follows from Eqns. (1) and (2).

**Property IV.** $\int_0^a f(x)\,dx = \int_0^a f(a-x)\,dx$

Put $x = a - t$, so that $dx = -dt$. Also when $x = 0, t = a$ ; when $x = a, t = 0$.

$$\therefore \int_0^a f(x)\,dx = -\int_a^0 f(a-t)\,dt = \int_0^a f(a-t)\,dt = \int_0^a f(a-x)\,dx \quad \text{[Prop. II]}$$

**Example 6.12.** (a) $\int_0^{\pi/2} \frac{\sqrt{\sin x}\,dx}{\sqrt{\sin x} + \sqrt{\cos x}}$ *(Mysore, 2019)* (b) $\int_0^{\pi/4} \log(1 + \tan\theta)d\theta$ *(DU, 2019)*

**Solution :** (a) Let $I = \int_0^{\pi/2} \frac{\sqrt{\sin x}\,dx}{\sqrt{\sin x} + \sqrt{\cos x}}$ ...(i)

$$= \int_0^{\pi/2} \frac{\sqrt{\sin\left(\frac{\pi}{2} - x\right)}\cdot dx}{\sqrt{\sin\left(\frac{\pi}{2} - x\right)} + \sqrt{\cos\left(\frac{\pi}{2} - x\right)}}$$

$$\therefore \quad I = \int_0^{\pi/2} \frac{\sqrt{\cos x}\cdot dx}{\sqrt{\cos x} + \sqrt{\sin x}} \quad ...(ii)$$

Now adding Eqns. (i) & (ii), we have $2I = \int_0^{\pi/2} \frac{\sqrt{\sin x} + \sqrt{\cos x}}{\sqrt{\sin x} + \sqrt{\cos x}}dx = \int_0^{\pi/2} dx\,[x]_0^{\pi/2} = \frac{\pi}{2}$

$$I = \frac{\pi}{4}$$

(b)
$$I = \int_0^{\pi/4} \log(1 + \tan\theta)d\theta = \int_0^{\pi/4} \log\left\{1 + \tan\left(\frac{\pi}{4} - \theta\right)\right\}d\theta$$

$$= \int_0^{\pi/4} \log\left(1 + \frac{1 - \tan\theta}{1 + \tan\theta}\right)d\theta = \int_0^{\pi/4} \log\left(\frac{2}{1 + \tan\theta}\right)d\theta$$

$$= \int_0^{\pi/4} \log 2\theta - \int_0^{\pi/4} \log(1 + \tan\theta)\,d\theta = \log 2[\theta]_0^{\pi/4} - I$$

$$2I = \log 2\cdot\frac{\pi}{4}$$

$$I = \frac{\pi}{8}\log 2$$

**Example 6.13.** *If* $I_n = \int_0^a (a^2 - x^2)^n dx$ *show that*

$$I_n = \left(\frac{2a^2 n}{2n+1}\right) I_{n-1}, \quad n > 0$$

*(Marathwada, 2019)*

**Solution :** Taking unity as second function and integrating w.r.t. $x$, we have

$$I_n = \int_0^a (a^2 - x^2)^n 1 \cdot dx$$

$$= \left[(a^2 - x^2)^n x\right] - \int_0^a n(a^2 - x^2)^{n-1}(-2x) \cdot x \cdot dx$$

$$= [0 - 0] + 2n\int_0^a (a^2 - x^2)^{n-1} x^2\, dx, \quad n > 0$$

$$= -2n\int_0^a (a^2 - x^2)^{n-1}(a^2 - x^2 - a^2)\, dx$$

$$I_n = -2n\int_0^a \left\{(a^2 - x^n)^n - a^2(a^2 - x^2)^{n-1}\right\} dx$$

$$= -2n\int_0^a (a^2 - x^2)^n dx + 2na^2\int_0^a (a^2 - x^2)^{n-1} dx$$

$$I_n = -2nI_n + 2na^2 I_{n-1}$$

$$I_n = \frac{2na^2}{2n+1} I_{n-1}, \quad n > 0$$

**Example 6.14.** *Evaluate*

*(a)* $\int_0^{\pi/2} \log \sin x \cdot dx$ *(b)* $\int_0^{\pi/2} \theta \sin^6 \theta \cos^4 \theta \, d\theta$ *(Marathwada, 2019)*

**Solution:** *(a)* $I = \int_0^{\pi/2} \log \sin x$ ...(*i*)

$$I = \int_0^{\pi/2} \left[\log \sin\left(\frac{\pi}{2} \quad x\right) dx\right]$$

$$I - \int_0^{\pi/2} \log \cos x\, dx \quad \text{...(ii)}$$

Now adding Eqns. (*i*) and (*ii*), we have

$$2I = \int_0^{\pi/2} (\log \sin x + \log \cos x) \cdot dx = \int_0^{\pi/2} \log\left(\sin \frac{2x}{2}\right) dx$$

$$= \int_0^{\pi/2} (\log \sin 2x - \log 2)\, dx = \int_0^{\pi/2} \log \sin 2x\, dx - \log 2 \cdot \frac{\pi}{2}$$

$$= I_1 - \frac{\pi}{2} \log_2 \quad \text{...(iii)}$$

Now let $2x = t$, $2dx = dt$, we have

$$I_1 = \int_0^{\pi} \log \sin t \frac{dt}{2} = \frac{1}{2}\int_0^{\pi} \log \sin t\; dt$$

$$I_1 = \frac{1}{2} \cdot 2\int_0^{\pi/2} \log \sin t\, dt = \int_0^{\pi/2} \log \sin x \cdot dx = I$$

From Eqn. (*iii*), we get

$$2I = I - \frac{\pi}{2} \log 2$$

$$I = -\frac{\pi}{2} \log 2 = \frac{\pi}{2} \log \frac{1}{2}$$

(*b*) Let

$$I = \int_0^{\pi} \theta \sin^6 \theta \cos^4 \theta \cdot d\theta \quad \text{...(i)}$$

$$= \int_0^{\pi} (\pi - \theta) \sin^6(\pi - \theta) \cos^4(\pi - \theta) d\theta$$

$$I = \int_0^{\pi} (\pi - \theta) \sin^6 \theta \cos^4 \theta \, d\theta \quad ...(ii)$$

Now adding Eqns. (*i*) and (*ii*) $\quad 2I = \int_0^{\pi} \pi \sin^6 \theta \cos^4 \theta \, d\theta$

$$= 2\pi \int_0^{\pi/2} \sin^6 \theta \cos^4 \theta \cdot d\theta = 2\pi \cdot \frac{5}{10} \cdot \frac{3}{8} \cdot \frac{1}{6} \cdot \frac{3}{4} \cdot \frac{1}{2} \cdot \frac{\pi}{2} = \frac{3\pi^2}{256}$$

$$I = \frac{3\pi^2}{512}$$

**Property V.** $\int_{-a}^{a} \mathbf{f(x)\,dx} = 2\int_0^a \mathbf{f(x)\,dx}$**, if f(x) is an even function,**

**= 0 if f(x) is an odd function.** (*Bhopal, 2008*)

$$\int_{-a}^{a} f(x)\,dx = \int_{-a}^{0} f(x)\,dx + \int_0^a f(x)\,dx \quad ...(1) \quad \text{[Prop. I]}$$

In $\int_{-a}^{0} f(x)\,dx$, put $x = -t$, so that $dx = -dt$

$$\therefore \quad \int_{-a}^{0} f(x)\,dx = -\int_a^0 f(-t)\,dt = \int_0^a f(-t)\,dt = \int_0^a f(-x)\,dx \quad \text{[Prop. II]}$$

Substituting in Eqn. (1), we get

$$\int_{-a}^{a} f(x)\,dx = \int_0^a f(-x)\,dx + \int_0^a f(x)\,dx \quad ...(2)$$

(*i*) *If $f(x)$ is an even function, $f(-x) = f(x)$.*

$\therefore$ from Eqn. (2), $\int_{-a}^{a} f(x)\,dx = \int_0^a f(x)\,dx + \int_0^a f(x)\,dx = 2\int_0^a f(x)\,dx$

(*ii*) *If $f(x)$ is an odd function, $f(-x) = -f(x)$.*

$\therefore$ from Eqn. (2), $\int_{-a}^{a} f(x)\,dx = -\int_0^a f(x)\,dx + \int_0^a f(x)\,dx = 0.$

**Property VI.** $\int_0^{2a} \mathbf{f(x)\,dx} = 2\int_0^a \mathbf{f(x)\,dx}$**, if f(2a – x) = f(x)**

**= 0, if f(2a – x) = – f(x)**

$$\int_0^{2a} f(x)\,dx = \int_0^a f(x)\,dx + \int_a^{2a} f(x)\,dx \quad ...(1) \quad \text{[Prop. III]}$$

In $\int_0^{2a} f(x)\,dx$, put $x = 2a - t$, so that $dx = -dt$

Also when $x = a$, $t = a$ ; when $x = 2a$, $t = 0$.

$$\therefore \quad \int_0^{2a} f(x)\,dx = -\int_a^0 f(2a - t)\,dt = \int_0^a f(2a - t)\,dt = \int_0^a f(2a - x)\,dx \quad \text{[Prop. II]}$$

Substituting in Eqn. (1), we get

$$\int_0^{2a} f(x)\,dx = \int_0^a f(x)\,dx + \int_0^a f(2a - x)\,dx \quad ...(2)$$

(*i*) If $f(2a - x) = f(x)$, then from Eqn. (2)

$$\int_0^{2a} f(x)\,dx = \int_0^a f(x)\,dx + \int_0^a f(x)\,dx = 2\int_0^a f(x)\,dx$$

(*ii*) If $f(2a - x) = -f(x)$, then from Eqn. (2)

$$\int_0^{2a} f(x)\,dx = \int_0^a f(x)\,dx - \int_0^a f(x)\,dx = 0.$$

**Cor. 1.** *If n is even,* $\int_0^{\pi} \sin^m x \cos^n x\,dx = 2\int_0^{\pi/2} \sin^m x \cos^n x\,dx$ *and if n is odd,* $\int_0^{\pi} \sin^m x \cos^n x\,dx = 0.$

**Cor. 2.** *If m is odd,* $\int_0^{2\pi} \sin^m x \cos^n x\, dx = 0$

and *if m is even,* $\int_0^{2\pi} \sin^m x \cos^n x\, dx = 2\int_0^{\pi} \sin^m x \cos^n x\, dx$

$$= 4\int_0^{\pi/2} \sin^m x \cos^n x\, dx \text{, if } n \text{ is even} = 0\text{, if } n \text{ is odd.}$$

**Example 6.15.** *Evaluate* $\int_0^{\pi} \theta \sin^2 \theta \cos^4 \theta\, d\theta.$ (*V.T.U., 2015*)

**Solution:** Let $I = \int_0^{\pi} \theta \sin^2 \theta \cos^4 \theta\, d\theta$

Then $I = \int_0^{\pi} (\pi - \theta) \sin^2 (\pi - \theta) \cos^4 (\pi - \theta)\, d\theta = \pi\int_0^{\pi} \sin^2 \theta \cos^4 \theta\, d\theta - I$ [Prop. IV]

or $2I = \pi\int_0^{\pi} \sin^2 \theta \cos^4 \theta\, d\theta = 2\pi\int_0^{\pi/2} \sin^2 \theta \cos^4 \theta\, d\theta$ [Prop. VI Cor. 2]

$$= 2\pi \cdot \frac{1.3.1}{6.4.2} = \frac{\pi}{2} = \frac{\pi^2}{16}$$

Hence $I = \frac{\pi^2}{32}$

**Example 6.16.** Prove that

$$\int_0^{\pi} \sin^3 \theta(1 + 2\cos\theta)(1 + \cos\theta)^2 d\theta = \frac{8}{3}$$ (*Anna, 2019*)

**Solution:** Let $I = \int_0^{\pi} \sin^3 \theta(1 + 2\cos\theta)(1 + \cos\theta)^2 d\theta = \frac{8}{3}$

$$= \int_0^{\pi} \sin^3 \theta(1 + 4\cos\theta + 5\cos^2\theta + 2\cos^3\theta) \cdot d\theta$$

$$= \int_0^{\pi} \sin^3 \theta\, d\theta + 4\int_0^{\pi} \sin^{3\theta} \cos\theta \cdot d\theta$$

$$+ 5\int_0^{\pi} \sin^3 \theta \cdot \cos^2\theta\, d\theta + 2\int_0^{\pi} \sin^3 \theta \cdot \cos^3\theta \cdot d\theta$$

$$= I_1 + I_2 + I_3 + I_4$$

Now $I_1 = \int_0^{\pi} \sin^3 \theta\, d\theta = 2\int_0^{\pi/2} \sin^3 \theta \cdot d\theta$

$$= 2 \cdot \frac{2}{3} = \frac{4}{3}$$

$$I_2 = 4\int_0^{\pi} \sin^3 \theta \cos\theta \cdot d\theta = 0$$

Since $f(\pi - 0) = -f(0)$

$$f(0) = \sin^2 \theta \cos\theta$$

$$f(\pi - \theta) = \sin^2 (\pi - \theta) \cos(\pi - \theta) = \sin^3 \theta \cos\theta$$

Also $I_3 = 5\int_0^{\pi} \sin^3 \theta \cdot \cos^2 \theta \cdot d\theta = 5 \times 2\int_0^{\pi/2} \sin^3 \theta \cos^3 \theta \cdot d\theta$

$$= 10 \cdot \frac{2.1}{5.3.1} = \frac{4}{3}$$

and $I_4 = 2\int_0^{\pi} \sin\theta \cos^3 \theta \cdot d\theta$

$$I = 0$$

$$\therefore \quad I = \frac{4}{3} + 0 + \frac{4}{3} + 0$$

$$I = \frac{8}{3}.$$

**Example 6.17.** *Evaluate* $\int_0^1 \frac{\sin^{-1} x}{x} dx$.

**Solution.** Put $\sin^{-1} x = \theta$ or $x = \sin\theta$ so that $dx = \cos\theta \, d\theta$

Also when $x = 0, \theta = 0$ ; when $x = 1, \theta = \pi/2$.

$$\therefore \quad \int_0^1 \frac{\sin^{-1} x}{x} dx = \int_0^{\pi/2} \theta . \frac{\cos\theta}{\sin\theta} d\theta \qquad \text{[Integrate by parts]}$$

$$= [\theta . \log \sin\theta]_0^{\pi/2} - \int_0^{\pi/2} 1 . \log \sin\theta \, d\theta$$

$$= -\int_0^{\pi/2} \log \sin\theta \, d\theta = -\left(-\frac{\pi}{2} \log 2\right) = \frac{\pi}{2} \log 2 \qquad \left[\because \underset{x \to 0}{\text{Lt}} (x \log x) = 0\right]$$

**Example 6.18 :** *If* $U_n = \int_0^{\pi/2} x^n \sin x \, dx$ *and* $n > 1$*, Prove that* $U_n + n(n-1)U_{n-2} = n\left(\frac{\pi}{2}\right)^{n-1}$

*(Osmania, 2019)*

**Solution :** Integrating by parts, we get

$$U_n = \left[x^n(-\cos x)\right]_0^{\pi/2} - \int_0^{\pi/2} nx^{n-1}(-\cos x)dx$$

$$= n\int_0^{\pi/2} x^{n-1} \cos x \cdot dx$$

Integrating again by parts, we get

$$U_n = n\left\{\left[x^{n-1} \sin x\right]_0^{\pi/2} - \int_0^{\pi/2} (n-1)x^{n-2} \sin x \, dx\right\}$$

$$= n\left\{\left(\frac{\pi}{2}\right)^{n-1} - (n-1)U_{n-2}\right\}$$

$$= U_n + n(n-1).$$

## PROBLEMS 6.4

Prove that :

1. (*i*) $\int_0^{\pi/2} \log \tan x \, dx = 0$ (*ii*) $\int_0^{\pi/2} \sin 2x \log \tan x \, dx = 0$

2. (*i*) $\int_0^{\infty} \frac{x^7 (1 - x^{12})}{(1 + x)^{28}} dx = 0$ (*ii*) $\int_0^{\pi/4} \log (1 + \tan\theta) \, d\theta = \frac{\pi}{8} \log_e 2$

3. (*i*) $\int_0^{\pi/2} \frac{\sin x}{\sin x + \cos x} dx = \frac{\pi}{4}$ (*ii*) $\int_0^{a} \frac{dx}{x + \sqrt{(a^2 + x^2)}} = \frac{\pi}{4}$

4. (*i*) $\int_0^{\pi/2} \frac{x \tan x}{\sec x + \cos x} dx = \frac{\pi^2}{4}$. (*ii*) $\int_0^{\pi} \frac{x}{1 + \sin x} dx = \pi$

5. (*i*) $\int_0^{\pi} \frac{x \tan x}{\sec x + \tan x} = \frac{1}{2} \pi (\pi - 2)$. (*ii*) $\int_0^{\pi/2} \frac{x \, dx}{\sin x + \cos x} = \frac{\pi}{2\sqrt{2}} \log (\sqrt{2} + 1)$.

Evaluate :

6. (*i*) $\int_0^{\pi} \sin^4 x \, dx$ (*ii*) $\int_{-\pi/2}^{\pi/2} \cos^7 x \, dx$ *(C.S.V.T.U., 2015)*

(iii) $\int_0^{\pi} \sin^6 x \cos^4 x\, dx$ (iv) $\int_0^{2\pi} \sin^4 x \cos^6 x\, dx$

7. (i) $\int_0^{\pi/2} \frac{\sin^2 x\, dx}{\sin x + \cos x}$ (ii) $\int_0^{\pi/2} \frac{x\, dx}{\sin x + \cos x}$

Prove that :

8. (i) $\int_0^1 \frac{\log(1+x)}{1-x^2} dx = \frac{\pi}{8} \log 2$ (ii) $\int_0^{\pi} (1+\cos\theta)(1+\cos^2\theta)\sin^3\theta\, d\theta = \frac{8}{5}$ (PTU, 2019)

9. (i) $\int_0^{\pi} \frac{x\, dx}{a^2 - \cos^2 x} = \frac{\pi^2}{2a\sqrt{(a^2-1)}}$ $(a > 1)$ (ii) $\int_0^{\pi} \frac{x\, dx}{1+\sin^2 x} = \frac{\pi^2}{2\sqrt{2}}$.

10. $\int_0^{\pi} \log(1+\cos\theta)\, d\theta = -\pi \log_e 2$

11. (i) $\int_0^{\infty} \frac{\log(1+x^2)}{1+x^2} dx = \pi \log_e 2$ (ii) $\int_0^{\infty} \frac{\log(x+1/x)}{1+x^2} dx = \pi \log_e 2$.

12. Evaluate

(i) $\int_0^{\frac{\pi}{2}} \frac{\sin^n x\, dx}{\sin^n x \cos^n x}$ (ii) $\int_0^{\pi} \frac{x \tan x}{\sec x + \tan x}$

Prove that :

13. (i) $\int_0^{\infty} \log\left(x + \frac{1}{x}\right)\frac{dx}{1+x^2} = \pi \log 2$ (ii) $\int_0^{\infty} \log(1+\cos\theta) d\theta = -\pi \log 2$

14. (i). $\int_{-\pi/2}^{\pi/2} \cos^2\theta (1+\sin\theta)^2\, d\theta = \frac{8}{5}$ (ii) $\int_0^{\pi/2} \sin 2x \log \tan x\, dx = 0$

15. (i) $\int_0^{\pi/2} \left(\frac{\sin x - \cos x}{1 + \sin x \cos x}\right) dx = 0$ (ii) $\int_0^{\pi/2} \left(\frac{\sin^2\phi\, d\phi}{a - b\cos\phi}\right) = \frac{2\pi}{b^2}\left[a - \sqrt{(a^2-b^2)}\right]$, $a > b > 0$

(Annamalai, 2018)

## 6.9 (1) INTEGRAL AS THE LIMIT OF A SUM

We have so far considered integration as inverse of differentiation. We shall now define the definite integral as the limit of a sum :

**Def.** *If $f(x)$ is continuous and single valued in the interval $[a, b]$, then the definite integral of $f(x)$ between the limits a and b is defined by the equation*

$$\int_a^b \mathbf{f(x)\, dx} = \underset{h \to 0}{\text{Lt}}\ \mathbf{h\,[f(a) + f(a+h) + f(a+2h) + \ldots + f(a + \overline{n-1}\,h)\,]},$$

*where $nh = b - a$.* ...(1)

**(2) Evaluation of limits of series**

The summation definition of a definite integral enables us to express the limits of sums of certain types of series as definite integrals which can be easily evaluated. We rewrite Eqn. (1) as follows :

$$\int_a^b f(x)\, dx = \underset{n \to \infty}{\text{Lt}}\ h \sum_{r=0}^{n-1} f(a + rh), \text{ where } nh = b - a.$$

Putting $a = 0$ and $b = 1$, so that $h = 1/n$, we get

$$\underset{n\to\infty}{\mathbf{Lt}}\ \frac{\mathbf{1}}{\mathbf{n}}\sum_{r=0}^{n-1}\mathbf{f}\left(\frac{\mathbf{r}}{\mathbf{n}}\right)=\int_0^1 \mathbf{f(x)\ dx}$$

*Thus to express a given series as definite integral:*

(*i*) Write the general term ($T_r$ or $T_{r+1}$ whichever involves $r$)

*i.e.*, $f(r/n)\,.\,1/n$

(*ii*) Replace $r/n$ by $x$ and $1/n$ *by* $dx$,

(*iii*) Integrate the resulting expression, taking

the lower limit $= \underset{n\to\infty}{\text{Lt}}\ (r/n)$ where $r$ is as in the first term,

and the upper limit $= \underset{n\to\infty}{\text{Lt}}\ (r/n)$ where $r$ is as in the last term.

**Example 6.19.** *Find the limit, when $n \to \infty$, of the series*

$$\frac{n}{n^2}+\frac{n}{n^2+1^2}+\frac{n}{n^2+2^2}+\ldots+\frac{n}{n^2+(n-1)^2}$$

**Solution.** Here the general term $(= T_{r+1}) = \dfrac{n}{n^2+r^2}=\dfrac{n}{1+(r/n)^2}\cdot\dfrac{1}{n}$

$$=\frac{1}{1+x^2}\,dx \qquad \text{[Putting } r/n = x \text{ and } 1/n = dx]$$

Now for the first term $r = 0$ and for the last term $r = n - 1$

$\therefore$ the lower limit of integration $= \underset{n\to\infty}{\text{Lt}}\left(\dfrac{0}{n}\right)=0$

and the upper limit of integration $= \underset{n\to\infty}{\text{Lt}}\left(\dfrac{n-1}{n}\right)=\underset{n\to\infty}{\text{Lt}}\left(1-\dfrac{1}{n}\right)=1.$

Hence, the required limit $=\displaystyle\int_0^1\frac{dx}{1+x^2}=\Big|\tan^{-1}x\Big|_0^1=\tan^{-1}(1)-\tan^{-1}(0)=\pi/4.$

**[To find limit of a product by integration :**

Let $P = \underset{n\to\infty}{\text{Lt}}$ (given product)

Take logs of both sides, so that

$\log P = \underset{n\to\infty}{\text{Lt}}$ (a series) $= k$ (say). Then $P = e^k$.]

**Example 6.20.** *Evaluate* $\underset{n\to\infty}{Lt}\left\{\left(1+\dfrac{1}{n}\right)\left(1+\dfrac{2}{n}\right)\ldots\left(1+\dfrac{n}{n}\right)\right\}^{1/n}.$ *(Bhopal, 2018)*

**Solution.** Let $P = \underset{n\to\infty}{\text{Lt}}\left\{\left(1+\dfrac{1}{n}\right)\left(1+\dfrac{2}{n}\right)\ldots\left(1+\dfrac{n}{n}\right)\right\}^{1/n}.$

Taking logs of both sides,

$$\log P=\underset{n\to\infty}{\text{Lt}}\ \frac{1}{n}\left\{\log\left(1+\frac{1}{n}\right)+\log\left(1+\frac{2}{n}\right)+\ldots+\log\left(1+\frac{n}{n}\right)\right\}$$

Its general term $=\log\left(1+\dfrac{r}{n}\right).\dfrac{1}{n}=\log(1+x)\,.\,dx$ [Putting $r/n = x$ and $1/n = dx$]

Also for first term $r = 1$ and for the last term $r = n$.

$\therefore$ The lower limit of integration $= \underset{n\to\infty}{\text{Lt}}\ (1/n)=0$ and the upper limit $= \underset{n\to\infty}{\text{Lt}}\ (n/n) = 1$

Hence $\log P=\displaystyle\int_0^1\log(1+x)\,dx=\int_0^1\log(1+x)\,.\,1\,dx$ [Integrate by parts]

$$=\Big|\log(1+x)\,.\,x\Big|_0^1-\int_0^1\frac{1}{1+x}\,.\,x\,dx$$

$$= \log 2 - \int_0^1 \frac{1+x-1}{1+x}\,dx = \log 2 - \int_0^1 dx + \int_0^1 \frac{dx}{1+x}$$

$$= \log 2 - \left| x \right|_0^1 + \left| \log(1+x) \right|_0^1 = \log 2 - 1 + \log 2$$

$$= \log 2^2 - \log_e e = \log(4/e). \text{ Hence, } P = 4/e.$$

**PROBLEMS 6.5**

Find the limit, as $n \to \infty$, of the series :

1. $\dfrac{1}{n+1} + \dfrac{1}{n+2} + \dfrac{1}{n+3} + \ldots + \dfrac{1}{2n}$.

2. $\dfrac{1}{n^3+1} + \dfrac{4}{n^3+8} + \dfrac{9}{n^3+27} + \ldots + \dfrac{r^2}{n^3+r^3} + \ldots + \dfrac{1}{2n}$.

3. $\underset{n\to\infty}{\text{Lt}} \dfrac{\sqrt{n}}{\sqrt{n^3}} + \dfrac{\sqrt{n}}{\sqrt{(n+3)^3}} + \dfrac{\sqrt{n}}{\sqrt{(n+6)^3}} + \ldots + \dfrac{\sqrt{n}}{\sqrt{\{n+3(n-1)\}^3}}$.

Evaluate :

4. $\underset{n\to\infty}{\text{Lt}} \displaystyle\sum_{r=1}^{n-1} \frac{1}{\sqrt{(n^2-r^2)}}$

5. $\underset{n\to\infty}{\text{Lt}} \dfrac{[(n+1)(n+2)\ldots(n+n)]^{1/n}}{n}$.

6. $\underset{n\to\infty}{\text{Lt}} \left\{\left(1+\dfrac{1}{n^2}\right)\left(1+\dfrac{2^2}{n^2}\right)\left(1+\dfrac{3^2}{n^2}\right)\ldots\left(1+\dfrac{n^2}{n^2}\right)\right\}^{1/n}$

## 6.10 AREAS OF CARTESIAN CURVES

**(1)** *Area bounded by the curve $y = f(x)$, the x-axis and the ordinates $x = a$, $x = b$ is* $\int_a^b$ **y dx.**

Let $AB$ be the curve $y = f(x)$ between the ordinates $LA$ $(x = a)$ and $MB$ $(x = b)$. (Fig. 6.1)

Let $P(x, y)$, $P'(x + \delta x, y + \delta y)$ be two neighbouring points on the curve and $NP$, $N'P'$ be their respective ordinates.

Let the area $ALNP$ be $A$, which depends on the position of $P$ whose abscissa is $x$. Then the area $PNN'P' = \delta A$.

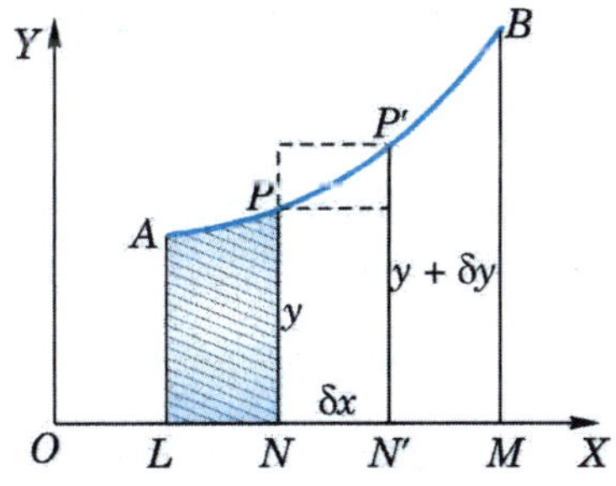

Fig. 6.1

Complete the rectangles $PN'$ and $P'N$.

Then the area $PNN'P'$ lies between the areas of the rectangles $PN'$ and $P'N$.

*i.e.*, $\delta A$ lies between $y\delta x$ and $(y + \delta y)\,\delta x$

$\therefore$ $\dfrac{\delta A}{\delta x}$ lies between $y$ and $y + \delta y$.

Now taking limits as $P' \to P$ *i.e.*, $\delta x \to 0$ (and $\therefore$ $\delta y \to 0$),

$$dA/dx = y$$

Integrating both sides between the limits $x = a$ and $x = b$, we have

$$\left| A \right|_a^b = \int_a^b y\,dx$$

or (value of $A$ for $x = b$) – (value of $A$ for $x = a$) $= \int_a^b y\,dx$

Thus area $ALMB = \int_a^b y\,dx$.

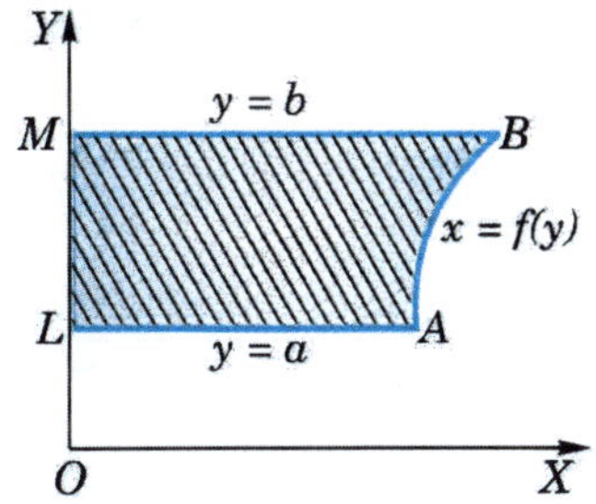

Fig. 6.2

**(2)** Interchanging $x$ and $y$ in the above formula, we see that *the area bounded by the curve $x = f(y)$, the y-axis and the abscissae $y = a$, $y = b$ is* $\int_a^b$ **x dy.** (Fig. 6.2)

**Obs. 1.** The area bounded by a curve, the $x$-axis and two ordinates is called the **area under the curve.** The process of finding the area of plane curves is often called **quadrature.**

**Obs. 2. Sign of an area.** *An area whose boundary is described in the anti-clockwise direction is considered positive and an area whose boundary is described in the clockwise direction is taken as negative.*

In Fig. 6.3, the area $ALMB \left(= \int_a^b y\, dx\right)$ which is described in the anti-clockwise direction and lies above the $x$-axis, will give a positive result.

In Fig. 6.4, the area $ALMB \left(= \int_a^b y\, dx\right)$ which is described in the clockwise direction and lies below the $x$-axis, will give a negative result.

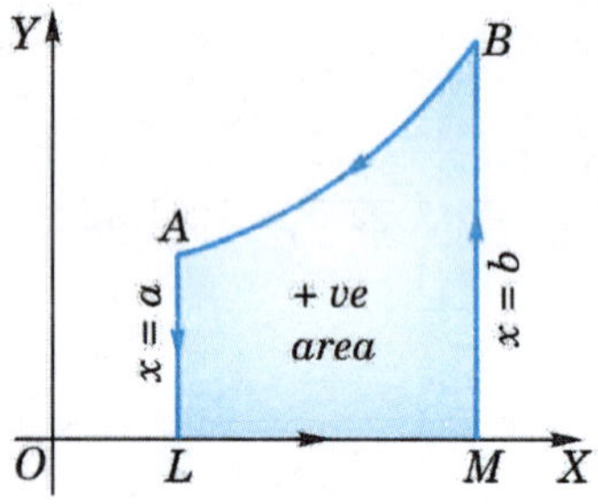


Fig. 6.3

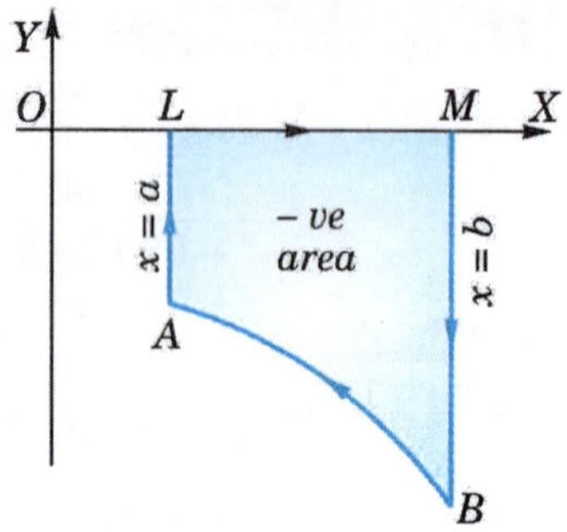


Fig. 6.4

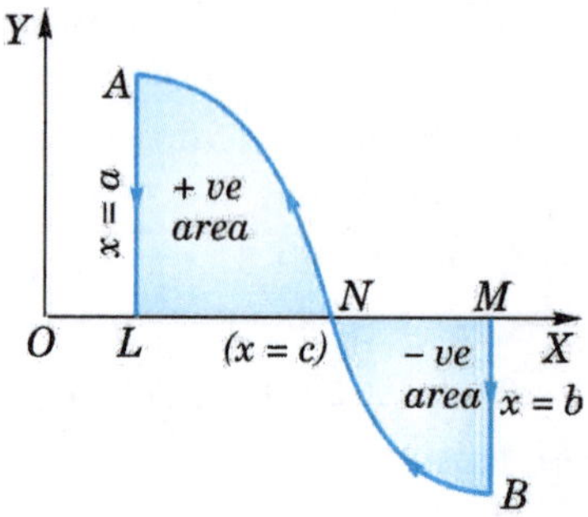


Fig. 6.5

In Fig. 6.5, the area $ALMB \left(= \int_a^b y\, dx\right)$ will not consist of the sum of the area $ALN \left(= \int_a^c y\, dx\right)$ and the area $NMB$ $\left(= \int_c^b y\, dx\right)$, but their difference.

Thus to find the total area in such cases the numerical value of the area of each portion must be evaluated separately and their results added afterwards.

**Example 6.21.** *Find the area of the loop of the curve $ay^2 = x^2 (a - x)$.* (V.T.U., 2012)

**Solution.** Let us trace the curve roughly to get the limits of integration.

(*i*) The curve is symmetrical about $x$-axis.

(*ii*) It passes through the origin. The tangents at the origin are $ay^2 = ax^2$ or $y = \pm x$. $\therefore$ Origin is a node.

(*iii*) The curve has no asymptotes.

(*iv*) The curve meets the $x$-axis at (0, 0) and ($a$, 0). It meets the $y$-axis at (0, 0) only.

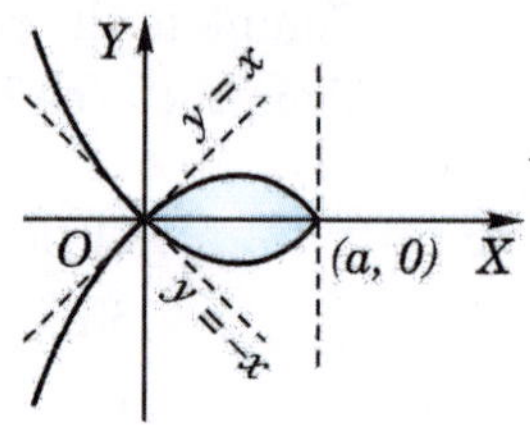


Fig. 6.6

From the equation of the curve, we have $y = \dfrac{x}{\sqrt{a}} \sqrt{(a - x)}$

For $x > a$, $y$ is imaginary. Thus no portion of the curve lies to the right of the line $x = a$. Also $x \to -\infty$, $y \to \infty$.

Thus the curve is as shown in Fig. 6.6.

$\therefore$ Area of the loop = 2 (area of upper half of the loop)

$$= 2\int_0^a y\, dx = 2\int_0^a x\sqrt{\left(\frac{a-x}{a}\right)}\, dx = \frac{2}{\sqrt{a}}\int_0^a [a - (a - x)]\sqrt{(a - x)}\, dx$$

$$= \frac{2}{\sqrt{a}}\int_0^a [a(a - x)^{1/2} - (a - x)^{3/2}]\, dx = 2\sqrt{a}\left|\frac{(a - x)^{3/2}}{-3/2}\right|_0^a - \frac{2}{\sqrt{a}}\left|\frac{(a - x)^{5/2}}{-5/2}\right|_0^a$$

$$= -\frac{4}{3}\sqrt{a(0 - a^{3/2})} + \frac{4}{5\sqrt{a}}(0 - a^{5/2}) = \frac{4}{3}a^2 - \frac{4}{5}a^2 = \frac{8}{15}a^2.$$

**Example 6.22.** *Find the area included between the curve $y^2(2a - x) = x^3$ and its asymptote.*

**Solution.** The curve is as shown in Fig. 4.23.

Area between the curve and the asymptote

$$= 2\int_0^{2a} y\, dx = 2\int_0^{2a} \sqrt{\left(\frac{x^3}{2a - x}\right)}\, dx$$

Put $x = 2a \sin^2 \theta$
so that $dx = 4a \sin\theta \cos\theta\, d\theta$

$$= 2\int_0^{\pi/2} \sqrt{\left\{\frac{(2a\sin^2\theta)^3}{2a\cos^2\theta}\right\}} \cdot 4a\sin\theta\cos\theta\, d\theta$$

$$= 16a^2\int_0^{\pi/2} \sin^4\theta\, d\theta = 16a^2 \cdot \frac{3\cdot 1}{4\cdot 2}\cdot\frac{\pi}{2} = 3\pi a^2.$$

**Example 6.23.** *Show that the area of the curve $a^2x^2 = y^3(2a - y)$ is equal to that of the circle whose radius is a.* (PTU, 2018)

**Solution:** The equation of the curve is $a^2x^2 = y^3(2a - y)$

Now trace the curve and curve is symmetrical about $y$-axis.

The curve passes through the origin. The tangents at the origin are $a^2x^2 = 0$ or $x = 0$

Now, two tangents at the origin are real and coincident, hence the origin is a cusp.

Now, from the equationputting $y = 2a\sin^2\theta$

$$x = \frac{y}{a}\sqrt{y(2a - y)}$$

$\therefore \quad dy = 4a\sin\theta\cos\theta\cdot d\theta$

when $y = 0, \quad \theta = 0$

when $y = 2a, \quad \theta = \frac{\pi}{2}$

Now,The area of the whole curve = 2 × Area $OAB$

Now,Area $OAB = \int_0^{2a} x\cdot dy$

$$= \int_0^{2a}\frac{y}{a}\sqrt{y(2a-y)}\, dy$$

$$= \int_0^{\pi/2}\frac{2a\sin^2\theta}{a}\sqrt{2a\sin^2\theta(2a - 2a\sin^2\theta)} \times 4a\sin\theta\cos\theta\cdot d\theta$$

$$= 16a^2\int_0^{\pi/2}\sin^4\theta\cos^2\theta\, d\theta$$

$$= 16a^2\cdot\frac{3\cdot 1}{6\cdot 4\cdot 2}\times\frac{\pi}{2} = \frac{\pi a^2}{2}$$

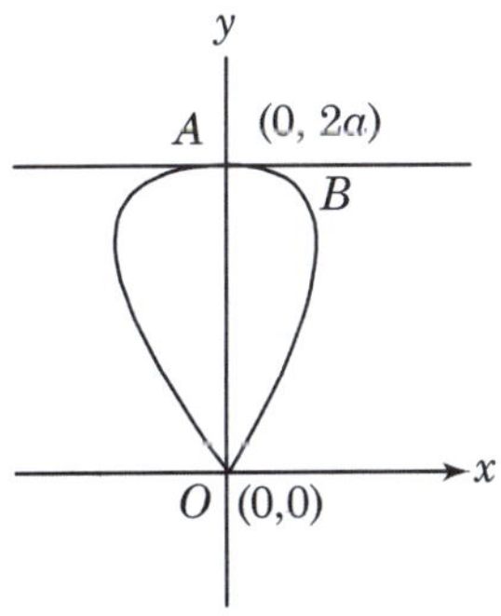

Fig. 6.7.

So, the area of the whole curve = 2 × area $OAB = \frac{2\times\pi a^2}{2}$

Hence the required area = $\pi a^2$.

**Example 6.24 :** *Find the area of the ellipse $\frac{x^2}{a^2} + \frac{y^2}{b^2} = 1$* (PTU, 2019)

**Solution:** Trace the ellipse roughly to find the limit of integration (*i*) The curve is symmetrical about both the axes. (*ii*) The curve does not pass through the origin. (*iii*) The curve meets the $x$ axis at $(\pm a, 0)$ and $y$ axis at $(0, \pm b)$ (*iv*) The curve has no asymptotes. (*v*) From the equation of the curve

$$y = \frac{b}{a}\sqrt{a^2 - x^2} \quad \text{(Taking +ve sign)}$$

Since the curve is symmetrical about both the axes.

area of the ellipse = 4 × area $OAB$

Now, area $OAB = \int_0^a y\, dx$

From the equation of the curve

$$y = \frac{b}{a}\sqrt{a^2 - x^2}$$

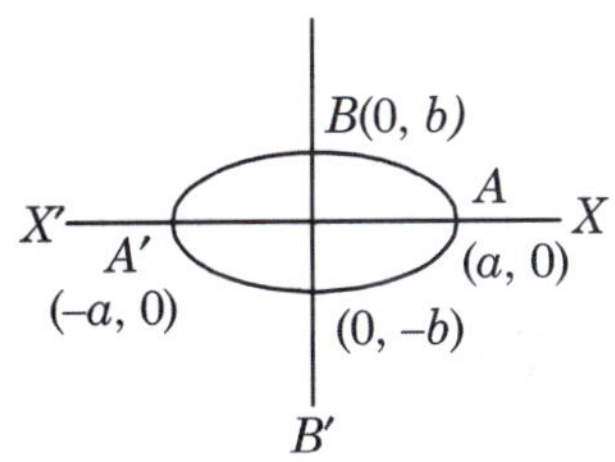

Fig. 6.8.

$\therefore$ area $OAB = \frac{b}{a}\int_0^{\pi/2} a^2\cos^2\theta\, d\theta$ putting $x = a\sin\theta, dx = a\cos\theta\cdot d\theta$

$$= ab\left[\frac{1}{2}\cdot\frac{\pi}{2}\right] = \frac{\pi ab}{4}$$

when $x = 0, \theta = 0$

$\therefore$ area of the ellipse $= 4 \times \dfrac{\pi ab}{4} = \pi ab$ when $x = a, \theta = \dfrac{\pi}{2}$

**Example 6.25 :** *Find the area of the segment cut off from the parabola $x^2 = 8y$ by the line $x - 2y + 8 = 0$* (*Bhilai, 2019; PTU, 2018*)

**Solution:** The equation of the parabola is

$$x^2 = 8y \quad ...(i)$$

and that of the straight line is

$$x - 2y + 8 = 0 \quad ..(ii)$$

from Eqns. (*i*) and (*ii*)

$$(2y-8)^2 = 8y$$
$$4y^2 - 40y + 64 = 0$$
$$y^2 - 10y + 16 = 0$$
$$(y-2)(y-8) = 0$$
$$y = 2, \quad y = 8,$$
$$y = 2, \quad x = -4 \text{ and when } y = 8, \quad x = 8$$

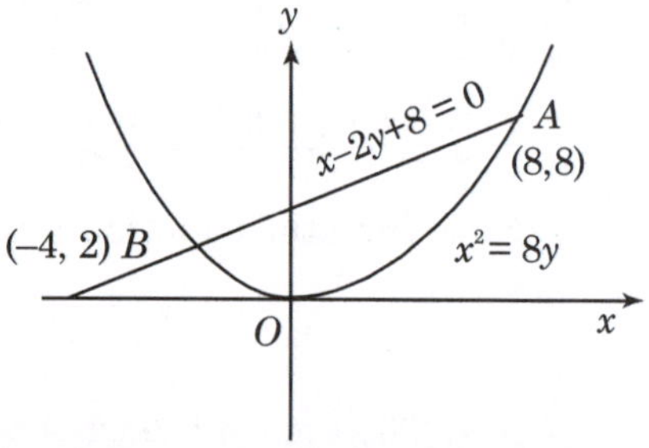

Fig. 6.9.

Hence the points of intersection (–4, 2) and (8, 8)

$$\text{Area } BOA = \int_{-4}^{8}(y_2 - y_1) = \int_{-4}^{8}(y_2 - y_1)dx = \int_{-4}^{8}\left(\frac{x+8}{2} - \frac{x^2}{8}\right)dx$$

$$= \left[\frac{1}{2}\left(\frac{x^2}{2} + 8x\right) - \frac{x^3}{24}\right]_{-4}^{8} = \left[\left\{\frac{1}{2}(32+64) - \frac{512}{24}\right\} - \left\{\frac{1}{2}\left(\frac{16}{2} - 32\right) - \left(\frac{-64}{24}\right)\right\}\right]$$

$$= 48 - \frac{64}{3} + 12 - \frac{8}{3} = 60 - \frac{72}{3} = 60 - 24 = 36$$

**Example 6.26 :** *Prove that the area common to the two paralbolas $x^2 = 4ay$ and $y^2 = 4ax$ is $\dfrac{16a^2}{3}$* (*Kurukshetra, 2019*)

**Solution:** The equations of the parabolas are

$$x^2 = 4ay \quad ...(i)$$
$$y^2 = 4ax \quad ...(ii)$$

Intersection of the curve Eqns. (*i*) and (*ii*)

$$\frac{x^4}{16a^2} = 4ax$$
$$x(x^3 - 64a^3) = 0$$
$$x = 0 \quad \text{or} \quad x = 4a$$

when
$$x = 0, \quad y = 0$$
$$x = 4a, \quad y = 4a$$

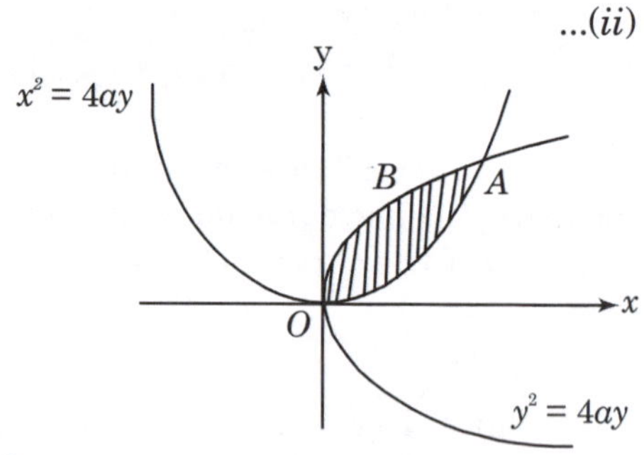

Fig. 6.10.

Hence the parabola is intersect at (0, 0) and (4*a*, 4*a*)

Required area = Area *ABO*

$$= \int_0^{4a}(y_2 - y_1)\,dx = \int_0^{4a}\left(\sqrt{4\,ax} - \frac{x^2}{4a}\right)dx = \left[2\sqrt{a}\,\frac{2}{3}x^{3/2} - \frac{3}{12a}\right]_0^{4a}$$

$$= \left[\left\{\frac{4\sqrt{a}}{3}(4a)^{3/2} - \frac{(4a)^3}{12a}\right\} - 0\right] = \frac{4\sqrt{a}}{3}\cdot 8a^{3/2} - \frac{16a^3}{3}$$

$$= \frac{32}{3}a^2 - \frac{16}{3}a^2 = \frac{16}{3}a^2$$

$$\text{Required area} = \frac{16}{3}a^2$$

**Example 6.27 :** *Show that the larger of the two areas, into which the circle $x^2 + y^2 = 64a^2$ is divided by the parabola $y^2 = 12\,ax$ is $\frac{16}{3}a^2(8\pi - \sqrt{3})$* (VTU, 2019)

**Solution:** The equation of the circle

$$x^2 + y^2 = 64a^2 \qquad ...(i)$$

and that of the parabola is

$$y^2 = 12\,ax \qquad ...(ii)$$

$$x^2 + 12ax - 64a^2 = 0$$

$$(x + 16a)(x - 4a) = 0$$

$$x = -16a \quad \text{and} \quad x = 4a$$

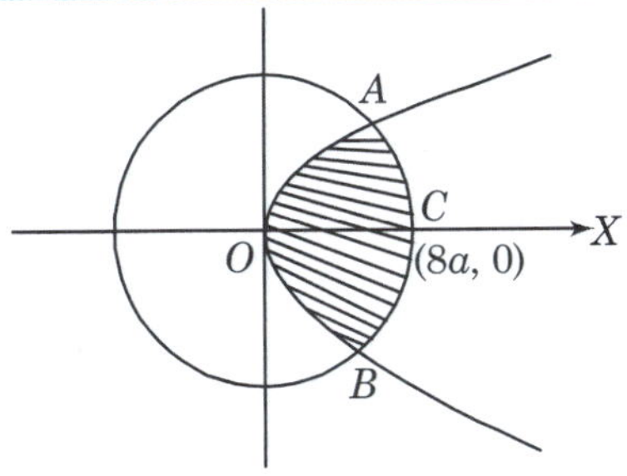

Fig. 6.11.

Rejecting $x = -16a$ being inadmissible as $x \not< 0$

$$x = 4a$$

$$y = \pm 4\sqrt{3}\,a$$

Hence the point of intersection are $(4a, 4\sqrt{3}\,a)$ and $(4a, -4\sqrt{3}\,a)$

$$\text{Area } OCA = \int_0^{4\sqrt{3}} (x_2 - x_1)\,dy = \int_0^{4\sqrt{3}\,a}\left[\sqrt{64\,a^2 - y^2} - \frac{y^2}{12a}\right]dy$$

$$= \left[\frac{y\sqrt{64a^2 - y^2}}{2} + \frac{64a^2}{2}\sin^{-1}\frac{y}{8a}\right]_0^{4\sqrt{3}\,a} - \left[\frac{y^3}{36a}\right]_0^{4\sqrt{3}\,a}$$

$$= 8\sqrt{3}\,a^2 + \frac{32\pi a^2}{3} - \frac{16\sqrt{3}\,a^2}{3} = \frac{32\pi a^2}{3} + \frac{8\sqrt{3}\,a^2}{3}$$

Now area of the semi circle $= \dfrac{\pi(8a)^2}{2} = 32\pi a^2$

So the unshaded area above the $x$-axis = Area of the semicircle – Area $OCA$

$$= 32\pi a^2 - \frac{32\pi a^2}{3} - \frac{8\sqrt{3}a^2}{3} = \frac{64\pi a^2}{3} - \frac{8\sqrt{3}a^2}{3} = \frac{8a^2}{3}(8\pi - \sqrt{3})$$

Required area = 2 × unshaded area above $x$-axis

$$= 2 \times \frac{8a^2}{3}(8\pi - \sqrt{3}) = \frac{16a^2}{3}(8\pi - \sqrt{3}).$$

**Example 6.28 :** *Find the area between the curve $x^2y^2 = a^2(y^2 - x^2)$ and its asymptotes.* (Kurukshetra, 2019)

**Solution:** Equation of the curve is $x^2y^2 = a^2(y^2 - x^2)$

Trace the curve roughly to find the limits of integration and the curve passes through the origin,

$$y^2 - x^2 = 0 \quad \text{or} \quad y = \pm x$$

The curve does not meet the axes at any other point except the origin.

$$y^2 - x^2 = 0 \quad \text{or} \quad y = \pm x$$

From the equation of the curve $y^2 = \dfrac{a^2x^2}{a^2 - x^2}$

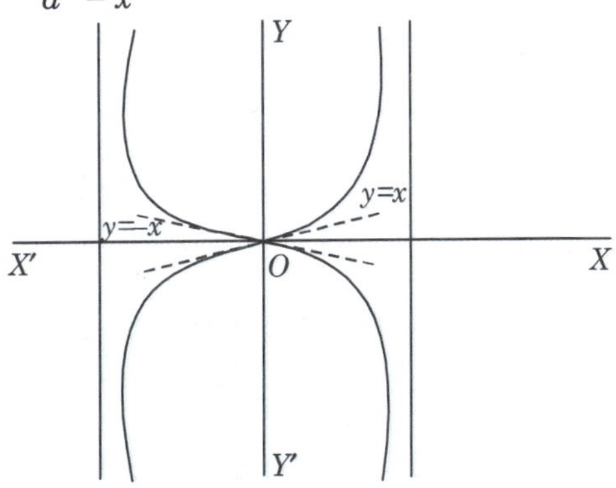

Fig. 6.12.

Area between the curve and its asymptotes

$$= 4 \times \text{Area between the curve}$$

$$= \int_0^a y\,dx = \int_0^a \frac{ax}{\sqrt{a^2 - x^2}}\,dx \qquad \text{putting } x = a\sin\theta \quad dx = a\cos\theta \cdot d\theta$$

$$= \int_0^{\pi/2} \frac{a \cdot a \cdot \sin\theta \cdot a\cos\theta \cdot d\theta}{a\cos\theta} \qquad \text{when } x = 0, \theta = 0$$

$$= a^2 \int_0^{\pi/2} \sin\theta \cdot d\theta \qquad \text{when } x = a, \theta = \frac{\pi}{2}$$

$$= a^2[-\cos\theta]_0^{\pi/2} = a^2$$

$\therefore$ The required area $= 4 \times a^2 = 4a^2$.

**Example 6.29 :** *Find the area enclosed by the curves*

$$x^2 = 4ay \quad and \quad y = \frac{8a^3}{x^2 + 4a^2}$$

**Solution:** The equation of the parabola is $x^2 = 4ay$

The curve does not meet the axes at any point except the origin.

In this curve has not asymptotes, As $y$ increases, $x$ also increases when $y \to \infty$, $x \to \infty$

For the shape of the second curve $y = \dfrac{8a^2}{x^2 + 4a^2}$

$$x^2 = 4ay \qquad ...(i)$$

$$y = \frac{8a^3}{x^2 + 4a^2} \qquad ...(ii)$$

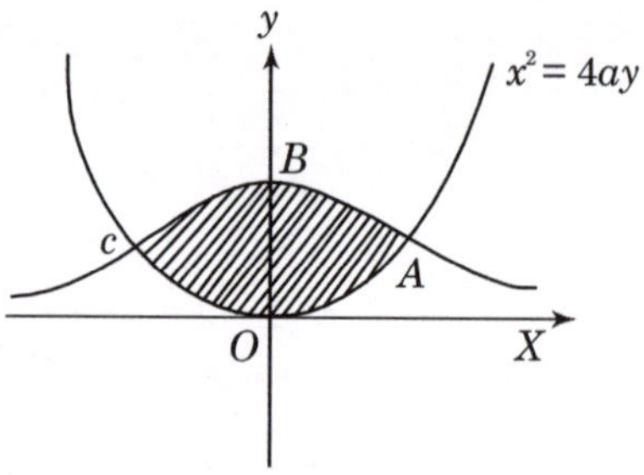


Fig. 6.13.

Eliminating $y$ between Eqns. $(i)$ and $(ii)$

$$x^2 = 4a \cdot \frac{8a^3}{x^2 + 4a^2}$$

$$x^2 + 4a^2x^2 - 32a^4 = 0$$

$$(x^2 + 8a^2)(x^2 - 4a^2) = 0$$

Rejecting $x^2 + 8a^2 = 0$ as it gives imaginary values of $x$,

we have $\qquad x^2 - 4a^2 = 0, \quad x = \pm 2a$

Thus the curves intersect at the points $(2a, 0)$ and $(-2a, a)$

The area between the two curves = 2 × Area $OAB$

$$\text{Area } OAB = \int_0^{2a} (y_2 - y_2)\,dx$$

$$= \int_0^{2a} \left( \frac{8a^3}{x^2 + 4a^2} - \frac{x^2}{4a} \right) dx = \left[ 8a^3 \cdot \frac{1}{2a} \cdot \tan^{-1}\frac{x}{2a} - \frac{x^3}{12a} \right]_0^{2a}$$

$$= \left[ \left( 4a^2 \tan^{-1} 1 - \frac{8a^2}{12} \right) - 0 \right] = \left( 4a^2 \cdot \frac{\pi}{4} - \frac{2}{3}a^2 \right) = a^2\left( \pi - \frac{2}{3} \right)$$

So, $\qquad$ the required area = 2 × Area $OAB = 2a^2\left(\pi - \dfrac{2}{3}\right)$

**Example 6.30 :** *Prove that the area of the loop of the curve*

$$3\,ay^2 = x(x - a)^2 \text{ is } \frac{8a^2}{15\sqrt{3}}$$

*Trace the curve roughly to find the limits and curve meets the x-axis at (a, 0) and does not meet the y-axis, the curve has no asymptotes parallel to the axes.*

**Solution:** When $x$ is increasing, $y$ also increase when $x \to \infty$, $y \to \pm\infty$

The shape of the curve is shown

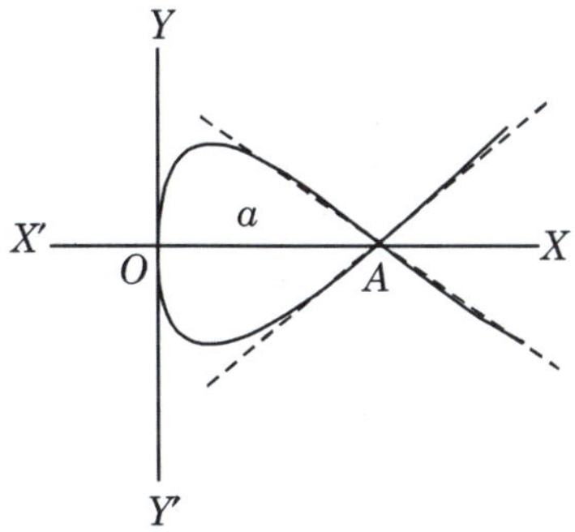

Fig. 6.14.

The curve is symmetrical about the $x$-axis.

$\therefore$ Area of the loop of the curve = 2 × Area under the arc $OA$

$$\text{Area under the arc } OA = \int_0^a y\,dx = \int_0^a \frac{\sqrt{x}(x-a)}{\sqrt{3}\,a}\,dx$$

$$= \frac{1}{\sqrt{3}a}\int_0^a \left(x^{3/2} - ax^{1/2}\right)dx$$

$$= \frac{1}{\sqrt{3}a}\left[\frac{2}{5}x^{5/2} - \frac{2}{3}ax^{3/2}\right]_0^a = \frac{2a^{5/2}}{\sqrt{3}\,a}\left[\frac{1}{5} - \frac{1}{3}\right] = \frac{-4a^2}{15\sqrt{3}}$$

Hence, the area of the loop of the curve = 2 × Area under the arc $OA$

$$= 2 \times \frac{4a^2}{15\sqrt{3}} = \frac{8a^2}{15\sqrt{3}}$$

## PROBLEMS 6.6

1. Show that the area included between the four infinite branches of the curve $x^2y^2 = a(x^2 + y^2)$ and its asymptotes is $4a^2$ *(JNTU, 2019)*
2. Find the area bounded by the curve $y = x(x-3)(x-5)$ and the $x$-axis.
3. Find the area included between the curve $ay^2 = x^3$, the $x$-axis and the ordinates $x = a$.
4. Find the area of the loop of the curve :
   (*i*) $3ay^2 = x(x-a)^2$ (*ii*) $x(x^2 + y^2) = a(x^2 - y^2)$ *(P.T.U., 2010)*
5. Find the whole area of the curve :
   (*i*) $a^2x^2 = y^3(2a - y)$ (*ii*) $8a^2y^2 = x^2(a^2 - x^2)$ *(V.T.U., 2006)*
6. Show that the area of the segment cut off from the parabola $y^2 = 2x$ by the straight line $y = 4x - 1$ is $\frac{9}{32}$. *(CSVTU, 2018)*
7. Show that the area cut off from a parabola by any double ordinate is two thirds of the corresponding rectangle contained by the double ordinate and its distance from the vertex.
8. Show that the area of the loop of the curve $y^2(a + x) = x^2(3a - x)$ is equal to the area between the curve and its asymptote.
9. Find the whole area of the *astroid* $x^{2/3} + y^{2/3} = a^{2/3}$ or $x = a\cos^3\theta$, $y = a\sin^3\theta$. *(C.S.V.T.U., 2014)*
10. Find the area bounded by the curve $\sqrt{x} + \sqrt{y} = \sqrt{a}$ and the coordinate axes. *(Delhi, 2012)*
11. Find the area included between the *cycloid* $x = a(\theta + \sin\theta)$, $y = a(1 - \cos\theta)$ *and its base.* Also find the area between the curve and the $x$-axis.
12. Find the area common to the circles $x^2 + y^2 = 4$ and $x^2 + y^2 = 4x$.
13. Find the area included between the circle $x^2 + y^2 = 2ax$ and the parabola $y^2 = ax$.
14. Find the area of the segment cut off from the parabola $y^2 = 4x$ by straight line $2x - 3y + 4 = 0$ *(Marathwada, 2018)*

(2) **Areas of polar curves.** *Area bounded by the curve $r = f(\theta)$ and the radii vectors*

$$\theta = \alpha,\ \theta = \beta \textit{ is } \frac{1}{2}\int_\alpha^\beta \mathbf{r^2}\,\mathbf{d\theta}.$$

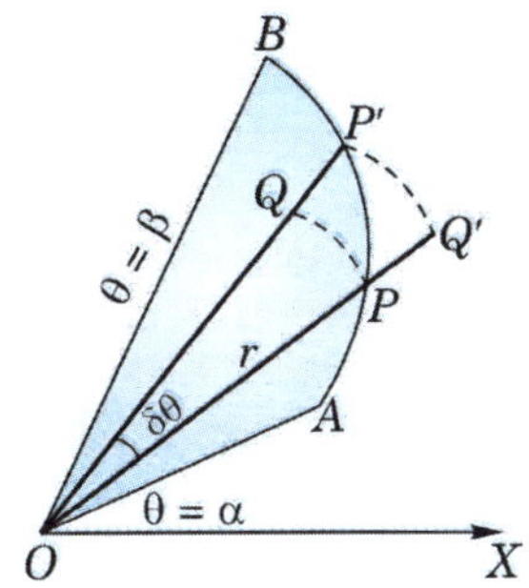

Fig. 6.15.

Let $AB$ be the curve $r = f(\theta)$ between the radii vectors $OA\,(\theta = \alpha)$ and $OB\,(\theta = \beta)$. Let $P(r, \theta)$, $P'(r + \delta r, \theta + \delta\theta)$ be any two neighbouring points on the curve. (Fig. 6.15)

Let the area $OAP = A$ which is a function of $\theta$. Then the area $OPP' = \delta A$. Mark circular arcs $PQ$ and $P'Q'$ with centre $O$ and radii $OP$ and $OP'$.

Evidently area $OPP'$ lies between the sectors $OPQ$ and $OP'Q'$ *i.e.*, $\delta A$ lies between $\frac{1}{2}r^2\,\delta\theta$ and $\frac{1}{2}(r+\delta r)^2\,\delta\theta$.

$\therefore \quad \dfrac{\delta A}{\delta\theta}$ lies between $\frac{1}{2}r^2$ and $\frac{1}{2}(r+\delta r)^2$.

Now taking limits as $\delta\theta \to 0$ $\qquad (\because \delta r \to 0),\ \dfrac{dA}{d\theta} = \dfrac{1}{2}r^2$

Integrating both sides from $\theta = \alpha$ to $\theta = \beta$, we get $\left| A \right|_\alpha^\beta = \int_\alpha^\beta \frac{1}{2}r^2\,d\theta$

or (value of $A$ for $\theta = \beta$) – (value of $A$ for $\theta = \alpha$) $= \dfrac{1}{2}\displaystyle\int_\alpha^\beta r^2\,d\theta$

Hence the required area $OAB = \dfrac{1}{2}\displaystyle\int_\alpha^\beta r^2\,d\theta$.

**Example 6.31.** *Find the area of the cardioid $r = a(1 - \cos\theta)$.*

**Solution.** The curve is as shown in Fig. 6.16. Its upper half is traced from $\theta = 0$ to $\theta = \pi$.

$$\therefore \quad \text{Area of the curve} = 2 \cdot \frac{1}{2}\int_0^\pi r^2\,d\theta = a^2\int_0^\pi (1-\cos\theta)^2\,d\theta$$

$$= a^2\int_0^\pi (2\sin^2\theta/2)^2\,d\theta = 4a^2\int_0^\pi \sin^4\theta/2 \cdot d\theta$$

$$= 8a^2\int_0^{\pi/2} \sin^4\phi\,d\phi, \text{ putting } \theta/2 = \phi \text{ and } d\theta = 2d\phi.$$

$$= 8a^2 \cdot \frac{3\cdot 1}{4\cdot 2}\cdot\frac{\pi}{2} = \frac{3\pi a^2}{2}.$$

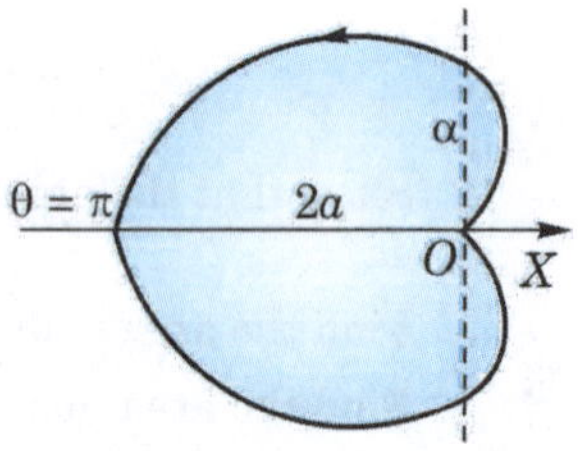

Fig. 6.16

**Example 6.32.** *Find the area of a loop of the curve $r = a\sin 3\theta$.* (V.T.U., 2013)

**Solution.** The curve is as shown in Fig. 6.17. It consists of three loops.

Putting $r = 0$, $\sin 3\theta = 0$ $\quad \therefore \quad 3\theta = 0$ or $\pi$ *i.e.*, $\theta = 0$ or $\pi/3$ which are the limits for the first loop.

$$\therefore \quad \text{Area of a loop} = \frac{1}{2}\int_0^{\pi/3} r^2\,d\theta = \frac{1}{2}a^2\int_0^{\pi/3}\sin^2 3\theta\,d\theta = \frac{a^2}{4}\int_0^{\pi/3}(1-\cos 6\theta)\,d\theta$$

$$= \frac{a^2}{4}\left|\theta - \frac{\sin 6\theta}{6}\right|_0^{\pi/3} = \frac{a^2}{4}\left(\frac{\pi}{3} - 0\right) = \frac{\pi a^2}{12}.$$

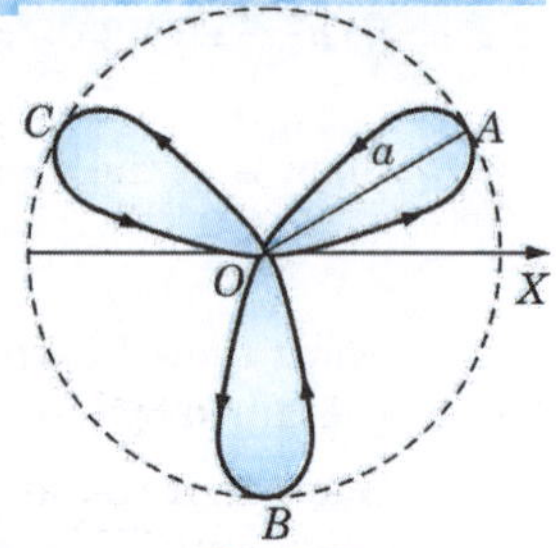

Fig. 6.17

**Obs.** The limits of integration for a loop of $r = a\sin n\theta$ or $r = a\cos n\theta$ are the two consecutive values of $\theta$ when $r = 0$.

**Example 6.33.** *Find the whole area of the hypocycloid $x = a\cos^3 t$, $y = b\sin^3 t$* (JNTU 2017)

**Solution:** The whole area of the curve = 4 × area $OAB$

$$\text{Area } OAB = \int_0^{\pi/2} y\cdot\frac{dx}{dt}\cdot dt$$

$$= \int_0^{\pi/2} b\sin^3 t(-3a\cos^2 t\sin t)\cdot dt$$

$$= -3ab\int_0^{\pi/2}\sin^4 t\cos^2 t = -3ab\cdot\frac{3\cdot 1}{6\cdot 4\cdot 2}\times\frac{\pi}{2}$$

$$= -\frac{3}{32}\pi ab = \frac{3}{32}\pi ab = 4\times\frac{3}{32}\pi ab = \frac{3}{8}\pi ab$$

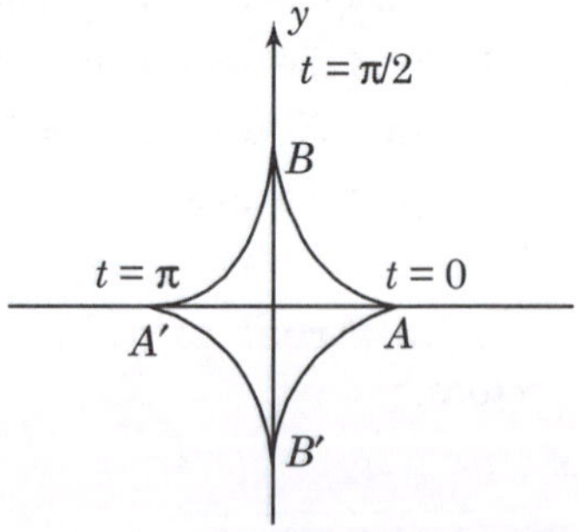

Fig. 6.18

**Example 6.34.** *Find the area common to the circles*

$$r = a\sqrt{2} \text{ and } r = 2a\cos\theta$$ (D.T.U., 2011)

**Solution.** The equations of the circles are $r = a\sqrt{2}$ ...(i) and $r = 2a\cos\theta$ ...(ii)

(i) represents a circle with centre at (0, 0) and radius $a\sqrt{2}$. (ii) represents a circle symmetrical about $OX$, with centre at $(a, 0)$ and radius $a$.

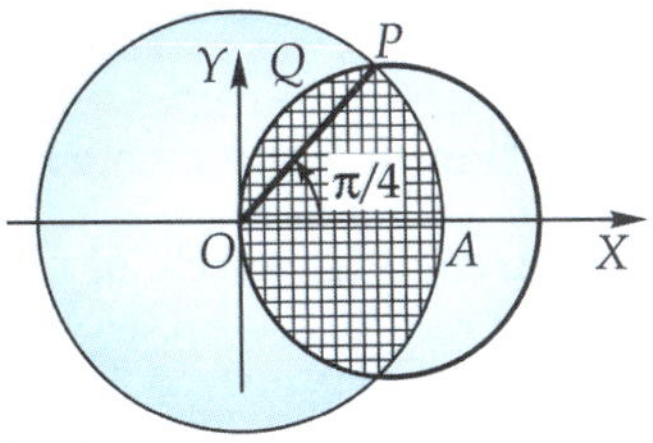

Fig. 6.19.

The circles are shown in Fig. 6.19. At their point of intersection $P$, eliminating $r$ from Eqns. (i) and (ii),

$$a\sqrt{2} = 2a\cos\theta \; i.e., \cos\theta = 1\sqrt{2}$$

or $$\theta = \pi/4$$

$\therefore$ Required area = 2 × area $OAPQ$ (By symmetry)

$= 2(\text{area } OAP + \text{area } OPQ)$

$$= 2\left[\frac{1}{2}\int_0^{\pi/4} r^2\,d\theta, \text{ for (i)} + \frac{1}{2}\int_{\pi/4}^{\pi/2} r^2\,d\theta, \text{ for (ii)}\right]$$

$$= \int_0^{\pi/4} (a\sqrt{2})^2\,d\theta + \int_{\pi/4}^{\pi/2} (2a\cos\theta)^2\,d\theta = 2a^2 \left|\,\theta\,\right|_0^{\pi/4} + 4a^2\int_{\pi/4}^{\pi/2} \frac{1+\cos 2\theta}{2}\,d\theta$$

$$= 2a^2(\pi/4 - 0) + 2a^2\left|\,\theta + \frac{\sin 2\theta}{2}\right|_{\pi/4}^{\pi/2} = \frac{\pi a^2}{2} + 2a^2\left(\frac{\pi}{2} - \frac{\pi}{4} - \frac{1}{2}\right) = a^2(\pi - 1).$$

**Example 6.35.** *Find the area common to the cardioids $r = a(1 + \cos\theta)$ and $r = a(1 - \cos\theta)$.* (D.T.U., 2015)

**Solution.** The cardioid $r = a(1 + \cos\theta)$ is $ABCOB'A$ and the cardioid $r = a(1 - \cos\theta)$ is $OC'BA'B'O$. Both the cardioids are symmetrical about the initial line $OX$ and intersect at $B$ and $B'$ (Fig. 6.20)

$\therefore$ Required area (shaded) = 2 area $OC'BCO$

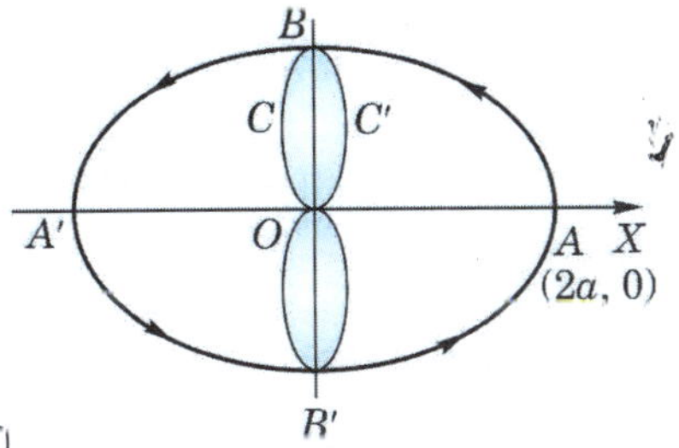

Fig. 6.20.

$= 2$ [area $OC'BO$ + area $OBCO$]

$$= 2\left[\left\{\int_0^{\pi/2} \frac{1}{2}r^2\,d\theta\right\}_{r=a(1-\cos\theta)} + \left\{\int_{\pi/2}^{\pi} \frac{1}{2}r^2\,d\theta\right\}_{r=a(1+\cos\theta)}\right]$$

$$= a^2\int_0^{\pi/2} (1-\cos\theta)^2\,d\theta + a^2\int_{\pi/2}^{\pi} (1+\cos\theta)^2\,d\theta$$

$$= a^2\left\{\int_0^{\pi/2} (1 - 2\cos\theta + \cos^2\theta)\,d\theta + \int_{\pi/2}^{\pi} [1 + 2\cos\theta + \cos^2\theta]\,d\theta\right\}$$

$$= a^2\left\{\int_0^{\pi} (1 + \cos^2\theta)\,d\theta - 2\int_0^{\pi/2} \cos\theta\,d\theta + 2\int_{\pi/2}^{\pi} \cos\theta\,d\theta\right\}$$

$$= a^2\left\{\int_0^{\pi} \left(1 + \frac{1+\cos 2\theta}{2}\right)d\theta - 2\left|\sin\theta\right|_0^{\pi/2} + 2\left|\sin\theta\right|_{\pi/2}^{\pi}\right\}$$

$$= a^2\left\{\left|\frac{3}{2}\theta + \frac{\sin 2\theta}{4}\right|_0^{\pi} - 2(1-0) + 2(0-1)\right\} = \left(\frac{3\pi}{2} - 4\right)a^2.$$

**Example 6.36 :** *Show that the whole area between the four infinite branches of the tractrix*

$$x = a\cos t + \frac{1}{2}a\log\tan^2\frac{t}{2},\; y = a\sin t$$

*is equal to the area of the circle whose is a.* (CSVTU, 2018)

**Solution :** The whole area under the curve

= 4 × Area under the curve in the first quadrant

So, = area in the first quadrant

$$= \int_0^{\pi/2} y\frac{dx}{dt}\,dt = \int_0^{\pi/2} a\sin t\left(-a\sin t + \frac{a}{\sin t}\right)dt$$

$$= a^2\int_0^{\pi/2} \sin t \cdot \frac{(1-\sin^2 t)}{\sin t}\, dt = a^2\int_0^{\pi/2} \cos^2 t \cdot dt = \frac{\pi a^2}{4}$$

Therefore the whole area under the curve $= 4 \times \frac{\pi a^2}{4} = \pi a^2$

Area of the circle of radius is $\pi a^2$.

## PROBLEMS 6.7

1. Find the whole area of
   (*i*) the *cardioid* $r = a\,(1 + \cos\theta)$ (*C.S.V.T.U., 2015*) (*ii*) the *lemniscate* $r^2 = a^2 \cos 2\theta$ ;
2. Show that the area of the curve
   $$x = \frac{a(1-t^2)}{1+t^2},\ y = \frac{2at}{1+t^2}$$
   is equal to the area of a circle of radius $a$. (*PTU, 2019*)
3. Show that the area of the loop of the curve $x = (1 - t^2)$,
   $$y = t(1-t^2),\ -1 \le t \le 1 \text{ is } \frac{8}{15}.$$ (*VTU, 2018*)
4. Find the area bounded by the parabola
   $x = at^2,\ y = 2at$ and the line $x = a$ (*Kerela, 2019*)
5. Find the area of the loop of the curve
   $$x = \frac{3at^2}{1+t^3},\ y = \frac{3at^2}{1+t^3}$$ (*PTU, 2018*)
6. Find the area of bounded by one branch of the curve
   $x = at - b\sin t \quad (0 < b \le a)$
   $y = a - b\cos t$
   and $a$ tangent to it at its lower points. (*Marathawada, 2019*)
7. Show that the area between the cissoid
   $$r = \frac{a\sin^2\theta}{\cos\theta} \text{ and its asymptote is } \frac{3\pi a^2}{4}$$
8. Find the area outside the *circle* $r = 2a\cos\theta$ and inside the *cardioid* $r = a(1 + \cos\theta)$. (*Delhi., 2015*)

## 6.11 LENGTHS OF CURVES

(**1**) *The length of the arc of the curve $y = f(x)$ between the points where $x = a$ and $x = b$ is*

$$\int_a^b \sqrt{\left[1 + \left(\frac{dy}{dx}\right)^2\right]}\, dx$$

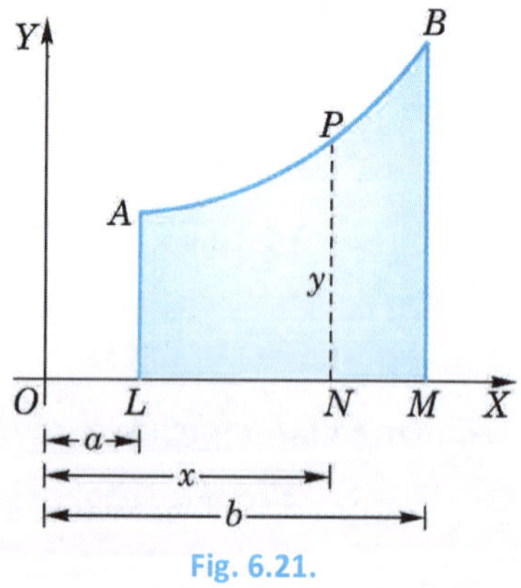

Fig. 6.21.

Let $AB$ be the curve $y = f(x)$ between the points $A$ and $B$ where $x = a$ and $x = b$ (Fig. 6.21)

Let $P(x, y)$ be any point on the curve and arc $AP = x$ so that it is a function of $x$.

Then
$$\frac{ds}{dx} = \sqrt{\left[1 + \left(\frac{dy}{dx}\right)^2\right]}$$

$\therefore$
$$\int_a^b \sqrt{\left[1 + \left(\frac{dy}{dx}\right)^2\right]}\, dx = \int_a^b \frac{ds}{dx} \cdot dx = |\, s\,|_{x=a}^{x=b}$$

= (value of $s$ for $x = b$) – (value of $s$ for $x = a$) = arc $AB$ – 0

Hence, the arc $AB = \int_a^b \sqrt{\left[1 + \left(\frac{dy}{dx}\right)^2\right]}\, dx.$

(**2**) *The length of the arc of the curve $x = f(y)$ between the points where $y = a$ and $y = b$, is*

$$\int_a^b \sqrt{\left[1 + \left(\frac{\mathbf{dx}}{\mathbf{dy}}\right)^2\right]}\, \mathbf{dy}$$

(**3**) *The length of the arc of the curve $x = f(t)$, $y = \phi(t)$ between the points where $t = a$ and $t = b$, is*

$$\int_a^b \sqrt{\left[\left(\frac{\mathbf{dx}}{\mathbf{dt}}\right)^2 + \left(\frac{\mathbf{dy}}{\mathbf{dt}}\right)^2\right]}\, \mathbf{dt}.$$

(**4**) *The length of the arc of the curve $r = f(\theta)$ between the points where $\theta = \alpha$* and *$\theta = \beta$, is*

$$\int_\alpha^\beta \sqrt{\left[\mathbf{r}^2 + \left(\frac{\mathbf{dr}}{\mathbf{d\theta}}\right)^2\right]}\, \mathbf{d\theta}.$$

(**5**) *The length of the arc of the curve $\theta = f(r)$ between the points where $r = a$ and $r = b$, is*

$$\int_a^b \sqrt{\left[1 + \left(\mathbf{r}\frac{\mathbf{d\theta}}{\mathbf{dr}}\right)^2\right]}\, \mathbf{dr}$$

**Example 6.37.** *Find the length of the arc of the parabola $x^2 = 4ay$ measured from the vertex to one extremity of the latus-rectum.*

**Solution.** Let $A$ be the vertex and $L$ an extremity of the latus-rectum so that at $A$, $x = 0$ and at $L$, $x = 2a$. (Fig. 6.22).

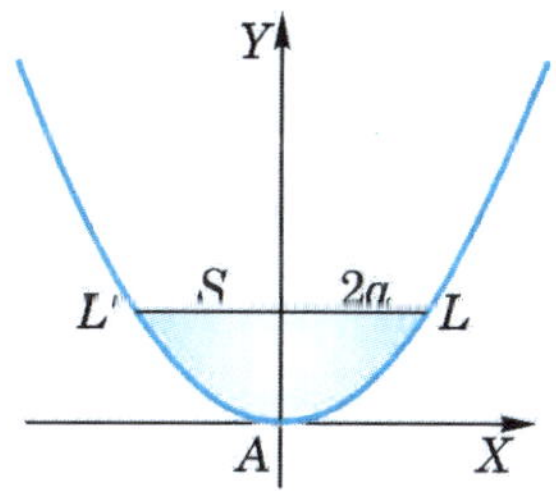

Fig. 6.22.

Now $\qquad y = x^2/4a$ so that $\frac{dy}{dx} = \frac{1}{4a} \cdot 2x - \frac{x}{2a}$

$$\therefore \qquad \text{arc } AL = \int_0^{2a} \sqrt{\left[1 + \left(\frac{dy}{dx}\right)^2\right]}\, dx$$

$$= \int_0^{2a} \sqrt{\left[1 + \left(\frac{x}{2a}\right)^2\right]}\, dx = \frac{1}{2a}\int_0^{2a} \sqrt{[(2a)^2 + x^2]}\, dx$$

$$= \frac{1}{2a}\left[\frac{x\sqrt{[(2a)^2 + x^2]}}{2} + \frac{(2a)^2}{2}\sinh^{-1}\frac{x}{2a}\right]_0^{2a} = \frac{1}{2a}\left[\frac{2a\sqrt{(8a)^2}}{2} + 2a^2\sinh^{-1} 1\right]$$

$$= a[\sqrt{2} + \sinh^{-1} 1] = a[\sqrt{2} + \log(1 + \sqrt{2})] \qquad [\because \ \sinh^{-1}x = \log[x + \sqrt{(1 + x^2)}]]$$

**Example 6.38.** *Find the perimeter of the loop of the curve $3ay^2 = x(x - a)^2$.* *(D.T.U., 2015)*

**Solution.** The curve is symmetrical about the $x$-axis and the loop lies between the limits $x = 0$ and $x = a$. (Fig. 6.23).

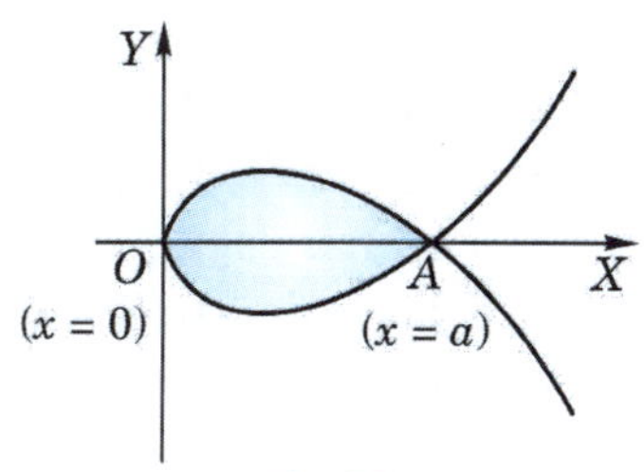

Fig. 6.23.

We have $\qquad y = \frac{\sqrt{x}(x - a)}{\sqrt{(3a)}}$

$$\therefore \qquad \frac{dy}{dx} = \frac{1}{\sqrt{(3a)}}\left[\frac{3}{2}x^{1/2} - \frac{a}{2}\cdot x^{-1/2}\right] = \frac{1}{2\sqrt{(3a)}}\,\frac{3x - a}{\sqrt{x}}$$

$$\therefore \quad \text{Perimeter of the loop} = 2\int_0^a \sqrt{\left[1 + \left(\frac{dy}{dx}\right)^2\right]}\, dx \qquad \text{(By symmetry)}$$

$$= 2\int_0^a \sqrt{\left[1 + \frac{(3x-a)^2}{12ax}\right]}\,dx = 2\int_0^a \frac{\sqrt{(9x^2+6ax+a^2)}}{\sqrt{(12ax)}}\,dx$$

$$= \frac{1}{\sqrt{(3a)}}\int_0^a \frac{3x+a}{\sqrt{x}}\,dx = \frac{1}{\sqrt{(3a)}}\int_0^a (3x^{1/2} + ax^{-1/2})\,dx$$

$$= \frac{1}{\sqrt{(3a)}}\left|\frac{3x^{3/2}}{3/2} + a\frac{x^{1/2}}{1/2}\right|_0^a = \frac{1}{\sqrt{(3a)}}(4a^{3/2}) = \frac{4a}{\sqrt{3}}.$$

**Example 6.39.** *Find the length of one arch of the cycloid*
$x = a(t - \sin t),\ y = a(1 - \cos t).$ *(V.T.U., 2013)*

**Solution.** As a point moves from one end $O$ to the other end of its first arch, the parameter $t$ increases from 0 to $2\pi$. [see Fig. 6.24]

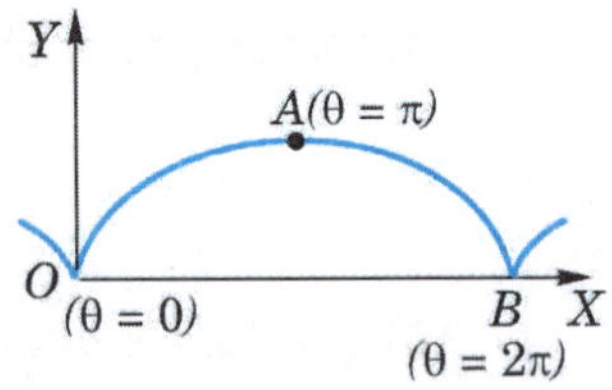

Fig. 6.24.

Also $\qquad \dfrac{dx}{dt} = a(1 - \cos t),\ \dfrac{dy}{dt} = a \sin t.$

$$\therefore \text{ Length of an arch } = \int_0^{2\pi} \sqrt{\left[\left(\frac{dx}{dt}\right)^2 + \left(\frac{dy}{dt}\right)^2\right]}\,dx$$

$$= \int_0^{2\pi} \sqrt{\{[a(1-\cos t)]^2 + (a\sin t)^2\}}\,dt = a\int_0^{2\pi} \sqrt{[2(1-\cos t)]}\,dt$$

$$= 2a\int_0^{2\pi} \sin t/2\,dt = 2a\left|-\frac{\cos t/2}{1/2}\right|_0^{2\pi} = 4a[(-\cos\pi) - (-\cos 0)] = 8a.$$

**Example 6.40.** *Find the entire length of the cardioid $r = a(1 + \cos\theta)$.*
*(V.T.U., 2013 ; P.T.U., 2010)*
*Also show that the upper half is bisected by $\theta = \pi/3$.* *(D.T.U., 2013 ; P.T.U., 2013)*

**Solution.** The cardioid is symmetrical about the initial line and for its upper half, $\theta$ increases from 0 to $\pi$ (Fig. 6.25)

Also $\qquad \dfrac{dr}{d\theta} = -a\sin\theta.$

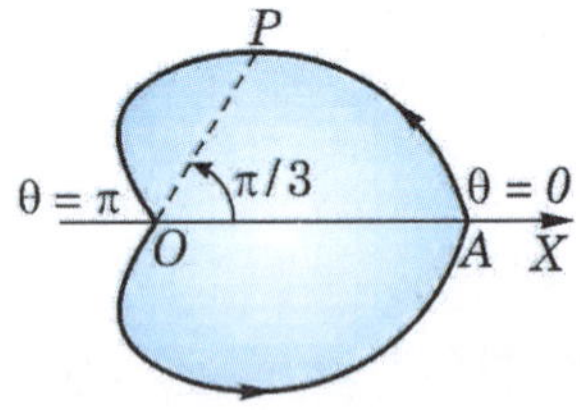

Fig. 6.25.

$$\therefore \text{ Length of the curve } = 2\int_0^{\pi} \sqrt{\left[r^2 + \left(\frac{dr}{d\theta}\right)^2\right]}\,d\theta$$

$$= 2\int_0^{\pi} \sqrt{\{[a(1+\cos\theta)]^2 + (-a\sin\theta)^2\}}\,d\theta = 2a\int_0^{\pi} \sqrt{[2(1+\cos\theta)]}\,d\theta$$

$$= 4a\int_0^{\pi} \cos\theta/2\,d\theta = 4a\left|\frac{\sin\theta/2}{1/2}\right|_0^{\pi} = 8a(\sin\pi/2 - \sin 0) = 8a.$$

$\therefore$ Length of upper half of the curve is $4a$. Also length of the arc $AP$ from 0 to $\pi/3$.

$$= a\int_0^{\pi/3} \sqrt{[2(1+\cos\theta)]}\,d\theta = 2a\int_0^{\pi/3} \cos\theta/2 \cdot d\theta$$

$$= 4a\left|\sin\theta/2\right|_0^{\pi/3} = 2a = \text{half the length of upper half of the cardioid.}$$

## PROBLEMS 6.8

1. Find the length of the arc of the *semi-cubical parabola* $ay^2 = x^3$ from the vertex to the ordinate $x = 5a$.
2. Find the length of the curve (*i*) $y = \log \sec x$ from $x = 0$ to $x = \pi/3$. *(V.T.U., 2010 S)*
   (*ii*) $y = \log[(e^x - 1)/(e^x + 1)]$ from $x = 1$ to $x = 2$. *(C.S.V.T.U., 2015)*
3. Find the length of the arc of the parabola $y^2 = 4ax$ (*i*) from the vertex to one end of the latus-rectum.
   (*ii*) cut off by the line $3y = 8x$. *(C.S.V.T.U., 2014)*

4. Find the perimeter of the loop of the following curves :
(*i*) $ay^2 = x^2(a - x)$ (*ii*) $9y^2 = (x - 2)(x - 5)^2$.
5. Find the length of the curve $y^2 = (2x - 1)^2$ cut off by the line $x = 4$.
6. Show that the whole length of the curve $x^2(a^2 - x^2) = 8a^2y^2$ is $\pi a\sqrt{2}$.
7. Show that the length of an arc of the curve $x^2 = a^2(1 - e^{y/x})$ measured from the origin to any point $(x, y)$ is $a \log \dfrac{a + x}{a - x} - x$. (*Delhi, 2012*)
8. (*a*) Find the length of an arch of the cycloid $x = a(\theta + \sin\theta)$, $y = a(1 - \cos\theta)$.
(*b*) By finding the length of the curve show that the curve $x = a(\theta - \sin\theta)$, $y = a(1 - \cos\theta)$, is divided in the ratio 1 : 3 at $\theta = 2\pi/3$.
9. Find the whole length of the curve $x = a\cos^3 t$, $y = a\sin^3 t$ *i.e.*, $x^{2/3} + y^{2/3} = a^{2/3}$. (*V.T.U., 2015*)
Also show that the line $\theta = \pi/3$ divides the length of this *astroid* in the first quadrant in the ratio 1 : 3.
10. Find the length of the loop of the curve $x = t^2$, $y = t - t^3/3$.
11. For the curve $r = ae^{\theta}\cot\alpha$, prove that $s/r$ = constant, $s$ being measured from the origin.
12. Find the perimeter of the *cardioid* $r = a(1 - \cos\theta)$. Also show that the upper half of the curve is bisected by the line $\theta = 2\pi/3$.
13. Find the whole length of the *lemniscate* $r^2 = a^2\cos 2\theta$.
14. Find the length of the parabola $r(1 + \cos\theta) = 2a$ as cut off by the latus-rectum.

## 6.12 (1) VOLUMES OF REVOLUTION

**Definition.** *A plane area when revolved about a line lying in its own plane, generates a solid of revolution.* The surface generated by the boundary of the plane area is called the **surface of revolution.** The fixed line is called the **axis of revolution.**

*For example* (*i*) A rectangle revolved about one of its sides generates a *right circular cylinder* (Fig. 6.26).

(*ii*) A right angled triangle revolved about one of its sides (along the right angle) generates a *right circular cone* (Fig. 6.27).

(*iii*) A semi-circle revolved about its diameter generates a *sphere* (Fig. 6.28).

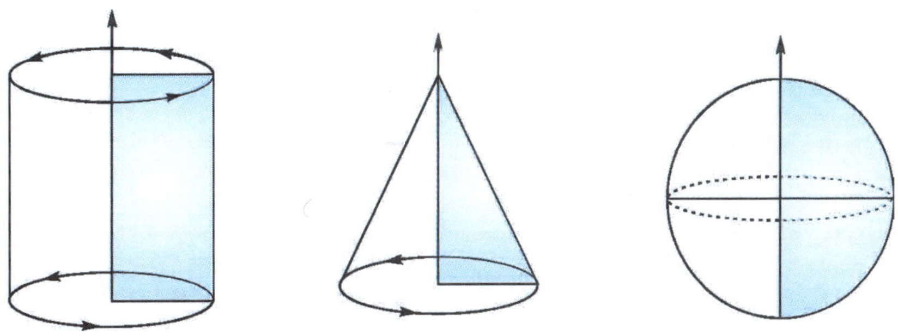

Fig. 6.26 Fig. 6.27 Fig. 6.28

(*iv*) A parabola revolved about $x$-axis generates a *paraboloid of revolution* (Fig. 6.29).

(*v*) A circle (errec)t revolved about an axis $ZZ'$ (not touching it) generates a *torus e.g.*, life-belt or inflated tube (Fig. 6.30).

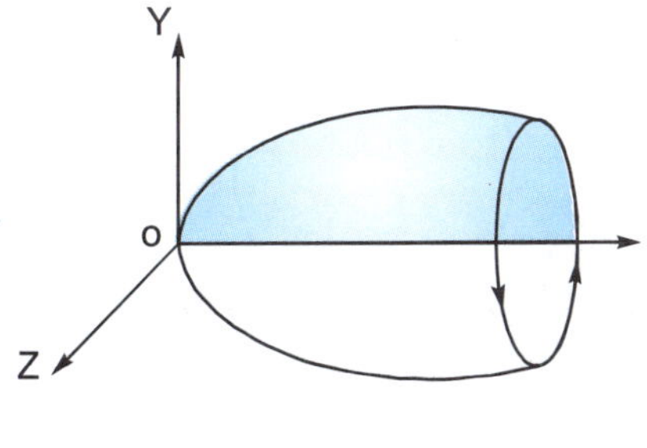


Fig. 6.29.

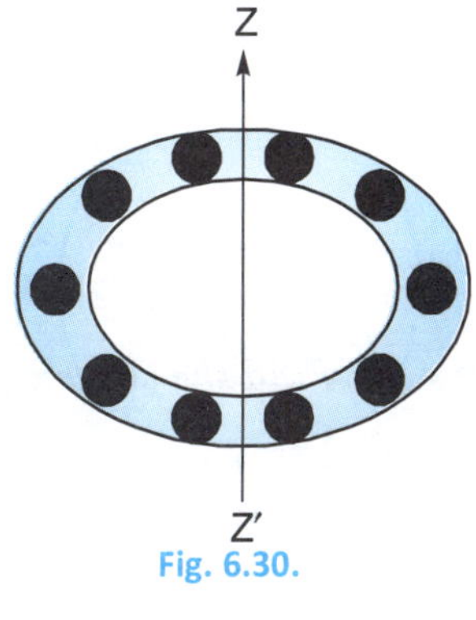


Fig. 6.30.

(a) **Revolution about x-axis.** *The volume of the solid generated by the revolution about the x-axis, of the area bounded by the curve $y = f(x)$, the x-axis and the ordinates $x = a$, $x = b$ is*

$$\int_a^b \pi \mathbf{y}^2\, \mathbf{dx}.$$

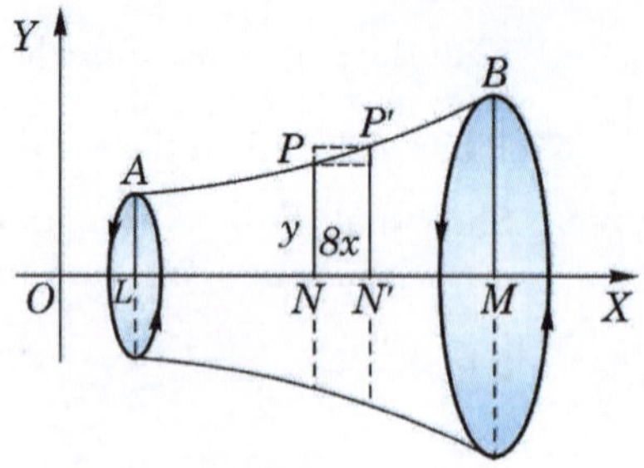

Fig. 6.31.

Let $AB$ be the curve $y = f(x)$ between the ordinates $LA(x = a)$ and $MB(x = b)$.

Let $P(x, y)$, $P'(x + \delta x, y + \delta y)$ be two neighbouring points on the curve and $NP$, $N'P'$ be their respective ordinates (Fig. 6.31).

Let the volume of the solid generated by the revolution about $x$-axis of the area $ALNP$ be $V$, which is clearly a function of $x$. Then the volume of the solid generated by the revolution of the area $PNN'P'$ is $\delta V$. Complete the rectangles $PN'$ and $P'N$.

The $\delta V$ lies between the volumes of the right circular cylinders generated by the revolution of rectangles $PN'$ and $P'N$,

*i.e.,* $\delta V$ lies between $\pi y^2\, \delta x$ and $\pi(y + \delta y)^2\, \delta x$.

$\therefore$ $\dfrac{\delta V}{\delta x}$ lies between $\pi y^2$ and $\pi(y + \delta y)^2$.

Now taking limits as $P' \to P$, *i.e.*, $\delta x \to 0$ (and $\therefore$ $\delta y \to 0$), $\dfrac{dV}{dx} = \pi y^2$

$$\therefore \quad \int_a^b \frac{dV}{dx}\, dx = \int_a^b \pi y^2\, dx \quad \text{or} \quad |\, V\,|_{x=a}^{b} = \int_a^b \pi y^2\, dx$$

or (value of $V$ for $x = b$) – (value of $V$ for $x = a$)

*i.e.*, volume of the solid obtained by the revolution of the area $ALMB = \displaystyle\int_a^b \pi y^2\, dx$.

**Example 6.41.** *Find the volume of the solid obtained by revolving about the x-axis the area enclosed by one arch of the cycloid*

$$x = a(\theta + \sin \theta)$$
$$y = a(1 + \cos \theta) \text{ and the x-axis}$$ (VTU, 2019)

**Solution:** The shape of the curve

For half of the cycloid θ varies from 0 to π

Therefore the required volume

$$= 2\int_0^\pi \pi y^2 \frac{dx}{d\theta} d\theta = 2\pi \int_0^\pi a^2 (1 + \cos\theta)^2 \cdot a(1 + \cos\theta) d\theta$$

$$= 2\pi a^3 \int_0^\pi (1 + \cos\theta)^3 d\theta = 2\pi a^3 \int_0^\pi \left(2\cos^2 \frac{\theta}{2}\right)^3 d\theta$$

$$= 2\pi a^3 \int_0^\pi 8 \cos^6 \frac{\theta}{2} d\theta \qquad \text{putting } \theta = 2t,\ d\theta = 2dt$$

$$= 32\pi a^3 \int_0^{\pi/2} \cos^6 t\, dt \qquad \text{when } \theta = 0,\ t = 0$$

$$= 32\pi a^3 \cdot \frac{5 \cdot 3 \cdot 1}{6 \cdot 4 \cdot 2} \times \frac{\pi}{2} \qquad \text{when } \theta = \pi,\ t = \frac{\pi}{2}$$

$$= 5\pi a^3$$

**Example 6.42.** *Find the volume formed by the revolution of loop of the curve $y^2(a + x) = x^2(3a - x)$, about the x-axis.* (P.T.U., 2011 S)

**Solution.** The curve is symmetrical about the $x$-axis, and for the upper half of its loop $x$ varies from 0 to $3a$ (Fig. 6.32)

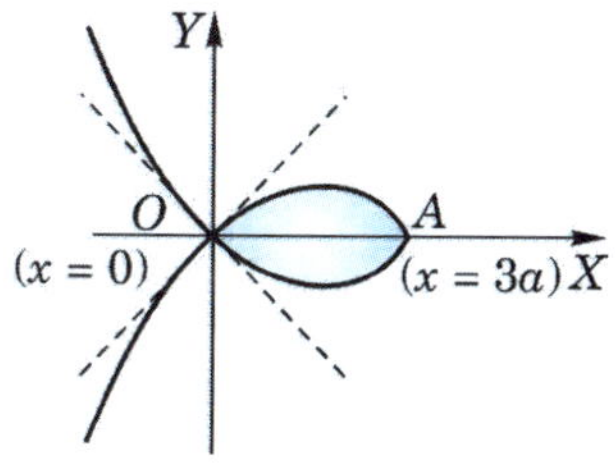

Fig. 6.32.

$$\therefore \quad \text{Volume of the loop} = \int_0^{3a} \pi y^2\, dx = \pi \int_0^{3a} \frac{x^2(3a-x)}{a+x}\, dx$$

$$= \pi \int_0^{3a} \frac{-x^3 + 3ax^2}{x+a}\, dx$$

[Divide the numerator by the denominator]

$$= \pi \int_0^{3a} \left[ -x^2 + 4ax - 4a^2 + \frac{4a^3}{x+a} \right] dx = \pi \left| -\frac{x^3}{3} + 4a \cdot \frac{x^2}{2} - 4a^2 x + 4a^3 \log(x+a) \right|_0^{3a}$$

$$= \pi \left[ \frac{-27a^3}{3} + 2a \cdot 9a^2 - 4a^2 \cdot 3a + 4a^3 \log 4a - (4a^3 \log a) \right]$$

$$= \pi a^3(-3 + 4 \log 4) = \pi a^3(8 \log 2 - 3).$$

**Example 6.43.** *Prove that the volume of the reel formed by the revolution of the cycloid $x = a(\theta + \sin\theta)$, $y = a(1 - \cos\theta)$ about the tangent at the vertex is $\pi^2 a^3$.*

**Solution.** The arch $AOB$ of the cycloid is symmetrical about the $y$-axis and the tangent at the vertex is the $x$-axis. For half the cycloid $OA$, $\theta$ varies from 0 to $\pi$. (Fig. 4.31).

Hence the required volume

$$= 2\int_{\theta=0}^{\theta=\pi} \pi y^2\, dx = 2\pi \int_0^{\pi} a^2(1-\cos\theta)^2 \,.\, a(1+\cos\theta)\, d\theta$$

$$= 2\pi a^3 \int_0^{\pi} (2\sin^2\theta/2)^2 \cdot (2\cos^2\theta/2)\, d\theta$$

$$= 16\pi a^3 \int_0^{\pi} \sin^4\theta/2 \cdot \cos^2\theta/2 \cdot d\theta \qquad [\text{Put } \theta/2 = \phi,\ d\theta = 2d\phi]$$

$$= 32\pi a^3 \int_0^{\pi/2} \sin^4\phi \cos^2\phi\, d\phi$$

$$= 32\pi a^3 \cdot \frac{3 \,.\, 1 \times 1}{6 \,.\, 4 \,.\, 2} \cdot \frac{\pi}{2} = \pi^2 a^3.$$

**Example 6.44.** *Find the volume of the solid formed by revolving about x-axis, the area enclosed by the parabola $y^2 = 4ax$, its evolute $27ay^2 = 4(x-2a)^3$ and the x-axis.*

**Solution.** The curve $27ay^2 = 4(x-2a)^3$ ...(*i*)
is symmetrical about $x$-axis and is a semi-cubical parabola with vertex at $A(2a, 0)$. The parabola $y^2 = 4ax$ and (*i*) intersect at $B$ and $C$ where $27a(4ax) = 4(x-2a)^3$ or $x^3 - 6ax^2 - 15a^2x - 8a^3 = 0$ which gives $x = -a, -a, 8a$. Since $x$ is not negative, therefore we have $x = 8a$ (Fig. 6.33).

$\therefore$ Required volume = Volume obtained by revolving the shaded area $OAB$ about $x$-axis = Vol. obtained by revolving area $OMBO$ – Vol. obtained by revolving area $ADBA$

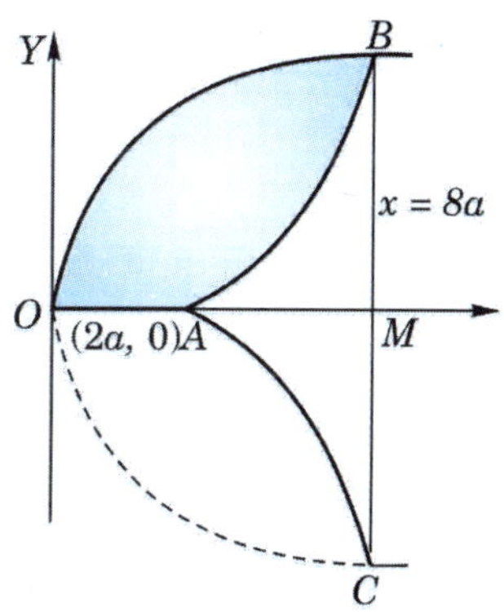

Fig. 6.33.

$$= \int_0^{8a} \pi y^2 (= 4ax)\, dx - \int_{2a}^{8a} \pi y^2 \,[\text{for } (i)]\, dx$$

$$= 4a\pi \left| \frac{x^2}{2} \right|_0^{8a} - \frac{4\pi}{27a} \int_{2a}^{8a} (x-2a)^3\, dx$$

$$= 128\pi a^3 - \frac{4\pi}{27a}\left|\frac{(x-2a)^4}{4}\right|_{2a}^{8a}$$

$$= 128\pi a^3 - \frac{\pi}{27a}(6a)^4 = 80\pi a^3.$$

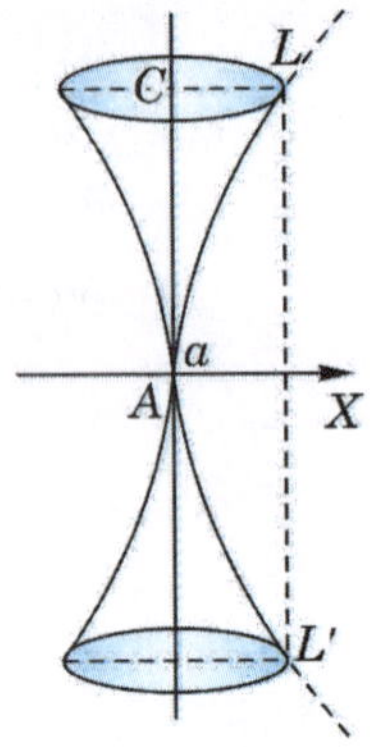

Fig. 6.34.

(*b*) **Revolution about the y-axis.** Interchanging $x$ and $y$ in the above formula, we see that the *volume of the solid generated by the revolution about y-axis, of the area, bounded by the curve $x = f(y)$, the y-axis and the abscissae $y = a, y = b$ is*

$$\int_a^b \pi \mathbf{x}^2 \, \mathbf{dy}.$$

**Example 6.45.** *Find the volume of the reel-shaped solid formed by the revolution about the y-axis, of the part of the parabola $y^2 = 4ax$ cut off by the latus-rectum.* (D.T.U., 2015)

**Solution.** Given parabola is $x = y^2/4a$.

Let $A$ be the vertex and $L$ one extremity of the latus-rectum. For the arc $AL$, $y$ varies from 0 to $2a$ (Fig. 6.34).

$\therefore$ required volume = 2 (volume generated by the revolution about the $y$-axis of the area $ALC$)

$$= 2\int_0^{2a} \pi x^2 \, dy = 2\pi \int_0^{2a} \frac{y^4}{16a^2} \cdot dy = \frac{\pi}{8a^2}\left|\frac{y^5}{5}\right|_0^{2a} = \frac{\pi}{40a^2}(32a^5 - 0) = \frac{4\pi a^3}{5}.$$

(*c*) **Revolution about any axis.** *The volume of the solid generated by the revolution about any axis LM of the area bounded by the curve AB, the axis LM and the perpendiculars AL, BM on the axis, is*

$$\int_{OL}^{OM} \pi(\mathbf{PN})^2 \, \mathbf{d(ON)}$$

*where O is a fixed point in LM and PN is perpendicular from any point P of the curve AB on LM.*

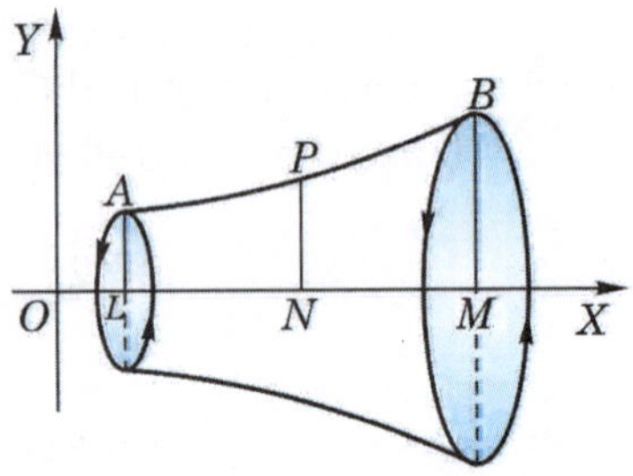

Fig. 6.35.

With $O$ as origin, take $OLM$ as the $x$-axis and $OY$, perpendicular to it as the $y$-axis (Fig. 6.35).

Let the coordinates of $P$ be $(x, y)$ so that $x = ON$, $y = NP$

If $OL = a$, $OM = b$, then required volume $= \int_a^b \pi y^2 \, dx = \int_{OL}^{OM} \pi(PN)^2 \, d(ON)$.

**Example 6.46 :** *Find the volume of the solid formed by the revolution of cissoid $y^2(a^2 - x) = a^2x$ about its asymptote.* (Kerala, 2019)

**Solution :** The asymptote of the curve is the line $x = a$

Let $P(x, y)$ be any point on the curve and $PL$ the perpendicular on the asymptote

$$\text{Required volume} = 2\int_0^{\infty} \pi(PL)^2 d(mL)$$

$$= 2\pi\int_0^{\infty} (a-x)^2 dy \qquad [\because PL = a - x, ML = y]$$

$$= 2\pi\int_0^{\infty}\left[a - \frac{ay^2}{y^2+a^2}\right]^2 dy$$

From the equation of the curve $\quad x = \dfrac{ay^2}{y^2+a^2}$

$$= 2\pi a^2\int_0^{\infty}\left[\frac{y^2+a^2-ay^2}{y^2+a^2}\right]^2 dy$$

$$= 2\pi a^6\int_0^{\infty}\frac{dy}{(y^2+a^2)^2} \quad \text{putting } y = a\tan\theta$$

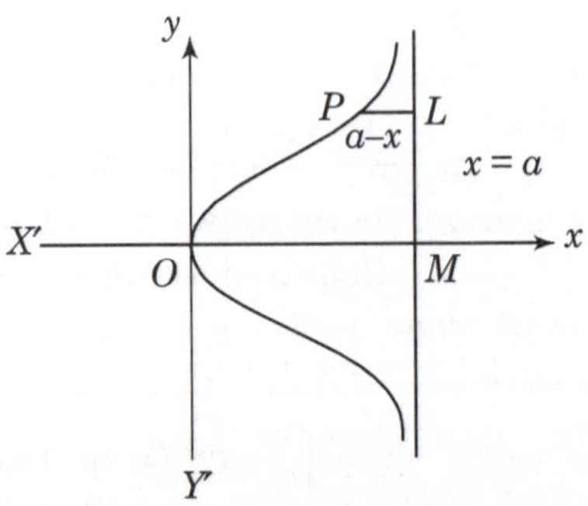

Fig. 6.36.

$$= 2\pi a^6 \int_0^{\pi/2} \frac{a \sec^2\theta \cdot d\theta}{a^4 \sec^4\theta} \qquad dy = a\sec^2\theta\, d\theta$$

$$= 2\pi a^3 \int_0^{\pi/2} \cos^2\theta \cdot d\theta \qquad \text{when } y = 0, \theta = 0$$

$$= 2\pi a^3 \cdot \frac{1}{2} \times \frac{\pi}{2} = \frac{1}{2}\pi^2 a^3 \qquad \text{when } y = \infty, \theta = \frac{\pi}{2}$$

**Example 6.47.** *The area cut off from the parabola $y^2 = 4ax$ by the chord joining the vertex to an end of the latus rectum is revolved through four right angles about the chord. Find the volume of the solid generated.*

*(Marathawada, 2019)*

**Solution:**

The equation of the curve is

$$y^2 = 4ax \qquad \text{...}(i)$$

The line LSL′ is latus rectum and $O$ the vertex of the parabola, the co-ordinate of $L$ are $(a, 2a)$, the equation of $OL$ is

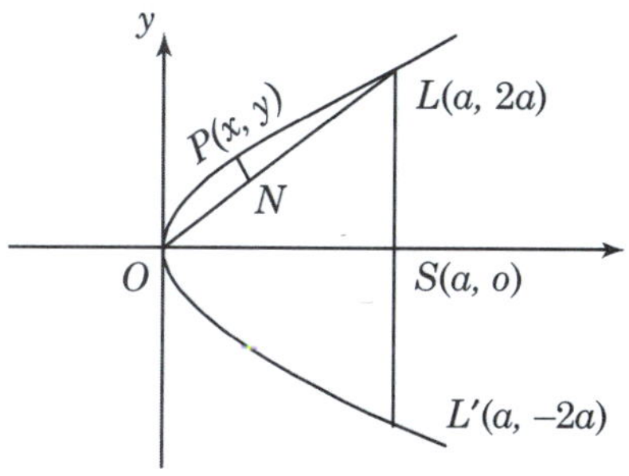


Fig. 6.37.

$$y - 0 = \frac{2a - 0}{a - 0}(x - 0)$$

$$ay = 2ax$$

$$y = 2x$$

$$2x - y = 0 \qquad \text{...}(ii)$$

Now, $PN$ = perpendicular distance of from the line $(ii)$

$$PN = \frac{2x - y}{\sqrt{2^2 + 1^2}} = \frac{2x - y}{\sqrt{5}}$$

$$PN = \frac{2\left(x - \sqrt{a}\sqrt{x}\right)}{\sqrt{5}} \qquad \left(\because\ y = 2\sqrt{a}\sqrt{x}\right)$$

$$= \frac{2\sqrt{x}\left(\sqrt{x} - \sqrt{a}\right)}{\sqrt{5}} \qquad \text{...}(iii)$$

$$ON^2 = OP^2 - PN^2 = x^2 + y^2 - \frac{(2x - y)^2}{5} = \frac{5x^2 + 5y^2 - 4x^2 - y^2 + 4xy}{5}$$

$$= \frac{x^2 + 4y^2 + 4xy}{5} = \frac{(x + 2y)^2}{5} = \frac{\left(x + 2\cdot 2\sqrt{ax}\right)^2}{5} = \frac{\left(x + 4\sqrt{a}\sqrt{x}\right)^2}{5}$$

$$ON = d\left[\frac{\left(x + 4\sqrt{a}\sqrt{x}\right)}{\sqrt{5}}\right]$$

$$d(ON) = \frac{x + 4\sqrt{a}\sqrt{x}}{\sqrt{5}} = \frac{\left[1 + 4\sqrt{a}\ 1/2\sqrt{x}\right]}{\sqrt{5}}$$

$$= \frac{\sqrt{x} + 2\sqrt{a}}{\sqrt{5}\sqrt{x}} \qquad \text{...}(iv)$$

$$\text{Required volume} = \int_0^a \pi(PN)^2 d(ON) = \pi\int_0^a \frac{4x}{5}\left(\sqrt{x} - \sqrt{a}\right)^2 \frac{1}{\sqrt{5}} \frac{\sqrt{x} + 2\sqrt{a}}{\sqrt{x}}\, dx$$

$$= \frac{4\pi}{5\sqrt{5}} \int_0^a \sqrt{x}\left(\sqrt{x} - \sqrt{a}\right)^2\left(\sqrt{x} + 2\sqrt{a}\right) dx$$

$$= \frac{4\pi}{5\sqrt{5}} \int_0^a 5x\left(x + a - 2\sqrt{a}\sqrt{x}\right)\left(\sqrt{x} + 2\sqrt{a}\right) dx$$

$$= \frac{4\pi}{5\sqrt{5}} \int_0^a (x^2 - 3ax + 2a^{3/2}x^{1/2})dx$$

$$= \frac{4\pi}{5\sqrt{5}}\left[\frac{x^3}{3} - \frac{3ax^2}{2} + 2\frac{2}{3}\cdot a^{3/2}\cdot x^{3/2}\right]_0^a = \frac{4\pi}{5\sqrt{5}}\left[\frac{a^3}{3} - \frac{3a^3}{2} + \frac{4}{3}a^3\right]$$

$$= \frac{4\pi}{5\sqrt{5}}\cdot\frac{a^3}{6} = \frac{2\pi a^3}{15\sqrt{15}}$$

**Example 6.48.** *Find the volume of the solid generated by the revolution of the cardioid $r = a(1 + \cos\theta)$ about the initial line.* *(Andhra., 2014 ; V.T.U., 2011 ; Kerala, 2010)*

**Solution.** The cardioid is symmetrical about the initial line and for its upper half $\theta$ varies from 0 to $\pi$. [Fig. 6.25].

$$\therefore \text{ required volume} = \int_0^\pi \frac{2}{3}\pi r^3 \sin\theta\, d\theta = \frac{2\pi}{3}\int_0^\pi a^3(1+\cos\theta)^3 \sin\theta\, d\theta$$

$$= -\frac{2\pi a^3}{3}\int_0^\pi (1+\cos\theta)^3\cdot(-\sin\theta)\,d\theta = -\frac{2\pi a^3}{3}\left|\frac{(1+\cos\theta)^4}{4}\right|_0^\pi = -\frac{\pi a^3}{6}[0-16] = \frac{8}{3}\pi a^3.$$

**Example 6.49 :** *Find the volume of the solid formed by the revolution of the loop of the curve $y^2(a-x) = x^2(a+x)$ above the x-axis.*

**Solution:** The curve can easily be traced and the shape is shown in the Fig. 6.38.

The equation of the curve is

$$y^2 = \frac{x^2(a+x)}{a-x} \quad \ldots(i)$$

The required volume = volume of revolution of the area under the arc $OAB$

For the upper half of the loop $x$ varies from $-a$ to 0

Therefore the required volume

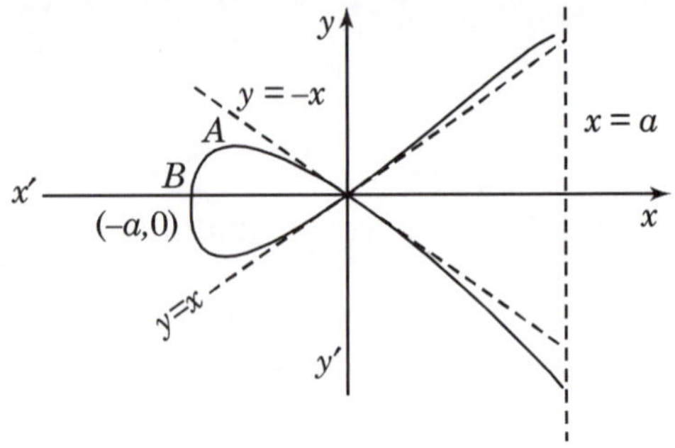

Fig. 6.38

$$= \pi\int_{-a}^0 y^2 dx = \pi\int_{-a}^0 \frac{x^2(a+x)}{a-x}dx = \int_{-a}^0 \frac{ax^2+x^3}{a-x}dx$$

$$= \pi\int_{-a}^0\left[-x^2 - 2ax - 2a^2 + \frac{2a^3}{(a-x)}\right]dx$$

$$= \pi\left[-\frac{x^3}{3} - \frac{2ax^2}{2} - 2a^2x + 2a^3\log(a-x)\right]_{-a}^0$$

$$= \pi\left[2a^3\log a - \left\{-\frac{a^3}{3} + a^3 - 2a^3 + 2a^3\log 2a\right\}\right]$$

$$= -\pi\left[2a^3\log\frac{a}{2a} + \frac{4a^3}{3}\right] = -2\pi a^3\left[\log\frac{1}{2} + \frac{2}{3}\right]$$

$$= 2\pi a^3\left[\log 2 - \frac{2}{3}\right]$$

## PROBLEMS 6.9

1. Find the volume generated by the revolution of the area bounded by $x$-axis, the *catenary* $y = c\cosh x/c$ and the ordinates $x = \pm c$, about the axis of $x$.
2. Find the volume of a spherical segment of height $h$ cut off from a sphere of radius $a$.
3. The area enclosed by the curve $y = \sin x$, $y = \cos x$ and the $x$-axis, between $x = 0$ and $x = \frac{\pi}{2}$, is revolved about the $x$-axis. Show that the volume of solid generated is $\frac{\pi}{4}(\pi - 2)$. *(Bhopal, 2019)*

4. Find the volume generated by revolving the area bounded by the curve $\sqrt{x} + \sqrt{y} = \sqrt{a}$, $x = 0$, $y = 0$ about the $x$-axis.
5. Find the volume of the solid generated by revolving the ellipse $x^2/a^2 + y^2/b^2 = 1$.
   (*i*) about the major axis. (*Bhopal, 2002 S*) (*ii*) about the minor axis.
   (*C.S.V.T.U., 2014 ; Rohtak, 2012*)
6. Obtain the volume of the frustrum of a right circular cone whose lower base has radius $R$, upper base is of radius $r$ and altitude is $h$.
7. Find the volume generated by the revolution of the curve $27ay^2 = 4(x - 2a)^3$ about the $x$-axis.
8. Find the volume of the solid formed by the revolution, about the $x$-axis, of the loop of the curve :
   (*i*) $y^2(a - x) = x^2 (a + x)$ (*ii*) $2ay^2 = x (x - a)^2$ (*iii*) $y^2 = x (2x - 1)^2$.
9. Find the volume obtained by revolving one arch of the cycloid
   (*i*) $x = a (t - \sin t), y = a (1 - \cos t)$, about its base.
   (*ii*) $x = a (\theta + \sin \theta), y = a (1 + \cos \theta)$, about the $x$-axis. (*Hazaribag., 2015*)
10. Find the volume of the spindle-shaped solid generated by the revolution of the *astroid* $x^{2/3} + y^{2/3} = a^{2/3}$ about the $x$-axis. (*P.T.U., 2010*)
11. The arc of cardioid $r = a (1 + \cos \theta)$ included between $\frac{-\pi}{2} \le \theta \le \frac{\pi}{2}$ is rotated about the line $\theta = \frac{\pi}{2}$. Find the area of the surface generated. (*VTU, 2018*)
12. Prove that the volume of the solid formed by the revolution of the curve $(a^2 + x^2) = a^3$, about its asymptote is $\frac{1}{2}\pi^2$ $a^3$.
13. Find the volume generated by the revolution about the initial line of
   (*i*) $r = 2a \cos \theta$ (*ii*) $r = a (1 - \cos \theta)$. (*P.T.U., 2011*)
14. A circular arc revolves about its chord. Prove that the area of the surface generated is $4\pi a^2 (\sin \alpha - a \cos \alpha)$ where $a$ is the radius and $2\alpha$ the angle subtended by the arc at the centre. (*Bhopal, 2019*)
15. Find the volume of the solid formed by revolving a loop of the lemniscate $r^2 = a^2 \cos 2\theta$ about the initial line.
16. Prove that the volume of the solid generated by the revolution about the $x$-axis of the loop of the curve.

$$x = t^2, \quad y = t - \frac{1}{3}t^3 \text{ is } \frac{3\pi}{4}.$$

17. The curve $r = a(1 + \cos \theta)$ revolves about the initial line. Find the surface area of the figure so formed. (*VTU, 2019*)

## 6.13 SURFACE AREAS OF REVOLUTION

(*a*) **Revolution about $x$-axis.** *The surface area of the solid generated by the revolution about x-axis, of the arc of the curve $y = f(x)$ from $x = a$ to $x = b$, is*

$$\int_{x=a}^{x=b} 2\pi y \, ds.$$

Let $AB$ be the curve $y = f(x)$ between the ordinates $LA$ $(x = a)$ and $MB$ $(x = b)$. Let $P(x, y)$, $P'$ $(x + \delta xy + \delta y)$ be two neighbouring points on the curve and $NP$, $N'P'$ be their respective ordinates (Fig. 6.19).

Let the arc $AP = s$ so that arc $PP' = \delta s$. Let the surface-area generated by the revolution about $x$-axis of the arc $AP$ be $S$ and that generated by the revolution of the arc $PP'$ be $\delta S$.

Since $\delta s$ is small, the surface area $\delta S$ may be regarded as lying between the curved surfaces of the right cylinders of radii $PN$ and $P'N'$ and of same thickness $\delta s$.

Thus $\delta S$ lies between $2\pi y \, \delta s$ and $2\pi (y + \delta y) \, \delta s$

$\therefore$ $\frac{\delta S}{\delta s}$ lies between $2\pi y$ and $2\pi (y + \delta y)$

Taking limits as $P' \to P$, *i.e.*, $\delta s \to 0$ and $\delta y \to 0$, $dS/dx = 2\pi y$

$$\therefore \quad \int_{x=a}^{x=b} \frac{dS}{ds} ds = \int_{x=a}^{x=b} 2\pi y \, ds \quad \text{or} \quad | S |_{x=a}^{x=b} = \int_{x=a}^{x=b} 2\pi y \, ds$$

or (value of $S$ for $x = b$) – (value of $S$ for $x = a$) = $\int_{x=a}^{x=b} 2\pi y \, dx$

or surface area generated by the revolution of the arc $AB - 0 = \int_{x=a}^{x=b} 2\pi y \, ds.$

Hence, the required surface area $= \int_{x=a}^{x=b} 2\pi y \, ds.$

**Obs. Practical forms of the formula S = $\int$ 2πy ds.**

*(i) Cartesian form* [*for the curve* $y = f(x)$]

$$S = \int 2\pi y \frac{ds}{dx} dx, \text{ where } \frac{ds}{dx} = \sqrt{\left[1 + \left(\frac{dy}{dx}\right)^2\right]}$$

*(ii) Parametric form* [*for the curve* $x = f(t), y = \phi(t)$]

$$S = \int 2\pi y \frac{ds}{dt} dt, \text{ where } \frac{ds}{dt} = \sqrt{\left[\left(\frac{dx}{dt}\right)^2 + \left(\frac{dy}{dt}\right)^2\right]}$$

*(iii) Polar form* [*for the curve* $r = f(\theta)$]

$$S = \int 2\pi y \frac{ds}{d\theta} d\theta, \text{ where } y = r \sin\theta, \ \frac{ds}{d\theta} = \sqrt{\left[r^2 + \left(\frac{dr}{d\theta}\right)^2\right]}.$$

**Example 6.50.** *Find the surface of the solid formed by revolving the cardioid $r = a(1 + \cos\theta)$ about the initial line.*

**Solution.** The cardioid is symmetrical about the initial line and for its upper half, $\theta$ varies from 0 to $\pi$ (Fig. 6.18).

Also
$$\frac{ds}{d\theta} = \sqrt{\left[r^2 + \left(\frac{dr}{d\theta}\right)^2\right]} = \sqrt{[a^2(1+\cos\theta)^2 + a^2 \sin^2\theta]}$$

$$= a\sqrt{[2(1+\cos\theta)]} = a\sqrt{[2.2\cos^2\theta/2]} = 2a\cos\theta/2$$

$$\therefore \quad \text{required surface} = \int_0^{\pi} 2\pi y \frac{ds}{d\theta} d\theta = 2\pi \int_0^{\pi} r\sin\theta \cdot 2a\cos\theta/2 \, d\theta$$

$$= 4\pi a \int_0^{\pi} a(1+\cos\theta)\sin\theta \cdot \cos\theta/2 \, d\theta = 4\pi a^2 \int_0^{\pi} 2\cos^2\frac{\theta}{2} . 2\sin\frac{\theta}{2}\cos\frac{\theta}{2} . \cos\frac{\theta}{2} d\theta$$

$$= 16\pi a^2 \int_0^{\pi} \cos^4\frac{\theta}{2}\sin\frac{\theta}{2} d\theta = 16\pi a^2(-2)\int_0^{\pi} \cos^4\frac{\theta}{2}\left(-\sin\frac{\theta}{2} . \frac{1}{2}\right) d\theta$$

$$= -32\pi a^2 \left|\frac{\cos^5\theta/2}{5}\right|_0^{\pi} = \frac{-32\pi a^2}{5}(0-1) = \frac{32\pi a^2}{5}.$$

(*b*) **Revolution about y-axis.** Interchanging $x$ and $y$ in the above formula, we see that *the surface of the solid generated by the revolution about y-axis, of the arc of the curve $x = f(y)$ from $y = a$ to $y = b$ is*

$$\int_{y=a}^{y=b} 2\pi x \, ds.$$

**Example 6.51.** *Find the surface area of the solid generated by the revolution of the astroid $x = a\cos^3 t$, $y = a\sin^3 t$, about the y-axis.*

**Solution.** The astroid is symmetrical about the $x$-axis, and for its portion in the first quadrant $t$ varies from 0 to $\pi/2$. (Fig. 4.29).

Also
$$\frac{dx}{dt} = -3a\cos^2 t\sin t, \quad \frac{dy}{dt} = 3a\sin^2 t\cos t.$$

$$\frac{ds}{dt} = \sqrt{\left[\left(\frac{dx}{dt}\right)^2 + \left(\frac{dy}{dt}\right)^2\right]} = \sqrt{[9a^2\cos^4 t\sin^2 t + 9a^2\sin^4 t\cos^2 t]}$$

$$= 3a\sin t\cos t\sqrt{(\cos^2 t + \sin^2 t)} = 3a\sin t\cos t$$

$$\therefore \quad \text{required surface} = 2\int_0^{\pi/2} 2\pi x \frac{ds}{dt} \cdot dt = 4\pi \int_0^{\pi/2} a\cos^3 t \cdot 3a\sin t\cos t \, dt$$

$$= 12\pi a^2 \int_0^{\pi/2} \sin t \cos^4 t \, dt = 12\pi a^2 \frac{3.1}{5.3.1} = \frac{12\pi a^2}{5}.$$

**Example 6.52 :** *Find the area included between the cycloid* *(VTU, 2018)*

$$x = a(t + \sin t), \quad y = a(1 - \cos t)$$

*Also find the area between the curve and x-axis.*

**Solution:** The shape of the curve is shown in the figure.

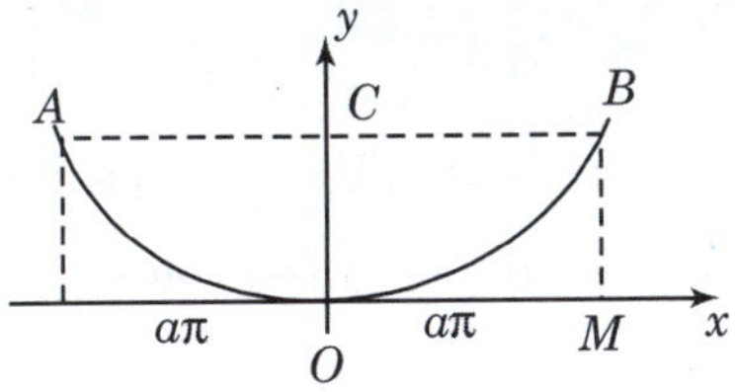

Fig. 6.39.

$$\text{Area } AOB = 2 \times \text{Area } OBC$$

$$= \int_0^{\pi} x \cdot \frac{dy}{dt} dt$$

$$= \int_0^{\pi} a(t + \sin t) \cdot a \sin t \cdot dt$$

$$= a^2 \int_0^{\pi} (t \sin t + \sin^2 t) \, dt$$

$$= a^2 \left[ |t(-\cos t)|_0^{\pi} + \int_0^{\pi} \cos t \cdot dt + \int_0^{\pi} \frac{1 - \cos 2t}{2} dt \right]$$

$$= a^2 \left[ (\pi - 0) + \{\sin t\}_0^{\pi} + \frac{1}{2} \left\{ t - \frac{\sin 2t}{2} \right\}_0^{\pi} \right]$$

$$= a^2 \left[ \pi + 0 + \frac{1}{2}\pi \right] = \frac{3}{2}\pi a^2$$

$$\therefore \quad \text{Area } AOB = 2 \times \frac{3}{2}\pi a^2 = 3\pi a^2$$

(*ii*) The area between the *x*-axis and cycloid

$$= 2 \times \text{Area } OBM$$

$$\text{Area } OBM = \text{Area } OCBM - \text{Area } OBC$$

$$= a\pi \cdot 2a - \frac{3}{2}\pi a^2$$ [$\because$ *OCBM* is a rectangular with sides $a\pi$ and $2a$]

$$= \frac{\pi a^2}{2}$$

Therefore the required area $= \dfrac{2 \times \pi a^2}{2} = \pi a^2$

**Example 6.53 :** *The circle $x^2 + y^2 = a^2$ is revolved about x-axis. Find the volume of sphere so formed.*

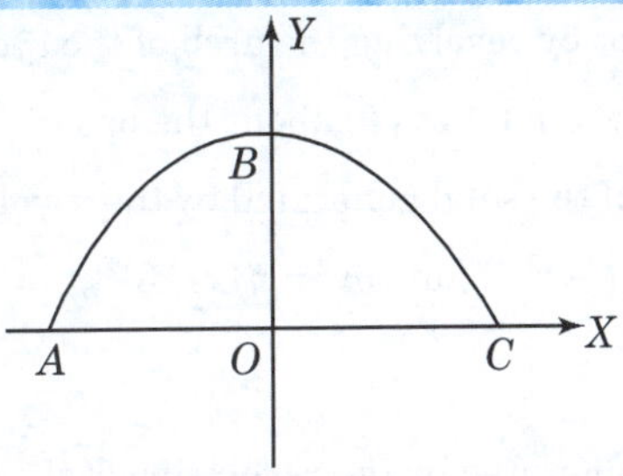

Fig. 6.40.

**Solution:** A semi-circular area *ABC* when revolved about its bounding diameter *AOC* will generate a sphere. equation of the circle, is $x^2 + y^2 = a^2$

$$y^2 = a^2 - x^2 \qquad ...(i)$$

So the required volume $= \int_{-a}^{a} \pi y^2 dx$

$$= \pi\int_{-a}^{a} (a^2 - x^2)\, dx \quad \text{from } (i) = 2\pi\int_0^a (a^2 - x^2)\, dx = 2\pi\left[a^2 x - \frac{x^3}{3}\right]_0^a$$

$$= 2\pi\left[a^3 - \frac{a^3}{3}\right] = \frac{4\pi a^3}{3}$$

**Example 6.54 :** *Find the volume of the spindle shaped solid generated by revolution of the astroid*

$$x = a\cos^3 t$$
$$y = a\sin^3 t \text{ about the y-axis}$$

**Solution:** Hear the part of the curve in the first quadrant $t$ varies from 0 to $\frac{\pi}{2}$

So, the required volume

$$= 2\int_0^{\pi/2} \pi x^3 \frac{dy}{dt}\, dt = 2\pi\int_0^{\pi/2} a^2 \cos^6 t \cdot 3a\sin^2 t \cdot \cos t \cdot dt$$

$$= 6\pi a^3 \int_0^{\pi/2} \sin^2 t \cdot \cos^7 t \cdot dt = 6\pi a^3 \frac{6\cdot 4\cdot 2}{9\cdot 7\cdot 5\cdot 3\cdot 1} = \frac{32}{105}\pi a^3$$

## PROBLEMS 6.10

1. Find the area of the surface generated by revolving the arc of the catenary $y = c\cosh x/c$ from $x = 0$ to $x = c$ about the $x$-axis.
2. Find the area of the surface formed by the revolution of $y^2 = 4ax$ about its axis, by the arc from the vertex to one end of the latus-rectum.
3. Find the surface of the solid generated by the revolution of the *ellipse* $x^2/a^2 + y^2/b^2 = 1$ about the $x$-axis.
4. Find the volume and surface of the *right circular cone* formed by the revolution of a right-angled triangle about a side which contains the right angle.
5. Obtain the surface area of a *sphere* of radius $a$.
6. Show that the surface area of the solid generated by the revolution of the curve $x = a\cos^3 t$, $y = a\sin^3 t$ about the $x$-axis, is $12\pi^2/5$. *(V.T.U., 2014)*
7. The arc of the curve $x^{2/3} + y^{2/3} = a^{2/3}$ in the first quadrant revolves about $x$-axis. Show that the area of the surface generated is $6\pi a^2/5$.
8. Find the surface area of the solid generated by revolving the *cycloid* $x = a(t - \sin t)$, $y = a(1 - \cos t)$ about the base. *(V.T.U., 2016 ; Marathwada, 2008 ; Cochin, 2005)*
9. Find the surface area of the solid got by revolving the arch of the *cycloid*
$x = a(\theta + \sin\theta)$, $y = a(1 + \cos\theta)$ about the base. *(V.T.U., 2010 S)*
10. Prove that the surface and volume of the solid generated by the revolution about the $x$-axis, of the loop of the curve
$x = t^2$, $y = t - t^3/3$, [or $9y^2 = x(x - 3)^2$],
are respectively $3\pi$ and $3\pi/4$.
11. Prove that the surface of the solid generated by the revolution of the *tractrix* $x = a\cos t + \frac{a}{2}\log\tan^2 t/2$, $y = a\sin t$, about $x$-axis is $4\pi a^2$.
12. Find the surface area of the solid of revolution of the curve $r = 2a\cos\theta$ about the initial line.
13. The part of *parabola* $y^2 = 4ax$ cut off by the latus-rectum revolves about the tangent at the vertex. Find the curved surface of the reel thus formed.

## 6.14 OBJECTIVE TYPE OF QUESTIONS

### PROBLEMS 6.11

Choose the correct answer or fill up the blanks in the following problems :

1. If $f(x) = f(2a - x)$, then $\int_0^{2a} f(x)\, dx$ is equal to

(a) $\int_a^0 f(2a-x)\, dx$ (b) $2\int_0^a f(x)\, dx$ (c) $-2\int_0^a f(x)\, dx$ (d) 0.

2. $\int_0^{\pi/2} \frac{\sqrt{\sin x}}{\sqrt{\sin x}+\sqrt{\cos x}}\, dx$ is equal to

(a) 0 (b) 1 (c) $\frac{\pi}{4}$ (d) $\frac{\pi}{2}$.

3. The value of definite integral $\int_{-a}^{a} |x|\, dx$ is equal to

(a) $a$ (b) $a^2$ (c) 0 (d) $2a$.

4. $\lim_{n\to\infty}\left[\frac{n}{n^2}+\frac{n}{n^2+1^2}+\frac{n}{n^2+2^2}+\ldots+\frac{n}{n^2+(n-1)^2}\right]$ is equal to

(a) $-\frac{\pi}{4}$ (b) 0 (c) $\frac{\pi}{4}$ (d) $\frac{\pi}{3}$.

5. $\int_0^{\pi/2} \frac{\cos 2x}{\cos x + \sin x}\, dx$ equals

(a) – 1 (b) 0 (c) 1 (d) 2.

6. $\lim_{n\to\infty}\left(\frac{1}{n}+\frac{1}{n+1}+\frac{1}{n+2}+\ldots+\frac{1}{3n}\right)$ equals

(a) $\log_e 2$ (b) $2\log_e 2$ (c) $\log_e 3$ (d) $2\log_e 3$.

7. $\int_0^{\pi} \sin^5\left(\frac{x}{2}\right)$ is equal to

(a) $\frac{16}{15}$ (b) $\frac{15}{16}\pi$ (c) $\frac{16}{15}\pi^2$ (d) $\frac{15}{16}$.

8. $\int_0^{\pi/2} \sin^{99} x \cos x\, dx$ is equal to

(a) $\frac{1}{99}$ (b) $\frac{\pi}{100}$ (c) $\frac{99}{100}$ (d) None of these.

9. The value of $\int_{-\pi/2}^{\pi/2} \cos^7 x\, dx$ is

(a) $\frac{32\pi}{35}$ (b) $\frac{32}{35}$ (c) zero.

10. The length of the arc of the equiangular spiral $r = ae^{\theta \cot \alpha}$ between the points for which the radii vectors are $r_1$ and $r_2$ is

(a) $(r_2 - r_1)$ cosec $\alpha$ (b) $(r_2 - r_1)\cos\alpha$ (c) $(r_2 - r_1)\sin\alpha$ (d) $(r_2 - r_1)\sec\alpha$.

11. The area of the region in the first quadrant bounded by the $y$-axis and the curves $y = \sin x$ and $y = \cos x$ is

(a) $\sqrt{2}$ (b) $\sqrt{2}+1$ (c) $\sqrt{2}-1$ (d) $2\sqrt{2}-1$.

12. The value of $\int_0^1 x^{3/2}(1-x)^{3/2}\, dx$ is

(a) $\pi/32$ (b) $-\pi/32$ (c) $3\pi/128$ (d) $-3\pi/128$. (*V.T.U., 2010*)

13. The entire length of the cardioid $r = 5(1+\cos\theta)$ is

(a) 40 (b) 30 (c) 20 (d) 5.

14. The area of the cardioid $r = a(1-\cos\theta)$ is .......

15. If $S_1$ and $S_2$ are surface areas of the solids generated by revolving the ellipses $\frac{x^2}{a^2}+\frac{y^2}{b^2}=1$ and $\frac{x^2}{b^2}+\frac{y^2}{a^2}=1$ about the $y$-axis, then
(a) $S_1 > S_2$ (b) $S_1 < S_2$ (c) $S_1 = S_2$ (d) can't predict.
16. The area of the loop of the curve $r = a \sin 3\theta$ is .......
17. If $I_n = \int_0^{\pi/4} \tan^n \theta \, d\theta$, then $n(I_{n-1} + I_{n+1}) =$ .......
18. $\int_0^2 x^3 \sqrt{(2x - x^2)}\, dx =$ .......
19. $\int_0^{\pi/2} \sin 2\theta \log \tan \theta \, d\theta$ is equal to
(a) 1 (b) –1 (c) 0 (d) $\pi/2$.
20. The area of the loop of the folium of Descartes $x^3 + y^3 - 3xy = 0$ is
(a) $\pi$ (b) $\pi/2$ (c) 1.5 (d) 3.
21. The volume of the frustrum of a right circular cone whose lower base has radius $r_1$ and upper base has radius $r_2$ and altitude is $h =$ .......
22. The length of the arc of the curve $y = \log \sec x$ from $x = 0$ to $x = \pi/4$ is
(a) $\log_e 2$ (b) $\log_e 3$ (c) $\log_e (1+\sqrt{2})$ (d) $\log_e (1+\sqrt{3})$.
23. If $v_1$ = volume of the solid generated by revolving the area included between $x$-axis and $x^2 + y^2 = a^2$ about $x$-axis ; $v_2$ = volume of the solid generated by revolving the entire area of the circle $x^2 + y^2 = a^2$ about $x$-axis, then
(a) $v_1 = v_2$ (b) $v_2 = 2v_1$ (c) $v_2 = 4v_1$ (d) $v_2 = 16v_1$.
24. If $f(r, \theta) = f(-r, \theta)$, then the curve is symmetrical about the .......
(a) initial line (b) pole (c) origin (d) tangential line. (*V.T.U., 2010*)
25. The volume generated by the revolution of the curve $y = a^3 (a^2 + x^2)^{-1}$ about its asymptote is
(a) $\pi^2 a^3/2$ (b) $\pi a^3/2$ (c) $\pi a^2/2$ (d) $\pi a/2$. (*V.T.U., 2010*)

CHAPTER 7

# Multiple Integrals and Beta, Gamma Functions



## 7.1 DOUBLE INTEGRALS

The definite integral $\int_a^b f(x)\,dx$ is defined as the limit of the sum

$$f(x_1)\,\delta x_1 + f(x_2)\,\delta x_2 + \ldots + f(x_n)\,\delta x_n,$$

where $n \to \infty$ and each of the lengths $\delta x_1, \delta x_2, \ldots$ tends to zero. A double integral is its counterpart in two dimensions.

Consider a function $f(x, y)$ of the independent variables $x, y$ defined at each point in the finite region $R$ of the $xy$-plane. Divide $R$ into $n$ elementary areas $\delta A_1, \delta A_2, \ldots, \delta A_n$. Let $(x_r, y_r)$ be any point within the $r$th elementary area $\delta A_r$. Consider the sum

$$f(x_1, y_1)\,\delta A_1 + f(x_2, y_2)\,\delta A_2 + \ldots + f(x_n, y_n)\,\delta A_n, \text{ i.e., } \sum_{r=1}^{n} f(x_r, y_r)\,\delta A_r$$

The limit of this sum, if it exists, as the number of sub-divisions increases indefinitely and area of each sub-division decreases to zero, is defined as the *double integral of f(x, y) over the region R* and is written as $\iint_R f(x, y)\,dA$.

Thus $$\iint_R f(x, y)\,dA = \underset{\substack{n \to \infty \\ \delta A \to 0}}{\text{Lt}} \sum_{r=1}^{n} f(x_r, y_r)\,\delta A_r \qquad \ldots(1)$$

The utility of double integrals would be limited if it were required to take limit of sums to evaluate them. However, there is another method of evaluating double integrals by successive single integrations.

For purpose of evaluation, Eqn. (1) is expressed as the repeated integral $\int_{x_1}^{x_2} \int_{y_1}^{y_2} f(x, y)\,dxdy$.

Its value is found as follows :

(*i*) *When $y_1, y_2$ are functions of x and $x_1, x_2$ are constants,* $f(x, y)$ is first integrated w.r.t. $y$ keeping $x$ fixed between limits $y_1, y_2$ and then resulting expression is integrated w.r.t. $x$ within the limits $x_1, x_2$ *i.e.,*

$$\boxed{I_1 = \int_{x_1}^{x_2} \boxed{\int_{y_1}^{y_2} f(x, y)\,dy}\,dx}$$

where integration is carried from the inner to the outer rectangle.

Fig. 7.1 illustrates this process. Here $AB$ and $CD$ are the two curves whose equations are $y_1 = f_1(x)$ and $y_2 = f_2(x)$. $PQ$ is a vertical strip of width $dx$.

Then the inner rectangle integral means that the integration is along one edge of the strip $PQ$ from $P$ to $Q$ ($x$ remaining constant), while the outer rectangle integral corresponds to the sliding of the edge from $AC$ to $BD$.

Thus the whole region of integration is the area $ABDC$.

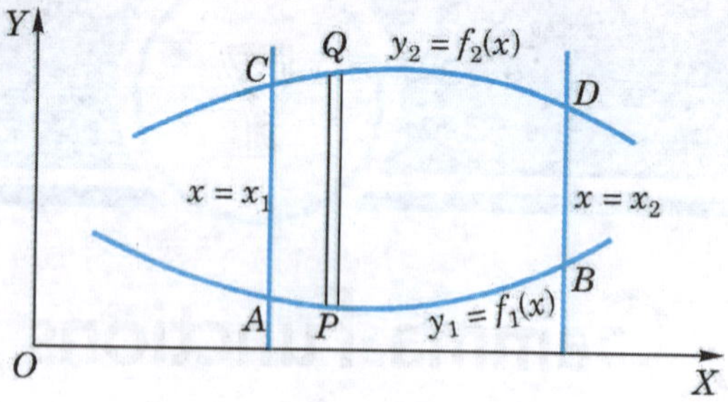


Fig. 7.1

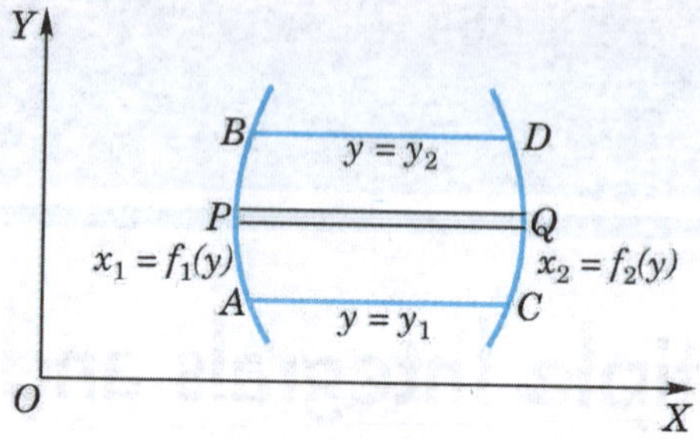


Fig. 7.2

(*ii*) *When $x_1, x_2$ are functions of $y$ and $y_1, y_2$ are constants, $f(x, y)$* is first integrated w.r.t. $x$ keeping $y$ fixed, within the limits $x_1, x_2$ and the resulting expression is integrated w.r.t. $y$ between the limits $y_1, y_2$, *i.e.*,

$$I_2 = \boxed{\int_{y_1}^{y_2} \boxed{\int_{x_1}^{x_2} f(x, y)\, dx}\, dy}$$ which is geometrically illustrated by Fig. 7.2.

Here $AB$ and $CD$ are the curves $x_1 = f_1(y)$ and $x_2 = f_2(y)$. $PQ$ is a horizontal strip of width $dy$.

Then inner rectangle indicates that the integration is along one edge of this strip from $P$ to $Q$ while the outer rectangle corresponds to the sliding of this edge from $AC$ to $BD$.

Thus the whole region of integration is the area $ABDC$.

(*iii*) *When both pairs of limits are constants,* the region of integration is the rectangle $ABDC$ (Fig. 7.3).

In $I_1$, we integrate along the vertical strip $PQ$ and then slide it from $AC$ to $BD$.

In $I_2$, we integrate along the horizontal strip $P'Q'$ and then slide it from $AB$ to $CD$.

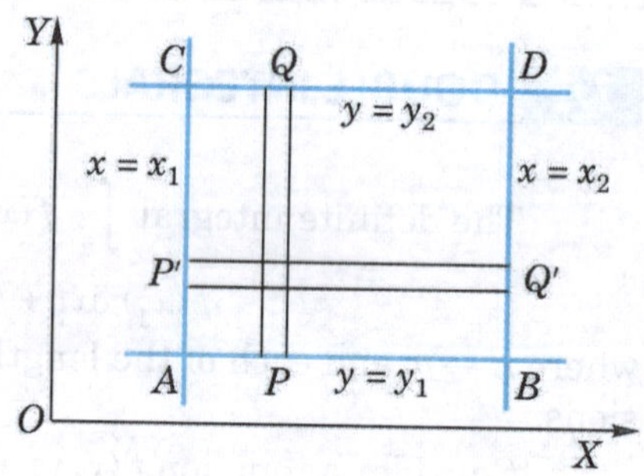


Fig. 7.3

Here obviously $I_1 = I_2$.

Thus *for constant limits, it hardly matters whether we first integrate w.r.t. x and then w.r.t. y or vice versa.*

**Example 7.1.** *Evaluate* $\int_0^5 \int_0^{x^2} x(x^2 + y^2)\, dxdy$.

**Solution.**
$$I = \int_0^5 dx \int_0^{x^2} (x^3 + xy^3)\, dy = \int_0^5 \left[x^3 y + x \cdot \frac{y^3}{3}\right]_0^{x^2} dx = \int_0^5 \left[x^3 \cdot x^2 + x \cdot \frac{y^6}{3}\right] dx$$
$$= \int_0^5 \left(x^5 + \frac{x^7}{3}\right) dx = \left|\frac{x^6}{6} + \frac{x^8}{24}\right|_0^5 = 5^6\left[\frac{1}{6} + \frac{5^2}{24}\right] = 18880.2 \text{ nearly.}$$

**Example 7.2.** *Evaluate* $\iint_A xy\, dx\, dy$, *where A is the domain bounded by x-axis, ordinate x = 2a and the curve $x^2 = 4ay$.* (V.T.U., 2016)

**Solution.** The line $x = 2a$ and the parabola $x^2 = 4ay$ intersect at $L(2a, a)$. Figure 7.4 shows the domain $A$ which is the area $OML$.

Integrating first over a vertical strip $PQ$, *i.e.*, w.r.t. $y$ from $P(y = 0)$ to $Q(y = x^2/4a)$ on the parabola and then w.r.t. $x$ from $x = 0$ to $x = 2a$, we have

$$\iint_A xy\, dx\, dy = \int_0^{2a} dx \int_4^{x^2/4a} xy\, dy = \int_0^{2a} x\left[\frac{y^2}{2}\right]_0^{x^2/4a} dx$$

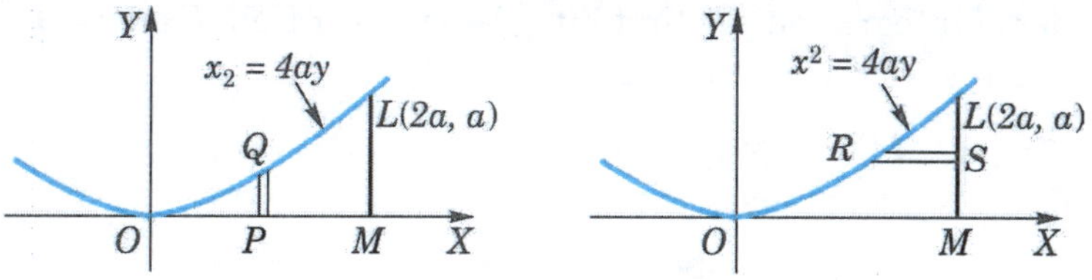


Fig. 7.4

$$= \frac{1}{32a^2}\int_0^{2a} x^5\, dx = \frac{1}{32a^2}\left|\frac{x^6}{6}\right|_0^{2a} = \frac{a^4}{3}.$$

**Otherwise** integrating first over a horizontal strip $RS$, *i.e.*, w.r.t. $x$ from, $R$ ($x = 2\sqrt{ay}$) on the parabola to $S(x = 2a)$ and then w.r.t. $y$ from $y = 0$ to $y = a$, we get

$$\iint_A xy\, dx\, dy = \int_0^a dx \int_{2\sqrt{(ay)}}^{2a} xy\, dx = \int_0^a y\left[\frac{x^2}{2}\right]_{2\sqrt{(ay)}}^{2a} dy$$

$$= 2a\int_0^a (ay - y^2)\, dy = 2a\left[\frac{ay^2}{2} - \frac{y^3}{3}\right]_0^a = \frac{a^4}{3}.$$

**Example 7.3.** *Evaluate* $\iint_R x^2\, dx\, dy$ *where R is the region in the first quadrant bounded by the lines* $x = y$, $y = 0$, $x = 8$ *and the curve* $xy = 16$.

**Solution.** The line $AL$ ($x = 8$) intersects the hyperbola $xy = 16$ at $A$ (8, 2) while the line $y = x$ intersects this hyperbola at $B$ (4, 4). Figure 7.5 shows the region $R$ of integration which is the area $OLAB$. To evaluate the given integral, we divide this area into two parts $OMB$ and $MLAB$.

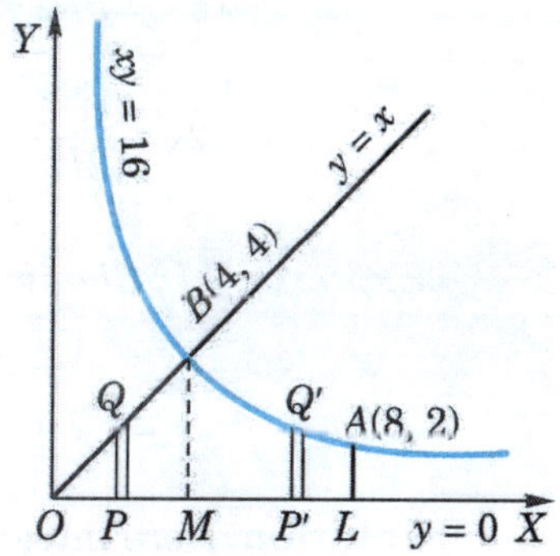


Fig. 7.5

$$\therefore \quad \iint_R x^2\, dxdy = \int_{x\text{ at }0}^{x\text{ at }M}\int_{y\text{ at }P}^{y\text{ at }Q} x^2\, dxdy + \int_{x\text{ at }M}^{x\text{ at }L}\int_{y\text{ at }P'}^{y\text{ at }Q'} x^2\, dxdy$$

$$= \int_0^4\int_0^x x^2\, dxdy + \int_4^8\int_0^{16/x} x^2\, dxdy$$

$$= \int_0^4 x^2\, dx\left|y\right|_0^x + \int_4^8 x^2\, dx\left|y\right|_0^{16/x}$$

$$= \int_0^4 x^3\, dx + \int_4^8 16x\, dx = \left|\frac{x^4}{4}\right|_0^4 + 16\left|\frac{x^2}{2}\right|_4^8 = 448$$

## 7.2 CHANGE OF ORDER OF INTEGRATION

In a double integral with variable limits, the change of order of integration changes the limit of integration. While doing so, sometimes it is required to split up the region of integration and the given integral is expressed as the sum of a number of double integrals with changed limits. To fix up the new limits, it is always advisable to draw a rough sketch of the region of integration.

The change of order of integration quite often facilitates the evaluation of a double integral. The following examples will make these ideas clear.

**Example 7.4.** *Evaluate* $\int_0^1\int_{e^x}^{e} dydx/\log y$ *by changing the order of integration.*

**Solution.** Here the integration is first w.r.t. $y$ from $P$ on $y = e^x$ to $Q$ on the line $y = e$. Then the integration is w.r.t. $x$ from $x = 0$ to $x = 1$, giving the shaded region $ABC$ (Fig. 7.6).

On changing the order of integration, we first integrate w.r.t. $x$ from $R$ on $x = 0$ to $S$ on $x = \log y$ and then w.r.t. $y$ from $y = 1$ to $y = e$.

Thus $$\int_0^1 \int_{e^x}^{e} \frac{dydx}{\log y} = \int_1^e \int_0^{\log y} \frac{dxdy}{\log y}$$

$$= \int_1^e \frac{dy}{\log y} \left| x \right|_0^{\log y}$$

$$= \int_1^e dy = \left| y \right|_1^e = e - 1.$$

$$= e - 1.$$

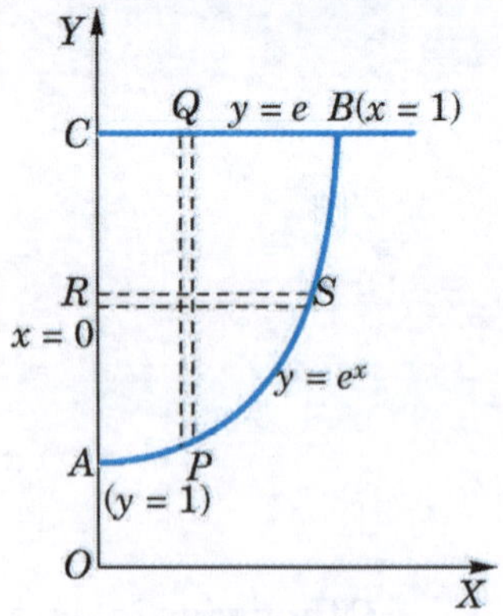

Fig. 7.6

**Example 7.5.** *Change the order of integration in* $I = \int_0^{4a} \int_{x^2/4a}^{2\sqrt{ax}} dydx$ *and hence evaluate.*

*(Andhra., 2021, 2016 ; Delhi., 2016 ; Nagpur, 2009)*

**Solution.** Here integration is first w.r.t. $y$ and $P$ on the parabola $x^2 = 4ay$ to $Q$ on the parabola $y^2 = 4ax$ and then w.r.t. $x$ from $x = 0$ to $x = 4a$ giving the shaded region of integration (Fig. 7.7).

On changing the order of integration, we first integrate w.r.t. $x$ from $R$ to $S$, then w.r.t. $y$ from $y = 0$ to $y = 4a$

$$\therefore \quad I = \int_0^{4a} \int_{y^2/4a}^{2\sqrt{ay}} dxdy = \int_0^{4a} dy \left| x \right|_{y^2/4a}^{2\sqrt{ay}} = \int_0^{4a} (2\sqrt{ay} - y^2/4a)\, dy$$

$$= \left| 2\sqrt{a}\, \frac{y^{3/2}}{3/2} - \frac{y^3}{12a} \right|_0^{4a} = \frac{32a^2}{3} - \frac{16a^2}{3} = \frac{16a^2}{3}.$$

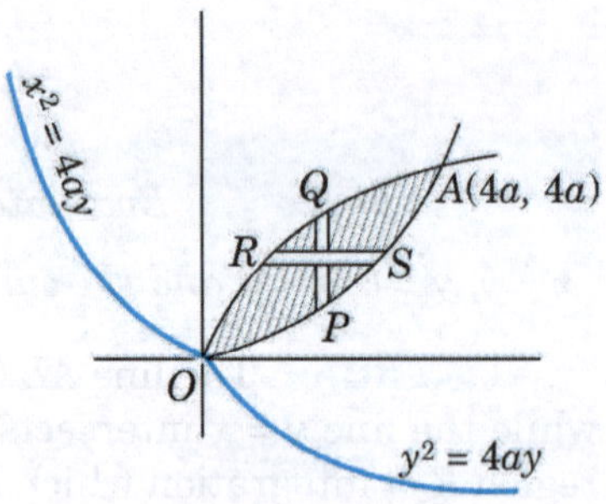

Fig. 7.7

**Example 7.6.** *Change the order of integration and hence evaluate*

$$I = \int_0^a \int_{\sqrt{ax}}^{a} \frac{y^2\, dxdy}{\sqrt{(y^4 - a^2x^2)}}$$

*(C.S.V.T.U., 2011)*

**Solution.** Here integration is first w.r.t. $y$ from $P$ on the parabola $y^2 = ax$ to $Q$ on the line $y = a$, then w.r.t. $x$ from $x = 0$ to $x = a$, giving the shaded region $OAB$ of integration (Fig. 7.8).

On changing the order of integration, we first integrate w.r.t. $x$ from $R$ to $S$, then w.r.t. $y$ from $y = 0$ to $y = a$.

$$\therefore \quad I = \int_0^a \int_0^{y^2/a} \frac{y^2\, dy}{\sqrt{(y^4 - a^2x^2)}}\, dx = \frac{1}{a}\int_0^a \int_0^{y^2/a} y^2\, dy \frac{dx}{\sqrt{[(y^2/a)^2 - x^2]}}\, dx$$

$$= \frac{1}{a}\int_0^a y^2 dy \left| \sin^{-1}\left(\frac{xa}{y^2}\right) \right|_0^{y^2/a} = \frac{1}{a}\int_0^a y^2 dy\ [\sin^{-1}(1) - \sin^{-1}(0)]$$

$$= \frac{\pi}{2a}\int_0^a y^2 dy = \frac{\pi}{2a}\left| \frac{y^3}{3} \right|_0^a = \frac{\pi a^2}{6}.$$

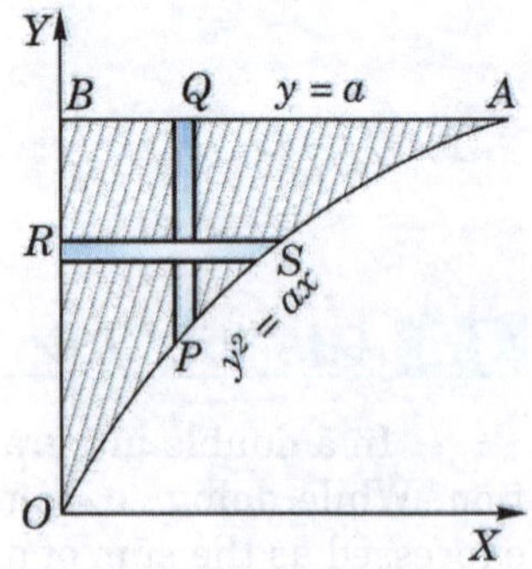

Fig. 7.8

**Example 7.7 :** *Evaluate* $\int_0^1 \int_{x^2}^{2-x} xy\, dxdy$ *(PU, 2018)*

**Solution :** Since the limits of $y$ are function of $x$, the integration will be performed w.r.t.

$$I = \int_0^1 \left[\frac{xy^2}{2}\right]_{x^2}^{2-x} dx = \int_0^1 \frac{x}{2}\left[(2-x)^2 - x^4\right] dx$$

$$= \frac{1}{2}\int_0^1 (4x - 4x^2 + x^3 - x^5)\cdot dx$$

$$= \frac{1}{2}\left[2x^2 - \frac{4}{3}x^3 + \frac{x^4}{4} - \frac{x^6}{6}\right]_0^1 = \frac{3}{8}$$

**Example 7.8 :** *Evaluate* $\iint (x^2 + y^2)\, dx\, dy$ *over the region in the positive quadrant for which* $x + y \le 1$

*(DU, 2019)*

**Solution:** The region of integration $R$ is the area bounded by the straight line $x + y = 1$.

$$x = 0, \quad x = 1 - y, \quad y = 0 \quad \text{and} \quad y = 1$$

$$= \iint_R (x^2 + y^2)\, dx\, dy = \int_0^1 \int_0^{1-y} (x^2 + y^2)\, dx\, dy$$

$$= \int_0^1 \left[\frac{x^3}{3} + xy^2\right]_0^{1-y} dy$$

$$= \int_0^1 \left[\frac{(1-y)^3}{3} + (1-y)y^2\right]\cdot dy$$

$$= \left[-\frac{(1-y)^4}{12} + \frac{y^3}{3} - \frac{y^4}{4}\right]_0^l$$

$$= \left(\frac{1}{3} - \frac{1}{4}\right) - \left(-\frac{1}{12}\right) = \frac{1}{6}$$

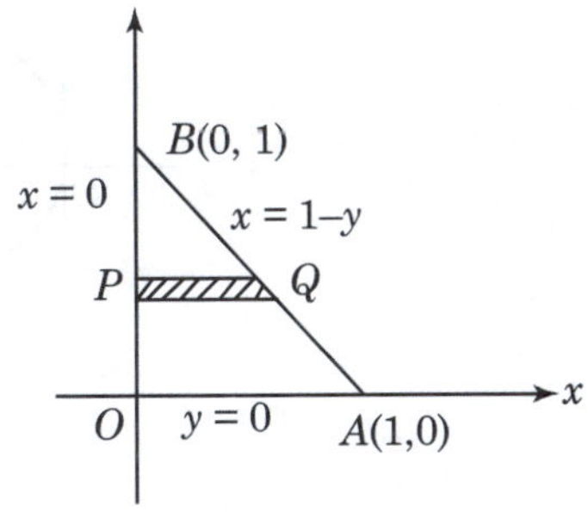


Fig. 7.9.

## 7.3 DOUBLE INTEGRALS IN POLAR COORDINATES

*To evaluate* $\int_{\theta_1}^{\theta_2} \int_{r_1}^{r_2} f(r, \theta)\, dr\, d\theta$, we first integrate w.r.t. $r$ between limits $r = r_1$ and $r = r_2$ keeping $\theta$ fixed and the resulting expression is integrated w.r.t. $\theta$ from $\theta_1$ to $\theta_2$. In this integral, $r_1, r_2$ are functions of $\theta$ and $\theta_1, \theta_2$ are constants.

*Figure 7.10 illustrates the process geometrically.*

Here $AB$ and $CD$ are the curves $r_1 = f_1(\theta)$ and $r_2 = f_2(\theta)$ bounded by the lines $\theta = \theta_1$ and $\theta = \theta_2$. $PQ$ is a wedge of angular thickness $\delta\theta$.

Then $\int_{r_1}^{r_2} f(r, \theta)\, dr$ indicates that the integration is along $PQ$ from $P$ to $Q$ while the integration w.r.t. $\theta$ corresponds to the turning of $PQ$ from $AC$ to $BD$.

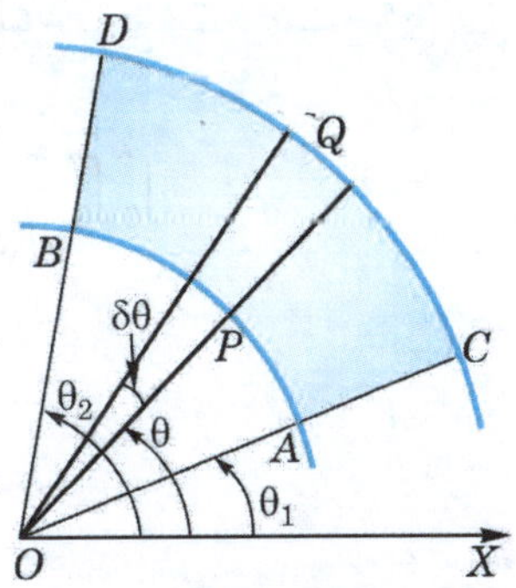


Fig. 7.10.

Thus the whole region of integration is the area $ACDB$. The order of integration may be changed with appropriate changes in the limits.

**Example 7.9.** *Evaluate* $\iint r \sin\theta\, dr\, d\theta$ *over the cardioid* $r = a\,(1 - \cos\theta)$ *above the initial line.*

*(Andhra., 2016 ; Kerala, 2005)*

**Solution.** To integrate first w.r.t. $r$, the limits are from 0 ($r = 0$) to $P$ [$r = a\,(1 - \cos\theta)$] and to cover the region of integration $R$, $\theta$ varies from 0 to $\pi$ (Fig. 7.11).

$$\therefore \quad \iint_R r \sin\theta\, dr d\theta = \int_0^{\pi} \sin\theta \left[\int_0^{r = a(1-\cos\theta)} r dr\right] d\theta$$

$$= \int_0^{\pi} \sin\theta\, d\theta \left|\frac{r^2}{2}\right|_0^{a(1-\cos\theta)} = \frac{a^2}{2}\int_0^{\pi} (1 - \cos\theta)^2 . \sin\theta\, d\theta$$

$$= \frac{a^2}{2}\left|\frac{(1-\cos\theta)^3}{3}\right|_0^{\pi} = \frac{a^2}{2}\cdot\frac{8}{3} = \frac{4a^2}{3}.$$

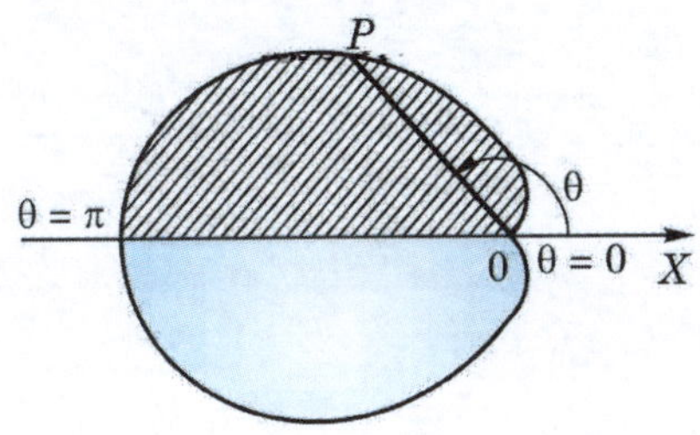


Fig. 7.11.

**Example 7.10.** *Evaluate* $I = \int_0^{2a}\int_0^{\sqrt{2ax-x^2}} x^2 dy\, dx$ *, by changing to polar coordinates* (PTU, 2019)

**Solution:** The limits for $y$ are from 0 to $\sqrt{2ax - x^2}$ and those for $x$ are from 0 to $2a$. Thus

$$y = 0, \quad y = \sqrt{2ax - x^2}$$
$$y = 0, \quad y^2 = 2ax - x^2$$
$$x^2 + y^2 = 2ax$$
$$I = \int_0^{\pi/2}\int_0^{2a\cos\theta} r^2 \cos^2\theta\, dr.d\theta$$

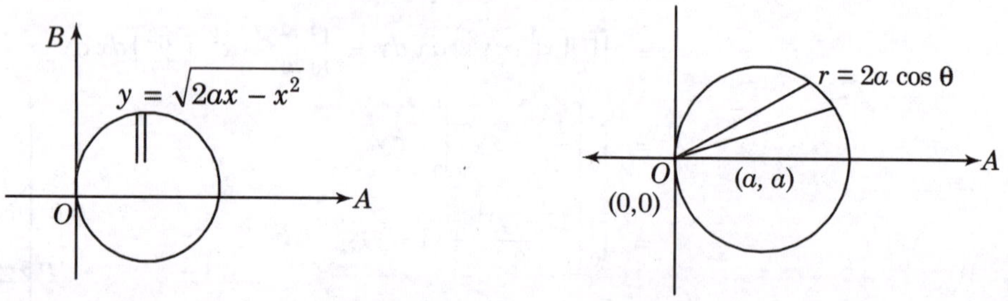


Fig. 7.12

$$= \int_0^{\pi/2}\int_0^{2a\cos\theta} r^3 \cos^2\theta\, dr.d\theta = \int_0^{\pi/2}\left[\frac{r^4}{4}\right]_0^{2a\cos\theta} \cos^2\theta \cdot d\theta$$

$$= 4a^4\int_0^{\pi/2}\cos^6\theta \cdot d\theta = 4a^4\,\frac{5\cdot 3\cdot 1}{6\cdot 4\cdot 2}\cdot\frac{\pi}{2} = \frac{5}{8}\pi a^4$$

**Example 7.11 :** Evaluate $\int_0^b\int_0^a \frac{dxdy}{\sqrt{(a^2-x^2)(b^2-y^2)}}$

**Solution:** $\int_0^b\int_0^a \frac{dxdy}{\sqrt{(a^2-x^2)(b^2-y^2)}} = \int_0^b \frac{1}{\sqrt{b^2-y^2}}\cdot\left[\sin^{-1}\frac{x}{a}\right]_0^a dy$

$$= \int_0^b \frac{1}{\sqrt{b^2-y^2}}\cdot\frac{\pi}{2}\cdot dy = \frac{\pi}{2}\int_0^b \frac{1}{\sqrt{b^2-a^2}} dy = \frac{\pi}{2}\left[\sin^{-1}\frac{y}{b}\right]_0^b$$

$$= \frac{\pi}{2}\cdot\frac{\pi}{2} = \frac{\pi^2}{4}$$

**Example 7.12 :** *Evaluate the integral*

$$\int_0^{\pi/2}\int_{a(1-\cos\theta)}^{a} r^2 dr\,.d\theta.$$

*Show the region of integration also.*

**Solution:** $\int_0^{\pi/2}\int_{a(1-\cos\theta)}^{a} r^2 dr\, d\theta = \int_0^{\pi/2}\left[\frac{r^3}{3}\right]_{a(1-\cos\theta)}^{a} d\theta$

$$= \frac{1}{3}\int_0^{\pi/2}\left[a^3 - a^3(1-\cos\theta)^3\right]d\theta$$

$$= \frac{a^3}{3}\int_0^{\pi/2}\left(3\cos\theta - 3\cos^2\theta + \cos^3\theta\right)\cdot d\theta$$

$$= \frac{a^3}{3}\left[\{3\sin\theta\}_0^{\pi/2} - 3\cdot\frac{1}{2}\cdot\frac{\pi}{2} + \frac{2}{3.1}\right]$$

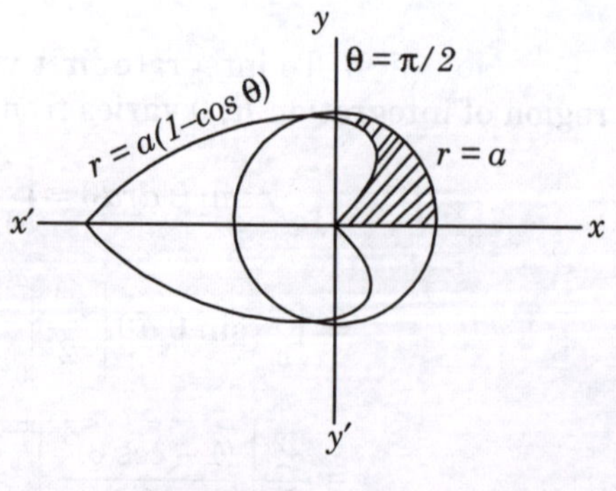


Fig. 7.13

$$= \frac{a^3}{3}\left[\left(3\sin\frac{\pi}{2} - 3\sin 0\right) - \frac{3\pi}{4} + \frac{2}{3}\right]$$

$$= \frac{a^3}{3}\left[3 - \frac{3\pi}{4} + \frac{2}{3}\right] = \frac{a^3}{36}(44 - 9\pi)$$

**Example 7.15 :** *Evaluate* $\iint r^3 dr\, d\theta$ *over the area included between the circles* $r = 2a\cos\theta$ *and* $r = 2b\cos\theta$ *($b < a$). In the region between the two circles be denoted by R. In the region R, of varies from* $-\frac{\pi}{2}$ *to* $\frac{\pi}{2}$.

**Solution:** Thus given integral

$$\iint_R r^3 dr\, d\theta = \int_{-\pi/2}^{\pi/2}\int_{2b\cos\theta}^{2a\cos\theta} r^3 dr\, d\theta$$

Here

$$= \int_{-\pi/2}^{\pi/2}\left[\frac{r^4}{4}\right]_{2b\cos\theta}^{2a\cos\theta} d\theta$$

$$= \frac{1}{4}\int_{-\pi/2}^{\pi/2}\left[(2a\cos\theta)^4 - (2b\cos\theta)^4\right]d\theta$$

$$= 4\left(a^4 - b^4\right)\int_{-\pi/2}^{\pi/2}\cos^4\theta\cdot d\theta$$

$$= 8\left(a^4 - b^4\right)\frac{3.1}{4.2}\cdot\frac{\pi}{2} = \frac{3}{2}\pi\left(a^4 - b^4\right).$$

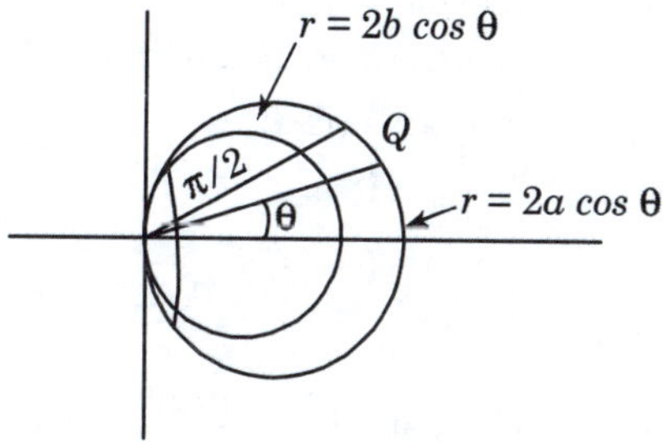


Fig. 7.14.

## PROBLEMS 7.1

Evaluate the following integrals (1–7) :

1. $\int_1^2\int_y^{y^2} dxdy$ (*K.T.U., 2016*)

2. Evaluate the following integrals $\int_a^b\int_\alpha^\beta r^2 \sin\theta\, d\theta\, dr$ (*PTU, 2019*) (**Ans.** $\left(\frac{b^3 - a^2}{3}\right)(\cos\alpha - \cos\beta)$ )

3. Solve $\int_0^\pi\int_0^\infty e^{-br^2} r\, dr\, d\theta$ (*VTU, 2018*) (**Ans.** $\frac{\pi}{2b}$ )

4. Solve $\int_0^{\pi/2}\int_{a(1+\cos\theta)}^{a} r\, dr\, d\theta$ (*JNTU, 2019*) (**Ans.** $\frac{a^2(8+\pi)}{8}$ )

5. $\iint_S xy\, dxdy$ over the positive quadrant of the circle $x^2 + y^2 = a^2$. (*Anna., 2016 ; Rohtak, 2011*

6. $\iint (x + y)^2 dxdy$ over the area bounded by the ellipse $x^2/a^2 + y^2/b^2 = 1$. (*U.K.T.U., 2013*)

7. $\iint xy(x + y)\, dxdy$ over the area between $y = x^2$ and $y = x$. (*V.T.U., 2015*)

Evaluate the following integrals by changing the order of integration (8–14) :

8. $\int_0^3\int_1^{\sqrt{(4-y)}} (x + y)\, dxdy$. (*Andhra., 2021, 2016*)

9. Solve $\int_0^{2a}\int_0^{\sqrt{2ax - x^2}} \frac{\phi'(y)\left(x^2 + y^2\right)x\, dx\, dy}{\sqrt{4a^2x^2 - \left(x^2 + y^2\right)^2}}$ (*PTU, 2019*)

10. $\int_0^1\int_x^{\sqrt{(2-x^2)}} \frac{x\, dydx}{\sqrt{(x^2 + y^2)}}$. (*Kurukshetra, 2013 ; P.T.U., 2010*)

**11.** $\int_0^{a/\sqrt{2}} \int_y^{\sqrt{(a^2-y^2)}} \log(x^2+y^2)\, dxdy\ (a > 0).$ *(C.S.V.T.U., 2015)*

**12.** Show that $\iint r^2 \sin\theta \cdot dr\, d\theta = \frac{2a^3}{3}$ where $R$ is the region bounded by the semicircle $r = 2a\cos\theta$, above the initial line. *(Andhra, 2022)*

**13.** $\int_0^{\infty} \int_x^{\infty} \frac{e^{-y}}{y}\, dydx.$ *(Andhra, 2023; C.S.V.T.U., 2012 ; P.T.U., 2011)*

**14.** $\int_0^{\infty} \int_0^{x} xe^{-x^2/y}\, dydx.$ *(U.P.T.U., 2013 ; Rohtak, 2012 )*

**15.** Sketch the region of integration of the following integrals and change the order of integrations,

(*i*) $\int_0^{2a} \int_{\sqrt{(2ax-x^2)}}^{\sqrt{(2ax)}} f(x, y)\, dxdy$ *(J.N.T.U., 2015)* (*ii*) $\int_0^{ae^{\pi/4}} \int_{2\log(r/a)}^{\pi/2} f(r, \theta)\, r\, drd\theta.$

**16.** $\int_0^{\infty}\int_0^{\infty} e^{-(x^2+y^2)} dx\, dy$ *(KTU, 2019)*

**17.** Evaluate $\iint \frac{r\, drd\theta}{\sqrt{a^2+r^2}}$ over one loop of the lemniscate $r^2 = a^2\cos 2\theta$.

**18.** Evaluate $\iint r^3\, drd\theta$ over the area bounded between the circles $r = 2\cos\theta$ and $r = 4\cos\theta$. *(Andhra, 2023)*

## 7.4 AREA ENCLOSED BY PLANE CURVES

**(1)** *Cartesian coordinates.*

Consider the area enclosed by the curves $y = f_1(x)$ and $y = f_2(x)$ and the ordinates $x = x_1$, $x = x_2$ [Fig. 7.15 (*a*)].

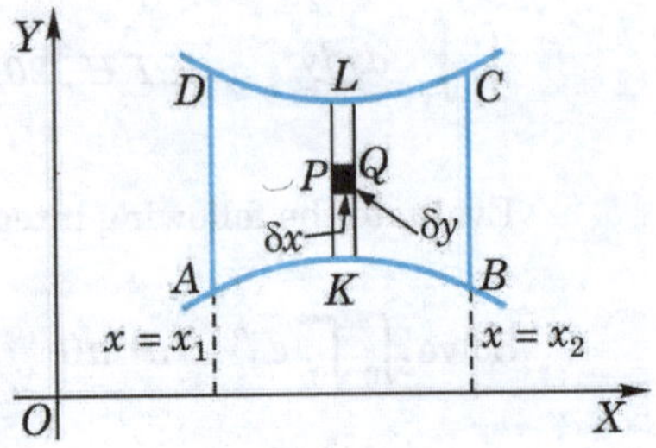

Fig. 7.15(*a*)

Divide this area into vertical strips of width $\delta x$. If $P(x, y)$, $Q(x + \delta x, y + \delta y)$ be two neighbouring points, then the area of the small rectangle $PQ = \delta x \delta y$.

$\therefore$ area of strip $KL = \underset{\delta y \to 0}{\text{Lt}} \Sigma\, \delta x\, \delta y$.

Since for all rectangles in this strip $\delta x$ is the same and $y$ varies from $y = f_1(x)$ to $y = f_2(x)$.

$$\therefore \quad \text{area of the strip } KL = \delta x \underset{\delta y \to 0}{\text{Lt}} \sum_{f_1(x)}^{f_2(x)} dy = \delta x \int_{f_1(x)}^{f_2(x)} dy.$$

Now adding up all such strips from $x = x_1$ to $x = x_2$, we get the area $ABCD$

$$= \underset{\delta x \to 0}{\text{Lt}} \sum_{x_1}^{x_2} \delta x \,.\int_{f_1(x)}^{f_2(x)} dy = \int_{x_1}^{x_2} dx \int_{f_1(x)}^{f_2(x)} dy = \int_{x_1}^{x_2} \int_{f_1(x)}^{f_2(x)} dxdy$$

Similarly, dividing the area $A'B'C'D$ [Fig. 7.15(*b*)] into horizontal strips of width $\delta y$, we get the area $A'B'C'D'$.

$$= \int_{y_1}^{y_2} \int_{f_1(y)}^{f_2(y)} dxdy$$

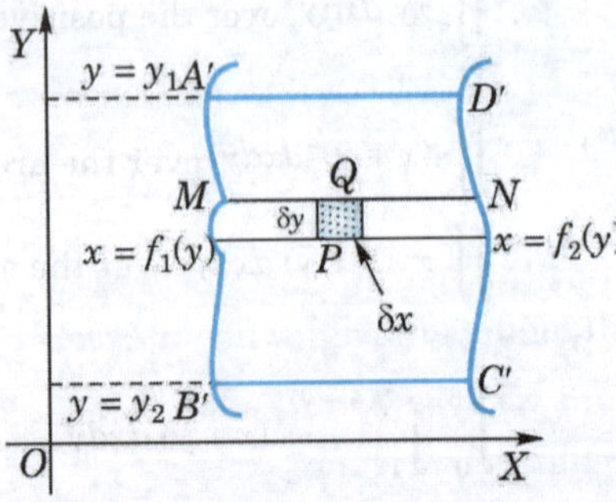

Fig. 7.15(*b*)

**(2)** *Polar coordinates.*

Consider an area $A$ enclosed by a curve whose equation is in polar coordinates.

Let $P(r, \theta)$, $Q(r + \delta r, \theta + \delta\theta)$ be two neighbouring points. Mark circular areas of radii $r$ and $r + \delta r$ meeting $OQ$ in $R$ and $OP$ (produced) in $S$ (Fig. 7.16).

Since arc $PR = r\delta\theta$ and $PS = \delta r$.

$\therefore$ area of the curvilinear rectangle $PRQS$ is approximately $= PR \,.\, PS = r\delta\theta \,.\, \delta r$.

If the whole area is divided into such curvilinear rectangles, the sum $\Sigma\Sigma r\delta\theta\delta r$ taken for all these rectangles, gives in the limit the area $A$.

Hence
$$A = \underset{\substack{\delta r \to 0 \\ \delta\theta \to 0}}{\text{Lt}} \Sigma\Sigma r\delta\theta\delta r = \iint r d\theta dr$$

where the limits are to be so chosen as to cover the entire area.

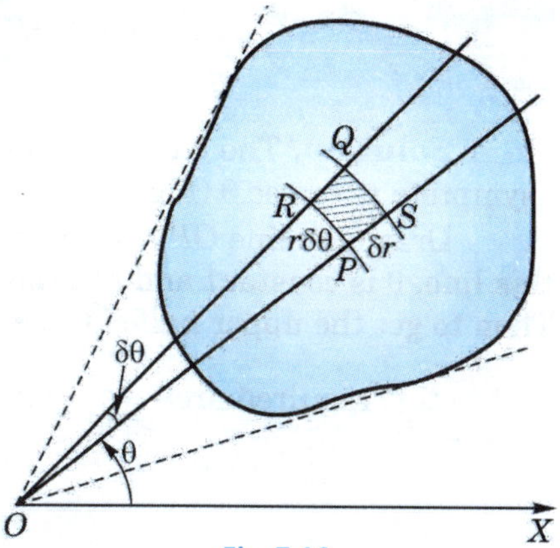

Fig. 7.16.

**Example 7.14.** *Find the area of a plate in the form of a quadrant of the ellipse*
$$\frac{x^2}{a^2} + \frac{y^2}{b^2} = 1.$$
*(Anna, 2012)*

**Solution.** Dividing the area into vertical strips of width $\delta x$, $y$ varies from $K(y = 0)$ to $L[y = b\sqrt{(1 - x^2/b^2)}]$ and then $x$ varies from 0 to $a$ (Fig. 7.17).

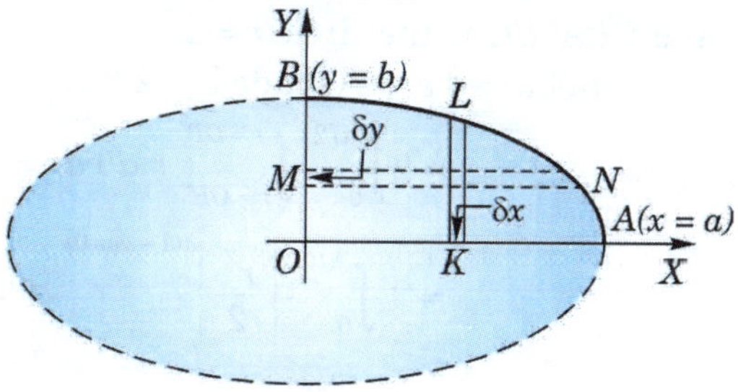

Fig. 7.17.

$\therefore$ required area

$$= \int_0^a dx \int_0^{b\sqrt{(1 - x^2/a^2)}} dy = \int_0^a dx\, [y]_0^{b\sqrt{(1 - x^2/a^2)}}$$

$$= \frac{b}{a}\int_0^a \sqrt{(a^2 - x^2)}\, dx = \pi ab/4.$$

**Otherwise,** dividing this area into horizontal strips of width $\delta y$, $x$ varies from $M(x = 0)$ to $N[x = a\sqrt{(1 - y^2/b^2)}\,]$ and then $y$ varies from 0 to $b$.

$\therefore$ required area
$$= \int_0^b dy \int_0^{a\sqrt{(1 - y^2/b^2)}} dx = \int_0^b dy\, [x]_0^{a\sqrt{(1 - y^2/b^2)}}$$

$$= \frac{a}{b}\int_0^b \sqrt{(b^2 - y^2)}\, dy = \pi ab/4.$$

**Obs.** The change of the order of integration does not in any way affect the value of the area.

**Example 7.15.** *Show that the area between the parabolas $y^2 = 4ax$ and $x^2 = 4ay$ is $\frac{16}{3}a^2$.*

*(Andhra, 2022 ; V.T.U., 2016 ; Anna, 2013)*

**Solution.** Solving the equations $y^2 = 4ax$ and $x^2 = 4ay$, it is seen that the parabolas intersect at $O$ (0, 0) and $A$ ($4a$, $4a$). As such for the shaded area between these parabolas (Fig. 7.18) $x$ varies from 0 to $4a$ and $y$ varies from $P$ to $Q$ *i.e.*, from $y = x^2/4a$ to $y = 2\sqrt{(ax)}$. Hence the *required area*

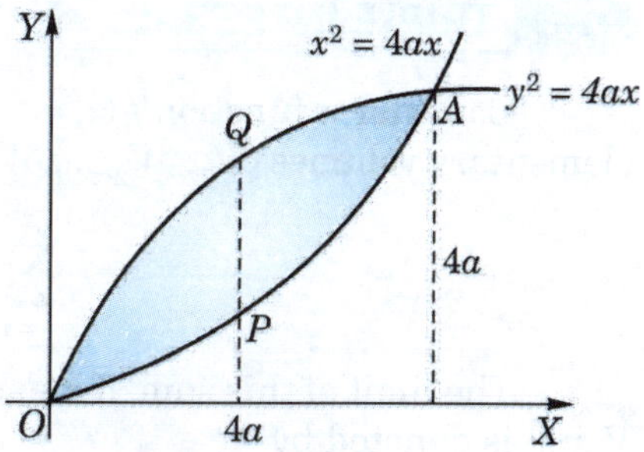

Fig. 7.18.

$$= \int_0^{4a}\int_{x^2/4a}^{2\sqrt{(ax)}} dydx = \int_0^{4a} (2\sqrt{(ax)} - x^2/4a)\, dx$$

$$= \left| 2\sqrt{a} \,.\, \frac{2}{3}x^{3/2} - \frac{1}{4a}\cdot\frac{x^3}{3} \right|_0^{4a} = \frac{32}{3}a^2 - \frac{16}{3}a^2 = \frac{16}{3}a^2.$$

**Example 7.16.** *Calculate the area included between the curve r = a (sec θ + cos θ) and its asymptote.* *(Andhra, 2022)*

**Solution.** The curve is symmetrical about the initial line and has an asymptote $r = a \sec \theta$ (Fig. 7.19).

Draw any line $OP$ cutting the curve at $P$ and its asymptote at $P'$. Along this line, θ is constant and $r$ varies from $a \sec \theta$ at $P'$ to $a (\sec \theta + \cos \theta)$ at $P$. Then to get the upper half of the area, θ varies from 0 to π/2.

$$\therefore \quad \text{required area} = 2\int_0^{\pi/2}\int_{a\sec\theta}^{a(\sec\theta+\cos\theta)} r\,dr d\theta$$

$$= 2\int_0^{\pi/2}\left[\frac{r^2}{2}\right]_{a\sec\theta}^{a(\sec\theta+\cos\theta)} d\theta$$

$$= a^2\int_0^{\pi/2}(2+\cos^2\theta)\,d\theta = 5\pi a^2/4.$$

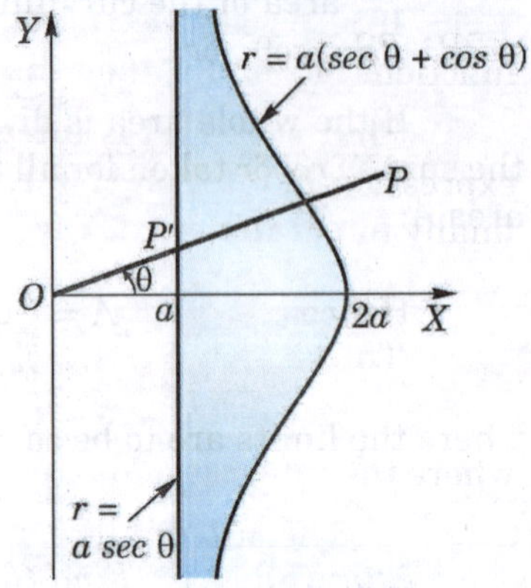


Fig. 7.19.

**Example 7.17.** *Find the area lying inside the cardioid r = a(1 + cos θ) and outside the circle r = a.*

**Solution.** In Fig. 7.20, $ABODA$ represents the cardioid $r = a(1 + \cos \theta)$ and $CBA'DC$ is the circle $r = a$.

Required area (shaded) = 2 (area $ABCA$)

$$= 2\int_0^{\pi/2}\int_{r=OP'}^{r=OP} r\,d\theta\,dr = 2\int_0^{\pi/2}\int_a^{a(1+\cos\theta)}(r dr)\,d\theta$$

$$= 2\int_0^{\pi/2}\left|\frac{r^2}{2}\right|_a^{a(1+\cos\theta)} d\theta = a^2\int_0^{\pi/2}[(1+\cos\theta)^2-1]\,d\theta$$

$$= a^2\int_0^{\pi/2}(\cos^2\theta+2\cos\theta)\,d\theta = a^2\left(\frac{1}{2}\cdot\frac{\pi}{2}+2\right)=\frac{a^2}{4}(\pi+8).$$

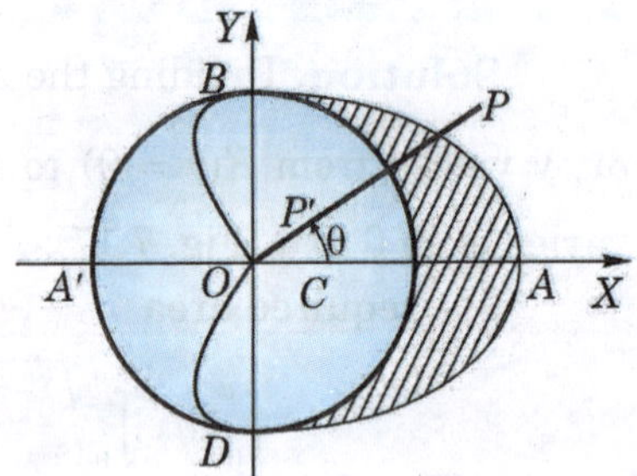


Fig. 7.20.

## PROBLEMS 7.2

1. Find, by double integration, the area lying between the parabola $y = 4x - x^2$ and the line $y = x$. *(J.N.T.U., 2015 ; C.S.V.T.U., 2014)*
2. Find the area lying between the parabola $y = x^2$ and the line $x + y - z = 0$. *(Andhra, 2022)*
3. By double integration, find the whole area of the curve $a^2x^2 = y^3(2a - y)$.
4. Find, by double integration, the area enclosed by the curves $y = 3x/(x^2 + 2)$ and $4y = x^2$. *(Andhra., 2013 ; Kurukshetra, 2013)*
5. Find, by double integration, the area of the lemniscate $r^2 = a^2 \cos 2\theta$. *(Andhra., 2021, 2016)*
6. Find, by double integration, the area lying inside the circle $r = a \sin \theta$ and outside the cardioid $r = a(1 - \cos \theta)$. *(Andhra., 2015 ; Rohtak, 2012 ; P.T.U., 2011)*
7. Find the area lying inside the cardioid $r = 1 + \cos \theta$ and outside the parabola $r (1 + \cos \theta) = 1$.
8. Find the area common to the circles $r = a \cos \theta$, $r = a \sin \theta$ by double integration.

## 7.5 TRIPLE INTEGRALS

Consider a function $f(x, y, z)$ defined at every point of the 3-dimensional finite region $V$. Divide $V$ into $n$ elementary volumes $\delta V_1, \delta V_2, \ldots, \delta V_n$. Let $(x_r, y_r, z_r)$ be any point within the $r$th sub-division $\delta V_r$. Consider the sum

$$\sum_{r=1}^{\infty} f(x_r, y_r, z_r)\delta V_r .$$

The limit of this sum, if it exists, as $n \to \infty$ and $\delta V_r \to 0$ is called the *triple integral of f(x, y, z) over the region V* and is denoted by

$$\iiint f(x, y, z)\, dV.$$

For purposes of evaluation, it can also be expressed as the repeated integral

$$\int_{x_1}^{x_2}\int_{y_1}^{y_2}\int_{z_1}^{z_2} f(x, y, z)\,dxdydz.$$

If $x_1$, $x_2$ are constants ; $y_1$, $y_2$ are either constants or functions of $x$ and $z_1$, $z_2$ are either constants or functions of $x$ and $y$, then this integral is evaluated as follows :

First $f(x, y, z)$ is integrated w.r.t. $z$ between the limits $z_1$ and $z_2$ keeping $x$ and $y$ fixed. The resulting expression is integrated w.r.t. $y$ between the limits $y_1$ and $y_2$ keeping $x$ constant. The result just obtained is finally integrated w.r.t. $x$ from $x_1$ to $x_2$.

Thus $$I = \boxed{\int_{x_1}^{x_2}\boxed{\int_{y_1(x)}^{y_2(x)}\boxed{\int_{z_1(x,y)}^{z_2(x,y)} f(x, y, z)dz}\,dy}\,dx}$$

where the integration is carried out from the innermost rectangle to the outermost rectangle.

The order of integration may be different for different types of limits.

**Example 7.18 :** *Evaluate the integral*

$$\int_0^{\log 2}\int_0^{x}\int_0^{x+\log y} e^{x+y+z}dz\,dy\,dx$$ *(PU, 2019)*

**Solution:** Let $$I = \int_0^{\log 2}\int_0^{x}\int_0^{x+\log y} e^{x+y+z}dz\,dy\,dx = \int_0^{\log 2}\int_0^{x} e^{x+y}\left[e^z\right]_0^{x+\log_y} dy\,dx$$

$$= \int_0^{\log 2}\int_0^{x} e^{x+y}.\left(e^{x+\log y}-1\right)dy\,dx = \int_0^{\log 2}\left[e^{2x}.ye^y - e^x.e^y\right]dy\,dx$$

$$= \int_0^{\log 2}\left[e^{2x}\left\{y\cdot e^y - e^y\right\}_0^x - e^x\left\{e^y\right\}_0^x\right] = \int_0^{\log 2}\left[e^{2x}\left(x\,e^x - e^x + 1\right) - e^x\left(e^x - 1\right)\right]dx$$

$$= \int_0^{\log 2}\left(xe^{3x} - e^{3x} + e^x\right)dx = \int_0^{\log 2}\left[(x-1)e^{3x} + e^x\right]dx$$

$$= \left[(x-1)\frac{e^{3x}}{3} - \frac{e^{3x}}{9} + e^x\right]_0^{\log 2} = (\log 2 - 1)\cdot\frac{e^{3\log 2}}{3} - \frac{e^{3\log 2}}{9} + e^{\log 2} - \frac{5}{9}$$

$$= (\log 2 - 1)\cdot\frac{e^{\log 2^3}}{3} - \frac{e^{\log 2^3}}{9} + e^{\log 2} - \frac{5}{9} = \frac{8}{3}(\log 2 - 1) - \frac{8}{9} + 2 - \frac{5}{9} = \frac{8}{3}\log 2 - \frac{19}{2}$$

**Example 7.19 :** *Evaluate the integral* $\iiint \frac{dx\,dy\,dz}{\sqrt{a^2 - x^2 - y^2 - z^2}}$ *taken throughout the volume of the sphere* $x^2 + y^2 + z^2 = a^2$.

**Solution:** Here the region of integration $V$ is the sphere

$$x^2 + y^2 + z^2 = a^2 \quad ...(i)$$

We can express any of the 3 variables

$$z = \pm\sqrt{a^2 - x^2 - y^2}$$

On this circle $y$ varies

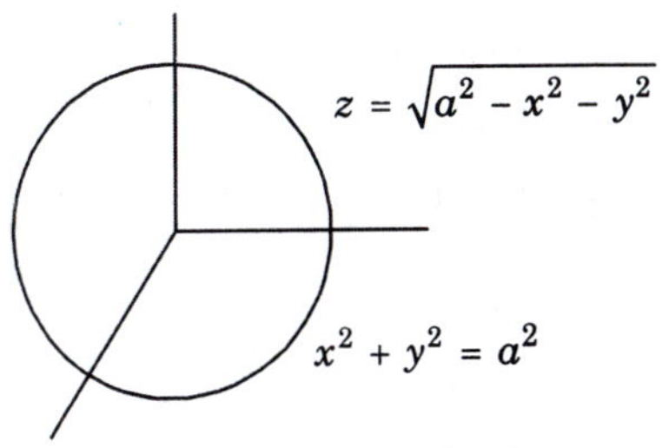


Fig. 7.21.

Let $$I = \iiint_V \frac{dx\,dy\,dz}{\sqrt{a^2 - x^2 - y^2 - z^2}}$$

$$= \int_{-a}^{a}\int_{-\sqrt{a^2-x^2}}^{\sqrt{a^2-x^2}}\int_{-\sqrt{a^2-x^2-y^2}}^{\sqrt{a^2-x^2-y^2}} \frac{dz\,dy\,dx}{\sqrt{a^2 - x^2 - y^2 - z^2}}$$

$$= 8\int_0^{a}\int_0^{\sqrt{a^2-x^2}}\int_0^{\sqrt{a^2-x^2-y^2}} \frac{dz\,dy\,dx}{\sqrt{a^2 - x^2 - y^2 - z^2}}$$

$$= 8\int_0^{a}\int_0^{\sqrt{a^2-x^2}}\left[\sin^{-1}\frac{z}{\sqrt{a^2 - x^2 - y^2}}\right]_0^{\sqrt{a^2-x^2-y^2}} dy\,dx = 8\int_0^{a}\int_0^{\sqrt{a^2-x^2}}\left[\sin^{-1}(1) - \sin^{-1}(0)\right]dy\,dx$$

$$= \int_0^a \int_0^{\sqrt{a^2-x^2}} \frac{\pi}{2} dy\, dx = 4\pi \int_0^a [y]_0^{\sqrt{a^2-x^2}} dx = 4\pi \int_0^a \sqrt{a^2-x^2}\, dx$$

$$= 4\pi \left[ \frac{x\sqrt{a^2-x^2}}{2} + \frac{a^2}{2} \sin^{-1} \frac{x}{a} \right]_0^a = 4\pi \left[ \frac{a^2}{2} \cdot \frac{\pi}{2} \right] = \pi^2 a^2$$

**Example 7.20.** *Evaluate* $\int_{-1}^{1} \int_0^z \int_{x-z}^{x+z} (x+y+z)\, dxdydz.$

*(Andhra, 2022 ; V.T.U., 2016 ; J.N.T.U., 2006)*

**Solution.** Integrating first w.r.t. $y$ keeping $x$ and $z$ constant, we have

$$I = \int_{-1}^{1} \int_0^z \left| xy + \frac{y^2}{2} + yz \right|_{x-z}^{x+z} dx\, dz = \int_{-1}^{1} \int_0^z \left[ (x+z)(2z) + \frac{1}{2} 4xz \right] dx\, dz$$

$$= 2 \int_{-1}^{1} \left| \frac{x^2 z}{2} + z^2 x + \frac{x^2}{2} z \right|_0^z dz = 2 \int_{-1}^{1} \left( \frac{z^3}{2} + z^3 + \frac{z^3}{2} \right) dz = 4 \left| \frac{z^4}{4} \right|_{-1}^{1} = 0.$$

**Example 7.21.** *Evaluate* $\int_0^1 \int_0^{\sqrt{(1-x^2)}} \int_0^{\sqrt{(1-x^2-y^2)}} xyz\, dxdydz.$

*(Andhra, 2020 ; Delhi., 2016 ; V.T.U., 2013)*

**Solution.** We have

$$I = \int_0^1 x \left[ \int_0^{\sqrt{(1-x^2)}} y \left\{ \int_0^{\sqrt{(1-x^2-y^2)}} z\, dz \right\} dy \right] dx = \int_0^1 x \left\{ \int_0^{\sqrt{(1-x^2)}} y \cdot \left| \frac{z^2}{2} \right|_0^{\sqrt{(1-x^2-y^2)}} dy \right\} dx$$

$$= \int_0^1 x \left\{ \int_0^{\sqrt{(1-x^2)}} y \cdot \frac{1}{2}(1-x^2-y^2)\, dy \right\} dx = \frac{1}{2} \int_0^1 x \left| (1-x^2)\frac{y^2}{2} - \frac{y^4}{4} \right|_0^{\sqrt{(1-x^2)}} dx$$

$$= \frac{1}{8} \int_0^1 [(1-x^2)^2 \cdot 2x - (1-x^2)^4 \cdot x]\, dx = \frac{1}{8} \int_0^1 (x - 2x^3 + x^5)\, dx$$

$$= \frac{1}{8} \left| \frac{x^2}{2} - \frac{2x^4}{4} + \frac{x^6}{6} \right|_0^1 = \frac{1}{8} \left( \frac{1}{2} - \frac{1}{2} + \frac{1}{6} \right) = \frac{1}{48}.$$

## PROBLEMS 7.3

Evaluate the following integrals :

1. $\int_0^3 \int_0^2 \int_0^1 (x+y+z)\, dz\, dy\, dx$ *(C.S.V.T.U., 2015)*
2. $\int_{-c}^{c} \int_{-b}^{b} \int_{-a}^{a} (x^2+y^2+z^2)\, dx\, dy\, dz$ *(V.T.U., 2016 )*
3. $\int_0^a \int_0^x \int_0^{x+y} e^{x+y+z}\, dz\, dy\, dx.$ *(V.T.U., 2012 ; Rohtak, 2011 S)*
4. $\int_1^e \int_1^{\log y} \int_1^{e^x} \log z\, dz\, dx\, dy.$ *(Andhra., 2016)*
5. Evaluate the following integrals $\int_0^a \int_0^x \int_0^{x+y} e^{x+y+z} dz\, dy\, dx$ *(Mysore, 2019)*
6. Solve $\int_0^1 \int_{y^2}^1 \int_0^{1-x} x\, dz\, dx\, dy$ *(PU, 2019)*
7. Solve $\int_0^1 \int_0^{\sqrt{1-x^2}} \int_0^{\sqrt{1-x^2-y^2}} \frac{1}{\sqrt{1-x^2-y^2-z^2}} dz\, dy\, dx$ *(Marathawada, 2019)*

8. Evaluate $\iiint (x+y+z)\,dx\,dy\,dz$ over the tetrahedron bunded by the planes $x = 0, y = 0, z = 0$ and $x + y + z = 1$ (*KTU, 2017*)

9. Evaluate $\int_0^{\pi}\int_0^{2}\int_0^{\sqrt{4-z^2}} z \sin y\, dx\, dy\, dz$. (*UPTU, 2018*)

10. Evaluate $\int_0^{\pi}\int_0^{\sin\theta}\int_0^{\pi\sin\theta} \pi\cos^2\theta\, dz\, dr\, d\theta$.

11. Compute $\int_0^{\pi}\int_0^{\pi/2}\int_0^{1} z\sin x + z\cos y\, dz.dy.dx$. (*Kashmir, 2017*)

## 7.6 VOLUMES OF SOLIDS

**(1) Volumes as double integrals.** Consider a surface $z = f(x, y)$. Let the orthogonal projection on $XY$-plane of its portion $S'$ be the area $S$ (Fig. 7.22).

Divide $S$ into elementary rectangles of area $\delta x\ \delta y$ by drawing lines parallel to $X$ and $Y$-axes. With each of these rectangles as base, erect a prism having its length parallel to $OZ$.

$\therefore$ volume of this prism between $S$ and the given surface $z = f(x, y)$ is $z\delta x\delta y$.

Hence the volume of the solid cylinder on $S$ as base, bounded by the given surface with generators parallel to the $Z$-axis.

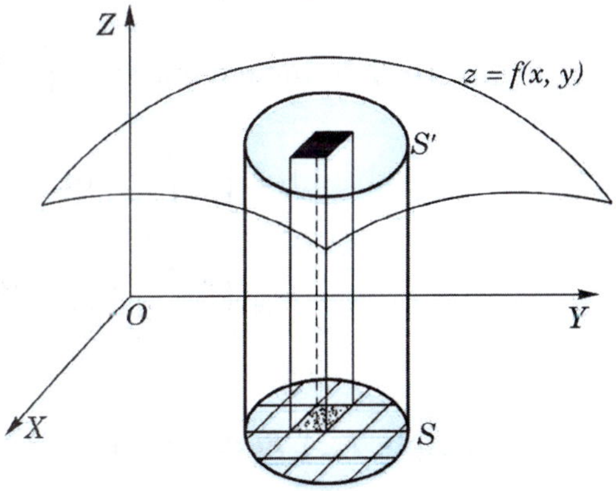

Fig. 7.22.

$$= \underset{\substack{\delta x \to 0 \\ \delta y \to 0}}{\text{Lt}} \Sigma\Sigma\, z\delta x\delta y$$

$$= \iint z\, dx\, dy \quad \text{or} \quad \iint f(x, y)\, dx\, dy$$

where the integration is carried over the area $S$.

**Obs.** While using polar coordinates, divide $S$ into elements of area $r\delta\theta\ \delta r$.

$\therefore$ replacing $dxdy$ by $r\delta\theta\delta r$, we get the required volume $= \iint zr\, d\theta\, dr$.

**Example 7.22.** *Find the volume bounded by the cylinder $x^2 + y^2 = 4$ and the planes $y + z = 4$ and $z = 0$.* (*Andhra, 2021 ; Kurukshetra, 2013*)

**Solution.** From Fig. 7.23, it is self-evident that $z = 4 - y$ is to be integrated over the circle $x^2 + y^2 = 4$ in the $XY$-plane. To cover the shaded half of this circle, $x$ varies from 0 to $\sqrt{(4-y^2)}$ and $y$ varies from $-2$ to 2.

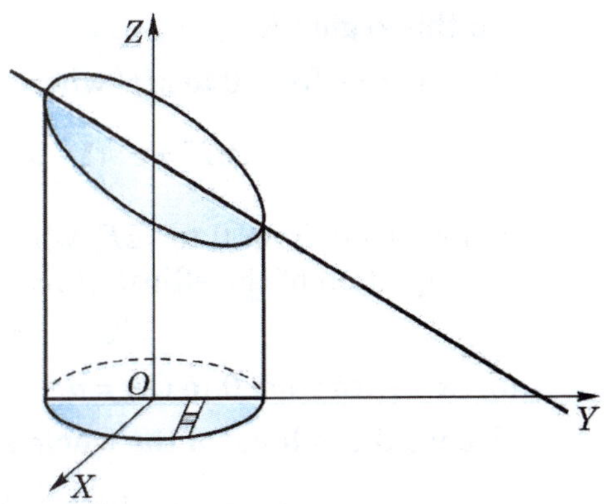

Fig. 7.23.

$\therefore$ Required volume

$$= 2\int_{-2}^{2}\int_0^{\sqrt{(4-y^2)}} z\, dx\, dy = 2\int_{-2}^{2}\int_0^{\sqrt{(4-y^2)}} (4-y)\, dx\, dy$$

$$= 2\int_{-2}^{2} (4-y)\,[x]_0^{\sqrt{(4-y^2)}}\, dy = 2\int_{-2}^{2} (4-y)\sqrt{(4-y^2)}\, dy$$

$$= 2\int_{-2}^{2} 4\sqrt{(4-y^2)}\, dy - 2\int_{-2}^{2} y\sqrt{(4-y^2)}\, dy$$

$$= 8\int_{-2}^{2} \sqrt{(4-y^2)}\, dy$$ [The second term vanishes as the integrand is an odd function.]

$$= 8 \left| \frac{y\sqrt{(4-y)^2}}{2} + \frac{4}{2}\sin^{-1}\frac{y}{2} \right|_{-2}^{2} = 16\pi.$$

**(2) Volume as triple integral**

Divide the given solid by planes parallel to the coordinate planes into rectangular parallelopipeds of volume $\delta x\, \delta y\, \delta z$ (Fig. 7.24).

$$\therefore \quad \text{the total volume} = \underset{\substack{\delta x \to 0 \\ \delta y \to 0 \\ \delta z \to 0}}{\text{Lt}} \Sigma\Sigma\Sigma\, \delta x\, \delta y\, \delta z$$

$$= \iiint dx\, dy\, dz$$

with appropriate limits of integration.

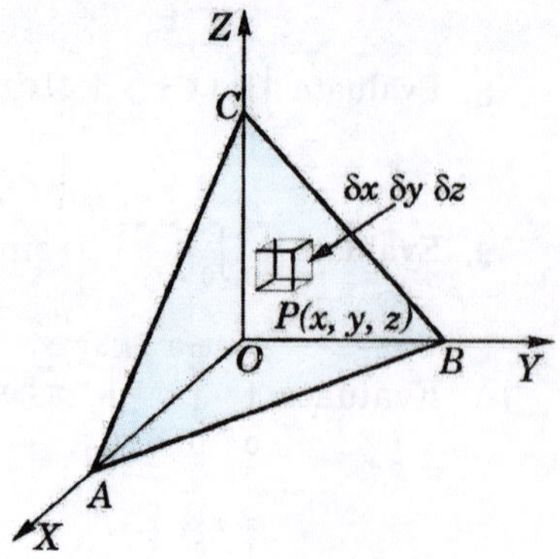


Fig. 7.24.

**Example 7.23.** *Calculate the volume of the solid bounded by the planes x = 0, y = 0, x + y + z = a and z = 0.*

**Solution.** Volume required $= \int_0^a \int_0^{a-x} \int_0^{a-x-y} dz\, dy\, dx$

$$= \int_0^a \int_0^{a-x} (a - x - y)\, dy\, dx = \int_0^a \left| (a-x)y - \frac{y^2}{2} \right|_0^{a-x} dx$$

$$= \int_0^a \left\{ (a-x)^2 - \frac{(a-x)^2}{2} \right\} dx = \frac{1}{2}\int_0^a (a-x)^2\, dx = \frac{1}{2}\left| -\frac{(a-x)^3}{3} \right|_0^a = \frac{a^3}{6}.$$

**Example 7.24.** *Find the volume of the ellipsoid* $\frac{x^2}{a^2} + \frac{y^2}{b^2} + \frac{z^2}{c^2} = 1.$

*(Trivendrum, 2019 ; Rohtak, 2012)*

**Solution.** Let $OABC$ be the positive octant of the given ellipsoid which is bounded by the planes $OAB$ ($z = 0$), $OBC$ ($x = 0$), $OCA$ ($y = 0$) and the surface $ABC$, *i.e.*,

$$\frac{x^2}{a^2} + \frac{y^2}{b^2} + \frac{z^2}{c^2} = 1.$$

Divide this region $R$ into rectangular parallellopipeds of volume $\delta x \delta y \delta z$. Consider such an element at $P(x, y, z)$. (Fig. 7.24)

$$\therefore \quad \text{the required volume} = 8 \iiint_R dx\, dy\, dz.$$

In this region $R$,

(*i*) $z$ varies from 0 to $MN$ where

$$MN = c\sqrt{(1 - x^2/a^2 - y^2/b^2)}.$$

(*ii*) $y$ varies from 0 to $EF$, where $EF = b\sqrt{(1 - x^2/a^2)}$ from the equation of the ellipse $OAB$, *i.e.*,

$$x^2/a^2 + y^2/b^2 = 1.$$

(*iii*) $x$ varies from 0 to $OA = a$.

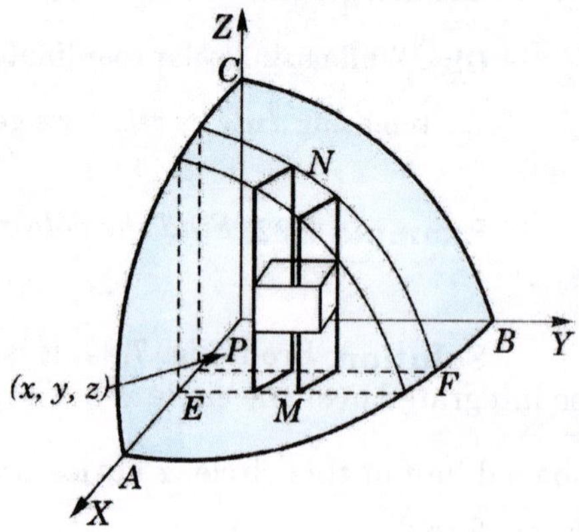


Fig. 7.25

Hence the volume of the whole ellipsoid

$$= 8 \int_0^a \int_0^{b\sqrt{(1-x^2/a^2)}} \int_0^{c\sqrt{(1-x^2/a^2-y^2/b^2)}} dx\, dy\, dz = 8\int_0^a dx \int_0^{b\sqrt{(1-x^2/a^2)}} dy \left| z \right|_0^{c\sqrt{(1-x^2/a^2-y^2/b^2)}}$$

$$= 8c \int_0^a dx \int_0^{b\sqrt{(1-x^2/a^2)}} \sqrt{(1 - x^2/a^2 - y^2/b^2)}\, dy$$

$$= \frac{8c}{b}\int_0^a dx \int_0^{\rho} \sqrt{(\rho^2 - y^2)}\, dy \quad \text{when } \rho = b\sqrt{(1 - x^2/a^2)}.$$

$$= \frac{8c}{b}\int_0^a dx \left[\frac{y\sqrt{(\rho^2 - y^2)}}{2} + \frac{\rho^2}{2}\sin^{-1}\frac{y}{\rho}\right]_0^{\rho} = \frac{8c}{b}\int_0^a \frac{b^2}{2}\left(1 - \frac{x^2}{a^2}\right)\frac{\pi}{2}\, dx$$

$$= 2\pi\, bc \int_0^a \left(1 - \frac{x^2}{a^2}\right) dx = 2\pi\, bc \left| x - \frac{x^3}{3a^2} \right|_0^a = \frac{4\pi abc}{3}.$$

**(3) Volumes of solids of revolution**

Consider an elementary area $\delta x \delta y$ at the point $P(x, y)$ of a plane area $A$. (Fig. 7.25)

As this elementary area revolves about $x$-axis, we get a ring of volume

$$= \pi[(y + \delta y)^2 - y^2]\, \delta x = 2\pi y\, \delta x\, \delta y,$$

nearly to the first powers of $\delta y$.

Hence the total volume of the solid formed by the revolution of the area $A$ about $x$-axis.

$$= \iint_A 2\pi y\, dxdy.$$

In polar coordinates, the above formula for the volume becomes

$$\iint_A 2\pi r \sin\theta \,.\, rd\theta dr, \text{ i.e. } \iint_A 2\pi r^2 \sin\theta\, d\theta dr$$

Similarly, the volume of the solid formed by the revolution of the area $A$ about $y$-axis $= \iint_A 2\pi x\, dx\, dy$.

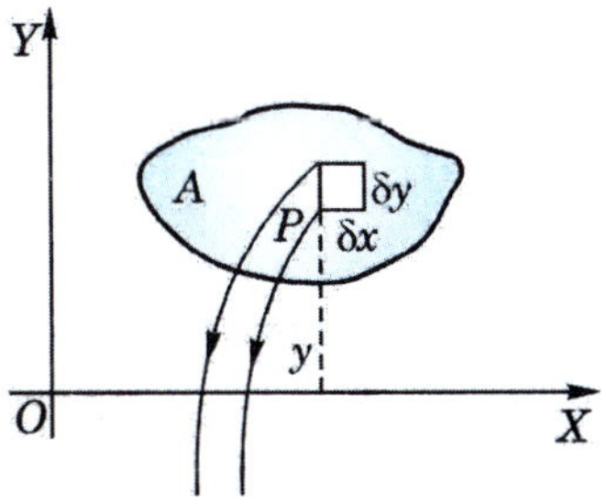


Fig. 7.26.

**Example 7.25.** *Prove that the volume enclosed between the cylinders*

$$x^2 + y^2 = 2ax \quad \text{and} \quad z^2 = 2ax \text{ is } \frac{128a^3}{15}$$

*(Andhra, 2019 ; Kerala, 2018)*

**Solution:** The required volume $V$ is that volume of the cylinder

$$x^2 + y^2 = 2ax \qquad \text{...}(i)$$

which in bounded by the cylinder

$$z^2 = 2ax \qquad \text{...}(ii)$$

only half of the volume is shown in the figure

$$x^2 + y^2 = 2ax$$

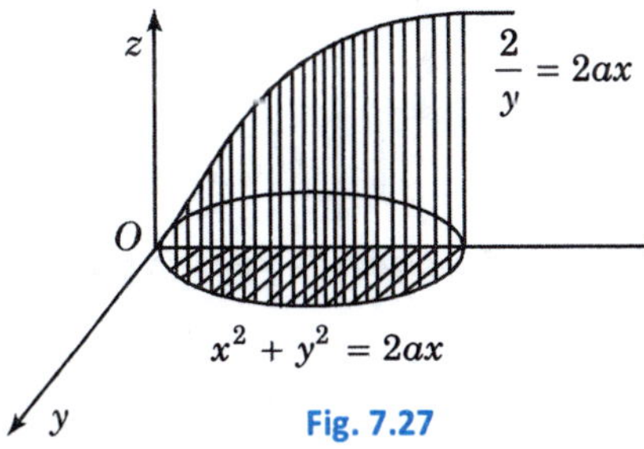


Fig. 7.27

Thus on this circle $y$ varies from $-\sqrt{2ax - x^2}$ to $\sqrt{2ax - x^2}$ and $x$ varies from 0 to $2a$.

$$V = 2\int_0^{2a}\int_{-\sqrt{2ax-x^2}}^{\sqrt{2ax-x^2}} z dy\, dx$$

$$= 4\int_0^{2a}\int_{\sqrt{2ax}}^{\sqrt{2ax-x^2}} dy\, dx = 4\sqrt{20}\int_0^{2a}\sqrt{x}\,[y]_0^{\sqrt{2ax-x^2}}\, dx$$

$$= 4\cdot\sqrt{2a}\int_0^{2a}\sqrt{x}\sqrt{2ax - x^2}\, dx$$

$$x = 2a\sin^2\theta \; dx = 4a\sin\theta\cos\theta \cdot d\theta$$

put $\qquad x = 2a, \quad \theta = \dfrac{\pi}{2}$

and when $\qquad x = 0, \quad \theta = 0$

$$V = 4\sqrt{2a}\int_0^{\pi/2}\sqrt{2a\sin^2\theta}\cdot\sqrt{4a^2\sin^2\theta - 4a^2\sin^4\theta}\,.\,4a\sin\theta\cos\theta\cdot d\theta$$

$$= 64a^3 \int_0^{\pi/2} \sin\theta \sin\theta \cdot \cos\theta \cdot \sin\theta \cos\theta \cdot d\theta$$

$$= 64a^3 \int_0^{\pi/2} \sin^3\theta \cos^2\theta \, d\theta = 64a^3 \cdot \frac{2}{5\cdot 3\cdot 1} = \frac{128a^3}{15}.$$

## 7.7 CHANGE OF VARIABLES

An appropriate choice of co-ordinates quite often facilitates the evaluation of a double or a triple integral. By changing the variables, a given integral can be transformed into a simpler integral involving the new variables.

**(1) In a double integral,** let the variables $x, y$ be changed to the new variables $u, v$ by the transformation.

$$x = \phi(u, v), \; y = \psi(u, v)$$

where $\phi(u, v)$ and $\psi(u, v)$ are continuous and have continuous first order derivatives in some region $R'_{uv}$ in the $uv$-plane which corresponds to the region $R_{xy}$ in the $xy$-plane. Then

$$\iint_{R_{xy}} f(x, y)\, dxdy = \iint_{R'_{uv}} f[\phi(u, v), \psi(u, v)] \,|\, J \,|\, dudv \quad ...(1)$$

where $J = \dfrac{\partial(x, y)}{\partial(u, v)} \; (\neq 0)$

is the *Jacobian of transformation** from $(x, y)$ to $(u, v)$ coordinates.

**(2) For triple integrals,** the formula corresponding to (1) is

$$\iiint_{R_{xyz}} f(x, y, z)\, dxdydz = \iiint_{R'_{uvw}} f[x(u, v, w), y(u, v, w), z(u, v, w)] \,|\, J \,|\, dudvdw$$

where $J = \dfrac{\partial(x, y, z)}{\partial(u, v, w)} \; (\neq 0)$

is the *Jacobian of transformation* from $(x, y, z)$ to $(u, v, w)$ coordinates.

**Particular cases :**

*(i)* ***To change cartesian coordinates (x, y) to polar coordinates (r, θ),*** *we have* $x = r\cos\theta$, $y = r\sin\theta$ *and*

$$J = \frac{\partial(x, y)}{\partial(r, \theta)} = r$$

$$\therefore \quad \iint_{R_{xy}} f(x, y)\, dx\, dy = \iint_{R'_{r\theta}} f(r\cos\theta, r\sin\theta) \,.\, r\, dr\, d\theta.$$

*(ii)* ***To change rectangular coordinates (x, y, z) to cylindrical coordinates*** $(\rho, \phi, z)$ — *Fig.* 8.27, *we have*

$$x = \rho\cos\phi, \; y = \rho\sin\phi, \; z = z$$

and $J = \dfrac{\partial(x, y, z)}{\partial(\rho, \phi, z)} = \rho$

Then $\displaystyle\iiint_{R_{xyz}} f(x, y, z)\, dxdydz = \iiint_{R'_{\rho\phi z}} f(\rho\cos\phi, \rho\sin\phi, z) \,.\, \rho d\rho d\phi dz.$

*(iii)* ***To change rectangular coordinates (x, y, z) to spherical polar coordinates (r,*** $\theta, \phi$)—*Fig.* 8.28, *we have*

$$x = r\sin\theta\cos\phi, \; y = r\sin\theta\sin\phi, \; z = r\cos\theta$$

and $J = \dfrac{\partial(x, y, z)}{\partial(r, \theta, \phi)} = r^2 \sin\theta$

Then $\displaystyle\iiint_{R_{xyz}} f(x, y, z)\, dxdydz = \iiint_{R'_{r\theta\phi}} f(r\sin\theta\cos\phi, r\sin\theta, \sin\phi, r\cos\theta) \,.\, r^2 \sin\theta\, drd\theta d\phi$

**Example 7.26.** *Find the volume of the region bounded by the cylinder $x^2 + y^2 = 4$ and the cone $x^2 + y^2 = z^2 \tan^2 \alpha$ $(0 < \alpha < \pi)$* *(Bhilai, 2019)*

**Solution :** The equation of the cone is

$$x^2 + y^2 = z^2 \tan^2 \alpha \qquad ...(i)$$

$$\Rightarrow \qquad z = \sqrt{(x^2 + y^2)\cot^2 \alpha}.$$

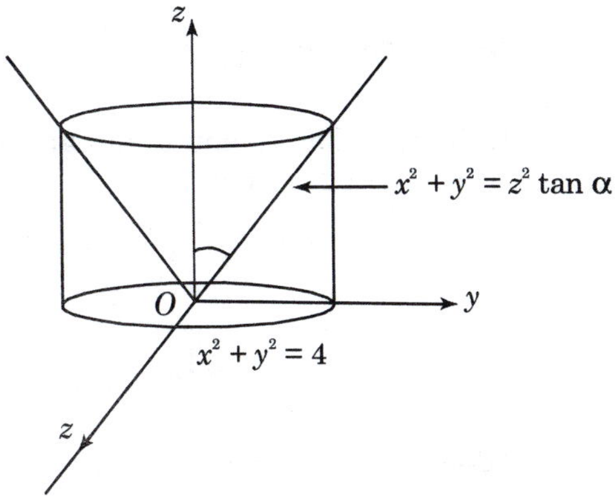

Fig. 7.28.

Only one eighth of the volume is shown in the figure,

Changing to polar co-ordinates by substituting

$x = r\cos\theta$, $y = r\sin\theta$, we get $z = r\cot\alpha$

The equation of the circle become $r = 2$.

$\therefore$ The required volume $= 8\int_0^{\pi/2}\int_0^2 zr\,dr\cdot d\theta$

$$= 8\int_0^{\pi/2}\int_0^2 r\cot\alpha\cdot r\cdot dr\cdot d\theta$$

$$= 8\cot\alpha\int_0^{\pi/2}\left[\frac{r^3}{3}\right]_0^2 d\theta$$

$$= \frac{8}{3}\cot\alpha\cdot\int_0^{\pi/2} 8\cdot d\theta$$

$$= \frac{64}{3}\cot\alpha\cdot\frac{\pi}{2} = \frac{32}{3}\pi\cot\alpha$$

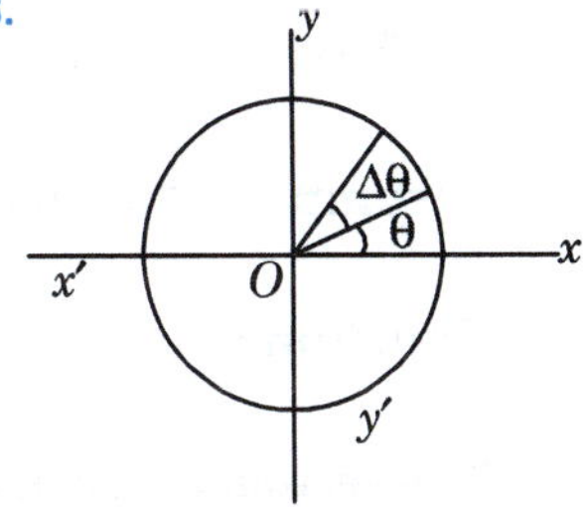

Fig. 7.29.

**Example 7.27 :** *Find the volume under the paraboloid $x^2 + y^2 + z^2 = 16$, over the rectangle $x = \pm a$, $y = \pm b$ in the xy-plane (where $a, b < 4$)* *(Mysore, 2019)*

**Solution:**

The equation of the parabolied is

$$x^2 + y^2 + z^2 = 16$$

$$z = 16 - x^2 - y^2 \qquad ...(i)$$

Its section by the $x - y$ plane $(z = 0)$ is the circle

$$x^2 + y^2 = 16 \qquad ...(ii)$$

(only one fourth of the volume is shown in the figure).

The required volume $V$ is obtained by integrating $z$ over this rectangle $x = \pm a, y = \pm b$

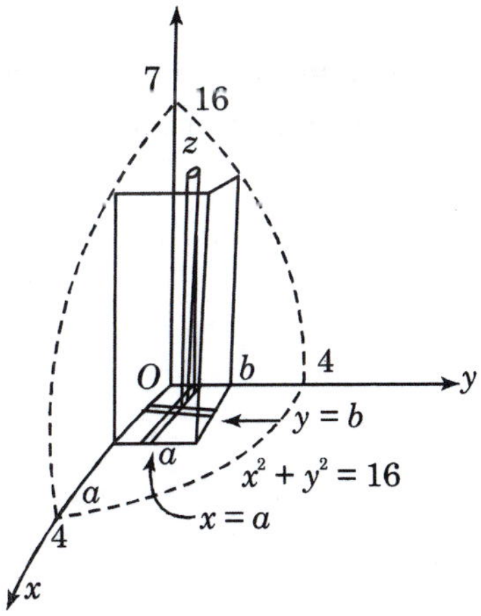

Fig. 7.30.

$$\because \qquad V = 4\int_0^a\int_0^b z\,dy\,dz$$

$$= 4\int_0^a\int_0^b (16 - x^2 - y^2)\,dy\,dz$$

$$= 4\int_0^a\left[16y - x^2y - \frac{y^3}{3}\right]_0^b dx$$

$$= 4\int_0^a\left(16b - x^2b - \frac{b^3}{3}\right)dx$$

$$= 4\left[16bx - \frac{x^3}{3}b - \frac{b^3x}{3}\right]_0^a = 4\left[16ab - \frac{a^3b}{3} - \frac{ab^3}{3}\right]$$

$$= \frac{4}{3}ab(48 - a^2 - b^2).$$

**Example 7.28.** *Evaluate* $\int_0^\infty \int_0^\infty e^{-(x^2+y^2)}\, dxdy$ *by changing to polar coordinates.*

(*Andhra., 2022, 2016 ; Anna., 2016 ; V.T.U., 2016 ; P.T.U., 2011*)

*Hence show that* $\int_0^\infty e^{-x^2}\, dx = \sqrt{\pi}/2$.

**Solution.** The region of integration being the first quadrant of the *xy*-plane, *r* varies from 0 to ∞ and θ varies from 0 to π/2. Hence,

$$I = \int_0^\infty \int_0^\infty e^{-(x^2+y^2)}\, dxdy = \int_{\theta=0}^{\pi/2} \int_{r=0}^{\infty} e^{-r^2}\, rdrd\theta$$

$$= -\frac{1}{2}\int_0^{\pi/2} \left\{\int_0^\infty e^{-r^2}(-2r)\,dr\right\} d\theta = -\frac{1}{2}\int_0^{\pi/2} \left| e^{-r^2} \right|_0^\infty d\theta = \frac{1}{2}\int_0^{\pi/2} d\theta = \frac{\pi}{4}. \qquad ...(i)$$

Also $$I = \int_0^\infty e^{-x^2}\, dx \times \int_0^\infty e^{-y^2}\, dy = \left\{\int_0^\infty e^{-x^2}\, dx\right\}^2 \qquad ...(ii)$$

Thus, from Eqns. (*i*) and (*ii*), we have $\int_0^\infty e^{-x^2} = \sqrt{\pi}/2$. ...(*iii*)

**Example 7.29.** *Find the volume bounded by xy plane, the cylinder* $x^2 + y^2 = 1$ *and the plane* $x + y + z = 3$. (*Rohtak, 2019*)

**Solution :**

The section of the plane $x + y + z = 3$ by the *x-y* plane is the straight line

$$x + y + z = 3$$

Let *V* be the volume of the solid figure which is bounded laterally by the cylinder

$$x^2 + y^2 = 1$$

and on the top by the plane $x + y + z = 3$

Thus limits of *z* are from 0 to $3 - x - y$ and *x* or *y* vary on the circle $x^2 + y^2 = 1$

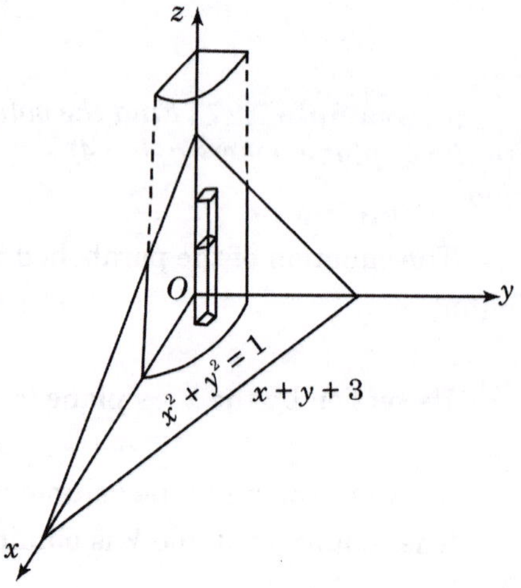


Fig. 7.31.

∴ $$V = \int_{-1}^{1}\int_{-\sqrt{1-x^2}}^{\sqrt{1-x^2}}\int_0^{3-x-y} dz\, dy\, dx$$

$$= \int_{-1}^{1}\int_{-\sqrt{1-x^2}}^{\sqrt{1-x^2}} [z]_0^{3-x-y}\, dydx$$

$$= \int_{-1}^{1}\int_{-\sqrt{1-x^2}}^{\sqrt{1-x^2}} (3 - x - y)dydx$$

$$= \int_{-1}^{1}\left(3y - xy - \frac{y^2}{2}\right)_{-\sqrt{1-x^2}}^{\sqrt{1-x^2}} dx$$

$$= \int_{-1}^{1}\left(6\sqrt{1-x^2} - 2x\sqrt{1-x^2}\right) dx$$

put $$x = \sin\theta$$

then $$dx = \cos\theta \cdot d\theta$$

Here θ varies from $-\frac{\pi}{2}$ to $\frac{\pi}{2}$

∴ $$V = \int_{-\pi/2}^{\pi/2}\left(6\cdot\sqrt{1-\sin^2\theta} - 2\sin\theta\sqrt{1-\sin^2\theta}\right)\cos\theta\cdot d\theta$$

$$= \int_{-\pi/2}^{\pi/2}(6\cos^2\theta - 2\sin\theta\cos^2\theta)\, d\theta$$

$$= 12\int_0^{\pi/2} \cos^2\theta\, d\theta - \int_{-\pi/2}^{\pi/2} 2\sin\theta\cos^2\theta\cdot d\theta$$

$$= 12\cdot\frac{1}{2}\cdot\frac{\pi}{2} + 2\left[\frac{\cos^3\theta}{3}\right]_{-\pi/2}^{\pi/2} = 3\pi$$

**Example 7.30.** *Find, by triple integration, the volume of the sphere $x^2 + y^2 + z^2 = a^2$.* (Andhra., 2016)

**Solution.** Changing to polar spherical coordinates by putting

$$x = r\sin\theta\cos\phi,\ y = r\sin\theta\sin\phi,\ z = r\cos\theta$$

we have $dx\, dy\, dz = r^2 \sin\theta\, dr\, d\theta\, d\phi$.

Also the volume of the sphere is 8 times the volume of its portion in the positive octant for which $r$ varies from 0 to $a$, $\theta$ varies from 0 to $\pi/2$ and $\phi$ varies from 0 to $\pi/2$.

∴ volume of the sphere

$$= 8\int_0^a \int_0^{\pi/2} \int_0^{\pi/2} r^2 \sin\theta\, dr\, d\theta\, d\phi = 8\int_0^a r^2\, dr \,.\int_0^{\pi/2} .\sin\theta\, d\theta \,.\int_0^{\pi/2} d\phi$$

$$= 8\,.\left|\frac{r^3}{3}\right|_0^a .\left|-\cos\theta\right|_0^{\pi/2} .\frac{\pi}{2} = 4\pi\,.\frac{a^3}{3}\,.(-0+1) = \frac{4}{3}\pi a^3.$$

**Example 7.31.** *Find the volume of the portion of the sphere $x^2 + y^2 + z^2 = a^2$ lying inside the cylinder $x^2 + y^2 = ay$.* (Andhra, 2020 ; U.P.T.U., 2012)

**Solution.** The required volume is easily found by changing to cylindrical coordinates $(\rho, \phi, z)$. Therefore, we have

$$x = \rho\cos\phi,\ y = \rho\sin\phi,\ z = z$$

and $$J = \frac{\partial(x, y, z)}{\partial(\rho, \phi, z)} = \rho.$$

Then the equation of the sphere becomes $\rho^2 + z^2 = a^2$ and that of cylinder becomes $\rho = a\sin\phi$.

The volume inside the cylinder bounded by the sphere is twice the volume shown shaded in the Fig. 7.32 for which $z$ varies from 0 to $\sqrt{(a^2 - \rho^2)}$, $\rho$ varies from 0 to $a\sin\phi$ and $\phi$ varies from 0 to $\pi$.

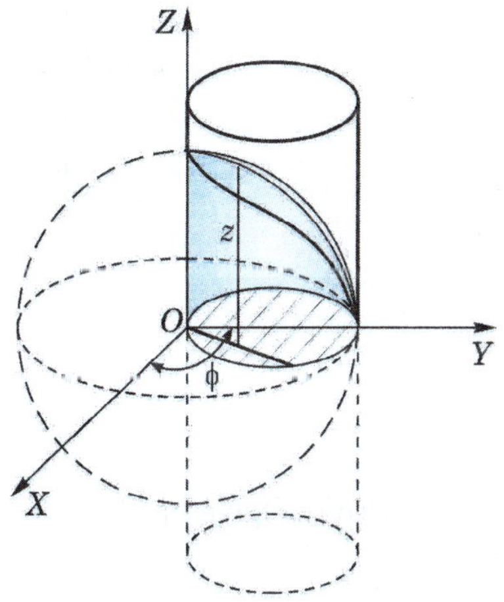

Fig. 7.32.

Hence the required volume $= 2\int_0^{\pi}\int_0^{a\sin\phi}\int_0^{\sqrt{(a^2-\rho^2)}} \rho dz d\rho d\phi$

$$= 2\int_0^{\pi}\int_0^{a\sin\phi} \rho\sqrt{(a^2-\rho^2)}\, d\rho d\phi = 2\int_0^{\pi}\left|-\frac{1}{3}(a^2-\rho^2)^{3/2}\right|_0^{a\sin\phi} d\phi$$

$$= \frac{2a^3}{3}\int_0^{\pi}(1-\cos^3\phi)\, d\phi = \frac{2a^3}{9}(3\pi - 4).$$

**Example 7.32.** *Find, by double integration, the volume of the solid generated by revolving the ellipse $\frac{x^2}{a^2} + \frac{y^2}{b^2} = 1$ about the x-axis.* (Kottayam, 2019)

**Solution.** Here we shall consider the revolution of the only upper half of the ellipse.

$$V = \int_{-a}^{a}\int_0^{b/a\sqrt{a^2-x^2}} 2\pi y\, dydx = 2\pi\int_{-a}^{a}\left[\frac{y^2}{2}\right]_0^{(b/a)\sqrt{a^2-x^2}} dx$$

$$= \pi \int_{-a}^{a} \frac{b^2}{a^2}(a^2 - x^2)\,dx$$

$$= \frac{2\pi b^2}{a^2}\int_0^a (a^2 - x^2)\,dx$$

$$= \frac{2\pi b^2}{a^2}\left[a^2 x - \frac{x^3}{3}\right]_0^a$$

$$= \frac{2\pi b^2}{a^2}\left[a^3 - \frac{a^3}{3}\right] = \frac{4\pi a b^2}{3}$$

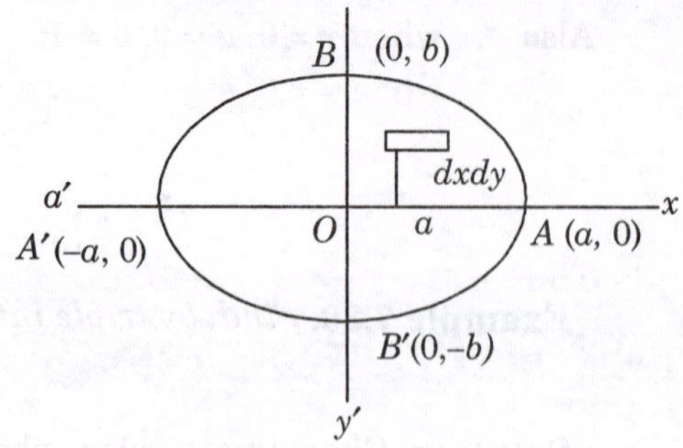


Fig. 7.33.

**Example 7.33 :** *Evaluate* $\iint_R (x + y)^2\, dxdy$, *where R is the parallelogram in the xy-plane with vertices (1, 0), (3, 1), (2, 2), (0, 1) using the transformation u = x + y and v = x – 2y.*

**Solution.** The region $R$, *i.e.*, parallelogram $ABCD$ in the $xy$-plane becomes the region $R'$, *i.e.*, rectangle $A'B'C'D'$ in the $uv$-plane as shown in Fig. 7.34, by taking

$$u = x + y \quad \text{and} \quad v = x - 2y \qquad \ldots(i)$$

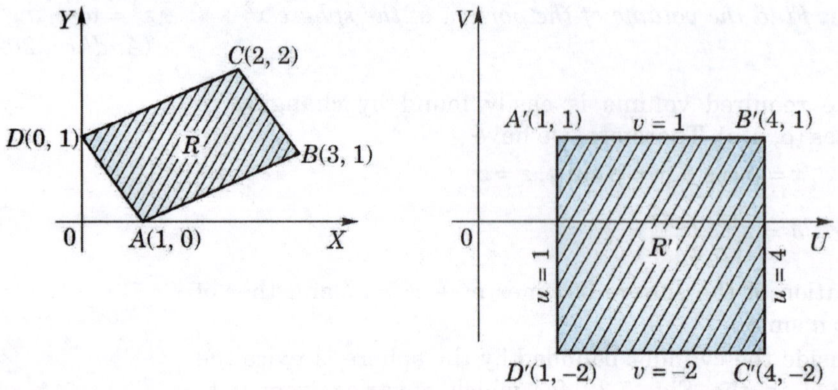


Fig. 7.34.

From (*i*), we have $\quad x = \frac{1}{3}(2u + v),\ y = \frac{1}{3}(u - v)$

$$\therefore \quad \frac{\partial(x, y)}{\partial(u, v)} = \begin{vmatrix} \frac{\partial x}{\partial u} & \frac{\partial y}{\partial u} \\ \frac{\partial x}{\partial v} & \frac{\partial y}{\partial v} \end{vmatrix} = -\frac{1}{3}$$

Hence, the given integral

$$= \iint_{R'} u^2 \,|\,J\,|\, dudv = \int_1^4 \int_{-2}^{1} u^2 \cdot \frac{1}{3} \cdot dudv = \frac{1}{3}\left|\frac{u^3}{3}\right|_1^4 \cdot |\,v\,|_{-2}^{1} = 21.$$

**Example 7.34 :** *Evaluate* $\iint_D xy\sqrt{(1 - x - y)}\, dx\, dy$ *where D is the region bounded by x = 0, y = 0 and x + y = 1 using the transformation x + y = u, y = uv.*

**Solution.** We have $x = u - uv,\ y = uv$

$$\therefore \quad J = \frac{\partial(x, y)}{\partial(u, v)} = \begin{vmatrix} \partial x/\partial u & \partial y/\partial u \\ \partial x/\partial v & \partial y/\partial v \end{vmatrix} = \begin{vmatrix} 1 - v & v \\ -u & u \end{vmatrix} = u.$$

Also when $x = 0, u = 0, v = 1$;
when $y = 0, u = 0, v = 0$ and
when $x + y = 1, u = 1$

$\therefore$ the limits of $u$ are from 0 to 1 and limits of $v$ are from 0 to 1.

Thus $$\iint_D xy\sqrt{(1-x-y)}\,dxdy = \int_0^1\int_0^1 u(1-v)\,uv\,(1-u)^{1/2}\,|J|\,dudv$$

$$= \int_0^1\int_0^1 u^3(1-u)^{1/2}\,v(1-v)\,du\,dv$$

where $u = \sin^2\theta$
$du = 2\sin\theta\cos\theta\,d\theta$
$u = 0, \theta = 0$
$u = 1, \theta = \pi/2$

$$= \int_0^1 u^3(1-u)^{1/2}\,du \times \int_0^1 v(1-v)\,dv$$

$$= \int_0^{\pi/2} \sin^6\theta\cos\theta\,.\,2\sin\theta\cos\theta\,d\theta \times \left|\frac{v^2}{2}-\frac{v^3}{3}\right|_0^1$$

$$= 2\int_0^{\pi/2} \sin^7\theta\cos^2\theta\,d\theta\left(\frac{1}{6}\right) = \frac{1}{3}\cdot\frac{6\,.\,1}{9\,.\,7\,.\,5\,.\,3} = \frac{2}{945}.$$

**Example 7.35 :** *Find the volume of the region bounded above by the sphere $x^2 + y^2 + z^2 = a^2$ and below by the cone $x^2 + y^2 = z^2$.* *(Marathwada, 2019)*

**Solution:** The equation of the sphere is

$$x^2 + y^2 + z^2 = a^2 \quad ...(i)$$

and that of the cone is

$$x^2 + y^2 = z^2 \quad ...(ii)$$

$$x = r\sin\theta\cos\phi$$

$$y = r\sin\theta\sin\phi, \quad z = r\cos\theta$$

Thus the equations (*i*) and (*ii*) polar spherical co-ordinates are respectively $r = a$ and $\theta = \pm\frac{\pi}{4}$. The volume shown in the figure is one-fourth only.

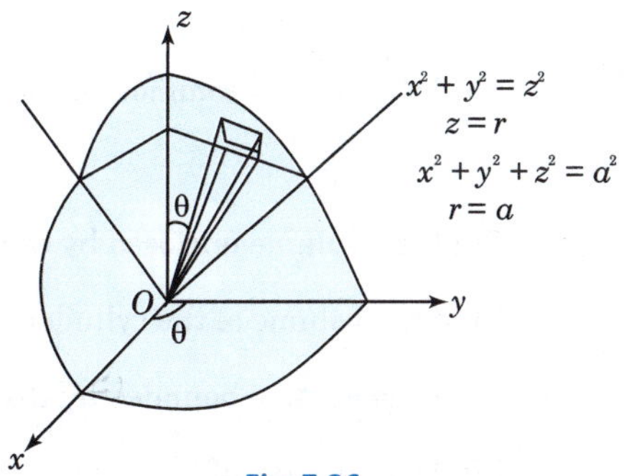

Fig. 7.36

$$V = 4\int_0^{\pi/2}\int_0^{\pi/4}\int_0^a r^2\sin\theta\,dr\,d\theta\,d\phi$$

$$= 4\int_0^{\pi/2}\int_0^{\pi/4}\left[\frac{r^3}{3}\right]_0^a \sin\theta\,d\theta\,d\phi$$

$$= \frac{4}{3}a^3\int_0^{\pi/2}[-\cos\theta]_0^{\pi/4}\,d\phi$$

$$= \frac{4}{3}a^3\left(1-\frac{1}{\sqrt{2}}\right)\int_0^{\pi/2}d\phi = \frac{4}{3}a^3\left(1-\frac{1}{\sqrt{2}}\right)[\phi]_0^{\pi/2} = \frac{2}{3}\pi a^3\left(1-\frac{1}{\sqrt{2}}\right)$$

## PROBLEMS 7.4

Evaluate the following integrals by changing to polar co-ordinates :

1. $\int_0^1\int_0^{\sqrt{(1-y^2)}} (x^2+y^2)\,dy\,dx.$ *(U.P.T.U., 2012 ; P.T.U., 2010)*

2. $\int_0^2\int_0^{\sqrt{(2x-x^2)}} \frac{x\,dx\,dy}{x^2+y^2}$

3. $\int_0^{4a}\int_{y^2/4a}^{y} \frac{x^2-y^2}{x^2+y^2}\,dxdy$ (*J.N.T.U., 2015 ; Andhra., 2013*)

4. Transform the following to cartesian form and hence evaluate $\int_0^{\pi}\int_0^{a} r^3 \sin\theta\cos\theta\, dr d\theta$.

5. By using the transformation $x + y = u, y = uv$, show that $\int_0^1\int_0^{1-x} e^{y/(x+y)}\,dydx = \frac{1}{2}(e-1)$.

6. Transform $\int_0^{\pi/2}\int_0^{\pi/2} \sqrt{\frac{\sin\phi}{\sin\theta}}\,d\phi\, d\theta$ by the substitution $x = \sin\phi\cos\theta, y = \sin\phi\sin\theta$ and show that its value is $\pi$.

Evaluate the following integrals by changing to spherical coordinates :

7. $\iiint_V \frac{dx\,dy\,dz}{x^2+y^2+z^2}$ where $V$ is the volume of the sphere $x^2 + y^2 + z^2 = a^2$.

8. Evaluate $\iiint \frac{dx\,dy\,dz}{(1+x+y+z)^3}$ over the volume of the tetrahedron $x = 0, y = 0, z = 0, x + y + z = 1$.

(*Anna., 2016; C.S.V.T.U., 2014*)

9. Show that $\iiint \frac{dx\,dy\,dz}{\sqrt{(a^2-x^2-y^2-z^2)}} = \frac{\pi^2a^3}{8}$, the integral being extended for all the values of the variables for which the expression is real. (*U.K.T.U., 2010*)

10. Find the volume bounded by the $xy$-plane, the cylinder $x^2 + y^2 = 1$ and the plane $x + y + z = 3$.

(*D.T.U., 2015 ; C.S.V.T.U., 2015*)

11. Find the volume bounded by the $xy$-plane, the paraboloid $2z = x^2 + y^2$ and the cylinder $x^2 + y^2 = 4$.

(*Andhra, 2023, 2015*)

12. Find the volume common to the cylinders $x^2 + y^2 = a^2$ and $x^2 + z^2 = a^2$.

(*Andhra., 2023, 2016 ; V.T.U., 2016 ; C.S.V.T.U., 2014*)

13. Find the volume enclosed by the cylinders $x^2 + y^2 = 2ax$ and $z^2 = 2ax$. (*J.N.T.U., 2015*)

14. Find the volume of the cylinder $x^2 + y^2 - 2ax = 0$, intercepted between the paraboloid $x^2 + y^2 = 2az$ and the $xy$-plane.

15. Find the volume bounded by the cylinder $x^2 + y^2 = 4$ and the hyperboloid $x^2 + y^2 - z^2 = 1$.

16. Find the volume of the region bounded by $z = x^2 + y^2, z = 0, x = -a, x = a$ and $y = -a, y = a$.

17. Prove, by using a double integral that the volume generated by the revolution of the cardioid $r = a(1 + \cos\theta)$ about its axis is $8\pi a^3/3$.

18. Find the volume of the tetrahedron bounded by the coordinate planes and the plane $\frac{x}{a}+\frac{y}{b}+\frac{z}{c} = 1$.

(*Andhra, 2022 ; U.P.T.U., 2012*)

19. Find the volume generated by the revolution of the curve $y^2(2a - x) = x^3$, about its asymptote through four right angles.

(*Bangalore, 2018*)

20. An trigangular prism is formed by the planes whose equations are $ay = bx, y = 0$ and $x = a$, prove that the volume of this prism between the plane $z = 0$ and the surface $z = c + xy$ is $\left(\frac{ab}{8}\right)(4c + ab)$ (*Kottayam, 2019*)

21. Find the volume of the cylindrical column standing on the area common to the parabolas $y^2 = x, x = y^2$ and cut off by the surface $z = 12 + y - x^2$ (*Bhilai, 2019*)

22. Find the volume bounded by the cylinder $x^2 + y^2 = 4$ and the planes $y + z = 4$ and $z = 0$ (*Gorakhpur, 2018*) (**Ans.** $16\pi$)

23. Find the volume cut off from the sphere $x^2 + y^2 + z^2 = a^2$ by the cylinder $x^2 + y^2 = ax$ (*Ranchi, 2017*)

24. By using double integral, show that the volume generated by revolution of the cardiod $r = a(1 + \cos\theta)$ about the initial line is $\frac{8}{3}\pi a^3$ (*Marathawada, 2018*)

25. The area bounded by the parabola $y^2 = 4x$ and the straight lines $x = 1$ and $y = 0$, in the first quadrant is revolved about the line $y = 2$. Find by double integration the volume of the solid generated. (*Bhilai, 2017*)

26. Show that the double integration that the area between the parabolas $y^2 = 4ax$ and $x^2 = 4ay$ is $\frac{16}{3}a^3$. (*Kanpur, 2019*)

27. Using double integration, find the mass of a plane in the shape of the curve $\left(\frac{x}{a}\right)^{2/3} + \left(\frac{y}{b}\right)^{2/3} = 1$, the density being given by $\rho = kxy$. (*Kerala, 2018*)

28. Calculate by double integration the volume generated by the revolution of the cardiod $r = a(1 - \cos\theta)$ about its axis. (*Bangalore, 2019*)

## 7.8 AREA OF A CURVED SURFACE

Consider a point $P$ of the surface $S : z = f(x, y)$. Let its projection on the $xy$-plane be the region $A$. Divide it into area elements by drawing lines parallel to the axes of $X$ and $Y$. (Fig. 7.37).

On the element $\delta x\delta y$ as base, erect a cylinder having generators parallel to $OZ$ and meeting the surface $S$ in an element of area $\delta S$.

As $\delta x\delta y$ is the projection of $\delta S$ on the $xy$-plane,

$\therefore$ $\delta x\delta y = \delta S \,.\, \cos\gamma$, where $\gamma$ is the angle between the $xy$-plane and the tangent plane to $S$ at $P$, *i.e.*, it is the angle between the $Z$-axis and the normal to $S$ at $P(= \angle Z'PN)$.

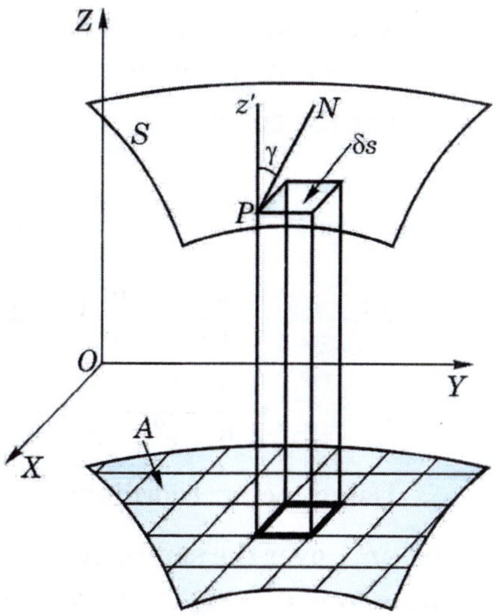

Fig. 7.37.

Now since the direction cosines of the normal to the surface $F(x, y, z) = 0$ proportional to

$$\frac{\partial F}{\partial x}, \frac{\partial F}{\partial y}, \frac{\partial F}{\partial z}.$$

$\therefore$ the direction cosines of the normal to $S[F = f(x, y) - z]$ are proportional to $-\frac{\partial z}{\partial x}, -\frac{\partial z}{\partial y}$, 1 and those of the $z$-axis are 0, 0, 1.

Hence $$\cos\gamma = \frac{1}{\sqrt{\left[\left(\frac{\partial z}{\partial x}\right)^2 + \left(\frac{\partial z}{\partial y}\right)^2 + 1\right]}}$$

$$\therefore \qquad \delta S = \frac{\delta x\, \delta y}{\cos\gamma} = \sqrt{\left[\left(\frac{\partial z}{\partial x}\right)^2 + \left(\frac{\partial z}{\partial y}\right)^2 + 1\right]}\, \delta x\, \delta y$$

Hence $S = \underset{\delta S \to 0}{\text{Lt}} \sum \delta S = \iint_A \sqrt{\left[\left(\frac{\partial z}{\partial x}\right)^2 + \left(\frac{\partial z}{\partial y}\right)^2 + 1\right]}\, dxdy$

Similarly, if $B$ and $C$ be the projections of $S$ on the $yz$-and $zx$-planes respectively, then

$$S = \iint_B \sqrt{\left[\left(\frac{\partial x}{\partial y}\right)^2 + \left(\frac{\partial x}{\partial z}\right)^2 + 1\right]}\, dydz$$

and $$S = \iint_C \sqrt{\left[\left(\frac{\partial y}{\partial z}\right)^2 + \left(\frac{\partial y}{\partial x}\right)^2 + 1\right]}\, dzdx.$$

**Example 7.36.** *Find the area of the portion of the cylinder $x^2 + z^2 = 4$ lying inside the cylinder $x^2 + y^2 = 4$.*

**Solution.** Figure 7.38 shows one-eighth of the required area. Its projection on the $xy$-plane is a quadrant circle $x^2 + y^2 = 4$.

For the cylinder $x^2 + z^2 = 4$, ...(*i*)

we have $\frac{\partial z}{\partial x} = -\frac{x}{z}, \frac{\partial z}{\partial y} = 0.$

so that $\left(\frac{\partial z}{\partial x}\right)^2 + \left(\frac{\partial z}{\partial y}\right)^2 + 1 = \frac{x^2 + z^2}{z^2} = \frac{4}{4 - x^2}.$

*Hence the required surface area* = 8 (surface area of the upper portion of Eqn. (*i*) lying within the cylinder $x^2 + y^2 = 4$ in the positive octant)

$$= 8\int_0^2 \int_0^{\sqrt{(4-x^2)}} \frac{2}{\sqrt{(4 - x^2)}}\, dxdy = 16\int_0^2 dx = 32 \text{ sq. units.}$$

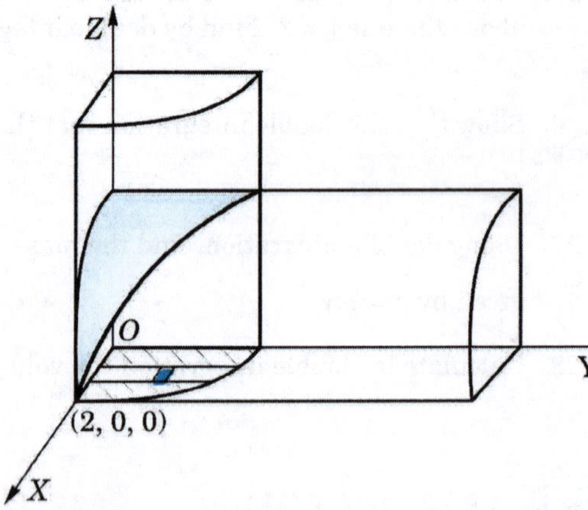

Fig. 7.38

**Example 7.37.** *Find the area of the portion of the sphere $x^2 + y^2 + z^2 = 9$ lying inside the cylinder $x^2 + y^2 = 3y$.*

**Solution.** Figure 7.39 shows one-fourth of the required area. Its projection on the $xy$-plane is the semi-circle $x^2 + y^2 = 3y$ bounded by the $Y$-axis.

For the sphere

$$x^2 + y^2 + z^2 = 9, \frac{\partial z}{\partial x} = -\frac{x}{z} \text{ and } \frac{\partial z}{\partial y} = -\frac{y}{z}$$

$$\therefore \quad \left(\frac{\partial z}{\partial x}\right)^2 + \left(\frac{\partial z}{\partial y}\right)^2 + 1 = (x^2 + y^2 + z^2)/z^2$$

$$= \frac{9}{9 - x^2 - y^2} = \frac{9}{9 - r^2} \quad \text{when } x = r\cos\theta, y = r\sin\theta.$$

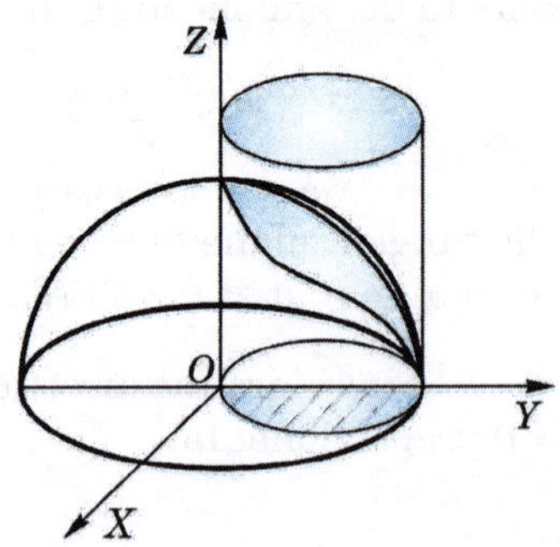

Fig. 7.39

Using polar coordinates, the required area is found by integrating $3/\sqrt{(9 - r^2)}$ over the semi-circle $r = 3\sin\theta$, for which $r$ varies from 0 to $3\sin\theta$ and $\theta$ varies from 0 to $\pi/2$.

*Hence the required surface area*

$$= 4\int_0^{\pi/2} \int_0^{3\sin\theta} \frac{3}{\sqrt{(9 - r^2)}}\, r d\theta\, dr = -6\int_0^{\pi/2} \left| \frac{\sqrt{(9 - r^2)}}{1/2} \right|_0^{3\sin\theta} d\theta$$

$$= 36\int_0^{\pi/2} (1 - \cos\theta)\, d\theta = 36 \left| \theta - \sin\theta \right|_0^{\pi/2} = 18(\pi - 2) \text{ sq. units.}$$

## PROBLEMS 7.5

1. Show that the surface area of the sphere $x^2 + y^2 + z^2 = a^2$ is $4\pi a^2$.
2. Find the area of the portion of the cylinder $x^2 + y^2 = 4y$ lying inside the sphere $x^2 + y^2 + z^2 = 16$.
3. Find the area of the portion of the sphere $x^2 + y^2 + z^2 = a^2$ lying inside the cylinder $x^2 + y^2 = ax$.
4. Find the area of the surface of the cone $x^2 + y^2 = z^2$ cut off by the surface of the cylinder $x^2 + y^2 = a^2$ above the $xy$-plane.
5. Compute the area of that part of the plane $x + y + z = 2a$ which lies in the first octant and is bounded by the cylinder $x^2 + y^2 = a^2$.

## 7.9 CALCULATION OF MASS

(*a*) **For a plane lamina,** if the surface density at the point $P(x, y)$ be $\rho = f(x, y)$ then the elementary mass at $P = \rho\delta x\delta y$.

$$\therefore \quad \text{total mass of the lamina} = \iint \rho\, dx\, dy \qquad ...(i)$$

with integrals embracing the whole area of the lamina.

In polar coordinates, taking $\rho = \phi(r, \theta)$ at the point $P(r, \theta)$,

$$\text{total mass of the lamina} = \iint \rho r\, d\theta\, dr \qquad ...(ii)$$

(*b*) **For a solid,** if the density at the point $P(x, y, z)$ be $\rho = f(x, y, z)$, then

*total mass of the solid* $= \iiint \rho\, dx\, dy\, dz$ with appropriate limits of integration.

**Example 7.38 :** *Find the mass of a plate in the form of one loop of the leminscate $r^2 - a^2 \sin^2 \theta$, where $\rho = kr^2$* (Kottayam, 2019)

**Solution:** Let $M$ be the required mass. Then

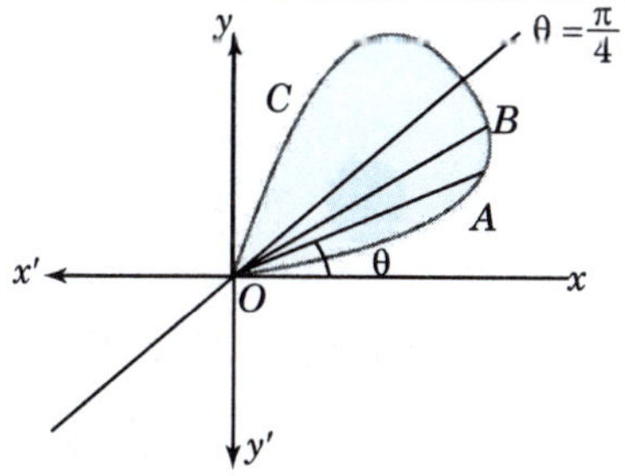


Fig. 7.40

$$M = 2 \times \text{mass of the area } OAB$$
$$= 2\iint_R \rho r\, dr\, d\theta$$

where $R$ is the area of the region $OAB$.

For the area $OAB$, $r$ varies from 0 to $a\sqrt{\sin^2\theta}$ and $\theta$ varies from 0 to $\frac{\pi}{4}$.

Thus
$$M = 2\int_0^{\pi/4}\int_0^{a\sqrt{\sin^2\theta}} kr^2 \cdot r \cdot dr \cdot d\theta$$
$$= 2k\int_0^{\pi/4}\left[\frac{r^4}{4}\right]_0^{a\sqrt{2}\sin^2\theta} d\theta = \frac{k}{2}\int_0^{\pi/4} r^4 \sin^2 2\theta\, d\theta$$
$$= \frac{ka^4}{2}\int_0^{\pi/4}\left(\frac{1-\cos 4\theta}{2}\right)d\theta = \frac{ka^4}{4}\left[\theta - \frac{\sin 4\theta}{4}\right]_0^{\pi/4} = \frac{ka^4\pi}{16}$$

## 7.10 CENTRE OF GRAVITY

(*a*) **To find the C.G. $(\bar{x}, \bar{y})$ of a plane lamina,** take the element of mass $\rho\delta x\delta y$ at the point $P(x, y)$. Then

$$\bar{x} = \frac{\iint x\rho\, dxdy}{\iint \rho\, dxdy}, \quad \bar{y} = \frac{\iint y\, \rho dxdy}{\iint \rho\, dxdy}$$ with integrals embracing the whole lamina.

While using polar coordinates, take the elementary mass as $\rho r\delta\theta\delta r$ at the point $P(r, \theta)$ so that $x = r\cos\theta$, $y = r\sin\theta$.

$$\therefore \quad \bar{x} = \frac{\iint r\cos\theta\,\rho r\,d\theta dr}{\iint \rho\, r d\theta dr},\ \bar{y} = \frac{\iint r\sin\theta\,\rho\, r d\theta dr}{\iint \rho\, r d\theta dr}$$

(*b*) **To find the C.G. $(\bar{x}, \bar{y}, \bar{z})$ of a solid,** take an element of mass $\rho\delta x\delta y\delta z$ enclosing the point $P(x, y, z)$. *Then*

$$\bar{x} = \frac{\iiint x\rho\, dxdydz}{\iiint \rho\, dxdydz},\ \bar{y} = \frac{\iiint y\rho\, dxdydz}{\iiint \rho\, dxdydz} \text{ and } \bar{z} = \frac{\iiint z\rho\, dxdydz}{\iiint \rho\, dxdydz}.$$

**Example 7.39.** *Find by double integration, the centre of gravity of the area of the cardioid $r = a(1 + \cos\theta)$.*

**Solution.** The cardioid being symmetrical about the initial line, its C.G. lies on $OX$, *i.e.*, $\bar{y} = 0$ (Fig. 7.35).

$$\therefore \quad \bar{x} = \frac{\iint \rho\, r\cos\theta\,.\,rd\theta dr}{\iint \rho\, rd\theta dr} = \frac{\int_{-\pi}^{\pi}\int_0^{a(1+\cos\theta)} \cos\theta.\, r^2 dr\,.\,d\theta}{\int_{-\pi}^{\pi}\int_0^{a(1+\cos\theta)} rdr\,.\,d\theta}$$

$$= \frac{\int_{-\pi}^{\pi} \cos\theta \left|\frac{r^3}{3}\right|_0^{a(1+\cos\theta)} d\theta}{\int_{-\pi}^{\pi} \left|\frac{r^2}{2}\right|_0^{a(1+\cos\theta)} d\theta} = \frac{2a}{3}\cdot\frac{\int_{-\pi}^{\pi}\cos\theta\,(1+\cos\theta)^3\,d\theta}{\int_{-\pi}^{\pi}(1+\cos\theta)^2\,d\theta}$$

$$= \frac{2a}{3}\cdot\frac{2.\int_0^{\pi}(3\cos^2\theta + \cos^4\theta)\,d\theta}{2.\int_0^{\pi}(1+\cos^2\theta)\,d\theta} \qquad \left\{\because \int_{-\pi}^{\pi}\cos^n\theta\,d\theta = 2\int_0^{\pi}\cos^n\theta\,d\theta \text{ or } 0 \text{ according as } n \text{ in even or odd.}\right\}$$

$$= \frac{2a}{3}\cdot\frac{2.\int_0^{\pi/2}(3\cos^2\theta + \cos^4\theta)\,d\theta}{2.\int_0^{\pi/2}(1+\cos^2\theta)\,d\theta} \text{ (as the powers of } \cos\theta \text{ are even)} = \frac{2a}{3}\cdot\frac{3.\frac{1}{2}.\frac{\pi}{2} + \frac{3.1}{4.2}.\frac{\pi}{2}}{\frac{\pi}{2} + \frac{1}{2}.\frac{\pi}{2}} = \frac{5a}{6}$$

Hence the C.G. of the cardioid is at $G(5a/6, 0)$.

**Example 7.40.** **(CG):** *A plate in the form of quadrant of the ellipse $\left(\frac{x}{a}\right)^2 + \left(\frac{y}{b}\right)^2 = 1$ is of the small but varying thickness, the thickness at any point proportional to the product of the distances of that point from the major and the minor axes. Find the co-ordinates of the centre of the gravity of the plate.* (Kottayam, 2019)

**Solution :** Let $OAB$ be the quadrant of the ellipse $\frac{x^2}{a^2} + \frac{y^2}{b^2} = 1$ with $OA = b$ and $OB = b$.

If $P(x, y)$ be a point on the plate then

$$\rho = Kxy$$

Let $(\bar{x}, \bar{y})$ be the coordinates of the centre of gravity then

$$\bar{x} = \frac{\iint_A x\rho\, dx\, dy}{\iint_A \rho\, dx\, dy} = \frac{K\iint_A x^2 y\, dx\, dy}{K\iint_A xy\, dx\, dy}$$

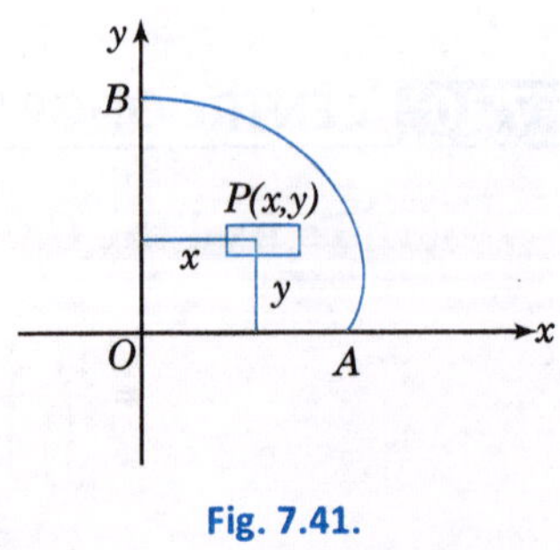

Fig. 7.41.

For the area $A$, $y$ varies from 0 to $\frac{b}{a}\sqrt{a^2 - x^2}$ and $x$ varies from 0 to $a$.

$$\therefore \quad \bar{x} = \frac{\int_0^a \int_0^{b/a\sqrt{a^2-x^2}} x^2 y \, dxdy}{\int_0^a \int_0^{b/a\sqrt{a^2-x^2}} xy \, dxdy} = \frac{\int_0^a x^2 \left[\frac{y^2}{2}\right]_0^{b/a\sqrt{a^2-x^2}} dx}{\int_0^a x \left[\frac{y^2}{2}\right]_0^{b/a\sqrt{a^2-x^2}} dx}$$

$$= \int_0^a \frac{(a^2x^2 - x^4)\,dx}{(a^2x - x^3)\,dx} = \frac{\left[\frac{a^2x^3}{3} - \frac{x^5}{5}\right]_0^a}{\left[\frac{a^2x^2}{2} - \frac{x^4}{4}\right]_0^a} = \frac{8}{15}a$$

By symmetry of the C.G. are $\left(\frac{8}{15}a, \frac{8}{15}b\right)$

## 7.11 CENTRE OF PRESSURE

Consider plane area $A$ immersed vertically in a homogeneous liquid. Take the line of intersection of the given plane with the free surface of the liquid as the $x$-axis and any line lying in this plane and perpendicular to it downwards as the $y$-axis (Fig. 7.42).

If $p$ be the pressure at the point $P(x, y)$ of the area $A$, then the pressure on an elementary area $\delta x \delta y$ at $P$ is $p\delta x\delta y$ which is normal to the plane.

$\therefore$ the resultant pressure on $A = \iint p dxdy$.

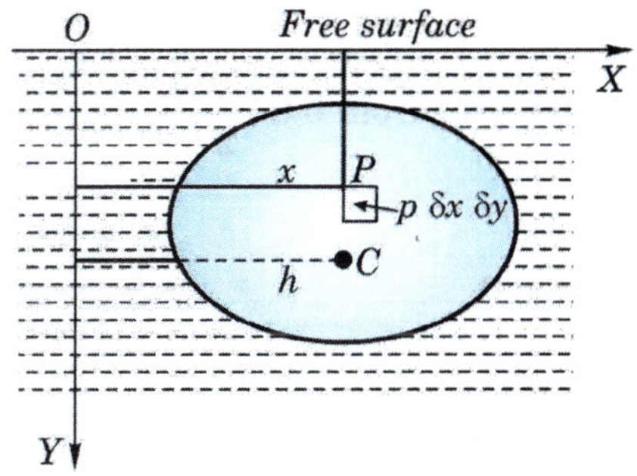


Fig. 7.42.

If this resultant pressure acting at $C(h, k)$ is equivalent to pressure at various points such as $p\delta x\delta y$ distributed over the whole area $A$, then $C$ is called the *centre of pressure*.

$\therefore$ taking the moment of the resultant pressure at $C$ and the sum of the moments of the individual pressures such as $p\delta x\delta y$ at $P(x, y)$ about the $y$-axis, we get

$$h \iint p\, dxdy = \iint x \,.\, p dxdy, \quad i.e., \; h = \iint x \,.\, dxdy \Big/ \iint p dxdy$$

Similarly, taking moments about $x$-axis, we have

$$k = \iint y \,.\, pdxdy \Big/ \iint p dxdy$$ with integrals embracing the whole of the area $A$.

While using polar coordinates, replace $x$ by $r \cos\theta$, $y$ by $r \sin\theta$ and $dxdy$ by $r\, d\theta dr$ in the above formulae.

**Example 7.41.** *A quadrant of a circle of radius a just immersed vertically in a homogeneous liquid withone edge in the surface. Determine the co-ordinate of the centre of pressure.* (Ranchi, 2019)

**Solution:** Let the equation of circle be $x^2 + y^2 = a^2$
Take the co ordinate axes an shown in the figure

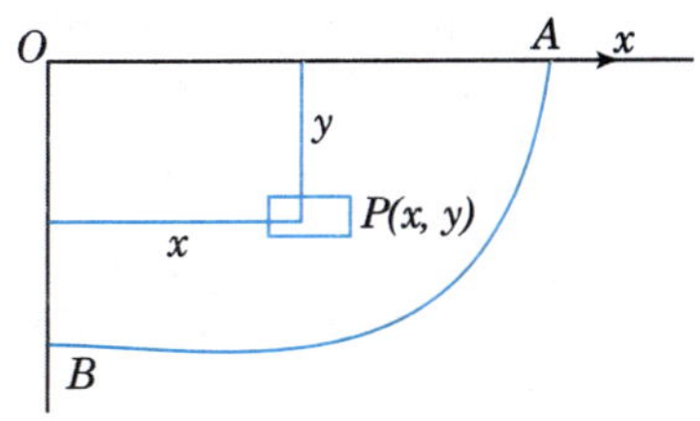


Fig. 7.43.

$$\bar{x} = \frac{\int_0^a \int_0^{\sqrt{a^2-x^2}} \rho x \cdot dy\, dx}{\int_0^a \int_0^{\sqrt{a^2-x^2}} \rho\, dy\, dx}$$

$$= \frac{\int_0^a \int_0^{\sqrt{a^2-x^2}} \rho y\, g \cdot x\, dy\, dx}{\int_0^a \int_0^{\sqrt{a^2-x^2}} \rho y\, g\, dy\, dx}$$

$$= \frac{\int_0^a \int_0^{\sqrt{a^2-x^2}} yx\, dy\, dx}{\int_0^a \int_0^{\sqrt{a^2-x^2}} y\, dy\, dx}$$

$$= \frac{\int_0^a \left[\frac{y^2}{2}\right]_0^{\sqrt{a^2-x^2}} x \cdot dx}{\int_0^a \left[\frac{y^2}{2}\right]_0^{\sqrt{a^2-x^2}} dx} = \frac{\int_0^a (a^2 - x^2) x \cdot dx}{\int_0^a (a^2 - x^2) dx} = \frac{\left[\frac{a^2x^2}{2} - \frac{x^4}{4}\right]_0^a}{\left[a^2x - \frac{x^3}{3}\right]_0^a} = \frac{\left(\frac{a^4}{2} - \frac{a^4}{4}\right)}{\left(a^3 - \frac{a^3}{3}\right)} = \frac{3}{8}a$$

Similarly, $$y = \frac{\int_0^a \int_0^{\sqrt{a^2-x^2}} y^2 dy dx}{\int_0^a \int_0^{\sqrt{a^2-x^2}} y\, dy dx}$$

$$\therefore \quad = \frac{\int_0^a \left[\frac{y^3}{3}\right]_0^{\sqrt{a^2-x^2}} dx}{\int_0^a \left[\frac{y_2}{2}\right]_0^{\sqrt{a^2-x^2}} dx} = \frac{\frac{2}{3}\int_0^a (a^2 - x^2)^{3/2} dx}{\int_0^a (a^2 - x^2) dx}$$

Let $x = a \sin \theta$, $dx = a \cos \theta \cdot d\theta$

$\theta$ varies from 0 to $\frac{\pi}{2}$

Then, $$y = \frac{2}{3} \frac{\int_0^a a^4 \cos^4 \theta\, d\theta}{\int_0^a a^3 \cos^3 \theta\, d\theta} = \frac{2}{3} \frac{\frac{3\pi}{16}}{\frac{2}{3}} a = \frac{3\pi a}{16}$$

Hence, the coordinates of centre of pressure are $\left(\frac{3a}{8}, \frac{3\pi a}{16}\right)$.

## PROBLEMS 7.6

1. A plate has for its edge the curve $y = e^x$, the line $x = 1$, and the co-ordinate axes. If the density varies as the square of the distance from the origin, find the mass of the plate. *(Andra, 2018)* (**Ans.** 2.84 $k$)
2. Find the mass of an elliptic plate $\frac{x^2}{a^2} + \frac{y^2}{b^2} = 1$, if the surface density at any point $P(x, y)$ on it is $\mu xy$. *(Nagpur, 2017)*
3. A thin plate is bounded by an arc of the parabola $y = 2x - x^2$ in the interval $0 \le x \le 2$ and the $x$-axis. Show that its mass is $\frac{26}{3} - \frac{15}{2}\log 3$, if the density at a point is $\frac{(1-y)}{(1+x)}$. *(PTU, 2017)*
4. Find the centroid of the area enclosed by the parabola $y^2 = 4ax$, the axis of $x$ and its latus-rectum. *(Andhra., 2016)*
5. The density at any point $(x, y)$ of a lamina is $\sigma(x + y)/a$ where $\sigma$ and $a$ are constants. The lamina is bounded by the lines $x = 0, y = 0, x = a, y = b$. Find the position of its centre of gravity.
6. Find the centroid of a loop of the lemniscate $r^2 = a^2 \cos 2\theta$.
7. A plane in the form of a quadrant of the ellipse $(x/a)^2 + (y/b)^2 = 1$ is of small but varying thickness, the thickness at any point being proportional to the product of the distances of that point from the axes ; show that the coordinates of the centroid are $(8a/15, 8b/15)$. *(Andhra., 2015 ; Nagpur, 2009)*
8. Find the co-ordinates of the centre of gravity a laminal in the shape of quadrant of the curve $x = a \cos^3 t$, $y = b \sin^3 t$, the density being given by $\rho = kxy$. *(Marathwada, 2018)*
9. Find the centroid of the tetrahedron bounded by the coordinate planes and the plane $x + y + z = 1$, the density at any point varying as its distance from the face $z = 0$.
10. Find $\bar{x}$ where $(\bar{x}, \bar{y}, \bar{z})$ is the centroid of the region $R$ bounded by the parabolic cylinder $z = 4 - x^2$ and the planes $x = 0, y = 0, y = 6, z = 0$. (Assume that the density is constant).

**11.** Find the mass of the lamina which is in the form of Leminscate of Bernoulli $r^2 = a^2 \cos^2\theta$, the density at a point being proportional to the square of its distance from the pole. *(Kottayam, 2019)*

**12.** Find the mass of the tetrahedron bounded by the coordinate planes and the palne $\frac{x}{a}+\frac{y}{b}+\frac{z}{c}=1$, the valiable density $\rho = \mu xyz$. *(Andhra, 2023)*

**13.** A horizontal boiler has a flat bottom and its ends consist of a square 1 metre wide surmounted by an isosceles triangle of height 0.5 metre. Determine the depth of the centre of pressure of either end when the boiler is just full.

**14.** Find the centre of gravity of the uniform solid in the form of $\left(\frac{x}{a}\right)^{2/3}+\left(\frac{y}{b}\right)^{2/3}+\left(\frac{z}{c}\right)^{2/3}=1$ contained in the first octant. *(PTU, 2018)*

**15.** Find the depth of the centre of pressure of a square lamina immersed in the liquid, with one vertex in the surface and the diagonal vertical.

**16.** Find the centre of pressure of a triangular lamina immersed in a homogeneous liquid with one side in the free surface.

**17.** A uniform semi-circular is lamina immersed in a fluid with its plane vertical and its boundary diameter on the free surface. If the density at any point of the fluid varies as the depth of the point below the free surface, find the position of the centre of pressure of the lamina.

**18.** Find the C.G. of the volume cut off from the cylinder $x^2 + y^2 = 2ax$ by the planes $z = mx$ and $z = nx$.

**19.** Determine the mass of a slab in the shape of a circle of radius $a$, if the density at any point $P$ on it, is proportional to $OP$, where is $O$ is a point on the axis of the cylinder. *(Mumbai, 2018)*

**20.** Find by double integration the centre of gravity of the area of the cardiod $r = a(1 + \cos\theta)$ *(Kolkata, 2019)*

**21.** Find the mass of the solid bounded by the planes $y = 0$, $z = 0$ and $z = h(h > 0)$ and the cylinder $x^2 + y^2 = a^2$ $(x > 0)$, if $\rho = Kyz$. *(Nagpur, 2019)*

## 7.12 (1) MOMENT OF INERTIA

If a particle of mass $m$ of a body be at a distance $r$ from a given line, then $mr^2$ is called the *moment of inertia of the particle about the given line* and the sum of similar expressions taken for all the particles of the body, *i.e.*, $\Sigma mr^2$ is called the *moment of inertia of the body about the given line* (Fig. 7.44).

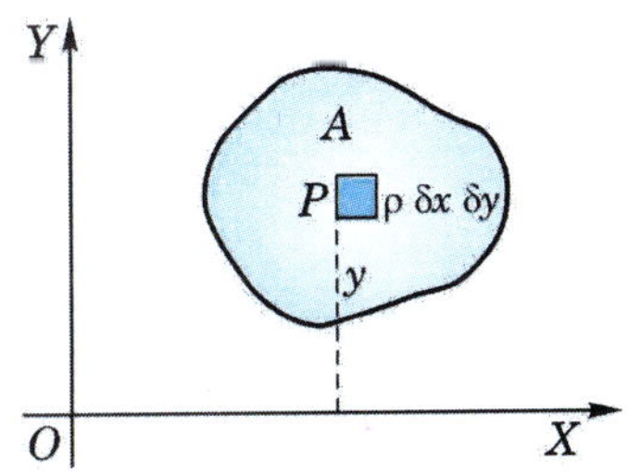

Fig. 7.44.

If $M$ be the total mass of the body and we write its moment of inertia $= Mk^2$, then $k$ is called the *radius of gyration* of the body about the axis.

**(2) M.I. of plane lamina.** Consider the elementary mass $\rho\delta x\delta y$ at the point $P(x, y)$ of a plane area $A$ so that its M.I. about $x$-axis $= \rho\delta x\delta y\, y^2$.

$\therefore$ M.I. of the lamina about $x$-axis, *i.e.* $I_x = \iint_A \rho\, y^2\, dxdy$.

Similarly, M.I. of the lamina about $y$-axis' *i.e.*, $I_y = \iint_A \rho\, x^2\, dxdy$.

Also M.I. of the lamina about an axis perpendicular to the $xy$-plane, *i.e.*,

$$I_z = \iint_A \rho(x^2 + y^2)\, dxdy.$$

**(3) M.I. of a solid.** Consider an elementary mass $\rho\delta x\delta y\delta z$ enclosing a point $P(x, y, z)$ of a solid of volume $V$.

Distance of $P$ from the $x$-axis $= \sqrt{(y^2 + z^2)}$.

$\therefore$ M.I. of this element about the $x$-axis $= \rho\, \delta x\delta y\delta z\, (y^2 + z^2)$.

Thus *M.I. of this solid about x-axis, i.e.,* $I_x = \iiint_V \rho(y^2 + z^2)\, dxdydz$.

Similarly, *its M.I. about y-axis, i.e.,* $I_y = \iiint_V \rho(z^2 + x^2)\, dxdydz$

and M.I. *about z-axis, i.e.*, $I_z = \iiint_V \rho(x^2 + y^2)\, dxdydz.$

(4) Sometimes we require the moment of inertia of a body about axes other than the principal axes. The following theorems prove useful for this purpose :

**I. Theorem of perpendicular axis.** *If the moment of inertia of a lamina about two perpendicular axes OX, OY in its plane are* $I_x$ *and* $I_y$*, then the moment of inertia of the lamina about an axis OZ, perpendicular to it is given by* $I_z = I_x + I_y$.

Its proof follows from the relations giving $I_x$, $I_y$ and $I_z$ for a plane lamina [(2) above].

**II. Steiner's theorem*.** *If the moment of inertia of a body of mass M about an axis through its centre of gravity is I, then I′, moment of inertia about a parallel axis at a distance d from the first axis, is given by* $I' = I + Md^2$.

Its proof will be found in any text book on Dynamics of a Rigid Body.

**Example 7.42.** *Find the M.I. of the area bounded by the curve* $r^2 = a^2 \cos 2\theta$ *about its axis.*

**Solution.** Given curve is symmetrical about the pole and for half of the loop in the first quadrant $\theta$ varies from 0 to $\pi/4$ (Fig. 7.45).

Elementary area at $P(r, \theta) = r\, d\theta dr$.

If $\rho$ be the surface density, then elementary mass

$$= \rho r\, d\theta dr \quad \ldots(i)$$

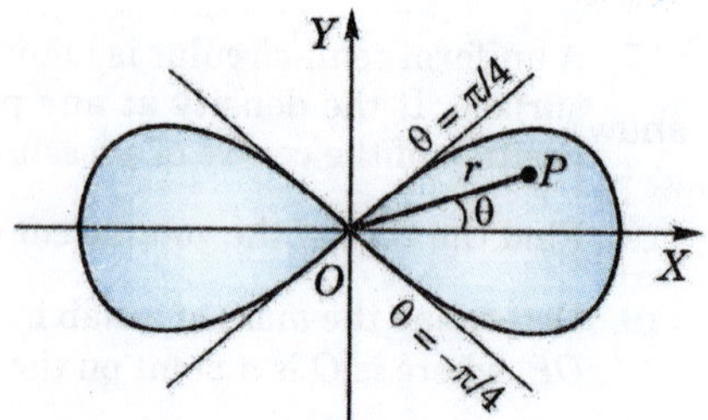

Fig. 7.45.

$$\therefore \text{ its total mass } M = 4\int_0^{\pi/4}\int_0^{a\sqrt{(\cos 2\theta)}} \rho r\, dr d\theta$$

$$= 2\rho a^2 \int_0^{\pi/4} \cos 2\theta\, d\theta = \rho a^2 \quad \ldots(ii)$$

Now M.I. of the elementary mass Eqn. (*i*) about the *x*-axis.

$$= \rho r\, d\theta dr \cdot y^2 = \rho r d\theta\, dr\, (r \sin\theta)^2 = \rho r^3 \sin^2\theta\, dr d\theta$$

Hence the M.I. of the whole area

$$= 4\int_0^{\pi/4}\int_0^{a\sqrt{(\cos 2\theta)}} \rho r^3 \sin^2\theta\, dr d\theta = 4\rho\int_0^{\pi/4} \sin^2\theta \cdot \left| r^4/4 \right|_0^{a\sqrt{(\cos 2\theta)}} d\theta$$

$$= \rho a^2 \int_0^{\pi/4} \cos^2 2\theta \cdot \sin^2\theta\, d\theta = \rho a^4 \int_0^{\pi/2} \cos^2\phi \cdot \sin^2\frac{\phi}{2} \cdot \frac{d\phi}{2} \quad [\text{Put } 2\theta = \phi,\ d\theta = d\phi/2]$$

$$= \frac{\rho a^4}{4}\int_0^{\pi/2} (\cos^2\phi - \cos^3\phi)\, d\phi = \frac{\rho a^4}{48}(3\pi - 8) = \frac{Ma^2}{48}(3\pi - 8). \quad [\text{By Eqn. } (ii)]$$

**Example 7.43.** *A thin plate of uniform thickness and density is in the form of a lamina bounded by the parabola* $x^2 = y$ *and the line* $y = x + 2$*. Find the moment of inertia of the lamina about the co-ordinate axes.* (Kottayam, 2019)

**Solution:** Let $A$ be the area, $M$ the mass and $\rho$ the density of the plate. The parabola $y = x^2$ and the line $y = x + 2$, intersect at points (–1, 1) and (2, 4)

$$M = \iint_A \rho\, dydx = \int_{-1}^{2}\int_{x^2}^{x+2} dy\, dx = \rho\int_{-1}^{2} [y]_{x^2}^{x+2} dx$$

$$= \rho\int_{-1}^{2}(x + 2 - x^2)\, dx = \rho\left[\frac{x^2}{2} + 2x - \frac{x^3}{3}\right]_{-1}^{2}$$

$$= \rho\left[\left(2 + 4 - \frac{8}{3}\right) - \left(\frac{1}{2} - 2 + \frac{1}{3}\right)\right] = \frac{9}{2}\rho \quad \ldots(i)$$

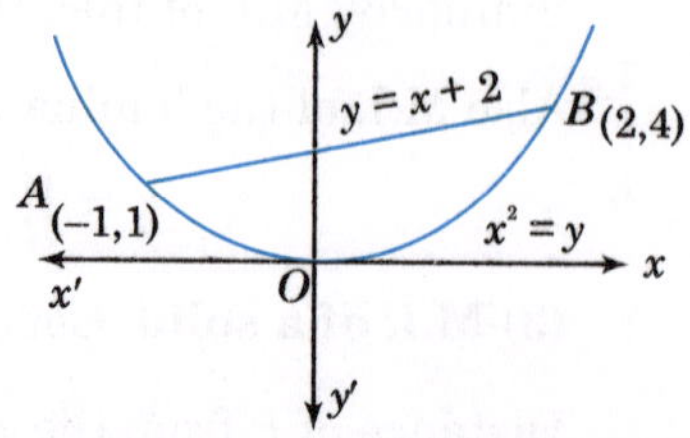

Fig. 7.46

Now,

$$I_x = \iint_A \rho y^2\, dy\, dx = \int_{-1}^{2}\int_{x^2}^{x+2} \rho y^2\, dy\, dx$$

$$= \int_{-1}^{2}\left[\frac{y^3}{3}\right]_{x^2}^{x+2} dx = \frac{\rho}{3}\int_{-1}^{2}\left[(x+2)^3 - (x^2)^3\right] dx = \frac{\rho}{3}\left[\frac{(x+2)^4}{4} - \frac{x^7}{7}\right]_{-1}^{2}$$

*Named after a Swiss geometrer *Jacob Steiner* (1796–1863) who was a professor at Berlin University.

$$= \frac{\rho}{3}\left[\left(4^3 - \frac{128}{7}\right) - \left(\frac{1}{4} + \frac{1}{7}\right)\right] = \frac{423}{28}\rho = \frac{47}{14}M$$

$$I_y = \iint_A \rho x^2 dy\, dx$$

$$= \rho\int_{-1}^{2}\int_{x^2}^{x+2} x^2\, dy\, dx = \rho\int_{-1}^{2} x^2[y]_{x^2}^{x+2} dx$$

$$= \rho\int_{-1}^{2} x^2(x + 2 - x^2)\, dx$$

$$= \rho\left[\frac{x^4}{4} + \frac{2x^3}{3} - \frac{x^5}{5}\right]_{-1}^{2} = \rho\left[\left(4 + \frac{16}{3} - \frac{32}{5}\right) - \left(\frac{1}{4} - \frac{2}{3} + \frac{1}{5}\right)\right]$$

$$= \frac{63}{20}\rho = \frac{7}{10}M$$

**Example 7.44.** *Find the moment of inertia of a solid right circular cylinder about its axis, and about of a diameter of the base.* *(Andhra, 2018)*

**Solution:** Let $r$ and $h$ be the radius and height of the cylinder respectively. Take the coordinate axes as shown in the figure. The equation of the base of the cylinder is $x^2 + y^2 = a^2$

Let $V$ be the volume, $\rho$ the density and $M$ the mass of the cylinder.

Then,

$$M = \iiint_V \rho\, dx\, dy\, dz$$

$$= \int_{-r}^{r}\int_{y=-\sqrt{r^2-x^2}}^{\sqrt{r^2-x^2}}\int_{z=0}^{h} \rho\, dz\, dy\, dx$$

$$= 4\rho\int_0^r\int_0^{\sqrt{r^2-x^2}} [z]_0^h\, dy\, dx$$

$$= 4\rho h\int_0^r [y]_0^{\sqrt{r^2-x^2}} dx$$

$$= 4\rho h\int_0^r \sqrt{r^2 - x^2}$$

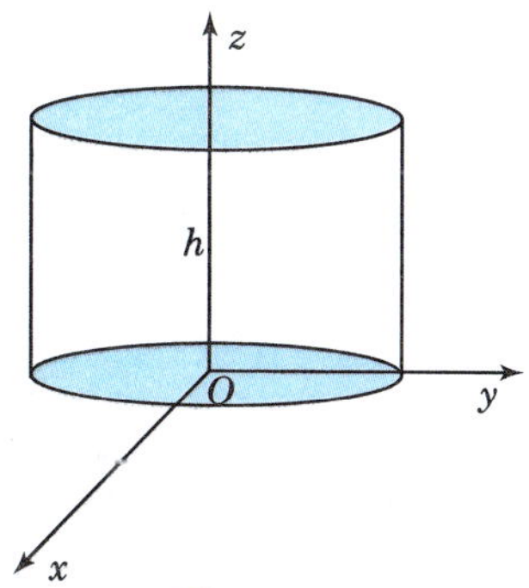


Fig. 7.47.

Let $x = r\sin\theta$, $dr = r\cos\theta\, d\theta$ $\quad\therefore\ M = 4\rho h\int_0^{\pi/2} r^2\cos^2\theta\, d\theta = 4\rho r^2 h\cdot\frac{\pi}{4} = \pi r^2 h\rho$

moment of inertia $I_x$ of cylinder about its axis *i.e.* about $x$-axis is given by

$$I_x = \iiint_V \rho(x^2 + y^2)\, dz\, dy\, dx = 4\rho\int_0^r\int_0^{\sqrt{r^2-y^2}}\int_0^h (x^2 + y^2)\, dz\, dy\, dx$$

$$= 4\rho\int_0^r\int_0^{\sqrt{r^2-x^2}} (x^2 + y^2)\cdot[z]_0^h\, dydx = 4\rho h\int_0^r\left[x^2 y + \frac{y^3}{3}\right]_0^{\sqrt{r^2-x^2}} dx$$

$$= 4\rho h\int_0^r\left[x^2\sqrt{r^2 - x^2} + \frac{1}{3}(r^2 - x^2)^{3/2}\right]dx$$

Let $x = r\sin\theta$, $dx = r\cos\theta$, $d\theta$

$\theta$ varies from 0 to $\frac{\pi}{2}$

$$I_x = 4\rho h\int_0^{\pi/2}\left[r^3\sin^2\theta\cos\theta + \frac{1}{3}r^3\cos^3\theta\right]r\cos\theta\cdot d\theta$$

$$= 4\rho h r^4\int_0^{\pi/2}\left(\sin^2\theta\cos^2\theta + \frac{1}{3}\cos^4\theta\right)d\theta$$

$$= 4\rho h r^4\left[\frac{1}{4}\cdot\frac{1}{2}\times\frac{\pi}{2} + \frac{1}{3}\cdot\frac{3}{4}\cdot\frac{1}{2}\cdot\frac{\pi}{2}\right]$$

$$= 4\rho h r^4\frac{\pi}{8} = \frac{\pi r^4 h\rho}{2} = (\pi r^2 h\rho)\frac{r^2}{2} = \frac{Mr^2}{2}$$

Now, diameter of the base is $y$-axis, hence moment of inertia $I_y$ of the cylinder, about $y$-axis is given by

$$I_y = \iiint_V \rho\left(y^2 + z^2\right) dz\, dy\, dx$$

Proceeding like previous part, we have

$$= 4\rho\int_0^r \int_0^{\sqrt{r^2-x^2}} \int_0^h \left(y^2 + z^2\right) dz\, dy\, dz$$

$$= 4\rho\int_0^r \int_0^{\sqrt{r^2-x^2}} \left[y^2 z + \frac{z^3}{3}\right]_0^h dy\, dx$$

$$= 4\rho\int_0^r \int_0^{\sqrt{r^2-x^2}} \left[y^2 h + \frac{h^3}{3}\right] dy\, dx = 4\rho h\int_0^r \int_0^{\sqrt{r^2-x^2}} \left(y^2 + \frac{h^2}{3}\right) dy\, dx$$

$$= 4\rho h\int_0^h \left[\left(r^2 - x^2\right)^{3/2} + h^2\sqrt{r^2 - x^2}\right] dx$$

put $x = r \sin\theta$, $dx = r\cos\theta \cdot d\theta$

$\theta$ varies from 0 to $\frac{\pi}{2}$ $\therefore$

$$I_y = \frac{4\rho h}{3}\int_0^{\pi/2} \left(r\cos^3\theta + h^2 \cdot r\cos\theta\right) r\cos\theta \cdot d\theta$$

$$= \frac{4\rho h r^2}{3}\int_0^{\pi/2} \left(r^2\cos^4\theta + h^2\cos^2\theta\right) d\theta = \frac{4\rho h r^2}{3}\left[r^2 \cdot \frac{3}{4} \cdot \frac{1}{2} \cdot \frac{\pi}{2} + h^2\frac{\pi}{4}\right]$$

$$= \frac{4\rho h r^2 \pi}{48}\left(3r^2 + 4h^2\right) = \frac{\pi r^2 \cdot h\rho}{12}\left(3r^2 + 4h^2\right)$$

$$= \frac{M}{12}\left(3r^2 + 4h^2\right)$$

**Example 7.45.** *A hemisphere of radius R has a cylindrical hole of radius a drilled through it, the axis of the hole being along the radius normal to the plane face of the hemisphere. Find its radius of gyration about a diameter of this face.*

**Solution.** M.I. of the given solid about $x$-axis

$$= \iiint \rho(y^2 + z^2)\, dx\, dy\, dz$$

The limits of integration for $z$ are from 0 to $z = \sqrt{(R^2 - x^2 - y^2)}$ found from the equation of the sphere $x^2 + y^2 + z^2 = R^2$. The limits for $x$ and $y$ are to be such as to cover the shaded area $A$ in the $xy$-plane between the concentric circles of radii $a$ and $R$ (Fig. 7.48).

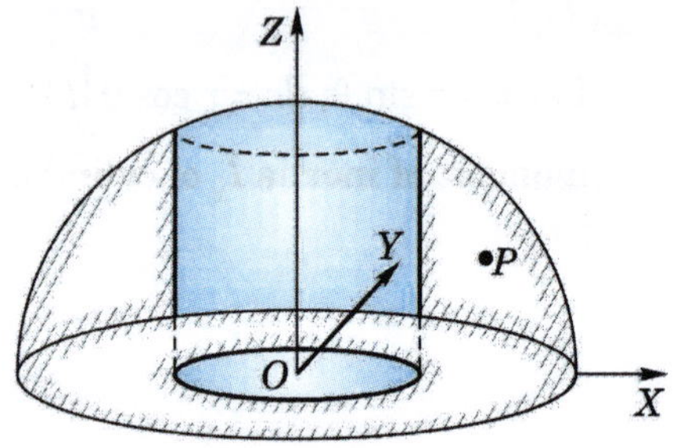

Fig. 7.48.

Thus the required M.I. about $x$-axis

$$= \rho\iint_A \int_0^{\sqrt{(R^2 - x^2 - y^2)}} (y^2 + z^2)\, dz\, dx\, dy$$

$$= \rho\iint_A \left| y^2 z + z^3/3 \right|_0^{\sqrt{(R^2 - x^2 - y^2)}} dx\, dy = \rho\iint_A \left[y^2(R^2 - x^2 - y^2)^{1/2} + \frac{1}{3}(R^2 - x^2 - y^2)^{3/2}\right] dx\, dy.$$

Now changing to polar coordinates, we have $x = r\cos\theta$, $y = r\sin\theta$ and $dxdy = rd\theta dr$.

Also to cover the area $A$, $r$ varies from $a$ to $R$ and $\theta$ varies from 0 to $2\pi$.

Hence the required M.I. about $x$-axis

$$= \rho\int_a^R \int_0^{2\pi} \left[r^2\sin^2\theta \cdot (R^2 - r^2)^{1/2} + \frac{1}{3}(R^2 - r^2)^{3/2}\right] r d\theta dr$$

$$= \rho\int_a^R \int_0^{2\pi} \left[\frac{1}{2}r^2(1 - \cos 2\theta) + \frac{1}{3}(R^2 - r^2)\right] d\theta \cdot r(R^2 - r^2)^{1/2}\, dr$$

$$= \rho\int_a^R \left| \frac{r^2}{2}\left(\theta - \frac{\sin 2\theta}{2}\right) + \frac{1}{3}(R^2 - r^2)\,\theta \right|_0^{2\pi} \cdot r(R^2 - r^2)^{1/2}\, dr$$

$$= \rho \int_a^R 2\pi \left( \frac{r^2}{2} + \frac{R^2 - r^2}{3} \right) . r(R^2 - r^2)^{1/2} \, dr$$

$$= \frac{\pi\rho}{3} \int_a^R (2R^2 + r^2)(R^2 - r^2)^{1/2} \cdot rdr \qquad \text{[Put } r^2 = t \text{ and } rdr = dt/2]$$

$$= \frac{\pi\rho}{6} \int_{a^2}^{R^2} (2R^2 + t)(R^2 - t)^{1/2} \, dt \qquad \text{[Integrate by parts]}$$

$$= \frac{\pi\rho}{9} \left[ (2R^2 + a^2)(R^2 - a^2)^{3/2} + \frac{2}{5}(R^2 - a^2)^{5/2} \right] = \frac{2\pi\rho}{3}(R^2 - a^2)^{3/2} \times \frac{1}{10}(4R^2 + a^2)$$

$$\left[ \because \text{ Mass} = \rho \int_0^{2\pi} \int_a^R \int_0^{\sqrt{(R^2 - r^2)}} dz \cdot rdr \cdot d\theta = \frac{2\pi\rho}{3}(R^2 - a^2)^{3/2} \right]$$

Hence, the radius of gyration = $[(4R^2 + a^2)/10]^{1/2}$.

## 7.13 (1) PRODUCT OF INERTIA

If a particle of mass $m$ of a body be at distances $x$ and $y$ from two given perpendicular lines, then $\Sigma mxy$ is called the *product of inertia* of the body about the given lines.

Consider an elementary mass $\delta x \delta y \delta z$ enclosing the point $P(x, y, z)$ of solid of volume $V$. Then the product of inertia (P.I.) of this element about the axes of $x$ and $y = \rho \delta x \delta y \delta z \; xy$.

$\therefore$ P.I. of the solid about $x$ and $y$-axes, *i.e.*, $P_{xy} = \iiint_V \rho xy \, dxdydz$

Similarly, $P_{yz} = \iiint_V \rho yz \, dxdydz$ and $P_{zx} = \iiint_V \rho zx \, dxdydz$.

In particular, for a plane lamina of surface density $\rho$ and covering a region $A$ in the $xy$-plane,

$$P_{xy} = \iint_A \rho xy \, dxdy \text{ whereas } P_{yz} = P_{zx} = 0. \qquad [\because \; z = 0]$$

**(2) Principal axes.** *The principal axes* of a lamina at a given point are that pair of axes in its plane through the given point, about which the product of inertia of the lamina vanishes.

Let $P(x, y)$ be a point of the plane area $A$ referred to rectangular axes $OX$, $OY$. Let $(x', y')$ be the coordinates of $P$ referred to another pair of rectangular axes $OX'$, $OY'$ in the same plane and inclined at an angle $\theta$ to the first pair (Fig. 7.49).

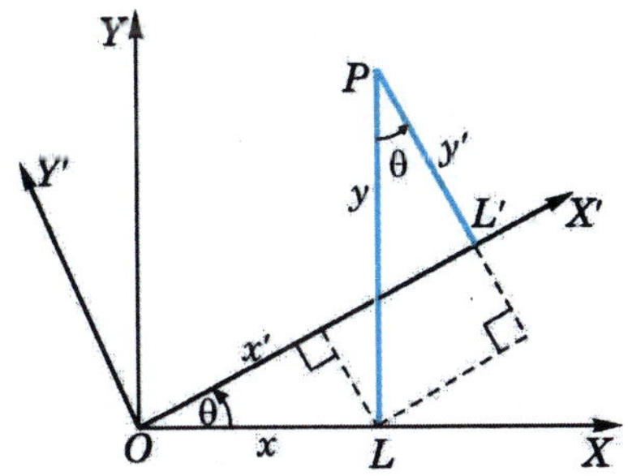

Fig. 7.49.

Then $\qquad x' = x \cos\theta + y \sin\theta$

$\qquad\qquad y' = y \cos\theta - x \sin\theta$

If $I_x$, $I_y$ be the moments of inertia of the area $A$ about $OX$ and $OY$ and $P_{xy}$ be its product of inertia about these axes, then

$$I_x = \iint_A \rho y^2 \, dA, \; I_y = \iint_A \rho x^2 \, dA, \; P_{xy} = \iint \rho xy \, dA.$$

$\therefore$ the product of inertia $P'_{xy}$ about $OX'$ and $OY'$ is given by

$$P'_{xy} = \int_A \rho x'y' \, dA = \iint_A \rho(x \cos\theta + y \sin\theta)(y \cos\theta - x \sin\theta) \, dA$$

$$= \sin\theta \cos\theta \iint_A \rho(y^2 - x^2) \, dA + (\cos^2\theta - \sin^2\theta) \iint_A \rho xy \, dA$$

$$= 1/2 \sin 2\theta \cdot (I_x - I_y) + \cos 2\theta \, P_{xy}.$$

Now $OX'$, $OY'$ will be the principal axes of the area $A$ if $P'_{xy}$ vanishes.

*i.e.*, If $\qquad 1/2 \sin 2\theta \, (I_x - I_y) + \cos 2\theta \, P_{xy} = 0$

*i.e.*, If $\qquad \mathbf{\tan 2\theta = 2P_{xy}/(I_y - I_x)}$.

This gives two values of $\theta$ differing by $\pi/2$.

**Example 7.46.** *Show that the principal axes at the node of a half-loop of the lemniscate $r^2 = a^2 \cos 2\theta$ are inclined to the initial line at angles*

$$\frac{1}{2}\tan^{-1}\frac{1}{2} \text{ and } \frac{\pi}{2}+\frac{1}{2}\tan^{-1}\frac{1}{2}.$$ (Nagpur, 2019)

**Solution.** Let the element of mass at $P(r, \theta)$ be $\rho r d\theta dr$.

Then $I_x = \rho\int_0^{\pi/4}\int_0^{a\sqrt{(\cos 2\theta)}} r^2 \sin^2\theta \cdot r d\theta dr$ [See Fig. 7.45]

$$= \frac{\rho a^4}{4}\int_0^{\pi/4} \sin^2\theta \cos^2 2\theta \, d\theta = \frac{\rho a^4}{16}\left(\frac{\pi}{4}-\frac{2}{3}\right)$$

$$I_y = \rho\int_0^{\pi/4}\int_0^{a\sqrt{(\cos 2\theta)}} r^2 \cos^2\theta \cdot r d\theta dr = \frac{\rho a^4}{16}\left(\frac{\pi}{4}+\frac{2}{3}\right)$$

and $$P_{xy} = \rho\int_0^{\pi/4}\int_0^{a\sqrt{(\cos 2\theta)}} r^2 \sin\theta\cos\theta \cdot r d\theta dr = \frac{\rho a^4}{48}.$$

Hence the required direction of the principal axes at $O$ are given by

$$\tan 2\theta = \frac{2P_{xy}}{I_y - I_x} = \frac{\rho a^4/24}{(\rho a^4/16)\times(4/3)} = \frac{1}{2}$$

or by $$\theta = \frac{1}{2}\tan^{-1}\frac{1}{2} \text{ and } \frac{\pi}{2}+\frac{1}{2}\tan^{-1}\frac{1}{2}.$$

**Example 7.47 :** *The lengths AB and AD of the sides of rectangle ABCD are 2a and 2b. Show that the inclination to AB of one of the principal axes at A is $\frac{1}{2}\tan^{-1}\left\{\frac{3ab}{2(a^2-b^2)}\right\}$.* (Bangalore, 2019)

**Solution:**

Take $A$ as the origin and $AB$, $AD$ the $x$ and $y$ axis respectively.

Let the principal axis through the point $A$ be inclined at an angle $\theta$ to $AB$.

Let $R$ be the area and $\rho$ the density of the rectangle

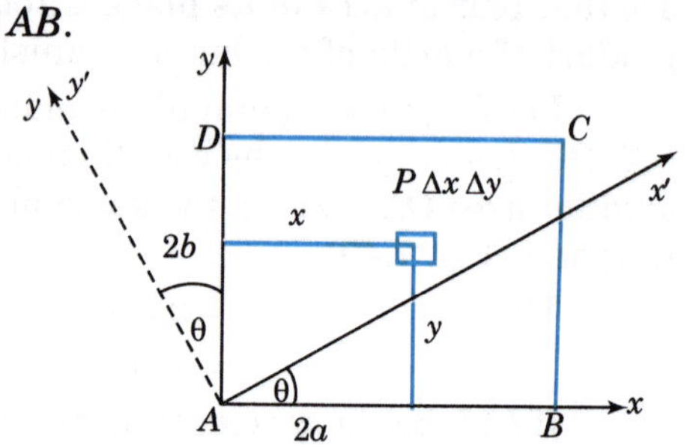

Fig. 7.50.

$$I_x = \int_0^{2b}\int_0^{2a} \rho y^2 dx\, dy = \rho\int_0^{2b} y^2 \cdot [x]_0^{2a} dy$$

$$= 2a\rho\int_0^{2b} y^2 dy = 2a\rho\left[\frac{y^3}{3}\right]_0^{2b} = \frac{16ab^3\rho}{3}$$

$$I_y = \int_0^{2b}\int_0^{2a} \rho x^2 dx\, dy$$

Proceeding like previous part, we get

$$I_y = \frac{16a^3b\rho}{3}$$

Now, $$\rho_{xy} = \int_0^{2b}\int_0^{2a} \rho xy\, dx\, dy = \rho\int_0^{2b}\left[\frac{x^2}{2}\right]_0^{2a} y\, dy = 2a^2\rho\int_0^{2b} y dy = 2a^2\rho\left[\frac{y^2}{2}\right]_0^{2b} = 4a^2b^2\rho$$

Hence, $$\tan\theta = \frac{2\rho_{xy}}{I_y - I_x} = 2\cdot\frac{4a^2b^2\rho}{\left(\frac{16}{3}a^3b - \frac{16}{3}ab^3\right)\rho}$$

$$\tan 2\theta = \frac{3ab}{2(a^2-b^2)}$$

$$\theta = \frac{1}{2}\tan^{-1}\left\{\frac{3ab}{2(a^2-b^2)}\right\}$$

## PROBLEMS 7.7

1. Show that the principal axes at the origin of the triangle enclosed by $x = 0, y = 0, \frac{x}{a} + \frac{y}{b} = 1$ are inclined at an angle $\alpha$ and $\frac{\pi}{2} + \alpha$ to the $x$-axis, where $\alpha = \frac{1}{2}\tan^{-1}\left(\frac{4ab}{a^2 - b^2}\right)$ *(Marathawada, 2019)*
2. Find the moment of inertia of a circular plate about a tangent.
3. Find the moment of inertia of the area $y = \sin x$ from $x = 0$ to $x = 2\pi$ about $OX$.
4. Find the product of inertia of quadrant of an ellipse $\frac{x^2}{a^2} + \frac{y^2}{b^2} = 1$, and show that at the centre the principal axes are inclined at an angle of $\frac{1}{2}\tan^{-1}\left(\frac{4}{\pi}\frac{ab}{a^2 - b^2}\right)$ to the axes. *(Mumbai, 2017)*
5. Find the moment of inertia about the initial line of the cardioid $r = a(1 + \cos\theta)$. *(Andhra., 2015)*
6. Find the moment of inertia of a uniform spherical ball of mass $M$ and radius $R$ about a diameter.
7. Find the moment of inertia of a solid right circular cylinder about its axis.. *(Nagpur, 2019)*
8. Find the M.I. of a solid right circular cone having base-radius $r$ and height $h$, about (*i*) its axis, (*ii*) an axis through the vertex and perpendicular to its axis, (*iii*) a diameter of its base.
9. Find the moment of inertia of a hollow sphere about a diameter, its external and internal radii being 51 metres and 49 metres.
10. Find the moment of inertia about $z$-axis of a homogeneous tetrahedron bounded by the planes $x = 0, y = 0, z = x + y$ and $z = 1$.
11. Find the moment of inertia of an octant of the ellipsoid $(x/a)^2 + (y/b)^2 + (z/c)^2 = 1$, about the $x$-axis.
12. Find the product of inertia of a quadrant of the ellipse $(x/a)^2 + (y/b)^2 = 1$, about the coordinate axes.
13. Show that the principal axes at the origin of the triangle enclosed by $x = 0, y = 0, (x/a) + (y/b) = 1$ are inclined to the $x$-axis at angles $\alpha$ and $\alpha + \pi/2$, where $\alpha = \frac{1}{2}\tan^{-1}[ab/(a^2 - b^2)]$
14. The lengths $AB$ and $AD$ of the sides of a rectangle $ABCD$ are $2a$ and $2b$. Show that the inclination to $AB$ of one of the principal axes at $A$ is $\frac{1}{2}\tan^{-1}\left\{\frac{3ab}{2(a^2 - b^2)}\right\}$.
15. A uniform lamina is in the form of an arc of the parabola $y^2 = 4ax$ bounded by double ordinate at a distance $h$ from the vertex. If $h = \frac{a\sqrt{7}}{3}(\sqrt{7} + 4)$, show that the principal axes at the end of a latus rectum are the tangent and normal at that point.
16. Find the product of inertia of equilateral triangle about two perpendicular axes in its plane at a vertex, one of the axes being along a side.
17. If an area is bounded by two concentric semicircles with their common bounding diameter in a free surface, Prove that the depth of the centre of pressure is $\frac{3}{10}\pi\frac{(a + b)(a^2 + b^2)}{a^2 + ab + b^2}$.
18. A rectangular lamina of length $2a$ and breadth $2b$ is completely immersed in a vertical plane in a fluid so that it centre is at depth $h$ and the side $2a$ makes an angle $\alpha$ with the horizontal. Find the co-ordinates of the centre of pressure. *(Kottayam, 2019)*

19. A horizontal boiler has a flat bottom and its ends are plane and semicircular. If it is just full of water, show that the depth of centre of pressure of either end is 0.7 × total depth (approximately).

20. Compute the moment of inertia of a right circular cone whose altitude is $h$ and base radius $r$ about (*i*) the axes of symmetry (*ii*) the diameter of the base.

## 7.14 BETA FUNCTION

*The beta function is defined as*

$$\beta(m, n) = \int_0^1 x^{m-1}(1-x)^{n-1}\, dx \begin{cases} m > 0 \\ n > 0 \end{cases} \qquad ...(1)$$

Putting $x = 1 - y$ in (1), we get $\beta(m, n) = -\int_1^0 (1-y)^{m-1}\, y^{n-1}\, dy$

$$= \int_0^1 y^{n-1}\,(1-y)^{m-1}\, dy = \beta(n, m)$$

Thus $\beta(m, n) = \beta(n, m)$ ...(2)

Putting $x = \sin^2\theta$ so that $dx = 2\sin\theta\cos\theta\, d\theta$, (1) becomes

$$\beta(m, n) = \int_0^{\pi/2} (\sin^2\theta)^{m-1}\,(\cos^2\theta)^{n-1}\, 2\sin\theta\cos\theta\, d\theta$$

$$= 2\int_0^{\pi/2} \sin^{2m-1}\theta\,\cos^{2n-1}\theta\, d\theta \qquad ...(3)$$

which expresses $\beta(m, n)$ in terms of trigonometric functions.

This function is also *Euler's integral of the first kind**.

## 7.15 (1) GAMMA FUNCTION

*The gamma function is defined as*

$$\Gamma(n) = \int_0^\infty e^{-x}\, x^{n-1}\, dx \;(n > 0) \qquad ...(i)$$

This integral is also known as *Euler's integral of the second kinds.* It defines a function of $n$ for positive values of $n$.

In particular, $\Gamma(1) = \int_0^\infty e^{-x}\, dx = \left| -e^{-x} \right|_0^\infty$

$= 1.$ ...(*ii*)

**(2) Reduction formula for Γ(n).**

Since $\Gamma(n+1) = \int_0^\infty e^{-x}\, x^n\, dx$ [Integrating by parts]

$$= \left| -x^n e^{-x} \right|_0^\infty + n\int_0^\infty e^{-x}\, x^{n-1} dx$$

$\therefore \quad \Gamma(n+1) = n\Gamma(n)$ ...(*iii*)

which is the reduction formula for $\Gamma(n)$. From this formula, it is clear that if $\Gamma(n)$ is known throughout a unit interval say : $1 < n \le 2$, then the values of $\Gamma(n)$ throughout the next unit interval $2 < n \le 3$ are found, from which the values of $\Gamma(n)$ for $3 < n \le 4$ are determined and so on. In this way, *the values of $\Gamma(n)$ for all positive values of $n > 1$ may be found by successive application of Eqn. (iii).*

Also using Eqn. (*iii*) in the form

$$\Gamma(n) = \frac{\Gamma(n+1)}{n} \qquad ...(iv)$$

*After an enormously creative Swiss mathematician *Leonhard Euler (1707–1783).* He studied under *John Bernoulli* and became a professor of mathematics in St. Petersburg, Russia. Even after becoming totally blind in 1771, he contributed to almost all branches of mathematics.

We can define $\Gamma(n)$ for values of $n$ for which the definition (1) fails. It gives the value of $\Gamma(n)$ for $0 < n \le 1$ in terms of the values of $\Gamma(n)$ for $1 < n \le 2$. Thus we can define $\Gamma(n)$ for all $n < 0$ provided its value for $1 < n \le 2$ is known. Also if $-1 < n < 0$, (4) gives $\Gamma(n)$ in terms of its values for $0 < n < 1$. Then we may find, $\Gamma(n)$ for $-2 < n < -1$ and so on.

*Thus (i) and (iv) together give a complete definition of $\Gamma(n)$ for all values of n except when n is zero or a negative integer and its graph is as shown in Fig. 7.51.* The values of $\Gamma(n)$ for $1 < n \le 2$ are given in (Table I-Appendix 2) from which the values of $\Gamma(n)$ for values of $n$ outside the interval $1 < n \le 2$ ($n \ne 0, -1, -2, -3$ .....) may be found.

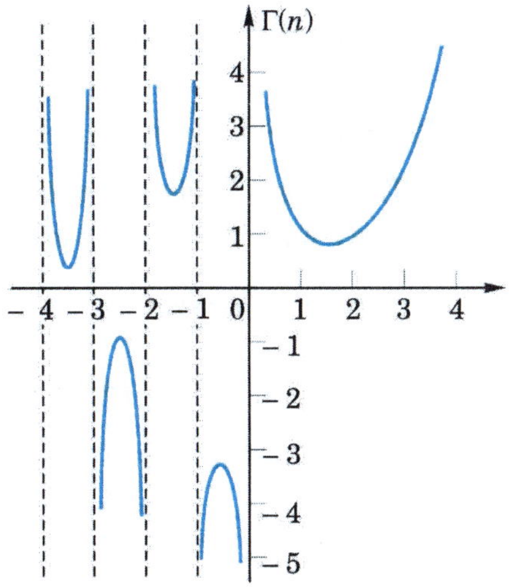


Fig. 7.51.

**(3) Value of Γ(n) in terms of factorial.**

Using $\Gamma(n + 1) = n\Gamma(n)$ successively, we get

$$\Gamma(2) = 1 \times \Gamma(1) = 1\,!$$
$$\Gamma(3) = 2 \times \Gamma(2) = 2 \times 1 = 2\,!$$
$$\Gamma(4) = 3 \times \Gamma(3) = 3 \times 2\,! = 3\,!$$
........ ....... .......

In general $\Gamma(n + 1) = n\,!$ provided $n$ is a positive integer ...(v)

Taking $n = 0$, it defines $0\,! = \Gamma(1) = 1$.

**(4) Value of $\Gamma\left(\frac{1}{2}\right)$.** We have

$$\Gamma\left(\tfrac{1}{2}\right) = \int_0^\infty e^{-x} x^{-1/2}\, dx \qquad \text{[Put } x = y^2 \text{ so that } dx = 2y\, dy]$$

$$= 2\int_0^\infty e^{-y^2}\, dy \text{ which is also } = 2\int_0^\infty e^{-x^2}\, dx$$

$$\therefore \quad \left[\Gamma\left(\tfrac{1}{2}\right)\right]^2 = 4\int_0^\infty \int_0^\infty e^{-(x^2+y^2)}\, dx\, dy = 4\int_0^{\pi/2}\int_0^\infty e^{-r^2}\, r\, dr\, d\theta$$

$$= 4 \cdot \frac{\pi}{2}\int_0^\infty e^{-r^2}\, r dr = 2\pi\left[\left(-\frac{1}{2}\right)e^{-r^2}\right]_0^\infty = \pi$$

whence $\qquad \Gamma\left(\tfrac{1}{2}\right) = \sqrt{\pi} = 1.772$ ...(vi) *(V.T.U., 2016 ; U.K.T.U., 2013 ; Delhi, 2012)*

## 7.16 RELATION BETWEEN BETA AND GAMMA FUNCTIONS

$$\beta(m, n) = \frac{\Gamma(m)\,\Gamma(n)}{\Gamma(m+n)} \qquad \text{...(1)}$$

We have $\qquad \Gamma(m) = \int_0^\infty e^{-t}\, t^{m-1} \qquad$ [Put $t = x^2$ so that $dt = 2x\, dx$

$$= 2\int_0^\infty e^{-x^2} x^{2m-1}\, dx \qquad \text{...(2)}$$

Similarly, $\qquad \Gamma(n) = 2\int_0^\infty e^{-y^2} y^{2n-1}\, dy$

$$\therefore \quad \Gamma(m)\Gamma(n) = 4\int_0^\infty e^{-x^2} x^{2m-1}\, dx \int_0^\infty e^{-y^2} y^{2n-1}\, dy$$

$$= 4\int_0^\infty \int_0^\infty e^{-(x^2+y^2)} x^{2m-1} y^{2n-1}\, dx\, dy \qquad \text{...(3)}$$

[∵ the limits of integration are constant.]

Now change to polar coordinates by writing $x = r\cos\theta$, $y = r\sin\theta$ and $dx\, dy = rd\theta dr$. To cover the region in Eqn. (3) which is the entire first quadrant, $r$ varies from 0 to ∞ and θ from 0 to π/2. Thus Eqn. (3) becomes

$$\Gamma(m)\Gamma(n) = 4\int_0^\infty \int_0^{\pi/2} e^{-r^2} r^{2(m+n)-1} \cos^{2m-1}\theta \sin^{2n-1}\theta\, d\theta dr$$

$$= \left[2\int_0^{\pi/2} \cos^{2m-1}\theta \sin^{2n-1}\theta\, d\theta\right] \times \left[2\int_0^\infty e^{-r^2} r^{2(m+n)-1}\, dr\right] \qquad \text{...(4)}$$

But by Eqn. (2), $2\int_0^\infty e^{-r^2} r^{2(m+n)-1}\, dr = \Gamma(m+n)$

and by Eqn. (3) of § 7.14, $2\int_0^{\pi/2} \cos^{2m-1}\theta \sin^{2n-1}\theta\, d\theta = \beta(m, n)$.

Thus Eqn. (4) gives $\Gamma(m)\Gamma(n) = \beta(m, n)\,\Gamma(m+n)$ *(Kurukshetra, 2013 ; V.T.U., 2013 ; Delhi, 2012 S)*

This follows Eqn. (1) which is extremely useful for evaluating definite integrals in terms of gamma functions.

**Cor.** *Rule to evaluate* $\int_0^{\pi/2} \sin^p x \cos^q x\, dx$.

$$\int_0^{\pi/2} \sin^p x \cos^q x\, dx = \frac{1}{2}\beta\left(\frac{p+1}{2}, \frac{q+1}{2}\right) \quad \text{[By (3) of § 7.14]}$$

$$= \frac{\Gamma\left(\frac{p+1}{2}\right)\Gamma\left(\frac{q+1}{2}\right)}{2\Gamma\left(\frac{p+q+2}{2}\right)} \quad ...(5)$$

In particular, when $q = 0$, and $p = n$, we have

$$\left.\begin{aligned} \int_0^{\pi/2} \sin^n x\, dx &= \frac{\Gamma\left(\frac{n+1}{2}\right)}{\Gamma\left(\frac{n+2}{2}\right)} \cdot \frac{\sqrt{\pi}}{2} \\ \text{Similarly,} \quad \int_0^{\pi/2} \cos^n x\, dx &= \frac{\Gamma\left(\frac{n+1}{2}\right)}{\Gamma\left(\frac{n+2}{2}\right)} \cdot \frac{\sqrt{\pi}}{2} \end{aligned}\right\} \quad ...(6)$$

**Example 7.48.** *Evaluate (i)* $\Gamma(-7/2)$ *(ii)* $\beta(7/2, -1/2)$.

**Solution.** (*i*) When $n$ is negative, we have $\Gamma(n) = \Gamma(n+1)/n$.

$$\therefore \quad \Gamma(-7/2) = \frac{\Gamma(-7/2+1)}{-7/2} = \frac{2}{7}\Gamma\left(\frac{-5}{2}\right),$$

$$\Gamma\left(-\frac{5}{2}\right) = \frac{\Gamma(-3/2)}{(-5/2)} = -\frac{2}{5}\Gamma\left(-\frac{3}{2}\right),$$

$$\Gamma\left(-\frac{3}{2}\right) = \frac{\Gamma(-1/2)}{-3/2} = -\frac{2}{3}\Gamma\left(-\frac{1}{2}\right) \text{ and } \Gamma\left(-\frac{1}{2}\right) = \frac{\Gamma(1/2)}{-1/2} = -2\sqrt{\pi}.$$

Hence
$$\Gamma\left(-\frac{7}{2}\right) = \left(-\frac{2}{7}\right)\left(-\frac{2}{5}\right)\left(-\frac{2}{3}\right)(-2)\sqrt{\pi} = \frac{16}{105}\sqrt{\pi}$$

(*ii*) We know that $\beta(m, n) = \dfrac{\Gamma(m)\Gamma(n)}{\Gamma(m+n)}$

$$\therefore \quad \beta\left(\frac{7}{2}, -\frac{1}{2}\right) = \frac{\Gamma\left(\frac{7}{2}\right)\Gamma\left(-\frac{1}{2}\right)}{\Gamma\left(\frac{7}{2}-\frac{1}{2}\right)} = \frac{\Gamma\left(\frac{7}{2}\right)\Gamma\left(-\frac{1}{2}\right)}{\Gamma(3)}$$

Also
$$\Gamma\left(\frac{7}{2}\right) = \frac{5}{2}\Gamma\left(\frac{5}{2}\right) \quad [\because \Gamma(n+1) = n\Gamma(n)]$$

$$\Gamma\left(\frac{5}{2}\right) = \frac{3}{2}\Gamma\left(\frac{3}{2}\right); \Gamma\left(\frac{3}{2}\right) = \frac{1}{2}\Gamma\left(\frac{1}{2}\right) = \frac{1}{2}\sqrt{\pi}$$

$\therefore \qquad \Gamma\left(\frac{7}{2}\right) = \left(\frac{5}{2}\right)\left(\frac{3}{2}\right)\left(\frac{1}{2}\right)\sqrt{\pi} = \frac{15}{8}\sqrt{\pi}$

Now $\qquad \Gamma\left(-\frac{1}{2}\right) = \frac{\Gamma\left(-\frac{1}{2}+1\right)}{-\frac{1}{2}} = -2\Gamma\left(\frac{1}{2}\right) = -2\sqrt{\pi}$

and $\qquad \Gamma(3) = 2\Gamma(2) = 2\times 1\times \Gamma(1) = 2 \qquad [\because \Gamma(1) = 1]$

Hence $\qquad \beta\left(\frac{7}{2}, -\frac{1}{2}\right) = \frac{(15/8)\sqrt{\pi}\times(-2)\sqrt{\pi}}{2} = -\frac{15}{8}\pi$

**Example 7.49.** *Show that*

*(a)* $\Gamma(n) = \int_0^1 \left(\log\frac{1}{y}\right)^{n-1} dy \ (n > 0).$

*(b)* *$\beta(p, q) = \int_0^\infty \frac{y^{q-1}}{(1+y)^{p+q}} dy$ *(D.T.U., 2011)*

$= \int_0^1 \frac{x^{p-1}+x^{q-1}}{(1+x)^{p+q}} dx.$ *(U.P.T.U., 2012 ; Kerala, 2010)*

**Solution.** *(a)* $\Gamma(n) = \int_0^\infty x^{n-1} e^{-x} dx \ (n > 0)$

$$= \int_1^0 \left(\log\frac{1}{y}\right)^{n-1} y\left(-\frac{1}{y}dy\right) = \int_0^1 \left(\log\frac{1}{y}\right)^{n-1} dy.$$

| Put $y = e^{-x}$ *i.e.,* $x = \log(1/y)$ so that $dx = -(1/y)\,dy$

*(b)* $\qquad \beta(p, q) = \int_0^1 x^{p-1}(1-x)^{q-1} dx$

$$= \int_\infty^0 \frac{1}{(1+y)^{p-1}}\left(\frac{y}{1+y}\right)^{q-1} \frac{-1}{(1+y)^2} dy$$

| Put $x = \frac{1}{1+y}$ *i.e.,* $y = \frac{1}{x} - 1$ so that $dx = \frac{-1}{(1+y)^2} dy$

$$= \int_0^\infty \frac{y^{q-1}}{(1+y)^{p+q}} dy = \int_0^1 \frac{y^{q-1}}{(1+y)^{p+q}} dy + \int_1^\infty \frac{y^{q-1}}{(1+y)^{p+q}} dy$$

Now substituting $y = 1/z$ in the second integral, we get

$$\int_1^\infty \frac{y^{q-1}}{(1+y)^{p+q}} dy = \int_1^0 \frac{1}{z^{q-1}} \cdot \frac{1}{(1+1/z)^{p+q}}\left(-\frac{1}{z^2}\right) dz$$

$$= \int_0^1 \frac{z^{p-1}}{(1+z)^{p+q}}$$

Hence $\qquad \beta(p, q) = \int_0^1 \frac{y^{q-1}}{(1+y)^{p+q}} dy + \int_0^1 \frac{z^{p-1}}{(1+z)^{p+q}} dz$

$$= \int_0^1 \frac{x^{p-1}+x^{q-1}}{(1+x)^{p+q}}$$

*Here Beta function is expressed as an *improper integral.*

**Example 7.50.** *Express the following integrals in terms of gamma functions :*

(a) $\int_0^1 \frac{dx}{\sqrt{(1-x^4)}}$ (V.T.U., 2014 ; Rohtak, 2011 S) (b) $\int_0^{\pi/2} \sqrt{(\tan\theta)}\, d\theta$. (Andhra, 2016 ; Madras, 2006)

(c) $\int_0^\infty \frac{x^c}{c^x}\, dx$ (U.P.T.U., 2006) (d) $\int_0^\infty a^{-bx^2}\, dx$.

(e) $\int_0^1 x^5 [\log(1/x)]^3\, dx$

**Solution.** (a) $\int_0^1 \frac{dx}{\sqrt{(1-x^4)}}$ | Put $x^2 = \sin\theta$, *i.e.*, $x = \sin^{1/2}\theta$ so that $dx = 1/2 \sin^{-1/2}\theta\cos\theta\, d\theta$

$$= \int_0^{\pi/2} \frac{1}{2} \cdot \frac{\sin^{-1/2}\theta \,.\, \cos\theta\, d\theta}{\sqrt{(1-\sin^2\theta)}} = \frac{1}{2}\int_0^{\pi/2} \sin^{-1/2}\theta\, d\theta = \frac{1}{2}\cdot\frac{\sqrt{\pi}}{2}\cdot\frac{\Gamma\left(\frac{-\frac{1}{2}+1}{2}\right)}{\Gamma\left(\frac{-\frac{1}{2}+2}{2}\right)} = \frac{\sqrt{\pi}}{4}\,\frac{\Gamma\left(\frac{1}{4}\right)}{\Gamma\left(\frac{3}{4}\right)}$$

(b) $\int_0^{\pi/2} \sqrt{(\tan\theta)}\, d\theta = \int_0^{\pi/2} \sin^{1/2}\theta \cos^{-1/2}\theta\, d\theta$

$$= \frac{\Gamma\left(\frac{\frac{1}{2}+1}{2}\right)\Gamma\left(\frac{-\frac{1}{2}+1}{2}\right)}{2\Gamma\left(\frac{\frac{1}{2}-\frac{1}{2}+2}{2}\right)} = \frac{\Gamma\left(\frac{3}{4}\right)\Gamma\left(\frac{1}{4}\right)}{2\Gamma(1)} = \frac{1}{2}\Gamma\left(\frac{1}{4}\right)\Gamma\left(\frac{3}{4}\right)$$

(c) $\int_0^\infty \frac{x^c}{c^x}\, dx = \int_0^\infty \frac{x^c}{e^{x\log c}}\, dx$ $[\because c^x = e^{\log c^x} = e^{x\log c}]$

$= \int_0^\infty e^{-x\log c}\, x^c\, dx$ [Put $x\log c = t$ so that $dx = dt/\log c$]

$$= \int_0^\infty e^{-t}\left(\frac{t}{\log c}\right)^c \frac{dt}{\log c} = \frac{1}{(\log c)^{c+1}}\int_0^\infty t^c e^{-t}\, dt = \Gamma(c+1)/(\log c)^{c+1}$$

(d) $\int_0^\infty a^{-bx^2}\, dx = \int_0^\infty e^{-bx^2\log a}\, dx$ [Put $(b\log a)x^2 = t$ so that $dx = dt/2\sqrt{(b\log a)}$]

$$= \frac{1}{2\sqrt{(b\log a)}}\int_0^\infty e^{-t}\, t^{-1/2}\, dt = \frac{\Gamma\left(\frac{1}{2}\right)}{2\sqrt{(b\log a)}} = \frac{\sqrt{\pi}}{2\sqrt{(b\log a)}}$$

(e) $\int_0^1 x^4 [\log(1/x)]^3\, dx = \frac{1}{625}\int_0^\infty e^{-t} \,.\, t^3\, dt$ [Put $x = e^{-t/5}$ so that $\log(1/x) = t/5$, $dx = -\frac{1}{5}e^{-t/5}\, dt$]

$$= \frac{\Gamma(4)}{625} = \frac{6}{625}.$$

**Example 7.51.** *Evaluate* $\int_0^\infty e^{-ax} x^{m-1} \sin bx\, dx$ *in terms of Gamma function.* *(Andhra, 2022)*

**Solution.** We have $\Gamma(m) = \int_0^\infty e^{-x} x^{m-1} dx$ [Put $x = ay, dx = ady$]

$$= \int_0^\infty e^{-ay} a^m y^{m-1} dy \quad \text{or} \quad \int_0^\infty e^{-ay} y^{m-1} dy = \Gamma(m)/a^m \qquad ...(i)$$

Then
$$I = \int_0^\infty e^{-ax} x^{m-1} \sin bx\, dx$$

$$= \int_0^\infty e^{-ax} x^{m-1} (\text{Imaginary part of } e^{ibx})\, dx$$

$$= \text{I.P. of} \int_0^\infty e^{-(a-ib)x} x^{m-1}\, dx$$

$$= \text{I.P. of } \{\Gamma(m)/(a-ib)^m \qquad \text{[By (i)]}$$

$$= \text{I.P. of } [\Gamma(m)/(r^m (\cos 0 - i \sin 0)^m] \qquad \text{where } a = r\cos\theta, b = r\sin\theta$$

$$= \text{I.P. of } \Gamma(m)/(r^m (\cos m\theta - i \sin m\theta)] \qquad \text{(Using Demoivere's theorem §19.5)}$$

$$= \text{I.P. of} \left\{\frac{\Gamma(m).(\cos m\theta + i\sin m\theta)}{r^m (\cos m\theta + i \sin m\theta)(\cos m\theta - i\sin m\theta)}\right\}$$

$$= \frac{\Gamma(m)}{r^m} \sin m\theta \qquad \text{where } r\ \sqrt{(a^2+b^2)},\ \theta = \tan^{-1} b/a.$$

**Example 7.52.** *Prove that* $\int_0^1 \frac{x^2 dx}{\sqrt{(1-x^4)}} \times \int_0^1 \frac{dx}{\sqrt{(1+x^4)}} = \frac{\pi}{4\sqrt{2}}$ *(Andhra, 2020)*

**Solution.** $\int_0^1 \frac{x^2 dx}{\sqrt{(1-x^4)}} \times \int_0^{\pi/2} \frac{\sin\theta}{\cos\theta} \cdot \frac{\cos\theta}{2\sqrt{(\sin\theta)}} d\theta$ $\left[\text{Putting } x^2 = \sin\theta, dx = \frac{\cos\theta\, d\theta}{2\sqrt{(\sin\theta)}}\right.$

$$= \frac{1}{2}\int_0^{\pi/2} \sqrt{(\sin\theta)}\, d\theta = \frac{1}{4}\beta\left(\frac{3}{4}, \frac{1}{2}\right) = \frac{1}{4}\frac{\Gamma(3/4)\Gamma(1/2)}{\Gamma(5/4)}$$

$$= \frac{\Gamma(3/4)\Gamma(1/2)}{\Gamma(1/4)}$$

$$\int_0^1 \frac{dx}{\sqrt{(1+x^4)}} = \int_0^{\pi/4} \frac{\sec^2\theta\, d\theta}{2\sqrt{(\tan\theta)}\sec\theta} \qquad \left[\text{Putting } x^2 = \tan\theta, dx = \frac{\sec^2\theta\, d\theta}{2\sqrt{(\tan\theta)}}\right.$$

$$= \frac{1}{\sqrt{2}}\int_0^{\pi/4} \frac{d\theta}{\sqrt{(\sin 2\theta)}} = \frac{1}{2\sqrt{2}}\int_0^{\pi/2} \sin^{-1/2}\phi\, d\phi \qquad \left[\text{Putting } 2\theta = \phi, d\theta = \frac{1}{2}d\phi\right.$$

$$= \frac{1}{4\sqrt{2}}\beta\left(\frac{1}{4}, \frac{1}{2}\right) = \frac{1}{4\sqrt{2}}\frac{\Gamma(1/4)\Gamma(1/2)}{\Gamma(3/4)}$$

$$\therefore \quad \int_0^1 \frac{x^2\, dx}{\sqrt{(1-x^4)}} \times \int_0^1 \frac{dx}{\sqrt{(1+x^4)}} = \frac{1}{4\sqrt{2}}\left\{\Gamma\left(\frac{1}{2}\right)\right\}^2 = \frac{\pi}{4\sqrt{2}}.$$

**Example 7.53.** *Prove that (i)* $\beta(m, 1/2) = 2^{2m-1}\beta(m, m)$ *(V.T.U., 2016)*

*(ii)* $\Gamma(m)\Gamma\left(m+\frac{1}{2}\right) = \frac{\sqrt{\pi}}{2^{2m-1}}\Gamma(2m)$ *(Legendre's Duplication Formula)*

*(V.T.U., 2010)*

*Hence evaluate* $\Gamma\left(\frac{1}{4}\right)\Gamma\left(\frac{3}{4}\right)$.

**Solution.** (*i*) We know that $\beta(m, n) = 2\int_0^{\pi/2} \sin^{2m-1}\theta \cos^{2n-1}\theta \, d\theta$ ...(1)

Putting $n = \frac{1}{2}$, we have $\beta\left(m, \frac{1}{2}\right) = 2\int_0^{\pi/2} \sin^{2m-1}\theta \, d\theta$ ...(2)

Again putting $n = m$ in Eqn. (1), we get $\beta(m, m) = 2\int_0^{\pi/2} (\sin\theta \cos\theta)^{2m-1} d\theta$

$$= \frac{1}{2^{2m-2}} \int_0^{\pi/2} \sin^{2m-1} 2\theta \, d\theta$$

$$= \frac{1}{2^{2m-1}} \int_0^{\pi} \sin^{2m-1}\phi \, d\phi, \text{ putting } 2\theta = \phi$$

$$= \frac{1}{2^{2m-1}} \cdot 2\int_0^{\pi/2} \sin^{2m-1}\phi \, d\phi$$

or $$2^{2m-1}\beta(m, m) = 2\int_0^{\pi/2} \sin^{2m-1}\theta \, d\theta = \beta\left(m, \frac{1}{2}\right) \quad \text{[by (2)]}$$

(*ii*) Rewriting the above result in terms of $\Gamma$ functions, we get

$$2^{2m-1}\frac{\Gamma(m)\Gamma(m)}{\Gamma(m+m)} = \frac{\Gamma(m)\Gamma\left(\frac{1}{2}\right)}{\Gamma\left(m+\frac{1}{2}\right)} \quad \left[\because \Gamma\left(\frac{1}{2}\right) = \sqrt{\pi}\right]$$

or $$\Gamma(m)\Gamma\left(m+\frac{1}{2}\right) = \frac{\sqrt{\pi}\,\Gamma(2m)}{2^{2m-1}}. \quad ...(\alpha)$$

Putting $m = \frac{1}{4}$ in $(\alpha)$, we get $\Gamma\left(\frac{1}{4}\right)\Gamma\left(\frac{3}{4}\right) = \frac{\sqrt{\pi}\,\Gamma(1/2)}{2^{-1/2}} = \sqrt{2}\sqrt{\pi}\sqrt{\pi} = \sqrt{2}\pi.$

**Example 7.54.** *Prove that*

(*a*) $\iint_D x^{l-1} y^{m-1} dx\,dy = \frac{\Gamma(l)\Gamma(m)}{\Gamma(l+m+1)} h^{l+m}$ *where D is the domain* $x \geq 0, y \geq 0$ *and* $x + y \leq h$.

(*b*) $\iiint_V x^{l-1} y^{n-1} z^{n-1} dxdydz = \frac{\Gamma(l)\Gamma(m)\Gamma(n)}{\Gamma(l+m+n+1)}$

*where V is the region* $x \geq 0, y \geq 0, z \geq 0$ *and* $x + y + z \leq 1$. *This important result is known as* **Dirichlet's integral*.**

**Solution.** (*a*) Putting $x/h = X$ and $y/h = Y$, we see that the given integral

$$= \iint_{D'} (hX)^{l-1}(hY)^{m-1} h^2 \, dXdY \text{ where } D' \text{ is the domain } X \geq 0, Y \geq 0 \text{ and } X + Y \leq 1.$$

$$= h^{l+m}\int_0^1\int_0^{1-X} X^{l-1} Y^{m-1} dY\,dX = h^{l+m}\int_0^1 X^{l-1}\left|\frac{Y^m}{m}\right|_0^{1-X} dX$$

$$= \frac{h^{l+m}}{m}\int_0^1 X^{l-1}(1-X)^m dX = \frac{h^{l+m}}{m}\beta(l, m+1) = \frac{h^{l+m}}{m}\cdot\frac{\Gamma(l)\Gamma(m+1)}{\Gamma(l+m+1)}$$

$$= h^{l+m}\frac{\Gamma(l)\Gamma(m)}{\Gamma(l+m+1)} \quad ...(i)\ [\because \Gamma(m+1)/m = \Gamma(m)]$$

(*b*) Taking $y + z \leq 1 - x (= h\text{: say})$, the triple integral

$$= \int_0^1\int_0^{1-x}\int_0^{1-x-y} x^{l-1} y^{m-1} z^{n-1} dzdydx$$

$$= \int_0^1 x^{l-1}\left[\int_0^h\int_0^{h-y} y^{m-1} z^{n-1} dzdy\right] dx = \int_0^1 x^{l-1}\frac{\Gamma(m)\Gamma(n)}{\Gamma(m+n+1)} h^{m+n} dx \quad ...\text{[By }(i)]$$

*Named after a German mathematician *Peter Gustav Lejeune Dirichlet* (1805–1859) who studied under Cauchy and succeeded Gauss at Gottingen. He is known for his contributions to Fourier series and number theory.

$$= \frac{\Gamma(m)\Gamma(n)}{\Gamma(m+n+1)} \int_0^1 x^{l-1}(1-x)^{m+n}\, dx = \frac{\Gamma(m)\Gamma(n)}{\Gamma(m+n+1)}\, \beta(l, m+n+1)$$

$$= \frac{\Gamma(m)\Gamma(n)}{\Gamma(m+n+1)} \cdot \frac{\Gamma(l)\Gamma(m+n+1)}{\Gamma(l+m+n+1)} = \frac{\Gamma(l)\Gamma(m)\Gamma(n)}{\Gamma(l+m+n+1)}.$$

**Example 7.55.** *Evaluate the integral* $\iiint x^{l-1} y^{m-1} z^{n-1}\, dxdydz$ *where x, y, z are all positive with condition,* $(x/a)^p + (y/b)^q + (z/c)^r \le 1$. *(U.P.T.U., 2011)*

**Solution.** Put $(x/a)^p = u$, *i.e.*, $x = au^{1/p}$ so that $dx = \frac{a}{p} u^{1/p-1}\, du$

$(y/b)^q = v$, *i.e.*, $y = bv^{1/q}$ so that $dy = \frac{b}{q} v^{1/q-1}\, dv$

and $(z/c)^r = w$, *i.e.*, $z = cw^{1/r}$ so that $dz = \frac{c}{r} w^{1/r-1}\, dw$

Then $\iiint x^{l-1} y^{m-1} z^{n-1}\, dxdydz$

$$= \iiint (au^{1/p})^{l-1} (bv^{1/q})^{m-1} (cw^{1/r})^{n-1} \left(\frac{a}{p}\right) u^{1/p-1} \left(\frac{b}{q}\right) v^{1/q-1} \left(\frac{c}{r}\right) w^{1/r-1}\, dudvdw$$

$$= \frac{a^l b^m c^n}{pqr} \iiint u^{l/p-1} v^{m/q-1} w^{n/r-1}\, du\, dv\, dw \text{ where } u+v+w \le 1.$$

$$= \frac{a^l b^m c^n}{pqr} \frac{\Gamma(l/p)\Gamma(m/q)\Gamma(n/r)}{\Gamma(l/p + m/q + n/r + 1)}$$ [By Dirichlet's integral]

**Example 7.56.** *The plane* $\frac{x}{a} + \frac{y}{b} + \frac{z}{c} = 1$ *meets the axes in A, B and C. Apply Dirichlet's integral to find the volume of the tetrahedron OABC.* *(Andhra, 2016)*

*Also find its mass if the density at any point is kxyz.*

**Solution.** Put $x/a = u$, $y/b = v$, $z/c = w$ then the tetrahedron $OABC$ has $u \ge 0$, $v \ge 0$, $w \ge 0$ and $u + v + w \le 1$.

$\therefore$ volume of this tetrahedron $= \iiint_D dx\, dy\, dz$

$$= \iiint_D abc\, du\, dv\, dw$$ $\left[\begin{array}{l} a\ dx = adu,\ dy = bdv,\ dz = cdw \\ \text{for } D' = u \ge 0, v \ge 0, w \ge 0\ \&\ u+v+w \le 1. \end{array}\right.$

$$= abc \iiint_D u^{1-1} v^{1-1} w^{1-1}\, du\, dv\, dw$$

$$= abc \frac{\Gamma(1)\Gamma(1)\Gamma(1)}{\Gamma(1+1+1+1)} = \frac{abc}{6}$$ [By Dirichlet's integral]

$$\text{Mass} = \iiint kxyzdx\, dy\, dz = \iiint k\,(au)(bv)(cw)\,abc\, du\, dv\, dw$$

$$= ka^2b^2c^2 \iiint u^{2-1} v^{2-1} w^{2-1}\, du\, dv\, dw$$

$$= ka^2b^2c^2 \frac{\Gamma(2)\Gamma(2)\Gamma(2)}{\Gamma(2+2+2+1)} ka^2b^2c^2 \cdot \frac{1}{6!} = \frac{k}{720} a^2b^2c^2.$$

## PROBLEMS 7.8

1. Compute :

(i) $\Gamma(4.5)$ (ii) $\Gamma\left(-\frac{3}{2}\right)$ *(U.P.T.U., 2013)*

(iii) $\beta\left(\frac{1}{2}, \frac{1}{2}\right)$ (iv) $\beta(2.5, 1.5)$ (v) $\beta\left(\frac{1}{2}, \frac{7}{2}\right)$.

2. Express the following integrals in terms of gamma functions :

(i) $\int_0^\infty e^{-x^2}\, dx$ (ii) $\int_0^\infty x^6 e^{-3x}\, dx$ (V.T.U., 2016)

(iii) $\int_0^\infty \sqrt{x}\, e^{-x^3}\, dx$ (iv) $\int_0^\infty \frac{dx}{x^{p+1}\,.\,(x-1)^q}\ (-p < q < 1)$

3. Show that :

(i) $\int_0^\infty \frac{x^4}{4^x}\, dx = \frac{\Gamma(5)}{(\log 4)^5}$

(ii) $\int_0^{\pi/2} \sqrt{(\cot\theta)}\, d\theta = \frac{1}{2}\Gamma\left(\frac{1}{4}\right)\Gamma\left(\frac{3}{4}\right)$ (Andhra., 2022, 2016)

(iii) $\int_0^{\pi/2} [\sqrt{(\tan\theta)} + \sqrt{(\sec\theta)}]\, d\theta = \frac{1}{2}\Gamma\left(\frac{1}{4}\right)\left\{\Gamma\left(\frac{3}{4}\right) + \sqrt{\pi}/\Gamma\left(\frac{3}{4}\right)\right\}$ (D.T.U., 2015)

(iv) $\int_0^{\pi/2} \sqrt{(\sin\theta)}\, d\theta \times \int_0^{\pi/2} \frac{d\theta}{\sqrt{(\sin\theta)}} = \pi.$ (Andhra, 2021; V.T.U., 2015)

(v) $\int_0^\infty e^{x^{-4}}\, dx \times \int_0^\infty x^2 e^{-x^4}\, dx = \frac{\pi}{8\sqrt{2}}$ (W.B.T.U., 2015)

(vi) $\int_0^1 \sqrt[3]{[x \log(1/x)]}\, dx = \left(\frac{3}{4}\right)^{4/3}\Gamma(4/3).$

4. Given $\int_0^\infty \frac{x^{n-1}}{1+x}\, dx = \frac{\pi}{\sin n\pi}$, show that $\Gamma(n)\,\Gamma(1-n) = \frac{\pi}{\sin n\pi}$. (C.S.V.T.U., 2012)

Hence evaluate $\int_0^\infty \frac{dy}{1+y^4}$. (U.P.T.U., 2012)

5. Prove that :

(i) $\int_0^1 \frac{x\,dx}{\sqrt{(1-x^5)}} = \frac{1}{5}\beta\left(\frac{2}{5}, \frac{1}{2}\right)$ (Andhra., 2022, 2016) (ii) $\int_0^1 \frac{dx}{\sqrt{(1+x^4)}} = \frac{1}{4\sqrt{2}}\beta\left(\frac{1}{4}, \frac{1}{2}\right)$. (U.P.T.U., 2011)

(iii) $\int_0^1 x^3 (1-\sqrt{x})^5\, dx = 2\beta(8, 6)$. (Andhra, 2019)

6. Show that (i) $\int_a^b (x-a)^{m-1}(b-x)^{n-1}\, dx = (b-a)^{m+n-1}\,\beta(m, n)$ (Andhra., 2014 ; P.T.U., 2010 )

(ii) $\int_0^\infty \frac{x^{m-1}}{(a+bx)^{m+n}}\, dx = \frac{1}{a^n, b^m}\beta(m, n)$ (Nagpur, 2009) (iii) $\int_0^\infty \frac{x^{10} - x^{18}}{(1+x)^{30}}\, dx = 0$

(iv) $\int_0^1 \frac{(1-x^4)^{3/4}}{(1+x^4)^2}\, dx = \frac{1}{2^{9/2}}\beta\left(\frac{7}{4}, \frac{1}{4}\right)$

7. Prove that $\int_0^1 x^m (\log x)^n\, dx = \frac{(-1)^n\, n!}{(m+1)^{n+1}}$, where $n$ is a positive integer and $m > -1$.

Hence evaluate $\int_0^1 x(\log x)^3\, dx$.

8. Show that $\int_0^1 y^{q-1}\left(\log\frac{1}{y}\right)^{p-1} dy = \frac{\Gamma(p)}{q^p}$, where $p > 0, q > 0$.

9. Express $\int_0^1 x^m (1-x^n)^p\, dx$ in terms of gamma functions (J.N.T.U, 2015)

Hence evaluate : (i) $\int_0^1 x^5 (1-x^3)^{10}\, dx$. (Andhra, 2023 ; Delhi, 2012) (ii) $\int_0^1 \frac{dx}{\sqrt{(1-x^n)}}$

10. Show that $\int_{-1}^2 (1+x)^{p-1}(1-x)^{q-1} dx = 2^{p+q-1}\beta(p, q)$. (Delhi., 2015)

11. Prove that $\int_0^\infty \frac{dx}{(e^x + e^{-x})^n} = \frac{1}{4}\beta\left(\frac{n}{2}, \frac{n}{2}\right)$ and hence evaluate $\int_0^\infty \operatorname{sech}^8 x\, dx$.

**12.** Prove that $\beta\left(n+\frac{1}{2}, n+\frac{1}{2}\right) = \frac{\Gamma(n+1/2)\sqrt{\pi}}{2^{2n}\,\Gamma(n+1)}$. Hence show that $2^n\,\Gamma(n+1/2) = 1 \cdot 3 \cdot 5 \ldots (2n-1)\sqrt{\pi}$

*(Andhra, 2021)*

**13.** Prove that :

(*i*) $\frac{\beta(m+1, n)}{\beta(m, n)} = \frac{m}{m+n}$ *(D.T.U., 2015)*

(*ii*) $\beta(n, n) = \frac{\sqrt{\pi}\,\Gamma(n)}{2^{2n-1}\,\Gamma\left(n+\frac{1}{2}\right)}$

(*iii*) $\Gamma\left(n+\frac{1}{2}\right) = \frac{\Gamma(2n+1)\sqrt{\pi}}{2^{2n}\,.\,\Gamma(n+1)}$

(*iv*) $\beta(m+1) + \beta(m, n+1) = \beta(m, n)$

**14.** Evaluate $\int_0^1\int_0^{1-y} xy\, dx\, dy$ using Drichlets integral. *(Andhra., 2016)*

**15.** Show that $\iint x^{m-1} y^{n-1}\, dx\, dy$ over the positive quadrant of the ellipse

$$\left(\frac{x}{a}\right)^2 + \left(\frac{y}{b}\right)^2 = 1 \text{ is } \frac{a^m b^n}{2n}\,\beta\left(\frac{m}{2}, \frac{n}{2}+1\right).$$

**16.** Show that the area in the first quadrant enclosed by the curve $(x/a)^\alpha + (y/b)^\beta = 1$, $\alpha > 0$, $\beta > 0$, is given by

$$\frac{ab}{\alpha+\beta}\,\frac{\Gamma(1/\alpha)\,\Gamma(1/\beta)}{\Gamma(1/\alpha + 1/\beta)}.$$

**17.** Find the mass of an octant of the ellipsoid $(x/a)^2 + (y/b)^2 + (z/c)^2 = 1$, the density at any point being $\rho = kxyz$.

## 7.17 (1) ELLIPTIC INTEGRALS

In Applied Mathematics, we often come across integrals of the form $\int_0^1 e^{-x^2}\, dx$ or $\int_0^1 \sin x^2\, dx$ which cannot be evaluated by any of the standard methods of integration. In such cases, we may find the value to any desired degree of accuracy by expanding their integrands as power series. An important class of such integrals is the *elliptic integrals*.

**Def.** The integral $F(k, \phi) = \int_0^\phi \frac{dx}{\sqrt{(1-k^2\sin^2 x)}}\ (k^2 < 1)$ ...(*i*)

which is a function of the two variables $k$ and $\phi$, is called the *elliptic integral of the first kind with modulus k* and *amplitude* $\phi$.

The integral $E(k, \phi) = \int_0^\phi \sqrt{(1-k^2\sin^2 x)}\, dx\ (k^2 < 1)$ ...(*ii*)

is called the *elliptic integral of the second kind* with *modulus k and amplitude* $\phi$.

The name *elliptic integral* arose from its original application in finding the length of an elliptic arc (Fig. 7.52), For instance, consider the ellipse

$$x = a\cos\phi, \qquad y = b\sin\phi, \qquad (a < b)$$

Then length of its arc

$$AP = \int_0^\phi \sqrt{\left[\left(\frac{dx}{d\phi}\right)^2 + \left(\frac{dy}{d\phi}\right)^2\right]}\, d\phi = \int_0^\phi \sqrt{[(-a\sin\phi)^2 + (b\cos\phi)^2]}\, d\phi$$

$$= \int_0^\phi \sqrt{[(b^2 + (a^2-b^2)\sin^2\phi]}\, d\phi = b\int_0^\phi \sqrt{\left[1-\left(1-\frac{a^2}{b^2}\right)\sin^2\phi\right]}\, d\phi$$

$= bE(k, \phi)$ for $k^2 = 1 - a^2/b^2 \le 1$.

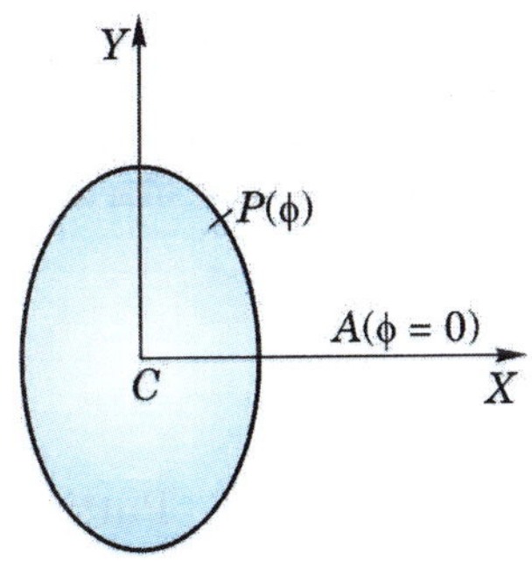

Fig. 7.52.

Also the perimeter of the ellipse

$$= 4b\int_0^{\pi/2} \sqrt{(1-k^2\sin^2\phi)}\, d\phi = 4bE(k, \pi/2).$$

This particular integral with upper limit $\phi = \pi/2$ is called the *complete elliptic integral of the second kind and is denoted by E(k).*

Thus $$E(k) = \int_0^{\pi/2} (1 - k^2 \sin^2 \phi)\, d\phi \; (k^2 < 1) \qquad ...(iii)$$

Similarly, the *complete elliptic integral of first kind* is

$$K(k) = \int_0^{\pi/2} \frac{d\phi}{\sqrt{(1 - k^2 \sin^2 \phi)}} \; (k^2 < 1) \qquad ...(iv)$$

*To evaluate it,* we expand the integral in the form

$$(1 - k^2 \sin^2 \phi)^{-1/2} = 1 + \frac{k^2}{2} \sin^2 \phi + \frac{3k^4}{4} \sin^4 \phi + ...$$

This series can be shown to be uniformly convergent for all $k$, and may, therefore, be integrated term by term [See § 9.19-II]. Then we have

$$K(k) = \int_0^{\pi/2} \left(1 + \frac{k^2}{2} \sin^2 \phi + \frac{3k^4}{8} \sin^4 \phi + \frac{5k^6}{16} \sin^6 \phi + ...\right) d\phi$$

$$= \frac{\pi}{2}\left[1 + \left(\frac{1}{2}\right)^2 k^2 + \left(\frac{1.3}{2.4}\right)^2 k^4 + \left(\frac{1.3.5}{2.4.6.}\right)^2 k^6 + ...\right] \qquad ...(v)$$

This series may be used to compute $K$ for various values of $k$. In particular, if $k = \sin 10°$; we have

$$K = \frac{\pi}{2}(1 + 0.00754 + 0.00012 + ...) = 1.5828 \qquad ...(vi)$$

In this way tables of the elliptic integrals are constructed. Values of $F(k, \phi)$ and $E(k, \phi)$ are readily available for $0 \le \phi \le \pi/2$, $0 < k < 1$. (*See Peirce's short tables*).

**Example 7.57.** *Express* $\int_0^{\pi/6} \frac{dx}{\sqrt{(\sin x)}}$ *in terms of elliptic integral.*

**Solution.** Put $\cos x = \cos^2 \phi$ and $dx = \dfrac{2 \cos \phi\, d\phi}{\sqrt{(1+ \cos^2 \phi)}}$

Then $$I = \int_0^{\pi/2} \frac{2 \cos^2 \phi}{\sqrt{(1+ \cos^2 \phi)}} d\phi = 2\int_0^{\pi/2} \frac{(1 + \cos^2 \phi) - 1}{\sqrt{(1 + \cos^2 \phi)}} d\phi$$

$$= 2\left\{\int_0^{\pi/2} \sqrt{(1+ \cos^2 \phi)}\, d\phi - \int_0^{\pi/2} \frac{d\phi}{\sqrt{(1 + \cos^2 \phi)}}\right\} = 2\left\{\int_0^{\pi/2} \sqrt{(2 - \sin^2 \phi)}\, d\phi - \int_0^{\pi/2} \frac{d\phi}{\sqrt{(2 - \sin^2 \phi)}}\right\}$$

$$= 2\sqrt{2}\int_0^{\pi/2} \sqrt{(1 - 1/2 \sin^2 \phi)}\, d\phi - \sqrt{2}\int_0^{\pi/2} \frac{d\phi}{\sqrt{(1 - 1/2 \sin^2 \phi)}} = 2\sqrt{2}\, E\left(\frac{1}{\sqrt{2}}\right) - \sqrt{2}\, K\left(\frac{1}{\sqrt{2}}\right)$$

(2) **Elliptic functions.** By putting $\sin x = t$ and $\sin \phi = z$, $(i)$ becomes

$$u = \int_0^{z} \frac{dt}{\sqrt{[(1 - t^2)(1 - k^2t^2)]}} \; (k^2 < 1) \qquad ...(vii)$$

This is known as *Jacobi's form of the elliptic integral of first kind** whereas Eqn.$(i)$ is the *Legendre's form*†.

If $k = 0$, Eqn. $(vii)$ gives $u = \sin^{-1} z$. By analogy, we denote Eqn. $(vii)$ $sn^{-1} z$ for a fixed non-zero value of $k$. This leads to the functions $sn\, u = z = \sin \phi$ and $cn\, u = \cos \phi$, which are called the *Jacobi's elliptic functions.*

The elliptic functions $sn\, u$ and $cn\, u$ are periodic with a period depending on $k$ and an amplitude equal to unity. These behave somewhat like $\sin u$ and $\cos u$. For instance

$$sn(0) = 0,\; cn(1) = 1 \quad \text{and} \quad sn(-u) = -sn(u),\; cn(-u) = cn(u).$$

**Example 7.58.** *Show that* $\int_0^{a/2} \frac{dx}{\sqrt{(2ax - x^2)}\sqrt{(a^2 - x^2)}} = \frac{2}{3a} K\left(\frac{1}{3}\right)$.

**Solution.** Putting $x = \dfrac{a}{2}(1 - \sin \theta),\; dx = -\dfrac{a}{2} \cos \theta\, d\theta,$

$$2ax - x^2 = \frac{a^2}{4}(1 - \sin \theta)(3 + \sin \theta) \text{ and } a^2 - x^2 = \frac{a^2}{4}(1 + \sin \theta)(3 - \sin \theta)$$

* See footnote p. 239.

† A French mathematician *Adrien Marie Legendre* (1752–1833) who made important contributions to number theory, special functions, calculus of variations and elliptic integrals.

Also when $x = 0, \theta = \pi/2$; when $x = a/2, \theta = 0$.

Thus the given integral

$$= \frac{4}{a^2}\int_{\pi/2}^{0} \frac{-(a/2)\cos\theta\, d\theta}{\sqrt{[(1-\sin^2\theta)(2-\sin^2\theta)]}} = \frac{2}{3a}\int_0^{\pi/2} \frac{d\theta}{\sqrt{[(1-(1/3)^2\sin^2\theta]}} = \frac{2}{3a} K\left(\frac{1}{3}\right).$$

## 7.18 (1) ERROR FUNCTION OR PROBABILITY INTEGRAL

*The error function or the probability integral is defined as*

$$erf(x) = \frac{2}{\sqrt{\pi}}\int_0^x e^{-t^2}\, dt.$$

This integral arises in the solution of certain partial differential equations of applied mathematics and occupies an important position in the probability theory.

*The complementary error function erfc (x) is defined as* $erfc(x) = 1 - erf(x)$.

**(2) Properties :** (i) $erf(-x) = -erf(x)$; (ii) $erf(0) = 0$

$$(iii)\ erf(\infty) = \frac{2}{\sqrt{\pi}}\int_0^\infty e^{-t^2} dt = \frac{2}{\sqrt{\pi}}\frac{\sqrt{\pi}}{2} = 1$$

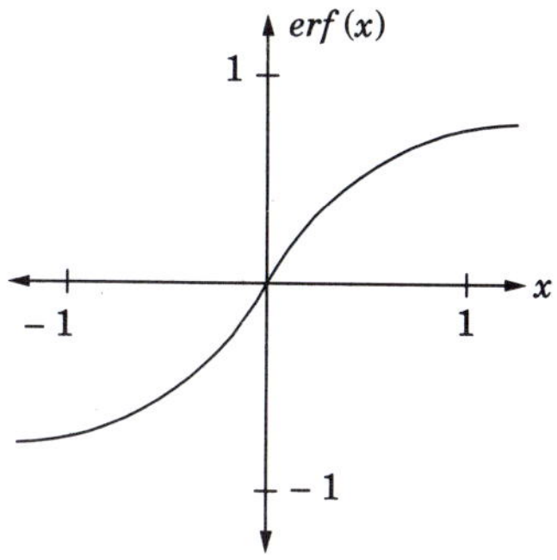

Fig. 7.53.

This proves that the total area under the *Normal* or *Gaussian error function curve* is unity – § 26.16.

**Example 7.59 :** *Evaluate* $\int_0^{\pi/2}\int_x^y \cos 2y\sqrt{1-a^2\sin^2 x}\, dx\, dy$ *(Kottayam, 2017)*

**Solution:** Here it is difficult to integrate w.r.t. $x$ first changing the order of integration.

New limits are $y = x$ to $y = \frac{\pi}{2}$ and $x = 0$ to $x = \frac{\pi}{2}$

$$I = \int_0^{\pi/2}\int_x^{\pi/2} \cos 2y\sqrt{1-a^2\sin^2 x}\, dx\, dy$$

$$I = \int_0^{\pi/2}\sqrt{1-a^2\sin^2 x}\left[\frac{\sin^2 2y}{2}\right]_x^{\pi/2} dx = -\frac{1}{2}\int_0^{\pi/2}\sqrt{1-a^2}\sin^2 x\sin 2x\, dx$$

put $\sin^2 x = t$

$2\sin x\cos x\, dx = dt$

$\sin 2x\, dx = dt$

when $x = 0, t = 0, x = \frac{\pi}{2}, t = 1$

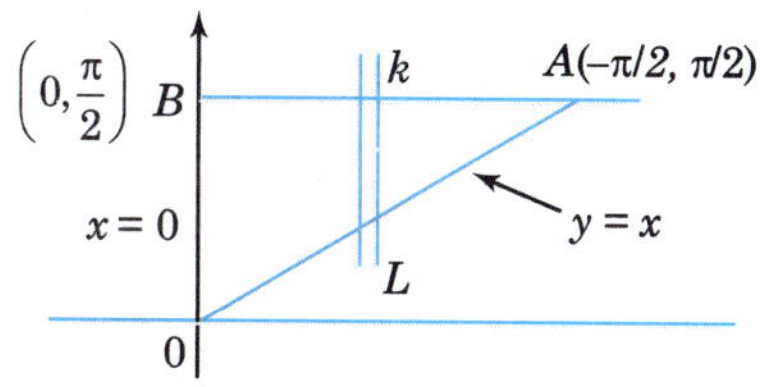

Fig. 7.54.

$$= -\frac{1}{2}\int_0^1\sqrt{1-a^2 t}\, dt = -\frac{1}{2}\left[\frac{(1-a^2t)^{3/2}}{(-a)^2\frac{3}{2}}\right]_0^1$$

$$= \frac{1}{3a^2}\left[(1-a^2t)^{3/2}\right]_0^1 = \frac{1}{3a^2}\left[(1-a^2)^{3/2} - 1\right]$$

**Example 7.60 :** *Evaluate* $\int_0^1\int_{1-\sqrt{1-y}}^{1+\sqrt{1-y}} \frac{dx\, dy}{(x^2-2x+y-3)^2}$ *(Kerala, 2018)*

**Solution:** Region of integration is bounded by

$x = 1-\sqrt{1-y}$ $(x-1)^2 = 1-y$

$x = 1+\sqrt{1+y}$ $(x-1)^2 = 1-y$

$y = 0$ to $y = 1$

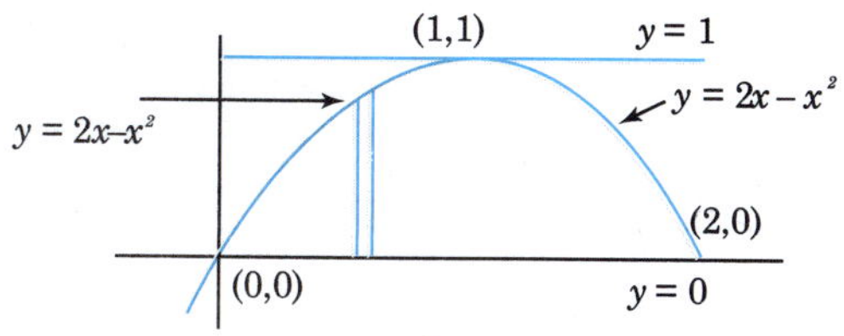

Fig. 7.55.

$$I = \int_0^2 \int_0^{2x-x^2} \left[\frac{dy\,dx}{(x^2-2x+y-3)^2}\right] = \int_0^2 \left[-\frac{1}{(x^2-2x+y-3)}\right]_0^{2x-x^2} dx$$

$$I = \int_0^2 \left[-\frac{1}{(x^2-2x+2x-x^2-3)} + \frac{1}{x^2-2x+0-3}\right] dx$$

$$= \int_0^2 \left[\frac{1}{3} + \frac{1}{(x-3)(x+1)}\right] dx = \int_0^2 \left[\frac{1}{3} + \frac{1}{4}\frac{1}{(x-3)} - \frac{1}{4}\frac{1}{(x+1)}\right] dx$$

$$= \int_0^2 \left[\frac{1}{3} - \frac{1}{4}\frac{1}{(3-x)} - \frac{1}{4}\frac{1}{(x+1)}\right] dx$$

$$= \left[\frac{1}{3}x + \frac{1}{4}\log(3-x) - \frac{1}{4}\log(x+1)\right]_0^2 = \left[\frac{2}{3} - \frac{1}{4}\log 3 - \frac{1}{4}\log 3\right]$$

$$= \frac{2}{3} - \frac{1}{2}\log 3.$$

**Example 7.61 :** *Change the order of integration in the double integration* $\int_0^5 \int_{2-x}^{2+x} f(x, y)\, dy\, dx$.

**Solution:** Let $I = \int_0^5 \left\{\int_{2-x}^{2+x} (x, y)\, dy\right\} dx$

Region of integration by $x = 0, x = 5, y = 2 - x$ *i.e.* $x + y = 2$; $y = 2 + x$.

Points of intersection are (0, 2) (5, –3), (5, 7) and the Region $R$ is shaded in Fig. 7.56.

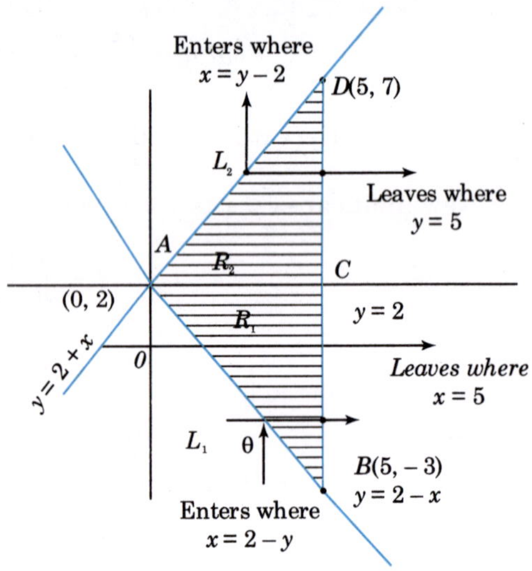


Fig. 7.56.

$$I = \int\left\{\int f(x, y) dx\right\} dy$$

$\therefore$ We have to imagine a horizontal strip and also we note that point $A(0, 2)$ the left end of horizontal strip changes its $x$-value from the straight line

$$x = 2 - y \quad \text{to} \quad x = y - 2$$

Let $R_1$ is $ABC$ and $R_2$ is $ACD$

$$I = \iint_{ABCD} f(x, y)\, dx\, dy$$

$$= \iint_{ABC} f(x, y)\, dx\, dy + \iint_{ACD} f(x, y)\, dx\, dy$$

Now to find the new limits for $R_1$, $R_2$, we will imagine two horizontal strip, in region $R_1$ and $R_2$

By changing

$$I = \int_{-3}^{2}\left\{\int_{2-y}^{5} f(x, y) dx\right\} dy + \int_2^7 \left\{\int_{y-2}^{5} f(x, y) dx\right\} dy.$$

## PROBLEMS 7.9

1. By means of the substitution $k \sin x = \sin z$, show that

(i) $\int_0^{\pi} \frac{dx}{\sqrt{(1 - k^2 \sin^2 x)}} = \frac{1}{k} F\left(\frac{1}{k}, \phi'\right)$;

(ii) $\int_0^{\phi} \sqrt{(1 - k^2 \sin^2 x)}\, dx = \left(\frac{1}{k} - k\right) F\left(\frac{1}{k}, \phi'\right) + kE\left(\frac{1}{k}, \phi'\right)$

where $k > 1$ and $\phi' = \sin^{-1} (k \sin \phi)$.

Express the following integrals in terms of elliptic integrals :

2. $\int_0^{\pi/2} \frac{dx}{\sqrt{(1+3\sin^2 x)}}$.
3. $\int_0^{\pi/2} \frac{dx}{\sqrt{(2-\cos x)}}$.
4. $\int_0^{\pi/2} \sqrt{(\cos x)}\, dx$.
5. Prove that $\int_0^1 \int_0^{\sqrt{1-x^2}} \frac{y\,dy\,dx}{(1+y^2)\sqrt{1-x^2-y^2}} = \frac{\pi}{4}\log 2$ (PTU, 2017)
6. Compute (i) $erf(0.3)$, (ii) $erf(0.5)$, correct to three decimal places.
7. Show that (i) $erf(x) + erf(-x) = 0$ (ii) $erfc(x) + erfc(-x) = 2$
8. Prove that

(i) $\frac{d}{dx}[erf(ax)] = \frac{2a}{\sqrt{\pi}} e^{-a^2x^2}$ (ii) $\frac{d}{dx}[erfc(ax)] = -\frac{2a}{\sqrt{\pi}} e^{-a^2x^2}$.

9. Prove that $\int_0^1 \int_0^{\sqrt{1-y^2}} \frac{\cos^{-1} x \cdot dxdy}{\sqrt{(1-x^2-y^2)(1-x^2)}} = \frac{\pi^3}{16}$
10. Prove that $\int_0^\infty \int_0^x xe^{-x^2/y}\, dydx = \frac{1}{2}$
11. Prove that $\int_0^a \int_0^x x\sqrt{(a^2-y^2)(x^2-y^2)}\, dy\, dx = \frac{8a^5}{45}$
12. Prove that $\int_0^a \int_0^y \frac{dx\,dy}{\sqrt{(a^2+x^2)(a-y)(y-x)}} = \pi \log(1+\sqrt{2})$

Express the following double integrals as single term integrals

13. $\int_0^1 \int_{-\sqrt{y}}^{\sqrt{y}} dx\,dy + \int_{-1}^2 \int_{-1}^1 dx\,dy$ and evaluate
14. $\int_0^1 \int_0^y (x^2+y^2)\, dx\,dy + \int_1^2 \int_0^{2-y} (x^2+y^2)\, dx\,dy$ and evaluate
15. $\int_{-3}^2 \int_{2-y}^5 f(x,y)dx\,dy + \int_2^7 \int_{y-2}^5 f(x,y)\,dx\,dy$
16. $\int_1^2 \int_{\sqrt{x}}^x f(x,y)dy\,dx + \int_2^4 \int_{\sqrt{x}}^2 f(x,y)\,dy\,dx$
17. $\int_0^1 \int_0^y f(x,y)dx\,dy + \int_1^\infty \int_0^{1/y} f(x,y)\,dx\,dy$
18. Prove that $\int_0^1 \int_{4y}^4 e^{x^2} dxdy = \frac{e^{16}-1}{8}$

## 7.19 OBJECTIVE TYPE OF QUESTIONS

### PROBLEMS 7.10

*Fill up the blanks or choose the correct answer from the following problems :*

1. $\int_1^a \int_1^b \frac{dxdy}{xy}$ = ..... (K.T.U., 2016 ; Anna, 2013)
2. $\int_0^1 \int_0^{1-x} dx\,dy$ = .......
3. $\int_0^\infty e^{-x^2} dx$ = ....... (U.P.T.U., 2011)
4. $\beta\left(\frac{1}{2}, \frac{1}{2}\right)$ = ....... (V.T.U., 2010)
5. $\Gamma(3.5)$ = .......
6. The surface area of the sphere $x^2+y^2+z^2+2x-4y+8z-2=0$ is .......
7. $\int_0^2 \int_1^3 \int_1^2 xy^2z\, dzdydx$ = .......
8. If $u = x + y$ and $v = x - 2y$, then the area-element $dxdy$ is replaced by ....... $dudv$.
9. In terms of Beta function $\int_0^{\pi/2} \sin^7\theta\sqrt{\cos\theta}\, d\theta$ = .......
10. The value of $\beta(2, 1) + \beta(1, 2)$ is .......
11. $\int_1^2 \int_1^3 xy^2\, dy\, dx$ = ....... (C.S.V.T.U., 2012)
12. Volume bounded by $x \geq 0, y \geq 0, z \geq 0$ and $x^2+y^2+z^2=1$ as a triple integral integral.

13. Value of $\int_0^1 \int_0^{x^2} xe^y \, dy \, dx$ is equal to (*a*) $e/2$ (*b*) $e - 1$ (*c*) $1 - e$ (*d*) $e/2 - 1$.

14. $\iint x^2 y^3 \, dxdy$ over the rectangle $0 \le x \le 1$ and $0 \le y \le 3$ is .......

15. $\int_0^{\pi} \int_0^{a \sin \theta} r \, dr \, d\theta =$ .......

16. $\int_{x=0}^{x=3} \int_{y=0}^{y=1/x} ye^{xy} \, dxdy =$ .......

17. $\int_0^{\pi/2} \int_0^{\infty} \frac{r \, dr \, d\theta}{(r^2 + a^2)} =$ .......

18. $\int_0^1 \int_0^{x^2} (x^2 + y^2) \, dx \, dy =$ ....... (*Anna, 2013*)

19. To change cartesian coordinates $(x, y, z)$ to spherical polar coordinate $(r, \theta, \phi)$; $dx\, dy\, dz$ is replaced by .......

20. $\int_0^2 \int_0^{x^2} e^{y/x} \, dydx =$ .......

21. $\iint (x + y)^2 \, dxdy$ over the area bounded by the ellipse $\frac{x^2}{a^2} + \frac{y^2}{b^2} = 1$, is .......

22. $\int_0^1 \int_0^{\sqrt{1+x^2}} \frac{dx\, dy}{1 + x^2 + y^2} =$ .......

23. $\iint xy\,(x + y) \, dx \, dy$ over the area between $y + x^2$ and $y = x$, is .......

24. Value of $\int_0^1 \int_x^{\sqrt{x}} xy \, dx \, dy$ is (*a*) zero (*b*) $-1/24$ (*c*) $1/24$ (*d*) 24. (*V.T.U., 2010*)

25. $\iint dx \, dy$ over the area bounded by $x = 0$, $y = 0$, $x^2 + y^2 = 1$ and $5y = 3$, is .......

26. $\iint_R y \, dx \, dy$ where $R$ is the region bounded by the parabolas $y^2 = 4x$ and $x^2 = 4y$, is .......

27. $\iint (x^2 + y^2) \, dx \, dy$ in the positive quadrant for which $x + y \le 1$, is .......

28. Area between the parabolas $y^2 = 4x$ and $x^2 = 4y$ is .......

29. Changing the order of integration in $\int_{-a}^{a} \int_0^{\sqrt{a^2 - y^2}} f(x, y) \, dx \, dy =$ .......

30. $\Gamma(1/4)\Gamma(3/4) =$ ....... (*V.T.U., 2011*)

31. $\beta(5/2, 7/2) =$ .......

32. $\int_0^{\infty} \int_0^x x e^{-x^2/y} \, dydx =$ .......

33. On changing to polar coordinates $\int_0^{2a} \int_0^{\sqrt{(2ax - x^2)}} dx \, dy$ becomes .......

34. A square lamina is immersed in the liquid with one vertex in the surface and the diagonal of length vertical. Its centre of pressure is at a depth .......

35. The centroid of the area enclosed by the parabola $y^2 = 4x$, $x$-axis and its latus-rectum is .......

36. The moment of inertia of a uniform spherical ball of mass 10 gm and radius 2 cm about a diameter is .......

37. M.I. of a solid right circular cone (base-radius $r$ and height $h$) about its axis is .......

38. $erf_c(-x) - erf(x) =$ .......

39. $\int_0^1 \frac{x - 1}{\log x} \, dx =$ .......

40. $\Gamma\left(\frac{3}{2}\right) =$ .......

41. Value of $\int_0^a \int_0^b \int_0^c x^2 y^2 z^2 \, dx \, dy \, dz$ is (*a*) $\frac{abc}{3}$ (*b*) $\frac{a^2b^2c^2}{27}$ (*c*) $\frac{a^3b^3c^3}{27}$ (*d*) $\frac{a^2b^2c^2}{9}$.

42. The integral $\int_0^1 \int_0^{\sqrt{(1-x^2)}} (x + y) \, dy \, dx$ after changing the order of integration.

(*a*) $\int_0^2 \int_0^{\sqrt{(1-y^2)}} (x + y) \, dx \, dy$  (*b*) $\int_0^1 \int_0^{\sqrt{(1-y^2)}} (x + y) \, dx \, dy$

(*c*) $\int_0^1 \int_0^{\sqrt{(1+y^2)}} (x + y) \, dx \, dy$  (*d*) $\int_0^{-1} \int_0^{\sqrt{(1-y^2)}} (x + y) \, dx \, dy$. (*V.T.U., 2011*)

43. Change the order of integration in $\int_0^a \int_x^a f(x, y) \, dxdy$. (*Anna, 2012*)

CHAPTER 8

# Vector Calculus and its Applications

1. Differentiation of vectors. 2. Curves in space. 3. Velocity and acceleration, Tangential and normal accelerations, Relative velocity and acceleration. 4. Scalar and vector point functions—Vector operator del. 5. Del applied to scalar point functions—Gradient. 6. Del applied to vector point functions—Divergence and Curl. 7. Physical interpretations of div **F** and curl **F**. 8. Del applied twice to point functions. 9. Del applied to products of point functions. 10. Integration of vectors. 11. Line integral—Circulation—Work. 12. Surface integral—Flux. 13. Green's theorem in the plane. 14. Stoke's theorem. 15. Volume integral. 16. Divergence theorem. 17. Green's theorem. 18. Irrotational and Solenoidal fields. 19. Orthogonal curvilinear coordinates, Del applied to functions in orthogonal curvilinear coordinates. 20. Cylindrical coordinates. 21. Spherical polar coordinates. 22. Objective Type of Questions.

## 8.1 (1) DIFFERENTIATION OF VECTORS

If a vector **R** varies continuously as a scalar variable $t$ changes, then **R** is said to be a function of $t$ and is written as $\mathbf{R} = \mathbf{F}(t)$.

Just as in scalar calculus, *we define* **derivative** *of a vector function* $\mathbf{R} = \mathbf{F}(t)$ *as*

$$\underset{\delta t \to 0}{\text{Lt}} \frac{\mathbf{F}(t + \delta t) - \mathbf{F}(t)}{\delta t} \text{ and write it as } \frac{d\mathbf{R}}{dt} \text{ or } \frac{d\mathbf{F}}{dt} \text{ or } \mathbf{F}'(t).$$

(2) **General rules of differentiation** are similar to those of ordinary calculus *provided the order of factors in vector products is maintained.* Thus, if $\phi$, **F, G, H** are scalar and vector functions of a scalar variable $t$, we have

(*i*) $\dfrac{d}{dt}(\mathbf{F} + \mathbf{G} - \mathbf{H}) = \dfrac{d\mathbf{F}}{dt} + \dfrac{d\mathbf{G}}{dt} - \dfrac{d\mathbf{H}}{dt}$  (*ii*) $\dfrac{d}{dt}(\mathbf{F}\phi) = \mathbf{F}\dfrac{d\phi}{dt} + \dfrac{d\mathbf{F}}{dt}\phi$

(*iii*) $\dfrac{d}{dt}(\mathbf{F} \,.\, \mathbf{G}) = \mathbf{F} \,.\, \dfrac{d\mathbf{G}}{dt} + \dfrac{d\mathbf{F}}{dt} \,.\, \mathbf{G}$  (*iv*) $\dfrac{d}{dt}(\mathbf{F} \times \mathbf{G}) = \mathbf{F} \times \dfrac{d\mathbf{G}}{dt} + \dfrac{d\mathbf{F}}{dt} \times \mathbf{G}$

(*v*) $\dfrac{d}{dt}[\mathbf{FGH}] = \left[\dfrac{d\mathbf{F}}{dt}\mathbf{GH}\right] + \left[\mathbf{F}\dfrac{d\mathbf{G}}{dt}\mathbf{H}\right] + \left[\mathbf{FG}\dfrac{d\mathbf{H}}{dt}\right]$

(*vi*) $\dfrac{d}{dt}[(\mathbf{F} \times \mathbf{G}) \times \mathbf{H}] = \left(\dfrac{d\mathbf{F}}{dt} \times \mathbf{G}\right) \times \mathbf{H} + \left(\mathbf{F} \times \dfrac{d\mathbf{G}}{dt}\right) \times \mathbf{H} + (\mathbf{F} \times \mathbf{G}) \times \dfrac{d\mathbf{H}}{dt}$

As an illustration, let us prove two of these, while the others can be proved similarly :

(*iv*) $$\frac{d}{dt}(\mathbf{F} \times \mathbf{G}) = \underset{\delta t \to 0}{\text{Lt}} \frac{(\mathbf{F} + \delta\mathbf{F}) \times (\mathbf{G} + \delta\mathbf{G}) - \mathbf{F} \times \mathbf{G}}{\delta t} = \underset{\delta t \to 0}{\text{Lt}} \frac{\mathbf{F} \times \delta\mathbf{G} + \delta\mathbf{F} \times \mathbf{G} + \delta\mathbf{F} \times \delta\mathbf{G}}{\delta t}$$

$$= \underset{\delta t \to 0}{\text{Lt}} \left[\mathbf{F} \times \frac{\delta\mathbf{G}}{\delta t} + \frac{\delta\mathbf{F}}{\delta t} \times \mathbf{G} + \frac{\delta\mathbf{F}}{\delta t} \times \delta\mathbf{G}\right] = \mathbf{F} \times \frac{d\mathbf{G}}{dt} + \frac{d\mathbf{F}}{dt} \times G \qquad [\because \ \delta\mathbf{G} \to 0 \text{ as } \delta t \to 0]$$

(v) $\frac{d}{dt}(\mathbf{FGH}) = \frac{d}{dt}[\mathbf{F}\,.\,(\mathbf{G}\times\mathbf{H})] = \mathbf{F}\,.\,\frac{d}{dt}(\mathbf{G}\times\mathbf{H}) + \frac{d\mathbf{F}}{dt}\,.\,(\mathbf{G}\times\mathbf{H})$

$$= \mathbf{F}\,.\left[\mathbf{G}\times\frac{d\mathbf{H}}{dt} + \frac{d\mathbf{G}}{dt}\times\mathbf{H}\right] + \frac{d\mathbf{F}}{dt}\,.\,(\mathbf{G}\times\mathbf{H})$$

$$= \mathbf{F}\,.\left[\mathbf{G}\times\frac{d\mathbf{H}}{dt}\right] + \mathbf{F}\,.\left[\frac{d\mathbf{G}}{dt}\times\mathbf{H}\right] + \frac{d\mathbf{F}}{dt}\,.\,(\mathbf{G}\times\mathbf{H})$$

$$= \left[\mathbf{FG}\frac{d\mathbf{H}}{dt}\right] + \left[\mathbf{F}\frac{d\mathbf{G}}{dt}\mathbf{H}\right] + \left[\frac{d\mathbf{F}}{dt}\mathbf{GH}\right]$$

$$= \frac{d\mathbf{F}}{dt}\mathbf{GH} + \left[\mathbf{F}\frac{d\mathbf{G}}{dt}\mathbf{H}\right] + \left[\mathbf{FG}\frac{d\mathbf{H}}{dt}\right]$$ (V.T.U., 2013)

**Obs. 1.** *If* $\mathbf{F}(t)$ *has a constant magnitude, then* $\mathbf{F}\,.\,\frac{d\mathbf{F}}{dt} = 0$

For $\mathbf{F}(t)\,.\,\mathbf{F}(t) = [\mathbf{F}(t)]^2$ = constant

$$\therefore \qquad \mathbf{F}\,.\,\frac{d\mathbf{F}}{dt} = 0, \quad i.e., \quad \frac{d\mathbf{F}}{dt} \perp \mathbf{F}.$$

**Obs. 2.** *If* $\mathbf{F}(t)$ *has constant (fixed) direction, then* $\mathbf{F}\times\frac{d\mathbf{F}}{dt} = \mathbf{0}$

Let $\mathbf{G}(t)$ be a unit vector in the direction of $\mathbf{F}(t)$ so that

$$\mathbf{F}(t) = f(t)\,\mathbf{G}(t) \quad \text{where} \quad f(t) = |\,\mathbf{F}(t)\,|.$$

$$\therefore \qquad \frac{d\mathbf{F}}{dt} = f\frac{d\mathbf{G}}{dt} + \frac{df}{dt}\mathbf{G} \quad \text{and} \quad \mathbf{F}\times\frac{d\mathbf{F}}{dt} = f\,\mathbf{G}\times\left[f\frac{d\mathbf{G}}{dt} + \frac{df}{dt}\mathbf{G}\right]$$

$$= f^2\,\mathbf{G}\times\frac{d\mathbf{G}}{dt} = \mathbf{0}.$$ [since $\mathbf{G}$ is constant, $d\mathbf{G}/dt = \mathbf{0}$.]

**Example 8.1 :** *Find the first and second derivatives of*

(i) $\left[\bar{r}\,\overrightarrow{\frac{dr}{dt}}\cdot\overrightarrow{\frac{d^2r}{dt^2}}\right]$ (ii) $\vec{r}\times\left(\overrightarrow{\frac{dr}{dt}}\times\frac{d^2r}{dt^2}\right)$ (Bhopal, 2019)

**Solution:** We denote $\frac{d\bar{r}}{dt}$ by $\dot{\bar{r}}$ and $\frac{d^2r}{dt^2}$ by $\ddot{\bar{r}}$ and soon

(i)
$$\frac{d}{dt}\left[\bar{r}\,\dot{\bar{r}}\,\ddot{\bar{r}}\right] = \left[\frac{d\bar{r}}{dt}\dot{\bar{r}}\,\ddot{\bar{r}}\right] + \left[\bar{r}\frac{d\dot{\bar{r}}}{dt}\ddot{\bar{r}}\right] + \left[\bar{r}\,\dot{\bar{r}}\frac{d\ddot{\bar{r}}}{dt}\right] = \left[\dot{\bar{r}}\,\dot{\bar{r}}\,\ddot{\bar{r}}\right] + \left[\bar{r}\frac{d\dot{\bar{r}}}{dt}\ddot{\bar{r}}\right] + \left[\bar{r}\,\dot{\bar{r}}\frac{d\ddot{\bar{r}}}{dt}\right]$$

$$= \left[\dot{\bar{r}}\,\dot{\bar{r}}\,\ddot{\bar{r}}\right] + \left[\bar{r}\,\ddot{\bar{r}}\,\ddot{\bar{r}}\right] + \left[\bar{r}\,\dot{\bar{r}}\,\dddot{\bar{r}}\right] = 0 + 0 + \left[\bar{r}\,\dot{\bar{r}}\,\dddot{\bar{r}}\right], \text{ since } \left[\bar{a}\,\bar{a}\,\bar{b}\right] = 0$$

$$= \left[\bar{r}\frac{d\bar{r}}{dt}\frac{d^3\bar{r}}{dt^2}\right] \qquad ...(i)$$

To get
$$\frac{d^2}{dt^2}\left[\bar{r}\,\dot{\bar{r}}\,\ddot{\bar{r}}\right] = \frac{d}{dt}\left\{\frac{d}{dt}\left[\bar{r}\,\dot{\bar{r}}\,\ddot{\bar{r}}\right]\right\} = \frac{d}{dt}\left[\bar{r}\,\dot{\bar{r}}\,\dddot{\bar{r}}\right] = \left[\dot{\bar{r}}\,\dot{\bar{r}}\,\dddot{\bar{r}}\right] + \left[\bar{r}\,\ddot{\bar{r}}\,\dddot{\bar{r}}\right] + \left[\bar{r}\,\dot{\bar{r}}\,\ddddot{\bar{r}}\right]$$

$$= \left[r\frac{d^2\bar{r}}{dt^2}\frac{d^3\bar{r}}{dt^3}\right] + \left[\bar{r}\frac{d\bar{r}}{dt}\cdot\frac{d^4\bar{r}}{dt^4}\right]$$

(ii) $\frac{d}{dt}\{\bar{r}\times(\dot{\bar{r}}\times\ddot{\bar{r}})\} = \dot{\bar{r}}\times(\dot{\bar{r}}\times\ddot{\bar{r}}) + \bar{r}\times(\ddot{\bar{r}}\times\ddot{\bar{r}}) + \bar{r}\times(\dot{\bar{r}}\times\dddot{\bar{r}})$

$$= \frac{d\dot{\bar{r}}}{dt}\times\left(\frac{d\bar{r}}{dt}\times\frac{d^2\bar{r}}{dt^2}\right) + \bar{r}\times\left(\frac{d\bar{r}}{dt}\times\frac{d^3\bar{r}}{dt^3}\right)$$

Similarly proceeding, it can be seen that

$$= \frac{d^2}{dt^2}\left[\bar{r}\times\left(\frac{d\bar{r}}{dt}\times\frac{d^2\bar{r}}{dt^2}\right)\right] = \frac{d}{dt}\left[\dot{\bar{r}}\times(\dot{\bar{r}}\times\ddot{\bar{r}}) + (\dot{\bar{r}}\times\bar{r}\times\dddot{\bar{r}})\right]$$

$$= \frac{d\bar{r}}{dt}\times\left(\frac{d\bar{r}}{dt}\times\frac{d^3\bar{r}}{dt^3}\right) + \bar{r}\times\left(\frac{d^2\bar{r}}{dt^2}\times\frac{d^3\bar{r}}{dt^2}\right) + \bar{r}\times\left(\frac{d\bar{r}}{dt}\times\frac{d^4\bar{r}}{dt^4}\right)$$

## 8.2 CURVES IN SPACE

(1) **Tangent.** Let $\mathbf{R}(t) = x(t)\mathbf{I} + y(t)\mathbf{J} + z(t)\mathbf{K}$ be the position vector of a point $P$. Then as the scalar parameter $t$ takes different values, the point $P$ traces out a *curve in space* (Fig. 8.1). If the neighbouring point $Q$ corresponds to $t + \delta t$, then $\delta\mathbf{R} = \mathbf{R}(t + \delta t) - \mathbf{R}(t)$ or $\delta\mathbf{R}/\delta t$ is directed along the chord $PQ$. As $\delta t \to 0$, Lt $\delta\mathbf{R}/\delta t$ becomes the tangent (vector) to the curve at $P$ whenever it exists and is not zero.

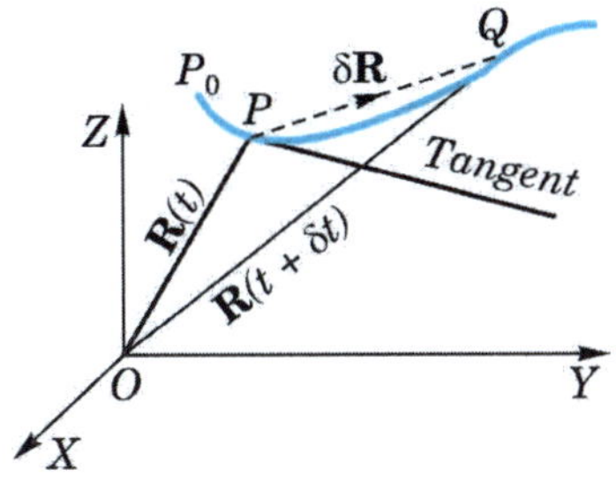

Fig. 8.1

Thus the vector $\mathbf{R}' = d\mathbf{R}/dt$ is a tangent to the space curve $\mathbf{R} = \mathbf{F}(t)$.

Let $P_0$ be a fixed point of this curve corresponding to $t = t_0$. If $s$ be the length of the arc $P_0P$, then

$$\frac{\delta s}{\delta t} = \frac{\delta s}{|\delta\mathbf{R}|} \cdot \frac{|\delta\mathbf{R}|}{\delta t} = \frac{\text{arc } PQ}{\text{chord } PQ}\left|\frac{\delta\mathbf{R}}{\delta t}\right|$$

As $Q \to P$ along the curve $QR$ *i.e.*, $\delta t \to 0$, then arc $PQ$/chord $PQ \to 1$ and

$$\frac{ds}{dt} = \left|\frac{d\mathbf{R}}{dt}\right| \quad \text{or} \quad |\mathbf{R}'(t)|.$$

If $\mathbf{R}'(t)$ is continuous, then arc $P_0P$ is given by

$$s = \int_{t_0}^{t} |\mathbf{R}'|\, dt = \int_{t_0}^{t} \sqrt{(x')^2 + (y')^2 + (z')^2}\, dt$$

If we take $s$ the parameter in place of $t$ then the magnitude of the tangent vector, *i.e.*, $|d\mathbf{R}/ds| = 1$. Thus denoting the unit tangent vector by $\mathbf{T}$, we have

$$\mathbf{T} = \frac{d\mathbf{R}}{ds} \qquad ...(1)$$

(2) **Principal normal.** Since $\mathbf{T}$ is a unit vector, we have

$$d\mathbf{T}/ds \,.\, \mathbf{T} = 0$$

*i.e.*, $d\mathbf{T}/ds$ is perpendicular to $\mathbf{T}$. Or else $d\mathbf{T}/ds = 0$, in which case $\mathbf{T}$ is a constant vector w.r.t. the arc length $s$ and so has a fixed direction, *i.e.*, the curve is a straight line.

If we denote a unit normal vector to the curve at $P$ by $\mathbf{N}$ then $d\mathbf{T}/ds$ is in the direction of $\mathbf{N}$ which is known as the *principal normal* to the space curve at $P$. The plane of $\mathbf{T}$ and $\mathbf{N}$ is called the *osculating plane* of the curve at $P$ (Fig. 8.2).

(3) **Binormal.** A third unit vector $\mathbf{B}$ defined by $\mathbf{B} = \mathbf{T} \times \mathbf{N}$, is called the *binormal at P*. Since $\mathbf{T}$ and $\mathbf{N}$ are unit vectors, $\mathbf{B}$ is also a unit vector perpendicular to both $\mathbf{T}$ and $\mathbf{N}$ and hence normal to the *osculating plane at P*.

Thus at each point $P$ of a space curve there are three mutually perpendicular unit vectors, $\mathbf{T}, \mathbf{N}, \mathbf{B}$ which form a moving trihedral such that

$$\mathbf{T} = \mathbf{N} \times \mathbf{B},\ \mathbf{N} = \mathbf{B} \times \mathbf{T},\ \mathbf{B} = \mathbf{T} \times \mathbf{N} \qquad ...(2)$$

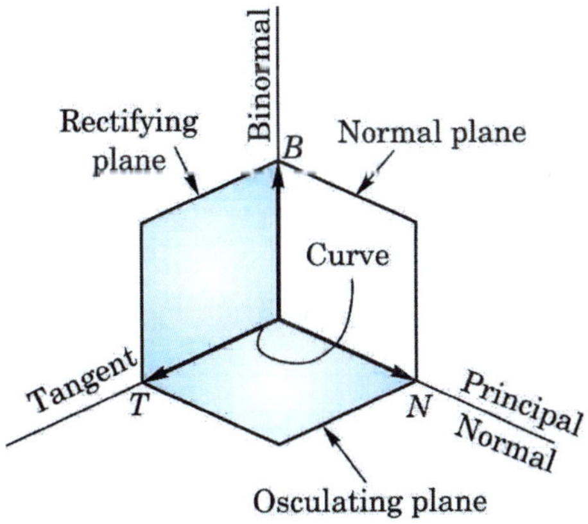

Fig. 8.2

This moving trihedral determines the following three fundamental planes at each point of the curve :

(*i*) The osculating plane containing $\mathbf{T}$ and $\mathbf{N}$

(*ii*) The normal plane containing $\mathbf{N}$ and $\mathbf{B}$

(*iii*) The rectifying plane containing $\mathbf{B}$ and $\mathbf{T}$.

(4) **Curvature.** The arc rate of turning of the tangent (*i.e.*, the magnitude of $d\mathbf{T}/ds$) is called the *curvature* of the curve and is denoted by $k$.

Since $d\mathbf{T}/ds$ is in the direction of the principal normal $\mathbf{N}$, therefore,

$$\frac{d\mathbf{T}}{ds} = k\mathbf{N} \qquad ...(3)$$

(5) **Torsion.** Since $\mathbf{B}$ is a unit vector, we have $\dfrac{d\mathbf{B}}{ds} \,.\, \mathbf{B} = 0$

Also $\mathbf{B} \,.\, \mathbf{T} = 0$, therefore $\dfrac{d\mathbf{B}}{ds} \,.\, \mathbf{T} + \mathbf{B} \,.\, \dfrac{d\mathbf{T}}{ds} = 0.$

or $\dfrac{d\mathbf{B}}{ds} \cdot \mathbf{T} + \mathbf{B} \cdot (k\mathbf{N}) = 0,$ *i.e.,* $\dfrac{d\mathbf{B}}{ds} \cdot \mathbf{T} = 0$ $[\because \mathbf{B} \cdot \mathbf{N} = 0]$

Hence $d\mathbf{B}/ds$ is perpendicular to both **B** and **T** and is, therefore, parallel to **N**.

The arc rate of turning of the binormal (*i.e.*, the magnitude of $d\mathbf{B}/ds$) is called *torsion* of the curve and is denoted by $\tau$. We may, therefore, write

$$\frac{d\mathbf{B}}{ds} = -\tau\mathbf{N} \qquad ...(4)$$

(The negative sign indicates that for $\tau > 0$, $d\mathbf{B}/ds$ has direction of $-\mathbf{N}$).

Finally to find $d\mathbf{N}/ds$, we differentiate $\mathbf{N} = \mathbf{B} \times \mathbf{T}$.

$$\therefore \quad \frac{d\mathbf{N}}{ds} = \frac{d\mathbf{B}}{ds} \times \mathbf{T} + \mathbf{B} \times \frac{d\mathbf{T}}{ds} = -\tau\mathbf{N} \times \mathbf{T} + \mathbf{B} \times k\mathbf{N}$$

Using the relation Eqn. (2), it reduces to $\dfrac{d\mathbf{N}}{ds} = \tau\mathbf{B} - k\mathbf{T}$ ...(5)

The Eqns. (3), (4) and (5) constitute the well-known *Frenet formulae** for space curves.

**Obs. 1.** $\rho = 1/k$ and $\sigma = 1/\tau$ are respectively called the radii of curvature and torsion.

**2.** For a plane curve $\tau = 0$.

**Example 8.2.** *Find the angle between the tangents to the curve* $\mathbf{R} = t^2\mathbf{I} + 2t\mathbf{J} - t^3\mathbf{K}$ *at the point* $t = \pm 1$. *(V.T.U., 2010)*

**Solution.** The tangent at any point '$t$' is given by

$$\frac{d\mathbf{R}}{dt} = 2t\mathbf{I} + 2\mathbf{J} - 3t^2\mathbf{K}$$

$\therefore$ the tangents $\mathbf{T}_1$, $\mathbf{T}_2$ at $t = 1$ and $t = -1$ are respectively given by

$$\mathbf{T}_1 = 2\mathbf{I} + 2\mathbf{J} - 3\mathbf{K} \; ; \; \mathbf{T}_2 = -2\mathbf{I} + 2\mathbf{J} - 3\mathbf{K},$$

Then the required $\angle\theta$ is given by $T_1T_2 \cos\theta = \mathbf{T}_1 \cdot \mathbf{T}_2 = 2(-2) + 2 \cdot 2 + (-3)(-3)$

*i.e.,* $\sqrt{17}\,\sqrt{17} \cos\theta = 9 \quad \therefore \quad \theta = \cos^{-1}(9/17).$

**Example 8.3.** *Find the curvature and torsion of the curve $x = a \cos t$, $y = a \sin t$, $z = bt$.*

(This curve is drawn on a circular cylinder cutting its generators at a constant angle and is known as a *circle helix*).

**Solution.** The vector equation of the curve is $\mathbf{R} = a \cos t\mathbf{I} + a \sin t\mathbf{J} + bt\mathbf{K}$

$$\therefore \quad d\mathbf{R}/dt = -a \sin t\mathbf{I} + a \cos t\mathbf{J} + b\mathbf{K}$$

Its arc length from $P_0$ $(t = 0)$ to any point $P(t)$ (Fig. 8.3) is given by

$$s = \int_0^t |d\mathbf{R}/dt|\, dt = \sqrt{(a^2 + b^2)}t$$

$$\therefore \quad \frac{ds}{dt} = \sqrt{(a^2 + b^2)}$$

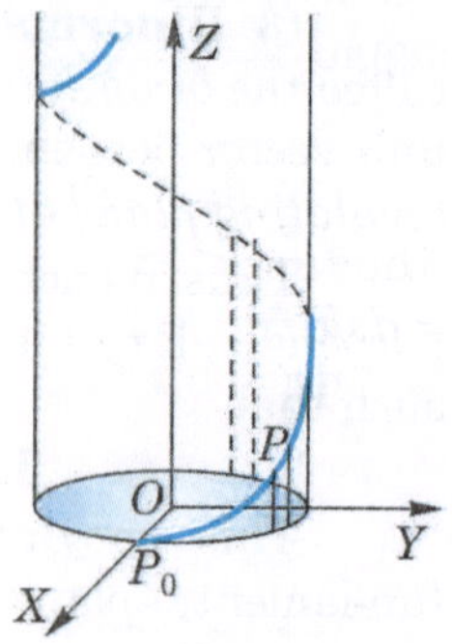

Fig. 8.3

Then $\mathbf{T} = \dfrac{d\mathbf{R}}{ds} = \dfrac{d\mathbf{R}}{dt} \Big/ \dfrac{ds}{dt} = \dfrac{-a \sin t\mathbf{I} + a \cos t\mathbf{J} + b\mathbf{K}}{\sqrt{(a^2 + b^2)}}$

and $\dfrac{d\mathbf{T}}{ds} = \dfrac{d\mathbf{T}}{dt} \Big/ \dfrac{ds}{dt} = \dfrac{-a(\cos t\mathbf{I} + \sin t\mathbf{J})}{a^2 + b^2}$

Thus $k = \left|\dfrac{d\mathbf{T}}{ds}\right| = \dfrac{a}{a^2 + b^2}$ ...(*i*) and $\mathbf{N} = -(\cos t\mathbf{I} + \sin t\mathbf{J})$

Also $\mathbf{B} = \mathbf{T} \times \mathbf{N} = (b \sin t\mathbf{I} - b \cos t\mathbf{J} + a\mathbf{K})/\sqrt{(a^2 + b^2)}$

$\therefore$ $\dfrac{d\mathbf{B}}{ds} = \dfrac{d\mathbf{B}}{dt} \Big/ \dfrac{ds}{dt} = b(\cos t\mathbf{I} + \sin t\mathbf{J})/(a^2 + b^2) = -\tau\mathbf{N} = \tau(\cos t\mathbf{I} + \sin t\mathbf{J})$

Hence $\tau = \dfrac{b}{a^2 + b^2}$. ...(*ii*)

* Named after a French mathematician *Jean-Frederic Frenet* (1816–1900).

## PROBLEMS 8.1

1. Show that, if $\mathbf{R} = \mathbf{A} \sin \omega t + \mathbf{B} \cos \omega t$, where $\mathbf{A}, \mathbf{B}, \omega$ are constants, then (*i*) $\dfrac{d^2\mathbf{R}}{dt^2} = -\omega^2\mathbf{R}$ *(Bhopal, 2007 S)*

   (*ii*) $\mathbf{R} \times \dfrac{d\mathbf{R}}{dt} = -\omega\mathbf{A} \times \mathbf{B}$. *(Rohtak, 2012)*
2. Given $\mathbf{R} = t^m \mathbf{A} + t^n \mathbf{B}$, where $\mathbf{A}, \mathbf{B}$ are constant vectors, show that, if $\mathbf{R}$ and $d^2\mathbf{R}/dt^2$ are parallel vectors, then $m + n = 1$, unless $m = n$.
3. If $\mathbf{P} = 5t^2\mathbf{I} + t^3\mathbf{J} - t\mathbf{K}$ and $\mathbf{Q} = 2\mathbf{I} \sin t - \mathbf{J} \cos t + 5t\mathbf{K}$, find (*i*) $\dfrac{d}{dt}(\mathbf{P} \cdot \mathbf{Q})$ ; (*ii*) $\dfrac{d}{dt}(\mathbf{P} \times \mathbf{Q})$.
4. If $\dfrac{d\mathbf{U}}{dt} = \mathbf{W} \times \mathbf{U}$ and $\dfrac{d\mathbf{V}}{dt} = \mathbf{W} \times \mathbf{V}$, prove that $\dfrac{d}{dt}(\mathbf{U} \times \mathbf{V}) = \mathbf{W} \times (\mathbf{U} \times \mathbf{V})$.
5. If $\mathbf{A} = x^2yz\mathbf{I} - 2xz^3\mathbf{J} + xz^2\mathbf{K}$ and $\mathbf{B} = 2z\mathbf{I} + y\mathbf{J} - x^2\mathbf{K}$, find $\dfrac{\partial^2}{\partial x \partial y}(\mathbf{A} \times \mathbf{B})$ at $(1, 0, -2)$.
6. If $\mathbf{R} = (a \cos t)\,\mathbf{I} + (a \sin t)\,\mathbf{J} + (at \tan \alpha)\,\mathbf{K}$, find the value of

   (*i*) $\left|\dfrac{d\mathbf{R}}{dt} \times \dfrac{d^2\mathbf{R}}{dt^2}\right|$ (*ii*) $\left|\dfrac{d\mathbf{R}}{dt}, \dfrac{d^2\mathbf{R}}{dt^2}, \dfrac{d^3\mathbf{R}}{dt^3}\right|$ *(P.T.U., 2010)*

   Also find the unit tangent vector at any point $t$ of the curve.
7. Find the unit tangent vector at any point on the curve $x = t^2 + 2, y = 4t - 5, z = 2t^2 - 6t$, where $t$ is any variable. Also determine the unit tangent vector at the point $t = 2$.
8. Find the equation of the tangent line to the curve $x = a \cos \theta, y = a \sin \theta, z = a\theta \tan \alpha$ at $\theta = \pi/4$.
9. Find the curvature of the (*i*) ellipse $\mathbf{R}(t) = a \cos t\mathbf{I} + b \sin t\mathbf{J}$ ; (*ii*) parabola $\mathbf{R}(t) = 2t\mathbf{I} + t^2\mathbf{J}$ at the point $t = 1$.
10. Find the education of the osculating plane and binormal to the curve

    (*i*) $x = 2 \cosh (t/2), y = 2 \sinh (t/2), z = 2t$ at $t = 0$ ; (*ii*) $x = e^t \cos t, y = e^t \sin t, z = e^t$ at $t = 0$.
11. A circular helix is given by the equation $\mathbf{R}(t) = (2 \cos t)\,\mathbf{I} + (2 \sin t)\,\mathbf{J} + \mathbf{K}$. Find the curvature and torsion of the curve at any point and show that they are constant.
12. Show that for the curve $\mathbf{R} = a(3t - t^3)\,\mathbf{I} + 3at^2\,\mathbf{J} + a(3t + t^2)\,\mathbf{K}$, the curvature equals torsion.

## 8.3 (1) VELOCITY AND ACCELERATION

Let the position of a particle $P$ at time $t$ on a path (curve) $C$ be $\mathbf{R}(t)$. At time $t + \delta t$, let the particle be at $Q$ (Fig. 8.1), then $\delta\mathbf{R} = \mathbf{R}(t + \delta t) - \mathbf{R}(t)$ or $\delta\mathbf{R}/\delta t$ is directed along $PQ$. As $Q \to P$ along $C$, the line $PQ$ becomes the tangent at $P$ to $C$.

$$\therefore \qquad \underset{\delta t \to 0}{\text{Lt}} \frac{\delta\mathbf{R}}{\delta t} = \frac{d\mathbf{R}}{dt} = \mathbf{V}$$

is the tangent vector of $C$ at $P$ which is the *velocity* (vector) $\mathbf{V}$ of the motion and its magnitude is the *speed* $v = ds/dt$, where $s$ is the arc length of $P$ from a fixed point $P_0$ $(s = 0)$ on $C$.

The derivative of the velocity vector $\mathbf{V}(t)$ is called the *acceleration* (vector) $\mathbf{A}(t)$, which is given by

$$\mathbf{A}(t) = \frac{d\mathbf{V}}{dt} = \frac{d^2\mathbf{R}}{dt^2}.$$

**(2) Tangential and normal accelerations.** It is important to note that the magnitude of acceleration is not always the rate of change of $|\mathbf{V}|$ because $\mathbf{A}(t)$ is not always tangential to the path $C$. Infact

$$\mathbf{V}(t) = \frac{d\mathbf{R}}{dt} = \frac{d\mathbf{R}}{ds} \cdot \frac{ds}{dt}. \text{ where } d\mathbf{R}/ds \text{ is a unit tangent vector of } C.$$

$$\therefore \qquad \mathbf{A}(t) = \frac{d\mathbf{V}}{dt} = \frac{d}{dt}\left[\frac{ds}{dt} \cdot \frac{d\mathbf{R}}{ds}\right] = \frac{d^2s}{dt^2} \cdot \frac{d\mathbf{R}}{ds} + \left(\frac{ds}{dt}\right)^2 \frac{d^2\mathbf{R}}{ds^2}$$

Now since $d\mathbf{R}/dt \cdot d^2\mathbf{R}/dt^2 = 0$, $d^2\mathbf{R}/dt^2$ is perpendicular to $d\mathbf{R}/dt$. Hence the acceleration $\mathbf{A}(t)$ is comprised of (*i*) the tangential component $d^2s/dt^2 \cdot d\mathbf{R}/ds$, called the *tangential acceleration, and*

(*ii*) the normal component $(ds/dt)^2 \cdot d^2\mathbf{R}/ds^2$, called the *normal acceleration.*

**Obs.** The acceleration is the time rate change of $|\mathbf{V}| = ds/dt$, if the normal acceleration is zero, for then

$$|A| = \left|\frac{d^2s}{dt^2}\right| \cdot \left|\frac{d\mathbf{R}}{ds}\right| = \left|\frac{d^2s}{dt^2}\right|.$$

**(3) Relative velocity and acceleration.** Let two particles $P_1$ and $P_2$ moving along the curves $C_1$ and $C_2$ have position vectors $\mathbf{R}_1$ and $\mathbf{R}_2$ at time $t$, (Fig. 8.4), so that

$\mathbf{R} = \overrightarrow{P_1P_2} = \mathbf{R}_2 - \mathbf{R}_1$

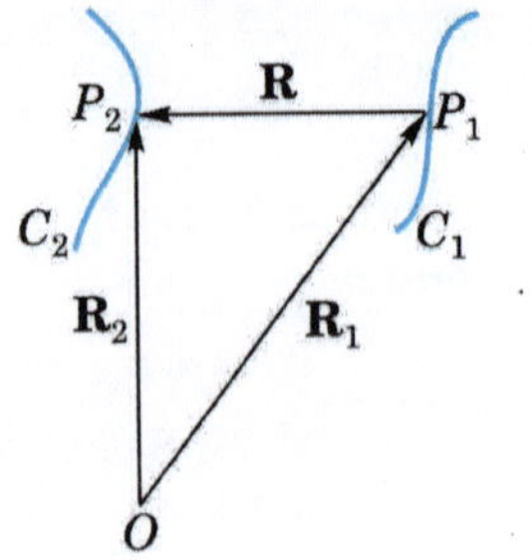

Fig. 8.4

Differentiating w.r.t. $t$, we get $\dfrac{d\mathbf{R}}{dt} = \dfrac{d\mathbf{R}_2}{dt} - \dfrac{d\mathbf{R}_1}{dt}$ ...(*iii*)

This defines the *relative velocity* (vector) of $P_2$ w.r.t. $P_1$ and states that the *velocity (vector) of $P_2$ relative to $P_1$ = velocity (vector) of $P_2$ – velocity (vector) of $P_1$*.

Again differentiating Eqn. (*iii*), we have $\dfrac{d^2\mathbf{R}}{dt^2} = \dfrac{d^2\mathbf{R}_2}{dt^2} - \dfrac{d^2\mathbf{R}_1}{dt^2}$ ...(*iv*)

*i.e.,* *acceleration (vector) of $P_2$ relative to $P_1$ = acceleration (vector) of $P_2$ – acceleration (vector) of $P_1$*.

**Example 8.4.** *A particle moves along the curve $x = t^3 + 1$, $y = t^2$, $z = 2t + 3$ where $t$ is the time. Find the components of its velocity and acceleration at $t = 1$ in the direction* **I + J + 3K.** (*V.T.U., 2016*)

**Solution.** *Velocity* $= \dfrac{d\mathbf{R}}{dt} = \dfrac{d}{dt}\,[(t^3 + 1)\,\mathbf{I} + t^2\mathbf{J} + (2t + 3)\mathbf{K}]$

$= 3t^2\,\mathbf{I} + 2t\mathbf{J} + 2\mathbf{K} = 3\mathbf{I} + 2\mathbf{J} + 2\mathbf{K}$ at $t = 1$

and *acceleration* $= \dfrac{d^2\mathbf{R}}{dt^2} = 6t\mathbf{I} + 2\mathbf{J} + 0\mathbf{K} = 6\mathbf{I} + 2\mathbf{J}$ at $t = 1$.

Now unit vector in the direction of **I + J + 3K** is $\dfrac{\mathbf{I} + \mathbf{J} + 3\mathbf{K}}{\sqrt{(1^2 + 1^2 + 3^2)}} = \dfrac{1}{\sqrt{11}}\,(\mathbf{I} + \mathbf{J} + 3\mathbf{K})$.

$\therefore$ component of velocity at $t = 1$ in the direction **I + J + 3K**

$$= \frac{(3\mathbf{I} + 2\mathbf{J} + 2\mathbf{K})\,.\,(\mathbf{I} + \mathbf{J} + 3\mathbf{K})}{\sqrt{(11)}} = \frac{3 + 2 + 6}{\sqrt{(11)}} = \sqrt{(11)}$$

and component of acceleration at $t = 1$ in the direction

$$\mathbf{I} + \mathbf{J} + 3\mathbf{K} = (6\mathbf{I} + 2\mathbf{J})\,.\,(\mathbf{I} + \mathbf{J} + 3\mathbf{K})/\sqrt{(11)} = \frac{6 + 2}{\sqrt{(11)}} = \frac{8}{\sqrt{(11)}}.$$

**Example 8.5 :** *If $\vec{R}$ be a unit vector function, show that $\dfrac{d\vec{R}}{d\theta}$ is a unit vector normal to $\vec{R}$, where $\theta$ is the angle which $\vec{R}$ makes with the axis of $x$.* (*KTU, 2018*)

**Solution:** Let $\overline{OP} = \vec{R}$ and $\overline{OQ} = \vec{R} + \delta\vec{R}$

Then $\overline{PQ} = \delta\vec{R}$

Since $\dfrac{d\vec{R}}{d\theta} = \lim\limits_{\delta\theta \to 0} \dfrac{\delta\vec{R}}{\delta\theta}$,

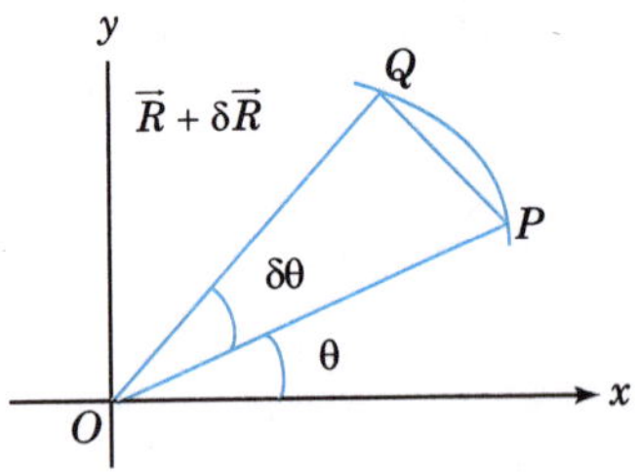

Fig. 8.5

We observe that $\dfrac{d\vec{R}}{d\theta}$ is normal to $\vec{R}$ as $\delta\vec{R}$ is normal $\vec{R}$ as $\delta\vec{R}$ is normal to $\vec{R}$ in the limiting position.

Further, $\left|\dfrac{d\vec{R}}{d\theta}\right| = \lim\limits_{\delta\theta \to 0} \left|\dfrac{d\vec{R}}{d\theta}\right|$

Hence $\dfrac{d\vec{R}}{d\theta}$ is a unit vector normal to $\vec{R}$

**Example 8.6.** *A particle moves along the curve* $\mathbf{R} = (t^3 - 4t)\,\mathbf{I} + (t^2 + 4t)\mathbf{J} + (8t^2 - 3t^3)\mathbf{K}$ *where $t$ denotes time. Find the magnitudes of acceleration along the tangent and normal at time $t = 2$.* (*Rohtak, 2011*)

**Solution.** Velocity $\dfrac{d\mathbf{R}}{dt} = (3t^2 - 4)\mathbf{I} + (2t + 4)\mathbf{J} + (16t - 9t^2)\mathbf{K}$

and acceleration $\dfrac{d^2\mathbf{R}}{dt^2} = 6t\mathbf{I} + 2\mathbf{J} + (16 - 18t)\mathbf{K}$

$\therefore$ at $t = 2$, velocity $\mathbf{V} = 8\mathbf{I} + 8\mathbf{J} - 4\mathrm{K}$ and acceleration $\mathbf{A} = 12\mathbf{I} + 2\mathbf{J} - 20\mathbf{K}$.

Since the velocity is along the tangent to the curve, therefore, the component of **A** along the tangent

$$= \mathbf{A} \cdot \frac{\mathbf{V}}{|\mathbf{V}|} = (12\mathbf{I} + 2\mathbf{J} - 20\mathbf{K}) \cdot \frac{8\mathbf{I} + 8\mathbf{J} - 4\mathbf{K}}{\sqrt{(64 + 64 + 16)}}$$

$$= \frac{12 \times 8 + 2 \times 8 + (-20) \times (-4)}{12} = 16.$$

Now the component of **A** along the normal

$$= |\, \mathbf{A} - \text{Resolved part of } \mathbf{A} \text{ along the tangent} \,|$$

$$= \left| 12\mathbf{I} + 2\mathbf{J} - 20\mathbf{K} - 16\frac{8\mathbf{I} + 8\mathbf{J} - 4\mathbf{K}}{12} \right| = \frac{1}{3}\,|\, 4\mathbf{I} - 26\mathbf{J} - 44\mathbf{K} \,| = 2\sqrt{73}.$$

**Example 8.7.** *Show that the radial and transverse velocities of a particle moving in a plane are respectively $\dfrac{dr}{d\theta}$ and $r\dfrac{d\theta}{dt}$ and accelerations in the same directions are respectively*

$$\frac{d^2r}{dt^2} - r\left(\frac{d\theta}{dt}\right)^2 = \frac{1}{r}\frac{1}{dt}\left(r^2\frac{d\theta}{dt}\right)$$ *(Agra, 2019)*

**Solution.** Let $\vec{r}$ be the position vector of a point $P$ and let $\hat{R}$ and $\hat{T}$ be unit vectors in the radial and transverse directions respectively as shown in Fig. 8.6.

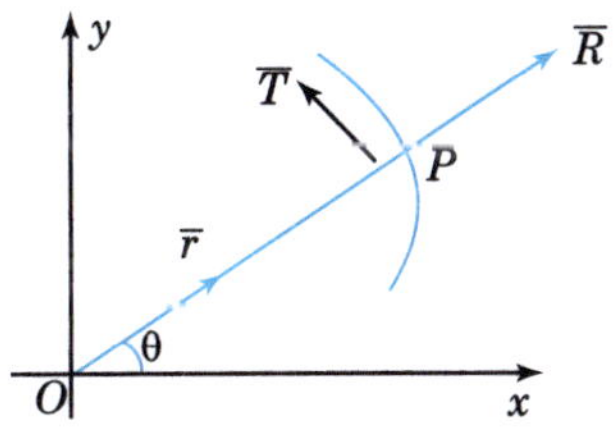

Fig. 8.6.

$$\vec{r} = r\hat{R}, \quad \text{where } r = |\vec{r}|$$

$$\therefore \quad \frac{d\vec{r}}{dt} = \frac{d}{dt}r\hat{R} = \frac{dr}{dt}\hat{R} + r\frac{d\hat{R}}{dt}$$

$$= \frac{dr}{dt}\hat{R} + r\frac{d\hat{R}}{d\theta}\frac{d\theta}{dt} = \frac{dr}{dt}\hat{R} + r\frac{d\theta}{dt}\hat{T}$$

Since $\qquad \dfrac{d\hat{R}}{d\theta} = \hat{T}$

Thus the velocity has components $\dfrac{dr}{dt}$ and $r\dfrac{d\theta}{dt}$ in the radial and transverse directions

$$\frac{d^2r}{dt^2}\hat{R} = \frac{d}{dt}\left(\frac{dr}{dt}\hat{R} + r\frac{d\theta}{dt}\hat{T}\right)$$

$$= \left(\frac{d^2r}{dt^2}\hat{R} + \frac{dr}{dt}\frac{d\hat{R}}{dt}\right) + \left(\frac{dr}{dt}\frac{d\theta}{dt}\hat{T} + r\frac{d^2\theta}{dt^2}\hat{T} + r\frac{d\theta}{dt}\frac{d\hat{T}}{dt}\right)$$

$$= \left(\frac{d^2r}{dt^2}\hat{R} + \frac{dr}{dt}\frac{d\hat{R}}{d\theta}\frac{d\theta}{dt}\right) + \left(\frac{dr}{dt}\frac{d\theta}{dt} + r\frac{d^2\theta}{dt^2}\right)\hat{T} + r\frac{d\theta}{dt}\frac{d\hat{T}}{d\theta}\cdot\frac{d\theta}{dt}$$

$$= \left(\frac{d^2r}{dt^2}\hat{R} + \frac{dr}{dt}\cdot\frac{d\theta}{dt}\hat{T}\right) + \left(\frac{dr}{dt}\frac{d\theta}{dt} + r\frac{d^2\theta}{dt^2}\right)\hat{T} - r\left(\frac{d\theta}{dt}\right)^2\hat{R}$$

Since $\qquad \dfrac{d\hat{R}}{d\theta} = \hat{T}$ and $\dfrac{d\hat{T}}{d\theta} = -\hat{R}$

$$\therefore \quad \frac{d^2r}{dt^2} = \left\{\frac{d^2r}{d^2t} - r\left(\frac{d\theta}{dt}\right)^2\hat{R} + \left\{2 - \frac{dr}{dt} - \frac{d\theta}{dt} + r\frac{d^2\theta}{dt^2}\right\}\right\}\hat{T}$$

Hence the radial and transvers components of the acceleration are $\left(\ddot{r} - r\,\theta^2\right)$ and $\left(2\dot{r}\theta - r\,\ddot{\theta}\right)$

**Example 8.8.** *A person going eastwards with a velocity of 4 km per hour, finds that the wind appears to blow directly from the north. He doubles his speed and the wind seems to come from north-east. Find the actual velocity of the wind.*

**Solution.** Let the actual velocity of the wind be $x\mathbf{I} + y\mathbf{J}$, where **I**, **J** represent velocities of 1 km per hour towards the east and north respectively. As the person is going eastwards with a velocity of 4 km per hour, his actual velocity is $4\mathbf{I}$.

Then the velocity of the wind relative to the man is $(x\mathbf{I} + y\mathbf{J}) - 4\mathbf{I}$, which is parallel to $-\mathbf{J}$, as it appears to blow from the north. Hence $x = 4$. ...(*i*)

When the velocity of the person becomes $8\mathbf{I}$, the velocity of the wind relative to man is $(x\mathbf{I} + y\mathbf{J}) - 8\mathbf{I}$. But this is parallel to $-(\mathbf{I} + \mathbf{J})$.

$\therefore$ $(x - 8)/y = 1$, which by Eqn. (*i*) gives $y = -4$.

Hence the actual velocity of the wind is $4(\mathbf{I} - \mathbf{J})$, *i.e.*, $4\sqrt{2}$ km. per hour towards the south-east.

**Example 8.9 :** *Find by vector methods the tangential and normal velocities and accelerations of a point moving in a plane curve.* (*CES, 2019*)

**Solution:** Let $\vec{r}$ be the positive vector of the point $P$. Its velocity

$$\vec{V}\frac{d\vec{r}}{dt} = \frac{d\vec{r}}{ds}\cdot\frac{ds}{dt}$$

Since
$$\frac{d\vec{r}}{ds} = \lim_{\delta s \to 0}\frac{\delta\vec{r}}{\delta s} = \hat{t}$$

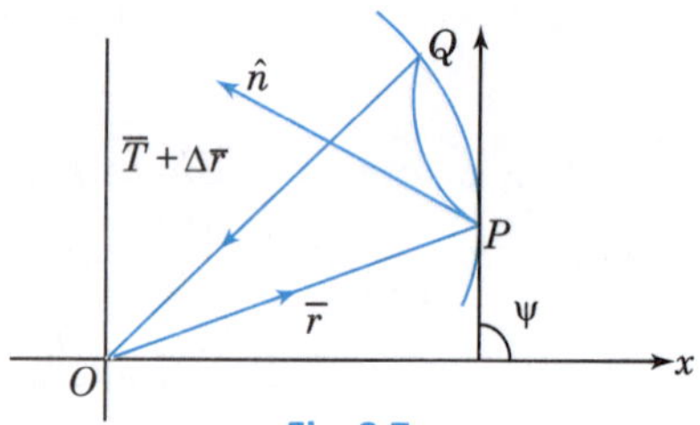

Fig. 8.7.

has a unit magnitude and direction that of tangent at $P$, we have

$$\vec{V} = v\hat{t}, \quad \text{where} \quad \frac{ds}{dt} = v$$

$$\frac{d\vec{V}}{dt} = \frac{dv}{dt}\hat{t} + v\frac{d\hat{t}}{dt} \qquad ...(i)$$

Now,
$$\frac{d\hat{t}}{dt} = \frac{d\hat{t}}{d\psi}\frac{d\psi}{dt} = \hat{\eta}\frac{d\psi}{dt}$$

$$= \hat{\eta}\frac{d\psi}{\rho s}\frac{ds}{dt} = \frac{v}{\rho}\hat{\eta} \qquad ...(ii)$$

Since
$$\rho = \frac{ds}{\rho\psi} \text{ or } \frac{1}{\rho} = \frac{d\psi}{ds}$$

from (*i*) and (*ii*), we have

$$\frac{\overline{dV}}{dt} = \frac{dv}{dt}\hat{t} + \frac{v^2}{\rho}\hat{\eta}$$

Hence the tangential acceleration of $\rho$ is $\frac{dv}{dt}$ or $\frac{d^2s}{dt^2}$ and normal acceleration is $\frac{v^2}{\rho}$.

## PROBLEMS 8.2

1. A particle moves along a circular helix in 3-space so that its position vector at time $t$ is $\mathbf{R} = 4(\cos \pi t)\mathbf{I} + 4(\sin \pi t)\mathbf{J} + t\mathbf{K}$. Find the displacement of the particle during the interval $1 \leq t \leq 5$. (*K.T.U., 2016*)
2. A particle moves along a curve $x = e^{-t}$, $y = 2\cos 3t$, $z = 2\sin 3t$, where $t$ is the time variable. Determine its velocity and acceleration vectors and also the magnitudes of velocity and acceleration at $t = 0$.
3. The position vector of a particle at time $t$ is $\mathbf{R} = \cos(t-1)\,\mathbf{I} + \sinh(t-1)\,\mathbf{J} + at^3\mathbf{K}$. Find the condition imposed on $a$ by requiring that at time $t = 1$, the acceleration is normal to the position vector.
4. A particle moves on the curve $x = 2t^2$, $y = t^2 - 4t$, $z = 3t - 5$, where $t$ is the time. Find the components of velocity and acceleration at time $t = 1$ in the direction $\mathbf{I} - 3\mathbf{J} + 2\mathbf{K}$. (*V.T.U., 2015*)
5. A particle moves so that its position vector is given by $\mathbf{R} = \mathbf{I}\cos\omega t + \mathbf{J}\sin\omega t$. Show that the velocity $\mathbf{V}$ of the particle is perpendicular to $\mathbf{R}$ and $\mathbf{R} \times \mathbf{V}$ is a constant vector.

6. A particle (position vector **R**) is moving in a circle with constant angular velocity ω. Show by vector methods, that the acceleration is equal to $-\omega^2\mathbf{R}$.

7. Show that if $\vec{r} = \vec{a}\sin\omega t + \vec{b}\cos\omega t$, where $\vec{a}, \vec{b}, \omega$ are constants, then $\frac{d^2\vec{r}}{dt^2} = -\omega^2\vec{r}$ and $\vec{r}\times\frac{d\vec{r}}{dt} = -\omega\,\vec{a}\times\vec{b}$ *(Mumbai, 2018)*

8. The position vector of a moving particle at a time $t$ is $\mathbf{R} = 3\cos t\,\mathbf{I} + 3\sin t\mathbf{J} + 4t\mathbf{K}$. Find the tangent and normal components of its acceleration at $t = 1$.

9. Find the angle between the tangents to the curve $\vec{r} = t^2i + 2tj - t^3k$ at the points $t = \pm 1$ *(Andra, 2018)*

10. If $\frac{d\vec{u}}{dt} = \vec{w}\times\vec{u}$ and $\frac{d\vec{v}}{dt} = \vec{w}\times\vec{v}$, prove that $\frac{d}{dt}(\vec{u}\times\vec{v}) = \vec{w}\times(\vec{u}\times\vec{v})$ *(Mysore, 2019)*

11. If $\vec{r} = a\cos t\,\hat{i} + a\sin t\,\hat{j} + at\tan\alpha\,\hat{k}$, show that $\left[\frac{d\vec{r}}{dt}\cdot\frac{d^2\vec{r}}{dt^2}\cdot\frac{d^3\vec{r}}{dt^3}\right] = a^3\tan\alpha$ *(IETE, 2019)*

12. Find the unit vector tangential at any point on the curve $x = t^2 + 1, y = 4t - 3, z = 2t^2 - 6t$.

Also find the unit tangential vector at the point $t = 2$,

13. Show that $\frac{d}{dt}\left[\left(\vec{a}\times\frac{d\vec{a}}{dt}\right)\times\frac{d^2\vec{a}}{dt^2}\right] = \left(\vec{a}\times\frac{d\vec{a}}{dt}\right)\times\frac{d^3\vec{a}}{dt^3} + \left(\vec{a}\times\frac{d^2\vec{a}}{dt^2}\right)\times\frac{d^2\vec{a}}{dt}$

14. If $\vec{F} = \frac{\vec{r}}{r}$, show that $\vec{F}\times\frac{d\vec{F}}{dt} = \frac{\vec{r}}{r^2}\times\frac{d\vec{r}}{dt}$

## 8.4 SCALAR AND VECTOR POINT FUNCTIONS

**(1)** *If to each point P*(**R**) *of a region E in space there corresponds a definite scalar denoted by f* (**R**), then *f* (**R**) *is called a* **scalar point function** *in E. The region E so defined is called a* **scalar field.**

The temperature at any instant, density of a body and potential due to gravitational matter are all examples of scalar point functions.

**(2)** *If to each point P*(**R**) *of a region E in space there corresponds a definite vector denoted by* **F**(**R**), *then it is called the* **vector point function** *in E. The region E so defined is called a* **vector field.**

The velocity of a moving fluid at any instant, the gravitational intensity of force are examples of vector point functions.

Differentiation of vector point functions follows the same rules as those of ordinary calculus. Thus if $\mathbf{F}(x, y, z)$ be a vector point function, then

$$\frac{d\mathbf{F}}{dt} = \frac{\partial\mathbf{F}}{\partial x}\frac{dx}{dt} + \frac{\partial\mathbf{F}}{\partial y}\frac{dy}{dt} + \frac{\partial\mathbf{F}}{\partial z}\frac{dz}{dt}$$

and
$$d\mathbf{F} = \frac{\partial\mathbf{F}}{\partial x}dx + \frac{\partial\mathbf{F}}{\partial y}dy + \frac{\partial\mathbf{F}}{\partial z}dz = \left(\frac{\partial}{\partial x}dx + \frac{\partial}{\partial y}dy + \frac{\partial}{\partial z}dz\right)\mathbf{F} \quad ...(i)$$

**(3) Vector operator del.** The operator on the right side of the Eqn. (*i*) is in the form of a scalar product of $\mathbf{I}\frac{\partial}{\partial x} + \mathbf{J}\frac{\partial}{\partial y} + \mathbf{K}\frac{\partial}{\partial z}$ and $\mathbf{I}dx + \mathbf{J}dy + \mathbf{K}dz$.

If ∇ (*read as del*) *be defined by the equation* $\nabla = \mathbf{I}\frac{\partial}{\partial x} + \mathbf{J}\frac{\partial}{\partial y} + \mathbf{K}\frac{\partial}{\partial z}$ ...(*ii*)

then Eqn. (*i*) may be written as $d\mathbf{F} = (\nabla\,.\,d\mathbf{R})\,\mathbf{F}$ for when $\mathbf{R} = x\mathbf{I} + y\mathbf{J} + z\mathbf{K}$, $d\mathbf{R} = \mathbf{I}dx + \mathbf{J}dy + \mathbf{K}dz$.

## 8.5 DEL APPLIED TO SCALAR POINT FUNCTIONS—GRADIENT

**(1) Def.** *The vector function $\nabla f$ is defined as the gradient of the scalar point function f and is written as grad f.*

Thus $$\text{grad } f = \nabla f = \mathbf{I}\frac{\partial f}{\partial x} + \mathbf{J}\frac{\partial f}{\partial y} + \mathbf{K}\frac{\partial f}{\partial z}$$

**(2) Geometrical interpretation.** Consider the scalar point function $f(\mathbf{R})$, where $\mathbf{R} = x\mathbf{I} + y\mathbf{J} + z\mathbf{K}$.

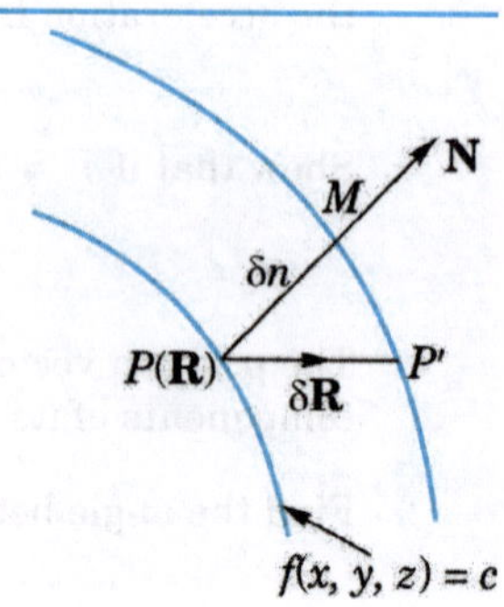

Fig. 8.8.

If a surface $f(x, y, z) = c$ be drawn through any point $P(\mathbf{R})$ such that at each point on it, the function has the same value as at $P$, then such a surface is called a *level surface* of the function $f$ through $P$, *e.g.*, equipotential or isothermal surface (Fig. 8.8).

Let $P'(\mathbf{R} + \delta\mathbf{R})$ be a point on a neighbouring level surface $f + \delta f$. Then

$$\nabla f . \delta\mathbf{R} = \left[\mathbf{I}\frac{\partial f}{\partial x} + \mathbf{J}\frac{\partial f}{\partial y} + \mathbf{K}\frac{\partial f}{\partial z}\right] . (\mathbf{I}\delta x + \mathbf{J}\delta y + \mathbf{K}\delta z) = \frac{\partial f}{\partial x}\delta x + \frac{\partial f}{\partial y}\delta y + \frac{\partial f}{\partial z}\delta z = \delta f.$$

Now if $P'$ lies on the same level surface as $P$, then $\delta f = 0$, *i.e.*, $\nabla f . \delta\mathbf{R} = 0$. This means that $\nabla f$ is perpendicular to every $\delta\mathbf{R}$ lying on this surface. Thus $\nabla f$ *is normal to the surface* $f(x, y, z) = c$.

$$\therefore \qquad \nabla f = |\nabla f| \mathbf{N}$$

where $\mathbf{N}$ is a unit vector normal to this surface. If the perpendicular distance $PM$ between the surfaces through $P$ and $P'$ be $\delta n$, then the rate of change of $f$ normal to the surface through $P$

$$= \frac{\partial f}{\partial n} = \underset{\delta n \to 0}{\text{Lt}} \frac{\delta f}{\delta n} = \underset{\delta n \to 0}{\text{Lt}} \nabla f . \frac{\delta \mathbf{R}}{\delta n}$$

$$= |\nabla f| \underset{\delta n \to 0}{\text{Lt}} \frac{\mathbf{N} . \delta\mathbf{R}}{\delta n} = |\nabla f|. \qquad [\because \mathbf{N} . \delta\mathbf{R} = |\delta\mathbf{R}| \cos\theta = \delta n]$$

Hence the magnitude of $\nabla f = \partial f/\partial n$.

Thus *grad f is a vector normal to the surface f = constant and has a magnitude equal to the rate of change of f along this normal.*

**(3) Directional derivative.** If $\delta r$ denotes the length $PP'$ and $\mathbf{N}'$ is a unit vector in the direction $PP'$, then the limiting value of $\delta f/\delta r$ as $\delta r \to 0$ (*i.e.*, $\partial f/\partial r$) is known as the *directional derivative of f at P along the direction PP'*.

Since $$\delta r = \delta n/\cos\alpha = \delta n/\mathbf{N} . \mathbf{N}'$$

$$\therefore \qquad \frac{\partial f}{\partial r} = \underset{\delta r \to 0}{\text{Lt}} \left[\mathbf{N} . \mathbf{N}' \frac{\delta f}{\delta n}\right] = \mathbf{N}' . \frac{\partial f}{\partial n}\mathbf{N} = \mathbf{N}' . \nabla f$$

Thus the directional derivation of $f$ in the direction of $\mathbf{N}'$ is the resolved part of $\nabla f$ in the direction $\mathbf{N}'$.

Since $$\nabla f . \mathbf{N}' = |\nabla f| \cos\alpha \le |\nabla f|$$

It follows that $\nabla f$ *gives the maximum rate of change of f.*

**Example 8.10.** *Prove that* $\nabla r^n = nr^{n-2}\mathbf{R}$, *where* $\mathbf{R} + x\mathbf{I} + y\mathbf{J} + z\mathbf{K}$.

*(W.B.T.U., 2016)*

**Solution.** We have $f(x, y, z) = r^n = (x^2 + y^2 + z^2)^{n/2}$

$$\therefore \qquad \frac{\partial f}{\partial x} = \frac{n}{2}(x^2 + y^2 + z^2)^{n/2-1}. 2x = nxr^{n-2}. \text{ Similarly, } \frac{\partial f}{\partial y} = nyr^{n-2} \text{ and } \frac{\partial f}{\partial z} = nzr^{n-2}$$

Thus $$\nabla r^n = \mathbf{I}\frac{\partial f}{\partial x} + \mathbf{J}\frac{\partial f}{\partial y} + \mathbf{K}\frac{\partial f}{\partial z} = nr^{n-2}(x\mathbf{I} + y\mathbf{J} + z\mathbf{K}) = nr^{n-2}\mathbf{R}.$$

*Otherwise :* The level surfaces for $f$ = constant, *i.e.*, $r^n$ = constant are concentric spheres with centre $O$ and hence unit normal $\mathbf{N}$ to the level surface through $P$ is along the radius $\mathbf{R}$

*i.e.,* $$\mathbf{N} = \hat{\mathbf{R}}.$$

$$\therefore \qquad \nabla f = \frac{\partial f}{\partial n} . \mathbf{N} = \frac{df}{dr}\hat{\mathbf{R}} = nr^{n-1}\hat{\mathbf{R}} \qquad [\because f = r^n]$$

$$= nr^{n-1}(\mathbf{R}/r) = nr^{n-2}\mathbf{R}.$$

**Example 8.11.** *If $\vec{r} = xi + yj + zk$, $|\vec{r}| = r$ and $\hat{r} = \dfrac{\vec{r}}{r}$ show that*

(i) grad $r = \hat{r}$ (ii) grad$\left(\dfrac{1}{r}\right) = -\dfrac{\hat{r}}{r^2}$ (iii) grad$\left(e^{r^2}\right) = 2\vec{r}e^{r^2}$

(*JNTUK, 2022 ; Kottayam, 2019; Delhi, 2013 ; UPTU, 2011*)

**Solution.** If $\vec{r} = xi + yj + zk$, $|\vec{r}| = r = \sqrt{x^2 + y^2 + z^2}$

or $r^2 = x^2 + y^2 + z^2$, we have on

$$2r\frac{\delta r}{\delta x} = 2x \quad \text{or} \quad \frac{\delta r}{\delta x} = \frac{x}{r}$$

So, $$\frac{\delta r}{\delta y} = \frac{y}{r} \text{ and } \frac{\delta r}{\delta z} = \frac{z}{r}$$

(i) grad $r = \nabla r = \left(i\dfrac{\partial}{\partial x} + j\dfrac{\partial}{\partial y} + k\dfrac{\partial}{\partial z}\right)r$

$$= i\frac{\partial r}{\partial x} + j\frac{\partial r}{\partial y} + k\frac{\partial r}{dz} = i\left(\frac{x}{r}\right) + j\left(\frac{y}{r}\right) + k\left(\frac{z}{r}\right) = \frac{xi + yj + zk}{r} = \frac{\vec{r}}{r} = \hat{r}$$

(ii) grad$\left(\dfrac{1}{r}\right) = \nabla\left(\dfrac{1}{r}\right) = \left(i\dfrac{\partial}{\partial x} + j\dfrac{\partial}{\partial y} + k\dfrac{\partial}{\partial z}\right)\left(\dfrac{1}{r}\right)$

$$= i\left(-\frac{1}{r^2}\frac{dr}{dx}\right) + j\left(-\frac{1}{r^2}\frac{\delta r}{dy}\right) + k\left(-\frac{1}{r^2}\frac{\delta r}{\delta z}\right)$$

$$= -\frac{ix}{r^3} - \frac{jy}{r^3} - \frac{kz}{r^3} = -\frac{1}{r^3}(x\hat{i} + y\hat{j} + \hat{k}z) = -\frac{\vec{r}}{r^3} = -\frac{\hat{r}}{r^2}$$

(iii) grad $(e^{r^2}) = \nabla(e^{r^2})$

$$= \left(i\frac{\delta}{\delta x} + j\frac{\delta}{\delta y} + k\frac{\delta}{\delta z}\right)e^{r^2} = i\left(e^{r^2}2r\frac{\delta r}{\delta x}\right) + j\left(e^{r^2}2r\frac{\delta r}{\delta y}\right) + k\left(e^{r^2}2r\frac{\delta r}{\delta z}\right)$$

$$= 2e^{r^2}(ix + jy + kz) = 2\vec{r}e^{r^2}$$

**Example 8.12.** *If $u = x + y + z$, $v = x^2 + y^2 + z^2$, $w = yz + zx + xy$, prove that grad u, grad v and grad w are coplanar.* (*U.K.T.U., 2010*)

**Solution.** grad $u = \left(\mathbf{I}\dfrac{\partial}{\partial x} + \mathbf{J}\dfrac{\partial}{\partial y} + \mathbf{K}\dfrac{\partial}{\partial z}\right)(x + y + z) = \mathbf{I} + \mathbf{J} + \mathbf{K}$

grad $v = 2x\mathbf{I} + 2y\mathbf{J} + 2z\mathbf{K}$, grad $w = (y + z)\,\mathbf{I} + (z + x)\,\mathbf{J} + (z + x)\,\mathbf{K}$

We know that three vectors are coplanar if their scalar triple product is zero.

Here [grad $u$, grad $v$, grad $w$]

$$= \begin{vmatrix} 1 & 1 & 1 \\ 2x & 2y & 2z \\ y+z & z+x & x+y \end{vmatrix}$$

$$= 2\begin{vmatrix} 1 & 1 & 1 \\ x & y & z \\ y+z & z+x & x+y \end{vmatrix} = 2\begin{vmatrix} 1 & 1 & 1 \\ x+y+z & y+z+x & z+x+y \\ y+z & z+x & x+y \end{vmatrix} \quad [\text{Operate } R_2 + R_3]$$

$$= 2(x+y+z)\begin{vmatrix} 1 & 1 & 1 \\ 1 & 1 & 1 \\ y+z & z+x & x+y \end{vmatrix} = 0.$$

Hence grad $u$, grad $v$ and grad $w$ are coplanar.

**Example 8.13.** *Find a unit vector normal to the surface $xy^3z^2 = 4$ at the point (–1, –1, 2).*

**Solution.** A vector normal to the given surface is $\nabla(xy^3z^2)$

$$= \mathbf{I}\frac{\partial}{\partial x}(xy^3z^2) + \mathbf{J}\frac{\partial}{\partial y}(xy^3z^2) + \mathbf{K}\frac{\partial}{\partial z}(xy^3z^2) = \mathbf{I}(y^3z^2) + \mathbf{J}(3xy^2z^2) + \mathbf{K}(2xy^3z)$$

$$= -4\mathbf{I} - 12\mathbf{J} + 4\mathbf{K} \text{ at the point } (-1, -1, 2).$$

Hence the desired unit normal to the surface

$$= \frac{-4\mathbf{I} - 12\mathbf{J} + 4\mathbf{K}}{\sqrt{[(-4)^2 + (-12)^2 + 4^2]}} = -\frac{1}{\sqrt{11}}(\mathbf{I} + 3\mathbf{J} - \mathbf{K}).$$

**Example 8.14.** *If* $\nabla\phi = (y^2 - 2xyz^3)i + (3 + 2xy - x^2z^3)j + (6z^3 - 3x^2yz^2)k$ *find* $\phi$.

**Solution.** Since $\nabla\phi = \frac{\partial\phi}{\partial x}i + \frac{\partial\phi}{\partial y}j + \frac{\partial\phi}{\partial z}k$, we have

$$\frac{\partial\phi}{\partial x} = (y^2 - 2xyz^3) \quad \text{or} \quad \phi = y^2x - x^2yz^3 + c_1(y, z)$$

$$\frac{\partial\phi}{\partial y} = (3 + 2xy - x^2z^3) \quad \text{or} \quad \phi = 3y + xy^2 - x^2yz^2 + c_2(x, z)$$

$$\frac{\partial\phi}{\partial z} = (6z^3 - 3x^2yz^2) \quad \text{or} \quad \phi = \frac{3}{2}z^4 - x^2yz^3 + c_3(x, y)$$

Now comparing the three values of $\phi$.

$$c_1(y, z) = 3y + \frac{3}{2}z^4, \quad c_2(x, z) = \frac{3}{2}z^4,$$

$$c_3(x, y) = 3y + xy^2$$

Hence
$$\phi = xy^2 - x^2yz^3 + 3y + \frac{3}{2}z^4 + c$$

**Example 8.15.** *Find the directional derivative of $f = x^2 - y^2 + 2z^2$ at the point P (1, 2, 3) in the direction of the line PQ where Q is the point (5, 0, 4). Also calculate the magnitude of the maximum directional derivative.*

*(CSVTU, 2019 ; V.T.U., 2016 ; U.K.T.U., 2011)*

**Solution.** We have $\nabla f = \left(\mathbf{I}\frac{\partial}{\partial x} + \mathbf{J}\frac{\partial}{\partial y} + \mathbf{K}\frac{\partial}{\partial z}\right)(x^2 - y^2 + 2z^2) = 2x\mathbf{I} - 2y\mathbf{J} + 4z\mathbf{K}$

$$= 2\mathbf{I} - 4\mathbf{J} + 12\mathbf{K} \text{ at } P(1, 2, 3)$$

Also
$$\vec{PQ} = \vec{OQ} - \vec{OP} = (5\mathbf{I} + 0\mathbf{J} + 4\mathbf{K}) - (\mathbf{I} + 2\mathbf{J} + 3\mathbf{K}) = 4\mathbf{I} - 2\mathbf{J} + \mathbf{K} = \mathbf{A} \text{ (say)}$$

$\therefore$ unit vector of $\mathbf{A} = \hat{A} = \frac{\mathbf{A}}{a} = \frac{4\mathbf{I} - 2\mathbf{J} + \mathbf{K}}{\sqrt{(16 + 4 + 1)}} = \frac{4\mathbf{I} + 2\mathbf{J} + \mathbf{K}}{\sqrt{21}}$

Thus the directional derivative of $f$ in the direction of $\vec{PQ}$

$$\nabla f . \hat{A} = (2\mathbf{I} - 4\mathbf{J} + 12\mathbf{K}) . (4\mathbf{I} - 2\mathbf{J} + \mathbf{K})/\sqrt{21}$$

$$= (8 + 8 + 12)/\sqrt{21} = 28/\sqrt{21}$$

The directional derivative of its maximum in the direction of the normal to the surface *i.e.*, in the direction of $\nabla f$.

Hence maximum value of this directional derivative

$$= |\nabla f| = |2\mathbf{I} - 4\mathbf{J} + 12\mathbf{K}|$$

$$= (4 + 16 + 144) = \sqrt{164} .$$

**Example 8.16.** *Find a unit vector normal to the surface $xy^3z^2 = 4$ at the point $(-1, -1, 2)$.* (*Mumbai, 2008*)

**Solution.** A vector normal to the given surface is $\nabla (xy^3z^2)$

$$= \mathbf{I}\frac{\partial}{\partial x}(xy^3z^2) + \mathbf{J}\frac{\partial}{\partial y}(xy^3z^2) + \mathbf{K}\frac{\partial}{\partial z}(xy^3z^2)$$

$$= \mathbf{I}(y^3 z^2) + \mathbf{J}(3xy^2z^2) + \mathbf{K}(2xy^3z)$$

$$= -4\mathbf{I} - 12\mathbf{J} + 4\mathbf{K} \text{ at the point } (-1, -1, 2).$$

Hence the desired unit normal to the surface

$$= \frac{-4\mathbf{I} - 12\mathbf{J} + 4\mathbf{K}}{\sqrt{[(-4)^2 + (-12)^2 + 4^2]}} = -\frac{1}{\sqrt{11}}(\mathbf{I} + 3\mathbf{J} - \mathbf{K}).$$

**Example 8.17.** *Find the angle between the surfaces $x^2 + y^2 + z^2 = 9$ and $z = x^2 + y^2 - 3$ at the point $(2, -1, 2)$.* (*V.T.U., 2016*)

**Solution.** Let $f_1 = x^2 + y^2 + z^2 - 9 = 0$ and $f_2 = x^2 + y^2 - z - 3 = 0$

Then $N_1 = \nabla f_1$ at $(2, -1, 2) = (2x\mathbf{I} + 2y\mathbf{J} + 2z\mathbf{K})$ at $(2, -1, 2)$
$= 4\mathbf{I} - 2\mathbf{J} + 4\mathbf{K}$

and $N_2 = \nabla f_2$ at $(2, -1, 2) = (2x\mathbf{I} + 2y\mathbf{J} - \mathbf{K})$ at $(2, -1, 2)$
$= 4\mathbf{I} - 2\mathbf{J} - \mathbf{K}$

Since the angle $\theta$ between the two surfaces at a point is the angle between their normals at that point and $\mathbf{N}_1, \mathbf{N}_2$ are the normals at $(2, -1, 2)$ to the given surfaces, therefore

$$\cos\theta = \frac{\mathbf{N}_1 \cdot \mathbf{N}_2}{n_1 n_2} = \frac{(4I - 2J + 4K).(4I - 2J - K)}{\sqrt{(16+4+16)}\sqrt{(16+4+1)}}$$

$$= \frac{4(4) + (-2)(-2) + 4(-1)}{6\sqrt{21}} = \frac{16}{6\sqrt{21}}$$

Hence the required angle $\theta = \cos^{-1}\left(\frac{8}{3\sqrt{21}}\right)$.

**Example 8.18.** *Find the values of a and b such that the surface $ax^2 - byz = (a + 2)x$ and $4x^2y + z^3 = 4$ cut orthogonally at $(1, -1, 2)$.* (*Hazaribag., 2015*)

**Solution.** Let $f_1 = ax^2 - byz - (a + 2)x = 0$ ...(*i*)

and $f_2 = 4x^2y + z^3 - 4 = 0$ ...(*ii*)

Then $\nabla f_1 = (2ax - a - 2)\mathbf{I} - 4z\mathbf{J} - by\mathbf{K} = (a - 2)\mathbf{I} - 2b\mathbf{J} + b\mathbf{K}$ at $(1, -1, 2)$.

$\nabla f_2 = 8xy\mathbf{I} + 4x^2\mathbf{J} + 3z^2\mathbf{K} = -8\mathbf{I} + 4\mathbf{J} + 12\mathbf{K}$ at $(1, -1, 2)$.

The surfaces Eqns. (*i*) and (*ii*) will cut orthogonally if $\nabla f_1 \cdot \nabla f_2 = 0$, *i.e.*, $-8(a - 2) - 8b + 12b = 0$

or $-2a + b + 4 = 0$ ...(*iii*)

Also since the point $(1, -1, 2)$ lies on Eqns. (*i*) and (*ii*),

$\therefore$ $a + 2b - (a + 2) = 0$ or $b = 1$

From Eqn. (*iii*), $-2a + 5 = 0$ or $a = 5/2$.

Hence $a = 5/2$ and $b = 1$.

**Example 8.19 :** *If $\theta$ is the acute angle between the surfaces $xy^2z = 3x + z^2$ and $3x^2 - y^2 + 2z = 1$ at the point $(1, -2, 1)$, show that $\cos\theta = \frac{3}{7\sqrt{6}}$*

**Solution:** Angle between two surfaces at a point is the angle between the normals to the surfaces at the point.

$\phi_1 = xy^2z - 3x - z^2 = 0$ and $\phi_2 = 3x^2 - y^2 + 2z = 1$

Then $n_1 = \text{grad}\,\phi_1 = (y^2z - 3)i + (2xyz)j + (xy^2 - 2z)k = i - 4j + 2k$

and $\overrightarrow{n_2} = \text{grad}\,\phi_2 = 6xi - 2yj + 2k$

$$= 6i + 4j + 2k$$

Since $\quad \vec{n_1} \cdot \vec{n_2} = |\vec{n_1}||\vec{n_2}| \cos\theta$, we have

$$\cos\theta = \frac{|n_1 n_2|}{|\vec{n_1}||\vec{n_2}|} = \left|\frac{6 - 16 + 4}{\sqrt{21}\sqrt{56}}\right| = \left|-\frac{6}{14\sqrt{6}}\right| = \frac{3}{7\sqrt{6}}$$

## PROBLEMS 8.3

1. (*a*) Find $\nabla\phi$, if $\phi = \log(x^2 + y^2 + z^2)$. *(Delhi, 2013 ; U.P.T.U., 2011)*
2. Find a unit vector normal to the surface $x^3 + y^3 + 3xyz = 3$ at the point (1, 2, –1).
3. Find the directional derivative of (*i*) $f(x, y) = e^x \sec y$ at $P(0, \pi/4)$ in the direction of $PQ$. whose $Q$ is the origin. *(K.T.U., 2016)*

   (*ii*) $\phi = x^2yz + 4xz^2$ at the point (1, –2, 1) in the direction of the vector $2\mathbf{I} - \mathbf{J} - 2\mathbf{K}$. *(JNTUK, 2022 ; V.T.U., 2015 ; C.S.V.T.U., 2012 ; Rohtak, 2012)*
4. What is the directional derivative of $\phi = xy^2 + yz^3$ at the point (2, –1, 1) in the direction of the normal to the surface $x \log z - y^2 = -4$ at (–1, 2, 1) ? *(C.S.V.T.U., 2014 S)*
5. Find the values of constants $a, b, c$ so that the directional derivative of $p = axy^2 + byz + cz^2x^3$ at (1, 2, –1) has a maximum magnitude 64 in the direction parallel to the $z$-axis. *(Anna., 2016 ; Andhra., 2015)*
6. If $\phi = 2xz^4 - x^2y$, show that $\nabla\phi = 10i - 4j - 16k$ and $|\nabla\phi| = 2\sqrt{93}$ at the point (2, –2, –1) *(Kerela, 2019)*
7. If the directional derivative of $\phi = ax^2y + by^2z + cz^2x$ at the point (1, 1, 1) has maximum magnitude 15 in the direction parallel to the line $\frac{x-1}{2} = \frac{y-3}{-2} = z$, find the values of $a$, $b$ and $c$. *(Rohtak, 2010)*
8. In what direction from (3, 1, –2) is the directional derivative of $\phi = x^2y^2z^4$ maximum ? Find also the magnitude of this maximum. *(Delhi, 2012)*
9. If $\vec{r} = \vec{xi} + \vec{yj} = \vec{zk}$ and $r = |\vec{r}|$, show that (*i*) $\text{grad}\,|\vec{r}|^3 = 3r\vec{r}$ (*ii*) $\nabla\phi(r) = \frac{\phi'(r)(\vec{r})}{r}$ *(Poona, 2019)*
10. The temperature of points in space is given by $T(x, y, z) = x^2 + y^2 - z$. A mosquito located at (1, 1, 2) desires to fly in such a direction that it will get warm as soon as possible. In what direction should it move ? *(Hydrabad, 2018; C.S.V.T.U., 2011)*
11. Calculate the angle between the normals to the surface $xy = z^2$ at the points (4, 1, 2) and (3, 3, –3). *(U.K.T.U., 2012)*
12. If $\phi(x, y) = \log_e \sqrt{x^2 + y^2}$, show that $\text{grad}\,\phi = \frac{\vec{r} - (\vec{k}\cdot\vec{r})\vec{k}}{\{\vec{r} - (\vec{k}\cdot\vec{r})\}\cdot\{\vec{r} - (\vec{k}\cdot\vec{r})\cdot\vec{k}\}}$ *(Rohtak, 2018)*
13. If $f$ and **G** are point functions, prove that the components of the latter normal and tangential to the surface $f = 0$ are $\frac{\nabla f \times (\mathbf{G}.\nabla f)}{(\nabla f)^2}$ and $\frac{\nabla f \times (\mathbf{G}.\nabla f)}{(\nabla f)^2}$
14. (*i*) If $\phi = (\vec{r} \times \vec{a})\cdot(\vec{r} \times \vec{b})$, show that $\nabla\phi = \vec{b} \times (\vec{r} \times \vec{a}) + \vec{a} \times (\vec{r} \times \vec{b})$

    (*ii*) If $\vec{u} = u(x, y, z, t)$ show that $\frac{d\vec{u}}{dt} = \frac{d\vec{u}}{dt} + \left(\frac{d\vec{r}}{dt}\cdot\nabla\right)\vec{u}$ (*iii*) $(\vec{u}\cdot\nabla)\vec{r} = \vec{u}$

## 8.6 DEL APPLIED TO VECTOR POINT FUNCTIONS

**(1) Divergence.** *The divergence of a continuously differentiable vector point function* **F** *is denoted by div* **F** *and is defined by the equation*

$$\textbf{div F} = \nabla\,.\,\mathbf{F} = \mathbf{I}\,.\,\frac{\partial \mathbf{F}}{\partial x} + \mathbf{J}\,.\,\frac{\partial \mathbf{F}}{\partial y} + \mathbf{K}\,.\,\frac{\partial \mathbf{F}}{\partial z}$$

If $\qquad \mathbf{F} = f\mathbf{I} + \phi\mathbf{J} + \psi\mathbf{K}$

then $\qquad \text{div}\,\mathbf{F} = \nabla\,.\,\mathbf{F} = \left(\mathbf{I}\frac{\partial}{\partial x} + \mathbf{J}\frac{\partial}{\partial y} + \mathbf{K}\frac{\partial}{\partial z}\right).(f\mathbf{I} + \phi\mathbf{J} + \psi\mathbf{K}) = \frac{\partial f}{\partial x} + \frac{\partial \phi}{\partial y} + \frac{\partial \psi}{\partial z}$

**(2) Curl.** *The curl of a continuously differentiable vector point function* **F** *is defined by the equation*

$$\textbf{curl F} = \nabla \times \mathbf{F} = \mathbf{I} \times \frac{\partial \mathbf{F}}{\partial x} + \mathbf{J} \times \frac{\partial \mathbf{F}}{\partial y} + \mathbf{K} \times \frac{\partial \mathbf{F}}{\partial z}$$

If $\mathbf{F} = f\mathbf{I} + \phi\mathbf{J} + \psi\mathbf{K}$ then curl $\mathbf{F} = \nabla \times \mathbf{F} = \left(\mathbf{I}\frac{\partial}{\partial x} + \mathbf{J}\frac{\partial}{\partial y} + \mathbf{K}\frac{\partial}{\partial z}\right) \times (f\mathbf{I} + \phi\mathbf{J} + \psi\mathbf{K})$

$$= \begin{vmatrix} \mathbf{I} & \mathbf{J} & \mathbf{K} \\ \frac{\partial}{\partial x} & \frac{\partial}{\partial y} & \frac{\partial}{\partial z} \\ f & \phi & \psi \end{vmatrix} = \mathbf{I}\left(\frac{\partial \psi}{\partial y} - \frac{\partial \phi}{\partial z}\right) + \mathbf{J}\left(\frac{\partial f}{\partial z} - \frac{\partial \psi}{\partial x}\right) + \mathbf{K}\left(\frac{\partial \phi}{\partial x} - \frac{\partial f}{\partial y}\right).$$

**Example 8.20.** *If $r = xi + yj + zk$ and $r = |\vec{r}|$*

*(i) $\text{div}\,\vec{r} = 3$* *(ii) $\text{div}\left(\frac{\vec{r}}{r^3}\right) = 0$* *(Varanasi–2018)*

**Solution.** Here $\vec{r} = \sum xi$

Hence $\frac{\delta \vec{r}}{\delta x} = i, \quad \frac{\delta \vec{r}}{\delta y} = j, \quad \frac{\delta \vec{r}}{\delta z} = k$

Also $r = |\vec{r}| = \sqrt{(x^2 + y^2 + z^2)}$

$$r^2 = x^2 + y^2 + z^2 = \sum x^2$$

$$2r\frac{\delta r}{\delta x} = 2x \quad \text{or} \quad \frac{\delta r}{\delta x} = \frac{x}{r}$$

Similarly $\frac{\delta r}{\delta y} = \frac{y}{r} \quad \text{and} \quad \frac{\delta r}{\delta z} = \frac{z}{r}$

(i) $\text{div}\,\vec{r} = \nabla\vec{r} = \left(i\frac{\delta}{\delta x} + j\frac{\delta}{\delta y} + k\frac{\delta}{\delta z}\right)\cdot(xi + yj + zk)$

$$= \frac{\delta x}{\delta x} + \frac{\delta y}{\delta y} + \frac{\delta z}{\delta z} = 1 + 1 + 1 = 3$$

(ii) $\text{div}\left(\frac{\vec{r}}{r^3}\right) = \nabla\cdot\left(\frac{\vec{r}}{r^3}\right) = \sum i\frac{\delta}{\delta x}\cdot\left(\frac{\vec{r}}{r^3}\right) = \sum i\frac{\delta}{\delta x}\left(\vec{r}\,\vec{r}^{-3}\right) = \sum i\left\{\frac{\delta \vec{r}}{\delta x}r^{-3} + \vec{r}\cdot\left(-3r^{-4}\frac{\delta r}{\delta x}\right)\right\}$

$$= \sum i\left\{\frac{i}{r^3} - \frac{3\vec{r}}{r^4}\left(\frac{x}{r}\right)\right\} = \sum\frac{1}{r^3} - 3\sum\frac{i\vec{r}x}{r^5}$$

$$= \frac{3}{r^3} - \frac{3}{r^5}\sum x^2 = \frac{3}{r^3} - \sum\frac{3r^2}{r^5} = 0$$

**Example 8.21.** If $r = xi + yj + zk$ and $r = |\vec{r}|$ show that $\text{div}(r\,\phi) = 3\phi + r\cdot\text{grad}\cdot\phi$ *(Calicut, 2018)*

**Solution.** $\vec{r} = \sum xi$

Hence $\frac{\delta \vec{r}}{\delta x} = i, \quad \frac{\delta \vec{r}}{\delta y} = j, \quad \frac{\delta \vec{r}}{\delta z} = k$

Also $r = |\vec{r}| = \sqrt{x^2 + y^2 + z^2}$

$$r = x^2 + y^2 + z^2 = \sum x^2$$

Hence $2r\frac{\delta r}{\delta x} = 2x \quad \text{or} \quad \frac{\delta r}{\delta x} = \frac{x}{r}$

Similarly $\frac{\delta r}{\delta y} = \frac{y}{r} \quad \text{and} \quad \frac{\delta r}{\delta z} = \frac{z}{r}$

Now, $\text{div}(\vec{r}\,\phi) = \nabla\cdot(\vec{r}\,\phi) = \sum i\frac{\delta}{\delta x}\cdot(\vec{r}\,\phi) = \sum i\frac{\delta}{\delta x}(\vec{r}\,\phi)$

$$= \sum i\left(\frac{\delta \vec{r}}{\delta x}\phi + \vec{r}\frac{\delta \phi}{\delta x}\right) = \sum i\left(i\phi + \vec{r}\frac{\delta \phi}{\delta x}\right) = \sum(i\cdot i)\phi + \sum\vec{r}\cdot i\frac{\delta \phi}{\delta x}$$

$$= \sum\phi + \vec{r}\cdot\sum i\frac{\delta \phi}{\delta x} = 3\phi + \vec{r}\cdot\text{grad}\,\phi$$

**Example 8.22 :** *Find div* **F** *and curl* **F**, *where* **F** = *grad* $(x^3 + y^3 + z^3 - 3xyz)$.

(*JNTUK, 2022 ; V.T.U., 2008 ; Kurukshetra, 2006*)

**Solution.** If $u = x^3 + y^3 + z^3 - 3xyz$, then

$$\mathbf{F} = \nabla u = \mathbf{I}\frac{\partial u}{\partial x} + \mathbf{J}\frac{\partial u}{\partial y} + \mathbf{K}\frac{\partial u}{\partial z}$$

$$= \mathbf{I}(3x^2 - 3yz) + \mathbf{J}(3y^2 - 3zx) + \mathbf{K}(3z^2 - 3xy)$$

$$\therefore \quad \text{div } \mathbf{F} = \frac{\partial}{\partial x}(3x^2 - 3yz) + \frac{\partial}{\partial y}(3y^2 - 3zx) + \frac{\partial}{\partial z}(3z^2 - 3xy) = 6(x + y + z)$$

$$\text{curl } \mathbf{F} = \begin{vmatrix} \mathbf{I} & \mathbf{J} & \mathbf{K} \\ \frac{\partial}{\partial x} & \frac{\partial}{\partial y} & \frac{\partial}{\partial z} \\ 3(x^2 - yz) & 3(y^2 - zx) & 3(z^2 - xy) \end{vmatrix}$$

$$= \mathbf{I}(-3x + 3x) - \mathbf{J}(-3y + 3y) + \mathbf{K}(-3z + 3z) = \mathbf{0}.$$

## 8.7 (1) PHYSICAL INTERPRETATION OF DIVERGENCE

Consider the motion of the fluid having velocity $\mathbf{V} = v_x\mathbf{I} + v_y\mathbf{J} + v_z\mathbf{K}$ at a point $P(x, y, z)$. Consider a small parallelopiped with edges $\delta x$, $\delta y$, $\delta z$ parallel to the axes in the mass of fluid, with one of its corners at $P$ (Fig. 8.9).

$\therefore$ the amount of fluid entering the face $PB'$ in unit time $= v_y \delta z \delta x$ and the amount of fluid leaving the face $P'B$ in unit time

$$= v_{y+\delta y}\delta z \delta x = \left(v_y + \frac{\partial v_y}{\partial y}\delta y\right)\delta z \delta x \quad \text{nearly}$$

$\therefore$ the net decrease of the amount of fluid due to flow across these two faces $= \dfrac{\partial v_y}{\partial y}\delta x \delta y \delta z$.

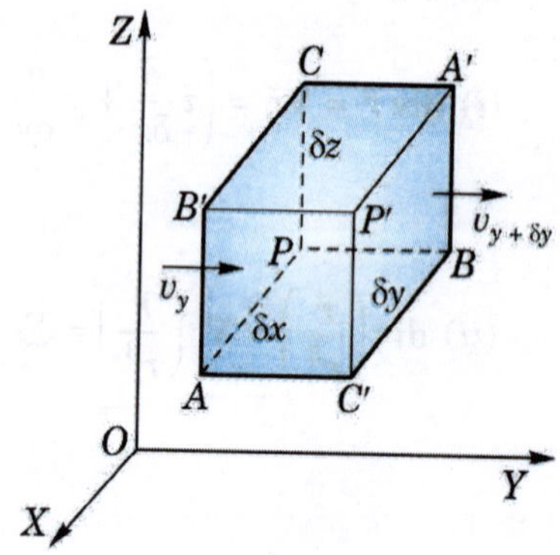

Fig. 8.9.

Finding similarly the contributions of other two pairs of faces, we have the total decrease of amount of fluid inside the parallelopiped per unit time $= \left(\dfrac{\partial v_x}{\partial x} + \dfrac{\partial v_y}{\partial y} + \dfrac{\partial v_z}{\partial z}\right)\delta x \delta y \delta z$.

Thus the rate of loss of fluid per unit volume

$$= \frac{\partial v_x}{\partial x} + \frac{\partial v_y}{\partial y} + \frac{\partial v_z}{\partial z} = \text{div } \mathbf{V}.$$

Hence *div* **V** *gives the rate at which fluid is originating at a point per unit volume.*

Similarly, if **V** represents an electric flux, div **V** is the amount of flux which diverges per unit volume in unit time. If **V** represents heat flux, div **V** is the rate at which heat is issuing from a point per unit volume. In general, the divergence of a vector point function representing any physical quantity gives at each point, the rate per unit volume at which the physical quantity is issuing from that point. This explains the justification for the name *divergence of a vector point function.*

If the fluid is incompressible, there can be no gain or loss in the volume element. Hence div **V** = 0, which is known in Hydrodynamics as the **equation of continuity** for incompressible fluids.

**Def.** *If the flux entering any element of space is the same as that leaving it, i.e., div* **V** = *0 everywhere then such a point function is called a* **solenoidal vector function.**

**(2) Physical interpretation of curl.** Consider the motion of a rigid body rotating about a fixed axis through $O$. If $\Omega$ be its angular velocity, then the velocity **V** of any particle $P(\mathbf{R})$ of the body is given by $\mathbf{V} = \Omega \times \mathbf{R}$.

If $\qquad \Omega = \omega_1\mathbf{I} + \omega_2\mathbf{J} + \omega_3\mathbf{K}$

and $\qquad \mathbf{R} = x\mathbf{I} + y\mathbf{J} + z\mathbf{K}$

then
$$\mathbf{V} = \Omega \times \mathbf{R} = \begin{vmatrix} \mathbf{I} & \mathbf{J} & \mathbf{K} \\ \omega_1 & \omega_2 & \omega_3 \\ x & y & z \end{vmatrix}$$
$$= \mathbf{I}(\omega_2 z - \omega_3 y) + \mathbf{J}(\omega_3 x - \omega_1 z) + \mathbf{C}(\omega_1 y - \omega_2 x)$$

$\therefore$
$$\text{curl } \mathbf{V} = \begin{vmatrix} \mathbf{I} & \mathbf{J} & \mathbf{K} \\ \dfrac{\partial}{\partial x} & \dfrac{\partial}{\partial y} & \dfrac{\partial}{\partial z} \\ \omega_2 z - \omega_3 y, & \omega_3 x - \omega_1 z, & \omega_1 y - \omega_2 x \end{vmatrix}$$
$$= \mathbf{I}(\omega_1 + \omega_1) + \mathbf{J}(\omega_2 + \omega_2) + \mathbf{K}(\omega_3 + \omega_3) \qquad [\because \ \omega_1, \omega_2, \omega_3 \text{ are constants.}]$$
$$= 2(\omega_1\mathbf{I} + \omega_2\mathbf{J} + \omega_3\mathbf{K}) = 2\Omega.$$

Hence $\Omega = \dfrac{1}{2}\text{curl } \mathbf{V}$

Thus the angular velocity of rotation at any point is equal to half the curl of the velocity vector which justifies the name *rotation* used for curl.

In general, the curl of any vector point function gives the measure of the angular velocity at any point of the vector field.

**Def.** *Any motion in which the curl of the velocity vector is zero is said to be* **irrotational,** *otherwise* **rotational.**

**Example 8.23** *Prove that div* $(r^n\mathbf{R}) = (n + 3)\, r^n$. *Hence show that* $\mathbf{R}/r^3$ *is solenoidal.*

**Solution.** We have $\mathbf{R} = x\mathbf{I} + y\mathbf{J} + z\mathbf{K}$ and $r = \sqrt{(x^2 + y^2 + z^2)}$

$\therefore$
$$div\ (r^n\mathbf{R}) = \nabla \,.\, (x^2 + y^2 + z^2)^{n/2}\, (x\mathbf{I} + y\mathbf{J} + z\mathbf{K})$$
$$= \frac{\partial}{\partial x}[x(x^2 + y^2 + z^2)^{n/2}] + \frac{\partial}{\partial y}[y(x^2 + y^2 + z^2)^{n/2}] + \frac{\partial}{\partial z}[z(x^2 + y^2 + z^2)^{n/2}]$$
$$= \Sigma\left\{1\,.\,(x^2 + y^2 + z^2)^{n/2} + x\,.\,\frac{n}{2}(x^2 + y^2 + z^2)^{\frac{n}{2}-1}\,.\,2x\right\}$$
$$= \Sigma r^n + n\,\Sigma x^2 (x^2 + y^2 + z^2)^{\frac{n}{2}-1} = 3r^n + nr^2\,.\,r^{n-2}$$

Thus $div\ (r^n\mathbf{R}) = (n + 3)\, r^n$

When $n = -3$, $div\ (\mathbf{R}/r^3) = 0$ *i.e.,* $\mathbf{R}/r^3$ is solenoidal.

**Example 8.24.** *Show that* $r^\alpha\mathbf{R}$ *is any irrotational vector for any value of* $\alpha$ *but is solenoidal if* $\alpha + 3 = 0$ *where* $\mathbf{R} = x\mathbf{I} + y\mathbf{J} + z\mathbf{K}$ *and r is the magnitude of* $\mathbf{R}$.

**Solution.** Let $\mathbf{A} = r^\alpha\mathbf{R} = (x^2 + y^2 + z^2)^{\alpha/2}\, (x\mathbf{I} + y\mathbf{J} + z\mathbf{K}) = \Sigma x\, (x^2 + y^2 + z^2)^{\alpha/2}\, \mathbf{I}$

$\therefore$
$$\text{curl } \mathbf{A} = \begin{vmatrix} \mathbf{I} & \mathbf{J} & \mathbf{K} \\ \dfrac{\partial}{\partial x} & \dfrac{\partial}{\partial y} & \dfrac{\partial}{\partial z} \\ x(x^2 + y^2 + z^2)^{\alpha/2} & y(x^2 + y^2 + z^2)^{\alpha/2} & z(x^2 + y^2 + z^2)^{\alpha/2} \end{vmatrix}$$
$$= \Sigma\, \mathbf{I}\left\{\frac{\alpha z}{2}(x^2 + y^2 + z^2)^{\alpha/2-1}\,(2y) - \frac{\alpha y}{2}(x^2 + y^2 + z^2)^{\alpha/2-1}\,.\,2z\right\} = 0$$

Hence $\mathbf{A}$ is irrotational for any value of $\alpha$.

But $\qquad div\ \mathbf{A} = \nabla\,.\,(r^\alpha\,\mathbf{R}) = (\alpha + 3)\, r^\alpha$

which is zero for $\alpha + 3 = 0$, *i.e.,* $\mathbf{A}$ is solenoidal if $\alpha + 3 = 0$.

## 8.8 DEL APPLIED TWICE TO POINT FUNCTIONS

$\nabla f$ and $\nabla \times \mathbf{F}$ being vector point functions, we can form their divergence and curl whereas $\nabla . \mathbf{F}$ being a scalar point function, we can have its gradients only. Thus we have the following five formulae :

**(1)** *div grad* $f = \nabla^2 f = \dfrac{\partial^2 f}{\partial x^2} + \dfrac{\partial^2 f}{\partial y^2} + \dfrac{\partial^2 f}{\partial z^2}$

**(2)** *curl grad* $f = \nabla \times \nabla f = \mathbf{0}$

**(3)** *div curl* $\mathbf{F} = \nabla . \nabla \times \mathbf{F} = 0$

**(4)** *curl curl* $\mathbf{F} =$ *grad div* $\mathbf{F} - \nabla^2 \mathbf{F}$, *i.e.*, $\nabla \times (\nabla \times \mathbf{F}) = \nabla(\nabla . \mathbf{F}) - \nabla^2 \mathbf{F}$

**(5)** *grad div* $\mathbf{F} =$ *curl curl* $\mathbf{F} + \nabla^2 \mathbf{F}$, *i.e.*, $\nabla(\nabla . \mathbf{F}) = \nabla \times (\nabla \times \mathbf{F}) + \nabla^2 \mathbf{F}$.

*Proofs.* (1) $\nabla^2 f = \nabla . \nabla f = \nabla . \left( \mathbf{I} \dfrac{\partial f}{\partial x} + \mathbf{J} \dfrac{\partial f}{\partial y} + \mathbf{K} \dfrac{\partial f}{\partial z} \right)$

$$= \frac{\partial}{\partial x}\left(\frac{\partial f}{\partial x}\right) + \frac{\partial}{\partial y}\left(\frac{\partial f}{\partial y}\right) + \frac{\partial}{\partial z}\left(\frac{\partial f}{\partial z}\right) = \frac{\partial^2 f}{\partial x^2} + \frac{\partial^2 f}{\partial y^2} + \frac{\partial^2 f}{\partial z^2} = \left(\frac{\partial^2}{\partial x^2} + \frac{\partial^2}{\partial y^2} + \frac{\partial^2}{\partial z^2}\right) f$$

$\nabla^2 = \dfrac{\partial^2}{\partial x^2} + \dfrac{\partial^2}{\partial y^2} + \dfrac{\partial^2}{\partial z^2}$ is called the *Laplacian operator* and $\nabla^2 f = 0$ is called the *Laplace's equation.*

$$\text{(2)}\quad \nabla \times \nabla f = \nabla \times \left( \mathbf{I} \frac{\partial f}{\partial x} + \mathbf{J} \frac{\partial f}{\partial y} + \mathbf{K} \frac{\partial f}{\partial z} \right) = \begin{vmatrix} \mathbf{I} & \mathbf{J} & \mathbf{K} \\ \frac{\partial}{\partial x} & \frac{\partial}{\partial y} & \frac{\partial}{\partial z} \\ \frac{\partial f}{\partial x} & \frac{\partial f}{\partial y} & \frac{\partial f}{\partial z} \end{vmatrix} = \Sigma \mathbf{I} \left( \frac{\partial^2 f}{\partial y \partial z} - \frac{\partial^2 f}{\partial z \partial y} \right) = \mathbf{0}$$

(*V.T.U., 2016 ; Calicut, 2013*)

$$\text{(3)}\quad \nabla \cdot \nabla \times \mathbf{F} = \left( \Sigma \mathbf{I} \frac{\partial}{\partial x} \right) \cdot \left( \mathbf{I} \times \frac{\partial \mathbf{F}}{\partial x} + \mathbf{J} \times \frac{\partial \mathbf{F}}{\partial y} + \mathbf{K} \times \frac{\partial \mathbf{F}}{\partial z} \right)$$

$$= \Sigma \mathbf{I} \cdot \left( \mathbf{I} \times \frac{\partial^2 \mathbf{F}}{\partial x^2} + \mathbf{J} \times \frac{\partial^2 \mathbf{F}}{\partial x \partial y} + \mathbf{K} \times \frac{\partial^2 \mathbf{F}}{\partial x \partial z} \right)$$

$$= \Sigma \left( \mathbf{I} \times \mathbf{I} \cdot \frac{\partial^2 \mathbf{F}}{\partial x^2} + \mathbf{I} \times \mathbf{J} \cdot \frac{\partial^2 \mathbf{F}}{\partial x \partial y} + \mathbf{I} \times \mathbf{K} \cdot \frac{\partial^2 \mathbf{F}}{\partial x \partial z} \right) = \Sigma \left( \mathbf{K} \cdot \frac{\partial^2 \mathbf{F}}{\partial x \partial y} - \mathbf{J} \cdot \frac{\partial^2 \mathbf{F}}{\partial x \partial z} \right) = 0.$$

(*V.T.U., 2016 ; U.P.T.U., 2011*)

$$\text{(4)}\quad \nabla \times (\nabla \times \mathbf{F}) = \left( \Sigma \mathbf{I} \frac{\partial}{\partial x} \right) \times \left( \mathbf{I} \times \frac{\partial \mathbf{F}}{\partial x} + \mathbf{J} \times \frac{\partial \mathbf{F}}{\partial y} + \mathbf{K} \times \frac{\partial \mathbf{F}}{\partial z} \right)$$

$$= \Sigma \mathbf{I} \times \left( \mathbf{I} \times \frac{\partial^2 \mathbf{F}}{\partial x^2} + \mathbf{J} \times \frac{\partial^2 \mathbf{F}}{\partial x \partial y} + \mathbf{K} \times \frac{\partial^2 \mathbf{F}}{\partial x \partial z} \right)$$

$$= \Sigma \left[ \left\{ \left( \mathbf{I} \cdot \frac{\partial^2 \mathbf{F}}{\partial x^2} \right) \mathbf{I} - (\mathbf{I} \cdot \mathbf{I}) \frac{\partial^2 \mathbf{F}}{\partial x^2} \right\} + \left\{ \left( \mathbf{I} \cdot \frac{\partial^2 \mathbf{F}}{\partial x \partial y} \right) \mathbf{J} - (\mathbf{I} \cdot \mathbf{J}) \frac{\partial^2 \mathbf{F}}{\partial x \partial y} \right\} + \left\{ \left( \mathbf{I} \cdot \frac{\partial^2 \mathbf{F}}{\partial x \partial z} \right) \mathbf{K} - (\mathbf{I} \cdot \mathbf{K}) \frac{\partial^2 \mathbf{F}}{\partial x \partial z} \right\} \right]$$

$$= \Sigma \left[ \left( \mathbf{I} \cdot \frac{\partial^2 \mathbf{F}}{\partial x^2} \right) \mathbf{I} + \left( \mathbf{I} \cdot \frac{\partial^2 \mathbf{F}}{\partial x \partial y} \right) \mathbf{J} + \left( \mathbf{I} \cdot \frac{\partial^2 \mathbf{F}}{\partial x \partial z} \right) \mathbf{K} \right] - \Sigma \frac{\partial^2 \mathbf{F}}{\partial x^2}$$

$$= \Sigma \mathbf{I} \frac{\partial}{\partial x} \left( \mathbf{I} \cdot \frac{\partial \mathbf{F}}{\partial x} + \mathbf{J} \cdot \frac{\partial \mathbf{F}}{\partial y} + \mathbf{K} \cdot \frac{\partial \mathbf{F}}{\partial z} \right) - \Sigma \frac{\partial^2 \mathbf{F}}{\partial x^2} = \nabla(\nabla \cdot \mathbf{F}) - \nabla^2 \mathbf{F}.$$

(5) is just another way of writing (4) above. (*Anna., 2016 ; V.T.U., 2014 ; Rohtak, 2012*)

**Obs.** *Interpretation of* $\nabla$ *as a vector according to rules of vector products leads to correct results so far so the repeated application of* $\nabla$ *is concerned.*

*e.g.*,
1. $\nabla . \nabla f = \nabla^2 f$ $\quad (\because \nabla . \nabla = \nabla^2)$
2. $\nabla \times \nabla f = \mathbf{0}$ $\quad (\because \nabla \times \nabla = \mathbf{0})$
3. $\nabla . \nabla \times \mathbf{F} = 0$ $\quad (\because [\nabla \nabla \mathbf{F}] = 0)$
4. $\nabla \times (\nabla \times \mathbf{F}) = \nabla(\nabla . \mathbf{F}) - \nabla^2 \mathbf{F}$ by expanding it as a vector triple product.

## 8.9 DEL APPLIED TO PRODUCTS OF POINT FUNCTIONS

*To prove that*

(1) grad $(fg) = f(\text{grad } g) + g(\text{grad } f)$   *i.e.* $\nabla(fg) = f\nabla g + g\nabla f$.

(2) *div* $(f\mathbf{G}) = (\text{grad } f) \cdot \mathbf{G} + f(\text{div } \mathbf{G})$   *i.e.* $\nabla(f\mathbf{G}) = \nabla f \cdot \mathbf{G} + f\nabla \cdot \mathbf{G}$

(3) *curl* $(f\mathbf{G}) = (\textit{grad } f) \times \mathbf{G} + f(\textit{curl } \mathbf{G})$   *i.e.* $\nabla \times (f\mathbf{G}) = \nabla f \times \mathbf{G} + f\nabla \times \mathbf{G}$

(4) *grad* $(\mathbf{F} \cdot \mathbf{G}) = (\mathbf{F} \cdot \nabla)\mathbf{G} + (\mathbf{G} \cdot \nabla)\mathbf{F} + \mathbf{F} \times \textit{curl } \mathbf{G} + \mathbf{G} \times \textit{curl } \mathbf{F}$

*i.e.,* $\nabla(\mathbf{F} \cdot \mathbf{G}) = (\mathbf{F} \cdot \nabla)\mathbf{G} + (\mathbf{G} \cdot \nabla)\mathbf{F} + \mathbf{F} \times (\nabla \times \mathbf{G}) + \mathbf{G} \times (\nabla \times \mathbf{F})$

(5) *div* $(\mathbf{F} \times \mathbf{G}) = \mathbf{G} \cdot (\textit{curl } \mathbf{F}) - \mathbf{F} \cdot (\textit{curl } \mathbf{G})$   *i.e.,* $\nabla \cdot (\mathbf{F} \times \mathbf{G}) = \mathbf{G} \cdot (\nabla \times \mathbf{F}) - \mathbf{F} \cdot (\nabla \times \mathbf{G})$   *(JNTUK, 2022)*

(6) *curl* $(\mathbf{F} \times \mathbf{G}) = \mathbf{F}(\textit{div } \mathbf{G}) - \mathbf{G}(\textit{div } \mathbf{F}) + (\mathbf{G} \cdot \nabla)\mathbf{F} - (\mathbf{F} \cdot \nabla)\mathbf{G}$

*i.e.,* $\nabla \times (\mathbf{F} \times \mathbf{G}) = \mathbf{F}(\nabla \cdot \mathbf{G}) - \mathbf{G}(\nabla \cdot \mathbf{F}) + (\mathbf{G} \cdot \nabla)\mathbf{F} - (\mathbf{F} \cdot \nabla)\mathbf{G}$

*Proofs* (1) $\nabla(fg) = \Sigma \mathbf{I} \cdot \frac{\partial}{\partial x}(fg) = \Sigma \mathbf{I}\left(f\frac{\partial g}{\partial x} + g\frac{\partial f}{\partial x}\right) = f\Sigma \mathbf{I}\frac{\partial g}{\partial x} + g\Sigma \mathbf{I}\frac{\partial f}{\partial x} = f\nabla g + g\nabla f$

(2) $$\nabla \cdot (f\mathbf{G}) = \Sigma \mathbf{I} \cdot \frac{\partial}{\partial x}(f\mathbf{G}) = \Sigma \mathbf{I} \cdot \left(\frac{\partial f}{\partial x}\mathbf{G} + f\frac{\partial \mathbf{G}}{\partial x}\right)$$

$$= \left(\Sigma \mathbf{I}\frac{\partial f}{\partial x}\right) \cdot \mathbf{G} + f\left(\Sigma \mathbf{I} \cdot \frac{\partial \mathbf{G}}{\partial x}\right) = \nabla f \cdot \mathbf{G} + f\nabla \cdot \mathbf{G}$$ *(V.T.U., 2016)*

(3) $$\nabla \times (f\mathbf{G}) = \Sigma \mathbf{I} \times \frac{\partial}{\partial x}(f\mathbf{G}) = \Sigma \mathbf{I} \times \left(f\frac{\partial \mathbf{G}}{\partial x} + \frac{\partial f}{\partial x}\mathbf{G}\right)$$

$$= f\Sigma \mathbf{I} \times f\frac{\partial \mathbf{G}}{\partial x} + \Sigma \mathbf{I}\frac{\partial f}{\partial x} \times \mathbf{G} = f\nabla \times \mathbf{G} + \nabla f \times \mathbf{G}$$ *(V.T.U. 2016)*

(4) $$\nabla(\mathbf{F} \cdot \mathbf{G}) = \Sigma \mathbf{I}\frac{\partial}{\partial x}(\mathbf{F} \cdot \mathbf{G}) = \Sigma \mathbf{I}\left(\frac{\partial \mathbf{F}}{\partial x} \cdot \mathbf{G} + \mathbf{F} \cdot \frac{\partial \mathbf{G}}{\partial x}\right) = \Sigma \mathbf{I}\frac{\partial \mathbf{F}}{\partial x} \cdot \mathbf{G} + \Sigma \mathbf{I}\left(\mathbf{F} \cdot \frac{\partial \mathbf{G}}{\partial x}\right) \quad ...(i)$$

Now $\mathbf{G} \times \left(\mathbf{I} \times \frac{\partial \mathbf{F}}{\partial x}\right) = \left(\mathbf{G} \cdot \frac{\partial \mathbf{F}}{\partial x}\right)\mathbf{I} - (\mathbf{G} \cdot \mathbf{I})\frac{\partial \mathbf{F}}{\partial x}$

or $$\left(\mathbf{G} \cdot \frac{\partial \mathbf{F}}{\partial x}\right)\mathbf{I} = \mathbf{G} \times \left(\mathbf{I} \times \frac{\partial \mathbf{F}}{\partial x}\right) + (\mathbf{G} \cdot \mathbf{I})\frac{\partial \mathbf{F}}{\partial x}$$

$$\therefore \quad \Sigma\left(\mathbf{G} \cdot \frac{\partial \mathbf{F}}{\partial x}\right)\mathbf{I} = \mathbf{G} \times \Sigma \mathbf{I} \times \frac{\partial \mathbf{F}}{\partial x} + \Sigma(\mathbf{G} \cdot \mathbf{I})\frac{\partial \mathbf{F}}{\partial x} = \mathbf{G} \times (\nabla \times \mathbf{F}) + (\mathbf{G} \cdot \nabla)\mathbf{F} \quad ...(ii)$$

Interchanging **F** and **G**, $\Sigma\left(\mathbf{F} \cdot \frac{\partial \mathbf{G}}{\partial x}\right)\mathbf{I} = \mathbf{F} \times (\nabla \times \mathbf{G}) + (\mathbf{F} \cdot \nabla)\mathbf{G} \quad ...(iii)$

Substituting in Eqn. (*i*) from Eqns.(*ii*) and (*iii*), we get

$$\nabla(\mathbf{F} \cdot \mathbf{G}) = (\mathbf{F} \cdot \nabla)\mathbf{G} + (\mathbf{G} \cdot \nabla)\mathbf{F} + \mathbf{F} \times (\nabla \times \mathbf{G}) + \mathbf{G} \times (\nabla \times \mathbf{F})$$

(5) $$\nabla \cdot (\mathbf{F} \times \mathbf{G}) = \Sigma \mathbf{I} \cdot \frac{\partial}{\partial x}(\mathbf{F} \times \mathbf{G}) = \Sigma \mathbf{I}\left(\frac{\partial \mathbf{F}}{\partial x} \times \mathbf{G} + F \times \frac{\partial \mathbf{G}}{\partial x}\right) = \Sigma \mathbf{I} \cdot \frac{\partial \mathbf{F}}{\partial x} \times \mathbf{G} - \Sigma \mathbf{I} \cdot \left(\frac{\partial \mathbf{G}}{\partial x} \times \mathbf{F}\right)$$

$$= \Sigma\left(I \times \frac{\partial \mathbf{F}}{\partial x}\right) \cdot \mathbf{G} - \Sigma\left(I \times \frac{\partial \mathbf{G}}{\partial x}\right) \cdot \mathbf{F}$$ $[\because \ \mathbf{A} \cdot (\mathbf{B} \times \mathbf{C}) = (A \times B) \cdot C]$

$$= \mathbf{G} \cdot (\nabla \times \mathbf{F}) - \mathbf{F} \cdot (\nabla \times \mathbf{G})$$

(6) $$\nabla \times (\mathbf{F} \times \mathbf{G}) = \Sigma \mathbf{I} \times \frac{\partial}{\partial x}(\mathbf{F} \times \mathbf{G}) = \Sigma \mathbf{I} \times \left(\frac{\partial \mathbf{F}}{\partial x} \times \mathbf{G} + \mathbf{F} \times \frac{\partial \mathbf{G}}{\partial x}\right)$$

$$= \Sigma\left[(\mathbf{I} \cdot \mathbf{G})\frac{\partial \mathbf{F}}{\partial x} - \left(\mathbf{I} \cdot \frac{\partial \mathbf{F}}{\partial x}\right)\mathbf{G}\right] + \Sigma\left[\left(\mathbf{I} \cdot \frac{\partial \mathbf{G}}{\partial x}\right)\mathbf{F} - (\mathbf{I} \cdot \mathbf{F})\frac{\partial \mathbf{G}}{\partial x}\right]$$

$$= \Sigma(\mathbf{G} \cdot \mathbf{I})\frac{\partial \mathbf{F}}{\partial x} - \mathbf{G}\Sigma \mathbf{I} \cdot \frac{\partial \mathbf{F}}{\partial x} + \mathbf{F}\Sigma \mathbf{I} \cdot \frac{\partial \mathbf{G}}{\partial x} - \Sigma(\mathbf{F} \cdot \mathbf{I})\frac{\partial \mathbf{G}}{\partial x}$$

$$= \mathbf{F}\left(\Sigma \mathbf{I} \cdot \frac{\partial \mathbf{G}}{\partial x}\right) - \mathbf{G}\Sigma\left(\mathbf{I} \cdot \frac{\partial \mathbf{F}}{\partial x}\right) + \Sigma(\mathbf{G} \cdot \mathbf{I})\frac{\partial \mathbf{F}}{\partial x} - \Sigma(\mathbf{F} \cdot \mathbf{I})\frac{\partial \mathbf{G}}{\partial x}$$

$$= \mathbf{F}(\nabla \cdot \mathbf{G}) - \mathbf{G}(\nabla \cdot \mathbf{F}) + (\mathbf{G} \cdot \nabla)\mathbf{F} - (\mathbf{F} \cdot \nabla)\mathbf{G}$$ *(P.T.U., 2010)*

**Rule to reproduce the above formulae easily :**

(*i*) *Treating each of the factors as constants separately, express the results of ∇-operation as a sum of the two terms.*

(*ii*) *Transform each of the two terms, noting that ∇ always appears before a function and keeping in mind whether the result of operation is a scalar or a vector.* To carry out the simplification, we may sometimes, employ the properties of triple products.

(*iii*) *Restore the change of treating the functions as constants.*

Let us illustrate the application of this rule to (**2**), (**4**) and (**6**) above :

(**2**) $\nabla \cdot (f\mathbf{G}) = \nabla \cdot (f_c\,\mathbf{G} + f\,\mathbf{G}_c) = f_c\,\nabla \cdot \mathbf{G} + \mathbf{G}_c \cdot \nabla f = f\,\nabla \cdot \mathbf{G} + \mathbf{G} \cdot \nabla f$

(**4**) $\nabla(\mathbf{F} \cdot \mathbf{G}) = \nabla(\mathbf{F}_c \cdot \mathbf{G}) + \nabla(\mathbf{F} \cdot \mathbf{G}_c)$

$$= [\mathbf{F}_c \times (\nabla \times \mathbf{G}) + (\mathbf{F}_c \cdot \nabla)\mathbf{G}] + [\mathbf{G}_c \times (\nabla \times \mathbf{F}) + (\mathbf{G}_c \cdot \nabla)\mathbf{F}]$$

$$= \mathbf{F} \times (\nabla \times \mathbf{G}) + (\mathbf{F} \cdot \nabla)\,\mathbf{G} + \mathbf{G} \times (\nabla \times \mathbf{F}) + (\mathbf{G} \cdot \nabla)\mathbf{F}$$

(**6**) $\nabla \times (\mathbf{F} \times \mathbf{G}) = \nabla \times (\mathbf{F}_c \times \mathbf{G}) + \nabla \times (\mathbf{F} \times \mathbf{G}_c) = [\nabla \cdot \mathbf{G}\mathbf{F}_c - (\mathbf{F}_c \cdot \nabla)\mathbf{G}] + (\mathbf{G}_c \cdot \nabla)\,\mathbf{F} - \nabla \cdot \mathbf{F}\mathbf{G}_c]$

$$= \mathbf{F}(\nabla \cdot \mathbf{G}) - (\mathbf{F} \cdot \nabla)\,\mathbf{G} + (\mathbf{G} \cdot \nabla)\mathbf{F} - \mathbf{G}(\nabla \cdot \mathbf{F}).$$

**Example 8.25.** *Show that div (grad $r^n$) = $\nabla^2(r^n) = n(n+1)r^{n-2}$ in particular $\nabla^2(1/r) = 0$*

*(JNTUK, 2022 ; Kurukshetra, 2013 ; D.T.U., 2012 ; U.K.T.U., 2012)*

**Solution.** $\nabla^2 r^n = \nabla \cdot (\nabla r^n)$

$$= \nabla \cdot \left(nr^{n-1}\frac{\mathbf{R}}{r}\right) = n\nabla \cdot (r^{n-2}\mathbf{R}) = n[(\nabla r^{n-2}) \cdot \mathbf{R} + r^{n-2}(\nabla \cdot \mathbf{R})] \qquad \text{[By § 8.9 (2)]}$$

$$= n\left[(n-2)r^{n-3}\frac{\mathbf{R}}{r} \cdot \mathbf{R} + r^{n-2}\,(3)\right] \qquad \text{[Using Ex. 8.18 (i)]}$$

$$= n[(n-2)r^{n-4}(r^2) + 3r^{n-2}] = n(n+1)\,r^{n-2} \qquad [\because\ \mathbf{R} \cdot \mathbf{R} = r^2]$$

*Otherwise :* $\nabla^2(r^n) = \dfrac{\partial^2(r^n)}{\partial x^2} + \dfrac{\partial^2(r^n)}{\partial y^2} + \dfrac{\partial^2(r^n)}{\partial z^2}$ [By § 8.8 (1)] ...(*i*)

Now $\dfrac{\partial(r^n)}{\partial x} = nr^{n-1}\dfrac{\partial r}{\partial x} = nr^{n-1}\dfrac{x}{r} = nr^{n-2}x$ $[\because\ r^2 = x^2 + y^2 + z^2]$

$$\therefore \quad \frac{\partial^2(r^n)}{\partial x^2} = n\left[r^{n-2} + (n-2)r^{n-3}\frac{\partial r}{\partial x}x\right] = n\left[r^{n-2} + (n-2)r^{n-3}\frac{x}{r}x\right]$$

$$= n\left[r^{n-2} + (n-2)r^{n-4}\,x^2\right] \qquad ...(ii)$$

Similarly, $\dfrac{\partial^2(r^n)}{\partial y^2} = n\left[r^{n-2} + (n-2)r^{n-4}\,y^2\right]$ ...(*iii*)

$$\frac{\partial^2(r^n)}{\partial z^2} = n\left[r^{n-2} + (n-2)r^{n-4}\,z^2\right] \qquad ...(iv)$$

Adding Eqn. (*ii*), Eqn. (*iii*) and (*iv*), Eqn. (*i*) gives

$$\nabla^2(r^n) = n\,[3r^{n-2} + (n-2)r^{n-4}(x^2 + y^2 + z^2)]$$

$$= n\,[3r^{n-2} + (n-2)r^{n-4}r^2] = n(n+1)r^{n-2}.$$

In particular $\nabla^2(1/r) = 0$.

**Example 8.26.** *If u***F** *= ∇v, where u, v are scalar fields and* **F** *is a vector field, show that* **F***. curl* **F** *= 0.*

*(Calicut, 2012)*

**Solution.** Since $\mathbf{F} = \dfrac{1}{u}\nabla v \quad \therefore \quad \text{curl}\,\mathbf{F} = \nabla \times \left(\dfrac{1}{u}\nabla v\right)$

or $\text{curl}\,\mathbf{F} = \nabla\dfrac{1}{u} \times \nabla v + \dfrac{1}{u}\nabla \times (\nabla v)$ [By § 8.9 (3)]

$$= \nabla\frac{1}{u} \times \nabla v \qquad [\because\ \nabla \times \nabla v = 0$$

Hence $\mathbf{F} \cdot \text{curl}\,\mathbf{F} = \dfrac{1}{u}\nabla v \cdot \left(\nabla\dfrac{1}{u} \times \nabla v\right) = 0$, for it is a scalar triple product in which two factors are equal.

**Example 8.27 :** If $\vec{V} = xy^2 i + 2x^2 y z j - 3y z^2 k$, *find curl* $\vec{V}$ *at the point (1, –1, 1)*

**Solution:**

$$\text{Curl}\,\vec{V} = \nabla \times \vec{U} = \begin{vmatrix} i & j & k \\ \frac{\delta}{\delta x} & \frac{\delta}{\delta y} & \frac{\delta}{\delta z} \\ xy^2 & 2x^2yz & -3yz^2 \end{vmatrix}$$

$$= i\left\{\frac{\delta}{\delta y}(-3yz^2) - \frac{\delta}{\delta z}(2x^2yz)\right\} - i\left\{\frac{\delta}{\delta x}(-3yz^2) - \frac{\delta}{\delta z}(xy^2)\right\} + k\left\{\frac{\delta}{\delta x}(2x^2yz) - \frac{\delta}{\delta y}(xy^2)\right\}$$

$$= i(-3z^2 - 2x^2y) - j(0 - 0) + k(4xy\,y - 2xy)$$

$$= -(i + 2k) \text{ at the point } (1, -1, 1)$$

**Example 8.28.** *If r and* **R** *have their usual meanings and* **A** *is a constant vector, prove that* $\nabla \times \left(\frac{\mathbf{A} \times \mathbf{R}}{r^n}\right) = \frac{2-n}{r^n}\mathbf{A} + \frac{n(\mathbf{A} \cdot \mathbf{R})}{r^{n+2}}\mathbf{R}.$ *(Andhra., 2015 ; V.T.U., 2012)*

**Solution.** $\nabla \times [r^{-n}(\mathbf{A} \times \mathbf{R})] = r^{-n}[\nabla \times (\mathbf{A} \times \mathbf{R})] + \nabla r^{-n} \times (\mathbf{A} \times \mathbf{R})]$ [By § 8.9 (3)]

$$= r^{-n}[(\nabla \cdot \mathbf{R})\mathbf{A} - (\mathbf{A} \cdot \nabla)\mathbf{R}] + (-nr^{-(n+1)}\mathbf{R}/r) \times (\mathbf{A} \times \mathbf{R})$$

$$= r^{-n}(3\mathbf{A} - \mathbf{A}) - nr^{-(n+2)}\mathbf{R} \times (\mathbf{A} \times \mathbf{R}) \qquad [\because \nabla \cdot \mathbf{R} = 3, (\mathbf{A} \cdot \nabla)\mathbf{R} = \mathbf{A}]$$

$$= 2\mathbf{A}r^{-n} - nr^{-(n+2)}[(\mathbf{R} \cdot \mathbf{R})\mathbf{A} - (\mathbf{A} \cdot \mathbf{R})\mathbf{R}]$$

$$= \frac{2\mathbf{A}}{r^n} - \frac{n}{r^{n+2}}[r^2\mathbf{A} - (\mathbf{A} \cdot \mathbf{R})\mathbf{R}] = \frac{2-n}{r^n}\mathbf{A} + \frac{n(\mathbf{A} \cdot \mathbf{R})}{r^{n+2}}\mathbf{R}.$$

**Example 8.29.** *If* $\nabla \cdot \vec{E} = 0$, $\nabla \vec{H} = 0$, $\nabla \times \vec{E} = \frac{-\delta \vec{H}}{\delta t}$, $\nabla \times \vec{H} = \frac{\delta \vec{E}}{\delta t}$, *show that* $\vec{E}$ *and* $\vec{H}$ *satisfy the wave equation* $\nabla^2 u = \frac{\delta^2 u}{\delta t^2}$. *(Calicut, 2019)*

**Solution.** Consider

$$\nabla \times (\nabla \times \vec{E}) = \nabla \times \left(\frac{-\delta \vec{H}}{\delta t}\right) = \frac{-\delta}{\delta t}(\nabla \times \vec{H})$$

$$= \frac{-\delta}{\delta t}\left(\frac{\delta \vec{E}}{\delta t}\right) = -\frac{\delta^2 E}{\delta t^2} \qquad ...(i)$$

Also we have,

$$\nabla \times (\nabla \times \vec{E}) = \nabla(\nabla \cdot \vec{E}) - \nabla^2 \vec{E}$$

$$= -\nabla^2 E \qquad ...(ii)$$

From Eqns. (*i*) and (*ii*),

$$\frac{\delta^2 E}{\delta t^2} = \nabla^2 \vec{E}$$

Similarly,

$$\nabla \times (\nabla \times \vec{H}) = \nabla \times \left(\frac{\delta \vec{E}}{\delta t}\right) = \frac{\delta}{\delta t}(\nabla \times \vec{E})$$

$$= \frac{\delta}{\delta t}\left(\frac{-\delta \vec{H}}{\delta t}\right) = \frac{-\delta^2 \vec{H}}{\delta t^2}$$

and

$$\nabla \times (\nabla \times \vec{H}) = \nabla(\nabla \cdot \vec{H}) - \nabla^2 \vec{H}$$

$$\therefore \qquad \frac{\delta^2 \vec{H}}{\delta t^2} = \nabla^2 \vec{H}$$

Thus $\vec{E}$ and $\vec{H}$ both satisfy the equation

$$\nabla^2 u = \frac{\delta^2 u}{\delta t^2}.$$

**Example 8.30.** *In electromagnetic theory, we have* $\nabla \cdot \mathbf{D} = \rho$, $\nabla \cdot \mathbf{H} = 0$, $\nabla \times D = -\frac{1}{c}\frac{\partial \mathbf{H}}{\partial t}$,

$\nabla \times \mathbf{H} = \frac{1}{c}\left(\rho\mathbf{V} + \frac{\partial \mathbf{D}}{\partial t}\right)$. *Prove that*

(i) $\nabla^2\mathbf{D} - \frac{1}{c^2}\frac{\partial^2 \mathbf{D}}{\partial t^2} = \nabla\rho + \frac{1}{c^2}\frac{\partial}{\partial t}(\rho\mathbf{V})$ (ii) $\nabla^2\mathbf{H} - \frac{1}{c^2}\frac{\partial^2 \mathbf{H}}{\partial t^2} = -\frac{1}{c}\nabla \times \rho\mathbf{V}$

**Solution.** (i) We have $\frac{1}{c^2}\left\{\frac{\partial^2 \mathbf{D}}{\partial t^2} + \frac{\partial}{\partial t}(\rho\mathbf{V})\right\} = \frac{1}{c^2}\frac{\partial}{\partial t}\left(\frac{\partial \mathbf{D}}{\partial t} + \rho\mathbf{V}\right)$

$= \frac{1}{c}\frac{\partial}{\partial t}(\nabla \times \mathbf{H}) = \frac{1}{c}\nabla \times \frac{\partial \mathbf{H}}{\partial t}$ $\left[\because \nabla \times \mathbf{H} = \frac{1}{c}\left(\rho\mathbf{V} + \frac{\partial \mathbf{D}}{\partial t}\right)\right]$

$= -\nabla \times (\nabla \times \mathbf{D})$ $\left[\because \nabla \times \mathbf{D} = -\frac{1}{c}\frac{\partial \mathbf{H}}{\partial t}\right]$

$= -[\nabla(\nabla \cdot \mathbf{D}) - \nabla^2\mathbf{D}]$ [Using § 8.8 (4)]

$= -\nabla\rho + \nabla^2\mathbf{D}$

Hence $\nabla^2\mathbf{D} - \frac{1}{c^2}\frac{\partial^2 \mathbf{D}}{\partial t^2} = \nabla\rho + \frac{1}{c^2}\frac{\partial}{\partial t}(\rho\mathbf{V})$

(ii) L.H.S. $= \nabla^2 H - \frac{1}{c^2}\frac{\partial^2 \mathbf{H}}{\partial t^2} = \nabla^2\mathbf{H} + \frac{1}{c}\frac{\partial}{\partial t}\left(-\frac{1}{c}\frac{\partial \mathbf{H}}{\partial t}\right)$

$= \nabla^2\mathbf{H} + \frac{1}{c}\frac{\partial}{\partial t}(\nabla \times \mathbf{D})$ $\left[\because \nabla \times \mathbf{D} = -\frac{1}{c}\frac{\partial \mathbf{H}}{\partial t}\right]$

$= \nabla^2\mathbf{H} + \frac{1}{c}\left(\nabla \times \frac{\partial \mathbf{D}}{\partial t}\right)$ $\left[\because \nabla \times \mathbf{H} = \frac{1}{c}\left(\rho V + \frac{\partial \mathbf{D}}{\partial t}\right)\right]$

$= \nabla^2\mathbf{H} + \nabla \times \left(\nabla \times \mathbf{H} - \frac{1}{C}\rho\mathbf{V}\right) = \nabla^2\mathbf{H} + \nabla \times (\nabla \times \mathbf{H}) - \frac{1}{c}\nabla \times (\rho\mathbf{V})$

$= \nabla^2\mathbf{H} + \nabla(\nabla \cdot \mathbf{H}) - \nabla^2\mathbf{H} - \frac{1}{c}\nabla \times (\rho\mathbf{V})$, [Using § 8.9 (4)]

$= \nabla(\nabla \cdot \mathbf{H}) - \frac{1}{c}\nabla \times (\rho\mathbf{V})$ $[\because \nabla \cdot \mathbf{H} = 0]$

$= -\frac{1}{c}\nabla \times \rho V =$ R.H.S.

## PROBLEMS 8.4

1. Find the divergence and curl of the vectors
   (i) $\vec{F} = (x^2 + yz)i + (y^2 + zx)j + (z^2 + xy)k$ *(Kerela, 2017)*
   (ii) $\vec{A} = x^2y\, i - 2xy\, j + 2yz\, k$ *(Rajasthan, 2018)*
   (iii) $\vec{V} = (x^2 - y^2)i + 2xyj + (x^2 - xy)k$ *(Rajasthan, 2019)*
2. If $\mathbf{F} = (x + y + 1)\,\mathbf{I} + \mathbf{J} - (x + y)\mathbf{K}$, show that $\mathbf{F}$. curl $\mathbf{F} = 0$. *(V.T.U., 2014)*
3. Find the value of $a$ if the vector $(ax^2y + yz)\,\mathbf{I} + (xy^2 - xz^2)\,\mathbf{J} + (2xyz - 2x^2y^2)\,\mathbf{K}$ has zero divergence. Find the curl of the above vector which has zero divergence.
4. Show that each of following vectors are solenoidal :
   (i) $(-x^2 + yz)\,\mathbf{I} + (4y - z^2x)\,\mathbf{J} + (2xz - 4z)\,\mathbf{K}$ *(P.T.U., 2010)*
   (ii) $3y^4z^2\mathbf{I} + 4x^3z^2\mathbf{J} + 3x^2y^2\mathbf{K}$ *(U.P.T.U., 2012)*
   (iii) Show that the vector $\vec{V} = (x + 3y)i + (y - 3z)j + (x - 2z)k$, is solenoidal *(Mysore, 2019)*
5. If **A** and **B** are irrotational, prove that **A** × **B** is solenoidal. *(V.T.U., 2016)*

6. If $u = x^2 + y^2 + z^2$ and $\mathbf{V} = x\mathbf{I} + y\mathbf{J} + z\mathbf{K}$, show that div $(u\mathbf{V}) = 5u$.
7. If $\mathbf{R} = x\mathbf{I} + y\mathbf{J} + z\mathbf{K}$ and $r \neq 0$, show that (*i*) $\nabla/(1/r^2) = -2\mathbf{R}/r^4$ ; $\nabla . (\mathbf{R}/r^2) = 1/r^2$

   (*ii*) div $(r^n \mathbf{R}) = (n + 3)r^n$ (*Rohtak, 2012*) (*iii*) curl $(r^n \mathbf{R}) = \mathbf{0}$

   (*iv*) $grad\left(\text{div}\,\dfrac{\mathbf{R}}{r}\right) = -\dfrac{2\mathbf{R}}{r^3}$. (*V.T.U., 2015*)

8. If $\vec{V} = \dfrac{xi + yj + zk}{\sqrt{x^2 + y^2 + z^2}}$ ; show that

   $\nabla \cdot \vec{V} = \dfrac{2}{\sqrt{x^2 + y^2 + z^2}}$ and $\nabla \times \vec{V} = 0$ (*Ranchi, 2019*)

9. Show that the vector field $\vec{A}$, where $\vec{A} = (x^2 - y^2 + x)i - (2xy + y)j$, a irrotational, and find a scalar $\phi$ such that $\vec{A}$ = grad $\phi$

   (*Kottayam, 2019*)

   (*i*) $\nabla^2 f(r) = f''(r) + \dfrac{2}{r} f'(r)$ (*Rohtak, 2012 ; C.S.V.T.U., 2011 S ; U.K.T.U., 2010*)

   (*ii*) $\nabla . (\phi\nabla\psi - \psi\nabla\phi) = \phi\nabla^2\psi - \psi\nabla^2\phi$.

10. If $\mathbf{A}$ is a constant vector and $\mathbf{R} = x\mathbf{I} + y\mathbf{J} + z\mathbf{K}$, prove that

    (*i*) grad $(\mathbf{A} . \mathbf{R}) = \mathbf{A}$ (*ii*) div $(\mathbf{A} \times \mathbf{R}) = 0$

    (*iii*) curl $(\mathbf{A} \times \mathbf{R}) = 2\mathbf{A}$ (*Delhi, 2013 ; P.T.U., 2010*) (*iv*) curl $[(\mathbf{A} . \mathbf{R})\mathbf{R}] = \mathbf{A} \times \mathbf{R}$.

11. Prove that (*i*) $\nabla \mathbf{A}^2 = 2(\mathbf{A} . \nabla)\,\mathbf{A} + 2\mathbf{A} \times (\nabla \times \mathbf{A})$, where $\mathbf{A}$ is a constant vector.

    (*ii*) $\nabla \times (\mathbf{R} \times \mathbf{U}) = \mathbf{R}(\nabla . \mathbf{U}) - 2\mathbf{U} - (\mathbf{R} . \nabla)\mathbf{U}$.

12. A vector field is given by $\vec{A} = (x^2 + xy^2)i + (y^2 + x^2y)j$. Show that the field is irrotational, and find the scalar potential. (*Dehradun, 2018*)
13. If $f = (x^2 + y^2 + z^2)^{-n}$, find div grad $f$ and determine $n$ if div grad $f = 0$. (*C.S.V.T.U., 2014*)
14. Find the constant $a, b, c$ so that $\vec{F} = (x + 2y + az)i + (bx - 3y - z)j + (4x + cy + 2z)k$ is irrotational. If $\vec{F}$ grad $\phi$, show that

    $\phi = \dfrac{x^2}{2} - \dfrac{3y^2}{2} + z^2 + 2xy + 4xz - yz$. (*Bhilai, 2019*)

15. If $u = x^2yz$, $v = xy - 3z^2$, find (*i*) $\nabla(\nabla u . \nabla v)$ ; (*ii*) $\nabla . (\nabla u \times \nabla v)$.
16. Find the directional derivative of $\nabla . (\nabla\phi)$ at the point $(1, -2, 1)$ in the direction of the normal to the surface $xy^2z = 3x + z^2$, where $\phi = 2x^3y^2z^4$.
17. Show that

    (*i*) $\nabla \times \left(\dfrac{\vec{a} \times \vec{r}}{r^n}\right) = \left(\dfrac{2-n}{r^n}\right)\vec{a} + nr^{-(n+2)}(\vec{a}\cdot\vec{r})\vec{r}$ (*ii*) $\nabla \cdot \left(\dfrac{\vec{a} \times \vec{r}}{r^n}\right) = 0$

    (*iii*) $\nabla\left(\dfrac{\vec{a}\cdot\vec{r}}{r^n}\right) = \dfrac{\vec{a}}{r^n} - \dfrac{n(\vec{a}\cdot\vec{r})\vec{r}}{r^{n+2}}$ (*Marathwada, 2019*)

18. If $\vec{a}$ is constant vector and $\vec{r}$ is the radius vector; show that

    (*i*) $\nabla(\vec{a}\cdot\vec{r}) = \vec{a}$ (*ii*) $\nabla\cdot(\vec{a}\times\vec{r}) = 0$ (*iii*) $\text{Curl}\left[(\vec{a}\cdot\vec{r})\vec{r}\right] = 0$

19. Show that the vector field defined by $\vec{F} = 2xy\,z^3\,i + x^2z^3\,j + 3x^2y\,z^3k$is irrotational. Find scalar potential $u$ such that $\vec{F}$ = grad $u$.
20. Show that if $\phi(x, y, z)$ is any solution of Laplace equation $\nabla^2\phi = 0$, then $\nabla\phi$ is a vector which is both solenoidal and irrotational.
21. Given $\nabla \times \vec{H} = \dfrac{1}{c}\dfrac{\delta\vec{E}}{\delta t}$, $\nabla \times \vec{E} = -\dfrac{1}{c}\dfrac{\delta\vec{H}}{\delta t}$, $\nabla\cdot\vec{H} = 0$, $\nabla\cdot\vec{E} = 4\pi\rho$, where $\rho = \rho(x, y, z)$ and $c$ = constant,

    Show that

(i) $\vec{E} = -\nabla\phi - \frac{1}{c}\frac{\delta \vec{A}}{\delta t}$ (ii) $\vec{H} = \nabla \times \vec{A}$

where $\vec{A}$ and $\phi$ satisfy the equations

$$\nabla \cdot \vec{A} + \frac{1}{c}\frac{\delta\phi}{\delta t} = 0, \quad \nabla^2\phi - \frac{1}{c^2}\frac{\delta^2\phi}{\delta t^2} = -4\pi\rho,$$

$$\nabla^2 A = \frac{1}{c^2}\frac{d^2\vec{A}}{\delta t^2}$$

22. Show that $\nabla^2\phi(r) = \frac{\delta^2\phi}{\delta r^2} + \frac{2}{r}\frac{\delta\phi}{\delta t}$, if $\nabla^2\phi(r) = 0$, show that $\phi(r) = A + \frac{B}{r}$ where $A$ and $B$ are arbitrary constants.

## 8.10 INTEGRATION OF VECTORS

If two vector functions $\mathbf{F}(t)$ and $\mathbf{G}(t)$ be such that

$$\frac{d\mathbf{G}(t)}{dt} = \mathbf{F}(t),$$

then $\mathbf{G}(t)$ is called an integral of $\mathbf{F}(t)$ with respect to the scalar variable $t$ and we write

$$\int \mathbf{F}(t)\, dt = \mathbf{G}(t).$$

If $\mathbf{C}$ be an arbitrary constant vector, we have

$$\mathbf{F}(t) = \frac{d\mathbf{G}(t)}{dt} = \frac{d}{dt}[\mathbf{G}(t) + \mathbf{C}] \quad \text{then} \quad \int \mathbf{F}(t)\, dt = \mathbf{G}(t) + \mathbf{C}$$

This is called the *indefinite integral of* $\mathbf{F}(t)$ and its *definite integral is*

$$\int_a^b \mathbf{F}(t)\, dt = \left[\mathbf{G}(t) + \mathbf{C}\right]_a^b = \mathbf{G}(b) - \mathbf{G}(a).$$

**Example 8.31.** *Given* $\mathbf{R}(t) = 3t^2\,\mathbf{I} + t\mathbf{J} - t^3\mathbf{K}$, *evaluate* $\int_0^1 (R \times d^2\mathbf{R}/dt^2)\, dt$.

**Solution.** $$\frac{d}{dt}\left(\mathbf{R} \times \frac{d\mathbf{R}}{dt}\right) = \frac{d\mathbf{R}}{dt} \times \frac{d\mathbf{R}}{dt} + \mathbf{R} \times \frac{d^2\mathbf{R}}{dt^2} = \mathbf{R} \times \frac{d^2\mathbf{R}}{dt^2}$$

$$\therefore \quad \int \left(\mathbf{R} \times \frac{d^2\mathbf{R}}{dt^2}\right) dt = \mathbf{R} \times \frac{d\mathbf{R}}{dt}$$

$$= (3t^2\mathbf{I} + t\mathbf{J} - t^3\mathbf{K}) \times (6t\mathbf{I} + \mathbf{J} - 3t^2\mathbf{K})$$

$$= \begin{vmatrix} \mathbf{I} & \mathbf{J} & \mathbf{K} \\ 3t^2 & t & -t^3 \\ 6t & 1 & -3t^2 \end{vmatrix}$$

$$= -2t^3\mathbf{I} + 3t^4\mathbf{J} - 3t^2\mathbf{K}$$

Thus $$\int_0^1 \left(\mathbf{R} \times \frac{d^2\mathbf{R}}{dt^2}\right) dt = \left| -2t^3\mathbf{I} + 3t^4\mathbf{J} - 3t^2\mathbf{K} \right|_0^1$$

$$= -2\mathbf{I} + 3\mathbf{J} - 3\mathbf{K}$$

### PROBLEMS 8.5

1. Given $\mathbf{F}(t) = (5t^2 - 3t)\mathbf{I} + 6t^3\mathbf{J} - 7t\,\mathbf{K}$, evaluate $\int_{t=2}^{t=4} \mathbf{F}(t)\,dt$.
2. Find $y(t)$ where $y''(t) = 12t^2\mathbf{I} - 2t\mathbf{J}$, $y(0) = 2\mathbf{I} - 4\mathbf{J}$, $y'(0) = 0$. *(K.T.U., 2016)*
3. If $\dfrac{d^2\mathbf{P}}{dt^2} = 6t\mathbf{I} - 12t^2\mathbf{J} + 4\cos t\mathbf{K}$, find $\mathbf{P}$, Given that $\dfrac{d\mathbf{P}}{dt} = -1 - 3\mathbf{K}$ and $\mathbf{P} = 2\mathbf{I} + \mathbf{J}$, when $t = 0$.
4. The acceleration of a particle at any time $t \geq 0$ is given by $12\cos 2t\mathbf{I} - 8\sin 2t\mathbf{J} + 16t\mathbf{K}$, the velocity and acceleration are initially zero. Find the velocity and displacement at any time.
5. If $\mathbf{R}(t) = \begin{cases} 2\mathbf{I} - \mathbf{J} + 2\mathbf{K} & \text{when } t = 1 \\ 3\mathbf{I} - 2\mathbf{J} + 4\mathbf{K} & \text{when } t = 2, \end{cases}$ show that $\int_1^2 \left(\mathbf{R}\cdot\dfrac{d\mathbf{R}}{dt}\right) dt = 10$.

## 8.11 (1) LINE INTEGRAL

Consider a continuous vector function $\mathbf{F}(\mathbf{R})$ which is defined at each point of curve $C$ in space. Divide $C$ into $n$ parts at the points $A = P_0, P_1, ..., P_{i-1}, P_i ..., P_n = B$ (Fig. 8.10). Let their position vectors be $\mathbf{R}_0, \mathbf{R}_1, ..., \mathbf{R}_{i-1}, \mathbf{R}_i, ..., \mathbf{R}_n$. Let $\mathbf{U}_i$, be the position vector of any point on the arc $P_{i-1}P_i$.

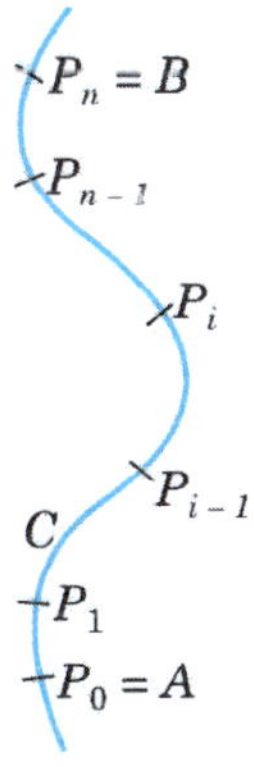

Fig. 8.10.

*Now consider the sum* $S = \sum_{i=0}^{n} \mathbf{F}(\mathbf{U}_i)\,.\,\delta\mathbf{R}_i$, *where* $d\mathbf{R}_i = \mathbf{R}_i - \mathbf{R}_{i-1}$.

*The limit of this sum as* $n \to \infty$ *in such a way that* $|\delta\mathbf{R}_i| \to 0$, *provided it exists, is called the* **tangential line integral** of $\mathbf{F}(\mathbf{R})$ *along* $C$ and is symbolically written as

$$\int_C \mathbf{F}(\mathbf{R})\,.\,d\mathbf{R} \quad \text{or} \quad \int_C \mathbf{F}\,.\,\frac{d\mathbf{R}}{dt}\,dt.$$

When the path of integration is a closed curve, this fact is denoted by using $\oint$ in place of $\int$.

If $\qquad \mathbf{F}(\mathbf{R}) = \mathbf{I}f(x, y, z) + \mathbf{J}\phi(x, y, z) + \mathbf{K}\psi(x, y, z)$

and $\qquad d\mathbf{R} = \mathbf{I}dx + \mathbf{J}dy + \mathbf{K}dz$

then $\qquad \int_C \mathbf{F}(\mathbf{R})\,.\,d\mathbf{R} = \int_C (f\,dx + \phi dy + \psi dz).$

Two other types of line integrals are $\int_C \mathbf{F}\times d\mathbf{R}$ and $\int_C f d\mathbf{R}$ which are both vectors.

(**2**) **Circulation.** *If* $\mathbf{F}$ *represents the velocity of a fluid particle then the line integral* $\int_C \mathbf{F}\,.\,d\mathbf{R}$ *is called the* **circulation of F** *around the curve. When the circulation of* **F** *around every closed curve in a region E vanishes,* **F** *is said to be* **irrotational** *in E.*

(**3**) **Work.** If **F** represents the force acting on a particle moving along an arc $AB$ then the work done during the small displacement $\delta\mathbf{R} = \mathbf{F}\,.\,\delta\mathbf{R}$.

$\therefore$ *the total work done by* **F** *during the displacement from A to B is given by the line integral* $\int_A^B \mathbf{F}.\,d\mathbf{R}$.

**Example 8.32.** *Given that* $\vec{r}(t) = 2i - j + 2k$, *when* $t = 2$, $r = (t) = 4i - 2j + 3k$, *when* $t = 3$

*Show that* $\int_2^3 \vec{r}\cdot\dfrac{\overrightarrow{dr}}{dt}\,dt = 10$. *(CSVTU, 2019)*

**Solution.** $\dfrac{d}{dt}(\vec{r}) = \dfrac{d}{dt}(\vec{r}\cdot\vec{r}) = 2r\cdot\dfrac{\overrightarrow{dr}}{dt}$, we have

$$\int_2^3 \frac{d}{dt}\left(\vec{r}\cdot\frac{\overrightarrow{dr}}{dt}\right)dt = \frac{1}{2} = [\vec{r}\cdot\vec{r}]_2^3 = \frac{1}{2}[\vec{r}(3)\cdot\vec{r}(3) - \vec{r}(2)\cdot\vec{r}(2)]$$

$$= \frac{1}{2}[(4i - 2j + 3k)\cdot(4i - 2j + 3k) - (2i - j + 2k)\cdot(2i - j + 2k)]$$

$$= \frac{1}{2}[(16 + 4 + 9) - (4 + 1 + 4)] = 10$$

**Example 8.33.** *Given that $\vec{r}(t) = 5t^2 i + tj - t^3 k$, then find the value of $\int_1^2 \left(r \times \frac{d^2r}{dt^2}\right) dt$.* *(Marathwada, 2019)*

**Solution.** We know that

$$\int \left(r \times \frac{d^2r}{dt^2}\right) dt = \left(r \times \frac{dr}{dt}\right) dt + c \qquad ...(i)$$

$$\therefore \qquad \int_1^2 \left(r \times \frac{d^2r}{dt^2}\right) dt = \left[\left(r \times \frac{dr}{dt}\right) + c\right]_1^2 = \left[r \times \frac{dr}{dt}\right]_1^2 .$$

Now as given,

$$r(t) = 5t^2 i + tj - t^3 k$$

$$\frac{dr}{dt} = 10ti + j - 3t^2 k$$

$$r \times \frac{dr}{dt} = \begin{vmatrix} i & j & k \\ 5t^2 & t & -t^3 \\ 10t & 1 & -3t^2 \end{vmatrix} = -2t^3\hat{i} + 5t^4\hat{j} - 5t^2\hat{k}$$

using this result, in equation (1) gives

$$\int_1^2 \left(r \times \frac{d^2r}{dt}\right) dt = [-2t^3 i + 5t^4 j - 5t^2 k]_1^2$$

$$= [-2t^3]_1^2 \, i + [5t^4]_1^2 \, j - [5t^2]_1^2 \, k$$

$$= -14i + 75j - 15k$$

**Example 8.34 :** *Evaluate $\vec{r}(t) = \int_0^1 ti + (t^2 - 2t)j + (3t^2 + 3t^3)k \cdot dt$.*

**Solution:** $\vec{r}(t) = \int_0^1 ti + (t^2 - 2t)j + (3t^2 + 3t^3)\, dt$

$$\vec{r}(t) = i\left[\frac{t^2}{2}\right]_0^1 + j\left[\frac{t^3}{3} - t^2\right]_0^1 + k\left[t^3 + \frac{3t^4}{4}\right]_0^1$$

$$= \frac{i}{2} - \frac{2}{3}j + \frac{7}{4}k$$

**Example 8.35 :** *Show that $\vec{F} = (2xy + z^3)\, i + x^2 j + 3z^2 k$ is a conservative field. Find its scalar potential and also the work done in moving a particle from (1, –2, 1) to (3, 1, 4).* *(Marathwada, 2019)*

**Solution :** 
$$\text{Curl } \vec{F} = \begin{vmatrix} i & j & k \\ \frac{\delta}{\delta x} & \frac{\delta}{\delta y} & \frac{\delta}{\delta z} \\ 2xy + z^3 & x^2 & 3z^2 x \end{vmatrix} = i(0) - j(3z^3 - 3z^2) + k(2x - 2x) = 0$$

So, $\vec{F}$ is a conservative field.

Let

$$\vec{F} = \text{grad}\, \phi = i\frac{\delta\phi}{\delta x} + j\frac{\delta\phi}{\delta y} + k\frac{\delta\phi}{\delta z} \text{ then}$$

$$\frac{\delta\phi}{\delta x} = 2xy + z^3, \quad \frac{\delta\phi}{\delta y} = x^2 \text{ and } \frac{\delta\phi}{\delta z} = 3z^2 x$$

Since

$$d\phi = \frac{\delta\phi}{\delta x}dx + \frac{\delta\phi}{\delta y}dy + \frac{\delta\phi}{\delta z}dz$$

$$= (2xy + z^3)\, dx + x^2 dy + 3z^2 x\, dz$$

$$= (2xy\, dx + x^2\, dy) + (z^3\, dx + 3z^2 x \cdot dz)$$

$$= d(x^2 y) + d(z^3 x)$$

on integrating, we have $\phi = x^2y + z^3x + c$,

which is the scalar potential of the vector field $\vec{F}$

$$= \int_A^B \vec{F}\cdot\overline{dr} = [\phi]_A^B = \left[x^2y + z^3 + c\right]_{(1,-2,1)}^{(3,1,4)}$$

$$= (9 + 192) - (-2 + 1) = 202 \text{ units.}$$

**Example 8.36 :** *A vector field is given by $\vec{F} = (\sin y)\,(i + x)\,(1 + \cos y)j$*
*Evaluate the line integral over the circular path $x^2 + y^2 = a^2, z = 0$. (Kurukshetra, 2019)*

**Solution:** Let $x = a\cos t, y = a\sin t, z = 0$, be the parametric equations of the circular path. Then

$$\oint_c \vec{F}\cdot d\vec{r} = \oint_c \left[\sin y i + x(1+\cos y)j\right]\cdot(dxi + dyj)$$

$$= \oint_c \left[\sin y dx + x(1+\cos y)dy\right]$$

$$= \oint_c (\sin y\, dx + x\cos y\, dy) + \oint_c x\, dy$$

$$= \oint_c [d(x\sin y)] + \oint_c x\, dy$$

$$= \int_0^{2\pi}\left[d\left\{a\cos t\sin(a\sin t)\right\}\right] + \int_0^{2\pi} a\cos t\cdot a\cos t\cdot dt$$

$$= \left[a\cos t\sin(a\sin t)\right]_0^{2\pi} + a^2\int_0^{2\pi}\cos^2 t\, dt$$

$$= 0 + a^2(\pi) = \pi a^2$$

## PROBLEMS 8.6

1. Find the arc length parametrization of the line $x = 1 + t, y = 3 - 2t, z = 4 + 2t$, that has the same direction as the given line and has reference point (1, 3, 4). *(K.T.U., 2016)*

2. Evaluate the line integral $\int_C [(x^2 + xy)dx + (x^2 + y^2)dy]$ where $C$ is the square formed by the lines $y = \pm 1$ and $x = \pm 1$.

3. Show that $\iint_c \vec{F}\cdot d\vec{r} = 3\pi$, given that $\vec{F} = zi + xi + yk$ and $c$ being the arc of the curve $\vec{r} = \cos i + \sin tj + tk$ from $t = 0$ to $t = 2\pi$. *(Agra, 2019)*

4. Compute the line integral $\int_C (y^2dx - x^2dy)$ about the triangle whose vertices are (1, 0), (0, 1) and (–1, 0). *(Rohtak, 2011 S)*

5. If $A = (3x^2 + 6y)\mathbf{I} - 14yz\mathbf{J} + 20xz^2\mathbf{K}$, evaluate $\int \mathbf{A}\,.\,d\mathbf{R}$ from (0, 0, 0) to (1, 1, 1) along the path $x = t, y = t^2, z = t^3$. *(C.S.V.T.U., 2014)*

6. Find the work done in moving a particle once round the circle $x^2 + y^2 = 9$ in the $xy$ plane if the field of force is $\vec{F} = (2x - y - z)i + (x + y - z^2)j + (3x - 2y + 4z)k$. Also find its scalar potential. *(Calicut, 2018)*

7. Find the total work done by the force $\mathbf{F} = 3xy\mathbf{I} - y\mathbf{J} + 2zx\mathbf{K}$ in moving a particle around the circle $x^2 + y^2 = 4$. *(V.T.U., 2010)*

8. Find the total work done in moving a particle in a force field given by $\mathbf{F} = 3xy\mathbf{I} - 5z\mathbf{J} + 10x\mathbf{K}$ along the curve $x = t^2 + 1, y = 2t^2, z = t^3$ from $t = 1$ to $t = 2$.

9. Using the line integral, compute the work done by the force $\mathbf{F} = (2y + 3)\mathbf{I} + xz\mathbf{J} + (yz - x)\mathbf{K}$ when it moves a particle from the point (0, 0, 0) to the point (2, 1, 1) along the curve $x = 2t^2, y = t, z = t^3$. *(U.P.T.U., 2011)*

10. Find the value of $\int_c \vec{F}\times\overline{dr}$, along the curve $x = \cos t, y = \sin t, z = 2\cos t$ from $t = 0$ to $t = \frac{\pi}{2}$, given that $\vec{F} = 2yi - zi + zk$ *(Roorkee, 2018)*

11. Find the value of $\int_c \left[(2x^2y + y + z^2)i + 2(1 + yz^3)j + (2z + 3y^2z^2)k\right] d\vec{r}$ along the curve $y^2 + z^2 = a^2$, $x = 0$ (*Poona, 2019*) (**Ans.** 0)

12. The acceleration of particle at any time $t$ is given by $e^t i + e^{2t} j + k$. Find the velocity $\vec{V}$, given that $\vec{V} = i + j$ when $t = 0$.

13. Show that (*i*) $\oint \vec{r} \cdot d\vec{r} = 0$ (*ii*) $\int d\vec{r} = 0$ (*iii*) $\oint \vec{r} \times d\vec{r} = 2\int ds$

14. Find the circulation of $\vec{F}$ around the curve $C$, where $\vec{F} = e^x \sin y\, i + e^x \cos y \cdot j$ and $C$ is the rectangle where vertices are $(0, 0), (1, 0)\ \left(1, \frac{\pi}{2}\right), \left(0, \frac{\pi}{2}\right)$.

## 8.12 (1) SURFACES

As seen in § 5.8, a surface $S$ may be represented by $F(x, y, z) = 0$.

The *parametric representation of* $S$ is of the form $\mathbf{R}(u, v) = x(u, v)\,\mathbf{I} + y(u, v)\,\mathbf{J} + z(u, v)\mathbf{K}$ and the continuous functions $u = \phi(t)$ and $v = \psi(t)$ of a real parameter $t$ represent a curve $C$ on this surface $S$.

For example, the parametric representation of the circular cylinder $x^2 + y^2 = a^2, -1 \le z \le 1$, (radius $a$ and height 2), is

$$\mathbf{R}(u, v) = a \cos u\mathbf{I} + a \sin u\mathbf{J} + v\mathbf{K}$$

where the parameters $u$ and $v$ vary in the rectangle $0 \le u \le 2\pi$ and $-1 \le v \le 1$. Also $u = t, v = bt$ represent a *circular helix* (Fig. 8.3) on this circular cylinder. The equation of the circular helix is $\mathbf{R} = a \cos t\mathbf{I} + a \sin t\mathbf{J} + bt\mathbf{K}$.

Differentiating $\mathbf{R} = \mathbf{R}(u, v)$, w.r.t. $t$, we get $\frac{d\mathbf{R}}{dt} = \frac{\partial \mathbf{R}}{\partial u} \cdot \frac{du}{dt} + \frac{\partial \mathbf{R}}{\partial v} \cdot \frac{dv}{dt}$

The vectors $\partial\mathbf{R}/\partial u$ and $\partial\mathbf{R}/\partial v$ are tangential to $S$ at $P$ and determine the tangent plane of $S$ at $P$. $\mathbf{N} = \partial\mathbf{R}/\partial u \times \partial\mathbf{R}/\partial v\ (\neq 0)$ gives a normal vector $\mathbf{N}$ of $S$ at $P$.

**Def.** *If S has a unique normal at each of its points whose direction depends continuously on the points of S, then the surface S is called a* **smooth surface**. If $S$ is not smooth but can be divided into finitely many smooth portions, then it is called a **piecewise smooth surface.**

For instance, the surface of a sphere is *smooth* while the surface of a cube is *piecewise smooth*.

**Def.** *A surface S is said to be* **orientable or two sided** *if the positive normal direction at any point P of S can be continued in a unique and continuous way to the entire surface*. If the positive direction of the normal is reversed as we move around a curve on $S$ passing through $P$, then the surface is **non-orientable** (*i.e.*, **one-sided**).

An example of a non-orientable surface is the *Mobius strip**. If we take a long rectangular strip of paper and giving a half-twist join the shorter sides so that the two points $A$ and the two points $B$ in Fig. 8.11. coincide, then the surface generated is non-orientable. Such a surface is a model of a Mobius strip.

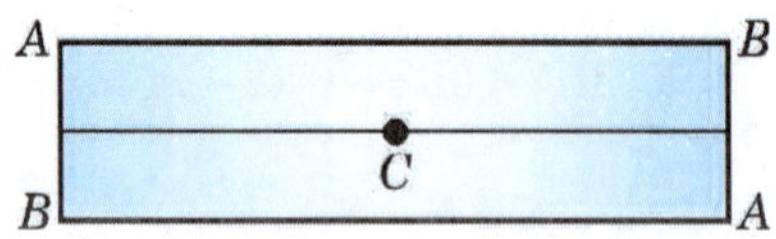

Fig. 8.11.

**(2) Surface integral.** Consider a continuous function $\mathbf{F}(\mathbf{R})$ and a surface $S$. Divide $S$ into a finite number of sub-surfaces. Let the surface element surrounding any point $P(\mathbf{R})$ be $\delta\mathbf{S}$ which can be regarded as a vector ; its magnitude being the area and its direction that of the outward normal to the element.

*Consider the sum* $\Sigma\mathbf{F}(\mathbf{R}) \cdot \delta\mathbf{S}$, *where the summation extends over all the sub-surfaces. The limit of this sum as the number of sub-surfaces tends to infinity and the area of each sub-surface tends to zero, is called the* **normal surface integral** *of* $\mathbf{F}(\mathbf{R})$ *over S and is denoted by*

$$\int_S \mathbf{F} \cdot d\mathbf{S} \quad \text{or} \quad \int_S \mathbf{F} \cdot \mathbf{N}ds \text{ where } \mathbf{N} \text{ is a unit outward normal at } P \text{ to } S.$$

Other types of surface integrals are $\int_S \mathbf{F} \times d\mathbf{S}$ and $\int_S f\, d\mathbf{S}$ which are both vectors.

*Named after a German mathematician *August Ferdinand Möbius (1790–1868)* who was a student of Gauss and professor of astronomy at Leipzig. His important contributions are in projective geometry, theory of surfaces and mechanics.

*Notation* : Only one integrals sign is used when there is one differential (say $d\mathbf{R}$ or $d\mathbf{S}$) and two (or three) signs when there are two (or three) differentials.

(3) **Flux across a surface.** *If* **F** *represent the velocity of a fluid particle then the total outward flux of* **F** *across a closed surface S is the surface integral* $\int_S \mathbf{F} \,.\, d\mathbf{S}$.

*When the flux of* **F** *across every closed surface S in a region E vanishes,* **F** *is said to be a* **solenoidal vector point function** *in E.*

It may be noted that **F** could equally well be taken as any other physical quantity *e.g.*, gravitational force, electric force and magnetic force.

**Example 8.37.** *What is the value of* $\iint_S \vec{A}\,\hat{n}\,ds$ *where* $\vec{A} = 18zi - 12j + 3yk$ *and S is the part of the plane* $2x + 3y + 6z = 12$*, which is in the first octant.*

**Solution.** Let $\phi = 2x + 3y + 6z - 12$, then

$$\text{grad}\,\phi = 2i + 3j + 6k,\; a\hat{n} = \frac{(2i+3j+6k)}{7} \text{ and } \hat{n}\cdot k = \frac{6}{7}$$

$$\vec{A}\cdot\hat{n} = (18zi - 12j + 3yk)\cdot\frac{(2i+3j+6k)}{7}$$

$$= \frac{36z - 36 + 18y}{7} = \frac{36 - 12x}{7}$$

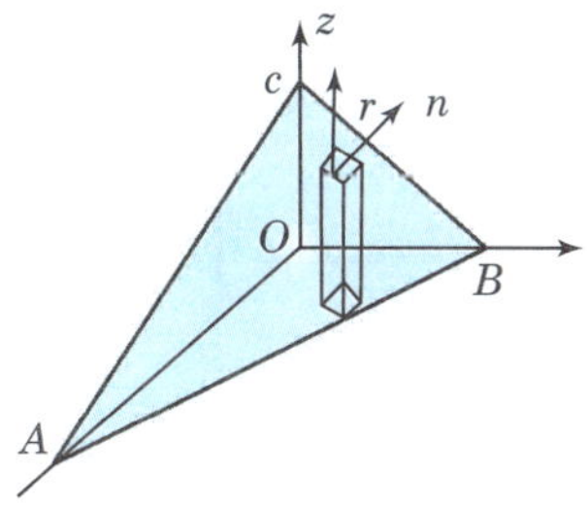

Fig. 8.12.

on using $z = \frac{1}{6}(12 - 2x - 3y)$ from the equation of the plane

we get
$$\iint_S \vec{A}\cdot\hat{n}\,ds = \iint_R \vec{A}\cdot\hat{n}\frac{dx\,dy}{\hat{n}\cdot k}$$

$$= \iint_R\left(\frac{36-12x}{7}\right)\frac{dx\,dy}{6/7} = \iint_R (6 - 2x)dx\,dy$$

where $R$ is the triangle $AOB$ in the $xy$-plane.

Hence we have to evaluate the double integral $AOB$.

$$\iint_S \vec{A}\hat{n}\,ds = \int_{x=0}^{\frac{6(12-2x)}{3}} (6 - 2x)dxdy = \int_{x=0}^{6} (6 - 2x)\left[y\right]_0^{\left(\frac{12-2x}{3}\right)} dx$$

$$= \frac{1}{3}\int_0^6 (6 - 2x)(12 - 2x)dx = \frac{1}{3}\int_0^6 (72 - 36x + 4x^2)\,dx = 24$$

**Example 8.38 :** *Evaluate* $\iint_S \vec{A}\hat{n}\,ds$ *where* $\vec{A} = zi + xj - 3y^2 zk$ *and S is the surface of the cylinder* $x^2 + y^2 = 16$ *in cluded in the first octant between* $z = 0$*, and* $z = 5$. *(Bikaner, 2019)*

**Solution:** Let projection of $S$ surface on the $yz$ plane be $R$, then

$$\iint_S \vec{A}\hat{n}\,ds = \iint_R \vec{A}\cdot\hat{n}\frac{dy\,dz}{\hat{n}\cdot\hat{i}}$$

Here $R$ is the region $OB\,B'O'$ enclosed by $y = 0$ to $y = 4$ and $z = 0$ to $z = 5$

Let $\phi = x^2 + y^2 - 16$. Then $\text{grad}\,\phi = 2xi + 2yj$

and
$$\hat{n} = \frac{2xi + 2yj}{\sqrt{4x^2 + 4y^2}} = \frac{xi + yj}{4}$$

Because on the surface of the cylinder $x^2 + y^2 = 16$.

Now,
$$\vec{A}\cdot\hat{n} = \left(zi + xj - 3y^2 zk\right)\cdot\frac{(xi + yj)}{4} = \frac{xz + xy}{4}$$

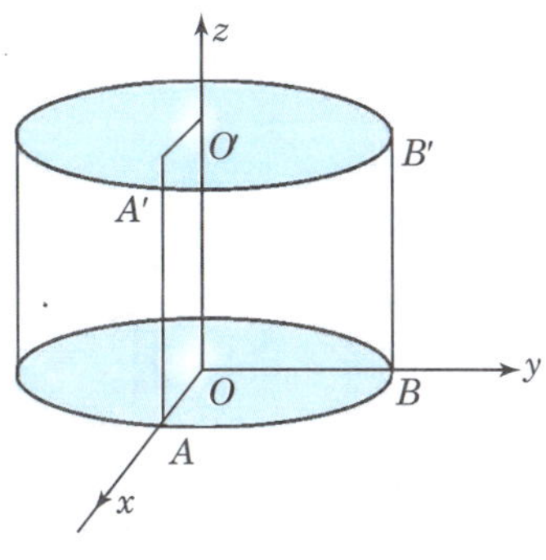

Fig. 8.13

Also $\hat{n}\cdot i = \left(\dfrac{xi + yj}{4}\right) = \dfrac{x}{4}$

$$\therefore \quad \iint_S \vec{A}\hat{n}\, ds = \iint_R \frac{x(y+z)}{4}\frac{dy\cdot dz}{x/4}, \text{ since } ds = \frac{dydz}{\hat{n}\cdot i}$$

$$= \int_{z=0}^{5}\int_{y=0}^{4}(y+z)dydz = \int_0^5\left[\frac{y^2}{2}+2y\right]_0^4 dz$$

$$= \int_0^5 (8+4z)dz = \left[8z + 2z^2\right]_0^5 = 90$$

## PROBLEMS 8.7

1. Show that $\iint_S \vec{F}\,\hat{n}\,ds = \dfrac{3}{2}$ where $\vec{F} = 4xzi - y^2j + yzk$ and $S$ is the surface of the cube bounded by the planes $x = 0,\ x = 1,\ y = 0,\ y = 1,\ z = 0,\ z = 1$ *(Calicut, 2019)*
2. Evaluate $\int_S \mathbf{F}\,.\,d\mathbf{S}$ where $\mathbf{F} = x\mathbf{I} + (z^2 - zx)\,\mathbf{J} - xy\mathbf{K}$ and $S$ is the triangular surface with vertices (2, 0, 0), (0, 2, 0) and (0, 0, 4).
3. Evaluate $\int_S \mathbf{F}\,.\,\mathbf{N}\,ds$ where $\mathbf{F} = 18z\mathbf{I} - 12\mathbf{J} + 3y\mathbf{K}$ and $S$ is the portion of the plane $2x + 3y + 6z = 12$ in the first octant. *(U.P.T.U., 2012)*
4. Prove that $\iint_S \vec{A}\hat{n}\,ds = \iint_R \vec{A}\hat{n}\dfrac{dx\,dy}{|\hat{n}\cdot k|}$ where $R$ is the projection of $S$ on the $x$-$y$ plane. *(Marathwada, 2018)*
5. If $\vec{A} = xyi + z^2j + 2yzk$, Find the the value of $\int_S \vec{A}\hat{n}\,ds$ where $S$ is the surface of tetrahedron bounded by the planes $x = 0, y = 0, z = 0$ and $x + y + z = 1$.
6. Evaluate $\iint_S \vec{A}\vec{n}\,ds$, where $\vec{A} = zi + xj + 3y^2zk$ and $S$ is the surface of the cylinder $x^2 + y^2 = 1$ included in the first octant between $z = 0$ and $z = 2$.

## 8.13 GREEN'S THEOREM IN THE PLANE*

*If $\phi(x, y)$, $\psi(x, y)$, $\phi_y$ and $\psi_x$ be continuous in a region E of the xy-plane bounded by a closed curve C, then*

$$\int_C (\phi dx + \psi dy) = \iint_E \left(\frac{\partial\psi}{\partial x} - \frac{\partial\phi}{\partial y}\right) dxdy \quad ...(1)$$

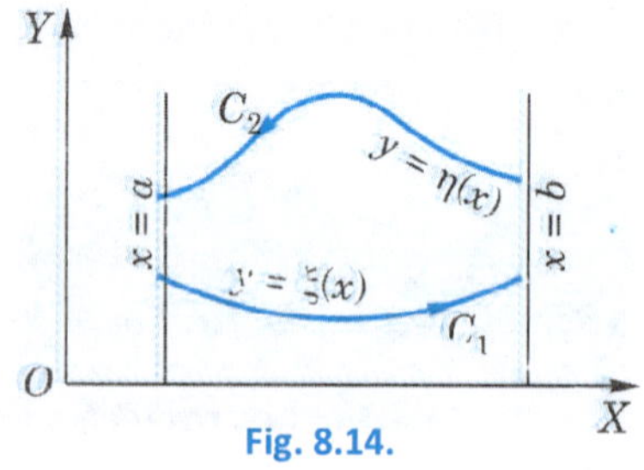

Fig. 8.14.

Consider the region $E$ bounded by a single closed curve $C$ which is cut by any line parallel to the axes at the most in two points.

Let $E$ be bounded by $x = a$, $y = \xi(x)$, $x = b$ and $y = \eta(x)$, where $\eta \geq \xi$, so that $C$ is divided into curves $C_1$ and $C_2$ (Fig. 8.14).

$$\therefore \quad \iint_E \frac{\partial\phi}{\partial y}dxdy = \int_a^b dx\left[\int_\xi^\eta \frac{\partial\phi}{\partial y}dy\right] = \int_a^b dx\,|\,\phi\,|_\xi^\eta$$

$$= \int_a^b [\phi(x,\eta) - \phi(x,\xi)]\,dx = -\int_{C_2}\phi(x,y)\,dx - \int_{C_1}\phi(x,y)\,dx$$

$$= -\int_C \phi(x,y)\,dx \quad ...(2)$$

Similarly, it can be shown that

$$\iint_E \frac{\partial\psi}{\partial x}dxdy = \int_C \psi(x,y)\,dy \quad ...(3)$$

*Named after the English mathematician *George Green* (1793–1841) who taught at Cambridge and is known for his work on potential theory in connection with waves, vibrations, elasticity, electricity and magnetism.

On subtracting Eqn. (2) from Eqn. (3), we get Eqn. (1).

This result can be extended to regions which may be divided into a finite number of sub-regions such that the boundary of each is cut at the most in two points by any line parallel to either axis. Applying Eqn. (1) to each of these sub-regions and adding the results, the surface integrals combine into an integral over the whole region; the line integrals over the common boundaries cancel (for each is covered twice but in opposite directions), whereas the remaining line integrals combine into the line integral over the external curve $C$.

**Obs.** This theorem converts a line integral around a closed curve into a double integral and is a special case of Stoke's theorem.

**Example 8.39.** *Verify Green's theorem for* $\int_C [(xy + y^2)\,dx + x^2 dy]$, *where C is bounded by* $y = x$ *and* $y = x^2$.

*(V.T.U., 2016 ; C.S.V.T.U., 2012 ; Delhi., 2012)*

**Solution.** Here $\phi = xy + y^2$ and $\psi = x^2$

$$\therefore \quad \int_C (\phi\, dx + \psi\, dy) = \int_{C_1} + \int_{C_2}$$

Along $C_1$, $y = x^2$ and $x$ varies from 0 to 1 (Fig. 8.15)

$$\therefore \quad \int_{C_1} = \int_0^1 [\{x(x)^2 + (x^2)^2\}]\,dx + x^2 d(x^2)]$$

$$= \int_0^1 (3x^3 + x^4)\,dx = \frac{19}{20}$$

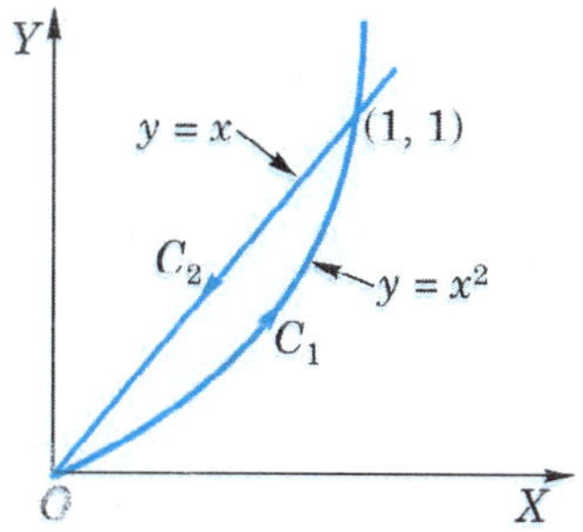

Fig. 8.15.

Along $C_2$, $y = x$ and $x$ varies from 1 to 0.

$$\therefore \quad \int_{C_2} = \int_1^0 [\{x(x) + (x)^2\}\,dx + x^2\, d(x)] = \int_1^0 3x^2\, dx = -1.$$

$$\text{Thus} \quad \int_C (\phi\, dx + \psi\, dy) = \frac{19}{20} - 1 = -\frac{1}{20} \qquad ...(i)$$

$$\text{Also} \quad \iint_E \left(\frac{\partial \psi}{\partial x} - \frac{\partial \phi}{\partial y}\right) dxdy = \iint_E \left[\frac{\partial}{\partial x}(x^2) - \frac{\partial}{\partial y}(xy + y^2)\right] dxdy$$

$$= \int_0^1 \int_{x^2}^x (2x - x - 2y)\, dydx = \int_0^1 \left[xy - y^2\right]_{x^2}^x dx = \int_0^1 (x^4 - x^3)\,dx = -\frac{1}{20} \qquad ...(ii)$$

Hence, Green theorem is verified from the equality of Eqns. $(i)$ and $(ii)$.

**Example 8.40.** *If C is a simple closed curve in the xy-plane not enclosing the origin, show that*

$$\int_C \mathbf{F} \,.\, d\mathbf{R} = 0, \quad \text{where} \quad \mathbf{F} = \frac{y\mathbf{I} - x\mathbf{J}}{x^2 + y^2}$$

$$\textbf{Solution.} \quad \int_C \mathbf{F} \,.\, d\mathbf{R} = \int_C \frac{y\mathbf{I} - x\mathbf{J}}{x^2 + y^2}(dx\mathbf{I} + dy\mathbf{J}) \qquad [\because \mathbf{R} = x\mathbf{I} + y\mathbf{J}]$$

$$= \int_C \frac{ydx - xdy}{x^2 + y^2} = \int_C (\phi dx + \psi dy)$$

where $\phi = \dfrac{y}{x^2 + y^2}$, $\psi = \dfrac{-x}{x^2 y^2}$

$$= \iint_S \left(\frac{\partial \psi}{\partial x} - \frac{\partial \phi}{\partial y}\right) dxdy \qquad \text{[By Green's theorem]}$$

$$= \iint_S \left[\frac{-(x^2 + y^2) + x(2x)}{(x^2 + y^2)^2} - \frac{(x^2 + y^2) - y(2y)}{(x^2 + y^2)^2}\right] dxdy$$

$$= \iint \left[\frac{x^2 - y^2}{(x^2 + y^2)^2} - \frac{x^2 - y^2}{(x^2 + y^2)^2}\right] dxdy = 0.$$

**Example 8.41.** *Using Green's theorem, evaluate* $\int_C [(y - \sin x)\, dx + \cos x\, dy]$ *where C is the plane triangle enclosed by the lines* $y = 0$, $x = \pi/2$ *and* $y = \frac{2}{\pi}x$.

(*C.S.V.T.U., 2014 S ; V.T.U., 2013 ; Rohtak, 2012 ; J.N.T.U., 2005*)

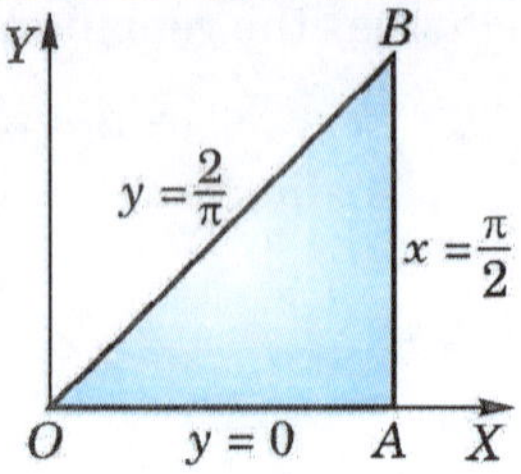

Fig. 8.16.

**Solution.** Here $\phi = y - \sin x$ and $\psi = \cos x$.

By Green's theorem $\int_C [(y - \sin x)\, dx + \cos x\, dy]$

$$= \iint_R \left(\frac{\partial \psi}{\partial x} - \frac{\partial \phi}{\partial y}\right) dx\, dy$$

$$= \int_{x=0}^{x=\pi/2} \int_{y=0}^{y=2x/\pi} (-\sin x - 1)\, dy\, dx = -\int_0^{\pi/2} (\sin x + 1) \Big| y \Big|_0^{2x/\pi} dx$$

$$= -\frac{2}{\pi}\int_0^{\pi/2} x(\sin x + 1)\, dx = -\frac{2}{\pi}\left\{\Big| x(-\cos x + x)\Big|_0^{\pi/2} - \int_0^{\pi/2} 1 . (-\cos x + x)\, dx\right\}$$

$$= -\frac{2}{\pi}\left\{\frac{\pi^2}{4} - \Big| -\sin x + \frac{x^2}{2}\Big|_0^{\pi/2}\right\} = -\frac{\pi}{2} + \frac{2}{\pi}\left(-1 + \frac{\pi^2}{8}\right) = -\left(\frac{\pi}{4} + \frac{2}{\pi}\right)$$

**Example 8.42 :** *Evaluate by Green's theorem* $\int_C e^{-x}(\sin y\, dx + \cos y\, dy)$ *where C is the rectangle with vertices* $(0, 0)\ (\pi, 0)\left(\pi, \frac{\pi}{2}\right), \left(0, \frac{\pi}{2}\right)$

**Solution:** Here $u = e^{-x} \sin y$ and $v = e^{-x} \cos y$

Hence $\frac{\delta u}{\delta y} = e^{-x} \cos y$ and $\frac{\delta v}{\delta x} - e^{-x} \cos y$

Now using Green's theorem, the given integral reduces to

$$= \int_{x=0}^{\pi} \int_{y=0}^{\pi/2} \left(-2e^{-x} \cos y\right) dx\, dy$$

$$= -2[-e^{-x}]_0^{\pi}\ [\sin y]_0^{\pi/2} = 2(e^{-x} - 1)$$

**Example 8.43 :** *If* $\vec{F} = (2x^2 - 3z)\, i - 2xy\, j - 4x\, k$, *then evaluate* $\iiint_V \nabla \cdot \vec{F}\, dV$, *where V is bounded by the plane* $x = 0, y = 0, z = 0$ *and* $2x + zy + z = 4$. (*Puducherry, 2018*)

**Solution:** We have

$$\nabla \cdot \vec{F} = \frac{\delta}{dx}(2x^2 - 3z) + \frac{\delta}{\delta y}(-2xy) + \frac{\delta}{\delta z}(-4x)$$

$$= 4x - 2x + 0 = 2x$$

The limits of integration for $z$ are from 0 to $4 - 2x - 2y$ for $y$ from 0 to $2 - x$ and $x$ from 0 to 2. Hence

$$\iiint_V \nabla\vec{F} \cdot dV = \int_0^2 \int_0^{2-x} \int_0^{4-2x-2y} 2x\, dx\, dy\, dz = 2\int_0^2 \int_0^{2-x} x(4 - 2x - 2y)\, dxdy$$

$$= 2\int_0^2 x\left(4y - 2xy - y^2\right)_0^{2-x} dx = 2\int_0^2 x(2 - x)(4 - 2x - 2 + x)dx$$

$$= 2\int_0^2 \left(4x - 4x^2 + x^3\right) dx = 2\left(2x^2 - \frac{4}{3}x^3 + \frac{x^4}{4}\right)_0^2 = \frac{8}{3}$$

**Example 8.44 :** If $\vec{A} = 2xzi - xj + y^2k$, *evaluate* $\iiint_V \vec{A} \cdot dV$ *where V is the region bounded by the surface. Given* $x = 0, y = 0, x = 2, y = 6, z = x^2, z = 4$.

**Solution:** Now $x$ and $y$ fixed and integrate from $z = x^2$ to $z = 4$ (from $P$ to $Q$ ), then keep $x$ fixed and integrate from $y = 0$ to $y = 6$ (from $R$ to $S$) and finally from $x = 0$ to $x = 2$

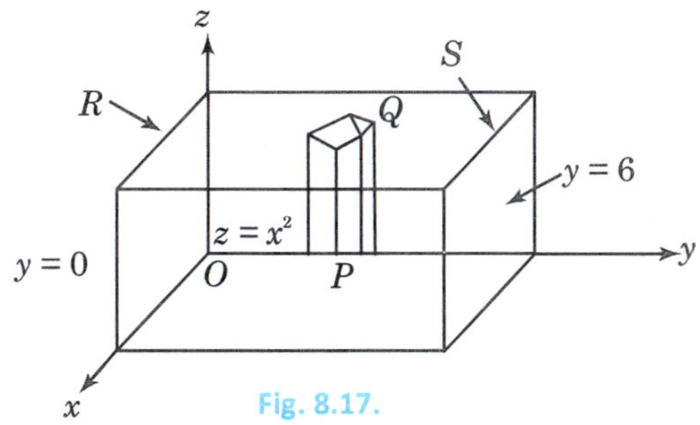

Fig. 8.17.

$$\therefore \quad \iiint_V \vec{A}\, dV = \int_{x=0}^{2}\int_{y=0}^{6}\int_{z=x^2}^{4} (2xz\,i - xj + y^2k)\, dz\, dy\, dx$$

$$= i\int_0^2\int_0^6\int_{x^2}^4 2xz\, dz\, dy\, dx - j\int_0^2\int_0^6\int_{x^2}^4 xz\, dz\, dy\, dx + k\int_0^2\int_0^6\int_{x^2}^4 y^2\, dz\, dy\, dx$$

$$= 128i - 24j + 384k$$

**Example 4.45 :** *Apply Green's theorem to evaluate* $\int_C [(2x^2 - y^2)dx + (x^2 + y^2)dy]$, *where C is the boundary of the area enclosed by the x-axis and the upper-half of the circle* $x^2 + y^2 = a^2$. *(U.P.T.U., 2005)*

**Solution.** By Green's theorem

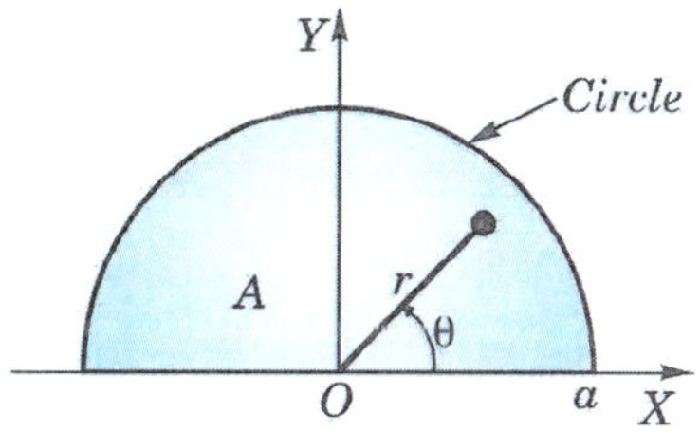

Fig. 8.18.

$$\int_C [(2x^2 - y^2)dx + (x^2 + y^2)dy]$$

$$= \iint_A \left[\frac{\partial}{\partial x}(x^2 + y^2) - \frac{\partial}{\partial y}(2x^2 - y^2)\right] dxdy$$

$$= 2\iint_A (x + y)\, dxdy, \text{ where } A \text{ is the region of Fig. 8.18.}$$

$$= 2\int_0^a\int_0^\pi r\,(\cos\theta + \sin\theta)\cdot r d\theta\, dr$$

[Changing to polar coordinates $(r, \theta)$, $r$ varies from 0 to $a$ and $\theta$ varies from 0 to $\pi$]

$$= 2\int_0^a r^2\, dr \int_0^\pi (\cos\theta + \sin\theta)\, d\theta$$

$$= 2\cdot\frac{a^3}{3}\cdot(1 + 1) = \frac{4a^3}{3}.$$

## PROBLEMS 8.8

1. Verify Green's theorem for $\mathbf{F} = (x^2 + y^2)$, $\boldsymbol{I} - 2xy\mathbf{J}$ taken round the rectangle bounded by the lines $x = \pm a$, $y = 0$ and $y = 6$. *(Anna., 2016)*
2. Verify Green's theorem for $\int_C [(3x - 8y^2)\, dx + (4y - 6xy)\, dy]$ where $C$ is the boundary of the region bounded by $x = 0$, $y = 0$ and $x + y = 1$. *(JNTUK, 2022 ; Calicut, 2013 ; C.S.V.T.U., 2011)*
3. Verify Green's theorem for $\int (x^2 ydx + x^2\, dy)$ where $C$ is the boundary described counter clockwise of triangle with vertices (0, 0), (1, 0), (1, 1). *(U.K.T.U., 2010)*
4. Apply Green's theorem to prove that the area enclosed by a plane curve is $\frac{1}{2}\int_C (xdy - ydx)$. *(W.B.T.U., 2012)*
Hence find the area of an ellipse whose semi-major and semi-minor axes are of lengths $a$ and $b$. *(V.T.U., 2014)*

5. Find the area between the parabola $y^2 = 4x$ and $x^2 = 4y$ with the help of Green's theorem in a plane.(*V.T.U., 2015*)

6. Evaluate $\int_C [(x^2 + xy)dx + (x^2 + y^2)dy]$, where $C$ is the square formed by the lines $x = \pm 1, y = \pm 1$. (*Andhra., 2016*)

7. Using Green's theorem in a plane, evaluate $\int_C [x^2(1+y)dx + (x^3+y^3)dy]$ where $C$ is the square formed by $x = \pm 1$ and $y = \pm 1$. (*Anna., 2016*)

8. Evaluate $\int_C [(x^2 - 2xy)dx + (x^2y + 3)dy]$, around the boundary of the region defined by $y^2 = 8x$ and $x = 2$.

9. Use Green's theorem in a plane to evaluate the integral $\int_C [(2x^2 - y^2)dx + (x^2 + y^2)dy]$, where $C$ is the boundary in the $x$-$y$ plane of the area enclosed by the $x$-axis and the semi-circle $x^2 + y^2 = 1$ in the upper half $x - y$ plane.

## 8.14 STOKE'S THEOREM* (*Relation between line and surface integrals*)

*If S be an open surface bounded by a closed curve C and* $\mathbf{F} = f_1\mathbf{I} + f_2\mathbf{J} + f_3\mathbf{K}$ *be any continuously differentiable vector point function, then*

$$\int_C \mathbf{F} \cdot d\mathbf{R} = \int_S curl\ \mathbf{F} \cdot \mathbf{N} ds$$

*where,* $\mathbf{N} = \cos\alpha\mathbf{I} + \cos\beta\mathbf{J} + \cos\gamma\mathbf{K}$ *is a unit external normal at any point of S.*

Writing $d\mathbf{R} = dx\mathbf{I} + dy\mathbf{J} + dz\mathbf{K}$, it may be reduced to the form

$$\int_C (f_1 dx + f_2 dy + f_3 dz) = \int_S \left[\left(\frac{\partial f_3}{\partial y} - \frac{\partial f_2}{\partial z}\right)\cos\alpha + \left(\frac{\partial f_1}{\partial z} - \frac{\partial f_3}{\partial x}\right)\cos\beta + \left(\frac{\partial f_2}{\partial x} - \frac{\partial f_1}{\partial y}\right)\cos\gamma\right] ds \quad ...(1)$$

Let us first prove that

$$\oint_C f_1 dx = \int_S \left(\frac{\partial f_1}{\partial z}\cos\beta - \frac{\partial f_1}{\partial y}\cos\gamma\right) ds \quad ...(2)$$

Let $z = g(x, y)$ be the equation of the surface $S$ whose projection on the $xy$-plane is the region $E$. Then the projection of $C$ on the $xy$-plane is the curve $C'$ enclosing region $E$.

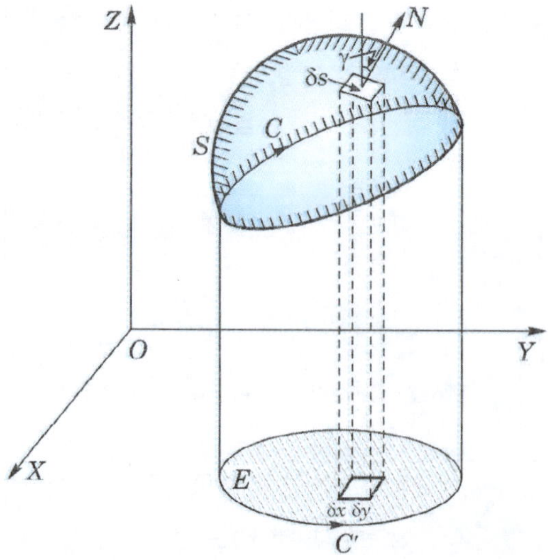

Fig. 8.19.

$$\therefore \quad \int_C f_1(x, y, z)dx = \int_{C'} f_1\{x, y, g(x, y)\}dx$$

$$= -\iint_E \frac{\partial}{\partial y} f_1(x, y, g)dxdy, \text{ by Green's theorem}$$

$$= -\iint_E \left(\frac{\partial f_1}{\partial y} + \frac{\partial f_1}{\partial z}\frac{\partial g}{\partial y}\right)dxdy \quad ...(3)$$

The direction cosines of the normal to the surface $z = g(x, y)$ are given by

$$\frac{\cos\alpha}{-\partial g/\partial x} = \frac{\cos\beta}{-\partial g/\partial y} = \frac{\cos\gamma}{1} \quad \text{(See p. 245)} \quad ...(4)$$

Moreover

$$dxdy = \text{projection of } ds \text{ on the } xy\text{-plane}$$

$$= ds\cos\gamma, \ i.e., \ ds = dxdy/\cos\gamma.$$

$\therefore$ right side of Eqn. (2)

* Named after an Irish mathematician *Sir George Gabriel Stokes* (1819–1903) who became professor in Cambridge. His important contributions are to infinite series, geodesy and theory of viscous fluids.

$$= \iint_E \left( \frac{\partial f_1}{\partial z} \frac{\cos \beta}{\cos \gamma} - \frac{\partial f_1}{\partial y} \right) dxdy = - \iint \left( \frac{\partial f_1}{\partial y} + \frac{\partial f_1}{\partial z} \cdot \frac{\partial g}{\partial y} \right) dxdy \qquad \left[ \because \frac{\cos \beta}{\cos \gamma} = - \frac{\partial g}{\partial y} \text{ by (4)} \right]$$

= Left side of Eqn. (2) by Eqn. (3)

Thus we have proved Eqn. (2). Similarly, we can prove the other corresponding relations for $f_2$ and $f_3$. Adding these three results, we get Eqn. (1).

**Cor. Green's theorem in a plane as a special case of Stokes theorem.** Let $\mathbf{F} = \phi \mathbf{I} + \psi \mathbf{J}$ be a vector function which is continuously differentiable in a region $S$ of the $xy$-plane bounded by a closed curve $C$. Then

$$\int_C \mathbf{F} \,.\, dR = \int_C (\phi \mathbf{I} + \psi \mathbf{J}) \cdot (dx\,\mathbf{I} + dy\,\mathbf{J}) = \int_C (\phi\, dx + \psi\, dy)$$

and
$$\text{curl } \mathbf{F} \cdot \mathbf{N} = \begin{vmatrix} \mathbf{I} & \mathbf{J} & \mathbf{K} \\ \partial/\partial x & \partial/\partial y & 0 \\ \phi & \psi & 0 \end{vmatrix} \cdot \mathbf{K} = \frac{\partial \psi}{\partial x} - \frac{\partial \phi}{\partial y}$$

Hence *Stoke's theorem* takes the form $\int_C (\phi\, dx + \psi\, dy) = \int_C \left( \frac{\partial \psi}{\partial x} - \frac{\partial \phi}{\partial y} \right) dxdy$ which is *Green's theorem* in a plane.

**Example 8.46.** *Verify Stoke's theorem for $\mathbf{F} = (x^2 + y^2)\mathbf{I} - 2xy\mathbf{J}$ taken around the rectangle bounded by the lines $x = \pm a$, $y = 0$, $y = b$.* *(C.S.V.T.U., 2014)*

**Solution.** Let $ABCD$ be the given rectangle as shown in Fig. 8.20.

$$\int_{ABCD} \mathbf{F} \cdot d\mathbf{R} = \int_{AB} \mathbf{F} \cdot d\mathbf{R} + \int_{BC} \mathbf{F} \cdot d\mathbf{R} + \int_{CD} \mathbf{F} \cdot d\mathbf{R} + \int_{DA} \mathbf{F} \cdot d\mathbf{R}$$

and
$$\mathbf{F} \cdot d\mathbf{R} = [(x^2 + y^2)\mathbf{I} - 2xy\mathbf{J}] \cdot (\mathbf{I}dx + \mathbf{J}dy) = (x^2 + y^2)dx - 2xydy$$

Along $AB$, $x = a$ (*i.e.*, $dx = 0$) and $y$ varies from 0 to $b$.

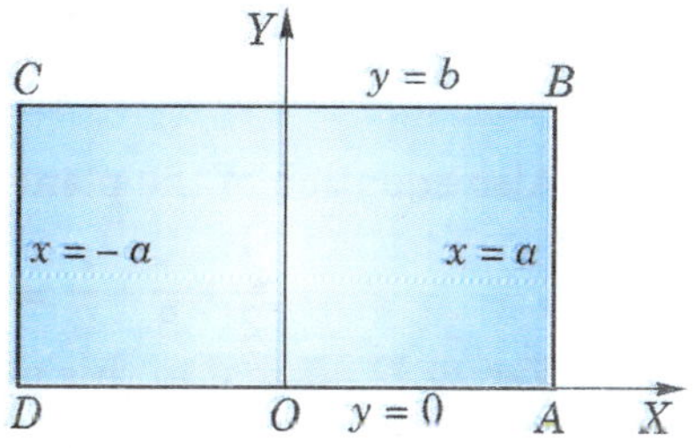

Fig. 8.20.

$$\therefore \quad \int_{AB} \mathbf{F} \cdot d\mathbf{R} = -2a \int_0^b y\, dy = -2a \cdot \frac{b^2}{2} = -ab^2 .$$

Similarly,
$$\int_{BC} \mathbf{F} \cdot d\mathbf{R} = \int_a^{-a} (x^2 + b^2) dx = -\frac{2a^3}{3} - 2ab^2 .$$

$$\int_{CD} \mathbf{F} \cdot d\mathbf{R} = 2a \int_b^0 y\, dy = -ab^2$$

and
$$\int_{DA} \mathbf{F} \cdot d\mathbf{R} = \int_{-a}^{a} x^2\, dx = \frac{2a^3}{3} .$$

Thus
$$\int_{ABCD} \mathbf{F} \cdot d\mathbf{R} = -4ab^2 \qquad ...(i)$$

Also since $\quad$ curl $\mathbf{F} = -4\mathbf{K}y$

$$\therefore \quad \int_S \text{curl } \mathbf{F} \cdot \mathbf{N}\, ds = \int_0^b \int_{-a}^{a} -4\mathbf{K}y \cdot \mathbf{K}\, dxdy = -4 \int_0^b \int_{-a}^{a} y\, dxdy$$

$$= -4 \int_0^b |x|_{-a}^{a}\, y\, dy = -8a \left| \frac{y^2}{2} \right|_0^b = -4ab^2 \qquad ...(ii)$$

Hence Stoke's theorem is verified from the equality of (*i*) and (*ii*).

**Example 8.47.** *Verify Stokc's theorem for the vector field $F = (2x - y)\,\mathbf{I} - yz^2\,\mathbf{J} - y^2z\,\mathbf{K}$ over the upper half surface of $x^2 + y^2 + z^2 = 1$, bounded by its projection on the xy-plane.*
*(W.B.T.U., 2016 ; C.S.V.T.U., 2015 ; V.T.U., 2014 ; P.T.U., 2013)*

**Solution.** The projection of the upper half of given sphere on the $xy$-plane ($z = 0$) is the circle $c[x^2 + y^2 = 1]$ (Fig. 8.17).

$$\oint_c \mathbf{F}.\, d\mathbf{R} = \oint_c [(2x - y)dx - yz^2\, dy - y^2\, zdz] = \oint_c (2x - y)dx \qquad [z = 0 \text{ in the } xy\text{-plane}]$$

$$= \int_{\theta=0}^{2\pi} (2 \cos \theta - \sin \theta)(- \sin \theta\, d\theta) \qquad [\text{Putting } x = \cos \theta,\ y = \sin \theta]$$

$$= \int_0^{2\pi} (-\sin 2\theta + \sin^2\theta)\, d\theta = \times 0 + 4\int_0^{\pi/2} \sin^2\theta\, d\theta = \pi. \quad ...(i)$$

Now $\quad \text{curl } \mathbf{F} = \begin{vmatrix} \mathbf{I} & \mathbf{J} & \mathbf{K} \\ \frac{\partial}{\partial x} & \frac{\partial}{\partial y} & \frac{\partial}{\partial z} \\ 2x - y & -yz^2 & -y^2z \end{vmatrix}$

$$= (-2yz + 2yz)\,\mathbf{I} + 0\,\mathbf{J} + \mathbf{K} = \mathbf{K}$$

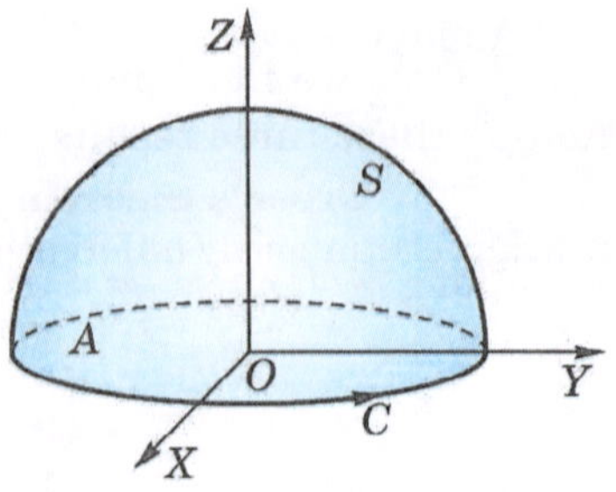

Fig. 8.21.

$$\therefore \quad \int \text{curl } \mathbf{F}.\, \mathbf{N}ds = \int_S \mathbf{K}.\mathbf{N}\, ds = \int_A \mathbf{K}.\mathbf{N} \frac{dxdy}{|\mathbf{N}.\mathbf{K}|}$$

where $A$ is the projection of $S$ on $xy$-plane and $ds = dxdy / \mathbf{N}.\mathbf{K}$

$$= \int_A dx\, dy = \text{area of circle } C = \pi \quad ...(ii)$$

Hence, the Stokes theorem is verified from the equality of Eqns. (*i*) and (*ii*).

**Example 8.48.** *Apply Stoke's theorem to evaluate* $\int_C [(x + y)dx + (2x - z)dy + (y + z)dz]$ *where C is the boundary of the triangle with vertices (2, 0, 0), (0, 3, 0) and (0, 0, 6).*

*(UPTU, 2018 ; V.T.U., 2016 ; P.T.U., 2010)*

**Solution.** Here $\quad \mathbf{F} = (x + y)\,\mathbf{I} + (2x - z)\,\mathbf{J} + (y + z)\,\mathbf{K}$

$$\therefore \quad \text{curl } \mathbf{F} = \begin{vmatrix} \mathbf{I} & \mathbf{J} & \mathbf{K} \\ \frac{\partial}{\partial x} & \frac{\partial}{\partial y} & \frac{\partial}{\partial z} \\ x + y & 2x - z & y + z \end{vmatrix} = 2\mathbf{I} + \mathbf{K}$$

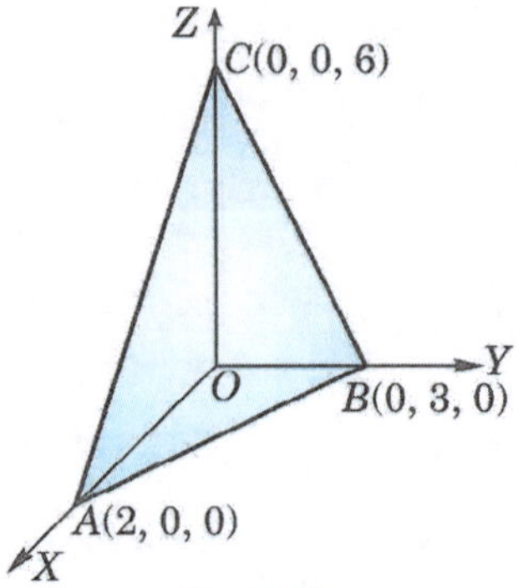

Fig. 8.22.

Also equation of the plane through $A, B, C$ (Fig. 8.22) is

$$\frac{x}{2} + \frac{y}{3} + \frac{z}{6} = 1 \text{ or } 3x + 2y + z = 6$$

Vector $\mathbf{N}$ normal to this plane is

$$\nabla (3x + 2y + z - 6) = 3\mathbf{I} + 2\mathbf{J} + \mathbf{K}$$

$$\therefore \quad \hat{\mathbf{N}} = \frac{3\mathbf{I} + 2\mathbf{J} + \mathbf{K}}{\sqrt{(9 + 4 + 1)}} = \frac{1}{\sqrt{14}}(3\mathbf{I} + 2\mathbf{J} + \mathbf{K})$$

Hence $\int_C [(x + y)dx + (2x - z)dy + (y + z)dz] = \int_C \mathbf{F}.\, d\mathbf{R}$

$$= \int_S \text{curl } \mathbf{F}.\hat{\mathbf{N}}\, ds \qquad \text{where } S \text{ is the triangle } ABC$$

$$= \int_S (2\mathbf{I} + \mathbf{K}).\left(\frac{3\mathbf{I} + 2\mathbf{J} + \mathbf{K}}{\sqrt{14}}\right) ds = \frac{1}{\sqrt{14}}(6 + 1)\int_S ds$$

$$= \frac{7}{\sqrt{14}} (\text{Area of } \Delta ABC) = \frac{7}{\sqrt{14}}.\, 3\sqrt{14} = 21.$$

**Example 8.49 :** *Verify Stokes's theorem for the function* $\vec{F} = x^2 i + xyj$ *integrated round the square whose sides are* $x = 0, y = 0, x = a$ *and* $y = a$ *in the plane* $z = 0$ *(Bhopal, 2019)*

**Solution:** Let $C$ denote the boundary of the square then

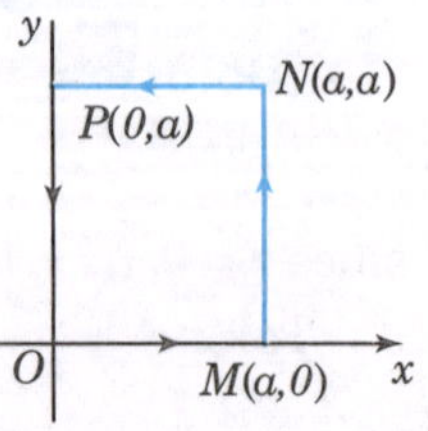

Fig. 8.23.

$$\int_C \vec{F} \cdot \vec{dr} = \int_C (x^2 i + xy\, j)(i\, dx + j\, dy)$$

$$= \int_C (x^2 dx + xy\, dy)$$

The curve $C$ consists of four line $OM$, $MN$, $NP$ and $PO$

We evaluate the integral separately along each path.

Along $OM$, $y = 0$, $dy = 0$, Hence

$$\int_{OM} (x^2dx + xydy) = \int_{OM} x^2dx = \int_0^a x^2dx = \frac{a^3}{3} \quad ...(i)$$

Along $MN$, $x = a$, $dx = 0$, Hence

$$\int_{MN} (x^2dx + xydy) = \int_0^a ay\,dy = \frac{1}{2}a^3 \quad ...(ii)$$

Along $NP$, $y = 0$, $ay = 0$, Hence

$$\int_{NP} (x^2dx + xydy) = \int_0^a x^2dx = \frac{-a^3}{3} \quad ...(iii)$$

Along $PO$, $x = 0$, $dx = 0$, Hence

$$\int_{PO} (x^2dx + xydy) = 0 \quad ...(iv)$$

$$\int_C (x^2dx + xy\,dy) = \frac{a^3}{3} + \frac{1}{2}a^3 - \frac{a^3}{3} - 0 = \frac{1}{2}a^3 \quad ...(v)$$

Again, $$\text{curl}\,\vec{F} = \begin{vmatrix} i & j & k \\ \frac{\delta}{\delta x} & \frac{\delta}{\delta y} & \frac{\delta}{\delta z} \\ x^2 & xy & 0 \end{vmatrix} = yk$$

For the open surface $S$, we choose the square $OMNP$, so that

$$\text{curl}\,\vec{F}\cdot\hat{n} = yk \quad k = y$$

and $$\iiint_S \text{curl}\,\vec{F}\cdot\hat{n} = yk \cdot k = y$$

and $$\iiint_S \text{curl}\,\vec{F}\cdot\hat{n}\,ds = \int_0^a\int_0^a y\,dx\cdot dy = [x]_0^a\left[\frac{y^2}{2}\right]_0^a = \frac{a^3}{2} \quad ...(vi)$$

From Eqns. $(v)$ and $(vi)$

$$\int_C \vec{F}\cdot d\vec{r} = \iint_S \text{curl}\,\vec{F}\cdot\hat{n}\cdot ds$$

This verifies the Stoke's theorem.

**Example 8.50.** *Verify Stoke's theorem when $\vec{F} = yi + zj + xk$ and surface $S$ is the part of the sphere $x^2 + y^2 + z^2 = 1$ above the $x - y$ plane.*

**Solution:** Here $\vec{F}\cdot\overrightarrow{dr} = (yi + zj + xk)\cdot(i\,dx + j\,dy + k\,dz)$

$$= y\,dx + z\,dy + x\,dz$$

The equation of circle $C$ in the $x$–$y$ plane is $x^2 + y^2 = 1$

$$\int_c \vec{F}\cdot\overrightarrow{dr} = \int[(yi + zi + xk)\cdot idx + jdy + kdz]$$

$$= \int(y\,dx + z\,dy + x\,dz) = \int_c y\,dx$$

since $z = 0$, $dz = 0$, in the $xy$ plane

Taking $x = \cos\phi$, $y = \sin\phi$

$$dx = -\sin\phi\,d\phi, \text{ we get}$$

$$\int_c \vec{F}\cdot\overrightarrow{dr} = \int_c y\,dx = \int_0^{\pi/2} \sin\phi(-\sin\phi)d\phi$$

$$= -4\int_0^{\pi/2} \sin^2\phi\, d\phi = -4\left(\frac{\pi}{4}\right) = -\pi \qquad ..(i)$$

Next

$$\text{curl}\,\vec{F} = \begin{vmatrix} i & j & k \\ \frac{\delta}{\delta x} & \frac{\delta}{\delta y} & \frac{\delta}{\delta z} \\ y & z & x \end{vmatrix} = -(i + j + k)$$

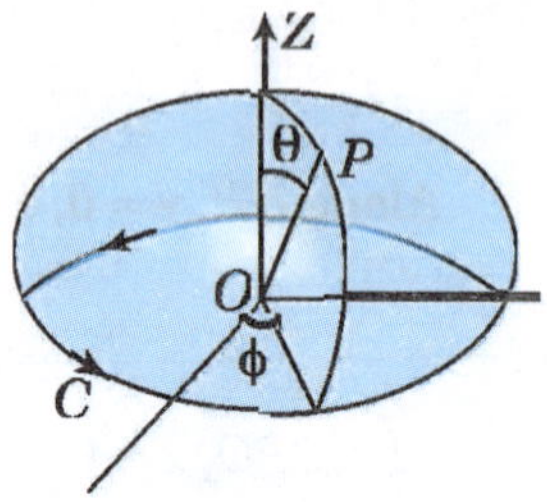

Fig. 8.24.

$$\hat{n} = \frac{\nabla\phi}{|\nabla\phi|} = \frac{2xi + 2yj + 2zk}{\sqrt{4x^2 + 4y^2 + 4z^2}}$$

$$= xi + yj + zk$$

using the spherical polar co-ordinates we get

$$\hat{n} = \sin\theta\cos\phi + i\sin\theta\sin\phi\, j + \cos\theta\, k$$

$\therefore$

$$\text{curl}\,\vec{F}\cdot\hat{n} = -(\sin\theta\cos\phi + \sin\theta\sin\phi\, j + \cos\theta\, k)$$

Now,

$$\iint \text{curl}\, F.\hat{n}\,dS = -\int_{\theta=0}^{\pi/2}\int_{\theta=0}^{2\pi} (\sin\theta\cos\phi + \sin\theta\sin\phi + \cos\theta)\sin\theta d\theta d\phi$$

$$= -\int_{\theta=0}^{\pi/2} [\sin\theta\sin\phi - \sin\theta\cos\phi + \cos\theta]_0^{2\pi} \sin\theta d\theta$$

$$= -2\pi\int_0^{\pi/2} \cos\theta\sin\theta\, d\theta = \pi\left[\cos\frac{2\theta}{2}\right]_0^{\pi/2} = -\pi \qquad ...(ii)$$

From Eqns. (*i*) and (*ii*) Stoke's theorem is verified.

**Example 8.51:** *f* $\mathbf{F} = 3y\mathbf{I} - xz\mathbf{J} + yz^2\mathbf{K}$ *and S is the surface of the paraboloid* $2z = x^2 + y^2$ *bounded by* $z = 2$*, evaluate* $\iint_S (\nabla \times \mathbf{F}) . d\mathbf{S}$ *using Stoke's theorem.*

**Solution.** By Stokes theorem,

$$I = \iint_S (\nabla \times \mathbf{F}) . d\mathbf{S}$$

$$= \oint_C \mathbf{F} . d\mathbf{R}$$

where $S$ is the surface $2z = x^2 + y^2$ bounded by $z = 2$.

$\therefore$

$$I = \oint_C \mathbf{F} . d\mathbf{R} = \oint_C (3yI - xzJ + yz^2K) . (dxI + dyJ + dzK)$$

$$= \oint_C (3ydx - xzdy + yz^2dz)$$

$\because\ S \equiv x^2 + y^2 = 4,\ z = 2$
$\therefore$ Put $x = 2\cos\theta,\ y = 2\sin\theta$
$C \equiv x^2 + y^2 = 4,\ \theta = 0$ to $2\pi$.

$$= \int_0^{2\pi} [6\sin\theta\,(-2\cos\theta\, d\theta) - 4\cos\theta\,(2\cos\theta\, d\theta) + 8\sin\theta\,(0)]$$

$$= -4\int_0^{2\pi} (12\sin^2\theta + 8\cos^2\theta)\, d\theta$$

$$= -4\left(12 . \frac{1}{2}\frac{\pi}{2} + 8 . \frac{1}{2} . \frac{\pi}{2}\right)$$

$$= -20\pi.$$

**Example 8.52.** *If S be any closed surface, prove that $\int_S$ curl* $\mathbf{F} . d\mathbf{S} = 0$.

**Solution.** Cut the surface $S$ by any plane and let $S_1, S_2$ denote its upper and lower portions. Let $C$ be the common curve bounding both these portions.

$$\therefore \quad \int_S \text{curl } \mathbf{F} . d\mathbf{S} = \int_{S_1} \text{curl } \mathbf{F} . d\mathbf{S} + \int_{S_2} \text{curl } \mathbf{F} . d\mathbf{S} = \int_C \mathbf{F} . d\mathbf{R} - \int_C \mathbf{F} . d\mathbf{R} = 0,$$

on applying Stoke's theorem. The second integral is negative because it is traversed in a direction opposite to that of the first.

## PROBLEMS 8.9

1. Apply Stoke's theorem to prove that

$$\int_C (ydx + zdy + xdz) = -2\sqrt{2}\,\pi a^2,$$

where $C$ is the curve given by $x^2 + y^2 + z^2 - 2ax - 2ay = 0$, $x + y = 2a$ and begins at the point $(2a, 0, 0)$ and goes at first below the $z$-plane.

*(Kottayan, 2019)*

2. Verify Stoke's theorem for a vector field defined by

(*i*) $\mathbf{F} = x^2\mathbf{I} + xy\mathbf{J}$ integrated round the square in the plane $z = 0$ and bounded by the lines $x = 0$, $y = 0$ and $x = a$, $y = a$.

*(D.T.U., 2015)*

(*ii*) $\mathbf{F} = -y^3\mathbf{I} + x^3\mathbf{J}$, in the region $x^2 + y^2 \leq 1$, $z = 0$. *(V.T.U., 2014)*

3. Apply Stoke's theorem to evaluate $\int_S \text{curl}\, \vec{F}.\hat{n}\, dS$ where $\vec{F} = xi + xj + yk$ and $S$ is the sphere $x^2 + y^2 + z^2 = 1$ above the $xy$ plane. Also verify Stoke's theorem. *(Agra, 2018)*

4. Verity Stoke's theorem for $\mathbf{F} = (y - z + 2)\mathbf{I} + (yz + 4)\mathbf{J} - xz\mathbf{K}$ where $S$ is the surface of the cube $x = 0$, $y = 0$, $z = 0$, $x = 2$, $y = 2$, $z = 2$ above the $xy$-plane.

5. Evaluate $\int_C \mathbf{F} . d\mathbf{R}$ where $\mathbf{F} = y\mathbf{I} + xz^3\mathbf{J} - zy^3\mathbf{K}$, $C$ is the circle $x^2 + y^2 = 4$, $z = 1.5$.

6. Verify Stoke's theorem for $\vec{F} = (x^2 + y^2)\, i - 2xy\, j$ taken round the rectangle bounded by the lines $x = \pm a$, $y = 0$, $y = b$. *(Jiwaji, 2019)*

7. If $S$ be the surface of the sphere $x^2 + y^2 + z^2 = 1$, prove that $\int_S \text{curl } \mathbf{F} . d\mathbf{S} = 0$.

8. Prove that $\int_C A \times \mathbf{R} . d\mathbf{R} = 2\mathbf{A} . \int d\mathbf{S}$, $\mathbf{A}$ being any constant vector, and deduce that $\oint_C \mathbf{R} \times d\mathbf{R}$ is twice the vector area of the surface enclosed by $C$.

9. If $\phi$ is a scalar point function, use Stoke's theorem to prove that (*i*) curl (grad $\phi$) = 0. (*ii*) div curl $\mathbf{F} = 0$.

10. Apply Stoke's theorem to evaluate $\int_C (ydx + zdy + xdz)$, where $C$ is the curve of intersection of $x^2 + y^2 + z^2 = a^2$ and $x + z = a$.

11. Use Stoke's theorem to evaluate $\int (\nabla \times \mathbf{F}) . \mathbf{N}\, ds$, where $\mathbf{F} = y\mathbf{I} + (x - 2xz)\,\mathbf{J} - xy\mathbf{K}$ and $S$ is the surface of the sphere $x^2 + y^2 + z^2 = a^2$ above the $xy$-plane. *(V.T.U., 2013)*

12. Evaluate $\int_S \text{curl}\, \vec{F}.\hat{n}\, dS$, where $\vec{F} = yi + zj + xk$ and $S$ is the portion of the surface $x^2 + y^2 - 2ax + az = 0$ above the $x - y$ plane. Also verify the Stoke's theorem.

13. Find the value of $\iint_S (\nabla \times \vec{F}) d\vec{S}$ taken over the upper portion of the surface $x^2 + y^2 - 2ax + az = 0$ and the bounding curve lies in the plane $z = 0$, when

$$\vec{F} = (y^2 + z^2 - x^2)i + (z^2 - x^2 - y^2)j + (x^2 + y^2 - z^2)k$$

14. If $\vec{F} = yi + (x - 2xz)j - xy\,k$, evalute $\iint_S (\nabla \times \vec{F}).\hat{n}\,dS$, where $S$ is the surface of the sphere $x^2 + y^2 + z^2 = a^2$ above the $xy$-plane.

15. Verify Stoke's theorem for $\vec{F} = (x^2 + y^2)\,i - 2xy\,j$ taken round the rectangle bounded by the lines $x = \pm a$, $y = 0, y = b$. *(Jiwaji, 2019)*

## 8.15 VOLUME INTEGRAL

Consider a continuous vector function $\mathbf{F}(\mathbf{R})$ and surface $S$ enclosing the region $E$. Divide $E$ into finite number of sub-regions $E_1, E_2, ..., E_n$. Let $\delta v_i$ be the volume of the sub-region $E_i$ enclosing any point whose position vector is $\mathbf{R}_i$.

*Consider the sum* $\quad \mathbf{V} = \sum_{i=1}^{n} \mathbf{F}(\mathbf{R}_i)\,\delta v_i$

*The limit of this sum as $n \to \infty$ in such a way that $\delta v_i \to 0$, is called the volume integral of $\mathbf{F}(\mathbf{R})$ over $E$ and is symbolically written as $\int_E \mathbf{F}\,dv$.*

If $\mathbf{F}(\mathbf{R}) = f(x, y, z)\mathbf{I} + \phi(x, y, z)\mathbf{J} + \psi(x, y, z)\mathbf{K}$ so that $\delta v = \delta x \delta y \delta z$, then

$$\int_E \mathbf{F}\,dv = \mathbf{I}\iiint_E f dxdydz + \mathbf{J}\iiint_E \phi dxdydz + \mathbf{K}\iiint_E \psi dxdydz.$$

## 8.16 GAUSS DIVERGENCE THEOREM* (*Relation between surface and volume integrals*)

If $\mathbf{F}$ is a continuously differentiable vector function in the region $E$ bounded by the closed surface $S$, then

$$\int_S \mathbf{F}\,.\,\mathbf{N}ds = \int_E div\,\mathbf{F}\,dv$$

*where $\mathbf{N}$ is the unit external normal vector.*

If $\mathbf{F}(\mathbf{R}) = f(x, y, z)\mathbf{I} + \phi(x, y, z)\mathbf{J} + \psi(x, y, z)\mathbf{K}$ then it is required to prove that

$$\iint_S (fdydz + \phi dzdx + \psi dxdy)$$

$$= \iiint_E \left(\frac{\partial f}{\partial x} + \frac{\partial \phi}{\partial y} + \frac{\partial \psi}{\partial z}\right) dxdydz \quad ...(1)$$

*Firstly consider such a surface $S$ that a line parallel to z-axis cuts it in two points; say $P_1(x, y, z_1)$ and $P_2(x, y, z_2)$ $(z_1 \le z_2)$* (Fig. 8.26).

If $S$ projects into the area $A_z$ on the $xy$-plane, then

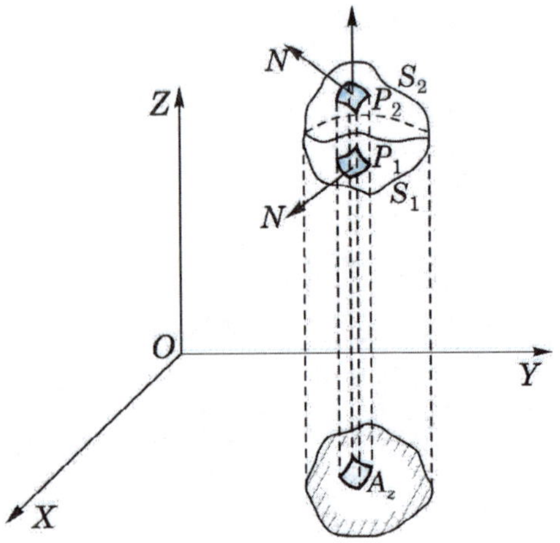

Fig. 8.26.

$$\iiint_E \frac{\partial \psi}{\partial z} dxdydz = \iint_{A_z} dxdy \int_{z_1}^{z_2} \frac{\partial \psi}{\partial z} dz$$

$$= \iint_{A_z} [\Psi(x, y, z_2) - \Psi(x, y, z_1)]\,dxdy = \iint_{A_z} \Psi(x, y, z_2)\,dxdy - \iint_{A_z} \Psi(x, y, z_1)\,dxdy \quad ...(2)$$

Let $S_1, S_2$ be the lower and upper parts of the surface $S$ corresponding to the points $P_1$ and $P_2$ respectively and $\mathbf{N}$ be the unit external normal vector at any point of $S$. As the external normal at any point of $S_2$ makes an acute angle with the positive direction of $z$-axis and that at any point of $S_1$ an obtuse angle, therefore

*See footnote p. 39.

$$\iint_{A_z} \Psi(x, y, z_2)\, dxdy = \int_{S_2} \psi \mathbf{N} \cdot \mathbf{K} ds \qquad ...(3)$$

$$\iint_{A_z} \Psi(x, y, z_1)\, dxdy = -\int_{S_1} \psi \mathbf{N} \cdot \mathbf{K} ds \qquad ...(4)$$

Using Eqns. (3) and (4), Eqns. (2) now becomes

$$\iiint_E \frac{\partial \psi}{\partial z} dxdydz = \int_{S_2} \psi \mathbf{N} \cdot \mathbf{K} ds + \int_{S_1} \psi \mathbf{N} \cdot \mathbf{K} ds = \int_S \psi \mathbf{N} \cdot \mathbf{K} ds \qquad ...(5)$$

Similarly, we have

$$\iiint_E \frac{\partial f}{\partial x} dxdydz = \int_S f \mathbf{N} \cdot \mathbf{I} ds \qquad ...(6)$$

$$\iiint_E \frac{\partial \phi}{\partial y} dxdydz = \int_S \phi \mathbf{N} \cdot \mathbf{J} ds \qquad ...(7)$$

Addition of Eqns. (5), Eqns. (6) and (7) gives

$$\iiint_E \left( \frac{\partial f}{\partial x} + \frac{\partial \phi}{\partial y} + \frac{\partial \psi}{\partial z} \right) dxdydz = \int_S (f\mathbf{I} + \phi \mathbf{J} + \psi \mathbf{K}) \cdot \mathbf{N} ds \text{ which is same as (1).}$$

*Secondly, consider a general region E.* Assume that it can be split up into a finite number of sub-regions each of which is met by a line parallel to any axis in only two points. Applying Eqns. (1) to each of these sub-regions and adding the results, the volume integrals will combine to give the volume integral over the whole region $E$. Also the surface integrals over the common boundaries of two sub-regions cancel because each occurs twice and having corresponding normals in opposite directions whereas the remaining surface integrals combine to give the surface integral over the entire surface $S$.

*Finally consider a region E bounded by two closed surfaces $S_1$, $S_2$ ($S_1$ being within $S_2$).* Noting that outward normal at points of $S_1$ is directed inwards (*i.e.*, away from $S_2$) and introducing an additional surface cutting $S_1$, $S_2$ so that all parts of $E$ are bounded by a single closed surface, the truth of the theorem follows as before. Thus theorem also holds for regions enclosed by several surfaces.

Hence the theorem is completely established.

**Example 8.53.** *Verify Divergence theorem for* $\mathbf{F} = (x^2 - yz)\mathbf{I} + (y^2 - zx)\mathbf{J} + (z^2 - xy)\mathbf{K}$ *taken over the rectangular parallelepiped* $0 \le x \le a$, $0 \le y \le b$, $0 \le z \le c$. *(Anna., 2016 ; Rohtak, 2012 ; P.T.U., 2010)*

**Solution.** As

$$\text{div } \mathbf{F} = \frac{\partial}{\partial x}(x^2 - yz) + \frac{\partial}{\partial y}(y^2 - zx) + \frac{\partial}{\partial z}(z^2 - xy) = 2(x + y + z)$$

$$\therefore \quad \int_R \text{div } \mathbf{F}\, dv = 2\int_0^c \int_0^b \int_0^a (x + y + z)\, dxdydz$$

$$= 2\int_0^c dz \int_0^b dy \left( \frac{a^2}{2} + ya + za \right)$$

$$= 2\int_0^c dz \left( \frac{a^2}{2} b + \frac{ab^2}{2} + abz \right)$$

$$= 2\left( \frac{a^2 b}{2} c + \frac{ab^2}{2} c + ab\frac{c^2}{2} \right)$$

$$= abc\,(a + b + c) \qquad ...(i)$$

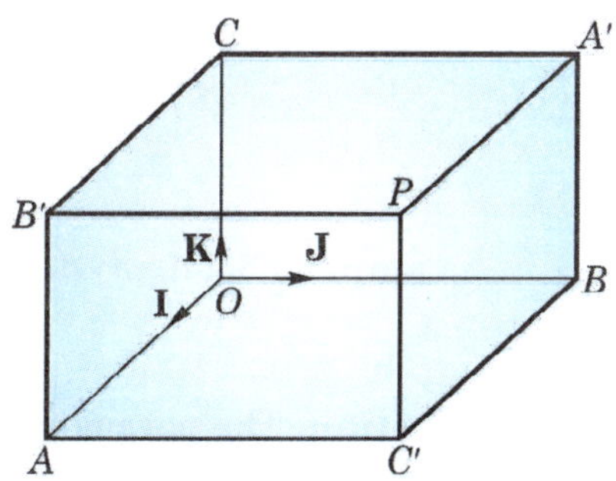


Fig. 8.27.

Also
$$\int_S \mathbf{F} \cdot \mathbf{N} ds = \int_{S_1} \mathbf{F} \cdot \mathbf{N} ds + \int_{S_2} \mathbf{F} \cdot \mathbf{N} ds + ... + \int_{S_6} \mathbf{F} \cdot \mathbf{N} ds$$

where $S_1$ in the face $OAC'B$, $S_2$ the face $CB'PA'$, $S_3$ the face $OBA'C$, $S_4$ the face $AC'PB'$, $S_5$ the face $OCB'A$ and $S_6$ the face $BAP'C'$ (Fig. 8.27).

Now $\int_{S_1} \mathbf{F} . \mathbf{N} ds = \int_{S_1} \mathbf{F} . (-\mathbf{K})\, ds = -\int_0^b \int_0^a (0 - xy)\, dxdy = \frac{a^2b^2}{4}$

$$\int_{S_2} \mathbf{F} . \mathbf{N} ds = \int_{S_2} \mathbf{F} . \mathbf{K} ds = \int_0^b \int_0^a (c^2 - xy)\, dxdy = abc^2 - \frac{a^2b^2}{4}$$

Similarly, $\int_{S_3} \mathbf{F} . \mathbf{N} ds = \frac{b^2c^2}{4}, \int_{S_4} \mathbf{F} . \mathbf{N} ds = a^2bc - \frac{b^2c^2}{4},$

$$\int_{S_5} \mathbf{F} . \mathbf{N} ds = \frac{c^2a^2}{4} \text{ and } \int_{S_6} \mathbf{F} . \mathbf{N} ds = ab^2c - \frac{c^2a^2}{4}$$

Thus $\int_S \mathbf{F} . \mathbf{N} ds = abc(a + b + c)$ ...(*ii*)

Hence the theorem is verified from the equality of Eqns. (*i*) and (*ii*).

**Example 8.54.** *Evaluate* $\int_S \mathbf{F} . \mathbf{ds}$ *where* $\mathbf{F} = 4x\mathbf{I} - 2y^2\mathbf{J} + z^2\mathbf{K}$ *and S is the surface bounding the region* $x^2 + y^2 = 4$, $z = 0$ *and* $z = 3$. (*V.T.U., 2016 ; C.S.V.T.U., 2015 ; Kurukshetra., 2013*)

**Solution.** By divergence theorem,

$$\int_S \mathbf{F} . \mathbf{ds} = \int_V div\, \mathbf{F}\, dv$$

$$= \int_V \left[\frac{\partial}{\partial x}(4x) + \frac{\partial}{\partial y}(-2y^2) + \frac{\partial}{\partial z}(z^2)\right] dv$$

$$= \iiint_V ((4 - 4y + 2z)\, dxdydz$$

$$= \int_{-2}^{2} \int_{\sqrt{(4-x^2)}}^{\sqrt{(4-x^2)}} \int_0^3 (4 - 4y + 2z)\, dzdydx$$

$$= \int_{-2}^{2} \int_{-\sqrt{(4-x^2)}}^{\sqrt{(4-x^2)}} \left| 4z - 4yz + z^2 \right|_0^3 dydx$$

$$= \int_{-2}^{2} \int_{-\sqrt{(4-x^2)}}^{\sqrt{(4-x^2)}} (12 - 12y + 9)\, dydx$$

$$= \int_{-2}^{2} \left| 21y - 6y^2 \right|_{-\sqrt{(4-x^2)}}^{\sqrt{(4-x^2)}} dx$$

$$= 42 \int_{-2}^{2} \sqrt{(4 - x^2)}\, dx = 84 \int_0^2 \sqrt{(4 - x^2)}\, dx = 84 \left| \frac{x\sqrt{(4 - x^2)}}{2} + \frac{4}{2} \sin^{-1} \frac{x}{2} \right|_0^2 = 84\pi.$$

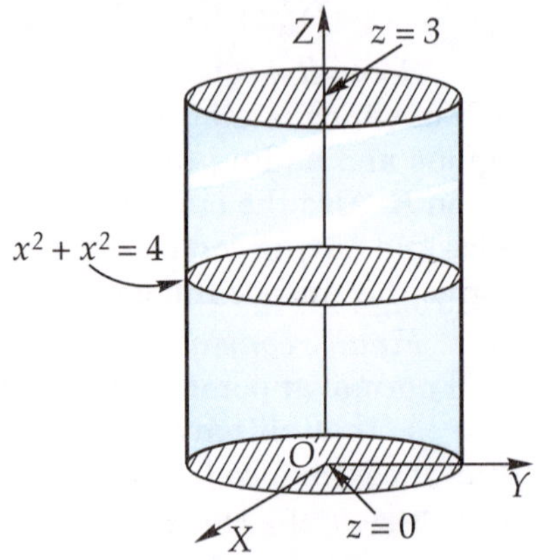

Fig. 8.28.

**Example 8.55.** *Apply divergence theorem to evaluate* $\int (lx^2 + my^2 + nz^2)\, ds$ *taken over the sphere* $(x - a)^2 + (y - b)^2 + (z - c)^2 = \rho^2$; *l, m, n being the direction cosines of the external normal to the sphere.*

**Solution.** The parametric equations of the sphere are $x = a + \rho \sin\theta \cos\phi$, $y = b + \rho \sin\theta \sin\phi$, $z = c + \rho \cos\theta$ and to cover the whole sphere, $r$ varies from 0 to $\rho$, $\theta$ varies from 0 to $\pi$ and $\phi$ from 0 to $2\pi$.

$$\therefore \int_S (lx^2 + my^2 + nz^2)\, ds = \int_S (x^2\mathbf{I} + y^2\mathbf{J} + z^2\mathbf{K}) . \mathbf{N} ds$$

$$= \int_V \text{div}\, (x^2\mathbf{I} + y^2\mathbf{J} + z^2\mathbf{K}) dv = 2 \int_V (x + y + z) dv$$

$$= 2 \int_0^{2\pi} \int_0^{\pi} \int_0^{\rho} [(a + b + c) + \rho(\sin\theta \cos\phi + \sin\theta \sin\phi + \cos\theta)] \times \rho^2 \sin\theta\, drd\theta d\phi$$

$$= 2(a + b + c) \frac{\rho^3}{3} \left| -\cos\theta \right|_0^{\pi} . 2\pi = \frac{8\pi}{3} (a + b + c)\, \rho^3.$$

**Example 8.56.** *Evaluate* $\int_S (a^2x^2 + b^2y^2 + c^2z^2)^{-1/2}\, dS$, *where S is the surface of the ellipsoid* $ax^2 + by^2 + cz^2 = 1$. *(U.P.T.U., 2012)*

**Solution.** Taking $\phi = ax^2 + by^2 + cz^2 - 1 = 0$, $\nabla \phi = 2ax\mathbf{I} + 2by\mathbf{J} + 2cz\mathbf{K}$

$\therefore$ Unit vector normal to the ellipsoid $= \hat{N} = \dfrac{\nabla\phi}{|\nabla\phi|} = \dfrac{ax\mathbf{I} + by\mathbf{J} + cz\mathbf{K}}{\sqrt{(a^2x^2 + b^2y^2 + c^2z^2)}}$

Since $\mathbf{F} \cdot \hat{\mathbf{N}} = (a^2x^2 + b^2y^2 + c^2z^2)^{-1/2}$, $\quad \therefore \mathbf{F} \cdot (ax\mathbf{I} + by\mathbf{J} + cz\mathbf{K}) = 1$

Obviously $\quad \mathbf{F} = x\mathbf{I} + y\mathbf{J} + z\mathbf{K}$ $\quad [\because ax^2 + by^2 + cz^2 = 1$

$\therefore$ By Divergence theorem,

$$\int_S \mathbf{F} \cdot d\mathbf{S} = \int_V \text{div } \mathbf{F} dv = \int_V \left[\frac{\partial}{\partial x}(x) + \frac{\partial}{\partial y}(y) + \frac{\partial}{\partial z}(z)\right] dv = 3\int_V dv = 3V$$

$$= 3 \cdot \frac{4\pi}{3} \frac{1}{\sqrt{(abc)}} = \frac{4\pi}{\sqrt{(abc)}}. \qquad \left[\because \text{ Vol. of ellipsoid} = \frac{4\pi}{3}\frac{1}{\sqrt{(abc)}}\right]$$

**Example 8.57.** *If the position vector of any point (x, y, z) within a closed surface S, be* $\mathbf{R}$ *measured from an origin O, then show that*

$$\iint_S \frac{\mathbf{R}}{r^3} \cdot \mathbf{N}\, ds = \begin{cases} 0, \text{ if } O \text{ lies outside } S \\ 4\pi, \text{ if } O \text{ lies inside } S \end{cases}$$

**Solution.** (*a*) *When O is outside S.* Here $\mathbf{F} = \mathbf{R}/r^3$ is continuously differentiable throughout the volume $V$ enclosed by $S$. Hence by Divergence theorem, we have

$$\iint_S \frac{\mathbf{R}}{r^3} \cdot \mathbf{N} ds = \iiint_V \text{div}\left(\frac{\mathbf{R}}{r^3}\right) dV = 0 \qquad \left[\because \text{div}\left(\frac{\mathbf{R}}{r^3}\right) = 0\right]$$

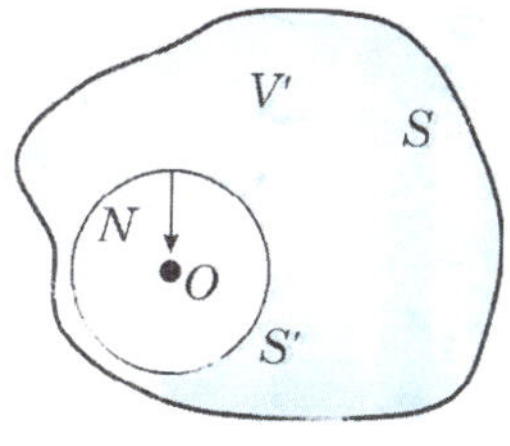


Fig. 8.29.

(*b*) *When O is inside S.* Hence $F = \mathbf{R}/r^3$ has a point of discontinuity at $O$ and as such Divergence theorem cannot be applied to the region $V$ enclosed by $S$. To remove this point of discontinuity, we enclose $O$ by a small sphere $S'$ of radius $\rho$.

Now $\mathbf{F}$ is continuously differentiable throughout the region $V'$ enclosed between $S$ and $S'$. Therefore applying Divergence theorem to region $V'$, we get

$$\iint_S \frac{\mathbf{R}}{r^3} \cdot \mathbf{N}\, ds + \iint_{S'} \frac{\mathbf{R}}{r^3} \cdot \mathbf{N}\, ds' = \iiint_{V'} \text{div}\left(\frac{\mathbf{R}}{r^3}\right) dV' = 0 \qquad \left[\because \text{div}\left(\frac{\mathbf{R}}{r^3}\right) = 0\right]$$

$$\therefore \qquad \iint_S \frac{\mathbf{R}}{r^3} \cdot \mathbf{N}\, ds = -\iint_{S'} \frac{\mathbf{R}}{r^3} \cdot \mathbf{N}\, ds' \qquad ...(i)$$

Now the outward normal $\mathbf{N}$ on the sphere $S'$ is directed towards the centre $O$. Therefore $\mathbf{N} = -\mathbf{R}/\rho$ on $S'$ (Fig. 8.29).

$$\therefore \qquad -\iint_{S'} \frac{\mathbf{R}}{r^3} \cdot \mathbf{N}\, ds' = -\iint_{S'} \frac{\mathbf{R}}{\rho^3} \cdot \left(-\frac{\mathbf{R}}{\rho}\right) ds' \qquad [\because \text{ on } S', r = \rho]$$

$$= \iint_{S'} \frac{r^2}{\rho^4} ds' = \iint_{S'} \frac{\rho^2}{\rho^4} ds' = \frac{1}{\rho^2}\iint_{S'} ds' = \frac{1}{\rho^2} \cdot 4\pi\rho^2 = 4\pi$$

Hence from (*i*), $\qquad \iint_S \frac{\mathbf{R}}{r^3} \cdot \mathbf{N}\, ds = 4\pi.$

## 8.17 GREEN'S THEOREM*

*If* $\phi$ *and* $\psi$ *are scalar point functions possessing continuous derivatives of first and second orders, then*

$$\int_E (\phi\nabla^2\psi - \psi\nabla^2\phi) dv = \int_S \left(\phi\frac{\partial\psi}{\partial n} - \psi\frac{\partial\psi}{\partial n}\right) ds \qquad ...(1)$$

*See footnote p. 809

*where ∂/∂n denotes differentiation in the direction of the external normal to the bounding surface S enclosing the region E.*

Applying Divergence theorem : $\int_S \mathbf{F}\,.\,\mathbf{N}ds = \int_S \nabla\,.\,\mathbf{F}dv$ to the function $\phi\nabla\psi$, we get

$$\int_S \phi\nabla\psi\,.\,\mathbf{N}ds = \int_E \nabla\,.\,(\phi\nabla\psi)\,dv = \int_E (\nabla\phi\,.\,\nabla\psi + \phi\nabla^2\psi)\,dv$$

$$= \int_E \nabla\phi\,.\,\nabla\psi dv + \int_E \phi\nabla^2\psi dv \quad \text{...(2)}$$

Interchanging $\phi$ and $\psi$, (*ii*) gives

$$\int_S \psi\nabla\phi\,.\,\mathbf{N}ds = \int_E \nabla\psi\,.\,\nabla\phi dv + \int_E \psi\nabla^2\phi dv \quad \text{...(3)}$$

Subtracting (3) from (2), we have $\int_S (\phi\nabla\psi - \psi\nabla\phi)\,.\,\mathbf{N}ds = \int_E (\phi\nabla^2\psi - \psi\nabla^2\phi)dv$

But $\nabla\psi\,.\,\mathbf{N} = \partial\psi/\partial n$ the directional derivative of $\psi$ along the external normal at any point of $S$. Hence

$$\int_S \left(\phi\frac{\partial\psi}{\partial n} - \psi\frac{\partial\phi}{\partial n}\right)ds = \int_E (\phi\nabla^2\psi - \psi\nabla^2\phi)\,dv$$ which is the required result (1).

Obs. **Harmonic function.** *A scalar point function $\phi$ satisfying the Laplace's equation $\nabla^2\phi = 0$ at every point of a region E, is called a harmonic function in E.*

If $\phi$ and $\phi$ be both harmonic functions in E, (1) gives

$$\int_S \phi\frac{\partial\psi}{\partial n}ds = \int_S \psi\frac{\partial\phi}{\partial n}ds$$ which is known as **Green's reciprocal theorem.**

## PROBLEMS 8.10

1. Verify divergence theorem for $\mathbf{F}$ taken over the cube bounded by $x = 0, x = 1\,; y = 0, y = 1\,; z = 0, z = 1$ where
   (*i*) $\mathbf{F} = 4xz\mathbf{I} - y^2\mathbf{J} + yz\mathbf{K}$ (*Delhi, 2013 ; Anna, 2012*) (*ii*) $x^2\mathbf{I} + z\mathbf{J} + yz\mathbf{K}$
   (*JNTUK, 2022 ; U.P.T.U., 2013 ; Calicut, 2013*)
2. Verify Gauss divergence theorem for the function $\mathbf{F} = y\mathbf{I} + x\mathbf{J} + z^2\mathbf{K}$ over the cylindrical region bounded by $x^2 + y^2 = 9$, $z = 0$ and $z = 2$. (*W.B.T.U., 2012*)
3. Using divergence theorem, prove that
   (*i*) $\int_S \mathbf{R}\,.\,d\mathbf{S} = 3V$ (*Delhi, 2013*) (*ii*) $\int_S \nabla r^2\,.\,d\mathbf{S} = 6V$ (*Calicut, 2012*)
   where $S$ is any closed surface enclosing a volume $V$ and $r^2 = x^2 + y^2 + z^2$.
4. Evaluate $\iint \mathbf{F}.\mathbf{N}ds$ given $\mathbf{F} = x\mathbf{I} + y\mathbf{J} + z\mathbf{K}$ over the sphere $x^2 + y^2 + z^2 = a^2$ by using Gauss Divergence theorem. (*V.T.U., 2015*)
5. If $S$ is any closed surface enclosing a volume $V$ and $\mathbf{F} = ax\mathbf{I} + by\mathbf{J} + cz\mathbf{K}$, prove that
   $$\int_S \mathbf{F}\,.\,\mathbf{N}ds = (a + b + c)V$$
6. For any closed surface $S$, prove that $\int [x(y - z)\mathbf{I} + y(z - x)\mathbf{J} + z(x - y)\mathbf{K}]\,.\,ds = 0$.
7. Use divergence theorem to evaluate $\int_S \mathbf{F}\,.\,d\mathbf{S}$, where
   (*i*) $\mathbf{F} = x^3\mathbf{I} + y^3\mathbf{J} + z^3\mathbf{K}$, and $S$ is the surface of the sphere $x^2 + y^2 + z^2 = a^2$. (*P.T.U., 2010*)
   (*ii*) $\mathbf{F} = [e^x, e^y, e^z]$ and $S$ is the surface of the cube $|\,x\,| \le 1, |\,y\,| \le 1, |\,z\,| \le 1$.
8. Evaluate $\iint (xdydz + ydzdx + zdxdy)$ over the surface of a sphere of radius $a$.
9. Evaluate $\int_S \mathbf{F}\,.\,d\mathbf{S}$ where $F = y^2z^2\mathbf{I} + z^2x^2\mathbf{J} + x^2y^2\mathbf{K}$ and $S$ is the upper part of the sphere $x^2 + y^2 + z^2 = a^2$ above $XOY$ plane. (*C.S.V.T.U., 2011 S*)
10. By transforming to triple integral, evaluate $\iint_S (x^3dydz + x^2ydzdx + x^2zdxdy)$ where $S$ is the closed surface consisting of the cylinder $x^2 + y^2 = a^2$ and the circular discs $z = 0$ and $z = b$. (*V.T.U., 2013*)

11. Evaluate $\int_S (\nabla \times \mathbf{F}) \cdot d\mathbf{S}$, where $S$ is the surface of the paraboloid $x^2 + y^2 + z = 4$ above the $xy$-plane, and $\mathbf{F} = (x^2 + y - 4)\mathbf{I} + 3xy\mathbf{J} + (2xz + z^2)\mathbf{K}$.

12. If $\mathbf{F} = (2x^2 - 3z)\mathbf{I} - 2xy\mathbf{J} - 4x\mathbf{K}$, then evaluate $\iiint_V \nabla \cdot \mathbf{F}\, dv$, where $V$ is bounded by $x = y = z = 0$ and $2x + 2y + z = 4$.

13. If $\mathbf{F} = \text{grad } \phi$ and $\nabla^2\phi = -4\pi\rho$, prove that $\int_S \mathbf{F} \cdot \mathbf{N} ds = -4\pi\rho \int_V dV$ where the symbol have their usual meanings.

## 8.18 (1) IRROTATIONAL FIELDS

*An irrotational field* **F** *is characterised by any one of the following conditions* :

(*i*) $\Delta \times \mathbf{F} = \mathbf{0}$.  (*ii*) *Circulation* $\int \mathbf{F} \cdot d\mathbf{R}$ *along every closed surface is zero.*

(*iii*) $\mathbf{F} = \Delta\phi$, *if the domain is simply connected.**

If $\nabla \times \mathbf{F} = \mathbf{0}$, then by Stoke's theorem,

$$\int_C \mathbf{F} \cdot d\mathbf{R} = \int_S \nabla \times \mathbf{F} \cdot d\mathbf{S} = \mathbf{0},$$ *i.e.*, the circulation along every closed surface is zero.

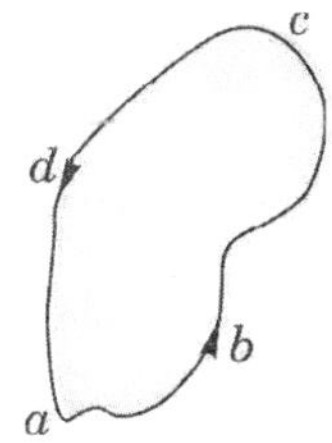

Fig. 8.30.

Again since $\nabla \times \nabla\phi = 0$

$\therefore$ in an irrotational field for which $\Delta \times \mathbf{F} = 0$, the vector **F** can always be expressed as the gradient of a scalar function $\phi$ provided the domain is simply connected. Thus

$$\mathbf{F} = \nabla\phi.$$

Such a scalar function $\phi$ is called the *potential*. In a rotational field, **F** cannot be expressed as the gradient of a scalar potential.

**Obs. 1.** *In an irrotational field, the line integral* **F** *between two points is independent of the path of integration and is equal to the potential difference between these points.*

If *aced* be any closed contour in an irrotational field F (Fig. 8.30), then

$$\int_{abcd} \mathbf{F} \cdot d\mathbf{R} = \int_{abc} \mathbf{F} \cdot d\mathbf{R} + \int_{cda} \mathbf{F} \cdot d\mathbf{R} = 0$$

or
$$\int_{abc} \mathbf{F} \cdot d\mathbf{R} = \int_{abc} \mathbf{F} \cdot d\mathbf{R}$$

*i.e.* the value of the line integral is independent of the path joining the end points.

Further, substituting $\mathbf{F} = \nabla\phi$, we have

$$\int_a^c \mathbf{F} \cdot d\mathbf{R} = \int_a^c \nabla\phi \cdot d\mathbf{R} = \int_a^c \left(\mathbf{I}\frac{\partial o}{\partial x} + \mathbf{J}\frac{\partial\phi}{\partial y} + \mathbf{K}\frac{\partial\phi}{\partial z}\right) \cdot (\mathbf{I}dx + \mathbf{J}dy + \mathbf{K}dz)$$

$$= \int_a^c \left(\frac{\partial\phi}{\partial x}dx + \frac{\partial\phi}{\partial y}dy + \frac{\partial\phi}{\partial z}dz\right) = \int_a^c d\phi = \phi_c - \phi_a.$$

**Obs. 2.** If **F** is a vector force acting on a particle, then $\oint_C \mathbf{F} \cdot d\mathbf{R}$ represents the work done in moving the particle around a closed path.

When $\oint_C \mathbf{F} \cdot d\mathbf{R} = 0$, *the field is said to be* **conservative,** *i.e.*, no work is done in displacement from a point $a$ to another point in the field and back to $a$ and the mechanical energy is conserved.

Thus *every irrotational field is conservative.*

**Obs. 3.** *The well-known equations of the Poisson and Laplace hold good for every irrotational field.*

Suppose $\nabla \cdot \mathbf{F} = f(x, y, z)$. Then $\nabla \cdot \nabla\phi = f(x, y, z)$ *i.e.*, $\nabla^2\phi = f(x, y, z)$ ...(*i*)

which is known as *Poisson's equation*. Its solutions for electrostatic fields enable us to determine the potential $\phi$ as a function of the charge distribution $f(x, y, z)$.

If $f(x, y, z) = 0$ then (*i*) reduces to $\nabla^2\phi = 0$ which is the *Laplace's equation*. The solutions of this equation are of great importance in modern engineering and physics, some of which we'll study in § 18.11 and 18.12.

*A domain $D$ is said to be *simply connected* if every closed curve in $D$ can be shrunk to any point within $D$.

(2) **Solenoidal fields.** *A solenoidal field* **F** *is characterised by any one of the following conditions :*

(*i*) $\nabla \cdot \mathbf{F} = 0$. (*ii*) *flux* $\int \mathbf{F} \cdot \mathbf{N}\, ds$ *across every closed surface is zero.* (*iii*) $\mathbf{F} = \nabla \times \mathbf{V}$.

If $\nabla \cdot \mathbf{F} = 0$ then by the Divergence theorem,

$$\int_S \mathbf{F} \cdot \mathbf{N} ds = \int_V \nabla \cdot \mathbf{F} dv = 0, \text{ i.e., the flux across every closed surface is zero.}$$

Again since $\nabla \cdot \nabla \times \mathbf{V} = 0$.

$\therefore$ in a solenoidal field for which $\nabla \cdot \mathbf{F} = 0$, the vector **F** can always be expressed as the curl of a vector function **V**; thus $\mathbf{F} = \nabla \times \mathbf{V}$.

**Example 8.58.** *A vector field is given by* $\mathbf{F} = (x^2 - y^2 + x)\mathbf{I} - (2xy + y)\mathbf{J}$.
*Show that the field is irrotational and find its scalar potential.* *(P.T.U., 2010)*
*Hence evaluate the line integral from (1, 2) to (2, 1).*

**Solution.** Since $\nabla \times \mathbf{F} = \begin{vmatrix} \mathbf{I} & \mathbf{J} & \mathbf{K} \\ \dfrac{\partial}{\partial x} & \dfrac{\partial}{\partial y} & \dfrac{\partial}{\partial z} \\ x^2 - y^2 + x & -(2xy + y) & 0 \end{vmatrix} = \mathbf{0}$

$\therefore$ this field is *irrotational* and the vector **F** can be expressed as the gradient of a scalar potential,

*i.e.*, $(x^2 - y^2 + x)\mathbf{I} - (2xy + y)\mathbf{J} = \nabla\phi = \dfrac{\partial\phi}{\partial x}\mathbf{I} + \dfrac{\partial\phi}{\partial y}\mathbf{J}$

whence
$$\partial\phi/\partial x = x^2 - y^2 + x \qquad ...(i)$$
$$\partial\phi/\partial y = -(2xy + y) \qquad ...(ii)$$

Integrating Eqn. (*i*) w.r.t. $x$, keeping $y$ constant, we get $\phi = \dfrac{x^3}{3} - y^2 x + \dfrac{x^2}{2} + f(y)$ ...(*iii*)

Similarly integrating Eqn. (*ii*) w.r.t. $y$, keeping $x$ constant, we obtain $\phi = -xy^2 - \dfrac{y^2}{2} + g(x)$ ...(*iv*)

Equating Eqns. (*iii*) and (*iv*), we get $\dfrac{x^3}{3} - y^2 x + \dfrac{x^2}{2} + f(y) = -xy^2 - \dfrac{y^2}{2} + g(x)$

$\therefore \qquad f(y) = -\dfrac{y^2}{2}$ and $g(x) = \dfrac{x^3}{3} + \dfrac{x^2}{2}$

Hence $\qquad \phi = \dfrac{x^3}{3} - xy^2 + \dfrac{x^2}{2} - \dfrac{y^2}{2}$

Since the field is irrotational,

$\therefore \quad \int \mathbf{F} \cdot d\mathbf{R}$ from (1, 2) to (2, 1) $= \phi_{1,2} - \phi_{2,1} = \left(\dfrac{1}{3} - 1 \times 4 + \dfrac{1}{2} - \dfrac{4}{2}\right) - \left(\dfrac{8}{3} - 2 \times 1 + \dfrac{4}{2} - \dfrac{1}{2}\right) = -7\dfrac{1}{3}$.

**Example 8.59.** *A fluid motion is given by* $\mathbf{V} = (y + z)\mathbf{I} + (z + x)\mathbf{J} + (x + y)\mathbf{K}$.
(*a*) *Is this motion irrotational? If so, find the velocity potential.*
(*b*) *Is the motion possible for an incompressible fluid?*

**Solution.** We have $\nabla \times \mathbf{V} = \begin{vmatrix} \mathbf{I} & \mathbf{J} & \mathbf{K} \\ \dfrac{\partial}{\partial x} & \dfrac{\partial}{\partial y} & \dfrac{\partial}{\partial z} \\ y + z & z + x & x + y \end{vmatrix} = \mathbf{I}(1 - 1) - \mathbf{J}(1 - 1) + \mathbf{K}(1 - 1) = \mathbf{0}.$

$\therefore$ this motion is irrotational and if $\phi$ is the velocity potential then $\mathbf{V} = \nabla\phi$.

*i.e.*, $\qquad (y + z)\mathbf{I} + (z + x)\mathbf{J} + (x + y)\mathbf{K} = \dfrac{\partial\phi}{\partial x}\mathbf{I} + \dfrac{\partial\phi}{\partial y}\mathbf{J} + \dfrac{\partial\phi}{\partial z}\mathbf{K}$

$\therefore$ $$\frac{\partial\phi}{\partial x} = y + z, \frac{\partial\phi}{\partial y} = z + x, \frac{\partial\phi}{\partial z} = x + y$$

Integrating these, we get $\phi = (y + z)x + f_1(y, z)$ ...(*i*)

$\phi = (z + x)y + f_2(z, x)$ ...(*ii*)

and $\phi = (x + y)z + f_3(x, y)$ ...(*iii*)

Equality of Eqns. (*i*), (*ii*) and (*iii*), requires that

$$f_1(y, z) = yz, f_2(z, x) = zx, f_3(x, y) = xy.$$

Hence $\phi = yz + zx + xy.$

(*b*) The fluid motion is possible if **V** satisfies the equation of continuity which for an incompressible fluid is $\nabla \cdot \mathbf{V} = 0$. [See § 8.7 (1)

Here $$\nabla \cdot \mathbf{V} = \frac{\partial}{\partial x}(y + z) + \frac{\partial}{\partial y} + (z + x) + \frac{\partial}{\partial z}(x + y) = 0.$$

Hence, the fluid motion is possible.

**Example 8.60.** *Find whether* $\int_C [2xyz^2 dx + (x^2z^2 + z \cos yz)\, dy + (2x^2yz + y \cos yz)\, dz]$ *is independent of the path joining* $(0, \pi/2, 1)$ *and* $(1, 0, 1)$*. If so, evaluate this line integral.*

**Solution.** The line integral of **F** is independent of path of integration if $\nabla \times \mathbf{F} = \mathbf{0}$.

$$= \int_C [2xyz^2\,\mathbf{I} + (x^2z^2 + z \cos yz)\mathbf{J} + (2x^2yz + y \cos yz)\mathbf{K}] \cdot (\mathbf{I}dx + \mathbf{J}dy + \mathbf{K}dz) = \int_C \mathbf{F} \cdot d\mathbf{R}$$

and $$\nabla \times \mathbf{F} = \begin{vmatrix} \mathbf{I} & \mathbf{J} & \mathbf{K} \\ \dfrac{\partial}{\partial x} & \dfrac{\partial}{\partial y} & \dfrac{\partial}{\partial z} \\ 2xyz^2 & x^2z^2 + z \cos yz & 2x^2yz + y \cos yz \end{vmatrix}$$

$$= \mathbf{I}\,[2x^2z + \cos yz - yz \sin yz - (2x^2z + \cos yz - yz \sin yz)]$$
$$- \mathbf{J}[4xyz - 4xyz] + \mathbf{K}[2xz^2 - 2xz^2] = \mathbf{0}$$

$\therefore$ the given integral is independent of the path $C$.

Now let $\mathbf{F} = \nabla\phi$

*i.e.*, $$(2xyz^2)\mathbf{I} + (x^2z^2 + z \cos yz)\mathbf{J} + (2x^2yz + y \cos yz)\,\mathbf{K} = \mathbf{I}\frac{\partial\phi}{\partial x} + \mathbf{J}\frac{\partial\phi}{\partial y} + \mathbf{K}\frac{\partial\phi}{\partial z}$$

$\therefore$ $$2xyz^2 = \frac{\partial\phi}{\partial x},\ x^2z^2 + z \cos yz = \frac{\partial\phi}{\partial y},\ 2x^2yz + y \cos yz = \frac{\partial\phi}{\partial z}$$

Integrating first w.r.t. $x$ partially, we get

$\phi = x^2y^2z^2 + \Psi_1(y, z)$ ...(*i*)

Integrating second w.r.t. $y$ partially, we get

$\phi = x^2yz^2 + \sin yz + \Psi_2(z, x)$ ...(*ii*)

Integrating third w.r.t. $z$ partially, we get

$\phi = x^2yz^2 + \sin yz + \Psi_3(x, y)$ ...(*iii*)

Comparing Eqns. (*i*), (*ii*), (*iii*), we have

$\Psi_1(y, z)$ = terms in $\phi$ independent of $x = \sin yz$

$\Psi_2(z, x)$ = terms in $\phi$ independent of $y = 0$

$\Psi_3(z, x)$ = terms in $\phi$ independent of $z = 0$

Thus $\phi = x^2yz^2 + \sin yz$

Hence the value of the given integral $= |\phi|_{(0,\pi/2,1)}^{(1,0,1)}$

$$= (0 + 0) - (0 + \sin \pi/2) = -1.$$

## PROBLEMS 8.11

1. If $\phi$ is a solution of the Laplace equation, prove that $\nabla\phi$ is both solenoidal and irrotational.
2. Show that the vector field defined by $\mathbf{F} = (x^2 + xy^2)\mathbf{I} + (y^2 + x^2y)\mathbf{J}$ is conservative and find the scalar potential. *(C.S.V.T.U., 2015)*

   Hence evaluate $\int \mathbf{F} \cdot d\mathbf{R}$ from (0, 1) to (1, 2). *(D.T.U., 2012)*
3. Find the work done by the variable force $\mathbf{F} = 2y\mathbf{I} + xy\mathbf{J}$ on a particle when it is displaced from the origin to the point $\mathbf{R} = 4\mathbf{I} + 2\mathbf{J}$ along the parabola $y^2 = x$.
4. Show that the vector field given by $\mathbf{A} = 3x^2y\mathbf{I} + (x^3 - 2yz^2)\mathbf{J} + (3z^2 - 2y^2z)\mathbf{K}$ is irrotational but not solenoidal. Also find $\phi(x, y, z)$ such that $\nabla\phi = \mathbf{A}$.
5. Show that the following vectors are irrotational and find the scalar potential in each case :

   (*i*) $(x^2 - yz)\mathbf{I} + (y^2 - zx)\mathbf{J} + (z^2 - xy)\mathbf{K}$ *(JNTUK., 2022)*

   (*ii*) $2xy\mathbf{I} + (x^2 + 2yz)\mathbf{J} + (y^2 + 1)\mathbf{K}$ *(V.T.U., 2013)*

   (*iii*) $(6xy + z^3)\mathbf{I} + (3x^2 - z)\mathbf{J} + (3xz^2 - y)\mathbf{K}$ *(W.B.T.U., 2012)*

   (*iv*) $(x^2 - 2yz)\mathbf{I} + (3y^2 - 2zx)\mathbf{J} + (3z^2 - 2xy)\mathbf{K}$. *(V.T.U., 2016)*
6. If $\nabla\phi = 2xyz^3\mathbf{I} + x^2yz^3\mathbf{J} + 3x^2yz^2\mathbf{K}$, find $\phi(x, y, z)$ given that $\phi(1, -2, 2) = 4$. *(Anna., 2016 ; D.T.U., 2015)*
7. Fluid motion is given by $\mathbf{V} = ax\mathbf{I} + ay\mathbf{J} - 2az\mathbf{K}$.

   (*i*) Is it possible to find out the velocity potential ? If so, find it.

   (*ii*) Is the motion possible for an incompressible fluid ?
8. Show that the vector field defined by $\mathbf{F} = (y \sin z - \sin x)\mathbf{I} + (x \sin z + 2yz)\mathbf{J} + (xy \cos z + y^2)\mathbf{K}$ is irrotational and find its velocity potential.
9. Show that $\mathbf{F} = (2xy + z^3)\mathbf{I} + x^2\mathbf{J} + 3xz^2\mathbf{K}$ is a conservative vector field and find a function $\phi$ such that $\mathbf{F} = \nabla\phi$. Also find the work done in moving an object in this field from (1, –2, 1) to (3, 1, 4). *(D.T.U., 2013)*
10. If $\mathbf{F} = (x + y + az)\mathbf{I} + (bx + 2y - z)\mathbf{J} + (x + cy + 2z)\mathbf{K}$, find $a, b, c$ such that curl $\mathbf{F} = \mathbf{0}$, then find $\phi$ such that $\mathbf{F} = \nabla\phi$. *(V.T.U., 2013)*
11. Find the constant $a$ so that $\mathbf{V}$ is a conservative vector field, where

$$\mathbf{V} = (axy - z^3)\mathbf{I} + (a - 2)x^2\mathbf{J} + (1 - a)xz^2\mathbf{K}.$$

   Calculate its scalar potential and work done in moving a particle from (1, 2, –3) to (1, –4, 2) in the field.

## 8.19 (1) ORTHOGONAL CURVILINEAR COORDINATES

Let the rectangular coordinates $(x, y, z)$ of any point be expressed as functions of $u, v, w$ so that

$$x = x(u, v, w),\ y = y(u, v, w),\ z = z(u, v, w) \quad ...(1)$$

Suppose that (1) can be solved for $u, v, w$ in terms of $x, y, z$, so that

$$u = u(x, y, z),\ v = v(x, y, z),\ w = w(x, y, z) \quad ...(2)$$

We assume that the functions in Eqns. (1) and (2) are single-valued and have continuous partial derivatives so that the correspondence between $(x, y, z)$ and $(u, v, w)$ is unique. Then $(u, v, w)$ are called *curvilinear coordinates* of $(x, y, z)$.

Each of $u, v, w$ has a level surface through an arbitrary point. The surfaces $u = u_0$, $v = v_0$, $w = w_0$ are called *coordinate surfaces* through $P(u_0, v_0, w_0)$. Each pair of these coordinate surfaces intersect in curves called the *coordinate curves*. The curve of intersection of $u = u_0$ and $v = v_0$ will be called the $w$-curve, for only $w$ changes along this curve. Similarly we define $u$ and $v$-curves.

In vector notation, Eqn. (1) can be written as $\mathbf{R} = x(u, v, w)\mathbf{I} + y(u, v, w)\mathbf{J} + z(u, v, w)\mathbf{K}$

$$\therefore \quad d\mathbf{R} = \frac{\partial \mathbf{R}}{\partial u} du + \frac{\partial \mathbf{R}}{\partial v} dv + \frac{\partial \mathbf{R}}{\partial w} dw \quad ...(3)$$

Then $\partial \mathbf{R}/\partial u$ is a tangent vector to the $u$-curve at $P$. If $\mathbf{T}_u$ is a unit vector at $P$ in this direction, then $\partial \mathbf{R}/\partial u = h_1 \mathbf{T}_u$, where $h_1 = |\partial \mathbf{R}/\partial u|$.

Similarly if $\mathbf{T}_v$ and $\mathbf{T}_w$ be unit tangent vectors to $v$- and $w$-curves at $P$, then

$$\frac{\partial \mathbf{R}}{\partial v} = h_2 \mathbf{T}_v \text{ and } \frac{\partial \mathbf{R}}{\partial w} = h_3 \mathbf{T}_w$$

where $h_2 = |\partial \mathbf{R}/\partial v|$ and $h_3 = |\partial \mathbf{R}/\partial w|$. [$h_1, h_2, h_3$ are called scalar factors.]

Then Eqn. (3) can be written as

$$d\mathbf{R} = h_1 du\, \mathbf{T}_u + h_2 dv\, \mathbf{T}_v + h_3\, dw\, \mathbf{T}_w \qquad ...(4)$$

Since $\nabla u$ is normal to the surface $u = u_0$ at $P$, therefore, a unit vector in this direction is given by $\mathbf{N}_u = \dfrac{\nabla u}{|\nabla u|}$.

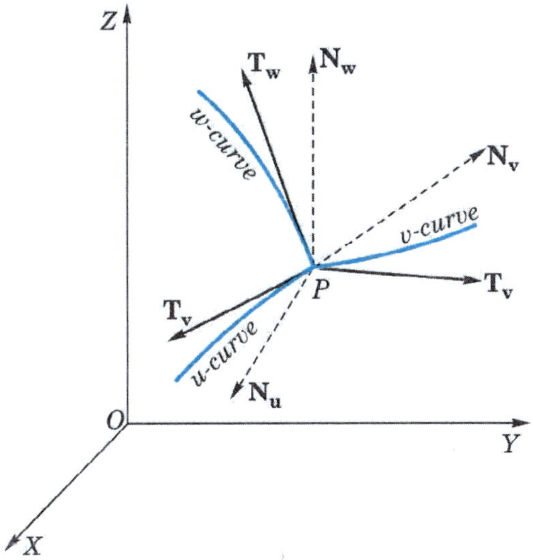


Fig. 8.31.

Similarly, the unit vectors $\mathbf{N}_v = \dfrac{\nabla v}{|\nabla v|}$ and $\mathbf{N}_w = \dfrac{\nabla w}{|\nabla w|}$ are normal to the surfaces $v = v_0$ and $w = w_0$ at $P$ respectively. Thus at each point $P$ of a curvilinear coordinate system there exist two triads of unit vectors : $\mathbf{T}_u$, $\mathbf{T}_v$, $\mathbf{T}_w$ tangents to $u$, $v$, $w$-curves and $\mathbf{N}_u$, $\mathbf{N}_v$, $\mathbf{N}_w$ normals to the co-ordinates surfaces (Fig. 8.31).

In particular, when the coordinate surfaces intersect a right angles, the three coordinate curves are also mutually orthogonal and $u$, $v$, $w$ are called the *orthogonal curvilinear coordinates*. In this case $\mathbf{T}_u$, $\mathbf{T}_v$, $\mathbf{T}_w$ and $\mathbf{N}_u$, $\mathbf{N}_v$, $\mathbf{N}_w$ are mutually perpendicular unit vector triads and hence become identical. Henceforth, we shall refer to orthogonal curvilinear coordinates only.

Multiplying Eqn. (3) scalarly by $\nabla u$, we get

$$\nabla u \,.\, d\mathbf{R} = du = \left(\nabla u \,.\, \frac{\partial \mathbf{R}}{\partial u}\right) du + \left(\nabla u \,.\, \frac{\partial \mathbf{R}}{\partial v}\right) dv + \left(\nabla u \,.\, \frac{\partial \mathbf{R}}{\partial w}\right) dw$$

whcncc
$$\nabla u \,.\, \frac{\partial \mathbf{R}}{\partial u} = 1,\ \nabla u \,.\, \frac{\partial \mathbf{R}}{\partial v} = 0,\ \nabla u \,.\, \frac{\partial \mathbf{R}}{\partial w} = 0$$

Similarly,
$$\nabla v \,.\, \frac{\partial \mathbf{R}}{\partial u} = 0,\ \nabla v \,.\, \frac{\partial \mathbf{R}}{\partial v} = 1,\ \nabla v \,.\, \frac{\partial \mathbf{R}}{\partial w} = 0$$

and
$$\nabla w \,.\, \frac{\partial \mathbf{R}}{\partial u} = 0,\ \nabla w \,.\, \frac{\partial \mathbf{R}}{\partial v} = 0,\ \nabla w \,.\, \frac{\partial \mathbf{R}}{\partial w} = 1.$$

These relations show that the sets $\partial \mathbf{R}/\partial u$, $\partial \mathbf{R}/\partial v$, $\partial \mathbf{R}/\partial w$ and $\nabla u$, $\nabla v$, $\nabla w$ constitute reciprocal system of vectors.

$$\therefore \qquad \nabla u = \frac{\dfrac{\partial \mathbf{R}}{\partial v} \times \dfrac{\partial \mathbf{R}}{\partial w}}{\left[\dfrac{\partial \mathbf{R}}{\partial u} \cdot \dfrac{\partial \mathbf{R}}{\partial v} \times \dfrac{\partial \mathbf{R}}{\partial w}\right]} = \frac{(h_2 \mathbf{T_v}) \times (h_3 \mathbf{T_w})}{[(h_1 \mathbf{T_u}) \,.\, (h_2 \mathbf{T_v}) \times (h_3 T_\mathbf{w})]}$$

$$= \frac{h_2 h_3\, \mathbf{T}_v \times \mathbf{T}_w}{h_1 h_2 h_3 [\mathbf{T}_u \mathbf{T}_v \mathbf{T}_w]} = \frac{\mathbf{T}_u}{h_1} \qquad [\because\ \mathbf{T}_u \mathbf{T}_v \mathbf{T}_w = 1]$$

or $\qquad \mathbf{T}_v = h_1\, \nabla u$

Similarly $\qquad \mathbf{T}_v = h_2 \nabla v$ and $\mathbf{T}_w = h_3\, \nabla w$ $\qquad ...(5)$

Also $\qquad = \mathbf{T}_v \times \mathbf{T}_w = h_2 h_3\, \nabla v \times \nabla w$

Similarly $\qquad \mathbf{T}_v = h_3 h_1\, \nabla w \times \nabla u$ and $\mathbf{T}_w = h_1 h_2\, \nabla u \times \nabla v$ $\qquad ...(6)$

**Arc, area and volume elements**

(*i*) *Arc element*. The element of arc length $ds$ is determined from Eqn. (4).

$$\therefore \qquad ds^2 = d\mathbf{R} \,.\, d\mathbf{R} = h_1^2 du^2 + h_2^2 dv^2 + h_3^2 dw^2 \qquad ...(7)$$

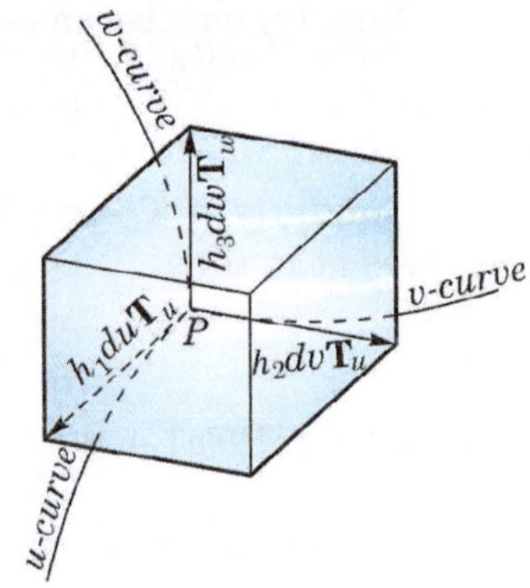


Fig. 8.32.

The arc length $ds_1$ along $u$-curve at $P$ is $h_1du$ for $v$ and $w$ are constants. Therefore the vector arc element along the $u$-curve is $d\mathbf{u} = h_1du\mathbf{T}_u$. Similarly vector arc elements along $v$ and $w$ curves at $P$ are $d\mathbf{v} = h_2dv\ \mathbf{T}_v$ and $dw = h_3dw\ \mathbf{T}_w$. The arc element $ds$ therefore corresponds to the length of the diagonal of the rectangular parallelopiped of Fig. 8.32.

(*ii*) *Area elements.* The area of the parallelogram formed by $d\mathbf{u}$ and $d\mathbf{v}$ is called the area element on the $uv$ surface which is perpendicular to $w$-curve and we denote it by $dS_w$. Hence, $dS_w = |du \times dv| = h_1h_2dudv$. Similarly, $dS_u = h_2h_3\,dvdw$, $dS_v = h_3h_1dwdu$.

(*iii*) *Volume element* is the volume of the parallelopiped formed by $d\mathbf{u}$, $d\mathbf{v}$, $d\mathbf{w}$.

$$\therefore \quad dV = [h_1du\mathbf{T}_u) \,.\, (h_2dv\mathbf{T}_v) \times (h_3dw\mathbf{T}_w)]$$
$$= h_1h_2h_3\,dudvdw \qquad ...(8)\ [\because [\mathbf{T}_u\mathbf{T}_v\mathbf{T}_w] = 1$$

This can also be written as

$$dV = \frac{\partial \mathbf{R}}{\partial u} \cdot \frac{\partial \mathbf{R}}{\partial v} \times \frac{\partial \mathbf{R}}{\partial w}\,dudvdw = \frac{\partial(x, y, z)}{\partial(u, v, w)}\,dudvdw \qquad ...(9)$$

where $\partial(x, y, z)/\partial(u, v, w)$ is called the *Jacobian of the transformation* from $(x, y, z)$ to $(u, v, w)$ coordinates.

### (2) Del applied to Functions in Orthogonal Curvilinear coordinates

*To prove that*

(**1**) $\nabla f = \dfrac{\mathbf{T}_u}{h_1}\dfrac{\partial f}{\partial u} + \dfrac{\mathbf{T}_v}{h_2}\dfrac{\partial f}{\partial v} + \dfrac{\mathbf{T}_w}{h_3}\dfrac{\partial f}{\partial w}$

(**2**) $\nabla \,.\, \mathbf{F} = \dfrac{1}{h_1h_2h_3}\left[\dfrac{\partial}{\partial u}(h_2h_3f_1) + \dfrac{\partial}{\partial v}(h_3h_1f_2) + \dfrac{\partial}{\partial w}(h_1h_2f_3)\right]$

(**3**) $\nabla \times \mathbf{F} = \begin{vmatrix} \dfrac{\mathbf{T}_u}{h_2h_3} & \dfrac{\mathbf{T}_v}{h_3h_1} & \dfrac{\mathbf{T}_w}{h_1h_2} \\ \dfrac{\partial}{\partial u} & \dfrac{\partial}{\partial v} & \dfrac{\partial}{\partial w} \\ h_1f_1 & h_2f_2 & h_3f_3 \end{vmatrix}$ where $\mathbf{F} = f_1\mathbf{T}_u + f_2\mathbf{T}_v + f_3\mathbf{T}_w$.

(**1**) Let $f(u, v, w)$ be any scalar point function in terms of $u, v, w$, the orthogonal curvilinear coordinates. Taking $u, v, w$ as functions of $x, y, z$, we have

$$\frac{\partial f}{\partial x} = \frac{\partial f}{\partial u}\frac{\partial u}{\partial x} + \frac{\partial f}{\partial v}\frac{\partial v}{\partial x} + \frac{\partial f}{\partial w}\frac{\partial w}{\partial x} \qquad ...(i)$$

$$\frac{\partial f}{\partial y} = \frac{\partial f}{\partial u}\frac{\partial u}{\partial y} + \frac{\partial f}{\partial v}\frac{\partial v}{\partial y} + \frac{\partial f}{\partial w}\frac{\partial w}{\partial y} \qquad ...(ii)$$

and
$$\frac{\partial f}{\partial z} = \frac{\partial f}{\partial u}\frac{\partial u}{\partial z} + \frac{\partial f}{\partial v}\frac{\partial v}{\partial z} + \frac{\partial f}{\partial w}\frac{\partial w}{\partial z} \qquad ...(iii)$$

Multiplying Eqn. (*i*) by **I**, Eqn.(*ii*) by **J**, Eqn. (*iii*) by **K** and adding, we have

$$\nabla f = \frac{\partial f}{\partial u}\nabla u + \frac{\partial f}{\partial v}\nabla v + \frac{\partial f}{\partial w}\nabla w \qquad ...(iv)$$

$$= \frac{\mathbf{T}_u}{h_1}\frac{\partial f}{\partial u} + \frac{\mathbf{T}_v}{h_2}\frac{\partial f}{\partial v} + \frac{\mathbf{T}_w}{h_3}\frac{\partial f}{\partial w} \qquad \text{[By (5) p. 413]}$$

which is the required result.

(**2**) Let $\mathbf{F}(u, v, w)$ be a vector point function such that

$$\mathbf{F} = f_1\mathbf{T}_u + f_2\mathbf{T}_v + f_3\mathbf{T}_w = \Sigma f_1h_2h_3\,\nabla v \times \nabla w \qquad \text{[By (6) p. 413]}$$

$$\therefore \quad \nabla \,.\, \mathbf{F} = \Sigma\nabla \,.\, \{(f_1h_2h_3)(\nabla v \times \nabla w)\}$$
$$= \Sigma[(f_1h_2h_3)\,\nabla \,.\, (\nabla v \times \nabla w) + (\nabla v \times \nabla w)\,\nabla(f_1h_2h_3)] \qquad ...(v)$$

Now $\nabla \,.\, (\nabla v \times \nabla w) = \nabla w \,.\, \nabla \times (\nabla v) - \nabla v \,.\, \nabla \times (\nabla w) = 0$ [By (5) p. 367]

and
$$\nabla(f_1h_2h_3) = \frac{\partial(f_1h_2h_3)}{\partial u}\nabla u + \frac{\partial(f_1h_2h_3)}{\partial v}\nabla v + \frac{\partial(f_1h_2h_3)}{\partial w}\nabla w \qquad \text{[By (iv) above]}$$

$\therefore$ Eqn. (*v*) now becomes

$$\nabla \cdot \mathbf{F} = \sum (\nabla v \times \nabla w) \cdot \left\{ \frac{\partial (f_1 h_2 h_3)}{\partial u} \nabla u + \frac{\partial (f_1 h_2 h_3)}{\partial v} \nabla v + \frac{\partial (f_1 h_2 h_3)}{\partial w} \nabla w \right\}$$

$$= [\nabla u, \nabla v, \nabla w] \sum \frac{\partial (f_1 h_2 h_3)}{\partial u} = \frac{1}{h_1 h_2 h_3} \sum \frac{\partial (f_1 h_2 h_3)}{\partial u}$$ which is the required result.

(*V.T.U., 2013*)

**Cor. Laplacian.** $\nabla^2 f = \nabla \cdot (\nabla f)$

$$= \nabla \cdot \left( \frac{\mathbf{T}_u}{h_1} \frac{\partial f}{\partial u} + \frac{\mathbf{T}_v}{h_2} \frac{\partial f}{\partial v} + \frac{\mathbf{T}_w}{h_3} \frac{\partial f}{\partial w} \right) = \frac{1}{h_1 h_2 h_3} \sum \frac{\partial}{\partial u} \left( \frac{1}{h_1} \frac{\partial f}{\partial u} h_2 h_3 \right)$$

(3) Let $\mathbf{F}(u, v, w)$ be a vector point function such that

$$\mathbf{F} = f_1 \mathbf{T}_u + f_2 \mathbf{T}_v + f_3 \mathbf{T}_w = f_1 h_1 \nabla u + f_2 h_2 \nabla v + f_3 h_3 \nabla w$$ [By Eqn. (5) p. 413]

$$\nabla \times \mathbf{F} - \sum \nabla \times (f_1 h_1 \nabla u)$$ [Using Eqn. (3) p. 412]

$$= \sum [\nabla (f_1 h_1) \times \nabla u + f_1 h_1 \, . \, \nabla \times \nabla u] = \sum \left[ \frac{\partial (f_1 h_1)}{\partial u} \nabla u + \frac{\partial (f_1 h_1)}{\partial v} \nabla v + \frac{\partial (f_1 h_1)}{\partial w} \nabla w \right] \times \nabla u$$

$$= \sum \left[ \frac{\partial (f_1 h_1)}{\partial v} \nabla v \times \nabla u + \frac{\partial (f_1 h_1)}{\partial w} \nabla w \times \nabla u \right]$$

$$= \sum \left[ \frac{\partial (f_1 h_1)}{\partial v} \left( - \frac{\mathbf{T}_u \times \mathbf{T}_v}{h_1 h_2} \right) + \frac{\partial (f_1 h_1)}{\partial w} \left( \frac{\mathbf{T}_w \times \mathbf{T}_u}{h_3 h_1} \right) \right]$$

$$= - \frac{\partial (f_1 h_1)}{\partial v} \frac{\mathbf{T}_w}{h_1 h_2} + \frac{\partial (f_1 h_1)}{\partial w} \frac{\mathbf{T}_v}{h_3 h_1} - \frac{\partial (f_2 h_2)}{\partial w} \frac{\mathbf{T}_u}{h_2 h_3} + \frac{\partial (f_2 h_2)}{\partial u} \frac{\mathbf{T}_w}{h_1 h_2} - \frac{\partial (f_3 h_3)}{\partial u} \frac{\mathbf{T}_v}{h_3 h_1} + \frac{\partial (f_3 h_3)}{\partial v} \frac{\mathbf{T}_u}{h_2 h_3}$$

$$= \frac{\mathbf{T}_u}{h_2 h_3} \left[ \frac{\partial (f_3 h_3)}{\partial v} - \frac{\partial (f_2 h_2)}{\partial w} \right]$$ + two similar terms, whence follows the required result.

(*V.T.U., 2013*)

## TWO SPECIAL CURVILINEAR SYSTEMS

## 8.20 (1) CYLINDRICAL COORDINATES

Any point $P(x, y, z)$ whose projection on the $xy$-plane is $Q(x, y)$ has the *cylindrical coordinates* $(\rho, \phi, z)$, where $\rho = OQ$, $\phi = \angle XOQ$ and $z = QP$.

The level surfaces $\rho = \rho_0$, $\phi = \phi_0$, $z = z_0$ are respectively cylinders about the $Z$-axis; planes through the $Z$-axis and planes perpendicular to the $Z$-axis.

The coordinate curves for $\rho$ are rays perpendicular to the $Z$-axis; for $\phi$, horizontal circles with centres on the $Z$-axis; for $z$, lines parallel to the $Z$-axis.

From Fig. 8.33, we have

$$x = \rho \cos \phi, \; y = \rho \sin \phi, \; z = z$$

(*i*) *Arc element.*

$\therefore$ $$(ds)^2 = (dx)^2 + (dy)^2 + (dz)^2 = (d\rho)^2 + \rho^2 (d\phi)^2 + (dz)^2$$

so that the scale factors are $h_1 = 1$, $h_2 = \rho$, $h_3 = 1$.

(*ii*) *Area elements* $dS_\rho = \rho d\phi dz$, $dS_\phi = dz d\rho$, $dS_z = \rho d\rho d\phi$ where $dS_\rho$ is the area element $\perp$ to $\rho$-direction, etc.

(*iii*) *Volume element* $dV = \rho d\rho d\phi dz$.

**(2) Cylindrical co-ordinate system is orthogonal**

At any point $P$, we have $x = \rho \cos \phi$, $y = \rho \sin \phi$, $z = z$,

so that $$\mathbf{R} = \rho \cos \phi \mathbf{I} + \rho \sin \phi \mathbf{J} + z\mathbf{K}$$

If $\mathbf{T}_\rho$, $\mathbf{T}_\phi$, $\mathbf{T}_z$ be the unit vectors at $P$ in the directions of the tangents to the $\rho$, $\phi$, $z$-curves respectively, then

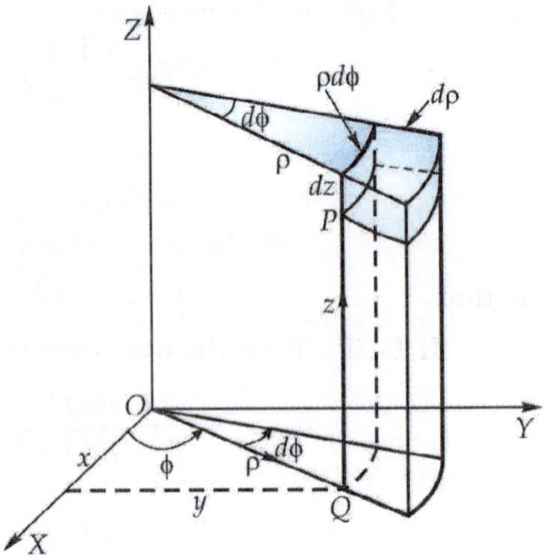

Fig. 8.33.

$$\mathbf{T}_\rho = \frac{\partial \mathbf{R}/\partial \rho}{|\partial \mathbf{R}/\partial \rho|} = \frac{\cos\phi \mathbf{I} + \sin\phi \mathbf{J}}{\sqrt{(\cos^2\phi + \sin^2\phi)}}$$
$$= \cos\phi \mathbf{I} + \sin\phi \mathbf{J}$$
$$\mathbf{T}_\phi = \frac{\partial \mathbf{R}/\partial \phi}{|\partial \mathbf{R}/\partial \phi|} = \frac{-\rho\sin\phi \mathbf{I} + \rho\cos\phi \mathbf{J}}{\sqrt{[(-\rho\sin\phi)^2 + (\rho\cos\phi)^2]}}$$
$$= -\sin\phi \mathbf{I} + \cos\phi \mathbf{J}$$

and $$\mathbf{T}_z = \frac{\partial \mathbf{R}/\partial z}{|\partial \mathbf{R}/\partial z|} = \mathbf{K}$$

Now $\mathbf{T}_\rho \cdot \mathbf{T}_\phi = (\cos\phi \mathbf{I} + \sin\phi \mathbf{J}) \cdot (-\sin\phi \mathbf{I} + \cos\phi \mathbf{J}) = -\cos\phi \sin\phi + \sin\phi\cos\phi = 0,$

$\mathbf{T}_\phi \cdot \mathbf{T}_z = (-\sin\phi \mathbf{I} + \cos\phi \mathbf{J}) \cdot \mathbf{K} = 0$, and $\mathbf{T}_z \cdot \mathbf{T}_\rho = \mathbf{K} \cdot (\cos\phi \mathbf{I} + \sin\phi \mathbf{J}) = 0.$

Hence the cylindrical coordinate system is orthogonal.

Also $\quad \mathbf{T}_\rho \times \mathbf{T}_\phi = (\cos\phi \mathbf{I} + \sin\phi \mathbf{J}) \times (-\sin\phi \mathbf{I} + \cos\phi \mathbf{J}) = (\cos^2\phi + \sin^2\phi)\, \mathbf{I} \times \mathbf{J} = \mathbf{K} = T_z$

$\mathbf{T}_\phi \times \mathbf{T}_z = (-\sin\phi \mathbf{I} + \cos\phi \mathbf{J}) \times \mathbf{K} = \sin\phi \mathbf{J} + \cos\phi \mathbf{I} = \mathbf{T}_\rho$

$\mathbf{T}_z \times \mathbf{T}_\rho = \mathbf{K} \times (\cos\phi \mathbf{I} + \sin\phi \mathbf{J}) = \cos\phi \mathbf{J} - \sin\phi \mathbf{I} = \mathbf{T}_\phi$

These conditions satisfied by $T_\rho$, $T_\phi$, and $T_z$, show that the *cylindrical coordinates system is a right handed orthogonal coordinate system.* *(V.T.U., 2016)*

**(3) Del applied to functions in Cylindrical coordinates**

We have $u = \rho$, $v = \phi$, $w = z$ and $h_1 = 1$, $h_2 = \rho$, $h_3 = 1$.

Let $T_\rho$, $T_\phi$, $T_z$ be the unit vectors in the directions of the tangents to the $\rho$, $\phi$, $z$ curves.

*(i) Expression for grad f.*

Since $$\nabla f = \frac{\mathbf{T}_u}{h_1}\frac{\partial f}{\partial u} + \frac{\mathbf{T}_v}{h_2}\frac{\partial f}{\partial v} + \frac{\mathbf{T}w}{h_3}\frac{\partial f}{\partial w}$$

$\therefore$ $$\nabla f = \frac{\partial f}{\partial \rho}\mathbf{T}_\rho + \frac{1}{\rho}\frac{\partial f}{\partial \phi}\mathbf{T}_\phi + \frac{\partial f}{\partial z}\mathbf{T}_z$$

*(ii) Expression for div* $\mathbf{F}$ *where* $\mathbf{F} = f_1\mathbf{T}_u + f_2\mathbf{T}_v + f_3\mathbf{T}_w$

Since $$\nabla \cdot \mathbf{F} = \frac{1}{h_1h_2h_3}\left[\frac{\partial}{\partial u}(h_2h_3f_1) + \frac{\partial f}{\partial v}(h_3h_1f_2) + \frac{\partial}{\partial w}(h_1h_2f_3)\right]$$

$\therefore$ $$\nabla \cdot \mathbf{F} = \frac{1}{\rho}\left\{\frac{\partial}{\partial \rho}(\rho f_1) + \frac{\partial f_2}{\partial \phi} + \frac{\partial}{\partial z}(\rho f_3)\right\}$$

*(iii) Expression for curl* $\mathbf{F}$ *where* $\mathbf{F} = f_1\mathbf{T}_u + f_2\mathbf{T}_v + f_3\mathbf{T}_w$

Since $$\nabla \times \mathbf{F} = \begin{vmatrix} \dfrac{\mathbf{T}_u}{h_2h_3} & \dfrac{\mathbf{T}_v}{h_3h_1} & \dfrac{\mathbf{T}_w}{h_1h_2} \\ \dfrac{\partial}{\partial u} & \dfrac{\partial}{\partial v} & \dfrac{\partial}{\partial w} \\ h_1f_1 & h_2f_2 & h_3f_3 \end{vmatrix} = \begin{vmatrix} \mathbf{T}_\rho/\rho & \mathbf{T}_\phi & \mathbf{T}_z/\rho \\ \dfrac{\partial}{\partial \rho} & \dfrac{\partial}{\partial \phi} & \dfrac{\partial}{\partial z} \\ f_1 & \rho f_2 & f_3 \end{vmatrix}$$

$$= \mathbf{T}_\rho\left(\frac{1}{\rho}\frac{\partial f_3}{\partial \phi} - \frac{\partial f_2}{\partial z}\right) + \mathbf{T}_\phi\left(\frac{\partial f_1}{\partial z} - \frac{\partial f_3}{\partial \rho}\right) + \mathbf{T}_z\left(\frac{\partial f_2}{\partial \rho} - \frac{1}{\rho}\frac{\partial f_1}{\partial \phi}\right)$$

*(iv) Expression for* $\nabla^2 f$

Since $$\nabla^2 f = \frac{1}{h_1h_2h_3}\left\{\frac{\partial}{\partial u}\left(\frac{1}{h_1}\frac{\partial f}{\partial u}h_2h_3\right) + \frac{\partial}{\partial v}\left(\frac{1}{h_2}\frac{\partial f}{\partial v}h_3h_1\right) + \frac{\partial}{\partial w}\left(\frac{1}{h_3}\frac{\partial f}{\partial w}h_1h_2\right)\right\}$$

$$\therefore \quad \nabla^2 f = \frac{1}{\rho}\left\{\frac{\partial}{\partial\rho}\left(\rho\frac{\partial f}{\partial\rho}\right)+\frac{\partial}{\partial\phi}\left(\frac{1}{\rho}\frac{\partial f}{\partial\phi}\right)+\frac{\partial}{\partial z}\left(\rho\frac{\partial f}{\partial z}\right)\right\} = \frac{\partial^2 f}{\partial\rho^2}+\frac{1}{\rho}\frac{\partial f}{\partial\rho}+\frac{1}{\rho^2}\frac{\partial^2 f}{\partial\phi^2}+\frac{\partial^2 f}{\partial z^2}.$$

**Example 8.61.** *Express the vector $z\mathbf{I} - 2x\mathbf{J} + y\mathbf{K}$ in cylindrical coordinates.* (V.T.U., 2016)

**Solution.** We have $x = \rho \cos \phi$, $y = \rho \sin \phi$ and $z = z$.

so that $\mathbf{R} = x\mathbf{I} + y\mathbf{J} + z\mathbf{K} = \rho \cos \phi\mathbf{I} + \rho \sin \phi\mathbf{J} + z\mathbf{K}$

If $\mathbf{T}_\rho$, $\mathbf{T}_\phi$, $\mathbf{T}_z$ be the unit vectors along the tangents to $\rho$, $\phi$ and $z$ curves respectively, then

$$\mathbf{T}_\rho = \frac{\partial \mathbf{R}/\partial\rho}{|\partial \mathbf{R}/\partial\rho|} = \frac{\cos\phi\mathbf{I}+\sin\phi\mathbf{J}}{\sqrt{(\cos^2\phi+\sin^2\phi)}} = \cos\phi\mathbf{I}+\sin\phi\mathbf{J}$$

$$\mathbf{T}_\phi = \frac{\partial \mathbf{R}/\partial\phi}{|\partial \mathbf{R}/\partial\phi|} = \frac{-\rho\sin\phi\mathbf{I}+\rho\cos\phi\mathbf{J}}{\sqrt{[(-\rho\sin\phi)^2+(\rho\cos\phi)^2]}} = -\sin\phi\mathbf{I}+\cos\phi\mathbf{J}$$

$$\mathbf{T}_z = \frac{\partial \mathbf{R}/\partial z}{|\partial \mathbf{R}/\partial z|} = \mathbf{K}$$

Let the expression for $\mathbf{F} = z\mathbf{I} - 2x\mathbf{J} + y\mathbf{K}$ in cylindrical coordinates be

$$\mathbf{F} = f_1\mathbf{T}_\rho + f_2\mathbf{T}_\phi + f_3\mathbf{T}_z \quad \text{...}(i)$$

Then $f_1 = \mathbf{F}\,.\,\mathbf{T}_\rho = z\cos\phi - 2x\sin\phi$

$f_2 = \mathbf{F}\,.\,\mathbf{T}_\phi = -z\sin\phi - 2x\cos\phi$

$f_3 = \mathbf{F}\,.\,\mathbf{T}_z = y$

Substituting the values of $f_1, f_2, f_3$ in (*i*), we get

$$\mathbf{F} = (z\cos\phi - 2x\sin\phi)\,\mathbf{T}_\rho - (z\sin\phi + 2x\cos\phi)\,\mathbf{T}_\phi + y\mathbf{T}_z$$

$$= (z\cos\phi - \rho\sin 2\phi)\,\mathbf{T}_\rho - (z\sin\phi + 2\rho\cos^2\phi)\,\mathbf{T}_\phi + \rho\sin\phi\,\mathbf{T}_z$$

**Example 8.62.** *Show that $\nabla$ (log $\rho$) and $\nabla\phi$, $\rho \neq 0$, $\phi \neq 0$ are solenoidal vectors.*

**Solution.** (*i*) $f = \log\rho$ is a function of $\rho$ only. We have to prove that $\nabla\,.\,(\nabla f)$, *i.e.*, $\nabla^2 f = 0$

$$\nabla^2 f = \frac{\partial^2 f}{\partial\rho^2}+\frac{1}{\rho}\frac{\partial f}{\partial\rho}+\frac{1}{\rho^2}\frac{\partial^2 f}{\partial\phi^2}+\frac{\partial^2 f}{\partial z^2} = \frac{\partial^2}{\partial\rho^2}(\log\rho)+\frac{1}{\rho}\frac{\partial(\log\rho)}{\partial\rho}+0+0 = -\frac{1}{\rho^2}+\frac{1}{\rho^2} = 0$$

Hence $\nabla(\log\rho)$ is a solenoidal vector.

(*ii*) $f = \nabla\phi$ is a function of $\phi$ only. We have to show that $\nabla\,.\,(\nabla f)$, *i.e.*, $\nabla^2 f = 0$.

$$\nabla^2 f = \frac{\partial^2 f}{\partial\rho^2}+\frac{1}{\rho}\frac{\partial f}{\partial\rho}+\frac{1}{\rho^2}\frac{\partial^2 f}{\partial\phi^2}+\frac{\partial^2 f}{\partial z^2} = 0+0+\frac{1}{\rho^2}\frac{\partial^2\phi}{\partial\phi^2}+0 = 0.$$

Hence the result.

## 8.21 (1) SPHERICAL POLAR COORDINATES

Let $P(x, y, z)$ be any point whose projection on the $XY$-plane is $Q(x, y)$. Then the spherical polar co-ordinates of $P$ are $(r, \theta, \phi)$ such that $r = OP$, $\theta = \angle ZOP$ and $\phi = \angle XOQ$.

The level surfaces $r = r_0$, $\theta = \theta_0$, $\phi = \phi_0$ are respectively spheres about $O$, cones about the $Z$-axis with vertex at $O$ and planes through the $Z$-axis.

The co-ordinate curves for $r$ are rays from the origin; for $\theta$, vertical circles with centre at $O$ (called *meridians*); for $\phi$, horizontal circles with centres on the $Z$-axis

From Fig. 8.34, we have

$$x = OQ\cos\phi = OP\cos(90° - \theta)\cos\phi = r\sin\theta\cos\phi,$$

$$y = OQ\sin\phi = r\sin\theta\sin\phi;\ z = r\cos\theta.$$

(*i*) *Arc element*

$$\therefore \qquad (ds)^2 = (dx)^2 + (dy)^2 + (dz)^2 = (dr)^2 + r^2 (d\theta)^2 + (r \sin \theta)^2 (d\phi)^2$$

so that the scale factors are

$$h_1 = 1, h_2 = r, h_3 = r \sin \theta.$$

(*ii*) *Area elements*

$$dS_r = r^2 \sin \theta \, d\theta d\phi, \; dS_\theta = r \sin \theta d\phi dr, \; dS_\phi = r dr d\theta$$

where, $dS_r$ is the area element perpendicular to the $r$-direction, etc.

(*iii*) *Volume element* $dV = r^2 \sin \theta \, dr d\theta d\phi$.

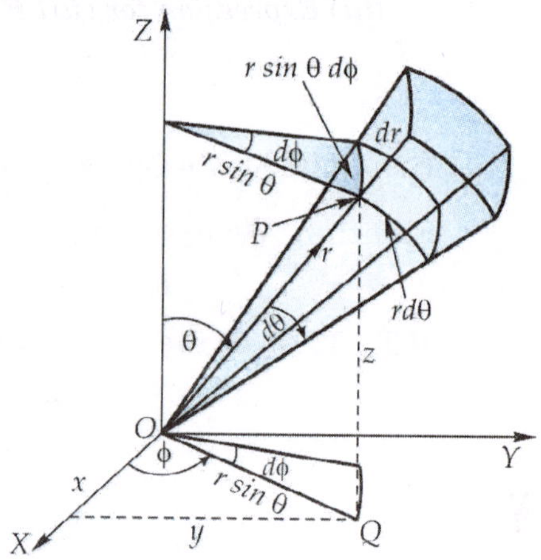


Fig. 8.34

**(2) Spherical polar coordinate system is orthogonal**

At any point $P$, we have $x = r \sin \theta \cos \phi, y = r \sin \theta \sin \phi, z = r \cos \theta$,

so that $\qquad \mathbf{R} = r \sin \theta \cos \phi \mathbf{I} + r \sin \theta \sin \phi \mathbf{J} + r \cos \theta \mathbf{K}$

If $\mathbf{T}_r, \mathbf{T}_\theta, \mathbf{T}_\phi$ be the unit vectors at $P$ in the directions of the tangents to the $r, \theta, \phi$-curves respectively, then

$$\mathbf{T}_r = \frac{\partial \mathbf{R}/\partial r}{|\partial \mathbf{R}/\partial r|} = \frac{\sin \theta \cos \phi \mathbf{I} + \sin \theta \sin \phi \mathbf{J} + \cos \theta \mathbf{K}}{\sqrt{(\sin^2 \theta \cos^2 \phi + \sin^2 \theta \sin^2 \phi + \cos^2 \theta)}}$$

$$= \sin \theta \cos \phi \mathbf{I} + \sin \theta \sin \phi \mathbf{J} + \cos \theta \mathbf{K}$$

$$\mathbf{T}_\theta = \frac{\partial \mathbf{R}/\partial \theta}{|\partial \mathbf{R}/\partial \theta|} = \frac{r \cos \theta \cos \phi \mathbf{I} + r \cos \theta \sin \phi \mathbf{J} - r \sin \theta \mathbf{K}}{r\sqrt{(\cos^2 \theta \cos^2 \phi + \cos^2 \theta \sin^2 \phi + \sin^2 \theta)}}$$

$$= \cos \theta \cos \phi \mathbf{I} + \cos \theta \sin \phi \mathbf{J} - \sin \theta \mathbf{K}$$

and

$$\mathbf{T}_\phi = \frac{\partial \mathbf{R}/\partial \phi}{|\partial \mathbf{R}/\partial \phi|} = \frac{-r \sin \theta \sin \phi \mathbf{I} + r \sin \theta \cos \phi \mathbf{J}}{r \sin \theta} = -\sin \phi \mathbf{I} + \cos \phi \mathbf{J}$$

Now $\qquad \mathbf{T}_r \cdot \mathbf{T}_\theta = \sin \theta \cos \theta \cos^2 \phi + \sin \theta \cos \theta \sin^2 \phi - \sin \theta \cos \theta = 0$

$$\mathbf{T}_\theta \cdot \mathbf{T}_\phi = -\cos \theta \cos \phi \sin \phi + \cos \theta \sin \phi \cos \phi = 0$$

$$\mathbf{T}_\phi \cdot \mathbf{T}_r = -\sin \theta \cos \phi \sin \phi + \sin \theta \sin \phi \cos \phi = 0$$

Also $\qquad \mathbf{T}_r \times \mathbf{T}_\theta = \sin \theta \cos \phi \cos \theta \sin \phi \mathbf{K} + \sin^2 \theta \cos \phi \mathbf{J} - \sin \theta \sin \phi \cos \theta \cos \phi \mathbf{K}$

$$-\sin^2 \theta \sin \phi \mathbf{I} + \cos^2 \theta \cos \phi \mathbf{J} - \cos^2 \theta \sin \phi \mathbf{I}$$

$$= -\sin \phi \mathbf{I} + \cos \phi \mathbf{J} = \mathbf{T}_\phi$$

$$\mathbf{T}_\theta \times \mathbf{T}_\phi = \cos \theta \cos^2 \phi \mathbf{K} + \sin^2 \phi \cos \theta \mathbf{K} + \sin \theta \sin \phi \mathbf{J} + \sin \theta \cos \phi \mathbf{I} = \mathbf{T}_r$$

and $\qquad \mathbf{T}_\phi \times \mathbf{T}_r = -\sin \theta \sin^2 \phi \mathbf{K} + \sin \phi \cos \theta \mathbf{J} - \sin \theta \cos^2 \phi \mathbf{K} + \cos \phi \cos \theta \mathbf{I} = \mathbf{T}_\theta$

The above conditions satisfied by $\mathbf{T}_r, \mathbf{T}_\theta$, and $\mathbf{T}_\phi$ show that *the spherical polar coordinate system is a right handed orthogonal coordinate system.* (*V.T.U., 2016*)

**(3) Del applied to functions in spherical polar coordinates**

We have $u = r, v = \theta, w = \phi$ and $h_1 = 1, h_2 = r, h_3 = r \sin \theta$.

Let $\mathbf{T}_r, \mathbf{T}_\theta, \mathbf{T}_\phi$ be the unit vectors in the directions of the tangents to the $r, \theta, \phi$-curves.

(*i*) *Expression for grad f*

Since

$$\nabla f = \frac{\mathbf{T}_u}{h_1}\frac{\partial f}{\partial u} + \frac{\mathbf{T}_v}{h_2}\frac{\partial f}{\partial v} + \frac{\mathbf{T}_w}{h_3}\frac{\partial f}{\partial w}$$

$$\therefore \qquad \nabla f = \frac{\partial f}{\partial r}\mathbf{T}_r + \frac{1}{r}\frac{\partial f}{\partial \theta}\mathbf{T}_\theta + \frac{1}{r \sin \theta}\frac{\partial f}{\partial \phi}\mathbf{T}_\phi$$

(*ii*) *Expression for div* **F** *where* $\mathbf{F} = f_1 \mathbf{T}_u + f_2 \mathbf{T}_v + f_3 \mathbf{T}_w$

Since

$$\nabla \cdot \mathbf{F} = \frac{1}{h_1 h_2 h_3}\left[\frac{\partial}{\partial u}(h_2 h_3 f_1) + \frac{\partial}{\partial v}(h_3 h_1 f_2) + \frac{\partial}{\partial w}(h_1 h_2 f_3)\right]$$

$$\therefore \qquad \nabla \cdot \mathbf{F} = \frac{1}{r^2 \sin \theta}\left[\frac{\partial}{\partial r}(r^2 \sin \theta \, f_1) + \frac{\partial}{\partial \theta}(r \sin \theta \, f_2) + \frac{\partial}{\partial \phi}(r f_3)\right]$$

$$= \frac{1}{r^2}\frac{\partial}{\partial r}(f_1 r^2) + \frac{1}{r \sin \theta}\frac{\partial}{\partial \theta}(f_2 \sin \theta) + \frac{1}{r \sin \theta}\frac{\partial f_3}{\partial \phi}$$

*(iii) Expression for curl* **F** *where* $\mathbf{F} = f_1\mathbf{T}_u + f_2\mathbf{T}_v + f_3\mathbf{T}_w$

Since
$$\nabla \times \mathbf{F} = \begin{vmatrix} \dfrac{\mathbf{T}_u}{h_2h_3} & \dfrac{\mathbf{T}_v}{h_3h_1} & \dfrac{\mathbf{T}_w}{h_1h_2} \\ \dfrac{\partial}{\partial u} & \dfrac{\partial}{\partial v} & \dfrac{\partial}{\partial w} \\ h_1f_1 & h_2f_2 & h_3f_3 \end{vmatrix}$$

∴
$$\nabla \times \mathbf{F} = \begin{vmatrix} \dfrac{\mathbf{T}_r}{r^2 \sin\theta} & \dfrac{\mathbf{T}_\theta}{r\sin\theta} & \dfrac{\mathbf{T}\phi}{r} \\ \dfrac{\partial}{\partial r} & \dfrac{\partial}{\partial \theta} & \dfrac{\partial}{\partial \phi} \\ f_1 & rf_2 & r\sin\theta f_3 \end{vmatrix}$$

$$= \frac{\mathbf{T}_r}{r^2\sin\theta}\left\{\frac{\partial}{\partial\theta}(r\sin\theta f_3) - \frac{\partial}{\partial\phi}(rf_2)\right\} - \frac{\mathbf{T}_\theta}{r\sin\theta}\left\{\frac{\partial}{\partial r}(r\sin\theta f_3) - \frac{\partial f_1}{\partial\phi}\right\} + \frac{\mathbf{T}_\phi}{r}\left\{\frac{\partial}{\partial r}(rf_2) - \frac{\partial f_1}{\partial\theta}\right\}$$

$$= \frac{\mathbf{T}_r}{r\sin\theta}\left\{\frac{\partial}{\partial\theta}(f_3\sin\theta) - \frac{\partial f_2}{\partial\phi}\right\} + \frac{\mathbf{T}_\theta}{r}\left\{\frac{1}{\sin\theta}\frac{\partial f_1}{\partial\phi} - \frac{\partial}{\partial r}(rf_3)\right\} + \frac{\mathbf{T}_\phi}{r}\left\{\frac{\partial}{\partial r}(rf_2) - \frac{\partial f_1}{\partial\theta}\right\}$$

*(iv) Expression for* $\nabla^2 f$.

Since
$$\nabla^2 f = \frac{1}{h_1h_2h_3}\left\{\frac{\partial}{\partial u}\left(\frac{h_2h_3}{h_1}\frac{\partial f}{\partial u}\right) + \frac{\partial}{\partial v}\left(\frac{h_3h_1}{h_2}\frac{\partial f}{\partial v}\right) + \frac{\partial}{\partial w}\left(\frac{h_1h_2}{h_3}\frac{\partial f}{\partial w}\right)\right\}$$

∴
$$\nabla^2 f = \frac{1}{r^2\sin\theta}\left\{\frac{\partial}{\partial r}\left(r^2\sin\theta\frac{\partial f}{\partial r}\right) + \frac{\partial}{\partial\theta}\left(\frac{r\sin\theta}{r}\frac{\partial f}{\partial\theta}\right) + \frac{\partial}{\partial\phi}\left(\frac{r}{r\sin\theta}\frac{\partial f}{\partial\phi}\right)\right\}$$

$$= \frac{1}{r^2}\frac{\partial}{\partial r}\left(r^2\frac{\partial f}{\partial r}\right) + \frac{1}{r^2\sin\theta}\frac{\partial}{\partial\theta}\left(\sin\theta\frac{\partial f}{\partial\theta}\right) + \frac{1}{r^2\sin^2\theta}\frac{\partial^2 f}{\partial\phi^2}$$

$$= \frac{\partial^2 f}{\partial r^2} + \frac{1}{r^2}\frac{\partial^2 f}{\partial\theta^2} + \frac{1}{r^2\sin^2\theta}\frac{\partial^2 f}{\partial\phi^2} + \frac{2}{r}\frac{\partial f}{\partial r} + \frac{\cot\theta}{r^2}\frac{\partial f}{\partial\theta}.$$

**Example 8.63.** *Express the vector field* $2y\mathbf{I} - z\mathbf{J} + 3x\mathbf{K}$ *in spherical polar coordinate system.*

**Solution.** We have $x = r\sin\theta\cos\phi,\ y = r\sin\theta\sin\phi,\ z = r\cos\theta$

so that $\mathbf{R} = r\sin\theta\cos\phi\mathbf{I} + r\sin\theta\sin\phi\mathbf{J} + r\cos\theta\mathbf{K}$.

If $\mathbf{T}_r, \mathbf{T}_\theta, \mathbf{T}_\phi$ be the unit vectors along the tangents to $r, \theta, \phi$, curves respectively, then

$$\mathbf{T}_r = \frac{\partial\mathbf{R}/\partial r}{|\partial\mathbf{R}/\partial r|} = \frac{\sin\theta\cos\phi\mathbf{I} + \sin\theta\sin\phi\mathbf{J} + \cos\theta\mathbf{K}}{\sqrt{[(\sin\theta\cos\phi)^2 + (\sin\theta\sin\phi)^2 + \cos^2\theta]}}$$
$$= \sin\theta\cos\phi\mathbf{I} + \sin\theta\sin\phi\mathbf{J} + \cos\theta\mathbf{K}$$

$$\mathbf{T}_\theta = \frac{\partial\mathbf{R}/\partial\theta}{|\partial\mathbf{R}/\partial\theta|} = \frac{r\cos\theta\cos\phi\mathbf{I} + r\cos\theta\sin\phi\mathbf{J} - r\sin\theta\mathbf{K}}{\sqrt{[(r\cos\theta\cos\phi)^2 + (r\cos\theta\sin\phi)^2 + (-r\sin\theta)^2]}}$$
$$= \cos\theta\cos\phi\mathbf{I} + \cos\theta\sin\phi\mathbf{J} - \sin\theta\mathbf{K}$$

$$\mathbf{T}_\phi = \frac{\partial\mathbf{R}/\partial\phi}{|\partial\mathbf{R}/\partial\phi|} = \frac{-r\sin\theta\sin\phi\mathbf{I} + r\sin\theta\cos\phi\mathbf{J}}{\sqrt{[(-r\sin\theta\sin\phi)^2 + (r\sin\theta\cos\phi)^2]}} = -\sin\phi\mathbf{I} + \cos\phi\mathbf{J}.$$

Let the expression for $\mathbf{F} = 2y\mathbf{I} - z\mathbf{J} + 3x\mathbf{K}$ in spherical polar coordinates be

$$\mathbf{F} = f_1\mathbf{T}_r + f_2\mathbf{T}_\theta + f_3\mathbf{T}_\phi \qquad \text{...}(i)$$

Then $f_1 = \mathbf{F}\,.\,\mathbf{T}_r = (2r\sin\theta\sin\phi\mathbf{I} - r\cos\theta\mathbf{J} + 3r\sin\theta\cos\phi\mathbf{K})\,.\,(\sin\theta\cos\phi\mathbf{I} + \sin\theta\sin\phi\mathbf{J} + \cos\theta\mathbf{K})$

$= 2r\sin^2\theta\sin\phi\cos\phi - r\sin\theta\cos\theta\sin\phi + 3r\sin\theta\cos\theta\cos\phi$

$$f_2 = \mathbf{F} \,.\, \mathbf{T}_\theta = (2r \sin\theta \sin\phi \mathbf{I} - r\cos\theta \mathbf{J} + 3r\sin\theta\cos\phi\mathbf{K}) \,.\, (\cos\theta\cos\phi\mathbf{I} + \cos\theta\sin\phi\mathbf{J} - \sin\theta\mathbf{K})$$
$$= 2r\sin\theta\cos\theta\sin\phi\cos\phi - r\cos^2\theta\sin\phi - 3r\sin^2\theta\cos\phi.$$

and
$$f_3 = \mathbf{F} \,.\, \mathbf{T}_\phi = (2r\sin\theta\sin\phi\mathbf{K} - r\cos\theta\mathbf{J} + 3r\sin\theta\cos\phi\mathbf{K}).\, (-\sin\phi\mathbf{I} + \cos\phi\mathbf{J})$$
$$= -2r\sin\theta\sin^2\phi - r\cos\theta\cos\phi$$

Substituting the values of $f_1, f_2, f_3$ in $(i)$, we get the desired expression.

**Example 8.64.** *Prove that $\nabla(\cos\theta) \times \nabla\phi = \nabla\,(1/r).\ r \neq 0.$*

**Solution.** In spherical polar coordinates,

$$\nabla f = \frac{\partial f}{\partial r}\mathbf{T}_r + \frac{1}{r}\frac{\partial f}{\partial\theta}\mathbf{T}_\theta + \frac{1}{r\sin\theta}\frac{\partial f}{\partial\phi}\mathbf{T}_\phi$$

$$\therefore \quad \nabla(\cos\theta) = \frac{1}{r}\frac{\partial}{\partial\theta}(\cos\theta)\mathbf{T}_\theta = -\frac{1}{r}\sin\theta\,\mathbf{T}_\theta \qquad ...(i)$$

$$\nabla\phi = \frac{1}{r\sin\theta}\frac{\partial}{\partial\phi}(\phi)\mathbf{T}_\phi = \frac{1}{r\sin\theta}\mathbf{T}_\phi \qquad ...(ii)$$

and
$$\nabla\left(\frac{1}{r}\right) = \frac{\partial}{\partial r}(r^{-1})\mathbf{T}_r = -\frac{1}{r^2}\mathbf{T}_r$$

Now from Eqns. $(i)$ and $(ii)$, we get

$$\nabla(\cos\theta) \times \nabla\phi = -\frac{1}{r^2}\mathbf{T}_\theta \times \mathbf{T}_\phi = -\frac{1}{r^2}\mathbf{T}_r = \nabla\left(\frac{1}{r}\right).$$

**Example 8.65.** *If $F = r^2\cos\theta\ \mathbf{T}_r - \frac{1}{r}\mathbf{T}_\theta + \frac{1}{r\sin\theta}\mathbf{T}_\phi$, find the value of $\mathbf{F} \times$ curl $\mathbf{F}$.*

**Solution.** In spherical coordinates,

$$\text{curl}\,\mathbf{F} = \begin{vmatrix} \mathbf{T}_r/r^2\sin\theta & \mathbf{T}_\theta/r\sin\theta & \mathbf{T}_\phi/r \\ \dfrac{\partial}{\partial r} & \dfrac{\partial}{\partial\theta} & \dfrac{\partial}{\partial\phi} \\ f_1 & rf_2 & r\sin\theta f_3 \end{vmatrix}$$

Here $f_1 = r^2\cos\theta,\ f_2 = -1/r,\ f_3 = 1/r\sin\theta.$

$$\therefore \quad \text{curl}\,\mathbf{F} = \frac{2}{r^2\sin\theta}\begin{vmatrix} \mathbf{T}_r & r\mathbf{T}_\theta & r\sin\theta\,\mathbf{T}_\phi \\ \dfrac{\partial}{\partial r} & \dfrac{\partial}{\partial\theta} & \dfrac{\partial}{\partial\phi} \\ r^2\cos\theta & -1 & 1 \end{vmatrix} = r\sin\theta\mathbf{T}_\phi$$

$$\therefore \quad \mathbf{F} \times \text{curl}\,\mathbf{F} = \left(r^2\cos\theta\mathbf{T}_r - \frac{1}{r}\mathbf{T}_\theta + \frac{1}{r\sin\theta}\mathbf{T}_\phi\right) \times (r\sin\theta\mathbf{T}_\phi) = -(r^3\sin\theta\cos\theta\mathbf{T}_\theta + \sin\theta\mathbf{T}_r).$$

**Example 8.66 :** *What is the value of $\iint_S \vec{F}\cdot ds$, where $\vec{F} = x^3\,i + y^3 j + z^3 k$ and S is the surface of the sphere $x^2 + y^2 + z^2 = a^2$* (Ranchi, 2019)

**Solution:**
$$\text{Div}\,\vec{F} = \frac{\delta}{\delta x}(x^3) + \frac{\delta}{\delta y}(y^3) + \frac{\delta}{\delta z}(z^3)$$
$$= 3x^2 + 3y^2 + 3z^2 = 3(x^2 + y^2 + z^2) = 3r^2$$

By Divergence theorem, we have

$$\iint_S \vec{F}\cdot\vec{ds} = \iiint_V \text{div}\,\vec{F}\cdot dV$$
$$= \int_0^{2\pi}\int_0^{\pi}\int_0^{a} 3r^2\left(r^2\sin\theta\,dr\,d\theta\,d\phi\right) = 3\int_0^{2\pi}\int_0^{\pi}\int_0^{a} r^4\sin\theta\,d\theta\,d\phi\,dr$$

Since limits of all the varibles are independent of each other.

We get
$$\iint_S \vec{F}\cdot\vec{ds} = 3\int_0^{2\pi}d\phi\int_0^{\pi}\sin\theta\cdot d\theta\int_0^{a}r^4 dr = 3[\phi]_0^{2\pi}[-\cos\theta]_0^{\pi}\left[\frac{r^5}{5}\right]_0^a = \frac{12}{5}\pi a^5$$

## PROBLEMS 8.12

1. Equation of a surface in spherical coordinates is $r = \sin\theta \sin\phi$. Find the equation of this surface in rectangular coordinates. *(K.T.U., 2016)*
2. Express the following vectors in cylindrical coordinates
   (*i*) $2y\mathbf{I} - z\mathbf{J} + 3x\mathbf{K}$ (*ii*) $2x\mathbf{I} - 3y^2\mathbf{J} + zx\mathbf{K}$
3. Express the following vectors in spherical polar coordinates
   (*i*) $x\mathbf{I} + 2y\mathbf{J} + yz\mathbf{K}$ (*ii*) $xy\mathbf{I} + yz\mathbf{J} + zx\mathbf{K}$
4. Find the cylinderical and spherical coordinates of the point that has rectangular coordinates $(x, y, z) = (4, -4, 4\sqrt{6})$. *(K.T.U., 2016)*
5. Evaluate $\nabla\phi = xyz$ in cylindrical coordinates.
6. Show that $\nabla(r/\sin\theta) \times \nabla\theta = \nabla\phi$.
7. Prove that $\mathbf{V} = \dfrac{\cos\theta}{r^3}(\mathbf{T}_r/\sin\theta - \mathbf{T}_\theta/\cos\theta + r^4\mathbf{T}_\phi)$ is solenoidal.
8. Show that (*i*) $\nabla^2(\log r) = 1/r^2$ (*ii*) $\nabla \times [(\cos\theta)(\nabla\phi)] = \nabla(1/r)$.
9. Prove that $\mathbf{V} = \rho z \sin 2\phi \left[\mathbf{T}_\rho + \cot 2\phi\mathbf{T}_\phi + \dfrac{\rho}{2z}\mathbf{T}_z\right]$ is irrotational.
10. If $u, v, w$ are orthogonal curvilinear coordinates with $h_1, h_2, h_3$ as scale factors, prove that
$$\left[\frac{\partial\mathbf{R}}{\partial u}, \frac{\partial\mathbf{R}}{\partial v}, \frac{\partial\mathbf{R}}{\partial w}\right] = \frac{1}{[\nabla u, \nabla v, \nabla w]} = h_1h_2h_3.$$

## 8.22 OBJECTIVE TYPE OF QUESTIONS

## PROBLEMS 8.13

*Fill up the blanks or choose the correct answer from the following problems :*

1. A unit tangent vector to the surface $x = t, y = t^2, z = t^3$ at $t = 1$ is .......
2. The equation of the normal to the surface $2x^2 + y^2 + 2z = 3$ at $(2, 1, -3)$ is .......
3. If $u = u(x, y)$ and $v = v(x, y)$, then the area-element $dudv$ is related to the area element $dxdy$ by the relation .......
4. If $A = 2x^2\mathbf{I} - 3yz\mathbf{J} + xz^2\,\mathbf{K}$, then $\nabla . \mathbf{A}$ = .......
5. div curl $\mathbf{F}$ = .......
6. Area bounded by a simple closed curve $C$ is ....... *(Rohtak, 2011)*
7. If $S$ is a closed surface enclosing a volume $V$ and if $\mathbf{R} = x\mathbf{I} + y\mathbf{J} + z\mathbf{K}$, then $\int_S \mathbf{R}\,.\,\mathbf{N}\,ds$ = .......
8. (*i*) grad $(1/r)$ = ...... ; (*ii*) div $\mathbf{R}$ = .......; (*iii*) curl $\mathbf{R}$ = ....... *(Anna., 2016)*
9. If $\mathbf{A}$ is such that $\nabla \times \mathbf{A} = 0$, then $\mathbf{A}$ is called .......
10. If $\nabla . \mathbf{F} = 3$, then $\int_S \mathbf{F}\,.\,\mathbf{N}\,ds$ where $S$ is a surface of a unit sphere, is .......
11. If $\nabla . \mathbf{F} = 0$, then $\mathbf{F}$ is called.......
12. The directional derivative of $\phi(x, y, z) = x^2yz + 4xz^2$ at the point $(1, -2, -1)$ in the direction $PQ$ where $P = (1, 2, -1)$ and $Q = (-1, 2, 3)$ is .......
13. If $u = x^2yz$, $v = xy - 3z^2$, then $\nabla . (\nabla u \times \nabla v)$ =.......
14. curl $(xy\mathbf{I} + yz\mathbf{J} + zx\mathbf{K})$ = .......
15. If $\mathbf{F} = f_1\mathbf{I} + f_2\mathbf{J} + f_3\mathbf{K}$, then $\nabla . \mathbf{F}$ = .......
16. If $\mathbf{F}$ is a conservative force field then curl $\mathbf{F}$ is .......
17. If $\phi = 3x^2y - y^3z^2$, grad $\phi$ at the point $(1, -2, -1)$ is....
18. curl $(x\mathbf{I} + y\mathbf{J} + z\mathbf{K})$ = ........
19. Workdone by a particle along the square formed by the lines $y = \pm 1$ and $x = \pm 1$ under the force $\mathbf{F} = (x^2 + xy)\mathbf{I} + (x^2 + y^2)\mathbf{J}$ is.......
20. Curl (grad $\phi$) = .......
21. If $\mathbf{A}$ is a constant vector, then *div* $(\mathbf{A} \times \mathbf{R})$ = .......
22. If $\mathbf{R} = x\mathbf{I} + y\mathbf{J} + z\mathbf{K}$, then $\nabla^2 \log r$ = ....... ; $\nabla(r^n)$ = ........ *(Anna., 2016)*
23. A level surface is defined as .......
24. Unit normal vector to the surface $z = 2xy$ at the point $(2, 1, 4)$ is .......
25. If the directional derivative of $f = ax + by + cz$ at $(1, 1, 1)$ has maximum magnitude 4 in direction parallel to $x$-axis, then the values of $a, b, c$ are .......
26. Maximum value of the directional derivative of $\phi = x^2 - 2y^2 + 4z^2$ at the point $(1, 1, -1)$ is .......

27. If $r^2 = x^2 + y^2 + z^2$, then $\nabla \cdot (\mathbf{R}/r) =$ ......
28. Directional derivative of $f = xyz$ at the point $(1, -1, -2)$ in the direction of the vector $2\mathbf{I} - 2\mathbf{J} + \mathbf{K}$ is .......
29. If $\mathbf{V} = x^2\mathbf{I} + xye^x\mathbf{J} + \sin z\mathbf{K}$, then $\nabla \cdot (\nabla \times F) =$ .......
30. If $f = \tan^{-1}(y/x)$ then div (grad $f$) is equal to
(a) 1 (b) $-1$ (c) 0 (d) 2.
31. The value of curl (grad $f$), where $f = 2x^2 - 3y^2 + 4z^2$ is
(a) $4x - 6y + 8z$, (b) $4x\mathbf{I} - 6y\mathbf{J} + 8z\mathbf{K}$ (c) 0 (d) 3.
32. The value of $\int \text{grad}\,(x + y - z)\, d\mathbf{R}$ from $(0, 1, -1)$ to $(1, 2, 0)$ is
(a) 0 (b) 3 (c) $-1$ (d) not obtainable.
33. If $\mathbf{F} = ax\mathbf{I} + by\mathbf{J} + cz\mathbf{K}$, then $\int_S \mathbf{F} \cdot d\mathbf{S}$, $S$ being the surface of a unit sphere, is
(a) $(4/3)\,\pi\,(a + b + c)^2$ (b) 0 (c) $4\pi/3\,(a + b + c)$ (d) none of these.
34. A necessary and sufficient condition that the line integral $\int_L \mathbf{F} \cdot d\mathbf{R}$ for every closed $C$ vanishes, is
(a) curl $\mathbf{F} = 0$ (b) div $\mathbf{F} = 0$ (c) curl $\mathbf{F} \neq 0$ (d) div $\mathbf{F} \neq 0$.
35. The value of $\iint_S (yzdydz + zxdz\,dx + xy\,dxdy)$, where $S$ is the surface of unit sphere $x^2 + y^2 + z^2 = 1$ is
(a) 0 (b) $4\pi$ (c) $4\pi/3$ (d) $10\pi$.
36. If $u = x^2 + y^2 + z^2$ and $\mathbf{V} = x\mathbf{I} + y\mathbf{J} + z\mathbf{K}$, then $\nabla(u\mathbf{V}) =$ ........
37. For any scalar function $\psi$, $\nabla \times \nabla \psi =$ .......
38. $\int_C \mathbf{F} \cdot d\mathbf{R}$ is independent of the path joining any two points if and only if it is .......
39. The value of the line integral $\int_C (y^2dx + x^2dy)$ where $C$ is the boundary of the square $-1 \le x \le 1, -1 \le y \le 1$ is
(a) 0 (b) $2(x + y)$ (c) 4 (d) 4/3. *(V.T.U., 2010)*
40. If $\mathbf{V}$ is the instantaneous velocity vector of the moving fluid at a point $P$, then div $\mathbf{V}$ represents .......
41. The spherical coordinate system is
(a) Orthogonal (b) Coplanar (c) Non-coplanar (d) Not orthogonal. *(V.T.U., 2010)*
42. Physical interpretation of $\nabla \phi$ is that ........
43. The magnitude of the vector drawn perpendicular to the surface $x^2 + 2y^2 + z^2 = 7$ at the point $(1, -1, 2)$ is
(a) 2/3 (b) 3/2 (c) 3 (d) 6.
44. The value of $\lambda$ so that the vector $(x + 3y)\mathbf{I} + (y - 2z)\mathbf{J} + (x + \lambda z)\mathbf{K}$ is a solenoidal vector, is
(a) $-2$ (b) 3 (c) 1 (d) none of these.
45. The work done by the force $\mathbf{F} = yz\mathbf{I} + zx\mathbf{J} + xy\mathbf{K}$, in moving a particle from the point $(1, 1, 1)$ to the point $(3, 3, 2)$ along the path $c$ is
(a) 17 (b) 10 (c) 0 (d) cannot be found.
46. Value of $\int_c (y^2dx + x^2dy)$ where $c$ is the boundary of the square $-1 \le x \le 1, -1 \le y \le 1$, is
(a) 4 (b) 0 (c) $2(x + y)$ (d) 4/3.
47. The directional derivative of $f(x, y) = (x^2 - y^2)/xy$ at $(1, 1)$ is zero along a ray making an angle with the positive direction of $x$-axis :
(a) 45° (b) 60° (c) 135° (d) none of these.
48. The vector $\mathbf{V} = e^x \sin y\mathbf{I} + e^x \cos y\mathbf{J}$, is
(a) solenoidal (b) irrotational (c) rotational.
49. In orthogonal curvilinear coordinates the value of $\dfrac{\partial(x, y, z)}{\partial(u, v, w)}$ is
(a) $h_1h_2h_3$ (b) $1/h_1h_2h_3$ (c) $h_1/h_2h_3$ (d) $h_1h_2/h_3$. *(V.T.U., 2015)*
50. If $u = 1/r$ where $r^2 = x^2 + y^2$, then $\nabla^2 u = 0$. (True or False)
51. $\mathbf{F} = (x + 3y)\,\mathbf{I} + (z - 3y)\mathbf{J} + (x + 2z)\mathbf{K}$ is a solenoidal vector function. (True or False)
52. $\mathbf{F} = yz\mathbf{I} + zx\mathbf{J} + xy\mathbf{K}$ is irrotational. (True or False)
53. $\mathbf{F} = (y^2 - z^2 + 3yz - 2x)\,\mathbf{I} + (3xz + 2xy)\,\mathbf{J} + (3xy - 2xz + 2z)\,\mathbf{K}$ is both solenoidal and irrotational. (True or False)
*(Calicut, 2013)*

CHAPTER 9

# Infinite Series

1. Introduction. 2. Sequences. 3. Series : Convergence. 4. General properties. 5. Series of positive terms— 6. Comparison tests. 7. Integral test. 8. Comparison of ratios. 9. D'Alembert's ratio test. 10. Raabe's test, Logarithmic test. 11. Cauchy's root test. 12. Alternating series ; Leibnitz's rule. 13. Series of positive or negative terms. 14. Power series. 15. Convergence of Exponential, Logarithmic and Binomial series. 16. Procedure for testing a series for convergence. 17. Uniform convergence. 18. Weierstrass's M-test. 19. Properties of uniformly convergent series. 20. Objective Type of Questions.

## 9.1 INTRODUCTION

Infinite series occur so frequently in all types of problems that the necessity of studying their convergence or divergence is very important. Unless a series employed in an investigation is convergent, it may lead to absurd conclusions. Hence it is essential that the students of engineering begin by acquiring an intelligent grasp of this subject.

## 9.2 SEQUENCES

**(1)** An ordered set of real numbers, $a_1, a_2, a_3, ..., a_n$ is called a *sequence* and is denoted by $(a_n)$. If the number of terms is unlimited, then the sequence is said to be an *infinite sequence* and $a_n$ is its *general term*.

For instance (*i*) 1, 3, 5, 7, ..., $(2n - 1)$, ..., (*ii*) 1, 1/2, 1/3, ..., $1/n$, ...,

(*iii*) 1, –1, 1, –1, ..., $(-1)^{n-1}$, ... are infinite sequences.

**(2) Limit.** A sequence is said to tend to a limit $l$, if for every $\varepsilon > 0$, a value $N$ of $n$ can be found such that $|a_n - l| < \varepsilon$ for $n \geq N$.

We then write $\underset{n \to \infty}{\text{Lt}} (a_n) = l$ or simply $(a_n) \to l$ as $n \to \infty$.

**(3) Convergence.** *If a sequence $(a_n)$ has a finite limit, it is called a* **convergent sequence.** If $(a_n)$ is not convergent, it is said to be **divergent.**

In the above examples, (*ii*) is convergent, while (*i*) and (*iii*) are divergent.

**(4) Bounded sequence.** A sequence $(a_n)$ is said to be bounded, if there exists a number $k$ such that $a_n < k$ for every $n$.

**(5) Monotonic sequence.** The sequence $(a_n)$ is said to increase steadily or to decrease steadily according as $a_{n+1} \geq a_n$ or $a_{n+1} \leq a_n$, for all values of $n$. Both increasing and decreasing sequences are called *monotonic sequences.*

A monotonic sequence always tends to a limit, finite or infinite. Thus, *a sequence which is monotonic and bounded is* **convergent.**

**(6) Convergence, Divergence and Oscillation.** If $\underset{n \to \infty}{\text{Lt}} (a_n) = l$ is finite and unique then the sequence is said to be *convergent.*

If $\underset{n \to \infty}{\text{Lt}} (a_n)$ is infinite ($\pm \infty$), the sequence is said to be *divergent*.

If $\underset{n \to \infty}{\text{Lt}} (a_n)$ is not unique, then $(a_n)$ is said to be *oscillatory*.

**Example 9.1.** *Examine the following sequences for convergence :*

*(i)* $a_n = \dfrac{n^2 - 2n}{3n^2 + n}$ *(ii)* $a_n = 2^n$ *(iii)* $a_n = 3 + (-1)^n$.

**Solution.** *(i)* $\underset{n \to \infty}{\text{Lt}} \left( \dfrac{n^2 - 2n}{3n^2 + n} \right) = \underset{n \to \infty}{\text{Lt}} \dfrac{1 - 2/n}{3 + 1/n} = 1/3$ which is finite and unique. Hence the sequence $(a_n)$ is convergent.

*(ii)* $\underset{n \to \infty}{\text{Lt}} (2^n) = \infty$. Hence the sequence $(a_n)$ is divergent.

*(iii)* $\underset{n \to \infty}{\text{Lt}} [3 + (-1)^n] = 3 + 1 = 4$ when $n$ is even

$= 3 - 1 = 2$, when $n$ is odd

*i.e.*, this sequence doesn't have a unique limit. Hence it oscillates.

## PROBLEMS 9.1

Examine the convergence of the following sequences :

1. $a_n = \dfrac{3n - 1}{1 + 2n}$
2. $a_n = 1 + 2/n$
3. $a_n = [n + (-1)^n]^{-1}$
4. $a_n = \sin n$
5. $a_n = 1/2n$
6. $a_n = 1 + (-1)^n/n$
7. $\left( \dfrac{n}{n - 1} \right)^2$
8. $a_n = 2n$.

## 9.3 SERIES

**(1) Def.** *If $u_1, u_2, u_3, \ldots, u_n, \ldots$ be an infinite sequence of real numbers, then*

$$u_1 + u_2 + u_3 + \ldots + u_n + \ldots \infty$$

*is called an infinite series.* An infinite series is denoted by $\Sigma u_n$ and the sum of its first $n$ terms is denoted by $s_n$.

**(2) Convergence, divergence and oscillation of a series.**

Consider the infinite series $\Sigma u_n = u_1 + u_2 + u_3 + \ldots + u_n + \ldots \infty$

and let the sum of the first $n$ terms be $s_n = u_1 + u_2 + u_3 + \ldots + u_n$

Clearly, $s_n$ is a function of $n$ and as $n$ increases indefinitely three possibilities arise :

*(i)* If $s_n$ tends to a finite limit as $n \to \infty$, the series $\Sigma u_n$ is said to be *convergent*.

*(ii)* If $s_n$ tends to $\pm \infty$ as $n \to \infty$, the series $\Sigma u_n$ is said to be *divergent*.

*(iii)* If $s_n$ does not tend to a unique limit as $n \to \infty$, then the series $\Sigma u_n$ is said to be *oscillatory* or *non-convergent*.

**Example 9.2.** *Examine for convergence the series (i) $1 + 2 + 3 + \ldots + n + \ldots \infty$.*
*(ii) $5 - 4 - 1 + 5 - 4 - 1 + 5 - 4 - 1 + \ldots \infty$*

**Solution.** *(i)* Here $s_n = 1 + 2 + 3 + \ldots + n = \dfrac{n(n + 1)}{2}$

$\therefore$ $\underset{n \to \infty}{\text{Lt}} s_n = \dfrac{1}{2} \underset{n \to \infty}{\text{Lt}} n(n + 1) \to \infty$. Hence this series is *divergent*.

*(ii)* Here $s_n = 5 - 4 - 1 + 5 - 4 - 1 + 5 - 4 - 1 + \ldots n$ terms

$= 0, 5$ or $1$ according as the number of terms is $3m$, $3m + 1$, $3m + 2$.

Clearly in this case, $s_n$ does not tend to a unique limit. Hence the series is *oscillatory*.

**Examples 9.3. Geometric series.** *Show that the series $1 + r + r^2 + r^3 + \dots \infty$ (i) converges if $|r| < 1$, (ii) diverges if $r \geq 1$, and (iii) oscillates if $r \leq -1$.* (U.K.T.U., 2012; Rohtak, 2010 S)

**Solution.** Let $s_n = 1 + r + r^2 + \dots + r^{n-1}$

*Case I.* When $|r| < 1$, $\underset{n\to\infty}{\text{Lt}}\ r^n = 0$.

Also $$s_n = \frac{1-r^n}{1-r} = \frac{1}{1-r} - \frac{r^n}{1-r} \text{ so that } \underset{n\to\infty}{\text{Lt}}\ s_n = \frac{1}{1-r}$$

$\therefore$ the series is convergent.

*Case II.* (i) When $r > 1$, $\underset{n\to\infty}{\text{Lt}}\ r^n \to \infty$.

Also $$s_n = \frac{r^n-1}{r-1} = \frac{r^n}{r-1} - \frac{1}{r-1} \text{ so that } \underset{n\to\infty}{\text{Lt}}\ s_n \to \infty$$

$\therefore$ the series is divergent.

(ii) When $r = 1$, then $s_n = 1 + 1 + 1 \dots + 1 = n$

and $\underset{n\to\infty}{\text{Lt}}\ s_n \to \infty$ $\therefore$ The series is divergent.

*Case III.* (i) When $r = -1$, then the series becomes $1 - 1 + 1 - 1 + 1 - 1 \dots$ which is an oscillatory series.

(ii) When $r < -1$, let $r = -\rho$ so that $\rho > 1$. Then $r^n = (-1)^n \rho^n$

and $$s_n = \frac{1-r^n}{1-r} = \frac{1-(-1)^n\rho^n}{1+\rho} \text{ as } \underset{n\to\infty}{\text{Lt}}\ \rho^n \to \infty.$$

$\therefore$ $\underset{n\to\infty}{\text{Lt}}\ s_n \to -\infty$ or $+\infty$ according as $n$ is even or odd. Hence the series oscillates.

**PROBLEMS 9.2**

Examine the following series for convergence :

1. $1 + \frac{1}{2} + \frac{1}{4} + \frac{1}{8} + \dots \infty$.
2. $1 - \frac{1}{3} + \frac{1}{3^2} - \frac{1}{3^3} + \frac{1}{3^4} - \dots \infty$.
3. $6 - 10 + 4 + 6 - 10 + 4 + 6 - 10 + 4 + \dots \infty$.
4. $\frac{1}{1\cdot 2} + \frac{1}{2.3} + \frac{1}{3.4} + \dots \infty$.
5. A ball is dropped from a height $h$ metres. Each time the ball hits the ground, it rebounds a distance $r$ times the distance fallen where $0 < r < 1$. If $h = 3$ metres and $r = 2/3$, find the total distance travelled by the ball.

## 9.4 GENERAL PROPERTIES OF SERIES

The truth of the following properties is self-evident and these may be regarded as axioms :

1. *The convergence or divergence of an infinite series remains unaffected by the addition or removal of a finite number of its terms ;* for the sum of these terms being the finite quantity does not on addition or removal alter the nature of its sum.

2. *If a series in which all the terms are positive is convergent, the series remains convergent even when some or all of its terms are negative ;* for the sum is clearly the greatest when all the terms are positive.

3. *The convergence or divergence of an infinite series remains unaffected by multiplying each term by a finite number.*

## 9.5 SERIES OF POSITIVE TERMS

1. *An infinite series in which all the terms after some particular terms are positive, is a positive term series.* *e.g.,* $-7 - 5 - 2 + 2 + 7 + 13 + 20 + \dots$ is a positive term series as all its terms after the third are positive.

2. *A series of positive terms either converges or diverges to $+\infty$ ;* for the sum of its first $n$ terms, omitting the negative terms, tends to either a finite limit or $+\infty$.

3. **Necessary condition for convergence.** *If a positive term series $\Sigma u_n$ is convergent, then* $\underset{n\to\infty}{Lt}\, u_n = 0$.

Let $s_n = u_1 + u_2 + u_3 + \ldots + u_n$. Since $\Sigma u_n$ is given to be convergent.

$\therefore$ $\underset{n\to\infty}{\text{Lt}}\, s_n$ = a finite quantity $k$ (say). Also $\underset{n\to\infty}{\text{Lt}}\, s_{n-1} = k$

But $u_n = s_n - s_{n-1}$ $\therefore$ $\underset{n\to\infty}{\text{Lt}}\, u_n = \underset{n\to\infty}{\text{Lt}}\,(s_n - s_{n-1}) = 0.$

Hence the result.

**Obs. 1.** *It is important to note that the converse of this result is not true.*

Consider, for instance, the series $1 + \frac{1}{\sqrt{2}} + \frac{1}{\sqrt{3}} + \ldots + \frac{1}{\sqrt{n}} + \ldots \infty$

Since the term go on descending,

$$\therefore \quad s_n = 1 + \frac{1}{\sqrt{2}} + \frac{1}{\sqrt{3}} + \ldots + \frac{1}{\sqrt{n}} > \frac{1}{\sqrt{n}} + \frac{1}{\sqrt{n}} + \frac{1}{\sqrt{n}} + \ldots + \frac{1}{\sqrt{n}} = \frac{n}{\sqrt{n}} \quad i.e., \ \sqrt{n}$$

$$\therefore \quad \underset{n\to\infty}{\text{Lt}}\, s_n = \underset{n\to\infty}{\text{Lt}}\, \sqrt{n} \to \infty$$

Thus the series is divergent even though $\underset{n\to\infty}{\text{Lt}}\, u_n = \underset{n\to\infty}{\text{Lt}}\, \frac{1}{\sqrt{n}} = 0$

*Hence* $\underset{n\to\infty}{Lt}\, u_n = 0$ *is a necessary but not sufficient condition for convergence of* $\Sigma n_n$.

**Obs. The above result leads to a simple test for divergence :**

*If* $\underset{n\to\infty}{Lt}\, u_n \neq 0$, *the series* $\Sigma u_n$ *must be divergent.*

## 9.6 COMPARISON TESTS

*I. If two positive term series* $\Sigma u_n$ *and* $\Sigma v_n$ *be such that*

(*i*) $\Sigma v_n$ *converges,* (*ii*) $u_n \le v_n$ *for all values of n, then* $\Sigma u_n$ *also converges.*

*Proof.* Since $\Sigma v_n$ is convergent,

$$\therefore \quad \underset{n\to\infty}{\text{Lt}}\,(v_1 + v_2 + v_3 + \ldots + v_n) = \text{a finite quantity } k \text{ (say)}$$

Also since $u_1 \le v_1, u_2 \le v_2, \ldots, u_n \le v_n$

$\therefore$ Adding, $u_1 + u_2 + \ldots + u_n \le v_1 + v_2 + \ldots + v_n$

$$\therefore \quad \underset{n\to\infty}{\text{Lt}}\,(u_1 + u_2 + \ldots + u_n) \le \underset{n\to\infty}{\text{Lt}}\,(v_1 + v_2 + \ldots + v_n) = k.$$

Hence the series $\Sigma u_n$ also converges.

**Obs.** If, however, the relation $u_n \le v_n$ holds for values of $n$ greater than a fixed number $m$, then the first $m$ terms of both the series can be ignored without affecting their convergence or divergence.

*II. If two positive term series* $\Sigma u_n$ *and* $\Sigma v_n$ *be such that :*

(*i*) $\Sigma v_n$ *diverges,* (*ii*) $u_n \ge v_n$ *for all values of n, then* $\Sigma u_n$ *also diverges.*

Its proof is similar to that of Test *I*.

*III.* **Limit form**

*If two positive term series* $\Sigma u_n$ *and* $\Sigma v_n$ *be such that*

$$\underset{n\to\infty}{Lt}\, \frac{u_n}{v_n} = \textit{finite quantity } (\neq 0), \textit{ then } \Sigma u_n \textit{ and } \Sigma v_n \textit{ converge or diverge together.}$$

*Proof.* Since $\underset{n\to\infty}{\text{Lt}}\, \frac{u_n}{v_n} = l$, a finite number ($\neq 0$)

By definition of a limit, there exists a positive number $\varepsilon$, however small, such that

$$\left| \frac{u_n}{v_n} - l \right| < \varepsilon \qquad \text{for } n \ge m$$

or $$-\varepsilon < \frac{u_n}{v_n} - l < \varepsilon \qquad \text{for } n \geq m$$

or $$l - \varepsilon < \frac{u_n}{v_n} < l + \varepsilon \qquad \text{for } n \geq m$$

Omitting the first $m$ terms of both the series, we have

$$l - \varepsilon < \frac{u_n}{v_n} < l + \varepsilon \qquad \text{for all } n \qquad ...(1)$$

*Case I. When $\Sigma v_n$ is convergent, then*

$$\underset{n \to \infty}{\text{Lt}} (v_1 + v_2 + ... + v_n) = k, \text{ a finite number} \qquad ...(2)$$

Also from Eqn. (1), $\frac{u_n}{v_n} < l + \varepsilon$, *i.e.*, $u_n < (l + \varepsilon)v_n$ for all $n$.

$$\therefore \quad \underset{n \to \infty}{\text{Lt}} (u_1 + u_2 + ... + u_n) < (l + \varepsilon) \underset{n \to \infty}{\text{Lt}} (v_1 + v_2 + ... + v_n) = (l + \varepsilon)k \qquad \text{[By Eqn. (2)]}$$

Hence $\Sigma u_n$ is also convergent.

*Case II. When $\Sigma v_n$ is divergent,* then

$$\underset{n \to \infty}{\text{Lt}} (v_1 + v_2 + ... + v_n) \to \infty \qquad ...(3)$$

Also from (1) $\quad l - \varepsilon < \frac{u_n}{v_n}$ or $u_n > (l - \varepsilon)v_n$ for all $n$

$$\therefore \quad \underset{n \to \infty}{\text{Lt}} (u_1 + u_2 + ... + u_n) > (l - \varepsilon) \underset{n \to \infty}{\text{Lt}} (v_1 + v_2 + ... + v_n) \to \infty \qquad \text{[By Eqn. (3)]}$$

Hence $\Sigma u_n$ is also divergent.

## 9.7 INTEGRAL TEST

*A positive term series $f(1) + f(2) + ... + f(n) + ...$, where $f(n)$ decreases as $n$ increases, converges or diverges according as the integral*

$$\int_1^\infty f(x)\, dx \qquad ...(1) \text{ is finite or infinite.}$$

The area under the curve $y = f(x)$, between any two ordinates lies between the set of inscribed and escribed rectangles formed by ordinates at $x = 1, 2, 3, ...$ as in Fig. 9.1. Then

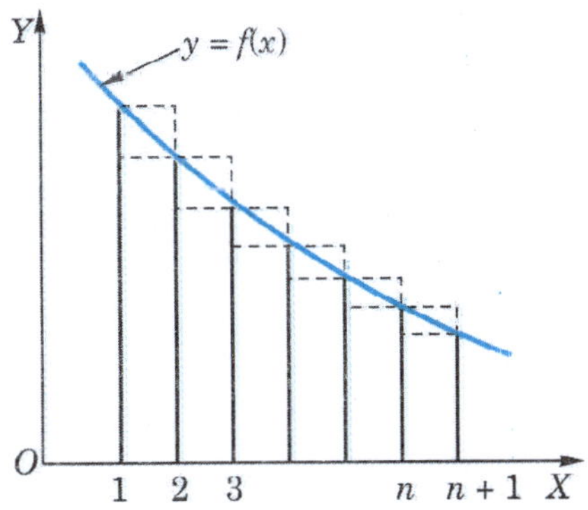

Fig. 9.1

$$f(1) + f(2) + ... + f(n) \geq \int_1^{n+1} f(x)\, dx \geq f(2) + f(3) + ... + f(n + 1)$$

or $$s_n \geq \int_1^{n+1} f(x)\, dx \geq s_{n+1} - f(1)$$

Taking limits as $n \to \infty$, we find from the second inequality that Lt $s_{n+1} \leq \int_1^\infty f(x)\, dx + f(1)$.

Hence if integral (1) is finite, so is Lt $s_{n+1}$. Similarly, from the first inequality, we see that if the integral (1) is infinite, so is Lt $s_n$. But the given series either converges or diverges to $+\infty$, *i.e.*, Lt $s_n$ is either finite or infinite as $n \to \infty$.

Hence the result follows.

**Example 9.4. Test for Comparison.** *Show that the p-series* *(D.T.U., 2015)*

$$\sum_{n=1}^{\infty} \frac{1}{n^p} = \frac{1}{1^p} + \frac{1}{2^p} + \frac{1}{3^p} + ... \infty$$

*(i) converges for $p > 1$ (ii) diverges for $p \leq 1$.* *(W.B.T.U., 2016)*

**Solution.** By the above test, this series will converge or diverge according as $\int_1^\infty \frac{dx}{x^p}$ is finite or infinite.

If $p \neq 1$, $\int_1^\infty \frac{dx}{x^p} = \underset{m\to\infty}{\text{Lt}} \int_1^m \frac{dx}{x^p} = \underset{m\to\infty}{\text{Lt}} \left(\frac{m^{1-p}-1}{1-p}\right)$

$= \frac{1}{p-1}$, *i.e.* finite for $p > 1$

$\to \infty$ for $p < 1$

If $p = 1$, $\int_1^\infty \frac{dx}{x} = \int_1^\infty \log x \to \infty$, this proves the result.

**Obs. Application of comparison tests.** Of all the above tests the 'limit form' is the most useful. To apply this comparison test to a given series $\Sigma u_n$, the auxiliary series $\Sigma v_n$ must be so chosen that Lt $(u_n/v_n)$ is non-zero and finite. To do this, we take $v_n$ equal to that term of $u_n$ which is of the highest degree in $1/n$ and the convergence or divergence of $v_n$ is known with the help of the above series.

**Example 9.5.** *Test for convergence the series*

(*i*) $\frac{1}{1.2.3} + \frac{3}{2.3.4} + \frac{5}{3.4.5} + \dots \infty$ (*Andhra, 2018 ; Anna, 2016; U.K.T.U., 2012*)

(*ii*) $\frac{1}{4.7.10} + \frac{4}{7.10.13} + \frac{9}{10.13.16} + \dots \infty$ (*Andhra, 2018 ; V.T.U., 2010*)

(*iii*) $\frac{1}{1+2} + \frac{1}{1+2^2} + \frac{1}{1+2^3}$

**Solution.** (*i*) We have $u_n = \frac{2n-1}{n(n+1)(n+2)} = \frac{1}{n^2}\frac{2-1/n}{(1+1/n)(1+2/n)}$

Take $v_n = 1/n^2$; then

$$\underset{n\to\infty}{\text{Lt}} \frac{u_n}{v_n} = \underset{n\to\infty}{\text{Lt}} \frac{2-1/n}{(1+1/n)(1+2/n)} = \frac{2-0}{(1+0)(1+0)}$$

= 2, which is finite and non-zero

∴ both $\Sigma u_n$ and $\Sigma v_n$ converge or diverge together.

But $\Sigma v_n = \Sigma 1/n^2$ is known to be convergent.

Hence $\Sigma u_n$ is also convergent.

(*ii*) Here $u_n = \frac{n^2}{(3n+1)(3n+4)(3n+7)} = \frac{1}{n\left(3+\frac{1}{n}\right)\left(3+\frac{4}{n}\right)\left(3+\frac{7}{n}\right)}$

Taking $v_n = \frac{1}{n}$, we find that

$$\underset{n\to\infty}{\text{Lt}} \frac{u_n}{v_n} = \underset{n\to\infty}{\text{Lt}} \frac{1}{\left(3+\frac{1}{n}\right)\left(3+\frac{4}{n}\right)\left(3+\frac{7}{n}\right)} = \frac{1}{27} \neq 0$$

Now since $\Sigma v_n$ is divergent, therefore $\Sigma u_n$ is also divergent.

(*iii*) Here $U_n = \frac{n}{1+2^n}$

and $U_{n+1} = \frac{n+1}{1+2^{n+1}}$

$$\frac{U_n}{U_{n+1}} = \frac{n}{1+2^n} \cdot \frac{1+2^{n+1}}{n+1} = \frac{n \cdot 2^{n+1}(1+1/2^{n+1})}{2^n(1+1/2^n)\cdot n(1+1/n)} = \frac{2(1+1/2^{n+1})}{(1+1/2^n)(1+1/n)}$$

$$\lim \frac{U_n}{U_{n+1}} = 2\frac{(1+0)}{(1+0)(1+0)} = 2 \text{ which is} > 1.$$

Hence, by ratio test series $\Sigma U_n$ is convergent.

**Example 9.6.** *Test the convergence of the series :*

(i) $\sum_{n=1}^{\infty} \frac{1}{\sqrt{n}+\sqrt{(n+1)}}$ (Delhi, 2012) (ii) $\frac{\sqrt{2}-1}{3^3-1}+\frac{\sqrt{3}-1}{4^3-1}+\frac{\sqrt{4}-1}{5^3-1}+\dots\infty$ (Delhi, 2012 S)

(iii) $\sum \frac{1}{n}\sin\frac{1}{n}$ (Anna, 2016; Rohtak, 2012 S)

**Solution.** (i) We have $u_n = \frac{\sqrt{(n+1)}-\sqrt{n}}{[\sqrt{(n+1)}+\sqrt{n}][\sqrt{(n+1)}-\sqrt{n}]} = \sqrt{(n+1)}-\sqrt{n}$

$$= \sqrt{n}\,[(1+1/n)^{1/2}-1] \qquad \text{(Expanding by Binomial Theorem)}$$

$$= \sqrt{n}\left\{\left(1+\frac{1}{2n}-\frac{1}{8n^2}+\dots\right)-1\right\} = \sqrt{n}\left(\frac{1}{2n}-\frac{1}{8n^2}+\dots\right) = \frac{1}{\sqrt{n}}\left(\frac{1}{2}-\frac{1}{8n}+\dots\right)$$

Taking $v_n = 1/\sqrt{n}$, we have

$$\underset{u\to\infty}{\text{Lt}} \frac{u_n}{v_n} = \underset{n\to\infty}{\text{Lt}}\left(\frac{1}{2}-\frac{1}{8n}+\dots\right) = \frac{1}{2}, \text{ which is finite and non-zero.}$$

$\therefore$ both $\Sigma u_n$ and $\Sigma v_n$ converge or diverge together.

But $\Sigma v_n = \Sigma 1/\sqrt{n}$ is known to be divergent. Hence $\Sigma u_n$ is also divergent.

(ii) We have $u_n = \frac{\sqrt{(n+1)}-1}{(n+2)^3-1} = \frac{\sqrt{n}}{n^3}\left[\frac{\sqrt{1+1/n}-1/\sqrt{n}}{(1+2/n)^3-1/n^3}\right]$

Taking $v_n = \frac{1}{n^{5/2}}$, we find that

$$\underset{n\to\infty}{\text{Lt}} \frac{u_n}{v_n} = \underset{n\to\infty}{\text{Lt}} \frac{[\sqrt{(1+1/n)}-1/\sqrt{n}]}{[(1+2/n)^3-1/n^3]} = 1 \neq 0$$

Since $\Sigma v_n$ is convergent, therefore $\Sigma u_n$ is also convergent.

(iii) Here $u_n = \frac{1}{n}\sin\frac{1}{n} = \frac{1}{n}\left[\frac{1}{n}-\frac{1}{3!\,n^3}+\frac{1}{5!\,n^5}-\dots\right] = \frac{1}{n^2}\left[1-\frac{1}{3!\,n^2}+\frac{1}{5!\,n^4}-\dots\right]$

Taking $v_n = \frac{1}{n^2}$, we have

$$\underset{n\to\infty}{\text{Lt}} \frac{u_n}{v_n} = \underset{n\to\infty}{\text{Lt}}\left[1-\frac{1}{3!\,n^2}+\frac{1}{5!\,n^4}\dots\right] = 1 \neq 0$$

Since $\Sigma v_n$ is convergent, therefore $\Sigma u_n$ is also convergent.

**Example 9.7.** *Determine the nature of the series :*

(i) *Test for convergence the series* $\frac{1}{a\cdot 1^2+b}+\frac{2}{a\cdot 2^2+b}+\frac{3}{a\cdot 3^2+b}$

(ii) *Test the convergence of the series* $\frac{1}{(m+1)(m+2)}+\frac{1}{(m+2)(m+3)}+\frac{1}{(m+3)(m+4)}+\dots\dots$

(iii) $\sum_{1}^{\infty}\frac{(\log n)^2}{n^{3/2}}$ (iv) $\sum_{2}^{\infty}\frac{1}{n(\log n)^p}\ (p>0)$ (P.T.U., 2010)

**Solution.** (i) Here $\Sigma U_n = \frac{1}{a\cdot 1^2+b}+\frac{2}{a\cdot 2^2+b}+\frac{3}{a\cdot 3^2+b} = \Sigma\left(\frac{n}{an^2+b}\right)$

$$U_n = \frac{n}{an^2+b} \qquad \therefore \text{ Let } V_n = \frac{n}{n^2} = \frac{1}{n}$$

Now,
$$\lim \frac{U_n}{V_n} = \lim\left(\frac{n}{an^2+b}\cdot n\right)$$

$$= \lim\left(\frac{1}{an^2+bn}\right) = \frac{1}{a} \qquad \text{[which is finite and non-zero]}$$

But the auxiliary series $\Sigma V_n = \Sigma\frac{1}{n}$ is divergent because here $p = 1$.

Hence by comparison test the given series is also divergent.

(*ii*)
$$U_n = \frac{1}{(m+n)(m+n+1)} = \frac{1}{m+n} - \frac{1}{m+n+1}$$

$$S_n = U_1 + U_2 + .... U_n$$

$$= \left(\frac{1}{m+1} - \frac{1}{m+2}\right) + \left(\frac{1}{m+2} - \frac{1}{m+3}\right) + \left(\frac{1}{m+3} - \frac{1}{m+4}\right)$$

$$+..... + \left(\frac{1}{m+n-1} - \frac{1}{m+n}\right) + \left(\frac{1}{m+n} - \frac{1}{m+n+1}\right)$$

$$S_n = \frac{1}{m+1} - \frac{1}{m+n+1}$$

$$\lim_{Sn} = \lim\left(\frac{1}{m+1} - \frac{1}{m+n+1}\right) = \frac{1}{m+1}$$

which is a finite quantity.

Hence the given series is convergent.

(*iii*) We have $\underset{n\to\infty}{\text{Lt}} \frac{(\log n)^2}{n^{1/4}} = 0$, *i.e.*, $\frac{(\log n)^2}{n^{1/4}} < 1$ or $(\log n)^2 < n^{1/4}$

$$\therefore \qquad u_n = \frac{(\log n)^2}{n^{3/2}} < \frac{n^{1/4}}{n^{3/2}} = \frac{1}{n^{5/4}}$$

Since $\Sigma 1/n^{5/4}$ converges by $p$-series. $(\because \ p = 5/4 > 1)$

Hence by comparison test, $\Sigma u_n$ also converges.

(*iv*) Let
$$f(n) = \frac{1}{n(\log n)^p} \text{ so that } f(x) = \frac{(\log x)^{-p}}{x}$$

$$\therefore \qquad f'(x) = \frac{-p}{x}(\log x)^{-p-1}\cdot\frac{1}{x} + (\log x)^{-p}\cdot\left(-\frac{1}{x^2}\right) = -\frac{1}{x^2}\left\{\frac{p}{(\log x)^{p+1}} + \frac{1}{(\log x)^p}\right\} < 0$$

*i.e.*, $f(x)$ is a decreasing function.

Also
$$\int_2^\infty f(x)\,dx = \int_2^\infty \frac{dx}{x(\log x)^p} = \left|\frac{(\log x)^{-p+1}}{-p+1}\right|_2^\infty$$

*If* $p > 1$, then $\qquad p - 1 = k(\text{say}) > 0$

$$\therefore \qquad \int_2^\infty f(x)\,dx = \left|\frac{(\log x)^{-k}}{-k}\right|_2^\infty = \frac{1}{k}[0 + (\log 2)^{-k}] \text{ which is finite}$$

Thus by integral test, *the given series converges for $p > 1$.*

*If* $p < 1$, then $1 - p > 0$ and $(\log x)^{1-p} \to \infty$ as $x \to \infty$.

$$\therefore \qquad \int_2^\infty f(x)\,dx \to \infty.$$

Thus *the given series diverges for p < 1.*

*If p = 1,* then $\int_2^\infty f(x)\,dx = \int_2^\infty \frac{dx}{x \log x} = |\log(\log x)|_2^\infty \to \infty$

Thus *the given series diverges for* $p = 1$.

**Example 9.8 :** *Test the convergence or divergence of the series* (P.T.U., 2014 S)

$$\frac{2}{1^p} + \frac{3}{2^p} + \frac{4}{3^p} + \frac{5}{4^p} + \ldots\ldots \infty$$

**Solution:** Here $U_n = \dfrac{n+1}{n^p}$

$\therefore$ Now let $\sum u_n = \sum \dfrac{1}{n^{p-1}}$ be the auxiliary series

$\therefore$ $u_n = \dfrac{1}{n^{p-1}}$

$$\frac{u_n}{v_n} = \frac{n+1}{n^{10}} \cdot n^{p-1} = \frac{n+1}{n} = \left(1 + \frac{1}{n}\right)$$

$\therefore$ $\underset{\bar{n}\to\infty}{\text{Lt}} \dfrac{u_n}{v_n} = \underset{\bar{n}\to\infty}{\text{Lt}} \left(1 + \dfrac{1}{n}\right)$

$= 1$ [a non-zero finite quantity]

Hence $\Sigma u_n$ and $\Sigma v_n$ converage or diverge together.

But $\sum u_n = \Sigma \dfrac{1}{n^{p-1}}$ is

(*i*) Convergent if $p - 1 > 1$ *i.e.*, $p > 2$

(*ii*) Divergent if $p - 1 \le 1$ *i.e.*, $p \le 2$

$\therefore$ $\sum u_n$ Even converges if $p > 2$ and diverges if $p \le 2$.

**Example 9.9 :** *Discuss the nature of the series.*

$$\sum U_n = \sum\left(\frac{2}{5n+1}\right)$$

**Solution:** Here $Un = \dfrac{2}{5n+1}$

Let $\sum v_n = \sum \dfrac{1}{n}$

$v_n = \dfrac{1}{n}$

Now, $\dfrac{u_n}{v_n} = \dfrac{2n}{5n+1} = \dfrac{2}{\left(5 + \dfrac{1}{n}\right)}$

$\therefore$ $\underset{n\to\infty}{\text{Lt}} \dfrac{u_n}{v_n} = \lim\limits_{n\to\infty} \dfrac{2}{(5 + 1/n)} = \dfrac{2}{5}$ [a non-zero finite quantity]

Hence $\Sigma u_n$ and $\Sigma v_n$ converge or divergent together. But $\sum v_n = \sum \frac{1}{n}$ is known to be divergent, therefore the given series is also divergent.

**Example 9.10 :** *Test the convergence of the series* $1+\frac{2^2}{2!}+\frac{3^2}{3!}+\frac{4^2}{3!}+....$

**Solution:** Here,
$$U_n=\frac{n^2}{n!}$$

and
$$U_{n+1}=\frac{(n+1)^2}{(n+1)!}$$

Now,
$$\frac{U_n}{U_{n+1}}=\frac{n^2}{n!}\cdot\frac{(n+1)!}{(n+1)^2}=\frac{n^2(n+1)}{(n+1)^2}=\frac{n^2}{n+1}=\frac{n}{1+\frac{1}{n}}$$

$\therefore$
$$\lim\frac{U_n}{U_{n+1}}=\lim\frac{n}{(1+1/n)}$$
$$=\infty,\text{ which is} > 1.$$

Hence by the ratio test the series $\Sigma U_n$ is convergent.

**Example 9.11 :** *Test for convergence the series* $\frac{1}{1\cdot 2\cdot 3}+\frac{3}{2\cdot 3\cdot 4}+\frac{5}{3\cdot 4\cdot 5}+....$

**Solution:** Here
$$\Sigma U_n=\frac{1}{1\cdot 2\cdot 3}+\frac{3}{2\cdot 3\cdot 4}+\frac{5}{3\cdot 4\cdot 5}$$
$$\Sigma U_n=\Sigma\frac{(2n-1)}{n(n+1)(n+2)}$$
$$U_n=\frac{(2n-1)}{n(n+1)(n+2)}$$

$\therefore$ Let
$$V_n=\frac{n}{n\cdot n:n}=\frac{1}{n^2}$$
$$\lim\frac{U_n}{V_n}=\lim\frac{(2n-1)n^2}{n(n+1)(n+2)}=\lim\frac{(2-1)/n}{(1+1/n)(1+2/n)}$$
$$\lim\frac{U_n}{V_n}=2,\text{ which is finite and no-zero.}$$

But the auxiliary series $\Sigma V_n=\Sigma\frac{1}{n^2}$ is convergent because $p = 2.\ e > 1$.

Hence by comparison test, the given series is also convergent.

## PROBLEMS 9.3

Test the following series for convergence :

1. $1-\frac{1}{3}+\frac{1}{3^2}-\frac{1}{3^3}+\frac{1}{3^4}-...\infty$
2. Test for convergence of the series. $\sqrt{\frac{1}{2}}+\sqrt{\frac{2}{3}}+\sqrt{\frac{3}{4}}+....+\sqrt{\frac{n}{n+1}}+....$ (*Bhilai-2019*)
3. $\frac{1}{2^2}\cdot\frac{1}{3^2}+\frac{5}{3^2\cdot 4^2}+\frac{7}{4^2\ 5^2}+.....$ (*KTU 2018*)
4. $\frac{1}{1.3}+\frac{2}{3.5}+\frac{3}{5.7}+...\infty$
5. $\frac{3}{5}+\frac{4}{5^2}+\frac{3}{5^3}+\frac{4}{5^4}....\infty$
6. $\frac{1}{1.2}+\frac{2}{3.4}+\frac{3}{5.6}+...$

7. $\frac{1}{1^2}+\frac{1+2}{1^2+2^2}+\frac{1+2+3}{1^2+2^2+3^2}+\dots\infty$

8. $\frac{1}{1.3.5}+\frac{2}{3.5.7}+\frac{3}{5.7.9}+\dots\infty$

9. $\frac{3}{1}+\frac{4}{8}+\frac{5}{27}+\frac{6}{64}+\dots\infty$

10. $\sum\frac{\sqrt{n}}{n^2+1}$

11. $\sum_{n=0}^{\infty}\frac{2n^3+5}{4n^5+1}$

12. $\sum\frac{(n+1)(n+2)}{n^2\sqrt{n}}$

13. $\sum_{n=1}^{\infty}[\sqrt{(n^2+1)}-n]$ (*JNTUK, 2020 ; D.T.C., 2013*)

14. $\sum[\sqrt[3]{(n^3+1)}-n]$ (*JNTUK, 2020 ; Andhra, 2016*)

15. $\sum[\sqrt{(n^4+1)}-\sqrt{(n^4-1)}]$ (*JNTUK, 2020 ; Rohtak, 2012 S*)

16. $\sum\frac{1}{\sqrt{n}}\tan\frac{1}{n}$ (*Delhi, 2012*)

17. $\sum\frac{\tan^{-1}n}{n^{1.1}}$ (*P.T.U., 2013*)

18. $\sum_{n=2}^{\infty}\frac{1}{\sqrt{n\log n}}$

19. $\sum_{n=1}^{\infty}\frac{\sqrt{(n+1)}-1}{(n+2)^3-1}$

20. $\frac{1}{2}+\frac{\sqrt{2}}{5}+\frac{\sqrt{3}}{10}+\frac{\sqrt{4}}{17}+\dots$

21. $x+\frac{3}{5}x^2+\frac{8}{10}x^3+\dots\frac{n^2-1}{n^2+1}x^n+\dots x>0.$

22. $1+\frac{2p}{2!}+\frac{3p}{3!}+\frac{4p}{4!}+\dots+\infty$

23. $1^2+2^2x+3^2x^2+\dots+(0<x<1).$

24. $\frac{1}{e}+\frac{8}{e^3}+\frac{27}{e^5}+\dots$

## 9.8 COMPARISON OF RATIOS

*If $\Sigma u_n$ and $\Sigma v_n$ be two positive term series, then $\Sigma u_n$ converges if (i) $\Sigma v_n$ converges, and (ii) from and after some particular term,*

$$\frac{u_{n+1}}{u_n}<\frac{u_{n+1}}{v_n}$$

Let the two series beginning from the particular term be $u_1+u_2+u_3+\dots$ and $v_1+v_2+v_3+\dots$

If $\frac{u_2}{u_1}<\frac{v_2}{v_1},\frac{u_3}{u_2}<\frac{v_3}{v_2},\dots$

then $\quad u_1+u_2+u_3+\dots=u_1\left(1+\frac{u_2}{u_1}+\frac{u_3}{u_1}+\dots\right)$

$$=u_1\left(1+\frac{u_2}{u_1}+\frac{u_2}{u_1}\cdot\frac{u_3}{u_2}+\dots\right)<u_1\left(1+\frac{v_2}{v_1}+\frac{v_2}{v_1}\cdot\frac{v_3}{v_2}+\dots\right)<\frac{u_1}{v_1}(v_1+v_2+v_3+\dots).$$

Hence, if $\Sigma v_n$ converges, $\Sigma u_n$ also converges.

**Obs.** A more convenient form of the above test to apply is as follows :

*$\Sigma u_n$ converges if (i) $\Sigma v_n$ converges and (ii) from and after a particular term* $\frac{u_n}{u_{n+1}}>\frac{v_n}{v_{n+1}}$.

*Similarly, $\Sigma u_n$ diverges, if (i) $\Sigma v_n$ diverges and (ii) from and after a particular term* $\frac{u_n}{u_{n+1}}<\frac{v_n}{v_{n+1}}$.

## 9.9 D'ALEMBERT'S RATIO TEST*

*In a positive term series $\Sigma u_n$, if*

*Called after the French mathematician *Jean le-Rond d'Alembert* (1717–1783), who also made important contributions to mechanics.

$\underset{n \to \infty}{Lt} \dfrac{u_{n+1}}{u_n} = \lambda$, *then the series converges for* $\lambda < 1$ *and diverges for* $\lambda > 1$.

***Case I.** When* $\underset{n \to \infty}{Lt} \dfrac{u_{n+1}}{u_n} = \lambda < 1$.

By definition of a limit, we can find a positive number $r$ (< 1) such that $\dfrac{u_{n+1}}{u_n} < r$ for all $n > m$

Leaving out the first $m$ terms, let the series be $u_1 + u_2 + u_3 + \ldots$

so that $\dfrac{u_2}{u_1} < r, \dfrac{u_3}{u_2} < r, \dfrac{u_4}{u_3} < r, \ldots$ and so on. Then $u_1 + u_2 + u_3 + \ldots \infty$

$$= u_1\left(1 + \frac{u_2}{u_1} + \frac{u_3}{u_2}\cdot\frac{u_2}{u_1} + \frac{u_4}{u_3}\cdot\frac{u_3}{u_2}\cdot\frac{u_2}{u_1} + \ldots \infty\right) < u_1(1 + r + r^2 + r^3 + \ldots \infty)$$

$$= \frac{u_1}{1-r}, \text{ which is finite quantity. Hence } \Sigma u_n \text{ is convergent.} \qquad [\because \; r < 1]$$

***Case II.** When* $\underset{n \to \infty}{Lt} \dfrac{u_{n+1}}{u_n} = \lambda > 1$

By definition of limit, we can find $m$, such that $\dfrac{u_{n+1}}{u_n} \geq 1$ for all $n \geq m$.

Leaving out the first $m$ terms, let the series be

$$u_1 + u_2 + u_3 + \ldots \text{ so that } \frac{u_2}{u_1} \geq 1, \quad \frac{u_3}{u_2} \geq 1, \quad \frac{u_4}{u_3} \geq 1 \text{ and so on.}$$

$$\therefore \qquad u_1 + u_2 + u_3 + u_4 + \ldots + u_n = u_1\left(1 + \frac{u_2}{u_1} + \frac{u_3}{u_2}\cdot\frac{u_2}{u_1} + \ldots\right)$$

$$\geq u_1(1 + 1 + 1 + \ldots \text{ to } n \text{ terms}) = nu_1$$

$$\therefore \qquad \underset{n \to \infty}{\text{Lt}}\,(u_1 + u_2 + \ldots + u_n)$$

$$\geq \underset{n \to \infty}{\text{Lt}}\,(nu_1), \text{ which tends to infinity. Hence } \Sigma u_n \text{ is divergent.}$$

**Obs. 1.** *Ratio test fails when* $\lambda = 1$. *Consider, for instance, the series* $\Sigma u_n = \Sigma 1/n^p$.

Here $$\lambda = \underset{n \to \infty}{\text{Lt}} \frac{u_{n+1}}{u_n} = \underset{n \to \infty}{\text{Lt}} \left[\frac{1}{(n+1)^p}\cdot\frac{n^p}{1}\right] = \underset{n \to \infty}{\text{Lt}} \frac{1}{(1+1/n)^p} = 1.$$

Then for all values of $p$, $\lambda = 1$ ; whereas $\Sigma 1/n^p$ converges for $p > 1$ and diverges for $p < 1$.

Hence $\lambda = 1$ both for convergence and divergence of $\Sigma u_n$, which is absurd.

**Obs. 2.** *It is important to note that this test makes no reference to the magnitude of* $u_{n+1}/u_n$ *but concerns only with the limit of this ratio.*

For instance in the series $1 + \dfrac{1}{2} + \dfrac{1}{3} + \dfrac{1}{4} + \ldots + \dfrac{1}{n} + \ldots$, the ratio $\dfrac{u_{n+1}}{u_n} = \dfrac{n}{n+1} < 1$ for all finite values of $n$, but tends to unity as $n \to \infty$. Hence the Ratio test fails although this series is divergent.

**Practical form of Ratio test.** Taking reciprocals, the ratio test can be stated as follows :

*In the positive term series* $\Sigma u_n$, *if* $\underset{n \to \infty}{Lt} \dfrac{u_n}{u_{n+1}} = k$, *then the series converges for* $k > 1$ *and diverges for* $k < 1$ *but fails for* $k = 1$.

**Example. 9.12.** *Test for convergence the series*

(i) $\frac{1}{2\sqrt{1}}+\frac{x^2}{3\sqrt{2}}+\frac{x^4}{4\sqrt{3}}+\frac{x^6}{5\sqrt{4}}+\ldots\infty.$ *(Andhra, 2016, 2014; P.T.U., 2013 S;)*

(ii) $1+\frac{2}{5}x+\frac{6}{9}x^2+\frac{14}{17}x^3+\ldots+\frac{2^n-2}{2^n+1}x^{n-1}+\ldots(x>0).$

**Solution.** (i) We have $u_n=\frac{x^{2n-2}}{(n+1)\sqrt{n}}$ and $u_{n+1}=\frac{x^{2n}}{(n+2)\sqrt{(n+1)}}$

$$\therefore \qquad \underset{n\to\infty}{\text{Lt}}\frac{u_n}{u_{n+1}}=\underset{n\to\infty}{\text{Lt}}\frac{x^{2n-2}}{(n+1)\sqrt{n}}\cdot\frac{(n+2)\sqrt{(n+1)}}{x^{2n}}$$

$$=\underset{n\to\infty}{\text{Lt}}\left[\frac{n+2}{n+1}\left(\frac{n+1}{n}\right)^{1/2}\right]x^{-2}=\underset{n\to\infty}{\text{Lt}}\left[\frac{1+2/n}{1+1/n}\cdot\sqrt{(1+1/n)}\right]x^{-2}=x^{-2}.$$

Hence $\Sigma u_n$ converges if $x^{-2}>1$, *i.e.*, for $x^2<1$ and diverges for $x^2>1$.

If $x^2=1$, then, $u_n=\frac{1}{(n+1)\sqrt{n}}=\frac{1}{n^{3/2}}\cdot\frac{1}{1+1/n}$

Taking $v_n=\frac{1}{n^{3/2}}$, we get $\underset{n\to\infty}{\text{Lt}}\frac{u_n}{v_n}=\underset{n\to\infty}{\text{Lt}}\frac{1}{1+1/n}=1$, a finite quantity.

$\therefore$ Both $\Sigma u_n$ and $\Sigma v_n$ converge or diverge together. But $\Sigma v_n=\sum\frac{1}{n^{3/2}}$ is a convergent series.

$\therefore$ $\Sigma u_n$ is also convergent. Hence the given series converges if $x^2\le 1$ and diverges if $x^2>1$.

(ii) Here $$\frac{u_n}{u_{n+1}}=\frac{2^n-2}{2^n+1}x^{n-1}\cdot\frac{2^{n+1}+1}{2^{n+1}-2}\frac{1}{x^n}=\frac{1-\frac{2}{2^n}}{1+\frac{1}{2^n}}\cdot\frac{2+\frac{1}{2^n}}{2-\frac{2}{2^n}}\frac{1}{x}$$

$$\therefore \qquad \underset{n\to\infty}{\text{Lt}}\frac{u_n}{u_{n+1}}=\frac{1-0}{1+0}\cdot\frac{2+0}{2-0}\frac{1}{x}=\frac{1}{x}$$

Thus by Ratio test, $\Sigma u_n$ converges for $x^{-1}>1$ *i.e.*, for $x<1$ diverges for $x>1$. But it fails for $x=1$.

When $x=1$, $$\underset{n\to\infty}{\text{Lt}}u_n=\underset{n\to\infty}{\text{Lt}}\frac{2^n-2}{2^n+1}=\underset{n\to\infty}{\text{Lt}}\frac{1-\frac{2}{2^n}}{1+\frac{1}{2^n}}=1\neq 0$$

$\therefore$ $\Sigma u_n$ diverges for $x=1$. Hence the given series converges for $x<1$ and diverges for $x\ge 1$.

**Example 9.13.** *Discuss the convergence of the series* $\frac{1}{3}x+\frac{1\cdot 2}{3\cdot 5}x^2+\frac{1\cdot 2\cdot 3}{3\cdot 5\cdot 7}x^3+\ldots.+\infty(x>0).$

**Solution.** Here $$U_n=\frac{1\cdot 2\cdot 3\ldots..n}{3\cdot 5\cdot 7\ldots..(2n+1)x^n}$$

and $$U_{n+1}=\frac{1\cdot 2\cdot 3\ldots..n(n+1)}{3\cdot 5\cdot 7\ldots..(2n+1)(2n+3)}x^{n+1}$$

$$\therefore \qquad \frac{U_n}{U_{n+1}}=\frac{1\cdot 2\cdot 3\ldots..n}{3\cdot 5\cdot 7\ldots.(2n+1)}x^n\times\frac{3\cdot 5\cdot 7(2n+1)(2n+3)}{1\cdot 2\cdot 3\ldots.n(n+1)}\times\frac{1}{x^{n+1}}$$

$$=\frac{(2n+3)}{(n+1)}\cdot\frac{1}{x}=\frac{(2+3/n)}{(1+1/n)}\cdot\frac{1}{x}$$

$$\therefore \qquad \underset{n\to\infty}{\text{Lt}}\frac{U}{U_{n+1}}=\underset{n\to\infty}{\text{Lt}}\frac{(2+3/n)}{(1+1/n)}\cdot\frac{1}{x}=\frac{2}{x}$$

The given series converges of the $2/x > 1$, *i.e.*, $2 > x$ or $x < 2$ and diverges if $\frac{2}{x} < 1$ or $2 < x$ or $x > 2$.

The test fails at $x = 2$.

When $$x = 2$$

$$\frac{U_n}{U_{n+1}} = \frac{2n+1}{(n+1)} \cdot \frac{1}{2}$$

$$\therefore \quad n\left(\frac{u_n}{u_{n+1}} - 1\right) = 2\left[\frac{(2n+3)}{2(n+1)} - 1\right] = \frac{n}{2n+2} = \frac{1}{\left(2 + \frac{2}{n}\right)}$$

$$\therefore \quad \underset{n\to\infty}{\text{Lt}}\, n\left(\frac{u_n}{u_{n+1}} - 1\right) = \underset{n\to\infty}{\text{Lt}} \frac{1}{\left(2 + \frac{2}{n}\right)} = \frac{1}{2} (< 1)$$

Thus the series diverges when $x = 2$.

Hence the given series converges when $x < 2$ and diverges for $x \geq 2$.

**Example 9.14.** *Examine the convergence of the series :*

(*i*) $\frac{x}{1+x} + \frac{x^2}{1+x^2} + \frac{x^3}{1+x^3} + \ldots \infty$

(*ii*) $1 + \frac{a+1}{b+1} + \frac{(a+1)(2a+1)}{(b+1)(2b+1)} + \frac{(a+1)(2a+1)(3a+1)}{(b+1)(2b+1)(3b+1)} + \ldots \infty$

**Solution.** (*i*) Here $u_n = \frac{x^n}{1+x^n}$ and $u_{n+1} = \frac{x^{n+1}}{1+x^{n+1}}$

$$\therefore \quad \underset{n\to\infty}{\text{Lt}} \frac{u_n}{u_{n+1}} = \underset{n\to\infty}{\text{Lt}} \left(\frac{x^n}{x^{n+1}} \cdot \frac{1+x^{n+1}}{1+x^n}\right) = \underset{n\to\infty}{\text{Lt}} \left(\frac{1+x^{n+1}}{x+x^{n+1}}\right)$$

$$= \frac{1}{x}, \text{ if } x < 1. \qquad [\because \; x^{n+1} \to 0 \text{ as } n \to \infty]$$

Also $$\underset{n\to\infty}{\text{Lt}} \frac{u_n}{u_{n+1}} = \underset{n\to\infty}{\text{Lt}} \left(\frac{1+1/x^{n+1}}{1+x/x^{n+1}}\right) = 1 \text{ if } x > 1.$$

$\therefore$ by *Ratio test*, $\Sigma u_n$ converges for $x < 1$ and fails for $x \geq 1$.

When $x = 1$, $\Sigma u_n = \frac{1}{2} + \frac{1}{2} + \frac{1}{2} + \ldots + \infty$, which is divergent.

Hence the given series converges for $x < 1$ and diverges for $x \geq 1$.

(*ii*) Neglecting the first term, we have

$$u_{n+1} = u_n \cdot \frac{n_{a+1}}{n_{b+1}}$$

$$\therefore \quad \underset{n\to\infty}{\text{Lt}} \frac{u_n}{u_{n+1}} = \underset{n\to\infty}{\text{Lt}} \frac{n_{b+1}}{n_{a+1}} = \underset{n\to\infty}{\text{Lt}} \frac{b+1/n}{a+1/n} = \frac{b}{a}.$$

By *Ratio test*, $\Sigma u_n$ converges for $b/a > 1$ or $a < b$, and diverges for $a > b$.

When $a = b$, the series becomes $1 + 1 + 1 + \ldots \infty$, which is divergent.

Hence the given series converges for $0 < a < b$ and diverges for $0 < b \leq a$.

**Example 9.15:** *Test for convergence the series.*

$$\frac{2^2}{3 \cdot 4} + \frac{2^2 \cdot 4^2}{3 \cdot 4 \cdot 5 \cdot 6} + \frac{2^2 \cdot 4^2 \cdot 6^2}{3 \cdot 4 \cdot 5 \cdot 6 \cdot 7 \cdot 8} + \ldots .\infty$$

**Solution:** Here
$$U_n = \frac{2^2 \cdot 4^2 \cdot 6^2 .....(2n)^2}{3 \cdot 4 \cdot 5 \cdot \quad (2n+2)}$$

and
$$U_{n+1} = \frac{2^2 \cdot 4^2 \cdot 6^2 .....(2n)^2 (2n+2)^2}{3 \cdot 4 \cdot 5 \cdot \quad (2n+2)(2n+3)(2n+4)}$$

$$\therefore \quad \frac{U_n}{U_{n+1}} = \frac{2^2 \cdot 4^2 \cdot 6^2 ....(2n)^2}{3 \cdot 4 \cdot 5 ....(2n+2)} \times \frac{3 \cdot 4 \cdot 5(2n+2)(2n+3)(2n+4)}{2^2 \cdot 4^2 \cdot 6^2 ...(2n)^2 (2n+2)^2}$$

$$\frac{U_n}{U_{n+1}} = \frac{(2n+3)(2n+4)}{(2n+2)^2} = \frac{(2+3/n)(2+4/n)}{(2+2/n)^2}$$

$$\therefore \quad \underset{n\to\infty}{\text{Lt}} \frac{U_n}{U_{n+1}} = \underset{n\to\infty}{\text{Lt}} \frac{(2+3/n)(2+4/n)}{(2+2/n)^2} = 1$$

Hence the ratio test fails and we apply higher test.

Now
$$n\left(\frac{U_n}{U_n+1} - 1\right) = n\left[\frac{(2n+3)(2n+4)}{(2n+2)^2}\right]$$

$$= \frac{6n^2 + 8n}{(2n+2)^2} = \frac{(3+4/n)}{2(1+1/n)^2}$$

$$\therefore \quad \underset{n\to\infty}{\text{Lt}}\, n\left(\frac{u_n}{u_{n+1}} - 1\right) = \underset{n\to\infty}{\text{Lt}}\, n \frac{(3+4/n)}{(1+1/n)^2} = \frac{3}{2} (> 1)$$

Hence the given convergence.

**Example 9.16:** *Test the convergence of the series.*
$$\Sigma \frac{(n-1)n^{-1}}{n^n}$$

**Solution:** Here $U_n = \dfrac{(n-1)^{n-1}}{n^n}$

$$U_{n+1} = \frac{n^n}{(n+1)^{n+1}}$$

$$\therefore \quad \frac{U_n}{U_{n+1}} = \frac{(n-1)^{n-1}}{n^n} \cdot \frac{(n+1)^{n+1}}{n^n}$$

$$= \frac{n^{n-1}\left(1 - \frac{1}{n}\right)^{n-1} \cdot n^{n+1}\left(1 + \frac{1}{n}\right)^{n+1}}{n^{2n}}$$

$$= \left(1 - \frac{1}{n}\right)^{n-1} \cdot \left(1 + \frac{1}{n}\right)^{n+1}$$

$$\frac{U_n}{U_{n+1}} = \left(1 - \frac{1}{n}\right)^{n-1} \cdot \left(1 + \frac{1}{n}\right)^{n} \cdot \left(1 + \frac{1}{n}\right)$$

$$\therefore \quad n \log \frac{u_n}{u_{n+1}} = n\left[(n-1)\log\left(1 - \frac{1}{n}\right) + \log(1 + 1/n)^n + \log\left(1 + \frac{1}{n}\right) + \log\left(1 + \frac{1}{n}\right)\right]$$

$$= n\left[(n-1)\left(-\frac{1}{n} - \frac{1}{2}n^2 - \frac{1}{3}n^2 + .....\right) + \log\left(1 + \frac{1}{n}\right)^n + \left(\frac{1}{n} - \frac{1}{2}n^2 + .....\right)\right]$$

$$= \left[ n\left(-1+\frac{1}{n}-\frac{1}{2n}+....\right)+\log\left(1+\frac{1}{n}\right)^n+\left(\frac{1}{n}-\frac{1}{2n^2}+....\right)^n+\left(\frac{1}{n}-\frac{1}{2n^2}+....\right)^2 \right]$$

$$\therefore \quad \underset{n\to\infty}{\text{Lt}}\ n\log\frac{u_n}{u_{n+1}} = \underset{n\to\infty}{\text{Lt}}\ n\left[(-1)+\frac{1}{n}-\frac{1}{2}n+....)+\log\left(1+\frac{1}{n}\right)^n+\left(\frac{1}{n}-\frac{1}{2}n^2+....1\right)\right]$$

$$= \underset{n\to\infty}{\text{Lt}}\ n\left[\left(-1+\frac{1}{2^n}.......\right)+\log e+\left(\frac{1}{n}-\frac{1}{2n^2}+....\right)\right]$$

$$= \underset{n\to\infty}{\text{Lt}}\ n\left[-1+\frac{3}{2n}+1+....\right]$$

$$= \frac{3}{2} > 1 \qquad [\because \log e = 1]$$

Hence the given series converges.

**Example 9.17:** *Test the convergence of the series* $1+\frac{2x}{2!}+\frac{3^2x^2}{3!}+\frac{4^3x^3}{4!}+\frac{5^4x^4}{5!}+.....$ *(JNTUK, 2020)*

**Solution:** Here $U_n = \dfrac{n^{n-1}x^{n-1}}{n!}$

$$U^{n+1} = \frac{(n+1)^n x^n}{n!}$$

$$U^{n+1} = \frac{(n+1)^n x^n}{(n+1)!}$$

$$\therefore \quad \frac{U_{n+1}}{U_n} = \frac{(n+1)^n}{n^{(n-1)}}\ \frac{x}{n+1}$$

$$= \left(1+\frac{1}{n}\right)^n \cdot \frac{n}{n+1}x$$

Thus
$$\underset{n\to\infty}{\text{Lt}}\frac{U_{n+1}}{U_n} = \underset{n\to\infty}{\text{Lt}}\left[(1+1/n)^n\frac{1}{1+1/n}\cdot x\right] = e^x$$

So, the series is convergent when $e^x < 1$ and divergent when $e^x > 1$. When $e^x = 1$ or $x = \frac{1}{e}$ we apply further test.

$$= \underset{n\to\infty}{\text{Lt}}\ n\log\left[\frac{e}{\left(1+\frac{1}{n}\right)^{n-1}}\right] = \underset{n\to\infty}{\text{Lt}}\ n\left\{\log e-(n-1)\log\left(1+\frac{1}{n}\right)\right\}$$

$$= \underset{n\to\infty}{\text{Lt}}\left\{1-(n-1)\left(\frac{1}{n}-\frac{1}{2n^2}+\frac{1}{3n^3}\right)\right\}$$

$$= \underset{n\to\infty}{\text{Lt}}\left\{n-(n-1)+\frac{(n-1)}{2n}-\frac{(n-1)}{3n^2}+.....\right\}$$

$$= \underset{n\to\infty}{\text{Lt}}\left\{\frac{3}{2}-\frac{1}{2n}-\frac{(n-1)}{3n^2}+....\right\} = \frac{3}{2} > 1$$

As such the given series is convergent when $x = \frac{1}{e}$.

## PROBLEMS 9.4

Test for convergence the following series :

1. $\sum_{n=1}^{\infty} \frac{n^2}{2n}$ (P.T.U., 2015)
2. $\sum \frac{1}{(n)^n}$ (Andhra, 2016)
3. $\sum_{n=1}^{\infty} \frac{n!\,3^n}{n^n}$
4. $\sum_{n=1}^{\infty} \left(\frac{n^2}{2^n} + \frac{1}{n^2}\right)$
5. $\sum \frac{z^n}{n(n+2)}$ (P.T.U., 2015 S)
6. $\sum_{n=2}^{\infty} \frac{x^n}{n(n-1)(n-2)}$
7. $\sum_{n=1}^{\infty} \frac{x^n}{(2n)!}$ (Delhi, 2012)
8. $\sum_{n=1}^{\infty} \frac{\sqrt{n}}{\sqrt{(n^2+1)}} x^n$
9. $\sqrt{\frac{1}{2}}\,x + \sqrt{\frac{2}{3}}\,x^2 + \sqrt{\frac{3}{4}}\,x^3 + \ldots \infty$
10. $1 + \frac{x}{2} + \frac{x^2}{5} + \frac{x^3}{10} + \ldots + \frac{x^n}{n^2+1} \ldots \infty$ (Andhra, 2018)
11. $1 + \frac{2^p}{2!} + \frac{3^p}{3!} + \frac{4^p}{4!} + \ldots \infty$
12. $\sum_{n=1}^{\infty} \frac{n^3 + a}{2^n + a}$
13. $\frac{2}{3.4} + \frac{2.4}{3.5.6} + \frac{2.4.6}{3.5.7.8} + \ldots$ (V.T.U., 2010)
14. $\left(\frac{1}{3}\right)^2 + \left(\frac{1.2}{3.5}\right)^2 + \left(\frac{1.2.3}{3.5.7}\right)^2 + \ldots$
15. $1 + \frac{1^2.2^2}{1.3.5} + \frac{1^2.2^2.3^2}{1.3.5.7.9} + \ldots \infty$ (Rohtak, 2011 S)
16. $\frac{1}{1^p} + \frac{x}{3^p} + \frac{x^2}{5^p} + \ldots + \frac{x^{n-1}}{(2n-1)^p} + \ldots \infty$
17. $\sum_{n=1}^{\infty} \frac{3.6.9\ldots 3n}{4.7.10\ldots(3n+1)} \cdot \frac{5^n}{3n+2}$
18. $1 + \frac{1+\alpha}{1+\beta} + \frac{(1+\alpha)(1+2\alpha)}{(1+\beta)(1+2\beta)} + \frac{(1+\alpha)(1+2\alpha)(1+3\alpha)}{(1+\beta)(1+2\beta)(1+3\beta)} + \ldots$

## 9.10 FURTHER TESTS OF CONVERGENCE

*When the Ratio test fails, we apply the following tests :*

**(1) Raabe's test*.** *In the positive term series* $\Sigma u_n$, if $\underset{n\to\infty}{\text{Lt}}\; n\left(\frac{u_n}{u_{n+1}} - 1\right) = k$, *then the series converges for k > 1 and diverges for k < 1,* but the test fails for $k = 1$.

When $k > 1$, choose a number $p$ such that $k > p > 1$, and compare $\Sigma u_n$ with the series $\sum \frac{1}{n^p}$ which is convergent since $p > 1$.

$\therefore$ $\Sigma u_n$ will converge, if from and after some term,

$$\frac{u_n}{u_{n+1}} > \frac{(n+1)^p}{n^p} \text{ or } \left(1 + \frac{1}{n}\right)^p \qquad \text{or if, } \quad \frac{u_n}{u_{n+1}} > 1 + \frac{p}{n} + \frac{p(p-1)}{2n^2} + \ldots$$

or if, $$n\left(\frac{u_n}{u_{n+1}} - 1\right) > p + \frac{p(p-1)}{2n} + \ldots \quad \text{or if, } \quad \underset{n\to\infty}{\text{Lt}}\; n\left(\frac{u_n}{u_{n+1}} - 1\right) > \underset{n\to\infty}{\text{Lt}}\left[p + \frac{p(p-1)}{2n} + \ldots\right]$$

*i.e.*, if $k > p$, which is true. Hence $\Sigma u_n$ is convergent.

The other case when $k < 1$ can be proved similarly.

**(2) Logarithmic test.** *In the positive term series* $\Sigma u_n$ *if* $\underset{n\to\infty}{Lt}\left(n \log \frac{u_n}{u_{n+1}}\right) = k$, *then the series converges for k > 1, and diverges for k < 1,* but the test fails for $k = 1$.

*Called after the Swiss mathematician *Joseph Ludwig Raabe* (1801–1859).

Its proof is similar to that of Raabe's test.

**Obs.** *Logarithmic test is a substitute for Raabe's test and should be applied when* **either** *n occurs as an exponent in* $u_n/u_{n+1}$, **or** *evaluation of* $\underset{n\to\infty}{\text{Lt}}$ *becomes easier on taking logarithm of* $u_n/u_{n+1}$.

**(3) Gauss Test.** In a positive term series $\Sigma u_n$, if $\dfrac{u_n}{u_{n+1}}$ can be expanded in the form $1+\dfrac{\lambda}{n}+0\left(\dfrac{1}{n^2}\right)$, then $\Sigma u_n$ converges for $\lambda > 1$ and diverges for $\lambda \le 1$.

**Obs. 1.** This test never fails, and is applied when all other tests fail.

**Obs. 2. If $u_n/u_{n+1}$ does not involve n as an exponent or a logarithm, the series $\Sigma u_n$ diverges.**

**Example 9.18.** *Test for convergence the series*

(*i*) $\sum \dfrac{4\cdot 7 \ldots (3n+1)}{1\cdot 2 \ldots n} x^n$ (*JNTUK, 2020*) (*ii*) $\sum \dfrac{(n\,!)^2}{(2n)\,!} x^{2n}$. (*Calicut, 2012*)

**Solution.** (*i*) Here $\dfrac{u_n}{u_{n+1}} = \dfrac{4\,.\,7 \ldots (3n+1)}{1\,.\,2 \ldots n} x^n \div \dfrac{4\,.\,7 \ldots (3n+4)}{1\,.\,2 \ldots (n+1)} x^{n+1} = \dfrac{n+1}{3n+4}\cdot\dfrac{1}{x} = \left[\dfrac{1+1/n}{3+4/n}\right]\dfrac{1}{x}$

$$\therefore \qquad \underset{n\to\infty}{\text{Lt}} \frac{u_n}{u_{n+1}} = \frac{1}{3x}.$$

Thus by *Ratio test*, the series converges for $\dfrac{1}{3x} > 1$, *i.e.*, for $x < \frac{1}{3}$ and diverges for $x > \frac{1}{3}$. But it fails for $x = \frac{1}{3}$. $\therefore$ Let us try the *Raabe's test.*

Now $\qquad \dfrac{u_n}{u_{n+1}} = \left(1+\dfrac{1}{n}\right)\left(1+\dfrac{4}{3n}\right)^{-1}$ [*Expand by Binomial Theorem*]

$$= \left(1+\frac{1}{n}\right)\left(1-\frac{4}{3n}+\frac{16}{9n^2}-\ldots\right) = 1-\frac{1}{3n}+\frac{4}{9n^2}+\ldots$$

$\therefore \qquad n\left(\dfrac{u_n}{u_{n+1}} - 1\right) = -\dfrac{1}{3}+\dfrac{4}{9n}+\ldots \qquad \therefore \quad \underset{n\to\infty}{\text{Lt}}\, n\left(\dfrac{u_n}{u_{n+1}}-1\right) = -\dfrac{1}{3}$ which $< 1$.

Thus by *Raabe's test*, the series diverges.

Hence the given series converges for $x < \frac{1}{3}$ and diverges for $x \ge \frac{1}{3}$.

(*ii*) Here $\qquad \dfrac{u_n}{u_{n+1}} = \left(\dfrac{n\,!}{(n+1)\,!}\right)^2 \dfrac{[2(n+1)]\,!}{(2n)\,!}\cdot\dfrac{x^{2n}}{x^{2(n+1)}} = \dfrac{(2n+1)(2n+2)}{(n+1)^2}\dfrac{1}{x^2} = \dfrac{2(2n+1)}{n+1}\dfrac{1}{x^2}$

$$\therefore \qquad \underset{n\to\infty}{\text{Lt}} \frac{u_n}{u_{n+1}} = \underset{n\to\infty}{\text{Lt}} \frac{2(2+1/n)}{1+1/n}\cdot\frac{1}{x^2} = \frac{4}{x^2}$$

Thus by *Ratio Test*, the series converges for $x^2 < 4$ and diverges for $x^2 > 4$. But fails for $x^2 = 4$.

When $x^2 = 4$, $\qquad n\left(\dfrac{u_n}{u_{n+1}}-1\right) = n\left(\dfrac{2n+1}{2n+2}-1\right) = -\dfrac{n}{2n+2}$

$$\therefore \qquad \underset{n\to\infty}{\text{Lt}}\, n\left(\frac{u_n}{u_{n+1}}-1\right) = -\frac{1}{2} < 1$$

Thus by *Raabe's test*, the series diverges.

Hence the given series converges for $x^2 < 4$ and diverges for $x^2 \ge 4$.

**Example 9.19.** *Discuss the convergence of the series*

$$x+\frac{2^2x^2}{2\,!}+\frac{3^3x^3}{3\,!}+\frac{4^4x^4}{4\,!}+\frac{5^5x^5}{5\,!}+\ldots\infty$$ (*JNTUK, 2020 ; D.T.U., 2013 ; Calicut, 2012*)

**Solution.** Here $\dfrac{u_n}{u_{n+1}} = \dfrac{n^n x^n}{n!} \div \dfrac{(n+1)^{n+1}\, x^{n+1}}{(n+1)!} = \dfrac{n^n}{(n+1)^n\, x} = \dfrac{1}{(1+1/n)^n} \cdot \dfrac{1}{x}$

$$\therefore \qquad \underset{n\to\infty}{\text{Lt}} \frac{u_n}{u_{n+1}} = \frac{1}{ex}.$$

Thus by Ratio test, the series converges for $x < 1/e$ and diverges for $x > 1/e$. But it fails for $x = 1/e$. Let us try the *log-test*.

Now $$\frac{u_n}{u_{n+1}} = \frac{e}{(1+1/n)^n}$$

$$\therefore \qquad \log \frac{u_n}{u_{n+1}} = \log_e e - n \log\left(1+\frac{1}{n}\right) = 1 - n\left(\frac{1}{n} - \frac{1}{2n^2} + \frac{1}{3n^3} - \cdots\right) = \frac{1}{2n} - \frac{1}{3n^2} + \ldots$$

$\therefore \qquad \underset{n\to\infty}{\text{Lt}} \left(n \log \dfrac{u_n}{u_{n+1}}\right) = \dfrac{1}{2}$, which $< 1$. Thus by the *log-test*, the series diverges.

Hence the given series converges for $x < 1/e$ and diverges for $x \geq 1/e$.

**Example 9.20.** *Discuss the convergence of the* ***hypergeometric series***

$$1 + \frac{\alpha\,.\,\beta}{1\,.\,\gamma}x + \frac{\alpha(\alpha+1)\,\beta(\beta+1)}{1\,.\,2\,.\,\gamma(\gamma+1)}x^2 + \frac{\alpha(\alpha+1)(\alpha+2)\,\beta(\beta+1)(\beta+2)}{1\,.\,2\,.\,3\,.\,\gamma(\gamma+1)(\gamma+2)}x^3 + \ldots \infty.$$ *(Rohtak, 2012)*

**Solution.** Neglecting the first term, we have

$$u_{n+1} = u_n \frac{(\alpha+n)(\beta+n)}{(n+1)(\gamma+n)}x$$

$$\therefore \qquad \underset{n\to\infty}{\text{Lt}} \frac{u_n}{u_{n+1}} = \underset{n\to\infty}{\text{Lt}} \frac{(n+1)(\gamma+n)}{(\alpha+n)(\beta+n)} \cdot \frac{1}{x} = \underset{n\to\infty}{\text{Lt}} \frac{(1+1/n)(1+\gamma/n)}{(1+\alpha/n)(1+\beta/n)} \cdot \frac{1}{x} = \frac{1}{x}$$

$\therefore$ by *Ratio test,* the series converges for $1/x > 1$, *i.e.*, for $x < 1$, and diverges for $x > 1$. But it fails for $x = 1$.

$\therefore$ let us try the *Raabe's test*.

$$\underset{n\to\infty}{\text{Lt}}\; n\left(\frac{u_n}{u_{n+1}} - 1\right) = \underset{n\to\infty}{\text{Lt}}\; n\left\{\frac{(n+1)(n+\gamma)}{(n+\alpha)(n+\beta)} - 1\right\} = \underset{n\to\infty}{\text{Lt}}\; n\left\{\frac{n(1+\gamma-\alpha-\beta)+\gamma-\alpha\beta}{n^2+n(\alpha+\beta)+\alpha\beta}\right\}$$

$$= \underset{n\to\infty}{\text{Lt}} \left\{\frac{(1+\gamma-\alpha-\beta)+(\gamma-\alpha\beta)\dfrac{1}{n}}{1+(\alpha+\beta)\dfrac{1}{n}+\alpha\beta\,.\,\dfrac{1}{n^2}}\right\} = 1+\gamma-\alpha-\beta$$

Thus the series converges for $1 + \gamma - \alpha - \beta > 1$, *i.e.*, for $\gamma > \alpha + \beta$ and diverges for $\gamma < \alpha + \beta$. But it fails for $\gamma = \alpha + \beta$. Since $u_n/u_{n+1}$ does not involve $n$ as an exponent or a logarithm, the series $\Sigma u_n$ diverges for $\gamma = \alpha + \beta$.

Hence the series converges for $x < 1$ and diverges for $x > 1$. When $x = 1$, the series converges for $\gamma > \alpha + \beta$ and diverges for $\gamma \leq \alpha + \beta$.

## PROBLEMS 9.5

Test the following series for convergence :

1. $\dfrac{x}{1.2} + \dfrac{x^2}{3.4} + \dfrac{x^3}{5.6} + \dfrac{x^4}{7.8} + \ldots \infty \; (x > 0)$
2. $1 + \dfrac{1}{2}x + \dfrac{1.3}{2.4}x^2 + \dfrac{1.3.5}{2.4.6}x^3 + \ldots \infty \; (x > 0)$
3. $1 + \dfrac{2}{3}x + \dfrac{2.3}{3.5}x^2 + \dfrac{2.3.4}{3.5.7}x^3 + \ldots \infty$
4. $1 + \dfrac{x}{2} + \dfrac{2!}{3^2}x^2 + \dfrac{3!}{4^3}x^3 + \dfrac{4!}{5^4}x^4 + \ldots \infty$ *(P.T.U., 2015 S)*

5. $1 + \frac{3}{7}x + \frac{3.6}{7.10}x^2 + \frac{3.6.9}{7.10.13}x^3 + \frac{3.6.9.12}{7.10.13.16}x^4 + \ldots \infty$ (*Delhi, 2015*)

6. $\frac{x}{1} + \frac{1}{2}\cdot\frac{x^3}{3} + \frac{1.3}{2.4}\cdot\frac{x^5}{5} + \frac{1.3.5}{2.4.6}\cdot\frac{x^7}{7} + \ldots \infty. (x > 0)$ (*Andhra, 2018 ; Rohtak, 2012 S*)

7. $\frac{1^2}{2^2} + \frac{1^2.3^2}{2^2.4^2} + \frac{1^2.3^2.5^2}{2^2.4^2.6^2} + \ldots \infty$ (*Rohtak, 2010 S*)

8. $\frac{a+x}{1!} + \frac{(a+2x)^2}{2!} + \frac{(a+3x)^3}{3!} + \ldots \infty$

9. $x^2 (\log 2)^q + x^3 (\log 3)^q + x^4 (\log 4)^q + \ldots \infty$ (*Rohtak, 2011 S*)

10. $\frac{1}{1^2} + \frac{1+2}{1^2+2^2} + \frac{1+2+3}{1^2+2^2+3^2} + \ldots$

11. $1 + \frac{a}{b}x + \frac{a(a+1)}{b(b+1)}x^2 + \frac{a(a+1)(a+2)}{b(b+1)(b+2)}x^3 + \ldots \infty\ (a, b > 0, x > 0).$ (*Calicut, 2013*)

12. $\frac{1}{3}x + \frac{1\cdot 2}{3\cdot 5}x^2 + \frac{1\cdot 2\cdot 3}{3\cdot 5\cdot 7}x^3 + \ldots\ldots\infty \quad (x > 0)$ (*UPTU* 2011)

13. $\frac{2}{5}x + \frac{2\cdot 4}{5\cdot 8}x^2 + \frac{2\cdot 4\cdot 6}{5\cdot 8\cdot 11}x^3 + \ldots\ldots\infty \quad (x > 0)$ (*Marthwada 2019*)

14. $1 + \frac{\alpha\cdot\beta}{1\cdot\gamma} - x + \frac{\alpha(\alpha+\gamma)\beta(\beta+1)}{1\cdot 2\cdot\gamma(\gamma+1)}x^2 + \frac{\alpha(\alpha+1)(\alpha+2)\beta(\beta+1)(\beta+2)}{1\cdot 2\cdot 3\ \gamma\ (\gamma+1)(\gamma+2)x^3} + \ldots.$ (*Annamalai, 2019*)

15. $\frac{3}{7}x + \frac{3\cdot 6}{7\cdot 10}x^2 + \frac{3\cdot 6\cdot 9}{7\cdot 10\cdot 13}x^3 + \ldots\ldots\infty\,(x > 0)$

16. $\frac{1}{2\cdot}x^3 + \frac{1\cdot 3}{2\cdot 4}\frac{x^5}{5} + \frac{1\cdot 3\cdot 5}{2\cdot 4\cdot 6}\frac{x^7}{7} + \ldots.\infty$

17. $\frac{1}{2} + \frac{1\cdot 3}{2\cdot 4}\cdot\frac{1}{2} + \frac{1\cdot 3\cdot 5}{2\cdot 4\cdot 6}\cdot\frac{1}{3} + \ldots \infty$

## 9.11 CAUCHY'S ROOT TEST

*In a positive series $\Sigma u_n$, if $\underset{n\to\infty}{\text{Lt}} (u_n)^{1/n} = \lambda$,*

*then the series converges for $\lambda < 1$, and diverges for $\lambda > 1$.*

*Case I. When* $\underset{n\to\infty}{\text{Lt}} (u_n)^{1/n} = \lambda < 1$.

By definition of a limit, we can find a positive number $r$ $(\lambda < r < 1)$ such that

$$(u_n)^{1/n} < r \text{ for all } n > m, \quad \text{or} \quad u_n < r^n \text{ for all } n > m.$$

Since $r < 1$, the geometric series $\Sigma r^n$ is convergent. Hence, by comparison test, $\Sigma u_n$ is also convergent.

*Case II. When* Lt $(u_n)^{1/n} = \lambda > 1$.

By definition of a limit, we can find a number $m$, such that

$$(u_n)^{1/n} > 1 \text{ for all } n > m, \quad \text{or} \quad u_n > 1 \text{ for all } n > m.$$

Omitting the first $m$ terms, let the series be $u_1 + u_2 + u_3 + \ldots.$ so that $u_1 > 1, u_2 > 1, u_3 > 1$ and so on.

$$\therefore \quad u_1 + u_2 + u_3 + \ldots + u_n > n \quad \text{and} \quad \underset{n\to\infty}{\text{Lt}} (u_1 + u_2 + \ldots + u_n) \to \infty$$

Hence the series $\Sigma u_n$ is divergent.

**Obs.** *Cauchy's root test fails when $\lambda = 1$.*

**Example 9.21.** *Test for convergence the series*

(i) $\Sigma \dfrac{n^3}{3^n}$ (ii) $\Sigma (\log n)^{-2n}$ (iii) $\Sigma (1 + 1/\sqrt{n})^{-n^{3/2}}$ *(Calicut, 2013 ; U.K.T.U., 2013)*

**Solution.** (i) We have $u_n = n^3/3^n$.

$$\therefore \quad \underset{n\to\infty}{\text{Lt}} (u_n)^{1/n} = \underset{n\to\infty}{\text{Lt}} \left(\frac{n^{3/n}}{3}\right) = \underset{n\to\infty}{\text{Lt}} \frac{(n^{1/n})^3}{3} = \frac{1}{3} (< 1) \qquad \left[\because \underset{n\to\infty}{\text{Lt}} n^{1/n} = 1\right]$$

Hence the given series converges by Cauchy's root test.

(ii) Here $\qquad u_n = (\log n)^{-2n}$

$$\therefore \quad \underset{n\to\infty}{\text{Lt}} (u_n)^{1/n} = \underset{n\to\infty}{\text{Lt}} (\log n)^{-2} = 0 \ (< 1) \qquad [\because \underset{n\to\infty}{\text{Lt}} \log n = 0]$$

Hence, by Cauchy's root test, the given series converges.

(iii) Here $\qquad u_n = (1 + 1/\sqrt{n})^{-n^{3/2}}$

$$\therefore \quad (u_n)^{1/n} = \left[\frac{1}{(1+1/\sqrt{n})^{n^{3/2}}}\right]^{1/n} = \frac{1}{(1+1/\sqrt{n})^{\sqrt{n}}}$$

$$\therefore \quad \underset{n\to\infty}{\text{Lt}} (u_n)^{1/n} = \underset{n\to\infty}{\text{Lt}} \frac{1}{(1+1/\sqrt{n})^{\sqrt{n}}} = \frac{1}{e},$$

which is < 1. Hence the given series is convergent.

**Example 9.22.** *Discuss the nature of the following series :*
(i) *Test the series for convergence and divergence.* *(Bhilai, 2019)*

$$\frac{1}{2} \cdot x + \left(\frac{2}{3}\right)^4 \cdot x^2 + \left(\frac{3}{4}\right)^9 \cdot x^3 + \ldots . \left(\frac{n}{n+1}\right)^n x^n + \ldots .$$

(ii) $\Sigma \dfrac{(n+1)^n x^n}{n^{n+1}}$ *(Rohtak, 2012)*

**Solution.**

(i) Given, $\qquad u_n = \left(\dfrac{n}{n+1}\right) n^2 \cdot x^n$

$$\therefore \quad (u_n)^{1/n} = \left(\frac{1}{1+\dfrac{1}{n}}\right)^n \cdot x = \frac{x}{\left(1+\dfrac{1}{n}\right)^n}$$

$$\therefore \quad \underset{n\to\infty}{\text{Lt}} (u_n)^{1/n} = \underset{n\to\infty}{\text{Lt}} \frac{x}{\left(1+\dfrac{1}{n}\right)^n} = \frac{x}{e}$$

The given series converges if $\dfrac{x}{e} < 1$, *i.e.*, $x < e$ and divergence if $x/e > 1$ *i.e.*, $x > e$

The above test fails at $x = e$.

when $x = e$ $\qquad u_n = \left(\dfrac{n}{n+1}\right) n^2 \cdot e^n$

Now, $\underset{n\to\infty}{\text{Lt}} u_n \neq 0$, so the given series diverges at $x = e$.

So the given series converges for $x < e$ and diverges for $x \geq e$.

(ii) Here $(u_n^{1/n}) = \dfrac{n+1}{n^{1+1/n}}x$

$\therefore \quad \underset{n\to\infty}{\text{Lt}}\,(u_n)^{1/n} = \underset{n\to\infty}{\text{Lt}}\,\dfrac{n+1}{n}\cdot\dfrac{1}{n^{1/n}}x = \underset{n\to\infty}{\text{Lt}}\left(1+\dfrac{1}{n}\right)\left(\dfrac{1}{n^{1/n}}\right)x = x \qquad \left[\because \underset{n\to\infty}{\text{Lt}}\, n/n = 1\right]$

$\therefore$ The given series converges for $x < 1$ and diverges for $x > 1$.

When $x = 1$, $u_n = \dfrac{(n+1)^n}{n^{n+1}} = \dfrac{1}{n}\left(1+\dfrac{1}{n}\right)^n$

Taking $\quad v_n = \dfrac{1}{n}$, $\underset{n\to\infty}{\text{Lt}}\left(\dfrac{u_n}{v_n}\right) = \underset{n\to\infty}{\text{Lt}}\left(1+\dfrac{1}{n}\right)^n = e \neq 0$ and finite.

$\therefore$ By comparison test both $\Sigma u_n$ and $\Sigma v_n$ behave alike.

But $\Sigma v_n = \Sigma\dfrac{1}{n}$ is divergent ($\because \; p = 1$). $\therefore \; \Sigma u_n$ also diverges. Hence the given series converges for $x < 1$ and diverges for $x \geq 1$.

(iii) Neglecting the first term 1, we have

$$u_n = \frac{(2n+1)}{n^3+1}x^n \text{ and } u_{n+1} = \frac{2n+3}{(n+1)^3+1}x^{n+1}$$

$$\frac{u_n}{u_{n+1}} = \frac{2n+1}{n^3+1}\cdot\frac{(n+1)^3+1}{2n+3}\cdot\frac{1}{x} = \frac{\left(2+\dfrac{1}{n}\right)}{\left(1+\dfrac{1}{n^3}\right)}\cdot\frac{\left(1+\dfrac{1}{n}\right)^3}{\left(2+\dfrac{3}{n}\right)}\cdot\frac{1}{x}$$

$$\lim\frac{u_n}{u_{n+1}} = \frac{2}{1}\cdot\frac{1+0}{2}\cdot\frac{1}{x} = \frac{1}{x}$$

$\therefore$ By ratio test the series is convergent if $\dfrac{1}{x} > 1$ *i.e.*, $x < 1$ and the test fails when $\dfrac{1}{x} = 1$, when $x = 1$.

When $x = 1$, $u_n = \dfrac{2n+1}{n^3+1} \qquad v_n = \dfrac{n}{n^3} = \dfrac{1}{n^2}$

$$\lim\frac{u_n}{v_n} = \lim\left(\frac{2n+1}{n^3+1}\cdot n^2\right) = \lim\left(\frac{2+\dfrac{1}{n}}{1+\dfrac{1}{n^3}}\right) = 2$$

which is finite and non-zero.

But the series $\Sigma v_n = \Sigma\dfrac{1}{n^2}$ is convergent as here $p = 2 > 1$.

So the comparison test the given series is also convergent. When $x = 1$ so the series is convergent if $x \leq 1$ and divergent if $x > 1$.

**Example 9.23:** *Discuss the convergence of the series*

(a) $\Sigma\left(1+\dfrac{1}{n}\right)^{n^2}$ $\qquad$ (b) $1+\dfrac{1}{2^2}+\dfrac{1}{3^3}+\ldots\ldots\dfrac{1}{n^n}+\ldots\ldots$

**Solution:** (a) Here, $U_n = \left(1+\dfrac{1}{n}\right)^{n^2}$

$$(U_n)^{1/n} = \left(1+\frac{1}{n}\right)^n$$

$$\therefore \qquad \underset{n\to\infty}{\text{Lt}} (u_n)^{1/n} = \underset{n\to\infty}{\text{Lt}} \left(1+\frac{1}{n}\right)^n = e > 1$$

So the given series is divergent.

(*ii*) Given, $1 + \dfrac{1}{2^2} + \dfrac{1}{3^2} + .... + \dfrac{1}{n^n} + ....$

$$\text{So,} \qquad U_n = \frac{1}{n^n} \text{ or } (U_n)^{1/n} = \frac{1}{n}$$

$$\therefore \qquad \underset{n\to\infty}{\text{Lt}} (u_n)^{1/n} = \underset{n\to\infty}{\text{Lt}} \frac{1}{n} = 0 < 1$$

Hence the given series is convergent.

## PROBLEMS 9.6

Discuss the convergence of the following series :

1. $\sum_{n=1}^{\infty} \dfrac{n^2}{3^n}$ (*Calicut, 2013*)
2. $\sum \dfrac{1}{(\log n)^n}$
3. $\sum \left(\dfrac{n}{n+1}\right)^{n^2}$ (*D.T.U., 2013 ; P.T.U., 2010*)
4. $1 + \dfrac{x}{2} + \dfrac{x^2}{3^2} + \dfrac{x^3}{4^3} + .. + (x > 0)$
5. $\sum \left(\dfrac{n+2}{n+3}\right)^n x^n$
6. $\sum \dfrac{[(2n+1)\,x]^n}{n^{n+1}}, x > 0$
7. $\dfrac{3}{4}x + \left(\dfrac{4}{5}\right)^2 x^2 + \left(\dfrac{5}{6}\right)^3 x^3 + ... -\infty \; (x > 0)$

## 9.12 ALTERNATING SERIES

(1) **Def.** *A series in which the terms are alternately positive or negative is called an alternating series.*

(2) **Leibnitz's series.** *An alternating series $u_1 - u_2 + u_3 - u_4 + ...$ converges if (i) each term is numerically less than its preceding term, and (ii) $\underset{n\to\infty}{Lt} u_n = 0$.*

*If $\underset{n\to\infty}{Lt} u_n \neq 0$, the given series is oscillatory.*

The given series is $\quad u_1 - u_2 + u_3 - u_4 + ...$

Suppose $\quad u_1 > u_2 > u_3 > u_4 ... > u_{n+1} ...$ ...(1)

and $\quad \underset{n\to\infty}{\text{Lt}} u_n = 0$ ...(2)

Consider the sum of $2n$ terms. It can be written as

$$s_{2n} = (u_1 - u_2) + (u_3 - u_4) + ... + (u_{2n-1} - u_{2n}) \quad ...(3)$$

or as

$$s_{2n} = u_1 - (u_2 - u_3) - (u_4 - u_5) ... - u_{2n} \quad ...(4)$$

By virtue of Eqn. (1), the expressions within the brackets in Eqns. (3) and (4) are all positive.

$\therefore$ It follows from Eqn. (3) that $s_{2n}$ is positive and increases with $n$.

Also from Eqn. (4), we note that $s_{2n}$ always remains less than $u_1$.

Hence $s_{2n}$ must tend to a finite limit.

Moreover $\quad \underset{n\to\infty}{\text{Lt}} s_{2n+1} = \underset{n\to\infty}{\text{Lt}} (s_{2n} + u_{2n+1}) = \underset{n\to\infty}{\text{Lt}} s_{2n} + 0$ [by (2)

Thus $\underset{n\to\infty}{\text{Lt}} s_n$ tends to the same finite limit whether $n$ is even or odd.

Hence the given series is convergent.

When $\underset{n\to\infty}{\text{Lt}}\, u_n \neq 0$, $\underset{n\to\infty}{\text{Lt}}\, s_{2n} \neq \underset{n\to\infty}{\text{Lt}}\, s_{2n+1}$. $\therefore$ The given series is oscillatory.

**Example 9.24.** *Discuss the convergence of the series*

(*i*) $1 - \dfrac{1}{\sqrt{2}} + \dfrac{1}{\sqrt{3}} - \dfrac{1}{\sqrt{4}} + \ldots$ (*ii*) $\dfrac{5}{2} - \dfrac{7}{4} + \dfrac{9}{6} - \dfrac{11}{8} + \ldots$

(*iii*) $\dfrac{1}{\log 2} - \dfrac{1}{\log 3} + \dfrac{1}{\log 4} - \dfrac{1}{\log 5} + \ldots$ (*P.T.U., 2010*)

**Solution.** (*i*) The terms of the given series are alternately positive and negative ; each term is numerically less than its preceding term $\left[\because \quad u_n - u_{n-1} = \dfrac{1}{\sqrt{n}} - \dfrac{1}{\sqrt{n-1}} < 0\right]$

Also $\underset{n\to\infty}{\text{Lt}}\, u_n = \underset{n\to\infty}{\text{Lt}}\, (1/\sqrt{n}) = 0$. Hence by Leibnitz's rule, the given series is convergent.

(*ii*) The terms of the given series are alternately positive and negative and

$$u_n - u_{n-1} = \frac{2n+3}{2n} - \frac{2n+1}{2n-2} = \frac{-6}{4n(n-1)} < 0 \text{ for } n > 1.$$

*i.e.*, $u_n < u_{n-1}$ for $n > 1$. Also $\underset{n\to\infty}{\text{Lt}}\, u_n = \underset{n\to\infty}{\text{Lt}}\, \dfrac{2n+3}{2n} = 1 \neq 0$

Hence by Leibnitz's rule, the given series is oscillatory.

(*iii*) The terms of the given series are alternately positive and negative.

Also $n + 2 > n + 1$, *i.e.*, $\log (n + 2) > \log (n + 1)$

*i.e.*, $\dfrac{1}{\log (n+2)} < \dfrac{1}{\log (n+1)}$, *i.e.*, $u_{n+1} < u_n$.

and $\underset{n\to\infty}{\text{Lt}}\, u_n = \underset{n\to\infty}{\text{Lt}}\, \dfrac{1}{\log (n+1)} = 0$

Hence the given series is convergent.

**Example 9.25.** *Examine the character of the series*

(*i*) $\displaystyle\sum_{n=1}^{\infty} \frac{(-1)^{n-1}\, n}{2n-1}$. (*ii*) $\displaystyle\sum_{n=2}^{\infty} \frac{(-1)^{n-1}\, x^n}{n(n-1)}$, $0 < x < 1$. (*JNTUK, 2023*)

**Solution.** (*i*) The terms of the given series are alternately positive and negative ; each term is numerically less than its preceding term.

$$\left[\because \quad u_n - u_{n-1} = \frac{n}{2n-1} - \frac{n-1}{2n-3} = \frac{-1}{(2n-1)(n-3)} < 0\right.$$

But $\underset{n\to\infty}{\text{Lt}}\, u_n = \underset{n\to\infty}{\text{Lt}}\, \dfrac{n}{2n-1} = \underset{n\to\infty}{\text{Lt}}\, \dfrac{1}{2-1/n} = \dfrac{1}{2}$ which is not zero.

Hence the given series is oscillatory.

(*ii*) The terms of the given series are alternately positive and negative

$$u_n - u_{n-1} = \frac{x^n}{n(n-1)} - \frac{x^{n-1}}{(n-1)(n-2)} = \frac{x^{n-1}[(n-2)x - n]}{n(n-1)(n-2)} < 0 \quad \text{for} \quad n \geq 2, \qquad (\because \quad 0 < x < 1)$$

*i.e.*, $u_n < u_{n-1}$ for $n \geq 2$. Also $\underset{n\to\infty}{\text{Lt}}\, u_n = \underset{n\to\infty}{\text{Lt}}\, \dfrac{x^n}{n(n-1)} = 0$ $(\because \quad 0 < x < 1)$

Hence the given series is convergent.

## PROBLEMS 9.7

**Discuss the convergence of the following series :**

1. $1-\frac{1}{2}+\frac{1}{3}-\frac{1}{4}+\ldots\infty.$

2. $1-\frac{1}{5}+\frac{1}{9}-\frac{1}{13}+\ldots\infty.$ *(V.T.U., 2010)*

3. $\frac{1}{1\cdot 2}-\frac{1}{3\cdot 4}+\frac{1}{5\cdot 6}-\frac{1}{7\cdot 8}+\ldots\infty$

4. $\frac{1}{6}-\frac{2}{11}+\frac{3}{16}-\frac{4}{21}+\frac{5}{26}-\ldots\infty.$

5. $1-2x+3x^2-4x^3+\ldots+\infty,\ \left(x<\frac{1}{2}\right).$

6. $\sum_{n=1}^{\infty}\frac{(-1)^{n+1}}{n^2}.$ *(JNTUK, 2022 ; P.T.U, 2015)*

7. $\sum_{n=1}^{\infty}(-1)^{n-1}\frac{n}{n^2+1}.$

8. $\sum_{n=0}^{\infty}\frac{(-1)^n}{n!}$

9. $\sum_{n=1}^{\infty}\frac{\cos n\pi}{n^2+1}.$ *(JNTUK, 2023)*

10. $\frac{x}{1+x}-\frac{x^2}{1+x^2}+\frac{x^3}{1+x^3}-\frac{x^4}{1+x^4}+\ldots\infty\ (0<x<1).$ *(Anna, 2016)*

11. $\left(\frac{1}{2}-\frac{1}{\log 2}\right)-\left(\frac{1}{2}-\frac{1}{\log 3}\right)+\left(\frac{1}{2}-\frac{1}{\log 4}\right)-\left(\frac{1}{2}-\frac{1}{\log 5}\right)+\ldots\infty.$

## 9.13 SERIES OF POSITIVE AND NEGATIVE TERMS

The series of positive terms and the alternating series are special types of these series with arbitrary signs.

**Def.** (**1**) *If the series of arbitrary terms* $u_1+u_2+u_3+\ldots+u_n+\ldots$

*be such that the series* $|u_1|+|u_2|+|u_3|+\ldots+|u_n|+\ldots$

*is convergent, then the series* $\Sigma u_n$ *is said to be* **absolutely convergent.**

(**2**) *If* $\Sigma\,|u_n|$ *is divergent but* $\Sigma u_n$ *is convergent, then* $\Sigma u_n$ *is said to be* **conditionally convergent.**

For instance, the series $1-\frac{1}{2^2}+\frac{1}{3^2}-\frac{1}{4^2}+\frac{1}{5^2}-\ldots$ *is absolutely convergent,* since the series

$1+\frac{1}{2^2}+\frac{1}{3^2}+\frac{1}{4^2}+\frac{1}{5^2}+\ldots$ is known to be convergent.

Again, since the alternating series $1-\frac{1}{2}+\frac{1}{3}-\frac{1}{4}+\frac{1}{5}-\ldots$ is convergent, and the series of absolute values $1+\frac{1}{2}+\frac{1}{3}+\frac{1}{4}+\frac{1}{5}+\ldots$ is divergent, so the original series is conditionally convergent.

**Obs. 1.** *An absolutely convergent series is necessarily convergent but not conversely.*

Let $\Sigma u$ be an absolutely convergent series.

Clearly $u_1+u_2+u_3+\ldots+u_n+\ldots$

$\le |u_1|+|u_2|+|u_3|+\ldots+|u_n|+\ldots$ which is known to be convergent.

Hence the series $\Sigma u_n$ is also convergent.

**Obs. 2.** As the series $\Sigma\,|u_n|$ is of positive terms, the tests already established for positive term series can be applied to examine $\Sigma u_n$ for its absolute convergence. For instance, Ratio test can be restated as follows :

*The series* $\Sigma\,u_n$ *is absolutely convergent if* $\underset{n\to\infty}{\text{Lt}}\frac{|u_{n+1}|}{|u_n|}<1,$

*and is divergent if* $\underset{n\to\infty}{\text{Lt}}\frac{|u_{n+1}|}{|u_n|}>1.$ This test fails when the limit is unity.

**Example 9.26.** *Examine the following series for convergence :*

$$(i)\ \frac{1}{2^3} - \frac{1}{3^3}(1+2) + \frac{1}{4^3}(1+2+3) - \frac{1}{5^3}(1+2+3+4) + \ldots \infty.$$

**Solution.** (*i*) Here $u_n = (-1)^{n-1} \dfrac{(1+2+3+\ldots+n)}{(n+1)^3}$

$$= (-1)^{n-1} \frac{n(n+1)}{2(n+1)^3} = (-1)^{n-1} \frac{n}{2(n+1)^2} = (-1)^{n-1} a_n \text{ (Say)}.$$

Then $$a_n - a_{n+1} = \frac{1}{2}\left[\frac{n}{(n+1)^2} - \frac{n+1}{(n+2)^2}\right] = \frac{1}{2} \frac{n^2+n-1}{(n+1)^2 (n+2)^2} > 0.$$

*i.e.,* $a_{n+1} < a_n$. Also $\underset{n\to\infty}{\text{Lt}}\ a_n = \dfrac{1}{2} \underset{n\to\infty}{\text{Lt}} \dfrac{n}{(n+1)^2} = 0.$

Thus by Leibnitz's rule, $\Sigma a_n$ and therefore $\Sigma u_n$ is convergent.

Also $|u_n| = \dfrac{1}{2} \dfrac{n}{n^2+1}$. Taking $v_n = \dfrac{1}{n}$, we note that

$$\underset{n\to\infty}{\text{Lt}} \frac{|u_n|}{v_n} = \frac{1}{2} \underset{n\to\infty}{\text{Lt}} \frac{n^2}{n^2+1} = \frac{1}{2} \neq 0$$

Since $\Sigma v_n$ is divergent, therefore $\Sigma\, |u_n|$ is also divergent.

*i.e.,* $\Sigma u_n$ is convergent but $\Sigma\, |u_n|$ is divergent.

Thus the given series $\Sigma u_n$ is conditionally convergent.

**Example 9.27.** (*i*) *Test the convergence of the series* $\displaystyle\sum_{n=1}^{\infty} \frac{1}{n^2+4n+3}$. (*KTU, 2018*)

(*ii*) *Test the convergence of the series* $\displaystyle\sum_{n=1}^{\infty} \frac{\sqrt{n^2+n+1} - \sqrt{n^2-n+1}}{n}$ (*Marathwada, 2018*)

**Solution.** (*i*) Given, $U_n = \dfrac{1}{n^2+4n+3}$

$$= \frac{1}{(n+1)(n+3)} = \frac{(n+3)-(n+1)}{2(n+1)(n+3)}$$

$$= \frac{1}{2}\left[\frac{1}{n+1} - \frac{1}{n+3}\right]$$

$$S_n = u_1 + u_2 + u_3 + u_n$$

$$= \frac{1}{2}\left[\left(\frac{1}{2}-\frac{1}{4}\right) + \left(\frac{1}{3}-\frac{1}{5}\right) + \left(\frac{1}{4}-\frac{1}{6}\right) + \left(\frac{1}{5}-\frac{1}{7}\right) + \ldots\ldots + \left(\frac{1}{n}-\frac{1}{n+2}\right) + \left(\frac{1}{n+1}-\frac{1}{n+3}\right)\right]$$

$$= \frac{1}{2}\left(\frac{1}{2}+\frac{1}{3}-\frac{1}{n+2}-\frac{1}{n+3}\right) = \frac{1}{2}\left(\frac{5}{6}-\frac{1}{n+2}-\frac{1}{n+3}\right)$$

$$\lim_{n\to\infty} Sn = \lim_{n\to\infty} \frac{1}{2}\left(\frac{5}{6}-\frac{1}{n+2}-\frac{1}{n+3}\right)$$

$$= \frac{5}{12} \text{ (finite)}$$

So the series $\Sigma u_n$ is convergents.

(*ii*) Let $$U_n = \frac{\sqrt{n^2+n+1} - \sqrt{n^2-n+1}}{n}$$

$$U_n = \left[1 + \left(\frac{1}{n} + \frac{1}{n^2}\right)\right]^{1/2} - \left[1 + \left(-\frac{1}{n} + \frac{1}{n^2}\right)\right]^{1/2}$$

Expand using binomial expasion

$$U_n = \left[1 + \frac{1}{2}\left(\frac{1}{n} + \frac{1}{n^2}\right) + \frac{\frac{1}{2}\left(\frac{1}{2}-1\right)}{2!}\left(\frac{1}{n}+\frac{1}{n^2}\right)^2 + \ldots\right] - \left[1 + \frac{1}{2}\left(\frac{-1}{n} + \frac{1}{n^2}\right) + \frac{\frac{1}{2}\left(\frac{1}{2}-1\right)}{2!} - \left(\frac{1}{n}+\frac{1}{n^2}\right)^2 + \ldots\right]$$

$$= \left[1 + \frac{1}{2n} + \frac{1}{2n^2} - \frac{1}{8}\left(\frac{1}{n^2} + \frac{2}{n^3} + \frac{1}{n^4}\right) + \ldots\right] - \left[1 - \frac{1}{2n} + \frac{1}{2n^2} - \frac{1}{8}\left(\frac{1}{n^2} - \frac{2}{n^3} + \frac{1}{n^4}\right) + \ldots\right]$$

$$= \frac{1}{n} - \frac{1}{2n^3} + \ldots = \frac{1}{n}\left(1 - \frac{1}{2n^2} + \ldots\right)$$

Let $$u_n = \frac{1}{n}$$

$$\lim_{n\to\infty} \frac{u_n}{v_n} = \lim_{n\to\infty}\left(1 - \frac{1}{2n^2} + \ldots\right)$$

and $\Sigma u_n = 1$ [finite and non zero]

and $\Sigma v_n = \sum \frac{1}{n}$ is divergent

Hence by comparison test, $\Sigma u_n$ is also divergent.

## 9.14 POWER SERIES

**(1) Def.** *A series of the form* $a_0 + a_1x + a_2x^2 + \ldots + a_nx^n + \ldots$ ...(*i*)
*where the a's are independent of x, is called a* **power series** *in x.* Such a series may converge for some or all values of *x*.

**(2) Interval of convergence**

In the power series (*i*), $u_n = a_nx^n$.

$$\therefore \quad \underset{n\to\infty}{\text{Lt}} \frac{u_{n+1}}{u_n} = \underset{n\to\infty}{\text{Lt}} \frac{a_{n+1}x^{n+1}}{a_nx^n} = \underset{n\to\infty}{\text{Lt}}\left(\frac{a_{n+1}}{a_n}\right) . x$$

If $\underset{n\to\infty}{\text{Lt}}\left(\frac{a_{n+1}}{a_n}\right) = l$, then by Ratio test, the series (*i*) converges, when $lx$ is numerically less than 1, *i.e.*, when $|x| < 1/l$ and diverges for other values.

Thus the power series (*i*) has an interval $-1/l < x < 1/l$ within which it converges and diverges for values of $x$ outside this interval. Such an interval is called the *interval of convergence of the power series.*

**Example 9.28.** *State the values of x for which the following series converge :*

(*i*) $x - \frac{x^2}{2} + \frac{x^3}{3} - \frac{x^4}{4} + \frac{x^5}{5} - \ldots \infty$, (*JNTUK, 2022; Calicut, 2013*)

(*ii*) *Test the convergence of the given series* (*Kottayam, 2019*)

$$1 + \frac{3}{7}x + \frac{3\cdot 6}{7\cdot 10}x^2 + \frac{3\cdot 6\cdot 9}{7\cdot 10\cdot 13}x^3 + \frac{3\cdot 6\cdot 9\cdot 12}{7\cdot 10\cdot 13\cdot 16}x^4 + \ldots$$ (*JNTUK, 2022*)

**Solution.** (*i*) Here $u_n = (-1)^{n-1}\dfrac{x^n}{n}$ and $u_{n+1} = (-1)^n \dfrac{x^{n+1}}{n+1}$

$$\therefore \qquad \frac{u_{n+1}}{u_n} = -\frac{n}{n+1}x \quad \text{and} \quad \underset{n\to\infty}{\text{Lt}}\left|\frac{u_{n+1}}{u_n}\right| = \left(\underset{n\to\infty}{\text{Lt}}\frac{1}{1+1/n}\right)|x| = |x|$$

$\therefore$ by Ratio test the given series converges for $|x| < 1$ and diverges for $|x| > 1$.

Let us examine the series for $x = \pm 1$.

*For* $x = 1$, the series reduces to $1 - \dfrac{1}{2} + \dfrac{1}{3} - \dfrac{1}{4} + \dfrac{1}{5} + \ldots$

which is an alternating series and is convergent.

*For* $x = -1$, the series becomes $-\left(1 + \dfrac{1}{2} + \dfrac{1}{3} + \dfrac{1}{4} + \dfrac{1}{5} + \ldots\right)$

which is a divergent series as can be seen by comparison with $p$-series when $p = 1$.

Hence the given series converges for $-1 < x \le 1$.

(*ii*) Leaving the first term, we have $n$th term of the sequence 3, 6, 9, .... is $3 + (n-1)3n$ and $n^{\text{th}}$ term of the sequence 7, 10, 13, ..... is $7 + (n-1) = 3n + 4$

$$u_n = \frac{3\cdot 6\cdot 9.....3n}{7\cdot 10\cdot 13.....(3n+4)}x^n$$

$$u_{n+1} = \frac{3\cdot 6\cdot 9.....3n(3n+3)}{7\cdot 10\cdot 13.....(3n+4)(3n+7)}x^{n-1}$$

$$\frac{u_n}{u_{n+1}} = \frac{(3n+7)}{(3n+3)}\cdot\frac{1}{x} = \frac{\left(3+\frac{7}{n}\right)}{\left(3+\frac{3}{n}\right)}\cdot\frac{1}{x}$$

$$\lim\frac{u_n}{u_{n+1}} = \lim\left(\frac{3+\frac{7}{n}}{3+\frac{3}{n}}\right)\cdot\frac{1}{x} = \frac{3}{3}\cdot\frac{1}{x} = \frac{1}{x}$$

$\therefore$ By ratio Diversion test, the series is convergent if $\dfrac{1}{x} > 1$ *i.e.*, if $x < 1$, divergent $\dfrac{1}{x} < 1$.

If $x > 1$ and thus fails if $\dfrac{1}{x} = 1$ *i.e.*, if $x = 1$.

When $x = 1$, we have $\qquad \dfrac{u_n}{u_{n+1}} = \dfrac{3n+7}{3n+3}$

Applying Raab's test we have

$$\lim n\left(\frac{u_n}{u_{n+1}} - 1\right) = \lim n\left(\frac{3n+7}{3n+3} - 1\right) = \lim\frac{4n}{3n+3}$$

$$= \lim\frac{4}{3+\frac{3}{n}} = \frac{4}{3} > 1$$

$\therefore$ The series is convergent when $x = 1$.

Hence, the given series convergent if $x \le 1$ and divergent if $x > 1$.

**Example 9.29.** *Test the series* $\dfrac{x}{\sqrt{3}} - \dfrac{x^2}{\sqrt{5}} + \dfrac{x^3}{\sqrt{7}} - \ldots$ *for absolute convergence and conditional convergence.*

*(V.T.U., 2010)*

**Solution.** We have $u_n = (-1)^{n-1} \dfrac{x^n}{\sqrt{(2n+1)}}$ and $u_{n+1} = \dfrac{(-1)^n x^{n+1}}{\sqrt{(2n+3)}}$

$$\therefore \quad \underset{n\to\infty}{\text{Lt}} \left|\frac{u_{n+1}}{u_n}\right| = \underset{n\to\infty}{\text{Lt}} \left| \frac{\dfrac{(-1)^n x^{n+1}}{\sqrt{(2n+3)}} \cdot \sqrt{(2n+1)}}{(-1)^{n-1} x^n} \right| = \underset{n\to\infty}{\text{Lt}} \left| (-1) \sqrt{\left(\frac{2n+1}{2n+3}\right)} x \right|$$

$$= \underset{n\to\infty}{\text{Lt}} \left| \sqrt{\left(\frac{2+1/n}{2+3/n}\right)} x \right| = |x|$$

Hence *the given series is absolutely convergent for* $|x| < 1$ *and is divergent for* $|x| > 1$ and the test fails for $|x| = 1$.

*For* $x = 1$, $\quad u_n = \dfrac{(-1)^{n-1}}{\sqrt{(2n+1)}}$. Since $2n+1 < 2n+3$ or $(2n+1)^{-1/2} > (2n+3)^{-1/2}$

*i.e.,* $\quad u_n > u_{n+1}$. Also $\underset{n\to\infty}{\text{Lt}}\, u_n = \underset{n\to\infty}{\text{Lt}} \dfrac{1}{\sqrt{(2n+1)}} = 0$.

$\therefore$ the series is convergent by Leibnitz's test.

But $\quad \dfrac{1}{\sqrt{3}} + \dfrac{1}{\sqrt{5}} + \dfrac{1}{\sqrt{7}} + \ldots$ has $u_n = \dfrac{1}{\sqrt{(2n+1)}} = \dfrac{1}{\sqrt{n}} \cdot \dfrac{1}{\sqrt{(2+1/n)}}$

On comparing it with $v_n = \dfrac{1}{\sqrt{n}}$, $\Sigma u_n$ is divergent.

Hence the *given series is conditionally convergent for* $x = 1$.

*For* $x = -1$, the series becomes $-\left(\dfrac{1}{\sqrt{3}} + \dfrac{1}{\sqrt{5}} + \dfrac{1}{\sqrt{7}} + \ldots\right)$

But we have seen that the series $\dfrac{1}{\sqrt{3}} + \dfrac{1}{\sqrt{5}} + \dfrac{1}{\sqrt{7}} + \ldots$ is divergent.

Hence, *the given series is divergent when* $x = -1$.

**Example 9.30 :** *Test for convergence the series* $\Sigma\left\{\dfrac{(n+1)^3}{n^k + (n+2)^k}\right\}$.

**Solution:**

$$U_n = \frac{(n+1)^3}{n^k + (n+2)^k}$$

To apply comparision test, take

$$V_n = \frac{n^3}{n^k} = \frac{1}{n^{k-3}}$$

Now,

$$\frac{\lim U_n}{V_n} = \lim \frac{(n+1)^3}{n^k + (n+2)^k} \cdot n^{k-3}$$

$$= \lim \frac{(1+1/n)^3}{1 + \left(1 + 2/n\right)^k} = \frac{1}{1+1} = \frac{1}{2}$$

which is finite and non-zero

But the auxiliary series $\quad \Sigma V_n = \Sigma \dfrac{1}{n^{k-3}}$

is convergent if $K - 3 > 1$ *i.e.,* if $K > 4$ and divergent if $K - 3 \le 1$ if $K \le 4$.

Hence by comparsion test the given series is convgt. if $k \ge 4$ and divergent if $k \le 4$.

**Example 9.31:** *Test for convergence the series* $\sum_{n=1}^{\infty} \frac{3n+1}{4n+3} x^n > 0$.

**Solution:** Here,

$$U_n = \frac{3n+1}{4n+3} x^n$$

$$U_{n+1} = \frac{3(n+1)+1}{4(n+1)+3} x^{n+1}$$

$$= \frac{3n+4}{4n+7} x^{n+1}$$

$$\therefore \quad \frac{U_n}{U_{n+1}} = \frac{(3n+1)(4n+7)}{(4n+3)(3n+4)} \cdot \frac{x^n}{x^{n+1}}$$

$$= \lim \frac{\left(3+\frac{1}{n}\right)\left(4+\frac{7}{n}\right)}{\left(4+\frac{3}{n}\right)\left(3+\frac{4}{n}\right)} \cdot \frac{1}{x} = \frac{3 \cdot 4}{4 \cdot 3} \cdot \frac{1}{x} = \frac{1}{x}$$

$\therefore$ By ratio test, the series $\Sigma U_n$ is convergent if $\frac{1}{x} > 1$.

*i.e.*, $x < 1$ divergent if $\frac{1}{x} < 1$ *i.e.*, if $x > 1$ and test fails if $\frac{1}{x} = 1$. *i.e.*, if $x = 1$.

When $\quad x = 1, U_n = \frac{3n+1}{4n+3}$

In case, $\quad \lim U_n = \lim \frac{3n+1}{4n+3} = \lim \frac{n\left(3+\frac{1}{n}\right)}{n\left(4+\frac{3}{n}\right)} = \lim \frac{3+\frac{1}{n}}{4+\frac{3}{n}}$

$$= \frac{3}{4}, \text{ which is } x > 0.$$

$\therefore$ $\Sigma u_n$ is divergent when $x = 1$.

Hence $\Sigma U_n$ is convergent if $x < 1$ and divergent if $x \geq 1$.

**Example 9.32:** *Test the convergence of the series* $1 + \frac{2^2}{3 \cdot 4} + \frac{2^2 \cdot 4^2}{3 \cdot 4 \cdot 5 \cdot 6} + \frac{2^2 \cdot 4^2 \cdot 6^2}{3 \cdot 4 \cdot 5 \cdot 6 \cdot 7 \cdot 8}$.

**Solution:** Neglecting the first term 1, we have

$$U_n = \frac{2^2 \cdot 4^2 \cdot 6^2 .....(2n)^2}{3 \cdot 4 \cdot 5 \cdot 6 \cdot 7 \cdot 8 .....(2n)(2n+1)(2n+1)}$$

$$U_{n+1} = \frac{2^2 \cdot 4^2 \cdot 6^2 .....(2n)^2 \ (2n+2)^2}{3 \cdot 4 \cdot 5 \cdot 6 \cdot 7 \cdot 8 .....(2n+2)(2n+3)(2n+4)}$$

$$\frac{U_n}{U_{n+1}} = \frac{(2n+3)(2n+4)}{(2n+2)^2}$$

$$= \frac{\left(2+\frac{3}{n}\right)\left(2+\frac{4}{n}\right)}{\left(2+\frac{2}{n}\right)^2}$$

$$\lim \frac{U_n}{U_{n+1}} = \frac{2 \cdot 2}{(2)^2} = 1. \text{ So that the ratio test fails.}$$

Apply Raab's test we have

$$\lim n\left(\frac{U_n}{U_{n+1}} - 1\right) = \lim n\left[\frac{(2n+3)(2n+4)}{(2n+2)^2} - 1\right]$$

$$= \lim n\left[\frac{6n+8}{(2n+2)^2}\right] = \lim\left[\frac{6+\frac{8}{n}}{\left(2+\frac{2}{n}\right)^2}\right]$$

$$= \frac{6}{2^2} = \frac{3}{2} > 1$$

Hence the series is convergents.

**Example 9.33 :** *Investigate the convergence of the series* $\sum_{n=1}^{\infty} \frac{2^n + 5}{3^n}$.

**Solution:** Given $\sum_{n=1}^{\infty} \frac{2^n+5}{3^n} = \sum_{n=1}^{\infty}\left[\left(\frac{2}{3}\right)^n + 5\left(\frac{1}{3^n}\right)^n\right] = \sum_{n=1}^{\infty}\left(\frac{2}{3}\right)^n + \sum_{n=1}^{\infty} 5\left(\frac{1}{3}\right)^n$

$$= \sum_{n=1}^{\infty}\left(\frac{2}{3}\right)^n + \sum_{n=1}^{\infty} 5\left(\frac{1}{3}\right)^n$$

$$= \sum_{n=1}^{\infty}\left(\frac{2}{3}\right)\left(\frac{2}{3}\right)^{n-1} + \sum_{n=1}^{\infty}\left(\frac{5}{3}\right)\left(\frac{1\cdot}{3}\right)^{n-1}$$

Both the series and geometric series with $r_1 = \frac{2}{3}$ and $r_2 = \frac{1}{3}$ respectively.

$$|r_1| < 1 \text{ and } |r_{2\cdot}| < 1$$

Hence, both the series are convergent

$$S_1 = \frac{a_1}{1-r_1} = \frac{\frac{2}{3}}{1-\frac{2}{3}} = \frac{\frac{2}{3}}{\frac{1}{3}} = 2$$

$$S_2 = \frac{a_2}{1-r_2} = \frac{\frac{5}{3}}{1-\frac{1}{3}} = \frac{\frac{5}{3}}{\frac{2}{3}} = \frac{5}{2}$$

Hence $\sum_{n=1}^{\infty} \frac{2^n+5}{3^n} = 2 + \frac{5}{2} = \frac{9}{2}$

**Example 6.34:** *Test the convergence of the series*

$$\sum_{n=1}^{\infty} \frac{5^n + a}{3^n + b},\ a > 0,\ b > 0$$

**Solution:** Let $u_n = \frac{5^n + a}{3^n + b}$

$$u_{n+1} = \frac{5^{n+1} + a}{3^{n+1} + b}$$

$$\lim_{n\to\infty} \frac{u_n}{u_n + 1} = \lim_{n\to\infty} \frac{5^n + a}{3^n + b} \cdot \frac{3^{n+1} + b}{5^{n+1} + a}$$

$$= \lim_{n\to\infty} \frac{1 + \frac{a}{5}n}{1 + \frac{b}{3}} \cdot \frac{3 + \frac{b}{3}n}{5 + \frac{a}{5}n}$$

$$= \frac{3}{5} < 1$$

Hence, by *D'* Alembert, ratio test the series is divergent.

**Example 9.35:** *Test the convergence of the series whose nth term is* $\left[\sqrt{(n^2 + 1)} - n\right]x^{2n}$. *(JNTUK, 2023)*

**Solution:**

$$u_n = \left[\sqrt{(n^2 + 1)} - n\right]x^{2n}$$

$$= n\left[\left(1 + \frac{1}{n^2}\right)^{1/2} - 1\right]x^{2n}$$

$$= n\left[1 + \frac{1}{2}\cdot\frac{1}{n^2} + \frac{\frac{1}{2}\left(\frac{1}{2} - 1\right)}{2!}\frac{1}{n^4} + \dots - 1\right]x^{2n}$$

$$= \frac{1}{n}\left[\frac{1}{2} - \frac{1}{8n^2} + \dots\right]x^{2n} \qquad \dots(i)$$

$$\therefore \qquad U_{n+1} = \left(\frac{1}{n+1}\right)\left[\frac{1}{2} - \frac{1}{8(n+1)^2} + \dots\right]x^{2n+2}$$

$$\therefore \qquad \frac{U_n}{U_{n+1}} = \left(1 + \frac{1}{n}\right)\frac{\left[\frac{1}{2} - \frac{1}{8}n^2 + \dots\right]}{\left[\frac{1}{2} - \frac{1}{8}(n+1)^2 + \dots\right]}\frac{-1}{x^2}$$

$$\therefore \qquad \lim \frac{u_n}{u_n + 1} = \frac{1}{x^2}$$

$\therefore$ By *D'* Alembert's ratio test, the given series is convergent if $\frac{1}{x^2} > 1$. *i.e.*, $x^2 < 1$ and divergent if $\frac{1}{x^2} < 1$. *i.e.*, $x^2 > 1$.

If $\frac{1}{x^2} = 1$. if $x^2 = 1$, the ratio test fails.

Now, when $x^2 = 1$. from (*i*), we have

$$U_n = \frac{1}{n}\left[\frac{1}{2} - \frac{1}{8n^2} - \dots\right]$$

In this case to apply comparision test, we take $V_n = \frac{1}{n}$.

Then $\frac{u_n}{v_v} = \frac{1}{2}$, which is finite and non-zero.

By comparision test $\Sigma u_n$ and $\Sigma v_n$ are either both convergent or both divergent.But for $V_n$.
Hence, the given series is convergent if $x^2 < 1$ and divergent if $x^2 \geq 1$.

**Example 9.36:** *Test the convergence of series* $1+\frac{1}{2}\cdot\frac{x^2}{4}+\frac{1\cdot 3\cdot 5}{2\cdot 4\cdot 6}\frac{x^4}{8}+\frac{1\cdot 3\cdot 5\cdot 7\cdot 9}{2\cdot 4\cdot 6\cdot 8\cdot 10}\frac{x^6}{12}+.....$ *(Bhilai, 2019)*

**Solution:** Neglecting the first term 1, we have $n$th term of the sequence $1\cdot 3\cdot 5\cdot 9.....$
$1+(n-1)\,4 = 4n-3$ and $n$th term of the sequence 4, 8, 12, is

$$4+(n-1)4 = 4n$$

$$U_n = \frac{1\cdot 3\cdot 5.....(4n-3)}{2\cdot 4\cdot 6.....(4n-2)}\frac{x^{2n}}{4n}$$

$$U_{n+1} = \frac{1\cdot 3\cdot 5.....(4n-3)(4n-1)(4n+1)}{2\cdot 4\cdot 6.....(4n-2)(4n)(4n+2)}\cdot\frac{x^{2n+2}}{4n+4}$$

Now,

$$\frac{U_n}{U_{n+1}} = \frac{4n(4n+2)}{(4n-1)(4n+1)}\cdot\frac{4n+4}{4n}\frac{1}{x^2}$$

$$= \frac{\left(4+\frac{2}{n}\right)\left(4+\frac{4}{n}\right)}{\left(4-\frac{1}{n}\right)\left(4+\frac{1}{n}\right)}\cdot\frac{1}{x^2}$$

$$\lim\frac{U_n}{U_{n+1}} = \frac{4\cdot 4}{4\cdot 4}\frac{1}{x^2} = \frac{1}{x^2}$$

$\therefore$ By ratio test the series $\Sigma U_n$ is convergent if $\frac{1}{x^2} > 1$.

If $x^2 < 1$, divergent if $\frac{1}{x^2} < 1$. *i.e.*, $x^2 > 1$ and test $\frac{1}{x^2} < 1$. *i.e.*, when $x^2 = 1$

When

$$x^2 = 1\ \frac{U_n}{U_{n+1}} = \frac{(4n+2)(4n+4)}{(4n-1)(4n+1)}$$

$$= \frac{16n^2+24n+8}{16n^2-1}$$

Applying Raab's test we have

$$\lim n\left(\frac{U_n}{U_{n+1}}-1\right) = \lim n\left(\frac{16n^2+24n+8}{16n^2-1}-1\right)$$

$$\lim\frac{n(24n+9)}{16n^2-1} = \lim\frac{24+\frac{9}{n}}{16-\frac{1}{n^2}} = \frac{24}{16} = \frac{3}{2} > 1$$

$\therefore$ The series is convergent when $x^2 = 1$.
Hence the given series is convergent if $x^2 \leq 1$ and divergent if $x^2 > 1$.

**Example 9.37:** *Test the convergence of the series*

*(JNTUK, 2020)*

$$x-\frac{x^2}{2^2}+\frac{x^3}{3^2}-\frac{x^4}{4^2}+.....$$

**Solution:** Here we shall apply '*D*' Alembert test.

$$U_n = (-1)^{n-1}\frac{x^n}{n^2}$$

$$U_{n+1} = (-1)^n \frac{x^{n+1}}{(n+1)^2}$$

$$\frac{U_{n+1}}{U_n} = \frac{-x^{n+1}}{(n+1)^2}\cdot\frac{n^2}{x^n} = -\frac{1}{\left(1+\frac{1}{n}\right)^2}\cdot x$$

$$\left|\frac{U_{n+1}}{U_n}\right| = \left|\frac{-1}{\left(1+\frac{1}{n}\right)^2}\cdot x\right| = \frac{1}{\left(1+\frac{1}{n}\right)^2}|x|$$

$$\therefore \quad \underset{n\to\infty}{\text{Lt}}\left|\frac{U_{n+1}}{U_n}\right| = \underset{n\to\infty}{\text{Lt}}\left[\frac{1}{\left(1+\frac{1}{n}\right)^2}\right]|x| = |x|$$

So, the given series convergence, when $|x| < 1$ and diverges for $|x| > 1$.

When $|x| = 1$. *D*' Alembert's ratio test fails.

Now, when $|x| = 1$, the series can be shown to be convergent by Leibnitz's test.

Hence the given series is convergent for $|x| \le 1$ and divergent for $|x| > 1$.

## 9.15 (1) CONVERGENCE OF EXPONENTIAL SERIES

*The series* $1 + x + \frac{x^2}{2!} + \ldots + \frac{x^n}{n!} + \ldots \infty$ *is convergent for all values of x.*

Here $\quad \underset{n\to\infty}{\text{Lt}}\frac{u_{n+1}}{u_n} = \underset{n\to\infty}{\text{Lt}}\left[\frac{x^n}{n!}\div\frac{x^{n-1}}{(n-1)!}\right] = \underset{n\to\infty}{\text{Lt}}\frac{x}{n} = 0$

Hence the series converges, whatever be the value of $x$.

**(2) Convergence of logarithmic series**

*The series* $x - \frac{x^2}{2} + \frac{x^3}{3} - \ldots + (-1)^n\frac{x^n}{n} + \ldots \infty$ *is convergent for* $-1 < x \le 1$.

Here $\quad \underset{n\to\infty}{\text{Lt}}\frac{u_{n+1}}{u_n} = \underset{n\to\infty}{\text{Lt}}\frac{(-1)^{n+1}x^{n+1}}{n+1}\cdot\frac{n}{(-1)^n x^n} = -x\underset{n\to\infty}{\text{Lt}}\frac{n}{n+1} = -x\underset{n\to\infty}{\text{Lt}}\left\{\frac{1}{1+1/n}\right\} = -x.$

Hence the series converges for $|x| < 1$ and diverges for $|x| > 1$.

When $x = 1$, the series being $1 - \frac{1}{2} + \frac{1}{3} - \frac{1}{4} + \ldots$, is convergent.

When $x = -1$, the series being $-\left(1 + \frac{1}{2} + \frac{1}{3} + \frac{1}{4} + \ldots\right)$, is divergent.

Hence the series converges for $-1 < x \le 1$.

**(3) Convergence of binomial series**

*The series* $1 + nx + \frac{n(n-1)}{2!}x^2 + \ldots + \frac{n(n-1)\ldots(n-r+1)}{r!}x^r + \ldots \infty$

*converges for* $|x| < 1$.

Here $\quad u_r = \frac{n(n-1)\ldots(n-r)}{(r-1)!}x^{r-1}$ and $u_{r+1} = \frac{n(n-1)\ldots(n-r+1)}{r!}x^r$

$$\therefore \quad \underset{r\to\infty}{\text{Lt}}\frac{u_{r+1}}{u_r} = \underset{r\to\infty}{\text{Lt}}\frac{n-r+1}{r}x = \underset{r\to\infty}{\text{Lt}}\left(\frac{n+1}{r}-1\right)x = -x \text{ for } r > n+1.$$

Hence, the series converges for $|x| < 1$.

**PROBLEMS 9.8**

1. Test the following series for conditional convergence : (*i*) $\sum \frac{(-1)^{n-1}}{\sqrt{n}}$ (*ii*) $\sum \frac{(-1)^{n-1} n}{n^2+1}$ (*iii*) $\sum_{n=1}^{\infty} (-1)^{n+1} \frac{n+1}{2n^2}$. (*Delhi, 2012*)

2. Prove that the series $\frac{\sin x}{1^3} - \frac{\sin 2x}{2^3} + \frac{\sin 3x}{3^3} - \ldots$ converges absolutely. (*Andhra, 2016; D.T.U., 2015; Rohtak, 2012*)

3. Test the following series for conditional convergence :

   (*i*) $1 - \frac{1}{2^p} + \frac{1}{3^p} - \frac{1}{4^p} + \ldots \infty$ (*Andhra, 2015*) (*ii*) $1 - \frac{1}{2} + \frac{1.3}{2.4} - \frac{1.3.5}{2.4.6} + \ldots \infty$

   (*iii*) $\frac{1}{2(\log 2)^p} - \frac{1}{3(\log 3)^p} + \frac{1}{4(\log 4)^p} \ldots\ldots$ (*Rohtak, 2012 S*)

4. Discuss the absolute convergence of (*i*) $\sum_{n=0}^{\infty} \frac{(-1)^n x^n}{n+1}$ (*Andhra, 2016 ; D.T.U., 2013*)

   (*ii*) $x - \frac{x^2}{4} + \frac{x^3}{9} - \frac{x^4}{16} + \ldots \infty$ (*D.T.U., 2012*) (*iii*) $\frac{1}{\sqrt{(1^3+1)}} - \frac{1}{\sqrt{(2^3+1)}} x + \frac{1}{\sqrt{(3^3+1)}} x^2 - \ldots \infty$

5. Find the nature of the series $\frac{x}{1 \cdot 2} - \frac{x^2}{2 \cdot 3} + \frac{x^3}{3 \cdot 4} - \frac{x^4}{4 \cdot 5} + \ldots \infty$

6. For what values of $x$ are the following series convergent :

   (*i*) $x - \frac{x^2}{\sqrt{2}} + \frac{x^3}{\sqrt{3}} - \frac{x^4}{\sqrt{4}} + \ldots \infty$ (*Anna, 2016 ; P.T.U., 2009 S ; V.T.U., 2008*) (*ii*) $x \quad \frac{x^2}{2^2} + \frac{x^3}{3^2} - \frac{x^4}{4^2} + \ldots \infty$

7. Find the radius of convergence of the series $\sum \frac{n!}{n^n} x^n$.

8. Prove that $\frac{1}{a} + \frac{1}{a+1} - \frac{1}{a+2} + \frac{1}{a+3} + \frac{1}{a+4} - \frac{1}{a+5} + \ldots$ is a divergent series.

9. Test the series $1 - \frac{1}{2\sqrt{2}} + \frac{1}{3\sqrt{3}} - \frac{1}{4\sqrt{4}}$ for

   (*i*) absolute convergence and (*ii*) conditional convergence. (*D.T.U., 2013*)

10. Find the radius of convergence and interval of convergence of the series $\sum_{k=1}^{\infty} \frac{(x-5)^k}{k^2}$. (*K.T.U., 2016*)

## 9.16 PROCEDURE FOR TESTING A SERIES FOR CONVERGENCE

First see whether the given series is

(*i*) a series with terms alternately positive and negative ;

(*ii*) a series of positive terms excluding power series ;

or (*iii*) a power series.

*For alternating series* (*i*), apply the Leibnitz's rule (§ 9.12).

*For series* (*ii*), first find $u_n$ and if possible evaluate Lt $u_n$. If Lt $u_n \neq 0$, the series is divergent. If Lt $u_n = 0$, compare $\Sigma u_n$ with $\Sigma 1/n^p$ and apply the comparison tests (§ 9.6).

If the comparison tests are not applicable, apply the Ratio test (§ 9.9). If Lt $u_n/u_{n+1} = 1$, *i.e.*, the ratio test fails, apply Raabe's test (§ 9.10). If Raabe's test fails for a similar reason, apply Logarithmic test (§ 9.10). If this also fails, apply Cauchy's root test (§ 9.11).

*For the power series* (*iii*), apply the Ratio test as in § 9.14. If the Ratio test fails, examine the series as in case (*ii*) above.

## PROBLEMS 9.9

Test the convergence of the following series :

1. $\frac{1}{1.2} - \frac{1}{2.2^2} + \frac{1}{3.2^3} - \ldots.$ *(Delhi, 2015)*

2. $\sum \left( \frac{1}{\sqrt{n}} - \sqrt{\frac{n}{n+1}} \right)$.

3. $1 + \frac{1}{2^2} + \frac{2^2}{3^3} + \frac{3^3}{4^4} + \ldots$

4. $\sum_{n=1}^{\infty} \sqrt{\left\{ \frac{2^n + 1}{3^n + 1} \right\}}$

5. $\frac{1}{1+\sqrt{2}} + \frac{2}{1+2\sqrt{3}} + \frac{3}{1+3\sqrt{4}} + \ldots \infty.$

6. $\frac{x}{1+\sqrt{1}} + \frac{x^2}{2+\sqrt{2}} + \frac{x^3}{3+\sqrt{3}} + \ldots \infty.$

7. $1 + \frac{2^2}{3^2} x + \frac{2^2 . 4^2}{3^2 . 5^2} x^2 + \frac{2^2 . 4^2 . 6^2}{3^2 . 5^2 . 7^2} x^3 + \ldots \infty.$

8. $\sum_{n=1}^{\infty} \frac{2^n - 2}{2^n + 1} . x^{n-1}$ $(x > 0)$.

9. $\sum_{1}^{\infty} \frac{n x^n}{(n+1)(n+2)}$ $(x > 0)$.

10. $\sum_{n=1}^{\infty} \frac{(2n)!}{(n!)^2} x^n$.

11. $\sum_{n=1}^{\infty} \frac{x^n}{(2n-1)^2 \, 2^n}$.

12. $\sum_{0}^{\infty} \frac{(3x+5)^n}{(n+1)!}$.

13. $\sum_{n=1}^{\infty} \frac{(x+2)^n}{3^n n}$.

14. $\sum_{n=2}^{\infty} \frac{(-1)^n}{\log n}$.

15. $\sum \frac{(-1)^{n-1} \sin nx}{n^3}$.

16. $\frac{\sqrt{2}-1}{3^3-1} + \frac{\sqrt{3}-1}{4^3-1} + \frac{\sqrt{4}-1}{5^3-1} + \ldots \infty.$

17. $\sum_{n=2}^{\infty} \frac{1}{(n \log n)(\log \log n)^p}$.

18. $\sum_{n=1}^{\infty} \frac{x^{n+1}}{(n+1)^{\sqrt{n}}}$ *(Rohtak, 2012 S)*

## 9.17 UNIFORM CONVERGENCE

Let $$u_1(x) + u_2(x) + \ldots \infty = \sum_{n=1}^{\infty} u_n(x) \qquad \ldots(1)$$

be an infinite series of functions each of which is defined in the interval $(a, b)$. Let $s_n(x)$ be the sum of its first $n$ terms, *i.e.*, $s_n(x) = u_1(x) + u_2(x) + \ldots + u_n(x)$

At some point $x = x_1$, if $\underset{n \to \infty}{\text{Lt}} s_n(x_1) = s(x_1)$,

then the series (1) is said to converge to sum $s(x_1)$ at that point. This means at $x = x_1$ given a positive number $\varepsilon$, we can find a number $N$ such that $|s(x_1) - s_n(x_1)| < \varepsilon$ for $n > N$ ...(2)

Evidently $N$ will depend on $\varepsilon$ but generally it will also depend on $x_1$. Now if we keep the same $\varepsilon$ but take some other value $x_2$ of $x$ for which (1) is convergent, then we may have to change $N$ for the inequality (2) to hold. If we wish to approximate the sum $s(x)$ of the series by its partial sums $s_n(x)$, we shall require different partial sums at different points of the interval and the problem will become quite complicated. If, however, we choose an $N$ which is independent of the values of $x$, the problem becomes simpler. Then the partial sum $s_n(x)$, $(n > N)$ approximates to $s(x)$ for all values of $x$ in the interval $(a, b)$ and $\varepsilon$ is uniform throughout this interval. Thus we have

**Definition.** *The series $\Sigma u_n(x)$ is said to be uniformly convergent in the interval $(a, b)$, if for a given $\varepsilon > 0$, a number $N$ can be found* **independent of x,** *such that for every $x$ in the interval $(a, b)$,*

$$|s(x) - s_n(x)| < \varepsilon \text{ for all } n > N.$$

**Example 9.38.** *Examine the geometric series $1 + x + x^2 + \ldots + x^{n-1} + \ldots \infty$ for uniform convergence in the interval $\left(-\frac{1}{2}, \frac{1}{2}\right)$.*

**Solution.** We have $s_n(x) = 1 + x + x^2 + \dots + x^{n-1} = \frac{1 - x^n}{1 - x}$.

and $s(x) = \underset{n \to \infty}{\text{Lt}} \frac{1 - x^n}{1 - x} = \frac{1}{1 - x}$ for $|x| < 1$

$\therefore$ $|s(x) - s_n(x)| = \left| \frac{x^n}{1 - x} \right| = \frac{|x^n|}{1 - x} = \frac{|x|^n}{1 - x}$ which will be $< \varepsilon$, if $|x|^n < \varepsilon (1 - x)$.

Choose $N$ such that $|x|^N = \varepsilon (1 - x)$

or $$N = \log [\varepsilon \ (1 - x)] / \log |x| \qquad ...(i)$$

Evidently $N$ increases with the increase of $|x|$ and in the interval $-\frac{1}{2} \le x \le \frac{1}{2}$, it assumes a maximum value $N' = \log (\varepsilon/2)/\log \frac{1}{2}$ at $x = \frac{1}{2}$ for a given $\varepsilon$.

Thus $|s(x) - s_n(x)| < \varepsilon$ for all $n \ge N'$ for every value of $x$ in the interval $\left(-\frac{1}{2}, \frac{1}{2}\right)$.

Hence the geometric series converges uniformly in the interval $\left(-\frac{1}{2}, \frac{1}{2}\right)$.

**Obs.** *The geometric series though convergent in the interval (–1, 1), is not uniformly convergent in this interval,* since we cannot find a fixed number $N$ for every $x$ in this interval
($\because$ $N$ given by $(i) \to \infty$ as $|x| \to 1$).

## 9.18 WEIERSTRASS'S M-TEST*

*A series $\Sigma u_n(x)$ is uniformly convergent in an interval (a, b), if there exists a convergent series $\Sigma M_n$ of positive constants such that $|u_n(x)| \le M_n$ for all values of x in (a, b).*

Since $\Sigma M_n$ is convergent, therefore, for a given $\varepsilon > 0$, we can find a number $N$, such that $|s - s_n| < \varepsilon$ for every $n > N$,

where $s = M_1 + M_2 + \dots + M_n + M_{n+1} + \dots$ and $s_n = M_1 + M_2 + \dots + M_n$

This implies that $|M_{n+1} + M_{n+2} + \dots| < \varepsilon$ for every $n > N$.

Since $|u_n(x)| \le M_n$

$\therefore$ $$|u_{n+1}(x)| + u_{n+2}(x) + \dots | \le | u_{n+1}(x)| + |u_{n+2}(x)| + \dots$$
$$\le M_{n+1} + M_{n+2} + \dots < \varepsilon \text{ for every } n > N.$$

*i.e.*, $|s(x) - s_n(x)| < \varepsilon$ for every $n > N$, where $s(x)$ is the sum of the series $\Sigma u_n(x)$.

Since $N$ does not depend on $x$, the series $\Sigma u_n(x)$ converges uniformly in $(a, b)$.

**Obs.** *$\Sigma u_n(x)$ is also absolutely convergent for every x, since $|u_n(x)| \le M_n$.*

**Example 9.39.** *Show that the following series converges uniformly in any interval :*

*(i)* $\sum \frac{\cos nx}{n^p}$ *(ii)* $\sum \frac{1}{n^3 + n^4 x^2}$.

**Solution.** (*i*) $\left| \frac{\cos nx}{n^p} \right| = \frac{|\cos nx|}{n^p} \le \frac{1}{n^p}$ $(= M_n)$ for all values of $x$.

Since $\sum_{n=1}^{\infty} M_n = \sum_{n=1}^{\infty} \frac{1}{n^p}$ converges for p > 1,

$\therefore$ By $M$-test, the given series converges uniformly for all real values of $x$ and $p > 1$.

(*ii*) For all values of $x$, $n^3 + n^4x^2 > n^3$

$\therefore$ $\left| \frac{1}{n^3 + n^4x^2} \right| < \frac{1}{n^3}$ $(= M_n)$. But $\Sigma M_n$ being $p$-series with $p > 1$, is convergent.

$\therefore$ By $M$-test, the given series converges uniformly in any interval.

---

* Named after the great German mathematician *Karl Weierstrass* (1815–1897) who made basic contributions to Calculus, Approximation theory, Differential geometry and Calculus of variations. He was also one of the founders of Complex analysis.

**Example 9.40.** *Examine the following series for uniform convergence :*

*(i)* $\sum_{n=1}^{\infty} \frac{\sin(nx + x^2)}{n(n+2)}$ *(ii)* $\sum_{n=1}^{\infty} \frac{1}{n^p + n^q x^2}$

**Solution.** *(i)* $\left|\frac{\sin(nx + x^2)}{n(n+2)}\right| = \frac{|\sin(nx + x^2)|}{n^2 + 2n} \le \frac{1}{n^2}$ $(= M_n)$ for all real $x$.

Since $\sum_{n=1}^{\infty} M_n = \sum_{n=1}^{\infty} \frac{1}{n^2}$ is convergent, therefore, by $M$-test, the given series is uniformly convergent for all real values of $x$.

*(ii)* For all real values of $x$, $x^2 \ge 0$, *i.e.*, $n^q x^2 \ge 0$

*i.e.*, $$n^p + n^q x^2 \ge n^p \quad \text{or} \quad \frac{1}{n^p + n^q x^2} \le \frac{1}{n^p} \; (= M_n)$$

Since $\sum_{n=1}^{\infty} M_n = \sum_{n=1}^{\infty} \frac{1}{n^p}$ is convergent for $p > 1$,

$\therefore$ by $M$-test, the given series is uniformly convergent for all real values of $x$ and $p > 1$.

## 9.19 PROPERTIES OF UNIFORMLY CONVERGENT SERIES

*I. If the series $\Sigma u_n(x)$ converges uniformly to sum $s(x)$ in the interval $(a, b)$ and each of the functions $u_n(x)$ is continuous in this interval, then the sum $s(x)$ is also continuous in $(a, b)$.*

*II. If the series $\Sigma u_n(x)$ converges uniformly in the interval $(a, b)$ and each of the functions $u_n(x)$ is continuous in this interval, then the series can be integrated term by term*

*i.e.*, $$\int_a^b [u_1(x) + u_2(x) + \ldots]\, dx = \int_a^b u_1(x)\, dx + \int_a^b u_2(x)\, dx + \ldots$$

*III. If $\Sigma u_n(x)$ is a convergent series having continuous derivatives of its terms, and the series $\Sigma u_n(x)$ converges uniformly, then the series can be differentiated term by term*

$$\frac{d}{dx}[u_1(x) + u_2(x) + \ldots] = u_1'(x) + u_2'(x) + \ldots.$$

**Example 9.41.** *Prove that* $\int_0^1 \left(\Sigma \frac{x^n}{n^2}\right) dx = \sum_{n=1}^{\infty} \frac{1}{n^2(n+1)}.$ *(U.K.T.U., 2011)*

**Solution.** $|x^n| \le 1$ for $0 \le x \le 1$

$\therefore$ $\left|\frac{x^n}{n^2}\right| \le \frac{1}{n^2}$ $(= M_n)$ for $0 \le x \le 1$. But $\Sigma M_n$ is a convergent series.

$\therefore$ by $M$-test, the series $\Sigma(x^n/n^2)$ is uniformly convergent in $0 \le x \le 1$. Also $x^n/n^2$ is continuous in this interval.

$\therefore$ the series $\Sigma(x^n/n^2)$ can be integrated term by term in the interval $0 \le x \le 1$.

*i.e.*, $$\int_0^1 \left(\Sigma \frac{x^n}{n^2}\right) dx = \Sigma\left(\int_0^1 \frac{x^n}{n^2}\, dx\right) = \Sigma\left(\frac{1}{n^2}\int_0^1 x^n\, dx\right) = \Sigma \frac{1}{n^2(n+1)}.$$

**Imp. Obs.** *There is no relation between absolute and uniform convergence.* In fact, a series may converge absolutely but not uniformly while another series may converge uniformly but not absolutely.

For instance, the series

$\frac{1}{x^2+1} - \frac{1}{x^2+2} + \frac{1}{x^2+3} - \ldots$ can be seen to converge uniformly but not absolutely, while the series

$x^2 + \frac{x^2}{x^2+1} + \frac{x^2}{(x^2+1)^2} + \frac{x^2}{(x^2+1)^3} + \ldots$ can be shown to converge absolutely but not uniformly.

**Example 9.42 :** *Test the convergence of the series* $\Sigma \dfrac{x^n}{n+\sqrt{1+n^2}}$.

**Solution:**

$$u_n = \frac{x^n}{n+\sqrt{(1+n^2)}}$$

$$u_{n+1} = \frac{x^{n+1}}{n+1+\sqrt{\{(1+(n+1)^2\}}}$$

Then,
$$\frac{u_n}{u_{n+1}} = \frac{(n+1)+\sqrt{1+(n+1)^2}}{n+\sqrt{(1+n^2)}}\,\frac{1}{x} = \frac{\left(1+\dfrac{1}{n}\right)+\sqrt{\left\{\dfrac{1}{n^2}+\left(1+\dfrac{1}{n^2}\right)^2\right\}}}{1+\sqrt{1+\dfrac{1}{n^2}}+1}\cdot\frac{1}{x}$$

$$\lim\frac{u_n}{u_{n+1}} = \frac{(1+0)+1}{1+1}\cdot\frac{1}{x} = \frac{1}{x}$$

$\therefore$ By ratio test the series is convergent if $\dfrac{1}{x} > 1$ if $x < 1$, divergent if $\dfrac{1}{x} < 1$, *i.e.*, if $x > 1$ and the test fails when $\dfrac{1}{x} = 1$.

When $x = 1$, we have $u_n = \dfrac{1}{n+\sqrt{(1+n)^2}}$

To apply comparison test, we take $V_n = \dfrac{1}{n}$

$$\frac{u_n}{v_n} = \lim\frac{n}{n+\sqrt{1+n^2}} = \lim\frac{1}{1+\sqrt{\left(\dfrac{1}{n^2}+1\right)}}$$

$= 1$, which is finite and non-zero.

Since the auxiliary series $\Sigma V_n = \Sigma\dfrac{1}{n}$ is divergent as $p = 1$. So it is divergent when $x = 1$.

Hence, the given series is convergent $x > 1$, divergent if $x > 1$.

## PROBLEMS 9.10

Test for uniform convergence the series :

1. $\displaystyle\sum_{n=1}^{\infty}\frac{x^n}{n^{3/2}}$.

2. $\displaystyle\sum\frac{\cos nx}{2^n}$.

3. $\dfrac{\sin x}{1^2}+\dfrac{\sin 2x}{2^2}+\dfrac{\sin 3x}{3^2}+\dfrac{\sin 4x}{4^2}+\ldots\infty$.

4. $\sin x-\dfrac{\sin 2x}{2\sqrt{2}}+\dfrac{\sin 3x}{3\sqrt{3}}-\dfrac{\sin 4x}{4\sqrt{4}}+\ldots\infty$. *(Andhra, 2014)*

5. $\dfrac{\cos x}{1^2}+\dfrac{\cos 3x}{3^2}+\dfrac{\cos 5x}{5^2}+\ldots\infty$.

6. $\dfrac{ax}{2}+\dfrac{a^2x^2}{5}+\dfrac{a^3x^3}{10}+\ldots+\dfrac{a^nx^n}{n^2+1}+\ldots\infty$.

7. Show that the series $\Sigma r^n\sin n\theta$ and $\Sigma r^n\cos n\theta$ converge uniformly for all real values of $\theta$ if $0 < r < 1$.

8. Show that $\dfrac{1}{1+x^2}-\dfrac{1}{2+x^2}+\dfrac{1}{3+x^2}-\dfrac{1}{4+x^2}+\ldots$ converges uniformly in the interval $x \geq 0$ but not absolutely.

9. Prove that $\displaystyle\sum\frac{x}{n(1+nx^2)}$ is uniformly convergent for all real values of $x$.

10. Examine the following series for uniform convergence :

(i) $\sum_{n=1}^{\infty} \frac{1}{n^4 + n^3x^2}$ (ii) $\sum_{n=1}^{\infty} \frac{\cos(x^2 + n^2x)}{n(n^2+2)}$.

11. Show that

(i) $\int_0^1 \left(\sum \frac{\sin x}{x}\right) dx = 1 - \frac{1}{3.3!} + \frac{1}{5.5!} - \frac{1}{7.7!} + \ldots \infty$; (ii) $\int_0^{\pi} \left(\sum \frac{\sin n\theta}{n^3}\right) d\theta = 2\sum_{n=1}^{\infty} \frac{1}{(2n-1)^4}$.

Find out which of the following series converge absolutely or conditionally.

12. $1 - \frac{1}{3^2} + \frac{1}{5^2} - \frac{1}{7^2} + \frac{1}{9^2} - \ldots\ldots$

13. $\frac{1}{\log 2} - \frac{1}{\log 3} - \frac{1}{\log 4} - \ldots\ldots$

14. $\frac{1}{2} - \frac{1}{2} - \frac{1}{2^2} - \frac{1}{3} - \frac{1}{2^3} -$

15. $\frac{1}{5\sqrt{2}} - \frac{1}{5\sqrt{3}} + \frac{1}{5\sqrt{4}} - \ldots\ldots + (-1)^n \frac{1}{5\sqrt{n}} + \ldots\ldots$

16. $\sum \frac{(-1)^{n-1}}{\sqrt{n}}$.

17. $\sum \frac{(-1)^{n-1} n}{n^2 + 1}$

18. $1 - \frac{1}{\sqrt{2}} + \frac{1}{\sqrt{3}} - \frac{1}{\sqrt{4}} + \ldots\ldots$

## 9.20 OBJECTIVE TYPE OF QUESTIONS

### PROBLEMS 9.11

*Choose the correct answer or fill up the blanks in each of the following problems :*

1. The series $\frac{1}{1^p} + \frac{1}{2^p} + \frac{1}{3^p} + \ldots$ converges if

(a) $p > 0$ (b) $p < 1$ (c) $p > 1$ (d) $p \leq 1$.

2. The series $\sum_{n=0}^{\infty} (2x)^n$ converges if

(a) $-1 \leq x \leq 1$ (b) $-\frac{1}{2} < x < \frac{1}{2}$ (c) $-2 < x < 2$ (d) $-\frac{1}{2} \leq x \leq \frac{1}{2}$.

3. The series $\frac{2}{1^2} - \frac{3}{2^2} + \frac{4}{3^2} - \frac{5}{4^2} + \ldots$ is

(a) conditionally convergent (b) absolutely convergent
(c) divergent (d) none of the above.

4. Which one of the following series is *not* convergent ?

(a) $\frac{1}{2\sqrt{2}} + \frac{1}{3\sqrt{3}} + \frac{1}{4\sqrt{4}} + \ldots \infty$ (b) $1\frac{1}{2} - 1\frac{1}{3} + 1\frac{1}{4} - 1\frac{1}{5} + \ldots \infty$

(c) $\frac{1}{2} - \frac{1}{3} + \frac{1}{4} - \frac{1}{5} + \ldots \infty$ (d) $x + x^2 + x^3 + x^4 + \ldots \infty$ where $|x| < 1$.

5. The sum of the alternating harmonic series $1 - \frac{1}{2} + \frac{1}{3} - \frac{1}{4} + \ldots$ is

(a) zero (b) infinite (c) log 2
(d) not defined as the series is not convergent.

6. Let $\Sigma u_n$ be a series of positive terms. Given that $\Sigma u_n$ is convergent and also

$$\underset{n\to\infty}{\text{Lt}} \frac{u_{n+1}}{u_n} \text{ exists, then the said limit is}$$

(a) necessarily equal to 1 (b) necessarily greater than 1
(c) may be equal to 1 or less than 1 (d) necessarily less than 1.

7. $\sum \left(1+\frac{1}{n}\right)^{-n^2}$ is

(a) convergent (b) oscillatory (c) divergent.

8. $1-\frac{1}{\sqrt{2}}+\frac{1}{\sqrt{3}}-\frac{1}{\sqrt{4}}+\ldots$ is

(a) oscillatory (b) conditionally convergent
(c) divergent (d) absolutely convergent.

9. $1-\frac{1}{2^2}-\frac{1}{3^2}+\frac{1}{4^2}+\frac{1}{5^2}-\frac{1}{6^2}-\frac{1}{7^2}+\frac{1}{8^2}+\ldots\infty$ is

(a) conditionally convergent (b) convergent
(c) oscillatory (d) divergent. (Anna, 2016)

10. $\int_0^1 \left(\sum \frac{x^n}{n^2}\right) dx =$

(a) $\sum_{n=0}^{\infty} \frac{1}{n(n+1)}$ (b) $\sum_{n=1}^{\infty} \frac{1}{n^2(n-1)}$ (c) $\sum_{n=0}^{\infty} \frac{1}{n(n-1)}$ (d) $\sum_{n=1}^{\infty} \frac{1}{n^2(n+1)}$.

11. If $\Sigma u_n$ is a convergent series of positive terms, then $\underset{n\to\infty}{\text{Lt}}\ u_n$ is

(a) 1 (b) ± 1 (c) 0 (d) 0. (V.T.U., 2010)

12. Geometric series $1+x+x^2+\ldots+x^{n-1}+\ldots\infty$

(a) converges in the interval ....... (b) converges uniformly in the interval .......

13. The series $x-\frac{x^2}{2}+\frac{x^3}{3}-\frac{x^4}{4}+\ldots\infty$ converges in the interval .......

14. If $\underset{n\to\infty}{\text{Lt}} \frac{u_n}{u_{n+1}} = k$, then $\Sigma u_n$ converges for $k$ .......

15. A sequence $(a_n)$ is said to be bounded, if there exists a number $k$ such that for every $n$, $a_n$ is .......

16. The series $2-5+3+2-5+3-5+\ldots\infty$ is ....... (Convergent etc.)

17. The series $1+\frac{x}{1!}+\frac{x^2}{2!}+\frac{x^3}{3!}+\frac{x^4}{4!}+\ldots\infty$ converges for .......

18. If $\underset{n\to\infty}{\text{Lt}}\ n\left\{\frac{u_n}{u_{n+1}}-1\right\} = k$, then $\Sigma u_n$ diverges for $k$ .......

19. A sequence which is monotonic and bounded is .......

20. The series $\frac{1}{1.2}+\frac{2}{3.4}+\frac{3}{5.6}+\ldots\infty$ is ....... (Convergent etc.)

21. The series $\frac{2^p}{1^q}+\frac{3^p}{2^q}+\frac{4^p}{3^q}+\ldots\infty$ converges for .......

22. The series $\sqrt{\frac{1}{4}}+\sqrt{\frac{2}{6}}+\sqrt{\frac{3}{8}}+\ldots\infty$ is ....... (Convergent etc.)

23. The series $\sqrt{\left(\frac{2^n-1}{3^n-1}\right)}$ is ... (Convergent etc.)

24. The series $1-\frac{1}{2}(x-2)+\frac{1}{4}(x-2)^2\ldots+\left(-\frac{1}{2}\right)^n(x-2)^n+\ldots\infty$ converges in the interval ......

25. Is the series $\sum_{n=1}^{\infty} \frac{8\tan^{-1} n}{1+n^2}$ convergent ?

26. The exponential series $1 + \frac{x}{1!} + \frac{x^2}{2!} + \ldots + \frac{x^n}{n!} + \ldots \infty$ is absolutely convergent. (True/False)

27. The series $\frac{1}{1.2} + \frac{1}{2.3} + \frac{1}{3.4} + \ldots + \frac{1}{n(n+1)} + \ldots \infty$, is ....... (Convergent/divergent/oscillatory)

28. Is the series $\sum n \tan 1/n$ convergent ?

29. The series $\sum \frac{1}{nx^n}$ converges for $x$ ......

30. The series $\sum_{n=1}^{\infty} \frac{x^n}{n^3}$ converges uniformly when $x$ lies in the interval .......

31. Every absolutely convergent series is necessarily
(*a*) divergent (*b*) convergent
(*c*) conditionally convergent (*d*) none of these.

32. The convergence of the series $1 - \frac{1}{3} + \frac{1}{5} - \frac{1}{7} + \ldots$ is tested by
(*a*) Ratio test (*b*) Raabe's test (*c*) Leibnitz's (*d*) Cauchy root test.

33. The series $\sum \frac{x^n}{(n+1)^n}, x > 0$ is
(*a*) divergent (*b*) convergent (*c*) oscillatory (*d*) none of these. (*V.T.U., 2010*)

34. $\sum \sin\left(\frac{1}{n}\right)$ is
(*a*) convergent (*b*) divergent (*c*) oscillatory (*d*) none of these. (*Delhi, 2012*)

35. $\sum_{n=2}^{\infty} \frac{1}{(\log n)^n}$ is convergent. (True or False)

36. Show that the series $\sum_{k=1}^{\infty} \frac{\cos k}{k^2}$ is convergent. (*K.T.U., 2016*)

CHAPTER 10

# Fourier Series and Harmonic Analysis

**1.** Introduction. **2.** Euler's Formulae. **3.** Conditions for a Fourier expansion. **4.** Functions having points of discontinuity. **5.** Change of interval. **6.** Odd and even function—Expansions of odd or even periodic functions. **7.** Half-range series. **8.** Typical wave-forms. **9.** Parseval's formula. **10.** Complex form of F-series. **11.** Practical Harmonic Analysis. **12.** Objective Type of Questions.

## 10.1 INTRODUCTION

In many engineering problems, especially in the study of periodic phenomenae* in conduction of heat, electro-dynamics and acoustics, it is necessary to express a function in a series of sines and cosines. Most of the single-valued functions which occur in applied mathematics can be expressed in the form.

$$\tfrac{1}{2}a_0{}^{\dagger} + a_1 \cos x + a_2 \cos 2x + \ldots\ldots{}^{\dagger}$$
$$+ b_1 \sin x + b_2 \sin 2x + \ldots\ldots$$

within a desired range of values of the variable. Such a series is known as the **Fourier series$**.

## 10.2 EULER'S FORMULAE

*The Fourier series for the function f(x) in the interval* $\alpha < x < \alpha + 2\pi$ *is given by*

$$f(x) = \frac{a_0}{2} + \sum_{n=1}^{\infty} a_n \cos nx + \sum_{n=1}^{\infty} b_n \sin nx$$

where

$$\left.\begin{aligned} a_0 &= \frac{1}{\pi}\int_{\alpha}^{\alpha+2\pi} f(x)\,dx \\ a_n &= \frac{1}{\pi}\int_{\alpha}^{\alpha+2\pi} f(x)\cos nx\,dx \\ b_n &= \frac{1}{\pi}\int_{\alpha}^{\alpha+2\pi} f(x)\sin nx\,dx \end{aligned}\right\} \quad \ldots(1)$$

These values of $a_0$, $a_n$, $b_n$ are known as *Euler's formulae*.

---

***Periodic functions.** If at equal intervals of abscissa $x$, the value of each ordinate $f(x)$ repeats itself, *i.e.*, $f(x) = f(x + a)$, for all $x$, then $y = f(x)$ is called a *periodic function* having **period** $\alpha$, *e.g.*, $\sin x$, $\cos x$ are periodic functions having a period $2\pi$.

† To write $a_0/2$ instead of $a_0$ is a conventional device to be able to get more symmetric formulae for the coefficients.

$ Named after the French mathematician and physicist *Jacques Fourier* (1768–1830) who was first to use Fourier series in his memorable work *'Theorie Analytique de la Chaleur'* in which he developed the theory of heat conduction. These series had a deep influence in the further development of mathematics and mathematical physics.

To establish these formulae, the following definite integrals will be required :

1. $\int_{\alpha}^{\alpha+2\pi} \cos nx\, dx = \left| \frac{\sin nx}{n} \right|_{\alpha}^{\alpha+2\pi} = 0$ $\qquad (n \neq 0)$

2. $\int_{\alpha}^{\alpha+2\pi} \sin nx\, dx = -\left| \frac{\cos nx}{n} \right|_{\alpha}^{\alpha+2\pi} = 0$ $\qquad (n \neq 0)$

3. $\int_{\alpha}^{\alpha+2\pi} \cos mx \cos nx\, dx$ $\qquad (n \neq 0)$

$$= \frac{1}{2}\int_{\alpha}^{\alpha+2\pi} [\cos (m+n)x + \cos (m-n)x]\, dx$$

$$= \frac{1}{2}\left| \frac{\sin (m+n)x}{m+n} + \frac{\sin (m-n)x}{m-n} \right|_{\alpha}^{\alpha+2\pi} = 0 \qquad (m \neq n)$$

4. $\int_{\alpha}^{\alpha+2\pi} \cos^2 nx\, dx = \left| \frac{x}{2} + \frac{\sin 2nx}{4n} \right|_{\alpha}^{\alpha+2\pi} = \pi$ $\qquad (n \neq 0)$

5. $\int_{\alpha}^{\alpha+2\pi} \sin mx \cos nx\, dx = -\frac{1}{2}\left[ \frac{\cos (m-n)x}{m-n} + \frac{\cos (m+n)x}{m+n} \right] = 0$ $\qquad (m \neq n)$

6. $\int_{\alpha}^{\alpha+2\pi} \sin nx \cos nx\, dx = \left| \frac{\sin^2 nx}{2n} \right|_{\alpha}^{\alpha+2\pi} = 0$ $\qquad (n \neq 0)$

7. $\int_{\alpha}^{\alpha+2\pi} \sin mx \sin nx\, dx = \frac{1}{2}\left| \frac{\sin (m-n)x}{m-n} - \frac{\sin (m+n)x}{m+n} \right|_{\alpha}^{\alpha+2\pi} = 0$ $\qquad (m \neq n)$

8. $\int_{\alpha}^{\alpha+2\pi} \sin^2 nx\, dx = \left| \frac{x}{2} - \frac{\sin 2nx}{4n} \right|_{\alpha}^{\alpha+2\pi} = \pi.$ $\qquad (n \neq 0)$

*Proof.* Let $f(x)$ be represented in the interval $(\alpha, \alpha + 2\pi)$ by the Fourier series :

$$f(x) = \frac{a_0}{2} + \sum_{n=1}^{\infty} a_n \cos nx + \sum_{n=1}^{\infty} b_n \sin nx \qquad ...(i)$$

To find the coefficients $a_0, a_n, b_n$, we assume that the series Eqn. (*i*) can be integrated term by term from $x = \alpha$ to $x = \alpha + 2\pi$.

*To find* $a_0$, integrate both sides of Eqn. (*i*) from $x = \alpha$ to $x = \alpha + 2\pi$. Then

$$\int_{\alpha}^{\alpha+2\pi} f(x)\, dx = \frac{1}{2} a_0 \int_{\alpha}^{\alpha+2\pi} dx + \int_{\alpha}^{\alpha+2\pi} \left( \sum_{n=1}^{\infty} a_n \cos nx \right) dx + \int_{\alpha}^{\alpha+2\pi} \left( \sum_{n=1}^{\infty} b_n \sin nx \right) dx$$

$$= \frac{1}{2} a_0(\alpha + 2\pi - \alpha) + 0 + 0 = a_0\pi \qquad \text{[By integrals (1) and (2) above]}$$

Hence $\qquad a_0 = \frac{1}{\pi}\int_{\alpha}^{\alpha+2\pi} f(x)\, dx\,.$

*To find* $a_n$, multiply each side of Eqn. (*i*) by $\cos nx$ and integrate from $x = \alpha$ to $x = \alpha + 2\pi$. Then

$$\int_{\alpha}^{\alpha+2\pi} f(x) \cos nx\, dx = \frac{1}{2} a_0 \int_{\alpha}^{\alpha+2\pi} \cos nx\, dx + \int_{\alpha}^{\alpha+2\pi} \left( \sum_{n=1}^{\infty} a_n \cos nx \right) \cos nx\, dx$$

$$+ \int_{\alpha}^{\alpha+2\pi} \left( \sum_{n=1}^{\infty} b_n \sin nx \right) \cos nx\, dx$$

$$= 0 + \pi a_n + 0 \qquad \text{[By integrals (1), (3), (4), (5) and (6)]}$$

Hence $\qquad a_n = \frac{1}{\pi}\int_{\alpha}^{\alpha+2\pi} f(x) \cos nx\, dx\,.$

*To find* $b_n$, multiply each side of Eqn. (*i*) by $\sin nx$ and integrate from $x = \alpha$ to $x = \alpha + 2\pi$. Then

$$\int_{\alpha}^{\alpha+2\pi} f(x)\sin nx\,dx = \frac{1}{2}a_0\int_{\alpha}^{\alpha+2\pi}\sin nx\,dx + \int_{\alpha}^{\alpha+2\pi}\left(\sum_{n=1}^{\infty} a_n\cos nx\right)\sin nx\,dx$$

$$+\int_{\alpha}^{\alpha+2\pi}\left(\sum_{n=1}^{\infty} b_n\sin nx\right)\sin nx\,dx$$

$$= 0 + 0 + \pi b_n \qquad \text{[By integrals (2), (5), (6), (7) and (8)]}$$

Hence $$b_n = \frac{1}{\pi}\int_{\alpha}^{\alpha+2\pi} f(x)\sin nx\,dx.$$

**Cor. 1.** Making $\alpha = 0$, *the interval becomes* $0 < x < 2\pi$, and the formulae (I) reduce to

$$\left.\begin{aligned} a_0 &= \frac{1}{\pi}\int_0^{2\pi} f(x)\,dx \\ a_n &= \frac{1}{\pi}\int_0^{2\pi} f(x)\cos nx\,dx \\ b_n &= \frac{1}{\pi}\int_0^{2\pi} f(x)\sin nx\,dx \end{aligned}\right\} \qquad \text{...(II)}$$

**Cor. 2.** Putting $\alpha = -\pi$, *the interval becomes* $-\pi < x < \pi$ and the formulae (I) take the form :

$$\left.\begin{aligned} a_0 &= \frac{1}{\pi}\int_{-\pi}^{\pi} f(x)\,dx \\ a_n &= \frac{1}{\pi}\int_{-\pi}^{\pi} f(x)\cos nx\,dx \\ b_n &= \frac{1}{\pi}\int_{-\pi}^{\pi} f(x)\sin nx\,dx \end{aligned}\right\} \qquad \text{...(III)}$$

**Example 10.1.** *Obtain the Fourier series for $f(x) = e^{-x}$ in the interval $0 < x < 2\pi$.*

*(Andhra, 2023 ; Calicut, 2013)*

**Solution.** Let $$e^{-x} = \frac{a_0}{2} + \sum_{n=1}^{\infty} a_n\cos nx + \sum_{n=1}^{\infty} b_n\sin nx \qquad \text{...}(i)$$

Then $$a_0 = \frac{1}{\pi}\int_0^{2\pi} f(x)\,dx = \frac{1}{\pi}\int_0^{2\pi} e^{-x}\,dx = \frac{1}{\pi}\Big|-e^{-x}\Big|_0^{2\pi} = \frac{1-e^{-2\pi}}{\pi}$$

and $$a_n = \frac{1}{\pi}\int_0^{2\pi} f(x)\cos nx\,dx = \frac{1}{\pi}\int_0^{2\pi} e^{-x}\cos nx\,dx$$

$$= \frac{1}{\pi(n^2+1)}\Big|e^{-x}(-\cos nx + n\sin nx)\Big|_0^{2\pi} = \left(\frac{1-e^{-2\pi}}{\pi}\right)\cdot\frac{1}{n^2+1}$$

$\therefore$ $$a_1 = \left(\frac{1-e^{-2\pi}}{\pi}\right)\frac{1}{2},\ a_2 = \left(\frac{1-e^{-2\pi}}{\pi}\right)\cdot\frac{1}{5} \text{ etc.}$$

Finally, $$b_n = \frac{1}{\pi}\int_0^{2\pi} f(x)\sin nx\,dx = \frac{1}{\pi}\int_0^{2\pi} e^{-x}\sin nx\,dx$$

$$= \frac{1}{\pi(n^2+1)}\Big|e^{-x}(-\sin nx - n\cos nx)\Big|_0^{2\pi} = \left(\frac{1-e^{-2\pi}}{\pi}\right)\cdot\frac{n}{n^2+1}$$

$\therefore$ $$b_1 = \frac{1-e^{-2\pi}}{\pi}\cdot\frac{1}{2},\ b_2 = \left(\frac{1-e^{-2\pi}}{\pi}\right)\cdot\frac{2}{5} \text{ etc.}$$

Substituting the values of $a_0, a_n, b_n$ in (*i*), we get

$$e^{-x} = \frac{1-e^{-2\pi}}{\pi}\left\{\frac{1}{2} + \left(\frac{1}{2}\cos x + \frac{1}{5}\cos 2x + \frac{1}{10}\cos 3x + \dots\right) + \left(\frac{1}{2}\sin x + \frac{2}{5}\sin 2x + \frac{3}{10}\sin 3x + \dots\right)\right\}.$$

**Example 10.2.** *Obtian the Fourier series to represent* $f(x) = \frac{1}{4}(\pi - x)^2,\ 0 < x < 2\pi$ (Calicut, 2018)

**Solution:** Let $\frac{1}{4}(\pi - x)^2 = \frac{1}{2}a_0 + \Sigma_{a_n} \cos n x + \Sigma b_n \sin nx$

Then $$a_0 = \frac{1}{\pi}\int_0^{2\pi} \frac{1}{4}(\pi - x)^2\, dx = \frac{1}{4\pi}\left[-\frac{1}{3}(\pi - x)^3\right]_0^{2\pi} = -\frac{1}{12\pi}\left[(-\pi)^3 - \pi^3\right] = \frac{\pi^2}{6};$$

$$a_n = \frac{1}{\pi}\int_0^{2\pi} \frac{1}{4}(\pi - 2)^2 \cos nx\, dx\ *$$

$$= \frac{1}{4\pi}\left[(\pi - x)^2 \cdot \sin\frac{nx}{n} - 2(\pi - x)(-1)\left(-\frac{\cos nx}{n^2}\right) + 2(-1)^2\left(-\frac{\sin nx}{n^3}\right)\right]_o^{2\pi}$$

$$= \frac{1}{4\pi}\left[\left(0 + \frac{2\pi \cos 2n\pi}{n^2} + 0\right) - \left(0 - \frac{2\pi \cos 0}{n^2} + 0\right)\right] = \frac{1}{4\pi}\left[\frac{2\pi}{n^2} + \frac{2\pi}{n^2}\right] = \frac{1}{n^2}$$

Since $\cos 2n\pi = \cos 0 = 1$

$$b_n = \frac{1}{\pi}\int_0^{2\pi} \frac{1}{4}(\pi - n)^2 \sin n x\, dx$$

$$= \frac{1}{4\pi}\left[(\pi - x)^2 \cdot \left(-\frac{\cos nx}{n}\right) - 2(\pi - x)(-1)\left(\frac{-\sin nx}{n^2}\right) + 2\left(\frac{\cos nx}{n^3}\right)\right]_0^{2\pi}$$

$$= \frac{1}{4\pi}\left[\left(-\frac{\pi^2 \cos 2n\pi}{n} - 0 + 2\frac{\cos 2n\pi}{n^3}\right) - \left(\frac{\pi^2}{n}\cos 0 - 0 + \frac{2\cos 0}{n^3}\right)\right]$$

$$= \frac{1}{4\pi}\left[\left(-\frac{\pi^2}{n} + \frac{2}{n^3}\right) - \left(\frac{-\pi^2}{n} + \frac{2}{n^3}\right)\right] = 0$$

$$\therefore \quad \frac{1}{4}(\pi - x)^2 = \frac{1}{2}\cdot\frac{\pi^2}{6} + \Sigma\frac{1}{n^2}\cos nx = \frac{\pi^2}{12} + \frac{\cos x}{1^2} + \frac{\cos 2x}{2^2} + \frac{\cos 3x}{3^2} + \ldots..$$

**Obs.** Putting $x = 0$, we find another interesting series $0 = -\frac{\pi^2}{3} + 4\left(\frac{1}{1^2} - \frac{1}{2^2} + \frac{1}{3^2} - \frac{1}{4^2} + \ldots\right)$

*i.e.,* $$\frac{1}{1^2} - \frac{1}{2^2} + \frac{1}{3^2} - \frac{1}{4^2} + \ldots = \frac{\pi^2}{12}.$$ (V.T.U., 2013)

**Note.** In the above example, we have used the results $\sin n\pi = 0$ and $\cos n\pi = (-1)^n$

Also $\sin\left(n + \frac{1}{2}\right)\pi = (-1)^n$ and $\cos\left(n + \frac{1}{2}\right)\pi = 0$. *The reader should remember these results.*

**Example 10.3.** *Expand* $f(x) = x \sin x$ *as a Fourier series in the interval* $0 < x < 2\pi$.
(Coimbatore, 2019; D.T.U., 2016; J.N.T.U., 2015; Rohtak, 2012)

---

* Apply the *general rule of integration by parts* which states that if $u$, $v$ be two functions of $x$ and dashes denote differentiations and suffixes integrations w.r.t. $x$, then

$$\int uv\ dx = uv_1 - u'v_2 + u''v_3 - u'''\ v_4 + \ldots$$

In other words : *Integral of the product of two functions*
*= 1st function × integral of 2nd – go on differentiating 1st, integrating 2nd signs alternately +ve and –ve.*

**Solution.** Let $$f(x) = \frac{a_0}{2} + \sum_{n=1}^{\infty} a_n \cos nx + \sum_{n=1}^{\infty} b_n \sin nx \quad \text{...(i)}$$

Then $$a_0 = \frac{1}{\pi}\int_0^{2\pi} x \sin x \, dx = \frac{1}{\pi}\left| x(-\cos x) - 1.(-\sin x)\right|_0^{2\pi} = -2.$$

$$a_n = \frac{1}{\pi}\int_0^{2\pi} x \sin x \cos nx \, dx = \frac{1}{2\pi}\int_0^{2\pi} x(2\cos nx \sin x)\, dx$$

$$= \frac{1}{2\pi}\int_0^{2\pi} x[\sin(n+1)x - \sin(n-1)x]\, dx$$

$$= \frac{1}{2\pi}\left[x\left\{\frac{-\cos(n+1)x}{n+1} + \frac{\cos(n-1)x}{n-1}\right\} - 1\left\{-\frac{\sin(n+1)x}{(n+1)^2} + \frac{\sin(n-1)x}{(n-1)^2}\right\}\right]_0^{2\pi}$$

$$= \frac{1}{2\pi}\left[2\pi\left\{-\frac{\cos 2(n+1)\pi}{n+1} + \frac{\cos 2(n-1)\pi}{n-1}\right\}\right] = \frac{2}{n^2-1}. \ (n \neq 1)$$

When $n = 1$, $$a_1 = \frac{1}{\pi}\int_0^{2\pi} x \sin x \cos x \, dx = \frac{1}{2\pi}\int_0^{2\pi} x \sin 2x \, dx$$

$$= \frac{1}{2\pi}\left[x\left(\frac{-\cos 2x}{2}\right) - 1.\left(-\frac{\sin 2x}{4}\right)\right]_0^{2\pi} = -\frac{1}{2}.$$

Finally, $$b_n = \frac{1}{\pi}\int_0^{2\pi} x \sin x \sin nx \, dx = \frac{1}{2\pi}\int_0^{2\pi} x[\cos(n-1)x - \cos(n+1)x]dx$$

$$= \frac{1}{2\pi}\left[x\left\{\frac{\sin(n-1)x}{n-1} - \frac{\sin(n+1)}{n+1}\right\} - 1.\left\{-\frac{\cos(n-1)x}{(n-1)^2} + \frac{\cos(n+1)x}{(n+1)^2}\right\}\right]_0^{2\pi}$$

$$= \frac{1}{2\pi}\left[\frac{\cos 2(n-1)\pi}{(n-1)^2} - \frac{\cos 2(n+1)\pi}{(n+1)^2} - \frac{1}{(n-1)^2} + \frac{1}{(n+1)^2}\right] = 0 \ (n \neq 1)$$

When $n = 1$, $$b_1 = \frac{1}{\pi}\int_0^{2\pi} x \sin x \sin x \, dx = \frac{1}{2\pi}\int_0^{2\pi} x(1 - \cos 2x)\, dx$$

$$= \frac{1}{2\pi}\left[x\left(x - \frac{\sin 2x}{2}\right) - 1.\left(\frac{x^2}{2} + \frac{\cos 2x}{4}\right)\right]_0^{2\pi} = \pi$$

Substituting the values of $a$'s and $b$'s, in Eqn. (*i*), we get

$$x \sin x = -1 + \pi \sin x - \frac{1}{2}\cos x + \frac{2}{2^2-1}\cos 2x + \frac{2}{3^2-1}\cos 3x + \ldots$$

**Example 10.4.** *Find the Fourier series to represent the function $e^{ax}$ from $x = -\pi$ to $x = \pi$.* (Ranchi, 2019)

**Solution:** Let $$e^{ax} = \frac{1}{2}a_0 + \Sigma a_n \cos nx + \Sigma b_n \sin nx$$

Then $$a_0 = \frac{1}{\pi}\int_{-\pi}^{\pi} e^x dx = \frac{1}{\pi}\left[\frac{e^{ax}}{a}\right]_{-\pi}^{\pi} = \frac{1}{\pi a}(e^{ax} - e^{-ax}) = \frac{2\sin h\, a\pi}{a\pi}$$

$$a_n = \int_{-\pi}^{\pi} e^{ax}\cos nx \, dx$$

$$= \frac{1}{\pi}\left\{\left[e^{ax}\cdot\frac{\sin nx}{n} - ae^{ax}\cdot\left(\frac{-\cos nx}{n^2}\right)\right]_{-\pi}^{\pi} + \int_{-\pi}^{\pi} a^2 e^{ax}\cdot\left(\frac{-\cos nx}{n^2}\right)dx\right]\right\}$$

$$a_n = \frac{1}{\pi}\left[0 + \frac{a}{n^2}(e^{4x} - e^{-ax})\cos^{n\pi}\right] - \frac{a^2}{n^2}\cdot a^n$$

$$\left(1+\frac{a^2}{n^2}\right)a_n = \frac{2a}{\pi n^2}\sin h\, a\pi \cos n\pi$$

$$a_n = \frac{2a(-1)^n \sin h\, a\pi}{\pi(n^2+a^2)}, \text{ since } \cos n\pi\ (-1)^n$$

Similarly,

$$b_n = \frac{2n(-1)^n \sin h\, a\pi}{\pi(n^2+a^2)}$$

$$\therefore \quad e^{ax} = \frac{\sum \sinh a\pi}{\pi a} + \sum \frac{2a(-1)^n \sinh a\pi}{\pi(n^2+a^2)}\cos nx + \frac{\sum 2n(-1)^n \sin h\, a\pi}{\pi(n^2+a^2)}\sin nx$$

$$e^{ax} = \frac{2\sinh ax}{\pi}\left[\frac{1}{2a} - \frac{a\cos x}{a^2+1^2} + \frac{a\cos 2x}{a^2+2^2} - \frac{a\cos 3x}{a^2+3^2} + \ldots\ldots - \left(\frac{\sin x}{a^2+1^2} - \frac{2\sin 2x}{a^2+2^2} + \frac{3\sin 3x}{a^2+3^2} - \ldots\ldots\right)\right]$$

**Example 10.5:** *Find the Fourier series for the function $f(x) = x + x^2$ in the interval $-\pi < x < \pi$. Hence show that*

$$\frac{\pi^2}{6} = 1 + \frac{1}{2^2} + \frac{1}{3^2} + \frac{1}{4^2} + \ldots\ldots$$ *(JNTUK, 2022 ; Marathwada, 2018)*

**Solution:** Let $x + x^2 = \frac{1}{2}a_0 + \Sigma a_n \cos nx + \Sigma b_n \sin nx$

Then

$$a_0 = \frac{1}{\pi}\int_{-\pi}^{\pi}(x+x^2)dx = \frac{1}{\pi}\left[\int_{-\pi}^{\pi} x\cdot dx + \int_{-\pi}^{\pi} x^2\,dx\right] = \frac{1}{\pi}\left[0 + 2\int_0^{\pi} x^2\,dx\right]$$

since $x$ is an odd function and $x^2$ is an even function

$$\therefore \quad a_0 = \frac{2}{\pi}\left[\frac{x^3}{3}\right]_0^{\pi} = \frac{2\pi^2}{3}$$

$$a_n = \frac{1}{\pi}\left[(x+x^2)\cos nx\cdot dx\right] = \frac{1}{\pi}\cdot 2\int_0^{\pi} x^2\cos nx\,dx;\ \text{since } \int_{-\pi}^{\pi} x\cos nx\,dx = 0$$

$$= \frac{2}{\pi}\left[x^2\frac{\sin nx}{n} - 2x\left(\frac{\cos nx}{n^2}\right) + 2\left(-\frac{\sin nx}{3}\right)\right]_0^{\pi}$$

$$= \frac{2}{\pi}\left[\left(0 + 2\pi\frac{\cos n\pi}{n^2} + 0\right) - (0+0+0)\right] = \frac{4(-1)^n}{n^2}$$

$$b_n = \frac{1}{\pi}\int_{-\pi}^{\pi}(x+x)^2 \sin nx\cdot dx$$

$$= \frac{2}{\pi}\int_0^{\pi} x\cdot \sin nx\,dx,\ \text{since } \int_{-\pi}^{\pi} x^2 \sin nx\,dx = 0$$

$$= \frac{2}{\pi}\left[x\left(-\frac{\cos nx}{n}\right) - 1\left(\frac{-\sin nx}{n^2}\right)\right]_0^{\pi} = \frac{2}{\pi}\left[\left(-\pi\frac{\cos n\pi}{n} - 0\right) - (0-0)\right] = \frac{-2(-1)^n}{n}$$

Hence,

$$x + x^2 = \frac{\pi^2}{3} + \sum\frac{4(-1)^n}{n^2}\cos nx + \sum\frac{(-2)(-1)^n}{n}\sin nx$$

$$= \frac{\pi^2}{3} - \pi\left(\frac{\cos x}{1^2} - \frac{\cos 2x}{2^2} + \frac{\cos 3x}{3^2}\ldots\ldots\right) + 2\left(\sin x - \frac{\sin 2x}{2} + \frac{\sin 3x}{3}\ldots\ldots\right)$$

which is true for all values of $x$ in the interval $-\pi < x < \pi$ but it does not represent.

$f(x) = x + x^2$ at $x = \pm\pi$ being the points of discontinuity of $f(x)$.

At $x = \pi$, the sum of the series is

$$= \frac{1}{2}[f(\pi+0) + f(\pi-0)] = \frac{1}{2}(\pi^2 - \pi + \pi^2 + \pi) = \pi^2$$

So, $$\pi^2 = \frac{\pi^2}{3} - 4\left(\frac{\cos\pi}{1^2} - \frac{\cos 2\pi}{2^2} + \frac{\cos 3\pi}{3^2} .... + 2(0)\right)$$

$$\pi^2 - \frac{\pi^2}{3} = 4\left(\frac{1}{1^2} + \frac{1}{2^2} + \frac{1}{2^3} + .....\right) 1 + \frac{1}{2^2} + \frac{1}{3^2} + .... \frac{\pi^2}{6}$$

**Example 10.6:** *Obtain Fourier series for $f(x) = x\cos x$, $0 < x < 2\pi$.* *(Coimbatore, 2019)*

**Solution:** Let $x\sin x = a_0 + \Sigma a_n \cos nx - \Sigma b_n \sin nx$

Then $$a_0 = \frac{1}{\pi}\int_0^{2\pi} x\cos x \cdot dx = \frac{1}{\pi}[x(-\sin x) - 1(-\cos x)]_0^{2\pi}$$

$$= \frac{1}{\pi}[-2\pi \sin 2\pi + 0] = -2;$$

$$a_n = \frac{1}{\pi}\int_0^{2\pi} x\sin x \cos nx\, dx = \frac{1}{2\pi}\int_0^{2\pi} x\{\cos x(n+1)x - \cos x(n-1)x\}\, dx$$

$\therefore$ $$a_n = \frac{1}{2\pi}\left[x\left\{\frac{-\cos(n+1)x}{n+1} + \frac{\cos(n-1)x}{n-1}\right\}\right] - 1\left\{-\frac{\cos(n+1)x}{(n+1)^2} + \frac{\cos(n-1)x}{(n-1)^2}\right\}\Bigg]_0^{2\pi}$$

$$= \frac{1}{2\pi}\left[2\pi^2\left\{\frac{-\cos 2(n+1)\pi}{n+1} + \frac{\cos^2(n-1)\pi}{n-1}\right\}\right] = -\frac{1}{n+1} + \frac{1}{n-1} = \frac{2}{n^2-1}, n \neq 1$$

Since $\cos(2n+1)\pi = \cos 2(n-1)\pi = 0$

If $$n = 1, a_1 = \frac{1}{\pi}\int_0^{2\pi} \cdot x \cdot \sin x \cdot \cos x\, dx$$

$$a_1 = \frac{1}{2\pi}\int_0^{2\pi} x \cdot \sin 2x \cdot dx = \frac{1}{2\pi}\left[x \cdot \left(\frac{-\cos 2x}{2}\right) - 1\left(\frac{-\sin 2x}{2}\right)\right]_0^{2\pi} = -\frac{1}{2};$$

$$b_n = \frac{1}{\pi}\int_0^{2\pi} x\sin x \cdot \sin nx\, dx = \frac{1}{2\pi}\int_0^{2\pi} x \cdot \{\cos(n-1)x - \cos(n+1)x\}\, dx$$

provided that $n \neq 1$.

when $n = 1$, we have $$b_1 = \frac{1}{\pi}\int_0^{2\pi} x \cdot \cos x \cdot \cos x \cdot dx = \frac{1}{2}\int_0^{2\pi} x(1 - \sin x)dx$$

$$= \frac{1}{2\pi}\left[x \cdot \left(x - \frac{\cos 2x}{2}\right) - 1\left(\frac{x^2}{2} + \frac{\sin 2x}{4}\right)\right]_0^{2\pi} = \pi$$

$\therefore$ $$x\sin x = -1 + \left(\frac{-1}{2}\sin x + \frac{2}{2^2-1}\sin 2x\right) + \frac{2}{3^2-1}\sin 3x + ..... + \pi\cos x + 0$$

$$= -1 + \pi\cos x - \frac{1}{2}\sin x + \frac{2}{2^2-1}\sin 2x + \frac{1}{3^2-1}\sin 3x + .....$$

## PROBLEMS 10.1

1. Obtain a Fourier series to represent $e^x$ from $x = -\pi$ to $x = \pi$. Hence derive series for $\pi/\sinh \pi$. *(Andhra, 2020 ; J.N.T.U., 2015)*

2. Prove that $x^2 = \dfrac{\pi^2}{3} + 4\sum_{n=1}^{\infty} (-1)^n \dfrac{\cos nx}{n^2}$, $-\pi < x < \pi$. *(Anna, 2013 ; U.P.T.U., 2012; Rohtak, 2011)*

   Hence show that (*i*) $\sum \dfrac{1}{n^2} = \dfrac{\pi^2}{6}$.

   (*ii*) $\sum \dfrac{1}{(2n-1)^2} = \dfrac{\pi^2}{8}$ (*iii*) $\dfrac{1}{1^2} - \dfrac{1}{2^2} + \dfrac{1}{3^2} - \dfrac{1}{4^2} + \ldots = \dfrac{\pi^2}{12}$ *(U.P.T.U., 2012)*

   (*iv*) $\sum \dfrac{1}{n^4} = \dfrac{\pi^4}{90}$. *(Anna, 2013)*

3. If $f(x) = \left(\dfrac{\pi - x}{2}\right)^2$ in the range 0 to $2\pi$, show that $f(x) = \dfrac{\pi^2}{12} + \sum_{n=1}^{\infty} \dfrac{\cos nx}{n^2}$. *(Kurukshetra, 2013 ; Anna, 2012)*

4. Prove that in the range $-\pi < x < \pi$, $\cosh ax$

$$= \frac{2a^2}{\pi} \sinh a\pi \left[\frac{1}{2a^2} + \sum_{n=1}^{\infty} \frac{(-1)^n}{n^2 + a^2} \cos nx\right].$$

## 10.3 CONDITIONS FOR A FOURIER EXPANSION

The reader must not be misled by the belief that the Fourier expansion of $f(x)$ in each case shall be valid. The above discussion has merely shown that if $f(x)$ has an expansion, then the coefficients are given by Euler's formulae. The problems concerning the possibility of expressing a function by Fourier series and convergence of this series are many and cumbersome. Such questions should be left to the curiosity of a pure-mathematician. However, almost all engineering applications are covered by the following well-known **Dirichlet's conditions:**

*Any function f(x) can be developed as a Fourier series* $\dfrac{a_0}{2} + \sum_{n=1}^{\infty} a_n \cos nx + \sum_{n=1}^{\infty} b_n \sin nx$ *where* $a_0, a_n, b_n$ *are constants, provided :*

(*i*) *$f(x)$ is periodic, single-valued and finite;*

(*ii*) *$f(x)$ has a finite number of discontinuties in any one period;*

(*iii*) *$f(x)$ has at the most a finite number of maxima and minima.*

In fact the problem of expressing any function $f(x)$ as a Fourier series depends upon the evaluation of the integrals.

$$\frac{1}{\pi}\int f(x) \cos nx \, dx \; ; \; \frac{1}{\pi}\int f(x) \sin nx \, dx$$

within the limits $(0, 2\pi)$, $(-\pi, \pi)$ or $(\alpha, \alpha + 2\pi)$ according as $f(x)$ is defined for every value of $x$ in $(0, 2\pi)$, $(-\pi, \pi)$ or $(\alpha, \alpha + 2\pi)$.

## PROBLEMS 10.2

State giving reasons whether the following functions can be expanded in Fourier series in the interval $-\pi \le x \le \pi$.

1. cosec $x$
2. sin $1/x$
3. $1/(3-t)$ in the interval $0 < t < 2\pi$. *(Delhi, 2012)*

## 10.4 FUNCTIONS HAVING POINTS OF DISCONTINUITY

In deriving the Euler's formulae for $a_0$, $a_n$, $b_n$, it was assumed that $f(x)$ was continuous. Instead a function may have a finite number of points of finite discontinuity *i.e.*, its graph may consist of a finite number of different curves given by different equations. Even then such a function is expressible as a Fourier series.

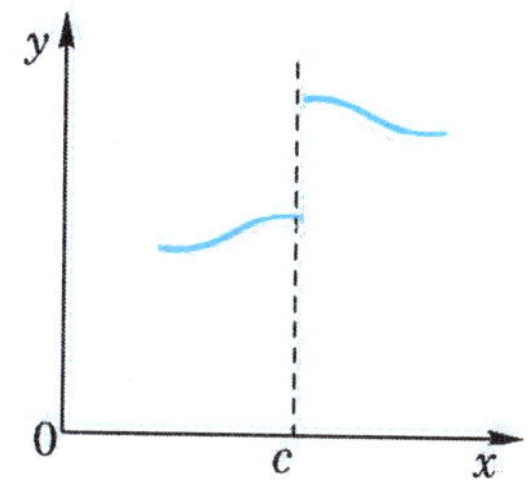

Fig. 10.1

For instance, if in the interval $(\alpha, \alpha + 2\pi)$, $f(x)$ is defined by

$$f(x) = \phi(x), \ \alpha < x < c.$$
$$= \psi(x), \ c < x < \alpha + 2\pi, \text{ i.e., } c \text{ is the point of discontinuity, then}$$

$$a_0 = \frac{1}{\pi}\left[\int_\alpha^c \phi(x)\,dx + \int_c^{\alpha+2\pi} \psi(x)\,dx\right]$$

$$a_n = \frac{1}{\pi}\left[\int_\alpha^c \phi(x)\cos nx\,dx + \int_c^{\alpha+2\pi} \psi(x)\cos nx\,dx\right]$$

and $$b_n = \frac{1}{\pi}\left[\int_\alpha^c \phi(x)\sin nx\,dx + \int_c^{\alpha+2\pi} \psi(x)\sin nx\,dx\right]$$

At a point of finite discontinuity $x = c$, there is a finite jump in the graph of function (Fig. 10.1). Both the limit on the left [*i.e.*, $f(c-0)$] and the limit on the right [*i.e.*, $f(c+0)$] exist and are different. At such a point, Fourier series gives the value of $f(x)$ as the arithmetic mean of these two limits,

*i.e.*, at $x = c$, $\quad f(x) = \frac{1}{2}[f(c-0) + f(c+0)]$.

**Example 10.7.** *Find the Fourier series expansion for $f(x)$, if*

$$f(x) = -\pi, \ -\pi < x < 0$$
$$x, \ 0 < x < \pi.$$

*(V.T.U., 2013 ; U.K.T.U., 2012)*

*Deduce that* $\frac{1}{1^2} + \frac{1}{3^2} + \frac{1}{5^2} + \ldots\ldots = \frac{\pi^2}{8}$. *(Andhra, 2021 ; Calicut, 2013)*

**Solution.** Let $$f(x) = \frac{a_0}{2} + \sum_{n=1}^{\infty} a_n \cos nx + \sum_{n=1}^{\infty} b_n \sin nx \quad \ldots(i)$$

Then $$a_0 = \frac{1}{\pi}\left[\int_{-\pi}^0 (-\pi)\,dx + \int_0^\pi x\,dx\right] = \frac{1}{\pi}\left[-\pi \left| x \right|_{-\pi}^0 + \left| x^2/2 \right|_0^\pi\right] = \frac{1}{\pi}\left(-\pi^2 + \frac{\pi^2}{2}\right) = -\frac{\pi}{2};$$

$$a_n = \frac{1}{\pi}\left[\int_{-\pi}^0 (-\pi)\cos nx\,dx + \int_0^\pi x\cos nx\,dx\right]$$

$$= \frac{1}{\pi}\left[-\pi\left|\frac{\sin nx}{n}\right|_{-\pi}^0 + \left|\frac{x\sin nx}{n} + \frac{\cos nx}{n^2}\right|_0^\pi\right]$$

$$= \frac{1}{\pi}\left[0 + \frac{1}{n^2}\cos n\pi - \frac{1}{n^2}\right] = \frac{1}{\pi n^2}(\cos n\pi - 1)$$

$\therefore$ $$a_1 = \frac{-2}{\pi\,.\,1^2}, \ a_2 = 0, \ a_3 = -\frac{2}{\pi\,.\,3^2}, \ a_4 = 0, \ a_5 = -\frac{2}{\pi\,.\,5^2} \text{ etc.}$$

Finally, $$b_n = \frac{1}{\pi}\left[\int_{-\pi}^0 (-\pi)\sin nx\,dx + \int_0^\pi x\sin nx\,dx\right]$$

$$= \frac{1}{\pi}\left[\left|\frac{\pi \cos nx}{n}\right|_{-\pi}^{0} + \left| -x\frac{\cos nx}{n} + \frac{\sin nx}{n^2}\right|_0^{\pi}\right]$$

$$= \frac{1}{\pi}\left[\frac{\pi}{n}(1 - \cos n\pi) - \frac{\pi}{n}\cos n\pi\right] = \frac{1}{n}(1 - 2\cos n\pi)$$

$\therefore$ $$b_1 = 3, b_2 = -\frac{1}{2}, b_3 = 1, b_4 = -\frac{1}{4}, \text{ etc.}$$

Hence substituting the values of $a$'s and $b$'s in Eqn. (*i*), we get

$$f(x) = -\frac{\pi}{4} - \frac{2}{\pi}\left(\cos x + \frac{\cos 3x}{3^2} + \frac{\cos 5x}{5^2} + \ldots\ldots\right) + 3\sin x - \frac{\sin 2x}{2} + \frac{3\sin 3x}{3} - \frac{\sin 4x}{4} + \ldots \quad \ldots(ii)$$

which is the required result.

Putting $x = 0$ in Eqn. (*ii*), we obtain $f(0) = -\frac{\pi}{4} - \frac{2}{\pi}\left(1 + \frac{1}{3^2} + \frac{1}{5^2} + \ldots\ldots \infty\right)$ ...(*iii*)

Now $f(x)$ is discontinuous at $x = 0$. As a matter of fact

$$f(0-0) = -\pi \text{ and } f(0+0) = 0 \quad \therefore \quad f(0) = \frac{1}{2}[f(0-0) + f(0+0)] = -\pi/2.$$

Hence Eqn. (*iii*) takes the form $-\frac{\pi}{2} = -\frac{\pi}{4} - \frac{2}{\pi}\left[\frac{1}{1^2} + \frac{1}{3^2} + \frac{1}{5^2} + \ldots\right]$ whence follows the result.

**Example 10.8.** *If* $f(x) = \begin{cases} 0, & -\pi \le x \le 0 \\ \sin x, & 0 \le x \le \pi \end{cases}$, *prove that* $f(x) = \frac{1}{\pi} + \frac{\sin x}{2} - \frac{2}{\pi}\sum_{n=1}^{\infty}\frac{\cos 2nx}{4n^2 - 1}$.

*(Andhra, 2016 ; W.B.T.U., 2015)*

*Hence show that* $\frac{1}{1.3} - \frac{1}{3.5} + \frac{1}{5.7} - \ldots - \infty = \frac{1}{4}(\pi - 2)$ *(Anna, 2016)*

**Solution.** Let $$f(x) = \frac{a_0}{2} + \sum_{n=1}^{\infty} a_n \cos nx + \sum_{n=1}^{\infty} b_n \sin nx$$

Then $$a_0 = \frac{1}{\pi}\int_{-\pi}^{\pi} f(x)\,dx = \frac{1}{\pi}\left[\int_{-\pi}^{0} 0\,dx + \int_0^{\pi} \sin x\,dx\right] = \frac{2}{\pi}$$

$$a_n = \frac{1}{\pi}\int_{-\pi}^{\pi} f(x)\cos nx\,dx = \frac{1}{\pi}\left[\int_{-\pi}^{0} 0\,.\,dx + \int_0^{\pi}\sin x\cos nx\,dx\right]$$

$$= \frac{1}{2\pi}\int_0^{\pi}[\sin(n+1)x - \sin(n-1)x]\,dx$$

$$= \frac{1}{2\pi}\left[-\frac{\cos(n+1)x}{n+1} + \frac{\cos(n-1)x}{n-1}\right]_0^{\pi}$$

$$= \frac{1}{2\pi}\left[-\frac{\cos(n+1)\pi}{n+1} + \frac{\cos(n-1)\pi}{n-1} + \frac{1}{n+1} - \frac{1}{n-1}\right] \quad (n \ne 1)$$

$$= \frac{1}{2\pi}\left\{\frac{1-(-1)^{n+1}}{n+1} - \frac{(-1)^{n-1}-1}{n-1}\right\} = 0, \text{ when } n \text{ is odd}$$

$$= -\frac{2}{\pi(n^2-1)}, \text{ when } n \text{ is even.}$$

When $n = 1$, $a_1 = \dfrac{1}{\pi}\displaystyle\int_0^{\pi} \sin x \cos x \, dx = \dfrac{1}{2\pi}\int_0^{\pi} \sin 2x \, dx = \dfrac{1}{2\pi}\left| -\dfrac{\cos 2x}{2}\right|_0^{\pi} = 0$

Finally, $b_n = \dfrac{1}{\pi}\displaystyle\int_{-\pi}^{\pi} f(x) \sin nx \, dx = \dfrac{1}{\pi}\left[\int_{-\pi}^{0} 0 \, . \, dx + \int_0^{\pi} \sin x \sin nx \, dx\right]$

$$= \frac{1}{2\pi}\int_0^{\pi} [\cos \overline{n-1}x - \cos \overline{n+1}x] \, dx = \frac{1}{2\pi}\left[\frac{\sin \overline{n-1}x}{n-1} - \frac{\sin \overline{n+1}x}{n+1}\right]_0^{\pi}$$

$= 0 \quad (n \neq 1)$

When $n = 1$, $b_1 = \dfrac{1}{\pi}\displaystyle\int_0^{\pi} \sin x \sin x \, dx = \dfrac{1}{2\pi}\int_0^{\pi} (1 - \cos 2x) \, dx = \dfrac{1}{2\pi}\left[x - \dfrac{\sin 2x}{2}\right]_0^{\pi} = \dfrac{1}{2}$

Hence $f(x) = \dfrac{1}{\pi} - \dfrac{2}{\pi}\left[\dfrac{\cos 2x}{2^2 - 1} + \dfrac{\cos 4x}{4^2 - 1} + \dfrac{\cos 6x}{6^2 - 1} + \ldots\right] + \dfrac{1}{2}\sin x$ ...(*i*)

Putting $x = \dfrac{\pi}{2}$ in Eqn. (*i*), we get $1 = \dfrac{1}{\pi} - \dfrac{2}{\pi}\left(-\dfrac{1}{1.3} + \dfrac{1}{3.5} - \dfrac{1}{5.7} + \ldots \infty\right) + \dfrac{1}{2}$

Whence $\dfrac{1}{1.3} - \dfrac{1}{3.5} + \dfrac{1}{5.7} - \ldots \infty = \dfrac{1}{4}(\pi - 2).$

**Example 10.9.** *Find the Fourier series for the function*

$$f(t) = \begin{cases} -1 & \text{for} \quad -\pi < t < -\pi/2 \\ 0 & \text{for} \quad -\pi/2 < t < \pi/2 \\ 1 & \text{for} \quad \pi/2 < t < \pi \end{cases}$$ *(U.P.T.U., 2012)*

**Solution.** Let $f(t) = \dfrac{a_0}{2} + \displaystyle\sum_{n=1}^{\infty} a_n \cos nt + \sum_{n=1}^{\infty} b_n \sin nt$ ...(*i*)

Then $a_0 = \dfrac{1}{\pi}\left\{\displaystyle\int_{\pi}^{-\pi/2} (-1) dt + \int_{-\pi/2}^{\pi/2} (0) dt + \int_{\pi/2}^{\pi} (1) \, dt\right\}$

$$= \frac{1}{\pi}\left\{\left| -x \right|_{-\pi}^{-\pi/2} + \left| x \right|_{\pi/2}^{\pi}\right\} = \frac{1}{\pi}(\pi/2 - \pi + \pi - \pi/2) = 0$$

$$a_n = \frac{1}{\pi}\left\{\int_{-\pi}^{-\pi/2} (-1) \cos nt \, dt + \int_{-\pi/2}^{\pi/2} (0) \cos nt \, dt + \int_{\pi/2}^{\pi} (1) \cos nt \, dt\right\}$$

$$= \frac{1}{\pi}\left\{\left| -\frac{\sin nt}{n}\right|_{-\pi}^{-\pi/2} + \left|\frac{\sin nt}{n}\right|_{\pi/2}^{\pi}\right\} = \frac{1}{n\pi}\left(\frac{\sin n\pi}{2} - \frac{\sin n\pi}{2}\right) = 0$$

and $b_n = \dfrac{1}{\pi}\left\{\displaystyle\int_{-\pi}^{-\pi/2} (-1) \sin nt \, dt + \int_{-\pi/2}^{\pi/2} (0) \sin nt \, dt + \int_{\pi/2}^{\pi} (1) \sin nt \, dt\right\}$

$$= \frac{1}{\pi}\left\{\left|\frac{\cos nt}{n}\right|_{-\pi}^{-\pi/2} + \left| -\frac{\cos nt}{n}\right|_{\pi/2}^{\pi}\right\}$$

$$= \frac{2}{n\pi}\left(\cos \frac{n\pi}{2} - \cos n\pi\right)$$

$\therefore \quad b_1 = \frac{2}{\pi}, b_2 = -\frac{2}{\pi}, b_3 = \frac{2}{3\pi}$ etc.

Hence substituting the values of $a$'s and $b$'s in (*i*),

we get $f(t) = \frac{2}{\pi}\left( \sin t - \sin 2t + \frac{1}{3}\sin 3t + ... \right)$.

## PROBLEMS 10.3

1. Find the Fourier series to represent the function $f(x)$ given by

$$f(x) = x \text{ for } 0 \le x \le \pi, \text{ and } = 2\pi - x \text{ for } \pi \le x \le 2\pi.$$ (*JNTUK, 2022*)

Deduce that $\frac{1}{1^2} + \frac{1}{3^2} + \frac{1}{5^2} + ... \infty = \frac{\pi^2}{8}$.

2. Find the Fourier series to represent the function $f(x) = x$ in the interval $(0, 2\pi)$. (*Kottayan 2018*)

3. Draw the graph of the function $f(x) = 0, -\pi < x < 0$

$$= x^2, 0 < x < \pi.$$

If $f(2\pi + x) = f(x)$, obtain Fourier series of $f(x)$. (*U.P.T.U., 2011*)

4. Find the Fourier series to represent $x - x^2$ in the interval $(-\pi, \pi)$.

5. Find the Fourier series for the function $f(x)$ in the interval $(-\pi, \pi)$ where $f(x) = \begin{cases} \pi + x, & -\pi < x < 0 \\ \pi - x, & 0 < x < \pi \end{cases}$ (*Delhi, 2012 S*)

6. Find a Fourier series for the function defined by

$$f(x) = \begin{cases} -1, \text{ for } -\pi < x < 0 \\ 0, \text{ for } x = 0 \\ 1, \text{ for } 0 < x < \pi \end{cases}$$ (*Andhra, 2016*)

Hence prove that $1 - \frac{1}{3} + \frac{1}{5} - \frac{1}{7} + ... \infty = \frac{\pi}{4}$.

7. Obtain the Fourier series for $f(x) = e^{-x}$ in the interval $0 \le x \le 2\pi$. (*Calicut-2019*)

8. Find the Fourier series to represent the function $f(x) = x \sin x, -\pi < x < \pi$.

Hence deduce that $\frac{\pi}{4} = \frac{1}{2} + \frac{1}{1 \cdot 3} - \frac{1}{3 \cdot 5} + .....$

## 10.5 CHANGE OF INTERVAL

In many engineering problems, the period of the function required to be expanded is not $2\pi$ but some other interval, say : $2c$. In order to apply the foregoing discussion to functions of period $2c$, this interval must be converted to the length $2\pi$. This involves only a proportional change in the scale.

Consider the periodic function $f(x)$ defined in $(\alpha, \alpha + 2c)$. To change the problem to period $2\pi$

put $\quad z = \pi x/c \quad$ or $\quad x = cz/\pi$ ...(1)

so that when $\quad x = \alpha, \quad z = \alpha\pi/c = \beta$ (say)

when $\quad x = \alpha + 2c, \quad z = (\alpha + 2c)\pi/c = \beta + 2\pi$.

Thus the function $f(x)$ of period $2c$ in $(\alpha, \alpha + 2c)$ is transformed to the function $f(cz/\pi)$ [$= F(z)$ say] of period $2\pi$ in $(\beta, \beta + 2\pi)$. Hence $f(cz/\pi)$ can be expressed as the Fourier series

$$f\left(\frac{cz}{\pi}\right) = \frac{a_0}{2} + \sum_{n=1}^{\infty} a_n \cos nz + \sum_{n=1}^{\infty} b_n \sin nz \quad ...(2)$$

where,
$$\left.\begin{aligned} a_0 &= \frac{1}{\pi}\int_{\beta}^{\beta+2\pi} f\left(\frac{cz}{\pi}\right)dz \\ a_n &= \frac{1}{\pi}\int_{\beta}^{\beta+2\pi} f\left(\frac{cz}{\pi}\right)\cos nz\,dz \\ b_n &= \frac{1}{\pi}\int_{\beta}^{\beta+2\pi} f\left(\frac{cz}{\pi}\right)\sin nz\,dz \end{aligned}\right\} \quad ...(3)$$

Making the inverse substitutions $z = \pi x/c$, $dz = (\pi/c)\,dx$ in Eqns. (2) and (3) *the Fourier expansion of $f(x)$ in the interval $(\alpha, \alpha + 2c)$ is given by*

$$f(x) = \frac{a_0}{2} + \sum_{n=1}^{\infty} a_n \cos\frac{n\pi x}{c} + \sum_{n=1}^{\infty} b_n \sin\frac{n\pi x}{c}$$

where
$$\left.\begin{aligned} a_0 &= \frac{1}{c}\int_{\alpha}^{\alpha+2c} f(x)\,dx \\ a_n &= \frac{1}{c}\int_{\alpha}^{\alpha+2c} f(x)\cos\frac{n\pi x}{c}dx \\ b_n &= \frac{1}{c}\int_{\alpha}^{\alpha+2c} f(x)\sin\frac{n\pi x}{c}dx \end{aligned}\right\} \quad ...(4)$$

**Cor.** Putting $\alpha = 0$ in (4), we get the results for the interval $(0, 2c)$ and putting $\alpha = -c$ in (4), we get results for the interval $(-c, c)$.

**Example 10.10.** *Expand $f(x) = e^{-x}$ as a Fourier series in the interval $(-l, l)$.*
*(K.T.U., 2016 ; Rohtak, 2010 S)*

**Solution.** The required series is of the form

$$e^{-x} = \frac{a_0}{2} + \sum_{n=1}^{\infty} a_n \cos\frac{n\pi x}{l} + \sum_{n=1}^{\infty} b_n \sin\frac{n\pi x}{l} \quad ...(i)$$

Then $$a_0 = \frac{1}{l}\int_{-l}^{l} e^{-x}\,dx = \frac{1}{l}\left|-e^{-x}\right|_{-l}^{l} = \frac{1}{l}(e^{l} - e^{-l}) = \frac{2\sinh l}{l}$$

and $$a_n = \frac{1}{l}\int_{-l}^{l} e^{x}\cos\frac{n\pi x}{l}dx \qquad \left[\because \int e^{ax}\cos bx\,dx = \frac{e^{ax}}{a^2+b^2}(a\cos bx + b\sin bx)\right]$$

$$= \frac{1}{l}\left|\frac{e^{-x}}{1+(n\pi/l)^2}\left(-\cos\frac{n\pi x}{l} + \frac{n\pi}{l}\sin\frac{n\pi x}{l}\right)\right|_{-l}^{l} = \frac{2l(-1)^n\sinh l}{l^2+(n\pi)^2} \qquad [\because \cos n\pi = (-1)^n]$$

$$\therefore \quad a_1 = \frac{-2l\sinh l}{l^2+\pi^2},\ a_2 = \frac{2l\sinh l}{l^2+2^2\pi^2},\ a_3 = \frac{2l\sinh l}{l^2+3^2\pi^2} \text{ etc.}$$

Finally, $$b_n = \frac{1}{l}\int_{-l}^{l} e^{-x}\sin\frac{n\pi x}{l}dx \qquad \left[\because \int e^{ax}\sin bx\,dx = \frac{e^{ax}}{a^2+b^2}(a\sin bx - b\cos bx)\right]$$

$$= \frac{1}{l}\left|\frac{e^{-x}}{1+(n\pi/l)^2}\left(-\sin\frac{n\pi x}{l} - \frac{n\pi}{l}\cos\frac{n\pi x}{l}\right)\right|_{-l}^{l} = \frac{2n\pi(-1)^n\sinh l}{l^2+(n\pi)^2}$$

$$\therefore \quad b_1 = \frac{-2\pi\sinh l}{l^2+\pi^2},\ b_2 = \frac{4\pi\sinh l}{l^2+2^2\pi^2},\ b_2 = \frac{-6\pi\sinh l}{l^2+3^2\pi^2} \text{ etc.}$$

Substituting the values of $a$'s and $b$'s in $(i)$, we get

$$e^{-x} = \sinh l\left\{\frac{1}{l} - 2l\left(\frac{1}{l^2+\pi^2}\cos\frac{\pi x}{l} - \frac{1}{l^2+2^2\pi^2}\cos\frac{2\pi x}{l} + \frac{1}{l^2+3^2\pi^2}\cos\frac{3\pi x}{l} - ...\right)\right.$$
$$\left. - 2\pi\left(\frac{1}{l^2+\pi^2}\sin\frac{n\pi}{l} - \frac{2}{l^2+2^2\pi^2}\sin\frac{2\pi x}{l} + \frac{3}{l^2+3^2\pi^2}\sin\frac{3\pi x}{l} - ...\right)\right\}$$

**Example 10.11.** *Find the Fourier series expansion of $f(x) = 2x - x^2$ in (0, 3) and hence deduce that* $\frac{1}{1^2} - \frac{1}{2^2} + \frac{1}{3^2} - \frac{1}{4^2} + \ldots - \infty = \frac{\pi}{12}$.

**Solution.** The required series is of the form

$$2x - x^2 = \frac{a_0}{2} + \sum_{n=1}^{\infty} a_n \cos\frac{n\pi x}{l} + \sum_{n=1}^{\infty} b_n \sin\frac{n\pi}{l}x \quad \text{where } l = 3/2. \quad \ldots(i)$$

Then $$a_0 = \frac{1}{l}\int_0^{2l} (2x - x^2)dx = \frac{2}{3}\left| x^2 - \frac{x^3}{3} \right|_0^3 = 0$$

$$a_n = \frac{1}{l}\int_0^{2l} (2x - x^2)\cos\frac{n\pi x}{l}dx = \frac{2}{3}\int_0^3 (2x - x^2)\cos\frac{2n\pi x}{3}dx$$

$$= \frac{2}{3}\left[(2x - x^2)\frac{\sin 2n\pi x/3}{2n\pi/3} - (2 - 2x)\frac{-\cos 2n\pi x/3}{(2n\pi/3)^2} + (-2)\frac{-\sin 2n\pi x/3}{(2n\pi/3)^3}\right]_0^3$$

$$= \frac{2}{3}\cdot\frac{9}{4n^2\pi^2}[(2 - 6)\cos 2n\pi - 2] = -\frac{9}{n^2\pi^2}$$

$$b_n = \frac{1}{l}\int_0^{2l} (2x - x^2)\sin\frac{n\pi x}{l}dx = \frac{2}{3}\int_0^3 (2x - x^2)\sin\frac{2n\pi x}{3}dx$$

$$= \frac{2}{3}\left[(2x - x^2)\frac{-\cos 2n\pi x/3}{2n\pi/3} - (2 - 2x)\frac{-\sin 2n\pi x/3}{(2n\pi/3)^2} + (-2)\frac{\cos 2n\pi x/3}{(2n\pi/3)^3}\right]_0^3$$

$$= \frac{2}{3}\left\{-\frac{6}{n^2\pi^2}\cos 2n\pi - \frac{27}{4n^3\pi^3}(\cos 2n\pi - 1)\right\} = \frac{3}{n\pi}$$

Substituting the values of $a_0$, $a_n$, $b_n$ in (*i*), we get

$$2x - x^2 = -\sum_{n=1}^{\infty}\frac{9}{n^2\pi^2}\cos\frac{2n\pi x}{3} + \sum_{n=1}^{\infty}\frac{3}{n\pi}\sin\frac{2n\pi x}{3}$$

Putting $x = 3/2$, we get

$$3 - \frac{9}{4} = -\sum_{n=1}^{\infty}\frac{9}{n^2\pi^2}\cos n\pi \quad \text{or} \quad -\sum_{n=1}^{\infty}\frac{\cos n\pi}{n^2} = \frac{\pi^2}{9}\cdot\frac{3}{4}$$

or $$\frac{1}{1^2} - \frac{1}{2^2} + \frac{1}{3^2} - \frac{1}{4^2} + \frac{1}{5^2} - \ldots \infty = \frac{\pi^2}{12}.$$

**Example 10.12.** *Obtain Fourier series for the function*

$$f(x) = \begin{cases} \pi x, & 0 \le x \le 1 \\ \pi(2 - x), & 1 \le x \le 2 \end{cases}$$ *(Andhra, 2016 ; D.T.U., 2016 ; Rohtak, 2013 ; V.T.U., 2011)*

*Deduce that* $\frac{1}{1^2} + \frac{1}{3^2} + \frac{1}{5^2} + \ldots \infty = \frac{\pi^2}{8}$. *(Andhra, 2022)*

**Solution.** The required series is of the form

$$f(x) = \frac{a_0}{2} + \sum_{n=1}^{\infty} a_n \cos n\pi x + \sum_{n=1}^{\infty} b_n \sin n\pi x$$

Then $$a_0 = \int_0^1 \pi x dx + \int_1^2 \pi(2 - x)dx = \pi\left|\frac{x^2}{2}\right|_0^1 + \pi\left|2x - \frac{x^2}{2}\right|_1^2 = \pi\left(\frac{1}{2}\right) + \pi\left\{(4 - 2) - \left(2 - \frac{1}{2}\right)\right\} = \pi$$

$$a_n = \int_0^1 \pi x \cos n\pi x dx + \int_1^2 \pi(2 - x)\cos n\pi x dx$$

$$= \left| \pi x \cdot \frac{\sin n\pi x}{n\pi} - \pi\left(-\frac{\cos n\pi x}{n^2\pi^2}\right)\right|_0^1 + \left|\pi(2-x)\frac{\sin n\pi x}{n\pi} - (-\pi)\left(-\frac{\cos n\pi x}{n^2\pi^2}\right)\right|_1^2$$

$$= \left(\frac{\cos n\pi}{n^2\pi} - \frac{1}{n^2\pi^2}\right) - \left(\frac{\cos 2n\pi}{n^2\pi} - \frac{\cos n\pi}{n^2\pi}\right) = \frac{2}{n^2\pi}[(-1)^n - 1]$$

$$= 0 \text{ when } n \text{ is even}; -\frac{4}{n^2\pi} \text{ when } n \text{ is odd.}$$

$$b_n = \int_0^1 \pi x \sin n\pi x \, dx + \int_1^2 \pi(2-x)\sin n\pi x \, dx$$

$$= \left|\pi x\left(-\frac{\cos n\pi x}{n\pi}\right) - \pi\left(-\frac{\sin n\pi x}{n^2\pi^2}\right)\right|_0^1 + \left|\pi(2-x)\left(\frac{-\cos n\pi x}{n\pi}\right) - (-\pi)\left(-\frac{\sin n\pi x}{n^2\pi^2}\right)\right|_1^2$$

$$= \left(-\frac{\cos n\pi}{n}\right) + \left(\frac{\cos n\pi}{n}\right) = 0$$

Hence $\quad f(x) = \frac{\pi}{2} - \frac{4}{\pi}\left(\frac{\cos \pi x}{1^2} + \frac{\cos 3\pi x}{3^2} + \frac{\cos 5\pi x}{5^2} + \ldots \infty\right)$

Putting $x = 2$, $0 = \frac{\pi}{2} - \frac{4}{\pi}\left(\frac{\cos 2\pi}{1^2} + \frac{\cos 6\pi}{3^2} + \frac{\cos 10\pi}{5^2} + \ldots \infty\right)$

Whence $\frac{1}{1^2} + \frac{1}{3^2} + \frac{1}{5^2} + \ldots \infty = \frac{\pi^2}{8}$.

**Example 10.13.** *Find the Fourier series for*

$$f(t) = 0, \ -2 < t < -1$$
$$= 1 + t, \ -1 < t < 0$$
$$= 1 - t, \ 0 < t < 1$$
$$= 0, \quad 1 < t < 2.$$

**Solution.** Let $\quad f(t) = \frac{a_0}{2} + \sum_{n=1}^{\infty} a_n \cos\frac{n\pi t}{2} + \sum_{n=1}^{\infty} b_n \sin\frac{n\pi t}{2} \quad \ldots(i)$

[$\because$ $2c = 2 - (-2)$ so that $c = 2$]

Then $\quad a_0 = \frac{1}{2}\left\{\int_{-2}^{-1}(0)dt + \int_{-1}^{0}(1+t)dt + \int_0^1 (1-t)dt + \int_1^2 (0)dt\right\} = \frac{1}{2}\left\{\left|t + \frac{t^2}{2}\right|_{-1}^0 + \left|t - \frac{t^2}{2}\right|_0^1\right\}$

$$= \frac{1}{2}\left\{-\left(-1+\frac{1}{2}\right) + \left(1 - \frac{1}{2}\right)\right\} = \frac{1}{2}$$

$$a_n = \frac{1}{2}\left\{\int_{-1}^0 (1+t)\cos\frac{n\pi t}{2}dt + \int_0^1 (1-t)\cos\frac{n\pi t}{2}dt\right\} \quad \text{[Integrate by parts]}$$

$$= \frac{1}{2}\left\{\left|(1+t)\left(\sin\frac{n\pi t}{2}\right)\frac{2}{n\pi} - (1)\left(-\cos\frac{n\pi t}{2}\right)\frac{4}{n^2\pi^2}\right|_{-1}^0\right.$$
$$\left. + \left|(1-t)\left(\sin\frac{n\pi t}{2}\right)\frac{2}{n\pi} - (-1)\left(-\cos\frac{n\pi t}{2}\right)\frac{4}{n^2\pi^2}\right|_0^1\right\}$$

$$= \frac{4}{n^2\pi^2}(1 - \cos n\pi/2)$$

$$b_n = \frac{1}{2}\left\{\int_{-1}^0 (1+t)\sin\frac{n\pi t}{2}dt + \int_0^1 (1-t)\sin\frac{n\pi t}{2}dt\right\}$$

$$= \frac{1}{2}\left\{ \left| (1+t)\left(-\cos\frac{n\pi t}{2}\right)\frac{2}{n\pi} - 1\left(-\sin\frac{n\pi t}{2}\right)\frac{4}{n^2\pi^2} \right|_{-1}^{0} + \left| (1-t)\left(-\cos\frac{n\pi t}{2}\right)\frac{2}{n\pi} - (-1)\left(-\sin\frac{n\pi t}{2}\right)\frac{4}{n^2\pi^2} \right|_{0}^{1} \right\}$$

$$= \frac{1}{2}\left\{\left(\frac{4}{n^2\pi^2}\sin\frac{n\pi}{2}\right) - \left(\frac{4}{n^2\pi^2}\sin\frac{n\pi}{2}\right)\right\} = 0$$

Substituting the values of $a$'s and $b$'s in ($i$), we get

$$f(t) = \frac{1}{4} + \sum_{n=1}^{\infty}\frac{4}{n^2\pi^2}\left(1-\cos\frac{n\pi}{2}\right)\cos\frac{n\pi t}{2}.$$

**Example 10.14:** *Obtain the Fourier series for the function $f(x) = x^2$, $-\pi < x < \pi$ Sketch the group of $f(x)$. Hence show that*

(i) $1 + \frac{1}{2^2} + \frac{1}{3^2} + \frac{1}{4^2} + \ldots\ldots \frac{\pi^2}{6}$ (ii) $1 - \frac{1}{2^2} + \frac{1}{3^2} + \frac{1}{4^2} + \ldots\ldots \frac{\pi^2}{12}$

(iii) $1 + \frac{1}{3^2} + \frac{1}{5^2} + \ldots\ldots \frac{\pi^2}{8}$

**Solution:** Since $f(x) = x^2$ is an even function $b_n = 0$.

Let $x^2 = a_0 + \Sigma a_n \cos nx$. Then

$$a_0 = \frac{2}{\pi}\int_0^{\pi} x^2 dx = \frac{2}{\pi}\left[\frac{x^3}{3}\right]_0^{\pi}$$

and $$a_n = \frac{2}{\pi}\int_0^{\pi} x^2\cos nx\, dx = \frac{2}{\pi}\left[x^2\left(\frac{\sin nx}{n}\right) - 2x\left(-\frac{\cos nx}{n^2}\right) + 2\left(-\frac{\sin nx}{n^2}\right)\right]_0^{\pi}$$

$$= \frac{2}{\pi}\left[\frac{2\pi\cos n\pi}{n^2}\right] = \frac{4}{n^2}(-1)^n \qquad [\because \cos n\,\pi(-1)^n]$$

$$\therefore \quad x^2 = \frac{\pi^2}{3} + \frac{\sum 4(-1)^n}{n^2}\cos nx$$

$$\therefore \quad x^2 = \frac{\pi^2}{3} - \left(\cos x - \frac{\cos 2x}{2^2} + \frac{\cos 3x}{3^2} - \frac{\cos 4x}{4^2} + \ldots\ldots\right)$$

The graph of $f(x) = x^2$ is shown

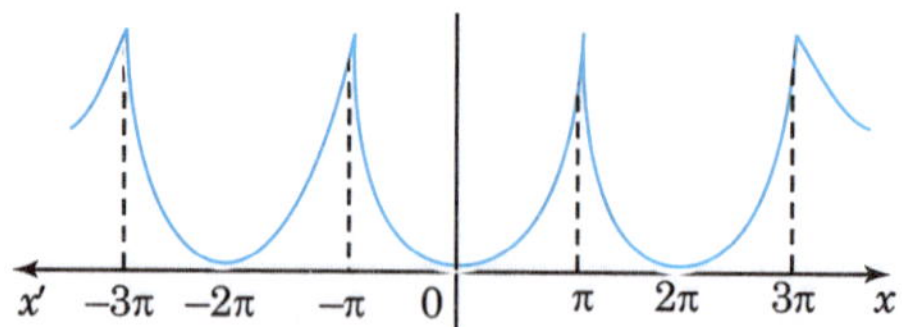


Fig. 10.2.

Now, obtain ($i$), ($ii$) and ($iii$) let $x = \pi$ and $x = 0$, on that conditions.

**Example 10.15:** *Express $f(x) = |x|, -\pi < x < \pi$ as a Fourier series. Draw a graph.* (Kottyam 2018)

**Solution:** Since $|x|$ is an even function of $x$, $b_n = 0$.

$$a_0 = \frac{2}{\pi}\int_0^{\pi}|x|\, dx = \frac{2}{\pi}\int_0^{\pi} x\, dx = \frac{2}{\pi}\left[\frac{x^2}{2}\right]_0^{\pi} = \pi$$

$$a_n = \frac{2}{\pi}\int_0^{\pi} |x| \cos nx \, dx = \frac{2}{\pi}\int_0^{\pi} x \cos nx \cdot dx$$

$$= \frac{2}{\pi}\left[x \cdot \frac{\sin nx}{n} - 1 \cdot \left(\frac{-\cos nx}{n^2}\right)\right]_0^{\pi}$$

$$= \frac{2}{\pi}\left[\frac{\cos nx}{n^2} - \frac{1}{n^2}\right] = \frac{2}{\pi n^2}\left[(-1)^n - 1\right]$$

$$= 0, \text{ if } n \text{ is even} = \frac{-4}{\pi n^2}, \text{ if } n \text{ is odd.}$$

$$\therefore \quad |x| = \frac{4}{2} - \frac{4}{\pi}\left(\cos x + \frac{\cos 3x}{3^2} + \frac{\cos 5x}{5^2} + \ldots\ldots\right)$$

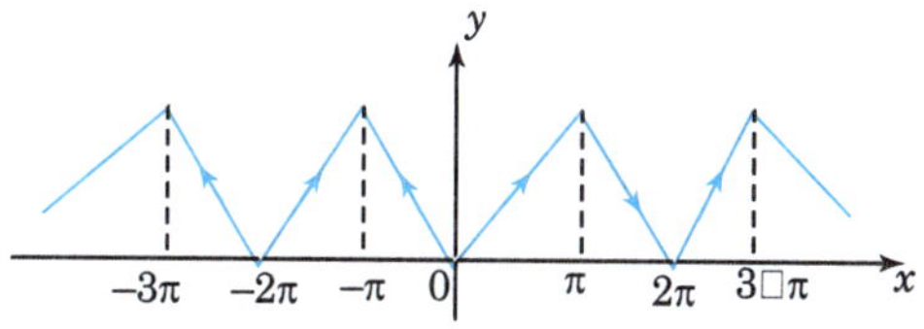


Fig. 10.3.

**Example 10.16:** *Find the Fourier series to represent the function $f(x) = |\sin x|, -\pi < x < \pi$.*

**Solution:** Since $|\sin x|$ is an even function, $a_n = 0$.

Let $$|\sin x| = a_0 + \Sigma a_n \cos nx$$

$$a_0 = \frac{2}{\pi}\int_0^{\pi} |\sin x| \, dx$$

$$= \frac{2}{\pi}\int_0^{\pi} \sin x dx = \frac{2}{\pi}[-\cos x]_0^{\pi} = \frac{4}{11}$$

$$a_n = \frac{2}{\pi}\int_0^{\pi} |\sin x| \cos nx \, dx$$

$$= \frac{1}{\pi}\int_0^{\pi} \{[\sin(n+1)x - \sin(n-1)x]\} \cdot dx$$

$= \dfrac{-4}{(n^2-1)\pi}$ or according as $n$ is even or odd provided that $-n \neq 1$. If $n$ = 1, then

$$a_1 = \frac{2}{\pi}\int_0^{\pi} \sin x \cos x \, dx = \int_0^{\pi} \sin 2n \, dx = 0$$

$$\therefore \quad |\sin x| = \frac{2}{\pi} - \frac{2}{\pi}\left(\frac{\cos 2x}{3} + \frac{\cos 4x}{15} + \ldots + \frac{\cos 2nx}{4n^2-1}\ldots\ldots\right)$$

**Example 10.17:** *For $-\pi < x < \pi$, show that $x \cos x = \dfrac{-1}{2}\sin x + 2\sum_{n=2}^{\infty}\dfrac{(-1)^n}{n^2-1}\sin nx$.*

**Solution:** Since $x \cos x$ is an odd function of $x$,

$$a_o = a_n = 0$$

Let $$x \cos x = \Sigma b_n \sin nx, \text{ where}$$

$$b_n = \frac{2}{\pi}\int_0^{\pi} x\cos x \cdot \sin nx \cdot dx = \frac{1}{\pi}\int_0^{\pi} x\{\sin(n+1)x + \sin(n-1)x\}\, dx$$

$$= \frac{1}{\pi}\left[x \cdot \left\{\frac{-\cos(n+1)x}{(n+1)} - \cos\frac{(n-1)x}{n-1}\right\} - 1\left\{\frac{-\sin(n+1)x}{(n+1)^2} - \sin\frac{(n-1)x}{(n-1)^2}\right\}\right]_0^{\pi}$$

$$= \frac{1}{\pi}\left[\pi \cdot \left\{\frac{-\cos(n+1)\pi}{n+1} - \frac{\cos(n-1)\pi}{n-1}\right\}\right], n \neq 1$$

$$= \left\{-\frac{1}{(n+1)} - \frac{1}{(n-1)}\right\} = \frac{-2n}{n^2-1}, \text{ if } n \text{ is odd (or } n \pm 1 \text{ are even)}$$

But
$$b_n = \left\{\frac{1}{(n+1)} + \frac{1}{(n-1)}\right\}$$

$$b_n = \frac{2n}{n^2-1}, \text{ if } n \text{ is even}$$

If $n = 1$
$$b_1 = \frac{2}{\pi}\int_0^{\pi} x\cos x \cdot \sin x \cdot dx$$

$$b_1 = \frac{1}{\pi}\int_0^{\pi} x \cdot \sin 2x \cdot dx = \frac{1}{\pi}\left[x \cdot \left(-\frac{\cos 2x}{2}\right) - 1\left(\frac{\sin 2x}{4}\right)\right]_0^{\pi} = -\frac{1}{2}$$

$\therefore$
$$x\cos x = b_1 \sin x + b_2 \sin 2x + b_3 \sin 3x + \ldots\ldots$$

$$= -\frac{1}{2}\sin x + \frac{4\sin 2x}{2^2-1} - \frac{6\sin 3x}{3^2-1} + \ldots\ldots$$

**Example 10.18 :** *Express $f(x) = \cos wx, -\pi < x < \pi$, where $w$ is a fraction as Fourier series.* (VTU, 2018)

**Solution:** Since $\cos wx$ is an even function, $bn = 0$

Let
$$\cos wx = a_0 + \Sigma a_n \cos nx$$

$$a_0 = \frac{2}{\pi}\int_0^{\pi} \cos wx\, dx = \frac{2}{\pi}\left[\sin\frac{wx}{w}\right]_0^{\pi} = \frac{2}{\pi w}\sin w\pi$$

$$a_n = \frac{2}{\pi}\int_0^{\pi} \cos wx \cos nx\, dx$$

$$a_n = \frac{1}{\pi}\int_0^{\pi} [\cos(n+w)x + \cos(n-w)x]\, dx = \frac{1}{\pi}\left[\frac{\sin(n+w)\pi}{n+w} + \frac{\sin(n-w)\pi}{n-w}\right]$$

$$= \frac{1}{\pi}\left[(-1)^n\frac{\sin w\pi}{n+w} - (-1)^n\frac{\sin w\pi}{n-w}\right] = \frac{(-1)^n \sin w\pi}{\pi}\left[\frac{1}{n+w} - \frac{1}{n-w}\right]$$

$$= (-1)^{n+1}\frac{2w\sin w\pi}{\pi(n^2-w^2)}$$

$\therefore$
$$\cos wx = \frac{2w\sin w\pi}{\pi}\left(\frac{1}{2w^2} + \frac{\cos x}{1-w^2} - \frac{\cos 2x}{4-w^2} + \frac{\cos 3x}{9-w^2} + \ldots\ldots\right)$$

**Example 10.19:** *Find the Fourier series for f(x) in the interval $(-\pi, -\pi)$, when $f(x) = \pi + x, -\pi < x < 0$*
*$= \pi - x, 0 < x < \pi$.*

**Solution:** From the graph of $f(x)$, it can be easily seen it is an even function. This can also be seen from

$$f(-x) = \pi + (-x) = \pi - x = f(x)$$

Hence $b_h = 0,$

Now let $f(x) = a_0 + \Sigma a_n \cos nx$

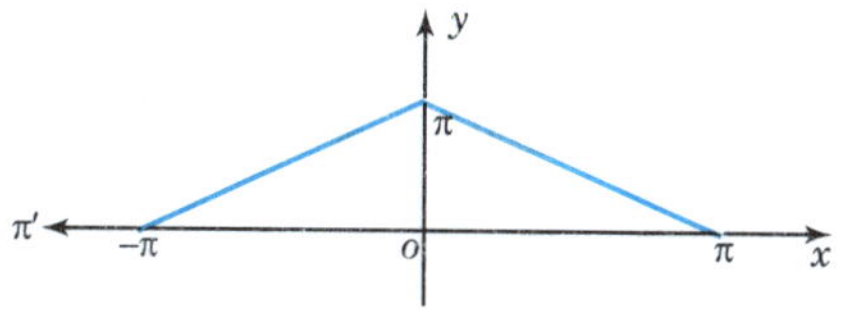


Fig. 10.4.

where
$$a_0 = \frac{2}{\pi}\int_0^{\pi} f(x)\cdot dx$$

$$= \frac{2}{\pi}\int_0^{\pi}(\pi - n)dx = \frac{2}{\pi}\left[\pi x - \frac{x^2}{2}\right]_0^{\pi} = \pi$$

$$a_n = \frac{2}{\pi}\int_0^{\pi} f(x)\cos nx$$

$$= \frac{2}{\pi}\int_0^{\pi}(\pi - x)\cos nx = \frac{2}{\pi}\left[(\pi - \overline{x})\frac{\sin nx}{n} - (-1)\left(\frac{-\cos nx}{n^2}\right)\right]_0^{\pi}$$

$$= \frac{2(1-\cos n\pi)}{\pi n^2} = 0, \quad \text{or} \quad \frac{4}{\pi n^2} \text{ according as } n \text{ is even or odd.}$$

$$\therefore \quad f(x) = \frac{\pi}{2} + \frac{4}{\pi}\left(\frac{\cos x}{1^2} + \frac{\cos 3x}{3^2} + \frac{\cos 5x}{5^4} + .....\right)$$

## PROBLEMS 10.4

1. Obtain the Fourier series for $f(x) = \left(\frac{\pi - x}{2}\right)$ in $0 \le x \le 2$. *(U.P.T.U., 2011)*
2. *(i)* Find the Fourier series to represent $x^2$ in the interval $(0, a)$. *(Andhra, 2016)*
   *(ii)* Find a Fourier series for $f(t) = 1 - t^2$ when $-1 \le t \le 1$.
3. If $f(x) = 2x - x^2$ in $0 \le x \le 2$, show that $f(x) = \frac{2}{3} - \sum_{n=1}^{\infty}\frac{4}{n^2\pi^2}\cos n\pi x$.
4. Find the Fourier series of $f(x) = \begin{cases} x, & 0 < x < 1 \\ 1 - x, & 1 < x < 2 \end{cases}$ *(K.T.U.,2016)*
5. Find the Fourier series for $f(x) = \begin{cases} x & \text{in } 0 \le x \le 3 \\ 6 - x & \text{in } 3 \le x \le 6 \end{cases}$
6. A sinusoidal voltage $E \sin \omega t$ is passed through a half-wave rectifier which clips the negative portion of the wave. Develop the resulting periodic function

$$U(t) = 0 \qquad \text{when } -T/2 < t < 0$$
$$= E\sin\omega t \qquad \text{when } 0 < t < T/2,$$

and $T = 2\pi/\omega$, in a Fourier series.

7. Find the Fourier series of the function $f(x) = \begin{cases} \pi x, & 0 < x < 1 \\ 0, & x = 1 \\ \pi(x-2), & 1 < x. < 2 \end{cases}$

Hence show that $\frac{\pi}{4} = \frac{1}{1} - \frac{1}{3} + \frac{1}{5} - \frac{1}{7} + ...$

8. Prove that for all values of $x$ between $-\pi$ and $\pi$. $\frac{1}{2}x = \sin x - \frac{1}{2}\sin 2x + \frac{1}{3}\sin 3x - \frac{1}{4}\sin 4x.....$

9. Show that for $-\pi < x < \pi$. $\sin a_x = \frac{2\sin ax}{\pi}\left[\frac{\sin x}{1-a^2} - \frac{2\sin 2x}{2^2-a^2} + \frac{3\sin 3x}{3^2-a^2}.....\right]$ (*JNTU, 2019*)

10. Prove that for $-\pi < x < \pi$ $\frac{x(\pi^2 - x^2)}{12} = \frac{\sin x}{1^3} - \frac{\sin 2x}{2^3} + \frac{\sin 3x}{3^3} - \frac{\sin 4x}{4^3} + .....$

11. Find the value of $\sum_{n=1}^{\infty}\frac{1}{n^2}$ using Fourier series. (*Andhra, 2017*)

12. When $x$ lies in the interval. $(-\pi, \pi)$ and $w$ is not an integer Show that.

$$\cos wx = \frac{2\sin w\pi}{\pi}\left(\frac{1}{2w} + \frac{w\cos x}{1-w^2} - \frac{w\cos 2x}{2^2-w^2} + \frac{w\cos 3x}{3^2-w^2}.....\right)$$ (*Annamalai, 2019*)

13. Prove that– $(-\pi < x < \pi)$

(*i*) $\frac{\pi\sin hax}{2\sin h\,a\pi} = \frac{\sin x}{a^2+1^2} - \frac{2\sin 2x}{a^2+2^2} + \frac{3\sin 3x}{a^2+3^2}$ (*ii*) $\frac{\pi\cos hax}{2\sin h\,a\pi} = \frac{1}{2a} - \frac{a\cos x}{a^2+1^2} + \frac{a\cos 2x}{a^2+2^2} - \frac{a\cos 3x}{a^2+3^2}$

## 10.6 (1) EVEN AND ODD FUNCTIONS

*A function $f(x)$ is said to be* **even** *if* $f(-x) = f(x)$,

*e.g.*, $\cos x$, $\sec x$, $x^2$ are all even functions. Graphically an even function is symmetrical about the $y$-axis.

*A function $f(x)$ is said to be* **odd** *if* $f(-x) = -f(x)$, (Fig. 10.5)

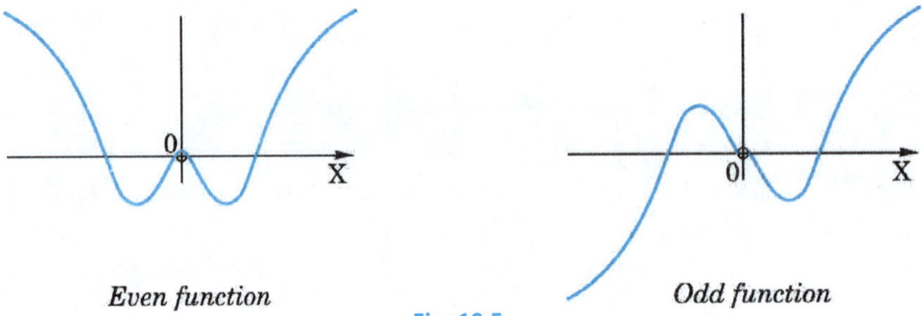


Fig. 10.5.

*e.g.* $\sin x$, $\tan x$, $x^3$ are odd functions. Graphically, an odd function is symmetrical about the origin.

We shall be using the following property of definite integrals in the next paragraph :

$$\int_{-c}^{c} \mathbf{f(x)dx} = 2\int_{0}^{c} \mathbf{f(x)dx}, \text{ when } f(x) \text{ is an even function.}$$

$$= \mathbf{0}, \text{ when } f(x) \text{ is an odd function.}$$

**(2) Expansions of even or odd periodic functions.** We know that a periodic function $f(x)$ defined in $(-c, c)$ can be represented by the Fourier series

$$f(x) = \frac{a_0}{2} + \sum_{n=1}^{\infty} a_n \cos\frac{n\pi x}{c} + \sum_{n=1}^{\infty} b_n \sin\frac{n\pi x}{c},$$

where $$a_0 = \frac{1}{c}\int_{-c}^{c} f(x)dx,\ a_n = \frac{1}{c}\int_{-c}^{c} f(x)\cos\frac{n\pi x}{c}dx,\ b_n = \frac{1}{c}\int_{-c}^{c} f(x)\sin\frac{n\pi x}{c}dx.$$

**Case I.** *When f(x) is an even function* $a_0 = \frac{1}{c}\int_{-c}^{c} f(x)dx = \frac{2}{c}\int_{0}^{c} f(x)dx.$

Since $f(x)\cos\frac{n\pi x}{c}$ is also an even function,

$\therefore \qquad a_n = \frac{1}{c}\int_{-c}^{c} f(x)\cos\frac{n\pi x}{c}dx = \frac{2}{c}\int_0^c f(x)\cos\frac{n\pi x}{c}dx$

Again since $f(x)\sin\frac{n\pi x}{c}$ is an odd function, $\therefore \; b_n = \frac{1}{c}\int_{-c}^{c} f(x)\sin\frac{n\pi x}{c}dx = 0.$

Hence, *if a periodic function f (x) is even, its Fourier expansion contains only cosine terms, and*

$$\left.\begin{aligned} a_0 &= \frac{2}{c}\int_0^c f(x)dx \\ a_n &= \frac{2}{c}\int_0^c f(x)\cos\frac{n\pi x}{c}dx \end{aligned}\right\} \qquad ...(1)$$

**Case II.** *When f(x) is an odd function,* $a_0 = \frac{1}{c}\int_{-c}^{c} f(x)dx = 0,$

Since $\cos\frac{n\pi x}{c}$ is an even function, therefore, $f(x)\cos\frac{n\pi x}{c}$ is an odd function.

$\therefore \qquad a_n = \frac{1}{c}\int_{-c}^{c} f(x)\cos\frac{n\pi x}{c} = 0$

Again since $\sin\frac{n\pi x}{c}$ is an odd function, therefore, $f(x)\sin\frac{n\pi x}{c}$ is an even function.

$\therefore \qquad b_n = \frac{1}{c}\int_{-c}^{c} f(x)\sin\frac{n\pi x}{c}dx = \frac{2}{c}\int_0^c f(x)\sin\frac{n\pi x}{c}dx$

Thus, *if a periodic function f (x) is odd, its Fourier expansion contains only sine terms and*

$$b_n = \frac{2}{c}\int_0^c f(x)\sin\frac{n\pi x}{c}dx \qquad ...(2)$$

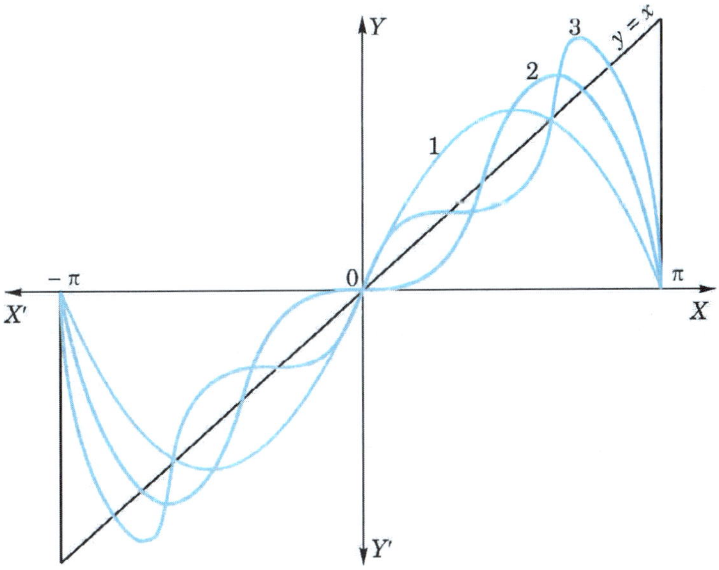


Fig. 10.6

**Example 10.20.** *Express f(x) = x/2 as a Fourier series in the interval –π < x < π.*

*(Andhra, 2016)*

**Solution.** Since $\qquad f(-x) = -x/2 = -f(x).$

$\therefore \qquad f(x)$ is an odd function and hence $f(x) = \sum_{n=1}^{\infty} b_n \sin nx$

where $\qquad b_n = \frac{2}{\pi}\int_0^{\pi} f(x)\sin nx\,dx = \frac{2}{\pi}\int_0^{\pi} \frac{x}{2}\sin nx\,dx$

$$= \frac{1}{\pi}\left[x\left(-\frac{\cos nx}{n}\right)-\left(\frac{-\sin nx}{n^2}\right)\right]_0^{\pi} = -\frac{\cos n\pi}{n}.$$

$\therefore$ $b_1 = 1/1, b_2 = -1/2, b_3 = 1/3, b_4 = -1/4$, etc.

Hence the series is $x/2 = \sin x - \frac{1}{2}\sin 2x + \frac{1}{3}\sin 3x - \frac{1}{4}\sin 4x + \ldots$ ...(*i*)

**Obs.** The graphs of $y = 2\sin x$, $y = 2(\sin x - \frac{1}{2}\sin 2x)$ and $y = 2(\sin x - \frac{1}{2}\sin 2x + \frac{1}{3}\sin 3x)$ are shown in Fig. 10.6, by the curves 1, 2 and 3 respectively. These illustrate the manner in which the successive approximations to the series (*i*) approach more and more closely to $y = x$ for all values of $x$ in $-\pi < x < \pi$, but not for $x = \pm\pi$.

As the series has a period $2\pi$, it represents the discontinuous function, called *saw-toothed waveform*, shown in Fig. 10.7. It is important to note that the given function $y = x$ is continuous and each term of the series (*i*) is continuous, but the function represented by the series (*i*) has finite discontinuities at $x = \pm\pi, \pm 3\pi, \pm 5\pi$ etc.

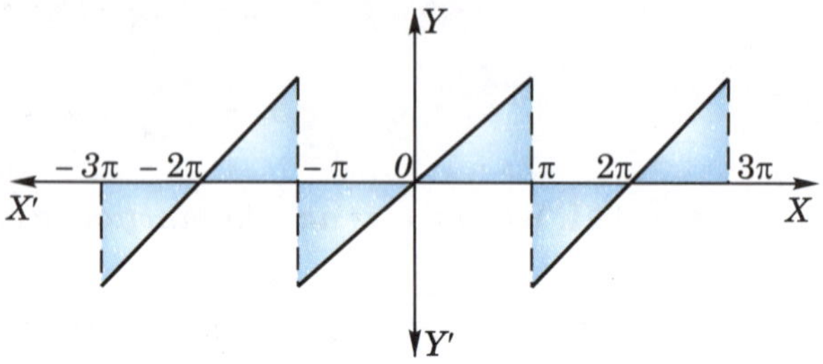

Fig. 10.7.

**Example 10.21.** *If f(x) = | cos x |, expand f(x) as a Fourier series in the interval (–π, π).*
*(Anna, 2016 ; Hazaribagh, 2015 ; Rohtak, 2010 S)*

**Solution.** As $f(-x) = |\cos(-x)| = |\cos x| = f(x)$, $|\cos x|$ is an even function.

$\therefore$
$$f(x) = \frac{a_0}{2} + \Sigma a_n \cos nx$$

where
$$a_0 = \frac{2}{\pi}\int_0^{\pi} |\cos x|\, dx = \frac{2}{\pi}\int_0^{\pi/2}\cos x\, dx + \int_{\pi/2}^{\pi}(-\cos x)\, dx$$

[$\because$ $\cos x$ is –ve when $\pi/2 < x < \pi$]

$$= \frac{2}{\pi}\left\{|\sin x|_0^{\pi/2} - |\sin x|_{\pi/2}^{\pi}\right\} = \frac{2}{\pi}[(1-0)-(0-1)] = \frac{4}{\pi}$$

and
$$a_n = \frac{2}{\pi}\int_0^{\pi}|\cos x|\cos nx\, dx = \frac{2}{\pi}\left[\int_0^{\pi/2}\cos x\cos nx\, dx + \int_{\pi/2}^{\pi}(-\cos x)\cos nx\, dx\right]$$

$$= \frac{1}{\pi}\left\{\int_0^{\pi/2}[\cos(n+1)x + \cos(n-1)x]\, dx - \int_{\pi/2}^{\pi}[\cos(n+1)x + \cos(n-1)x]\, dx\right\}$$

$$= \frac{1}{\pi}\left\{\left|\frac{\sin(n+1)x}{n+1} + \frac{\sin(n-1)x}{n-1}\right|_0^{\pi/2} - \left|\frac{\sin(n+1)x}{n+1} + \frac{\sin(n-1)x}{n-1}\right|_{\pi/2}^{\pi}\right\}$$

$$= \frac{1}{\pi}\left[\left\{\frac{\sin(n+1)\pi/2}{n+1} + \frac{\sin(n-1)\pi/2}{n-1}\right\} + \left\{\frac{\sin(n+1)\pi/2}{n+1} + \frac{\sin(n-1)\pi/2}{n-1}\right\}\right]$$

$$= \frac{2}{\pi}\left(\frac{\cos n\pi/2}{n+1} - \frac{\cos n\pi/2}{n-1}\right) = \frac{-4\cos n\pi/2}{\pi(n^2-1)}\ (n \neq 1)$$

In particular $a_1 = \frac{2}{\pi}\left[\int_0^{\pi/2}\cos^2 x\, dx - \int_{\pi/2}^{\pi}\cos^2 x\, dx\right] = 0$

Hence $|\cos x| = \frac{2}{\pi} + \frac{4}{\pi}\left\{\frac{1}{3}\cos 2x - \frac{1}{15}\cos 4x + \ldots\right\}.$

**Example 10.22.** *Obtain Fourier series for the function f(x) given by*

$$f(x) = 1 + 2x/\pi, \quad -\pi \le x \le 0,$$
$$= 1 - 2x/\pi, \quad 0 \le x \le \pi.$$

*Deduce that* $\frac{1}{1^2} + \frac{1}{3^2} + \frac{1}{5^2} + \ldots = \frac{\pi^2}{8}$. *(J.N.T.U.K., 2022, 2015 ; Kurukshetra, 2013 ; V.T.U., 2010)*

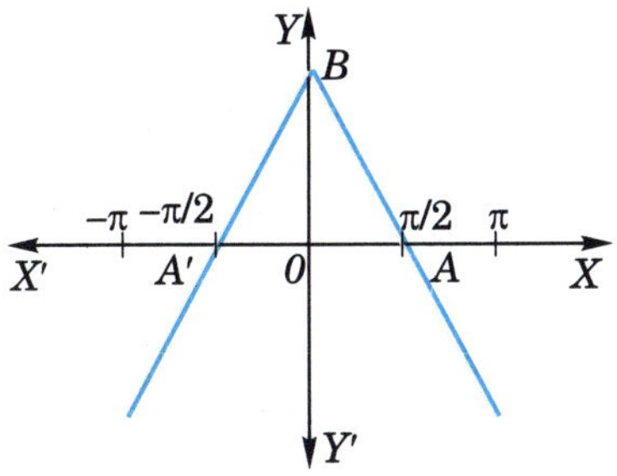

Fig. 10.8

**Solution.** Since $f(-x) = 1 - \frac{2x}{\pi}$ in $(-\pi, 0) = f(x)$ in $(0, \pi)$

and $f(-x) = 1 + \frac{2x}{\pi}$ in $(0, \pi) = f(x)$ in $(-\pi, 0)$

$\therefore$ $f(x)$ is an even function in $(-\pi, \pi)$. This is also clear from its graph $A'BA$ (Fig. 10.8) which is symmetrical about the $y$-axis.

$$\therefore \quad f(x) = \frac{a_0}{2} + \sum_{n=1}^{\infty} a_n \cos nx \qquad ...(i)$$

where $a_0 = \frac{2}{\pi}\int_0^{\pi} f(x)dx = \frac{2}{\pi}\int_0^{\pi}\left(1 - \frac{2x}{\pi}\right)dx = \frac{2}{\pi}\left(x - \frac{x^2}{\pi}\right)_0^{\pi} = 0$

and $a_n = \frac{2}{\pi}\int_0^{\pi} f(x)\cos nx\, dx = \frac{2}{\pi}\int_0^{\pi}\left(1 - \frac{2x}{\pi}\right)\cos nx\, dx$

$$= \frac{2}{\pi}\left|\left(1 - \frac{2x}{\pi}\right)\frac{\sin nx}{n} - \left(-\frac{2}{\pi}\right)\left(-\frac{\cos nx}{n^2}\right)\right|_0^{\pi} = \frac{2}{\pi}\left(-\frac{2\cos n\pi}{\pi n^2} + \frac{2}{\pi n^2}\right) = \frac{4}{n^2\pi^2}[1 - (-1)^n]$$

$\therefore$ $a_1 = 8/\pi^2, a^3 = 8/3^2\pi^2, a_5 = 8/5^2\pi^2, ...$

and $a_2 = a_4 = a_6 = ... = 0.$

Thus substituting the values of $a$'s in Eqn. $(i)$, we get

$$f(x) = \frac{8}{\pi^2}\left[\frac{\cos x}{1^2} + \frac{\cos 3x}{3^2} + \frac{\cos 5x}{5^2} + ...\right] \qquad ...(ii)$$

as the required Fourier expansion

Putting $x = 0$ in Eqn. $(ii)$, we get $1 = f(0) = \frac{8}{\pi^2}\left(\frac{1}{1^2} + \frac{1}{3^2} + \frac{1}{5^2} + ...\right)$

whence follows the desired result.

**Example 10.23:** *Find the Fourier series to represent the function $f(x)$ given by $f(x) = x,\ 0 \le x \le \pi$ and $f(x) = 2\pi - x,\ \pi \le x \le 2\pi$. Draw sketch of $f(x)$.* (Osmania 2018)

**Solution:** The given function is represented by two straight line.

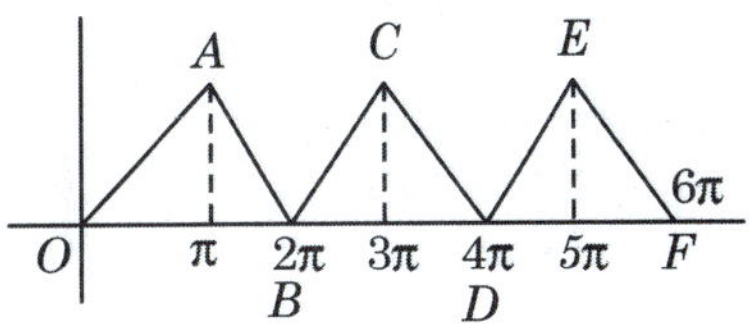

Fig. 10.9.

$OA$ and $AB$ in the interval $(0, 2\pi)$, since $f(x)$ is periodic, it is represented by $BC$ and $CD$ in the interval $(2\pi, 4\pi)$.

$$f(x) = \frac{1}{2}a_o + \Sigma a_n \cos nx + \Sigma b_n \sin nx$$

Then $a_0 = \frac{1}{\pi}\int_0^{2\pi} f(x)\,dx = \frac{1}{\pi}\left[\int_0^{\pi} x\,dx + \int_0^{2\pi}(2\pi - x)dx\right]$

$$= \frac{1}{\pi}\left\{\left[\frac{x^2}{2}\right]_0^{\pi} + \left[2\pi x - \frac{x^2}{2}\right]_{\pi}^{2\pi}\right\} = \frac{1}{\pi}\left\{\left(\frac{x^2}{2} - 0\right) + (4\pi^2 - 2\pi^2) - \left(2\pi^2 - \frac{\pi^2}{2}\right)\right\} = \pi$$

$$a_n = \frac{1}{\pi}\int_0^{2\pi} f(x)\cos nx\,dx = \frac{1}{\pi}\left[\int_0^{\pi} x\cos nx\,dx + \int_0^{2\pi}(2\pi - x)\cos nx\,dx\right]$$

$$= \frac{1}{\pi}\left\{\left(\frac{\cos n\pi}{n^2} - \frac{\cos o}{n^2}\right) + \left(\frac{-\cos^2 n\pi}{n^2} + \frac{\cos n\pi}{n^2}\right)\right\} = \frac{2}{\pi n^2}\{(-1)^n - 1\} = \frac{-4}{\pi n^2}, \text{ if } n \text{ is odd}$$

and $$a_n = \frac{2}{\pi n^2}[1-1] = 0, \text{ if } n \text{ is even}$$

$\therefore$ $$a_1 = \frac{-4}{\pi},\ a_3 = \frac{-4}{\pi \cdot 3^2},\ a_5 = \frac{-4}{\pi\, 5^2}$$

and $$a_2 = a_4 = a_6 \ldots.\ 0$$

$$b_n = \frac{1}{\pi}\int_0^{2\pi} f(x)\sin nx\, dx = \frac{1}{\pi}\left\{\int_0^{\pi} x\sin nx\, dx + \int_{\pi}^{2\pi}(2\pi - x)\sin nx\, dx\right\}$$

$$= \frac{1}{\pi}\left[\left\{x\cdot\left(\frac{-\cos nx}{n}\right) - 1\left(\frac{-\sin nx}{n^2}\right)_0^{\pi} + (2\pi - x)\left(\frac{-\cos x}{n}\right) - (-1)\left(\frac{-\sin nx}{n^2}\right)\right\}\right]_{\pi}^{2\pi}$$

$$= \frac{1}{\pi}\left[\left(\frac{-\pi\cos n\pi}{n} - 0\right) + \left(0 + \frac{\pi\cos n\pi}{n}\right)\right] = 0$$

$\therefore$ $$f(x) = \frac{\pi}{2} - \frac{4}{\pi}\left(\cos x + \frac{\cos 3x}{3^2} + \frac{\cos 5x}{5^2} + \ldots..\right).$$

**Example 10.24:** *An alternating current after passing through a rectifier has the form*

$$i = i_o \sin x, \text{ for } 0 \le x \le \pi = 0$$
$$i = 0 \text{ or } \pi \le x \le 2\pi$$

*where $i_o$ is the maximum current and the period is $2\pi$. Express i as Fourier series.* (Bhopal, 2019)

**Solution:** Let $i = a_o + \Sigma a_n \cos nx + \Sigma b_n \sin nx$

Then $$a_o = \frac{1}{\pi}\int_o^{2\pi} f(x)\, dx = \frac{1}{\pi}\int_o^{2\pi} i\, dx = \frac{1}{\pi}\left\{\int_o^{\pi} I_o \sin x\, dx + \int_o^{2\pi} 0\, dx\right\}$$

$$= \frac{I_o}{\pi}[-\cos x]_o^{\pi} + 0 = \frac{I_o}{\pi}(-\cos\pi + \cos o) = \frac{2I_o}{\pi}$$

$$a_n = \frac{1}{\pi}\int_o^{2\pi} i\cos nx\, dx = \frac{1}{\pi}\left\{\int_o^{\pi} I_o \sin x\cos nx\, dx + \int_{\pi}^{2\pi} 0\cdot\cos nx\cdot dx\right\}$$

$$= \frac{I_o}{2\pi}\int_o^{\pi}\{\sin(n+1)x - \sin(n-1)x\}\, dx$$

$$= \frac{I_o}{2\pi}\left[\frac{-\cos(n+1)\pi}{n+1} + \frac{\cos(n-1)x}{n-1}\right]_o^{\pi}, n \ne 1$$

$$= \frac{I_o}{2\pi}\left[\frac{-\cos(n+1)\pi}{n+1} + \frac{\cos(n-1)\pi}{n-1} + \frac{1}{n+1} - \frac{1}{n-1}\right]$$

If $n$ is odd then $(n+1)$ as well as $(n-1)$ are even and hence $\cos(n+1)\pi = \cos(n-1)\pi = 1$ Then $a_n = 0$.

$$a_3 = a_5 = 0$$

If $n$ is even $(n+1)$ and $(n-1)$ are odd. As such

$$\cos(n+1)\pi = \cos(n-1)\pi = -1. \text{ then.}$$

$$a_n = \frac{J_o}{2\pi}\left(\frac{2}{n+1} - \frac{2}{n-1}\right) = \frac{2I_o}{\pi(n^2-1)}$$

$$a_2 = \frac{2I_o}{\pi(2^2-1)}, a_4 = \frac{2I_o}{\pi(4^2-1)}, a_6 = \frac{2I_o}{\pi(6^2-1)}$$

Since the above result does not hold good when $\pi = 1$.

when $n \neq 1$, we have

$$b_n = \frac{1}{\pi}\int_o^{\pi} I_o \sin x \cdot \sin nx \cdot dx + \int_{\pi}^{2\pi} 0\, dx$$

$$= \frac{I_o}{2\pi}\int_o^{\pi}\{\cos(n-1)x - \cos(n+1)x\}\, dx$$

$$= \frac{I_o}{2\pi}\left[\frac{\sin(n-1)x}{n-1} - \sin\frac{(n+1)x}{n+1}\right]_o^{\pi} = 0$$

$$b_2 = b_3 = b_4 = 0$$

$$\therefore \quad i = \frac{I_o}{\pi} + \frac{I_o}{2}\sin x - \frac{2I_o}{\pi}\left(\frac{\cos 2x}{2^2-1} + \frac{\cos 4x}{4^2-1} + \frac{\cos 6x}{6^2-1} + .....\right)$$

**Example 10.25:** *If* $f(x) = 0, \quad -\pi \le x \le 0$

$= \sin x, \ 0 \le x \le \pi$

*show that* $$f(x) = \frac{1}{\pi} - \frac{2}{\pi}\sum_{n=1}^{\infty}\frac{\cos 2nx}{4n^2-1} + \frac{1}{2}\sin x$$

*Hence show that (i)* $\frac{1}{1.3} + \frac{1}{3.5} + \frac{1}{5.7} + .... = \frac{1}{2}$

*(ii)* $\frac{1}{1.3} - \frac{1}{3.5} + \frac{1}{5.7} - \frac{1}{7.9} + .... = \frac{\pi - 2}{4}$ *(Madurai, 2019)*

**Solution:** Let $$f(x) = \frac{1}{2}a_0 + \Sigma_{a_n}\cos nx + \Sigma_{b_n}\sin nx, \text{ then}$$

$$a_o = \frac{1}{\pi}\int_{-\pi}^{\pi} f(x)dx = \frac{1}{\pi}\left\{\int_{-\pi}^{0} 0\, dx + \int_o^{\pi}\sin x\, dx\right\} \quad ...(i)$$

$$= \frac{1}{\pi}[-\cos x]_0^{\pi} = \frac{1}{\pi}[-\cos\pi + \cos 0] = \frac{2}{\pi}$$

$$a_n = \frac{1}{\pi}\int_{-\pi}^{\pi} f(x)\cos nx\, dx = \frac{1}{\pi}\left\{0 + \int_o^{\pi}\sin x\cos nx\, dx\right\}$$

$$= \frac{1}{2}\int_o^{\pi}\{\sin(n+1)x - \sin(n-1)x\}\, dx$$

$$= \frac{1}{2\pi}\left[-\frac{\cos(x+1)x}{x+1} + \frac{\cos(x-1)x}{x-1}\right]_o^{\pi}, \text{ where } n \neq 1$$

$= 0$, when $x$ is odd *i.e.*, $n = 3, 5, 7....$

$$= -\frac{2}{\pi(n^2-1)}, \text{ when } n \text{ is even}$$

In case $n = 1$, we have
$$a_1 = \frac{1}{\pi}\int_0^{\pi} \sin x \cos x \cdot dx$$

$$= \frac{1}{2\pi}\int_0^{\pi} \sin 2x \, dx = \frac{1}{2\pi}\left[-\frac{\cos 2x}{2}\right]_0^{\pi} = 0$$

$$b_1 = \frac{1}{\pi}\int_0^{\pi} \sin x \cdot \sin x \cdot dx$$

$$= \frac{1}{2\pi}\int_0^{\pi} (1 - \cos 2x) dx = \frac{1}{2\pi}\left[x - \frac{\sin 2x}{2}\right]_0^{\pi} = \frac{1}{2}$$

If $n \neq 1$, we have

$$b_n = \frac{1}{\pi}\int_0^{\pi} \sin x \sin nx \, dx$$

$$= \frac{1}{2\pi}\int_0^{\pi} \{\cos(n-1)x - \cos(n+1)x\} \, dx = 0$$

$\therefore$
$$f(x) = \frac{1}{\pi} - \frac{2}{\pi}\left\{\frac{\cos 2x}{2^2 - 1} + \frac{\cos 4x}{4^2 - 1} + \frac{\cos 6x}{6^2 - 1} + \ldots\ldots + \frac{1}{2}\sin x\right\}$$

$$= \frac{1}{\pi} - \frac{2}{\pi}\sum_{n=1}^{\infty} \frac{\cos 2nx}{(2n)^2 - 1} + \frac{1}{2}\sin x \quad \ldots(i)$$

(*i*) Let $x = 0$, then $f(x) = 0$, Hence from Eqn. (*i*), we get

$$0 = \frac{1}{\pi} - \frac{2}{\pi}\sum_{n=1}^{\infty} \frac{1}{4n^2 - 1} \text{ or } \sum \frac{1}{4n^2 - 1} = \frac{1}{2}$$

$$\frac{1}{2} = \sum_{n=1}^{\infty} \frac{1}{(2n-1)(2n+1)}$$

$$= \frac{1}{1 \cdot 3} + \frac{1}{3 \cdot 5} + \frac{1}{5 \cdot 7} + \ldots\ldots$$

(*ii*) Let $x = \frac{\pi}{2}$, then $f(x) = \sin\frac{\pi}{2} = 1$. Hence from

$$I = \frac{1}{\pi} - \frac{2}{\pi}\sum_{n=1}^{\infty} \frac{\cos n\pi}{4n^2 - 1} + \frac{1}{2}$$

$$I = \frac{1}{\pi} - \frac{2}{\pi}\sum_{n=1}^{\infty} \frac{(-1)^n}{(2n-1)(2n+1)} + \frac{1}{2}$$

or
$$\frac{1}{1 \cdot 3} - \frac{1}{3 \cdot 5} + \frac{1}{5 \cdot 7} \ldots\ldots = \frac{\pi - 2}{4}$$

## PROBLEMS 10.5

1. Find the Fourier series to represent the following functions. Also sketch the graphs of the functions.

$$f(x) = -\pi, \quad -\pi < x < 0$$
$$f(x) = x \quad 0 < x < \pi$$

Hence show that $1 + \frac{1}{3^2} + \frac{1}{5^2} + .... \frac{\pi^2}{8}$. (*Ranchi, 2019*)

2. Show that for $-\pi < x < \pi$, $\sin ax = \frac{2 \sin a\pi}{\pi}\left(\frac{\sin x}{1^2 - a^2} - \frac{2 \sin 2x}{2^2 - a^2} + \frac{3 \sin 3x}{3^2 - a^2} - ...\right)$ (*Calicut, 2012*)

3. Find the Fourier series to represent the following functions. Also sketch the graphs of the functions.

$$f(x) = \cos x, \quad \frac{-\pi}{2} < x < \frac{\pi}{2}$$
$$= 0, \quad \frac{\pi}{2} < x < \frac{3\pi}{2}$$ (*Ranchi, 2019*)

4. Find the Fourier series to represent the function. Also sketch the graph of the functions. (*Sambalpur, 2019*)

$$f(x) = x^2, 0 \le x \le \pi$$
$$= -x^2, \pi \le x \le 2\pi$$

5. For a function $f(x)$ defined by $f(x) = |x|, -2 < x < 2$, obtain a Fourier series. (*J.N.T.U., 2015*)

Deduce that $\frac{1}{1^2} + \frac{1}{3^2} + \frac{1}{5^2} + \frac{1}{7^2} + ... = \frac{\pi^2}{8}$. (*Anna, 2013 ; Rohtak, 2013*)

6. Find the Fourier series to represent the function

(*i*) $f(x) = |\sin x|, -\pi < x < \pi$. (*Anna, 2015 ; Delhi, 2015*)

(*ii*) $f(x) = |\cos (\pi x/l)|$ in the interval $(-1, 1)$.

7. Given $f(x) = \begin{cases} -x + 1 \text{ for } -\pi \le x \le 0, \\ x + 1 \text{ for } 0 \le x \le \pi. \end{cases}$

Is the function even or odd? Find the Fourier series for $f(x)$ and deduce the value of $\frac{1}{1^2} + \frac{1}{3^2} + \frac{1}{5^2} + ...$

8. Find the Fourier series to represent the following functions. Also sketch the graphs of the functions.

$$f(x) = -K, -\pi < x < 0$$
$$= K, 0 < x < \pi$$

Also deduce that $\frac{\pi}{4} = 1 - \frac{1}{3} + \frac{1}{5} - \frac{1}{7} + .....$

9. A function is defined as follows :

$$f(x) = -x \text{ when } -\pi < x \le 0 = x \quad \text{when } 0 < x < \pi.$$

Show that $f(x) = \frac{\pi}{2} - \frac{4}{\pi}\left[\frac{1}{1^2} \cos x + \frac{1}{3^2} \cos 3x + \frac{1}{5^2} \cos 5x + ...\right]$

Deduce that $\sum_{n=1}^{\infty} \frac{1}{(2n-1)^2} = \frac{\pi^2}{8}$.

10. If $f(x) = 0, -\pi \le x \le 0 = \frac{\pi x}{4}, \quad 0 < x < \pi$ (*VTU, 2017*)

Show that $$f(x) = \frac{\pi^2}{16} + \frac{\sum\{(-1)^n - 1\}}{4n^2} \cos nx - \frac{\sum(-1)^n \pi}{4n} \sin nx$$

Hence show that $$\frac{\pi^2}{8} = 1 + \frac{1}{3^2} + \frac{1}{5^2} + .....$$

11. Given, $f(x) = 1 + \frac{2x}{\pi}$, when $-\pi \le x \le 0 = 1 - \frac{2x}{\pi}$, when $0 \le x \le \pi$ (*Hazaribagh, 2018*)

Show that $$f(x) = \frac{8}{\pi^2}\left[\frac{\cos x}{1^2} + \frac{\cos 3x}{3^2} + \frac{\cos 5x}{5^2} + ....\right]$$

12. A Sinusoidal voltage $E \sin t$ is passed through a half wave rectifier which clips the negative portion of the wave. The resulting periodic function $\phi$ is defined by $\phi(t) = \begin{cases} E \sin t, & \text{for } 0 \le t \le \pi \\ 0 & \text{for } \pi \le t \le 2\pi \end{cases}$ show that

$$\phi(t) = \frac{E}{\pi} + \frac{1}{2} E \sin t - \frac{2E}{\pi} \sum \frac{\cos 2nt}{4n^2 - 1}$$

## 10.7 HALF RANGE SERIES

Many a time is is required to obtain a Fourier expansion of a function $f(x)$ for the range $(0, c)$ which is half the period of the Fourier series. As it is immaterial whatever the function may be outside the range $0 < x < c$, we extend the function to cover the range $-c < x < c$ so that the new function may be odd or even. The Fourier expansion of such a function of half the period, therefore, consists of sine or cosine terms only. In such cases the graphs for the values of $x$ in $(0, c)$ are the same but outside $(0, c)$ are different for odd or even functions. That is why we get different forms of series for the same function as is clear from the examples 10.26 and 10.27.

**Sine series.** If it be required to expand $f(x)$ as a sine series in $0 < x < c$ ; then we extend the function reflecting it in the origin, so that $f(x) = -f(-x)$.

Then the extended function is odd in $(-c, c)$ and the expansion will give the desired Fourier sine series :

$$\left.\begin{aligned} f(x) &= \sum_{n=1}^{\infty} b_n \sin \frac{n\pi x}{c} \\ \text{where} \quad b_n &= \frac{2}{c}\int_0^c f(x) \sin \frac{n\pi x}{c}\, dx \end{aligned}\right\} \quad ...(1)$$

**Cosine series.** If it be required to express $f(x)$ as a cosine series in $0 < x < c$, we extend the function reflecting it in the $y$-axis, so that $f(-x) = f(x)$.

Then the extended function is even in $(-c, c)$ and its expansion will give the required Fourier cosine series :

$$\left.\begin{aligned} f(x) &= \frac{a_0}{2} + \sum_{n=1}^{\infty} a_n \cos \frac{n\pi x}{c} \\ \text{where} \quad a_0 &= \frac{2}{c}\int_0^c f(x)\, dx \\ \text{and} \quad a_n &= \frac{2}{c}\int_0^c f(x) \cos \frac{n\pi x}{c}\, dx \end{aligned}\right\} \quad ...(2)$$

**Example 10.26.** *Find the half range sine series for the function $f(x) = e^x$, for $0 < x < 1$.* (Ranchi. 2019)

**Solution:** Let
$$e^x = \sum b_n \sin n\pi x, \text{ since } l = 1$$

Now
$$b_n = 2\int_0^1 e^x \sin n\pi x \, dx \; ..... \qquad ...(i)$$

$$= 2\left\{\left(e^x\left(\frac{-\cos n\pi x}{n\pi}\right) - e^x\left(\frac{-\sin n\pi x}{n^2\pi^2}\right)\right)_0^1 + \int_0^1 e^x\left(\frac{-\sin n\pi x}{n^2\pi^2}\right)dx\right\}$$

$$= 2\left[e\left(\frac{-\cos n\pi}{n\pi}\right) + e^0\frac{\cos 0}{n\pi}\right] - \frac{2}{n^2\pi^2}\int_0^1 e^x \sin n\pi x \cdot dx$$

$$= \frac{2}{n\pi}\left[-e(-1)^n + 1\right] - \frac{1}{n^2\pi^2}[b_n], \text{ from } (i)$$

So,
$$b_n = \frac{2n\pi}{n^2\pi^2 + 1}\left[1 - (-1)^n e\right]$$

So,
$$e^x = 2\pi\left[\frac{1+e}{\pi^2+1}\sin \pi x + \frac{2(1-e)}{4\pi^2+1}\sin 2\pi x + \frac{3(1+e)}{9\pi^2+1}\sin 3\pi x + .....\right]$$

**Example 10.27.** *Express $f(x) = x$ as a half-range cosine series in $0 < x < 2$.*

*(Andhra, 2016 ; Delhi, 2015 ; V.T.U., 2013)*

**Solution.** The graph of $f(x) = x$ in $(0, 2)$ is the line $OA$. Let us extend the function $f(x)$ in the interval $(-2, 0)$ shown by the line $OB'$ so that the new function is symmetrical about the $y$-axis and, therefore, *represents an even function in $(-2, 2)$*. (Fig. 10.10)

Hence the Fourier series for $f(x)$ over the full period $(-2, 2)$ will contain only cosine terms given by

$$f(x) = \frac{a_0}{2} + \sum_{n=1}^{\infty} a_n \cos\frac{n\pi x}{2}$$

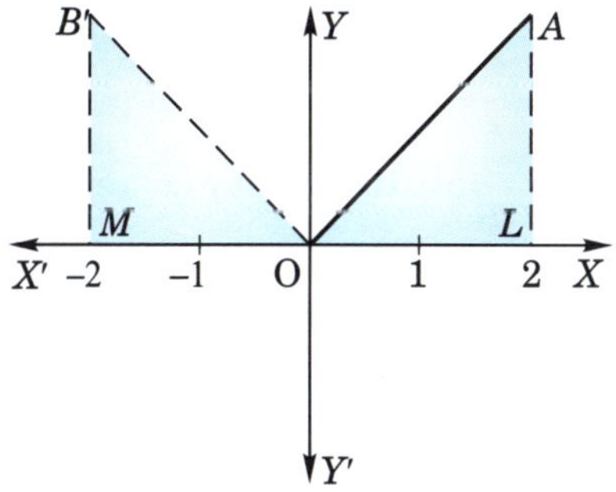

Fig. 10.10.

where
$$a_0 = \frac{2}{2}\int_0^2 f(x)\,dx = \int_0^2 x\,dx = 2$$

and
$$a_n = \frac{2}{2}\int_0^2 f(x)\cos\frac{n\pi x}{2}\,dx = \int_0^2 x\cos\frac{n\pi x}{2}\,dx$$

$$= \left|\frac{2x}{n\pi}\sin\frac{n\pi x}{2} + \frac{4}{n^2\pi^2}\cos\frac{n\pi x}{2}\right|_0^2 = \frac{4}{n^2\pi^2}[(-1)^n - 1]$$

Thus $a_1 = -8/\pi^2$, $a_2 = 0$, $a_3 = -8/3^2\pi^2$, $a_4 = 0$, $a_5 = -8/5^2\pi^2$ etc.

Hence the desired Fourier series for $f(x)$ over the half-range $(0, 2)$ is

$$f(x) = 1 - \frac{8}{\pi^2}\left[\frac{\cos \pi x/2}{1^2} + \frac{\cos 3\pi x/2}{3^2} + \frac{\cos 5\pi x/2}{5^2} + ...\right]$$

**Important Obs.** It must be clearly understood that we expand a function in $0 < x < c$ as a series of sines or cosines, merely looking upon it as an odd or even function of period $2c$. It hardly matters whether the function is odd or even or neither.

**Example 10.28.** *Show that the series $\frac{l}{\pi}\sum_{n=1}^{\infty}\frac{1}{n}\sin\frac{2n\pi x}{l}$ represents $\frac{1}{2}l - x$, when $0 < x < l$.*

*(Shri Venkateswara, 2019)*

**Solution.** Infact the given series is half range sine series for the function $f(x) = \frac{1}{2}l - x$ in the interval $(0, l)$.

Let $$\frac{1}{2}l - x = \sum b_n \sin\frac{n\pi x}{l}$$

$$b_n = \frac{2}{l}\int_0^l \left(\frac{1}{2}l - x\right)\cdot \sin\frac{n\pi x}{l}\,dx$$

$$= \frac{2}{l}\left[\frac{1}{2}(l-x)\cdot\frac{l}{r\cdot\pi}\left(-\cos\frac{n\pi x}{l}\right) - (-1)\frac{l^2}{n^2\pi^2}\left(-\sin\frac{n\pi x}{l}\right)\right]_0^l$$

$$= \frac{2}{l}\left[\frac{l^2}{2n\pi}\cos n\pi + \frac{l^2}{2n\pi}\cos 0\right] = \frac{l}{n\pi}\left[(-1)^n + 1\right] = 2\frac{l}{n\pi} \text{ or } 0$$

according as $n$ is even or odd

$$b_1 = b_3 = b_5 = 0$$

$$\frac{1}{2}l - x = \frac{l}{\pi}\sin\frac{2\pi x}{l} + \frac{1}{2\pi}\sin\frac{4\pi x}{l} + \frac{l}{3\pi}\sin\frac{6\pi x}{l} + \ldots..$$

$$= \frac{l}{\pi}\left[\sin\frac{2\pi x}{l} + \frac{1}{2\pi}\frac{\sin 2(2\pi x)}{l} + \frac{1}{3}\sin\frac{3(2\pi x)}{l} + \ldots..\right]$$

$$= \frac{l}{\pi}\cdot\sum_{n=1}^{\infty}\frac{1}{n}\sin\frac{n(2\pi x)}{l} = \frac{l}{\pi}\sum_{n=1}^{\infty}\frac{1}{n}\sin\frac{2n\pi x}{l}$$

**Example 10.29.** *Find the half range sine* $f(x) = \begin{cases} \frac{1}{4} - x, & 0 < x < \frac{1}{2} \\ x - \frac{3}{4} & \frac{1}{2} < x < 1 \end{cases}$

*(Andhra, 2019 ; Sambapur, 2019)*

**Solution.** Let $f(x)$ represent an odd function in $(-1, 1)$ so that $f(x) = \sum_{n=1}^{\infty} b_n \sin n\pi x$

where $$b_n = \frac{2}{1}\int_0^1 f(x)\sin n\pi x\,dx$$

$$= 2\left[\int_0^{\frac{1}{2}}\left(\frac{1}{4} - x\right)\sin n\pi x\,dx + \int_{\frac{1}{2}}^1\left(x - \frac{3}{4}\right)\sin n\pi x\,dx\right]$$

$$= 2\left|-\left(\frac{1}{4} - x\right)\frac{\cos n\pi x}{n\pi} - \frac{\sin n\pi x}{n^2\pi^2}\right|_0^{\frac{1}{2}} + 2\left|-\left(x - \frac{3}{4}\right)\frac{\cos n\pi x}{n\pi} + \frac{\sin n\pi x}{n^2\pi^2}\right|_{\frac{1}{2}}^1$$

$$= 2\left[\frac{1}{4n\pi}\cos\frac{n\pi}{2} + \frac{1}{4n\pi} - \frac{\sin n\pi/2}{n^2\pi^2}\right] + 2\left[-\frac{1}{4n\pi}\cos n\pi - \frac{1}{4n\pi}\cos\frac{n\pi}{2} - \frac{\sin n\pi/2}{n^2\pi^2}\right]$$

$$= \frac{1}{2n\pi}[1 - (-1)^n] - \frac{4\sin n\pi/2}{n^2\pi^2}$$

Thus $$b_1 = \frac{1}{\pi} - \frac{4}{\pi^2};\ b_2 = 0$$

$$b_3 = \frac{1}{3\pi} + \frac{4}{3^2\pi^2};\ b_4 = 0$$

$$b_5 = \frac{1}{5\pi} - \frac{4}{5^2\pi^2};\ b_6 = 0 \text{ etc.}$$

Hence $$f(x) = \left(\frac{1}{\pi} - \frac{4}{\pi^2}\right)\sin\pi x + \left(\frac{1}{3\pi} + \frac{4}{3^2\pi^2}\right)\sin 3\pi x + \left(\frac{1}{5\pi} - \frac{4}{5^2\pi^2}\right)\sin 5\pi x + \cdots$$

## PROBLEMS 10.6

1. Obtain cosine and sine series for $f(x) = x$ in the interval $0 \le x \le \pi$. Hence show that

$$\frac{1}{1^2} + \frac{1}{3^2} + \frac{1}{5^2} + \ldots = \frac{\pi^2}{8}.$$

*(Andhra, 2023 ; J.N.T.U., 2015 ; Anna, 2012 ; Rohtak, 2010 S)*

2. Find the Fourier series to represent the function

$$f(x) = 0, \quad \text{for} \quad -2 < x < 0$$
$$= 1 \quad \text{for} \quad 0 < x < 2$$

*(Alagappa University, 2017)*

3. Find the Fourier cosine series of the function $f(x) = \pi - x$ in $0 < x < \pi$. Hence show that

$$\sum_{r=0}^{\infty} \frac{1}{(2r+1)^2} = \frac{\pi^2}{8}$$

*(W.B.T.U., 2015)*

4. Find the half-range cosine series for the function $f(x) = (x - 1)^2$ in the interval $0 < x < 1$.

*(Andhra, 2013 ; Anna, 2013 ; V.T.U., 2010)*

Hence show that (*i*) $\frac{1}{1^2} + \frac{1}{2^2} + \frac{1}{3^2} + \ldots = \frac{\pi^2}{6}$ (*ii*) $\pi^2 = 8\left(\frac{1}{1^2} + \frac{1}{3^2} + \frac{1}{5^2} + \ldots\right)$

5. Find the half-range sine series for the function $f(t) = t - t^2$, $0 < t < 1$. *(U.P.T.U., 2011)*

6. Find the half range sine series for $f(x) = x \cos x$ in $(0, \pi)$. *(K.T.U., 2016)*

7. Obtain the half-range sine series for $e^x$ in $0 < x < 1$.

8. Find the Fourier cosine series of $f(x) = x(\pi - x)$ in $0 < x < \pi$. *(Andhra, 2021 ; K.T.U., 2016)*

9. Find the half range Fourier sine series of $f(x) = x(\pi - x)$, $0 \le x \le \pi$ *(V.T.U., 2015 S ; J.N.T.U., 2015)*

(*i*) $\sum \frac{1}{n^4} = \frac{\pi^4}{90}$ (*ii*) $\frac{1}{1^3} - \frac{1}{3^3} + \frac{1}{5^3} - \frac{1}{7^3} + \ldots = \frac{\pi^3}{32}$. *(J.N.T.U., 2015)*

(*iii*) $\sum_{n=1}^{\infty} \frac{1}{(2n-1)^6} = \frac{\pi^6}{960}$

10. If $f(x) = x, \quad 0 < x < \pi/2$

$$= \pi - x, \quad \pi/2 < x < \pi,$$

show that (*i*) $f(x) = \frac{4}{\pi}\left[\sin x - \frac{1}{3^2}\sin 3x + \frac{1}{5^2}\sin 5x - \ldots\right]$ *(J.N.T.U., 2015)*

(*ii*) $f(x) = \frac{\pi}{4} - \frac{2}{\pi}\left[\frac{1}{12}\cos 2x + \frac{1}{3^2}\cos 6x + \frac{1}{5^2}\cos 10x + \ldots\right]$ *(JNTUK, 2022 ; Rohtak, 2012 S ; V.T.U., 2011)*

11. Find the half-range cosine series expansion of the function $f(x) = \begin{cases} 0, & 0 \le x \le l/2 \\ l - x, & l/2 \le x \le l \end{cases}$. *(P.T.U., 2010)*

12. If $f(x) = \sin x$ for $0 \le x \le \pi/4$

$$= \cos x \text{ for } \pi/4 \le x \le \pi/2,$$ expand $f(x)$ in a series of sines.

13. For the function defined by the graph *OAB* in Fig. 10.11, find the half-range Fourier sine series.

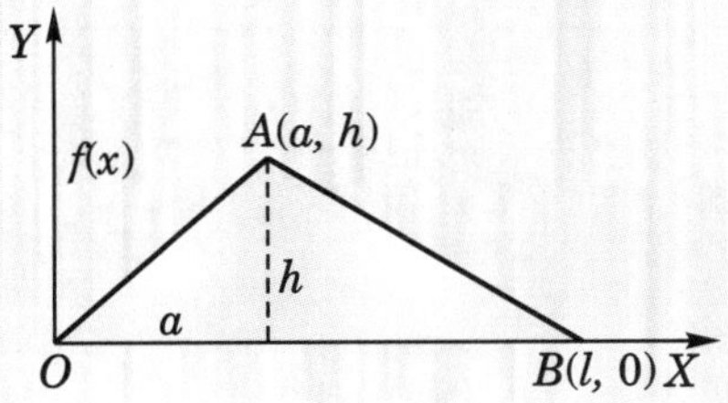

Fig. 10.11.

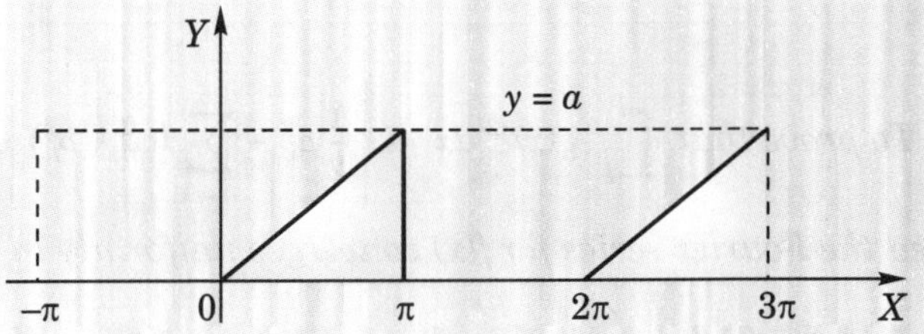

Fig 10.12.

14. If $f(x) = \pi x$ for $0 \le x < 1$, for $(x) = 0$ for $x = 1$ and

$f(x) = \pi(x - 2)$ for $1 < x \le 2$, show that

$$f(x) = 2\left(\sin \pi x - \frac{1}{2}\sin^2 \pi x + \frac{1}{3}\sin 3\pi x - \frac{1}{4}\sin 4\pi x + .....\right)$$

15. Find the Fourier series to represent the following function

$$f(x) = \pi x, \text{ from } x = -c \text{ to } x = c.$$

16. Find the Fourier series to represent the following function.

$$f(x) = e^{-x}, \text{ from } x = -a \text{ to } x = a$$

17. Find the Fourier series to represent the function. *(Andhra, 2020)*

$$f(x) = x^2 - 2, \text{ from } x = -2 \text{ to } x = 2.$$

## 10.8 TYPICAL WAVEFORMS

We give below six typical waveforms usually met with in communication engineering :

(1) *Square waveform* is an extension of the function of the square form.

(2) *Saw-toothed waveform* (Fig. 10.7) is an extension of the function in Ex. 10.20, page 485.

(3) *Modified saw-toothed waveform* (Fig. 10.12) is extension of the function

$$f(x) = 0, \qquad -\pi < x \le 0$$
$$= x, \qquad 0 \le x < \pi,$$

Its Fourier expansion is

$$f(x) = \frac{a}{4} - \frac{2a}{\pi^2}\left(\cos x + \frac{\cos 3x}{3^2} + \frac{\cos 5x}{5^2} + ...\right) + \frac{\alpha}{\pi}\left(\sin x - \frac{\sin 2x}{2} + \frac{\sin 3x}{3} + ...\right)$$

(4) *Triangular waveform* (Fig. 10.8) is an extension of the function of Ex. 10.22, page 486.

(5) *Half-wave rectifier* (Fig. 10.2) is an extension of the function of Problem 2, page 491.

(6) *Full-wave rectifier* (Fig. 10.13) is an extension of the function $f(x) = a \sin x$, $0 \le x \le \pi$. Its Fourier expansion is

$$f(x) = \frac{4a}{\pi}\left\{\frac{1}{2} - \frac{1}{1.3}\cos 2x - \frac{1}{3.5}\cos 4x - \frac{1}{5.7}\cos 6x - ...\right\}$$

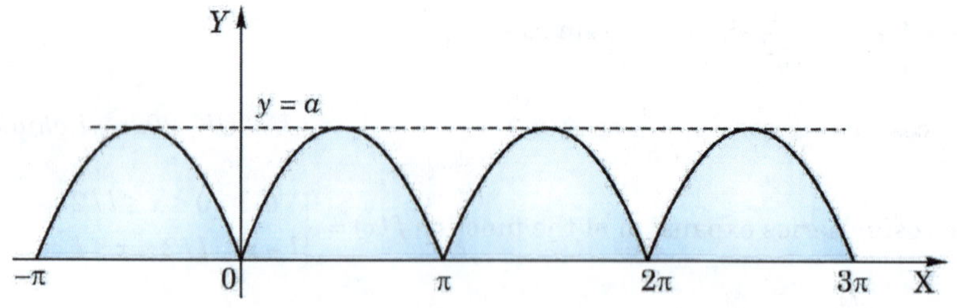


Fig. 10.13.

## 10.9 (1) PARSEVAL'S FORMULA*

*To prove that* $\int_{-l}^{l} [f(x)]^2\, dx = l\left\{\frac{1}{2}a_0^2 + \sum_{n=1}^{\infty}(a_n^2 + b_n^2)\right\}$,

*provided the Fourier series for f(x) converges uniformly in (–l, l).*

*Named after the French mathematician *Marc Antoine Parseval* (1755–1836).

The Fourier series for $f(x)$ in $(-l, l)$ is $f(x) = \frac{a_0}{2} + \sum_{n=1}^{\infty}\left(a_n \cos\frac{n\pi x}{l} + b_n \sin\frac{n\pi x}{l}\right)$ ...(1)

Multiplying both sides of Eqn. (1) by $f(x)$ and integrating term by term from $-l$ to $l$ [which is justified as the series (1) is uniformly convergent], we get

$$\int_{-l}^{l}[f(x)]^2\,dx = \frac{a_0}{2}\int_{-l}^{l} f(x)\,dx + \sum_{n=1}^{\infty}\left\{a_n\int_{-l}^{l} f(x)\cos\frac{n\pi x}{l}\,dx + b_n\int_{-l}^{l} f(x)\sin\frac{n\pi x}{l}\,dx\right\} \quad ...(2)$$

Now $\int_{-l}^{l} f(x)\,dx = la_0,$

$$\int_{-l}^{l} f(x)\cos\frac{n\pi x}{l}\,dx = la_n \text{ and } \int_{-l}^{l} f(x)\sin\frac{n\pi x}{l}\,dx = lb_n,$$

$\therefore$ Eqn. (2) takes the form $\int_{-l}^{l}[f(x)]^2\,dx = l\left\{\frac{a_0^2}{2} + \sum_{n=1}^{\infty}(a_n^2 + b_n^2)\right\}$ ...(3)

which is the desired *Parseval's formula.*

**Cor. 1.** *If* $f(x) = \frac{a_0}{2} + \sum_{n=1}^{\infty}\left(a_n \cos\frac{n\pi x}{l} + b_n \sin\frac{n\pi x}{l}\right)$ *in* $(0, 2l)$, then

$$\int_0^{2l}[f(x)]^2\,dx = l\left\{\frac{a_0^2}{2} + \sum_{n=1}^{\infty}(a_n^2 + b_n^2)\right\} \quad ...(4)$$

**Cor. 2.** *If the half-range cosine series is (0, l) for f(x) is*

$$f(x) = \frac{a_0}{2} + \sum_{n=1}^{\infty} a_n \cos\left(\frac{n\pi x}{l}\right), \text{ then}$$

$$\int_0^{l}[f(x)]^2\,dx = \frac{l}{2}\left(\frac{a_0^2}{2} + a_1^2 + a_2^2 + a_3^2 + \ldots\infty\right) \quad ...(5)$$

**Cor. 3.** *If the half-range sine series in (0, l) for f(x) is* $f(x) = \sum_{n=1}^{\infty} b_n \sin\left(\frac{n\pi x}{l}\right)$, then

$$\int_0^{l}[f(x)]^2 = \frac{l}{2}(b_1^2 + b_2^2 + b_3^2 + \ldots\infty) \quad ...(6)$$

**(2) Root mean square (rms) value.** *The root mean square value of the function f(x) over an interval (a, b) is defined as*

$$[f(x)]_{\text{rms}} = \sqrt{\left\{\frac{\int_a^b [f(x)]^2\,dx}{b-a}\right\}} \quad ...(7)$$

The use of root mean square value of a periodic function is frequently made in the theory of mechanical vibrations and in electric circuit theory. The r.m.s. value is also known as the effective value of the function.

**Example 10.30.** *Obtain the Fourier series for* $y = x^2$ *in* $-\pi < x < \pi$. *Using the two values of y, show that*

$$\frac{1}{1^4} + \frac{1}{2^4} + \frac{1}{3^4} + \frac{1}{4^4} + \ldots = \frac{\pi^4}{90}$$

**Solution.** Let $y = \frac{a_0}{2} + \sum_{n=1}^{\infty} a_n \cos nx + \sum_{n=1}^{\infty} b_n \sin nx$

We have $a_0 = 2\frac{n^2}{3}, a_n = \frac{4}{n^2}(-1)^n, b_n = 0$ for all $n$

If $\bar{y}$ be the r.m.s. value of $y$ in $(-\pi, \pi)$, then

$$(\bar{y})^2 = \frac{\pi}{2\pi}\left[\frac{a_0^2}{2} + \sum_{n=1}^{\infty}(a_n^2 + b_n^2)\right] \qquad \text{[By (3) and (7) §10.9]}$$

$$= \frac{1}{4}\left(\frac{2\pi^2}{3}\right)^2 + \frac{1}{2}\sum_{n=1}^{\infty}\left[\frac{16}{n^4}(-1)^{2n} + 0\right] = \frac{\pi^4}{9} + 8\sum_{1}^{\infty}\frac{1}{n^4}$$

Also by definition,

$$(\bar{y})^2 = \frac{1}{2\pi}\int_{-\pi}^{\pi} y^2\, dx = \frac{1}{2\pi}\int_{-\pi}^{\pi} x^4\, dx = \frac{\pi^4}{5}$$

Equating the two values of $(\bar{y})^2$, we get

$$\frac{\pi^4}{9} + 8\sum_{1}^{\infty}\frac{1}{n^4} = \frac{\pi^4}{5} \text{ i.e., } \sum_{1}^{\infty}\frac{1}{n^4} = \frac{\pi^4}{90}.$$

## PROBLEMS 10.7

1. By using the sine series for $f(x) = 1$ in $0 < x < \pi$, show that $\frac{\pi^2}{8} = 1 + \frac{1}{3^2} + \frac{1}{5^2} + \frac{1}{7^2} + \dots$ *(Andhra, 2019)*

2. Prove that in $0 < x < l$, $x = \frac{l}{2} - \frac{4l}{\pi^2}\left(\cos\frac{\pi x}{l} + \frac{1}{3^2}\cos\frac{3\pi x}{l} + \frac{1}{5^2}\cos\frac{5\pi x}{l} + \dots\right)$

   and deduce that $$\frac{1}{1^4} + \frac{1}{3^4} + \frac{1}{5^4} + \dots\dots = \frac{\pi^4}{96}.$$

3. If $\frac{a_0}{2} + \sum a_n \cos\frac{n\pi x}{l}$ is the half-range cosine series of $f(x)$ of period $2l$ in $(0, l)$, then show that the mean square value of $f(x)$ in $(0, l)$ is $\frac{l}{2}\left\{\frac{a_0^2}{2} + \sum_{n=1}^{\infty} a_n^2\right\}$

   Use this result to evaluate $1^{-4} + 3^{-4} + 5^{-4} + \dots\dots$ from the half-range cosine series of the function $f(x)$ of period 4 defined in $(0, 2)$ by

   $$f(x) = \begin{cases} \pi x, & 0 < x < 1 \\ \pi(2 - x), & 1 < x < 2 \end{cases}.$$

## 10.10 COMPLEX FORM OF FOURIER SERIES

The Fourier series of a periodic function $f(x)$ of period $2l$, is

$$f(x) = \frac{a_0}{2} + \sum_{n=1}^{\infty}\left(a_n \cos\frac{n\pi x}{l} + b_n \sin\frac{n\pi x}{l}\right) \qquad \dots(1)$$

Since $\cos\theta = \frac{1}{2}(e^{i\theta} + e^{-i\theta})$ and $\sin\theta = \frac{1}{2i}(e^{i\theta} - e^{-i\theta})$,

therefore, we can express Eqn. (1) as

$$f(x) = \frac{a_0}{2} + \sum_{n=1}^{\infty}\left\{a_n\left(\frac{e^{in\pi x/l} + e^{-in\pi x/l}}{2}\right) + b_n\left(\frac{e^{in\pi x/l} - e^{-in\pi x/l}}{2i}\right)\right\}$$

$$= c_0 + \sum_{n=1}^{\infty}\left\{c_n e^{in\pi x/l} + c_{-n} e^{-in\pi x/l}\right\} \qquad \dots(2)$$

where $c_0 = \frac{1}{2}a_0,\ c_n = \frac{1}{2}(a_n - ib_n),\ c_{-n} = \frac{1}{2}(a_n + ib_n)$

Now
$$c_n = \frac{1}{2l}\left\{\int_{-l}^{l} f(x)\cos\frac{n\pi x}{l}dx - i\int_{-l}^{l} f(x)\sin\frac{n\pi x}{l}dx\right\}$$
$$= \frac{1}{2l}\int_{-l}^{l} f(x)\left(\cos\frac{n\pi x}{l} - i\sin\frac{n\pi x}{l}\right)dx = \frac{1}{2l}\int_{-l}^{l} f(x)e^{-in\pi x/l}\,dx$$

and
$$c_{-n} = \frac{1}{2l}\int_{-l}^{l} f(x)\left(\cos\frac{n\pi x}{l} - i\sin\frac{n\pi x}{l}\right)dx = \frac{1}{2l}\int_{-l}^{l} f(x)e^{in\pi x/l}\,dx$$

Combining these, we have $c_n = \dfrac{1}{2l}\displaystyle\int_{-l}^{l} f(x)e^{-in\pi x/l}\,dx$

where $n = 0, \pm 1, \pm 2, \pm 3, ......$ ...(3)

Then the series (2) can be compactly written as :
$$f(x) = \sum_{n=-\infty}^{\infty} c_n\, e^{in\pi x/l}$$
which is the *complex form of Fourier series* and its coefficients are given by Eqn. (3).

**Obs.** The complex form of a Fourier series is especially useful in problems on electrical circuits having impressed periodic voltage.

**Example 10.31.** *Find the half range sine series for the function $f(x) = e^x$, for $0 < x < 1$.* *(Ranchi. 2019)*

**Solution:** Let $e^x = \sum b_n \sin n\pi x$, since $l = 1$

Now
$$b_n = 2\int_0^1 e^x \sin n\pi x\, dx \,..... \qquad ...(i)$$
$$= 2\left\{\left(e^x\left(\frac{-\cos n\pi x}{n\pi}\right) - e^x\left(\frac{-\sin n\pi x}{n^2\pi^2}\right)\right)_0^1 + \int_0^1 e^x\left(\frac{-\sin n\pi x}{n^2\pi^2}\right)dx\right\}$$
$$= 2\left[e\left(\frac{-\cos n\pi}{n\pi}\right) + e^0\frac{\cos 0}{n\pi}\right] - \frac{2}{n^2\pi^2}\int_0^1 e^x \sin n\pi x \cdot dx$$
$$= \frac{2}{n\pi}\left[-e(-1)^n + 1\right] - \frac{1}{n^2\pi^2}[b_n], \text{ from Eqn. } (i)$$

So,
$$b_n = \frac{2n\pi}{n^2\pi^2 + 1}\left[1 - (-1)^n e\right]$$

So,
$$e^x = 2\pi\left[\frac{1+e}{\pi^2+1}\sin\pi x + \frac{2(1-e)}{4\pi^2+1}\sin 2\pi x + \frac{3(1+e)}{9\pi^2+1}\sin 3\pi x + .....\right]$$

**Example 10.32.** *Expand $x(\pi - x)$ in a half range sine series in the interval $(0, \pi)$ upto first three terms.*

**Solution:** Let $x(\pi - x) = \Sigma b_n \sin nx$

where
$$b_n = \frac{2}{\pi}\int_0^{\pi} (\pi x - x^2)\sin nx\, dx$$
$$= \frac{2}{\pi}\left[(\pi x - x^2)\left(\frac{-\cos nx}{n}\right) - (\pi - 2x)\left(\frac{-\sin nx}{n^2}\right) + (-2)\left(\frac{\cos nx}{n^3}\right)\right]_0^{\pi}$$
$$= \frac{2}{\pi}\left[\left(0 - 0 - \frac{-2\cos n\pi}{n^3}\right) - \left(0 - 0 - \frac{2\cos 0}{n^3}\right)\right]$$
$$= \frac{4}{\pi n^3}[1 - (-1)^n] = 0 \text{ or } \frac{8}{\pi n^3} \text{ according as } n \text{ is even or odd.}$$

$$a_2 = a_4 = a_6 = 0\ldots \text{ and } a_1 = \frac{8}{\pi},\ a_3 = \frac{8}{\pi 3^2}$$

$$\therefore \qquad x(\pi - x) = \frac{8}{\pi}\left(\sin x + \frac{\sin 3x}{3^3} + \frac{\sin 5x}{5^3} + \ldots\right).$$

**Example 10.33:** *Given that $f(x) = kx$, when $0 \le x \le \frac{l}{2}$.*

*$= k(l - x)$, when $\frac{l}{2} \le x \le l$. Expand $f(x)$ as a Fourier series.* (Pondicherry, 2018)

**Solution:** Since interval is $(0, l)$, we can expand $f(x)$ either as a sine or a cosine series.

For cosine series, let

$$f(x) = \frac{an}{2} + \sum a_n \cos\frac{n\pi x}{l}$$

where $$a_o = \frac{2}{l}\int_0^l f(x)\,dx = \frac{2}{l}\left[\int_0^{l/2} kx \cdot dx \int_{l/2}^l k(l-n)dx\right] = \frac{1}{2}kl$$

and $$a_n = \frac{2}{l}\int_0^{l/2}\left[k(x)\cos\frac{n\pi x}{l}dx + \int_{l/2}^l k(l-x)\cos\frac{n\pi x}{l}dx\right]$$

$$= \frac{2}{l}\left[kx \cdot \frac{l}{n\pi}\sin\frac{n\pi x}{l} + k\frac{l^2}{n^2\pi^2}\cos\frac{n\pi x}{l}\right]_0^{l/2} + \frac{2}{l}\left[k(l-x)\frac{l}{n\pi}\sin\frac{n\pi x}{l} - \frac{kl^2}{n^2\pi^2}\cdot\cos\frac{n\pi x}{l}\right]_{l/2}^{l}$$

$$= \frac{2}{l}\left[\frac{kl^2}{2n\pi}\sin\frac{n\pi}{2} + \frac{kl^2}{n^2\pi^2}\left(\cos\frac{n\pi}{2} - 1\right)\right] + \frac{2}{l}\left[\left(0 - \frac{kl^2}{n^2\pi^2}\cos n\pi\right)\right]$$

$$= \frac{2kl}{n^2\pi^2}\left(2\cos\frac{n\pi}{2} - 1 - \cos n\pi\right)$$

If $$n = 1,\ a_1 = \frac{2kl}{\pi^2}\left(2\cos\frac{\pi}{2} - 1 - \cos\pi\right) - \left(\frac{kl^2}{2n\pi}\sin\frac{n\pi}{2} - \frac{kl^2}{n^2\pi^2}\cos\frac{n\pi}{2}\right) = \frac{2kl}{\pi^2}(0 - 1 + 1) = 0$$

Thus $$a_3 = a_4 = a_5 = a_7 = a_8 = a_9 = \ldots = 0.$$

Again $$a_2 = \frac{2kl}{4\pi^2}[2\cos\pi - 1 - \cos 2\pi]$$

$$= \frac{2kl}{4\pi^2}(-4) = \frac{-8kl}{2^2\pi^2}$$

$$a_6 = \frac{-8kl}{6^2\pi^2},\ a_{10} = \frac{-8kl}{10^2\pi^2}$$

$$\therefore \qquad f(x) = \frac{kl}{4} - \frac{8kl}{\pi^2}\left(\frac{1}{2^2}\cos\frac{2\pi x}{l} + \frac{1}{6^2}\cos\frac{6\pi x}{l} + \frac{1}{10^2}\cos\frac{10\pi x}{l} + \ldots\right)$$

**Example 10.34:** *Let $f(x) = \omega x$, when $0 \le x \le \frac{l}{2}$*

*$= \omega(l - x)$. When $\frac{1}{2} \le x \le l$ show that $f(x)$*

$$= \frac{4\omega l}{\pi^2}\sum_{n=0}^{\infty}\frac{(-1)^n}{(2n+1)^2}\sin\frac{(2n+1)\pi x}{l}.$$ *Hence obtain the sum of series* $1 + \frac{1}{3^2} + \frac{1}{5^2} + \ldots$

(Assam, 2018)

**Solution.** Let $f(x) = \omega x$, when $0 \le x \le \frac{l}{2}$

$$= \omega(l - x), \text{ when } \frac{l}{2} \le x \le l$$

$$f(x) = \sum bx \sin \frac{n\pi x}{l}$$

$$b_n = \frac{2}{l}\int_0^l f(x)\sin\frac{n\pi x}{l}\,dx = \frac{2}{l}\left[\int_0^{l/2} wx \cdot \sin\frac{n\pi x}{l}\,dx + \int_{1/2}^{l} w(l - x)\sin\frac{n\pi x}{l}\,dx\right]$$

$$= \frac{2\omega}{l}\left[x \cdot \left(-\frac{l}{n\pi}\cos\frac{n\pi x}{l}\right) - 1\left(\frac{l^2}{n^2\pi^2}\cdot\sin\frac{n\pi x}{l}\right)\right]_0^{l/2} +$$

$$\frac{2\omega}{l}\left[(l - x)\left(\frac{-l}{n\pi}\cdot\cos\frac{n\pi x}{l}\right) - (-1)\left(\frac{l^2}{n^2\pi^2}\cdot\sin\frac{n\pi x}{l}\right)\right]_{l/2}^{l}$$

$$= \frac{2\omega}{l}\left[\frac{-l}{2n\pi}\cos\frac{n\pi}{2} + \frac{l^2}{n^2\pi^2}\sin\frac{n\pi}{2} - 0\right] + \frac{2\omega}{l}\left[0 - \left(\frac{-l^2}{2n\pi}\frac{\cos n\pi}{2} - \frac{l^2}{n^2\pi}\sin\frac{n\pi}{2}\right)\right]$$

$$= \frac{2\omega}{l}\left(\frac{2l^2}{n^2\pi^2}\sin\frac{n\pi}{2}\right) = \frac{4wl}{n^2\pi^2}\sin\frac{n\pi}{2}$$

when $n$ is even, $b_n = 0$

$$b_2 = b_4 = b_4 = 0$$

Also $$b_1 = \frac{4\omega l}{\pi^2}\sin\frac{\pi}{2} = \frac{4\omega l}{\pi^2}$$

$$b_3 = 4\frac{\omega l}{3^2\pi^2}\sin\frac{3\pi}{2} = -\frac{4\omega l}{3^2\pi^2}$$

$$b_5 = 4\frac{\omega l}{5^2\pi^2}\sin\frac{5\pi}{2} = \frac{4\,\omega l}{5^2\pi^2}, \text{ and so on}$$

$\therefore$ $$f(x) = b_1\sin\frac{\pi x}{l} + b_3\frac{\sin 3\pi x}{l} + b_5\sin\frac{5\pi x}{l} + \ldots..$$

$$= \frac{4\omega l}{\pi^2}\left[\sin\frac{\pi x}{l} - \frac{1}{3^2}\sin\frac{3\pi x}{l} + \frac{1}{5^2}\sin\frac{5\pi x}{l}\ldots.\right]$$

$$= \frac{4\omega l}{\pi^2}\sum_{n=1}^{\infty}\frac{(-1)^n}{(2n+1)^2}\sin\frac{(2n+1)\pi x}{l} \quad \ldots(i)$$

Substituting $x = \frac{l}{2}$ in Eqn. $(i)$

$$= 1 + \frac{1}{3^2} + \frac{1}{5^2} + \ldots = \frac{\pi^2}{8}.$$

## PROBLEMS 10.8

Find the complex form of the Fourier series of the following periodic functions :

1. $f(x) = e^{ax}, -l < x < l.$
2. $f(t) = \sin t, 0 < t < \pi$
3. $f(x) = \cos ax, -\pi < x < \pi$ *(Anna, 2013)*
4. Find the half range sine series for the function $f(x) = x^2$ in the range $(0, \pi)$. *(Madurai, 2019)*
5. $f(x) = \begin{cases} 0 \text{ when } 0 < x < l \\ a \text{ when } l < x < 2l \end{cases}.$
6. Express $f(x) = x$ as a sine and cosine half range series in the interval $0 \le x \le \pi$. *(Ranchi, 2018)*
7. Obtain the half range sine and cosine series for $f(t) = t - t^2, 0 < t < 1$.
8. Find the half range sine and cosine series for $f(x) = 2x - 1$ for $0 < x < 1$.
9. Find the half range sine series for

$$f(x) = \frac{1}{4} - x, \quad 0 < x < \frac{1}{2} = x - \frac{3}{4}, \quad \frac{1}{2} < x < 1.$$

## 10.11 PRACTICAL HARMONIC ANALYSIS

We have discussed at length, the problem of expanding $y = f(x)$ as Fourier series :

$$f(x) = \frac{a_0}{2} + \sum_{n=1}^{\infty} (a_n \cos nx + b_n \sin nx) \quad ...(1)$$

where
$$a_0 = \frac{1}{\pi} \int_0^{2\pi} f(x)\, dx$$

$$a_n = \frac{1}{\pi} \int_0^{2\pi} f(x) \cos nx\, dx \quad ...(2)$$

$$b_n = \frac{1}{\pi} \int_0^{2\pi} f(x) \sin nx\, dx$$

So far, the function has always been defined by an explicit function of an independent variable. In practice, however, the function is often given not by a formula but by a graph or by a table of corresponding values. In such cases, the integrals in Eqn. (2) cannot be evaluated and instead, the following alternative forms of Eqn. (2) are employed.

Since the mean value of a function $y = f(x)$ over the range $(a, b)$ is $\frac{1}{b-a} \int_a^b f(x)\, dx$.

$\therefore$ the Eqn. (2) give,

$$\mathbf{a_0} = 2 \times \frac{1}{2\pi} \int_0^{2\pi} f(x)\, dx = \mathbf{2[mean\ value\ of\ f(x)\ in\ (0, 2\pi)]}$$

$$\mathbf{a}_n = 2 \times \frac{1}{2\pi} \int_0^{2\pi} f(x) \cos nx\, dx = \mathbf{2[mean\ value\ of\ f(x)\ cos\ nx\ in\ (0, 2\pi)]} \quad ...(3)$$

$$\mathbf{b}_n = 2 \times \frac{1}{2\pi} \int_0^{2\pi} f(x) \sin nx\, dx = \mathbf{2[mean\ value\ of\ f(x)\ sin\ nx\ in\ (0, 2\pi)]}$$

There are numerous other methods of finding the value of $a_0$, $a_n$, $b_n$ which constitute the field of **harmonic analysis.**

In Eqn. (1), *the term* ($a_1 \cos x + b_1 \sin x$) *is called the* **fundamental or first harmonic,** *the term* ($a_2 \cos 2x + b_2 \sin 2x$) **the second harmonic** *and so on.*

**Example 10.35.** *The displacement y of a part of a mechanism is tabulated with corresponding angular movement x° of the crank. Express y as a Fourier series neglecting the harmonic above the third :*

| $x°$ | 0 | 30 | 60 | 90 | 120 | 150 | 180 | 210 | 240 | 270 | 300 | 330 |
|---|---|---|---|---|---|---|---|---|---|---|---|---|
| $y$ | 1.80 | 1.10 | 0.30 | 0.16 | 1.50 | 1.30 | 2.16 | 1.25 | 1.30 | 1.52 | 1.76 | 2.00 |

(*V.T.U., 2013*)

**Solution.** Let the Fourier series upto the third harmonic representing $y$ in $(0, 2\pi)$ be

$$y = \tfrac{1}{2}a_0 + a_1 \cos x + a_2 \cos 2x + a_3 \cos 3x + b_1 \sin x + b_2 \sin 2x + b_3 \sin 3x \qquad ...(i)$$

To evaluate the coefficients, we form the following table.

| $x°$ | $\sin x$ | $\cos x$ | $\sin 2x$ | $\cos 2x$ | $\sin 3x$ | $\cos 3x$ | $y$ | $y \sin x$ | $y \cos x$ | $y \sin 2x$ | $y \cos 2x$ | $y \sin 3x$ | $y \cos 3x$ |
|---|---|---|---|---|---|---|---|---|---|---|---|---|---|
| 0 | 0 | 1 | 0 | 1 | 0 | 1 | 1.80 | 0.00 | 1.80 | 0.00 | 1.80 | 0.00 | 1.80 |
| 30 | 0.50 | 0.87 | 0.87 | 0.50 | 1 | 0 | 1.10 | 0.55 | 0.96 | 0.96 | 0.55 | 1.10 | 0.00 |
| 60 | 0.87 | 0.50 | 0.87 | –0.50 | 0 | –1 | 0.30 | 0.26 | 0.15 | 0.26 | –0.15 | 0.00 | –0.30 |
| 90 | 1.00 | 0 | 0 | –1.00 | –1 | 0 | 0.16 | 0.16 | 0.00 | 0.00 | –0.16 | –0.16 | 0.00 |
| 120 | 0.87 | –0.50 | –0.87 | –0.50 | 0 | 1 | 0.50 | 0.43 | –0.25 | –0.43 | –0.25 | 0.00 | 0.50 |
| 150 | 0.50 | –0.87 | –0.87 | –0.50 | 1 | 0 | 1.30 | 0.65 | –1.13 | –1.13 | 0.65 | 1.30 | 0.00 |
| 180 | 0 | –1.00 | 0 | 1.00 | 0 | –1 | 2.16 | 0.00 | –2.16 | –0.00 | 2.16 | 0.00 | –2.16 |
| 210 | –0.50 | –0.87 | 0.87 | 0.50 | –1 | 0 | 1.25 | –0.63 | –1.09 | 1.09 | 0.63 | –1.25 | 0.00 |
| 240 | –0.87 | –0.50 | 0.87 | –0.50 | 0 | 1 | 1.30 | –1.13 | –0.65 | 1.13 | –0.65 | 0.00 | 1.30 |
| 270 | –1.00 | 0 | 0 | –1.00 | 1 | 0 | 1.52 | –1.52 | 0.00 | 0.00 | –1.52 | 1.52 | 0.00 |
| 300 | –0.87 | 0.50 | –0.87 | –0.50 | 0 | –1 | 1.76 | –1.53 | 0.88 | –1.53 | –0.88 | 0.00 | –1.76 |
| 330 | –0.50 | 0.87 | –0.87 | 0.50 | –1 | 0 | 2.00 | –1.00 | 1.74 | –1.74 | 1.00 | –2.00 | 0.00 |
| | | | | | | $\Sigma =$ | 15.15 | –3.76 | 0.25 | –1.39 | 3.18 | 0.51 | –0.62 |

$$\therefore \qquad a_0 = 2 \cdot \frac{\Sigma y}{12} = \frac{15.15}{6} = 2.53 \;;\; a_1 = \frac{1}{6}\Sigma y \cos x = \frac{0.25}{6} = 0.04$$

$$a_2 = \frac{1}{6}\Sigma y \cos 2x = \frac{3.18}{6} = 0.53 \;;\; a_3 = \frac{1}{6}\Sigma y \cos 3x = \frac{-0.62}{6} = -0.1$$

$$b_1 = \frac{1}{6}\Sigma y \sin x = \frac{-3.76}{6} = -0.63\;;$$

$$b_2 = \frac{1}{6}\Sigma y \sin 2x = \frac{-1.39}{6} = -0.23$$

$$b_3 = \frac{1}{6}\Sigma y \sin 3x = \frac{0.51}{6} = 0.085$$

Substituting the values of $a$'s and $b$'s in Eqn. ($i$), we get

$$y = 1.26 + 0.04 \cos x + 0.53 \cos 2x - 0.1 \cos 3x - 0.63 \sin x - 0.23 \sin 2x + 0.085 \sin 3x.$$

**Example 10.36.** *The following table gives the variations of periodic current over a period.*

| *t sec* | 0 | *T/6* | *T/3* | *T/2* | *2T/3* | *5T/6* | *T* |
|---|---|---|---|---|---|---|---|
| *A amp.* | *1.98* | *1.30* | *1.05* | *1.30* | *–0.88* | *–0.25* | *1.98* |

*Show that there is a direct current part of 0.75 amp in the variable current and obtain the amplitude of the first harmonic.* (*Andhra, 2015 ; Delhi, 2015*)

**Solution.** Here length of the interval is $T$, *i.e.* $C = T/2$ (§ 10.5)

Then $$A = \frac{a_0}{2} + a_1 \cos\frac{2\pi t}{T} + b_1 \sin\frac{2\pi t}{T} + a_2 \cos\frac{4\pi t}{T} + b_2 \sin\frac{4\pi t}{T} + \ldots$$

The desired values are tabulated as follows :

| $t$ | $2\pi t/T$ | $\cos 2\pi t/T$ | $\sin 2\pi t/T$ | $A$ | $A \cos 2\pi t/T$ | $A \sin 2\pi t/T$ |
|---|---|---|---|---|---|---|
| 0 | 0 | 1.0 | 0.000 | 1.98 | 1.980 | 0.000 |
| $T/6$ | $\pi/3$ | 0.5 | 0.866 | 1.30 | 0.650 | 1.126 |
| $T/3$ | $2\pi/3$ | –0.5 | 0.866 | 1.05 | –0.525 | 0.909 |
| $T/2$ | $\pi$ | –1.0 | 0.000 | 1.30 | –1.300 | 0.000 |
| $2T/3$ | $4\pi/3$ | –0.5 | –0.866 | –0.88 | 0.440 | 0.762 |
| $5T/6$ | $5\pi/3$ | 0.5 | –0.866 | –0.25 | –0.125 | 0.217 |
| | | | $\Sigma =$ | 4.5 | 1.12 | 3.014 |

$$\therefore \quad a_0 = 2 \cdot \frac{1}{6}\Sigma A = \frac{1}{3}(4.5) = 1.5$$

$$a_1 = 2 \cdot \frac{1}{6}\Sigma A \cos\frac{2\pi t}{T} = \frac{1}{3}(1.12) = 0.373$$

$$b_1 = 2 \cdot \frac{1}{6}\Sigma A \sin\frac{2\pi t}{T} = \frac{1}{3}(3.014) = 1.005$$

Thus the direct current part in the variable current = $a_0/2 = 0.75$ and amplitude of the first harmonic

$$= \sqrt{(a_1^2 + b_1^2)} = \sqrt{[(0.373)^2 + (1.005)^2]} = 1.072.$$

**Example 10.37.** *Obtain the first three coefficients in the Fourier cosine series for y, where y is given in the following table :*

| $x$ : | 0 | 1 | 2 | 3 | 4 | 5 |
|---|---|---|---|---|---|---|
| $y$ : | 4 | 8 | 15 | 7 | 6 | 2 |

*(Anna, 2016)*

**Solution.** Taking the interval as 60°, we have

| $\theta =$ | 0° | 60° | 120° | 180° | 240° | 300° |
|---|---|---|---|---|---|---|
| $x =$ | 0 | 1 | 2 | 3 | 4 | 5 |
| $y =$ | 4 | 8 | 15 | 7 | 6 | 2 |

$\therefore$ Fourier cosine series in the intervals $(0, 2\pi)$ is

$$y = \frac{a_0}{2} + a_1 \cos\theta + a_2 \cos 2\theta + a_3 \cos 3\theta + \ldots$$

| $\theta°$ | $\cos\theta$ | $\cos 2\theta$ | $\cos 3\theta$ | $y$ | $y\cos\theta$ | $y\cos 2\theta$ | $y\cos 3\theta$ |
|---|---|---|---|---|---|---|---|
| 0° | 1 | 1 | 1 | 4 | 4 | 4 | 4 |
| 60° | $\frac{1}{2}$ | $-\frac{1}{2}$ | –1 | 8 | 4 | –4 | –8 |
| 120° | $-\frac{1}{2}$ | $-\frac{1}{2}$ | 1 | 15 | –7.5 | –7.5 | 15 |
| 180° | –1 | 1 | –1 | 7 | –7 | 7 | –7 |
| 240° | $-\frac{1}{2}$ | $-\frac{1}{2}$ | 1 | 6 | –3 | –3 | 6 |
| 300° | $\frac{1}{2}$ | $-\frac{1}{2}$ | –1 | 2 | 1 | –1 | –2 |
| | | | $\Sigma =$ | 42 | –8.5 | –4.5 | 8 |

Hence $$a_0 = 2 \cdot \frac{42}{6} = 14,\ a_1 = 2\left(\frac{-8.5}{6}\right) = -2.8,\ a_2 = 2\left(\frac{-4.5}{6}\right) = -1.5,$$

$$a_3 = 2\left(\frac{8}{6}\right) = 2.7.$$

**Example 10.38.** *A circuit analyse the current i (amp) given by the table below into its constituent harmonic upto to third.*

| θ° | 0 | 30 | 60 | 90 | 120 | 150 | 180 | 210 | 240 | 270 | 300 | 330 |
|---|---|---|---|---|---|---|---|---|---|---|---|---|
| i° | 0 | 24 | 33 · 5 | 27 · 5 | 18 · 5 | 13 | 0 | –24 | –33.5 | –27 · 5 | –18 · 2 | –13 |

*Find the current flowing.* (Bangalore, 2019)

**Solution:** Since the values of $i$ reverse in sign at $\theta = \pi$, only the odd harmonies $a_1, a_2, a_5, ...., b_1, b_3, b_5$.

Let $i = a_1 \cos\theta + b_1 \sin\theta + a_3 \cos 3\theta + b_3 \sin 3\theta$.

Now we prepare the table

| $\theta$ | $i$ | $\cos\theta$ | $\sin\theta$ | $\cos 3\theta$ | $\sin 3\theta$ |
|---|---|---|---|---|---|
| 0 | 0 | 1 | 0 | 1 | 0 |
| 30° | 24° | 0.8666 | 0.5 | 0 | 1 |
| 60° | 33.5 | 0.5 | 0.866 | –1 | 0 |
| 90° | 27.5 | 0 | 1 | 0 | –1 |
| 120° | 18.2 | –0.5 | 0.866 | 1 | 0 |
| 150° | 13 | –0.866 | 0.5 | 0 | 1 |

$$a_1 = \frac{2}{n}\sum i\cos\theta = \frac{2}{6}[0\cdot866(24-13)+0\cdot5(33\cdot5-18\cdot2)]$$

$$= \frac{1}{3}(9\cdot526+7\cdot65) = 5\cdot7253$$

$$b_1 = \frac{2}{n}\sum i\sin\theta = \frac{2}{6}[0.5(24+13)+0.866(33.5+18.2)+27.5]$$

$$= \frac{1}{3}(90\cdot77) = 30\cdot26$$

$$a_3 = \frac{2}{n}\sum i\cos 3\theta = \frac{2}{6}[-33\cdot5+18\cdot2] = -5\cdot1$$

$$b_3 = \frac{2}{n}\sum i\sin 3\theta = \frac{2}{6}[24-27\cdot5+13] = 3\cdot17$$

$$\therefore \quad i = 5.7253\cos\theta + 30.26\sin\theta - 5.1\cos 3\theta + 3.17\sin 3\theta.$$

**Example 10.39:** *In a machine the displacement y of a given point for a certain angle θ is observed is follows:*

| θ° : 0 | 30 | 60 | 90 | 120 | 150 | 180 | 210 | 240 | 270 | 300 | 330 | |
|---|---|---|---|---|---|---|---|---|---|---|---|---|
| y : 7.9 | 8 | 7 · 2 | 5 · 6 | 3 · 6 | 1 · 7 | 0 · 5 | 0 · 2 | 0 · 9 | 2 · 5 | 4 · 7 | 6 · 8 | |

*Compute the coefficient of sin 2θ in the Fourier series representing the above variation.* (Rajasthan, 2018)

**Solution:** Tabulating the values of $\sin 2\theta$, for the given values of $\theta$, we have

$\sin 2\theta$ : 0 | 0 · 866 | 0 · 866 | 0 | –0 · 866 | –0 · 866 | 0 | 0 · 866 | 0 · 866 | 0 | –0 · 866 | –0 · 866

$$\therefore \quad = \frac{2}{n}\sum y\sin 2x$$

$$= \frac{1}{12}[0{\cdot}866(8+0{\cdot}72-3{\cdot}6-1{\cdot}7+0{\cdot}2+0{\cdot}9-4{\cdot}7-6{\cdot}8]$$

$$= \frac{1}{6}\times 0\cdot866\,(16\cdot3-16\cdot8) = -0.072.$$

**Example 10.40:** *The following table gives the variations of a periodic current over a period.*

| $t$(sec) : | 0 | $\frac{T}{6}$ | $\frac{T}{3}$ | $\frac{T}{2}$ | $\frac{2T}{2}$ | $\frac{5T}{6}$ | $T$ |
|---|---|---|---|---|---|---|---|
| $A(amp)$ : | 1.98 | 1.30 | 1.05 | 1.30 | –0.88 | –0.25 | 1.98 |

*Show by numerical analysis, that there is a direct current part of 0.75 amp in the variable current and obtain the amplitude of the first harmonic.* (*Nagarjuna, 2019*)

**Solution:** Let $A = \frac{1}{2}a_0 + a_1 \cos \frac{2\pi t}{T} + b_n \sin \frac{2\pi t}{T} + ....$ where, $1/2a_0$ represents the direct current part and $\sqrt{a_1^2 + b_1^2}$ gives the amplitude of the first harmonic.

| $t$ : | 0 | $T/6$ | $T/3$ | $T/2$ | $2T/3$ | $5T/6$ |
|---|---|---|---|---|---|---|
| $A$ : | 1.98 | 1.3 | 1.05 | 1.3 | –0.88 | –0.25 |
| $\cos\left(\frac{2\pi t}{T}\right)$ : | 1 | 0.5 | –0.5 | –1 | –0.5 | 0.5 |
| $\sin\left(\frac{2\pi t}{T}\right)$ : | 0 | 0.866 | 0.866 | 0 | –0.866 | –0.866 |

Now with the help of the table, we have

$$a_0 = \frac{2}{n}\sum A = \frac{2}{6}[1.98 + 1.3 + 1.05 + 1.3 - 0.88 - 0.25]$$

$$= \frac{1}{3}(5.63 - 1.13) = \frac{1}{3}(4.5) = 1.50$$

$$a_1 = \frac{2}{n}\sum A \cos\left(\frac{2\pi t}{T}\right)$$

$$= \frac{2}{6}[(1.98 - 1.3) + 0.5(1.3 - 1.05 + 0.88 - 0.25)]$$

$$= \frac{1}{3}(0.68 + 0.5 \times 0.88) = \frac{1}{3}(1.12) = 0.373$$

$$b_1 = \frac{2}{n}\sum A \sin\left(\frac{2\pi t}{T}\right) = \frac{2}{6}[0 + 0.866(1.3 + 1.05 + 0.88 + 0.25)]$$

$$= \frac{1}{3} \times 0.866 \times 3.48 = 1.005$$

$$\therefore \quad A = 0.75 + 0.373 \cos\left(\frac{2\pi t}{T}\right) + 1.005 \times \sin\left(\frac{2\pi t}{T}\right) + .....$$

Hence $A$ has a direct current part of 0.75 amp.
The amplitude of the first harmonic is given by

$$\sqrt{(0.373)^2 + (1.005)^2} = \sqrt{1.1491} = 1.072$$

**Example 10.41:** *The turning moment T units of the crank shaft of a steam engine is given for a series of values of the crank angle θ in degrees.* (*Marathwada, 2019*)

| θ : | 0° | 30° | 60° | 90° | 120° | 150° | 180° |
|---|---|---|---|---|---|---|---|
| $T$ : | 0 | 5224 | 8097 | 7850 | 5499 | 2626 | 0 |

*Find the first four terms in a series of sines to represent T. Also calculate T when θ = 75°.*

**Solution:** Since the sine series represent on odd periodic function. We have to obtain half range sine series

$T = b_1 \sin \theta + b_2 \sin 2\theta + b_3 \sin 3\theta + b_4 \sin 4\theta$

To obtain $b_1, b_2, b_3, b_4$.

| θ° | i | cos θ | sin θ | cos 3θ | sin 3θ |
|---|---|---|---|---|---|
| 0° | 0 | 0 | 0 | 0 | 0 |
| 30° | 5224 | 0.5 | 0.866 | 1 | 0.866 |
| 60° | 8097 | 0.866 | 0.866 | 0 | –0.866 |
| 90° | 7850 | 1 | 0 | –1 | 0 |
| 120° | 8499 | 0.866 | –0.866 | 0 | 0.866 |
| 150° | 2626 | 0.5 | –0.866 | 1 | –0.866 |

Now,
$$b_1 = \frac{2}{n}\sum y \sin\theta$$
$$= \frac{2}{6}[0.5(5224 + 2626) + 0.866(8097 + 5499) + 7850]$$
$$= \frac{1}{3}[3925 + 11774 + 7850] = \frac{1}{3}(23549) = 7850$$
$$b_2 = \frac{2}{n}\sum y \sin 2\theta = \frac{1}{3}[0.866(5224 + 8097 - 5499 - 2626]$$
$$= \frac{1}{3} \times 0.866 \times 5196 = 1500$$
$$b_3 = \frac{2}{n}\sum y \sin 3\theta = \frac{2}{6}(5224 - 7850 + 2626) = 0$$
$$b_4 = \frac{2}{n}\sum y \sin 4\theta = \frac{2}{6}[0.866(5224 - 8097) + 5499 - 2626] = 0$$

∴
$$T = 7850 \sin\theta + 1500 \sin 2\theta$$
$$= \theta = 75°, \text{ then } T = 7850 \sin 75° + 1500 \sin 150°$$
$$= 7850 \times 0.9659 + 1500 \times 0.5 = 8332.$$

## PROBLEMS 10.9

1. The turning moment $T$ on the crank-shaft of a steam engine for the crank angle θ (degress) is recorded as follows:

| θ | 0° | 15° | 30° | 45° | 60° | 75° | 90° | 105° | 120° | 135° | 150° | 165° | 180° |
|---|---|---|---|---|---|---|---|---|---|---|---|---|---|
| T | 0 | 2.7 | 5.2 | 7 | 8.1 | 8.3 | 7.9 | 6.8 | 6.6 | 4.1 | 2.6 | 1.2 | 0 |

Expand $T$ is a series of sines up to second harmonics. *(UTU, 2019)*

2. Compute the first two harmonics of the Fourier series of $f(x)$ given in the following table :

| $x$ : | 0 | π/3 | 2π/3 | π | 4π/3 | 5π/3 | 2π |
|---|---|---|---|---|---|---|---|
| $f(x)$ : | 1.0 | 1.4 | 1.9 | 1.7 | 1.5 | 1.2 | 1.0 |

*(Anna, 2015)*

3. Obtain the constant term and the coefficients of the first sine and cosine terms in the Fourier expansion of $y$ as given in the following table :

| $x$ : | 0 | 1 | 2 | 3 | 4 | 5 |
|---|---|---|---|---|---|---|
| $y$ : | 9 | 18 | 24 | 28 | 26 | 20 |

*(V.T.U., 2016 ; Anna, 2012 ; Delhi, 2012 ; U.P.T.U., 2012)*

4. In a machine the displacement $y$ of a given point is given for a certain angle $\theta$ as follows :

| $\theta°$ : | 0 | 30 | 60 | 90 | 120 | 150 | 180 | 210 | 240 | 270 | 300 | 330 |
|---|---|---|---|---|---|---|---|---|---|---|---|---|
| $y$ : | 7.9 | 8.0 | 7.2 | 5.6 | 3.6 | 1.7 | 0.5 | 0.2 | 0.9 | 2.5 | 4.7 | 6.8 |

Find the coefficient of sin $2\theta$ in the Fourier series representing the above variation. *(V.T.U., 2015 S)*

5. Obtain $a_0$, $a_1$ and $b_1$ in the Fourier expansion of $y$ from the following table

| $x$ : | 0 | 1 | 2 | 3 | 4 | 5 |
|---|---|---|---|---|---|---|
| $y$ : | 9 | 18 | 24 | 28 | 26 | 20 |

*(Ranchi, 2019)*

6. Find the Fourier series up to third harmonic in given table.

| $x$ | 0° | 30° | 60° | 90° | 120° | 150° | 180° | 210° | 240° | 270° | 300° | 330° |
|---|---|---|---|---|---|---|---|---|---|---|---|---|
| $y$ | 25 | 40 | 50.5 | 57.5 | 61.5 | 63.3 | 68.2 | 59.2 | 52.2 | 44.2 | 35.8 | 28.7 |

*(M.E. Annamalai, 2019)*

7. A part of a machine has an osscillatory motion. The displacement $y$ at a time $t$ is given below:
The displacement $y$ at a time $t$ is given below:

| $t$ : | 0.02 | 0.04 | 0.06 | 0.08 | 0.1 | 0.12 | 0.14 | 0.16 | 0.18 | 0.2 |
|---|---|---|---|---|---|---|---|---|---|---|
| $y$ : | 0.64 | 1.13 | 1.34 | 0.95 | 0.0 | –0.92 | –1.33 | –1.166 | –0.66 | 0 |

Find the constant in the equation

$$y = A \sin (10 \pi t + \alpha_1) + \beta \sin (20\pi t + \alpha_2)$$

8. Find the Fourier series upto third hormonic

| $x$ | 0 | 30 | 60 | 90 | 120 | 150 | 180 | 210 | 240 | 270 | 300 | 330 |
|---|---|---|---|---|---|---|---|---|---|---|---|---|
| $y$ | 1.8 | 1.1 | 0.3 | 0.16 | 0.5 | 1.3 | 2.16 | 1.25 | 1.3 | 1.52 | 1.76 | 2 |

## 10.12 OBJECTIVE TYPE OF QUESTIONS

### PROBLEMS 10.10

*Fill up the blanks or choose the correct answer in each of the following problems :*

1. The period of cos $3x$ is $x$ = ......
2. If $x = c$ is a point of discontinuity then the Fourier series of $f(x)$ at $x = c$ gives $f(x)$ = ......
3. A function $f(x)$ defined for $0 < x < 1$ can be extended to an odd periodic function in ......
4. The mathematical function representing the following graph is ......
5. Fourier expansion of an odd function has only ...... terms.
6. Formulae for evaluation of Fourier coefficients for a given set of points $(x_i, y_i)$ : $i = 0, 1, 2, ...... n$ are......
7. If $f(x) = x^4$ in $(-1, 1)$, then the Fourier coefficient $b_n$ = ......
8. The period of a constant function is ......
9. If $f(t) = \begin{cases} -1, & -1 < t < 0 \\ 1, & 0 < t < 1 \end{cases}$, then $f(t)$ is an ......
10. Fourier expansion of an even function $f(x)$ in $(-\pi, \pi)$ has only ...... terms.
11. If $f(x) = \begin{cases} -x, & -\pi < x < 0 \\ x, & 0 < x < \pi \end{cases}$, then $f(x)$ is an ...... function in $(-\pi, \pi)$.
12. The smallest period of the function $\sin\left(\dfrac{2n\pi x}{k}\right)$ is ......
13. In the Fourier series expansion of $f(x) = |\sin x|$ in $(-\pi, \pi)$, the value of $b_n$ = ......
14. In the Fourier series for $f(x) = x$ in $(-\pi \le x \le \pi)$, the ...... terms are absent.

15. If $f(x)$ is an even function in $(-l, l)$, then the value of $b_n$ = ......
16. If $f(x) = x^2$ in $-2 < x < 2$, $f(x + 4) = f(x)$, then $a_n$ is ......
17. If $f(x)$ is a periodic function with period $2T$, then the value of the Fourier coefficient $b_n$ = ......
18. Dirichlet conditions for the expansion of a function as a Fourier series in the interval $c_1 \le x \le c_2$ are ......
19. If $f(x) = x \sin x$ in $(-\pi, \pi)$, then the value of $b_n$ = ......
20. The formulae for finding the half range cosine series for the function $f(x)$ in $(0, l)$ are ......
21. The half-range sine series for 1 in $(0, \pi)$, is ......
22. Period of $|\sin t|$ is ......
23. The value of $bn$ in the Fourier series of $f(x) = |x|$ in $(-\pi, \pi)$ = ......
24. If $f(x)$ is defined in $(0, l)$ then the period of $f(x)$ to expand it as a half range sine series is ......
25. The complex form of Fourier series for $e^{-x}$ in $(-1, 1)$ is ......
26. $f(x)$ is an odd function in $(-\pi, \pi)$, then the graph of $f(x)$ is symmetric about the $x$-axis. (True or False)
27. $f(x) = \begin{cases} -\pi, & -\pi < x < 0 \\ x, & 0 < x < \pi, \end{cases}$ then $f(0)$ = ......
28. If $f(x) = \begin{cases} \pi x \text{ in } & 0 \le x \le 1 \\ \pi(2 - x) \text{ in } & 1 \le x \le 2, \end{cases}$ then it is ...... function. (odd or even)
29. If $f(x)$ is an odd function in $(-l, l)$, then the values of $a_0$ and $a_n$ are......
30. The root mean square value of $f(t) = 3 \sin 2t + 4 \cos 2t$ over the range $0 \le t \le \pi$ is......
31. In the Fourier series expansion of the function

$$f(x) = \begin{cases} -(\pi + x), & -\pi < x < 0 \\ -(\pi - x), & 0 < x < \pi, \end{cases}$$ the value of $b_n$ is ...... *(P.T.U., 2010)*

32. Let $f(x)$ be defined in $(0, 2\pi)$ by

$$f(t) = \begin{cases} \dfrac{1 + \cos x}{\pi - x}, 0 < x < \pi \\ \cos x, \pi < x < 2\pi, \end{cases}$$ $f(x) + 2\pi = f(x)$. The value of $f(\pi)$ is ......

33. The mean value of $f(x) \cos nx$ in $(0, 2\pi)$ = ......
34. Using sine series for $f(x) = 1$ in $0 < x < \pi$, show that $1 + \frac{1}{3^2} + \frac{1}{5^2} + \frac{1}{7^2} + \ldots \infty$ = .....
35. Fourier series representing $f(x) = |x|$ in $-\pi < x < \pi$, is ......
36. Fourier series of $f(x) = \cos^4 x$ in $(0, 2\pi)$ is ......
37. If $f(x) = x^2 + x$ in $(0, l)$, then the even extension of $f(x)$ in $(-l, 0)$ is ......
38. If $f(x) = x(l - x)$ in $(0, l)$, then the extension of $f(x)$ in $(l, 2l)$ so as to get sine series is ......
39. A function $f(x)$ defined in $(-\pi, \pi)$ can be expanded into Fourier series containing both sine and cosine terms. (True or False)
40. The function $f(x) = \begin{cases} 1 - x \text{ in } & -\pi < x < 0 \\ 1 + x \text{ in } & 0 < x < \pi, \end{cases}$ is an odd function. (True or False)
41. If $f(x) = x^2$ in $(-\pi, \pi)$, then the Fourier series of $f(x)$ contains only sine terms. (True or False)
42. $\sin(1/x)$ can be expanded into Fourier series in $(-\pi, \pi)$ (True or False) *(Delhi, 2015)*

CHAPTER 11

# Differential Equations of First Order

**1.** Definitions. **2.** Practical approach to differential equations. **3.** Formation of a differential equation. **4.** Solution of a differential equation—Geometrical meaning. **5.** Equations of the first order and first degree. **6.** Variables separable. **7.** Homogeneous equations. **8.** Equations reducible to homogeneous form. **9.** Linear equations. **10.** Bernoulli's equation. **11.** Exact equations. **12.** Equations reducible to exact equations. **13.** Equations of the first order and higher degree. **14.** Clairut's equation. **15.** Objective Type of Questions.

## 11.1 DEFINITIONS

**(1)** *A* **differential equation** *is an equation which involves differential coefficients or differentials.*

Thus (*i*) $e^x\,dx + e^y\,dy = 0$ (*ii*) $\dfrac{d^2x}{dt^2} + n^2x = 0$

(*iii*) $y = x\dfrac{dy}{dx} + \dfrac{x}{dy/dx}$ (*iv*) $\left[1+\left(\dfrac{dy}{dx}\right)^2\right]^{3/2} \Big/ \dfrac{d^2y}{dx^2} = c$

(*v*) $\dfrac{dx}{dt} - wy = a\cos pt,\ \dfrac{dy}{dt} + wx = a\sin pt$ (*vi*) $x\dfrac{\partial u}{\partial x} + y\dfrac{\partial u}{\partial y} = 2u$

(*vii*) $\dfrac{\partial^2 y}{\partial t^2} = c^2\dfrac{\partial^2 y}{\partial x^2}$ are all examples of differential equations.

**(2)** *An* **ordinary differential equation** *is that in which all the differential coefficients have reference to a single independent variable.* Thus the equations (*i*) to (*v*) are all ordinary differential equations.

*A* **partial differential equation** *is that in which there are two or more independent variables and partial differential coefficients with respect to any of them.* Thus the equations (*vi*) and (*vii*) are partial differential equations.

**(3)** *The* **order** *of a differential equation is the order of the highest derivative appearing in it.*

*The* **degree** *of a differential equation is the degree of the highest derivative occurring in it, after the equation has been expressed in a form free from radicals and fractions as far as the derivatives are concerned.*

Thus, from the examples above,

(*i*) is of the first order and first degree ; (*ii*) is of the second order and first degree ;

(*iii*) written as $y\dfrac{dy}{dx} = x\left(\dfrac{dy}{dx}\right)^2 + x$ is clearly of the first order but of second degree ;

and (*iv*) written as $\left[1+\left(\dfrac{dy}{dx}\right)^2\right]^3 = c^2\left(\dfrac{d^2y}{dx^2}\right)^2$ is of the second order and second degree.

## 11.2 PRACTICAL APPROACH TO DIFFERENTIAL EQUATIONS

Differential equations arise from many problems in oscillations of mechanical and electrical systems, bending of beams, conduction of heat, velocity of chemical reactions etc., and as such play a very important role in all modern scientific and engineering studies.

The approach of an engineering student to the study of differential equations has got to be practical unlike that of a student of mathematics, who is only interested in solving the differential equations without knowing as to how the differential equations are formed and how their solutions are physically interpreted.

*Thus for an applied mathematician, the study of a differential equation consists of three phases :*

(*i*) *formulation of differential equation from the given physical situation, called* **modelling.**

(*ii*) *solutions of this differential equation, evaluating the arbitrary constants from the given conditions,* and

(*iii*) *physical interpretation of the solution.*

## 11.3 FORMATION OF A DIFFERENTIAL EQUATION

An ordinary differential equation is formed in an attempt to eliminate certain arbitrary constant from a relation in the variables and constants. It will, however, be seen later that the partial differential equations may be formed by the elimination of either arbitrary constants or arbitrary functions. In applied mathematics, every geometrical or physical problem when translated into mathematical symbols gives rise to a differential equation.

**Example 11.1.** *Form the differential equation of simple harmonic motion given by* $x = A \cos (nt + \alpha)$.

**Solution.** To eliminate the constants $A$ and $\alpha$ differentiating it twice, we have

$$\frac{dx}{dt} = -nA \sin (nt + \alpha) \quad \text{and} \quad \frac{d^2x}{dt^2} = -n^2A \cos (nt + \alpha) = -n^2x$$

Thus $$\frac{d^2x}{dt^2} + n^2x = 0$$

is the desired differential equation which states that the acceleration varies as the distance from the origin.

**Example 11.2.** *Obtain the differential equation of all circles of radius a and centre* $(h, k)$.

**Solution.** Such a circle is $(x - h)^2 + (y - k)^2 = a^2$ ...(*i*)

where $h$ and $k$, the coordinates of the centre, and $a$ are the constants.

Differential it twice, we have

$$x - h + (y - k)\frac{dy}{dx} = 0 \quad \text{and} \quad 1 + (y - k)\frac{d^2y}{dx^2} + \left(\frac{dy}{dx}\right)^2 = 0$$

Then $$y - k = -\frac{1 + (dy/dx)^2}{d^2y/dx^2}$$

and $$x - h = -(y - k)\, dy/dx = \frac{\frac{dy}{dx}\left[1 + \left(\frac{dy}{dx}\right)^2\right]}{d^2y/dx^2}$$

Substituting these in Eqn. (*i*) and simplifying, we get $[1 + (dy/dx)^2]^3 = a^2 (d^2y/dx^2)^2$ ...(*ii*)

as the required differential equation

Writing Eqn. (*ii*) in the form $\dfrac{[1 + (dy/dx)^2]^{3/2}}{d^2y/dx^2} = a$,

it states that the radius of curvature of a circle at any point is constant.

**Example 11.3.** *Obtain the differential equation of the coaxial circles of the system* $x^2 + y^2 + 2ax + c^2 = 0$ *where c is a constant and a is a variable.*

**Solution.** We have $x^2 + y^2 + 2ax + c^2 = 0$ ...(*i*)

Differentiating w.r.t. $x$, $2x + 2y dy/dx + 2a = 0$

or
$$2a = -2\left(x + y\frac{dy}{dx}\right)$$

Substituting in Eqn. (*i*), $x^2 + y^2 - 2(x + y\, dy/dx)x + c^2 = 0$

or
$$2xy\, dy/dx = y^2 - x^2 + c^2$$

which is the required differential equation.

## 11.4 (1) SOLUTION OF A DIFFERENTIAL EQUATION

*A solution (or integral) of a differential equation is a relation between the variables which satisfies the given differential equation.*

For example,
$$x = A\cos(nt + \alpha) \quad ...(1)$$

is a solution of
$$\frac{d^2x}{dt^2} + n^2x = 0 \text{ [Example 11.1]} \quad ...(2)$$

The **general** (*or* **complete**) **solution** *of a differential equation is that in which the number of arbitrary constants is equal to the order of the differential equation.* Thus Eqn. (1) is a general solution Eqn. (2) as the number of arbitrary constants ($A$, $\alpha$) is the same as the order of Eqn. (2).

A **particular solution** *is that which can be obtained from the general solution by giving particular values to the arbitrary constants.*

For example,
$$x = A\cos(nt + \pi/4)$$

is the particular solution of the Eqn. (2) as it can be derived from the general solution Eqn. (1) by putting $\alpha = \pi/4$.

A differential equation may sometimes have an additional solution which cannot be obtained from the general solution by assigning a particular value to the arbitrary constant. Such a solution is called a **singular solution** and is not of much engineering interest.

**Linearly independent solution.** *Two solutions $y_1(x)$ and $y_2(x)$ of the differential equation*
$$\frac{d^2y}{dx^2} + a_1(x)\frac{dy}{dx} + a_2(x)\, y = 0 \quad ...(3)$$

*are said to be linearly independent if $c_1y_1 + c_2y_2 = 0$ such that $c_1 = 0$ and $c_2 = 0$*

If $c_1$ and $c_2$ are not both zero, then the two solutions $y_1$ and $y_2$ are said to be *linearly dependent.*

*If $y_1(x)$ and $y_2(x)$ any two solutions of Eqn. (3), then their linear combination $c_1y_1 + c_2y_2$ where $c_1$ and $c_2$ are constants, is also a solution of Eqn. (3).*

**Example 11.4.** *Find the differential equation whose set of independent solutions is $[e^x, xe^x]$.*

**Solution.** Let the general solution of the required differential equation be $y = c_1e^x + c_2xe^x$ ...(*i*)

Differentiating Eqn. (*i*) w.r.t. $x$, we get
$$y_1 = c_1e^x + c_2(e^x + xe^x)$$
$$\therefore \quad y - y_1 = c_2e^x \quad ...(ii)$$

Again differentiating Eqn. (*ii*) w.r.t. $x$, we obtain
$$y_1 - y_2 = c_2e^x \quad ...(iii)$$

Subtracting Eqn. (*iii*) from Eqn. (*ii*), we get
$$y - y_1 - (y_1 - y_2) = 0 \quad \text{or} \quad y - 2y_1 + y_2 = 0$$

which is the desired differential equation.

**(2) Geometrical meaning of a differential equation.** Consider any differential equation of the first order and first degree
$$\frac{dy}{dx} = f(x, y) \quad ...(1)$$

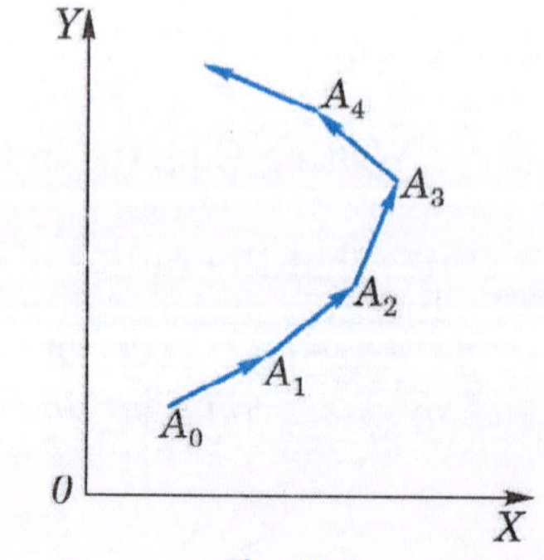

Fig. 11.1

If $P(x, y)$ be any point, then Eqn. (1) can be regarded as an equation giving the value of $dy/dx$ (= $m$) when the values of $x$ and $y$ are known (Fig. 11.1).

Let the value of $m$ at the point $A_0(x_0, y_0)$ derived from Eqn. (1) be $m_0$. Take a neighbouring point $A_1(x_1, y_1)$ such that the slope of $A_0A_1$ is $m_0$. Let the corresponding value of $m$ at $A_1$ be $m_1$. Similarly take a neighbouring point $A_2(x_2, y_2)$ such that the slope of $A_1A_2$ is $m_1$ and so on.

If the successive points $A_0, A_1, A_2, A_3$ ... are chosen very near one another, the broken curve $A_0A_1A_2A_3$ ... approximates to a smooth curve $C[y = \phi(x)]$ which is a solution of Eqn. (1) associated with the initial point $A_0$ $(x_0, y_0)$. Clearly the slope of the tangent to $C$ at any point and the coordinates of that point satisfy Eqn. (1).

A different choice of the initial point will, in general, give a different curve with the same property. *The equation of each such curve is thus a* **particular solution** *of the differential Eqn. (1). The equation of the whole family of such curves is the general solution of* Eqn. (1). The slope of the tangent at any point of each member of this family and the co-ordinates of that point satisfy Eqn. (1).

Such a simple geometric interpretation of the solutions of a second (or higher) order differential equation is not available.

## PROBLEMS 11.1

Form the differential equations from the following equations :

1. $y = ax^3 + bx^2$.
2. $x = a \sin (at + b)$ *(Andhra, 2016)*
3. $y = C_1 \cos 2x + C_2 \sin 2x$
4. $xy = Ae^x + Be^{-x} + x^2$. *(U.P.T.U., 2011)*
5. $y = e^x (A \cos x + B \sin x)$.
6. $y = ae^{2x} + be^{-3x} + ce^x$.

Find the differential equations of :

7. A family of circles passing through the origin and having centres on the $x$-axis.
8. All circles of radius 5, with their centres on the $y$-axis.
9. All parabolas with $x$-axis as the axis and $(a, 0)$ as focus.
10. If $y_1(x) = \sin 2x$ and $y_2(x) = \cos 2x$ are two solutions of $y'' + 4y = 0$, show that $y_1(x)$ and $y_2(x)$ are linearly independent solutions.
11. Determine the differential equation whose set of independent solutions is $[e^x, xe^x, x^2 e^x]$

## 11.5 EQUATIONS OF THE FIRST ORDER AND FIRST DEGREE

It is not possible to solve such equations in general. We shall, however, discuss some special methods of solution which are applied to the following types of equations :

(*i*) Equations where variables are separable, (*ii*) Homogeneous equations,

(*iii*) Linear equations, (*iv*) Exact equations.

In other cases, the particular solution may be determined numerically (Chapter 31).

## 11.6 VARIABLES SEPARABLE

If in an equation it is possible to collect all functions of $x$ and $dx$ on one side and all the functions of $y$ and $dy$ on the other side, then the *variables are said to be separable.* Thus the general form of such an equation is $f(y)\, dy = \phi(x)\, dx$

Integrating both sides, we get $\int f(y)\, dy = \int \phi(x)\, dx + c$ as its solution.

**Example 11.5.** *Solve* $dy/dx = \dfrac{x(2 \log x + 1)}{\sin y + y \cos y}$. *(V.T.U., 2012)*

**Solution.** Given equation is $x (2 \log x + 1)\, dx = (\sin y + y \cos y)\, dy$

Integrating both sides, $2\int (\log x \,.\, x + x)\, dx = \int \sin y\, dy + \int y \cos y\, dy + c$

or $$2\left[\left(\log x \,.\, \frac{x^2}{2} - \int \frac{1}{x} \cdot \frac{x^2}{2}\, dx\right) + \frac{x^2}{2}\right] = -\cos y + \left[y \sin y - \int \sin y \,.\, 1\, dy + c\right]$$

or $$2x^2 \log x - \frac{x^2}{2} + \frac{x^2}{2} = -\cos y + y \sin y + \cos y + c$$

Hence the solution is $2x^2 \log x - y \sin y = c$.

**Example 11.6.** *Solve* $\frac{dy}{dx} = e^{3x-2y} + x^2 e^{-2y}$.

**Solution.** Given equation is $\frac{dy}{dx} = e^{-2y}(e^{3x} + x^2)$ or $e^{2y}\,dy = (e^{3x} + x^2)\,dx$

Integrating both sides, $\int e^{2y}\,dy = \int (e^{3x} + x^2)\,dx + c$

or $$\frac{e^{2y}}{2} = \frac{e^{3x}}{3} + \frac{x^3}{3} + c \quad \text{or} \quad 3e^{2y} = 2(e^{3x} + x^3) + 6c.$$

**Example 11.7.** *Solve* $(x + y + 1)^2 \frac{dy}{dx} = 1$. *(Marathwada, 2019)*

**Solution:** Given equation is

$$(x + y + 1)^2 \frac{dy}{dx} = 1 \qquad \ldots(i)$$

$$x + y + 1 = z$$

$$1 + \frac{dy}{dx} = \frac{dz}{dx}$$

$$\frac{dy}{dx} = \frac{dz}{dx} - 1$$

Substituting these value in Eqn. $(i)$

$$z^2\left(\frac{dz}{dx} - 1\right) = 1$$

$$\frac{dz}{dx} = 1 + \frac{1}{z^2} = \frac{z^2 + 1}{z^2}$$

$$\frac{z^2}{z^2 + 1}\,dz = dx$$

Integrating both side, we have

$$\int \frac{z^2}{z^2 + 1}\,dz = \int dx + C$$

$$z - \tan^{-1} z = x + c$$

Substituting the value of $z$, the required solution is

$$(x + y + 1) - \tan^{-1}(x + y + z) = x + c$$

$$y = \tan^{-1}(x + y + z) + c - 1$$

**Example 11.8.** *Solve* $dy/dx = (4x + y + 1)^2$, *if* $y(0) = 1$.

**Solution.** Putting $4x + y + 1 = t$, we get $\frac{dy}{dx} = \frac{dt}{dx} - 4$.

$\therefore$ the given equation becomes $\frac{dt}{dx} - 4 = t^2$ or $\frac{dt}{dx} = 4 + t^2$

Integrating both sides, we get $\int \frac{dt}{4 + t^2} = \int dx + c$

or $\qquad \dfrac{1}{2}\tan^{-1}\dfrac{t}{2} = x + c \quad$ or $\quad \dfrac{1}{2}\tan^{-1}\left[\dfrac{1}{2}(4x + y + 1)\right] = x + c.$

or $\qquad 4x + y + 1 = 2\tan 2(x + c)$

When $x = 0, y = 1 \quad \therefore \quad \dfrac{1}{2}\tan^{-1}(1) = c$ *i.e.* $c = \pi/8$.

Hence the solution is $4x + y + 1 = 2\tan(2x + \pi/4)$.

**Example 11.9.** *Solve* $\dfrac{y}{x}\dfrac{dy}{dx} + \dfrac{x^2 + y^2 - 1}{2(x^2 + y^2) + 1} = 0.$ *(Calicut, 2013)*

**Solution.** Putting $x^2 + y^2 = t$, we get $2x + 2y\dfrac{dy}{dx} = \dfrac{dt}{dx}$ or $\dfrac{y}{x}\dfrac{dy}{dx} = \dfrac{1}{2x}\dfrac{dt}{dx} - 1.$

Therefore the given equation becomes $\dfrac{1}{2x}\dfrac{dt}{dx} - 1 + \dfrac{t-1}{2t+1} = 0$

or $\qquad \dfrac{1}{2x}\dfrac{dt}{dx} = 1 - \dfrac{t-1}{2t+1} = \dfrac{t+2}{2t+1} \quad$ or $\quad 2x\,dx = \dfrac{2t+1}{t+2}dt$

or $\qquad 2x\,dx = \left(2 - \dfrac{3}{t+2}\right)dt$

Integrating, we get $\qquad x^2 = 2t - 3\log(t + 2) + c$

or $\qquad x^2 + 2y^2 - 3\log(x^2 + y^2 + 2) + c = 0 \qquad [\because \; t = x^2 + y^2]$

which is the required solution.

**Example 11.10:** *Solve the differential equation* $\sin y' = x$.

**Solution:** Solving the given equation for $y'$, we get

$$y' = n\pi + (-1)^n \sin{}^{1}(x)$$

$$\frac{dy}{dx} = \pi n + ( 1)^n \sin^{-1} x$$

Separating the variables and integrating, we have

$$\int dy = \int\left[n\pi + (-1)^n \sin^{-1} x\right]dx + c$$

$$y = n\pi x + (-1)^n \int \sin^{-1} x\,dx + c \qquad \text{...}(i)$$

To evaluate $\int \sin^{-1} x\,dx$, we take 1 as second function

$$\int \sin^{-1} x\,dx = \int \sin^{-1} x \cdot 1 \cdot dx = \sin^{-1} x \cdot x - \int \frac{x}{\sqrt{1-x^2}}dx$$

$$= x\sin^{-1} x + \sqrt{1-x^2}$$

Substituting in Eqn. (*i*)

$$y = n\pi x + (-1)^n (x\sin^{-1} x + \sqrt{(1-x^2)}) + c$$

**Example 11.11:** *A particle falling in a medium whose resistance is proportional to the square of velocity. Satisfies the equation* $\dfrac{du}{dt} = g - ku^2$ *where k is constant and g the acceleration due to gravity. Given that when* $t = 0, u = 0$. *Solve the equation and show that the value of u will never exceed* $\sqrt{\dfrac{g}{k}}$. *(Kurukshetra, 2019)*

**Solution.** The equation of motion is

$$\frac{du}{dt} = g - ku^2 = k\left(\frac{g}{k} - v^2\right)$$

separating the variables and integrating, we get

$$\int \frac{du}{\frac{g}{k} - u^2} = k\int dt + c$$

$$\therefore \qquad \frac{1}{2}\sqrt{\frac{k}{g}} \log\left(\frac{\sqrt{\frac{g}{k}} + u}{\sqrt{\frac{g}{k}} - u}\right) = kt + c \qquad ...(i)$$

initially, when $t = 0, v = 0.$ $\therefore c = 0$ $(\because \log 1 = 0)$

Substituting $c = 0$ in Eqn. $(i)$ and simplifying

$$\frac{\sqrt{\frac{g}{k}} + u}{\frac{\sqrt{g}}{k} - u} = e^{2\sqrt{gk}\,t}$$

$$\frac{u}{\sqrt{g/k}} = \frac{e^{2\sqrt{gkt}} + 1}{e^{2\sqrt{gkt}} + 1}$$

Applying dividendo and componendo

$$u = \sqrt{g/k}\,\frac{\left(e^{2\sqrt{gkt}} - 1\right)}{\left(e^{2\sqrt{gkt}} + 1\right)}\,;\ u = \sqrt{\frac{g}{k}}\left(\frac{1 - e^{2\sqrt{gkt}}}{1 + e^{2\sqrt{gkt}}}\right)$$

Now as $t \to \infty$, $v = \sqrt{\frac{g}{k}}$

Hence $u$ can never exceed $\sqrt{\frac{g}{k}}$.

## PROBLEMS 11.2

Solve the following differential equations :

1. $y\sqrt{(1 - x^2)}\,dy + x\sqrt{(1 - y^2)}\,dx = 0.$
2. $(x^2 - yx^2)\dfrac{dy}{dx} + y^2 + xy^2 = 0.$
3. $\sec^2 x \tan y\, dx + \sec^2 y \tan x\, dy = 0.$
4. $\dfrac{y}{x}\dfrac{dy}{dx} = \sqrt{(1 + x^2 + y^2 + x^2y^2)}.$ (V.T.U., 2011)
5. $e^x \tan y\, dx + (1 - e^x)\sec^2 y\, dy = 0.$
6. $\dfrac{dy}{dx} = xe^{y - x^2}$, if $y = 0$ when $x = 0.$ (J.N.T.U., 2006)
7. $x\dfrac{dy}{dx} + \cot y = 0$ if $y = \pi/4$ when $x = \sqrt{2}.$
8. $e^x(\cos y\, dx - \sin y\, dy) = 0$, when $y(0) = 0$ (P.T.U., 2015)
9. $(xy^2 + x)\,dx + (yx^2 + y)\,dy = 0.$
10. $\dfrac{dy}{dx} = e^{2x - 3y} + 4x^2 e^{-3y}.$
11. $y - x\dfrac{dy}{dx} = a\left(y^2 + \dfrac{dy}{dx}\right).$
12. $(x + 1)\dfrac{dy}{dx} + 1 = 2e^{-y}.$
13. $(x - y)^2\dfrac{dy}{dx} = a^2.$
14. $(x + y + 1)^2\dfrac{dy}{dx} = 1.$
15. $\sin^{-1}(dy/dx) = x + y$ (V.T.U., 2010)
16. $\dfrac{dy}{dx} = \cos(x + y + 1)$
17. $\dfrac{dy}{dx} - x\tan(y - x) = 1.$
18. $x^4\dfrac{dy}{dx} + x^3y + \operatorname{cosec}(xy) = 0.$

## 11.7 HOMOGENEOUS EQUATIONS

are of the form $\frac{dy}{dx} = \frac{f(x, y)}{\phi(x, y)}$

where $f(x, y)$ and $\phi(x, y)$ are homogeneous functions of the same degree in $x$ and $y$ (see page 226).

*To solve a homogeneous Eqn.* (*i*) Put $y = vx$, then $\frac{dy}{dx} = v + x\frac{dv}{dx}$,

(*ii*) Separate the variables $v$ and $x$, and integrate.

**Example 11.12.** *Solve* $(x^2 - y^2)\, dx - xy\, dy = 0.$

**Solution.** Given equation is $\frac{dy}{dx} = \frac{x^2 - y^2}{xy}$ which is homogeneous in $x$ and $y$. ...(*i*)

Put $y = vx$, then $\frac{dy}{dx} = v + x\frac{dv}{dx}$. $\therefore$ (*i*) becomes $v + x\frac{dv}{dx} = \frac{1 - v^2}{v}$

or $$x\frac{dv}{dx} = \frac{1 - v^2}{v} - v = \frac{1 - 2v^2}{v}.$$

Separating the variables, $\frac{v}{1 - 2v^2}\, dv = \frac{dx}{x}$

Integrating both sides, $\int \frac{v\, dv}{1 - 2v^2} = \int \frac{dx}{x} + c$

or $$-\frac{1}{4}\int \frac{-4v}{1 - 2v^2}\, dv = \int \frac{dx}{x} + c \quad \text{or} \quad -\frac{1}{4}\log(1 - 2v^2) = \log x + c$$

or $$4\log x + \log(1 - 2v^2) = -4c \quad \text{or} \quad \log x^4(1 - 2v^2) = -4c$$ [Put $v = y/x$]

or $$x^4(1 - 2y^2/x^2) = e^{-3c} = c'$$

Hence the required solution is $x^2(x^2 - 2y^2) = c'$.

**Example 11.13.** *Solve* $(x \tan y/x - y \sec^2 y/x)\, dx - x \sec^2 y/x\, dy = 0.$ *(V.T.U., 2013)*

**Solution.** The given equation may be rewritten as

$$\frac{dy}{dx} = \left(\frac{y}{x}\sec^2\frac{y}{x} - \tan\frac{y}{x}\right)\cos^2 y/x \quad ...(i)$$

which is a homogeneous equation. Putting $y = vx$, (*i*) becomes $v + x\frac{dv}{dx} = (v\sec^2 v - \tan v)\cos^2 v$

or $$x\frac{dv}{dx} = v - \tan v \cos^2 v - v$$

Separating the variables $\frac{\sec^2 v}{\tan v}\, dv = -\frac{dx}{x}$

Integrating both sides $\log \tan v = -\log x + \log c$

or $$x \tan v = c \quad \text{or} \quad x \tan y/x = c.$$

**Example 11.14.** *Solve* $(1 + e^{x/y})\, dx + e^{x/y}(1 - x/y)\, dy = 0.$ *(V.T.U., 2014 S)*

**Solution.** The given equation may be rewritten as

$$\frac{dx}{dy} = -\frac{e^{x/y}(1 - x/y)}{1 + e^{x/y}} \quad ...(i)$$

which is a homogeneous equation. Putting $x = vy$ so that Eqn. (*i*) becomes

$$v + y\frac{dv}{dy} = -\frac{e^v(1-v)}{1+e^v} \quad \text{or} \quad y\frac{dv}{dy} = -\frac{e^v(1-v)}{1+e^v} - v = -\frac{v+e^v}{1+e^v}$$

Separating the variables, we get

$$-\frac{dy}{y} = \frac{1+e^v}{v+e^v}\,dv = \frac{d(v+e^v)}{v+e^v}$$

Integrating both sides, $-\log y = \log(v + e^v) + c$

or $\quad y(v + e^v) = e^{-c} \quad \text{or} \quad x + ye^{x/y} = c' \quad \text{(say)}$

which is the required solution.

## PROBLEMS 11.3

Solve the following differential equations :

1. $xydx - (x^2 + 2y^2)\,dy = 0$ *(J.N.T.U., 2015)*
2. $x^2y\,dx - (x^3 + y^3)\,dy = 0.$ *(V.T.U., 2010)*
3. $(x^2y - 2xy^2)\,dx - (x^3 - 3x^2y)\,dy = 0.$ *(D.T.U., 2011)*
4. $(y^3 - 3x^2y)\,dx - (x^3 - 3xy^2)\,dy = 0$ *(V.T.U., 2015)*
5. $\left(y - \frac{dy}{dx}\right) = a\left(y^2 + \frac{dy}{dx}\right)$ *(Bhopal, 2018)*
6. $\tan y\left(\frac{dy}{dx}\right) = \sin(x+y) + \sin(x-y)$ *(Bhilai, 2018)*
7. Solve the following differential equation.
   $\cos y \log(\sec x + \tan x)\,dx - \cos x \log(\sec x + \tan y)\,dy = 0.$ *(Chennai, 2018)*

[Equations solvable like homogeneous equations : When a differential equation contains $y/x$, a number of times, solve it like a homogeneous equation by putting $y/x = v$].

8. $\frac{dy}{dx} = \frac{y}{x} + \sin\frac{y}{x}.$
9. $ye^{x/y}\,dx = (xe^{x/y} + y^2)\,dy.$
10. $xy(\log x/y)\,dx + [y^2 - x^2\log(x/y)]\,dy = 0.$
11. $x\,dx + \sin^2(y/x)(ydx - xdy) = 0.$
12. $x\cos\frac{y}{x}(ydx + xdy) = y\sin\frac{y}{x}(xdy - ydx).$
13. $[x\cos(y/x) + y\sin(y/x)]\,y - [y\sin(y/x) - x\cos(y/x)]\,x\frac{dy}{dx} = 0.$ *(V.T.U., 2013)*
14. $3e^x\tan y\,dx + (1 - e^x)\sec^2 y\,dy = 0$ *(Bundel Khand, 2018)*
15. $(a^2 + y^2)\,dx + 2x\sqrt{ax + x^2}\,dy = 0$, given $y = 0$ when $x = 0$.
16. $(x - y)^2\frac{dy}{dx} = a^2.$
17. Aero space agency has established the law "The velocity $v$ of a projectile covering a distance $x$ in the barrel of a gun is given by $v = \frac{ax^n}{b + x^n}$; $n < 1$".What is the relationship between the time $t$ for which the projectile moved in the barrel and the distance $x$ covered by it. *(Rohtak, 2018)*

## 11.8 EQUATIONS REDUCIBLE TO HOMOGENEOUS FORM

The equations of the form $\dfrac{dy}{dx} = \dfrac{ax + by + c}{a'x + b'y + c'}$ ...(1)

can be reduced to the homogeneous form as follows :

**Case I. When $\dfrac{\mathbf{a}}{\mathbf{a'}} \neq \dfrac{\mathbf{b}}{\mathbf{b'}}$**

Putting $x = X + h,\ y = Y + k$, ($h, k$ being constants)

so that $dx = dX,\ dy = dY$, Eqn. (1) becomes

$$\frac{dY}{dX} = \frac{aX + bY + (ah + bk + c)}{a'X + b'Y + (a'h + b'k + c')} \quad ...(2)$$

Choose $h, k$ so that Eqn. (2) may become homogeneous.

Put $ah + bk + c = 0$, and $a'h + b'k + c' = 0$

so that $$\frac{h}{bc' - b'c} = \frac{k}{ca' - c'a} = \frac{1}{ab' - ba'}$$

or $$h = \frac{bc' - b'c}{ab' - b'a},\ k = \frac{ca' - c'a}{ab' - ba'} \quad ...(3)$$

Thus when $ab' - ba' \neq 0$, Eqn. (2) becomes $\dfrac{dY}{dX} = \dfrac{aX + bY}{a'X + b'Y}$ which is homogeneous in $X, Y$ and can be solved by putting $Y = vX$.

**Case II. When $\dfrac{\mathbf{a}}{\mathbf{a'}} = \dfrac{\mathbf{b}}{\mathbf{b'}}$.**

*i.e.*, $ab' - b'a = 0$, the above method fails as $h$ and $k$ become infinite or indeterminate.

Now $\dfrac{a}{a'} = \dfrac{b}{b'} = \dfrac{1}{m}$ (say)

$\therefore$ $a' = am,\ b' = bm$ and Eqn. (1) becomes

$$\frac{dy}{dx} = \frac{(ax + by) + c}{m(ax + by) + c'} \quad ...(4)$$

Put $ax + by = t$, so that $a + b\dfrac{dy}{dx} = \dfrac{dt}{dx}$

or $$\frac{dy}{dx} = \frac{1}{b}\left(\frac{dt}{dx} - a\right) \quad \therefore \quad (4) \text{ becomes } \frac{1}{b}\left(\frac{dt}{dx} - a\right) = \frac{t + c}{mt + c'}$$

or $$\frac{dt}{dx} = a + \frac{bt + bc}{mt + c'} = \frac{(am + b)t + ac' + bc}{mt + c'}$$

so that the variables are separable. In this solution, putting $t = ax + by$, we get the required solution of Eqn. (1).

**Example 11.15.** *Solve* $\dfrac{dy}{dx} = \dfrac{y + x - 2}{y - x - 4}$.

**Solution.** Given equation is $\dfrac{dy}{dx} = \dfrac{y + x - 2}{y - x - 4}$ $\left[\text{Case } \dfrac{a}{a'} \neq \dfrac{b}{b'}\right]$ ...(*i*)

Putting $x = X + h$ , $y = Y + k$, ($h, k$ being constants) so that $dx = dX,\ dy = dY$, Eqn. (*i*) becomes

$$\frac{dY}{dX} = \frac{Y + X + (k + h - 2)}{Y - X + (k - h - 4)} \quad ...(ii)$$

Put $k + h - 2 = 0$ and $k - h - 4 = 0$ so that $h = -1,\ k = 3$.

$\therefore$ Eqn. (*ii*) becomes $\dfrac{dY}{dX} = \dfrac{Y + X}{Y - X}$ which is homogeneous in $X$ and $Y$. ...(*iii*)

$\therefore$ put $Y = vX$, then $\dfrac{dY}{dX} = v + X\dfrac{dv}{dX}$

$\therefore$ Eqn. (*iii*) becomes $v + X\dfrac{dv}{dX} = \dfrac{v+1}{v-1}$ or $X\dfrac{dv}{dX} = \dfrac{v+1}{v-1} - v = \dfrac{1+2v-v^2}{v-1}$

or $$\frac{v-1}{1+2v-v^2}\,dv = \frac{dX}{X}.$$

Integrating both sides, $-\dfrac{1}{2}\displaystyle\int \frac{2-2v}{1+2v-v^2}\,dv = \int \frac{dX}{X} + c.$

or $$-\frac{1}{2}\log(1+2v-v^2) = \log X + c$$

or $$\log\left(1+\frac{2Y}{X}-\frac{Y^2}{X^2}\right) + \log X^2 = -2c$$

or $$\log(X^2+2XY-Y^2) = -2c \quad \text{or} \quad X^2+2XY-Y^2 = e^{-2c} = c' \qquad \ldots(iv)$$

Putting $X = x - h = x + 1$, $Y = y - k = y - 3$,

Eqn. (*iv*) becomes $(x+1)^2 + 2(x+1)(y-3) - (y-3)^2 = c'$

or $x^2 + 2xy - y^2 - 4x + 8y - 14 = c'$ which is the required solution.

**Example 11.16 :** *Solve $(3y + 2x + 4)\,dx - (4x + 6y + 5)\,dy = 0$.* (V.T.U., 2013)

**Solution.** Given equation is $\dfrac{dy}{dx} = \dfrac{(2x+3y)+4}{2(2x+3y)+5}$ ...(*i*)

Putting $2x + 3y = t$ so that $2 + 3\dfrac{dy}{dx} = \dfrac{dt}{dx}$ $\therefore$ Eqn. (*i*) becomes $\dfrac{1}{3}\left(\dfrac{dt}{dx} - 2\right) = \dfrac{t+4}{2t+5}$

or $$\frac{dt}{dx} = 2 + \frac{3t+12}{2t+5} = \frac{7t+22}{2t+5} \quad \text{or} \quad \frac{2t+5}{7t+22}\,dt = dx$$

Integrating both sides, $\displaystyle\int \frac{2t+5}{7t+22}\,dt = \int dx + c$

or $$\int\left(\frac{2}{7} - \frac{9}{7}\cdot\frac{1}{7t+22}\right)dt = x + c \quad \text{or} \quad \frac{2}{7}t - \frac{9}{49}\log(7t+22) = x + c$$

Putting $t = 2x + 3y$, we have $14(2x + 3y) - 9\log(14x + 21y + 22) = 49x + 49c$

or $21x - 42y + 9\log(14x + 21y + 22) = c'$ which is the required solution.

**Example 11.17:** *Solve the differential equation $(3y - 7x + 7)\,dx + (7y - 3x + 3)\,dy = 0$.*

**Solution :** $$\frac{dy}{dx} = \frac{-3y-7x+7}{7y-3x+3} \qquad \ldots(i)$$

Here, $$\frac{a}{b} \neq \frac{a^1}{b^1}$$

Let $- x = X + h, y = Y + k$

$$\therefore \quad \frac{dy}{dx} = \frac{dY}{dX}$$

Substituting these values in Eqn. (*i*), we get

$$\frac{dy}{dx} = \frac{3(Y+k) - 7(X+h) + 7}{7(Y+k-3)(X+h)+3}$$

$$= \frac{-3(Y+7X)+(-7h+3k+7)}{(7Y-3X)+(-3h+7k+3)} \qquad \ldots(ii)$$

Now choosing $h$ and $k$ is such manner,

$$-7h + 3k + 7 = 0$$

$$-3h + 7k + 3 = 0$$

Solving these equations, we get $h = 1, k = 0$

These value of $h$ and $k$, reduce the Eqn. $(ii)$ to the form

$$\frac{dY}{dX} = \frac{-(3Y - 7X)}{(7Y - 3X)} \quad ...(iii)$$

$$Y = VX$$

$$\therefore \quad \frac{dY}{dX} = V + X\frac{dV}{dX}$$

Substituting these values in Eqn. $(ii)$; we get

$$V + X\frac{dV}{dX} = \frac{-(3VX - 7X)}{7VX - 3X} = \frac{7 - 3V}{7V - 3}$$

$$X\frac{dV}{dX} = \frac{7 - 3V}{7V - 3} - V = \frac{-7(V^2 - 1)}{7V - 3}$$

Separating the variables and integrating.

$$\text{we get} \int \frac{7V - 3}{V^2 - 1}\, dV = -7\int \frac{dX}{X} + c$$

or
$$\int \left(\frac{2}{V - 1} + \frac{5}{V + 1}\right) dV = -7 \log X + c$$

$$2 \log (V - 1) + 5 \log (V + 1) = -7 \log X + c$$

$$\log [(V - 1)^2 (V + 1)^5] + 7 \log X = c$$

or
$$\log [X^7 (V - 1)^2 (V + 1)^5] = c$$

Replacing $V$ by $Y/X$, we get

$$\log \left[ X^7 \left(\frac{Y}{X} - 1\right)^2 \left(\frac{Y}{X} + 1\right)^5 \right] = c$$

$$(Y - X)^2 (Y + X)^5 = C^6 = K$$

Substituting $\quad X = x - h = x - 1$ and $Y = y - K = y$

So the solution is $(y - x + 1)^2 (y + x - 1)^5 = K$.

## PROBLEMS 11.4

Solve the following differential equations :

1. $(ax + by - a)dx = (bx + ay - b)dy$ *(Mysore, 2018)*
2. $(2x + y - 3)\, dy = (x + 2y - 3)\, dx$. *(V.T.U., 2009S)*
3. $(2x + 5y + 1)\, dx - (5x + 2y - 1)\, dy = 0$.
4. $\dfrac{dy}{dx} + \dfrac{ax + hy + g}{hx + by + f} = 0$.
5. $\dfrac{dy}{dx} = \dfrac{2x + 3y + 4}{4x + 6y + 5}$ *(Mysore 2018)*
6. $(4x - 6y - 1)\, dx + (3y - 2x - 2)\, dy = 0$.
7. $(x + 2y)(dx - dy) = dx + dy$.
8. $\dfrac{dy}{dx} = \dfrac{x + y - 2}{y - x - 4}$

## 11.9 LINEAR EQUATIONS

*A differential equation is said to be linear if the dependent variable and its differential coefficients occur only in the first degree and not multiplied together.*

*Thus* **the standard form of a linear equation of the first order,** *commonly known as* **Leibnitz's linear equation,** *is*

$$\frac{dy}{dx} + Py = Q \quad \text{where, } P, Q \text{ are the functions of } x. \qquad ...(1)$$

To solve the equation, multiply both sides by $e^{\int P dx}$ so that we get

$$\frac{dy}{dx} \cdot e^{\int P dx} + y\,(e^{\int P dx} P) = Qe^{\int P dx} \quad \text{i.e.,} \quad \frac{d}{dx}(ye^{\int P dx}) = Qe^{\int P dx}$$

Integrating both sides, we get $ye^{\int P dx} = \int Q e^{\int P dx} dx + c$ as the required solution.

**Obs.** The factor $e^{\int P dx}$ on multiplying by which the left-hand side of Eqn. (1) becomes the differential coefficient of a single function, is called the **integrating factor (I.F.)** of the linear Eqn. (1).

*It is important to remember that* **I.F.** $= e^{\int P dx}$

*and the solution is* $y$ **(I.F.)** $= \int Q(\text{I.F.})\, dx + c$.

**Example 11.18.** *Solve* $(x + 1)\dfrac{dy}{dx} - y\, e^{3x} (x + 1)^2$. *(V.T.U., 2014)*

**Solution.** Dividing throughout by $(x + 1)$, given equation becomes

$$\frac{dy}{dx} - \frac{y}{x+1} = e^{3x}(x + 1) \text{ which is Leibnitz's equation.} \qquad ...(i)$$

Here $\quad P = -\dfrac{1}{x+1}$ and $\int P dx = -\int \dfrac{dx}{x+1} = -\log(x + 1) = \log(x + 1)^{-1}$

$$\therefore \quad \text{I.F.} = e^{\int P dx} = e^{\log (x+1)^{-1}} = \frac{1}{x+1}$$

Thus the solution of Eqn. (*i*) is $y(\text{I.F.}) = \int [e^{3x}(x+1)](\text{I.F.})\, dx + c$

or $\quad \dfrac{y}{x+1} = \int e^{3x} dx + c = \dfrac{1}{3} e^{3x} + c \quad$ or $\quad y = \left(\frac{1}{3}e^{3x} + c\right)(x + 1).$

**Example 11.19.** *Solve* $\left(\dfrac{e^{-2\sqrt{x}}}{\sqrt{x}} - \dfrac{y}{\sqrt{x}}\right)\dfrac{dx}{dy} = 1.$

**Solution.** Given equation can be written as $\dfrac{dy}{dx} + \dfrac{y}{\sqrt{x}} = \dfrac{e^{-2\sqrt{x}}}{\sqrt{x}}$ $\qquad ...(i)$

$$\therefore \quad \text{I.F.} = e^{\int x^{1/2} dx} = e^{2\sqrt{x}}$$

Thus solution of Eqn. (*i*) is $y\,(\text{I.F.}) = \int \dfrac{e^{-2\sqrt{x}}}{\sqrt{x}} (\text{I.F.})\, dx + c$

or $$ye^{2\sqrt{x}} = \int \frac{e^{-2\sqrt{x}}}{\sqrt{x}} \cdot e^{2\sqrt{x}} dx + c$$

or $$ye^{2\sqrt{x}} = \int x^{-1/2} dx + c \quad \text{or} \quad ye^{2\sqrt{x}} = 2\sqrt{x} + c.$$

**Example 11.20.** *Solve* $3x(1 - x^2)\, y^2 \dfrac{dy}{dx} + (2x^2 - 1)\, y^3 = ax^3$

**Solution.** Putting $y^3 = z$ and $3y^2 \dfrac{dy}{dx} = \dfrac{dz}{dx}$, the given equation becomes

$$x(1 - x^2)\frac{dz}{dx} + (2x^2 - 1)\, z = ax^3, \quad \text{or} \quad \frac{dz}{dx} + \frac{2x^2 - 1}{x - x^3} z = \frac{ax^3}{x - x^3} \qquad \ldots(i)$$

which is Leibnitz's equation in $z$

$$\therefore \qquad \text{I.F.} = \exp\left(\int \frac{2x^2 - 1}{x - x^3} dx\right)$$

Now, $$\int \frac{2x^2 - 1}{x - x^3} dx = \int \left(-\frac{1}{x} - \frac{1}{2}\frac{1}{1+x} + \frac{1}{2}\cdot\frac{1}{1-x}\right) dx = -\log x - \frac{1}{2}\log(1 + x) - \frac{1}{2}\log(1 - x)$$

$$= -\log\,[x\sqrt{(1 - x^2)}]$$

$$\therefore \qquad \text{I.F.} = e^{-\log [x\sqrt{(1 - x^2)}]} = [x\sqrt{(1 - x^2)}]^{-1}$$

Thus the solution of $(i)$ is

$$z(\text{I.F.}) = \int \frac{ax^3}{x - x^3} (\text{I.F.})\, dx + c$$

or $$\frac{z}{[x\sqrt{(1 - x^2)}]} = a\int \frac{x^3}{x(1 - x^2)} \cdot \frac{1}{x\sqrt{(1 - x^2)}} dx + c = a\int x(1 - x^2)^{-3/2}\, dx$$

$$= -\frac{a}{2}\int (-2x)(1 - x^2)^{-3/2}\, dx + c = a\,(1 - x^2)^{-1/2} + c$$

Hence the solution of the given equation is

$$y^3 = ax + cx\sqrt{(1 - x^2)}. \qquad [\because \ z = y^3]$$

**Example 11.21.** *Solve* $y\,(\log y)\, dx + (x - \log y)\, dy = 0.$

**Solution.** We have $$\frac{dx}{dy} + \frac{x}{y\log y} = \frac{1}{y} \qquad \ldots(i)$$

which is a Leibnitz's equation in $x$

$$\therefore \qquad \text{I.F.} = e^{\int \frac{1}{y\log y} dy} = e^{\log(\log y)} = \log y$$

Thus the solution of Eqn. $(i)$ is $x\,(\text{I.F.}) = \displaystyle\int \frac{1}{y}(\text{I.F.})\, dy + c$

$$x\log y = \int \frac{1}{y}\log y\, dy + c = \frac{1}{2}(\log y)^2 + c$$

*i.e.,* $$x = \frac{1}{2}\log y + c\,(\log y)^{-1}.$$

**Example 11.22.** *Solve* $(1 + y^2)\, dx = (\tan^{-1} y - x)\, dy.$ *(P.T.U., 2013 S)*

**Solution.** This equation contains $y^2$ and $\tan^{-1} y$ and is, therefore, not a linear in $y$; but since only $x$ occurs, it can be written as

$$(1 + y^2)\frac{dx}{dy} = \tan^{-1} y - x \quad \text{or} \quad \frac{dx}{dy} + \frac{x}{1 + y^2} = \frac{\tan^{-1} y}{1 + y^2}$$

which is a Leibnitz's equation in $x$.

$$\therefore \qquad \text{I.F.} = e^{\int P dy} = e^{\int \frac{1}{1 + y^2} dy} = e^{\tan^{-1} y}$$

Thus the solution is $x\,(\text{I.F.}) = \displaystyle\int \frac{\tan^{-1} y}{1 + y^2} (\text{I.F.})\, dy + c$

or $$xe^{\tan^{-1}y} = \int \frac{\tan^{-1}y}{1+y^2} \cdot e^{\tan^{-1}y}\, dy + c$$ $\left[\begin{array}{l}\text{Put } \tan^{-1} y = t \\ \therefore \dfrac{dy}{1+y^2} = dt\end{array}\right]$

$$= \int te^t\, dt + c = t \cdot e^t - \int 1 \cdot e^t\, dt + c$$ (Integrating by parts)

$$= t \cdot e^t - e^t + c = (\tan^{-1} y - 1)\, e^{\tan^{-1} y} + c$$

or $$x = \tan^{-1} y - 1 + ce^{-\tan^{-1} y}.$$

**Example 11.23.** *Solve $r \sin \theta\, d\theta + (r^3 - 2r^2 \cos \theta + \cos \theta)\, dr = 0$.*

**Solution.** Given equation can be rewritten as

$$\sin \theta \frac{d\theta}{dr} + \frac{1}{r}(1 - 2r^2) \cos \theta = - r^2 \qquad ...(i)$$

Put $\cos \theta = y$ so that $- \sin \theta\, d\theta/dr = dy/dr$

Then Eqn. (*i*) becomes $-\dfrac{dy}{dr} + \left(\dfrac{1}{r} - 2r\right) y = - r^2$ or $\dfrac{dy}{dr} + \left(2r - \dfrac{1}{r}\right) y = r^2$

which is a Leibnitz's equation $\therefore$ I.F. $= e^{\int (2r - 1/r)\, dr} = e^{r^2 - \log r} = \dfrac{1}{r} e^{r^2}$

Thus, its solution is $y\left(\dfrac{1}{r} e^{r^2}\right) = \int r^2 \cdot e^{r^2} \cdot \dfrac{1}{r}\, dr + c$

or $$ye^{r^2}/r = \frac{1}{2}\int e^{r^2}\, 2r\, dr + c = \frac{1}{2} e^{r^2} + c$$

or $$2e^{r^2} \cos \theta = re^{r^2} + 2\, cr \quad \text{or} \quad r(1 + 2\, ce^{-r^2}) = 2 \cos \theta.$$

**Example 11.24:** *Solve $(1 + y^2) + (x - e^{\tan^{-1} y}) \dfrac{dy}{dx} = 0$.* *(VTU, 2018)*

**Solution:** The given equation involves $y^2$. So, it can't be linear in $y$. however it may be written as

$$\frac{dx}{dy} + \frac{x}{1+y^2} = \frac{e^{-\tan^{-1} y}}{1+y^2} \qquad ...(i)$$

Here $$\text{'}P\text{'} = \frac{1}{1+y^2},\ \text{'}Q\text{'} = \frac{e^{-\tan^{-1} y}}{1+y^2} \qquad ...(i)$$

$\therefore$ $$I.F = e^{\int \frac{1}{1+y^2} dy} = e^{\tan^{-1} y}$$

So the Eqn. (*i*)

$$x(I.F.) = \int \frac{e^{\tan^{-1} y} e^{-\tan^{-1} y}}{1+y^2}\, dy + c$$

or $$x \cdot e^{\tan^{-1} y} = \int \frac{1}{1+y^2}\, dy + c$$

or $$x\left(e^{\tan^{-1} y}\right) = \tan^{-1} y + c$$

or $$x = e^{\tan^{-1} y} + (e^{-\tan^{-1} y})$$

which is required solution.

**PROBLEMS 11.5**

Solve the following differential equations :

1. $\cos^2 x \dfrac{dy}{dx} + y = \tan x.$
2. $x \log x \dfrac{dy}{dx} + y = \log x^2.$ *(V.T.U., 2011)*
3. $2y' \cos x + 4y \sin x = \sin 2x$, given $y = 0$ when $x = \pi/3$.
4. $\cosh x \dfrac{dy}{dx} + y \sinh x = 2 \cosh^2 x \sinh x$ .
5. $(1 - x^2) \dfrac{dy}{dx} - xy = 1$ *(V.T.U., 2010)*
6. $(1 - x^2) \dfrac{dy}{dx} + 2xy = x \sqrt{(1 - x^2)}$
7. $\dfrac{dy}{dx} = -\dfrac{x + y \cos x}{1 + \sin x}.$
8. $dr + (2r \cot \theta + \sin 2\theta) d\theta = 0.$ *(J.N.T.U., 2015)*
9. $\dfrac{dy}{dx} + 2xy = 2e^{-x^2}$
10. $(x + 2y^3) \dfrac{dy}{dx} = y.$
11. $y(1 + x - y^2) \dfrac{dy}{dx} = 1$ *(P.T.U., 2014)*
12. $\sqrt{(1 - y^2)}\, dx = (\sin^{-1} y - x)\, dy.$ *(V.T.U., 2015)*
13. $(1 + y^2)\, dx + (x - e^{-\tan^{-1} y})\, dy = 0.$ *(V.T.U., 2016)*
14. $dy/dx + x e^x y = e^{(1-x)e^x}$
15. $\dfrac{dy}{dx} = \dfrac{y}{2y \log y + y - x}$
16. $\dfrac{dy}{dx} = \dfrac{1}{x \cos y + \sin 2y}$
17. $(2x - 10y^3) \dfrac{dy}{dx} + y = 0$
18. $\dfrac{dy}{dx} - y \tan x = 3e^{-\sin x}$, $y = 4$, when $x = 0$

## 11.10 BERNOULLI'S EQUATION

*The equation* $\dfrac{dy}{dx} + Py = Qy^n$ ...(1)

*where P, Q are functions of x, is reducible to the Leibnitz's linear equation and is usually called the Bernoulli's equation*.*

*To solve* Eqn. (1), divide both sides by $y^n$, so that $y^{-n} \dfrac{dy}{dx} + Py^{1-n} = Q$ ...(2)

Put $y^{1-n} = z$ so that $(1 - n)\, y^{-n} \dfrac{dy}{dx} = \dfrac{dz}{dx}$.

$\therefore$ Eqn. (2) becomes $\dfrac{1}{1-n} \dfrac{dz}{dx} + Pz = Q$ or $\dfrac{dz}{dx} + P(1 - n)\, z = Q(1 - n)$,

which is Leibnitz's linear in $z$ and can be solved easily.

**Example 11.25.** *Solve* $x \dfrac{dy}{dx} + y = x^3y^6.$ *(V.T.U., 2016)*

**Solution.** Dividing throughout by $xy^6$, $y^{-6} \dfrac{dy}{dx} + \dfrac{y^{-5}}{x} = x^2$ ...(*i*)

Put $y^{-5} = z$, so that $-5y^{-6} \dfrac{dy}{dx} = \dfrac{dz}{dx}$ $\therefore$ Eqn. (*i*) becomes $-\dfrac{1}{5} \dfrac{dz}{dx} + \dfrac{z}{x} = x^2$

*Named after the Swiss mathematician *Jacob Bernoulli* (1654–1705) who is known for his basic work in probability and elasticity theory. He was professor at Basel and had amongst his students his youngest brother *Johann Bernoulli* (1667–1748) and his nephew *Niklaus Bernoulli* (1687–1759). Johann is known for his basic contributions to Calculus while Niklaus had profound influence on the development of Infinite series and probability. His son *Daniel Bernoulli* (1700–1782) is known for his contributions to kinetic theory of gases and fluid flow.

or $$\frac{dz}{dx} - \frac{5}{x} z = -5x^2 \text{ which is Leibnitz's linear in } z. \quad ...(ii)$$

$$\text{I.F.} = e^{-\int (5/x)dx} = e^{-5 \log x} = e^{\log x^{-5}} = x^{-5}$$

$\therefore$ the solution of Eqn. (*ii*) is $z\,(\text{I.F.}) = \int (-5x^2)(\text{I.F.})\,dx + c$ or $zx^{-5} = \int (-5x^2)\, x^{-5}\, dx + c$

or $$y^{-5}x^{-5} = -5.\,\frac{x^{-2}}{-2} + c \qquad [\because z = y^{-5}]$$

Dividing throughout by $y^{-5}x^{-5}$, $1 = (2.5 + cx^2)\, x^3y^5$ which is the required solution.

**Example 11.26.** *Solve* $xy\,(1 + xy^2)\dfrac{dy}{dx} = 1.$ *(Andhra, 2015)*

**Solution.** Rewriting the given equation as

$$\frac{dx}{dy} - yx = y^3x^2$$

and dividing by $x^2$, we have

$$x^{-2}\frac{dx}{dy} - yx^{-1} = y^3 \qquad ...(i)$$

Putting $x^{-1} = z$ so that $-x^{-2}\dfrac{dx}{dy} = \dfrac{dz}{dy}$ Eqn. (*i*) becomes

$$\frac{dz}{dy} + yz = -y^3 \text{ which is Leibnitz's linear in } z.$$

Here $$\text{I.F.} = e^{\int y\,dy} = e^{y^2/2}$$

$\therefore$ the solution is $z\,(\text{I.F.}) = \int (-y^3)(\text{I.F.})\,dy + c$

or $$ze^{y^2/2} = -\int y^2 .\, e^{\frac{1}{2}y^2} .\, y dy + c \qquad \left|\begin{array}{l}\text{Put } \frac{1}{2}y^2 = t \\ \text{so that } y\,dy = dt\end{array}\right.$$

$$= -2\int t .\, e^t\, dt + c \qquad \text{[Integrate by parts]}$$

$$= -2\,[t .\, e^t - \int 1.\, e^t\, dt] + c = -2\,[te^t - e^t] + c = (2 - y^2)\, e^{y^2/2} + c$$

or $$z = (2 - y^2) + ce^{-\frac{1}{2}y^2} \text{ or } 1/x = (2 - y^2) + ce^{-\frac{1}{2}y^2}.$$

**Note.** *General equation reducible to Leibnitz's linear is* $f'(y)\dfrac{dy}{dx} + Pf(y) = Q$ ...(A)

*where P, Q are functions of x. To solve it, put* $f(y) = z$.

**Example 11.27.** *Solve* $\dfrac{dy}{dx} + x \sin 2y = x^3 \cos^2 y.$ *(Andhra, 2021 ; Jaipur, 2017 ; V.T.U., 2016)*

**Solution.** Dividing throughout by $\cos^2 y$, $\sec^2 y\,\dfrac{dy}{dx} + 2x\,\dfrac{\sin y \cos y}{\cos^2 y} = x^3$

or $$\sec^2 y\,\frac{dy}{dx} + 2x \tan y = x^3 \text{ which is of the form (A) above.} \qquad ...(i)$$

$\therefore$ put $\tan y = z$ so that $\sec^2 y\,\dfrac{dy}{dx} = \dfrac{dz}{dx}$ $\therefore$ Eqn. (*i*) becomes $\dfrac{dz}{dx} + 2xz = x^3$.

This is Leibnitz's linear equation in $z$. $\therefore$ $\text{I.F.} = e^{\int 2x\,dx} = e^{x^2}$

$\therefore$ the solution is $ze^{x^2} = \int e^{x^2}\, x^3 dx + c = \dfrac{1}{2}(x^2 - 1)\, e^{x^2} + c.$

Replacing $z$ by $\tan y$, we get $\tan y = \dfrac{1}{2}(x^2 - 1) + ce^{-x^2}$ which is the required solution.

**Example 11.28.** $\frac{dy}{dx} + \frac{x}{1-x^2} y = x\sqrt{y}$ *(Alagappa, 2018)*

**Solution** Given, $\frac{dy}{dx} + \frac{x}{1-x^2} y = x\sqrt{y}$

dividing both sides by $\sqrt{y}$,

$$y^{-1/2}\frac{dy}{dx} + \frac{x}{1-x^2} y^{1/2} = x \qquad ...(i)$$

put $y^{1/2} = z, \quad \therefore \frac{1}{2} y^{-1/2}\frac{dy}{dx} = \frac{dz}{dx}$

$$y^{-1/2}\frac{dy}{dx} = 2\frac{dz}{dx}$$

Substituting these value in eqn. Eqn. (*i*)

$$\frac{2\,dz}{dx} + \frac{x}{1-x^2} z = x$$

$$\frac{dz}{dx} + \frac{x}{2(1-x^2)} z = \frac{x}{2} \qquad ...(ii)$$

$$I.F. = e^{\frac{1}{2}\int \frac{x}{(1-x^2)}dx} = e^{-1/4\log(1-x^2)} = (1-x^2)^{-1/4} = \frac{1}{(1-x^2)^{1/4}}$$

So, the solution of Eqn. (*ii*) is

$$Z(I.F) = \int \frac{x}{2}(I.F.)dx + c$$

$$\frac{Z}{(1-x^2)^{1/4}} = \int \frac{x}{2}\cdot\frac{1}{(1-x^2)^{1/4}}\,dx + c = -\frac{1}{3}(1-x^2)^{3/2} + C$$

$$\sqrt{y} = \frac{-1}{3}(1-x^2) + c(1-x^2)^{1/4} \qquad \{\because z = \sqrt{y}\}$$

**Example 11.29:** *Solve* $\frac{dy}{dx} - \frac{\tan y}{1+x} = (1+x)e^x \sec y$ *(Bhilai, 2019)*

**Solution:** Given, $\frac{dy}{dx} - \frac{\tan y}{1+x} = (1+x)e^x \sec y$

dividing throughout by sec $y$, the equation $\cos y\frac{dy}{dx} - \frac{\sin y}{1+x} = (1+x)\,e^x$ ...(*i*)

put $\sin y = z$

$$\therefore \quad \cos y\frac{dy}{dx} = \frac{dz}{dx}$$

Substituting these values in Eqn. (*i*)

$$\frac{dz}{dx} - \frac{z}{1+x} = (1+x)e^x \qquad ...(ii)$$

The Eqn. (*ii*) is linear in $z$,

$$I.F. = e^{-\int \frac{dx}{1+x}} = e^{-\log(1+x)} = \frac{1}{1+x}$$

So, the solution of Eqn. (*ii*) is

$$Z(I.F.) = \int (1+x)e^x \cdot (I.F.)dx + C$$

$$Z \cdot \frac{1}{1+x} = \int (1+x)e^x \cdot \frac{1}{1+x} dx + C$$

and $$Z \cdot \frac{1}{1+x} = \int e^x dx + c = e^x + c$$

$$\sin y = (e^x + c)(1+x), \text{ since } Z = \sin y$$

## PROBLEMS 11.6

Solve the following equations :

1. $\dfrac{dy}{dx} = y \tan x - y^2 \sec x.$ (*Andhra, 2016 ; V.T.U., 2015*)
2. $r \sin\theta - \cos\theta \dfrac{dr}{d\theta} = r^2.$ (*V.T.U., 2015*)
3. $2xy' = 10x^3y^5 + y.$
4. $y' + 4xy + xy^3 = 0.$ (*P.T.U., 2015*)
5. $\dfrac{dy}{dx} + x(x+y) = x^3(x+y)^3 - 1$
6. $\left(\dfrac{dz}{dx}\right) + \left(\dfrac{z}{x}\right)\log z = \dfrac{z}{x}(\log z)^2.$
7. $\dfrac{dy}{dx} + \dfrac{1}{x} = \dfrac{e^y}{x^2}$
8. $x(x-y)\,dy + y^2\,dx = 0.$
9. $(x^3y^2 + xy)\,dx = dy$
10. $e^y\left(\dfrac{dy}{dx} + 1\right) = e^x.$
11. $\sec^2 y \dfrac{dy}{dx} + x \tan y = x^3.$
12. $\tan y \dfrac{dy}{dx} + \tan x = \cos y \cos^2 x.$
13. $\dfrac{dy}{dx} = \dfrac{y}{x + \sqrt{(xy)}}.$ (*V.T.U., 2011*)
14. $\dfrac{dy}{dx} + \dfrac{y}{x} = x^3y^2$ (*Andhra, 2022*)

## 11.11 EXACT DIFFERENTIAL EQUATIONS

**(1) Def.** *A differential equation of the form* $\mathbf{M}(x, y)\,\mathbf{dx} + \mathbf{N}(x, y)\,\mathbf{dy} = 0$ *is said to be* ***exact*** *if its left hand member is the exact differential of some function* $u(x, y)$ *i.e.,* $du = Mdx + Ndy = 0$. Its solution, therefore, is $u(x, y) = c$.

**(2) Theorem.** The necessary and sufficient condition for the differential equation $Mdx + Ndy = 0$ to be exact is

$$\frac{\partial \mathbf{M}}{\partial \mathbf{y}} = \frac{\partial \mathbf{N}}{\partial \mathbf{x}}$$

*Condition is necessary :*

The equation $Mdx + Ndy = 0$ will be exact, if

$$Mdx + Ndy \equiv du \qquad ...(1)$$

where $u$ is some function of $x$ any $y$.

But $$du = \frac{\partial u}{\partial x} dx + \frac{\partial u}{\partial y} dy \qquad ...(2)$$

$\therefore$ Equating coefficients of $dx$ and $dy$ in Eqns. (1) and (2), we get $M = \frac{\partial u}{\partial x}$ and $N = \frac{\partial u}{\partial y}$

$$\therefore \quad \frac{\partial M}{\partial y} = \frac{\partial^2 u}{\partial y \partial x} \text{ and } \frac{\partial N}{\partial x} = \frac{\partial^2 u}{\partial x \partial y}.$$

But $$\frac{\partial^2 u}{\partial y \partial x} = \frac{\partial^2 u}{\partial x \partial y} \qquad \text{(Assumption)}$$

$\therefore$ $\frac{\partial M}{\partial y} = \frac{\partial N}{\partial x}$ which is the necessary condition for exactness.

*Condition is sufficient : i.e.,* if $\frac{\partial M}{\partial y} = \frac{\partial N}{\partial x}$, then $Mdx + Ndy = 0$ is exact.

Let $\int Mdx = u$, where $y$ is supposed constant while performing integration.

Then $$\frac{\partial}{\partial x}\left(\int Mdx\right) = \frac{\partial u}{\partial x}, \quad i.e., \; M = \frac{\partial u}{\partial x} \qquad \left\{ \frac{\partial M}{\partial y} = \frac{\partial N}{\partial x} \text{ (given)} \text{ and } \frac{\partial^2 u}{\partial y \partial x} = \frac{\partial^2 u}{\partial x \partial y} \right. \qquad ...(3)$$

$$\therefore \quad \frac{\partial M}{\partial y} = \frac{\partial^2 u}{\partial y \partial x} \quad \text{or} \quad \frac{\partial N}{\partial x} = \frac{\partial^2 u}{\partial x \partial y} = \frac{\partial}{\partial x}\left(\frac{\partial u}{\partial y}\right)$$

Interacting both sides w.r.t. $x$ (taking $y$ as constant).

$$N = \frac{\partial u}{\partial y} + f(y), \text{ where } f(y) \text{ is a function of } y \text{ alone.} \qquad ...(4)$$

$$\therefore \quad Mdx + Ndy = \frac{\partial u}{\partial x} dx + \left\{\frac{\partial u}{\partial y} + f(y)\right\} dy \qquad \text{[By (3) and (4)]}$$

$$= \left\{\frac{\partial u}{\partial x} dx + \frac{\partial u}{\partial y} dy\right\} + f(y)\, dy = du + f(y)\, dy = d\left[u + \int f(y) dy\right] \qquad ...(5)$$

which shows that $Mdx + Ndy = 0$ is exact.

**(3) Method of solution.** By Eqn. (5), the equation $Mdx + Ndy = 0$ becomes $d[u + \int f(y) dy] = 0$

Integrating $\quad u + \int f(y) dy = 0.$

But $\quad u = \int_{y \text{ constant}} Mdx$ and $f(y)$ = terms of $N$ not containing $x$.

$\therefore$ *The solution of $Mdx + Ndy = 0$ is*

$$\int_{(y \text{ cons.})} \mathbf{M\, dx} + \int \textbf{(terms of N not containing x) dy} = \mathbf{c}$$

*provided* $$\frac{\partial M}{\partial y} = \frac{\partial N}{\partial x}.$$

**Example 11.30.** *Solve* $(y^2 e^{xy^2} + 4x^3)\, dx + (2xy\, e^{xy^2} - 3y^2)\, dy = 0.$

**Solution.** Here $\quad M = y^2 e^{xy^2} + 4x^3$ and $N = 2xy\, e^{xy^2} - 3y^2$

$$\therefore \quad \frac{\partial M}{\partial y} = 2y\, e^{xy^2} + y^2\, e^{xy^2} . 2xy = \frac{\partial N}{\partial x}$$

Thus the equation is exact and its solution is

$$\int_{(y \text{ const.})} Mdx + \int (\text{terms of } N \text{ not containing } x)\, dy = c$$

*i.e.,* $\int_{(y\text{ const.})} (y^2 e^{y^2 x} + 4x^3)\,dx + \int (-3y^2)\,dy = c$ or $e^{xy^2} + x^4 - y^3 = c.$

**Example 11.31.** *Solve* $\left\{y\left(1+\frac{1}{x}\right)+\cos y\right\} dx + (x + \log x - x \sin y)\, dy = 0.$

**Solution.** Here $M = y(1 + 1/x) + \cos y$ and $N = x + \log x - x \sin y$

$$\therefore \quad \frac{\partial M}{\partial y} = 1 + 1/x - \sin y = \frac{\partial N}{\partial x}$$

Then the equation is exact and its solution is

$$\int_{(y\text{ const})} M dx + \int (\text{terms of } N \text{ not containing } x)\, dy = c$$

$$\int_{(y\text{ const})} \left\{\left(1+\frac{1}{x}\right)y + \cos y\right\} dx = c \quad \text{or} \quad (x + \log x)\,y + x \cos y = c.$$

**Example 11.32.** *Solve* $(1 + 2xy \cos x^2 - 2xy)\, dx + (\sin x^2 - x^2)\, dy = 0.$

**Solution.** Here $M = 1 + 2xy \cos x^2 - 2xy$ and $N = \sin x^2 - x^2$

$$\therefore \quad \frac{\partial M}{\partial y} = 2x \cos x^2 - 2x = \frac{\partial N}{\partial x}$$

Thus the equation is exact and its solution is

$$\int_{(y\text{ const})} M dx + \int (\text{terms of } N \text{ not containing } x) = c$$

*i.e.,* $\int_{(y\text{ const})} (1 + 2xy \cos x^2 - 2xy)\, dx = c$ or $x + y\left[\int \cos x^2 \,.\, 2x dx - \int 2x\, dx\right] = c$

or $x + y \sin x^2 - yx^2 = c.$

**Example 11.33.** *Solve* $\frac{dy}{dx} + \frac{y \cos x + \sin y + y}{\sin x + x \cos y + x} = 0.$ *(Andhra, 2016 ; V.T.U., 2016 ; Rohtak, 2011)*

**Solution.** Given equation can be written as

$$(y \cos x + \sin y + y)\, dx + (\sin x + x \cos y + x)\, dy = 0.$$

Here, $M = y \cos x + \sin y + y$ and $N = \sin x + x \cos y + x.$

$$\therefore \quad \frac{\partial M}{\partial y} = \cos x + \cos y + 1 = \frac{\partial N}{\partial x}.$$

Thus the equation is exact and its solution is

$$\int_{(y\text{ const.})} M dx + \int (\text{terms of } N \text{ not containing } x)\, dy = c$$

*i.e.,* $\int_{(y\text{ const.})} (y \cos x + \sin y + y)\, dx + \int (0)\, dx = c$ or $y \sin x + (\sin y + y)\, x = c.$

**Example 11.34.** *Solve* $(2x^2 + 3y^2 - 7)\, xdx - (3x^2 + 2y^2 - 8)\, ydy = 0.$

**Solution.** Given equation can be written as

$$\frac{ydy}{xdx} = \frac{2x^2 + 3y^2 - 7}{3x^2 + 2y^2 - 8}$$

or $$\frac{ydy + xdx}{ydy - xdx} = \frac{5(x^2 + y^2 - 3)}{-x^2 + y^2 + 1}$$ [By componendo & dividendo]

or $$\frac{xdx + ydy}{x^2 + y^2 - 3} = 5 \cdot \frac{xdx - ydy}{x^2 - y^2 - 1}$$

Integrating both sides, we get

$$\int \frac{2xdx + 2ydy}{x^2 + y^2 - 3} = 5\int \frac{2xdx - 2ydy}{x^2 - y^2 - 1} + c$$

or $$\log(x^2 + y^2 - 3) = 5\log(x^2 - y^2 - 1) + \log c'$$ [Writing $c = \log c'$]

or $$x^2 + y^2 - 3 = c'(x^2 - y^2 - 1)^5$$

which is the required solution.

## PROBLEMS 11.7

Solve the following equations :

1. $(x^2 - ay)\,dx = (ax - y^2)\,dy$. *(Calicut, 2012)*
2. $(x^2 + y^2 - a^2)\,xdx + (x^2 - y^2 - b^2)\,ydy = 0$
3. $(x^2 - 4xy - 2y^2)\,dx + (y^2 - 4xy - 2x^2)\,dy = 0$. *(Andhra, 2023)*
4. $(x^4 - 2xy^2 + y^4)\,dx - (2x^2y - 4xy^3 + \sin y)\,dy = 0$
5. $(5x^4 + 3x^2y^2 - 2xy^3)\,dx + (2x^3y - 3x^2y^2 - 5y^4)\,dy = 0$
6. $(3x^2 + 6xy^2)\,dx + (6x^2y + 4y^3)\,dy = 0$
7. $\frac{2x}{y^3}dx + \frac{y^2 - 3x^2}{y^4}dy = 0$
8. $y\sin 2x\,dx - (1 + y^2 + \cos^2 x)\,dy = 0$ *(D.T.U., 2011)*
9. $ye^{xy}dx + (xe^{xy} + 2y)\,dy = 0$
10. $(x - y)(dx - dy) = dx + dy$ *(U.K.T.U., 2012)*
11. $(\sec x\tan x\tan y - e^x)\,dx + \sec x\sec^2 y\,dy = 0$
12. $(2xy + y - \tan y)\,dx + (x^2 - x\tan^2 y + \sec^2 y)\,dy = 0$.

## 11.12 EQUATIONS REDUCIBLE TO EXACT EQUATIONS

Sometimes a differential equation which is not exact, can be made so on multiplication by a suitable factor called an *integrating factor*. The rules for finding integrating factors of the equation $Mdx + Ndy = 0$ are as follows :

**(1) I.F. found by inspection.** In a number of cases, the integrating factor can be found after regrouping the terms of the equation and recognizing each group as being a part of an exact differential. In this connection the following integrable combinations prove quite useful :

$$xdy + ydx = d(xy)$$

$$\frac{xdy - ydx}{x^2} = d\left(\frac{y}{x}\right);\ \frac{xdy - ydx}{xy} = d\left[\log\left(\frac{y}{x}\right)\right]$$

$$\frac{xdy - ydx}{y^2} = -d\left(\frac{x}{y}\right);\ \frac{xdy - ydx}{x^2 + y^2} = d\left(\tan^{-1}\frac{y}{x}\right)$$

$$\frac{xdy - ydx}{x^2 - y^2} = d\left(\frac{1}{2}\log\frac{x + y}{x - y}\right).$$

**Example 11.35.** *Solve $y(2xy + e^x)\,dx = e^x\,dy$.*

**Solution.** It is easy to note that the terms $ye^xdx$ and $e^xdy$ should be put together.

$\therefore$ $$(ye^xdx - e^xdy) + 2xy^2\,dx = 0$$

Now we observe that the term $2xy^2\,dx$ should not involve $y^2$. This suggests that $1/y^2$ may be I.F. Multiplying throughout by $1/y^2$, it follows

$$\frac{ye^x\,dx - e^x\,dy}{y^2} + 2xdx = 0 \quad \text{or} \quad d\left(\frac{e^x}{y}\right) + 2xdx = 0$$

Integrating, we get $\frac{e^x}{y} + x^2 = c$ which is the required solution.

**(2) I.F. of a homogeneous equation.** *If $Mdx + Ndy = 0$ be a homogeneous equation in $x$ and $y$, then $1/(Mx + Ny)$ is an integrating factor $(Mx + Ny \neq 0)$.*

**Example 11.36.** *Solve $(x^2y - 2xy^2)\,dx - (x^3 - 3x^2y)\,dy = 0$.*

**Solution.** This equation is homogeneous in $x$ and $y$.

$$\therefore \qquad \text{I.F.} = \frac{1}{Mx + Ny} = \frac{1}{(x^2y - 2xy^2)\,x - (x^3 - 3x^2y)y} = \frac{1}{x^2y^2}$$

Multiplying throughout by $1/x^2y^2$, the equation becomes

$$\left(\frac{1}{y} - \frac{2}{x}\right)dx - \left(\frac{x}{y^2} - \frac{3}{y}\right)dy = 0, \text{ which is exact.}$$

$$\therefore \text{ the solution is } \int_{(y\text{ const})} Mdx + \int (\text{terms of } N \text{ not containing } x)\,dy = c \text{ or } \frac{x}{y} - 2\log x + 3\log y = c.$$

**(3) I.F. for an equation of the type $f_1(xy)ydx + f_2(xy)xdy = 0$.**
*If the equation $Mdx + Ndy = 0$ be of this form, then $1/(Mx - Ny)$ is an integrating factor $(Mx - Ny \neq 0)$.*

**Example 11.37.** *Solve $(1 + xy)\,ydx + (1 - xy)\,xdy = 0$.* *(C.S.V.T.U., 2014 ; V.T.U., 2014)*

**Solution.** The given equation is of the form $f_1(xy)\,ydx + f_2(xy)\,xdy = 0$

Here $\qquad M = (1 + xy)\,y,\ N = (1 - xy)\,x.$

$$\therefore \qquad \text{I.F.} = \frac{1}{Mx - Ny} = \frac{1}{(1 + xy)yx - (1 - xy)xy} = \frac{1}{2x^2y^2}$$

Multiplying throughout by $1/2x^2y^2$, it becomes

$$\left(\frac{1}{2x^2y} + \frac{1}{2x}\right)dx + \left(\frac{1}{2xy^2} - \frac{1}{2y}\right)dy = 0, \text{ which is an exact equation.}$$

$$\therefore \text{ the solution is } \int_{(y\text{ const})} Mdx + \int (\text{terms of } N \text{ not containing } x)\,dy = c$$

or
$$\frac{1}{2y}\left(-\frac{1}{x}\right) + \frac{1}{2}\log x - \frac{1}{2}\log y = c \quad \text{or} \quad \log\frac{x}{y} - \frac{1}{xy} = c'.$$

**(4) In the equation $Mdx + Ndy = 0$,**

(*a*) *if* $\dfrac{\dfrac{\partial M}{\partial y} - \dfrac{\partial N}{\partial x}}{N}$ *be a function of $x$ only $= f(x)$ say, then $e^{\int f(x)dx}$ is an integrating factor.*

(*b*) *if* $\dfrac{\dfrac{\partial N}{\partial x} - \dfrac{\partial M}{\partial y}}{M}$ *be a function of $y$ only $= F(y)$ say, then $e^{\int F(y)dy}$ is an integrating factor.*

**Example 11.38.** *$(\sec x \tan x \tan y - e^x)\,dx + \sec x \sec^2 y\,dy = 0$* *(Amravati, 2018)*

**Solution.** Here,
$$M = \sec x \tan x \tan y - e^x$$
$$N = \sec x \sec^2 x$$
$$\frac{\delta M}{\delta y} = \sec x \tan x \sec^2 y = \frac{\delta N}{\delta x}$$

So, the given equation is exact and its solution is

$$\int M dx + \int \text{terms of } N \text{ that do not contain } x) dy = C$$

$$\int (\sec x \tan x \tan y - e^x) dx + \int dy = C$$

$$\sec x \tan y - e^x = C.$$

**Example 11.39.** *Solve* $(xy^3 + y)\, dx + 2\,(x^2y^2 + x + y^4)\, dy = 0.$ *(V.T.U., 2015 ; P.T.U., 2014 ; D.T.U., 2013)*

**Solution.** Here $M = xy^3 + y$, $N = 2(x^2y^2 + x + y^4)$

$$\therefore \quad \frac{1}{M}\left(\frac{\partial N}{\partial x} - \frac{\partial M}{\partial y}\right) = \frac{1}{y(xy^2+1)}(4xy^2 + 2 - 3xy^2 - 1) = \frac{1}{y}, \text{ which is a function of } y \text{ alone.}$$

$$\therefore \quad \text{I.F.} = e^{\int 1/y\, dy} = e^{\log y} = y$$

Multiplying throughout by $y$, it becomes $(xy^4 + y^2)\, dx + (2x^2y^3 + 2xy + 2y^5)\, dy = 0$, which is an exact equation.

$$\therefore \quad \text{its solution is } \int_{(y \text{ const})} (M dx) + \int (\text{terms of } N \text{ not containing } x)\, dy = 0$$

or $$\int_{(y \text{ const})} (xy^4 + y^2) dx + \int 2y^5\, dy = c \quad \text{or} \quad \frac{1}{2}x^2y^4 + xy^2 + \frac{1}{3}y^6 = c.$$

**Example 11.40.** *Solve* $(y \log y)\, dx + (x - \log y)\, dy = 0$

**Solution.** Here $M = y \log y$ and $N = x - \log y$

$$\therefore \quad \frac{1}{M}\left(\frac{\partial N}{\partial x} - \frac{\partial M}{\partial y}\right) = \frac{1}{y \log y}(1 - \log y - 1) = -\frac{1}{y}, \text{ which is a function of } y \text{ alone.}$$

$$\therefore \quad \text{I.F.} = e^{-\int \frac{1}{y} dy} = e^{-\log y} = \frac{1}{y}$$

Multiplying the given equation throughout by $1/y$, it becomes

$$\log y\, dx + \frac{1}{y}(x - \log y)\, dy = 0$$

which is an exact equation $\left[\because \frac{\partial}{\partial y}(\log y) = \frac{\partial}{\partial x}\left(\frac{x - \log y}{y}\right)\right.$

$$\therefore \quad \text{its solution is } \int_{(y \text{ const})} (M dx) + \int (\text{terms of } N \text{ not containing } x)\, dy = c$$

or $$\log y \int dx + \int \left(\frac{-\log y}{y}\right) dy = c \quad \text{or} \quad x \log y - \frac{1}{2}(\log y)^2 = c.$$

**(5) For the equation of the type**

$$x^a y^b\, (m y dx + n x dy) + x^{a'} y^{b'}\, (m' y dx + n' x dy) = 0,$$

*an integrating factor is* $x^h y^k$

*where* $$\frac{a + h + 1}{m} = \frac{b + k + 1}{n}, \quad \frac{a' + h + 1}{m'} = \frac{b' + k + 1}{n'}.$$

**Example 11.41.** *Solve* $y(xy + 2x^2y^3)\, dx + x\,(xy - x^2y^2)\, dy = 0.$ *(Rohtak, 2012)*

**Solution.** Rewriting the equation as $xy\,(ydx + xdy) + x^2y^2\,(2ydx - xdy) = 0$ and comparing with

$$x^a y^b\, (m y dx + n x dy) + x^{a'} y^{b'}\, (m' y dx + n' x dy) = 0,$$

we have $a = b = 1$, $m = n = 1$; $a' = b' = 2$, $m' = 2$, $n' = -1$.

$$\therefore \quad \text{I.F.} = x^h y^k.$$

where $$\frac{a + h + 1}{m} = \frac{b + k + 1}{n}, \quad \frac{a' + h + 1}{m'} = \frac{b' + k + 1}{n'}$$

i.e. $$\frac{1+h+1}{1}=\frac{1+k+1}{1}, \frac{2+h+1}{2}=\frac{2+k+1}{-1}$$

or $$h-k=0,\ h+2k+9=0$$

Solving these, we get $h = k = -3$. $\therefore$ I.F. $= 1/x^3y^3$.

Multiplying throughout by $1/x^3y^3$, it becomes

$$\left(\frac{1}{x^2y}+\frac{2}{x}\right)dx+\left(\frac{1}{xy^2}-\frac{1}{y}\right)dx=0, \text{ which is an exact equation.}$$

$\therefore$ The solution is $\int_{(y\ const)} Mdx + \int (\text{terms of } N \text{ not containing } x)\, dy = c$

or $$\frac{1}{y}\left(-\frac{1}{x}\right)+2\log x-\log y=c \quad \text{or} \quad 2\log x-\log y-1/xy=c.$$

## PROBLEMS 11.8

Solve the following equations :

1. $xdy - ydx + a(x^2 + y^2)\, dx = 0$. (*C.S.V.T.U., 2012*)
2. $xdx + ydy = \dfrac{a^2(xdy - ydx)}{x^2 + y^2}$. (*Kurukshetra, 2011*)
3. $ydx - xdy + \log x\, dx = 0$.
4. $\dfrac{dy}{dx} = \dfrac{x^3 + y^3}{xy^2}$.
5. $(x^3y^2 + x)\, dy + (x^2y^3 - y)\, dx = 0$. (*C.S.V.T.U., 2012*)
6. $(x^2y^2 + xy + 1)ydx + (x^2y^2 - xy + 1)\, xdy = 0$.
7. $(y^4 + 2y)\, dx + (xy^3 + 2y^4 - 4x)\, dy = 0$.
8. $(1 + e^{x/y})dx + \left(1 - \dfrac{x}{y}\right)e^{x/y}dy = 0$. (*Gondwana, 2019*)
9. $(5x^3 + 12x^2 + 6y^2)\, dx + 6xydy = 0$ (*P.T.U., 2015*)
10. $x^4 \dfrac{dy}{dx} + x^3y + \text{cosec}\,(xy) = 0$.
11. $(y - xy^2)\, dx - (x + x^2y)\, dy = 0$ (*P.T.U., 2014*)
12. $ydx - xdy + 3x^2y^2 e^{x^3} dx = 0$. (*Calicut, 2013*)
13. $x\, dx + ydy = \dfrac{a^2(x\, dy - y\, dx)}{x^2 + y^2}$. (*Shiva Ji, 2018*)
14. $(y^2 e^{xy^2} + 4x^3)dx + (2xy\, e^{xy^2} - 3y^2)\, dy = 0$
15. $(2xy \cos^2 x + 2xy + 1)\, dx + (\sin x^2 - x^2)\, dy = 0$
16. $\{\cos \tan y + \cos (x + y)\}\, dx + \{\sin x \sec^2 y + \cos (x + y)\}\, dy = 0$
17. $(xy^2 - e^{1/x^2})\, dx - x^2ydy = 0$ (*Andhra, 2021*)

## 11.13 EQUATIONS OF THE FIRST ORDER AND HIGHER DEGREE

As $dy/dx$ will occur in higher degrees, it is convenient to denote $dy/dx$ by $p$. Such equations are of the form $f(x, y, p) = 0$. Three cases arise for discussion :

**Case. I. Equation solvable for p.** A differential equation of the first order but of the $n$th degree is of the form

$$p^n + P_1 p^{n-1} + P_2 p^{n-2} + \ldots + P_n = 0 \qquad \ldots(1)$$

where $P_1, P_2, \ldots, P_n$ are functions of $x$ and $y$.

Splitting up the left hand side of Eqn. (1) into $n$ linear factors, we have

$$[p - f_1(x, y)]\,[p - f_2(x, y)] \ldots [p - f_n(x, y)] = 0.$$

Equating each of the factors to zero,

$$p = f_1(x, y),\ p = f_2(x, y),\ \ldots,\ p = f_n(x, y)$$

Solving each of these equations of the first order and first degree, we get the solutions

$$F_1(x, y, c) = 0,\ F_2(x, y, c) = 0,\ \ldots,\ F_n(x, y, c) = 0.$$

These $n$ solutions constitute the general solution of (1).

*Otherwise*, the general solution of Eqn. (1) may be written as

$$F_1(x, y, c) \,.\, F_2(x, y, c) \ldots\ldots\, F_n(x, y, c) = 0.$$

**Example 11.42.** *Solve* $\frac{dy}{dx} - \frac{dx}{dy} = \frac{x}{y} - \frac{y}{x}$. (V.T.U., 2016 ; P.T.U., 2014)

**Solution.** Given equation is $p - \frac{1}{p} = \frac{x}{y} - \frac{y}{x}$ where $p = \frac{dy}{dx}$ or $p^2 + p\left(\frac{y}{x} - \frac{x}{y}\right) - 1 = 0$.

Factorising $(p + y/x)(p - x/y) = 0$.

Thus we have $p + y/x = 0$ ...(i) and $p - x/y = 0$ ...(ii)

From Eqn. (i), $\frac{dy}{dx} + \frac{y}{x} = 0$ or $xdy + ydx = 0$

i.e., $d(xy) = 0$. Integrating, $xy = c$.

From Eqn. (ii), $\frac{dy}{dx} - \frac{x}{y} = 0$ or $xdx - ydx = 0$

Integrating, $x^2 - y^2 = c$. Thus $xy = c$ or $x^2 - y^2 = c$, constitute the required solution.

**Otherwise,** combining these into one, the required solution can be written as

$$(xy - c)(x^2 - y^2 - c) = 0.$$

**Example 11.43.** *Solve* $p^2 + 2py \cot x = y^2$. (C.S.V.T.U., 2015 ; V.T.U., 2012)

**Solution.** We have $p^2 + 2py \cot x + (y \cot x)^2 = y^2 + y^2 \cot^2 x$

or $$p + y \cot x = \pm y \operatorname{cosec} x$$

i.e., $$p = y(-\cot x + \operatorname{cosec} x) \quad ...(i)$$

or $$p = y(-\cot x - \operatorname{cosec} x) \quad ...(ii)$$

From Eqn. (i), $\frac{dy}{dx} = y(-\cot x + \operatorname{cosec} x)$ or $\frac{dy}{y} = (\operatorname{cosec} x - \cot x)\, dx$

Integrating, $\log y = \log \tan \frac{x}{2} - \log \sin x + \log c = \log \frac{c \tan x/2}{\sin x}$

or $$y = \frac{c}{2 \cos x^2/2} \text{ or } y(1 + \cos x) = c \quad ...(iii)$$

From Eqn. (ii), $\frac{dy}{dx} = -y(\cot x + \operatorname{cosec} x)$ or $\frac{dy}{y} = -(\cot x + \operatorname{cosec} x)\, dx$

Integrating, $\log y = -\log \sin x - \log \tan \frac{x}{2} + \log c = \log \frac{c}{\sin x \tan \frac{x}{2}}$

or $$y = \frac{c}{2 \sin^2 \frac{x}{2}} \text{ or } y(1 - \cos x) = c \quad ...(iv)$$

Thus combining Eqns. (iii) and (iv), the required general solution is

$$y(1 \pm \cos x) = c.$$

## PROBLEMS 11.9

Solve the following equations :

1. $y\left(\frac{dy}{dx}\right)^2 + (x - y)\frac{dy}{dx} - x = 0.$
2. $xp^2 - (2x + 3y)p + 6y = 0.$ (V.T.U., 2015)
3. $p(p + y) = x(x + y).$ (V.T.U., 2011)
4. $y = x[p + \sqrt{(1 + p^2)}].$
5. $xy\left(\frac{dy}{dx}\right)^2 - (x^2 + y^2)\frac{dy}{dx} + xy = 0.$
6. $p^3 + 2xp^2 - y^2p^2 - 2xy^2p = 0.$

**Case II. Equations solvable for y.** If the given equation, on solving for $y$, takes the form

$$y = f(x, p). \quad ...(1)$$

then differentiation with respect to $x$ gives an equation of the form

$$p = \frac{dy}{dx} = \phi\left(x, p, \frac{dp}{dx}\right).$$

Now it may be possible to solve this new differential equation in $x$ and $p$.

Let its solution be $F(x, p, c) = 0$. ...(2)

The elimination of $p$ from Eqns. (1) and (2) gives the required solution.

In case elimination of $p$ is not possible, then we may solve Eqns. (1) and (2) for $x$ and $y$ and obtain

$$x = F_1(p, c), y = F_2(p, c)$$

as the required solution, where $p$ is the parameter.

**Obs.** This method is especially useful for equations which do not contain $x$.

**Example 11.44.** *Solve $y - 2px = \tan^{-1}(xp^2)$.* *(V.T.U., 2014 ; P.T.U., 2013)*

**Solution.** Given equation is $y = 2px + \tan^{-1}(xp^2)$ ...(*i*)

Differentiating both sides with respect to $x$, $\dfrac{dy}{dx} = p = 2\left(p + x\dfrac{dp}{dx}\right) + \dfrac{p^2 + 2xp\dfrac{dp}{dx}}{1 + x^2p^4}$

or $$p + 2x\frac{dp}{dx} + \left(p + 2x\frac{dp}{dx}\right) \cdot \frac{p}{1 + x^2p^4} = 0 \text{ or } \left(p + 2x\frac{dp}{dx}\right)\left(1 + \frac{p}{1 + x^2p^4}\right) = 0$$

This gives $p + 2x\, dp/dx = 0$.

Separating the variables and integrating, we have $\displaystyle\int \frac{dx}{x} + 2\int \frac{dp}{p}$ = a constant

or $\log x + 2 \log p = \log c$ or $\log xp^2 = \log c$

whence $xp^2 = c$ or $p = \sqrt{(c/x)}$ ...(*ii*)

Eliminating $p$ from Eqns. (*i*) and (*ii*), we get $y = 2\sqrt{(c/x)}x + \tan^{-1} c$

or $y = 2\sqrt{(cx)} + \tan^{-1} c$ which is the general solution of Eqn. (*i*).

**Obs.** The significance of the factor $1 + p/(1 + x^2p^4) = 0$ which we didn't consider, will not be considered here as it concerns 'singular solution' of Eqn. (*i*) whereas *we are interested only in finding general solution.*

**Caution.** Sometimes Eqn. (*i*) is tempted to write Eqn. (*ii*) as

$$\frac{dy}{dx} = \sqrt{\left(\frac{c}{x}\right)}$$

and integrating it to say that the required solution is $y = 2\sqrt{(cx)} + c'$. Such a reasoning is *incorrect*.

**Example 11.45.** *Solve $y + px = x^4 p^2$* *(Patna, 2018)*

**Solution:** The given equation is $y = x^4 p^2 - px$ ...(*i*)

Differentiating Eqn. (*i*) with respect to $x$.

$$\frac{dy}{dx} = 4x^3p^2 + 2x^4p\frac{dp}{dx} - p - x\frac{dp}{dx}$$

verify $\dfrac{dy}{dx} = p$, we get

$$p = 4x^3p^2 + 2x^4p\frac{dp}{dx} - p - x\frac{dp}{dx}$$

$$(2x^4p - x)\frac{dp}{dx} + 4x^3p^2 - 2p = 0$$

$$x(2x^3 - p - 1)\frac{dp}{dx} + 2p(2x^3 p - 1) = 0$$

$$(2x^3 p - 1)\left(x\frac{dp}{dx} + 2p\right) = 0$$

Consider the factor

$$x\frac{dp}{dx} + 2p = 0 \quad ...(ii)$$

$$\frac{dp}{p} + \frac{2dx}{x} = 0$$

Integrating $\log p + 2\log x = \log c$

$$px^2 = c$$

$$p = c/x^2 \quad ...(iii)$$

$$y + \frac{xc}{x^2} = x^4 \cdot \frac{c^2}{x^4}$$

$$y + \frac{c}{x} = c^2$$

$xy + c = c^2 x$, which is the solution the significance of the factor.

$$2x^3 p - 1 = 0$$

**PROBLEMS 11.10**

Solve the following equations :

1. $y = 2px - p^2$. (*P.T.U., 2014*)
2. $y + 2px = p^n$.
3. $y = x + a\tan^{-1} p$.
4. $y - 3x + \log p$. (*V.T.U., 2015*)
5. $xp^2 + x = 2yp$. (*V.T.U., 2012*)
6. $x^2\left(\frac{dy}{dx}\right)^4 + 2x\frac{dy}{dx} - y = 0$.
7. $y = p\sin p + \cos p$. (*V.T.U., 2013*)
8. $y^2 \log y = xpy + p^2$

**Case III. Equations solvable for x.** If the given equation on solving for $x$, takes the form

$$x = f(y, p) \quad ...(1)$$

then differentiation with respect to $y$ gives an equation of the form

$$\frac{1}{p} = \frac{dx}{dy} = \phi\left(y, p, \frac{dp}{dy}\right) \quad ...(2)$$

Now it may be possible to solve the new differential equation in $y$ and $p$. Let its solution be $F(y, p, c) = 0$.

The elimination of $p$ from Eqns. (1) and (2) gives the required solution. In case the elimination is not feasible, Eqns. (1) and (2) may be expressed in terms of $p$ and $p$ may be regarded as a parameter.

**Obs.** This method is especially useful for equations which do not contain $y$.

**Example 11.46.** *Solve $y = 2px + y^2p^3$.* (*C.S.V.T.U., 2012*)

**Solution.** Given equation, on solving for $x$, takes the form $x = \dfrac{y - y^2p^3}{2p}$

Differentiating with respect to $y$, $\dfrac{dx}{dy}\left(=\dfrac{1}{p}\right) = \dfrac{1}{2} \cdot \dfrac{p\left(1 - 2y \cdot p^3 - y^2 3p^2\dfrac{dp}{dy}\right) - (y - y^2p^3)\dfrac{dp}{dy}}{p^2}$

or $$2p = p - 2yp^4 - 3y^2p^3\frac{dp}{dy} - y\frac{dp}{dy} + y^2p^3\frac{dp}{dy}$$

or $$p + 2yp^4 + 2y^2p^3\frac{dp}{dy} + y\frac{dp}{dy} = 0 \text{ or } p(1 + 2yp^3) + y\frac{dp}{dy}(1 + 2yp^3) = 0.$$

or $\left(p + y\dfrac{dp}{dy}\right)(1 + 2py^3) = 0$ This gives $p + y\dfrac{dp}{dy} = 0.$ or $\dfrac{d}{dy}(py) = 0.$

Integrating $py = c.$ ...(*i*)

Thus eliminating from the given equation and Eqn. (*i*), we get $y = 2\dfrac{c}{y}x + \dfrac{c^3}{y^3}y^2$ or $y^2 = 2cx + c^3$

which is the required solution.

**PROBLEMS 11.11**

Solve the following equations :

1. $p^3 - 4xyp + 8y^2 = 0.$ (*V.T.U., 2016*)
2. $p^3y + 2px = y.$
3. $x - yp = ap^2.$
4. $p = \tan\left(x - \dfrac{p}{1+p^2}\right).$ (*V.T.U., 2016 ; C.S.V.T.U., 2015*)

## 11.14 CLAIRAUT'S EQUATION*

An equation of the form $y = px + f(p)$ is known as Clairaut's equation ...(1)

Differentiating with respect to $x$, we have $p = p + x\dfrac{dp}{dx} + f'(p)\dfrac{dp}{dx}$

or $[x + f'(p)]\dfrac{dp}{dx} = 0$ $\therefore \dfrac{dp}{dx} = 0$, or $x + f'(p) = 0$

$$\frac{dp}{dx} = 0, \text{ gives } p = c \qquad ...(2)$$

Thus eliminating $p$ from Eqns. (1) and (2), we get $y = cx + f(c)$ ...(3)

as the general solution of Eqn. (1).

*Hence the solution of the Clairaut's equation is obtained on replacing p* by *c*.

**Obs.** If we eliminate $p$ from $x + f'(p) = 0$ and Eqn. (1), we get an equation involving no constant. *This is the* **singular solution** of Eqn. (1) *which gives the envelope of the family of straight lines* Eqn. (3).

*To obtain the singular solution, we proceed as follows :*

(*i*) Find the general solution by replacing $p$ by $c$ *i.e.,* Eqn. (3)

(*ii*) Differentiate this w.r.t. $c$ giving $x + f(c) = 0.$ ...(4)

(*iii*) Eliminate c from Eqns. (3) and (4) which will be the *singular solution*.

**Example 11.47.** *Solve p = sin (y – xp).* (*P.T.U., 2014*)
*Also find its singular solutions.* (*V.T.U., 2014*)

**Solution.** Given equation can be written as

$$\sin^{-1} p = y - xp \text{ or } y = px + \sin^{-1} p$$ which is the *Clairaut's equation*.

$\therefore$ its solution is $y = cx + \sin^{-1} c.$

*To find the singular solution,* differentiate (*i*) w.r.t. $c$ giving

$$0 = x + \frac{1}{\sqrt{1-c^2}} \qquad ...(ii)$$

To eliminate $c$ from (*i*) and (*ii*), we rewrite (*ii*) as

$$c = N(x^2 - 1)/x$$

Now substituting this value of $c$ in (*i*), we get

$$y = N(x^2 - 1) + \sin^{-1}\{N(x^2 - 1)/x\}$$

which is the desired singular solution.

*After the name of a youthful prodigy *Alexis Claude Clairaut* (1713–1765) who first solved this equation. A French mathematician who is also known for his work in Astronomy and Geodesy.

**Obs. Equations reducible to Clairaut's form.** Many equations of the first order but of higher degree can be easily reduced to the Clairaut's form by making suitable substitutions.

**Example 11.48.** *Solve*

$$y + \log p = x$$ (*Bhilai, 2018*)

**Solution:** The given equation is

$$y = x - \log p \quad ...(i)$$

Differentiating Eqn. $(i)$ $w.r.t.$ $x$, we get

$$\frac{dy}{dx} = 1 - \frac{1}{p}\frac{dp}{dx}$$

Let $\dfrac{dy}{dx} = p$, we get

$$p = 1 - \frac{1}{p}\frac{dp}{dx}$$

$$\frac{dp}{p(1-p)} = dx$$

$$\left[\frac{1}{p} + \frac{1}{1-p}\right]dp = dx$$

Integrating, $\log p - \log(1-p) = x + c$

$$\log \frac{p}{1-p} = x + c$$

$$x = \log \frac{p}{1-p} - c$$

Substituting this value of $x$ in Eqn. $(i)$ $y = \log\left(\dfrac{p}{1-p}\right) - c - \log p \quad ...(iii)$

$$y = -\log(1-p) - c$$

**Example 11.49:** *Solve $p^2 - 2p \sinh x - 1 = 0$*

**Solution:** The given equation is quadratic in $p$

$$\therefore \quad p = \frac{2\sin h\, x \pm \sqrt{4\sin h^2 x + 4}}{2} = \sin h\, x \pm \cos h\, x$$

$$= \left(\frac{e^x - e^{-x}}{2}\right) \pm \left(\frac{e^x + e^{-x}}{2}\right)$$

$$\therefore \quad \frac{dy}{dx} = e^x \text{ or } \frac{dy}{dx} = -e^{-x}$$

First we consider $\dfrac{dy}{dx} = e^x$

Integrating, we get $y = e^x + c$

$$y - e^x - c = 0$$

when $\dfrac{dy}{dx} = -e^{-x} \quad ...(i)$

Integrating, we get $y = e^{-x} + c$

$$y - e^{-x} - c = 0 \quad ...(ii)$$

From Eqns. $(i)$ and $(ii)$

$$(y - e^x - c)(y - e^{-x} - c) = 0$$

### PROBLEMS 11.12

1. Find the general and singular solution of the equations :
   (i) $(y + px)^2 = px^2$ (Goa, 2018)
   (ii) $p = \log (px - y)$. (P.T.U., 2013)
   (iii) $xp^3 - yp^2 + 1 = 0$ (V.T.U., 2015)
   (iv) $\sin px \cos y = \cos px \sin y + p$ (W.B.T.U., 2015 ; C.S.V.T.U., 2015)

Solve the following equations :

2. $y + 2\left(\dfrac{dy}{dx}\right)^2 = (x+1)\dfrac{dy}{dx}$.
3. $(y - px)(p - 1) = p$.
4. $(x - a)\left(\dfrac{dy}{dx}\right)^2 + (x - y)\dfrac{dy}{dx} - y = 0$.
5. $x^2(y - px) = yp^2$. (Andhra, 2013 ; C.S.V.T.U., 2015 ; Andhra, 2014)
6. $(px + y)^2 = py^2$.
7. $(px - y)(x + py) = 2p$. (V.T.U., 2012)

## 11.15 OBJECTIVE TYPE OF QUESTIONS

### PROBLEMS 11.13

*Fill up the blanks or choose the correct answer in the following problems :*

1. $y = cx - c^2$, is the general solution of the differential equation
   (i) $(y')^2 - xy' + y = 0$ (ii) $y'' = 0$ (iii) $y' = c$ (iv) $(y')^2 + xy' + y = 0$.
2. The differential equation having a basis for its solution as sinh $6x$ and cosh $6x$ is
   (i) $y'' + 36y = 0$ (ii) $y'' - 36y = 0$ (iii) $y'' + 6y = 0$ (iv) none of these.
3. The differential equation $(dx/dy)^2 + 5y^{1/3} = x$ is
   (i) linear of degree 3 (ii) non-linear of order 1 and degree 6
   (iii) non-linear of order 1 and degree 2.
4. The differential equation $ydx/dy + 1 = y$, $y(0) = 1$, has
   (i) a unique solution (ii) two solutions
   (iii) infinite number of solutions (iv) no solution
5. Solution of $(x^2 + y^2)\,dy = xy\,dx$ is .......
6. Solution of $(3x - 2y)\,dx = xdy$ is .......
7. Solution of $dy/dx - y = 2xy^2 e^{-x}$ is .......
8. The differential equation $(y^2 e^{xy^2} + 6x)\,dx + (2xye^{xy^2} - 4y)\,dy = 0$ is
   (i) linear, homogeneous and exact (ii) non-linear, homogeneous and exact
   (iv) non-linear, non-homogeneous and exact (iv) non-linear, non-homogeneous and inexact.
9. Solution of $xdx + ydy + \dfrac{xdy - ydx}{x^2 + y^2}$ is .......
10. Solution of $dy/dx = \dfrac{x^3 + y^3}{xy^2}$ is ......
11. The differential equation $(x + x^8 + ay^2)\,dx + (y^8 - y + bxy)\,dy = 0$ is exact if
    (i) $b = 2a$ (ii) $a = b$ (iii) $a \neq 2b$ (iv) $a = 1, b = 3$.
12. Solution of $xy\,(1 + xy^2)\,dy = dx$ is .......
13. Solution of $xp^2 - yp + a = 0$ is .......
14. The general solution of the differential equation $p = \log (px - y)$ is
    (a) $y = px + e^p$ (b) $y = px - e^p$ (c) $y = px - e^c$ (d) $y = cx - e^c$ (V.T.U., 2013)
15. Solution of $\dfrac{dy}{dx} - \dfrac{dx}{dy} = \dfrac{x}{y} - \dfrac{y}{x}$ is .......
16. The order of the differential equation $(1 + y_1^2)^{3/2}/y_2 = c$ is .......
17. The general solution of $\dfrac{1}{x^2y^2}(xdy + ydx) = 0$ is .......

18. Integrating factor of the differential equation $\frac{dx}{dy} + \frac{3x}{y} = \frac{1}{y^2}$ is

(a) $e^{y^3}$ (b) $y^3$ (c) $x^3$ (d) $-y^3$.

19. Solution of the equation $\frac{dy}{dx} = \frac{y}{x} - \text{cosec}\,\frac{y}{x}$ is

(a) $\cos(y/x) - \log x = c$ (b) $\cos(y/x) + \log x = c$

(c) $\cos^2(y/x) + \log x = c$ (d) $\cos^2(y/x) - \log x = c$. (V.T.U., 2010)

20. Solution of $x\sqrt{(1+x^2)} + y\sqrt{(1+y^2)}\,dy/dx = 0$ is ........

21. Solution of $dy/dx + y = 0$ given $y(0) = 5$ is .......

22. The substitution that transforms the equation $\frac{dy}{dx} = \frac{x+y+1}{2x+2y+3}$ to homogeneous form is .......

23. Integrating factor of $xy' + y = x^3y^6$ is .......

24. Solution of the exact differential equation $Mdx + Ndy = 0$ is .......

25. Solution of $(2x^3y^2 + x^4)\,dx + (x^4y + y^4)\,dy = 0$ is ........

26. The general solution of the differential equation $\frac{dy}{dx} + \frac{y}{x} = \tan 2x$ is .......

27. Degree of the differential equation $\left(\frac{d^2y}{dx^2}\right)^2 + x\left(\frac{dy}{dx}\right)^5 x^2y = 0$ is

(a) 2 (b) 0 (c) 3 (d) 5

28. Integrating factor of the differential equation $\frac{dy}{dx} + y\cos x = \frac{\sin 2x}{2}$ is

(a) $e^{\sin^2 x}$ (b) $e^{\sin^3 x}$ (c) $e^{\sin x}$ (d) $\sin x$

29. The differential equation of the family of circles with centre as origin is ......

30. Solution of $x\,e^{-x^2}\,dx + \sin y\,dy = 0$ is ........

31. Solution of $p = \sin(y - xp)$ is

(a) $y = \frac{c}{x} + \sin^{-1} c$ (b) $y = cx + \sin c$ (c) $y = cx + \sin^{-1} c$ (d) $y = x + \sin^{-1} c$. (V.T.U., 2011)

32. Differential equation obtained by eliminating $A$ and $B$ form $y = A\cos x + B\sin x$ is $d^2y/dx^2 - y = 0$. (True or False)

33. $(x^3 - 3xy^2)\,dx + (y^3 - 2x^2y)\,dy = 0$ is an exact differential equation. (True or False)

34. The singular solution of the equation $y = px - \log p$ is

(a) $y = 1 - \log x$ (b) $y = 1 - \log(1/x)$ (c) $y = \log x - 2x$ (d) none of these. (V.T.U., 2012)

35. If $M$ and $N$ in $M(x, y)\,dx + N(x, y)\,dy = 0$ satisfy the equation $\frac{\partial M}{\partial y} + \frac{3}{y}M = \frac{\partial N}{\partial x}$, then show that $y^3$ is an integrating factor. (U.P.T.U., 2012)

36. $y = (c_1 + c_2x + c_3x^2)\,e^x$ is the solution of the differential equation ........, where $c_1, c_2, c_3$ are constants. (U.P.T.U., 2012)

37. Solution of $y(2xy + e^x)\,dx - e^x\,dy = 0$ is ...... . (Delhi, 2012S)

38. For what values of $a$ and $b$ the differential equation $(y + x)^3\,dx + (ax + by^3) = 0$ is exact. (P.T.U., 2015)

CHAPTER 12

# Applications of Differential Equations of First Order



## 12.1 INTRODUCTION

In this chapter, we shall consider only such practical problems which give rise to differential equations of the first order. The fundamental principles required for the formation of such differential equations are given in each case and are followed by illustrative examples.

## 12.2 GEOMETRIC APPLICATIONS

(*a*) *Cartesian coordinates.* Let $P(x, y)$ be any point on the curve $f(x, y) = 0$ (Fig. 12.1), then [as per 4.6 §(1) and 4.11(1) and (4)], we have

(*i*) slope of the tangent at $P(= \tan \psi) = dy/dx$

(*ii*) equation of the tangent at $P$ is

$$Y - y = \frac{dy}{dx}(X - x)$$

so that its $x$-intercept $(= OT)$

$$= x - y \, . \, dx/dy$$

and $y$-intercept $(= OT') = y - x \, . \, dy/dx$

(*iii*) equation of the normal at $P$ is $Y - y = -\dfrac{dx}{dy}(X - x)$

(*iv*) length of the tangent $(= PT) = y\sqrt{[1 + (dx/dy)^2]}$

(*v*) length of the normal $(= PN) = y\sqrt{[1 + (dy/dx)^2]}$

(*vi*) length of the sub-tangent $(= TM) = y \, . \, dx/dy$

(*vii*) length of the sub-normal $(= MN) = y \, . \, dy/dx$

(*viii*) $\dfrac{ds}{dx} = [1 + (dy/dx)^2]$; $\dfrac{ds}{dy} = \sqrt{[1 + (dx/dy)^2]}$

(*ix*) differential of the area $= ydx$ or $xdy$

(*x*) $\rho$, radius of curvature at $P = \dfrac{[1 + (dy/dx)^2]^{3/2}}{d^2y/dx^2}$

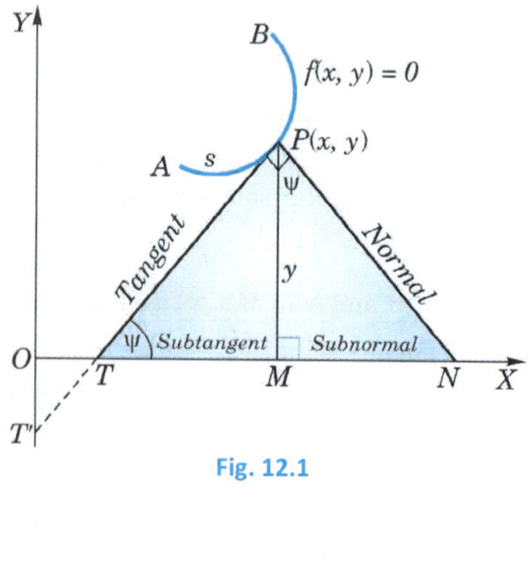

Fig. 12.1

(*b*) *Polar coordinates*. Let $P(r, \theta)$ be any point on the curve $r = f(\theta)$ (Fig. 12.2), then [as per § 4.7, 4.9 (2) and 4.11 (4)], we have

(*i*) $\psi = \theta + \phi$

(*ii*) $\tan \phi = r d\theta/dr$, $p = r \sin \phi$

(*iii*) $$\frac{1}{p^2} = \frac{1}{r^2} + \frac{1}{r^4}\left(\frac{dr}{d\theta}\right)^2$$

(*iv*) polar sub-tangent ($= OT$) $= r^2 d\theta/dr$

(*v*) polar sub-normal ($ON$) $= dr/d\theta$

(*vi*) $$\frac{ds}{dr} = \sqrt{\left[1 + \left(r\frac{d\theta}{dr}\right)^2\right]};\ \frac{ds}{d\theta} = \sqrt{\left[r^2 + \left(\frac{dr}{d\theta}\right)^2\right]}$$

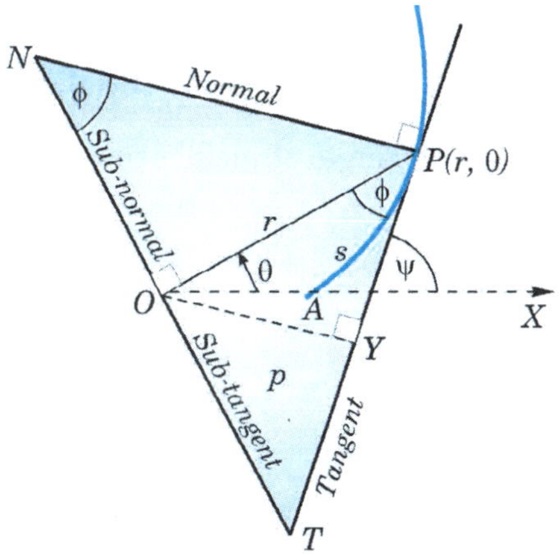

Fig. 12.2

**Example 12.1.** *Show that the curve in which the portion of the tangent included between the co-ordinates axes is bisected at the point of contact is a rectangular hyperbola.*

**Solution.** Let the tangent at any point $P(x, y)$ of a curve cut the axes at $T$ and $T'$ (Fig. 12.3).

We know that its $x$-intercept ($= OT$) $= x - y \, . \, dx/dy$

and $y$-intercept ($= OT'$) $= y - x \, . \, dy/dx$

$\therefore$ the co-ordinates of $T$ and $T'$ are

$$(x - y \, . \, dx/dy, 0), (0, y - x \, dy/dx)$$

Since $P$ is the mid-point of $TT'$

$$\therefore \quad \frac{[x - y.dx/dy] + 0}{2} = x$$

or $$x - y \, . \, dx/dy = 2x \text{ or } x\,dy + y\,dx = 0$$

or $$d(xy) = 0 \text{ Integrating, } xy = c$$

which is the equation of a rectangular hyperbola, having $x$ and $y$ axes as its asymptotes.

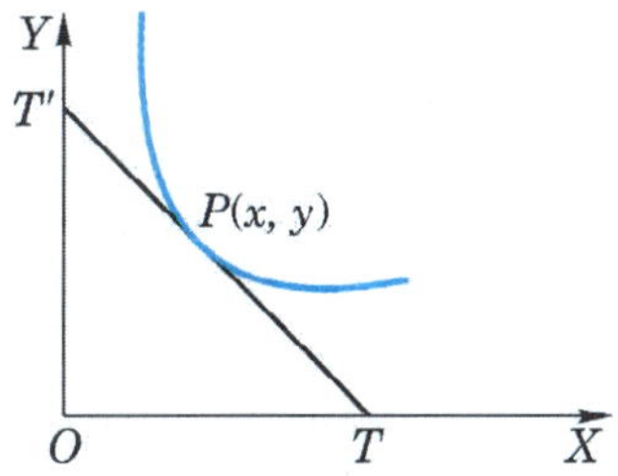

Fig. 12.3

**Example 12.2.** *Find the curve for which the normal makes equal angles with the radius vector and the initial line.*

**Solution.** Let $PT$ and $PN$ be the tangent and normal at $P(r, \theta)$ of the curve so that

$$\tan \phi = r \, d\theta/dr$$

By the condition of the problem,

$$\angle OPN = 90° - \phi = \angle ONP \text{ (Fig. 12.4).}$$

$$\therefore \quad \theta = \angle PON = 180° - (180° - 2\phi) = 2\phi$$

or $$\theta/2 = \phi \quad \therefore \quad \tan \frac{\theta}{2} = \tan \phi = r\frac{d\theta}{dr}.$$

Here the variables are separable.

$$\therefore \quad \frac{dr}{r} = \frac{\cos \theta/2}{\sin \theta/2} d\theta$$

Integrating both sides $\log r = 2 \log \sin \theta/2 + \log c$

or $$r = c \sin^2 \theta/2 = \frac{1}{2} c(1 - \cos \theta)$$

Thus the curve is the cardioid $r = a(1 - \cos \theta)$.

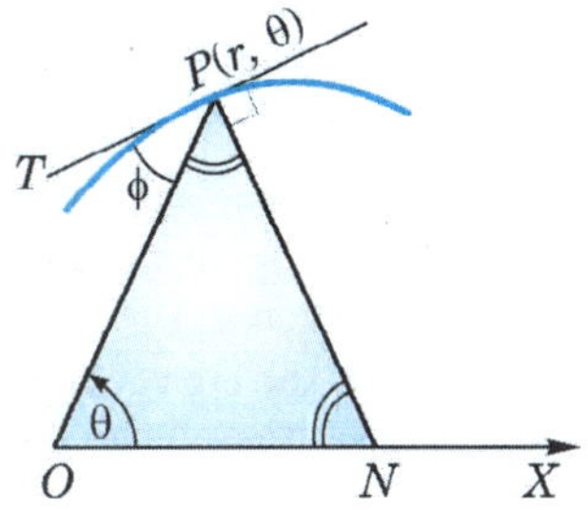

Fig. 12.4

**Example 12.3.** *Find the shape of a reflector such that light coming from a fixed source is reflected in parallel rays.*

**Solution.** Taking the fixed source of light as the origin and the $X$-axis parallel to the reflected rays; the reflector will be a surface generated by the revolution of a curve $f(x, y) = 0$ about $X$-axis (Fig. 12.5).

In the $XY$-plane, let $PP'$ be the reflected ray, where $P$ is the point $(x, y)$ on the curve $f(x, y) = 0$.

If $TPT'$ be the tangent at $P$, then

$\because$ angle of incidence = angle of reflection,

$\therefore$ $\phi = \angle OPT = \angle P'PT' = \angle OTP = \psi$

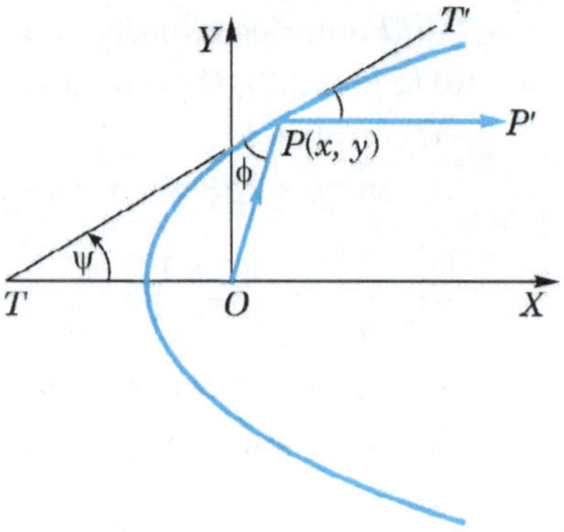

Fig. 12.5

i.e., $$p = \frac{dy}{dx} = \tan \angle XOP = \tan 2\phi$$

$$= \frac{2 \tan \phi}{1 - \tan^2 \phi} = \frac{2p}{1 - p^2}$$

or $$2x = \frac{y}{p} - yp \text{ which is solvable for } x \quad ...(i)$$

$\therefore$ differentiating $(i)$ w.r.t. $y$, $\dfrac{2}{p} = \dfrac{1}{p} - \dfrac{y}{p^2}\dfrac{dp}{dy} - p - y\dfrac{dp}{dy}$

i.e., $$\left(\frac{1}{p} + p\right) + \left(\frac{1}{p^2} + 1\right) y \frac{dp}{dy} = 0 \text{ or } \left(\frac{1}{p} + p\right)\left(1 + \frac{y}{p}\frac{dp}{dy}\right) = 0$$

This gives $dp/p = -dy/y$

Integrating, $\log p = \log c - \log y$, i.e., $p = c/y$ ...(ii)

Thus eliminating $p$ from $(i)$ and $(ii)$, we have family of curves $y^2 = 2cx + c^2$.

Hence the reflector is a member of the family of paraboloids of revolution $y^2 + z^2 = 2cx + c^2$.

## PROBLEMS 12.1

1. Find the equation of the curve which passes through
   (i) the point (3, –4) and has the slope $2y/x$ at the point $(x, y)$ on it.
   (ii) the origin and has the slope $x + 3y - 1$.
2. At every point on a curve the slope is the sum of the abscissa and the product of the ordinate and the abscissa, and the curve passes through (0, 1). Find the equation of the curve.
3. A curve is such that the length of the perpendicular from origin on the tangent at any point $P$ of the curve is equal to the abscissa of $P$. Prove that the differential equation of the curve is
   $$y^2 - 2xy\frac{dy}{dx} - x^2 = 0, \text{ and hence find the curve.}$$
4. A plane curve has the property that the tangents from any point on the $y$-axis to the curve are of constant length $a$. Find the differential equation of the family to which the curve belongs and hence obtain the curve.
5. Determine the curve whose sub-tangent is twice the abscissa of the point of contact and passes through the point (1, 2).
6. Determine the curve in which the length of the sub-normal is proportional to the square of the ordinate.
7. The tangent at any point of a certain curve forms with the coordinate axes a triangle of constant area $A$. Find the equation to the curve.
8. Find the curve which passes through the origin and is such that the area included between the curve, the ordinate and the $x$-axis is twice the cube of that ordinate.
9. Find the curve whose (i) polar sub-tangent is constant.
   (ii) polar sub-normal is proportional to the sine of the vectorial angle.
10. Determine the curve for which the angle between the tangent and the radius vector is twice the vectorial angle.
11. Find the curve for which the tangent at any point $P$ on it bisects the angle between the ordinate at $P$ and the line joining $P$ to the origin.
12. Find the curve for which the tangent, the radius vector $r$ and the perpendicular from the origin on the tangent form a triangle of area $kr^2$.

## 12.3 (1) ORTHOGONAL TRAJECTORIES

*Two families of curves such that every member of either family cuts each member of the other family at right angles are called* **orthogonal trajectories** *of each other (Fig. 12.6).*

The concept of the orthogonal trajectories is of wide use in applied mathematics especially in field problems. For instance, in an electric field, the paths along which the current flows are the orthogonal trajectories of the equipotential curves and *vice versa*. In fluid flow, the stream lines and the equipotential lines (lines of constant velocity potential) are orthogonal trajectories. Likewise, the lines of heat flow for a body are perpendicular to the isothermal curves. The problem of finding the orthogonal trajectories of a given family of curves depends on the solution of the first order differential equations.

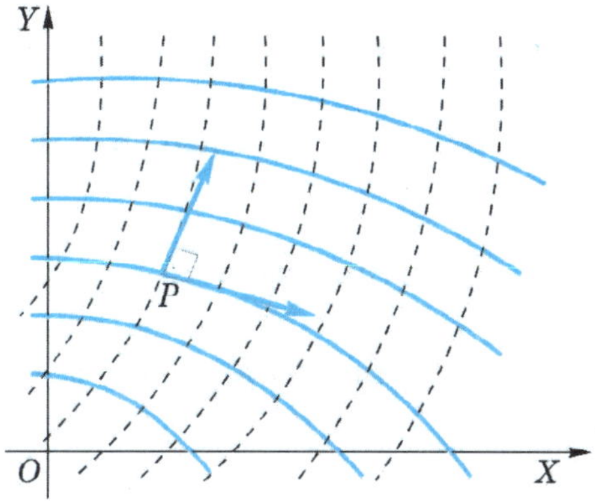

Fig. 12.6

**(2) To find the orthogonal trajectories of the family of curves F(x, y, c) = 0.**

(*i*) *Form its differential equation in the form f(x, y, dy/dx) = 0 by eliminating c.*

(*ii*) *Replace, in this differential equation, dy/dx by –dx/dy*, (so that the product of their slopes at each point of intersection is – 1).

(*iii*) *Solve the differential equation of the orthogonal trajectories i.e., f (x, y, – dx/dy) = 0.*

**Example 12.4.** *If the stream lines (paths of fluid particles) of a flow around a corner are xy = constant find their orthogonal trajectories (called equipotential lines-§ 20.6)*

**Solution.** Taking the axes as the walls, the stream lines of the flow around the corner of the walls is

$$xy = c \quad \text{...}(i)$$

Differentiating, we get, $$x\frac{dy}{dx} + y = 0 \quad \text{...}(ii)$$

as the differential equation of the given family (*i*).

Replacing $\frac{dy}{dx}$ by $-\frac{dx}{dy}$ in Eqn. (*ii*), we obtain $x\left(-\frac{dx}{dy}\right) + y = 0$

or $$xdx - ydy = 0 \quad \text{...}(iii)$$

as the differential equation of the orthogonal trajectories.

Integrating Eqn. (*iii*), we get $x^2 - y^2 = c'$ as the required orthogonal trajectories of Eqn. (*i*) *i.e.*, the *equipotential lines*, shown dotted in Fig. 12.7.

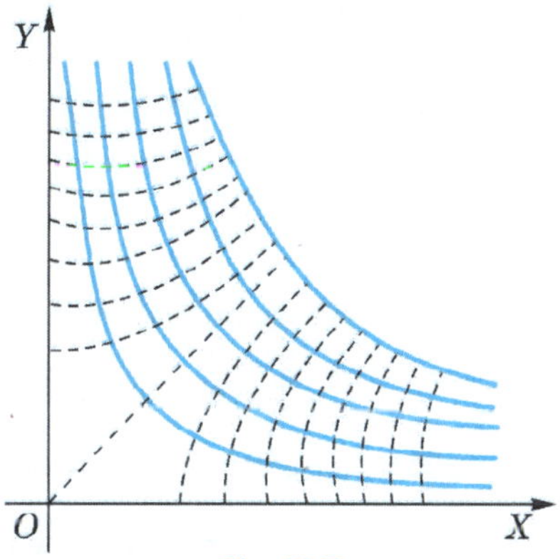

Fig. 12.7

**Example 12.5.** *Find the orthogonal trajectories of the family of confocal conics $\frac{x^2}{a^2} + \frac{y^2}{a^2+\lambda} = 1$, where λ is the parameter.* (*Andhra, 2017*)

**Solution.** Differentiating the given equation, we get $\frac{2x}{a^2} + \frac{2y}{a^2+\lambda}\frac{dy}{dx} = 0$

or $$\frac{y}{a^2+\lambda} = -\frac{x}{a^2(dy/dx)} \quad \text{or} \quad \frac{y^2}{a^2+\lambda} = \frac{-xy}{a^2(dy/dx)}$$

Substituting this in the given equation, we get

$$\frac{x^2}{a^2} - \frac{xy}{a^2(dy/dx)} = 1 \quad \text{or} \quad (x^2 - a^2)\frac{dy}{dx} = xy \quad \text{...}(i)$$

which is the differential equation of the given family.

Changing $dy/dx$ to $-dx/dy$ in Eqn. (*i*), we get $(a^2 - x^2)\,dx/dy = xy$ as the differential equation of the orthogonal trajectories.

Separating the variables and integrating, we obtain

$$\int y\,dy = \int \frac{a^2 - x^2}{x}dx + c \quad \text{or} \quad \frac{1}{2}y^2 = a^2 \log x - \frac{1}{2}x^2 + c$$

or $$x^2 + y^2 = 2a^2 \log x + c' \quad [c' = 2c]$$

which is the equation of the required orthogonal trajectories.

**Example 12.6.** *Find the orthogonal trajectories of a system of confocal and coaxial parabolas.*
*(JNTUK, 2023 ; Andhra, 2015 ; V.T.U., 2015)*

**Solution.** The equation of the family of confocal parabolas having $x$-axis as their axis, is of the form

$$y^2 = 4a(x + a) \quad \text{...}(i)$$

Differentiating, $$y\frac{dy}{dx} = 2a \quad \text{...}(ii)$$

Substituting the value of $a$ from Eqn. $(ii)$ in Eqn. $(i)$, we get $y^2 = 2y\dfrac{dy}{dx}\left(x + \dfrac{1}{2}y\dfrac{dy}{dx}\right)$

*i.e.,* $$y\left(\frac{dy}{dx}\right)^2 + 2x\frac{dy}{dx} - y = 0 \text{ as the differential equation of the family.} \quad \text{...}(iii)$$

Replacing $\dfrac{dy}{dx}$ by $-\dfrac{dx}{dy}$ in Eqn. $(iii)$, we obtain $y\left(\dfrac{dx}{dy}\right)^2 - 2x\dfrac{dx}{dy} - y = 0$

or $$y\left(\frac{dy}{dx}\right)^2 + 2x\frac{dy}{dx} - y = 0 \text{ which is the same as } (iii).$$

Thus we see that a system of confocal and coaxial parabolas is *self-orthogonal, i.e.*, each member of the family Eqn. $(i)$ cuts every other member of the same family orthogonally.

**(3) To find the orthogonal trajectories of the curves F(r, θ, c) = 0.**

$(i)$ *Form its differential equation in the form $f(r, \theta, dr/d\theta) = 0$ by eliminating c.*

$(ii)$ *Replace in this differential equation,*

$$\frac{dr}{d\theta} \text{ by } -r^2\frac{d\theta}{dr}$$

[∵ for the given curve through $P(r, \theta)$ $\tan\phi = r d\theta/dr$

and for the orthogonal trajectory through $P$

$$\tan\phi' = \tan(90° + \phi) = -\cot\phi = -\frac{1}{r}\frac{dr}{d\theta}$$

Thus for getting the differential equation of the orthogonal trajectory

$$r\frac{d\theta}{dr} \text{ is to be replaced by } -\frac{1}{r}\frac{dr}{d\theta}$$

or $$\frac{dr}{d\theta} \text{ is to be replaced by } -r^2\frac{d\theta}{dr}.]$$

$(iii)$ *Solve the differential equation of the orthogonal trajectories*

*i.e.,* $$f(r, \theta, -r^2 d\theta/dr) = 0.$$

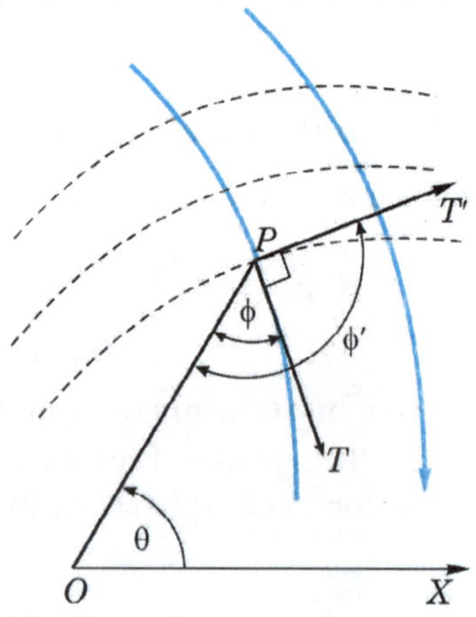

Fig. 12.8

**Example 12.7.** *Find the orthogonal trajectory of the cardioids $r = a(1 - \cos\theta)$.*
*(JNTUK, 2022 ; V.T.U., 2014)*

**Solution.** Differentiating $$r = a(1 - \cos\theta). \quad \text{...}(i)$$

with respect to θ, we get $$\frac{dr}{d\theta} = a\sin\theta \quad \text{...}(ii)$$

Eliminating $a$ from Eqns. $(i)$ and $(ii)$, we obtain

$$\frac{dr}{d\theta}\cdot\frac{1}{r} = \frac{\sin\theta}{1-\cos\theta} = \cot\frac{\theta}{2} \text{ which is the differential equation of the given family.}$$

Replacing $dr/d\theta$ by $-r^2\, d\theta/dr$, we obtain

$$\frac{1}{r}\left(-r^2\frac{d\theta}{dr}\right) = \cot\frac{\theta}{2} \quad \text{or} \quad \frac{dr}{r} + \tan\frac{\theta}{2}\,d\theta = 0$$

as the differential equation of orthogonal trajectories. It can be rewritten as

$$\frac{dr}{r} = -\frac{(\sin\theta/2)\,d\theta}{\cos\theta/2}$$

Integrating, $\log r = 2 \log \cos \theta/2 + \log c$

or $$r = c \cos^2 \theta/2 = \frac{1}{2} c(1 + \cos\theta) \quad \text{or} \quad r = a'(1 + \cos\theta)$$

which is the required orthogonal trajectory.

**Example 12.8.** *Find the orthogonal trajectory of the family of curves $r^n = a \sin n\theta$.* (*V.T.U., 2016*)

**Solution.** We have $n \log r = \log a + \log \sin n\theta$.

Differentiating w.r.t. $\theta$, we have

$$\frac{n}{r}\frac{dr}{d\theta} = \frac{n \cos n\theta}{\sin n\theta} \quad \text{or} \quad \frac{1}{r}\frac{dr}{d\theta} = \cot n\theta$$

Replacing $dr/d\theta$ by $-r^2\, d\theta/dr$, we obtain

$$\frac{1}{r}\left(-r^2 \frac{d\theta}{dr}\right) = \cot n\theta \quad \text{or} \quad \tan n\theta \cdot d\theta - \frac{dr}{r} = 0$$

Integrating, $$\int \frac{dr}{r} + \int \frac{\sin n\theta}{\cos n\theta}\, d\theta = c,$$

*i.e.,* $$\log r - \frac{1}{n} \log \cos n\theta = c \quad \text{or} \quad \log (r^n/\cos n\theta) = nc = \log b. \text{ (say)}$$

or $r^n = b \cos n\theta$, which is the required orthogonal trajectory.

## PROBLEMS 12.2

Find the orthogonal trajectories of the family of :

1. Parabolas $y^2 = 4ax$.
2. Parabolas $y = ax^2$. (*Rohtak, 2011*)
3. Semi-cubical parabolas $ay^2 = x^3$. (*Andhra, 2016*)
4. Coaxial circles $x^2 + y^2 + 2\lambda x + c = 2$, $\lambda$ being the parameter.
5. Confocal conics $\dfrac{x^2}{a^2 + \lambda} + \dfrac{y^2}{b^2 + \lambda} = 1$, $\lambda$ being the parameter. (*JNTUK, 2023; V.T.U., 2016*)
6. $x^{2/3} + y^{2/3} = a^{2/3}$ (*J.N.T.U., 2022*)
7. Cardioids $r = a\,(1 + \cos\theta)$.
8. Family of cardioids $r = a \cos^2(\theta/2)$ (*V.T.U., 2015*)
9. $r = 2a\,(\cos\theta + \sin\theta)$ (*V.T.U., 2010 S*)
10. Confocal and coaxial parabolas $r = 2a/(1 + \cos\theta)$. (*JNTUK, 2020*)
11. Curves $r^2 = a^2 \cos 2\theta$.
12. $r^n \cos n\theta = a^n$. (*V.T.U. 2011*)
13. $r^n \sin n\theta = a^n$ (*J.N.T.U., 2015*)
14. Show that the family of parabolas $x^2 = 4a\,(y + a)$ is self orthogonal. (*Andhra, 2018*)
15. Show that the family of curves $r^n = a \sec n\theta$ and $r^n = b \operatorname{cosec} n\theta$ are orthogonal.
16. The *electric lines of force* of two opposite charges of the same strength at $(\pm 1, 0)$ are circles (through these points) of the form $x^2 + y^2 - ay = 1$. Find their *equipotential lines* (orthogonal trajectories).

   [**Isogonal trajectories.** *Two families of curves such that every member of either family cuts each member of the other family at a constant angle $\alpha$ (Say), are called* ***isogonal trajectories*** *of each other.* The slopes $m$, $m'$ of the tangents to the corresponding curves at each point, are connected by the relation $\dfrac{m - m'}{1 + mm'} = \tan a = \text{const.}$]
17. Find the isogonal trajectories of the family of circles $x^2 + y^2 = a^2$ which intersect at 45°.

## 12.4 PHYSICAL APPLICATIONS

(1) Let a body of mass $m$ start moving from $O$ along the straight line $OX$ under the action of a force $F$. After any time $t$, let it be moving at $P$ where $OP = x$, then

(*i*) its velocity $(v) = \dfrac{dx}{dt}$ (*ii*) its acceleration $(a) = \dfrac{dv}{dt}$ or $\dfrac{vdv}{dx}$ or $\dfrac{d^2x}{dt^2}$

If, however, the body be moving along a curve, then

(*i*) its velocity $(v) = ds/dt$ and

(*ii*) its acceleration $(a) = \dfrac{dv}{dt}, v\dfrac{dv}{ds}$ or $\dfrac{d^2s}{dt^2}$.

The quantity $mv$ is called the *momentum.*

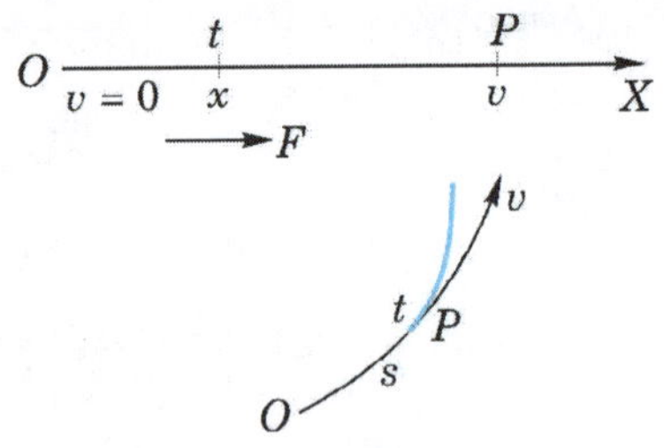

Fig. 12.9

**(2) Newton's second law** *states that* $F = \dfrac{d}{dt}(mv)$.

If $m$ is constant, then $F = m\dfrac{dv}{dt} = ma$, *i.e., net force = mass × acceleration.*

**(3) Hooke's law*** *states that tension of an elastic string (or a spring) is proportional to extension of the string (or the spring) beyond its natural length.*

Thus $\quad\quad$ $\mathbf{T = \lambda e/1}$,

where $e$ is the extension beyond the natural length $l$ and $\lambda$ is the *modulus of elasticity.*

Sometimes for a spring, we write $\mathbf{T = ke}$,

where $e$ is the extension beyond the natural length and $k$ is the *stiffness of the spring.*

**(4) Systems of units**

*I.* **F.P.S.** [foot (*ft.*) pound (*lb.*), second (*sec.*)] **system**. If mass $m$ is in *pounds* and acceleration ($a$) is in *ft/sec*², then the force $F(= ma)$ is in *poundals.*

*II.* **C.G.S.** [centimetre (*cm.*), gram (*g*), second (*sec*)] **system**. If mass $m$ is in *grams* and acceleration $a$ is in *cm/sec*² then the force $F(= ma)$ is *dynes.*

*III.* **M.K.S.** [metre (*m*), kilogram (*kg.*), second (*sec*)] **system**. If mass $m$ is in kilograms and acceleration $a$ in *m/sec*², then the force $F(= ma)$ is in *newtons* (*nt*).

These are called *absolute units.* If $g$ is the acceleration due to gravity and $w$ is the weight of the body, then $w/g$ is the mass of the body in *gravitational units.*

$$g = 32 \text{ ft/sec}^2 = 980 \text{ cm/sec}^2 = 9.8 \text{ m/sec}^2 \text{ approx.}$$

**Example 12.9. Motion of a boat across a stream.** *A boat is rowed with a velocity u directly across a stream of width a. If the velocity of the current is directly proportional to the product of the distances from the two banks, find the path of the boat and the distance down stream to the point where it lands.*

**Solution.** Taking the origin at the point from where the boat starts, let the axes be chosen as in Fig. 12.10.

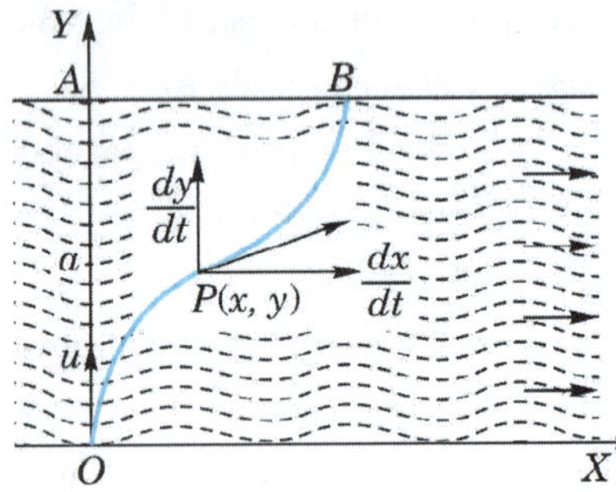

Fig. 12.10

At any time $t$ after its start from $O$, let the boat be at $P(x, y)$, so that

$dx/dt$ = velocity of the current $= ky(a - y)$

$dy/dt$ = velocity with which the boat is being rowed $= u$.

$$\therefore \quad \frac{dy}{dx} = \frac{dy}{dt} \div \frac{dx}{dt} = \frac{u}{ky(a-y)} \quad \quad ...(i)$$

This gives the direction of the resultant velocity of the boat which is also the direction of the tangent to the path of the boat.

Now Eqn. (*i*) is of variables separable form and we can write it as

$$y(a - y)dy = \frac{u}{k}\,dx$$

Integrating, we get $\quad \dfrac{ay^2}{2} - \dfrac{y^3}{3} = \dfrac{u}{k}x + c$

Since $y = 0$ when $x = 0$, $\therefore\ c = 0$.

Hence the equation to the path of the boat is $x = \dfrac{k}{6u}y^2(3a - 2y)$

Putting $y = a$, we get the distance $AB$, down stream where the boat lands $= ka^3/6u$.

*Named after an English physicist *Robert Hooke* (1635–1703) who had discovered the law of gravitation earlier than Newton.

**Example 12.10. Resisted motion.** *A moving body is opposed by a force per unit mass of value cx and resistance per unit of mass of value $bv^2$ where x and v are the displacement and velocity of the particle at that instant. Find the velocity of the particle in terms of x, if it starts from rest.*

**Solution.** By Newton's second law, the equation of motion of the body is $v\dfrac{dv}{dx} = -cx - bv^2$

or $$v\frac{dv}{dx} + bv^2 = -cx \quad \text{...(i)}$$

This is Bernoulli's equation. $\therefore$ Put $v^2 = z$ and $2v\, dv/dx = dz/dx$, so that (*i*) becomes

$$\frac{dz}{dx} + 2bz = -2cx \quad \text{...(ii)}$$

This is Leibnitz's linear equation and I.F. $= e^{2bx}$.

$\therefore$ the solution of (*ii*) is $ze^{2bx} = -\int 2cxe^{2bx}\,dx + c'$ [Integrate by parts]

$$= -2c\left[x \cdot \frac{e^{2bx}}{2b} - \int 1 \cdot \frac{e^{2bx}}{2b}\,dx\right] + c' = -\frac{cx}{b}e^{2bx} + \frac{c}{2b^2}e^{2bx} + c'$$

or $$v^2 = \frac{c}{2b^2} + c'e^{-2bx} - \frac{cx}{b} \quad \text{...(iii)}$$

Initially $v = 0$ when $x = 0$ $\therefore$ $0 = c/2b^2 + c'$.

Thus, substituting $c' = -c/2b^2$ in (*iii*), we get $v^2 = \dfrac{c}{2b^2}(1 - e^{-2bx}) - \dfrac{cx}{b}$.

**Example 12.11. Resisted vertical motion.** *A particle falls under gravity in a resisting medium whose resistance varies with velocity. Find the relation between distance and velocity if initially the particle starts from rest.*

**Solution.** After falling a distance $s$ in time $t$ from rest, let $v$ be velocity of the particle. The forces acting on the particle are its weight $mg$ downwards and resistance $m\lambda v$ upwards.

$\therefore$ equating of motion is $m\dfrac{dv}{dt} = mg - m\lambda v$

or $$\frac{dv}{dt} = g - \lambda v \quad \text{or} \quad \frac{dv}{g - \lambda v} = dt$$

Integrating, $\displaystyle\int \frac{dv}{g - \lambda v} = \int dt + c$ or $-\dfrac{1}{\lambda}\log(g - \lambda v) = t + c$

Since $v = 0$ when $t = 0$, $\therefore$ $c = -\dfrac{1}{\lambda}\log g$

Thus $$\frac{1}{\lambda}\log\left[\frac{g}{g - \lambda v}\right] = t \quad \text{or} \quad \frac{g - \lambda v}{g} = e^{-\lambda t}$$

or $$\frac{ds}{dt} = v = \frac{g}{\lambda}(1 - e^{-\lambda t}) \quad \text{...(i)}$$

Integrating, $$s = \frac{g}{\lambda}\int (1 - e^{-\lambda t})dt + c' \quad \text{or} \quad s = \frac{g}{\lambda}\left(t + \frac{1}{\lambda}e^{-\lambda t}\right) + c'$$

Since $s = 0$ when $t = 0$, $\therefore$ $c' = -g/\lambda^2$

Thus $$s = \frac{g}{\lambda}t + \frac{g}{\lambda^2}(e^{-\lambda t} - 1) \quad \text{...(ii)}$$

Eliminating $t$ from Eqns. (*i*) and (*ii*), we get

$$s = \frac{g}{\lambda^2}\log\left(\frac{g}{g - \lambda v}\right) - \frac{v}{\lambda}$$

which is the desired relation between $s$ and $v$.

**Example 12.12.** *A body of mass m, falling from rest is subject to the force of gravity and an air resistance proportional to the square of the velocity (i.e., $kv^2$). If it falls through a distance x and possesses a velocity v at that instant, prove that*

$$\frac{2kx}{m} = \log \frac{a^2}{a^2 - v^2}, \text{ where } mg = ka^2. \qquad (U.P.T.U., 2011)$$

**Solution.** If the body be moving with the velocity $v$ after having fallen through a distance $x$, then its equation of motion is

$$mv \frac{dv}{dx} = mg - kv^2 \quad \text{or} \quad mv \frac{dv}{dx} = k(a^2 - v^2). \qquad [\because mg = ka^2] \quad ...(i)$$

$\therefore$ separating the variables and integrating, we get $\int \frac{vdv}{a^2 - v^2} = \int \frac{k}{m} dx + c$

or
$$-\frac{1}{2} \log (a^2 - v^2) = \frac{kx}{m} + c \qquad ...(ii)$$

Initially, when $x = 0$, $v = 0$. $\therefore -\frac{1}{2} \log a^2 = c$ ...(iii)

Subtracting Eqn. (*iii*) from (*ii*), we have $\frac{1}{2}[\log a^2 - \log (a^2 - v^2)] = kx/m$

or
$$\frac{2kx}{m} = \log\left(\frac{a^2}{a^2 - v^2}\right).$$

**Obs.** *When the resistance becomes equal to the weight, the acceleration becomes zero and particle continues to fall with a constant velocity, called the* **limiting** *or* **terminal** *velocity.* From Eqn. (*i*), it follows that the acceleration will become zero when $v = a$. *Thus, the limiting velocity, i.e., the maximum velocity which the particle can attain is a.*

**Example 12.13. Velocity of escape from the earth.** *Find the initial velocity of a particle which is fired in radial direction from the earth's centre and is supposed to escape from the earth. Assume that it is acted upon by the gravitational attraction of the earth only.*

**Solution.** According to Newton's law of gravitation, the acceleration $\alpha$ of the particle is proportional to $1/r^2$, where $r$ is the variable distance of the particle from the earth's centre. Thus

$$\alpha = v \frac{dv}{dr} = -\frac{\mu}{r^2}$$

where, $v$ is the velocity when at a distance $r$ from the earth's centre. The acceleration is negative because $v$ is decreasing. When $r = R$, the earth's radius then $\alpha = -g$, the acceleration of gravity at the surface.

*i.e.,*
$$-g = -\mu/R^2, \text{ i.e., } \mu = gR^2 \quad \therefore \quad v \frac{dv}{dr} = -\frac{gR^2}{r^2}$$

Separating the variables and integrating, we obtain $\int vdv = -gR^2 \int \frac{dr}{r^2} + c$

*i.e.,*
$$v^2 = \frac{2gR^2}{r} + 2c \qquad ...(i)$$

On the earth's surface $r = R$ and $v = v_0$ (say), the initial velocity. Then

$$v_0^2 = 2gR + 2c, \quad \text{i.e.,} \quad 2c = v_0^2 - 2gR$$

Inserting this value of $c$ in Eqn. (*i*), we get $v^2 = \frac{2gR^2}{r} + v_0^2 - 2gR$

When $v$ vanishes, the particle stops and the velocity will change from positive to negative and the particle will return to the earth. Thus the velocity will remain positive, if and only if $v_0^2 \geq 2gR$ and then the particle projected from the earth with this velocity will escape from the earth. Hence the minimum such velocity of projection $v_0 = \sqrt{(2gR)}$ *is called the velocity of escape* from the earth.

**Example 12.14. Rotating cylinder containing liquid.** *A cylindrical tank of radius r is filled with water to a depth h. When the tank is rotated with angular velocity ω about its axis, centrifugal force tends to drive the water outwards from the centre of the tank. Under steady conditions of uniform rotation, show that the section of the free surface of the water by a plane through the axis, is the curve*

$$y = \frac{\omega^2}{2g}\left(x^2 - \frac{r^2}{2}\right) + h.$$

**Solution.** Let the figure represent an axial section of the cylindrical tank. Forces acting on a particle of mass $m$ at $P(x, y)$ on the curve, cut out from the free surface of water, are :

(*i*) the weight $mg$ acting vertically downwards,

(*ii*) the centrifugal force $m\omega^2 x$ acting horizontally outwards.

As the motion is steady, $P$ moves just on the surface of the water and, therefore, there is no force along the tangent to the curve. Thus the resultant $R$ of $mg$ and $m\omega^2 x$ is along the outward normal to the curve.

$\therefore \quad R \cos \psi = mg$ and $R \sin \psi = m\omega^2 x$

whence $$\frac{dy}{dx} = \tan \psi = \frac{m\omega^2 x}{mg} = \frac{\omega^2 x}{g} \qquad ...(i)$$

This is the differential equation of the surface of the rotating liquid.

Integrating Eqn. (*i*), we get

$$\int dy = \frac{\omega^2}{g} \int x\,dx + c$$

*i.e.,* $$y = \frac{\omega^2 x^2}{2g} + c \qquad ...(ii)$$

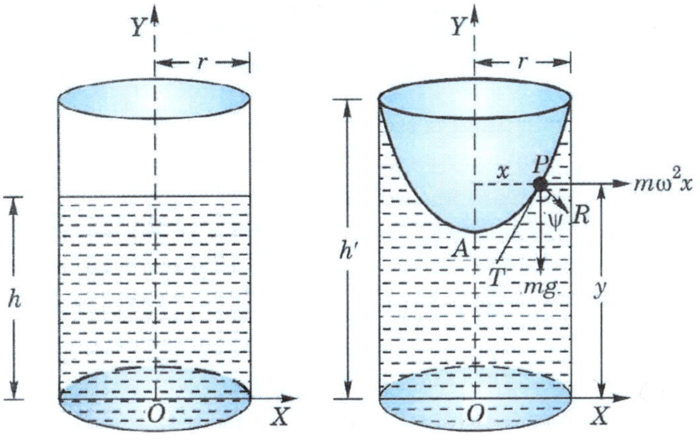


Fig 12.11

*To find c, we note that the volume of the liquid remains the same in both cases (Fig. 12.11).*

When $x = 0$ in Eqn. (*ii*), $OA\ (= y) = c$. When

$x = r$ in Eqn. (*ii*), $$h'\ (= y) = \frac{\omega^2 r^2}{2g} + c \qquad ...(iii)$$

Now the volume of the liquid in the non-rotational case $= \pi r^2 h$, and the volume of the liquid in the rotational case

$$= \pi r^2 h' - \int_{OA}^{h'} \pi x^2 dy = \pi r^2 h' - \frac{2\pi g}{\omega^2}\int_c^{h'} (y - c)\, dy \qquad \text{[From Eqn. (ii)]}$$

$$= \pi r^2 h' - \frac{\pi g}{\omega^2}(h' - c)^2 = \pi r^2 \left(\frac{\omega^2 r^2}{4g} + c\right) \qquad \text{[By Eqn. (iii)]}$$

Thus $$\pi r^2 h = \pi r^2 \left(\frac{\omega^2 r^2}{4g} + c\right) \text{ whence } c = h - \frac{\omega^2 r^2}{4g}$$

$\therefore$ Eqn. (*ii*) becomes, $$y = \frac{\omega^2 x^2}{2g} + h - \frac{\omega^2 x^2}{4g} \quad \text{or} \quad y = \frac{\omega^2}{2g}\left(x^2 - \frac{r^2}{2}\right) + h$$

which is the desired equation of the curve.

**Example 12.15. Discharge of water through a small hole.** *If the velocity of flow of water through a small hole is $0.6\sqrt{2gy}$ where g is the gravitational acceleration and y is the height of water level above the hole, find the time required to empty a tank having the shape of a right circular cone of base radius a and height h filled completely with water and having a hole of area $A_0$ in the base.*

**Solution.** At any time $t$, let the height of the water level be $y$ and radius of its surface be $r$ (Fig. 12.12) so that

$$\frac{h - y}{r} = \frac{h}{a} \quad \text{or} \quad r = a(h - y)/h$$

$\therefore$ surface area of the liquid $= \pi r^2 = \pi a^2 (1 - y/h)^2$

Volume of water drained through the hole per unit time

$$= 0.6\sqrt{(2gy)}\,A_0 = 4.8\sqrt{y}\,A_0 \qquad [\because\ g = 32]$$

$\therefore$ rate of fall of liquid level $= 4.8\,A_0\sqrt{y} \div \pi a^2(1 - y/h)^2$

*i.e.,* $$\frac{dy}{dt} = -\frac{4.8\,A_0\sqrt{y}}{\pi a^2(1 - y/h)^2} \quad \text{(–ve is taken since the water level decreases)}$$

Hence time to empty the tank $(= t)$

$$= -\int_h^0 \frac{\pi a^2 (1 - y/h)^2}{4.8\,A_0\sqrt{y}}\,dy = \frac{\pi a^2}{4.8\,A_0}\int_0^h (y^{-1/2} - 2y^{1/2}/h + y^{3/2}/h^2)\,dy$$

$$= \frac{\pi a^2}{4.8\,A_0}\left|2y^{1/2} - \frac{4}{3h}y^{3/2} + \frac{2}{5h^2}y^{5/2}\right|_0^h = 0.2\,\pi a^2\sqrt{h}/A_0.$$

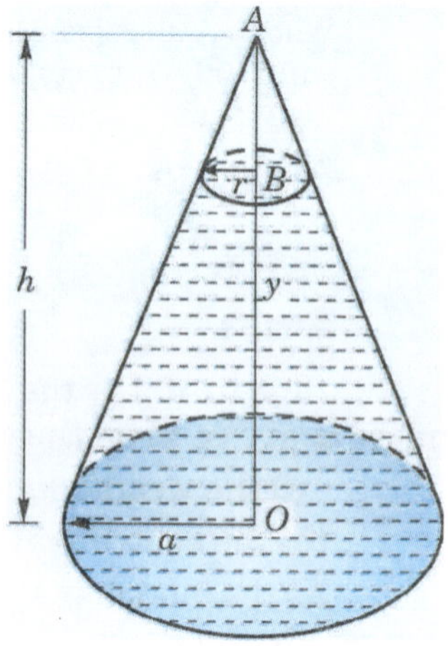

Fig. 12.12

**Example 12.16. Atmospheric pressure.** *Find the atmospheric pressure p lb. per ft. at a height z ft. above the sea-level, both when the temperature is constant or variable.*

**Solution.** Take a vertical column of air of unit cross-section.

Let $p$ be the pressure at a height $z$ above the sea-level and $p + \delta p$ at height $z + \delta z$. Let $\rho$ be the density at a height $z$. (Fig. 12.13)

Now since the thin column $\delta z$ of air is being pressured upwards with pressure $p$ and downwards with $p + \delta p$, we get by considering its equilibrium;

$$p = p + \delta p + g\rho\delta z.$$

Taking the limit, we get $\quad dp/dz = -g\rho \qquad ...(i)$

which is the differential equation giving the atmospheric pressure at height $z$.

(*i*) *When the temperature is constant,* we have by Boyle's law*, $p = k\rho \qquad ...(ii)$

$\therefore$ Substituting the value of $\rho$ from Eqn. (*ii*) in Eqn. (*i*), we get

$$\frac{dp}{dz} = -gp/k \quad \text{or} \quad \int\frac{dp}{p} = -\frac{g}{k}\int dz + c \quad \text{or} \quad \log p = -\frac{g}{k}z + c$$

At the sea-level, where $z = 0$, $p = p_0$ (say) then $c = \log p_0$

$$\therefore \qquad \log p - \log p_0 = -\frac{g}{k}z \ \textit{i.e.,}\ \log p/p_0 = -gz/k$$

Hence $p$ is given by $p = p_0 e^{-gz/k}$.

(*ii*) *When the temperature varies,* we have $p = k\rho^n$.

Proceeding as above, we shall find that $p$ is given by $\frac{n}{n-1}(p_0^{1-1/n} - p^{1-1/n}) = gk^{-1/n}z$.

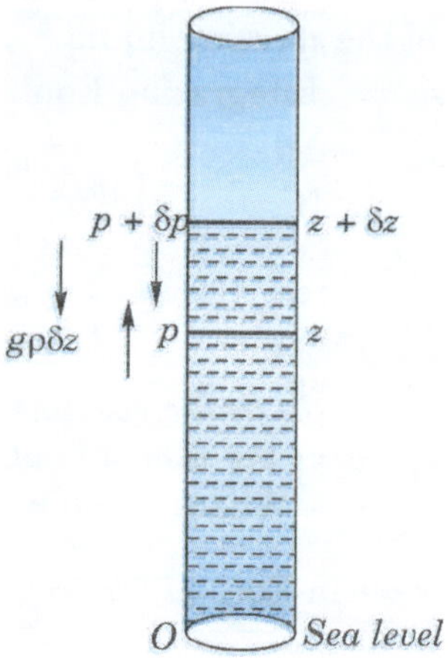

Fig. 12.13

## PROBLEMS 12.3

1. A particle of mass $m$ moves under gravity in a medium whose resistance is $k$ times its velocity, where $k$ is a constant. If the particle is projected vertically upwards with a velocity $v$, show that the time to reach the highest point is $\frac{m}{k}\log_e\left(1 + \frac{kv}{mg}\right)$.
2. A body of mass $m$ falls from rest under gravity and air resistance proportional to square of velocity. Find velocity as function of time.
3. A body of mass $m$ falls from rest under gravity in a field whose resistance is $mk$ times the velocity of the body. Find the terminal velocity of the body and also the time taken to acquire one half of its limiting speed.
4. A particle is projected with velocity $v$ along a smooth horizontal plane in the medium whose resistance per unit mass is $\mu$ times the cube of the velocity. Show that the distance it has described in time $t$ is $\frac{1}{\mu v}(\sqrt{1 + 2\mu v^2 t} - 1)$.

*Named after the English physicist *Robert Boyle* (1627–1691) who was one of the founders of the Royal Society.

5. When a bullet is fired into a sand tank, its retardation is proportional to the square root of its velocity. How long will it take to come to rest if it enters the sand bank with velocity $v_0$?
6. A particle of mass $m$ is attached to the lower end of a light spring (whose upper end is fixed) and is released. Express the velocity $v$ as a function of the stretch $x$ feet.
7. A chain coiled up near the edge of a smooth table just starts to fall over the edge. The velocity $v$ when a length $x$ has fallen is given by $xv \dfrac{dv}{dx} + v^2 = gx$.

   Show that $v = 8\sqrt{(x/3)}$ ft/sec.
8. A toboggan weighing 200 lb., descends from rest on a uniform slope of 5 in 13 which is 15 yards long. If the coefficient of friction is 1/10 and the air resistance varies as the square of the velocity and is 3 lb. weight when the velocity is 10 ft/sec.; prove that its velocity at the bottom is 38.6 ft/sec and show that however long, the slope is the velocity cannot exceed 44 ft per sec. [**Hint.** Fig. 12.14. Equation of motion is

$$\frac{W}{g} \cdot v \frac{dv}{dx} = -\mu R - kv^2 + W \sin \alpha]$$

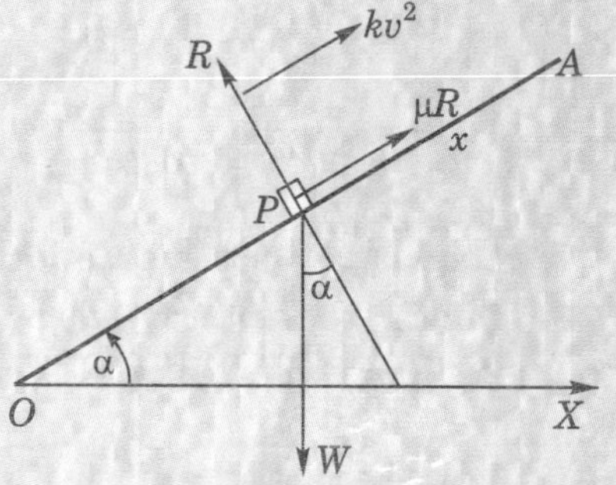

Fig. 12.14

9. Show that a particle projected from the earth's surface with a velocity of 7 miles/sec. will not return to the earth. [Take earth's radius = 3960 miles and $g = 32.17$ ft/sec²].
10. A cylindrical tank 1.5 m. high stands on its circular base of diameter 1 m. and is initially filled with water. At the bottom of the tank there is a hole of diameter 1 cm., which is opened at some instant, so that the water starts draining under gravity. Find the height of water in the tank at any time $t$ sec. Find the times at which the tank is one-half full, one quarter full, and empty.

    [**Hint.** Take $g = 980$ cm/sec² in $v = 0.6\sqrt{(2gy)}$]
11. The rate at which water flows from a small hole at the bottom of a tank is proportional to the square root of the depth of the water. If half the water flows from a cylindrical tank (with vertical axis) in 5 minutes, find the time required to empty the tank.
12. A conical cistern of height $h$ and semi-vertical angle $\alpha$ is filled with water and is held in vertical position with vertex downwards. Water leaks out from the bottom at the rate of $kx^2$ cubic cms per second, $k$ is a constant and $x$ is the height of water level from the vertex. Prove that the cistern will be empty in $(\pi h \tan^2 \alpha)/k$ seconds.
13. Upto a certain height in the atmosphere, it is found that the pressure $p$ and the density $\rho$ are connected by the relation $p = k\rho^n$ $(n > 1)$. If this relation continued to hold upto any height, show that the density would vanish at a finite height.

## 12.5 SIMPLE ELECTRIC CIRCUITS

We shall consider circuits made up of

(*i*) three passive elements—resistance, inductance, capacitance and

(*ii*) an active element—voltage source which may be a battery or a generator.

**(1) Symbols**

| *Element* | *Symbol* | *Unit** |
|---|---|---|
| 1. *Quantity of electricity* | $q$ | coulomb |
| 2. *Current* (= time rate flow of electricity) | $i$ | ampere (A) |
| 3. *Resistance, R* | 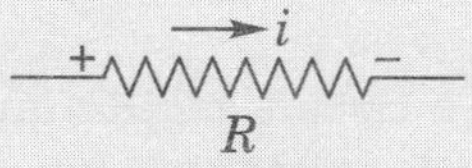 | ohm (Ω) |
| 4. *Inductance, L* | 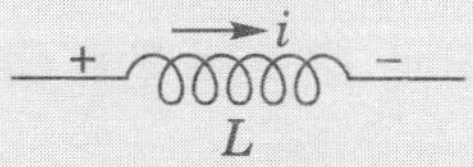 | henry (H) |
| 5. *Capacitance, C* | 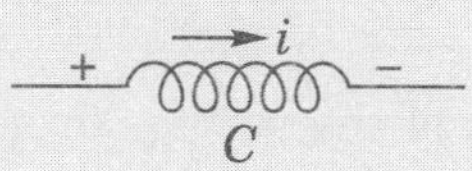 | farad (F) |

*These units are respectively named after the French engineer and physicist *Charles Augustin de Coulomb* (1736–1806); French physicist *Andre Marie Ampere* (1775–1836); German physicist *George Simon Ohm* (1789–1854); Italian physicist *Joseph Henry* (1797–1878); American physicist *Michael Faraday* (1791–1867) and the Italian physicist *Alessandro Volta* (1745–1827).

| *Element* | *Symbol* | *Unit* |
|---|---|---|
| 6. *Electromotive force (e.m.f.) or voltage, E* | 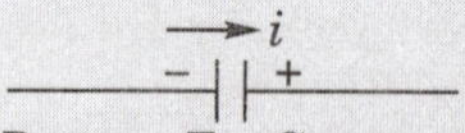 *Battery, E = Constant* 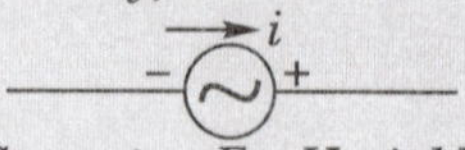 *Generator, E = Variable* | volt (V) |

7. *Loop* is any closed path formed by passing through two or more elements in series.
8. *Nodes* are the terminals of any of these elements.

**(2) Basic relations**

(*i*) $i = \dfrac{dq}{dt}$ or $q = \int i dt$ [∵ current is the rate of flow of electricity]

(*ii*) Voltage drop across resistance $R = Ri$ [Ohm's Law]

(*iii*) Voltage drop across inductance $L = L\dfrac{di}{dt}$

(*iv*) Voltage drop across capacitance $C = \dfrac{q}{C}$.

**(3) Kirchhoff's laws*.** The formulation of differential equations for an electrical circuit depends on the following two Kirchhoff's laws which are of cardinal importance :

**I.** *The algebraic sum of the voltage drops around any closed circuit is equal to the resultant electromotive force in the circuit.*

**II.** *The algebraic sum of the currents flowing into (or from) any node is zero.*

**(4) Differential equations**

(*i*) *R, L series circuit.* Consider a circuit containing resistance $R$ and inductance $L$ in series with a voltage source (battery) $E$. (Fig. 12.15).

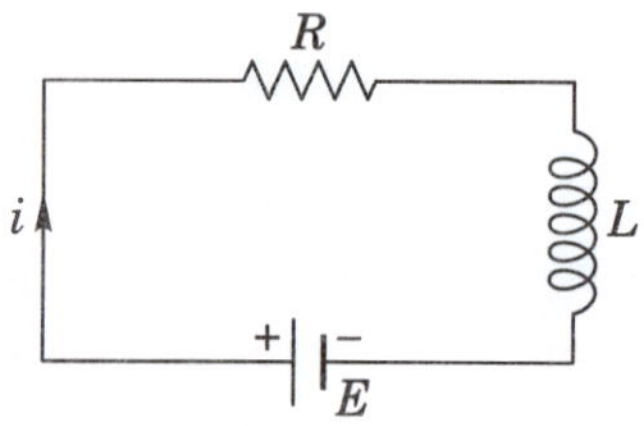

Fig. 12.15

Let $i$ be the current flowing in the circuit at any time $t$. Then by Kirchhoff's first law, we have sum of voltage drops across $R$ and $L = E$

*i.e.,* $$Ri + L\frac{di}{dt} = E \quad \text{or} \quad \frac{di}{dt} + \frac{R}{L}i = \frac{E}{L} \qquad ...(1)$$

This is a Leibnitz's linear equation.

$$\text{I.F.} = e^{\int \frac{R}{L}dt} = e^{Rt/L} \text{ and therefore, its solution is } i(\text{I.F.}) = \int \frac{E}{L}(\text{I.F.})dt + c$$

or $$i \cdot e^{Rt/L} = \int \frac{E}{L} e^{Rt/L}\, dt + c = \frac{E}{L} \cdot \frac{L}{R} \cdot e^{Rt/L} + c \text{ whence } i = \frac{E}{R} + ce^{-Rt/L} \qquad ...(2)$$

If initially there is no current in the circuit, *i.e.*, $i = 0$, when $t = 0$, we have $c = -E/R$.

Thus Eqn. (2) becomes $i = \dfrac{E}{R}\left(1 - e^{-Rt/L}\right)$ which shows that $i$ increases with $t$ and attains the maximum value $E/R$.

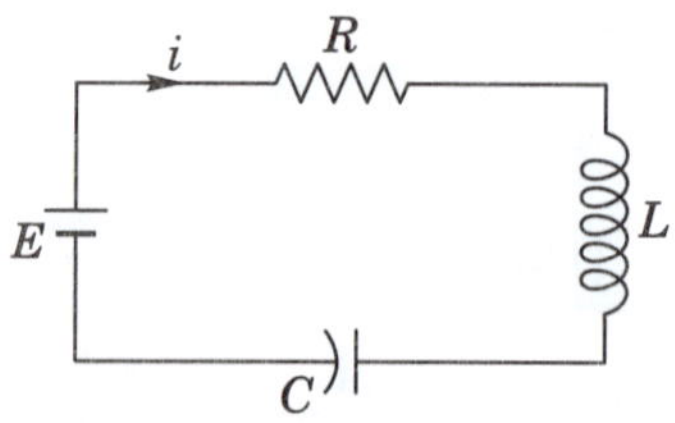

Fig. 12.16

(*ii*) *R, L, C series circuit.* Now consider a circuit containing resistance $R$, inductance $L$ and capacitance $C$ all in series with a constant e.m.f. $E$ (Fig. 12.16)

If $i$ be the current in the circuit at time $t$, then the charge $q$ on the condenser $= \int i\, dt$, *i.e.*, $i = \dfrac{dq}{dt}$.

Applying Kirchhoff's law, we have, sum of the voltage drops across $R$, $L$ and $C = E$.

*i.e.,* $$Ri + L\frac{di}{dt} + \frac{q}{C} = E$$

*Named after the German physicist *Gustav Robert Kirchhoff* (1824–1887).

or $$L\frac{d^2q}{dt^2} + R\frac{dq}{dt} + \frac{q}{C} = E.$$

This is the desired differential equation of the circuit and will be solved in § 14.5.

**Example 12.17.** *Show that the differential equation for the current i in an electrical circuit containing an inductance L and a resistance R in series and acted on by an electromotive force E sin ωt satisfies the equation L di/dt + Ri = E sin ωt.*

*Find the value of the current at any time t, if initially there is no current in the circuit.*

*(C.S.V.T.U., 2012 ; Kurukshetra, 2005)*

**Solution.** By Kirchhoff's first law, we have sum of voltage drops across $R$ and $L = E \sin \omega t$

*i.e.,* $$Ri + L\frac{di}{dt} = E \sin \omega t.$$

This is the required differential equation which can be written as $\frac{di}{dt} + \frac{R}{L}i = \frac{E}{L}\sin \omega t$

This is a *Leibnitz's equation*. Its I.F. $= e^{\int \frac{R}{L}dt} = e^{\frac{Rt}{L}}$

$\therefore$ the solution is $i(\text{I.F.}) = \int \frac{E}{L}\sin \omega t \cdot (\text{I.F.})dt + c$

or $$ie^{Rt/L} = \frac{E}{L}\int e^{Rt/L}\sin \omega t\, dt + c = \frac{E}{L}\frac{e^{Rt/L}}{\sqrt{[(R/L)^2 + \omega^2]}}\sin\left(\omega t - \tan^{-1}\frac{L\omega}{R}\right) + c$$

or $$i = \frac{E}{\sqrt{(R^2 + \omega^2 L^2)}}\sin(\omega t - \phi) + ce^{-Rt/L} \quad \text{where } \tan\phi = L\omega/R \qquad ...(i)$$

Initially when $t = 0$ ; $i = 0$. $\therefore$ $0 = \frac{E\sin(-\phi)}{\sqrt{(R^2 + \omega^2 L^2)}} + c$, *i.e.*, $c = \frac{E\sin\phi}{\sqrt{(R^2 + \omega^2 L^2)}}$

Thus Eqn. (*i*) takes the form $i = \frac{E\sin(\omega t - \phi)}{\sqrt{(R^2 + \omega^2 L^2)}} + \frac{E\sin\phi}{\sqrt{(R^2 + \omega^2 L^2)}} \cdot e^{-Rt/L}$

or $$i = \frac{E}{\sqrt{R^2 + \omega^2 L^2}}[\sin(\omega t - \phi) + \sin\phi \,.\, e^{-Rt/L}]$$ which gives the current at any time $t$.

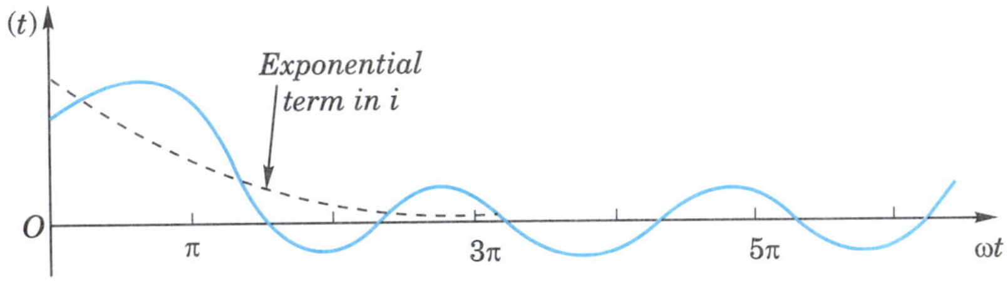


Fig. 12.17

**Obs.** As $t$ increases indefinitely, the exponential term will approach zero. This implies that after sometime the current $i(t)$ will execute nearly harmonic oscillations only (Fig. 12.17). *(C.S.V.T.U., 2012)*

## PROBLEMS 12.4

1. When a switch is closed in a circuit containing a battery $E$, a resistance $R$ and an inductance $L$, the current $i$ builds up at a rate given by $L\, di/dt + Ri = E$.
   Find $i$ as a function of $t$. How long will it be, before the current has reached one-half its final value if $E = 6$ volts, $R = 100$ ohms and $L = 0.1$ henry ?
2. When a resistance $R$ ohms is connected in series with an inductance $L$ henries with an e.m.f. of $E$ volts, the current $i$ amperes at time $t$ is given by $L\, di/dt + Ri = E$.
   If $E = 10 \sin t$ volts and $i = 0$ when $t = 0$, find $i$ as a function of $t$. *(C.S.V.T.U., 2015)*

3. A constant e.m.f. $E$ volts is applied to a circuit containing a constant resistance $R$ ohms in series and a constant inductance $L$ henries. If the initial current is zero, show that the current builds upto half its theoretical maximum in $(L \log 2)/R$ seconds. (*Rohtak, 2012*)
4. A resistance of 100 Ω, an inductance of 0.5 henry are connected in series with a battery of 20 volts. Find the current in the circuit at $t = 0.5$ sec, if $i = 0$ at $t = 0$. (*JNTUK, 2020*)
5. The equation of electromotive force in terms of current $i$ for an electrical circuit having resistance $R$ and condenser of capacity $C$ in series, is

$$E = Ri + \int \frac{idt}{C}$$

Find the current $i$ at any time $t$ when $E = E_m \sin \omega t$. (*C.S.V.T.U., 2014*)
6. A resistance $R$ in series with inductance $L$ is shunted by an equal resistance $R$ with capacity $C$. An alternating e.m.f. $E \sin pt$ produces currents $i_1$ and $i_2$ in two branches. If initially there is no current, determine $i_1$ and $i_2$ from the equations

$$L\frac{di_1}{dt} + Ri_1 = E \sin pt \quad \text{and} \quad \frac{i_2}{C} + R\frac{di_2}{dt} = pE \cos pt.$$

Verify that if $R^2C = L$, the total current $i_1 + i_2$ will be $(E \sin pt)/R$.

## 12.6 NEWTON'S LAW OF COOLING*

According to this law, *the temperature of a body changes at a rate which is proportional to the difference in temperature between that of the surrounding medium and that of the body itself.*

If $\theta_0$ is the temperature of the surroundings and $\theta$ that of the body at any time $t$, then

$$\frac{d\theta}{dt} = -k(\theta - \theta_0), \text{ where } k \text{ is a constant.}$$

**Example. 12.18:** *Newton's law of cooling states that the temperature of an object changes at a rate proportional to the difference of temperature between the object and the surroundings. Supposing water a temperature 100°C cools to 80°C in ten minutes, in a room maintained at temperature of 30°C. Find when the temperature of water will become 40°C.* (*Marathwada, 2018*)

**Solution:** Let unit of time be 1 minute. If $T$ is temperature of object at any instant $t$.

$$\frac{dT}{dt} = K(T - 30) \qquad ...(i)$$

the negative sign shows that temperature decrease with time.

Separating the variables in Eqn. $(i)$

$$\int \frac{dT}{T-30} = -K\int dt + \log c$$

$$\log(T - 30) = -Kt + \log c$$

$$T - 30 = ce^{-kt} \qquad ...(ii)$$

Initially, when $t = 0$, $T = 100°$ C

∴ $C = 70$

Hence from Eqn. $(ii)$ $T = 30 + 70\,(e^{-10k})t/10$ ...(iii)

Given when $t = 10$, $T = 80$

Now from Eqn. $(iii)$, we get $e^{-10k} = \frac{5}{7}$

$$T = 30 + 70\left(\frac{5}{7}\right)^{t/10} \qquad ...(iv)$$

Now, when $T = 40$°C, we get from Eqn. $(iv)$

*Named after the great English mathematician and physicist *Sir Issac Newton* (1642–1727) whose contributions are of utmost importance. He discovered many physical laws, invented Calculus alongwith *Leibnitz* (see footnote p. 154) and created analytical methods of investigating physical problems. He became professor at Cambridge in 1699, but his '*Mathematical Principles of Natural Philosophy*' containing development of classical mechanics had been completed in 1687.

$$40 = 30 + 70\left(\frac{5}{7}\right)^{t/10}$$

$$\frac{1}{7} = \left(\frac{5}{7}\right)^{t/10}$$

$$\frac{t}{10} = \frac{\log_{10}\left(\frac{1}{7}\right)}{\log_{10}\left(\frac{5}{7}\right)} = \frac{0 \cdot 8451}{0 \cdot 1461} = 5.79$$

$$t = 57.9 \text{ minutes.}$$

## 12.7 HEAT FLOW

*The fundamental principles involved in the problems of heat conduction are :*

(*i*) *Heat flows from a higher temperature to the lower temperature.*

(*ii*) *The quantity of heat in a body is proportional to its mass and temperature.*

(*iii*) *The rate of heat-flow across an area is proportional to the area and to the rate of change of temperature with respect to its distance normal to the area.*

If $q$ (cal./sec.) be the quantity of heat that flows across a slab of area $\alpha$ (cm$^2$) and thickness $\delta x$ in one second, where the difference of temperature at the faces is $\delta T$, then by Eqn. (*iii*) above

$$q = -k\alpha dT/dx \qquad ...(A)$$

where $k$ is a constant depending upon the material of the body and is called the *thermal conductivity*.

**Example 12.19.** *A pipe 20 cm in diameter contains steam at 150°C and is protected with a covering 5 cm thick for which k = 0.0025. If the temperature of the outer surface of the covering is 40°C, find the temperature half-way through the covering under steady state conditions.*

**Solution.** Let $q$ cal./sec. be the constant quantity of heat flowing out radially through a surface of the pipe having radius $x$ cm. and length 1 cm (Fig. 12.18). Then the area of the lateral surface (belt) = $2\pi x$.

$\therefore$ the equation (*A*) above gives

$$q = -k \cdot 2\pi x \cdot \frac{dT}{dx} \quad \text{or} \quad dT = -\frac{q}{2\pi k} \cdot \frac{dx}{x}$$

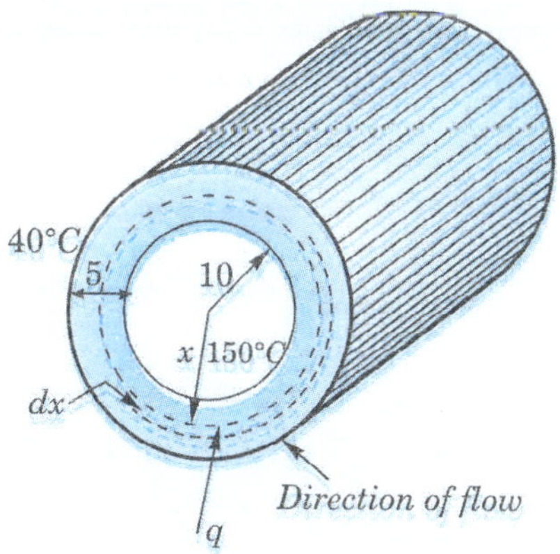


Fig. 12.18

Integrating, we have

$$T = -\frac{q}{2\pi k}\log_e x + c$$

Since $T = 150$, when $x = 10$. $\therefore$ $150 = -\frac{q}{2\pi k}\log_e 10 + c \qquad ...(i)$

Again since $T = 40$, when $x = 15$, $40 = -\frac{q}{2\pi k}\log_e 15 + c \qquad ...(ii)$

Subtracting Eqns. (*ii*) from (*i*), $110 = \frac{q}{2\pi k}\log_e 1.5 \qquad ...(iii)$

Let $T = t$, when $x = 12.5$ $\therefore$ $t = -\frac{q}{2\pi k}\log_e 12.5 + c \qquad ...(iv)$

Subtracting Eqns. (*i*) from (*iv*), $t - 150 = -\frac{q}{2\pi k}\log_e 1.25 \qquad ...(v)$

Dividing Eqns. (*v*) by (*iii*), $\frac{t-150}{110} = -\frac{\log_e 1.25}{\log_e 1.5}$, whence $t = 89.5$°C.

## PROBLEMS 12.5

1. If the temperature of the air is 30°C and the substance cools from 100°C to 70°C in 15 minutes, find when the temperature will be 40°C. *(JNTUK, 2020 ; V.T.U., 2015 ; Rohtak, 2010)*

2. If the air is maintained at 30°C and the temperature of the body cools from 80°C to 60°C in 12 minutes, find the temperature of the body after 24 minutes. *(Andhra, 2022 ; JNTUK, 2022)*

3. In a certain chemical rection, the rate of conversion of a substance at time $t$ is proportional to the quantity of substance at time $t$ is proportional to the quantity of substance still untransformed at that instant. The amount of substance remaining untransformed at the end of the hour and at the end of four hour are 60 gm and 21 gms respectively. How may grams of substance were present initially. *(Guwahati, 2018)*

4. A pipe 20 cm. in diameter contains steam at 200°C. It is covered by a layer of insulations 6 cm thick and thermal conductivity 0.0003. If the temperature of the outer surface is 30°C, find the heat loss per hour from two metre length of the pipe. *(C.S.V.T.U., 2015)*

5. The temperature of a cups of coffee is 92°C. When freshly poured, the room temperature being 24°C. In one minute it has cooled to 88°C. How long a period must elapse, before the temperature of the cup becomes 65° C. *(Tezpur University 2018)*

6. The quantity of light absorbed in passing through a thin layer of water is proportional to the quantity of incident light and to the thickness of the layer. If one half of the original quantity of light is absorbed is passing through 3 meter thick layer of water, what part of this quantity will reach a depth of 30 meters.

7. Neutrons in an atomic pile increase at a rate proportional to the number of neutrons. present at any instant. If $N_0$ neutrons are initially present, and $N_1$ and $N_2$ neutrons are present at time $T_1$ and $T_2$ respectively show that

$$\left(\frac{N_2}{N_0}\right)^{T_1} = \left(\frac{N_1}{N_0}\right)^{T_2}$$

8. A body originally at 80°C coolsdown to 60°C in 20 minutes, the temperature of the air being 40°C. What will be the temperature of the body after 40 minutes from the original? *(Andhra, 2022)*

## 12.8 RATE OF DECAY OF RADIO-ACTIVE MATERIALS

*This law states that disintegration at any instant is proportional to the amount of material present.*

If $u$ is the amount of material at any time $t$, then $\dfrac{du}{dt} = -ku$, where $k$ is a constant.

**Example 12.20.** *Uranium disintegrates at a rate proportional to the amount present at any instant. If $m_1$ and $m_2$ grams of uranium are present at times $T_1$ and $T_2$ respectively, show that the half life of uranium is*

$$\frac{(T_2 - T_1)\log_e 2}{\log_e(m_1 / m_2)}$$

*(Andhra, 2022; Baddi University, 2018)*

**Solution:** If $m$ grams of uranium are present at any time $t$, then $\dfrac{dm}{dt} = -km$,

where $k$ is a constant.

Separating the variables and integrating

$$\int \frac{dm}{m} = -k\int dt + \log c$$

$$\log m = -kt + \log c$$

$$\log\left(\frac{m}{c}\right) = -kt$$

$$m = ce^{-kt} \qquad \ldots(i)$$

Initially, when $t = 0$, let $m = m_o$

From Eqn. ($i$), we get. $m_o = c$

Substituting in Eqn ($i$), we get $m = m_o\, e^{-kt}$ ...($ii$)

Now. when $t = T_1$, $m = m_1$, and when $t = T_2$, $m = m_2$.

Hence from Eqn. ($ii$), we get $m_1 = m_o\, e^{-kT_1}$ ...($iii$)

$$m_2 = m_o\, e^{-kT_2} \quad ...(iv)$$

Dividing Eqn. ($iii$) by Eqn. ($iv$), we get

$$\frac{m_1}{m_2} = e^{k}(T_2 - T_1)$$

$$k = \frac{1}{(T_2 - T)} \log_c (m_1 / m_2)$$

Let $T$ be the half life of uranium *i.e.*,

when $$t = T_1,\ m = \frac{1}{2} m_o$$

Hence from Eqn. ($ii$), we get $\frac{1}{2} m_o = m_o e^{-kT}$

$$-kT = \log_e \left(\frac{1}{2}\right) = -\log_e(2) \quad ...(v)$$

Substituting the value $k$ in Eqn. ($v$), we get

$$T = \frac{(T_2 - T_1)\log_e 2}{\log_e (m_1 / m_2)}$$

## 12.9 CHEMICAL REACTIONS AND SOLUTIONS

A type of problems which are especially important to chemical engineers are those concerning either chemical reactions or chemical solutions. These can be best explained through the following example :

**Example 12.21:** *A tank contains 100 liters of solution having 50 gm of salt dissolved in it. Saltwater containing 1 gm of salt per litre runs into the tank at a rate of 2 litres per minute and the well stirred mixture runs out of it at the rate of 3 litres per minute. Find the amount of the salt present in the tank after 40 minutes.*

*(Mumbai, 2018)*

**Solution:** Let $x$ gm of salt be present in the tank after $t$ minute. Now, After $t$ minute the volume of solution in the tank is $(100 - t)$ litres containing $x$ gm of salt. The solution leaves the tank at the rate of 3 litres per minute decreasing the salt content by $\left(\frac{3x}{100 - t}\right)$ gm per minute.

Thus change in salt concentration in one minute

$$= \left(2 - \frac{3x}{100 - t}\right) \text{gm}$$

Hence, $$\frac{dx}{dt} = 2 - \frac{3x}{100 - t}$$

$$\frac{dx}{dt} + \frac{3}{100 - t} x = 2 \quad ...(i)$$

Hence, $$I.F = e^{\int \frac{3}{100-t} dt} = e^{-3\log(100-t)} = e^{\log(100-t)^{-3}} = \frac{1}{(100-t)^3}$$

The solution of Eqn. ($i$) is

$$x \cdot \frac{1}{(100-t)^3} = 2\int \frac{1}{(100-t)^3}\, dt + c$$

$$\frac{x}{(100-t)^3} = \frac{1}{(100-t)^2} + c.... \qquad ...(ii)$$

Initially, when $t = 0$, $x = 50$ from Eqn. $(ii)$, we get

$$\frac{50}{(100)^3} = \frac{1}{(100)^2} + c$$

$$c = \frac{-50}{(100)^3}$$

Substituting the value of $c$ in Eqn. $(ii)$

$$x = (100 - t) - \frac{50}{(100)^3}(100-t)^3$$

when $t = 40$,

$$x = (100 - 40) - \frac{50}{(100)^3} \times (60)^3$$

$$x = 60 - 10.8 = 49.2 \text{ gm.}$$

**Example 12.22:** *In a culture of yeast at each instant the time rate change of active ferment is proportional to the amount present. If the active ferment doubles in two hours, how much can be expected at the end of 8 hours at the same rule of growth. Find also, how much time will elapse, before the active ferment grows to eight times of its initial values.* (*Sri Sai. Uni., 2018*)

**Solution:** Let $y$ be the amount of active ferment present at any time $t$ and $y_o$, the amount present initially.

Then,
$$\frac{dy}{dt} = ky$$

separating the variables and integrating

$$\int \frac{dy}{y} = K\int dt + \log c$$

where $\log c$ is constant of integration

$$\log y = kt + \log c$$

$$\log\left(\frac{y}{c}\right) = kt$$

$$y = ce^{kt} \qquad ...(i)$$

Initially when $t = 0$, $y = y_o$

From Eqn. $(i)$ we get $\qquad c = y_o$

Hence the Eqn. $(i)$ become $\qquad y = y_o\, e^{kt}$

Also when $\qquad t = 2, y = 2y_o \qquad ...(ii)$

Therefore from Eqn. $(ii)$ $\qquad 2y_o = y_o\, e^{2k}$

$$e^{2k} = 2$$

Now, Eqn. $(ii)$ $\qquad y = y_o\,(e^k)^t = y_o(2)^{t/2}$

Now, when $\qquad t = 8$, let $y = y_1$

From Eqn. $(ii)$, we get

$$y_1 = y_o(2)^4 = 16y_o$$

Hence after 8 hours the amount of active ferment becomes, 16 times of the original amount.

Also, $\qquad y = 8y_o$, let $t = t_1$, then from Eqn. $(ii)$

$$8y_o = y_o(2)^{\frac{t_1}{2}}$$

$$t_1 = 6$$

**Example 12.23:** *In the inversion of raw sugar, the time rate change varies as the amount of raw sugar remaining. If after 10 hours, 8000 kg of raw sugar has been reduced to 800 kg, how much raw sugar will remain at the end of 24 hours.* (Assam, 2018)

**Solution:** Let $m$ kg of raw sugar be present at the end of $t$ hours. Taking unit of time as 1 hour, we have

$$\frac{dm}{dt} = -\text{km}$$

separating the variables and integrating

$$\int \frac{dm}{dt} = -k\int dt + \log c$$

$$\log m = -kt + \log c$$

$$m = ce^{-kt} \qquad ...(i)$$

Initially, when $t = 0,\ m = 8000$

$$c = 8000$$

Hence, Eqn. $(i)$ becomes $m = 8000\, e^{-\lambda t}$ ...(ii)

Also, when $t = 10$, $m = 800$, so from Eqn. $(ii)$

we get, $800 = 8000\, e^{-10k}$

$$e^{-10k} = \frac{1}{10}$$

Substituting this value in Eqn. $(ii)$ $m = 8000(e^{-10k})^{t/10}$

$$m = 8000\left(\frac{1}{10}\right)^{t/10}$$

$t = 24$, we have

$$m = 8000\left(\frac{1}{10}\right)^{2.4}$$

$$m = 31.848 \text{ kg}$$

**Example 12.24:** *The air in a hall 20 m × 10 m × 5 m is tested and found to contain 0.3% of carbon dioxide ($CO_2$). Fresh air containing 0.05% of carbon di-oxide was admitted through the ventilators at the rate of room in per minute. Determine the percent of carbondioxide after 50 minutes.* (Chennai, 2018)

**Solution:** Let $x$ cum of $CO_2$ be present in the hall after $t$ minutes

Concentration of $CO_2$ in the hall $= \dfrac{x}{20 \times 10 \times 5} = \dfrac{x}{1000} = \dfrac{x}{10^3}$

Rate of $CO_2$ entering the hall $= 100 \times \dfrac{5}{10000} = \dfrac{1}{20}$ cum/min

Rate of $CO_2$ leaving the hall $= 100 \times \dfrac{x}{1000} = \dfrac{x}{10}$ cum/min

Thus we have

$$\frac{dx}{dt} = \frac{1}{20} - \frac{x}{10}$$

$$\frac{dx}{dt} + \frac{1}{10}x = \frac{1}{20} \qquad ...(i)$$

The Eqn. $(i)$ is linear $I.F. = e^{\int \frac{1}{10}dt} = e^{t/10}$

Hence solution of Eqn. (*i*), is

$$x \cdot e^{t/10} = \frac{1}{20}\int e^{t/10} dt + c$$

$$x \cdot e^{t/10} = \frac{1}{2} e^{t/10} + c$$

$$x = \frac{1}{2} + ce^{-t/10} \qquad ...(ii)$$

Initially, when $t = 0$, $x = 1000 \times \frac{3}{1000} = 3$ from Eqn. (*ii*), we get

$$3 + \frac{1}{2} + C = 0$$

$$C = \frac{5}{2}$$

Substituting value of $C$ in Eqn. (*iii*) $x = \frac{1}{2}(1 + 5e^{-t/10})$

when $t = 50$, from Eqn. (*iii*) $x = \frac{1}{2}(1 + 5e^{-5}) = \frac{1}{2}(1 + 5 \times 0 \cdot 00674) = 0.517.$

This % of $CO_2$ after 50 min. $= \frac{0.517}{1000} \times 100 = 0.0517\ \%$

## PROBLEMS 12.6

1. Calculate the amount of heat passing 1 sq. cm of a refrigerator wall, which is 30 cm thick. The temperature inside the refrigerator being 0°C while outside it is 20°C. Assum $k = 0.002$ cgs units. (*Calicut, 2018*)
2. The rate at which bacteria multiply is proportional to the instantaneous number present. If the original number doubles in 2 hours, in how many hours will it triple ?
3. Radium decomposes at a rate proportional to the amount present. If a fraction $p$ of the original amount disappears in 1 year, how much will remain at the end of 21 years ?
4. If 30% of radio active substance disappeared in 10 days. How long will it take for 90% of it to disappear ? (*Andhra, 2016*)
5. A pipe 10 cm in diameter contains steam at 100°C. It is covered with asbestos, 5 cm thick, for which $k = 0.0006$ cgs units and the outside surface is at 30°C. Find the heat loss per hour from 9 metre long pipe. (*Rajasthan, 2018*)
6. A long hollow pipe has an inner diameter of 10 cm and outer diameter of 20 cm. The inner surface is kept at 200°C and the outer diameter is kept at 50°C. The thermal conductivity is 0.12 cgs units. How much heat is lost per minute from a portion of pipe 20 meters long? Find the temperature at a distance of 7.5 cm from the common axis of the cylinder? (*Marathwada, 2019*)
7. A tank contains 1000 gallons of brine in which 500 lt. of salt are dissolved. Fresh water runs into the tank at the rate of 10 gallons /minute and the mixture kept uniform by stirring, runs out at the same rate. How long will it be before only 50 lt. of salt is left in the tank ?

   [**Hint.** If $u$ be the amount of salt after $t$ minutes, then $du/dt = -10u/1000$.]
8. A thin rod is in the form of cone, the radius of one end being twice that of the other. The rod has a length $l$ and weight $w$ per unit volume. It is suspended from the smaller end. The extension $y$ at a distance $x$ below this end is given by a equation $E\left(\frac{l+x}{l^2}\right) \cdot \frac{dy}{dx} = \frac{8wl}{3}\left[1 - \frac{(l+x)^3}{8l^3}\right]$ where $E$ is the Young's modulus of the rod. Show that the extension at the lower end is $\frac{5wl^2}{6E}$. (*Maharashtra, 2018*)

## 12.10 OBJECTIVE TYPE OF QUESTIONS

### PROBLEMS 12.7

*Fill up the blanks or choose the correct answer in the following problems :*

1. If a coil having a resistance of 15 ohms and an inductance of 10 henries is connected to 90 volts supply then the current after 2 secs is .....
2. A tennis ball dropped from a height of 6 m, rebounds infinitely often. If it rebounds 80% of the distance that it falls, then the total distance for these bounces is .....
3. Radium decomposes at a rate proportional to the amount present. If 5% of the original amount disappears in 50 years then ......% will remain after 100 years.
4. The curve whose polar subtangent is constant is ......
5. The curve in which the length of the subnormal is proportional to the square of the ordinate, is ......
6. The curve in which the portion of the tangent between the axes is bisected at the point of contact, is ......
7. If the stream lines of a flow around a corner are $xy = c$, then the equipotential lines are ......
8. The orthogonal trajectories of a system of confocal and coaxial parabolas is ......
9. When a bullet is fired into a sand tank, its retardation is proportional to √(velocity). If it enters the sand tank with velocity $v_o$, it will come to rest after ...... seconds.
10. The rate at which bacteria multiply is proportional to the instantaneous number present. If the original number doubles in two hours, then it will triple after ..... hours.
11. Ram and Sunil order coffee and receive cups simultaneously at equal temperature. Ram adds a spoon of cold cream but doesn't drink for 10 minutes, Sunil waits for 10 minutes and adds a spoon of cold cream and begins to drink. Who drinks the hotter coffee ?
12. The equation $y - 2x = c$ represents the orthogonal trajectories of the family
    (*i*) $y = ae^{-2x}$ (*ii*) $x^2 + 2y^2 = a$ (*iii*) $xy = a$ (*iv*) $x + 2y = a$.
13. In order to keep a body in air above the earth for 12 seconds, the body should be thrown vertically up with a velocity of
    (*a*) $\sqrt{6}$ g m/sec (*b*) $\sqrt{12}$ g m/sec (*c*) 6 g m/sec (*d*) 12g m/sec.
14. The orthogonal trajectory of the family $x^2 + y^2 = c^2$ is
    (*a*) $x + y = c$ (*b*) $xy = c$ (*c*) $x^2 + y^2 = x + y$ (*d*) $y = cx$. (*V.T.U., 2010*)
15. If a thermometer is taken outdoors where the temperature is 0°C, from a room having temperature 21°C and the reading drops to 10°C in 1 minute then its reading will be 5°C after ......minutes.
16. The equation of the curve for which the angle between the tangent and the radius vector is twice the vectorial angle is $r^2 = 2a \sin 2\theta$. This satisfies the differential equation
    (*a*) $r\dfrac{dr}{d\theta} = \tan 2\theta$ (*b*) $r\dfrac{dr}{d\theta} = \cos 2\theta$ (*c*) $r\dfrac{d\theta}{dr} = \tan 2\theta$ (*d*) $r\dfrac{d\theta}{dr} = \cos 2\theta$.
17. Two balls of $m_1$ and $m_2$ grams are projected vertically upwards such that the velocity of projection of $m_1$ is double that of $m_2$. If the maximum height to which $m_1$ and $m_2$ rise be $h_1$ and $h_2$ respectively then
    (*a*) $h_1 = 2h_2$ (*b*) $2h_1 = h_2$ (*c*) $h_1 = 4h_2$ (*d*) $4h_1 = h_2$.
18. Two balls are projected simultaneously with same velocity from the top of a tower, one vertically upwards and the other vertically downwards. If they reach the ground in times $t_1$ and $t_2$, then the height of the tower is
    (*a*) $\dfrac{1}{2} gt_1t_2$ (*b*) $\dfrac{1}{2} g(t_1^2 + t_2^2)$ (*c*) $\dfrac{1}{2} g(t_1^2 - t_2^2)$ (*d*) $\dfrac{1}{2} g(t_1 + t_2)^2$.
19. A particle projected from the earth's surface with a velocity of 7 miles/sec will return to the earth. (Taking $g = 32.17$ and earth's radius = 3960 miles) (True/False)
20. If a particle falls under gravity with air resistance $k$ times its velocity, then its velocity cannot exceed $g/k$. (True/False)

CHAPTER 13

# Linear Differential Equations



## 13.1 DEFINITIONS

**Linear differential equations** *are those in which the dependent variable and its derivatives occur only in the first degree and are not multiplied together.* Thus the general linear differential equation of the $n$th order is of the form

$$\frac{d^n y}{dx^n} + p_1 \frac{d^{n-1} y}{dx^{n-1}} + p_2 \frac{d^{n-2} y}{dx^{n-2}} + ... + p_n y = X,$$

where $p_1, p_2, ..., p_n$ and $X$ are functions of $x$ only.

**Linear differential equations with constant co-efficients** are of the form

$$\frac{d^n y}{dx^n} + k_1 \frac{d^{n-1} y}{dx^{n-1}} + k_2 \frac{d^{n-2} y}{dx^{n-2}} + ... + k_n y = X$$

where $k_1, k_2, ..., k_n$ are constants. Such equations are most important in the study of electro-mechanical vibrations and other engineering problems.

## 13.2 (1) THEOREM

*If $y_1, y_2$ are only two solutions of the equation*

$$\frac{d^n y}{dx^n} + k_1 \frac{d^{n-1} y}{dx^{n-1}} + k_2 \frac{d^{n-2} y}{dx^{n-2}} + ... + k_n y = 0 \quad ...(1)$$

*then $c_1 y_1 + c_2 y_2$ $(= u)$ is also its solution.*

Since $y = y_1$ and $y = y_2$ are solutions of Eqn. (1).

$$\therefore \quad \frac{d^n y_1}{dx^n} + k_1 \frac{d^{n-1} y_1}{dx^{n-1}} + k_2 \frac{d^{n-2} y_1}{dx^{n-2}} + ... + k_n y_1 = 0 \quad ...(2)$$

and

$$\frac{d^n y_2}{dx^n} + k_1 \frac{d^{n-1} y_2}{dx^{n-1}} + k_2 \frac{d^{n-2} y_2}{dx^{n-2}} + ... + k_n y_2 = 0 \quad ...(3)$$

If $c_1, c_2$ be two arbitrary constants, then

$$\frac{d^n (c_1 y_1 + c_2 y_2)}{dx^n} + k_1 \frac{d^{n-1} (c_1 y_1 + c_2 y_2)}{dx^{n-1}} + ... + k_n (c_1 y_1 + c_2 y_2)$$

$$= c_1\left(\frac{d^n y_1}{dx^n} + k_1 \frac{d^{n-1} y_1}{dx^{n-1}} + \dots + k_n y_1\right) + c_2\left(\frac{d^n y_2}{dx^n} + k_1 \frac{d^{n-1} y_2}{dx^{n-1}} + \dots + k_n y_2\right)$$

$$= c_1(0) + c_2(0) = 0 \qquad \text{[By (2) and (3)]}$$

*i.e.,* $$\frac{d^n u}{dx^n} + k_1 \frac{d^{n-1} u}{dx^{n-1}} + \dots + k_n y = 0 \qquad \dots(4)$$

This proves the theorem.

**(2)** Since the general solution of a differential equation of the $n$th order contains $n$ arbitrary constants, it follows, from above, that if $y_1, y_2, y_3, \dots, y_n$, are $n$ independent solutions of Eqn. (1), then $c_1y_1 + c_2y_2 + \dots + c_ny_n$ $(= u)$ is its complete solution.

**(3)** If $y = v$ be any particular solution of

$$\frac{d^n y}{dx^n} + k_1 \frac{d^{n-1} y}{dx^{n-1}} + \dots + k_n y = X \qquad \dots(5)$$

then $$\frac{d^n v}{dx^n} + k_1 \frac{d^{n-1} v}{dx^{n-1}} + \dots + k_n v = X \qquad \dots(6)$$

Adding Eqns. (4) and (6), we have $\frac{d^n (u+v)}{dx^n} + k_1 \frac{d^{n-1}(u+v)}{dx^{n-1}} + \dots + k_n(u+v) = X$

This shows that $y = u + v$ is the complete solution of Eqn. (5).

*The part u is called the* **complementary function (C.F.)** *and the part v is called the* **particular integral (P.I.)** *of Eqn. (5).*

$\therefore$ the complete solution (C.S.) of Eqn. (5) is **y = C.F. + P.I.**

*Thus in order to solve the question Eqn. (5), we have to first find the C.F., i.e., the complete solution of Eqn. (1), and then the P.I., i.e. a particular solution of Eqn. (5).*

## 13.3 OPERATOR D

Denoting $\frac{d}{dx}, \frac{d^2}{dx^2}, \frac{d^3}{dx^3}$ etc. by $D, D^2, D^3$ etc., so that

$\frac{dy}{dx} = Dy$, $\frac{d^2y}{dx^2} = D^2y$, $\frac{d^3x}{dx^3} = D^3y$ etc., the Eqn. (5) above can be written in the *symbolic form* $(D^n + k_1D^{n-1} + \dots + k_n)y = X$, *i.e.,* $f(D)y = X$,

where $f(D) = D^n + k_1D^{n-1} + \dots + k_n$, *i.e.*, a polynomial in $D$.

Thus *the symbol D stands for the operation of differentiation and can be treated much the same as an algebraic quantity i.e., f (D) can be factorised by ordinary rules of algebra and the factors may be taken in any order.* For instance

$$\frac{d^2y}{dx^2} + 2\frac{dy}{dx} - 3y = (D^2 + 2D - 3)\,y = (D + 3)(D - 1)y \text{ or } (D - 1)(D + 3)\,y.$$

## 13.4 RULES FOR FINDING THE COMPLEMENTARY FUNCTION

*To solve the equation* $$\frac{d^n y}{dx^n} + k_1 \frac{d^{n-1} y}{dx^{n-1}} + k_2 \frac{d^{n-2} y}{dx^{n-2}} + \dots + k_n y = 0 \qquad \dots(1)$$

*where k's are constants.*

The Eqn. (1) in *symbolic form is*

$$(D^n + k_1D^{n-1} + k_2D^{n-2} + \dots + k_n)y = 0 \qquad \dots(2)$$

Its symbolic co-efficient equated to zero *i.e.*

$$D^n + k_1D^{n-1} + k_2D^{n-2} + \dots + k_n = 0$$

is called the *auxiliary equation* (A.E.). Let $m_1, m_2, \dots, m_n$ be its roots.

**Case I.** *If all the roots be real and different*, then Eqn. (2) is equivalent to

$$(D - m_1)(D - m_2) \dots (D - m_n)y = 0 \qquad \dots(3)$$

Now Eqn. (3) will be satisfied by the solution of $(D - m_n)y = 0$, *i.e.*, by $\frac{dy}{dx} - m_n y = 0$.

This is a Leibnitz's linear and I.F. = $e^{-m_n x}$

$\therefore$ its solution is $y e^{-m_n x} = c_n$, *i.e.*, $y = c_n e^{m_n x}$

Similarly, since the factors in Eqn. (3) can be taken in any order, it will be satisfied by the solutions of $(D - m_1) y = 0$, $(D - m_2)y = 0$ etc. *i.e.*, by $y = c_1 e^{m_1 x}$, $y = c_2 e^{m_2 x}$ etc.

*Thus the complete solution of Eqn. (1) is* $y = c_1 e^{m_1 x} + c_2 e^{m_2 x} + \ldots + c_n e^{m_n x}$ ...(4)

**Case II.** *If two roots are equal (i.e.,* $m_1 = m_2$*)*, then Eqn. (4) becomes

$$y = (x_1 + xc_2) e^{m_1 x} + c_3 e^{m_3 x} + \ldots + c_n e^{m_n x}$$

$$y = C e^{m_1 x} + c_3 e^{m_3 x} + \ldots + c_n e^{m_n x} \qquad [\because \ c_1 + c_2 = \text{one arbitrary constant } C]$$

It has only $n - 1$ arbitrary constants and is, therefore, not the complete solution of Eqn. (1). In this case, we proceed as follows :

The part of the complete solution corresponding to the repeated root is the complete solution of $(D - m_1)(D - m_1)y = 0$

Putting $(D - m_1)y = z$, it becomes $(D - m_1) z = 0$ or $\frac{dz}{dx} - m_1 z = 0$

This is a Leibnitz's linear in $z$ and I.F. = $e^{-m_1 x}$. $\therefore$ its solution is $ze^{-m_1 x} = c_1$ or $z = c_1 e^{m_1 x}$

Thus $\qquad (D - m_1)y = z = c_1 e^{m_1 x}$ or $\frac{dy}{dx} - m_1 y = c_1 e^{m_1 x}$ ...(5)

Its I.F. being $e^{-m_1 x}$, the solution of Eqn. (5) is

$$y e^{-m_1 x} = \int c_1 e^{m_1 x} \, dx + c_2 = c_1 x + c_2 \text{ or } y = (c_1 x + c_2) e^{m_1 x}$$

Thus the complete solution of Eqn. (1) is $y = (c_1 x + c_2) e^{m_1 x} + c_3 e^{m_3 x} + \ldots + c_n e^{m_n x}$

If, however, the A.E. has three equal roots (*i.e.*, $m_1 = m_2 = m_3$), then the complete solution is

$$y = (c_1 x^2 + c_2 x + c_3) e^{m_1 x} + c_4 e^{m_4 x} + \ldots + c_n e^{m_n x}$$

**Case III.** *If one pair of roots be imaginary, i.e.,* $m_1 = \alpha + i\beta$, $m_2 = \alpha - i\beta$, then the complete solution is

$$y = c_1 e^{(\alpha + i\beta)x} + c_2 e^{(\alpha - i\beta)x} + c_3 e^{m_3 x} + \ldots + c_n e^{m_n x}$$

$$= e^{\alpha x}(c_1 e^{i\beta x} + c_2 e^{-i\beta x}) + c_3 e^{m_3 x} + \ldots + c_n e^{m_n x}$$

$$= e^{\alpha x}[c_1(\cos \beta x + i \sin \beta x) + c_2 (\cos \beta x - i \sin \beta x)] + c_3 e^{m_3 x} + \ldots + c_n e^{m_n x}$$

[$\because$ by *Euler's Theorem*, $e^{i\theta} = \cos \theta + i \sin \theta$]

$$= e^{\alpha x} (C_1 \cos \beta x + C_2 \sin \beta x) + c_3 e^{m_3 x} + \ldots + c_n e^{m_n x}$$

where $C_1 = c_1 + c_2$ and $C_2 = i(c_1 - c_2)$.

**Case IV.** *If two points of imaginary roots be equal i.e.,* $m_1 = m_2 = \alpha + i\beta$, $m_3 = m_4 = \alpha - i\beta$, then by case II, the complete solution is

$$y = e^{\alpha x}[(c_1 x + c_2) \cos \beta x + (c_3 x + c_4) \sin \beta x] + \ldots + c_n e^{m_n x}.$$

**Example 13.1.** *Solve* $\frac{d^2 x}{dt^2} + 5\frac{dx}{dt} + 6x = 0$, *given* $x(0) = 0$, $\frac{dx}{dt}(0) = 15$. *(V.T.U., 2010)*

**Solution.** Given equation in *symbolic form is* $(D^2 + 5D + 6) x = 0$.

Its A.E. is $D^2 + 5D + 6 = 0$, *i.e.*, $(D + 2)(D + 3) = 0$ whence $D = -2, -3$.

$\therefore \qquad$ C.S. is $x = c_1 e^{-2t} + c_2 e^{-3t}$ and $\frac{dx}{dt} = -2ae^{-2t} - 3c_2 e^{-3t}$

When $t = 0$, $x = 0$. $\quad \therefore \quad 0 = c_1 + c_2$ ...(*i*)

When $t = 0$, $dx/dt = 15$ $\quad \therefore \quad 15 = -2c_1 - 3c_2$ ...(*ii*)

Solving Eqn. (*i*) and (*ii*), $c_1 = 15, c_2 = -15$.

Hence the required solution is $x = 15\,(e^{-2t} - e^{-3t})$.

**Example 13.2.** *Solve* $\dfrac{d^2x}{dt^2} + 6\dfrac{dx}{dt} + 9x = 0.$

**Solution.** Given equation in *symbolic form is* $(D^2 + 6D + 9) = 0$

$\therefore$ A.E. is $D^2 + 6D + 9 = 0$, *i.e.*, $(D + 3)^2 = 0$ whence $D = -3, -3$.

Hence the C.S. is $x = (c_1 + c_2 t)\, e^{-3t}$.

**Example 13.3.** *Solve* $(D^3 + D^2 + 4D + 4) = 0.$

**Solution.** Here the A.E. is $D^3 + D^2 + 4D + 4 = 0$ *i.e.*, $(D^2 + 4)(D + 1) = 0$ $\therefore$ $D = -1, \pm 2i$.

Hence the C.S. is $y = c_1 e^{-x} + e^{0x}(c_2 \cos 2x + c_3 \sin 2x)$

*i.e.*, $y = c_1 e^{-x} + c_2 \cos 2x + c_3 \sin 2x$.

**Example 13.4.** *Solve* (*i*) $(D^4 - 4D + 4)\, y = 0$

(*ii*) $(D^2 + 1)^3\, y = 0$ *where* $D \equiv d/dx$.

**Solution.** (*i*) The A.E. equation is $D^4 - 4D^2 + 4 = 0$ or $(D^2 - 2)^2 = 0$

$\therefore$ $D^2 = 2, 2$ *i.e.*, $D = \pm \sqrt{2}, \pm \sqrt{2}$.

Hence the C.S. is $((c_1 + c_2 x)\, e^{\sqrt{2}x} + (c_3 + c_4 x)\, e^{-\sqrt{2}x})$ [Roots being repeated]

(*ii*) The A.E. equation is $(D^2 + 1)^3 = 0$

$\therefore$ $D = \pm i, \pm i, \pm i$.

Hence the C.S. is $y = e^{ox}\,[(c_1 + c_2 x + c_3 x^2)\cos x + (c_4 + c_5 x + c_6 x^2)\sin x]$

*i.e.*, $y = (c_1 + c_2 + c_3 x^2)\cos x + (c_4 + c_5 x + c_6 x^2)\sin x$.

**Example 13.5.** *Solve* $\dfrac{d^4x}{dt^4} + 4x = 0.$

**Solution.** Given equation in *symbolic form is* $(D^4 + 4)\, x = 0$

$\therefore$ A.E. is $D^4 + 4 = 0$ or $(D^4 + 4D^2 + 4) - 4D^2 = 0$ or $(D^2 + 2)^2 - (2D)^2 = 0$

or $(D^2 + 2D + 2)(D^2 - 2D + 2) = 0$

$\therefore$ either $D^2 + 2D + 2 = 0$ or $D^2 - 2D + 2 = 0$

whence $D = \dfrac{-2 \pm \sqrt{(-4)}}{2}$ and $\dfrac{2 \pm \sqrt{(-4)}}{2}$ *i.e.*, $D = -1 \pm i$ and $1 \pm i$.

Hence the required solution is $x = e^{-t}(c_1 \cos t + c_2 \sin t) + e^{t}(c_3 \cos t + c_4 \sin t)$.

## PROBLEMS 13.1

Solve :

1. $y'' - 4y' - 5y = 0.$ (*P.T.U., 2015*)
2. $y'' - 2y' + 5y = 0, y(0) = -3, y'(0) = 1.$ (*K.T.U., 2016*)
3. $\dfrac{d^2x}{dt^2} - 4\dfrac{dx}{dt} + 13x = 0, x(0) = \dfrac{dx(0)}{dt} = 2.$
4. $\dfrac{d^3y}{dx^3} + y = 0.$ (*C.S.V.T.U., 2012*)
5. $\dfrac{d^3y}{dx^3} + \dfrac{d^2y}{dx^2} - \dfrac{dy}{dx} - y = 0.$ (*Calicut, 2013*)
6. $\dfrac{d^3y}{dx^3} - 6\dfrac{d^2y}{dx^2} + 11\dfrac{dy}{dx} - 6y = 0$ (*V.T.U., 2016*)
7. $\dfrac{d^3y}{dx^3} - 3\dfrac{d^2y}{dx^2} + 3\dfrac{dy}{dx} - y = 0.$ (*V.T.U., 2014*)
8. $\dfrac{d^4y}{dx^4} - 4y = 0$ (*C.S.V.T.U., 2015*)
9. $(D^4 + 18D^2 + 81)y = 0$ (*J.N.T.U., 2015*)
10. $\dfrac{d^4y}{dx^4} + 8\dfrac{d^2y}{dx^2} + 16y = 0.$
11. $(4D^4 - 8D^3 - 7D^2 + 11D + 6)y = 0$
12. $(D^2 + 1)^2 (D - 1)y = 0.$
13. $(D^2 + D + 1)^2 (D - 2)\, y = 0.$ (*Delhi, 2012*)
14. If $\dfrac{d^4x}{dt^4} = m^4 x$, show that $x = c_1 \cos mt + c_2 \sin mt + c_3 \cosh mt + c_4 \sinh mt$.

## 13.5 INVERSE OPERATOR

**(1) Definition.** $\frac{1}{f(D)}X$ is that function of $x$, not containing arbitrary constants which when operated upon by $f(D)$ gives $X$.

*i.e.,* $$f(D)\left\{\frac{1}{f(D)}X\right\} = X$$

Thus $\frac{1}{f(D)}X$ *satisfies the equation* $f(D)y = X$ *and is, therefore, its particular integral.*

Obviously, $f(D)$ and $1/f(D)$ are inverse operators.

**(2)** $\mathbf{\frac{1}{D}X = \int X dx}$

Let $$\frac{1}{D}X = y \quad \text{...(i)}$$

Operating by $D$, $$D\frac{1}{D}X = Dy \quad i.e., \; X = \frac{dy}{dx}$$

Integrating both sides w.r.t. $x$, $y = \int X\,dx$, no constant being added as Eqn. (*i*) does not contain any constant.

Thus $$\frac{1}{D}X = \int X\,dx.$$

**(3)** $\mathbf{\frac{1}{D-a}X = e^{ax}\int Xe^{-ax}dx.}$

Let $$\frac{1}{D-a}X = y \quad \text{...(ii)}$$

Operating by $D - a$, $(D-a)\cdot\frac{1}{D-a}X = (D-a)y.$

or $$X = \frac{dy}{dx} - ay, \; i.e., \; \frac{dy}{dx} - ay = X$$ which is a Leibnitz's linear equation.

$\therefore$ I.F. being $e^{-ax}$, its solution is

$$ye^{-ax} = \int Xe^{-ax}dx,$$ no constant being added as Eqn. (*ii*) doesn't contain any constant.

Thus $$\frac{1}{D-a}X = y = e^{ax}\int Xe^{-ax}dx.$$

## 13.6 RULES FOR FINDING THE PARTICULAR INTEGRAL

Consider the equation $\frac{d^n y}{dx^n} + k_1\frac{d^{n-1}y}{dx^{n-1}} + k_2\frac{d^{n-2}y}{dx^{n-2}} + \ldots + k_n y = X$

which is *symbolic form of* $(D^n + k_1D^{n-1} + k_2D^{n-2} + \ldots + k_n)y = X.$

$$\therefore \quad \text{P.I.} = \frac{1}{D^n + k_1D^{n-1} + k_2D^{n-2} + \ldots + k_n}X.$$

**Case I. When X = e^ax^**

Since
$$De^{ax} = ae^{ax}$$
$$D^2e^{ax} = a^2e^{ax}$$
........................
........................
$$D^ne^{ax} = a^ne^{ax}$$

$$\therefore \quad (D^n + k_1D^{n-1} + \ldots + k_n)e^{ax} = (a^n + k_1a^{n-1} + \ldots + k_n)e^{ax}, \quad i.e., \quad f(D)e^{ax} = f(a)e^{ax}$$

Operating on both sides by $\frac{1}{f(D)}, \frac{1}{f(D)} f(D) e^{ax} = \frac{1}{f(D)} f(a) e^{ax}$ or $e^{ax} = f(a) \frac{1}{f(D)} e^{ax}$

$\therefore$ dividing by $f(a)$,

$$\frac{1}{\mathbf{f(D)}} \mathbf{e^{ax}} = \frac{1}{\mathbf{f(a)}} \mathbf{e^{ax}} \quad \textit{provided } f(a) \neq 0 \qquad \text{...(1)}$$

**If f(a) = 0,** *the above rule fails and we proceed further.*

Since $a$ is a root of A.E. $f(D) = D^n + k_1 D^{n-1} + ... + k_n = 0$.

$\therefore$ $D - a$ is a factor of $f(D)$. Suppose $f(D) = (D - a)\,\phi(D)$, where $\phi(a) \neq 0$. Then

$$\frac{1}{f(D)} e^{ax} = \frac{1}{D-a} \cdot \frac{1}{\phi(D)} e^{ax} = \frac{1}{D-a} \cdot \frac{1}{\phi(a)} e^{ax} \qquad \text{[By Eqn. (1)]}$$

$$= \frac{1}{\phi(a)} \cdot \frac{1}{D-a} e^{ax} = \frac{1}{\phi(a)} \cdot e^{ax} \int e^{ax} . e^{-ax} dx \qquad \text{[By §13.5 (3)]}$$

$$= \frac{1}{\phi(a)} e^{ax} \int dx - x \frac{1}{\phi(a)} e^{ax} \quad i.e., \quad \frac{1}{\mathbf{f(D)}} \mathbf{e^{ax}} = \mathbf{x} \frac{1}{\mathbf{f'(a)}} \mathbf{e^{ax}} \qquad \text{...(2)}$$

$$\left[ \begin{array}{l} \because \; f'(D) = (D-a)\phi'(D) + 1 . \phi(D) \\ \therefore \; f'(a) = 0 \times \phi'(a) + \phi(a) \end{array} \right.$$

If $f'(a) = 0$, then applying (2) again, we get $\frac{1}{\mathbf{f(D)}} \mathbf{e^{ax}} = \mathbf{x^2} \frac{1}{\mathbf{f''(a)}} \mathbf{e^{ax}}$, *provided* $f''(a) \neq 0$ ...(3)

and so on.

**Example 13.6.** *Find the P.I. of $(D^2 + 5D + 6)y = e^x$.*

**Solution.** $$\text{P.I.} = \frac{1}{D^2 + 5D + 6} e^x \; [\text{Put } D = 1] \; = \frac{1}{1^2 + 5.1 + 6} e^x = \frac{e^x}{12}.$$

**Example 13.7.** *Find the P.I. of $(D + 2)(D - 1)^2 y = e^{-2x} + 2 \sinh x$.*

**Solution.** $$\text{P.I.} = \frac{1}{(D+2)(D-1)^2} [e^{-2x} + 2 \sinh x] - \frac{1}{(D+2)(D-1)^2} [e^{-2x} + e^x - e^{-x}]$$

Let us evaluate each of these terms separately.

$$\frac{1}{(D+2)(D-1)^2} e^{-2x} = \frac{1}{D+2} \cdot \left[ \frac{1}{(D-1)^2} e^{-2x} \right]$$

$$= \frac{1}{D+2} \cdot \frac{1}{(-2-1)^2} e^{-2x} = \frac{1}{9} \cdot \frac{1}{D+2} e^{-2x}$$

$$= \frac{1}{9} \cdot x \cdot \frac{1}{1} e^{-2x} = \frac{x}{9} e^{-2x} \qquad \left[ \because \frac{d}{dD}(D+2) = 1 \right]$$

$$\frac{1}{(D+2)(D-1)^2} e^x = \frac{1}{1+2} \cdot \frac{1}{(D-1)^2} e^x = \frac{1}{3} \cdot x^2 \cdot \frac{1}{2} e^x = \frac{x^2}{6} e^x \qquad \left[ \because \frac{d^2}{dD^2}(D-1)^2 = 2 \right]$$

and $$\frac{1}{(D+2)(D-1)^2} e^{-x} = \frac{1}{(-1+2)(-1-1)^2} e^{-x} = \frac{e^{-x}}{4}$$

Hence, P.I. $= \frac{x}{9} e^{-2x} + \frac{x^2}{6} e^x + \frac{1}{4} e^{-x}$.

**Case II. When X = sin (ax + b) or cos (ax + b).**

Since $$D \sin(ax + b) = a \cos(ax + b)$$
$$D^2 \sin(ax + b) = -a^2 \sin(ax + b)$$
$$D^3 \sin(ax + b) = -a^3 \cos(ax + b)$$

$$D^4 \sin(ax+b) = a^4 \sin(ax+b)$$

*i.e.*, $$D^2 \sin(ax+b) = (-a^2)\sin(ax+b)$$

$$(D^2)^2 \sin(ax+b) = (-a^2)^2 \sin(ax+b)$$

In general $$(D^2)^r \sin(ax+b) = (-a^2)^r \sin(ax+b)$$

$\therefore$ $$f(D^2)\sin(ax+b) = f(-a^2)\sin(ax+b)$$

Operating on both sides $1/f(D^2)$,

$$\frac{1}{f(D^2)} \cdot f(D^2)\sin(ax+b) = \frac{1}{f(D^2)} f(-a^2)\sin(ax+b)$$

or $$\sin(ax+b) = f(-a^2)\frac{1}{f(D^2)}\sin(ax+b)$$

$\therefore$ Dividing by $f(-a^2)$, $\mathbf{\dfrac{1}{f(D^2)}\sin(ax+b) = \dfrac{1}{f(-a^2)}\sin(ax+b)}$ *provided* $f(-a^2) \neq 0$ ...(4)

**If $\mathbf{f(-a^2) = 0}$,** *the above rule fails and we proceed further.*

Since $\cos(ax+b) + i\sin(ax+b) = e^{i(ax+b)}$ [Euler's theorem]

$\therefore$ $$\frac{1}{f(D^2)}\sin(ax+b) = \text{I.P. of } \frac{1}{f(D^2)}e^{i(ax+b)}$$ [Since $f(-a^2) = 0$ $\therefore$ by (2)]

$$= \text{I.P. of } x\frac{1}{f'(D^2)}e^{i(ax+b)}$$ where $D^2 = -a^2$

$\therefore$ $\mathbf{\dfrac{1}{f(D^2)}\sin(ax+b) = x\,\dfrac{1}{f'(-a^2)}\sin(ax+b)}$ *provided* $f'(-a^2) \neq 0$ ...(5)

*If* $f'(-a^2) = 0$, $\mathbf{\dfrac{1}{f(D^2)} \cdot \sin(ax+b) = x^2\,\dfrac{1}{f''(-a^2)}\sin(ax+b)}$, *provided* $f''(-a^2) \neq 0$, and so on.

Similarly, $\mathbf{\dfrac{1}{f(D^2)}\cos(ax+b) = \dfrac{1}{f(-a^2)}\cos(ax+b)}$, *provided* $f(-a^2) \neq 0$

*If* $f(-a^2) = 0$, $\mathbf{\dfrac{1}{f(D^2)}\cos(ax+b) = x \cdot \dfrac{1}{f'(-a^2)}\cos(ax+b)}$, *provided* $f'(-a^2) \neq 0$.

*If* $f'(-a^2) = 0$, $\mathbf{\dfrac{1}{f(D^2)}\cos(ax+b) = x^2\,\dfrac{1}{f''(-a^2)}\cos(ax+b)}$, *provided* $f''(-a^2) \neq 0$ and so on.

**Example 13.8.** *Find the P.I. of $(D^3+1)y = \cos(2x-1)$.*

**Solution.** $$\text{P.I.} = \frac{1}{D^3+1}\cos(2x-1)$$ [Put $D^2 = -2^2 = -4$]

$$= \frac{1}{D(-4)+1}\cos(2x-1)$$ [Multiply and divide by $1+4D$]

$$= \frac{(1+4D)}{(1-4D)(1+4D)}\cos(2x-1) = (1+4D) \cdot \frac{1}{1-16D^2}\cos(2x-1)$$ [Put $D^2 = -2^2 = -4$]

$$= (1+4D)\frac{1}{1-16(-4)}\cos(2x-1) = \frac{1}{65}[\cos(2x-1) + 4D\cos(2x-1)]$$

$$= \frac{1}{65}[\cos(2x-1) - 8\sin(2x-1)].$$

**Example 13.9.** *Find the P.I. of $\dfrac{d^3y}{dx^3} + 4\dfrac{dy}{dx} = \sin 2x$.* *(C.S.V.T.U., 2011)*

**Solution.** Given equation in *symbolic form* is $(D^3+4D)y = \sin 2x$

$\therefore$ $$\text{P.I.} = \frac{1}{D(D^2+4)}\sin 2x \qquad [\because D^2+4=0 \text{ for } D^2=-2^2, \therefore \text{ Apply (5) 570}]$$

$$= x\frac{1}{3D^2+4}\sin 2x \qquad \left[\because \frac{d}{dD}[D^3+4D]=3D^2+4\right]$$

[Put $D^2=-2^2=-4$]

$$= x\frac{1}{3(-4)+4}\sin 2x = -\frac{x}{8}\sin 2x.$$

**Case III. When $\mathbf{X = x^m}$.**

Here $$\text{P.I.} = \frac{1}{f(D)}x^m = [f(D)]^{-1}x^m.$$

Expand $[f(D)]^{-1}$ in ascending powers of $D$ as far as the term in $D^m$ and operate on $x^m$ term by term. Since the $(m+1)$th and higher derivatives of $x^m$ are zero, we need not consider terms beyond $D^m$.

**Example 13.10.** *Find the P.I. of* $\dfrac{d^2y}{dx^2}+\dfrac{dy}{dx}=x^2+2x+4.$

**Solution.** Given equation in *symbolic form* is $(D^2+D)y = x^2+2x+4$.

$\therefore$ $$\text{P.I.} = \frac{1}{D(D+1)}(x^2+2x+4) = \frac{1}{D}(1+D)^{-1}(x^2+2x+4)$$

$$= \frac{1}{D}(1-D+D^2-\ldots)(x^2+2x+4) = \frac{1}{D}[x^2+2x+4-(2x+2)+2]$$

$$= \int(x^2+4)dx = \frac{x^3}{3}+4x.$$

**Case IV. When $\mathbf{X = e^{ax}V}$,** *V being a function of x.*

If $u$ is a function of $x$, then

$$D(e^{ax}u) = e^{ax}Du + ae^{ax}u + e^{ax}(D+a)u$$
$$D^2(e^{ax}u) = a^{ax}D^2u + 2ae^{ax}Du + a^2e^{ax}u = e^{ax}(D+a)^2u$$

and in general, $$D^n(e^{ax}u) = e^{ax}(D+a)^nu$$

$\therefore$ $$f(D)(e^{ax}u) = e^{ax}f(D+a)u$$

Operating both sides by $1/f(D)$,

$$\frac{1}{f(D)}\cdot f(D)(e^{ax}u) = \frac{1}{f(D)}[e^{ax}f(D+a)u]$$

$$e^{ax}u = \frac{1}{f(D)}[e^{ax}f(D+a)u]$$

Now put $f(D+a)u = V$, *i.e.*, $u = \dfrac{1}{f(D+a)}V$, so that $e^{ax}\dfrac{1}{f(D+a)}V = \dfrac{1}{f(D)}(e^{ax}V)$

*i.e.*, $$\mathbf{\frac{1}{f(D)}(e^{ax}V) = e^{ax}\frac{1}{f(D+a)}V.} \qquad \ldots(6)$$

**Example 13.11.** *Find P.I. of* $(D^2-2D+4)y = e^x\cos x$.

**Solution.** $$\text{P.I.} = \frac{1}{D^2-2D+4}e^x\cos x \qquad [\text{Replace } D \text{ by } D+1]$$

$$= e^x\frac{1}{(D+1)^2-2(D+1)+4}\cos x = e^x\frac{1}{D^2+3}\cos x \qquad [\text{Put } D^2=-1^2=-1]$$

$$= e^x\frac{1}{-1+3}\cos x = \frac{1}{2}e^x\cos x.$$

**Case V. When X is any other function of x.**

Here $\quad \text{P.I.} = \dfrac{1}{f(D)} X.$

If $\quad f(D) = (D - m_1)(D - m_2) \ldots (D - m_n)$, resolving into partial fractions,

$$\frac{1}{f(D)} = \frac{A_1}{D - m_1} + \frac{A_2}{D - m_2} + \ldots + \frac{A_n}{D - m_n}$$

$$\therefore \quad \text{P.I.} = \left[\frac{A_1}{D - m_1} + \frac{A_2}{D - m_2} + \ldots + \frac{A_n}{D - m_n}\right] X$$

$$= A_1 \frac{1}{D - m_1} X + A_2 \frac{1}{D - m_2} X + \ldots + A_n \frac{1}{D - m_n} X$$

$$= A_1 . e^{m_1 x} \int X e^{-m_1 x} dx + A_2 . e^{m_2 x} \int X e^{-m_2 x} dx + \ldots + A_n . e^{m_n x} \int X e^{-m_n x} dx \qquad \text{[By §13.5 ...(3)]}$$

**Obs.** This method is a general one and can, therefore, be employed to obtain a particular integral in any given case.

## 13.7 WORKING PROCEDURE TO SOLVE THE EQUATION

$$\frac{d^n y}{dx^n} + k_1 \frac{d^{n-1} y}{dx^{n-1}} + \ldots + k_{n-1} \frac{dy}{dx} + k_n y = X$$

of which the *symbolic form* is

$$(D^n + k_1 D^{n-1} + \ldots + k_{n-1} D + k_n) y = X .$$

**Step I. To find the complementary function**

(*i*) *Write the A.E.*

*i.e.*, $\quad D^n + k_1 D^{n-1} + \ldots + k_{n-1} D + k_n = 0$ and solve it for $D$.

(*ii*) Write the *C.F.* as follows :

| *Roots of A.E.* | *C.F.* |
|---|---|
| 1. $m_1, m_2, m_3 \ldots$ (real and different roots) | $c_1 e^{m_1 x} + c_2 e^{m_2 x} + c_3 e^{m_3 x} + \ldots$ |
| 2. $m_1, m_1, m_3 \ldots$ (two real and equal roots) | $(c_1 + c_2 x) e^{m_1 x} + c_3 e^{m_3 x} + \ldots$ |
| 3. $m_1, m_1, m_1, m_4 \ldots$ (three real and equal roots) | $(c_1 + c_2 x + c_3 x^2) e^{m_1 x} + c_4 e^{m_4 x} + \ldots$ |
| 4. $\alpha + i\beta, \alpha - i\beta, m_3 \ldots$ (a pair of imaginary roots) | $e^{\alpha x}(c_1 \cos \beta x + c_2 \sin \beta x) + c_3 e^{m_3 x} + \ldots$ |
| 5. $\alpha \pm i\beta, \alpha \pm i\beta, m_5 \ldots$ (2 pairs of equal imaginary roots) | $e^{\alpha x}[(c_1 + c_2 x) \cos \beta x + (c_3 + c_4 x) \sin \beta x] + c_5 e^{m_5 x} + \ldots$ |

**Step II. To find the particular integral**

*From symbolic form P.I.* $= \dfrac{1}{D^n + k_1 D^{n-1} + \ldots + k_{n-1} D + k_n} X = \dfrac{1}{f(D)}$ or $\dfrac{1}{\phi(D^2)} X$

(*i*) *When* $X = e^{ax}$

$$\text{P.I.} = \frac{1}{f(D)} e^{ax}, \quad \text{put } D = a, \qquad [f(a) \neq 0]$$

$$= x \frac{1}{f'(D)} e^{ax}, \quad \text{put } D = a, \qquad [f(a) = 0, f'(a) \neq 0]$$

$$= x^2 \frac{1}{f''(D)} e^{ax}, \quad \text{put } D = a, \qquad [f'(a) = 0, f''(a) \neq 0]$$

and so on.

where $\quad f'(D) =$ diff. coeff. of $f(D)$ w.r.t. $D$

$f''(D) =$ diff. coeff. of $f'(D)$ w.r.t. $D$, etc.

(*ii*) *When* $X = \sin(ax + b)$ or $\cos(ax + b)$.

$$\text{P.I.} = \frac{1}{\phi(D^2)} \sin(ax + b) \text{ [or } \cos(ax + b)], \quad \text{put } D^2 = -a^2 \quad [\phi(-a^2) \neq 0]$$

$$= x \frac{1}{\phi'(D^2)} \sin(ax + b) \text{ [or } \cos(ax + b)], \quad \text{put } D^2 = -a^2 \quad [\phi(-a^2) = 0, \phi'(-a^2) \neq 0]$$

$$= x^2 \frac{1}{\phi''(D^2)} \sin(ax + b) \text{ [or } \cos(ax + b)], \quad \text{put } D^2 = -a^2 \quad [\phi'(-a^2) \neq 0, \phi''(-a^2) \neq 0]$$

and so on.

where $\phi'(D^2)$ = diff. coeff. of $\phi(D^2)$ w.r.t. $D$,

$\phi''(D^2)$ = diff. coeff. of $\phi'(D^2)$ w.r.t. $D$, etc.

(*iii*) *When* $X = x^m$, $m$ being a positive integer.

$$\text{P.I.} = \frac{1}{f(D)} x^m = [f(D)]^{-1} x^m$$

To evaluate it, expand $[f(D)]^{-1}$ in ascending powers of $D$ by Binomial theorem as far as $D^m$ and operate on $x^m$ term by term.

(*iv*) *When* $X = e^{ax}V$, where $V$ is a function of $x$.

$$\text{P.I.} = \frac{1}{f(D)} e^{ax} V = e^{ax} \frac{1}{f(D + a)} V$$

and then evaluate $\frac{1}{f(D+a)} V$ as in (*i*), (*ii*), and (*iii*).

(*v*) *When X is any function of x*.

$$\text{P.I.} = \frac{1}{f(D)} X$$

Resolve $\frac{1}{f(D)}$ into partial fractions and operate each partial fraction on $X$ remembering that

$$\frac{1}{D - a} X = e^{ax} \int X e^{-ax} \, dx.$$

**Step III. To find the complete solution**

Then the C.S. is $y$ = C.F. + P.I.

**Example 13.12.** *Solve* $\frac{d^2y}{dx^2} + \frac{dy}{dx} + y = (1 - e^x)^2$.

**Solution.** Given equation in *symbolic form* is $(D^2 + D + 1)y = (1 - e^x)^2$

(*i*) *To find C.F.*

Its A.E. is $D^2 + D + 1 = 0$, $\therefore \quad D = \frac{1}{2}(-1 \pm \sqrt{3}i)$

Thus C.F. $= e^{-x/2} \left( c_1 \cos \frac{\sqrt{3}}{2} x + c_2 \sin \frac{\sqrt{3}}{2} x \right)$

(*ii*) *To find P.I.*

$$\text{P.I.} = \frac{1}{D^2 + D + 1} (1 - 2e^x + e^{2x}) = \frac{1}{D^2 + D + 1} (e^{0x} - 2e^x + e^{2x})$$

$$= \frac{1}{0^2 + 0 + 1} e^{0x} - 2 \cdot \frac{1}{1^2 + 1 + 1} e^x + \frac{1}{2^2 + 2 + 1} e^{2x} = 1 - \frac{2}{3} e^x + \frac{e^{2x}}{7}$$

(*iii*) Hence the C.S. is $y = e^{-x/2} \left( c_1 \cos \frac{\sqrt{3}}{2} x + c_2 \sin \frac{\sqrt{3}}{2} x \right) + 1 - \frac{2}{3} e^x + \frac{e^{2x}}{7}$.

**Example 13.13.** *Solve $y'' + 4y' + 4y = 3 \sin x + 4 \cos x$, $y(0) = 1$ and $y'(0) = 0$.*

**Solution.** Given equation in *symbolic form* is $(D^2 + 4D + 4)y = 3 \sin x + 4 \cos x$

(*i*) *To find C.F.*

Its A.E. is $(D + 2)^2 = 0$ where $D = -2, -2$ $\therefore$ C.F. $= (c_1 + c_2 x)e^{-2x}$.

(*ii*) *To find P.I.*

$$\text{P.I.} = \frac{1}{D^2 + 4D + 4}(3 \sin x + 4 \cos x) = \frac{1}{-1 + 4D + 4}(3 \sin x + 4 \cos x)$$

$$= \frac{4D - 3}{16D^2 - 9}(3 \sin x + 4 \cos x) = \frac{(4D - 3)}{-16 - 9}(3 \sin x + 4 \cos x)$$

$$= \frac{-1}{25}\{3(4 \cos x - 3 \sin x) + 4(-4 \sin x - 3 \cos x)\} = \sin x$$

(*iii*) C.S. is $y = (c_1 + c_2)\, e^{-2x} + \sin x$

When $x = 0, y = 1,$ $\therefore$ $1 = c_1$

Also $y' = c_2 e^{-2x} + (c_1 + c_2 x)(-2)e^{-2x} + \cos x$.

When $x = 0, y' = 0,$ $\therefore$ $0 = c_2 - 2c_1 + 1$, *i.e.*, $c_2 = 1$.

Hence the required solution is $y = (1 + x)\, e^{-2x} + \sin x$.

**Example 13.14.** *Solve $(D - 2)^2 = 8(e^{2x} + \sin 2x + x^2)$.* (*V.T.U., 2012*)

**Solution.** (*i*) *To find C.F.*

Its A.E. is $(D - 2)^2 = 0$, $\therefore$ $D = 2, 2$.

Thus C.F. $= (c_1 + c_2 x)\, e^{2x}$.

(*ii*) *To find P.I.*

$$\text{P.I.} = 8\left[\frac{1}{(D-2)^2}e^{2x} + \frac{1}{(D-2)^2}\sin 2x + \frac{1}{(D-2)^2}x^2\right]$$

Now $\dfrac{1}{(D-2)^2}e^{2x} = x^2 \dfrac{1}{2(1)}e^{2x}$ [$\because$ by putting $D = 2$, $(D - 2)^2 = 0$, $2(D - 2) = 0$]

$$= \frac{x^2 e^{2x}}{2}.$$

$$\frac{1}{(D-2)^2}\sin 2x = \frac{1}{D^2 - 4D + 4}\sin 2x = \frac{1}{(-2^2) - 4D + 4}\sin 2x$$

$$= -\frac{1}{4}\int \sin 2x\, dx = -\frac{1}{4}\left(\frac{-\cos 2x}{2}\right) = \frac{1}{8}\cos 2x$$

and
$$\frac{1}{(D-2)^2}x^2 = \frac{1}{4}\left(1 - \frac{D}{2}\right)^{-2}x^2 = \frac{1}{4}\left[1 + (-2)\left(\frac{D}{2}\right) + \frac{(-2)(-3)}{2!}\left(-\frac{D}{2}\right)^2 + \ldots\right]x^2$$

$$= \frac{1}{4}\left(1 + D + \frac{3D^2}{4} + \ldots\right)x^2 = \frac{1}{4}\left(x^2 + 2x + \frac{3}{2}\right)$$

Thus P.I. $= 4x^2 e^{2x} + \cos 2x + 2x^2 + 4x + 3$.

(*iii*) Hence the C.S. is $y = (c_1 + c_2 x)\, e^{2x} + 4x^2 e^{2x} + \cos 2x + 2x^2 + 4x + 3$.

**Example 13.15.** *Find the complete solution of $y'' - 2y' + 2y = x + e^x \cos x$.*

**Solution.** Given equation in *symbolic form* is $(D^2 - 2D + 2)y = x + e^x \cos x$

(*i*) *To find C.F.*

Its A.E. is $D^2 - 2D + 2 = 0$ $\therefore$ $D = \dfrac{2 \pm \sqrt{(4 - 8)}}{2} = 1 \pm i$.

Thus C.F. $= e^x\,(c_1 \cos x + c_2 \sin x)$

(*ii*) *To find P.I.*

$$\text{P.I.} = \frac{1}{D^2 - 2D + 2}(x) + \frac{1}{D^2 - 2D + 2}(e^x \cos x)$$

$$= \frac{1}{2}\left[1 - \left(D - \frac{D^2}{2}\right)\right]^{-1}(x) + e^x \frac{1}{(D+1)^2 - 2(D+1) + 2}(\cos x)$$

$$= \frac{1}{2}\left(1 + D - \frac{D^2}{2}\right)x + e^x \frac{1}{D^2 + 1}\cos x \qquad \text{[Case of failure]}$$

$$= \frac{1}{2}(x + 1 - 0) + e^x \cdot x\frac{1}{2D}\cos x = \frac{1}{2}(x+1) + \frac{xe^x}{2}\int \cos x\, dx = \frac{1}{2}(x+1) + \frac{xe^x}{2}\sin x$$

(*iii*) Hence the C.S. is $y = e^x(c_1 \cos x + c_2 \sin x) + \frac{1}{2}(x+1) + \frac{xe^x}{2}\sin x$.

**Example 13.16.** *Solve* $\frac{d^2y}{dx^2} - 3\frac{dy}{dx} + 2y = xe^{3x} + \sin 2x.$ *(Andhra, 2022)*

**Solution.** Given equation in *symbolic form* is $(D^2 - 3D + 2)y = xe^{3x} + \sin 2x$

(*i*) *To find C.F.*

Its A.E. is $D^2 - 3D + 2 = 0$ or $(D-2)(D-1) = 0$ whence $D = 1, 2$.

Thus C.F. $= c_1e^x + c_2e^{2x}$

(*ii*) *To find P.I.*

$$\text{P.I.} = \frac{1}{D^2 - 3D + 2}(xe^{3x} + \sin 2x) = \frac{1}{D^2 - 3D + 2}(e^{3x} \cdot x) + \frac{1}{D^2 - 3D + 2}(\sin 2x)$$

$$= e^{3x} \cdot \frac{1}{(D+3)^2 - 3(D+3) + 2}(x) + \frac{1}{-4 - 3D + 2}(\sin 2x)$$

$$= e^{3x} \cdot \frac{1}{D^2 + 3D + 2}(x) - \frac{3D - 2}{9D^2 - 4}(\sin 2x) = \frac{e^{3x}}{2} \cdot \left[1 + \left\{\frac{3D + D^2}{2}\right\}\right]^{-1} x - \frac{(3D - 2)}{9(-4) - 4}(\sin 2x)$$

$$= \frac{e^{3x}}{2}\left(1 - \frac{3D}{2} \ldots\right)x + \frac{1}{40}(6\cos 2x - 2\sin 2x) = \frac{e^{3x}}{2}\left(x - \frac{3}{2}\right) + \frac{1}{20}(3\cos 2x - \sin 2x)$$

(*iii*) Hence the C.S. is $y = c_1e^x + c_2e^{2x} + e^{3x}\left(\frac{x}{2} - \frac{3}{4}\right) + \frac{1}{20}(3\cos 2x - \sin 2x)$.

**Example 13.17.** *Solve* $\frac{d^2y}{dx^2} - 4y = x \sinh x.$

**Solution.** Given equation in *symbolic form* is $(D^2 - 4)y = x \sinh x$.

(*i*) *To find C.F.*

Its A.E. is $D^2 - 4 = 0$, whence $D = \pm 2$.

Thus C.F. $= c_1e^{2x} + c_2e^{-2x}$

(*ii*) *To find P.I.*

$$\text{P.I.} = \frac{1}{D^2 - 4}x \sinh x = \frac{1}{D^2 - 4}x\left(\frac{e^x - e^{-x}}{2}\right) = \frac{1}{2}\left[\frac{1}{D^2 - 4}e^x \cdot x - \frac{1}{D^2 - 4}e^{-x} \cdot x\right]$$

$$= \frac{1}{2}\left[e^x \frac{1}{(D+1)^2 - 4}x - e^{-x}\frac{1}{(D-1)^2 - 4}x\right] = \frac{1}{2}\left[e^x \frac{1}{D^2 + 2D - 3}x - e^{-x}\frac{1}{D^2 - 2D - 3}x\right]$$

$$= \frac{1}{2}\left[\frac{e^x}{-3}\left\{1-\left(\frac{2D}{3}+\frac{D^2}{3}\right)\right\}^{-1}.x - \frac{e^{-x}}{-3}\left\{1+\left(\frac{2D}{3}-\frac{D^2}{3}\right)\right\}^{-1}\cdot x\right]$$

$$= -\frac{1}{6}\left[e^x\left(1+\frac{2D}{3}+...\right)x - e^{-x}\left(1-\frac{2D}{3}+...\right)x\right] = -\frac{1}{6}\left[e^x\left(x+\frac{2}{3}\right)-e^{-x}\left(x-\frac{2}{3}\right)\right]$$

$$= -\frac{x}{3}\left(\frac{e^x - e^{-x}}{2}\right) - \frac{2}{9}\left(\frac{e^x + e^{-x}}{2}\right) = -\frac{x}{3}\sinh x - \frac{2}{9}\cosh x\,.$$

(*iii*) Hence the C.S. is $y = c_1e^{2x} + c_2e^{-2x} - \frac{x}{3}\sinh x - \frac{2}{9}\cosh x$.

**Example 13.18.** *Solve* $(D^2 - 1)y = x \sin 3x + \cos x.$ *(C.S.V.T.U., 2015)*

**Solution.** (*i*) *To find C.F.*

Its A.E. is $D^2 - 1 = 0$, whence $D = \pm 1$. $\therefore$ C.F. $= c_1e^x + c_2e^{-x}$

(*ii*) *To find P.I.*

$$\text{P.I.} = \frac{1}{D^2-1}(x\sin 3x + \cos x) = \frac{1}{D^2-1}x\,(\text{I.P. of } e^{3ix}) + \frac{1}{D^2-1}\cos x$$

$$= \text{I.P. of } \frac{1}{D^2-1}e^{3ix}.x + \frac{1}{(-1)^2-1}\cos x = \text{I.P. of}\left[e^{3ix}\frac{1}{(D+3i)^2-1}x\right] - \frac{\cos x}{2}$$

[Replacing $D$ by $D + 3i$]

$$= \text{I.P. of}\left[e^{3ix}\frac{1}{D^2+6iD-10}x\right] - \frac{\cos x}{2}$$

$$= \text{I.P. of}\left[e^{3ix}.\frac{1}{-10}\left(1-\frac{3iD}{5}-\frac{D^2}{10}\right)^{-1}x\right] - \frac{\cos x}{2}$$ [Expand by Binomial theorem]

$$= \text{I.P. of}\left[e^{3ix}.\frac{1}{-10}\left(1+\frac{3iD}{5}+......\right)x\right] - \frac{\cos x}{2} = \text{I.P. of}\left[-\frac{e^{3ix}}{10}\left(x+\frac{3i}{5}\right)\right] - \frac{\cos x}{2}$$

$$= \text{I.P. of}\left[\frac{-1}{10}(\cos 3x + i\sin 3x)\left(x+\frac{3i}{5}\right)\right] - \frac{\cos x}{2}$$

$$= -\frac{1}{10}\text{I.P. of}\left[\left(x\cos 3x - \frac{3\sin 3x}{5}\right) + i\left(x\sin 3x + \frac{3}{5}\cos 3x\right)\right] - \frac{\cos x}{2}$$

$$= -\frac{1}{10}\left(x\sin 3x + \frac{3}{5}\cos 3x\right) - \frac{\cos x}{2}.$$

(*iii*) Hence the C.S. is $y = c_1e^x + c_2e^{-x} - \frac{1}{50}(5x\sin 3x + 3\cos 3x + 25\cos x)$.

**Example 13.19.** *Solve* $\frac{d^2y}{dx^2} - 2\frac{dy}{dx} + y = xe^x \sin x.$

*(Andhra, 2014 ; V.T.U., 2013 ; C.S.V.T.U., 2011 S)*

**Solution.** Given equation in *symbolic form* is $(D^2 - 2D + 1)\,y = xe^x \sin x$

(*i*) *To find C.F.*

Its A.E. is $D^2 - 2D + 1 = 0$, *i.e.*, $(D - 1)^2 = 0$

$\therefore$ $D = 1, 1$. Thus C.F. $= (c_1 + c_2x)e^x$

(*ii*) *To find P.I.*

$$\text{P.I.} = \frac{1}{(D-1)^2} e^x \,.\, x \sin x = e^x \,.\, \frac{1}{(D+1-1)^2} x \sin x$$

$$= e^x \frac{1}{D^2} x \sin x = e^x \frac{1}{D} \int x \sin x \, dx \qquad \text{[Integrate by parts]}$$

$$= e^x \frac{1}{D}\left[x(-\cos x) - \int 1 \,.\, (-\cos x)\, dx\right] = e^x \int [-x \cos x + \sin x]\, dx$$

$$= e^x\left[-\left\{x \sin x - \int 1 \,.\, \sin x \, dx\right\} - \cos x\right] = e^x[-x \sin x - \cos x - \cos x]$$

$$= -e^x(x \sin x + 2 \cos x).$$

(*iii*) Hence the C.S. is $y = (c_1 + c_2x)\, e^x - e^x(x \sin x + 2 \cos x)$.

**Example 13.20.** *Solve* $(D^3 - 4D^2 + 4D)y = 0$

*where,* $D = \dfrac{d}{dx}$.

**Solution:**

$$A.E. \text{ is } D^3 - 4D^2 + D = 0$$
$$D(D^2 - 4D + 1) = 0$$
$$D = 0, 2, 2.$$

Hence the solution of the given equation is

$$y = C_1e^x + (C_2x + C_3)e^{2x}$$

or

$$y = C_1 + (C_2x + C_3)e^{2x}$$

**Example 13.21.** *Solve* $(D^2 - 4D + 4)y = 8x^2e^{2x} \sin 2x.$

(*V.T.U., 2013 ; Calicut, 2012*)

**Solution.** (*i*) *To find C.F.*

Its A.E. is $D^2 - 4D + 4 = 0$ *i.e.*, $(D - 2)^2 = 0$. $\therefore \quad D = 2, 2$

$\therefore \quad$ C.F. $= (c_1 + c_2x)\, e^{2x}$

(*ii*) *To find P.I.*

$$\text{P.I.} = \frac{1}{(D-2)^2}(8x^2e^{2x} \sin 2x) = 8e^{2x} \frac{1}{(D+2-2)^2}(x^2 \sin 2x)$$

$$= 8e^{2x} \frac{1}{D^2}(x^2 \sin 2x) = 8e^{2x} \,.\, \frac{1}{D} \int x^2 \sin 2x \, dx$$

$$= 8e^{2x} \,.\, \frac{1}{D}\left\{x^2\left(\frac{-\cos 2x}{2}\right) - \int 2x \left(\frac{-\cos 2x}{2}\right) dx\right\}$$

$$= 8e^{2x} \frac{1}{D}\left\{-\frac{x^2}{2}\cos 2x + x\frac{\sin 2x}{2} - \int 1 \,.\, \frac{\sin 2x}{2}\, dx\right\}$$

$$= 8e^{2x} \int \left\{-\frac{x^2}{2}\cos 2x + \frac{x}{2}\sin 2x + \frac{\cos 2x}{4}\right\} dx$$

$$= 8e^{2x}\left[\left\{\frac{-x^2}{2}\frac{\sin 2x}{2} - \int (-x)\frac{\sin 2x}{2} dx\right\} + \left\{\int \frac{x}{2} \sin 2x \, dx\right\} + \frac{\sin 2x}{8}\right]$$

$$= 8e^{2x}\left[\left(\frac{-x^2}{4} + \frac{1}{8}\right)\sin 2x + \int x \sin 2x \, dx\right]$$

$$= 8e^{2x}\left[\left(\frac{1}{8} - \frac{x^2}{4}\right)\sin 2x + x\left(\frac{-\cos 2x}{2}\right) - \int 1 \,.\, \left(\frac{-\cos 2x}{2}\right) dx\right]$$

$$= 8e^{2x}\left[\left(\frac{1}{8}-\frac{x^2}{4}\right)\sin 2x - \frac{x\cos 2x}{2} + \frac{\sin 2x}{4}\right] = e^{2x}[(3-2x^2)\sin 2x - 4x\cos 2x]$$

(*iii*) Hence the C.S. is $y = e^{2x}[c_1 + c_2x + (3-2x^2)\sin 2x - 4x\cos 2x]$.

**Example 13.22.** *Solve the differential equation*

$$y'' + 4y' + 29y = 0$$

*Given that at* $x = 0, y = 0, \dfrac{dy}{dx} = 15$

**Solution:** $\dfrac{d}{dx} = D$, the given equation can be

$$(D^2 + 4D + 29)y = 0$$

A.E. $\quad D^2 + 4D + 29 = 0$

or $\quad D = \dfrac{-4 \pm \sqrt{16 - 4 \times 29}}{2} = -2 \pm 5i$

Hence the solution of given equation, $y$

$$y = e^{-2x}(C_1\cos 5x + C_2\sin 5x) \quad ...(i)$$

The arbitrary constants $C_1$ and $C_2$ are to be determined from the given conditions

when $\quad x = 0, y = 0,$

therefore from Eqn. (*i*) $\quad C_1 = 0$

Differentiating Eqn. (*i*) *w.r.t.* $x$

$$\frac{dy}{dx} = e^{-2x}(-5C_1\sin 5x + 5C_2\cos 5x) - 2e^{-2x}(C_1\cos 5x + C_2\sin 5x)$$

Substituting $\quad x = 0, \dfrac{dy}{dx} = 15$ and using $C_1 = 0$, we get

$$15 = 5C_2$$

$$C_2 = 3$$

Substituting the value of $C_1$ and $C_2$ in Eqn. (*i*) the solution

$$y = 3e^{-2x}\sin 5x.$$

**Example 13.23:** *If* $\dfrac{d^4x}{dt^4} = m^4x$. *Show that* $x = c_1\cos mt + c_2\sin mt + c_3\cos h\, mt + c_4\sin h\, mt$ (*VTU, 2018*)

**Solution:** Denoting $\dfrac{d}{dt}$ by $D$, the given equation can be written as

$$(D^4 - m^4)x = 0$$

A.E is $\quad D^4 - m^4 = 0$

$$(D^2 + m^2)(D^2 - m^2) = 0$$

$$(D^2 + m^2)(D - m)(D + m) = 0$$

$$D = \pm mi, \pm m$$

Hence the solution of the given equation is

$$x = c_1\cos mt + c_2\sin mt + k_1e^{mt} + k_2e^{-mt} \quad ...(i)$$

$$e^{mt} = \cos h\, mt + \sin h.\, mt$$

$$e^{-mt} = \cos h\, mt - \sin h\, mt$$

From Eqn. (*i*) becomes $\quad x = c_1\cos mt + c_2\sin mt + k_1(\cos h\, mt + \sin h\, mt) + k_2(\cosh mt - \sin hmt)$

$$= c_1\cos mt + c_2\sin mt + (k_1 + k_2)\cos h\, mt + (k_1 - k_2)\sin h\, mt$$

$$x = c_1\cos mt + c_2\sin mt + c_3\cos h\, mt + c_4\sin h\, mt$$

where $\quad k_1 + k_2 = c_3$

and $\quad k_1 - k_2 = c_4$

### PROBLEMS 13.2

Solve :

1. $\dfrac{d^2y}{dx^2}-6\dfrac{dy}{dx}+9y=6e^{3x}+7e^{-2x}-\log 2$

2. $\dfrac{d^2y}{dx^2}+4\dfrac{dy}{dx}+5y=-2\cosh x$. Also find $y$ when $y=0$, $\dfrac{dy}{dx}=1$ at $x=0$.

3. $\dfrac{d^2y}{dx^2}-4y=\cos h\,(2x-1)+3^x$ (V.T.U., 2014)

4. $\dfrac{d^2y}{dx^2}+\dfrac{dy}{dx}=\cos x$. (Rohtak, 2012)

5. $\dfrac{d^2y}{dx^2}+3\dfrac{dy}{dx}+2y=4\cos^2 x$.

6. $(D^2-D+2)y=\sin 2x$. (V.T.U., 2013)

7. $\dfrac{d^3y}{dx^3}+2\dfrac{d^2y}{dx^2}+\dfrac{dy}{dx}=e^{-x}+\cos 2x$. (Anna, 2016)

8. $(D^2-4D+4)y=x^{2x}+\cos 2x+4$ (V.T.U., 2015)

9. $(D^3-5D^2+7D-3)y=e^{2x}\cosh x$.

10. $\dfrac{d^2y}{dx^2}-y=e^x+x^2e^x$. (Andhra, 2022)

11. $(D^3-D)y=2x+1+4\cos x+2e^x$. (JNTUK, 2021)

12. $(D^3-6D^2+11D-6)y=2^x+\cos 2x$. (V.T.U., 2013)

13. $(D^2+1)^2\,y=x^4+2\sin x\cos 3x$.

14. $\dfrac{d^2y}{dx^2}+5\dfrac{dy}{dx}+6y=e^{-2x}\sin 2x$.

15. $(D^2-3D+2)y=\sin(e^x)$ (U.P.T.U., 2012)

16. $\dfrac{d^2y}{dx^2}+2\dfrac{dy}{dx}+3y=e^x\cos x$. (V.T.U., 2010)

17. $(D^2+4D+3)y=e^{-x}\sin x+xe^{3x}$.

18. $\dfrac{d^2y}{dx^2}+2y=x^2e^{3x}+e^x\cos 2x$.

19. $\dfrac{d^4y}{dx^4}-y=\cos x\cosh x$.

20. $(D^2+1)y=x^2e^x+\sin x$ (K.T.U., 2016)

21. $\dfrac{d^2y}{dx^2}+16y=x\sin 3x$. (V.T.U., 2010 S)

22. $(D^2+2D+1)y=x\cos x$.

23. $(D^2-1)y=x\sin x+(1+x^2)e^x$.

24. $(D^2+2D+1)y=xe^{-x}\cos x$. (Anna, 2016 ; J.N.T.U., 2015)

25. $\dfrac{d^2y}{dx^2}+2\dfrac{dy}{dx}+y=x^2e^{-x}\cos x$ (U.P.T.U., 2012)

26. $\dfrac{d^2y}{dx^2}+3\dfrac{dy}{dx}+2y=e^{e^x}$. (C.S.V.T.U., 2011)

27. $(D^2+a^2)y=\tan ax$.

## 13.8 TWO OTHER METHODS OF FINDING P.I.

**I. Method of variation of parameters.** *This method is quite general and applies to equations of the form*

$$y''+py'+qy=X \qquad ...(1)$$

*where p, q, and X are functions of x. It gives* P.I. $=-y_1\int\dfrac{y_2X}{W}dx+y_2\int\dfrac{y_1X}{W}dx$ ...(2)

*where $y_1$ and $y_2$ are the solutions of* $y''+py'+qy=0$ ...(3)

*and* $W=\begin{vmatrix} y_1 & y_2 \\ y_1' & y_2' \end{vmatrix}$ *is called the Wronskian* of $y_1, y_2$.*

*Proof.* Let the C.F. of Eqn. (1) be $y=c_1y_1+c_2y_2$

Replacing $c_1, c_2$ (regarded as *parameters*) by unknown functions $u(x)$ and $v(x)$, let the P.I. be

$$y=uy_1+vy_2 \qquad ...(4)$$

Differentiating Eqn. (4) w.r.t. $x$, we get $y'=uy'_1+vy'_2+u'y_1+v'y_2$

*Named after the Polish mathematician and philosopher *Hoene Wronsky* (1778–1853).

$$= uy_1' + vy_2' \qquad ...(5)$$

on assuming that $u'y_1 + v'y_2 = 0$ ...(6)

Differentiate Eqn. (4) and substitute in Eqn. (1). Then noting that $y_1$ and $y_2$, satisfy Eqn. (3), we obtain

$$u'y_1' + v'y_2' = X \qquad ...(7)$$

Solving Eqns. (6) and (7), we get

$$u' = -\frac{y_2 X}{W}, v' = \frac{y_1 X}{W} \qquad \text{where, } W = y_1 y_2' - y_2 y_1'$$

Integrating $u = -\int \frac{y_2 X}{W} dx, v = \int \frac{y_1 X}{W} dx$. Substituting these in Eqn. (4), we get Eqn. (2).

**Example 13.24.** *Using the method of variation of parameters, solve*

$$\frac{d^2y}{dx^2} + 4y = \tan 2x.$$

*(Andhra, 2022, 2016 ; K.T.U., 2016 ; V.T.U., 2016)*

**Solution.** Given equation in *symbolic form* is $(D^2 + 4)y = \tan 2x$.

*(i) To find C.F.*

Its A.E. is $D^2 + 4 = 0$, $\therefore$ $D = \pm 2i$

Thus C.F. is $y = c_1 \cos 2x + c_2 \sin 2x$.

*(ii) To find P.I.*

Here $y_1 = \cos 2x$, $y_2 = \sin 2x$ and $X = \tan 2x$

$$\therefore \quad W = \begin{vmatrix} y_1 & y_2 \\ y_1' & y_2' \end{vmatrix} = \begin{vmatrix} \cos 2x & \sin 2x \\ -2\sin 2x & 2\cos 2x \end{vmatrix} = 2$$

Thus,

$$\text{P.I.} = -y_1 \int \frac{y_2 X}{W} dx + y_2 \int \frac{y_1 X}{W} dx$$

$$= -\cos 2x \int \frac{\sin 2x \tan 2x}{2} dx + \sin 2x \int \frac{\cos 2x \tan 2x}{2} dx$$

$$= -\frac{1}{2}\cos 2x \int (\sec 2x - \cos 2x) dx + \frac{1}{2}\sin 2x \int \sin 2x\, dx$$

$$= -\frac{1}{4}\cos 2x[\log(\sec 2x + \tan 2x) - \sin 2x] - \frac{1}{4}\sin 2x \cos 2x$$

$$= -\frac{1}{4}\cos 2x \log(\sec 2x + \tan 2x)$$

Hence the C.S. is $y = c_1 \cos 2x + c_2 \sin 2x - \frac{1}{4}\cos 2x \log(\sec 2x + \tan 2x)$.

**Example 13.25.** *Solve, by the method of variation of parameters, $d^2y/dx^2 - y = 2/(1 + e^x)$.*

*(D.T.U., 2016 ; U.P.T.U. 2011)*

**Solution.** Given equation is $D^2 - 1 = 2/(1 + e^x)$

A.E. is $D^2 - 1 = 0, D = \pm 1$, $\therefore$ C.F. $= c_1 e^x + c_2 e^{-x}$

Here $y_1 = e^x, y_2 = e^{-x}$ and $X = 2/(1 + e^x)$

$$\therefore \quad W = \begin{vmatrix} y_1 & y_2 \\ y_1' & y_2' \end{vmatrix} = \begin{vmatrix} e^x & e^{-x} \\ e^x & -e^{-x} \end{vmatrix} = -e^x e^{-x} - e^x e^{-x} = -2.$$

Thus

$$\text{P.I.} = -y_1 \int \frac{y_2 X}{W} dx + y_2 \int \frac{y_1 X}{W} dx = -e^x \int \frac{e^{-x}}{-2} \cdot \frac{2}{1+e^x} dx + e^{-x} \int \frac{e^x}{-2} \cdot \frac{2}{1+e^x} dx$$

$$= e^x \int \left(\frac{1}{e^x} - \frac{1}{1+e^x}\right) dx - e^{-x} \log(1+e^x) = e^x \left[e^{-x} - \int \frac{e^{-x}}{e^{-x}+1} dx\right] - e^{-x}\log(1+e^x)$$

$$= e^x[-e^{-x} + \log(e^{-x}+1)] - e^{-x}\log(1+e^x) = -1 + e^x \log(e^{-x}+1) - e^{-x}\log(e^x+1)$$

Hence C.S. is $y = c_1 e^x + c_2 e^{-x} - 1 + e^x \log(e^{-x}+1) - e^{-x}\log(e^x+1)$.

**Example 13.26.** *Solve by the method of variation of parameters $y'' - 6y' + 9y = e^{3x}/x^2$.*

*(Andhra, 2021 ; D.T.U., 2013 ; P.T.U., 2013S)*

**Solution.** Given equation is $(D^2 - 6D + 9)y = e^{3x}/x^2$

A.E. is $D^2 - 6D + 9 = 0$ *i.e.* $(D - 3)^2 = 0$ $\therefore$ C.F. $= (c_1 + c_2x)e^{3x}$

Here $y_1 = e^{3x}$, $y_2 = xe^{3x}$ and $X = e^{3x}/x^2$

$$\therefore \qquad W = \begin{vmatrix} y_1 & y_2 \\ y_1' & y_2' \end{vmatrix} = \begin{vmatrix} e^{3x} & xe^{3x} \\ 3e^{3x} & e^{3x} + 3xe^{3x} \end{vmatrix} = e^{6x}.$$

Thus
$$\text{P.I.} = -y_1 \int \frac{y_2 X}{W} dx + y_2 \int \frac{y_1 X}{W} dx = -e^{3x} \int \frac{xe^{3x}}{e^{6x}} \frac{e^{3x}}{x^2} dx + xe^{3x} \int \frac{e^{3x}}{e^{6x}} \cdot \frac{e^{3x}}{x^2} dx$$

$$= -e^{3x} \int \frac{dx}{x} + xe^{3x} \int x^{-2} dx = -e^{3x} (\log x + 1)$$

Hence C.S. is $\quad y = (c_1 + c_2x)e^{3x} - e^{3x} (\log x + 1).$

**Example 13.27.** *Solve, by the method of variation of parameters, $y'' - 2y' + y = e^x \log x$.*

*(JNUTK, 2022 ; V.T.U., 2014 ; Rohtak, 2011)*

**Solution.** Given equation in symbolic form is $(D^2 - 2D + 1)\, y = e^x \log x$

*(i) To find C.F.*

Its A.E. is $(D - 1)^2 = 0$, $\therefore$ $D = 1, 1$

Thus C.F. is $y = (c_1 + c_2x)e^x$

*(ii) To find P.I.*

Here $y_1 = e^x$, $y_2 = xe^x$ and $X = e^x \log x$

$$\therefore \qquad W = \begin{vmatrix} y_1 & y_2 \\ y_1' & y_2' \end{vmatrix} = \begin{vmatrix} e^x & xe^x \\ e^x & (1 + x)\, e^x \end{vmatrix} = e^{2x}$$

Thus
$$\text{P.I.} = -y_1 \int \frac{y_2 X}{W} dx + y_2 \int \frac{y_1 X}{W} dx$$

$$= -e^x \int \frac{xe^x \,.\, e^x \log x}{e^{2x}} dx + xe^x \int \frac{e^x \,.\, e^x \log x}{e^{2x}} dx = -e^x \int x \log x \, dx + xe^x \int \log x \, dx$$

$$= -e^x \left( \frac{x^2}{2} \log x - \int \frac{1}{x} \cdot \frac{x^2}{2} dx \right) + x \,.\, e^x \left( x \log x - \int \frac{1}{x} \,.\, x \, dx \right)$$

$$= -e^x \left( \frac{x^2}{2} \log x - \frac{x^2}{4} \right) + x \; e^x (x \log x - x) = \frac{1}{4} x^2 e^x (2 \log x - 3)$$

Hence C.S. is $y = (c_1 + c_2x)\, e^x + \dfrac{1}{4} x^2 e^x (2 \log x - 3).$

## II. Method of undetermined coefficients

To find the P.I. of $f(D)\, y = X$, we assume a trial solution containing unknown constants which are determined by substitution in the given equation. The trial solution to be assumed in each case, depends on the form of $X$. Thus choose P.I. from the following table depending on the nature of $X$.

| *S.No.* | *Function X* | *Choice of P.I.* |
|---|---|---|
| 1. | $ke^{lx}$ | $ce^{lx}$ |
| 2. | $k \sin (ax + b)$ or $k \cos (ax + b)$ | $c_1 \sin (ax + b) + c_2 \cos (ax + b)$ |
| 3. | $ke^{lx} \sin (ax + b)$ or $ke^{lx} \cos (ax + b)$ | $c_1e^{lx} \sin (ax + b) + c_2e^{lx} \cos (ax + b)$ |
| 4. | $kx^{mx}$ where $m = 0, 1, 2, \ldots$ | $c_0 + c_1x + c_2x + \ldots + c_mx^m$ |

**Obs. 1.** When $X$ is a sum of functions in col. 2 of the above table, then trial P.I. is chosen as a sum of corresponding functions in col. 3.

**Obs. 2.** The above method holds so long as no term in the trial solution appears in the C.F. If any term of the trial solution appears in the C.F., then the choice from the above table is modified as under :

If a term $u$ of $X$ is also a term of the C.F., then the choice from the table corresponding to $u$ should be multiplied by

(*i*) $x$ if $u$ corresponds to a simple root of C.F.

(*ii*) $x^2$ if $u$ corresponds to a double root of C.F. and so on.

**Obs. 3.** When $X$ is $\tan x$ or $\sec x$ etc. this method fails since the number of terms obtained by differentiating such functions is infinite.

**Example 13.28.** *Solve* $\dfrac{d^2y}{dx^2} + 2\dfrac{dy}{dx} + 4y = 2x^2 + 3e^{-x}$.

**Solution.** Here $\quad$ C.F. $= e^{-x}(c_1 \cos \sqrt{3}\,x + c_2 \sin \sqrt{3}\,x)$

Assume P.I. as $\quad y = a_1x^2 + a_2x + a_3 + a_4e^{-x}$

$\therefore \quad Dy = 2a_1x + a_2 - a_4e^{-x}$ and $D^2y = 2a_1 + a_4e^{-x}$

Substituting these in the given equation, we get

$$4a_1x^2 + (4a_1 + 4a_2)\,x + (2a_1 + 2a_2 + 4a_3) + 3a_4e^{-x} = 2x^2 + 3e^{-x}$$

Equating corresponding coefficients on both sides, we get

$$4a_1 = 2,\ 4a_1 + 4a_2 = 0,\ 2a_1 + 2a_2 + 4a_3 = 0,\ 3a_4 = 3$$

Then $\quad a_1 = \dfrac{1}{2},\ a_2 = -\dfrac{1}{2},\ a_3 = 0,\ a_4 = 1.$ Thus P.I. $= \dfrac{1}{2}x^2 - \dfrac{1}{2}x + e^{-x}$

$\therefore$ C.S. is $y = e^{-x}(c_1 \cos \sqrt{3x} + c_2 \sin \sqrt{3x}) + \dfrac{1}{2}x^2 - \dfrac{1}{2}x + e^{-x}$.

**Example 13.29.** *Solve* $(D^2 + 1)\,y = \sin x$.

**Solution.** Here C.F. $= c_1 \cos x + c_2 \sin x$

We would normally assume a trial solution as $a_1 \cos x + a_2 \sin x$.

However, since these terms appear in the C.F., we multiply by $x$ and assume the trial P.I. as

$$y = x\,(a_1 \cos x + a_2 \sin x)$$

$\therefore \quad Dy = (a_1 + a_2x) \cos x + (a_2 - a_1x) \sin x$ and $D^2y = (2a_2 - a_1x) \cos x - (2a_1 + a_2x) \sin x$

Substituting these in the given equation, we get $2a_1 \cos x - 2a_2 \sin x = \sin x$

Equating corresponding coefficients,

$$2a_1 = 0, \quad -2a_2 = 1 \quad \text{so that } a_1 = 0,\ a_2 = -\frac{1}{2}. \quad \text{Thus P.I.} = -\frac{1}{2}\,x \sin x$$

$\therefore$ C.S. is $y = c_1 \cos x + c_2 \sin x - \dfrac{1}{2}\,x \sin x$.

**Example 13.30.** *Solve by the method of undetermined coefficients,*

$$\frac{d^2y}{dx^2} - y = e^{3x} \cos 2x - e^{2x} \sin 3x.$$

**Solution.** Its A.E. is $D^2 - 1 = 0, \quad \therefore \quad D = \pm 1$.

Thus C.F. $= c_1e^x + c_2e^{-x}$

Assume P.I. as $y = e^{3x}(c_1 \cos 2x + c_2 \sin 2x) - e^{2x}(c_3 \cos 3x + c_4 \sin 3x)$

$$\therefore \quad \frac{dy}{dx} = e^{3x}\{(3c_1 + 2c_2) \cos 2x + (3c_2 - 2c_1) \sin 2x\} - e^{2x}\{(2c_3 + 3c_4) \cos 3x + (2c_4 - 3c_3) \sin 3x\}$$

and
$$\frac{d^2y}{dx^2} = e^{3x}\{(5c_1 + 12c_2) \cos 2x + (5c_2 - 12c_1) \sin 2x\} - e^{2x}\{(12c_4 - 5c_3) \cos 3x - (5c_4 + 12c_3) \sin 3x\}$$

Substituting these in the given equation, we get

$$e^{3x}\{(4c_1 + 12c_2) \cos 2x + (4c_2 - 12c_1) \sin 2x\} - e^{2x}\{(12c_4 - 6c_3) \cos 3x - (6c_4 + 12c_3) \sin 3x\}$$
$$= e^{3x} \cos 2x - e^{2x} \sin 3x$$

Equating corresponding coefficients,

$$4c_1 + 12c_2 = 1,\ 4c_2 - 12c_1 = 0\ ;\ 12c_4 - 6c_3 = 0,\ 6c_4 + 12c_3 = -1$$

whence $c_1 = 1/40$, $c_2 = 3/40$, $c_3 = -1/15$, $c_4 = -1/30$

Thus P.I. $= \dfrac{1}{40} e^{3x} (\cos 2x + 3 \sin 2x) + \dfrac{1}{30} e^{2x} (2 \cos 3x + \sin 3x)$

Hence C.S. is $y = c_1 e^x + c_2 e^{-x} + \dfrac{1}{30} e^{2x} (2 \cos 3x + \sin 3x) + \dfrac{1}{40} e^{3x} (\cos 2x + 3 \sin 2x)$.

## PROBLEMS 13.3

Solve by the method of variation of parameters :

1. $\dfrac{d^2y}{dx^2} + a^2 y = \text{cosec}\, ax.$ *(Anna, 2012)*
2. $\dfrac{d^2y}{dx^2} + y = \sec x.$ *(V.T.U., 2014 ; D.T.U., 2013)*
3. $\dfrac{d^2y}{dx^2} + y = \tan x.$ *(Andhra, 2022 ; Anna, 2016 ; C.S.V.T.U., 2014 ; U.K.T.U., 2011)*
4. $\dfrac{d^2y}{dx^2} + y = x \sin x.$
5. $y'' - 4y' + 4y = e^{-2x}$ *(P.T.U., 2015S)*
6. $\dfrac{d^2y}{dx^2} - 2\dfrac{dy}{dx} + y = e^x / x.$ *(V.T.U., 2013)*
7. $(D^2 + 1)^3 (D^2 + D + 1)^2 y = 0$ *(Pune, 2018)*
8. $y'' - 2y' + 2y = e^x \tan x.$ *(V.T.U., 2010)*
9. $\dfrac{d^2y}{dx^2} - 2\dfrac{dy}{dx} = e^x \sin x.$
10. $\dfrac{d^2y}{dx^2} + y = \dfrac{1}{1 + \sin x}.$
11. $\dfrac{d^2y}{dx^2} - 2\dfrac{dy}{dx} + y = e^x \log_e x.$ *(Delhi, 2015)*

Solve by the method of undetermined coefficients :

12. $\dfrac{d^2y}{dx^2} + y = 2 \cos x.$
13. $\dfrac{d^2y}{dx^2} + \dfrac{dy}{dx} - 2y = x + \sin x.$ *(V.T.U., 2010)*
14. $(D^2 - 2D + 3)\, y = x^3 + \cos x.$
15. $(D^2 - 2D)\, y = e^x \sin x.$
16. $L\dfrac{d^2Q}{dt^2} + R\dfrac{dQ}{dt} + \dfrac{Q}{C} = 0$, given that $Q = Q_0$, $\dfrac{dQ}{dt} = 0$, when $t = 0$ and $CR^2 < 4L = 0$ *(Marathwada, 2019)*
17. If $\dfrac{d^2y}{dx^2} - a^2 y = 0$. Show that $y = c_1 \cos h\, ax + c_2 \sin h\, ax$. *(Mumbai, 2018)*
18. $2y'' + y' + 2 \sin^2 30° \cos^2 30°\, y = 0.$

## 13.9 EQUATIONS REDUCIBLE TO LINEAR EQUATIONS WITH CONSTANT COEFFICIENTS

Now we shall study two such forms of linear differential equations with variable coefficients which can be reduced to linear differential equations with constant coefficients by suitable substitutions.

**I. Cauchy's homogeneous linear equation.** An equation of the form

$$x^n \frac{d^n y}{dx^n} + k_1 x^{n-1} \frac{d^{n-1}y}{dx^{n-1}} + \ldots + k_{n-1} x \frac{dy}{dx} + k_n y = X \quad \ldots(1)$$

where, $X$ is a function of $x$, is called *Cauchy's homogeneous linear equation*.

Such equations can be reduced to linear differential equations with constant coefficients, by putting

$$x = e^t \quad \text{or} \quad t = \log x. \text{ Then if } D = \frac{d}{dt}$$

$$\frac{dy}{dx} = \frac{dy}{dt} \cdot \frac{dt}{dx} = \frac{dy}{dt} \cdot \frac{1}{x}, \quad \textit{i.e.,} \quad x\frac{dy}{dx} = Dy.$$

$$\frac{d^2y}{dx^2} = \frac{d}{dx}\left(\frac{1}{x}\frac{dy}{dt}\right) = -\frac{1}{x^2}\frac{dy}{dt} + \frac{1}{x}\frac{d}{dt}\left(\frac{dy}{dt}\right)\frac{dt}{dx} = -\frac{1}{x^2}\frac{dy}{dt} + \frac{1}{x}\frac{d^2y}{dt^2}\frac{dt}{dx} = \frac{1}{x^2}\left(\frac{d^2y}{dt^2} - \frac{dy}{dt}\right)$$

*i.e.,* $\quad x^2\dfrac{d^2y}{dx^2} = D(D-1)\,y$. Similarly, $x^3 \dfrac{d^3y}{dx^3} = D(D-1)(D-2)\,y$ and so on.

After making these substitutions in Eqn. (1), there results a linear equation with constant coefficients, which can be solved as before.

**Example 13.31.** *Solve* $\dfrac{d^2y}{dx^2} - 5\dfrac{dy}{dx} + 6y = e^x$. (*Osmania, 2018*)

**Solution:** Denoting $\dfrac{d}{dx}$ by $D$, the given equation can be written as

$$(D^2 - 5D + 6)y = e^x$$

A.E is $\quad D^2 - 5D + 6$ or $(D-2)(D-3) = 0$

$$D = 2, D = 3$$

Hence $\quad C.F = c_1 e^{2x} + c_2 e^{3x}$

$$P.I = \frac{1}{D^2 - 5D + 6} e^x = \frac{1}{1^2 - 5.1 + 6} e^x = \frac{e^x}{2}$$

Hence the complete solution is $\quad y = CG + P.I$

$$y = C_1 e^{2x} + C_2 e^{3x} + \frac{e^x}{2}$$

**Example 13.32.** *Solve* $x^2\dfrac{d^2y}{dx^2} + 3x\dfrac{dy}{dx} + y = \dfrac{1}{(1-x)^2}$. (*JNTUK, 2021 ; Andhra 2015 ; D.T.U., 2015*)

**Solution.** Put $x = e^t$ *i.e.,* $t = \log x$ so that $x\, dy/dx = Dy$, $x^2\dfrac{d^2y}{dx^2} = D(D-1)\,y$

Then the given equation becomes

$$[D(D-1) + 3D + 1]\, y = \frac{1}{(1-e^t)^2} \quad \text{or} \quad (D^2 + 2D + 1)\, y = \frac{1}{(1-e^t)^2}$$

Its A.E. is $\quad D^2 + 2D + 1 = 0 \quad$ or $\quad (D+1)^2 = 0$ *i.e.,* $D = -1, -1$.

$$\therefore \quad \text{C.F.} = (c_1 + c_2 x)\, e^{-t} = (c_1 + c_2 \log x)\, \frac{1}{x}$$

$$\text{P.I.} = \frac{1}{(D+1)^2}\frac{1}{(1-e^t)^2} = \frac{1}{D+1}\, u, \text{ where } u = \frac{1}{D+1} \cdot \frac{1}{(1-e^t)^2} \textit{ i.e. } \frac{du}{dt} + u = (1-e^t)^{-2}$$

which is Leibnitz's linear equation having I.F. = $e^t$

$$\therefore \quad \text{its solution is} \quad ue^t = \int \frac{e^t}{(1-e^t)^2}\, dt = \frac{1}{1-e^t} \quad \text{or} \quad u = \frac{e^{-t}}{1-e^t}$$

$\therefore$ $$\text{P.I.} = \frac{1}{D+1}\left(\frac{e^{-t}}{1-e^t}\right) = e^{-t}\int \frac{1}{1-e^t}\,dt = \frac{1}{x}\int \frac{dx}{x(1-x)}$$

$$= \frac{1}{x}\int\left(\frac{1}{x}+\frac{1}{1-x}\right)dx = \frac{1}{x}\,[\log x - \log(1-x)] = \frac{1}{x}\log\frac{x}{1-x}$$

Hence the solution is $$y = \left\{c_1 + c_2\log x + \log\frac{x}{1-x}\right\}\frac{1}{x}.$$

**Example 13.33.** *Solve* $x^2\dfrac{d^2y}{dx^2} + x\dfrac{dy}{dx} + y = \log x \sin(\log x).$

*(Andhra, 2018 ; C.S.V.T.U., 2011 S)*

**Solution.** Putting $x = e^t$ *i.e.* $t = \log x$, the given equation becomes

$$[D(D-1) + D + 1]\,y = t\sin t \quad i.e. \quad (D^2+1)\,y = t\sin t \qquad \text{...}(i)$$

Its A.E. is $D^2 + 1 = 0$ *i.e.* $D = \pm i$.

$\therefore$ $$\text{C.F.} = c_1\cos t + c_2\sin t$$

and $$P.I. = \frac{1}{D^2+1}\,t\sin t = \frac{1}{D^2+1}\,t\,(\text{I.P. of } e^{it})$$

$$= \text{I.P. of } e^{it}\frac{1}{(D+i)^2+1}\,t = \text{I.P. of } e^{it}\,.\,\frac{1}{D^2+2iD}\,t$$

$$= \text{I.P. of } e^{it}\frac{1}{2iD\,(1+D/2i)}\,t = \text{I.P. of } \frac{1}{2i}e^{it}\frac{1}{D}\left(1-\frac{iD}{2}\right)^{-1}t$$

$$= \text{I.P. of } \frac{1}{2i}e^{it}\frac{1}{D}\left(1+\frac{iD}{2}+\ldots\right)t = \text{I.P. of } \frac{1}{2i}e^{it}\frac{1}{D}\left(t+\frac{i}{2}\right)$$

$$= \text{I.P. of } \frac{e^{it}}{2i}\int\left(t+\frac{i}{2}\right)dt = \text{I.P. of } \frac{e^{it}}{2i}\left(\frac{t^2}{2}+\frac{it}{2}\right)$$

$$= \text{I.P. of } e^{it}\left(-\frac{i}{4}t^2+\frac{t}{4}\right) = \text{I.P. of } (\cos t + i\sin t)\left(-\frac{it^2}{4}+\frac{t}{4}\right) = -\frac{t^2}{4}\cos t + \frac{t}{4}\sin t$$

Hence the C.S. of $(i)$ is $y = c_1\cos t + c_2\sin t - \dfrac{t^2}{4}\cos t + \dfrac{t}{4}\sin t$

or $$y = c_1\cos(\log x) + c_2\sin(\log x) - \frac{1}{4}(\log x)^2\cos(\log x) + \frac{1}{4}\log(\log x)\sin(\log x)$$

which is the required solution.

**Example 13.34.** *Solve* $x^2\dfrac{d^2y}{dx^2} + 4x\dfrac{dy}{dx} + 2y = e^{x^x}.$ *(JNTUK, 2022 ; D.T.U., 2015)*

**Solution.** Putting $x = e^t$, *i.e.*, $t = \log x$, the given equation becomes

$$[D(D-1) + 4D + 2]\,y = e^{e^t} \quad i.e., (D^2+3D+2)\,y = e^{e^t}$$

Its A.E. is $D^2 + 3D + 2 = 0$ whence $D = -1, -2$.

$\therefore$ C.F. $= c_1e^{-t} + c_2e^{-2t} = c_1x^{-1} + c_2x^{-2}$

and $$\text{P.I.} = \frac{1}{(D^2+3D+2)}e^{e^t} = \frac{1}{(D+1)(D+2)}e^{e^t} = \left(\frac{1}{D+1}-\frac{1}{D+2}\right)e^{e^t}$$

Now $$\frac{1}{D+1}e^{e^t} = \frac{1}{D+1}e^{-t}\,.\,e^te^{e^t} = e^{-t}\frac{1}{(D-1)+1}e^te^{e^t}$$

$$= e^{-t}\frac{1}{D}e^t e^{e^t} = e^{-t}\int e^{e^t} d(e^t) = x^{-1}\int e^x dx = x^{-1}e^x$$

$$\frac{1}{D+2}e^{e^t} = \frac{1}{D+2}e^{-2t}\,.\,e^{2t}e^{e^t} = e^{-2t}\frac{1}{(D-2)+2}e^{2t}e^{e^t}$$

$$= e^{-2t}\frac{1}{D}e^{e^t}e^{2t} = e^{-2t}\int e^{e^t}e^t\, d(e^t)$$

$$= x^{-2}\int e^x\, x\, dx \qquad [\because \ e^t = x]$$

$$= x^{-2}(xe^x - e^x) \qquad \text{[Integrating by parts]}$$

$\therefore$ P.I. $= x^{-1}e^x - x^{-2}(xe^x - e^x) = x^{-2}e^x$

Hence the required solution is $y = c_1 x^{-1} + x^{-2}(c_2 + e^x)$.

**II. Legendre's linear equation*.** An equation of the form

$$(ax+b)^n\frac{d^n y}{dx^n} + k_1(ax+b)^{n-1}\frac{d^{n-1}y}{dx^{n-1}} + \ldots + k_n y = X \qquad \ldots(2)$$

where $k$'s are constants and $X$ is a function of $x$, is called *Legendre's linear equation.*

Such equations can be reduced to linear equations with constant coefficients by the substitution $ax + b = e^t$, *i.e.*, $t = \log(ax + b)$.

Then, if $D = \dfrac{d}{dt}$, $\dfrac{dy}{dx} = \dfrac{dy}{dt}\cdot\dfrac{dt}{dx} = \dfrac{a}{ax+b}\cdot\dfrac{dy}{dt}$ *i.e.* $(ax+b)\dfrac{dy}{dx} = a\,Dy$

$$\frac{d^2y}{dx^2} = \frac{d}{dx}\left(\frac{a}{ax+b}\frac{dy}{dt}\right) = \frac{-a^2}{(ax+b)^2}\frac{dy}{dt} + \frac{a}{ax+b}\frac{d}{dt}\left(\frac{dy}{dt}\right)\frac{dt}{dx} = \frac{a^2}{(ax+b)^2}\left(\frac{d^2y}{dt^2} - \frac{dy}{dt}\right)$$

*i.e.*, $(ax+b)^2\dfrac{d^2y}{dx^2} = a^2 D(D-1)\,y$. Similarly, $(ax+b)^3\dfrac{d^3y}{dx^3}$ $a^3D(D-1)(D-2)\,y$ and so on.

After making these replacements in Eqn. (2), there results a linear equation with constant coefficients.

**Example 13.35:** *Solve, $(D^2 - a^2)y = e^{ax} + e^{-ax}$* *(Jaipur, 2019)*

**Solution:** A.E. is

$$D^2 - a^2 = 0 \text{ or}$$

$$(D-a)(D+a) = 0$$

$$\therefore \quad D = a,\ D = -a$$

$$\therefore \quad \text{C.F.} = c_1e^{ax} + c_2e^{-ax}$$

$$\text{P.I.} = \frac{1}{(D^2-a^2)}(e^{4x} + e^{-ax}) = \frac{1}{(D^2-a^2)}e^{ax} + \frac{1}{D^2-a^2}e^{-ax}$$

$$= \frac{1}{D^2-a^2}e^{ax} + \frac{1}{D^2-a^2}e^{-ax} = \frac{x}{2D}e^x - \frac{x}{2D}e^{-ax}$$

$$= \frac{x}{2a}(e^{ax} - e^{-ax})$$

Hence the complete solution is $y$ = C.F. + P.I.

$$y = c_1e^{ax} + c_2e^{-ax} + \frac{x}{2a}(e^{ax} - e^{ax})$$

$$y = \left(c_1 + \frac{x}{2a}\right)e^{ax} + \left(c_2 - \frac{x}{2a}\right)e^{-ax}$$

* A French mathematician *Adrien Marie Legender* (1752 – 1833) who made important contributions to number theory, special functions, calculus of variations and elliptic integrals.

**Example 13.36.** *Solve* $(1+x)^2 \dfrac{d^2y}{dx^2} + (1+x)\dfrac{dy}{dx} + y = 2\sin[\log(1+x)]$. ...(*i*)

(*K.T.U., 2016 ; C.S.V.T.U., 2014 ; P.T.U., 2014*)

**Solution.** This is a Legendre's linear equation.

$\therefore$ put $1 + x = e^t$, *i.e.*, $t = \log(1+x)$, so that $(1+x)\dfrac{dy}{dx} = Dy$

and $$(1+x)^2 \frac{d^2y}{dx^2} = D(D-1)\,y, \text{ where } D = \frac{d}{dt}$$

Then Eqn. (*i*) becomes $D(D-1)\,y + Dy + y = 2\sin t$

or $$(D^2+1)\,y = 2\sin t \quad ...(ii)$$

This is a linear equation with constant co-efficients

Its A.E. is $D^2 + 1 = 0$, whence $D = \pm i$ $\therefore$ C.F. $= c_1 \cos t + c_2 \sin t$

and $$\text{P.I.} = 2\frac{1}{D^2+1}\sin t = 2t \cdot \frac{1}{2D}\sin t$$

$$= t\int \sin t\, dt = -t\cos t \qquad [\because \text{ on replacing } D^2 \text{ by } -1^2,\ D^2+1=0]$$

Hence the solution of Eqn. (*ii*) is $y = c_1\cos t + c_2 \sin t - t\cos t$ and on replacing $t$ by $\log(1+x)$, we get $y = c_1 \cos[\log(1+x)] + c_2 \sin[\log(1+x)] - \log(1+x)\cos[\log(1+x)]$ as the required solution.

**Example 13.37.** *Solve* $\dfrac{d^2y}{dx^2} - 3\dfrac{dy}{dx} + 2y = xe^{3x} + \sin 2x$ (*C.C.S., 2018*)

**Solution:** Denoting $\dfrac{d}{dx}$ by $D$, the given equation can be written as $(D^2 - 3D + 2)y = xe^{3x} + \sin 2x$

$$\text{A.E.} = (D^2 - 3D + 2) = 0$$
$$\text{A.E.} = (D-2)(D-1) = 0$$
$$D = 1, D = 2$$
$$C.F. = C_1 e^x + C_2 e^{2x}$$

$$\text{P.I.} = \frac{1}{D^2 - 3D + 2} xe^{3x} + \frac{1}{D^2 - 3D + 2}\sin 2x$$

$$= e^{3x}\frac{1}{(D+3)^2 - 3(D+3) + 2}x + \frac{1}{(-2)^2 - 3D + 2}\sin 2x$$

$$= e^{3x}\frac{1}{D^2 + 3D + 2}x - \frac{1}{3D+2}\sin 2x$$

$$= \frac{e^{3x}}{2}\left[1 + \frac{3D}{2} + \frac{D^2}{2}\right]^{-1} x - \frac{3D-2}{9D^2 - 4}\sin 2x$$

$$= \frac{e^{3x}}{2}\left[1 - \frac{3D}{2} - ....\right]x - \frac{3D-2}{9(-2)^2 - 4}\sin 2x$$

$$= \frac{e^{3x}}{2}\left(x - \frac{3}{2}\right) + \frac{1}{40}(3.2\cos 2x - 2\sin 2x)$$

$$= \frac{e^{3x}}{2}(2x-3) + \frac{1}{20}(3\cos 2x - \sin 2x)$$

Hence the complete solution is $y = \text{C.F.} + \text{P.I.}$

$$\therefore \quad y = C_1 e^x + C_2 e^{2x} + \frac{e^{3x}}{4}(2x-3) + \frac{1}{20}(\cos 2x - \sin 2x)$$

**Example 13.38 :** *Solve* $\frac{d^4y}{dx^4} - m^4y = \cos mx$

**Solution:** The equation can be written in the symbolic form is

$$(D^2 - m^2)y = \cos mx$$

$$\text{A.E. is } D^2 - m^4 = 0 \text{ or } (D^2 - m^2)(D^2 + m^2) = 0$$

$$D = \pm m, D = \pm mi$$

$$\therefore \quad \text{C.F.} = C_1e^{mx} + C_2\, e^{-mx} + C_3 \cos mx + C_4 \sin mx$$

$$\text{P.I.} = \frac{1}{D^4 - m^4}\cos mx = \frac{1}{(D^2 + m^2)(D^2 - m^2)}\cos mx$$

$$= -\frac{1}{2m^2}\frac{1}{D^2 + m^2}\cos mx = -\frac{1}{2m^2}\cdot\frac{x}{2D}\cos mx$$

$$= -\frac{1}{2m^2}\cdot\frac{x}{2}\int \cos mx \cdot dx \quad \left[\because \frac{1}{D}X = \int X\,dx\right]$$

$$= -\frac{1}{4m^2x}\frac{\sin mx}{m} = \frac{-1}{4m^3}x \cdot \sin mx$$

$$y = \text{C.F} + \text{P.I.}$$

$$y = C_1\, e^{mx} + C_2\, e^{-mx} + C_3 \cos mx + C_4 \sin mx - \left|\frac{-x}{4m^3}\right|\sin mx.$$

**Example 13.39:** *Solve* $(D^2 + a^2)\, y = \tan ax$

**Solution:**

$$\text{A.E. } (D^2 + a^2) = 0$$

$$D = \pm\, ai$$

$$\text{C.F.} = C_1 \cos ax + C_2 \sin ax$$

$$\text{P.I.} = \frac{1}{(D^2 + a^2)}\tan ax = \frac{1}{(D + ai)(D - ai)}\tan ax$$

$$= \frac{1}{2ia}\left[\frac{1}{(D - ai)} - \frac{1}{D + ai}\right]\tan ax \qquad \text{...}(i)$$

Now

$$\frac{1}{D - ai}\tan ax = e^{\tan x}\int \tan ax\, e^{\tan x} dx \qquad \text{...}(ii)$$

$$= e^{\tan x}\int \tan ax(\cos ax - i\sin ax)dx = e^{\tan x}\left[\int \sin ax - \frac{\sin^2 ax}{\cos ax}\right]dx$$

$$= e^{\tan x}\int\left[\sin ax - i\left(\frac{1 - \cos^2 ax}{\cos ax}\right)\right]dx$$

$$= e^{\tan x}\int [\sin ax - i(\sec ax - \cos ax)]dx$$

$$= e^{\tan x}\left[\frac{-\cos ax}{a} - \frac{i}{a}\log(\sec ax + \tan ax) + \frac{i\sin ax}{a}\right]$$

$$= \frac{-1}{a}e^{iax}\left[(\cos ax - i\sin ax) + i\log(\sec ax + \tan ax)\right]$$

$$= \frac{-1}{a}e^{iax}\left[e^{-iax} + i\log(\sec ax + \tan ax)\right]$$

$$= -\frac{1}{a}\left[1 + i\, e^{iax}\log(\sec ax + \tan ax)\right]$$

changing $i$ to $-i$ in Eqn. ($i$), we get

$$\frac{1}{D+ia}\tan ax = -\frac{1}{a}[1 - i\,e^{-\tan x}\log(\sec ax + \tan ax)]$$

Thus from Eqn. ($i$)

$$P.I. = \frac{1}{2ia}\left[-\frac{1}{a}\{1 + i\,e^{iax}\log(\sec ax + \tan ax)\} + \frac{1}{a}\{1 - e^{-iax}\log(\sec ax + \tan ax\}\right]$$

$$= -\frac{1}{a^2}\log(\sec x + \tan ax)\left(\frac{e^{iax} + e^{-iax}}{2}\right)$$

$$= -\frac{1}{a^2}\log(\sec ax + \tan ax)\cdot\cos ax = -\frac{1}{a^2}\cos ax\log(\sec ax + \tan ax)$$

Hence the complete solution

$$y = C.F + P.I$$

$$y = C_1\cos ax + C_2\sin ax - \frac{1}{a^2}\cos ax\log(\sec ax + \tan ax)$$

## PROBLEMS 13.4

Solve :

1. $x^2y'' + xy' + y = x.$ (*P.T.U., 2015*)
2. $x^2\frac{d^2y}{dx^2} - 2x\frac{dy}{dx} - 4y = x^4.$
3. $x^2\frac{d^2y}{dx^2} - 3x\frac{dy}{dx} + 4y = (1 + x^2).$
4. $x\frac{d^2y}{dx^2} - \frac{2y}{x} = x + \frac{1}{x^2}.$
5. The radial displacement $u$ in a rotating disc at a distance $r$ from the axis is given by $r^2\frac{d^2u}{dr^2} + r\frac{du}{dr} - u + kr^3 = 0,$ where $k$ is a constant. Solve the equation under the conditions $u = 0$, when $r = 0$, $u = 0$ when $r = a$.

Solve :

6. $x^2\frac{d^2y}{dx^2} + 4x\frac{dy}{dx} + 2y = \log x.$
7. $x^3\frac{d^3y}{dx^3} + 3x^2\frac{d^2y}{dx^2} + x\frac{dy}{dx} + y = x + \log x$ (*K.T.U., 2016*)
8. $x^2y'' + xy' + y = 2\cos^2(\log x).$ (*V.T.U., 2011*)
9. $x^2\frac{d^2y}{dx^2} + 4x\frac{dy}{dx} + 2y = e^x.$ (*U.P.T.U., 2012*)
10. $x^3\frac{d^3y}{dx^3} + 2x^2\frac{d^2y}{dx^2} + 2y = 10\left(x + \frac{1}{x}\right)$
11. $x^2\frac{d^2y}{dx^2} + 3x\frac{dy}{dx} + y = \frac{1}{(1-x)^2}.$ (*JNTUK, 2022*)
12. $x^2\frac{d^2y}{dx^2} + 5x\frac{dy}{dx} + 4y = x\log x.$ (*C.S.V.T.U., 2012*)
13. $(x^2D^2 - xD - 2)\,y = x^2\log x.$ (*Anna, 2016*)
14. $x^2\frac{d^2y}{dx^2} + 2x\frac{dy}{dx} - 12y = x^3\log x.$
15. $x^2y'' - xy' + 2y = x\sin(\log x)$ (*V.T.U., 2013*)
16. $(2x + 3)^2\frac{d^2y}{dx^2} - (2x + 3)\frac{dy}{dx} - 12y = 6x.$ (*JNTUK, 2022 ; Andhra, 2022*)
17. $(2x + 7)^2y'' - 6(2x + 7)\,y' + 8y = 8x.$ (*Anna, 2016*)
18. $(1+x)^2\frac{d^2y}{dx^2} + (1+x)\frac{dy}{dx} + y = 4\cos\log(1+x)$ (*P.T.U., 2015*)
19. $(1 + x)^2\frac{d^2y}{dx^2} + (1 + x)\frac{dy}{dx} + y = \sin[2\log(1 + x)]$ (*JNTUK, 2022*)
20. $(3x + 2)^2\frac{d^2y}{dx^2} + 5(3x + 2)\frac{dy}{dx} - 3y = x^2 + x + 1.$
21. $(x - 1)^3\frac{d^3y}{dx^3} + 2(x - 1)^2\frac{d^2y}{dx^2} - 4(x - 1)\frac{dy}{dx} + 4y = 4\log(x - 1).$

## 13.10 (1) LINEAR DEPENDENCE OF SOLUTIONS

Consider the initial value problem consisting of the homogeneous linear equation

$$y'' + py' + qy = 0 \qquad ...(1)$$

with variable coefficients $p(x)$ and $q(x)$ and two initial conditions $y(x_0) = k_0$, $y'(x_0) = k_1$ ...(2)

Let its general solution be $y = c_1 y_1 + c_2 y_2$ ...(3)

which is made up of two linearly dependent solutions $y_1$ and $y_2$.*

*If $p(x)$ and $q(x)$ are continuous functions on some open interval I and $x_0$ is any fixed point on I, then the* above initial value problem has a **unique solution** *$y(x)$ on the interval I.*

**(2) Theorem.** *If $p(x)$ and $q(x)$ are continuous on an open interval I, then the solutions $y_1$ and $y_2$ of Eqn. (1) are linearly dependent in I if and only if the Wronskian* $W(y_1, y_2) = \begin{vmatrix} y_1 & y_2 \\ y_1' & y_2' \end{vmatrix} = 0$ for some $x_0$ on $I$. If there is an $x = x_1$ in $I$ at which $W(y_1, y_2) \neq 0$, then $y_1, y_2$ *are linearly independent on I.*

*Proof.* If $y_1, y_2$ are linearly dependent solutions of Eqn. (1) then there exist two constants $c_1, c_2$ not both zero, such that

$$c_1 y_1 + c_2 y_2 = 0 \qquad ...(4)$$

Differentiating w.r.t. $x$, $c_1 y_1' + c_2 y_2' = 0$ ...(5)

Eliminating $c_1, c_2$ from Eqns. (4) and (5), we get

$$W(y_1, y_2) = \begin{vmatrix} y_1 & y_2 \\ y_1' & y_2' \end{vmatrix} = 0$$

Conversely, suppose $W(y_1, y_2) = 0$ for some $x = x_0$ on $I$ and show that $y_1, y_2$ are linearly dependent.

Consider the equation

$$\left.\begin{aligned} c_1 y_1(x_0) + c_2 y_2(x_0) = 0 \\ c_1 y_1'(x_0) + c_2 y_2'(x_0) = 0 \end{aligned}\right\} \qquad ...(6)$$

which, on eliminating $c_1, c_2$, give $W(y_1, y_2) = \begin{vmatrix} y_1(x_0) & y_2(x_0) \\ y'_1(x_0) & y'_2(x_0) \end{vmatrix} = 0$

Hence the system has a solution in which $c_1, c_2$ are not both zero.

Now introduce the function $\bar{y}(x) = c_1 y_1(x) + c_2 y_2(x)$

Then $y(x)$ is a solution of Eqn. (1) on $I$. By Eqn. (6), this solution satisfies the initial conditions $y(x_0) = 0$ and $y'(x_0) = 0$. Also since $p(x)$ and $q(x)$ are continuous on $I$, this solution must be unique. But $\mathbf{y = 0}$ is obviously another solution of Eqn. (1) satisfying the given initial conditions. Hence $\bar{\mathbf{y}} = y$ *i.e.*, $c_1 y_1 + c_2 y_2 = 0$ in $I$. Now since $c_1, c_2$ are not both zero, it implies that $y_1$ and $y_2$ are linearly dependent on $I$.

**Example 13.40.** *Show that the two functions sin 2x, cos 2x are independent solutions of $y'' + 4y = 0$.*

**Solution.** Substituting $y_1 = \sin 2x$ (or $y_2 = \cos 2x$) in the given equation we find that $y_1, y_2$ are its solutions.

Also $$W(y_1, y_2) = \begin{vmatrix} \sin 2x & \cos 2x \\ 2\cos 2x & -2\sin 2x \end{vmatrix} = -2 \neq 0$$

for any value of $x$. Hence the solutions $y_1, y_2$ are linearly independent.

### PROBLEMS 13.5

Solve :

1. Show that the functions $y_1 = x^{1/4}$, $y_2 = x^{5/4}$ are linearly independent and form a set of fundamental solutions of the differential equation $16x^3y'' - 8xy' + 5y = 0$ $(x > 0)$. *(Birla, Ranchi, 2016)*
2. Show that $e^{-x}$, $xe^{-x}$ are independent solutions of $y'' + 2y' + y = 0$ in any interval.
3. Show that $e^x \cos x$, $e^x \sin x$ are independent solutions of the equation $xy'' - 2y' = 0$.
4. If $y_1, y_2$ be two solutions of $y'' + p(x)y' + q(x)y = 0$, show that the *Wronskian* can be expressed as $W(y_1, y_2) = ce^{-\int p dx}$.

---

* As in §2.12, $y_1, y_2$ are said to be *linearly dependent* in an interval $I$, if and only if there exist numbers $\lambda_1, \lambda_2$ not both zero such that $\lambda_1 y_1 + \lambda_2 y_2 = 0$ for all $x$ in $I$.

If no such numbers other than zero exist, then $y_1, y_2$ are said to be *linearly independent*.

## 13.11 SIMULTANEOUS LINEAR EQUATIONS WITH CONSTANT COEFFICIENTS

Quite often we come across linear differential equations in which there are two or more dependent variables and a single independent variable. Such equations are known as *simultaneous linear equations*. Here we shall deal with systems of linear equations with constant coefficients only. Such a system of equations is solved by eliminating all but one of the dependent variables and then solving the resulting equations as before. Each of the dependent variables is obtained in a similar manner.

**Example 13.41.** *Solve the simultaneous equations :*

$$\frac{dx}{dt} + 5x - 2y = t, \quad \frac{dy}{dt} + 2x + y = 0$$

*being given $x = y = 0$ when $t = 0$.* (Delhi, 2016, C.S.V.T.U., 2012)

**Solution.** Taking $d/dt = D$, the given equations become

$$(D + 5)\, x - 2y = t \quad \text{...(i)}$$
$$2x + (D + 1)\, y = 0 \quad \text{...(ii)}$$

Eliminate $x$ as if $D$ were an ordinary algebraic multiplier. Multiplying Eqn. (*i*) by 2 and operating on Eqn. (*ii*) by $D + 5$ and then subtracting, we get

$$[-4 - (D + 5)(D + 1)]\, y = 2t \text{ or } (D^2 + 6D + 9)\, y = -2t$$

Its auxiliary equation is $D^2 + 6D + 9 = 0$, *i.e.*, $(D + 3)^2 = 0$

whence $D = -3, -3 \quad \therefore \quad \text{C.F.} = (c_1 + c_2 t)\, e^{-3t}$

and $$\text{P.I.} = \frac{1}{(D+3)^2}(-2t) = -\frac{2}{9}\left(1 + \frac{D}{3}\right)^{-2} t = -\frac{2}{9}\left(1 - \frac{2D}{3} + \dots\right) t = -\frac{2t}{9} + \frac{4}{27}$$

Hence $$y = (c_1 + c_2 t)\, e^{-3t} - \frac{2t}{9} + \frac{4}{27} \quad \text{...(iii)}$$

*Now to find* $x$, either eliminate $y$ from Eqns. (*i*) and (*ii*) and solve the resulting equation or substitute the value of $y$ in Eqn. (*ii*). Here, it is more convenient to adopt the latter method.

From Eqn. (*iii*), $Dy = c_2 e^{-3t} + (c_1 + c_2 t)(-3)\, e^{-3t} - \frac{2}{9}$

$\therefore$ Substituting for $y$ and $Dy$ in (*ii*), we get

$$x = -\frac{1}{2}[Dy + y] = \left[\left(c_1 - \frac{1}{2}c_2\right) + c_2 t\right] e^{-3t} + \frac{t}{9} + \frac{1}{27} \quad \text{...(iv)}$$

Hence Eqn. (*iii*) and (*iv*) constitute the solutions of the given equations.

Since $x = y = 0$ when $t = 0$, the Eqn. (*iii*) and (*iv*) give

$$0 = c_1 + \frac{4}{27} \text{ and } c_1 - \frac{1}{2}c_2 + \frac{1}{27} = 0 \text{ whence } c_1 = -\frac{4}{27},\ c_2 = -\frac{2}{9}.$$

Hence the desired solutions are

$$x = -\frac{1}{27}(1 + 6t)\, e^{-3t} + \frac{1}{27}(1 + 3t),\ y = -\frac{2}{27}(2 + 3t)\, e^{-3t} + \frac{2}{27}(2 - 3t).$$

**Example 13.42.** *Solve the simultaneous equations* $\frac{dx}{dt} + 2y + \sin t = 0,\ \frac{dy}{dt} - 2x - \cos t = 0$ *given that $x = 0$ and $y = 1$ when $t = 0$.* (V.T.U., 2016)

**Solution.** Given equations are

$$Dx + 2y = -\sin t \quad \text{...(i)}$$
$$-2x + Dy = \cos t \quad \text{...(ii)}$$

Eliminating $x$ by multiplying Eqn. (*i*) by 2 and Eqn. (*ii*) by $D$ and then adding, we get

$$4y + D^2 y = -2 \sin t - \sin t \text{ or } (D^2 + 4)y = -3 \sin t$$

Its A.E. is $D = \pm 2i$

$\therefore$ $\text{C.F.} = c_1 \cos 2t + c_2 \sin 2t$

$$\text{P.I.} = -3\,\frac{1}{D^2+4}\sin t = -3\,\frac{1}{-1+4}\sin t = -\sin t$$

$\therefore$ $$y = c_1 \cos 2t + c_2 \sin 2t - \sin t \quad ...(iii)$$

and $$dy/dt = -2 \sin 2t + 2c_2 \cos 2t - \cos t \quad ...(iv)$$

Substituting Eqn. (*iii*) in (*ii*), we get

$$2x = Dy - \cos t = -2c_1 \sin 2t + 2c_2 \cos 2t - 2\cos t$$

or $$x = -c_1 \sin 2t + c_2 \cos 2t - \cos t \quad ...(v)$$

When $t = 0, x = 0, y = 1$, Eqns. (*iii*) and (*v*) give $1 = c_1, 0 = c_2 - 1$

Hence $x = \cos 2t - \sin 2t - \cos t, y = \cos 2t + \sin 2t - \sin t$.

**Example 13.43.** *Solve the simultaneous equations*

$$\frac{dx}{dt} + \frac{dy}{dt} - 2y = 2\cos t - 7\sin t, \quad \frac{dx}{dt} - \frac{dy}{dt} + 2x = 4\cos t - 3\sin t. \qquad \text{(U.K.T.U., 2011)}$$

**Solution.** Given equations are

$$Dx + (D-2)y = 2\cos t - 7\sin t \quad ...(i)$$

$$(D+2)\,x - Dy = 4\cos t - 3\sin t \quad ...(ii)$$

Eliminate $y$ by operating on Eqn. (*i*) by $D$ and Eqn. (*ii*) by $(D-2)$ and then adding, we get

$$D^2x + (D-2)(D+2)x = -2\sin t - 7\cos t + 4(-\sin t - 2\cos t) - 3(\cos t - 2\sin t)$$

or $$2(D^2 - 2)\,x = -18\cos t \text{ or } (D^2-2)\,x = -9\cos t$$

Its A.E. is $$D^2 - 2 = 0 \text{ or } D = \pm\sqrt{2}, \quad \therefore \text{ C.F.} = c_1 e^{\sqrt{st}} + c_2 e^{-\sqrt{2}t}$$

$$\text{P.I.} = (-9)\,\frac{1}{D^2-2}\cos t = \frac{-9\cos t}{-1-2} = 3\cos t.$$

Hence $x = c_1 e^{\sqrt{2}t} + c_2 e^{-\sqrt{2}t}\;\; 3\cos t$.

Now substituting this value of $x$ in Eqn. (*ii*), we get

$$Dy = (D+2)\,(c_1 e^{\sqrt{2}t} + c_2 e^{-\sqrt{2}t} + 3\cos t) - 4\cos t + 3\sin t$$

$$= c_1\sqrt{2}\,e^{\sqrt{2}t} + 2c_1 e^{\sqrt{2}t} + c_2(-\sqrt{2}\,e^{-2t}) + 2c_2\, e^{-\sqrt{2}t} - 3\sin t + 6\cos t - 4\cos t + 3\sin t$$

$$= (2+\sqrt{2})\,c_1 e^{\sqrt{2}t} + (2-\sqrt{2})\,c_2 e^{-\sqrt{2}t} + 2\cos t$$

Hence $$y = (\sqrt{2}+1)\,c_1 e^{\sqrt{2}t} - (\sqrt{2}-1)\,c_2 e^{-\sqrt{2}t} + 2\sin t + c_3.$$

**Example 13.44.** *The small oscillations of a certain system with two degrees of freedom are given by the equations*

$$D^2x + 3x - 2y = 0$$

$$D^2x + D^2y - 3x + 5y = 0$$

*where $D = d/dt$. If $x = 0, y = 0, Dx = 3, Dy = 2$ when $t = 0$, find $x$ and $y$ when $t = 1/2$.*

**Solution.** Given equations are $(D^2+3)\,x - 2y = 0$ ...(*i*)

$$(D^2-3)x + (D^2+5)y = 0 \quad ...(ii)$$

To eliminate $x$, operate these equations by $D^2 - 3$ and $D^2 + 3$ respectively and subtract Eqn. (*i*) from Eqn. (*ii*). Then

$$[(D^2+3)(D^2+5) + 2(D^2-3)]\,y = 0 \quad \text{or} \quad (D^4 + 10D^2 + 9)y = 0$$

Its auxiliary equation is $D^4 + 10D^2 + 9 = 0$ whence $D = \pm i, \pm 3i$

Thus $$y = c_1 \cos t + c_2 \sin t + c_3 \cos 3t + c_4 \sin 3t \quad ...(iii)$$

*To find x*, we eliminate $y$ from Eqns. (*i*) and (*ii*).

$\therefore$ operating Eqn. (*i*) by $D^2 + 5$ and multiplying Eqn. (*ii*) by 2 and adding, we get

$$(D^4 + 10D^2 + 9)\,x = 0. \text{ Thus } x = k_1 \cos t + k_2 \sin t + k_3 \cos 3t + k_4 \sin 3t \quad ...(iv)$$

To find the relations between the constants in Eqns. (*iii*) and (*iv*), substitute these values of $x$ and $y$ either of the given equations, say Eqn. (*i*). This gives

$$2(k_1 - c_1)\cos t + 2(k_2 - c_2)\sin t - 2(3k_3 + c_3)\cos 3t - 2(3k_4 + c_4)\sin 3t = 0$$

which must hold for all values of $t$.

$\therefore$ Equating to zero the coefficients of $\cos t$, $\sin t$, $\cos 3t$ and $\sin 3t$, we get

$$k_1 = c_1,\ k_2 = c_2,\ k_3 = -c_3/3,\ k_4 = -c_4/3$$

Thus $$x = c_1 \cos t + c_2 \sin t - \frac{1}{3}(c_3 \cos 3t + c_4 \sin 3t) \qquad ...(v)$$

Hence Eqns. (*iii*) and (*iv*) constitute the solutions of Eqns. (*i*) and (*ii*).

Since $x = y = 0$, when $t = 0$; $\therefore$ Eqns. (*iii*) and (*v*) give

$$0 = c_1 + c_3 \text{ and } c_1 - \frac{1}{3}c_3 = 0 \text{ i.e. } c_1 = c_3 = 0$$

Thus Eqns. (*iii*) and (*v*) reduce to

$$\left.\begin{aligned} y &= c_2 \sin t + c_4 \sin 3t \\ \text{and} \quad x &= c_2 \sin t - \frac{c_4}{3}\sin 3t \end{aligned}\right\} \qquad ...(vi)$$

$\therefore$ $Dx = c_2 \cos t - c_4 \cos 3t$ and $Dy = c_2 \cos t + 3c_4 \cos 3t$.

Since $Dx = 3$ and $Dy = 2$ when $t = 0$

$\therefore$ $3 = c_2 - c_4$ and $2 = c_2 + 3c_4$, whence $c_2 = \dfrac{11}{4}, c_4 = -\dfrac{1}{4}$.

Hence Eqn. (*vi*) becomes $x = \dfrac{1}{4}\left(11 \sin t + \dfrac{1}{3}\sin 3t\right), y = \dfrac{1}{4}(11 \sin t - \sin 3t)$ ...(*vii*)

$\therefore$ when $t = 1/2$, $x = \dfrac{1}{4}\left[11 \sin(0.5) + \dfrac{1}{3}\sin(1.5)\right] = \dfrac{1}{4}\left[[11(0.4794 + \dfrac{1}{3}(0.9975)\right] = 1.4015$

and $$y = \frac{1}{4}[11 \sin(0.5) - \sin(1.5)] = 1.069.$$

**Example 13.45.** *Solve the simultaneous equations:* $\dfrac{dx}{dt} = 2y, \dfrac{dy}{dt} = 2z, \dfrac{dz}{dt} = 2x$. (*C.S.V.T.U., 2011S*)

**Solution.** Differentiating first equation w.r.t. $t$, $\dfrac{d^2x}{dt^2} = 2\dfrac{dy}{dt} = 2(2z)$

Again differentiating w.r.t. $t$, $\dfrac{d^3x}{dt^2} = 4\dfrac{dz}{dt} = 4(2x)$ ...(*i*)

or $$(D^3 - 8)x = 0$$

Its A.E. is $$D^3 - 8 = 0 \quad \text{or} \quad (D - 2)(D^2 + 2D + 4) = 0$$

or $$D = 2, -1 \pm i\sqrt{3}$$

$\therefore$ the solution of Eqn. (*i*) is $$x = c_1e^{2t} + e^{-t}(c_2 \cos\sqrt{3}t + c_3 \sin\sqrt{3}t) \qquad ...(ii)$$

From the first equation, we have $y = \dfrac{1}{2}\dfrac{dx}{dt}$

$$\therefore \quad y = \frac{1}{2}[2c_1e^{2t} + (-1)e^{-t}(c_2 \cos\sqrt{3}t + c_3 \sin\sqrt{3}t) + e^{t}(-\sqrt{3}\,c_2 \sin\sqrt{3}t + \sqrt{3}\,c_3 \cos\sqrt{3}t)]$$

or $$y = c_1e^{2t} + \frac{1}{2}e^{-t}\left\{(\sqrt{3}c_3 - c_2)\cos\sqrt{3}t - (c_3 + \sqrt{3}c_2)\sin\sqrt{3}t\right\} \qquad ...(iii)$$

From the second equation, we have $z = \dfrac{1}{2}\dfrac{dy}{dt}$

$$\therefore \quad z = \frac{1}{2}2c_1e^{2t} + \frac{1}{4}\Big[(-1)e^{-t}\left\{(\sqrt{3}c_3 - c_2)\cos\sqrt{3}t - (c_3 + \sqrt{3}c_2)\sin\sqrt{3}t\right\}$$
$$+ e^{-t}\left\{\sqrt{3}(c_2 - \sqrt{3}c_3)\sin\sqrt{3}t - \sqrt{3}(c_3 + \sqrt{3}c_2)\cos\sqrt{3}t\right\}\Big]$$
$$= c_1e^{2t} + \frac{1}{4}e^{-t}\left\{(-2c_2 - 2\sqrt{3}c_3)\cos\sqrt{3}t - (2\sqrt{3}\,c_2 - 2c_3)\sin\sqrt{3}t\right\}$$

or $$z = c_1e^{2t} - \frac{1}{2}e^{-t}\left\{(\sqrt{3}\,c_2 - c_3)\sin\sqrt{3t} + (c_2 + \sqrt{3}c_3)\cos\sqrt{3t}\right\} \quad ...(iv)$$

Hence the Eqns. (*ii*), (*iii*) and (*iv*) taken together give the required solution.

**Example 13.46:** *If* $\frac{d^4x}{dt^4} = m^4x$. *Show that* $x = c_1 \cos mt + c_2 \sin mt + c_3 \cosh mt + c_4 \sinh mt$ *(VTU, 2018)*

**Solution:** Denoting $\frac{d}{dt}$ by $D$, the given equation can be written as

$$(D^4 - m^4)x = 0$$

A.E. is
$$D^4 - m^4 = 0$$
$$(D^2 + m^2)(D^2 - m^2) = 0$$
$$(D^2 + m^2)(D - m)(D + m) = 0$$
$$D = \pm mi, \pm m$$

Hence the solution of the given equation is

$$x = c_1 \cos mt + c_2 \sin mt + k_1 e^{mt} + k_2 e^{-mt} \quad ...(i)$$
$$e^{mt} = \cosh mt + \sinh.\, mt$$
$$e^{-mt} = \cosh mt - \sinh mt$$

From Eqn. (*i*) becomes

$$x = c_1 \cos mt + c_2 \sin mt + k_1(\cosh mt + \sinh mt) + k_2(\cosh mt - \sinh mt)$$
$$= c_1 \cos mt + c_2 \sin mt + (k_1 + k_2)\cosh mt + (k_1 - k_2)\sinh mt$$
$$x = c_1 \cos mt + c_2 \sin mt + c_3 \cosh mt + c_4 \sinh mt$$

where $k_1 + k_2 = c_3$

and $k_1 - k_2 = c_4$

## PROBLEMS 13.6

Solve the following simultaneous equations :

1. $\frac{dx}{dt} = 5x + y, \frac{dy}{dt} = y - 4x.$ *(Andhra, 2018 ; P.T.U., 2014)*
2. $\frac{dx}{dt} - 7x + y = 0, \frac{dy}{dt} - 2x - 5y = 0.$ *(Andhra, 2018 ; V.T.U., 2015)*
3. $\frac{dx}{dt} + y = \sin t, \frac{dy}{dt} + x = \cos t$; given that $x = 2$ and $y = 0$ when $t = 0$. *(Anna, 2012 ; C.S.V.T.U., 2011)*
4. $\frac{dx}{dt} + 2x + 3y = 0, 3x + \frac{dy}{dt} + 2y = 2e^{2t}.$
5. $\frac{dx}{dt} + x - 2y = 0, \frac{dy}{dt} - x + 4y = 0, x(0) = y(0) = 1.$ *(U.P.T.U., 2011)*
6. $\frac{dx}{dt} + 2y = e^t, \frac{dy}{dt} - 2x = e^{-t}.$
7. $\frac{dx}{dt} + 2x - 3y = t; \frac{dy}{dt} - 3x + 2y = e^{2t}.$ *(U.K.T.U., 2012)*
8. $(D - 1)x + Dy = 2t + 1, (2D + 1)x + 2Dy = t.$
9. $(D + 1)x + (2D + 1)y = e^t, (D - 1)x + (D + 1)y = 1.$
10. $Dx + Dy + 3x = \sin t, Dx + y - x = \cos t.$
11. $t\frac{dx}{dt} + y = 0, t\frac{dy}{dt} + x = 0$ given $x(1) = 1, y(-1) = 0.$ *(C.S.V.T.U., 2015)*
12. $\frac{dx}{dt} + x = y + e^t, \frac{dy}{dt} + y = x + e^t.$ *(Delhi, 2012 S)*
13. $\frac{dx}{dt} + \frac{dy}{dt} + 3x = \sin t, \frac{dx}{dt} + y - x = \cos t.$ *(D.T.U., 2013)*
14. $\frac{d^2x}{dt^2} - 3x - 4y = 0, \frac{d^2y}{dt^2} + x + y = 0.$
15. $y'' = x, x'' = y.$ *(Anna, 2016)*

16. $\frac{d^2x}{dt^2} + y = \sin t, \frac{d^2y}{dt^2} + x = \cos t.$

17. A mechanical system with two degrees of freedom satisfies the equations

$$2\frac{d^2x}{dt^2} + 3\frac{dy}{dt} = 4; 2\frac{d^2y}{dt^2} - 3\frac{dx}{dt} = 0.$$

Obtain expression for $x$ and $y$ in terms of $t$, given $x, y, dx/dt, dy/dt$ all vanish at $t = 0$.

18. $x^2 \frac{d^2y}{dx^2} - x \cdot \frac{dy}{dx} - 3y = x^2 \log x$

19. $(x^2 D^2 + 4xD + 2) y = x + \sin x$

20. $x^4 \frac{d^4y}{dy^4} + 2x^3 \frac{d^3y}{dx^3} + x^2 \frac{d^2y}{dx^2} - x\frac{dy}{dx} + y = \log x$

21. $x^4 \frac{d^3y}{dx^3} + 2x^3 \frac{d^2y}{dx^2} - x^2 \frac{dy}{dx} + xy = 1$

22. $x^3 \frac{d^3y}{dx^3} + 2x^2 \frac{d^2y}{dx^2} + 3x\frac{dy}{dx} - 3y = x^2 + x$

23. $x^2 \frac{d^2y}{dx^2} + 9x\frac{dy}{dx} + 25y = 50$

24. $x^2 \frac{d^2y}{dx^2} - 2y = x^2 + \frac{1}{x}$

## 13.12 OBJECTIVE TYPE OF QUESTIONS

### PROBLEMS 13.7

*Fill up the blanks or choose the correct answer in the following problems :*

1. The solution of $\frac{d^3y}{dt^2} - x = 0$ is ........ *(J.N.T.U., 2015)*
2. P.I. of the differential equation $(D^2 + D + 1) y = \sin 2x$ is .......
3. P.I. of $y'' - 3y' + 2y = 12$ is .......
4. The Wronskian of $x$ and $e^x$ is .......
5. The C.F. of $y'' - 2y' + y = xe^x \sin x$ is
   (a) $C_1e^x + C_2e^{-x}$ (b) $(C_1x + C_2)e^x$ (c) $(C_1 + C_2x)e^{-x}$ (d) None of these. *(V.T.U., 2010)*
6. The general solution of the differential equation $y^{iv} + 6y^{iii}\ ay^{ii}$ is ....... *(K.T.U., 2016)*
7. The particular integral of $(D^2 + a^2) y = \sin ax$ is
   (a) $-\frac{x}{2a}\cos ax$ (b) $\frac{x}{2a}\cos ax$ (c) $-\frac{ax}{2}\cos ax$ (d) $\frac{ax}{2}\cos ax$.
8. The solution of the differential equation $(D^2 - 2D + 5)^2 y = 0$, is .......
9. The solution of the differential equation $y'' + y = 0$ satisfying the conditions $y(0) = 1$ and $y(\pi/2) = 2$, is .......
10. $e^{-x} (c_1 \cos \sqrt{3}x + c_2 \sin \sqrt{3}x) + c_3e^{2x}$ is the general solution of
    (a) $d^3y/dx^3 + 4y = 0$ (b) $d^3y/dx^3 - 8y = 0$
    (c) $d^3y/dx^3 + 8y = 0$ (d) $d^3y/dx^3 - 2d^2y/dx^2 + dy/dx - 2 = 0$.
11. The solution of the differential equation $(D^2 + 1)^2 y = 0$ is ....... .
12. The particular integral of $d^2y/dx^2 + y = \cos h\ 3x$ is ....... .
13. The solution of $x^2y'' + xy' = 0$ is .......
14. The general solution of $(D^2 - 2)^2, y = 0$ is.......
15. P.I. of $(D - 1)^2 y = e^x \sin x$ is ........ *(Anna, 2012)*
16. If $f(D) = D^2 - 2, \frac{1}{f(D)} e^{2x} =$ .......
17. If $f(D) = D^2 + 5, \frac{1}{f(D)} \sin 2x =$ ........
18. The particular integral of $(D^2 - 4D + 4) y = 2^x$ is ....... *(Anna, 2013)*

19. The general solution of $(4D^3 + 4D^2 + D)\,y = 0$ is .......

20. P.I. of $(D^2 + 4)y = \cos 2x$ is

(a) $\frac{1}{2}\sin 2x$ (b) $\frac{1}{2}x \sin 2x$ (c) $\frac{1}{4}x \sin 2x$ (d) $\frac{1}{2}x \cos 2x$.

21. By the method of undetermined coefficients $y_p$ of $y'' + 3y' + 2y = 12x^2$ is of the form

(a) $a + bx + cx^2$ (b) $a + bx$ (c) $ax + bx^2 + cx^3$ (d) None of these. (V.T.U., 2010)

22. In the equation $\frac{dx}{dt} + y = \sin t + 1$, $\frac{dy}{dt} + x = \cos t$ if $y = \sin t + 1 + e^{-t}$, then $x =$ .......

23. $(x^2D^2 + xD + 7)y = 2/x$ converted to a linear differential equation with constant coefficients is .......

24. P.I. of $\frac{d^2y}{dx^2} + \frac{dy}{dx} = x^2 + 2x + 4$ is

(a) $\frac{x^2}{3} + 4x$ (b) $\frac{x^3}{3} + 4$ (c) $\frac{x^3}{3} + 4x$ (d) $\frac{x^3}{3} + 4x^2$.

25. The solution of the differential equation $\frac{d^2y}{dx^2} - 3\frac{dy}{dx} + 2y = e^{3x}$ is given by

(a) $y = C_1e^{x} + C_2e^{2x} + \frac{1}{2}e^{3x}$ (b) $y = C_1e^{-x} + C_2e^{-2x} + \frac{1}{2}e^{3x}$

(c) $y = C_1e^{-x} + C_2e^{2x} + \frac{1}{2}e^{3x}$ (d) $y = C_1e^{-x} + C_2e^{2x} + \frac{1}{2}e^{-3x}$.

26. The particular integral of the differential equation $(D^3 - D)y = e^x + e^{-x}$, $D = \frac{d}{dx}$ is

(a) $\frac{1}{2}(e^x + e^{-x})$ (b) $\frac{1}{2}x\,(e^x + e^{-x})$ (c) $\frac{1}{2}x^2\,(e^x + e^{-x})$ (d) $\frac{1}{2}x^2\,(e^x - e^{-x})$.

27. The complementary function of the differential equation $x^2y'' - xy' + y = \log x$ is ........

28. The homogeneous linear differential equation whose auxiliary equation has roots 1, –1 is ......

29. The particular integral of $(D^2 - 6D + 9)y = \log 2$ is ....... (V.T.U., 2011)

30. To transform $x\frac{d^2y}{dx^2} + \frac{dy}{dx} = \frac{1}{x}$ into a linear differential equation with constant coefficients, put $x =$ ........

31. The P.I. of $(D - 2)^2\,y = \sin 2x$ is ........ (P.T.U., 2014)

32. The P.I. of $(D^2 - 2D + 1)y = xe^x$ is ........ (K.T.U., 2016)

33. The particular integral of $(D^2 - 4)y = \sin 3x$ is

(a) 1/4 (b) –1/13 (c) 1/5 (d) None of these. (V.T.U., 2010)

34. The differential equation whose auxiliary equation has the roots 0, –1, –1 is ......

35. Complementary function of $x^2\frac{d^2y}{dx^2} - x\frac{dy}{dx} - y = 2x\log x$ is

(a) $(C_1 + C_2\,x)e^x$ (b) $(C_1 + C_2\log x)x$ (d) $(C_1 + C_2\,x)\log x$ (d) $(C_1 + C_2\log x)e^x$.

36. The general solution of $(D^2 - D - 2)\,x = 0$ is $x = c_1e^t + c_2e^{-2t}$ (True or False)

37. $\frac{1}{f(D)}(x^2e^{ax}) = \frac{1}{f(D+a)}(e^{ax}x^2)$. (True or False)

CHAPTER 14

# Applications of Linear Differential Equations



## 14.1 INTRODUCTION

The linear differential equations with constant coefficients find their most important applications in the study of electrical, mechanical and other linear systems. In fact such equations play a dominant role in unifying the theory of electrical and mechanical oscillatory systems.

We shall begin by explaining the types of oscillations of the mechanical systems and the equivalent electrical circuits. Then we shall study at some length the slightly less striking applications such as deflection of beams and whirling of shafts. At the end, we'll take up some of the applications of simultaneous linear differential equations.

## 14.2 SIMPLE HARMONIC MOTION

When the acceleration of a particle is proportional to its displacement from a fixed point and is always directed towards it, then the motion is said to be *simple harmonic*.

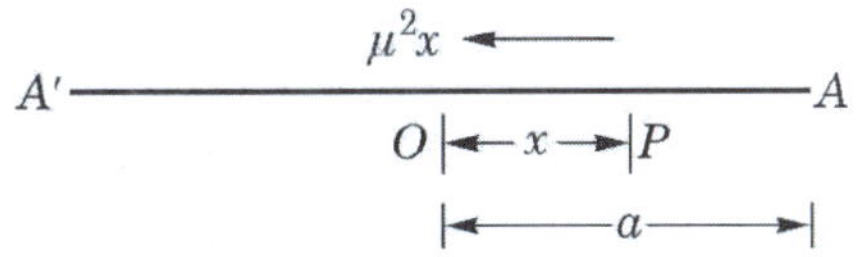


Fig. 14.1

If the displacement of the particle at any time $t$, from fixed point $O$ is $x$ (Fig. 14.1), then

$$\frac{d^2x}{dt^2} = -\mu^2 x \quad \text{or} \quad (D^2 + \mu^2)x = 0, \qquad ...(1)$$

$\therefore$ its solution is $\quad x = c_1 \cos \mu t + c_2 \sin \mu t \qquad ...(2)$

$\therefore$ its velocity at $\quad P = \frac{dx}{dt} = \mu(-c_1 \sin \mu t + c_2 \cos \mu t) \qquad ...(3)$

If the particle starts from rest at $A$, where $OA = a$,

then from Eqn. (2), (at $t = 0, x = a$) $\quad a = c_1$

and from Eqn. (3), (at $t = 0, dx/dt = 0$) $\quad 0 = c_2$.

Thus $\quad x = a \cos \mu t \qquad ...(4)$

and $\quad \frac{dx}{dt} = -a\mu \sin \mu t = -\sqrt{(a^2 - x^2)} \qquad ...(5)$

which give the displacement and the velocity of the particle at any time $t$.

**Nature of motion.** The particle starts from $A$ towards $O$ under acceleration directed towards $O$ which gradually decreases until it vanishes at $O$, when the particle has acquired the maximum velocity. On passing through $O$, retardation begins and the particle comes to an instantaneous rest at $A'$, where $OA' = OA$. It then retraces its path and goes on oscillating between $A$ and $A'$.

*The* **amplitude** *or maximum displacement from the centre is a.*

*The* **periodic time,** *i.e., the time of complete oscillation is* $2\pi/\mu$, for when $t$ is increased by $2\pi/\mu$, the values of $x$ and $dx/dt$ remain unaltered.

*The* **frequency** *or the number of oscillations per second is*

$$1/\text{periodic time, i.e., } \mu/2\pi.$$

**Example 14.1.** *In the case of a stretched elastic horizontal string which has one end fixed and a particle of mass m attached to the other, find the equation of motion of the particle given that l is the natural length of the string and e is its elongation due to weight mg. Also find the displacement s of particle when initially s = 0, v = 0.*

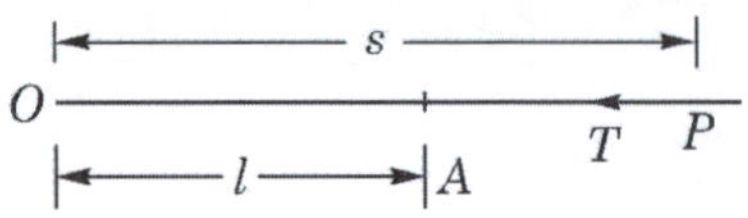

Fig. 14.2

**Solution.** Let $OA(= l)$ be the elastic horizontal string with the end $O$ fixed and having a particle of mass $m$ attached to the end $A$. (Fig. 14.2)

At any time $t$, let the particle be at $P$ where $OP = s$ ; so that the elongation $AP = s - l$.

Since for the elongation $e$, tension $= mg$

$\therefore$ for the elongation $s - l$, tension $T = \dfrac{mg(s-l)}{e}$

Tension being the only horizontal force, the equation of motion is

$$m\frac{d^2s}{dt^2} = -T \quad \text{or} \quad \frac{d^2s}{dt^2} = -\frac{T}{m} = -\frac{g(s-l)}{e} \qquad ...(i)$$

which is the required equation of motion.

Now Eqn. $(i)$ can be written as $(D^2 + g/e)s = gl/e$, where $D = d/dt$ ...$(ii)$

$\therefore$ the auxiliary equation is $D^2 + g/e = 0$ or $D = \pm i\sqrt{(g/e)}$

$$\therefore \qquad \text{C.F.} = c_1 \cos\sqrt{(g/e)}\,t + c_2 \sin\sqrt{(g/e)}\,t$$

and
$$\text{P.I.} = \frac{1}{D^2 + g/e}\cdot\frac{gl}{e} = \frac{gl}{e}\cdot\frac{l}{D^2 + g/e}e^{0t} = l$$

Thus the solution of Eqn. $(ii)$ is

$$s = c_1 \cos\sqrt{(g/e)}\,t + c_2 \sin\sqrt{(g/e)}\,t + l \qquad ...(iii)$$

When $t = 0$, $s = s_0$, $\quad\therefore\quad s_0 = c_1 + 0 + l$ *i.e.,* $c_1 = s_0 - l$

Again from Eqn. $(iii)$, $\quad \dfrac{ds}{dt} = -c_1\sqrt{(g/e)}\sin\sqrt{(g/e)}\,t + c_2\sqrt{(g/e)}\cos\sqrt{(g/e)}\,t$

When $t = 0$, $ds/dt = 0$ ; $\quad\therefore\quad 0 = c_2$.

Substituting the values of $c_1$ and $c_2$ in Eqn. $(iii)$, we have

$$s = (s_0 - l)\cos\sqrt{(g/e)}\,t + l \text{ which is the required result.}$$

**Example 14.2.** *Two particles of masses $m_1$ and $m_2$ are tied to the ends of an elastic string of natural length a and modulus λ. They are placed on a smooth table so that the string is just taut and $m_2$ is projected with any velocity directly away from $m_1$. Show that the string will become slack after the lapse of time* $\pi\sqrt{[am_1m_2/\lambda(m_1+m_2)]}$.

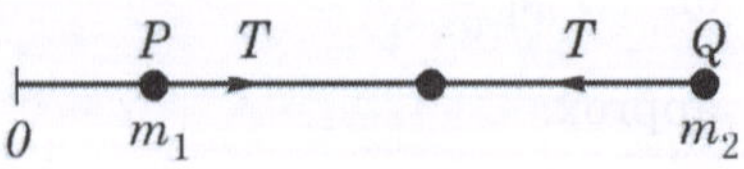

Fig. 14.3

**Solution.** Taking $O$ as fixed point of reference, let particle $m_1$ be at $O$ and $m_2$ at a distance $a$ from $m_1$ at time $t = 0$ Fig. 14.3. At any time $t$, let $m_1$ be of a distance $x$ from $O$ and $m_2$ be at a distance $y$ from $O$. Then the equation of motion of $m_1$ is

$$m_1 d^2x/dt^2 = T \qquad ...(i)$$

and equation of motion of $m_2$ is $m_2 d^2y/dt^2 = -T$ ...$(ii)$

where $T = \lambda(y - x)/a$

From Eqns. (*i*) and (*ii*) $d^2y/dt^2 - d^2x/dt^2 = -\dfrac{T}{m_2} - \dfrac{T}{m_1}$

or $$\frac{d^2(y-x)}{dt^2} = -\left(\frac{1}{m_1}+\frac{1}{m_2}\right)\frac{\lambda(y-x)}{a} \quad \text{or} \quad \frac{d^2u}{dt^2} = -\frac{\lambda(m_1+m_2)u}{m_1 m_2 a} \quad \text{where } u = y - x$$

This is S.H.M. with periodic time $\tau = 2\pi\sqrt{\left\{\dfrac{am_1m_2}{\lambda(m_1+m_2)}\right\}}$

The string will acquire its original length (*i.e.* become slack) after time $\tau_1$ of $m_2$ moving towards $m_1$ such that

$$\tau_1 = \frac{\tau}{4}+\frac{\tau}{4} = \frac{\tau}{2} = \pi\sqrt{\left\{\frac{am_1m_2}{\lambda(m_1+m_2)}\right\}}.$$

**Example 14.3.** *A particle of mass m executes S.H.M. in the line joining the points A and B, on a smooth table and is connected with these points by elastic strings whose tensions in equilibrium are each T. If l, l′ be the extensions of the strings beyond their natural lengths, find the time of an oscillation.* (*P.T.U., 2014*)

**Solution.** In the equilibrium position, let the particle be at $C$ so that $AC = a + l$ and $BC = a' + l'$, where $a, a'$ are natural lengths of the strings (Fig. 14.4). Then the tensions (at this time) are given by

$$T = \lambda l/a = \lambda' l'/a' \quad \ldots(i)$$

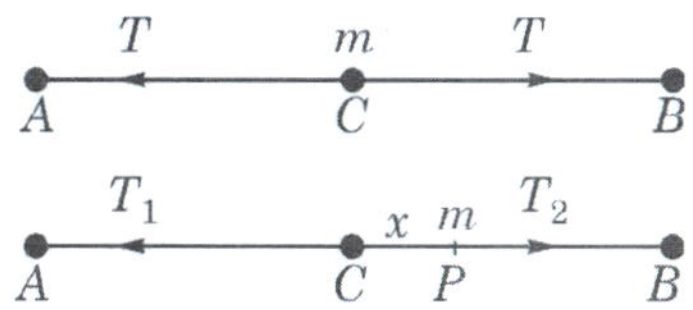

Fig. 14.4

At any time $t$, let the particle be at $P$, so that $CP = x$. Then

$$T_1 = \lambda\frac{l+x}{a} \text{ and } T_2 = \lambda'\frac{l'-x}{a'}$$

$\therefore$ the equation of motion is $m\dfrac{d^2x}{dt^2} = T_2 - T_1 = \lambda'\dfrac{l'-x}{a'} - \lambda\dfrac{l+x}{a}$

$$= \left(\frac{\lambda' l'}{a'} - \frac{\lambda l}{a}\right) - \left(\frac{\lambda'}{a'}+\frac{\lambda}{a}\right)x = -\left(\frac{T}{l'}+\frac{T}{l}\right)x \quad \text{[By } (i)]$$

or $$\frac{d^2x}{dt^2} = -\mu x \quad \text{where } \mu = \frac{l+l'}{ll'}\cdot\frac{T}{m}$$

Hence the periodic time $= \dfrac{2\pi}{\sqrt{\mu}} = 2\pi\sqrt{\left\{\dfrac{mll'}{(l+l')T}\right\}}.$

## 14.3 (1) SIMPLE PENDULUM

A heavy particle attached by a light string to a fixed point and oscillating under gravity constitutes a *simple pendulum*.

Let $O$ be the fixed point, $l$ be the length of the string and $A$ be the position of the bob initially (Fig. 14.5). If $P$ be the position of the bob at any time $t$, such that arc $AP = s$ and $\angle AOP = \theta$, then $s = l\theta$.

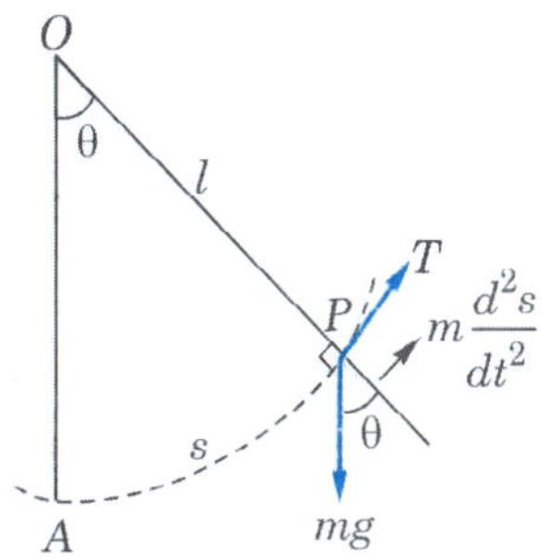

Fig. 14.5

$\therefore$ the equation of motion along $PT$ is $m\dfrac{d^2s}{dt^2} = -mg\sin\theta$

*i.e.*, $$\frac{d^2(l\theta)}{dt^2} = -g\sin\theta$$

or $$\frac{d^2\theta}{dt^2} = -\frac{g}{l}\sin\theta = -\frac{g}{l}\left(\theta - \frac{\theta^3}{3!}+\ldots\right) = -\frac{g\theta}{l}$$ to a first approx.

Here the auxiliary equation being $D^2 + g/l = 0$, we have $D = \pm\sqrt{(g/l)}\,i$

$\therefore$ its solution is $\theta = c_1\cos\sqrt{(g/l)}\,t + c_2\sin t$.

*Thus the motion of the bob is simple harmonic and the time of an oscillation is* $2\pi\sqrt{(l/g)}$.

**Obs.** The movement of the bob from one end to the other constitutes half an oscillation and is called a *beat* or *a swing*.

Thus the time of one beats = $\pi\sqrt{(l/g)}$.

*A seconds pendulum beats 86400 times a day* for there are 86,400 seconds in 24 hours.

**(2) Gain or loss of oscillations.** Let a pendulum of length $l$ make $n$ beats in time $T$, so that

$$T = \text{time of } n \text{ beats} = n\pi\sqrt{(l/g)} \quad \text{or} \quad n = \frac{T}{\pi}(g/l)^{1/2}$$

Taking logs, $\log n = \log (T/\pi) + \frac{1}{2}(\log g - \log l)$.

Taking differentials of both sides, we get $\frac{dn}{n} = \frac{1}{2}\left(\frac{dg}{g} - \frac{dl}{l}\right)$.

If only $g$ changes, $l$ remaining constant, $\frac{\mathbf{dn}}{\mathbf{n}} = \frac{\mathbf{dg}}{\mathbf{2g}}$ ...(1)

If only $l$ changes, $g$ remaining constant, $\frac{\mathbf{dn}}{\mathbf{n}} = -\frac{\mathbf{dl}}{\mathbf{2l}}$. ...(2)

**Example 14.4.** *Find how many seconds a clock would lose per day if the length of its pendulum were increased in the ratio 900 : 901.* (Rohtak, 2010)

**Solution.** If the original length $l$ of the string be increased to $l + dl$, then

$$\frac{l + dl}{dl} = \frac{901}{900}. \quad \therefore \quad \frac{dl}{l} = \frac{901}{900} - 1 = \frac{1}{900}.$$

$\therefore$ using Eqn. (2) above, we have $\frac{dn}{n} = -\frac{dl}{2l} = -\frac{1}{1800}$

*i.e.,*
$$dn = -\frac{n}{1800} = -\frac{86400}{1800} = -48.$$

Since $dn$ is negative, the clock will lose 4 seconds per day.

**Example 14.5.** *A simple pendulum of length $l$ is oscillating through a small angle $\theta$ in a medium in which the resistance is proportional to the velocity. Find the differential equation of its motion. Discuss the motion and find the period of oscillation.*

**Solution.** Let the position of the bob (of mass $m$), at any time $t$ be $P$ and $O$ be the point of suspension such that $OP = l$, $\angle AOP = \theta$ and therefore, arc $AP = s = l\theta$. (Fig. 14.6)

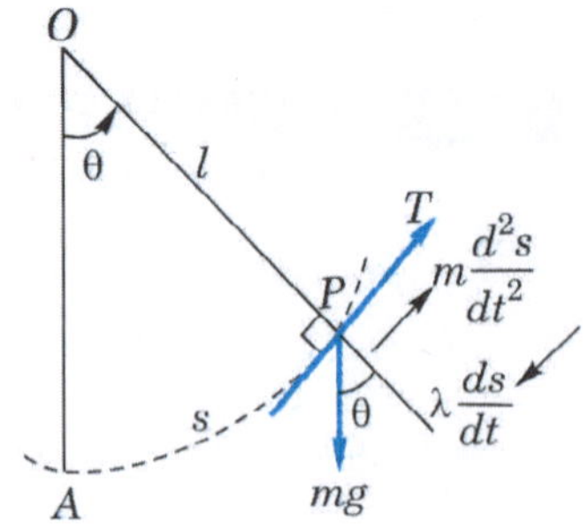

Fig. 14.6

$\therefore$ the equation of motion along the tangent $PT$ is

$$m\frac{d^2s}{dt^2} = -mg \sin\theta - \lambda\frac{ds}{dt} \text{ where } \lambda \text{ is a constant.}$$

or
$$\frac{d^2(l\theta)}{dt^2} + \frac{\lambda}{m}\frac{d(l\theta)}{dt} + g\sin\theta = 0$$

Replacing $\sin\theta$ by $\theta$ since it is small and writing $\lambda/m = 2k$, we get

$$\frac{d^2\theta}{dt^2} + 2k\frac{d\theta}{dt} + \frac{g\theta}{l} = 0 \quad ...(i)$$

which is the required differential equation.

Its auxiliary equation has roots $D = k \pm \sqrt{(k^2 - w^2)}$ where $w = g/l$.

*The oscillatory motion of the bob is only possible when $k < w$.*

Then the roots of the auxiliary equation are $-k \pm i\sqrt{(w^2 - k^2)}$.

$\therefore$ the solution of Eqn. $(i)$ is

$$\theta = e^{-kt}$$

which gives a vibratory motion of period $2\pi/\sqrt{(w^2 - k^2)}$.

**Example 14.6.** *A pendulum of length $l$ has one end of the string fastened to a peg on a smooth plane inclined to the horizon at an angle $\alpha$. With the string and the weight on the plane, its time of oscillation is t sec. If the pendulum of length $l'$ oscillates in one sec. when suspended vertically, prove that $\alpha = \sin^{-1}\left(\frac{l}{l't^2}\right)$.*

**Solution.** At any time $t$, let the bob of mass $m$ be at $P$ and $O$ be the point of suspension so that $OP = l$ and $\angle AOP = \theta$ (Fig. 14.7).

The component of weight along the plane being $mg \sin \alpha$, the equation of motion of the bob along the tangent at $P$ is

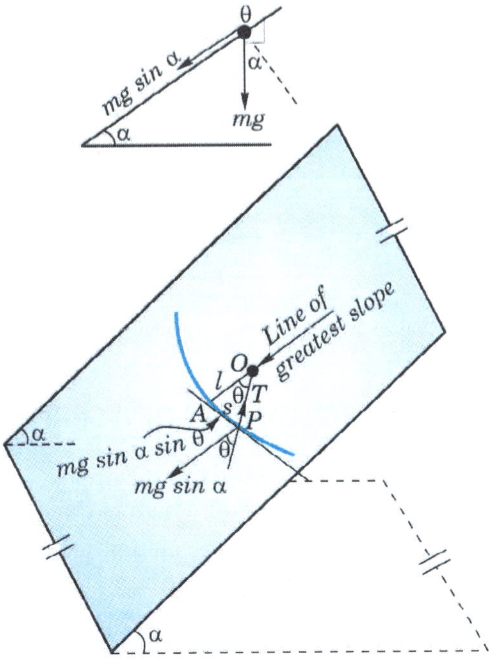

Fig. 14.7

$$m\frac{d^2s}{dt^2} = -mg \sin \alpha \sin \theta$$

or $$\frac{d^2(l\theta)}{dt^2} = -g \sin \alpha \sin \theta \qquad [\because \; s = l\theta]$$

or $$\frac{d^2\theta}{dt^2} = -g \sin \alpha\left(\theta - \frac{\theta^3}{3!} + \ldots\right)$$

or $$\frac{d^2\theta}{dt^2} = -\mu\theta \quad \text{where } \mu = \frac{g \sin \alpha}{l}, \text{ to a first approximation.}$$

$\therefore$ the motion being simple harmonic, the time of oscillation $t$.

$$= \frac{2\pi}{\mu} = 2\pi\sqrt{\frac{l}{g \sin \alpha}} \qquad \ldots(i)$$

We know that for a pendulum of length $l'$ when suspended vertically, the time of oscillation

$$1 = 2\pi\sqrt{l'/g} \qquad \ldots(ii)$$

$\therefore$ dividing Eqns. $(i)$ by $(ii)$, we have $t = \sqrt{\left(\frac{l}{l' \sin \alpha}\right)}$

or $$t^2 = l/l' \sin \alpha \quad \text{or} \quad \alpha = \sin^{-1}(l/l't^2).$$

## PROBLEMS 14.1

1. A particle is executive simple harmonic motion with amplitude 20 cm and time 4 seconds. Find the time required by the particle in passing between points which are at distances 15 cm and 5 cm from the centre of force and are on the same side of it.
2. At the ends of three successive seconds, the distances of a point moving with S.H.M. from its mean position are $x_1$, $x_2$, $x_3$. Show that the time of a complete oscillation is
$$2\pi / \cos^{-1}\left(\frac{x_1 + x_3}{2x_2}\right).$$
3. An elastic string of natural length $2a$ and modulus $\lambda$ is stretched between two points $A$ and $B$ distant $4a$ apart on a smooth horizontal table. A particle of mass $m$ is attached to the middle of the string. Show that it can vibrate in line $AB$ with period $2\pi/\omega$, where $\omega^2 = 2\lambda/am$.
4. A particle of mass $m$ moves in a straight line under the action of force $mn^2(OP)$, which is always directed towards fixed point $O$ in the line. If the resistance to the motion is $2\lambda\, mnv$, where $v$ is the speed and $0 < \lambda < 1$, find the displacement $x$ in terms of the time $t$ given that when $t = 0$, $x = 0$ and $dx/dt = u$ where $OP = x$. (U.P.T.U., 2010)
5. A point moves in a straight line towards the centre of force $\mu/(\text{distance}^3)$ starting from rest at a distance $a$ from the centre of force, show that the time of reaching a point $b$ from the centre of force is $a\sqrt{(a^2 - b^2)}/\sqrt{\mu}$ and that its velocity then is $\frac{\sqrt{\mu}}{ab}\sqrt{(a^2 - b^2)}$.

**6.** A clock loses five seconds a day, find the alteration required in the length of its pendulum in order that it may keep correct time.

**7.** A clock with a seconds pendulum loses 10 seconds per day at a place where $g = 32$ ft/sec$^2$. What change in the gravity is necessary to make it accurate ?

**8.** A seconds pendulum which gains 10 seconds per day at one place loses 10 seconds per day at another; compare the acceleration due to gravity at the two places. *(Kurukshetra, 2011)*

**9.** Show that the free oscillations of a galvanometer needle, as affected by the viscosity of the surrounding air which varies directly as the angular velocity of the needle, are determined by the equation $\frac{d^2\theta}{dt^2} + K\frac{d\theta}{dt} + \mu\theta = 0$, where $k$ is the co-efficient of viscosity and $\theta$ is the angular deflection of the needle at time $t$. Obtain $\theta$ in terms of $t$ and discuss the different cases that can arise.

**10.** If $I = \frac{d^2\theta}{dt^2} = -\,mgl \sin\theta$, where $I, m, g, l$ are constant, given that at $t = 0$, $\theta = 0$ and $d\theta/dt = \omega_0 = m\sqrt{(mgl)}/I$, then show that $t = \frac{2}{\omega_0}\log\frac{\pi+\theta}{4}$.

## 14.4 OSCILLATIONS OF A SPRING

(*i*) **Free oscillations.** Suppose a mass $m$ is suspended from the end $A$ of a light spring, the other end of which is fixed at $O$. (Fig. 14.8)

Let $e$ (= $AB$) be the elongation produced by the mass $m$ hanging in equilibrium. If $k$ be the restoring force per unit stretch of the spring due to elasticity, then for the equilibrium at $B$,

$$mg = T = ke \qquad \text{...(1)}$$

At any time $t$, after the motion ensues, let the mass be at $P$, where $BP = x$. Then the equation of motion of $m$ is

$$m\frac{d^2x}{dt^2} = mg - k(e + x) = -\,kx \qquad \text{[By (1)]}$$

Or writing $k/m = \mu^2$, it becomes

$$\frac{d^2x}{dt^2} + \mu^2 x = 0 \qquad \text{...(2)}$$

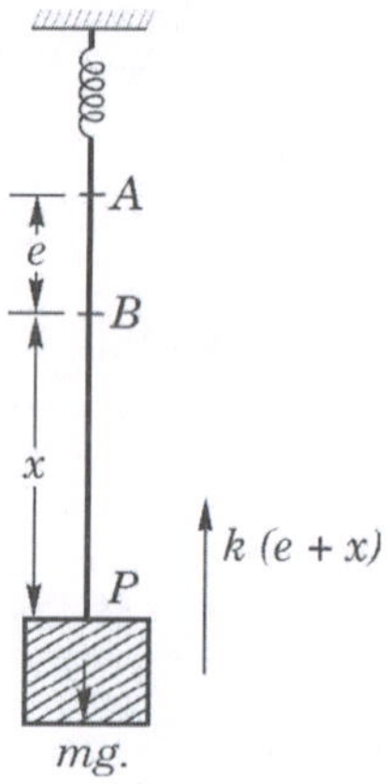

Fig. 14.8

This equation represents the free vibrations of the spring which are of the simple harmonic form having centre of oscillation at $B$—its equilibrium position and the *period of oscillation*

$$= \frac{2\pi}{\mu} = 2\pi\sqrt{\left(\frac{e}{g}\right)}. \qquad \left[\because \frac{1}{\mu} = \sqrt{\left(\frac{m}{k}\right)} = \sqrt{\left(\frac{e}{g}\right)}, \text{[By (1)]}\right]$$

(*ii*) **Damped oscillations.** If the mass $m$ be subjected to do damping force proportional to velocity (say : $r\,dx/dt$) (Fig. 14.9), then the equation of motion becomes

$$m\frac{d^2x}{dt^2} = mg - k(e + x) - r\frac{dx}{dt}$$

$$= -\,kx - r\frac{dx}{dt} \qquad \text{[By (1)]}$$

Or wriing $r/m = 2\lambda$ and $k/m = \mu^2$, it becomes

$$\frac{d^2x}{dt^2} + 2\lambda\frac{dx}{dt} + \mu^2 x = 0 \qquad \text{...(3)}$$

$\therefore$ its auxiliary equation is

$$D^2 + 2\lambda D + \mu^2 = 0 \quad \text{whence} \quad D = -\,\lambda \pm .$$

**Case I.** *When* $\lambda > \mu$, the roots of the auxiliary equation are real and distinct (say $\gamma_1, \gamma_2$).

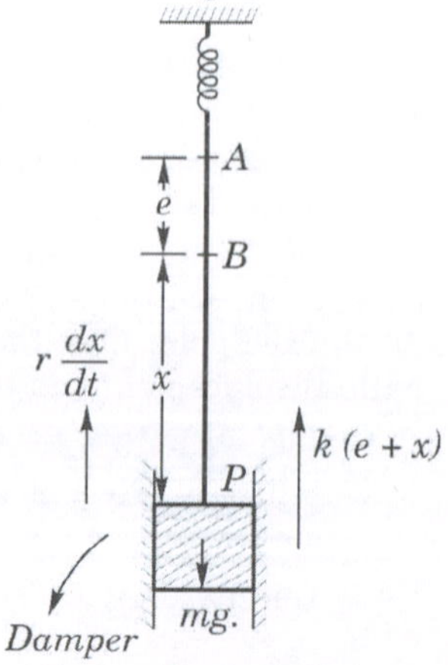

Fig. 14.9

$\therefore$ the solution of Eqn. (3) is $x = c_1 e^{\gamma_1 t} + c_2 e^{\gamma_2 t}$ ...(4)

To determine $c_1, c_2$ let the spring be stretched to a length $x = l$ and then released so that

$$x = l \text{ and } dx/dt = 0 \text{ at } t = 0.$$

$\therefore$ from Eqn. (4), $l = c_1 + c_2$

Also from $\dfrac{dx}{dt} = c_1\gamma_1 e^{\gamma_1 t} + c_2\gamma_2 e^{\gamma_2 t}$, we get

$$0 = c_1\gamma_1 + c_2\gamma_2$$

whence $c_1 = \dfrac{-l\gamma_2}{\gamma_1 - \gamma_2}$ and $c_2 = \dfrac{l\gamma_1}{\gamma_1 - \gamma_2}$

Hence the solution of Eqn. (3) is

$$x = \frac{l}{\gamma_1 - \gamma_2}(\gamma_1 e^{\gamma_2 t} - \gamma_2 e^{\gamma_1 t}) \quad ...(5)$$

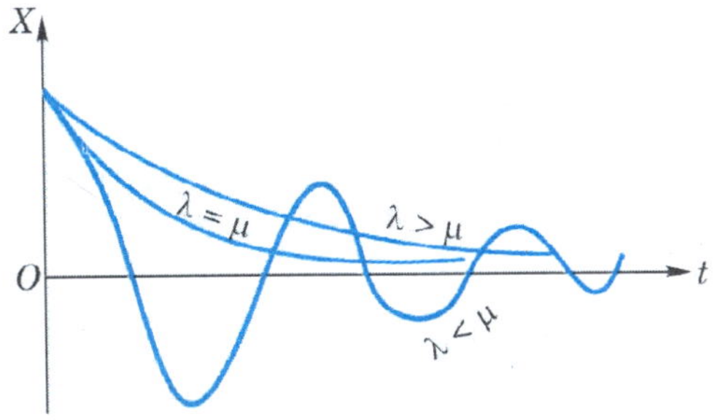

Fig. 14.10

which shows that $x$ is always positive and decreases to zero as $t \to \infty$ (Fig. 14.10).

The restoring force, in this case, is so great that the motion is non-oscillatory and is, therefore, referred to as *over-damped* or *dead-beat* motion.

**Case II.** *When* $\lambda = \mu$, the roots of the auxiliary equation are real and equal, (each being $= -\lambda$).

$\therefore$ The general solution of Eqn. (3) becomes $x = (c_1 + c_2 t)e^{-\lambda t}$.

As in case I, if $x = l$ and $dx/dt = 0$ at $t = 0$, then $c_1 = l$ and $c_2 = \lambda l$.

Hence the solution of Eqn. (3) is $x = l(1 + \lambda t)e^{-\lambda t}$ which also shows that $x$ is always positive and decreases to zero as $t \to \infty$ (Fig. 14.10).

The nature of motion is similar to that of the previous case and is called the *critically damped motion* for it separates the non-oscillatory motion of case I from the most interesting oscillatory motion of case III.

**Case III.** *When* $\lambda < \mu$, the roots of the auxiliary equation are imaginary, *i.e.* $D = -\lambda \pm i\alpha$, where $\alpha^2 = \mu^2 - \lambda^2$.

$\therefore$ the solution of Eqn. (3) is $x = e^{-\lambda t}(c_1 \cos \alpha t + c_2 \sin \alpha t)$ ...(6)

As in case I, $x = l$, $dx/dt = 0$ at $t = 0$, then $c_1 = l$ and $c_2 = \lambda l/\alpha$

Thus the solution of Eqn. (3) becomes $x = le^{-\lambda t}\left(\cos \alpha t + \dfrac{\lambda}{\alpha}\sin \alpha l\right)$.

which can be put in the form $x = l\sqrt{\left\{1 + \left(\dfrac{\lambda}{\alpha}\right)^2\right\}} e^{-\lambda t} \cos\left\{\alpha - \tan^{-1}\dfrac{\lambda}{\alpha}\right\}$ ...(7)

Here the presence of the trigonometric factor in Eqn. (7) shows that the *motion is oscillatory*, having

(*a*) the variable amplitude $= l\sqrt{[1 + (\lambda/\alpha)^2]}e^{-\lambda t}$ which decreases with time,

(*b*) the periodic time $T = 2\pi/\alpha$.

But the periodic time of free oscillations is $T' = 2\pi/\mu$.

As $$\alpha = \sqrt{(\mu^2 - \lambda^2)} < \mu$$

$$\therefore \quad \frac{2\pi}{\alpha} > \frac{2\pi}{\mu}, \quad i.e. \quad T > T'.$$

This shows that the *effect of damping is to increase the period of oscillation and the motion ultimately dies away*. Such a motion is termed as *damped oscillatory motion*.

(*iii*) **Forced oscillations** (*without damping*). If the point of the support of the spring is also vibrating with some external periodic force, then the resulting motion is called the *forced oscillatory motion*.

Taking the external periodic force to be $mp \cos nt$, the equation of motion is

$$m\frac{d^2x}{dt^2} = mg - k(e + x) + mp \cos nt$$

$$= -kx + mp \cos nt \qquad [\because \; mg = ke] \quad ...(8)$$

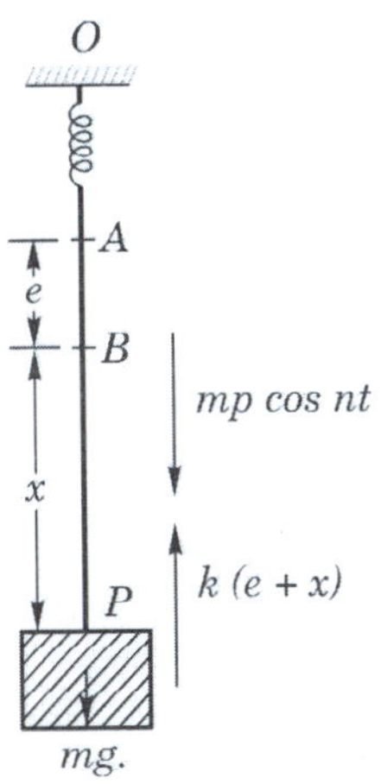

Fig. 14.11

Or writing $k/m = \mu^2$, Eqn. (8) takes the form

$$\frac{d^2x}{dt^2} + \mu^2 x = p \cos nt \quad ...(9)$$

Its C.F. $= c_1 \cos \mu t + c_2 \sin \mu t$ and P.I. $= p \frac{1}{D^2 + \mu^2} \cos nt.$

*New two cases arise :*

**Case I.** *When* $\mu \neq n$.

$$\text{P.I.} = \frac{p}{\mu^2 - n^2} \cos nt.$$

$\therefore$ the complete solution of Eqn. (9) is $x = c_1 \cos \mu t + c_2 \sin \mu t + \frac{p}{\mu^2 - n^2} \cos nt.$

On writing $c_1 \cos \mu t + c_2 \sin \mu t$ as $r \cos (\mu t + \phi)$, we have

$$x = r \cos (\mu t + \phi) + \frac{p}{\mu^2 - n^2} \cos nt \quad ...(10)$$

*This shows that the motion is compounded of two oscillatory motions : the first* (due to the C.F.) *gives free oscillations of period* $2\pi/\mu$, *and the second* (due to the P.I.) *gives forced oscillations of period* $2\pi/n$.

Also we observe that if the frequency of free oscillations is very high (*i.e.*, $\mu$ is large), then the amplitude of forced oscillations is small.

**Case II.** *When* $\mu = n$.

$$\text{P.I.} = pt \cdot \frac{1}{2D} \cos \mu t = \frac{pt}{2} \int \cos \mu t \, dt = \frac{pt}{2\mu} \sin \mu t$$

$\therefore$ the complete solution of (9) is $x = c_1 \cos \mu t + c_2 \sin \mu t + \frac{pt}{2\mu} \sin \mu t$

$$= \left(c_2 + \frac{pt}{2\mu}\right) \sin \mu t + c_1 \cos \mu t.$$

Putting $c_2 + pt/2\mu = \rho \cos \psi$ and $c_1 = \rho \sin \psi$, we get

$$x = \rho \sin (\mu t + \psi) \quad ...(11)$$

This shows that the oscillations are of period $2\pi/\mu$ and amplitude $\rho = \sqrt{[(c_2 + pt/2\mu)^2 + c_1^2]}$, which clearly increases with time (Fig. 14.12).

Thus the amplitude of the oscillations may become abnormally large causing over-strain and consequently breakdown of the system. In practice, however, collapse rarely occurs, though the amplitudes may become dangerously large since there is always some resistance present in the system.

*This phenomenon of the impressed frequency becoming equal to the natural frequency of the system, is reffered to as* **resonance.**

Thus, while designing a machine or a structure, the occurrence of resonance should always be avoided to check the rupture of the system at any stage. That is why, the soldiers break step while marching over a bridge for the fear that their steps may not be in rhym with the natural frequency of the bridge causing its collapse due to 'resonance'.

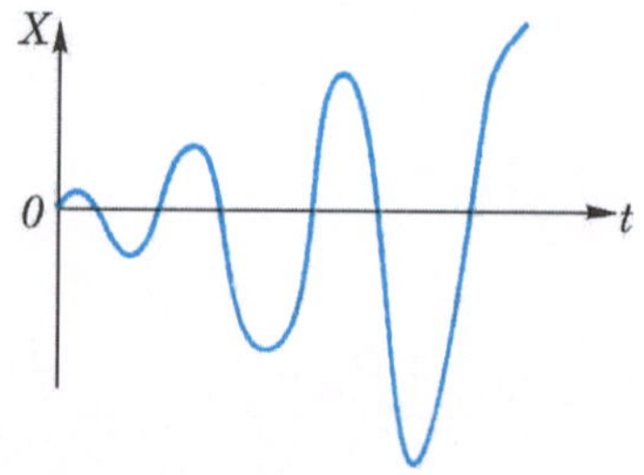

Fig. 14.12

(*iv*) **Forced oscillations** (*with damping*). If, in addition, there is a damping force proportional to velocity (say : $r\,dx/dt$) (Fig. 14.13), then the equation (8) becomes

$$m\frac{d^2x}{dt^2} = mg - k(e + x) + mp \cos nt - r\frac{dx}{dt}$$

$$= -kx + mp \cos nt - r\frac{dx}{dt} \quad [\because mg = ke]$$

On writing $r/m = 2\lambda$ and $k/m = \mu^2$, it takes the form

$$\frac{d^2x}{dt^2} + 2\lambda \frac{dx}{dt} + \mu^2 x = p \cos nt \qquad ...(12)$$

Its auxiliary equation is $D^2 + 2\lambda D + \mu^2 = 0$ whence $D = -\lambda \pm \sqrt{\lambda^2 - \mu^2}$.

$$\therefore \qquad \text{C.F.} = e^{-\lambda t}[c_1 e^{t\sqrt{(\lambda^2-\mu^2)}} + c_2 e^{-t\sqrt{(\lambda^2-\mu^2)}}].$$

It represents the free oscillations of the system which die out as $t \to \infty$.

Also the P.I.

$$= p\frac{1}{D^2 + 2\lambda D + \mu^2}\cos nt = p\frac{1}{-n^2 + 2\lambda D + \mu^2}\cos nt$$

$$= p\frac{(\mu^2 - n^2) - 2\lambda D}{(\mu^2 - n^2)^2 - 4\lambda^2 D^2}\cos nt = p\frac{(\mu^2 - n^2)^2 \cos nt + 2\lambda n \sin nt}{(\mu^2 - n^2)^2 + 4\lambda^2 n^2}$$

Putting $\mu^2 - n^2 = R\cos\theta$ and $2\lambda n = R \sin\theta$, we get

$$\text{P.I.} = \frac{p}{\sqrt{[(\mu^2 - n^2)^2 + 4\lambda^2 n^2]}}\cos(nt - \theta)$$

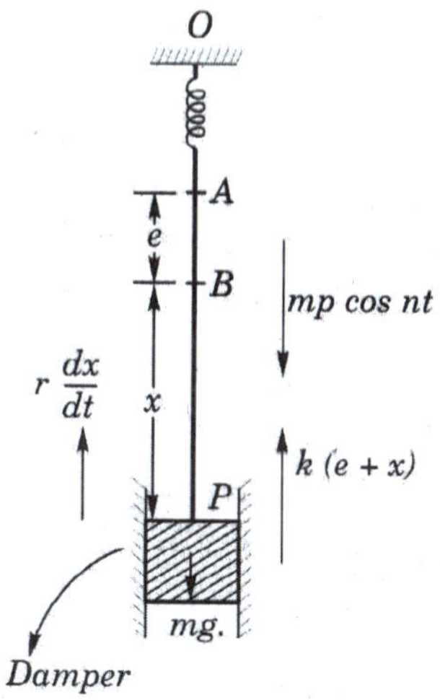

Fig. 14.13

which represents the forced oscillations of the system having

(*a*) a constant amplitude

$$= p/\sqrt{[(\mu^2 - n^2)^2 + 4\lambda^2 n^2]}$$

and (*b*) the period $= 2\pi/n$ which is the same as that of the impressed force.

Thus *with the increase of time, the free oscillations die away while the forced oscillations continue giving the steady state motion.*

**Example 14.7.** *In a system, the amplitude of motion is 15 meters and the period is 4 seconds. Find the time required by the particle is passing between the points which are at a distances of 4 metres and 2 meters from the centre of the force and are on same side of it. Also find the velocities at these points.* (*Annamalai, 2018*)

**Solution:** The equation of the simple harmonic motion is

$$\frac{d^2x}{dt^2} = -\mu^2 x \; .... \qquad ...(i)$$

Time period $= \dfrac{2\pi}{\mu}$

Since time period is 4 seconds $\dfrac{2\pi}{\mu} = 4$

$$\mu = \frac{\pi}{2}$$

Hence solution of equation Eqn. (*i*) is $x = 5\cos\dfrac{\pi}{2}t$ ...(*ii*)

{$\because$ $a$ = Amplitude = 5 meters and $x = a\cos\mu t$}

Let $t_1$ seconds and $t_2$ seconds be the times when the particle is at a distance 4 meters and 2 metres

$$4 = 5\cos\frac{\pi}{2}t_1$$

$$t_1 = \frac{2}{\pi}\cos^{-1}\left(\frac{4}{5}\right)$$

$$2 = 5\cos\frac{\pi}{2}t_2$$

$$t_2 = \frac{2}{\pi}\cos^{-1}\left(\frac{2}{5}\right)$$

Time required in passing through these points

$$t_2 - t_1 = \frac{2}{\pi}\left(\cos^{-1}\frac{2}{5} - \cos^{-1}\frac{4}{5}\right)$$
$$= 0.33 \text{ sec}$$

Differentiating Eqn. (*ii*) *w.r.t. t*

$$\frac{dx}{dt} = \frac{-5\pi}{2} \sin\frac{\pi}{2} t$$

$$= \frac{-5\pi}{2}\sqrt{1 - \cos^2\frac{\pi}{2} t} = \frac{-5\pi}{2}\sqrt{1 - \frac{\pi^2}{25}} \qquad \text{from Eqn. } (iii)$$

$$v = -\frac{\pi}{2}\sqrt{25 - x^2}$$

When $x = 4$ meters, $v = -\frac{\pi}{2}\sqrt{25 - 4^2}$

$$v = -\frac{3\pi}{2} = 4.71 \text{ m/sec}$$

When $x = 2$ meters, $v = -\frac{\pi}{2}\sqrt{25 - 2^2}$

$$v = \frac{\sqrt{21}\pi}{2} = \text{met/sec}.$$

**Example 14.8.** *Three successive seconds, the distance of a point moving in simple harmonic motion from its mean position are $x_1$, $x_2$, $x_3$ respectively. Show that the time of a complete oscillation is* $2\pi / \cos^{-1}\left(\frac{x_1 + x_3}{2x_2}\right)$.

*(AU, 2018)*

**Solution:** Let the equation of SHM be

$$x_1 = a \cos \mu t \qquad ...(i)$$
$$x_2 = a \cos [\mu(t + 1)] \qquad ...(ii)$$
$$x_3 = a \cos [\mu(t + 2)] \qquad ...(iii)$$

From Eqns. (*ii*) and (*iii*) $x_1 + x_3 = a[\cos \mu t + \cos \{\mu(t + 2)\}$

$$= 2a \cos\frac{\mu t + \mu t + 2\mu}{2} \cos\frac{\mu t - \mu t - 2\mu}{2}$$

$$= 2a \cos \{\mu(t + 1)\} \cos \mu \qquad \{\because \cos(-\mu) = \cos \mu\}$$

$$x_1 + x_3 = 2x_2 \cos \mu$$

$$\mu = \cos^{-1}\left(\frac{x_1 + x_3}{2x_2}\right)$$

Period of complete oscillation $T = \frac{2\pi}{\mu} = \frac{2\pi}{\cos^{-1}\left(\frac{x_1 + x_3}{2x_2}\right)}$.

**Example 14.9.** *A spring which stretches by an amount e under a force $m\lambda^2 e$ is suspended from a support O and has a mass m at the lower end. Initially the mass is at rest in its equilibrium position at a point A below O. A vertical oscillation is now given to the support O such that at any time (t > 0) its displacement below its initial position is a sin nt. Show that the displacement x of the mass below A is given by*

$$d^2x/dt^2 + \lambda^2 x = \lambda^2 a \sin nt.$$

*Hence show that if $n \neq \lambda$, the displacement is given by $x = \lambda a (\lambda \sin nt - n \sin \lambda t)/(\lambda^2 - n^2)$. What happens when $n = \lambda$ ?*

**Solution.** If $k$ be the stiffness of the spring then $m\lambda^2 e = ke$ *i.e.,* $k = m\lambda^2$.

Also in equilibrium $mg = ke$ ...(*i*)

Initially the mass is in equilibrium at $A$ (Fig. 14.7). At time $t$, the support $P$ is given a downward displacement $a \sin nt$. If the mass is displaced through a further distance $x$ from $A$, then the equation of motion of the mass is given by

$$m\frac{d^2x}{dt^2} = mg - k(x+e) + ka \sin nt$$

$$= -kx + ka \sin nt \qquad \text{[By Eqn. } (i)]$$

or $$\frac{d^2x}{dt^2} + \lambda^2 x = \lambda^2 a \sin nt \qquad [\because \; k = m\lambda^2]$$

or $$(D^2 + \lambda^2)x = \lambda^2 a \sin nt \qquad ...(ii)$$

Its A.E. $= c_1 \cos \lambda t + c_2 \sin \lambda t$

$$\text{P.I.} = \frac{1}{D^2+\lambda^2}\lambda^2 a \sin nt.$$

Now two cases arise :

**Case I.** *When* $n \neq \lambda$

$$\text{P.I.} = \lambda^2 a \frac{1}{n^2+\lambda^2}\sin nt$$

$\therefore$ the complete solution of Eqn. (*ii*) is $x = c_1 \cos \lambda t + c_2 \sin \lambda t + \dfrac{\lambda^2 a}{\lambda^2 - n^2}\sin nt$ ...(*iii*)

$$\therefore \qquad \frac{dx}{dt} = -c_1\lambda \sin \lambda t + c_2 \lambda \cos \lambda t + \frac{\lambda^2 an}{\lambda^2-n^2}\cos nt$$

Initially when $t = 0$, $x = 0$ and $dx/dt = 0$.

$\therefore \quad c_1 = 0$ and $0 = c_2\lambda + \lambda^2 an/(\lambda^2 - n^2)$ *i.e.*, $c_2 = \lambda an/(\lambda^2 - n^2)$

Thus, substituting the values of $c_1$ and $c_2$ in Eqn. (*iii*), we have

$$x = -\frac{\lambda an}{\lambda^2-n^2}\sin \lambda t + \frac{\lambda^2 a}{\lambda^2-n^2}\sin nt = \frac{\lambda a}{\lambda^2-n^2}(\lambda \sin nt - n \sin \lambda t)$$

**Case II.** *When* $n = \lambda$

$$\text{P.I.} = \lambda^2 a \frac{1}{D^2+\lambda^2}\sin nt = \lambda^2 at \cdot \frac{1}{2D}\sin \lambda t = \frac{\lambda^2 at}{2}\int \sin \lambda t \, dt = -\frac{\lambda at}{2}\cos \lambda t$$

$\therefore$ the complete solution is

$$x = c_1 \cos \lambda t + c_2 \sin \lambda t - \frac{\lambda at}{2}\cos \lambda t \qquad ...(iv)$$

$$\therefore \qquad \frac{dx}{dt} = -c_1\lambda \sin \lambda t + c_2\lambda \cos \lambda t + \frac{\lambda^2 at}{2}\sin \lambda t - \frac{\lambda a}{2}\cos \lambda t$$

When $t = 0$, $x = 0$ and $dx/dt = 0$

$\therefore \quad 0 = c_1$ and $0 = c_2\lambda - \lambda a/2$ *i.e.*, $c_2 = a/2$.

Thus, substituting the values of $c_1$ and $c_2$ in Eqn. (*iv*), we get

$$x = \frac{a}{2}\sin \lambda t - \frac{\lambda at}{2}\cos \lambda t$$

$$= \frac{a}{2}(\sin \lambda t - \lambda t \cos \lambda t) \qquad [\text{Put } 1 = r \cos \phi \text{ and } \lambda t = r \sin \phi]$$

$$= \frac{ar}{2}\sin(\lambda t - \phi)$$

Its amplitude $\left(\dfrac{ar}{2}\right) = \dfrac{a}{2}\sqrt{(1+\lambda^2 t^2)}$, which increases with time. Hence the phenomenon of *resonance* occurs.

**Example 14.10.** *A particle is moving in a straight line and when the distance of the particle from the equilibrium position has values $x_1$ and $x_2$. The corresponding velocities are $u_1$ and $u_2$ respectively. Show that the period of oscillation is $2\pi\{(x_2^2 - x_1^2)/(u_1^2 - u_2^2)\}^{1/2}$ find (i) maximum velocity (ii) amplitude.*

(*Marathwada 2018*)

**Solution:** Let the equation of simple harmonic motion be $x = a \cos \mu t$ ...(*i*)

If $u$ is the velocity of the particle at any time $t$, then

$$u = \frac{dx}{dt} = -\mu a \sin \mu t \quad \text{...(ii)}$$

from Eqns. (*i*) and (*ii*)

$$\frac{x}{a} = \cos \mu t$$

$$\frac{\mu}{\mu a} = -\sin \mu t$$

Squaring and adding

$$\frac{x^2}{a^2} + \frac{\mu^2}{\mu^2 a^2} = 1 \quad \text{...(iii)}$$

when $x = x_1$, $u = u_1$ and when $x = x_2$, $u = u_2$ substituting these values in Eqn. (*iii*)

$$\frac{x_1^2}{a^2} + \frac{u_1^2}{\mu^2 a^2} = 1 \quad \text{...(iv)}$$

and

$$\frac{x_2^2}{a^2} + \frac{u_2^2}{\mu^2 a^2} = 1 \quad \text{...(v)}$$

Subtracting Eqn. (*v*) from Eqn. (*iv*)

$$\frac{x_1^2 - x_2^2}{a^2} + \frac{u_1^2 - u_2^2}{\mu^2 a^2} = 0$$

$$\mu^2 = \frac{u_1^2 - u_2^2}{x_2^2 - x_1^2}$$

$$\mu = \left(\frac{u_1^2 - u_2^2}{x_2^2 - x_1^2}\right)^{1/2}$$

Hence time period,

$$T = \frac{2\pi}{\mu} = 2\pi \Big/ \left(\frac{u_1^2 - u_2^2}{x_2^2 - x_1^2}\right)^{1/2} = 2\pi\left\{(x_2^2 - x_1^2) - (u_1^2 - u_2^2)\right\}^{1/2}$$

From Eqn. (*iv*), the amplitude $a$ is give by

$$a = \left(x_1^2 + \frac{u_1^2}{\mu^2}\right)^{1/2}$$

Substituting value of $\mu^2$, we get

$$a = \left\{x_1^2 + u_1^2\left(\frac{x_2^2 - x_1^2}{u_1^2 - u_2^2}\right)\right\}^{1/2}$$

$$a = \left\{\frac{u_1^2 x_2^2 - u_2^2 x_1^2}{u_1^2 - u_2^2}\right\}^{1/2}$$

Maximum velocity can be obtained from Eqn. (*ii*) when $\sin \mu t = 1$

$$= \left(\frac{u_1^2 - u_2^2}{x_2^2 - x_1^2}\right)^{1/2} \cdot \left(\frac{u_1^2 x_2^2 - u_2^2 x_1^2}{u_1^2 - u_2^2}\right)^{1/2} = \left(\frac{u_1^2 x_2^2 - u_2^2 x_1^2}{x_1^2 - x_2^2}\right)^{1/2}$$

## PROBLEMS 14.2

1. An elastic string of natural length $a$ is fixed at one end and a particle of mass $m$ hangs freely from the other end. The modulus of elasticity is $mg$. The particle is pulled down a further distance $l$ below its equilibrium position and released from rest. Show that the motion of the particle is simple harmonic and find the pariodicity.
2. A mass of 4 lb suspended from a light elastic string of natural length 3 feet extends it to a distance 2 feet. One end of the string is fixed and a mass of 2 lb is attached to other. The mass is held so that the string is just unstretched and is then let go. Find the amplitude, the period and the maximum velocity of the ensuing simple harmonic motion.
3. A light elastic string of natural length $l$ has one extremity fixed at a point $A$ and the other end attached to a stone, the weight of which in equilibrium would extend the string to a depth $l_1$. Show that if the stone be dropped from rest at $A$, it will come to instantaneous rest at a depth $\sqrt{(l_1^2 - l^2)}$ below the equilibrium position.
4. A 4 lb weight on a string stretches it 6 in. Assuming that a damping force in lb wt. equal to $\lambda$ times the instantaneous velocity in ft/sec. acts on the weight, show that the motion is over damped, critically damped or oscillatory according as $\lambda > = < 2$. Find the period of oscillation when $\lambda = 1.5$.
5. A mass of 200 gm is tied at the end of a spring which extends to 4 cm under a force 196,000 dynes. The spring is pulled 5 cm and released. Find the displacement $t$ seconds after release if there be a damping force of 2000 dynes per cm per second.
6. A body weighing 16 lb is suspended by a spring in a fluid whose resistance in lb wt. is twice the speed of the body in ft/sec. A pull of 25 lb wt. would stretch the spring 3 inches. The body is drawn 3 inches below the equilibrium position in the fluid and then released. Find the period of oscillations and the time required for the damping factor to be reduced to one-tenth of its initial value.
7. A mass $M$ suspended from the end of a helical spring is subjected to a periodic force $f = F \sin \omega t$ in the direction of its length. The force $f$ is measured positive vertically downwards and at zero time $M$ is at rest. If the spring stiffness is $S$, prove that the displacement of $M$ at time $t$ from the commencement of motion is given by
$$x = \frac{F}{M(p^2 - \omega^2)}\left[\sin \omega t - \frac{\omega}{p}\sin pt\right]$$
where, $p^2 = S/M$ and damping effects are neglected.
8. A vertical spring having 4.5 lb/ft. has 16 lb wt. suspended from it. An external force of 24 sin $9t$ ($t \geq 0$) lb wt. is applied. A damping force given numerically in lb. wt. by four times its velocity in ft/sec, is assumed to act. Initially the weight is at rest at its equilibrium position. Determine the position of the weight at any time. Also find the amplitude, period and the frequency of the steady-state solution.
9. A body weighing 4 lb hangs at rest on a spring producing in the spring an extension of 1ft. The upper end of the spring in now made to execute a vertical simple harmonic oscillation $x = \sin 4t$, $x$ being measured vertically downwards in feet. If the body is subject to a frictional resistance whose magnitude in lb wt. is one-quarter of its velocity in feet per second, obtain the differential equation for the motion of the body and find the expression for its displacement at time $t$, when $t$ is large.
10. A body executes damped forced vibrations given by the equation
$$\frac{d^2x}{dt^2} + 2k\frac{dx}{dt} + b^2x = e^{-kt}\sin nt.$$
Solve the equation for both the cases, when $n^2 \neq b^2 - k^2$ and $n^2 = b^2 - k^2$.

## 14.5 OSCILLATORY ELECTRICAL CIRCUIT

(*i*) **L-C circuit**

Consider an electrical circuit containing an inductance $L$ and capacitance $C$ (Fig. 14.14).

Let $i$ be the current and $q$ the charge in the condenser plate at any time $t$, so that the voltage drop across

$$L = L\frac{di}{dt} = L\frac{d^2q}{dt^2}$$

and the voltage drop across $C = q/C$.

As there is no applied e.m.f. in the circuit, therefore, by Kirchhoff's first law, we have

$$L\frac{d^2q}{dt^2} + \frac{q}{C} = 0.$$

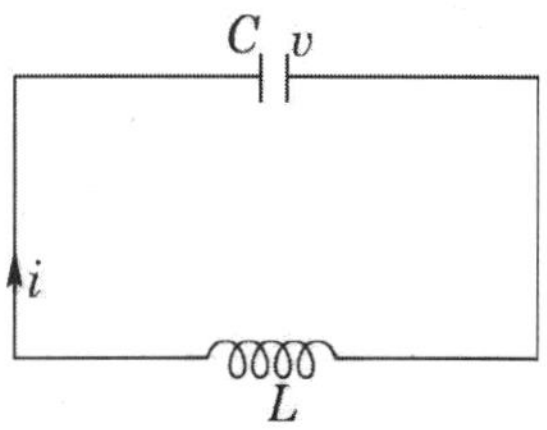

Fig. 14.14.

Or dividing by $L$ and writing $1/LC = \mu^2$, we get $\dfrac{d^2q}{dt^2} + \mu^2 q = 0$ ...(1)

This equation is precisely same as (2) on page 525 and, therefore, *it represents free electrical oscillations of the current having period* $2\pi/\mu = 2\pi\sqrt{LC}$.

*Thus the discharging of a condenser through an inductance L is same as the motion of the mass m at the end of a spring.*

(*ii*) **L–C–R circuit**

Now consider the discharge of a condenser $C$ through an inductance $L$ and the resistance $R$ (Fig. 14.15). Since the voltage drop across $L$, $C$ and $R$ are respectively

$$L\frac{d^2q}{dt^2}, \frac{q}{C} \text{ and } R\frac{dq}{dt}.$$

$\therefore$ by Kirchhoff's law, we have $L\dfrac{d^2q}{dt^2} + R\dfrac{dq}{dt} + \dfrac{q}{C} = 0$ ...(2)

Or writing $R/L = 2\lambda$ and $1/LC = \mu^2$, we have $\dfrac{d^2q}{dt^2} + 2\lambda\dfrac{dq}{dt} + \mu^2 q = 0$

This equation is same as Eqn. (3) and, therefore has the same solution as for the mass $m$ on a spring with a damper.

Thus *the charging or discharging of a condenser through the resistance R and an inductance L is an electrical analogue of the damped oscillations of mass m on a spring.*

(*iii*) **L–C circuit with e.m.f.** = *p cos nt*.

The Eqn. (1) for an $L$–$C$ circuit (Fig. 14.16), now becomes $L\dfrac{d^2q}{dt^2} + \dfrac{q}{C} = p \cos nt$.

Or writing $1/LC = \mu^2$, we have $\dfrac{d^2q}{dt^2} + \mu^2 q = \dfrac{p}{L}\cos nt$ ...(3)

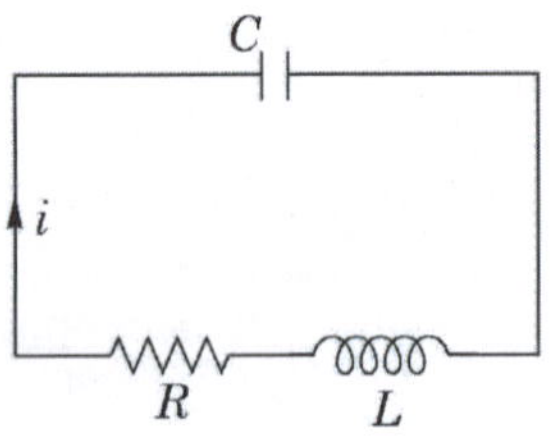


Fig. 14.15

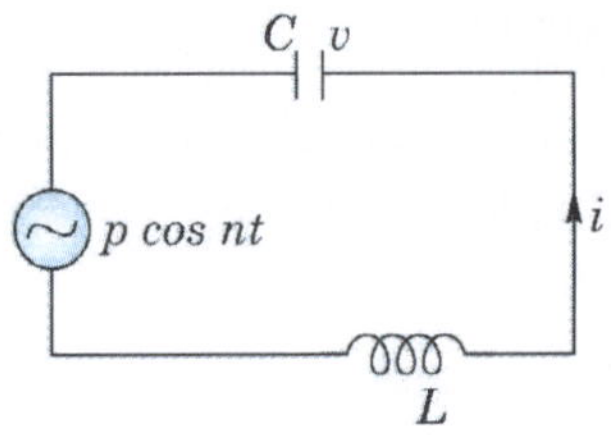


Fig. 14.16

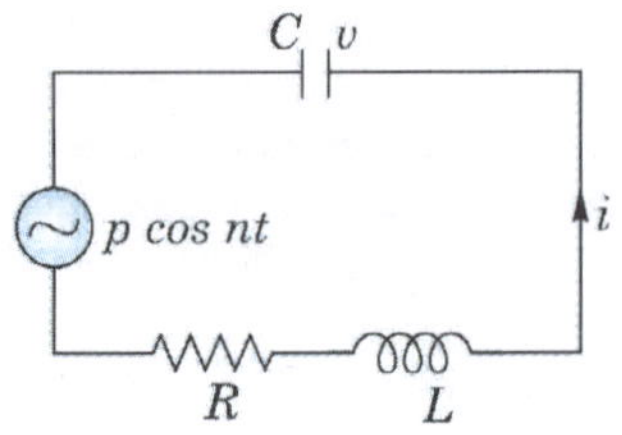


Fig. 14.17

This equation is of the same form as Eqn. (9) and, therefore, has the solution as for the motion of a mass $m$ on a spring with external periodic force $p \cos nt$ acting on it.

*Thus the condenser placed in series with source of e.m.f. (= p cos nt) and discharging through a coil containing inductance L is an electrical analogue of the forced oscillations of the mass m on a spring.*

An electrical instance of *resonance* phenomena occurs while tuning a radio-station, for the natural frequency of the tuning of $L$–$C$ circuit is made equal to the frequency of the desired radio-station, giving the maximum output of the receiver at the said receiving station.

(*iv*) **L–C–R circuit with e.m.f.** = *p cos nt*.

The equation of Eqn. (2) above, now becomes $L\dfrac{d^2q}{dt^2} + R\dfrac{dq}{dt} + \dfrac{q}{C} = p \cos nt$. (Fig. 14.17)

Or writing $R/L = 2\lambda$ and $1/LC = \mu^2$ as before, we have

$$\frac{d^2q}{dt^2} + 2\lambda\frac{dq}{dt} + \mu^2 q = \frac{p}{L}\cos nt \qquad ...(4)$$

This equation is exactly same as Eqn. (12) on and, therefore, its C.F. represents the free oscillations of the circuit whereas the P.I. represents the forced oscillations.

Here also as $t$ increases, the free oscillations die out while the forced oscillations persist giving steady motion.

*Thus the L–C–R circuit with a source of alternating e.m.f. is an electrical equivalent of the mechanical phenomena of forced oscillations with resistance.*

## 14.6 ELECTRO-MECHANICAL ANALOGY

We have just seen, how merely by renaming the variables, the differential equation representing the oscillation of a weight on a spring represents an analogous electrical circuit. As electrical circuits are easy to assemble and the currents and voltages are accurately measured with ease, this affords a practical method of studying the oscillations of complicated mechanical systems which are expensive to make and unwieldy to handle by considering an equivalent electrical circuit. While making an electrical equivalent of a mechanical system, the following correspondences between the elements should be kept in mind, noting that the circuit may be in series or in parallels :

| *Mech. System* | *Series circuit* | *Parallel circuit* |
|---|---|---|
| Displacement | Current $i$ | Voltage $E$ |
| Force or couple | Voltage $E$ | Current $i$ |
| Mass $m$ or $M.I.$ | Inductance $L$ | Capacitance $C$ |
| Damping force | Resistance $R$ | Conductance $1/R$ |
| Spring modulus | Elastance $1/C$ | Susceptance $1/L$ |

**Example 14.11.** *An uncharged condenser of capacity C is charged by applying an e.m.f. E sin* $t/\sqrt{(LC)}$, *through leads of self-inductance L and negligible resistance. Prove that at any time t, the charge on one of the plates is* $\frac{EC}{2}\left\{\sin\frac{t}{\sqrt{(LC)}}-\frac{t}{\sqrt{(LC)}}\cos\frac{t}{\sqrt{(LC)}}\right\}$

**Solution.** If $q$ be the change on the condenser, the differential equation of the circuit is

$$L\frac{d^2q}{dt^2}+\frac{q}{C}=E\sin\frac{t}{\sqrt{(LC)}} \qquad ...(i)$$

Its A.E. is $\quad LD^2+1/C=0$ or $D=\pm 1/\sqrt{(LC)}$

$\therefore \quad$ C.F. $= c_1\cos t/\sqrt{LC}+c_2\sin t/\sqrt{(LC)}$

and $\quad$ P.I. $=\dfrac{1}{LD^2+\dfrac{1}{C}}E\sin\dfrac{t}{\sqrt{(LC)}}$ $\qquad \left[\text{Putting } D^2=-\dfrac{1}{LC},\text{ denom.}=0\right]$

$$=Et\frac{1}{2LD}\sin\frac{t}{\sqrt{(LC)}}=\frac{Et}{2L}\int\sin\frac{t}{\sqrt{(LC)}}dt=-\frac{Et}{2L}\sqrt{(LC)}\cos\frac{t}{\sqrt{(LC)}}=-\frac{Et}{2}\sqrt{\frac{C}{L}}\cos\frac{t}{\sqrt{LC}}$$

Thus the C.S. of (*i*) is $q=c_1\cos\dfrac{t}{\sqrt{LC}}+c_2\sin\dfrac{t}{\sqrt{LC}}-\dfrac{Et}{2}\sqrt{\dfrac{C}{L}}\cos\dfrac{t}{\sqrt{LC}}$

When $t=0, q=0, c_1=0$

$$\therefore \quad q=c_2\sin\frac{t}{\sqrt{LC}}-\frac{Et}{2}\sqrt{\frac{C}{L}}\cos\frac{t}{\sqrt{LC}} \qquad ...(ii)$$

Differentiating Eqn. (*ii*) w.r.t. $t$, we get

$$\frac{dq}{dt}=\frac{c_2}{\sqrt{LC}}\cos\frac{t}{\sqrt{LC}}-\frac{E}{2}\sqrt{\frac{C}{L}}\left\{\cos\frac{t}{\sqrt{LC}}-\frac{t}{\sqrt{LC}}\sin\frac{t}{\sqrt{LC}}\right\}$$

Also when $t=0, dq/dt=i=0$,

$$\therefore \quad \frac{c_2}{\sqrt{LC}}-\frac{E}{2}\sqrt{\frac{C}{L}}=0 \quad\text{or}\quad c_2=\frac{EC}{2}.$$

Substituting the value of $c_2$ in Eqn. (*ii*), $q$ at any time $t$ is given by

$$q=\frac{EC}{2}\left\{\sin\frac{t}{\sqrt{LC}}-\frac{t}{\sqrt{LC}}\cos\frac{t}{\sqrt{LC}}\right\}.$$

**Example 14.12.** *In an L—C—R circuit, the charge q on a plate of a condenser is given by*

$$L\frac{d^2q}{dt^2}+R\frac{dq}{dt}+\frac{q}{C}=E\sin pt.$$

*The circuit is tuned to resonance so that* $p^2=1/LC$. *If initially the current i and the charge q be zero, show that, for small values of R/L, the current in the circuit at time t is given by*

$(Et/2L)\sin pt.$ $\qquad$ (*JNTUK, 2021 ; P.T.U., 2014 S*)

**Solution.** Given differential equation is $(LD^2 + RD + 1/C)q = E \sin pt$ ...(i)

Its auxiliary equation is $LD^2 + RD + 1/C = 0$,

which gives $$D = \frac{1}{2L}\left[-R \pm \sqrt{\left(R^2 - \frac{4L}{C}\right)}\right] = -\frac{R}{2L} + \sqrt{\left(\frac{R^2}{4L^2} - \frac{1}{LC}\right)}$$

As $R/L$ is small, therefore, to the first order in $R/L$,

$$D = -\frac{R}{2L} \pm i\frac{1}{\sqrt{(LC)}} = -\frac{R}{2L} \pm ip \qquad \left[\because\ p^2 = \frac{1}{LC}\right]$$

$\therefore$ C.F. $= e^{-(Rt/2L)}(c_1 \cos pt + c_2 \sin pt)$

$= (1 - Rt/2L)(c_1 \cos pt + c_2 \sin pt)$ rejecting terms in $(R/L)^2$ etc.

and $$\text{P.I.} = \frac{1}{LD^2 + RD + 1/C} E \sin pt = E\frac{1}{-Lp^2 + RD + 1/C} \sin pt$$

$$= \frac{E}{R}\int \sin pt\, dt = -\frac{E}{Rp} \cos pt \qquad \left[\because p^2 = \frac{1}{LC}\right]$$

Thus the complete solution of Eqn. (i) is $q = \left(1 - \frac{Rt}{2L}\right)(c_1 \cos pt + c_2 \sin pt) - \frac{E}{Rp}\cos pt$ ...(ii)

$$\therefore \quad i = \frac{dq}{dt} = \left(1 - \frac{Rt}{2L}\right)(-c_1 \sin pt + c_2 \cos pt)\, p - \frac{R}{2L}(c_1 \cos pt + c_2 \sin pt) + \frac{E}{R}\sin pt \quad ...(iii)$$

Initially, when $t = 0, q = 0, i = 0$ $\therefore$ from Eqn. (ii), $0 = c_1 - E/Rp$ $\therefore$ $c_1 = E/Rp$ and from Eqn. (iii),

$0 = c_2 p - Rc_1/2L$ $\therefore$ $c_2 = Rc_1/2Lp = E/2Lp^2$

Thus, substituting these values of $c_1$ and $c_2$ in Eqn. (iii), we get

$$i = \left(1 - \frac{Rt}{2L}\right)\left(-\frac{E}{Rp}\sin pt + \frac{E}{2Lp^2}\cos pt\right)p - \frac{R}{2L}\left(\frac{E}{Rp}\cos pt + \frac{E}{2Lp^2}\sin pt\right) + \frac{E}{R}\sin pt$$

$$= \frac{Et}{2L}\sin pt. \qquad [\because\ R/L \text{ is small}]$$

## PROBLEMS 14.3

1. Show that the frequency of free vibrations in a closed electrical circuit with inductance $L$ and capacity $C$ in series is $\frac{30}{\pi\sqrt{(LC)}}$ per minute. *(Rohtak, 2011)*
2. An inductance of 2 heneries and a resistance of 20 ohms are connected in series with e.m.f. $E$ volts. If the current is zero when $t = 0$, find the current at the end of 0.01 seconds if $E = 100$ volts. *(P.T.U., 2015)*
3. The differential equation for a circuit in which self-inductance and capacitance neutralize each other is $L\frac{d^2i}{dt^2} + \frac{i}{C} = 0$. Find the current $i$ as a function of $t$ given that $I$ is the maximum current, and $i = 0$ when $t = 0$.
4. A constant e.m.f. $E$ at $t = 0$ is applied to a circuit consisting of inductance $L$, resistance $R$ and capacitance $C$ in series. The initial values of the current and the charge being zero, find the current at any time $t$, if $CR^2 < 4L$. Show that the amplitudes of the successive vibrations are in geometrical progression.
5. The damped LCR circuit is governed by the equation $L\frac{d^2q}{dt^2} + R\frac{dq}{dt} + \frac{q}{C} = 0$ where, $L, R, C$ are positive constants. Find the conditions under which the circuit is over damped, under damped and critically damped. Find also the critical resistance.
6. A condenser of capacity $C$ discharged through an inductance $L$ and resistance $R$ in series and the charge $q$ at time $t$ satisfies the equation $L\frac{d^2q}{dt^2} + R\frac{dq}{dt} + \frac{q}{C} = 0$. Given that $L = 0.25$ heneries, $R = 250$ ohms, $C = 2 \times 10^{-6}$ farads, and that when $t = 0$, charge $q$ is 0.002 coulombs and the current $dq/dt = 0$, obtain the value of $q$ in terms of $t$.
7. An e.m.f. $E \sin pt$ is applied at $t = 0$ to a circuit containing a capacitance $C$ and inductance $L$. The current $i$ satisfies the equation $L\frac{di}{dt} + \frac{1}{C}\int i\, dt = E \sin pt$. If $p^2 = 1/LC$ and initially the current $i$ and the charge $q$ are zero, show that the current at time $t$ is $(Et/2L) \sin pt$, where $i = dq/dt$. *(P.T.U., 2014 S)*

8. For an $L$—$R$—$C$ circuit, the charge $q$ on a plate of the condenser is given by $L\frac{d^2q}{dt^2}+R\frac{dq}{dt}+\frac{q}{C}=E\sin\omega t$, where $i=\frac{dq}{dt}$. The circuit is tuned to resonance so that $\omega^2 = 1/LC$.

If $CR^2 < 4L$ and initially $q = 0$, $i = 0$, show that $q=\frac{E}{R\omega}\left[e^{-Rt/2C}\left(\cos pt+\frac{R}{2Lp}\sin pt\right)-\cos\omega t\right]$

where $$p^2=\frac{1}{LC}-\frac{R^2}{4L^2}.$$

9. An alternating E.M.F. $E\sin pt$ is applied to a circuit at $t = 0$. Given the equation for the current $i$ as $L\frac{d^2i}{dt^2}+R\frac{di}{dt}+\frac{i}{C}=pE\cos pt$, find the current $i$ when (*i*) $CR^2 > 4L$, (*ii*) $CR^2 < 4L$.

10. The equation of motion of a point in the end of a tunning fork for small amplitude of motion is given by

$$\frac{d^2x}{dt}+p^2x=0$$

The fork can be driven by a varying magnetic field so that the equation of forced vibration will be of the form

$$\frac{d^2x}{dt^2}+p^2x=k\sin\omega t$$

Solve the equation for forced vibration when $p \neq \omega$ and also for $p = \omega$. Discuss the physical significance of the result.

(*Shimla University, 2018*)

11. In the case of stretched horizontal string which has one end fixed and a particle of mass $m$ attached to the other end.

Prove that the equation of motion is $e\frac{d^2S}{dt^2}=-g(S-l)$.

where $l$ is the natural length of the string and $e$ is elongation due to a weight mg. (*Kerala, Univ. 2013*)

Also find $s$ given that $S = S_o$, $u = o$, when $t = 0$.

12. A particle of mass $m$ is attracted by each of two centres with a force proportional to the distance. Find the law of motion if the distance between the centre is $2b$. Initially, the particle was located on the line connecting the centres at a distance $c$ from the midpoint and had a velocity zero.

## 14.7 DEFLECTION OF BEAMS

Consider a uniform beam as made up of fibres running lengthwise. We have to find its deflection under given loadings.

In the bent form, the fibres of the lower half are stretched and those of upper half are compressed. In between these two, there is a layer of unstrained fibres called the *neutral surface*. The fibre which was initially along the $x$-axis (the central horizontal axis of the beam) now lies in the neutral surface, in the form of a curve called the *deflection curve* or the *elastic curve*. We shall encounter differential equations while finding the equation of this curve.

Consider a cross-section of the beam cutting the elastic curve in $P$ and the neutral surface in the line $AA'$— called the neutral axis of this section (Fig. 14.18).

It is well-known from mechanics that the bending moment $M$ about $AA'$, of all forces acting on either side of the two portions of the beam separated by this cross-section, is given by the *Bernoulli-Euler law*

$$M = EI/R$$

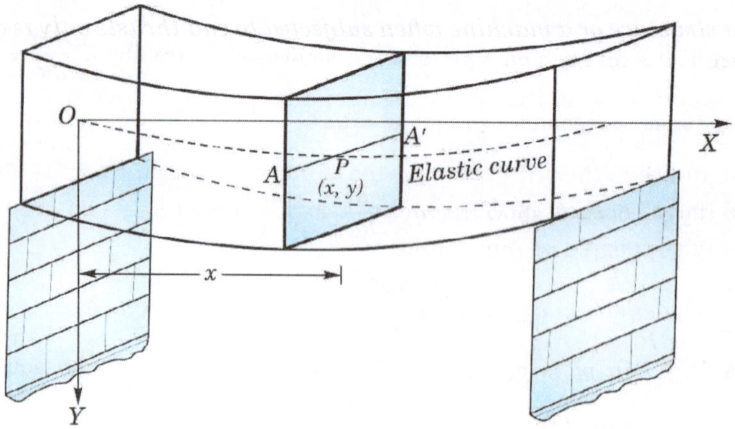


Fig. 14.18.

where, $E$ = modulus of elasticity of the beam,

$I$ = moment of inertia of the cross-section about $AA'$,

and $R$ = radius of curvature of the elastic curve at $P(x, y)$.

If the deflection of the beam is small, the slope of the elastic curve is also small so that we may neglect $(dy/dx)^2$ in the formula,

$$R = \left[1 + \left(\frac{dy}{dx}\right)^2\right]^{3/2} \bigg/ \frac{d^2y}{dx^2}$$. Thus for small deflections, $R = 1/(d^2y/dx^2)$.

Hence (1) *Bending moment* $\mathbf{M} = \mathbf{EI}\,\dfrac{\mathbf{d^2y}}{\mathbf{dx^2}}$

(2) *Shear force* $\left(= \dfrac{dM}{dx}\right) = EI\,\dfrac{d^3y}{dx^3}$;

(3) *Intensity of loading* $\left(= \dfrac{d^2M}{dx^2}\right) = EI\,\dfrac{d^4y}{dx^4}$

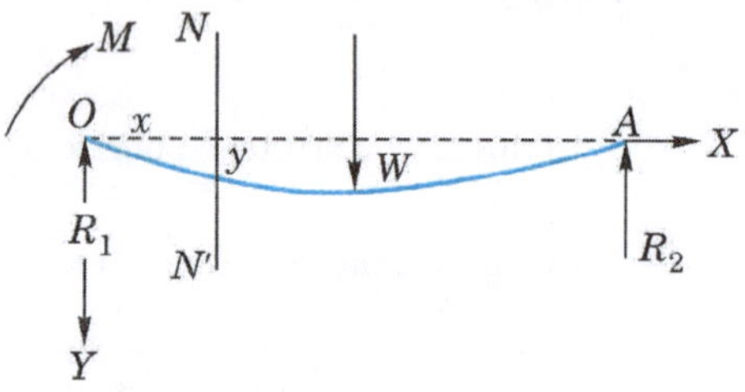


Fig. 14.19.

(4) *Convention of signs.* The sum of the moments about a section $NN'$ due to external forces on the left of the section, if anti-clockwise is taken as positive and if clockwise (as in Fig. 14.19) is taken as negative.

The deflection $y$ downwards and length $x$ to the right are taken as positive. The slope $dy/dx$ will be positive if downwards in the direction of $x$-positive.

(5) *End conditions.* The arbitrary constants appearing in the solution of the differential Eqn. (1) for a given problem are found from the following end conditions :

(*i*) *At a freely supported end* (Fig. 14.20), there being no deflection and no bending moment, we have $y = 0$ and $d^2y/dx^2 = 0$.

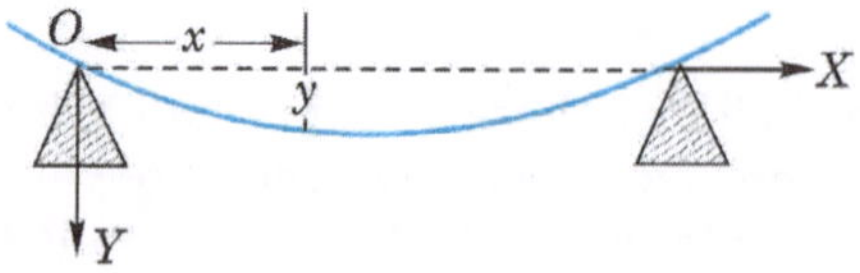


Fig. 14.20.

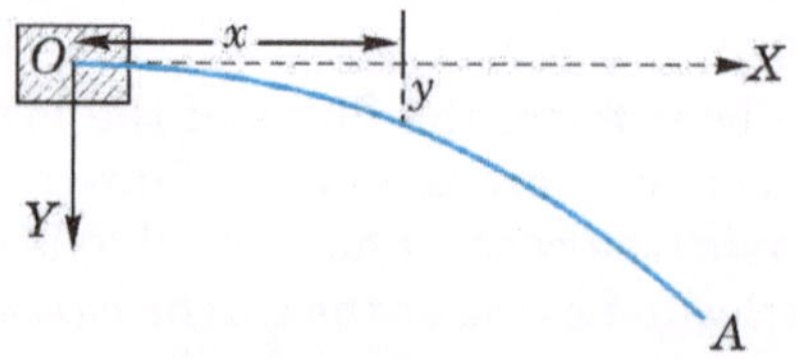


Fig. 14.21.

(*ii*) *At a (horizontal) fixed end* (Fig. 14.21), the deflection and the slope of the beam being both zero, we have

$$y = 0 \text{ and } dy/dx = 0.$$

(*ii*) *At a perfectly free end* (*A* in Fig. 14.24), there being no bending moment or shear force, we have

$$\frac{d^2y}{dx^2} = 0 \quad \text{and} \quad \frac{d^3y}{dx^3} = 0$$

(6) *A member of a structure or a machine when subjected to end thrusts only is called a* **strut** *and a vertical strut is called a* **column**.

There are four possible ways of the end fixation of a strut:

(*i*) Both ends fixed, called a *built-in* or *encastre* strut.

(*ii*) One end fixed and the other freely supported, hinged or pin-jointed.

(*iii*) One end fixed and the other end free, called a *cantilever.*

(*iv*) Both ends freely supported or pin-jointed.

**Example 14.13.** *The deflection of a strut of length l with one end (x = 0) built-in and the other supported and subjected to end thrust P, satisfies the equation*

$$\frac{d^2y}{dx^2} + a^2 y = \frac{a^2 R}{P}(l - x).$$

*Prove that the deflection curve is* $y = \frac{R}{P}\left(\frac{\sin ax}{a} - l\cos ax + l - x\right)$, *where* $al = \tan al$.

**Solution.** Given differential equation is $(D^2 + a^2)y = \frac{a^2R}{P}(l - x)$ ...(*i*)

Its auxiliary equation is $D^2 + a^2 = 0$, whence $D = \pm\ ai$.

$$\therefore \qquad C.F. = \frac{1}{D^2 + a^2}\frac{a^2R}{P}(l - x) = \frac{R}{P}\left(1 + \frac{D^2}{a^2}\right)^{-1}(l - x)$$

$$= \frac{R}{P}\left(1 - \frac{D^2}{a^2} + \ldots\right)(l - x) = \frac{R}{P}(l - x)$$

Thus the complete solution of (*i*) is $y = c_1 \cos ax + c_2 \sin ax + \frac{R}{P}(l - x)$ ...(*ii*)

Also $\qquad \frac{dy}{dx} = -c_1 a \sin ax + c_2 a \cos ax - \frac{R}{P}$ ...(*iii*)

Now as the end $O$ is built in (Fig. 14.22). $\therefore\ y = dy/dx = 0$ at $x = 0$.

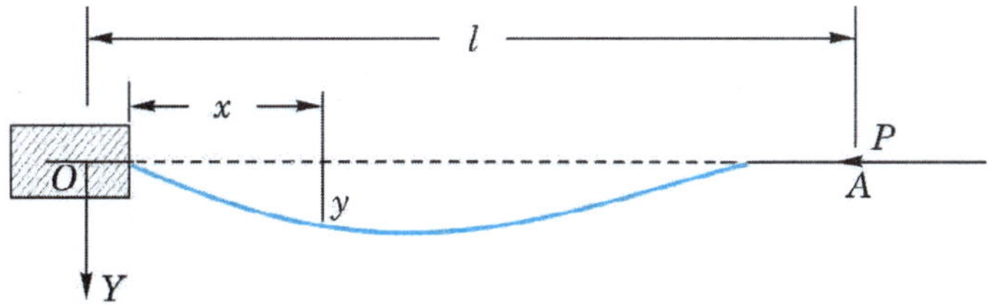


Fig. 14.22.

$\therefore$ from Eqns. (*ii*) and (*iii*), we have

$$0 = c_1 + Rl/P \text{ and } 0 = c_2 a - R/P$$

whence $\qquad c_1 = -Rl/P \quad \text{and} \quad c_2 = R/aP$

Thus Eqn. (*ii*) becomes $\quad y = \frac{R}{P}\left(\frac{\sin ax}{a} - l\cos ax + l - x\right)$ ...(*iv*)

which is the desired equation of the deflection curve.

The end $A$ being freely supported $y = 0$ when $x = l$ (We don't need the other condition $d^2y/dx^2 = 0$].

$\therefore$ Eqn. (*iv*) gives $\qquad 0 = \frac{R}{P}\left(\frac{\sin al}{a} - l\cos al\right)$ whence $al = \tan al$.

**Example 14.14.** *A horizontal tie-rod is freely pinned at each end. It carries a uniform load w lb per unit length and has a horizontal pull P. Find the central deflection and the maximum bending moment, taking the origin at one of its ends.*

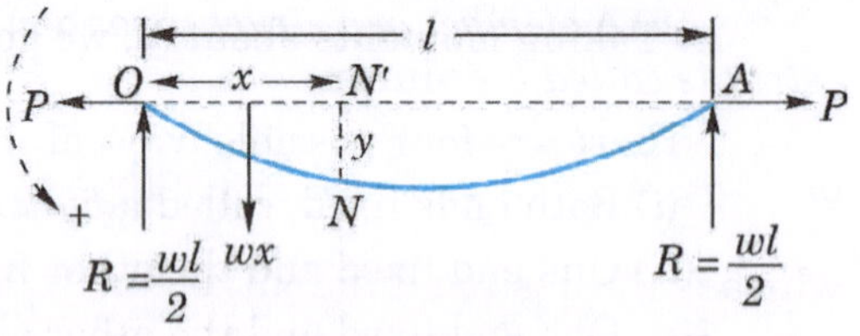

Fig. 14.23.

**Solution.** Let $OA$ be the given beam of length $l$ (Fig.14.23).

At each end there is a vertical reaction $R = wl/2$.

The external forces acting to the left of the section $NN'$ are :

(*i*) the horizontal pull $P$, (*ii*) the reaction $R = wl/2$ and (*iii*) the weight of the portion $ON = wx$ acting mid-way.

Taking moments about, $N$, we have

$$EI\frac{d^2y}{dx^2} = Py - \frac{wl}{2}x + wx.\frac{x}{2}$$

or $$EI\frac{d^2y}{dx^2} - Py = \frac{w}{2}(x^2 - lx) \quad \text{or} \quad \frac{d^2y}{dx^2} - a^2y = \frac{w}{2EI}(x^2 - lx), \text{ where } a^2 = \frac{P}{EI} \qquad ...(i)$$

This is the differential equation of the elastic curve. Its auxiliary equation is $D^2 - a^2 = 0$, whence $D = \pm a$.

$\therefore$ $$\text{C.F.} = c_1 \cosh ax + c_2 \sinh ax$$

and $$\text{P.I.} = \frac{1}{D^2 - a^2}\frac{w}{2EI}(x^2 - lx) = \frac{-w}{2EIa^2}\left(1 - \frac{D^2}{a^2}\right)^{-1}(x^2 - lx)$$

$$= -\frac{w}{2P}\left(1 + \frac{D^2}{a^2}...\right)(x^2 - lx) = -\frac{w}{2P}\left(x^2 - lx + \frac{2}{a^2}\right).$$

Thus the complete solution of Eqn. (*i*) is $y = c_1 \cosh ax + c_2 \sinh ax - \frac{W}{2P}\left(x^2 - lx + \frac{2}{a^2}\right)$ ...(*ii*)

At the end $O$, $y = 0$ when $x = 0$, [We don't need the other condition $d^2y/dx^2 = 0$]

$\therefore$ Eqn. (*ii*) gives $0 = c_1 - w/Pa^2$, or $c_1 = w/Pa^2$ ...(*iii*)

At the end $A$, $y = 0$ when $x = l$, [We don't need the other condition $d^2y/dx^2 = 0$]

$\therefore$ Eqn. (*ii*) gives $0 = c_1 \cosh al + c_2 \sinh al - w/Pa^2$ or $c_2 \sinh al = \frac{W}{Pa^2}(1 - \cosh al)$

whence $$c_2 = -\frac{w}{Pa^2}\tanh\frac{al}{2} \qquad ...(iv)$$

Substituting these values of $c_1$ and $c_2$ in (*ii*), we get

$$y = \frac{w}{Pa^2}\left(\cosh ax - \tanh\frac{al}{2}\sinh ax\right) - \frac{w}{2P}\left(x^2 - lx + \frac{2}{a^2}\right)$$

which gives the deflection of the beam at $N$.

*Thus the central deflection* $= y$ (at $x = l/2$)

$$= \frac{w}{Pa^2}\left(\cosh\frac{al}{2} - \tanh\frac{al}{2}\sinh\frac{al}{2} - 1\right) + \frac{wl^2}{8P} = \frac{w}{Pa^2}\left(\text{sech}\frac{al}{2} - 1\right) + \frac{wl^2}{8P}$$

Also the bending moment is maximum at the point of maximum deflection ($x = l/2$).

$\therefore$ *The maximum bending moment*

$$= EI\frac{d^2y}{dx^2}(\text{at } x = l/2) = Py + \frac{w}{2}(x^2 - lx)\ (\text{at } x = l/2) = \frac{w}{a}\left(\text{sech}\frac{al}{2} - 1\right)$$

**Example 14.15.** *A cantilever beam of length l and weighing w lb/unit is subjected to a horizontal compressive force P applied at the free end. Taking the origin at the free end and y-axis upwards, establish the differential equation of the beam and hence find the maximum deflection.*

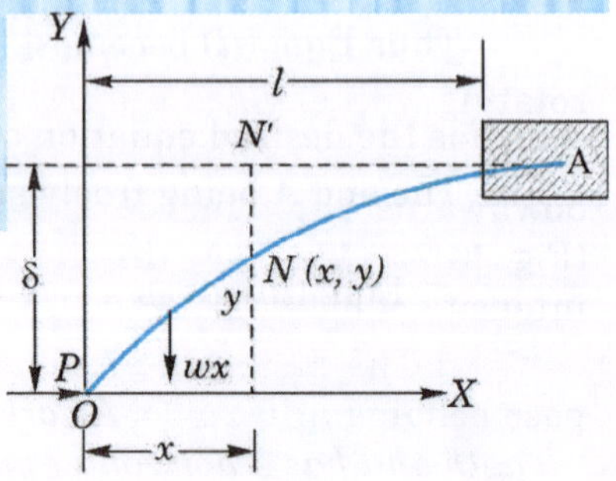

Fig. 14.24

**Solution.** Let $N(x, y)$ be any point of the beam referred to axes through the free end as shown (Fig. 14.24).

The external forces acting to the left of the section $NN'$, are

(*i*) the compressive force $P$,

(*ii*) the weight of the portion $ON = wx$ acting midway.

$\therefore$ Taking moments about $N$, we get $EI \dfrac{d^2y}{dx^2} = -Py - wx \cdot \dfrac{x}{2}$

or $$EI \frac{d^2y}{dx^2} + Py = -\frac{wx^2}{2} \quad \text{...(i)}$$

which is the desired differential equation.

Dividing by $EI$ and taking $P/EI = n^2$, we get

$$\frac{d^2y}{dx^2} + n^2y = -\frac{wn^2}{2P} \cdot x^2$$

Its auxiliary equation is $D^2 + n^2 = 0$, whence $D = \pm ni$.

$$\text{C.F.} = c_1 \cos nx + c_2 \sin nx$$

$$\therefore \quad \text{P.I.} = \frac{1}{D^2 + n^2}\left(-\frac{wn^2}{2P}x^2\right) = -\frac{w}{2P}\left(1 + \frac{D^2}{n^2}\right)^{-1} x^2 = -\frac{w}{2P}\left(1 - \frac{D^2}{n^2} + \ldots\right)x^2 = \frac{w}{2P}\left(\frac{2}{n^2} - x^2\right)$$

Thus the complete solution of Eqn. (*i*) is $y = c_1 \cos nx + c_2 \sin nx + \dfrac{w}{2P}\left(\dfrac{2}{n^2} - x^2\right)$ ...(*ii*)

The boundary conditions at the fixed end are

$x = l$, $y = \delta$, the maximum deflection and $dy/dx = 0$.

Using the first condition (*i.e.* $y = \delta$, when $x = l$), Eqn. (*ii*) gives

$$\delta = c_1 \cos nl + c_2 \sin nl + \frac{w}{2P}\left(\frac{2}{n^2} - l^2\right) \quad \text{...(iii)}$$

Differentiating Eqn. (*ii*), we get $\dfrac{dy}{dx} = n(-c_1 \sin nx + c_2 \cos nx) - \dfrac{wx}{P}$.

Applying the second condition, it gives $0 = n(-c_1 \sin nl + c_2 \cos nl) - wl/P$ ...(*iv*)

Also imposing the boundary condition for the free end (*i.e.* $x = 0$, $d^2y/dx^2 = 0$) on

$$\frac{d^2y}{dx^2} = -n^2(c_1 \cos nx + c_2 \sin nx) - \frac{w}{P},$$

we get $$0 = -n^2c_1 - w/P, \text{ i.e., } c_1 = -w/Pn^2.$$

Substituting this value of $c_1$ in Eqn. (*iv*), we get $c_2 = \dfrac{wl}{Pn} \sec nl - \dfrac{w}{Pn^2} \tan nl$

Thus, substituting the values of $c_1$ and $c_2$ in Eqn. (*iii*), we get

the *maximum deflection* $\delta = \dfrac{w}{Pn^2}\left(1 - \dfrac{l^2n^2}{2} - \sec nl + nl \tan nl\right)$.

## 14.8 WHIRLING OF SHAFTS

**(1) Critical or Whirling speeds.** A shaft seldom rotates about its geometrical axis for there is always some non-symmetrical crookedness in the shaft. In fact, the dead weight of the shaft causes some deflection which tends to become large at certain speeds. Such speeds at which the deflection of the shaft reaches a stage, where the shaft will fracture unless the speed is lowered are called the *critical or whirling speeds* of the shaft.

**(2) Differential equation of the rotating shaft.** Consider a shaft of weight $W$ per unit length which is rotating with angular velocity $\omega$.

Take its original horizontal position and the vertical downwards through the end $O$ as the axes of $x$ and $y$ (Fig. 14.25). We know that for a uniformly loaded beam, the intensity of loading at $P(x, y) = EI\, d^4y/dx^4$.

$\therefore$ The restoring force (*i.e.* the internal action to oppose bending at $P(x, y) = EI\, d^4y/dx^4$.

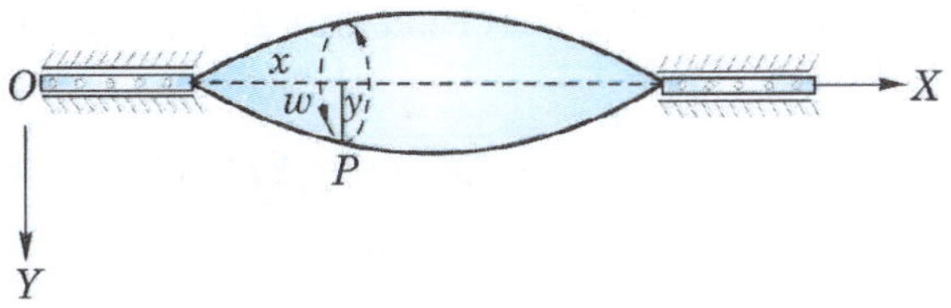


Fig. 14.25

Also the centrifugal force per unit length at $P = mr\omega^2$, *i.e.* $\frac{Wy}{g}\omega^2$.

As the restoring force arising out of the rigidity or stiffness of the shaft balances the centrifugal force which causes further deflection.

$$\therefore \quad EI\frac{d^4y}{dx^4} = \frac{W}{g}y\omega^2 \quad \text{or} \quad \frac{d^4y}{dx^4} - a^4y = 0, \text{ where } a^4 = \frac{W\omega^2}{gEI}$$

which is the desired differential equation.

Its auxiliary equation being $D^4 - a^4 = 0$, we have

$$D = \pm a, \pm ai.$$

Hence its solution is $y = c_1e^{ax} + c_2e^{-ax} + c_3 \cos ax + c_4 \sin ax$ which may be put in the form

$$y = A \cosh ax + B \sinh ax + C \cos ax + D \sin ax.$$

**(3) End conditions.** To determine the arbitrary constants $A, B, C, D$ we use the following end conditions :

(*i*) *At an end in a short or flexible bearings* (Fig. 14.26), there being no deflection and also no bending moment, we have

$$y = 0 \text{ and } \frac{d^2y}{dx^2} = 0.$$

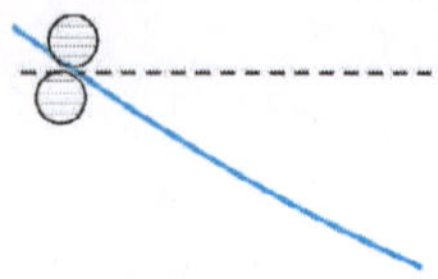

Fig. 14.26.

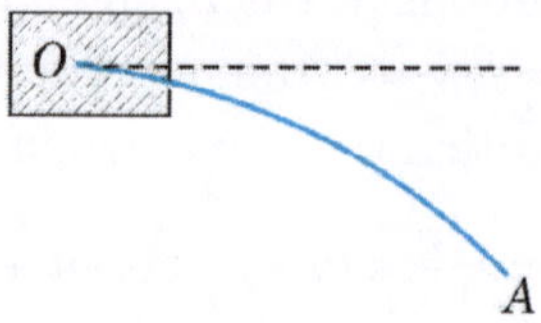


Fig. 14.27.

(*ii*) *At an end in long or fixed bearings* (Fig 14.27), the deflection and the slope of the shaft being both zero, we have

$$y = 0 \text{ and } \frac{dy}{dx} = 0.$$

(*ii*) *At a perfectly free end* (such as *A* in Fig. 14.27), there being no bending moment and no shear force, we have

$$\frac{d^2y}{dx^2} = 0 \text{ and } \frac{d^3y}{dx^3} = 0.$$

**Example 14.16.** *The differential equation for the displacement y of a whirling shaft when the weight of the shaft is taken into account is*

$$EI\frac{d^4y}{dx^4} - \frac{W\omega^2}{g}y = W.$$

*Taking the shaft of length 2l with the origin at the centre and short bearings at both ends, show that the maximum deflection of the shaft is*

$$\frac{g}{2\omega^2}(\text{sech } al + \sec al - 2).$$

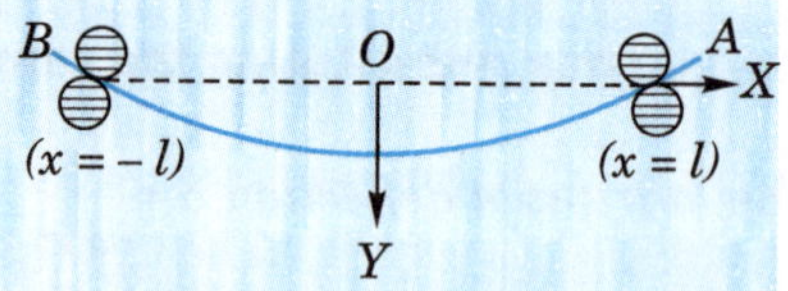


Fig. 14.28.

**Solution.** Given differential equation can be written as

$$\frac{d^4y}{dx^4} - a^4y = \frac{W}{EI}, \text{ where } a^4 = \frac{W\omega^2}{EIg} \quad ...(i)$$

Its $\quad$ C.F. $= A \cosh ax + B \sinh ax + C \cos ax + D \sin ax$

and $\quad$ $$\text{P.I.} = \frac{1}{D^4 - a^4}\cdot\frac{W}{EI} = \frac{W}{EI}\cdot\frac{1}{D^4 - a^4}e^{0.x} = -\frac{W}{EIa^4} = -\frac{g}{\omega^2}$$

Thus the complete solution of Eqn. (*i*) is

$$y = A \cosh ax + B \sinh ax + C \cos ax + D \sin ax - \frac{g}{\omega^2} \quad ...(ii)$$

Differentiating it twice, we get

$$\frac{1}{a}\frac{dy}{dx} = A \sinh ax + B \cosh ax - C \sin ax + D \cos ax$$

$$\frac{1}{a^2}\frac{d^2y}{dx^2} = A \cosh ax + B \sinh ax - C \cos ax - D \sin ax \quad ...(iii)$$

*As the end A* of the shaft is in short bearings (Fig. 14.28)

∴ when $x = l$; $y = 0$, $d^2y/dx^2 = 0$

∴ from Eqns. (*ii*) and (*iii*), we have

$$0 = A \cosh al + B \sinh al + C \cos al + D \sin al - \frac{g}{\omega^2} \quad ...(iv)$$

$$0 = A \cosh al + B \sinh al - C \cos al - D \sin al \quad ...(v)$$

Similarly *at the end B*, $x = -l$, $y = 0$, $d^2y/dx^2 = 0$.

∴ from Eqns. (*ii*) and (*iii*), we get

$$0 = A \cosh al - B \sinh al + C \cos al - D \sin al - \frac{g}{\omega^2} \quad ...(vi)$$

$$0 = A \cosh al - B \sinh al - C \cos al + D \sin al \quad ...(vii)$$

Adding Eqns. (*iv*) and (*vi*), and eqns. (*v*) and (*vii*), we get

$$A \cosh al + C \cos al = \frac{g}{\omega^2} \quad \text{and} \quad A \cosh al - C \cos al = 0.$$

whence $$A = \frac{g}{2\omega^2 \cosh al} \text{ and } C = \frac{g}{2\omega^2 \cos al}$$

Again subtracting Eqn. (*vi*) from Eqn. (*iv*) and Eqn. (*vii*) from Eqn. (*v*), we get

$B \sinh al + D \sin al = 0$ and $B \sinh al - D \sin al = 0$, whence $B = 0$ and $D = 0$.

Substituting the values of *A, B, C* and *D* in (*ii*), we get

$$y = \frac{g}{2\omega^2}\left[\frac{\cosh ax}{\cosh al} + \frac{\cos ax}{\cos al} - 2\right]$$

*Thus the maximum deflection = value of y at the centre (x = 0)*

$$= \frac{g}{2\omega^2}(\text{sech } al + \sec al - 2).$$

**Example 14.17.** *The whirling speed of a shaft of length l is given by*

$$\frac{d^4y}{dx^4} - m^4y = 0 \text{ where } m^4 = \frac{W\omega^2}{gEI},$$

*and y is the displacement at distance x from one end. If the ends of the shaft are constrained in long bearings, show that the shaft will whirl when cos ml cosh ml = 1.*

**Solution.** The solution of the given differential equation is

$$y = A \cosh mx + B \sinh mx + C \cos mx + D \sin mx \quad ...(i)$$

which on differentiation gives,

$$\frac{1}{m}\frac{dy}{dx} = A \sinh mx + B \cosh mx - C \sin mx + D \cos mx \quad ...(ii)$$

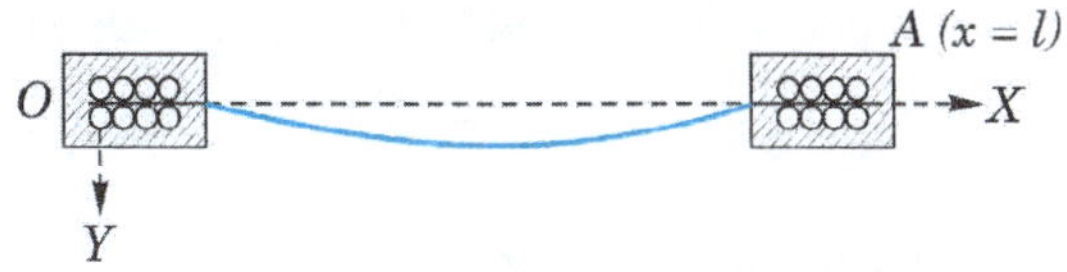

Fig. 14.29.

As the end *O* of the shaft is fixed in long bearings (Fig. 14.29).

∴ when $x = 0$, $y = 0$, $dy/dx = 0$,

∴ from Eqns. (*i*) and (*ii*), we have

$$0 = A + C \quad \text{or} \quad C = -A \quad ...(iii)$$

and $\quad 0 = B + D \quad$ or $\quad D = -B \quad$ ...(iv)

Similarly, at the end $A$, $x = l$, $y = 0$, $dy/dx = 0$.

$\therefore$ From Eqns. (i) and (ii), we have

$$0 = A \cosh ml + B \sinh ml + C \cos ml + D \sin ml \quad ...(v)$$

$$0 = A \sinh ml + B \cosh ml - C \sin ml + D \cos ml \quad ...(vi)$$

Substituting the values of $C$ and $D$ in Eqn. (v) and (vi), we get

$$A(\cosh ml - \cos ml) + B(\sinh ml - \sin ml) = 0$$

and
$$A(\sinh ml + \sin ml) + B(\cosh ml - \cos ml) = 0$$

Eliminating $A$ and $B$ from these equations, we get

$$\frac{\cosh ml - \cos ml}{\sinh ml - \sin ml} = -\frac{B}{A} = \frac{\sinh ml + \sin ml}{\cosh ml - \cos ml}$$

or
$$\cosh^2 ml - 2\cosh ml \cos ml + \cos^2 ml = \sinh^2 ml - \sin^2 ml$$

or
$$-2\cosh ml \cos ml + 2 = 0 \text{ or } \cos ml \cosh ml = 1$$

which must be satisfied when the shaft whirls.

The solution of this equation gives $ml = 4.73 = 3\pi/2$ radians approximately.

$$\therefore \quad \omega\sqrt{\left(\frac{W}{gEI}\right)}\, l^2 = m^2 l^2 = \frac{9\pi^2}{4}$$

Thus the *whirling speed of a shaft with ends in long bearings.*

$$= \omega = \frac{9\pi^2}{4l^2}\sqrt{\left(\frac{gEI}{W}\right)} \text{ approximately.}$$

**Obs. 1.** *When the shaft has one long bearing and the other short bearing*, the condition to be satisfied is $\tan ml = \tanh ml$, of which the solution is $ml = 3.927$

or
$$\omega\sqrt{\left(\frac{W}{gEI}\right)} \,.\, l^2 = m^2 l^2 = (3.927)^2 = 15.4 \text{ nearly.}$$

Thus *the whirling speed* $= \omega = \dfrac{15.4}{l^2}\sqrt{\left(\dfrac{gEI}{W}\right)}$

**Obs. 2.** *When the shaft has both short bearings*, the condition to be satisfied is $\sin ml = 0$ *i.e.* $ml = \pi$ (least non-zero value).

$$\therefore \quad \omega\sqrt{\left(\frac{W}{gEI}\right)} \,.\, l^2 = m^2 l^2 = \pi^2. \text{ Thus the whirling speed } = \omega = \frac{\pi^2}{l^2}\sqrt{\left(\frac{gEI}{W}\right)}.$$

**Obs. 3.** *When the shaft has one long bearing*, the condition to the satisfied is $\cos ml \cosh ml = -1$.

Its solution gives $\quad ml = 1.865$

$$\therefore \quad \omega\sqrt{\left(\frac{W}{gEI}\right)} \,.\, l^2 = m^2 l^2 = (1.865)^2 = 3.5 \text{ nearly. Thus the whirling speed } \omega = \frac{3.5}{l^2}\sqrt{\left(\frac{gEI}{W}\right)}.$$

## PROBLEMS 14.4

1. A horizontal tie-rod of length $2l$ with concentrated load $W$ at the centre and ends freely hinged, satisfies the differential equation $EI\dfrac{d^2y}{dx^2} = Py - \dfrac{W}{2}x$. With conditions $x = 0, y = 0$ and $x = l, dy/dx = 0$, prove that the deflection $\delta$ and the bending moment $M$ at the centre $(x = l)$ are given by $\delta = \dfrac{W}{2Pn}(nl - \tanh nl)$ and $M = -\dfrac{W}{2n}\tanh nl$, where $n^2EI = P$.

2. A light horizontal strut $AB$ is freely pinned at $A$ and $B$. It is under the action of equal and opposite compressive forces $P$ at its ends and it carries a load $W$ at its centre. Then for $0 < x < l/2$, $EI\dfrac{d^2y}{dx^2} + Py + \dfrac{1}{2}Wx = 0$. Also $y = 0$ at $x = 0$ and $dy/dx = 0$ at $x = l/2$.

   Prove that $y = \dfrac{W}{2P}\left(\dfrac{\sin nx}{n\cos nl/2} - x\right)$ where $n^2 = \dfrac{P}{EI}$.

3. A uniform horizontal strut of length $l$ freely supported at both ends, carries a uniformly distributed load W per unit length. If the thrust at each end is, $P$, prove that the maximum deflection is $\frac{W}{Pa^2}\left(\sec\frac{al}{2}-1\right)-\frac{Wl^2}{8P}$, where $\frac{P}{EI}=a^2$. Prove also that the maximum bending moment is of the magnitude $\frac{W}{a^2}\left(\sec\frac{al}{2}-1\right)$.
4. The shape of a strut of length $l$ subjected to an end thrust $P$ and lateral load $w$ per unit length, when the ends are built in, is given by $EI\frac{d^2y}{dx^2}+Py=\frac{wx^2}{2}-\frac{wlx}{2}+M$, where $M$ is the moment at a fixed end. Find $y$ in terms of $x$, given that $y = 0$, $dy/dx = 0$ at $x = 0$ and $dy/dx = 0$ at $x = l/2$.
5. A light horizontal strut of length $l$ is clamped at one end carries a vertical load $W$ at the free end. If the horizontally thrust at the free end is $P$, show that the strut satisfies the differential equation

   $El\frac{d^2y}{dx^2}=(\delta-y)P+W(l-x)$, where $y$ is the displacement of a point at a distance $x$ from the fixed end and $\delta$, the deflection at the free end.

   Prove that the deflection at the free end is given by $\frac{W}{nP}(\tan nl - nl)$, where $n^2EI = P$.
6. A long column fixed at one end ($x = 0$) and hinged at the other ($x = l$) is under the action of axial load $P$. If a force $F$ is applied laterally at the hinge to prevent lateral movement, show that it satisfies the equation $\frac{d^2y}{dx^2}+n^2y=\frac{En^2}{P}$ $(l-x)$, where $EIn^2 = P$. Hence determine the equation of the deflection curve.
7. A long column of length $l$ is fixed at one end and is completely free at the other end. If $y$ is the lateral deflection at a point distance $x$ from the fixed end, when load $P$ is axially applied, find the differential equation satisfied by $x$ and $y$. Show that the deflection curve is given by $y = a\{1-\cos\sqrt{P/EI})\,x\}$ and find the least value of the critical load ($a$ is the lateral deflection of the free end).
8. The differential equation for the displacement $y$ of a heavy whirling shaft is $\frac{d^4y}{dx^4}=a^4\left(y+\frac{g}{\omega^2}\right)$, where $a^4=\frac{W\omega^2}{gEI}$. If both ends are in short bearings, the ends being $x = 0$ and $x = l$, find the bending moment of the centre of the shaft.

## 14.9 APPLICATIONS OF SIMULTANEOUS LINEAR EQUATIONS

So far we have considered engineering systems having only one degree of freedom. The analysis of a system having more than one degree of freedom depends on the solution of simultaneous linear equations. In fact such equations form the basis of the theory of projectiles and the coupled circuits having self and mutual inductance. The details of such applications are best explained through the following examples :

**Example 14.18. Projectile with resistance.** *Find the path of a particle projected with a velocity $v$ at an angle $\alpha$ to the horizon in a medium whose resistance, apart from gravity, varies as velocity. Also find the greatest height attained.*

**Solution.** Let the axes of $x$ and $y$ be respectively horizontal and vertical with origin at the point of projection (Fig. 14.30).

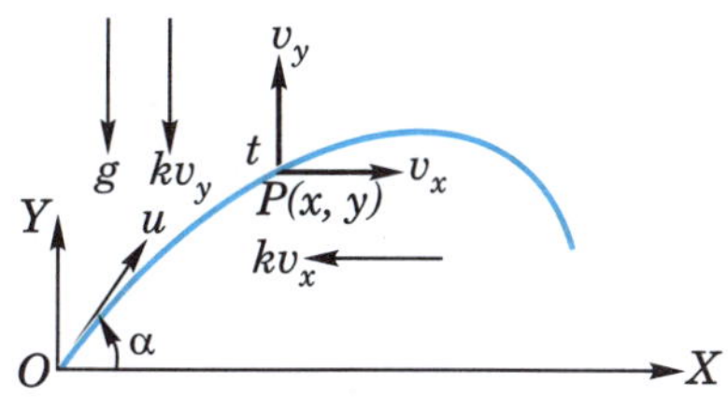

Fig. 14.30

Let $P(x, y)$ be the position of the projectile at the time $t$, where the velocity components parallel to the axes are

$$v_x=\frac{dx}{dt},\ v_y=\frac{dy}{dt}$$

$\therefore$ the equations of motion are:

| *Parallel to x-axis* | *Parallel to y-axis* |
|---|---|
| $m\frac{dv_x}{dt}=-mkv_x$ | $m\frac{dv_y}{dt}=-mg-mkv_y$ |
| or $\frac{dv_x}{dt}=-kv_x$ | $\frac{dv_y}{dt}=-(g+kv_y)$ |

Separating the variables and integrating, we have

$$\int \frac{dv_x}{v_x} = -k\int dt + c_1 \qquad \Bigg| \qquad \int \frac{dv_y}{g + kv_y} = -\int dt + c_2$$

or $\log v_x = -kt + c_1 \qquad \Bigg| \qquad \frac{1}{k}\log (g + kv_y) = -t + c_2.$

Initially when $t = 0$, $v_x = u \cos \alpha$, $v_y = u \sin \alpha$.

$$\log u \cos \alpha = c_1 \qquad \Bigg| \qquad \frac{1}{k}\log (g + ku \sin \alpha) = c_2$$

Subtracting,

$$\log\left(\frac{v_x}{u\cos\alpha}\right) = -kt \qquad \Bigg| \qquad \frac{1}{k}\log\left(\frac{g + kv_y}{g + ku\sin\alpha}\right) = -t$$

or $\frac{dx}{dt} = v_x = u \cos \alpha e^{-kt} \quad ...(i) \qquad \Bigg| \qquad \frac{dy}{dt} = v_y = \frac{1}{k}[(g + ku \sin \alpha)e^{-kt} - g] \quad ...(ii)$

Again integrating, we get

$$x = \frac{u\cos\alpha}{-k}e^{-kt} + c_3,\ y = -\frac{1}{k}\left(\frac{g}{k} + u\sin\alpha\right)e^{-kt} - \frac{g}{k}t + c_4$$

Initially when $t = 0$, $x = 0$, $y = 0$,

$$\therefore \qquad 0 = \frac{u\cos\alpha}{k} + c_3,\ 0 = -\frac{1}{k}\left(\frac{g}{k} + u\sin\alpha\right) + c_4$$

Subtracting, we get $\qquad x = \frac{u\cos\alpha}{k}(1 - e^{-kt}) \qquad ...(iii)$

$$y = \frac{1}{k}\left(\frac{g}{k} + u\sin\alpha\right)(1 - e^{-kt}) - \frac{gt}{k} \qquad ...(iv)$$

Eliminating $t$ from Eqns. (*iii*) and (*iv*), we obtain $y = \left(\frac{g}{k} + u\sin\alpha\right)\frac{x}{u\cos\alpha} + \frac{g}{k^2}\log\left(1 - \frac{kx}{u\cos\alpha}\right)$

which is the required equation of the trajectory.

The projectile will attain the greatest height when $dy/dt = 0$.

*i.e.*, when $e^{-kt} = g/(g + ku \sin \alpha)$, *i.e.*, at time $t = \frac{1}{k}\log\left(1 + \frac{ku\sin\alpha}{g}\right)$. [From (*ii*)]

*Substituting the value of t in* Eqn. (*iv*), we get the greatest height attained

$$(= y) = \frac{u\sin\alpha}{k} - \frac{g}{k^2}\log\left(1 + \frac{ku\sin\alpha}{g}\right).$$

**Example 14.19.** *Newtons law of cooling states that the temperature of an object changes at a rate proportional to the difference of temperature between the object and the surroundings. Supposing water at a temperature 100°C cools to 80°C in 10 minutes, in a room maintained at a temperature of 30°C. Find when the temperature of water will become 40°C.* *(Kumaun, 2018)*

**Solution:** Let unit of time be 1 minute. If $T$ is the temperature of the object at any instant $t$, then by Newtons law of cooling.

$$\frac{dT}{dt} = -k(T - 30) \qquad ...(i)$$

the negative sign shown that temperature decreases

$$\int \frac{dT}{T - 30} = -k\int dt + \log C$$

$$\log(T-30) = -kt + \log C$$

$$T - 30 = ce^{-kt} \qquad \text{...(ii)}$$

Initially, when $t = 0$, $T = 100°\ C$ $\quad C = 70$

Hence from Eqn. (*ii*) $\quad T - 30 = 70e^{-kt}$

Also, when $t = 10$, $T = 80°C$ $\quad 80 - 30 = 70e^{-10k}$

$$e^{-10k} = \frac{5}{7}$$

$$T = 30 + 70(e^{-10k})^{t/10}$$

$$T = 30 + 70\left(\frac{5}{7}\right)^{t/10}$$

$T = 40°$ C, we get from given equation

$$40 = 30 + 70\left(\frac{5}{7}\right)^{t/10}$$

$$\frac{1}{7} = \left(\frac{5}{7}\right)^{t/10}$$

$$\frac{t}{10} = \frac{\log_{10}\left(\frac{1}{7}\right)}{\log_{10}\left(\frac{5}{7}\right)} = \frac{0\cdot 8451}{0\cdot 1461} = 5.79$$

$t = 57.9$ minutes.

**Example 14.20:** *In the inversion of raw sugar, the time rate change varies as the amount of raw sugar remaining. If after 10 hours, 8000 kg of raw sugar has been reduced to 800 kg, how much raw sugar will remain at the end of 24 hours.* *(Marathwada, 2018)*

**Solution:** Let $m$ kg of raw sugar be present at the end of $t$ hours. Taking unit of time as 1 hour.

$$\frac{dm}{dt} = km$$

Separating the variables and integrating

$$\int \frac{dm}{dt} = -k\int dt + \log c$$

$$\log m = -kt + \log c$$

$$m = ce^{-kt} \qquad \text{...(i)}$$

Initially, when $t = 0$, $m = 8000$ $\quad c = 8000$

Hence, Eqn. (*i*) becomes $\quad m = 8000\, e^{-\lambda t} \qquad \text{...(ii)}$

Also, when $t = 10$, $m = 800$ $\quad 800 = 8000e^{-10k}$

$$e^{-10k} = \frac{1}{10}$$

Substituting this value in Eqn. (*ii*) $\quad m = 8000\,(e^{-10k})t/10$

$$= 8000\left(\frac{1}{10}\right)^{t/10}$$

$t = 24.$

$$m = 8000\left(\frac{1}{10}\right)^{2.4}$$

$$m = 31.84$$

the required amount of raw sugar, after 24 hours.

## PROBLEMS 14.5

1. A particle is projected with velocity $u$, at an elevation $\alpha$. Neglecting air resistance, show that the equation to its path is the parabola $y = x \tan \alpha - \frac{gx^2}{2u^2 \cos^2 \alpha}$. Also find the time of flight and range on the horizontal plane.
2. An inclined plane makes angle $\alpha$ with the horizontal. A projectile is launched from the bottom of the inclined plane with speed $V$ in a direction making angle $\beta$ with the horizontal. Set up the differential equations and find (*i*) the range on the incline, (*ii*) the maximum range up the incline.
3. The radioactive isotope Thorium 234 disintegrates at a rate proportional to the amount present. If 100 milligrams of this material is reduced to 82.04 milligrams in 7 days, find an expression for the amount present at any time. Also find the time interval that must elapse before the mass reduces to one half of its original value. *(Timkur Uni., 2018)*
4. A particle moving in a plane is subjected to a force directed towards a fixed point $O$ and proportional to the distance of the particle from $O$. Show that the differential equations of motion are of the form $\frac{d^2x}{dt^2} = -k^2x$, $\frac{d^2y}{dt^2} = -k^2y$.
   Find the cartesian equation of the path of the particle if $x = 1, y = 0, \frac{dx}{dt} = 0$ and $dy/dt = 2$, when $t = 0$.
5. The currents $i_1$ and $i_2$ in mesh are given by the differential equations
   $$\frac{di_1}{dt} - \omega i_2 = a \cos pt, \quad \frac{di_2}{dt} + \omega i_1 = a \sin pt.$$ Find the currents $i_1$ and $i_2$ if $i_1 = i_2 = 0$ at $t = 0$.
6. Neutrons in an atomic pile increase at a rate proportional to the number of neutrons present at any instant. If $N_0$ neutrons are initially present and $N_1$ and $N_2$ neutrons are present at time $T_1$ and $T_2$ respectively, show that
   $$\left(\frac{N_2}{N_0}\right)^{T_1} = \left(\frac{N_1}{N_0}\right)^{T_2}.$$ *(Kannada Uni., 2018)*
7. The motion of a particle is governed by the equations
   $$\frac{d^2x}{dt^2} - n\frac{dy}{dt} = 0, \quad \frac{d^2y}{dt^2} + n\frac{dx}{dt} = n^2a,$$ when $x = y = \frac{dx}{dt} = \frac{dy}{dt} = 0$ at $t = 0$. Find $x$ and $y$ in terms of $t$.
8. Under certain conditions, the motion of an electron is given by the equations $m\frac{d^2x}{dt^2} + eH\frac{dy}{dt} = eE$ and $m\frac{d^2y}{dt^2} - eH\frac{dx}{dt} = 0$. Find the path of the electron, if it started from rest at the origin.
9. The voltage $V$ and the current $i$ at a distance $x$ from the source satisfy the equations $-dV/dt = Ri$, $-di/dx = GV$, where $R$, $G$ are constants. If $V = V_0$ at $x = 0$ and $V = 0$ at the receiving end $x = l$, show that $V = V_0 \sinh n(l - x)/\sinh nl$, $i = V_0/(G/R). \cosh n(l - x)/\sinh nl$, where $n^2 = RG$.
10. If 30% of a radioactive substance disappeared in 10 days, how long will it take for 90% of it to disappear.
11. In a certain chemical reaction, the rate of conversion of a substance at time $t$ is proportional to the quantity of substance still untransformed at that instant. The amount of substance remaining untransformed at the end of one hour and at the end of four hours are 60 gms and 21 gms respectively. How many grams of substance were present initially. *(Gulbarge, 2018)*

## 14.10 OBJECTIVE TYPE OF QUESTIONS

## PROBLEMS 14.6

*Fill up the blanks or choose the correct answer in the following problems:*

1. A particle executing simple harmonic motion of amplitude 5 cm has a speed of 8 cm/sec when at a distance of 3 cm from the centre of the path. The period of the motion of the particle will be
   (*a*) $\pi/2$ sec (*b*) $\pi$ sec (*c*) $2\pi$ sec (*d*) $4\pi$ sec.
2. A ball of mass $m$ is suspended from a fixed point $O$ by a light string of natural length $l$ and modulus of elasticity $\lambda$. If the ball is displaced vertically, its motion will be S.H.M. of period

(a) $2\pi\sqrt{(m/\lambda l)}$ (b) $2\pi\sqrt{(ml/\lambda)}$ (c) $2\pi\sqrt{(l/m\lambda)}$ (d) $2\pi\sqrt{(\lambda m/l)}$.

3. The periodic time of the motion described by the differential equation $\frac{d^2x}{dt^2} + 4x = 0$ is

(a) $\pi/2$ (b) $\pi$ (c) $2\pi$.

4. A particle is projected with a velocity $u$ at an angle of 60° to the horizontal. The time of flight of the projectile is equal to

(a) $\sqrt{3}u/2g$ (b) $\sqrt{3}u/g$ (c) $u/g$ (d) $u/2g$.

5. A body of 6.5 kg is suspended by two strings of lengths 5 and 12 metres attached to two points in the same horizontal line whose distance apart is 13 meters. The tension of the strings are

(a) 2 kg & 6.5 kg (b) 2.5 kg & 6 kg (c) 2.25 kg & 6.25 kg (d) 3 kg & 5.5 kg.

6. A particle is projected at an angle of 30° to the horizontal with a velocity of 1962 cm/sec then the time of flight is

(a) 1 sec (b) 2 sec (c) 2.5 sec (d) 3 sec.

7. A point moves with S.H.M. whose period is 4 seconds. If it starts from rest at a distance of 4 meters from the centre of its path, then the time it takes, before it has described 2 metres is

(a) $\frac{1}{3}$ second (b) $\frac{2}{3}$ second (c) $\frac{3}{4}$ second (d) $\frac{4}{5}$ second.

8. If the length of the pendulum of a clock be increased in the ratio 720 : 721, it would loose ....... seconds per day.

9. The frequency of free vibrations in a closed circuit with inductance $L$ and capacity $C$ in series is ....... per minute.

10. If a clock with a seconds pendulum loses 10 seconds per day at a place having $g = 32$ ft/sec$^2$, $g$ should be increased by ....... ft/sec$^2$, to keep correct time.

11. The soldiers break step while marching over a bridge for the fear that their steps may not be in rhym with the natural frequency of the bridge causing its collapse due to .......

12. A horizontal tie-rod is freely pinned at each end. If it carries a uniform load $w$ lb per unit length and has a horizontal pull $P$, then the differential equation of the elastic curve is .......

13. The conditions for an end of a whirling shaft to be in fixed bearings are ....... and .......

CHAPTER 15

# Differential Equations of Other Types

**1.** Introduction. **2.** Equations of the form $d^2y/dx^2 = f(x)$. **3.** Equations of the form $d^2y/dx^2 = f(y)$. **4.** Equations which do not contain *y*. **5.** Equations which do not contain *x*. **6.** Equations whose one solution is known. **7.** Equations which can be solved by changing the independent variable. **8.** Total differential equation : $Pdx + Qdy + Rdz = 0$. **9.** Simultaneous total differential equations. **10.** Equations of the form $dx/P = dy/Q = dz/R$.

## 15.1 INTRODUCTION

In this chapter, we propose to study some other important types of ordinary differential equations which require special methods for their solution and have varied applications as illustrated side by side.

## 15.2 EQUATIONS OF THE FORM $d^2y/dx^2 = f(x)$

Integrating with respect to $x$, we have $\frac{dy}{dx} = \int f(x)dx + c = F(x)$. (say)

Again integrating, we get $y = \int F(x)dx + c'$ as the required solution.

In general, the solution of the equations of the form $\frac{d^n y}{dx^n} = f(x)$ is obtained by integrating it $n$ times successively.

**Example 15.1.** *Solve* $\frac{d^2 y}{dx^2} = xe^x$.

**Solution.** Integrating, we get $\frac{dy}{dx} = xe^x - \int e^x\,dx + c_1 = (x-1)\,e^x + c_1$

Again integrating, we get

$$y = (x-1)e^x - \int e^x dx + c_1 x + c_2 = (x-2)e^x + c_1 x + c_2.$$

**PROBLEMS 15.1**

Solve :

1. $\frac{d^2y}{dx^2} = x^2 \sin x.$
2. $\frac{d^3y}{dx^3} = x + \log x.$

3. A beam of length $2l$ with uniform load $w$ per unit length is freely supported at both ends. Prove that the maximum deflection of the beam is $\frac{5wl^4}{24EI}$.

[**Hint.** Taking the origin at the left end, we have $EI\, d^4y/dx^4 = w$. At each end, $y = 0$ and $d^2y/dx^2 = 0$.]

4. For a cantilever beam of length $l$ with a uniform load of $w$ per unit length, show that the maximum deflection at the free end is $wl^4/EI$, where the symbols have the usual meaning.

## 15.3 EQUATIONS OF THE FORM $d^2y/dx^2 = f(y)$

Multiplying both sides by $2dy/dx$, we have $2\frac{dy}{dx}\cdot\frac{d^2y}{dx^2} = 2f(y)\frac{dy}{dx}$

Integrating with respect to $x$, $\left(\frac{dy}{dx}\right)^2 = 2\int f(y)\, dy + c = F(y)$ (say)

or
$$\frac{dy}{dx} = \sqrt{[F(y)]}$$

Separating the variables and integrating, we get $\int \frac{dy}{\sqrt{[F(y)]}} = x + c$, whence follows the desired solution.

*Such equations occur quite frequently in Dynamics.*

**Example 15.2.** *Solve $d^2y/dx^2 = 2(y^3 + y)$ under the conditions $y = 0$, $dy/dx = 1$, when $x = 0$.*

**Solution.** Multiplying by $2\, dy/dx$, the given equation becomes

$$2\frac{dy}{dx}\cdot\frac{d^2y}{dx^2} = 4(y^3 + y)\frac{dy}{dx}$$

Integrating w.r.t. $x$, $\left(\frac{dy}{dx}\right)^2 = 4\left(\frac{y^4}{4} + \frac{y^2}{2}\right) + c = y^4 + 2y^2 + c$ ...(*i*)

As $dy/dx = 1$ for $y = 0$, $\therefore\ c = 1$

$\therefore$ Eqn. (*i*) takes the form $(dy/dx)^2 = y^4 + 2y^2 + 1 = (y^2 + 1)^2$ or $dy/dx = y^2 + 1$

Separating the variables and integrating, we have $\int \frac{dy}{1+y^2} = \int dx + c'$

or
$$\tan^{-1} y = x + c' \quad ...(ii)$$

Thus Eqn. (*ii*) becomes $\tan^{-1} y = x$ or $y = \tan x$ which is the required solution.

**Example 15.3.** *A point moves in a straight line towards a centre of force $\mu/(\text{distance})^3$, starting from rest at a distance 'a' from the centre of force, show that the time of reaching a point distant 'b' from the centre of force is $\frac{a}{\sqrt{\mu}}\sqrt{(a^2 - b^2)}$ and that its velocity is $\frac{\sqrt{\mu}}{ab}\sqrt{(a^2 - b^2)}$.*

**Solution.** Let $O$ be the centre of force and $A$ the point of start so that $OA = a$. At any time $t$, let the point be at $P$ where $OP = x$ so that

$$\frac{d^2x}{dt^2} = \frac{-\mu}{x^3} \quad ...(i)$$

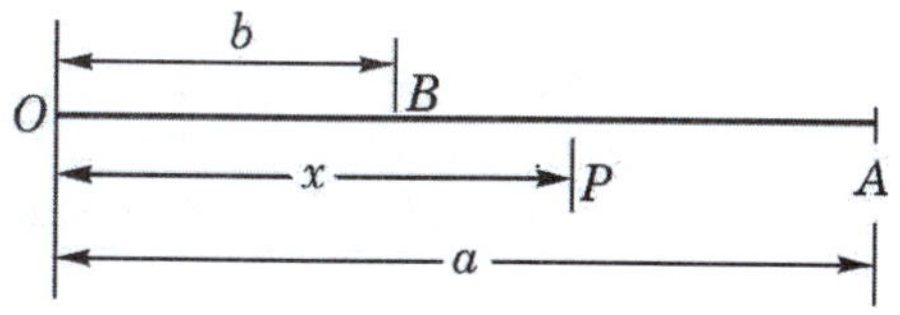

Fig. 15.1

Multiplying both sides by $2\, dx/dt$, we get

$$\frac{2dx}{dt}\cdot\frac{d^2x}{dt^2} = -\frac{\mu}{x^3}\cdot\frac{2dx}{dt}$$

Integrating both sides, we obtain

$$\left(\frac{dx}{dt}\right)^2 = -\mu \int \frac{2}{x^3}\frac{dx}{dt}\,.\,dt + c = +\frac{\mu}{x^2} + c$$

When $x = a$, velocity $dx/dt = 0$. $\therefore\ c = -\mu/a^2$.

$$\therefore \qquad \left(\frac{dx}{dt}\right)^2 = \mu\left(\frac{1}{x^2} - \frac{1}{a^2}\right) = \frac{\mu(a^2 - x^2)}{a^2x^2} \qquad ...(ii)$$

At $B$ $(x = b)$, velocity towards $O = \dfrac{\sqrt{\mu(a^2 - b^2)}}{ab}$

Again Eqn. $(ii)$ can be rewritten as $\dfrac{-ax\,dx}{\sqrt{(a^2 - x^2)}} = \sqrt{\mu}\,dt$ [–ve is taken since point is moving towards $O$]

Integrating both sides, we get

$$\sqrt{\mu}\int dt = -\int \frac{ax\,dx}{\sqrt{(a^2 - x^2)}} + c' \quad \text{or} \quad \sqrt{\mu}t = a\sqrt{(a^2 - x^2)} + c' \qquad ...(iii)$$

Since $t = 0$ at $x = a$, $\therefore\ c' = 0$

Thus Eqn. $(iii)$ gives $t = \dfrac{a}{\sqrt{\mu}}\sqrt{(a^2 - x^2)}$

Hence at $B$ $(x = b)$ $t = \dfrac{a}{\sqrt{\mu}}\sqrt{(a^2 - b^2)}$.

### PROBLEMS 15.2

Solve :

1. $d^2y/dx^2 = 3\sqrt{y}$ given that $y = 1$, $dy/dx = 2$ when $x = 0$.
2. $\dfrac{d^2y}{dx^2} = \dfrac{36}{y^2}$, given that when $x = 0$, $\dfrac{dy}{dx} = 0$, $y = 8$.
3. If $d^2r/dt^2 = \omega^2 r$, find the value of $r$ in terms of $t$, if $r = a$ and $dr/dt = v$, when $t = 0$.
4. The motion of a particle let fall from a point outside the earth is given by $d^2x/dt^2 = -ga^2/x^2$. Given that $x = h$ and $dx/dt = 0$, when $t = 0$, find $t$ in terms of $x$.
5. A particle is acted upon by a force $\mu\,(x + a^4/x^3)$ per unit mass towards the origin, where $x$ is the distance from the origin at time $t$. If it starts from rest at a distance $a$, show that it will arrive at the origin in time $\pi/(4\sqrt{\mu})$.

## 15.4 EQUATIONS WHICH DO NOT CONTAIN y

A second order equation of this form is

$$f(d^2y/dx^2, dy/dx, x) = 0$$

On putting $dy/dx = p$ and $d^2y/dx^2 = dp/dx$, it becomes

$$f(dp/dx, p, x) = 0.$$

This is an equation of the first order in $x$ and $p$ and can, therefore, be solved easily.

If its solution is $(p =)\ dy/dx = \phi\,(x)$, then $y = \int \phi\,(x)dx + c$ is the required solution.

**Obs.** This method may be used to reduce any such equation of the $n$th order to one of the $(n - 1)$th order. If, however, the lowest derivative in such an equation is $d^ry/dx^r$

$(i)$ put $d^ry/dx^r = p$ ; $(ii)$ find $p$ and therefrom find $y$, (*See* Ex. 15.5).

**Example 15.4.** *Solve* $x\dfrac{d^2y}{dx^2} = \sqrt{\left[1 + \left(\dfrac{dy}{dx}\right)^2\right]}$

**Solution.** Putting $dy/dx = p$ and $d^2y/dx^2 = dp/dx$, the given equation becomes

$$x dp/dx = \sqrt{(1+p^2)}.$$

Separating the variables and integrating, we get

$$\int \frac{dp}{\sqrt{(1+p^2)}} = \int \frac{dx}{x} + \text{constant}$$

or $$\log\left[p + \sqrt{(1+p^2)}\right] = \log x + \log c = \log cx.$$

$\therefore$ $$p + \sqrt{(1+p^2)} = cx \quad \text{or} \quad 1 + p^2 = (cx - p)^2$$

or $$(p =)\ \frac{dy}{dx} = \frac{1}{2}\left(cx - \frac{1}{cx}\right).$$

$\therefore$ integrating again, we have $y = \frac{1}{2}\left(c\frac{x^2}{2} - \frac{1}{c}\log x\right) + c'$ as the required solution.

**Example 15.5.** *Solve* $\frac{d^4y}{dx^4} \cdot \frac{d^3y}{dx^3} = 1$.

**Solution.** Putting $d^3y/dx^3 = p$ and $d^4y/dx^4 = dp/dx$, the given equation becomes $\frac{dp}{dx} p = 1$.

Integrating w.r.t. $x$, $\int p dp = x + c_1$, *i.e.* $p^2/2 = x + c_1$ or $(p =)\ d^3y/dx^3 = \sqrt{2}(x + c_1)^{1/2}$.

Integrating thrice successively, we get

$$\frac{d^2y}{dx^2} = \sqrt{2}\frac{(x+c_1)^{3/2}}{3/2} + c_2, \quad \frac{dy}{dx} = \frac{2\sqrt{2}}{3} \cdot \frac{(x+c_1)^{5/2}}{5/2} + c_2 x + c_3$$

$$y = \frac{4\sqrt{2}}{15} \frac{(x+c_1)^{7/2}}{7/2} + c_2\frac{x^2}{2} + c_3 x + c_4$$

Hence $y = \frac{8\sqrt{2}}{105}(x + c_1)^{7/2} + \frac{1}{2}c_2 x^2 + c_3 x + c_4$ is the desired solution.

## PROBLEMS 15.3

Solve the following equations :

1. $x\frac{d^2y}{dx^2} + 2\frac{dy}{dx} + 6x = 0.$
2. $(1 + x^2)\frac{d^2y}{dx^2} + 1 + \left(\frac{dy}{dx}\right)^2 = 0.$
3. $2x\frac{d^3y}{dx^3} \cdot \frac{d^2y}{dx^2} = \left(\frac{d^2y}{dx^2}\right)^2 - a^2.$
4. $\left[1 + \left(\frac{dy}{dx}\right)^2\right]^{3/2} = a\frac{d^2y}{dx^2}.$
5. A particle of mass $m$ grammes is constrained to move in a horizontal circular path of radius $a$ cm and is subjected to a resistance proportional to the square of the speed at any instant. Show that the differential equation of motion is of the form $m\frac{d^2\theta}{dt^2} + \mu a\left(\frac{d\theta}{dt}\right)^2 = 0$. If the particle starts with an angular velocity $\omega$, find its angular displacement $\theta$ at time $t$ sec.
6. When the inner of two concentric spheres of radii $r_1$ and $r_2 (r_1 < r_2)$ carries an electric charge, the differential equation for the potential $v$ at any point between two spheres at a distance $r$ from their common centre is $\frac{d^2v}{dt^2} + \frac{2}{r}\frac{dv}{dr} = 0$. Solve for $v$ given $v = v_1$ when $r = r_1$ and $v = v_2$ when $r = r_2$.

## 15.5 EQUATIONS WHICH DO NOT CONTAIN x

A second order equation of this form is

$$f(d^2y/dx^2, dy/dx, y) = 0.$$

On putting $\dfrac{dy}{dx} = p$ and $\dfrac{d^2y}{dx^2} = \dfrac{dp}{dx} = \dfrac{dp}{dy} \cdot \dfrac{dy}{dx} = p\dfrac{dp}{dy}$, it becomes

$$f(p\,dp/dy, p, y) = 0.$$

This is an equation of the first order in $y$ and $p$ and can, therefore, be solved easily.

**Example 15.6.** *Solve* $y\dfrac{d^2y}{dx^2} + \dfrac{dy}{dx}\left(\dfrac{dy}{dx} - 2y\right) = 0.$

**Solution.** On putting $dy/dx = p$ and $d^2y/dx^2 = p\,dp/dy$, the given equation becomes

$$yp\frac{dp}{dy} + p(p - 2y) = 0.$$

This gives *either* $p = 0$, of which the solution is $y = c$ ;

or $$\left(y\frac{dp}{dy} + p\right) - 2y = 0 \quad i.e., \quad (ydp + pdy) = 2ydy \text{ i.e., } d(py) = 2ydy.$$

Integrating, $\quad py = 2\int ydy + c_1 = y^2 + c_1.$

Separating the variables and integrating, we get

$$\int \frac{ydy}{y^2 + c_1} = \int dx + c_2 \quad \text{or} \quad \frac{1}{2}\log(y^2 + c_1) = x + c_2 \text{ whence } y^2 + c_1 = c_3e^{2x}$$

Hence the required solutions are $y = c$ and $y^2 + c_1 = c_3\,e^{2x}$.

**Example 15.7.** *Find the curve in which the radius of curvature is twice the normal and in the opposite direction.*

**Solution.** At any point $P(x, y)$ of a curve, the radius of curvature

$$\rho = \left[1 + \left(\frac{dy}{dx}\right)^2\right]^{3/2} \Bigg/ \frac{d^2y}{dx^2}$$

and the length of the normal $(PN)$

$$= y\sqrt{[1 + (dy/dx)^2]}.$$

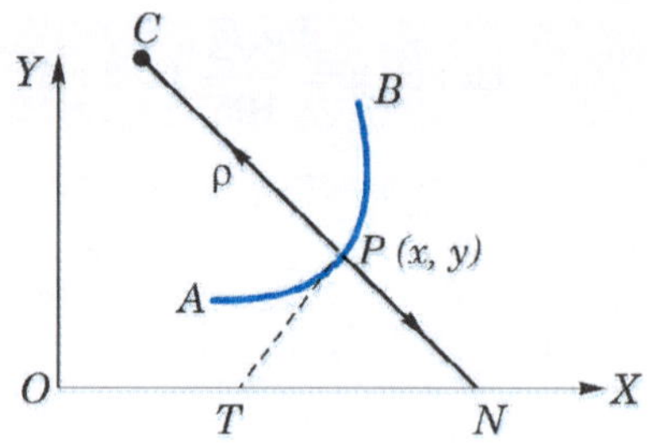

Fig. 15.2

Also we know that $\rho$ is measured inwards and the normal is measured outwards, *i.e.*, both of them are positive when measured in opposite directions. So the sign will be positive (or negative) according as $\rho$ and the normal run in the opposite (or same) directions.

Thus for the given curve $\left[1 + \left(\dfrac{dy}{dx}\right)^2\right]^{3/2} \div \dfrac{d^2y}{dx^2} = 2y\sqrt{\left[1 + \left(\dfrac{dy}{dx}\right)^2\right]}$

or $$1 + \left(\frac{dy}{dx}\right)^2 = 2y\frac{d^2y}{dx^2}.$$

On putting $dy/dx = p$ and $d^2y/dx^2 = p\,dp/dy$, the given equation becomes

$$1 + p^2 = 2y \,.\, p\,dp/dy.$$

$\therefore$ separating variables and integrating, we have

$$\int \frac{2pdp}{1 + p^2} = \int \frac{dy}{y} + \text{constant}$$

or $$\log(1 + p^2) = \log y + \log a = \log ay$$

$\therefore$ $$1 + p^2 = ay \text{ or } (p =)\, dy/dx = \sqrt{(ay - 1)}$$

$\therefore$ separating the variables and integrating, we get

$$\int dx + b = \int (ay - 1)^{-1/2} dy$$

or
$$x + b = \frac{2}{a}(ay - 1)^{1/2} \quad \text{or} \quad a^2(x + b)^2 = 4(ay - 1)$$

which is required equation of the curve and represents a system of parabolas having axes parallel to $y$-axis.

## PROBLEMS 15.4

Solve the following equations :

1. $2\dfrac{d^2y}{dx^2} - \left(\dfrac{dy}{dx}\right)^2 + 4 = 0.$
2. $y\dfrac{d^2y}{dx^2} + \left(\dfrac{dy}{dx}\right)^2 = 1.$
3. $y\dfrac{d^2y}{dx^2} - \left(\dfrac{dy}{dx}\right)^2 = y^2 \log y.$
4. $y(1 - \log y)\dfrac{d^2y}{dx^2} + (1 + \log y)\left(\dfrac{dy}{dx}\right)^2 = 0.$
5. Find the curve in which the radius of curvature is equal to the normal and is in the same direction.

## 15.6 EQUATIONS WHOSE ONE SOLUTION IS KNOWN

Consider the equation $d^2y/dx^2 + P\,dy/dx + Q = R$, where $P$, $Q$ and $R$ are functions of $x$ only. If $y = u(x)$ is a known solution of this equation, then put $y = uv$ in it. It *reduces the differential equation to one of first order in $dv/dx$ which can be completely solved**.

One integral belonging to the C.F. can be found by inspection as follows ;

(*i*) If $1 + P + Q = 0$, then $y = e^x$ is a solution,

(*ii*) If $I - P + Q = 0$, then $y = e^{-x}$ is a solution,

(*iii*) If $P + Qx = 0$, then $y = x$ is a solution.

**Example 15.8.** *Solve* $x\dfrac{d^2y}{dx^2} - (2x - 1)\dfrac{dy}{dx} + (x - 1)y = 0.$

**Solution.** The given equation is $\dfrac{d^2y}{dx^2} - \left(2 - \dfrac{1}{x}\right)\dfrac{dy}{dx} + \left(1 - \dfrac{1}{x}\right)y = 0$ ...(*i*)

Here $1 + P + Q = 1 - (2 - 1/x) + (1 - 1/x) = 0$

$\therefore$ $y = e^x$ is a part of C.F. of (*i*)

Now let $y = e^x v$ ...(*ii*)

so that $\dfrac{dy}{dx} = e^x v + e^x \dfrac{dv}{dx}$ ...(*iii*) and $\dfrac{d^2y}{dx^2} = e^x v + 2e^x\dfrac{dv}{dx} + e^x\dfrac{d^2v}{dx^2}$ ...(*iv*)

Substituting Eqn. (*iv*), Eqns. (*iii*) and (*ii*) in Eqn. (*i*), we get

$$x\left(e^x v + 2e^x\frac{dv}{dx} + e^x\frac{d^2v}{dx^2}\right) - (2x - 1)\left(e^x v + e^x\frac{dv}{dx}\right) + (x - 1)e^x v = 0$$

or cancelling $e^x$, it becomes $x\dfrac{d^2v}{dx^2} + \dfrac{dv}{dx} = 0$ or $x\dfrac{dp}{dx} + p = 0$, where $p = \dfrac{dv}{dx}$.

Integrating, we get $\displaystyle\int \frac{dp}{p} = -\int \frac{dx}{x} + c$ or $\log p = -\log x + \log c_1$

*i.e.*,
$$p = \frac{c_1}{x} \quad \text{or} \quad \frac{dv}{dx} = \frac{c_1}{x}.$$

Again integrating, we obtain $v = c_1 \log x + c_2$

Hence the complete solution of (*i*) is $y = e^x(c_1 \log x + c_2)$.

*This method is due to **Lagrange** (See p. 158)

**Example 15.9.** *Solve $(1 - x^2) y'' - 2xy' + 2y = 0$ given that $y = x$ is a solution.*

**Solution.** Let $y = xv$ so that $y' = v + x \dfrac{dv}{dx}$

and
$$y'' = x\frac{d^2v}{dx^2} + 2\frac{dv}{dx}$$

Substituting these in the given equation, we get

$$(1 - x^2)\left(x\frac{d^2v}{dx^2} + 2\frac{dv}{dx}\right) - 2x\left(v + x\frac{dv}{dx}\right) + 2xv = 0$$

or
$$(x - x^3)\frac{d^2v}{dx^2} + (2 - 4x^2)\frac{dv}{dx} = 0$$

or
$$(x - x^3)\frac{dp}{dx} + (2 - 4x^2)\, p = 0 \quad \text{where } p = \frac{dv}{dx}$$

Integrating, we get $\displaystyle\int \frac{dp}{p} + \int \frac{2 - 4x^2}{x - x^3}\, dx = c$

or
$$\log p + \int \frac{2}{x}dx - \int \frac{dx}{1 - x} - \int \frac{dx}{1 + x} = c$$

or
$$\log p + 2 \log x + \log (1 - x) - \log (1 + x) = \log c_1$$

$$px^2 (1 - x)/(1 + x) = c_1 \text{ or } \frac{dv}{dx} = \frac{c_1(1 + x)}{x^2 (1 - x)}$$

Again integrating, $v = c_1 \displaystyle\int \left(\frac{2}{x} + \frac{1}{x^2} + \frac{2}{1 - x}\right) dx + c_2$

or
$$v = c_1 \{2 \log (x/1 - x) - 1/x\} + c_2$$

Hence the required complete solution is $y = x\, [c_1 \{\log (x/1 - x)^2 - 1/x\} + c_2]$

**Obs.** Here $P + Qx = 0$. That is why $y = x$ is a solution of the given equation.

## PROBLEMS 15.5

1. If $y = e^{x^2}$ is a solution of $y'' - 4xy' + (4x^2 - 2)\, y = 0$, find a second independent solution.
2. Solve $x^2y'' - (x^2 + 2x)\, y' + (x + 2)y = x^3 e^x$.
3. Solve $x \dfrac{d^2y}{dx^2} - (2x - 1) \dfrac{dy}{dx} + (x - 1)\, y = e^x$ given that $y = e^x$ is one integral.
4. Solve $\sin^2 x \dfrac{d^2y}{dx^2} = 2y$, given that $y = \cot x$ is a solution.
5. Solve $\dfrac{d^2y}{dx^2} - \cot x \dfrac{dy}{dx} - (1 - \cot x)\, y = e^x \sin x$.

## 15.7 EQUATIONS WHICH CAN BE SOLVED BY CHANGING THE INDEPENDENT VARIABLE

Consider the equation
$$\frac{d^2y}{dx^2} + P\frac{dy}{dx} + Qy = R \qquad ...(1)$$

To change the independent variable $x$ to $z$, let $z = f(x)$

Then
$$\frac{dy}{dx} = \frac{dy}{dz} \cdot \frac{dz}{dx} \qquad ...(2)$$

$$\frac{d^2y}{dx^2} = \frac{d}{dx}\left(\frac{dy}{dz} \cdot \frac{dz}{dx}\right) = \frac{dy}{dz} \cdot \frac{d^2z}{dx^2} + \left(\frac{dz}{dx}\right)^2 \frac{d^2y}{dx^2} \qquad ...(3)$$

Substituting Eqns. (2) and (3) in Eqn. (1), we get $\frac{d^2y}{dz^2} + P_1 \frac{dy}{dz} + Q_1 y = R_1$ ...(4)

where $$P_1 = \left(\frac{d^2z}{dx^2} + P\frac{dz}{dx}\right)\Big/\left(\frac{dz}{dx}\right)^2, Q_1 = Q\Big/\left(\frac{dz}{dx}\right)^2, R_1 = R\Big/\left(\frac{dz}{dx}\right)^2$$

Now Eqn. (4) can be solved by taking $Q_1$ = a constant.

**Example 15.10.** *Solve, by changing the independent variable,* $x\frac{d^2y}{dx^2} - \frac{dy}{dx} + 4x^3y = x^5$.

**Solution.** Given equation is $\frac{d^2y}{dx^2} - \frac{1}{x}\frac{dy}{dx} + 4x^2y = x^4$ ...(*i*)

Here $P = -1/x$, $Q = 4x^2$ and $R = x^4$.

Choose $z$ so that $Q/(dz/dx)^2$ = const. or $(dz/dx)^2 = 4x^2$ (say)

or $$\frac{dz}{dx} = 2x \quad \text{or} \quad z = x^2$$

Changing the independent variable $x$ to $z$ by $z = x^2$, we get

$$\frac{d^2y}{dz^2} + P \cdot \frac{dy}{dz} + Q_1 y = R_1 \qquad ...(ii)$$

where, $P_1 = \left(\frac{d^2z}{dx^2} + P\frac{dz}{dx}\right)\Big/\left(\frac{dz}{dx}\right)^2 = [2 + (-x^{-1})\,2x]/4x^2 = 0$

$$Q_1 = \frac{Q}{(dz/dx)^2} = \frac{4x^2}{4x^2} = 1, R_1 = \frac{R}{(dz/dx)^2} = \frac{x^4}{4x^2} = \frac{x^2}{4} = \frac{z}{4}$$

$\therefore$ Eqn. (*ii*) takes the form $\frac{d^2y}{dz^2} + y = \frac{z}{4}$ or $(D^2 + 1)\,y = \frac{z}{4}$

Its A.E. is $D^2 + 1 = 0$, *i.e.*, $D = \pm i$

$$\text{C.F.} = c_1 \cos z + c_2 \sin z$$

$$\text{P.I.} = \frac{1}{D^2+1}\frac{z}{4} = \frac{1}{4}(1 + D^2)^{-1}z = \frac{1}{4}(1 - D^2 \,.........)\, z = \frac{z}{4}.$$

Hence the complete solution of Eqn. (*i*) is

$$y = c_1 \cos z + c_2 \sin z + \frac{z}{4} \quad \text{or} \quad y = c_1 \cos x^2 + c_2 \sin x^2 + \frac{x^2}{4}.$$

**Example 15.11.** *Solve* $\frac{d^2y}{dx^2} + \cot x\frac{dy}{dx} + 4y \operatorname{cosec}^2 x = 0.$ ...(*i*)

**Solution.** Here $P = \cot x, Q = 4 \operatorname{cosec}^2 x$

Choosing $z$ so that $Q\Big/\left(\frac{dz}{dx}\right)^2$ = const. or $\left(\frac{dz}{dx}\right)^2 = \operatorname{cosec}^2 x$ (say)

$$dz/dx = \operatorname{cosec} x \text{ or } z = \int \operatorname{cosec} x\, dx = \log \tan x/2$$

Changing the independent variable $x$ to $z$, we get

$$\frac{d^2y}{dz^2} + P_1\frac{dy}{dz} + Q_1 y = R_1 \qquad ...(ii)$$

where $P_1 = \left(\frac{d^2z}{dx^2} + P\frac{dz}{dx}\right)\Big/\left(\frac{dz}{dx}\right)^2 = (-\operatorname{cosec} x \cot x + \cot x \operatorname{cosec} x)/\operatorname{cosec}^2 x = 0$

$$Q_1 = Q\Big/\left(\frac{dz}{dx}\right)^2 = \frac{4 \operatorname{cosec}^2 x}{\operatorname{cosec}^2 x} = 4, R_1 = 0$$

$\therefore$ Eqn. (*ii*) takes the form $\dfrac{d^2y}{dz^2} + 4y = 0$

Its solution is $y = c_1 \cos(2z) + c_2 \sin(2z)$

*i.e.,* $$y = c_1 \cos(2 \log \tan x/2) + c_2 \sin(2 \log \tan x/2)$$

This is the required complete solution of Eqn. (*i*).

## PROBLEMS 15.6

Solve the following equations (by changing the independent variable) :

1. $\dfrac{d^2y}{dx^2} + \tan x \dfrac{dy}{dx} + y \cos^2 x = 0.$
2. $\dfrac{d^2y}{dx^2} + \dfrac{2}{x}\dfrac{dy}{dx} + \dfrac{y}{x^4} = 0.$
3. $\dfrac{d^2y}{dx^2} - \cot x \dfrac{dy}{dx} - \sin^2 xy = 0.$
4. $x^4 \dfrac{d^2y}{dx^2} + 2x^3 \dfrac{dy}{dx} + n^2 y = 0$ (*U.K.T.U., 2011*)
5. $x\dfrac{d^2y}{dx^2} + (4x^2 - 1)\dfrac{dy}{dx} + 4x^3y = 2x^3.$
6. $\cos x \dfrac{d^2y}{dx^2} + \sin x \dfrac{dy}{dx} - 2y \cos^3 x = 2 \cos^5 x.$
7. $\dfrac{d^2y}{dx^2} + 2x\dfrac{dy}{dx} + (x^2 - 8)y = x^2 e^{-x^2/2}.$ (*U.P.T.U., 2013*)

## 15.8 TOTAL DIFFERENTIAL EQUATIONS

**(1)** An ordinary differential equation of the first order and first degree involving three variables is of the form

$$P + Q\frac{dy}{dx} + R\frac{dz}{dx} = 0 \qquad ...(1)$$

where $P, Q, R$ are functions of $x, y, z$ and $x$ is the independent variable.

In terms of differentials, eqn. (1) can be written as

$$Pdx + Qdy + Rdz = 0 \qquad ...(2)$$

which is integrable only if

$$P\left(\frac{\partial Q}{\partial z} - \frac{\partial R}{\partial y}\right) + Q\left(\frac{\partial R}{\partial x} - \frac{\partial P}{\partial z}\right) + R\left(\frac{\partial P}{\partial y} - \frac{\partial Q}{\partial x}\right) = 0 \qquad ...(3)$$

**(2) Rule** *to solve* $Pdx + Qdy + Rdz = 0$

*If the condition of integrability is satisfied, consider one of the variables say : z, as constant so that dz = 0. Then integrate the equation Pdx + Qdy = 0. Replace the arbitrary constant appearing in its integral by ϕ(z). Now differentiate the integral just obtained with respect to x, y, z. Finally, compare this result with the given differential equation to determine ϕ(z).*

**Example 15.12.** *Solve* $(y^2 + yz)dx + (z^2 + zx)dy + (y^2 - xy)dz = 0.$

**Solution.** Here $P = y^2 + yz,\ Q = z^2 + zx,\ R = y^2 - xy.$

$$\therefore \quad P\left(\frac{\partial Q}{\partial z} - \frac{\partial R}{\partial y}\right) + Q\left(\frac{\partial R}{\partial x} - \frac{\partial P}{\partial z}\right) + R\left(\frac{\partial P}{\partial y} - \frac{\partial Q}{\partial x}\right)$$

$$= (y^2 + yz)\,[2z + x - (2y - x)] + (z^2 + zx)\,[-y - y] + (y^2 - xy)\,[(2y + z) - z] = 0$$

Hence the condition of integrability is satisfied.

Considering $z$ as constant, the given equation becomes

$$(y^2 + yz)dx + (z^2 + zx)dy = 0, \quad \text{or} \quad \frac{dx}{z(z + x)} + \frac{dy}{y(y + z)} = 0$$

Integrating and noting that $z$ is a constant, we get

$$\frac{1}{z}\int \frac{dx}{z + x} + \frac{1}{z}\int \left(\frac{1}{y} - \frac{1}{y + z}\right) dy = \text{constant}$$

*i.e.,* $$\log(z + x) + \log y - \log(y + z) = \text{constant.}$$

*i.e.,* $$\frac{y(z + x)}{y + z} = \text{constant} = \phi(z), \text{ say} \qquad ...(i)$$

or $$y(z + x) - (y + z)\,\phi(z) = 0$$

Differentiating w.r.t. $x, y, z$, we obtain

$$y(dz + dx) + (z + x)dy - [(y + z)\phi'(z)dz + (dy + dz)\phi(z)] = 0$$

or $$ydx + [z + x - \phi(z)]dy + [y - (y + z)\,\phi'(z) - \phi(z)]\,dz = 0 \qquad ...(ii)$$

Comparing Eqn. (*ii*) with the given differential equation, we get

$$\frac{y^2 + yz}{y} = \frac{z^2 + zx}{z + x - \phi(z)} = \frac{y^2 - xy}{y - (y + z)\phi'(z) - \phi(z)}.$$

The relation $\dfrac{y^2 + yz}{y} = \dfrac{z^2 + zx}{z + x - \phi(z)}$ reduces to Eqn. (*i*). $\therefore$ it gives no information about $\phi(z)$.

Taking $\dfrac{y^2 + yz}{y} = \dfrac{y^2 - xy}{y - (y + z)\phi'(z) - \phi(z)}$, we get

$$y^2 - xy = (y + z)\,[y - (y + z)\phi'(z) - \phi(z)] = y^2 + yz - (y + z)^2\,\phi'(z) - (y + z)\,\phi(z)$$

$$= y^2 + yz - (y + z)^2\,\phi'(z) - y(z + x) \qquad \text{[From Eqn. (i)]}$$

$$= y^2 - xy - (y + z)^2\,\phi'(z)$$

*i.e.*, $(y + z)^2\,\phi'(z) = 0$, *i.e.*, $\phi'(z) = 0$ so that $\phi(z) = c$

Hence the required solution is $y(z + x) = (y + z)\,c$. [From Eqn. (*i*)]

**Obs.** *Sometimes the integral is readily obtained by simply regrouping the terms in the given equation* as is illustrated below.

**Example 15.13.** *Solve* $xdx + zdy + (y + 2z)\,dz = 0$.

**Solution.** Regrouping the terms, we can write the given equation as

$$xdx + (ydz + zdy) + 2z\,dz = 0$$

of which the integral is $\dfrac{x^2}{2} + yz + z^2 = c$.

**Example 15.14:** $y'' + y'\cos y = 0$

**Solution:** Let $z = y' = \dfrac{dy}{dx}$

thus, $$y'' = \frac{dz}{dx} = \frac{dz}{dy}\frac{dy}{dx} = \frac{dy}{dy}y' = \frac{dz}{dy}z$$

Above equation becomes first order differential equation of $z$ (dependent variable) with respect to $y$ (independent variable)

$$\frac{dz}{dy}z + z^3\cos y = 0$$

which can be solved by separation of variables

$$\frac{-dz}{z^2} = \cos y\,dy \text{ or } \frac{1}{z} = \sin y + c_1$$

or $$z = y' = \frac{dy}{dx} = \frac{1}{\sin y + c_1}$$

which can be solved by separation of variables again,

$$(\sin y + c_1)\,dy = dx$$

$$\Rightarrow \quad -\cos y + c_1 y + c_2 = 0$$

**Example 15.15. :** *Solve* $\dfrac{dy}{dx} + 3y = e^{-x}$

**Solution:** Given, $$\frac{dy}{dx} + 3y = e^{-x}$$

We multiply both side of by $e^{3x}$

We obtain $$e^{3x}\frac{dy}{dx} + 3e^{3x}y = e^{2x}$$

We note that the given terms are the derivative of the product $ye^{3x}$

Thus $$\frac{d}{dx}\left[y^{e\,3x}\right] = e^{2x}$$

Now, Integrate $$ye^{3x} = \frac{1}{2}e^{2x} + C$$

and $$y = \frac{1}{2}e^{-x} + ce^{-3x}$$

**Example 15.16:** *Solve for $x > 0$*

$$\frac{dy}{dx} + \frac{y}{x} = x$$

*Subject to $y = 4$, when $x = 3$.* (*Chennai, 2018*)

**Solution:** Given $$\frac{dy}{dx} + \frac{y}{x} = x, \quad x > 0,$$

$$P(x) = \frac{1}{x} \text{ and } Q(x) = x$$

Our integrating factor $$V(x) = e^{\int P(x)\,dx} = e^{\int \frac{1}{x}dx} = e^{\ln|x|} \qquad \{\because |x| = x\}$$

$$V(x) = x \qquad \text{...}(i)$$

Now, multiplication by $x$ in Eqn. ($i$) and differentiate

$$x\frac{dy}{dx} + y = x^2$$

and $$\frac{d}{dx}[x\,y] = x^2$$

Integration $$xy = \frac{1}{3}x^3 + c$$

and $$y = \frac{1}{3}x^2 + cx^{-1}$$

The I.C. gives $4 = y(3) = \dfrac{9}{3} + \dfrac{C}{3}$

and so $c = 3$.

Thus, $y(x) = \dfrac{1}{3}x^2 + 3x^{-1}, x > 0;$

## PROBLEMS 15.7

Solve :

1. $(mz - ny)dx + (nx - lz)dy + (ly - mx)dz = 0.$
2. $(y^2 + z^2 - x^2)dx - 2xydy - 2xzdz = 0.$
3. $yzdx - 2zxdy - 3xydz = 0.$
4. $(2xz - yz)dx + (2yz - zx)dy - (x^2 - xy + y^2)dz = 0.$
5. $x\,dx + y\,dy - \sqrt{a^2 - x^2 - y^2}\,dz = 0.$ (*Poona, 2019*)

6. $(yz + xyz)dx + (zx + xyz)\,dy + (xy + xyz)\,dz = 0.$
7. $yz\,dx - 2xz\,dy + (xy - zy^3)\,dz = 0$
8. $(e^x y + e^x)\,dx + (e^y z + e^x)\,dy + (e^y - e^x y - e^y. z).\,dz = 0$
9. $(y^2 + yz + z^2)\,dx + (z^2 + zx + x^2)dy + (x^2 + xy + y^2)dz = 0.$

## 15.9 SIMULTANEOUS TOTAL DIFFERENTIAL EQUATIONS

These equations in three variables are given by

$$\left.\begin{aligned} Pdx + Qdy + Rdz = 0 \\ P'dx + Q'dy + R'dz = 0 \end{aligned}\right\} \quad ...(1)$$

where $P$, $Q$, $R$ and $P'$, $Q'$, $R'$ are any functions of $x, y, z$.

(*a*) If each of these equations is integrable and have solutions $f(x, y, z) = c$ and $Y(x, y, z) = c¢$ respectively, then these taken together constitute the solution of the simultaneous Eqn. (1).

(*b*) If one or both the Eqn. (1) is not integrable, then we write these as follows :

$$\frac{dx}{QR' - Q'R} = \frac{dy}{RP' - R'P} = \frac{dz}{PQ' - P'Q}$$

and solve these by the methods explained below.

## 15.10 EQUATIONS OF THE FORM dx/P = dy/Q = dz/R

**(1) Method of grouping**

See if it is possible to take two fractions $dx/P = dz/R$ from which $y$ can be cancelled or is absent, leaving equations in $x$ and $z$ only.

If so, integrate it by giving $\phi(x, z) = c$. ...(1)

Again see if one variable say : $x$ is absent or can be removed may be with the help of (1), from the equation $dy/Q = dz/R$.

Then integrate it by giving $\psi(y, z) = c'$ ...(2)

These two independent solutions (1) and (2) taken together constitute the complete solution required.

**Example 15.17.** *Solve* $\dfrac{dx}{z^2 y} = \dfrac{dy}{z^2 x} = \dfrac{dz}{y^2 x}$.

**Solution.** Taking the first two fractions and cancelling $z^2$, we get

$$\frac{dx}{y} = \frac{dy}{x} \quad \text{or} \quad xdx - ydy = 0$$

which on integration gives $x^2 - y^2 = c$. ...(*i*)

Again taking the second and third fractions and cancelling $x$, we have

$$\frac{dy}{z^2} = \frac{dz}{y^2}, \text{ i.e., } y^2dy - z^2dz = 0.$$

Its integral is $y^3 - z^3 = c'$. ...(*ii*)

Thus Eqns. (*i*) and (*ii*) taken together constitute the required solution of the given equations.

**(2) Method of multipliers**

By a proper choice of the multipliers $l, m, n$ which are not necessarily constants, we write

$$\frac{dx}{P} = \frac{dy}{Q} = \frac{dz}{R} = \frac{ldx + mdy + ndz}{lP + mQ + nR} \text{ such that } lP + mQ + nR = 0.$$

Then $ldx + mdy + ndz = 0$ can be solved giving the integral $\phi(x, y, z) = c$ ...(1)

Again search for another set of multipliers $\lambda, \mu, \gamma$

so that $\lambda P + \mu Q + \gamma R = 0$

giving $\lambda dx + \mu dy + \gamma dz = 0,$

which on integration gives the solution $\psi(x, y, z) = c'$ ...(2)

These two solutions Eqns. (1) and (2) taken together constitute the required solution.

**Example 15.18.** *Solve* $\dfrac{dx}{x(y^2 - z^2)} = \dfrac{dy}{-y(z^2 + x^2)} = \dfrac{dz}{z(x^2 + y^2)}$.

**Solution.** Using the multipliers $x, y, z$

each fraction $= \dfrac{xdx + ydy + zdz}{x^2(y^2 - z^2) - y^2(z^2 + x^2) + z^2(x^2 + y^2)} = \dfrac{xdx + ydy + zdz}{0}$

$\therefore$ $xdx + ydy + zdz = 0$, which on integration gives the solution $x^2 + y^2 + z^2 = c$ ...(*i*)

Again using the multipliers $1/x, -1/y, -1/z$

each fraction $= \dfrac{\frac{1}{x}dx - \frac{1}{y}dy - \frac{1}{z}dz}{(y^2 - z^2) + (z^2 + x^2) - (x^2 + y^2)} = \dfrac{\frac{1}{x}dx - \frac{1}{y}dy - \frac{1}{z}dz}{0}$ so that $\dfrac{dx}{x} - \dfrac{dy}{y} - \dfrac{dz}{z} = 0$

which on integration gives $\log x - \log y - \log z =$ constant or $yz = c'x$. ...(*ii*)

Hence the solution of the given equation is $x^2 + y^2 + z^2 = c$ ; $yz = c'x$.

## PROBLEMS 15.8

**Solve :**

1. $\dfrac{xdx}{y^2z} = \dfrac{dy}{xz} = \dfrac{dz}{y^2}$.
2. $\dfrac{dx}{mz - ny} = \dfrac{dy}{nx - lz} = \dfrac{dx}{ly - mx}$.
3. $\dfrac{dx}{x^2 - yz} = \dfrac{dy}{y^2 - zx} = \dfrac{dz}{z^2 - xy}$.
4. $\dfrac{dx}{y - zx} = \dfrac{dy}{yz + x} = \dfrac{dz}{x^2 + y^2}$.
5. $\dfrac{dx}{x(y^2 - z^2)} = \dfrac{dy}{y(z^2 - x^2)} = \dfrac{dz}{z(x^2 - y^2)}$.
6. $\dfrac{dx}{x^2 - y^2 - z^2} = \dfrac{dy}{2xy} = \dfrac{dz}{2xz}$. (*Calicut, 2012*)

CHAPTER 16

# Series Solution of Differential Equations and Special Functions

1. Introduction. 2. Validity of series solution. 3. Series solution when $x = 0$ is an ordinary point. 4. Frobenius method. 5. Bessel's equation. 6. Recurrence formulae for $J_n(x)$. 7. Expansions for $J_0$ and $J_1$. 8. Value of $J_{1/2}$. 9. Generating function for $J_n(x)$. 10. Equations reducible to Bessel's equation. 11. Orthogonality of Bessel functions; Fourier-Bessel expansion of $f(x)$. 12. Ber and Bei functions. 13. Legendre's equation. 14. Rodrigue's formula, Legendre polynomials. 15. Generating function for $P_n(x)$. 16. Recurrence formulae for $P_n(x)$. 17. Orthogonality of Legendre polynomials, Fourier-Legendre expansion for $f(x)$. 18. Other special functions. 19. Strum-Liouville problem, Orthogonality of eigen functions. 20. Objective Type of Questions.

## 16.1 INTRODUCTION

Many differential equations arising from physical problems are linear but have variable coefficients and do not permit a general solution in terms of known functions. Such equations can be solved by numerical methods (Chapter 28), but in many cases it is easier to find a solution in the form of an infinite convergent series.

The series solution of certain differential equations give rise to special functions such as Bessel's function, Legendre's polynomial, Lagurre's polynomial, Hermite's polynomial, Chebyshev polynomials. Strum-Lioville problem based on the orthogonality of functions is also included which shows that Bessel's, Legendre's and other equations can be considered from a common point of view. These special functions have many applications in engineering. For instance, the solutions of Wave equation, Heat equation with cylinderical symmetry can be expressed in terms of Bessel functions and with spherical symmetry in terms of Legendre polynomials. Also Fourier-Bessel series, Fourier-Legendre series and Chebyshev polynomials are useful in approximation theory.

## 16.2 VALIDITY OF SERIES SOLUTION OF THE EQUATION

$$P_0(x)\frac{d^2y}{dx^2} + P_1(x)\frac{dy}{dx} + P_2(x)y = 0 \qquad ...(i)$$

can be determined with the help of the following theorems :

**Def. 1.** If $P_0(a) \neq 0$, then $x = a$ is called and **ordinary point** of Eqn. (*i*), otherwise a **singular point.**

**2.** *A singular point $x = a$ of (1) is called* **regular** *if, when* Eqn. (*i*) *is put in the form*

$$\frac{d^2y}{dx^2} + \frac{Q_1(x)}{x-a}\frac{dy}{dx} + \frac{Q_2(x)}{(x-a)^2}y = 0,$$

*$Q_1(x)$ and $Q_2(x)$ possess derivatives of all orders in the neighbourhood of a.*

**3.** *A singular point which is not regular is called an* **irregular singular point.**

**Theorem I.** *When $x = a$ is an ordinary point of Eqn. (i), its every solution can be expressed in the form*

$$y = a_0 + a_1(x-a) + a_2(x-a)^2 + \dots \qquad ...(ii)$$

**Theorem II.** *When $x = a$ is a regular singularity of Eqn. (i), at least one of the solutions can be expressed as*

$$y = (x - a)^m [a_0 + a_1(x - a) + a_2(x - a)^2 + ...] \quad ...(iii)$$

**Theorem III.** *The series* Eqns. *(ii) and (iii) are convergent at every point within the circle of convergence at a.* A solution in series will be valid only if the series is convergent.

## 16.3 SERIES SOLUTION WHEN X = 0 IS AN ORDINARY POINT OF THE EQUATION

$$P_0 \frac{d^2y}{dx^2} + P_1 \frac{dy}{dx} + P_2 y = 0 \quad ...(1)$$

*where P's are polynomials in x and $P_0 \neq 0$ at $x = 0$.*

(*i*) Assume its solution to be of the form $y = a_0 + a_1x + a_2x^2 + ... + a_nx^n + ...$ ...(2)

(*ii*) Calculate $dy/dx$, $d^2y/dx^2$ from Eqn. (2) and substitute the values of $y$, $dy/dx$, $d^2y/dx^2$ in Eqn. (1).

(*iii*) Equate to zero the coefficients of the various powers of $x$ and determine $a_2, a_3, a_4$ ... in terms of $a_0, a_1$. (The result obtained by equating to zero is the coefficient of $x^n$ that is called the *recurrence relation*).

(*iv*) Substituting the values of $a_2, a_3, a_4$ ... in Eqn. (2), we get the desired series solution having $a_0, a_1$ as its arbitrary constants.

**Example 16.1.** *Solve in series the equation* $\frac{d^2y}{dx^2} + xy = 0.$ *(V.T.U., 2010)*

**Solution.** Here $x = 0$ is an ordinary point since coefficient of $y'' \neq 0$ at $x = 0$.

Assume its solution is $y = a_0 + a_1x + a_2x^2 + a_3x^3 + ... + a_nx^n + ...$ ...(*i*)

Then $$\frac{dy}{dx} = a_1 + 2a_2x + 3a_3x^2 + ... + na_nx^{n-1} + ...$$

and $$\frac{d^2y}{dx^2} = 2 . 1a_2 + 3 . 2a_3x + 4 . 3a_4x^2 + ... + n(n-1)a_nx^{n-2} + ...$$

Substituting in the given differential equation

$$1 . 1a_2 + 3 . 2a_3x + 4 . 3a_4x^2 + ... + n(n-1)a_nx^{n-2} + ... + x(a_0 + a_1x + a_2x^2 + ... + a_nx^n + ...) = 0$$

or $$2. 1a_2 + (3 . 2a_3 + a_0)x + (4 . 3a_4 + a_1)x^2 + (5 . 4a_5 + a_2)x^3 + ... + [(n+2)(n+1)a_{n+2} + a_{n-1}]x^n + ... = 0.$$

Equating to zero the co-efficients of the various powers of $x$,

$a_2 = 0,$ [Coeff. of $x^0 = 0$]

$3 . 2a_3 + a_0 = 0$, *i.e.*, $a_3 = -\frac{a_0}{3!}$ [Coeff. of $x = 0$]

$4 . 3a_4 + a_1 = 0$, *i.e.*, $a_4 = -\frac{2a_1}{4!}$ [Coeff. of $x^2 = 0$]

$5 . 4a_5 + a_2 = 0$, *i.e.*, $a_5 = -\frac{a_2}{5 . 4} = 0$ and so on. [Coeff. of $x^3 = 0$]

In general, $(n+2)(n+1)a_{n+2} + a_{n-1} = 0$ [Coeff. of $x^n = 0$]

*i.e.*, $$a_{n+2} = \frac{-a_{n-1}}{(n+2)(n+1)} \quad ..(ii)$$

which is the *recurrence relation.*

Putting $n = 4, 5, 6, ...$ in Eqn. (*ii*) successively, $a_6 = -\frac{a_3}{6 . 5} = \frac{4a_0}{6!}$; $a_7 = -\frac{a_4}{7 . 6} = \frac{5 . 2a_1}{7!}$

$$a_8 = -\frac{a_5}{8 . 7} = 0; \; a_9 = -\frac{a_6}{9 . 8} = -\frac{7 . 4a_0}{9!} \text{ and so on.}$$

Substituting these values in Eqn. (*i*), we get

$$y = a_0 \left(1 - \frac{x^3}{3!} + \frac{1 . 4x^6}{6!} - \frac{1 . 4 . 7x^9}{9!} + ...\right) + a_1 \left(x - \frac{2x^4}{4!} + \frac{2 . 5x^7}{7!} - ...\right)$$

which is the required solution.

**Example 16.2.** *Solve in series* $(1-x^2)\frac{d^2y}{dx^2} - x\frac{dy}{dx} + 4y = 0.$

**Solution.** Here $x = 0$ is an ordinary point since coefficient of $y'' \neq 0$ at $x = 0$.

Assume the solution of the given equation to be

$$y = a_0 + a_1x + a_2x^2 + a_3x^3 + \dots + a_nx^n + \dots \qquad \dots(i)$$

Then $$\frac{dy}{dx} = a_1 + 2a_2x + 3a_3x^2 + \dots + na_nx^{n-1} + \dots$$

and $$\frac{d^2y}{dx^2} = 2a_2 + 3.2a_3x + 4.3a_4x^2 + \dots + n(n-1)a_nx^{n-2} + \dots$$

Substituting in the given equation, we get

$$(1-x^2)[2a_2 + 3.2a_3x + 4.3a_4x^2 + \dots + n(n-1)a_nx^{n-2} + \dots]$$
$$-x[a_1 + 2a_2x + 3a_3x^2 + \dots + na_nx^{n-1} + \dots] + 4[a_0 + a_1x + a_2x^2 + \dots + a_nx^n + \dots] = 0$$

Equating to zero the coefficients of the various powers of $x$,

$$2a_2 + 4a_0 = 0 \quad i.e., \quad a_2 = -2a_0 \qquad [\text{coeff. of } x^0 = 0]$$

$$3.2a_3 - a_1 + 4a_1 = 0 \quad i.e., \quad a_3 = -\frac{1}{2}a_1 \qquad [\text{coeff. of } x^1 = 0]$$

$$4.3a_4 - 2a_2 - 2a_2 + 4a_2 = 0 \quad i.e., \quad a_4 = 0 \qquad [\text{coeff. of } x^2 = 0]$$

$$5.4a_5 - 3.2a_3 - 3a_3 + 4a_3 = 0 \qquad [\text{coeff. of } x^3 = 0]$$

*i.e.,* $$20a_5 - 5a_3 = 0 \quad i.e., \quad a_5 = -\frac{a_1}{8} \text{ and so on.}$$

In general, $(n+2)(n+1)a_{n+2} - n(n-1)a_n - na_n + 4a_n = 0$

or $$a_{n+2} = \frac{n-2}{n+1}a_n \qquad \dots(ii)$$

which is the *recurrence relation*

Putting $n = 4, 5, 6, 7\dots$ in Eqn. $(ii)$ successively,

$$a_6 = 0; \quad a_7 = \frac{3}{6}a_5 = -\frac{3}{6}\frac{a_1}{8}; \; a_8 = 0; \; a_9 = \frac{5.3}{8.6}\cdot\frac{a_1}{8}\dots$$

Substituting these values in Eqn. $(i)$, we get

$$y = a_0(1-2x^2) + a_1x\left(1 - \frac{x^2}{2} - \frac{x^4}{8} - \frac{3}{6}\cdot\frac{x^6}{8} - \frac{5.3}{8.6}\cdot\frac{x^8}{8} - \dots\right).$$

**PROBLEMS 16.1**

Solve the following equations in series :

1. $\frac{d^2y}{dx^2} + y = 0$, given $y(0) = 0$. *(V.T.U., 2016)*
2. $\frac{d^2y}{dx^2} + x^2y = 0$.
3. $y'' + xy' + y = 0$. *(V.T.U., 2015 ; Andhra, 2013)*
4. $(1-x^2)y'' + 2y = 0$, given $y(0) = 4$, $y'(0) = 5$. *(Andhra, 2015)*
5. $(1+x^2)\frac{d^2y}{dx^2} + x\frac{dy}{dx} - y = 0$. *(V.T.U., 2013)*
6. $(1-x^2)\frac{d^2y}{dx^2} - x\frac{dy}{dx} + 4y = 0$. *(U.P.T.U., 2012)*

## 16.4 FROBENIUS* METHOD : Series solution when x = 0 is a regular singularity of the equation

$$\mathbf{P_0\frac{d^2y}{dx^2} + P_1\frac{dy}{dx} + P_2y = 0} \qquad \dots(1)$$

*A German mathematician *F.G. Frobenius* (1849–1917) who is known for his contributions to the theory of matrices and groups.

(*i*) Assume the solution to be $y = x^m(a_0 + a_1x + a_2x^2 + a_3x^3 + ... + a_nx^n + ...)$ ...(2)

(*ii*) Substitute from Eqn. (2) for $y$, $dy/dx$, $d^2y/dx^2$ in Eqn. (1) as before.

(*iii*) Equate to zero the coefficient of the lowest degree term in $x$. It gives a quadratic equation known as the *indicial equation*.

(*iv*) Equating to zero the coefficients of the other powers of $x$, find the values of $a_1, a_2, a_3,...$ in terms of $a_0$.

*The complete solution depends on the nature of roots of the indicial equation.*

**Case I. When roots of the indicial equation are distinct and do not differ by an integer,** *the complete solution is*

$$y = c_1(y)_{m_1} + c_2(y)_{m_2}$$

where $m_1, m_2$ are the roots.

**Example 16.3.** *Solve in series the equation* $9x(1-x)\dfrac{d^2y}{dx^2} - 12\dfrac{dy}{dx} + 4y = 0.$

**Solution.** Here $x = 0$ is a singular point since coefficient of $y'' = 0$ at $x = 0$.

Substituting $y = a_0x^m + a_1x^{m+1} + a_2x^{m+2} + ...$

$$\therefore \quad \frac{dy}{dx} = ma_0x^{m-1} + (m+1)a_1x^m + (m+2)a_2x^{m+1} + ...$$

and
$$\frac{d^2y}{dx^2} = m(m-1)a_0x^{m-2} + (m+1)ma_1x^{m-1} + (m+2)(m+1)a_2x^m + ...$$

in the given equation, we obtain

$$9x(1-x)\,[m(m-1)a_0x^{m-2} + (m+1)ma_1x^{m-1} + (m+2)(m+1)a_2x^m + ...]$$
$$- 12[ma_0x^{m-1} + (m+1)a_1x^m + (m+2)a_2x^{m+1} + ...] + 4[a_0x^m + a_1x^{m+1} + a_2x^{m+2} + ...] = 0.$$

The lowest power of $x$ is $x^{m-1}$. Its coefficient equated to zero gives

$$a_0(9m(m-1) - 12m] = 0, \text{ i.e., } m(3m-7) = 0 \quad \text{as} \quad a_0 \neq 0.$$

Thus the roots of the *indicial equation* are $m = 0, 7/3$. *i.e., Roots are distinct and do not differ by an integer.*

The coefficient of $x^m$ equated to zero gives $a_1\{9(m+1)m - 12(m+1)\} + a_0\{4 - 9m(m-1)\} = 0$

*i.e.,* $\quad 3a_1(3m-4)(m+1) - a_0(3m-4)(3m+1) = 0$

*i.e.,* $\quad 3a_1(m+1) = a_0(3m+1).$

Similarly $3a_2(m+2) = a_1(3m+4)$, $3a_3(m+3) = a_2(5m+7)$ and so on.

$$\therefore \quad a_1 = \frac{3m+1}{3(m+1)}a_0,\ a_2 = \frac{(3m+4)a_1}{3(m+2)} = \frac{(3m+4)(3m+1)}{3^2(m+2)(m+1)}a_0,\ a_3 = \frac{(3m+7)(3m+4)(3m+1)}{3^3(m+3)(m+2)(m+1)}a_0 \text{ etc.}$$

*When* $m = 0$, $a_1 = \dfrac{1}{3}a_0$, $a_2 = \dfrac{1.4}{3.6}a_0$, $a_3 = \dfrac{1.4.7}{3.6.9}a_0$ etc. giving the particular solution

$$y_1 = a_0\left[1 + \frac{1}{3}x + \frac{1.4}{3.6}x^2 + \frac{1.4.7}{3.6.9}x^3 + ...\right]$$

*When* $m = 7/3$, the particular solution is

$$y_2 = a_0x^{7/3}\left[1 + \frac{8}{10}x + \frac{8.11}{10.13}x^2 + \frac{8.11.14}{10.13.16}x^3 + ...\right]$$

Thus the complete solution is $y = c_1y_1 + c_2y_2$

*i.e.,*
$$y = C_1\left[1 + \frac{1}{3}x + \frac{1.4}{3.6}x^2 + \frac{1.4.7}{3.6.9}x^3 + ...\right]$$
$$+ C_2x^{7/3}\left[1 + \frac{8}{10}x + \frac{8.11}{10.13}x^2 + \frac{8.11.14}{10.13.16}x^3 + ...\right], \quad \text{where } C_1 = c_1a_0,\ C_2 = c_2a_0.$$

**Case II. When roots of the indicial equation are equal** *the complete solution is*

$$y = c_1(y)_{m_1} + c_2\left(\frac{\partial y}{\partial m}\right)_{m_1}$$

where $m_1, m_1$ are the roots.

**Example 16.4.** *Solve in series the equation* $x\dfrac{d^2y}{dx^2} + \dfrac{dy}{dx} - xy = 0.$ *(U.P.T.U. 2012)*

**Solution.** Substituting $y = a_0x^m + a_1x^{m+1} + a_2x^{m+2} + \ldots$ ...(*i*)

$$\frac{dy}{dx} = ma_0x^{m-1} + (m+1)a_1x^m + (m+2)a_2x^{m+1} + \ldots$$

and $$\frac{d^2y}{dx^2} = m(m-1)a_0x^{m-2} + (m+1)ma_1x^{m-1} + (m+2)(m+1)a_2x^m + \ldots$$

in the given equation, we obtain

$$x[m(m-1)a_0x^{m-2} + (m+1)ma_1x^{m-1} + (m+2)(m+1)a_2x^m + \ldots]$$
$$+ [ma_0x^{m-1} + (m+1)a_1x^m + (m+2)a_2x^{m+1} + \ldots]$$
$$- x[a_0x^m + a_1x^{m+1} + a_2x^{m+2} + \ldots] = 0.$$

The lowest power of $x$ is $x^{m-1}$. Its coefficient equated to zero gives $a_0[m(m-1) + m] = 0$.

*i.e.,* $m^2 = 0$ as $a_0 \neq 0$. $\therefore$ $m = 0, 0$.

The coefficients of $x^m, x^{m+1}, \ldots$ equated to zero give

$$a_1[(m+1)m + m + 1] = 0, \quad i.e., \quad a_1 = 0$$

$$a_2(m+2)^2 - a_0 = 0,\ a_3(m+3)^2 + a_1 = 0,\ a_4(m+4)^2 + a_2 = 0 \text{ and so on.}$$

Clearly $a_3 = a_5 = a_7 \ldots\ldots = 0$.

Also $$a_2 = \frac{a_0}{(m+2)^2},\ a_4 = \frac{a_2}{(m+4)^2} = \frac{a_0}{(m+2)^2(m+4)^2} \text{ etc.}$$

$$\therefore \quad y = a_0x^m\left[1 + \frac{x^2}{(m+2)^2} + \frac{x^4}{(m+2)^2(m+4)^2} + \frac{x^6}{(m+2)^2(m+4)^2(m+6)^2} + \ldots\ldots\right] \quad \ldots(ii)$$

Putting $m = 0$, the first solution is

$$y_1 = a_0\left[1 + \frac{x^2}{2^2} + \frac{x^4}{2^2 . 4^2} + \frac{x^6}{2^2 . 4^2 . 6^2} + \ldots\ldots\right] \quad \ldots(iii)$$

This gives only one solution instead of two. To get the second solution, differentiate (*ii*) partially w.r.t. $m$.

$$\frac{\partial y}{\partial m} = y \log x + a_0x^m\left\{\frac{-x^2}{(m+2)^2}\frac{2}{m+2} + \frac{x^4}{(m+2)^2(m+4)^2}\left[\frac{2}{m+2} + \frac{2}{m+4}\right] + \ldots\right\}$$

$\therefore$ the second solution is $y_2 = \left(\dfrac{\partial y}{\partial m}\right)_{m=0}$

$$= y_1 \log x + a_0\left\{\frac{-1}{2^2}x^2 - \frac{1}{2^2 . 4^2}\left(1 + \frac{1}{2}\right)x^4 + \frac{1}{2^2 . 4^2 . 6^2}\left(1 + \frac{1}{2} + \frac{1}{3}\right)x^6 - \ldots\right\} \quad \ldots(iv)$$

Hence the complete solution is $y = c_1y_1 + c_2y_2$. [From Eqn. (*iii*) and (*iv*)]

*i.e.,* $$y = (C_1 + C_2 \log x)\left[1 + \frac{1}{2^2}x^2 + \frac{1}{2^2 . 4^2}x^4 + \frac{1}{2^2 . 4^2 . 6^2}x^6 + \ldots\right]$$
$$- C_2\left\{\frac{1}{2^2}x^2 + \frac{1}{2^2 . 4^2}\left(1 + \frac{1}{2}\right)x^4 + \frac{1}{2^2 . 4^2 . 6^2}\left(1 + \frac{1}{2} + \frac{1}{3}\right)x^6 - \ldots\right\}$$

where $C_1 = a_0c_1,\ C_2 = a_0c_2$.

**Obs.** The above differential equation is called *Bessel's equation of order zero*, $y_1$ is called *Bessel function of the first kind of order zero* and is denoted by $J_0(x)$. It is absolutely convergent for all values of $x$ whether real or complex.

$y_2$ is called the *Bessel function of the second kind of order zero or the Neumann function* and is denoted by $Y_0(x)$.

Thus the complete solution of the *Bessel's equation of order zero is* $y = AJ_0(x) + BY_0(x)$.

**Case III. When roots of indicial equation are distinct and differ by an integer, making a coefficient of y infinite.**

Let $m_1$ and $m_2$ be the roots such that $m_1 < m_2$. If some of the coefficients of $y$ series become infinite when $m = m_1$, we modify the form of $y$ by replacing $a_0$ by $b_0(m - m_1)$. Then the complete solution is

$$y = C_1(y)_{m_2} + C_2\left(\frac{\partial y}{\partial m}\right)_{m_1}$$

**Obs. 1.** Two independent solution can also be obtained by putting $m = m_1$ (lesser of the two roots) in the modified form of $y$ and $\partial y/\partial m$.

**Obs. 2.** *If one of the coefficients* (say : $a_1$) *becomes* **indeterminate** *when* $m = m_2$, *the complete solution is given by putting* $m = m_2$ *in y which contains two arbitrary constants.*

**Example 16.5.** *Obtain the series solution of the equation*

$$x(1-x)\frac{d^2y}{dx^2} - (1+3x)\frac{dy}{dx} - y = 0. \qquad \textit{(Calicut, 2012)}$$

**Solution.** Here $x = 0$ is a singular point, since coefficient of $y''$ is zero at $x = 0$.

$\therefore$ substituting $\quad y = a_0x^m + a_1x^{m+1} + a_2x^{m+2} + \ldots \qquad \ldots(i)$

$$\frac{dy}{dx} = ma_0x^{m-1} + (m+1)a_1x^m + (m+2)\,a_2\,x^{m+1} + \ldots$$

and
$$\frac{d^2y}{dx^2} = m(m-1)\,a_0x^{m-2} + (m+1)ma_1\,x^{m-1} + (m+2)(m+1)\,a_2x^m + \ldots$$

in the given equation, we obtain

$$x(1-x)[m(m-1)\,a_0x^{m-2} + (m+1)ma_1\,x^{m-1} + (m+2)(m+1)a_2x^m + \ldots]$$
$$-(1+3x)[ma_0x^{m-1} + (m+1)a_1x^m + (m+2)a_2x^{m+1} + \ldots] - [a_0x^m + a_1x^{m+1} + a_2x^{m+2} + \ldots] = 0$$

Equating to zero the coefficients of the lowest power of $x$, we get $a_0[m(m-1) - m] = 0$, $(a_0 \neq 0)$,

*i.e.,* $\quad m(m-2) = 0$, *i.e.* $m = 0, 2$ *i.e.*, the two roots are distinct and differ by an integer.

Equating to zero the coefficients of successive powers of $x$, we get

$$(m-1)a_1 = (m+1)a_0,\; ma_2 = (m+2)a_1,\; (m+1)a_3 = (m+3)\,a_2 \text{ and so on.}$$

*i.e.,*
$$a_1 = \frac{m+1}{m-1}a_0,\; a_2 = \frac{(m+1)(m+2)}{(m-1)m}a_0,\; a_3 = \frac{(m+1)(m+2)(m+3)}{(m-1)m(m+1)}a_0 \text{ etc.}$$

Thus Eqn. $(i)$ becomes

$$y = a_0x^m\left[1 + \frac{m+1}{m-1}x + \frac{(m+1)(m+2)}{(m-1)m}x^2 + \frac{(m+1)(m+2)(m+3)}{(m-1)m(m+1)}x^3 + \ldots\right] \qquad \ldots(ii)$$

Putting $m = 2$ (greater of the two roots) in Eqn. $(ii)$, the first solution is

$$y_1 = a_0x^2\left[1 + 3x + \frac{3.4}{2}x^2 + \frac{4.5}{2}x^3 + \ldots\right]$$

If we put $m = 0$ in Eqn. $(ii)$, the coefficients become infinite.

To obviate this difficulty, put $a_0 = b_0(m - 0)$ so that

$$y = b_0x^m\left[m + \frac{m(m+1)}{m-1}x + \frac{(m+1)(m+2)}{m-1}x^2 + \frac{(m+1)(m+2)(m+3)}{(m-1)(m+1)}x^3 + \ldots\right]$$

$$\therefore \quad \frac{\partial y}{\partial m} = b_0x^m\log x\left[m + \frac{m(m+1)x}{m-1} + \frac{(m+1)(m+2)}{m-1}x^2 + \frac{(m+1)(m+2)(m+3)}{(m-1)(m+1)}x^3 + \ldots\right]$$
$$+ b_0x^m\left[1 + \frac{m^2-2m-1}{(m-1)^2}x + \frac{m^2-m-5}{(m-1)^2}x^2 + \frac{m^2-2m-11}{(m-1)^2}x^3 + \ldots\right]$$

$\therefore$ the second solution is $y_2 = \left(\frac{\partial y}{\partial m}\right)_{m=0}$

$$= b_0 \log x\,[-1.2x^2 - 2.3x^3 - 3.4x^4 - ...] + b_0[1 - x - 5x^2 - 11x^3 - ...]$$

Hence the complete solution is $y = c_1y_1 + c_2y_2$

*i.e.*, $$y = \frac{1}{2}\,c_1a_0[1.2x^2 + 2.3x^3 + 3.4x^4 + ...] - b_0c_2 \log x\,[1.2x^2 + 2.3x^3 + 3.4x^4 + ...]$$
$$- b_0c_2\,[-1 + x + 5x^2 + 11x^3 + ...]$$

*i.e.*, $$y = (C_1 + C_2 \log x)\,(1.2x^2 + 2.3x^3 + 3.4x^4 + ...) + C_2\,(-1 + x + 5x^2 + 11x^3 + ...)$$

where $C_1 = \frac{1}{2}\,c_1a_0,\ C_2 = -\,b_0c_2.$

**Example 16.6.** *Find the first four terms in each portion of the series solution around $x_0 = 0$ for the following differential equation.*

$$(x^2 + 1)y'' - 4xy' + 6y = 0$$ *(Kurukshetra., 2018)*

**Solution:** We have a differential equation that does not have a constant coefficient for the second derivative

$$P(x) = x^2 + 1$$
$$P(0) = 1 \neq 0$$

So, $x_0 = 0$, is an ordinary point for this differential equation.

$$y(x) = \sum_{n=0}^{\infty} a_nx^n, \quad y'(x) = \sum_{n=1}^{\infty} na_nx^{n-1}$$

$$y''(x) = \sum_{n=2}^{\infty} n(n-1)a_n\,x^{n-2}$$

$$(x^2+1)\cdot\sum_{n=2}^{\infty} n(n-1)a_nx^{n-2} - 4x\sum_{n=1}^{\infty} na_nx^{n-1} + 6\sum_{n=0}^{\infty} a_nx^n = 0$$

Now, break up the first term into two

$$\sum_{n=2}^{\infty} n(n-1)a_nx^n + \sum_{n=2}^{\infty} n(n-1)a_nx^{n-2} - \sum_{n=1}^{\infty} 4na_nx^n + \sum_{n=0}^{\infty} 6a_nx^n = 0$$

We will only need to shift the second series down by two get all the exponents.

$$\sum_{n=2}^{\infty} n(n-1)a_nx^n + \sum_{n=0}^{\infty} (n+2)(n+1)a_n + 2x^n - \sum_{n=1}^{\infty} 4na_nx^n + \sum_{n=0}^{\infty} 6a_nx^n = 0$$

$$\sum_{n=0}^{\infty}\left[n(n-1)a_n + (n+2)(n+1)a_{n+2} - 4na_n + 6a_n\right]x^n = 0$$

$$\sum_{n=0}^{\infty}\left[n^2 - 5n + 6)a_n + (n+2)(n+1)a_{n+2}\right]x^n = 0$$

$$\sum_{n=0}^{\infty}\left[(n-2)(n-3)a_n + (n+2)(n+1)a_{n+1}\right]x^n = 0$$

$$(n-2)\,(n-3)a_n + (n+2)(n+1)a_{n+2} = 0,\ n = 0, 1, 2, ...$$

$$a_{n+2} = \frac{-(n-2)(n-3)a_n}{(n+2)(n+1)}, \quad n = 0, 1, 2, ...$$

$$n = 0 \ : \ a_2 = -\,3a_0$$

$$n = 1 \ : \ a_3 = -\frac{1}{3}a_1$$

$$n = 2 \quad : \quad a_4 = -\frac{0}{12}a_2 = 0$$

$$n = 3 \quad : \quad a_5 = \frac{-0}{20}a_3 = 0$$

Now, from this point on all the coefficients are zero.

So, $$y(x) = a_0 \{1 - 3x^2\} + a_1 \left\{x - \frac{1}{3}x^3\right\}.$$

**Example 16.7.** *Solve the equation in series* $\frac{d^2y}{dx^2} + x^2y = 0$ *(UPTU, 2019)*

**Solution:** Let $y = a_0 + a_1x + a_2x^2 + a_3x^3 + \dots anx^n + \dots$ ...(*i*)

$$\frac{d^2y}{dx^2} = 2a_2 + 6a_3x + 24a_4x^2 + \dots n(n + 1)a_nx^{n+2}$$

Substituting these expressions for $y$ and $\frac{d^2y}{dx^2}$ in the given equation

$[2a_2 + 6a_3x + 12a_4x^2 + \dots + n(n-1)\, a_nx^{n-2} + \dots] + x^2[a_0 + a_1x + a_2x^2 + a_3x^3 + \dots a_n\, x^n + \dots] = 0$

$= 2a_2 + a_3x + (12a_4 + a_0)x^2 + \{n(n-1)a_n + a_{n-4}\}x^{n-2} + \dots = 0$

Since this is an identify being true for every of $x$, equality to zero the coefficients of various powers of $x$.

$$2a_2 = 0$$
$$6a_3 = 0$$
$$12a_4 + a_0 = 0$$

- - - - - - - - - - - - - -

- - - - - - - - - - - - - -

- - - - - - - - - - - - - -

$$n(n-1)\, a_n + a_{n-4} = 0$$

These equations $Q$ is

$$a_2 = 0$$
$$a_3 = 0$$
$$a_4 = -\frac{a_0}{12}$$

- - - - - - - - - - - - - - - - - - - - - - - -

- - - - - - - - - - - - - - - - - - - - - - - -

- - - - - - - - - - - - - - - - - - - - - - - -

$$a_n = -\frac{a_{n-4}}{(n-1)n}$$

The last of these is called a recurrent relation and is used to determine the various values of $a$'s by taking different values of $n$.

on substituting $n = 4, 5, 6, \dots$

$$a_4 = \frac{-a_0}{3 \cdot 4}$$

$$a_5 = \frac{-a_1}{4 \cdot 5}$$

$$a_6 = \frac{-a_2}{5 \cdot 6} = 0$$

$$a_7 = \frac{-a_3}{6 \cdot 7} = 0$$

$$a_8 = \frac{-a_4}{7 \cdot 8} = \frac{a_0}{3 \cdot 4 \cdot 7 \cdot 8}$$

$$a_9 = -\frac{a_5}{89} = \frac{a_1}{4 \cdot 5 \cdot 8 \cdot 9} \text{ etc}$$

Now, $\quad a_2 = a_3 = a_6 = a_7 = a_{10} = a_{11} = a_{14} = a_{15} = 0$

Substituting these values of $a_2$, $a_3$, $a_4$ in Eqn. (*i*), we have

$$y = a_0 + a_1 x - \frac{a_0 x^4}{3 \cdot 4} - \frac{a_1 x^5}{4 \cdot 5} + \frac{a_o x^8}{3 \cdot 4 \cdot 7 \cdot 8} + \frac{a_1 x^9}{4 \cdot 5 \cdot 8 \cdot 9} \cdots$$

$$y = a_0 \left(1 - \frac{x_4}{3 \cdot 4} + \frac{x^8}{3 \cdot 4 \cdot 7 \cdot 8} \cdots\right) + a_1 \left(x - \frac{x^5}{4 \cdot 5} + \frac{x^9}{4 \cdot 5 \cdot 8 \cdot 9} \cdots\right)$$

which is the required solution containing two arbitrary constant $a_0$ and $a_1$.

**Example 16.8:** Solve $(e^{x+y} - y)\, dx + (xe^{x+y} + 1)\, dy = 0$.

**Solution:** Given

$$(e^{x+y} - y)dx + (xe^{x+y} + 1)dy = 0 \qquad \ldots(i)$$

Let

$$M = e^{x+y} - 1$$

$$N = x\, e^{x+y} + 1$$

$$\frac{\delta M}{\delta y} = e^{x+y} - 1$$

$$\frac{\delta N}{\delta x} = e^{x+y} + x\, e^{x+y} + 0$$

$$\frac{\dfrac{\delta M}{\delta y} - \dfrac{\delta N}{\delta x}}{N} = \frac{e^{x+y} - 1 - e^{x+y} - xe^{x+y}}{xe^{x+y} + 1} = \frac{-(1 + xe^{x+y})}{(xe^{x+y} + 1)} = -1$$

$$\text{I.F.} = e^{\int (-1) dx} = e^{-x}$$

Multiply Eqn. (*i*) by I.F.

$$e^{-x}(e^{x+y} - y)dx + e^{-x}(x\, e^{x+y} + 1)dy = 0$$

or,

$$(e^y - ye^{-x})\, dx + (x\, e^y + e^{-x})\, dy = 0 \qquad \ldots(ii)$$

Now Eqn. (*ii*) became

diff$^n$. eq is given by

$$\because \qquad \frac{\delta}{\delta y}(e^y - ye^{-x}) = e^y - e^{-x} = \frac{\delta}{\delta x}(xe^y + e^{-x})$$

Now, Integrate

$$\int (e^y - ye^{-x})dy + \int 0 \cdot dx = C$$

keeping $y$ is constant.

or

$$\int (e^y - ye^{-x})dx + 0 = C$$

or,

$$e^y x - \frac{ye^{-x}}{(-1)} + C = C$$

$$\therefore \qquad xe^y + ye^{-x} = C$$

**Example 16.9:** *Solve in series the equations*

$$(1-x^2)y'' + 2xy' + y = 0$$ *(Rajasthan, 2018)*

**Solution:** Let $y = \Sigma a_k x^{m+k}$, $a_0 \neq 0$

to be a solution of the given differential equation

Put the value of $y$, $y'$ and $y''$ and

$\Sigma (m+k)(m+k-1) a_k x^{m+k-2} - \Sigma[-(m+k)(m+k-3)+1]a_k x^{m+k} = 0$

Equating the co-efficient of $x^{m-2}$. We get the individual equation $m(m-1)a_0 = 0 \Rightarrow m = 0, \quad m = 1$

thus the roots differ by the integer 1.

Equating to zero the coefficient of $x^{m+k-2}$

$$a_k = \left[\frac{(m+k-2)(m+k-5)-1}{(m+k)(m+k-1)}\right]a_{k-2}$$

Find the value of $a_1$, equate the co-efficient of $x^{m-1}$ to zero, getting

$$m(m+1)a_1 = 0$$

Which is satisfies for $m = 0$ and so we can take $a_1$ to arbitrary.

$$a_k = \frac{[(k-2)(k-5)-1]}{k(k-1)}a_{k-2}$$

Now, let $k = 2, 3, 4, 5 \ldots$ $\quad a_2 = -\frac{1}{2}a_0, \; a_3 = -\frac{1}{2}a_1, \; a_4 = \frac{1}{8}a_0, \; a_5 = \frac{1}{40}a_1$

Hence the general solution is $\quad y = a_0\left(1 - \frac{x^2}{2} + \frac{x^4}{8}\ldots\right) + a_1\left(x - \frac{x^3}{2} + \frac{x^5}{40}\ldots\right).$

## PROBLEMS 16.2

Solve the following equations in power series :

1. $4x\frac{d^2y}{dx^2} + 2\frac{dy}{dx} + y = 0.$ *(V.T.U., 2012)*
2. $x(1-x)y'' - 3xy' - y = 0$ *(Gujarat University, 2018)*
3. $x\frac{d^2y}{dx^2} + \frac{dy}{dx} - y = 0.$ *(V.T.U., 2014)*
4. $3x\frac{d^2y}{dx^2} + (1-x)\frac{dy}{dx} - y = 0.$
5. $x\frac{d^2y}{dx^2} + (1+x)\frac{dy}{dx} + 2y = 0.$
6. $2x^2y'' + xy' - (x+1)y = 0.$ *(U.P.T.U., 2011)*
7. $8x^2\frac{d^2y}{dx^2} + 10x\frac{dy}{dx} - (1+x)y = 0.$
8. $2x(1-x)\frac{d^2y}{dx^2} + (1-x)\frac{dy}{dx} + 3y = 0.$ *(D.T.U., 2016; U.P.T.U., 2010)*
9. $x(1-x)\frac{d^2y}{dx^2} - 3\frac{dy}{dx} + 2y = 0.$
10. $xy'' + y' + x^2y = 0$ *(Andhra University, 2018)*
11. $x^2y'' + x(x-1)y' + (1-x)y = 0$ about $x = 0$. *(U.P.T.U., 2011)*
12. $xy'' + y' - y = 0$ *(Timkur, 2017)*
13. $x\frac{d^2y}{dx^2} + \frac{dy}{dx} + xy = 0$

## 16.5 BESSEL'S EQUATION*

One of the most important differential equations in applied mathematics is

$$x^2\frac{d^2y}{dx^2} + x\frac{dy}{dx} + (x^2 - n^2)y = 0 \qquad ...(1)$$

which is known as *Bessel's equation of order n*. Its particular solutions are called *Bessel functions or cylinderical functions of order n*. Many physical problems involving vibrations or heat conduction in cylindrical regions give rise to this equation.

Substituting $y = a_0x^m + a_1x^{m+1} + a_2x^{m+2} + ...$

(1) takes the form

$$a_0(m^2 - n^2)\,x^m + a_1\{(m+1)^2 - n^2\}x^{m+1} + \{a_2[(m+2)^2 - n^2] + a_0\}x^{m+2} + ... = 0.$$

Equating to zero the coefficient of $x^m$, we obtain the indicial equation $m^2 - n^2 = 0$ (as $a_0 \neq 0$) where $m = n$ or $-n$.

$$a_1 = a_3 = a_5 = a_7 = ... = 0$$

and $$a_2 = -\frac{a_0}{(m+2)^2 - n^2},\ a_4 = -\frac{a_2}{(m+4)^2 - n^2} \text{ etc.}$$

These give $$y = a_0x^m\left(1 - \frac{1}{(m+2)^2 - n^2}x^2 + \frac{1}{[(m+2)^2 - n^2][(m+4)^2 - n^2]}x^4 - \cdots\right)$$

For $m = n$, we get

$$y_1 = a_0x^n\left\{1 - \frac{1}{4(n+1)}x^2 + \frac{1}{4^2\cdot 2!\,(n+1)(n+2)}x^4 - \frac{1}{4^3\cdot 3!\,(n+1)(n+2)(n+3)}x^6 + \cdots\right\} \qquad ...(2)$$

and for $m = -n$, we have

$$y_2 = a_0x^{-n}\left\{1 - \frac{1}{4(-n+1)}x^2 + \frac{1}{4^2\cdot 2!\,(-n+1)(-n+2)}x^4 - \frac{1}{4^3\cdot 3!\,(-n+1)(-n+2)(-n+3)}x^6 + \cdots\right\} \qquad ...(3)$$

**Case I.** *When n is not integral or zero,* the complete solution of Eqn. (1) is $y = c_1y_1 + c_2y_2$.

If we take $a_0 = 1/2^n\,\Gamma(n+1)$, then the solution given by Eqn. (2) is called the *Bessel function of the first kind of order n* and is denoted by $J_n(x)$. Thus

$$J_n(x) = \left(\frac{x}{2}\right)^n\left\{\frac{1}{\Gamma(n+1)} - \frac{1}{1!\,\Gamma(n+2)}\left(\frac{x}{2}\right)^2 + \frac{1}{2!\,\Gamma(n+3)}\left(\frac{x}{2}\right)^4 - \frac{1}{3!\,\Gamma(n+4)}\left(\frac{x}{2}\right)^6 + \cdots\right\}\ (n > 0)$$

*i.e.* $$J_n(x) = \sum_{r=0}^{\infty}(-1)^r\left(\frac{x}{2}\right)^{n+2r}\frac{1}{r!\,\Gamma(n+r+1)} \qquad ...(4)$$

and corresponding to Eqn. (3), we have $$J_{-n}(x) = \sum_{r=0}^{\infty}(-1)^r\left(\frac{x}{2}\right)^{-n+2r}\frac{1}{r!\,\Gamma(-n+r+1)} \qquad ...(5)$$

which is called the *Bessel function of the first kind of order – n.*

Hence complete solution of the Bessel's Eqn. (1) may be expressed in the form.

$$y = AJ_n(x) + BJ_{-n}(x). \qquad ...(6)$$ (*V.T.U., 2015*)

**Case II.** *When n is zero,* $y_1 = y_2$ and the complete solution of Eqn. (1), which reduces to the *Bessel's equation of order zero,* is obtained as in Example 16.4.

---

* Named after the German mathematician and astronomer *Friederich Wilhelm Bessel* (1784–1846) whose paper on Bessel functions appeared in 1826. He studied Astronomy of his own and became director of Königsberg observatory.

**Case III.** *When n is integral, $y_2$ fails to give a solution for positive values of n and $y_1$ fails to give a solution for negative values.* Thus another independent integral of the Bessel's Eqn. (1) is needed to form its general solution. We now proceed to find an independent solution of Eqn. (1), when $n$ is an integer.

Let $y = u(x)J_n(x)$ be a solution of Eqn. (1). Substituting the values of $y$, $y' = u'J_n + uJ_n'$ and $y'' = u''J_n + 2u'J_n' + uJ_n''$ in (1), we obtain

$$x^2(u''J_n + 2u'J_n' + uJ_n'') + x(u'J_n + uJ_n') + (x^2 - n^2)uJ_n = 0$$

or
$$u\{x^2J_n'' + xJ_n' + (x^2 - n^2)J_n\} + x^2u''J_n + 2x^2u'J_n' + xu'J_n = 0. \quad ...(7)$$

Now since $J_n$ is a solution of Eqn. (1), therefore, $x^2J_n'' + xJ_n' + (x^2 - n^2)J_n = 0$

$\therefore$ Eqn. (7) reduces to $x^2u''J_n + 2x^2u'J_n' + xu'J_n = 0$.

Dividing throughout by $x^2u'J_n$, it becomes $\dfrac{u''}{u'} + 2\dfrac{J_n'}{J_n} + \dfrac{1}{x} = 0$

*i.e.,*
$$\frac{d}{dx}(\log u') + 2\frac{d}{dx}(\log J_n) + \frac{d}{dx}(\log x) = 0 \text{ or } \frac{d}{dx}\{\log(u'J_n^2x)\} = 0.$$

Integrating, $\log(u'J_n^2x) = \log B$, whence $xu'J_n^2 = B$.

$$\therefore \quad u' = \frac{B}{xJ_n^2} \text{ or } u = B\int\frac{dx}{xJ_n^2} + A.$$

Thus
$$y = AJ_n(x) + BJ_n(x)\int\frac{dx}{x[J_n(x)]^2}.$$

Hence the complete solution of the Bessel's Eqn. (1) is

$$y = AJ_n(x) + BY_n(x) \quad ...(8) \quad (V.T.U., 2012)$$

where
$$Y_n(x) = J_n(x)\int\frac{dx}{x[J_n(x)]^2} \quad ...(9)$$

$Y_n(x)$ is called the *Bessel function of the second kind of order n or Neumann function**.

**Obs.** Putting $k = -n + r$, *i.e.* $r = k + n$, and noting that $\Gamma(k+1) = k!$ where $k$ is an integer, Eqn. (5) may be written as

$$J_{-n}(x) = \sum_{k=0}^{\infty}\frac{(-1)^{k+n}(x/2)^{2k+n}}{(k+n)!\,k!} = (-1)^n\sum_{k=0}^{\infty}\frac{(-1)^k(x/2)^{n+2k}}{k!\,\Gamma(k+n+1)}$$

Hence $J_{-n}(x) = (-1)^nJ_n(x)$. ...(10)

## 16.6 RECURRENCE FORMULAE FOR $J_n(x)$

The following recurrence formulae can easily be derived from the series expression for $J_n(x)$ :

(1) $\dfrac{d}{dx}[x^nJ_n(x)] = x^nJ_{n-1}(x)$.

(2) $\dfrac{d}{dx}[x^{-n}J_n(x)] = -x^{-n}J_{n+1}(x)$.

(3) $J_n(x) = \dfrac{x}{2n}[J_{n-1}(x) + J_{n+1}(x)]$. *(Andhra 2018)*

(4) $J_n'(x) = \dfrac{1}{2}[J_{n-1}(x) - J_{n+1}(x)]$.

(5) $J_n'(x) = \dfrac{n}{x}J_n(x) - J_{n+1}(x)$.

(6) $J_{n+1}(x) = \dfrac{2n}{x}J_n(x) - J_{n-1}(x)$.

*These formulae are very useful in the solution of boundary value problems and in establishing the various properties of Bessel functions.*

*Proofs.* (1) Multiplying (4) of page 884 by $x^n$, we have

$$x^nJ_n(x) = x^n\sum_{r=0}^{\infty}\frac{(-1)^r(x/2)^{n+2r}}{r!\,\Gamma(n+r+1)} = \sum_{r=0}^{\infty}\frac{(-1)^rx^{2(n+r)}}{2^{n+2r}\,r!\,\Gamma(n+r+1)}$$

$$\therefore \quad \frac{d}{dx}[x^nJ_n(x)] = \sum_{r=0}^{\infty}\frac{(-1)^r\,2(n+r)x^{2(n+r)-1}}{2^{n+2r}r!\,\Gamma(n+r+1)} = x^n\sum_{r=0}^{\infty}\frac{(-1)^r(x/2)^{n-1+2r}}{r!\,\Gamma(n-1+r+1)} = x^nJ_{n-1}(x).$$

*(D.T.U., 2013)*

* Named after the German mathematician and physicist *Carl Neumann* (1832–1925) whose work on potential theory gave impetus for development of integral equations by *Volterra* of Rome, *Fredholm* of Stockholm and *Hilbert* of Gottingen.

(**2**) Multiplying (4) of page 884 by $x^{-n}$, we have

$$x^{-n}J_n(x) = \sum_{r=0}^{\infty} \frac{(-1)^r x^{2r}}{2^{n+2r} r!\,\Gamma(n+r+1)}$$

$$\therefore \quad \frac{d}{dx}[x^{-n}J_n(x)] = \sum_{r=0}^{\infty} \frac{(-1)^r 2r\, x^{2r-1}}{2^{n+2r} r!\,\Gamma(n+r+1)} = -x^{-n}\sum_{r=1}^{\infty} \frac{(-1)^{r-1} x^{n+1+2(r-1)}}{2^{n+1+2(r-1)}(r-1)!\,\Gamma(n+r+1)}$$

$$= -x^{-n}\sum_{k=0}^{\infty} \frac{(-1)^k (x/2)^{n+1+2k}}{k!\,\Gamma(n+1+k+1)} = -x^{-n}J_{n+1}(x), \text{ where } k = r-1.$$

(**3**) From (1), we have $x^n J_n'(x) + nx^{n-1}J_n(x) = x^n J_{n-1}(x)$

or dividing by $x^n$, $\quad J_n'(x) + (n/x)J_n(x) = J_{n-1}(x)$ ...(*i*)

Similarly from (2), we get $x^{-n}J_n'(x) - nx^{-n-1}J_n(x) = -x^{-n}J_{n+1}(x)$

or $\quad -J_n'(x) + \frac{n}{x}J_n(x) = J_{n+1}(x)$ ...(*ii*)

Adding Eqns. (*i*) and (*ii*), we obtain $\frac{2n}{x}J_n(x) = J_{n-1}(x) + J_{n+1}(x)$

*i.e.,* $\quad J_n(x) = \frac{x}{2n}[J_{n-1}(x) + J_{n+1}(x)]$

(**4**) Subtracting Eqn. (*ii*) from Eqn. (*i*), we get $2J_n'(x) = J_{n-1}(x) - J_{n+1}(x)$

*i.e.,* $\quad J_n'(x) = \frac{1}{2}[J_{n-1}(x) - J_{n+1}(x)].$

(**5**) is another way of writing Eqn. (*ii*). (*D.T.U., 2012*)

(**6**) is another way of writing (3).

## 16.7 (1) EXPANSIONS FOR $J_0$ AND $J_1$

We have from (4) of page 884,

$$J_0(x) = 1 - \frac{1}{(1!)}\left(\frac{x}{2}\right)^2 + \frac{1}{(2!)^2}\left(\frac{x}{2}\right)^4 - \frac{1}{(3!)^2}\left(\frac{x}{2}\right)^6 + \cdots \quad ...(1)$$

and

$$J_1(x) = \frac{x}{2}\left[1 - \frac{1}{1!2!}\left(\frac{x}{2}\right)^2 + \frac{1}{2!3!}\left(\frac{x}{2}\right)^4 - \frac{1}{3!4!}\left(\frac{x}{2}\right)^6 + \cdots\right] \quad ...(2)$$

Because of their special importance, the values of $J_0(x)$ and $J_1(x)$ are given in Appendix 2 : Table II to four decimal places at intervals of 0.1. With the help of these values, the graphs of $J_0(x)$ and $J_1(x)$ can be drawn as shown in Fig. 16.1, for $x > 0$. Their close resemblance to graphs of $\cos x$ and $\sin x$ is interesting.

**Obs.** The roots of the equation $J_0(x) = 0$ are useful in some physical problems. This equation has no complex roots but an infinite number of real roots. Its first four roots are $x = 2.4, 5.52, 8.65, 11.79$ approximately.

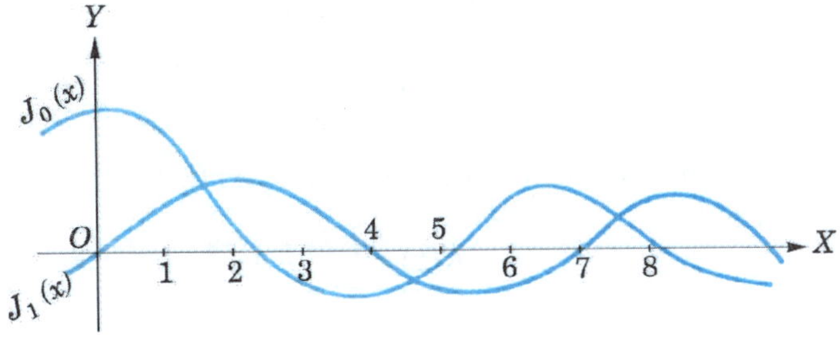

Fig. 16.1

## 16.8 VALUE OF $J_{1/2}$*

We may think that $J_0(x)$ is the simplest of the $J$'s but actually $J_{1/2}(x)$ is simpler, for it can be expressed in a finite form. Taking $n = \frac{1}{2}$ in (4) of page 884, we have

*Bessel functions $J_n(x)$ of order $n = \pm\frac{1}{2}, \pm\frac{3}{2}, \pm\frac{5}{2}$ etc. are Elementary Bessel functions and can be expressed in terms of sines, cosines and powers of $x$.

$$J_{1/2}(x) = \left(\frac{x}{2}\right)^{1/2}\left\{\frac{1}{\Gamma\left(\frac{3}{2}\right)} - \frac{1}{1!\,\Gamma\left(\frac{5}{2}\right)}\left(\frac{x}{2}\right)^2 + \frac{1}{2!\,\Gamma\left(\frac{7}{2}\right)}\left(\frac{x}{2}\right)^4 - \ldots\right\}$$

$$= \left(\frac{x}{2}\right)^{1/2}\left\{\frac{1}{\frac{1}{2}\Gamma\left(\frac{1}{2}\right)} - \frac{1}{\frac{3}{2}\cdot\frac{1}{2}\Gamma\left(\frac{1}{2}\right)}\left(\frac{x}{2}\right)^2 + \frac{1}{2\cdot\frac{5}{2}\cdot\frac{3}{2}\cdot\frac{1}{2}\Gamma\left(\frac{1}{2}\right)}\left(\frac{x}{2}\right)^4 - \ldots\right\}$$

$$= \frac{\sqrt{x}}{\sqrt{2\Gamma\left(\frac{1}{2}\right)}}\left\{\frac{2}{1!} - \frac{2x^2}{3!} + \frac{2x^4}{5!} - \ldots\right\}$$

Now multiplying the series by $x/2$ and outside by $2/x$, we get

$$J_{1/2}(x) = \frac{\sqrt{2}}{\sqrt{x}\sqrt{\pi}}\left\{\frac{x}{1!} - \frac{x^3}{3!} + \frac{x^5}{5!} - \ldots\right\} = \sqrt{\left(\frac{2}{\pi x}\right)}\sin x. \qquad \ldots(3)$$ (*Andhra, 2015 ; V.T.U., 2013*)

Similarly taking $n = \frac{1}{2}$ in (5) of page 847, it can be shown that

$$J_{-\frac{1}{2}}(x) = \sqrt{\left(\frac{2}{\pi x}\right)}\cos x. \qquad \ldots(4)$$

(*Andhra, 2021 ; Delhi, 2012 ; Calicut, 2012*)

**Example 16.10.** *Express $J_5(x)$ in terms of $J_0(x)$ and $J_1(x)$.*

**Solution.** We know that

$$J_n(x) = \frac{x}{2n}[J_{n-1}(x) + J_{n+1}(x)] \text{ i.e. } J_{n+1}(x) = \frac{2n}{x}J_n(x) - J_{n-1}(x)$$

Putting $n = 1, 2, 3, 4$ successively, $J_2(x) = \frac{2}{x}J_1(x) - J_0(x) \quad \ldots(i)$ $\qquad J_3(x) = \frac{4}{x}J_2(x) - J_1(x) \quad \ldots(ii)$

$$J_4(x) = \frac{6}{x}J_3(x) - J_2(x) \quad \ldots(iii) \qquad J_5(x) = \frac{8}{x}J_4(x) - J_3(x) \quad \ldots(iv)$$

Substituting the value of $J_2(x)$ in Eqn. (*ii*), we have

$$J_3(x) = \frac{4}{x}\left\{\frac{2}{x}J_1(x) - J_0(x)\right\} - J_1(x) = \left(\frac{8}{x^2} - 1\right)J_1(x) - \frac{4}{x}J_0(x) \qquad \ldots(v)$$

Now substituting the values of $J_3(x)$ from Eqn. (*v*) and $J_2(x)$ from Eqn. (*i*) in Eqn. (*iii*), we get

$$J_4(x) = \left(\frac{48}{x^3} - \frac{8}{x}\right)J_1(x) + \left(1 - \frac{24}{x^2}\right)J_0(x) \qquad \ldots(vi)$$

Finally putting the values of $J_4(x)$ from Eqn. (*vi*) and $J_3(x)$ from Eqn. (*v*) in Eqn. (*iv*), we obtain

$$J_5(x) = \left(\frac{384}{x^4} - \frac{72}{x^2} - 1\right)J_1(x) + \left(\frac{12}{x} - \frac{192}{x^3}\right)J_0(x).$$

**Example 16.11.** *Prove that* $J_{5/2}(x) = \sqrt{\left(\frac{2}{\pi x}\right)}\left\{\frac{3 - x^2}{x^2}\sin x - \frac{3}{x}\cos x\right\}.$

**Solution.** We know that $J_{n+1}(x) = \frac{2n}{x}J_n(x) - J_{n-1}(x) \qquad \ldots(i)$

Putting $n = \frac{1}{2}$, we get $\quad J_{3/2}(x) = \frac{1}{x}J_{1/2}(x) - J_{-1/2}(x) = \sqrt{\left(\frac{2}{\pi x}\right)}\left(\frac{\sin x}{x} - \cos x\right)$

Again putting $n = \frac{3}{2}$ in Eqn. (*i*), we get $J_{5/2}(x) = \frac{3}{x}J_{3/2}(x) - J_{1/2}(x)$

$$= \frac{3}{x}\left[\sqrt{\left(\frac{2}{\pi x}\right)}\left(\frac{\sin x}{x} - \cos x\right)\right] - \sqrt{\left(\frac{2}{\pi x}\right)}\sin x = \sqrt{\left(\frac{2}{\pi x}\right)}\left[\frac{3-x^2}{x^2}\sin x - \frac{3}{x}\cos x\right]$$

which is the required result.

**Example 16.12.** *Prove that*

(a) $J_1''(x) = -J_1(x) + \frac{1}{x}J_2(x)$. *(U.P.T.U., 2011)*

(b) $J_n''(x) = \frac{1}{4}[J_{n-2}(x) - 2J_n(x) + J_{n+2}(x)]$. *(Andhra., 2021, 2016 ; Delhi., 2015)*

(c) $\frac{d}{dx}[xJ_n(x)J_{n+1}(x)] = x[J_n^2(x) - J_{n+1}^2(x)]$. *(Andhra., 2016 ; Delhi., 2015 ; Calicut, 2012)*

**Solution.** (a) We know that $x^2\frac{d^2y}{dx^2} + x\frac{dy}{dx}(x^2 - n^2) = 0$ ...(i)

Since $J_n(x)$ is a solution of Eqn. (i), we have $x^2 J_n'' + xJ_n + (x^2 - n^2) J_n = 0$ ...(ii)

We know that $xJ_n' = nJ_n - xJ_{n+1}$ ...(iii)

Now substituting the value of $xJ_n'$ from Eqns. (iii) in (ii), we get $x^2J_n'' = -nJ_n + xJ_{n+1} + (n^2 - x^2) J_n$

or $$x^2J_n''(x) = (n^2 - n - x^2) J_n(x) + xJ_{n+1}(x)$$

Putting $n = 1$, we get $x^2J_1''(x) = -x^2J_1(x) + xJ_2(x)$

Dividing by $x^2$, we get $J_1''(x) = -J_1(x) + \frac{1}{x}J_2(x)$

(b) We know that $J_n'(x) = \frac{1}{2}\{J_{n-1}(x) - J_{n+1}(x)\}$ ...(i)

Differentiating both sides, we get $J_n''(x) = \frac{1}{2}\{J_{n-1}'(x) - J_{n+1}'(x)\}$ ...(ii)

Changing $n$ to $n - 1$ in Eqn. (i), we obtain $J_{n-1}'(x) = \frac{1}{2}\{J_{n-2}(x) - J_n(x)\}$ ...(iii)

Changing $n$ to $n + 1$ in Eqn. (i), we have $J_{n+1}'(x) = \frac{1}{2}\{J_n(x) - J_{n+2}(x)\}$ ...(iv)

Substituting the values of $J_{n-1}'(x)$ and $J_{n+1}'(x)$ from Eqn. (iii) and (iv) in Eqn. (ii), we get

$$J_n'' = \frac{1}{4}[J_{n-2}(x) - 2J_n(x) + J_{n+2}(x)]$$

(c) $\frac{d}{dx}[xJ_n(x) J_{n+1}(x)] = J_n(x)J_{n+1}(x) + x[J_n(x)J'_{n+1}(x) + J_n'(x) J_{n+1}(x)]$ ...(i)

From (5) of § 16.6, we have $J_n'(x) = \frac{n}{x}J_n(x) - J_{n+1}(x)$ ...(ii)

Changing $n$ to $n + 1$ in Eqn. (i) of page 650, we get $J'_{n+1}(x) = J_n(x) - \frac{n+1}{x}J_{n+1}(x)$ ...(iii)

Now substituting from Eqn. (iii) and (ii) in Eqn. (i), we get

$$\frac{d}{dx}[xJ_n(x) J_{n+1}(x)] = J_n(x) J_{n+1}(x) + x\left[J_n(x)\left\{J_n(x) - \frac{n+1}{x}J_{n+1}(x)\right\} + \left\{\frac{n}{x}J_n(x) - J_{n+1}(x)\right\}J_{n+1}(x)\right]$$

$$= x\{J_n^2(x) - J_{n+1}^2(x)\}.$$

**Example 16.13.** *Prove that :*

(a) $\int J_3(x)dx = c - J_2(x) - \frac{2}{x}J_1(x)$.

(b) $\int xJ_0^2(x)dx = \frac{1}{2}x^2\{J_0^2(x) + J_1^2(x)\}$. *(U.P.T.U., 2012 ; D.T.U., 2011)*

**Solution.** (*a*) We know that $\frac{d}{dx}[x^{-n}J_n(x)] = -x^{-n}J_{n+1}(x)$ [§ 16.6 (2)] ...(*i*)

or $$\int x^{-n}J_{n+1}(x)dx = -x^{-n}J_n(x) \quad ...(ii)$$

$\therefore$ $$\int J_3(x)dx = \int x^2 \cdot x^{-2}J_3(x)dx + c \quad \text{[Integrate by parts]}$$

$$= x^2 . \int x^{-2}J_3(x)dx - \int 2x\left[\int x^{-2}J_3(x)dx\right]dx + c$$

$$= x^2[-x^{-2}J_2(x)] - \int 2x[-x^{-2}J_2(x)]\,dx + c \quad \text{[By Eqn. (ii) when } n = 2]$$

$$= c - J_2(x) + \int \frac{2}{x}J_2(x)\,dx = c - J_2(x) - \frac{2}{x}J_1(x) \quad \text{[By Eqn. (ii) when } n = 1]$$

(*b*) $$\int xJ_0^2(x)dx = \int J_0^2(x)\cdot xdx \quad \text{[Integrate by parts]}$$

$$= J_0^2(x)\cdot\frac{1}{2}x^2 - \int 2J_0(x)J_0'(x)\cdot\frac{1}{2}x^2\,dx$$

$$= \frac{1}{2}x^2J_0^2(x) + \int x^2J_0(x)J_1(x)\,dx \quad \text{[By Eqn. (i) when } n = 0]$$

$$= \frac{1}{2}x^2J_0^2(x) + \int xJ_1(x)\cdot\frac{d}{dx}[xJ_1(x)]\,dx \quad \left[\because \frac{d}{dx}[xJ_1(x)] = xJ_0(x) \text{ by § 16.6 (1)}\right]$$

$$= \frac{1}{2}x^2J_0^2(x) + \frac{1}{2}[xJ_1(x)]^2 = \frac{1}{2}x^2[J_0^2(x) + J_1^2(x)].$$

## 16.9 GENERATING FUNCTION FOR $J_n(x)$

*To prove that* $e^{\frac{1}{2}x(t-1/t)} = \sum_{n=-\infty}^{\infty} t^nJ_n(x)$. We have $e^{\frac{1}{2}x(t-t^{-1})} = e^{xt/2} \times e^{-x/2t}$

$$= \left[1 + \left(\frac{xt}{2}\right) + \frac{1}{2!}\left(\frac{xt}{2}\right)^2 + \frac{1}{3!}\left(\frac{xt}{2}\right)^3 + \cdots\right] \times \left[1 - \left(\frac{x}{2t}\right) + \frac{1}{2!}\left(\frac{x}{2t}\right)^2 - \frac{1}{3!}\left(\frac{x}{2t}\right)^3 + \cdots\right]$$

The coefficient of $t^n$ in this product

$$= \frac{1}{n!}\left(\frac{x}{2}\right)^n - \frac{1}{(n+1)!}\left(\frac{x}{2}\right)^{n+2} + \frac{1}{2!(n+2)!}\left(\frac{x}{2}\right)^{n+4} - \ldots = J_n(x).$$

As all the integral powers of $t$, both positive and negative occur, we have

$$e^{\frac{1}{2}x(t-t^{-1})} = J_0(x) + tJ_1(x) + t^2J_2(x) + t^3J_3(x) + \ldots + t^{-1}J_{-1}(x) + t^{-2}J_{-2}(x) + t^{-3}J_{-3}(x) + \ldots$$

$$= \sum_{n=-\infty}^{\infty} t^nJ_n(x)$$ *(Calicut, 2012 ; Delhi, 2012)*

This shows that Bessel functions of various orders can be derived as coefficients of different powers of $t$ in the expansion of $e^{\frac{1}{2}x(t-1/t)}$. For this reason, it is known as the *generating function of Bessel functions.*

**Example 16.14.** *Show that*

(*a*) $J_0(x) = \frac{1}{\pi}\int_0^{\pi}\cos(x\cos\phi)\,d\phi.$

(*b*) $J_0^2 + 2J_1^2 + 2J_2^2 + 2J_3^2 + \ldots = 1.$

(*c*) $J_{5/2}(x) = \sqrt{\left(\frac{2}{\pi x}\right)}\left(\frac{3-x^2}{x^2}\sin x - \frac{3\cos x}{x}\right)$

(*d*) $J_4(x) = \left(\frac{48}{x^3} - \frac{8}{x}\right)J_1(x) + \left(1 - \frac{24}{x^2}\right)J_0(x)$ *(Sambalpur, 2019)*

**Solution.**

(*a*) Changing $\theta$ to $\frac{1}{2}\pi - \phi$ in § 16.6 (3), we get

$$\cos(x\cos\phi) = J_0 + 2J_2\cos(\pi - 2\phi) + 2J_4\cos(2\pi - 4\phi) + \ldots$$
$$= J_0 - 2J_2\cos 2\phi + 2J_4\cos 4\phi - \ldots$$

Integrating both sides w.r.t. $\phi$ from 0 to $\pi$, we get

$$\int_0^{\pi} \cos(x\cos\phi)\,d\phi = \int_0^{\pi} \{J_0(x) - 2J_2(x)\cos 2\phi + 2J_4(x)\cos 4\phi - \cdots\}\,d\phi$$

$$= \left| J_0(x)\cdot\phi - 2J_2(x)\cdot\frac{1}{2}\sin 2\phi + 2J_4(x)\cdot\frac{1}{4}\sin 4\phi - \cdots \right|_0^{\pi} = J_0(x)\,.\,\pi \text{ whence follows the result.}$$

(*b*) Squaring § 16.6 (3) and (4) and integrating w.r.t. $\phi$ from 0 to $\pi$ and noting that ($m$, $n$ being integers),

$$\int_0^{\pi} \cos m\theta \cos n\theta\,d\theta = \int_0^{\pi} \sin m\theta \sin n\theta\,d\theta = 0, \qquad (m \neq n)$$

and $$\int_0^{\pi} \cos^2 n\theta\,d\theta = \int_0^{\pi} \sin^2 n\theta\,d\theta = \pi/2, \text{ we obtain}$$

$$[J_0(x)]^2\,\pi + 4\,[J_2(x)]^2\frac{\pi}{2} + 4\,[J_4(x)]^2\,\frac{\pi}{2} + \ldots = \int_0^{\pi} \cos^2(x\sin\theta)\,d\theta$$

$$4\,[J_1(x)]^2\,\frac{\pi}{2} + 4\,[J_3(x)]^2\,\frac{\pi}{2} + \ldots = \int_0^{\pi} \sin^2(x\sin\theta)\,d\theta$$

Adding, $$\pi\,[J_0^2 + 2J_1^2 + 2J_2^2 + 2J_3^2 + \cdots] = \int_0^{\pi} d\theta = \pi$$

Hence $$J_0^2 + 2J_1^2 + 2J_2^2 + 2J_3^2 + \cdots = 1.$$

(*c*) Writing recurrence relation

$$J_{n+1}(x) = \frac{2n}{x}J_n(x) - J_{n-1}(x) \qquad \ldots(i)$$

Now on substituting $n = \frac{3}{2}$ in Eqn. (*i*)

$$J_{5/2}(x) = \frac{3}{x}J_{3/2}(x) - J_{1/2}(x)$$

and taking $n = \frac{1}{2}$ $$J_{3/2}(x) = \frac{1}{x}J_{1/2}(x) - J_{-1/2}(x)$$

$$\therefore \quad J_{5/2}(x) = \frac{3}{2}\left\{\frac{1}{x}J_{1/2}(x) - J_{-1/2}(x)\right\} - J_{1/2}(x) = \left(\frac{3}{x^2} - 1\right)J_{1/2}(x) - \frac{3}{2}J_{-1/2}(x)$$

$$= \sqrt{\frac{2}{\pi x}}\left\{\frac{3 - x^2}{x^2}\sin x - \frac{3\cos x}{x}\right\}$$

on putting the values of $J_{1/2}(x)$ and $J_{-1/2}(x)$

(*d*) Let $n = 1, 2, 3$ successively in Eqn. (*i*), we have

$$J_2(x) = \frac{2}{x}J_1(x) - J_0(x)$$

$$J_3(x) = \frac{4}{x}J_2(x) - J_1(x) = \frac{4}{x}\left\{\frac{2}{x}J_1(x) - J_0(x)\right\} - J_1(x)$$

and $$J_4(x) = \frac{6}{x}J_3(x) - J_2(x)$$

$$= \frac{6}{x}\left[\frac{4}{x}\left\{\frac{2}{x}J_1(x) - J_0(x)\right\} - J_1(x)\right] - \left\{\frac{2}{x}J_1(x) - J_0(x)\right\}$$

$$= \left(\frac{48}{x^3} - \frac{8}{x}\right) J_1(x) + \left(1 - \frac{24}{x^2}\right) J_0(x)$$

**Example 16.15:** *Prove that*

(*a*) $\frac{d}{dx}\left(J_n^2 + J_{n+1}^2\right) = 2\left(\frac{n}{x} J_n^2 - \frac{n+1}{x} J_{n+1}^2\right)$ *and hence show that*

(*b*) $J_0^2 + 2(J_1^2 + J_2^2 + J_3^2 + ...) = 1$ (*Calicut, 2018*)

**Solution:** We have $\frac{d}{dx}(J_n^2 + J_{n+1}^2) = 2J_n \cdot J_n' + 2J_{n+1} \cdot J'_{n+1}$ ...(*i*)

We know that $J_n' = \frac{n}{x} J_n - J_{n+1}$

and $J_n' = J_{n-1} - \frac{n}{x} J_n$

$$J'_{n+1} = J_n - \left(\frac{n+1}{2}\right) J_{n+1},$$

Now replacing $n$ by $(n + 1)$,

Substituting the value of $J_n'$ and $J'_{n+1}$ in Eqn. (*i*) and simplifying, we get the result

(*b*) Let $n = 0, 1, 2, 3$ ...

$$\frac{d}{dx}\left(J_0^2 + J_1^2\right) = 2\left(0 - \frac{1}{x} J_1^2\right)$$

$$\frac{d}{dx}\left(J_1^2 + J_2^2\right) = 2\left(\frac{1}{x} J_1^2 - \frac{2}{x} J_2^2\right)$$

$$\frac{d}{dx}\left(J_2^2 + J_3^2\right) = 2\left(\frac{2}{x} J_2^2 - \frac{3}{x} J_3^2\right)$$

..................................................

..................................................

..................................................

..................................................

..................................................

..................................................

Adding them column wise and recalling that $J_n \to 0$

$$\frac{d}{dx}\left[J_0^2 + 2(J_1^2 + J_2^2 + J_3^2 + ...)\right] = 0$$

Now integrating w.r.t $x$, we get

$$[J_0^2 + 2(J_1^2 + J_2^2 + J_3^2 ...)] = c \qquad ...(ii)$$

Replacing $x$ by 0 in Eqn. (*ii*) and recalling $J_0(0) = 1$ and $J_n(0) = 0$ for $n \geq 1$, we get $c = 1$

Hence $[J_0 + 2(J_1^2 + J_2^2 + J_3^2 ...)] = 1$

**Example 16.16:** *Show that*

(*a*) $4J_0''(x) + 3J_0'(x) + J_3(x) = 0$ (*b*) $J_3(x)dx + J_2(x) + \frac{2}{x} J_1(x) = 0$ (*Marathwada, 2018*)

**Solution:** (*a*) We know that

$$J_0'(x) = J_1(x)$$

Differentiating it, we get

$$J''_0(x) = -J'_1(x) \quad ...(i)$$

Now taking $n = 1$ in the relation

$$2J'_n(x) = J_{n-1}(x) - J_{n+1}(x) \quad ...(ii)$$

We have

$$2J'_1(x) = J_0(x) - J_2(x) \quad ...(iii)$$

from Eqns. (*i*) and (*iii*), we have

$$2J''_0(x) = J_2(x) - J_0(x) \quad ...(iv)$$

Differentiating Eqn. (*iv*), we have

$$2J'''_0(x) = J_2'(x) - J_1'(x) \quad ...(v)$$

Once again from Eqn. (*iii*), on taking $n = 2$ we have,

$$2J'_2(x) = J_1(x) - J_3(x) \quad ...(vi)$$

Now, from Eqns. (*v*) and (*vi*), we have

$$4J'''(x) = [J_1(x) - J_3(x)] - 2J_0'(x) = [-J_0'(x) - J_3(x)] - J_0'(x)$$

$$4J'''(x) + 3J_0'(x) - J_3(x) = 0$$

(*b*) We have already show that

$$\frac{d}{dx}\left(x^{-n}J_n(x)\right) = -x^{-n}J_{n+1}(x)$$

So, $$\int x^{-n}J_{n+1}dx = -x^{-n}J_n \quad ...(vii)$$

on taking $n = 2$, in Eqn. (*vii*) we get

$$\int x^{-2}J_3\,dx = -x^{-2}J_2 \quad ...(viii)$$

Now, $$\int J_3\,dx = \int x^2(x^{-2}J_3)\,dx$$

$$= x^2\left(-x^{-2}J_2\right) - \int 2x(-x^{-2}J_2)\,dx$$

on integrating by parts and using the result Eqn. (*i*)

$\therefore$ $$\int J_3\,dx = -J_2 + 2\int x^{-1}J_2\,dx$$

$$= -J_2 + 2(-x^{-1}J_1)$$

on using Eqn. (*vii*) again when $n = 1$

So, $$\int J_3\,dx + J_2 + \frac{2}{x}J_1 = 0$$

## PROBLEMS 16.3

1. Compute $J_0(2)$, $J_1(1)$ correct to three decimal places.
2. Show that (*i*) $J_3(x) = \left(\frac{8}{x^2} - 1\right)J_1(x) - \frac{4}{x}J_0(x)$ (*U.P.T.U., 2012*)

   (*ii*) $J_4(x) = \left(\frac{48}{x^3} - \frac{8}{x}\right)J_1(x) + \left(1 - \frac{24}{x^4}\right)J_0(x)$. (*D.T.U., 2013*) (*iii*) $J_{3/2}(x) = \sqrt{\left(\frac{2}{\pi x}\right)}\left(\frac{\sin x}{x} - \cos x\right)$ (*Guwahati, 2019*)
3. Show that

   (*i*) $J_{-1/2}(x) = J_{1/2}(x)\cot x$. (*ii*) $J'_{1/2}(x)\,J_{-1/2}(x) - J'_{-1/2}(x)\,J_{1/2}(x) = 2/\pi x$

   (*iii*) $J_{-3/2}(x) = -\sqrt{\left(\frac{2}{\pi x}\right)}\left(\sin x + \frac{\cos x}{x}\right)$ (*W.B.T.U., 2015*) (*iv*) $J_{-5/2}(x) = \sqrt{\left(\frac{2}{\pi x}\right)}\left(\frac{3}{x}\sin x + \frac{3 - x^2}{x^2}\cos x\right)$
4. Prove that (*i*) $\frac{d}{dx}J_0(x) = -J_1(x)$. (*ii*) $\frac{d}{dx}[xJ_1(x)] = xJ_0(x)$. (*Andhra, 2014*)

(iii) $\frac{d}{dx}[x^n J_n(ax)] = ax^n J_{n-1}(ax)$.  (iv) $J'_n(x) = -\frac{n}{2} J_n(x) + J_{n-1}(x)$

5. Show by the use of recurrence formula, that

(i) $J_0''(x) = \frac{1}{2}[J_2(x) - J_0(x)]$  (ii) $4J_0'''(x) + 3J_0'(x) + J_3(x) = 0$.

6. Prove that

(i) $\frac{d}{dx}\{J_n^2(x)\} = \frac{x}{2n}[J_{n-1}^2(x) - J_{n+1}^2(x)]$

(ii) $\frac{d}{dx}\{J_n^2(x) + J_{n+1}^2(x)\} = 2\left\{\frac{n}{2}J_n^2(x) - \frac{n+1}{x}J_{n+1}^2(x)\right\}$.  *(Andhra, 2016 ; Calicut, 2012)*

7. Prove that (i) $\int_0^{\pi/2} \sqrt{(\pi x)}\, J_{1/2}(2x)\, dx = 1$.  *(U.P.T.U., 2012)*

(ii) $\int_0^r xJ_0(ax)dx = \frac{r}{a}J_1(ar)$.  (iii) $\int x^2 J_1(x)dx = x^2 J_2(x)$.  *(U.P.T.U., 2011)*

8. Prove that (i) $\int J_0(x)\, J_1(x)\, dx = -\frac{1}{2}[J_0(x)]^2$.  (ii) $\int_0^\infty e^{-ax} J_0(bx)dx = \frac{1}{\sqrt{a^2+b^2}}$.

9. Starting with the series of §16.9, prove that

$$2nJ_n(x) = x\{J_{n-1}(x) + J_{n+1}(x)\} \text{ and } xJ_n'(x) = nJ_n(x) - xJ_{n+1}(x).$$

10. Establish the *Jacobi series*

$$\cos(x\cos\theta) = J_0 - 2J_2\cos 2\theta + 2J_4\cos 4\theta - \ldots$$
$$\sin(x\cos\theta) = 2[J_1\cos\theta - J_3\cos 3\theta + J_5\cos 5\theta - \ldots]$$

11. Prove that (i) $\sin x = 2[J_1 - J_3 + J_5 - \ldots]$

(ii) $\cos x = J_0 - 2J_2 + 2J_4 - 2J_6 + \ldots$

(iii) $1 = J_0 + 2J_2 + 2J_4 + 2J_6 + \ldots$

12. Show that

(i) $J_{-3/2}(x) = -\sqrt{\left(\frac{2}{\pi x}\right)}\left(\frac{\cos x}{x} + \sin x\right)$  (ii) $J_{-5/2}(x) = \sqrt{\left(\frac{2}{\pi x}\right)}\left(\frac{3-x^2}{x^2}\cos x + \frac{3}{x}\sin x\right)$

13. Show that

(i) $2J_0'' = J_2 - J_0$  (ii) $4J_n'' = J_{n-2} - 2J_n + J_{n+2}$

14. Show that $\frac{d}{dx}\left[x^n J_n(ax)\right] = ax^n J_{n-1}(ax)$ and hence deduce, $\frac{d}{dx}[xJ_1(x)] = xJ_0(x)$

15. Show that $x^n J_n(x)$ is the solution of $xy'' + (1-2n)y' + xy = 0$.

## 16.10 EQUATIONS REDUCIBLE TO BESSEL'S EQUATION

In many problems, we come across such differential equations which can easily be reduced to Bessel's equation and, therefore, can be solved by means of Bessel functions.

**(1)** *To reduce the differential equation* $x^2\frac{d^2y}{dx^2} + x\frac{dy}{dx} + (k^2x^2 - n^2)y = 0$ *to Bessel form.*

Put $t = kx$, so that $\frac{dy}{dx} = k\frac{dy}{dt}$ and $\frac{d^2y}{dx^2} = k^2\frac{d^2y}{dt^2}$.

Then (**1**) becomes $t^2\frac{d^2y}{dt^2} + t\frac{dy}{dt} + (t^2 - n^2)y = 0$

$\therefore$ its solution is $y = c_1J_n(t) + c_2J_{-n}(t)$, $n$ is non-integral,

or $y = c_1J_n(t) + c_2Y_n(t)$, $n$ is integral.

Hence the solution of (**1**) is

$$y = c_1J_n(kx) + c_2J_{-n}(kx),\ n \text{ is non-integral}$$

or $$y = c_1J_n(kx) + c_2Y_n(kx),\ n \text{ is integral.}$$

**(2)** *To reduce the differential equation* $x\dfrac{d^2y}{dx^2} + a\dfrac{dy}{dx} + k^2xy = 0$ *to Bessel's equation,*

put $$y = x^nz,$$

so that $$\frac{dy}{dx} = x^n\frac{dz}{dx} + nx^{n-1}z \text{ and } \frac{d^2y}{dx^2} = x^n\frac{d^2z}{dx^2} + 2nx^{n-1}\frac{dz}{dx} + n(n-1)x^{n-2}z$$

Then (**2**) takes the form $x^{n+1}\dfrac{d^2z}{dx^2} + (2n + a)x^n\dfrac{dz}{dx} + [k^2x^2 + n^2 + (a-1)n]x^{n-1}z = 0.$

Dividing throughout by $x^{n-1}$ and putting $2n + a = 1$,

$$x^2\frac{d^2z}{dx^2} + x\frac{dz}{dx} + (k^2x^2 - n^2)z = 0.$$

Its solution by (**1**) is $z = c_1J_n(kx) + c_2J_{-n}(kx)$, $n$ is non-integral

or $$z = c_1J_n(kx) + c_2Y_n(kx),\ n \text{ is integral}$$

Hence the solution of (**2**) is $y = x^n\,[c_1J_n(kx) + c_2J_{-n}(kx)]$, $n$ is non-integral

or $$y = x^n\,[c_1J_n(kx) + c_2Y_n(kx)],\ n \text{ is integral, where } n = (1-a)/2.$$

**(3)** *To reduce the differential equation* $x\dfrac{d^2y}{dx^2} + c\dfrac{dy}{dx} + k^2x^ry = 0$ *to Bessel form*, put $x = t^m$, *i.e.* $t = x^{1/m}$,

so that $$\frac{dy}{dx} = \frac{dy}{dt}\cdot\frac{dt}{dx} = \frac{1}{m}t^{1-m}\frac{dy}{dt}$$

and $$\frac{d^2y}{dx^2} = \frac{d}{dt}\left(\frac{1}{m}t^{1-m}\frac{dy}{dt}\right)\cdot\frac{1}{m}t^{1-m} = \frac{1}{m^2}t^{2-2m}\frac{d^2y}{dt^2} + \frac{1-m}{m^2}t^{1-2m}\frac{dy}{dt}$$

Then (**3**) takes the form $\dfrac{1}{m^2}t^{2-m}\dfrac{d^2y}{dt^2} + \dfrac{1-m+cm}{m^2}t^{1-m}\dfrac{dy}{dt} + k^2t^{mr}y = 0$

or multiplying throughout by $m^2/t^{1-m}$, $t\dfrac{d^2y}{dt^2} + (1 - m + cm)\dfrac{dy}{dt} + (km)^2t^{mr+m-1}y = 0.$

In order to reduce it to (**2**), we set $mr + m - 1 = 1$, *i.e.*, $m = 2/(r+1)$

and $$a = 1 - m + cm = (r + 2c - 1)/(r + 1).$$

Thus it reduces to $t\dfrac{d^2y}{dt^2} + a\dfrac{dy}{dt} + (km)^2ty = 0$ which is similar to (**2**).

Hence the solution of (**3**) is $y = x^{n/m}\{c_1J_n(k_mx^{1/m}) + c_2J_{-n}(kmx^{1/m})\}$, $n$ is a fraction

or $$y = x^{n/m}\,\{c_1J_n(kmx^{1/m}) + c_2Y_n(kmx^{1/m})\},\ n \text{ is an integer}$$

where $$n = \frac{1-a}{2} = \frac{1-c}{1+r} \text{ and } m = \frac{2}{1+r}.$$

**Example 16.17.** *Solve the differential equations :*

(*i*) $y'' + \dfrac{y'}{x} + \left(8 - \dfrac{1}{x^2}\right)y = 0.$ (*U.P.T.U., 2012*)

(*ii*) $4y'' + 9xy = 0.$ (*iii*) $xy'' + y' + \dfrac{1}{4}y = 0.$

**Solution.** (*i*) Rewriting the given equation as $x^2\dfrac{d^2y}{dx^2} + x\dfrac{dy}{dx} + (8x^2 - 1)y = 0,$

and comparing with (**1**) above, we see that $n = 1$ and $k = 2\sqrt{2}$.

∴ The solution of the given equation is

$$y = c_1 J_n(kx) + c_2 Y_n(kx)$$

*i.e.,*
$$y = c_1 J_1(2\sqrt{2x}) + c_2 Y_1(2\sqrt{2x}).$$

(*ii*) Rewriting the given equation as $x\dfrac{d^2y}{dx^2} + \dfrac{9}{4}x^2 y = 0$ ...(α)

and comparing with (**3**) above, we find that $c = 0$, $k = 3/2$ and $r = 2$.

∴
$$n = \frac{1-c}{1+r} = \frac{1}{3},$$

$$m = \frac{2}{1+r} = \frac{2}{3}$$

and
$$\frac{n}{m} = \frac{1}{2}.$$

Hence the solution of (α) is $y = x^{n/m}\{c_1 J_n(kmx^{1/m}) + c_2 Y_{-n}(kmx^{1/m})\}$

$$y = \sqrt{x}\,[c_1 J_{1/3}(x^{3/2}) + c_2 J_{-1/3}(x^{3/2})].$$

(*iii*) Multiplying by $x$, the given equation becomes

$$x^2y'' + xy' + \frac{1}{4}xy = 0 \qquad \text{...(α)}$$

Comparing with (**3**) above, we get $c = 1$, $k = 1/2$ & $r = 0$

∴
$$m = \frac{2}{1+r} = 2,\ n = \frac{1-c}{1+r} = 0 \ \&\ \frac{n}{m} = 0$$

Hence the solution of (α)

$$y = x^{n/m}\{c_1 J_n(kmx^{1/m}) + c_2 Y_n(kmx^{1/m})\}$$

$$= x^0\left\{c_1 J_0\left(\frac{1}{2}2x^{1/2}\right) + c_2 Y_0\left(\frac{1}{2}.2x^{1/2}\right)\right\}$$

*i.e.,*
$$y = c_1 J_0(\sqrt{x}) + c_2 Y_0(\sqrt{x})$$

## 16.11 (1) ORTHOGONALITY OF BESSEL FUNCTIONS

*We shall prove that*

$$\int_0^1 xJ_n(\alpha x)J_n(\beta x)\,dx = \begin{cases} 0, & \alpha \neq \beta \\ \dfrac{1}{2}[J_{n+1}(\alpha)]^2, & \alpha = \beta \end{cases},$$

*where* α, β *are the roots of* $J_n(x) = 0$.

We know that the solution of the equation

$$x^2u'' + xu' + (\alpha^2x^2 - n^2)u = 0 \qquad \text{...(1)}$$

and
$$x^2v'' + xv' + (\beta^2x^2 - n^2)v = 0 \qquad \text{...(2)}$$

are
$u = J_n(\alpha x)$ and $v = J_n(\beta x)$ respectively.

Multiplying Eqn. (1) by $v/x$ and Eqn. (2) by $u/x$ and subtracting, we get

$$x(u''v - uv'') + (u'v - uv') + (\alpha^2 - \beta^2)xuv = 0$$

or
$$\frac{d}{dx}\{x(u'v - uv')\} = (\beta^2 - \alpha^2)\,xuv.$$

Now integrating both sides from 0 to 1,

$$(\beta^2 - \alpha^2)\int_0^1 xuv\,dx = \left[x(u'v - uv')\right]_0^1 = (u'v - uv')_{x=1} \quad ...(3)$$

Since $\quad u = J_n(\alpha x),$

$$\therefore \quad u' = \frac{d}{dx}[J_n(\alpha x)] = \frac{d}{d(\alpha x)}[J_n(\alpha x)].\frac{d(\alpha x)}{dx} = \alpha J_n'(\alpha x)$$

Similarly, $v = J_n(\beta x)$ and $v' = \beta J_n'(\beta x)$. Substituting these values in Eqn. (3), we get

$$\int_0^1 xJ_n(\alpha x)J_n(\beta x)\,dx = \frac{\alpha J_n'(\alpha)J_n(\beta) - \beta J_n(\alpha)J_n'(\beta)}{\beta^2 - \alpha^2} \quad ...(4)$$

If $\alpha$ and $\beta$ are distinct roots of $J_n(x) = 0$, then $J_n(\alpha) = J_n(\beta) = 0$, and Eqn. (4) reduces to

$$\int_0^1 xJ_n(\alpha x)J_n(\beta x)\,dx = 0 \quad ...(5)$$

This is known as the *orthogonality relation of Bessel functions.* *(D.T.U., 2016 ; V.T.U., 2016 )*

When $\beta = \alpha$, the right side of Eqn. (4) is of 0/0 form. Its value can be found by considering $\alpha$ as a root of $J_n(x) = 0$ and $\beta$ as a variable approaching $\alpha$. Then Eqn. (4) gives

$$\underset{\beta\to\alpha}{\text{Lt}}\int_0^1 xJ_n(\alpha x)J_n(\beta x)\,dx = \underset{\beta\to\alpha}{\text{Lt}}\frac{\alpha J_n'(\alpha)J_n(\beta)}{\beta^2 - \alpha^2}$$

or by L'Hospital's rule, $\displaystyle\int_0^1 xJ_n^2(\alpha x)\,dx = \underset{\beta\to\alpha}{\text{Lt}}\frac{\alpha J_n'(\alpha)J_n'(\beta)}{2\beta} = \frac{1}{2}[J_n'(\alpha)]^2 = \frac{1}{2}[J_{n+1}(\alpha)]^2$ ...(6) [By (5)]

**Obs.** If however, the interval be from 0 to 1, it can be shown that

$$\int_0^1 xJ_n^2(\alpha x)\,dx = \frac{1}{2}[J_n'(\alpha)]^2 \quad \text{where } \alpha \text{ is the root of } J_n(x) = 0. \quad ...(7)$$

**(2) Fourier-Bessel expansion.** If $f(x)$ is a continuous function having finite number of oscillations in the interval $(0, a)$, then we can write

$$f(x) = c_1J_n(\alpha_1 x) + c_2J_n(\alpha_2 x) + ... + c_nJ_n(\alpha_n x) + ... \quad ...(8)$$

where $\alpha_1, \alpha_2$ ... are the positive roots of $J_n(x) = 0$.

To determine the coefficients $c_n$, multiply both sides of Eqn. (8) by $xJ_n(\alpha_n x)$ and integrate from 0 to $a$. Then all integrals on the right of Eqn. (1) vanish by Eqn. (5), except the term in $c_n$. This gives

$$\int_0^a xf(x)J_n(\alpha_n x)\,dx = c_n\int_0^a xJ_n^2(\alpha_n x)\,dx = c_n\frac{a^2}{2}J_{n+1}^2(a\alpha_n) \quad \text{[By (7)]}$$

$$\therefore \quad c_n = \frac{2}{a^2J_{n+1}^2(a\alpha_n)}\int_0^a xf(x)J_n(\alpha_n x)\,dx$$

Equation (8) is known as the *Fourier-Bessel expansion of f (x).*

**Example 16.18.** *If $\alpha_1, \alpha_2, ... \alpha_n$ are the positive roots of $J_0(x) = 0$, show that*

$$\frac{1}{2} = \sum_{n-1}^{\infty}[J_0(\alpha_n x)/\alpha_n J_1(\alpha_n)].$$

**Solution.** If $\quad f(x) = c_1J_n(\alpha_1 x) + c_2J_n(\alpha_2 x) + ... + c_rJ_n(\alpha_n x) + ...$ ...(*i*)

then $$c_r = \frac{2}{a^2J_{n+1}^2(a\alpha_r)}\int_0^a xf(x)J_n(\alpha_n x)\,dx$$

Taking $f(x) = 1$, $a = 1$ and $n = 0$, we get

$$c_r = \frac{2}{J_1^2(\alpha_r)}\int_0^1 xJ_0(\alpha_r x)\,dx = \frac{2}{J_1^2(\alpha_r)}\left|\frac{xJ_1(\alpha_r x)}{\alpha_r}\right|_0^1 = \frac{2}{\alpha_r J_1(\alpha_r)}$$

From Eqn. (*i*), $$1 = \sum_{r=1}^{\infty}\frac{2}{\alpha_r J_1(\alpha_r)}J_0(\alpha_r x) \quad \text{or} \quad \frac{1}{2} = \sum_{n=1}^{\infty}\frac{J_0(\alpha_n x)}{\alpha_n J_1(\alpha_n)}.$$

**Example 16.19.** *Expand $f(x) = x^2$ in the interval $0 < x < 2$ in terms of $J_2(\alpha_n x)$, where $\alpha_n$ are determined by $J_2(2\alpha_n) = 0$.*

**Solution.** Let the Fourier-Bessel expansion of $f(x)$ be $x^2 = \sum_{n=1}^{\infty} c_n J_2(\alpha_n x)$.

Multiplying both sides by $xJ_2(\alpha_n x)$ and integrating w.r.t. $x$ from 0 to 2, we get

$$\int_0^2 x^3 J_2(\alpha_n x)dx = c_n \int_0^2 xJ_2^2(\alpha_n x)dx = c_n \frac{(2)^2}{2} J_3^2(2\alpha_n) \qquad \text{[By (7) Page 661]}$$

or
$$\left| \frac{x^3 J_3(\alpha_n x)}{\alpha_n} \right|_0^2 = 2c_n J_3^2(2\alpha_n), \; c_n = \frac{4}{\alpha_n J_3(2\alpha_n)}$$

Hence
$$x^2 = 4 \sum_{n=1}^{\infty} \frac{J_2(\alpha_n x)}{\alpha_n J_3(2\alpha_n)}.$$

## 16.12 BER AND BEI FUNCTIONS

Consider the differential equation

$$x\frac{d^2y}{dx^2} + \frac{dy}{dx} - ixy = 0 \qquad \text{...(1)}$$

which occurs in certain problems of electrical engineering. This is Eqn. (1) of §16.10 with $n = 0$ and $k^2 = -i$, so that its particular solution is

$$y = J_0(kx) = J_0[(-i)^{1/2} x] = J_0(i^{3/2} x)$$

Replacing $i^{3/2} x$ in the series for $J_0(x)$ [§16.8], we get

$$y = 1 - \frac{i^3 x^2}{2^2} + \frac{i^6 x^4}{(2\,!)^2 2^4} - \frac{i^9 x^6}{(3\,!)^2 2^6} + \frac{i^{12} x^8}{(4!)^2 2^8} - \ldots$$

$$= \left[1 - \frac{x^4}{2^2 \cdot 4^2} + \frac{x^8}{2^2 \cdot 4^2 \cdot 6^2 \cdot 8^2} - \ldots\right] + i\left[\frac{x^2}{2^2} - \frac{x^6}{2^2 \cdot 4^2 \cdot 6^2} + \frac{x^{10}}{2^2 \cdot 4^2 \cdot 6^2 \cdot 8^2 \cdot 10^2} - \ldots\right] \qquad \text{...(2)}$$

which is complex for $x$ real. The series in the above brackets are taken to define *Bessel-real (or ber)* and *Bessel-imaginary (or bei) functions.*

Thus
$$ber\, x = 1 + \sum_{m=1}^{\infty} (-1)^m \frac{x^{4m}}{2^2 \cdot 4^2 \cdot 6^2 \ldots (4m)^2} \qquad \text{...(3)}$$

and
$$bei\, x = -\sum_{m=1}^{\infty} (-1)^m \frac{x^{4m-2}}{2^2 \cdot 4^2 \cdot 6^2 \ldots (4m-2)^2} \qquad \text{...(4)}$$

so that $y = ber\, x + i\, bei\, x$ is a solution of Eqn. (1).

Tables giving numerical values of *ber x* and *bei x* are also available.

**Example 16.20.** *Prove that* (i) $\frac{d}{dx}(x\, ber'\, x) = -x\, bei\, x$ (ii) $\frac{d}{dx}(x\, bei'\, x) = x\, ber\, x$.

(*Delhi, 2012 S*)

**Solution.** We have $x\, ber' x = x \sum_{m=1}^{\infty} (-1)^m \frac{4mx^{4m-1}}{2^2 \cdot 4^2 \cdot 6^2 \ldots (4m)^2}$

$$= \sum_{m=1}^{\infty} (-1)^m \cdot \frac{x^{4m}}{2^2 \cdot 4^2 \cdot 6^2 \ldots (4m-2)^2\, 4m} = -\int_0^{\infty} x\, bei\, x\, dx$$

or
$$\frac{d}{dx}(x\, ber'\, x) = -x\, bei\, x$$

Again $\int_0^x x \, ber \, x \, dx = \frac{x^2}{2} + \sum_{m=1}^{\infty} (-1)^m \frac{x^{4m+2}}{2^2 \cdot 4^2 \cdot 6^2 ...(4m)^2(4m+2)}$

$$= - \sum_{m=1}^{\infty} (-1)^m \frac{x^{4m-2}}{2^2 \cdot 4^2 \cdot 6^2 ...(4m-4)^2(4m-2)} = x \, bei' \, x \quad \text{or} \quad \frac{d}{dx}(x \, bei' \, x) = x \, ber \, x.$$

## PROBLEMS 16.4

Obtain the solutions of the following differential equations in terms of Bessel functions :

1. $y'' + \frac{y'}{x} + \left(1 - \frac{1}{9x^2}\right) y = 0.$
2. $y'' + \frac{y'}{2} + \left(1 - \frac{1}{6.25x^2}\right) y = 0.$
3. $xy'' + \alpha y' + k^2 xy = 0.$ (*V.T.U., 2010*)
4. $x^2 y'' - xy' + 4x^2 y = 0.$
5. $xy'' + y = 0.$
6. Show that (*i*) $x^n J_n(x)$ is a solution of the equation $xy'' + (1 - 2n) y' + xy = 0.$
   (*ii*) $x^{-n} J_n(x)$ is a solution of the equation $xy'' + (1 + 2n)y' + xy = 0.$
7. Show that under the transformation $y = u/\sqrt{x}$, Bessel equation becomes
   $$u'' + \left(1 + \frac{1 - 4n^2}{4x^2}\right) u = 0.$$ Hence find the solution of this equation.
8. By the use of substitution $y = u/\sqrt{x}$, show that the solution of the equation $x^2 \frac{d^2y}{dx^2} + x\frac{dy}{dx} + \left(x^2 - \frac{1}{4}\right) y = 0$ can be written in the form $y = c_1 \frac{\sin x}{\sqrt{x}} + c_2 \frac{\cos x}{\sqrt{x}}.$
9. Show that $\int_0^p x(ber^2 x + bei^2 x) \, dx = p(ber \, p \, bei' \, p - bei \, p \, ber' p).$
10. If $\alpha_1, \alpha_2, ..., \alpha_n$ are the positive roots of $J_0(x) = 0$, prove that
    $$x^2 = 2 \sum_{n=1}^{\infty} \frac{\alpha_n^2 - 4}{\alpha_n^3 J_1(\alpha_n)} J_0(\alpha_n x).$$
11. Expand $f(x) = x^3$ in the interval $0 < x < 3$ in terms of functions $J_1(\alpha_n x)$ where $\alpha_n$ are determined by $J_1(3\alpha) = 0.$
12. By the use the substitution $y = \frac{u}{\sqrt{x}}$, show that the solution of the equation $x^2 y'' + xy' + \left(x^2 - \frac{1}{4}\right) y = 0$ can be written in the form
    $$y = c_1 \frac{\sin x}{\sqrt{x}} + c_2 \frac{\cos x}{\sqrt{x}}$$
13. Show that under the transformation on $y = \frac{u}{\sqrt{x}}$, Bessel's equation reduces to
    $$u'' + \left(1 + \frac{1 - 4n^2}{4x^2}\right) u = 0$$
    Hence find the solution of this equation **[Ans.** $c_1 \sqrt{x} J_n(x) + c_2 \sqrt{x} \, J_{-n}(x)$**]**
14. Prove that
    (*i*) $\cos (x \cos \phi) = J_0 - 2J_2 \cos 2\phi + 2J_4 \cos 4\phi - ...............$
    (*ii*) $\sin (x \cos \phi) = 2[J_1 \cos \phi - J_3 \cos 3\phi + J_5 \cos 5\phi ...]$
    Hence, show that $$J_0(x) = \frac{1}{\pi} \int_0^{\pi} \cos (x \cos \phi) d\phi$$ (*Jiwaji University, 2019*)
15. Show that $\frac{d}{dx}(x \, J_n J_{n+1}) = x(J_n^{\,2} - J_{n+1}^{\,2})$ and hence show that
    $$x = 2J_0 J_1 + 6J_1 J_2 + 10J_2 J_3 + ...$$

## 16.13 LEGENDRE'S EQUATION

Another differential equation of importance in Applied Mathematics, particularly in boundary value problems for spheres, is *Legendre's equation,*

$$(1-x^2)\frac{d^2y}{dx^2} - 2x\frac{dy}{dx} + n(n+1)y = 0 \quad ...(1)$$

Here $n$ is a real number. But in most applications only integral values of $n$ are required.

Substituting $\quad y = a_0x^m + a_1x^{m+1} + a_2x^{m+2} + ... \ (a_0 \neq 0),$

Equation (1) takes the form

$$a_0(m)(m-1)x^{m-2} + a_1(m+1)mx^{m-1} + ... + [a_{r+2}(m+r+2)(m+r+1)$$
$$-\{(m+r)(m+r+1) - n(n+1)\}a_r]x^{m+r} + ... = 0$$

Equating to zero the coefficient of the lowest power of $x$, *i.e.*, of $x^{m-2}$, we get

$$a_0m(m-1) = 0,\ m = 0, 1 \qquad [\because\ a_0 \neq 0]$$

Equating to zero the coefficients of $x^{m-1}$ and $x^{m+r}$, we get $a_1(m+1)m = 0$ ...(2)

$$a_{r+2}(m+r+2)(m+r+1) - \{(m+r)(m+r+1) - n(n+1)\}a_r = 0 \quad ...(3)$$

*When* $m = 0$, Eqn. (2) is satisfied and therefore, $a_1 \neq 0$. Then Eqn. (3) gives, taking $r = 0, 1, 2, ...$ in turn,

$$a_2 = -\frac{n(n+1)}{2!}a_0, \quad a_3 = -\frac{(n-1)(n+2)}{3!}a_1$$

$$a_4 = \frac{-(n-2)(n+3)}{4.3}a_2 = \frac{n(n-2)(n+1)(n+3)}{4!}a_0,\ a_5 = -\frac{(n-3)(n+4)}{5.4}a_3 = \frac{(n-1)(n-3)(n+2)(n+4)}{5!}a_1, \text{ etc.}$$

Hence for $m = 0$, there are two independent solutions of Eqn. (1) :

$$y_1 = a_0\left\{1 - \frac{n(n+1)}{2!}x^2 + \frac{(n-2)n(n+1)(n+3)}{4!}x^4 - ......\right\} \quad ...(4)$$

$$y_2 = a_1\left\{x - \frac{(n-1)(n+2)}{3!}x^3 + \frac{(n-3)(n-1)(n+2)(n+4)}{5!}x^5 - ......\right\} \quad ...(5)$$

*When* $m = 1$, Eqn. (2) shows that $a_1 = 0$. Therefore, Eqn. (3) gives

$$a_3 = a_5 = a_7 = ... = 0 \quad \text{and} \quad a_2 = -\frac{(n-1)(n+2)}{3!}a_0,\ a_4 = \frac{(n-3)(n-1)(n+2)(n+4)}{5!}a_0, \text{ etc.}$$

Thus for $m = 1$, we get the solution Eqn. (5) again. Hence $y = y_1 + y_2$ is the general solution of Eqn. (1).

If $n$ is a positive even integer, the series Eqn. (4) terminates at the term in $x^n$ and $y_1$ becomes a polynomial. Similarly if $n$ is an odd integer, Eqn. (5) becomes a polynomial of degree $n$. Thus, whenever $n$ is a positive integer, the general solution of Eqn. (1) consists of a polynomial solution and an infinite series solution.

These polynomial solutions, with $a_0$ or $a_1$ so chosen that the value of the polynomial is 1 for $x = 1$, are called *Legendre polynomials* of order $n$ and are denoted by $P_n(x)$. The infinite series solution with ($a_0$ or $a_1$ properly chosen) is called *Legendre function of the second kind* and is denoted by $Q_n(x)$.

## 16.14 (1) RODRIGUE'S FORMULA*

*We shall prove that* $P_n(x) = \dfrac{1}{n!\,2^n}\dfrac{d^n}{dx^n}(x^2-1)^n$ ...(1)

Let $\quad v = (x^2-1)^n$. Then $v_1 = \dfrac{dv}{dx} = 2nx(x^2-1)^{n-1}$

*i.e.*, $\quad (1-x^2)v_1 + 2nxv = 0$ ...(2)

Differentiating Eqn. (2), $(n+1)$ times by Leibnitz's theorem

$$(1-x^2)v_{n+2} + (n+1)(-2x)v_{n+1} + \frac{1}{2!}(n+1)n(-2)v_n + 2n[xv_{n+1} + (n+1)v_n] = 0$$

or

$$(1-x^2)\frac{d^2(v_n)}{dx^2} - 2x\frac{d(v_n)}{dx} + n(n+1)v_n = 0$$

* Named after the French mathematician and economist *Olinde Rodrigue* (1794–1851).

which is Legendre's equation and $cv_n$ is its solution. Also its finite series solution is $P_n(x)$.

$$\therefore \qquad P_n(x) = cv_n = c\frac{d^n}{dx^n}(x^2-1)^n \qquad ...(3)$$

To determine the constant $c$, put $x = 1$ in Eqn. (3). Then

$$1 = c\left[\frac{d^n}{dx^n}\{(x-1)^n(x+1)^n\}\right]_{x=1}$$

$= c[n\,!\,(x+1)^n]$

$+$ terms containing $(x-1)$ and its powers$]_{x=1}$

$= c\,.\,n\,!\,2^n$, *i.e.* $c = 1/n\,!\,2^n$.

Substituting this value of $c$ in Eqn. (3), we get Eqn. (1), which is known as the *Rodrigue's formula*.

(*Andhra, 2016 ; D.T.U., 2015 ; V.T.U., 2015 ; W.B.T.U., 2015*)

**Obs.** All roots of $P_n(x) = 0$ are real and lie between $-1$ and $+1$.

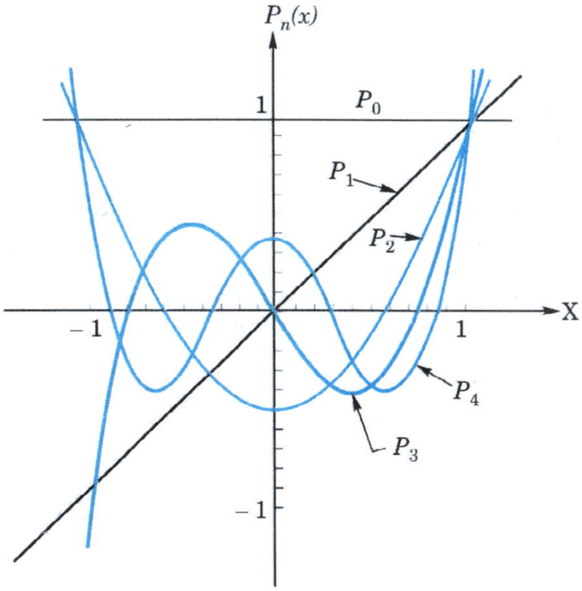


Fig. 16.2. Legendre polynomials.

**(2) Legendre polynomials.** Using Eqn. (1), we get

$P_0(x) = 1, \qquad P_1(x) = x,$

$P_2(x) = \frac{1}{2}(3x^2-1), \qquad P_3(x) = \frac{1}{2}(5x^3-3x),$

$$P_4(x) = \frac{1}{8}(35x^4 - 30x^2 + 3),\ P_5(x) = \frac{1}{8}(63x^5 - 70x^3 + 15x), \text{ etc.}$$

In general, we have $P_n(x) = \sum_{r=0}^{N} \frac{(-1)^r\,(2n-2r)!}{2^n r\,!\,(n-r)!\,(n-2r)!} x^{n-2r}$ ...(4)

where $N = \frac{1}{2}n$ or $\frac{1}{2}(n-1)$ according as $n$ is even or odd.

Let us derive Eqn. (4) from Eqn. (1).

By Binomial theorem, $(x^2-1)^n = \sum_{r=0}^{n} {}^nC_r(x^2)^{n-r}(-1)^r = \sum_{r=0}^{n}(-1)^r \frac{n\,!}{r\,!\,(n-r)!}x^{2n-2r}$

$$\therefore \text{ by Eqn. (1), } P_n = \frac{1}{n\,!\,2^n}\sum_{n=0}^{n}\frac{(-1)^r\,n\,!}{r\,!\,(n-r)!}\frac{d^n\,(x^{2n-2r})}{dx^n} = \sum_{r=0}^{N}\frac{(-1)^r\,(2n-2r)!}{2^n r\,!\,(n-r)!\,(n-2r)!}x^{n-2r}$$

This is same as Eqn. (4), and the last term ($r = N$) is such that the power of $x$ (*i.e.*, $n-2r$) for this term is either 0 or 1.

**Example 16.21.** *Express $f(x) = x^4 + 3x^3 - x^2 + 5x - 2$ in terms of Legendre polynomials.*

(*V.T.U., 2015 ; U.P.T.U., 2013*)

**Solution.** Since $P_4(x) = \frac{35}{8}x^4 - \frac{15}{4}x^2 + \frac{3}{8}\cdot \quad \therefore x^4 = \frac{8}{35}P_4(x) + \frac{6}{7}x^2 - \frac{3}{35}$

$$\therefore \qquad f(x) = \left[\frac{8}{35}P_4(x) + \frac{6}{7}x^2 - \frac{3}{35}\right] + 3x^3 - x^2 + 5x - 2$$

$$= \frac{8}{35}P_4(x) + 3x^3 - \frac{1}{7}x^2 + 5x - \frac{73}{35} \qquad \left[\because\ x^3 = \frac{2}{5}P_3(x) + \frac{3}{5}x\,;\ x^2 = \frac{2}{3}P_2(x) + \frac{1}{3}\right]$$

$$= \frac{8}{35}P_4(x) + 3\left[\frac{2}{5}P_3(x) + \frac{3}{5}x\right] - \frac{1}{7}\left[\frac{2}{3}P_2(x) + \frac{1}{3}\right] + 5x - \frac{73}{35}$$

$$= \frac{8}{35}P_4(x) + \frac{6}{5}P_3(x) - \frac{2}{21}P_2(x) + \frac{34}{5}x - \frac{224}{105} \qquad [\because\ x = P_1(x),\ 1 = P_0(x)]$$

$$= \frac{8}{35} P_4(x) + \frac{6}{5} P_3(x) - \frac{2}{21} P_2(x) + \frac{34}{5} P_1 x - \frac{224}{105} P_0(x).$$

**Example 16.22.** *Show that for any function f (x), for which the nth derivative is continuous,*

$$\int_{-1}^{1} f(x)\, P_n(x) dx = \frac{1}{2^n\, n!} \int_{-1}^{1} (1 - x^2)^n\, f^n(x)\, dx.$$

**Solution.** Using Rodrigue's formula : $P_n(x) = \dfrac{1}{2^n\, n!} \dfrac{d^n}{dx^n} (x^2 - 1)^n$,

$$\int_{-1}^{1} f(x)\, P_n(x)\, dx = \frac{1}{2^n\, n!} \int_{-1}^{1} f(x) \frac{d^n (x^2 - 1)^n}{dx^n}\, dx \qquad \text{[Integrate by parts]}$$

$$= \frac{1}{2^n\, n!} \left[ \left| f(x) \cdot \frac{d^{n-1}(x^2 - 1)^n}{dx^{n-1}} \right|_{-1}^{1} - \int_{-1}^{1} f'(x) \cdot \frac{d^{n-1}(x^2 - 1)^n}{dx^{n-1}} dx \right]$$

$$= \frac{(-1)}{2^n\, n!} \int_{-1}^{1} f'(x) \frac{d^{n-1}}{dx^{n-1}} (x^2 - 1)^n\, dx$$

$$= \frac{(-1)^2}{2^n\, n!} \int_{-1}^{1} f''(x) \frac{d^{n-2}}{dx^{n-2}} (x^2 - 1)^n\, dx \qquad \text{[Again integrating by parts]}$$

$$= \frac{(-1)^n}{2^n\, n!} \int_{-1}^{1} f^n(x)\ (x^2 - 1)^n\, dx \qquad \text{[Integrating by parts } (n-2) \text{ times]}$$

$$= \frac{(-1)^{2n}}{2^n\, n!} \int_{-1}^{1} f^n(x)\, (1 - x^2)^n\, dx = \frac{1}{2^n\, n!} \int_{-1}^{1} f^n(x)\, (1 - x^2)^n\, dx$$

## 16.15 GENERATING FUNCTION FOR $P_n(x)$

*To show that* $(1 - 2xt + t^2)^{-\frac{1}{2}} = \sum_{n=0}^{\infty} t^n P_n(x)$. *(Andhra, 2015)*

Since
$$(1 - z)^{-\frac{1}{2}} = 1 + \frac{1}{2} z + \frac{\frac{1}{2} \cdot \frac{3}{2}}{2!} z^2 + \frac{\frac{1}{2} \cdot \frac{3}{2} \cdot \frac{5}{2}}{3!} z^3 + \ldots$$

$$= 1 + \frac{2!}{(1!)^2\, 2^2} z + \frac{4!}{(2!)^2\, 2^4} z^2 + \frac{6!}{(3!)^2\, 2^6} z^3 + \ldots$$

$\therefore$
$$[1 - t(2x - t)]^{-\frac{1}{2}} = 1 + \frac{2!}{(1!)^2\, 2^2} t(2x - t) + \frac{4!}{(2!)^2\, 2^4} t^2 (2x - t)^2 + \ldots$$

$$+ \frac{(2n - 2r)!}{[(n - r)!]^2\, 2^{2n - 2r}} t^{n-r} (2x - t)^{n-r} + \ldots + \frac{(2n)!}{(n!)^2\, 2^{2n}} t^n (2x - t)^n + \ldots \qquad \ldots(1)$$

The term in $t^n$ from the term containing $t^{n-r} (2x - t)^{n-r}$

$$= \frac{(2n - 2r)!}{[(n - r)!]^2\, 2^{2n - 2r}} t^{n-r} \cdot {}^{n-r}C_r (-t)^r (2x)^{n-2r}$$

$$= \frac{(2n - 2r)!}{[(n - r)!]^2\, 2^{2n - 2r}} \times \frac{(n - r)!}{r!\,(n - 2r)!} (-1)^r t^n \cdot (2x)^{n-2r} = \frac{(-1)^r\, (2n - 2r)!}{2^n\, r!\,(n - r)!\,(n - 2r)!} x^{n-2r} \cdot t^n.$$

Collecting all terms in $t^n$ which will occur in the term containing $t^n (2x - t)^n$ and the preceding terms, we see that terms in $t^n$.

$$= \sum_{r=0}^{N} \frac{(-1)^r\, (2n - 2r)!}{2^n\, r!\,(n - r)!\,(n - 2r)!} x^{n-2r} \cdot t^n = P_n(x) t^n$$

where $N = \frac{1}{2}n$ or $\frac{1}{2}(n-1)$ according as $n$ is even or odd.

Hence (1) may be written as $[1 - t(2x - t)]^{-1/2} = \sum_{n=0}^{\infty} P_n(x)\, t^n$ ...(2)

This shows that $P_n(x)$ is the coefficient of $t^n$ in the expansion of $(1 - 2xt + t^2)^{-1/2}$. That is why, it is known as the *generating function of Legendre polynomials*.

**Cor. 1.** *$P_n(1) = 1$.* *(Andhra, 2016 ; W.B.T.U., 2015 ; D.T.U., 2012 S)*

Taking $x = 1$ in (2), we have $(1 - 2t + t^2)^{-1/2} = \sum_{n=0}^{\infty} P_n(1)\, t^n$

*i.e.,*
$$\sum_{n=0}^{\infty} P_n(1)\, t^n = (1-t)^{-1} = 1 + t + t^2 + \dots + t^n + \dots$$

Equating coefficients of $t^n$, we get $P_n(1) = 1$.

**Cor. 2.** *$P_n(-1) = (-1)^n$.* *(Andhra, 2016)*

Taking $x = -1$ in Eqn. (2), we have

$$\sum_{n=0}^{\infty} P_n(-1)\, t^n = (1+t)^{-1} = 1 - t + t^2 - \dots + (-1)^n t^n + \dots$$

Equating coefficients of $t^n$, we get the desired result.

**Cor. 3.** $P_n(0) = \begin{cases} (-1)^{n/2} \dfrac{1 \times 3 \times 5 \dots (n-1)}{2 \times 4 \times 6 \times \dots n}, & \text{when } n \text{ is even} \\ 0, & \text{when } n \text{ is odd} \end{cases}$

Putting $x = 0$ in Eqn. (2), we get $\sum_{n=0}^{\infty} P_n(0)\; t^n = (1 + t^2)^{-1/2}$

$$= 1 - \frac{1}{2}t^2 + \frac{1.3}{2.4}t^4 \dots + (-1)^r \frac{1 \cdot 3 \cdot 5 \cdots (2r+1)}{2 \cdot 4 \cdot 6 \cdots 2r} t^{2r} + \dots$$

Equating coefficient of $t^{2m}$, we get $P_{2m}(0) = (-1)^m \dfrac{1 \cdot 3 \cdot 5 \cdots (2m-1)}{2 \cdot 4 \cdot 6 \cdots 2m}$

Similarly equating coefficients of $t^{2m+1}$, we have $P_{2m+1}(0) = 0$.

**Cor. 4.** *$P'_n(1) = \frac{1}{2} n(n+1)$*

Since $P_n(x)$ is a solution of Legendre's eduation, $(1 - x^2) P''_n(x) - 2xP'_n(x) + n(n+1) P_n(x) = 0$

Putting $x = 1$, $-2P'_n(1) + n(n+1) P_n(1) = 0$ or $P'_n(1) = \frac{1}{2} n(n+1)$ $[\because\ P_n(1) = 1]$

## 16.16 RECURRENCE FORMULAE FOR $P_n(x)$

The following recurrence formulae can be easily derived from the generating function for $P_n(x)$ :

**(1)** $(n+1) P_{n+1}(x) = (2n+1)xP_n(x) - nP_{n-1}(x)$

**(2)** $nP_n(x) = xP'_n(x) - P'_{n-1}(x)$

**(3)** $(2n+1)P_n(x) = P'_{n+1}(x) - P'_{n-1}(x)$

**(4)** $P'_n(x) = xP'_{n-1}(x) + nP_{n-1}(x)$.

**(5)** $(1 - x^2)P'_n(x) = n\,[P_{n-1}(x) - xP_n(x)]$.

*Proofs.* **(1)** We know that $(1 - 2xt + t^2)^{-1/2} = \sum_{n=0}^{\infty} P_n(x) t^n$ ...(*i*)

Differentiating partially w.r.t. $t$, we get

$$-\frac{1}{2}(1 - 2xt + t^2)^{-3/2}(-2x + 2t) = \Sigma n P_n(x)\, t^{n-1}$$

or
$$(x - t)(1 - 2xt + t^2)^{-1/2} = (1 - 2xt + t^2)\, \Sigma n P_n(x)\, t^{n-1}$$

or $$(x-t)\,\Sigma P_n(x)\,t^n = (1-2xt+t^2)\,\Sigma nP_n(x)\,t^{n-1}$$

Equating coefficients of $t^n$ from both sides, we get

$$xP_n(x) - P_{n-1}(x) = (n+1)\,P_{n+1}(x) - 2nxP_n(x) + (n-1)\,P_{n-1}(x)$$

whence follows the required result. (*U.P.T.U., 2012*)

(**2**) Differentiating Eqn. (*i*) partially w.r.t. $x$,

$$-\frac{1}{2}\,(1-2xt+t^2)^{-3/2}\,.\,(-2t) = \Sigma P_n'(x)t^n$$

i.e., $$t(1-2tx+t^2)^{-3/2} = \Sigma P_n'(x)\,t^n \qquad ...(ii)$$

Again differentiating Eqn. (*i*) partially w.r.t. $t$, we have

$$(x-t)\,(1-2tx+t^2)^{-3/2} = \Sigma nP_n(x)t^{n-1} \qquad ...(iii)$$

Dividing Eqn. (*iii*) by Eqn. (*ii*), we get $\dfrac{x-t}{t} = \dfrac{\Sigma nP_n(x)t^{n-1}}{\Sigma P_n'(x)t^n}$

i.e., $$\Sigma nP_n(x)\,t^n = (x-t)\,\Sigma P_n'(x)\,t^n$$

Equating coefficients of $t^n$ from both sides, we get (**2**).

(**3**) Differentiating (**1**) w.r.t. $x$, we get

$$(n+1)P'_{n+1}(x) = (2n+1)P_n(x) + (2n+1)xP'_n(x) - nP'_{n-1}(x) \qquad ...(iv)$$

Substituting for $xP_n'(x)$ from (**2**) in Eqn. (*iv*), we obtain

$$(n+1)\,P'_{n+1}(x) = (2n+1)\,P_n(x) + (2n+1)\,[nP_n(x) + P'_{n-1}(x)] - nP'_{n-1}(x)$$

or $$(2n+1)\,P_n(x) = P'_{n+1}(x) - P'_{n-1}(x)$$

(**4**) Rewriting Eqn. (*iv*) as

$$\begin{aligned}(n+1)\,P'_{n+1}(x) &= (2n+1)\,P_n(x) + (n+1)\,xP_n'(x) + n[xP_n'(x) - P'_{n-1}(x)]\\ &= (2n+1)\,P_n(x) + (n+1)\,xP_n'(x) + n^2P_n(x) \qquad \text{[by (2)]}\\ &= (n+1)xP_n'(x) + (n^2+2n+1)P_n(x)\end{aligned}$$

or $$P'_{n+1}(x) = xP_n'(x) + (n+1)\,P_n(x)$$

Replacing $n$ by $(n-1)$, we get (**4**).

(**5**) Rewriting (**2**) and (**4**) as (*U.P.T.U., 2010*)

$$xP_n'(x) - P'_{n-1}(x) = nP_n(x) \qquad ...(v)$$

and $$P_n'(x) - xP'_{n-1}(x) = nP_{n-1}(x) \qquad ...(vi)$$

Multiplying Eqn. (*v*) by $x$ and substracting from Eqn. (*vi*), we get

$$(1-x^2)P_n'(x) = n[P_{n-1}(x) - xP_n(x)].$$

**Example 16.23.** *Prove that* $(2n+1)\,(1-x^2)\,P_n'(x) = n\,(n+1)\,[P_{n-1}(x) - P_{n+1}(x)]$.

**Solution.** We have the recurrence formula

$$(n+1)\,P_{n+1}(x) = (2n+1)\,xP_n(x) - nP_{n-1}(x)$$

or $$(\overline{n+1}+n)\,xP_n(x) = (n+1)\,P_{n+1}(x) + nP_{n-1}(x)$$

or $$\begin{aligned}(n+1)\,[xP_n(x) - P_{n+1}(x)] &= n\,[P_{n-1}(x) - xP_n(x)]\\ &= (1-x^2)\,P_n'(x) \quad [\because\ (1-x^2)\,P_n'(x) = n\,[P_{n-1}(x) - xP_n(x)] \qquad ...(i)\end{aligned}$$

or $$xP_n(x) = P_{n+1}(x) + \frac{(1-x^2)\,P_n'(x)}{n+1} \qquad ...(ii)$$

Also from Eqn. (*i*) $$xP_n(x) = P_{n-1}(x) - \frac{(1-x^2)\,P_n'(x)}{n} \qquad ...(iii)$$

From Eqns. (*ii*) and (*iii*), $$P_{n-1}(x) - \frac{(1-x^2)\,P_n'(x)}{n} = P_{n+1}(x) + \frac{(1-x^2)\,P_n'(x)}{n+1}$$

or $$n\,(n+1)\,P_{n-1}(x) - (n+1)\,(1-x^2)\,P_n'(x) = n\,(n+1)\,P_{n+1}(x) + n\,(1-x^2)\,P_n'(x)$$

or $$(2n+1)\,(1-x^2)\,P_n'(x) = n\,(n+1)\,[P_{n-1}(x) - P_{n+1}(x)]$$

## 16.17 (1) ORTHOGONALITY OF LEGENDRE POLYNOMIALS

*We shall prove that,* $$\int_{-1}^{1} P_m(x)\, P_n(x)\, dx = \begin{cases} 0, & m \neq n \\ \dfrac{2}{2n+1}, & m = n \end{cases}$$

We know that the solutions of

$$(1-x)^2 u'' - 2xu' + m(m+1)\, u = 0 \qquad ...(1)$$

and $$(1-x^2)v'' - 2xv' + n(n+1)\, v = 0 \qquad ...(2)$$

are $P_m(x)$ and $P_n(x)$ respectively.

Multiplying Eqn. (1) by $v$ and Eqn. (2) by $u$ and subtracting, we get

$$(1-x^2)\,(u''v - uv'') - 2x(u'v - uv') + [m(m+1) - n(n+1)]uv = 0$$

or $$\frac{d}{dx}\{(1-x^2)\,(u'v - uv')\} + (m-n)\,(m+n+1)\, uv = 0.$$

Now integrating from –1 to 1, we get

$$(m-n)\,(m+n+1)\int_{-1}^{1} uv\, dx = \Big| (1-x^2)\,(uv' - u'v) \Big|_{-1}^{1} = 0.$$

Hence $$\int_{-1}^{1} P_m(x)\, P_n(x)\, dx = 0. \qquad (m \neq n) \qquad ...(3)$$

This is known as the *orthogonality property of Legendre polynomials.*

(*Hamirpur, 2015 ; U.P.T.U., 2013*)

When $m = n$, we have from Rodrigue's formula,

$$(n\,!\,2^n)^2 \int_{-1}^{1} P_n^2(x)\, dx = \int_{-1}^{1} D^n(x^2-1)^n \cdot D^n(x^2-1)^n\, dx \qquad \text{[Integrate by parts]}$$

$$= \Big| D^n(x^2-1)^n \cdot D^{n-1}(x^2-1)^n \Big|_{-1}^{1} - \int_{-1}^{1} D^{n+1}(x^2-1)^n \,.\, D^{n-1}(x^2-1)^n\, dx$$

Since $D^{n-1}(x^2-1)^n$ has $x^2 - 1$ as a factor, the first term on the right vanishes for $x = \pm 1$. Thus

$$(n\,!\,2^n)^2 \int_{-1}^{1} P_n^2(x)\, dx = -\int_{-1}^{1} D^{n+1}(x^2-1)^n \,.\, D^{n-1}(x^2-1)^n\, dx$$

[Integrate by parts $(n-1)$ times]

$$= (-1)^n \int_{-1}^{1} D^{2n}(x^2-1)^n \,.\, (x^2-1)^n\, dx = (-1)^n \int_{-1}^{1} (2n)\,!\,(x^2-1)^n\, dx$$

$$= 2(2n)\,! \int_{0}^{1} (1-x^2)^n\, dx \qquad \text{[Put } x = \sin\theta\text{]}$$

$$= 2(2n)\,! \int_{0}^{\pi/2} \cos^{2n+1}\theta\, d\theta = 2(2n)\,!\, \frac{2n(2n-2)\cdots 4\cdot 2}{(2n+1)\,(2n-1)\cdots 2\,.\,1}$$

$$= 2(2n)\,!\,[2n(2n-2)...\,4\,.\,2]^2/(2n+1)\,! = \frac{2}{2n+1}\,(2^n n\,!)^2$$

Hence $$\int_{-1}^{1} P_n^2(x) dx = 2/(2n+1). \qquad ...(4) \quad (D.T.U., 2013 ; V.T.U., 2012)$$

**(2) Fourier-Legendre expansion of f(x).** If $f(x)$ be a function defined from $x = -1$ to $x = 1$, we can write

$$f(x) = \sum_{n=0}^{\infty} c_n P_n(x) \qquad ...(5)$$

To determine the coefficient $c_n$, multiply both sides by $P_n(x)$ and integrate from –1 to 1. Then Eqn. (3) and Eqn. (4) give

$$\int_{-1}^{1} f(x) P_n(x) dx = c_n \int_{-1}^{1} P_n^2(x)\, dx = \frac{2c_n}{2n+1} \quad \text{or} \quad c_n = \left(n + \frac{1}{2}\right)\int_{-1}^{1} f(x) P_n(x)\, dx$$

Equation (5) is known as *Fourier-Legendre expansion of f (x).*

**Example 16.24.** *Show that*

*(a)* $P_n(1) = 1$

*(b)* $P_n(-x) = (-1)^n p_n(x)$ *(BHU, 2018)*

**Solution:** (*a*) We have

$$\sum z^n \cdot P_n(x) = (1 - 2xz + z^2)^{-1/2} \quad ...(i)$$

$$x = 1$$

then

$$\sum z^n P_n(1) = (1 - 2z + z^2)^{-1/2} = (1 - z)^{-1}$$

$$= 1 + z + z^2 + ... z^n + ...$$

$$= \sum z^n$$

Equating the co-efficients of $z^n$ from both sides, we get

$$P_n(1) = 1$$

(*b*) Replacing $z$ and $x$ by $(-z)$ and $(-x)$ and $(-x)$ respectively in Eqn. (*i*), we get

$$\sum(-z)^n p_n(-x) = (1 - 2xz + z^2)^{-1/2}$$

$$\sum z^n (-1)^n p_n(-x) = \sum z^n P_n(x) \quad ...(ii)$$

Equating the co-efficients of $z^n$ on both side of (*ii*)

$$((-1)^n p'_n - x) = P_n(x) \quad ...(iii)$$

Now multiplying both sides of (*iii*) by $(-1)^n$, we have

$$P_n(-x) = (-1)^n p_n(x)$$

$$(-1)^{2n} = 1$$

**Example 16.25:** *Show that*

$$P_{2n+1}(0) = 0$$

*and*

$$P_{2n}(0) = (-1)^n \cdot \frac{1.3.5...(2n-1)}{2.4.6...2n}$$

*(Mysore, 2018)*

**Solution:** Let $x = 0$

$$\sum z^n P_n(x) = (1 - 2xz + z^2)^{-1/2}, \quad \text{then}$$

$$\sum z^n P_n(0) = (1 + z^2)^{-1/2}$$

$$= [1 - (1 - z^2)]^{-1/2}$$

$$\sum z^n P_n(0) = 1 + \frac{1}{2}(-z^2) + \frac{1.3}{2.4}(-z^2) + \frac{1.3.5}{2.4.6}(-z^2)^3 + ... + ... \frac{1.3.5...(2k-1)}{2.4.. + ...2k}(-z^2)^k + ...$$

Equating the co-efficients of $z^{2n+1}$ on the two sides of the above result.

$$P_{2n+1}(0) = 0$$

Now equating the co-efficients of $z^{2n}$ on the two sides $P$.

$$P_{2n}(0) = \frac{1.3.5...(2n-1)}{2.4.. + ...2n}(-1)^n$$

$$= (-1)^n \frac{(2n)!}{(2^n \cdot n!)^2}$$

**Example 16.26.** *Show that*

*(a)* $P_n(1) = 1$

*(b)* $P_n(-x) = (-1)^n p_n(x)$ *(BHU, 2018)*

**Solution:** (*a*) We have

$$\sum z^n \cdot P_n(x) = (1 - 2xz + z^2)^{-1/2} \quad ...(i)$$

$$x = 1$$

then

$$\sum z^n P_n(1) = (1 - 2z + z^2)^{-1/2} = (1 - z)^{-1}$$

$$= 1 + z + z^2 + ... z^n + ...$$

$$= \Sigma z^n$$

Equating the co-efficients of $z^n$ from both sides, we get

$$P_n(1) = 1$$

(*b*) Replacing $z$ and $x$ by $(-z)$ and $(-x)$ and $(-x)$ respectively in Eqn. (*i*), we get

$$\Sigma(-z)^n p_n(-x) = (1 - 2xz + z^2)^{-1/2}$$

$$\Sigma z^n (-1)^n p_n(-x) = \Sigma z^n P_n(x) \quad \text{...(ii)}$$

Equating the co-efficients of $z^n$ on both side of Eqn. (*ii*)

$$((-1)^n p'_n - x) = P_n(x) \quad \text{...(iii)}$$

Now multiplying both sides of Eqn. (*iii*) by $(-1)^n$, we have

$$P_n(-x) = (-1)^n p_n(x)$$

$$(-1)^{2n} = 1$$

**Example 16.27.** *Show that* $\int_{-1}^{1} x^2 \, P_{n-1} P_{n+1} \, dx = \dfrac{2n(n+1)}{(2n-1)(2n+1)(2n+3)}$.

*(Delhi, 2015 ; Calicut, 2012)*

**Solution.** We have from the recurrence relation (1),

$$(2n + 1) x P_n = (n + 1) P_{n+1} + n P_{n-1}$$

$$\therefore \quad xP_{n-1} = \frac{1}{2n-1} \{nP_n + (n-1) P_{n-2}\}$$

and

$$xP_{n+1} = \frac{1}{2n+3} \{(n+2) P_{n+2} + (n+1) P_n\}P$$

$$\therefore \quad x^2 P_{n-1} P_{n+1} = \frac{1}{(2n-1)(2n+3)} \{n(n+2) P_n P_{n+2} + n(n+1) P_n^2$$
$$+ (n-1)(n+2) P_{n-2} P_{n+2} + (n^2 - 1) P_n P_{n-2}\}$$

Integrating both sides from –1 to 1 and using orthogonality of Legendre polynomials, we get

$$\int_{-1}^{1} x^2 P_{n-1} P_{n+1} \, dx = \frac{n(n+1)}{(2n-1)(2n+3)} \int_{-1}^{1} P_n^2 \, dx = \frac{2n(n+1)}{(2n-1)(2n+1)(2n+3)}.$$

**Example 16.28.** *If* $f(x) = 0, \quad -1 < x \le 0$
$\quad\quad = x, \quad 0 < x < 1,$

*show that* $f(x) = \frac{1}{4} P_0(x) + \frac{1}{2} P_1(x) + \frac{5}{16} P_2(x) - \frac{3}{32} P_4(x) + \ldots\ldots$

**Solution.** Let $\quad f(x) = \sum_{n=0}^{\infty} c_n P_n(x)$

Then $c_n$ is given by $c_n = \left(n + \frac{1}{2}\right) \int_{-1}^{1} f(x) P_n(x) \, dx$

$$= \left(n + \frac{1}{2}\right) \left[\int_{-1}^{0} 0 \,.\, P_n(x) \, dx + \int_0^1 xP_n(x) \, dx\right] = \left(n + \frac{1}{2}\right) \int_0^1 x \, P_n(x) \, dx$$

$$\therefore \quad c_0 = \frac{1}{2} \int_0^1 xP_0(x) \, dx = \frac{1}{2} \int_0^1 x \, dx = \frac{1}{4}$$

$$c_1 = \frac{3}{2} \int_0^1 xP_1(x) \, dx = \frac{3}{2} \int_0^1 x^2 \, dx = \frac{1}{2}$$

$$c_2 = \frac{5}{2}\int_0^1 xP_2(x)\,dx = \frac{5}{2}\int_0^1 x\,.\,\frac{3x^2-1}{2}\,dx = \frac{5}{4}\left|\frac{3x^4}{4}-\frac{x^2}{2}\right|_0^1 = \frac{5}{16}$$

$$c_3 = \frac{7}{2}\int_0^1 xP_3(x)\,dx = \frac{7}{2}\int_0^1 x\,.\,\frac{5x^3-3x}{2}\,dx = \frac{7}{4}\left|5\frac{x^5}{5}-3\frac{x^3}{3}\right|_0^1 = 0$$

$$c_4 = \frac{9}{2}\int_0^1 xP_4(x)\,dx = \frac{9}{2}\int_0^1 x\,.\,\frac{35x^4-30x^2+3}{8}\,dx$$

$$= \frac{9}{16}\left|35\frac{x^6}{6}-35\frac{x^4}{4}+3\frac{x^2}{2}\right|_0^1 = -\frac{3}{32} \text{ and so on.}$$

Hence $$f(x) = \frac{1}{4}P_0(x) + \frac{1}{2}P_1(x) + \frac{5}{16}P_2(x) - \frac{3}{32}P_4(x) + \dots .$$

**Example16.29:** *Prove that*

*(i)* $\int_{-1}^{1} P_n(x)dx = 0,$ *if* $n \neq 0$ *(ii)* $\int_{-1}^{1} x^m P_n(x)dx = 0,$ *if* $m < n$

*m and n are positive integer*

**Solution:** (*i*)

$$\int_{-1}^{1} P_n(x)dx = \frac{1}{2^n(n!)}\int_{-1}^{1} D^n(x^2-1)^n\,dx$$

$$= \frac{1}{2^n(n!)}\left[D^{n-1}(x^2-1)^n\right]_{-1}^{1}$$

But

$$D^{n-1}[(x^2-1)^n] = D^{n-1}[(x+1)^n(x-1)^n]$$

$$= D^{n-1}(x+1)^n(x-1)^n + nc_1 D^{n-2}(x+1)^n - D(x-1)^n$$

$$+ \dots (x+1)^n \cdot D^{n-1}(x-1)^n$$

$$= n!\,(x+1)(x-1)^n + \dots + (x+1)^n\,.\,n!\,(n-1)$$

$$= 0, \quad \text{when} \quad x = -1, \quad \text{or} \quad 1,$$

since each term contains both $(x-1)$ and $(x+1)$

$\therefore$

$$\int_{-1}^{1} P_{(n)}dx = 0$$

(*ii*)

$$\int_{-1}^{1} x^m P_n(x) = \frac{1}{2^n(n!)}\int_{-1}^{1} x^m\cdot D^n(x^2-1)^n\cdot dx$$

$$= \frac{1}{2^n(n!)}\left[\left\{x^m\cdot D^{n-1}(x^2-1)^n\right\}_{-1}^{1} - \int_{-1}^{1} mx^{m-1}\cdot D^{n-1}(x^2-1)^n\,dx\right]$$

$$= \frac{(-1)m}{2^n(n!)}\int_{-1}^{1} x^{m-1}D^{n-1}(x^2-1)^n\,dx$$

Since $[D^{n-1}(x^2-1)^n]_{-1}^{1} = 0$

Proceeding similar

$$\int_{-1}^{1} x^m P_n(x)dx = \frac{(-1)^m m!}{2^n(n!)}\int_{-1}^{1} x^0 D^{n-m}(x^2-1)^n\,dx$$

$$= \frac{(-1)^m m!}{2^n(n!)}\left[D^{n-m-1}(x^2-1)^n\right]_{-1}^{1}$$

$$= 0, \quad \text{if } n \geq m+1$$

## PROBLEMS 16.5

1. Show that (*i*) $P_n(-x) = (-1)^n P_n(x)$. (*Hazaribag, 2015 ; Calicut, 2012*)

(*ii*) $P_{2n}(0) = \dfrac{(-1)^n (2n)!}{2^{2n} (n!)^2}$. (*Calicut, 2012*) (*iii*) $P_{2n+1}(0) = 0$. (*Calicut, 2012*)

2. Prove that (*i*) $P_{2n}'(0) = 0$ (*ii*) $P_{2n+1}'(0) = \dfrac{(-1)^n (2n+1)!}{2^{2n} (n!)^2}$.

3. Express the following in terms of Legendre polynomials : (*i*) $1 + 2x + x^2$ (*Calicut, 2012*)

(*ii*) $x^3 + 2x^2 - x - 3$, (*D.T.U., 2013*)

(*iii*) $x^3 + 5x^2 + 6x + 1$ (*Delhi, 2012*)

4. Prove that (*i*) $(1 - x^2) P_n'(x) = (n+1)[xP_n(x) - P_{n+1}(x)]$,

(*ii*) $P_n(x) - P'_{n+1}(x) - 2xP_n'(x) + P'_{n-1}(x)$. (*iii*) $P_n(x) P_{n+1/2}(x) = \dfrac{\sqrt{\pi}}{2^{2n+1}} P_{2m}(x)$

5. Prove that (*i*) $\displaystyle\int_{-1}^{1} [P_2(x)]^2 \, dx = \frac{2}{5}$. (*Andhra, 2013*) (*ii*) $\displaystyle\int_0^1 P_{2n}(x) \, dx = 0$.

6. Prove that $\displaystyle\int_{-1}^{1} P_n(x)(1 - 2hx + h^2)^{-1/2} \, dx = \frac{2h^n}{2n+1}$.

7. Using Rodrigue's formula, show that $P_n(x)$ satisfies the differential equation

$$\frac{d}{dx}\left[(1 + x^2) \frac{d}{dx}[P_n(x)]\right] + n(n+1) P_n(x) = 0.$$ (*Andhra, 2014*)

8. Expand the following functions in terms of Legendre polynomials in the interval $-1 < x < 1$ :

(*i*) $f(x) = x^3 + 2x^2 - x - 3$ (*ii*) $f(x) = x^4 + x^3 + 2x^2 - x - 3$.

9. If $f(x) = 0, \quad -1 < x < 0$

$= 1, \quad 0 < x < 1$, show that $f(x) = \frac{1}{2}P_0(x) + \frac{3}{4}P_1(x) - \frac{7}{16}P_3(x) + \ldots\ldots$ (*Hazaribag, 2015*)

10. Show that

(*i*) $x^2 = \dfrac{1}{3}[P_0(x) + 2P_2(x)]$ (*ii*) $x^3 = \dfrac{1}{5}[3P_1(x) + 2P_3(x)]$

(*iii*) $P_6(x) = \dfrac{1}{16}[231x^6 - 315x^4 + 105x^2 - 5]$

11. Using Rodrigue's formula, show that $P_n(x)$ satisfies the different equation

$$\frac{d}{dx}\left[(1 - x^2) \frac{d}{dx}\{P_n(x)\}\right] + n(n+1)P_n(x) = 0$$

12. Show that

(*i*) $P_n'(x) = xP'_{n-1}(x) + n P_{n-1}(x)$ (*ii*) $P_n(x) = P'_{n+1}(x) - 2x P_n'(x) + P'_{n-1}(x)$

13. Prove that

(*i*) $\displaystyle\int_{-1}^{1} x P_n P_n' dx = \frac{2n}{2n+1}$ (*ii*) $\displaystyle\int_{-1}^{1} x^2 P_{n+1} P_{n-1} \, dx = \frac{2n(n+1)}{(2n-1)(2n+1)(2n+3)}$

14. Show that

(*i*) $\displaystyle\int_{-1}^{1} (1 - x)\left(\frac{dP_n}{dx}\right)^2 dx = \frac{2n(n+1)}{2n+1}$ (*ii*) $\displaystyle\int_{-1}^{1} (1 - x^2) P_n' P_m'' dx = 0, m \neq n$

## 16.18 OTHER SPECIAL FUNCTIONS

The following special functions occur in numerous engineering problems. We state below their important properties which can be verified by similar methods :

(1) **Laguerre's polynomials***. These are the solutions of *Laguerre's differential equation*

$$xy'' + (1 - x)y' + ny = 0 \quad ...(1)$$

These polynomials $L_n(x)$, are given by the corresponding Rodrigue's formula

$$L_n(x) = e^x \frac{d^n}{dx^n}(x^n e^{-x}) \quad ...(2)$$

In particular, $L_0(x) = 1$ ; $L_1(x) = 1 - x$, $L_2(x) = 2 - 4x + x^2$ ; $L_3(x) = 6 - 18x + 9x^2 - x^3$.

Their *generating function* is given by

$$\frac{e^{-xt/(1-t)}}{1-t} = \sum_{n=0}^{\infty} \frac{L_n(x)}{n!} t^n \quad ...(3)$$

The *orthogonal property* for these polynomials is

$$\int_0^{\infty} e^{-x} L_m(x) L_n(x)\, dx = \begin{cases} 0, & m \neq n \\ (n!)^2, & m = n \end{cases} \quad ...(4)$$

(2) **Hermite's polynomials†**. These are the solutions of Hermite's differential equation

$$y'' - 2xy' + 2ny = 0 \quad ...(5)$$

These polynomials $H_n(x)$, are given by the Rodrigue's formula

$$H_n(x) = (-1)^n e^{x^2} \frac{d^{2n}}{dx^n}(e^{-x^2}) \quad ...(6)$$

In particular, $H_0(x) = 1$ ; $H_1(x) = 2x$ ; $H_2(x) = 4x^2 - 2$ ; $H_3(x) = 8x^3 - 12x$.

Their *generating function is* given by

$$e^{2tx - t^2} = \sum_{n=0}^{\infty} \frac{H_n(x)}{n!} t^n \quad ...(7)$$

The *orthogonal property* of these polynomials is

$$\int_{-\infty}^{\infty} e^{-x^2} H_m(x) H_n(x)\, dx = \begin{cases} 0, & m \neq n \\ 2^n n! \sqrt{\pi}, & m = n \end{cases} \quad ...(8)$$

(3) **Chebyshev polynomials****. These polynomials denoted by $T_n(x)$, are the solutions of the differential equation

$$(1 - x^2)y'' - xy' + n^2 y = 0 \quad ...(9)$$

Their *generating function* is

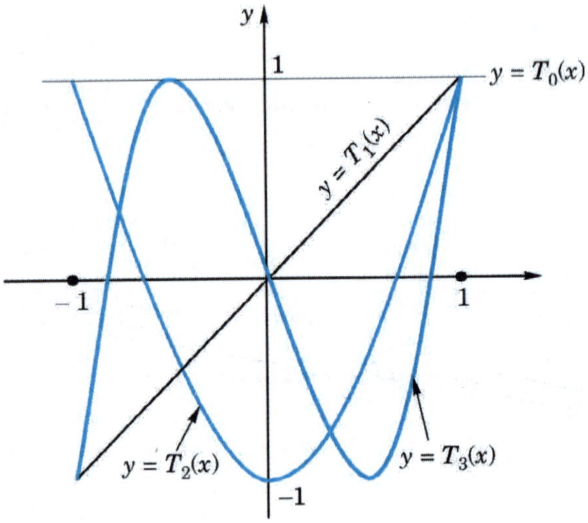

Fig. 16.3. Graphs of $T_0(x)$, $T_1(x)$, $T_2(x)$, $T_3(x)$.

$$\frac{1 - xt}{1 - 2xt + t^2} = \sum_{n=0}^{\infty} T_n(x) t^n \quad ...(10)$$

and $$T_n(x) = \frac{n}{2} \sum_{r=0}^{N} (-1)^r \frac{(n - r - 1)!}{r!(n - 2r)!} (2x)^{n-2r} \quad ...(11)$$

where $N = \frac{n}{2}$, if $n$ is even and $N = \frac{1}{2}(n - 1)$, if $n$ is odd.

---

* Named after the French mathematician *Edmond Laguerre* (1834–86) who is known for his work in infinite series and geometry.

† See footnote p. 80.

** Named after the Russian mathematician *Pafnuti Chebyshev* (1821–1894) who is known for his work in the theory of numbers and approximation theory.

In particular, $T_0(x) = 1$, $T_1(x) = x$, $T_2(x) = 2x^2 - 1$, $T_3(x) = 4x^3 - 3x$. Also, we have the recurrence relation

$$T_{n+1}(x) = 2x\, T_n(x) - T_{n-1}(x) \qquad \text{...(12)}$$

which defines $T_{n+1}$ in terms of $T_n$ and $T_{n-1}$.

Their *orthogonal property* is

$$\int_{-1}^{1} (1 - x^2)^{-1/2}\, T_m(x) T_n(x)\, dx = \begin{cases} 0, & m \neq n \\ \dfrac{\pi}{2}, & m = n \\ \pi, & m = n = 0 \end{cases} \qquad \text{...(13)}$$

**Example 16.30.** *Prove that* $\int_0^\infty e^{-x} L_m(x) L_n(x)\, dx = 0,\ m \neq n.$

**Solution.** Since $L_m(x)$ and $L_n(x)$ are the solutions of the Laguerre's differential Eqn. (1).

$\therefore$
$$xL_m'' + (1 - x)L_m' + mL_m = 0 \qquad \text{...(i)}$$
$$xL_n'' + (1 - x)L_n' + nL_n = 0 \qquad \text{...(ii)}$$

Multiplying Eqn. (*i*) by $L_n$ and Eqn. (*ii*) by $L_m$ and subtracting, we get

$$x(L_nL_m'' - L_mL_n'') + (1 - x)(L_nL_m' - L_mL_n') = (n - m)L_mL_n$$

or
$$\frac{d}{dx}(L_nL_m' - L_mL_n') + \frac{1 - x}{x}(L_nL_m' - L_mL_n') = \frac{(n - m)L_mL_n}{x}$$

This is Leibnitz's linear equation and its

$$\text{I.F.} = e^{\int\left(\frac{1}{x} - 1\right)dx} = e^{\log x - x} = xe^{-x}.$$

$\therefore$ Its solution is $\left|(L_nL_m' - L_mL_n')\, xe^{-x}\right|_0^\infty = \int_0^\infty \frac{(n - m)L_mL_n}{x}\, xe^{-x}\, dx$

or
$$\int_0^\infty e^{-x} L_mL_n\, dx = \left|\frac{(L_nL_m' - L_mL_n')xe^{-x}}{n - m}\right|_0^\infty = 0 \text{ which proves the result.}$$

**Example 16.31.** *Prove that* $H_n(x) = (-1)^n e^{x^2} \frac{d^{2n}}{dx^n}(e^{-x^2})$.

**Solution.** The generating function for $H_n(x)$ is $e^{2tx - t^2} = e^{x^2} . e^{-(t - x)^2} = \sum_{n=0}^\infty H_n(x) . \frac{t^n}{n!}$

Then
$$\left[\frac{\partial^n}{\partial t^n}(e^{2tx - t^2})\right]_{t=0} = H_n(x) \qquad \text{...(i)}$$

Also
$$\left[\frac{\partial^n}{\partial t^n}(e^{2tx - x^2})\right]_{t=0} = e^{x^2}\left[\frac{\partial^n}{\partial t^n}\{e^{-(t - x)^2}\}\right]_{t=0}$$

$$= e^{x^2}\left[\frac{\partial^n}{\partial(-x)^n}\{e^{-(t - x)^2}\}\right]_{t=0} = (-1)^n \frac{d^n}{dx^n}(e^{-x^2}) \qquad \text{...(ii)}$$

Equating Eqns. (*i*) and (*ii*), we get the desired result.

## PROBLEMS 16.6

1. Using the generating function (3) page 674, obtain the recurrence formula
$$L_{n+1}(x) = (2n + 1 - x)\, L_n(x) - n^2L_{n-1}(x).$$
2. Show that (*i*) $nL_{n-1}(x) = nL'_{n-1}(x) - L'_n(x)$. (*ii*) $L'_n(x) = L'_{n-1}(x) - L_{n-1}(x)$.
3. Show that (*i*) $H_{2n}(0) = (-1)^n \frac{2n!}{n!}$ (*ii*) $H_{2n+1}(0) = 0$

4. Prove that (i) $H_n'(x) = 2n\, H_{n-1}(x)$ (ii) $\dfrac{d^m}{dx^m}[H_n(x)] = \dfrac{2^m . n!}{(n-m)!} H_{n-m}(x), m < n.$
5. Using the generating function (7) page 603, obtain the recurrence formula $2xH_n(x) = 2nH_{n-1}(x) + H_{n+1}(x)$.
6. Prove that (i) $\int_{-\infty}^{\infty} e^{-x^2} H_2(x) H_3(x) = 0,$ (ii) $\int_{-\infty}^{\infty} e^{-x^2} \{H_2(x)\}^2 = 8\sqrt{\pi}.$
7. Express $x^3$ in terms of Chebyshev polynomials $T_1$ and $T_3$.
8. Show that (i) $T_5 = 16x^5 - 20x^3 + 5x$.
   (ii) $(1 - x^2)T_n' = nT_{n-1}(x) - nxT_n(x)$.
9. Prove that $\dfrac{1-t^2}{1-2xt+t^2} = T_0(x) + 2\sum_{n=1}^{\infty} T_n(x)\, t^n.$

## 16.19 (1) STRUM*-LIOUVILLE† PROBLEM

Legendre's equation $(1 - x^2)y'' - 2xy' + n(n + 1)\, y = 0$ ...(i)

can be written as, $[(1 - x^2)y']' + \lambda y = 0$ $[\lambda = n(n + 1)]$

Bessel's equation $X^2 \dfrac{d^2y}{dx^2} + X\dfrac{dy}{dx} + (X^2 - n^2)y = 0$ can be transformed by putting $X = kx$ (so that $\dfrac{dy}{dx} = \dfrac{dy}{dx}\cdot\dfrac{dx}{dX} = \dfrac{y'}{k}, \dfrac{d^2y}{dX^2} = \dfrac{y''}{k^2}$ to the form

$$x^2y'' + xy' + (k^2x^2 - n^2)\, y = 0$$

or
$$(xy'' + y') + (\lambda x - n^2/x)\, y = 0 \qquad [\lambda = k^2]$$

or
$$(xy')' + (\lambda x - n^2/x)\, y = 0 \qquad ...(ii)$$

Both the Eqns. (i) and (ii) are of the form

$$[r(x)\, y']' + [\lambda p(x) + q(x)]y = 0 \qquad ...(1)$$

which is known as the *Strum-Liouville equation*. Similarly Laguerre's, Hermite's equations etc. can also be reduced to Eqn. (1). Thus all the above equations of engineering utility can be considered with a common approach by means of Strum-Liouville's equation.

Eqution (1) considered on some interval $a \le x \le b$, satisfying the conditions

$$\alpha_1 y(a) + \alpha_2 y'(a) = 0, \beta_1 y(b) + \beta_2 y'(b) = 0 \qquad ...(2)$$

with the real constants : $\alpha_1, \alpha_2$ not both zero and $\beta_1, \beta_2$ not both zero. The conditions Eqn. (2) at the end points are called *boundary condtions.*

*A differential equation together with the boundary conditions, is called a* **boundary value problem.** *Equation (1) together with boundary conditions Eqn. (2) is called a* **Strum-Liouville problem†.**

Obviously $y = 0$ is a solution of the problem for any value of the parameter $\lambda$ which is a trivial solution and as such is of no practical utility. Any other solution of Eqn. (1) satisfying Eqn. (2) is called an *eigen function* of the problem and the corresponding value of $\lambda$ is called an eign value of the problem.

**A special case.** Taking $r = p = 1$ and $q = 0$ in Eqn. (1), we get

$$y'' + \lambda y = 0 \qquad ...(3)$$

Also if $\alpha_1 = \beta_1 = 1$ and $\alpha_2 = \beta_2 = 0$, then the boundary conditions Eqn. (2) become

$$y(a) = 0, \quad y(b) = 0 \qquad ...(4)$$

Thus Eqns. (3) and (4) constitute the *simplest form of Strum-Liouville problem.*

**(2) Orthogonality.** Of the various properties of eigen functions of Strum-Liouville problem the orthogonality is of special importance.

* Named after the Swiss mathematician *J.C.F. Strum* (1803–1855) who later became Poisson's successor at Sorbonne university, Paris.

† Named after the French professor *Joseph Liouville* (1809–1882) who is known for his important contributions to complex analysis, special functions, number theory and differential geometry.

**Def.** *Two functions $y_m(x)$ and $y_n(x)$ defined on some interval $a \le x \le b$, are said to be orthogonal on this interval w.r.t. the weight function $p(x) > 0$, if*

$$\int_a^b p(x)\, y_m(x) y_n(x)\, dx = 0 \text{ for } m \neq n.$$

*The* **norm** *of $y_m$, denoted by $\| y_m \|$, is defined to be the non-negative square root of* $\int_a^b p(x)\, [y_m(x)]^2\, dx$.

*Thus*

$$\| y_m \| = \sqrt{\left\{ \int_a^b p(x)\, [y_m(x)]^2\, dx \right\}}$$

The functions which are orthogonal on $a \le x \le b$ and have norm equal to 1, are called **orthonormal** on this interval.

**(3) Orthogonality of eigen functions.**

**Theorem.** *If (i) the functions p, q, r and r′ in the Strum-Liouville Eqn. (1) be continuous in $a \le x \le b$ ;*

*(ii) $y_m(x)$, $y_n(x)$ be two eigen functions of the Strum-Liouville problem corresponding to eigen values $\lambda_m$ and $\lambda_n$ respectively ;*

*then $y_m(x)$ and $y_n(x)$ $(m \neq n)$ are orthogonal on that interval w.r.t. the weight function $p(x)$.*

*Proof.* Since $y_m$ and $y_n$ satisfy (1) above

$$(ry'_m)' + (\lambda_m p + q)\, y_m = 0$$
$$(ry'_n)' + (\lambda_n p + q)\, y_n = 0$$

Multiplying the first equation by $y_n$ and the second by $-y_m$ and adding, we get

$$(\lambda_m - \lambda_n)\, p y_m y_n = y_m (ry'_n) - y_n (ry'_m)$$
$$= \frac{d}{dx} [(ry'_n)\, y_m - (ry'_m) y_n], \text{ after differentiation.}$$

Now integrating both sides w.r.t. $x$ from $a$ to $b$, we obtain

$$(\lambda_m - \lambda_n) \int_a^b p y_m y_n\, dx = \left[ (ry'_n) y_m - (ry'_m) y_n \right]_a^b$$
$$= r(b)\, [y'_n(b)\, y_m(b) - y'_m(b)\, y_n(b)] - r(a)\, [y'_n(a)\, y_m(a) - y'_m(a)\, y_n(a)] \qquad \text{...(A)}$$

The R.H.S. will vanish if the boundary conditions are of one of the following forms :

I. $y(a) = y(b) = 0$ ; II. $y'(a) = y'(b) = 0$ ; III. $\alpha_1 y(a) + \alpha_2 y'(a) = 0$, $\beta_1 y(b) + \beta_2 y'(b) = 0$ where either $\alpha_1$ and $\alpha_2$ is not zero and either $\beta_1$ or $\beta_2$ is not zero.

Thus in each case (A) reduces to $\int_a^b p y_m y_n\, dx = 0 \quad (m \neq n)$

which shows that the eigen functions $y_m$ and $y_n$ are orthogonal on $a \le x \le b$ w.r.t. the weight function $p(x) = 0$.

**Obs.** The third form of the boundary conditions in fact contains the first two forms as special cases.

**Cor. 1. Orthogonality of Legendre polynomials** has already been established directly in § 16.17. But if follows at once from the above theorem.

We have already seen in Eqn. (*i*) so that Legendre's equation is Strum-Liouville equation

$$[(1 - x^2)\, y']' + \lambda y = 0 \qquad [\lambda = n(n + 1)]$$

with $r(x) = 1 - x^2$, $p(x) = 1$ and $q(x) = 0$.

Since $y(-1) = y(1) = 0$ and for $n = 0, 1, 2, ..., \lambda = 0, 1.2, 2.3, ...$, the Legendre polynomials are the solutions of the problem *i.e.*, these are the eigen functions. Thus it follows by the above theorem, that they are orthogonal on $-1 \le x \le 1$.

**Cor. 2. Orthogonality of Bessel functions** has also been established directly in § 16.11. But it can easily be seen to follow from the above theorem.

In Eqn. (*i*), we transformed the Bessel's equation

$$X^2 \frac{d^2 J_n}{dx^2} + X \frac{dJ_n}{dx} + (X^2 - n^2)\, J_n(x) = 0$$

into $[xJ'_n(kx)]' + (k^2 x - n^2/x)\, J_n(kx) = 0$ which is Strum-Liouville equation with $r(x) = x$, $p(x) = x$, $q(x) = -n^2/x$ and $\lambda = k^2$. Since $r(0) = 0$, it follows from the above theorem that those solutions of $J_n(kx)$ which are zero at $x = 0$ form an orthogonal set on $0 \le x \le R$ with weight function $p(x) = x$.

**Example 16.32.** *For the Strum-Liouville problem $y'' + \lambda y = 0$, $y(0) = 0$, $y(l) = 0$, find the eigen functions and show that they are orthogonal.*

**Solution.** For $\lambda = -\gamma^2$, the general solution of the equation is $y(x) = c_1 e^{\gamma x} + c_2 e^{-\gamma x}$

The above boundary conditions give $c_1 = c_2 = 0$ and $y = 0$ which is not an eigen function.

*For* $\lambda = \gamma^2$, the general solution is $y(x) = A \cos \gamma x + B \sin \gamma x$

The first boundary condition gives $y(0) = A = 0$ and the second boundary condition gives $y(l) = B \sin \gamma l = 0$, $\gamma = 0, \pm \pi/l, \pm 2\pi/l, \ldots$ Thus the eigen values are $\lambda = 0, \pi^2/l^2, 4\pi^2/l^2, \ldots$ and taking $B = 1$, the corresponding eigen functions are

$$y_n(x) = \sin (n\pi x/l) \qquad n = 0, 1, 2, \ldots$$

From the above theorem, it follows that the said eigen functions are orthogonal on the interval $0 \leq x \leq l$.

**Obs.** This problem concerns an elastic string stretched between fixed points $x = 0$ and $x = l$ and allowed to vibrate. Here $y(x)$ is the space function of the deflection $u(x, t)$ of the string where $t$ is the time. (See §18.4).

## PROBLEMS 16.7

Find the eigen functions of each of the following *Strum-Liouville problems* and verify their orthogonality :

1. $y'' + \lambda y = 0, y(0) = 0, y(\pi) = 0.$
2. $y'' + \lambda y = 0, y(0) = 0, y'(l) = 0.$
3. $y'' + \lambda y = 0, y'(0) = 0, y'(\pi) = 0.$
4. $y'' + \lambda y = 0, y(\pi) = y(-\pi), y'(\pi) = y'(-\pi).$
5. $(xy')' + \lambda x^{-1} y = 0, y(1) = 0, y'(e) = 0.$

Transform each of the following equations to the *Strum-Liouville equations* indicating the weight function :

6. *Laguerre's equation* : $xy'' + (1 - x) y' + ny = 0.$
7. Hermite's equation : $y'' - 2xy' + 2ny = 0.$

## 16.20 OBJECTIVE TYPE OF QUESTIONS

## PROBLEMS 16.8

*Fill up the blanks or choose the correct answer in the following problems :*

1. In terms of Legendre polynomials $2 - 3x + 4x^2$ is ......
2. $J_{-1/2} =$ .........
3. $\int_{-1}^{1} P_n^2(x)\, dx =$ .........
4. $P_{2n+1}(0) =$ .........
5. $\int_{-1}^{1} x^m P_n(x)\, dx =$ ......($m$ being an integer $< n$)
6. The recurrence relation connecting $J_n(x)$ to $J_{n-1}(x)$ and $J_{n+1}(x)$ is .........
7. Orthogonality relation for Bessel functions is ..........
8. Bessel's equation of order zero is .........
9. $J_{1/2} =$ .........
10. $\frac{d}{dx}[x^n J_n(x)] =$ .........
11. Value of $P_2(x)$ is .........
12. $\int_{-1}^{1} P_3(x) P_4(x)\, dx =$ .........
13. $P_n(-1) = (-1)^n$ (True or False)
14. Rodrigue's formula for $P_n(x)$ is .........
15. $\int_0^1 x J_n(\alpha x) J_n(\beta x)\, dx = 0$, if .........
16. Expansion of $5x^3 + x$ interms of Legendre polynomials is .........
17. Generating function of $P_n(x)$ is .........
18. $\frac{d}{dx}[J_0(x)] =$ .........
19. Bessel equation of order 4 is $x^2 y'' + xy' + (x^2 - 4) y = 0.$ (True or False)
20. $\frac{d}{dx}[x^2 J_2(x)] = x^2 J_1(x).$ (True or False)
21. Legendre's polynomial of first degree $= x.$ (True or False)

22. If $\alpha$ is a root of $P_n(x) = 0$, then $P_{n+1}(\alpha)$ and $P_{n-1}(\alpha)$ are of opposite signs. (True or False)

23. $x = 0$ is a regular singular point of $2x^2y'' + 3xy' + (x^2 - 4)y = 0$. (True or False)

24. $\cos x = 2J_1 - 2J_3 + 2J_5 - \ldots\ldots$ (True or False)

25. If $J_0$ and $J_1$ are Bessel functions, then $J_1'(x)$ is given by

(*a*) $-J_0$ (*b*) $J_0(x) - 1/x\, J_1(x)$ (*c*) $J_0(x) + \frac{1}{x} J_1(x)$.

26. If $J_n(x)$ is the Bessel function of first kind, then $\int_0^\pi [J_{-2}(x) - J_2(x)]\, dx =$

(*a*) 2 (*b*) –2 (*c*) 0 (*d*) 1.

27. If $J_{n+1}(x) = \frac{2}{x} J_n(x) - J_0(x)$, then $n$ is

(*a*) 0 (*b*) 2 (*c*) –1 (*d*) none of these.

28. The series $x - \frac{x^3}{2^2(1!)^2} + \frac{x^5}{2^4(2!)^2} - \frac{x^7}{2^6(3!)^2} + \ldots\ldots \infty$ equals

(*a*) $J_{1/2}(x)$ (*b*) $J_0(x)$ (*c*) $xJ_0(x)$ (*d*) $xJ_{1/2}(x)$.

29. If $\int_{-1}^{1} P_n(x)\, dx = 2$, then $n$ is

(*a*) 0 (*b*) 1 (*c*) –1 (*d*) none of these.

30. The value of $\int_{-1}^{1} (2x+1)P_3(x)\, dx$ where $P_3(x)$ is the third degree Legendre polynomial, is

(*a*) 1 (*b*) –1 (*c*) 2 (*d*) 0.

31. The value of the integral $\int_{-1}^{1} x^3 P_3(x)\, dx$, where $P_3(x)$ is a Legendre polynomial of degree 3, is

(*a*) 0 (*b*) $\frac{2}{35}$ (*c*) $\frac{4}{35}$ (*d*) $\frac{11}{35}$.

32. The polynomial $2x^2 + x + 3$ in terms of Legendre polynomials is

(*a*) $\frac{1}{3}(4P_2 - 3P_1 + 11P_0)$ (*b*) $\frac{1}{3}(4P_2 + 3P_1 - 11P_0)$

(*c*) $\frac{1}{3}(4P_2 + 3P_1 + 11P_0)$ (*d*) $\frac{1}{3}(4P_2 - 3P_1 - 11P_0)$.

33. If $P_n(x)$ be the Legendre polynomial, then $P_n'(-x)$ is equal to

(*a*) $(-1)^n P_n(x)$ (*b*) $(-1)^n P_n'(x)$ (*c*) $(-1)^{n+1} P_n'(x)$ (*d*) $P_n''(x)$.

34. Legendre polynomial $P_5(x) = \lambda(63x^5 - 70x^3 + 15x)$ where $\lambda$ is equal to

(*a*) 1/2 (*b*) 1/5 (*c*) 1/8 (*d*) 1/10.

35. $\int_{-1}^{1} (1+x)\, P_n(x)\, dx$, $(n > 1)$, is equal to

(*a*) $\frac{1}{2n+1}$ (*b*) $\frac{2}{2n+1}$ (*c*) $\frac{n}{2n+1}$ (*d*) 0.

36. The singular points of the differential equation $x^3(x-1)y'' + 2(x-1)y' + y - 0$ are .........

37. Show that $x^4 = \frac{1}{35}[8P_4(x) + 20P_2(x) + 7P_0(x)]$ (*Calicut, 2012*)

38. The roots of the initial equation for the power series solution of the differential equation

$2x^2y'' + xy' + (x^2 - 3)y = 0$ are ....... . (*U.P.T.U., 2011*)

39. Show that $\sum_{n=0}^{\infty} P_n(x) = \frac{1}{\sqrt{[2(1-x)]}}$. (*U.P.T.U., 2012*)

40. Classify the singular points of the differential equation $x^2 \frac{d^2y}{dx^2} + ax\frac{dy}{dx} + by = 0$. (*U.P.T.U., 2012*)

CHAPTER 17

# Partial Differential Equations



## 17.1 INTRODUCTION

The reader has, already been introduced to the notion of partial differential equations. Here, we shall begin by studying the ways in which partial differential equations are formed. Then we shall investigate the solutions of special types of partial differential equations of the first and higher orders.

In what follows $x$ and $y$ will, usually be taken as the independent variables and $z$, the dependent variable so that $z = f(x, y)$ and we shall employ the following notation :

$$\frac{\partial z}{\partial x} = p, \frac{\partial z}{\partial y} = q, \frac{\partial^2 z}{\partial x^2} = r, \frac{\partial^2 z}{\partial x \partial y} = s, \frac{\partial^2 z}{\partial y^2} = t.$$

## 17.2 FORMATION OF PARTIAL DIFFERENTIAL EQUATIONS

Unlike the case of ordinary differential equations which arise from the elimination of arbitrary constants; *the partial differential equations can be formed either by the elimination of arbitrary constants or by the elimination of arbitrary functions from a relation involving three or more variables.* The method is best illustrated by the following examples :

**Example 17.1.** *Derive a partial differential equation (by eliminating the constants) from the equation*

$$2z = \frac{x^2}{a^2} + \frac{y^2}{b^2}. \qquad ...(i) \quad (JNTUK, 2022)$$

**Solution.** Differentiating Eqn. $(i)$ partially with respect to $x$ and $y$, we get

$$2\frac{\partial z}{\partial x} = \frac{2x}{a^2} \quad \text{or} \quad \frac{1}{a^2} = \frac{1}{x}\frac{\partial z}{\partial x} = \frac{p}{x}$$

and

$$\frac{2\partial z}{\partial y} = \frac{2y}{b^2} \quad \text{or} \quad \frac{1}{b^2} = \frac{1}{y}\frac{\partial z}{\partial y} = \frac{q}{y}$$

Substituting these values of $1/a^2$ and $1/b^2$ in $(i)$, we get

$$2z = xp + yq$$

as the desired partial differential equation of the first order.

**Example 17.2.** *Form the partial differential equations (by eliminating the arbitrary functions) from*
*(a)* $z = (x + y)\,\phi\,(x^2 - y^2)$
*(b)* $z = f(x + at) + g(x - at)$ *(c)* $f\,(x^2 + y^2, z - xy) = 0$

**Solution.** (*a*) We have $z = (x + y)\,\phi\,(x^2 - y^2)$

Differentiating $z$ partially with respect to $x$ and $y$,

$$p = \frac{\partial z}{\partial x} = (x + y)\,\phi'\,(x^2 - y^2)\,.\,2x + \phi\,(x^2 - y^2), \quad ...(i)$$

$$q = \frac{\partial z}{\partial y} = (x + y)\,\phi'\,(x^2 - y^2)\,.\,(-2y) + \phi\,(x^2 - y^2) \quad ...(ii)$$

From Eqn. (*i*), $p - \dfrac{z}{x + y} = 2x\,(x + y)\,\phi'\,(x^2 - y^2)$

From Eqn. (*ii*), $q - \dfrac{z}{x + y} = -2y\,(x + y)\,\phi'\,(x^2 - y^2)$

Division gives $\dfrac{p - z/(x + y)}{q - z/(x + y)} = -\dfrac{x}{y}$

*i.e.*, $[p(x + y) - z]y + [q\,(x + y) - z]x$

*i.e.*, $(x + y)\,(py + qx) - z(x + y) = 0$

Hence $py + qz = z$ is required equation.

(*b*) We have $z = f\,(x + at) + g(x - at)$ ...(*i*)

Differentiating $z$ partially with respect to $x$ and $t$,

$$\frac{\partial z}{\partial x} = f'\,(x + at) + g'(x - at),\ \frac{\partial^2 z}{\partial x^2} = f''\,(x + at) + g''(x - at) \quad ...(ii)$$

$$\frac{\partial z}{\partial t} = af'\,(x + at) - ag'(x - at),\ \frac{\partial^2 z}{\partial t^2} = a^2 f''\,(x + at) + a^2 g''(x - at) = a^2\frac{\partial^2 z}{\partial x^2} \quad \text{[By Eqn. (ii)]}$$

Thus the desired partial differential equation is $\dfrac{\partial^2 z}{\partial t^2} = a^2\,\dfrac{\partial^2 z}{\partial x^2}$

which is an equation of the second order and Eqn. (*i*) is its solution.

(*c*) Let $x^2 + y^2 = u$ and $z - xy = v$ so that $f\,(u, v) = 0$.

Differentiating partially w.r.t. $x$ and $y$, we have

$$\frac{\partial f}{\partial u}\left(\frac{\partial u}{\partial x} + \frac{\partial u}{\partial z}p\right) + \frac{\partial f}{\partial v}\left(\frac{\partial v}{\partial x} + \frac{\partial v}{\partial z}p\right) = 0$$

or $$\frac{\partial f}{\partial u}(2x) + \frac{\partial f}{\partial v}(-y + p) = 0 \quad ...(i)$$

and $$\frac{\partial f}{\partial u}\left(\frac{\partial u}{\partial y} + \frac{\partial u}{\partial z}q\right) + \frac{\partial f}{\partial v}\left(\frac{\partial v}{\partial y} + \frac{\partial v}{\partial z}q\right) = 0 \quad \text{or} \quad \frac{\partial f}{\partial u}(2y) + \frac{\partial f}{\partial v}(-x + q) = 0 \quad ...(ii)$$

Eliminating $\dfrac{\partial f}{\partial u}$ and $\dfrac{\partial f}{\partial v}$ from (*i*) and (*ii*), we get

$$\begin{vmatrix} 2x & -y + p \\ 2y & -x + q \end{vmatrix} = 0 \quad \text{or} \quad xq - yp = x^2 - y^2.$$

**Example 17.3.** *Form a P.D.E. by eliminating the function f from the relation* $z = y^2 + 2f\left(\dfrac{1}{x} + \log y\right)$

(*Rohtak, 2019*)

**Solution:** Given, $$z = y^2 + 2f\left(\frac{1}{x} + \log y\right)$$

Differentiating the given relation partially w.r.t. only

$$1 = \frac{\delta z}{\delta x} = 2f'\left(\frac{1}{x} + \log y\right)\left(-\frac{1}{x^2}\right)$$

$$-px^2 = 2f'\left(\frac{1}{x} + \log y\right) \quad ...(i)$$

$$q = \frac{\delta z}{\delta y} = 2y + 2f'\left(\frac{1}{x} + \log y\right)\left(\frac{1}{y}\right)$$

$$qy - 2y^2 = 2f'\left(\frac{1}{x} + \log y\right) \quad ...(ii)$$

From Eqn. (*i*) and Eqn. (*ii*), we get $-px^2 = qy - 2y^2$

$$x^2p + yq = 2y^2$$

which is P.D.E. of first order and first degree.

**Example 17.4:** *Obtain a P.D.E by eliminating functions involved in the relation given below.*

$$\phi = f_1(x + iy) + f_2(x - iy)$$

**Solution:** Differentiating $\phi$ twice partially w.r.t and $y$ we

$$\frac{\delta\phi}{\delta x} = f_1'(x + iy) + f_2'(x - iy)$$

$$\frac{\delta^2\phi}{\delta x^2} = f_1'(x + iy) + f_2''(x - iy) \quad ...(i)$$

$$\frac{\delta\phi}{\delta y} = if_1'(x + iy) - if_2'(x - iy)$$

$$\frac{\delta^2\phi}{\delta y^2} = -f_1''(x + iy) + f_2''(x - iy) \quad ...(ii)$$

Adding Eqn. (*i*) and (*ii*) $\quad \dfrac{\delta^2\phi}{\delta x^2} + \dfrac{\delta^2\phi}{\delta y^2} = 0$

which is a linear P.D.E. of second order.

## PROBLEMS 17.1

From the partial differential equation (by eliminating the arbitrary constants from :

1. $(x - a)^2 + (y - b)^2 + z^2 = c^2$. (*Anna, 2013 ; Delhi, 2012*)
2. $\dfrac{x^2}{a^2} + \dfrac{y^2}{b^2} + \dfrac{z^2}{c^2} = 1$ (*JNTUK, 2022 ; K.T.U., 2016*)
3. $(x - a)^2 + (y - b)^2 = z^2 \cot^2 \alpha$
4. $y = xy + y\sqrt{x + a} + b$. (*JNTUK, 2022, 2015*)
5. $z = a \log \left\{\dfrac{b(y-1)}{1-x}\right\}$ (*JNTUK, 2022*)
6. Find the differential equation of all spheres of fixed radius having their centres in the $xy$-plane.
7. Find the differential equation of all spheres whose centres lie on the $z$-axis. (*J.N.T.U., 2015*)

Form the partial differential equations (by eliminating the arbitrary functions) from :

8. $z = f(x^2 - y^2)$
9. $z = f\left(\dfrac{xy}{z}\right)$ (*V.T.U., 2013 ; D.T.U., 2011*)
10. $z = f(x^2 + y^2) + x + y$
11. $z = yf(x) + xg(y)$. (*K.T.U., 2016*)
12. $z = x^2 f(y) + y^2 g(x)$.
13. $z = f(x) + e^y g(x)$. (*JNTUK, 2022; V.T.U., 2013*)
14. $xyz = \phi(x + y + z)$.
15. $z = f_1(x) f_2(y)$.
16. $z = e^{my} \phi(x - y)$.
17. $z = y^2 + 2f\left(\dfrac{1}{x} + \log y\right)$. (*V.T.U., 2010 ; J.N.T.U., 2010*)

18. $lx + my + nz = \phi(x^2 + y^2 + z^2)$ (*JNTUK, 2022; Delhi, 2012S*)

19. $z = f_1(y + 2x) + f_2(y - 3x)$.

20. $v = \frac{1}{r}[f(r - at) + F(r + at)]$.

21. $z = xf_1(x + t) + f_2(x + t)$.

22. $F(xy + z^2, x + y + z) = 0$. (*J.N.T.U 2015 ; V.T.U., 2015*)

23. $F(x + y + z, x^2 + y^2 + z^2) = 0$. (*Rohtak, 2012*)

24. If $u = f(x^2 + 2yz, y^2 + 2zx)$, prove that $(y^2 - zx)\frac{\partial u}{\partial x} + (x^2 - yz)\frac{\partial u}{\partial y} + (z^2 - xy)\frac{\partial u}{\partial z} = 0$.

## 17.3 SOLUTIONS OF A PARTIAL DIFFERENTIAL EQUATION

It is clear from the above examples that a partial differential equation can result both from elimination of arbitrary constants and from the elimination of arbitrary functions.

The solution $$f(x, y, z, a, b) = 0 \quad \ldots(1)$$
of a first order partial differential equation which contains two arbitrary constants is called a *complete integral*.

A solution obtained from the complete integral by assigning particular values to the arbitrary constants is called a particular integral.

If we put $b = \phi(a)$ in Eqn. (1) and find the envelope of the family of surfaces $f[x, y, z, \phi(a)] = 0$, then we get a solution containing an arbitrary function $\phi$, which is called the *general integral*.

The envelope of the family of surfaces Eqn. (1), with parameters $a$ and $b$, if it exists, is called a *singular integral*. The singular integral differs from the particular integral in that it is not obtained from the complete integral by giving particular values to the constants.

## 17.4 EQUATIONS SOLVABLE BY DIRECT INTEGRATION

We now consider such partial differential equations which can be solved by direct integration. In place of the usual constants of integration, we must, however, use arbitrary functions of the variable held fixed.

**Example 17.5.** *Solve* $\frac{\partial^3 z}{\partial x^2 \partial y} + 18xy^2 + \sin(2x - y) = 0.$ (*V.T.U., 2010*)

**Solution.** Integrating twice with respect to $x$ (keeping $y$ fixed),

$$\frac{\partial^2 z}{\partial x \partial y} + 9x^2y^2 - \frac{1}{2}\cos(2x - y) = f(y)$$

$$\frac{\partial z}{\partial y} + 3x^3y^2 - \frac{1}{4}\sin(2x - y) = xf(y) + g(y).$$

Now integrating with respect to $y$ (keeping $x$ fixed)

$$z + x^3y^3 - \frac{1}{4}\cos(2x - y) = x\int f(y)dy + \int g(y)dy + w(x)$$

The result may be simplified by writing

$$\int f(y)dy = u(y) \text{ and } \int g(y)dy = v(y).$$

Thus $z = \frac{1}{4}\cos(2x - y) - x^3y^3 + xu(y) + v(y) + w(x)$ where $u, v, w$ are arbitrary functions.

**Example 17.6.** *Solve* $\frac{\partial^2 z}{\partial x^2} + z = 0$, *given that when* $x = 0$, $z = e^y$ *and* $\frac{\partial z}{\partial x} = 1$. (*V.T.U., 2012*)

**Solution.** If $z$ were function of $x$ alone, the solution would have been $z = A\sin x + B\cos x$, where $A$ and $B$ are constants. Since $z$ is a function of $x$ and $y$, $A$ and $B$ can be arbitrary functions of $y$. Hence the solution of the given equation is $z = f(y)\sin x + \phi(y)\cos x$

$$\therefore \quad \frac{\partial z}{\partial x} = f(y)\cos x - \phi(y)\sin x$$

When $x = 0$; $z = e^y$, $\therefore e^y = \phi(y)$. When $x = 0$, $\dfrac{\partial z}{\partial x} = 1$, $\therefore 1 = f(y)$.

Hence the desired solution is $z = \sin x + e^y \cos x$.

**Example 17.7.** *Solve* $\dfrac{\partial^2 z}{\partial x \partial y} = \sin x \sin y$, *for which* $\dfrac{\partial z}{\partial y} = -2 \sin y$ *when* $x = 0$ *and* $z = 0$ *when* $y$ *is an odd multiple of* $\pi/2$. *(V.T.U., 2016)*

**Solution.** Given equation is $\dfrac{\partial^2 z}{\partial x \partial y} = \sin x \sin y$

Integrating w.r.t. $x$, keeping $y$ constant, we get

$$\frac{\partial z}{\partial y} = -\cos x \sin y + f(y) \qquad ...(i)$$

When $x = 0$, $\dfrac{\partial z}{\partial y} = -2 \sin y$, $\therefore -2 \sin y = -\sin y + f(y)$ or $f(y) = -\sin y$

$\therefore$ Eqn. $(i)$ becomes $\dfrac{\partial z}{\partial y} = -\cos x \sin y - \sin y$

Now integrating w.r.t. $y$, keeping $x$ constant, we get

$$z = \cos x \cos y + \cos y + g(x) \qquad ...(ii)$$

When $y$ is an odd multiple of $\pi/2$, $z = 0$.

$\therefore$ $0 = 0 + 0 + g(x)$ or $g(x) = 0$ $[\because \cos(2n + 1)\pi/2 = 0]$

Hence from Eqn. $(ii)$, the complete solution is $z = (1 + \cos x) \cos y$.

## PROBLEMS 17.2

Solve the following equations :

1. $\dfrac{\partial^2 z}{\partial x \partial y} = \dfrac{x}{y} + a$.
2. $\dfrac{\partial^2 z}{\partial y \partial x} = xy$. *(Rohtak, 2012)*
3. $\dfrac{\partial^2 u}{\partial x \partial t} = e^{-t} \cos x$.
4. $\dfrac{\partial^3 z}{\partial x^2 \partial y} = \cos(2x + 3y)$. *(V.T.U., 2014)*
5. $\dfrac{\partial^2 z}{\partial y^2} = z$, gives that when $y = 0$, $z = e^x$ and $\dfrac{\partial z}{\partial y} = e^{-x}$
6. $\dfrac{\partial^2 z}{\partial x^2} = a^2 z$ given that when $x = 0$, $\dfrac{\partial z}{\partial x} = a \sin y$ and $\dfrac{\partial z}{\partial y} = 0$. *(Rohtak, 2011)*

## 17.5 LINEAR EQUATIONS OF THE FIRST ORDER

*A linear partial differential equation of the first order,* commonly known as *Lagrange's Linear equation*, is of the form

$$Pp + Qq = R \qquad ...(1)$$

where $P$, $Q$ and $R$ are functions of $x, y, z$. This equation is called a *quasi-linear equation*. When $P$, $Q$ and $R$ are independent of $z$ it is known as *linear equation*.

Such an equation is obtained by eliminating an arbitrary function $\phi$ from $\phi(u, v) = 0$ ...(2)
where $u, v$ are some functions of $x, y, z$.

Differentiating Eqn. (2) partially with respect to $x$ and $y$.

$$\frac{\partial \phi}{\partial u}\left(\frac{\partial u}{\partial x} + \frac{\partial u}{\partial z} p\right) + \frac{\partial \phi}{\partial v}\left(\frac{\partial v}{\partial x} + \frac{\partial v}{\partial z} p\right) = 0$$

and $$\frac{\partial \phi}{\partial u}\left(\frac{\partial u}{\partial y} + \frac{\partial u}{\partial z} q\right) + \frac{\partial \phi}{\partial v}\left(\frac{\partial v}{\partial y} + \frac{\partial v}{\partial z} q\right) = 0.$$

Eliminating $\frac{\partial \phi}{\partial u}$ and $\frac{\partial \phi}{\partial v}$, we get $\begin{vmatrix} \frac{\partial u}{\partial x} + \frac{\partial u}{\partial z} p & \frac{\partial v}{\partial x} + \frac{\partial v}{\partial z} p \\ \frac{\partial u}{\partial y} + \frac{\partial u}{\partial z} q & \frac{\partial u}{\partial y} + \frac{\partial v}{\partial z} q \end{vmatrix} = 0$

which simplifies to $\left(\frac{\partial u}{\partial y}\frac{\partial v}{\partial z} - \frac{\partial u}{\partial z}\frac{\partial v}{\partial y}\right) p + \left(\frac{\partial u}{\partial z}\frac{\partial v}{\partial x} - \frac{\partial u}{\partial x}\frac{\partial v}{\partial z}\right) q = \left(\frac{\partial u}{\partial x}\frac{\partial v}{\partial y} - \frac{\partial u}{\partial y}\frac{\partial v}{\partial x}\right)$ ...(3)

This is of the same form as Eqn. (1).

Now suppose $u = a$ and $v = b$, where $a, b$ are constants, so that

$$\frac{\partial u}{\partial x} dx + \frac{\partial u}{\partial y} dy + \frac{\partial u}{\partial z} dz = du = 0$$

$$\frac{\partial v}{\partial x} dx + \frac{\partial v}{\partial y} dy + \frac{\partial v}{\partial z} dz = dv = 0.$$

By cross-multiplication, we have

$$\frac{dx}{\frac{\partial u}{\partial y}\frac{\partial v}{\partial z} - \frac{\partial u}{\partial z}\frac{\partial v}{\partial y}} = \frac{dy}{\frac{\partial u}{\partial z}\frac{\partial v}{\partial x} - \frac{\partial u}{\partial x}\frac{\partial v}{\partial z}} = \frac{dz}{\frac{\partial u}{\partial x}\frac{\partial v}{\partial y} - \frac{\partial u}{\partial y}\frac{\partial v}{\partial x}}.$$

or $$\frac{dx}{P} = \frac{dy}{Q} = \frac{dz}{R}$$ ...(4) [By virtue of Eqn. (1) and Eqn. (3)]

The solutions of these equations are $u = a$ and $v = b$.

$\therefore$ $\phi(u, v) = 0$ is the required solution of Eqn. (1).

*Thus to solve the equation $Pp + Qq = R$.*

(*i*) *form the subsidiary equations* $\frac{dx}{P} = \frac{dy}{Q} = \frac{dz}{R}$.

(*ii*) *solve these simultaneous equations by the method of § 16.10 giving $u = a$ and $v = b$ as its solutions.*

(*iii*) *write the complete solution as $\phi(u, v) = 0$* or $u = f(v)$.

**Example 17.8.** *Solve* $(z^2 - 2yz - y^2)p + (xy + zx)q = xy - zx$ (*Rohtak, 2018*)

**Solution:** The subsidiary equations are

$$\frac{dx}{z^2 - 2yz - y^2} = \frac{dy}{xy + zx} = \frac{dz}{xy - zx}$$

$$\frac{x\,dx}{z^2 - 2yz - y^2} = \frac{dy}{y + z} = \frac{dz}{y - z} \quad ...(i)$$

using Eqn. (*i*), $y, z$ as multipliers, we get

$$x\,dx + y\,dy + z\,dz = 0$$

$$\therefore \quad x^2 + y^2 + z^2 = c_1 \quad ...(ii)$$

Taking the last two members of Eqn. (*i*) we get

$$(y - z)dy = (y + z)\,dz$$

$$y\,dy - (z\,dy + y\,dz) - z\,dz = 0$$

$$y\,dy - d(yz) - z\,dz = 0$$

$$\therefore \quad \frac{y^2}{2} - yz - \frac{z^2}{2} = c_2 \quad \text{or} \quad y^2 - 2yz - z^2 = c_1 \quad ...(iii)$$

From Eqns. (*ii*) and (*iii*), we get the solution as

$$\phi(x^2 + y^2 + z^2, y^2 - 2yz - z^2) = 0$$

**Example 17.9:** *Solve* : $(P - q)z = z^2 + (x + y)^2$

**Solution:** The subsidiary equations are

$$\frac{dx}{z} = \frac{dy}{-z} = \frac{dz}{z^2 + (x+y)^2}$$

Taking the first two members, we have

$$dx = -dy \quad \text{or} \quad dx + dy = 0$$

$$x + y = c_1 \qquad ...(i)$$

Taking the first and the last members, we

$$dx = \frac{z\,dz}{z^2 + c_1^2}, \quad \text{since } x + y = c_1$$

$$\therefore \qquad c_2 = x - \frac{1}{2}\log\left[z^2 + (x+y)^2\right] \qquad ...(ii)$$

From Eqn. (*i*) and (*ii*), we get the general solution is

$$\phi\left[x + y,\ x - \frac{1}{2}\log\left(x^2 + y^2 + z^2 + 2xy\right)\right] = 0$$

**Example 17.10.** *Solve* $(mz - ny)\dfrac{\partial z}{\partial x} + (nx - lz)\dfrac{\partial z}{\partial y} = ly - mx.$

*(JNTUK, 2022; U.P.T.U., 2012 ; Rohtak, 2011 ; V.T.U., 2010)*

**Solution.** Here the subsidiary equations are $\dfrac{dx}{mz - ny} = \dfrac{dy}{mx - lz} = \dfrac{dz}{ly - mx}$

Using multipliers $x$, $y$, and $z$, we get each fraction $= \dfrac{xdx + ydy + zdz}{0}$

$\therefore \qquad xdx + ydy + zdz = 0$ which on integration gives $x^2 + y^2 + z^2 = a \qquad ...(i)$

Again using multipliers $l$, $m$ and $n$, we get each fraction $= \dfrac{ldx + mdy + ndz}{0}$

$\therefore \qquad ldx + mdy + ndx = 0$ which on integration gives $lx + my + nz = b \qquad ...(ii)$

Hence from Eqn. (*i*) and (*ii*), the required solution is $x^2 + y^2 + z^2 = f(lx + my + nz)$.

**Example 17.11.** *Solve* $(x^2 - y^2 - z^2)\,p + 2xyq = 2xz.$

*(JNTUK, 2022; Delhi 2016 ; V.T.U., 2013)*

**Solution.** Here the subsidiary equations are $\dfrac{dx}{x^2 - y^2 - z^2} = \dfrac{dy}{2xy} = \dfrac{dz}{2xz}$

From the last two fractions, we have $\dfrac{dy}{y} = \dfrac{dz}{z}$

which on integration gives $\log y = \log z + \log a$ or $y/z = a \qquad ...(i)$

Using multipliers $x$, $y$ and $z$, we have

$$\text{each fraction} = \frac{xdx + ydy + zdz}{x(x^2 + y^2 + z^2)} \qquad \therefore \qquad \frac{2xdx + 2ydy + 2zdz}{x^2 + y^2 + z^2} = \frac{dz}{z}$$

which on integration gives $\log(x^2 + y^2 + z^2) = \log z + \log b$

$$\text{or} \qquad \frac{x^2 + y^2 + z^2}{z} = b \qquad ...(ii)$$

Hence from Eqn. (*i*) and (*ii*), the required solution is $x^2 + y^2 + z^2 = zf(y/z)$.

**Example 17.12.** *Solve* $x^2(y - z)\,p + y^2(z - x)\,q = z^2(x - y).$

*(Anna, 2016 ; U.P.T.U., 2012)*

**Solution.** Here the subsidiary equations are

$$\frac{dx}{x^2(y - z)} = \frac{dy}{y^2(z - x)} = \frac{dz}{z^2(x - y)}$$

Using the multipliers $1/x$, $1/y$ and $1/z$, we have

$$\text{each fraction} = \frac{\frac{1}{x}dx + \frac{1}{y}dy + \frac{1}{z}dz}{0}$$

$\therefore$ $$\frac{dx}{x} + \frac{dy}{y} + \frac{dz}{z} = 0 \text{ which on integration gives}$$

$$\log x + \log y + \log z = \log a \quad \text{or} \quad xyz = a \qquad ...(i)$$

Using the multipliers $\frac{1}{x^2}, \frac{1}{y^2}$ and $\frac{1}{z^2}$, we get

$$\text{each fraction} = \frac{\frac{1}{x^2}dx + \frac{1}{y^2}dy + \frac{1}{z^2}dz}{0}$$

$\therefore$ $$\frac{dx}{x^2} + \frac{dy}{y^2} + \frac{dz}{z^2} = 0, \text{ which on integrating gives}$$

$$\frac{1}{x} + \frac{1}{y} + \frac{1}{z} = 0 \qquad ...(ii)$$

Hence from Eqns. (*i*) and (*ii*), the complete solution is

$$xyz = f\left(\frac{1}{x} + \frac{1}{y} + \frac{1}{z}\right).$$

**Example 17.13.** *Solve* $(x^2 - yz)\,p + (y^2 - zx)\,q = z^2 - xy$. *(JNTUK, 2022; V.T.U., 2016 ; Anna, 2015)*

**Solution.** Here the subsidiary equations are

$$\frac{dx}{x^2 - yz} = \frac{dy}{y^2 - zx} = \frac{dz}{z^2 - xy} \qquad ...(i)$$

$$\text{Each of these equations} = \frac{dx - dy}{x^2 - y^2 - (y - x)z} = \frac{dy - dz}{y^2 - z^2 - x(z - y)}$$

*i.e.*, $$\frac{d(x - y)}{(x - y)(x + y + z)} = \frac{d(y - z)}{(y - z)(x + y + z)} \quad \text{or} \quad \frac{d(x - y)}{x - y} = \frac{d(y - z)}{y - z}$$

Integrating, $$\log(x - y) = \log(y - z) + \log c \quad \text{or} \quad \frac{x - y}{y - z} = c \qquad ...(ii)$$

$$\text{Each of the subsidiary Eqn. } (i) = \frac{xdx + ydy + zdz}{x^3 + y^3 + z^3 - 3xyz}$$

$$= \frac{xdx + ydy + zdz}{(x + y + z)(x^2 + y^2 + z^2 - yz - zx - xy)} \qquad ...(iii)$$

$$\text{Also each of the subsidiary equations} = \frac{dx + dy + dz}{x^2 + y^2 + z^2 - yz - zx - xy} \qquad ...(iv)$$

Equating Eqns. (*iii*) and (*iv*) and cancelling the common factor, we get

$$\frac{xdx + ydy + zdz}{x + y + z} = dx + dy + dz$$

or $$\int (xdx + ydy + zdz) = \int (x + y + z)d(x + y + z) + c'$$

or $$x^2 + y^2 + z^2 = (x + y + z)^2 + 2c' \quad \text{or} \quad xy + yz + zx + c' = 0 \qquad ...(v)$$

Combining Eqn. (*ii*) and (*v*), the general solution is

$$\frac{x - y}{y - z} = f(xy + yz + zx).$$

## PROBLEMS 17.3

Solve the following equations :

1. $xp + yq = 3z$.
2. $p\sqrt{x} + q\sqrt{y} = \sqrt{z}$.
3. $(z - y)p + (x - z)q = y - x$.
4. $p \cos(x + y) + q \sin(x + y) = z$. (*J.N.T.U., 2015*)
5. $\left(\frac{b-c}{a}\right)yzp + \left(\frac{c-a}{b}\right)zxq = \left(\frac{a-b}{c}\right)xy$. (*Andhra, 2015*)
6. $p \tan x + q \tan y = \tan z$.
7. $p - q = \log(x + y)$.
8. $xp - yq = y^2 - x^2$ (*J.N.T.U., 2015 S ; V.T.U., 2014*)
9. $(y + z)p - (z + x)q = x - y$. (*Rohtak, 2012*)
10. $x(y - z)p + y(z - x)q = z(x - y)$.
11. $x(y^2 - z^2)p + y(z^2 - x^2)q - z(x^2 - y^2) = 0$. (*JNTUK, 2022; K.T.U., 2016 ; Anna, 2013*)
12. $x(y^2 + z)p - y(x^2 + z)q = z(x^2 - y^2)$ (*V.T.U., 2014*)
13. $y^2p - xyq = x(z - 2y)$.
14. $(y^2 + z^2)p - xyq + zx = 0$. (*K.T.U., 2016 ; U.P.T.U., 2011*)
15. $(z^2 - 2yz - y^2)p + (xy + zx)q = xy - zx$.
16. $px(z - 2y^2) = (z - qy)(z - y^2 - 2x^3)$.

## 17.6 NON-LINEAR EQUATIONS OF THE FIRST ORDER

Those equations in which $p$ and $q$ occur other than in the first degree are called *non-linear partial differential equations of the first order*. The *complete solution* of such an equation contains only two arbitrary constants (*i.e.*, equal to the number of independent variables involved) and the particular integral is obtained by giving particular values to the constants.]

Here we shall discuss four standard forms of these equations.

**Form I. f(p, q) = 0,** *i.e., equations containing p and q only.*

Its complete solution is $z = ax + by + c$ ...(1)

where $a$ and $b$ are connected by the relation $f(a, b) = 0$ ...(2)

[Since from Eqn. (1), $p = \frac{\partial z}{\partial x} = a$ and $q = \frac{\partial z}{\partial y} = b$, which when substituted in Eqn. (2) give $f(p, q) = 0$].

Expressing Eqn. (2) as $b = \phi(a)$ and substituting this value of $b$ in Eqn. (1), we get the required solution as $z = ax + \phi(a)y + c$ in which $a$ and $c$ are arbitrary constants.

**Example 17.14.** *Solve $p - q = 1$.*

**Solution.** The complete solution is $z = ax + by + c$ where $a - b = 1$

Hence $z = ax + (a - 1)y + c$ is the desired solution.

**Example 17.15.** *Solve $x^2p^2 + y^2q^2 = z^2$.* (*Rohtak, 2011*)

**Solution.** Given equation can be reduced to the above form by writing it as

$$\left(\frac{x}{z}\cdot\frac{\partial z}{\partial x}\right)^2 + \left(\frac{y}{z}\cdot\frac{\partial z}{\partial y}\right)^2 = 1 \qquad ...(i)$$

and setting $\frac{dx}{x} = du, \frac{dy}{y} = dv, \frac{dz}{z} = dw$ so that $u = \log x, v = \log y, w = \log z$.

Then Eqn. (*i*) becomes $\left(\frac{\partial w}{\partial u}\right)^2 + \left(\frac{\partial w}{\partial v}\right)^2 = 1$

*i.e.,* $P^2 + Q^2 = 1$ where $P = \frac{\partial w}{\partial u}$ and $Q = \frac{\partial w}{\partial v}$.

Its complete solution is $w = au + bv + c$ ...(*ii*)

where $a^2 + b^2 = 1$ or $b = \sqrt{(1 - a^2)}$.

$\therefore$ Eqn. (*ii*) becomes $w = au + \sqrt{(1 - a^2)}\, v + c$

or $\log z = a \log x + \sqrt{(1 - a^2)} \log y + c$ which is the required solution.

**Form II. f(z, p, q) = 0,** *i.e., equations not containing x and y.*

As a trial solution, assume that $z$ is a function of $u = x + ay$, where $a$ is an arbitrary constant.

$$\therefore \qquad p = \frac{\partial z}{\partial x} = \frac{dz}{du} \cdot \frac{\partial u}{\partial x} = \frac{dz}{du} \qquad\qquad q = \frac{\partial z}{\partial y} = \frac{dz}{du} \cdot \frac{\partial u}{\partial y} = a\frac{dz}{du}$$

Substituting the values of $p$ and $q$ in $f(z, p, q) = 0$, we get

$f\left(z, \dfrac{\partial z}{\partial u}, a\dfrac{dz}{du}\right) = 0$ which is an ordinary differential equation of the first order.

Rewriting it as $\dfrac{dz}{du} = \phi(z, a)$ it can be easily integrated giving

$F(z, a) = u + b$, or $x + ay + b = F(z, a)$ which is the desired complete solution.

*Thus to solve $f(z, p, q) = 0$,*

*(i) assume $u = x + ay$ and substitute $p = dz/du$, $q = a\, dz/du$ in the given equation;*

*(ii) solve the resulting ordinary differential equation in z and u;*

*(iii) replace u by $x + ay$.*

**Example 17.16.** *Solve $p(1 + q) = qz$.* *(J.N.T.U., 2015 ; Rohtak, 2012)*

**Solution.** Let $u = x + ay$, so that $p = dz/du$ and $q = a\, dz/du$.

Substituting these values of $p$ and $q$ in the given equation, we have

$$\frac{dz}{du}\left(1 + a\frac{dz}{du}\right) = az\frac{dz}{du} \text{ or } a\frac{dz}{du} = az - 1 \qquad \text{or} \qquad \int \frac{a\,dz}{az - 1} = \int du + b$$

or $\log(az - 1) = u + b$ or $\log(az - 1) = x + ay + b$

which is the required complete solution.

**Example 17.17.** *Solve $q^2 = z^2p^2(1 - p^2)$.*

**Solution.** Setting $u = y + ax$ and $z = f(u)$, we get

$$p = \frac{\partial z}{\partial x} = \frac{dz}{du} \cdot \frac{\partial u}{\partial x} = a\frac{dz}{du} \text{ and } q = \frac{dz}{du} \cdot \frac{\partial u}{\partial y} = \frac{dz}{du}$$

$\therefore$ The given equation becomes $\left(\dfrac{dz}{du}\right)^2 = a^2z^2\left(\dfrac{dz}{du}\right)^2\left\{1 - a^2\left(\dfrac{dz}{du}\right)^2\right\}$ ...(*i*)

or $$a^4z^2\left(\frac{dz}{du}\right)^2 = a^2z^2 - 1$$

or $$\frac{dz}{du} = \frac{\sqrt{(a^2z^2 - 1)}}{a^2z}$$

Integrating, $$\int \frac{a^2z}{\sqrt{(a^2z^2 - 1)}}\,dz = \int du + c$$

or $$(a^2z^2 - 1)^{1/2} = u + c$$

*i.e.,* $$a^2z^2 = (y + ax + c)^2 + 1 \qquad [\because u = y + ax]$$

The second factor in Eqn. (*i*) is $dz/du = 0$. Its solution is $z = c'$.

**Example 17.18.** *Solve $z^2(p^2x^2 + q^2) = 1$.*

**Solution.** Given equation can be reduced to the above form by writing it as

$$z^2\left[\left(x\frac{\partial z}{\partial x}\right)^2 + \left(\frac{\partial z}{\partial y}\right)^2\right] = 1 \qquad ...(i)$$

Putting $X = \log x$, so that $x\dfrac{\partial z}{\partial x} = \dfrac{\partial z}{\partial X}$, Eqn. (*i*) takes the standard form

$$z^2\left[\left(\frac{\partial z}{\partial X}\right)^2 + \left(\frac{\partial z}{\partial y}\right)^2\right] = 1 \qquad ...(ii)$$

Let $u = X + ay$ and put $\dfrac{\partial z}{\partial X} = \dfrac{dz}{du}$ and $\dfrac{\partial z}{\partial y} = a\dfrac{dz}{du}$ in Eqn. (*ii*), so that

$$z^2\left[\left(\frac{dz}{du}\right)^2 + a^2\left(\frac{dz}{du}\right)^2\right] = 1 \quad \text{or} \quad \sqrt{(1+a^2)}\, zdz = \pm\, du$$

Integrating, $\sqrt{(1+a^2)z^2} = \pm\, 2u + b = \pm\, 2(X + ay) + b$

or $z^2\sqrt{(1+a^2)} = \pm\, 2\,(\log x + ay) + b$

which is the complete solution required.

**Form III. f(x, p) = F(y, q),** *i.e., equations in which z is absent and the terms containing x and p can be separated from those containing y and q.*

As a trial solution assume that $f(x, p) = F(y, q) = a$, say

Then solving for $p$, we get $p = \phi(x)$

and solving for $q$, we get $q = \psi(y)$

Since $dz = \dfrac{\partial z}{\partial x}dx + \dfrac{\partial z}{\partial y}dy = pdx + qdy$

$\therefore$ $dz = \phi(x)dx + \psi(y)dy$

Integrating, $z = \int \phi(x)dx + \int \psi(y)dy + b$

which is the desired complete solution containing two constants $a$ and $b$.

**Example 17.19.** *Solve* $p^2 + q^2 = x + y$.

**Solution.** Given equation is $p^2 - x = y - q^2 = a$, say

$\therefore$ $p^2 - x = a$ gives $p = \sqrt{(a+x)}$

and $y - q^2 = a$ gives $q = \sqrt{(y-x)}$

Substituting these values of $p$ and $q$ in $dz = pdx + qdy$, we get

$$dz = \sqrt{(a+x)}\, dx + \sqrt{(y-a)}\, dy$$

$\therefore$ integrating gives, $z = \dfrac{2}{3}(a+x)^{3/2} + \dfrac{2}{3}(y-a)^{3/2} + b$

which is the required complete solution.

**Example 17.20.** *Solve* $z^2(p^2 + q^2) = x^2 + y^2$.

**Solution.** The equation can be reduced to the above form by writing it as

$$\left(z\frac{\partial z}{\partial x}\right)^2 + \left(z\frac{\partial z}{\partial y}\right)^2 = x^2 + y^2 \qquad ...(i)$$

and putting $zdz = dZ$, *i.e.*, $Z = \dfrac{1}{2}z^2$

$$\therefore \quad \frac{\partial Z}{\partial x} = \frac{\partial Z}{\partial z}\cdot\frac{\partial z}{\partial x} = z\frac{\partial z}{\partial x} = P$$

and $$\frac{\partial Z}{\partial y} = \frac{\partial Z}{\partial z}\cdot\frac{\partial z}{\partial y} = z\frac{\partial z}{\partial y} = Q$$

$\therefore$ Eqn. (*i*) becomes $\quad P^2 + Q^2 = x^2 + y^2$

or $\quad P^2 - x^2 = y^2 - Q^2 = a$, say.

$$\therefore \quad P = \sqrt{(x^2 + a)} \text{ and } Q = \sqrt{(y^2 + a)}\,..$$

$$\therefore \quad dZ = Pdx + Qdy \text{ gives}$$

$$dZ = \sqrt{(x^2 + a)}\, dx + \sqrt{(y^2 - a)}\, dy$$

Integrating, we have

$$Z = \frac{1}{2} x \sqrt{(x^2 + a)} + \frac{1}{2} a \log [x + \sqrt{(x^2 + a)}] + \frac{1}{2} y \sqrt{(y^2 - a)} - \frac{1}{2} a \log [y + \sqrt{(y^2 - a)}] + b$$

or

$$z^2 = x\sqrt{(x^2 + a)} + y\sqrt{(y^2 - a)} + a \log \frac{x + \sqrt{(x^2 + a)}}{y + \sqrt{(y^2 - a)}} + 2b$$

which is the required complete solution.

**Example 17.21.** *Solve* $(x + y)(p + q)^2 + (x - y)(p - q)^2 = 1.$

**Solution.** This equation can be reduced to the form $f(x, q) = F(y, q)$ by putting $u = x + y$, $v = x - y$ and taking $z = z(u, v)$.

Then

$$p = \frac{\partial z}{\partial x} = \frac{\partial z}{\partial u}\cdot\frac{\partial u}{\partial x} + \frac{\partial z}{\partial v}\cdot\frac{\partial v}{\partial x} = P + Q$$

and

$$q = \frac{\partial z}{\partial y} = \frac{\partial z}{\partial u}\cdot\frac{\partial u}{\partial y} + \frac{\partial z}{\partial v}\cdot\frac{\partial v}{\partial y} = P - Q, \text{ where } P = \frac{\partial z}{\partial u}, Q = \frac{\partial z}{\partial v}$$

Substituting these, the given equation reduces to

$$u(2P)^2 + v(2Q)^2 = 1 \quad \text{or} \quad 4P^2u = 1 - 4Q^2v = a \text{ (say)}$$

$$P = \pm \frac{1}{2}\sqrt{\frac{a}{u}},\ Q = \pm \frac{1}{2}\sqrt{\frac{1-a}{v}}$$

$$\therefore \quad dz = \frac{\partial z}{\partial u} du + \frac{\partial z}{\partial v} dv = Pdu + Qdv$$

$$= \pm \frac{\sqrt{a}}{2}\frac{du}{\sqrt{u}} \pm \frac{\sqrt{1-a}}{2}\frac{dv}{\sqrt{v}}$$

Integrating, we have

$$z = \pm \sqrt{a}\sqrt{u} \pm \sqrt{1-a}\sqrt{v} + b$$

or

$$z = \pm \sqrt{a(x + y)} \pm \sqrt{(1 - a)(x - y)} + b$$

which is the required complete solution.

**Form IV. z = px + qy + f(p, q):** *an equation analogous to the Clairaut's equation (§11.14).*

Its complete solution is $z = ax + by + f(a, b)$ which is obtained by writing $a$ for $p$ and $b$ for $q$ in the given equation.

**Example 17.22.** *Solve* $z = px + qy + \sqrt{(1 + p^2 + q^2)}$. *(Anna, 2016)*

**Solution.** Given equation is of the form $z = px + qy + f(p, q)$ where $f(p, q) = \sqrt{(1 + p^2 + q^2)}$

$\therefore$ Its complete solution is $z = ax + by + \sqrt{(1 + a^2 + b^2)}$.

## PROBLEMS 17.4

Obtain the complete solution of the following equations :

1. $p + q = pq$. *(Anna, 2013)*
2. $p^2 + q^2 = 1$.
3. $z = p^2 + q^2$. *(Hazaribag, 2015 ; Calicut, 2012)*
4. $(x^2 - y^2 - z^2)p + 2xy\, q = 2xz$ *(ITE, 2019)*

5. $yp + xq + pq = 0.$

6. $p + q = \sin x + \sin y.$

7. $(y^3 - 2x^3)xp + (2y^3 - x^3)yq = 9z(x^3 - y^3)$ *(Andra, 2017)*

8. $\sqrt{p} + \sqrt{q} = x + y.$

9. $p^2 + q^2 = x^2 + y^2.$ *(JNTUK, 2022)*

10. $z = px + qy + \sin(p + q).$ *(Calicut, 2012)*

11. $\cos(x + y)p + \sin(x + y)q = z$ *(Bhopal, 2018)*

12. $z = px + qy + p^2q^2.$ *(Anna., 2015)*

13. $(pq - p - q)(z - px - qy) = pq.$ *(Calicut, 2012)*

14. $(x - y)(px - qy) = (p - q)^2.$ **[Hint.** Use $x + y = u$, $xy = v$]

15. $(x + 2z)p + (4zx - y)q = 2x + y$

16. $x(y^2 + z)p - y(x^2 + z) = z(x^2 - y^2)$

17. $(x^2 + 3y^2)xp + (y^2 + 3x^2)yq = 2(x^2 + y^2)z$

## 17.7 CHARPIT'S METHOD*

We now explain a general method for finding the complete integral of a non-linear partial differential equation which is due to Charpit's.

Consider the equation

$$f(x, y\, z, p, q) = 0 \qquad \text{...(1)}$$

Since $z$ depends on $x$ and $y$, we have

$$dz = \frac{\partial z}{\partial x}dx + \frac{\partial z}{\partial y}dy = pdx + qdy \qquad \text{...(2)}$$

Now if we can find another relation involving $x, y, z, p, q$ such as $\phi(x, y, z, p, q) = 0$ ...(3)
then we can solve Eqns. (1) and (3) for $p$ and $q$ and substitute in Eqn. (2). This will give the solution provided Eqn. (2) is integrable.

To determine $\phi$, we differentiate Eqns. (1) and (3) with respect to $x$ and $y$ giving

$$\frac{\partial f}{\partial x} + \frac{\partial f}{\partial z}p + \frac{\partial f}{\partial p}\frac{\partial p}{\partial x} + \frac{\partial f}{\partial q}\frac{\partial q}{\partial x} = 0 \qquad \text{...(4)}$$

$$\frac{\partial \phi}{\partial x} + \frac{\partial \phi}{\partial z}p + \frac{\partial \phi}{\partial p}\frac{\partial p}{\partial x} + \frac{\partial \phi}{\partial q}\frac{\partial q}{\partial x} = 0 \qquad \text{...(5)}$$

$$\frac{\partial f}{\partial y} + \frac{\partial f}{\partial z}q + \frac{\partial f}{\partial p}\frac{\partial p}{\partial y} + \frac{\partial f}{\partial q}\frac{\partial q}{\partial y} = 0 \qquad \text{...(6)}$$

$$\frac{\partial \phi}{\partial y} + \frac{\partial \phi}{\partial z}q + \frac{\partial \phi}{\partial p}\frac{\partial p}{\partial y} + \frac{\partial \phi}{\partial q}\frac{\partial q}{\partial y} = 0 \qquad \text{...(7)}$$

Eliminating $\frac{\partial p}{\partial x}$ between the equations Eqns. (4) and (5), we get

$$\left(\frac{\partial f}{\partial x}\frac{\partial \phi}{\partial p} - \frac{\partial \phi}{\partial x}\frac{\partial f}{\partial p}\right) + \left(\frac{\partial f}{\partial z}\frac{\partial \phi}{\partial p} - \frac{\partial \phi}{\partial z}\frac{\partial f}{\partial p}\right)p + \left(\frac{\partial f}{\partial q}\frac{\partial \phi}{\partial p} - \frac{\partial \phi}{\partial q}\frac{\partial f}{\partial p}\right)\frac{\partial q}{\partial x} = 0 \qquad \text{...(8)}$$

Also eliminating $\frac{\partial q}{\partial y}$ between the equations Eqns. (6) and (7), we obtain

$$\left(\frac{\partial f}{\partial y}\frac{\partial \phi}{\partial q} - \frac{\partial \phi}{\partial y}\frac{\partial f}{\partial q}\right) + \left(\frac{\partial f}{\partial z}\frac{\partial \phi}{\partial q} - \frac{\partial \phi}{\partial z}\frac{\partial f}{\partial q}\right)q + \left(\frac{\partial f}{\partial p}\frac{\partial \phi}{\partial q} - \frac{\partial \phi}{\partial p}\frac{\partial f}{\partial q}\right)\frac{\partial p}{\partial y} = 0 \qquad \text{...(9)}$$

*Charpit's memoir containing this method was presented to the Paris Academy of Sciences in 1784.

Adding Eqns. (8) and (9) and using $\frac{\partial q}{\partial x} = \frac{\partial^2 z}{\partial x \partial y} = \frac{\partial p}{\partial y}$,

we find that the last terms in both cancel and the other terms, on rearrangement, give

$$\left(\frac{\partial f}{\partial x} + p\frac{\partial f}{\partial z}\right)\frac{\partial \phi}{\partial p} + \left(\frac{\partial f}{\partial y} + q\frac{\partial f}{\partial z}\right)\frac{\partial \phi}{\partial q} + \left(-p\frac{\partial f}{\partial p} - q\frac{\partial f}{\partial q}\right)\frac{\partial \phi}{\partial z} + \left(-\frac{\partial f}{\partial p}\right)\frac{\partial \phi}{\partial x} + \left(-\frac{\partial f}{\partial q}\right)\frac{\partial \phi}{\partial y} = 0 \quad ...(10)$$

*i.e.,*

$$\left(-\frac{\partial f}{\partial p}\right)\frac{\partial \phi}{\partial x} + \left(-\frac{\partial f}{\partial q}\right)\frac{\partial \phi}{\partial y} + \left(-p\frac{\partial f}{\partial p} - q\frac{\partial f}{\partial q}\right)\frac{\partial \phi}{\partial z} + \left(\frac{\partial f}{\partial x} + p\frac{\partial f}{\partial z}\right)\frac{\partial \phi}{\partial p} + \left(\frac{\partial f}{\partial y} + q\frac{\partial f}{\partial z}\right)\frac{\partial \phi}{\partial q} = 0 \quad ...(11)$$

This is Lagrange's linear equation (§ 17.5) with $x, y, z, p, q$ as independent variables and $\phi$ as the dependent variable. Its solution will depend on the solution of the subsidiary equations

$$\frac{dx}{-\frac{\partial f}{\partial p}} = \frac{dy}{-\frac{\partial f}{\partial q}} = \frac{dz}{-p\frac{\partial f}{\partial p} - q\frac{\partial f}{\partial q}} = \frac{dp}{\frac{\partial f}{\partial x} + p\frac{\partial f}{\partial z}} = \frac{dq}{\frac{\partial f}{\partial y} + q\frac{\partial f}{\partial z}} = \frac{\partial \phi}{0}$$

An integral of these equations involving $p$ or $q$ or both, can be taken as the required relation Eqn. (3), which alongwith Eqn. (1) will give the values of $p$ and $q$ to make Eqn. (2) integrable. Of course, we should take the simplest of the integrals so that it may be easier to solve for $p$ and $q$.

**Example 17.23.** *Solve* $(p^2 + q^2)y = qz$.

**Solution.** Let $f(x, y, z, p, q) = (p^2 + q^2)y - qz = 0$ ...(*i*)

Charpit's subsidiary equations are

$$\frac{dx}{-2py} = \frac{dy}{z - 2qy} = \frac{dz}{-qz} = \frac{dp}{-pq} = \frac{dq}{p^2}$$

The last two of these give $pdp + qdq = 0$

Integrating, $p^2 + q^2 = c^2$ ...(*ii*)

Now to solve Eqn. (*i*) and Eqn. (*ii*), put $p^2 + q^2 = c^2$ in Eqn. (*i*), so that $q = c^2y/z$

Substituting this value of $q$ in Eqn. (*ii*), we get $p = c\sqrt{(z^2 - c^2y^2)}/z$

Hence $$dz = pdx + qdy = \frac{c}{z}\sqrt{(z^2 - c^2y^2)}\,dx + \frac{c^2y}{z}dy$$

or $$zdz - c^2y\,dy = c\sqrt{(z^2 - c^2y^2)}\,dx \quad \text{or} \quad \frac{\frac{1}{2}d(z^2 - c^2y^2)}{\sqrt{(z^2 - c^2y^2)}} = c\,dx$$

Integrating, we get $\sqrt{(z^2 - c^2y^2)} = cx + a$ or $z^2 = (a + cx)^2 + c^2y^2$ which is the required complete integral.

**Example 17.24.** *Solve* $(t + y + z)\frac{\delta t}{dx} + (t + z + x)\frac{\delta t}{\delta y} + (t + x + y)\frac{\delta t}{\delta z} = x + y + z$

**Solution:** The subsidiary equations are

$$\frac{dx}{t + y + z} = \frac{dy}{t + z + x} = \frac{dz}{t + x + y} = \frac{dt}{x + y + z}$$

$$\therefore \quad \frac{dx + dy + dz + dt}{3(x + y + z + t)} = \frac{dz - dt}{-(x - t)} = \frac{dy - dt}{-(y - t)} = \frac{dz - dt}{-(z - t)}$$

Hence, $(x + y + z + t)^{1/3}(x - t) = c_1$ ...(*i*)

$(x + y + z + t)^{1/3}(y - t) = c_2$ ...(*ii*)

$(x + y + z + t)^{1/3}(z - t) = c_3$ ...(*iii*)

From Eqns. (*i*), (*ii*), and (*iii*), the general solutions can be given by

$\phi[(x + y + z + t)^{1/3}(x - t), (x + y + z + t)^{1/3}(y - t), (x + y + z + t)^{1/3}(z - t) = 0$

**Example 17.25.** *Solve* $2z + p^2 + qy + 2y^2 = 0$. (*Rohtak, 2011*)

**Solution.** Let $f(x, y, z, p, q) = 2z + p^2 + qy + 2y^2$

Charpit's subsidiary equations are

$$\frac{dx}{-2p} = \frac{dy}{-y} = \frac{dz}{-(2p^2 + qy)} = \frac{dp}{2p} = \frac{dq}{4y + 3q}$$

From first and fourth ratios, $dp = -dx$ or $p = -x + a$

Substituting $p = a - x$ in the given equation, we get

$$q = \frac{1}{y}[-2z - 2y^2 - (a - x)^2]$$

$$\therefore \qquad dz = pdx + qdy = (a - x)dx - \frac{1}{y}[2z + 2y^2 + (a - x)^2]dy$$

Multiplying both sides by $2y^2$,

$$2y^2dz + 4yz\,dy = 2y^2(a - x)dx - 4y^3dy - 2y(a - x)^2dy$$

Integrating $\qquad 2zy^2 = -[y^2(a - x)^2 + y^4] + b$

or $\qquad y^2[(x - a)^2 + 2z + y^2] = b$, which is the desired solution.

## PROBLEMS 17.5

Solve the following equations :

1. $z = p^2x + q^2x$.
2. $z^2 = pq\,xy$. (*Hazaribag, 2015*)
3. $1 + p^2 = qz$.
4. $(p^2 + q^2)y = qz$ (*Osmania, 2018*)
5. $p(p^2 + 1) + (b - z)q = 0$.
6. $z^2 = pq\,xy$ (*Calicut 2019*)

**Example 17.26:** *Solve* : $\frac{\delta^3u}{\delta x^3} - \frac{\delta^3u}{dy^3} = x^3y^3$ (*Calicut, 2019*)

**Solution:** We have $(D^3 - D'^3)u = x^3y^3$

Auxiliary equation is $m^3 - 1 = 0$

$$(m - 1)(m - \omega)(m - \omega^2) = 0 \Rightarrow m = 1, \omega, \omega^2$$

when $\omega$ is one of the cube roots of unity

$\therefore$ The C.F. is given by $\qquad z = \phi_1(y + x) + \phi_2(y + \omega x) + \phi_3(y + \omega^2x)$

The P.I. is given by

$$z = \frac{1}{(D^3 - D'^3)}x^3y^3 = \frac{1}{D^3}\left(1 - \frac{D'^3}{D^3}\right)^{-1}.x^3y^3$$

$$= \frac{1}{D^3}\left(1 + \frac{D'^3}{D^3} + \ldots\right)x^3y^3 = \frac{1}{D^3}\cdot x^3y^3 + \frac{D'^3}{D^6}\cdot x^3y^3$$

$$= \frac{x^6y^3}{4\cdot5\cdot6} + \frac{1}{D^6}(x^3 \times 6) = \frac{x^6y^3}{120} + \frac{x^9}{10080}$$

Hence the complete solution is given by

$$z = \phi_1(y + x) + \phi_2(y + \omega x) + \phi_3(y + \omega^2x) + \frac{x^6y^3}{120} + \frac{x^9}{10080}$$

## 17.8 HOMOGENEOUS LINEAR EQUATIONS WITH CONSTANT COEFFICIENTS

An equation of the form

$$\frac{\partial^n z}{\partial x^n} + k_1\frac{\partial^n z}{\partial x^{n-1}\partial y} + \ldots + k_n\frac{\partial^n z}{\partial y^n} = F(x, y) \qquad \ldots(1)$$

in which $k$'s are constants, is called a *homogeneous linear partial differential equation of the nth order with constant coefficients.* It is called homogeneous because all terms contain derivatives of the same order.

On writing, $\frac{\partial^r}{\partial x^r} = D^r$ and $\frac{\partial^r}{\partial y^r} = D'^r$. Equation (1) becomes $(D^n + k_1 D^{n-1} D'^r + D' + ... + k_n D'^n)\, z = F(x, y)$

or briefly $$f(D, D')z = F(x, y) \quad ...(2)$$

As in the case of ordinary linear equations with constant coefficients the complete solution of Eqn. (1) consists of two parts, namely : the *complementary function* and *the particular integral.*

The complementary function is the complete solution of the equation $f(D, D')z = 0$, which must contain $n$ arbitrary functions. The particular integral is the particular solution of Eqn. (2).

## 17.9 RULES FOR FINDING THE COMPLEMENTARY FUNCTION

Consider the equation $\frac{\partial^2 z}{\partial x^2} + k_1 \frac{\partial^2 z}{\partial x \partial y} + k_2 \frac{\partial^2 z}{\partial y^2} = 0$ ...(1)

which in *symbolic form is* $(D^2 + k_1 DD' + k_2 D'^2)z = 0$ ...(2)

Its symbolic operator equated to zero, *i.e.*, $D^2 + k_1 DD' + k_2 D'^2 = 0$ is called the *auxiliary equation* (A.E.)

Let its root be $D/D' = m_1, m_2$.

**Case I.** *If the roots be real and distinct* then Eqn. (2) is equivalent to

$$(D - m_1 D')(D - m_2 D')z = 0 \quad ...(3)$$

It will be satisfied by the solution of

$$(D - m_2 D')z = 0, \text{ i.e., } p - m_2 q = 0.$$

This is a Lagrange's linear and the subsidiary equations are

$$\frac{dx}{1} = \frac{dy}{-m_2} = \frac{dz}{0}, \text{ whence } y + m_2 x = a \text{ and } z = b.$$

$\therefore$ its solution is $z = \phi(y + m_2 x)$.

Similarly Eqn. (3) will also be satisfied by the solution of

$$(D - m_1 D')z = 0, \quad i.e., \quad \text{by } z = f(y + m_1 x)$$

Hence the complete solution of Eqn. (1) is $z = f(y + m_1 x) + \phi(y + m_2 x)$.

**Case II.** *If the roots be equal* (*i.e.*, $m_1 = m_2$), then Eqn. (2) is equivalent to

$$(D - m_1 D')^2 z = 0 \quad ...(4)$$

Putting $(D - m_1 D')z = u$, it becomes $(D - m_1 D')u = 0$ which gives

$$u = \phi(y + m_1 x)$$

$\therefore$ Eqn. (4) takes the form $(D - m_1' D)z = \phi(y + m_1 x)$ or $p - m_1 q) = \phi(y + m_1 x)$

This is again Lagrange's linear and the subsidiary equations are

$$\frac{dx}{1} = \frac{dy}{-m_1} = \frac{dz}{\phi(y + m_1 x)}$$

giving $$y + m_1 x = a \text{ and } dz = \phi(a)\, dx, \text{ i.e., } z = \phi(a)x + b$$

Thus the complete solution of (1) is

$$z - x\phi(y + m_1 x) = f(y + m_1 x). \quad i.e., \quad z = f(y + m_1 x) + x\phi(y + m_1 x).$$

**Example 17.27.** *Solve* $2\frac{\partial^2 z}{\partial x^2} + 5\frac{\partial^2 z}{\partial x \partial y} + 2\frac{\partial^2 z}{\partial y^2} = 0.$

**Solution.** Given equation in *symbolic form* is $(2D^2 + 5DD' + 2D'^2)z = 0$.

Its *auxiliary equation* is $2m^2 + 5m + 2 = 0$, where $m = D/D'$.

which gives $$m = -2, -1/2.$$

Here the complete solution is $z = f_1(y - 2x) + f_2(y - \frac{1}{2}x)$

which may be written as $z = f_1(y - 2x) + f_2(2y - x)$.

**Example 17.28.** *Solve* $4r + 12s + 9t = 0.$ *(P.T.U., 2010)*

**Solution.** Given equation in *symbolic form* is $(4D^2 + 12DD' + 9D'^2)z = 0$

for $$r = \frac{\partial^2 z}{\partial x^2} = D^2 z,\ s = \frac{\partial^2 z}{\partial x \partial y} = DD'z \text{ and } t = \frac{\partial^2 z}{\partial y^2} = D'^2 z.$$

$\therefore$ Its *auxiliary equation* is $4m^2 + 12m + 9 = 0$, whence $m = -3/2, -3/2$

Hence the complete solution is $z = f_1(y - 1.5x) + xf_2(y - 1.5x)$.

**Example 17.29 :** *Solve : $2zx - px^2 - 2q\,xy + pq = 0$*

**Solution:** Let $$F = 2zx - px^2 - 2q\,xy + pq \quad ...(i)$$

The Charpit's subsidiary equations are

$$\frac{dp}{2z - 2qy} = \frac{dq}{0} = \frac{dz}{px^2 - pq + 2xy\,q - pq} = \frac{dx}{x^2 - q} = \frac{dy}{2xy - p} = \frac{dF}{0}$$

$$\Rightarrow \quad dq = 0, \quad \text{or} \quad q = a \quad ...(ii)$$

Substituting these values of $p$ and $q$ in

$$dz = pdx + qdy$$

$$dz = \frac{2x(z - ay)}{x^2 - a}dx + ady$$

$$dz - a\,dy = \frac{2x(z - ay)}{x^2 - a}dx$$

$$\frac{dz - a\,dy}{z - ay} = \frac{2x}{x^2 - a}dx$$

Integrating, $$\log(z - ay) = \log(x^2 - a) + \log b$$

$$z = ay + b(x^2 - a)$$

**Example 17.30:** *Find the complete integral of the equation*

$$2(z + xp + yq) = y\,p^2$$ *(Hissar, 2019 M.E.)*

**Solution:** Hence $$F = 2(z + xp + yq) - yq^2 = 0 \quad ...(i)$$

The Charpit's subsidiary equations are

$$\frac{dp}{2p + 2q} = \frac{dq}{2p - p^2 + 2q} = \frac{dz}{-p(2x - 2yp) - 2qy} = \frac{dx}{-2xy + 2yp} = \frac{dy}{-2y}$$

Taking the first and the last member, we have

$$\frac{dp}{p} + 2\frac{dy}{y} = 0 \Rightarrow py^2 = a$$

substituting $p = \dfrac{a}{y^2}$ in Eqn. (*i*), we get, $q = \dfrac{-z}{y} - \dfrac{ax}{y^3} + \dfrac{a^2}{2y^4}$

Hence, $$dz = \left(\frac{a}{y^2}\right)dx + \left(-\frac{z}{y} - \frac{ax}{y^3} + \frac{a^2}{2y^4}\right)dy$$

or $$y\,dz + z\,dy = a\left(\frac{y\,dx - x\,dy}{y^2}\right) + \frac{a^2}{2y^3}dy$$

integrating $$yz = \frac{ax}{y} - \frac{a^2}{4y^2} + b$$

$$z = \frac{ax}{y^2} - \frac{a^2}{4y^3} + \frac{b}{y}$$

## 17.10 RULES FOR FINDING THE PARTICULAR INTEGRAL

Consider the equation $(D^2 + k_1DD' + k_2D'^2)z = F(x, y)$ *i.e.*, $f(D, D')z = F(x, y)$.

$$\therefore \quad \text{P.I.} = \frac{1}{f(D, D')} F(x, y)$$

**Case I.** *When* $F(x, y) = e^{ax + by}$

Since $De^{ax+by} = ae^{ax+by}$ ; $D'e^{ax+by} = be^{ax+by}$

$\therefore$ $D^2e^{ax+by} = a^2e^{ax+by}$ ; $DD'e^{ax+by} = abe^{ax+by}$

and $D'^2e^{ax+by} = b^2e^{ax+by}$

$\therefore$ $(D^2 + k_1DD' + k_2D'^2)e^{ax+by} = (a^2 + k_1ab + k_2b^2)\, e^{ax+by}$

*i.e.*, $f(D, D')e^{ax+by} = f(a, b)\, e^{ax+by}$

Operating both sides by $1/f(D, D')$, we get

$$\text{P.I.} = \frac{1}{f(D, D')} e^{ax+by} = \frac{1}{f(a, b)} e^{ax+by}$$

**Case II.** *When* $F(x, y) = \sin(mx + ny)$ or $\cos(mx + ny)$

Since $D^2 \sin(mx + ny) = -m^2 \sin(mx + ny)$

$DD' \sin(mx + ny) = -mn \sin(mx + ny)$

and $D'^2 \sin(mx + ny) = -n^2 \sin(mx + ny)$.

$\therefore$ $f(D^2, DD', D'^2) \sin(mx + ny) = f(-m^2, -mn, -n^2) \sin(mx + ny)$

Operating both sides by $1/f(D^2, DD', D'^2)$, we get

$$\text{P.I.} = \frac{1}{f(D^2, DD', D'^2)} \sin(mx + ny) = \frac{1}{f(-m^2 - mn, -n^2)} \sin(mx + ny)$$

Similarly about the P.I. for $\cos(mx + ny)$.

**Case III.** *When* $F(x, y) = x^my^n$, $m$ and $n$ being constants.

$$\therefore \quad \text{P.I.} = \frac{1}{f(D, D')} x^m y^n = [f(D, D')]^{-1} x^my^n.$$

To evaluate it, we expand $[f(D, D')]^{-1}$ in ascending powers of $D$ or $D'$ by Binomial theorem and then operate on $x^my^n$ term by term.

**Case IV.** *When* $F(x, y)$ *is any function of* $x$ *and* $y$.

$$\therefore \quad \text{P.I.} = \frac{1}{f(D, D')} F(x, y)$$

To evaluate it, we resolve $1/f(D, D')$ into partial fractions treating $f(D, D')$ as a function of $D$ alone and operate each partial fraction on $F(x, y)$ remembering that

$$\frac{1}{D - mD'} F(x, y) = \int F(x, c - mx)\, dx$$

where $c$ is replaced by $y + mx$ after integration.

## 17.11 WORKING PROCEDURE TO SOLVE THE EQUATION

$$\frac{\partial^n z}{\partial x^n} + k_1 \frac{\partial^n z}{\partial x^{n-1} \partial y} + \ldots + k^n \frac{\partial^n z}{\partial y^n} = F(x, y).$$

Its *symbolic form* is $(D^n + k_1D^{n-1} D' + \ldots + k_n D'^n)z = F(x, y)$

or briefly $f(D, D')z = F(x, y)$

**Step I. To find the C.F.**

(*i*) *Write the A.E.*

*i.e.*, $m^n + k_1m^{n-1} + \ldots + k_n = 0$ *and solve it for* $m$.

*(ii) Write the C.F. as follows*

| *Roots of A.E.* | *C.F.* |
|---|---|
| 1. $m_1, m_2, m_3$ ... (distinct roots) | $f_1(y + m_1x) + f_2(y + m_2x) + f_3(y + m_3x) + ...$ |
| 2. $m_1, m_1, m_3$ ... (two equal roots) | $f_1(y + m_1x) + xf_2(y + m_1x) + f_3(y + m_3x) + ...$ |
| 3. $m_1, m_1, m_1$ ... (three equal roots) | $f_1(y + m_1x) + xf_2(y + m_1x) + x^2f_3(y + m_1x) + ...$ |

**Step II. To find the P.I.**

From the *symbolic form*, P.I. $= \dfrac{1}{f(D, D')} F(x, y)$.

*(i) When* $F(x, y) = e^{ax + by}$ P.I. $= \dfrac{1}{f(D, D')} e^{ax + by}$ [Put $D = a$ and $D' = b$]

*(ii) When* $F(x, y) = \sin(mx + ny)$ or $\cos(mx + ny)$

$$\text{P.I.} = \frac{1}{f(D^2, DD', D'^2)} \sin \text{ or } \cos(mx + ny) \qquad [\text{Put } D^2 = -m^2, DD' = -mn, D'^2 = -n^2]$$

*(iii) When* $F(x, y) = x^m y^n$, P.I. $= \dfrac{1}{f(D, D')} x^m y^n = [f(D, D')]^{-1} x^m y^n$.

*Expand* $[f(D, D')]^{-1}$ *in ascending powers of D or D′ and operate on* $x^m y^n$ *term by term.*

*(iv) When F(x, y) is any function of x and y* P.I. $= \dfrac{1}{f(D, D')} F(x, y)$.

*Resolve* $1/f(D, D')$ *into partial fractions considering* $f(D, D')$ *as a function of D alone and operate each partial fraction on* $F(x, y)$ *remembering that*

$$\frac{1}{D - mD'} F(x, y) = \int F(x, c - mx)\, dx \text{ where } c \text{ is replaced by } y + mx \text{ after integration.}$$

**Example 17.31.** *Solve* $(D^2 + 4DD' - 5D'^2)z = \sin(2x + 3y)$. *(JNTUK, 2022)*

**Solution.** A.E. of the given equation is $m^2 + 4m - 5 = 0$ *i.e.*, $m = 1, -5$

$\therefore$ C.F. $= f_1(y + x) + f_2(y - 5x)$

$$\text{P.I.} = \frac{1}{D^2 + 4DD' - 5D'^2} \sin(2x + 3y) \qquad [\text{Put } D^2 = -2^2, DD' = -2 \times 3, D'^2 = -3^2]$$

$$= \frac{1}{-4 + 4(-6) - 5(-9)} \sin(2x + 3y) = \frac{1}{17} \sin(2x + 3y).$$

Hence, the C.S. is $z = f_1(y + x) + f_2(y - 5x) + \dfrac{1}{17} \sin(2x + 3y)$.

**Example 17.32.** *Solve :* $\dfrac{\delta^2 z}{\delta x^2} + 3\dfrac{\delta^2 z}{\delta x\, \delta y} + 2\dfrac{\delta^2 z}{\delta y^2} = x + y$ *(I.T.E, 2018)*

**Solution:** We have $(D^2 + 3DD' + 2D'^2)z = x + y$

The auxiliary equation is

$$m^2 + 3m + 2 \Rightarrow m = -1, -2,$$

$\therefore$ The C.F. is $z = \phi_1(y - x) + \phi_2(y - 2x)$

Now to obtain P.I. $z = \dfrac{1}{D^2 + 3DD' + 2D'^2}(x + y) = \dfrac{1}{D^2\left(1 + \dfrac{3D'}{D} + \dfrac{2D'^2}{D^2}\right)}(x + y)$

$$= \frac{1}{D^2}\left[1 + \left(\frac{3D'}{D} + \frac{2D'^2}{D^2}\right)\right]^{-1}(x + y) = \frac{1}{D^2}\left[1 - \frac{3D'}{D} + ...\right](x + y)$$

$$= \frac{1}{D^2}\left[x + y - \frac{3}{D}(1)\right] = \frac{1}{D^2}(x + y - 3x)$$

$$= \frac{1}{D^2}(y - 2x) = \frac{1}{D^2}(y) - \frac{1}{D^2}(2x) = \frac{x^2 y}{2} - \frac{x^3}{3}$$

Hence the complete solution is $\quad z = \phi_1(y - x) + \phi_2(y - 2x) + \frac{x^2 y}{2} - \frac{x^3}{3}$

**Example 17.33.** *Find the surface satisfying $t = 6x^3y$ containing the two lines $y = 0 = z$ and $y = 1 = z$.* *(Kottayem, 2020)*

**Solution.** The given 2nd order PDE is

$$t = 6x^3 y$$

$\Rightarrow$ $$\frac{\delta t}{\delta y} = 6x^3 y$$

Integrating w.r.t. $y$, we have $\quad t = \frac{6x^3y^2}{2} + f(x)$

$$\frac{\partial z}{\delta y} = 3x^3 y^2 + f(x)$$

Integrating w.r.t. $y$ $\quad z = \frac{3x^3y^3}{3} + y\, f(x) + F(x) \qquad ...(i)$

Using given condition $\quad y = 0 = z$ in eqn. $(i)$, we have

$$0 = 0 + 0 + F(x)$$

$$F(x) = 0 \qquad ...(ii)$$

Also using $\quad y = 1 = z$ in eqn. $(i)$, we get

$$1 = x^3 + f(x) + F(x)$$

Using Eqn. $(ii)$, we get $\quad 1 = x^3 + f(x) + 0$

$$f(x) = 1 - x^3 \qquad ...(iii)$$

Using Eqns. $(ii)$ and $(iii)$ in Eqn. $(i)$

$$Z = x^3y^3 + y\,(1 - x^3)$$

which is the required surface containing the two lines.

**Example 17.34.** *Solve* $\frac{\delta^2 z}{\delta x^2} + \frac{\delta^2 z}{\delta y^2} = \cos m\,x \cos xy + 30\,(2x + y)$

**Solution:** The C.F. is easily obtained by

$$z = \phi_1(y + ix) + \phi_2(y - ix)$$

To obtain P.I., we have $\quad z = \frac{1}{D^2 + D'^2}[\cos mx \cos ny + 30(2x + y)]$

$$= \frac{1}{2}\cdot\frac{\cos(mx + ny)}{D^2 + D'^2} + \frac{1}{2}\frac{\cos(mx - ny)}{D^2 + D'^2} + \frac{30(2x + y)}{D^2 + D'^2}$$

$$= \frac{1}{2}\cdot\frac{-\cos(mx + ny)}{m^2 + n^2} + \frac{1}{2} - \frac{\cos(mx - ny)}{m^2 + n^2} + \frac{30}{2^2 + 1^2}\iint t\,dt\,dt, \quad \text{where } t = 2x + y$$

$$= -\frac{1}{2}\cdot\frac{\cos(mx + ny) + \cos(mx - ny)}{m^2 + n^2} + \frac{30}{5}\times\frac{t^3}{6}$$

$$= -\frac{\cos mx \cos ny}{m^2 + n^2} + (2x + y)^3$$

Hence the complete solution is

$$z = \phi_1(y + ix) + \phi_2(y - ix) - \frac{\cos mx \cos ny}{m^2 + n^2} + (2x + y)^3$$

**Example 17.35.** *Solve* $r - 4s + 4t = e^{2x+y}$. *(Delhi, 2015., Hazaribag, 2015)*

**Solution.** Given equation is $\dfrac{\partial^2 z}{\partial x^2} - 4\dfrac{\partial^2 z}{\partial x \partial y} + 4\dfrac{\partial^2 z}{\partial y^2} = e^{2x+y}$.

*i.e.,* in *symbolic form* $(D^2 - 4DD' + 4D'^2)z = e^{2x+y}$.

Its A.E. is $(m-2)^2 = 0$, whence $m = 2, 2$.

$\therefore$ $\quad$ C.F. $= f_1(y + 2x) + xf_2(y + 2x)$

$$\text{P.I.} = \frac{1}{(D-2D')^2} e^{2x+y}$$

The usual rule fails because $(D - 2D')^2 = 0$ for $D = 2$ and $D' = 1$.

$\therefore$ to obtain the P.I., we find from $(D - 2D')\,u = e^{2x+y}$, the solution

$$u = \int F(x, c - mx)\, dx = \int e^{2x + (c-2x)}\, dx = xe^c = xe^{2x+y} \qquad [\because\ y = c - mx = c - 2x]$$

and from $(D - 2D')$ $\quad z = u = xe^{2x+y}$, the solution

$$z = \int xe^{2x+(c-2x)} dy = \frac{1}{2}x^2 e^c = \frac{1}{2}x^2 e^{2x+y} \qquad [\because\ y = c - mx = c - 2x]$$

Hence, the C.S. is $\quad z = f_1(y + 2x) + xf_2(y + 2x) + \dfrac{1}{2}x^2 e^{2x+y}$.

**Example 17.36.** *Solve* $\dfrac{\partial^3 z}{\partial x^3} - 2\dfrac{\partial^3 z}{\partial x^2 \partial y} = 2e^{2x} + 3x^2 y$. *(JNTUK, 2022; Anna, 2016 ; C.S.V.T.U., 2007)*

**Solution.** Given equation in *symbolic form* is

$$(D^3 - 2D^2 D')z = 2e^{2x} + 3x^2 y$$

Its A.E. is $m^3 - 2m^2 = 0$, whence $m = 0, 0, 2$.

$\therefore$ $\quad$ C.F. $= f_1(y) + xf_2(y) + f_3(y + 2x)$

$$\text{P.I.} = \frac{1}{D^3 - 2D^2 D'}(2e^{2x} + 3x^2 y) = 2\frac{1}{D^3 - 2D^2 D'}e^{2x} + 3\frac{1}{D^3(1 - 2D'/D)}x^2 y$$

$$= 2\frac{1}{2^3 - 2 \,.\, 2^2(0)}e^{2x} + \frac{3}{D^3}(1 - 2D'/D)^{-1}\, x^2 y$$

$$= \frac{1}{4}e^{2x} + \frac{3}{D^3}\left(1 + \frac{2D'}{D} + \frac{4D'^2}{D^2} + \ldots\right)x^2 y$$

$$= \frac{1}{4}e^{2x} + \frac{3}{D^3}\left(x^2 y + \frac{2}{D}x^2 \,.\, 1\right) = \frac{1}{4}e^{2x} + \frac{3}{D^3}\left(x^2 y + \frac{2}{3}x^3\right) \qquad \left[\because\ \frac{1}{D}f(x) = \int f(x)\, dx\right]$$

$$= \frac{1}{4}e^{2x} + 3y\frac{x^5}{3 \,.\, 4 \,.\, 5} + 2 \,.\, \frac{x^6}{4 \,.\, 5 \,.\, 6} \qquad \left[\because\ \frac{1}{D^3}f(x) = \int\left[\int\left(\int f(x)\, dx\right)dx\right]dx\right]$$

$$= \frac{e^{2x}}{4} + \frac{x^5 y}{20} + \frac{x^6}{60}$$

Hence the C.S. is $z = f_1(y) + xf_2(y) + f_3(y + 2x) + \dfrac{e^2x}{4} + \dfrac{x^5 y}{20} + \dfrac{x^6}{60}$

**Example 17.37.** *Solve* $\dfrac{\partial^2 z}{\partial x^2} + \dfrac{\partial^2 z}{\partial x \partial y} - 6\dfrac{\partial^2 z}{\partial y^2} = y \cos x$. *(Anna, 2013)*

*or* $\quad r + s - 6t = y \cos x$.

**Solution.** Its *symbolic form* is $(D^2 + DD' - 6D'^2)\, z = y \cos x$

and the A.E. is $m^2 + m - 6 = 0$, whence $m = -3, 2$.

$\therefore$ $\quad$ C.F. $= f_1(y - 3x) + f_2(y + 2x)$

$$\text{P.I.} = \frac{1}{(D - 2D')(D + 3D')} y \cos x = \frac{1}{D - 2D'}\left[\int (c + 3x) \cos x \, dx\right]_{c \to y - 3x}$$

$[\because\ y = c - mx = c + 3x]$

$$= \frac{1}{D - 2D'}[(c + 3x) \sin x + 3 \cos x]_{c \to y - 3x}$$ [Integrating by parts]

$$= \frac{1}{D - 2D'}(y \sin x + 3 \cos x) = \left[\int \{(c - 2x) \sin x \ 3 \cos x\} dx\right]_{c \to y - 2x}$$

$$= [(c - 2x)(- \cos x) - (-2)(- \sin x) + 3 \sin x]_{c \to y + 2x}$$

$$= - y \cos x + \sin x$$

Hence the C.S. is $z = f_1(y - 3x) + f_2(y + 2x) + \sin x - y \cos x$.

**Example 17.38.** *Solve* $4\dfrac{\partial^2 z}{\partial x^2} - 4\dfrac{\partial^2 z}{\partial x \partial y} + \dfrac{\partial^2 z}{\partial y^2} = 16 \log (x + 2y)$.

**Solution.** Its *symbolic form* is $4D^2 - 4DD' + D'^2 = 16 \log (x + 2y)$
and the A.E. is $4m^2 - 4m + 1 = 0$, $m = 1/2, 1/2$.

$\therefore$ $\quad$ C.F. $= f_1\left(y + \dfrac{1}{2}x\right) + x f_2\left(y + \dfrac{1}{2}x\right)$

$$\text{P.I.} = \frac{1}{(2D - D')^2} 16 \log (x + 2y) = 4\frac{1}{\left(D - \frac{1}{2}D'\right)}\left\{\frac{1}{D - \frac{1}{2}D'} \log (x + 2y)\right\}$$

$$= 4\frac{1}{D - \frac{1}{2}D'}\left[\int \log\left\{x + 2\left(c - \frac{x}{2}\right)\right\} dx\right]_{c \to y + x/2}$$ $[\because\ y = c - mx = c - x/2]$

$$= 4\frac{1}{D - \frac{1}{2}D'}\left[\int \log (2c) \, dx\right]_{c \to y + x/2} = 4\frac{1}{D - \frac{1}{2}D'}[x \log (x + 2y)]$$

$$= 4\left[\int \left\{x \log\left[x + 2\left(c - \frac{x}{2}\right)\right]\right\} dx\right]_{c \to y + x/2} = 4\left[\log 2c \int x \, dx\right]_{c \to y + x/2} = 2x^2 \log (x + 2y)$$

Hence the C.S. is $z = f_1\left(y + \dfrac{x}{2}\right) + x f_2\left(y + \dfrac{x}{2}\right) + 2x^2 \log (x + 2y)$.

## PROBLEMS 17.6

Solve the following equations :

1. $\dfrac{\partial^3 z}{\partial x^3} - 4\dfrac{\partial^3 x}{\partial x^2 \partial y} + 4\dfrac{\partial^3 z}{\partial x \partial y^2} = 0.$

2. $\dfrac{\partial^2 z}{\partial x^2} + \dfrac{\partial^2 z}{\partial x \partial y} + 4\dfrac{\partial^2 z}{\partial y^2} = e^{2x + y}.$ *(K.T.U., 2016)*

3. $(D^2 - 2DD' + D'^2)z = e^{x + y}.$

4. $\dfrac{\partial^3 z}{\partial x^3} - 4\dfrac{\partial^3 z}{\partial x^2 \partial y} + 5\dfrac{\partial^3 z}{\partial x \partial y^2} - 2\dfrac{\partial^3 z}{\partial y^3} = e^{2x + y}.$

5. $\dfrac{\partial^2 z}{\partial x^2} - 2\dfrac{\partial^2 z}{\partial x \partial y} + \dfrac{\partial^2 z}{\partial y^2} = \sin x.$

6. $\dfrac{\partial^2 y}{\partial t^2} - a^2\dfrac{\partial^2 y}{\partial x^2} = E \sin pt.$

7. $\dfrac{\partial^3 z}{\partial x^3} - \dfrac{4\partial^3 z}{\partial z^2 \partial y} + 4\dfrac{\partial^3 z}{\partial x \partial y^2} = \cos (2x + y).$ *(K.T.U., 2016)*

8. $(D^3 - 7DD'^2 - 6D'^3)z = \sin (2x + y)$. *(Anna, 2013)*

9. $\frac{\partial^2 z}{\partial x^2} - 3\frac{\partial^2 z}{\partial x \partial y} + 2\frac{\partial^2 z}{\partial y^2} = e^{2e-y} + e^{x+y} + \cos(x + 2y).$

10. $(D^2 + DD' - 6D'^2)z = y \sin x.$ (U.P.T.U., 2012)

11. $\frac{\partial^2 z}{\partial x^2} - \frac{\partial^2 z}{\partial x \partial y} = \sin x \cos 2y.$ (JNTUK, 2022; Hazaribag, 2015 ; Delhi, 2012)

12. $(D^2 - DD')z = \cos 2y (\sin x + \cos x).$

13. $(D^2 - D'^2)z = e^{x-y} \sin(x + 2y).$

14. $(D^2 + (a + b)DD' + abD'^2)z = xy.$ (U.P.T.U., 2013)

15. $\frac{\partial^2 z}{\partial x^2} + 2\frac{\partial^2 z}{\partial x \partial y} + \frac{\partial^2 z}{\partial y^2} = x^2 + xy + y^2.$

16. $(D^2 - DD' - 2D'^2)z = (y - 1)e^x.$ (Delhi, 2012S)

17. $(D^3 + D^2D' - DD'^2 - D'^3)z = e^x \cos 2y.$

18. $(D^2 + 2DD' + D'^2)z = 2\cos y - x \sin y.$

## 17.12 NON-HOMOGENEOUS LINEAR EQUATIONS

If in the equation $f(D, D')z = F(x, y)$ ...(1)

the polynomial expression $f(D, D')$ is not homogeneous, then Eqn. (1) is a non-homogeneous linear partial differential equation. As in the case of homogeneous linear partial differential equations, its complete solution = C.F. + P.I.

The methods to find P.I. are the same as those for homogeneous linear equations.

To find the C.F., we factorize $f(D, D')$ into factors of the form $D - mD' - c$. To find the solution of $(D - mD' - c)z = 0$, we write it as $p - mq = cz$ ...(2)

The subsidiary equations are

$$\frac{dx}{1} = \frac{dy}{-m} = \frac{dz}{cz}$$

Its integrals are $y + mx = a$ and $z = be^{cx}$.

Taking $b = \phi(a)$, we get $z = e^{cx}\phi(y + mx)$

as the solution of Eqn. (2). The solution corresponding to various factors added up, give the C.F. of Eqn. (1).

**Example 17.39.** *Solve* $(D^2 + 2DD' + D'^2 - 2D - 2D')z = \sin(x + 2y).$ (Anna, 2016)

**Solution.** Here $f(D, D') = (D + D')(D + D' - 2)$

Since the solution corresponding to the factor $D - mD' - c$ is known to be

$$z = e^{cx}\phi(y + mx)$$

$$\therefore \quad \text{C.F.} = \phi_1(y - x) + e^{2x} f_2(y - x)$$

$$\therefore \quad \text{P.I.} = \frac{1}{D^2 + 2DD' + D'^2 - 2D - 2D'}\sin(x + 2y)$$

$$= \frac{1}{-1 + 2(-2) + (-4) - 2D - 2D'}\sin(x + 2y)$$

$$= -\frac{1}{2(D + D') + 9}\sin(x + 2y)$$

$$= -\frac{2(D + D') - 9}{4(D^2 + 2DD' + D'^2) - 81}\sin(x + 2y)$$

$$= \frac{1}{39}[2\cos(x + 2y) - 3\sin(x + 2y)]$$

Hence the complete solution is

$$z = \phi_1(y - x) + e^{2x}\phi_2(y - x) + \frac{1}{39}[2\cos(x + 2y) - 3\sin(x + 2y)].$$

## PROBLEMS 17.7

Solve the following equations :

1. $\frac{\partial^2 z}{\partial x^2}+\frac{\partial^2 z}{\partial x \partial y}+\frac{\partial z}{\partial y}-z=e^{-x}$.
2. $(D-D'-1)(D-D'-2)z=e^{2x-y}$.
3. $(D+D'-1)(D+2D'-3)z=4+3x+6y$.
4. $(D^2-D'^2-3D+3D')z=xy+e^{x+2y}$. (J.N.T.U., 2015)
5. $\frac{\partial^2 z}{\partial x^2}-\frac{\partial^2 z}{\partial x \partial y}+\frac{\partial z}{\partial y}=x^2+y^2$.
6. $\frac{\delta^2 w}{\delta x^2}-2a\frac{\delta^2 w}{\delta x\, \delta y}+a^2\frac{\delta^2 w}{dy^2}=4(y+ax)$ (CSVTU, 2018)
7. $(2DD'+D'^2-3D')z=3\cos(3x-2y)$.
8. $r-t=\tan x \tan y(\tan^2 x-\tan^2 y)$ (VIT, Vellore, 2019)
9. $(D^2-2DD'+D'^2)\phi=2\cos y-x\sin y$
10. $(D^2-6DD'+9D'^2)w=6x-2y$
11. $(D^3-2D^2D'+DD'^2+DD'^3)=e^{x+y}$

## 17.13 NON-LINEAR EQUATIONS OF THE SECOND ORDER

We now give a method due to *Monge**, for integrating the equation $Rr+Ss+Tt=V$ ...(1)
in which $R, S, T, V$ are functions of $x, y, z, p$ and $q$.

Since $$dp=\frac{\partial p}{\partial x}dx+\frac{\partial p}{\partial y}dy=rdx+tdy, \text{ and } dq=sdx+tdy,$$

we have $$r=(dp-tdy)/dx \text{ and } t=(dq-sdx)/dy.$$

Substituting these values of $r$ and $t$ in Eqn. (1), and rearranging the terms, we get

$$(Rdpdy+Tdqdx-Vdxdy)-s(Rdy^2-Sdydx+Tdx^2)=0 \quad ...(2)$$

Let us consider the equations

$$Rdy^2-Sdydx+Tdx^2=0 \quad ...(3)$$

$$Rdpdy+Tdqdx-Vdxdy=0 \quad ...(4)$$

which are known as *Monge's equations.*

Since Eqn. (3) can be factorised, we obtain its integral first. In case the factors are different, we may get two distinct integrals of Eqn. (3). Either of these together with Eqn. (4) will give an integral of Eqn. (4). If need be, we may also use the relation $dz=pdx+qdy$ while solving Eqns. (3) and (4).

Let $u(x, y, z, p, q)=a$ and $v(x, y, z, p, q)=b$ be the integrals of Eqns. (3) and (4) respectively. Then $u=a$, $v=b$ evidently constitute a solution of Eqn. (2) and therefore, of Eqn. (1) also. Taking $b=\phi(a)$, we find a general solution of Eqn. (1) to be $v=\phi(u)$, which should be further integrated by methods of first order equations.

**Example 17.40.** *Solve $(D^2-DD'-2D'^2)z=(y-1)e^x$* (AMIE, 2018)

**Solution.** The C.F. can be easily obtained

$$z=\phi_1(y+2x)+\phi_2(y-x)$$

To get P.I., we proceed as follows
$$z=\frac{1}{D^2-DD'-2D'^2)}(y-1)e^x=\frac{1}{(D+D')(D-2D')}(y-1)e^x$$

$$=\frac{1}{(D+D')}\int(c-2x-1)e^x dx \quad \text{since} \quad y+2x=c$$

$$=\frac{1}{(D+D')}\cdot(c-2x+1)e^x=\frac{1}{D+D'}(y+1)e^x$$

on replacing $c$ by $y+2x$

Again $$z=\int(c'+x+1)e^x dx \quad \text{as} \quad y-x=c'=(c'+x)e^x=(y-x+x)e^x=ye^x$$

so the complete solution is $$z=\phi_1(y+2x)+\phi_2(y-x)+y\,e^x$$

---

*Named after Gaspard Monge (1746-1818), Professor at Paris.

**Example 17.41.** *Solve $(r - s)\, y + (s - t)\, x + q - p = 0$* *(Jadavpur, 2018)*

**Solution:** The equation can be written as

$$ry + s(x - y) - tx + q - p = 0$$

$$y\, dy\, dp - y\, dx\, dq + (q - p)\, dx\, dy = 0 \qquad ...(i)$$

$$y\, dy^2 - (x - y)\, dx\, dy - x\, dx^2 = 0 \qquad ...(ii)$$

from Eqn. (*ii*) we have

$$dy + dx = 0 \qquad ...(iii)$$

$$y\, dy - x\, dx = 0 \qquad ...(iv)$$

Integrating these we get

$$y + x = c_1 \qquad ...(v)$$

$$y^2 - x^2 = c_2 \qquad ...(vi)$$

Combining Eqns. (*i*) and (*iii*), we have $y\, dp + x\; dq + q\, dx + p\, dy = 0$

$$(y\, dp + p\, dy) + (x\, dq + q\, dx) = 0$$

$$Py + xq = \text{constant} = f(y + x) \text{ from Eqn. } (v)$$

This is of Lagrays form

$$\frac{dx}{y} = \frac{dy}{x} = \frac{dz}{f(x + y)}$$

From first two members, we get

$$x\, dx - y\, dy = 0$$

$$x^2 - y^2 = c$$

Also

$$\frac{dx}{y} = \frac{dz}{f(x + y)} \quad \text{or} \quad dz = \frac{f\left\{x + \sqrt{x^2 - c}\right\}}{\sqrt{x^2 - c}}$$

since

$$y^2 = x^2 - c \quad \text{or} \quad y = \sqrt{(x^2 - c)}$$

Now integrating, we have

$$z = F_1\left[x + \sqrt{(x^2 - c)}\right] + \text{constant}$$

$$= F_1(x + y) + F_2\,(x^2 - y^2)$$

**Example 17.42:** *Solve the equations*

$$\frac{\delta^2 u}{dx^2} - \frac{\delta^2 u}{\delta x\, \delta y} - 2\frac{\delta^2 u}{\delta y^2} + 2\frac{\delta u}{\delta x} + 2\frac{\delta u}{\delta y} = e^{2x + 3y} + \sin\,(2x + y) + xy$$

*(Rohtak, 2018)*

**Solution:** The given equation is

$$(D^2 - DD' - 2D'^2 + 2D + 2D')u = e^{2x + 3y} + \sin\,(2x + y) + xy$$

$$(D + D')\,(D - 2D' + 2)u = e^{2x + 3y} + \sin\,(2x + y) + xy$$

Hence C.F.

$$u = \phi_1(y - x) + e^{-2x}\,\phi_2\,(2x + y)$$

P.I. corresponding to $e^{2x + 3y}$

$$u = \frac{1}{(D + D')(D - 2D' + 2)}\, e^{2x+3y} = \frac{1}{(2 + 3)(2 - 6 + 2)}\, e^{2x+3y} = \frac{-1}{10}\, e^{2x+3y}$$

Next P.I. corresponding to $\sin\,(2x + y)$ is given

$$u = \frac{1}{(D^2 - D' - 2D'^2 + 2D + 2D')}\sin(2x + y)$$

$$= \frac{1}{-2^2 + 2.1 - 2(-1)^2 + 2D + 2D'}\sin(2x + y)$$

$$= \frac{1}{2(D + D')}\sin(2x + y) = \frac{1}{2}\cdot\frac{D - D'}{D^2 - D'^2}\sin(2x + y)$$

$$= \frac{1}{2}\cdot\frac{D - D'}{-2^2 - (-1)^2}\sin(2x + y) = -\frac{1}{6}\cos(2x + y)$$

Finally P.I. corresponding to $xy$

$$= \frac{1}{(D + D')(D - 2D' + 2)}\, xy$$

$$= \frac{1}{2D}\left(1+\frac{D'}{D}\right)\left[1-\frac{1}{2}(D-2D')+\frac{1}{4}(D-2D)^2+...\right]xy$$

$$= \frac{1}{2D}\left(1+\frac{D'}{D}\right)\left(xy-\frac{1}{2}y+x-1\right) = \frac{1}{2D}\left(xy-\frac{1}{2}y-\frac{1}{2}x^2+\frac{3}{2}x-1\right)$$

$$= \frac{1}{2}\left(\frac{1}{2}x^2y-\frac{1}{2}xy-\frac{1}{6}x^3+\frac{3}{4}x^2-x\right)$$

Hence the complete solution is

$$u = \phi_1(y-x)+e^{-2x}\,\phi_2(2x+y)-\frac{1}{10}e^{2x+3y}-\frac{1}{6}\cos(2x+y)+\frac{1}{4}\left(x^2y-xy-\frac{1}{3}x^3+\frac{3}{2}x^2-2x\right)$$

**Example 17.43:** *Solve*

$$\frac{1}{x^2}\frac{\delta^2 z}{\delta x^2}-\frac{1}{x^3}\frac{\delta z}{\delta x} = \frac{1}{y^2}\frac{\delta^2 z}{\delta y^2}-\frac{1}{y^3}\frac{\delta z}{\delta y} \qquad \textit{(Mysore, 2018)}$$

**Solution:** Let $\frac{x^2}{2} = u$ and $\frac{y^2}{2} = v$

Then $x\,dx = du$ and $y\,dy = dv$

Now, $\frac{\delta z}{\delta u} = \frac{\delta z}{\delta x}\cdot\frac{\delta x}{\delta u} = \frac{1}{x}\cdot\frac{\delta z}{\delta x}$

$$\frac{\delta^2 z}{\delta u^2} = \frac{\delta}{\delta x}\left(\frac{\delta z}{\delta u}\right)\cdot\frac{\delta x}{\delta u} = \frac{\delta}{\delta x}\left(\frac{1}{x}\cdot\frac{\delta z}{\delta x}\right)\cdot\frac{1}{x} = \frac{-1}{x^3}\frac{\delta z}{\delta x}+\frac{1}{x^2}\frac{\delta^2 z}{\delta x^2}$$

Similarly $\frac{\delta^2 z}{\delta v^2} = -\frac{1}{y^3}\frac{\delta z}{\delta y}+\frac{1}{y^2}\frac{\delta^2 z}{\delta y^2}$

Thus the given equation $= \frac{\delta^2 z}{\delta u^2}-\frac{\delta^2 z}{\delta v^2} = 0$ or $(D^2-D'^2)_z = 0$

$$(D-D')(D+D')_z = 0$$

$\therefore$ The solution is given by $z = \phi_1(v+u)+\phi_2(v-u) = \phi_1\left(\frac{y^2}{2}+\frac{x^2}{2}\right)+\phi_2\left(\frac{y^2}{2}-\frac{x^2}{2}\right)$

$$= f_1(y^2+x^2)+f_2(y^2-x^2)$$

## PROBLEMS 17.8

Solve :

1. $(q+1)s = (p+1)t$.
2. $r - t\cos^2 x + p\tan x = 0$.
3. $(1+q)^2 r - 2(1+p+q+pq)s + (1+p)^2 t = 0$ *(Guwahati, 2018)*
4. $xy(t-r)+(x^2-y^2)(s-2) = py - qx$.
5. $q^2r - 2pqs + p^2t = pq^2$.
6. $(1+q)^2r - 2(1+p+q+pq)s + (1+p)^2t = 0$.
7. $(q+1)s = (p+1)t$
8. $(x-y)(xr - xs - ys - yt) = (x+y)(p-q)$
9. $y^2r - 2yz + t = p + 6y$
10. $x^2r + 2xys + y^2 t = 0$
11. $pt - qs = q^3$

## 17.14 OBJECTIVE TYPE OF QUESTIONS

## PROBLEMS 17.9

*Fill up the blanks or choose the correct answer in each of the following problems :*

1. The equation $\frac{\partial^2 z}{\partial x^2}+2xy\left(\frac{\partial z}{\partial x}\right)^2+\frac{\partial z}{\partial y} = 5$ is of order ....... and degree .......
2. The complementary function of $(D^2-4DD'+4D'^2)z = x+y$ is ......

3. The solution of $\frac{\partial^2 z}{\partial y^2} = \sin(xy)$ is ......
4. A solution of $(y - z)p + (z - x)q = x - y$ is .......
5. The particular integral of $(D^2 + DD')z = \sin(x + y)$ is .......
6. The partial differential equation obtained from $z = ax + by + ab$ by eliminating $a$ and $b$ is ......
7. Solution of $\sqrt{p} + \sqrt{q} = 1$ is ......
8. Solution of $p\sqrt{x} + q\sqrt{y} = \sqrt{z}$ is ......
9. Solution of $p - q = \log(x + y)$.
10. The order of the partial differential equation obtained by eliminating $f$ from $z = f(x^2 + y^2)$, is ......
11. The solution of $x\, \partial z/\partial x = 2x + y$ is
12. By eliminating $a$ and $b$ from $z = a(x + y) + b$, the p.d.e. formed is ......
13. The solution of $[D^3 - 3D^2D' + 2DD'^2]z = 0$ is ......
14. By eliminating the arbitrary constants from $z = a^2x + ay^2 + b$, the partial differential equation formed is ......
15. A solution of $u_{xy} = 0$ is of the form .......
16. If $u = x^2 + t^2$ is a solution of $c^2 \frac{\partial^2 u}{\partial x^2} = \frac{\partial^2 u}{\partial t^2}$, then $c$ = ......
17. The general solution of $u_{xx} = xy$ is .......
18. The complementary function of $r - 7s + 6t = e^{x+y}$ is .......
19. The solution of $xp + yq = z$ is

    (i) $f(x^2, y^2) = 0$ (ii) $f(xy, yz)$ (iii) $f(x, y) = 0$ (iv) $f\left(\frac{x}{y}, \frac{y}{z}\right) = 0$.

20. The solution of $(y - z)p + (z - x)q = x - y$, is

    (i) $f(x^2 + y^2 + z^2) = xyz$ (ii) $f(x + y + z) = xyz$

    (iii) $f(x + y + z) = x^2 + y^2 + z^2$ (iv) $f(x^2 + y^2 + z^2, xyz) = 0$.

21. The partial differential equation from $z = (c + x)^2 + y$ is

    (i) $z = \left(\frac{\partial z}{\partial x}\right)^2 + y$ (ii) $z = \left(\frac{\partial z}{\partial y}\right)^2 + y$ (iii) $z = \frac{1}{4}\left(\frac{\partial z}{\partial x}\right)^2 + y$ (iv) $z = \frac{1}{4}\left(\frac{\partial z}{\partial y}\right)^2 + y$.

22. The solution of $p + q = z$ is

    (i) $f(xy, y \log z) = 0$ (ii) $f(x + y, y + \log z) = 0$

    (iii) $f(x - y, y - \log z) = 0$ (iv) None of these.

23. Particular integral of $(2D^2 - 3DD' + D'^2)z = e^{x+2y}$ is

    (i) $\frac{1}{2}e^{x+2y}$ (ii) $-\frac{x}{2}e^{x+2y}$ (iii) $xe^{x+2y}$ (iv) $x^2e^{x+2y}$.

24. The solution of $\frac{\partial^3 z}{\partial x^3} = 0$ is

    (i) $z = (1 + x + x^2) f(y)$ (ii) $z = (1 + y + y^2) f(x)$

    (iii) $z = f_1(x) + y f_2(x) + y^2 f_3(x)$ (iv) $z = f_1(y) + x f_2(y) + x^2 f_3(y)$.

25. Particular integral of $(D^2 - D'^2)z = \cos(x + y)$ is

    (i) $x\cos(x + y)$ (ii) $\frac{x}{2}\cos(x + y)$ (iii) $x\sin(x + y)$ (iv) $\frac{x}{2}\sin(x + y)$

26. The solution of $\partial^2 z/\partial x^2 = \partial^2 z/\partial y^2$ is

    (i) $z = f_1(y + x) + f_2(y - x)$ (ii) $z = f_1(y + x) + f_1(y - x)$

    (iii) $z = f(x^2 - y^2)$ (iv) $z = f(x^2 + y^2)$.

27. Solution of $(D - 1)(D - D' + 1)z = 0$ is ...... *(Anna, 2012)*
28. Solution of partial differential equation $(p - q)(z - px - qy) = 1$ is ...... *(Rohtak, 2011)*
29. $xu_x + yu_y = u^2$ is a non-linear partial differential equation. (True or False)
30. $xu_x + u_{xx} = 0$ is a non-linear partial differential equation. (True or False)
31. $u = x^2 - y^2$ is a solution of $u_{xx} + u_{yy} = 0$. (True or False)
32. $u = e^{-t}\sin x$ is a solution of $\frac{\partial^2 u}{\partial t^2} + \frac{\partial^2 u}{\partial x^2} = 0$. (True or False)
33. $x\frac{\partial u}{\partial x} + t\frac{\partial u}{\partial t} = 2u$ is an ordinary differential equation. (True or False)

CHAPTER 18

# Applications of Partial Differential Equations



## 18.1 INTRODUCTION

In physical problems, we always seek a solution of the differential equation which satisfies some specified conditions known as the boundary conditions. The differential equation together with these boundary conditions, constitute a *boundary value problem.*

In problems involving ordinary differential equations, we may first find the general solution and then determine the arbitrary constants from the initial values. But the same process is not applicable to problems involving partial differential equations for the general solution of a partial differential equation contains arbitrary functions which are difficult to adjust so as to satisfy the given boundary conditions. Most of the boundary value problems involving linear partial differential equations can be solved by the following method.

## 18.2 METHOD OF SEPARATION OF VARIABLES

It involves a solution which breaks up into a product of functions each of which contains only one of the variables. The following example explains this method :

**Example 18.1.** (*a*) *Solve (by the method of separation of variables) :*

$$\frac{\partial^2 z}{\partial x^2} - 2\frac{\partial z}{\partial x} + \frac{\partial z}{\partial y} = 0. \qquad \text{(V.T.U., 2015)}$$

(*b*) *Solve the above equation when* $z(0, y) = 0$ *and* $\frac{\partial z}{\partial x}(0, y) = e^{-3y}$. (*Delhi, 2016*)

**Solution.** (*a*) Assume the trial solution $z = X(x)Y(y)$ ...(*i*)

where $X$ is a function of $x$ alone and $Y$ that of $y$ alone.

Substituting this value of $z$ in the given equation, we have

$$X''Y - 2X'Y + XY' = 0 \qquad \text{where } X' = \frac{dX}{dx},\ Y' = \frac{dY}{dy} \text{ etc.}$$

Separating the variables, we get $\dfrac{X'' - 2X'}{X} = -\dfrac{Y'}{Y}$ ...(*ii*)

Since $x$ and $y$ are independent variables, therefore, Eqn. (*ii*) can only be true if each side is equal to the same constant, $a$ (say).

$$\therefore \quad \frac{X'' - 2X'}{X} = a, \text{ i.e. } X'' - 2X' - aX = 0 \qquad ...(iii)$$

and $$-Y'/Y = a, \quad i.e., \quad Y' + aY = 0 \qquad ...(iv)$$

To solve the ordinary linear Eqn. (*iii*), the auxiliary equation is

$$m^2 - 2m - a = 0, \quad \text{whence } m = 1 \ \sqrt{(1+a)}\,.$$

$\therefore$ the solution of Eqn. (*iii*) is $X = c_1 e^{[1+\sqrt{(1+a)}]x} + c_2 e^{[1+\sqrt{(1+a)}]x}$

and the solution of Eqn. (*iv*) is $Y = c_3 e^{-ay}$.

Substituting these values of $X$ and $Y$ in Eqn. (*i*), we get

$$z = \{c_1 e^{[1+\sqrt{(1+a)}]x} + c_2 e^{[1-\sqrt{(1+a)}]x}\} \,.\, c_3 e^{-ay}$$

*i.e.,* $$z = \{k_1 e^{[1+\sqrt{(1+a)}]x} + k_2 e^{[1-\sqrt{(1+a)}]x}\}\, e^{-ay} \qquad ...(v)$$

which is the required complete solution. ...(*vi*)

(*b*) When $z(0, y) = 0$, (*v*) given $k_1 + k_2 = 0$.

Also from Eqn. (*v*), $\dfrac{\partial z}{\partial x} = \{k_1[1+\sqrt{(1+a)}]e^{[1+\sqrt{(1+a)}]x} + k_2[1-\sqrt{(1+a)}]\, e^{[1-\sqrt{(1+a)}]x}\}\, e^{-ay}$

When $\dfrac{\partial z}{\partial x}(0, y) = e^{-3y}$, $e^{-3y} = \{k_1\,[1+\sqrt{(1+a)}] + k_2\,[1-\sqrt{(1+a)}]\}e^{-ay}$

or $$k_1 + k_2 + k_1\sqrt{(1+a)} - k_2\sqrt{(1+a)} = e^{a-3y}.$$

Using Eqn. (*vi*), $k_1 - k_2 = e^{a-3y}/\sqrt{(1+a)}$ ... (*vii*)

Solving Eqns. (*vi*) and (*vii*), we get $k_1 = \dfrac{1}{2\sqrt{1+a}}e^{a-3y}$, $k_2 = -\dfrac{1}{2\sqrt{1+a}}e^{a-3y}$

Substituting these values of $k_1$ and $k_2$ in Eqn. (*v*), we get the result.

**Obs.** In practical problems, the unknown constants $a$, $k_1$, $k_2$ are determined from the given boundary conditions.

**Example 18.2.** *Using the method of separation of variables, solve* $\dfrac{\partial u}{\partial x} = 2\dfrac{\partial u}{\partial t} + u$ *where* $u(x, 0) = 6e^{-3x}$.

*(JNTUK, 2022; V.T.U., 2013 ; Calicut, 2012)*

**Solution.** Assume the solution $u(x, t) = X(x)T(t)$

Substituting in the given equation, we have

$$X'T = 2XT' + XT \quad \text{or} \quad (X' - X)\,T = 2XT'$$

or $$\frac{X' - X}{2X} = \frac{T'}{T} = k \text{ (say)}$$

$\therefore$ $$X' - X - 2kX = 0 \quad \text{or} \quad \frac{X'}{X} = 1 + 2k \quad ...(i) \quad \text{and} \quad \frac{T'}{T} = k \qquad ...(ii)$$

Solving Eqn. (*i*), $\log X = (1 + 2k)x + \log c$ or $X = ce^{(1+2k)x}$

From Eqn. (*ii*), $\log T = kt + \log c'$ or $T = c'e^{kt}$

Thus $u(x, t) = XT = cc'\, e^{(1+2k)x}\, e^{kt}$ ...(*iii*)

Now $6e^{-3x} = u(x, 0) = cc'\, e^{(1+2k)x}$ $\therefore$ $cc' = 6$ and $1 + 2k = -3$ or $k = -2$

Substituting these values in Eqn. (*iii*), we get

$$u = 6e^{-3x}\, e^{-2t} \quad i.e., \quad u = 6e^{-(3x+2t)}$$ which is the required solution.

## PROBLEMS 18.1

Solve the following equations by the method of *separation of variables* :

1. $py^3 + qx^2 = 0$. (V.T.U., 2014)
2. $x^2\dfrac{\partial u}{\partial x} + y^2\dfrac{\partial u}{\partial y} = 0$.
3. $\dfrac{\partial u}{\partial x} = 4\dfrac{\partial u}{\partial y}$, given that $u(0, y) = 8e^{-3y}$. *(JNTUK, 2022)*
4. $4\dfrac{\partial u}{\partial x} + \dfrac{\partial u}{\partial y} = 3u$, given $u = 3e^{-y} - e^{-5y}$, when $x = 0$. *(Hazaribag., 2015 ; J.N.T.U., 2015 ; Delhi 2012)*
5. $3\dfrac{\partial u}{\partial x} + 2\dfrac{\partial u}{\partial y} = 0$, $u(x, 0) = 4e^{-x}$. *(JNTUK, 2022; V.T.U., 2014)*
6. Find a solution of the equation $\dfrac{\partial^2 u}{\partial x^2} = \dfrac{\partial u}{\partial y} + 2u$ in the form $u = f(x)\,g(y)$. Solve the equation subject to the conditions $u = 0$ and $\partial u/\partial x = 1 + e^{-3y}$, when $x = 0$ for all values of $y$. *(Delhi, 2015)*

## 18.3 PARTIAL DIFFERENTIAL EQUATIONS OF ENGINEERING

A number of problems in engineering give rise to the following well-known partial differential equations :

(*i*) *Wave equation :* $\dfrac{\partial^2 y}{\partial t^2} = c^2 \dfrac{\partial^2 y}{\partial x^2}$.

(*ii*) *One dimensional heat flow equation :* $\dfrac{\partial u}{\partial t} = c^2 \dfrac{\partial^2 u}{\partial x^2}$.

(*iii*) *Two dimensional heat flow equation* which in steady state becomes the two dimensional *Laplace's equation :* $\dfrac{\partial^2 u}{\partial x^2} + \dfrac{\partial^2 u}{\partial y^2} = 0$.

(*iv*) *Transmission line equations.*

(*v*) *Vibrating membrane.* Two dimensional wave equation.

(*vi*) *Laplace's equation* in three dimensions.

Besides these, the partial differential equations frequently occur in the theory of Elasticity and Hydraulics.

Starting with the method of separation of variables, we find their solutions subject to specific boundary conditions and the combination of such solution gives the desired solution. Quite often a certain condition is not applicable. In such cases, the most general solution is written as the sum of the particular solutions already found and the constants are determined using Fourier series so as to satisfy the remaining conditions.

## 18.4 (1) VIBRATIONS OF A STRETCHED STRING—WAVE EQUATION

Consider a tightly stretched elastic string of length $l$ and fixed ends $A$ and $B$ and subjected to constant tension $T$ (Fig. 18.1). The tension $T$ will be considered to be large as compared to the weight of the string so that the effects of gravity are negligible.

Let the string be released from rest and allowed to vibrate. We shall study the subsequent motion of the string, with no external forces acting on it, assuming that each point of the string makes small vibrations at right angles to the equilibrium position $AB$, of the string entirely in one plane.

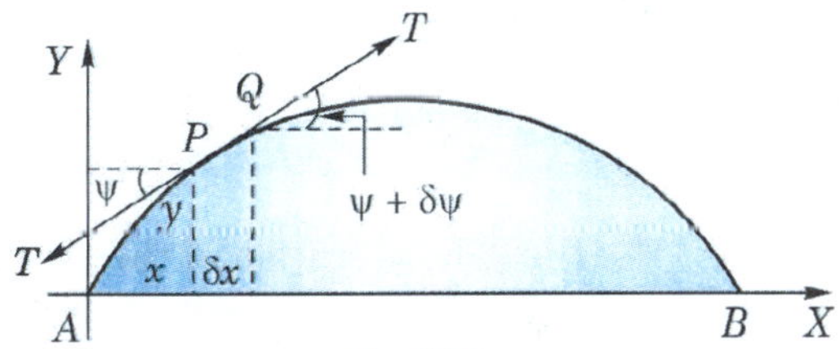

Fig. 18.1

Taking the end $A$ as the origin, $AB$ as the $x$-axis and $AY$ perpendicular to it as the $y$-axis ; so that the motion takes place entirely in the $xy$-plane. Figure 18.1 shows the string in the position $APB$ at time $t$. Consider the motion of the element $PQ$ of the string between its points $P(x, y)$ and $Q(x + \delta x, y + \delta y)$, where the tangents make angles $\psi$ and $\psi + \delta\psi$ with the $x$-axis. Clearly the element is moving upwards with the acceleration $\partial^2 y/\partial t^2$. Also the vertical component of the force acting on this element.

$$= T \sin (\psi + \delta\psi) - T \sin \psi = T[\sin (\psi + \delta\psi) - \sin \psi]$$

$$= T\,[\tan (\psi + \delta\psi) - \tan \psi], \text{ since } \psi \text{ is small} = T\left[\left\{\frac{\partial y}{\partial x}\right\}_{x+\delta x} - \left\{\frac{\partial y}{\partial x}\right\}_x\right]$$

If $m$ be the mass per unit length of the string, then by Newton's second law of motion, we have

$$m\delta x \,.\, \frac{\partial^2 y}{\partial t^2} = T\left[\left\{\frac{\partial y}{\partial x}\right\}_{x+\delta x} - \left\{\frac{\partial y}{\partial x}\right\}_x\right] \quad i.e., \quad \frac{\partial^2 y}{\partial t^2} = \frac{T}{m}\left[\frac{\left\{\frac{\partial y}{\partial x}\right\}_{x+\delta x} - \left\{\frac{\partial y}{\partial x}\right\}_x}{\delta x}\right]$$

Taking limits as $Q \to P$ *i.e.*, $dx \to 0$, we have $\dfrac{\partial^2 y}{\partial t^2} = c^2 \dfrac{\partial^2 y}{\partial x^2}$, where $c^2 = \dfrac{T}{m}$ ...(1)

This is the partial differential equation giving the transverse vibrations of the string. It is also called the one dimensional *wave equation*.

**(2) Solution of the wave equation.** Assume that a solution of Eqn. (1) is of the form

$z = X(x)T(t)$ where $X$ is a function of $x$ and $T$ is a function of $t$ only.

Then $\dfrac{\partial^2 y}{\partial t^2} = X \,.\, T''$ and $\dfrac{\partial^2 y}{\partial x^2} = X''\,.\,T$

Substituting these in Eqn. (1), we get $XT'' = c^2X''T$ *i.e.*, $\dfrac{X''}{X} = \dfrac{1}{c^2}\dfrac{T''}{T}$ ...(2)

Clearly the left side of Eqn. (2) is a function of $x$ only and the right side is a function of $t$ only. Since $x$ and $t$ are independent variables, Eqn. (2) can hold good if each side is equal to a constant $k$ (say). Then Eqn. (2) leads to the ordinary differential equations :

$$\frac{d^2X}{dx^2} - kX = 0 \quad ...(3) \qquad \text{and} \qquad \frac{d^2T}{dt^2} - kc^2T = 0 \quad ...(4)$$

Solving Eqns. (3) and (4), we get

(*i*) *When k is positive* and $= p^2$, say $X = c_1e^{px} + c_2e^{-px}$ ; $T = c_3e^{cpt} + c_4e^{-cpt}$.

(*ii*) *When k is negative* and $= -p^2$ say $X = c_5 \cos px + c_6 \sin px$ ; $T = c_7 \cos cpt + c_8 \sin cpt$.

(*iii*) *When k is zero.* $X = c_9x + c_{10}$ ; $T = c_{11}t + c_{12}$.

Thus the various possible solutions of wave-equation Eqn. (1) are

$$y = (c_1e^{px} + c_2e^{-px})(c_3e^{cpt} + c_4e^{-cpt}) \quad ...(5)$$

$$y = (c_5 \cos px + c_6 \sin px)(c_7 \cos cpt + c_8 \sin cpt) \quad ...(6)$$

$$y = (c_9x + c_{10})(c_{11}t + c_{12}) \quad ...(7)$$

Of these three solutions, we have to choose that solution which is consistent with the physical nature of the problem. As we will be dealing with problems on vibrations, $y$ must be a periodic function of $x$ and $t$. Hence their solution must involve trigonometric terms. Accordingly the solution given by Eqn. (6), *i.e.*, of the form

$$y = (C_1 \cos px + C_2 \sin px)(C_3 \cos cpt + C_4 \sin cpt) \quad ...(8)$$

is the only suitable solution of the *wave equation*. (*K.T.U., 2016*)

**Example 18.3.** *A string is stretched and fastened to two points l apart. Motion is started by displacing the string in the form $y = a \sin(\pi x/l)$ from which it is released at time $t = 0$. Show that the displacement of any point at a distance x from one end at time t is given by*

*$y(x, t) = a \sin(\pi x/l). \cos(\pi ct/l)$.*

(*V.T.U., 2013 ; Rohtak, 2012 ; U.K.T.U., 2012*)

**Solution.** The vibration of the string is given by $\dfrac{\partial^2 y}{\partial t^2} = c^2 \dfrac{\partial^2 y}{\partial x^2}$ ...(*i*)

As the end points of the string are fixed, for all time,

$$y(0, t) = 0 \quad ...(ii) \qquad \text{and} \qquad y(l, t) = 0 \quad ...(iii)$$

Since the initial transverse velocity of any point of the string is zero,

therefore, $$\left(\frac{\partial y}{\partial t}\right)_{t=0} = 0 \quad ...(iv)$$

Also $$y(x, 0) = a \sin(\pi x/l) \quad ...(v)$$

Now we have to solve Eqn. (*i*) subject to the *boundary conditions Eqns.* (*ii*) and (*iii*) and *initial conditions* Eqns. (*iv*) and (*v*). Since the vibration of the string is periodic, therefore, the solution of Eqn. (*i*) is of the form

$$y(x, t) = (C_1 \cos px + C_2 \sin px)(C_3 \cos cpt + C_4 \sin cpt) \quad ...(vi)$$

By Eqn. (*ii*), $y(0, t) = C_1(C_3 \cos cpt + C_4 \sin cpt) = 0$

For this to be true for all time, $C_1 = 0$.

Hence $$y(x, t) = C_2 \sin px (C_3 \cos cpt + C_4 \sin cpt) \quad ...(vii)$$

and $$\frac{\partial y}{\partial t} = C_2 \sin px \{C_3(-cp \,.\, \sin cpt) + C_4(cp \,.\, \cos cpt)\}$$

$\therefore$ By Eqn. (*iv*), $\left(\dfrac{\partial y}{\partial t}\right)_{t=0} = C_2 \sin px \,.\, (C_4\, cp) = 0$, whence $C_2C_4cp = 0$.

If $C_2 = 0$, Eqn. (*vii*) will lead to the trivial solution $y(x, t) = 0$,

$\therefore$ the only possibility is that $C_4 = 0$.

Thus Eqn. (*vii*) becomes $y(x, t) = C_2C_3 \sin px \cos cpt$ ...(*viii*)

$\therefore$ By Eqn. (*iii*), $y(l, t) = C_2C_3 \sin pl \cos cpt = 0$ for all $t$.

Since $C_2$ and $C_3 \neq 0$, we have $\sin pl = 0$. $\therefore$ $pl = n\pi$, *i.e.*, $p = n\pi/l$, where $n$ is an integer.

Hence Eqn. (*i*) reduces to $y(x, t) = C_2C_3 \sin \frac{n\pi x}{l} \cos \frac{n\pi ct}{l}$.

[These are the solutions of Eqn. (*i*) satisfying the boundary conditions. These functions are called the **eigen functions** corresponding to the **eigen values** $\lambda_n = cn\pi/l$ of the vibrating string. The set of values $\lambda_1, \lambda_2, \lambda_3, \ldots$ is called its **spectrum.**]

Finally, imposing the last condition Eqn. (*v*), we have $y(x, 0) = C_2C_3 \sin \frac{n\pi x}{l} = a \sin \frac{\pi x}{l}$

which will be satisfied by taking $C_2C_3 = a$ and $n = 1$.

Hence the required solution is $y(x, t) = a \sin \frac{\pi x}{l} \cos \frac{\pi ct}{l}$ ...(*ix*)

**Obs.** We have from Eqn. (*ix*) $\frac{\partial^2 y}{\partial t^2} = -a\left(\frac{\pi c}{l}\right)^2 \sin \frac{\pi x}{l} \cos \frac{\pi ct}{l} = -\left(\frac{\pi c}{l}\right)^2 y.$

This shows that the motion of each point $y(x, t)$ of the string is simple harmonic with period $= 2\pi/(\pi c/l)$, *i.e.*, $2l/c$.

Thus we can look upon Eqn. (*ix*) as a sine wave $y = y_0 \sin(\pi x/l)$ of wave length $l$, wave-velocity $c$ and amplitude $y_0 = a \cos(\pi ct/l)$ which varies harmonically with time $t$. Whatever $t$ may be, $y = 0$ when $x = 0, l, 2l, 3l$ etc. and these points called *nodes*, remain undisturbed during wave motion. Thus Eqn. (*ix*) represents a *stationary sine wave* of varying amplitudes whose frequency is $c/2l$. Such waves often occur in electrical and mechanical vibratory systems.

**Example 18.4.** *A tightly stretched string with fixed end points $x = 0$ and $x = l$ is initially in a position given by $y = y_0 \sin^3(\pi x/l)$. If it is released from rest from this position, find the displacement $y(x, t)$.*

*(Delhi, 2015 ; Anna, 2012)*

**Solution.** The equation of the vibrating string is $\frac{\partial^2 y}{\partial t^2} = c^2 \frac{\partial^2 y}{\partial x^2}$ ...(*i*)

The boundary conditions are $y(0, t) = 0, y(l, t) = 0$ ...(*ii*)

Also the initial conditions are $y(x, 0) = y_0 \sin^3\left(\frac{\pi x}{l}\right)$ ...(*iii*)

and $\left(\frac{\partial y}{\partial t}\right)_{t=0} = 0$ ...(*iv*)

Since the vibration of the string is periodic, therefore, the solution of Eqn. (*i*) is of the form

$$y(x, t) = (c_1 \cos px + c_2 \sin px)(c_3 \cos cpt + c_4 \sin cpt)$$

By Eqn. (*ii*), $y(0, t) = c_1(c_3 \cos cpt + c_4 \sin cpt) = 0$

For this to be true for all time, $c_1 = 0$.

$\therefore$ $y(x, t) = c_2 \sin px\,(c_3 \cos cpt + c_4 \sin cpt)$

Also by Eqn. (*ii*), $y(l, t) = c_2 \sin pl\,(c_3 \cos cpt + c_4 \sin cpt) = 0$ for all $t$.

This gives $pl = n\pi$ or $p = n\pi/l$, $n$ being an integer.

Thus $y(x, t) = c_2 \sin \frac{n\pi x}{l}\left(c_3 \cos \frac{cn\pi t}{l} + c_4 \sin \frac{cn\pi t}{l}\right)$ ...(*v*)

$$\frac{\partial y}{\partial t} = \left(c_2 \sin \frac{n\pi x}{l}\right)\frac{cn\pi}{l}\left(-c_3 \sin \frac{cn\pi t}{l} + c_4 \cos \frac{cn\pi t}{l}\right)$$

By Eqn. (*iv*), $\left(\frac{\partial y}{\partial t}\right)_{t=0} = \left(c_2 \sin \frac{n\pi x}{l}\right)\frac{cn\pi}{l} \cdot c_4 = 0$, *i.e.* $c_4 = 0$.

Thus Eqn. (*v*) becomes $y(x, t) = c_2c_3 \sin \frac{n\pi x}{l} \cos \frac{n\pi ct}{l} = b_n \sin \frac{n\pi x}{l} \cos \frac{n\pi ct}{l}$

Adding all such solutions the general solution of Eqn. (*i*) is

$$y(x, t) = \sum_{n=1}^{\infty} b_n \sin \frac{n\pi x}{l} \cos \frac{n\pi ct}{l} \qquad \ldots(vi)$$

$\therefore$ from (*iii*), $$y_0 \sin^3 \frac{\pi x}{l} = \sum_{n=1}^{\infty} b_n \sin \frac{n\pi x}{l}$$

or $$y_0 \left\{ \frac{3 \sin \frac{\pi x}{l} - \sin \frac{3\pi x}{l}}{4} \right\} = b_1 \sin \frac{\pi x}{l} + b_2 \sin \frac{2\pi x}{l} + b_3 \sin \frac{3\pi x}{l} + \ldots$$

Comparing both sides, we have

$$b_1 = 3\, y_0/4,\ b_2 = 0,\ b_3 = -\,y_0/4,\ b_4 = b_5 = \ldots = 0.$$

Hence from Eqn. (*vi*), the desired solution is

$$y(x, t) = \frac{3y_0}{4} \sin \frac{\pi x}{l} \cos \frac{\pi ct}{l} - \frac{y_0}{4} \sin \frac{3\pi x}{l} \cos \frac{3\pi ct}{l}.$$

**Example 18.5.** *A tightly stretched flexible string has its ends fixed at x = 0 and x = l. At time t = 0, the string is given a shape defined by F(x) = μx(l – x), where μ is a constant, and then released. Find the displacement of any point x of the string at any time t > 0.*

(*Anna, 2015*)

**Solution.** The equation of the string is $\frac{\partial^2 y}{\partial t^2} = c^2 \frac{\partial^2 y}{\partial x^2}$ ...(*i*)

The boundary conditions are $y(0, t) = 0,\ y(l, t) = 0$ ...(*ii*)

Also the initial conditions are $y(x, 0) = \mu x(l - x)$ ...(*iii*)

and $$\left(\frac{\partial y}{\partial t}\right)_{t=0} = 0 \quad \ldots(iv)$$

The solution of Eqn. (*i*) is of the form

$$y(x, t) = (c_1 \cos px + c_2 \sin px)\,(c_3 \cos cpt + c_4 \sin cpt)$$

By Eqn. (*ii*), $$y(0, t) = c_1(c_3 \cos cpt + c_4 \sin cpt) = 0$$

For this to be true for all time, $c_1 = 0$.

$\therefore$ $$y(x, t) = c_2 \sin px\,(c_3 \cos cpt + c_4 \sin cpt)$$

Also by Eqn. (*ii*) $$y(l, t) = c_2 \sin pl(c_3 \cos cpt + c_4 \sin cpt) = 0 \text{ for all } t.$$

This gives $pl = n\pi$ or $p = n\pi/l$, $n$ being an integer.

Thus $$y(x, t) = c_2 \sin \frac{n\pi x}{l} \left( c_3 \cos \frac{n\pi ct}{l} + c_4 \sin \frac{n\pi ct}{l} \right) \quad \ldots(v)$$

$$\frac{\partial y}{\partial t} = \left( c_2 \sin \frac{n\pi x}{l} \right) \frac{n\pi c}{l} \left( -\,c_3 \sin \frac{n\pi ct}{l} + c_4 \cos \frac{n\pi ct}{l} \right)$$

$\therefore$ by Eqn. (*iv*) $$\left(\frac{\partial y}{\partial t}\right)_{t=0} = \left( c_2 \sin \frac{n\pi x}{l} \right) \frac{n\pi c}{l} \cdot c_4 = 0$$

Thus Eqn. (*v*) becomes $y(x, t) = c_2 c_3 \sin \frac{n\pi x}{l} \cos \frac{n\pi ct}{l} = b_n \sin \frac{n\pi x}{l} \cos \frac{n\pi ct}{l}$

Adding all such solutions, the general solution of Eqn. (*i*) is

$$y(x, t) = \sum_{n=1}^{\infty} b_n \sin \frac{n\pi x}{l} \cos \frac{n\pi ct}{l} \quad \ldots(vi)$$

From Eqn. (*iii*), $$\mu(lx - x^2) = y(x, 0) = \sum_{n=1}^{\infty} b_n \sin \frac{n\pi x}{l}$$

where $$b_n = \frac{2}{l} \int_0^l \mu(lx - x^2) \sin \frac{n\pi x}{l}\, dx, \text{ by Fourier half-range sine series}$$

$$= \frac{2\mu}{l} \left\{ \left| (lx - x^2) \left( -\frac{\cos n\pi x/l}{n\pi/l} \right) \right|_0^l - \int_0^l (l - 2x) \left( -\frac{\cos n\pi x/l}{n\pi/l} \right) dx \right\}$$

$$= \frac{2\mu}{l} \cdot \frac{1}{n\pi} \left\{ \int_0^l (l - 2x) \frac{\cos n\pi x}{l} dx \right\} = \frac{2\mu}{n\pi} \left\{ \left| (l - 2x) \frac{\sin n\pi x/l}{n\pi/l} \right|_0^l - \int_0^l (-2) \frac{\sin n\pi x/l}{n\pi/l} dx \right\}$$

$$= \frac{2\mu}{n\pi} \cdot \frac{2l}{n\pi} \int_0^l \sin \frac{n\pi x}{l} dx = \frac{4\mu l}{n^2\pi^2} \left| \frac{-\cos n\pi x/l}{n\pi/l} \right|_0^l = \frac{4\mu l^2}{n^3\pi^3} \{1 - (-1)^n\}$$

Hence from Eqn. (*vi*), the desired solution is

$$y(x, t) = \frac{4\mu l^2}{\pi^3} \sum_{n=1}^{\infty} \frac{1 - (-1)^n}{n^3} \sin \frac{n\pi x}{l} \cos \frac{n\pi ct}{l}$$

$$= \frac{8\mu l^2}{\pi^3} \sum_{m=1}^{\infty} \frac{1}{(2m-1)^3} \sin \frac{(2m-1)\pi}{l} x \cos \frac{(2m-1)\pi ct}{l}.$$

**Example 18.6.** *When a particle is moving in a straight line and when the distance of the particle from the equilibrium position has the values $x_1$ and $x_2$ the corresponding velocities are $u_1$ and $u_2$ respectively. Show that the period of oscillation is* $2\pi \left\{ (x_2^2 - x_1^2) \big/ (u_1^2 - u_2^2) \right\}^{1/2}$.

*find also Eqn. (i) the maximum velocity Eqn. (ii) amplitude* (*Marathwada, 2019*)

**Solution:** Let the equation of SHM be

$$x = a \cos \mu t \qquad \ldots(i)$$

If $u$ is the velocity of the particle at any time $t$.

$$u = \frac{dx}{dt} = -\mu a \sin \mu t \qquad \ldots(ii)$$

From Eqns. (*i*) and (*ii*)

$$\frac{x}{a} = \cos \mu t$$

$$\frac{u}{\mu a} = -\sin \mu t$$

Squaring and adding

$$\frac{x^2}{a^2} + \frac{u^2}{\mu^2 a^2} = 1 \qquad \ldots(iii)$$

when $x = x_1$, $u = u_1$ and
when $x = x_2$, $u = u_2$

Substituting these values in Eqn. (*iii*), we get

$$\frac{x_1^2}{a^2} + \frac{u_1^2}{\mu^2 a^2} = 1 \qquad \ldots(iv)$$

and

$$\frac{x_2^2}{a^2} + \frac{u_2^2}{\mu^2 a^2} = 1 \qquad \ldots(v)$$

Subtraction Eqn. (*v*) from Eqn. (*iv*)

$$\frac{x_1^2 - x_2^2}{a^2} + \frac{u_1^2 - u_2^2}{\mu^2 a^2} = 0$$

$$\mu^2 = \left( \frac{u_1^2 - u_2^2}{x_2^2 - x_1^2} \right)^{1/2}$$

Hence time period

$$T = \frac{2\pi}{\mu} = 2\pi \Big/ \left( \frac{u_1^2 - u_2^2}{x_2^2 - x_1^2} \right)^{1/2}$$

$$= 2\pi\{(x_2^2 - x_1^2)/(u_1^2 - u_2^2)\}^{1/2}$$

from Eqn. (*iv*), the amplitude a is given by

$$a = \left(x_1^2 + \frac{u_1^2}{\mu^2}\right)^{1/2}$$

Substituting value of $\mu^2$, we get $\quad a = \left\{x_1^2 + u_1^2\left(\frac{x_2^2 - x_1^2}{u_1^2 - u_2^2}\right)\right\}^{1/2} = \left\{\frac{u_1^2 x_2^2 - u_2^2 x_1^2}{u_1^2 - u_2^2}\right\}^{1/2}$

Maximum velocity can be obtained from Eqn. (*ii*) when $\sin \mu t = 1$

Henec maximum velocity $= \mu a$

$$= \left(\frac{u_1^2 - u_2^2}{x_2^2 - x_1^2}\right)^{1/2}\left(\frac{u_1^2 x_2^2 - u_2^2 x_1^2}{u_1^2 - u_2^2}\right) = \left(\frac{u_1^2 x_2^2 - u_2^2 x_1^2}{x_2^2 - x_1^2}\right)^{1/2}$$

**Example 18.7.** *A tightly stretched string with fixed end points $x = 0$ and $x = l$ is initially at rest in its equilibrium position. If each of its points is given a velocity $\lambda x(l - x)$, find the displacement of the string at any distance x from one end at any time t.* *(K.T.U., 2016 ; Hamirpur, 2015 ; Anna, 2013)*

**Solution.** The equation of the vibrating string is $\dfrac{\partial^2 y}{\partial t^2} = c^2 \dfrac{\partial^2 y}{\partial x^2}$ ...(*i*)

The boundary conditions are $\quad y(0, t) = 0, y(l, t) = 0$ ...(*ii*)

Also the initial conditions are $\quad y(x, 0) = 0$ ...(*iii*)

and $\quad \left(\dfrac{\partial y}{\partial t}\right)_{t=0} = \lambda x(l - x)$ ...(*iv*)

As in example 18.6, the general solution of Eqn. (*i*) satisfying the conditions Eqns. (*ii*) and (*iii*) is

$$y(x, t) = \sum_{n=1}^{\infty} b_n \sin \frac{n\pi x}{l} . \sin \frac{n\pi ct}{l} \quad ...(v)$$

$$\frac{\partial y}{\partial t} = \sum_{n=1}^{\infty} b_n \sin \frac{n\pi x}{l} . \cos \frac{n\pi ct}{l} . \left(\frac{n\pi c}{l}\right)$$

By Eqn. (*iv*), $\quad \lambda x(l - x) = \left(\dfrac{\partial y}{\partial t}\right)_{t=0} = \dfrac{\pi c}{l}\displaystyle\sum_{n=1}^{\infty} nb_n \sin \frac{n\pi x}{l}$

$$\therefore \quad \frac{\pi c}{l} nb_n = \frac{2}{l}\int_0^l \lambda x(l - x) \sin \frac{n\pi x}{l}\, dx$$

$$= \frac{2\lambda}{l}\left| (lx - x^2)\left(-\frac{l}{n\pi}\cos\frac{n\pi x}{l}\right) - (l - 2x)\left(-\frac{l^2}{n^2\pi^2}\sin\frac{n\pi x}{l}\right) + (-2)\left(\frac{l^3}{n^3\pi^3}\cos\frac{n\pi x}{l}\right)\right|_0^l$$

$$= \frac{4\lambda l^2}{n^3\pi^3}(1 - \cos n\pi) = \frac{4\lambda l^2}{n^3\pi^3}\,[1 - (-1)^n]$$

or $\quad b_n = \dfrac{4\lambda l^3}{c\pi^4 n^4}[1 - (-1)^n] = \dfrac{8\lambda l^3}{c\pi^4 (2m - 1)^4}$ taking $n = 2m - 1$.

Hence, from Eqn. (*v*), the desired solution is

$$y = \frac{8\lambda l^3}{c\pi^4}\sum_{m=1}^{\infty}\frac{1}{(2m - 1)^4}\sin\frac{(2m - 1)\pi x}{l}\sin\frac{(2m - 1)\pi ct}{l}.$$

**Example 18.8.** *At the end of three successive seconds, the distances of a point moving in simple harmonic motion from its mean position are $x_1$, $x_2$, $x_3$ respectively. Show that the time of a complete oscillation is $2\pi\Big/\cos^{-1}\left(\dfrac{x_1 + x_3}{2x_3}\right)$.* *(Bhilai, 2018)*

**Solution:** Let the equation of SHM be

$$x = a \cos \mu t \quad ...(i)$$

If the distance of the moving point from the mean positive are $x_1$, $x_2$ and $x_3$ at the end of $t$, $t + 1$, $t + 2$ seconds respectively, then from Eqn. (*i*)

$$x_1 = a \cos \mu t \quad ...(ii)$$
$$x_2 = a \cos [\mu (t + 1)] \quad ...(iii)$$
$$x_3 = a \cos [\mu (t + 2)] \quad ...(iv)$$

From Eqns. (*ii*) and (*iv*) $\quad x_1 + x_3 = a[\cos \mu t + \cos \{\mu (t + 2)\}]$

$$= 2a \cos \frac{\mu t + \mu t + 2\mu}{2} \cos \frac{\mu t - \mu t - 2\mu}{2}$$

$$= 2a \cos \{\mu (t + 1)\} \cos \mu \qquad [\because \cos (-\mu) = \cos \mu]$$

$$= 2.\ x_2 \cos \mu$$

$$\mu = \cos^{-1}\left(\frac{x_1 + x_2}{2x_3}\right)$$

Hence period of a complete oscillation

$$T = \frac{2\pi}{\mu} = \frac{2\pi}{\cos^{-1}\left(\frac{x_1 + x_2}{2x_3}\right)}$$

**(3) D'Alembert's solution of the wave equation**

$$\frac{\partial^2 y}{\partial t^2} = c^2 \frac{\partial^2 y}{\partial x^2} \quad ...(1)$$

Let us introduce the new independent variables $u = x + ct$, $v = x - ct$ so that $y$ becomes a function of $u$ and $v$.

Then $\frac{\partial y}{\partial x} = \frac{\partial y}{\partial u} + \frac{\partial y}{\partial v}$

and
$$\frac{\partial^2 y}{\partial x^2} = \frac{\partial}{\partial x}\left(\frac{\partial y}{\partial u} + \frac{\partial y}{\partial v}\right) = \frac{\partial}{\partial u}\left(\frac{\partial y}{\partial u} + \frac{\partial y}{\partial v}\right) + \frac{\partial}{\partial v}\left(\frac{\partial y}{\partial u} + \frac{\partial y}{\partial v}\right) = \frac{\partial^2 y}{\partial u^2} + 2\frac{\partial^2 y}{\partial u \partial v} + \frac{\partial^2 y}{\partial v^2}$$

Similarly,
$$\frac{\partial^2 y}{\partial t^2} = c^2\left(\frac{\partial^2 y}{\partial u^2} - 2\frac{\partial^2 y}{\partial u \partial v} + \frac{\partial^2 y}{\partial v^2}\right)$$

Substituting in Eqn. (1), we get $\frac{\partial^2 y}{\partial u \partial v} = 0 \quad ...(2)$

Integrating Eqn. (2) w.r.t. $v$, we get $\frac{\partial y}{\partial u} = f(u) \quad ...(3)$

where $f(u)$ is an arbitrary function of $u$. Now integrating Eqn. (3) w.r.t. $u$, we obtain

$$y = \int f(u)\, du + \psi(v)$$

where $\psi(v)$ is an arbitrary function of $v$. Since the integral is a function of $u$ alone, we may denote it by $\phi(u)$. Thus

$$y = \phi(u) + \psi(v)$$

*i.e.*
$$y(x, t) = \phi(x + ct) + \psi(x - ct) \quad ...(4)$$

This is the *general solution of the wave Eqn.* (1).

Now to determine $\phi$ and $\psi$, suppose initially $u(x, 0) = f(x)$ and $\partial y(x, 0)/\partial t = 0$.

Differentiating Eqn. (4) w.r.t. $t$, we get $\frac{\partial y}{\partial t} = c\phi'(x + ct) - c\psi'(x - ct)$

At $t = 0$, $\quad \phi'(x) = \psi'(x) \quad ...(5)$

and $\quad y(x, 0) = \phi(x) + \psi(x) = f(x) \quad ...(6)$

Eqn. (5) gives, $\quad \phi(x) = \psi(x) + k$

$\therefore$ Eqn. (6) becomes $\quad 2\psi(x) + k = f(x)$

or $$\psi(x) = \frac{1}{2}[f(x) - k] \text{ and } \phi(x) = \frac{1}{2}[f(x) + k]$$

Hence the solution of Eqn. (4) takes the form

$$y(x, t) = \frac{1}{2}[f(x + ct) + k] + \frac{1}{2}[f(x - ct) - k] = f(x + ct) + f(x - ct) \qquad ...(7)$$

which is the *d'Alembert's solution* of the wave Eqn.* (1) *(V.T.U., 2016 ; Calicut, 2012)*

**Obs.** The above solution gives a very useful method of solving partial differential equations by change of variables.

**Example 18.9.** *Find the deflection of a vibrating string of unit length having fixed ends with initial velocity zero and initial deflection $f(x) = k(\sin x - \sin 2x)$.* *(V.T.U., 2011)*

**Solution.** By D'Alembert's method, the solution is

$$y(x, t) = \frac{1}{2}[f(x + ct) + f(x - ct)]$$

$$= \frac{1}{2}[k\{\sin(x + ct) - \sin 2(x + ct)\} + k\{\sin(x - ct) - \sin 2(x - ct)\}]$$

$$= k[\sin x \cos ct - \sin 2x \cos 2ct]$$

Also $$y(x, 0) = k(\sin x - \sin 2x) = f(x)$$

and $$\partial y(x, 0)/\partial t = k(-c \sin x \sin ct + 2c \sin 2x \sin 2ct)_{t=0} = 0$$

*i.e.*, the given boundary conditions are satisfied.

## PROBLEMS 18.2

1. Solve completely the equation $\partial^2 y/\partial t^2 = c^2 \partial^2 y/\partial x^2$, representing the vibrations of a string of length $l$, fixed at both ends, given that $y(0, t) = 0$ ; $y(l, t) = 0$ ; $y(x, 0) = f(x)$ and $\partial y(x, 0)/\partial t = 0$, $0 < x < l$.

2. Solve the wave equation $\frac{\partial^2 u}{\partial t^2} = c^2 \frac{\partial^2 u}{\partial x^2}$ under the conditions $u(0, t) = 0$, $u(l, t) = 0$ for all $t$ ; $u(x, 0) = f(x)$ and $\left(\frac{\partial u}{\partial t}\right)_{t=0} = g(x)$, $0 < x < l$. *(JNTUK, 2022, 2015)*

3. Solve the wave equation $\frac{\partial^2 u}{\partial t^2} = \frac{a^2 \partial^2 v}{\partial x^2}$ under the conditions $u = 0$ when $x = 0$ and $x = \pi$, $\frac{\partial u}{\partial t} = 0$ when $t = 0$ and $u(x, 0) = x$, $0 < x < \pi$. *(D.T.U., 2012)*

4. The equation of motion of a point on the end of a tunning fork for small amplitude of motion is given by
$$\frac{d^2 x}{dt^2} + p^2 x = 0$$
The fork can be driven by a varying magnetic field so that the equation of forced vibration will be of the form
$$\frac{d^2 x}{dt^2} + p^2 x = k \sin \omega t$$
Solve the equation for forced vibration when $p \neq \omega$ and also for $p = \omega$. Discuss the physical significance of the result. *(Bhopal, 2019)*

5. A tightly stretched string of length $l$ has its ends fastened at $x = 0$, $x = l$. The mid-point of the string is then taken to height $h$ and then released from rest in that position. Find the lateral displacement of a point of the string at time $t$ from the instant of release.

6. A tightly stretched string with fixed end points at $x = 0$ and $x = 1$, is initially in a position given by
$$f(x) = \begin{cases} x, & 0 \le x \le \frac{1}{2} \\ 1 - x, & \frac{1}{2} \le x \le 1 \end{cases}$$

*See in page 433.

If it is released from this position with velocity $a$, perpendicular to the $x$-axis, show that the displacement $u(x, t)$ at any point $x$ of the string at any time $t > 0$, is given by

$$u(x, t) = \frac{4\sqrt{2}}{\pi^2}\left[\sum_{n=1}^{\infty}\left\{\frac{\sin[(4\pi-3)\pi x]\cos[(4\pi-3)\pi at-\pi/4]}{(4n-3)^2} - \frac{\sin[(4\pi-1)\pi x]\cos[(4\pi-1)\pi at-\pi/4]}{(4n-1)^2}\right\}\right]$$

7. Using D'Alembert's method, find the deflection of a vibrating string of unit length having fixed ends, with initial velocity zero and initial deflection :

(i) $f(x) = a(x - x^2)$ (*Calicut, 2012 ; Kerala, M. Tech., 2005*) (ii) $f(x) = a \sin^2 \pi x$.

8. In the case of stretched horizontal string which has one end fixed and a particle of mass $m$ attached to the other end, prove that the equation of motion is

$$e\frac{d^2s}{dt^2} = -g(s - l)$$

where $l$ is the natural length of the string and $e$ is elongation due to a weight $mg$, also find $s$ given that $S = 50$, $v = 0$, when $t = 0$.

## 18.5 (1) ONE-DIMENSIONAL HEAT FLOW

Consider a homogeneous bar of uniform cross-section $\alpha$(cm$^2$). Suppose that the sides are covered with a material impervious to heat so that the stream lines of heat-flow are all parallel and perpendicular to the area $\alpha$. Take one end of the bar as the origin and the direction of flow as the positive $x$-axis (Fig. 18.2). Let $\rho$ be the density (gr/cm$^3$), $s$ the specific heat (cal./gr. deg.) and $k$ the thermal conductivity (cal./cm. deg. sec.).

Let $u(x, -t)$ be the temperature at a distance $x$ from $O$. If $\delta u$ be the temperature change in a slab of thickness $\delta x$ of the bar, then by §12.7 Eqn. (*ii*), the quantity of heat in this slab = $s\rho\alpha\, \delta x\delta u$. Hence the rate of increase of heat in this slab, *i.e.*,

$s\rho\alpha\delta x \dfrac{\partial u}{\partial t} = R_1 - R_2$, where $R_1$ and $R_2$ are respectively the rate (cal./sec.) of inflow and outflow of heat.

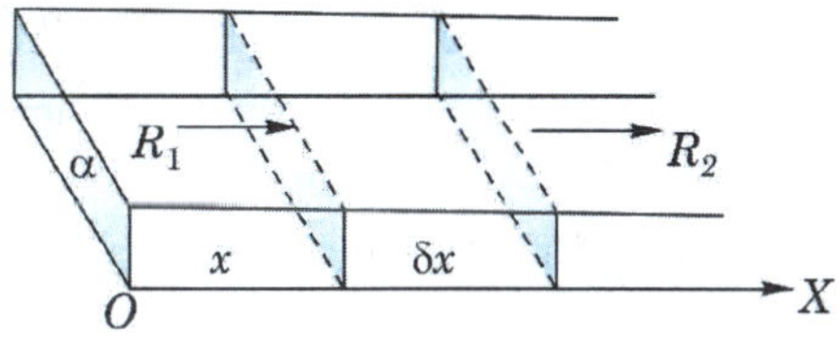

Fig. 18.2.

Now $R_1 = -k\alpha\left(\dfrac{\partial u}{\partial x}\right)_x$ and $R_2 = -k\alpha\left(\dfrac{\partial u}{\partial x}\right)_{x+\delta x}$

the negative sign appearing as a result of Eqn. (*i*) on p. 556.

Hence $s\rho\alpha\delta x \dfrac{\partial u}{\partial t} = -k\alpha\left(\dfrac{\partial u}{\partial x}\right)_x + k\alpha\left(\dfrac{\partial u}{\partial x}\right)_{x+\delta x}$ *i.e.*, $\dfrac{\partial u}{\partial t} = \dfrac{k}{s\rho}\left\{\dfrac{(\partial u/\partial x)_{x+\delta x} - (\partial u/\partial x)_x}{\delta x}\right\}$

Writing $k/s\rho = c^2$, called the *diffusivity* of the substance (cm$^2$/sec.), and taking the limit as $\delta x \to 0$, we get

$$\frac{\partial \mathbf{u}}{\partial \mathbf{t}} = \mathbf{c}^2 \frac{\partial^2 \mathbf{u}}{\partial \mathbf{x}^2} \quad ...(1)$$

This is the *one-dimensional heat-flow equation.* (*V.T.U., 2013 ; Calicut, 2012*)

**(2) Solution of the heat equation.** Assume that a solution of Eqn. (1) is of the form

$$u(x, t) = X(x) \,.\, T(t)$$

where $X$ is a function of $x$ alone and $T$ is a function of $t$ only.

Substituting this in Eqn. (1), we get

$$XT' = c^2X''T, \text{ i.e., } X''/X = T'/c^2T \quad ...(2)$$

Clearly the left side of Eqn. (2) is a function of $x$ only and the right side is a function of $t$ alone. Since $x$ and $t$ are independent variables, Eqn. (2) can hold good if each side is equal to a constant $k$ (say). Then Eqn. (2) leads to the ordinary differential equations

$$\frac{d^2X}{dx^2} - kX = 0 \quad ...(3) \qquad \text{and} \qquad \frac{dT}{dt} - kc^2T = 0 \quad ...(4)$$

Solving Eqns. (3) and (4), we get

(*i*) *When k is positive* than $k = p^2$, say :

$$X = c_1e^{px} + c_2e^{-px},\ T = c_3\, e^{c^2p^2t}\ ;$$

(*ii*) *When k is negative* than $k = -p^2$, say :

$$X = c_4 \cos px + c_5 \sin px,\ T = c_6\, e^{-c^2p^2t}\ ;$$

(*iii*) *When k is zero :*

$$X = c_7x + c_8,\ T = c_9.$$

Thus the various possible solutions of the heat-equation Eqn. (1) are

$$u = (c_1e^{px} + c_2e^{-px})\, c_3e^{c^2p^2t} \qquad ...(5)$$

$$u = (c_4 \cos px + c_5 \sin px)\, c_6e^{-c^2p^2t} \qquad ...(6)$$

$$u = (c_7x + c_8)c_9 \qquad ...(7)$$

Of these three solutions, we have to choose that solution which is consistent with the physical nature of the problem. As we are dealing with problems on heat conduction, it must be a transient solution, *i.e.*, $u$ is to decrease with the increase of time $t$. Accordingly, the solution given by Eqn. (6), *i.e.*, of the form

$$u = (C_1 \cos px + C_2 \sin px)\, e^{-c^2p^2t} \qquad ...(8)$$

is the only suitable solution of the heat equation. (*V.T.U., 2016 ; J.N.T.U., 2015*)

**Example 18.10.** *Solve the equation* $\dfrac{\partial u}{\partial t} = \dfrac{\partial^2 u}{\partial x^2}$ *with boundary conditions* $u(x, 0) = 3 \sin n\pi x$, $u(0, t) = 0$ *and* $u(1, t) = 0$, *where* $0 < x < 1$, $t > 0$. (*K.T.U., 2016*)

**Solution.** The solution of the equation $\dfrac{\partial u}{\partial t} = \dfrac{\partial^2 u}{\partial x^2}$ ...(*i*)

is $\quad u(x, t) = (c_1 \cos px + c_2 \sin px)\ e^{-p^2t}$ ...(*ii*)

When $x = 0$, $\quad u(0, t) = c_1e^{-p^2t} = 0 \quad i.e.,\quad c_1 = 0.$

$\therefore$ Eqn. (*ii*) becomes $\quad u(x, t) = c_2 \sin pxe^{-p^2t}$ ...(*iii*)

When $x = 1$, $\quad u(1, t) = c_2 \sin p \,.\, e^{-p^2t} = 0$ or $\sin p = 0$

*i.e.*, $\quad p = n\pi.$

$\therefore$ Eqn. (*iii*) reduces to $\quad u(x, t) = b_n e^{-(n\pi)^2 t} \sin n\pi x$ where $b_n = c_2$

Thus the general solution of Eqn. (*i*) is $u(x, t) = \sum b_n e^{-n^2\pi^2t} \sin n\pi x$ ...(*iv*)

When $t = 0$, $3 \sin n\pi x = u(0, t) = \sum\limits_{n=1}^{\infty} b_n e^{-n^2\pi^2t} \sin n\pi x$

Comparing both sides, $b_n = 3$

Hence from Eqn. (*iv*), the desired solution is

$$u(x, t) = 3 \sum_{n=1}^{\infty} e^{-n^2\pi^2t} \sin n\pi x.$$

**Example 18.11.** *Solve the differential equation* $\dfrac{\partial u}{\partial t} = \alpha^2 \dfrac{\partial^2 u}{\partial x^2}$ *for the conduction of heat along a rod without radiation, subject to the following conditions :*

(*i*) *u is not infinite for* $t \to \infty$, (*ii*) $\dfrac{\partial u}{\partial x} = 0$ *for* $x = 0$ *and* $x = l$,

(*iii*) $u = lx - x^2$ *for* $t = 0$, *between* $x = 0$ *and* $x = l$. (*J.N.T.U., 2015*)

**Solution.** Substituting $u = X(x)T(t)$ in the given equation, we get

$$XT' = \alpha^2X''T \quad i.e.,\quad X''/X = \frac{T'}{\alpha^2T} = -k^2 \text{ (say)}$$

$$\therefore \qquad \frac{d^2X}{dx^2} + k^2X = 0 \text{ and } \frac{dT}{dt} + k^2\alpha^2T = 0 \qquad ...(1)$$

Their solutions are $\qquad X = c_1 \cos kx + c_2 \sin kx,\ T = c_3 e^{-k^2\alpha^2 t}$ ...(2)

If $k^2$ is changed to $-k^2$, the solutions are

$$X = c_4 e^{kx} + c_5 e^{-kx},\ T = c_6 e^{k^2\alpha^2 t} \qquad ...(3)$$

If $k^2 = 0$, the solutions are $\qquad X = c_7 x + c_8,\ T = c_9$ ...(4)

In (3), $T \to \infty$ for $t \to \infty$ therefore, $u$ also $\to \infty$ *i.e.*, the given condition condition (*i*) is not satisfied. So we reject the solutions Eqn. (3) while Eqns. (2) and (4), satisfy this condition.

Applying the condition (*ii*) to (4), we get $c_7 = 0$.

$$\therefore \qquad u = XT = c_8c_9 = a_0 \quad \text{(say)} \qquad ...(5)$$

From (2), $\qquad \frac{\partial u}{\partial x} = (-c_1 \sin kx + c_2 \cos kx)\, kc_3 e^{-k^2\alpha^2 t}$

Applying the condition (*ii*), we get $c_2 = 0$ and $-c_1 \sin kl + c_2 \cos kl = 0$

*i.e.*, $\qquad c_2 = 0 \quad \text{and} \quad kl = n\pi$ ($n$ an integer)

$$\therefore \qquad u = c_1 \cos kx \,.\, c_3 e^{-k^2\alpha^2 t}$$

$$= a_n \cos\left(\frac{n\pi x}{l}\right)\frac{e^{-n^2\pi^2\alpha^2 t}}{l^2} \qquad ...(6)$$

Thus the general solution being the sum of Eqns. (5) and (6), is

$$u = a_0 + \Sigma a_n \cos(n\pi x/l)\, e^{-n^2\pi^2\alpha^2 t/l^2} \qquad ...(7)$$

Now using the condition (*iii*), we get

$$lx - x^2 = a_0 + \Sigma a_n \cos(n\pi x/l)$$

This being the expansion of $lx - x^2$ as a half-range cosine series in $(0, l)$, we get

$$a_0 = \frac{1}{l}\int_0^l (lx - x^2)\,dx = \frac{1}{l}\left|\frac{lx^2}{2} - \frac{x^3}{3}\right|_0^l = \frac{l^2}{6}$$

and

$$a_n = \frac{2}{l}\int_0^l (lx - x^2)\cos\frac{n\pi x}{l}\,dx = \frac{2}{l}\Bigg|(lx - x^2)\left(\frac{l}{n\pi}\sin\frac{n\pi x}{l}\right)$$

$$-(l - 2x)\left(-\frac{l^2}{n^2\pi^2}\cos\frac{n\pi x}{l}\right) + (-2)\left(-\frac{l^3}{n^3\pi^3}\sin\frac{n\pi x}{l}\right)\Bigg|_0^l$$

$$= \frac{2}{l}\left\{0 - \frac{l^3}{n^2\pi^2}(\cos n\pi + 1) + 0\right\} = -\frac{4l^2}{n^2\pi^2} \text{ when } n \text{ is even, otherwise } 0.$$

Hence taking $n = 2m$, the required solution is

$$u = \frac{l^2}{6} - \frac{l^2}{\pi^2}\sum_{m=1}^{\infty}\frac{1}{m^2}\cos\left(\frac{2m\pi x}{l}\right)e^{-4m^2\pi^2\alpha^2 t/l^2}.$$

**Example 18.12.** (*a*) *An insulated rod of length l has its ends A and B maintained at 0°C and 100°C respectively until steady state conditions prevail. If B is suddenly reduced to 0°C and maintained at 0°C, find the temperature at a distance x from A at time t.* *(JNTUK, 2022; K.T.U., 2016 ; J.N.T.U., 2015)*

(*b*) *Solve the above problem if the change consists of raising the temperature of A to 20°C and reducing that of B to 80°C.*

**Solution.** (*a*) Let the equation for the conduction of heat be

$$\frac{\partial u}{\partial t} = c^2\frac{\partial^2 u}{\partial x^2} \qquad ...(i)$$

Prior to the temperature change at the end $B$, when $t = 0$, the heat flow was independent of time (*steady state condition*). When $u$ depends only on $x$, Eqn. (*i*) reduces to $\partial^2 u/\partial x^2 = 0$.

Its general solution is $u = ax + b$ ...(*ii*)

Since $u = 0$ for $x = 0$ and $u = 100$ for $x = l$, therefore, Eqn. (*ii*) gives $b = 0$ and $a = 100/l$.

Thus the *initial condition* is expressed by $u(x, 0) = \dfrac{100}{l} x$ ...(*iii*)

Also the *boundary conditions* for the subsequent flow are

$u(0, t) = 0$ for all values of $t$ ...(*iv*)

and $u(l, t) = 0$ for all values of $t$ ...(*v*)

Thus we have to find a temperature function $u(x, t)$ satisfying the differential Eqn. (*i*) subject to the initial condition Eqn. (*iii*) and the boundary conditions Eqns. (*iv*) and (*v*).

Now the solution of Eqn. (*i*) is of the form

$$u(x, t) = (C_1 \cos px + C_2 \sin px)\, e^{-c^2p^2t} \quad ...(vi)$$

By Eqn. (*iv*), $u(0, t) = C_1 e^{-c^2p^2t} = 0$, for all values of $t$.

Hence $C_1 = 0$ and Eqn. (*vi*) reduces to $u(x, t) = C_2 \sin px \,.\, e^{-c^2p^2t}$ ...(*vii*)

Applying Eqn. (*v*), Eqn. (*vii*) gives $u(l, t) = C_2 \sin pl \,.\, e^{-c^2p^2t} = 0$, for all values of $t$.

This requires $\sin pl = 0$ *i.e.*, $pl = n\pi$ as $C_2 \neq 0$. $\therefore$ $p = n\pi/l$, where $n$ is any integer.

Hence Eqn. (*vii*) reduces to $u(x, t) = b_n \sin \dfrac{n\pi x}{l} \cdot e^{-c^2n^2\pi^2t/l^2}$, where $b_n = C_2$.

[These are the solutions of Eqn. (*i*) satisfying the boundary conditions Eqns. (*iv*) and (*v*). These are the **eigen functions** corresponding to the **eigen values** $\lambda_n = cn\pi/l$, of the problem.]

Adding all such solutions, the most general solution of (*i*) satisfying the boundary conditions Eqns. (*iv*) and (*v*) is

$$u(x, t) = \sum_{n=1}^{\infty} b_n \sin \frac{n\pi x}{l} \cdot e^{-c^2n^2\pi^2t/l^2} \quad ...(viii)$$

Putting $t = 0$, $$u(x, 0) = \sum_{n=1}^{\infty} b_n \sin \frac{n\pi x}{l} \quad ...(ix)$$

In order that the condition Eqn. (*iii*) may be satisfied, Eqns. (*iii*) and (*ix*) must be same. This requires the expansion of $100x/l$ as a half-range Fourier sine series in $(0, l)$. Thus

$$\frac{100x}{l} = \sum_{n=1}^{\infty} b_n \sin \frac{n\pi x}{l} \text{ where } b_n = \frac{2}{l}\int_0^l \frac{100x}{l} \cdot \sin \frac{n\pi x}{l}\, dx$$

$$= \frac{200}{l^2}\left[x\left\{-\frac{\cos(n\pi x/l)}{(n\pi/l)}\right\} - (1)\left\{-\frac{\sin(n\pi x/l)}{(n\pi/l)^2}\right\}\right]_0^l = \frac{200}{l^2}\left(-\frac{l^2}{n\pi}\cos n\pi\right) = \frac{200}{n\pi}(-1)^{n+1}$$

Hence Eqn. (*viii*) gives $u(x, t) = \dfrac{200}{\pi}\displaystyle\sum_{n=1}^{\infty} \frac{(-1)^{n+1}}{n} \sin \frac{n\pi x}{l} \cdot e^{-(cn\pi/l)^2 t}$

(*b*) Here the initial condition remains the same as Eqn. (*iii*) above, and the boundary conditions are

$u(0, t) = 20$ for all values of $t$ ...(*x*)

$u(l, t) = 80$ for all values of $t$ ...(*xi*)

In part (*a*), the boundary values (*i.e.*, the temperature at the ends) being zero, we were able to find the desired solution easily. Now the boundary values being non-zero, we have to modify the procedure.

We split up the temperature function $u(x, t)$ into two parts as

$$u(x, t) = u_s(x) + u_t(x, t) \quad ...(xii)$$

where $u_s(x)$ is a solution of Eqn. (*i*) involving $x$ only and satisfying the boundary conditions Eqns. (*x*) and (*xi*) ; $u_t(x, t)$ is then a function defined by Eqn. (*xii*). Thus $u_s(x)$ is a steady state solution of the form Eqn. (*ii*) and $u_t(x, t)$ may be regarded as a transient part of the solution which decreases with increase of $t$.

Since $u_s(0) = 20$ and $u_s(l) = 80$, therefore, using Eqn. (*ii*) we get

$$u_s(x) = 20 + (60/l)x \qquad \text{...(xiii)}$$

Putting $x = 0$ in Eqn. (*xii*), we have by (*x*),

$$u_t(0, t) = u(0, t) - u_s(0) = 20 - 20 = 0 \qquad \text{...(xiv)}$$

Putting $x = l$ in Eqn. (*xii*), we have by Eqn. (*xi*),

$$u_t(l, t) = u(l, t) - u_s(l) = 80 - 80 = 0 \qquad \text{...(xv)}$$

Also $\quad u_t(x, 0) = u(x, 0) - u_s(x) = \dfrac{100x}{l} - \left(\dfrac{60x}{l} + 20\right)$ [by Eqns. (*iii*) and (*xiii*)

$$= \frac{40x}{l} - 20 \qquad \text{...(xvi)}$$

Hence Eqns. (*xiv*) and (*xv*) give the boundary conditions and Eqn. (*xvi*) gives the initial condition relative to the transient solution. Since the boundary values given by Eqns. (*xiv*) and (*xv*) are both zero, therefore, as in part (*a*), we have $u_t(x, t) = (C_1 \cos px + C_2 \sin px)\, e^{-c^2p^2t}$

By Eqn. (*xiv*), $\quad u_t(0, t) = C_1 e^{-c^2p^2t} = 0$, for all values of $t$.

Hence $C_1 = 0$ and $\quad u_t(x, t) = C_2 \sin px \,.\, e^{-c^2p^2t} \qquad$ ...(*xvii*)

Applying Eqn. (*xv*), it gives $u_t(l, t) = C_2 \sin\, pl e^{-c^2p^2t} = 0$ for all values of $t$.

This requires $\quad \sin pl = 0$, *i.e.* $pl = n\pi$ as $C_2 \neq 0$. $p = n\pi/l$, when $n$ is any integer.

Hence Eqn. (*xvii*) reduces to $u_t(x, t) = b_n \sin \dfrac{n\pi x}{l} e^{-c^2n^2\pi^2t/l^2}$ where $b_n = C_2$.

Adding all such solutions, the most general solution of Eqn. (*xvii*) satisfying the boundary conditions Eqn. (*xiv*) and Eqn. (*xv*) is

$$u_t(x, t) = \sum_{n=1}^{\infty} b_n \sin \frac{n\pi x}{l} e^{-c^2n^2\pi^2t/l^2} \qquad \text{...(xviii)}$$

Putting $t = 0$, we have $\quad u_t(x, 0) = \displaystyle\sum_{n=1}^{\infty} b_n \sin \frac{n\pi x}{l} \qquad$ ...(*xix*)

In order that the condition Eqn. (*xvi*) may be satisfied, Eqns. (*xvi*) and (*xix*) must be same. This requires the expansion of $(40/l)\,x - 20$ as a half-range Fourier sine series in $(0, l)$. Thus

$$\frac{40x}{l} - 20 = \sum_{n=1}^{\infty} b_n \sin \frac{n\pi x}{l} \quad \text{where} \quad b_n = \frac{2}{l}\int_0^l \left(\frac{40x}{l} - 20\right) \sin \frac{n\pi x}{l}\, dx = -\frac{40}{n\pi}(1 + \cos n\pi)$$

*i.e.*, $b_n = 0$, when $n$ is odd ; $= -80/n\pi$, when $n$ is even

Hence Eqn. (*xviii*) becomes $u_t(x, t) = \displaystyle\sum_{n=2,4\ldots}^{\infty} \left(\frac{-80}{n\pi}\right) \sin \frac{n\pi x}{l} \cdot e^{-c^2n^2\pi^2t/l^2}$ [Take $n = 2m$]

$$= -\frac{40}{\pi} \sum_{m=1}^{\infty} \frac{1}{m} \sin \frac{2m\pi x}{l} \cdot e^{-4c^2m^2\pi^2t/l^2} \qquad \text{...(xx)}$$

Finally combining Eqns. (*xiii*) and (*xx*), the required solution is

$$u(x, t) = \frac{40x}{l} + 20 - \frac{40}{\pi} \sum_{m=1}^{\infty} \frac{1}{m} \sin \frac{2m\pi x}{l} \cdot e^{-4c^2m^2\pi^2t/l^2}.$$

**Example 18.13.** *The ends A and B of a rod 20 cm long have the temperature at 30°C and 80°C until steady-state prevails. The temperature of the ends are changed to 40°C and 60°C respectively. Find the temperature distribution in the rod at time t.*

**Solution.** Let the heat equation be $\dfrac{\partial u}{\partial t} = c^2 \dfrac{\partial^2 u}{\partial x^2} \qquad$ ...(*i*)

In steady state condition, $u$ is independent of time and depends on $x$ only, (*i*) reduces to

$$\partial^2 u / \partial x^2 = 0. \qquad \text{...(ii)}$$

Its solution is $u = a + bx$

Since $u = 30$ for $x = 0$ and $u = 80$ for $x = 20$, therefore $a = 30$, $b = (80 - 30)/20 = 5/2$

Thus the initial conditions are expressed by

$$u\,(x, 0) = 30 + \frac{5}{2}x \qquad ...(iii)$$

The boundary conditions are $u\,(0, t) = 40$, $u\,(20, t) = 60$

Using Eqn. $(ii)$, the steady state temperature is

$$u\,(x, 0) = 40 + \frac{60 - 40}{20}\,x = 40 + x \qquad ...(iv)$$

To find the temperature $u$ in the intermediate period,

$$u\,(x, t) = u_s\,(x) + u_t\,(x, t)$$

where $u_s\,(x)$ is the steady state temperature distribution of the form Eqn. $(iv)$ and $u_t\,(x, t)$ is the transient temperature distribution which decreases to zero as $t$ increases.

Since $u_t\,(x, t)$ satisfies one dimensional heat equation

$$\therefore \quad u(x, t) = 40 + x + \sum_{n=1}^{\infty} (a_n \cos px + b_n \sin px)\, e^{-p^2 t} \qquad ...(v)$$

$$u(0, t) = 40 = 40 + \sum_{n=1}^{\infty} a_n e^{-p^2 t} \quad \text{where } a_n = 0.$$

$$\therefore \text{ Eqn. } (v) \text{ reduces to } u\,(x, t) = 40 + x + \sum_{n=1}^{\infty} b_n \sin pxe^{-p^2 t} \qquad ...(vi)$$

Also $$u\,(20, t) = 60 = 40 + 20 + \sum_{n=1}^{\infty} b_n \sin 20\, pe^{-p^2 t}$$

or $$\sum_{n=1}^{\infty} b_n \sin 20\, pe^{-p^2 t} = 0 \; i.e., \sin 20p = 0 \; i.e., p = n\pi/20$$

Thus Eqn. $(vi)$ becomes $$u\,(x, t) = 40 + x + \sum_{n=1}^{\infty} b_n \sin \frac{n\pi x}{20}\, e^{-n\pi t/20} \qquad ...(vii)$$

Using Eqn. $(iii)$, $$30 + \frac{5}{2}x = u\,(0, t) = 40 + x + \sum_{n=1}^{\infty} b_n \sin \frac{n\pi x}{20}$$

or $$\frac{3x}{2} - 10 = \sum_{n=1}^{\infty} b_n \sin \frac{n\pi x}{20}$$

where $b_n = \dfrac{2}{20}\displaystyle\int_0^{20} \left(\frac{3x}{2} - 10\right) \sin \frac{n\pi x}{20}\, dx = -\frac{20}{n\pi}\,(1 + 2\cos n\pi)$

Hence from Eqn. $(vii)$, the desired solution is

$$u = 40 + x - \frac{20}{\pi} \sum_{n=1}^{\infty} \frac{1 + 2\cos n\pi}{n} \sin \frac{n\pi x}{20}\, e^{-(n\pi/20)^2 t}.$$

**Example 18.14. Bar with insulated ends.** *A bar 100 cm long, with insulated sides, has its ends kept at 0°C and 100°C until steady state conditions prevail. The two ends are then suddenly insulated and kept so. Find the temperature distribution.* *(D.T.U., 2011)*

**Solution.** The temperature $u\,(x, t)$ along the bar satisfies the equation

$$\frac{\partial u}{\partial t} = c^2 \frac{\partial^2 u}{\partial x^2} \qquad ...(i)$$

By law of heat conduction, the rate of heat flow is proportional to the gradient of the temperature. Thus, if the ends $x = 0$ and $x = l$ (= 100 cm) of the bar are insulated (Fig. 18.4) so that no heat can flow through the ends, the boundary conditions are

$$\frac{\partial u\,(0, t)}{\partial x} = 0, \frac{\partial u\,(l, t)}{\partial x} = 0 \text{ for all } t \qquad ...(ii)$$

Initially, under steady state conditions, $\dfrac{\partial^2 u}{\partial x^2} = 0$. Its solution is $u = ax + b$.

Since $u = 0$ for $x = 0$ and $u = 100$ for $x = l$ $\quad\therefore\quad b = 0$ and $a = 1$.

Thus the initial condition is $u\,(x, 0) = x \quad 0 < x < l$. ...(iii)

Now the solution of Eqn. (*i*) is of the form $u\,(x, t) = (c_1 \cos px + c_2 \sin px)\, e^{-c^2p^2t}$ ...(iv)

Differentiating partially w.r.t. $x$, we get

$$\frac{\partial u}{\partial x} = (-c_1 p \sin px + c_2 p \cos px)\, e^{-c^2p^2t} \qquad ...(v)$$

Putting $x = 0$, $\quad \left(\dfrac{\partial u}{\partial x}\right)_0 = c_2 p e^{-c^2p^2t} = 0$ for all $t$. [By (*ii*)]

$\therefore \quad c_2 = 0$

Putting $x = l$ in (*v*), $\quad \left(\dfrac{\partial u}{\partial x}\right)_l = -c_1 p \sin pl e^{-c^2p^2t}$ for all $t$. [By (*ii*)]

$\therefore \quad c_1 p \sin pl = 0$ *i.e.*, $p$ being $\neq 0$, either $c_1 = 0$ or $\sin pl = 0$.

When $c_1 = 0$, Eqn. (*iv*) gives $u\,(x, t) = 0$ which is a trivial solution, therefore $\sin pl = 0$.

or $\quad pl = n\pi \quad$ or $\quad p = n\pi/l, \quad n = 0, 1, 2, ....$

Hence Eqn. (*iv*) becomes $u\,(x, t) = c_1 \cos \dfrac{n\pi x}{l} e^{-c^2n^2\pi^2t/l^2}$.

$\therefore$ the most general solution of Eqn. (*i*) satisfying the boundary conditions Eqn. (*ii*) is

$$u(x, t) = \sum_{n=0}^{\infty} A_n \cos \frac{n\pi x}{l}\, e^{-c^2n^2\pi^2t/l^2} = A_0 + \sum_{n=1}^{\infty} A_n \cos \frac{n\pi x}{l}\, e^{-c^2n^2\pi^2t/l^2} \quad (\text{where } A_n = c_1) \quad ...(vi)$$

Putting, $t = 0$, $u\,(x, 0) - A_0 + \displaystyle\sum_{n=1}^{\infty} A_n \cos \frac{n\pi x}{l} = x$ [by Eqn. (*iii*)]

This requires the expansion of $x$ into a half range cosine series in $(0, l)$.

Thus $\quad x = \dfrac{a_0}{2} + \displaystyle\sum_{n=1}^{\infty} a_n \cos n\pi x/l \quad$ where $a_0 = \dfrac{2}{l}\displaystyle\int_0^l x dx = l$

and

$$a_n = \frac{2}{l}\int_0^l x \cos \frac{n\pi x}{l} dx = \frac{2l}{n^2\pi^2}(\cos n\pi - 1)$$

$= 0$, when $n$ is even ; $= -\,4l/n^2\pi^2$, when $n$ is odd.

$\therefore \quad A_0 = \dfrac{a_0}{2} = l/2$, and $A_n = a_n = 0$ for $n$ even ; $= -\,4l/n^2\pi^2$ for $n$ odd.

Hence (*vi*) takes the form

$$u(x, t) = \frac{l}{2} + \sum_{n=1,3,...}^{\infty} \frac{4l}{n^2\pi^2} \cos \frac{n\pi x}{l}\, e^{-c^2n^2\pi^2t/l^2}$$

$$= \frac{l}{2} - \frac{4l}{\pi^2}\sum_{1}^{\infty} \frac{1}{(2n-1)^2} \cos \frac{(2n-1)\pi x}{l}\, e^{-c^2(2n-1)^2\pi^2t/l^2} \qquad ...(vii)$$

This is the required temperature at a point $P_1$ distant $x$ from end $A$ at any time $t$.

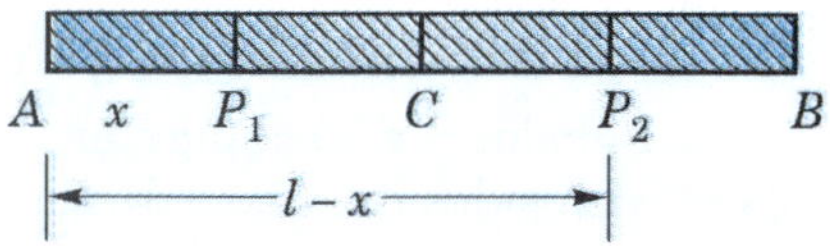


Fig. 18.3.

**Obs.** *The sum of the temperatures at any two points equidistant from the centre is always 100°C, a constant.*

Let $P_1$, $P_2$ be two points equidistant from the centre $C$ of the bar so that $CP_1 = CP_2$ (Fig. 18.3).

If $\quad AP_1 = BP_2 = x$ (say), then $AP_2 = l - x$.

$\therefore$ Replacing $x$ by $l - x$ in (*vii*), we get the temperature at $P_2$ as

$$u(l-x,t) = \frac{l}{2} - \frac{4l}{\pi^2}\sum_1^\infty \frac{1}{(2n-1)^2}\cos\frac{(2n-1)\pi(l-x)}{l}\, e^{\frac{-c^2(2n-1)^2\pi^2 t}{l^2}}$$

$$= \frac{l}{2} + \frac{4l}{\pi^2}\sum_1^\infty \frac{1}{(2n-1)^2}\cos\frac{(2n-1)\pi x}{l}\, e^{\frac{-c^2(2n-1)^2\pi^2 t}{l^2}} \qquad ...(viii)$$

$$\left\{\because \cos\frac{(2n-1)\pi(l-x)}{l} = \cos\left[2n\pi - \pi - \frac{(2n-1)\pi x}{l}\right] = -\cos\frac{(2n-1)\pi x}{l}\right.$$

Adding Eqns. (*vii*) and (*viii*), we get $u(x,t) + u(l-x,t) = l = 100°C$.

## PROBLEMS 18.3

1. Solve $\frac{\partial u}{\partial t} = a^2\frac{\partial^2 u}{\partial x^2}$ subject to the conditions $u(0,t) = 0, = u(l,t), t > 0$;

$$u(x,0) = \begin{cases} x, 0 < x < l/2 \\ l-x, l/2 \le x \le l. \end{cases}$$
(*K.T.U., 2016 ; Anna, 2015*)

2. A homogeneous rod of conducting material of length 100 cm has its ends kept at zero temperature and the temperature initially is

$$u(x,0) = x, \qquad 0 \le x \le 50$$
$$= 100 - x, \qquad 50 \le x \le 100.$$

Find the temperature $u(x,t)$ at any time.

3. A bar of 10 cm long, with insulated sides has its ends $A$ and $B$ maintained at temperatures 50°C and 100°C respectively, until steady-state conditions prevail. The temperature $A$ is suddenly raised to 90°C and at the same time that at $B$ is lowered to 60°C. Find the temperature distribution in the bar at time $t$. (*P.T.U., 2010*)

Show that the temperature at the middle point of the bar remains unaltered for all time, regardless of the material of the bar.

4. The temperatures at one end of a bar, 50 cm long with insulated sides, is kept at 0°C and that the other end is kept at 100°C until steady-state conditions prevail. The two ends are then suddenly insulated, so that the temperature gradient is zero at each end thereafter. Find the temperature distribution.

5. Find the solution of $\frac{\partial V}{\partial t} = k\frac{\partial^2 V}{\partial x^2}$ having given that $V = V_0 \sin nt$ when $x = 0$ for all values of $t$ and $V = 0$ when $x$ is very large.

## 18.6 TWO-DIMENSIONAL HEAT FLOW

Consider the flow of heat in a metal plate of uniform thickness α (cm), density ρ (gr/cm$^3$), specific heat $s$ (cal/gr deg) and thermal conductivity $k$ (cal/cm sec deg). Let $XOY$ plane be taken in one face of the plate (Fig. 18.4). If the temperature at any point is independent of the $z$-coordinate and depends only on $x, y$ and time $t$, then the flow is said to be two-dimensional. In this case, the heat flow is in the $XY$-plane only and is zero along the normal to the $XY$-plane.

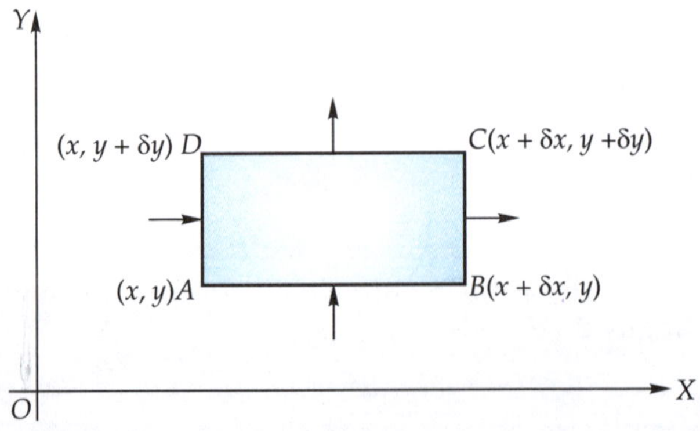

Fig. 18.4.

Consider a rectangular element $ABCD$ of the plane with sides $\delta x$ and $\delta y$. By the amount of heat entering the element in 1 sec. from the side $AB$

$$= -k\alpha\delta x\left(\frac{\partial u}{\partial y}\right)_y$$

and the amount of heat entering the element in 1 second from the side $AD = -k\alpha\delta y\left(\frac{\partial u}{\partial x}\right)_x$

The quantity of heat flowing out through the side $CD$ per sec. $= -k\alpha\delta x\left(\frac{\partial u}{\partial y}\right)_{y+\delta y}$

and the quantity of heat flowing out through the side $BC$ per second $= -k\alpha\delta y\left(\frac{\partial u}{\partial x}\right)_{x+\delta x}$

Hence the total gain of heat by the rectangular element $ABCD$ per second

$$= -k\alpha\delta x\left(\frac{\partial u}{\partial y}\right)_y - k\alpha\delta y\left(\frac{\partial u}{\partial x}\right)_x + k\alpha\delta x\left(\frac{\partial u}{\partial y}\right)_{y+\delta y} + k\alpha\delta y\left(\frac{\partial u}{\partial x}\right)_{x+\delta x}$$

$$= k\alpha\delta x\left[\left(\frac{\partial u}{\partial y}\right)_{y+\delta y} - \left(\frac{\partial u}{\partial y}\right)_y\right] + k\alpha\delta y\left[\left(\frac{\partial u}{\partial x}\right)_{x+\delta x} - \left(\frac{\partial u}{\partial x}\right)_x\right]$$

$$= k\alpha\delta x\delta y\left[\frac{\left(\frac{\partial u}{\partial x}\right)_{x+\delta x} - \left(\frac{\partial u}{\partial x}\right)_x}{\partial x} + \frac{\left(\frac{\partial u}{\partial y}\right)_{y+\delta y} - \left(\frac{\partial u}{\partial y}\right)_y}{\partial y}\right] \quad \text{...(1)}$$

Also the rate of gain of heat by the element

$$= \rho\delta x\delta y\alpha s\,\frac{\partial u}{\partial t} \quad \text{...(2)}$$

Thus equating Eqns. (1) and (2),

$$k\alpha\delta x\delta y\left[\frac{\left(\frac{\partial u}{\partial x}\right)_{x+\delta x} - \left(\frac{\partial u}{\partial x}\right)_x}{\delta x} + \frac{\left(\frac{\partial u}{\partial y}\right)_{y+\delta y} - \left(\frac{\partial u}{\partial y}\right)_y}{\delta y}\right] = \rho\delta x\delta y\alpha s\,\frac{\partial u}{\partial t}.$$

Dividing both sides by $\alpha\delta x\delta y$ and taking limits as $\delta x \to 0$, $\delta y \to 0$, we get

$$k\left(\frac{\partial^2 u}{\partial x^2} + \frac{\partial^2 u}{\partial y^2}\right) = \rho s\frac{\partial u}{\partial t}$$

*i.e.*,
$$\frac{\partial \mathbf{u}}{\partial \mathbf{t}} = \mathbf{c}^2\left(\frac{\partial^2 \mathbf{u}}{\partial \mathbf{x}^2} + \frac{\partial^2 \mathbf{u}}{\partial \mathbf{y}^2}\right)$$

where $c^2 = k/\rho s$ is the *diffusivity*. ...(3)

Hence the Eqn. (3) gives the temperature distribution of the plane in the *transient state*.

**Cor.** In the *steady state*, $u$ is independent of $t$, so that $\partial u/\partial t = 0$ and the above equation reduces to,

$$\frac{\partial^2 \mathbf{u}}{\partial \mathbf{x}^2} + \frac{\partial^2 \mathbf{u}}{\partial \mathbf{y}^2} = \mathbf{0}$$

which is the well known **Laplace's equation** *in two dimensions.*

**Obs.** When the stream lines are curves in space, *i.e.*, the heat flow is three dimensional, we shall similarly arrive at the equation

$$\frac{\partial u}{\partial t} = c^2\left(\frac{\partial^2 u}{\partial x^2} + \frac{\partial^2 u}{\partial y^2} + \frac{\partial^2 u}{\partial z^2}\right)$$

In a *steady state*, it reduces to $\frac{\partial^2 \mathbf{u}}{\partial \mathbf{x}^2} + \frac{\partial^2 \mathbf{u}}{\partial \mathbf{y}^2} + \frac{\partial^2 \mathbf{u}}{\partial \mathbf{z}^2} = \mathbf{0}$

which is the *three dimensional* **Laplace's equation.**

## 18.7 SOLUTION OF LAPLACE'S EQUATION

$$\frac{\partial^2 \mathbf{u}}{\partial \mathbf{x}^2} + \frac{\partial^2 \mathbf{u}}{\partial \mathbf{y}^2} = \mathbf{0} \quad \text{...(1)}$$

Let $u = X(x)Y(y)$ be a solution of Eqn. (1).

Substituting it in Eqn. (1), we get $\frac{d^2X}{dx^2}Y + X\frac{\partial^2 Y}{\partial y^2} = 0$

or separating the variables, $\frac{1}{X}\frac{d^2X}{dx^2} = -\frac{1}{Y}\frac{d^2Y}{dy^2}$ ...(2)

Since $x$ and $y$ are independent variables, Eqn. (2) can hold good only if each side of Eqn. (2) is equal to a constant $k$ (say). Then Eqn. (2) leads to the ordinary differential equations

$$\frac{d^2X}{dx^2} - kX = 0$$

and $$\frac{d^2Y}{dy^2} + kY = 0.$$

Solving these equations, we get

(*i*) *When k is positive* and is equal to $p^2$, say

$$X = c_1e^{px} + c_2e^{-px}, \qquad Y = c_3 \cos py + c_4 \sin py$$

(*ii*) *When k is negative*, and is equal to $-p^2$, say

$$X = c_5 \cos px + c_6 \sin px, \qquad Y = c_7e^{py} + c_8e^{-py}$$

(*iii*) *When k is zero* ; $X = c_9x + c_{10}$, $Y = c_{11}y + c_{12}$.

Thus the various possible solutions of Eqn. (1) are

$$u = (c_1e^{px} + c_2e^{-px})(c_3 \cos py + c_4 \sin py) \quad ...(3)$$

$$u = (c_5 \cos px + c_6 \sin px)(c_7e^{py} + c_8e^{-py}) \quad ...(4)$$

$$u = (c_9x + c_{10})(c_{11}y + c_{12}) \quad ...(5)$$

Of these we take that solution which is consistent with the given boundary conditions.

(*V.T.U., 2016 ; Hazaribag, 2015*)

## Temperature distribution in long plates

**Example 18.15.** *An infinitely long plane uniform plate is bounded by two parallel edges and an end at right angles to them. The breadth is π ; this end is maintained at a temperature $u_0$ at all points and other edges are at zero temperature. Determine the temperature at any point of the plate in the steady-state.*

**Solution.** In the steady state (Fig. 18.5), the temperature $u(x, y)$ at any point $P(x, y)$ satisfies the equation

$$\frac{\partial^2 u}{\partial x^2} + \frac{\partial^2 u}{\partial y^2} = 0 \quad ...(i)$$

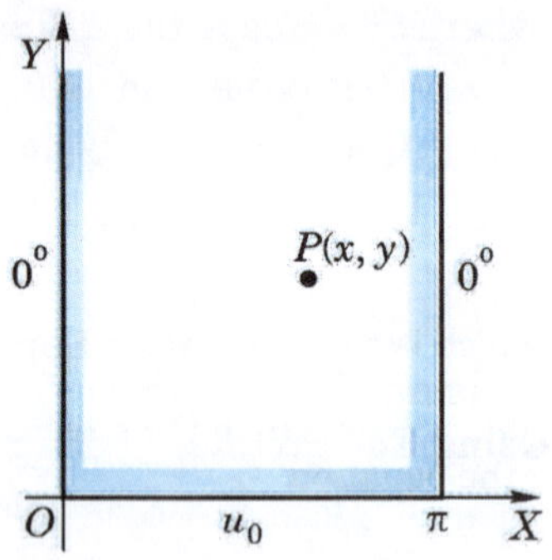

Fig. 18.5.

The boundary conditions are $u(0, y) = 0$ for all values of $y$ ...(*ii*)

$u(\pi, y) = 0$ for all values of $y$ ...(*iii*)

$u(x, \infty) = 0$ in $0 < x < \pi$ ...(*iv*)

$u(x, 0) = u_0$ in $0 < x < \pi$ ...(*v*)

Now the three possible solutions of (*i*) are

$$u = (c_1e^{px} + c_2e^{-px})(c_3 \cos py + c_4 \sin py) \quad ...(vi)$$

$$u = (c_5 \cos px + c_6 \sin px)(c_7 e^{py} + c_8e^{-py}) \quad ...(vii)$$

$$u = (c_9x + c_{10})(c_{11}y + c_{12}) \quad ...(viii)$$

Of these, we have to choose that solution which is consistent with the physical nature of the problem. The solution Eqn. (*vi*) cannot satisfy the condition Eqn. (*ii*) for $u \neq 0$ for $x = 0$, for all values of $y$. The solution Eqn. (*viii*) cannot satisfy the condition Eqn. (*iv*). Thus the only possible solution is Eqn. (*vii*), *i.e.* of the form

$$u(x, y) = (C_1 \cos px + C_2 \sin px)(C_3 e^{py} + C_4 e^{-py}) \quad ...(ix)$$

By Eqn. (*ii*), $u(0, y) = C_1(C_3e^{py} + C_4e^{-py}) = 0$ for all $y$.

Hence $C_1 = 0$ and (*ix*) reduces to

$$u(x, y) = C_2 \sin px\,(C_3e^{py} + C_4e^{-py}) \quad ...(x)$$

By Eqn. (*iii*), $u(\pi, y) = C_2 \sin p\pi(C_3e^{py} + C_4e^{-py}) = 0$, for all $y$.

This requires $\sin p\pi = 0$, *i.e.* $p\pi = n\pi$ as $C_2 \neq 0$. $\therefore$ $p = n$, an integer.

Also to satisfy the condition Eqn. (*iv*), *i.e.*, $u = 0$ as $y \to \infty$, $C_3 = 0$.

Hence Eqn. $(x)$ takes the form $u(x, y) = b_n \sin nx \cdot e^{-ny}$, where $b_n = C_2C_4$.

∴ the most general solution satisfying Eqns. $(ii)$, $(iii)$ and $(iv)$ is of the form

$$u(x, y) = \sum_{n=1}^{\infty} b_n \sin nx \cdot e^{-ny} \quad ...(xi)$$

Putting $y = 0$,
$$u(x, 0) = \sum_{n=1}^{\infty} b_n \sin nx \quad ...(xii)$$

In order that the condition Eqn. $(v)$ may be satisfied, Eqns. $(v)$ and $(xii)$ must be same. This requires th expansion of $u$ as a half-range Fourier sine series in $(0, \pi)$. Thus

$$u = \sum_{n=1}^{\infty} b_n \sin nx \text{ where } b_n = \frac{2}{\pi}\int_0^{\pi} u_0 \sin nx\, dx = \frac{2u_0}{n\pi}[1 - (-1)^n]$$

*i.e.*, $b_n = 0$, if $n$ is even ; $= 4u_0/n\pi$, if $n$ is odd.

Hence Eqn. $(xi)$ becomes $u(x, y) = \dfrac{4u_0}{\pi}\left[e^{-y}\sin x + \dfrac{1}{3}e^{-3y}\sin 3x + \dfrac{1}{5}e^{-5y}\sin 5x + ...\right]$.

## Temperature distribution in finite plates

**Example 18.16.** *Solve the Laplace equation* $\dfrac{\partial^2 u}{\partial x^2} + \dfrac{\partial^2 u}{\partial y^2} = 0$ *subject to the conditions* $u(0, y) = u(l, y) = u(x, 0) = 0$ *and* $u(x, a) = \sin n\pi x/l$. *(U.P.T.U., 2012 ; V.T.U., 2011)*

**Solution.** The three possible solutions of

$$\frac{\partial^2 u}{\partial x^2} + \frac{\partial^2 u}{\partial y^2} = 0 \quad ...(i)$$

are
$$u = (c_1e^{px} + c_2e^{-px})(c_3 \cos py + c_4 \sin py) \quad ...(ii)$$
$$u = (c_5 \cos px + c_6 \sin px)(c_7e^{py} + c_8e^{-py}) \quad ...(iii)$$
$$u = (c_9 x + c_{10})(c_{11}y + c_{12}) \quad ...(iv)$$

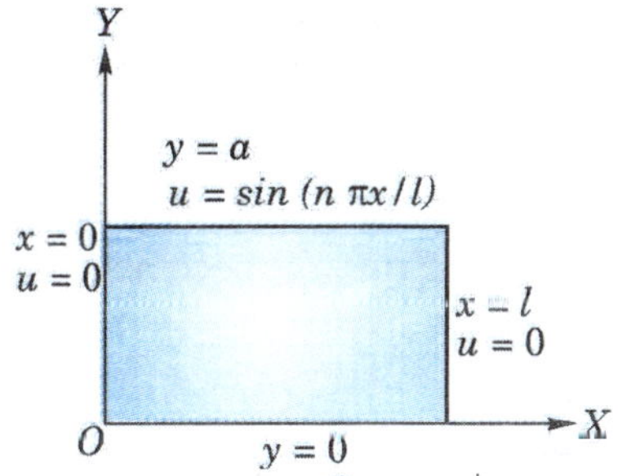

Fig. 18.6.

We have to solve Eqn. $(i)$ satisfying the following boundary conditions

$$u(0, y) = 0 \quad ...(v) \qquad u(l, y) = 0 \quad ...(vi)$$
$$u(x, 0) = 0 \quad ...(vii) \qquad u(x, a) = \sin n\pi x/l \quad ...(viii)$$

Using Eqns. $(v)$ and $(vi)$ in Eqn. $(ii)$, we get

$$c_1 + c_2 = 0, \quad \text{and} \quad c_1e^{pl} + c_2e^{-pl} = 0$$

Solving these equations, we get $c_1 = c_2 = 0$ which lead to trivial solution. Similarly, we get a trivial solution by using Eqns. $(v)$ and $(vi)$ in Eqn. $(iv)$. Hence the suitable solution for the present problem is solution Eqn. $(iii)$.

Using Eqn. $(v)$ in Eqn. $(iii)$, we have $c_5(c_7e^{py} + c_8e^{-py}) = 0$ *i.e.*, $c_5 = 0$

∴ Eqn. $(iii)$ becomes $\quad u = c_6 \sin px(c_7e^{py} + c_8e^{-py}) \quad ...(ix)$

Using Eqn. $(vi)$, we have $c_6 \sin pl\,(c_7e^{py} + c_8e^{-py}) = 0$

∴ either $c_6 = 0$ or $\sin pl = 0$

If we take $c_6 = 0$, we get a trivial solution.

Thus $\sin pl = 0$ whence $pl = n\pi$ or $p = n\pi/l$ where $n = 0, 1, 2, ......$

∴ Eqn. $(ix)$ becomes $u = c_6 \sin(n\pi x/l)(c_7e^{n\pi y/l} + c_8e^{-n\pi y/l}) \quad ...(x)$

Using Eqn. $(vii)$, we have $0 = c_6 \sin n\pi x/l \cdot (c_7 + c_8)$ *i.e.*, $c_8 = -c_7$.

Thus the solution suitable for this problem is

$$u(x, y) = b_n \sin\frac{n\pi x}{l}(e^{n\pi y/l} - e^{-n\pi y/l}) \quad \text{where} \quad b_n = c_6c_7$$

Now using the condition Eqn. $(viii)$, we have

$$u(x, a) = \sin\frac{n\pi x}{l} = b_n \sin\frac{n\pi x}{l}(e^{n\pi a/l} - e^{-n\pi a/l}),$$

we get $$b_n = \frac{1}{(e^{n\pi a/l} - e^{-n\pi a/l})}$$

Hence the required solution is

$$u(x, y) = \frac{e^{n\pi y/l} - e^{-n\pi y/l}}{e^{n\pi a/l} - e^{-n\pi a/l}} \sin \frac{n\pi x}{l} = \frac{\sinh (n\pi y/l)}{\sinh (n\pi a/l)} \sin \frac{n\pi x}{l}.$$

**Example 18.17.** *The function v (x, y) satisfies the Laplace's equation in rectangular coordinates (x, y) and for points within the rectangle x = 0, x = a, y = 0, y = b, it satisfies the conditions v (0, y) = v (a, y) = v (x, b) = 0 and v (x, 0) = x(a – x), 0 < x < a. Show that v (x, y) is given by*

$$v(x, y) = \frac{8a^2}{\pi^3} \sum_{n=0}^{\infty} \frac{\sin (2n+1)\, \pi x/a}{(2n+1)^3} \frac{\sinh (2n+1)\, \pi(b-y)/a}{\sinh (2n+1)\, \pi b/a}$$

**Solution.** The only possible solution of

$$\frac{\partial^2 v}{\partial x^2} + \frac{\partial^2 v}{\partial y^2} = 0 \qquad \text{...(i)}$$

is of the form $$v(x, y) = (c_1 \cos px + c_2 \sin px)(c_3 e^{py} + c_4 e^{-py}) \qquad \text{...(ii)}$$

The boundary conditions are

$$v(0, y) = 0\ ; \quad v(a, y) = 0 \qquad \text{...(iii)}$$
$$v(x, b) = 0 \qquad \text{...(iv)}$$
$$v(x, 0) = x(a - x),\ 0 < x < a. \qquad \text{...(v)}$$

Using Eqn. (*iii*) $v(0, y) = c_1(c_3 e^{py} + c_4 e^{-py}) = 0$ *i.e.*, $c_1 = 0$.

∴ Eqn. (*ii*) becomes $$v(x, y) = c_2 \sin px\,(c_3 e^{py} + c_4 e^{-py}) \qquad \text{...(vi)}$$

Again using Eqn. (*iii*), $v(a, y) = c_2 \sin pa\,(c_3 e^{py} + c_4 e^{-py}) = 0$.

*i.e.*, $\sin pa = 0$, *i.e.* $pa = n\pi$ or $p = n\pi/a$

∴ Eqn. (*vi*) becomes $$v(x, y) = c_2 \sin \frac{n\pi x}{a}\left(c_3 e^{\frac{n\pi y}{a}} + c_4 e^{-\frac{n\pi y}{a}}\right)$$

or $$v(x, y) = \sin \frac{n\pi x}{a}\,(Ae^{n\pi y/a} + Be^{-n\pi y/a}) \quad \text{where } A = c_2c_3,\ B = c_2c_4 \qquad \text{...(vii)}$$

Now using Eqn. (*iv*), $$v(x, b) = \sin \frac{n\pi x}{a}\left(Ae^{\frac{n\pi b}{a}} + Be^{-\frac{n\pi b}{a}}\right) = 0$$

*i.e.*, $$Ae^{n\pi b/a} + Be^{-n\pi b/a} = 0 \quad \text{or} \quad Ae^{n\pi b/a} - Be^{-n\pi b/a} = -\frac{1}{2} b_n \text{ (say)}$$

Thus eqn. (*vii*) becomes $$v(x, y) = \sin \frac{n\pi x}{a} \cdot \frac{1}{2} b_n \left\{e^{n\pi(b-y)/a} - e^{-n\pi(b-y)/a}\right\}$$

$$= b_n \sin \frac{n\pi x}{a} \sinh \frac{n\pi(b-y)}{a}$$

∴ the most general solution of Eqn. (*i*) is

$$v(x, y) = \sum_{n=1}^{\infty} b_n \sin \frac{n\pi x}{a} \sinh \frac{n\pi(b-y)}{a} \qquad \text{...(viii)}$$

Using the condition Eqn. (*v*), we have

$$x(a-x) = v(x, 0) = \sum_{n=1}^{\infty} b_n \sinh \frac{n\pi b}{a} \sin \frac{n\pi x}{a}$$

where $$b_n \sinh \frac{n\pi b}{a} = \frac{2}{a}\int_0^a x(a-x) \sin \frac{n\pi x}{a}\, dx$$

$$= \frac{2}{a}\left| (ax - x^2)\left(\frac{-\cos n\pi x/a}{n\pi/a}\right) - (a-2x)\left(-\frac{\sin n\pi x/a}{(n\pi/a)^2}\right) + (-2)\left\{\frac{\cos n\pi x/a}{(n\pi/a)^3}\right\}\right|_0^a$$

$$= 0 - 0 + \frac{4a^2}{n^3\pi^3}(1 - \cos n\pi)$$

$$= \frac{8a^2}{n^3\pi^3} \text{ when } n \text{ is odd, otherwise zero when } n \text{ is even.}$$

Hence from (*viii*), the required solution is

$$v(x, y) = \frac{8a^2}{\pi^3} \sum_{n=1,3,5,\ldots}^{\infty} \frac{\sinh n\pi(b-y)/a}{n^3 \sinh n\pi b/a} \sin \frac{n\pi x}{a}$$

or
$$v(x, y) = \frac{8a^2}{\pi^3} \sum_{n=0}^{\infty} \frac{\sinh (2n+1)\pi(b-y)/a}{(2n+1)^3 \sinh (2n+1)\pi b/a} \sin \frac{(2n+1)\pi x}{a}.$$

## PROBLEMS 18.4

1. A long rectangular plate of width $\pi$ cm. with insulated surface has its temperature $u$ equal to zero on both the long sides and one of the short sides so that $u(0, y) = 0$, $u(\pi, y) = 0$, $u(x, \infty) = 0$, $u(x, 0) = kx$. Show that the steady-state temperature within the plate is

$$u(x, y) = 2k \sum_{n=1}^{\infty} \frac{(-1)^{n+1}}{n} e^{-ny} \sin nx..$$ (*Delhi, 2016*)

2. A rectangular plate with insulated surface is 8 cm. wide and so long compared to its width that it may be considered infinite in length without introducing an appreciable error. If the temperature along one short edge $y = 0$ is given by

$$u(x, 0) = 100 \sin (\pi x/8), \quad 0 < x < 8;$$

while the two long edges $x = 0$ and $x = 8$ as well as the other short edge are kept at 0°C, show that the steady-state temperature at any point of the plane is given by

$$u(x, y) = 100e^{-\pi y/8} \sin (\pi x/8).$$ (*JNTUK, 2022; Calicut, 2012*)

3. A rectangular plate with insulated surface is 10 cm. wide and so long compared to its width that it may be considered infinite in length without introducing an appreciable error. If the temperature of the short edge $y = 0$ is given by

$$u = 20\,x \quad \text{for } 0 \le x \le 5$$

and
$$u = 20(10 - x) \quad \text{for } 5 \le x \le 10$$

and the two long edges $x = 0$, $x = 10$ as well as the other short edge are kept at 0°C, prove that the temperature $u$ at any point $(x, y)$ is given by

$$u = \frac{40}{\pi^2} \sum_{1}^{\infty} \frac{(-1)^{n+1}}{(2n-1)^2} \sin \frac{(2n-1)\pi x}{10} \cdot e^{-(2n-1)\pi y/10}$$ (*V.T.U., 2015 S ; Anna, 2013*)

4. Solve $\frac{\partial^2 u}{\partial x^2} + \frac{\partial^2 u}{\partial y^2} = 0$ for $0 < x < \pi$, $0 < y < \pi$, with conditions given: $u(0, y) = u(\pi, y) = u(x, \pi) = 0$, $u(x, 0) = \sin^2 x$.

5. A square plate is bounded by the lines $x = 0$, $y = 0$, $x = 20$ and $y = 20$. Its faces are insulated. The temperature along the upper horizontal edge is given by

$$u(x, 20) = x(20 - x), \text{ when } 0 < x < 20,$$

while other three edges are kept at 0°C. Find the steady state temperature in the plate.

6. The temperature $u$ is maintained at 0° along three edges of a square plate of length 100 cm. and the fourth edge is maintained at 100° until steady-state conditions prevail. Find an expression for the temperature $u$ at any point $(x, y)$. Hence show that the temperature at the centre of the plate

$$= \frac{200}{\pi}\left[\frac{1}{\cosh \pi/2} - \frac{1}{3\cosh 3\pi/2} + \frac{1}{5\cosh 5\pi/2} - \ldots\right].$$

7. A square thin metal plate of side $a$ is bounded by the lines $x = 0$, $x = a$, $y = 0$, $y = a$. The edges $x = 0$, $y = a$ are kept at zero temperature, the edge $y = 0$ is insulated and the edge $x = a$ is kept at constant temperature $T_0$. Show that in the steady state conditions, the temperature $u(x, y)$ at the point $(x, y)$ is given by

$$u(x, y) = \frac{4T_0}{\pi} \sum_{n=1}^{\infty} \frac{(-1)^{n-1} \sinh \frac{(2n-1)\pi x}{2a} \cos \frac{(2n-1)\pi y}{2a}}{(2n-1) \sinh \frac{(2n-1)\pi}{2}}.$$

8. A rectangular plate has sides $a$ and $b$. Taking the side of length $a$ as $OX$ and that of length $b$ as $OY$ and other sides to be $x = a$ and $y = b$, the sides $x = 0, x = a, y = b$ are insulated and the edge $y = 0$ is kept at temperature $u_0 \cos \frac{\pi x}{a}$. Find the temperature $u(x, y)$ in the steady-state.

## 18.8 (1) LAPLACE'S EQUATION IN POLAR COORDINATES

In the study of steady-state temperature distribution in a rectangular plate, it is usually convenient to employ Cartesian coordinates as hitherto done. Sometimes Polar coordinates $(r, \theta)$ are found to be more useful and the Cartesian form of Laplace's equation is replaced by its polar form :

$$r^2 \frac{\partial^2 u}{\partial r^2} + r \frac{\partial u}{\partial r} + \frac{\partial^2 u}{\partial \theta^2} = 0$$

**(2) Solution of Laplace's equation**

$$r^2 \frac{\partial^2 u}{\partial r^2} + r \frac{\partial u}{\partial r} + \frac{\partial^2 u}{\partial \theta^2} = 0 \qquad ...(1)$$

Assume that a solution of Eqn. (1) is of the form $u = R(r) . \phi(\theta)$ where $R$ is a function of $r$ alone and $\phi$ is a function of $\theta$ only.

Substituting it in Eqn. (1), we get $r^2R''\phi + rR'\phi + R\phi'' = 0$ or $\phi(r^2R'' + rR') + R\phi'' = 0$.

Separating the variables $\frac{r^2R'' + rR'}{R} = -\frac{\phi''}{\phi}$ ...(2)

Clearly the left side of Eqn. (2) is a function of $r$ only and the right side is a function of $\theta$ alone. Since $r$ and $\theta$ are independent variables, Eqn. (2) can hold good only if each side is equal to a constant $k$ (say). Then Eqn. (2) leads to the ordinary differential equations

$$r^2 \frac{d^2R}{dr^2} + r\frac{dR}{dr} - kR = 0 \quad ...(3) \qquad \text{and} \qquad \frac{d^2\phi}{d\theta^2} + k\phi = 0 \quad ...(4)$$

Putting $r = e^z$, Eqn. (3) reduces to $\frac{d^2R}{dz^2} - kR = 0$ ...(5)

Solving Eqns. (5) and (4), we get

(i) *When k is positive* and $= p^2$, say :

$$R = c_1e^{pz} + c_2e^{-pz} = c_1r^p + c_2r^{-p}, \ \phi = c_3 \cos p\theta + c_4 \sin p\theta$$

(ii) *When k is negative* and $= -p^2$, say

$$R = c_5 \cos pz + c_6 \sin pz = c_5 \cos (p \log r) + c_6 \sin (p \log r), \ \phi = c_7e^{p\theta} + c_8e^{-p\theta}$$

(iii) *When k is zero* :

$$R = c_9z + c_{10} = c_9 \log r + c_{10}, \ \phi = c_{11}\theta + c_{12}$$

Thus the three possible solutions of (1) are

$$u = (c_1r^p + c_2r^{-p}) (c_3 \cos p\theta + c_4 \sin p\theta) \qquad ...(6)$$

$$u = [c_5 \cos (p \log r) + c_6 \sin (p \log r)] (c_7e^{p\theta} + c_8e^{-p\theta}) \qquad ...(7)$$

$$u = (c_9 \log r + c_{10}) (c_{11}\theta + c_{12}) \qquad ...(8)$$

Of these solutions, we have to take that solution which is consistent with the physical nature of the problem. The general solution will consist of a sum of terms of type Eqns. (6), (7) or (8). *(Andhra, 2015)*

**Example 18.18.** *The diameter of a semi-circular plate of radius a is kept at 0°C and the temperature at the semi-circular boundary is T°C. Show that the steady state temperature in the plate is given by*

$$u(r, \theta) = \frac{4T}{\pi} \sum_{n=1}^{\infty} \frac{1}{2n-1}\left(\frac{r}{a}\right)^{2n-1} \sin (2n-1)\theta.$$

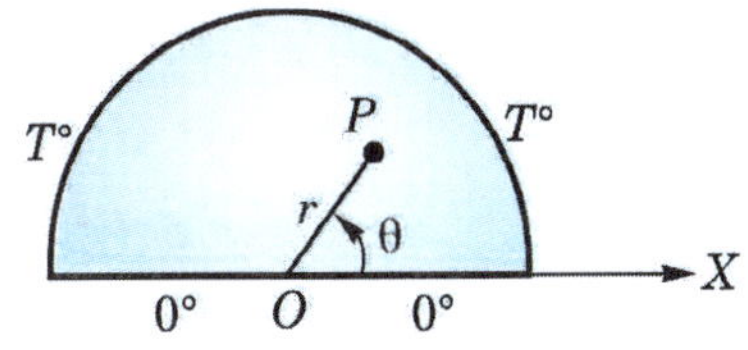

Fig. 18.7.

**Solution.** Take the centre of the circle as the pole and bounding diameter as the initial line as in Fig. 18.7. Let the steady state temperature at any point $P(r, \theta)$ be $u(r, \theta)$, so that $u$ satisfies the equation

$$r^2 \frac{\partial^2 u}{\partial r^2} + r\frac{\partial u}{\partial r} + \frac{\partial^2 u}{\partial \theta^2} = 0 \quad ...(i)$$

The boundary conditions are :

$$u(r, 0) = 0 \quad \text{in } 0 \le r \le a \quad ...(ii)$$

$$u(r, \pi) = 0 \quad \text{in } 0 \le r \le a \quad ...(iii)$$

and

$$u(a, \theta) = T \quad ...(iv)$$

The three possible solutions of (*i*) are

$$u = (c_1 r^p + c_2 r^{-p})(c_3 \cos p\theta + c_4 \sin p\theta) \quad ...(v)$$

$$u = [c_5 \cos (p \log r) + c_6 \sin (p \log r)](c_7 e^{p\theta} + c_8 e^{-p\theta}) \quad ...(vi)$$

$$u = (c_9 \log r + c_{10})(c_{11}\theta + c_{12}) \quad ...(vii)$$

From Eqns. (*ii*) and (*iii*), $u = 0$ when $r = 0$ *i.e.*, $u$ must be finite at the origin. Thus the solutions Eqns. (*vi*) and (*vii*) are to be rejected. Hence the only suitable solution is Eqn. (*v*).

By Eqn. (*ii*), $u(r, \theta) = (c_1 r^p + c_2 r^{-p})\, c_3 = 0$

Hence $c_3 = 0$ and Eqn. (*v*) becomes $u(r, \theta) = (c_1 r^p + c_2 r^{-p})\, c_4 \sin p\theta$ ...(*viii*)

By Eqn. (*iii*), $u(r, \pi) = (c_1 r^p + c_2 r^{-p})\, c_4 \sin p\pi = 0.$

As $c_4 \ne 0$, $\sin p\pi = 0$, *i.e.*, $p = n$, where $n$ is any integer.

Hence Eqn. (*viii*) reduces to $u(r, \theta) = (c_1 r^n + c_2 r^{-n})\, c_4 \sin n\theta$ ...(*ix*)

Since $u = 0$, when $r = 0$, $\therefore$ $c_2 = 0$ and (*ix*) becomes

$u(r, \theta) = b_n r^n \sin n\theta$, where $b_n = c_1 c_4$.

$\therefore$ the most general solution of Eqn. (*i*) is of the form

$$u(r, \theta) = \sum_{n=1}^{\infty} b_n r^n \sin n\theta \quad ...(x)$$

Putting $r = a$,

$$u(a, \theta) = \sum_{n=1}^{\infty} b_n a^n \sin n\theta \quad ...(xi)$$

In order that Eqn. (*iv*) may be satisfied, Eqns. (*iv*) and (*xi*) must be same. This requires the expansion of $T$ as a half-range Fourier sine series in $(0, \pi)$. Thus

$$T = \sum_{n=1}^{\infty} B_n \sin n\theta, \text{ where } B_n = \frac{2}{\pi}\int_0^{\pi} T \sin n\theta \, d\theta = \frac{2T}{n\pi}(1 - \cos n\pi) \text{ and } B_n = b_n a^n$$

$$\therefore \quad b_n = \frac{B_n}{a^n} = \frac{2T}{n\pi a^n}(1 - \cos n\pi)$$

*i.e.*, $b_n = 0$, if $n$ is even

$= \dfrac{4T}{n\pi a^n}$, if $n$ is odd.

Hence Eqn. (*x*) gives $u(r, \theta) = \dfrac{4T}{\pi}\left\{\dfrac{(r/a)}{1}\sin\theta + \dfrac{(r/a)^3}{3}\sin 3\theta + \dfrac{(r/a)^5}{5}\sin 5\theta + \ldots\right\}$.

**Example 18.19.** *The differential equation*

$$\frac{d^2x}{dt^2} + 2k\frac{dx}{dt} + n^2 x = 0;\ (k < n)$$

*represents the damped harmonic oscillations of a particle. Solve this equation and show that the ratio of the amplitude of any oscillation to that of the preceding one is constant i.e. its amplitudes form a G.P.*

*(Ranchi, 2019)*

**Solution:** The given equation

$$\frac{d^2x}{dt^2} + 2k\frac{dx}{dt} + n^2x = 0 \quad ...(i)$$

is a linear differential equation with constant, co-efficients

Hence $D^2 + 2kD + n^2 = 0$

$$D = \frac{-k \pm \sqrt{4k^2 - 4n^2}}{2} = -k \pm i\mu$$

where $\mu = \sqrt{n^2 - k^2}$

Thus, the solution of Eqn. (*i*)

$$x = e^{-kt}(c_1 \cos \mu t + c_2 \sin \mu t) \quad ...(ii)$$

Let $c_1 = r \cos \phi, c_2 = r \sin \phi$

the Eqn. (*ii*) reduces to $x = re^{-kt} \cos(\mu t - \theta) = \sqrt{c_1^2 + c_2^2}\, e^{-kt} \cos\left(\mu t - \tan^{-1}\frac{c_2}{c_1}\right)$

which shows the motion is oscillatory, with time

$$T = \frac{2\pi}{\mu}$$

The amplitude of the motion is $a = \sqrt{c_1^2 + c_2^2}\, e^{-kt}$

The particle will be the extreme position

Thus, $\frac{2\pi}{\mu}, \frac{4\pi}{\mu}, \frac{6\pi}{\mu}$

Let the amplitude corresponding to these times be $a_1\, a_2\, a_3$

when $t = \frac{2\pi}{\mu}, a_1 = \sqrt{c_1^2 + c_2^2}\, e^{-2\pi t} / \mu$

$$t = \frac{4\pi}{\mu}, a_2 = \sqrt{c_1^2 + c_2}\, e^{-4\pi k} / \mu$$

$$t = \frac{6\pi}{\mu}, a_3 = \sqrt{c_1^2 + c_2^2}\, e^{-6\pi k} / \mu$$

The ratio of successive amplitude $a_1, a_2, a_3...$ is constant and equal to $e^{-2\pi k/\mu}$.

So, the successive amplitude from a G.P.

**Example 18.20:** *A particle is projected with velocity u, making an angle α with the horizontal. Neglecting air resistance, show that the equation of its path is the parabola*

$$y = x \tan \alpha - \frac{g}{2u^2 \cos^2 \alpha}.x^2$$

*find the time of flight and range on the horizontal plane.* (*Karnataka, 2019*)

**Solution:** Let the particle be projected with velocity u making an angle α with the horizontal through *O* as *x* axis and *OY* perpendicular to *OX* as the *y* axis. The particle falls to the ground at the point *L*. The distance OL along the *x*-axis is called the range of the projectile. Let *p*(*x*, *y*) be the position of the particle at time *t*.

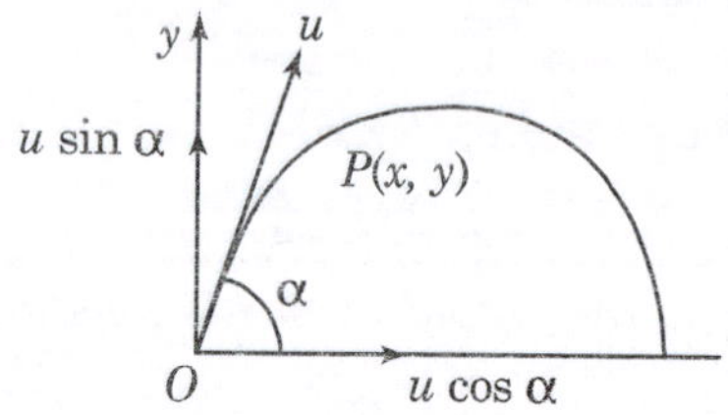


Fig. 18.8.

Resolving the velocity $u$ in its component along $OX$ and $OY$

If there is no force along the horizontal acceleration in this direction is zero. $\frac{d^2x}{dt^2} = 0$ ...(*i*)

In verticle direction $\frac{d^2y}{dt^2} = -g$ ...(*ii*)

Integrating equation Eqn. (*i*) w.r.t. $t \frac{dx}{dt} = c_1$ ...(*iii*)

$$\frac{dx}{dt} = u \cos \alpha$$

Integrating w.r.t. $t$ $x = (u \cos \alpha)\, t + c_2$ ...(*iv*)

Initially at $t = 0, x = 0$

$$c_2 = 0$$

Hence from Eqn. (*i*) $x = (u \cos \alpha)\; t$ ...(*v*)

Integrating Eqn. (*ii*) $\frac{dy}{dt} = -gt + c_3$ ...(*vi*)

Initially $t = 0$, $\frac{dy}{dt} = u \sin \alpha$

Thus $c_3 = u \sin \alpha$

Substituting the value of $c_3$ in Eqn. (*vi*)

$$\frac{dy}{dt} = u \sin \alpha - gt$$

Integrating w.r.t. $t$ $y = (u \sin \alpha)\, t - 1/2\, gt^2 + c_4$ ...(*vii*)

Initially at $t = 0, y = 0\ c_4 = 0$

The Eqn. (*vii*), $y = (u \sin \alpha)\, t - 1/2\, gt^2$ ...(*viii*)

from Eqns. (*v*) and (*viii*) $t = \frac{x}{u \cos \alpha}$ [from (*v*)]

Substituting this value of $t$ $y = (u \sin \alpha).\frac{x}{u \cos \alpha} - \frac{1}{2} g \left( \frac{x^2}{u^2 \cos^2 \alpha} \right)$

$$y = x \tan \alpha - \frac{g}{2u^2 \cos^2 \alpha} x^2 \qquad ...(ix)$$

which is the equation of the path of the projectile and the path is a parabola.

At the point $L$, $y = 0$ there from Eqn. (*viii*)

$$0 = (u \sin \alpha)\, t - 1/2\, gt^2$$

$$t = \frac{2u \sin \alpha}{g}$$

The range $R(OL)$ is the distance covered by the particle in time $t$, velocity $u \cos \alpha$

$$R = u \cos \alpha \times t$$

$$R = u \cos \alpha \times \frac{2u \sin \alpha}{g} = \frac{u^2 \sin 2\alpha}{g}$$

## PROBLEMS 18.5

1. A semi-circular plate of radius $a$ has its circumference kept at temperature $u(a, \theta), = k\theta(\pi - \theta)$ while the boundary diameter is kept at zero temperature. Find the steady state temperature distribution $u(r, \theta)$ of the plate assuming the lateral surfaces of the plate to be insulated.
2. A semi-circular plate of radius 10 cm has insulated faces and heat flows in plane curves. The bounding diameter is kept at 0°C and on the circumference the temperature distribution maintained is $u(10, \theta) = (400/\pi)(\pi\theta - \theta^2), 0 \le \theta \le \pi$. Determine the temperature distribution $u(r, \theta)$ at any point on the plate.
3. A plate in the shape of truncated quadrant of a circle, is bounded by $r = a, r = b$ and $\theta = 0, \theta = \pi/2$. It has its faces insulated and heat flows in plane curves. It is kept at temperature 0°C along three of the edges while along the edge $r = a$, it is kept at temperature $\theta(\pi/2 - \theta)$. Determine the temperature distribution.
4. Determine the steady state temperature at the points on the sector $0 \le \theta \le \pi/4, 0 \le r \le a$ of a circular plate, if the temperature is maintained at 0°C along the side edges and at a constant temperature $k$°C along the curved edges.
5. Find the steady-state temperature in a circular plate of radius $a$ which has one-half of its circumference at 0°C and the other half at 60°C.
6. If the radii of the inner and outer boundaries of a circular annulus area 10 cm and 20 cm and
$$u(10, \theta) = 15 \cos \theta, u(20, \theta) = 30 \sin \theta,$$
find the value of $u(r, \theta)$ in the annulus. [$u(r, \theta)$ satisfies Laplace equation in the interior of the annulus.]
7. A plate in the form of a ring is bounded by the lines $r = 2$ and $r = 4$. Its surfaces are insulated and the temperature along the boundaries are
$$u(2, \theta) = 10 \sin \theta + 6 \cos \theta, u(4, \theta) = 17 \sin \theta + 15 \cos \theta$$
Find the steady-state temperature $u(r, \theta)$ in the ring.

## 18.9 (1) VIBRATING MEMBRANE—TWO DIMENSIONAL WAVE EQUATION

We shall now derive the equation for the vibrations of a tightly stretched membrane, such as the membrane of a drum. We shall assume that the membrane is uniform and the tension $T$ in it per unit length is the same in all directions at every point.

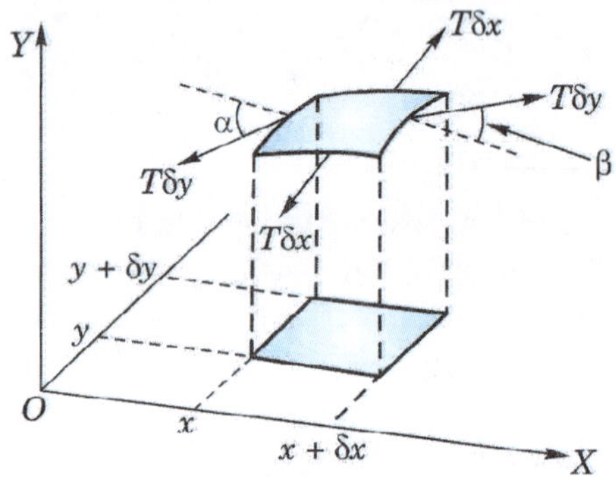

Fig. 18.9

Consider the forces acting on an element $\delta x \delta y$ of the membrane (Fig. 18.9). Forces $T\delta x$ and $T\delta y$ act on the edges along the tangent to the membrane. Let $u$ be its small displacement perpendicular to the $xy$-plane, so that the forces $T\delta y$ on its opposite edges of length $\delta y$ make angles α and β to the horizontal. So their vertical component

$= T\delta y \sin \beta - T\delta y \sin \alpha$

$= T\delta y (\tan \beta - \tan \alpha)$ approximately, since α and β are small

$$= T\delta y \left\{ \left( \frac{\partial u}{\partial x} \right)_{x + \delta x} - \left( \frac{\partial u}{\partial x} \right)_x \right\} = T\delta y \frac{\partial^2 u}{\partial x^2} \delta x, \text{ up to a first order of approximation.}$$

Similarly, the vertical component of the force $T\delta x$ acting on the edges of length $\delta x$

$$= T\delta x \frac{\partial^2 u}{\partial y^2} \delta y$$

If $m$ be the mass per unit area of the membrane, then the equation of motion of the element is

$$m\delta x\delta y \frac{\partial^2 u}{\partial t^2} = T\left[ \frac{\partial^2 u}{\partial x^2} + \frac{\partial^2 u}{\partial y^2} \right] \delta x\delta y \quad \text{or} \quad \frac{\partial^2 \mathbf{u}}{\partial \mathbf{t}^2} = \mathbf{c}^2 \left[ \frac{\partial^2 \mathbf{u}}{\partial \mathbf{x}^2} + \frac{\partial^2 \mathbf{u}}{\partial \mathbf{y}^2} \right] \quad \text{where} \quad c^2 = T/m \qquad ...(1)$$

This is the wave equation in two dimensions.

**(2) Solution of the two-dimensional wave equation - Rectangular membrane.** Assume that a solution of Eqn. (1) is of the form $u = X(x)Y(y)T(t)$

Substituting this in Eqn. (1) and dividing by $XYT$, we get

$$\frac{1}{c^2 T} \frac{d^2 T}{dt^2} = \frac{1}{X} \frac{d^2 X}{dx^2} + \frac{1}{Y} \frac{d^2 Y}{dy^2}$$

This can hold good if each member is a constant. Choosing the constants suitably, we have

$$\frac{d^2X}{dx^2} + k^2X = 0, \quad \frac{d^2Y}{dy^2} + l^2Y = 0 \quad \text{and} \quad \frac{d^2T}{dt^2} + (k^2 + l^2)\, c^2T = 0$$

Hence a solution of (1) is

$$u = (c_1 \cos kx + c_2 \sin kx)\,(c_3 \cos ly + c_4 \sin ly) \times [c_5 \cos \sqrt{(k^2 + l^2)}ct + c_6 \sin \sqrt{(k^2 + l^2)}ct\,] \qquad ...(2)$$

Now suppose the membrane is rectangular and is stretched between the lines $x = 0, x = a, y = 0, y = b$. Then the condition $u = 0$ when $x = 0$ gives

$$0 = c_1(c_3 \cos ly + c_4 \sin ly)[c_5 \cos \sqrt{(k^2 + l^2)ct} + c_6 \sin \sqrt{(k^2 + l^2)ct}\,] \quad i.e., \quad c_1 = 0.$$

Then putting $c_1 = 0$ in Eqn. (2) and applying the condition $u = 0$ when $x = a$, we get $\sin ka = 0$ or $k = m\pi/a$. ($m$ being an integer)

Similarly, applying the conditions $u = 0$, when $y = 0$ and $y = b$, we obtain

$$c_3 = 0 \quad \text{and} \quad l = n\pi/b \qquad (n \text{ being an integer})$$

Thus the solution Eqn. (2) becomes

$$u(x, y, t) = c_2c_4 \sin \frac{m\pi x}{a} \sin \frac{n\pi y}{b} (c_5 \cos p_{mn}t + c_6 \sin p_{mn}t)$$

where $p_{mn} = \pi c\sqrt{[(m/a)^2 + (n/b)^2]}$ ...(3)

[These are the solutions of the wave Eqn. (1) which are zero on the boundary of the rectangular membrane. These functions are called **eigen functions** and the numbers $p_{mn}$ are the **eigen values** of the vibrating membrane.]

Choosing the constants $c_2$ and $c_4$ so that $c_2c_4 = 1$, we can write the general solution of the equation (1) as

$$u = \sum_{m=1}^{\infty} \sum_{n=1}^{\infty} \sin \frac{m\pi x}{a} \sin \frac{n\pi y}{b} (A_{mn} \cos pt + B_{mn} \sin pt) \qquad ...(4)$$

If the membrane starts from rest from the initial position $u = f(x, y)$, *i.e.*, $\frac{\partial u}{\partial t} = 0$ when $t = 0$, then Eqn. (3) gives $B_{mn} = 0$.

Also using the condition $u = f(x, y)$ when $t = 0$, we get

$$f(x, y) = \sum_{m=1}^{\infty} \sum_{n=1}^{\infty} A_{mn} \sin \frac{m\pi x}{a} \sin \frac{n\pi y}{b}$$

This is *double Fourier series*. Multiplying both sides by $\sin (m\pi x/a) \sin (n\pi y/b)$ and integrating from $x = 0$ to $x = a$ and $y = 0$ to $y = b$, every term on the right except one, becomes zero. Hence we obtain

$$\int_0^a \int_0^b f(x, y) \sin \frac{m\pi x}{a} \sin \frac{n\pi y}{b}\, dydx = \frac{ab}{4} A_{mn} \qquad ...(5)$$

which gives the coefficients in the solution and is called the **generalised Euler's formula.**

## Rectangular Membranes

**Example 18.21.** *Find the deflection $u(x, y, t)$ of the square membrane with $a = b = 1$ and $c = 1$, if the initial velocity is zero and the initial deflection is $f(x, y) = A \sin \pi x \sin 2\pi y$.*

**Solution.** Taking $a = b = 1$ and $f(x, y) = A \sin \pi x \sin 2\pi y$, in Eqn. (5), we get

$$A_{mn} = 4\int_0^1 \int_0^1 A \sin \pi x \sin 2\pi y \sin m\pi x \sin n\pi y\, dydx$$

$$= 4A\int_0^1 \sin \pi x \sin m\pi x\, dx \left(\int_0^1 \sin 2\pi y \sin n\pi y\, dy\right) = 0, \quad \text{for } m \neq 1$$

$$= 4A\left(\frac{1}{2}\right)\int_0^1 \sin 2\pi y \sin n\pi y\, dy, \text{ for } m = 1 \qquad \left[\because \int_0^1 \sin \pi x \sin \pi x\, dx = \tfrac{1}{2}\right]$$

*i.e.,* $$A_{mn} = 2A\int_0^1 \sin 2\pi y \sin n\pi y\, dy = 0, \text{ for } n \neq 2$$

$$= 2A\left(\frac{1}{2}\right), \text{ for } n = 2.$$

$\therefore$ $A_{12} = A$. Also from Eqn. (3), $p_{mn} = \pi\sqrt{(m^2 + n^2)}$ $[\because\ a = b = 1 = c]$

$\therefore$ $p_{12} = \pi\sqrt{(1^2 + 2^2)} = \sqrt{5}\,\pi.$

Hence from Eqn. (4), the required solution is $u(x, y, t) = A \sin \pi x \sin 2\pi y \cos (\sqrt{5}\,\pi t)$.

**Example 18.22.** *Find the vibration u(x, y, t) of a rectangular membrane (0 < x < a, 0 < y < b) whose boundary is fixed given that it starts from rest and u(x, y, 0) = hxy(a − x)(b − y).*

**Solution.** Proceeding as in § 18.9 (**2**), we have from Eqn. (4),

$$u = \sum_{m=1}^{\infty}\sum_{n=1}^{\infty} \sin\frac{m\pi x}{a}\sin\frac{n\pi y}{b}(A_{mn}\cos pt + B_{mn}\sin pt) \text{ where } p = \pi c\sqrt{[(m/a)^2 + (n/b)^2}$$

Since the membrane starts from rest $\partial u/\partial t = 0$ when $t = 0$,

$$\therefore \quad \sum_{m=1}^{\infty}\sum_{n=1}^{\infty} \sin\frac{m\pi x}{a}\sin\frac{n\pi y}{b}(-A_{mn}\,p\sin pt + pB_{mn}\cos pt) = 0 \text{ when } t = 0$$

This gives $B_{mn} = 0$

$$\therefore \quad u(x, y, t) = \sum_{m=1}^{\infty}\sum_{n=1}^{\infty} A_{mn}\sin\frac{m\pi x}{a}\sin\frac{n\pi y}{b}\cos pt \qquad ...(i)$$

Then $$hxy(a - x)(b - y) = u(x, y, 0) = \sum_{m=1}^{\infty}\sum_{n=1}^{\infty} A_{mn}\sin\frac{m\pi x}{a}\sin\frac{n\pi y}{b}$$

where $$A_{mn} = \frac{2}{a}\cdot\frac{2}{b}\int_0^a\int_0^b hxy(a-x)(b-y)\,.\sin\frac{m\pi x}{a}\sin\frac{n\pi y}{b}\,dydx$$

$$= \frac{4h}{ab}\left\{\int_0^a x(a-x)\sin\frac{m\pi x}{a}dx\right\}\left\{\int_0^b y(b-y)\sin\frac{n\pi y}{b}dy\right\}$$

$$= \frac{4h}{ab}\left|(ax - x^2)\left(\frac{-\cos m\pi x/a}{m\pi/a}\right) - (a - 2x)\left\{\frac{-\sin\frac{m\pi x}{a}}{(m\pi/a)^2}\right\} + (-2)\frac{\cos m\pi x/a}{(m\pi/a)^3}\right|_0^a$$

$$\times\left|(by - y^2)\left(\frac{-\cos n\pi y/b}{n\pi/b}\right) - (b - 2y)\left\{\frac{-\sin n\pi/b}{(n\pi/b)^2}\right\} + (-2)\frac{\cos n\pi y/b}{(n\pi/b)^3}\right|_0^b$$

$$= \frac{4h}{ab}\frac{2a^3}{m^3\pi^3}\cdot\frac{2b^3}{n^3\pi^3}(1 - \cos m\pi)(1 - \cos n\pi)$$

Hence from Eqn. (*i*), we get

$$u(x, y, t) = \sum_{m=1}^{\infty}\sum_{n=1}^{\infty} A_{mn}\sin\frac{m\pi x}{a}\sin\frac{n\pi y}{b}\cos pt$$

where $A_{mn} = \dfrac{16ha^2b^2}{m^3n^3\pi^6}(1 - \cos m\pi)(1 - \cos n\pi)$ and $p = \pi c\sqrt{[(m/a)^2 + (n/b)^2]}$

## Circular Membranes*

**Example 18.23.** *A circular membrane of unit radius fixed along its boundary starts vibrating from rest and has initial deflection u(r, 0) = f(r). Show that the deflection u(r, t) of the membrane at any instant is given by*

$$u(r, t) = \sum_{m=1}^{\infty} A_m \cos(c\alpha_m t)\,.\,J_o(\alpha_m r) \text{ where } A_m = \frac{2}{J_1^2(\alpha_m)}\int_0^1 r f(r) J_0(\alpha_m r)\,dr,$$

*and $\alpha_m (m = 1, 2, ...)$ are the positive roots of the Bessel function $J_0(k) = 0$.*

*Drums, telephones and microphones provide examples of circular membrane and as such are quite useful in engineering.

**Solution.** The vibrations of a plane circular membrane are governed by 2-dimensional wave equation in polar coordinates *i.e.*,

$$\frac{\partial^2 u}{\partial t^2} = c^2 \left( \frac{\partial^2 u}{\partial r^2} + \frac{1}{r}\frac{\partial u}{\partial r} + \frac{1}{r^2}\frac{\partial^2 u}{\partial \theta^2} \right)$$

For a radially symmetric membrane (in which $u$ does not depend on $\theta$) the above equation reduces to

$$\frac{\partial^2 u}{\partial t^2} = c^2 \left( \frac{\partial^2 u}{\partial r^2} + \frac{1}{r}\frac{\partial u}{\partial r} \right) \quad \text{...}(i)$$

For the given membrane fixed along its boundary, the boundary condition is

$$u(1, t) = 0 \quad \text{for all } t \geq 0 \quad \text{...}(ii)$$

For solutions not depending on $\theta$,

initial deflection $u(r, 0) = f(r)$ ...(*iii*)

and initial velocity $\left(\frac{\partial u}{\partial t}\right)_{t=0} = 0$ ...(*iv*)

which are the initial conditions. We find the solutions $u(r, t) = R(r)T(t)$ ...(*v*)

satisfying the boundary condition Eqn. (*ii*).

Differentiating and substituting Eqn. (*v*) in Eqn. (*i*), we get

$$\frac{\partial^2 T/\partial t^2}{c^2 T} = \frac{1}{R}\left( \frac{\partial^2 R}{\partial r^2} + \frac{1}{r}\frac{\partial R}{\partial r} \right) = -k^2 \text{ (say)}$$

This leads to $\frac{\partial^2 T}{\partial t^2} + p^2 T = 0$ where $p = ck$ ...(*vi*)

and

$$\frac{\partial^2 R}{\partial r^2} + \frac{1}{r}\frac{\partial R}{\partial r} + k^2 R = 0 \quad \text{...}(vii)$$

Now putting $s = kr$, Eqn. (*vii*) transforms to $\frac{d^2 R}{ds^2} + \frac{1}{s}\frac{dR}{ds} + R = 0$ which is Bessel's equation. Its general solution $R = aJ_0(s) + bY_0(s)$ where $J_0$ and $Y_0$ are Bessel's functions of the first and second kind of order zero.

Since the deflection of the membrane is always finite, we must have $b = 0$. Then taking $a = 1$, we get

$$R(r) = J_0(s) = J_0(kr)$$

On the boundary of the circular membrane, we must have $J_0(k) = 0$, which is satisfied for

$$k = \alpha_m, \; m = 1, 2, \ldots$$

Thus the solutions of Eqn. (*vii*) are $R(r) = J_0(\alpha_m r)$, $m = 1, 2, \ldots$ and the corresponding solutions of (*vi*) are $T(t) = A_m \cos p_m t + B_m \sin p_m t$, where $p_m = ck_m = c\alpha_m$.

Hence the general solution of Eqn. (*i*) satisfying Eqn. (*ii*) are

$$u(r, t) = (A_m \cos p_m t + B_m \sin p_m t)\, J_0(\alpha_m r)$$

which are the *eigen functions* of the problem and the corresponding *eigen values* are $p_m$.

To find that solution which also satisfies the initial conditions Eqns. (*iii*) and (*iv*), consider the series

$$u(r, t) = \sum_{m=1}^{\infty} (A_m \cos p_m t + B_m \sin p_m t)\, J_0(\alpha_m r)$$

Putting $t = 0$ and using Eqn. (*iii*), we get $u(r, 0) = \sum_{m=1}^{\infty} A_m J_0(\alpha_m r) = f(r)$

Here, the constants $A_m$ must be the coefficients of Fourier-Bessel series with $m = 0$, *i.e.*,

$$A_m = \frac{2}{J_1^2(\alpha_m)} \int_0^1 r\, f(r)\, J_0(\alpha_m r)\, dr$$

Using Eqn. (*iv*), we get $B_m = 0$. Hence the result.

## PROBLEMS 18.6

1. A mass $M$ suspended from the end of a helical spring is subjected to a periodic force $f = F \cos \omega t$ in the direction of its length. The force $f$ is measured positive vertically downwards and at zero time $M$ is at rest. If the spring stiffness is $S$, Prove that the displacement of $M$ at time $t$ from the commencement of motion is given by

$$x = \frac{F}{M(p^2 - \omega^2)} (\cos \omega t - \cos pt)$$

where $p^2 = S/m$ and damping effects are neglected.

2. A spring is loaded with 9.8 kg weight, its point of support has a motion given by $\sin \omega t$ and the resistance to the motion of the load is ten times the instantaneous speed. If the tension of the spring be 50 times the stretch and its mass neglected, show that the motion is given by

$$\frac{d^2x}{dt^2} + 10\frac{dx}{dt} + 50x = 50 \sin \omega t$$

Find the amplitude of steady motion when $\omega^2 = 50$. *(Periyauni, Salem, 2019)*

3. A body executes damped forced vibrations given by the equation

$$\frac{d^2x}{dt^2} + 2k\frac{dx}{dt} + b^2x = e^{-kt} \sin \omega t$$

Solve the equation for both the cases when $\omega^2 \neq b^2 - k^2$ and when $\omega^2 = b^2 - k^2$. *(KTU, 2018)*

## 18.10 TRANSMISSION LINE

Consider a cable $l$ km in length, carrying an electric current with resistance $R$ ohms/km, inductance $L$ henries/km ; capacitance $C$ farads/km and leakance $G$ mhos/km (Fig. 18.10).

Let the instantaneous voltage and current at any point $P$, distant $x$ km from the sending end $O$, and at time $t$ sec be $v(x, t)$ and $i(x, t)$ respectively. Consider a small length $PQ(= \delta x)$ of the cable.

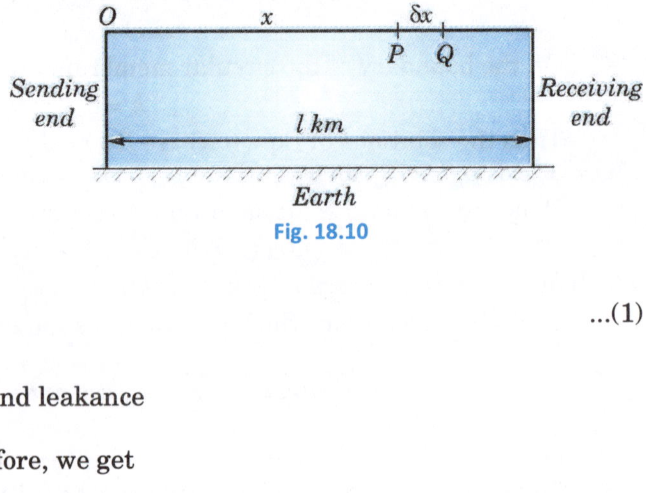

Fig. 18.10

Now since the voltage drop across the segment $\delta x$

= voltage drop due to resistance + voltage drop due to inductance

$$\therefore \quad -\delta v = iR\delta x + L\delta x \cdot \frac{\partial i}{\partial t}$$

and dividing by $\delta x$ and taking limits as $\delta x \to 0$, we get

$$-\frac{\partial v}{\partial x} = Ri + L\frac{\partial i}{\partial t} \quad \text{...(1)}$$

Similarly the current loss between $P$ and $Q$

= current lost due to capacitance and leakance

$$\therefore \quad -\delta i = C\frac{\partial v}{\partial t}\delta x + Gv\delta x$$ from which as before, we get

$$-\frac{\partial i}{\partial x} = C\frac{\partial v}{\partial t} + Gv \quad \text{...(2)}$$

Rewriting the simultaneous partial differential Eqns. (1) and (2) as

$$\left(R + L\frac{\partial}{\partial t}\right)i + \frac{\partial v}{\partial x} = 0 \quad \text{...(3)}$$

and
$$\frac{\partial i}{\partial x} + \left(C\frac{\partial}{\partial t} + G\right)v = 0, \quad \text{...(4)}$$

we shall eliminate $i$ and $v$ in turn.

$\therefore$ operating Eqn. (3) by $\frac{\partial}{\partial x}$ and Eqn. (4) by $\left(R + L\frac{\partial}{\partial t}\right)$ and subtracting

$$\frac{\partial^2 v}{\partial x^2} - \left(R + L\frac{\partial}{\partial t}\right)\left(C\frac{\partial}{\partial t} + G\right)v = 0$$

or
$$\frac{\partial^2 v}{\partial x^2} = LC\frac{\partial^2 v}{\partial t^2} + (LG + RC)\frac{\partial v}{\partial t} + RGv \qquad ...(5)$$

Again operating Eqn. (3) by $\left(C\frac{\partial}{\partial t} + G\right)$ and Eqn. (4) by $\frac{\partial}{\partial x}$ and subtracting

$$\left(C\frac{\partial}{\partial t} + G\right)\left(R + L\frac{\partial}{\partial t}\right)i - \frac{\partial^2 i}{\partial x^2} = 0$$

or
$$\frac{\partial^2 i}{\partial x^2} = LC\frac{\partial^2 i}{\partial t^2} + (LG + RC)\frac{\partial i}{\partial t} + RGi \qquad ...(6)$$

which is Eqn. (5) with $v$ replaced by $i$. The Eqns. (5) and (6) are called the *telephone equations.*

**Cor. 1.** *If* $L = G = 0$, the Eqns. (5) and (6) become

$$\frac{\partial^2 v}{\partial x^2} = RC\frac{\partial v}{\partial t} \qquad ...(7) \qquad\qquad \frac{\partial^2 i}{\partial x^2} = RC\frac{\partial i}{\partial t} \qquad ...(8)$$

which are known as the *telegraph equations.*

Rewriting Eqn. (7) as $\frac{\partial v}{\partial t} = \frac{1}{RC}\frac{\partial^2 v}{\partial x^2}$, we observe that it is similar to the heat equation.

**Cor. 2.** *If* $R = G = 0$, the Eqns. (5) and (6) become

$$\frac{\partial^2 v}{\partial x^2} = LC\frac{\partial^2 v}{\partial t^2} \qquad ...(9) \qquad\qquad \frac{\partial^2 i}{\partial x^2} - LC\frac{\partial^2 i}{\partial t^2} \qquad ...(10)$$

which are called the *radio equations.*

Rewriting Eqn. (9) as $\frac{\partial^2 v}{\partial t^2} = k^2\frac{\partial^2 v}{\partial x^2}$ where $k^2 = \frac{1}{LC}$,

its general solution is $v(x, t) = f_1(x + kt) + f_2(x - kt)$.

Similarly from Eqn. (10), $i(x, t) = F_1(x + kt) + F_2(x - kt)$.

Thus the voltage $v(x, t)$ for the current $i(x, t)$ at any point along the lossless transmission line can be obtained by the superposition of a progressive wave and a receding wave travelling with equal velocities ($k$). This is the case of oscillations of $v(x, t)$ and $i(x, t)$ at high frequencies.

**Cor. 3.** *If* $L = C = 0$, *e.g.*, in the case of a submarine cable, then (5) gives

$$\frac{\partial^2 v}{\partial x^2} = GRv, \text{ i.e. } (D^2 - GR)v = 0$$

$\therefore$
$$v(x) = A\cosh(\sqrt{GR}\,.\,x) + B\sinh(\sqrt{GR}\,.\,x) \qquad ...(11)$$

Since by Eqn. (1), $\quad Ri = -\frac{\partial v}{\partial x} = -\sqrt{GR}\,[A\sinh(\sqrt{GR}\,.\,x) + B\cosh(\sqrt{GR}\,.\,x)]$

$\therefore$
$$i(x) = -\sqrt{G/R}\,[A\sinh(\sqrt{GR}\,.\,x) + B\cosh(\sqrt{GR}\,.\,x)] \qquad ...(12)$$

If $v(0) = v_0$ and $i(0) = i_0$, then $v_0 = A$ and $i_0 = -\sqrt{G/R}\ B$.

Hence writing $\sqrt{GR} = \gamma$ and $\sqrt{R/G} = z_0$, Eqn. (11) and Eqn. (12) give

$$v(x) = v_0\cosh\gamma x - i_0 z_0\sinh\gamma x \qquad ...(13)$$

and
$$i(x) = i_0\cosh\gamma x - \frac{v_0}{z_0}\sinh\gamma x. \qquad ...(14)$$

**Obs. Steady-state solutions.** We have so far considered the transient state solutions only. The steady-state solutions of transmission lines are however, obtained by assuming $v = Ve^{iwt}$ and $i = Ie^{iwt}$, where $V$ and $I$ are complex functions of $x$ only. Substituting these in Eqns. (5) and (6), we get two ordinary linear differential equations of the second order which can be solved at once.

**Example 18.24.** *Neglecting R and G, find the e.m.f. v(x, t) in a line of length l, t seconds after the ends were suddenly grounded, given that* $i(x, 0) = i_0$ *and* $v(x, 0) = e_1 \sin \frac{\pi x}{l} + e_5 \sin \frac{5\pi x}{l}$.

*(U.P.T.U., 2012)*

**Solution.** Since $R$ and $G$ are negligible, we use the *Radio equation* $\frac{\partial^2 v}{\partial x^2} = LC \frac{\partial^2 v}{\partial t^2}$ ...(*i*)

Since the ends are suddenly grounded, we have the boundary conditions

$$v(0, t) = 0, \; v(l, t) = 0 \quad ...(ii)$$

Also the initial conditions are $i(x, 0) = i_0$

and
$$v(x, 0) = e_1 \sin \pi x/l + e_5 \sin 5\pi x/l \quad ...(iii)$$

$$\therefore \quad \frac{\partial i}{\partial x} = -c\frac{\partial v}{\partial t} \quad \text{gives} \quad \frac{\partial v}{\partial t}(x, 0) = 0 \quad ...(iv)$$

Let $v = X(x)T(t)$ be the solution of (*i*).

$\therefore$ Eqn. (*i*) gives $\quad X''T = LCXT''$

$$\frac{X''}{X} = LC\frac{T''}{T} = -k^2 \text{ (say)}$$

$$\therefore \quad X'' + k^2 X = 0 \quad \text{and} \quad T'' + (k^2/LC)T = 0$$

Solving these equations, we get

$$v = (c_1 \cos kx + c_2 \sin kx)\left(c_3 \cos \frac{k}{\sqrt{LC}} . t + c_4 \sin \frac{k}{\sqrt{LC}} t\right)$$

Using the boundary conditions Eqn. (*ii*), we get

$$c_1 = 0 \quad \text{and} \quad k = n\pi/l.$$

$$\therefore \quad v = \sin \frac{n\pi x}{l}\left(a_n \cos \frac{n\pi}{l\sqrt{LC}} t + b_n \sin \frac{n\pi}{l\sqrt{LC}} t\right)$$

Using the initial Eqn. (*iv*), we get $b_n = 0$

$$\therefore \quad v = a_n \sin \frac{n\pi x}{l} \cos \frac{n\pi}{l\sqrt{LC}} t$$

Thus the most general solution of Eqn. (*i*) is

$$v = \sum_{n=1}^{\infty} a_n \sin \frac{n\pi x}{l} \cos \frac{n\pi t}{l\sqrt{LC}}$$

Finally by the initial condition Eqn. (*iii*), we have

$$e_1 \sin \frac{\pi x}{l} + e_5 \sin \frac{5\pi x}{l} = \sum a_n \sin \frac{n\pi x}{l}$$

$\therefore \quad a_1 = e_1$ and $a_5 = e_5$ while all other $a$'s are zero.

Hence
$$v = e_1 \sin \frac{\pi x}{l} \cos \frac{\pi t}{l\sqrt{LC}} + e_5 \sin \frac{5\pi x}{l} \cos \frac{5\pi t}{l\sqrt{LC}}$$

which is the required solution.

**Example 18.25.** *A telephone line 3000 km. long has a resistance of 4 ohms/km. and a capacitance of* $5 \times 10^{-7}$ *farad/km. Initially both the ends are grounded so that the line is uncharged. At time t = 0, a constant e.m.f. E is applied to one end, while the other end is left grounded. Assuming the inductance and leakance to be negligible, show that the steady state current of the grounded end at the end of 1 sec. is 5.3%.*

**Solution.** Since $L = 0$, $G = 0$, we use the telegraph equation

$$\frac{\partial^2 v}{\partial x^2} = RC\frac{\partial v}{\partial t}$$

Let $v = X(x)T(t)$ be its solution so that

$$TX'' = RCXT' \quad \text{or} \quad \frac{X''}{X} = RC\frac{T'}{T} = -k^2 \text{ (say)}$$

$\therefore$ $$X'' + k^2X = 0 \text{ and } T' + (k^2/RC)T = 0$$

Solving these equations, we get

$$X = c_1 \cos kx + c_2 \sin kx, \; T = c_3 e^{-k^2t/RC}$$

giving $$v = (c_1 \cos kx + c_2 \sin kx)\, c_3 e^{-k^2t/RC} \qquad ...(i)$$

When $t = 0$ ; $v = 0$ at $x = 0$ and $v = 0$ at $x = l$

$\therefore$ $$0 = c_1c_3 \;;\; 0 = (c_1 \cos kl + c_2 \sin kl)c_3$$

*i.e.,* $$c_1c_3 = 0 \text{ and } kl = n\pi \;(n \text{ an integer})$$

Putting these in Eqn. (*i*) and adding a linear term, we have

$$v = a_0x + b_0 + \sum_{n=1}^{\infty} \sin\frac{n\pi x}{l}\, e^{-n^2\pi^2t/RCl^2} \qquad ...(ii)$$

The end conditions of the problem are

$$v = 0 \text{ at } x = 0 \text{ and } v = E \text{ at } x = l$$

Using these, Eqn. (*ii*) gives $b_0 = 0$ and $a_0 = E/l$

Then Eqn. (*ii*) becomes $\quad v = \dfrac{E}{l}x + \sum b_n \sin\dfrac{n\pi x}{l}\, e^{-n^2\pi^2t/RCl^2}$

Also $v = 0$ when $t = 0$, we get $-Ex/l = \sum b_n \sin n\pi x/l$

This requires the expansion of $(-Ex/l)$ as a half-range sine series in $(0, l)$.

$\therefore$ $$b_n = \frac{2}{l}\int_0^l \left(\frac{-Ex}{l}\right)\sin\left(\frac{n\pi x}{l}\right)dx$$

$$= \frac{2}{l}\left[\left(\frac{-Ex}{l}\right)\left(-\frac{l}{n\pi}\cos\frac{n\pi x}{l}\right) - \left(\frac{-E}{l}\right)\left(-\frac{l^2}{n^2\pi^2}\sin\frac{n\pi x}{l}\right)\right]_0^l = \frac{2}{l}\left(\frac{El}{n\pi}\cos n\pi\right) = \frac{2E}{n\pi}(-1)^n.$$

Thus $$v = \frac{Ex}{l} + \frac{2E}{\pi}\sum_{n=1}^{\infty} \sin\frac{n\pi x}{l}\, e^{-n^2\pi^2t/RCl^2} \qquad ...(iii)$$

Also when $L = 0$, $\dfrac{-\partial v}{\partial x} = Ri$

*i.e.,* $$i = -\frac{1}{R}\frac{\partial v}{\partial x} = -\frac{E}{lR} - \frac{2E}{lR}\sum_{n=1}^{\infty}(-1)^n \cos\frac{n\pi x}{l}\, e^{-n^2\pi^2t/RCl^2}$$

At the grounded end $(x = 0)$, the current is

$$i = -\frac{E}{lR} - \frac{2E}{lR}\sum_{n=1}^{\infty}(-1)^n e^{-n^2\pi^2t/RCl^2}$$

When $t = 1$ sec, $$i = -\frac{E}{lR}\left(1 - 2e^{\pi^2/RCl^2} + 2e^{-4\pi^2/RCl^2} - ...\right) \qquad ...(iv)$$

Since $$\frac{\pi^2}{RCl^2} = \frac{(3.14)^2}{4(5\times10^{-7})(3000)^2} = 0.548$$

$\therefore$ $$e^{-\pi^2/RCl^2} = e^{-0.548} = 0.578$$

When $\quad t \to \infty, i \to -E/lR$

Hence from (*iv*), we have

$$i = -\frac{E}{lR}\{1 - 2(0.578) + 2(0.578)^4 - 2(0.578)^9 + ...\}$$

$$= -\frac{E}{lR}\{1 - 1.156 + 0.223 - 0.014 + ...\}$$

$$= i_\infty(0.053) = 5.3\% \text{ of } i_\infty.$$

**Example 18.26.** *A transmission line 1000 kilometers long is initially under steady-state conditions with potential 1300 volts at the sending end (x = 0) and 1200 volts at the receiving end (x = 1000). The terminal end of the line is suddenly grounded, but the potential at the source is kept at 1300 volts. Assuming the inductance and leakance to be negligible, find the potential v(x, t).*

**Solution.** The equation of the telegraph line is

$$\frac{\partial^2 v}{\partial x^2} = RC\frac{\partial v}{\partial t} \quad \text{or} \quad \frac{\partial v}{\partial t} = \frac{1}{RC}\frac{\partial^2 v}{\partial x^2} \qquad ...(i)$$

$v_s$ = initial steady voltage satisfying $\frac{\partial^2 v}{\partial x^2} = 0 = 1300 - x/10 = v(x, 0)$ ...(*ii*)

$v_s'$ = steady voltage (after grounding the terminal end) when steady conditions are ultimately reached = $1300 - 1.3x$

$\therefore \quad v(x, t) = v_s' + v_t(x, t)$, where $v_t(x, t)$ is the transient part

$$= 1300 - 1.3x + \sum_{n=1}^{\infty} b_n e^{-(n^2\pi^2 t)/(l^2 RC)} \sin\frac{n\pi x}{l} \qquad \text{[By Eqn. } (viii)\text{, p. 722]} \qquad ...(iii)$$

where, $l$ = 1000 kilometers.

Putting $t = 0$, we have from Eqns. (*ii*) and (*iii*)

$$1300 - 0.1x = v(x, 0) = 1300 - 1.3x + \sum_{n=1}^{\infty} b_n \sin\frac{n\pi x}{l}$$

*i.e.*

$$1.2x = \sum_{n=1}^{\infty} b_n \sin\frac{n\pi x}{l} \text{ where } b_n = \frac{2}{l}\int_0^1 1.2 \sin\frac{n\pi x}{l}\,dx = \frac{2400}{\pi}\cdot\frac{(-1)^{n+1}}{n}$$

Hence

$$v(x, t) = 1300 - 1.3x + \frac{2400}{\pi}\sum_{n=1}^{\infty}\frac{(-1)^{n+1}}{n} e^{-(n^2\pi^2 t)/(l^2 RC)} \sin\frac{n\pi x}{1000}.$$

## PROBLEMS 18.7

1. Find the current $i$ and voltage $e$ in a line of length $l$, $t$ seconds after the ends are suddenly grounded, given that $i(x, 0) = i_0$, $e(x, 0) = e_0 \sin(\pi x/l)$.
Also $R$ and $G$ are negligible.
2. Show that a transmission line with negligible resistance and leakage propagates waves of current and potential with a velocity equal to $l/\sqrt{(LC)}$, where $L$ is the self-inductance and $C$ is the capacitance.
3. In a telephone wire of length $l$, a steady voltage distribution of 20 volts at the source end and 12 volts at the terminal end is maintained. At time $t = 0$, the receiving end is grounded. Assuming $L = 0$, $G = 0$, determine the voltage and current $t$ sec later where symbols have their usual meanings. (*U.P.T.U., 2011*)
4. Obtain the solution of the *radio equation*

$$\frac{\partial^2 v}{\partial x^2} = LC\frac{\partial^2 v}{\partial t^2}$$

appropriate to the case when a periodic e.m.f. $V_0$ cos$t$ $pt$ is applied at the end $x = 0$ of the line.

## 18.11 LAPLACE'S EQUATION IN THREE DIMENSIONS

We have seen that the three dimensional heat flow equation in steady state reduces to

$$\nabla^2 u = \frac{\partial^2 u}{\partial x^2} + \frac{\partial^2 u}{\partial y^2} + \frac{\partial^2 u}{\partial z^2} = 0 \qquad ...(1)$$

which is the *Laplace's equation in three dimensions.*

Laplace's equation also arises in the study of gravitational potential at $(x, y, z)$ of a particle of mass $m$ situated at $(\xi, \eta, \zeta)$ given by

$$\frac{Gm}{r} \text{ where } r = \sqrt{[(x-\zeta)^2 + (y-\eta)^2 + (z-\zeta)^2]}$$

This function is called the *potential of the gravitational field* and satisfies the Laplace's equation.

If a mass of density $\rho$ at $(\xi, \eta, \zeta)$ is distributed throughout a region $R$, then the gravitational potential $u$ at an external point $(x, y, z)$ is given by

$$u(x, y, z) = G \iiint_R \frac{\rho}{r}\, d\xi\, d\eta\, d\zeta \qquad ...(2)$$

Since $\nabla^2(1/r) = 0$ and $\rho$ is independent of $x$, $y$ and $z$, we get

$$\nabla^2 u = \iiint_R \rho\, \nabla^2 (1/r)\, d\xi\, d\eta\, d\zeta = 0.$$

This shows that the gravitational potential defined by Eqn. (2) also obeys Laplace's equation.

Thus Laplace's Eqn. (1) is one of the most important equations arising in connection with numerous problems of physics and engineering. *The theory of its solutions is called the* **potential theory** *and its solutions are called the* **harmonic functions.**

In most of the problems leading to Laplace's equation, it is required to solve the equation subject to certain boundary conditions. A proper choice of coordinate system makes the solution of the problem simpler. Now we proceed to take up the solutions of Eqn. (1) in its other forms.

## 18.12 SOLUTIONS OF THREE DIMENSIONAL LAPLACE'S EQUATION

(1) **Cartesian form of $\nabla^2 u = 0$** is $\dfrac{\partial^2 u}{\partial x^2} + \dfrac{\partial^2 u}{\partial y^2} + \dfrac{\partial^2 u}{\partial z^2} = 0$ ...(1)

Let $u = X(x)Y(y)Z(z)$ ...(2)

be a solution of Eqn. (1). Substituting Eqn. (2) in Eqn. (1) and dividing by $XYZ$, we obtain

$$\frac{1}{X}\frac{d^2X}{dx^2} + \frac{1}{Y}\frac{d^2Y}{dy^2} + \frac{1}{Z}\cdot\frac{d^2Z}{dz^2} = 0 \qquad ...(3)$$

which is of the form $F_1(x) + F_2(y) + F_3(z) = 0$.

As $x, y, z$ are independent, this will hold good only if $F_1, F_2, F_3$ are constants. Assuming these constants to be $k^2, l^2, -(k^2 + l^2)$ respectively, Eqn. (3) leads to the equations

$$\frac{d^2X}{dx^2} - k^2 X = 0, \quad \frac{d^2Y}{dy^2} - l^2 Y = 0, \quad \frac{d^2Z}{dz^2} + (k^2 + l^2)Z = 0$$

Their solutions are $X = c_1 e^{kx} + c_2 e^{-kx}$, $Y = c_3 e^{ly} + c_4 e^{-ly}$

$$Z = c_5 \cos \sqrt{(k^2 + l^2)}z + c_6 \sin \sqrt{(k^2 + l^2)}z$$

Thus a possible solution of Eqn. (1) is

$$u = (c_1 e^{kx} + c_2 e^{-kx})(c_3 e^{ly} + c_4 e^{-ly})\{c_5 \cos \sqrt{(k^2 + l^2)}z + c_6 \sin \sqrt{(k^2 + l^2)}z\}.$$

Since the three constants could have been taken as $-k^2, -l^2$ and $k^2 + l^2$, an alternative solution of Eqn. (1) will be

$$u = (c_1 \cos kx + c_2 \sin kx)(c_3 \cos ly + c_4 \sin ly)\{c_5 e^{\sqrt{(k^2 + l^2)}z} + c_6 e^{-\sqrt{(k^2 + l^2)}z}\}.$$

(2) **Cylindrical form of $\nabla^2 u = 0$** is $\dfrac{\partial^2 u}{\partial \rho^2} + \dfrac{1}{\rho}\dfrac{\partial u}{\partial \rho} + \dfrac{1}{\rho^2}\dfrac{\partial^2 u}{\partial \phi^2} + \dfrac{\partial^2 u}{\partial z^2} = 0$ ...(1)

Let $u = R(\rho)\, H(\phi)\, Z(z)$ [(iv)]

be a solution of Eqn. (1). Substituting it in Eqn. (1), and dividing by $RHZ$, we get

$$\frac{1}{R}\left(\frac{d^2R}{d\rho^2} + \frac{1}{\rho}\frac{dR}{d\rho}\right) + \frac{1}{\rho^2 H}\frac{d^2H}{d\phi^2} + \frac{1}{Z}\frac{d^2Z}{dz^2} = 0 \qquad ...(2)$$

Assuming that $\dfrac{d^2H}{d\phi^2} = -n^2H$ and $\dfrac{d^2Z}{dz^2} = k^2Z,$ ..(3)

Eqn. (2) reduces to $\dfrac{1}{R}\left(\dfrac{d^2R}{d\rho^2} + \dfrac{1}{\rho}\dfrac{dR}{d\rho}\right) - \dfrac{n^2}{\rho^2} + k^2 = 0$

or $$\rho^2\frac{d^2R}{d\rho^2} + \rho\frac{dR}{d\rho} + (k^2\rho^2 - n^2)R = 0.$$

This is Bessel's equation [§ 16.10 (1)] and its solution is $R = c_1J_n(k\rho) + c_2Y_n(k\rho)$. (*V.T.U., 2016*)

Also the solutions of Eqn. (3) are

$$H = c_3 \cos n\phi + c_4 \sin n\phi,\ Z = c_5e^{kz} + c_6e^{-kz}$$

Thus a solution of (1) is

$$u = [c_1J_n(k\rho) + c_2Y_n(k\rho)]\,[c_3 \cos n\phi + c_4 \sin n\phi]\,[c_5\, e^{kz} + c_6e^{-kz}]$$

which is known as a *cylindrical harmonic.*

**(3) Spherical form of $\nabla^2 u = 0$** is

$$\frac{\partial^2 u}{\partial r^2} + \frac{2}{r}\frac{\partial u}{\partial r} + \frac{1}{r^2}\frac{\partial^2 u}{\partial \theta^2} + \frac{\cot\theta}{r^2}\frac{\partial u}{\partial \theta} + \frac{1}{r^2\sin^2\theta}\frac{\partial^2 u}{\partial \phi^2} = 0 \qquad ...(1)$$

Let $u = R(r)\, G(\theta)\, H(\phi)$ be a solution of Eqn. (1).

Then $$\frac{1}{R}\left(r^2\frac{d^2R}{dr^2} + 2r\frac{dR}{dr}\right) + \frac{1}{G}\left(\frac{d^2G}{d\theta^2} + \cot\theta\frac{dG}{d\theta}\right) + \frac{1}{H\sin^2\theta}\frac{d^2H}{d\phi^2} = 0$$

Putting $\dfrac{1}{R}\left(r^2\dfrac{d^2R}{dr^2} + 2r\dfrac{dR}{dr}\right) = n(n+1)$ ...(2) and $\dfrac{1}{H}\dfrac{d^2H}{d\phi^2} = -m^2,$ ...(3)

the above equation takes the form

$$\frac{d^2G}{d\theta^2} + \cot\theta\frac{dG}{d\theta} + [n(n+1) - m^2 \operatorname{cosec}^2\theta]\,G = 0 \qquad ...(4)$$

Now differentiating the *Legendre's equation* (§ 16.13)

$$(1 - x^2)y'' - 2xy' + n(n+1)y = 0,$$

$m$ times with respect to $x$ and writing $u = d^my/dx^m$, we get

$$(1 - x^2)u'' - 2(m+1)xu' + (n - m)(n + m + 1)u = 0 \qquad ...(5)$$

Now putting $G = (1 - x^2)^{m/2}\, u$ in Eqn. (5), we get

$$(1 - x^2)\frac{d^2G}{dx^2} - 2x\frac{dG}{dx} + \left[n(n+1) - \frac{m^2}{1 - x^2}\right]G = 0 \qquad ...(6)$$

Now putting $x = \cos\theta$ in Eqn. (6), it reduces to Eqn. (4) and its solution is

$$G = c_1P_n^m(\cos\theta) + c_2Q_n^m(\cos\theta)$$

The solution of Eqn. (3) is $H = c_3 \cos m\phi + c_4 \sin m\phi$

To solve Eqn. (2), write $R = r^k$, so that $k(k-1) + 2k = n(n+1)$ which gives $k = n$ or $-(n+1)$

Thus $$R = c_5r^n + c_6r^{-n-1}$$

Hence the general solution of Eqn. (1) is

$$u = \sum_{n=0}^{\infty}\sum_{m=0}^{\infty} \{c_1P_n^m(\cos\theta) + c_2Q_n^m(\cos\theta)\}(c_3 \cos m\phi + c_4 \sin m\phi) \times (c_5r^n + c_6r^{-n-1})$$

Any solution of Eqn. (1) is known as a *spherical harmonic.*

**Example 18.27.** *Find the potential in the interior of a sphere of unit radius when the potential on the surface is $f(\theta) = \cos^2 \theta$.*

**Solution.** Take the origin at the centre of the given sphere $S$. Since the potential is independent of $\phi$ on $S$, so also is the potential at any point. Therefore, the Laplace's equation in spherical co-ordinates reduces to

$$\frac{\partial^2 u}{\partial r^2} + \frac{2}{r}\frac{\partial u}{\partial r} + \frac{1}{r^2}\frac{\partial^2 u}{\partial \theta^2} + \frac{\cot\theta}{r^2}\frac{\partial u}{\partial \theta} = 0 \qquad ...(i)$$

Putting $u(r, \theta) = R(r)\, G(\theta)$ in Eqn. (*i*) and proceeding as in § 18.12 (3), we obtain the equations

$$\frac{\partial^2 G}{d\theta^2} + \cot\theta \frac{dG}{d\theta} + n(n+1)G = 0 \qquad ...(ii)$$

and

$$\frac{1}{R}\left(r^2 \frac{d^2R}{dr^2} + 2r\frac{dR}{dr}\right) = n(n+1) \qquad ...(iii)$$

Putting $\cot\theta = v$, Eqn. (*ii*) takes the form

$$(1 - v^2)\frac{d^2G}{dv^2} - 2v\frac{dG}{dv} + n(n+1)G = 0$$

which is Legendre's equation. Its solutions are

$$G = P_n(v) = P_n(\cos\theta) \text{ for } n = 0, 1, 2,...$$

The solutions of Eqn. (*iii*) are $R_n(r) = r^n$, $\overline{R_n}(r) = 1/r^{n+1}$.

Hence the equation (*i*) has the following two sets of solutions

$$u_n(r, \theta) = c_n r^n P_n(\cos\theta) \text{ and } \bar{u}_n(r, \theta) = c_n P_n(\cos\theta)/r^{n+1}, \text{ where } n = 0, 1, 2,...$$

For points inside $S$, we have the general equation $u(r, \theta) = \sum_{n=0}^{\infty} c_n r^n P_n(\cos\theta)$ ...(*iv*)

On the boundary of $S$, $u(1, \theta) = f(\theta)$ $\therefore$ $f(\theta) = \sum_{n=0}^{\infty} c_n P_n(\cos\theta)$

which is Fourier-Legendre expansion of $f(\theta)$. Hence by Eqn. (5) p. 661, we have

$$c_n = \left(n + \frac{1}{2}\right)\int_{-1}^{1} f(\theta)\, P_n(x)\, dx \text{ where } x = \cos\theta.$$

$$= \left(n + \frac{1}{2}\right)\int_{-1}^{1} x^2 P_n(x) dx \qquad [\because f(\theta) = \cos^2\theta]$$

$$= \left(n + \frac{1}{2}\right)\int_{-1}^{1} \left[\frac{2}{3}P_2(x) + \frac{1}{3}P_0(x)\right] P_n(x)\, dx \qquad [\because P_2(x) = \tfrac{1}{2}(3x^2 - 1)]$$

Using the orthogonality of Legendre polynomials, we get

$c_n = 0$, except for $n = 0, 2$. Hence

$$c_0 = \frac{1}{2}\cdot\frac{1}{3}\int_{-1}^{1} P_0^2(x)\, dx = \frac{1}{3},\ c_2 = \frac{5}{2}\cdot\frac{2}{3}\int_{-1}^{0} P_2^2(x)\, dx = \frac{2}{3}.$$

Substituting in Eqn. (*iv*), we get $u(r, \theta) = \frac{1}{3} + \frac{2}{3}r^2 P_2(\cos\theta)$ or $u(r, \theta) = \frac{1}{3} + r^2(\cos^2\theta - \frac{1}{3})$.

**Example 18.28.** *In a system, the amplitude of motion is 5 meters and the period is 4 seconds. Find the time required by the particle in passing between the points which are at distance of 4 meters and 2 meters from the centre of the force and are on same side of it. Also find the velocities at these points.* (Calicut, 2019)

**Solution:** The equation of the SHM is

$$\frac{d^2x}{dt^2} = -\mu^2 x \qquad ...(i)$$

Time period $= \dfrac{2\pi}{\mu}$

Since time period is 4 seconds $\dfrac{2\pi}{\mu} = 4 \quad ; \quad \mu = \dfrac{\pi}{2}$

Hence solution of Eqn. (*i*) is $x = 5 \cos \dfrac{\pi}{2} t$

Let $t_1$ seconds and $t_2$ seconds by the times, when the particle is at a distance 4 meters and 2 meters respectively, from the centre of motion.

Then from Eqn. (*ii*) $4 = 5 \cos \dfrac{\pi}{2} t_1 \quad ; \quad t_1 = \dfrac{2}{\pi} \cos^{-1}\left(\dfrac{4}{5}\right)$

$$2 = 5 \cos \frac{\pi}{2} t_2 \quad ; \quad t_2 = \frac{2}{\pi} \cos^{-1}\left(\frac{2}{5}\right)$$

Time required in passing through these points $t_2 - t_1 = \dfrac{2}{\pi}\left(\cos^{-1}\dfrac{2}{5} - \cos^{-1}\dfrac{4}{5}\right) = 0.33$ seconds

Differentiating Eqn. (*ii*)

$$\frac{dx}{dt} = -\frac{5\pi}{2} \sin \frac{\pi}{2} t = -\frac{5\pi}{2} \sqrt{1 - \cos^2 \frac{\pi}{2} t} = -\frac{5\pi}{2} \sqrt{1 - \frac{\pi^2}{25}} \qquad \text{from eq}^n \ (ii)$$

$$v = -\frac{\pi}{2} \sqrt{25 - x^2}$$

When $x = 4$ meters $v = -\dfrac{\pi}{2} \sqrt{25 - 4^2} \ ; \ v = -\dfrac{3\pi}{2} = 4.71$ met/sec

When $x = 2$ meters, $v = -\dfrac{\pi}{2} \sqrt{25 - 2^2} \ ; \ v = \dfrac{\sqrt{21}\pi}{2}$ m/sec

The negative sign with velocity indicates that it is directed towards centre of the motion.

## PROBLEMS 18.8

1. Show that a solution of Laplace's equation in cylindrical co-ordinates, which remains finite at $r = 0$, may be expressed in the form

$$u = \sum_{n=0}^{\infty} J_n(kr) \{e^{kz} (A_n \cos n\theta + B_n \sin n\theta) + e^{-kz} (C_n \cos n\theta + D_n \sin n\theta)\}.$$

2. The potential on the surface of a unit sphere is $f(\theta) = \cos 2\theta$. Show that the potential at all points of space is given by

$$u(r, \theta) = 2r^2(\cos^2 \theta - 1/3) - \frac{1}{3} \text{ for } r < 1,$$

and
$$u(r, \theta) = 2r^{-3}(\cos^2 \theta - 1/3) - r^{-1}/3 \text{ for } r > 1.$$

3. Show that in spherical polar coordinates $(r, \theta, \phi)$, Laplace's equation possesses solutions of the form

$$(Ar^n + B/r^{n+1})P_n(\mu)e^{\pm im\phi},$$

where $\mu = \cos \theta$, $A, B, m, n$ are constants and $P_n(\mu)$ satisfies Legendre's equation

$$(1 - \mu^2) \frac{d^2P_n}{d\mu^2} - 2\mu \frac{dP_n}{d\mu} + \left\{n(n+1) - \frac{m^2}{1 - \mu^2}\right\} P_n = 0.$$

4. Initially tank $A$ contains 400 liters of brine and 100 kg of salt and tank $B$ contains 200 liters of fresh water. Brine from tank $A$ runs in to tank $B$ at 12 liters/minute and from tank $B$ into tank $A$ at 8 litres/min. If each tank is kept well stirred how much salt does tank $A$ contains after 50 minutes. **(Ans.** 34.315 kg)

5. In Thomson's experimental determination of the ratio $(m/e)$ of mass to charge of an electron, in which the electrons were subjected to an electric field of intensity $E$ and a magnetic field of intensity $H$, the equations

$$m \frac{d^2x}{dt^2} + He \frac{dy}{dt} = Ee \quad \text{and} \quad m \frac{d^2y}{dt^2} - He \frac{dx}{dt} = 0$$

were obtained. If $x = 0, y = 0$ and $\frac{dx}{dt} = \frac{dy}{dt} = 0$ at $t = 0$, find the parametric equations of the motion of an electron. *(Ujjain, 2019)*

6. Two coils of a transformer are identical with resistance $R$, inductance $L$, mutual inductance $M$, and a voltage E is impressed on primary. Determine the currents in the coils at any instant, assuming that there is no current initially.

## 18.13 OBJECTIVE TYPE OF QUESTIONS

### PROBLEMS 18.9

*Fill up the blanks in each of the following questions :*

1. The radio equations for the potential and current are .........
2. The partial differential equation representing variable heat flow in three dimensions, is .........
3. Temperature gradient is defined as .........
4. The differential equation $f_{xx} + 2f_{xy} + 4f_{yy} = 0$ is classified as .........
5. The partial differential equation of the transverse vibrations of a string is .........
6. The steady state temperature of a rod of length $l$ whose ends are kept at 30° and 40° is .........
7. The equation $u_t = c^2 u_{xx}$ is classified as ........
8. The two dimensional steady state heat flow equation in polar coordinates is .......
9. The mathematical function of the initial form of the string given by the following graph is .......
10. When a vibrating string fastened to two points $l$ apart, has an initial velocity $u_0$, its initial conditions are .......
11. In two dimensional heat flow, the temperature along the normal to the $xy$-plane is .......
12. If a square plate has its faces and the edge $y = 0$ insulated, its edges $x = 0$ and $x = a$ are kept at zero temperature and the fourth edge is kept at temperature $u$, then the boundary conditions for this problem are .........
13. If the ends $x = 0$ and $x = l$ are insulated in one dimensional heat flow problems, then the boundary conditions are .........

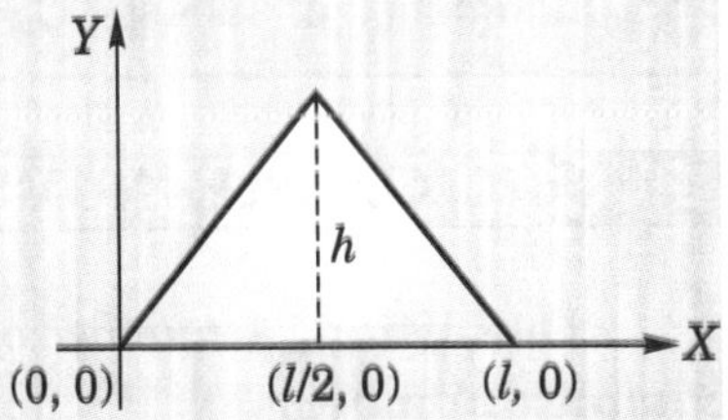

Fig. 18.11.

14. D'Alembert's solution of the wave equation is .........
15. The partial differential equation of 2-dimensional heat flow in .........
16. A rod 50 cm long with insulated sides has its end $A$ and $B$ kept at 20° and 70°C respectively. The steady state temperature distribution of the rod is .........
17. The three possible solutions of Laplace equation in polar coordinates are .........
18. Solution of $\frac{\partial u}{\partial x} = 4\frac{\partial u}{\partial y}$, given $u(0, y) = 8e^{-3y}$, is .........
19. Solution of $\frac{\partial z}{\partial x} + 4z = \frac{\partial z}{\partial t}$, given $z(x, 0) = 4e^{-3x}$, is .......
20. In the equation $\frac{\partial u}{\partial t} = \alpha^2 \frac{\partial^2 u}{\partial x^2}, \alpha^2$ represents .......
21. The telegraph equations for potential and current are ......
22. The general solution of one-dimensional heat flow equation when both ends of the bar are kept at zero temperature, is of the form ........
23. The three possible solutions of Laplace equation $\frac{\partial^2 v}{\partial x^2} + \frac{\partial^2 v}{\partial y^2} = 0$ are .......

CHAPTER 19

# Complex Numbers and Functions

**1.** Complex Numbers. **2.** Argand's diagram. **3.** Geometric representation of $z_1 \pm z_2$ ; $z_1z_2$ and $z_1/z_2$. **4.** De Moivre's theorem. **5.** Roots of a complex number. **6.** To expand sin $n\theta$, cos $n\theta$ and tan $n\theta$ in powers of sin $\theta$, cos $\theta$ and tan $\theta$ respectively; Addition formulae for any number of angles; To expand $\sin^m \theta$, $\cos^n \theta$ and $\sin^m \theta \cos^n \theta$ in a series of sines or cosines of multiples of $\theta$. **7.** Complex function: Definition. **8.** Exponential function of a complex variable. **9.** Circular functions of a complex variable. **10.** Hyperbolic functions. **11.** Inverse hyperbolic functions. **12.** Real and imaginary parts of circular and hyperbolic functions. **13.** Logarithmic functions of a complex variable. **14.** Summation of series – '$C + iS$' method. **15.** Approximations and Limits. **16.** Objective Type of Questions.

## 19.1 (1) COMPLEX NUMBERS

**Definition.** *A number of the form $x + iy$, where x and y are real numbers and $i = \sqrt{(-1)}$, is called a* **complex number.**

$x$ is called the *real part* of $x + iy$ and is written as $R(x + iy)$ and $y$ is called the *imaginary part* and is written as $I(x + iy)$.

*A pair of complex numbers $x + iy$ and $x - iy$ are said to be conjugate of each other.*

**Properties : (1)** *If $x_1 + iy_1 = x_2 + iy_2$, then $x_1 - iy_1 = x_2 - iy_2$.*

**(2)** *Two complex numbers $x_1 + iy_1 =$ and $x_2 + iy_2$ are said to be equal when*

$$R(x_1 + iy_1) = R(x_2 + iy_2),\ i.e.,\ x_1 = x_2$$

and

$$I(x_1 + iy_1) = I(x_2 + iy_2),\ i.e.,\ y_1 = y_2.$$

**(3)** *Sum, difference, product and quotient of any two complex numbers is itself a complex number.*

If $x_1 + iy_1$ and $x_2 + iy_2$ be two given complex numbers, then

(*i*) their sum $= (x_1 + iy_1) + (x_2 + iy_2) = (x_1 + x_2) + i(y_1 + y_2)$

(*ii*) their difference $= (x_1 + iy_1) - (x_2 + iy_2) = (x_1 - x_2) + i(y_1 - y_2)$

(*iii*) their product $= (x_1 + iy_1) + (x_2 + iy_2) = x_1x_2 - y_1y_2 + i(x_1y_2 + x_2y_1)$

and (*iv*) their quotient $= \dfrac{x_1 + iy_1}{x_2 + iy_2} = \dfrac{(x_1 + iy_1)(x_2 - iy_2)}{(x_2 + iy_2)(x_2 - iy_2)} = \dfrac{x_1x_2 + y_1y_2}{x_2^2 + y_2^2} + i\,\dfrac{x_2y_1 - x_1y_2}{x_2^2 + y_2^2}$

**Obs.** $i^2 = -1, i^3 = -i, i^4 = 1$. In general $i^{4n} = 1, i^{4n+1} = i, i^{4n+2} = -1, i^{4n+3} = -i$.

**(4)** *Every complex number $x + iy$ can always be expressed in the form $r$ (cos $\theta$ + i sin $\theta$).*

Put $R(x + iy)$, *i.e.*, $x = r \cos \theta$ ...(*i*)

and $I(x + iy)$, *i.e.*, $y = r \sin \theta$ ...(*ii*)

Squaring and adding, we get $x^2 + y^2 = r^2$ *i.e.* $r = \sqrt{(x^2 + y^2)}$ (taking positive square root only)

Dividing Eqn. (*ii*) by Eqn. (*i*), we get $y/x = \tan \theta$ *i.e.* $\theta = \tan^{-1} (y/x)$.

Thus $x + iy = r (\cos \theta + i \sin \theta)$ where $r = \sqrt{(x^2 + y^2)}$ and $\theta = \tan^{-1} (y/x)$.

**Definitions.** *The number* $r = +\sqrt{(x^2 + y^2)}$ *is called the* **modulus** *of* $x + iy$ *and is written as mod* $(x + iy)$ *or* $|x + iy|$.

*The angle* $\theta$ *is called the amplitude or argument of* $x + iy$ *and is written as amp* $(x + iy)$ *or arg* $(x + iy)$.

Evidently the amplitude $\theta$ has an infinite number of values. The value of $\theta$ which lies between $-\pi$ and $\pi$ is called the *principal value of the amplitude.* Unless otherwise specified, we shall take amp $(z)$ to mean the principal value.

**Note.** *$\cos\theta + i\sin\theta$ is briefly written as cis $\theta$ (pronounced as 'sis $\theta$')*

**(5)** *If the conjugate of* $z = x + iy$ be $\bar{z}$, *then*

(*i*) $R(z) = \frac{1}{2}(z + \bar{z}),\ I(z) = \frac{1}{2i}(z - \bar{z})$

(*ii*) $|z| = \sqrt{R^2(z) + I^2(z)} = |\bar{z}|$

(*iii*) $z\bar{z} = |z|^2$

(*iv*) $\overline{z_1 + z_2} = \bar{z}_1 + \bar{z}_2$

(*v*) $\overline{z_1 z_2} = \bar{z}_1 \cdot \bar{z}_2$

(*vi*) $\overline{(z_1/z_2)} = \bar{z}_1/\bar{z}_2$, where $\bar{z}_2 \neq 0$.

**Example 19.1.** *Reduce* $1 - \cos\alpha + i\sin\alpha$ *to the modulus amplitude form.*

**Solution.** Put $1 - \cos\alpha = r\cos\theta$ and $\sin\alpha = r\sin\theta$

$$\therefore \quad r = (1 - \cos\alpha)^2 + \sin^2\alpha = 2 - 2\cos\alpha = 4\sin^2\alpha/2$$

*i.e.,*
$$r = 2\sin\alpha/2$$

and
$$\tan\theta = \frac{\sin\alpha}{1 - \cos\alpha} = \frac{2\sin\alpha/2\cos\alpha/2}{2\sin^2\alpha/2} = \cot\alpha/2$$

$$= \tan\left(\frac{\pi}{2} - \frac{\alpha}{2}\right) \quad \therefore \theta = \frac{\pi - \alpha}{2}.$$

Thus $1 - \cos\alpha + i\sin\alpha = 2\sin\frac{\alpha}{2}\left[\cos\frac{\pi - \alpha}{2} + i\sin\frac{\pi - \alpha}{2}\right]$.

**Example 19.2.** *Find the complex number z if arg* $(z + 1) = \pi/6$ *and arg* $(z - 1) = 2\pi/3$.

**Solution.** Let $z = x + iy$ so that $z + 1 = (x + 1) + iy$ and $(z - 1) = (x - 1) + iy$

Since $\quad arg\ (z + 1) = \pi/6, \quad \therefore \tan^{-1}\left(\frac{y}{x+1}\right) = 30°$

*i.e.,*
$$\frac{y}{x+1} = \tan 30° = 1/\sqrt{3}, \text{ or } \sqrt{3y} = x + 1 \quad ...(i)$$

Also since $\quad \arg(z - 1) = 2\pi/3, \quad \therefore \tan^{-1}\left(\frac{y}{x+1}\right) = 120°$

*i.e.,*
$$\frac{y}{x-1} = \tan 120° = -\sqrt{3}, \text{ or } y = -\sqrt{3}x + \sqrt{3} \text{ or } \sqrt{3}y = -3x + 3 \quad ...(ii)$$

Subtracting Eqn. (*ii*) from Eqn. (*i*), we get $4x - 2 = 0$ *i.e.,* $x = 1/2$

From Eqn. (*i*), $\sqrt{3}\,y = 1/2 + 1$, *i.e.,* $y = \sqrt{3}/2$

Hence $z = \frac{1}{2} + \frac{\sqrt{3}}{2}i$.

**Example 19.3.** *Find the real values of x, y so that* $-3 + ix^2y$ *and* $x^2 + y + 4i$ *may represent complex conjugate numbers.*

**Solution.** If $z = -3 + ix^2y$, then $\bar{z} = x^2 + y + 4i$

so that
$$z = (x^2 + y) - 4i$$

$$\therefore \quad -3 + ix^2y = x^2 + y - 4i$$

Equating real and imaginary parts from both sides, we get

$$-3 = x^2 + y, x^2y = -4$$

Eliminating $x, (y + 3)y = -4$

or $y^2 + 3y - 4 = 0$ *i.e.*, $y = 1$ or $-4$

When $y = 1$, $x^2 = -3 - 1$ or $x = + 2i$ which is not feasible

When $y = -4$, $x^2 = 1$ or $x = \pm 1$

Hence $x = 1$, $y - 4$ or $x = -1, y = -4$.

## 19.2 (1) GEOMETRIC REPRESENTATION OF IMAGINARY NUMBERS

Let all the real numbers be represented along $X'OX$, the positive real numbers being along $OX$ and negative ones along $OX'$. Let $OA$ be equal to one unit of measurement (Fig. 19.1).

Take a point $L$ on $OX$ such that $OL = x\ (OA)$.

Then $L$ on $OX$ represents the positive real number $x$ and $i.ix = i^2x = -x$ is represented by a point $L'$ on $OX'$ distant $OL$ from $O$.

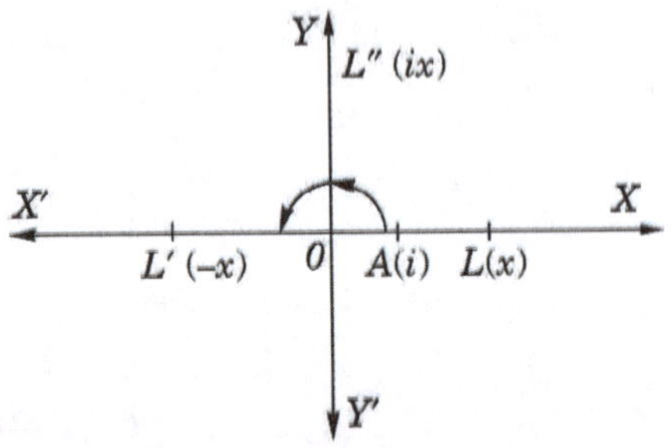

Fig. 19.1

From this we infer that the multiplication of the real number $x$ by $i$ twice amounts to the rotation of $OL$ through two right angles to the position $OL'$.

Thus it naturally follows that the multiplication of a real number by $i$ is equivalent to the rotation of $OL$ through one right angle to the position $OL''$.

*Hence, if $Y'OY$ be a line perpendicular to the real axis $X'OX$, then all imaginary numbers are represented by points on $Y'OY$, called the* **imaginary axis**, *the positive ones along $OY$ and negative ones along $OY'$.**

**Obs. Geometric interpretation of i*.** From the above, it is clear that $i$ is an operation which when multiplied to any real number makes it imaginary and rotates its direction through a right angle on the complex plane.

**(2) Geometric representation of complex numbers†**

Consider two lines $X'OX$, $Y'OY$ at right angles to each other.

Let all the real numbers be represented by points on the line $X'OX$ (called the *real axis*), positive real numbers being along $OX$ and negative ones along $OX'$. Let the point $L$ on $OX$ represent the real number $x$ (Fig. 19.2).

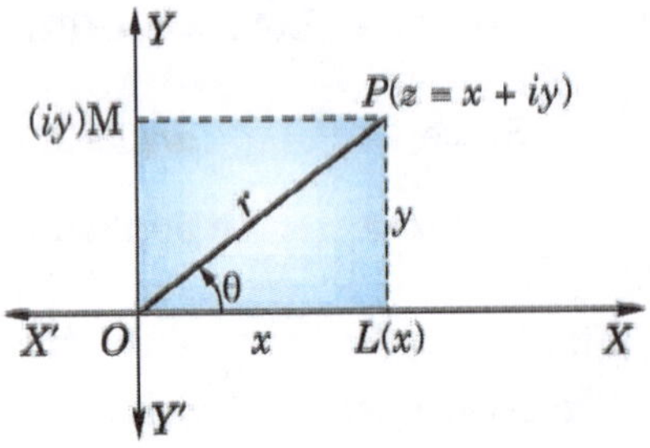

Fig. 19.2

Since the multiplication of a real number by $i$ is equivalent to the rotation of its direction through a right angle. Therefore, let all the imaginary numbers be represented by points on the line $Y'OY$ (called the *imaginary axis*), the positive ones along $OY$ and negative ones along $OY'$. Let the point $M$ on $OY$ represent the imaginary number $iy$.

Complete the rectangle $OLPM$. *Then the point whose cartesian coordinates are $(x, y)$ uniquely represents the complex number $z = x + iy$ on the complex plane $z$.* The diagram in which this representation is carried out is called the **Argand's diagram**.

If $(r, \theta)$ be the polar coordinates of $P$, then $r$ is the modulus of $z$ and $\theta$ is its amplitude.

**Obs.** Since a complex number has magnitude and direction, therefore, it can be represented like a vector. *Hereafter we shall often refer to the complex number $z = x + iy$ as*

(*i*) *the point $z$ whose co-ordinates are $(x, y)$ or* (*ii*) *the vector $z$ from $O$ to $P(x, y)$.*

**Example 19.4.** *The centre of a regular hexagon is at the origin and one vertex is given by $\sqrt{3} + i$ on the Argand diagram. Determine the other vertices.*

* The first mathematician to propose a geometric representation of imaginary number $i$ was *Kuhn of Denzig* (1750–51).

† The geometric representation of complex numbers came into mathematics through the memoire of *Jean Robert Argand*, Paris 1806.

**Solution.** Let $\vec{OA} = \sqrt{3} + i$ so that

$OA = 2$ and $\angle XOA = \tan^{-1} 1/\sqrt{3} = 30°$. (Fig. 19.3)

Being a regular hexagon, $OB = OC = 2$

$$\angle XOB = 30° + 60° = 90°$$

and $$\angle XOC = 30° + 120° = 150°$$

$$\therefore \quad \vec{OB} = 2(\cos 90° + i \sin 90°) = 2i$$

$$\vec{OC} = 2(\cos 150° + i \sin 150°) = -\sqrt{3} + i$$

Since $\vec{AD}, \vec{BE}, \vec{CF}$ are bisected at $O$,

$$\therefore \quad \vec{OD} = -\vec{OA} = -\sqrt{3} - i$$

$$\vec{OE} = -\vec{OB} = -2i \text{ and } \vec{OF} = -\vec{OC} = \sqrt{3} - i.$$

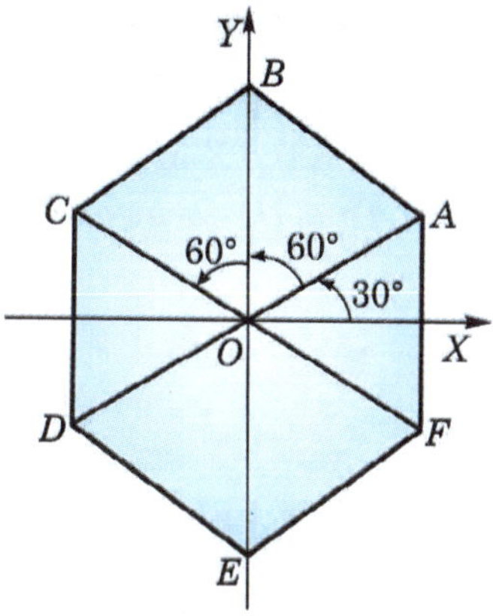

Fig. 19.3

## 19.3 (1) GEOMETRIC REPRESENTATION OF $z_1 + z_2$

Let $P_1, P_2$ represent the complex numbers $z_1 = x_1 + iy_1$ and $z_2 = x_2 + iy_2$. (Fig. 19.4)

Complete the parallelogram $OP_1PP_2$. Draw $P_1L$, $P_2M$ and $PN \perp$ s to $OX$. Also draw $P_1K \perp PN$.

Since $\quad ON = OL + LN = OL + OM = x_1 + x_2 \quad [\because LN = P_1K = OM]$

and $\quad NP = NK + KP = LP_1 + MP_2 = y_1 + y_2$.

The coordinates of $P$ are $(x_1 + x_2, y_1 + y_2)$ and it represents the complex number

$$z = x_1 + x_2 + i(y_1 + y_2) = (x_1 + iy_1) + (x_2 + iy_2) = z_1 + z_2.$$

Thus the point $P$ which is the extremity of the diagonal of the parallelogram having $OP_1$ and $OP_2$ as adjacent sides, represents the sum of the complex numbers $P_1(z_1)$ and $P_2(z_2)$ such that

$$|z_1 + z_2| = OP \text{ and amp } (z_1 + z_2) = \angle XOP.$$

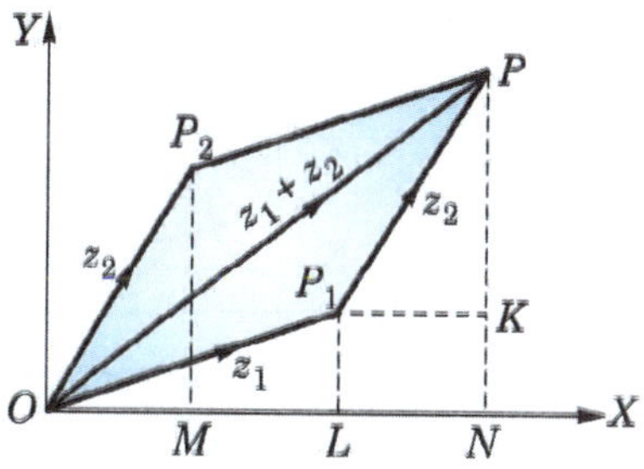

Fig. 19.4

**Obs.** Vectorially, we have $\vec{OP_1} + \vec{P_1P} = \vec{OP}$.

**(2) Geometric representation of $z_1 - z_2$**

Let $P_1, P_2$ represent the complex numbers $z_1 = x_1 + iy_1$ and $z_2 = x_2 + iy_2$ (Fig. 19.5). Then the subtraction of $z_2$ from $z_1$ may be taken as addition of $z_1$ to $-z_2$.

Produce $P_2O$ backwards to $R$ such that $OR = OP_2$. Then the coordinates of $R$ are evidently $(-x_2, -y_2)$ and so it corresponds to the complex number $-x_2 - iy_2 = -z_2$.

Complete the parallelogram $ORQP_1$, then the sum of $z_1$ and $(-z_2)$ is represented by $OQ$ *i.e.*, $z_1 - z_2 = \vec{OQ} = \vec{P_2P_1}$.

*Hence the complex number $z_1 - z_2$ is represented by the vector $P_2P_1$.*

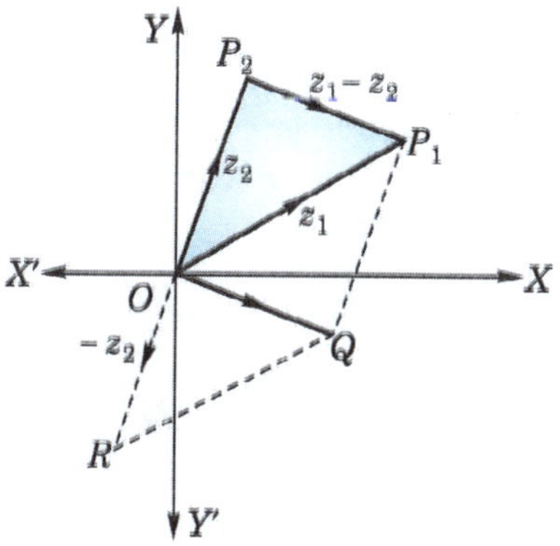

Fig. 19.5

**Obs.** By means of the relation $\vec{P_2P_1} = \vec{OP_1} - \vec{OP_2}$, any vector $\vec{P_2P_1}$ may be referred to the origin.

**Example 19.5.** *Find the locus of P(z) when*
(*i*) $|z - a| = k$;
(*ii*) *amp* $(z - a) = \alpha$, where $k$ and $\alpha$ are constants.

**Solution.** Let $a, z$ be represented by $A$ and $P$ in the complex plane, $O$ being the origin (Fig. 19.6).

Then $z - a = \vec{OP} - \vec{OA} = \vec{AP}$

(*i*) $|z - a| = k$ means that $AP = k$.

*Thus the locus of P(z) is a circle whose centre is A(a) and radius k.*

(*ii*) amp $(z - a)$, *i.e.*, amp $(\vec{AP}) = \alpha$, means that $AP$ always makes a constant angle with the $X$-axis.

*Thus the locus of P(z) is a straight line through A(a) making an ∠α with OX.*

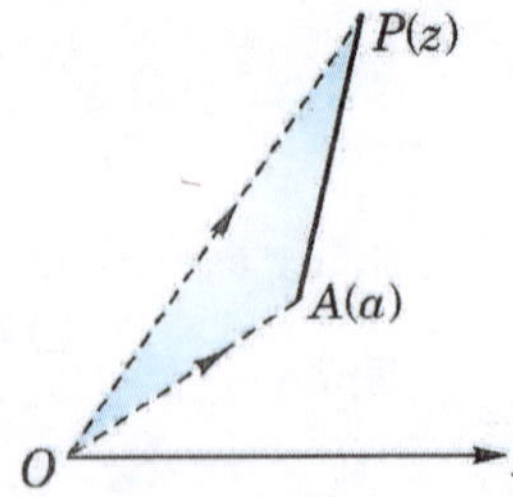


Fig. 19.6

**Example 19.6.** *Determine the region in the z-plane represented by*
(*i*) $1 < |z + 2i| \le 3$ (*ii*) $R(z) > 3$ (*iii*) $\pi/6 \le amp\ (z) \le \pi/3$.

**Solution.** (*i*) $|z + 2i| = 1$ is a circle with centre $(-2i)$ and radius 1 and $|z + 2i| = 3$ is a circle with the same centre and radius 3.

Hence $1 < |z + 2i| \le 3$ represents the region outside the circle $|z + 2i| = 1$ and inside (including circumference of) the circle $|z + 2i| = 3$ [Fig. 19.7].

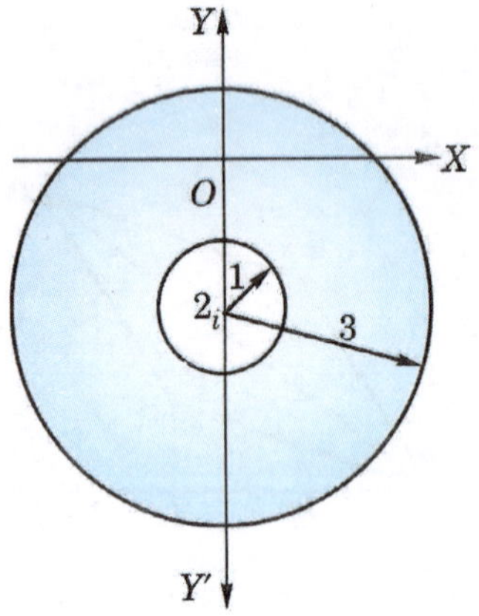


Fig. 19.7

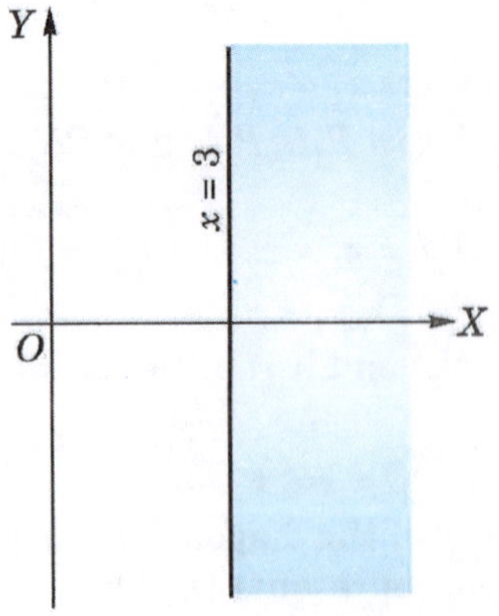


Fig. 19.8

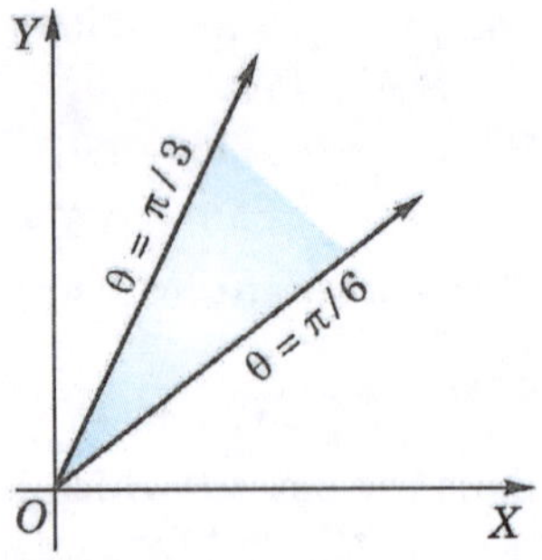


Fig. 19.9

(*ii*) $R(z) > 3$, defines all points $(z)$ whose real part is greater than 3. Hence it represents the region of the complex plane to the right of the line $x = 3$ [Fig. 19.8].

(*iii*) If $z = r(\cos\theta + i\sin\theta)$, then amp $(z) = \theta$.

$\therefore$ $\pi/6 \le$ amp $(z) \le \pi/3$ defines the region bounded by and including the lines $\theta = \pi/6$ and $\theta = \pi/3$. [Fig. 19.9].

**Example 19.7.** *If $z_1$, $z_2$ be any two complex numbers, prove that*
(*i*) $|z_1 + z_2| \le |z_1| + |z_2|$ [*i.e., the modulus of the sum of two complex numbers is less than or at the most equal to the sum of their moduli*].
(*ii*) $|z_1 - z_2| \ge |z_1| - |z_2|$ [*i.e., the modulus of the difference of two complex numbers is greater than or at the most equal to the difference of their moduli*].

**Solution.** Let $P_1$, $P_2$ represent the complex numbers $z_1$, $z_2$ (Fig. 19.10). Complete the parallelogram $OP_1PP_2$, so that

$$|z_1| = OP_1,\ |z_2| = OP_2 = P_1P,$$

and $$|z_1 + z_2| = OP.$$

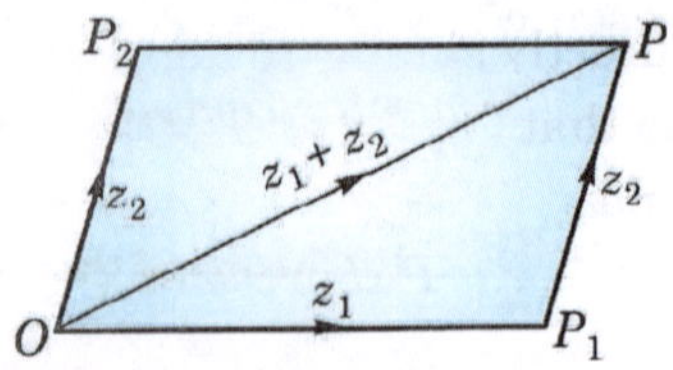


Fig. 19.10

Now from $\Delta OP_1P$, $OP \le OP_1 + P_1P$, the sign of equality corresponding to the case when $O$, $P_1$, $P$ are collinear.

Hence $$|z_1 + z_2| \le |z_1| + |z_2| \quad ...(i)$$

Again $$|z_1| = |(z_1 - z_2) + z_2| \le |z_1 - z_2| + |z_2| \quad \text{[By (i)]}$$

Thus $$|z_1 - z_2| \ge |z_1| - |z_2| \quad ...(ii)$$

**Obs.** $|z_1 + z_2 + z_3| \le |z_1| + |z_2| + |z_3|$.

In general, $|z_1 + z_2 + ... + z_n| \le |z_1| + |z_2| + ... + |z_n|$.

**Example 19.8.** *If* $|z_1 + z_2| = |z_1 - z_2|$, *prove that the difference of amplitudes of* $z_1$ *and* $z_2$ *is* $\pi/2$.

**Solution.** Let $z_1 + z_2 = r(\cos\theta + i\sin\theta)$ and $z_1 - z_2 = r(\cos\phi + i\sin\phi)$

Then $$2z_1 = r[(\cos\theta + \cos\phi) + i(\sin\theta + \sin\phi)]$$

$$= r\left\{2\cos\frac{\theta+\phi}{2}\cos\frac{\theta-\phi}{2} + 2i\sin\frac{\theta+\phi}{2}\cos\frac{\theta-\phi}{2}\right\}$$

or $$z_1 = r\cos\frac{\theta-\phi}{2}\left(\cos\frac{\theta+\phi}{2} + i\sin\frac{\theta+\phi}{2}\right) \text{ i.e., amp } (z_1) = \frac{\theta+\phi}{2} \quad ...(i)$$

Also $$2z_2 = r(\cos\theta - \cos\phi) + i(\sin\theta - \sin\phi)$$

$$= 2r\sin\frac{\theta-\phi}{2}\left(-\sin\frac{\theta+\phi}{2} + i\cos\frac{\theta+\phi}{2}\right)$$

or $$z_2 = r\sin\frac{\theta-\phi}{2}\left\{\cos\left(\frac{\pi}{2} + \frac{\theta+\phi}{2}\right) + i\sin\left(\frac{\pi}{2} + \frac{\theta+\phi}{2}\right)\right\}$$

*i.e.,* $$\text{amp } (z_2) = \frac{\pi}{2} + \frac{\theta+\phi}{2} \quad ...(ii)$$

Hence [Eqn. (*ii*) – Eqn. (*i*)], gives amp $(z_2)$ – amp $(z_1) = \frac{\pi}{2}$.

**Example 19.9.** *Show that the equation of the ellipse having foci at* $z_1$, $z_2$ *and major axis 2a, is* $|z - z_1| + |z - z_2| = 2a$.

*Also find its eccentricity.*

**Solution.** Let $P(z)$ be any point on the given ellipse (Fig. 19.11) having foci at $S(z_1)$ and $S'(z_2)$ so that $SP = |z - z_1|$ and $S'P = |z - z_2|$.

We know that $SP + S'P = AA'$ $(= 2a)$

*i.e.,* $$|z - z_1| + |z - z_2| = 2a$$

which is the desired equation of the ellipse.

Also we know that $SS' = 2ae$, $e$ being the eccentricity.

or $$\left|\vec{OS'} - \vec{OS}\right| = 2ae \quad \text{or} \quad |z_2 - z_1| = 2ae$$

or $$|z_1 - z_2| = 2ae \text{ whence } e = |z_1 - z_2|/2a.$$

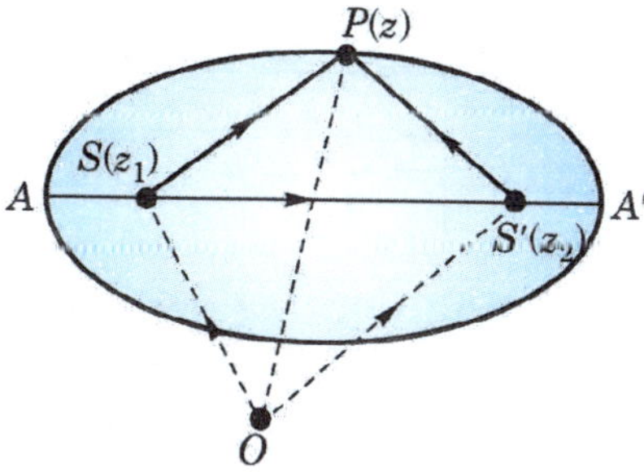

Fig. 19.11

**(3) Geometric Representation of $z_1z_2$.** Let $P_1$, $P_2$ represent the complex numbers

$$z_1 = x_1 + iy_1 = r_1(\cos\theta_1 + i\sin\theta_1)$$

and $$z_2 = x_2 + iy_2 = r_2(\cos\theta_2 + i\sin\theta_2)$$

Measure off $OA = 1$ along $OX$ (Fig. 19.12). Construct $\Delta OP_2P$ on $OP_2$ directly similar to $\Delta OAP_1$,

so that $$OP/OP_1 = OP_2/OA \text{ i.e., } OP = OP_1 \cdot OP_2 = r_1r_2$$

and $$\angle AOP = \angle AOP_2 + \angle P_2OP = \angle AOP_2 + \angle AOP_1 = \theta_2 + \theta_1$$

∴ $P$ represents the number

$$r_1r_2[\cos(\theta_1 + \theta_2) + i\sin(\theta_1 + \theta_2)].$$

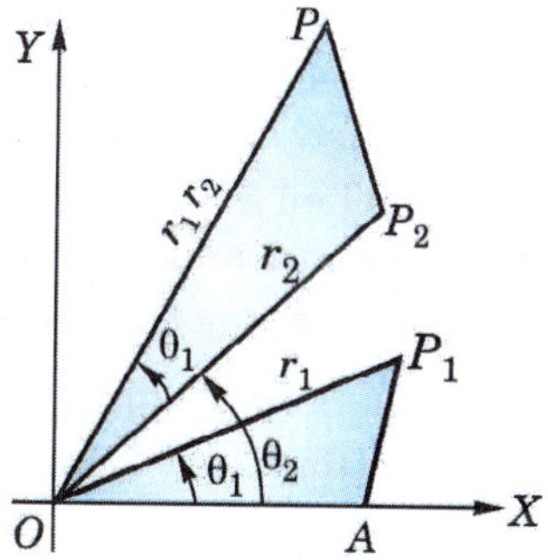

Fig. 19.12

Hence *the product of two complex numbers* $z_1$, $z_2$ *is represented by the point P, such that* Eqn. (*i*) $|z_1z_2| = |z_1| \cdot |z_2|$.

(*ii*) **amp $(z_1z_2)$ = amp $(z_1)$ + amp $(z_2)$.**

**Cor.** The effect of multiplication of any complex number $z$ by $\cos\theta + i\sin\theta$ is to rotate its direction through an angle $\theta$, for the modulus of $\cos\theta + i\sin\theta$ is unity.

**(4) Geometric representation of $z_1/z_2$.**

Let $P_1$, $P_2$ represent the complex numbers

$$z_1 = x_1 + iy_1 = r_1 (\cos \theta_1 + i \sin \theta_1)$$

and $$z_2 = x_2 + iy_2 = r_2 (\cos \theta_2 + i \sin \theta_2)$$

Measure off $OA = 1$, construct triangle $OAP$ on $OA$ directly similar to the triangle $OP_2P_1$ (Fig. 19.13), so that

$$\frac{OP}{OA} = \frac{OP_1}{OP_2} \quad i.e., \quad OP = \frac{OP_1}{OP_2} = \frac{r_1}{r_2}$$

and $$\angle XOP = \angle P_2OP_1 = \angle AOP_1 - \angle AOP_2 = \theta_1 - \theta_2.$$

$\therefore$ $P$ represents the number

$$(r_1/r_2) [\cos (\theta_1 - \theta_2) + i \sin (\theta_1 - \theta_2)].$$

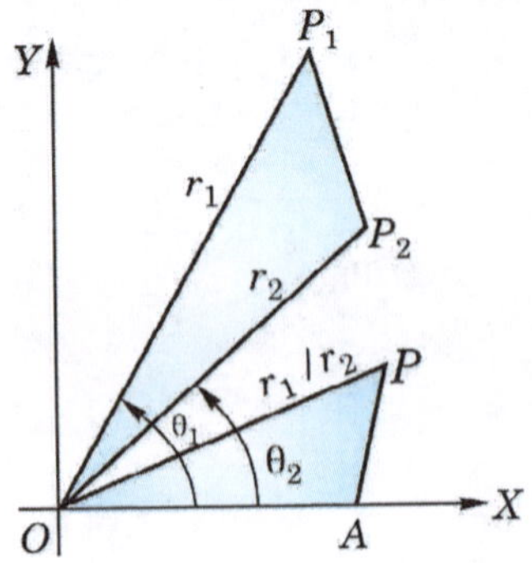

Fig. 19.13

*Hence the complex number $z_1/z_2$ is represented by the point P, such that*

(*i*) $|\mathbf{z_1/z_2}| = |\mathbf{z_1}|/|\mathbf{z_2}|$

(*ii*) **amp** $(\mathbf{z_1/z_2})$ = **amp** $(\mathbf{z_1})$ – **amp** $(\mathbf{z_2})$.

**Note.** *If $P_1(z_1)$, $P_2(z_2)$ and $P_3(z_3)$ be any three points, then*

$$\text{amp}\left(\frac{z_3 - z_2}{z_1 - z_2}\right) = \angle P_1P_2P_3.$$

Join $O$, the origin, to $P_1$, $P_2$, and $P_3$. Then from the Fig. 19.14, we have

$$\vec{P_2P_1} = z_1 - z_2 \quad \text{and} \quad \vec{P_2P_3} = z_3 - z_2$$

$$\therefore \quad \text{amp}\left(\frac{z_3 - z_2}{z_1 - z_2}\right) = \text{amp}\left[\frac{\vec{P_2P_3}}{\vec{P_2P_1}}\right]$$

$$= \text{amp}\,(\vec{P_2P_3}) - \text{amp}\,(\vec{P_2P_1}) = \beta - \alpha = \angle P_1P_2P_3.$$

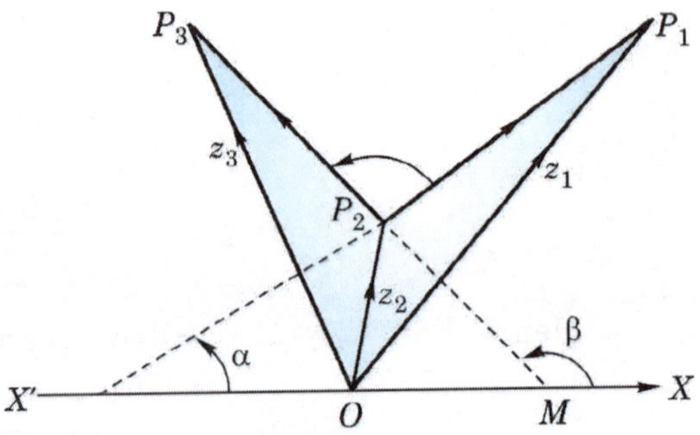

Fig. 19.14

**Example 19.10.** *Find the locus of the point z, when*

(*i*) $\left|\frac{z-a}{z-b}\right| = k$ (*ii*) *amp* $\left(\frac{z-a}{z-b}\right) = \alpha$ *where k and α are constants.*

**Solution.** Let $A(a)$ and $B(b)$ be any two fixed points on the complex plane and let $P(z)$ be any variable point (Fig. 19.15).

(*i*) Since $|z - a| = AP$ and $|z - b| = BP$.

$\therefore$ The point $P$ moves so that $\left|\frac{z-a}{z-b}\right| = \frac{|z-a|}{|z-b|} = \frac{AP}{BP} = k$

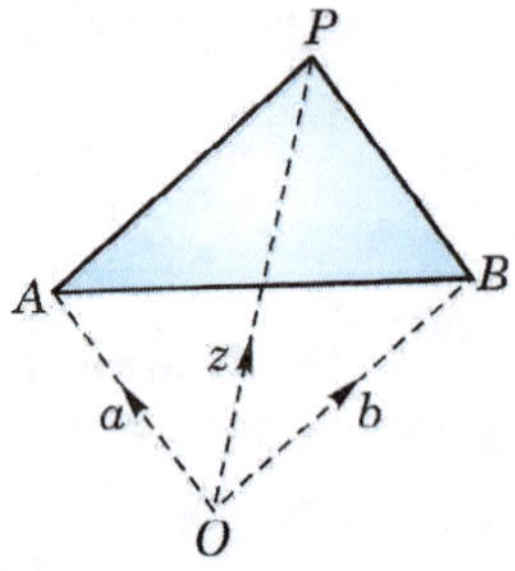

Fig. 19.15

*i.e.*, $P$ moves so that its distances from two fixed points are in a constant ratio, which is obviously the *Appollonius* circle.

When $k = 1$, $BP = AP$ *i.e.*, $P$ moves so that its distance from two fixed points are always equal and thus the locus of $P$ is the right bisector of $AB$.

*Hence the locus of P(z) is a circle (unless k = 1, when the locus is the right bisector of AB).*

**Obs.** For different values of $k$, the equation represents family of non-intersecting coaxial circles having $A$ and $B$ as its limiting points.

(*ii*) From the figure 19.16, we have amp $\left(\frac{z-a}{z-b}\right) = \angle APB = \alpha$.

*Hence the locus of P(z) is the arc APB of the circle which passes through the fixed points A and B.*

If, however, $P'(z')$ be a point on the lower arc $AB$ of this circle, then

amp $\left(\frac{z'-a}{z'-b}\right) = \angle BP'A = \alpha - \pi$, which shows that the locus of $P'$ is the arc $AP'B$ of the same circle.

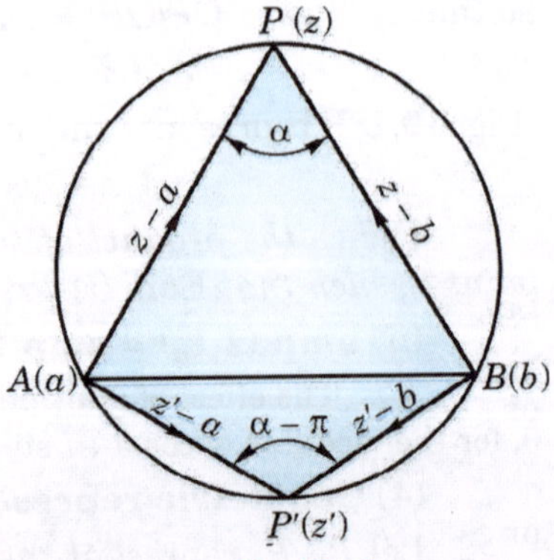

Fig. 19.16

**Obs.** For different values of $\alpha$ from $-\pi$ to $\pi$, the equation represents a family of intersecting coaxial circles having $AB$ as their common radical axis.

**Example 19.11.** *Prove that the area of the triangle whose vertices are the points $z_1, z_2, z_3$ on the Argand diagram is*

$$\sum\left\{\frac{(z_2 - z_3)|z_1|^2}{4iz_1}\right\}$$ (Ambala, 2018)

**Solution:** Let $z_1 = x_1 + iy_1$, $z_2 = x_2 + iy_2$ and $z_3 = x_3 + iy_3$

Area of the triangle $\Delta = \frac{1}{2}\begin{vmatrix} x_1 & y_1 & 1 \\ x_2 & y_2 & 1 \\ x_3 & y_3 & 1 \end{vmatrix} = \frac{1}{2i}\begin{vmatrix} x_1 & iy_1 & 1 \\ x_2 & iy_2 & 1 \\ x_3 & iy_3 & 1 \end{vmatrix}$ multiplying $c_2$ by $i$

$= \frac{1}{2i}\begin{vmatrix} x_1 & x_1 + iy_1 & 1 \\ x_2 & x_2 + iy_2 & 1 \\ x_3 & x_3 + iy_3 & 1 \end{vmatrix}$ Adding $c_1$ to $c_2 = \frac{1}{2i}\sum x.(z_2 - z_3) = \frac{1}{2i}\sum\left(\frac{z_1 + \bar{z}_2}{2}\right)(z_2 - z_3)$

$$= \frac{1}{4i}\left[\sum z_1(z_2 - z_3) + \sum \bar{z}_1(z_2 - z_3)\right]$$

Now $\sum z_1(z_2 - z_3) = 0$ $\quad \Delta = \frac{1}{4i}\sum \bar{z}_1(z_2 - z_3) = \frac{1}{4i}\sum\frac{\bar{z}_1(z_2 - z_3)z_1}{z_1} = \sum\frac{(z_1.\bar{z}_1)(z_2 - z_3)}{4iz_1} = \sum\frac{|z_1|^2(z_2 - z_3)}{4iz_1}$

$$= \sum\frac{(z_2 - z_3)|z_1|^2}{4iz_1} \qquad (\because z_1\bar{z}_1 = z_1^2)$$

**Example 19.12.** *If $z_1, z_2, z_3$ be the vertices of an isosceles triangle, right angled at $z_2$, prove that*

$$z_1^2 + z_3^2 + 2z_2^2 = 2z_2(z_1 + z_3).$$

**Solution.** Let $A(z_1)$, $B(z_2)$, $C(z_3)$ be the vertices of $\Delta ABC$ such that $AB = BC$ and $\angle ABC = \pi/2$. (Fig. 19.17)

Then $|z_1 - z_2| = |z_3 - z_2| = r$ (say).

If amp $(z_1 - z_2) = \theta$ then amp $(z_3 - z_2) = \pi/2 + \theta$

$\therefore$ $z_1 - z_2 = r(\cos\theta + i\sin\theta)$,

and $z_3 - z_2 = r\left[\cos\left(\frac{\pi}{2} + \theta\right) + i\sin\left(\frac{\pi}{2} + \theta\right)\right] = r(-\sin\theta + i\cos\theta)$

*i.e.,* $z_3 - z_2 = ir(\cos\theta + i\sin\theta) = i(z_1 - z_2)$

or $(z_3 - z_2)^2 = -(z_1 - z_2)^2$ or $z_1^2 + z_3^2 + 2z_2^2 = 2z_3(z_1 + z_2)$.

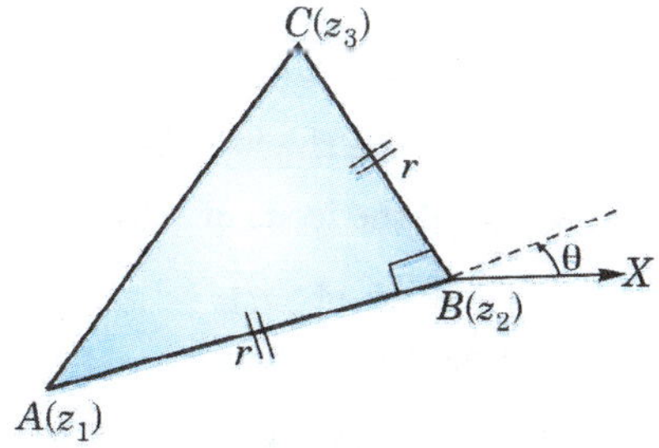

Fig. 19.17

**Example 19.13.** *If $z_1, z_2, z_3$ be the vertices of an equilateral triangle, prove that*

$$z_1^2 + z_2^2 + z_3^2 = z_1z_2 + z_2z_3 + z_3z_1.$$

**Solution.** Since $\Delta ABC$ is equilateral, therefore, $BC$ when rotated through 60° coincides with $BA$ (Fig. 19.18). But to turn the direction of a complex number through an $\angle\theta$, we multiply it by $\cos\theta + i\sin\theta$.

$\therefore$ $\vec{BC}(\cos\pi/3 + i\sin\pi/3) = \vec{BA}$

*i.e.,* $(z_3 - z_2)\left(\frac{1 + i\sqrt{3}}{2}\right) = z_1 - z_2$

or $i\sqrt{3}(z_3 - z_2) = 2z_1 - z_2 - z_3$

Squaring, $-3(z_3 - z_2)^2 = (2z_1 - z_2 - z_3)^2$

or $4(z_1^2 + z_2^2 + z_3^2 - z_1z_2 - z_2z_3 - z_3z_1) = 0$

whence follows the required condition.

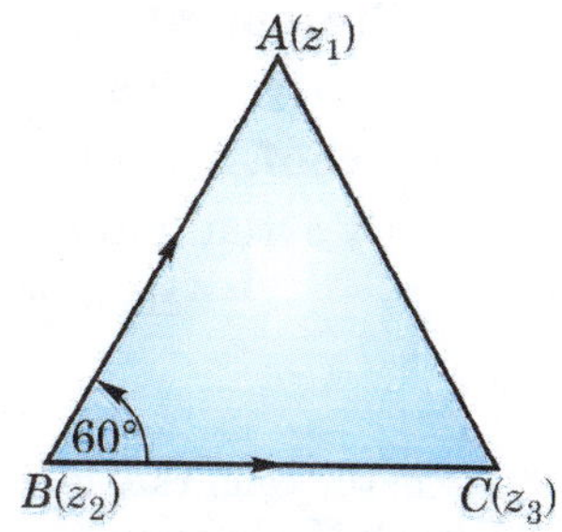

Fig. 19.18

## PROBLEMS 19.1

1. Express the following in the form $a + ib$: $\frac{3}{1+i} - \frac{1}{2-i} + \frac{1}{1-i}$. Also find its conjugate. (*V.T.U., 2016*)
2. Find $|z|$ and arg $(z)$, when $z = \frac{(2-3i)(1+i)}{2+i}$. (*P.T.U., 2015*)
3. Express the following in the modulus-amplitude form:

   (*i*) $1 + \cos\theta + i\sin\theta$ (*V.T.U., 2013*) (*ii*) $\frac{1}{(2+i)^2} - \frac{1}{(2-i)^2}$. (*V.T.U., 2011 S*)
4. If $x$ and $y$ are real, solve the equation $\frac{iy}{ix+1} - \frac{3y+4i}{3x+y} = 0$.
5. If $\alpha - i\beta = \frac{1}{a-ib}$, prove that $(\alpha^2 + \beta^2)(a^2 + b^2) = 1$.
6. Find what curve $z\bar{z} + (1+i)z + (1-i)\bar{z} = 0$ represents?
7. In an Argand diagram, show that $9 + i$, $4 + 13i$, $-8 + 8i$ and $-3 - 4i$ form a square.
8. If $|z_1| = |z_2|$ and amp $(z_1)$ + amp $(z_2) = 0$, then show that $z_1$ and $z_2$ are conjugate complex numbers.
9. A rectangle is constructed in the complex plane and its sides parallel to the axes and its centre is situated at the origin. If one of the vertices of the rectangle is $1 + i\sqrt{3}$, find the complex numbers representing the other three vertices of the rectangle. Find also the area of the rectangle.
10. The centre of a regular hexagon is at the origin and one vertex is given by $1 + i$ on the Argand diagram. Find the remaining vertices.
11. What domain of the $z$-plane is represented by

    (*i*) $2 \le |z+3| < 4$ (*ii*) $I(z) > 2$

    (*iii*) $\pi/3 <$ amp $(z) < \pi/2$ (*iv*) $|z+2| + |z-2| < 4$.
12. If $|z^2 - 1| = |z|^2 + 1$, prove that $z$ lies on the imaginary axis.
13. What are the loci given by (*i*) $|z-1| + |z+1| = 3$ (*ii*) $|z-3| = k\,|z+1|$ for $k = 1$ and 2.
14. Find the locus of $z$ given by :

    (*i*) $|z| = |z-2|$. (*ii*) $|3z-1| = |z-3|$.
15. Find the locus of $z$ :

    (*i*) when $\frac{z+i}{z+2}$ is real, (*ii*) when $\frac{z-i}{z-2}$ is purely imaginary.
16. Interpret geometrically the following loci.

    (*i*) $\left(\frac{z-i}{z+i}\right) \ge 2$ (*ii*) Arg $\left(\frac{z-1}{z+1}\right) = \frac{\pi}{2}$

## 19.4 DE MOIVRE'S THEOREM*

**Statement :** *If n be (i) an integer, positive or negative* $(\cos\theta + i\sin\theta)^n = \cos n\theta + i\sin n\theta$;
*(ii) a fraction, positive or negative, one of the values of* $(\cos\theta + i\sin\theta)^n$ is $\cos n\theta + i\sin n\theta$.

**Proof. Case I.** *When n is a positive integer.*

By actual multiplication

$$\text{cis}\,\theta_1\,\text{cis}\,\theta_2 = (\cos\theta_1\cos\theta_2 - \sin\theta_1\sin\theta_2) + i(\sin\theta_1\cos\theta_2 + \cos\theta_1\sin\theta_2)$$
$$= \cos(\theta_1+\theta_2) + i\sin(\theta_1+\theta_2), \text{ i.e., } \text{cis}\,(\theta_1+\theta_2)$$

*One of the remarkable theorems in mathematics; called after the name of its discoverer *Abraham De Moivre* (1667–1754), a French Mathematician.

Similarly cis $\theta_1$ cis $\theta_2$ cis $\theta_3$ = cis $(\theta_1 + \theta_2)$ cis $\theta_3$ = cis $(\theta_1 + \theta_2 + \theta_3)$

Proceeding in this way,

$$\text{cis } \theta_1 \text{ cis } \theta_2 \text{ cis } \theta_3 \ldots \text{ cis } \theta_n = \text{cis } (\theta_1 + \theta_2 + \theta_3 + \ldots + \theta_n)$$

Now putting $\theta_1 = \theta_2 = \theta_3 = \ldots = \theta_n = \theta$, we obtain $(\text{cis } \theta)^n = \text{cis } n\theta$.

**Case II.** *When n is a negative integer.*

Let $n = -m$, where $m$ is a + ve integer.

$$\therefore \quad (\text{cis } \theta)^n = (\text{cis } \theta)^{-m} = \frac{1}{(\text{cis } \theta)^m} = \frac{1}{\text{cis } m\theta} \quad \text{(By case I)}$$

$$= \frac{\cos m\theta - i \sin m\theta}{(\cos m\theta + i \sin m\theta)(\cos m\theta - i \sin m\theta)}$$

[Multiplying the num. and denom. by $(\cos m\theta - i \sin m\theta)$]

$$= \frac{\cos m\theta - i \sin m\theta}{\cos^2 m\theta + \sin^2 m\theta} = \cos m\theta - i \sin m\theta$$

$$= \cos(-m\theta) + i \sin(-m\theta) = \text{cis } (-m\theta) = \text{cis } n\theta \quad [\because \ -m = n]$$

**Case III.** *When n is a fraction, positive or negative.*

Let $n = p/q$, where $q$ is a + ve integer and $p$ is any integer +ve or –ve

Now $(\text{cis } \theta/q)^q = \text{cis } (q \cdot \theta/q) = \text{cis } \theta$

$\therefore$ Taking $q$th root of both sides cis $(\theta/q)$ is one of the $q$ values of $(\text{cis } \theta)^{1/q}$, *i.e.*, one of the values of $(\text{cis } \theta)^{1/q} = \text{cis } \theta/p$

Raise both sides to power $p$, then one of the values of $(\text{cis } \theta)^{p/q} = (\text{cis } \theta/q)^p = \text{cis } (p/q)\, \theta$ *i.e.*, one of the values of $(\text{cis } \theta)^n = \text{cis } n\theta$. (By case I and II)

Thus *the theorem is completely established for all rational values of n.*

**Cor.**
1. *cis* $\theta_1 \cdot$ *cis* $\theta_2 \ldots$ *cis* $\theta_n$ = *cis* $(\theta_1 + \theta_2 + \ldots + \theta_n)$
2. $(\cos \theta - i \sin \theta)^n = \cos n\theta - i \sin n\theta = (\cos \theta + i \sin \theta)^{-n}$
3. $(\text{cis } m\theta)^n = \text{cis } mn\theta = (\text{cis } n\theta)^m$.

**Example 19.14.** *Simplify* $\dfrac{(\cos\theta + i \sin\theta)^5}{(\sin\theta + i\cos\theta)^7}$. *(Puducherry; 2018)*

**Solution.** $\dfrac{(\cos\theta + i\sin\theta)^5}{(\sin\theta + i\cos\theta)^7} = (\cos\theta + i \sin\theta)^5 (\sin\theta + i\cos\theta)^{-7}$

$$= (\cos\theta + i\sin\theta)^5 \cdot \left[\cos\left\{\frac{\pi}{2} - \theta\right\} + i\sin\left\{\frac{\pi}{2} - \theta\right\}\right]^{-7}$$

By De Moivre's theorem

$$= (\cos 5\theta + i\sin 5\theta)\left[\cos\left\{\frac{-7\pi}{2} + 7\theta\right\} + i\sin\left\{-\frac{7\pi}{2} + 7\theta\right\}\right]$$

$$= \cos\left(5\theta - \frac{7\pi}{2} + 7\theta\right) + i\sin\left(5\theta - \frac{7\pi}{2} + 7\theta\right)$$

$$= \cos\left(\frac{7\pi}{2} - 12\theta\right) - i\sin\left(\frac{7\pi}{2} - 12\theta\right) = -\sin 12\,\theta + i\cos 12\,\theta$$

**Example 19.15.** *Prove that*

$$(1 + \cos\theta + i\sin\theta)^n + (1 + \cos\theta - i\sin\theta)^n = 2^{n+1}\cos^n(\theta/2) \cdot (\cos n\theta/2).$$

**Solution.** Put $1 + \cos\theta = r\cos\alpha$, $\sin\theta = r\sin\alpha$.

$$\therefore \quad r^2 = (1 + \cos\theta)^2 + \sin^2\theta = 2 + 2\cos\theta = 4\cos^2\theta/2 \qquad i.e., \quad r = 2\cos\theta/2$$

and

$$\tan\alpha = \frac{\sin\theta}{1 + \cos\theta} = \frac{2\sin\theta/2 \cdot \cos\theta/2}{2\cos^2\theta/2} = \tan\theta/2 \qquad i.e., \quad \alpha = \theta/2.$$

$$\therefore \quad \text{L.H.S.} = [r(\cos\alpha + i\sin\alpha)]^n + [r(\cos\alpha - i\sin\alpha)]^n$$

$= r^n[(\cos\alpha + i\sin\alpha)^n + (\cos\alpha - i\sin\alpha)^n] = r^n(\cos n\alpha + i\sin n\alpha + \cos n\alpha - i\sin n\alpha)$

$= r^n \cdot 2\cos n\alpha$ [Substituting the values of $r$ and $\alpha$]

$= 2^{n+1}\cos^n(\theta/2)\cos(n\theta/2)$.

**Example 19.16.** *Prove that* $\dfrac{(1+\sin\phi+i\cos\phi)^n}{(1+\sin\phi-i\cos\theta)^n} = \cos\left(\dfrac{n\pi}{2}-n\phi\right)+i\sin\left(\dfrac{n\pi}{2}-n\phi\right)$. *(Cochin, Uni., 2018)*

**Solution.** Let $1+\sin\phi = r\cos\theta$ and $\cos\phi = r\sin\theta$

$$\therefore \quad \tan\theta = \frac{r\sin\theta}{r\cos\theta} = \frac{\cos\phi}{1+\sin\phi} = \frac{\sin\left(\frac{\pi}{2}-\phi\right)}{1+\cos\left(\frac{\pi}{2}-\phi\right)} = \frac{2\sin\left(\frac{\pi}{4}-\frac{\phi}{2}\right)\cos\left(\frac{\pi}{4}-\frac{\phi}{2}\right)}{2\cos^2\left(\frac{\pi}{4}-\frac{\phi}{2}\right)}$$

$$\tan\theta = \tan\left(\frac{\pi}{4}-\frac{\phi}{2}\right) = \tan\left(\frac{\pi}{4}-\frac{\phi}{2}\right)$$

$$\theta = \frac{\pi}{4}-\frac{\phi}{2}$$

$$\text{L.H.S} \quad \frac{[r(\cos\theta+i\sin\theta)]^n}{[r(\cos\theta-i\sin\theta)]^n} = (\cos\theta+i\sin\theta)^n(\cos\theta-i\sin\theta)^{-n}$$

$$= (\cos n\theta + i\sin n\theta)(\cos n\theta + i\sin n\theta) = \cos 2n\theta + i\sin 2n\theta$$

$$= \cos 2n\left(\frac{\pi}{4}-\frac{\phi}{2}\right)+i\sin 2n\left(\frac{\pi}{4}-\frac{\phi}{2}\right) \quad \left[\because \theta = \frac{\pi}{4}-\frac{\phi}{2}\right]$$

$$= \cos\left(\frac{n\pi}{2}-n\phi\right)+i\sin\left(\frac{n\pi}{2}-n\phi\right) \text{ R.H.S}$$

**Example 19.17:** *If the complex numbers $z_1$, $z_2$ and $z_3$ be the vertices of on equilateral triangle and $z_0$ be the circumcentre of the triangle. Then show that $z_1^2 + z_2^2 + z_3^2 = 3z_0^2$.*

**Solution:** Consider $A(z_1)$, $B(z_2)$ and $C(z_3)$, be the vertices of the equilateral triangle and $D(z_0)$ its circumcentre on the Argand diagram. Let each side of the $\Delta ABC$ is $2a$.

$$z_1 = OA = -a$$

$$z_3 = OC = a$$

$$z_2 = i.\, OB$$

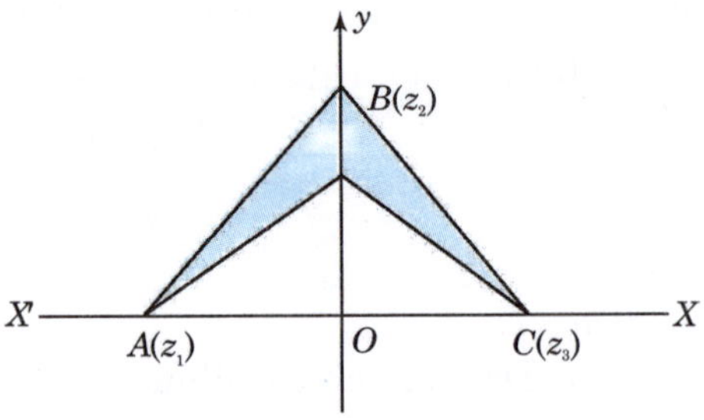


Fig. 19.19.

From $\Delta OBC$, $\dfrac{OB}{OC} = \tan 60° = \sqrt{3}$

$$z_2 = OB = a\sqrt{3}$$

From $\Delta ODC$, $\dfrac{OD}{OC} = \tan 30 = \dfrac{1}{\sqrt{3}}$

$$OD = \frac{a}{\sqrt{3}},\ z_0 = i,\ OD = i.\ \frac{a}{\sqrt{3}}$$

$$z_1^2+z_2^2+z_3^2 = (-a)^2+\left(ia\sqrt{3}\right)^2+(a)^2 = -a^2 = 3\left(\frac{ia}{\sqrt{3}}\right)^2$$

$$z_1^2+z_2^2+z_3^2 = 3z_0^{\,2}$$

**Example 19.18.** *If $\sin\alpha+\sin\beta+\sin\gamma = \cos\alpha+\cos\beta+\cos\gamma = 0$,*

*prove that* (i) $\sin 2\alpha + \sin 2\beta + \sin 2\gamma = 0$

(ii) $\sin 3\alpha + \sin 3\beta + \sin 3\gamma = 3\sin(\alpha+\beta+\gamma)$

(iii) $\sin 4\alpha + \sin 4\beta + \sin 4\gamma = 2\Sigma\sin 2(\alpha+\beta)$

(iv) $\sin(\alpha+\beta)+\sin(\beta+\gamma)+\sin(\gamma+\alpha) = 0$.

**Solution.** Let $a = \text{cis}\,\alpha,\ b = \text{cis}\,\beta$ and $c = \text{cis}\,\gamma$.

Then $a + b + c = (\cos\alpha + \cos\beta + \cos\gamma) + i(\sin\alpha + \sin\beta + \sin\gamma) = 0$ ...(1)

(*i*) $\frac{1}{a} + \frac{1}{b} + \frac{1}{c} = (\cos\alpha + i\sin\alpha)^{-1} + (\cos\beta + i\sin\beta)^{-1} + (\cos\gamma + i\sin\gamma)^{-1}$

$$= \sum \frac{\cos\alpha - i\sin\alpha}{\cos\alpha - i\sin\alpha} \cdot \frac{1}{\cos\alpha + i\sin\alpha} = \sum(\cos\alpha - i\sin\alpha)$$

$= (\cos\alpha + \cos\beta + \cos\gamma) - i(\sin\alpha + \sin\beta + \sin\gamma) = 0$ (Given)

or $bc + ca + ab = 0$ ...(2)

$\therefore\ a^2 + b^2 + c^2 = (a + b + c)^2 - 2(bc + ca + ab) = 0$ [By Eqn. (1) & Eqn. (2) ...(3)]

or $(cis\,\alpha)^2 + (cis\,\beta)^2 + (cis\,\gamma)^2 = cis\,2\alpha + cis\,2\beta + cis\,2\gamma = 0$

Equating imaginary parts from both sides, we get

$$\sin 2\alpha + \sin 2\beta + \sin 2\gamma = 0$$

(*ii*) Since $a + b + c = 0$, $\therefore\ a^3 + b^3 + c^3 = 3abc$

or $(cis\,\alpha)^3 + (cis\,\beta)^3 + (cis\,\gamma)^3 = 3\,cis\,\alpha\,cis\,\beta\,cis\,\gamma$

or $cis\,3\alpha + cis\,3\beta + cis\,3\gamma = 3\,cis\,(\alpha + \beta + \gamma)$

Equating imaginary parts from both sides, we get

$$\sin 3\alpha + \sin 3\beta + \sin 3\gamma = 3\sin(\alpha + \beta + \gamma)$$

(*iii*) From Eqn. (1), $a + b = -c$ or $(a + b)^2 = c^2$ or $a^2 + b^2 - c^2 = -2ab$

Again squaring, $a^4 + b^4 + c^4 + 2a^2b^2 - 2b^2c^2 - 2c^2a^2 = 4a^2b^2$

*i.e.*, $a^4 + b^4 + c^4 = 2(a^2b^2 + b^2c^2 + c^2a^2)$

or $(cis\,\alpha)^4 + (cis\,\beta)^4 + (cis\,\gamma)^4 = 2\sum (cos\,\alpha)^2\ (cis\,\beta)^2$

or $cis\,4\alpha + cis\,4\beta + cis\,4\gamma = 2\sum cis\,2\alpha\,cis\,2\beta = 2\sum cis\,2\,(\alpha + \beta)$

Equating imaginary parts from both sides, we get

$$\sin 4\alpha + \sin 4\beta + \sin 4\gamma = 2\sum \sin 2(\alpha + \beta)$$

(*iv*) From Eqn.(2), $ab + bc + ca = 0$

or $cis\,\alpha\,cis\,\beta + cis\,\beta\,cis\,\gamma + cis\,\gamma\,cis\,\alpha = 0$

or $cis\,(\alpha + \beta) + cis\,(\beta + \gamma) + cis\,(\gamma + \alpha) = 0$

Equating imaginary parts from both sides, we get

$$\sin(\alpha + \beta) + \sin(\beta + \gamma) + \sin(\gamma + \alpha) = 0$$

**Example 19.19:** *If $a = \cos 2\alpha + i\sin 2\alpha$, $b = \cos 2\beta + i\sin 2\beta$; $c = \cos 2\gamma + i\sin 2\gamma$ and $d = \cos 2\delta + i\sin 2\delta$. Then prove that $\sqrt{\frac{ab}{c}} + \sqrt{\frac{c}{ab}} = 2\cos(\alpha + \beta - \gamma)$*

**Solution:** We have $\frac{ab}{c} = \frac{(\cos 2\alpha + i\sin 2\alpha)(\cos 2\beta + i\sin 2\beta)}{\cos 2\gamma + i\sin 2\gamma} = \cos 2(\alpha + \beta - \gamma) + i\sin 2(\alpha + \beta - \gamma)$

$\therefore\ \sqrt{\frac{ab}{c}} = [\cos 2(\alpha + \beta - \gamma) + i\sin 2(\alpha + \beta - \gamma)]^{1/2} = \cos 2.\frac{1}{2}(\alpha + \beta - \gamma) + i\sin 2.\frac{1}{2}(\alpha + \beta - \gamma)$

$= \cos(\alpha + \beta - \gamma) + i\sin(\alpha + \beta - \gamma)$ ...(*i*)

$\therefore\ \sqrt{\frac{c}{ab}} = \left[\frac{1}{\cos(\alpha + \beta - \gamma) + i\sin(\alpha - \beta - \gamma)}\right]$ [By Eqn.(*i*)]

$= [\cos(\alpha + \beta - \gamma) + i\sin(\alpha + \beta - \gamma)]^{-1} = \cos(\alpha + \beta - \gamma) - i\sin(\alpha + \beta - \gamma)$ ...(*ii*)

By De Moivre's theorem

From Eqns.(*i*) and (*ii*) we have

$$\sqrt{\frac{ab}{c}} + \sqrt{\frac{c}{ab}} = 2\cos(\alpha + \beta - \gamma)$$

## PROBLEMS 19.2

1. Prove that (i) $\dfrac{(\cos 5\theta - i \sin 5\theta)^2 (\cos 7\theta + i \sin 7\theta)^{-3}}{(\cos 4\theta - i \sin 4\theta)^9 (\cos \theta + i \sin \theta)^5} = 1$

(ii) $\dfrac{(\cos \alpha + i \sin \alpha)^4}{(\sin \beta + i \cos \beta)^5} = \sin (4\alpha + 5\beta) - i \cos (4\alpha + 5\beta)$. (iii) $\left(\dfrac{\cos \theta + i \sin \theta)^4}{\sin \theta + i \cos \theta}\right) = \cos 8\theta + i \sin 8\theta$. *(C.S.V.T.U., 2015)*

2. If $a = \text{cis } 2\alpha$, $b = \text{cis } 2\beta$, $c = \text{cis } 2\gamma$ and $d = \text{cis } 2\delta$, prove that

(i) $\sqrt{\dfrac{ab}{c}} + \sqrt{\dfrac{c}{ab}} = 2 \cos (\alpha + \beta - \gamma)$ (ii) $\sqrt{\dfrac{ab}{cd}} + \sqrt{\dfrac{cd}{ab}} = 2 \cos (\alpha + \beta - \gamma - \delta)$.

3. Find the general value of $\theta$ which satisfies the equation

$$(\cos \theta + i \sin \theta)(\cos 2\theta + i \sin 2\theta) \ldots (\cos n\theta + i \sin n\theta) = 1.$$

4. Prove that (i) $(a + ib)^{m/n} + (a - ib)^{m/n} = 2(a^2 + b^2)^{m/2n} \cos \left(\dfrac{m}{n} \tan^{-1} \dfrac{b}{a}\right)$. *(C.S.V.T.U., 2014)*

(ii) $(1 + i)^n + (1 - i)^n = 2^{n/2 + 1} \cos n\pi/4$.

5. Simplify $[\cos \alpha - \cos \beta + i (\sin \alpha - \sin \beta)]^n + [\cos \alpha - \cos \beta - i (\sin \alpha - \sin \beta)]^n$

6. Prove that (i) $(1 + \sin \theta + i \cos \theta)^n + (1 + \sin \theta - i \cos \theta)^n = 2^{n+1} \cos^n \left(\dfrac{\pi}{4} - \dfrac{\theta}{2}\right) \cos \left(\dfrac{n\pi}{4} - \dfrac{n\theta}{2}\right)$.

*(P.T.U., 2014 ; C.S.V.T.U., 2011 S)*

(ii) $\left[\dfrac{1 + \sin \alpha + i \cos \alpha}{1 + \sin \alpha - i \cos \alpha}\right]^n = \cos \left(\dfrac{n\pi}{2} - n\alpha\right) + i \sin \left(\dfrac{n\pi}{2} - n\alpha\right)$. *(P.T.U., 2013)*

7. If $2 \cos \theta = x + 1/x$ and $2 \cos \phi = y + 1/y$, show that one of the values of

(i) $x^m y^n + \dfrac{1}{x^m y^n}$ is $2 \cos (m\theta + n\phi)$. *(C.S.V.T.U., 2012)*

(ii) $\dfrac{x^m}{y^n} + \dfrac{y^n}{x^m}$ is $2 \cos (m\theta - n\phi)$.

8. If $\alpha, \beta$ be the roots of $x^2 - 2x + 4 = 0$, prove that $\alpha^n + \beta^n = 2^{n+1} \cos n\pi/3$.

9. If $x^2 - 2x \cos \theta + 1 = 0$, show that $x^{2n} - 2x^n \cos n\theta + 1 = 0$.

10. If $x = \cos \alpha + i \sin \alpha$, $y = \cos \beta + i \sin \beta$, $z = \cos \gamma + i \sin \gamma$ and $x + y + z = 0$, then prove that

$$x^{-1} + y^{-1} + z^{-1} = 0.$$

11. If $\sin \theta + \sin \phi + \sin \psi = 0 = \cos \theta + \cos \phi + \cos \psi$, prove that

(i) $\cos 2\theta + \cos 2\phi + \cos 2\psi = 0$

(ii) $\cos 3\theta + \cos 3\phi + \cos 3\psi = 3 \cos (\theta + \phi + \psi)$

(iii) $\cos 4\theta + \cos 4\phi + \cos 4\psi = 2 \sum \cos 2(\phi + \psi)$.

12. If $\cos \alpha + \cos \beta + \cos \gamma = 0$ and $\sin \alpha + \sin \beta + \sin \gamma = 0$, prove that

(i) $\sin^2 \alpha + \sin^2 \beta + \sin^2 \gamma = \cos^2 \alpha + \cos^2 \beta + \cos^2 \gamma = 3/2$

(ii) $\cos (\alpha + \beta) + \cos (\beta + \gamma) + \cos (\gamma + \alpha) = 0$

13. If $\sin \alpha + 2 \sin \beta + 3 \sin \gamma = 0$, $\cos \alpha + 2 \cos \beta + 3 \cos \gamma = 0$, prove that $\sin 3\alpha + 8 \sin 3\beta + 27 \sin 3\gamma = 18 \sin (\alpha + \beta + \gamma)$ and $\cos 3\alpha + 8 \cos 3\beta + 27 \sin 3\gamma = 18 \cos (\alpha + \beta + \gamma)$. *(C.S.V.T.U., 2014 S)*

14. Show that $\{(\cos \theta - \cos \phi) + i(\sin \theta - \sin \phi)\}^n + \{\cos \theta - \cos \phi) - i(\sin\theta - \sin\phi\}^n = 2^{n+1} \sin n \dfrac{\theta - \phi}{2} \cos n\left(\dfrac{\pi + \theta + \phi}{2}\right)$.

15. If $x = \cos \theta + i \sin \theta$, show that

(i) $x^m + \dfrac{1}{x^m} = 2 \cos m\theta$ (ii) $x^m - \dfrac{1}{x^m} = 2 \text{ i} \sin m\,\theta$

16. Show that $(a + bi)^{m/n} + (a - ib)^{m/n} = 2(a^2 + b^2)^{m/2n} \cos\left(\dfrac{m}{n} \tan^{-1} b/a\right)$

17. Prove that $(1 + i)^n + (1 - i)^n = 2^{\frac{n+2}{2}} \cos \dfrac{n\pi}{4}$.

## 19.5 ROOTS OF A COMPLEX NUMBER

*There are q and only q distinct values of* $(\cos\theta + i\sin\theta)^{1/q}$, *q being an integer.*

Since $\cos\theta = \cos(2n\pi + \theta)$ and $\sin\theta = \sin(2n\pi + \theta)$, where $n$ is any integer.

$\therefore$ $\text{cis}\,\theta = \text{cis}\,(2n\pi + \theta)$.

By De Moivre's theorem one of the values of

$$(\text{cis}\,\theta)^{1/q} = [\text{cis}\,(2n\pi + \theta)]^{1/q} = \text{cis}\,(2n\pi + \theta)/q \qquad \ldots(1)$$

Giving $n$ the values 0, 1, 2, 3, ..., $(q-1)$ successively, we get the following $q$ values of $(\text{cis}\,\theta)^{1/q}$;

$$\left.\begin{array}{ll} \text{cis}\,\theta/q & (\text{for } n = 0) \\ \text{cis}\,(2\pi + \theta)/q & (\text{for } n = 1) \\ \text{cis}\,(4\pi + \theta)/q & (\text{for } n = 2) \\ \ldots\ldots\ldots\ldots & \ldots\ldots\ldots\ldots \\ \text{cis}\,[2(q-1)\pi + \theta]/q & (\text{for } n = q-1) \end{array}\right\} \qquad \ldots(2)$$

Putting $n = q$ in Eqn. (1), we get a value of $(\text{cis}\,\theta)^{1/q} = \text{cis}\,(2\pi + \theta/q) = \text{cis}\,\theta/q$, which is the same as the value of $n = 0$.

Similarly for $n = q + 1$, we get a value of $(\text{cis}\,\theta)^{1/q}$ to be $\text{cis}\,(2\pi + \theta)/q$, which is the same as the value for $n = 1$ and so on.

Thus, the values of $(\text{cis}\,\theta)^{1/q}$ for $n = q, q+1, q+2$ etc. are the mere repetition of the $q$ values obtained in Eqn. (2).

Moreover, the $q$ values given by Eqn.(2) are clearly distinct from each other, for no two of the angles involved therein are equal or differ by a multiple of $2\pi$.

Hence $(\text{cis}\,\theta)^{1/q}$ has $q$ and only $q$ distinct values given by Eqn. (2).

**Obs.** $(\text{cis}\,\theta)^{p/q}$ *where p/q is a rational fraction in its lowest terms, has also q and only q distinct values*; which are obtained by putting $n = 0, 1, 2, \ldots, q-1$ successively in $\text{cis}\,p\,(2n\pi + \theta)/q$.

Note that $(\text{cis}\,\theta)^{6/15}$ has only 5 distinct values and not 15; because 6/15 in its lowest terms = 2/5

$\therefore$ *In order to find the distinct values of* $(\text{cis}\,\theta)^{p/q}$ *always see that* $p/q$ *is in its lowest terms.*

**Note.** The above discussion can usefully be employed for extracting any assigned root of a given quantity. We have only to express it in the form $r(\cos\theta + i\sin\theta)$ and proceed as above.

**Example 19.20.** *Find the cube roots of unity and show that they form an equilateral triangle in the Argand diagram.*

**Solution.** If $x$ be a cube root of unity, then

$$x = (1)^{1/3} = (\cos 0 + i\sin 0)^{1/3} = (\text{cis}\,0)^{1/3} = (\text{cis}\,2n\pi)^{1/3} = \text{cis}\,2n\pi/3$$

where $n = 0, 1, 2$.

$\therefore$ the three vlaues of $x$ are $\text{cis}\,0 = 1$,

$$\text{cis}\,2\pi/3 = \cos 120° + i\sin 120° = -\frac{1}{2} + i\frac{\sqrt{3}}{2},$$

and
$$\text{cis}\,4\pi/3 = \cos 240° + i\sin 240° = -\frac{1}{2} - i\frac{\sqrt{3}}{2}.$$

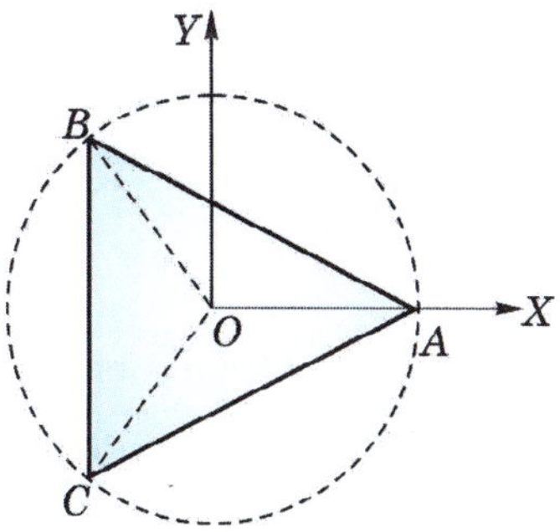

Fig. 19.20.

These three cube roots are represented by the points $A, B, C$ on the Argand diagram such that $OA = OB = OC$ and $\angle AOB = 120°$, $\angle AOC = 240°$ (Fig. 19.20).

$\therefore$ these points lie on a circle with centre $O$ and unit radius such that $\angle AOB = \angle BOC = \angle COA = 120°$ *i.e.*, $AB = BC = CA$.

Hence $A, B, C$ form an equilateral triangle.

**Example. 19.21.** *Find all the values of* $\left(\frac{1}{2} + \frac{\sqrt{3}i}{2}\right)^{3/4}$. (*D.T.U., 2013*)

*Also show that the continued product of these values is 1.*

**Solution.** Put $1/2 = r\cos\theta$ and $\sqrt{3}/2 = r\sin\theta$ so that $r = 1$ and $\theta = \pi/3$

$\therefore \quad (1/2 + \sqrt{3}i/2)^{3/4} = [(\cos \pi/3 + i \sin \pi/3)^3]^{1/4} = (\text{cis } \pi)^{1/4}$

$= [\text{cis } (2n + 1)\pi]^{1/4} = \text{cis } (2n + 1)\, \pi/4$ where $n = 0, 1, 2, 3.$

Hence the required values are cis $\pi/4$, cis $3\pi/4$, cis $5\pi/4$ and cis $7\pi/4$.

$\therefore$ their continued product $= \text{cis}\left(\frac{\pi}{4} + \frac{3\pi}{4} + \frac{5\pi}{4} + \frac{7\pi}{4}\right) = \text{cis } 4\pi = 1.$

**Example 19.22.** *Use De Moivre's theorem to solve the equation.*

$$x^4 - x^3 + x^2 - x + 1 = 0.$$

**Solution.** '$x^4 - x^3 + x^2 - x + 1$' is a G.P. with common ratio $(-x)$, therefore

$$\frac{1-(-x)^5}{1-(-x)} = 0, \qquad x \neq -1 \quad \text{or} \quad x^5 + 1 = 0$$

*i.e.,* $x^5 = -1 = \text{cis } \pi = \text{cis } (2n + 1)\pi$

$\therefore \quad x = [\text{cis } (2n + 1)\pi]^{1/5} = \text{cis } (2n + 1)\pi/5$, where $n = 0, 1, 2, 3, 4$

Hence the values are cis $\pi/5$, cis $3\pi/5$, cis $\pi$, cis $7\pi/5$, cis $9\pi/5$

or $\cos\frac{\pi}{5} + i \sin\frac{\pi}{5}, \cos\frac{3\pi}{5} + i \sin\frac{3\pi}{5}, -1, \cos\frac{3\pi}{5} - i \sin\frac{3\pi}{5}, \cos\frac{\pi}{5} - i \sin\frac{\pi}{5}$

Rejecting the value $-1$ which corresponds to the factor $x + 1$, the required roots are :

$$\cos \pi/5 \pm i \sin \pi/5, \cos 3\pi/5 \pm i \sin 3\pi/5.$$

**Example 19.23.** *Show that the roots of the equation $(x - 1)^n = x^n$, n being a positive integer are $\frac{1}{2}(1 + i \cot r\pi/n)$, where r has the values 1, 2, 3, ..., n – 1.*

**Solution.** Given equation is $\left(\frac{x-1}{x}\right)^n = 1$ or $1 - \frac{1}{x} = (1)^{1/n}$

or $\frac{1}{x} = 1 - (1)^{1/n} = 1 - \text{cis}\frac{2r\pi}{n}, r = 0, 1, 2, \ldots (n - 1).$ $[\because \quad 1 = \text{cis } 2\pi r]$

or $= \left(1 - \cos\frac{2r\pi}{n}\right) - i \sin\frac{2r\pi}{n} = 2 \sin^2\frac{r\pi}{n} - 2i \sin\frac{r\pi}{n} \cos\frac{r\pi}{n}$

$$\therefore \quad x = \frac{1}{2\sin\frac{r\pi}{n}} \cdot \frac{1}{\left(\sin\frac{r\pi}{n} - i\cos\frac{r\pi}{n}\right)} = \frac{\sin\frac{r\pi}{n} + i\cos\frac{r\pi}{n}}{2\sin\frac{r\pi}{n}}$$

$= \frac{1}{2}\left(1 + i\cot\frac{r\pi}{n}\right), r = 1, 2, \ldots (n - 1).$ $[\because \quad \cot 0 \to \infty]$

Hence the roots of the given equation are $\frac{1}{2}(1 + i \cot r\pi/n)$ where $r = 1, 2, 3, \ldots (n - 1)$.

**Example 19.24.** *Find the $7^{th}$ root of unity and prove that the sum of their $n^{th}$ power always vanishes unless n be a multiple of 7, n being an integer and then the sum is 7.* (Bhopal, 2019)

**Solution.** $1 = \cos 0 + i \sin 0 = \cos n\,\pi + i \sin^2 n\pi$

$\therefore \quad (1)^{1/7} = (\cos 2n\pi + i \sin 2n\pi)^{1/7}$

Apply De-Moivres theorem

$$(1)^{1/7} = \cos\frac{2n\pi}{7} + i\sin\frac{2n\pi}{7}$$

$$n = 0, 1, 2, 3 \ldots 6$$

So, the roots are $\cos 0 + i \sin 0 = 1$

$$\cos\frac{2\pi}{7} + i\sin\frac{2\pi}{7} = \alpha$$

$$\cos\frac{4\pi}{7}+i\sin\frac{4\pi}{7} = \alpha^2 \text{ and so on upto } \alpha^6$$

Let the sum of $n^{\text{th}}$ power

$$S = 1 + \alpha^n + (\alpha^2)^n + ... + (\alpha^6)^n = 1 + \alpha^n + (\alpha^n)^2 + ... + (\alpha^n)^6$$

$$S = \frac{1-(\alpha^n)^7}{1-\alpha^n} \qquad ...(i)$$

when $n$ is not a multiple of 7

$$(\alpha^n)^7 = \left(\cos\frac{2\pi}{7}+i\sin\frac{2\pi}{7}\right)^{7n} = \cos\frac{2\pi.7n}{7}+i\sin\frac{2\pi.7n}{7} = \cos n\pi + \text{i} \sin n\pi = 1$$

$\therefore$ $$1-(\alpha^n)^7 = 1-1 = 0$$

$$1-\alpha^n = 1-\left(\cos\frac{2n\pi}{7}+i\sin\frac{2n\pi}{7}\right) \neq 0$$

as $n$ is not a multiple of 7

$$S = 0, \text{ from (i)}$$

$\therefore$ $$\alpha^n = \left(\cos\frac{2\pi}{7}+i\sin\frac{2\pi}{7}\right)^{7m} = \cos 2\pi m + i \sin 2\pi m = 1$$

Similarly $\alpha^{2n}$, $\alpha 4n$ etc each term is 1

$\therefore$ $$S = 1 + \alpha^n + \alpha^{2n} + \ldots \alpha^{6n} = 1 + 1 + 1 \ldots + 1 = 7$$

**Example 19.25:** *Find all the values of* $(1+\sqrt{3}i)^{1/3}$. *(UPTU, 2018)*

**Solution:** Let

$$1 = r\cos\theta$$

$$\sqrt{3} = r\sin\theta.$$

$$r = 2$$

$$\theta = \frac{\pi}{3}$$

$\therefore$ $$1+\sqrt{3}i = 2\left(\cos\frac{\pi}{3}+i\sin\frac{\pi}{3}\right) = 2\left[\cos\left(2n\pi+\frac{\pi}{3}\right)+i\sin\left(2n\pi+\frac{\pi}{3}\right)\right]$$

$$= 2\left[\cos(6n+1)\frac{\pi}{3}+i\sin(6n+1)\frac{\pi}{3}\right]$$

$\therefore$ $$(1+\sqrt{3}i)^{1/3} = (2)^{1/3}\left[\cos(6n++1)\frac{\pi}{3}+i\sin(6n+1)\frac{\pi}{3}\right]^{1/3} = (2)^{1/3}\left[\cos(6n+1)\frac{\pi}{9}+i\sin(6n+1)\frac{\pi}{9}\right]$$

for $n = 0, 1, 2$

Hence the values are $$(2)^{1/3}\left(\cos\frac{\pi}{9}+i\sin\frac{\pi}{9}\right), (2)^{1/3}\left(\cos\frac{7\pi}{9}+i\sin\frac{7\pi}{9}\right), (2)^{1/3}\left(\cos\frac{13\pi}{9}+i\sin\frac{13\pi}{9}\right)$$

Thus the required values are

$$(2)^{1/3}\left(\cos\frac{r\pi}{9}+i\sin\frac{r\pi}{9}\right), \text{ for } r = 1, 7, 13$$

**Example 19.26.** *Find the equation whose roots are $2\cos\pi/7$, $2\cos 3\pi/7$, $2\cos 5\pi/7$.*

**Solution.** Let $y = \cos\theta + i\sin\theta$, where $\theta = \pi/7, 3\pi/7, ..., 13\pi/7$.

Then $$y^7 = (\cos\theta + i\sin\theta)^7 = \cos 7\theta + i\sin 7\theta = -1 \quad \text{or} \quad y^7 + 1 = 0$$

or $$(y + 1)(y^6 - y^5 + y^4 - y^3 + y^2 - y + 1) = 0$$

Leaving the factor $y + 1$ which corresponds to $\theta = \pi$,

We get $$y^6 - y^5 + y^4 - y^3 + y^2 - y + 1 = 0 \qquad ...(i)$$

Its roots are $y = \text{cis}\,\theta$ where $\theta = \pi/7, 3\pi/7, 5\pi/7, 9\pi/7, 11\pi/7, 13\pi/7$.

Dividing (*i*) by $y^3$, $(y^3 + 1/y^3) - (y^2 + 1/y^2) + (y + 1/y) - 1 = 0$

or $$\{(y + 1/y)^3 - 3(y + 1/y)\} - \{(y + 1/y)^2 - 2\} - (y + 1/y) - 1 = 0$$

or $$x^3 - x^2 - 2x + 1 = 0 \quad \text{...(ii)}$$

where $x = y + 1/y = 2\cos\theta$.

Now since $\cos 13\pi/7 = \cos \pi/7$, $\cos 11\pi/7 = \cos 3\pi/7$, $\cos 9\pi/7 = \cos 5\pi/7$

Hence the roots of (*ii*) are $2\cos\frac{\pi}{7}, 2\cos\frac{3\pi}{7}, 2\cos\frac{5\pi}{7}$.

## PROBLEMS 19.3

1. Find all the values of

   (*i*) $(1 - i\sqrt{3})^{1/4}$ (*V.T.U., 2016*) (*ii*) $(-1 + i)^{2/5}$

   (*iii*) $(-1 + i\sqrt{3})^{3/2}$ (*iv*) $(1 + i\sqrt{3})^{1/3} + (1 - i\sqrt{3})^{1/3}$.
2. If $\omega$ is a complex cube root of unity, prove that $1 + \omega + \omega^2 = 0$.
3. Find all the values of $(-1)^{1/6}$.
4. Mark by points on the Argand diagram, all the values of $(1 + i\sqrt{3})^{1/5}$ and verify that they form a pentagon.
5. Use De Moivre's theorem to solve the following equations :

   (*i*) $(z - 1)^3 = 8$ (*P.T.U., 2015 S*) (*ii*) $x^7 + x^4 + x^3 + 1 = 0$

   (*iii*) $x^9 + x^5 - x^4 - 1 = 0$ (*iv*) $(x - 1)^5 + x^5 = 0$.
6. Find the roots common to the equations $x^4 + 1 = 0$ and $x^6 - i = 0$.
7. Solve the equation $x^{12} - 1 = 0$ and find which of its roots satisfy the equation $x^4 + x^2 + 1 = 0$.
8. Show that the roots of $(x + 1)^7 = (x - 1)^7$ are given by $\pm i \cot r\pi/7$, $r = 1, 2, 3$.
9. Find the equation whose roots are $2\cos 2\pi/7$, $2\cos 4\pi/7$, $2\cos 6\pi/7$.
10. Prove that the *n*th roots of unity form a geometric progression.

    Also show that the sum of these *n* roots is zero and their product is $(-1)^{n-1}$. (*P.T.U., 2013 S*)

## 19.6 (1) TO EXPAND sin nθ, cos nθ AND tan nθ IN POWERS OF sin θ, cos θ AND tan θ RESPECTIVELY (n BEING A POSITIVE INTEGER)

We have, $\cos n\theta + i\sin n\theta = (\cos\theta + i\sin\theta)^n$ (By De Moivre's theorem)

$$= \cos^n\theta + {}^nC_1\cos^{n-1}\theta\,(i\sin\theta) + {}^nC_2\cos^{n-2}\theta\,(i\sin\theta)^2 + {}^nC_3\cos^{n-3}\theta\,(i\sin\theta)^3 + \dots$$

(By Binomial theorem)

$$= (\cos^n\theta - {}^nC_2\cos^{n-2}\theta\sin^2\theta + \dots) + i\,({}^nC_1\cos^{n-1}\theta\sin\theta - {}^nC_3\cos^{n-3}\theta\sin^3\theta + \dots)$$

Equating real and imaginary parts from both sides, we get

$$\cos n\theta = \cos^n\theta - {}^nC_2\cos^{n-2}\theta\sin^2\theta + {}^nC_4\cos^{n-4}\theta\sin^4\theta - \dots \quad \text{...(1)}$$

$$\sin n\theta = {}^nC_1\cos^{n-1}\theta\sin\theta - {}^nC_3\cos^{n-3}\theta\sin^3\theta + {}^nC_5\cos^{n-5}\theta\sin^5\theta - \dots \quad \text{...(2)}$$

Replacing every $\sin^2\theta$ by $1 - \cos^2\theta$ in Eqn. (1) and every $\cos^2\theta$ by $1 - \sin^2\theta$ in Eqn. (2), we get the desired expansions of $\cos n\theta$ and $\sin n\theta$.

Dividing Eqn. (2) by Eqn. (1),

$$\tan n\theta = \frac{{}^nC_1\cos^{n-1}\theta\sin\theta - {}^nC_3\cos^{n-3}\theta\sin^3\theta + {}^nC_5\cos^{n-5}\theta\sin^5\theta - \dots}{\cos^n\theta - {}^nC_2\cos^{n-2}\theta\sin^2\theta + {}^nC_4\cos^{n-4}\theta\sin^4\theta - \dots}$$

and dividing numerator and denominator by $\cos^n\theta$, we get

$$\tan n\theta = \frac{{}^nC_1\tan\theta - {}^nC_3\tan^3\theta + {}^nC_5\tan^5\theta - \dots}{1 - {}^nC_2\tan^2\theta + {}^nC_4\tan^4\theta - \dots}.$$

**Example 19.27.** *Express cos 6θ in terms of cos θ.*

**Solution.** We know that $\cos n\theta = \cos^n\theta - {}^nC_2\cos^{n-2}\theta\sin^2\theta + {}^nC_4\cos^{n-4}\theta\sin^4\theta - \dots$

Put $n = 6$, then $\cos 6\theta = \cos^6\theta - {}^6C_2 \cos^4\theta \sin^2\theta + {}^6C_4 \cos^2\theta \sin^4\theta - {}^6C_6 \sin^6\theta$

$$= \cos^6\theta - 15\cos^4\theta(1-\cos^2\theta) + 15\cos^2\theta(1-\cos^2\theta)^2 - (1-\cos^2\theta)^3$$

$$= 32\cos^6\theta - 48\cos^4\theta + 18\cos^2\theta - 1.$$

**(2) Addition formulae for any number of angles**

We have, $\cos(\theta_1 + \theta_2 + \ldots + \theta_n) + i\sin(\theta_1 + \theta_2 + \ldots + \theta_n)$

$$= (\cos\theta_1 + i\sin\theta_1)(\cos\theta_2 + i\sin\theta_2)\ldots(\cos\theta_n + i\sin\theta_n)$$

Now $\cos\theta_1 + i\sin\theta_1 = \cos\theta_1(1 + i\tan\theta_1)$, $\cos\theta_2 + i\sin\theta_2 = \cos\theta_2(1 + i\tan\theta_2)$ and so on.

$\therefore$ $\cos(\theta_1 + \theta_2 + \ldots + \theta_n) + i\sin(\theta_1 + \theta_2 + \ldots + \theta_n)$

$$= \cos\theta_1\cos\theta_2\ldots\cos\theta_n(1 + i\tan\theta_1)(1 + i\tan\theta_2)\ldots(1 + i\tan\theta_n)$$

$$= \cos\theta_1\cos\theta_2\ldots\cos\theta_n[1 + i(\tan\theta_1 + \tan\theta_2 + \ldots + \tan\theta_n) + i^2(\tan\theta_1\tan\theta_2 + \tan\theta_2\tan\theta_3 + \ldots) + i^3(\tan\theta_1\tan\theta_2\tan\theta_3 + \ldots) + \ldots + \ldots]$$

$$= \cos\theta_1\cos\theta_2\ldots\cos\theta_n(1 + is_1 - s_2 - is_3 + s_4 + \ldots)$$

where $s_1 = \tan\theta_1 + \tan\theta_2 + \ldots + \tan\theta_n$, $s_2 = \Sigma\tan\theta_1\tan\theta_2$, $s_3 = \Sigma\tan\theta_1\tan\theta_2\tan\theta_3$ etc.

Equating real and imaginary parts, we have

$$\cos(\theta_1 + \theta_2 + \ldots + \theta_n) = \cos\theta_1\cos\theta_2\ldots\cos\theta_n(1 - s_2 + s_4 - \ldots)$$

$$\sin(\theta_1 + \theta_2 + \ldots + \theta_n) = \cos\theta_1\cos\theta_2\ldots\cos\theta_n(s_1 - s_3 + s_5 - \ldots)$$

and by division, we get $\mathbf{\tan(\theta_1 + \theta_2 + \ldots + \theta_n) = \dfrac{s_1 - s_3 + s_5 - \ldots}{1 - s_2 + s_4 - s_6 + \ldots}}$.

**Example 19.28.** *If $\tan^{-1}x + \tan^{-1}y + \tan^{-1}z = \pi/2$, show that $xy + yz + zx = 1$.*

**Solution.** Let $\tan^{-1}x = \alpha$, $\tan^{-1}y = \beta$, $\tan^{-1}z = \gamma$ so that $x = \tan\alpha$, $y = \tan\beta$, $z = \tan\gamma$

We know that $$\tan(\alpha + \beta + \gamma) = \frac{\tan\alpha + \tan\beta + \tan\gamma - \tan\alpha\tan\beta\tan\gamma}{1 - \tan\alpha\tan\beta - \tan\beta\tan\gamma - \tan\gamma\tan\alpha}$$

$\therefore$ $$\tan\pi/2 = \frac{x + y + z - xyz}{1 - xy - yz - zx} \quad \text{or} \quad 1 - xy - yz - zx = 0$$

Hence $xy + yz + zx = 1$.

**Example 19.29.** *If $\theta_1, \theta_2, \theta_3$ be three values of $\theta$ which satisfy the equation $\tan 2\theta = \lambda\tan(\theta + \alpha)$ and such that no two of them differ by a multiple of $\pi$, show that $\theta_1 + \theta_2 + \theta_3 + \alpha$ is a multiple of $\pi$.*

**Solution.** Given equation can be written as $\dfrac{2t}{1-t^2} = \lambda\dfrac{t + \tan\alpha}{1 - t\,.\tan\alpha}$ where $t = \tan\theta$

or $$\lambda t^3 + (\lambda - 2)\tan\alpha\,.\,t^2 + (2 - \lambda)t - \lambda\tan\alpha = 0$$

$\therefore$ $\tan\theta_1, \tan\theta_2, \tan\theta_3$, being its roots, we have

$$s_1 = \Sigma\tan\theta_1 = -\frac{\lambda - 2}{\lambda}\tan\alpha \qquad \text{[By § 1.3]}$$

$$s_2 = \Sigma\tan\theta_1\tan\theta_2 = \frac{2 - \lambda}{\lambda} \quad \text{and} \quad s_3 = \tan\alpha$$

$\therefore$ $$\tan(\theta_1 + \theta_2 + \theta_3) = \frac{s_1 - s_3}{1 - s_2} = \frac{(-1 + 2/\lambda)\tan\alpha - \tan\alpha}{1 - (2/\lambda - 1)}$$

$$= -\tan\alpha = \tan(n\pi - \alpha)$$

Thus $\theta_1 + \theta_2 + \theta_3 = n\pi - \alpha$, whence follows the result.

**(3) To expand $\sin^m\theta$, $\cos^n\theta$ or $\sin^m\theta\cos^n\theta$ in a series of sines or cosines of multiples of $\theta$**

If $z = \cos\theta + i\sin\theta$ then $1/z = \cos\theta - i\sin\theta$.

By De Moivre's theorem, $z^p = \cos p\theta + i\sin p\theta$ and $1/z^p = \cos p\theta - i\sin p\theta$

$\therefore$ $z + 1/z = 2\cos\theta$, $z - 1/z = 2i\sin\theta$ ; $z^p + 1/z^p = 2\cos p\theta$, $z^p - 1/z^p = 2i\sin p\theta$

These results are used to expand the powers of sin θ or cos θ or their products in a series of sines or cosines of multiples of θ.

**Example 19.30.** *Expand $\cos^8\theta$ in a series of cosines of multiples of θ.*

**Solution.** Let $z = \cos\theta + i\sin\theta$, so that $z + 1/z = 2\cos\theta$ and $z^p + 1/z^p = 2\cos p\theta$.

$\therefore (2\cos\theta)^8 = (z + 1/z)^8$

$$= z^8 + {}^8C_1 z^7 \cdot \frac{1}{z} + {}^8C_2 z^6 \cdot \frac{1}{z^2} + {}^8C_3 z^5 \cdot \frac{1}{z^3} + {}^8C_4 z^4 \cdot \frac{1}{z^4} + {}^8C_5 z^3 \cdot \frac{1}{z^5} + {}^8C_6 z^2 \cdot \frac{1}{z^6} + {}^8C_7 z \cdot \frac{1}{z^7} + \frac{1}{z^8}$$

$$= (z^8 + 1/z^8) + {}^8C_1(z^6 + 1/z^6) + {}^8C_2(z^4 + 1/z^4) + {}^8C_3(z^2 + 1/z^2) + {}^8C_4$$

$$= (2\cos 8\theta) + 8(2\cos 6\theta) + 28(2\cos 4\theta) + 56(2\cos 2\theta) + 70.$$

Hence $$\cos^8\theta = \frac{1}{128}[\cos 8\theta + 8\cos 6\theta + 28\cos 4\theta + 56\cos 2\theta + 35].$$

**Example 19.31.** *Expand $\sin^7\theta\cos^3\theta$ in a series of sines of multiples of θ.*

**Solution.** Let $z = \cos\theta + i\sin\theta$

so that $z + 1/z = 2\cos\theta,\ z - 1/z = 2i\sin\theta$ and $z^p - 1/z^p = 2i\sin p\theta$.

$$\therefore\quad (2i\sin\theta)^7(2\cos\theta)^3 = (z - 1/z)^7(z + 1/z)^3$$

$$= (z - 1/z)^4[(z - 1/z)(z + 1/z)]^3 = (z - 1/z)^4(z^2 - 1/z^2)^3$$

$$= \left(z^4 - 4z^2 + 6 - \frac{4}{z^2} + \frac{1}{z^4}\right)\left(z^6 - 3z^2 + \frac{3}{z^2} - \frac{1}{z^6}\right)$$

$$= \left(z^{10} - \frac{1}{z^{10}}\right) - 4\left(z^8 - \frac{1}{z^8}\right) + 3\left(z^6 - \frac{1}{z^6}\right) + 8\left(z^4 - \frac{1}{z^4}\right) - 14\left(z^2 - \frac{1}{z^2}\right)$$

$$= 2i\sin 10\theta - 4(2i\sin 8\theta) + 3(2i\sin 6\theta) + 8(2i\sin 4\theta) - 14(2i\sin 2\theta)$$

Since, $i^7 = -i$,

$$\therefore\quad \sin^7\theta\cos^3\theta = -\frac{1}{2^9}[\sin 10\theta - 4\sin 8\theta + 3\sin 6\theta + 8\sin 4\theta - 14\sin 2\theta].$$

**Obs.** The expansion of $\sin^m\theta\cos^n\theta$ is a series of sines or cosines of multiples of θ according as $m$ is odd or even.

## PROBLEMS 19.4

1. Express $\sin 6\theta/\sin\theta$ as a polynomial in $\cos\theta$?

Prove that (2–5) :

2. $\sin 7\theta/\sin\theta = 7 - 56\sin^2\theta + 112\sin^4\theta - 64\sin^6\theta$.
3. $\dfrac{1+\cos 7\theta}{1+\cos\theta} = (x^3 - x^2 - 2x + 1)^2$, where $x = 2\cos\theta$. *(P.T.U., 2013 S)*
4. $2(1 + \cos 8\theta) = (x^4 - 4x^2 + 2)^2$, where $x = 2\cos\theta$.
5. $\tan 5\theta = \dfrac{5t - 10t^3 + t^5}{1 - 10t^2 + 5t^4}$, where $t = \tan\theta$.
6. If $\tan^{-1}x + \tan^{-1}y + \tan^{-1}z = \pi$, show that $x + y + z = xyz$.
7. If α, β, γ be the roots of the equation $x^3 + px^2 + qx + p = 0$, prove that $\tan^{-1}\alpha + \tan^{-1}\beta + \tan^{-1}\gamma = n\pi$ radians except in one particular case.

Prove that (8-12) :

8. $\sin^5\theta = \dfrac{1}{10}(\sin 5\theta - 5\sin 3\theta + 10\sin\theta)$. *(D.T.U., 2013)*
9. $\cos^7\theta = \dfrac{1}{16}(\cos 7\theta + 7\cos 5\theta + 21\cos 3\theta + 35\cos\theta)$.
10. $\cos^6\theta - \sin^6\theta = \dfrac{1}{16}(\cos 6\theta + 15\cos 2\theta)$.
11. $\sin^8\theta = 2^{-7}(\cos 8\theta - 8\cos 6\theta + 28\cos 4\theta - 56\cos 2\theta + 35)$.
12. $32\sin^4\theta\cos^2\theta = \cos 6\theta - 2\cos 4\theta - \cos 2\theta + 2$.
13. $\sin^5\theta\cos^2\theta = \dfrac{1}{64}(\sin 7\theta - 3\sin 5\theta + \sin 3\theta + 5\sin\theta)$. *(P.T.U., 2014 S)*

14. Expand $\cos^5\theta \sin^7\theta$ in a series of sines of multiples of $\theta$ ?
15. If $\cos^5\theta = A\cos\theta + B\cos 3\theta + C\cos 5\theta$, find $\sin^5\theta$ in terms of $A, B, C$.
16. If $\sin^4\theta\cos^3\theta = A_1\cos\theta + A_3\cos 3\theta + A_5\cos 5\theta + A_7\cos 7\theta$, prove that
$$A_1 + 9A_3 + 25A_5 + 49A_7 = 0.$$

## 19.7 COMPLEX FUNCTION

**Definition.** *If for each value of the complex variable z (= x + iy) in a given region R, we have one or more values of w (= u + iv), then w is said to be a* **complex function** *of z* and we write $w = u(x, y) + iv(x, y) = f(z)$ where $u, v$ are real functions of $x$ and $y$.

If to each value of $z$, there corresponds one and only one value of $w$, then $w$ is said to be a *single-valued function of z* otherwise a *multi-valued function*. For example, $w = 1/z$ is a single-valued function and $w = \sqrt{z}$ is a multi-valued function of $z$. The former is defined at all points of the $z$-plane except at $z = 0$ and the latter assumes two values for each value of $z$ except at $z = 0$.

## 19.8 EXPONENTIAL FUNCTION OF A COMPLEX VARIABLE

**(1) Definition.** When $x$ is real, we are already familiar with the exponential function

$$e^x = 1 + \frac{x}{1!} + \frac{x^2}{2!} + \ldots + \frac{x^n}{n!} + \ldots \infty .$$

Similarly, *we define the exponential function of the complex variable $z = x + iy$, as*

$$\mathbf{e^z \text{ or } exp(z) = 1 + \frac{z}{1!} + \frac{z^2}{2!} + \ldots + \frac{z^n}{n!} + \ldots \infty} \qquad \ldots(i)$$

**(2) Properties :**

*I. Exponential form of $z = re^{i\theta}$*

Putting $x = 0$ in Eqn. $(i)$, we get

$$e^{iy} = 1 + \frac{iy}{1!} + \frac{(iy)^2}{2!} + \frac{(iy)^3}{3!} + \frac{(iy)^4}{4!} + \ldots \infty$$

$$= \left(1 - \frac{y^2}{2!} + \frac{y^4}{4!} - \ldots\right) + i\left(y - \frac{y^3}{3!} + \frac{y^5}{5!} - \ldots\right) = \cos y + i \sin y$$

Thus, $\quad e^z = e^x . e^{iy} = e^x(\cos y + i\sin y)$

Also $\quad x + iy = r(\cos\theta + i\sin\theta) = re^{i\theta}$. Thus, $z = re^{i\theta}$

*II. $e^z$ is periodic function having imaginary period $2\pi i$,* $\quad [\because \; e^{z+2n\pi i} = e^z . e^{2n\pi i} = e^z]$.

*III. $e^z$ is not zero for any value of z.*

Since $\quad e^z = e^{x+iy} = re^{i\theta}$ or $e^x . e^{iy} = re^{i\theta}$

$\therefore \quad r = e^x > 0, y = \theta, |e^{iy}| = 1,$

Thus $\quad |e^z| = |e^x| . |e^{iy}| = e^x \neq 0.$

*IV.* $\quad e^{\bar{z}} = \overline{e^z}$

Since $\quad e^{\bar{z}} = e^{x-iy} = e^x . e^{-iy} = e^x(\cos y - i\sin y)$

$$= \overline{e^x(\cos y + i\sin y)} = \overline{e^z}$$

## 19.9 CIRCULAR FUNCTIONS OF A COMPLEX VARIABLE

**(1) Definitions:**

Since $\quad e^{iy} = \cos y + i\sin y$ and $e^{-iy} = \cos y - i\sin y$.

$\therefore$ the circular functions of real angles can be written as

$$\sin y = \frac{e^{iy} - e^{-iy}}{2i}, \cos y = \frac{e^{iy} + e^{-iy}}{2} \text{ and so on.}$$

It is, therefore, natural to *define the circular functions of the complex variable z by the equations :*

$$\mathbf{\sin z = \frac{e^{iz} - e^{-iz}}{2i}, \cos z = \frac{e^{iz} + e^{-iz}}{2}, \tan z = \frac{\sin z}{\cos z}}$$

*with cosec z, sec z and cot z as their respective reciprocals.*

**(2) Properties :**

*I. Circular functions are periodic :* sin $z$, cos $z$ are periodic functions having real period $2\pi$ while tan $z$, cot $z$ have period $\pi$. [$a$ $\sin(z + 2n\pi) = \sin z$, $\tan(z + n\pi) = \tan z$ etc.]

*II. Even and odd functions : cos z, sec z are even functions while sin z, cosec z are odd functions.*

$$[\because \quad \cos z = \frac{e^{-iz} + e^{iz}}{2} = \cos z, \text{ and } \sin(-z) = \frac{e^{-iz} - e^{iz}}{2i} = \frac{e^{iz} - e^{-iz}}{2i} = -\sin z]$$

*III. Zeros of sin z are given by* $z = \pm 2n\pi$ *and zeros of cos z are given by* $z = \pm \frac{1}{2}(2n + 1)\pi$, $n = 0, 1, 2, ...$

*IV. All the formulae for real circular functions are valid for complex circular functions*

*e.g.*, $\sin^2 z + \cos^2 z = 1$, $\sin(z_1 \pm z_2) = \sin z_1 \cos z_2 \pm \cos z_1 \sin z_2$.

**(3) Euler's theorem $e^{iz} = \cos z + i \sin z$.**

By definition $\cos z + i \sin z = \dfrac{e^{iz} + e^{-iz}}{2} + i\dfrac{e^{iz} - e^{-iz}}{2i} = e^{iz}$ where $z = x + iy$.

Also we have shown that $e^{iy} = \cos y + i \sin y$, where $y$ is real.

Thus $e^{i\theta} = \mathbf{\cos\theta + \mathit{i} \sin\theta}$, *where θ is real or complex. This is called the Euler's theorem.*

**Cor.** *De Moivre's theorem for complex numbers*

Whether θ is real or complex, we have

$$(\cos\theta + i\sin\theta)^n = (e^{i\theta})^n = e^{in\theta} = \cos n\theta + i \sin n\theta$$

Thus De Moivre's theorem is true for all θ (real or complex).

**Example 19.32.** *Prove that (i)* $[\sin(\alpha + \theta) - e^{i\alpha}\sin\theta]^n = \sin^n\alpha\, e^{-in\theta}$
*(ii)* $\sin(\alpha - n\theta) + e^{-i\alpha}\sin n\theta = e^{-in\theta}\sin\alpha$.

**Solution.** (*i*) L.H.S. $= [\sin\alpha\cos\theta + \cos\alpha\sin\theta - (\cos\alpha + i\sin\alpha)\sin\theta]^n$

$$= (\sin\alpha\cos\theta - i\sin\alpha\sin\theta)^n$$

$$= \sin^n\alpha(\cos\theta - i\sin\theta)^n = \sin^n\alpha(e^{-i\theta})^n = \sin^n\alpha e^{-in\theta}$$

(*ii*) L.H.S. $= \sin\alpha\cos n\theta - \cos\alpha\sin n\theta + (\cos\alpha - i\sin\alpha)\sin n\theta$

$$= \sin\alpha\cos n\theta - i\sin\alpha\sin n\theta$$

$$= \sin\alpha(\cos n\theta - i\sin n\theta) = \sin\alpha \,.\, e^{-in\theta}.$$

**Example 19.33.** *Given* $\dfrac{1}{\rho} = \dfrac{1}{L\rho i} + C\rho i + \dfrac{1}{R}$, *where L, ρ, R are real, express ρ in the form* $Ae^{i\theta}$ *giving the values of A and θ.*

**Solution.** $$\frac{1}{\rho} = \frac{R + L\rho^2 CR(-1) + L\rho i}{L\rho Ri} = \frac{(R - L\rho^2 CR) + iLR}{L\rho Ri}$$

or $$\rho = L\rho\frac{Ri}{(R - L\rho^2 CR) + iL\rho} \times \frac{(R - L\rho^2 CR) - iL\rho}{(R - L\rho^2 CR) - iL\rho}$$

$$= \frac{L^2\rho^2 R + iL\rho R(R - L\rho^2 CR)}{(R - L\rho^2 CR)^2 + (L\rho)^2}$$

$$= A(\cos\theta + i\sin\theta), \text{ say}$$

Equating real and imaginary parts, we have

$$A \cos \theta = \frac{L^2\rho^2 R}{(R - L\rho^2 CR)^2 + (L\rho)^2} \qquad ...(i)$$

$$A \sin \theta = \frac{L\rho R(R - L\rho^2 CR)}{(R - L\rho^2 CR)^2 + (L\rho)^2} \qquad ...(ii)$$

Squaring and adding (*i*) and (*ii*),

$$A^2 = \frac{(L^2\rho^2 R)^2 + (L\rho R)^2 (R - L\rho^2 CR)^2}{[(R - L\rho^2 CR)^2 + (L\rho)^2]^2} \quad \text{or} \quad A = \frac{L\rho R}{\sqrt{[(R - L\rho^2 CR)^2 + (L\rho)]^2}} \qquad ...(iii)$$

Dividing Eqn. (*ii*) by Eqn. (*i*),

$$\tan \theta = \frac{R - L\rho^2 CR}{L\rho} \quad \text{or} \quad \theta = \tan^{-1}\left\{\frac{R(1 - LC\rho^2)}{L\rho}\right\} \qquad ...(iv)$$

Hence $\qquad P = A(\cos \theta + i \sin \theta) = Ae^{i\theta}$

where $A$ and $\theta$ are given by Eqns. (*iii*) and (*iv*).

## 19.10 HYPERBOLIC FUNCTIONS

**(1) Definitions:** *If x be real or complex,*

(*i*) $\dfrac{e^x - e^{-x}}{2}$ *is defined as* **hyperbolic sine of x** and is written as **sinh x.**

(*ii*) $\dfrac{e^x + e^{-x}}{2}$ *is defined as* **hyperbolic cosine of x** *and is written as* **cosh x.**

Thus $\sinh x = \dfrac{e^x - e^{-x}}{2}$ and $\cosh x = \dfrac{e^x + e^{-x}}{2}$

Also we define,

$$\tanh x = \frac{\sinh x}{\cosh x} = \frac{e^x - e^{-x}}{e^x + e^{-x}}; \ \coth x = \frac{1}{\tanh x} = \frac{e^x + e^{-x}}{e^x - e^{-x}}$$

$$\text{sech } x = \frac{1}{\cosh x} = \frac{2}{e^x + e^{-x}}; \ \text{cosech } x = \frac{1}{\sinh x} = \frac{2}{e^x - e^{-x}}$$

**(2) Properties**

*I. Periodic functions : sinh z and cosh z are periodic functions having imaginary period* $2\pi i$.

$[\because \ \sinh (z + 2\pi i) = \sinh z \ ; \ \cosh (z + 2\pi i) = \cosh z]$

*II. Even and odd functions : cosh z is an even function while sinh z is an odd function*

*III.* $\sinh 0 = 0$, $\cosh 0 = 1$, $\tanh 0 = 0$.

*IV.* **Relations between hyperbolic and circular functions.**

Since for all values of $\theta$, $\qquad \sin \theta = \dfrac{e^{i\theta} - e^{-i\theta}}{2i}$ and $\cos \theta = \dfrac{e^{i\theta} + e^{-i\theta}}{2}$

$\therefore$ Putting $\theta = ix$, we have $\qquad \sin ix = \dfrac{e^{-x} - e^x}{2i} = -\dfrac{e^x - e^{-x}}{2i}$ $\qquad [\because \ e^{i\theta} = e^{i \cdot ix} = e^{-x}]$

$$= i^2 \frac{e^x - e^{-x}}{2i} = i \cdot \frac{e^x - e^{-x}}{2} = i \sinh x$$

and, therefore, $\qquad \cos ix = \dfrac{e^{-x} + e^x}{2} = \cosh x$

Thus $\qquad$ **sin ix = i sinh x** $\qquad$ ...(*i*)

**cos ix = cosh x** $\qquad$ ...(*ii*)

and $\therefore$ $\qquad$ **tan ix = i tanh x** $\qquad$ ...(*iii*)

**Cor.** $\qquad$ **sinh ix = i sin x** $\qquad$ ...(*iv*)

**cosh ix = cos x** $\qquad$ ...(*v*)

**tanh ix = i tan x** $\qquad$ ...(*vi*)

**V. Formulae of hyperbolic functions**

*(a) Fundamental formulae*

(1) $\cosh^2 x - \sinh^2 x = 1$ (2) $\text{sech}^2 x + \tanh^2 x = 1$ (3) $\coth^2 x - \text{cosech}^2 x = 1$.

*(b) Addition formulae*

(4) $\sinh (x \pm y) = \sinh x \cosh y \pm \cosh x \sinh y$ (5) $\cosh (x \pm y) = \cosh x \cosh y \pm \sinh x \sinh y$

(6) $\tanh (x \pm y) = \dfrac{\tanh x \pm \tanh y}{1 \pm \tanh x \tanh y}$

*(c) Functions of 2x.*

(7) $\sinh 2x = 2 \sinh x \cosh x$

(8) $\cosh 2x = \cosh^2 x + \sinh^2 x = 2 \cosh^2 x - 1 = 1 + 2 \sinh^2 x$

(9) $\tanh 2x = \dfrac{2 \tanh x}{1 + \tanh^2 x}$

*(d) Functions of 3x*

(10) $\sinh 3x = 3 \sinh x + 4 \sinh^3 x$

(11) $\cosh 3x = 4 \cosh^3 x - 3 \cosh x$ (12) $\tanh 3x = \dfrac{3 \tanh x + \tanh^3 x}{1 + 3 \tanh^2 x}$

*(e) Functions of sin x + sin y*

(13) $\sinh x + \sinh y = 2 \sinh \dfrac{x+y}{2} \cosh \dfrac{x-y}{2}$ (14) $\sinh x - \sinh y = 2 \cosh \dfrac{x+y}{2} \sinh \dfrac{x-y}{2}$

(15) $\cosh x + \cosh y = 2 \cosh \dfrac{x+y}{2} \cosh \dfrac{x-y}{2}$ (16) $\cosh x - \cosh y = 2 \sinh \dfrac{x+y}{2} \sinh \dfrac{x-y}{2}$.

*Proofs.* (1) Since, for all values of θ, we have $\cos^2 \theta + \sin^2 \theta = 1$.

$\therefore$ putting $\theta = ix$, we get $\cos^2 ix + \sin^2 ix = 1$ or $\cosh^2 x - \sinh^2 x = 1$

*Otherwise* : $\cosh^2 x - \sinh^2 x = \left(\dfrac{e^x + e^{-x}}{2}\right)^2 - \left(\dfrac{e^x - e^{-x}}{2}\right)^2 = \dfrac{1}{4}[e^{2x} + e^{-2x} + 2 - e^{2x} - e^{-2x} + 2] = 1.$

Similarly we can establish the formulae Eqns. (2) and (3).

Eqn. (4) $\sinh (x + y) = (1/i) \sin i(x + y) = -i[\sin ix \cos iy + \cos ix \sin iy]$

$= -i[i \sinh x \,.\, \cosh y + \cosh x \,.\, i \sinh y] = \sinh x \cosh y + \cosh x \sinh y.$

*Otherwise :* $\sinh x \cosh y + \cosh x \sinh y$

$$= \frac{e^x - e^{-x}}{2} \cdot \frac{e^y + e^{-y}}{2} + \frac{e^x + e^{-x}}{2} \cdot \frac{e^y - e^{-y}}{2} = \frac{e^{x+y} - e^{-(x+y)}}{2} = \sinh (x + y)$$

Similarly we can establish the formulae Eqn. (5) and (6).

Eqn. (12) $\tan 3A = \dfrac{3 \tan A - \tan^3 A}{1 - 3 \tan^2 A}$

Putting $A = ix$, $\tan 3ix = \dfrac{3 \tan ix - \tan^3 ix}{1 - 3 \tan^2 ix}$ or $i \tanh 3x = \dfrac{3(i \tanh x) - (i \tanh x)^3}{1 - 3(i \tanh x)^2}$

$$\tanh 3x = \frac{3 \tanh x + \tanh^3 x}{1 + 3 \tanh^2 x}$$

Similarly, we can establish the formulae Eqns. (7) to (11).

Eqn. (16) $\cos C - \cos D = -2 \sin \dfrac{C+D}{2} \sin \dfrac{C-D}{2}$

Putting $C = ix$, and $D = iy$, $\cos ix - \cos iy = -2 \sin i \dfrac{x+y}{2} \sin i \dfrac{x-y}{2}$

or $\cosh x - \cosh y = -2\left(i \sinh \dfrac{x+y}{2}\right)\left(i \sinh \dfrac{x-y}{2}\right) = 2 \sinh \dfrac{x+y}{2} \sinh \dfrac{x-y}{2}$

Similarly, we can establish the formulae Eqns. (13) to (15).

## 19.11 INVERSE HYPERBOLIC FUNCTIONS

**(1) Definitions:** If *sinh* $u = z$, then $u$ is called the hyperbolic sine inverse of $z$ and is written as $u = \sinh^{-1} z$. Similarly we define $\cosh^{-1} z$, $\tanh^{-1} z$, etc.

The inverse hyperbolic functions like other inverse functions are many-valued, but we shall consider only their principal values.

**(2)** *To show that (i)* $\sinh^{-1} z = \log [z + \sqrt{(z^2 + 1)}]$

*(ii)* $\cosh^{-1} z = \log [z + \sqrt{(z^2 - 1)}]$, *(iii)* $\tanh^{-1} z = \frac{1}{2} \log \frac{1+z}{1-z}$.

(*i*) Let $\sinh^{-1} z = u$, then $z = \sinh u = \frac{1}{2}(e^u - e^{-u})$

or $$2z = e^u - 1/e^u \quad \text{or} \quad e^{2u} - 2ze^u - 1 = 0$$

This being a quadratic in $e^u$, we have

$$e^u = \frac{2z \pm \sqrt{(4z^2 + 4)}}{2} = z \pm \sqrt{(z^2 + 1)}$$

$\therefore$ Taking the positive sign only, we have

$$e^u = z + \sqrt{(z^2 + 1)} \quad \text{or} \quad u = \log [z + \sqrt{(z^2 + 1)}]$$

Similarly we can establish (*ii*)

(*iii*) Let $\tanh^{-1} z = u$, then $z = \tanh u$

*i.e.,* $$z = \frac{e^u - e^{-u}}{e^u + e^{-u}}.$$

Applying componendo and dividendo, we get $\frac{1+z}{1-z} = e^u/e^{-u} = e^{2u}$

or $$2u = \log \left(\frac{1+z}{1-z}\right) \text{ whence follows the result.}$$

**Example 19.34.** *If $u = \log \tan (\pi/4 + \theta/2)$, prove that*

*(i) $\tanh u/2 = \tan \theta/2$*

*(ii) $\theta = -i \log \tan\left(\frac{\pi}{4} + \frac{iu}{2}\right)$.*

**Solution.** We have $$e^u = \tan\left(\frac{\pi}{4} + \frac{\theta}{2}\right) \quad \text{or} \quad \frac{e^{u/2}}{e^{-u/2}} = \frac{1 + \tan \theta/2}{1 - \tan \theta/2}$$

By componendo and dividendo, we get

$$\frac{e^{u/2} - e^{-u/2}}{e^{u/2} + e^{-u/2}} = \tan \theta/2 \quad i.e., \quad \tanh \frac{u}{2} = \tan \frac{\theta}{2} \qquad ...(i)$$

or $$\frac{1}{i} \tan \frac{iu}{2} = \frac{1}{i} \tanh \frac{i\theta}{2} \quad \text{or} \quad \frac{i\theta}{2} = \tanh^{-1}\left(\tan \frac{iu}{2}\right) = \frac{1}{2} \log \frac{1 + \tan iu/2}{1 - \tan iu/2}$$

or $$\theta = \frac{1}{i} \log \tan\left(\frac{\pi}{4} + \frac{iu}{2}\right) = -i \log \tan\left(\frac{\pi}{4} + \frac{iu}{2}\right). \qquad ...(ii)$$

**Example 19.35.** *Show that $\tanh^{-1} (\cos \theta) = \cosh^{-1} (\text{cosec } \theta)$.*

**Solution.** Let $\tanh^{-1} (\cos \theta) = \phi$ so that $\cos \theta = \tanh \phi$

or $\tanh^2 \phi = \cos^2 \theta$ or $1 - \text{sech}^2 \phi = \cos^2 \theta$

or $\text{sech}^2 \phi = 1 - \cos^2 \theta = \sin^2 \theta$ or $\text{sech } \phi = \sin \theta$

or $\cosh \phi = \text{cosec } \theta$ or $\phi = \cosh^{-1} (\text{cosec } \theta)$.

**Example 19.36.** *Find tanh x, if 5 sinh x – cosh x = 5.*

**Solution.** We have $5(\sinh x - 1) = \cosh x$

or $25(\sinh x - 1)^2 = \cosh^2 x = 1 + \sinh^2 x$

or $24 \sinh^2 x - 50 \sinh x + 24 = 0$ or $12 \sinh^2 x - 25 \sinh x + 12 = 0$

or $(3 \sinh x - 4)(4 \sinh x - 3) = 0$ whence $\sinh x = 4/3$ or $3/4$.

$\therefore$ $\cosh x = \sqrt{(1 + \sinh^2 x)} = 5/3$ or $-5/4$ [$\because$ $\cosh x = 5/4$ doesn't satisfy (*i*)]

Hence $\tanh x = \dfrac{4}{5}$ or $-\dfrac{3}{5}$.

## PROBLEMS 19.5

1. Separate into real and imaginary parts

   (*i*) $\exp(z^2)$ where $z = x + iy$ (*ii*) $\exp(5 + i\pi/2)$ (*iii*) $\exp(5 + 3i)^2$.

2. From the definitions of sin *z* and cos *z*, prove that

   (*i*) $\cos 2z = 2\cos^2 z - 1$ (*ii*) $\dfrac{\sin 2z}{1 - \cos 2z} = \cot z$ (*iii*) $\sin 3z = 3 \sin z - 4 \sin^3 z$.

3. Prove that $[\sin(\alpha - \theta) + e^{-i\alpha} \sin\theta]^n = \sin^{n-1}\alpha\, \{\sin(\alpha - n\theta) + e^{-i\alpha} \sin n\theta\}$

4. If $z = e^{i\theta}$, show that $\dfrac{z^2 - 1}{z^2 + 1} = i \tan\theta$.

5. If $y = \log \tan x$, show that $\sinh ny = \dfrac{1}{2}(\tan^n x - \cot^n x)$.

6. If $\tan y = \tan\alpha \tanh\beta$ and $\tan z = \cot\alpha \tanh\beta$, prove that $\tan(y + z) = \sinh 2\beta \operatorname{cosec} 2\alpha$.

7. Prove that

   (*i*) $\cosh(\alpha + \beta) - \cosh(\alpha - \beta) = 2 \sinh\alpha \sinh\beta$

   (*ii*) $\sinh(\alpha + \beta)\cosh(\alpha - \beta) = \dfrac{1}{2}(\sinh 2\alpha + \sinh 2\beta)$.

8. Express $\cosh^7\theta$ in terms of hyperbolic cosines of multiples of θ.

9. If $\tan x/2 = \tanh u/2$, prove that

   (*i*) $\tan x = \sinh u$ and $\cos x \cosh u = 1$; (*ii*) $u = \log_e \tan(\pi/4 + x/2)$.

10. If $\cosh x = \sec\theta$, prove that

    (*i*) $\tanh^2 x/2 = \tan^2\theta/2$ (*ii*) $x = \log_e \tan(\pi/4 + \theta/2)$.

11. Show that $\tan^{-1} z = \dfrac{i}{2} \log \dfrac{i + z}{i - z}$.

12. Prove that

    (*i*) $\sinh^{-1} x = \cosh^{-1}\sqrt{(1 + x^2)} = \tanh^{-1}\dfrac{x}{\sqrt{(1 - x^2)}} = \dfrac{1}{2}\operatorname{cosech}^{-1}\dfrac{1}{2x\sqrt{(1 + x^2)}}$

    (*ii*) $\tanh^{-1} x = \sinh^{-1}\dfrac{x}{\sqrt{(1 - x^2)}}$.

13. Show that

    (*i*) $\sinh^{-1}(\tan\theta) = \log\tan(\pi/4 + \theta/2)$ (*ii*) $\operatorname{sech}^{-1}(\sin\theta) = \log\cot\theta/2$.

14. Find all the values of *z* which satisfy $e^z = 1 + i$. (*P.T.U., 2015*)

15. Solve the equation

    (*i*) $\sin z = 2$. (*P.T.U., 2015 S*) (*ii*) $7\cosh x + 8\sinh x = 1$, for real values of *x*.

16. Find tanh *x* if $\sinh x - \cosh x = 5$.

## 19.12 REAL AND IMAGINARY PARTS OF CIRCULAR AND HYPERBOLIC FUNCTIONS

(1) *To separate the real and imaginary parts of*

*(i) sin (x + iy); (ii) cos (x + iy); (iii) tan (x + iy); (iv) cot (x + iy); (v) sec (x + iy); (vi) cosec (x + iy).*

*Proofs.* (*i*) $\sin(x+iy) = \sin x \cos iy + \cos x \sin iy = \sin x \cosh y + i \cos x \sinh y.$

Similarly, $\cos(x+iy) = \cos x \cosh y - i \sin x \sinh y$

(*iii*) Let $\alpha + i\beta = \tan(x+iy)$ then $\alpha - i\beta = \tan(x-iy)$

Adding, $2\alpha = \tan(x+iy) + \tan(x-iy)$

*i.e.,* $$\alpha = \frac{\sin(x+iy+x-iy)}{2\cos(x+iy)\cos(x-iy)} = \frac{\sin 2x}{\cos 2x + \cos 2iy} = \frac{\sin 2x}{\cos 2x + \cosh 2y}$$

Subtracting, $2i\beta = \tan(x+iy) - \tan(x-iy)$

*i.e.,* $$i\beta = \frac{\sin 2iy}{2\cos(x+iy)\cos(x-iy)} = \frac{i\sinh 2y}{\cos 2x + \cosh 2y}$$

$\therefore$ $$\beta = \frac{\sinh 2y}{\cos 2x + \cosh 2y}$$

Similarly, $\cot(x+iy) = \dfrac{\sin 2x - i\sinh 2y}{\cosh 2y - \cos 2x}$.

(*v*) Let $\alpha + i\beta = \sec(x+iy)$ then $\alpha - i\beta = \sec(x-iy)$

Adding, $2\alpha = \sec(x+iy) + \sec(x-iy)$

*i.e.,* $$\alpha = \frac{\cos(x-iy)+\cos(x+iy)}{2\cos(x+iy)\cos(x-iy)} = \frac{2\cos x \cos iy}{\cos 2x + \cos 2iy} = \frac{2\cos x \cosh y}{\cos 2x + \cosh 2y}$$

Subtracting, $2i\beta = \sec(x+iy) - \sec(x-iy)$

*i.e.,* $$i\beta = \frac{\cos(x-iy)-\cos(x+iy)}{2\cos(x+iy)\cos(x-iy)} = \frac{2\sin x \sin iy}{\cos 2x + \cos 2iy} = \frac{2i\sin x \sinh y}{\cos 2x + \cosh 2y}$$

$\therefore$ $$\beta = \frac{2\sin x \sinh y}{\cos 2x + \cosh 2y}$$

Similarly, $\operatorname{cosec}(x+iy) = 2\,\dfrac{\sin x \cosh y - i\cos x \sinh y}{\cosh 2y - \cos 2x}$.

**(2)** *To separate the real and imaginary parts of*

*(i) sinh (x + iy); (ii) cosh (x + iy); (iii) tanh (x + iy).*

*Proofs.* (*i*) $\sinh(x+iy) = (1/i)\sin i(x+iy) = (1/i)\sin(ix-y)$

$= (1/i)[\sin ix \cos y - \cos ix \sin y] = (1/i)[i\sinh x \cos y - \cosh x \sin y]$

$= \sinh x \cos y + i\cosh x \sin y$

Similarly, $\cosh(x+iy) = \cosh x \cos y + i\sinh x \sin y.$

(*iii*) If $\alpha + i\beta = \tanh(x+iy) = (1/i)\tan(ix-y)$

then $\alpha - i\beta = \tanh(x-iy) = (1/i)\tan(ix+y)$

Adding, $2\alpha = (1/i)[\tan(ix-y) + \tan(ix+y)]$

$$\alpha = \frac{\sin(ix-y+ix+y)}{i\,.\,2\cos(ix-y)\cos(ix+y)} = \frac{(1/i)\sin 2ix}{\cos 2ix + \cos 2y} = \frac{\sinh 2x}{\cosh 2x + \cos 2y}.$$

Subtracting, $2i\beta = (1/i)[\tan(ix-y) - \tan(ix+y)]$

*i.e.,* $$i\beta = -\frac{\sin[(ix+y)-(ix-y)]}{i\,.\,2\cos(ix+y)\cos(ix-y)}$$

$\therefore$ $$\beta = \frac{\sin 2y}{\cos 2ix + \cos 2y} = \frac{\sin 2y}{\cosh 2x + \cos 2y}.$$

**Example 19.37.** *If cosh (u + iv) = x + iy, prove that*

$$\frac{x^2}{\cosh^2 u} + \frac{y^2}{\sinh^2 u} = 1 \quad \text{(P.T.U., 2009 S)} \qquad \frac{x^2}{\cos^2 v} - \frac{y^2}{\sin^2 v} = 1.$$

**Solution.** Since $x + iy = \cosh(u + iv) = \cos(iu - v)$

$$= \cos iu \cos v + \sin iu \sin v = \cosh u \cos v + i \sinh u \sin v.$$

$\therefore$ equating real and imaginary parts, we get $x = \cosh u \cos v$ ; $y = \sinh u \sin v$

*i.e.,*
$$\frac{x}{\cosh u} = \cos v \quad \text{and} \quad \frac{y}{\sinh u} = \sin v$$

Squaring and adding, we get the first result.

Again
$$\frac{x}{\cos v} = \cosh u \quad \text{and} \quad \frac{y}{\sin v} = \sinh u.$$

$\therefore$ squaring and subtracting, we get the second result.

**Example 19.38.** *If $\tan(\theta + i\phi) = e^{i\alpha}$, show that*

$$\theta = (n + 1/2)\,\pi/2 \quad \text{and} \quad \phi = \frac{1}{2}\log\tan(\pi/4 + \alpha/2).$$

**Solution.** Since $\tan(\theta + i\phi) = \cos\alpha + i\sin\alpha \quad \therefore \quad \tan(\theta - i\phi) = \cos\alpha - i\sin\alpha$

$\therefore$
$$\tan 2\theta = \tan[(\theta + i\phi) + (\theta - i\phi)]$$

$$= \frac{\tan(\theta + i\phi) + \tan(\theta - i\phi)}{1 - \tan(\theta + i\phi)\tan(\theta - i\phi)} = \frac{2\cos\alpha}{1 - (\cos^2\alpha + \sin^2\alpha)} = \frac{2\cos\alpha}{0} \to \infty$$

*i.e.,*
$$2\theta = n\pi + \pi/2 \quad \text{or} \quad \theta = (n + 1/2)\pi/2$$

Also
$$\tan 2i\phi = \tan[(\theta + i\phi) - (\theta - i\phi)] = \frac{\tan(\theta + i\phi) - \tan(\theta - i\phi)}{1 + \tan(\theta + i\phi)\tan(\theta - i\phi)}$$

or
$$i\tanh 2\phi = \frac{2i\sin\alpha}{1 + (\cos^2\alpha + \sin^2\alpha)} = i\sin\alpha \quad \text{or} \quad \frac{e^{2\phi} - e^{-2\phi}}{e^{2\phi} + e^{-2\phi}} = \frac{\sin\alpha}{1}$$

By componendo and dividendo, we get

$$\frac{e^{2\phi}}{e^{-2\phi}} = \frac{1 + \sin\alpha}{1 - \sin\alpha} = \frac{\cos^2\alpha/2 + \sin^2\alpha/2 + 2\sin\alpha/2\cos\alpha/2}{\cos^2\alpha/2 + \sin^2\alpha/2 - 2\sin\alpha/2\cos\alpha/2}$$

or
$$e^{4\phi} = \frac{(\cos\alpha/2 + \sin\alpha/2)^2}{(\cos\alpha/2 - \sin\alpha/2)^2} = \left(\frac{1 + \tan\alpha/2}{1 - \tan\alpha/2}\right)^2$$

or
$$e^{2\phi} = \frac{1 + \tan\alpha/2}{1 - \tan\alpha/2} = \tan\left(\frac{\pi}{4} + \frac{\alpha}{2}\right). \quad \text{Hence } \phi = \frac{1}{2}\log\tan(\pi/4 + \alpha/2).$$

**Example 19.39.** *Separate $\tan^{-1}(x + iy)$ into real and imaginary parts.* (P.T.U., 2013)

**Solution.** Let $\alpha + i\beta = \tan^{-1}(x + iy)$. Then $\alpha - i\beta = \tan^{-1}(x - iy)$

Adding,
$$2\alpha = \tan^{-1}(x + iy) + \tan^{-1}(x - iy)^* = \tan^{-1}\frac{(x + iy) + (x - iy)}{1 - (x + iy)(x - iy)}$$

$\therefore$
$$\alpha = \frac{1}{2}\tan^{-1}\frac{2x}{1 - x^2 - y^2}$$

Subtracting,
$$2i\beta = \tan^{-1}(x + iy) - \tan^{-1}(x - iy) = \tan^{-1}\frac{(x + iy) - (x - iy)}{1 + (x + iy)(x - iy)}$$

$$= \tan^{-1} i\frac{2y}{1 + x^2 + y^2} = i\tanh^{-1}\frac{2y}{1 + x^2 + y^2} \qquad [\because \tan^{-1} iz = i\tanh^{-1} z]$$

$\therefore$
$$\beta = \frac{1}{2}\tanh^{-1}\frac{2y}{1 + x^2 + y^2}.$$

**Example 19.40.** *Separate $\sin^{-1}(\cos\theta + i\sin\theta)$ into real and imaginary parts, where $\theta$ is a positive acute angle.* (P.T.U., 2014)

---

* $\tan^{-1} A \pm \tan^{-1} B = \tan^{-1}\dfrac{A \pm B}{1 \mp AB}$

**Solution.** Let $\sin^{-1}(\cos\theta + i\sin\theta) = x + iy$

Then $\cos\theta + i\sin\theta = \sin(x+iy) = \sin x\cosh y + i\cos x\sinh y$

$\therefore$ $\cos\theta = \sin x\cosh y$ ...(*i*) and $\sin\theta = \cos x\sinh y$ ...(*ii*)

Squaring and adding, we have

$$1 = \sin^2 x\cosh^2 y + \cos^2 x\sinh^2 y = \sin^2 x(1+\sinh^2 y) + \cos^2 x\sinh^2 y$$
$$= \sin^2 x + \sinh^2 y(\sin^2 x + \cos^2 x)$$

or $$1 - \sin^2 x = \sinh^2 y, \quad i.e. \quad \cos^2 x = \sinh^2 y.$$

Hence from Eqn. (*ii*), we have $\sin^2\theta = \cos^4 x$, *i.e.*, $\cos^2 x = \sin\theta$ because $\theta$ being a positive acute angle, $\sin\theta$ is positive.

As $x$ is to be between $-\pi/2$ and $\pi/2$, therefore, we have

$$\cos x = +\sqrt{(\sin\theta)} \quad \text{or} \quad x = \cos^{-1}\sqrt{(\sin\theta)}$$

The relation Eqn. (*ii*), then, gives $\sinh y = \sqrt{(\sin\theta)}$ so that $y = \log[\sqrt{(\sin\theta)} + \sqrt{(1+\sin\theta)}]$.

## PROBLEMS 19.6

**1.** If $\sin(A+iB) = x+iy$, prove that

(*i*) $\dfrac{x^2}{\cosh^2 B} + \dfrac{y^2}{\sinh^2 B} = 1$ (*D.T.U., 2013*) (*ii*) $\dfrac{x^2}{\sin^2 A} - \dfrac{y^2}{\cos^2 A} = 1.$ (*P.T.U., 2010*)

**2.** If $\cos(\alpha + i\beta) = r(\cos\theta + i\sin\theta)$, prove that (*i*) $e^{2\beta} = \dfrac{\sin(\alpha-\theta)}{\sin(\alpha+\theta)}$.

(*ii*) $\beta = \dfrac{1}{2}\log\dfrac{\sin(\alpha-\theta)}{\sin(\alpha+\theta)}$. (*D.T.U., 2012*)

**3.** If $\cos(\theta + i\phi) = \cos\alpha + i\sin\alpha$, prove that

(*i*) $\sin^2\theta = \pm\sin\alpha$ (*Madras, 2003*) (*ii*) $\cos 2\theta + \cosh 2\phi = 2$.

**4.** If $\tan(A+iB) = x+iy$, prove that

(*i*) $x^2 + y^2 + 2x\cot 2A = 1$. (*ii*) $x^2 + y^2 - 2y\coth 2B + 1 = 0$. (*iii*) $x\sinh 2B = y\sin 2A$.

**5.** If $\tan(\theta + i\phi) = \tan\alpha + i\sec\alpha$, prove that $e^{2\phi} = \pm\cot\alpha/2$ and $2\theta = \left(n+\dfrac{1}{2}\right)\pi + \alpha$.

**6.** If $\tan(x+iy) = \sin(u+iv)$, prove that $\dfrac{\sin 2x}{\sinh 2y} = \dfrac{\tan u}{\tan v}$. (*C.S.V.T.U., 2011S*)

**7.** If $\operatorname{cosec}(\pi/4 + ix) = u + iv$, prove that $(u^2+v^2) = 2(u^2 - v^2)$.

**8.** If $x = 2\cos\alpha\cosh\beta$, $y = 2\sin\alpha\sinh\beta$, prove that $\sec(\alpha+i\beta) + \sec(\alpha - i\beta) = \dfrac{4x}{x^2+y^2}$.

**9.** If $a + ib = \tanh(v + i\pi/4)$, prove that $a^2 + b^2 = 1$.

**10.** Reduce $\tan^{-1}(\cos\theta + i\sin\theta)$ to the form $a + ib$.

Hence show that $\tan^{-1}(e^{i\theta}) = \dfrac{n\pi}{2} + \dfrac{\pi}{4} - \dfrac{i}{2}\log\tan\left(\dfrac{\pi}{4} - \dfrac{\theta}{2}\right)$.

**11.** Separate $\cos^{-1}(\cos\theta + i\sin\theta)$ into real and imaginary parts, where $\theta$ is a positive acute angle.

**12.** If $\sin^{-1}(u+iv) = \alpha + i\beta$, prove that $\sin^2\alpha$ and $\cosh^2\beta$ are the roots of the equation

$$x^2 - x(1+u^2+v^2) + u^2 = 0.$$

**13.** If $\cos^{-1}(x+iy) = \alpha + i\beta$, show that

(*i*) $x^2\sec^2\alpha - y^2\operatorname{cosec}^2\alpha = 1$, (*ii*) $x^2\operatorname{sech}^2\beta + y^2\operatorname{cosech}^2\beta = 1$.

**14.** Prove that (*i*) $\sin^{-1}(ix) = 2n\pi + i\log(\sqrt{1+x^2} + x)$ (*ii*) $\sin^{-1}(\operatorname{cosec}\theta) = \pi/2 + i\log\cot\theta/2$.

**15.** If $\tan(x+iy) = i$, where $x$ and $y$ are real, prove that $x$ is indeterminate and $y$ is infinite. (*P.T.U., 2015*)

## 19.13 LOGARITHMIC FUNCTION OF A COMPLEX VARIABLE

**(1)** *Definition. If $z(= x+iy)$ and $w(= u+iv)$ be so related that $e^w = z$, then $w$ is said to be a logarithm of $z$ to the base e and is written as $w = \log_e z$.* ...(*i*)

Also $e^{w+2in\pi} = e^w \cdot e^{2in\pi} = z$ $[\because\ e^{2in\pi} = 1]$

$\therefore$ $\log z = w + 2in\pi$ ...(*ii*)

*i.e., the logarithm of a complex number has an infinite number of values and is, therefore, a multi-valued function.*

The general value of the logarithm of $z$ is written as $Log\ z$ (beginning with capital L) so as to distinguish it from its principal value which is written as $\log z$. This principal value is obtained by taking $n = 0$ in $Log\ z$.

Thus from Eqns. (*i*) and (*ii*), $Log\ (x + iy) = 2in\pi + \log (x + iy)$.

**Obs. 1.** If $y = 0$, then $Log\ x = 2in\pi + \log x$.

*This shows that the logarithm of a real quantity is also multi-valued. Its principal value is real while all other values are imaginary.*

**2.** *We know that the logarithm of a negative quantity has no real value. But we can now evaluate this.*

*e.g.*
$$\log_e(-2) = \log_e 2(-1) = \log_e 2 + \log_e(-1) = \log_e 2 + i\pi \qquad [\because\ -1 = \cos\pi + i\sin\pi = e^{i\pi}]$$
$$= 0.6931 + i\,(3.1416).$$

**(2)** *Real and imaginary parts of Log $(x + iy)$.*

$$Log\ (x+iy) = 2in\pi + \log(x+iy)$$
$$= 2in\pi + \log[r(\cos\theta + i\sin\theta)] \qquad \left\{\begin{array}{l}\text{Put } x = r\cos\theta,\ y = r\sin\theta \text{ so that}\\ r = \sqrt{(x^2+y^2)} \text{ and } \theta = \tan^{-1}(y/x)\end{array}\right.$$
$$= 2in\pi + \log(re^{i\theta})$$
$$= 2in\pi + \log r + i\theta = \log\sqrt{(x^2+y^2)} + i\,\{2n\pi + \tan^{-1}(y/x)\}$$

**(3)** *Real and imaginary parts of* $(\alpha + i\beta)^{x+iy}$

$$(\alpha+i\beta)^{x+iy} = e^{(x+iy)\,Log\,(\alpha+i\beta)} = e^{(x+iy)\,[2in\pi + \log(\alpha+i\beta)]} \qquad \left\{\begin{array}{l}\text{Put } \alpha = r\cos\theta,\ \beta = r\sin\theta \text{ so that}\\ r = \sqrt{(\alpha^2+\beta^2)} \text{ and } \theta = \tan^{-1}\beta/\alpha\end{array}\right.$$
$$= e^{(x+iy)\,[2in\pi + \log re^{i\theta}]} = e^{(x+iy)\,[\log r + i\,(2n\pi+\theta)]}$$
$$= e^{A+iB} = e^{A}(\cos B + i\sin B).$$

where, $A = x\log r - y\,(2n\pi + \theta)$ and $B = y\log r + x\,(2n\pi + \theta)$.

$\therefore$ the required real part $= e^A\cos B$ and the imaginary part $= e^A\sin B$.

**Example 19.41.** *Find the general value of log* $(-i)$. *(P.T.U., 2014 S)*

**Solution.**
$$Log\,(-i) = 2in\pi + \log[0 + i(-1)] \qquad \left\{\begin{array}{l}\text{Put } 0 = r\cos\theta,\ -1 = r\sin\theta 2\\ \text{so that } r = 1 \text{ and } \theta = -\pi/2\end{array}\right.$$
$$= 2in\pi + \log[r(\cos\theta + i\sin\theta)] = 2in\pi + \log(re^{i\theta})$$
$$= 2in\pi + \log r + i\theta = 2in\pi + \log 1 + i\,(-\pi/2) = i\left(2n - \frac{1}{2}\right)\pi.$$

**Example 19.42.** *Prove that (i)* $i^i = e^{-(4n+1)\pi/2}$ *and* $Log\ i^i = -\left(2n + \frac{1}{2}\right)\pi$.

*(ii)* $(\sqrt{i})^{\sqrt{i}} = e^{-\alpha}\,\text{cis}\,\alpha$, *where* $\alpha = \pi/4\sqrt{2}$.

**Solution.** (*i*) By definition, we have

$$i^i = e^{i\,Log\,i} = e^{i\,(2in\pi + \log i)} = e^{-2n\pi + i\log[\exp(i\pi/2)]} \qquad [\because\ i = \text{cis}\,\pi/2 = \exp(i\pi/2)]$$
$$= e^{-2n\pi + i(i\pi/2)} = e^{-(2n+1/2)\pi}$$

Taking logarithms, we get (*ii*)

(*ii*)
$$(\sqrt{i})^{\sqrt{i}} = e^{\sqrt{i}\log\sqrt{i}}$$

Now
$$\sqrt{i}\log\sqrt{i} = \frac{1}{2}\left(\cos\frac{\pi}{2} + i\sin\pi/2\right)^{1/2}\log\left(\cos\frac{\pi}{2} + i\sin\frac{\pi}{2}\right)$$
$$= \frac{1}{2}\left(\cos\frac{\pi}{4} + i\sin\frac{\pi}{4}\right)\log(e^{i\pi/2}) = \frac{1}{2}\left(\cos\frac{\pi}{4} + i\sin\frac{\pi}{4}\right)\frac{i\pi}{2}$$
$$= \frac{i\pi}{4}\left(\frac{i}{\sqrt{2}} + \frac{i}{\sqrt{2}}\right) = -\frac{\pi}{4\sqrt{2}} + i\frac{\pi}{4\sqrt{2}}$$

Hence $(\sqrt{i})^{\sqrt{i}} = e^{-\alpha + i\alpha}$ where $\alpha = \pi/4\sqrt{2}$

$= e^{-\alpha} . e^{i\alpha} = e^{-\alpha} (\cos \alpha + i \sin \alpha)$.

**Example 19.43.** *If $(a + ib)^b = m^{x+iy}$, prove that one of the values of $y/x$ is $2 \tan^{-1} (b/a) \div \log (a^2 + b^2)$.* (C.S.V.T.U., 2015)

**Solution.** Taking logarithms, $(a + ib)^p = m^{x+iy}$ gives $p \log (a + ib) = (x + iy) \log m$

or $$p\left(\frac{1}{2}\log (a^2 + b^2) + i \tan^{-1}\frac{b}{a}\right) = x \log m + iy \log m$$

Equating real and imaginary parts from both sides, we get

$$\frac{p}{2} \log (a^2 + b^2) = x \log m \quad ...(i), \qquad p \tan^{-1}\frac{b}{a} = y \log m \quad ...(ii)$$

Division of Eqn. (*ii*) by Eqn. (*i*) gives

$$y/x = 2 \tan^{-1} (b/a)/\log (a^2 + b^2).$$

**Example 19.44.** *If $i^{i^{i\ldots\infty}} = A + iB$, prove that $\tan \pi A/2 = B/A$ and $A^2 + B^2 = e^{-\pi/B}$.* (C.S.V.T.U., 2011)

**Solution.** $i^{i^{i\ldots\infty}} = A + iB$ i.e. $i^{A+iB} = A + iB$

or $$A + iB = e^{(A+iB)\log i} = e^{(A+iB)\log(\cos \pi/2 + i \sin \pi/2)}$$

$$= \exp [(A + iB) \log (e^{i\pi/2})] = e^{(A+iB)(i\pi/2)}$$

$$= e^{-B\pi/2} . e^{i\pi A/2} = e^{-B\pi/2}\left(\cos\frac{\pi A}{2} + i \sin \frac{\pi A}{2}\right)$$

Equating real and imaginary parts, we get

$$A = e^{-B\pi/2} \cos\frac{\pi A}{2} \quad ...(i) \qquad B = e^{-B\pi/2} \sin\frac{\pi A}{2} \quad ...(ii)$$

Division of Eqn. (*ii*) by Eqn. (*i*) gives $B/A = \tan \pi A/2$

Squaring and adding Eqn. (*i*) and Eqn. (*ii*), $A^2 + B^2 = e^{-B\pi}$.

**Example 19.45.** *Prove that $\log\left(\frac{a+ib}{a-ib}\right) = 2i \tan^{-1}\left(\frac{b}{a}\right)$. Hence evaluate $\cos\left[i \log\left(\frac{a+ib}{a-ib}\right)\right]$.*

**Solution.** Putting $a = r \cos \theta$, $b = r \sin \theta$ so that $\theta = \tan^{-1} b/a$, we have

$$\log\left(\frac{a+ib}{a-ib}\right) = \log\frac{r(\cos\theta + i \sin\theta)}{r(\cos\theta - i\sin\theta)} = \log (e^{i\theta} \div e^{-i\theta})$$

$$= \log e^{2i\theta} = 2i\theta = 2i \tan^{-1} b/a.$$

Thus $$\cos\left[i \log\left(\frac{a+ib}{a-ib}\right)\right] = \cos [i (2i\theta)] = \cos 2\theta = \frac{1-\tan^2\theta}{1+\tan^2\theta} = \frac{1-(b/a)^2}{1+(b/a)^2} = \frac{a^2-b^2}{a^2+b^2}.$$

**Example 19.46.** *Separate into real and imaginary parts $\log \sin (x + iy)$.*

**Solution.** $\log \sin (x + iy) = \log (\sin x \cos iy + \cos x \sin iy)$

$$= \log (\sin x \cosh y + i \cos x \sinh y) = \log r (\cos \theta + i \sin \theta),$$

where $r \cos \theta = \sin x \cosh y$ and $r \sin \theta = \cos x \sinh y$,

so that $$r = \sqrt{(\sin^2 x \cosh^2 y + \cos^2 x \sinh^2 y)}$$

$$= \sqrt{\frac{1-\cos 2x}{2}\cdot\frac{1+\cosh 2y}{2} + \frac{1+\cos 2x}{2}\cdot\frac{\cosh 2y - 1}{2}} = \sqrt{\left[\frac{1}{2}(\cosh 2y - \cos 2x)\right]}$$

and $\theta = \tan^{-1} (\cot x \tanh y)$.

Thus $\log \sin (x + iy) = \log (re^{i\theta}) = \log r + i\theta$

$$= \frac{1}{2}\log\left[\frac{1}{2}(\cosh 2y - \cos 2x)\right] + i \tan^{-1}(\cot x \tanh y).$$

**Example 19.47.** *Find all the roots of the equation*
*(i) sin z = cosh 4* *(ii) sinh z = i.*

**Solution.** (*i*) $\sin z = \cosh 4 = \cos 4i = \sin (\pi/2 - 4i)$

$\therefore$ $$z = n\pi + (-1)^n (\pi/2 - 4i) \quad \left\{\because \text{ If } \sin\theta = \sin\alpha \text{ then } \theta = n\pi + (-1)^n \alpha\right.$$

(*ii*) $$i = \sinh z = \frac{e^z - e^{-z}}{2}$$

or $$e^{2z} - 2ie^z - 1 = 0, \quad i.e. \quad (e^z - i)^2 = 0 \quad i.e., \quad e^z = i$$

or $$z = Log\, i = 2in\pi + \log i = 2in\pi + \log e^{i\pi/2} = 2in\pi + i\pi/2 = i\left(2n + \frac{1}{2}\right)\pi.$$

## PROBLEMS 19.7

1. Find the general value of
(*i*) $\log (6 + 8i)$ (*ii*) $\log (-1)$. (*iii*) $\log [(1 + i) \log i]$ (*P.T.U., 2015*)
2. Show that (*i*) $\log (1 + i \tan \alpha) = \log (\sec \alpha) + i\alpha$, where $\alpha$ is an acute angle.
(*ii*) $\text{Log}_e \dfrac{3 - i}{3 + i} = 2i\left(n\pi - \tan^{-1}\dfrac{1}{3}\right)$.
3. If $(a_1 + ib_1)(a_2 + ib_2)\ldots\ldots(a_n + ib_n) = A + iB$, prove that
(*i*) $(a_1^2 + b_1^2)(a_2^2 + b_2^2)\ldots\ldots(a_n^2 + b_n^2) = A^2 + B^2$
(*ii*) $\tan^{-1}\dfrac{b_1}{a_1} + \tan^{-1}\dfrac{b_2}{a_2} + \ldots\ldots + \tan^{-1}\dfrac{b_n}{a_n} = \tan^{-1}\dfrac{B}{A}$. (*D.T.U., 2012*)
4. Find the modulus and argument of (*i*) $(1 - i)^{1+i}$. (*P.T.U., 2010*) (*ii*) $i^{\log (1 + i)}$
5. If $i^{\alpha + i\beta} = \alpha + i\beta$, prove that $\alpha^2 + \beta^2 = e^{-(4n+1)\pi\beta}$. (*C.S.V.T.U., 2014*)
6. Prove that $\log \left\{\dfrac{\sin (x + iy)}{\sin (x - iy)}\right\} = 2i \tan^{-1} (\cot x \tanh y)$.
7. Prove that $\tan \left[i \log \left(\dfrac{a - ib}{a + ib}\right)\right] = \dfrac{2ab}{a^2 - b^2}$.
8. If $\tan \log (x + iy) = a + ib$ where $a^2 + b^2 \neq 1$, show that $\tan \log (x^2 + y^2) = \dfrac{2a}{1 - a^2 - b^2}$. (*C.S.V.T.U., 2015*)
9. If $\sin^{-1} (x + iy) = \log (A + iB)$, show that $\dfrac{x^2}{\sin^2 u} - \dfrac{y^2}{\cos^2 u} = 1$, where $A^2 + B^2 = e^{2u}$.
10. Separate into real and imaginary parts $\log \cos (x + iy)$.
11. Find all the roots of the equation, (*i*) $\cos z = 2$, (*ii*) $\tanh z + 2 = 0$.

## 19.14 SUMMATION OF SERIES – 'C + iS' METHOD

This is the most general method and is applied to find the sum of a series of the form

$$a_0 \sin \alpha + a_1 \sin (\alpha + \beta) + a_2 \sin (\alpha + 2\beta) + \ldots$$

or $$a_0 \cos \alpha + a_1 \cos (\alpha + \beta) + a_2 \cos (\alpha + 2\beta) + \ldots$$

**Procedure.** (*i*) Put the given series = *S* (or *C*) according as it is a series of sines (or cosines). Then write *C* (or *S*) = a similar series of cosines (or sines).

*e.g.*, If $$S = a_0 \sin \alpha + a_1 \sin (\alpha + \beta) + a_2 \sin (\alpha + 2\beta) + \ldots$$

then $$C = a_0 \cos \alpha + a_1 \cos (\alpha + \beta) + a_2 \cos (\alpha + 2\beta)\ldots$$

(*ii*) Multiply the series of sines by *i* and add to the series of cosines, so that

$$C + iS = a_0 [\cos \alpha + i \sin \alpha] + a_1 [\cos (\alpha + \beta) + i \sin (\alpha + \beta)] + \ldots$$
$$= a_0 e^{i\alpha} + a_1 e^{i(\alpha + \beta)} + a_2 e^{i(\alpha + 2\beta)} + \ldots$$

(*iii*) Sum up this last series using any of the following standard series :

**(1) Exponential series** *i.e.,* $1 + x + \frac{x^2}{2!} + \frac{x^3}{3!} + \ldots \infty = e^x$

**(2) Sine, cosine, sinh or cosh series**

*i.e.,* $x - \frac{x^3}{3!} + \frac{x^5}{5!} - \ldots \infty = \sin x$, $1 - \frac{x^2}{2!} + \frac{x^4}{4!} - \ldots \infty = \cos x$

$x + \frac{x^3}{3!} + \frac{x^5}{5!} + \ldots \infty = \sinh x$, $1 + \frac{x^2}{2!} + \frac{x^4}{4!} + \ldots \infty = \cosh x$

**(3) Logarithmic series**

*i.e.,* $x - \frac{x^2}{2} + \frac{x^3}{3} - \ldots \infty = \log(1 + x)$, $-\left(x + \frac{x^2}{2} + \frac{x^3}{3} + \ldots \infty\right) = \log(1 - x)$

**(4) Gregory's series**

*i.e.,* $x - \frac{x^3}{3} + \frac{x^5}{5} - \ldots \infty = \tan^{-1} x$, $x + \frac{x^3}{3} + \frac{x^5}{5} + \ldots \infty = \tanh^{-1} x = \frac{1}{2} \log \frac{1+x}{1-x}$

**(5) Binomial series**

*i.e.,* $1 + nx + \frac{n(n-1)}{1.2} x^2 + \frac{n(n-1)(n-2)}{1.2.3} x^3 + \ldots \infty = (1 + x)^n$

$1 - nx + \frac{n(n+1)}{2!} x^2 - \frac{n(n+1)(n+2)}{3!} x^3 + \ldots \infty = (1 + x)^{-n}$

$1 + nx + \frac{n(n+1)}{2!} x^2 + \frac{n(n+1)(n+2)}{3!} x^3 + \ldots \infty = (1 - x)^{-n}$

**(6) Geometric series**

*i.e.,* $a + ar + ar^2 + \ldots$ to $n$ terms $= a \frac{1 - r^n}{1 - r}$, $a + ar + ar^2 + \ldots \infty = \frac{a}{1 - r}, |r| < 1.$

(*iv*) Finally express the sum thus obtained in the form $A + iB$ so that by equating the real and imaginary parts, we get $C = A$ and $S = B$.

**Series depending on exponential series**

**Example 19.48.** *Sum the series* $\sin\alpha + \frac{1}{2}\sin 3\alpha + \frac{1.3}{2.4}\sin 5\alpha + \ldots\infty$ (*Kerala, 2019*)

**Solution.** $S = \sin\alpha + \frac{1}{2}\sin 3\alpha + \frac{1.3}{2.4}\sin 5\alpha + \ldots\infty$

$C = \cos\alpha + 1/2\cos 3\alpha + \frac{1.3}{2.4}\cos 5\alpha + \ldots\infty$

$\therefore$ $C + iS = (\cos\alpha + i\sin\alpha) + 1/2(\cos 3\alpha + i\sin 3\alpha) + \frac{1.3}{2.4}(\cos 5\alpha + i\sin 5\alpha) + \ldots\infty$

$= e^{i\alpha} + \frac{1}{2}e^{3i\alpha} + \frac{1.3}{2.4}e^{5i\alpha} + \ldots = e^{i\alpha}\left[1 + \frac{1}{2}e^{2i\alpha} + \frac{1.3}{2.4}e^{4i\alpha} + \ldots\right]$

$= e^{i\alpha}(1 - e^{2i\phi})^{-1/2}$

$= (\cos\alpha + i\sin\alpha)(1 - \cos 2\alpha - i\sin 2\alpha)^{-1/2}$

$= (\cos\alpha + i\sin\alpha)(2\sin^2\alpha - 2i\sin\alpha\cos\alpha)^{-1/2}$

$= (2\sin\alpha)^{-1/2}(\cos\alpha + i\sin\alpha).(\sin\alpha - i\cos\alpha)^{-1/2}$

$= 2(\sin\alpha)^{-1/2}(\cos\alpha + i\sin\alpha).\left[\cos\left(\frac{\pi}{2} - \alpha\right) - i\sin\left(\frac{\pi}{2} - \alpha\right)\right]^{-1/2}$

$= 2(\sin\alpha)^{-1/2}(\cos\alpha + i\sin\alpha).\left[\cos\left(\frac{\pi}{4} - \frac{2}{2}\right) + i\sin\left(\frac{\pi}{4} - \frac{\alpha}{2}\right)\right]$

$C + iS = (2\sin\alpha)^{-1/2}\left[\cos\left(\frac{\pi}{4} + \frac{\alpha}{2}\right) + i\sin\left(\frac{\pi}{4} + \frac{\alpha}{2}\right)\right]$

Equating imaginary parts on both sides

$S = (2\sin\alpha)^{-1/2}\sin\left(\frac{\pi}{4} + \frac{\alpha}{2}\right)$

### Series depending on logarithmic series

**Example 19.49.** *Sum the series*

$$\sin^2\theta - \frac{1}{2}\sin 2\theta \sin^2\theta + \frac{1}{3}\sin 3\theta \sin^3\theta - \frac{1}{4}\sin 4\theta \sin^4\theta + \ldots \infty.$$

*(P.T.U., 2010 )*

**Solution.** Let $S = \sin\theta \,.\, \sin\theta - \frac{1}{2}\sin 2\theta \,.\, \sin^2\theta + \frac{1}{3}\sin 3\theta \,.\, \sin^3\theta - \ldots \infty$

and $C = \cos\theta \,.\, \sin\theta - \frac{1}{2}\cos 2\theta \,.\, \sin^2\theta + \frac{1}{3}\cos 3\theta \,.\, \sin^3\theta - \ldots \infty$

$$\therefore \quad C + iS = e^{i\theta}\sin\theta - \frac{e^{2i\theta}\sin^2\theta}{2} + \frac{e^{3i\theta}\sin^3\theta}{3} - \ldots \infty$$
$$= \log(1 + e^{i\theta}\sin\theta) = \log[1 + (\cos\theta + i\sin\theta)\sin\theta]$$
$$= \log[1 + \cos\theta\sin\theta + i\sin^2\theta] \text{ [Put } 1 + \cos\theta\sin\theta = r\cos\alpha\,;\ \sin^2\theta = r\sin\alpha] \quad \ldots(i)$$
$$= \log r(\cos\alpha + i\sin\alpha) = \log re^{i\alpha} = \log r + i\alpha$$

Equating imaginary parts, we have $S = \alpha = \tan^{-1}\left(\dfrac{\sin^2\theta}{1 + \cos\theta\sin\theta}\right)$. [from $(i)$]

### Series depending on binomial series

**Example 19.50.** *Find the sum to infinity of the series*

$$1 - \frac{1}{2}\cos\theta + \frac{1.3}{2.4}\cos 2\theta - \frac{1.3.5}{2.4.6}\cos 3\theta + \ldots (-\pi < \theta < \pi).$$ *(P.T.U., 2014 S ; C.S.V.T.U., 2011 S)*

**Solution.** Let $C = 1 - \frac{1}{2}\cos\theta + \frac{1.3}{2.4}\cos 2\theta - \frac{1.3.5}{2.4.6}\cos 3\theta + \ldots \infty$

and $S = 0 - \frac{1}{2}\sin\theta + \frac{1.3}{2.4}\sin 2\theta - \frac{1.3.5}{2.4.6}\sin 3\theta + \ldots \infty$

$$\therefore \quad C + iS = 1 - \frac{1}{2}e^{i\theta} + \frac{1.3}{2.4}e^{2i\theta} - \frac{1.3.5}{2.4.6}e^{3i\theta} - \ldots$$
$$= 1 + \left(-\frac{1}{2}\right)e^{i\theta} + \frac{-\frac{1}{2}\left(-\frac{1}{2} - 1\right)}{1.2}e^{2i\theta} + \frac{-\frac{1}{2}\left(-\frac{1}{2} - 1\right)\left(-\frac{1}{2} - 2\right)}{1.2.3}e^{3i\theta} + \ldots$$
$$= (1 + e^{i\theta})^{-1/2} = (1 + \cos\theta + i\sin\theta)^{-1/2} = \left(2\cos^2\frac{\theta}{2} + i\,.\,2\sin\frac{\theta}{2}\cos\frac{\theta}{2}\right)^{-1/2}$$
$$= \left(2\cos\frac{\theta}{2}\right)^{-1/2}\left(\cos\frac{\theta}{2} + i\sin\frac{\theta}{2}\right)^{-1/2} = \left(2\cos\frac{\theta}{2}\right)^{-1/2}\left(\cos\frac{\theta}{4} - i\sin\frac{\theta}{4}\right).$$

Equating real parts, we have $C = (2\cos\theta/2)^{-1/2}\cos\theta/4$.

## PROBLEMS 19.8

Sum the following series :

1. $\cos\theta + \sin\theta\cos 2\theta + \dfrac{\sin^2\theta}{1.2}\cos 3\theta + \ldots \infty$. *(C.S.V.T.U., 2014 S)*
2. $\sin\alpha - \dfrac{\sin(\alpha + 2\beta)}{2!} + \dfrac{\sin(\alpha + 4\beta)}{4!} - \ldots \infty$.
3. Sum the series $1 + n\sin\theta + \dfrac{n(n-1)}{2!}\sin 2\theta + \dfrac{n(n-1)(n-2)}{3!}\sin 3\theta + \ldots \sin n\theta$ *(Marathwada, 2019)*
4. $\cos\theta - \dfrac{1}{2}\cos 2\theta + \dfrac{1}{3}\cos 3\theta \ldots \infty$.
5. $e^{\alpha}\cos\beta - \dfrac{e^{3\alpha}}{3}\cos 3\beta + \dfrac{e^{5\alpha}}{5}\cos 5\beta - \ldots \infty$. *(C.S.V.T.U., 2014)*
6. $c\sin\alpha + \dfrac{c^3}{3}\sin 3\alpha + \dfrac{c^5}{5}\sin 5\alpha + \ldots \infty$.
7. Sum the series $1 + {}^nC_1\cos\alpha + {}^nC_2\cos 2\alpha + {}^nC_3\cos 3\alpha + \ldots\infty$ *(Patna, 2018)*

8. $n \sin \alpha + \frac{n(n+1)}{1.2} \sin 2\alpha + \frac{n(n+1)(n+2)}{1.2.3} \sin 3\alpha + ... \infty$. *(P.T.U., 2013 S)*

9. Sum the series $x \sin \theta - x^2 \sin 2\theta + x^2 \sin 3\theta$ ... to $n$ *(Bhilai, 2019)*

10. $\cos \alpha + \cos (\alpha + \beta) + \cos (\alpha + 2\beta) + ...$ to $n$ terms. *(C.S.V.T.U., 2011)*

11. $\sin \alpha \sin \beta - \frac{1}{2} \sin^2 \alpha \sin 2\beta + \frac{1}{3} \sin^3 \alpha \sin 3\beta + ...\infty$. *(C.S.V.T.U., 2015)*

12. $1 + x \cos \theta + x^2 \cos 2\theta + ... + x^{n-1} \cos (n-1)\theta$.

13. Assuming the validity of the expression, show that $\sin\left(\frac{\pi}{4}+\theta\right) = \frac{1}{\sqrt{2}}\left(1+\theta-\frac{\theta^2}{2!}-\frac{\theta^3}{3!}+...\right)$. *(D.T.U., 2013)*

## 19.15 APPROXIMATIONS AND LIMITS

**Example 19.51.** *If $\frac{\sin\theta}{\theta} = \frac{599}{600}$, find an approximate value of $\theta$ in radians.*

**Solution.** Since $\frac{\sin\theta}{\theta} = 1 - \frac{1}{600}$ which is nearly equal to 1. $\therefore$ $\theta$ must be very small.

We know that $\sin\theta = \theta - \frac{\theta^3}{3!} + \frac{\theta^5}{5!} + ...$

$$\therefore \quad \frac{\sin\theta}{\theta} = 1 - \frac{\theta^2}{6} + \frac{\theta^4}{5!}$$

Omitting $\theta^4$ and higher powers, we have

$$\frac{\sin\theta}{\theta} = 1 - \frac{\theta^2}{6} = 1 - \frac{1}{600}$$

or $\theta^2 = \frac{1}{100}$. Hence $\theta = 0.1$ radians.

**Example 19.52.** *Solve approximately $\sin\left(\frac{\pi}{6}+\theta\right) = 0.51$.*

**Solution.** Since 0.51 is nearly equal to 1/2, which is the value of $\sin \pi/6$, so $\theta$ must be very small.

$$\therefore \quad \sin\left(\frac{\pi}{6}+\theta\right) = \sin\frac{\pi}{6}\cos\theta + \cos\frac{\pi}{6}\sin\theta = \frac{1}{2}\left(1-\frac{\theta^2}{2!}+...\right) + \frac{\sqrt{3}}{2}\left(\theta-\frac{\theta^3}{3!}+...\right)$$

$$= \frac{1}{2} + \frac{\sqrt{3}}{2}\theta, \text{ omitting } \theta^2 \text{ and higher powers of } \theta.$$

Hence the given equation becomes,

$$\frac{1}{2} + \frac{\sqrt{3}}{2}\theta = 0.51 \quad \text{or} \quad \theta = \frac{1}{50\sqrt{3}}$$

or $$\theta = \frac{1}{50\sqrt{3}} \text{ radian} = \frac{\sqrt{3}}{150} \times 57.29 \text{ degrees nearly} = 39.7'.$$

**Example 19.53.** *If $\sin(\alpha + i\beta) = x + iy$. Prove that*

*(i)* $\frac{x^2}{\cosh^2\beta} + \frac{y^2}{\sin h^2\beta} = 1$ *(ii)* $\frac{x^2}{\sin^2\alpha} + \frac{y^2}{\cos^2\alpha} = 1$

**Solution:** $\sin(\alpha + i\beta) = x + iy$

$\sin \alpha \cosh \beta + i \cos \alpha \sinh \beta = x + iy$

Equality real and imaginary parts, we have

$$x = \sin \alpha \cosh \beta$$
$$y = \cos \alpha \sinh \beta$$
$$\frac{x}{\cosh \beta} = \sin \alpha$$
$$\frac{y}{\sinh \beta} = \cos \alpha$$

$\therefore$
$$\frac{x^2}{\cosh^2 \beta} + \frac{y^2}{\sinh^2 \beta} = \sin^2 \alpha + \cos^2 \alpha$$
$$\frac{x^2}{\cosh^2 \beta} + \frac{y^2}{\sinh^2 \beta} = 1$$

Also
$$\frac{x}{\sin \alpha} = \cosh \beta$$
$$\frac{y}{\cos \alpha} = \sinh \beta$$
$$\frac{x^2}{\sin^2 \alpha} - \frac{y^2}{\cos^2 \alpha} = \cosh^2 \beta - \sinh^2 \beta$$

$\therefore$
$$\frac{x^2}{\sin^2 \alpha} - \frac{y^2}{\cos^2 \alpha} = 1$$

**Example 19.54:** *If $\tan(\theta + i\phi) = \tan \alpha + i \sec \alpha$, then show that*

$$2\theta = n\pi + \frac{\pi}{2} + \alpha$$
$$e^{2\phi} = \pm \cot(\alpha/2)$$

**Solution:** Here
$$\tan(\theta + i\phi) = \tan \alpha + i \sec \alpha$$

$\therefore$
$$\tan(\theta - i\phi) = \tan \alpha - i \sec \alpha$$

Now
$$\tan 2\theta = \tan[(\theta + i\phi) + (\theta - i\phi)]$$
$$= \frac{\tan(\theta + i\phi) + \tan(\theta - i\phi)}{1 - \tan(\theta + i\phi)\tan(\theta - i\phi)} = \frac{(\tan \alpha + i\sec \alpha) + (\tan \alpha - i\sec \alpha)}{1 - (\tan + i\sec \alpha)(\tan \alpha - i\sec \alpha)}$$
$$= \frac{2\tan \alpha}{1 - (\tan^2 \alpha + \sec^2 \alpha)} = \frac{-2\tan \alpha}{\tan^2 \alpha} = -\cot \alpha = \tan\left[\left(n\pi + \frac{\pi}{2}\right) + \alpha\right]$$

$\therefore$
$$2\theta = n\pi + \frac{\pi}{2}$$
$$\tan 2i\phi = \tan[(\theta + i\phi) - (\theta - i\phi)]$$
$$= \frac{\tan(\theta + i\phi) - \tan(\theta - i\phi)}{1 - \tan(\theta + i\phi)\tan(\theta - i\phi)} = \frac{(\tan \alpha + i\sec \alpha) - (\tan \alpha - i\sec \alpha)}{1 + (\tan \alpha + i\sec \alpha)(\tan \alpha - i\sec \alpha)}$$
$$= \frac{2i\sec \alpha}{1 + (\tan^2 \alpha + \sec^2 \alpha)} = \frac{2i\sec \alpha}{2\sec^2 \alpha} = i \cos \alpha$$

$\therefore$
$$i \tan 2h\, 2\phi = i \cos \alpha$$
$$\tan h\, 2\phi = \cos \phi$$

$$\frac{e^{2\phi}-e^{-2\phi}}{e^{2\phi}+e^{-2\phi}}=\frac{\cos\alpha}{1}$$

$$\frac{e^{2\phi}}{e^{-2\phi}}=\frac{1+\cos\alpha}{1-\cos\alpha}=\frac{2\cos^2\alpha/2}{2\sin^2\alpha/2}=\cot^2\frac{\alpha}{2}$$

$$e^{4\phi}=\cot^2\alpha/2$$

$$e^{2\phi}=\pm\cot\alpha/2$$

**PROBLEMS 19.9**

1. Given $\frac{\sin\theta}{\theta}=\frac{5045}{5046}$, show that $\theta$ is 1°58′ nearly.
2. If $\frac{\sin\theta}{\theta}=\frac{2165}{2166}$, find an approximate value of $\theta$ in radians.
3. If $\cos\theta=\frac{1681}{1682}$, find $\theta$ approximately.
4. Solve approximately the equation $\cos\left(\frac{\pi}{3}+\theta\right)=0.49$.

## 19.16 OBJECTIVE TYPE OF QUESTIONS

**PROBLEMS 19.10**

*Choose the correct answer or fill up the blanks in each of the following problems :*

1. If $x+iy=\sqrt{2}+3i$, then $x^2+y$ is
   (a) 7 (b) 5 (c) 13 (d) $\sqrt{2}+3$.
2. The real part of $(\sin x+i\cos x)^5$ is
   (a) $-\cos 5x$ (b) $-\sin 5x$ (c) $\sin 5x$ (d) $\cos 5x$.
3. The number $(i)^i$ is
   (a) a purely imaginary number (b) an irrational number
   (c) a rational number (d) an integer.
4. The relation $|3-z|+|3+z|=5$ represents
   (a) a circle (b) a parabola (c) an ellipse (d) a hyperbola.
5. $z$ is a complex number with $|z|=1$ and $\arg(z)=3\pi/4$. The value of $z$ is
   (a) $(1+i)/\sqrt{2}$ (b) $(-1+i)/\sqrt{2}$ (c) $(1-i)/\sqrt{2}$ (d) $(-1-i)/\sqrt{2}$.
6. If $f(z)=e^{2z}$, then the imaginary part of $f(z)$ is
   (a) $e^y\sin x$ (b) $e^x\cos y$ (c) $e^{2x}\cos 2y$ (d) $e^{2x}\sin 2y$.
7. Expansion of $\sin^m\theta\cos^n\theta$ is a series of sines of multiples of $\theta$ when $m$ is ......
8. Expansion of $\cos 6\theta$ in terms of $\cos\theta$ is ......
9. If $f(z)=3\bar{z}$, then the value of $f(z)$ at $z=2+4i$ is ......
10. If $x=\cos\theta+i\sin\theta$, then $x^n-1/x^n=$ ......
11. Imaginary part of $(2+i3)/(3-i4)$ is ......
12. Real part of $\cosh(x+iy)$ is ......
13. If $\frac{\sin\theta}{\theta}=\frac{2165}{2166}$, then $\theta=$ ...... approximately.
14. If $\tan x/2=\tanh y/2$, then $\cos x\cosh y=$ ......
15. Imaginary part of $\sin\bar{z}$ is ......
16. Modulus of $(\sqrt{i})^{\sqrt{i}}=$ ...........
17. If $\sin\alpha+\sin\beta+\sin\gamma=0=\cos\alpha+\cos\beta+\cos\gamma$, then $\cos 3\alpha+\cos 3\beta+\cos 3\gamma=3\cos$ (............)
18. $\log(-1)=$ ................
19. $(i)^i$ is purely real or imaginary
20. If $\sin\theta=\tanh\phi$, then $\tan\theta=$ ...............................
21. Imaginary part of $\tan(\theta+i\phi)=$ ...............................
22. $\cos 5\alpha=$ (.........) $\cos^5\alpha$ + (.....) $\cos^3\alpha$ + (.....) $\cos\alpha$.
23. Cube roots of unity form .................................. triangle.

24. If $|z_1 + z_2| = |z_1 - z_2|$ then amp $(z_1)$ – amp $(z_2)$ is ....................
25. If $-3 + ix^2y$ and $x^2 + y + 4i$ represent conjugate complex number then $x$ = .................. and $y$ = ......................
26. If $\left|\frac{z-a}{z-b}\right| = k\ (\neq 1)$, then the locus of $z$ is ................................
27. $(-i)^{-i}$ is purely real. (True or False)
28. The statements Re $z > 0$ and $|z - 1| < |z + 1|$ are equivalent. (True or False)
29. Hyperbolic functions are periodic. (True or False)
30. $n$th roots of unity form a G.P. (True or False)
31. $\sin ix = -i \sinh x$. (True or False)
32. If the sum and product of two complex numbers are real, then the two numbers must be either real or conjugate. (True or False)
33. The modulus of the sum of two complex numbers $\geq$ to the sum of their moduli. (True or False)

CHAPTER 20

# Calculus of Complex Functions



## 20.1 INTRODUCTION

In the previous chapter, we have dealt with some elementary complex functions—the exponential, logarithmic, circular and hyperbolic functions, evaluated at specific complex values. These functions are useful in the study of fluid mechanics, thermodynamics and electric fields. It, therefore, seems desirable to study the calculus of such functions.

## 20.2 (1) LIMIT OF A COMPLEX FUNCTION

A *function* $w = f(z)$ *is said to tend to* **limit** $l$ *as* $z$ *approaches a point* $z_0$, *if for every real* $\varepsilon$, *we can find a positive real* $\delta$ *such that*

$$| f(z) - l | < \epsilon \quad \text{for} \quad | z - z_0 | < \delta$$

*i.e.*, for every $z \neq z_0$ in the $\delta$–disc (dotted) of $z$-plane, $f(z)$ has a value lying in the $\varepsilon$-disc of $w$-plane (Fig. 20.1). In symbols, we write $\underset{z \to z_0}{\text{Lt}} f(z) = l$.

This definition of limit though similar to that in ordinary calculus, is quite different for in real calculus $x$ approaches $x_0$ only along the line whereas here *z approaches* $z_0$ *from any direction in the z-plane.*

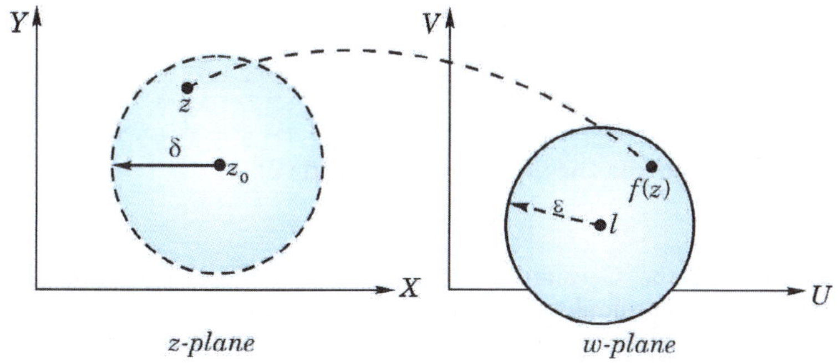


Fig. 20.1

**(2) Continuity of f(z).** *A function $w = f(z)$ is said to be* **continuous** *at $z = z_0$, if*

$$\underset{z \to z_0}{\text{Lt}} f(z) = f(z_0).$$

Further $f(z)$ is said to be continuous in any region $R$ of the $z$-plane, if it is continuous at every point of that region.

Also if $w = f(z) = u(x, y) + iv(x, y)$ is continuous at $z = z_0$, then $u(x, y)$ and $v(x, y)$ are also continuous at $z = z_0$, *i.e.*, at $x = x_0$ and $y = y_0$. Conversely if $u(x, y)$ and $v(x, y)$ are continuous at $(x_0, y_0)$, then $f(z)$ will be continuous at $z = z_0$. [cf. § 5.1 (3)].

## 20.3 (1) DERIVATIVE OF f(z)

Let $w = f(z)$ be a single-valued function of the variable $z = x + iy$. Then *the derivative of $w = f(z)$ is defined to be*

$$\frac{dw}{dz} = f'(z) = \underset{\delta z \to 0}{\text{Lt}} \frac{f(z + \delta z) - f(z)}{\delta z},$$

*provided the limit exists and has the same value for all the different ways in which $\delta z$ approaches zero.*

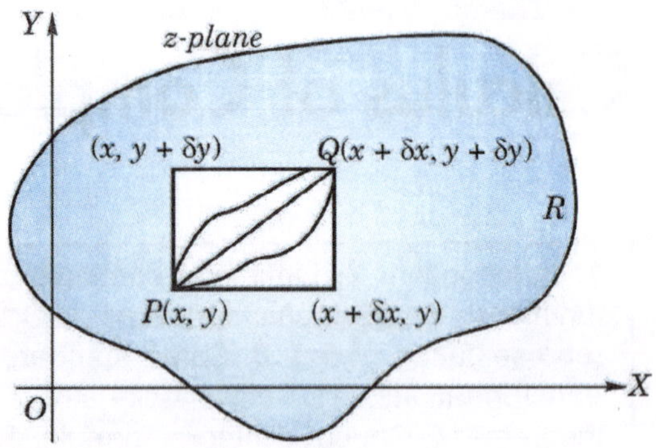

Fig. 20.2

Suppose $P(z)$ is fixed and $Q(z + \delta z)$ is a neighbouring point (Fig. 20.2). The point $Q$ may approach $P$ along any straight or curved path in the given region, *i.e.*, $\delta z$ may tend to zero in any manner and $dw/dz$ may not exist. It, therefore, becomes a fundamental problem to determine the necessary and sufficient conditions for $dw/dz$ to exist. The fact is settled by the following theorem.

**(2) Theorem.** *The necessary and sufficient conditions for the derivative of the function $w = u(x, y) + iv(x, y) = f(z)$ to exist for all values of $z$ in a region $R$, are*

(*i*) $\frac{\partial u}{\partial x}, \frac{\partial u}{\partial y}, \frac{\partial v}{\partial x}, \frac{\partial v}{\partial y}$ *are continuous functions of $x$ and $y$ in $R$;*

(*ii*) $\frac{\partial u}{\partial x} = \frac{\partial v}{\partial y}, \frac{\partial u}{\partial y} = -\frac{\partial v}{\partial x}$.

*The relations (ii) are known as* **Cauchy-Riemann* equations** or briefly *C-R* equations.

(*a*) *Condition is necessary.*

If $f(z)$ possesses a unique derivative at $P(z)$, then

$$f'(z) = \underset{\delta z \to 0}{\text{Lt}} \frac{f(z + \delta z) - f(z)}{\delta z}$$

$$= \underset{\substack{\delta x \to 0 \\ \delta y \to 0}}{\text{Lt}} \frac{\{u(x + \delta x, y + \delta y) + iv(x + \delta x, y + \delta y)\} - \{u(x, y) + iv(x, y)\}}{\delta x + i\delta y}$$

Since $\delta z$ can approach zero in any manner, we can first assume $\delta z$ to be wholly real and then wholly imaginary. When $\delta z$ is wholly real, then $\delta y = 0$ and $\delta z = \delta x$.

$$\therefore \quad f'(z) = \underset{\delta x \to 0}{\text{Lt}} \left( \frac{u(x + \delta x, y) - u(x, y)}{\delta x} + i \frac{v(x + \delta x, y) - v(x, y)}{\delta x} \right) = \frac{\partial u}{\partial x} + i \frac{\partial v}{\partial x} \quad ...(1)$$

When $\delta z$ is wholly imaginary, then $\delta x = 0$ and $\delta z = i\delta y$.

$$\therefore \quad f'(z) = \underset{\delta y \to 0}{\text{Lt}} \left( \frac{u(x, y + \delta y) - u(x, y)}{i\delta y} + i \frac{v(x, y + \delta y) - v(x, y)}{i\delta y} \right) = \frac{1}{i} \frac{\partial u}{\partial y} + \frac{\partial v}{\partial y} = \frac{\partial v}{\delta y} - i \frac{\partial u}{\partial y} \quad ...(2)$$

Now the existence of $f'(z)$ requires the equality of (1) and (2).

---

*Named after *Cauchy* (p. 584) and the German mathematician *Bernhard Riemann* (1826—1866) who along with *Weierstrass* (p. 95) laid the foundations of complex analysis. Riemann introduced the concept of integration and made basic contributions to number theory and mathematical analysis. He developed the Riemannian geometry which formed the mathematical base for Einstein's relativity theory.

$$\therefore \qquad \frac{\partial u}{\partial x} + i\frac{\partial v}{\partial x} = \frac{\partial v}{\partial y} - i\frac{\partial u}{\partial y}$$

On equating the real and imaginary parts from both sides, we get

$$\frac{\partial u}{\partial x} = \frac{\partial v}{\partial y}, \frac{\partial u}{\partial y} = -\frac{\partial v}{\partial x} \qquad ...(3)$$

*Thus the necessary conditions for the existence of the derivative of $f(z)$ is that the C-R equations should be satisfied.* (V.T.U., 2016)

(b) *Condition is sufficient.* Suppose $f(z)$ is a single-valued function possessing partial derivatives $\frac{\partial u}{\partial x}, \frac{\partial u}{\partial y}, \frac{\partial v}{\partial x}, \frac{\partial v}{\partial y}$ at each point of the region and the *C-R* Eqn. (3) are satisfied.

Then by Taylor's theorem for functions of two variables

$$f(z + \delta z) = u(x + \delta x, y + \delta y) + iv(x + \delta x, y + \delta y)$$

$$= u(x, y) + \left(\frac{\partial u}{\partial x}\delta x + \frac{\partial u}{\partial y}\delta y\right) + ... + i\left[v(x, y) + \left(\frac{\partial v}{\partial x}\delta x + \frac{\partial v}{\partial y}\delta y\right) + ...\right]$$

$$= f(z) + \left(\frac{\partial u}{\partial x} + i\frac{\partial v}{\partial x}\right)\partial x + \left(\frac{\partial u}{\partial y} + i\frac{\partial v}{\partial y}\right)\delta y$$

[Omitting terms beyond the first powers of $\delta x$ and $\delta y$]

or
$$f(z + \delta z) - f(z) = \left(\frac{\partial u}{\partial x} + i\frac{\partial v}{\partial x}\right)\partial x + \left(\frac{\partial u}{\partial y} + i\frac{\partial v}{\partial y}\right)\delta y.$$

Now using the *C-R* equation (3), replace $\frac{\partial u}{\partial y}$ and $\frac{\partial v}{\partial y}$ by $-\frac{\partial v}{\partial x}$ and $\frac{\partial u}{\partial x}$ respectively.

Then
$$f(z + \delta z) - f(z) = \left[\frac{\partial u}{\partial x} + i\frac{\partial v}{\partial x}\right]\delta x + \left[-\frac{\partial v}{\partial x} + i\frac{\partial u}{\partial x}\right]\delta y = \left[\frac{\partial u}{\partial x} + i\frac{\partial v}{\partial x}\right]\delta x + \left[i\frac{\partial v}{\partial x} + \frac{\partial u}{\partial x}\right]i\delta y$$

$$= \left[\frac{\partial u}{\partial x} + i\frac{\partial v}{\partial x}\right](\delta x + i\delta y) = \left[\frac{\partial u}{\partial x} + i\frac{\partial v}{\partial x}\right]\delta z$$

$$\therefore \qquad f'(z) = \underset{\delta z \to 0}{\text{Lt}} \frac{f(z + \delta z) - f(z)}{\delta z} = \frac{\partial u}{\partial x} + i\frac{\partial v}{\partial x} \quad \text{or} \quad \frac{\partial v}{\partial y} - i\frac{\partial u}{\partial y}$$

which by Eqns. (1) or (2) proves the sufficiency of conditions.

## 20.4 ANALYTIC FUNCTIONS

*A function $f(z)$ which is single-valued and possesses a unique derivative with respect to z at all points of a region R, is called an* **analytic function** *of z in that region. An analytic function is also called a regular function* or an *holomorphic function.*

*A function which is analytic everywhere in the complex plane, is known as an* **entire function.** As derivative of a polynomial exists at every point, a polynomial of any degree is an *entire function.*

*A point at which an analytic function ceases to possess a derivative is called a* **singular point** *of the function.*

Thus if $u$ and $v$ are real single-valued functions of $x$ and $y$ such that $\delta u/\delta x$, $\delta u/\delta y$, $\delta v/\delta x$, $\delta v/\delta y$ are continuous throughout a region $R$, then the Cauchy-Riemann equations

$$\frac{\partial \mathbf{u}}{\partial \mathbf{x}} = \frac{\partial \mathbf{v}}{\partial \mathbf{y}} \text{ and } \frac{\partial \mathbf{u}}{\partial \mathbf{y}} = -\frac{\partial \mathbf{v}}{\partial \mathbf{x}} \qquad ...(1)$$

are both necessary and sufficient conditions for the function $f(z) = u + iv$ to be analytic in $R$. The derivative of $f(z)$ is then, given by Eqn. (1) of p. 786 or (2) of p. 786.

The real and imaginary parts of an analytic function are called *conjugate functions*. The relation between two conjugate functions is given by *C-R* Eqn. (1).

**Example 20.1.** *If $w = \log z$, find $dw/dz$ and determine where w is non-analytic.*

**Solution.** We have $w = u + iv = \log (x + iy) = \frac{1}{2} \log (x^2 + y^2) + i \tan^{-1} y/x$ [By (**2**), p. 786]

so that $u = \frac{1}{2} \log (x^2 + y^2),\ v = \tan^{-1} y/x.$

$$\therefore \quad \frac{\partial u}{\partial x} = \frac{x}{x^2 + y^2} = \frac{\partial v}{\partial y},\ \frac{\partial u}{\partial y} = \frac{y}{x^2 + y^2} = -\frac{\partial v}{\partial x}.$$

Since the Cauchy-Riemann equations are satisfied and the partial derivatives are continuous except at (0, 0). Hence $w$ is analytic everywhere except at $z = 0$.

$$\therefore \quad \frac{dw}{dz} = \frac{\partial u}{\partial x} + i \frac{\partial v}{\partial x} = \frac{x}{x^2 + y^2} + i \frac{-y}{x^2 + y^2} = \frac{x - iy}{(x + iy)(x - iy)} = \frac{1}{x + iy} = \frac{1}{z} \ (z \neq 0).$$

**Obs.** The definition of the derivative of a function of complex variable is identical in form to that of the derivative of a function of real variable. Hence the rules of differentiation for complex functions are the same as those of real calculus. *Thus* **if, a complex function is once known to be analytic, it can be differentiated just in the ordinary way.**

**Example 20.2.** *If $f(z)$ is an analytic function with constant modulus, show that $f(z)$ is constant.* *(JNTUK, 2023; Andhra., 2022, 2015)*

**Solution.** If $f(z) = u + iv$ is an analytic function, then

$$|f(z)| = \sqrt{(u^2 + v^2)} \text{ is constant} = c \text{ (say) or } u^2 + v^2 = c^2 \quad ...(i)$$

Differentiating $(i)$ partially w.r.t. $x$ and $y$, we get

$$2u \frac{\partial u}{\partial x} + 2v \frac{\partial v}{\partial x} = 0; \qquad 2u \frac{\partial u}{\partial y} + 2v \frac{\partial v}{\partial y} = 0$$

or
$$u \frac{\partial u}{\partial x} + v \frac{\partial v}{\partial x} = 0 \quad ...(ii) \qquad u \frac{\partial u}{\partial y} + v \frac{\partial v}{\partial y} = 0 \quad ...(iii)$$

Since $\frac{\partial u}{\partial x} = \frac{\partial v}{\partial y};\ \frac{\partial u}{\partial y} = -\frac{\partial v}{\partial x}$ by C-R equations,

$\therefore$ Eqn. $(iii)$ becomes $-u \frac{\partial v}{\partial x} + v \frac{\partial u}{\partial x} = 0$ ...$(iv)$

Squaring and adding Eqns. $(ii)$ and $(iv)$, we obtain

$$u^2 \left(\frac{\partial u}{\partial x}\right)^2 + v^2 \left(\frac{\partial v}{\partial x}\right)^2 + u^2 \left(\frac{\partial v}{\partial x}\right)^2 + v^2 \left(\frac{\partial u}{\partial x}\right)^2 = 0$$

or
$$(u^2 + v^2)\left[\left(\frac{\partial u}{\partial x}\right)^2 + \left(\frac{\partial v}{\partial x}\right)^2\right] = 0 \text{ or } \left(\frac{\partial u}{\partial x}\right)^2 + \left(\frac{\partial v}{\partial x}\right)^2 = 0 \quad [\because u^2 + v^2 = c^2 \neq 0] \quad ...(v)$$

Now
$$f'(z) = \frac{\partial u}{\partial x} + i \frac{\partial v}{\partial x}$$

$$\therefore \quad |f'(z)|^2 = \left(\frac{\partial u}{\partial x}\right)^2 + \left(\frac{\partial v}{\partial x}\right)^2 = 0 \quad \text{[By } (v)\text{]}$$

or $f'(z) = 0.$ or $f(z) =$ constant.

**Example 20.3.** *Show that the function $f(z) = \sqrt{|xy|}$ is not analytic at the origin even though C.R. equations are satisfied thereof.* *(JNTUK, 2023; Andhra, 2021)*

**Solution.** If $f(z) = \sqrt{|xy|} = u(x, y) + iv(x, y)$, then $u(x, y) = \sqrt{|xy|}$, $v(x, y) = 0$

At the origin, we have

$$\frac{\partial u}{\partial x} = \underset{x \to 0}{\text{Lt}} \frac{u(x, 0) - u(0, 0)}{x} = \underset{x \to 0}{\text{Lt}} \frac{0 - 0}{x} = 0$$

$$\frac{\partial u}{\partial y} = \underset{y \to 0}{\text{Lt}} \frac{u(0, y) - u(0, 0)}{y} = \underset{x \to 0}{\text{Lt}} \frac{0 - 0}{y} = 0$$

$$\frac{\partial v}{\partial x} = \underset{x \to 0}{\text{Lt}} \frac{v(x, 0) - v(0, 0)}{x} = \underset{x \to 0}{\text{Lt}} \frac{0 - 0}{x} = 0$$

$$\frac{\partial v}{\partial y} = \underset{y \to 0}{\text{Lt}} \frac{v(0, y) - v(0, 0)}{y} = \underset{y \to 0}{\text{Lt}} \frac{0 - 0}{y} = 0$$

$$\therefore \qquad \frac{\partial u}{\partial x} = \frac{\partial v}{\partial y} \text{ and } \frac{\partial u}{\partial y} = -\frac{\partial v}{\partial x}$$

*i.e.*, C.R. equations are satisfied at the origin.

However $$f'(0) = \underset{z \to 0}{\text{Lt}} \frac{f(z) - f(0)}{z - 0} = \underset{z \to 0}{\text{Lt}} \frac{\sqrt{|xy|} - 0}{x + iy}$$

$$= \underset{x \to 0}{\text{Lt}} \frac{\sqrt{|mx^2|} - 0}{x(1 + im)}, \text{ when } z \to 0 \text{ along the line } y = mx$$

$$= \frac{\sqrt{|m|}}{1 + im} \text{ which is not unique.}$$

$\therefore$ $f'(0)$ does not exist. Hence $f(z)$ is not analytic at the origin.

**Example 20.4.** *Prove that the function $f(z)$ defined by*

$$f(z) = \frac{x^3(1 + i) - y^3(1 - i)}{x^2 + y^2} \, (z \neq 0), \, f(0) = 0$$

*is continuous and the Cauchy-Riemann equations are satisfied at the origin, yet $f'(0)$ does not exist.* *(JNTUK, 2023; W.B.T.U., 2015)*

**Solution.** $$\underset{z \to 0}{\text{Lt}} f(z) = \underset{\substack{x \to 0 \\ y \to 0}}{\text{Lt}} \frac{x^3(1 + i) - y^3(1 - i)}{x^2 + y^2} = \underset{y \to 0}{\text{Lt}} \frac{-y^3(1 - i)}{y^2} = \underset{y \to 0}{\text{Lt}} [-y(1 - i)] = 0$$

$$\underset{z \to 0}{\text{Lt}} f(z) = \underset{\substack{y \to 0 \\ x \to 0}}{\text{Lt}} \frac{x^3(1 + i) - y^3(1 - i)}{x^2 + y^2} = \underset{x \to 0}{\text{Lt}} \frac{x^3(1 + i)}{x^2} = \underset{x \to 0}{\text{Lt}} [x(1 + i)] = 0$$

Also $f(0) = 0$ (given).

Thus $\underset{z \to 0}{\text{Lt}} f(z) = f(0)$ when $x \to 0$ first and then $y \to 0$ and also vice-versa. Now let both $x$ and $y$ tend to zero simultaneously along the path $y = mx$. Then

$$\underset{z \to 0}{\text{Lt}} f(z) = \underset{\substack{y \to mx \\ x \to 0}}{\text{Lt}} \frac{x^3(1 + i) - y^3(1 - i)}{x^2 + y^2}$$

$$= \underset{x \to 0}{\text{Lt}} \frac{x^3(1 + i) - m^3x^3(1 - i)}{(1 + m^2)x^2} = \underset{x \to 0}{\text{Lt}} \frac{x[1 + i - m^3(1 - i)]}{1 + m^2} = 0$$

Hence $\underset{z \to 0}{\text{Lt}} f(z) = f(0)$, in whatever manner $z \to 0$. $\therefore$ $f(z)$ is continuous at the origin.

Now $$f(z) = \frac{x^3 - y^3}{x^2 + y^2} + i\frac{x^3 + y^3}{x^2 + y^2} = u(x, y) + iv(x, y).$$

Also $u(0, 0) = 0$, and $v(0, 0) = 0$ $\qquad [\because \ f(0) = 0]$

$$\therefore \qquad \left(\frac{\partial u}{\partial x}\right)_{0,0} = \underset{x \to 0}{\text{Lt}} \frac{u(x, 0) - u(0, 0)}{x} = \underset{x \to 0}{\text{Lt}} \frac{x}{x} = 1$$

$$\left(\frac{\partial u}{\partial y}\right)_{0,0} = \underset{y \to 0}{\text{Lt}} \frac{u(0, y) - u(0, 0)}{y} = \underset{y \to 0}{\text{Lt}} \frac{-y}{y} = -1$$

$$\left(\frac{\partial v}{\partial x}\right)_{0,0} = \underset{x \to 0}{\text{Lt}} \frac{v(x, 0) - v(0, 0)}{x} = \underset{x \to 0}{\text{Lt}} \frac{x}{x} = 1$$

and $$\left(\frac{\partial v}{\partial y}\right)_{0,0} = \underset{y \to 0}{\text{Lt}} \frac{v(0, y) - v(0, 0)}{y} = \underset{y \to 0}{\text{Lt}} \frac{y}{y} = 1.$$

Hence at (0, 0), $\frac{\partial u}{\partial x} = \frac{\partial v}{\partial y}$ and $\frac{\partial u}{\partial y} = -\frac{\partial v}{\partial x}$

Thus the *C-R* equations are satisfied at the origin.

But $$f'(0) = \underset{z \to 0}{\text{Lt}} \frac{f(z) - f(0)}{z} = \underset{z \to 0}{\text{Lt}} \frac{f(z)}{z} = \underset{z \to 0}{\text{Lt}} \frac{x^3 - y^3 + i(x^3 + y^3)}{(x^2 + y^2)(x + iy)}.$$

If $z \to 0$ along the path $y = mx$, then $f'(0) = \frac{1 - m^3 + i(1 + m^3)}{(1 + m^2)(1 + im)}$

which assumes different values as $m$ varies. So $f'(z)$ is not unique at (0, 0) *i.e.*, $f'(0)$ does not exist. Thus $f(z)$ is not analytic at the origin even though it is continuous and satisfies the *C-R* equations there at.

**Example 20.5.** *Show that polar form of Cauchy-Riemann equations are*

$$\frac{\partial u}{\partial r} = \frac{1}{r}\frac{\partial v}{\partial \theta}, \frac{\partial v}{\partial r} = -\frac{1}{r}\frac{\partial u}{\partial \theta}.$$ *(JNTUK, 2023; Andhra., 2015 ; Hazaribag., 2015)*

*Deduce that* $\frac{\partial^2 u}{\partial r^2} + \frac{1}{r}\frac{\partial u}{\partial r} + \frac{1}{r^2}\frac{\partial^2 u}{\partial \theta^2} = 0.$ *(V.T.U., 2012, 2012 S)*

**Solution.** If $(r, \theta)$ be the coordinates of a point whose cartesian coordinates are $(x, y)$, then $z = x + iy = re^{i\theta}$.

$$\therefore \quad u + iv = f(z) = f(re^{i\theta})$$

where $u$ and $v$ are now expressed in terms of $r$ and $\theta$.

Differentiating it partially w.r.t. $r$ and $\theta$, we have

$$\frac{\partial u}{\partial r} + i\frac{\partial v}{\partial r} = f'(re^{i\theta}) \,.\, e^{i\theta}$$

and $$\frac{\partial u}{\partial \theta} + i\frac{\partial v}{\partial \theta} = f'(re^{i\theta}) \,.\, ire^{i\theta} = ir\left(\frac{\partial u}{\partial r} + i\frac{\partial v}{\partial r}\right)$$

Equating real and imaginary parts, we get

$$\frac{\partial u}{\partial r} = \frac{1}{r}\frac{\partial v}{\partial \theta} \quad ...(i) \qquad \frac{\partial v}{\partial r} = -\frac{1}{r}\frac{\partial u}{\partial \theta} \quad ...(ii)$$

Differentiating Eqn. (*i*) partially w.r.t. $r$, we get

$$\frac{\partial^2 u}{\partial r^2} = -\frac{1}{r^2}\frac{\partial v}{\partial \theta} + \frac{1}{r}\frac{\partial^2 v}{\partial \theta \partial r} \quad ...(iii)$$

Differentiating Eqn. (*ii*) partially w.r.t. $\theta$, we have

$$\frac{\partial^2 u}{\partial \theta^2} = -r\frac{\partial^2 v}{\partial r \partial \theta} \quad ...(iv)$$

Thus using Eqns. (*i*), (*ii*) and (*iv*)

$$\frac{\partial^2 u}{\partial r^2} + \frac{1}{r}\frac{\partial u}{\partial r} + \frac{1}{r^2}\frac{\partial^2 u}{\partial \theta^2} = -\frac{1}{r^2}\frac{\partial v}{\partial \theta} + \frac{1}{r}\frac{\partial^2 v}{\partial \theta \partial r} + \frac{1}{r}\left(\frac{1}{r}\frac{\partial v}{\partial \theta}\right) + \frac{1}{r^2}\left(-r\frac{\partial^2 v}{\partial r \partial \theta}\right) = 0 \qquad \left[\because \frac{\partial^2 v}{\partial \theta \partial r} = \frac{\partial^2 v}{\partial r \partial \theta}\right]$$

## 20.5 (1) HARMONIC FUNCTIONS

If $f(z) = u + iv$ be an analytic function in some region of the $z$-plane, then the Cauchy-Riemann equations are satisfied.

*i.e.*, $$\frac{\partial u}{\partial x} = \frac{\partial v}{\partial y} \quad ...(1) \qquad \text{and} \qquad \frac{\partial u}{\partial y} = -\frac{\partial v}{\partial x}. \quad ...(2)$$

Differentiating Eqn. (1) with respect to $x$ and Eqn. (2) with respect to $y$, we obtain

$$\frac{\partial^2 u}{\partial x^2} = \frac{\partial^2 v}{\partial x \partial y} \quad ...(3) \qquad \text{and} \qquad \frac{\partial^2 u}{\partial y^2} = -\frac{\partial^2 v}{\partial y \partial x}. \quad ...(4)$$

Adding Eqns. (3) and (4) and assuming that $\frac{\partial^2 v}{\partial x \partial y} = \frac{\partial^2 v}{\partial y \partial x}$, we get

$$\frac{\partial^2 u}{\partial x^2} + \frac{\partial^2 u}{\partial y^2} = 0. \quad ...(5)$$

Similarly, by differentiating Eqn. (1) with respect to $y$ and Eqn. (2) with respect to $x$ and subtracting, we obtain

$$\frac{\partial^2 v}{\partial x^2} + \frac{\partial^2 v}{\partial y^2} = 0. \quad ...(6)$$

Thus both the functions $u$ and $v$ satisfy the Laplace's equation in two variables. For this reason, they are known as **harmonic functions** and their theory is called **potential theory.** *(D.T.U., 2012)*

**(2) Orthogonal system.** Consider the two families of curves

$$u(x, y) = c_1 \quad ...(7) \quad \text{and} \quad v(x, y) = c_2 \quad ...(8)$$

Differentiating Eqn. (7), we get $\frac{\partial u}{\partial x} + \frac{\partial u}{\partial y} \cdot \frac{dy}{dx} = 0$

or $$\frac{dy}{dx} = -\frac{\partial u / \partial x}{\partial u / \partial y} = \frac{\partial v / \partial y}{\partial v / \partial x} = m_1 \text{ (say)} \quad \text{[By (1) and (2)]}$$

Similarly Eqn. (8) gives $\frac{dy}{dx} = -\frac{\partial v / \partial x}{\partial v / \partial y} = m_2$ (say)

$\therefore$ $m_1 m_2 = -1$, *i.e.*, Eqns. (7) and (8) form an orthogonal system.

Hence *every analytic function $f(z) = u + iv$ defines two families of curves $u(x, y) = c_1$ and $v(x, y) = c_2$, which form an orthogonal system.*

## 20.6 APPLICATIONS TO FLOW PROBLEMS

As the real and imaginary parts of an analytic function are the solutions of the Laplace's equation in two variables, the conjugate functions provide solutions to a number of field and flow problems.

As an illustration, consider the irrotational motion of an incompressible fluid in two dimensions. Assuming the flow to be in planes parallel to the $xy$-plane, the velocity $\mathbf{V}$ of a fluid particle can be expressed as

$$\mathbf{V} = v_x \mathbf{I} + v_y \mathbf{J} \quad ...(1)$$

Since the motion is irrotational, therefore, by § 6.18 (1), there exist a scalar function $\phi(x, y)$ such that

$$\mathbf{V} = \nabla\phi(x, y) = \frac{\partial \phi}{\partial x}\mathbf{I} + \frac{\partial \phi}{\partial y}\mathbf{J} \quad ...(2)$$

[The function $\phi(x, y)$ is called the *velocity potential* and the curves $\phi(x, y) = c$ are known as *equipotential lines.*]

Thus from Eqns. (1) and (2), $v_x = \frac{\partial \phi}{\partial x}$ and $v_y = \frac{\partial \phi}{\partial y}$ ...(3)

Also the fluid being incompressible div $\mathbf{V} = 0$ [by § 8.7 (1)] *i.e.*, $\frac{\partial v_x}{\partial x} + \frac{\partial v_y}{\partial y} = 0.$

Substituting the values of $v_x$ and $v_y$ from Eqn. (3), we get $\frac{\partial^2 \phi}{\partial x^2} + \frac{\partial^2 \phi}{\partial y^2} = 0$

which shows that the velocity potential $\phi$ is *harmonic*. It follows that there must exist a conjugate harmonic function $\psi(x, y)$ such that $w(z) = \phi(x, y) + i\psi(x, y)$ ...(4)
is analytic.

Also the slope at any point of the curve $\psi(x, y) = c'$ is given by

$$\frac{dy}{dx} = -\frac{\partial\psi/\partial x}{\partial\psi/\partial y} = \frac{\partial\phi/\partial y}{\partial\phi/\partial x} \quad \text{[By } C\text{-}R \text{ equations]}$$

$$= v_y/v_x \quad \text{[By (3)]}$$

This shows that the velocity of the fluid particle is along the tangent to the curve $\psi(x, y) = c'$, *i.e.* the particle moves along this curve. Such curves are known as *stream lines* and $\psi(x, y)$ is called the *stream function.* Also the equipotential lines $\phi(x, y) = c$ and the stream lines $\psi(x, y) = c'$ cut orthogonally.

From Eqn. (4), $\frac{\partial w}{dz} = \frac{\partial \phi}{\partial x} + i\frac{\partial \psi}{\partial x} = \frac{\partial \phi}{\partial x} - i\frac{\partial \phi}{\partial y}$ [By *C-R* equations]

$= v_x - iv_y$ [By Eqn. (3)]

$\therefore$ The magnitude of the fluid velocity $= \sqrt{(v_x^2 + v_y^2)} = | dw/dz |$.

Thus the flow pattern is fully represented by the function $w(z)$ which is known as the **complex potential.**

Similarly the complex potential $w(z)$ can be taken to represent any other type of 2-dimensional steady flow. In electrostatics and gravitational fields, the curves $\phi(x, y) = c$ and $\psi(x, y) = c'$ are *equipotential lines* and *lines of force*. In heat flow problems, the curves $\phi(x, y) = c$ and $\psi(x, y) = c'$ are known as *isothermals* and *heat flow lines* respectively.

Given $\phi(x, y)$, we can find $\psi(x, y)$ and *vice-versa*.

**Example 20.6.** *If $w = \phi + i\psi$ respresents the complex potential for an electric field and $\psi = x^2 - y^2 + \frac{x}{x^2+y^2}$, determine the function $\phi$.* *(V.T.U., 2016)*

**Solution.** It is readily verified that $\psi$ satisfies the Laplace's equation.

$\therefore$ $\phi$ and $\psi$ must satisfy the Cauchy-Riemann equations :

$$\frac{\partial \phi}{\partial x} = \frac{\partial \psi}{\partial y} \quad ...(i) \qquad \frac{\partial \phi}{\partial y} = -\frac{\partial \psi}{\partial x} \quad ...(ii)$$

$\therefore$ by Eqn. (*i*), $\frac{\partial \phi}{\partial x} = \frac{\partial}{\partial y}\left[x^2 - y^2 + \frac{x}{x^2+y^2}\right] = -2y - \frac{2xy}{(x^2+y^2)^2}$

Integrating w.r.t. $x$, we get $\phi = -2xy + \frac{y}{x^2+y^2} + \eta(y)$ where $\eta(y)$ is an arbitrary function of $y$.

$\therefore$ Eqn. (*ii*) gives $-2x + \frac{x^2-y^2}{(x^2+y^2)^2} + \eta'(y) = -2x + \frac{x^2-y^2}{(x^2+y^2)^2}$

whence $\eta'(y) = 0$, *i.e.*, $\eta(y) = c$, an arbitrary constant.

Thus $$\phi = -2xy + \frac{y}{x^2+y^2} + c$$

*Otherwise* (Milne-Thomson's method*) :

We have $$\frac{dw}{dz} = \frac{\partial \phi}{\partial x} + i\frac{\partial \psi}{\partial x} = \frac{\partial \psi}{\partial y} + i\frac{\partial \psi}{\partial x} = \left[-2y - \frac{2xy}{(x^2+y^2)^2}\right] + i\left[2x + \frac{y^2-x^2}{(x^2+y^2)^2}\right]$$

By Milne-Thomson's method, we express $dw/dz$ in terms of $z$, on replacing $x$ by $z$ and $y$ by 0.

$$\therefore \quad \frac{dw}{dz} = i\left(2z - \frac{1}{z^2}\right)$$

Integrating w.r.t. $z$, we get $w = i\,(z^2 + 1/z) + A$, where $A$ is a complex constant.

---

* Since $z = x + iy$ and $\bar{z} = x - iy$, we have

$$x = \frac{1}{2}(z + \bar{z}), \qquad y = \frac{1}{2i}(z + \bar{z})$$

$$\therefore \quad f(z) = \phi(x, y) + i\psi(x, y) \quad ...(1)$$

$$= \phi\left[\frac{z+\bar{z}}{2}, \frac{z-\bar{z}}{2i}\right] + i\psi\left[\frac{z+\bar{z}}{2}, \frac{z-\bar{z}}{2i}\right]$$

Now considering this as a formal identity in the two independent variables $z$, $\bar{z}$ and putting $\bar{z} = z$, we get

$$f(z) = \phi(z, 0) + i\psi(z, 0) \quad ...(2)$$

$\therefore$ Eqn. (2) is the same as Eqn. (1), if we replace $x$ by $z$ and $y$ by 0.

Thus **to express any function in terms of z, replace x by z and y by 0.** *This provides an elegant method of finding f (z) when its real part or the imaginary part is given. It is due to Milne-Thomson.*

Hence $$\phi = R\left[i\left(z^2 + \frac{1}{z}\right) + A\right] = -2xy + \frac{y}{x^2 + y^2} + c.$$

**Example 20.7.** *Find the analytic function, whose real part is sin 2x/(cosh 2y – cos 2x).*
*(Andhra., 2015 ; V.TU., 2015)*

**Solution.** Let $f(z) = u + iv$, where $u = \sin 2x/(\cosh 2y - \cos 2x)$

$$\therefore \quad f'(z) = \frac{\partial u}{\partial x} + i\frac{\partial v}{\partial x} = \frac{\partial u}{\partial x} - i\frac{\partial u}{\partial y} \qquad \text{[By } C\text{-}R \text{ equations]}$$

$$= \frac{(\cosh 2y - \cos 2x)\,2\cos 2x - \sin 2x\,(2\sin 2x)}{(\cosh 2y - \cos 2x)^2} - i\,\frac{\sin 2x(-2\sinh 2y)}{(\cosh 2y - \cos 2x)^2}$$

$$= \frac{2\cos 2x\cosh 2y - 2}{(\cosh 2y - \cos 2x)^2} + i\,\frac{2\sin 2x\sinh 2y}{(\cosh 2y - \cos 2x)^2}$$

By Milne-Thomson's method, we express $f'(z)$ in terms of $z$ by putting $x = z$ and $y = 0$.

$$\therefore \quad f'(z) = \frac{2\cos 2z - 2}{(1 - \cos 2z)^2} + i(0) = \frac{-2}{1 - \cos 2z} = \frac{-2}{2\sin^2 z} = -\operatorname{cosec}^2 z$$

Integrating w.r.t. $z$, we get $f(z) = \cot z + ic$, taking the constant of integration as imaginary since $u$ does not contain any constant.

**Example 20.8.** *Determine the analytic function $f(z) = u + iv$, if $u - v = \dfrac{\cos x + \sin x - e^{-y}}{2(\cos x - \cosh y)}$ and $f(\pi/2) = 0$.*
*(JNTUK, 2023; Calicut, 2011S)*

**Solution.** We have $$u - v = \frac{\cos x + \sin x - e^{-y}}{2(\cos x - \cosh y)}$$

$$\therefore \quad \frac{\partial u}{\partial x} - \frac{\partial v}{\partial x} = \frac{(\sin x - \cos x)\cosh y + 1 - e^{-y}\sin x}{2(\cos x - \cosh y)^2} \qquad ...(i)$$

and $$\frac{\partial u}{\partial y} - \frac{\partial v}{\partial y} = \frac{(\cos x - \cosh y)\,e^{-y} + (\cos x + \sin x - e^{-y})\sinh y}{2(\cos x - \cosh y)^2}$$

or $$-\frac{\partial v}{\partial x} - \frac{\partial u}{\partial x} = \frac{(\sin x + \cos x)\sinh y + e^{-y}\,(\cos x - \cosh y - \sinh y)}{2(\cos x - \cosh y)^2} \qquad ...(ii)$$

Subtracting Eqns. (*ii*) from Eqn. (*i*), we get

$$2\frac{\partial u}{\partial x} = \frac{(\sin x - \cos x)\cosh y - (\sin x + \cos x)\sinh y + 1 - e^{-y}(\sin x + \cos x - \cosh y - \sinh y)}{2(\cos x - \cosh y)^2}$$

Adding Eqns. (*i*) and (*ii*), we have

$$-2\frac{\partial v}{\partial x} = \frac{(\sin x - \cos x)\cosh y + (\sin x + \cos x)\sinh y + 1 + e^{-y}(-\sin x + \cos x - \cosh y - \sinh y)}{2(\cos x - \cosh y)^2}$$

Thus $$f'(z) = \frac{\partial u}{\partial x} + i\frac{\partial v}{\partial x} = \frac{1 - \cos z}{2(1 - \cos z)^2} \qquad \text{[Putting } x = z \text{ and } y = 0]$$

$$= \frac{1}{2(1 - \cos z)} = \frac{1}{4\sin^2 z/2} = \frac{1}{4}\operatorname{cosec}^2\frac{z}{2} \quad \text{or} \quad f(z) = -\frac{1}{2}\cot\frac{z}{2} + c$$

Since $f(\pi/2) = 0$, $\quad 0 = -\dfrac{1}{2}\cot \pi/4 + c$, whence $c = \dfrac{1}{2}$

Hence $$f(z) = \frac{1}{2}\left(1 - \cot\frac{z}{2}\right).$$

**Example 20.9.** *Find the conjugate harmonic of $v(r, \theta) = r^2 \cos 2\theta - r \cos \theta + 2$. Show that v is harmonic.* *(JNTUK, 2023; Andhra, 2020)*

**Solution.** Let $f(z) = u + v$. Using C-R equations in polar coordinates (Ex. 20.5),

$$r\frac{\partial u}{\partial r} = \frac{\partial v}{\partial \theta} = -2r^2 \sin 2\theta + r \sin \theta \quad ...(i)$$

$$-\frac{1}{r}\frac{\partial u}{\partial \theta} = \frac{\partial v}{\partial r} = 2r \cos 2\theta - \cos \theta \quad ...(ii)$$

$\therefore$ (*i*) gives, $\frac{\partial u}{\partial r} = -2r \sin 2\theta + \sin \theta$

Integrating w.r.t., $r$

$$u = -r^2 \sin 2\theta + r \sin \theta + \phi(\theta)$$ where $\phi(\theta)$ is an arbitrary function.

$$\therefore \quad \frac{\partial u}{\partial \theta} = -2r^2 \cos 2\theta + r \cos \theta + \phi'(\theta) \quad ...(iii)$$

From Eqns. (*ii*) and (*iii*), we get

$$-2r^2 \cos 2\theta + r \cos \theta = \frac{\partial u}{\partial \theta} = -2r^2 \cos 2\theta + r \cos \theta + \phi'(\theta)$$

$$\therefore \quad \phi'(\theta) = 0 \quad \text{or} \quad \phi(\theta) = c$$

Thus $u = -r^2 \sin 2\theta + r \sin \theta + c$ is the conjugate harmonic of $v$.

Now $v$ will be harmonic if it satisfies the Laplace equation $\frac{\partial^2 v}{\partial r^2} + \frac{1}{r} + \frac{\partial v}{\partial r} + \frac{1}{r^2}\frac{\partial^2 v}{\partial \theta^2} = 0$

From Eqn. (*i*), $\frac{\partial^2 v}{\partial \theta^2} = -4r^2 \cos 2\theta + r \cos \theta$. From Eqn. (*ii*), $\frac{\partial^2 v}{\partial r^2} = 2 \cos 2\theta$

$$\therefore \quad \frac{\partial t^2 v}{\partial r^2} + \frac{1}{r}\frac{\partial v}{\partial r} + \frac{1}{r^2}\frac{\partial^2 v}{\partial \theta^2} = 2 \cos 2\theta + \frac{1}{r}(2r \cos 2\theta - \cos \theta) + \frac{1}{r^2}(-4r^2 \cos 2\theta + r \cos \theta)$$

$$= 4 \cos 2\theta - \frac{1}{r}\cos \theta - 4 \cos 2\theta + \frac{1}{r}\cos \theta = 0$$

Hence $v$ is harmonic.

**Example 20.10.** (*a*) *Find the orthogonal trajectories of the family of curves*

$$x^4 + y^4 - 6x^2y^2 = \text{constant}$$

(*b*) *Show that the curves $r^n = \alpha \sec n\theta$ and $r^n = \beta \operatorname{cosec} n\theta$ cut orthogonally.*

**Solution.** (*a*) Take $u(x, y) = x^4 + y^4 - 6x^2y^2$. Then the family of curves $v(x, y) =$ constant will be the required trajectories if $f(z) = u + iv$ is analytic.

Now $$\frac{\partial u}{\partial x} = 4x^3 - 12xy^2, \quad \frac{\partial u}{\partial y} = 4y^3 - 12x^2y$$

$$\therefore \quad \frac{\partial v}{\partial y} = \frac{\partial u}{\partial x} = 4x^3 - 12xy^2$$

Integrating, $$v = 4x^3y - 4xy^3 + c(x)$$

Differentiating partially w.r.t. $x$

$$12x^2y - 4y^3 + \frac{dc(x)}{dx} = \frac{\partial v}{\partial x} = -\frac{\partial u}{\partial y} = -4y^3 + 12x^2y$$

$$\therefore \quad \frac{dc(x)}{dx} = 0 \quad \text{or} \quad c = \text{constant}$$

Thus the required orthogonal trajectories are $v =$ constant or $x^3y - xy^3 =$ constant.

(*b*) Writing $u(r, \theta) = r^n \cos n\theta = \alpha$ and $v(r, \theta) = r^n \sin n\theta = \beta$,

we have $u(r, \theta) + iv(r, \theta) = \alpha + i\beta = r^n (\cos n\theta + i \sin n\theta) = r^n \cdot e^{in\theta} = (re^{i\theta})^n = z^n$

This is an analytic function.

Thus $f(z) = u + iv$, gives the curves $u = \alpha$ and $v = \beta$ which cut orthogonally.

**Example 20.11.** *Two concentric circular cylinders of radii $r_1, r_2$ $(r_1 < r_2)$ are kept at potentials $\phi_1$ and $\phi_2$ respectively. Using complex function $w = a \log z + c$, prove that the capacitance per unit length of the capacitor formed by them is $2\pi\lambda / \log (r_2/r_1)$, where $\lambda$ is the dielectric constant of the medium.*

**Solution.** We have $\phi + i\psi = a \log (re^{i\theta}) + c,$ where $z = x + iy = re^{i\theta}$

$\therefore$ $\phi = a \log r + c,$ and $\psi = a\theta$

so that $\phi_1 = a \log r_1 + c,$ $\phi_2 = a \log r_2 + c$

Thus the potential difference $= \phi_2 - \phi_1 = a (\log r_2 - \log r_1)$

Also the total charge (or flux) $= \int_0^{2\pi} d\psi = \int_0^{2\pi} a\, d\theta = 2\pi a.$

The capacitance being the charge required to maintain a unit potential difference ; the capacitance without dielectric

$$= \frac{\text{charge}}{\text{potential difference}} = \frac{2\pi a}{a(\log r_2 - \log r_1)} = \frac{2\pi}{\log (r_2/r_1)}.$$

A medium of dielectric constant $\lambda$ increases the potential difference to $\lambda$ times that in vacuum for the same charge. Thus the capacitance with dielectric $= 2\pi\lambda/\log (r_2/r_1)$.

**Example 20.12.** *If f(z) is a regular function of z, prove that*

$$\left(\frac{\partial^2}{\partial x^2} + \frac{\partial^2}{\partial y^2}\right) | f(z)|^2 = 4 \mid f'(z) \mid^2. \qquad \textit{(V.T.U., 2016 ; D.T.U., 2012)}$$

*or*

$$\nabla^2 \mid f(z) \mid^2 = 4 \mid f'(z) \mid^2$$

**Solution.** Let $f(z) = u(x, y) + iv(x, y)$ so that $\mid f(z) \mid^2 = u^2 + v^2 = \phi(x, y)$, (say).

$$\therefore \quad \frac{\partial \phi}{\partial x} = 2u \frac{\partial u}{\partial x} + 2v \frac{\partial v}{\partial x} \text{ and } \frac{\partial^2 \phi}{\partial x^2} = 2\left\{u \frac{\partial^2 u}{\partial x^2} + \left(\frac{\partial u}{\partial x}\right)^2 + v \frac{\partial^2 v}{\partial x^2} + \left(\frac{\partial v}{\partial x}\right)^2\right\}$$

Similarly,

$$\frac{\partial^2 \phi}{\partial y^2} = 2\left\{u \frac{\partial^2 u}{\partial y^2} + \left(\frac{\partial u}{\partial y}\right)^2 + v \frac{\partial^2 v}{\partial y^2} + \left(\frac{\partial v}{\partial y}\right)^2\right\}$$

Adding, we have

$$\frac{\partial^2 \phi}{\partial x^2} + \frac{\partial^2 \phi}{\partial y^2} = 2\left\{u\left(\frac{\partial^2 u}{\partial x^2} + \frac{\partial^2 u}{\partial y^2}\right) + v\left(\frac{\partial^2 v}{\partial x^2} + \frac{\partial^2 v}{\partial y^2}\right)\right\} + 2\left\{\left(\frac{\partial u}{\partial x}\right)^2 + \left(\frac{\partial u}{\partial y}\right)^2 + \left(\frac{\partial v}{\partial x}\right)^2 + \left(\frac{\partial v}{\partial y}\right)^2\right\} \quad ...(i)$$

Since $u, v$ have to satisfy Cauchy-Riemann equations and the Laplace's equation.

$$\therefore \quad \left(\frac{\partial u}{\partial x}\right)^2 = \left(\frac{\partial v}{\partial y}\right)^2 ; \left(\frac{\partial u}{\partial y}\right)^2 = \left(-\frac{\partial v}{\partial x}\right)^2 \text{ and } \frac{\partial^2 u}{\partial x^2} + \frac{\partial^2 u}{\partial y^2} = 0 = \frac{\partial^2 v}{\partial x^2} + \frac{\partial^2 v}{\partial y^2}.$$

Thus Eqn. $(i)$ takes the form $\dfrac{\partial^2 \phi}{\partial x^2} + \dfrac{\partial^2 \phi}{\partial y^2} = 4\left\{\left(\dfrac{\partial u}{\partial x}\right)^2 + \left(\dfrac{\partial v}{\partial x}\right)^2\right\}$

Hence

$$\left(\frac{\partial^2}{\partial x^2} + \frac{\partial^2}{\partial y^2}\right) | f(z)|^2 = 4 \, |f'(z)|^2 \quad \text{or} \quad \nabla^2 \, |f(z)|^2 = 4 \, |f'(z)|^2.$$

## PROBLEMS 20.1

1. If $f(z) = \begin{cases} x^3 y(y - ix)/(x^6 + y^2), & z \neq 0 \\ 0 & , z = 0 \end{cases}$ prove that $[f(z) - f(0)]/z \to 0$ as $z \to 0$ along any radius vector but not as $z \to 0$ along the curve $y = ax^3$. *(U.K.T.U., 2010)*

2. Show that (*a*) $f(z) = xy + iy$ is everywhere continuous but is not analytic.
(*b*) $f(z) = z + 2\bar{z}$ is not analytic anywhere in the complex plane.

3. If $f(z) = u + iv$ is analytic, then show that $|f'(z)|^2 = \begin{vmatrix} \frac{\partial u}{\partial x} & \frac{\partial u}{\partial y} \\ \frac{\partial v}{\partial x} & \frac{\partial v}{\partial y} \end{vmatrix}$.

4. Find the constants $a, b, c, d$ and $e$ if $f(z) = (ax^4 + bx^2y^2 + cy^4 + dx^2 - 2y^2) + i(4x^3y - exy^3 + 4xy)$ is analytic.

5. Prove that the following functions are analytic and find their derivative:
(*i*) $z^n$ (*V.T.U. 2010 S*) (*ii*) $\sinh z$ (*Delhi., 2016*)

6. Determine which of the following functions are analytic :
(*i*) $2xy + i(x^2 - y^2)$ (*ii*) $(x - iy)/(x^2 + y^2)$ (*iii*) $\cosh z$.

7. (*a*) Determine $p$ such that the function $f(z) = \frac{1}{2}\log_e(x^2 + y^2) + i\tan^{-1}(px/y)$ be an analytic function.
(*b*) Show that $u = \frac{1}{2}\log(x^2 + y^2)$ is harmonic and find its harmonic conjugate function. (*U.P.T.U., 2010*)

8. Show that each of the following functions is not analytic at any point :
(*i*) $\bar{z}$ (*ii*) $|z|^2$. (*W.B.T.U., 2015*)

9. Show that $u + iv = (x - iy)/(x - iy + a)$ where $a \neq 0$, is not an analytic function of $z = x + iy$ whereas $u - iv$ is such a function.

10. Show that $f(z) = \begin{cases} xy^2(x + iy) \div (x^2 + y^4), & z \neq 0 \\ 0, & z = 0 \end{cases}$
is not analytic at $z = 0$, although *C-R* equations are satisfied at the origin.

11. Verify if $f(z) = \dfrac{xy^2(x + iy)}{x^2 + y^4}, z \neq 0; f(0) = 0$ is analytic or not.

12. Examine the nature of the function $f(z) = \dfrac{x^3y^5(x + iy)}{x^6 + y^{10}}, z \neq 0; f(0) = 0$. (*U.P.T.U., 2011*)

13. For the function $f(z)$ defined by $f(z)^2 = (\bar{z})^2/z, z \neq 0; f(0) = 0$, show that the C-R equations are satisfied at (0, 0), but $f(z)$ is not differentiable at (0, 0). (*P.T.U., 2010*)

14. Determine the analytic function whose real part is
(*i*) $x^3 - 3xy^2 + 3x^2 - 3y^2$ (*W.B.T.U., 2015*) (*ii*) $\cos x \cosh y$
(*iii*) $y/(x^2 + y^2)$ (*iv*) $y + e^x \cos y$
(*v*) $e^{-x}(x \sin y - y \cos y)$ (*U.P.T.U., 2012*)
(*vi*) $e^{2x}(x \cos 2y - y \sin 2y)$
(*vii*) $x \sin x \cosh y - y \cos x \sinh y$
(*viii*) $e^x[(x^2 - y^2)\cos y - 2xy \sin y]$. (*Birla Ranchi., 2016 ; V.T.U., 2010 S*)

15. Find the regular function whose imaginary part is
(*i*) $(x - y)/(x^2 + y^2)$ (*ii*) $-\sin x \sinh y$ (*iii*) $e^x \sin y$ (*Andhra, 2021*)
(*iv*) $e^{-x}(x \sin y - y \cos y)$ (*v*) $e^{-x}(x \cos y + y \sin y)$ (*vi*) $\dfrac{2 \sin x \sin y}{\cos 2x + \cosh 2y}$.

16. Find the analytic function $z = u + iv$, if
(*i*) $u - v = (x - y)(x^2 + 4xy + y^2)$ (*V.T.U., 2014*)
(*ii*) $u - v = \dfrac{\cos x + \sin x - e^{-y}}{2\cos x - e^y - e^{-y}}$, when $f\left(\dfrac{\pi}{2}\right) = 0$
(*iii*) $u + v = \dfrac{2 \sin 2x}{e^{2y} - e^{-2y} - 2\cos 2x}$. (*JNTUK, 2023 ; Andhra, 2022*)

17. An electrostatic field in the *xy*-plane is given by the potential function $\phi = 3x^2y - y^3$, find the stream function.

18. If the potential function is $\log(x^2 + y^2)$, find the flux function and the complex potential function.

19. Prove that $u = x^2 - y^2$ and $v = \dfrac{y}{x^2 + y^2}$ are harmonic functions of $f(x, y)$ but are not harmonic conjugates. (*Anna., 2016*)

20. Show that the function $u = e^{-2xy}\sin(x^2 - y^2)$ is harmonic. Find the conjugate function $v$ and express $u + iv$ as an analytic function of $z$. (*Anna., 2016*)

21. For $w = \exp(z^2)$, find $u$ and $v$, and prove that the curves $u(x, y) = c_1$ and $v(x, y) = c_2$ where $c_1$ and $c_2$ are constants, cut orthogonally.
22. Find the orthogonal trajectories of the family of curves
(i) $x^3y - xy^3 = c$ (ii) $e^x \cos y - xy = c$
(iii) $r^2 \cos 2\theta = c$.
23. In a two dimensional fluid flow, the stream function $\psi$ is given, find the velocity potential $\phi$ :
(i) $\psi = -y/(x^2 + y^2)$ (ii) $\psi = \tan^{-1}(y/x)$. (Anna., 2016)
24. Find the analytic function $f(z) = u + iv$, given
(i) $u = a(1 + \cos\theta)$ (ii) $v = (r - 1/r)\sin\theta, r \neq 0$.
25. If $f(z)$ is an analytic function of $z$, show that
$$\left\{\frac{\partial}{\partial x}|f(z)|\right\}^2 + \left\{\frac{\partial}{\partial y}|f(z)|\right\}^2 = |f'(z)|^2.$$ (V.T.U., 2015)
26. If $f(z)$ is an analytic function of $z$, prove that
(i) $\left(\frac{\partial^2}{\partial x^2} + \frac{\partial^2}{\partial y^2}\right)\log|f(z)| = 0$ (ii) $\left(\frac{\partial^2}{\partial x^2} + \frac{\partial^2}{\partial y^2}\right)|R\, f(z)|^2 = 2\,|f'(z)|^2$ (Anna., 2016)
(iii) $\left(\frac{\partial^2}{\partial x^2} + \frac{\partial^2}{\partial y^2}\right)|f(z)|^p = p^2\,|f'(z)|^2\,|f(z)|^{p-2}$.
27. Prove that $\psi = \log[(x - 1)^2 + (y - 2)^2]$ is harmonic in every region which does not include the point (1, 2). Find a function $\phi$ such that $\phi + i\psi$ is an analytic function of the complex variable $z = x + iy$. Express $\phi + i\psi$ as a function of $z$.

## 20.7 GEOMETRICAL REPRESENTATION OF w = f(z)

To find the geometrical representation of a function of a complex variable, it requires a departure from the usual practice of cartesian plotting, where we associate a curve to a real function $y = f(x)$.

In the complex domain, the function $w = f(z)$

*i.e.,* $$u + iv = f(x + iy) \quad ...(1)$$

involves four real variables $x, y, u, v$. Hence a four dimensional region is required to plot (1) in the cartesian fashion. As it is not possible to have 4-dimensional graph papers, we make use of two complex planes, one for the variable $z = x + iy$, and the other for the variable $w = u + iv$. If the point $z$ describes some curve $C$ in the $z$-plane, the point $w$ will move along a corresponding curve $C'$ in the $w$-plane, since to each point $(x, y)$, there corresponds a point $(u, v)$ (Fig. 20.3). We then, say that *a curve C in the z-plane is mapped into the corresponding curve C' in the w-plane by the function w = f(z) which defines a* **mapping** *or* **transformation** *of the z-plane into the w-plane.*

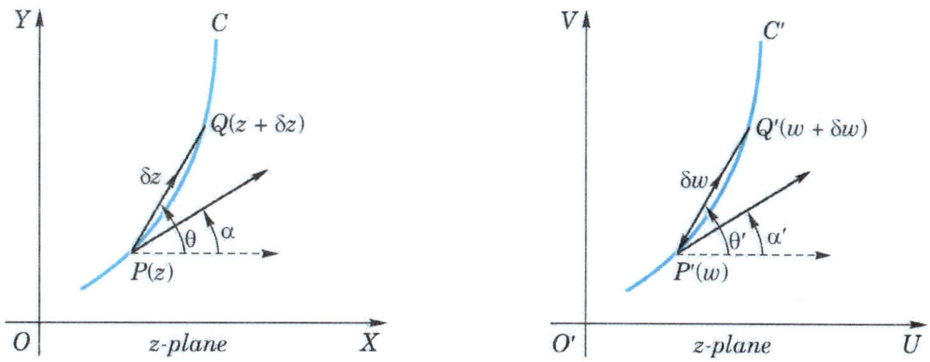


Fig. 20.3

## 20.8 SOME STANDARD TRANSFORMATIONS

**(1) Translation.** $w = z + c$, where $c$ is a complex constant.

If $z = x + iy$, $c = c_1 + ic_2$ and $w = u + iv$, then the transformation becomes $u + iv = x + iy + c_1 + ic_2$ whence $u = x + c_1$ and $v = y + c_2$, *i.e.* the point $P(x, y)$ in the $z$-plane is mapped onto the point $P'(x + c_1, y + c_2)$ in the $w$-plane.

Every point in the $z$-plane is mapped onto $w$-plane in the same way. Thus if the $w$-plane is superposed on the $z$-plane, figure is shifted through a distance given by the vector $c$. Accordingly, this transformation maps a figure in the $z$-plane into a figure in the $w$-plane of the same shape and size.

In particular, this transformation changes circles into circles.

**(2) Magnification and rotation. w = cz,** where $c$ is a complex constant.

If $c = \rho e^{i\alpha}$, $z = re^{i\theta}$ and $w = Re^{i\phi}$, then

$$Re^{i\phi} = \rho e^{i\alpha} \cdot re^{i\theta} = \rho re^{i(\theta + \alpha)}$$

whence $R = \rho r$ and $\phi = \theta + \alpha$, *i.e.* the point $P(r, \theta)$ in the $z$-plane is mapped onto the point $P'(\rho r, \theta + \alpha)$ in the $w$-plane. Hence the transformation consists of magnification (or contraction) of the radius vector of $P$ by $\rho = |c|$ and its rotation through an $\angle\alpha$ = amp ($c$). Accordingly any figure in the $z$-plane is transformed into a geometrically similar figure in the $w$-plane. In particular, this transformation maps circles into circles.

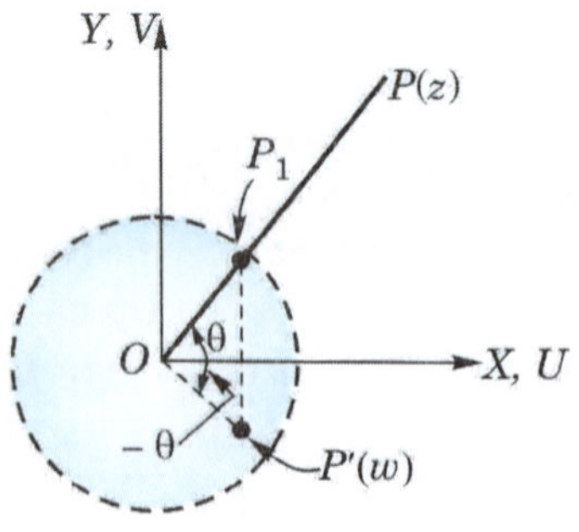

Fig. 20.4

**(3) Inversion and reflection. w = 1/z.**

Here it is convenient to think the $w$-plane as superposed on $z$-plane (Fig. 20.4).

If $z = re^{i\theta}$ and $w = Re^{i\phi}$, then $Re^{i\phi} = \dfrac{1}{r} e^{-i\theta}$

whence $R = 1/r$ and $\phi = -\theta$. Thus, if $P$ be $(r, \theta)$ and $P_1$ be $(1/r, \theta)$, *i.e.* $P_1$ is the inverse* of $P$ w.r.t. the unit circle with centre $O$, then the reflection $P'$ of $P_1$ in the real axis represents $w = 1/z$.

*Hence this transformation is an inversion of z w.r.t. the unit circle* $|z| = 1$ *followed by reflection of the inverse into the real axis.*

**Obs. 1.** Clearly the function $w = 1/z$ maps the interior of the unit circle $|z| = 1$ onto the exterior of the unit circle $|w| = 1$ and the exterior of $|z| = 1$ onto the interior of $|w| = 1$. In particular, the origin $z = 0$ corresponds to the improper point $w = \infty$, called the *point at infinity* and the image of the improper point $z = \infty$ is the origin $w = 0$.

**2.** *This transformation maps a circle onto a circle or to a straight line if the former goes through the origin.*

To prove this, we write $z = 1/w$ as $x + iy = \dfrac{1}{u + iv} = \dfrac{u - iv}{u^2 + v^2}$

so that $$x = \frac{u}{u^2 + v^2} \text{ and } y = \frac{-v}{u^2 + v^2}. \qquad ...(1)$$

Now the general equation of any circle in the $z$-plane is

$$x^2 + y^2 + 2gx + 2fy + c = 0 \qquad ...(2)$$

which on substituting from Eqn. (1), becomes $\dfrac{u^2}{(u^2 + v^2)^2} + \dfrac{v^2}{(u^2 + v^2)^2} + 2g\dfrac{u}{u^2 + v^2} + 2f\dfrac{-v}{u^2 + v^2} + c = 0$

or $$c(u^2 + v^2) + 2gu - 2fv + 1 = 0 \qquad ...(3)$$

This is the equation of a circle in the $w$-plane. If $c = 0$, the circle Eqn. (2) passes through the origin and its image, *i.e.*, Eqn. (3) reduces to a straight line. Hence the result.

Regarding a straight line as the limiting form of a circle with infinite radius, we conclude that the *transformation* $w = 1/z$ *always maps a circle into a circle.*

**(4) Bilinear transformation.** The transformation

$$w = \frac{az + b}{cz + d} \qquad ...(1)$$

where $a$, $b$, $c$ and $d$ are complex constants and $ad - bc \neq 0$ is known as the **bilinear transformation.**** The condition $ad - bc \neq 0$ ensures that $dw/dz \neq 0$, *i.e.*, the transformation is conformal. If $ad - bc = 0$ every point of the $z$-plane is a *critical point.*

The inverse mapping of Eqn. (1) is

$$z = \frac{-dw + b}{cw - a} \qquad ...(2)$$

which is also a bilinear transformation.

---

*The inverse of a point $A$ w.r.t. a circle with centre $O$ and radius $k$ is defined as the point $B$ on the line $OA$ such that $OA.OB = k^2$.

** First studied by *Möbius* (p. 392). Hence, sometimes called *Möbius transformation*

**Obs. 1.** From Eqn. (1), we see that each point in the $z$-plane except $z = -d/c$, corresponds a unique point in the $w$-plane. Similarly, Eqn. (2) shows that each point in the $w$-plane except $w = a/c$, maps into a unique point in the $z$-plane. Including the images of the two exceptional points as the infinite points in the two planes, it follows that *there is one to one correspondence between all points in the two planes.*

**Obs. 2. Invariant points of bilinear transformation.** If $z$ maps into itself in the $w$-plane (*i.e.*, $w = z$), then Eqn. (1) gives

$$z = \frac{az+b}{cz+d} \quad \text{or} \quad cz^2 + (d-a)z - b = 0$$

*The roots of this equation* (say : $z_1, z_2$) *are defined as the invariant or fixed points of the bilinear transformation Eqn. (1).* If however, the two roots are equal, the bilinear transformation is said to be *parabolic.*

**Obs. 3.** Dividing the numerator and denominator of the right side of Eqn. (1) by one of the four constants, it is clear that Eqn. (1) has only three essential arbitrary constants. Hence *three conditions are required to determine a bilinear transformation.* For instance, three distinct points $z_1, z_2, z_3$ can be mapped into any three specified points $w_1, w_2, w_3$.

**Two important properties :**

*I. A bilinear transformation maps circles into circles.* (*Calicut, 2012*)

By actual division, Eqn. (1) can be written as $w = \frac{a}{c} + \frac{bc-ad}{c^2} \cdot \frac{1}{z+d/c}$

which is a combination of the transformations

$$w_1 = z + d/c,\ w_2 = 1/w_1,\ w_3 = \frac{bc-ad}{c^2} w_2,\ w = \frac{a}{c} + w_3 .$$

By these transformations, we successively pass from $z$-plane to $w_1$-plane, from $w_1$-plane to $w_2$-plane, from $w_2$-plane to $w_3$-plane and finally from $w_3$-plane to $w$-plane. Now each of these transformations is one or other of the standard transformations $w = z + c$, $w = cz$, $w = 1/z$ and under each of these a circle always maps onto a circle. Hence the bilinear transformation maps circles into circles.

*II. A bilinear transformation preserves cross-ratio*† *of four points.*

Let the points $z_1, z_2, z_3, z_4$ of the $z$-plane map onto the points $w_1, w_2, w_3, w_4$ of the $w$-plane respectively under the bilinear transformation Eqn. (1). If these points are finite, then from Eqn. (1), we have

$$w_j - w_k = \frac{az_j+b}{cz_j+d} - \frac{az_k+b}{cz_k+d} = \frac{ad-bc}{(cz_j+d)(cz_k+d)}(z_j - z_k).$$

Using this relation for $j, k = 1, 2, 3, 4$, we get $\frac{(w_1-w_2)(w_3-w_4)}{(w_1-w_4)(w_3-w_2)} = \frac{(z_1-z_2)(z_3-z_4)}{(z_1-z_4)(z_3-z_2)}$

Thus the cross-ratio of four points is invariant under bilinear transformation.

This property is very useful in finding a bilinear transformation. If one of the points, say : $z_1 \to \infty$, the quotient of those two differences which contain $z_1$, is replaced by 1 *i.e.*,

$$\frac{(z_1-z_2)(z_3-z_4)}{(z_1-z_4)(z_3-z_2)} = \frac{z_3-z_4}{z_3-z_2}.$$

**Example 20.13.** *Find the bilinear transformation which maps the points $z = 1, i, -1$ onto the points $w = i, 0, -i$.*

*Hence find (a) the image of $|z| < 1$,* (*V.T.U., 2015 ; Calicut, 2012*)

*(b) the invariant points of this transformation.*

**Solution.** Let the points $z_1 = 1, z_2 = i, z_3 = -1$ and $z_4 = z$ map onto the points $w_1 = i, w_2 = 0, w_3 = -i$ and $w_4 = w$. Since the cross-ratio remains unchanged under a bilinear transformation.

$$\therefore \quad \frac{(1-i)(-1-z)}{(1-z)(-1-i)} = \frac{(i-0)(-i-w)}{(i-w)(-i-0)} \quad \text{or} \quad \frac{w+i}{w-i} = \frac{(z+1)(1-i)}{(z-1)(1+i)}$$

By componendo dividendo, we get $\frac{2w}{2i} = \frac{(z+1)(1-i)+(z-1)(1+i)}{(z+1)(1-i)-(z-1)(1+i)}$

†**Def.** *If $t_1, t_2, t_3, t_4$ be any four numbers, then* $\frac{(t_1-t_2)(t_3-t_4)}{(t_1-t_4)(t_3-t_2)}$ *is said to be their cross-ratio and is denoted* $(t_1, t_2, t_3, t_4)$.

$$\therefore \qquad w = \frac{1+iz}{1-iz} \qquad ...(i)$$

which is the required bilinear transformation.

(a) Rewriting Eqn. (i) as $z = i\,\dfrac{1-w}{1+w}$

$$\therefore \qquad \left|\frac{i(1-w)}{1+w}\right| = |z| < 1 \quad \text{or} \quad |i|\ |1-w| < |1+w|$$

or $\qquad |1-u-iv| < |1+u+iv| \qquad [\because \ |i| = 1]$

or $\qquad (1-u)^2 + v^2 < (1+u)^2 + v^2$ which reduces to $u > 0$.

Hence the interior of the circle $x^2 + y^2 = 1$ in the $z$-plane is mapped onto the entire half of the $w$-plane to the right of the imaginary axis.

(b) To find the invariant points of the transformation, we put $w = z$ in Eqn. (i).

$$\therefore \qquad z = \frac{1+iz}{1-iz} \quad \text{or} \quad iz^2 + (i-1)z + 1 = 0$$

or
$$z = \frac{1 - i \pm \sqrt{[(i-1)^2 - 4i]}}{2i} = -\tfrac{1}{2}\{1 + i \pm \sqrt{(6i)}\}$$

which are the required invariant points.

**Example 20.14.** *Show that $w = \dfrac{i-z}{i+z}$ maps the real axis of z-plane into the circle $|w| = 1$ and the half plane $y > 0$ into the interior of the unit circle $|w| = 1$ in the w-plane.*

**Solution.** Since $\qquad w = (i-z)/(i+z)$,

$\therefore \qquad |w| = 1$ becomes $|(i-z)/(i+z)| = 1$ or $|i-z| = |i+z|$

*i.e.,* $\qquad |i - x - iy| = |i + x + iy|$ or $|-x + i(1-y)| = |x + i(1+y)|$

$\therefore \qquad \sqrt{x^2 + (1-y)^2} = \sqrt{[x^2 + (1+y)^2}$ or $(1-y)^2 = (1+y)^2$

$\therefore \qquad 4y = 0$ or $y = 0$ which is the real axis.

Hence the real axis of the $z$-plane is mapped to the circle $|w| = 1$

Now, for the interior of the circle $|w| = 1$

$$|w| < 1 \quad i.e., \quad |i-z| < |i+z| \quad \text{or} \quad (1-y)^2 < (1+y)^2$$

$\therefore \qquad -4y < 0 \quad$ *i.e.,* $\quad y > 0$

Hence the half plane $y > 0$ is mapped into the interior of the circle $|w| = 1$.

## PROBLEMS 20.2

1. Find the invariant points of the transformation $w = (z-1)/(z+1)$.
2. Find the transformation which maps the points $-1, i, 1$ of the $z$-plane onto $1, i, -1$ of the $w$-plane respectively. Also find its *invariant points.* (Andhra, 2022; V.T.U., 2016)
3. Find the bilinear transformation which maps $1, i, -1$ to $2, i, -2$ respectively. (Andhra, 2021; Anna, 2016)
   Find the fixed and critical points of the transformation.
4. Determine the bilinear transformation that maps the points $1-2i, 2+i, 2+3i$ respectively into $2+2i, 1+3i, 4$.
5. Find the bilinear transformation which maps
   (i) the points $z = 1, i, -1$ into the points $w = 0, 1, \infty$ (V.T.U., 2014)
   (ii) the points $z = 0, 1, i$ into the points $w = 1+i, -i, 2-i$ (V.T.U., 2010S)
   (iii) $R(z) > 0$ into interior of unit circle so that $z = \infty, i, 0$ map into $w = -1, -i, 1$. (Andhra, 2020)
6. Under the transformation $w = \dfrac{z-1}{z+1}$, show that the map of the straight line $x = y$ is a circle and find its centre and radius.

7. Show that the bilinear transformation $w = (2z + 3)/(z - 4)$ maps the circle $x^2 + y^2 - 4x = 0$ into the line $4u + 3 = 0$.
8. Show that the condition for transformation $w = (az + b)/(cz + d)$ to make the circle $|w| = 1$, correspond to a straight line in the $z$-plane is $|a| = |c|$.
9. Show that the transformation $w = i(1 - z)/(1 + z)$ maps the circle $|z| = 1$ into the real axis of the $w$-plane and the interior of the circle $|z| < 1$ into the upper half of the $w$-plane. *(Andhra., 2015)*
10. If $z_0$ is the upper half of the $z$-plane, show that the bilinear transformation $w = e^{i\alpha}\left(\frac{z - z_0}{z - \bar{z}_0}\right)$ maps the upper half of the $z$-plane into the interior of the unit circle at the origin in the $w$-plane.

## 20.9 (1) CONFORMAL TRANSFORMATION

Suppose two curves $C$, $C_1$ in the $z$-plane intersect at the point $P$ and the corresponding curves $C'$ and $C_1'$ in the $w$-plane intersect at $P'$ (Fig. 20.5). *If the angle of intersection of the curves at P is the same as the angle of intersection of the curves at P′ in magnitude and sense, then the transformation is said to be* **conformal.**

**(2) Theorem.** *The transformation effected by an analytic function $w = f(z)$ is conformal at every point of the z-plane where $f'(z) \neq 0$.*

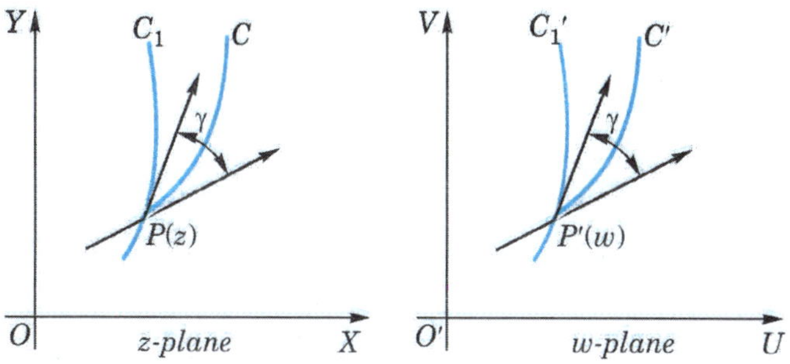

Fig. 20.5

Let $P(z)$ be a point in the region $R$ of the $z$-plane and $P'(w)$ the corresponding point in the region $R'$ of the $w$-plane (Fig. 20.3). Suppose $z$ moves on a curve $C$ and $w$ moves on the corresponding curve $C'$. Let $Q(z + \delta z)$ be a neighbouring point on $C$ and $Q'(w + \delta w)$ be the corresponding point on $C'$ so that $\vec{PQ} = \delta z$ and $\vec{P'Q'} = \delta w$. Then $\delta z$ is a complex number whose modulus $r$ is the length $PQ$ and amplitude $\theta$ is the angle which $PQ$ makes with the $x$-axis.

$$\therefore \qquad \delta z = re^{i\theta}$$

Similarly, if the modulus and amplitude of $\delta w$ be $r'$ and $\theta'$, then $\delta w = r'e^{i\theta'}$.

Hence, $$\frac{\delta w}{\delta z} = \frac{r'}{r} e^{i(\theta' - \theta)}$$

Now if the tangent at $P$ to the curve $C$ makes an $\angle\alpha$ with the $x$-axis and the tangent at $P'$ to $C'$ makes an $\angle\alpha'$ with the $u$-axis, then as $\delta z \to 0$, $\theta \to \alpha$ and $\theta' \to \alpha'$.

$$\therefore \qquad f'(z) = \frac{dw}{dz} = \left(\text{Lt}\,\frac{r'}{r}\right) . e^{i(\alpha' - \alpha)} \qquad ...(1)$$

If $\rho$ is the modulus and $\phi$ the amplitude of the function $f(z)$ which is supposed to be non-zero, then

$$f'(z) = \rho e^{i\phi} \qquad ...(2)$$

$\therefore$ from Eqns. (1) and (2), we have $\rho = \text{Lt}\,\dfrac{r'}{r}$ ...(3)

and $$\phi = \alpha' - \alpha. \qquad ...(4)$$

Now let $C_1$ be another curve through $P$ in the $z$-plane and $C_1'$ the corresponding curve through $P'$ in the $w$-plane. If the tangent at $P$ to $C_1$ makes an $\angle\beta$ with the $x$-axis and tangent at $P'$ to $C_1'$ makes an $\angle\beta'$ with the $u$-axis, then as in Eqn. (4),

$$\psi = \beta' - \beta \qquad ...(5)$$

Equating Eqns. (4) and (5), $\alpha' - \alpha = \beta' - \beta$ or $\beta - \alpha = \beta' - \alpha' = \gamma$ (Fig. 20.5)

Thus the angle between the curves before and after the mapping is preserved in magnitude and direction. Hence the mapping by the analytic function $w = f(z)$ is conformal at each point where $f'(z) \neq 0$.

**Obs. 1.** *A point at which $f'(z) = 0$ is called a* **critical point** *of the transformation.*

**Obs. 2.** The relation Eqn. (4), *i.e.*, $\alpha' = \alpha + \phi$ shows that the tangent at $P$ to the curve $C$ is rotated through an $\angle\phi$ = amp $\{f'(z)\}$ under the given transformation.

**Obs. 3.** The relation Eqn. (3) shows that in the transformation, elements of arc passing through $P$ in any direction are changed in the ratio $\rho : 1$, where $\rho = |f'(z)|$, *i.e., an infinitesimal length in the z-plane is magnified by the factor* $|f'(z)|$. Consequently *the infinitesimal areas are magnified by the factor* $|f'(z)|^2$ *in a conformal transformation.*

If $w = f(z)$ is analytic then $u$ and $v$ must satisfy $C$-$R$ equations.

$$\therefore \qquad J\left(\frac{u,v}{x,y}\right) = \begin{vmatrix} \frac{\partial u}{\partial x} & \frac{\partial u}{\partial y} \\ \frac{\partial v}{\partial x} & \frac{\partial v}{\partial y} \end{vmatrix} = \begin{vmatrix} \frac{\partial u}{\partial x} & \frac{-\partial v}{\partial x} \\ \frac{\partial v}{\partial x} & \frac{\partial u}{\partial x} \end{vmatrix} = \left(\frac{\partial u}{\partial x}\right)^2 + \left(\frac{\partial v}{\partial x}\right)^2 = \left|\frac{\partial u}{\partial x} + i\frac{\partial v}{\partial x}\right|^2 = |f'(z)|^2$$

Hence *in a conformal transformation, infinitesimal areas are magnified by the factor* $J\left(\frac{u,v}{x,y}\right)$.

*Also the condition of a conformal mapping is* $J\left(\frac{u,v}{x,y}\right) \neq 0$.

**Obs. 4.** *The angle preserving property of the conformal transformation has many important physical applications.* For instance, consider the flow of an incompressible fluid in a plane with velocity potential $\phi(x, y)$ and stream function $\psi(x, y)$. We know that $\phi$ and $\psi$ are real and imaginary parts of some analytic function $w = f(z)$. As $\phi$ = constant and $\psi$ = constant represent a system of orthogonal curves; these are transformed by the function $w = f(z)$ into a set of orthogonal lines in the $w$-plane and *vice-versa*.

Thus, *the conjugate functions $\phi$ and $\psi$ when subjected to conformal transformation remain conjugate functions, i.e., the solutions of Laplace's equation remain solutions of the Laplace's equation after the transformation*. This is the main reason for the great importance of the conformal transformation in applications.

## 20.10 SPECIAL CONFORMAL TRANSFORMATIONS

(1) *Transformation* $w = z^2$.

We have $\quad u + iv = (x + iy)^2 = x^2 - y^2 + 2ixy.$

$\therefore \quad u = x^2 - y^2$ and $v = 2xy$ ...(1)

If $u$ is constant (say, $a$), then $x^2 - y^2 = a$ which is a rectangular hyperbola. Similarly, if $v$ is constant (say, $b$), then $xy = b/2$ which also represents a rectangular hyperbola.

Hence a pair of lines $u = a$, $v = b$ parallel to the axes in the $w$-plane, map into pair of orthogonal rectangular hyperbolae in the $z$-plane as shown in Fig. 12.7 (p. 545). *(Anna., 2016 ; V.T.U., 2014)*

Again, if $x$ is constant (say, $c$), then $y = v/2c$ and $y^2 = c^2 - u$. Elimination of $y$ from these equations gives $v^2 = 4c^2(c^2 - u)$, which represents a parabola. Similarly, if $y$ is a constant (say, $d$), then elimination of $x$ from the equations Eqn. (1) gives $v^2 = 4d^2(d^2 + u)$ which is also a parabola.

Hence the pair of lines $x = c$ and $y = d$ parallel to the axes in the $z$-plane map into orthogonal parabolas in the $w$-plane as shown in Fig. 20.6.

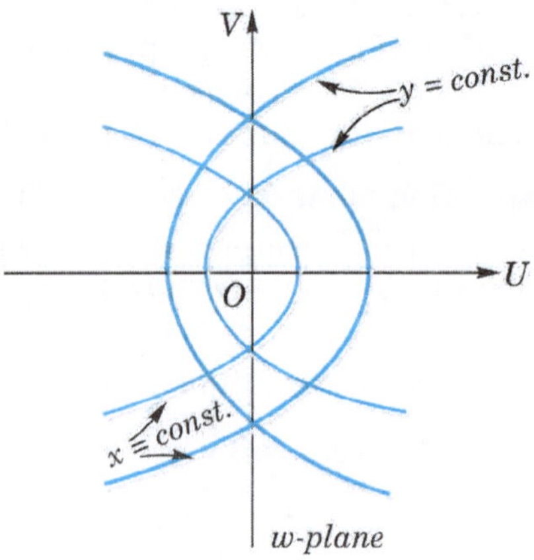

Fig. 20.6

Also since $\frac{dw}{dz} = 2z = 0$ for $z = 0$, therefore, it is a *critical point* of the mapping.

Taking $z = re^{i\theta}$ and $w = Re^{i\phi}$ then in polar form $w = z^2$ becomes $Re^{i\phi} = r^2e^{2i\theta}$.

*This shows that upper half of the z-plane $0 < \theta < \pi$ transforms into the entire w-plane $0 \leq \phi < 2\pi$. The same is true of the lower half.*

**Obs. 1.** Taking the axes to represent two walls, a single quadrant could be used to represent fluid flow at a corner wall. This transformation can also represent the electrostatic field in the vicinity of a corner conductor.

**Obs. 2.** *For the tansformation $w = z^n$, n being a positive integer, we have* $dw/dz = 0$ at $z = 0$.

Also $\quad Re^{i\phi} = (re^{i\theta})^n = r^ne^{in\theta}$

$\therefore \quad R = r^n$ and $\phi = n\theta$, when $0 < \theta < \pi/n$, correspondingly $0 < \phi < \pi$.

*Hence $w = z^n$ gives a conformal mapping of the z-plane everywhere except at the origin and that is fans out a sector of z-plane of central angle $\pi/n$ to cover the upper half of the w-plane.*

(2) *Joukowski's transformation** $w = z + 1/z$.

Since $\frac{dw}{dz} = \frac{(z+1)(z-1)}{z^2}$, the mapping is conformal except at the points $z = 1$ and $z = -1$ which correspond to the points $w = 2$ and $w = -2$ of the $w$-plane.

*Named after the Russian mathematician *Nikolai Jegorovich Joukowshi* (1847-1921).

Changing to polar coordinates,

$$w = u + iv = r(\cos\theta + i\sin\theta) + \frac{1}{r(\cos\theta + i\sin\theta)}$$

$$= r(\cos\theta + i\sin\theta) + \frac{1}{r}(\cos\theta - i\sin\theta)$$

$\therefore$ $$u = (r + 1/r)\cos\theta \text{ and } v = (r - 1/r)\sin\theta$$

Elimination of θ gives $$\frac{u^2}{(r+1/r)^2} + \frac{v^2}{(r-1/r)^2} = 1 \quad ...(1)$$

while the elimination of $r$ gives $$\frac{u^2}{4\cos^2\theta} - \frac{v^2}{4\sin^2\theta} = 1 \quad ...(2)$$

From Eqn. (1), it follows that the circles $r$ = constant of $z$-plane transform into a family of ellipses of the $w$-plane (Fig. 20.7). These ellipses are confocal for $(r + 1/r)^2 - (r - 1/r)^2 = 4$, *i.e.*, a constant.

In particular, the unit circle ($r = 1$) in the $z$-plane flattens out to become the segment $u = -2$ to $u = 2$ of the real axis in $w$-plane. Thus the exterior of the unit circle in the $z$-plane maps into the entire $w$-plane.

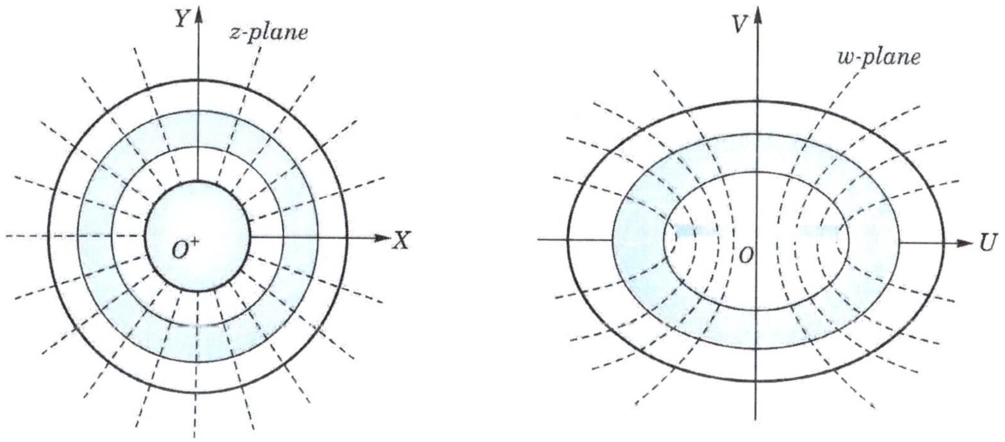


Fig. 20.7

From Eqn. (2), it is clear that the radial lines θ = constant of the $z$-plane transform into a family of hyperbolae which are also confocal (Fig. 20.7).

**Obs. 1.** $v = \left(r - \frac{1}{r}\right)\sin\theta = 0$ gives $r = \pm 1$ or $\theta = 0, \pi$, *i.e.*, this streamline consists of the unit circle $r = 1$ and the $x$-axis ($\theta = 0$ to $\theta = \pi$). For large $z$, the flow is nearly uniform and parallel to the $x$-axis. This can be interpreted as a flow around a circular cylinder of unit radius having two **stagnation points*** at $A(z = 1)$ and $B(z = -1)$. (Fig. 20.8)

[$\because$ $dw/dz = 0$ at $z = \pm 1$]

**Obs. 2.** This transformation is also used to map the exterior of the profile of an aeroplane wing on the exterior of a nearly circular region. These airfoils are known as *Joukowski airfoils*.

**(3)** *Transformation* $w = e^z$.

Writing $z = x + iy$ and $w = \rho e^{i\phi}$, we have $\rho e^{i\phi} = e^{x+iy} = e^x \cdot e^{iy}$

whence $\rho = e^x$ ...(1) and $\phi = y$ ...(2)

From (1), it is clear that the lines parallel to $y$-axis ($x$ = const.) map into circles ($\rho$ = const.) in the $w$-plane, their radii being less than or greater than 1 according as $x$ is less than or greater than 0 (Fig. 20.9). (*V.T.U., 2011*)

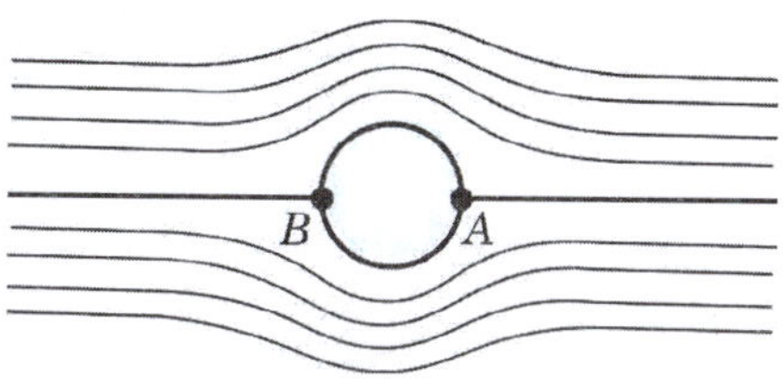


Fig. 20.8

Similarly, it follows from Eqn. (2) that the lines parallel to the $x$-axis ($y$ = const.) map into the radial lines ($\phi$ = const.) of the $w$-plane. Thus any horizontal strip of height $2\pi$ in the $z$-plane will cover once the entire $w$-plane.

* *Stagnation points* are those at which the fluid velocity is zero.

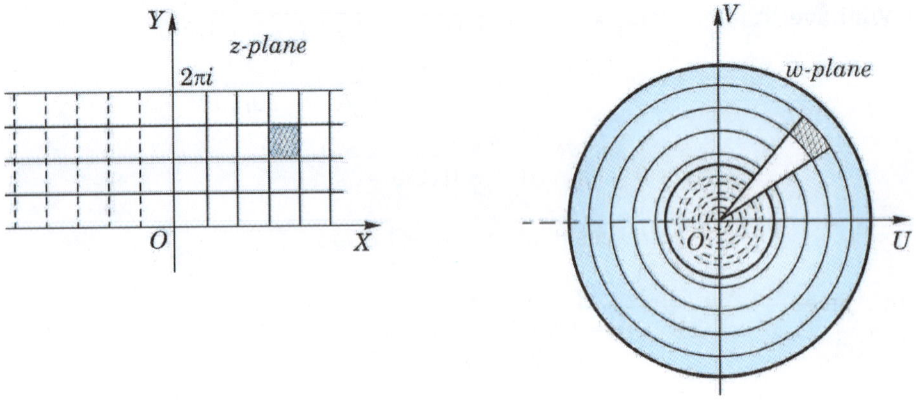


Fig. 20.9

The rectangular region $a_1 \le x \le a_2$, $b_1 \le y \le b_2$ in the $z$-plane (shown shaded) transforms into the region $e^{a_1} \le \rho \le e^{a_2}$, $b_1 \le \phi \le b_2$ in the $w$-plane bounded by circles and rays (shown shaded). (*V.T.U., 2012S*)

**Obs.** This transformation can be used to obtain the circulation of a liquid around a cylindrical obstacle, the electrostatic field due to a charged circular cylinder etc.

(**4**) *Transformation $w = \cosh z$.*

We have $$u + iv = \cosh(x + iy) = \cosh x \cos y + i \sinh x \sin y$$

so that $$u = \cosh x \cos y \text{ and } v = \sinh x \sin y.$$

Elimination of $x$ from these equations gives

$$\frac{u^2}{\cos^2 y} - \frac{v^2}{\sin^2 y} = 1 \quad ...(1)$$

while elimination of $y$ gives $$\frac{u^2}{\cosh^2 x} + \frac{v^2}{\sinh^2 z} = 1 \quad ...(2)$$

Eqn. (1) shows that the lines parallel to $x$-axis (*i.e.*, $y$ = const.) in the $z$-plane map into hyperbolae in the $w$-plane.

Eqn. (2) shows that the lines parallel to the $y$-axis (*i.e.*, $x$ = const.) in the $z$-plane map into ellipse in the $w$-plane (Fig. 20.10). The rectangular region $a_1, \le x \le a_2$, $b_1 \le y \le b_2$ in the $z$-plane (shown shaded) transforms into the shaded region in the $w$-plane bounded by the corresponding hyperbolae and ellipses.

**Obs.** This transformation can be used.

(*i*) to obtain the circulation of liquid around an elliptic cylinder;

(*ii*) to determine the electrostatic field due to a charged cylinder;

(*iii*) to determine the potential between two confocal elliptic (or hyperbolic) cylinders.

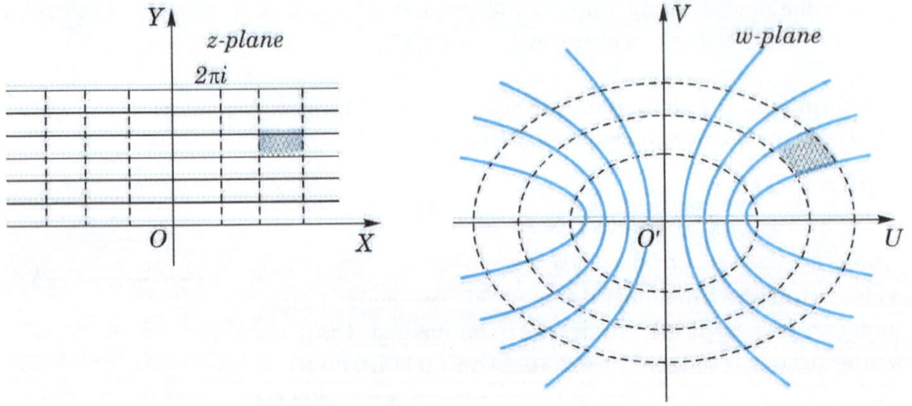


Fig. 20.10

**Example 20.15.** *Show that under the transformation $w = (z - i)/(z + i)$, real axis in the $z$-plane is mapped into the circle $|w| = 1$. Which portion of the $z$-plane corresponds to the interior of the circle?*

**Solution.** We have $$|w| = \left|\frac{z-i}{z+i}\right| = \frac{|z-i|}{|z+i|} = \frac{|x+i(y-1)|}{|x+i(y+1)|}$$

$$= \sqrt{\{x^2+(y-1)^2\}}/\sqrt{\{x^2+(y+1)^2\}}$$

Now the real axis in $z$-plane *i.e.*, $y = 0$, transforms into

$$|w| = \sqrt{(x^2+1)}/\sqrt{\{x^2+1)} = 1.$$

Hence the real axis in the $z$-plane is mapped into the circle $|w| = 1$.

The interior of the circle, *i.e.*, $|w| < 1$, gives

$$\{x^2+(y-1)^2\}/\{x^2+(y+1)^2\} < 1 \text{ i.e., } -4y < 0 \text{ or } y > 0.$$

Thus the upper half of the $z$-plane corresponds to the interior of the circle $|w| = 1$.

## PROBLEMS 20.3

1. Determine the region of the $w$-plane into which the following regions are mapped by the transformation $w = z^2$:
   (*i*) first quadrant of $z$-plane
   (*ii*) region bounded by $x = 1$, $y = 1$, $x + y = 1$ (*V.T.U., 2016*)
   (*iii*) the region $1 \le x \le 2$ and $1 \le y \le 2$
   (*iv*) circle $|z - 1| = 2$.
2. Find the transformation which maps the triangular region $0 \le \arg z \le \pi/3$ into the unit circle $w \le 1$.
3. Discuss the transformation $w = \sqrt{z}$. Is it conformal at the origin?
4. Under the transformation $w = 1/z$, find the image of
   (*a*) the circle $|z - 2i| = 2$
   (*b*) the straight line $y - x + 1 - 0$
   (*c*) the hyperbola $x^2 - y^2 = 1$.
5. Show that under the transformation $w = 1/z$, (*a*) circle $x^2 + y^2 - 6x = 0$ is transformed into a straight line in the $w$-plane.
   (*b*) the circle $(x - 3)^2 + y^2 = 2$ is transformed into a circle with centre (3/7, 0) and radius $\sqrt{2}/17$.
6. Show that the transformation $w = 1/z$ transforms all circles and straight lines into the circles and straight lines in the $w$-plane. Which circles in the $z$-plane become straight lines in the $w$-plane, and which straight lines are transformed into other straight lines ?
7. Show that the transformation $w = z + 1/z$ converts the straight line $\arg z = \alpha$ ($|\alpha| < a/2$) into a branch of hyperbola of eccentricity sec $\alpha$. (*V.T.U., 2015*)
8. Show that the transformation $w = z + (a^2 - b^2)/4z$ transforms the circle of radius $\frac{1}{2}(a + b)$, centre at the origin, in the $z$-plane into ellipse of semi-axes $a$, $b$ in the $w$-plane.
9. Show that the transformation $w = z + a^2/z$ transforms circles with origin at the centre in the $z$-plane into co-axial concentric, confocal ellipses in the $w$-plane. (*V.T.U., 2012*)
10. Show that the function $w = A(z + a^2/z)$ may be used to represent the flow of a perfect incompressible fluid past a circular cylinder. Also find the stagnation points.
11. Show that by the relation $u + iv = \cos(x + iy)$, the infinite strip bounded by $x = c$, $x = d$, where $c$ and $d$ lie between 0 and $\pi/2$, is mapped into the region between the two branches of the hyperbola lying in $u > 0$.
12. Prove that the transformation $w = \sin z$, maps the families of lines $x$ = constant and $y$ = constant into two families of confocal central conics.
13. Discuss the transformation $w = e^z$, and show that it transforms the region between the real azis and a line parallel to real axis at $y = \pi$, into the upper half of the $w$-plane.
14. Discuss fully the transformation $w = c \cosh z$, where $c$ is a real number. What physical problem can we study with the help of this transformation ?

## 20.11 SCHWARZ-CHRISTOFFEL TRANSFORMATION

*This transformation maps the interior of a polygon of the w-plane into the upper half of the z-plane and the boundary of the polygon into the real axis. The formula of this transformation is*

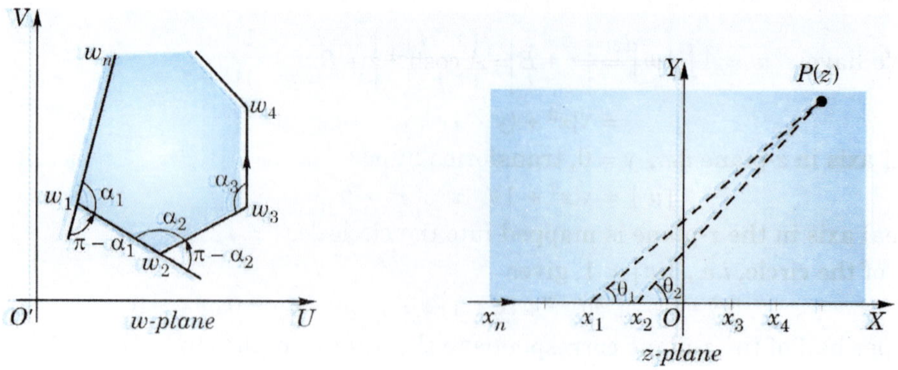


Fig. 20.11

$$\frac{dw}{dz} = A(z - x_1)^{\frac{\alpha_1}{\pi} - 1} (z - x_2)^{\frac{\alpha_2}{\pi} - 1} \ldots (z - x_n)^{\frac{\alpha_n}{\pi} - 1} \quad \ldots(1)$$

or
$$w = A \int (z - x_1)^{\frac{\alpha_1}{\pi} - 1} (z - x_2)^{\frac{\alpha_2}{\pi} - 1} \ldots (z - x_n)^{\frac{\alpha_n}{\pi} - 1} dz + B \quad \ldots(2)$$

where $\alpha_1, \alpha_2, \ldots, \alpha_n$ are the interior angles of the polygon having vertices $w_1, w_2, \ldots, w_n$ which map into the points $x_1, x_2, \ldots, x_n$ on the real-axis of the $z$-plane (Fig. 20.11). Also $A$ and $B$ are complex constants which determines the size and position of the polygon.

*Proof.* We have from Eqn. (1),

$$\text{amp}\left(\frac{dw}{dz}\right) = \text{amp}\ [(A) + \left(\frac{\alpha_1}{\pi} - 1\right) \text{amp}\ (z - x_1) + \left(\frac{\alpha_2}{\pi} - 1\right) \text{amp}\ (z - x_2)$$

$$\ldots. + \left(\frac{\alpha_n}{\pi} - 1\right) \text{amp}\ (z - x_n) \quad \ldots(3)$$

As $z$ moves along the real axis from the left towards $x_1$, suppose that $w$ moves along the side $w_n w_1$ of the polygon towards $w_1$. As $z$ crosses $x_1$ from left to right, $\theta_1 = \text{amp}\ (z - x_1)$ changes from $\pi$ to 0 while all other terms of Eqn. (3) remain unaffected. Hence only $\left(\frac{\alpha_1}{\pi} - 1\right)$ amp $(z - x_1)$ decreases by $\left(\frac{\alpha_1}{\pi} - 1\right) \pi = \alpha_1 - \pi$, *i.e.* increases by $\pi - \alpha_1$ in the anti-clockwise direction. In other words, amp $(dw/dz)$ increases by $\pi - \alpha_1$. Thus the direction of $w_1$ turns through the angle $\pi - \alpha_1$ and $w$ now moves along the side $w_1 w_2$ of the polygon.

Similarly when $z$ passes through $x_2$, $\theta_1 = \text{amp}\ (z - x_1)$ and $\theta_2 = \text{amp}\ (z - x_2)$ change from $\pi$ to 0 while all other terms remain unchanged. Hence the side $w_1 w_2$ turns through the angle $\pi - \alpha_2$. Proceeding in this way, we see that as $z$ moves along $x$-axis, $w$ traces the polygon $w_1\, w_2\, w_3 \ldots w_n$ and conversely.

**Example 20.16.** *Find the transformation which maps the semi-infinite strip in the w-plane (Fig. 20.12) into the upper half of the z-plane*

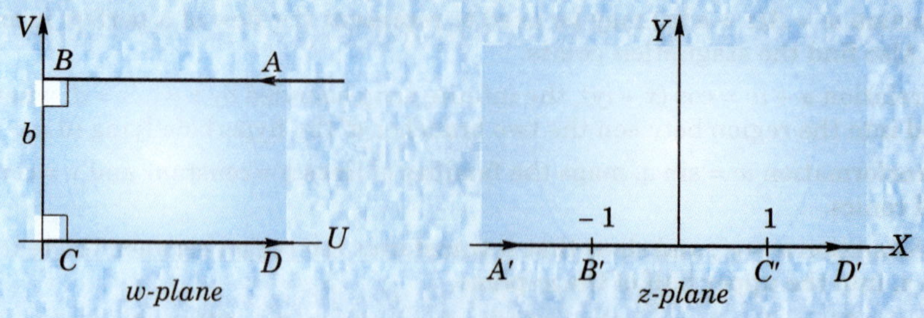


Fig. 20.12

**Solution.** Consider $ABCD$ as the limiting case of a triangle with two vertices $B$ and $C$ and the third vertex $A$ or $D$ at infinity. Let the vertices $B$ and $C$ map into the points $B'$ $(-1)$ and $C'$ $(1)$ of the $z$-plane. Since the interior angles at $B$ and $C$ are $\pi/2$, we have by the Schwarz-Christoffel transformation,

$$\frac{dw}{dz} = A(z + 1)^{\frac{\pi/2}{\pi} - 1} (z - 1)^{\frac{\pi/2}{\pi} - 1} = A / \sqrt{(z^2 - 1)}$$

$$\therefore \qquad w = A\int \frac{dz}{\sqrt{(z^2-1)}} + B = A\cosh^{-1} z + B$$

When $\quad z = 1, w = 0. \quad \therefore \quad 0 = A\cosh^{-1}(1) + B$, *i.e.*, $B = 0$.

When $z = -1, w = ib. \quad \therefore \quad ib = A\cosh^{-1}(-1) + 0$, *i.e.*, $\cosh(ib/A) = -1$

or $\quad \cos\dfrac{b}{A} = -1 = \cos\pi$. Thus $A = \dfrac{b}{\pi}$.

Hence $\quad w = \dfrac{b}{\pi}\cosh^{-1} z$ or $z = \cosh\dfrac{\pi w}{b}$.

## PROBLEMS 20.4

1. Find the transformation which maps the semi-infinite strip of width $\pi$ bounded by the lines $v = 0$, $v = \pi$ and $u = 0$ into the upper half of the $z$ plane.
2. Show how you will use Schwarz-Christoffel transformation to map the semi-infinite strip enclosed by the real axis and the lines $u = \pm 1$ of the $w$-plane into the upper half of the $z$-plane.
3. Find the mapping function which maps semi-infinite strip in the $z$-plane $-\pi/2 \le x \le \pi/2, y \ge 0$ into half $w$-plane for which $v \ge 0$, such that the points $(-\pi/2, 0)$, $(\pi/2, 0)$ in the $\pi$-plane are mapped into the points $(-1, 0)$, $(1, 0)$ respectively in $w$-plane.
4. Find the transformation which will map the interior of the infinite strip bounded by the lines $v = 0$, $v = \pi$ onto the upper half of the $z$-plane. *(Hazaribag., 2015)*

## 20.12 COMPLEX INTEGRATION

We have already discussed the concept of the line integral as applied to vector fields in § 8.11. Now we shall consider the line integral of a complex function.

Consider a continuous function $f(z)$ of the complex variable $z = x + iy$ defined at all points of a curve $C$ having end points $A$ and $B$. Divide $C$ into $n$ parts at the points

$$A = P_0(z_0), P_1(z_1), \ldots, P_i(z_i), \ldots P_n(z_n) = B.$$

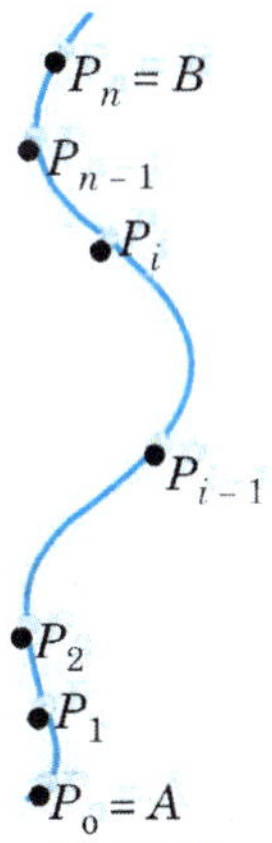

Fig. 20.13

Let $\delta z_i = z_i - z_{i-1}$ and $\zeta_i$ be any point on the arc $P_{i-1}P_i$. The limit of the sum $\sum_{i=1}^{n} f(\zeta_i)\delta z_i$ as $n \to \infty$ *in such a way that the length of the chord* $\delta z_i$ *approaches zero, is called the* **line integral of f (z)** *taken along the path C, i.e.,*

$$\int_C f(z)\, dz.$$

Writing $f(z) = u(x, y) + iv(x, y)$ and noting that $dz = dx + idy$,

$$\int_C f(z)\, dz = \int_C (udx - vdy) + i\int_C (vdx + udy)$$

which shows that the evaluation of the line integral of a complex function can be reduced to the evaluation of two line integrals of real functions.

**Obs.** *The value of the integral is independent of the path of integration when the integrand is analytic.*

**Example 20.17.** *Prove that*

*(i)* $\displaystyle\int_C \frac{dz}{z-a} = 2\pi i.$ $\qquad$ *(ii)* $\displaystyle\int_C (z-a)^n\, dz = 0$ [*n, any integer* $\neq -1$]

*where C is the circle* $|z - a| = r$.

**Solution.** The parametric equation of $C$ is $z - a = re^{i\theta}$, where $\theta$ varies from 0 to $2\pi$ as $z$ describes $C$ once in the positive (anti-clockwise) sense. (Fig. 20.14)

*(i)* $$\int_C \frac{dz}{z-a} = \int_0^{2\pi} \frac{1}{re^{i\theta}} \cdot ire^{i\theta}\, d\theta \qquad [\because dz = ire^{i\theta}\, d\theta]$$

$$= i\int_0^{2\pi} d\theta = 2\pi i$$

(ii) $\int_C (z-a)^n \, dz = \int_0^{2\pi} r^n e^{ni\theta} \,.\, ire^{i\theta} \, d\theta$

$$= ir^{n+1} \int_0^{2\pi} e^{(n+1)\theta i} \, d\theta = \frac{r^{n+1}}{n+1} \left| e^{(n+1)\theta i} \right|_0^{2\pi}, \text{ provided } n \neq -1$$

$$= \frac{r^{n+1}}{n+1} [e^{2(n+1)\pi i} - 1] = 0 \qquad [\because \; e^{2(n+1)\pi i} = 1]$$

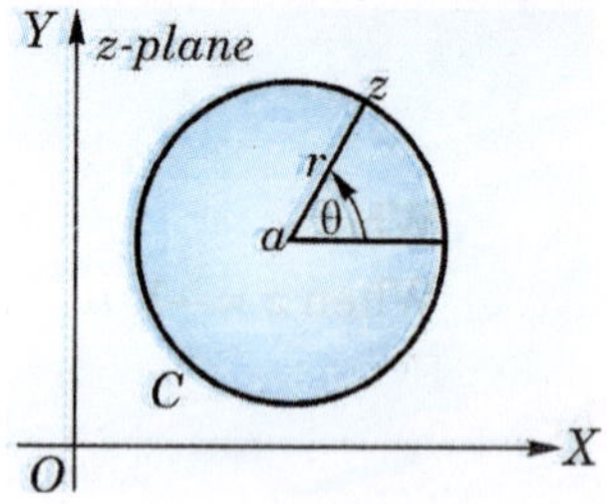

Fig. 20.14

**Example 20.18.** *Evaluate* $\int_0^{2+i} (\bar{z})^2 \, dz$, *along (i) the line* $y = x/2$, *(ii) the real axis to 2 and then vertically to* $2 + i$. (U.P.T.U., 2012)

**Solution.** (i) Along the line $OA$, $x = 2y$, $z = (2+i)y$, $\bar{z} = (2-i)y$ and $dz = (2+i)\,dy$ (Fig. 20.15)

$$\therefore \qquad I = \int_0^{2+i} (\bar{z})^2 \, dz = \int_0^1 (2-i)^2 y^2 \,.\, (2+i) \, dy$$

$$= 5(2-i) \left| \frac{y^3}{3} \right|_0^1 = \frac{5}{3}(2-i)$$

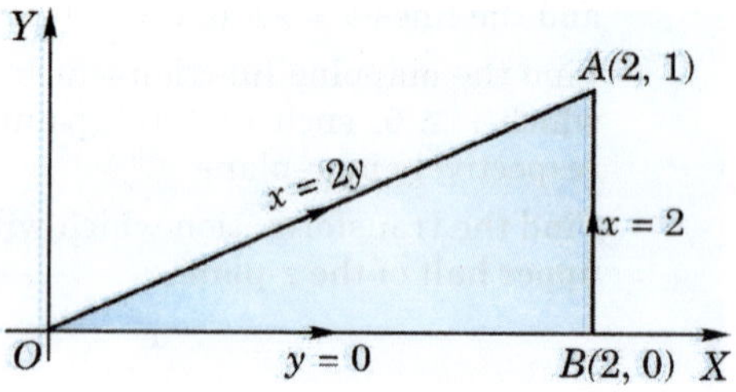

Fig. 20.15

(ii) $$I = \int_{OB} (\bar{z})^2 \, dz + \int_{BA} (\bar{z})^2 \, dz.$$

Now along $OB$, $z = x$, $\bar{z} = x$, $dz = dx$;

and along $BA$, $z = 2 + iy$, $\bar{z} = 2 - iy$, $dz = i\,dy$

$$\therefore \qquad I = \int_0^2 x^2 dx + \int_0^1 (2-iy)^2 \,.\, i\,dy = \left| \frac{x^3}{3} \right|_0^2 + \int_0^1 [\,4y + (4-y^2)i\,]\, dy$$

$$= \frac{8}{3} + 4 \,.\, \frac{1}{2} + \left(4.1 - \frac{1}{3}\right) i = \frac{1}{3}(14 + 11i).$$

**Example 20.19.** *Evaluate* $\int_C (z^2 + 3z + 2)\, dz$ *where C is the arc of the cycloid* $x = a\,(\theta + \sin\theta)$, $y = a\,(1 - \cos\theta)$ *between the points* $(0, 0)$ *and* $(\pi a, 2a)$.

**Solution.** $f(z) = z^2 + 3z + 2$ is analytic in the $z$-plane being a polynomial. As such, the line integral of $f(z)$ between $O$ and $A$ is independent of the path (Fig. 20.16). We therefore, take the path from $O$ to $L$ and $L$ to $A$ so that

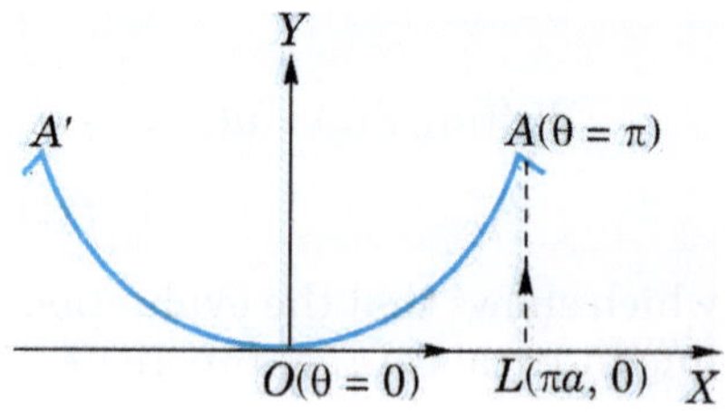

Fig. 20.16

$$\int_C f(z)\, dz = \int_{OL} f(z)\, dz + \int_{LA} f(z)\, dz \qquad ...(i)$$

$$\therefore \qquad \int_{OL} f(z)\, dz = \int_0^{\pi a} (x^2 + 3x + 2)\, dx$$

[$\because$ along $OL$, $y = 0$, $x = 0$ at $O$, $x = \pi a$ at $L$

$$= \left| \frac{x^3}{3} + \frac{3x^2}{2} + 2x \right|_0^{\pi a} = \frac{\pi a}{6}(2\pi^2 a^2 + 9\pi a + 12) \qquad ...(ii)$$

and $$\int_{LA} f(z)\, dz = \int_0^{2a} [(\pi a + iy)^2 + 3\,(\pi a + iy) + 2]\, i\,dy$$

[$\because$ along $LA$, $x = \pi a$, $z = \pi a + iy$, $dz = i\,dy$ and $y$ varies from 0 (at $L$) to $2a$ (at $A$)

$$= L \left| \frac{(\pi a + iy)^3}{3i} + 3\,\frac{(\pi a + iy)^2}{2i} + 2y \right|_0^{2\pi} = \frac{a^3}{3}(\pi + 2i)^3 + \frac{3a^2}{2}(\pi + 2i)^2 + 4ia \qquad ...(iii)$$

$\therefore$ substituting from Eqns. (ii) and (iii) in Eqn. (i), we get

$$\int_C f(z)\, dz = \frac{\pi a}{6}(2\pi^2 a^2 + 9\pi a + 12) + \frac{a^3}{3}(\pi + 2i)^3 + \frac{3a^2}{2}(\pi + 2i)^2 + 4ia$$

## PROBLEMS 20.5

1. Evaluate $\int_0^{1+i}(x^2 - iy)\,dz$ along the paths (a) $y = x$ and (b) $y = x^2$.

2. Evaluate $\int_{1-i}^{2+i}(2x + iy + 1)\,dz$, along the two paths:

   (i) $x = t + 1, y = 2t^2 - 1$ (ii) the straight line joining $1 - i$ and $2 + i$. (Andhra, 2020)

3. Evaluate $\int_{1-i}^{2+3i}(z^2 + z)\,dz$ along the line joining the points (1, –1) and (2, 3).

4. Show that for every path between the limits, $\int_{-2}^{-2+i}(2 + z)^2\,dz = -i/3$.

5. Show that $\oint_C (z + 1)\,dz = 0$, where $C$ is the boundary of the square whose vertices are at the points $z = 0, z = 1, z = 1 + i$ and $z = i$.

6. Evaluate $\int_C |z|\,dz$, where $C$ is the contour

   (i) straight line from $z = -i$ to $z = i$.

   (ii) left half of the unit circle $|z| = 1$ from $z = -i$ to $z = i$. (Andhra, 2022)

   (iii) circle given by $|z + 1| = 1$ described in the clockwise sense.

7. Find the value of $\int_0^{1+i}(x - y + ix^2)\,dz$

   (i) along the straight line from $z = 0$ to $z = 1 + i$

   (ii) along real axis from $z = 0$ to $z = 1$ and then along a line parallel to the imaginary axis from $z = 1$ to $z = 1 + i$.

8. Prove that $\int dz/z = -\pi i$ or $\pi i$, according as $C$ is the semi-circular arc $|z| = 1$ above or below the real axis.

9. Evaluate $\int_C (z - z^2)\,dz$, where $C$ is the upper half of the circle $|z| = 1$. (Andhra, 2022)

   What is the value of this integral if $C$ is the lower half of the above circle ?

## 20.13 CAUCHY'S THEOREM

*If $f(z)$ is an analytic function and $f'(z)$ is continuous at each point within and on a closed curve C, then $\int_C f(z)\,dz = 0$.*

Writing $f(z) = u(x, y) + iv(x, y)$ and noting that $dz = dx + idy$

$$\oint_C f(z)dz = \oint_C (udx - vdy) = i \oint_C (vdx + udy) \quad ...(1)$$

Since $f'(z)$ is continuous, therefore, $\frac{\partial u}{\partial x}, \frac{\partial u}{\partial y}, \frac{\partial v}{\partial x}, \frac{\partial v}{\partial y}$ are also continuous in the region $D$ enclosed by $C$. Hence the Green's theorem can be applied to (1), giving

$$\oint_C f(z)dz = -\iint_D \left[\frac{\partial v}{\partial x} + \frac{\partial v}{\partial y}\right] dxdy + i \iint_D \left[\frac{\partial u}{\partial x} - \frac{\partial v}{\partial y}\right] dxdy \quad ...(2)$$

Now $f(z)$ being analytic, $u$ and $v$ necessarily satisfy the Cauchy-Riemann equations and thus the integrands of the two double integrals in (2) vanish identically.

Hence $$\oint_C f(z)\,dz = 0.$$

**Obs. 1.** The Cauchy-Riemann equations are precisely the conditions for the two real integrals in Eqn. (1) to be independent of the path. *Hence the line integral of a function f(z) which is analytic in the region D, is independent of the path joining any two points of D.*

**Obs. 2. Extension of Cauchy's theorem.** *If f(z) is analytic in the region D between two simple closed curves C and* $C_1$, then $\oint_C f(z)dz = \oint_{C_1} f(z)dz$.

To prove this, we need to introduce the cross-cut $AB$. Then $\oint f(z)dz = 0$ where the path is as indicated by arrows in Fig. 20.17, *i.e.*, along $AB$—along $C_1$ in clockwise sense and along $BA$—along $C$ in anti-clockwise sense.

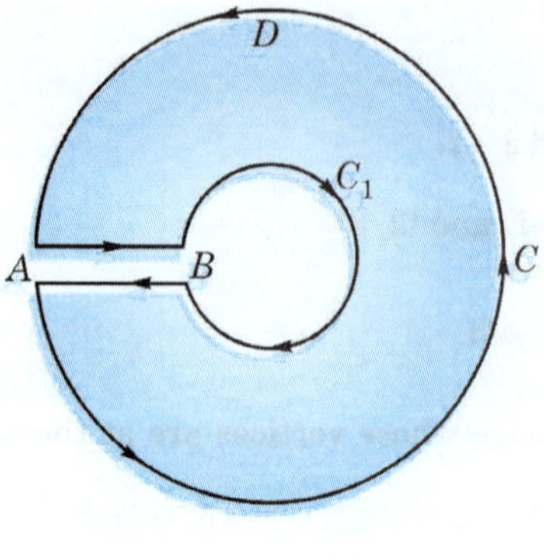


Fig. 20.17(*a*)

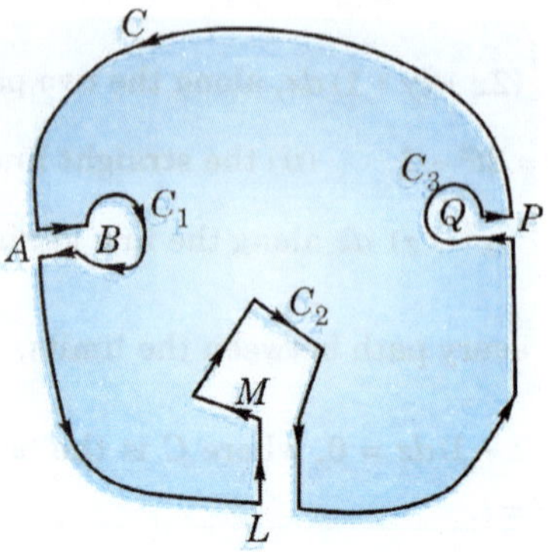


Fig. 20.17(*b*)

*i.e.*, $$\int_{AB} f(z)dz + \int_{C_1} f(z)dz + \int_{AB} f(z)dz + \int_{C} f(z)dz = 0$$

But, since the integrals along $AB$ and along $BA$ cancel, it follows that

$$\int_{C} f(z)dz + \int_{C_1} f(z)dz = 0$$

Reversing the direction of the integral around $C_1$ and transposing, we get

$$\int_{C} f(z)dz + \int_{C_1} f(z)dz$$ each integration being taken in the anti-clockwise sense.

If $C_1$, $C_2$, $C_3$, ... *be any number of closed curves within C* (*Fig. 20.17(b)*), *then*

$$\oint_C f(z)dz = \oint_{C_1} f(z)dz + \oint_{C_2} f(z)dz + \oint_{C_3} f(z)dz + \ldots$$

## 20.14 CAUCHY'S INTEGRAL FORMULA

*If f (z) is analytic within and on a closed curve and if a is any point within C, then*

$$f(a) = \frac{1}{2\pi i} = \oint_C \frac{f(z)dz}{z-a}$$

Consider the function $f(z)/(z-a)$ which is analytic at all points within $C$ except at $z = a$. With the point $a$ as centre and radius $r$, draw a small circle $C_1$ lying entirely within $C$.

Now $f(z)/(z-a)$ being analytic in the region enclosed by $C$ and $C_1$, we have by Cauchy's theorem,

$$\oint_C \frac{f(z)}{z-a}dz = \oint_{C_1} \frac{f(z)}{z-a}dz$$

$\left\{\begin{array}{l}\text{For any point on } C_1, \\ z - a = re^{i\theta} \text{ and } dz = ire^{i\theta}\, d\theta\end{array}\right.$

$$= \oint_C \frac{f(a+re^{i\theta})}{re^{i\theta}} . ire^{i\theta}\, d\theta = i \oint_{C_1} f(a+re^{i\theta})d\theta \quad \ldots(1)$$

In the limiting form, as the circle $C_1$ shrinks to the point $a$, *i.e.*, as $r \to 0$, the integral Eqn. (1) will approach to

$$\oint_C f(a)d\theta = if(a)\int_0^{2\pi} d\theta = 2\pi i f(a) \oint f(z)dz. \text{ Thus } \oint_C \frac{f(z)}{z-a}dz = 2\pi i f(a)$$

*i.e.*, $$f(a) = \frac{1}{2\pi i}\oint_C \frac{f(z)}{z-a}dz \quad \ldots(2)$$

which is the desired *Cauchy's integral formula.* (*Birla Ranchi., 2016 ; U.P.T.U., 2012 ; V.T.U., 2011 S*)

**Cor.** Differentiating both sides of Eqn. (2) w.r.t. $a$,

$$f'(a) = \frac{1}{2\pi i}\oint_C \frac{\partial}{\partial a}\left[\frac{f(z)}{z-a}\right]dz = \frac{1}{2\pi i}\oint_C \frac{f(z)}{(z-a)^2}dz \quad \ldots(3)$$

Similarly, $$f''(a) = \frac{2!}{2\pi i}\oint_C \frac{f(z)}{(z-a)^3}\oint_C dz \quad \ldots(4)$$

and in general, $$f^n(a) = \frac{n!}{2\pi i}\oint_C \frac{f(z)}{(z-a)^{n+1}}\,dz. \qquad ...(5)$$

Thus it follows from the results Eqns. (2) to (5) that *if a function f(z) is known to be analytic on the simple closed curve C then the values of the function and all its derivatives can be found at any point of C.* Incidently, we have established a remarkable fact that **an analytic function possesses derivatives of all orders and these are themselves all analytic.**

**Example 20.20.** *Evaluate* $\int_C \frac{z^2 - z + 1}{z - 1}\,dx$, *where C is the circle*

(*i*) $|z| = 1$, (*ii*) $|z| = \frac{1}{2}$. (JNTUK, 2023)

**Solution.** (*i*) Here $f(z) = z^2 - z + 1$ and $a = 1$.

Since $f(z)$ is analytic within and on circle $C : |z| = 1$ and $a = 1$ lies on $C$.

$\therefore$ by Cauchy's integral formula $\frac{1}{2\pi i}\int_C \frac{f(z)}{z-a} = f(a) = 1$ *i.e.*, $\int_C \frac{z^2 - z + 1}{z - 1}\,dz = 2\pi i.$

(*ii*) In this case, $a = 1$ lies outside the circle $C : |z| = 1/2$. So $(z^2 - z + 1)/(z - 1)$ is analytic everywhere within $C$.

$\therefore$ by Cauchy's theorem $\int_C \frac{z^2 - z + 1}{z - 1}\,dz = 0.$

**Example 20.21.** *Evaluate, using Cauchy's integral formula:*

(*i*) $\oint_C \frac{\sin \pi z^2 + \cos \pi z^2}{(z-1)(z-2)}\,dz$ *where C is the circle* $|z| = 3$ (V.T.U., 2012 ; Calicut, 2011 ; U.P.T.U., 2010)

(*ii*) $\oint_C \frac{\cos \pi z}{z^2 - 1}\,dz$ *around a rectangle with vertices* $2 \pm i, -2 \pm 1$ (JNTUK, 2023)

(*iii*) $\oint_C \frac{e^{tz}}{z^2 + 1}\,dz$ *where C is the circle* $|z| = 3$.

**Solution.** (*i*) $f(z) = \sin \pi z^2 + \cos \pi z^2$ is analytic within the circle $|z| = 3$ and the two singular points $z = 1$ and $z = 2$ lie inside this circle.

$$\therefore \quad \oint_C \frac{f(z)dz}{(z-1)(z-2)} = \oint_C (\sin \pi z^2 + \cos \pi z^2).\left(\frac{1}{z-2} - \frac{1}{z-1}\right)dz$$

$$= \oint_C \frac{(\sin \pi z^2 + \cos \pi z^2)}{z-2}\,dz - \oint_C \frac{(\sin \pi z^2 + \cos \pi z^2)}{z-1}\,dz$$

$$= 2\pi i\,[\sin \pi(2)^2 + \cos \pi(2)^2] - 2\pi i[\sin \pi(1)^2 + \cos \pi(1)^2]$$

[By Cauchy's integral formula]

$$= 2\pi i\,(0 + 1) - 2\pi i\,(0 - 1) = 4\pi i$$

(*ii*) $f(z) = \cos \pi z$ is analytic in the region bounded by the given rectangle and the two singular points $z = 1$ and $z = -1$ lie inside this rectangle. (Fig. 20.18)

$$\therefore \quad \oint_C \frac{\cos \pi z}{z^2 - 1}\,dz = \frac{1}{2}\oint_C \left(\frac{1}{z-1} - \frac{1}{z+1}\right)\cos \pi z\,dz$$

$$= \frac{1}{2}\oint_C \frac{\cos \pi z}{z-1}\,dz - \oint_C \frac{\cos \pi z}{z+1}\,dz$$

$$= \frac{1}{2}\{2\pi i \cos \pi\,(1)\} - \frac{1}{2}\{2\pi i \cos \pi\,(-1)\} = 0.$$

[By Cauchy's integral formula]

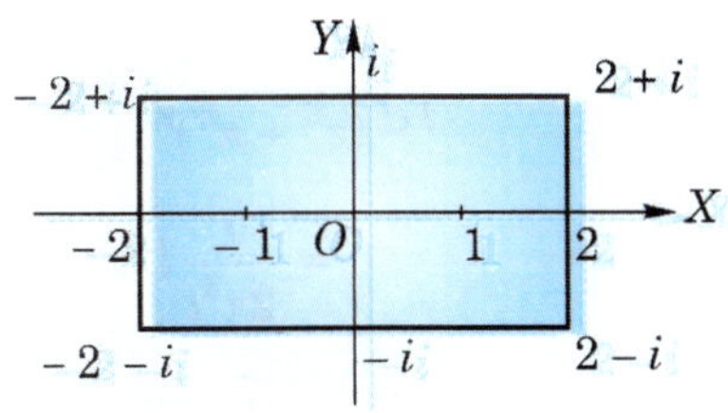


Fig. 20.18

(*iii*) $f(z) = e^{tz}$ is analytic within the circle $|z| = 3$.

The singular points are given by $z^2 + 1 = 0$ *i.e.*, $z = i$ and $z = -i$ which lie within this circle.

$$\therefore \quad \oint_C \frac{e^{tz}}{z^2+1} dz = \oint_C \frac{1}{2i}\left(\frac{1}{z-i} - \frac{1}{z+i}\right) e^{tz} dz = \frac{1}{2i}\left\{\oint_C \frac{e^{tz}}{z-i} dz - \oint_C \frac{e^{tz}}{z+i} dz\right\}$$

$$= \frac{1}{2i}\{2\pi i e^{t(i)} - 2\pi i e^{t(-i)}\} \qquad \text{[By Cauchy's integral formula]}$$

$$= 2\pi i \left(\frac{e^{it} - e^{-it}}{2i}\right) = 2\pi i \sin t.$$

**Example 20.22.** *Evaluate*

(*i*) $\oint_C \frac{\sin^2 z}{(z - \pi/6)^3} dz$, *where C is the circle* $|z| = 1$ *(JNTUK, 2023)*

(*ii*) $\oint_C \frac{e^{2z}}{(z+i)^4} dz$, *where C is the circle* $|z| = 3$

(*iii*) $\oint_C \frac{e^z}{(z^2+\pi^2)^2} dz$, *where C is* $|z| = 4$. *(JNTUK, 2023)*

**Solution.** (*i*) $f(z) = \sin^2 z$ is analytic inside the circle $C$: $|z| = 1$ and the point $a = \pi/6$ (= 0.5 approx.) lies within $C$.

$\therefore$ by Cauchy's integral formula $f''(a) = \frac{2!}{2\pi i}\oint_C \frac{f(z)}{(z-a)^3} dz$,

we get
$$\oint_C \frac{\sin^2 z}{(z-\pi/6)^3} dz = \pi i \left[\frac{d^2}{dz^2}(\sin^2 z)\right]_{z=\pi/6}$$

$$= \pi i (2\cos 2z)_{z=\pi/6} = 2\pi i \cos \pi/3 = \pi i.$$

(*ii*) $f(z) = e^{2z}$ is analytic within the circle $C$: $|z| = 3$. Also $z = -1$ lies inside $C$.

$\therefore$ By Cauchy's integral formula: $f'''(a) = \frac{3!}{2\pi i}\int_C \frac{f(z)dz}{(z-a)^4}$

we get
$$\int_C \frac{e^{2z}}{(z+1)^4} dz = \frac{2\pi i}{6}\left|\frac{d^3(e^{2z})}{dz^3}\right|_{z=-1} = \frac{\pi i}{3}[8e^{2z}]_{z=-1} = \frac{8\pi i}{3} e^{-2}$$

(*iii*) $\frac{e^z}{(z^2+\pi^2)^2} = \frac{e^z}{(z+\pi i)^2 (z-\pi i)^2}$ is not analytic at $z = \pm \pi i$.

However both $z = \pm \pi i$ lie within the circle $|z| = 4$.

Now
$$\frac{1}{(z+\pi i)^2(z-\pi i)^2} = \frac{A}{z+\pi i} + \frac{B}{(z+\pi i)^2} + \frac{C}{z-\pi i} + \frac{D}{(z-\pi i)^2}$$

where $A = 7/2\pi^3 i$, $C = -7/2\pi^3 i$, $B = D = -1/4\pi^2$

$$\therefore \quad \int_C \frac{e^z}{(z^2+\pi^2)^2} dz = \frac{7}{2\pi^3 i}\left\{\int_C \frac{e^z}{z+\pi i} dz - \int_C \frac{e^z}{z-\pi i} dz\right\} - \frac{1}{4\pi^2}\left\{\int_C \frac{e^z}{(z+\pi i)^2} dz + \int \frac{e^z}{(z-\pi i)^2} dz\right\}$$

$$= \frac{7}{2\pi^3 i}[2\pi i f(-\pi i) - 2\pi i f(\pi i)] - \frac{1}{4\pi^2}[2\pi i f'(-\pi i) + 2\pi i f'(\pi i)] \qquad \text{where } f(z) = e^z.$$

$$= \frac{7}{\pi^2}(e^{-\pi i} - e^{\pi i}) - \frac{i}{2\pi}(e^{-\pi i} + e^{\pi i}) = -\frac{14i}{\pi^2}\left(\frac{e^{\pi i} - e^{-\pi i}}{2i}\right) - \frac{i}{\pi}\left(\frac{e^{\pi i} + e^{-\pi i}}{2}\right) \qquad [\S\ 19.9]$$

$$= -\frac{14i}{\pi^2}\sin \pi - \frac{i}{\pi}\cos \pi = \frac{i}{\pi}.$$

**Example 20.23.** *Verify Cauchy's theorem by integrating $e^{iz}$ along the boundary of the triangle with the verticles at the points $1 + i$, $-1 + i$ and $-1 - i$.* (JNTUK, 2023; U.P.T.U., 2012)

**Solution.** The boundary of the given triangle consists of three lines $AB$, $BC$, $CA$. (Fig. 20.19).

$$\oint_{ABC} e^{iz}dz = \int_{AB} e^{iz}dz + \int_{BC} e^{iz}dz + \int_{CA} e^{iz}dz$$

Now $$\int_{AB} e^{iz}dz = \int_1^{-1} e^{i(x+i)}dx \quad \left| \begin{array}{l} \because \text{Along } AB : y = 1 \\ \therefore z = x + i \text{ and } dz = dx \end{array} \right.$$

$$= \int_1^{-1} e^{(ix-1)}\,dx = \left| \frac{e^{ix-1}}{i} \right|_1^{-1} = \frac{e^{-i-1} - e^{i-1}}{i}$$

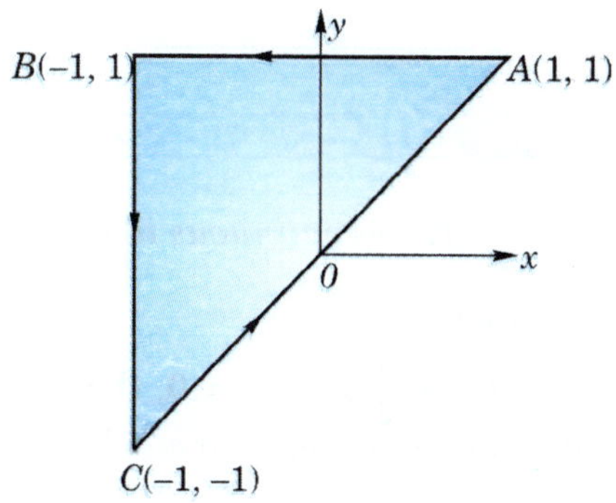

Fig. 20.19

$$\int_{BC} e^{iz}dz = \int_1^{-1} e^{i(-1+1iy)}\,idy$$

$$\left| \begin{array}{l} \because \text{Along } BC : x = -1 \\ \therefore z = -1 + iy,\ dz = idy \end{array} \right.$$

$$= i\int_1^{-1} e^{-i-y}\,dy = i\left| \frac{e^{-i-y}}{-1} \right|_1^{-1} = \frac{e^{-i+1} - e^{-i-1}}{i}$$

$$\int_{CA} e^{iz}dz = \int_{-1}^{1} e^{i(1+i)x}\,(1+i)dx \quad \left| \begin{array}{l} \because \text{Along } CA : y = x \\ \therefore z = (1 + i, x,\ dz = (1+i)\,dx \end{array} \right.$$

$$= (1+i)\,\frac{e^{(i-1)} - e^{-(i-1)}}{i(1+i)} = \frac{e^{i-1} - e^{-i+1}}{i}$$

Thus from Eqn. (*i*) $$\oint_{ABC} e^{iz}\,dz = \frac{e^{-i-1} - e^{i-1}}{i} + \frac{e^{-i+1} - e^{-i-1}}{i} + \frac{e^{i-1} - e^{-i+1}}{i} = 0 \qquad \ldots(ii)$$

Also since $f(x) = e^{iz}$ is analytic everywhere,

$\therefore$ by Cauchy's theorem $\oint f(z) = 0$ ...(*iii*)

Hence from Eqns. (*ii*) and (*iii*), $\oint_{ABC}$ Cauchy's theorem is verified.

**Example 20.24.** *If $F(\zeta) = \oint_C \dfrac{4z^2 + z + 5}{z - \zeta}\,dz$, where $C$ is the ellipse $(x/2)^2 + (y/3)^2 = 1$, find the value of (a) $F(3.5)$; (b) $F(i)$, $F''(-1)$ and $F''(-i)$.* (JNTUK, 2023)

**Solution.** (*a*) Since $\zeta = 3.5$ is the only singular point of $(4z^2 + z + 5)/(z - 3.5)$ and it lies outside the ellipse $C$, therefore, $(4z^2 + z + 5)/(z - 3.5)$ is analytic everywhere within $C$.

Hence by Cauchy's theorem,

$$\oint_C \frac{4z^2 + z + 5}{z - 3.5}\,dz = 0, \textit{ i.e., } F(3.5) = 0.$$

(*b*) Since $f(z) = 4z^2 + z + 5$ is analytic within $C$ and $\zeta = i$, $-1$ and $-i$ all lie within $C$, therefore, by Cauchy's integral formula

$$f(\zeta) = \frac{1}{2\pi i}\oint_C \frac{f(z)}{z - \zeta}\,dz$$

*i.e.,* $$\oint_C \frac{4z^2+z+5}{z-\zeta}dz = 2\pi i(4\zeta^2+\zeta+5)$$

*i.e.,* $$F(\zeta) = 2\pi i(4\zeta^2+\zeta+5)$$

$\therefore$ $$F'(\zeta) = 2\pi i\,(8\zeta+1) \text{ and } F''(\zeta) = 16\pi i$$

Thus $$F(i) = 2\pi i(-4+i+5) = 2\pi(i-1)$$

$$F'(-1) = 2\pi i[8(-1)+1] = -14\pi i \text{ and } F''(-i) = 16\pi i.$$

## 20.15 (1) CONVERSE OF CAUCHY'S THEOREM: MORERA'S THEOREM*

*If $f(z)$ is continuous in a region D and $\oint_C f(z)\,dz = 0$ around every simple closed curve C in D, then $f(z)$ is analytic in D.*

Since $\oint f(z)\,dz = 0$, then the line integral of $f(z)$ from a fixed point $z_0$ to a variable point $z$ must be independent of the path and hence must be a function of $z$ only. Thus

$$\int_{z_0}^{z} f(z)dz = \phi(z), \text{ (say)},$$

Let $$\phi(z) = U + iV \text{ and } f(z) = u + iv$$

Then $$U + iV = \int_{(x_0,\,y_0)}^{(x,y)} (u+iv)\,(dx+idy) = \int_{(x_0,\,y_0)}^{(x,y)} (udx - vdy) + i\int_{(x_0,\,y_0)}^{(x,y)} (vdx + udy)$$

$\therefore$ $$U = \int_{(x_0,\,y_0)}^{(x,y)} (udx - vdy),\ V = \int_{(x_0,\,y_0)}^{(x,y)} (vdx + udy)$$

Differentiating under the integral sign,

$$\frac{\partial U}{\partial x} = u,\ \frac{\partial U}{\partial y} = -v,\ \frac{\partial V}{\partial x} = v,\ \frac{\partial V}{\partial y} = u \quad \therefore \quad \frac{\partial U}{\partial x} = \frac{\partial V}{\partial y},\ \frac{\partial U}{\partial y} = -\frac{\partial V}{\partial x}$$

Thus $U$ and $V$ satisfy C-R equations.

Also, since $f(z)$ is given to be continuous, $u$ and $v$ and therefore, $\partial U/\partial x$, $\partial U/\partial y$, $\partial V/\partial x$, $\partial V/\partial y$, are also continuous.

$\therefore$ $\phi(z)$ is an analytic function and

$$\phi'(z) = \frac{\partial U}{\partial x} + i\frac{\partial V}{\partial x} = u + iv = f(z).$$

Thus, $f(z)$ is the derivative of an analytic function $\phi(z)$. Hence $f(z)$ is analytic by § 20.14 Cor.

(2) **Cauchy's inequality.** *If $f(z)$ is analytic within and on the circle C: $|z-a| = r$, then*

$$|f^n(a)| \le \frac{Mn!}{r^n} \qquad \text{...(I)}$$

*where M is the maximum value of $|f(z)|$ on C.*

From Eqn. (5) of § 20.14, we get

$$|f^n(a)| = \frac{n!}{2\pi}\left|\oint_C \frac{f(z)dz}{(z-a)^{n+1}}\right|$$

$$\le \frac{n!}{2\pi}\cdot\frac{M}{r^{n+1}}\oint_C |z| \qquad [\because\ |f(z)| < M]$$

$$= \frac{n!M}{2\pi r^{n+1}}\oint_C ds = \frac{Mn!}{2\pi r^{n+1}}\,2\pi r = \frac{Mn!}{r^n}$$

(3) **Liouville's theorem**[$]. *If $f(z)$ is analytic and bounded for all $z$ in the entire complex plane, then $f(z)$ is a constant.*

---

* Named after the Italian mathematician, *Giacinto Morera* (1856–1909) who worked in Turin.

$ See footnote p. 677.

Taking $n = 1$ and replacing $a$ by $z$, Eqn. (I) gives

$$|f'(z)| \le M/r$$

As $r \to \infty$, it gives $f'(z) = 0$ *i.e.*, $f(z)$ is constant for all $z$.

**(4) Poisson's integral formulae.** *If $f(z)$ is analytic within and on the circle C: $|z| = \rho$ and $z = re^{i\theta}$ is any point within C, then*

$$f(re^{i\theta}) = \frac{1}{2\pi}\int_0^{2\pi} \frac{\rho^2 - r^2}{\rho^2 - 2r\rho\cos(\theta - \phi) + r^2} f(re^{i\phi})d\phi$$

By Cauchy's integral formula, $f(z) = \dfrac{1}{2\pi i}\oint_C \dfrac{f(w)}{w - z}\, dw$ ...(1)

As the inverse of the point $x$ w.r.t. $C$ lies outside $C$ and is given by $\rho^2/\bar{z}$.

$\therefore$ by Cauchy's theorem,

$$0 = \frac{1}{2\pi i}\int \frac{f(w)}{w - \rho^2/\bar{z}}\, dw \quad ...(2)$$

Subtracting (2) from (1), $f(z) = \dfrac{1}{2\pi i}\displaystyle\int \left(\frac{1}{w - z} - \frac{1}{w - \rho^2/\bar{z}}\right) f(w)\, dw$

$$= \frac{1}{2\pi i}\,\%_C \frac{z\bar{z} - \rho^2}{\bar{z}w^2 - (z\bar{z} + \rho^2)\, w + z\rho^2)} f(w)\, dw \quad ...(3)$$

Taking $w = \rho e^{i\phi}$ and noting that $\bar{z} = re^{-i\theta}$, (3) gives

$$f(re^{i\theta}) = \frac{1}{2\pi i}\int_0^{2\pi} \frac{(r^2 - \rho^2)\, f(\rho e^{i\phi})\,.\, \rho i e^{i\phi}\, d\phi}{re^{-i\theta}\,.\,\rho^2 e^{2i\phi} - (r^2 + \rho^2)\,\rho e^{i\phi} + re^{i\theta}\rho^2}$$

$$= \frac{1}{2\pi}\int_0^{2\pi} \frac{(\rho^2 - r^2)\, f(\rho e^{i\phi})d\phi}{\rho^2 + r^2 - r\rho[e^{i(\theta-\phi)} + e^{-i(\theta-\phi)}]} = \frac{1}{2\pi}\int_0^{2\pi} \frac{(\rho^2 - r^2)\, f(\rho e^{i\phi})d\phi}{\rho^2 - 2r\rho\cos(\theta - \phi) + r^2} \quad ...(4)$$

This is called *Poisson's integral formula** for a circle. It expresses the values of a harmonic function within a circle in terms of its values on the boundary.

Writing $f(re^{i\theta}) = u(r, \theta) + iv(r, \theta)$ and $f(\rho e^{i\phi}) = u(\rho, \phi) + iv(\rho, \phi)$ in Eqn. (4) and equating real and imaginary parts from both sides, we get the formulae:

$$u(r, \theta) = \int_0^{2\pi} \frac{(\rho^2 - r^2)u(\rho, \phi)d\phi}{\rho^2 - 2r\rho\cos(\theta - \phi) + r^2} \quad ...(5)$$

and

$$v(r, \theta) = \int_0^{2\pi} \frac{(\rho^2 - r^2)v(\rho, \phi)d\phi}{\rho^2 - 2r\rho\cos(\theta - \phi) + r^2} \quad ...(6)$$

## PROBLEMS 20.6

1. Evaluate $\oint_C (z - a)^{-1} dz$, where $C$ is a simple closed curve and the point $z = a$ is (*i*) outside $C$, (*Calicut, 2012*)
(*ii*) inside $C$.
2. Evaluate $\oint_C \dfrac{dz}{(z - a)^n}$, $n = 2, 3, 4, \ldots$, where $C$ is a closed curve containing the point $z = a$.
3. Evaluate (*i*) $\oint_C \dfrac{e^z}{z^2 + 1} dz$, where $C$ is the circle $|z| = 1/2$.
(*ii*) $\oint_C \dfrac{e^{3iz}}{(z + \pi)^3} dz$, where $C$ is the circle $|z - \pi| = 3$.

* Named after the French mathematician *Simeon Denis Poisson* (1781–1840) who was a professor in Paris and made contributions to partial differential equations, potential theory and probability.

4. Use Cauchy's integral formula to calculate:

(i) $\oint_C \frac{e^{2z}}{(z+1)(z-2)}$, where, $C$ is $|z| = 3$. (*V.T.U., 2016*) (ii) $\oint_C \frac{z^2+1}{z(2z+1)} dz$, where, $C$ is $|z| = 1$. (*D.T.U., 2011*)

(iii) $\oint_C \frac{\sin \pi z + \cos \pi z}{(z-1)(z-2)} dz$ where $C$ is $|z| = 4$. (*JNTUK, 2023*)

5. Evaluate (i) $\oint_C \frac{z^3 - 2z + 1}{(z-i)^2}$ where $C$ is $|z| = 2$.

(ii) $\oint_C \frac{e^{-z}}{(z-1)(z-2)^2} dz$ where $C$ is $|z| = 3$.

(iii) $\oint_C \frac{e^{2z}}{(z+1)^4} dz$ where $C$ is $|z| = 2$. (*Anna., 2016*)

6. Evaluate, using Cauchy's integral formulae:

(i) $\oint_C \frac{z}{z^2 - 3z + 2} dz$, where $C$ is $|z-2| = \frac{1}{2}$.

(ii) $\oint_C \frac{e^z dz}{(z+1)^2}$, where $C$ is $|z-1| = 3$. (*Andhra, 2021*)

(iii) $\oint_C \frac{\log z}{(z-1)^3} dz$ where $C$ is $|z-1| = \frac{1}{2}$. (*JNTUK, 2023; Andhra, 2022*)

7. Evaluate $f(2)$ and $f(3)$ where $f(a) = \oint_C \frac{2z^2 - z - 2}{z - a} dz$ and $C$ is the circle $|z| = 2.5$.

8. If $\phi(\zeta) = \oint_C \frac{3z^2 + 7z + 1}{z - \zeta} dz$, where $C$ is the circle $|z| = 2$ find the values of

(i) $\phi(3)$, (ii) $\phi'(1-i)$ (iii) $\phi''(1-i)$.

9. Evaluate $\oint_C \frac{z^3 + z + 1}{z^2 - 7z + 2} dz$, where $C$ is the ellipse $4x^2 + 9y^2 = 1$.

10. Verify Cauchy's theorem for the integral of $z^3$ taken over the boundary of the (i) rectangle with vertices $-1, 1, 1+i, -1+i$; (ii) triangle with vertices (1, 2), (1, 4), (3, 2).

## 20.16 (1) SERIES OF COMPLEX TERMS

Let $$(a_1 + ib_1) + (a_2 + ib_2) + \ldots + (a_n + ib_n) + \ldots \quad \ldots(1)$$

be an infinite series of complex terms ; $a$'s and $b$'s being real numbers. *If the series $\Sigma a_n$ and $\Sigma b_n$ converge to the sums A and B, then series (1) is said to* **converge** *to the sum $A + iB$.* Also if (1) is a convergent series, then $\underset{n\to\infty}{\text{Lt}} (a_n + ib_n) = 0$.

*The series (1) is said to be* **absolutely convergent** *if the series*

$$|a_1 + ib_1| + |a_2 + ib_2| + \ldots + |a_n + ib_n| + \ldots$$

*is convergent.* Since $|a_n|$ and $|b_n|$ are both $\le |a_n + ib_n|$, it follows *that an absolutely convergent series is convergent.*

Next let the series of functions $u_1(z) + u_2(z) + \ldots + u_n(z) + \ldots\ldots$ ...(2)

converge to the sum $S(z)$ and $S_n(z)$ be the sum of its first $n$ terms. Then *the series (2) is said to be* **uniformly convergent** *in a region R, if corresponding to any positive number ε, there exists a positive number N, depending on ε, but not on z, such that for every z in R.*

$$|S(z) - S_n(z)| < \varepsilon \text{ for } n > N.$$

As in the case of real series **Weirstrass's M-test** holds for series of complex terms. So *the series (2) is uniformly convergent in a region R if there is a convergent series of positive constants $\Sigma M_n$ such that $|u_n(z)| \le M_n$ for all z in R.*

Also a *uniformly convergent series of continuous complex functions is itself continuous and can be integrated term by term.*

**Obs.** *If a power series $\Sigma a_n z^n$ converges for $z = z_1$, then it converges absolutely for $|z| < |z_1|$.*

Since $\Sigma a_n z_1^n$ converges, therefore, $\underset{n\to\infty}{\text{Lt}}\ a_1 z_1^n = 0$ and so we can find a number $k$ such that $|a_n z_1^n| < k$ for all $n$. Then $\Sigma a_n z^n = \Sigma |a_n z_1^n| \,.\, |z/z_1|^n < \Sigma k t^n$ where $t = |z/z_1|$.

But the series $\Sigma t^n$ converges for $t < 1$. Hence the series $\Sigma a_n z^n$ converges absolutely for $|z| < |z_1|$, *i.e.*, if a circle with centre at the origin and radius $|z|$ be drawn, then the given series converges absolutely at all points inside the circle.

Such a circle $|z| = R$ within which series $\Sigma a_n z^n$ converges, is called the *circle of convergence and R* is called the *radius of convergence*.

*A power series is uniformly convergent in any region which lies entirely within its circle of convergence.*

**(2) Taylor's series.** *If $f(z)$ is analytic inside a circle C with centre at a, then for z inside C,*

$$f(z) = f(a) + f'(a)(z-a) + \frac{f''(a)}{2!}(z-a)^2 + \ldots + \frac{f^n(a)}{n!}(z-a)^n + \ldots \qquad \ldots(i)$$

**Proof.** Let $z$ be any point inside $C$. Draw a circle $C_1$ with centre at $a$ enclosing $z$ (Fig. 20.20). Let $t$ be a point on $C_1$. We have

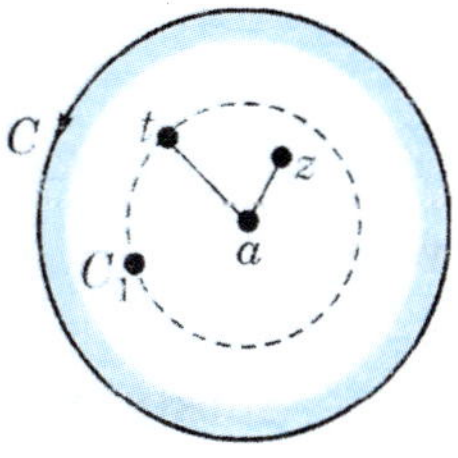

Fig. 20.20

$$\frac{1}{t-z} = \frac{1}{t-a-(z-a)} = \frac{1}{t-a}\left(1 - \frac{z-a}{t-a}\right)^{-1}$$

$$= \frac{1}{t-a}\left[1 + \frac{z-a}{t-a} + \left(\frac{z-a}{t-a}\right) + \ldots + \left(\frac{z-a}{t-a}\right)^n + \ldots\right] \qquad \ldots(ii)$$

As $|z-a| < |t-a|$, *i.e.* $|(z-a)/(t-a)| < 1$, this series converges uniformly. So, multiplying both sides of Eqn. (*ii*) by $f(t)$, we can integrate over $C_1$.

$$\therefore \quad \oint_{C_1} \frac{f(t)}{t-z}dz = \oint_{C_1} \frac{f(t)}{t-a}dz + (z-a)\oint_{C_1} \frac{f(t)}{(t-a)^2}dt + \ldots + (z-a)^n \,.\, \oint_{C_1} \frac{f(t)}{(t-a)^{n+1}}dt + \ldots \qquad \ldots(iii)$$

Since $f(t)$ is analytic on and inside $C_1$, therefore, applying the formulae (2) to (5) of p. 815 (*iii*), we get Eqn. (*i*) which is known as *Taylor's series*.

**Obs.** Another remarkable fact is that *complex analytic functions can always be represented by power series of the form* (*i*).

**(3) Laurent's series†.** *If f(z) is analytic in the ring-shaped region R bounded by two concentric circles C and $C_1$ of radii r and $r_1$ $(r > r_1)$ and with centre at a, then for all z in R*

$$f(z) = a_0 + a_1(z-a) + a_2(z-a)^2 + \ldots + a_{-1}(z-a)^{-1} + a_{-2}(z-a)^{-2} + \ldots$$

where $a_n = \dfrac{1}{2\pi i}\oint \dfrac{f(t)}{(t-a)^{n+1}}\,dt,$

$\Gamma$ *being any curve in R, encircling $C_1$* (as in Fig. 20.21).

**Proof.** Introduce cross-out $AB$, then $f(z)$ is analytic in the region $D$ bounded by $AB$, $C_1$ described clockwise, $BA$ and $C$ described anti-clockwise (see Fig. 20.17). Then if $z$ be any point in $D$, we have

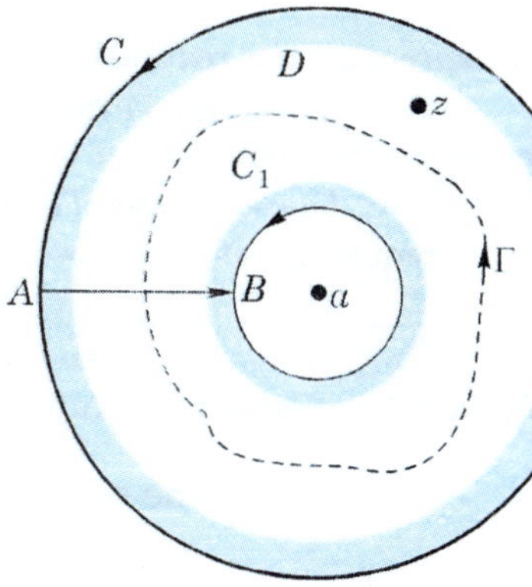

Fig. 20.21

$$f(z) = \frac{1}{2\pi i}\left[\int_{AB}\frac{f(t)}{t-z}dt + \oint_{C_1}\frac{f(t)}{t-z}dt + \int_{BA}\frac{f(t)}{t-z}dt + \oint_{C_1}\frac{f(t)}{t-z}dt\right]$$

$$= \frac{1}{2\pi i}\left[\oint_{C}\frac{f(t)}{t-z}dt - \oint_{C_1}\frac{f(t)}{t-z}dt\right] \qquad \ldots(i)$$

where both $C$ and $C_1$ are described anti-clockwise in Eqn. (*i*) and integrals along $AB$ and $BA$ cancel (Fig. 20.21).

For the first integral in Eqn. (*i*), expanding $1/(t-z)$ as in §20.16(2), we get

$$\frac{1}{2\pi i}\oint_C \frac{f(t)}{t-z}dt = \sum_{n=1}^{\infty}\frac{(z-a)^n}{2\pi i}\oint_C \frac{f(t)}{(t-a)^{n+1}}dt$$

† Named after the French engineer and mathematician *Pierre Alphonse Laurent* (1813–1854) who published this theorem in 1843.

$$= \Sigma a_n (z-a)^n \text{ where } a_n = \frac{1}{2\pi i}\oint_C \frac{f(t)}{(t-a)^{n+1}}dt \qquad ...(ii)$$

For the second integral in Eqn. (*i*), let $t$ lie on $C_1$. Then we write

$$\frac{1}{t-z} = \frac{1}{(t-a)-(z-a)} = -\frac{1}{z-a}\left(1-\frac{t-a}{z-a}\right)^{-1}$$

$$= -\frac{1}{z-a}\left[1+\frac{t-a}{z-a}+\left(\frac{t-a}{z-a}\right)^2+\ldots+\left(\frac{t-a}{z-a}\right)^{n-1}+\ldots\right]$$

As $|t-a| < |z-a|$, *i.e.*, $|(t-a)/(z-a)| < 1$, this series converges uniformly. So multiplying both sides by $f(t)$ and integrating over $C_1$, we get

$$-\frac{1}{2\pi i}\oint_C \frac{f(t)}{t-z}dt = \sum_{n=1}^{\infty}\frac{1}{(z-a)^n}\cdot\frac{1}{2\pi i}\oint_C (t-a)^{n-1} f(t)dt = \sum_{n=1}^{\infty} a_{-n}(z-a)^{-n} \qquad ...(iii)$$

where $$a_{-n} = \frac{1}{2\pi i}\oint_{C_1}\frac{f(t)}{(t-a)^{-n+1}}dt$$

Substituting from Eqns. (*ii*) and (*iii*) in Eqn. (*i*), we get

$$f(z) = \sum_{n=0}^{\infty} a_n (z-a)^n + \sum_{n=1}^{\infty} a_{-n}(z-a)^{-n}. \qquad ...(iv)$$

Now $f(t)/(t-a)^{n+1}$ being analytic in the region between $C$ and $\Gamma$, we can take the integral giving $a_n$ over $\Gamma$. Similarly we can take the integral giving $a_{-n}$ over $\Gamma$. Hence Eqn. (*iv*) can be written as

$$f(z) = \sum_{-\infty}^{\infty} a_n (z-a)^n \text{ where } a_n = \frac{1}{2\pi i}\oint_\Gamma \frac{f(t)}{(t-a)^{n+1}}dt$$

which is known as *Laurent's series.*

**Obs. 1.** *As $f(z)$ is not given to be analytic inside $\Gamma$,* $a_n = \frac{1}{2\pi i}\oint_\Gamma \frac{f(t)}{(t-a)^{n+1}}dt \neq \frac{f^n(a)}{n!}$

However, *if $f(z)$ is analytic inside $\Gamma$,* then $a_{-n} = 0$ ; $a_n = \frac{1}{2\pi i}\oint_\Gamma \frac{f(t)}{(t-a)^{n+1}}dt = \frac{f^n(a)}{n!}$

and *Laurent's series reduces to Taylor's series.*

**Obs. 2.** *To obtain Taylor's or Laurent's series, simply expand f(z) by binomial theorem* instead of finding $a_n$ by complex integration which is quite complicated.

**Obs. 3.** *Laurent series of a given analytic function f(z) in its annulus of convergence is unique.* There may be different Laurent series of $f(z)$ in two annuli with the same centre.

**Example 20.25.** *Show that the series $z(1-z) + z^2(1-z) + z^3(1-z) + \ldots\infty$ converges for $|z| < 1$. Determine whether it converges absolutely or not.*

**Solution.** Let the sum of the first $n$ terms of the series be $s_n$, so that

$$s_n = z - z^2 + z^2 - z^3 + z^3 - z^4 + \ldots + z^n - z^{n+1} = z - z^{n+1}$$

For $|z| < 1$, $z^{n+1} \to 0$ as $n \to \infty$.

$\therefore$ $\underset{n\to\infty}{\text{Lt}}(s_n) = z$, *i.e.*, the given series converges for $|z| < 1$.

$$|s_n(z)| = |z(1-z)| + |z^2(1-z)| + \ldots + |z^n(1-z)|$$
$$= |1-z|\{|z| + |z|^2 + |z|^3 + \ldots + |z|^n\}$$

For $|z| < 1$, $\underset{n\to\infty}{\text{Lt}}|s_n(z)| = |1-z|\frac{|z|}{1-|z|}$ [G.P.]

Hence the given series converges absolutely.

**Example 20.26.** *Expand sin z in a Taylor's series about z = 0 and determine the region of convergence.*

**Solution.** Given $f(z) = \sin z, f'(z) = \cos \theta, f''(z) = -\sin z, f''(z) = -\cos z, \ldots$

$\therefore$ $f(0) = 0, f'(0) = 1, f''(0) = 0, f'''(0) = 1$

By Taylor's series about $z = 0$, we have

$$f(z) = f(0) + \frac{(z-0)}{1!} f'(0) + \frac{(z-0)^2}{2!} f''(0) + \frac{(z-0)^3}{3!} f'''(0) + \ldots$$

*i.e.,*
$$\sin z = z - \frac{z^3}{3!} + \frac{z^5}{5!} - \ldots + (-1)^{n-1} \frac{z^{2n} - 1}{(2n-1)!} + \ldots$$

Hence
$$\sin z = \sum_{n=1}^{\infty} a_n (z-0)^{2n-1} \text{ where } a_n = \frac{(-1)^{n-1}}{(2n-1)!}$$

Since
$$\rho = \underset{n\to\infty}{\text{Lt}} \left| \frac{a_{n+1}}{a_n} \right| = \underset{n\to\infty}{\text{Lt}} \left| \frac{(2n-1)!}{(2n+1)!} \right| = 0$$

Thus the radius of convergence of $f(z) = 1/\rho = \infty$

*i.e.*, the region of convergence of $f(z)$ is all reals.

**Example 20.27.** *Find Taylor's expansion of* (JNTUK, 2022)

(*i*) $f(z) = \dfrac{1}{(z+1)^2}$ *about the point z = –i.*

(*ii*) $f(z) = \dfrac{2z^3+1}{z^2+z}$ *about the point z = i.*

**Solution.** (*i*) To expand $f(z)$ about $z = -i$, *i.e.*, in powers of $z + i$, put $z + i = t$. Then

$$f(z) = \frac{1}{(t-i+1)^2} = (1-i)^{-2}\,[1 + t/(1-i)]^{-2} = \frac{i}{2}\left[1 - \frac{2t}{1-i} + \frac{3t^2}{(1-i)^2} - \frac{4t^3}{(1-i)^3} + \ldots\right]$$

[Expanding by Binomial theorem]

$$= \frac{i}{2}\left[1 + \sum_{n=1}^{\infty} (-1)^n \frac{(n+1)(z+i)^n}{(1-i)^n}\right]$$

(*ii*)
$$f(z) = \frac{2z^3+1}{z(z+1)} = 2z - 2 + \frac{2z+1}{z(z+1)} = (2i-2) + 2(z-i) + \frac{1}{z} + \frac{1}{z+1} \quad \ldots(i)$$

[By partial fractions]

To expand $1/z$ and $1/(z + 1)$ about $z = i$, put $z - i = t$, so that

$$\frac{1}{z} = \frac{1}{(t+i)} = \frac{1}{i}\left(1 + \frac{t}{i}\right)^{-1}$$ [Expanding by Binomial theorem]

$$= \frac{1}{i}\left[1 - \frac{t}{i} + \frac{t^2}{i^2} - \frac{t^3}{i^3} + \frac{t^4}{i^4} - \ldots \infty\right] = \frac{1}{i} + \frac{t}{1} + \frac{t^2}{i^3} - \frac{t^3}{i^4} + \frac{t^4}{i^5} - \ldots \infty$$

$$= -i + (z-i) + \sum_{n=2}^{\infty} (-1)^n \frac{(z-i)^n}{i^{n+1}} \quad \ldots(ii)$$

and
$$\frac{1}{z+1} = \frac{1}{t+i+1} = \frac{1}{1+i}\left(1 + \frac{t}{1+i}\right)^{-1}$$ [Expanding by Binomial theorem]

$$= \frac{1}{1+i}\left[1 - \frac{t}{1+i} + \frac{t^2}{(1+t)^2} - \frac{t^3}{(1+i)^3} + \frac{t^4}{(1+i)^4} - \ldots \infty\right]$$

$$= \frac{1-i}{2} - \frac{t}{2i} + \left[\frac{t^2}{(1+i)^3} - \frac{t^3}{(1+i)^4} + \frac{t^4}{(1+i)^5} - \ldots \infty\right] = \frac{1}{2} - \frac{i}{2} - \frac{z-i}{2i} + \sum_{n=2}^{\infty} (-1)^n \frac{(z-i)^n}{(1+i)^{n+1}} \quad \ldots(iii)$$

Substituting from (*ii*) and (*iii*) in (*i*), we get

$$f(z) = \left(2i - 2 - i + \frac{1}{2} - \frac{i}{2}\right) + \left(2 + 1 - \frac{1}{2i}\right)(z - i) + \sum_{n=2}^{\infty} (-1)^n \left\{\frac{1}{i^{n+1}} + \frac{1}{(1+i)^{n+1}}\right\}(z-i)^n$$

$$= \left(\frac{i}{2} - \frac{3}{2}\right) + \left(3 + \frac{i}{2}\right)(z - i) + \sum_{n=2}^{\infty} (-1)^n \left\{\frac{1}{i^{n+1}} + \frac{1}{(1+i)^{n+1}}\right\}(z-i)^n.$$

**Example 20.28.** *Expand $f(z) = 1/[(z-1)(z-2)]$ in the region:*

*(a) $|z| < 1$, (W.B.T.U., 2015)* *(b) $1 < |z| < 2$,* *(JNTUK, 2023; Andhra., 2015)*

*(c) $|z| > 2$, (Hazaribag., 2015)* *(d) $0 < |z - 1| < 1$.* *(U.P.T.U., 2010 ; V.T.U., 2010)*

**Solution.** (*a*) By partial fractions $\dfrac{1}{(z-1)(z-2)} = \dfrac{1}{z-2} - \dfrac{1}{z-1}$ ...(*i*)

$$= -\frac{1}{2}\left(1 - \frac{z}{2}\right)^{-1} + (1 - z)^{-1} \quad ...(ii)$$

*For $|z| < 1$,* both $|z/2|$ and $|z|$ are less than 1. Hence Eqn. (*ii*) gives on expansion

$$f(z) = -\frac{1}{2}\left(1 + \frac{z}{2} + \frac{z^2}{4} + \frac{z^3}{8} + ...\right) + (1 + z + z^2 + z^3 + ...)$$

$$= \frac{1}{2} + \frac{3}{4}z + \frac{7}{8}z^2 + \frac{15}{16}z^3 + ...$$ which is a *Taylor's series.*

(*b*) *For $1 < |z| < 2$,* we write Eqn. (*i*) as

$$f(z) = -\frac{1}{2}\frac{1}{(1 - z/2)} - \frac{1}{z(1 - z^{-1})} \quad ...(iii)$$

and notice that both $|z/2|$ and $|z^{-1}|$ are less than 1. Hence Eqn. (*iii*) gives on expansion

$$f(z) = -\frac{1}{2}\left(1 + \frac{z}{2} + \frac{z^2}{4} + \frac{z^3}{8} + ...\right) - \frac{1}{z}(1 + z^{-1} + z^{-2} + z^{-3} + ...)$$

$$= ... - z^{-4} - z^{-3} - z^{-2} - z^{-1} - \frac{1}{2} - \frac{1}{4}z - \frac{1}{8}z^2 - \frac{1}{16}z^3 - ...$$

which is a *Laurent's series.*

(*c*) *For $|z| > 2$,* we write Eqn. (*i*) as

$$f(z) = \frac{1}{z(1 - 2z^{-1})} - \frac{1}{z(1 - z^{-1})}$$

$$= z^{-1}(1 + 2z^{-1} + 4z^{-2} + 8z^{-3} + ...) - z^{-1}(1 + z^{-1} + z^{-2} + z^{-3} + ...)$$

$$= ... + 7z^{-4} + 3z^{-3} + z^{-2} + ...$$

(*d*) *For $0 < |z - 1| < 1$,* we write Eqn. (*i*) as

$$f(z) = \frac{1}{(z-1) - 1} - \frac{1}{z - 1}$$

$$= -(z - 1)^{-1} - [1 - (z - 1)]^{-1}$$

$$= -(z - 1)^{-1} - [1 + (z - 1) + (z - 1)^2 + (z - 1)^3 + ...].$$

**Example 20.29.** *Find the Laurents' expansion of $f(z) = \dfrac{7z - 2}{(z + 1)z(z - 2)}$ in the region $1 < z + 1 < 3$.* *(JNTUK, 2022; Andhra, 2022)*

**Solution.** Writing $z + 1 = u$, we have

$$f(z) = \frac{7(u-1) - 2}{u(u-1)(u-1-2)} = \frac{7u - 9}{u(u-1)(u-3)}$$

$$= -\frac{3}{u} + \frac{1}{u-1} + \frac{2}{u-3}$$ (splitting into partial fraction)

$$= -\frac{3}{u} + \frac{1}{u(1 - 1/u)} - \frac{2}{3(1 - u/3)} = -\frac{3}{u} + \frac{1}{u}\left(1 - \frac{1}{u}\right)^{-1} - \frac{2}{3}\left(1 - \frac{u}{3}\right)^{-1}$$

Since $1 < u < 3$ or $1/u < 1$ and $u/3 < 1$, expanding by Binomial theorem,

$$f(z) = \frac{-3}{u} + \frac{1}{u}\left(1 + \frac{1}{u} + \frac{1}{u^2} + \frac{1}{u^3} + \dots \infty\right) - \frac{2}{3}\left(1 + \frac{u}{3} + \frac{u^2}{3^2} + \frac{u^3}{3^3} + \dots \infty\right)$$

$$= -\frac{2}{u} + \frac{1}{u^2} + \frac{1}{u^3} + \dots \infty - \frac{2}{3}\left(1 + \frac{u}{3} + \frac{u^2}{3^2} + \frac{u^3}{3^3} + \dots \infty\right)$$

Hence $f(z) = -\frac{2}{z+1} + \frac{1}{(z+1)^2} + \frac{1}{(z+1)^3} + \dots \infty - \frac{2}{3}\left[1 + \frac{z+1}{3} + \frac{(z+1)^2}{3^2} + \frac{(z+1)^3}{3^3} + \dots \infty\right]$

which is valid in the region $1 < z + 1 < 3$.

## 20.17 (1) ZEROS OF AN ANALYTIC FUNCTION

**Def.** *A zero of an analytic function f(z) is that value of z for which f(z) = 0.*

If $f(z)$ is analytic in the neighbourhood of a point $z = a$, then by Taylor's theorem

$$f(z) = a_0 + a_1(z - a) + a_2(z - a)^2 + \dots + a_n(z - a)^n + \dots \qquad \text{where } a_n = \frac{f^n(a)}{n!}.$$

If $a_0 = a_1 = a_2 = \dots = a_{m-1} = 0$ but $a_m \neq 0$, then $f(z)$ is said to have a *zero of order m* at $z = a$.

When $m = 1$, the zero is said to be *simple*. In the neighbourhood of zero ($z = a$) of order $m$,

$$f(z) = a_m(z - a)^m + a_{m+1}(z - a)^{m+1} + \dots \infty$$
$$= (z - a)^m \phi(z) \text{ where } \phi(z) = a_m + a_{m+1}(z - a) + \dots$$

Then $\phi(z)$ is analytic and non-zero in the neighbourhood of $z = a$.

**(2) Singularities of an analytic function**

We have already defined a *singular point of a function as the point at which the function ceases to be analytic.*

(*i*) **Isolated singularity.** *If $z = a$ is a singularity of f(z) such that f(z) is analytic at each point in its neighourhood* (*i.e.*, there exists a circle with centre $a$ which has no other singularity), *then $z = a$ is called an* **isolated singularity.**

In such a case, $f(z)$ can be expanded in a Laurent's series around $z = a$, giving

$$f(z) = a_0 + a_1(z - a) + a_2(z - a)^2 + \dots + a_{-1}(z - a)^{-1} + a_{-2}(z - a)^{-2} + \dots \qquad \dots(1)$$

*For example*, $f(z) = \cot(\pi/z)$ is not analytic where $\tan(\pi/z) = 0$ *i.e.* at the points $\pi/z = 4\pi$ or $z = 1/n$ ($n = 1, 2, 3, \dots$).

Thus $z = 1, 1/2, 1/3, \dots$ are all *isolated singularities* as there is no other singularity in their neighbourhood.

But when $n$ is large, $z = 0$ is such a singularity that there are infinite number of other singularities in its neighbourhood. Thus $z = 0$ is the *non-isolated singularity* of $f(z)$.

(*ii*) **Removable singularity.** If all the negative powers of $(z - a)$ in Eqn. (1) are zero, then $f(z) = \sum_{n=0}^{\infty} a_n(z-a)^n$. Here the singularity can be removed by defining $f(z)$ at $z = a$ in such a way that it becomes analytic at $z = a$. Such a singularity is called a *removable singularity.*

Thus *if* $\underset{z\to a}{\text{Lt}} f(z)$ *exists finitely, then $z = a$ is a removable singularity.*

(*iii*) **Poles.** *If all the negative powers of $(z - a)$ in (i) after the nth are missing, then the singularity at $z = a$ is called a* **pole of order n.**

A pole of first order is called a **simple pole.**

(*iv*) **Essential singularity.** *If the number of negative powers of $(z - a)$ in Eqn. (1) is infinite, then $z = a$ is called an essential singularity.* In this case, $\underset{z\to a}{\text{Lt}} f(z)$ does not exist.

**Example 20.30.** *Find the nature and location of singularities of the following functions:*

(*i*) $\frac{z - \sin z}{z^2}$ (*ii*) $(z + 1) \sin \frac{1}{z-2}$ (*iii*) $\frac{1}{\sin z - \cos z}$. (*JNTUK, 2022; W.B.T.U., 2015*)

**Solution.** (*i*) Here $z = 0$ is a singularity.

Also $$\frac{z - \sin z}{z^2} = \frac{1}{z^2}\left\{z - \left(z - \frac{z^3}{3!} + \frac{z^5}{5!} - \frac{z^7}{7!} + \ldots\right)\right\} = \frac{z}{3!} - \frac{z^3}{5!} + \frac{z^5}{7!} - \ldots$$

Since there are no negative powers of $z$ in the expansion, $z = 0$ *is a removable singularity.*

(*ii*) $$(z + 1)\sin\frac{1}{z-2} = (t + 2 + 1)\sin\frac{1}{t} \qquad \text{where } t = z - 2$$

$$= (t + 3)\left\{\frac{1}{t} - \frac{1}{3!t^3} + \frac{1}{5!t^5} - \ldots\right\} = \left(1 - \frac{1}{3!t^2} + \frac{1}{5!t^4} - \ldots\right) + \left(\frac{3}{t} - \frac{1}{2t^3} + \frac{3}{5!t^5} - \ldots\right)$$

$$= 1 + \frac{3}{t} - \frac{1}{6t^2} - \frac{1}{2t^3} + \frac{1}{120t^4} - \ldots = 1 + \frac{3}{z-2} - \frac{1}{6(z-2)^2} - \frac{1}{2(z-2)^3} + \ldots$$

Since there are infinite number of terms in the negative powers of $(z - 2)$, $z = 2$ *is an essential singularity.*

(*iii*) Poles of $f(z) = \dfrac{1}{\sin z - \cos z}$ are given by equating the denominator to zero, *i.e.*, by $\sin z - \cos z = 0$ or $\tan z = 1$ or $z = \pi/4$. Clearly $z = \pi/4$ is a *simple pole of f(z).*

**Example 20.31.** *What type of singularity have the following functions :*

(*i*) $\dfrac{1}{1-e^z}$ (*ii*) $\dfrac{e^{2z}}{(z-1)^4}$ (*iii*) $\dfrac{e^{1/z}}{z^2}$.

**Solution.** (*i*) Poles of $f(z) = 1/(1 - e^z)$ are found by equating to zero $1 - e^z = 0$ or $e^z = 1 = e^{2n\pi i}$

$\therefore$ $$z = 2n\pi i \ (n = 0, \pm 1, \pm 2, \ldots)$$

Clearly $f(z)$ *has a simple pole at* $z = 2\pi i$.

(*ii*) $$\frac{e^{2z}}{(z-1)^4} = \frac{e^{2(t+1)}}{t^4} = \frac{e^2}{t^4}\cdot e^{2t} \qquad \text{where } t = z - 1$$

$$= \frac{e^2}{t^4}\left\{1 + \frac{2t}{1!} + \frac{(2t)^2}{2!} + \frac{(2t)^3}{3!} + \frac{(2t)^4}{4!} + \frac{(2t)^5}{5!} + \ldots\right\} = e^2\left\{\frac{1}{t^4} + \frac{2}{t^3} + \frac{2}{t^2} + \frac{4}{3t} + \frac{2}{3} + \frac{4t}{15} + \ldots\right\}$$

$$= e^2\left\{\frac{1}{(z-1)^4} + \frac{2}{(z-1)^3} + \frac{2}{(z-1)^2} + \frac{4}{3(z-1)} + \frac{2}{3} + \frac{4}{15}(z-1) + \ldots\right\}$$

Since there are finite (4) number of terms containing negative powers of $(z - 1)$,

$\therefore$ $z = 1$ *is a pole of 4th order.*

(*iii*) $$f(z) = \frac{e^{1/z}}{z^2} = \frac{1}{z^2}\left\{1 + \frac{1}{1!z} + \frac{1}{2!z^2} + \frac{1}{3!z^3} + \ldots\right\} = z^{-2} + z^{-3} + \frac{z^{-4}}{2} + \ldots \infty$$

Since there are infinite number of terms in the negative powers of $z$, therefore $f(z)$ has *an essential singularity* at $z = 0$.

## PROBLEMS 20.7

1. Obtain the expansion of $(z - 1)/z^2$ in a Taylor's series in powers of $(z - 1)$ and determine the region of convergence.
2. Find the first three times of the Taylor's series expansion of $f(z) = 1/(z^2 + 4)$ about $z = -i$. Also find the region of convergence.
3. Expand in Taylor's series (*i*) $(z - 1)/(z + 1)$ about the point $z = 1$.

   (*ii*) $\cos z$ about the point $z = \pi/2$. (*iii*) $\dfrac{1}{z^2 - z - 6}$ about (*a*) $z = -1$ (*b*) $z = 1$
4. Expand the following functions in Laurent's series :

   (*i*) $f(z) = \dfrac{1}{z - z^2}$ for $1 < |z + 1| < 2$.

(ii) $f(z) = \dfrac{1}{(z-1)(z+3)}$ for $1 < |z| < 3$. *(Calicut, 2012)*

(iii) $f(z) = z/[(z-1)(z-3)]$ for $|z-1| < 2$. *(Andhra, 2021)*

5. Find the Laurent's expansion of (i) $\dfrac{e^z}{(z-1)^2}$, about $z = 1$.

(ii) $e^{2z}/(z-1)^3$ about the singularity $z = 1$.

6. Expand the following functions in Laurent series.

(i) $(z-1)/z^2$ for $|z-1| > 1$ (ii) $\dfrac{1-\cos z}{z^3}$, about $z = 0$.

7. Find the Laurent's series expansion of

(i) $\dfrac{z^2-1}{z^2+5z+6}$ about $z = 0$ in the region $2 < |z| < 3$ *(V.T.U., 2011S)*

(ii) $\dfrac{z^2-6z-1}{(z-1)(z-3)(z+2)}$ in the region $3 < |z+2| < 5$ *(Calicut, 2011)*

(iii) $\dfrac{7z^2-9z-18}{z^3-9z}$ in the region (a) $|z| > 3$ (b) $0 < |z-3| < 3$. *(V.T.U., 2010S)*

8. Find the Laurent's expansion of $1/[(z^2+1)(z^2+2)]$ for (a) $0 < |z| < 1$ ; (b) $1 < |z| < \sqrt{2}$ ; (c) $|z| > 2$. *(JNTUK, 2022)*

Find the nature and location of the singularities of the following functions :

9. $\dfrac{1}{z(2-z)}$. 10. $\sin(1/z)$. 11. $\tan\left(\dfrac{1}{z}\right)$.

12. $\dfrac{z^2-1}{(z-1)^3}$. 13. $\dfrac{e^z}{(z-1)^4}$. 14. $\dfrac{\cot \pi z}{(z-a)^2}$.

## 20.18 (1) RESIDUES

*The coefficient of $(z-a)^{-1}$ in the expansion of $f(z)$ around an isolated singularity is called the* **residue** *of $f(z)$ at that point.* Thus in the Laurent's series expansion of $f(z)$ around $z = a$ *i.e.*, $f(z) = a_0 + a_1(z-a) + a_2(z-a)^2 + \dots + a_{-1}(z-a)^{-1} + a_{-2}(z-a)^{-2} + \dots$, the residue of $f(z)$ at $z = a$ is $a_{-1}$.

$$\therefore \qquad Res\, f(a) = \frac{1}{2\pi i}\oint_C f(z)$$

*i.e.*,
$$\oint_C f(z)\,dz = 2\pi i \text{ Res } f(a). \qquad \dots(1)$$

**(2) Residue Theorem**

*If $f(z)$ is analytic in a closed curve C except at a finite number of singular points within C, then $\oint_C f(z)\,dz = 2\pi i \times$ (sum of the residues at the singular points within C).*

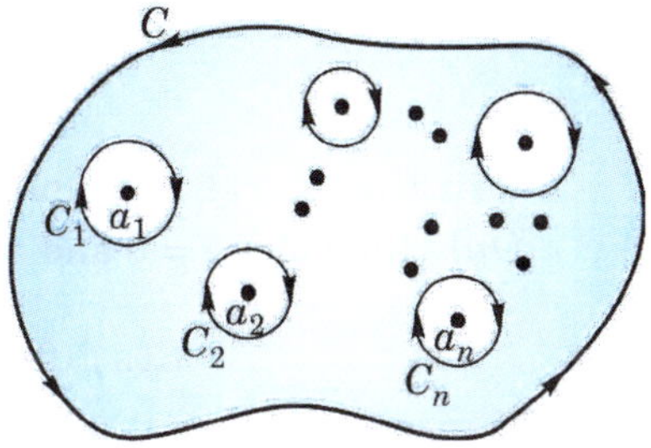

Fig. 20.22

Let us surround each of the singular points $a_1, a_2, \dots, a_n$ by a small circle such that it encloses no other singular point (Fig. 20.22). Then these circles $C_1, C_2, \dots, C_n$ together with C, form a multiply connected region in which $f(z)$ is analytic.

$\therefore$ applying Cauchy's theorem, we have

$$\oint_C f(z)\,dz = \oint_{C_1} f(z)dz + \oint_{C_2} f(z)dz + \dots + \oint_{C_n} f(z)dz \qquad \text{[by (1)}$$

$= 2\pi i\,[\text{Res } f(a_1) + \text{Res } f(a_2) + \dots + \text{Res } f(a_n)]$ which is the desired result.

## 20.19 CALCULATION OF RESIDUES

**(1)** *If $f(z)$ has a simple pole at $z = a$, then*

$$\textbf{Res f(a)} = \underset{z \to a}{\textbf{Lt}} \; [\mathbf{(z-a)f(z)}]. \qquad ...(1)$$

Laurent's series in this case is

$$f(z) = c_0 + c_1(z-a) + c_2(z-a)^2 \dots + c_{-1}(z-a)^{-1}$$

Multiplying throughout by $z - a$, we have

$$(z-a)f(z) = c_0(z-a) + c_1(z-a)^2 + \dots + c_{-1}.$$

Taking limits as $z \to a$, we get

$$\underset{z \to a}{\text{Lt}} \; [(z-a)f(z)] = c_{-1} = \text{Res}\, f(a).$$

**(2)** *Another formula for Res $f(a)$ :*

Let $f(z) = \phi(z)/\psi(z)$, where $\psi(z) = (z-a)\,F(z)$, $F(a) \neq 0$.

Then
$$\underset{z \to a}{\text{Lt}} \; [(z-a)\phi(z)/\psi(z)]$$

$$= \underset{z \to a}{\text{Lt}} \frac{(z-a)[\phi(a) + (z-a)\phi'(a) + \dots]}{\psi(a) + (z-a)\psi'(a) + \dots}$$

$$= \underset{z \to a}{\text{Lt}} \frac{\phi(a) + (z-a)\phi'(a) + \dots}{\psi'(a) + (z-a)\psi''(a) + \dots}, \quad \text{since } \psi(a) = 0$$

Thus
$$\textbf{Res f(a)} = \frac{\boldsymbol{\phi}\mathbf{(a)}}{\boldsymbol{\psi}'\mathbf{(a)}}.$$

**(3)** *If $f(z)$ has a pole of order $n$ at $z = a$, then*

$$\textbf{Res f(a)} = \frac{\mathbf{1}}{\mathbf{(n-1)!}} \left\{ \frac{\mathbf{d^{n-1}}}{\mathbf{dz^{n-1}}} [\mathbf{(z-a)^n f(z)}] \right\}_{\mathbf{z=a}}$$

Here
$$f(z) = c_0 + c_1(z-a) + c_2(z-a)^2 + \dots + c_{-1}(z-a)^{-1} + \dots + c_{-n}(z-a)^{-n}.$$

Multiplying throughout by $(z-a)^n$, we get

$$(z-a)^n f(z) = c_0(z-a)^n + c_1(z-a)^{n+1} + c_2(z-a)^{n+2} + \dots + c_{-1}(z-a)^{n-1} + c_{-2}(z-a)^{n-2} + \dots + c_{-n}.$$

Differentiating both sides w.r.t. $z$, $n - 1$ times and putting $z = a$, we get

$$\left\{ \frac{d^{n-1}}{dz^{n-1}} [(z-a)^n f(z)] \right\}_{z=a} = (n-1)!\, c_{-1} \text{ whence follows the result.}$$

**Obs.** In many cases, the residue of a pole $(z = a)$ can be found, by putting $z = a + t$ in $f(z)$ and expanding it in powers of $t$ where $|t|$ is quite small.

**Example 20.32.** *Find the sum of the residues of $f(z) = \dfrac{\sin z}{z \cos z}$ at its poles inside the circle $|z| = 2$.*

*(JNTUK, 2022)*

**Solution.** $f(z)$ has simple poles at $z = 0, \pm \pi/2, \pm 3\pi/2, - \dots$

Only the poles $z = 0$ and $z = \pm \pi/2$ lies inside $|z| = 2$.

$$\therefore \quad \text{Res}\, f(0) = \underset{z \to 0}{\text{Lt}} \; [z \,.\, f(z)] = \underset{z \to 0}{\text{Lt}} \left( \frac{\sin z}{\cos z} \right) = 0.$$

$$\text{Res}\, f(\pi/2) = \underset{z \to \pi/2}{\text{Lt}} \left[ \left( z - \frac{\pi}{2} \right) f(z) \right] = \underset{z \to \pi/2}{\text{Lt}} \left\{ \frac{(z - \pi/2)\sin z}{z \cos z} \right\}$$

$$= \underset{z \to \pi/2}{\text{Lt}} \frac{(z - \pi/2)\cos z + \sin z}{\cos z - z \sin z} \qquad \left[ \text{Being } \frac{0}{0} \text{ form} \right]$$

$$= \frac{1}{-\pi/2} = -\frac{2}{\pi}$$

and $$\text{Res}\, f(-\pi/2) = \underset{z\to-\pi/2}{\text{Lt}} \left\{\frac{(z+\pi/2)\sin z}{z\cos z}\right\} = \underset{z\to-\pi/2}{\text{Lt}} \frac{(z+\pi/2)\cos z + \sin z}{\cos z - z\sin z} = \frac{-1}{-\pi/2} = \frac{2}{\pi}$$

Hence sum of residues $= 0 - \frac{2}{\pi} + \frac{2}{\pi} = 0.$

**Example 20.33.** *Determine the poles of the function* (JNTUK, 2022)
*$f(z) = z^2/(z-1)^2(z+2)$ and the residue at each pole.*
*Hence evaluate $\oint_C f(z)dz$, where C is the circle $|z| = 2.5$.*

**Solution.** Since $\underset{z\to-2}{\text{Lt}} \{(z+2)f(z)\} = \underset{z\to-2}{\text{Lt}} \frac{z^2}{(z-1)^2} = \frac{4}{9}.$

which is finite and non-zero, the function has a simple pole at $z = -2$ and Res $f(-2) = 4/9$.

Also since $\underset{z\to1}{\text{Lt}} [(z-1)^2 f(z)]$ is finite and non-zero, $f(z)$ has a pole of order two at $z = 1$.

$$\therefore \quad \text{Res}\, f(1) = \frac{1}{1!}\left[\frac{d}{dz}\{(z-1)^2 f(z)\}\right]_{z=1} = \left[\frac{d}{dz}\left(\frac{z^2}{z+2}\right)\right]_{z=1} = \left[\frac{z^2+4z}{(z+2)^2}\right]_{z=1} = \frac{5}{9}.$$

[*Otherwise writing* $z = 1 + t$,

$$f(z) = \frac{(1+t)^2}{t^2(3+t)} = \frac{1}{3t^2}(1+t)^2(1+t/3)^{-1} = \frac{1}{3t^2}(1+t)^2\left(1 - \frac{t}{3} + \frac{t^2}{9} - \ldots\right)$$

$$= \frac{1}{3t^2}\left(1 + \frac{5}{3}t + \frac{4}{9}t^2 - \ldots\right) = \frac{1}{3t^2} + \frac{5}{9t} + \frac{4}{27} - \ldots \qquad \ldots(i)$$

$$\therefore \quad \text{Res}\, f(1) = \text{coefficient of } \frac{1}{t} \text{ in } (i) = \frac{5}{9}.]$$

Clearly $f(z)$ is analytic on $|z| = 2.5$ and at all points inside except the poles $z = -2$ and $z = 1$. Hence by residue theorem

$$\oint_C f(z)dz = 2\pi i\, [\text{Res}\, f(-2) + \text{Res}\, f(1)] = 2\pi i\left[\frac{4}{9} + \frac{5}{9}\right] = 2\pi i.$$

**Example 20.34.** *Find the residue of $f(z) = \dfrac{z^3}{(z-1)^4(z-2)(z-3)}$ at its poles and hence evaluate $\oint_C f(z)dz$ where C is the circle $|z| = 2.5$.* (JNTUK, 2023; Anna., 2016)

**Solution.** The poles of $f(z)$ are given by $(z-1)^4(z-2)(z-3) = 0$.

$\therefore$ $z = 1$ is a pole of order 4, while $z = 2$ and $z = 3$ are simple poles.

$$\text{Res}\, f(1) = \frac{1}{3!}\frac{d^3}{dz^3}\left\{(z-1)^4 \cdot \frac{z^3}{(z-1)^4(z-2)(z-3)}\right\}_{z=1} = \frac{1}{6}\frac{d^3}{dz^3}\left\{\frac{z^3}{(z-2)(z-3)}\right\}_{z=1}$$

$\therefore$ $z = 1$ is a pole of order 4, while $z = 2$ and $z = 3$ are simple poles.

$$\text{Res}\, f(1) = \frac{1}{3!}\frac{d^3}{dz^3}\left\{(z-1)^4 \cdot \frac{z^3}{(z-1)^4(z-2)(z-3)}\right\}_{z=1} = \frac{1}{6}\frac{d^3}{dz^3}\left\{\frac{z^3}{(z-2)(z-3)}\right\}_{z=1}$$

$$= \frac{1}{6}\frac{d^3}{dz^3}\left[z + 5 - \frac{8}{z-2} + \frac{27}{z-3}\right] = \frac{1}{6}\left[-8 \cdot \frac{(-1)^3\, 3!}{(z-2)^4} + \frac{27 \cdot (-1)^3\, 3!}{(z-2)^4}\right]_{z=1}$$

$$= -\left[-8 + \frac{27}{16}\right] = \frac{101}{16}.$$

$$\text{Res } f(2) = \underset{z\to 2}{\text{Lt}} \left\{(z-2)\,.\frac{z^3}{(z-1)^4(z-2)(z-3)}\right\} = \underset{z\to 2}{\text{Lt}} \left\{\frac{z^3}{(z-1)^4(z-3)}\right\} = \frac{8}{(1)^4(-1)} = -8$$

$$\text{Res } f(3) = \underset{z\to 3}{\text{Lt}} \left\{(z-3)\,.\frac{z^3}{(z-1)^4(z-2)(z-3)}\right\} = \frac{27}{(2)^4\,.\,1} = \frac{27}{16}$$

Now $\oint_C f(z)dz = 2\pi i\,[\text{Res } f(1) + \text{Res } f(2)]$ [∵ Pole $z = 3$ is outside $C$]

$$= 2\pi i\left(\frac{101}{16} - 8\right) = \frac{-27\pi i}{8}.$$

**Example 20.35.** *Evaluate*

$$\oint_C \frac{z-3}{z^2+2z+5}dz, \text{ where } C \text{ is the circle}$$ (JNTUK, 2023)

(i) $|z| = 1$, (ii) $|z+1-i| = 2$, (iii) $|z+1+i| = 2$.

**Solution.** The poles of $f(z) = \dfrac{z-3}{z^2+2z+5}$ are given by $z^2 + 2z + 5 = 0$

*i.e.,* by $$z = \frac{-2 \pm \sqrt{(4-20)}}{2} = -1 \pm 2i.$$

(i) Both the poles $z = -1 + 2i$ and $z = -1 - 2i$ lie outside the circle $|z| = 1$. Therefore, $f(z)$ is analytic everywhere within $C$.

Hence by Cauchy's theorem, $\oint_C \dfrac{z-3}{z^2+2z+5}dz = 0.$

(ii) Here only one pole $z = -1 + 2i$ lies inside the circle $C : |z + 1 - i| = 2$. Therefore, $f(z)$ is analytic within $C$ except at this pole.

$$\therefore \quad \text{Res } f(-1+2i) = \underset{z\to -1+2i}{\text{Lt}} [\{z-(-1+2i)\}\, f(z)] = \underset{z\to -1+2i}{\text{Lt}} \frac{(z+1-2i)(z-3)}{z^2+2z+5}$$

$$= \underset{z\to -1+2i}{\text{Lt}} \frac{z-3}{z+1+2i} = \frac{-4+2i}{4i} = i + 1/2.$$

Hence by residue theorem $\oint_C f(z)dz = 2\pi i \text{ Res } f(-1+2i) = 2\pi i(i + 1/2) = \pi(i-2).$

(iii) Here only the pole $z = -1 - 2i$ lies inside the circle $C : |z + 1 + i| = 2$. Therefore, $f(z)$ is analytic within $C$ except at this pole.

$$\therefore \quad \text{Res } f(-1-2i) = \underset{z\to -1-2i}{\text{Lt}} \frac{(z+1+2i)(z-3)}{z^2+2z+5}$$

$$= \underset{z\to -1-2i}{\text{Lt}} \frac{z-3}{z+1-2i} = \frac{-4-2i}{-4i} = \frac{1}{2} - i$$

Hence by residue theorem, $\oint_C f(z)dz = 2\pi i \text{ Res } f(-1-2i) = 2\pi i\left(\frac{1}{2} - i\right) = \pi(2+i).$

**Example 20.36.** *Evaluate* $\oint_C \dfrac{e^z}{\cos \pi z}dz$, *where C is the unit circle* $|z| = 1$. (JNTUK, 2022)

**Solution.** $f(z) = e^x/\cos \pi z$ has simple poles at $z = \pm 1/2, \pm 3/2, \pm 5/2, \ldots$

Out of these only the poles at $z = 1/2$ and $z = -1/2$ lie inside the given circle $|z| = 1$.

$$\therefore \quad \text{Res } f(1/2) = \underset{z\to 1/2}{\text{Lt}} \left[\left(z-\frac{1}{2}\right)f(z)\right] = \underset{z\to 1/2}{\text{Lt}} \left\{\frac{\left(z-\frac{1}{2}\right)e^z}{\cos \pi z}\right\} \quad \left[\frac{0}{0} \text{ form}\right]$$

$$= \underset{z \to 1/2}{\text{Lt}} \frac{e^z + \left(z - \frac{1}{2}\right)e^z}{-\pi \sin \pi z} = \frac{e^{1/2}}{-\pi}$$

and $$\text{Res } f(-1/2) = \underset{z \to -1/2}{\text{Lt}} \left\{ \frac{\left(z + \frac{1}{2}\right)e^z}{\cos \pi z} \right\} \qquad \left[\frac{0}{0} \text{ form}\right]$$

$$= \underset{z \to -1/2}{\text{Lt}} \frac{e^z + \left(z + \frac{1}{2}\right)e^z}{-\pi \sin \pi z} = \frac{e^{-1/2}}{\pi}$$

Hence $$\oint_C \frac{e^z}{\cos \pi z} dz = 2\pi i \left( \text{Res } f\left(\frac{1}{2}\right) + \text{Res } f\left(-\frac{1}{2}\right) \right\}$$

$$= 2\pi i \left( -\frac{e^{1/2}}{\pi} + \frac{e^{-1/2}}{\pi} \right) = -4i \left( \frac{e^{1/2} - e^{-1/2}}{2} \right) = -4i \sinh \frac{1}{2}.$$

**Example 20.37.** *Evaluate* $\oint_C \tan z \, dz$ *where C is the circle* $|z| = 2$. *(Andhra, 2022)*

**Solution.** The poles of $f(z) = \sin z/\cos z$ are given by $\cos z = 0$ *i.e.* $z = \pm \pi/2, \pm 3\pi/2, \pm 5\pi/2, \ldots$

Of these poles, $z = \pi/2$, and $-\pi/2$ only are within the given circle.

$$\therefore \quad \text{Res } f(\pi/2) = \underset{z \to \pi/2}{\text{Lt}} \frac{\sin z}{\frac{d}{dz}(\cos z)} = \underset{z \to \pi/2}{\text{Lt}} \left( \frac{\sin z}{-\sin z} \right) = -1 \qquad [\text{By § 20.19 (2)}]$$

Similarly $$\text{Res } f(-\pi/2) = \underset{z \to -\pi/2}{\text{Lt}} \frac{\sin z}{\frac{d}{dz}(\cos z)} = -1.$$

Hence by residue theorem,

$$\oint_C f(z)dz = 2\pi i \{\text{Res } f(\pi/2) + \text{Res } f(-\pi/2)\} = 2\pi i(-1-1) = -4\pi i.$$

**Example 20.38.** *Evaluate* $\oint_C \frac{\sin \pi z^2 + \cos \pi z^2}{(z-1)^2 (z-2)} dz$, *where C is the circle* $|z| = 3$.

*(JNTUK, 2022; V.T.U., 2015 ; Calicut, 2012)*

**Solution.** $$f(z) = \frac{\sin \pi z^2 + \cos \pi z^2}{(z-1)^2 (z-2)}$$

is analytic within the circle $|z| = 3$ excepting the poles $z = 1$ and $z = 2$.

Since $z = 1$ is a pole of order 2.

$$\therefore \quad \text{Res } f(1) = \frac{1}{1!}\left[\frac{d}{dz}\{(z-1)^2 f(z)\}\right]_{z=1} = \left[\frac{d}{dz}\left(\frac{\sin \pi z^2 + \cos \pi z^2}{(z-2)}\right)\right]_{z=1}$$

$$= \left[\frac{(z-2)(2\pi z \cos \pi z^2 - 2\pi z \sin \pi z^2) - (\sin \pi z^2 + \cos \pi z^2)}{(z-2)^2}\right]_{z=1}$$

$$= (-1)(-2\pi) - (-1) = 2\pi + 1$$

Also $$\text{Res } f(2) = \underset{z \to 2}{\text{Lt}} [(z-2) f(z)] = \underset{z \to 2}{\text{Lt}} \frac{\sin \pi z^2 + \cos \pi z^2}{(z-1)^2} = 1$$

Hence by residue theorem,

$$\oint_C f(z) \, dz = 2\pi i \, [\text{Res } f(1) + \text{Res } f(2)] = 2\pi i \, (2\pi + 1 + 1) = 4\pi(\pi + 1)i.$$

## PROBLEMS 20.8

1. Expand $f(z) = 1/[z^2 (z - i)]$ as a Laurent's series about $i$ and hence find the residue thereat.
2. Find the residue of (*i*) $ze^z/(z - 1)^3$ at its pole.
 (*ii*) $z^2/(z^2 + a^2)$ at $z = ai$.
3. Determine the poles of the following functions and the residue at each pole :

 (*i*) $\dfrac{z^2+1}{z^2-2z}$ (*ii*) $\dfrac{z^2-2z}{(z+1)^2(z^2+1)}$ (*iii*) $\dfrac{2z+4}{(z+1)(z^2+1)}$ (*Andhra, 2021*)
4. Find the residues of the following functions at each pole.
 (*i*) $(1 - e^{2z})/z^4$ (*D.T.U., 2013*) (*ii*) $ze^{iz}/(z^2 + 1)$ (*P.T.U., 2010*) (*iii*) $\cot z$.
5. $\oint_C \dfrac{z^2+4}{(z-2)(z+3)}\,dz$, where $C$ is (*i*) $|z + 1| = 2$ (*ii*) $|z - 2| = 2$.
6. Evaluate the following integrals :

 (*i*) $\oint_C \dfrac{1-2z}{z(z-1)(z-2)}\,dz$ where, $C$ is the circle $|z| = 1.5$. (*Hazaribag., 2015*)

 (*ii*) $\oint_C \dfrac{e^{2z}\,dz}{(z+2)(z+4)(z+7)}$ for $C$ as circle $|z| = 3$.

 (*iii*) $\oint_C \dfrac{4z^2-4z+1}{(z-2)(4+z^2)}\,dz$, $C: |z| = 1$

 (*iv*) $\oint_C \dfrac{3z^2+z+1}{(z^2-1)(z+3)}\,dz$, $C: |z| = 2$.
7. Evaluate

 (*i*) $\int_C \dfrac{2z+1}{(2z-1)^2}\,dz$, where $C$ is $|z| = 1$ (*ii*) $\oint_C \dfrac{z+4}{z^2+2z+5}\,dz$, where $C$ is $|z + 1 - i| = 2$

 (*iii*) $\int_C \dfrac{z^2-2z}{(z+1)^2(z^2+4)}\,dz$, where $C$ is the circle $|z| = 10$.
8. Evaluate :

 (*i*) $\oint_C \dfrac{z\,dz}{(z-1)(z-2)^2}$, $C: |z - 2| = \frac{1}{2}$. (*JNTUK, 2022; Andhra, 2020*)

 (*ii*) $\oint_C \dfrac{3z^2+2}{(z-1)(z^2+9)}\,dz$, $C: |z - 2| = 2$.

 (*iii*) $\oint_C \dfrac{dz}{(z^2+4)^2}$, $C: |z - i| = 2$.
9. Evaluate :

 (*i*) $\oint_C \dfrac{e^{-z}}{z^2}\,dz$, $C: |z| = 1$. (*ii*) $\oint_C z^2 e^{1/z}$, $C: |z| = 1$.

 (*iii*) $\oint_C \dfrac{e^z\,dz}{z^2+4}$, $C: |z - i| = 2$. (*iv*) $\oint_C \dfrac{e^{2z}\,dz}{(z+1)^4}$, $C: |z| = 2$.
10. Evaluate the following integrals : (*i*) $\oint_C \dfrac{\sin^6 z}{(z-\pi/6)^3}\,dz$, $C: |z| = 1$

 (*ii*) $\oint_C \dfrac{z\sec z}{(1-z)^2}\,dz$, $C: |z| = 3$ (*iii*) $\oint_C \dfrac{z\cos z}{(z-\pi/2)^3}\,dz$, $C: |z - 1| = 1$.
11. Evaluate $\oint_C \dfrac{dz}{\sinh 2z}$ where $C$ is the circle $|z| = 2$.
12. Obtain Laurent's expansion for the function $f(z) = 1/z^2 \sinh z$ and evaluate

 $\oint_C \dfrac{z}{z^2 \sinh z}$, where $C$ is the circle $|z - 1| = 2$.

## 20.20 EVALUATION OF REAL DEFINITE INTEGRALS

Many important definite integrals can be evaluated by applying the Residue theorem to properly chosen integrals. The contours chosen will consist of straight lines and circular arcs.

(*a*) **Integration around the unit circle.** An integral of the type $\int_0^{2\pi} f(\sin\theta, \cos\theta)\, d\theta$, where the integrand is a rational function of sin θ and cos θ can be evaluated by writing $e^{i\theta} = z$.

Since $\sin\theta = \frac{1}{2i}\left(z - \frac{1}{z}\right)$ and $\cos\theta = \frac{1}{2}\left(z + \frac{1}{z}\right)$, then integral takes the form $\int_C f(z)\,dz$, where $f(z)$ is a rational function of $z$ and $C$ is a unit circle $|z| = 1$.

Hence *the integral is equal to* $2\pi i$ *times the sum of the residues at those poles of* $f(z)$ *which are within C.*

**Example 20.39.** *Show that*

$$\int_0^{2\pi} \frac{\cos 2\theta\, d\theta}{1 - 2a\cos\theta + a^2} = \frac{2\pi a^2}{1 - a^2}, \ (a^2 < 1).$$

**Solution.** Putting $z = e^{i\theta}$, $d\theta = dz/iz$, $\cos\theta = \frac{1}{2}(z + 1/z)$ and $\cos 2\theta = \frac{1}{2}(e^{2i\theta} + e^{-2i\theta}) = \frac{1}{2}(z^2 + 1/z^2)$

∴ the given integral

$$I = \int_C \frac{\frac{1}{2}(z^2 + 1/z^2)}{1 - a(z + 1/z) + a^2} \cdot \frac{dz}{iz} = \frac{1}{2i}\int_C \frac{(z^4 + 1)\,dz}{z^2(z - az^2 - a + a^2 z)}$$

$$= \frac{1}{2i}\int_C \frac{(z^4 + 1)\,dz}{z^2(z - a)(1 - az)} = \int_C f(z)\,dz \qquad \text{where } C \text{ is the unit circle } |z| = 1.$$

Now $f(z)$ has simple poles at $z = a$, $1/a$ and the second order pole at $z = 0$, of which the poles at $z = 0$ and $z = a$ lie within the unit circle.

∴ $$\text{Res } f(a) = \underset{z \to a}{\text{Lt}}\, [(z - a) f(z)] = \frac{1}{2i}\underset{z \to a}{\text{Lt}}\left[\frac{z^4 + 1}{z^2(1 - az)}\right] = \frac{a^4 + 1}{2ia^2(1 - a^2)}$$

and $$\text{Res } f(0) = \underset{z \to 0}{\text{Lt}}\, \frac{d}{dz}[z^2 f(z)] = \frac{1}{2i}\underset{z \to 0}{\text{Lt}}\, \frac{d}{dz}\left[\frac{z^4 + 1}{z - az^2 - a + a^2 z}\right]$$

$$= \frac{1}{2i}\underset{z \to 0}{\text{Lt}}\, \frac{(z - az^2 - a + a^2 z)(4z^3) - (z^4 + 1)(1 - 2az + a^2)}{(z - az^2 - a + a^2 z)^2} = -\frac{1 + a^2}{2ia^2}$$

Hence $$I = 2\pi i\,[\text{Res } f(a) + \text{Res } f(0)] = 2\pi i\left[\frac{a^4 + 1}{2ia^2(1 - a^2)} - \frac{1 + a^2}{2ia^2}\right] = \frac{2\pi a^2}{1 - a^2}.$$

**Example 20.40.** *By integrating around a unit circle, evaluate*

$$\int_0^{2\pi} \frac{\cos 3\theta}{5 - 4\cos\theta}\, d\theta.$$ *(JNTUK, 2022; Andhra, 2022)*

**Solution.** Putting $z = e^{i\theta}$, $d\theta = dz/iz$, $\cos\theta = \frac{1}{2}(z + 1/z)$

and $$\cos 3\theta = \frac{1}{2}(e^{3i\theta} + e^{-3i\theta}) = \frac{1}{2}(z^3 + 1/z^3).$$

∴ the given integral $$I = \int_C \frac{\frac{1}{2}(z^3 + 1/z^3)}{5 - 2(z + 1/z)} \cdot \frac{dz}{iz}$$

$$= -\frac{1}{2i}\int_C \frac{z^6 + 1}{z^3(2z^2 - 5z + 2)}\,dz = -\frac{1}{2i}\int_C \frac{(z^6 + 1)\,dz}{z^3(2z - 1)(z - 2)}$$

$$= -\frac{1}{2i}\int f(z)\,dz, \quad \text{where } C \text{ is the unit circle } |z| = 1.$$

Now $f(z)$ has a pole of order 3 at $z = 0$ and simple poles at $z = \frac{1}{2}$ and $z = 2$. Of these only $z = 0$ and $z = 1/2$ lie within the unit circle.

$$\therefore \quad \text{Res } f(1/2) = \underset{z\to 1/2}{\text{Lt}} \frac{(z-1/2)(z^6+1)}{(2z-1)(z-2)} = \underset{z\to 1/2}{\text{Lt}} \left\{\frac{z^6+1}{2z^3(z-2)}\right\} = -\frac{65}{24}$$

$$\text{Res } f(0) = \frac{1}{(n-1)!}\left\{\frac{d^{n-1}}{dz^{n-1}}[(z-0)^n f(z)]\right\}_{z=0} \quad \text{where } n = 3$$

$$= \frac{1}{2}\left\{\frac{d^2}{dz^2}\left(\frac{z^6+1}{2z^2-5z+2}\right)\right\}_{z=0} = \frac{d}{dz}\left[\frac{(2z^2-5z+2)\,6z^5-(z^6+1)(4z-5)}{2(2z^2-5z+2)^2}\right] \text{ at } z=0$$

$$= \left\{\frac{d}{dz}\left[\frac{8z^7-25z^6+12z^5-4z+5}{2(2z^2-5z+2)^2}\right]\right\}_{z=0}$$

$$= \left[\frac{(2z^2-5z+2)^2(56z^6-150z^5+60z^4-4)-(8z^7-25z^6+12z^5-4z+5)2(2z^2-5z+2)(4z-5)}{2(2z^2-5z+2)^4}\right]_{z=0}$$

$$= \frac{4(-4)-5(-20)}{2\times 16} = \frac{84}{32} = \frac{21}{8}$$

Hence
$$I = \frac{-1}{2i}\{2\pi i\,[\text{Res } f(1/2) + \text{Res } f(0)]\} = -\pi\left[-\frac{65}{24}+\frac{21}{8}\right] = -\pi\left(-\frac{1}{12}\right) = \frac{\pi}{12}.$$

(*b*) **Integration around a small semi-circle.** To evaluate $\int_{-\infty}^{\infty} f(x)\,dx$, we consider $\int_C f(z)$, where $C$ is the contour consisting of the semi-circle $C_R : |z| = R$, together with the diameter that closes it.

Supposing that $f(z)$ has no singular point on the real axis, we have, by the Residue theorem,

$$\int_{C_R} f(z)\,dz + \int_{-R}^{R} f(x)\,dx = 2\pi i \sum \text{Res } f(a).$$

Finally making $R$ tend to $\infty$, we find the value of $\int_{-\infty}^{\infty} f(x)\,dx$, provided $\int_{C_R} f(z)\,dz \to 0$ as $R \to \infty$.

**Example 20.41.** *Evaluate* $\int_{-\infty}^{\infty} \frac{x^2\,dx}{(x^2+1)(x^2+4)}$. *(JNTUK, 2022)*

**Solution.** Consider $\int_C \frac{z^2 dz}{(z^2+1)(z^2+4)} = \int_C f(z)\,dz$ ...(*i*)

where $C$ is the contour consisting of the semi-circle $C_R$ of radius $R$ together with the part of the real axis from $-R$ to $R$ as shown in Fig. 20.23.

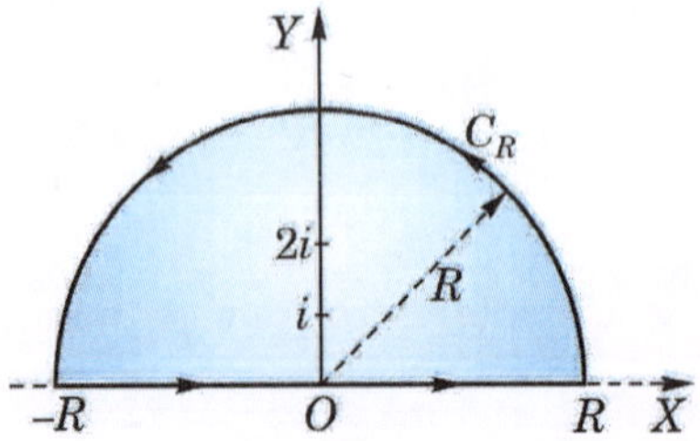

Fig. 20.23

The integrand has simple poles at $z = \pm i$, $z = \pm 2i$ of which $z = i, 2i$ only lie inside $C$.

$\therefore$ by the Residue theorem,

$$\int_C f(z)\,dz = 2\pi i\,[\text{Res } f(i) + \text{Res } f(2i)]$$

$$= 2\pi i\,[\underset{z\to i}{\text{Lt}}\,(z-i)\,f(z) + \underset{z\to 2i}{\text{Lt}}\,(z-2i)\,f(z)]$$

$$= 2\pi i\left[\frac{i^2}{2i(i^2+4)} + \frac{4i^2}{(4i^2+1)(4i)}\right] = 2\pi i\left(\frac{i}{6}-\frac{i}{3}\right) = \frac{\pi}{3} \quad \text{...(ii)}$$

Also $$\int_C f(z)\,dz = \int_{-R}^{R} f(x)\,dx + \int_{C_R} f(z)\,dz \qquad ...(iii)$$

Now let $R \to \infty$, so as to show that the second integral in Eqn. (*iii*) vanishes. For any point on $C_R$ as $|z| \to \infty$

$$f(z) = \frac{1}{z^2} \cdot \frac{1}{(1 + z^{-2})(1 + 4z^{-2})}$$

decreases as $1/z^2$ and tends to zero whereas the length of $C_R$ increases with $z$.

Consequently, $\underset{|z|\to\infty}{\text{Lt}} \int_{C_R} f(z)\,dz = 0.$

Hence from Eqns. (*i*), (*ii*) and (*iii*), we get $\int_{-\infty}^{\infty} \frac{x^2\,dx}{(x^2+1)(x^2+4)} = \frac{\pi}{3}.$

**Example 20.42.** *Evaluate* $\int_0^{\infty} \frac{\cos ax}{x^2+1}\,dx.$

**Solution.** Consider $\int_C \frac{e^{iaz}}{z^2+1}\,dz = \int_C f(z)\,dz$

where $C$ is the contour consisting of the semi-circle $C_R$ of radius $R$ together with the part of the real axis from $-R$ to $R$ as shown in Fig. 20.23.

The integrand has simple poles at $z = i$ amd $z = -i$, of which $z = i$ only lies inside $C$.

$\therefore$ by Residue theorem, $\int_C f(z)\,dz = 2\pi i \text{ Res } f(i) = 2\pi i \underset{z\to i}{\text{Lt}} [(z-i) f(z)]$

$$= 2\pi i \underset{z\to i}{\text{Lt}} \frac{(z-i)\,e^{iaz}}{z^2+1} = 2\pi i \underset{z\to i}{\text{Lt}} \frac{e^{iaz}}{z+i} = 2\pi i\,\frac{e^{-a}}{2i} = \pi e^{-a} \qquad ...(i)$$

Also $$\int_C f(z)\,dz = \int_{-R}^{R} f(x)\,dx + \int_{C_R} f(z)\,dz \qquad ...(ii)$$

Now $|z| = R$ on $C_R$ and $|z^2 + 1| \geq R^2 - 1$.

Also $|e^{iaz}| = |e^{ia(x+iy)}| = |e^{iax} . e^{-ay}| = e^{-ay} < 1 \qquad [\because \; y > 0]$

$$\therefore \quad \left|\frac{e^{iaz}}{z^2+1}\right| = |e^{iaz}| . \frac{1}{|z^2+1|} < 1 . \frac{1}{R^2-1}$$

Thus $\int_{C_R} f(z)\,dz = \left|\int_{C_R} \frac{e^{iaz}}{z^2+1}\,dz\right| < \int_{C_R} \frac{1}{R^2-1}|dz| < \frac{\pi R}{R^2-1}$ which $\to$ to 0 as $R \to \infty$. ...(*iii*)

Hence from Eqns. (*i*), (*ii*) and (*iii*), we get

$$\pi e^{-a} = \int_{-\infty}^{\infty} f(x)\,dx + 0 \quad \text{or} \quad \int_{-\infty}^{\infty} \frac{e^{iax}}{x^2+1}\,dx = \pi e^{-a}$$

Equating real parts from both sides, we obtain

$$\int_{-\infty}^{\infty} \frac{\cos ax}{x^2+1}\,dx = \pi e^{-a}$$

Since $\cos ax/(x^2 + 1)$ is an even function of $x$, we have

$$2\int_0^{\infty} \frac{\cos ax}{x^2+1}\,dx = \pi e^{-a} \quad \text{or} \quad \int_0^{\infty} \frac{\cos ax}{x^2+1}\,dx = \frac{\pi}{2} e^{-a}.$$

(*c*) **Integration around rectangular contours**

**Example 20.43.** *Evaluate* $\int_{-\infty}^{\infty} \frac{e^{ax}}{e^x+1}\,dx.$

**Solution.** Consider $\int_C \frac{e^{az}}{e^z+1}\,dz = \int_C f(z)\,dz$ where $C$ is the rectangle $ABCD$ with vertices at $(R, 0)$, $(R, 2\pi)$, $(-R, 2\pi)$ and $(-R, 0)$, $R$ being positive (Fig. 20.24).

$f(z)$ has finite poles given by

$$e^z = -1 = e^{(2n+1)\pi i}$$

or $$z = (2n+1)\pi i, \text{ where } n = 0, \pm 1, \pm 2, \pm 3, \ldots$$

The only pole inside the rectangle is $z = \pi i$.

$\therefore$ by Residue theorem,

$$\int_C f(z)dz = 2\pi i \text{ Res } f(\pi i)$$

$$= 2\pi i \left[ e^{az} \Big/ \frac{d}{dz}(e^z + 1) \right]_{z = \pi i}$$

$$= 2\pi i e^{a\pi i}/e^{\pi i} = -2\pi i e^{a\pi i} \qquad [\because \; e^{\pi i} = -1] \qquad ...(i)$$

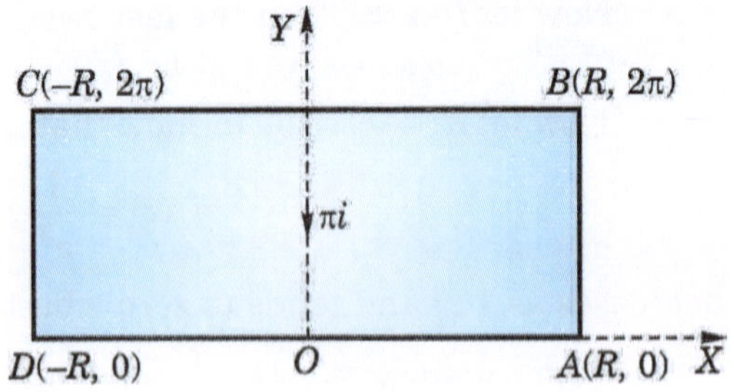

Fig. 20.24

Also $$\int_C f(z)dz = \int_{AB} f(z)dz + \int_{BC} f(z)dz + \int_{CD} f(z)dz + \int_{DA} f(z)dz$$

$$= \int_0^{2\pi} f(R+iy)idy + \int_R^{-R} f(x+2\pi i)dx + \int_{2\pi}^0 f(-R+iy)idy + \int_{-R}^R f(x)dx$$

[$\because$ $z = R + iy$ along $AB$, $z = x + 2\pi i$ along $BC$, $z = -R + iy$ along $CD$ and $z = x$ along $DA$.]

or $$\int_C f(z)dz = i\int_0^{2\pi} \frac{e^{a(R+iy)}}{e^{R+iy}+1}dy - \int_{-R}^R \frac{e^{a(x+2\pi i)}}{e^{x+2\pi i}+1}dx - i\int_0^{2\pi} \frac{e^{a(-R+iy)}}{e^{-R+iy}+1}dy + \int_{-R}^R \frac{e^{ax}}{e^x+1}dx \qquad ...(ii)$$

Now for any two complex numbers $z_1, z_2$

$$|z_1| \geq |z_2|, \text{ we have } |z_1 + z_2| \geq |z_1| - |z_2|$$

so that $|e^{R+iy} + 1| \geq e^R - 1$. Also $|e^{a(R+iy)}| = e^{aR}$

$\therefore$ for the integrand of first integral in Eqn. (*ii*), we have

$$\left| \frac{e^{a(R+iy)}}{e^{R+iy}+1} \right| \leq \frac{e^{aR}}{e^R - 1} \text{ which} \to 0 \text{ as } R \to \infty. \qquad [\because \; a > 1]$$

Similarly, for the integrand of the third integral in Eqn. (*ii*), we get

$$\left| \frac{e^{a(-R+iy)}}{e^{-R+iy}+1} \right| \leq \frac{e^{-aR}}{1 - e^{-R}} \text{ which also} \to 0 \text{ as } R \to \infty. \qquad [\because \; a < 0]$$

Hence as $R \to \infty$, since the first and third integrals in Eqn. (*ii*) approach zero, we get

$$\int_C f(z)dz = -e^{2a\pi i}\int_{-\infty}^{\infty} \frac{e^{ax}}{e^x+1}dx + \int_{-\infty}^{\infty} \frac{e^{ax}}{e^x+1}dx = \left(1 - e^{2a\pi i}\right)\int_{-\infty}^{\infty} \frac{e^{ax}}{e^x+1}dx \qquad ...(iii)$$

Thus from Eqns. (*i*) and (*iii*), we obtain $(1 - e^{2a\pi i}) \displaystyle\int_{-\infty}^{\infty} \frac{e^{ax}}{e^x+1}dx = -2\pi i e^{a\pi i}$

$\therefore$ equating real parts, we get $\displaystyle\int_{-\infty}^{\infty} \frac{e^{ax}}{e^x+1}dx = \frac{\pi}{\sin a\pi}$.

**Example 20.44.** *Show that* $\displaystyle\int_0^{\infty} e^{-x^2} \cos 2mx \, dx = \frac{1}{2}\sqrt{\pi}e^{-m^2}$.

**Solution.** Integrate $f(z) = e^{-z^2}$ along the rectangle ABCDA having vertices $A(-l)$, $B(l)$, $C(l + im)$, $D(-l + im)$ (Fig. 20.25). $f(z)$ has no poles inside this contour. As such

$$\int_{AB} f(z)dz + \int_{BC} f(z)dz + \int_{CD} f(z)dz + \int_{DA} f(z)dz = 0 \qquad ...(i)$$

On $AB : z = x$, on $BC : z = l + iy$, on $CD : z = x + im$ and on $DA : z = -l + iy$.

Therefore, Eqn. (*i*) becomes

$$\int_{-l}^{l} e^{-x^2} dx + \int_0^m e^{-(l+iy)^2} idy + \int_l^{-l} e^{-(x+im)^2} dx + \int_m^0 e^{-(-l+iy)^2} dy = 0$$

or $$\int_{-l}^{l} e^{-x^2} dx - \int_{-l}^{l} e^{-x^2 - 2imx + m^2} dx + \int_0^m e^{-l^2 - 2ily + y^2} . idy$$

$$- \int_0^m e^{-l^2 + 2ily + y^2} . idy = 0 \qquad ...(ii)$$

Now let $l \to \infty$. Then the last two integrals

$$= ie^{-l^2} \int_0^m e^{y^2} (e^{-2ily} - e^{2ily})\, dy = 2e^{-l^2} \int_0^m e^{y^2} \sin 2ly\, dy \to 0$$

[$\because$ As $l \to \infty$, $e^{-l^2} \to 0$ and $\sin 2ly$ is finite]

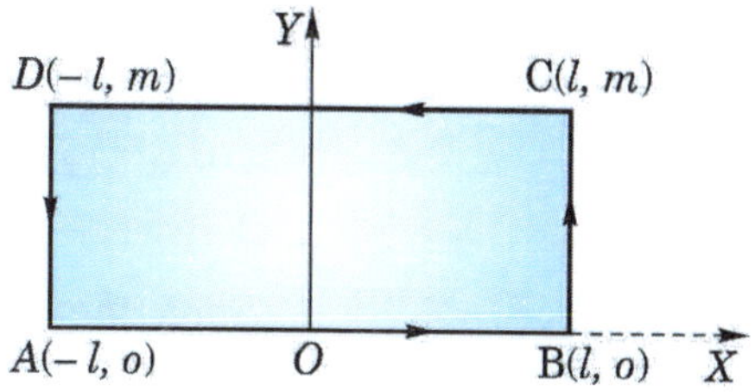

Fig. 20.25

Hence Eqn. (*ii*) reduces to

$$\int_{-\infty}^{\infty} e^{-x^2} dx - e^{m^2} \int_{-\infty}^{\infty} e^{-x^2} (\cos 2mx - i \sin 2mx)\, dx = 0$$

Equating real parts, we get

$$e^{m^2} \int_{-\infty}^{\infty} e^{-x^2} \cos 2mx\, dx = \int_{-\infty}^{\infty} e^{-x^2} dx = \sqrt{\pi}$$

or $$\int_0^{\infty} e^{-x^2} \cos 2mx\, dx = \frac{1}{2}\sqrt{\pi} e^{-m^2}$$

(*d*) **Indenting the contours having poles on the real axis.** So far we have considered such cases in which there is no pole on the real axis. When the integrand has a simple pole on the real axis, we delete it from the region by indenting the contour (*i.e.*, by drawing a small semi-circle having the pole for the centre). The method will be clear from the following example.

**Example 20.45.** *Evaluate* $\int_0^{\infty} \frac{\sin mx}{x} dx$, *when* $m > 0$. *(JNTUK, 2022; U.P.T.U., 2007)*

**Solution.** Consider the integral $\int_C \frac{e^{miz}}{z} dz = \int_C f(z) dz$ where $C$ consists of

(*i*) the real axis from $r$ to $R$,

(*ii*) the upper half of the circle $C_R : |z| = R$,

(*iii*) the real axis $-R$ to $-r$,

(*iv*) the upper half of the circle $C_r : |z| = r$ (Fig. 20.26).

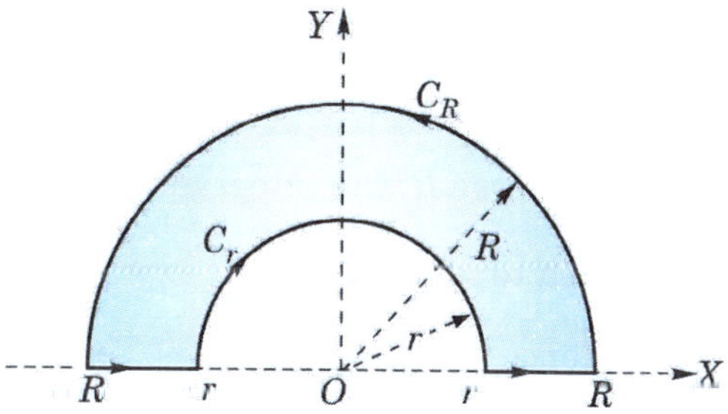

Fig. 20.26

Since $f(z)$ has no singularity inside $C$ (its only singular point being a simple pole at $z = 0$ which has been deleted by drawing $C_r$), we have by Cauchy's theorem :

$$\int_r^R f(x)dx + \int_{C_R} f(z)dz + \int_{-R}^{-r} f(x)dx + \int_{C_r} f(z)dz = 0 \quad ...(i)$$

Now $$\int_{C_R} f(z)dz = \int_0^{\pi} \frac{e^{imR(\cos\theta + i\sin\theta)}}{Re^{i\theta}} . Rie^{i\theta}\, d\theta \qquad [\because \; z = Re^{i\theta}]$$

$$= i\int_0^{\pi} e^{imR(\cos\theta + i\sin\theta)}\, d\theta$$

Since $|e^{imR(\cos\theta + i\sin\theta)}| = |e^{-mR\sin\theta + imR\cos\theta}| = e^{-mR\sin\theta}$

$\therefore$ $$\left|\int_{C_R} f(z)dz\right| \le \int_0^{\pi} e^{-mR\sin\theta}\, d\theta = 2\int_0^{\pi/2} e^{-mR\sin\theta}\, d\theta$$

$$= 2\int_0^{\pi/2} e^{-2mR\theta/\pi}\, d\theta \qquad [\because \; \text{for } 0 \le \theta \le \pi/2,\ \sin\theta/\theta \ge 2/\pi]$$

$$= \frac{\pi}{mR}\left(1 - e^{-mR}\right) \text{ which} \to 0 \text{ as } R \to \infty,$$

Also $$\int_{C_r} f(z)dz = i\int_{\pi}^{0} e^{imr(\cos\theta + i\sin\theta)}\, d\theta \to i\int_{\pi}^{0} d\theta \; i.e., -i\pi \text{ as } r \to 0.$$

Hence as $r \to 0$ and $R \to \infty$, we get from (*i*) $\int_0^{\infty} f(x)dx + 0 + \int_{-\infty}^{0} f(x)dx - i\pi = 0$

or $$\int_{-\infty}^{\infty} f(x)dx = i\pi \; i.e., \int_{-\infty}^{\infty} \frac{e^{imx}}{x} dx = i\pi \quad ...(ii)$$

Equating imaginary parts from both sides,

$$\int_{-\infty}^{\infty} \frac{\sin mx}{x} dx = \pi. \text{ Hence } \int_0^{\infty} \frac{\sin mx}{x} dx = \frac{\pi}{2}.$$

**Obs.** Equating real parts from both sides of (*ii*), we get

$$\int_{-\infty}^{\infty} \frac{\cos mx}{x} dx = 0.$$

**Example 20.46.** *Show that* $\int_0^\infty \frac{x^{p-1}}{1+x} dx = \frac{\pi}{\sin p\pi}, 0 < p < 1.$

**Solution.** Integrate $f(z) = \frac{z^{p-1}}{1+z}$ along the contour consisting of the circles α and γ of radii $a$ and $R$ and the lines $AB$ and $FG$ along $x$-axis (Fig. 20.27). There is a simple pole at $z = -1$ which is within the contour.

$$\therefore \quad \text{Res } f(-1) = \underset{z\to -1}{\text{Lt}} (1+z) \cdot \frac{z^{p-1}}{1+z} = \underset{z\to -1}{\text{Lt}} z^{p-1} = (-1)^{p-1} = e^{i\pi(p-1)}$$

Thus
$$\int_{AB} f(z)dz + \int_\gamma f(z)dz + \int_{FG} f(z)dz + \int_\alpha f(z)dz = 2\pi i \, e^{i\pi(p-1)} \quad ...(i)$$

On $AB$ : $z = x$ and on $FG$ : $z = xe^{2\pi i}$

$$\therefore \quad \int_{AB} f(z)dz + \int_{FG} f(z)dz = \int_a^R \frac{x^{p-1}}{1+x} dx + \int_R^a \frac{(xe^{2\pi i})^{p-1}}{1+xe^{2\pi i}} dx \, e^{2\pi i}$$

$$= \int_a^R \frac{x^{p-1}}{1+x} [1 - e^{2\pi i(p-1)}] \, dx$$

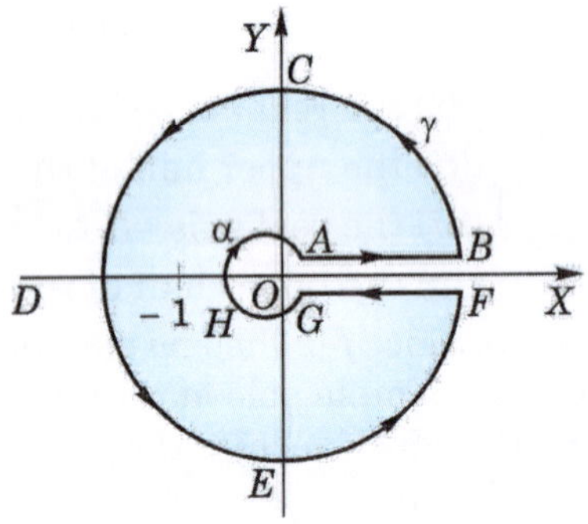

Fig. 20.27

On the circle $\gamma : z = Re^{i\theta}$. So

$$\int_\gamma f(z)dz = \int_0^{2\pi} \frac{(Re^{i\theta})^{p-1}}{1+Re^{i\theta}} Re^{i\theta} \, i \, d\theta$$

For large $R$, the integrand is of the order $\frac{R^{p-1} . R}{1+R}$ *i.e.* $R^{p-1}$ which tends to zero as $R \to \infty$. $(\because \quad p < 1)$

Hence $\int_\gamma f(z)dz \to 0$ as $R \to \infty$

On the circle $\alpha : z = ae^{i\theta}$. So

$$\int_\alpha f(z)dz = \int_{2\pi}^0 \frac{(ae^{i\theta})^{p-1}}{1+ae^{i\theta}} ae^{i\theta} \, i d\theta$$

For small α, the integrand is of the order $a^p$ which tends to zero as $a \to 0$. $(\because \quad p > 0)$

Thus on taking limits as $a \to 0$ and $R \to \infty$, (*i*) gives

$$\int_0^\infty \frac{x^{p-1}}{1+x} \{1 - e^{2\pi i(p-1)}\} dx = 2\pi i \, e^{i\pi(p-1)}$$

or
$$\int_0^\infty \frac{x^{p-1}}{1+x} dx = \frac{2\pi i \, e^{i\pi(p-1)}}{1 - e^{2\pi i(p-1)}} = \frac{2\pi i \, e^{ip\pi}(-1)}{1 - e^{2ip\pi}(1)} = \frac{2i \, . \, \pi}{e^{ip\pi} - e^{-ip\pi}} = \frac{\pi}{\sin p\pi}.$$

**Example 20.47.** *Prove that* $\int_0^\infty \sin x^2 \, dx = \int_0^\infty \cos x^2 \, dx = \frac{1}{2}\sqrt{\left(\frac{\pi}{2}\right)}.$

**Solution.** Consider $\int_C e^{-z^2} dz$ where $C$ consists of the real axis from $O$ to $A$, part of circle $AB$ of radius $R$ and the line $\theta = \frac{\pi}{4}$. (Fig. 20.28).

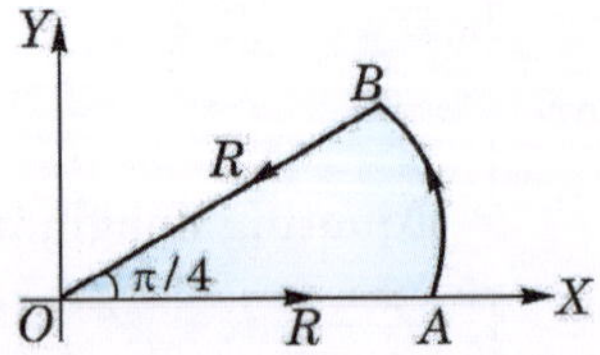

Fig. 20.28

$e^{-z^2}$ has no singularity within $C$.

$$\therefore \quad \int_{OA} e^{-z^2} dz + \int_{AB} e^{-z^2} dz + \int_{BO} e^{-z^2} dz = 0 \quad ...(i)$$

On $OA : z = x$, $\therefore \quad \int_{OA} e^{-z^2} dz = \int_0^R e^{-x^2} dx \to \sqrt{\pi/2}$ as $R \to \infty$

On $AB : z = Re^{i\theta}$,

$$\therefore \quad \int_{AB} e^{-z^2} dz = \int_0^{\pi/4} e^{-R^2(\cos 2\theta + i \sin 2\theta)} . Re^{i\theta} . id\theta \to 0 \quad \text{as} \quad R \to \infty.$$

[$\because$ integrand $\to 0$ as $R \to \infty$]

On $BO : z = re^{i\pi/4}$ and $z^2 = r^2 e^{i\pi/2} = ir^2$

$$\therefore \quad \int_{BO} e^{-z^2} dz = \int_R^0 e^{-ir^2} . e^{i\pi/4} dr = -\int_0^R e^{-ix^2} \frac{1+i}{\sqrt{2}} dx$$

$$\to -\int_0^\infty (\cos x^2 - i \sin x^2) \frac{1+i}{\sqrt{2}} dx \quad \text{when } R \to \infty$$

Substituting these in $(i)$, we get

$$\frac{1}{2}\sqrt{\pi} + 0 - \int_0^\infty (\cos x^2 - i \sin x^2) \left(\frac{1+i}{\sqrt{2}}\right) dx = 0$$

Equating real and imaginary parts, we obtain

$$\int_0^\infty (\cos x^2 + \sin x^2) dx = \frac{1}{2}\sqrt{(2\pi)} \quad \text{and} \quad \int_0^\infty (\cos x^2 - \sin x^2) dx = 0$$

Hence $\quad \int_0^\infty \sin x^2 \, dx = \int_0^\infty \cos x^2 \, dx = \frac{1}{2}\sqrt{\left(\frac{\pi}{2}\right)}$.

## PROBLEMS 20.9

Apply the calculus of residues, to prove that

1. $\int_0^{2\pi} \frac{d\theta}{1 - 2p \sin\theta + p^2} = \frac{2\pi}{1-p^2} (0 < p < 1)$. )
2. $\int_0^{2\pi} \frac{d\theta}{1 - 2r \cos\theta + r^2} = \frac{\pi}{1-r^2}$.
3. $\int_0^{2\pi} \frac{d\theta}{13 + 12\cos\theta} = \frac{\pi}{5}$. *(Anna., 2016)*
4. $\int_0^{2\pi} \frac{\cos 2\theta}{5 + 4\cos\theta} d\theta = \frac{\pi}{6}$. *(W.B.T.U., 2015 ; U.P.T.U., 2010)*
5. $\int_0^{2\pi} \frac{\sin^2\theta \, d\theta}{a + b\cos\theta} = \frac{2\pi}{b^2}[a - \sqrt{(a^2 - b^2)}], (0 < b < a)$. *(Andhra, 2021)*
6. $\int_0^{2\pi} \frac{a d\theta}{a^2 + \sin^2\theta} = \frac{2\pi}{\sqrt{(1+a^2)}}, (a > 0)$.
7. $\int_0^{2\pi} \frac{d\theta}{(5 - 3\cos\theta)^2} = \frac{5\pi}{32}$. *(Andhra, 2020)*
8. $\int_{-\infty}^{\infty} \frac{x^2 dx}{(x^2 + a^2)(x^2 + b^2)} = \frac{\pi}{a+b} (a, b > 0)$. *(Andhra., 2015)*
9. $\int_{-\infty}^{\infty} \frac{x^2 - x + 2}{x^4 + 10x^2 + 9} dx = \frac{5\pi}{12}$.
10. $\int_0^\infty \frac{dx}{x^4 + a^4} = \frac{\pi\sqrt{2}}{4a^3} (a > 0)$. *(Anna., 2016)*
11. $\int_0^\infty \frac{dx}{(1+x^2)^2} = \frac{\pi}{4}$. *(D.T.U., 2013)*
12. $\int_0^\infty \frac{dx}{x^6 + 1} = \frac{\pi}{3}$.
13. $\int_0^\infty \frac{x^2}{x^6 + 1} dx = \frac{\pi}{6}$.
14. $\int_{-\infty}^{\infty} \frac{\cos mx}{e^x + e^{-x}} dx = \frac{\pi}{2} \text{sech} \frac{m\pi}{2}$.
15. $\int_0^\infty \frac{1 - \cos x}{x^2} dx = \pi/2$.
16. $\int_0^\infty \frac{\sin x}{x} dx = \frac{\pi}{2}$. *(D.T.U., 2012)*
17. $\int_{-\infty}^{\infty} \frac{\sin x}{x^2 + 4x + 5} dx = \frac{-\pi \sin 2}{e}$.
18. $\int_{-\infty}^{\infty} \frac{\cos x \, dx}{(x^2 + a^2)(x^2 + b^2)} = \frac{\pi}{a^2 - b^2}\left(\frac{e^{-b}}{b} - \frac{e^{-a}}{a}\right) (a > b > 0)$.
19. $\int_0^\infty \frac{x \sin mx}{x^2 + a^2} dx = \frac{\pi}{2} e^{-ma} (m > 0)$. *(Anna., 2016)*

## 20.21. OBJECTIVE TYPE OF QUESTIONS

### PROBLEMS 20.10

*Select the correct answer or fill up the blanks in each of the following questions :*

1. The only function that is analytic from the following is
   (*i*) $f(z) = \sin z$ (*ii*) $f(z) = \bar{z}$ (*iii*) $f(z) = \text{Im}(z)$ (*iv*) $R(iz)$.
2. If $f(z) = u(x, y) + iv(x, y)$ is analytic, then $f'(z) =$
   (*i*) $\frac{\partial u}{\partial x} - i\frac{\partial v}{\partial x}$ (*ii*) $\frac{\partial u}{\partial x} - i\frac{\partial u}{\partial y}$ (*iii*) $\frac{\partial v}{\partial y} - i\frac{\partial v}{\partial x}$.
3. If $2x - x^2 + ay^2$ is to be harmonic, then $a$ should be
   (*a*) 1 (*b*) 2 (*c*) 3 (*d*) 0.
4. The analytic function which maps the angular region $0 \le \theta \le \pi/4$ onto the upper half plane is
   (*i*) $z^2$ (*ii*) $4z$ (*iii*) $z^4$ (*iv*) $2\theta$.
5. An angular domain in the complex plane is defined by $0 < \text{amp}(z) < \pi/4$. The mapping which maps this region onto the left half plane is
   (*i*) $w = z^4$ (*ii*) $w = iz^4$ (*iii*) $w = -z^4$ (*iv*) $w = -iz^4$.
6. The mapping $w = z^2 - 2z - 3$ is
   (*i*) conformal within $|z| = 1$ (*ii*) not conformal at $z = 1$
   (*iii*) not conformal at $z = -1$ and $z = 3$ (*iv*) conformal everywhere.
7. If $z = re^{i\theta}$, then the image of $\theta$ = constant under the mapping $w(z) = Re^{i\phi} = iz^3$ is
   (*i*) $\phi = 3\theta$ (*ii*) $\phi = 3\theta + \pi/2$ (*iii*) $\phi = 3\theta - \pi/2$ (*iv*) $\phi = \theta^3$.
8. The fixed points of the mapping $w = (5z + 4)/(z + 5)$ are
   (*i*) 2, 2 (*ii*) 2, –2 (*iii*) –2, –2 (*iv*) –4/5, 5.
9. The value of $\int_C (4x^3\,dx + 3y^2z^2\,dy + 2y^3 z\,dz)$ where $C$ is any path joining $A(-1, 1, 0)$ to $B(1, 2, 1)$ is
   (*i*) 0 (*ii*) 1 (*iii*) 8 (*iv*) –8.
10. The value of $\int_C \frac{3z^2 + 7z + 1}{z + 1} dz$ where $C$ is $|z| = 1/2$ is
    (*i*) $2\pi i$ (*ii*) 0 (*iii*) $\pi i$ (*iv*) $\pi i/2$.
11. The value of $\int_C \frac{3z + 4}{z(2z + 1)} dz$ where $C$ is the circle $|z| = 1$ is
    (*i*) $2\pi i$ (*ii*) $3\pi i$ (*iii*) 4 (*iv*) –4.
12. The residue of a function can be found if the pole is an isolated singularity :
    (*i*) True (*ii*) False (*iii*) Partially false (*iv*) none of these.
13. The value of $\int_C \frac{z\,dz}{\sin z}$ where $C : |z| = 4$ is
    (*i*) $2\pi i$ (*ii*) 0 (*iii*) $-2\pi i$ (*iv*) $4\pi i$.
14. The value of $\int_C \tanh z\,dz$, where $C : |z| = 3$, is
    (*i*) 0 (*ii*) $\pi i$ (*iii*) $2\pi i$ (*iv*) $4\pi i$.
15. The harmonic conjugate of the function $u(x, y) = 2x(1 - y)$ is ......
16. Harmonic conjugate of $x^3 - 3xy^2$ is .......
17. The curves $u(x, y) = c$ and $v(x, y) = c'$ are orthogonal if ......
18. The value of $\int_0^{1+i} z^2\,dz$ along the line $x = y$ is ......
19. Residue of $\frac{\cos z}{z}$ at $z = 0$ is ......
20. The critical point of the transformation $w^2 = (z - a)(z - b)$ is ......
21. Image of $|z + 1| = 1$ under the mapping $w = 1/z$ is......
22. The poles of $f(z) = (z^3 - 1)/(z^3 + 1)$ are $z =$ ......
23. $w = \log z$ is analytic everywhere except at $z =$ ......
24. If $f(z) = -\frac{1}{z - 1} - 2[1 + (z - 1) + (z - 1)^2 + \ldots\ldots]$, then the residue of $f(z)$ at $z = 1$ is ......
25. If $|z| < 1$ then Taylor's series expansion of $\log(1 + z)$ about $z = 0$ is ......

26. The value of $\int_C \frac{4z^2+z+5}{z-4}dz$ where $C$ is $9x^2 + 4y^2 = 36$, is ......

27. The value of $\int z^4 e^{1/z}\, dz$, where $C$ is $|z| = 1$, is

(i) $\pi i$ (ii) $\pi i/12$ (iii) $\pi i/60$ (iv) $-\pi i/60$.

28. If $f(z)$ has a pole of order three at $z = a$ Res $[f(a)]$ = ......

29. The value of $\int_C \frac{e^z\, dz}{(z-3)^2}$, $C$ being $|z| = 2$, is ......

30. The $CR$ equations for $f(z) = u(x, y) + iv(x, y)$ to be analytic are ......

31. If $f(z)$ is analytic in a simply connected domain $D$ and $C$ is any simple closed path then $\int_C f(z)dz$ = ......

32. The harmonic conjugate of $e^x \cos y$ is ......

33. The value of $\oint_C \cos z\, dz$ where $C$ is the circle $|z| = 1$, is ......

34. The singularity of $f(z) = z/(z-2)^3$ is ......

35. The function $f(z) = \bar{z}$ is analytic at ......

36. $C$-$R$ equations for a function to be analytic, in polar form, are ......

37. If $C$ is the circle $|z - a| = r$, $\int_C (z-a)^n\, dz$ [ $n$, any integer $\neq -1$] = ......

38. A simply connected region is that ......

39. A holomorphic function is that ......

40. The poles of the function $f(z) = \frac{z^2}{(z-1)^2(z+2)}$ are at $z$ = ......

41. The cross-ratio of four points $z_1, z_2, z_3, z_4$ is ......

42. The value of $\int_C |z|\, dz$, where $C$ is the contour represented by the straight line from $z = -i$ to $z = i$, is ......

43. Taylor's series expansion of $\left(\frac{1}{z-2} - \frac{1}{z-1}\right)$ in the region $|z| < 1$, is ......

44. The invariant points of the transformation $w = (1+z)/(1-z)$ are $z$ = ......

45. The residue at $z = 0$ of $\frac{1+e^z}{z\cos z + \sin z}$ is ......

46. The transformation $w = Cz$ consists of ......

47. The residue of $f(z)$ at a pole is ......

48. The value of $\int_C \frac{1}{z-1}dz$, $C$ being $|z| = 2$, is ......

49. If $C$ is $|z| = 1/2$, $\int_C \frac{z^2-z+1}{z-1}dz$ = ......

50. Singular points of $\frac{\cos \pi z}{(z-1)(z-2)}$ are ......

51. Taylor series expansion of $\frac{1}{z-2}$ in $|z| < 1$ is ......

52. $\underset{z\to\infty}{\text{Lt}} \frac{iz^3+iz-1}{(2z+3)(z-1)^2}$ = ........................

53. The poles of $\frac{(z-1)^2}{z(z-2)^2}$ are at $z$ = ......

54. Cauchy's integral theorem states that ......

55. The critical points of the transformation $w = z + 1/z$ are ......

56. $\int_c \frac{dz}{2z-3}$, where $|z| = 1$, is .......

57. The zeroes and singularities of $\frac{z^2+1}{1-z^2}$ are ......

58. Residue of $\tan z$ at $z = \pi/2$ is ......

59. Singularity of $e^{z^{-1}}$ at $z = 0$ is of the type .......

60. Res $(e^{1/z})_{z=0}$ = ......

61. Taylor's series expansion of $\sin z$ about $z = \pi/4$ is ......

62. Image of $|z| = 2$ under $w = z + 3 + 2i$ is ............

63. The poles of $\cot z$ are ...................

64. If $a$ is simple pole, then Res $[\phi(z)/\psi(z)]_{z=a}$ = .........

65. Bilinear transformation always transforms circles into ................

66. If $f(z)$ and $\overline{f(z)}$ are analytic functions, then $f(z)$ is constant. (True or False)

67. The function $u(x, y) = 2xy + 3xy^2 - 2y^3$ is a harmonic functions. (True or False)

68. The function $e^x \cos y$ is harmonic. (True or False)

69. $\int_C \frac{f(z)}{z-a} dz = 2\pi i f(a)$, if $z = a$ is a point within $C$. (True or False)

70. The transformation affected by an analytic function $w = f(z)$ is conformal at every point of the $z$-plane where $f'(z) = 0$. (True or False)

71. The function $\bar{z}$ is not analytic at any point. (True or False)

72. Under the transformation $w = 1/z$, circle $x^2 + y^2 - 6x = 0$ transforms into a straight line in the $w$-plane. (True or False)

73. If $w = f(z)$ is analytic, then $\frac{dw}{dz} = -i \frac{\partial w}{\partial y}$. (True or False)

74. An analytic function with constant imaginary part is constant. (True or False)

75. If $u + iv$ is analytic, then $v - iu$ is also analytic. (True or False)

76. $f(z) = I_m(z)$ is not analytic. (True or False)

77. The cross-ratio of four points is not invariant under bilinear transformation. (True or False)

78. $z = 0$ is not a critical point of the mapping $w = z^2$. (True or False)

79. $f(z) = \text{Re}(z^2)$ is analytic. (True or False)

80. An analytic function with constant modulus is constant. (True or False)

81. The function $|\bar{z}|^2$ is not analytic at any point. (True or False)

82. If $f(z) = z^2$, then the family of curves $x^2 - y^2 = C_1$, and $xy = C_2$ are orthogonal. (True or False)

83. $f(x) = x^2 + xy + y^2 + i(ax^2 + 2by + cy^2)$ is analytic, if $a = ......$, $b = ......$, $c = ......$ .

CHAPTER 21

# Laplace Transforms



## 21.1 INTRODUCTION

The knowledge of Laplace transforms has in recent years become an essential part of mathematical background required of engineers and scientists. This is because the transform methods provide an easy and effective means for the solution of many problems arising in engineering.

This subject originated from the operational methods applied by the English engineer Oliver Heaviside (1850–1925), to problems in electrical engineering. Unfortunately, Heaviside's treatment was unsystematic and lacked rigour, which was placed on sound mathematical footing by Bromwich and Carson during 1916–17. It was found that Heaviside's operational calculus is best introduced by means of a particular type of definite integrals called Laplace transforms.*

## 21.2 (1) DEFINITION

*Let f(t) be a function of t defined for all positive values of t. Then the* **Laplace transforms** *of f(t), denoted by* $L\{f(t)\}$ *is defined by*

$$L\{f(t)\} = \int_0^\infty e^{-st} f(t)\, dt \qquad ...(1)$$

*provided that the integral exists.* $s$ is a parameter which may be a real or complex number.

$L\{f(t)\}$ being clearly a function of $s$ is briefly written as $\bar{f}(s)$ *i.e.*, $L\{f(t)\} = \bar{f}(s)$,

which can also be written as $f(t) = L^{-1}\{\bar{f}(s)\}$.

Then $f(t)$ is called the **inverse Laplace transform** of $\bar{f}(s)$. The symbol $L$, which transforms $f(t)$ into $\bar{f}(s)$, is called the *Laplace transformation operator.*

**(2) Conditions for the existence**

*The Laplace transform of* $f(t)$ *i.e.,* $\int_0^\infty e^{-st} f(t)\, dt$ *exists for* $s > a$, *if*

*Pierre de Laplace (1749–1827) used such transforms, much earlier in 1799, while developing the theory of probability.

(*i*) *$f(t)$ is continuous* (*ii*) $\underset{t \to \infty}{\text{Lt}} \{e^{-at} f(t)\}$ *is finite.*

It should however, be noted that the above conditions are sufficient and not necessary. (*U.P.T.U., 2012*)

For example, $L(1/\sqrt{t}\,)$ exists, though $1/\sqrt{t}$ is infinite at $t = 0$.

**(3) Advantages**

(*i*) Particular solution of a differential equations is obtained directly without first finding the general solution and than finding the constants by substituting the initial conditions.

(*ii*) Non-homogenous differential equations can be solved without the necessity of first solving the corresponding homogeneous differential equations.

(*iii*) Applicable not to continuous functions but also to discontinuous functions, periodic functions, step functions, impulse functions.

(*iv*) Laplace transforms of various functions are readily available in the form of tables see Appendix 2 Table III which reduces the problem of solving differential equations to mere algebraic manupulations.

(*v*) Laplace transforms are very useful for finding the solutions of both ordinary and partial differential equations, solutions of simultaneous differential equations, solution of linear difference equations (Chap. 31), solution of integral equations and in evaluating complicated definite integrals.

## 21.3 TRANSFORMS OF ELEMENTARY FUNCTIONS

The direct application of the definition gives the following formulae :

**(1)** $\mathbf{L(1) = \frac{1}{s}}$ $(s > 0)$

**(2)** $\mathbf{L(t^n) = \frac{n!}{s^{n+1}}}$, when $n = 0, 1, 2, 3, \ldots$ $\left[\text{Otherwise } \frac{\Gamma(n+1)}{s^{n+1}}\right]$

**(3)** $\mathbf{L(e^{at}) = \frac{1}{s-a}}$ $(s > a)$

**(4)** $\mathbf{L(\sin at) = \frac{a}{s^2+a^2}}$ $(s > 0)$

**(5)** $\mathbf{L(\cos at) = \frac{s}{s^2+a^2}}$ $(s > 0)$

**(6)** $\mathbf{L(\sinh at) = \frac{a}{s^2-a^2}}$ $(s > |\,a\,|)$

**(7)** $\mathbf{L(\cosh at) = \frac{s}{s^2-a^2}}$ $(s > |\,a\,|)$

*Proofs.* **(1)** $$L(1) = \int_0^\infty e^{-st} . 1\, dt = \left| -\frac{e^{-st}}{s} \right|_0^\infty = \frac{1}{s} \text{ if } s > 0.$$

**(2)** $$L(t^n) = \int_0^\infty e^{-st} . t^n\, dt = \int_0^\infty e^{-p} . \left(\frac{p}{s}\right)^n \frac{dp}{s}, \text{ on putting } st = p$$

$$= \frac{1}{s^{n+1}} \int_0^\infty e^{-p} . p^n\, dp = \frac{\Gamma(n+1)}{s^{n+1}}, \text{ if } n > -1 \text{ and } s > 0.$$

In particular $L(t^{-1/2}) = \frac{\Gamma(1/2)}{s^{1/2}} = \sqrt{\frac{\pi}{s}}$ ; $L(t^{1/2}) = \frac{\Gamma(3/2)}{s^{3/2}} = \frac{\sqrt{\pi}}{2s^{3/2}}$

In $n$ be a positive integer, $\Gamma(n + 1) = n\,!$ [(*v*) p. 350],

therefore, $L(t^n) = n\,!/s^{n+1}$. (*V.T.U., 2013*)

**(3)** $$L(e^{at}) = \int_0^\infty e^{-st} . e^{at}\, dt = \int_0^\infty e^{-(s-a)t}\, dt = \left| \frac{e^{-(s-a)t}}{-(s-a)} \right|_0^\infty = \frac{1}{s-a}, \text{ if } s > a.$$

**(4)** $L(\sin at) = \int_0^\infty e^{-st} \sin at\, dt = \left| \frac{e^{-st}}{s^2+a^2} (-s \sin at - a \cos at) \right|_0^\infty = \frac{a}{s^2+a^2}$

Similarly, the reader should prove **(5)** himself.

**(6)** $L(\sinh at) = \int_0^\infty e^{-st} \sinh at\, dt = \int_0^\infty e^{-st} \left( \frac{e^{at}-e^{-at}}{2} \right) dt$

$$= \frac{1}{2}\left[\int_0^\infty e^{-(s-a)t}\, dt - \int_0^\infty e^{-(s+a)t}\, dt\right] = \frac{1}{2}\left[\frac{1}{s-a} - \frac{1}{s+a}\right] = \frac{a}{s^2-a^2} \text{ for } s > |a|.$$

Similarly, the reader should prove **(7)** himself.

## 21.4 PROPERTIES OF LAPLACE TRANSFORMS

**I. Linearity property.** *If a, b, c be any constants and f, g, h any functions of t, then*

$$\mathbf{L\,[af(t) + bg(t) - ch(t)] = aL\,\{f(t)\} + bL\,\{g(t)\} - cL\,\{h(t)\}}$$

For by definition,

$$\text{L.H.S.} = \int_0^\infty e^{-st}\,[af(t) + bg(t) - ch(t)]\, dt$$

$$= a\int_0^\infty e^{-st} f(t)\, dt + b\int_0^\infty e^{-st} g(t)\, dt - c\int_0^\infty e^{-st} h(t)\, dt = aL\{f(t)\} + bL\{g(t)\} - cL\{h(t)\}$$

This result can easily be generalised.

Because of the above property of $L$, it is called a *linear operator*.

**II. First shifting property.** *If* $L\{f(t)\} = \bar{f}(s)$, then

$$\mathbf{L\,\{e^{at} f(t)\} = \bar{f}(s-a)}.$$

By definition, $L\{e^{at} f(t)\} = \int_0^\infty e^{-st} e^{at} f(t)\, dt = \int_0^\infty e^{-(s-a)t} f(t)\, dt$

$$= \int_0^\infty e^{-rt} f(t)\, dt, \text{ where } r = s - a = \bar{f}(r) = \bar{f}(s-a).$$

*Thus, if we know the transform* $\bar{f}(s)$ *of* $f(t)$, *we can write the transform of* $e^{at} f(t)$ *simply replacing s by* $s-a$ *to get* $\bar{f}(s-a)$.

*Application of this property leads us to the following useful results :*

**(1)** $\mathbf{L\,(e^{at}) = \frac{1}{s-a}}$ $\left[\because L(1) = \frac{1}{s}\right]$

**(2)** $\mathbf{L\,(e^{at}t^n) = \frac{n!}{(s-a)^{n+1}}}$ $\left[\because L(t^n) = \frac{n!}{s^{n+1}}\right]$

**(3)** $\mathbf{L\,(e^{at} \sin bt) = \frac{b}{(s-a)^2+b^2}}$ $\left[\because L(\sin bt) = \frac{b}{s^2+b^2}\right]$

**(4)** $\mathbf{L\,(e^{at} \cos bt) = \frac{s-a}{(s-a)^2+b^2}}$ $\left[\because L(\cos bt) = \frac{s}{s^2+b^2}\right]$

**(5)** $\mathbf{L\,(e^{at} \sinh bt) = \frac{b}{(s-a)^2-b^2}}$ $\left[\because L(\sinh bt) = \frac{b}{s^2-b^2}\right]$

**(6)** $\mathbf{L\,(e^{at} \cosh bt) = \frac{s-a}{(s-a)^2-b^2}}$ $\left[\because L(\cosh bt) = \frac{s}{s^2-b^2}\right]$

where in each case $s > a$.

**Example 21.1.** *Find the Laplace transforms of*
*(i) sin 2t sin 3t* *(ii)* $\cos^2 2t$ *(iii)* $\sin^3 2t$. *(V.T.U., 2014 ; D.T.U., 2012)*

**Solution.** (*i*) Since $\sin 2t \sin 3t = \frac{1}{2}[\cos t - \cos 5t]$

$$\therefore \quad L(\sin 2t \sin 3t) = \tfrac{1}{2}[L(\cos t) - L(\cos 5t)] = \frac{1}{2}\left[\frac{s}{s^2+1^2} - \frac{s}{s^2+5^2}\right] = \frac{12s}{(s^2+1)(s^2+25)}$$

(*ii*) Since $\cos^2 2t = \frac{1}{2}(1 + \cos 4t)$

$$\therefore \quad L(\cos^2 2t) = \frac{1}{2}[L(1) + L\cos 4t] = \frac{1}{2}\left(\frac{1}{s} + \frac{s}{s^2+16}\right)$$

(*iii*) Since $\sin 6t = 3\sin 2t - 4\sin^3 2t$

or $\sin^3 2t = \frac{3}{4}\sin 2t - \frac{1}{4}\sin 6t$

$$\therefore \quad L(\sin^3 2t) = \tfrac{3}{4}L(\sin 2t) - \tfrac{1}{4}L(\sin 6t)$$

$$= \frac{3}{4}\cdot\frac{2}{s^2+2^2} - \frac{1}{4}\cdot\frac{2}{s^2+6^2} = \frac{48}{(s^2+4)(s^2+36)}.$$

**Example 21.2.** *Find the Laplace transform of*
*(i)* $e^{-t}\cos t - \cos 2t$ *(ii)* $t^2 e^t \sin 4t$

**Solution:** (*i*) Since $\cot t \cos 2t = \frac{1}{2}(\cos 3t + \cos t)$

$$L(e^{-t}\cos t\cos 2t) = \frac{1}{2}L\left[\left\{(e^{-t}\cos 3t) + \frac{1}{2}(e^{-t}\cos t)\right\}\right] = \frac{1}{2}\left[\frac{s+1}{(s+1)^2+9} + \frac{s+1}{(s+1)^2+1}\right]$$

$$= \frac{(s+1)(s^2+2s+6)}{(s^2+2s+10)(s^2+2s+2)}$$

(*ii*) Since $L(t^2) = \dfrac{2}{S^2}$, we have $L\{t^2 e^{i4t}\} = \dfrac{2}{(s-4i)^3} = \dfrac{2(s+4i)^3}{(s-4i)^3(s+4i)^3}$

or $$L\{t^2\cos 4t + it^2 \sin 4t\} = \frac{2(s^2 - 48s) + 8i(5s^2-16)}{(s^2+16)^3}$$

Equating imaging parts $$L(t^2 \sin 4t) = \frac{8(3s^2-16)}{(s^2+16)^3}$$

Now applying first shifting theorem, we get

$$L\{e^t t^2 \sin 4t\} = \frac{8\{3(s-1)^2 - 16]}{[(s-1)^2+16]^2} = \frac{8(3S^2 - 6S - 13)}{[(S^2 - 2S + 17)^3}$$

**Example 21.3.** *If* $L\,f(t) = \bar{f}(s)$, *show that*

$$L[(\sinh at)f(t)] = \frac{1}{2}[\bar{f}(s-a) - \bar{f}(s+a)]$$

$$L[(\cosh at)f(t)] = \frac{1}{2}[\bar{f}(s-a) + \bar{f}(s+a)]$$

*Hence evaluate (i) sinh 2t sin 3t (ii) cosh 3t cos 2t.*

**Solution.** We have $L[(\sinh at)f(t)] = L\left[\frac{1}{2}(e^{at} - e^{-at})f(t)\right] = \frac{1}{2}[L\{e^{at}f(t) - L\{e^{-at}f(t)\}]$

$$= \frac{1}{2}[\bar{f}(s-a) - \bar{f}(s+a)], \text{ by shifting property.}$$

Similarly, $$L[(\cosh at) f(t)] = \frac{1}{2}[L\{e^{at} f(t)\} + \{e^{-at} f(t)\}]$$

$$= \frac{1}{2}[\bar{f}(s-a) + \bar{f}(s+a)], \text{ by shifting property.}$$

(*i*) Since $L(\sin 3t) = \dfrac{3}{s^2+3^2}$, the first result gives

$$L(\sinh 2t \sin 3t) = \frac{1}{2}\left\{\frac{3}{(s-2)^2+3^2} - \frac{3}{(s+2)^2+3^2}\right\} = \frac{12s}{s^4+10s^2+169}$$

(*ii*) Since $L(\cos 2t) = \dfrac{s}{s^2+2^2}$, the second result gives

$$L(\cosh 3t \cos 2t) = \frac{1}{2}\left\{\frac{s-3}{(s-3)^2+2^2} + \frac{s+3}{(s+3)^2+2^2}\right\} = \frac{2s(s^2-5)}{s^4-10s^2+169}.$$

**Example 21.4.** *Show that*

(*i*) $L(t \sin at) = \dfrac{2as}{(s^2+a^2)^2}$ (*ii*) $L(t \cos at) = \dfrac{s^2-a^2}{(s^2+a^2)^2}$.

**Solution.** Since $L(t) = 1/s^2$. $\therefore$ $L(te^{iat}) = \dfrac{1}{(s-ia)^2} = \dfrac{(s+ia)^2}{[(s-ia)(s+ia)]^2}$

or $$L[t(\cos at + i \sin at)] = \frac{(s^2-a)^2 + i(2as)}{(s^2+a^2)^2}$$

Equating the real and imaginary parts from both sides, we get the desired results.

**Example 21.5.** *Find the Laplace transform of $f(t)$ defined as*

(*i*) $f(t) = t/\tau$, *when* $0 < t < \tau$
$\quad - 1$, *when* $t > \tau$.

(*ii*) $f(t) = \begin{cases} 1, & 0 < t \le 1 \\ t, & 1 < t \le 2 \\ 0, & t > 2 \end{cases}$

**Solution.** (*i*) $$Lf(t) = \int_0^\tau e^{-st} \cdot \frac{t}{\tau} dt + \int_\tau^\infty e^{-st} \cdot 1\, dt = \frac{1}{\tau}\left[\left| t \cdot \frac{e^{-st}}{-s} \right|_0^\tau - \int_0^\tau 1 \cdot \frac{e^{-st}}{-s} dt\right] + \left| \frac{e^{-st}}{-s} \right|_\tau^\infty$$

$$= \frac{1}{\tau}\left[\frac{\tau e^{-s\tau} - 0}{-s} - \left| \frac{e^{-st}}{s^2} \right|_0^\tau\right] + \frac{0 - e^{-s\tau}}{-s} = \frac{-e^{-s\tau}}{s} - \frac{e^{-s\tau}-1}{\tau s^2} + \frac{e^{-s\tau}}{s} = \frac{1-e^{-s\tau}}{\tau s^2}.$$

(*ii*) $$L\{f(t)\} = \int_0^1 e^{-st} \cdot 1\, dt + \int_1^2 e^{-st} \cdot t\, dt + \int_2^\infty e^{-st} \cdot (0)\, dt$$

$$= \left| \frac{e^{-st}}{-s} \right|_0^1 + \left| t \cdot \frac{e^{-st}}{-s} - \frac{e^{-st}}{s^2} \right|_1^2 = \frac{1-e^{-s}}{s} + \left\{\left(-\frac{2e^{-2s}}{s} - \frac{e^{-2s}}{s^2}\right) - \left(\frac{e^{-s}}{-s} - \frac{e^{-s}}{s^2}\right)\right\}$$

$$= \frac{1}{s} - \frac{2e^{-2s}}{s} + \frac{e^{-s}}{s^2} - \frac{e^{-2s}}{s^2}.$$

**Example 21.6.** *Find the Laplace transform of* (*i*) $\left(\sqrt{t} - \dfrac{1}{\sqrt{t}}\right)^3$. (*Rohtak, 2011*)

(*ii*) $\dfrac{\cos\sqrt{t}}{\sqrt{t}}$

**Solution.** (*i*) Since $(\sqrt{t} - 1/\sqrt{t})^3 = t^{3/2} - 3t^{1/2} + 3t^{-1/2} - t^{-3/2}$

$$\therefore \quad L(\sqrt{t} - 1/\sqrt{t}) = L(t^{3/2}) - 3L(t^{1/2}) + 3L(t^{-1/2}) - L(t^{-3/2})$$

$$= \frac{\Gamma(3/2+1)}{s^{3/2+1}} - 3\frac{\Gamma(1/2+1)}{s^{1/2+1}} + 3\frac{\Gamma(-1/2+1)}{s^{-1/2+1}} - \frac{\Gamma(-3/2+1)}{s^{-3/2+1}}$$

$$= \frac{\frac{3}{2}\Gamma\left(\frac{3}{2}\right)}{s^{5/2}} - 3\frac{\frac{1}{2}\Gamma\left(\frac{1}{2}\right)}{s^{3/2}} + 3\frac{\Gamma\left(\frac{1}{2}\right)}{s^{1/2}} - \frac{\Gamma\left(-\frac{1}{2}\right)}{s^{-1/2}}$$

$$= \frac{3}{4}\frac{\sqrt{\pi}}{s^{5/2}} - \frac{3}{2}\frac{\sqrt{\pi}}{s^{3/2}} + \frac{3\sqrt{\pi}}{s^{1/2}} + \frac{2\sqrt{\pi}}{s^{-1/2}} \qquad \left[\because \Gamma\left(\frac{1}{2}\right) = \sqrt{\pi}, \Gamma\left(-\frac{1}{2}\right) = -2\sqrt{\pi}\right]$$

$$= \frac{\sqrt{\pi}}{4}\left(\frac{3}{s^{5/2}} - \frac{6}{s^{3/2}} + \frac{12}{s^{1/2}} + \frac{8}{s^{-1/2}}\right).$$

(*ii*) We know that $\cos\theta = 1 - \frac{\theta^2}{2!} + \frac{\theta^4}{4!} - \frac{\theta^6}{6!} + \ldots \infty$

$$\therefore \quad \cos\sqrt{t} = 1 - \frac{t}{2!} + \frac{t^2}{4!} - \frac{t^3}{6!} + \ldots$$

and
$$\frac{\cos\sqrt{t}}{\sqrt{t}} = t^{-1/2} - \frac{t^{1/2}}{2!} + \frac{t^{3/2}}{4!} - \frac{t^{5/2}}{6!} + \ldots$$

and
$$L\left(\frac{\cos\sqrt{t}}{\sqrt{t}}\right) = \frac{\Gamma(1/2)}{s^{1/2}} - \frac{1}{2!}\frac{\Gamma(3/2)}{s^{3/2}} + \frac{1}{4!}\frac{\Gamma(5/2)}{s^{5/2}} - \frac{1}{6!}\frac{\Gamma(7/2)}{s^{7/2}} + \ldots$$

$$= \frac{\Gamma 1/2}{\sqrt{s}} - \frac{1}{2}\cdot\frac{1/2\,\Gamma 1/2}{s^{3/2}} + \frac{1}{4!}\frac{3/2\,.\,1/2\,.\,\Gamma(1/2)}{s^{5/2}} - \frac{1}{6!}\frac{5/2\,.\,3/2\,.\,1/2\,.\,\Gamma(1/2)}{s^{7/2}} + \ldots$$

$$= \sqrt{\left(\frac{\pi}{2}\right)}\left[1 - \frac{1}{(4s)} + \frac{1}{2!}\frac{1}{(4s)^2} - \frac{1}{3!}\frac{1}{(4s)^3}\ldots\right] = \sqrt{\left(\frac{\pi}{s}\right)}e^{-1/4s}.$$

**Example 21.7.** *Find the Laplace transform of the function*
*(i)* $f(t) = |t-1| + |t+1|, t \geq 0$
*(ii)* $f(t) = [t]$, *where* [ ] *stands for the greatest integer function.* *(Andhra, 2021 ; P.T.U., 2010)*

**Solution.** (*i*) Given function is equivalent to

$$f(t) = \begin{cases} 2, & 0 \leq t < 1 \\ 2t, & t \geq 1 \end{cases}$$

$$\therefore \quad Lf(t) = \int_0^1 e^{-st}(2)\,dt + \int_1^\infty e^{-st}(2t)\,dt = 2\left[\left|\frac{e^{-st}}{-s}\right|_0^1 + 2\left|\frac{t.e^{-st}}{-s}\right|_1^\infty - \left|\frac{e^{-st}}{(-s)^2}\right|_1^\infty\right]$$

$$= 2\left(\frac{e^{-s}}{-s} + \frac{1}{s}\right) + 2\left(\frac{0 - e^{-s}}{-s} - \frac{0 - e^{-s}}{s^2}\right) = \frac{2}{s}\left(1 + \frac{e^{-s}}{s}\right)$$

(*ii*) Given function is equivalent to

$$[t] = 0 \text{ in } (0, 1) + 1 \text{ in } (1, 2) + 2 \text{ in } (2, 3) + 3 \text{ in } (3, 4) + \ldots$$

$$\therefore \quad L[f(t)] = \int_0^\infty e^{-st}[f(t)]dt = \int_0^\infty e^{-st}[t]dt$$

$$= \int_0^1 e^{-st}(0)dt + \int_1^2 e^{-st}(1)dt + \int_2^3 e^{-st}(2)dt + \int_3^4 e^{-st}(3)dt + \ldots \infty$$

$$= 0 + \left|\frac{e^{-st}}{-s}\right|_1^2 + 2\left|\frac{e^{-st}}{-s}\right|_2^3 + 3\left|\frac{e^{-st}}{-s}\right|_3^4 + \ldots \infty$$

$$= -\frac{1}{s}[(e^{-2s} - e^{-s}) + 2(e^{-3s} - e^{-2s}) + 3(e^{-4s} - e^{-3s}) + \ldots \infty]$$

$$= \frac{1}{s}(e^{-s} + e^{-2s} + e^{-3s} + \dots \infty) = \frac{1}{s}\left(\frac{e^{-s}}{1-e^{-s}}\right) = \frac{1}{s(e^{s}-1)}.$$

**III. Change of scale property.** *If* $L\{f(t)\} = \bar{f}(s)$, *then* $L\{f(at)\} = \frac{1}{a}\bar{f}\left(\frac{s}{a}\right)$

$$L\{f(at)\} = \int_0^\infty e^{-st} f(at)\,dt = \int_0^\infty e^{-su/a} f(u)\,.\,du/a$$ | Put $at = u$, $dt = du/a$

$$= \frac{1}{a}\int_0^\infty e^{-su/a} f(u)du = \frac{1}{a}\bar{f}(s/a).$$

**Example 21.8.** *Find* $L\left\{\frac{\sin at}{t}\right\}$, *given that* $L\left\{\frac{\sin t}{t}\right\} = \tan^{-1}\left\{\frac{1}{s}\right\}$.

**Solution.** By the above property,

$$L\left\{\frac{\sin at}{at}\right\} = \frac{1}{a}\tan^{-1}\left\{\frac{1}{(s/a)}\right\} = \frac{1}{a}\tan^{-1}\left(\frac{a}{s}\right) \text{ i.e., } L\left\{\frac{\sin at}{t}\right\} = \tan^{-1}\left\{\frac{a}{s}\right\}.$$

## PROBLEMS 21.1

Find the Laplace transforms of

1. $e^{2t} + 4t^3 - 2\sin 3t + 3\cos 3t$.
2. $1 + 2\sqrt{t} + 3/\sqrt{t}$.
3. $\cos(at + b)$.
4. $(\sin t - \cos t)^2$. (V.T.U., 2016)
5. $\cos t \cos 2t \cos 3t$. (V.T.U., 2016)
6. $\sin t \sin 2t \sin 3t$. (V.T.U., 2013)
7. $\sin^5 t$.
8. $\cos^3 2t$.
9. $e^{-at}\sinh bt$
10. $e^{-t}\cos^2 t$. (V.T.U., 2016)
11. $e^{-t}\sin^2 3t$ (Anna., 2016)
12. $\cosh at \sin at$.
13. $\sinh 3t \cos^2 t$ (Andhra, 2022)
14. $t^2 e^{-2t}$. (J.N.T.U., 2015)
15. $wt \cos wt + \sin wt$. (Delhi., 2016)
16. $t\sqrt{(1 + \sin t)}$
17. $f(t) = \begin{cases} 4, & 0 \le t \le 1 \\ 3, & t > 1 \end{cases}$
18. $f(t) = \begin{cases} \sin t, & 0 < t < \pi \\ 0, & t > \pi. \end{cases}$ (JNTUK, 2022; Delhi, 2013)
19. $f(t) = \begin{cases} \cos(t - 2\pi/3), & t > 2\pi/3 \\ 0, & t < 2\pi/3 \end{cases}$
20. If $L[f(t)] = \frac{1}{s(s^2+1)}$, find $L[e^{-t} f(2t)]$.
21. Show that $L[te^{at}\sin at] = \frac{2a(s-a)}{(s^2 - 2as - 2a^2)^2}$ (AKTU, 2018)
22. Show that $L\left[e^{-at} - 2e^{at/2}\cos\frac{1}{2}\sqrt{3}\,at\right] = \frac{2a^2 - 2as - s^2}{s^2 + a^2}$ (Bhilai, 2018)
23. Show that $L\{\sin kt \sin ht\} = \frac{2k^2 s}{s^4 + 4k^4}$
24. Using the expression

$$\sin x = x - \frac{x^2}{3!} + \frac{x^5}{5!} - \frac{x^7}{7!};\quad \text{show that } L = \{\sin\sqrt{t}\}\frac{\sqrt{\pi}}{2S^{3/2}}e^{-1/4s}$$

25. If $L\{t(t)\} = \bar{f}(s)$, show that

(i) $L\{f(at)\} = \frac{1}{a}\bar{f}\left(\frac{S}{a}\right)$ (ii) $L\{f(t/a)\} = a\,\bar{f}\,(aS)$

## 21.5 TRANSFORMS OF PERIODIC FUNCTIONS

*If $f(t)$ is a periodic function with period $T$, i.e., $f(t+T)=f(t)$, then*

$$L\{f(t)\} = \frac{\int_0^T e^{-st} f(t)dt}{1-e^{-sT}}$$

We have $\quad L\{f(t)\} = \int_0^\infty e^{-st} f(t)dt = \int_0^T e^{-st} f(t)dt + \int_T^{2T} e^{-st} f(t)dt + \int_{2T}^{3T} e^{-st} f(t)dt + \ldots$

In the second integral put $t = u + T$, in the third integral put $t = u + 2T$, and so on. Then

$$L\{f(t)\} = \int_0^T e^{-st} f(t)dt + \int_0^T e^{-s(u+T)} f(u+T)du + \int_0^T e^{-s(u+2T)} f(u+2T)du + \ldots$$

$$= \int_0^T e^{-st} f(t)dt + e^{-sT}\int_0^T e^{-su} f(u)du + e^{-2sT}\int_0^T e^{-su} f(u)du + \ldots$$

$[\because \ f(u) = f(u+T) = f(u+2T)$ etc.$]$

$$= \int_0^T e^{-st} f(t)dt + e^{-sT}\int_0^T e^{-st} f(t)dt + e^{-2sT}\int_0^T e^{-st} f(t)dt + \ldots$$

$$= (1 + e^{-sT} + e^{-2sT} + \ldots \infty)\int_0^T e^{-st} f(t)dt$$

$$= \frac{1}{1-e^{-sT}}\int_0^T e^{-st} f(t)dt.$$

(*D.T.U., 2015 ; V.T.U., 2008 ; Mumbai, 2006*)

**Example 21.9.** *Find the Laplace transform of the function*

$$f(t) = E \sin \omega t, \quad 0 < t < \pi/\omega$$
$$= 0, \quad \pi/\omega < t < 2\pi/\omega.$$

(*Anna., 2016 ; Delhi ; 2016 ; V.T.U., 2013*)

**Solution.** Since $f(t)$ is a periodic function with period $2\pi/\omega$.

$$\therefore \quad L\{f(t)\} = \frac{1}{1-e^{-2\pi s/\omega}}\int_0^{2\pi/\omega} e^{-st} f(t)dt$$

$$= \frac{1}{1-e^{-2\pi s/\omega}}\left[\int_0^{\pi/\omega} e^{-st} E\sin\omega t\, dt + \int_{\pi/\omega}^{2\pi/\omega} e^{-st}.0\, dt\right]$$

$$= \frac{E}{1-e^{-2\pi s/\omega}}\left|\frac{e^{-st}(-s\sin\omega t - \omega\cos\omega t)}{s^2+\omega^2}\right|_0^{\pi/\omega} = \frac{E(\omega e^{-\pi s/\omega}+\omega)}{(1-e^{-2\pi s/\omega})(s^2+\omega^2)} = \frac{E\omega}{(1-e^{-\pi s/\omega})(s^2+\omega^2)}.$$

**Example 21.10.** *Draw the graph of the periodic function*

$$f(t) = \begin{cases} t, & 0 < t < \pi \\ \pi - t, & \pi < t < 2\pi. \end{cases}$$

*and find its Laplace transform.* (*V.T.U., 2012*)

**Solution.** Here the period of $f(t) = 2\pi$ and its graph is as in Fig. 21.1.

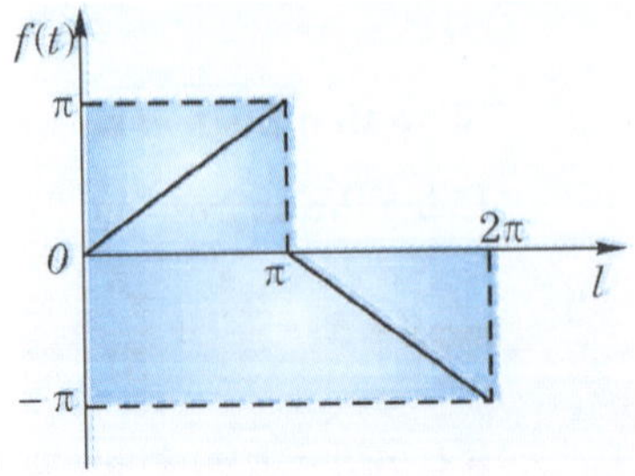


Fig. 21.1

$$\therefore \quad L f(t) = \frac{1}{1-e^{-2\pi s}}\left\{\int_0^\pi e^{-st}\, t\, dt + \int_\pi^{2\pi} e^{-st}(\pi - t)dt\right\}$$

$$= \frac{1}{1-e^{-2\pi s}}\left\{\left| t\left(\frac{e^{-st}}{-s}\right) - 1.\left(\frac{e^{-st}}{s^2}\right)\right|_0^\pi + \left|(\pi - t)\left(\frac{e^{-st}}{-s}\right) - (-1)\left(\frac{e^{-st}}{s^2}\right)\right|_\pi^{2\pi}\right\}$$

$$= \frac{1}{1-e^{-2\pi s}}\left\{\frac{-\pi e^{-\pi s}}{s} - \frac{e^{-\pi s}}{s^2} + \frac{1}{s^2} + \frac{\pi e^{-2\pi s}}{s} + \frac{e^{-2\pi s}}{s^2} - \frac{e^{-\pi s}}{s^2}\right\}$$

$$= \frac{1}{1-e^{-2\pi s}}\left\{\frac{\pi}{s}\left(e^{-2\pi s} - e^{-\pi s}\right) + \frac{1}{s^2}\left(1 + e^{-2\pi s} - 2e^{-\pi s}\right)\right\}.$$

## 21.6 TRANSFORMS OF SPECIAL FUNCTIONS

**(1) Transform of Bessel functions $J_0(x)$ and $J_1(x)$.**

We know that $$J_0(x) = 1 - \frac{x^2}{2^2} + \frac{x^4}{2^2 . 4^2} - \frac{x^6}{2^2 . 4^2 . 6^2} + \ldots$$

$$\therefore \quad L\{J_0(x)\} = \frac{1}{s} - \frac{1}{2^2}\frac{2!}{s^3} + \frac{1}{2^2 . 4^2} . \frac{4!}{s^5} - \frac{1}{2^2 . 4^2 . 6^2} . \frac{6!}{s^7} + \ldots$$

$$= \frac{1}{s}\left\{1 - \frac{1}{2}\left(\frac{1}{s^2}\right) + \frac{1 . 3}{2 . 4}\left(\frac{1}{s^4}\right) - \frac{1 . 3 . 5}{2 . 4 . 6}\left(\frac{1}{s^6}\right) + \ldots\right\}$$

$$= \frac{1}{s}\left(1 + \frac{1}{s^2}\right)^{-1/2} = \frac{1}{\sqrt{(s^2+1)}} \qquad \ldots(1)$$

Also since $J_0'(x) = -J_1(x)$.

$$\therefore \quad L\{J_1(x)\} = -L\{J_0'(x)\} = -[sL\{J_0(x)\} - 1] = 1 - \frac{s}{\sqrt{(s^2+1)}} \qquad \ldots(2)$$

**(2) Transform of Error function**

We know that $$erf(\sqrt{x}) = \frac{2}{\sqrt{\pi}}\int_0^{\sqrt{x}} e^{-t^2}\, dt$$

$$= \frac{2}{\sqrt{\pi}}\int_0^{\sqrt{x}}\left(1 - t^2 + \frac{t^4}{2!} - \frac{t^6}{3!} + \ldots\right)dt = \frac{2}{\sqrt{\pi}}\left(x^{1/2} - \frac{x^{3/2}}{3} + \frac{x^{5/2}}{5 . 2!} - \frac{x^{7/2}}{7 . 3!} + \ldots\right)$$

$$\therefore \quad L\{erf(\sqrt{x})\} = \frac{2}{\sqrt{\pi}}\left\{\frac{\Gamma(3/2)}{s^{3/2}} - \frac{\Gamma(5/2)}{3s^{5/2}} + \frac{\Gamma(7/2)}{5 . 2!\, s^{7/2}} - \frac{\Gamma(9/2)}{7 . 3!\, s^{9/2}} + \ldots\right\}$$

$$= \frac{1}{s^{3/2}} - \frac{1}{2}\frac{1}{s^{5/2}} + \frac{1 . 3}{2 . 4} . \frac{1}{s^{7/2}} - \frac{1 . 3 . 5}{2 . 4 . 6} . \frac{1}{s^{9/2}} + \ldots$$

$$= \frac{1}{s^{3/2}}\left\{1 - \frac{1}{2} . \frac{1}{s} + \frac{1 . 3}{2 . 4} . \frac{1}{s^2} - \frac{1 . 3 . 5}{2 . 4 . 6} . \frac{1}{s^3} + \ldots\right\}$$

$$= \frac{1}{s^{3/2}}\left[1 + \frac{1}{s}\right]^{-1/2} = \frac{1}{s\sqrt{(s+1)}}. \qquad \ldots(3)$$

**(3) Transform of Laguerre's polynomials $L_n(x)$**

We know that $$L_n(x) = e^x \frac{d^n}{dx^n}(x^n e^{-x})$$

$$L[L_n(t)] = \int_0^\infty e^{-st} e^t \frac{d^n}{dt^n}(t^n e^{-t})\, dt = \int_0^\infty e^{-(s-1)t} \frac{d^n}{dt^n}(e^{-t} t^n)\, dt$$

$$= \left| e^{-(s-1)t} \frac{d^{n-1}}{dt^{n-1}}(e^{-t} t^n)\right|_0^\infty + \int_0^\infty e^{-(s-1)t\,(s-1)}\,(s-1)\frac{d^{n-1}}{dt^{n-1}}(e^{-t} t^n)\, dt$$

$$= (s-1)\int_0^\infty e^{-(s-1)t} \frac{d^{n-1}}{dt^{n-1}}(e^{-t} t^n)\, dt. \qquad \text{(Integrating by parts)}$$

$$= (s-1)^n \int_0^\infty e^{-(s-1)t} . e^{-t} . t^n\, dt = (s-1)^n \int_0^\infty e^{-st} . t^n\, dt$$

$$= (s-1)^n L(t^n) = (s-1)^n . \frac{n!}{s^{n+1}}$$

Hence $$L[L_n(x)] = \frac{n!(s-1)^n}{s^n+1} \quad (s > 1).$$

**Example 21.11.** *Evaluate (i)* $L\{e^{-at} J_0(at)\}$ *(ii)* $L(erf\, 2\sqrt{t})$.

**Solution.** (*i*) We know that $L\{J_0(at)\} = \dfrac{1}{\sqrt{(s^2 + a^2)}}$

By shifting property, we get

$$L\{e^{-at} J_0(at)\} = \frac{1}{\sqrt{[(s+a)^2 + a^2]}} = \frac{1}{\sqrt{(s^2 + 2sa + 2a^2)}}$$

(*ii*) We know that $L(erf \sqrt{t}) = \dfrac{1}{s(s+1)}$

$$\therefore \qquad L(erf\; 2\sqrt{t}) = L[erf\sqrt{(4t)}] = \frac{1}{4} \cdot \frac{1}{\frac{s}{4}\sqrt{\left(\frac{s}{4}+1\right)}} = \frac{2}{s\sqrt{(s+4)}}.$$

**PROBLEMS 21.2**

1. Find the Laplace transform of the *saw-toothed wave* of period $T$, given $f(t) = t/T$ for $0 < t < T$.
2. Find the Laplace transform of the *full-wave rectifier*

$$f(t) = \sin\frac{\pi t}{a} \text{ such that } f(t+a) = f(t). \qquad \textit{(Anna., 2016)}$$

3. Find the Laplace transform of the *square-wave* (or *meander*) function of period $a$ defined as

$$f(t) = E, \quad \text{when } 0 < t < \alpha/2$$
$$= -E, \quad \text{when } \alpha/2 < t < \alpha. \qquad \textit{(Anna., 2016 ; V.T.U., 2015)}$$

4. Find the Laplace transform of the *triangular wave* of period $2a$ given by

$$f(t) = t, \quad 0 < t < a$$
$$= 2a - t, \quad a < t < 2a. \qquad \textit{(Andhra, 2022; V.T.U., 2016 ; Delhi, 2012)}$$

Find the Laplace transform of the following functions :

5. $J_0(ax)$. 6. $e^{-at} J_0(bt)$. 7. $e^{2t} erf(\sqrt{t})$.

## 21.7 TRANSFORMS OF DERIVATIVES

**(1)** *If $f'(t)$ be continuous and $L\{f(t)\} = \bar{f}(s)$, then $L\{f'(t)\} = s\bar{f}(s) - f(0)$.*

$$L\{f(t)\} = \int_0^\infty e^{-st} f'(t)dt \qquad \text{[Integrate by parts]}$$

$$= \left| e^{-st} f(t) \right|_0^\infty - \int_0^\infty (-s)e^{-st} \cdot f(t)dt.$$

Now assuming $f(t)$ to be such that $\underset{t\to\infty}{\text{Lt}}\; e^{-st} f(t) = 0$. When this condition is satisfied, $f(t)$ is said to be *exponential order s*.

Thus, $$L\{f'(t)\} = f(0) + s\int_0^\infty e^{-st} f(t)dt$$

whence follows the desired result.

**(2)** *If $f'(t)$ and its first $(n-1)$ derivatives be continuous, then*

$$\mathbf{L\{f^n(t)\} = s^n\, \bar{f}(s) - s^{n-1} f(0) - s^{n-2} f'(0) - \ldots - f^{n-1}(0).}$$

Using the general rule of integration by parts.

$$L\{f^n(t)\} = \int_0^\infty e^{-st} f^n(t)dt$$
$$= |e^{-st} f^{n-1}(t) - (-s)e^{-st} f^{n-2}(t) + (-s)^2 e^{-st} f^{n-3}(t) - \ldots$$
$$+ (-1)^{n-1}(-s)^{n-1} e^{-st} \cdot f(t)\,|_0^\infty + (-1)^n(-s)^n \int_0^\infty e^{-st} f(t)dt$$

$$= -f^{n-1}(0) - sf^{n-2}(0) - s^2 f^{n-3}(0) - \ldots - s^{n-1} f(0) + s^n \int_0^\infty e^{-st} f(t)dt$$

Assuming that $\underset{t\to\infty}{\text{Lt}}\; e^{-st} f^m(t) = 0$ for $m = 0, 1, 2, \ldots, n-1$.

This proves the required result.

## 21.8 TRANSFORMS OF INTEGRALS

*If* $L\{f(t)\} = \bar{f}(s)$, *then* $\mathbf{L\left\{\int_0^t f(u)\,du\right\} = \frac{1}{s}\bar{f}(s)}$.

Let $\phi(t) = \int_0^t f(u)\,du$, then $\phi'(t) = f(t)$ and $\phi(0) = 0$

$\therefore$ $L\{\phi'(t)\} = s\bar{\phi}(s) - \phi(0)$ [By § 21.7 (1)]

or $\bar{\phi}(s) = \frac{1}{s}L\{\phi'(t)\}$ *i.e.*, $L\left(\int_0^t f(u)\,du\right) = \frac{1}{s}\bar{f}(s)$. (*JNTUK, 2022*)

**Obs.** While Laplace transform of the derivative of a function corresponds to multiplication of the transform by $s$, Laplace transform of the integral of a function corresponds to division of the transform by $s$.

## 21.9 MULTIPLICATION BY $t^n$

*If* $L\{f(t)\} = \bar{f}(s)$, *then*

$$\mathbf{L\{t^n f(t)\} = (-1)^n \frac{d^n}{ds^n}[\bar{f}(s)]}, \textit{ where } n = 1, 2, 3 \ldots$$

We have $\int_0^\infty e^{-st} f(t)\,dt = \bar{f}(s)$.

Differentiating both sides with respect to $s$, $\frac{d}{ds}\left\{\int_0^\infty e^{-st} f(t)\,dt\right\} = \frac{d}{ds}\{\bar{f}(s)\}$

or By Leibnitz's rule for differentiation under the integral sign.

$$\int_0^\infty \frac{\partial}{\partial s}(e^{-st})\,f(t)\,dt = \frac{d}{ds}\{\bar{f}(s)\}$$

or $\int_0^\infty \{-te^{-st} f(t)\}\,dt = \frac{d}{ds}[\bar{f}(s)]$ or $\int_0^\infty e^{-st}\,[tf(t)]\,dt = -\frac{d}{ds}[\bar{f}(s)]$

which proves the theorem for $n = 1$.

Now assume the theorem to be true for $n = m$ (say), so that

$$\int_0^\infty e^{-st}\,[t^m f(t)]\,dt = (-1)^m \frac{d^m}{ds^m}[\bar{f}(s)]$$

Then $\frac{d}{ds}\left[\int_0^\infty e^{-st}\,t^m f(t)\,dt\right] = (-1)^m \frac{d^{m+1}}{ds^{m+1}}[\bar{f}(s)]$

or By Leibnitz's rule, $\int_0^\infty (-te^{-st})\,.\,t^m f(t)\,dt = (-1)^m \frac{d^{m+1}}{ds^{m+1}}[\bar{f}(s)]$

or $\int_0^\infty e^{-st}\,[t^{m+1} f(t)]\,dt = (-1)^{m+1} \frac{d^{m+1}}{ds^{m+1}}[\bar{f}(s)]$.

This shows that, if the theorem, is true for $n = m$, it is also true for $n = m + 1$. But it is true for $n = 1$. Hence it is true for $n = 1 + 1 = 2$, and $n = 2 + 1 = 3$ and so on.

Thus the theorem is true for all positive integral values of $n$. (*W.B.T.U., 2015 ; Andhra., 2014*)

**Example 21.12.** *Find the Laplace transforms of*

(*i*) $t \cos at$ (*ii*) $t^2 \sin at$

(*iii*) $te^{-t} \sin 3t$. (*V.T.U., 2016 ; Rohtak, 2012 ; D.T.U., 2011*)

**Solution.** (*i*) Since $L(\cos at) = s/(s^2 + a^2)$

$$\therefore \quad L(t \cos at) = -\frac{d}{ds}\left(\frac{s}{s^2 + a^2}\right) = -\frac{s^2 + a^2 - s\,.\,2s}{(s^2 + a^2)^2}$$

$$= \frac{s^2 - a^2}{(s^2 + a^2)^2}$$ [*cf.* Example 21.4]

(*ii*) Since $\sin at = \dfrac{a}{s^2 + a^2}$,

$$\therefore \quad L(t^2 \sin at) = (-1)^2 \frac{d^2}{ds^2}\left(\frac{a}{s^2 + a^2}\right) = \frac{d}{ds}\left\{\frac{-2as}{(s^2 + a^2)^2}\right\} = \frac{2a(3s^2 - a^2)}{(s^2 + a^2)^3}.$$

(*iii*) Since $L(\sin 3t) = \dfrac{3}{s^2 + 3^2}$, therefore $L(t \sin 3t) = -\dfrac{d}{ds}\left(\dfrac{3}{s^2 + 3^2}\right) = \dfrac{6s}{(s^2 + 9)^2}$

Now using the shifting property (§ 21.4 II), we get

$$L(e^{-t} t \sin 3t) = \frac{6(s + 1)}{[(s + 1)^2 + 9]^2} = \frac{6(s + 1)}{(s^2 + 2s + 10)^2}.$$

**Example 21.13.** *Evaluate (i) $L\{t J_0(at)\}$ (ii) $L\{t J_1(t)\}$ (iii) $L\{t \, erf \, 2\sqrt{t}\}$.*

**Solution.** (*i*) Since $L\{J_0(at)\} = \dfrac{1}{\sqrt{(s^2 + a^2)}}$

$$\therefore \quad L\{tJ_0(at)\} = -\frac{d}{ds}\{L J_0(at)\} = -\frac{d}{ds}\frac{1}{\sqrt{(s^2 + a^2)}} = \frac{s}{(s^2 + a^2)^{3/2}}$$

(*ii*) Since $L\{J_1(t)\} = 1 - \dfrac{s}{\sqrt{(s^2 + 1)}}$

$$\therefore \quad L\{t J_1(t)\} = -\frac{d}{ds}\{L J_0(t)\} = -\frac{d}{ds}\left\{1 - \frac{s}{\sqrt{(s^2 + 1)}}\right\} = \frac{1}{(s^2 + 1)^{3/2}}$$

(*iii*) Since $L(erf \sqrt{t}) = \dfrac{1}{s\sqrt{(s + 1)}}$

$$\therefore \quad L(erf \, 2\sqrt{t}) = L \, erf \sqrt{4(t)} = \frac{1}{4} \cdot \frac{1}{\frac{s}{4}\sqrt{\left(\frac{s}{4} + 1\right)}} = \frac{2}{s\sqrt{(s + 4)}}$$

Thus $$L(t \, erf \, 2\sqrt{t}) = -\frac{d}{ds}\left\{\frac{2}{s\sqrt{(s + 4)}}\right\} = -\frac{d}{ds}\left\{\frac{2}{\sqrt{(s^3 + 4s^2)}}\right\} = \frac{3s + 8}{s^2(s + 4)^{3/2}}$$

## 21.10 DIVISION BY *t*

*If* $L\{f(t)\} = \bar{f}(s)$, *then* $\mathbf{L}\left\{\dfrac{\mathbf{1}}{\mathbf{t}}\mathbf{f(t)}\right\} = \int_s^\infty \mathbf{\bar{f}(s)\,ds}$ *provided the integral exists.*

We have $\bar{f}(s) = \int_0^\infty e^{-st} f(t)\,dt$

Integrating both sides with respect to $s$ from $s$ to $\infty$.

$$\int_s^\infty \bar{f}(s)\,ds = \int_s^\infty \left[\int_0^\infty e^{-st} f(t)\,dt\right] ds = \int_0^\infty \int_s^\infty f(t)\,e^{-st}\,ds\,dt$$

[Changing the order of integration]

$$= \int_0^\infty f(t)\left[\int_s^\infty e^{-st}\,ds\right] dt$$ [$\because$ $t$ is independent of $s$]

$$= \int_0^\infty f(t)\left|\frac{e^{-st}}{-t}\right|_s^\infty dt = \int_0^\infty e^{-st} \cdot \frac{f(t)}{t}\,dt = L\left\{\frac{1}{t}f(t)\right\}.$$

**Example 21.14.** *Find the Laplace transform of (i) $(1 - e^t)/t$* *(D.T.U., 2012)*

*(ii)* $\dfrac{\cos at - \cos bt}{t}$. *(V.T.U., 2016 ; D.T.U., 2015 ; U.K.T.U., 2012)*

**Solution.** (*i*) Since $L(1 - e^t) = L(1) - L(e^t) = \dfrac{1}{s} - \dfrac{1}{s-1}$

$$\therefore \quad L\left(\frac{1-e^t}{t}\right) = \int_s^\infty \left(\frac{1}{s} - \frac{1}{s-1}\right) ds = \left| \log s - \log (s-1) \right|_0^\infty$$

$$= \left| \log\left(\frac{s}{s-1}\right)\right|_s^\infty = -\log\left[\frac{1}{1-1/s}\right] = \log\left(\frac{s-1}{s}\right)$$

(*ii*) Since $L(\cos at - \cos bt) = \dfrac{s}{s^2+a^2} - \dfrac{s}{s^2+b^2}$

$$\therefore \quad L\left(\frac{\cos at - \cos bt}{t}\right) = \int_s^\infty \left(\frac{s}{s^2+a^2} - \frac{s}{s^2+b^2}\right) ds$$

$$= \left| \frac{1}{2}\log(s^2+a^2) - \frac{1}{2}\log(s^2+b^2)\right|_s^\infty$$

$$= \frac{1}{2}\underset{s\to\infty}{\text{Lt}} \log\frac{s^2+a^2}{s^2+b^2} - \frac{1}{2}\log\frac{s^2+a^2}{s^2+b^2}$$

$$= \frac{1}{2}\log\left(\frac{1+0}{1+0}\right) - \frac{1}{2}\log\left(\frac{s^2+a^2}{s^2+b^2}\right) = \log\left(\frac{s^2+b^2}{s^2+a^2}\right)^{1/2}$$

$[\because \ \log 1 = 0]$

**Example 21.15.** *Evaluate (i)* $L\left\{e^{-t}\int_0^t \frac{\sin t}{t}\,dt\right\}$ *(Andhra, 2022)*

*(ii)* $L\left\{t\int_0^t \frac{e^{-t}\sin t}{t}\,dt\right\}$ *(Andhra, 2022)* *(iii)* $L\left\{\int_0^t\int_0^t\int_0^t (t\sin t)\,dt\,dt\,dt\right\}$.

**Solution.** (*i*) We know that $L(\sin t) = \dfrac{1}{s^2+1}$

$$L\left(\frac{\sin t}{t}\right) = \int_0^\infty \frac{1}{s^2+1}\,ds = \frac{\pi}{2} - \tan^{-1} s = \cot^{-1} s$$

$$\therefore \quad L\left\{\int_0^t \frac{\sin t}{t}\,dt\right\} = \frac{1}{s}\cot^{-1} s$$

Thus by shifting property, $L\left\{e^{-t}\left(\int_0^t \frac{\sin t}{t}\,dt\right)\right\} = \dfrac{1}{s+1}\cot^{-1}(s+1)$.

(*ii*) Since $L\left(\dfrac{\sin t}{t}\right) = \cot^{-1} s$

$$\therefore \quad L\left(e^{-t}.\frac{\sin t}{t}\right) = \cot^{-1}(s+1)$$ *(V.T.U., 2013)*

and $$L\left\{\int_0^t e^{-t}\frac{\sin t}{t}\,dt\right\} = \frac{1}{s}\cot^{-1}(s+1)$$

Hence $$L\left\{t.\int_0^t e^{-t}\frac{\sin t}{t}\,dt\right\} = -\frac{d}{ds}\left\{\frac{\cot^{-1}(s+1)}{s}\right\}$$

$$= -\frac{s \cdot \left[\dfrac{-1}{1+(s+1)^2}\right] - \cot^{-1}(s+1)}{s^2} = \frac{s + (s^2 + 2s + 2)\cot^{-1}(s+1)}{s^2(s^2 + 2s + 2)}$$

(*iii*) Since $\quad L(\sin t) = \dfrac{1}{s^2+1}$

$\therefore \quad L(t \sin t) = -\dfrac{d}{ds} \dfrac{1}{(s^2+1)} = \dfrac{2s}{(s^2+1)^2}$

Thus $\quad L\left\{\int_0^t \int_0^t \int_0^t (t \sin t)\, dt \,.\, dt \,.\, dt\right\} = \dfrac{1}{s^3} L(t \sin t) = \dfrac{1}{s^3} \cdot \dfrac{2s}{(s^2+1)^2} = \dfrac{2}{s^2(s^2+1)^2}$

## 21.11 EVALUATION OF INTEGRALS BY LAPLACE TRANSFORMS

**Example 21.16.** *Evaluate* (*i*) $\displaystyle\int_0^\infty te^{-2t} \sin t\, dt$ *(D.T.U., 2016)*

(*ii*) $\displaystyle\int_0^\infty \frac{\sin mt}{t} dt$ (*iii*) $\displaystyle\int_0^\infty e^{-t}\left(\frac{\cos at - \cos bt}{t}\right) dt$

(*iv*) $L\left\{\displaystyle\int_0^t \frac{e^{-t} \sin t}{t} dt\right\}$. *(Andhra, 2022)*

**Solution.** (*i*) $\quad \displaystyle\int_0^\infty te^{-2t} \sin t\, dt = \int_0^\infty e^{-st}(t \sin t)\, dt$ where $s = 2$

$= L(t \sin t)$, by definition.

$$= (-1)\frac{d}{ds}\left(\frac{1}{s^2+1}\right) = \frac{2s}{(s^2+1)^2} = \frac{2 \times 2}{(2^2+1)^2} = \frac{4}{25}.$$

(*ii*) Since $\quad L(\sin mt) = m/(s^2 + m^2) = f(s)$, say.

$\therefore$ Using $\quad$ § 21.10, $L\left(\dfrac{\sin mt}{t}\right) = \displaystyle\int_s^\infty f(s)\, ds = \int_0^\infty \frac{m\, ds}{s^2+m^2} = \left|\tan^{-1}\frac{s}{m}\right|_s^\infty$

or by Def., $\quad \displaystyle\int_0^\infty e^{-st} \frac{\sin mt}{t} dt = \frac{\pi}{2} - \tan^{-1}\frac{s}{m}$

Now $\quad \displaystyle\operatorname*{Lt}_{s \to 0} \tan^{-1}(s/m) = 0$ if $m > 0$ or $\pi$ if $m < 0$.

Thus taking limits as $s \to 0$, we get

$$\int_0^\infty \frac{\sin mt}{t} dt = \frac{\pi}{2} \text{ if } m > 0 \text{ or } -\pi/2 \text{ if } m < 0$$

(*iii*) We know that $\quad L(\cos at) = \dfrac{s}{s^2+a^2}$ and $L(\cos bt) = \dfrac{s}{s^2+b^2}$

$$\therefore \quad L\,\frac{\cos at - \cos bt}{t} = \int_s^\infty \left(\frac{s}{s^2+a^2} - \frac{s}{s^2+b^2}\right) ds$$

$$= \frac{1}{2}\left\{\log\left(\frac{s^2+a^2}{s^2+b^2}\right)\right\}_s^\infty = \frac{1}{2}\log\left(\frac{s^2+b^2}{s^2+a^2}\right)$$

This implies that $\displaystyle\int_0^\infty e^{-st}\left(\frac{\cos at - \cos bt}{t}\right) dt = \frac{1}{2}\log\left(\frac{s^2+b^2}{s^2+a^2}\right)$

Taking $s = 1$, we get $$\int_0^\infty \left(e^{-t}\,\frac{\cos at - \cos bt}{t}\right) dt = \frac{1}{2}\log\left(\frac{1+b^2}{1+a^2}\right)$$

(*iv*) Since $$L\left(\frac{\sin t}{t}\right) = \int_s^\infty \frac{ds}{s^2+1} = \tan^{-1} s = \frac{\pi}{2} - \tan^{-1} s = \cot^{-1} s.$$

$\therefore$ $$L\left\{e^t\left(\frac{\sin t}{t}\right)\right\} = \cot^{-1}(s-1), \text{ by shifting property (§ 21.4 } II).$$

Thus $$L\left[\int_0^t \left\{e^t\left(\frac{\sin t}{t}\right)\right\} dt\right] = \frac{1}{s}\cot^{-1}(s-1), \text{ by § 21.8.}$$

## PROBLEMS 21.3

1. Given $L[\sin(\sqrt{t})] = \dfrac{\sqrt{\pi}}{2s^{3/2}}\, e^{-1/4\pi}$, prove that $L\left[\dfrac{\cos(\sqrt{t})}{\sqrt{t}}\right] = \sqrt{\left(\dfrac{\pi}{s}\right)}\, e^{-1/4\pi}$. (*JNUTK, 2022*)

Find the Laplace transforms of the following functions :

2. $t \sin^2 t$
3. $t(\sin^3 t - \cos^3 t)$. (*W.B.T.U., 2015*)
4. $t^2 \cos at$.
5. $t^5 e^{4t} \cosh 3t$.
6. $te^{at} \sin at$. (*U.K.T.U., 2012*)
7. $te^{-2t} \sin 4t$. (*V.T.U., 2015*)
8. $t^2 e^{-t} \cos t$. (*Anna., 2016*)
9. $(e^{-at} - e^{-bt})/t$. (*J.N.T.U., 2011*)
10. $(\sin t)/t$. (*P.T.U., 2010*)
11. $(e^{at} - \cos bt)/t$.
12. $(e^{-t} \sin t)/t$.
13. $(1 - \cos t)/t^2$. (*J.N.T.U., 2015*)
14. $2^t + \dfrac{\cos 2t - \cos 3t}{t} + t \sin t$.
15. Evaluate (*i*) $\displaystyle\int_0^\infty \frac{\cos 6t - \cos 4t}{t}\, dT$ (*U.P.T.U., 2012*)

    (*ii*) $\displaystyle\int_0^\infty \frac{e^{-\sqrt{2}t} \sinh t \sin t}{t}\, dt$ (*iii*) $\displaystyle\int_0^\infty te^{-2t} \sin 3t\, dt$

    (*iv*) $\displaystyle\int_0^\infty t^3 e^{-t} \sin t\, dt$. (*Anna., 2016*)
16. Prove that (*i*) $\displaystyle\int_0^\infty \frac{e^{-at} - e^{-bt}}{t}\, dt = \log\frac{b}{a}$. (*U.K.T.U., 2012*)

    (*ii*) $\displaystyle\int_0^\infty \frac{e^{-t}\sin t}{t}\, dt = \frac{\pi}{4}$. (*iii*) $\displaystyle\int_0^\infty \frac{e^{-t}\sin^2 t}{t}\, dt = \frac{1}{4}\log 5$. (*Hazaribag., 2015*)

    (*iv*) $\displaystyle\int_{t=0}^\infty \int_{u=0}^t \frac{e^{-t}\sin u}{u}\, du dt = \frac{\pi}{4}$ (*U.P.T.U., 2011*)
17. Evaluate (*i*) $L\displaystyle\int_0^t e^t \sin t\, dt$ (*Delhi., 2016*)

    (*ii*) $L\displaystyle\int_0^t e^{-t} \cos t\, dt$. (*J.N.T.U., 2015*)

    (*iii*) $L\displaystyle\int_0^t \frac{e^t \sin t}{t}\, dt$. (*D.T.U., 2016*)

    (*iv*) $dt\displaystyle\int_0^t e^{-2t}\, t \sin^3 t\, dt$. (*D.T.U., 2012*)
18. Show that (*i*) $L[t J_0(at)] = \dfrac{s}{(s^2+a^2)^{3/2}}$ (*ii*) $\displaystyle\int_0^\infty te^{-3t} J_0(4t)\, dt = 3/125$.

## 21.12 INVERSE TRANSFORMS — METHOD OF PARTIAL FRACTIONS

Having found the Laplace transforms of a few functions, let us now determine the inverse transforms of given functions of $s$. We have seen that $L\{f(t)\}$ in each case, is a rational algebraic function. Hence *to find the* inverse transforms, we first express the given function of $s$ into partial fractions which will, then, be recognizable as one of the following standard forms :

(1) $L^{-1}\left[\frac{1}{s}\right] = 1.$ (2) $L^{-1}\left[\frac{1}{s-a}\right] = e^{at}.$

(3) $L^{-1}\left[\frac{1}{s^n}\right] = \frac{t^{n-1}}{(n-1)!}$, n = 1, 2, 3, ... (4) $L^{-1}\left[\frac{1}{(s-a)^n}\right] = \frac{e^{at}\, t^{n-1}}{(n-1)!}.$

(5) $L^{-1}\left(\frac{1}{s^2+a^2}\right) = \frac{1}{a}\sin at.$ (6) $L^{-1}\left(\frac{s}{s^2+a^2}\right) = \cos at.$

(7) $L^{-1}\left(\frac{1}{s^2-a^2}\right) = \frac{1}{a}\sinh at.$ (8) $L^{-1}\left(\frac{s}{s^2-a^2}\right) = \cosh at.$

(9) $L^{-1}\left[\frac{1}{(s-a)^2+b^2}\right] = \frac{1}{b}e^{at}\sin bt.$ (10) $L^{-1}\left[\frac{s-a}{(s-a)^2+b^2}\right] = e^{at}\cos bt.$

(11) $L^{-1}\left(\frac{s}{(s^2+a^2)^2}\right) = \frac{1}{2a}t\sin at.$ (12) $L^{-1}\left[\frac{1}{(s^2+a^2)^2}\right] = \frac{1}{2a^3}(\sin at - at\cos at).$

*The reader is strongly advised to commit these results to memory.* The results (1) to (10) follow at once from their corresponding results in § 21.3 and 21.4. As illustrations, we shall prove (11) and (12). Example 21.4 gives

$$L(t\sin at) = \frac{2as}{(s^2+a^2)^2} \text{ and } L(t\cos at) = \frac{s^2-a^2}{(s^2+a^2)^2}$$

$$\therefore \quad t\sin at = 2a\,L^{-1}\left[\frac{s}{(s^2+a^2)^2}\right], \text{ whence follows (11).}$$

Also,

$$t\cos at = L^{-1}\left[\frac{s^2-a^2}{(s^2+a^2)^2}\right] = L^{-1}\left[\frac{(s^2+a^2)-2a^2}{(s^2+a^2)^2}\right]$$

$$= L^{-1}\left[\frac{1}{s^2+a^2}\right] - 2a^2\,L^{-1}\left[\frac{1}{(s^2+a^2)^2}\right]$$

$$= \frac{1}{a}\sin at - 2a^2\,L^{-1}\left[\frac{1}{(s^2+a^2)^2}\right] \text{ whence follows (12).}$$

**Obs.** *Go through the note on the 'partial fractions' given in para 10 of 'useful information' in Appendix I.*

**Example 21.17.** *Find the inverse Laplace transform of*

(*i*) $\frac{s-a}{(s-a)^2+b^2}$ (*ii*) $\frac{s+8}{(s^2+4s+5)}$ (*iii*) $\frac{s}{(s^2-1)^2}$ .... (*UPTU, 2017*)

**Solution.** (*i*) We know that

$$L\{\cos bt\} = \frac{s}{s^2+b^2}$$

So,

$$L^{-1}\frac{s-a}{(s-a)^2+b^2} = e^{at}\cos bt$$

(*ii*) Here

$$\frac{s+8}{s^2+4s+5} = \frac{(s+2)+6}{(s+2)^2+1} = \frac{s+2}{(s+2)^2+1} + \frac{6}{(s+2)^2+1}$$

Now, $$L\{\cos t\} = \frac{s}{s^2+1}$$

$$L\{e^{-2t}\} = \frac{s+2}{(s+1)^2+1}$$

Also $$L^{-1}\left[\frac{s+2}{(s+2)^2+1}\right] = e^{-2t}\cos t \text{ and}$$

$$L^{-1}\left[\frac{1}{s^2+4s+s}\right] = e^{-2t}\sin t$$

Hence $$L^{-1}\left[\frac{s+8}{s^2+4s+5}\right] = L^{-1}\left[\frac{s+2}{(s+2)^2+1}\right] + L^{-1}\left[\frac{6}{(s+2)^2+1}\right] = e^{-2t}\cos t + 6e^{-2t}\sin t$$

$$= e^{-2t}(\cos t + 6\sin t).$$

(*iii*) Since $$L(t) = \frac{1}{S^2}, L[e^t\, t] = \frac{1}{(s-1)^2}$$

and $$L[e^{-t}.t] = \frac{1}{(s+1)^2}$$

Now, consider $$L[te^t - te^{-t}] = \frac{1}{(s-1)^2} - \frac{1}{(s+1)^2} = \frac{(s+1)^2-(s-1)^2}{(s-1)^2(s+1)^2}$$

$$L[t(e^t - e^{-t})] = \frac{4s}{(s^2-1)^2} \text{ or}$$

$$L[t\cos ht] = \frac{2s}{(s^2-1)^2}$$

$\therefore$ $$L^{-1}\left[\frac{2s}{(s^2-1)^2}\right] = t\cos ht$$

$$L^{-1}\left[\frac{s}{(s^2-1)^2}\right] = \frac{t}{2}\cos h\, t$$

**Example 21.18.** *Find the inverse transforms of*

(*i*) $\dfrac{2s^2-6s+5}{s^3-6s^2+11s-6}$ (*V.T.U., 2014*)

(*ii*) $\dfrac{4s+5}{(s-1)^2(s+2)}$ (*V.T.U., 2016*)

**Solution.** (*i*) Here the denominator = $(s-1)(s-2)(s-3)$.

So let $$\frac{2s^2-6s+5}{(s-1)(s-2)(s-3)} = \frac{A}{s-1} + \frac{B}{s-2} + \frac{C}{s-3}$$

Then $A = [2\,.\,1^2 - 6\,.\,1 + 5]/(1-2)(1-3) = \frac{1}{2}$

$B = [2\,.\,2^2 - 6\,.\,2 + 5]/(2-1)(2-3) = -1$

and $C = [2\,.\,3^2 - 6\,.\,3 + 5]/(3-1)(3-2) = \frac{5}{2}$.

$\therefore$ $$L^{-1}\left(\frac{2s^2-6s+5}{s^3-6s^2+11s-6}\right) = \frac{1}{2}L^{-1}\left(\frac{1}{s-1}\right) - L^{-1}\left(\frac{1}{s-2}\right) + \frac{5}{2}L^{-1}\left(\frac{1}{s-3}\right)$$

$$= \tfrac{1}{2}e^{t} - e^{2t} + \tfrac{5}{2}e^{3t}.$$

(*ii*) Let $$\frac{4s+5}{(s-1)^2\,(s+2)} = \frac{A}{s-1} + \frac{B}{(s-1)^2} + \frac{4(-2)+5}{(-2-1)^2(s+2)}$$

Multiplying both sides by $(s-1)^2\,(s+2)$, $4s + 5 = A\,(s-1)(s+2) + B\,(s+2) - \frac{1}{3}(s-1)^2$

Putting $s = 1$, $9 = 3B$, $\therefore\ B = 3$.

Equating the coefficients of $s^2$ from both sides,

$$0 = A - \tfrac{1}{3}, \quad \therefore\ A = \tfrac{1}{3}.$$

$$\therefore \quad L^{-1}\left[\frac{4s+5}{(s-1)^2\,(s+2)}\right] = \frac{1}{3}L^{-1}\left(\frac{1}{s-1}\right) + 3L^{-1}\left[\frac{1}{(s-1)^2}\right] - \frac{1}{3}L^{-1}\left(\frac{1}{s+2}\right)$$

$$= \tfrac{1}{3}e^{t} + 3te^{t} - \tfrac{1}{3}e^{-2t}.$$

**Example 21.19.** *Find the inverse transforms of*

(*i*) $\dfrac{5s+3}{(s-1)(s^2+2s+5)}$ *(Andhra., 2014 ; V.T.U., 2012)*

(*ii*) $\dfrac{1+2s}{(s+2)^2(s-1)^2}$ *(Trivendrum, 2018)*

**Solution.** (*i*) Let $$\frac{5s+3}{(s-1)(s^2+2s+5)} = \frac{5(1)+3}{(s-1)(1^2+2\,.\,1+5)} + \frac{As+B}{s^2+2s+5}$$

Multiplying both sides by $(s-1)\,(s^2+2s+5)$,

$$5s + 3 = 1\,.\,(s^2+2s+5) + (As+B)\,(s-1).$$

Equating the coefficients of $s^2$ from both sides,

$$0 = 1 + A, \quad \therefore\ A = -1.$$

Putting $s = 0$, $3 = 5 - B$, $\therefore\ B = 2$.

$$\therefore \quad L^{-1}\left[\frac{5s+3}{(s-1)(s^2+2s+5)}\right] = L^{-1}\left(\frac{1}{s-1}\right) + L^{-1}\left(\frac{-s+2}{s^2+2s+5}\right)$$

$$= L^{-1}\left(\frac{1}{s-1}\right) + L^{-1}\left[\frac{-(s+1)+3}{(s+1)^2+4}\right] = L^{-1}\left(\frac{1}{s-1}\right) - L^{-1}\left[\frac{s+1}{(s+1)^2+2^2}\right] + 3L^{-1}\left[\frac{1}{(s+1)^2+2^2}\right]$$

$$= e^{t} - e^{-t}\cos 2t + \tfrac{3}{2}\,e^{-t}\sin 2t.$$

(*ii*) $\dfrac{1+2s}{(s+2)^2\,(s-1)}$

Here denominator has repeated factors $\bar{f}(s) = \dfrac{1+2s}{(s+2)^2(s-1)^2} = \dfrac{A}{s+2} + \dfrac{B}{(s+2)^2} + \dfrac{C}{s-1} + \dfrac{D}{(s-1)^2}$ ...(*i*)

Now $B$ and $D$ are given by

$$B = \lim_{s\to -2}\left[(s+2)^2\bar{f}(s)\right] = \lim_{s\to -2}\left[\frac{1+2s}{(s-1)^2}\right] = \frac{1-4}{9} = \frac{-1}{3}$$

and $$D = \lim_{x\to 1}\left[(s-1)\bar{f}(s)\right] = \lim_{s\to 1}\left[\frac{1+2s}{(s+2)^2}\right] = \frac{1+2}{9} = \frac{1}{3}$$

To get $A$ and $C$ first put $s = 0$ in (1)

$$\frac{1+0}{(0+2)^2(0-1)^2} = \frac{A}{2} + \frac{B}{4} + \frac{C}{-1} + \frac{D}{1}$$

$$\frac{1}{4} = \frac{A}{2} - \frac{1}{1^2} - C + \frac{1}{3}$$

$$\frac{A}{2} - C = 0 \qquad ...(ii)$$

Next multiply Eqn. (*i*) by $S$ and let $S \to \infty$

and
$$\lim_{s\to\infty}\left[\frac{s+2s^2}{(s+2)^2(s-1)^2}\right] = \lim_{s\to\infty}\left[\frac{As}{s+2} + \frac{Bs}{(s+2)^2} + \frac{Cs}{(s-1)} + \frac{Ds}{(s-1)^2}\right]$$

$$0 = A + C \qquad ...(iii)$$

Solving Eqns. (*ii*) and (*iii*), we have

$$A = C = 0$$

$$\because \quad L^{-1}\{f(x)\} = L^{-1}\left\{-\frac{1}{3(s+2)^2} + \frac{1}{3(s-1)^2}\right\} = \frac{1}{3}t(e^{2t} - e^{-2t})$$

## PROBLEMS 21.4

Find the inverse Laplace transforms of :

1. $\frac{2s-5}{4s^2+25} + \frac{4s-18}{9-s^2}$.
2. $\frac{a(s^2-2a^2)}{s^4+4a^4}$
3. $\frac{s}{(2s-1)(3s-1)}$. (*V.T.U., 2010*)
4. $\frac{s}{(s+1)(s+2)}$. (*Calicut, 2013*)
5. $\frac{3s+2}{s^2-s-2}$. (*Andhra, 2018; V.T.U., 2010 S*)
6. $\frac{3s+7}{s^2-2s-3}$ (*U.K.T.U., 2014*)
7. $\frac{s^3+6s^2+14s}{(s+2)^2}$
8. Prove that $L^{-1}[e^{-\sqrt{s}}] = \frac{1}{2t\sqrt{\pi t}}e^{-1/4t}$
9. $\frac{3s+1}{(s+1)^4}$. (*Anna., 2016*)
10. $\frac{2p^2-6p+5}{p^3-6p^2+11p-6}$.
11. $\frac{1+2s}{(s+2)^2(s-1)^2}$. (*JNTUK, 2022*)
12. $\frac{s}{(s-3)(s^2+4)}$.
13. $\frac{s}{(s+1)^2(s^2+1)}$.
14. $\frac{1}{s^3-a^3}$.
15. $\frac{s^2+6}{(s^2+1)(s^2+4)}$.
16. $\frac{2s-3}{s^2+4s+13}$. (*D.T.U., 2011*)
17. $\frac{s^2+s}{(s^2+1)(s^2+2s+2)}$.
18. $\frac{s}{s^4+s^2+1}$. (*D.T.U., 2011*)
19. $\frac{a(s^2-2a^2)}{s^4+4a^4}$.
20. $\frac{1}{s^4-a^4}$
21. $(s^2+2s+5)^{-2}$
22. Show that $L^{-1}\left[\frac{4(s+3)}{(s^2+6s+13)^2}\right] = te^{-3t}\sin 2t$
23. Show that $L^{-1}\left[\frac{s^3}{s^4+64}\right] = \cosh 2t \cos 2t$
24. Show that $L^{-1}\left[\frac{s^4-s^3+2s^2-8}{s^3(s^2+4)}\right] = 1 - t^2 - \sin^2 t$

## 21.13 OTHER METHODS OF FINDING INVERSE TRANSFORMS

We have seen that the most effective method of finding the inverse transforms is by means of partial fractions. However, various other methods are available which depend on the following *important inversion formulae.*

**I. Shifting property for inverse Laplace transforms.**

*If* $L^{-1}[\bar{f}(s)] = f(t)$, *then*

$$L^{-1}[\bar{f}'(s-a)] = e^{at} f(t) = e^{at} L^{-1}[\bar{f}(s)].$$

**II.** *If* $L^{-1}\{\bar{f}(s)\} = f(t)$ *and* $f(0) = 0$, *then*

$$L^{-1}[s\bar{f}(s)] = \frac{d}{dt}\{f(t)\}$$

In general, $L^{-1}\{s^n \bar{f}(s)\} = \dfrac{d^n}{dt^n}\{f(t)\}$ provided $f(0) = f'(0) = \ldots = f^{n-1}(0) = 0$.

The above formulae at once follow from the results of § 21.7 (Transforms of derivatives).

**III.** *If* $L^{-1}\{\bar{f}(s)\} = f(t)$, *then*

$$L^{-1}\left\{\frac{\bar{f}(s)}{s}\right\} = \int_0^t f(t)\, dt$$

This result follows from § 21.8 (Transforms of integrals)

Also $$L^{-1}\left\{\frac{\bar{f}(s)}{s^2}\right\} = \int_0^t \left\{\int_0^t f(t)\, dt\right\} dt$$

$$L^{-1}\left\{\frac{\bar{f}(s)}{s^3}\right\} = \int_0^t \left\{\int_0^t \left(\int_0^t f(t)\, dt\right) dt\right\} dt \text{ and so on.}$$

**IV.** *If* $L^{-1}\{\bar{f}(s)\} = f(t)$, *then*

$$t f(t) = L^{-1}\left\{-\frac{d}{ds}[\bar{f}(s)]\right\}$$

This result follows from $L\{t f(t)\} = -\dfrac{d}{ds}\{\bar{f}(s)\}$ (§ 21.9)

**V. The formula of** § **21.10,** *i.e.*,

$$L\left\{\frac{f(t)}{t}\right\} = \int_s^\infty \bar{f}(s)\, ds$$

is useful in finding $f(t)$ when $f(s)$ is given, provided the inverse transform of $\int_s^\infty \bar{f}(s)\, ds$ can be conveniently calculated.

**Example 21.20.** *Find the inverse Laplace transforms of the following :*

(*i*) $\dfrac{s^2}{(s-2)^3}$ (*ii*) $\dfrac{s+2}{s^2-4s+13}$ (*V.T.U., 2016*) (*iii*) $\dfrac{(s+2)^2}{(s^2+4s+8)^2}$

**Solution.** (*i*) Since $s^2 = (s-2)^2 + 4(s-2) + 4$

$$\therefore \quad \frac{s^2}{(s-2)^3} = \frac{1}{s-2} + \frac{4}{(s-2)^2} + \frac{4}{(s-2)^3}$$

$$\therefore \quad L^{-1}\left\{\frac{s^2}{(s-2)^3}\right\} = L^{-1}\left\{\frac{1}{s-2}\right\} + 4L^{-1}\left\{\frac{1}{(s-2)^2}\right\} + 4L^{-1}\left\{\frac{1}{(s-2)^3}\right\}$$

$$= e^{2t} + 4e^{2t} t + 2e^{2t} t^2.$$ [using shifting property]

(*ii*) $$\frac{s+2}{s^2-4s+13} = \frac{s-2}{(s-2)^2+3^2} + \frac{4}{(s-2)^2+3^2}$$

$$\therefore \quad L^{-1}\left\{\frac{s+2}{s^2-4s+13}\right\} = L^{-1}\left\{\frac{s-2}{(s-2)^2+3^2}\right\} + \frac{4}{3}L^{-1}\left\{\frac{3}{(s-2^2)+3^2}\right\}$$

$$= e^{2t}\cos 3t + \frac{4}{3}e^{2t}\sin 3t. \qquad \text{[Using shifting property]}$$

(*iii*) $$L^{-1}\frac{(s+2)^2}{(s^2+4s+8)^2} = L^{-1}\frac{(s+2)^2}{(s^2+4s+4+4)^2} = L^{-1}\frac{(s+2)^2}{[(s+2)^2+4]^2}$$

$$= e^{-2t}L^{-1}\left\{\frac{s^2}{(s^2+4)^2}\right\} = e^{-2t}L^{-1}\left\{\frac{s^2+4-4}{(s^2+4)^2}\right\}$$

$$= e^{-2t}L^{-1}\left\{\frac{1}{s^2+4} - \frac{4}{(s^2+4)^2}\right\} = \frac{e^{-2t}\sin 2t}{2} - 4e^{-2t}L^{-1}\left\{\frac{1}{(s^2+4)^2}\right\}$$

$$= \frac{e^{-2t}\sin 2t}{2} - 4e^{-2t}\left\{\frac{1}{4}\left(\frac{\sin 2t}{4} - \frac{t\cos 2t}{2}\right)\right\}$$

$$= e^{-2t}\left\{\frac{\sin 2t}{2} - \frac{\sin 2t}{4} + \frac{t\cos 2t}{2}\right\} = e^{-2t}\left\{\frac{\sin 2t}{4} + \frac{t\cos 2t}{2}\right\}.$$

**Example 21.21.** *Find the inverse transform of (i) $1/s\,(s^2+a^2)$ (ii) $1/s\,(s+a)^3$.* (*Andhra, 2021*)

**Solution.** (*i*) Since $L^{-1}\left(\frac{1}{s^2+a^2}\right) = \frac{1}{a}\sin at.$

therefore, by formula III above,

$$L^{-1}\left\{\frac{1}{s(s^2+a^2)}\right\} = \int_0^t \frac{1}{a}\sin at\,dt = \frac{1}{a^2}[-\cos at]_0^t = (1-\cos at)/a^2$$

(*ii*) $$L^{-1}\left\{\frac{1}{s(s+a)^3}\right\} = L^{-1}\left\{\frac{1}{[(s+a)-a](s+a)^3}\right\} = e^{-at}L^{-1}\left\{\frac{1}{(s-a)s^3}\right\}$$

Now $$L^{-1}\left\{\frac{1}{s-a}\right\} = e^{at} \quad \therefore \quad L^{-1}\left\{\frac{1}{(s-a)s}\right\} = \int_0^t e^{at}\,dt = \frac{e^{at}}{a} - \frac{1}{a}, \text{ by III above}$$

$$\therefore \quad L^{-1}\left\{\frac{1}{(s-a)s^2}\right\} = \frac{1}{a}\int_0^t (e^{at}-1)\,dt = \frac{1}{a^2}(e^{at}-at-1)$$

$$L^{-1}\left\{\frac{1}{(s-a)\,s^3}\right\} = \frac{1}{a^2}\int_0^t (e^{at}-at-1)\,dt = \frac{1}{a^3}\left(e^{at} - \frac{a^2}{2}t^2 - at - 1\right)$$

Hence $$L^{-1}\left\{\frac{1}{s(s+a)^3}\right\} = e^{-at}\cdot\frac{1}{a^3}\left(e^{at} - \frac{a^2t^2}{2} - at - 1\right) = \frac{1}{a^3}\left(1 - e^{-at} - ate^{-at} - \frac{a^2}{2}t^2e^{-at}\right).$$

**Example 21.22.** *Find the inverse Laplace transforms of :*

(*i*) $\frac{s}{(s^2+a^2)^2}$ (*V.T.U., 2015*) (*ii*) $\frac{s^2}{(s^2+a^2)^2}$ (*iii*) $\frac{1}{(s^2+a^2)^2}$.

**Solution.** (*i*) If $f(t) = L^{-1}\frac{s}{(s^2+a^2)^2}$, then by formula V above,

$$L\left\{\frac{f(t)}{t}\right\} = \int_s^\infty \frac{s}{(s^2+a^2)^2}\,ds = \frac{1}{2}\int_s^\infty \frac{2s}{(s^2+a^2)^2}\,ds = -\frac{1}{2}\left(\frac{1}{s^2+a^2}\right)_s^\infty = \frac{1}{2}\cdot\frac{1}{s^2+a^2}$$

$$\therefore \quad \frac{f(t)}{t} = \frac{1}{2}L^{-1}\left(\frac{1}{s^2+a^2}\right) = \frac{\sin at}{2a}$$

Hence, $f(t) = \dfrac{1}{2a}\, t \sin at.$

**Otherwise :** Let $f(t) = L^{-1}\left(\dfrac{1}{s^2 + a^2}\right) = \dfrac{\sin at}{a}$ so that $\bar{f}(s) = \dfrac{1}{s^2 + a^2}$

Then by (IV) above, $t\, f(t) = L^{-1}\left\{-\dfrac{d}{ds}[\bar{f}(s)]\right\} = L^{-1}\left\{-\dfrac{d}{ds}\left(\dfrac{1}{s^2 + a^2}\right)\right\}$

or $$\frac{t \sin at}{a} = L^{-1}\left\{\frac{2s}{(s^2 + a^2)^2}\right\}. \text{ Hence } L^{-1}\left\{\frac{s}{(s^2 + a^2)^2}\right\} = \frac{1}{2a}\, t \sin at.$$

(*ii*) In (*i*), we have proved that

$$L^{-1}\left\{\frac{s}{(s^2 + a^2)^2}\right\} = \frac{1}{2a}\, t \sin at = f(t), \text{ say}$$

Since $f(0) = 0$, we get from formula II above, that

$$L^{-1}\left\{\frac{s^2}{(s^2 + a^2)^2}\right\} = L^{-1}\left\{s \cdot \frac{s}{(s^2 + a^2)^2}\right\} = \frac{d}{dt}\{f(t)\}$$

$$= \frac{d}{dt}\left(\frac{1}{2a}\, t \sin at\right) = \frac{1}{2a}(\sin at + at \cos at)$$

(*iii*) In (*i*), we have shown that

$$L^{-1}\left\{\frac{s}{(s^2 + a^2)^2}\right\} = \frac{1}{2a}(t \sin at) = f(t), \text{ say}$$

By formula III above, we have

$$L^{-1}\left\{\frac{1}{(s^2 + a^2)^2}\right\} = L^{-1}\left\{\frac{1}{s} \cdot \frac{s}{(s^2 + a^2)^2}\right\} = \int_0^t f(t)\, dt = \int_0^t \frac{t \sin at}{2a}\, dt$$

$$= \frac{1}{2a}\left\{\left| t \cdot \frac{-\cos at}{a}\right|_0^t - \int_0^t 1 \cdot \left(\frac{-\cos at}{a}\right) dt\right\}$$

$$= \frac{1}{2a}\left\{\frac{-t \cos at}{a} + \frac{\sin at}{a^2}\right\} = \frac{1}{2a^3}(\sin at - at \cos at).$$

**Example 21.23.** *Find the inverse Laplace transforms of*

(*i*) $\dfrac{s + 2}{s^2 (s + 1)(s - 2)}$ (*ii*) $\dfrac{s + 2}{(s^2 + 4s + 5)^2}$.

**Solution.** (*i*) $L^{-1}\left\{\dfrac{s + 2}{(s + 1)(s - 2)}\right\} = \dfrac{4}{3} L^{-1}\left(\dfrac{1}{s - 2}\right) - \dfrac{1}{3} L^{-1}\left(\dfrac{1}{s + 1}\right) = \dfrac{4}{3} e^{2t} - \dfrac{1}{3} e^{-t}$

By III above, $L^{-1}\left\{\dfrac{s + 2}{s(s + 1)(s - 2)}\right\} = \displaystyle\int_0^t L^{-1}\left(\dfrac{s + 2}{(s + 1)(s - 2)}\right) dt$

$$= \int_0^t \left(\frac{4}{3} e^{2t} - \frac{1}{3} e^{-t}\right) dt = \frac{2}{3} e^{2t} + \frac{1}{3} e^{-t} - 1$$

Again by III above, $L^{-1}\dfrac{s + 2}{s^2 (s + 1)(s - 2)} = \displaystyle\int_0^t L^{-1}\left\{\dfrac{s + 2}{s(s + 1)(s - 2)}\right\} dt$

$$= \int_0^t \left(\frac{2}{3} e^{2t} + \frac{1}{3} e^{-t} - 1\right) dt = \frac{1}{3}(e^{2t} - e^{-t} - t).$$

(ii) $$L^{-1}\left(\frac{1}{s^2+4s+5}\right) = L^{-1}\left\{\frac{1}{(s+2)^2+1}\right\} = e^{-2t}\sin t$$

By II above, $L^{-1}\left\{\frac{d}{ds}\left(\frac{1}{s^2+4s+5}\right)\right\} = (-1)^1 t \, . \, e^{-2t}\sin t$

*i.e.,* $$L^{-1}\left\{\frac{-(2s+4)}{(s^2+4s+5)^2}\right\} = -t\, . \, e^{-2t}\sin t$$

or $$L^{-1}\left\{\frac{s+2}{(s^2+4s+5)^2}\right\} = \frac{1}{2}t\, . \, e^{-2t}\sin t.$$

**Example 21.24.** *Find the inverse Laplace transforms of the following :*

*(i)* $\log\frac{s+1}{s-1}$ *(ii)* $\log\frac{s^2+1}{s(s+1)}$

*(JNTUK, 2022; Andhra., 2022, 2016 ; Delhi, 2013 ; V.T.U., 2013)*

*(iii)* $\cot^{-1}\left(\frac{s}{2}\right)$ *(Rohtak, 2011)* *(iv)* $\tan^{-1}\left(\frac{2}{s^2}\right)$. *(Hazaribag., 2015 ; V.T.U., 2011)*

**Solution.** (*i*) If $f(t) = L^{-1}\log\frac{s+1}{s-1}$, then by IV above,

$$t\,f(t) = L^{-1}\left\{-\frac{d}{ds}\log\left(\frac{s+1}{s-1}\right)\right\} = -L^{-1}\left\{\frac{d}{ds}\log(s+1)\right\} + L^{-1}\left\{\frac{d}{ds}\log(s-1)\right\}$$

$$= -L^{-1}\left(\frac{1}{s+1}\right) + L^{-1}\left(\frac{1}{s-1}\right) = -e^{-t} + e^{t} = 2\sinh t$$

Thus $f(t) = (2\sinh t)/t.$

(*ii*) If $f(t) = L^{-1}\log\frac{s^2+1}{s(s+1)}$, then by IV above,

$$t\,f(t) = L^{-1}\left\{-\frac{d}{ds}\log\left(\frac{s^2+1}{s(s+1)}\right)\right\} = -L^{-1}\left\{\frac{d}{ds}\log(s^2+1)\right\} + L^{-1}\left\{\frac{d}{ds}\log s\right\}$$

$$+ L^{-1}\left\{\frac{d}{ds}\log(s+1)\right\}$$

$$= -L^{-1}\left(\frac{2s}{s^2+1}\right) + L^{-1}\left(\frac{1}{s}\right) + L^{-1}\left(\frac{1}{s+1}\right) = -2\cos t + 1 + e^{-t}$$

Thus $f(t) = \frac{1}{t}(1 + e^{-t} - 2\cos t).$

(*iii*) If $f(t) = L^{-1}\cot^{-1}\left(\frac{s}{2}\right)$, then by IV above,

$$t\,f(t) = L^{-1}\left\{-\frac{d}{ds}\cot^{-1}\left(\frac{s}{2}\right)\right\} = L^{-1}\left(\frac{2}{s^2+2^2}\right) = \sin 2t$$

Thus $f(t) = (\sin 2t)/t.$

(*iv*) If $f(t) = L^{-1}\left(\tan^{-1}\frac{2}{s^2}\right)$, then by IV above,

$$t\,f(t) = L^{-1}\left\{-\frac{d}{ds}\tan^{-1}\left(\frac{2}{s^2}\right)\right\} = L^{-1}\left\{\frac{4s}{s^4+4}\right\}$$

$$= L^{-1}\left\{\frac{4s}{(s^2+2)^2-(2s)^2}\right\} = L^{-1}\left\{\frac{4s}{(s^2+2+2s)(s^2+2-2s)}\right\}$$

$$= L^{-1}\left\{\frac{1}{s^2-2s+2}-\frac{1}{s^2+2s+2}\right\} = L^{-1}\left\{\frac{1}{(s-1)^2+1}-\frac{1}{(s+1)^2+1}\right\}$$

$$= e^t \sin t - e^{-t}\sin t = 2\sinh t \sin t.$$

## 21.14 CONVOLUTION THEOREM

*If* $L^{-1}\{\bar{f}(s)\} = f(t)$, *and* $L^{-1}\{\bar{g}(s)\} = g(t)$,

*then* $L^{-1}\{\bar{f}(s)\,\bar{g}(s)\} = \int_0^t f(u)\,g(t-u)\,du = F * G$

[*$F * G$ is called the* **convolution** *or* **falting** *of F and G.*]

Let $\phi(t) = \int_0^t f(u)\,g(t-u)\,du$

$$L\{\phi(t)\} = \int_0^\infty e^{-st}\left\{\int_0^t f(u)\,g(t-u)\,du\right\}dt = \int_0^\infty \int_0^t e^{-st} f(u)\,g(t-u)\,du\,dt \qquad ...(1)$$

The domain of integration for this double integral is the entire area lying between the lines $u = 0$ and $u = t$ (Fig. 21.2).

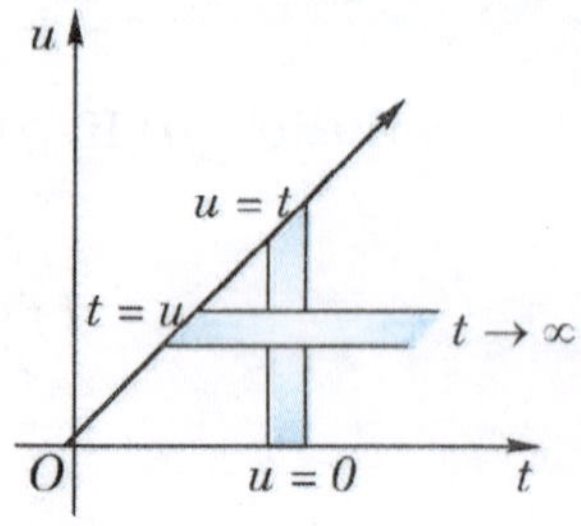

Fig. 21.2

On changing the order of integration, we get

$$L\{\phi(t)\} = \int_0^\infty \int_u^\infty e^{-st} f(u)\,g(t-u)\,dt\,du$$

$$= \int_0^\infty e^{-su} f(u)\left\{\int_0^\infty e^{-s(t-u)}\,g(t-u)\,dt\right\}du$$

$$= \int_0^\infty e^{-su} f(u)\left\{\int_0^\infty e^{-sv}\,g(v)\,dv\right\}du \text{ on putting } t-u=v$$

$$= \int_0^\infty e^{-su} f(u)\,\bar{g}(s)\,du = \int_0^\infty e^{-su} f(u)\,du\,.\,\bar{g}(s)$$

$= \bar{f}(s)\,.\,\bar{g}(s)$ whence follows the desired result. (*D.T.U., 2012*)

**Example 21.25.** *Apply Convolution theorem to evaluate*

(*i*) $L^{-1}\dfrac{s}{(s^2+a^2)^2}$. (*JNTUK, 2022; Andhra., 2022, 2015 ; W.B.T.U., 2015 ; Anna, 2012*)

(*ii*) $L^{-1}\dfrac{s^2}{(s^2+a^2)(s^2+b^2)}$. (*V.T.U., 2012*)

**Solution.** (*i*) Since $f(t) = L^{-1}\left(\dfrac{s}{s^2+a^2}\right) = \cos at$ and $g(t) = L^{-1}\left(\dfrac{s}{s^2+a^2}\right) = \dfrac{1}{a}\sin at$

∴ by Convolution theorem, we get

$$L^{-1}\left[\frac{s}{s^2+a^2}\cdot\frac{1}{s^2+a^2}\right] = \int_0^t \cos au\,\frac{\sin a(t-u)}{a}\,du \qquad \left[\because\ f(u) = \cos au,\quad g(t-u) = \frac{1}{a}\sin a(t-u)\right.$$

$$= \frac{1}{2a}\int_0^t [\sin at - \sin(2au - at)]\,dt = \frac{1}{2a}\left|u\sin at + \frac{1}{2a}\cos(2au-at)\right|_0^t = \frac{1}{2a}\,t\sin at$$

Hence $L^{-1}\left\{\dfrac{s}{(s^2+a^2)^2}\right\} = \dfrac{1}{2a}\,t\sin at.$

(*ii*) Since $f(t) = L^{-1}\left(\frac{s}{s^2+a^2}\right) = \cos at$ and $g(t) = L^{-1}\left(\frac{s}{s^2+b^2}\right) = \cos bt$,

$\therefore$ by Convolution theorem, we get

$$L^{-1}\left(\frac{s}{s^2+a^2}\cdot\frac{s}{s^2+b^2}\right) = \int_0^t \cos\ au \cos b(t-u)\,du \qquad [\because\ f(u) = \cos au,\ g(t-u) = \cos b(t-u)]$$

$$= \frac{1}{2}\int_0^t \{\cos\ [(a-b)\,u + bt] + \cos\ [(a+b)\,u - bt]\}\,du$$

$$= \frac{1}{2}\left|\frac{\sin[(a-b)u+bt]}{a-b} + \frac{\sin\,[(a+b)u-bt]}{a+b}\right|_0^t$$

$$= \frac{1}{2}\left\{\frac{\sin at - \sin bt}{a-b} + \frac{\sin at + \sin bt]}{a+b}\right\} = \frac{a\sin at - b\sin bt}{a^2-b^2}.$$

**Example 21.26.** *Evaluate* (*i*) $L^{-1}\dfrac{1}{(s^2+1)(s^2+9)}$ *(Andhra, 2021)*

(*ii*) $L^{-1}\dfrac{s}{(s^2+1)(s^2+4)(s^2+9)}$.

**Solution.** (*i*) Since $L^{-1}\left(\frac{1}{s^2+1}\right) = \sin t,\ L^{-1}\left(\frac{1}{s^2+9}\right) = \frac{\sin 3t}{3}$

$\therefore$ by Convolution theorem, we get

$$L^{-1}\left(\frac{1}{s^2+1}\cdot\frac{1}{s^2+9}\right) = \int_0^t \sin u\,.\,\frac{\sin 3(t-u)}{3}\,du$$

$$= \frac{1}{6}\int_0^t [\cos(4u-3t) - \cos(3t-2u)]\,du = \frac{1}{6}\left|\frac{\sin(4u-3t)}{4} - \frac{\sin(3t-2u)}{-2}\right|_0^t$$

$$= \frac{1}{6}\left\{\frac{1}{4}(\sin t + \sin 3t) + \frac{1}{2}(\sin t - \sin 3t)\right\} = \frac{1}{8}(\sin t - \frac{1}{3}\sin 3t)$$

(*ii*) Since $L^{-1}\left(\frac{s}{s^2+4}\right) = \cos 2t$ and $L^{-1}\left\{\frac{1}{(s^2+1)(s^2+9)}\right\} = \frac{1}{8}\left[\sin t - \frac{1}{3}\sin 3t\right]$ [By (*i*)]

$\therefore$ by Convolution theorem, we get

$$L^{-1}\frac{s}{(s^2+1)(s^2+4)(s^2+9)} = L^{-1}\left\{\frac{1}{(s^2+1)(s^2+9)}\cdot\frac{s}{s^2+4}\right\}$$

$$= \int_0^t \frac{1}{8}(\sin u - \frac{1}{3}\sin 3u)\,.\cos 2\,(t-u)\,du$$

$$= \frac{1}{8}\int_0^t [\sin u\cos 2(t-u) - \frac{1}{3}\sin 3u\cos 2(t-u)]du$$

$$= \frac{1}{8}\int_0^t \left[\frac{1}{2}\{\sin(2t-u) - \sin(3u-2t)\} - \frac{1}{6}\{\sin(u+2t) - \sin(5u-2t)\}\right]du$$

$$= \frac{1}{16}\left[\left|\frac{-\cos(2t-u)}{-1} + \frac{\cos(3u-2t)}{3}\right|_0^t\right] - \frac{1}{48}\left[\left|-\cos(u+2t) + \frac{\cos(5u-2t)}{5}\right|_0^t\right]$$

$$= \frac{1}{12}\cos t - \frac{1}{10}\cos 2t + \frac{1}{60}\cos 3t.$$

## PROBLEMS 21.5

Find the inverse transforms of :

1. $\dfrac{1}{s^2(s+5)}$.
2. $\dfrac{1}{s(s+2)^3}$.
3. $\dfrac{s}{a^2s^2+b^2}$.
4. $\dfrac{1}{s^2(s^2+a^2)}$.
5. $\dfrac{1}{s^3(s^2+1)}$.
6. $\dfrac{s+2}{(s^2+4s+8)^2}$.
7. $\dfrac{2as}{(s^2+a^2)^2}$.
8. $\dfrac{s^2}{(s+a)^3}$.
9. $\log\left(\dfrac{1+s}{s}\right)$.
10. $\log\left(\dfrac{s+a}{s+b}\right)$. (*V.T.U., 2016 ; U.K.T.U., 2012*)
11. $\log\left\{\dfrac{s+1}{(s+2)(s+3)}\right\}$.
12. $\dfrac{1}{2}\log\left(\dfrac{s^2+b^2}{s^2+a^2}\right)$.
13. $\log\left(1+\dfrac{1}{s^2}\right)$. (*J.N.T.U., 2015*)
14. $\log\dfrac{s^2+1}{(s-1)^2}$.
15. $\tan^{-1}\left(\dfrac{2}{s}\right)$
16. $\cot^{-1}(s)$. (*V.T.U., 2005*)
17. $s\log\dfrac{s-1}{s+1}$

Using Convolution theorem, evaluate :

18. $L^{-1}\left\{\dfrac{1}{(s+a)(s+b)}\right\}$.
19. $L^{-1}\dfrac{1}{(s^2+a^2)^2}$.
20. $L^{-1}\dfrac{1}{s^2(s^2+a^2)}$.
21. $L^{-1}\left\{\dfrac{1}{s^2(s+1)^2}\right\}$. (*Andhra, 2018; V.T.U., 2016 ; J.N.T.U., 2015*)
22. $L^{-1}\left\{\dfrac{1}{(s-2)(s+2)^2}\right\}$. (*JNTUK, 2022*)
23. $L^{-1}\left\{\dfrac{s}{(s-1)(s^2+1)}\right\}$. (*V.T.U., 2015*)
24. $L^{-1}\left\{\dfrac{1}{s^3(s^2+1)}\right\}$. (*U.P.T.U., 2012*)
25. $L^{-1}\dfrac{s}{(s^2+a^2)^3}$. (*U.P.T.U., 2012*)
26. $L^{-1}\dfrac{s}{(s^2+a^2)(s^2+b^2)}$ (*Andhra, 2018; Anna., 2016*)
27. $L^{-1}\left\{\dfrac{1}{(s^2+4s+13)^2}\right\}$.
28. Show that (*i*) $L^{-1}\left(\dfrac{1}{s}\sin\dfrac{1}{s}\right)=t-\dfrac{t^3}{(3\,!)^2}+\dfrac{t^5}{(5\,!)^2}-\dfrac{t^7}{(7\,!)^2}+\ldots$

    (*ii*) $L^{-1}\left(\dfrac{1}{s}\cos\dfrac{1}{s}\right)=1-\dfrac{t^2}{(2\,!)^2}+\dfrac{t^4}{(4\,!)^2}-\dfrac{t^6}{(6\,!)^2}+\ldots$

## 21.15 (1) APPLICATION TO DIFFERENTIAL EQUATIONS

The Laplace transform method of solving differential equations yields particular solutions without the necessity of first finding the general solution and then evaluating the arbitrary constants. This method is, in general, shorter than our earlier methods and is specially useful for solving linear differential equations with constant coefficients.

**(2) Working procedure** *to solve a linear differential equation with constant coefficients by transform method :*

1. *Take the Laplace transform of both sides of the differential equation using the formula of § 21.7, and the given initial conditions.*

2. *Transpose the terms with minus signs to the right.*

3. *Divide by the coefficient of $\bar{y}$, getting $\bar{y}$ as a known function of s.*

4. *Resolve this function of s into partial fractions and take the inverse transform of both sides.* This gives $y$ as a function of $t$ which is the desired solution satisfying the given conditions.

**Example 21.27.** *Solve the equations*
*$x''' - 2x'' - 5x' = 0$, $x = 0$, $x' = 1$ at $t = 0$ and $x = 1$ at $t = \pi/8$* (Rohtak. 2018)

**Solution:** We have

$$(D^3 - 2D^2 + 5D)x = 0$$

$$x = 0, x_1 = 1, \text{ and } x = 1 \text{ at } x = \frac{\pi}{8}$$

Let
$$\frac{d^2x}{dt^2} = x_2, \quad \text{when } t = 0$$

$$(s^3 - 2s^2 + 5s)\bar{x} = (s^2 - 2s).0 + (s - 2)\,1. + x_2$$

$$(s^3 - 2s^2 + 5s)\bar{x} = (s^2 - 2s).0 + (s - 2).1 + x_2$$

$$\bar{x} = \frac{s - 2 + x_2}{s(s^2 - 2s + 5)} = \frac{A}{s} = \frac{Bs + C}{s^2 - 2s + 5}$$

$$A = \frac{x_1 - 2}{5}, B = \frac{2 - x_2}{5} \text{ and } C = \frac{2x_2 + 1}{5}$$

So,
$$\bar{x} = \frac{x_2 - 2}{5s} + \frac{(2 - x_2)s + 2x_2 + 1}{5(s^2 - 2s + 5)}$$

$$= \left(\frac{x_2 - 2}{5}\right) \cdot \frac{1}{s} + \left(\frac{2 - x_2}{5}\right) \cdot \frac{s - 1}{(s - 1)^2 + (2)^2} + \left(\frac{x_2 + 3}{10}\right) \cdot \left(\frac{2}{(s - 1)^2 + 2^2}\right)$$

Now taking the Inverse transform, we get

$$x = \left(\frac{x_2 - 2}{5}\right) \cdot 1 + \left(\frac{2 - x_2}{5}\right) e^t \cos 2t + \left(\frac{x_2 + 3}{10}\right) e^t \sin 2t$$

Using the last condition $x = 1$, when $t = \pi/8 = \frac{1}{5}\left[(x_2 - 2) + (2 - x_2)e^{\pi/8} \cdot \frac{1}{\sqrt{2}} + \frac{(x_2 + 3)}{2} e^{\pi/8} \cdot \frac{1}{\sqrt{2}}\right]$

$$5 = x_2\left(1 - \frac{1}{\sqrt{2}} e^{\pi/8} + \frac{1}{2\sqrt{2}} e^{\pi/8}\right) - 2 + \frac{2}{\sqrt{2}} e^{\pi/8} + \frac{3}{2\sqrt{2}} e^{\pi/8}$$

$$= 7\left(1 - \frac{1}{2\sqrt{2}} e^{\pi/8}\right) = x_2\left(1 - \frac{1}{2\sqrt{2}} e^{\pi/8}\right)$$

$$x_2 = 7$$

Hence $1 - e^t(\cos 2t - \sin 2t)$

**Example 21.28.** *Use transform method to solve*
$$\frac{d^2x}{dt^2} - 2\frac{dx}{dt} + x = e^t \text{ with } x = 2, \frac{dx}{dt} = -1 \text{ at } t = 0.$$ (Andhra, 2021; V.T.U., 2016)

**Solution.** Taking the Laplace transforms of both sides, we get

$$[s^2\bar{x} - sx(0) - x'(0)] - 2\{s\bar{x} - x(0)\} + \bar{x} = \frac{1}{s - 1}$$

Using the given conditions, it reduces to

$$(s^2 - 2s + 1)\bar{x} = \frac{1}{s - 1} + 2s - 5 = \frac{2s^2 - 7s + 6}{s - 1}$$

$$\therefore \quad \bar{x} = \frac{2s^2 - 7s + 6}{(s - 1)^3} = \frac{2}{s - 1} - \frac{3}{(s - 1)^2} + \frac{1}{(s - 1)^3} \text{ on breaking into partial fractions.}$$

On inversion, we obtain $x = 2L^{-1}\left(\frac{1}{s-1}\right) - 3L^{-1}\frac{1}{(s-1)^2} + L^{-1}\frac{1}{(s-1)^3}$

$$= 2e^t - \frac{3e^t \,.\, t}{1!} + \frac{e^t \,.\, t^2}{2!} = 2e^t - 3te^t + \tfrac{1}{2}t^2e^t.$$

**Example 21.29.** *Solve* $(D^2 + n^2)\,x = a \sin(nt + \alpha)$, $x = Dx = 0$ *at* $t = 0$.

**Solution.** Taking the Laplace transforms of both sides, we get

$$[s^2\bar{x} - sx(0) - x'(0)] + n^2\bar{x} = aL\{\sin nt \,.\, \cos\alpha + \cos nt \,.\, \sin\alpha\}$$

On using the given conditions,

$$(s^2 + n^2)\,\bar{x} = a\cos\alpha \,.\, \frac{n}{s^2+n^2} + a\sin\alpha \,.\, \frac{s}{s^2+n^2}$$

$$\therefore \qquad \bar{x} = an\cos\alpha \,.\, \frac{1}{(s^2+n^2)^2} + a\sin\alpha \,.\, \frac{s}{(s^2+n^2)^2}$$

On inversion, we obtain

$$x = an\cos\alpha \,.\, \frac{1}{2n^3}(\sin nt - nt\cos nt) + a\sin\alpha \,.\, \frac{t}{2n}\sin nt \qquad \text{[By (11) and (12) p. 854]}$$

$$= a\{\sin nt\cos\alpha - nt\cos(nt+\alpha)\}/2n^2.$$

**Example 21.30.** *Solve* $(D^3 - 3D^2 + 3D - 1)\,y = t^2e^t$ *given that* $y(0) = 1$, $y'(0) = 0$, $y''(0) = -2$.

**Solution.** Taking the Laplace transforms of both sides, we get

$$[s^3\bar{y} - s^2y(0) - sy'(0) - y''(0)] - 3[s^2\bar{y} - sy(0) - y'(0)] + 3[s\bar{y} - y(0)] - \bar{y} = \frac{2}{(s-1)^3}$$

Using the given conditions, it reduces to

$$\bar{y} = \frac{s^2 - 3s + 1}{(s-1)^3} + \frac{2}{(s-1)^6} = \frac{(s-1)^2 - (s-1) - 1}{(s-1)^3} + \frac{2}{(s-1)^6}$$

$$= \frac{1}{s-1} - \frac{1}{(s-1)^2} - \frac{1}{(s-1)^3} + \frac{2}{(s-1)^6}$$

On inversion, we obtain $y = L^{-1}\left(\frac{1}{s-1}\right) - L^{-1}\frac{1}{(s-1)^2} - L^{-1}\frac{1}{(s-1)^3} + 2L^{-1}\frac{1}{(s-1)^6}$

$$= e^t\left(1 - t - \frac{1}{2}t^2 + \frac{1}{60}t^5\right).$$

**Example 21.31.** *Solve* $\frac{d^2x}{dt^2} + 9x = \cos 2t$, *if* $x(0) = 1$, $x(\pi/2) = -1$. *(D.T.U., 2015 ; J.N.T.U., 2015)*

**Solution.** Since $x'(0)$ is not given, we assume $x'(0) = a$.

Taking the Laplace transforms of both sides of the equation, we have

$$L(x'') + 9L(x) = L(\cos 2t) \text{ i.e., } [s^2\bar{x} - s\,x(0) - x'(0)] + 9\bar{x} = \frac{s}{s^2+4}$$

$$(s^2 + 9)\,\bar{x} = s + a + \frac{s}{s^2+4} \quad \text{or} \quad \bar{x} = \frac{s+a}{s^2+9} + \frac{s}{(s^2+4)(s^2+9)}$$

or
$$\bar{x} = \frac{a}{s^2+9} + \frac{1}{5}\cdot\frac{s}{s^2+4} + \frac{4}{5}\cdot\frac{s}{s^2+9}.$$

On inversion, we get $\quad x = \frac{a}{3}\sin 3t + \frac{1}{5}\cos 2t + \frac{4}{5}\cos 3t$

When $t = \pi/2$, $-1 = -\frac{a}{3} - \frac{1}{5}$ or $\frac{a}{3} = \frac{4}{5}$ $\qquad \left[\because\ x\left(\frac{\pi}{2}\right) = -1\right]$

Hence the solution is $x = \frac{1}{5}(\cos 2t + 4\sin 3t + 4\cos 3t)$.

**Obs.** *Laplace transform method can also be used for solving ordinary differential equations with variable coefficients of the form $t^m y^{(n)}(t)$ because $L[t^m y^{(n)}(t)] = (-1)^m \frac{d^m}{ds^m}[L\, y^{(n)}(t)]$.*

**Example 21.32.** *Solve $ty'' + 2y' + ty = \cos t$ given that $y(0) = 1$.*

**Solution.** Taking Laplace transform of both sides of the equation and noting that

$$L[t\{f(t)\}] = -\frac{d}{ds}[L\{f(t)\}], \text{ we get}$$

$$-\frac{d}{ds}[s^2\bar{y} - sy(0) - y'(0)] + 2[s\bar{y} - y(0)] - \frac{d}{ds}(\bar{y}) = \frac{s}{s^2+1}$$

or $$-\left(s^2\frac{d\bar{y}}{ds} + 2s\bar{y}\right) + y(0) + 0 + 2s\bar{y} - 2y(0) - \frac{d}{ds}(\bar{y}) = \frac{s}{s^2+1}$$

or $$(s^2+1)\frac{d\bar{y}}{ds} + 1 = -\frac{s}{s^2+1} \quad \text{or} \quad \frac{d\bar{y}}{ds} = -\frac{1}{s^2+1} - \frac{s}{(s^2+1)^2}$$

On inversion and noting that $L^{-1}\{\bar{f}'(s)\} = -t\, f(t)$, we get

$$-ty = -\sin t - \tfrac{1}{2}t\sin t$$ [See § 21.12 (11)]

or $$y = \frac{1}{2}\left(1 + \frac{2}{t}\right)\sin t$$ which is the desired solution.

**Example 21.33.** *Solve $x\frac{d^2y}{dx^2} + \frac{dy}{dx} + xy = 0$, $y(0) = 2$, $y'(0) = 0$.*

**Solution.** Taking Laplace transform of both sides of the equation, we get

$$L(xy'') + L(y') + L(xy) = 0$$

or $$-\frac{d}{ds}[s^2\bar{y} - sy(0) - y'(0)] + [s\bar{y} - y(0)] - \frac{d\bar{y}}{ds} = 0 \quad \text{or} \quad (s^2+1)\frac{d\bar{y}}{ds} + s\bar{y} = 0$$

Separating the variables, $\int \frac{d\bar{y}}{\bar{y}} + \int \frac{s\,ds}{s^2+1} = c$

or $$\log\bar{y} + \tfrac{1}{2}\log(s^2+1) = \log c' \quad \text{or} \quad \bar{y} = \frac{c'}{\sqrt{(s^2+1)}}$$

Inversion gives $y = c' J_0(x)$

To find $c'$, we have $y(0) = c' J_0(0)$, *i.e.*, $c' = 2$

Hence $y = 2J_0(x)$.

**Example 21.34.** *A condensor of capacity C is charged to potential E and discharged at t = 0, through an inductive resistance L, R, show that the charge q at time t is given by*

$$q = \frac{CE}{n}e^{-\mu t}(\mu \sin nt + n\cos nt)$$

$$\mu = \frac{R}{2L} \text{ and } n^2 = \left(\frac{1}{LC} - \frac{R^2}{4L^2}\right) > 0$$ *(Patna, 2018)*

**Solution:** Let $i$ be the current. Then the potential difference inductance $L$,

Since applied potential is zero.

$$L\frac{di}{dt} + Ri + \frac{q}{C} = 0$$

$$L\frac{d^2q}{dt^2} + R\frac{dq}{dt} + \frac{q}{c} = 0$$

$$i = \frac{dq}{dt}$$

The initial condition at $t = 0$, $q_0$ = initial charge = $EC$,

$$l_0 = \left(\frac{dq}{dt}\right)t = 0 = q = 0$$

Let $L[q] = \bar{q}$, then

$$L\left[s^2\bar{q} - sq_0 - \bar{q}_1\right] + R\left[\bar{s}q - q_0\right] + \frac{1}{c}\bar{q}^0 = 0$$

$$\bar{q} = \frac{EC\left(S + \frac{R}{L}\right)}{S^2 + \frac{R}{L}S + \frac{1}{LC}}$$

Substituting $\frac{R}{2} = 2\mu$ and $\frac{1}{LC} = R^2 + \frac{R^2}{4L^2} = n^2 + \mu^2$

We have $= \frac{EC(s + 2\mu)}{s^2 + 2\mu s + \mu^2 + n^2}$

Now taking inverse transforms, we have

$$q = EC\left[e^{-\mu t}\cos nt + \frac{\mu}{n}\cdot e^{-\mu t}\sin nt\right] = \frac{EC}{n}e^{-\mu t}(\mu\sin nt + \mu\cos nt)$$

**Exemple 21.35:** *Find the inverse Laplace transforms of*

*(i)* $\log\left(1 + \frac{1}{s^2}\right)$ *(ii)* $\log\left(\frac{s+1}{s-1}\right)$

**Solution:** (*i*) Let $L\{f(t)\} = \bar{f}(s) = \log\left(1 + \frac{1}{s^2}\right)$

$$L\{t(t)\} = \frac{-d}{ds}\left[\log\left(1 + \frac{1}{s^2}\right)\right] = -\left[\frac{-2/s^2}{1 + \frac{1}{s^2}}\right]$$

$$= \frac{2}{s(s^2+1)} = 2\left(\frac{1}{s} - \frac{s}{s^2+1}\right) = 2L[(1 - \cos t)] = L[2 - 2\cos t)]$$

Comparing both sides, we get $tf(t) = (2 - 2\cos t)$ or $f(t) = \frac{2(1 - \cos t)}{t}$

(*ii*) Let $L\{f(t)\} = \bar{f}(s) = \log\left(\frac{s+1}{s-1}\right) = \log(s+1) - \log(s-1)$

$\therefore$ $L\{t(f(t)\} = -\frac{d}{ds}[\log(s+1) - \log(s-1)]$

$$= -\left[\frac{1}{s+1} - \frac{1}{s-1}\right] = \left[\frac{1}{s-1} - \frac{1}{s+1}\right] = L[e^t - e^{-t}] = L[2\sin ht]$$

Comparing both side $tf(t) = 2\sin ht$

$$\text{or } f(t) = \frac{2\sin ht}{t}$$

**Example 21.36:** *Apply convolution theorem to evaluate*

$$L^{-1}\left\{\frac{s^2}{(s^2+a^2)(s^2+b^2)}\right\}$$

**Solution:** $L^{-1}\left\{\frac{s}{s^2+a^2}\right\} = \cos at$

and $L^{-1}\left\{\frac{s}{s^2+a^2}\cdot\frac{s}{s^2+b^2}\right\} = \int_o^t \cos au \cos b(t-4)du$

$$= \frac{1}{2}\int_o^t [\cos(au - bu + bt) + \cos(au + bu - bt]\,du$$

$$= \frac{1}{2}\left[\frac{\sin(au-bu+bt)}{a-b} + \sin\frac{(au+bu-bt)}{a+b}\right]_0^t$$

$$= \frac{1}{2}\left(\frac{\sin at - \sin bt}{a-b} + \frac{\sin at + \sin bt}{a+b}\right) = \frac{a\sin at - b\sin bt}{a^2-b^2}$$

## PROBLEMS 21.6

Solve the following equations by the transform method :

1. $\frac{dy}{dt} + y = \cos 2t, y(0) = 1.$ *(Delhi., 2016)*
2. $y''(t) + y(t) = 0, y(0) = 2, y(\pi/2) = 1.$ *(V.T.U., 2015)*
3. $y'' + 4y' + 3y = e^{-t}, y(0) = y'(0) = 1.$ *(Andhra, 2021)*
4. $y'' + y = t, y(0) = 1, y'(0) = 0.$ *(Andhra, 2023)*
5. $y'' - 3y' + 2y = e^{3t}$, when $y(0) = 1$ and $y'(0) = 0$. *(V.T.U., 2010)*
6. $(D^2 - 3D + 2)y = 4e^{2t}$ with $y(0) = -3, y(0) = 5$. *(Andhra, 2022)*
7. $y'' + 25y = 10\cos 5t$ given that $y(0) = 2, y'(0) = 0$. *(Andhra, 2022)*
8. $(D^2 + \omega^2)y = \cos \omega t, t > 0$, given that $y = 0$ and $Dy = 0$ at $t = 0$.
9. $\frac{d^2y}{dt^2} + 2\frac{dy}{dt} - 3y = \sin t, y = \frac{dy}{dt} = 0$ when $t = 0$. *(D.T.U., 2016)*
10. $\frac{d^4y}{dt^4} - k^4y = 0$, where $y(0) = 1, y'(0) = y''(0) = y'''(0) = 0$. *(JNTUK, 2022)*
11. $y''''(t) + 2y''(t) + y(t) = \sin t$, when $y(0) = y'(0) = y''(0) = y'''(0) = 0$.
12. $\frac{d^2x}{dt^2} + 2\frac{dx}{dt} + 5x = e^{-t}\sin t$. *(P.T.U., 2010)*
13. $y'' + y' = t^2 + 2t$ given $y = 4, y' = -2$, when $t = 0$. *(Anna., 2016)*
14. $\frac{d^3y}{dt^3} - 3\frac{d^2y}{dt^2} + 3\frac{dy}{dt} - y = t^2e^{2t}$, where $y = 1, \frac{dy}{dt} = 0, \frac{d^2y}{dt^2} = -2$ at $t = 0$. *(V.T.U., 2012)*
15. $(D^2 + 1)x = t\cos 2t, x = Dx = 0$ at $t = 0$. *(D.T.U., 2011)*
16. $ty'' + 2y' + ty = \sin t$, when $y(0) = 1$.
17. $ty'' + (1 - 2t)y' - 2y = 0$, when $y(0) = 1, y'(0) = 2$.
18. $y'' + 2ty' - y = t$, when $y(0) = 0, y'(0) = 1$.
19. $ty'' + y' + 4ty = 0$ when $y(0) = 3, y'(0) = 0$.
20. A voltage $Ee^{-at}$ is applied at $t = 0$ to a circuit of inductance $L$ and resistance $R$. Show (by the transform method) that the current at time $t$ is $\frac{E}{R-aL}(e^{-at} - e^{-Rt/L})$.
21. Workout example 12.17, p. 555 by the transform method.
22. Obtain the equation for the forced oscillation of a mass $m$ attached to the lower end of an elastic spring whose upper end is fixed and whose stiffness is $k$, when the driving force is $F_0 \sin at$. Solve this equation (using the Laplace transforms) when $a^2 \neq k/m$, given that initial velocity and displacement (from equilibrium position) are zero.

$\left[\textbf{Hint : } \text{The equation of motion is } \frac{d^2x}{dt^2} + \frac{k}{m}x = \frac{F_0}{m}\sin at \text{ and } x = \frac{dx}{dt} = 0 \text{ when } t = 0.\right]$

## 21.16 SIMULTANEOUS LINEAR EQUATIONS WITH CONSTANT COEFFICIENTS

The Laplace transform method can also be applied with advantage to the solution of simultaneous linear differential equations.

**Example 21.37.** *Solve the simultaneous equations* $\frac{dx}{dt} + 5x - 2y = t$, $\frac{dy}{dt} + 2x + y = 0$ *being given* $x = y = 0$ *when* $t = 0$. [Ex. 13.38]

**Solution.** Taking the Laplace transforms of the given equations, we get

$$[s\bar{x} - x(0)] + 5\bar{x} - 2\bar{y} = 1/s^2 \quad i.e., \quad (s+5)\bar{x} - 2\bar{y} = 1/s^2 \qquad ...(i)\ [\because\ x(0) = 0]$$

and
$$s\bar{y} - y(0) + 2\bar{x} + \bar{y} = 0 \quad i.e., \quad 2\bar{x} + (s+1)\bar{y} = 0 \qquad ...(ii)\ [\because\ y(0) = 0]$$

Solving (*i*) and (*ii*) for $\bar{x}$, we get

$$\bar{x} = \begin{vmatrix} 1/s^2 & -2 \\ 0 & s+1 \end{vmatrix} \div \begin{vmatrix} s+5 & -2 \\ 2 & s+1 \end{vmatrix} = \frac{s+1}{s^2(s+3)^2} = \frac{1}{27s} + \frac{1}{9s^2} - \frac{1}{27(s+3)} - \frac{2}{9(s+3)^2}$$

Substituting the value of $\bar{x}$ in (*ii*), we get

$$\bar{y} = -\frac{2}{s^2(s+3)^2} = \frac{4}{27s} - \frac{2}{9s^2} - \frac{4}{27(s+3)} - \frac{2}{9(s+3)^2}$$

On inversion, we get

$$x = \frac{1}{27} + \frac{t}{9} - \frac{1}{27}e^{-3t} - \frac{2}{9}te^{-3t},\ y = \frac{4}{27} - \frac{2t}{9} - \frac{4}{27}e^{-3t} - \frac{2}{9}te^{-3t}.$$

**Example 21.38.** *The coordinates* $(x, y)$ *of a particle moving along a plane curve at any time t, are given by* $dy/dt + 2x = \sin 2t$, $dx/dt - 2y = \cos 2t$, $(t > 0)$. *If at* $t = 0$, $x = 1$ *and* $y = 0$, *show by transforms, that the particle moves along the curve* $4x^2 + 4xy + 5y^2 = 4$.

**Solution.** Taking the Laplace transforms of the given equations and noting that $y(0) = 0$, $x(0) = 1$,

we get
$$[s\bar{y} - y(0)] + 2\bar{x} = \frac{2}{s^2 + 2^2} \quad \text{or} \quad 2\bar{x} + s\bar{y} = \frac{2}{s^2+4} \qquad ...(i)$$

and
$$[s\bar{x} - x(0)] - 2\bar{y} = \frac{s}{s^2 + 2^2} \quad \text{or} \quad s\bar{x} - 2\bar{y} = \frac{s}{s^2+4} + 1 \qquad ...(ii)$$

Multiplying Eqn. (*i*) by $s$ and Eqn. (*ii*) by 2 and subtracting, we get

$$(s^2+4)\bar{y} = -2 \quad \text{or} \quad \bar{y} = -2/(s^2+4)$$

On inversion,
$$y = -2L^{-1}\left[\frac{1}{s^2+4}\right] = -\sin 2t$$

From the given first equation,

$$2x = \sin 2t - dy/dt = \sin 2t - \frac{d}{dt}(-\sin 2t)$$

or
$$2x = \sin 2t + 2\cos 2t \quad \text{or} \quad 4x^2 = (\sin 2t + 2\cos 2t)^2 \qquad ...(iii)$$

Also
$$4xy = (\sin 2t + 2\cos 2t)(-2\sin 2t) = -2(\sin^2 2t + 2\sin 2t \cos 2t) \qquad ...(iv)$$

and
$$5y^2 = 5\sin^2 2t. \qquad ...(v)$$

Adding Eqns. (*iii*), (*iv*), and (*v*), we obtain

$$4x^2 + 4xy + 5y^2 = \sin^2 2t + 4\sin 2t\cos 2t + 4\cos^2 2t - 2\sin^2 2t$$
$$-4\sin 2t\cos 2t + 5\sin^2 2t = 4\sin^2 2t + 4\cos^2 2t = 4.$$

**Example 21.39.** *The small oscillations of a certain system with two degrees of freedom are given by the equations :* $D^2x + 3x - 2y = 0$, $D^2x + D^2y - 3x + 5y = 0$ *where* $D = d/dt$. *If* $x = 0$, $y = 0$, $x = 3$, $y = 2$ *when* $t = 0$, *find x and y when* $t = 1/2$.

**Solution.** Taking the Laplace transform of both the equations, we get

$$[s^2\bar{x} - sx(0) - x'(0)] + 3\bar{x} - 2\bar{y} = 0 \quad i.e., \quad (s^2+3)\bar{x} - 2\bar{y} = 3 \qquad ...(i)$$

and $$[s^2\bar{x} - sx(0) - x'(0)] + [s^2\bar{y} - sy(0) - y'(0)] - 3\bar{x} + 5\bar{y} = 0 \; i.e., (s^2-3)\bar{x} + (s^2+5)\bar{y} = 5 \qquad ...(ii)$$

Solving (*i*) and (*ii*) for $\bar{x}$ and $\bar{y}$, we get

$$\bar{x} = \begin{vmatrix} 3 & -2 \\ 5 & s^2+5 \end{vmatrix} \div \begin{vmatrix} s^2+3 & -2 \\ s^2-3 & s^2+5 \end{vmatrix} = \frac{3s^2+25}{(s^2+1)(s^2+9)}$$
$$= \frac{11}{4}\cdot\frac{1}{s^2+1} + \frac{1}{4}\cdot\frac{1}{s^2+9}$$

and $$\bar{y} = \begin{vmatrix} s^2+3 & 3 \\ s^2-3 & 5 \end{vmatrix} \div \begin{vmatrix} s^2+3 & -2 \\ s^2-3 & s^2+5 \end{vmatrix} = \frac{2s^2+24}{(s^2+1)(s^2+9)} = \frac{11}{4}\cdot\frac{1}{s^2+1} + \frac{3}{4}\cdot\frac{1}{s^2+9}.$$

On inversion, we get $x = \frac{11}{4}\sin t + \frac{1}{12}\sin 3t$ ; $y = \frac{11}{4}\sin t - \frac{1}{4}\sin 3t$

**Obs.** The student should compare the earlier solutions of the above examples with those given now and appreciate the superiority of the transform method over others.

## PROBLEMS 21.7

Solve the following simultaneous equations (by using Laplace transforms) :

1. $\frac{dx}{dt} - y = e^t,\ \frac{dy}{dt} + x = \sin t$, given $x(0) = 1, y(0) = 0$. (*Rohtak, 2011 ; U.K.T.U., 2011*)
2. $\frac{dx}{dt} + y = \sin t,\ \frac{dy}{dt} + x = \cos t$, given that $x = 2$ and $y = 0$ when $t = 0$. (*Hamirpur, 2015 ; Andhra., 2013 ; V.T.U., 2013*)
3. $\frac{d^2x}{dt^2} - x = y,\ \frac{d^2y}{dt^2} + y = -x$, given that at $t = 0$ ; $x = 2, y = -1, \frac{dx}{dt} = 0$ and $\frac{dy}{dt} = 0$.
4. $3\frac{dx}{dt} + \frac{dy}{dt} + 2x = 1,\ \frac{dx}{dt} + 4\frac{dy}{dt} + 3y = 0$ ; given $x = 0, y = 0$ when $t = 0$.
5. $(D-2)x - (D+1)y = 6e^{3t}$ ; $(2D-3)x + (D-3)y = 6e^{3t}$ given $x = 3, y = 0$ when $t = 0$.
6. The currents $i_1$ and $i_2$ in mesh are given by the differential equations ; $di_1/dt - \omega i_2 = a\cos pt,\ di_2/dt + \omega i_1 = a\sin pt$. Find the currents $i_1$ and $i_2$ by Laplace transform, if $i_1 = i_2 = 0$ at $t = 0$.

## 21.17 (1) UNIT STEP FUNCTION

At times, we come across such fractions of which the inverse transform cannot be determined from the formulae so far derived. In order to cover such cases, we introduce the *unit step function* (or *Heaviside's unit function**).

**Def.** *The unit step function $u(t-a)$ is defined as follows :*

$$u(t-a) = \begin{cases} 0 \text{ for } t < a \\ 1 \text{ for } t \geq a \end{cases}$$

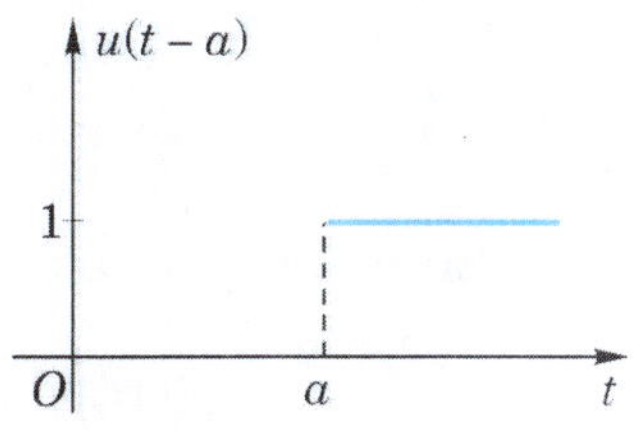

Fig. 21.3

*where, a is always positive (Fig. 21.3). It is also denoted as $H(t-a)$* or $U_a$

**(2) Transform of unit function.**

$$L\{u(t-a)\} = \int_0^\infty e^{-st}\,u(t-a)\,dt = \int_0^a e^{-st}.0\,dt + \int_a^\infty e^{-st}.1\,dt = 0 + \left|\frac{e^{-st}}{-s}\right|_a^\infty$$

Thus $$L\{u(t-a)\} = e^{-as}/s.$$

*Named after the British Electrical Engineer Oliver Heaviside (1850–1925).

The product $f(t)\, u(t-a) = \begin{cases} 0 & \text{for } t < a \\ f(t) & \text{for } t \ge a. \end{cases}$

The function $f(t-a)\,.\,u(t-a)$ represents the graph of $f(t)$ shifted through a distance $a$ to the right and is of special importance.

**Second shifting property.** If $L\{f(t)\} = \bar{f}(s)$, *then*

$$\mathbf{L\{f(t-a).\,u(t-a)\} = e^{-as}\,\bar{f}(s)}$$

$$L\{f(t-a)\,.\,u(t-a)\} = \int_0^\infty e^{-st} f(t-a)\; u(t-a)\, dt$$

$$= \int_0^a e^{-st} f(t-a)\,(0)\, dt + \int_a^\infty e^{-st} f(t-a)\, dt \qquad [\text{Put } t-a=u]$$

$$= \int_0^\infty e^{-s(u+a)} f(u)\, du = e^{-sa}\int_0^\infty e^{-su} f(u)\, du = e^{-as}\,\bar{f}(s).$$ *(Hamirpur., 2015)*

**Example 21.40.** *Express the following function (Fig. 21.4) in terms of unit step function and find its Laplace transform.* *(Andhra, 2021)*

**Solution.** We have $f(t) = \begin{cases} 0, & 0 < t < 1 \\ t-1, & 1 < t < 2 \\ 1, & t > 2 \end{cases}$

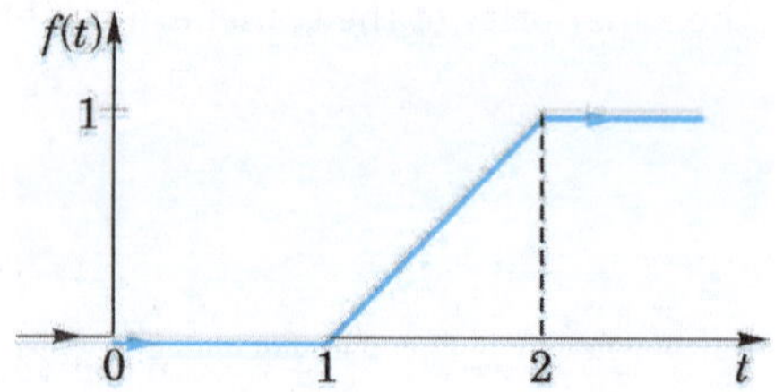

Fig. 21.4

or $$f(t) = (t-1)\,[u(t-1) - u(t-2)] + u(t-2)$$
$$= (t-1)\,u(t-1) - (t-2)\,u(t-2)$$

By second shifting property,

$$L[f(t-a)\,u(t-a)] = e^{-as}\,L[f(t)].$$

Also $$L[f(t)] = L(t) = 1/s^2.$$

$\therefore$ $$L[(t-1)\,u(t-1)]$$
$$= e^{-s}\,.\,\frac{1}{s^2} \text{ and } L\,[(t-2)\,u(t-2)] = e^{-2s}.\,\frac{1}{s^2}$$

Hence $$L\,[f(t)] = L\,[(t-1)\,u(t-1) - (t-2)\,u(t-2)] = \frac{e^{-s} - e^{-2s}}{s^2}.$$

**Example 21.41.** *Using unit step function, find the Laplace transform of*

$$f(t) = \begin{cases} \sin t, & 0 \le t < \pi \\ \sin 2t, & \pi \le t < 2\pi. \\ \sin 3t, & t \ge 2\pi \end{cases}$$

**Solution.** $$f(t) = \sin t\,[u(t-0) - u(t-\pi)] + \sin 2t\,[u(t-\pi) - u(t-2\pi)] + \sin 3t\,.\,u(t-2\pi)$$
$$= \sin t + (\sin 2t - \sin t)\,u(t-\pi) + (\sin 3t - \sin 2t)\,u(t-2\pi)$$

Since $L[f(t-a)\,u(t-a)] = e^{-as}\,\bar{f}(s)$ and $L(\sin at) = \dfrac{a}{s^2+a^2}$,

$$L[f(t)] = L(\sin t) + L[(\sin 2t - \sin t)\,.\,u(t-\pi)] + L[(\sin 3t - \sin 2t)\,.\,u(t-2\pi)]$$

$$= \frac{1}{s^2+1} + e^{-\pi s}\left(\frac{2}{s^2+4} - \frac{1}{s^2+1}\right) + e^{-2\pi s}\left(\frac{3}{s^2+9} - \frac{2}{s^2+4}\right).$$

**Example 21.42.** (*i*) *Express the function (Fig. 21.5) in terms of unit step function and find its Laplace transform.*

(*ii*) *Obtain the Laplace transform of* $e^{-t}[1 - u(t-2)]$.

**Solution.** (*i*) We have $f(t) = \begin{cases} t-1, & 1 < t < 2 \\ 3-t, & 2 < t < 3. \end{cases}$

or $$f(t) = (t-1)\{u(t-1) - u(t-2)\} + (3-t)\{u(t-2) - u(t-3)\}$$
$$= (t-1)\,u(t-1) - 2(t-2)\,u(t-2) + (t-3)\,u(t-3)$$

Since $L\{f(t-a)\,u(t-a)\} = e^{-as}\,\bar{f}(s)$ ...(*i*)

$$\therefore \quad L[f(t)] = e^{-s}\cdot\frac{1}{s^2} - 2e^{-2s}\cdot\frac{1}{s^2} + e^{-3s}\cdot\frac{1}{s^2} = \frac{e^{-s}(1-e^{-s})^2}{s^2} \qquad [\because\ f(t) = t]$$

(*ii*) $L\{e^{-t}[1-u(t-2)]\} = L(e^{-t}) - L\{e^{-t}u(t-2)\} = \dfrac{1}{s+1} - e^{-2}L\{e^{-(t-2)}u(t-2)]$

Taking $f(t) = e^{-t}$, $\bar{f}(s) = \dfrac{1}{s+1}$ and using (*i*) above,

$$L\{e^{-(t-2)}u(t-2)\} = e^{-2s}\cdot\frac{1}{s+1}$$

Hence $Le^{-t}\{1-u(t-2)\} = \{1-e^{-2(s+1)}\}/(s+1).$

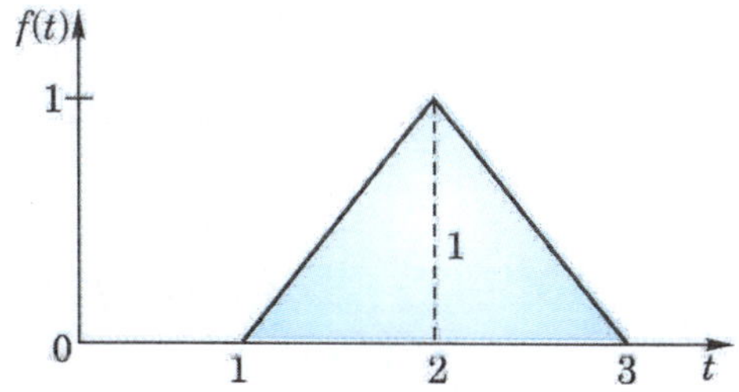

Fig. 21.5

**Example 21.43.** *Using Laplace transform, evaluate* $\int_0^\infty e^{-t}(1+2t-t^2+t^3)\,H(t-1)\,dt.$

**Solution.** We have $L\{(1+2t-t^2+t^3)\,H(t-1)\}$

$$= e^{-s}L[1+2(t+1)-(t+1)^2+(t+1)^3] = e^{-s}L(3+3t+2t^2+t^3)$$

$$= e^{-s}\left(3\cdot\frac{1}{s} + 3\cdot\frac{1}{s^2} + 2\cdot\frac{2!}{s^3} + \frac{3!}{s^4}\right) = e^{-s}\left(\frac{3}{s} + \frac{3}{s^2} + \frac{4}{s^3} + \frac{6}{s^4}\right)$$

By definition, this implies that

$$\int_0^\infty e^{-st}(1+2t-t^2+t^3)\,H(t-1)\,dt = e^{-s}\left(\frac{3}{s} + \frac{3}{s^2} + \frac{4}{s^3} + \frac{6}{s^4}\right)$$

Taking $s = 1$, we obtain

$$\int_0^\infty e^{-t}(1+2t-t^2+t^3)\,H(t-1)\,dt = e^{-1}(3+3+4+6) = 16/e.$$

**Example 21.44.** *Evaluate* (*i*) $L^{-1}\left\{\dfrac{e^{-s}-3e^{-3s}}{s^2}\right\}$ (*ii*) $L^{-1}\left\{\dfrac{se^{-as}}{s^2-w^2}\right\}, a > 0.$

**Solution.**
$$L^{-1}\left\{e^{-s}\cdot\frac{1}{s^2}\right\} = \begin{cases} t-1, & t>1 \\ 0, & t<1 \end{cases} = (t-1)\,u(t-1)$$

$$L^{-1}\left\{e^{-3s}\cdot\frac{1}{s^2}\right\} = \begin{cases} t-3, & t>3 \\ 0, & t<3 \end{cases} = (t-3)\,u(t-3)$$

$$\therefore \quad L^{-1}\left\{\frac{e^{-s}-3e^{-3s}}{s^2}\right\} = L^{-1}\left(\frac{e^{-s}}{s^2}\right) - 3L^{-1}\left(\frac{e^{-3s}}{s^2}\right) = (t-1)\,u(t-1) - 3(t-3)\,u(t-3)$$

(*ii*) We know that $L^{-1}\left(\dfrac{s}{s^2-w^2}\right) = \cosh wt$

$$\therefore \quad L^{-1}\left(\frac{se^{-as}}{s^2-w^2}\right) = \begin{cases} \cosh w(t-a), & t>a \\ 0, & t<a \end{cases}$$

$= \cosh w(t-a)\,u(t-a)$, by second shifting property.

**Example 21.45.** *Find the inverse Laplace transform of :*

(i) $\dfrac{se^{-s/2} + \pi e^{-s}}{s^2 + \pi^2}$  (ii) $\dfrac{e^{-cs}}{s^2(s+a)}$ $(c > 0)$.

**Solution.** (i) Since $L^{-1} \dfrac{s}{s^2 + \pi^2} = \cos \pi t$, $L^{-1}\left(\dfrac{\pi}{s^2+\pi^2}\right) = \sin \pi t$

and $$L^{-1}[e^{-as}\bar{f}(s)] = f(t-a) \,.\, u(t-a) \qquad ...(\lambda)$$

$$\therefore \quad L^{-1}\left\{\frac{se^{-s/2} + \pi e^{-s}}{s^2+\pi^2}\right\} = L^{-1}\left\{e^{-s/2} \cdot \frac{s}{s^2+\pi^2}\right\} + L^{-1}\left\{e^{-s} \cdot \frac{\pi}{s^2+\pi^2}\right\}$$

$$= \cos \pi (t - 1/2) \,.\, u(t - 1/2) + \sin \pi (t-1) \,.\, u(t-1)$$

$$= \sin \pi t \,.\, u(t - 1/2) - \sin \pi t \,.\, u(t-1) = \{u(t-1/2) - u(t-1)\} \sin \pi t$$

(ii) $$L^{-1}\left\{\frac{e^{-cs}}{s^2(s+a)}\right\} = L^{-1}\left\{e^{-cs}\left(-\frac{1}{a^2}\cdot\frac{1}{s} + \frac{1}{a}\cdot\frac{1}{s^2} + \frac{1}{a^2}\cdot\frac{1}{s+a}\right)\right\}$$

Using $(\lambda)$ above, we have

$$L^{-1}\left\{\frac{e^{-cs}}{s^2(s+a)}\right\} = -\frac{1}{a^2}\{1 \,.\, u(t-c)\} + \frac{1}{a}\{(t-c) \,.\, u(t-c)\} + \frac{1}{a^2}\{e^{-a(t-c)} \,.\, u(t-c)\}$$

$$= \frac{1}{a^2}\{a(t-c) - 1 + e^{-a(t-c)}\}\, u(t-c).$$

**Example 21.46.** *A particle of mass m can oscillate about the position of equilibrium under the effect of a restoring force $mk^2$ times the displacement. It started from rest by a constant force F which acts for time T and then ceases. Find the amplitude of the subsequent oscillation.*

**Solution.** The constant force $F$ acting from $t = 0$ to $t = T$ can be expressed as

$$F[1 - u(t-T)],\ 0 < t < T$$

$\therefore$ equation of motion of the particle is

$$m\frac{d^2x}{dt^2} = F[1 - u(t-T)] - mk^2x \qquad \text{or} \qquad \frac{d^2x}{dt^2} + k^2x = \frac{F}{m}[1 - u(t-T)]$$

Taking Laplace transform of both sides, we get

$$(s^2 + k^2)\,\bar{x} = \frac{F}{ms}(1 - e^{-sT}) \qquad [\because\ x = 0,\ \dot{x} = 0 \text{ at } t = 0]$$

or $$\bar{x} = \frac{F}{m}\cdot\frac{1 - e^{-sT}}{s(s^2+k^2)} = \frac{F}{m}(1 - e^{-sT})\cdot\frac{1}{k^2}\left(\frac{1}{s} - \frac{s}{s^2+k^2}\right)$$

$$= \frac{F}{mk^2}\left\{(1 - e^{-sT})\frac{1}{s} - (1 - e^{-sT})\cdot\frac{s}{s^2+k^2}\right\}$$

Taking inverse Laplace transform, we obtain

$$x = \frac{F}{mk^2}[(1 - \cos kt) - \{1 - \cos k(t-T)\}]\, u(t-T)$$

*i.e.,* $$x = \frac{F}{mk^2}(1 - \cos kt) \text{ for } 0 < t < T$$

and $$x = \frac{F}{mk^2}(1 - \cos kt) - \{1 - \cos k(t-T)\} \text{ for } t > T$$

$$= \frac{F}{mk^2}\{\cos k(t-T) - \cos kT\} \text{ for } t > T$$

or $$x = \frac{2F}{mk^2} \sin \frac{kT}{2} \ . \sin k\,(t - T/2) \text{ for } t > T$$

Hence the amplitude of subsequent oscillation (*i.e.*, for $t > T$) $= \dfrac{2F}{mk^2} \sin \dfrac{kT}{2}$.

**Example 21.47.** *In an electrical circuit with e.m.f. E(t), resistance R and inductance L, the current i builds up at the rate given by*

$$L\, di/dt + Ri = E(t). \qquad \text{...}(i)$$

*If the switch is connected at t = 0 and disconnected at t = a, find the current i at any instant.*

**Solution.** We have $i = 0$ at $t = 0$ and $E(t) = \begin{cases} E \text{ for } 0 < t < a \\ 0 \text{ for } t > a \end{cases}$

$\therefore$ taking the Laplace transform of both sides, Eqn. (*i*) becomes

$$(Ls + R)\bar{i} = \int_0^\infty e^{-st} \, E(t)\, dt = \int_0^a e^{-st} \, E dt = \frac{E}{s}(1 - e^{-as})$$

or $$\bar{i} = \frac{E}{s(Ls + R)} - \frac{Ee^{-as}}{s(Ls + R)}$$

On inversion, we get $$i = L^{-1}\left\{\frac{E}{s(Ls + R)}\right\} - L^{-1}\left\{\frac{Ee^{-as}}{s(Ls + R)}\right\} \qquad \text{...}(ii)$$

Now $$L^{-1}\left\{\frac{E}{s(Ls + R)}\right\} = \frac{E}{R}\left\{L^{-1}\left(\frac{1}{s}\right) - L^{-1}\left(\frac{1}{s + R/L}\right)\right\} = \frac{E}{R}\,(1 - e^{-Rt/L})$$

and $$L^{-1}\left\{\frac{Ee^{-as}}{s(Ls + R)}\right\} = \frac{E}{R}\,\{1 - e^{-R(t-a)/L}\}\, u\,(t - a) \qquad \text{[By the second shifting property]}$$

Thus Eqn. (*ii*) becomes $$i = \frac{E}{R}\{1 - e^{-Rt/L}\} - \frac{E}{R}\{1 - e^{-R(t-a)/L}\}\, u\,(t - a)$$

Hence $$i = \frac{E}{R}\{1 - e^{-Rt/L}\} \text{ for } 0 < t < a$$

and $$i = \frac{E}{R}[\{1 - e^{-Rt/L}\} - \{1 - e^{-R(t-a)/L}\}] = \frac{E}{R} e^{-Rt/L}\,(e^{-Ra/L} - 1) \text{ for } t > a.$$

**Example 21.48.** *Calculate the maximum deflection of an encastre beam 1 ft. long carrying a uniformly distributed load w lb./ft. on its central half length.*

**Solution.** Taking the origin at the end $A$, we have

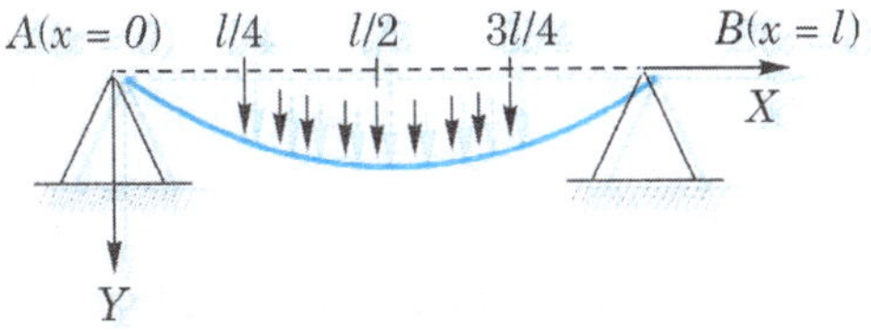


Fig. 21.6

$$EI \frac{d^4 y}{dx^4} = w(x)$$

where $w(x) = w\{u(x - l/4) - u(x - 3l/4)\}$

Taking the Laplace transform of both sides, (Fig. 21.6), we get

$$EI[s^4 \bar{y} - s^3 y(0) - s^2 y'(0) - sy''(0) - y'''(0)]$$

$$= w\left(\frac{e^{-ls/4}}{s} - \frac{e^{-3ls/4}}{s}\right)$$

Using the conditions $y(0) = y'(0) = 0$ and taking $y''(0) = c_1$ and $y'''(0) = c_2$, we have

$$EI\,\bar{y} = w\left(\frac{e^{-ls/4}}{s^5} - \frac{e^{-3ls/4}}{s^5}\right) + \frac{c_1}{s^3} + \frac{c_2}{s^4}$$

On inversion, we get $EIy = \dfrac{w}{24}[(x - l/4)^4 u(x - l/4) - (x - 3l/4)^4 u(x - 3l/4)] + \frac{1}{2}c_1 x^2 + \frac{1}{6}c_2 x^3 \qquad \text{...}(i)$

For $x > 3l/4$, $$EIy = \frac{w}{24}[(x - l/4)^2 - (x - 3l/4)^2] + \frac{1}{2}c_1 x^2 + \frac{1}{6}c_2 x^3$$

and $$EIy' = \frac{w}{6}[(x - l/4)^3 - (x - 3l/4)^3] + c_1x + \tfrac{1}{2}c_2x^2$$

Using the conditions $y(l) = 0$ and $y'(l) = 0$, we get $0 = \frac{w}{24}\left\{\left(\frac{3l}{4}\right)^4 - \left(\frac{l}{4}\right)^4\right\} + \tfrac{1}{2}c_1l^2 + \tfrac{1}{6}c_2l^3$

and $$0 = \frac{w}{6}\left\{\left(\frac{3l}{4}\right)^3 - \left(\frac{l}{4}\right)^3\right\} + c_1l + \tfrac{1}{2}c_2l^2$$

whence $c_1 = 11\,wl^2/192$ ; $c_2 = -\,wl/4$.

Thus for $l/4 < x < 3l/4$, (*i*) gives $EIy = \frac{w}{24}\left(x + \frac{1}{4}\right)^4 + \frac{11wl^2}{384}x^2 - \frac{wl}{24}x^3$

Hence the maximum deflection $= y(l/2) = \frac{13wl^4}{6144EI}$.

**Example 21.49:** *Solve Laplace transform* $\frac{d^2y}{dt^2} + 2\frac{dy}{dt} + 5y = e^{-t}\sin t,$

$$y(0) = 0$$
$$y_i(0) = 1$$

**Solution:** The given differential equation is

$$(D^2 + 2D + 5)y = e^{-t}\sin t \qquad \text{...(i)}$$

With initial conditions $y_0 = 0$ and $y_1 = 1$

Taking the Laplace transforms of both side of (*i*), we get

$$(s^2 + 2s + 5)y = \frac{1}{(s+1)^2 + 1} + (s + 2)y_0 + y_1$$

$$(s^2 + 2s + 5)y = \frac{1}{s^2 + 2s + 2} + 1$$

Since $y_0 = 0, y_1 = 1$ $$y = \frac{1}{(s^2 + 2s + 2)(s^2 + 2s + 5)} + \frac{1}{s^2 + 2s + 5}$$

$$y = \frac{1}{3}\left[\frac{1}{s^2 + 2s + 2} - \frac{1}{s^2 + 2s + 5}\right] + \frac{1}{s^2 + 2s + 5}$$

$$y = \frac{1}{3}\left[\frac{1}{(s+1)^2 + 1} + \frac{2}{(s+1)^2 + 2^2}\right]$$

Now taking the inverse Laplace $$y = \frac{1}{3}\left[e^{-t}\sin t + e^{-t}\sin 2t\right] = \frac{1}{3}\left[e^{-t}(\sin t + \sin 2t)\right]$$

**Example. 21.50:** *A particle of mass m can perform small oscillations about a position of equilibrium under a force $mn^2$ times the displacement. It is started from rest by a constant force F which acts for a time T and then ceases. Show that the amplitude of the subsequent oscillation is* $\frac{2F}{mn^2}\sin\frac{nT}{2}$. *(BPUT, Rourkela, 2019)*

**Solution:** Since the constent force $F$ acts for $0 < t < T$, and then ceases, it can be represented by

$$F[1 - u(t - T)]$$

Equation $$m\frac{d^2x}{dt^2} = -mn^2x + F[1 - u(t - T)]$$

Since $x_0 = x_1 = 0$

$$\bar{x} = \frac{F(1-e^{-st})}{ms(s^2+n^2)} \qquad ...(i)$$

$$\frac{1}{s(s^2+n^2)} = \frac{1}{n^2}\left[\frac{1}{s} - \frac{s}{s^2+n^2}\right]$$

$$L^{-1}\left[\frac{1}{s(s^2+n^2)}\right] = \frac{1}{n^2}\left[\frac{1}{n^2}(1-\cos nt)\right]$$

Second shifting theorem

$$L^{-1}\left\{\frac{e^{-st}}{s(s^2+n^2)}\right\} = \frac{1}{n^2}\left[1-\cos n(t-T)u(t-T)\right] \qquad ...(ii)$$

Now taking inverse transform $\quad x = \dfrac{F}{mn^2}\left[(1-\cos nt) - \{1-\cos n(t-T)u(t-T)\}\right]$

$$x = \frac{F}{mn^2}(1-\cos nt)\text{, for } 0 < t < T$$

for $t > T$ $\quad x = \dfrac{F}{mn^2}\{(1-\cos nt) - \{1-\cos n(t-T\}\}$

$$= \frac{F}{mn^2}\left[\cos n(t-T) - \cos nT\right] = \frac{2F}{mn^2}\sin\frac{nT}{2}\cdot\sin n\left(t-\frac{T}{2}\right)$$

Hence the amplitude of the subsequent $\dfrac{2F}{mn^2}\sin\dfrac{nT}{2}$.

## 21.18 (1) UNIT IMPULSE FUNCTION

The idea of a very large force acting for a very short time is of frequent occurrence in mechanics. To deal with such and similar ideas, we introduce the *unit impulse function* (also called *Dirac delta function**).

Thus unit impulse function is considered as the limiting form of the function (Fig. 21.7) :

$$\delta_\varepsilon(t-a) = 1/\varepsilon, \quad a \le t \le a+\varepsilon$$
$$= 0, \quad \text{otherwise}$$

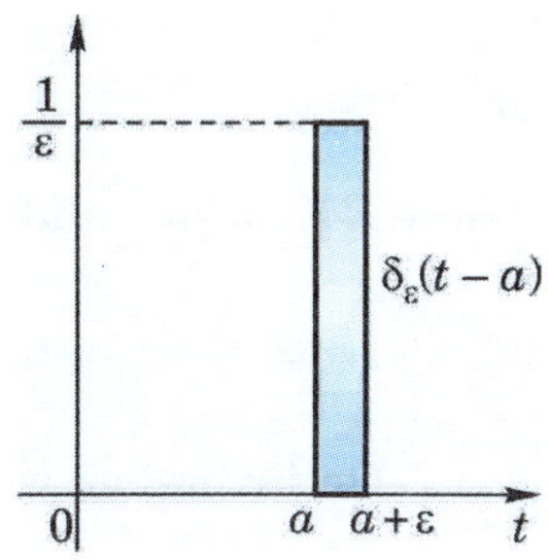

Fig. 21.7

as $\varepsilon \to 0$. It is clear from Fig. 21.7 that as $\varepsilon \to 0$, the height of the strip increases indefinitely and the width decreases in such a way that its area is always unity.

*Thus the unit impulse function* $\delta(t-a)$ *is defined as follows :*

$$\delta(t-a) = \infty \text{ for } t = a\ ; = 0 \text{ for } t \ne a,$$

such that $\quad \int_0^\infty \delta(t-a)\, dt = 1. \quad (a \ge 0)$

As an illustration, a load $w_0$ acting at the point $x = a$ of a beam may be considered as the limiting case of uniform loading $w_0/\varepsilon$ per unit length over the portion of the beam between $x = a$ and $x = a + \varepsilon$. Thus

$$w(x) = w_0/\varepsilon \quad a < x < a+\varepsilon,$$
$$= 0, \quad \text{otherwise}$$

*i.e.*, $\quad w(x) = w_0\delta(x+a).$

* After the English physicist *Paul Dirac* (1902-84), who was awarded the Nobel prize in 1933 for his work in Quantum mechanics.

**(2) Transform of unit impulse function.** If $f(t)$ be a function of $t$ continuous at $t = a$, then

$$\int_0^\infty f(t)\,\delta_\varepsilon(t-a)\,.\,dt = \int_0^{a+\varepsilon} f(t)\,.\,\frac{1}{\varepsilon}\,dt$$

$$= (a+\varepsilon-a)\,f(\eta)\,.\,\frac{1}{\varepsilon} = f(\eta), \quad \text{where } a < \eta < a+\varepsilon.$$

by Mean value theorem for integrals.

As $\varepsilon \to 0$, we get $\int_0^\infty f(t)\,\delta(t-a)\,dt = f(a)$. *(JNTUK, 2022, 2015; V.T.U., 2016)*

In particular, when $f(t) = e^{-st}$, we have $\mathbf{L\{\delta(t-a)\} = e^{-as}}$.

**Example 21.51.** *Evaluate (i)* $\int_0^\infty \sin 2t\,\delta(t-\pi/4)\,dt$ *(ii)* $L\frac{1}{t}\,\delta(t-a)$. *(Hazaribag., 2015)*

**Solution.** (*i*) We know that $\int_0^\infty f(t)\,\delta(t-a)\,dt = f(a)$

$$\therefore \quad \int_0^\infty \sin 2t\,\delta(t-\pi/4)\,dt = \sin(2\,.\,\pi/4) = 1$$

(*ii*) We know that $L\{\delta(t-a)\} = e^{-as}$

$$\therefore \quad L\left[\frac{1}{t}\,\delta(t-a)\right] = \int_s^\infty L[\delta(t-a)]\,ds = \int_s^\infty e^{-as}\,ds$$

$$= \left|\frac{e^{-as}}{-a}\right|_s^\infty = \frac{1}{a}\,e^{-as}.$$

**Example 21.52.** *An impulsive voltage $E\delta(t)$ is applied to a circuit consisting of L, R, C in series with zero initial conditions. If i be the current at any subsequent time t, find the limit of i as $t \to 0$ ?*

**Solution.** The equation of the circuit governing the current $i$ is

$$L\frac{di}{dt} + Ri + \frac{1}{C}\int_0^t i\,dt = E\delta(t) \qquad \text{where } i = 0, \text{ when } t = 0.$$

Taking Laplace transform of both sides, we get

$$L\,[s\bar{i} - i(0)] + R\bar{i} + \frac{1}{C}\frac{1}{s}\,\bar{i} = E \qquad [\text{Using § 21.7 and 21.8}]$$

or $$\left(s^2 + \frac{R}{L}s + \frac{1}{CL}\right)\bar{i} = \frac{E}{L}s \quad \text{or} \quad (s^2 + 2as + a^2 + b^2)\,\bar{i} = (E/L)\,s$$

where $R/L = 2a$ and $1/CL = a^2 + b^2$

or $$\bar{i} = \frac{E}{L}\,\frac{(s+a)-a}{(s+a)^2+b^2} = \frac{E}{L}\left\{\frac{s+a}{(s+a)^2+b^2} - a\,\frac{1}{(s+a)^2+b^2}\right\}$$

On inversion, we get

$$i = \frac{E}{L}\left\{e^{-at}\cos bt - \frac{a}{b}e^{-at}\sin bt\right\}$$

Taking limits as $t \to 0$, $i \to E/L$

Although the current $i = 0$ initially, yet a large current will develop instantaneously due to impulsive voltage applied at $t = 0$. In fact, we have determined the limit of this current which is $E/L$.

**Example 21.53.** *A beam is simply supported at its end $x = 0$ and is clamped at the other end $x = l$. It carries a load $w$ at $x = l/4$. Find the resulting deflection at any point.* *(Andhra., 2014)*

**Solution.** The differential equation for deflection is

$$\frac{d^4y}{dx^4} = \frac{w}{EI}\,\delta(x - l/4)$$

Taking the Laplace transform, we have $s^4\bar{y} - s^3y(0) - s^2y'(0) - sy''(0) - y'''(0) = \frac{w}{EI}e^{-ls/4}$

Using the conditions $y(0) = 0, y''(0) = 0$ and taking $y'(0) = c_1$ and $y'''(0) = c_2$, we get

$$\bar{y} = \frac{c_1}{s^2} + \frac{c_2}{s^4} + \frac{w}{EI}\frac{e^{-ls/4}}{s^4}.$$

On inversion, it gives $$y = c_1 x + c_2 \frac{x^3}{3!} + \frac{w}{EI}\frac{(x - l/4)^3}{3!}\, u(x - l/4)$$

*i.e.*, $$\left.\begin{aligned} y &= c_1 x + \tfrac{1}{6}c_2 x^3, & 0 < x < l/4 \\ \text{and} \quad y &= c_1 x + \tfrac{1}{6}c_2 x^3 + \frac{\omega}{6EI}(x - l/4)^3, & l/4 < x < l \end{aligned}\right\} \quad \ldots(i)$$

Using the conditions $y(l) = 0$ and $y'(l) = 0$, we get

$$0 = c_1 l + \tfrac{1}{6}c_2 l^3 + 9wl^3/128EI \quad \text{and} \quad 0 = c_1 + \tfrac{1}{2}c_2 l^2 + 9wl^2/32EI$$

whence $c_1 = 9wl^2/256\,EI, \qquad c_2 = -81w/128EI.$

Substituting the values of $c_1$ and $c_2$ in Eqn. (*i*), we get the deflection at any point.

## PROBLEMS 21.8

1. Express the function in term of unit step function and find Lpalce transform
$[D^2 + n^2]x = a \sin(nt + \alpha), x_o = x_1 = 0$ (*Kerala 2018*)

2. Sketch the graph of the following functions and express them in terms of unit step function. Hence find their Laplace transforms :
(*i*) $f(t) = 2t$ for $0 < t < \pi$, $f(t) = 1$ for $t > \pi$ (*V.T.U., 2012*)
(*ii*) $f(t) = \begin{cases} t^2 \text{ for } 0 < t \le 1 \\ 4t \text{ for } t > 1 \end{cases}$
(*iii*) $f(t) = \cos(wt + \phi)$ for $0 < t < T$, $f(t) = 0$ for $t > T$.

3. Express the following functions in terms of unit step function and hence find its Laplace transform.
(*i*) $f(t) = \begin{cases} \cos t, & 0 < t < \pi \\ 1, & \pi < t < 2\pi \\ \sin t, & t > 2\pi \end{cases}$ (*Andhra, 2022, V.T.U., 2016*)
(*ii*) $f(t) = \begin{cases} \cos t, & 0 < t < \pi \\ \cos 2t, & \pi < t < 2\pi \\ \cos 3t, & t > 2\pi \end{cases}$ (*V.T.U., 2014*)
(*iii*) $f(t) = \begin{cases} t^2, & 0 < t < 2 \\ 4t, & 2 < t < 4 \\ 8, & t > 4 \end{cases}$ (*V.T.U., 2014*)
(*iv*) $f(t) = \begin{cases} \sin 2t, & 2\pi < t < 4\pi \\ 0, & \text{otherwise} \end{cases}$ (*Andhra, 2017*)

4. Evaluate (*i*) $L\{e^{t-1} u(t - 1)\}$ (*ii*) $L\{(t - 1)^2 u(t - 1)\}$
(*iii*) $L(1 + 2t - 3t^2 + 4t^3) H(t - 2)$ (*iv*) $L\{t^2 u(t - 1) + \delta(t - 1)\}$.

5. Evaluate $\int_0^\infty e^{-t}(1 + 3t + t^2)\, u(t - 2)\, dt$.

6. Find the inverse Laplace transforms of :
(*i*) $\dfrac{e^{-\pi s}}{s^2 + 1}$ (*ii*) $\dfrac{e^{-s}}{(s + 1)^3}$ (*P.T.U., 2010*)
(*iii*) $\dfrac{e^{-s}}{\sqrt{(s + 1)}}$ (*Anna., 2016*) (*iv*) $\dfrac{3}{s} - 4\dfrac{e^{-s}}{s^2} + 4\dfrac{e^{-3s}}{s^2}$.

7. Solve using Laplace transforms $\dfrac{d^2y}{dt^2} + 4y = u(t - 2)$, where $u$ is a step function $y(0) = 0, y'(0) = 1$.
(*Andhra., 2013 ; Delhi, 2012 S*)

8. A mechanical system with two degree of freedom satisfies the equation

$$2D^2x + 3Dy = 4$$
$$2D^2y - 3Dx = 0$$

Find $x$ and $y$ at any time $t$ given that $x_0 = y_0 = x_1 = y_1 = 0$. *(Andra, 2019)*

9. A beam has its ends clamped at $x = 0$ and $x = l$. A concentrated load $W$ acts vertically downwards at the point $x = l/3$. Find the resulting deflection.

[**Hint.** The differential equation and the boundary conditions are $\dfrac{d^4y}{dx^2} = \dfrac{W}{EI}\,\delta\,(x - l/3)$ and

$y(0) = y'(0) = 0, y(l) = y'(l) = 0.$]

10. If currents $i_1$ and $i_2$ in a mesh are given by

$$Di_1 - \omega i_2 = a \cos pt$$
$$Di_2 + \omega i_1 = a \sin pt$$

$i_1 = i_2 = 0$ at $t = 0$ find $i_1$, $i_2$ at any $t$. *(FMU, Balasore, 2018)*

11. An impulse $I$ (kg-sec) is applied to a mass $m$ attached to a spring having a spring constant $k$. The system is damped with damping constant $\mu$. Derive expressions for displacement and velocity of the mass, assuming initial conditions $x(0) = x'(0) = 0$.

[**Hint.** The equation of motion is $m\dfrac{d^2x}{dt^2} = I\,\delta(x) - kx - \mu\dfrac{dx}{dt}$.]

12. $(D^2 - D - 2)x = 20 \sin 2t, x(0) = -1, x'(0) = 2$

13. $\dfrac{d^2y}{dt^2} + 2\dfrac{dy}{dt} + 2y = \dfrac{17}{2}\sin 5t$

$y = 2$ and $\dfrac{dy}{dt} = -4$ when $t = 0$

14. $\dfrac{dy}{dt} + 2y + \displaystyle\int_0^t y \cdot dt = \sin t.\ y(0) = 1$

15. $(D^2 - 3D + 2)x + (D - 1)y = 0, -(D - 1)x + (D^2 - 5D + 4)y = 0$ given that $x_0 = x_1 = y_1 = 0, y_0 = 1$.

16. $Dx - Dy - 2x + 2y = 1 - 2t, D^2x + 2Dy + x = 0$ given that $x_0 = y_0 = x_1 = 0$.

## 21.19 OBJECTIVE TYPE OF QUESTIONS

### PROBLEMS 21.9

*Fill up the blanks or choose the correct answer in each of the following problems :*

1. Laplace transform of $(t \sin t)$ = ........
2. $L\{\delta(t)\}$ = ........

   (a) 0 (b) $e^{-as}$ (c) $\infty$ (d) 1.
3. If $L\{f(t)\} = f(s)$, then $L\{e^{-at} f(t)\}$ is

   (a) $f(s - a)$ (b) $f(s + a)$ (c) $f(s)$ (d) none of these.
4. $L(e^{-3t} \cos 5t)$ = ........ *(Calicut, 2013)*
5. Inverse Laplace transform of $(s + 2)^{-2}$ is ........
6. Inverse Laplace transform of $1/(s^2 + 4s + 13)$ = ........
7. Laplace transform of $f'(t)$ = ........
8. $L^{-1}\left[\dfrac{s}{(2s+3)^2}\right]$ = ........
9. $L(\cosh^2 2t)$ = ........

10. $L(e^t) =$ ........

11. $L(e^{-t} t^k) =$ ........

12. $\int_0^\infty e^{-2t} \cos 3t \, dt =$ ........

13. $L[u(t-a)] =$ ........

14. $L^{-1}(\sqrt{t}) =$ ........

15. If $L\{F(t)\} = f(s)$, then $L\left\{\frac{d^2F(t)}{dt^2}\right\} =$ .......

16. $L(\cos^3 4t) =$ ........

17. $L\left(\frac{\sin at}{t}\right) =$ ........

18. $L^{-1}\left\{\frac{1}{(s+3)^5}\right\} =$ ........

19. $L \cos(2t+3) =$ ........

20. $L^{-1}(1/s^n)$ is possible only when $n$ is

(*a*) zero (*b*) –ve integer (*c*) +ve integer (*d*) negative rational.

21. If $L^{-1}[\phi(s)] = f(t)$, the $L^{-1}[e^{-as}\phi(s)] =$ ........

22. $L\{u(t+2)\} =$ ..............

(*a*) $e^{-2s}/s^2$ (*b*) $e^{2s}$ (*c*) $\frac{e^{2s}}{s}$ (*d*) $\frac{e^{-2s}}{s}$ (*V.T.U., 2011 S*)

23. $L^{-1}\left\{\frac{s^2-3s+4}{s^3}\right\} =$ ........ (*V.T.U., 2010 S*)

24. If $L\{f(t)\} = \bar{f}(s)$, then $L^{-1}\left\{\frac{\bar{f}(s)}{s}\right\} =$ ........

25. If $f(t)$ is a periodic function with period $T$, then $L[f(t)] =$ ........

26. If $y$ satisfies $y'' + 3y' + 2y = e^{-t}$ with $y(0) = y'(0) = 0$, then $L[y(t)] =$ ........

27. $L[e^{3t}(2\cos 5t + 3\sin 4t)] =$ ........

28. $L(4^t) =$ ........

29. $L^{-1}\left\{\frac{1}{\sqrt{(s+3)}}\right\} =$ ........

30. Laplace transform of $\sin 2t \, \delta(t-2)$ is

(*a*) $e^{2s} \sin 4$ (*b*) $e^{-2s} \sin 2$ (*c*) $e^{-4s} \sin 2$ (*d*) $e^{-2s} \sin 4$.

31. If $L^{-1}\left\{\frac{s}{(s+1)^2}\right\} = \frac{t \sin t}{2}$ then $L^{-1}\left\{\frac{8s}{(4s^2+1)^2}\right\}$

32. $L^{-1}\{e^{-as} F(s)\} =$

(*a*) $f(t)\, u(t)$ (*b*) $f(t-a)\, u(t)$ (*c*) $f(t-a)\, u(t-a)$ (*d*) None of these.

33. $L^{-1}\left\{\frac{1}{(s+a)^2}\right\} =$

(*a*) $e^{at}$ (*b*) $e^{-at}$ (*c*) $te^{-at}$

(*d*) $te^{at}$ (*e*) $-t$.

34. Laplace transform of $t^4 e^{-at}$ is

(*i*) $\frac{4!}{(s+a)^4}$ (*ii*) $\frac{4!}{(s+a)^5}$ (*iii*) $\frac{4!}{(s-a)^4}$ (*iv*) $\frac{5!}{(s-a)^5}$.

35. Laplace transform of $te^{at} \sin(at)$, $t > 0$, is

(*i*) $\frac{s-a}{(s-a)^2+a^2}$ (*ii*) $\frac{a(s-a)}{(s-a)^2+a^2}$ (*iii*) $\frac{2a(s-a)}{[(s-a)^2+a^2]^2}$ (*iv*) $\frac{(s-a)^2}{(s-a)^2+a^2}$.

36. $L^{-1} \frac{s^2}{(s^2+4)^2}$ is

(*i*) $\frac{1}{4}\sin 2t + t\cos 2t$ (*ii*) $\frac{1}{4}\sin 2t + \frac{1}{2}\cos 2t$ (*iii*) $\sin 2t + \frac{t}{2}\cos 2t$ (*iv*) $\frac{1}{2}\sin 2t + \frac{t}{4}\cos 2t$.

37. $L^{-1} \frac{1}{s(s^2+1)}$ is

(*i*) $1 + \sin t$ (*ii*) $1 - \sin t$ (*iii*) $1 + \cos t$ (*iv*) $1 - \cos t$.

38. $L [u(t - a)]$ where $u(t - a)$ is a unit step function, is

(i) $\frac{e^{-as}}{s}$ (ii) $\frac{e^{as}}{s}$ (iii) $\frac{e^{-as}}{s^2}$ (iv) $\frac{e^{as}}{s^2}$. (V.T.U., 2011)

39. For a periodic function of period $2\pi$, $\int_{a+2\pi}^{b+2\pi} f(x)\, dx =$ ........

40. $L [\delta(t - a)]$ where $\delta(t - a)$ is a unit impulse function, is

(i) $e^{as}$ (ii) $e^{-as}$ (iii) $e^{s}$ (iv) $e^{-as}/s$ (Andhra., 2016)

41. Laplace transform of $\sin^2 3t$ is

(i) $\frac{3}{s^2 + 36}$ (ii) $\frac{6}{(s^2 + 36)}$ (iii) $\frac{18}{s(s^2 + 36)}$ (iv) $\frac{18}{s^2 + 36}$ (V.T.U., 2010)

42. $[L(t^2 e^{-3t})] =$

(i) $\frac{1}{(s + 3)^3}$ (ii) $\frac{2}{(s + 3)^2}$ (iii) $\frac{3}{(s + 3)^3}$ (iv) $\frac{2}{(s + 3)^3}$ (V.T.U., 2011)

43. $L(\sin t \cos t) =$ ...... . (Delhi, 2012 S)

44. $L^{-1} \frac{1}{(s + 1)(s + 2)(s + 3)} =$ ...... . (V.T.U., 2013)

45. $\frac{d^2}{ds^2} [L f(t)] - L(t^2 f(t)] = 0.$ (True or False)

46. Laplace transform of $f(t)$ is defined for +ve and –ve values of $t$. (True or False)

47. If $L[f(t)] = \phi(s)$, then $L[t f(t)] = \frac{d}{ds} [\phi(s)]$. (True or False)

CHAPTER 22

# Fourier Transforms



## 22.1 INTRODUCTION

In the previous chapter, the reader has already been acquainted with the use of Laplace transforms in the solution of ordinary differential equations. In this chapter, the well-known Fourier transforms will be introduced and their properties will be studied which will be used in the solution of partial differential equations. The choice of a particular transform to be employed for the solution of an equation depends on the boundary conditions of the problem and the ease with which the transform can be inverted. A Fourier transform when applied to a partial differential equation reduces the number of its independent variables by one.

The theory of integral transforms afford mathematical devices through which solutions of numerous boundary value problems of engineering can be obtained *e.g.*, conduction of heat, transverse vibrations of a string, transverse oscillations of an elastic beam, free and forced vibrations of a membrane, transmission lines etc. Some of these applications will be illustrated in the last section.

## 22.2 DEFINITION

*The integral transform of a function f(x) denoted by I[f(x)], is defined by*

$$\bar{f}(s) = \int_{x_1}^{x_2} f(x) K(s, x)\, dx$$

where $K(s, x)$ is called the *kernel* of the transform and is a known function of $s$ and $x$. The function $f(x)$ is called the *inverse transform* of $\bar{f}(s)$.

Three simple examples of a kernel are as follows :

(*i*) *When* $K(s, x) = e^{-sx}$, *it leads to the* **Laplace transform** *of* $f(x)$, *i.e.*,

$$\bar{f}(s) = \int_0^\infty f(x) e^{-sx}\, dx. \qquad \text{[Chap. 21]}$$

(*ii*) *When* $K(s, x) = e^{isx}$, *we have the* **Fourier transform** *of* $f(x)$, *i.e.*,

$$F(s) = \int_{-\infty}^{\infty} f(x)\, e^{isx}\, dx.$$

(*iii*) When $K(s, x) = x^{s-1}$, it gives the *Mellin transform* of $f(x)$ *i.e.*,

$$M(s) = \int_0^\infty f(x)x^{s-1}\,dx.$$

Other special transforms arise when the kernel is a sine or a cosine function or a Bessel's function. These lead to *Fourier sine* or *cosine transforms* and the *Hankel transform* respectively.

In order to introduce the *Fourier transforms*, we shall first derive the Fourier integral theorem.

## 22.3 (1) FOURIER INTEGRAL THEOREM

Consider a function $f(x)$ which satisfies the Dirichlet's conditions (Art. 10.3) in every interval $(-c, c)$ so that, we have

$$f(x) = \frac{a_0}{2} + \sum_{n=1}^{\infty}\left(a_n \cos\frac{n\pi x}{c} + b_n \sin\frac{n\pi x}{c}\right) \qquad \text{...(1)}$$

where $a_0 = \frac{1}{c}\int_{-c}^{c} f(t)\,dt,\ a_n = \frac{1}{c}\int_{-c}^{c} f(t)\cos\frac{n\pi t}{c}\,dt,$ and $b_n = \frac{1}{c}\int_{-c}^{c} f(t)\sin\frac{n\pi t}{c}\,dt.$

Substituting the values of $a_0$, $a_n$ and $b_n$ in Eqn. (1), it takes the form

$$f(x) = \frac{1}{2c}\int_{-c}^{c} f(t)\,dt + \frac{1}{c}\sum_{n=1}^{\infty}\int_{-c}^{c} f(t)\cos\frac{n\pi(t-x)}{c}\,dt \qquad \text{...(2)}$$

If we assume that $\int_{-\infty}^{\infty} |f(x)|\,dx$ converges, the first term on the right side of Eqn. (2) approaches 0 as $c \to \infty$, since

$$\left|\frac{1}{2c}\int_{-c}^{c} f(t)\,dt\right| \le \frac{1}{2c}\int_{-\infty}^{\infty} |f(t)|\,dt$$

The second term on the right side of Eqn. (2) tends to

$$\underset{c\to\infty}{\text{Lt}}\ \frac{1}{c}\sum_{n=1}^{\infty}\int_{-\infty}^{\infty} f(t)\cos\frac{n\pi(t-x)}{c}\,dt$$

$$= \underset{\delta\lambda\to 0}{\text{Lt}}\ \frac{1}{\pi}\sum_{n=1}^{\infty}\delta\lambda\int_{-\infty}^{\infty} f(t)\cos n\delta\lambda\,(t-x)\,dt, \text{ on writing } \pi/c = \delta\lambda$$

This is of the form $\underset{\delta\lambda\to 0}{\text{Lt}}\ \sum_{n=1}^{\infty} F(n\delta\lambda)$, *i.e.*, $\int_0^\infty F(\lambda)\,d\lambda$

Thus as $c \to \infty$, Eqn. (2) becomes $f(x) = \frac{1}{\pi}\int_0^\infty\int_{-\infty}^{\infty} f(t)\cos\lambda(t-x)\,dt\,d\lambda$ ...(3)

which is known as the **Fourier integral** of $f(x)$.

**Obs.** We have given a heuristic demonstration of the Fourier integral theorem which simply helps in deriving the result (3). It cannot however, be taken as a rigorous proof for that would, involve a proof of the convergence of the Fourier integral which is beyond the scope of this book. When $f(x)$ satisfies the above-mentioned conditions, equation (3) holds good at a point of continuity. If however, $x$ is point of discontinuity, we replace $f(x)$ by $\frac{1}{2}[f(x+0) + f(x-0)]$ as in the case of Fourier series.

**(2) Fourier sine and cosine integrals.** Expanding $\cos\lambda(t - x)$, (3) may be written as

$$f(x) = \frac{1}{\pi}\int_0^\infty \cos\lambda x\int_{-\infty}^{\infty} f(t)\cos\lambda t\,dt\,d\lambda + \frac{1}{\pi}\int_0^\infty \sin\lambda x\int_{-\infty}^{\infty} f(t)\sin\lambda t\,dt\,d\lambda \qquad \text{...(4)}$$

If $f(x)$ is an odd function, $f(t)\cos\lambda t$ is also an odd function while $f(t)\sin\lambda t$ is even. Then the first term on the right side of (4) vanishes and, we get

$$f(x) = \frac{2}{\pi}\int_0^\infty \sin\lambda x\int_0^\infty f(t)\sin\lambda t\,dt\,d\lambda \qquad \text{...(5)}$$

which is known as the *Fourier sine integral*.

Similarly, if $f(x)$ is an even function, Eqn. (4) takes the form

$$f(x) = \frac{2}{\pi}\int_0^\infty \cos \lambda x \int_0^\infty f(t) \cos \lambda t \, dt \, d\lambda \quad ...(6)$$

which is known as the *Fourier cosine integral*.

**Obs.** A function $f(x)$ defined in the interval $(0, \infty)$ is expressed either as a Fourier sine integral or as a Fourier cosine integral, merely looking upon it as an odd or even function in $(-\infty, \infty)$ on the lines of half-range Fourier series.

(3) **Complex form of Fourier integrals.** Eqn. (3) can be written as

$$f(x) = \frac{1}{2\pi}\int_{-\infty}^\infty \int_{-\infty}^\infty f(t) \cos \lambda(t-x) dt d\lambda \quad ..(7)$$

because $\cos \lambda(t-x)$ is an even function of $\lambda$. Also since $\sin \lambda(t-x)$ is an odd function of $\lambda$, we have

$$0 = \frac{1}{2\pi}\int_{-\infty}^\infty \int_{-\infty}^\infty f(t) \sin \lambda(t-x) dt d\lambda \quad ...(8)$$

Now multiply Eqn. (8) by $i$ and add it to Eqn. (7), so that

$$f(x) = \frac{1}{2\pi}\int_{-\infty}^\infty \int_{-\infty}^\infty f(t)\, e^{i\lambda(t-x)} dt d\lambda \quad ...(9)$$

which is the *complex* form of the *Fourier integral*.

**(4) Fourier integral representation of a function**

Using Eqn. (4), *a function F(x) may be represented by a Fourier integral as*

$$F(x) = \frac{1}{\pi}\int_0^\infty [A(\lambda) \cos \lambda x + B(\lambda) \sin \lambda x] \, d\lambda$$

where $$A(\lambda) = \int_{-\infty}^\infty f(t) \cos \lambda t \, dt \, ; \, B(\lambda) = \int_{-\infty}^\infty f(t) \sin \lambda t \, dt \quad ...(10)$$

*If $f(x)$ is an odd function*, then [By (5)]

$$f(x) = \frac{1}{\pi}\int_0^\infty B(\lambda) \sin \lambda x \, d\lambda \text{ where } B(\lambda) = 2\int_0^\infty f(t) \sin \lambda t \, dt \quad ...(11)$$

*If $f(x)$ is an even function*, then [By (6)]

$$f(x) = \frac{1}{\pi}\int_0^\infty A(\lambda) \cos \lambda x \, d\lambda \text{ where } A(\lambda) = 2\int_0^\infty f(t) \cos \lambda t \, dt \quad ...(12)$$

**Example 22.1.** *Express $f(x) = 1$ for $0 \le x \le \pi$,*

*$= 0$ for $x > \pi$,*

*as a Fourier sine integral and hence evaluate*

$$\int_0^\infty \frac{1-\cos(\pi\lambda)}{\lambda} \sin(x\lambda) d\lambda.$$

*(JNTUK, 2022, 2015 ; Rohtak, 2013)*

**Solution.** The Fourier sine integral for $f(x) = \frac{2}{\pi}\int_0^\infty \sin(\lambda x)\, d\lambda \int_0^\infty f(t) \sin(\lambda t)\, dt$

$$= \frac{2}{\pi}\int_0^\infty \sin(\lambda x)\, d\lambda \int_0^\pi \sin(\lambda t)\, dt$$

$$= \frac{2}{\pi}\int_0^\infty \sin(\lambda x)\, d\lambda \left| \frac{-\cos(\lambda t)}{\lambda} \right|_0^\pi = \frac{2}{\pi}\int_0^\infty \frac{1-\cos(\lambda\pi)}{\lambda} \sin(\lambda x) d\lambda$$

$$\therefore \quad \int_0^\infty \frac{1-\cos(\lambda\pi)}{\lambda} \sin(\lambda x) d\lambda = \frac{\pi}{2} f(x) = \begin{cases} \pi/2 & \text{for } 0 \le x < \pi \\ 0 & \text{for } x > \pi \end{cases}$$

At $x = \pi$, which is a point of discontinuity of $f(x)$, the value of the above integral

$$= \frac{\pi}{2}\left[\frac{f(\pi-0)+f(\pi+0)}{2}\right] = \frac{\pi}{2} \cdot \frac{1+0}{2} = \frac{\pi}{4}.$$

## 22.4 (1) FOURIER TRANSFORMS

Rewriting Eqn. (9) of §22.3 as

$$f(x) = \frac{1}{2\pi}\int_{-\infty}^{\infty} e^{-isx}\, ds \int_{-\infty}^{\infty} f(t)e^{ist}\, dt,$$

it follows that if
$$F(s) = \int_{-\infty}^{\infty} f(t)\, e^{ist}\, dt \qquad ...(1)$$

then
$$f(x) = \frac{1}{2\pi}\int_{-\infty}^{\infty} F(s)\, e^{-isx}\, ds \qquad ...(2)$$

*The function F(s), defined by Eqn. (1), is called the* **Fourier transform** of $f(x)$. *Also the function $f(x)$, as given by Eqn. (2), is called the* **inverse Fourier transform** *of F(s).* Sometimes, we call *Eqn.* (2) as an *inversion formula* corresponding to Eqn. (1).

**(2) Fourier sine and cosine transforms.** From Eqn. (5) of § 22.3, it follows that if

$$F_s(s) = \int_0^{\infty} f(x) \sin sx\, dx \qquad ...(3)$$

then
$$f(x) = \frac{2}{\pi}\int_0^{\infty} F_s(s) \sin sx\, ds \qquad ...(4)$$

*The function $F_s(s)$, as defined by Eqn. (3), is known as the* **Fourier sine transform** *of $f(x)$ in $0 < x < \infty$. Also the function $f(x)$, as given by Eqn. (4) is called the* **inverse Fourier sine transform** *of $F_s(s)$.*

Similarly, if follows from Eqn. (6) of § 22.3 that if

$$F_c(s) = \int_0^{\infty} f(x) \cos sx\, dx \qquad ...(5)$$

then
$$f(x) = \frac{2}{\pi}\int_0^{\infty} F_c(s) \cos sx\, ds \qquad ...(6)$$

*The function $F_c(s)$ as defined by Eqn.* (5) *is known as the* **Fourier cosine transform** *of $f(x)$ in* $0 < x < \infty$. *Also the function $f(x)$, as given by Eqn.* (6), *is called the* **inverse Fourier cosine transform** *of $F_c(s)$.*

(3) **Finite Fourier sine and cosine transforms.** These transforms are useful for such a boundary-value problem in which at least two of the boundaries are parallel and separated by a finite distance.

*The* **finite Fourier sine transform** *of $f(x)$, in $0 < x < c$, is defined as*

$$F_s(n) = \int_0^{c} f(x) \sin \frac{n\pi x}{c}\, dx \qquad ...(7)$$

where $n$ is an integer.

*The function $f(x)$ is then called the* **inverse finite Fourier sine transform** *of $F_s(n)$ which is given by*

$$f(x) = \frac{2}{c}\sum_{n=1}^{\infty} F_s(n) \sin \frac{n\pi x}{c} \qquad ...(8)$$

*The* **finite Fourier cosine transform** *of $f(x)$, in $0 < x < c$, is defined as*

$$F_c(n) = \int_0^{c} f(x) \cos \frac{n\pi x}{c}\, dx \qquad ...(9)$$

where $n$ is an integer.

*The function $f(x)$ is then called the* **inverse finite Fourier cosine transform** *of $F_c(n)$ which is given by*

$$f(x) = \frac{1}{c} F_c(0) + \frac{2}{c}\sum_{n=1}^{\infty} F_c(n) \cos \frac{n\pi x}{c} \qquad ...(10)$$

**Obs.** The finite Fourier sine transform is useful for problems involving boundary conditions of heat distribution on two parallel boundaries, while the finite cosine transform is useful for problems in which the velocities normal to two parallel boundaries are among the boundary conditions.

## 22.5 PROPERTIES OF FOURIER TRANSFORMS

(1) **Linear property.** *If F(s) and G(s) are Fourier transforms of $f(x)$ and $g(x)$ respectively, then*

$$F[a f(x) + bg(x)] = a F(s) + bG(s)$$

*where a and b are constants.*

We have $$F(s) = \int_{-\infty}^{\infty} e^{isx} f(x)\,dx \text{ and } G(s) = \int_{-\infty}^{\infty} e^{isx} g(x)\,dx$$

$$\therefore \quad F[af(x) + bg(x)] = \int_{-\infty}^{\infty} e^{isx} [af(x) + bg(x)]\,dx = a\int_{-\infty}^{\infty} e^{isx} f(x)\,dx + b\int_{-\infty}^{\infty} e^{isx} g(x)dx$$
$$= aF(s) + bG(s)$$

**(2) Change of scale property.** *If F(s) is the complex Fourier transform of f(x), then*

$$F\{f(ax)\} = \frac{1}{a} F\left(\frac{s}{a}\right),\ a \neq 0$$

We have $$F(s) = \int_{-\infty}^{\infty} e^{isx} f(x)dx \qquad ...(i)$$

$$\therefore \quad F\{f(ax)\} = \int_{-\infty}^{\infty} e^{isx} f(ax)dx$$ Put $ax = t$ so that $dx = dt/a$

$$= \int_{-\infty}^{\infty} e^{ist/a} f(t)dt/a = \frac{1}{a}\int_{-\infty}^{\infty} e^{i(s/a)t} f(t)dt = \frac{1}{a} F\left(\frac{s}{a}\right) \qquad \text{[By (i)]}$$

**Cor.** *If $F_s(s)$ and $F_c(s)$ are the Fourier sine and cosine transforms of $f(x)$ respectively, then*

$$F_s\{f(ax)\} = \frac{1}{a} F_s\left(\frac{s}{a}\right) \quad \text{and} \quad F_c\{f(ax)\} = \frac{1}{a} F_s\left(\frac{s}{a}\right).$$

**(3) Shifting property.** *If F(s) is the complex Fourier transform of f(x), then*

$$F\{f(x-a)\} = e^{isa} F(s)$$

We have $$F(s) = \int_{-\infty}^{\infty} e^{isx} f(x)dx \qquad ...(i)$$

$$\therefore \quad F\{f(x-a)\} = \int_{-\infty}^{\infty} e^{isx} f(x-a)dx$$ Put $x - a = t$ so that $dx = dt$

$$= \int_{-\infty}^{\infty} e^{is(t+a)} f(t)dt = e^{isa}\int_{-\infty}^{\infty} e^{ist} f(t)dt = e^{isa} F(s) \qquad \text{[By (i)]}$$

**(4) Modulation theorem.** *If F(s) is the complex Fourier transform of f(x), then*

$$F\{f(x) \cos ax\} = \frac{1}{2}[F(s+a) + F(s-a)\}$$ *(JNTUK., 2022, 2015)*

We have $$F(s) = \int_{-\infty}^{\infty} e^{isx} f(x)dx \qquad ...(i)$$

$$\therefore \quad F\{f(x)\cos ax\} = \int_{-\infty}^{\infty} e^{isx} f(x)\cos ax\,dx = \int_{-\infty}^{\infty} e^{isx} \cdot f(x) \cdot \frac{e^{iax} + e^{-iax}}{2}\,dx$$

$$= \frac{1}{2}\left[\int_{-\infty}^{\infty} e^{i(s+a)x} f(x)dx + \int_{-\infty}^{\infty} e^{i(s-a)x} f(x)dx\right] = \frac{1}{2}[F(s+a) + F(s-a)].$$

**Cor.** *If $F_s(s)$ and $F_c(s)$ are Fourier sine and cosine transforms of $f(x)$ respectively, then*

(*i*) $F_s\{f(x)\cos ax\} = \frac{1}{2}[F_s(s+a) + F_s(s-a)]$

(*ii*) $F_c\{f(x)\sin ax\} = \frac{1}{2}[F_s(s+a) - F_s(s-a)]$ *(J.N.T.U., 2015)*

(*iii*) $F_s\{f(x)\sin ax\} = \frac{1}{2}[F_c(s-a) - F_c(s+a)]$

**Obs.** This theorem is of great importance in radio and television where the harmonic carrier wave is modulated by an envelope.

**Example 22.2.** *Find the Fourier transform of*

$$f(x) = \begin{cases} 1 \text{ for } |x| < 1 \\ 0 \text{ for } |x| > 1 \end{cases}.$$

*Hence evaluate* $\int_0^{\infty} \frac{\sin x}{x}\,dx.$ *(V.T.U., 2016 ; Anna., 2015 ; W.B.T.U., 2015)*

**Solution.** The Fourier transform of $f(x)$, *i.e.*,

$$F\{f(x)] = \int_{-\infty}^{\infty} f(x)e^{isx}\,dx = \int_{-1}^{1}(1)\,e^{isx}\,dx = \left|\frac{e^{isx}}{is}\right|_{-1}^{1} = \frac{e^{is}-e^{-is}}{is}$$

Thus $F\{f(x)\} = F(s) = 2\,\dfrac{\sin s}{s}$, $s \neq 0$. For $s = 0$, we have $F(s) = 2$.

Now by the inversion formula, we get

$$f(x) = \frac{1}{2\pi}\int_{-\infty}^{\infty} F(s)e^{-isx}ds, \text{ or } \frac{1}{2\pi}\int_{-\infty}^{\infty}\frac{2\sin s}{s}e^{-isx}ds = \begin{cases}1 \text{ for } |x|<1\\ 0 \text{ for } |x|>1\end{cases}$$

Putting $x = 0$, we get

$$\int_{-\infty}^{\infty}\frac{\sin s}{s}ds = \pi \quad \therefore \int_{0}^{\infty}\frac{\sin s}{s}ds = \frac{\pi}{2}, \text{ since the integrand is even.}$$

**Example 22.3.** *Find the Fourier transform of :*

$$f(x) = \begin{cases}1-x^2, & |x| \le 1\\ 0 & ,|x|>1\end{cases}$$

*Hence evaluate* $\displaystyle\int_0^{\infty}\frac{x\cos x-\sin x}{x^3}\cos\frac{x}{2}\,dx.$ *(Delhi., 2015 ; Hazaribag., 2015 ; J.N.T.U., 2015)*

**Solution.** $F\{f(x)\} = \displaystyle\int_{-\infty}^{\infty} f(x)\,e^{isx}\,dx = F(s)$, say

$$= \int_{-\infty}^{-1}(0)\,e^{isx}\,dx + \int_{-1}^{1}(1-x^2)\,e^{isx}\,dx + \int_{1}^{\infty}(0)\,e^{isx}\,dx = \left|(1-x^2)\frac{e^{isx}}{is} - (2x)\frac{e^{isx}}{(is)^2} + (-2)\frac{e^{isx}}{(is)^3}\right|_{-1}^{1}$$

$$= 2\left(\frac{e^{is}+e^{-is}}{-s^2}\right) - 2\left(\frac{e^{is}-e^{-is}}{-is^3}\right) = -\frac{4}{s^3}(s\cos s - \sin s)$$

Now by inversion formula, we have

$$f(x) = \frac{1}{2\pi}\int_{-\infty}^{\infty} F(s)e^{-isx}\,ds$$

or
$$-\frac{1}{2\pi}\int_{-\infty}^{\infty}\frac{4}{s^3}(s\cos s-\sin s)\,e^{-isx}\,ds = \begin{cases}1-x^2, & |x|\le 1\\ 0, & |x|>1\end{cases}$$

Putting $x = 1/2$, we obtain

$$-\frac{1}{2\pi}\int_{-\infty}^{\infty}\frac{4}{s^3}(s\cos s-\sin s)e^{-is/2}ds = \frac{3}{4}$$

or
$$\int_{-\infty}^{\infty}\frac{s\cos s-\sin s}{s^3}\left(\cos\frac{s}{2} - i\sin\frac{s}{2}\right)ds = -\frac{3\pi}{8}$$

or
$$\int_{-\infty}^{\infty}\frac{s\cos s-\sin s}{s^3}\,.\cos\frac{s}{2}\,ds = -\frac{3\pi}{8}$$

or
$$\int_{0}^{\infty}\frac{x\cos x-\sin x}{x^3}\,.\cos\frac{x}{2}\,dx = -\frac{3\pi}{16}, \text{ since the integral is even.}$$

**Example 22.4.** (*a*) *Find the Fourier transform of* $e^{-a^2x^2}$, *a < 0. Hence deduce that* $e^{-x^2/2}$ *is self reciprocal in respect of Fourier transform.* *(D.T.U., 2016 ; Anna, 2015 ; J.N.T.U., 2015)*

(*b*) *Find Fourier transform of* (*i*) $e^{-2(x-3)^2}$ (*ii*) $e^{-x^2}\cos 3x$.

**Solution.** (*a*) $\displaystyle F(e^{-a^2x^2}) = \int_{-\infty}^{\infty} e^{-a^2x^2}\,.\,e^{isx}\,dx = \int_{-\infty}^{\infty} e^{-a^2(x^2-isx/a^2)}dx$

$$= \int_{-\infty}^{\infty} e^{-a^2(x-is/2a^2)^2}\,.\,e^{-s^2/4a^2}\,dx$$

$$= \int_{-\infty}^{\infty} e^{-t^2} . e^{-s^2/4a^2} dt/a \qquad \text{[Putting } a(x - is/2a^2) = t, dx = dt/a$$

$$= \frac{e^{-s^2/4a^2}}{a} \int_{-\infty}^{\infty} e^{-t^2} dt = \frac{e^{-s^2/4a^2}}{a} \sqrt{\pi} \qquad \left[\because \int_{-\infty}^{\infty} e^{-t^2} dt = \sqrt{\pi}\right]$$

Hence $\qquad F(e^{-a^2x^2}) = \frac{\sqrt{\pi}}{a} e^{-s^2/4a^2}$

Taking $a^2 = 1/2$, we have

$$F(e^{-x^2/2}) = \frac{\sqrt{\pi}}{(1/\sqrt{2})} e^{-s^2/2} = \sqrt{2\pi} e^{-s^2/2} \qquad \textit{(Anna., 2016)}$$

*i.e.*, Fourier transform of $e^{-x^2/2}$ is a constant times $e^{-s^2/2}$. Also the functions $e^{-x^2/2}$ and $e^{-s^2/2}$ are the same. Hence it follows that $e^{-x^2/2}$ *is self-reciprocal under the Fourier transform.*

(*b*) Since $e^{-2x^2} = e^{-(2x)^2/2} = f(2x)$ where $f(x) = e^{-x^2/2}$

$\therefore$ by change of scale property, $F[f(2x)] = \frac{1}{2} F(s/2)$

*i.e.*, $$F(e^{-2x^2}) = F[e^{-(2x)^2/2}] = \sqrt{2\pi} e^{-(s/2)^2/2} = \sqrt{2\pi} e^{-s^2/8}$$

By shifting property $F f(x-3) = e^{i3s} F(3)$

$$\therefore \qquad F[e^{-2(x-3)^2}] = e^{3is} \sqrt{2\pi} e^{-s^2/8} = \sqrt{2\pi} e^{(3is - s^2/8)} \qquad ...(i)$$

Also by modulation theorem,

$$F[f(x) \cos 2x] = \frac{1}{2} [F(s+a) + F(s-a)]$$

$$F(e^{-x^2} \cos 3x) = \frac{1}{2} \sqrt{2\pi} [e^{-(s+3)^2/2} + e^{-(s-3)^2/2}]. \qquad ...(ii)$$

**Example 22.5.** *Find the Fourier cosine transform of* $e^{-x^2}$. *(W.B.T.U., 2015 ; V.T.U., 2010)*

**Solution.** We have $F_c(e^{-x^2}) = \int_0^{\infty} e^{-x^2} \cos sx \, dx = I$ (say)

Differentiating under the integral sign w.r.t. $s$,

$$\frac{dI}{ds} = -\int_0^{\infty} xe^{-x^2} \sin sx \, dx = \frac{1}{2}\int_0^{\infty} (\sin sx)(-2xe^{-x^2}) dx$$

$$= \frac{1}{2}\left\{\left|\sin sx . e^{-x^2}\right|_0^{\infty} - s\int_0^{\infty} \cos sx . e^{-x^2} dx\right\}$$

$$= -\frac{s}{2}\int_0^{\infty} e^{-x^2} \cos sx \, dx = -\frac{s}{2} I \qquad \text{or} \qquad \frac{dI}{I} = -\int \frac{s}{2} ds + \log c$$

or $$\log I = -\frac{s^2}{4} + \log c = \log (ce^{-s^2/4})$$

$$\therefore \qquad I = ce^{-s^2/4} \quad \text{or} \quad \int_0^{\infty} e^{-x^2} \cos sx \, dx = ce^{-s^2/4}$$

Putting $s = 0$, $$c = \int_0^{\infty} e^{-x^2} dx = \frac{\sqrt{\pi}}{2}. \; i.e. \; I = \frac{\sqrt{\pi}}{2} e^{-s^2/4}.$$

Hence $$F_c(e^{-x^2}) = \frac{\sqrt{\pi}}{2} e^{-s^2/4}.$$

**Example 22.6.** *Find the Fourier sine transform of* $e^{-|x|}$.

*Hence show that* $\int_0^{\infty} \frac{x \sin mx}{1+x^2} dx = \frac{\pi e^{-m}}{2}$, $m > 0$. *(V.T.U., 2016 ; D.T.U., 2015)*

**Solution.** $x$ being positive in the interval $(0, \infty)$, $e^{-|x|} = e^{-x}$

$\therefore$ Fourier sine transform of $f(x) = e^{-|x|}$ is given by

$$F_s\{f(x)\} = \int_0^\infty f(x) \sin sx \, dx = \int_0^\infty e^{-x} \sin sx \, dx$$

$$= \left| \frac{e^{-x}}{1+s^2} (-\sin sx - s \cos sx) \right|_0^\infty = \frac{s}{1+s^2}$$

Using Inversion formula for Fourier sine transforms, we get

$$f(x) = \frac{2}{\pi} \int_0^\infty F_s\{f(x)\} \sin sx \, dx \quad \text{or} \quad e^{-x} = \frac{2}{\pi} \int_0^\infty \frac{s}{1+s^2} \sin sx \, ds$$

or changing $x$ to $m$,
$$e^{-m} = \frac{2}{\pi} \int_0^\infty \frac{s \sin ms}{1+s^2} ds = \frac{2}{\pi} \int_0^\infty \frac{x \sin mx}{1+m^2} dx$$

Hence $\displaystyle \int_0^\infty \frac{x \sin mx}{1+m^2} dx = \frac{\pi e^{-m}}{2}$.

**Example 22.7.** *Find the Fourier cosine transform of* $f(x) = \begin{cases} x, \text{ for } 0 < x < 1 \\ 2 - x, \text{ for } 1 < x < 2 \\ 0, \text{ for } x > 2 \end{cases}$

**Solution.** Fourier cosine transform of $f(x)$ *i.e.*, $F_c[f(x)]$

$$= \int_0^\infty f_c(x) \cos sx \, dx = \int_0^1 x \cos sx \, dx + \int_1^2 (2-x) \cos sx \, dx + \int_2^\infty 0 \, . \, dx$$

$$= \left| x \frac{\sin sx}{s} - \left( \frac{-\cos sx}{s^2} \right) \right|_0^1 + \left| (2-x) \frac{\sin sx}{s} - (-1) \frac{-\cos sx}{s^2} \right|_1^2$$

$$= \left( \frac{\sin s}{s} + \frac{\cos s}{s^2} - \frac{1}{s^2} \right) + \left( -\frac{\cos 2s}{s^2} - \frac{\sin s}{s} + \frac{\cos s}{s^2} \right)$$

$$= \frac{2 \cos s}{s^2} - \frac{\cos 2s}{s^2} - \frac{1}{s^2}.$$

**Example 22.8.** *Find the Fourier sine transform of* $e^{-ax}/x$. *(Andhra., 2015 ; V.T.U., 2013)*

**Solution.** Let $\quad f(x) = e^{-ax}/x$, then its Fourier sine transform

*i.e.*
$$F_s\{f(x)\} = \int_0^\infty f(x) \sin sx \, dx = \int_0^\infty \frac{e^{-ax}}{x} \sin sx \, dx = F(s), \text{ say}$$

Differentiating both sides w.r.t. $s$, we get

$$\frac{d}{ds}\{F(s)\} = \int_0^\infty \frac{xe^{-ax} \cos sx}{x} dx = \int_0^\infty e^{-ax} \cos sx \, dx = \frac{a}{s^2 + a^2}$$

Integrating w.r.t. $s$, we obtain $F(s) = \displaystyle \int_0^\infty \frac{a}{s^2 + a^2} ds = \tan^{-1} \frac{s}{a} + c$

But $F(s) = 0$, when $s = 0$; $\quad \therefore \quad c = 0$. Hence $F(s) = \tan^{-1}(s/a)$.

**Example 22.9.** *Find the Fourier cosine transform of* $f(x) = 1/(1 + x^2)$.

*(Rohtak, 2011 ; V.T.U., 2011 S)*

*Hence derive Fourier sine transform of* $\phi(x) = x/(1 + x^2)$.

**Solution.**
$$F_c[f(x)] = \int_0^\infty \frac{\cos sx}{1+x^2} dx = I, \text{ say} \qquad ...(i)$$

$\therefore$
$$\frac{dI}{ds} = \int_0^\infty \frac{-x \sin sx}{1+x^2} dx = -\int_0^\infty \frac{x^2 \sin sx}{x(1+x^2)} dx \qquad ...(ii)$$

$$= -\int_0^\infty \frac{[(1+x^2)-1]\sin sx}{x(1+x^2)}dx = -\int_0^\infty \frac{\sin sx}{x}dx + \int_0^\infty \frac{\sin sx}{x(1+x^2)}dx$$

or $$\frac{dI}{ds} = -\frac{\pi}{2} + \int_0^\infty \frac{\sin sx}{x(1+x^2)}dx \quad ...(iii)$$

$\therefore$ $$\frac{d^2I}{ds^2} = \int_0^\infty \frac{x\cos sx}{x(1+x^2)}dx = I$$

or $$\frac{d^2I}{ds^2} - I = 0 \quad \text{or} \quad (D^2-1)I = 0, \text{ where } D = \frac{dI}{ds}$$

Its solution is $$I = c_1e^s + c_2e^{-s} \quad ...(iv)$$

$\therefore$ $$dI/ds = c_1e^s - c_2e^{-s} \quad ...(v)$$

When $s = 0$, Eqns. (*i*) and (*iv*) give $c_1 + c_2 = \int_0^\infty \frac{dx}{1+x^2} = \frac{\pi}{2}$

Also when $s = 0$, Eqns. (*iii*) and (*v*) give $c_1 - c_2 = -\pi/2$.

Solving these, $c_1 = 0$, $c_2 = \pi/2$.

Thus from Eqns. (*i*) and (*iv*), we have $F_c[f(x)] = I = (\pi/2)e^{-s}$

Now $$F_s[\phi(x)] = \int_0^\infty \frac{x\sin sx}{1+x^2}dx = -\frac{dI}{ds}, \text{ from Eqn. } (ii)$$

$$= (\pi/2)e^{-s}, \text{ from Eqn. } (v), \text{ with } c_1 = 0, c_2 = \pi/2.$$

**Example 22.10.** *Find the Fourier sine and cosine transform of $x^{n-1}$, $n > 0$ and prove that $\frac{1}{\sqrt{x}}$ is self-reciprocal under Fourier sine and cosine transforms.* *(Anna, 2012 ; Rohtak, 2011 S)*

**Solution.** (*i*) We know that $$F_s(x^{n-1}) = \int_0^\infty x^{n-1}\sin sx\,dx \quad ...(i)$$

and $$F_c(x^{n-1}) = \int_0^\infty x^{n-1}\cos sx\,dx \quad ...(ii)$$

$\therefore$ $$F_c(x^{n-1}) + i\,F_s(x^{n-1}) = \int_0^\infty (\cos sx + i\sin sx)x^{n-1}dx$$

$$= \int_0^\infty e^{isx}x^{n-1}dx = \int_0^\infty e^{-t}\left(-\frac{t}{is}\right)^{n-1}\left(-\frac{dt}{is}\right) \quad [\text{Where } isx = -t]$$

$$= \left(-\frac{1}{(i)}\right)^n \int_0^\infty e^{-t}t^{n-1}dt = \frac{(i)^{2n}}{(i)^n s^n}\Gamma(n) = \frac{(i)^n}{s^n}\Gamma(n)$$

$$= \left(\cos\frac{\pi}{2} + i\sin\frac{\pi}{2}\right)^n \Gamma(n)/s^n = \left(\cos\frac{n\pi}{2} + i\sin\frac{n\pi}{2}\right)\Gamma(n)/s^n$$

Equating real and imaginary parts, we get

$$F_c(x^{n-1}) = \frac{\Gamma(n)}{s^n}\cos\frac{n\pi}{2} \quad \text{and} \quad F_s(x^{n-1}) = \frac{\Gamma(n)}{s^n}\sin\frac{n\pi}{2}.$$

(*ii*) Putting $n = \frac{1}{2}$, $F_c\left(\frac{1}{\sqrt{x}}\right) = \frac{\Gamma\left(\frac{1}{2}\right)}{\sqrt{s}}\cos\frac{\pi}{4} = \frac{\sqrt{\pi}}{\sqrt{s}}\cdot\frac{1}{\sqrt{2}} = \sqrt{\left(\frac{\pi}{2}\right)}\cdot\frac{1}{\sqrt{s}}$ $\left[\because \sqrt{\left(\frac{1}{2}\right)} = \sqrt{\pi}\right]$

*i.e.* Fourier cosine tranform of $\frac{1}{\sqrt{x}}$ is constant times $\frac{1}{\sqrt{s}}$. Also $\frac{1}{\sqrt{x}}$ and $\frac{1}{\sqrt{s}}$ are the same. Hence it follows $\frac{1}{\sqrt{x}}$ is self-reciprocal under the Fourier cosine tranform.

Similarly putting $n = \frac{1}{2}$, $F_5\left(\frac{1}{\sqrt{x}}\right) = \sqrt{\left(\frac{\pi}{2}\right)}\frac{1}{\sqrt{s}}$, *i.e.*, $\frac{1}{\sqrt{x}}$ is self-reciprocal under Fourier sine tranform.

**Example 22.11.** *(a) Show that* $F_s[x\,f(x)] = -\dfrac{d}{ds}\{F_c(s)\};\ F_c[x\,f(x)] = \dfrac{d}{ds}\{F_s(s)\}$.

*(b) Find the Fourier sine and cosine transform of* $xe^{-ax}$

**Solution.** (*a*)
$$\frac{d}{ds}\{F_c(s)\} = \frac{d}{ds}\left\{\int_0^\infty f(x)\cos sx\,dx\right\} = \int_0^\infty f(x)\,(-x\sin sx)dx$$

$$= -\int_0^\infty \{x\,f(x)\}\ \sin sxdx = -F_s[x\,f(x)] \qquad \text{...(i)}$$

$$\frac{d}{ds}\{F_s(s)\} = \frac{d}{ds}\left\{\int_0^\infty f(x)\sin sx\,dx\right\} = \int_0^\infty f(x)\,(x\cos sx)dx$$

$$= \int_0^\infty \{x\,f(x)\}\cos sxdx = F_c\,[x\,f(x)] \qquad \text{...(ii)}$$

(*b*) We have
$$F_s(e^{-ax}) = \int_0^\infty e^{-ax}\ \sin sx\,dx = \frac{e^{-ax}}{a^2+s^2}\,|-a\sin sx - s\cos sx\,|_0^\infty$$

$$= \frac{s}{a^2+s^2} \qquad \text{...(iii)}$$

and
$$F_c(e^{-ax}) = \int_0^\infty e^{-ax}\ \cos sx\,dx = \frac{e^{-ax}}{a^2+s^2}\,|-a\cos sx + s\sin sx\,|_0^\infty$$

$$= \frac{a}{a^2+s^2} \qquad \text{...(iv)}$$

Now
$$F_c(xe^{-ax}) = -\frac{d}{ds}\ \{F_c(e^{-ax})\} \qquad \text{[by Eqn. (i)]}$$

$$= -\frac{d}{ds}\left(\frac{a}{a^2+s^2}\right) = \frac{2as}{(a^2+s^2)^2} \qquad \text{[by Eqn. (iv)]}$$

$$F_c(xe^{-ax}) = \frac{d}{ds}\{F_s(e^{-ax})\} \qquad \text{[by Eqn. (ii)]}$$

$$= \frac{d}{ds}\left(\frac{s}{a^2+s^2}\right) = \frac{(a^2+s^2)-s\,(2s)}{(a^2+s^2)^2} = \frac{a^2-s^2}{(a^2+s^2)^2}\,. \qquad \text{[by Eqn. (iii)]}$$

**Example 22.12.** *Find the Fourier transform of f(t)*

*Let* $f(t) = \begin{cases} 1 - t/a, \text{ for } & 0 < t < a \\ 1 + t/a, \text{ for } & -a < t < 0 \\ 0 & \text{otherwise} \end{cases}$ *find F[f(t)]?* *(GMU, Sambalpur, 2018)*

**Solution:**
$$F[f(t)] = \frac{1}{\sqrt{2\pi}}\int_{-\infty}^{\infty} f(t)\,e^{ist}\ dt = \frac{1}{\sqrt{2\pi}}\left[\int_{-a}^{0}(1+t/a)e^{ist}dt + \int_0^a\left(1-\frac{t}{a}\right)e^{ist}\,dt\right]$$

changing $t$ by $-t$ in the first intgral

$$F[f(t)] = \frac{1}{\sqrt{2\pi}}\left[\int_0^a\left(1-\frac{t}{a}\right)e^{-ist}dt + \int_0^a\left(1-\frac{t}{a}\right)e^{ist}\,dt\right]$$

$$= \frac{1}{\sqrt{2\pi}}\int_0^a\left(1-\frac{t}{a}\right)(e^{-ist}+e^{ist})dt$$

$$= \frac{2}{\sqrt{2\pi}} \int_0^a \left(1 - \frac{t}{a}\right) \cos t.dt = \sqrt{\frac{2}{\pi}} \left[\left(1 - \frac{t}{a}\right)\frac{\sin st}{s} - \left(-\frac{1}{a}\right)\left(-\frac{\cos st}{s^2}\right)\right]_0^a$$

$$= \sqrt{\frac{2}{\pi}} \left[-\frac{1}{as^2} \cos sa + \frac{1}{as^2}\right] = \frac{1}{as^2}\left[-\frac{1}{as^2} \cos sa + \frac{1}{as^2}\right]$$

$$= \frac{1}{as^2} \sqrt{\frac{2}{\pi}}(1 - \cos sa) = \frac{2\sqrt{2} \sin^2\left(\frac{sa}{2}\right)}{as^2\sqrt{\pi}}$$

**Example 22.13:** *Find Fourier transform of Delta function $\delta(t - a)$.* *(NOU, Bhubaneswer, 2018)*

**Solution:** The delta function possesses the shifting properly

$$\int_{-\infty}^{\infty} \phi(t) \cdot \delta(t - a)dt = \phi(a)$$

where $\phi(t)$ is any differentiable function

Now let $\phi(t) = e^{ist}$. then

$$\int_{-\infty}^{\infty} e^{ist} \cdot \delta(t - a)dt = e^{-isa}$$

$$\frac{1}{\sqrt{2}\,\pi} \int_{-\infty}^{\infty} \delta(t - a)e^{ist} dt = \frac{1}{\sqrt{2}\,\pi} e^{isa}$$

$$F[\delta(t - a)] = \frac{1}{\sqrt{2}\,\pi} e^{isa}$$

**Example 22.14.** *Solve the integral equation*

$$\int_0^{\infty} f(\theta) \cos \alpha\theta \, d\theta = \begin{cases} 1 - \alpha, & 0 \le a \le 1 \\ 0 & , \alpha > 1 \end{cases}$$ *(V.T.U., 2013)*

*Hence evaluate* $\int_0^{\infty} \frac{\sin^2 t}{t^2} dt.$ *(V.T.U., 2016)*

**Solution.** We have $\int_0^{\infty} f(\theta) \cos \alpha\theta \, d\theta = F_c(\alpha)$

$$\therefore \quad F_c(\alpha) = \begin{cases} 1 - \alpha, & 0 \le a \le 1 \\ 0 & , \alpha > 1 \end{cases} \quad ...(i)$$

By the inversion formula, we have

$$f(\theta) = \frac{2}{\pi} \int_0^{\infty} F_c(\alpha) \cos \alpha\theta \, d\alpha = \frac{2}{\pi} \int_0^1 (1 - \alpha) \cos \alpha\theta \, d\alpha \quad \text{[Integrating by parts]}$$

$$= \frac{2}{\pi}\left[\left|(1 - \alpha)\frac{\sin \alpha\theta}{\theta}\right|_0^1 - \int_0^1 (-1)\frac{\sin \alpha\theta}{\theta} d\alpha\right] = \frac{2}{\pi\theta}\left|-\frac{\cos \alpha\theta}{\theta}\right|_0^1 = \frac{2(1 - \cos \theta)}{\pi\theta^2}$$

Now $$F_c(\alpha) = \int_0^{\infty} f(\theta) \cos \alpha\theta \, d\theta = \int_0^{\infty} \frac{2(1 - \cos \theta)}{\pi\theta^2} \cos \alpha\theta \, d\theta \quad ...(ii)$$

$\therefore$ From Eqns. (*i*) and (*ii*), we have

$$\frac{2}{\pi} \int_0^{\infty} \frac{1 - \cos \theta}{\theta^2} \cos \alpha\theta \, d\theta = \begin{cases} 1 - \alpha, & 0 \le \alpha \le 1 \\ 0 & , \alpha > 1 \end{cases}$$

Now letting $\alpha \to 0$, we get $\frac{2}{\pi} \int_0^{\infty} \frac{1 - \cos \theta}{\theta^2} d\theta = 1$

or $$\int_0^\infty \frac{2\sin^2 \theta/2}{\theta^2}\, d\theta = \pi/2 \qquad \text{[Put } \theta/2 = t, \text{ so that } d\theta = 2dt]$$

$\therefore$ $$\int_0^\infty \frac{\sin^2 t}{t^2} = \pi/2.$$

## PROBLEMS 22.1

1. Express the function $f(x) = \begin{cases} 1 \text{ for } |x| \le 1 \\ 0 \text{ for } |x| > 1 \end{cases}$ as a Fourier integral.

   Hence evaluate $\int_0^\infty \frac{\sin \lambda \cos \lambda x}{\lambda}\, d\lambda$. *(Andhra., 2015 ; J.N.T.U., 2015)*

2. Find the Fourier integral representation for

   (*i*) $f(x) = \begin{cases} 1 - x^2, \text{ for } |x| \le 1 \\ 0, \text{ for } |x| > 1 \end{cases}$ (*ii*) $f(x) = \begin{cases} e^{ax}, \text{ for } x \le 0, a > 0 \\ e^{-ax}, \text{ for } x \ge 0\ a < 0 \end{cases}$

3. Using the Fourier integral representation, show that

   (*i*) $\int_0^\infty \frac{\omega \sin x\omega}{1+\omega^2}\, d\omega = \frac{\pi}{2} e^{-x}\ (x > 0)$ (*ii*) $\int_0^\infty \frac{\cos \omega x}{1+\omega^2}\, d\omega = \frac{\pi}{2}\, e^{-x}\quad (x \ge 0)$

   *(JNTUK, 2022; V.T.U., 2015 S)*

   (*iii*) $\int_0^\infty \frac{\sin \omega \cos x\omega}{\omega}\, d\omega = \frac{\pi}{2}$ when $0 \le x < 1$. (*iv*) $\int_0^\infty \frac{\sin \pi\alpha \sin \alpha\theta}{1-\alpha^2}\, d\alpha = \begin{cases} \frac{1}{2}\pi \sin \theta, & 0 \le \theta \le \pi \\ 0, & \theta > \pi \end{cases}$

4. Find the Fourier transforms of

   (*i*) $f(x) = \begin{cases} 1, |x| < a \\ 0, |x| > a \end{cases}$ *(Anna, 2013 ; Kurukshetra, 2013)*

   Hence evaluate $\int_{-\infty}^\infty \frac{\sin ax}{x}\, dx$

   (*ii*) $f(x) = \begin{cases} x^2, |x| < a \\ 0, |x| > a \end{cases}$ *(Kurukshetra, 2012S)*

5. Find the Fourier transform of $f(x) = \begin{cases} a^2 - x^2 \text{ for } |x| \le a \\ 0 \text{ for } |x| > a \end{cases}$

   Hence deduce that $\int_0^\infty \frac{\sin t - t\cos t}{t^3}\, dt = \frac{\pi}{4}$. *(JNTUK, 2022; Anna, 2012)*

6. Given $F(e^{-x^2}) = \sqrt{\pi} e^{-s^2/4}$, find the Fourier transform of

   (*i*) $e^{-x^2/3}$ (*ii*) $e^{-4(x-3)^2}$ (*iii*) $e^{-|x|}$ *(V.T.U., 2015S)*

7. Find the Fourier transform of the function $f(t) = u(t)e^{-t}$ where $u(t)$ is the unit step function. *(Delhi, 2012)*

8. Find the Fourier sine and cosine transforms of $f(x) = \begin{cases} 1, 0 \le x < 2 \\ 0, x \ge 2 \end{cases}$

9. Using the Fourier sine transform of $e^{-ax}$ $(a > 0)$, show that $\int_0^\infty \frac{x \sin kx}{a^2 + x^2}\, dx = \frac{\pi}{2} e^{-ak}\ (k > 0)$. *(D.T.U., 2016)*

   Hence obtain the Fourier sine transform of $x/(a^2 + x^2)$.

10. Find the Fourier sine transform of $e^{-ax}$ and evaluate $\int_0^\infty \frac{x \sin x}{x^2 + a^2}\, dx$. *(Anna., 2016)*

11. Find the Fourier cosine transform of $e^{-ax}$.

    Hence evaluate $\int_0^\infty \frac{\cos \lambda x}{x^2 + \alpha^2}\, dx$. *(J.N.T.U., 2015)*

12. If the Fourier sine transform of $f(x)$ is $e^{-as}/s$, find $f(x)$. Hence obtain the inverse Fourier sine transform of $1/s$.

13. Find the Fourier cosine transform of $e^{-x^2}$ and hence evaluate Fourier sine transform of $xe^{-x^2}$.
14. Find the Fourier cosine transform of $e^{-a^2x^2}$ for any $a > 0$ and hence prove that $e^{-x^2/2}$ is self-reciprocal under Fourier cosine transform. *(JNTUK, 2022)*
15. Find the Fourier sine transform of (*i*) $\dfrac{1}{x(x^2 + a^2)}$. *(Rohtak, 2013)*
(*ii*) $[e^{-ax}/x]$, $a > 0$
16. Obtain Fourier sine transform of

$$(i)\ f(x) = \begin{cases} \sin x, & 0 < x < a \\ 0, & x > a \end{cases} \qquad (ii)\ f(x) = \begin{cases} 4x, & \text{for } 0 < x < 1 \\ 4 - x, & \text{for } 1 < x < 4 \\ 0, & \text{for } x > 4 \end{cases}$$ *(V.T.U., 2013)*

17. Find the Fourier cosine transform of $(1 - x/\pi)^2$.
18. Find the finite Fourier sine and cosine transforms of $f(x) = 2x$, $0 < x < 4$. *(V.T.U., 2011)*
19. Find the finite sine transform of $f(x) = \begin{cases} -x, & x < c \\ \pi - x, & x > c \end{cases}$ where $0 \le c \le \pi$.
20. Show that the inverse finite Fourier sine transform of $F_s(n) = \dfrac{1}{\pi}\left\{1 + \cos n\pi - 2\cos\dfrac{n\pi}{2}\right\}$ is

$$f(x) = \begin{cases} 1, & 0 < x < \pi/2 \\ -1, & \pi/2 < x < \pi \end{cases}$$

21. Solve the integral equation $\displaystyle\int_0^\infty f(x)\sin tx\,dx = \begin{cases} 1, & 0 \le t < 1, \\ 2, & 1 \le t < 2, \\ 0, & t \ge 2 \end{cases}$ *(Rohtak, 2013)*
22. Solve the integral equation $\displaystyle\int_0^\infty f(x)\cos \alpha x\,dx = e^{-\alpha}$. *(V.T.U., 2015 S ; Rohtak, 2011 S)*

Hence show that $\displaystyle\int_0^\infty \frac{\cos \alpha x}{1+x^2}dx = \frac{\pi}{2}e^{-\alpha}$. *(Andhra., 2015)*

## 22.6 (1) CONVOLUTION

*The convolution of two functions $f(x)$ and $g(x)$ over the interval $(-\infty, \infty)$ is defined as*

$$f * g = \int_{-\infty}^{\infty} f(u)\, g(x - u)du = h(x).$$

(2) **Convolution theorem for Fourier transforms.** *The Fourier transform of the convolution of $f(x)$ and $g(x)$ is the product of their Fourier transforms, i.e.,*

$$F\{f(x) * g(x)\} = F\{f(x)\} \,.\, F\{g(x)\}$$

We have

$$F\{f(x) * g(x)\} = F\left\{\int_{-\infty}^{\infty} f(u)\, g(x - u)du\right\}$$

$$= \int_{-\infty}^{\infty}\left\{\int_{-\infty}^{\infty} f(u)\, g(x - u)du\right\} e^{isx}\, dx = \int_{-\infty}^{\infty} f(u)\left\{\int_{-\infty}^{\infty} g(x - u)\,.\, e^{isx}\, dx\right\}du$$

[Changing the order of integration]

$$= \int_{-\infty}^{\infty} f(u)\left\{\int_{-\infty}^{\infty} e^{is(x-u)}\,.\, g(x - u)\, d(x - u)\right\} e^{isu}\, du$$

$$= \int_{-\infty}^{\infty} e^{isu} f(u)\left\{\int_{-\infty}^{\infty} e^{ist}\, g(t)\, dt\right\} du \text{ where } x - u = t$$

$$= \int_{-\infty}^{\infty} e^{isu} f(u)\, du \,.\, F\{g(t)\} = \int_{-\infty}^{\infty} e^{isx} f(x)\, dx \,.\, F\{g(x)\} = F\{f(x)\}\,.\, F\{g(x)\}$$

## 22.7 PARSEVAL'S IDENTITY FOR FOURIER TRANSFORMS

*If the Fourier transforms of $f(x)$ and $g(x)$ are $F(s)$ and $G(s)$ respectively, then*

(*i*) $\displaystyle\frac{1}{2\pi}\int_{-\infty}^{\infty} F(s)\overline{G}(s)ds = \int_{-\infty}^{\infty} f(x)\overline{g}(x)dx$ (*ii*) $\displaystyle\frac{1}{2\pi}\int_{-\infty}^{\infty} [F(s)]^2\, ds = \int_{-\infty}^{\infty} |f(x)|^2\, dx$

*where bar implies the complex conjugate.*

(*i*) $\int_{-\infty}^{\infty} f(x)\,\bar{g}\,(dx) \int_{-\infty}^{\infty} f(x)\left\{\frac{1}{2\pi}\int_{-\infty}^{\infty} \bar{G}(s)\,e^{isx}\,ds\right\}dx$ [Using the inversion formula for Fourier transform]

$$= \frac{1}{2\pi}\int_{-\infty}^{\infty} \bar{G}(s)\left\{\int_{-\infty}^{\infty} f(x)\,e^{isx}\,dx\right\}ds \qquad \text{[Changing the order of integration]}$$

$$= \frac{1}{2\pi}\int_{-\infty}^{\infty} \bar{G}(s)F(s)ds, \text{ by definition of F-transform.}$$

(*ii*) Taking $g(x) = f(x)$, we get

$$\frac{1}{2\pi}\int_{-\infty}^{\infty} F(s)\,\bar{F}\,(s)ds = \int_{-\infty}^{\infty} f(x)\,\bar{f}(x)dx \text{ or } \frac{1}{2\pi}\int_{-\infty}^{\infty} |\,F(s)\,|^2\,ds = \int_{-\infty}^{\infty} |\,f(x)\,|^2\,dx$$

**Obs.** The following Parseval's identities for Fourier cosine and sine transforms can be proved as above :

(*i*) $\frac{2}{\pi}\int_0^{\infty} F_c(s)\,G_c(s)ds = \int_0^{\infty} f(x)\,g(x)dx$ (*ii*) $\frac{2}{\pi}\int_0^{\infty} F_s(s)\,G_s(s)ds = \int_0^{\infty} f(x)\,g(x)dx$

(*iii*) $\frac{2}{\pi}\int_0^{\infty} [F_c(s)]^2\,ds = \int_0^{\infty} [f(x)]^2\,dx$ (*iv*) $\frac{2}{\pi}\int_0^{\infty} [F_s(s)]^2\,ds = \int_0^{\infty} [f(x)]^2\,dx.$

**Example 22.15.** *Using Parseval's identities, prove that*

(*i*) $\int_0^{\infty} \frac{dt}{(a^2+t^2)(b^2+t^2)} = \frac{\pi}{2ab(a+b)}$

(*ii*) $\int_0^{\infty} \frac{t^2}{(t^2+1)^2}\,dt = \frac{\pi}{4}$ (*iii*) $\int_0^{\infty} \frac{\sin at}{t(a^2+t^2)}\,dt = \frac{\pi}{2}\cdot\frac{1-e^{-a^2}}{a^2}.$

**Solution.** (*i*) Let $f(x) = e^{-ax}$ and $g(x) = e^{-bx}$. Then $F_c(s) = \frac{a}{a^2+s^2}$, $G_c(s) = \frac{b}{b^2+s^2}$

Now using Parseval's identity for Fourier cosine transforms, *i.e.*,

$$\frac{2}{\pi}\int_0^{\infty} F_c(s)G_c(s)ds = \int_0^{\infty} f(x)\,g(x)dx \qquad ...(1)$$

We have $\frac{2}{\pi}\int_0^{\infty} \frac{ab}{(a^2+s^2)(b^2+s^2)}\,ds = \int_0^{\infty} e^{-(a+b)x}dx$

or $\frac{2ab}{\pi}\int_0^{\infty} \frac{ds}{(a^2+s^2)(b^2+s^2)} = \left|\frac{e^{-(a+b)x}}{-(a+b)}\right|_0^{\infty} = \frac{1}{a+b}$

Thus $\int_0^{\infty} \frac{dt}{(a^2+t^2)(b^2+t^2)} = \frac{\pi}{2ab(a+b)}$

(*ii*) Let $f(x) = \frac{x}{x^2+1}$ so that $F_s[f(x)] = \frac{\pi}{2}e^{-s}$

Now using Parseval's identity for sine transform, *i.e.*,

$$\frac{2}{\pi}\int_0^{\infty} [F_s\{f(x)\}]^2\,ds = \int_0^{\infty} [f(x)]^2\,dx$$

or $\int_0^{\infty}\left(\frac{x}{x^2+1}\right)^2 dx = \frac{2}{\pi}\int_0^{\infty}\left(\frac{\pi}{2}e^{-s}\right)^2 ds = \frac{\pi}{2}\left|\,e^{-2s}/-2\right|_0^{\infty} = \frac{\pi}{-4}(0-1) = \frac{\pi}{4}$

Hence $\int_0^{\infty} \frac{t^2}{(t^2+1)^2}\,dt = \frac{\pi}{4}$

(*iii*) Let $f(x) = e^{-ax}$ and $g(x) = \begin{cases}1, & 0 < x < a\\ 0, & x > a\end{cases}$. Then $F_c(s) = \frac{a}{a^2+s^2}$, $G_c(s) = \frac{\sin as}{s}$

Now using (1) above, we have $\dfrac{2}{\pi}\displaystyle\int_0^\infty \frac{a \sin as}{s(a^2+s^2)}\,ds = \int_0^a e^{-ax}\,.\,1\,dx = \frac{1-e^{-a^2}}{a}$

Thus $$\int_0^\infty \frac{\sin at}{t(a^2+t^2)}\,dt = \frac{\pi}{2a^2}(1-e^{-a^2}).$$

**Example 22.16.** *Find the Fourier transform of $f(x)$ given by* $f(x) = \begin{cases} a-|x|, & \text{for } |x|<a \\ 0, & \text{for } |x|>a \end{cases}$

*Hence show that* (i) $\displaystyle\int_0^\infty \left(\frac{\sin t}{t}\right)^2 dt = \frac{\pi}{2}$ (ii) $\displaystyle\int_0^\infty \left(\frac{\sin t}{t}\right)^4 dt = \pi/3.$ *(Anna, 2016)*

**Solution.** Fourier transform of $f(x)$ *i.e.* $F[f(x)] = \displaystyle\int_{-\infty}^\infty f(x)\,e^{isx}\,dx = \int_{-a}^a [a-|x|]e^{isx}\,dx$

$$= \int_{-a}^a [a-|x|](\cos sx + i\sin sx)\,dx$$

$$= 2\int_0^a (a-x)\cos sx\,dx + 0 \qquad \left\{\because \begin{array}{l}[a-|x|]\cos x \text{ in an even function} \\ [a-|x|]\sin x \text{ is an odd function}\end{array}\right.$$

$$= 2\left|(a-x)\frac{\sin sx}{s} - \frac{\cos sx}{s^2}\right|_0^a = 2\,\frac{1-\cos as}{s^2} = 4\,\frac{\sin^2 as/2}{s^2}$$

(*i*) By inversion formula,

$$f(x) = \frac{1}{2\pi}\int_{-\infty}^\infty F(s)\,e^{-isx}\,ds = \frac{1}{2\pi}\int_{-\infty}^\infty \frac{4\sin^2 as/2}{s^2}\,e^{-isx}\,ds$$

To evaluate $\displaystyle\int_0^\infty \left(\frac{\sin t}{t}\right)^2 dt$, put $x = 0$ and $a = 2$ so that

$$f(0) = \frac{2}{\pi}\int_{-\infty}^\infty \frac{\sin^2 s}{s^2}\,ds = \frac{4}{\pi}\int_0^\infty \left(\frac{\sin s}{s}\right)^2 ds \qquad \left[\because \frac{\sin s}{s} \text{ is an even function}\right]$$

$$\therefore \quad \int_0^\infty \left(\frac{\sin s}{s}\right)^2 ds = \frac{\pi}{4} f(0) = \frac{\pi}{2}. \qquad [\because f(0) = a = 2$$

(*ii*) Using Parseval's identity

$$\frac{1}{2\pi}\int_{-\infty}^\infty [F(s)]^2 ds = \int_{-\infty}^\infty |f(x)|^2\,dx$$

$$\frac{1}{2\pi}\int_{-\infty}^\infty \left(\frac{4\sin^2 as/2}{s^2}\right)^2 dx = \int_{-a}^a |[a-|x|]|^2\,dx$$

$$\frac{16}{\pi}\int_0^\infty \left(\frac{\sin as/2}{s}\right)^4 ds = 2\int_0^a (a-x)^2\,dx = 2\left|\frac{(a-x)^3}{-3}\right|_0^a = \frac{2}{3}a^3$$

Putting $t = as/2$ and $dt = ads/2$

$$\frac{16}{\pi}\int_0^\infty \left(\frac{\sin t}{2t/a}\right)\frac{2}{a}\,dt = \frac{2}{3}a^3 \quad \text{or} \quad \frac{2a^3}{\pi}\int_0^\infty \left(\frac{\sin t}{t}\right)^4 dt = \frac{2}{3}a^3$$

Hence $$\int_0^\infty \left(\frac{\sin t}{t}\right)^4 dt = \frac{\pi}{3}.$$ *(Calicut, 2011 S)*

## PROBLEMS 22.2

1. Verify Convolution theorem for $f(x) = g(x) = e^{-x^2}$. *(Anna, 2013)*

2. Use Convolution theorem to find the inverse Fourier transform of $\frac{i}{(1+s^2)^2}$, given that $\frac{2}{(1+s^2)}$ is the Fourier transform of $e^{-|x|}$. *(V.T.U., 2010 S)*

3. Using Parseval's identity, show that

   (i) $\int_0^{\infty} \frac{dt}{(t^2+1)^2} = \frac{\pi}{4}$ *(Andhra, 2015)* (ii) $\int_0^{\infty} \frac{t^2}{(4+t^2)(9+t^2)} dt = \frac{\pi}{10}$.

4. Find the Fourier transform of $f(x) = \begin{cases} 1 \text{ if } |x| < 1 \\ 0 \text{ if } |x| > 1 \end{cases}$.

   Hence deduce that $\int_0^{\infty} \left(\frac{\sin t}{t}\right)^2 dt = \frac{\pi}{2}$. *(Anna, 2015)*

5. Evaluate $\int_0^{\infty} \left(\frac{1-\cos x}{x}\right)^2 dx$.

6. Let $f(x) = \begin{cases} x & |x| < a \\ 0 & |x| > a \end{cases}$, show that $F[f(x)] = i\sqrt{\frac{2}{\pi}}(\sin as - as \cos as) / s^2$

7. If $\phi(x) = \begin{cases} x^2 & |x| < a \\ 0 & |x| > a \end{cases}$; Prove that $F[\phi(x)] = \frac{1}{s^3}\sqrt{\frac{2}{\pi}}\{a^2s^2 - 2) \sin as + 2as \cos as\}$

8. If $f(x) = \begin{cases} \frac{1}{2\varepsilon}, & |x| \le \varepsilon \\ 0 & |x| > \varepsilon \end{cases}$ Show that the Fourier transform of $f(x)$ is $\frac{\sin s\varepsilon}{s\varepsilon}$ *(Rohtak, ME. 2019)*

9. Use the integral $\int_{-\infty}^{\infty} e^{-y^2} dy = \sqrt{\pi}$ to prove that Fourier, transform of $\exp\left(-\frac{x^2}{2}\right)$ is $\exp\left(\frac{-s^2}{2}\right)$.

10. Let $\bar{f}(P) = \int_{-\infty}^{\infty} e^{ipx} f(x) \cdot dx$ and let $f(x) = \begin{cases} 1 - x^2 & |x| < 1 \\ 0 & |x| > 1 \end{cases}$, show that $\bar{f}(P) = \frac{4}{P^3}(P \cos p - \sin p)$

    Hence prove that $\int_{-\infty}^{\infty} \left(\frac{x \cos x - \sin x}{x^3}\right) \cdot \cos\frac{x}{2} dx = \frac{3\pi}{16}$.

## 22.8 RELATION BETWEEN FOURIER AND LAPLACE TRANSFORMS

*If*
$$f(t) = \begin{cases} e^{-xt} g(t), & t > 0 \\ 0 \quad , & t < 0 \end{cases} \quad \ldots(i)$$
*then*
$$F\{f(t)\} = L\{g(t)\}.$$

We have
$$F\{f(t)\} = \int_{-\infty}^{\infty} e^{ist} f(t) dt = \int_{-\infty}^{0} e^{ist} . 0 . dt + \int_0^{\infty} e^{ist} . e^{-xt} g(t) dt$$
$$= \int_0^{\infty} e^{(is-x)t} g(t)\, dt = \int_0^{\infty} e^{-pt} g(t) dt \quad \text{where } p = x - is$$

Hence *the Fourier transform of* $f(t)$ [defined by Eqn. $(i)$] is the *Laplace transform of* $g(t)$.

## 22.9 FOURIER TRANSFORMS OF THE DERIVATIVES OF A FUNCTION

The Fourier transform of the function $u(x, t)$ is given by
$$F[u(x, t)] = \int_{-\infty}^{\infty} ue^{isx} dx$$
*Then the Fourier transform of* $\partial^2u/\partial x^2$, *i.e.*

$$F\left[\frac{\partial^2 u}{\partial x^2}\right]=\int_{-\infty}^{\infty}\frac{\partial^2 u}{\partial x^2}e^{isx}\,dx=\left|e^{isx}\frac{\partial u}{\partial x}-is\,e^{isx}.u\right|_{-\infty}^{\infty}+(is)^2\int_{-\infty}^{\infty}ue^{isx}\,dx,$$

on applying the general rule of integration by parts. If $u$ and $\frac{\partial u}{\partial x}$ tend to zero as $x$ tends to $\pm\infty$, then

$$\mathbf{F}\left[\frac{\partial^2 \mathbf{u}}{\partial \mathbf{x}^2}\right]=-\mathbf{s}^2\,\mathbf{F}[\mathbf{u}] \qquad ...(1)$$

Similarly in the case of Fourier sine and cosine transforms, we have

$$\mathbf{F_s}\left[\frac{\partial^2 \mathbf{u}}{\partial \mathbf{x}^2}\right]=\mathbf{s}(\mathbf{u})_{\mathbf{x}=0}-\mathbf{s}^2\mathbf{F_s}[\mathbf{u}] \qquad ...(2)$$

and

$$\mathbf{F_c}\left[\frac{\partial^2 \mathbf{u}}{\partial \mathbf{x}^2}\right]=-\left(\frac{\partial \mathbf{u}}{\partial \mathbf{x}}\right)_{\mathbf{x}=0}-\mathbf{s}^2\,\mathbf{F_c}[\mathbf{u}] \qquad ...(3)$$

In general, the *Fourier transform of the nth derivative of* $f(x)$ is given by

$$\mathbf{F}\left[\frac{\mathbf{d^n f}}{\mathbf{dx^n}}\right]=(-\mathbf{is})^{\mathbf{n}}\,\mathbf{F}[\mathbf{f}(\mathbf{x})] \qquad \text{(J.N.T.U., 2015) ...(4)}$$

provided the first $n-1$ derivatives vanish as $x \to \pm\infty$.

For $\quad F[f^n(x)]=\int_{-\infty}^{\infty}f^n(x)\,e^{isx}\,dx$

$$=\left|e^{isx}f^{n-1}-ise^{isx}f^{n-2}+(is)^2e^{isx}f^{n-3}-\ldots\right|_{-\infty}^{\infty}+(-is)^n\int_{-\infty}^{\infty}f\,.\,e^{isx}\,dx$$

by the general rule of integration by parts, whence follows Eqn. (4).

## 22.10 INVERSE LAPLACE TRANSFORMS BY METHOD OF RESIDUES

Let the Laplace transform of $f(x)$ be $\bar{f}(s)$ so that

$$\bar{f}(s)=\int_0^{\infty}f(t)e^{-st}\,dt \qquad ...(1)$$

Multiply both sides by $e^{xs}$ and integrate w.r.t. $s$ within the limits $a-ir$ and $a+ir$. Then

$$\int_{a-ir}^{a+ix}e^{xs}\,\bar{f}(s)ds=\int_{a-ir}^{a+ir}e^{xs}\int_0^{\infty}f(t)e^{-st}\,dt\,ds \qquad \text{[Put } s=a-iu\text{]}$$

$$=\int_r^{-r}e^{x(a-iu)}\int_0^{\infty}f(t)e^{-(a-iu)t}\,dt\,(-idu)=ie^{ax}\int_{-r}^{r}e^{-ixu}\int_0^{\infty}[e^{-at}f(t)]e^{iut}\,dtdu$$

$$=ie^{ax}\int_{-r}^{r}e^{-ixu}\int_{-\infty}^{\infty}\phi(t)\,e^{iut}\,dt\,du$$

where $\phi(t)=\begin{cases}e^{-at}f(t) & \text{for } t\ge 0\\ 0 & \text{for } t<0\end{cases}$.

Proceeding to limits as $r\to\infty$, we get

$\int_{a-i\infty}^{a+i\infty}e^{xs}\,\bar{f}(s)\,ds=ie^{ax}\,.\,2\pi\phi(x)$, by (2) of § 22.4 $=2\pi ie^{ax}.e^{-ax}f(x)$ for $x>0$.

Hence $\quad f(x)=\int_{a-i\infty}^{a+i\infty}e^{xs}\,\bar{f}(s)ds\ (x>0) \qquad ...(2)$

which is called the *complex inversion formula*. It provides a direct means for obtaining the inverse Laplace transform of a given function.

The integration in Eqn. (2) is performed along a line $LM$ parallel to the imaginary axis in the complex plane $z=x+iy$ such that all the singularities of $\bar{f}(s)$ lie to its left* (Fig. 22.1). Let us take a contour $C$ which is composed of the

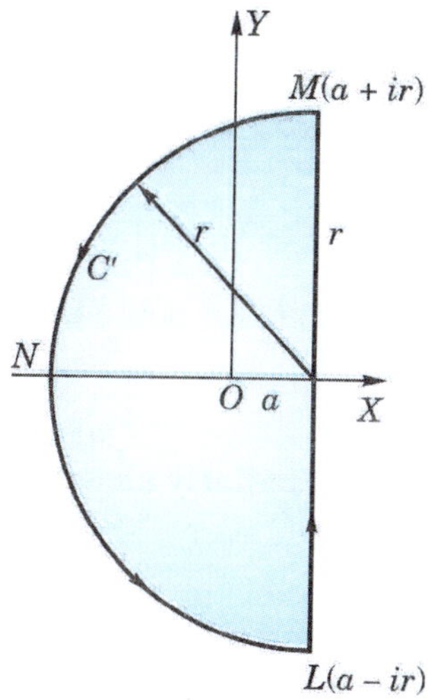

Fig. 22.1

* This has been so assumed simply to ensure the convergence of the integral (1).

line $LM$ and the semi-circle $C'$ (*i.e.*, $MNL$). Then from (2)

$$\frac{1}{2\pi i}\int_{LM} e^{xs}\,\bar{f}(s)\,ds = \frac{1}{2\pi i}\int_{C} e^{xs}\,\bar{f}(s)\,ds - \frac{1}{2\pi i}\int_{C'} e^{xs}\,\bar{f}(s)\,dx$$

The integral over $C'$ tends to zero as $r \to \infty$ (under certain conditions*). Therefore,

$$\mathbf{f(x)} = \underset{r\to\infty}{\mathbf{Lt}}\ \frac{1}{2\pi i}\int_{C} e^{xs}\,\bar{f}(s)\,dx$$

$$= \textbf{sum of the residues of } e^{xs}\bar{f}(s) \textbf{ at the poles of } f(s) \qquad ...(3)$$

[By §20.18]

**Example 22.17.** *Evaluate* $L^{-1}\left\{\dfrac{1}{(s-1)(s^2+1)}\right\}$ *by the method of residues.*

**Solution.** Since $\left|\dfrac{1}{(s-1)(s^2+1)}\right| \sim \left|\dfrac{1}{s^3}\right|$ for $|s| \to \infty$, therefore,

$$L^{-1}\left[\frac{1}{(s-1)(s^2+1)}\right] = \text{sum of Res}\left[\frac{e^{xs}}{(s-1)(s^2+1)}\right] \text{at the poles } s = 1, \pm i$$

Now
$$(\text{Res})_{s=1} = \underset{s\to 1}{\text{Lt}}\left[\frac{(s-1)\,.\,e^{xs}}{(s-1)(s^2+1)}\right] = \frac{e^x}{2} \qquad \text{[By § 20.19 (1)]}$$

$$(\text{Res})_{s=i} = \underset{s\to i}{\text{Lt}}\left[\frac{(s-i)\,.\,e^{xs}}{(s-1)(s^2+1)}\right] = \frac{e^{ix}}{(i-1)(i-1)} = -\frac{1}{2}\cdot\frac{e^{ix}}{1+i}$$

Changing $i$ to $-i$, we get $(\text{Res})_{s=-i} = -\dfrac{1}{2}\cdot\dfrac{e^{ix}}{1-i}$

$$\therefore \quad L^{-1}\left[\frac{1}{(s-1)(s^2+1)}\right] = \frac{e^x}{2} - \frac{1}{2}\left(\frac{e^{ix}}{1+i} + \frac{e^{-ix}}{1-i}\right) = \frac{1}{2}(e^x - \sin x - \cos x).$$

**Example 22.18.** *Prove that* $L^{-1}\left(\dfrac{e^{-c\sqrt{s}}}{s}\right) = 1 - \text{erf}\left(\dfrac{c}{\sqrt{2x}}\right)$.

**Solution.** By the complex inversion formula,

$$L^{-1}\left(\frac{e^{-c\sqrt{s}}}{s}\right) = \frac{1}{2\pi i}\int_{a-i\infty}^{a+i\infty} e^{xs}\,.\,\frac{e^{-c\sqrt{s}}}{s}\,ds.$$

Since $s = 0$ is a branch point of the integrand, we take a contour $LMNPQST$ as shown in Fig. 22.2, so that it doesn't include any singularity. Therefore, by Cauchy's theorem (§20.13), we have

$$\left\{\int_{LM} + \int_{MN} + \int_{NP} + \int_{PQS} + \int_{ST} + \int_{TL}\right\} \times e^{xs}\,\frac{e^{-c\sqrt{s}}}{s}\,ds = 0 \qquad ...(i)$$

If $ON = \rho$ and $OP = \varepsilon$, then along $NP$, $s = Re^{i\pi}$, therefore,

$$\int_{NP} = \int_{\rho}^{\epsilon} e^{-xR}\,\frac{e^{-ic\sqrt{R}}}{R}\,dR$$

Similarly along $ST$, $s = Re^{-i\pi}$, therefore,

$$\int_{ST} = \int_{\epsilon}^{\rho} e^{-xR}\,\frac{e^{c\sqrt{R}}}{R}\,dR$$

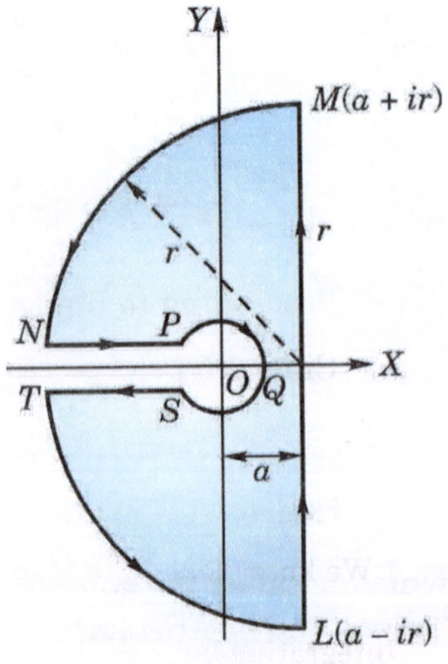

Fig. 22.2

* If positive constants $A$ and $k$ can be so found that $|\bar{f}(s)| < Ar^{-k}$ for every point on $C'$, then

$$\underset{r\to\infty}{\text{Lt}}\ \frac{1}{2\pi i}\int_{C'} e^{xs}\bar{f}(s)\,ds = 0. \qquad \textit{(Jordan's Lemma)}$$

Along the circle $PQS$, $s = \varepsilon\, e^{i\theta}$. Also $e^{xs}$ and $e^{-c\sqrt{\sigma}}$ are both approximately 1 since ε is small. Therefore,

$$\int_{PQS} = \int_{\pi}^{-\pi} \frac{1}{\varepsilon e^{i\theta}} \,.\, \varepsilon e^{i\theta}\, i d\theta = -2\pi i \text{ approximately.}$$

For $c > 0$, $| e^{-c\sqrt{s}}/s | < |s|^{-1}$.

But $\int_{MN}$ and $\int_{TL}$ both tend to zero as $r \to \infty$

Thus Eqn. (*i*) takes the form

$$\int_{a-ir}^{a+ir} \frac{e^{xs - c\sqrt{s}}}{s}\, ds + \int_{\varepsilon}^{\rho} e^{-xR}\, \frac{e^{ic\sqrt{R}} - e^{-ic\sqrt{R}}}{R}\, dR - 2\pi i = 0$$

Taking limits as $\varepsilon \to 0$ and $\rho \to \infty$, we get

$$\int_{a-i\infty}^{a+i\infty} \frac{e^{xs - c\sqrt{s}}}{s}\, ds = 2\pi i - 2i \int_0^{\infty} e^{-xR}\, \frac{\sin c\sqrt{R}}{R}\, dR$$

or $$\frac{1}{2\pi i}\int_{a-i\infty}^{a+i\infty} \frac{e^{xs - c\sqrt{s}}}{s}\, ds = 1 - \frac{2}{\pi}\int_0^{\infty} e^{-t^2}\, \frac{\sin (ct/\sqrt{x})}{t}\, dt^{*}, \text{ where } R = t^2/x$$

$$= 1 - \frac{2}{\pi}\cdot\frac{\pi}{2}\, erf\left(\frac{C}{2\sqrt{x}}\right) \text{ whence follows the result.}$$

**Example 22.19:** *Find the sine and cosine Fourier Transforms of $2e^{-5x} + 5e^{-2x}$*

**Solution:** Now use the following results

$$\int_0^{\infty} e^{-ax} \cos bx\, dx = \frac{a}{a^2 + b^2}$$

and $$\int_0^{\infty} e^{-ax} \sin bx\, dx = \frac{b}{a^2 + b^2}$$

Now $$F_e[2e^{-5x} + 5e^{-2x}] = 2F_e[e^{-5x}] + 5F_e[e^{-2x}]$$

$$= 2\sqrt{\frac{2}{\pi}}\int_0^{\infty} e^{-5x} \cos sx + 5\sqrt{\frac{2}{\pi}}\int_0^{\infty} e^{2x}.\cos sx\, dx$$

$$= 2\sqrt{\frac{2}{\pi}} \times \frac{5}{s^2 + 5^2} + 5\sqrt{\frac{2}{\pi}}\int_0^{\infty} e^{-2x}.\cos sx\, dx$$

$$= 2\sqrt{\frac{2}{\pi}} \times \frac{5}{s^2 + 5^2} + 5\sqrt{\frac{2}{\pi}} \times \frac{2}{s^2 + 2^2}$$

$$= 10\sqrt{\frac{2}{\pi}}\left(\frac{1}{s^2 + 25} + \frac{1}{s^2 + 4}\right)$$

Similarly, $$F_S[2e^{-5x} + 5e^{-2x}] = s\sqrt{\frac{2}{11}}\left(\frac{2}{s^2 + 25} + \frac{5}{s^2 + 4}\right)$$

---

* We know that $\int_0^{\infty} e^{-t^2} \cos 2mt\, dt = \frac{1}{2}\sqrt{\pi} e^{-m^2}$ [Example 20.44

Integrating both sides w.r.t. $m$ from 0 to $c/2\sqrt{x}$.

$$\int_0^{\infty} e^{-t^2} \left|\frac{\sin 2mt}{2t}\right|_0^{c/2\sqrt{x}} dt = \frac{1}{2}\sqrt{\pi}\int_0^{c/2\sqrt{\pi}} e^{-m^2}\, dm$$

or $$\int_0^{\infty} e^{-t^2}\, \frac{\sin (ct/\sqrt{x})}{t}\, dt = \frac{\pi}{2}\, erf\left(\frac{c}{2\sqrt{x}}\right).$$ [By § 7.18(1)]

**Example 22.20:** *Prove that*

(*i*) $F_C\,[te^{-at}] = \sqrt{\dfrac{2}{\pi}}\,\dfrac{a^2 - s^2}{(a^2 + s^2)^2}$ (*ii*) $F_S\,[te^{-at}] = \sqrt{\dfrac{2}{\pi}}\,\dfrac{2as}{(a^2 + s^2)^2}$

**Solution:** (*i*) We know that

$$F_c[e^{-at}] = \sqrt{\frac{2}{\pi}}\,\frac{a}{s^2 + a^2}$$

Now, differentiating both side

$$F_c[-te^{-at}] = \sqrt{\frac{2}{\pi}}\left[\frac{(s^2 + a^2).(1 - a)(2.a)}{(s^2 + a^2)^2}\right] = \sqrt{\frac{2}{\pi}}\,\frac{s^2 - a^2}{(s^2 + a^2)^2}$$

$$F_c\,[t_e{}^{-at}] = \sqrt{\frac{2}{\pi}}\,\frac{a^2 - s^2}{(s^2 + a^2)^2}$$

(*ii*) Similary.

$$F_e[e^{-at}] = \sqrt{\frac{2}{\pi}}\,\frac{s}{s^2 + a^2}$$

$$F_S[-te^{-at}] = \sqrt{\frac{2}{\pi}}\left[\frac{-2as}{(s^2 + a^2)^2}\right]$$

$$F_S[te^{-at}] = \sqrt{\frac{2}{\pi}}\,\frac{2as}{(s^2 + a^2)^2}.$$

## PROBLEMS 22.3

Using the method of residues, evaluate the inverse Laplace transform of each of the following :

1. $\dfrac{1}{(s+1)(s-2)^2}$.
2. $\dfrac{1}{(s-2)(s^2+1)}$.
3. $\dfrac{1}{s^2(s^2-a^2)}$.
4. $\dfrac{1}{(s-1)^2(s^2+1)}$.
5. $\dfrac{1}{(s^2+1)^2}$.

## 22.11 APPLICATION OF TRANSFORMS TO BOUNDARY VALUE PROBLEMS

In one dimensional boundary value problems, the partial differential equation can easily be transformed into an ordinary differential equation by applying a suitable transform. The required solution is then obtained by solving this equation and inverting by means of the complex inversion formula or by any other method. In two dimensional problems, it is sometimes required to apply the transforms twice and the desired solution is obtained by double inversion.

(*i*) If in a problem $u(x, t)_{x=0}$ is given then we use infinite sine transform to remove $\partial u^2/\partial x^2$ from the differential equation.

In case $[\partial u(x, t)/\partial x]_{x=0}$ is given then we employ infinite cosine transform to remove $\partial^2 u/\partial x^2$.

(*ii*) If in a problem $u(0, t)$ and $u(l, t)$ are given, then we use finite sine transform to remove $\partial^2 u/\partial x^2$ from the differential equation.

In case $(\partial u/\partial x)_{x=0}$ and $(\partial u/\partial x)_{x=l}$ are given, then we employ finite cosine transform to remove $\partial^2 u/\partial x^2$.

The method of solution is best explained through the following examples.

### Heat conduction

**Example 22.21.** *Determine the distribution of temperature in the semi-infinite medium $x \geq 0$, when the end $x = 0$ is maintained at zero temperature and the initial distribution of temperature is $f(x)$.*

**Solution.** Let $u(x, t)$ be the temperature at any point $x$ and at any time $t$. We have to solve the heat-flow equation (§ 18.5)

$$\frac{\partial u}{\partial t} = c^2 \frac{\partial^2 u}{\partial x^2} \ (x > 0, t > 0) \quad ...(i)$$

subject to the initial condition $u(x, 0) = f(x)$ ...(ii)

and the boundary condition $u(0, t) = 0$ ...(iii)

Taking Fourier sine transform of (1) and denoting $F_s[u(x, t)]$ by $\bar{u}_s$, we have

$$\frac{d\bar{u}_s}{dt} = c^2 [su(0, t) - s^2 \bar{u}_s] \quad \text{[By (2) of § 22.9]}$$

or $$\frac{d\bar{u}_s}{dt} + c^2 s^2 \bar{u}_s = 0 \quad \text{[By (iii)] ...(iv)}$$

Also the Fourier sine transform of Eqn. (ii) is $\bar{u}_s = \bar{f}(s)$ at $t = 0$. ...(v)

Solving Eqn. (iv) and using Eqn. (v), we get $\bar{u}_s = \bar{f}_s(s)e^{-c^2 s^2 t}$

Hence taking its inverse Fourier sine transform, we obtain

$$u(x, t) = \frac{2}{\pi}\int_0^\infty \bar{f}_s(s)e^{-c^2 s^2 t} \sin xs\, ds.$$

**Example 22.22.** *Solve $\partial u/\partial t = 2\partial^2 u/\partial x^2$, if $u(0, t) = 0$, $u(x, 0) = e^{-x}$ $(x > 0)$, $u(x, t)$ is bounded where $x > 0$, $t > 0$.*

**Solution.** Given $\partial u/\partial t = 2\partial^2 u/\partial x^2, x > 0, t > 0$ ...(i)

with boundary conditions : $u(0, t) = 0$, $u(x, t)$ is bounded ...(ii)

and initial condition $u(x, 0) = e^{-x}, x > 0$ ...(iii)

Since $u(0, t)$ is given, we take Fourier sine transform of both sides of Eqn. (i) so that

$$\int_0^\infty \frac{\partial u}{\partial t} \sin px\, dx = 2\int_0^\infty \frac{\partial^2 u}{\partial x^2} \sin px\, dx$$

or $$\frac{d}{dt}\int_0^\infty u(x, t) \sin px\, dx = 2\left[\left|\frac{\partial u}{\partial x}\sin px\right|_0^\infty - \int_0^\infty \frac{\partial u}{\partial x} . p\cos px\, dx\right] \quad \text{(Integrating by parts)}$$

or $$\frac{d\bar{u}_s}{dt} = -2p\int_0^\infty \frac{\partial u}{\partial x}\cos px\, dx, \text{ if } \frac{\partial u}{\partial x} \to \frac{\partial u}{\partial x} \to 0 \text{ as } x \to \infty \text{ where } \bar{u}_s(p, t) = \int_0^\infty u(x, t)\sin px\, dx$$

$$= -2p[|u(x, t)\cos px|_0^\infty - \int_0^\infty u(x, t) - (-p\sin px)\, dx] \quad \text{[Again integrating by parts]}$$

$$= -2p\,[0 - u(0, t) + p\int_0^\infty u(x, t)\sin px\, dx] \quad [\because \ u(x, t) \to 0 \text{ as } x \to \infty \text{ by (ii)}]$$

$$= 2pu(0, t) - 2p^2\bar{u}_s$$

or $$\frac{d\bar{u}_s}{dt} = -2p^2\bar{u}_s \quad \text{[By (ii)]}$$

Integrating $\int \frac{\partial \bar{u}_s}{\bar{u}_s} - \log c = -2p^2\int dt$ or $\log \bar{u}_s - \log c = -2p^2 t$

$\therefore$ $$\bar{u}_s(p, t) = ce^{-2p^2 t} \quad ...(iv)$$

Taking Fourier sine transform of both sides of Eqn. (iii), we get

$$\int_0^\infty u(x, 0)\sin px\, dx = \int_0^\infty e^{-x}\sin px\, dx$$

or $$\bar{u}_s(p, 0) = \left|\frac{e^{-x}}{1+p^2}(-\sin px - p\cos px)\right|_0^\infty = \frac{p}{1+p^2} \quad ...(v)$$

Putting $t = 0$ in Eqn. (iv) and using Eqn. (v), we obtain $p/(1 + p^2) = c$

Thus Eqn. (iv) becomes $\bar{u}_s(p, t) = \dfrac{p}{1+p^2}e^{-2p^2 t}$

Now taking inverse Fourier sine transform, we get

$$u(x, t) = \frac{2}{\pi}\int_0^\infty \frac{pe^{-2p^2t}}{1+p^2} \sin px \, dp.$$

**Example 22.23.** *Solve the equation* $\frac{\partial u}{\partial t} = \frac{\partial^2 u}{\partial x^2}$, $(x > 0, t > 0)$ *subject to the conditions*

*(i)* $u = 0$, *when* $x = 0, t > 0$ *(ii)* $u = \begin{cases} 1, & 0 \le x < 1 \\ 0, & x > 1, \text{ when } t = 0 \end{cases}$ *(iii)* $u(x, t)$ *is bounded.*

*(Kurukshetra, 2013)*

**Solution.** Since $u(0, t) = 0$, we take Fourier sine transform of both sides of the given equation, we get

$$\int_0^\infty \frac{\partial u}{\partial t} \sin sx \, dx = \int_0^\infty \frac{\partial^2 u}{\partial x^2} \sin sx \, dx$$

$$\frac{\partial}{\partial t}\int_0^\infty u \sin sx \, dx = -s^2 \bar{u}(s) + s\, u(0) \qquad [\because u = 0, \text{ when } x = 0]$$

or $$\frac{\partial \bar{u}}{\partial t} = -s^2\bar{u} \quad \text{or} \quad \frac{\partial \bar{u}}{\partial t} + s^2\bar{u} = 0 \quad \text{or} \quad (D^2 + s^2)\bar{u} = 0 \text{ i.e., } D = \pm is$$

$\therefore$ Its solution is $\bar{u}(s, t) = e^{-s^2t}$ ...(1)

Since $$\bar{u}(s, t) = \int_0^\infty u(x, t) \sin sx \, dx$$

$\therefore$ $$\bar{u}(s, 0) = \int_0^\infty u(x, 0) \sin sx \, dx = \int_0^1 1 . \sin sx \, dx \qquad [\text{By } (ii)]$$

$$= \frac{1 - \cos s}{s} \qquad ...(2)$$

From Eqns. (1) and (2), $$c = \bar{u}(s, 0) = \frac{1 - \cos s}{s}$$

Thus Eqn. (1) gives $$\bar{u}(s, t) = \frac{1 - \cos s}{s} e^{-s^2t}$$

Now taking inverse Fourier sine transform, we get

$$u(x, t) = \int_0^\infty \frac{1 - \cos s}{s} e^{-s^2t} \, ds$$

which is the desired solution.

**Example 22.24.** *Using finite Fourier transform, solve*

$$\frac{\partial u}{\partial t} = \frac{\partial^2 u}{\partial x^2}$$

*given* $u(0, t) = 0$, $u(4, t) = 0$ *and* $u(x, 0) = 2x$ *where* $0 < x < 4$, $t > 0$.

**Solution.** Since $u(0, t) = 0$, we take finite Fourier sine transform of both sides of the given equation

$$\int_0^4 \frac{\partial u}{\partial t} \sin \frac{n\pi}{4} x \, dx = \int_0^4 \frac{\partial^2 u}{\partial x^2} \sin \frac{n\pi}{4} x \, dx$$

or $$\frac{d}{dt}(\bar{u}s) = F_s\left(\frac{\partial^2 u}{\partial x^2}\right)$$

$$= -\frac{n^2\pi^2}{16}\bar{u}_s + \frac{n\pi}{4}[u(0, t) - (-1)^n u(4, t)]$$

$$= -\frac{n^2\pi^2}{16}\bar{u}_s \qquad [\because u(0, t) = 0, u(4, t) = 0.]$$

or $$\frac{d\bar{u}_s}{\bar{u}_s} = -\frac{n^2\pi^2}{16} dt$$

Integrating both sides, $\log \bar{u}_s = -\dfrac{n^2\pi^2}{16}t + c$

or $\bar{u}_s(x, 0) = \alpha e^{\frac{-n^2\pi^2 t}{16}}$ ...(i)

Putting $t = 0$, $a = \bar{u}_s(x, 0) = \int_0^4 u(x, 0) \sin \dfrac{n\pi x}{4} dx$ [$\because$ $u(x, 0) = 2x$]

$$= \int_0^4 2x \sin \frac{n\pi x}{4} dx = -\frac{32}{n\pi} \cos n\pi$$

Thus Eqn. (i) gives, $\bar{u}_s(x, 0) = -\dfrac{32}{n\pi} \cos n\pi e^{-n^2\pi^2 t/16} = -\dfrac{32}{n\pi}(-1)^n e^{-n^2\pi^2 t/16}$

Now taking inverse Fourier sine transform, we get

$$u(x, 0) = \frac{2}{4}\sum_{n-1}^{\infty} \frac{32}{n\pi}(-1)^{n+1} e^{-n^2\pi^2 t/16} \sin\left(\frac{n\pi x}{4}\right) = 16 \sum_{n=1}^{\infty} \frac{(-1)^{n+1}}{n\pi} e^{-n^2\pi^2 t/16} \sin\left(\frac{n\pi x}{4}\right).$$

**Example 22.25.** *If the initial temperature of an infinite bar is given by*

$$\theta(x) = \begin{cases} \theta_0 \text{ for } |x| < a \\ 0 \text{ for } |x| > a, \end{cases}$$

*determine the temperature at any point x and at any instant t.*

**Solution.** To determine the temperature $\theta(x, t)$ at any point at any time, we have to solve the equation

$$\frac{\partial \theta}{\partial t} = c^2 \frac{\partial^2 \theta}{\partial x^2} \quad (t > 0) \qquad ...(i)$$

subject to the initial condition $\theta(x, 0) = \begin{cases} \theta_0 \text{ for } |x| < a \\ 0 \text{ for } |x| > a \end{cases}$ ...(ii)

Taking Fourier transform of Eqn. (i) and denoting $F[\theta(x, t)]$ by $\bar{\theta}$, we find

$$\frac{d\bar{\theta}}{dt} = c^2 s^2 \bar{\theta} \qquad \text{[by (1) of § 22.9] ...(iii)}$$

Also the Fourier transform of (2) is

$$\bar{\theta}(s, 0) = \int_{-\infty}^{\infty} \theta(x, 0) e^{isx} dx = \int_{-a}^{a} \theta_0 e^{isx} dx = \theta_0 \frac{e^{isa} - e^{-isa}}{is} = 2\theta_0 \frac{\sin as}{s} \qquad ...(iv)$$

Solving Eqn. (iii) and using Eqn. (iv), we get $\bar{\theta} = \dfrac{2\theta_0 \sin as}{s} e^{-c^2 s^2 t}$

Hence taking its inverse Fourier transform, we get

$$\theta(x, t) = \frac{1}{2\pi}\int_{-\infty}^{\infty} \frac{2\theta_0 \sin as}{s} e^{-c^2 s^2 t} e^{-isx} ds = \frac{\theta_0}{\pi}\int_{-\infty}^{\infty} \frac{\sin as}{s} e^{-c^2 s^2 t} (\cos xs - i \sin xs) ds$$

$$= \frac{2\theta_0}{\pi}\int_0^{\infty} \frac{\sin as}{s} e^{-c^2 s^2 t} \cos xs \, ds \qquad \left\{\begin{array}{l}\text{The second integral vanishes as} \\ \text{its integrand is an odd function}\end{array}\right.$$

$$= \frac{\theta_0}{\pi}\int_0^{\infty} e^{-c^2 s^2 t} \frac{\sin (a + x)s + \sin (a - x)s}{s} ds$$

$$= \frac{\theta_0}{\pi}\int_0^{\infty} e^{-v^2} \left\{\sin \frac{(a + x)v}{c\sqrt{t}} + \sin \frac{(a - x)v}{c\sqrt{t}}\right\} \frac{dv}{v} \quad \text{where } v^2 = c^2 s^2 t$$

$$= \frac{\theta_0}{\pi}\left\{erf \frac{(a + x)}{2c\sqrt{t}} + erf \frac{(a - x)}{2c\sqrt{t}}\right\}.$$

**Example 22.26.** *A bar of length a is at zero temperature. At $t = 0$, the end $x = a$ is suddenly raised to temperature $u_0$ and the end $x = 0$ is insulated. Find the temperature at any point x of the bar at any time $t > 0$, assuming that the surface of the bar is insulated.*

**Solution.** Here we have to solve the differential equation

$$\frac{\partial u}{\partial t} = c^2 \frac{\partial^2 u}{\partial x^2} \quad (0 < x < a, t > 0) \qquad ...(i)$$

subject to the conditions

$$u(x, 0) = 0 \quad ...(ii)\ ; \quad u_x(0, t) = 0 \quad ...(iii) \quad \text{and} \quad u(a, t) = u_0 \qquad ...(iv)$$

The Laplace transform of Eqn. (*i*), if $L[u(x, t)] = \bar{u}(x, s)$, is

$$s\bar{u} - u(x, 0) = c^2 \frac{d^2\bar{u}}{dx^2}$$

Using Eqn. (*ii*), we get $\dfrac{d^2\bar{u}}{dx^2} - \dfrac{s}{c^2}\bar{u} = 0$ ...(*v*)

Similarly the Laplace transforms of Eqns. (*iii*) and (*iv*) are

$$\bar{u}_x(0, s) = 0 \quad ...(vi)\ ; \qquad \bar{u}(a, s) = \frac{u_0}{s} \qquad ...(vii)$$

Solving Eqn. (*v*), we have $\bar{u} = C_1 e^{x\sqrt{s}/c} + C_2 e^{-x\sqrt{s}/c}$

Using Eqn. (*vi*), we find $C_1 = C_2$ so that

$$\bar{u} = C_1 (e^{\sqrt{sx/c}} + e^{-\sqrt{sx/c}}) = 2C_1 \cosh(\sqrt{sx/c})$$

Now using Eqn. (*vii*), we have $\bar{u} = \dfrac{u_0 \cosh(\sqrt{sx/c})}{s \cosh(\sqrt{sa/c})}$

By the inversion formula (3) §22.10, we get

$$u(x, t) = \text{sum of the residues of } \left(\frac{e^{st} . u_0 \cosh(\sqrt{sx/c})}{s \cosh(\sqrt{sa/c})}\right) \text{ at all the poles which occur at } s = 0$$

and $\cosh(\sqrt{sa/c}) = 0$ *i.e.*, at $s = 0$, $\sqrt{sa/c} = \left(n - \dfrac{1}{2}\right)\pi i$, $n = 0, \pm 1, \pm 2, ...$

or at $s = 0$, $s\ (= s_n) = -\dfrac{(2n-1)^2 c^2 \pi^2}{4a^2} = 0, 1, 2, ...$

Now

$$(\text{Res})_{s=0} = \underset{s \to 0}{\text{Lt}} \left\{ s . \frac{u_0 e^{st} \cosh(\sqrt{sx/c})}{s \cosh(\sqrt{sa/c})} \right\} = u_0$$

$$(\text{Res})_{s=s_n} = u_0 \underset{s \to s_n}{\text{Lt}} \left\{ (s - s_n) . \frac{u_0 e^{st} \cosh(\sqrt{sx/c})}{s \cosh(\sqrt{sx/c})} \right\}$$

$$= u_0 \underset{s \to s_n}{\text{Lt}} \left\{ \frac{s - s_n}{\cosh(\sqrt{sa/c})} \right\} . \underset{s \to s_n}{\text{Lt}} \left\{ \frac{e^{st} \cosh(\sqrt{sa/c})}{s} \right\} \qquad \left[\frac{0}{0} \text{ form}\right]$$

$$= u_0 \underset{s \to s_n}{\text{Lt}} \frac{1}{\sinh(\sqrt{sa/c}) . (a/2\sqrt{s/c})} . \underset{s \to s_n}{\text{Lt}} \left\{ \frac{e^{st} \cosh(\sqrt{sx/c})}{s} \right\}$$

$$= \frac{4u_0 (-1)^n}{(2n-1)\pi} e^{-(2n-1)^2 \pi^2 c^2 t/4a^2} \cos \frac{(2n-1)\pi x}{2a}$$

Thus we get $u(x, t) = u_0 + \dfrac{4u_0}{\pi} \displaystyle\sum_{n=1}^{\infty} \dfrac{(-1)^n}{2n-1} e^{-(2n-1)^2 \pi^2 c^2 t/4a^2} \cos \dfrac{(2n-1)\pi x}{2a}$.

### Vibrations of a string

**Example 22.27.** *An infinite string is initially at rest and that the initial displacement is $f(x)$, $(-\infty < x < \infty)$. Determine the displacement $y(x, t)$ of the string.*

**Solution.** The equation for the vibration of the string is

$$\partial^2 y/\partial t^2 = c^2 \partial^2 y/\partial x^2 \quad \text{...(i)}$$

and the initial conditions are

$$(\partial y/\partial t)_{t=0} = 0 \; ; \; y(x, 0) = f(x) \quad \text{...(ii)}$$

Multiplying Eqn. (*i*) by $e^{isx}$ and integrating w.r.t. $x$ from $-\infty$ to $\infty$, we get

$$\partial^2 Y/\partial t^2 = c^2(-s^2 Y) \qquad \text{provided } y \text{ and } \partial y/\partial x \to 0 \text{ as } x \to \infty$$

$\therefore$ a solution of $d^2Y/dt^2 + c^2s^2Y = 0$ is $\quad Y = A_1 \cos cst + A_2 \sin cst$ ...(*iii*)

Also Fourier transforms of Eqn. (*ii*) are

$$\partial y/\partial t = 0 \quad \text{and} \quad Y = F(s) \text{ when } t = 0$$

Applying these to Eqn. (*iii*), we get

$$A_2 = 0 \quad \text{and} \quad A_1 = F(s)$$

Thus $\quad Y = F(s) \cos cst$

Now taking inverse Fourier transforms, we get

$$y(x, t) = \frac{1}{2\pi}\int_{-\infty}^{\infty} F(s) \cos cst \,.\, e^{-isx}\, dx$$

$$= \frac{1}{2\pi}\int_{-\infty}^{\infty} F(s) \frac{e^{icst} + e^{-icst}}{2} \,.\, e^{-isx}\, dx$$

$$= \frac{1}{4\pi}\int_{-\infty}^{\infty} [F(s) e^{-is(x-ct)} + F(s) e^{-is(x+ct)}]\, ds$$

$$= \frac{1}{2}[f(x-ct) + f(x+ct)] \qquad \left[\because\; f(x) = \frac{1}{2\pi}\int_{-\infty}^{\infty} F(s)\, e^{-isx}\, ds\right]$$

**Example 22.28.** *An infinitely long string having one end at $x = 0$, is initially at rest along the x-axis. The end $x = 0$ is given a transverse displacement $f(t)$, $t > 0$. Find the displacement of any point of the string at any time.*

**Solution.** Let $y(x, t)$ be the transverse displacement of any point $x$ of the string at any time $t$. Then we have to solve the wave equation (§ 18.4).

$$\frac{\partial^2 y}{\partial t^2} = c^2 \frac{\partial^2 y}{\partial x^2} \qquad (x > 0, t > 0) \quad \text{...(i)}$$

subject to the conditions $y(x, 0) = 0$, $y_t(x, 0) = 0$, $y(0, t) = f(t)$ and the displacement $y(x, t)$ is bounded.

The Laplace transform of Eqn. (*i*), writing $L[y(x, t)] = \bar{y}(x, s)$ is

$$s^2 \bar{y} - sy(x, 0) - \frac{\partial y(x, 0)}{\partial t} = c^2 \frac{\partial^2 \bar{y}}{\partial x^2}$$

Using the first two conditions, we have

$$\frac{\partial^2 \bar{y}}{\partial x^2} = \left[\frac{s}{c}\right]^2 \bar{y} \quad \text{...(ii)}$$

Similarly the Laplace transforms of the third and fourth conditions are

$$\bar{y}(0, s) = \bar{f}(s) \text{ at } x = 0 \quad \text{...(iii)} \quad \text{and} \quad \bar{y}(x, s) \text{ is bounded.} \quad \text{...(iv)}$$

Solving Eqn. (*ii*), we get

$$\bar{y}(x, s) = C_1 e^{sx/c} + C_2 e^{-sx/c}$$

To satisfy condition Eqn. (*iv*), we must have $C_1 = 0$

Using the condition Eqn. (*iii*), we get $C_2 = \bar{f}(s)$.

$$\therefore \quad \bar{y}(x, s) = \bar{f}(s) e^{-sx/c}$$

Using the complex inversion formula, we obtain

$$y = \frac{1}{2\pi i}\int_{a-i\infty}^{a+i\infty} e^{(t-x/c)s}\,\bar{f}(s)\,ds = f(t - x/c).$$

**Example 22.29.** *A tightly stretched flexible string has its ends fixed at x = 0 and x = l. At time t = 0, the string is given a shape defined by F(x) = μx(l – x), where μ is a constant and then released. Find the displacement of any point x of the string at any time t > 0.*

**Solution.** We have to solve the wave equation $\dfrac{\partial^2 y}{\partial t^2} = c^2 \dfrac{\partial^2 y}{\partial x^2}$ $\quad (x > 0, t > 0)$

subject to the conditions $y(0, t) = 0, y(l, t) = 0$

and $y(x, 0) = \mu x(l - x), y_t(x, 0) = 0$

Now taking Laplace transform, writing $L[y(x, t)] = \bar{y}(x, s)$, we get

$$s^2\bar{y} - s\bar{y}(x, 0) - \frac{\partial y(x, 0)}{\partial t} = c^2 \frac{\partial^2 \bar{y}}{\partial x^2} \qquad ...(i)$$

where $\bar{y}(0, s) = 0,\ \bar{y}(l, s) = 0$ ...(ii)

$\therefore$ (i) reduces to $\dfrac{\partial^2 \bar{y}}{\partial x^2} - \left(\dfrac{s}{c}\right)^2 \bar{y} = -\dfrac{\mu s x(l - x)}{c^2}$

Its solution is $\bar{y}(x, s) = c_1 \cosh(sx/c) + c_2 \sinh(sx/c) + \dfrac{\mu x(l - x)}{s} - \dfrac{2c^2\mu}{s^3}$

Applying the conditions Eqn. (ii), we get

$$c_1 = 2c^2\mu/s^2 \quad \text{and} \quad c_2 = \frac{2c^2\mu}{s^3}\left[\frac{1 - \cosh(sl/c)}{\sinh(sl/c)}\right] - \frac{2c^2\mu}{s^3}\tanh(s/2c)$$

Thus $$\bar{y}(x, s) = \frac{2c^2\mu}{s^3}\left[\frac{\cosh\{s(2x - l)/2c\}}{\cosh(sl/2c)}\right] + \frac{\mu x(l - x)}{s} - \frac{2c^2\mu}{s^3}$$

Now using the inversion formula (3) § 22.10, we get

$y(x, t)$ = sum of the residues of

$$2c^2\mu\left[e^{st}\frac{\cosh\{s(2x - l)/2c\}}{s^3\cosh(sl/2c)}\right] \text{ at all the poles } + \mu x(l - x) - c^2\mu t^2$$

Proceeding exactly as in Example 22.23, we have,

sum of the residues of $2c^2\mu\left[\dfrac{e^{st}\cosh\{s(2x - l)/2c\}}{s^3\cosh sl/2c}\right]$ at all the poles

$$= c^2\mu\left[t^2 + \left(\frac{2x - l}{2c}\right)^2 - \left(\frac{l}{2c}\right)^2\right]$$

$$- \frac{32c^2\mu}{\pi^3}\left(\frac{l}{2c}\right)^2 \sum_{n=1}^{\infty}\left[\frac{(-1)^n}{(2n - 1)^3}\cos\left\{\frac{(2n - 1)\pi(2x - l)}{2l}\right\}\cos\left\{\frac{(2n - 1)\pi ct}{l}\right\}\right]$$

$$= c^2\mu t^2 - \mu x(l - x) + \frac{8\mu l^2}{\pi^3}\sum_{n=1}^{\infty}\left[\frac{1}{(2n - 1)^3}\sin\frac{(2n - 1)\pi x}{l}\cos\frac{(2n - 1)\pi ct}{l}\right]$$

Hence $$y(x, t) = \frac{8\mu l^2}{\pi^3}\sum_{n=1}^{\infty}\left[\frac{1}{(2n - 1)^3}\sin\frac{(2n - 1)\pi x}{l}\cos\frac{(2n - 1)\pi ct}{l}\right].$$

## Transmission lines

**Example 22.30.** *A semi-infinite transmission line of negligible inductance and leakance per unit length has its voltage and current equal to zero. A constant voltage $v_0$ is applied at the sending end (x = 0) at t = 0. Find the voltage and current at any point (x > 0) and at any instant.*

**Solution.** Let $v(x, t)$ and $i(x, t)$ be the voltage and current at any point $x$ and at any time $t$. If $L = 0$ and $G = 0$, then the transmission line equations [(1) and (2) of § 18.10] become

$$\frac{\partial v}{\partial x} = -Ri, \frac{\partial i}{\partial x} = -C\frac{\partial v}{\partial t} \quad i.e., \quad \frac{\partial^2 v}{\partial x^2} = RC\frac{\partial v}{\partial t} \qquad ...(i)$$

The boundary conditions are $v(0, t) = v_0$ and $i(x, t)$ is finite for all $x$ and $t$.

The initial conditions are $v(x, 0) = 0, i(x, 0) = 0$. ...(ii)

Laplace transforms of Eqn. (i), are

$$\frac{d^2\bar{v}}{dx^2} = RC(s\bar{v} - 0) \quad \text{or} \quad \frac{d^2\bar{v}}{dx^2} - RCs\bar{v} = 0 \qquad ...(iii)$$

Laplace transforms of the conditions in Eqn. (ii), are

$$\bar{v}(0, s) = \frac{v_0}{s} \text{ at } x = 0 \qquad ...(iv)$$

and $\bar{v}(x, s)$ remains finite as $x \to \infty$ ...(v)

$\therefore$ the solution of Eqn. (iii) is

$$\bar{v}(x, s) = C_1 e^{\sqrt{RCs}\,x} + C_2 e^{-\sqrt{RCs}\,x}$$

To satisfy condition Eqn. (v), we must have $C_1 = 0$.

Using the condition Eqn. (iv), we get $C_2 = v_0/s$

Thus
$$\bar{v}(x, s) = \frac{v_0}{s} e^{-\sqrt{RCs}\,x}$$

Using the inversion formula, we obtain

$$v(x, t) = v_0 L^{-1}\left\{\frac{e^{-\sqrt{RC}\,x\,.\sqrt{s}}}{s}\right\} = v_0 \, erfc\left(x\frac{\sqrt{RC}}{2\sqrt{t}}\right) \qquad \text{[By Ex. 22.17]}$$

$$= v_0 \frac{x\sqrt{RC}}{2\sqrt{\pi}} \int_0^t u^{-3/2} e^{-(RCx^2/4u)}\, du$$

$\therefore$ since $i = -\dfrac{1}{R}\dfrac{\partial v}{\partial x}$, we obtain by differentiation,

$$i(x, t) = \frac{v_0 x}{2\sqrt{x}}\sqrt{\frac{C}{R}}\, t^{-3/2} e^{(-RCx^2/4t)}.$$

**Example 22.31.** *A transmission line of length l has negligible inductance and leakance. A constant voltage $v_0$ is applied at the sending end (x = 0) and is open circuited at the far end. Assuming the initial voltage and current to be zero, determine the voltage and current.*

**Solution.** For a transmission line with $L = G = 0$, the voltage $v$ and current $i$ are given by the equations

$$\frac{\partial^2 v}{\partial x^2} = RC\frac{\partial v}{\partial t} \text{ and } \frac{\partial v}{\partial x} + Ri = 0 \qquad ...(i)$$

The boundary conditions are (for $t > 0$)

$$v = v_0 \text{ at } x = 0 \text{ and } i = \frac{\partial v}{\partial x} = 0 \quad \text{at } x = l \qquad ...(ii)$$

The initial condition is $v = 0$ at $t = 0$ $(x > 0)$

Laplace transforms of Eqns. (i) and (ii) are

$$\frac{\partial^2\bar{v}}{\partial x^2} = RC(s\bar{v} - 0) \qquad ...(iii)$$

and
$$\bar{v} = v_0/s \text{ at } x = 0, \ \frac{\partial\bar{v}}{\partial x} = 0 \text{ at } x = l \qquad ...(iv)$$

∴ the solution of Eqn. (*iii*) is

$$\bar{v} = c_1 \cosh \sqrt{(RCs)\, x} + c_2 \sinh \sqrt{RCs\, x}$$

Applying conditions Eqn. (*iv*), it gives

$$v_0/s = c_1,\ 0 = c_1 \sinh \sqrt{(RCs)}\, l + c_2 \sinh \sqrt{(RCs)}\, l$$

$$\therefore \quad \bar{v} = \frac{v_0}{s}\left[\cosh \sqrt{(RCs)x} - \frac{\sinh \sqrt{(RCs)l}}{\cosh \sqrt{(RCs)l}} \sinh \sqrt{(RCs)x}\right]$$

$$= \frac{v_0}{s} \frac{\cosh pq\sqrt{s}}{\cosh p\sqrt{s}} \qquad \text{where } p = \sqrt{(RC)\,l} \text{ and } q = (l - x)/l$$

By the inversion formula (3) § 22.10, we get

$$v(x, t) = \text{sum of the residues of } (e^{st}\bar{v}) \text{ at all poles of } e^{st}\bar{v}. \qquad \text{...}(iv)$$

These poles are at $s = 0$ and $p\sqrt{s} = \pm i(2n - 1)\pi/2 = \pm ipk$ (say)

Now $\quad Res\,(e^{st}\bar{v})_{s=0} = \underset{s \to 0}{\text{Lt}} \dfrac{se^{st}\, v_0 \cosh pq\sqrt{s}}{s \cosh p\sqrt{s}} = v_0$

and

$$Res(e^{st}\bar{v})_{s=-k^2} = \underset{s \to -k^2}{\text{Lt}} \frac{(s + k^2)e^{st}\, v_0 \cosh pq\sqrt{s}}{s \cosh p\sqrt{s}} \qquad \left(\frac{0}{0}\text{ form}\right)$$

$$= \underset{s \to -k^2}{\text{Lt}} \frac{v_0 . e^{st} \cosh pq\sqrt{s} + (s + k^2)(...)}{\cosh p\sqrt{s} + s \sinh p\sqrt{s} \,.\, \frac{1}{2} ps^{-1/2}}$$

$$= \frac{v_0\, e^{-k^2 t} \cosh (ipqk) + 0}{0 + 1/2\,(ipk) \sinh (ipk)} = \frac{2v_0 e^{-k^2 t} \cos (pqk)}{-pk \sin pk}$$

Adding up all the residues, (*iv*) gives

$$v(x, t) = v_0 + \frac{4v_0}{\pi} \sum_{n=1}^{\infty} \frac{(-1)^n}{2n - 1} e^{-[(2n-1)^2\pi^2 t/4RCl^2]} \cos [(2n - 1)\, \pi(l - x)/2l]$$

$[\because \ pk = (2n - 1)\,\pi/2,\ -\sin pk = (-1)^n,\ pqk = \frac{1}{2}(2n - 1)\,\pi(l - x)/l,\ k^2 = (2n - 1)^2\pi^2/4RCl^2]$

Also $\quad i = -\dfrac{1}{R}\dfrac{\partial v}{\partial x}.$ [By Eqn. (*i*)]

**Example 22.32:** *Solve by Fourier transform,* $\dfrac{\delta^2 y}{dt^2} = C^2 \dfrac{\delta^2 y}{\delta x^2}$, *given* $\left(\dfrac{\delta y}{\delta t}\right)_{t=0} = 0$ *and* $y(x, 0) = f(x)$

*(Anna. 2018)*

**Solution:** Multiply both sides of the wave equation by $\dfrac{1}{\sqrt{2}\,\pi} e^{isx}$ and integrating w.r.t. $x$ from $-\infty$ to $\infty$

We have

$$\frac{1}{\sqrt{2}} \int_{-\infty}^{\infty} \frac{\delta^2 y}{\delta t^2} \cdot e^{isx}\, dx = \frac{1}{\sqrt{2x}} \cdot c^2 \int_{-\infty}^{\infty} \frac{\delta^2 y}{\delta x^2} e^{isx} dx$$

$$= \frac{\delta^2}{\delta t^2}\left[\frac{1}{\sqrt{2\pi}} \int_{-\infty}^{\infty} y \cdot e^{isx} dx\right] = c^2.[(is)^2 y]$$

$$\frac{\delta^2}{\delta t^2}[y] = -C^2 S^2\, y, \text{ where } y = y(s, t) = F[y, x, t]$$

we have assume that $y$ and $\dfrac{\delta y}{dx} \to 0$ as $x \to \infty$

So, $\quad \dfrac{\delta^2 y}{dt^2} + c^2 s^2 y = 0$

$\Rightarrow \qquad y = A \cos cst + B \sin cst \qquad ...(i)$

Taking Fourier Transform of the given initial

We get $\qquad \dfrac{\delta y}{dt} = 0$ and $y = \bar{f}(s)$, when $t = 0$

Using these condition in Eqn. $(i)$ $\quad A = \bar{f}(s)$ and $B = 0$

$$y = \bar{f}(s) \cos cst$$

Using Fourier Inversion Theorem

$$y(x, t) = \frac{1}{\sqrt{2}\,\pi}\int_{-\infty}^{\infty} \bar{f}(s)\cos cst\, c^{-isx} dx = \frac{1}{\sqrt{2}\,\pi}\int_{-\infty}^{\infty} \bar{f}(s)\cdot\left(\frac{e^{ist} + e^{-icst}}{2}\right)e^{-isx} dx$$

$$= \frac{1}{2\sqrt{2}\,\pi}\int_{-\infty}^{\infty}\left\{\bar{f}(s)\cdot e^{-is(x-ct)} + \bar{f}(s)e^{-is(x+ct)}\right\}dx$$

$$= \frac{1}{2}\{f(x - ct) + f(x + ct)\}$$

$$f(x) = \frac{1}{\sqrt{2\pi}}\int_{-\infty}^{\infty} f(x)e^{-is\,x} dx$$

## PROBLEMS 22.4

1. Solve the differential equation using Laplace transform method, $\dfrac{\partial y}{\partial t} = 3\dfrac{\partial^2 y}{\partial x^2}$
where $y(\pi/2, t) = 0$, $(\partial y/\partial x)_{x=0} = 0$ and $y(x, 0) = 30 \cos 5x$.

2. Using suitable transforms, solve the differential equation $\dfrac{\partial^2 V}{\partial x^2} = \dfrac{\partial V}{\partial t}$, $0 \le x \le \pi$, $t \ge 0$.
where $V(0, t) = 0 = V(\pi, t)$ and $V(x, 0) = V_0$ constant.

3. The initial temperature along the length of an infinite bar is given by $u(x, 0) = \begin{cases} 2, & |x| < 1 \\ 0, & |x| > 1 \end{cases}$. If the temperature $u(x, t)$ satisfies the equation $\dfrac{\partial u}{\partial t} = \dfrac{\partial^2 u}{\partial x^2}$, $-\infty < x < \infty$, $t > 0$, find the temperature at any point of the bar at any point $t$.

4. Solve the differential equation $\dfrac{\partial u}{\partial t} = \dfrac{\partial^2 u}{\partial t^2}$ subject to the conditions $\left(\dfrac{\partial u}{\partial t}\right)_{\text{at } x = 0} = 0$ and $u(x, 0) = \begin{cases} x, & a < x < 1 \\ 0, & x > 1 \end{cases}$
*(Delhi, 2012)*

5. Use the complex form of the Fourier transform to show that
$$V = \frac{1}{2\sqrt{(\pi t)}}\int_{-\infty}^{\infty} \bar{f}(u)e^{[-(x-u)^2/4t]} du$$
is the solution of the boundary value problem
$$\frac{\partial V}{\partial t} = \frac{\partial^2 V}{\partial x^2}, -\infty < x < \infty, t > 0\,;\ V = f(x) \text{ when } t = 0.$$

6. A semi-infinite solid ($x > 0$) is initially at temperature zero. At time $t = 0$, a constant temperature $\theta_0 > 0$ is applied and maintained at the face $x = 0$. Show that the temperature at any point $x$ and at any time $t$, is given by $\theta(x, t) = \theta_0 \, erfc(x/2c\sqrt{t})$.

7. A solid is initially at constant temperature $\theta_0$, while the ends $x = 0$ and $x = a$ are maintained at temperature zero. Determine the temperature at any point of the solid at any later time $t > 0$.

8. An infinite string is initially at rest along the $x$-axis. Its one end which is at $x = 0$, is given a periodic transverse displacement $a_0 \sin \omega t$, $t > 0$. Show that the displacement of any point of the string at any time is given by

$$y(x, t) = \begin{cases} a_0 \sin \omega(t - x/c), & t > x/c \\ 0, & t < x/c, \end{cases}$$

where $c$ is the wave velocity.

9. An infinite string has an initial transverse displacement $y(x, 0) = f(x), -\infty < x < \infty$, and is initially at rest. Show that

$$y(x, t) = \frac{1}{2} [f(x + ct) + f(x - ct)].$$

10. A semi-infinite transmission line has negligible inductance and leakance per unit length. A voltage $v$ is applied at the sending end ($x = 0$) which is given by

$$v(0, t) = \begin{cases} v_0, & 0 < t < \tau \\ 0, & t > \tau \end{cases}$$

Show that the voltage at any point $x > 0$ at any time $t > 0$ is given by

$$v(x, t) = v_0 \, erfc\left[\frac{x}{2}\sqrt{(RC/t)}\right].$$

## 22.12 OBJECTIVE TYPE OF QUESTIONS

### PROBLEMS 22.5

*Fill in the blanks or choose the correct answer in each of the following problems :*

1. Fourier cosine transform of $f(t)$ is .........
2. Fourier sine transform of $1/x$ is .........
3. Convolution theorem for Fourier transforms states that .........
4. If Fourier transform of $f(x)$ is $F(s)$, then the inversion formula is .........
5. $F[x^n f(x)] =$ .........
6. If $F\{f(x)\} = F(s)$, then $F\{f(x - a)\} =$ .........
7. Fourier sine integral representation of a function $f(x)$ is given by .........
8. If $F_c[f(ax)] = k\, F_c(s/a)$, then $k =$ .........
9. Fourier transform of second derivative of $u(x, t)$ is .........
10. If $f(x) = \begin{cases} 1, & 0 \le x \le \pi \\ 0, & x > \pi \end{cases}$, then Fourier sine integral of $f(x)$ is .........
11. Fourier sine transform of $f'(x)$ in the interval $(0, l)$ is .........
12. If $F(\lambda)$ is the Fourier transform of $f(x)$, then the Fourier transform of $f(ax)$ is .........
13. Inverse finite Fourier sine transform of $F_s(p) = \dfrac{1 - \cos p\pi}{(p\pi)^2}$ for $p = 1, 2, 3, \ldots$ and $0 < x < \pi$ is .........
14. If Fourier transform of $f(x) = F(s)$, then Fourier Transform of $f(2x)$ is .........
15. Fourier cosine transform of $e^{-x}$ is .........
16. $f(x) = 1, 0 < x < \infty$ cannot be represented by a Fourier integral. (True or False)
17. $\int_0^\infty |f(x)|^2 \, dx = \int_0^\infty |F_c(s)|^2 \, dx.$ (True or False)
18. Fourier transform is a linear operation. (True or False)
19. $F_s[x f(x)] = -\dfrac{d}{ds} F_c(s).$ (True or False)
20. Kernel of Fourier transform is $e^{sx}$. (True or False)
21. Finite Fourier cosine transform of $f(x) = 1$ in $(0, \pi)$ is zero. (True or False)
22. Fourier sine tansform of $\dfrac{1}{x} = \dfrac{\pi}{2}$. (True or False) ***(Delhi, 2012 S)***

CHAPTER 23

# Z-Transforms



## 23.1 INTRODUCTION

The development of communication branch is based on discrete analysis. Z-transform plays the same role in discrete analysis as Laplace transform in continuous systems. As such, Z-transform has many properties similar to those of the Laplace transform (§ 21.2). The main difference is that the Z-transform operates not on functions of continuous arguments but on sequences of the discrete integer-valued arguments, *i.e.* $n = 0, \pm 1, \pm 2$, ...... The analogy of Laplace transform to Z-transform can be carried further. For every operational rule of Laplace transforms, there is a corresponding operational rule of Z transforms and for every application of the Laplace transform, there is a corresponding application of Z-transform. A discrete system is expressible as a difference equation (§ 30.2) and its solutions are found using Z-transforms.

## 23.2 DEFINITION

*If the function $u_n$ is defined for discrete values ($n = 0, 1, 2, \ldots$) and $u_n = 0$ for $n < 0$, then its Z-transform is defined to be*

$$Z(u_n) = U(z) = \sum_{n=0}^{\infty} u_n z^{-n} \text{ whenever the infinite series converges.} \quad \ldots(i)$$

*The inverse Z-transform* is written as $Z^{-1}[U(z)] = u_n$.

If we insert a particular complex number $z$ into the power series Eqn. $(i)$, the resulting value of $Z(u_n)$ will be a complex number. Thus the Z-transform $U(z)$ is a complex valued function of a complex variable $z$.

## 23.3 SOME STANDARD Z-TRANSFORMS

*The direct application of the definition gives the following results :*

**(1)** $Z(a^n) = \dfrac{z}{z-a}$ **(2)** $Z(n^p) = -z \dfrac{d}{dz} Z(n^{p-1})$, *p being a +ve integer.*

*Proof.* **(1)** By definition, $Z(a^n) = \sum_{n=0}^{\infty} a^n z^{-n}$

$$= 1 + (a/z) + (a/z)^2 + (a/z)^3 + \ldots = \frac{1}{1-(a/z)} = \frac{z}{z-a}$$

**(2)** $$Z(n^p) = \sum_{n=0}^{\infty} n^p z^{-n} = z \sum_{n=0}^{\infty} n^{p-1} . n . z^{-(n+1)} \quad \ldots(i)$$

Changing $p$ to $p-1$, we get $Z(n^{p-1}) = \sum_{n=0}^{\infty} n^{p-1} . z^{-n}$

Differentiating it w.r.t. $z$,

$$\frac{d}{dz}[Z(n^{p-1})] = \sum_{n=0}^{\infty} n^{p-1} . (-n)\, z^{-(n+1)} \quad \ldots(ii)$$

Substituting Eqn. $(ii)$ in $(i)$, we obtain $Z(n^p) = -z\frac{d}{dz}[Z(n^{p-1})]$

which is the desired *recurrence formula.*

*In particular, we have the following formulae :*

**(3)** $Z(1) = \dfrac{z}{z-1}$ [Taking $a = 1$ in (1) **(4)** $Z(n) = \dfrac{z}{(z-1)^2}$ [Taking $p = 1$ in (2)]

**(5)** $Z(n^2) = \dfrac{z^2+z}{(z-1)^3}$ *(V.T.U., 2014)* **(6)** $Z(n^3) = \dfrac{z^3+4z^2+z}{(z-1)^4}$

**(7)** $Z(n^4) = \dfrac{z^4+11z^3+11z^2+z}{(z-1)^5}$.

## 23.4 LINEARITY PROPERTY

*If a, b, c be any constants and $u_n, v_n, w_n$ be any discrete functions, then*

$$Z(au_n + bv_n - cw_n) = aZ(u_n) + bZ(v_n) - cZ(w_n)$$

*Proof.* By definition, $Z(au_n + bv_n - cw_n) = \sum_{n=0}^{\infty} (au_n + bv_n - cw_n)z^{-n}$

$$= a\sum_{n=0}^{\infty} u_n z^{-n} + b\sum_{n=0}^{\infty} v_n z^{-n} - c\sum_{n=0}^{\infty} w_n z^{-n}$$

$$= aZ(u_n) + bZ(v_n) - cZ(w_n).$$

## 23.5 DAMPING RULE

*If $Z(u_n) = U(z)$, then $Z(a^{-n} u_n) = U(az)$*

*Proof.* By definition, $Z(a^{-n} u_n) = \sum_{n=0}^{\infty} a^{-n} u_n . z^{-n} = \sum_{n=0}^{\infty} u_n . (az)^{-n} = U(az)$.

**Cor.** $\mathbf{Z(a^n u_n) = U(z/a)}$

Obs. The geometric factor $a^{-n}$ when $|a| < 1$, damps the function $u_n$, hence the name *damping rule.*

## 23.6 SOME STANDARD RESULTS

*The application of the damping rule leads to the following standard results :*

**(1)** $Z(na^n) = \dfrac{az}{(z-a)^2}$ **(2)** $Z(n^2a^n) = \dfrac{az^2+a^2z}{(z-a)^3}$

(3) $Z(\cos n\theta) = \dfrac{z(z-\cos\theta)}{z^2 - 2z\cos\theta + 1}$ (4) $Z(\sin n\theta) = \dfrac{z\sin\theta}{z^2 - 2z\cos\theta + 1}$

(5) $Z(a^n \cos n\theta) = \dfrac{z(z - a\cos\theta)}{z^2 - 2az\cos\theta + a^2}$ (6) $Z(a^n \sin n\theta) = \dfrac{az\sin\theta}{z^2 - 2az\cos\theta + a^2}$.

*Proofs.* (1) We know that $Z(n) = \dfrac{z}{(z-1)^2}$. Applying damping rule, we have

$$Z(na^n) = U(a^{-1}z) = \frac{a^{-1}z}{(a^{-1}z-1)^2} = \frac{az}{(z-a)^2}.$$

(2) We know that $Z(n^2) = \dfrac{z^2+z}{(z-1)^3}$. Applying damping rule, we have

$$Z(n^2a^n) = U(a^{-1}z) = \frac{(a^{-1}z)^2 + a^{-1}z}{(a^{-1}z-1)^3} = \frac{a(z^2+az)}{(z-a)^3}.$$

(3) and (4) We know that $Z(1) = \dfrac{z}{z-1}$. Applying damping rule, we have

$$Z(e^{-in\theta}) = Z(e^{-i\theta})^n \cdot 1 = \frac{ze^{i\theta}}{ze^{i\theta}-1} = \frac{z}{z-e^{-i\theta}} = \frac{z(z-e^{i\theta})}{(z-e^{-i\theta})(z-e^{i\theta})}$$

$$= \frac{z(z-\cos\theta) - iz\sin\theta}{z^2 - z(e^{i\theta}+e^{-i\theta})+1} = \frac{z(z-\cos\theta) - iz\sin\theta}{z^2 - 2z\cos\theta + 1}$$

Equating real and imaginary parts, we get (3) and (4). *(V.T.U., 2013)*

(5) We know that $Z(\cos n\theta) = \dfrac{z(z-\cos\theta)}{z^2-2z\cos\theta+1}$. By damping rule, we have

$$Z(a^n\cos n\theta) = \frac{a^{-1}z(a^{-1}z-\cos\theta)}{(a^{-1}z)^2 - 2(a^{-1}z)\cos\theta + 1} = \frac{z(z-a\cos\theta)}{z^2-2az\cos\theta+a^2}$$

Similarly using (4) above, we get (6).

**Example 23.1.** *Find the Z-transform of the following :*
*(i)* $3n - 4\sin n\pi/4 + 5a$ *(ii)* $(n+1)^2$ *(V.T.U., 2010)*
*(iii)* $\sin(3n+5)$.

**Solution.** *(i)* $Z\left(3n - 4\sin\dfrac{n\pi}{4} + 5a\right) = 3Z(n) - 4Z\left(\sin\dfrac{n\pi}{4}\right) + 5a\,Z(1)$ [By Linearty property]

$$= 3\cdot\frac{z}{(z-1)^2} - 4\cdot\frac{z\sin n\pi/4}{z^2-2z\cos\pi/4+1} + 5a\cdot\frac{z}{z-1}$$ [Using formulae for $Z(1)$, $Z(n)$, $Z(\sin n\theta)$]

$$= \frac{(3-5a)z+5az^2}{(z-1)^2} - \frac{2\sqrt{2}z}{z^2-\sqrt{2}z+1}$$

*(ii)* $Z(n+1)^2 = Z(n^2+2n+1) = Z(n^2) + 2Z(n) + Z(1)$

$$= \frac{z^2+z}{(z-1)^3} + 2\frac{z}{(z-1)^2} + \frac{z}{z-1} = \frac{z^2(2z+1)}{(z-1)^3}$$

*(iii)* $Z[\sin(3n+5)] = Z(\sin 3n\cos 5 + \cos 3n\sin 5)$

$= \cos 5 . Z(\sin 3n) + \sin 5 . Z(\cos 3n)$ (using formulae for $Z(\sin n\theta)$, $Z(\cos n\theta)$

$$= \cos 5\cdot\frac{z\sin 3}{z^2-2z\cos 3+1} + \sin 5\cdot\frac{z(z-\cos 3)}{z^2-2z\cos 3+1} = z\cdot\frac{(z\sin 5 - \sin 2)}{z^2-2z\cos 3+1}.$$

**Example 23.2.** *Find the Z-transforms of the following*
*(i)* $e^{an}$ *(ii)* $ne^{an}$ *(iii)* $n^2e^{an}$.

**Solution.** *(i)* Let $u_n = 1$, $e^{an} = (e^{-a})^{-n} = k^{-n}$ where $k = e^{-a}$. By damping rule $Z(k^{-n}\, u_n) = U(kz)$,

$$\therefore \quad Z(e^{an}) = Z(k^{-n}\,.\,1) = U(kz) = \frac{kz}{kz-1} \qquad \left[\because \; U(z) = Z(1) = \frac{z}{z-1}\right]$$

$$= \frac{z}{z-1/k} = \frac{z}{z-e^a}$$

*(ii)* Let $u_n = n$, $e^{an} = (e^{-a})^{-n} = k^{-n}$ where $k = e^{-a}$

By damping rule, $Z(e^{an}\,.\,n) = Z(k^{-n}\,.\,n) = U(k\,z)$ where $U(z) = Z(n) = \dfrac{z}{(z-1)^2}$

$$\frac{kz}{(kz-1)^2} = \frac{z}{k(z-1/k)^2} = \frac{e^a z}{(z-e^a)^2}$$

*(iii)* Let $u_n = n^2$, $e^{an} = (e^{-a})^{-n} = k^{-n}$ where $k = e^{-a}$

By damping rule,

$$Z(e^{an}\,.\,n^2) = Z(k^{-n}\,.\,n^2) = U(kz) \qquad \text{where} \quad U(z) = Z(n^2) = \frac{z^2+z}{(z-1)^3}$$

$$= \frac{(kz)^2 + kz}{(kz-1)^3} = \frac{z(z+1/k)}{(z-1/(k)^3} = \frac{ze^a(z+e^a)}{(z-e^a)^3}.$$

**Example 23.3.** *Find the Z-transform of (i) cosh nθ.* (V.T.U., 2014) *(ii)* $a^n$ *cosh nθ.*

**Solution.** *(i)*

$$Z(\cosh n\theta) = Z\left(\frac{e^{n\theta}+e^{-n\theta}}{2}\right)$$

$$= \frac{1}{2}\left[Z\left\{(e^{-\theta})^{-n}\,.\,1\right\} + Z\left\{(e^{\theta})^{-n}\,.\,1\right\}\right]$$

Apply damping rule to both terms, taking $u_n = 1$.

$$Z(\cosh n\theta) = \frac{1}{2}\left[\frac{ze^{-\theta}}{ze^{-\theta}-1} + \frac{ze^{\theta}}{ze^{\theta}-1}\right] \qquad \left[\because \; z(1) = \frac{z}{z-1}\right]$$

$$= \frac{1}{2}\left\{\frac{2z^2 - z(e^{\theta}+e^{-\theta})}{z^2 - z(e^{\theta}+e^{-\theta})+1}\right\} = \frac{z^2 - z\cosh\theta}{z^2 - 2z\cosh\theta + 1}$$

*(ii)*

$$Z(a^n \cosh n\theta) = Z[(a^{-1})^{-n}\,.\,\cosh n\theta] \qquad \text{[Apply damping rule using (i)]}$$

$$= \frac{(a^{-1}z)^2 - (a^{-1}z)\cosh\theta}{(a^{-1}z)^2 - 2(a^{-1}z)\cosh\theta + 1} = \frac{z(z-a\cosh\theta)}{z^2 - 2az\cosh\theta + a^2}.$$

**Example 23.4.** *Find the Z-transforms of*
*(i)* $e^t \sin 2t$ *(ii)* $r^n \cos n\alpha$. $(n \geq 0)$ (Anna, 2015)

**Solution.** *(i)* We know that $Z(\sin 2t) = \dfrac{z\sin 2}{z^2 - 2z\cos 2 + 1}$ ...(A)

$$\therefore \quad Z(e^t \sin 2t) = Z[(e^{-1})^{-t}\,.\,\sin 2t] \qquad \text{[Apply damping rule, using (A)]}$$

$$= \frac{(e^{-1}z)\sin 2}{(e^{-1}z)^2 - 2(e^{-1}z)\cos 2 + 1} = \frac{ez\sin 2}{z^2 - 2ez\cos 2 + e^2}.$$

*(ii)* We know that $Z(\cos n\alpha) = \dfrac{z(z-\cos\alpha)}{z^2 - 2z\cos\alpha + 1}$ ...(B)

$$\therefore \quad Z(r^n \cos n\alpha) = Z[(r^{-1})^{-n}\,.\,\cos n\alpha] \qquad \text{[Apply damping rule, using (B)]}$$

$$= \frac{(r^{-1}z)[r^{-1}z - \cos\alpha]}{(r^{-1}z)^2 - 2(r^{-1}z)\cos\alpha + 1} = \frac{z(z - r\cos\alpha)}{z^2 - 2rz\cos\alpha + r^2}.$$

**Example 23.5.** *Find the Z-transforms of*

(*i*) $\cos\left(\frac{n\pi}{2}+\frac{\pi}{4}\right)$ (*Anna, 2012*) (*ii*) $\cosh\left(\frac{n\pi}{2}+\theta\right)$.

**Solution.** (*i*) $$Z\left[\cos\left(\frac{n\pi}{2}+\frac{\pi}{4}\right)\right]=Z\left(\cos\frac{n\pi}{2}\cos\frac{\pi}{4}-\sin\frac{n\pi}{2}\sin\frac{\pi}{4}\right)$$

$$=\cos\frac{\pi}{4}.Z\left(\cos\frac{n\pi}{2}\right)-\sin\frac{\pi}{2}.Z\left(\sin\frac{n\pi}{2}\right)$$ [Using formulae for $Z(\sin n\alpha)$ and $Z(\cos n\alpha)$]

$$=\frac{1}{\sqrt{2}}\left\{\frac{z(z-\cos\pi/2)}{z^2-2z\cos\pi/2+1}-\frac{z\sin\pi/2}{z^2-2z\cos\pi/2+1}\right\}=\frac{1}{\sqrt{2}}\left(\frac{z^2}{z^2+1}-\frac{z}{z^2+1}\right)=\frac{z(z-1)}{\sqrt{2}(z^2+1)}$$

(*ii*) $$Z\left[\cosh\left(\frac{n\pi}{2}+\theta\right)\right]=Z\left[\frac{e^{n\pi/2+\theta}+e^{-(n\pi/2+\theta)}}{2}\right]=\frac{1}{2}\left[e^{\theta}Z(e^{n\pi/2})+e^{-\theta}Z(e^{-n\pi/2})\right]$$

Since, $Z(a^n)=\frac{z}{z-a}$, $\therefore$ $Z(e^{n\pi/2})=Z(e^{\pi/2})^n=\frac{z}{z-e^{\pi/2}}$, $Z(e^{-n\pi/2})=\frac{z}{z-e^{-\pi/2}}$

Thus $$Z\left[\cosh\left(\frac{n\pi}{2}+\theta\right)\right]=\frac{1}{2}\left\{e^{\theta}.\frac{z}{z-e^{\pi/2}}+e^{-\theta}.\frac{z}{z-e^{-\pi/2}}\right\}$$

$$=\frac{z}{2}\left\{\frac{z(e^{\theta}+e^{-\theta})-[e^{(\pi/2-\theta)}+e^{-(\pi/2-\theta)}]}{z^2-z(e^{\pi/2}+e^{-\pi/2})+1}\right\}=\frac{z^2\cosh\theta-z\cosh\left(\frac{\pi}{2}-\theta\right)}{z^2-2z\cosh\left(\frac{\pi}{2}\right)+1}.$$

**Example 23.6.** *Find the Z-transform of*

(*i*) ${}^nC_p$ $(0\le p\le n)$ (*ii*) ${}^{n+p}C_p$.

**Solution.** (*i*) $$Z({}^nC_p)=\sum_{p=0}^{n}\left({}^nC_pz^{-p}\right)=1+{}^nC_1z^{-1}+{}^nC_2z^{-2}+\ldots+{}^nC_nz^{-n}=(1+z^{-1})^n$$

(*ii*) $$Z({}^{n+p}C_n)=\sum_{p=0}^{n}{}^{n+p}C_pz^{-p}$$

$$=1+{}^{n+1}C_1z^{-1}+{}^{n+2}C_2z^{-2}+{}^{n+3}C_3z^{-3}+\ldots\infty$$

$$=1+(n+1)z^{-1}+\frac{(n+2)(n+1)}{2!}z^{-2}+\frac{(n+3)(n+2)(n+1)}{3!}z^{-3}+\ldots\infty$$

$$=1+(-n-1)(-z^{-1})+\frac{(-n-1)(-n-2)}{2!}(-z^{-1})^2$$

$$+\frac{(-n-1)(-n-2)(-n-3)}{3!}(-z^{-1})^3+\ldots\infty$$

$$=(1-z^{-1})^{-n-1}.$$

**Example 23.7.** *Find the Z-transform of*

(*i*) *unit impulse sequence* $\delta(n)=\begin{cases}1 & n=0\\ 0 & n\neq 0\end{cases}$ (*ii*) *unit step sequence* $u(n)=\begin{cases}0 & n<0\\ 1 & n\geq 0\end{cases}$.

**Solution.** (*i*) $$Z[\delta(n)]=\sum_{n=0}^{\infty}\delta(n)z^{-n}=1+0+0+\ldots=1$$

(*ii*) $$Z[u(n)]=\sum_{n=0}^{\infty}u(n)z^{-n}=1+z^{-1}+z^{-2}+z^{-3}+\ldots=\frac{1}{1-z^{-1}}=\frac{z}{z-1}.$$

## 23.7 (1) SHIFTING $U_N$ TO THE RIGHT

If $Z(u_n) = U(z)$, *then* $Z(u_{n-k}) = z^{-k} U(z) \quad (k > 0)$

*Proof.* By definition,

$$Z(u_{n-k}) = \sum_{n=0}^{\infty} u_{n-k} z^{-n} = u_0 z^{-k} + u_1 z^{-(k+1)} + \ldots = z^{-k} \sum_{n=0}^{\infty} u_n z^{-n} = z^{-k} U(z)$$

**Obs.** This rule will be very useful in applications to difference equations.

(2) **Shifting $u_n$ to the left.** If $Z(u_n) = U(z)$, then

$$Z(u_{n+k}) = z^k[U(z) - u_0 - u_1 z^{-1} - u_2 z^{-2} - \ldots - u_{k-1} z^{-(k-1)}]$$

*Proof.* $Z(u_{n+k}) = \sum_{n=0}^{\infty} u_{n+k} z^{-n} = z^k \sum_{n=0}^{\infty} u_{n+k} z^{-(n+k)}$

$$= z^k \left[ \sum_{n=0}^{\infty} u_n z^{-n} - \sum_{n=0}^{k-1} u_n z^{-n} \right]$$

Hence $Z(u_{n+k}) = z^k[U(z) - u_0 - u_1 z^{-1} - u_2 z^{-2} - \ldots - u_{k-1} z^{-(k-1)}]$

*In particular, we have the following standard results :*

(1) $Z(u_{n+1}) = z[U(z) - u_0]$ ; (2) $Z(u_{n+2}) = z^2[U(z) - u_0 - u_1 z^{-1}]$

(3) $Z(u_{n+3}) = z^3[U(z) - u_0 - u_1 z^{-1} - u_2 z^{-2}]$.

**Example 23.8.** *Show that* $Z\left(\frac{1}{n!}\right) = e^{1/z}$.

*Hence evaluate* $Z[1/(n+1)!]$ *and* $Z[1/(n+2)!]$.

**Solution.** We have $Z\left(\frac{1}{n!}\right) = \sum_{n=0}^{\infty} \frac{1}{n!} z^{-n} = 1 + \frac{z^{-1}}{1!} + \frac{z^{-2}}{2!} + \frac{z^{-3}}{3!} + \ldots = e^{1/z}$.

Shifting $(1/n!)$ one unit to the left gives

$$Z\left[\frac{1}{(n+1)!}\right] = z\left[Z\left(\frac{1}{n!}\right) - 1\right] = z(e^{-1/z} - 1)$$

Similarly shifting $(1/n!)$ two units to the left gives

$$Z\left[\frac{1}{(n+2)!}\right] = z^2(e^{1/z} - 1 - z^{-1}).$$ *(Delhi, 2015)*

## 23.8 MULTIPLICATION BY n

If $Z(u_n) = u(z)$, then $Z(nu_n) = -z \dfrac{dU(z)}{dz}$

*Proof.* $Z(nu_n) = \sum_{n=0}^{\infty} n \, . \, u_n z^{-n} = -z \sum_{n=0}^{\infty} u_n (-n) z^{-n-1} = -z \sum_{n=0}^{\infty} u_n \frac{d}{dz}(z^{-n})$.

$$= -z \sum_{n=0}^{\infty} \frac{d}{dz}(u_n z^{-n}) = -z \frac{d}{dz}\left( \sum_{n=0}^{\infty} u_n z^{-n} \right) = -z \frac{d}{dz} U(z).$$

**Obs.** We have, $Z(n^2 u_n) = \left(-z \dfrac{d}{dz}\right)^2 u(z)$

In general, $Z(n^m u_n) = \left(-z \dfrac{d}{dz}\right)^m u(z)$.

**Example 23.9.** *Find the Z-transform of (i) $n \sin n\theta$ (ii) $n^2 e^{n\theta}$.*

**Solution.** (*i*) We know that $Z(nu_n) = -z \dfrac{dU(z)}{dz}$ and $Z(\sin n\theta) = \dfrac{z \sin \theta}{z^2 - 2z \cos \theta + 1}$

$$\therefore \quad Z(n \sin n\theta) = -z \frac{d}{dz}[Z(\sin n\theta)] = -z \frac{d}{dz}\left(\frac{z \sin \theta}{z^2 - 2z \cos \theta + 1}\right)$$

$$= -z \frac{\sin \theta - z^2 \sin \theta}{(z^2 - 2z \cos \theta + 1)^2} = \frac{z(z^2 - 1) \sin \theta}{(z^2 - 2z \cos \theta + 1)^2}$$

(*ii*) We know that $Z(e^{n\theta}) = \dfrac{z}{z - e^{\theta}}$

$$\therefore \quad Z(n^2 e^{n\theta}) = \left(-z\frac{d}{dz}\right)^2 (Ze^{n\theta}) = \left(-z\frac{d}{dz}\right)\left[-z\frac{d}{dz}\left(\frac{z}{z - e^{\theta}}\right)\right]$$

$$= \left(-z\frac{d}{dz}\right)\left\{-z\frac{(z - e^{\theta})(1) - z(1)}{(z - e^{\theta})^2}\right\} = -z\frac{d}{dz}\left\{\frac{ze^{\theta}}{(z - e^{\theta})^2}\right\}$$

$$= -ze^{\theta}\left\{\frac{(z - e^{\theta})^2(1) - z[2(z - e^{\theta})]}{(z - e^{\theta})^4}\right\} = -ze^{\theta}\frac{z - e^{\theta} - 2z}{(z - e^{\theta})^3} = \frac{z(z + e^{\theta})e^{\theta}}{(z - e^{\theta})^3}.$$

## 23.9 TWO BASIC THEOREMS

In applications, we often need the values of $u_n$ for $n = 0$ or as $n \to \infty$ without requiring complete knowledge of $u_n$. We can find this as the behaviour of $u_n$ for small values of $n$ is related to the behaviour of $U(z)$ as $z \to \infty$ and *vice-versa*. The precise relationship is given by the following *initial and final value theorems :*

**(1) Initial value theorem.** *If $Z(u_n) = U(z)$, then $u_0 = \underset{z \to \infty}{Lt} U(z)$* (*V.T.U., 2015*)

*Proof.* We know that $U(z) = Z(u_n) = u_0 + u_1 z^{-1} + u_2 z^{-2} + \ldots$

Taking limits as $z \to \infty$, we get $\underset{z \to \infty}{\text{Lt}} [U(z)] = u_0$, as required.

Similarly additional initial values can be found successively, giving :

$$u_1 = \underset{z \to \infty}{\text{Lt}} \{z[U(z) - u_0]\} \; ; \; u_2 = \underset{z \to \infty}{\text{Lt}} \{z^2[U(z) - u_0 - u_1 z^{-1}]\} \text{ and so on.}$$

**(2) Final value theorem.** *If $Z(u_n) = U(z)$, then*

$$\underset{n \to \infty}{Lt}(u_n) = \underset{z \to 1}{Lt}(z - 1) U(z)$$

*Proof.* By definition, $Z(u_{n+1} - u_n) = \sum_{n=0}^{\infty} (u_{n+1} - u_n) z^{-n}$

or $$Z(u_{n+1}) - Z(u_n) = \sum_{n=0}^{\infty} (u_{n+1} - u_n) z^{-n}$$

or $$z[U(z) - u_0] - U(z) = \sum_{n=0}^{\infty} (u_{n+1} - u_n) z^{-n}$$

or $$U(z)(z - 1) - u_0 z = \sum_{n=0}^{\infty} (u_{n+1} - u_n) z^{-n}$$

Taking limits of both sides as $z \to 1$, we get

$$\underset{z \to 1}{\text{Lt}} [(z - 1) U(z)] - u_0 = \sum_{n=0}^{\infty} (u_{n+1} - u_n) = \underset{n \to \infty}{\text{Lt}} [(u_1 - u_0) + (u_2 - u_1) + \ldots + (u_{n+1} - u_n)]$$

$$= \underset{n \to \infty}{\text{Lt}} [u_{n+1}] - u_0 = u_\infty - u_0$$

Hence $\quad u_\infty = \underset{z \to 1}{\text{Lt}} [(z-1) U(z)].$

**Example 23.10.** *If* $U(z) = \dfrac{2z^2 + 5z + 14}{(z-1)^4}$, *evaluate* $u_2$ *and* $u_3$.

**Solution.** Writing $\quad U(z) = \dfrac{1}{z^2} \cdot \dfrac{2 + 5z^{-1} + 14z^{-2}}{(1 - z^{-1})^4}$

By initial value theorem, $\quad u_0 = \underset{z \to \infty}{\text{Lt}} U(z) = 0$

Similarly, $\quad u_1 \;\underset{z \to \infty}{\text{Lt}} \{z [U(z) - u_0]\} = 0$

Now $\quad u_2 = \underset{z \to \infty}{\text{Lt}} \{z^2 [U(z) - u_0 - u_1 z^{-1}]\} = 2 - 0 - 0 = 2$

and $\quad u_3 = \underset{z \to \infty}{\text{Lt}} z^3[U(z) - u_0 - u_1 z^{-1} - u_2 z^{-2}] = \underset{z \to \infty}{\text{Lt}} z^3[U(z) - 0 - 0 - 2z^{-2}]$

$$= \underset{z \to \infty}{\text{Lt}} z^3 \left[\frac{2z^2 + 5z + 14}{(z-1)^4} - \frac{2}{z^2}\right] = \underset{z \to \infty}{\text{Lt}} z^3 \left\{\frac{13z^3 + 2z^2 + 8z - 2}{z^2 (z-1)^4}\right\} = 13.$$

## PROBLEMS 23.1

1. Find the Z-transforms of the following sequences :

   (*i*) $\dfrac{a^n}{n!}$ $(n \geq 0)$ (*Anna, 2012*) (*ii*) $\dfrac{1}{(n+1)!}$ (*iii*) $(\cos\theta + i \sin\theta)^n$.

2. Using the linearity property, find the Z-transforms of the following functions :

   (*i*) $2n + 5 \sin n\pi/4 - 3a^4$ (*J.N.T.U., 2015*) (*ii*) $\frac{1}{2}(n-1)(n+2)$

   (*iii*) $(n+1)(n+2)$ (*iv*) $(2n-1)^2$ (*V.T.U., 2011 S*)

3. Show that (*i*) $Z(\sinh n\theta) = \dfrac{z \sinh\theta}{z^2 - 2z\cosh\theta + 1}$ (*V.T.U., 2014*) (*ii*) $Z(a^n \sinh n\theta) = \dfrac{az\sinh\theta}{z^2 - 2az\cosh\theta + a^2}$.

4. Show that (*i*) $Z(e^{-an}\cos n\theta) = \dfrac{ze^a(ze^a - \cos\theta)}{z^2e^{2a} - 2ze^a\cos\theta + 1}$ ; (*ii*) $Z(e^{-an}\sin n\theta) = \dfrac{ze^a\sin\theta}{z^2e^{2a} - 2ze^a\cos\theta + 1}$

   Also evaluate $Z(e^{3n}\sin 2n)$.

5. Using $Z(n^2) = \dfrac{z^2 + z}{(z-1)^3}$, show that $Z(n+1)^2 = \dfrac{z^3 + z^2}{(z-1)^3}$.

6. Find the Z-transforms of (*i*) $\sin(n+1)\theta$ (*J.N.T.U., 2015*) (*ii*) $\cos\left(\dfrac{k\pi}{8} + \alpha\right)$. (*Hazaribag., 2015*)

7. Find the Z-transform of $\cos n\theta$ and hence find $Z(n\cos n\theta)$.

8. Find the Z-transform of (*i*) $\cos(n\pi/2)$. (*Anna, 2016 ; Calicut, 2012*) (*ii*) $a^n\cos(n\pi/2)$.

9. Find the Z-transforms of the following

   (*i*) $e^{-an}$ (*ii*) $e^{-2n}$ (*V.T.U., 2010 S*) (*iii*) $e^{-an}n^2$.

10. Show that (*i*) $Z[\delta(n+1)] = 1/z$ (*ii*) $(1/2)^n u(n) = \dfrac{2z}{2z-1}$.

11. Show that $Z({}^{n+p}C_p) = (1 - 1/z)^{-(p+1)}$. Using the damping rule, deduce that

    $$Z({}^{n+p}C_p\, a^n) = (1 + a/z)^{-(p+1)}.$$

12. If $Z(u_n) = \dfrac{z}{z-1} + \dfrac{z}{z^2+1}$, find the Z-transform of $u_{n+2}$. (*Andhra, 2015*)

13. If $U(z) = \dfrac{2z^2 + 3z + 12}{(z-1)^4}$, find the value of $u_2$ and $u_3$.

14. Given that $Z(u_n) = \dfrac{2z^2 + 3z + 4}{(z-3)^3}$, $|z| > 3$, show that $u_1 = 2$, $u_2 = 21$, $u_3 = 139$.

15. Show that (i) $Z\left(\dfrac{1}{n}\right) = z \log \dfrac{z}{z-1}$. (ii) $Z\left\{\dfrac{1}{n(n+1)}\right\}$. *(Anna, 2016 ; J.N.T.U., 2015)*

16. Using $Z(n) = \dfrac{z}{(z-1)^2}$, show that $Z(n \cos n\theta) = \dfrac{(z^3 + z)\cos\theta - 2z^2}{(z^2 - 2z\cos\theta + 1)^2}$.

## 23.10 SOME USEFUL Z-TRANSFORMS

| *Sr. No.* | *Sequence $u_n$ $(n \geq 0)$* | *Z-transform $U(z) = Z(u_n)$* |
|---|---|---|
| 1. | $k$ | $kz/(z-1)$ |
| 2. | $-k$ | $kz/(z+1)$ |
| 3. | $n$ | $z/(z-1)^2$ |
| 4. | $n^2$ | $(z^2+z)/(z-1)^3$ |
| 5. | $n^p$ | $-z\, d/dz\, [Z(n^{p-1})]$, $p$ + *ve* integer. |
| 6. | $\delta(n) = \begin{cases} 1 & n = 0 \\ 0 & n \neq 0 \end{cases}$ | 1 |
| 7. | $u(n) = \begin{Bmatrix} 0 & n < 0 \\ 1 & n \geq 0 \end{Bmatrix}$ | $z/(z-1)$ |
| 8. | $a^n$ | $z/(z-a)$ |
| 9. | $na^n$ | $az/(z-a)^2$ |
| 10. | $n^2a^n$ | $(az^2 + a^2z)/(z-a)^3$ |
| 11. | $\sin n\theta$ | $\dfrac{z \sin\theta}{z^2 - 2z\cos\theta + 1}$ |
| 12. | $\cos n\theta$ | $\dfrac{z(z - \cos\theta)}{z^2 - 2z\cos\theta + 1}$ |
| 13. | $a^n \sin n\theta$ | $\dfrac{az\sin\theta}{z^2 - 2az\cos\theta + a^2}$ |
| 14. | $a^n \cos n\theta$ | $\dfrac{z(z - a\cos\theta)}{z^2 - 2az\cos\theta + a^2}$ |
| 15. | $\sinh n\theta$ | $\dfrac{z\sinh\theta}{z^2 - 2z\cosh\theta + 1}$ |
| 16. | $\cosh n\theta$ | $\dfrac{z(z - \cosh\theta)}{z^2 - 2z\cosh\theta + 1}$ |
| 17. | $a^n \sinh n\theta$ | $\dfrac{az\sinh\theta}{z^2 - 2az\cosh\theta + a^2}$ |
| 18. | $a^n \cosh n\theta$ | $\dfrac{z(z - a\cosh\theta)}{z^2 - 2az\cosh\theta + a^2}$ |
| 19. | $a^n u_n$ | $U(z/a)$ |
| 20. | $u_{n+1}$ | $z[U(z) - u_0]$ |
| | $u_{n+2}$ | $z^2[U(z) - u_0 - u_1z^{-1}]$ |
| | $u_{n+3}$ | $z^3[U(z) - u_0 - u_1z^{-1} - u_2z^{-2}]$ |
| 21. | $u_{n-k}$ | $z^{-k}U(z)$ |
| 22. | $nu_n$ | $-z d/dz[U(z)]$ |
| 23. | $u_0$ | $\underset{z\to\infty}{\text{Lt}}\ U(z)$ |
| 24. | $\underset{n\to\infty}{\text{Lt}}\ (u_n)$ | $\underset{z\to 1}{\text{Lt}}\ [(z-1)\,U(z)]$ |

## 23.11 SOME USEFUL INVERSE Z-TRANSFORMS

| Sr. No. | $U(z)$ | Inverse Z-transform $u_n = z^{-1}[U(z)]$ |
|---|---|---|
| 1. | $\dfrac{1}{z-a}$ | $a^{n-1}$ |
| 2. | $\dfrac{1}{z+a}$ | $(-a)^{n-1}$ |
| 3. | $\dfrac{1}{(z-a)^2}$ | $(n-1)a^{n-2}$ |
| 4. | $\dfrac{1}{(z-a)^3}$ | $\dfrac{1}{2}(n-1)(n-2)\,a^{n-3}$ |
| 5. | $\dfrac{z}{z-a}$ | $a^n$ |
| 6. | $\dfrac{z}{z+a}$ | $(-a)^n$ |
| 7. | $\dfrac{z^2}{(z-a)^2}$ | $(n+1)a^n$ |
| 8. | $\dfrac{z^3}{(z-a)^3}$ | $\dfrac{1}{2!}(n+1)(n+2)\,a^n\,u(n)$ |

## 23.12 CONVOLUTION THEOREM

*If* $Z^{-1}[U(z)] = u_n$ *and* $Z^{-1}[V(z)] = v_n$*, then*

$$Z^{-1}[U(z)\,.\,V(z)] = \sum_{m=0}^{n} u_m\,.\,v_{n-m} = u_n * v_n$$

*where the symbol* $*$ *denotes the convolution operation.*

*Proof.* We have $U(z) = \sum_{n=0}^{\infty} u_n z^{-n}$, $V(z) = \sum_{n=0}^{\infty} v_n z^{-n}$

$$\therefore \quad U(z)\,V(z) = (u_0 + u_1 z^{-1} + u_2 z^{-2} + \ldots + u_n z^{-n} + \ldots \infty) \times (v_0 + v_1 z^{-1} + v_2 z^{-2} + \ldots + v_n z^{-n} + \ldots \infty)$$

$$= \sum_{n=0}^{\infty} (u_0 v_n + u_1 v_{n-1} + u_2 v_{n-2} + \ldots + u_n v_0) z^{-n} = Z(u_0 v_n + u_1 v_{n-1} + \ldots + u_n v_0)$$

whence follows the desired result.

**Obs.** The convolution theorem plays an important role in the solution of difference equations and in probability problems involving sums of two independent random variables.

**Example 23.11.** *Use convolution theorem to evaluate* $Z^{-1}\left\{\dfrac{z^2}{(z-a)(z-b)}\right\}$.

*(J.N.T.U., 2015 ; Anna, 2013)*

**Solution.** We know that $Z^{-1}\left\{\dfrac{z}{z-a}\right\} = a^n$ and $Z^{-1}\left\{\dfrac{z}{z-b}\right\} = b^n$

$$\therefore \quad Z^{-1}\left\{\frac{z^2}{(z-a)(z-b)}\right\} = Z^{-1}\left\{\frac{z}{z-a}\cdot\frac{z}{z-b}\right\} = a^n * b^n$$

$$= \sum_{m=0}^{n} a^m\,.\,b^{n-m} = b^n \sum_{m=0}^{n}\left(\frac{a}{b}\right)^m \text{ which is a G.P.}$$

$$= b^n\,.\,\frac{(a/b)^{n+1} - 1}{a/b - 1} = \frac{a^{n+1} - b^{n+1}}{a-b}.$$

## 23.13 CONVERGENCE OF Z-TRANSFORMS

Z-transform operation is performed on a sequence $u_n$ which may exist in the range of integers $-\infty < n < \infty$, and we write

$$U(z) = \sum_{n=-\infty}^{\infty} u_n z^{-n} \qquad \text{...(1)}$$

where $u_n$ represents a number in the sequence for $n$ = an integer. *The region of the z-plane in which (1) converges absolutely is known as the region of convergence (ROC) of U(z).*

We have so far discussed *one-sided Z-transform only for which* $n \geq 0$. Here the sequence is always *right-sided* and the *ROC* is always outside a prescribed circle say $|z| > |a|$ Fig. 23.1(*i*)]. For a left-handed sequence, the *ROC* is always inside any prescribed contour $|z| < |b|$. [Fig. 23.1(*ii*)].

## 23.14 TWO-SIDED Z-TRANSFORM OF $u_n$ IS DEFINED BY

$$U(z) = \sum_{n=-\infty}^{\infty} u_n z^{-n} \qquad \text{...(2)}$$

In this case, the sequence is two-sided and the region of convergence for (2) is the *annular region* $|b| < |z| < |c|$ [Fig. 23.1(*iii*)]. The inner circle bounds the terms in negative powers of $z$ and the outer circle bounds the terms in positive powers of $z$. *The shaded annulus of convergence is necessary for the two sided sequence and its Z-transform to exist.*

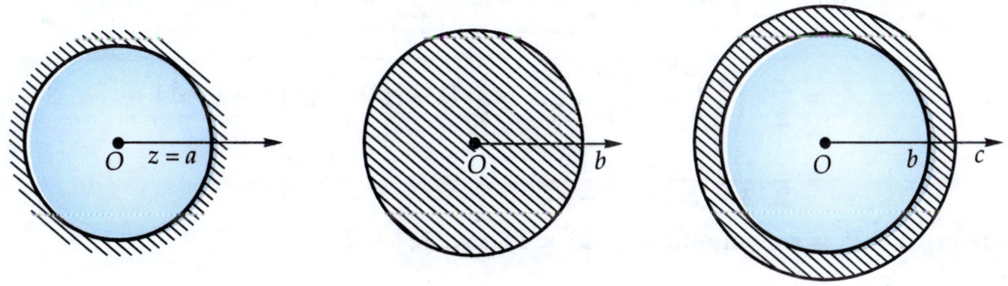


$|z| > |a|$ (*i*) $\qquad |z| < |b|$ (*ii*) $\qquad |b| < |z| < |c|$ (*iii*)

Fig. 23.1

**Example 23.12.** *Find the Z-transfrom and region of convergence of*

(*a*) $u(n) = \begin{cases} 4^n & \text{for } n < 0 \\ 2^n & \text{for } n \geq 0 \end{cases}$ $\qquad$ (*b*) $u(n) = {}^nc_k,\ n \geq k.$

**Solution.** By definition $\quad Z[u(n)] = \sum_{-\infty}^{\infty} u(n) Z^{-n} = \sum_{-\infty}^{-1} 4^n z^{-n} + \sum_{n=0}^{\infty} 2^n z^{-n}$

Putting $-n = m$ in the first series, we get

$$Z[u(n)] = \sum_{1}^{\infty} 4^{-m} z^m + \sum_{n=0}^{\infty} 2^n z^{-n}$$

$$\left\{\frac{z}{4} + \frac{z^2}{4^2} + \frac{z^3}{4^3} + \ldots\right\} + \left\{1 + \frac{2}{z} + \frac{2^2}{z^2} + \ldots\right\}$$

$$= \frac{z}{4}\left\{1 + \left(\frac{z}{4}\right) + \left(\frac{z}{4}\right)^2 + \ldots\right\} + \left\{1 + \left(\frac{2}{z}\right) + \left(\frac{2}{z}\right)^2 + \ldots\right\} \qquad \text{...}(i)$$

$$= \frac{z}{4} \cdot \frac{1}{1 - (z/4)} + \frac{1}{1 - (2/z)} = \frac{z}{4 - z} + \frac{z}{z - 2} = \frac{2z}{(4 - z)(z - 2)}$$

Now the two series in (*i*) being G.P. will be convergent if $|z/4| < 1$ and $|2/z| < 1$ *i.e.*, if $|z| < 4$ and $2 < |z|$. *i.e.* $2 < z < 4$.

Hence $Z[u(n)]$ is convergent if $z$ lies between the annulus as shown shaded in Fig. 23.2. Hence ROC is $2 < z < 4$.

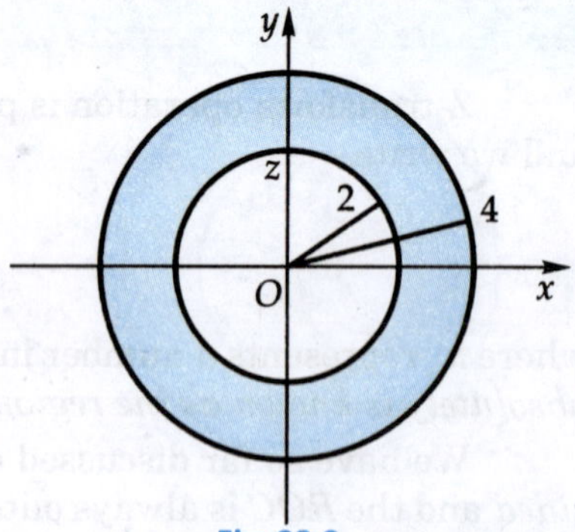


Fig. 23.2

(*b*) By definition, $Z[u(n)] = \sum_{-\infty}^{\infty} {}^nC_k z^{-n} = \sum_{n=k}^{\infty} {}^nC_k 2^n z^{-n}$

To find the sum of this series, we replace $n$ by $k + r$

$$\therefore \quad Z[u(n)] = \sum_{r=0}^{\infty} {}^{k+r}C_k z^{-(k+r)} = z^{-k} \sum_{r=0}^{\infty} {}^{k+r}C_r z^{-r} \qquad [\because \ {}^kC_r = {}^kC_{k-r}]$$

$$= z^{-k}[1 + {}^{k+1}C_1 z^{-1} + {}^{k+1}C_2 z^{-2} + ...]$$

$$= z^{-k}(1 - 1/z)^{-(k+1)}$$

This series is convergence for $|1/z| < 1$ *i.e.*, for $|z| > 1$.

Hence ROC is $|z| > 1$.

**Example 23.13.** *Find the Z-transform and the radius of convergence of*

(*a*) $f(n) = 2^n, n < 0$ (*b*) $f(n) = 5^n/n!, n \geq 0$.

**Solution.** (*a*) Assuming that $f(n) = 0$ for $n \geq 0$ we have

$$Z[f(n)] = \sum_{-\infty}^{\infty} f(n) z^{-n} = \sum_{n=-\infty}^{-1} 2^n z^{-n} = \sum_{m=1}^{\infty} 2^{-m} z^m \quad \text{where} \quad m = -n$$

$$= \frac{z}{2} + \left(\frac{z}{2}\right)^2 + \left(\frac{z}{2}\right)^3 + ...\infty = \frac{z}{2}\{1 + (z/2) + (z/2)^2 + ... \infty\}$$

$$= \frac{z}{2} \cdot \frac{1}{1-(z/2)} = \frac{z}{2-z}$$

This series being a G.P. is convergent if $|z/2| < 1$ *i.e.*, $|z| < 2$.

Hence ROC is $|z| < 2$.

(*b*) By definition, $Z[u(n)] = \sum_{n=0}^{\infty} \frac{5^n}{n!} \cdot z^{-n} = \sum_{0}^{\infty} \frac{(5/z)^n}{n!} = 1 + \left(\frac{5}{z}\right) + \frac{1}{2!}\left(\frac{5}{z}\right)^2 + \frac{1}{3!}\left(\frac{5}{z}\right)^3 + ... \infty$

$$= e^{5/z}$$

The above series is convergent for all values of $z$.

Hence ROC is the entire $z$-plane.

**PROBLEMS 23.2**

Find the Z-transform and its ROC in each of the following sequences :

1. $u(n) = 4^n, n \geq 0$.
2. $u(n) = 2^n, n < 0$.
3. $u(n) = 4^n$, for $n < 0$ and $= 3^n$ for $n \geq 0$.
4. $u(n) = n5^n, n \geq 0$.
5. $u(n) = 2^n/n, n > 1$.
6. $u(n) = 3^n/n!, n \geq 0$.
7. $u(n) = e^{an}, n \geq 0$.

## 23.15 EVALUATION OF INVERSE Z-TRANSFORMS

We can obtain the inverse Z-transforms using any of the following three methods :

I. **Power series method.** This is the simplest of all the methods of finding the inverse Z-transform. If $U(z)$ is expressed as the ratio of two polynomials which cannot be factorized, we simply divide the numerator by the denominator and take the inverse Z-transform of each term in the quotient.

**Example 23.14.** *Find the inverse Z-transform of log (z / z + 1) by power series method.*

**Solution.** Putting $z = 1/t$, $U(z) = \log\left(\frac{1/y}{1/y+1}\right) = -\log(1+y) = -y + \frac{1}{2}y^2 - \frac{1}{3}y^3 + \ldots$

$$= -z^{-1} + \frac{1}{2}z^{-2} - \frac{1}{3}z^{-3} + \ldots$$

Thus
$$u_n = \begin{cases} 0 & \text{for } n = 0 \\ (-1)^n/n & \text{otherwise} \end{cases}.$$

**Example 23.15.** *Find the inverse Z-transform of $z/(z+1)^2$ by division method.*

**Solution.**
$$U(z) = \frac{z}{z^2+2z+1} = z^{-1} - \frac{2+z^{-1}}{z^2+2z+1}, \text{ by actual division}$$

$$= z^{-1} - 2z^{-2} + \frac{3z^{-1}+2z^{-2}}{z^2+2z+1} = z^{-1} - 2z^{-2} + 3z^{-3} - \frac{4z^{-2}+3z^{-3}}{z^2+2z+1}$$

Continuing this process of division, we get an infinite series *i.e.*,

$$U(z) = \sum_{n=0}^{\infty} (-1)^{n-1}\, nz^{-n}$$

Thus $u_n = (-1)^{n-1} n.$

II. **Partial fractions method.** This method is similar to that of finding the inverse Laplace transforms using partial fractions. The method consists of decomposing $U(z)/z$ into partial fractions, multiplying the resulting expansion by $z$ and then inverting the same.

**Example 23.16.** *Find the inverse Z-transforms of*

(*i*) $\dfrac{2z^2+3z}{(z+2)(z-4)}$ (*V.T.U., 2016 ; Delhi., 2015*) (*ii*) $\dfrac{z^3-20z}{(z-2)^3(z-4)}$. (*Calicut, 2012 ; V.T.U., 2011*)

**Solution.** (*i*) We write $U(z) = \dfrac{2z^2+3z}{(z+2)(z-4)}$ as $\dfrac{U(z)}{z} = \dfrac{2z+3}{(z+2)(z-4)} = \dfrac{A}{z+2} + \dfrac{B}{z-4}$

where $A = 1/6$ and $B = 11/6$

$$\therefore \quad U(z) = \frac{1}{6}\frac{z}{z+2} + \frac{11}{6}\frac{z}{z-4}$$

On inversion, we have

$$u_n = \frac{1}{6}(-2)^n + \frac{11}{6}(4)^n \qquad \text{[Using § 23.10 (8)]}$$

(*ii*) We write
$$U(z) = \frac{z^3-20z}{(z-2)^3(z-4)}$$

as
$$\frac{U(z)}{z} = \frac{z^2-20}{(z-2)^3(z-4)} = \frac{A+Bz+Cz^2}{(z-2)^3} + \frac{D}{z-4}$$

Readily we get $D = 1/2$. Multiplying throughout by $(z-2)^3(z-4)$, we get

$$z^2 - 20 = (A + Bz + Cz^2)(z-4) + D(z-2)^3.$$

Putting $z = 0, 1, -1$ successively and solving the resulting simultaneous equations, we get $A = 6$, $B = 0$, $C = 1/2$.

Thus
$$U(z) = \frac{1}{2}\cdot\frac{12z+z^3}{(z-2)^3} - \frac{z}{z-4} = \frac{1}{2}\frac{z(z-2)^2+4z^2+8z}{(z-2)^3} - \frac{z}{z-4}$$

$$= \frac{1}{2}\left\{\frac{z}{z-2} + 2\frac{2z^2+4z}{(z-2)^3}\right\} - \frac{z}{z-4}$$

On inversion, we get $u_n = \frac{1}{2}(2^n + 2 \,.\, n^2 2^n) - 4^n$ [Using § 23.10 (8) & (10)]

$= 2^{n-1} + n^2 2^n - 4^n.$

**Example 23.17.** *Find the inverse Z-transform of*

*$2(z^2 - 5z + 6.5)/[(z-2)(z-3)^2]$, for $2 < |z| < 3$.*

**Solution.** Splitting into partial fractions, we obtain

$$U(z) = \frac{2(z^2 - 5z + 6.5)}{(z-2)(z-3)^2} = \frac{A}{z-2} + \frac{B}{z-3} + \frac{C}{(z-3)^2} \quad \text{where } A = B = C = 1$$

$$\therefore \quad U(z) = \frac{1}{z-2} + \frac{1}{z-3} + \frac{1}{(z-3)^2}$$

$$= \frac{1}{2}\left(1 - \frac{2}{z}\right)^{-1} - \frac{1}{3}\left(1 - \frac{z}{3}\right)^{-1} + \frac{1}{9}\left(1 - \frac{z}{3}\right)^{-2} \quad \text{so that } 2/z < 1 \text{ and } z/3 < 1$$

$$= \frac{1}{z}\left(1 + \frac{2}{z} + \frac{4}{z^2} + \frac{8}{z^3} + \ldots\right) - \frac{1}{3}\left(1 + \frac{z}{3} + \frac{z^2}{9} + \frac{z^3}{27} + \ldots\right) + \frac{1}{9}\left(1 + \frac{2z}{3} + \frac{3z^2}{9} + \frac{4z^3}{27} + \ldots\right)$$

where $2 < |z| < 3$.

$$= \left(\frac{1}{2} + \frac{2}{z^2} + \frac{2^2}{z^3} + \frac{2^3}{z^4} + \ldots\right) - \left(\frac{1}{3} + \frac{z}{3^2} + \frac{z^2}{3^3} + \frac{z^3}{3^4} + \ldots\right) + \left(\frac{1}{3^2} + \frac{2z}{3^3} + \frac{3z^2}{3^4} + \frac{4z^3}{3^5} + \ldots\right)$$

$$= \sum_{n=1}^{\infty} 2^{n-1} z^{-n} - \sum_{n=0}^{\infty} \left(\frac{1}{3}\right)^{n+1} z^n + \sum_{n=0}^{\infty} (n+1)\left(\frac{1}{3}\right)^{n+2} z^n$$

On inversion, we get $u_n = 2^{n-1}, n \geq 1$ and $u_n = -(n+2)3^{n-2}, n \leq 0$.

*III.* **Inversion integral method.** *The inverse Z-transform of U(z) is given by the formula*

$$u_n = \frac{1}{2\pi i}\int_c U(z) z^{n-1}\, dz$$

*= sum of residues of $U(z)\, z^{n-1}$ at the poles of U(z) which are inside the contour C drawn according to the ROC given.*

The following examples will illustrate the application of this formula :

**Example 23.18.** *Using the inversion integral method, find the inverse Z-transform of*

$$\frac{z}{(z-1)(z-2)}$$ *(V.T.U., 2010 S)*

**Solution.** Let $U(z) = \frac{z}{(z-1)(z-2)}$. Its poles are at $z = 1$ and $z = 2$.

Using $U(z)$ in the inversion integral, we have

$$u_n = \frac{1}{2\pi i}\int_C U(z) z^{n-1}\, dz,$$

where $C$ is a circle large enough to enclose both the poles of $U(z)$.

$= \text{sum of residues of } U(z)\, z^{n-1} \text{ at } z = 1 \text{ and } z = 2.$

Now $$Res\, [U(z)\, z^{n-1}]_{z=1} = \underset{z \to 1}{\text{Lt}} \left\{(z-1) \,.\, \frac{z^n}{(z-1)(z-2)}\right\} = -1$$

and $$Res\, [U(z)\, z^{n-1}]_{z=2} = \underset{z \to 2}{\text{Lt}} \left\{(z-2) \,.\, \frac{z^n}{(z-1)(z-2)}\right\} = 2^n$$

Thus the required inverse Z-transform $u_n = 2^n - 1, n = 0, 1, 2, \ldots$

**Example 23.19.** *Find the inverse Z-transform of $2z/[(z-1)(z^2+1)]$.*

**Solution.** Let $U(z) = \dfrac{2z}{(z-1)(z+i)(z-i)}$. It has three poles at $z = 1$, $z = \pm i$.

Using $U(z)$ in the inversion integral, we have

$$u_n = \frac{1}{2\pi i}\int_C U(z) \, . \, z^{n-1}\, dz \text{, where } C \text{ is a circle large enough to enclose the poles of } U(z).$$

$$= \text{sum of residues of } U(z) \, . \, z^{n-1} \text{ at } z = 1, z = \pm i.$$

Now $$Res\,[U(z)\, z^{n-1}]_{z=1} = \underset{z\to 1}{\text{Lt}} \left\{(z-1)\frac{2z^n}{(z-1)(z^2+1)}\right\} = 1$$

$$Res\,[U(z)\, z^{n-1}]_{z=i} = \underset{z\to i}{\text{Lt}} \left\{(z-i)\frac{2z^n}{(z-1)(z+i)(z-i)}\right\} = \frac{-(i)^n}{1+i}$$

$$Res\,[U(z)\, z^{n-1}]_{z=-i} = \underset{z\to -i}{\text{Lt}} \left\{(z+i)\frac{2z^n}{(z-1)(z+i)(z-i)}\right\} = \frac{(-i)^n}{i-1}$$

Hence $$u_n = 1 - \frac{(i)^n}{1+i} - \frac{(-i)^n}{1-i}.$$

## PROBLEMS 23.3

Using convolution theorem, evaluate the inverse Z-transforms of the following :

1. $\dfrac{z^2}{(z-1)(z-3)}$.
2. $\left(\dfrac{z}{z-a}\right)^2$ *(Anna, 2016)*
3. $\left(\dfrac{z}{z-1}\right)^3$.
4. Show that (*a*) $\dfrac{1}{n!} * \dfrac{1}{n!} = \dfrac{2^n}{n!}$ (*b*) $Z^{-1}\left(\dfrac{z^2}{(z+a)(z+b)}\right) = \dfrac{(-1)}{b-a}(b^{n+1} - a^{n+1})$.

Find the inverse Z-transforms of the following :

5. $\dfrac{4z}{z-a}$, $|z| > |a|$.
6. $\dfrac{5z}{(2-z)(3z-1)}$.
7. $\dfrac{z}{(z-1)^2}$.
8. $\dfrac{18z^2}{(2z-1)(4z+1)}$.
9. $\dfrac{2z}{(z-1)(z^2+1)}$. *(Delhi., 2015)*
10. $\dfrac{8z-z^3}{(4-z)^3}$.
11. $\dfrac{3z^2-18z+26}{(z-2)(z-3)(z-4)}$.
12. $\dfrac{4z^2-2z}{z^3-5z^2+8z-4}$. *(V.T.U., 2011 S)*
13. $\dfrac{z^3+3z}{(z-1)^2(z^2+1)}$.
14. $\dfrac{(1-e^{at})z}{(z-1)(z-e^{-at})}$.
15. Obtain $Z^{-1}\{1/[(z-2)(z-3)]\}$ for (*i*) $|z| < 2$ ; (*ii*) $2 < |z| < 3$; (*iii*) $|z| > 3$.
16. Evaluate $Z^{-1}\{(z-5)^{-3}\}$ for $|z| > 5$.

Using inversion integral, find the inverse Z-transform of the following functions :

17. $\dfrac{z+3}{(z+1)(z-2)}$.
18. $\dfrac{(2z-1)z}{2(z-1)(z+0.5)}$.
19. $\dfrac{1}{z(z-1)(z+0.5)}$.
20. $\dfrac{z^2+z}{(z-1)(z^2+1)}$.
21. $\dfrac{z}{z^2-2z+2}$. *(Anna., 2016)*
22. $\dfrac{2z(z^2-1)}{(z^2+1)^2}$.

## 23.16 (1) APPLICATION TO DIFFERENCE EQUATIONS

Just as the Laplace transforms method is quite effective for solving linear differential equations (§ 21.15), the *Z*-transforms are quite useful for solving linear difference equations.

The performance of discrete systems is expressed by suitable difference equations. Also *Z*-transform plays an important role in the analysis and representation of discrete-time systems. To determine the frequency response of such systems, the solution of difference equations is required for which *Z*-transform method proves useful.

**(2) Working procedure** *to solve a linear difference equation with constant coefficients by Z-transforms :*

1. *Take the Z-transform of both sides of the difference equations using the formulae of § 26.16 and the given conditions.*

2. *Transpose all terms without U(z) to the right.*

3. *Divide by the coefficient of U(z), getting U(z) as a function of z.*

4. *Express this function in terms of the Z-transforms of known functions and take the inverse Z-transform of both sides. This gives $u_n$ as a function of n which is the desired solution.*

**Example 23.20.** *Using the Z-transform, solve*

$$u_{n+2} + 4u_{n+1} + 3u_n = 3^n \text{ with } u_0 = 0,\ u_1 = 1.$$

*(Andhra., 2015 ; Delhi., 2015 ; J.N.T.U., 2015 ; V.T.U., 2012 S)*

**Solution.** If $\quad Z(u_n) = U(z)$, then $Z(u_{n+1}) = z[U(z) - u_0]$,

$$Z(u_{n+2}) = z^2[U(z) - u_0 - u_1 z^{-1}]$$

Also $\quad Z(2^n) = z/(z-2)$

∴ taking the *Z*-transforms of both sides, we get

$$z^2[U(z) - u_0 - u_1 z^{-1}] + 4z\,[U(z) - u_0] + 3U(z) = z/(z-3)$$

Using the given conditions, it reduces to

$$U(z)\,(z^2 + 4z + 3) = z + z/(z-3)$$

$$\therefore \quad \frac{U(z)}{z} = \frac{1}{(z+1)(z+3)} + \frac{1}{(z-3)(z+1)(z+3)} = \frac{3}{8}\frac{1}{z+1} + \frac{1}{24}\frac{1}{z-3} - \frac{5}{12}\frac{1}{z+3},$$

on breaking into partial fractions.

$$U(z) = \frac{3}{8}\frac{z}{z+1} + \frac{1}{24}\frac{z}{z-3} - \frac{5}{12}\frac{z}{z+3}$$

On inversion, we obtain

$$u_n = \frac{3}{8}Z^{-1}\left(\frac{z}{z+1}\right) + \frac{1}{24}Z^{-1}\left(\frac{z}{z-3}\right) - \frac{5}{12}Z^{-1}\left(\frac{z}{z+3}\right) = \frac{3}{8}(-1)^n + \frac{1}{24}3^n - \frac{5}{12}(-3)^n.$$

**Example 23.21.** *Solve $y_{n+2} + 6y_{n+1} + 9y_n = 2^n$ with $y_0 = y_1 = 0$, using Z-transforms.*

*(Hazaribag, 2015 ; V.T.U., 2014)*

**Solution.** If $Z(y_n) = Y(z)$, then $Z(y_{n+1}) = z\,[(Y(z) - y_0]$, $Z(y_{n+2}) = z^2[Y(z) - y_0 - y_1 z^{-1}]$

Also $\quad Z(2^n) = z/(z-2)$.

Taking *Z*-transforms of both sides, we get

$$z^2[Y(z) - y_0 - y_1 z^{-1}] + 6z[Y(z) - y_0] + 9Y(z) = z/(z-2)$$

Since $y_0 = 0$, and $y_1 = 0$, we have $Y(z)\,(z^2 + 6z + 9) = z/(z-2)$

or $$\frac{Y(z)}{z} = \frac{1}{(z-2)(z+3)^2} = \frac{1}{25}\left[\frac{1}{z-2} - \frac{1}{z+3} - \frac{5}{(z+3)^2}\right],$$ on splitting into partial fractions.

or $$Y(z) = \frac{1}{25}\left[\frac{z}{z-2} - \frac{z}{z+3} - 5\frac{z}{(z+3)^2}\right]$$

On taking inverse *Z*-transform of both sides, we obtain

$$y_n = \frac{1}{25}\left[Z^{-1}\left(\frac{z}{z-2}\right) - Z^{-1}\left(\frac{z}{z+3}\right) + \frac{5}{3}Z^{-1}\left(-\frac{3z}{(z+3)^2}\right)\right]$$

$$= \frac{1}{25}\{2^n - (-3)^n + \frac{5}{3}n(-3)^n\} \qquad \left[\because\ Z^{-1}\left\{\frac{az}{(z-a)^2}\right\} = na^n\right]$$

**Example 23.22.** *Find the response of the system $y_{n+2} - 5y_{n+1} + 6y_n = u_n$, with $y_0 = 0, y_1 = 1$ and $u_n = 1$ for $n = 0, 1, 2, 3, ....$ by Z-transform method.* (V.T.U., 2010)

**Solution.** Taking Z-transform of both sides of the given equation, we get

$$z^2(Y(z) - y_0 - y_1 z^{-1}) - 5z(Y(z) - y_0) + 6Y(z) = \frac{z}{z-1}$$

Substituting the values $y_0 = 0, y_1 = 1$, it reduces to

$$(z^2 - 5z + 6)\, Y(z) = \frac{z}{z-1} + z = \frac{z^2}{z-1}$$

or

$$\frac{Y(z)}{z} = \frac{z}{(z-1)(z-2)(z-3)}$$

$$= \frac{A}{z-1} + \frac{B}{z-2} + \frac{C}{z-3} \qquad \text{where } A = \frac{1}{2}, B = -2, C = \frac{3}{2}$$

so that

$$Y(z) = \frac{1}{2}\frac{z}{z-1} - 2\frac{z}{z-2} + \frac{3}{2}\frac{z}{z-3}$$

On inversion, we obtain $\quad y_n = \frac{1}{2} - 2(2)^n + \frac{3}{2}(3)^n$

**Obs.** The initial values given in the problem automatically appear in the generated sequence.

**Example 23.23.** *Solve the difference equation $y_n + \frac{1}{4}y_{n-1} = u_n + \frac{1}{3}u_{n-1}$ where $u_n$ is a unit step sequence.*

**Solution.** Taking Z-transform of both sides of the given equation, we get

$$Y(z) + \frac{1}{4}z^{-1}\, Y(z) = 1 + \frac{1}{3}z^{-1}$$

or

$$Y(z) = \left(1 + \frac{1}{3}z^{-1}\right)\Big/\left(1 + \frac{1}{4}z^{-1}\right) = \left(z + \frac{1}{3}\right)\Big/\left(z + \frac{1}{4}\right)$$

There being only one simple pole at $z = -1/4$, consider the contour $|z| > 1/4$.

$$\therefore \quad Res\,[Y(z)\, z^{n-1}]_{z=-1/4} = \underset{z \to -1/4}{\text{Lt}} \left\{\left(z + \frac{1}{4}\right) \cdot \left(z + \frac{1}{3}\right) z^{n-1} \Big/ \left(z + \frac{1}{4}\right)\right\}$$

$$= \underset{z \to -1/4}{\text{Lt}} \left(z + \frac{1}{3}\right) z^{n-1} = \left(-\frac{1}{4} + \frac{1}{3}\right)\left(-\frac{1}{4}\right)^{n-1} = \frac{1}{12} \cdot \left(-\frac{1}{4}\right)^{n-1}$$

Hence by inversion integral method, we have

$$y_n = \frac{1}{12}\left(-\frac{1}{4}\right)^{n-1}.$$

**Example 23.24.** *Using the Z-transform, solve $u_{n+2} - 2u_{n+1} + u_n = 3n + 5$.*

**Solution.** Given equation is $u_{n+2} - 2u_{n+1} + u_n = 3n + 5$.

Taking the Z-transforms of both sides, we get

$$z^2[U(z) - u_0 - u_1 z^{-1}] - 2z[U(z) - u_0] + U(z) = 3 \cdot \frac{z}{(z-1)^2} + 5 \cdot \frac{z}{z-1}$$

or

$$U(z)\,(z^2 - 2z + 1) = \frac{5z^2 - 2z}{(z-1)^2} + u_0(z^2 - 2z) + u_1 z$$

or

$$U(z) = \frac{5z^2 - 2z}{(z-1)^4} + u_0 \frac{z^2 - 2z}{(z-1)^2} + u_1 \frac{z}{(z-1)^2}$$

On inversion, we obtain

$$u_n = Z^{-1}\left\{\frac{5z^2 - 2z}{(z-1)^4}\right\} + u_0 Z^{-1}\left\{\frac{z^2 - 2z}{(z-1)^2}\right\} + u_1 Z^{-1}\left\{\frac{z}{(z-1)^2}\right\} \qquad ...(i)$$

Noting that $Z(1) = \dfrac{z}{z-1}, Z(n) = \dfrac{z}{(z-1)^2}$

$$Z(n^2) = \frac{z^2 + z}{(z-1)^3}, Z(n^3) = \frac{z^3 + 4z^2 + z}{(z-1)^4}$$

We write $\dfrac{5z^2 - 2z}{(z-1)^4} \equiv A\dfrac{z^3 + 4z^2 + z}{(z-1)^4} + B\dfrac{z^2 + z}{(z-1)^3} + C\dfrac{z}{(z-1)^2} + D\dfrac{z}{z-1}$

Equating coefficients of like powers of $z$, we find

$$A = \frac{1}{2}, B = 1, C = -\frac{3}{2}, D = 0$$

$$\therefore \qquad Z^{-1}\left\{\frac{5z^2 - 2z}{(z-1)^4}\right\} = \frac{1}{2}n^3 + n^2 - \frac{3}{2}n = \frac{1}{2}n(n-1)(n+3)$$

Also $\qquad Z^{-1}\left\{\dfrac{z^2 - 2z}{(z-1)^2}\right\} = Z^{-1}\left\{\dfrac{z}{z-1}\right\} - Z^{-1}\left\{\dfrac{z}{(z-1)^2}\right\} = 1 - n$

and $\qquad Z^{-1}\left\{\dfrac{z}{(z-1)^2}\right\} = n.$

Substituting these values in $(i)$ above, we get

$$u_n = \frac{1}{2}n(n-1)(n+3) + u_0(1-n) + u_1 n$$

$$= \frac{1}{2}n(n-1)(n+3) + c_0 + c_1 n. \qquad \text{where } c_0 = u_0, c_1 = u_1 - u_0$$

**Example 23.25.** *Using residue method, solve* $y_k + \frac{1}{9}y_{k-2} = \frac{1}{3^k}\cos\frac{k\pi}{2}, k \geq 0.$

**Solution.** Taking $Z$-transform of both sides of the given equation, we get

$$Z\left\{y_k + \frac{1}{9}y_{k-2}\right\} = Z\left\{\frac{1}{3^k}\cos\frac{k\pi}{2}\right\}$$

or $\qquad Y(z) + \dfrac{1}{9}z^{-2}\,Y(z) = \dfrac{z^2}{z^2 + 1/9} \qquad$ or $\qquad \left(1 + \dfrac{1}{9}z^{-2}\right)Y(z) = \dfrac{z^2}{z^2 + \frac{1}{9}}$

or $\qquad Y(z) = \dfrac{z^2}{\left(1 + \frac{1}{9}z^{-2}\right)\left(z^2 + \frac{1}{9}\right)} = \dfrac{z^4}{\left(z^2 + \frac{1}{9}\right)^2}$

There are two poles of second order at $z = i/3$ and $z = -i/3$.

$$\therefore \qquad \text{Residue at } (z = i/3) = \left[\frac{d}{dz}\left\{\left(\frac{z-i}{3}\right)^2 \frac{z^{k-1}\,z^4}{(z^2 + 1/9)^2}\right\}\right]$$

$$= \left[\frac{d}{dz}\left\{\frac{z^{k+3}}{(z + i/3)^2}\right\}\right]_{z=i/3} = \left[\frac{(z + i/3)^2\,(k+3)z^{k+2} - z^{(k+3)}\,.\,2(z + i/3)}{(z + i/3)^4}\right]_{z=i/3}$$

$$= \left[\frac{(z + i/3)\,(k+3)z^{k+2} - 2z^{k+3}}{(z + i/3)^3}\right]_{z=i/3} = \left(\frac{3}{2i}\right)^3\left[(2k+6)\left(\frac{i}{3}\right)^{k+3} - 2\left(\frac{i}{3}\right)^{k+3}\right]$$

$$= \frac{1}{8}(2k+4)\left(\frac{i}{3}\right)^k = \frac{1}{4}(k+2)\left(\frac{1}{3}\right)^k\left(\cos\frac{\pi}{2} + i\sin\frac{\pi}{2}\right)^k = \frac{1}{4}(k+2)\left(\frac{1}{3}\right)^k\left(\cos\frac{k\pi}{2} + i\sin\frac{k\pi}{2}\right) \quad ...(i)$$

Changing $i$ to $-i$ in Eqn. $(i)$, we have

$$\text{Residue at } (z = -i/3) = \frac{1}{4}(k+2)\left(\frac{1}{3}\right)^k\left(\cos\frac{k\pi}{2} - i\sin\frac{k\pi}{2}\right) \quad ...(ii)$$

Adding Eqns. $(i)$ and $(ii)$, we obtain $y_k = \frac{1}{2}(k+2)\left(\frac{1}{3}\right)^k \cos\frac{k\pi}{2}$.

## PROBLEMS 23.4

Solve the following difference equations using Z-transforms (1 – 8) :

1. $x(n+2) - 3x(n+1) + 2x(n) = 0$, given that $x(0) = 0$, $x(1) = 1$. *(Anna, 2015)*
2. $u(n+2) + 2u(n+1) + u(n) = n$, given that $u(0) = u(1) = 0$. *(V.T.U., 2016)*
3. $y_{n+2} - 4y_n = 0$ given that $y_0 = 0$, $y_1 = 2$.
4. $f(n) + 3f(n-1) - 4f(n-2) = 0$, $n \geq 2$, given that $f(0) = 3$, $f(1) = -2$.
5. $y_{(n+3)} - 3y(n+1) + 2y(n) = 0$, given that $y(0) = 4$, $y(1) = 0$ and $y(2) = 8$.
6. $y_{n+2} - 5y_{n+1} + 6y_n = 36$, given that $y(0) = y(1) = 0$.
7. $y_{n+2} - 6y_{n+1} + 9y_n = 3^n$.
8. $y_{n+2} - 5y_{n+1} + 6y_n = 5^n$. *(J.N.T.U., 2015)*
9. $y_{n+2} + y_n = 2$. *(Anna., 2016)*
10. $u_{x+2} + u_x = 5(2^x)$ given that $u_0 = 1$, $u_1 = 0$.
11. $y_{n+2} + 4y_{n+1} + 3y_n = 2^n$ with $y_0 = 0$, $y_1 = 1$.
12. $u_{n+2} - 2u_{n+1} + u_n = 2^n$ with $u_0 = 2$, $u_1 = 1$. *(V.T.U., 2015)*
13. $y_{n+2} - 6y_{n+1} + 8y_n = 2^n + 6n$.
14. $y_k + \frac{1}{25}y_{k-2} = \left(\frac{1}{5}\right)^k \cos\frac{k\pi}{2}$, $(k \geq 0)$.
15. Find the response of the system given by $y_n + 3y_{(n-1)} = u_n$ where $u_n$ is a unit step sequence and $y_{(-1)} = 1$.
16. Find the impulse response of a system described by $y_{(n+1)} + 2y_{(n)} = \delta_n$ ; $y_0 = 0$.

## 23.17 OBJECTIVE TYPE OF QUESTIONS

## PROBLEMS 23.5

Choose the correct answer or fill up the blanks in each of the following problems :

1. $Z(1) = ......$
2. If $u_n$ is defined for $n = 0, 1, 2, ......$ only, then $Z(u_n) = ....$
3. Z-transform of $n = ....$
4. $Z(na^n) = ......$
5. $Z(\sin n\theta) = ......$
6. Z-transform of $(1/n!)$ is *(Anna., 2016)*
7. $Z(n^2) = ......$
8. $Z(\cos n\theta + i\sin n\theta) = .....$ *(Anna., 2016)*
9. $Z^{-1}\left(\frac{1}{z-2}\right) = ....$
10. $Z^{-1}\left\{\frac{z}{(z+1)^2}\right\} = ......$
11. Initial value theorem on Z-transform states that .......
12. Z-transform is linear. (True or False)
13. If $Z(u_n) = u(z)$, then $\underset{n\to\infty}{\text{Lt}}(u_n) = \underset{z\to\infty}{\text{Lt}}(z-1)\,u(z)$. (True or False)
14. Z-transform of the sequence $\{2^k\}$, $k \geq 0$ is $z/(z-2)$. (True or False)
15. Z-transform of $\{a^k/k!\}$, $k \geq 0 = e^{a/z}$. (True or False)
16. Z-transform of $\{^nC_r\}$, $(0 \leq r \leq n)$ is $(1+z)^n$. (True or False)
17. Z-transform of unit impulse sequence $\delta(n) = \begin{cases} 1, & n < 0 \\ 0, & n \geq 0 \end{cases}$, is $z/z-1$. (True or False)
18. Z-transform of unit step sequence $u(n) = \begin{cases} 0, & n < 0, \\ 1, & n \geq 0, \end{cases}$ is 1. (True or False)

CHAPTER 24

# Empirical Laws and Curve-fitting



## 24.1 INTRODUCTION

In many branches of applied mathematics, it is required to express a given data, obtained from observations, in the form of a *Law* connecting the two variables involved. Such a *Law* inferred by some scheme, is known as *Empirical Law*. For example, it may be desired to obtain the law connecting the length and the temperature of a metal bar. At various temperatures, the length of the bar is measured. Then, by one of the methods explained below, a law is obtained that represents the relationship existing between temperature and length for the observed values. This relation can then be used to predict the length at an arbitrary temperature.

**(2) Scatter diagram.** To find a relationship between the set of paired observations $x$ and $y$ (say), we plot their corresponding values on the graph taking one of the variables along the $x$-axis and other along the $y$-axis *i.e.* $(x_1, y_1)$, $(x_2, y_2)$, $(x_n, y_n)$. The resulting diagram showing a collection of dots is called a *scatter diagram*. A smooth curve that approximates the above set of points is known as the *approximating curve*.

**(3) Curve fitting.** Several equations of different types can be obtained to express the given data approximately. But the problem is to find the equation of the curve of *'best fit'* which may be most suitable for predicting the unknown values. The process of finding such an equation of 'best fit' is known as *curve-fitting*.

If there are $n$ pairs of observed values then it is possible to fit the given data to an equation that contains $n$ arbitrary constants for we can solve $n$ simultaneous equations for $n$ unknowns. If it were desired to obtain an equation representing these data but having less than $n$ arbitrary constants, then we can have recourse to any of the four methods : *Graphical method, Method of Least squares, Method of Group averages* and *Method of Moments*. The graphical method fails to give the values of the unknowns uniquely and accurately while the other methods do. *The method of Least squares is, probably, the best to fit a unique curve to a given data*. It is widely used in applications and can be easily implemented on a computer.

## 24.2 GRAPHICAL METHOD

*When the curve representing the given data is a* **linear law** $y = mx + c$ ; *we proceed as follows :*

(*i*) *Plot the given points on the graph paper* taking a suitable *scale*.

(*ii*) *Draw the straight line of best fit* such that the points are evenly distributed about the line.

(*iii*) *Taking two suitable points* $(x_1, y_1)$ *and* $(x_2, y_2)$ *on the line, calculate m,* the slope of the line and $c$, its intercept on $y$-axis.

When the points representing the observed values do not approximate to a straight line, a smooth curve is drawn through them. From the shape of the graph, we try to infer the law of the curve and then reduce it to the form $y = mx + c$.

## 24.3 LAWS REDUCIBLE TO THE LINEAR LAW

We give below some of the laws in common use, indicating the way these can be reduced to the linear form by suitable substitutions :

**(1)** *When the law is* $y = mx^n + c$.

Taking $x^n = X$ and $y = Y$ the above law becomes $Y = mX + c$

**(2)** *When the law is* $y = ax^n$.

Taking logarithms of both sides, it becomes $\log_{10} y = \log_{10} a + n \log_{10} x$

Putting $\log_{10} x = X$ and $\log_{10} y = Y$, it reduces to the form $Y = nX + c$, where $c = \log_{10} a$.

**(3)** *When the law is* $y = ax^n + b \log x$.

Writing it as $\dfrac{y}{\log x} = a \dfrac{x^n}{\log x} + b$ and taking $x^n/\log x = X$ and $y/\log x = Y$,

the given law becomes, $Y = aX + b$.

**(4)** *When the law is* $y = ae^{bx}$

Taking logarithms, it becomes $\log_{10} y = (b \log_{10} e)\, x + \log_{10} a$

Putting $x = X$ and $\log_{10} y = Y$, it takes the form $Y = mX + c$ where $m = b \log_{10} e$ and $c = \log_{10} a$.

**(5)** *When the law is* $xy = ax + by$.

Dividing by $x$, we have $y = b \dfrac{y}{x} + a$.

Putting $y/x = X$ and $y = Y$, it reduces to the form $Y = bX + a$.

**Example 24.1.** *R is the resistance to maintain a train at speed V; find a law of the type* $R = a + bV^2$ *connecting R and V, using the following data :*

| *V (miles/hour)* | *10* | *20* | *30* | *40* | *50* |
|---|---|---|---|---|---|
| *R (lb/ton)* | *8* | *10* | *15* | *21* | *30* |

**Solution.** Given law is $R = a + bV^2$ ...(*i*)

Taking $V^2 = x$ and $R = y$, (*i*) becomes

$$y = a + bx \quad ...(ii)$$

which is a *linear law*.

Table for the values of $x$ and $y$ is as follows :

| $x$ | 100 | 400 | 900 | 1600 | 2500 |
|---|---|---|---|---|---|
| $y$ | 8 | 10 | 15 | 21 | 30 |

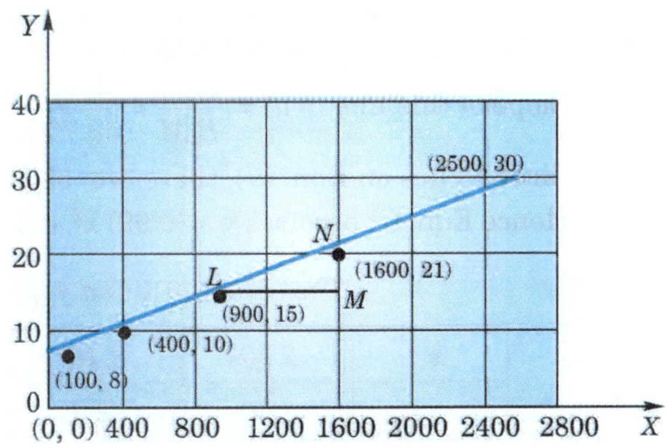


Fig. 24.1

Plot these points. Draw the straight line of best fit through these points (Fig. 24.1)

Slope of this line (= $b$)

$$= \frac{MN}{LM} = \frac{21 - 15}{1600 - 900} = \frac{6}{700} = 0.0085 \text{ nearly.}$$

Since $L$ (900, 15) lies on Eqn. (*ii*),

$\therefore$ $15 = a + 0.0085 \times 900$,

whence $a = 15 - 7.65 = 7.35$ nearly.

**Example 24.2.** *The following values of x and y are supposed to follow the law* $y = ax^2 + b \log_{10} x$. *Find graphically the most probable values of the constants a and b.*

| $x$ | *2.85* | *3.88* | *4.66* | *5.69* | *6.65* | *7.77* | *8.67* |
|---|---|---|---|---|---|---|---|
| $y$ | *16.7* | *26.4* | *35.1* | *47.5* | *60.6* | *77.5* | *93.4* |

**Solution.** Given law is $y = ax^2 + b \log_{10} x$

*i.e.* $$\frac{y}{\log_{10} x} = a\frac{x^2}{\log_{10} x} + b \qquad ...(i)$$

Taking $x^2/\log_{10} x = X$ and $y/\log_{10} x = Y$

Eqn. $(i)$ becomes $Y = aX + b \qquad ...(ii)$

This is a *linear law*. Table for the values of $X$ and $Y$ is as follows :

| $X = x^2/\log_{10} x$ | 17.93 | 25.56 | 32.49 | 42.87 | 53.75 | 67.80 | 80.83 |
|---|---|---|---|---|---|---|---|
| $Y = y/\log_{10} x$ | 35.59 | 44.83 | 52.50 | 62.90 | 73.65 | 87.04 | 99.56 |
| Points | $P_1$ | $P_2$ | $P_3$ | $P_4$ | $P_5$ | $P_6$ | $P_7$ |

Plot these points and draw the straight line of best fit through these points (Fig. 24.2).

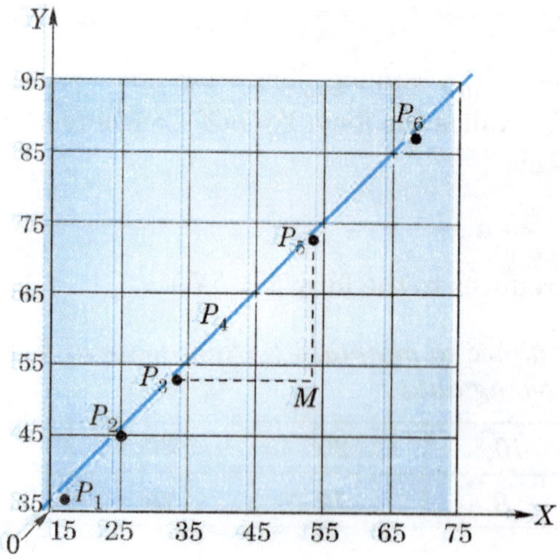


Fig. 24.2

Slope of this line $(= a) = \dfrac{MP_5}{P_3M} = \dfrac{73.65 - 52.50}{53.75 - 32.49} = \dfrac{21.15}{21.26} = 0.99$

Since $P_3$ lies on Eqn. $(ii)$, therefore, $52.50 = 0.99 \times 32.49 + b$ whence $b = 20.2$

Hence Eqn. $(i)$ becomes $y = (0.99)\, x^2 + (20.2) \log_{10} x$.

**Example 24.3.** *The values of x and y obtained in an experiment are as follows :*

| $x$ | 2.30 | 3.10 | 4.00 | 4.92 | 5.91 | 7.20 |
|---|---|---|---|---|---|---|
| $y$ | 33.0 | 39.1 | 50.3 | 67.2 | 85.6 | 125.0 |

*The probable law is $y = ae^{bx}$. Test graphically the accuracy of this law and if the law holds good, find the best values of the constants.*

**Solution.** Given law is $y = ae^{bx} \qquad ...(i)$

Taking logarithms to base 10, we have $\log_{10} y = \log_{10} a + (b \log_{10} e)\, x$

Putting $x = X$ and $\log_{10} y = Y$, it becomes $y = (b \log_{10} e)\, X + \log_{10} a \qquad ...(ii)$

Table for the values of $X$ and $Y$ is as under :

| $X = x$ | 2.30 | 3.10 | 4.00 | 4.92 | 5.91 | 7.20 |
|---|---|---|---|---|---|---|
| $Y = \log_{10} y$ | 1.52 | 1.59 | 1.70 | 1.83 | 1.93 | 2.1 |
| Points | $P_1$ | $P_2$ | $P_3$ | $P_4$ | $P_5$ | $P_6$ |

**Scale :** 1 small division along $x$–axis = 0.1

10 small divisions along $y$–axis = 0.1.

Plot these points and draw the line of best fit. As these points are lying almost along a straight line, the given law is nearly accurate (Fig. 24.3).

Now slope of this line ($= b \log_{10} e$)

$$= \frac{MN}{NM} = 0.12$$

whence $$b = \frac{0.12}{\log_{10} e} = 0.12 \times 2.303 = 0.276$$

Since the point $L$ (4, 1.71) lies on Eqn. (*ii*), therefore, $1.71 = 0.12 \times 4 + \log_{10} a$ whence $a = 17$ nearly.

Hence the curve of best fit is $y = 17\, e^{0.276x}$.

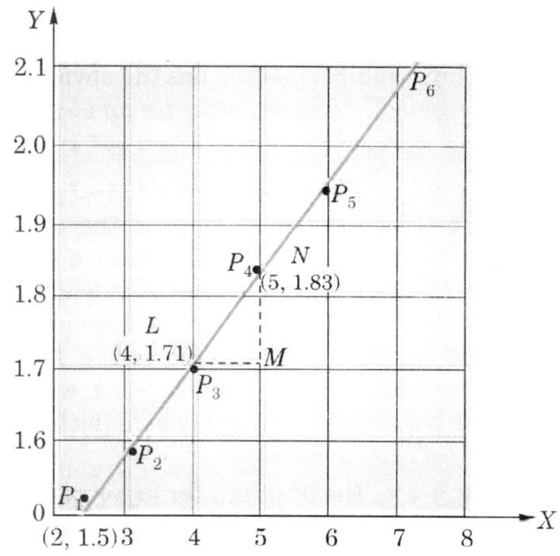


Fig. 24.3

## PROBLEMS 24.1

1. If $p$ is the pull required to lift the weight by means of a pulley block, find a linear law of the form $p = a + bw$, connecting $p$ and $w$, using the following data :

| $w$ (lb) : | 50 | 70 | 100 | 120 |
|---|---|---|---|---|
| $p$ (lb) : | 12 | 15 | 21 | 25 |

Compute $p$, when $w = 150$ lb.

2. The resistance $R$ of a carbon filament lamp was measured at various values of the voltage $V$ and the following observations were made :

| Voltage | V... | 62 | 70 | 78 | 84 | 92 |
|---|---|---|---|---|---|---|
| Resistance | R... | 73 | 70.7 | 69.2 | 67.8 | 66.3 |

Assuming a law of the form $R = \frac{a}{V} + b$, find by graphical method the best value of $a$ and $b$.

3. Verify if the values of $x$ and $y$, related as shown in the following table, obey the law $y = a + b\sqrt{x}$. If so, find graphically the values of $a$ and $b$.

| $x$ : | 500 | 1,000 | 2,000 | 4,000 | 6,000 |
|---|---|---|---|---|---|
| $y$ : | 0.20 | 0.33 | 0.38 | 0.45 | 0.51 |

4. The following values of $T$ and $l$ follow the law $T = al^n$. Test if this is so and find the best values of $a$ and $n$.

| $T = 1.0$ | 1.5 | 2.0 | 2.5 |
|---|---|---|---|
| $l = 25$ | 56.2 | 100 | 1.56 |

5. Find the best value of $a$ and $b$ if $y = ax + b \log_{10} x$ is the curve which represents most closely the observed values given below :

| $x$ : | 2 | 3 | 4 | 5 | 6 |
|---|---|---|---|---|---|
| $y$ : | 3.74 | 5.99 | 7.47 | 8.92 | 9.86 |

6. Fit the curve $y = ae^{bx}$ to the following data :

| $x$ : | 0 | 2 | 4 |
|---|---|---|---|
| $y$ : | 5.1 | 10 | 31.1 |

7. The following are the results of an experiment on friction of bearings. The speed being constant, corresponding values of the coefficient of friction and the temperature are shown in the table :

| $t$ : | 120 | 110 | 100 | 90 | 80 | 70 | 60 |
|---|---|---|---|---|---|---|---|
| $\mu$ : | 0.0051 | 0.0059 | 0.0071 | 0.0085 | 0.00102 | 0.00124 | 0.00148 |

If $\mu$ and $t$ are given by the law $\mu = ae^{bt}$, find the values of $a$ and $b$ by plotting the graph for $\mu$ and $t$.

## 24.4 PRINCIPLE OF LEAST SQUARES

The graphical method has the obvious drawback of being unable to give a unique curve of fit. *The principle of least squares, however, provides an elegant procedure for fitting a unique curve to a given data.*

Let the curve, $y = a + bx + cx^2 + ... + kx^{m-1}$ ...(1)

be fitted to the set of $n$ data points $(x_1, y_1), (x_2, y_2)...(x_n, y_n)$.

Now we have to determine the constants $a, b, c,...k$ such that it represents the curve of best fit. In case $n = m$, on substituting the values $(x_i, y_i)$ in (1), we get $n$ equations from which a unique set of $n$ constants can be found. But when $n > m$, we obtain $n$ equations which are more than the $m$ constants and hence cannot be solved for these constants. So we try to determine those values of $a, b, c,...k$ which satisfy all the equations as nearly as possible and thus may give the best fit. In such cases, we apply the *principle of least squares*.

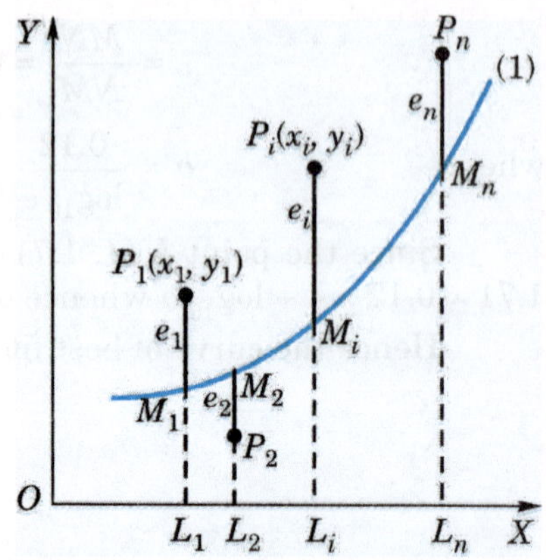

Fig. 24.4

At $x = x_i$, the *observed* (or *experimental*) *value* of the ordinate is $y_i = P_iL_i$ and the corresponding value on the fitting curve (1) is $a + bx_i + cx_i^2 + ... + kx_i^m = M_iL_i$ ($= \eta_i$, say) which is the *expected* (or *calculated*) *value* (Fig. 24.4). The difference of the observed and the expected values *i.e.* $y_i - \eta_i$ ($= e_i$) is called the *error* (or *residual*) at $x = x_i$. Clearly some of the errors $e_1, e_2, ... e_n$ will be positive and others negative. Thus to give equal weightage to each error, we square each of these and form their sum *i.e.* $E = e_1^2 + e_2^2 + ... e_n^2$.

*The curve of best fit is that for which e's are as small as possible i.e., E, the sum of the squares of the errors is a minimum.* This is known as the *principle of least squares* and was suggested by *Legendre* in 1806.

**Obs.** The principle of least squares does not help us to determine the form of the appropriate curve which can fit a given data. It only determines the best possible values of the constants in the equation when the form of the curve is known before hand. The selection of the curve is a matter of experience and practical considerations.

## 24.5 (1) METHOD OF LEAST SQUARES

For clarity, suppose it is required to fit the curve

$$y = a + bx + cx^2 \quad ...(1)$$

to a given set of observations $(x_1, y_1), (x_2, y_2), ..., (x_5, y_5)$. For any $x_i$, the observed value is $y_i$ and the expected value is $\eta_i = a + bx_i + cx_i^2$ so that the error $e_i = y_i - \eta_i$.

$\therefore$ the sum of the squares of these errors is

$$E = e_1^2 + e_2^2 + ... + e_5^2$$

$$= [y_1 - (a + bx_1 + cx_1^2)]^2 + [y_2 - (a + bx_2 + cx_2^2)]^2 + ... + [y_5 - (a + bx_5 + cx_5^2)]^2 \quad \text{[See § 5.12 (3)]}$$

For $E$ to be minimum, we have

$$\frac{\partial E}{\partial a} = 0 = 2[y_1 - (a + bx_1 + cx_1^2)] - 2[y_2 - (a + bx_2 + cx_2^2)] - ... - 2[y_5 - (a + bx_5 + cx_5^2)] \quad ...(2)$$

$$\frac{\partial E}{\partial b} = 0 = -2x_1[y_1 - (a + bx_1 + cx_1^2)] - 2x_2[y_2 - (a + bx_2 + cx_2^2)] - ... - 2x_5[y_5 - (a + bx_5 + cx_5^2)] \quad ...(3)$$

$$\frac{\partial E}{\partial c} = 0 = -2x_1^2[y_1 - (a + bx_1 + cx_1^2)] - 2x_2^2[y_2 - (a + bx_2 + cx_2^2)] - ... - 2x_5^2[y_5 - (a + bx_5 + cx_5^2)] \quad ...(4)$$

Equation (2) simplifies to

$$y_1 + y_2 + ... + y_5 = 5a + b(x_1 + x_2 + ... + x_5) + c(x_1^2 + x_2^2 + ... + x_5^2)$$

*i.e.,*
$$\Sigma y_i = 5a + b\Sigma x_i + c\,\Sigma x_i^2 \quad ...(5)$$

Equation (3) becomes

$$x_1y_1 + x_2y_2 + ... + x_5y_5 = a(x_1 + x_2 + ... + x_5) + b(x_1^2 + x_2^2 + ... + x_5^2) + c(x_1^3 + x_2^3 + ... + x_5^3)$$

*i.e.,* $$\Sigma x_i y_i = a\Sigma x_i + b\Sigma x_i^2 + c\Sigma x_i^3 \quad ...(6)$$

Similarly Eqn. (4) simplifies to $\Sigma x_i^2 y_i = a\Sigma x_i^2 + b\Sigma x_i^3 + c\Sigma x_i^4$ ...(7)

The Eqns. (5), (6) and (7) are known as *Normal equations* and can be solved as simultaneous equations in *a, b, c*. The values of these constants when substituted in Eqn. (1) give the desired curve of best fit.

**(2) Working procedure**

(*a*) **To fit the straight line $y = a + bx$**

(*i*) Substitute the observed set of *n* values in this equation.

(*ii*) Form normal equations for each constant

*i.e.,* $$\Sigma y = na + b\Sigma x, \ \Sigma xy = a\Sigma x + b\Sigma x^2$$

[The normal equation for the unknown *a* is obtained by multiplying the equations by the coefficient of *a* and adding. The normal equation for *b* is obtained by multiplying the equations by the coefficient of *b* (*i.e.*, *x*) and adding.]

(*iii*) Solve these normal equations as simultaneous equations for *a* and *b*.

(*iv*) Substitute the values of *a* and *b* in $y = a + bx$, which is the required line of best fit. (*Andhra., 2014*)

(*b*) **To fit the parabola : $y = a + bx + cx^2$**

(*i*) Form the normal equations $\Sigma y = na + b\Sigma x + c\Sigma x^2$

$$\Sigma xy = a\Sigma x + b\Sigma x^2 + c\Sigma x^3$$

and $$\Sigma x^2 y = a\Sigma x^2 + b\Sigma x^3 + c\Sigma x^4$$

[The normal equation for *c* has been obtained by multiplying the equations by the coefficient of *c* (*i.e.*, $x^2$) and adding.]

(*ii*) Solve these as simultaneous equations for *a, b, c*.

(*iii*) Substitute the values of *a, b, c* in $y = a + bx + cx^2$, which is the required parabola of best fit.

(*c*) In general, the curve $y = a + bx + cx^2 + ... + kx^{m-1}$ can be fitted to a given data by writing *m* normal equations.

**Example 24.4.** *If P is the pull required to lift a load W by means of a pulley block, find a linear law of the form P = mW + c connecting P and W, using the following data :*

| *P = 12* | *15* | *21* | *25* |
|---|---|---|---|
| *W = 50* | *70* | *100* | *120* |

*where P and W are taken in kg-wt. Compute P when W = 150 kg. wt.* (V.T.U., 2013)

**Solution.** The corresponding normal equations are

$$\left.\begin{aligned}\Sigma P &= 4c + m\Sigma W \\ \Sigma WP &= c\Sigma W + m\Sigma W^2\end{aligned}\right\} \quad ...(i)$$

The values of $\Sigma W$ etc. are calculated by means of the following table :

| W | P | W² | WP |
|---|---|---|---|
| 50 | 12 | 2500 | 600 |
| 70 | 15 | 4900 | 1050 |
| 100 | 21 | 10000 | 2100 |
| 120 | 25 | 14400 | 3000 |
| Total = 340 | 73 | 31800 | 6750 |

∴ The Eqn. (*i*) becomes $73 = 4c + 340m$ and $6750 = 340c + 31800m$

*i.e.,* $$2c + 170m = 365 \quad ...(ii)$$

and $$34c + 3180m = 675 \quad ...(iii)$$

Multiplying Eqn. (*ii*) by 17 and substracting from Eqn. (*iii*), we get

$$m = 0.1879 \quad \therefore \text{ from Eqn. } (ii), \quad c = 2.2785$$

Hence the line of best fit is

$$P = 2.2759 + 0.1879\,W$$

When $W = 150$ kg., $P = 2.2785 + 0.1879 \times 150 = 30.4635$ kg.

**Obs.** The calculations get simplified when the central values of $x$ is zero. It is therefore, advisable to make the central value zero, if it be not so. This is illustrated by the next example.

**Example 24.5.** *Fit a second degree parabola to the following data :*

| *x* | *0* | *1* | *2* | *3* | *4* |
|---|---|---|---|---|---|
| *y* | *1* | *1.8* | *1.3* | *2.5* | *6.3* |

**Solution.** Let $u = x - 2$ and $v = y$ so that the parabola of fit $y = a + bx + cx^2$ becomes

$$v = A + Bu + Cu^2 \qquad \text{...}(i)$$

The normal equations are

$$\Sigma v = 5A + B\Sigma u + C\Sigma u^2 \quad \text{or} \quad 12.9 = 5A + 10C$$

$$\Sigma uv = A\Sigma u + B\Sigma u^2 + C\Sigma u^2 \quad \text{or} \quad 11.3 = 10B$$

$$\Sigma u^2 v = A\Sigma u^2 + B\Sigma u^3 + C\Sigma u^4 \quad \text{or} \quad 33.5 = 10A + 34C$$

Solving these as simultaneous equations, we get

$$A = 1.48, \qquad B = 1.13, \qquad C = 0.55.$$

$\therefore$ Eqn. $(i)$ becomes, $\quad v = 1.48 + 1.13u + 0.55u^2$

or $\quad y = 1.48 + 1.13\,(x - 2) + 0.55\,(x - 2)^2$

Hence $y = 1.42 - 1.07x + 0.55x^2$.

**Example 24.6.** *Fit a second degree parabola to the following data :*

| *x = 1.0* | *1.5* | *2.0* | *2.5* | *3.0* | *3.5* | *4.0* |
|---|---|---|---|---|---|---|
| *y = 1.1* | *1.3* | *1.6* | *2.0* | *2.7* | *3.4* | *4.1* |

**Solution.** We shift the origin to (2.5, 0) and take 0.5 as the new unit. This amounts to changing the variable $x$ to $X$, by the relation $X = 2x - 5$.

Let the parabola of fit be $y = a + bX + cX^2$. The values of $\Sigma X$ etc., are calculated as below :

| $x$ | $X$ | $y$ | $Xy$ | $X^2$ | $X^2y$ | $X^3$ | $X^4$ |
|---|---|---|---|---|---|---|---|
| 1.0 | –3 | 1.1 | –3.3 | 9 | 9.9 | –27 | 81 |
| 1.5 | –2 | 1.3 | –2.6 | 4 | 5.2 | –8 | 16 |
| 2.0 | –1 | 1.6 | –1.6 | 1 | 1.6 | –1 | 1 |
| 2.5 | 0 | 2.0 | 0.0 | 0 | 0.0 | 0 | 0 |
| 3.0 | 1 | 2.7 | 2.7 | 1 | 2.7 | 1 | 1 |
| 3.5 | 2 | 3.4 | 6.8 | 4 | 13.6 | 8 | 16 |
| 4.0 | 3 | 4.1 | 12.3 | 9 | 36.9 | 27 | 81 |
| Total | 0 | 16.2 | 14.3 | 28 | 69.9 | 0 | 196 |

The normal equations are

$$7a + 28c = 16.2\,; \qquad 28b = 14.3\,; \qquad 28a + 196c = 69.9$$

Solving these as simultaneous equations, we get

$$a = 2.07,\ b = 0.511,\ c = 0.061$$

$\therefore \quad y = 2.07 + 0.511X + 0.061X^2$

Replacing $X$ by $2x - 5$ in the above equation, we get

$$y = 2.07 + 0.511\,(2x - 5) + 0.061\,(2x - 5)^2$$

which simplifies to $y = 1.04 - 0.198x + 0.244x^2$. This is the required parabola of best fit.

**Example 24.7.** *Fit a second degree parabola to the following data :*

| $x$ | 1989 | 1990 | 1991 | 1992 | 1993 | 1994 | 1995 | 1996 | 1997 |
|---|---|---|---|---|---|---|---|---|---|
| $y$ | 352 | 356 | 357 | 358 | 360 | 361 | 361 | 360 | 359 |

**Solution.** Taking $u = x - 1993$ and $v = y - 357$, the equation $y = a + bx + cx^2$ becomes

$$v = A + Bu + Cu^2 \quad \ldots(i)$$

| $x$ | $u = x - 1993$ | $y$ | $v = y - 357$ | $uv$ | $u^2$ | $u^2v$ | $u^2$ | $u^4$ |
|---|---|---|---|---|---|---|---|---|
| 1989 | – 4 | 352 | – 5 | 20 | 16 | – 80 | – 64 | 256 |
| 1990 | – 3 | 360 | – 1 | 3 | 9 | – 9 | – 27 | 81 |
| 1991 | – 2 | 357 | 0 | 0 | 4 | 0 | – 8 | 16 |
| 1992 | – 1 | 358 | 1 | – 1 | 1 | 1 | – 1 | 1 |
| 1993 | 0 | 360 | 3 | 0 | 0 | 0 | 0 | 0 |
| 1994 | 1 | 361 | 4 | 4 | 1 | 4 | 1 | 1 |
| 1995 | 2 | 361 | 4 | 8 | 4 | 16 | 8 | 16 |
| 1996 | 3 | 360 | 3 | 9 | 9 | 27 | 27 | 81 |
| 1997 | 4 | 359 | 2 | 8 | 16 | 32 | 64 | 256 |
| Total | $\Sigma u = 0$ | | $\Sigma v = 11$ | $\Sigma uv = 51$ | $\Sigma u^2 = 60$ | $\Sigma u^2v = -9$ | $\Sigma u^3 = 0$ | $\Sigma u^4 = 708$ |

The normal equations are

$$\Sigma v = 9A + B\Sigma u + C\Sigma u^2 \quad \text{or} \quad 11 = 9A + 60C$$

$$\Sigma uv = A\Sigma u + B\Sigma u^2 + C\Sigma u^3 \quad \text{or} \quad 51 = 60B \quad \text{or} \quad B = \frac{17}{20}$$

$$\Sigma u^2v = A\Sigma u^2 + B\Sigma u^3 + C\Sigma u^4 \quad \text{or} \quad -9 = 60A + 708C$$

On solving these equations, we get $A = \dfrac{694}{231}, B = \dfrac{17}{20}, C = -\dfrac{247}{924}$

$\therefore$ Eqn. ($i$) becomes $v = \dfrac{694}{231} + \dfrac{17}{20}u - \dfrac{247}{924}u^2$

or $$y - 357 = \frac{694}{231} + \frac{17}{20}(x - 1933) - \frac{247}{924}(x - 1933)^2$$

or $$y = \frac{694}{231} - \frac{32861}{20} - \frac{247}{924}(1933)^2 + \frac{17}{20}x + \frac{247 \times 3866}{924}x - \frac{247}{924}x^2$$

or $$y = 3 - 1643.05 - 998823.36 + 357 + 0.85x + 1033.44x - 0.267x^2$$

Hence $$y = -1000106.41 + 1034.29x - 0.267x^2.$$

## PROBLEMS 24.2

1. By the method of least squares, find the straight line that best fits the following data :

| $x$ : | 1 | 2 | 3 | 4 | 5 |
|---|---|---|---|---|---|
| $y$ : | 14 | 27 | 40 | 55 | 68 |

2. Fit a straight line to the following data :

| *Year x* : | 1961 | 1971 | 1981 | 1991 | 2001 | |
|---|---|---|---|---|---|---|
| *Production y* : | 8 | 10 | 12 | 10 | 16 | (in thousand tons) |

and find the expected production in 2006.

3. A simply supported beam carries a concentrated load $P$ (lb) at its mid-point. Corresponding to various values of $P$, the maximum deflection $Y$ (in) is measured. The data are given below :

| $P$: | 100 | 120 | 140 | 160 | 180 | 200 |
|---|---|---|---|---|---|---|
| $Y$: | 0.45 | 0.55 | 0.60 | 0.70 | 0.80 | 0.85 |

Find a law of the form $Y = a + bP$.

4. The results of measurement of electric resistance $R$ of a copper bar at various temperatures $t$°C are listed below :

| $t$: | 19 | 25 | 30 | 36 | 40 | 45 | 50 |
|---|---|---|---|---|---|---|---|
| $R$: | 76 | 77 | 79 | 80 | 82 | 83 | 85 |

If $R = a + bt$, find $a$ and $b$. *(Delhi, 2016)*

5. Find the best possible curve of the form $y = a + bx$, using method of least squares for the data :

| $x$: | 1 | 3 | 4 | 6 | 8 | 9 | 11 | 14 |
|---|---|---|---|---|---|---|---|---|
| $y$: | 1 | 2 | 4 | 4 | 5 | 7 | 8 | 9 |

*(V.T.U., 2011)*

6. Fit a straight line to the following data

| $x$: | 1 | 2 | 3 | 4 | 5 | 6 | 7 | 8 | 9 |
|---|---|---|---|---|---|---|---|---|---|
| $y$: | 9 | 8 | 10 | 12 | 11 | 13 | 14 | 16 | 5 |

7. Find the parabola of the form $y = a + bx + cx^2$ which fits most closely with the observations :

| $x$: | –3 | –2 | –1 | 0 | 1 | 2 | 3 |
|---|---|---|---|---|---|---|---|
| $y$: | 4.63 | 2.11 | 0.67 | 0.09 | 0.63 | 2.15 | 4.58 |

8. Fit a parabola $y = a + bx + cx^2$ to the following data :

| $x$: | 2 | 4 | 6 | 8 | 10 |
|---|---|---|---|---|---|
| $y$: | 3.07 | 12.85 | 31.47 | 57.38 | 91.29 |

9. Fit a second degree parabola to the following data :

| $x$: | 1 | 2 | 3 | 4 | 5 | 6 | 7 | 8 | 9 | 10 |
|---|---|---|---|---|---|---|---|---|---|---|
| $y$: | 124 | 129 | 140 | 159 | 228 | 289 | 315 | 302 | 263 | 210 |

10. The following table gives the results of the measurements of train resistances ; $V$ is the velocity in miles per hour. $R$ is the resistance in pounds per ton :

| $V$: | 20 | 40 | 60 | 80 | 100 | 120 |
|---|---|---|---|---|---|---|
| $R$: | 5.5 | 9.1 | 14.9 | 22.8 | 33.3 | 46.0 |

If $R$ is related to $V$ by the relation $R = a + bV + cV^2$, find $a$, $b$, and c. *(Delhi., 2015)*

11. The velocity $V$ of a liquid is known to vary with temperature according to a quadratic law $V = a + bT + cT^2$. Find the best values of $a$, $b$ and $c$ for the following table :

| $T$: | 1 | 2 | 3 | 4 | 5 | 6 | 7 |
|---|---|---|---|---|---|---|---|
| $V$: | 2.31 | 2.01 | 3.80 | 1.66 | 1.55 | 1.47 | 1.41 |

*(U.P.T.U., MCA, 2010)*

## 24.6 FITTING OF OTHER CURVES

**(1) $y = ax^b$**

Taking logarithms, $\log_{10} y = \log_{10} a + b \log_{10} x$

*i.e.*, $Y = A + bX$ where $X = \log_{10} x, Y = \log_{10} y$ and $A = \log_{10} a$. *(i)*

∴ The normal equations for (*i*) are : $\Sigma Y = nA + b\Sigma X$, $\Sigma XY = A\Sigma X + b\Sigma X^2$

from which $A$ and $b$ can be determined. Then $a$ can be calculated from $A = \log_{10} a$.

**(2) $y = ae^{bx}$** *(Exponential curve)*

Taking logarithms, $\log_{10} y = \log_{10} a + bx \log_{10} e$

*i.e.*, $Y = A + Bx$ where $Y = \log_{10} y$, $A = \log_{10} a$ and $B = b \log_{10} e$

Here the normal equations are : $\Sigma Y = nA + B\Sigma x$, $\Sigma xY = A\Sigma x + B\Sigma x^2$

from which $A$, $B$ can be found and consequently $a$, $b$ can be calculated.

**(3) $xy^n = b$ (or $pv^\gamma = k$)** *(Gas equation)*

Taking logarithms, $\log_{10} x + a \log_{10} y = \log_{10} b$ or $\log_{10} y = \frac{1}{a} \log_{10} b - \frac{1}{a} \log_{10} x$.

This is of the form $Y = A + BX$

where $X = \log_{10} x$, $Y = \log_{10} y$, $A = \frac{1}{a} \log_{10} b$, $B = -\frac{1}{a}$.

Here also the problem reduces to finding a straight line of best fit through the given data.

**Example 24.8.** *Find the least squares fit of the form $y = a_0 + a_1x^2$ to the following data :*

| $x$ | −1 | 0 | 1 | 2 |
|---|---|---|---|---|
| $y$ | 2 | 5 | 3 | 0 |

**Solution.** Putting $x^2 = X$, we have $y = a_0 + a_1X$

∴ the normal equations are : $\Sigma y = 4a + a_1\Sigma X$ ; $\Sigma Y = a_0\Sigma X + a_1\Sigma X^2$.

The values of $\Sigma X$, $\Sigma X^2$ etc. are calculated below :

| $x$ | $y$ | $X$ | $X^2$ | $XY$ |
|---|---|---|---|---|
| − 1 | 2 | 1 | 1 | 2 |
| 0 | 5 | 0 | 0 | 0 |
| 1 | 3 | 1 | 1 | 3 |
| 2 | 0 | 4 | 16 | 0 |
| | $\Sigma y = 10$ | $\Sigma X = 10$ | $\Sigma X^2 = 18$ | $\Sigma XY = 5$ |

∴ the normal equations become $10 = 4a_0 + 6a_1$ ; $5 = 6a_0 + 18a_1$

Solving these equations we get, $a_0 = 4.167$, $a_1 = -1.111$.

Hence the curve of best fit is $y = 4.167 - 1.111X$ *i.e.,* $y = 4.167 - 1.111x^2$.

**Example 24.9.** *An experiment gave the following values :*

| | | | | | |
|---|---|---|---|---|---|
| *v (ft/min)* | : | 350 | 400 | 500 | 600 |
| *t (min)* | : | 61 | 26 | 7 | 2.6 |

*It is known that v and t are connected by the relation $v = at^b$. Find the best possible values of a and b.* (V.T.U., 2013)

**Solution.** We have $\log_{10} v = \log_b a + b \log_{10} t$

or $y = A + 6X$, where $X = \log_{10} t$, $y = \log_{10} v$, $A = \log_{10} a$

∴ the normal equations are

$$\Sigma Y = 4A + b\Sigma X \quad ...(i)$$

$$\Sigma XY = A\Sigma X + b\Sigma X^2 \quad ...(ii)$$

Now $\Sigma X$ etc. are calculated as in the following table :

| $v$ | $t$ | $X = \log_{10} t$ | $y = \log_{10} v$ | $XY$ | $X^2$ |
|---|---|---|---|---|---|
| 350 | 61 | 1.7853 | 2.5441 | 4.542 | 3.187 |
| 400 | 26 | 1.4150 | 2.6021 | 3.682 | 2.002 |
| 500 | 7 | 0.8451 | 2.6990 | 2.281 | 0.714 |
| 600 | 2.6 | 0.4150 | 2.7782 | 1.153 | 0.172 |
| Total | | 4.4604 | 10.6234 | 11.658 | 6.075 |

∴ Equations (*i*) and (*ii*) become

$$4A + 4.46b = 10.623 \; ; \; 4.46A + 6.075b = 11.658$$

Solving these, $A = 2.845$, $b = -0.1697$

∴ $a = \text{antilog } A = \text{antilog } 2.845 = 699.8$.

**Example 24.10.** *Predict the mean radiation dose at an altitude of 3000 feet by fitting an exponential curve to the given data :*

| | | | | | | | | |
|---|---|---|---|---|---|---|---|---|
| *Altitude (x)* | : | 50 | 450 | 780 | 1200 | 4400 | 4800 | 5300 |
| *Dose of radiation (y)* | : | 28 | 30 | 32 | 36 | 51 | 58 | 69 |

**Solution.** Let $y = ab^x$ be the exponential curve.

Then $\log_{10} y = \log_{10} a + x \log_{10} b$

or $Y = A + Bx$ where $Y = \log_{10} y, A = \log_{10} a, B = \log_{10} b$

$\therefore$ the normal equations are

$$\Sigma Y = 7A + B\,\Sigma x \quad ...(i)$$

$$\Sigma x\,Y = A\,\Sigma x + B\,\Sigma x^2 \quad ...(ii)$$

Now $\Sigma x$ etc. are calculated as follows :

| $x$ | $y$ | $Y = \log_{10} y$ | $xY$ | $x^2$ |
|---|---|---|---|---|
| 50 | 28 | 1.447158 | 72.3579 | 2500 |
| 450 | 30 | 1.477121 | 664.7044 | 202500 |
| 780 | 32 | 1.505150 | 1174.0170 | 608400 |
| 1200 | 36 | 1.556303 | 1867.5636 | 1440000 |
| 4400 | 51 | 1.707570 | 7513.3080 | 19360000 |
| 4800 | 58 | 1.763428 | 8464.4544 | 23040000 |
| 5300 | 69 | 1.838849 | 9745.8997 | 28090000 |
| $\Sigma$ = 16980 | | 11.295579 | 29502.305 | 72743400 |

$\therefore$ equations (*i*) and (*ii*) become

$$11.295579 = 7A + 16980B$$

$$29502.305 = 16980A + 72743400B$$

Solving these equations, we get $A = 1.4521015$, $B = 0.0000666289$

$$\therefore \quad \log_{10} y = Y = 1.4521015 + 0.0000666289x$$

Hence $y$ (*at* $x = 3000$) $= 44.874$ *i.e.* 44.9 approx.

**Example 24.11.** *The pressure and volume of a gas are related by the equation* $pv^\gamma = k$, $\gamma$ *and* $k$ *being constants. Fit this equation to the following set of observations :*

| $p$ (kg/cm$^2$) : | 0.5 | 1.0 | 1.5 | 2.0 | 2.5 | 3.0 |
|---|---|---|---|---|---|---|
| $v$ (litres) : | 1.62 | 1.00 | 0.75 | 0.62 | 0.52 | 0.46 |

(V.T.U., 2011)

**Solution.** We have $\log_{10} p + \gamma \log_{10} v = \log_{10} k$

or $$\log_{10} v = \frac{1}{\gamma} \log_{10} k - \frac{1}{\gamma} \log_{10} p \quad \text{or} \quad Y = A + BX$$

where $X = \log_{10} p$, $Y = \log_{10} v$, $A = \frac{1}{\gamma} \log_{10} k$, $B = -\frac{1}{\gamma}$.

$\therefore$ the normal equations are

$$\Sigma Y = 6A + B\Sigma X \quad ...(i)$$

$$\Sigma XY = A\Sigma X + B\Sigma X^2 \quad ...(ii)$$

Now $\Sigma X$ etc. are calculated as follows :

| $p$ | $v$ | $X = \log_{10} p$ | $Y = \log_{10} v$ | $XY$ | $X^2$ |
|---|---|---|---|---|---|
| .5 | 1.62 | −0.3010 | 0.2095 | −0.0630 | 0.0906 |
| 1.0 | 1.00 | 0.0000 | 0.0000 | −0.0000 | 0.0000 |
| 1.5 | 0.75 | 0.1761 | −0.1249 | −0.0220 | 0.0310 |
| 2.0 | 0.62 | 0.3010 | −0.2076 | −0.0625 | 0.0906 |
| 2.5 | 0.52 | 0.3979 | −0.2840 | −0.1130 | 0.1583 |
| 3.0 | 0.46 | 0.4771 | −0.3372 | −0.1609 | 0.2276 |
| Total | | 1.0511 | −0.7442 | −0.4214 | 0.5981 |

$\therefore$ equations (*i*) and (*ii*) become

$$6A + 1.0511B = -0.7442$$

$$1.0511A + 0.5981B = -0.4214$$

Solving these, we get $A = 0.0132$, $B = -0.7836$.

$\therefore \quad \gamma = -1/B = 1.1276$ and $k = \text{antilog}(A\gamma) = \text{antilog}(0.0168) = 1.039$.

Hence the equation of best fit is $pv^{1.276} = 1.039$.

## PROBLEMS 24.3

1. If $V$ (km/hr) and $R$ (kg/ton) are related by a relation of the type $R = a + bV^2$, find by the method of least squares $a$ and $b$ with the help of the following table :

| $V$: | 10 | 20 | 30 | 40 | 50 |
|---|---|---|---|---|---|
| $R$: | 8 | 10 | 15 | 21 | 30 |

2. Using the method of least squares fit the curve $y = ax^2 + bx$ to following observations :

| $x$: | 1 | 2 | 3 | 4 | 5 | 6 | 7 | 8 |
|---|---|---|---|---|---|---|---|---|
| $y$: | 1 | 1.2 | 1.8 | 2.5 | 3.6 | 4.7 | 6.6 | 9.1 |

*(U.P.T.U., 2011)*

3. Fit the curve $y = ax + b/x$ to the following data :

| $x$: | 1 | 2 | 3 | 4 | 5 | 6 | 7 | 8 |
|---|---|---|---|---|---|---|---|---|
| $y$: | 5.4 | 6.3 | 8.2 | 10.3 | 12.6 | 14.9 | 17.3 | 19.5 |

*(U.P.T.U., 2010)*

4. Estimate $y$ at $x = 2.25$ by fitting the *indifference curve* of the form $xy = Ax + B$ to the following data :

| $x$: | 1 | 2 | 3 | 4 |
|---|---|---|---|---|
| $y$: | 3 | 1.5 | 6 | 7.5 |

5. Find the least square curve $y = ax + b/x$ for the following data :

| $x$: | 1 | 2 | 3 | 4 |
|---|---|---|---|---|
| $y$: | −1.5 | 0.99 | 3.88 | 7.66 |

6. Predict $y$ at $x = 3.75$, by fitting a *power curve* $y = ax^b$ to the given data :

| $x$: | 1 | 2 | 3 | 4 | 5 | 6 |
|---|---|---|---|---|---|---|
| $y$: | 298 | 4.26 | 5.21 | 6.10 | 6.80 | 7.50 |

7. Fit the curve of the form $y = ae^{bx}$ to the following data :

| $x$: | 77 | 100 | 185 | 239 | 285 |
|---|---|---|---|---|---|
| $y$: | 2.4 | 3.4 | 7.0 | 11.1 | 19.6 |

*(Andhra, 2020; V.T.U., 2011 S)*

8. Obtain the least squares fit of the form $f(t) = ae^{-3t} + be^{-2t}$ for the data :

| $x$ : | 0.1 | 0.2 | 0.3 | 0.4 |
|---|---|---|---|---|
| $f(t)$ : | 0.76 | 0.58 | 0.44 | 0.35 |

9. The voltage $v$ across a capacitor at time $t$ seconds is given by the following table :

| $t$ : | 0 | 2 | 4 | 6 | 8 |
|---|---|---|---|---|---|
| $v$ : | 150 | 63 | 28 | 12 | 5.6 |

Use the method of least squares to fit a curve of the form $v = ae^{kt}$ to this data.

10. Using method of least squares, fit a relation of the form $y = ab^x$ to the following data :

| $x$: | 2 | 3 | 4 | 5 | 6 |
|---|---|---|---|---|---|
| $y$: | 144 | 172.8 | 207.4 | 248.8 | 298.5 |

## 24.7 METHOD OF GROUP AVERAGES

Let the straight line, $\qquad y = a + bx \qquad$ ...(1)

fit the set of $n$ observations

$(x_1, y_1), (x_2, y_2), \ldots, (x_n, y_n)$ quite closely. (Fig. 24.5)

When $x = x_1$, the *observed* (*or experimental*) *value of* $y = y_1 = L_1P_1$ and from Eqn. (1),

$$y = a + bx_1 = L_1M_1,$$

which is known as the *expected* (or *calculated*) *value* of $y$ at $L_1$.

Then $e_1$ = observed value at $L_1$ – expected value at $L_1$

$$= y_1 - (a + bx_1) = M_1P_1,$$

which is called the *error* (or *residual*) at $x_1$. Similarly the errors for the other observations are

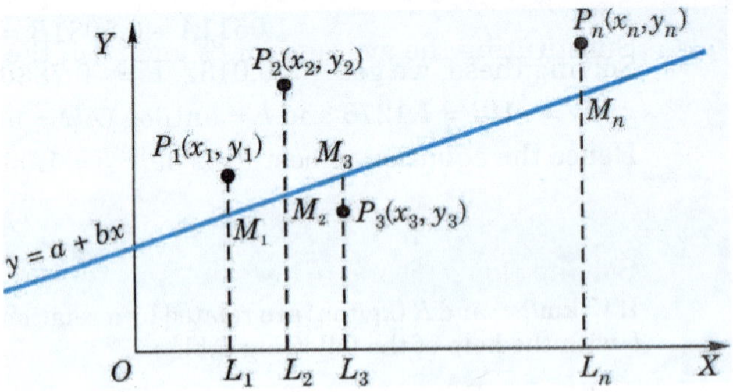

Fig. 24.5

$$e_2 = y_2 - (a + bx_2) = M_2P_2$$

..........................................

$$e_n = y_n - (a + bx_n) = M_nP_n$$

Some of these errors may be positive and others negative.

The method of group averages is based on the assumption that the sum of the residuals is zero. To find the constants $a$ and $b$ is Eqn. (1), we require two equations. As such we divide the data into two groups : the first containing $k$ observations

$$(x_1, y_1), (x_2, y_2) \ldots (x_k, y_k) ;$$

and the second group having the remaining $n - k$ observations

$$(x_{k+1}, y_{k+1}), (x_{k+2}, y_{k+2}), \ldots, (x_n, y_n).$$

Assuming that the sum of the errors in each group is zero, we get

$$\{y_1 - (a + bx_1)\} + \{y_2 - (a + bx_2)\} + \ldots \{y_k - (a + bx_k)\} = 0$$
$$\{y_{k+1} - (a + bx_{k+1})\} + \{y_{k+2} - (a + bx_{k+2})\} + \ldots + \{y_n - (a + bx_n)\} = 0$$

On simplification, we obtain

$$\frac{y_1 + y_2 + \ldots + y_k}{k} = a + b\,\frac{x_1 + x_2 + \ldots + x_k}{k} \qquad \ldots(2)$$

$$\frac{y_{k+1} + y_{k+2} + \ldots + y_n}{n-k} = a + b\,\frac{x_{k+1} + x_{k+2} + \ldots + x_n}{n-k} \qquad \ldots(3)$$

In (2), $\frac{1}{k}(x_1 + x_2 + \ldots + x_k)$ and $\frac{1}{k}(y_1 + y_2 + \ldots + y_k)$ are simply the average values of $x$'s and $y$'s of the first group. Hence the Eqns. (2) and (3) are obtained from Eqn. (1) by replacing $x$ and $y$ by their respective averages of the two groups. Solving Eqns. (2) and (3), we get $a$ and $b$.

**Obs.** The main drawback of this method is that a different grouping of the observations will give different values of $a$ and $b$. In practice, we divide the data in such a way that each group contains almost an equal number of observations.

**Example 24.12.** *The latent heat of vaporisation of steam r, is given in the following table at different temperatures t :*

| *t* : | 40 | 50 | 60 | 70 | 80 | 90 | 100 | 110 |
|---|---|---|---|---|---|---|---|---|
| *r* : | 1069.1 | 1063.6 | 1058.2 | 1052.7 | 1049.3 | 1041.8 | 1036.3 | 1030.8 |

*For this range of temperature, a relation of the form r = a + bt is known to fit the data. Find the values of a and b by the method of group averages.*

**Solution.** Let us divide the data into two groups each containing four readings. Then we have

| $t$ | $r$ | $t$ | $r$ |
|---|---|---|---|
| 40 | 1069.1 | 80 | 1049.3 |
| 50 | 1063.6 | 90 | 1041.8 |
| 60 | 1058.2 | 100 | 1036.3 |
| 70 | 1052.7 | 110 | 1030.8 |
| $\Sigma t = 220$ | $\Sigma r = 4243.6$ | $\Sigma t = 380$ | $\Sigma r = 4158.2$ |

Substituting the averages of $t$'s and $r$'s of the two groups in the given relation, we get

$$\frac{4243.6}{4} = a + b\,\frac{220}{4} \quad i.e.,\ 1060.9 = a + 55b \qquad ...(i)$$

$$\frac{4158.2}{4} = a + b\,\frac{380}{4} \quad i.e.,\ 1039.55 = a + 95b \qquad ...(ii)$$

Solving Eqns. (*i*) and (*ii*), we obtain

$$a = 1090.26,\ b = -0.534.$$

## 24.8 FITTING A PARABOLA

We have applied the method of averages to *linear law* involving two constants only. To fit the parabola

$$y = a + bx + cx^2 \qquad ...(1)$$

which contains three constants, to a set of observations, we proceed as follows :

Let $(x_1, y_1)$ be a point on Eqn. (1) satisfying the given data so that

$$y_1 = a + bx_1 + cx_1^2$$

Then $$y - y_1 = b\,(x - x_1) + c\,(x^2 - x_1^2)$$

or $$\frac{y - y_1}{x - x_1} = b + c\,(x + x_1)$$

Putting $x + x_1 = X$ and $(y - y_1)/(x - x_1) = Y$, it takes the linear form

$$Y = b + cX.$$

Now $b$ and $c$ can be found as before.

**Example 24.13.** *The corresponding values of x and y are given by the following table :*

| *x* : | *87.5* | *84.0* | *77.8* | *63.7* | *46.7* | *36.9* |
|---|---|---|---|---|---|---|
| *y* : | *292* | *283* | *270* | *235* | *197* | *181* |

**Solution.** Taking $x = 84$, $y = 283$ as a particular point on $y = a + bx + cx^2$,

we get $$283 = a + b\,(84) + c\,(84)^2 \qquad ...(i)$$

$$\therefore \quad y - 283 = b\,(x - 84) + c\,[x^2 - (84)^2]$$

or $$\frac{y - 283}{x - 84} = b + c\,(x + 84)$$

*i.e.,* $$Y = b + cX \qquad ...(ii)$$

where $X = x + 84$, $Y = (y - 283)/(x - 84)$.

Now we have the following table of values :

| $x$ | $y$ | $X = x + 84$ | $Y = (y - 283)/(x - 84)$ |
|---|---|---|---|
| 87.5 | 292 | 171.5 | 2.571 |
| 84.0 | 283 | — | — |
| 77.8 | 270 | 161.8 | 2.097 |
| | | $\Sigma X = 333.3$ | $\Sigma Y = 4.668$ |
| 63.7 | 235 | 147.7 | 2.364 |
| 46.7 | 197 | 130.7 | 2.306 |
| 36.9 | 181 | 120.9 | 2.166 |
| | | $\Sigma X = 399.3$ | $\Sigma Y = 6.836$ |

Substituting the averages of $X$ and $Y$ in Eqn. (*ii*), we get

$$\frac{4.668}{2} = b + c\,\frac{333.3}{2} \quad i.e.,\ 2.33 = b + 166.65\,c \qquad ...(iii)$$

$$\frac{6.836}{3} = b + c\,\frac{399.3}{3} \quad i.e.,\ 2.28 = b + 131.1\,c \qquad ...(iv)$$

Eqn. (*iv*)–(*iii*) gives $c = 0.0014$

and Eqn. (*iii*) gives $b = 2.0967$ *i.e.*, 2.1 nearly

From Eqn. (*i*), we get $a = 96.9988$ *i.e.*, 97 nearly.

Hence the parabola of fit is

$$y = 97 + 2.1x + .0014x^2.$$

**Example 24.14.** *The train resistance R (lbs / ton) is measured for the following values of its velocity V (km / hr) :*

| V | 20 | 40 | 60 | 80 | 100 |
|---|---|---|---|---|---|
| R | 5 | 9 | 14 | 25 | 36 |

*If R is related to V by the formula $R = a + bV^n$, find a, b, and n.*

**Solution.** To find $a$, we take the following three values of $v$ which are in G.P. :

$$v_1 = 20, \quad v_2 = 40, \quad v_3 = 80$$

Then $R_1 = 5, \quad R_2 = 9, \quad R_3 = 25$

$\therefore \quad (R_1 - a)(R_3 - a) = (R_2 - a)^2$

whence
$$a = \frac{R_1R_3 - R_2^2}{R_1 + R_3 - 2R_2} = 3.67$$

Thus $R - 3.67 = bV^n$ or $\log_{10}(R - 3.67) = \log_{10} b + n \log_{10} V$

*i.e.*, $$Y = k + nX \qquad ...(i)$$

where $X = \log_{10} V$, $Y = \log_{10}(R - 3.67)$, $k = \log_{10} b$.

Now we have the following table of values :

| V | R | $X = \log_{10} V$ | $Y = \log_{10}(R - 3.67)$ |
|---|---|---|---|
| 20 | 5 | 1.3010 | 0.1238 |
| 40 | 9 | 1.6021 | 0.7267 |
| 60 | 14 | 1.7782 | 1.0141 |
| | | $\Sigma X = 4.6813$ | $\Sigma Y = 1.8646$ |
| 80 | 25 | 1.9031 | 1.3290 |
| 100 | 36 | 2.0000 | 1.5096 |
| | | $\Sigma X = 3.9031$ | $\Sigma X = 2.8396$ |

Substituting the averages of $X$'s and $Y$'s in Eqn. (*i*), we obtain

$$\frac{1.8646}{2} = k + n\,\frac{4.6813}{2} \qquad i.e., \ 0.6215 = k + 1.5604\,n \qquad ...(ii)$$

$$\frac{2.8386}{2} = k + n\,\frac{3.9031}{2} \qquad i.e., \ 1.4193 = k + 1.9516\,n \qquad ...(iii)$$

Solving Eqn. (*ii*) and (*iii*), we get $n = 2.04$, $k = -2.56$ approx.

$$b = \text{antilog } k = \text{antilog }(-2.56) = 0.0028.$$

## PROBLEMS 24.4

1. Fit a straight line of the form $y = a + bx$ to the following data by the method of group averages :

| $x$ : | 0 | 5 | 10 | 15 | 20 | 25 |
|---|---|---|---|---|---|---|
| $y$ : | 12 | 15 | 17 | 22 | 24 | 30 |

2. The weights of a calf taken at weekly intervals are given below :

| Age : | 1 | 2 | 3 | 4 | 5 | 6 | 7 | 8 | 9 | 10 |
|---|---|---|---|---|---|---|---|---|---|---|
| Weight : | 52.5 | 58.7 | 65.0 | 70.2 | 75.4 | 81.1 | 87.2 | 95.5 | 102.2 | 108.4 |

Find a straight line of best fit.

3. Using the method of averages, fit a parabola $y = ax^2 + bx + c$ to the following data :

| $x$ : | 20 | 40 | 60 | 80 | 100 | 120 |
|---|---|---|---|---|---|---|
| $y$ : | 5.5 | 9.1 | 14.9 | 22.8 | 33.3 | 46.0 |

4. While testing a centrifugal pump, the following data is obtained. It is assumed to fit the equation $y = a + bx + cx^2$, where $x$ is the discharge in litre/sec and $y$, head in metres of water. Find the values of the constants $a, b, c$ by the method of group averages.

| $x$ : | 2 | 2.5 | 3 | 3.5 | 4 | 4.5 | 5 | 5.5 | 6 |
|---|---|---|---|---|---|---|---|---|---|
| $y$ : | 18 | 17.8 | 17.5 | 17 | 15.8 | 14.8 | 13.3 | 11.7 | 9 |

5. By the method of averages, fit a curve of the form $y = ae^{bx}$ to the following data :

| $x$ : | 5 | 15 | 20 | 30 | 35 | 40 |
|---|---|---|---|---|---|---|
| $y$ : | 10 | 14 | 25 | 40 | 50 | 62 |

## 24.9 METHOD OF MOMENTS

Let $(x_1, y_1), (x_2, y_2), \ldots (x_n, y_n)$ be the set of $n$ observations such that

$$x_2 - x_1 = x_3 - x_2 = \ldots = x_n - x_{n-1} = h \text{ (say)}$$

We define the *moments of the observed values* of $y$ as follows :

$m_1$, the 1st moment $= h\Sigma y$

$m_2$, the 2nd moment $= h\Sigma xy$

$m_3$, the 3rd moment $= h\Sigma x^2y$ and so on.

Let the curve fitting the given data be $y = f(x)$. Then the *moments of the calculated values* of $y$ are

$\mu_1$, the 1st moment $= \int y dx$

$\mu_2$, the 2nd moment $= \int xy\, dx$

$\mu_3$, the 3rd moment $= \int x^2 y\, dx$ and so on.

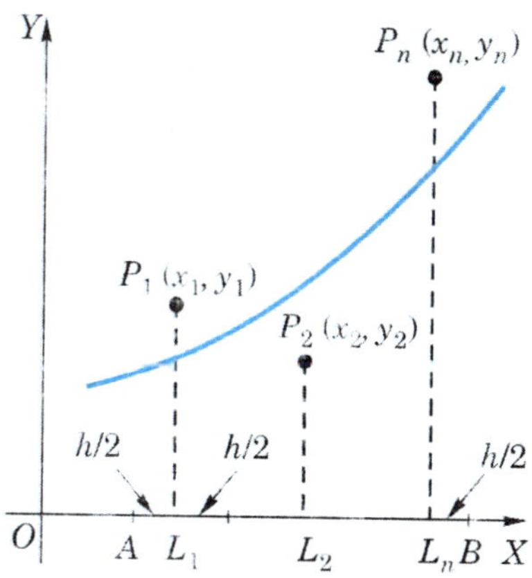

Fig. 24.6

*This method is based on the assumption that the moment of the observed values of y are respectively equal to the moments of the calculated values of y i.e.,* $m_1 = \mu_1, m_2 = \mu_2, m_3 = \mu_3$ *etc. These equations (known as observation equations)* are used to determine the constants in $f(x)$.

$m$'s are calculated from the tabulated values of $x$ and $y$ while $\mu$'s are computed as follows :

In Fig. 24.6, $y_1$ the ordinate of $P_1$ $(x = x_1)$, can be taken as the value of $y$ at the mid-point of the interval $(x_1 - h/2, x_1 + h/2)$. Similarly, $y_n$, the ordinate of $P_n$ $(x = x_n)$, can be taken as the value of $y$ at the mid-point of the interval $(x_n - h/2, x_n + h/2)$. If $A$ and $B$ be the points such that

$$OA = x_1 - h/2 \text{ and } OB = x_n + h/2,$$

then
$$\mu_1 = \int y\, dx = \int_{x_1 - h/2}^{x_n + h/2} f(x)\, dx$$

$$\mu_2 = \int_{x_1 - h/2}^{x_n + h/2} xf(x)\, dx$$

and
$$\mu_3 = \int_{x_1 - h/2}^{x_n + h/2} x^2 f(x)\, dx.$$

**Example 24.15.** *Fit a straight line $y = a + bx$ to the following data by the method of moments :*

| $x$ : | 1 | 2 | 3 | 4 |
|---|---|---|---|---|
| $y$ : | 16 | 19 | 23 | 26 |

**Solution.** Since only two constants $a$ and $b$ are to be found, it is sufficient to calculate the first two moments in each case. Here $h = 1$.

$$m_1 = h\Sigma y = 1\,(16 + 19 + 23 + 26) = 84$$
$$m_2 = h\Sigma xy = 1\,(1 \times 16 + 2 \times 19 + 3 \times 22 + 4 \times 26) = 227$$

To compute the moments of calculated values of $y = a + bx$, the limits of integration will be $1 - h/2$ and $4 + h/2$ *i.e.*, 0.5 to 4.5

$$\therefore \quad \mu_1 = 2\int_{0.5}^{4.5}(a + bx)\,dx = \left| ax + b\frac{x^2}{2} \right|_{0.5}^{4.5} = 4a + 10b$$

$$\mu_2 = \int_{0.5}^{4.5} x(a + bx)\,dx = 10a + \frac{91}{3}b.$$

Thus, the observation equations $m_r = \eta_r\,(r = 1, 2)$ are $4a + 10b = 84$ ; $10a + \dfrac{91}{3}b = 227$

Solving these, $a = 13.02$ and $b = 3.19$.

Hence the required equation is $y = 13.02 + 3.19x$.

**Example 24.16.** *Given the following data :*

| $x$ : | 0 | 1 | 2 | 3 | 4 |
|---|---|---|---|---|---|
| $y$ : | 1 | 5 | 10 | 22 | 38 |

*find the parabola of best fit by the method of moments.*

**Solution.** Let the parabola of best fit be $y = a + bx + cx^2$ ...(*i*)

Since three constants are to be found, we calculate the first three moments in each case. Here $h = 1$.

$$m_1 = h\Sigma y = 1\,(1 + 5 + 10 + 22 + 38) = 76$$
$$m_2 = h\Sigma xy = 1\,(0 + 5 + 20 + 66 + 152) = 243$$
$$m_3 = h\Sigma x^2y = 1\,(0 + 5 + 40 + 198 + 608) = 851$$

For computing the moments of calculated values of Eqn. (*i*), the limits of integration will be $0 - h/2$ and $4 + h/2$ *i.e.*, –0.5 and 4.5.

$$\therefore \quad \mu_1 = \int_{-0.5}^{4.5}(a + bx + cx^2)\,dx = 5a + 10b + 30.4c$$

$$\mu_2 = \int_{-0.5}^{4.5} x\,(a + bx + cx^2)\,dx = 10a + 30.4b + 102.5c$$

$$\mu_3 = \int_{-0.5}^{4.5} x^2\,(a + bx + cx^2)\,dx = 30.4a + 102.5b + 369.1c$$

Thus the observation equations $m_r = \mu_r\,(r = 1, 2, 3)$ are

$$5a + 10b + 30.4c = 76\ ;\ 10a + 30.4 + 102.5c = 243\ ;\ 30.4a + 102.5b + 369.1c = 851$$

Solving these equations, we get $a = 0.4$, $b = 3.15$, $c = 1.4$.

Hence the parabola of best fit is $y = 0.4 + 3.15x + 1.4x^2$.

## PROBLEMS 24.5

1. Use the method of moments to fit the straight line $y = a + bx$ to the data :

| $x$ : | 1 | 2 | 3 | 4 |
|---|---|---|---|---|
| $y$ : | 0.17 | 0.18 | 0.23 | 0.32 |

2. Fit a straight line to the following data, using the method of moments :

| $x$ : | 1 | 3 | 5 | 7 | 9 |
|---|---|---|---|---|---|
| $y$ : | 1.5 | 2.8 | 4.0 | 4.7 | 6.0 |

3. Fit a parabola of the form $y = a + bx + cx^2$ to the data :

| $x$ : | 1 | 2 | 3 | 4 |
|---|---|---|---|---|
| $y$ : | 1.7 | 1.8 | 2.3 | 3.2 |

by the method of moments.

4. By using the method of moments, fit a parabola to the following data :

| $x$ : | 1 | 2 | 3 | 4 |
|---|---|---|---|---|
| $y$ : | 0.30 | 0.64 | 1.32 | 5.40 |

## 24.10 OBJECTIVE TYPE OF QUESTIONS

### PROBLEMS 24.6

*Fill up the blanks or choose the correct answer in the following problems :*

1. The law $y = ax^2 + bx$ converted to linear form is .........
2. The gas equation $pv^r = k$ can be reduced to $y = a + bx$ where $a =$ ......... and $b =$ ......... .
3. The principle of *'least squares'* states that .........
4. $y = ax^b + c$ in linear form is .........
5. To fit the straight line $y = mx + c$ to $n$ observations, the normal equations are
   (*i*) $\Sigma y = n\ \Sigma x + \Sigma cm,\ \Sigma xy = c\ \Sigma x^2 + c\ \Sigma n.$
   (*ii*) $\Sigma y = m\ \Sigma x + nc,\ \Sigma xy = m\ \Sigma x^2 + c\ \Sigma x.$
   (*iii*) $\Sigma y = c\ \Sigma x + m\ \Sigma n,\ \Sigma\, xy = c\ \Sigma x^2 + m\ \Sigma x.$
6. To fit $y = ab^x$ by least square method, normal equations are .........
7. The observation equations for fitting a straight line by *method of moments* are .........
8. The *method of group averages* is based on the assumption that the sum of the residuals is .........
9. $y = ax^2 + b \log_{10} x$ reduced to linear law takes the form .........
10. Given $\begin{bmatrix} x: & 0 & 1 & 2 \\ y: & 0 & 1.1 & 2.1 \end{bmatrix}$ then the straight line of best fit is ......... .
11. The *method of moments* is based on the assumption that .........
12. In $y = a + bx$, $\Sigma x = 50$, $\Sigma y = 80$, $\Sigma xy = 1030$, $\Sigma x^2 = 750$ and $n = 10$, then $a =$ ........., $b =$ ......... .
13. $y = x/(ax + b)$ in linear form is .........
14. If $y = a + bx + cx^2$ and

| $x$ : | 0 | 1 | 2 | 3 | 4 |
|---|---|---|---|---|---|
| $y$ : | 1 | 1.8 | 1.3 | 2.5 | 7.3 |

then the first normal equation is :
(*a*) $15 = 5a + 10b + 29c$, (*b*) $15 = 5a + 10b + 31c$
(*c*) $12.9 = 5a + 10b + 30c$ (*d*) $34 = 5a + 10b + 27c$.

15. If $y = 2x + 5$ is the best fit for 8 pairs of values $(x, y)$ by the method of least squares and $\Sigma y = 120$, then $\Sigma X =$
(*a*) 35 (*b*) 40 (*c*) 45 (*d*) 30.

CHAPTER 25

# Statistical Methods



## 25.1 INTRODUCTION

Statistics deals with the methods for collection, classification and analysis of numerial data for drawing valid conclusions and making reasonable decisions. It has meaningful applications in production engineering, in the analysis of experimental data, etc. The importance of statistical methods in engineering is on the increase. As such we shall now introduce the student to this interesting field.

## 25.2 (1) COLLECTION OF DATA

The collection of data constitutes the starting point of any statistical investigation. Data may be collected for each and every unit of the whole lot (*population*), for it would ensure greater accuracy. But complete enumeration is prohibitively expensive and time consuming. As such out of a very large number of items, a few of them (a *sample*) are selected and conclusions drawn on the basis of this sample are taken to hold for the population.

(2) **Classification of data.** The data collected in the course of an inquiry is not in an easily assimilable form. As such, its proper classification is necessary for making intelligent inferences. The classification is done by dividing the raw data into a convenient number of groups according to the values of the variable and finding the frequency of the variable in each group.

Let us, for example, consider the raw data relating to marks obtained in Mechanics by a group of 64 students :

| | | | | | | | |
|---|---|---|---|---|---|---|---|
| 79 | 88 | 75 | 60 | 93 | 71 | 59 | 85 |
| 84 | 75 | 82 | 68 | 90 | 62 | 88 | 76 |
| 65 | 75 | 87 | 74 | 62 | 95 | 78 | 63 |
| 78 | 82 | 75 | 91 | 77 | 69 | 74 | 68 |
| 67 | 73 | 81 | 72 | 63 | 76 | 75 | 85 |
| 80 | 73 | 57 | 88 | 78 | 62 | 76 | 53 |
| 62 | 67 | 97 | 78 | 85 | 76 | 65 | 71 |
| 78 | 89 | 61 | 75 | 95 | 60 | 79 | 83 |

This data can conveniently be grouped and shown in a tabular form as follows :

| *Class* | *Frequency* | *Cumulative frequency* |
|---|---|---|
| 50—54 | 1 | 1 |
| 55—59 | 2 | 3 |
| 60—64 | 9 | 12 |
| 65—69 | 7 | 19 |
| 70—74 | 8 | 27 |
| 75—79 | 17 | 44 |
| 80—84 | 6 | 50 |
| 85—89 | 8 | 58 |
| 90—94 | 3 | 61 |
| 95—99 | 3 | 64 |
| | Total = 64 | |

It would be seen from the above table that there is one student getting marks between 50—54, two students getting marks between 55—59, nine students getting marks between 60—64 and so on. Thus the 64 figure have been put into only 10 groups, called the **classes**. The width of the class is called the **class interval** and the number in that interval is called the **frequency.** The mid-point or the mid-value of the class is called the **class mark.** The above table showing the classes and the corresponding frequencies is called a *frequency table. Thus a set of raw data summarised by distributing it into a number of classes alongwith their frequencies is known as a* **frequency distribution**.

While forming a frequency distribution, the number of classes should not ordinarily exceed 20, and should not, in general, be less than 10. As far as possible, the class intervals should be of equal width.

(3) **Cumulative frequency.** In some investigations, we require the number of items less than a certain value. We add up the frequencies of the classes upto that value and call this number as the *cumulative frequency*. In the above table, the third column shows the cumulative frequencies, *i.e.*, the number of students, getting less than 54 marks, less than 59 marks and so on.

## 25.3 GRAPHICAL REPRESENTATION

A convenient way of representing a sample frequency distribution is by means of graphs. It gives to the eyes the general run of the observations and at the same time makes the raw data readily intelligible. We give below the important types of graphs in use :

(1) **Histogram.** A histogram is drawn by erecting rectangles over the class intervals, such that the areas of the rectangles are proportional to the class frequencies. If the class intervals are of equal size, the height of the rectangles will be proportional to the class frequencies themselves (Fig. 25.1).

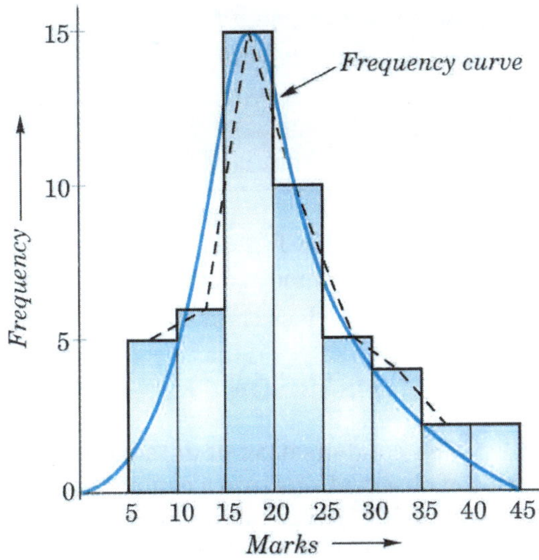

Fig. 25.1

(2) **Frequency polygon.** A frequency polygon for an ungrouped data can be obtained by joining points plotted with the variable values as the abscissae and the frequencies as the ordinates. For a grouped distribution, the abscissae of the points will be the mid-values of the class intervals. In case the intervals are equal, the frequency polygon can be obtained by joining the middle points of the upper sides of the rectangles of the histogram by straight lines (shown by dotted lines in Fig. 25.1). If the class intervals become very very small, the frequency polygon takes the form of a smooth curve called the *frequency curve*.

(3) **Cumulative frequency curve-Ogive.** Very often, it is desired to show in a diagrammatic form, not the relative frequencies in the various intervals, but the cumulative frequencies above or below a given value. For example, we may wish to read off from a diagram the number or proportions of people whose income is not less than any given amount, or proportion of people whose height does not exceed any stated value. Diagrams of

this type are known as *cumulative frequency curves* or *ogives*. These are of two kinds *'more than'* or *'less than'* and typically they look somewhat like a long drawn $S$ (Fig. 25.2).

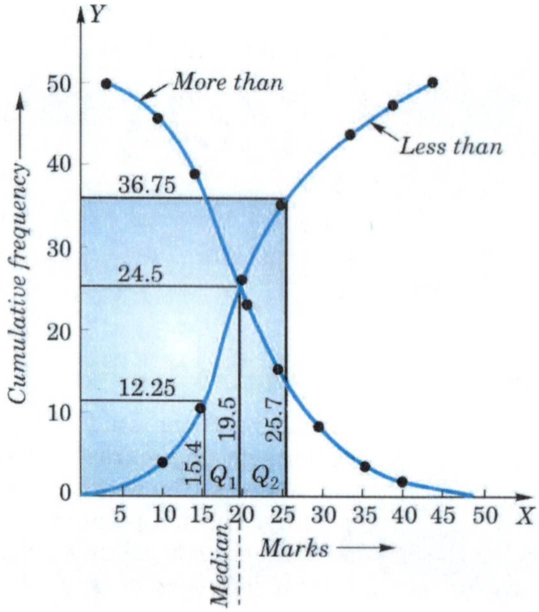


Fig. 25.2

**Example 25.1.** *Draw the histogram, frequency polygon, frequency curve and the ogive 'less than' and 'more than' from the following distribution of marks obtained by 49 students :*

| Class (Marks group) | Frequency (No. of students) | Cumulative frequency (Less than) | Cumulative frequency (More than) |
|---|---|---|---|
| 5—10 | 5 | 5 | 49 |
| 10—15 | 6 | 11 | 44 |
| 15—20 | 15 | 26 | 38 |
| 20—25 | 10 | 36 | 23 |
| 25—30 | 5 | 41 | 13 |
| 30—35 | 4 | 45 | 8 |
| 35—40 | 2 | 47 | 4 |
| 40—45 | 2 | 49 | 2 |

**Solution.** In Fig. 25.1, the rectangles show the *histogram*; the dotted polygon represents the *frequency polygon* and the smooth curve is the *frequency curve*.

The *ogives* 'less than' and 'more than' are shown in Fig. 25.2.

## 25.4 COMPARISON OF FREQUENCY DISTRIBUTIONS

The condensation of data in the form of a frequency distribution is very useful as far as it brings a long series of observations into a compact form. But in practice, we are generally interested in comparing two or more series. The inherent inability of the human mind to grasp in its entirety even the data in the form of a frequency distribution compels us to seek for certain constants which could concisely give an insight into the important characteristics of the series. The chief constants which summarise the fundamental characteristics of the frequency distributions are (*i*) *Measures of central tendency*, (*ii*) *Measures of dispersion and* (*iii*) *Measures of skewness*.

## 25.5 MEASURES OF CENTRAL TENDENCY

A frequency distribution in general, shows clustering of the data around some central value. Finding of this central value or the average is of importance, as it gives a most representative value of the whole group.

Different methods give different averages which are known as the *measures of central tendency*. The commonly used measures of central value are *Mean, Median, Mode, Geometric mean and Harmonic mean.*

**(1) Mean.** If $x_1, x_2, x_3, ..., x_n$ are a set of $n$ values of a variate, then the *arithmetic mean* (or simply *mean*) is given by

$$\bar{\mathbf{x}} = \frac{x_1 + x_2 + x_3 + ... + x_n}{n}, \text{ i.e. } \frac{\Sigma \mathbf{x}_i}{\mathbf{n}} \quad ...(1)$$

In a *frequency distribution*, if $x_1, x_2, ..., x_n$ be the mid-values of the class-intervals having frequencies $f_1, f_2, ..., f_n$ respectively, we have

$$\bar{\mathbf{x}} = \frac{f_1x_1 + f_2x_2 + ... + f_nx_n}{f_1 + f_2 + ... + f_n} = \frac{\Sigma \mathbf{f}_i\mathbf{x}_i}{\Sigma \mathbf{f}_i} \quad ...(2)$$

**Calculation of mean.** Direct method of computing especially when applied to grouped data involves heavy calculations and in order to avoid these, the following formulae are generally used :

*I. Short-cut method* $\quad \bar{\mathbf{x}} = \mathbf{A} + \dfrac{\Sigma \mathbf{f}_i d_i}{\Sigma \mathbf{f}_i} \quad ...(3)$

*II. Step-deviation method* $\bar{\mathbf{x}} = \mathbf{A} + \mathbf{h}\dfrac{\Sigma \mathbf{f}_i\mathbf{u}_i}{\Sigma \mathbf{f}_i} \quad ...(4)$

where $d = x - A$ and $u = (x - A)/h$, $A$ being an arbitrary origin and $h$ the equal class interval.

*Proof.* If $x_1, x_2, ..., x_n$ are the mid-values of the classes with frequencies $f_1, f_2, ..., f_n$, we have

$$\Sigma f_ix_i = \Sigma f_i (A + d_i) = A\Sigma f_i + \Sigma f_id_i$$

$$\therefore \quad \bar{x} = \frac{\Sigma f_ix_i}{\Sigma f_i} = A + \frac{\Sigma f_id_i}{\Sigma f_i}$$

Further $u_i = d_i/h$ or $d_i = hu_i$. Substituting this value in (3), we get (4).

**Obs.** *The algebraic sum of the deviations of all the variables from their mean is zero, for*

$$\Sigma f_i (x_i - \bar{x}) = \Sigma f_ix_i - \bar{x}\,\Sigma f_i = \Sigma f_ix_i - \frac{\Sigma f_ix_i}{\Sigma f_i}.\Sigma f_i = 0.$$

**Cor.** *If $\bar{x}_1, \bar{x}_2$ be the means of two samples of size $n_1$ and $n_2$, then the mean $\bar{x}$ of the combined sample of size $n_1 + n_2$ is given by*

$$\bar{\mathbf{x}} = \frac{\mathbf{n}_1\bar{\mathbf{x}}_1 + \mathbf{n}_2\bar{\mathbf{x}}_2}{\mathbf{n}_1 + \mathbf{n}_2}$$

For $\quad n_1\bar{x}_1$ = sum of all observations of the first sample,

and $\quad n_2\bar{x}_2$ = sum of all observations of the second sample.

$\therefore$ sum of the observations of the combined sample = $n_1\bar{x}_1 + n_2\bar{x}_2$.

Also number of the observations in the combined sample = $n_1 + n_2$.

$\therefore$ mean of the combined sample = $\dfrac{n_1\bar{x}_1 + n_2\bar{x}_2}{n_1 + n_2}$.

**Example 25.2.** *The following is the frequency distribution of a random sample of weekly earnings of 509 employees :*

| *Weekly earnings :* | 10 | 12 | 14 | 16 | 18 | 20 | 22 | 24 | 26 | 28 | 30 | 32 | 34 | 36 | 38 | 40 |
|---|---|---|---|---|---|---|---|---|---|---|---|---|---|---|---|---|
| *No. of employees :* | 3 | 6 | 10 | 15 | 24 | 42 | 75 | 90 | 79 | 55 | 36 | 26 | 19 | 13 | 9 | 7 |

*Calculate the average weekly earnings.*

**Solution.** The calculations are arranged in the following table. The arbitrary origin is generally taken as the value corresponding to the maximum frequency.

*By direct method, we have*

$$\text{Mean } \bar{x} = \frac{\Sigma fx}{\Sigma f} = \frac{13{,}315}{509} = 26.16$$

*By step-deviation method,* we have

$$\bar{x} = A + h\frac{\Sigma fu}{\Sigma f} = 25 + 2 \times \frac{295}{509}$$

$$= 25 + 1.16 = 26.16, \text{ which is same as found above.}$$

| *Weekly earnings* | *Mid value* $x$ | *No. of employees* $f$ | $f \times x$ | *Step deviations* $u = (x - 25)/2$ | $f \times u$ |
|---|---|---|---|---|---|
| 10—12 | 11 | 3 | 33 | –7 | –21 |
| 12—14 | 13 | 6 | 78 | –6 | –36 |
| 14—16 | 15 | 10 | 150 | –5 | –50 |
| 16—18 | 17 | 15 | 255 | –4 | –60 |
| 18—20 | 19 | 24 | 456 | –3 | –72 |
| 20—22 | 21 | 42 | 882 | –2 | –84 |
| 22—24 | 23 | 75 | 1725 | –1 | –75 |
| 24—26 | 25 | 90 | 2250 | 0 | –398 |
| 26—28 | 27 | 79 | 2133 | 1 | 79 |
| 28—30 | 29 | 55 | 1595 | 2 | 110 |
| 30—32 | 31 | 36 | 1116 | 3 | 108 |
| 32—34 | 33 | 26 | 858 | 4 | 104 |
| 34—36 | 35 | 19 | 665 | 5 | 95 |
| 36—38 | 37 | 13 | 481 | 6 | 78 |
| 38—40 | 39 | 9 | 351 | 7 | 63 |
| 40—42 | 41 | 7 | 287 | 8 | 56 |
| | | | | | + 693 |
| | | $\Sigma f = 509$ | $\Sigma fx = 13{,}315$ | | $\Sigma fu = 295$ |

**(2) Median.** *If the values of a variable are arranged in the ascending order of magnitude, the median is the middle item if the number is odd and is the mean of the two middle items if the number is even.* Thus the median is equal to the mid-value, *i.e.*, the value which divides the total frequency into two equal parts.

*For the grouped data,*

$$\textbf{Median} = \mathbf{L} + \frac{(\frac{1}{2}\mathbf{N} - \mathbf{C})}{\mathbf{f}} \times \mathbf{h}$$

where $L$ = lower limit of the median class, $N$ = total frequency,

$f$ = frequency of the median class, $h$ = width of the median class,

and $C$ = cumulative frequency upto the class preceding the median class.

**Quartiles.** Quartiles are those values which divide the frequency into four equal parts, when the values are arranged in the ascending order of magnitude. The **lower quartile** ($Q_1$) is mid-way between the lower extreme and the median. The **upper quartile** ($Q_3$) is midway between the median the upper extreme.

*For the grouped data,* these are calculated by the formulae :

$$\mathbf{Q_1} = \mathbf{L} + \frac{(\frac{1}{4}\mathbf{N} - \mathbf{C})}{\mathbf{f}} \times \mathbf{h}$$

and

$$\mathbf{Q_3} = \mathbf{L} + \frac{(\frac{3}{4}\mathbf{N} - \mathbf{C})}{\mathbf{f}} \times \mathbf{h}$$

where $L$ = lower limit of the class in which $Q_1$ or $Q_3$ lies, $f$ = frequency of this class, $h$ = width of the class

and $C$ = cumulative frequency upto the class preceding the class in which $Q_1$ or $Q_3$ lies.

*The difference between the upper and lower quartiles, i.e.,* $Q_3$—$Q_1$ *is called the* **inter-quartile range.**

**Obs.** The ogives give a ready method of marking on the curve the values of the median and the quartiles. The two ogives 'less than' and 'more than' cut each other at the median (Fig. 25.2).

**(3) Mode.** *The mode is defined as that value of the variable which occurs most frequently, i.e., the value of the maximum frequency.*

*For a grouped distribution,* it is given by the formula

$$\textbf{Mode} = \mathbf{L} + \frac{\Delta_1}{\Delta_1 + \Delta_2}\,\mathbf{h}$$

where $L$ = lower limit of the class containing the mode,

$\Delta_1$ = excess of modal frequency over frequency of preceding class,

$\Delta_2$ = excess of modal frequency over following class,

and $h$ = size of modal class.

For a frequency curve (Fig. 25.1), the abscissa of the highest ordinate determines the value of the mode. There may be one or more modes in a frequency curve. Curves having a single mode are termed as *unimodal*, those having two modes as *bi-modal* and those having more than two modes as *multi-modal*.

**Obs.** In a symmetrical distribution, the mean, median and mode coincide. For other distributions, however, they are different and are known to be connected by the empirical relationship :

**Mean – Mode = 3(Mean – Median).**

**Example 25.3.** *Calculate median and the lower and upper quartiles from the distribution of marks obtained by 49 students of example 25.1. Find also the semi-interquartile range and the mode.*

**Solution.** *Median* (or 49/2) falls in the class (15—20) and is given by

$$15 + \frac{(49/2) - 11}{15} \times 5 = 15 + \frac{13.5}{3} = 19.5 \text{ marks.}$$

*Lower quartile* $Q_1$ (or 49/4) = 12.25) also falls in the class 15—20.

$$\therefore \qquad Q_1 = 15 + \frac{(49/4) - 11}{15} \times 5 = 15 + \frac{12.5}{3} = 15.4 \text{ marks}$$

*Upper quartile* (or $\frac{3}{4} \times 49 = 36.75$) falls in the class 25—30.

$$\therefore \qquad Q_3 = 25 + \frac{36.75 - 36}{5} \times 5 = 25.75 \text{ marks.}$$

*Semi-interquartile range* = $\frac{1}{2}(Q_3 - Q_1) = \frac{25.75 - 15.4}{2} = \frac{10.35}{2} = 5.175.$

*Mode*. It is seen that the mode value falls in the class 15—20. Employing the formula for the grouped distribution, we have

$$\textit{Mode} = 15 + \frac{15 - 6}{(15-6) + (15-10)} \times 5 = 18.2 \text{ marks.}$$

**Obs.** In Fig. 25.2, the ogives meet at a point whose abscissa is 19.5 which is the *median* of the distribution. The values for the lower and upper quartiles are similarly seen to be 15.4 (for frequency 12.25) and 25.7 (for frequency 36.75).

**Example 25.4.** *Given below are the marks obtained by a batch of 20 students in a certain class test in Physics and Chemistry.*

| *Roll No. of students* | *Marks in Physics* | *Marks in Chemistry* | *Roll No. of students* | *Marks in Physics* | *Marks in Chemistry* |
|---|---|---|---|---|---|
| 1 | 53 | 58 | 11 | 25 | 10 |
| 2 | 54 | 55 | 12 | 42 | 42 |
| 3 | 52 | 25 | 13 | 33 | 15 |
| 4 | 32 | 32 | 14 | 48 | 46 |
| 5 | 30 | 26 | 15 | 72 | 50 |
| 6 | 60 | 85 | 16 | 51 | 64 |
| 7 | 47 | 44 | 17 | 45 | 39 |
| 8 | 46 | 80 | 18 | 33 | 38 |
| 9 | 35 | 33 | 19 | 65 | 30 |
| 10 | 28 | 72 | 20 | 29 | 36 |

*In which subject is the level of knowledge of the students higher ?*

**Solution.** The subject for which the value of the median is higher will be the subject in which the level of knowledge of the students is higher. To find the median in each case, we arrange the marks in ascending order of magnitude :

| Sr. No. | Marks in Physics | Marks in Chemistry | Sr. No. | Marks in Physics | Marks in Chemistry |
|---|---|---|---|---|---|
| 1 | 25 | 10 | 11 | 46 | 42 |
| 2 | 28 | 15 | 12 | 47 | 44 |
| 3 | 29 | 25 | 13 | 48 | 46 |
| 4 | 30 | 26 | 14 | 51 | 50 |
| 5 | 32 | 30 | 15 | 52 | 55 |
| 6 | 33 | 32 | 16 | 53 | 58 |
| 7 | 33 | 33 | 17 | 54 | 64 |
| 8 | 35 | 36 | 18 | 60 | 72 |
| 9 | 42 | 38 | 19 | 65 | 80 |
| 10 | 45 | 39 | 20 | 72 | 85 |

Median marks in Physics = A.M. of marks of 10th and 11th terms

$$= \frac{45 + 46}{2} = 45.5$$

Median marks in Chemistry = A.M. of marks of 10th and 11th items.

$$= \frac{39 + 42}{2} = 40.5$$

Since the median marks in Physics is greater than the median marks in Chemistry; the level of knowledge in Physics is higher.

**Example 25.5.** *An incomplete frequency distribution is given as below :*

| *Variable* : | *10—20* | *20—30* | *30—40* | *40—50* | *50—60* | *60—70* | *70—80* |
|---|---|---|---|---|---|---|---|
| *Frequency* : | *12* | *30* | *?* | *65* | *?* | *25* | *18* |

*Given that the total frequency is 229 and median is 46, find the missing frequencies.*

**Solution.** Let $f_1, f_2$ be the missing frequencies of the classes 30—40 and 50—60 respectively.

Since the median lies in the class 40—50,

$$\therefore \qquad 46 = 40 + \frac{229/2 - (12 + 30 + f_1)}{65} \times 10$$

which gives $f_1 = 33.5$ which can be taken as 34.

$$\therefore \qquad f_2 = 229 - (12 + 30 + 34 + 65 + 25 + 18) = 45.$$

**(4) Geometric mean.** If $x_1, x_2, ..., x_n$ are a set of $n$ observations, then the *geometric mean* is given by

$$\text{G.M.} = (x_1 x_2 \ldots x_n)^{1/n}$$

or
$$\log \text{G.M.} = \frac{1}{n}(\log x_1 + \log x_2 + \ldots + \log x_n) \qquad \ldots(1)$$

In a frequency distribution, let $x_1, x_2, ..., x_n$ be the central values with corresponding frequencies $f_1, f_2, ..., f_n$, we have

$$\text{G.M.} = \left[(x_1)^{f_1} . (x_2)^{f_2} \ldots (x_n)^{f_n}\right]^{1/n} \quad \text{where } n = \Sigma f_i.$$

or
$$\log \text{G.M.} = \frac{1}{n}[f_1 \log x_1 + f_2 \log x_2 + \ldots + f_n \log x_n] \qquad \ldots(2)$$

Hence Eqns. (1) and (2) show that logarithm of G.M. = A.M. of logarithms of the values.

**(5) Harmonic mean.** If $x_1, x_2, \ldots x_n$ be a set of $n$ observations, then the *harmonic mean is defined as the reciprocal of the* (*arithmetic*) *mean of the reciprocals of the quantities.* Thus

$$\text{H.M.} = \frac{1}{\frac{1}{n}\left(\frac{1}{x_1} + \frac{1}{x_2} + \ldots + \frac{1}{x_n}\right)}$$

In a frequency distribution, $\text{H.M.} = \dfrac{1}{\frac{1}{n}\left(\frac{f_1}{x_1} + \frac{f_2}{x_2} + \ldots + \frac{f_n}{x_n}\right)}$ where $n = \Sigma f_i$.

**Example 25.6.** *Three cities A, B, C are equidistant from each other. A motorist travels from A to B at 30 km/hr, from B to C at 40 km/hr, from C to A at 50 km/hr. Determine the average speed.*

**Solution.** Let $AB = BC = CA = s$ km

Time taken to travel from $A$ to $B = s/30$

Time taken to travel from $B$ to $C = s/40$

Time taken to travel from $C$ to $A = s/50$

$$\therefore \text{ average time taken } = \frac{1}{3}\left(\frac{s}{30}+\frac{s}{40}+\frac{s}{50}\right)$$

$$\text{Thus the average speed } = \frac{s}{\frac{1}{3}\left(\frac{s}{30}+\frac{s}{40}+\frac{s}{50}\right)}$$

*In other words, the average speed is the harmonic mean of 30, 40, 50 km/hr.*

$$\text{Hence the average speed } = \frac{1}{\frac{1}{3}\left(\frac{1}{30}+\frac{1}{40}+\frac{1}{50}\right)} = 38.3 \text{ km/hr.}$$

**Obs.** Of the various measures of central tendency, *the mean* is the most important for it can be computed easily. *The median,* though more easily calculable, cannot be applied with ease to theoretical analysis. Median is of advantage when there are exceptionally large and small values at the ends of the distribution.

*The mode,* though most easily calculated, has the least significance. It is particularly misleading in distributions which are small in numbers or highly unsymmetrical.

*The geometrical mean* though difficult to compute, finds application in cases like populations where we are concerned with a quantity whose changes tend to be directly proportional to the quantity itself.

*The harmonic mean* is useful in limited situations where time and rate or prices are involved.

## PROBLEMS 25.1

1. Draw the histogram and frequency polygon for the following distribution. Also calculate the arithmetic mean :

| Class interval | 0—99 | 100—199 | 200—299 | 300—399 | 400—499 | 500—599 | 600—699 | 700—799 |
|---|---|---|---|---|---|---|---|---|
| Frequency | 10 | 54 | 184 | 264 | 246 | 40 | 1 | 1 |

2. The following marks were given to a batch of candidates :

| | | | | | | | | | |
|---|---|---|---|---|---|---|---|---|---|
| 66 | 62 | 45 | 79 | 32 | 51 | 56 | 60 | 51 | 49 |
| 25 | 42 | 54 | 54 | 58 | 70 | 43 | 58 | 50 | 52 |
| 38 | 67 | 50 | 59 | 48 | 65 | 71 | 30 | 46 | 55 |
| 82 | 51 | 63 | 45 | 53 | 40 | 35 | 56 | 70 | 52 |
| 67 | 55 | 57 | 30 | 63 | 42 | 74 | 58 | 44 | 55 |

Draw a cumulative frequency curve.

Hence find the proportion of candidates securing more than 50 marks. Also mark off the median, the first and third quartiles.

3. Find the mean, median and mode for the following :

| Mid Value : | 15 | 20 | 25 | 30 | 35 | 40 | 45 | 50 | 55 |
|---|---|---|---|---|---|---|---|---|---|
| Frequency : | 2 | 22 | 19 | 14 | 3 | 4 | 6 | 1 | 1 |

4. Calculate mean, median and mode of the following data relating to weight of 120 articles :

| Weight (in gm) : | 0—10 | 10—20 | 20—30 | 30—40 | 40—50 | 50—60 |
|---|---|---|---|---|---|---|
| No. of articles : | 14 | 17 | 22 | 26 | 23 | 18 |

5. The population of a country was 300 million in 1971. It became 520 million in 1989. Calculate the percentage compound rate of growth per annum.
[**Hint.** Use $P_n = P_0(1 + r)^n$, $r$ being the growth rate.]
6. The number of divorces per 1000 marriages in the United States increased from 84 in 1970 to 108 in 1990. Find the annual increase of the divorce rate for the period 1970 to 1990.
7. An aeroplane flies along the four sides of a square at speeds of 100, 200, 300 and 400 km/hr. respectively. What is the average speed of the plane in its flight around the square.
8. A man having to drive 90 km. wishes to achieve an average speed of 30 km/hr. For the first half of the journey, he averages only 20 km/hr. What must be his average speed for the second half of the journey if his overall average is to be 30 km/hr.

9. Following table gives the cumulative frequency of the age of a group of 199 teachers. Find the mean and median age of the group.

| Age in years : | 20—25 | 25—30 | 30—35 | 35—40 | 40—45 | 45—50 | 50—55 | 55—60 | 60—65 | 65—70 |
|---|---|---|---|---|---|---|---|---|---|---|
| Cum. frequ. : | 21 | 40 | 90 | 130 | 146 | 166 | 176 | 186 | 195 | 199 |

10. Recast the following cumulative table in the form of an ordinary frequency distribution and determine the median and the mode :

| *No. of days absent* | *No. of students* | *No. of days absent* | *No. of students* |
|---|---|---|---|
| Less than 5 | 29 | Less than 30 | 644 |
| Less than 10 | 224 | Less than 35 | 650 |
| Less than 15 | 465 | Less than 40 | 653 |
| Less than 20 | 582 | Less than 45 | 655 |
| Less than 25 | 634 | | |

## 25.6 MEASURES OF DISPERSION

Although measures of central tendency do exhibit one of the important characteristics of a distribution, yet they fail to give any idea as to how the individual values differ from the central value, *i.e.*, whether they are closely packed around the central value or widely scattered away from it. Two distributions may have the same mean and the same total frequency, yet they may differ in the extent to which the individual values may be spread about the average (*See* Fig. 25.3). The magnitude of such a variation is called *dispersion*. The important measures of dispersion are given below :

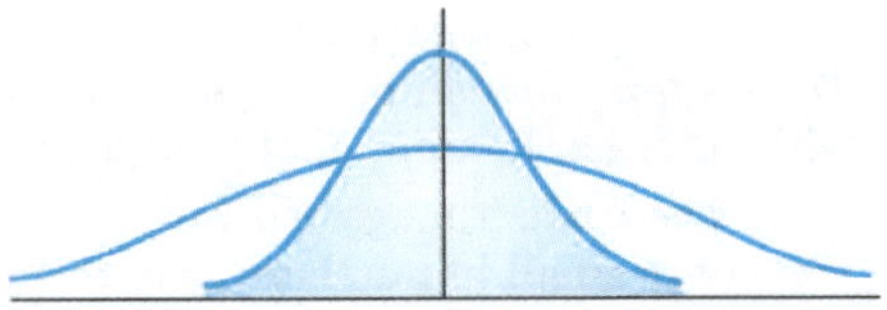

*Same mean different dispersion*

Fig. 25.3

**(1) Range.** *This is the simplest measure of dispersion and is given by the difference between the greatest and the least values in the distribution.* If the weekly wages of a group of labourers are

₹ 21 23 28 25 35 42 39 48

then *range* = Max. value – Min. value = 48 – 21 = ₹ 27.

**(2) Quartile deviation or semi-interquartile range.** One half of the interquartile range is called *quartile deviation,* or *semi-interquartile range.* If $Q_1$ and $Q_3$ are the first and third quartiles, the semi-interquartile range

$$Q = \tfrac{1}{2}(Q_3 - Q_1).$$

**(3) Mean deviation.** The mean deviation is the mean of the absolute differences of the values from the mean, median or mode. Thus *mean deviation (M.D.)*

$$= \frac{1}{n}\Sigma f_i\,|x_i - A|$$

where $A$ is either the mean or the median or the mode. As the positive and negative differences have equal effects, only the absolute value of differences is taken into account.

**(4) Standard deviation.** The most important and the most powerful measure of dispersion is the *standard deviation (S.D.)* : generally denoted by σ. It is computed as the square root of the mean of the squares of the differences of the *variate* values from their mean.

Thus *standard deviation (S.D.)*

$$\sigma = \sqrt{\left[\frac{\Sigma \mathbf{f}_i(\mathbf{x}_i - \bar{\mathbf{x}})^2}{\mathbf{N}}\right]} \qquad ...(1)$$

where $N$ is the total frequency $\Sigma f_i$.

If however, the deviations are measured from any other value, say $A$, instead of $\bar{x}$, it is called the *root-mean-square deviation.*

*The square of the standard deviation is known as the* **variance.**

**Calculation of S.D.** The change of origin and the change of scale considerably reduces the labour in the calculation of standard deviation. The formulae for the computation of σ are as follows :

*I. Short-cut method*

$$\sigma = \sqrt{\left[\frac{\Sigma \mathbf{f}_i \mathbf{d}_i^2}{\Sigma \mathbf{f}_i} - \left(\frac{\Sigma \mathbf{f}_i \mathbf{d}_i}{\Sigma \mathbf{f}_i}\right)^2\right]} \quad ...(2)$$

*II. Step-deviation method*

$$\sigma = \mathbf{h}\sqrt{\left\{\frac{\Sigma \mathbf{f}_i \mathbf{d}_i'^2}{\Sigma \mathbf{f}_i} - \left(\frac{\Sigma \mathbf{f}_i \mathbf{d}_i'}{\Sigma \mathbf{f}_i}\right)^2\right\}} \quad ...(3)$$

where $d_i = x_i - A$ and $d'_i = (x_i - A)/h$, being the assumed mean and $h$ the equal class interval.

*Proof.* We know that $x_i - \bar{x} = (x_i - A) - (\bar{x} - A)$

$$\therefore \quad \Sigma f_i(x_i - \bar{x})^2 = \Sigma f_i[d_i - (\bar{x} - A)]^2 = \Sigma f_i d_i^2 + (\bar{x} - A)^2 \Sigma f_i - 2(\bar{x} - A)\Sigma f_i d_i$$

$$= \Sigma f_i d_i^2 - \frac{(\Sigma f_i d_i)^2}{\Sigma f_i} \qquad \left[\because \bar{x} = A + \frac{\Sigma f_i d_i}{\Sigma f_i}\right]$$

Hence
$$\sigma^2 = \frac{\Sigma f_i (x - \bar{x})^2}{\Sigma f_i} = \frac{\Sigma f_i d_i^2}{\Sigma f_i} - \left(\frac{\Sigma f_i d_i}{\Sigma f_i}\right)^2$$

Further $d'_i = (x_i - A)/h = d_i/h$ or $d_i = hd'_i$, then substituting this value in Eqn. (2), we get Eqn. (3).

**Obs.** *The root mean square deviation is least when measured from the mean.*
The root mean square deviation is given by

$$s^2 = \frac{\Sigma f_i d_i^2}{\Sigma f_i} \quad \text{and} \quad \frac{\Sigma f_i d_i}{\Sigma f_i} = \left[A + \frac{\Sigma f_i d_i}{\Sigma f_i}\right] - A = \bar{x} - A$$

$\therefore$ from Eqn. (2), we have $s^2 = \sigma^2 + (\bar{x} - A)^2$ ...(4)

This shows that $s^2$ is always $> \sigma^2$ and the least value of $s^2 = \sigma^2$. This occurs when $A = \bar{x}$.

## 25.7 (1) COEFFICIENT OF VARIATION

The ratio of the standard deviation to the mean, is known as the *coefficient of variation*. As this is a ratio having no dimension, it is used for comparing the variations between the two groups with different means. It is often expressed as a percentage.

$\therefore$ *Coefficient of variation* $= \dfrac{\sigma}{\bar{x}} \times 100$

**(2) Relations between measures of dispersion**

(*i*) *Quartile deviation = 2/3 (standard deviation)*

(*ii*) *Mean deviation = 4/5 (standard deviation)*

## 25.8 STANDARD DEVIATION OF THE COMBINATION OF TWO GROUPS

*If $m_1, \sigma_1$ be the mean and S.D. of a sample of size $n_1$ and $m_2, \sigma_2$ be those for a sample of size $n_2$, then the S.D. $\sigma$ of the combined sample of size $n_1 + n_2$ is given by*

$$(\mathbf{n}_1 + \mathbf{n}_2)\sigma^2 = \mathbf{n}_1\sigma_1^2 + \mathbf{n}_2\sigma_2^2 + \mathbf{n}_1\mathbf{D}_1^2 + \mathbf{n}_2\mathbf{D}_2^2$$

*where $D_i = m_i - m$, $m$ being the mean of combined sample.*

From Eqn. (4), we have $ns^2 = n\sigma^2 + n(\bar{x} - A)^2$ where $n$ is the size of the sample.

*i.e.* sum of the squares of the deviations from $A = n\sigma^2 + n(\bar{x} - A)^2$.

Now let us apply this result to the first given sample taking $A$ at $m$. Then, sum of the squares of the deviations of $n_1$ items from $m = n_1\sigma_1^2 + n_1(m_1 - m)^2$ ...(5)

Similarly for the second given sample taking $A$ at $m$, sum of the squares of the deviations of $n_2$ items from $m = n_2\sigma_2^2 + n_2(m_2 - m)^2$ ...(6)

Adding Eqns. (5) and (6), sum of the squares of the deviations of $n_1 + n_2$ items from $m$

$$= n_1\sigma_1^2 + n_2\sigma_2^2 + n_1(m_1 - m)^2 + n_2(m_2 - m)^2$$

$$\therefore \quad (n_1 + n_2)\sigma^2 = n_1\sigma_1^2 + n_2\sigma_2^2 + n_1D_1^2 + n_2D_2^2$$

*This result can be extended to the combination of any number of samples, giving a result of the form*

$$(\Sigma n_i)\sigma^2 = \Sigma(n_i\sigma_i^2) + \Sigma(n_i D_i^2).$$

**Example. 25.7.** *Calculate the mean, median and standard deviation from the following table.*

| Marks Obtained | 1 – 10 | 11 – 20 | 21 – 30 | 31 – 40 | 41 – 50 | 51 – 50 |
|---|---|---|---|---|---|---|
| No. of Candidates | 3 | 16 | 26 | 31 | 16 | 8 |

(*Rajasthan, 2018*)

**Solution.** Let $A = 35.5$, then $u = \dfrac{x - A}{h} = \dfrac{x - 35.5}{10}$

$$\text{Now, mean } \bar{x} = A + h\frac{\sum fu}{N} = 35.5 + 10 \times \left(\frac{-35}{100}\right) = 35.5 - 3.5 = 32.0$$

$$\text{Further } \sigma = h\sqrt{\frac{fu}{N} - \left(\frac{\sum fu}{N}\right)^2} = 10\sqrt{\frac{165}{100} - \left(-\frac{35}{100}\right)^2} = 10\sqrt{1.53} = 12.4$$

| Class Interval | $x$ | $f$ | $C$ | $u$ | $fu$ | $fu^2$ |
|---|---|---|---|---|---|---|
| 1 – 10 | 5.5 | 3 | 3 | –3 | –9 | 27 |
| 11 – 20 | 15.5 | 16 | 19 | –2 | –32 | 64 |
| 21 – 30 | 25.5 | 26 | 45 | –1 | –26 | 26 |
| 31 – 40 | 35.5 | 31 | 76 | 0 | 0 | 0 |
| 41 – 50 | 45.5 | 16 | 92 | 1 | 16 | 16 |
| 51 – 60 | 55.5 | 8 | 100 | 2 | 16 | 32 |
| Total | | 100 | | | –35 | 165 |

The median class is 30.5 – 40.5. Hence on modifying the class interval as continuous we have

$$\text{median} = l + \frac{\left(\frac{N}{2} - C\right) \times h}{f} = 30.5 + \frac{(50 - 45)}{31} \times 10 = 30.5 + 1.8 = 32.3.$$

**Example 25.8.** *Calculate the mean and standard deviation of the following frequency distribution :*

| *Weekly wages is ₹* | *No. of men* |
|---|---|
| *4.5—12.5* | *4* |
| *12.5—20.5* | *24* |
| *20.5—28.5* | *21* |
| *28.5—36.5* | *18* |
| *36.5—44.5* | *5* |
| *44.5—52.5* | *3* |
| *52.5—60.5* | *5* |
| *60.5—68.5* | *8* |
| *68.5—76.5* | *2* |

**Solution.** The calculations are arranged in the table below :

| *Wages class* ₹ | *Mid value x* | *No. of men f* | *Step deviation* $d' = \dfrac{x - 32.5}{8}$ | $fd'$ | $fd'^2$ |
|---|---|---|---|---|---|
| 4.5—12.5 | 8.5 | 4 | –3 | –12 | 36 |
| 12.5—20.5 | 16.5 | 24 | –2 | –48 | 96 |
| 20.5—28.5 | 24.5 | 21 | –1 | –21 | 21 |
| 28.5—36.5 | 32.5 | 18 | 0 | 0 | 0 |
| 36.5—44.5 | 40.5 | 5 | 1 | 5 | 5 |
| 44.5—52.5 | 48.5 | 3 | 2 | 6 | 12 |
| 52.5—60.5 | 56.5 | 5 | 3 | 15 | 45 |
| 60.5—68.5 | 64.5 | 8 | 4 | 32 | 128 |
| 68.5—76.5 | 72.5 | 2 | 5 | 10 | 50 |
| | | $\Sigma f = 90$ | | $\Sigma fd' = -13$ | $\Sigma fd'^2 = 393$ |

$\therefore$ mean wage $= 32.5 + 8 \times \dfrac{\Sigma fd'}{\Sigma f} = 32.5 + 8\left(\dfrac{-13}{90}\right) = ₹\ 31.35$

Standard deviation $= 8\sqrt{\dfrac{\Sigma fd'^2}{\Sigma f} - \left(\dfrac{\Sigma fd'}{\Sigma f}\right)^2} = 8\sqrt{\dfrac{393}{90} - \left(\dfrac{-13}{90}\right)^2} = ₹\ 16.64.$

**Example 25.9.** *The following are scores of two batsmen A and B in a series of innings :*

| | | | | | | | | | | |
|---|---|---|---|---|---|---|---|---|---|---|
| *A :* | *12* | *115* | *6* | *73* | *7* | *19* | *119* | *36* | *84* | *29* |
| *B :* | *47* | *12* | *16* | *42* | *4* | *51* | *37* | *48* | *13* | *0* |

*Who is the better score getter and who is more consistent ?*

**Solution.** Let $x$ denote score of $A$ and $y$ that of $B$.

Taking 51 as the origin, we prepare the following table :

| $x$ | $d(= x - 51)$ | $d^2$ | $y$ | $\delta(= y - 51)$ | $\delta^2$ |
|---|---|---|---|---|---|
| 12 | –39 | 1521 | 47 | –4 | 16 |
| 115 | 64 | 4096 | 12 | –39 | 1521 |
| 6 | –45 | 2025 | 16 | –35 | 1225 |
| 73 | 22 | 484 | 42 | –9 | 81 |
| 7 | –44 | 1936 | 4 | –47 | 2209 |
| 19 | –32 | 1024 | 51 | 0 | 0 |
| 119 | 68 | 4624 | 37 | –14 | 196 |
| 36 | –15 | 225 | 48 | –3 | 9 |
| 84 | 33 | 1089 | 13 | –38 | 1444 |
| 29 | –22 | 484 | 0 | 51 | 2601 |
| Total | –10 | 17508 | | –240 | 9302 |

*For A,* A.M. $\bar{x} = 51 + \dfrac{\Sigma d}{n} = 51 - \dfrac{10}{10} = 50$

S.D. $\sigma_1 = \sqrt{\left\{\dfrac{\Sigma d^2}{n} - \left(\dfrac{\Sigma d}{n}\right)^2\right\}} = \sqrt{\{1750.8 - (-1)^2\}} = 41.8$

$\therefore$ coefficient of variation $= \dfrac{\sigma_1}{\bar{x}} \times 100 = \dfrac{41.8}{50} \times 100 = 83.6\%$

*For B,* A.M. $\bar{y} = 51 + \dfrac{\Sigma \delta}{n} = 51 - \dfrac{240}{10} = 27$

S.D. $\sigma_2 = \sqrt{\left\{\dfrac{\Sigma \delta^2}{n} - \left(\dfrac{\Sigma \delta}{n}\right)^2\right\}} = \sqrt{\{930.2 - (-24)^2\}} = 18.8$

$\therefore$ coefficient of variation $= \dfrac{\sigma^2}{\bar{y}} \times 100 = \dfrac{18.8}{27} \times 100 = 69.6\%$

Since the A.M. of $A >$ A.M. of $B$, it follows that $A$ is a better score getter (*i.e.*, more efficient) than $B$.

Since the coefficient of variation of $B <$ the coefficient of variation of $A$, it means that $B$ is more consistent than $A$. Thus even though $A$ is a better player, he is less consistent.

**Example 25.10.** *The numbers examined, the mean weight and S.D. in each group of examination by three medical examiners are given below. Find the mean weight and S.D. of the entire data when grouped together.*

| *Med. Exam.* | *No. Examined* | *Mean Wt. (lbs.)* | *S.D. (lbs.)* |
|---|---|---|---|
| A | *50* | *113* | *6* |
| B | *60* | *120* | *7* |
| C | *90* | *115* | *8* |

**Solution.** We have $n_1 = 50,\ \bar{x}_1 = 113,\ \sigma_1 = 6$

$n_2 = 60,\ \bar{x}_2 = 120,\ \sigma_2 = 7$

$n_3 = 90,\ \bar{x}_3 = 115,\ \sigma_3 = 8.$

If $\bar{x}$ is the mean of the entire data,

$$\bar{x} = \frac{n_1\bar{x}_1 + n_2\bar{x}_2 + n_3\bar{x}_3}{n_1 + n_2 + n_3} = \frac{50 \times 113 + 60 \times 120 + 90 \times 115}{50 + 60 + 90} = \frac{23200}{200} = 116 \text{ lb.}$$

If $\sigma$ is the S.D. of the entire data,

$$N\sigma^2 = n_1\sigma_1^2 + n_2\sigma_2^2 + n_3\sigma_3^2 + n_1D_1^2 + n_2D_2^2 + n_3D_3^2$$

where $N = n_1 + n_2 + n_3 = 200$, $D_1 = \bar{x}_1 - \bar{x} = -3$, $D_2 = \bar{x}_2 - \bar{x} = 4$ and $D_3 = \bar{x}_3 - \bar{x} = -1$.

$$\therefore \quad 200\sigma^2 = 50 \times 36 + 60 \times 49 + 90 \times 64 + 50 \times 9 + 60 \times 16 + 90 \times 1$$

$$= 1800 + 2940 + 5760 + 450 + 960 + 90$$

$$\sigma^2 = \frac{12000}{200} = 60. \quad \text{Hence } \sigma = \sqrt{60} = 7.746 \text{ lb.}$$

## PROBLEMS 25.2

1. The crushing strength of 8 cement concrete experimental blocks, in metric tonnes per sq. cm., was 4.8, 4.2, 5.1, 3.8, 4.4, 4.7, 4.1 and 4.5. Find the mean crushing strength and the standard deviation.
2. Show that the variance of the first $n$ positive integers is $\frac{1}{12}(n^2 - 1)$.
3. The mean of five items of an observation is 4 and the variance is 5.2. If three of the items are 1, 2 and 6, then find the other two.
4. For the distribution

| $x$: | 5 | 6 | 7 | 8 | 9 | 10 | 11 | 12 | 13 | 14 | 15 |
|---|---|---|---|---|---|---|---|---|---|---|---|
| $f$: | 18 | 15 | 34 | 47 | 68 | 90 | 80 | 62 | 35 | 27 | 11 |

find the mean, median and lower and upper quartiles, variance and the standard deviation.

5. The following table shows the marks obtained by 100 candidates in an examination. Calculate the mean, median and standard deviation :

| Marks obtained : | 1—10 | 11—20 | 21—30 | 31—40 | 41—50 | 51—60 |
|---|---|---|---|---|---|---|
| No. of candidates : | 3 | 16 | 26 | 31 | 16 | 8 |

6. Compute the quartile deviation and standard deviation for the following :

| $x$: | 100—109 | 110—119 | 120—129 | 130—139 | 140—149 | 150—159 | 160—169 | 170—179 |
|---|---|---|---|---|---|---|---|---|
| $f$: | 15 | 44 | 133 | 150 | 125 | 82 | 35 | 16 |

7. Calculate (*i*) mean deviation about the mean, (*ii*) mean deviation about the median for the following distribution :

| *Class* : | 3—4.9 | 5—6.9 | 7—8.9 | 9—10.9 | 11—12.9 | 13—14.9 | 15—16.9 |
|---|---|---|---|---|---|---|---|
| $f$ : | 5 | 8 | 30 | 82 | 45 | 24 | 6 |

8. Two observers bring the following two sets of data which represent measurements of the same quantity :

| I. | 105.1 | 103.4 | 104.2 | 104.7 | 104.8 | 105.0 | 104.9 |
|---|---|---|---|---|---|---|---|
| II. | 105.3 | 105.1 | 104.8 | 105.2 | 106.7 | 102.9 | 103.1 |

Calculate the standard deviation in each case. Which set of data is more reliable ? Can the same conclusion be reached by calculating the mean deviation ?

**Obs.** The smaller the coefficient of variation, the greater is the *reliability* or *consistency* in the data.

9. The heights and weights of the 10 armymen are given below. In which characteristics are they more variable ?

| Height in cm. | 170 | 172 | 168 | 177 | 179 | 171 | 173 | 178 | 173 | 179 |
|---|---|---|---|---|---|---|---|---|---|---|
| Weight in kg. | 75 | 74 | 75 | 76 | 77 | 73 | 76 | 75 | 74 | 75 |

10. The index number of prices of two articles $A$ and $B$ for six consecutive weeks are given below :

| $A$: | 314 | 326 | 336 | 368 | 404 | 412 |
|---|---|---|---|---|---|---|
| $B$: | 330 | 331 | 320 | 318 | 321 | 330 |

Find which has a more variable price ?

11. The scores of two golfers $A$ and $B$ in 12 rounds are given below. Who is the better player and who is the more consistent player ?

| $A$: | 74 | 75 | 78 | 72 | 78 | 77 | 79 | 81 | 79 | 76 | 72 | 71 |
|---|---|---|---|---|---|---|---|---|---|---|---|---|
| $B$: | 87 | 84 | 80 | 88 | 89 | 85 | 86 | 82 | 82 | 79 | 86 | 80 |

12. The scores obtained by two batsmen $A$ and $B$ in 10 matches are given below :

| $A$: | 30 | 44 | 66 | 62 | 60 | 34 | 80 | 46 | 20 | 38 |
|---|---|---|---|---|---|---|---|---|---|---|
| $B$: | 34 | 46 | 70 | 38 | 55 | 48 | 60 | 34 | 45 | 30 |

Calculating mean, S.D. and coefficient of variation for each batsman, determine who is more efficient and who is more consistent.

13. Find the mean and standard deviation of the following two samples put together :

| *Sample No.* | *Size* | *Mean* | *S.D.* |
|---|---|---|---|
| 1 | 50 | 158 | 5.1 |
| 2 | 60 | 164 | 4.6 |

14. A distribution consists of three components with frequencies 200, 250 and 300 having means 25, 10 and 15 and S.Ds. 3, 4 and 5 respectively. Show that the mean of the combined distribution is 16 and its S.D. is 7.2 approximately.

## 25.9 (1) MOMENTS

*The rth moment about the mean $\bar{x}$ of a distribution is denoted by $\mu_r$ and is given by*

$$\mu_r = \frac{1}{N} \Sigma f_i (x_i - \bar{x})^r \qquad ...(1)$$

The corresponding moment about any point $a$ is defined as

$$\mu_r' = \frac{1}{N} \Sigma f_i (x_i - a)^r \qquad ...(2)$$

In particular, we have $\mu_0 = \mu_0' = 1$ ...(3)

$$\mu_1 = \frac{1}{N} \Sigma f_i (x_i - \bar{x}) = 0; \ \mu_1' = \frac{1}{N} \Sigma f_i (x_i - a) = \bar{x} - a = d, \text{ say} \qquad ...(4)$$

$$\mu_2 = \frac{1}{N} \Sigma f_i (x_i - \bar{x})^2 = \sigma^2. \qquad ...(5)$$

**(2) Moments about the mean in terms of moments about any point.**

We have

$$\mu_r = \frac{1}{N} \Sigma f_i (x_i - \bar{x})^r = \frac{1}{N} \Sigma f [(x_i - a) - (\bar{x} - a)]^r$$

$$= \frac{1}{N} \Sigma f_i (X_i - d)^r \quad \text{where} \quad X_i = x_i - a, \ d = \bar{x} - a.$$

$$= \frac{1}{N} [\Sigma f_i X_i^r - {}^rC_1 d \Sigma f_i X_i^{r-1} + {}^rC_2 d^2 \Sigma f_i X_i^{r-2} - ...]$$

$$= \mu_r' - {}^rC_1 d \mu_{r-1}' + {}^rC_2 d^2 \mu_{r-2}'' - ...... \qquad ...(6)$$

In particular,

$$\mu_2 = \mu'_2 - \mu'^2_1 \qquad ...(7)$$

$$\mu_3 = \mu'_3 - 3\mu'_2\mu'_1 + 2\mu'^3_1 \qquad ...(8)$$

$$\mu_4 = \mu'_4 - 4\mu'_3\mu'_1 + 6\mu'_2\mu'^2_1 - 3\mu'^4_1 \qquad ...(9)$$

These three results should be committed to memory. It should be noted that in each of these relations, the sum of the coefficients of the various terms on the right side is zero. Also each term on the right side is of the same dimension as the term on the left.

## 25.10 SKEWNESS

*Skewness measures the degree of asymmetry or the departure from symmetry.* If the frequency curve has a longer 'tail' to the right, *i.e.*, the mean is to the right of the mode [as in Fig. 25.4 (*a*)], then the distribution is said to have *positive skewness*. If the curve is more elongated to the left, then it is said to have *negative skewness* [Fig. 25.4 (*b*)].

The following three measures of skewness deserve mention :

(*i*) *Pearson's* coefficient of skewness* $= \dfrac{\text{mean} - \text{mode}}{\sigma}$

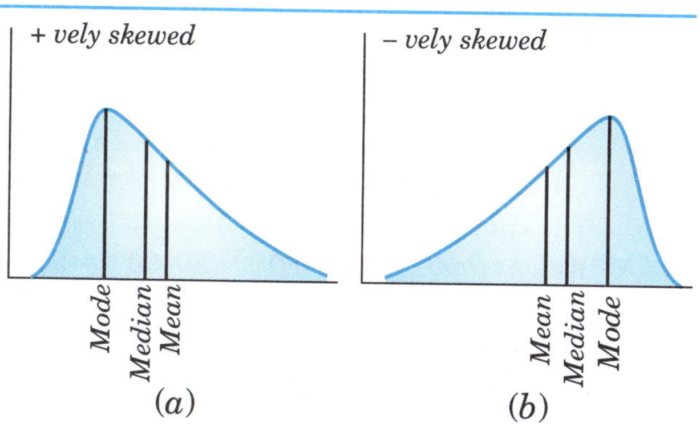

Fig. 23.4.

* After the English statistician and biologist *Karl Pearson* (1857–1936) who did pioneering work and found the English school of statistics.

(ii) *Quartile coefficient of skewness* $= \dfrac{Q_3 + Q_1 - 2Q_2}{Q_3 - Q_1}$

Its value always lies between –1 and +1.

(iii) *Coefficient of skewness based on third moment* $\gamma_1 = \sqrt{\beta_1}$.

where $\beta_1 = \mu_3^2/\mu_2^3$

Thus $\gamma_1 = \sqrt{\beta_1}$ gives the simplest measure of *skewness*.

## 25.11 KURTOSIS

Kurtosis measures the degree of peakedness of a distribution and is given by $\beta_2 = \mu_4/\mu_2^2$.

(*a*) *Leptokurtic* (*b*) *Platykurtic* (*c*) *Mesokurtic*

Fig. 25.5

$\gamma_2 = \beta_2 - 3$ gives the excess of *Kurtosis*. The curves with $\beta_2 > 3$ are called *Leptokurtic* and those with $\beta_2 < 3$ as *Platykurtic*. The normal curve for which $\beta_2 = 3$, is called *Mesokurtic* [Fig. 25.5].

**Example 25.11.** *The first four moments about the working mean 28.5 of a distribution are 0.294, 7.144, 42.409 and 454.98. Calculate the moments about the mean. Also evaluate $\beta_1$, $\beta_2$ and comment upon the skewness and kurtosis of the distribution.*

**Solution.** The first four moments about the arbitrary origin 28.5 are $\mu'_1 = 0.294$, $\mu'_2 = 7.144$, $\mu'_3 = 42.409$, $\mu'_4 = 454.98$.

$$\therefore \quad \mu'_1 = \frac{1}{N}\sum f_i(x_i - 28.5) = \frac{1}{N}\sum f_i x_i - 28.5 = \bar{x} - 28.5 = 0.294 \text{ or } \bar{x} = 28.794$$

$$\mu_2 = \mu'_2 - \mu'^2_1 = 7.144 - (0.294)^2 = 7.058$$

$$\mu_3 = \mu'_3 - 3\mu'_2\mu'_1 + 2\mu'^3_1 = 42.409 - 3(7.144)(0.294) + 2(0.294)^3 = 36.151.$$

$$\mu_4 = \mu'_4 - 4\mu'_3\mu'_1 + 6\mu'_2\mu'^2_1 - 3\mu'^4_1$$
$$= 454.98 - 4(42.409) \times (0.294) + 6(7.144)(0.294)^2 - 3(0.294)^4 = 408.738$$

Now $\quad \beta_1 = \mu_3^2/\mu_2^3 = (36.151)^2/(7.058)^3 = 3.717$

$\beta_2 = \mu_4/\mu_2^2 = 408.738/(7.058)^2 = 8.205.$

$\therefore \quad \gamma_1 = \sqrt{\beta_1} = 1.928$, which indicates considerable skewness of the distribution.

$\gamma_2 = \beta_2 - 3 = 5.205$ which shows that the distribution is *leptokurtic*.

**Example 25.12.** *Calculate the median, quartiles and the quartile coefficient of skewness from the following data :*

| *Weight (lbs)* : | 70–80 | 80–90 | 90–100 | 100–110 | 110–120 | 120–130 | 130–140 | 140–150 |
|---|---|---|---|---|---|---|---|---|
| *No. of persons* : | 12 | 18 | 35 | 42 | 50 | 45 | 20 | 8 |

**Solution.** Here total frequency $N = \Sigma f_i = 230$.

The cumulative frequency table is

| *Weight (lbs)* : | 70–80 | 80–90 | 90–100 | 100–110 | 110–120 | 120–130 | 130–140 | 140–150 |
|---|---|---|---|---|---|---|---|---|
| *f* : | 12 | 18 | 35 | 42 | 50 | 45 | 20 | 8 |
| *cum. f.* : | 12 | 30 | 65 | 107 | 157 | 202 | 222 | 230 |

Now $N/2 = 230/2 = 115$th item which lies in 110–120 group.

$$\therefore \quad \text{median or } Q_2 = L + \frac{N/2 - C}{f} \times h = 110 + \frac{115 - 107}{50} \times 10 = 111.6$$

Also $N/4 = 230/4 = 57.5$ *i.e.* $Q_1$ is 57.5th or 58th item which lies in 90–100 group.

$$\therefore \quad Q_1 = L + \frac{N/4 - C}{f} \times h = 90 + \frac{57.5 - 30}{35} \times 10 = 97.85$$

Similarly, $3N/4 = 172.5$ *i.e.* $Q_3$ is 173rd item which lies in 120–130 group.

$$\therefore \quad Q_3 = L + \frac{3N/4 - C}{f} \times h = 120 + \frac{172.5 - 157}{45} \times 10 = 123.44$$

Hence quartile coefficient of skewness $= \dfrac{Q_1 + Q_3 - 2Q_2}{Q_3 - Q_1}$

$$= \frac{97.85 + 123.44 - 2 \times 111.6}{123.44 - 97.85} = -\ 0.07 \text{ (approx.)}.$$

## PROBLEMS 25.3

1. Calculate the first four moments of the following distribution about the mean :

| $x$ : | 0 | 1 | 2 | 3 | 4 | 5 | 6 | 7 | 8 |
|---|---|---|---|---|---|---|---|---|---|
| $f$ : | 1 | 8 | 28 | 56 | 70 | 56 | 28 | 8 | 1 |

Also evaluate $\beta_1$ and $\beta_2$.

2. The following table gives the monthly wages of 72 workers in a factory. Compute the standard deviation, quartile deviation, coefficients of variation and skewness.

| *Monthly wages (in ₹)* | *No. of workers* | *Monthly wages (in ₹)* | *No. of workers* |
|---|---|---|---|
| 12.5—17.5 | 2 | 37.5—42.5 | 4 |
| 17.5—22.5 | 22 | 42.5—47.5 | 6 |
| 22.5—27.5 | 19 | 47.5—52.5 | 1 |
| 27.5—32.5 | 14 | 52.5—57.5 | 1 |
| 32.5—37.5 | 3 | | |

3. Find Pearson's coefficient of skewness for the following data :

| Class : | 10–19 | 20–29 | 30–39 | 40–49 | 50–59 | 60–69 | 70–79 | 80–89 |
|---|---|---|---|---|---|---|---|---|
| Frequency : | 5 | 9 | 14 | 20 | 25 | 15 | 8 | 4 |

4. Compute the quartile coefficient of skewness for the following distribution :

| $x$ : | 3–7 | 8–12 | 13–17 | 18–22 | 23–27 | 28–32 | 33–37 | 38–42 |
|---|---|---|---|---|---|---|---|---|
| $f$ : | 2 | 108 | 580 | 175 | 80 | 32 | 18 | 5 |

Also compute the measure of skewness based on the third moment.

5. The first three moments of a distribution about the value 2 of the variable are 1, 16 and –40. Show that the mean = 3, the variance = 15 and $\mu_3 = -86$.
6. Compute skewness and kurtosis, if the first four moments of a frequency distribution $f(x)$ about the value $x = 4$ are respectively 1, 4, 10 and 45.
7. The first four moments of a distribution about the value '4' of the variable are –1.5, 17, –30 and 108. State whether the distribution is leptokurtic or platykurtic.
8. The first four moments of a distribution about the value '0' are –0.20, 1.76, –2.36 and 10.88. Find the moments about the mean and the kurtosis.

## 25.12 CORRELATION

So far we have confined our attention to the analysis of observations on a single variable. There are, however, many phenomenae where the changes in one variable are related to the changes in the other variable. For instance, the yield of a crop varies with the amount of rainfall, the price of a commodity increases with the reduction in its supply and so on. Such a simultaneous variation, *i.e.* when the changes in one variable are associated or followed by changes in the other, is called *correlation.* Such a data connecting two variables is called *bivariate population.*

If an increase (or decrease) in the values of one variable corresponds to an increase (or decrease) in the other, the correlation is said to be *positive.* If the increase (or decrease) in one corresponds to the decrease (or increase) in the other, the correlation is said to be *negative.* If there is no relationship indicated between the variables, they are said to be *independent or uncorrelated.*

To obtain a measure of relationship between the two variables, we plot their corresponding values on the graph, taking one of the variables along the $x$-axis and the other along the $y$-axis. (Fig. 25.6).

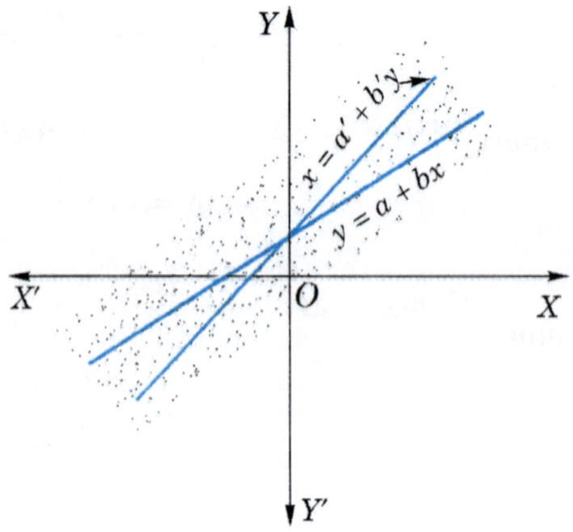

Fig. 25.6

Let the origin be shifted to $(\bar{x}, \bar{y})$, where $\bar{x}, \bar{y}$ are the means of $x$'s and $y$'s that the new co-ordinates are given by

$$X = x - \bar{x}, \qquad Y = y - \bar{y}.$$

Now the points $(X, Y)$ are so distributed over the four quadrants of $XY$-plane that the product $XY$ is positive in the first and third quadrants but negative in the second and fourth quadrants. The algebraic sum of the products can be taken as describing the trend of the dots in all the quadrants.

$\therefore$ (*i*) If $\Sigma XY$ is positive, the trend of the dots is through the first and third quadrants,

(*ii*) if $\Sigma XY$ is negative the trend of the dots is in the second and fourth quadrants, and

(*iii*) if $\Sigma XY$ is zero, the points indicate no trend *i.e.* the points are evenly distributed over the four quadrants.

The $\Sigma XY$ or better still $\frac{1}{n}\Sigma XY$, *i.e.*, the average of $n$ products may be taken as a measure of correlation. If we put $X$ and $Y$ in their units, *i.e.*, taking $\sigma_x$ as the unit for $x$ and $\sigma_y$ for $y$, then

$$\frac{1}{n}\Sigma\frac{X}{\sigma_x}\cdot\frac{Y}{\sigma_y}, \text{ i.e., } \frac{\Sigma XY}{n\sigma_x\sigma_y}$$

is the *measure of correlation.*

## 25.13 COEFFICIENT OF CORRELATION

The numerical measure of correlation is called the *coefficient of correlation* and is defined by the relation

$$r = \frac{\Sigma XY}{n\sigma_x\sigma_y}$$

where $X$ = deviation from the mean $\bar{x} = x - \bar{x}$, $Y$ = deviation from the mean $\bar{y} = y - \bar{y}$,

$\sigma_x$ = S.D. of $x$-series, $\sigma_y$ = S.D. of $y$-series and $n$ = number of values of the two variables.

**Methods of calculation :**

(*a*) *Direct method.* Substituting the value of $\sigma_x$ and $\sigma_y$ in the above formula, we get

$$r = \frac{\Sigma XY}{\sqrt{(\Sigma X^2 \Sigma Y^2)}} \qquad ...(1)$$

Another form of the formula (1) which is quite handy for calculation is

$$r = \frac{n\Sigma xy - \Sigma x\Sigma y}{\sqrt{[\{n\Sigma x^2 - (\Sigma x)^2\} \times \{n\Sigma y^2 - (\Sigma y)^2\}]}} \qquad ...(2)$$

(*b*) *Step-deviation method.* The direct method becomes very lengthy and tedious if the means of the two series are not integers. In such cases, use is made of assumed means. If $d_x$ and $d_y$ are step-deviations from the assumed means, then

$$r = \frac{n\Sigma d_x d_y - \Sigma d_x \Sigma d_y}{\sqrt{[\{n\Sigma d_x^2 - (\Sigma d_x)^2\} \times \{n\Sigma d_y^2 - (\Sigma d_y)^2\}]}} \qquad ...(3)$$

where $d_x = (x - a)/h$ and $d_y = (y - b)/k$.

**Obs.** The change of origin and units do not alter the value of the correlation coefficient since $r$ is a pure number.

(*c*) *Co-efficient of correlation for grouped data.* When $x$ and $y$ series are both given as frequency distributions, these can be represented by a two-way table known as the *correlation-table*. It is double-entry table with one series along the horizontal and the other along the vertical as shown on page 969. The co-efficient of correlation for such a *bivariate frequency distribution* is calculated by the formula.

$$r = \frac{n(\Sigma f d_x d_y) - (\Sigma f d_x)(\Sigma f d_y)}{\sqrt{[\{n\Sigma f d_x^2 - (\Sigma f d_x)^2\} \times \{n\Sigma f d_y^2 - (\Sigma f d_y)^2\}]}} \quad ...(4)$$

where, $d_x$ = deviation of the central values from the assumed mean of $x$-series,

$d_y$ = deviation of the central values from the assumed mean of $y$-series,

$f$ is the frequency corresponding to the pair $(x, y)$

and $n(= \Sigma f)$ is the total number of frequencies.

**Example 25.13.** *In two groups A and B, there are 6 winning posibilities are given below.*

| A | –1 | 1 | 2 | 4 | 6 | 7 |
|---|---|---|---|---|---|---|
| B | –1 | 2 | 3 | 3 | 5 | 8 |

*find the correlation of them.* (*Marathwada, 2019*)

**Solution.** Given,

| A | –1 | 1 | 2 | 4 | 6 | 7 |
|---|---|---|---|---|---|---|
| B | –1 | 2 | 3 | 3 | 5 | 8 |

, $n = 6$

| $A$ | $B$ | $A^2$ | $B^2$ | $AB$ |
|---|---|---|---|---|
| –1 | –1 | 1 | 1 | 1 |
| 1 | 2 | 1 | 4 | 2 |
| 2 | 3 | 4 | 9 | 6 |
| 4 | 3 | 16 | 9 | 12 |
| 6 | 5 | 36 | 25 | 30 |
| 7 | 8 | 49 | 64 | 56 |
| 19 | 20 | 107 | 112 | 109 |

$\Sigma A, \Sigma B, \Sigma A^2, \Sigma B^2, \Sigma AB$

$$A = \hat{a}x + \hat{b}$$

$$\hat{a} = \frac{(\Sigma A)(\Sigma B) - n\,\Sigma\,AB}{(\Sigma A)^2 - n\Sigma A^2} = \frac{(19)^2(20) - 6(107)}{(19)^2 - 6(107)} = \frac{-262}{-281} \approx 0.9324$$

$$B = \hat{a}A + b$$

$$\hat{b} = \frac{(\Sigma A)(\Sigma AB)(\Sigma B)(\Sigma A^2)}{(\Sigma A^2) - n\Sigma A^2} = \frac{(19)(107) - (20)(107)}{(19)^2 - 6(107)} = \frac{-107}{-281} = 0.3808$$

So, $$B = \hat{a}A + \hat{b}$$

$\therefore$ $$B = 0.9324A + 0.3808$$

$$S_A = \sqrt{\frac{\Sigma(A - \bar{A})^2}{n-1}} = \sqrt{\frac{\Sigma A^2 - \frac{1}{n}(\Sigma A)^2}{n-1}} = \sqrt{\frac{107 - \frac{1}{6}(19)^2}{5}} \approx 3.0605$$

$$S_B = \sqrt{\frac{112 - \frac{1}{6}(20)^2}{5}} = 3.0111$$

$$\delta = \left(\frac{S_A}{S_B}\right)\hat{a} \approx 0.9477 (\approx 1)$$

$$\delta = \frac{\Sigma_{AB} - \frac{1}{n}(\Sigma_A)(\Sigma_B)}{(n-1)S_A \cdot S_B} = 0.9477.$$

**Exercise 25.14:** *Find the mean deviation from the median for the data:*

| Marks | 0 – 10 | 10 – 20 | 20 – 30 | 30 – 40 | 40 – 50 |
|---|---|---|---|---|---|
| No. of Students | 5 | 8 | 15 | 16 | 6 |

**Solution.**

| Class Interval | $x$ (mid value) | $f$ | $c$ | $\lvert x - \bar{x}\rvert$ | $f\lvert x - \bar{x}\rvert$ |
|---|---|---|---|---|---|
| 0 – 10 | 5 | 5 | 5 | 23 | 115 |
| 10 – 20 | 15 | 8 | 13 | 13 | 104 |
| 20 – 30 | 25 | 15 | 28 | 3 | 45 |
| 30 – 40 | 35 | 16 | 44 | 7 | 112 |
| 40 – 50 | 45 | 6 | 50 | 17 | 102 |
| Total | | $\Sigma f = 50$ | | | $\Sigma f\lvert x - \bar{x}\rvert = 478$ |

Here $N = 50, \dfrac{N}{2} = 25$

Hence median class is 20 – 30.

Let $x$ be the median then

$$x = \frac{l + \left(\frac{N}{2} - C\right)}{f} \times h$$

$$= 20 + \frac{(25-13)}{15} \times 10 = 28$$

$\therefore$ mean deviation from the median

$$\bar{x} = \frac{1}{N}\left|\sum f(x - \bar{x})\right| = \frac{1}{50}(478) = 9.56 \text{ marks.}$$

**Example 25.15.** *Calculate the coefficient of correlation for the data:*

| | | | | | | | | |
|---|---|---|---|---|---|---|---|---|
| $X$ : | 55 | 56 | 57 | 57 | 58 | 59 | 60 | 62 |
| $Y$ : | 57 | 58 | 55 | 58 | 62 | 62 | 59 | 61 |

*(Bhilai, 2018)*

**Solution.**

| $X$ | $Y$ | $x$ | $y$ | $x^2$ | $y^2$ | $xy$ |
|---|---|---|---|---|---|---|
| 55 | 57 | –3 | –2 | 9 | 4 | 6 |
| 56 | 58 | –2 | –1 | 4 | 1 | 2 |
| 57 | 55 | –1 | –4 | 1 | 16 | 4 |
| 57 | 58 | –1 | –1 | 1 | 1 | 1 |
| 58 | 62 | 0 | 3 | 0 | 9 | 0 |
| 59 | 62 | 1 | 3 | 1 | 9 | 3 |
| 60 | 59 | 2 | 0 | 4 | 0 | 0 |
| 62 | 61 | 4 | 2 | 16 | 4 | 8 |
| $\Sigma x = 464$ | $\Sigma y = 472$ | $\Sigma x = 0$ | $\Sigma y = 0$ | $\Sigma x^2 = 36$ | $\Sigma y^2 = 44$ | $\Sigma xy = 24$ |

$$\bar{x} = \frac{\Sigma x}{h} = \frac{468}{8} = 58$$

$$\bar{y} = \frac{\Sigma y}{h} = \frac{472}{8} = 59$$

$$r = \frac{\Sigma xy}{\sqrt{\Sigma x^2 \, \Sigma y^2}} = \frac{24}{\sqrt{36 \times 44}} = 0.603$$

## 25.14 LINES OF REGRESSION

It frequently happens that the dots of the scatter diagram generally, tend to cluster along a well defined direction which suggests a linear relationship between the variables $x$ and $y$. Such a line of best-fit for the given distribution of dots is called the *line of regression* (Fig. 25.6). In fact there are two such lines, one giving the best possible mean values of $y$ for each specified value of $x$ and the other giving the best possible mean values of $x$ for given values of $y$. The former is known as the *line of regression of y on x* and the latter as the *line of regression of x on y*.

Consider first the line of regression of $y$ on $x$. Let the straight line satisfying the general trend of $n$ dots in a scatter diagram be

$$y = a + bx \qquad ...(1)$$

We have to determine the constants $a$ and $b$ so that Eqn. (1) gives for each value of $x$, the best estimate for the average value of $y$ in accordance with the *principle of least squares* (page 936), therefore, the normal equations for $a$ and $b$ are

$$\Sigma y = na + b\Sigma x \qquad ...(2)$$

and
$$\Sigma xy = a\Sigma x + b\Sigma x^2 \qquad ...(3)$$

Eqn. (2) gives $\quad \frac{1}{n}\Sigma y = a + b \,.\, \frac{1}{n}\Sigma x$ *i.e.*, $\bar{y} = a + b\bar{x}$.

This shows that $(\bar{x}, \bar{y})$, *i.e.*, the means of $x$ and $y$, lie on Eqn. (1).

Shifting the origin to $(\bar{x}, \bar{y})$, Eqn. (3) takes the form

$$\Sigma(x - \bar{x})(y - \bar{y}) = a\Sigma(x - \bar{x}) + b\Sigma(x - \bar{x})^2, \text{ but } a\Sigma(x - \bar{x}) = 0,$$

$$\therefore \qquad b = \frac{\Sigma(x - \bar{x})(y - \bar{y})}{\Sigma(x - \bar{x})^2} = \frac{\Sigma XY}{\Sigma X^2} = \frac{\Sigma XY}{n\sigma_x^2} = r\frac{\sigma_y}{\sigma_x} \qquad \left[\because r = \frac{\Sigma XY}{n\sigma_x \sigma_y}\right]$$

Thus the line of best fit becomes $\quad y - \bar{y} = r\frac{\sigma_y}{\sigma_x}(x - \bar{x}) \qquad ...(4)$

which is the equation of *the line of regression of y on x*. Its slope is called the *regression coefficient of y on x*.

Interchanging $x$ and $y$, we find that the line of regression of $x$ on $y$ is

$$x - \bar{x} = r\frac{\sigma_x}{\sigma_y}(y - \bar{y}) \qquad ...(5)$$

Thus the *regression coefficient of y on x* $= r\sigma_y/\sigma_x \qquad ...(6)$

and the *regression coefficient of x on y* $= r\sigma_x/\sigma_y \qquad ...(7)$

**Cor.** *The correlation coefficient r is the geometric mean between the two regression co-efficients.*

For $\qquad r\frac{\sigma_y}{\sigma_x} \times r\frac{\sigma_x}{\sigma_y} = r^2.$

**Example 25.16.** *Following records are available:*

*Regression equations:* $8x - 10y + 16 = 0$, $40x - 18y = 214$, $\sigma_x = 3$

*Find the (i) mean values of x and y (ii) coefficient of correlation* (*Mumbai (MCA). 2019*)

**Solution.** Since both the lines of regressian pass through the points $(\bar{x}, \bar{y})$, we have

$$8\bar{x} - 10\bar{y} + 66 = 0$$

$$40\bar{x} - 18\bar{y} - 214 = 0$$

Solving them $\bar{x} = 13$ and $\bar{y} = 17$ writing the regressian lines as

$$y = 0.8x + 6.6$$

and

$$x = 0.45y + 5.35$$

We get regression coefficients $\quad r\dfrac{\sigma y}{\sigma x} = 0.8$ and $r\dfrac{\sigma x}{\sigma y} = 0.45$

Multiplying then, $\quad r^2 = 0.8 \times 0.45 = 0.36$ or $r = 0.6$

Also, $\quad r\dfrac{\sigma_y}{xy} = 0.8, \quad \dfrac{0.6 \times \sigma_y}{3} = 0.8$

$$\sigma_y = 4.$$

**Example 25.17.** *In the following table are recorded data showing the test scores made by salesman on an intelligence test and their weekly sales :*

| *Salesman* | *1* | *2* | *3* | *4* | *5* | *6* | *7* | *8* | *9* | *10* |
|---|---|---|---|---|---|---|---|---|---|---|
| *Test scores* | *40* | *70* | *50* | *60* | *80* | *50* | *90* | *40* | *60* | *60* |
| *Sales (000)* | *2.5* | *6.0* | *4.5* | *5.0* | *4.5* | *2.0* | *5.5* | *3.0* | *4.5* | *3.0* |

*Calculate the regression line of sales on test scores and estimate the most probable weekly sales volume if a salesman makes a score of 70.*

**Solution.** With the help of the table below, we have

$$\bar{x} = \text{mean of } x \text{ (test scores)} = 60 + 0/10 = 60$$

$$\bar{y} = \text{mean of } y \text{ (sales)} = 4.5 + (-4.5)/10 = 4.05.$$

Regression line of sales ($y$) on scores ($x$) is given by

$$y - \bar{y} = r(\sigma_y / \sigma_x)(x - \bar{x})$$

where

$$r\frac{\sigma_y}{\sigma_x} = \frac{\Sigma XY}{\sigma_x \sigma_y} \times \frac{\sigma_y}{\sigma_x} = \frac{\Sigma XY}{(\sigma_x)^2} = \left[\Sigma d_x d_y - \frac{\Sigma d_x \Sigma d_y}{n}\right] \Big/ \left[\Sigma d_x^2 - (\Sigma d_x)^2 / n\right]$$

$$= \frac{140 - \dfrac{0 \times (-4.5)}{10}}{2400 - 0^2/10} = \frac{140}{2400} = 0.06$$

$\therefore$ the required regression line is

$$y - 4.05 = 0.06(x - 60) \quad \text{or} \quad y = 0.06x + 0.45.$$

For $x = 70, y = 0.06 \times 70 + 0.45 = 4.65$.

Thus the most probable weekly sales volume for a score of 70 is 4.65.

| *Test scores* $x$ | *Sales* $y$ | *Deviation of x from assumed mean (= 60)* $d_x$ | *Deviation of y from assumed average (= 4.5)* $d_y$ | $d_x \times d_y$ | $d_x^2$ | $d_y^2$ |
|---|---|---|---|---|---|---|
| 40 | 2.5 | −20 | −2 | 40 | 400 | 4 |
| 70 | 6.0 | 10 | 1.5 | 15 | 100 | 2.25 |
| 50 | 4.5 | −10 | 0 | 0 | 100 | 0 |
| 60 | 5.0 | 0 | 0.5 | 0 | 0 | 2.25 |
| 80 | 4.5 | 20 | 0 | 0 | 400 | 0 |
| 50 | 2.0 | −10 | −2.5 | 25 | 100 | 6.25 |
| 90 | 5.5 | 30 | 1 | 30 | 900 | 1.00 |
| 40 | 3.0 | −20 | −1.5 | 30 | 400 | 2.25 |
| 60 | 4.5 | 0 | 0 | 0 | 0 | 0 |
| 60 | 3.0 | 0 | −1.5 | 0 | 0 | 2.25 |
| | | $\Sigma d_x = 0$ | $\Sigma d_y = -4.5$ | $\Sigma d_x d_y = 140$ | $\Sigma d_x^2 = 2400$ | $\Sigma d_y^2 = 18.25$ |

**Example 25.18.** *If θ is the angle between the two regression lines, show that*

$$\tan\theta = \frac{1-r^2}{r} \cdot \frac{\sigma_x\sigma_y}{\sigma_x^2+\sigma_y^2}$$

*Explain the significance when r = 0 and r = ± 1.*

**Solution.** The equations to the line of regression of $y$ on $x$ and $x$ on $y$ are

$$y - \bar{y} = r\frac{\sigma_y}{\sigma_x}(x-\bar{x}) \text{ and } x - \bar{x} = r\frac{\sigma_x}{\sigma_y}(y-\bar{y})$$

∴ their slopes are $m_1 = r\sigma_y/\sigma_x$ and $m_2 = \sigma_y/r\sigma_x$

Thus $$\tan\theta = \frac{m_2-m_1}{1+m_1m_2} = \frac{\sigma_y/r\sigma_x - r\sigma_y/\sigma_x}{1+\sigma_y^2/\sigma_x^2} = \frac{1-r^2}{r}\cdot\frac{\sigma_x\sigma_y}{\sigma_x^2+\sigma_y^2}$$

*When* $r = 0$, *tan* $\theta \to \infty$ or $\theta = \pi/2$ *i.e. when the variables are independent, the two lines of regression are perpendicular to each other.*

*When* $r = \pm 1$, $\tan\theta = 0$ *i.e.,* $\theta = 0$ *or* $\pi$. *Thus the lines of regression coincide i.e., there is perfect correlation between the two variables.*

**Example 25.19.** *In a partially destroyed laboratory record, only the lines of regression of y on x and x on y are available as 4x – 5y + 33 = 0 and 20x – 9y = 107 respectively. Calculate* $\bar{x}, \bar{y}$ *and the coefficient of correlation between x and y.*

**Solution.** Since the regression lines pass through $(\bar{x}, \bar{y})$, therefore,

$$4\bar{x} - 5\bar{y} + 33 = 0, \quad 20\bar{x} - 9\bar{y} = 107.$$

Solving these equations, we get $\bar{x} = 13$, $\bar{y} = 17$.

Rewriting the line of regression of $y$ on $x$ as $y = \frac{4}{5}x + \frac{33}{5}$, we get

$$b_{yx} = r\frac{\sigma_y}{\sigma_x} = \frac{4}{5} \quad ...(i)$$

Rewriting the line of regression of $x$ on $y$ as $x = \frac{9}{20}y + \frac{107}{9}$, we get

$$b_{xy} = r\frac{\sigma_x}{\sigma_y} = \frac{9}{20} \quad ...(ii)$$

Multiplying Eqns. (*i*) and (*ii*), we get

$$r^2 = \frac{4}{5} \times \frac{9}{20} = 0.36 \quad \therefore \quad r = 0.6$$

Hence $r = 0.6$, the positive sign being taken as $b_{yx}$ and $b_{xy}$ both are positive.

**Example 25.20.** *Establish the formula* $r = \dfrac{\sigma_x^2 + \sigma_y^2 - \sigma_{x-y}^2}{2\sigma_x\sigma_y}$

*Hence calculate r from the following data :*

| *x* : | 21 | 23 | 30 | 54 | 57 | 58 | 72 | 78 | 87 | 90 |
|---|---|---|---|---|---|---|---|---|---|---|
| *y* : | 60 | 71 | 72 | 83 | 110 | 84 | 100 | 92 | 113 | 135 |

**Solution.** (*a*) Let $z = x - y$ so that $\bar{z} = \bar{x} - \bar{y}$.

∴ $$z - \bar{z} = (x-\bar{x}) - (y-\bar{y})$$

or $$(z-\bar{z})^2 = (x-\bar{x})^2 + (y-\bar{y})^2 - 2(x-\bar{x})(y-\bar{y})$$

Summing up for $n$ terms, we have

$$\Sigma(z-\bar{z})^2 = \Sigma(x-\bar{x})^2 + \Sigma(y-\bar{y})^2 - 2\Sigma(x-\bar{x})(y-\bar{y})$$

or $$\frac{\Sigma(z-\bar{z})^2}{n} = \frac{\Sigma(x-\bar{x})^2}{n} + \frac{\Sigma(y-\bar{y})^2}{n} - 2\frac{\Sigma(x-\bar{x})(y-\bar{y})}{n}$$

*i.e.,* $$\sigma_z^2 = \sigma_x^2 + \sigma_y^2 - 2r\sigma_x\sigma_y \qquad \left[\because\ r = \frac{\Sigma(x-\bar{x})(y-\bar{y})}{n\sigma_x\sigma_y}\right.$$

which is the required result.

(*b*) To fine *r*, we have to calculate $\sigma_x$, $\sigma_y$ and $\sigma_{x-y}$. We make the following table :

| $x$ | $X = x - 54$ | $X^2$ | $y$ | $Y = y - 100$ | $Y^2$ | $y - x$ | $(x - y)^2$ |
|---|---|---|---|---|---|---|---|
| 21 | –33 | 1089 | 60 | –40 | 1600 | 39 | 1521 |
| 23 | –31 | 961 | 71 | –29 | 841 | 48 | 2304 |
| 30 | –24 | 576 | 72 | –28 | 784 | 42 | 1764 |
| 54 | 0 | 0 | 83 | –17 | 289 | 29 | 841 |
| 57 | 3 | 9 | 110 | 10 | 100 | 53 | 2809 |
| 58 | 4 | 16 | 84 | –16 | 256 | 26 | 676 |
| 72 | 18 | 324 | 100 | 0 | 0 | 28 | 784 |
| 78 | 24 | 576 | 92 | –8 | 64 | 14 | 196 |
| 87 | 33 | 1089 | 113 | 13 | 169 | 26 | 676 |
| 90 | 36 | 1296 | 135 | 35 | 1225 | 45 | 2025 |
| Total | 30 | 5936 | | –80 | 5328 | 350 | 13596 |

$$\therefore \quad \sigma_x^2 = \frac{\Sigma X^2}{N} - \left(\frac{\Sigma X}{N}\right)^2 = \frac{5636}{10} - \left(\frac{30}{10}\right)^2 = 593.6 - 9 = 584.6$$

$$\sigma_y^2 = \frac{\Sigma Y^2}{N} - \left(\frac{\Sigma Y}{N}\right)^2 = \frac{5328}{10} - \left(\frac{-80}{10}\right)^2 = 532.8 - 64 = 468.8$$

$$\sigma_{x-y}^2 = \frac{\Sigma(x-y)^2}{N} - \left\{\frac{\Sigma(x-y)}{N}\right\}^2 = 1359.6 - 1225 = 134.6$$

From the above formula,

$$r = \frac{\sigma_x^2 + \sigma_y^2 - \sigma_{x-y}^2}{2\sigma_x\sigma_y} = \frac{584.6 + 468.8 - 134.6}{2 \times 24.18 \times 23.85} = 0.876.$$

**Example 25.21.** *While calculating correlation coefficient between two variables x and y from 25 pairs of observations, the following results were obtained : n = 25, Σx = 125, Σx² = 650, Σy = 100, Σy² = 460, Σxy = 508.*

*Later it was discovered at the time of checking that the pairs of values*

| x | y |
|---|---|
| 8 | 12 |
| 6 | 8 |

*were copied down as*

| x | y |
|---|---|
| 6 | 14 |
| 8 | 6 |

*Obtain the correct value of correlation coefficient.* (*V.T.U., 2011 S*)

**Solution.** To get the correct results, we subtract the incorrect values and add the corresponding correct values.

$\therefore$ The correct results would be

$$\Sigma n = 25,\ \Sigma x = 125 - 6 - 8 + 8 + 6 = 125,\ \Sigma x^2 = 650 - 6^2 - 8^2 + 8^2 + 6^2 = 650$$

$$\Sigma y = 100 - 14 - 6 + 12 + 8 = 100,\ \Sigma y^2 = 460 - 14^2 - 6^2 + 12^2 + 8^2 = 436$$

$$\Sigma xy = 508 - 6 \times 14 - 8 \times 6 + 8 \times 12 + 6 \times 8 = 520$$

$$r = \frac{n\Sigma xy - (\Sigma x)(\Sigma y)}{\sqrt{[\{n\Sigma x^2 - (\Sigma x)^2\}\{n\Sigma y^2 - (\Sigma y^2)\}]}} = \frac{25 \times 520 - 125 \times 100}{\sqrt{[\{25 \times 650 - (125)^2\}\{25 \times 436 - (100)^2\}]}}$$

$$= \frac{20}{\sqrt{(25 \times 36)}} = \frac{2}{3}.$$

## 25.15 STANDARD ERROR OF ESTIMATE

The sum of the squares of the deviations of the points from the line of regression of *y* on *x* is

$$\Sigma(y - a - bx)^2 = \Sigma(Y - bX)^2, \text{ where } X = x - \bar{x},\ Y = y - \bar{y}$$

$$= \sum \left( Y - r\frac{\sigma_y}{\sigma_x} X \right)^2 = \Sigma Y^2 - 2r(\sigma_y/\sigma_x)\, \Sigma XY + r^2(\sigma_y^2/\sigma_x^2)\, \Sigma X^2$$

$$= n\sigma_y^2 - 2r(\sigma_y/\sigma_x)\, r\,.\, n\sigma_x\sigma_y + r^2(\sigma_y^2/\sigma_x^2)\,.\, n\sigma_x^2 = n\sigma_y^2(1 - r^2).$$

Denoting this sum of squares by $nS_y^2$, we have $S_y = \sigma_y \sqrt{(1 - r^2)}$ ...(1)

Since $S_y$ is the root mean square deviation of the points from the regression line of $y$ on $x$, it is called the *standard error of estimate* of $y$. Similarly the standard error of estimate of $x$ is given by

$$S_x = \sigma_x \sqrt{(1 - r^2)} \qquad ...(2)$$

Since the sum of the squares of deviations cannot be negative, it follows that

$$r^2 \le 1 \quad \text{or} \quad -1 \le r \le 1.$$

*i.e., correlation coefficient lies between –1 and 1.*

If $r = 1$ or $-1$, the sum of the squares of deviations from either line of regression is zero. Consequently each deviation is zero and all the points lie on both the lines of regression. These two lines coincide and we say that the correlation between the variables is *perfect*. The nearer $r^2$ is to unity the closer are the points to the lines of regression. Thus *the departure of $r^2$ from unity is a measure of departure from linearity of the relationship between the variables.*

## 25.16 RANK CORRELATION*

A group of $n$ individuals may be arranged in order to merit with respect to some characteristic. The same group would give different orders for different characteristics. Considering the orders corresponding to two characteristics $A$ and $B$, the correlation between these $n$ pairs of ranks is called the *rank correlation* in the characteristics $A$ and $B$ for that group of individuals.

Let $x_i, y_i$ be the ranks of the $i$th individuals in $A$ and $B$ respectively. Assuming that no two individuals are bracketed equal in either case, each of the variables taking the values 1, 2, 3, ..., $n$, we have

$$\bar{x} = \bar{y} = \frac{1 + 2 + 3 + \ldots + n}{n} = \frac{n(n+1)}{2n} = \frac{n+1}{2}$$

If $X$, $Y$ be the deviations of $x$, $y$ from their means, then

$$\Sigma X_i^2 = \Sigma(x_i - \bar{x})^2 = \Sigma x_i^2 + n(\bar{x})^2 - 2\bar{x}\, \Sigma x_i = \Sigma n^2 + \frac{n(n+1)^2}{4} - 2\frac{n+1}{2}\,.\, \Sigma n$$

$$= \frac{n(n+1)(2n+1)}{6} + \frac{n(n+1)^2}{4} - \frac{n(n+1)^2}{2} = \frac{1}{12}(n^3 - n)$$

Similarly $\quad \Sigma Y_i^2 = \dfrac{1}{12}(n^3 - n)$

Now let $\quad d_i = x_i - y_i \quad$ so that $\quad d_i = (x_i - \bar{x}) - (y_i - \bar{y}) = X_i - Y_i$

$\therefore \quad \Sigma d_i^2 = \Sigma X_i^2 + \Sigma Y_i^2 - 2\Sigma X_i Y_i$

or $\quad \Sigma X_i Y_i = \frac{1}{2}(\Sigma X_i^2 + \Sigma Y_i^2 - \Sigma d_i^2) = \dfrac{1}{12}(n^3 - n) - \dfrac{1}{2}\Sigma d_i^2.$

Hence the correlation coefficient between these variables is

$$r = \frac{\Sigma X_i Y_i}{\sqrt{(\Sigma X_i^2 \Sigma Y_i^2)}} = \frac{\frac{1}{12}(n^3 - n) - \frac{1}{2}\Sigma d_i^2}{\frac{1}{12}(n^3 - n)} = 1 - \frac{6\, \Sigma d_i^2}{n^3 - n}$$

This is called the *rank correlation coefficient* and is denoted by $\rho$.

**Example 25.22.** *Ten participants in a contest are ranked by two judges as follows :*

| $x$ : | 1 | 6 | 5 | 10 | 3 | 2 | 4 | 9 | 7 | 8 |
|---|---|---|---|---|---|---|---|---|---|---|
| $y$ : | 6 | 4 | 9 | 8 | 1 | 2 | 3 | 10 | 5 | 7 |

*Calculate the rank correlation coefficient* $\rho$.

* It was developed by a psychologist Charles Edward Spearman in 1906.

**Solution.** If $d_i = x_i - y_i$, then $d_i = -5, 2, -4, 2, 2, 0, 1, -1, 2, 1$

$\therefore$ $\Sigma d_i^2 = 25 + 4 + 16 + 4 + 4 + 0 + 1 + 1 + 4 + 1 = 60$

Hence $\rho = 1 - \frac{6\Sigma d_i^2}{n^3 - n} = 1 - \frac{6 \times 60}{990} = 0.6$ nearly.

**Example 25.23.** *Three judges, A, B, C, give the following ranks. Find which pair of judges has common approach*

| | | | | | | | | | | |
|---|---|---|---|---|---|---|---|---|---|---|
| *A* : | 1 | 6 | 5 | 10 | 3 | 2 | 4 | 9 | 7 | 8 |
| *B* : | 3 | 5 | 8 | 4 | 7 | 10 | 2 | 1 | 6 | 9 |
| *C* : | 6 | 4 | 9 | 8 | 1 | 2 | 3 | 10 | 5 | 7 |

**Solution.** Here $n = 10$.

| | *Ranks by* | | $d_1$ | $d_2$ | $d_3$ | $d_1^2$ | $d_2^2$ | $d_3^2$ |
|---|---|---|---|---|---|---|---|---|
| $A (= x)$ | $B (= y)$ | $C (= z)$ | $x - y$ | $y - z$ | $z - x$ | | | |
| 1 | 3 | 6 | −2 | −3 | 5 | 4 | 9 | 25 |
| 6 | 5 | 4 | 1 | 1 | −2 | 1 | 1 | 4 |
| 5 | 8 | 9 | −3 | −1 | 4 | 9 | 1 | 16 |
| 10 | 4 | 8 | 6 | −4 | −2 | 36 | 16 | 4 |
| 3 | 7 | 1 | −4 | 6 | −2 | 16 | 36 | 4 |
| 2 | 10 | 2 | −8 | 8 | 0 | 64 | 64 | 0 |
| 4 | 2 | 3 | 2 | −1 | −1 | 4 | 1 | 1 |
| 9 | 1 | 10 | 8 | −9 | 1 | 64 | 81 | 1 |
| 7 | 6 | 5 | 1 | 1 | −2 | 1 | 1 | 4 |
| 8 | 9 | 7 | −1 | 2 | −1 | 1 | 4 | 1 |
| Total | | | 0 | 0 | 0 | 200 | 214 | 60 |

$\therefore$

$$\rho(x, y) = 1 - \frac{6\Sigma d_1^2}{n(n^2 - 1)} = 1 - \frac{6 \times 200}{10 \times 99} = -0.2$$

$$\rho(y, z) = 1 - \frac{6\Sigma d_2^2}{n(n^2 - 1)} = 1 - \frac{6 \times 214}{10 \times 99} = -0.3$$

$$\rho(z, x) = 1 - \frac{6\Sigma d_3^2}{n(n - 1)} = 1 - \frac{6 \times 60}{10 \times 99} = 0.6$$

Since $\rho(z, x)$ is maximum, the pair of judges $A$ and $C$ have the nearest common approach.

**Example 25.24:** *A simply supported beam carries a concentrated load P(kg) at its mid-points. The following table gives max. deflection y(cm) corresponding to various values of P.*

| | | | | | | |
|---|---|---|---|---|---|---|
| *P* : | 100 | 120 | 140 | 160 | 180 | 200 |
| *y* : | 0.45 | 0.55 | 0.60 | 0.70 | 0.80 | 0.85 |

*Find a low of the form $y = a + bP$*
*Also find the value of max-deflation when P = 150 kg.* (Calicut, 2019)

**Solution.** Let $A = 140$ and take $x = \frac{P - 140}{20}$, since $h = 20$.

| $P$ | $y$ | $x$ | $xy$ | $x^2$ |
|---|---|---|---|---|
| 100 | 0.45 | −2 | −9.0 | 4 |
| 120 | 0.55 | −1 | −0.55 | 1 |
| 140 | 0.60 | 0 | 0 | 0 |
| 160 | 0.70 | 1 | 0.70 | 1 |
| 180 | 0.80 | 2 | 1.60 | 4 |
| 200 | 0.85 | 3 | 2.55 | 9 |
| Total | 3.95 | 3 | 3.40 | 19 |

Here, $\Sigma y = 3.95$, $\Sigma x = 3$, $\Sigma xy = 3.40$, $\Sigma x^2 = 19$, and $x = 6$

Hence the normal equations.

$$3.95 = 6a + 3b$$
$$3.40 = 3a + 19b$$

Solving then $a = 0.6176$ and $b = 0.0814$

$\therefore$ The straight line of best fit is given by

$$y = a + bx$$
$$y = 0.6176 + 0.814\,x = 0.6176 + 0.0814\left(\frac{P-140}{20}\right)$$
$$y = 0.0478 + 0.00407\text{P}$$

when $P = 150, y = 0.0478$ to $0.00407 \times 150$

$$= 0.0478 + 0.6105 = 0.6583.$$

**Example 25.25:** *Fit a second degress parabola $y = a + bx + cx^2$ to the following data:*

| $x$ : | 1 | 3 | 5 | 7 | 9 |
|---|---|---|---|---|---|
| $y$ : | 2 | 7 | 10 | 11 | 9 |

*(AKTU, 2019)*

**Solution.** Let $X = \dfrac{x-5}{2}$ and $Y = y - 7$ and

$y = A + BX + CX^2$ be the equation of the parabola with respect to the new origin.

From the table, we note that $\Sigma X = 0$

$$\Sigma X^2 = 10, \quad \Sigma X^3 = 0$$
$$\Sigma X^4 = 34, \quad \Sigma Y = 4$$
$$\Sigma XY = 18, \quad \Sigma X^2Y = -8$$

| $x$ | $y$ | $X$ | $Y$ | $XY$ | $X^2$ | $X^2Y$ | $X^3$ | $X^4$ |
|---|---|---|---|---|---|---|---|---|
| 1 | 2 | –2 | –5 | 10 | 4 | –20 | –8 | 16 |
| 3 | 7 | –1 | 0 | 0 | 1 | 0 | –1 | 1 |
| 5 | 10 | 0 | 3 | 0 | 0 | 0 | 0 | 0 |
| 7 | 11 | 1 | 4 | 4 | 1 | 4 | 1 | 1 |
| 9 | 9 | 2 | 2 | 4 | 4 | 8 | 8 | 16 |
| Total | | 0 | 4 | 18 | 10 | –8 | 0 | 34 |

Hence the normal equations reduces to

$$5A + 0.B + 10.C = 40$$
$$0.A + 10B + 0.C = 18$$
$$10A + 0.B + 34C = -18$$

Solving them, we have $A = \dfrac{108}{35}$, $B = 1.8$, $C = \dfrac{-8}{7}$

$\therefore$ The equation of the parabola of best fit is

$$Y = \frac{108}{35} + 1.8X - \frac{8}{7}X^2$$

$$y - 7 = \frac{108}{35} + 1.8\left(\frac{x-5}{2}\right) - \frac{8}{7}\left(\frac{x-5}{2}\right)^2$$

$$y = -1.557 + 3.757x - 0.2857x^2.$$

**Example 25.26:** *Find the coefficient of correlation for the following table.*

| $x$ : | 10 | 14 | 18 | 22 | 26 | 30 |
|---|---|---|---|---|---|---|
| $y$ : | 18 | 12 | 24 | 6 | 30 | 36 |

*(Karnataka, 2019)*

**Solution.** Here the values of $x$ and $y$ are given that equal intervals

$$h = 4,\ k = 6,$$

Hence take
$$u = \frac{X - A}{h} = \frac{X - 22}{4} \text{ and } v = \frac{Y - B}{K} = \frac{Y - 24}{6}$$

| $X$ | $Y$ | $u$ | $v$ | $u^2$ | $v^2$ | $u$ |
|---|---|---|---|---|---|---|
| 10 | 18 | –3 | –1 | 9 | 1 | 3 |
| 14 | 12 | –2 | –2 | 4 | 4 | 4 |
| 18 | 24 | –1 | 0 | 1 | 0 | 0 |
| 22 | 6 | 0 | –3 | 0 | 9 | 0 |
| 26 | 30 | 1 | 1 | 1 | 1 | 1 |
| 30 | 36 | 2 | 2 | 4 | 4 | 4 |
| Total | | $\Sigma u = -3$ | $\Sigma v = -3$ | $\Sigma u^2 = 19$ | $\Sigma v^2 = 19$ | $\Sigma uv = 12$ |

$$\pi = \frac{\Sigma vu - \dfrac{\Sigma u\,\Sigma v}{h}}{\sqrt{\left\{\Sigma u^2 - \dfrac{(\Sigma u)^2}{h}\right\}}\sqrt{\left\{\Sigma v^2 - \dfrac{(\Sigma v^2)}{h}\right\}}} = \frac{12 - (-3)(-3)/6}{\sqrt{\left\{19 - \dfrac{(-3)^2}{6}\right\}}\sqrt{\left\{19 - \dfrac{(-3)^2}{6}\right\}}}$$

$$= \frac{12 - 1.5}{\sqrt{(19 - 1.5)}\sqrt{19 - 1.5}} = \frac{10.5}{17.5} = \frac{3}{5} = 0.6.$$

**Example 25.27:** *Calculate the coefficient of correlation for the following table.*

| $x/y$ | 0 – 4 | 4 – 8 | 8 – 12 | 12 – 16 |
|---|---|---|---|---|
| 0 – 5 | 7 | – | – | – |
| 5 – 10 | 6 | 8 | – | – |
| 10 – 15 | – | 5 | 3 | – |
| 15 – 20 | – | 7 | 2 | – |
| 20 – 25 | – | – | – | 9 |

*(UPTU (MCA), 2019)*

**Solution.** Replace the class interval for $X$ and $Y$ be their mid points and then let

$$u = \frac{X - 10}{4}, \qquad v = \frac{Y - 12.5}{5}$$

| $X / Y$ | 2 | 6 | 10 | 14 | $f$ | $v$ | $fv$ | $fv^2$ |
|---|---|---|---|---|---|---|---|---|
| 2.5 | $\boxed{4}$ 7 | | | | 7 | –2 | –14 | 28 |
| 7.5 | $\boxed{2}$ 6 | $\boxed{1}$ 8 | | | 14 | –1 | –14 | 14 |
| 12.5 | | $\boxed{0}$ 5 | $\boxed{0}$ 3 | | 8 | 0 | 0 | 0 |
| 17.5 | | $\boxed{-1}$ 7 | $\boxed{0}$ 2 | | 9 | 1 | 9 | 9 |
| 22.5 | | | | $\boxed{2}$ 9 | 9 | 2 | 18 | 36 |
| $f$ | 13 | 20 | 5 | 9 | 47 | Totals | –1 | 87 |
| $u$ | –2 | –1 | 0 | –1 | Totals | | | |
| $fu$ | –26 | –20 | 0 | 9 | –37 | | | |
| $fu^2$ | 52 | 20 | 0 | 9 | 81 | | | |

From the table $\Sigma fu = -37$, $\Sigma fu^2 = 81$, $\Sigma fv = -1$, $\Sigma fv^2 = 87$.

To obtain $\Sigma fuv$, first obtain $uv$ for each cell as shown in the table enclosed by the arc. Then

$$\Sigma fuv = 7 \times 4 + 6 \times 2 + 8 \times 1 + 5 \times 0 + 3 \times 0 + 7 \times (-1) + 2 \times 0 + 9 \times 2 = 59.$$

$$r = \frac{\Sigma fuv - \frac{1}{N}\Sigma fu\,\Sigma fv}{\sqrt{\left\{\Sigma fu^2 - \frac{(\Sigma fu)^2}{N}\right\}}\sqrt{\left\{\Sigma fv^2 - \frac{(\Sigma fv)^2}{N}\right\}}} = \frac{59 - \frac{1}{47}(-37)(-1)}{\sqrt{\left\{81 - \frac{(-37)^2}{47}\right\}}\sqrt{\left\{87 - \frac{(-1)^2}{47}\right\}}}$$

$$= \frac{2736}{3159.8} = 0.80.$$

**Example 25.28 :** *Find the correlation coefficient and the equations of regression lines for the following value of x and y:*

| $x$ : | 1 | 2 | 3 | 4 | 5 |
|---|---|---|---|---|---|
| $y$ : | 2 | 5 | 3 | 8 | 7 |

*(CSVTU, 2018)*

**Solution.** Let the assumed means $A = 3$ and $B = 5$.

| $x$ | $y$ | $u = x - 3$ | $v = y - 5$ | $u^2$ | $v^2$ | $uv$ |
|---|---|---|---|---|---|---|
| 1 | 2 | –2 | –3 | 4 | 9 | 6 |
| 2 | 5 | –1 | 0 | 1 | 0 | 0 |
| 3 | 3 | 0 | –2 | 0 | 4 | 0 |
| 4 | 8 | 1 | 3 | 1 | 9 | 3 |
| 5 | 7 | 2 | 2 | 4 | 4 | 4 |
| Total | | $\Sigma u = 0$ | $\Sigma v = 0$ | $\Sigma u^2 = 10$ | $\Sigma v^2 = 26$ | $\Sigma uv = 13$ |

Now
$$r = \frac{\Sigma uv - \frac{\Sigma u\,\Sigma v}{n}}{\sqrt{\left\{\Sigma u^2 \frac{(\Sigma u^2)}{n}\right\}}\sqrt{\left\{\Sigma v^2 - \frac{(\Sigma v)^2}{n}\right\}}} = \frac{13}{\sqrt{10}\sqrt{26}} = 8.0$$

Also
$$r\frac{\sigma y}{\sigma x} = \frac{\Sigma XY}{\Sigma X^2} = \frac{\Sigma uv - \frac{\Sigma u \Sigma v}{n}}{\left\{\Sigma u^2 - \frac{(\Sigma u)^2}{n}\right\}} = \frac{13}{10} = 1.3$$

and
$$r\frac{\sigma x}{\sigma y} = \frac{\Sigma XY}{\Sigma Y^2} = \frac{\Sigma uv - \frac{\Sigma u \Sigma v}{n}}{\left\{\Sigma v^2 - \frac{(\Sigma v)^2}{n}\right\}} = \frac{13}{26} = 0.5$$

Further
$$\bar{x} = A + \frac{\Sigma u}{n} = 3 + 0 = 3$$

$$\bar{y} = B + \frac{\Sigma v}{n} = 5 + 0 = 5$$

$\therefore$ Equation of the lines of regressions are

$$y - \bar{y} = r\frac{\sigma y}{\sigma x}(x - \bar{x})$$

$$y - 5 = 1.3(x - 3) \quad \text{or} \quad y = 1.3x + 1.1$$

$$x - \bar{x} = r\frac{\sigma x}{\sigma y}(y - \bar{y})$$

$$x - 3 = 0.5\,(y - 5) \text{ or } x = 0.5\,y + 0.5$$

## PROBLEMS 25.4

1. Find the correlation co-efficient and the regression lines of $y$ and $x$ and $x$ on $y$ for the following data :

(a)

| $x$ : | 1 | 2 | 3 | 4 | 5 |
|---|---|---|---|---|---|
| $y$ : | 2 | 5 | 3 | 8 | 7 |

(*Andhra, 2021; V.T.U., 2010*)

(b)

| $x$ : | 2 | 4 | 6 | 8 | 10 |
|---|---|---|---|---|---|
| $y$ : | 5 | 7 | 9 | 8 | 11 |

(*U.P.T.U., 2011*)

2. The following figures show $X$, the average concentration of smoke (mg/100 cum) at certain places in a state, and $Y$, the intensity of ultra-violet radiation (Standard natural ultra minutes per day)

| $x$ : | 24 | 13 | 27 | 12 | 31 | 42 | 13 | 29 | 17 | 11 |
|---|---|---|---|---|---|---|---|---|---|---|
| $y$ : | 24 | 25 | 21 | 25 | 22 | 29 | 24 | 20 | 25 | 26 |

Calculate the correlation coefficient and interpret the result. (*Roorkee M.E. 2018*)

3. Find the coefficient of correlation between industrial production ($X$ in crore tons) and export ($y$ in crore tons) in 2018, and comment on the result:

| $x$ : | 55 | 56 | 58 | 59 | 60 | 60 | 62 |
|---|---|---|---|---|---|---|---|
| $y$ : | 35 | 38 | 38 | 39 | 44 | 43 | 45 |

(*Madurai, 2019*)

4. Ten people of various heights as under, were requested to read the letters on a car at 25 yards distance. The number of letters correctly read is given below :

| Height (in feet) : | 5.1 | 5.3 | 5.6 | 5.7 | 5.8 | 5.9 | 5.10 | 5.11 | 6.0 | 6.1 |
|---|---|---|---|---|---|---|---|---|---|---|
| No. of letters : | 11 | 17 | 19 | 14 | 8 | 15 | 20 | 6 | 8 | 12 |

Is there any correlation between heights and visual power ?

5. Using the formula $r = \dfrac{\sigma_x^2 + \sigma_y^2 - \sigma_{x-y}^2}{2\sigma_x\sigma_y}$, find $r$ from the following data :

| $x$ : | 92 | 89 | 87 | 86 | 83 | 77 | 71 | 63 | 53 | 50 |
|---|---|---|---|---|---|---|---|---|---|---|
| $y$ : | 86 | 88 | 91 | 77 | 68 | 85 | 52 | 82 | 37 | 57 |

6. Find the correlation between $x$ (marks in Mathematics) and $y$ (marks in Engineering Drawing) given in the following data :

| $y$ \ $x$ | *10—40* | *40—70* | *70—100* | *Total* |
|---|---|---|---|---|
| 0—30 | 5 | 20 | — | 25 |
| 30—60 | — | 28 | 2 | 30 |
| 60—90 | — | 32 | 13 | 45 |
| Total | 5 | 80 | 15 | 100 |

7. The following table gives the data of rain fall and discharge in a certain river. Obtain the line of regression of $y$ on $x$.

| Rain fell : | 1.5 | 1.8 | 2.6 | 2.9 | 3.4 |
|---|---|---|---|---|---|
| Discharge : | 33 | 36 | 40 | 46 | 53 |

(*Annamalai, 2019*)

8. If the coefficient of correlation between two variables $x$ and $y$ is 0.5 and the acute angle between their lines of regression is $\tan^{-1}(3/8)$, show that $\sigma_x = \frac{1}{2}\sigma_y$.

9. For two random variables $x$ and $y$ with the same mean, the two regression lines are $y = ax + b$ and $x = \alpha y + \beta$. Show that $\dfrac{b}{\beta} = \dfrac{1-a}{1-\alpha}$. Find also the common mean. (*U.P.T.U., 2010*)

10. Two random variables have the regression lines with equations $3x + 2y = 26$ and $6x + y = 31$. Find the mean values and the correlation coefficient between $x$ and $y$.

11. The following results were obtained from records of age ($x$) and blood pressure ($y$) of a group of 10 men :

$$\left.\begin{array}{lcc} & x & y \\ \textit{Mean} & 53 & 142 \\ \textit{Variance} & 130 & 165 \end{array}\right\} \text{ and } \Sigma(x-\bar{x})(y-\bar{y}) = 1220.$$

Find the appropriate regression equation and use it to estimate the blood pressure of a man whose age is 45.

12. Compute the standard error of estimate $S_x$ for the respective heights of the following 12 couples :

| | | | | | | | | | | | | | |
|---|---|---|---|---|---|---|---|---|---|---|---|---|---|
| Height $x$ of husband (inches) | : | 68 | 66 | 68 | 65 | 69 | 66 | 68 | 65 | 71 | 67 | 68 | 70 |
| Height $y$ of wife (inches) | : | 65 | 63 | 67 | 64 | 68 | 62 | 70 | 66 | 68 | 67 | 69 | 71 |

13. Calculate the rank correlation coefficient from the following data showing ranks of 10 students in two subjects :

| | | | | | | | | | | | |
|---|---|---|---|---|---|---|---|---|---|---|---|
| Maths | : | 3 | 8 | 9 | 2 | 7 | 10 | 4 | 6 | 1 | 5 |
| Physics | : | 5 | 9 | 10 | 1 | 8 | 7 | 3 | 4 | 2 | 6 |

14. If $\theta$ is acute angle between the two regression lines in case of two variables $x$ and $y$, show that

$$\tan\theta = \frac{1-\pi^2}{\pi}\cdot\frac{\sigma_x\sigma_y}{\sigma_x^{\ 2}+\sigma_y^{\ 2}}$$

where $r$, $\sigma_x$, $\sigma_y$ have their usual meanings. (*Madurai. 2019*)

15. Find the correlation between $X$ and $Y$ given in the following table.

| | 10 – 40 | 40 – 70 | 70 – 100 |
|---|---|---|---|
| 0 – 30 | 5 | 20 | – |
| 30 – 60 | – | 28 | 2 |
| 60 – 90 | – | 32 | 13 |

(*Chennai. 2018*)

16. The two regression equation of the variables $x$ and $y$ are

$$x = 19.13 - 0.87y$$

and

$$y = 11.64 - 0.50\,x$$

(*i*) find $\bar{x}$ and $\bar{y}$

(*ii*) coefficient of correlation $r$. (*AMIE, 2018*)

17. Establish the formula

$$\sigma^2_{x-y} = \sigma_x^{\ 2} + \sigma_y^{\ 2} - r\sigma_x\sigma_y$$

where $r$ is the correlation coefficient between $x$ and $y$. (*KTU, 2019*)

## 25.17 OBJECTIVE TYPE OF QUESTIONS

### PROBLEMS 25.5

*Select the correct answer or fill up the blanks in each of the following questions :*

1. The median of the numbers 11, 10, 12, 13, 9 is

(*a*) 12.5 (*b*) 12 (*c*) 10.5 (*d*) 11.

2. The mode of the numbers 7, 7, 7, 9, 10, 11, 11, 11, 12 is

(*a*) 11 (*b*) 12 (*c*) 7 (*d*) 7 and 11.

3. S.D. is defined as

(*a*) $\sqrt{\dfrac{\Sigma f(x-\bar{x})^2}{\Sigma f}}$ (*b*) $\dfrac{\Sigma f(x-\bar{x})}{\Sigma f}$ (*c*) $\dfrac{\Sigma f(x-\bar{x})^2}{\Sigma f}$.

4. Coefficient of variation is

(*a*) $\dfrac{\sigma}{\bar{x}}\times 100$ (*b*) $\dfrac{\sigma}{\bar{x}}$ (*c*) $\sqrt{\dfrac{\sigma^2}{\bar{x}}}\times 100$.

5. Average scores of three batsman $A$, $B$, $C$ are respectively 40, 45 and 55 and their S.D.s are respectively 9, 11, 16. Which batsman is more consistent ?
(*a*) $A$ (*b*) $B$ (*c*) $C$.
6. The equations of regression lines are $y = 0.5x + a$ and $x = 0.4y + b$. The correlation coefficient is
(*a*) $\sqrt{0.2}$ (*b*) 0.45 (*c*) $-\sqrt{0.2}$.
7. If the correlation coefficient is 0, the two regression lines are
(*a*) parallel (*b*) perpendicular (*c*) coincident (*d*) inclined at 45° to each other.
8. If $r_1$ and $r_2$ are two regression coefficients, then signs of $r_1$ and $r_2$ depend on ........
9. Regression coefficient of $y$ on $x$ is 0.7 and that of $x$ on $y$ is 3.2. Is the correlation coefficient $r$ consistent ?
10. The standard deviation of the numbers 24, 48, 64, 36, 53 is ........
11. If $y = x + 1$ and $x = 3y - 7$ are the two lines of regression then $\bar{x}$ = ........, $\bar{y}$ = ........ and $r$ = ........
12. If the two regression lines are perpendicular to each other, then their coefficient of correlation is ........
13. Quartile deviation is defined as ........
14. The minimum value of correlation coefficient is ........
15. Prediction error of $Y$ is defined as ........
16. If $X$ and $Y$ are independent, then the correlation coefficient between $X$ and $Y$ is ........
17. The point of intersection of the two regression lines is ........
18. The smaller the coefficient of variation, the greater is the ........ in the data.
19. The moment coefficient of skewness is given by ........
20. Kurtosis measures the ........ of a distribution.
21. The equation of the line of regression of $y$ on $x$ is ........
22. Coefficient of variation = ........
23. The angle between two regression lines is given by ........
24. A frequency curve is said to be Mesokurtic when $\beta_2$ is ........
25. Correlation coefficient is the geometrical mean between ........
26. When the variables are independent, the two lines of regression are ........
27. Arithmetic mean of the coefficients of regression is ........ than the coefficient of correlation.
28. If two regression lines coincide then the coefficient of correlation is ........
29. The rank coefficient is given by ........
30. The ratio of the standard deviation to the mean is known as ........
31. The value of $\Sigma f(x - \bar{x})$ = ........
32. The value of coefficient of correlation lies between ........ and ........
33. If the two regression coefficients are – 0.4 and – 0.9, then the correlation coefficient is ........
34. A distribution with the following constants is positively skew : $Q_1 = 25.8$, median = 49.0, $Q_3 = 64.2$. (True or False)
35. Quartile coefficient of skewness is $\dfrac{Q_3 + Q_1 - 2Q_2}{Q_3 - Q_1}$. (True or False)
36. Skewness indicates peakedness of the frequency distribution. (True or False)

CHAPTER 26

# Probability and Distributions



## 26.1 (1) INTRODUCTION

We often hear such statements : 'It is likely to rain today', 'I have a fair chance of getting admission', and 'There is an even chance that in tossing a coin the head may come up'. In each case, we are not certain of the outcome, but we wish to assess the chances of our predictions coming true. The study of probability provides a mathematical framework for such assertions and is essential in every decision making process. Before defining probability, let us explain a few terms :

**(2) Principle of counting.** If an event can happen in $n_1$ ways and thereafter for each of these events a second event can happen in $n_2$ ways, and for each of these first and second events a third event can happen for $n_3$ ways and so on, then the number of ways these $m$ event can happen is given by the product $n_1 \cdot n_2 \cdot n_3 \ldots n_m$.

**(3) Permutations.** A permutation of a number of objects is their arrangement in some definite order. Given three letters $a, b, c$, we can permute them two at a time as "$bc, cb$ ; $ca, ac$; $ab, ba$" yielding 6 permutations. The combinations or groupings are only 3, *i.e.*, $bc, ca, ab$. Here the order is immaterial.

*The number of permutations of n different thing taken r at a time is*

$$n(n-1)(n-2)\ldots(n-r+1), \text{ which is denoted by } {}^nP_r.$$

Thus $${}^nP_r = n(n-1)(n-2)\ldots(n-r+1) = \frac{n!}{(n-r)!}$$

*Permutations with repetitions.* The number of permutations of $n$ objects of which $n_1$ are alike, $n_2$ are alike and $n_3$ are alike is $\dfrac{n!}{n_1!\, n_2!\, n_3!}$.

**(4) Combinations.** *The number of combinations of n different objects taken r at a time is denoted by* ${}^nC_r$. If we take any one of the combinations, its $r$ objects can be arranged in $r!$ ways. So the total number of arrangements which can be obtained from all the combinations is ${}^nP_r = {}^nC_r \cdot r!$.

Thus $${}^nC_r = \frac{{}^nP_r}{r!} = \frac{n!}{r!\,(n-r)!}$$

Also $${}^nC_{n-r} = {}^nC_r$$

*e.g.,* $${}^{25}P_4 = 25 \times 24 \times 23 \times 22;\ {}^{25}C_{21} = {}^{25}C_4 = \frac{25 \times 24 \times 23 \times 22}{4 \times 3 \times 2 \times 1}.$$

**Example 26.1.** *In how many ways can one make a first, second, third and fourth choice among 12 firms leasing construction equipment.*

**Solution.** First choice can be made from any of the 12 firms. Thereafter the second choice can be made from among the remaining 11 firms. Then the third choice can be made from the remaining 10 firms and the fourth choice can be made from the 9 firms.

Thus from the *principle of counting*, the number of ways in which first, second, third and fourth choice can be affected = 12 × 11 × 10 × 9 = 11880.

**Example 26.2.** *Find the number of permutations of all the letters of the word (i) Committee (ii) Engineering.*

**Solution.** (*i*) $n = 9, n_1\,(m, m) = 2, n_2\,(t, t) = 2, n_3\,(e, e) = 2$

$$\therefore \text{ no. of permutations } = \frac{n!}{n_1!\,.\,n_2!\,.\,n_3!} = \frac{9!}{2!\,.\,2!\,.\,2!} = 45360.$$

(*ii*) $n = 11, n_1(e\text{'s}) = 3, n_2\,(g, g) = 2, n_3\,(i, i) = 2, n_4\,(n\text{'s}) = 3$

$$\therefore \text{ no. of permutations } = \frac{11!}{3!\,2!\,2!\,3!} = 277200.$$

**Example 26.3.** *From six engineers and five architects a committee is to be formed having three engineers and two architects. How many different committees can be formed if (i) there is no restriction. (ii) two particular engineers must be included. (iii) one particular architect must be excluded.*

**Solution.** (*i*) Number of committees $^6C_3 \times {}^5C_2 = \frac{6\,.\,5\,.\,4}{3\,.\,2\,.\,1} \times \frac{5\,.\,4}{2\,.\,1} = 200.$

(*ii*) Here we have to choose one engineer from the remaining four engineers.

$$\therefore \text{ no. of committees } = {}^4C_1 \times {}^5C_2 = 4 \times \frac{5\,.\,4}{2\,.\,1} = 40$$

(*iii*) Here we have to choose two architects from the remaining four architects.

$$\therefore \text{ no. of committees } = {}^6C_3 \times {}^4C_2 = \frac{6\,.\,5\,.\,4}{3\,.\,2\,.\,1} \times \frac{4\,.\,3}{2\,.\,1} = 120.$$

### PROBLEMS 26.1

1. If a test consists of 12 true-false questions, in how many different ways can a student make the test paper with one answer to each question.
2. How many 4-digit numbers can be formed from the six digits 2, 3, 5, 6, 7 and 9, without repetition ? How many of these are less than 500 ?
3. A student has to answer 9 out of 12 questions. How many choices has he (*i*) if he must answer first two questions (*ii*) if he must answer at least four of the first five questions.
4. How many car number plates can be made if each plate contains two different letters followed by three different digits ? Solve the problem (*a*) with repetitions and (*b*) without repetitions.

## 26.2 (I) BASIC TERMINOLOGY

(*i*) **Exhaustive events.** A set of events is said to be *exhaustive*, if it includes all the possible events. For example, in tossing a coin there are two exhaustive cases either head or tail and there is no third possibility.

(*ii*) **Mutually exclusive events.** If the occurrence of one of the events procludes the occurrence of all other, then such a set of events is said to be *mutually exclusive*. Just as tossing a coin, either head comes up or the tail and both can't happen at the same time, *i.e.*, these are two mutually exclusive cases.

(*iii*) **Equally likely events.** If one of the events cannot be expected to happen in preference to another then such events are said to be *equally likely*. For instance, in tossing a coin, the coming of the head or the tail is equally likely.

Thus when a *die** is thrown, the turning up of the six different faces of the die are exhaustive, mutually exclusive and equally likely.

(*iv*) **Odds in favour of an event.** If the number of ways favourable to an event $A$ is $m$ and the number of ways not favourable to $A$ is $n$ then *odds in favour of* $A = m/n$ and *odds against* $A = n/m$.

**(2) Definition of probability.** *If there are n exhaustive, mutually exclusive and equally likely cases of which m are favourable to an event A, then probability (p) of the happening of A is*

$$\mathbf{P(A) = m/n.}$$

As there are $n - m$ cases in which $A$ will not happen (denoted by $A'$), *the chance of A not happening is q or P (A′) so that*

$$q = \frac{n-m}{n} = 1 - \frac{m}{n} = 1 - p$$

*i.e.*, $P(A') = 1 - P(A)$ so that $\mathbf{P(A) + P(A') = 1}$,

*i.e., if an event is certain to happen then its probability is unity, while if it is certain not to happen, its probability is zero.*

**Obs.** This definitions of probability fails when

(*i*) number of outcomes is infinite (not exhaustive) and (*ii*) outcomes are not equally likely.

**(3) Statistical (or Empirical) definition of probability.** *If in n trials, an event A happens m times, then the probability (p) of happening of A is given by*

$$p = P(A) = \underset{n \to \infty}{Lt} \frac{m}{n}$$

**Example 26.4.** *Find the chance of throwing (a) four, (b) an even number with an ordinary six faced die.*

**Solution.** (*a*) There are six possible ways in which the die can fall and of these there is only one way of throwing 4. Thus the required chance $= \frac{1}{6}$.

(*b*) There are six possible ways in which the die can fall. Of these there are only 3 ways of getting 2, 4 or 6. Thus the required chance $= 3/6 = \frac{1}{2}$.

**Example 26.5.** *What is the chance that a leap year selected at random will contain 53 Sundays ?*

**Solution.** A leap year consists of 366 days, so that there are 52 full weeks (and hence 52 Sundays) and two extra days. These two days can be (*i*) Monday, Tuesday (*ii*) Tuesday, Wednesday, (*iii*) Wednesday, Thursday (*iv*) Thursday, Friday (*v*) Friday, Saturday (*vi*) Saturday, Sunday (*vii*) Sunday, Monday.

Of these 7 cases, the last two are favourable and hence the required probability $= \frac{2}{7}$.

**Example 26.6.** *A five figure number is formed by the digits 0, 1, 2, 3, 4 without repetition. Find the probability that the number formed is divisible by 4.*

**Solution.** The five digits can be arranged in 5! ways, out of which 4! will begin with zero.

$\therefore$ total number of 5-figure numbers formed = 5! – 4! = 96.

Those numbers formed will be divisible by 4 which will have two extreme right digits divisible by 4, *i.e.*, numbers ending in 04, 12, 20, 24, 32, 40.

Now numbers ending in 04 = 3! = 6, numbers ending in 12 = 3! – 2! = 4,
numbers ending in 20 = 3! = 6, numbers ending in 24 = 3! – 2! = 4,
numbers ending in 32 = 3! – 2! = 4, and numbers ending in 40 = 3! = 6.

[The numbers having 12, 24, 32 in the extreme right are (3! – 2!) since the numbers having zero on the extreme left are to excluded.]

* *Die* is a small cube. Dots 1, 2, 3, 4, 5, 6 are marked on its six faces. The outcome of throwing a die is the number of dots on its upper face.

$\therefore$ total number of favourable ways = 6 + 4 + 6 + 4 + 4 + 6 = 30.

Hence the required probability $= \frac{30}{96} = \frac{5}{16}$.

**Example 26.7.** *A bag contains 40 tickets numbered 1, 2, 3, ... 40, of which four are drawn at random and arranged in ascending order* $(t_1 < t_2 < t_3 < t_4)$. *Find the probability of* $t_3$ *being 25?*

**Solution.** Here exhaustive number of cases $= {}^{40}C_4$

If $t_3 = 25$, then the tickets $t_1$ and $t_2$ must come out of 24 tickets numbered 1 to 24. This can be done in ${}^{24}C_2$ ways.

Then $t_4$ must come out of the 15 tickets (numbering 25 to 40) which can be done in ${}^{15}C_1$ ways.

$\therefore$ favourable number of cases $= {}^{24}C_2 \times {}^{15}C_1$

Hence the probability of $t_3$ being 25 $= \frac{{}^{24}C_2 \times {}^{15}C_1}{{}^{40}C_4} = \frac{414}{9139}$.

**Example 26.8.** *An urn contains 5 red and 10 black balls. Eight of them are placed in another urn. What is the chance that the latter then contains 2 red and 6 black balls ?*

**Solution.** The number of ways in which 8 balls can be drawn out of 15 is ${}^{15}C_8$.

The number of ways of drawing 2 red balls is ${}^5C_2$ and corresponding to each of these ${}^5C_2$ ways of drawing a red ball, there are ${}^{10}C_6$ ways of drawing 6 black balls.

$\therefore$ the total number of ways in which 2 red and 6 black balls can be drawn is ${}^5C_2 \times {}^{10}C_6$.

$\therefore$ the required probability $= \frac{{}^5C_2 \times {}^{10}C_6}{{}^{15}C_8} = \frac{140}{429}$.

**Example 26.9.** *A committee consists of 9 students two of which are from 1st year, three from 2nd year and four from 3rd year. Three students are to be removed at random. What is the chance that (i) the three students belong to different classes, (ii) two belong to the same class and third to the different class, (iii) the three belong to the same class ?* (V.T.U., 2014)

**Solution.** (*i*) The total number of ways of choosing 3 students out of 9 is ${}^9C_3$, *i.e.*, 84.

A student can be removed from 1st year students in 2 ways, from 2nd year in 3 ways and from 3rd year in 4 ways, so that the total number of ways of removing three students, one from each group is 2 × 3 × 4.

Hence the required chance $= \frac{2 \times 3 \times 4}{{}^9C_3} = \frac{24}{84} = \frac{2}{7}$.

(*ii*) The number of ways of removing two from 1st year students and one from others

$$= {}^2C_2 \times {}^7C_1.$$

The number of ways of removing two from 2nd year students and one from others

$$= {}^3C_2 \times {}^6C_1.$$

The number of ways of removing 2 from 3rd year students and one from others

$$= {}^4C_2 \times {}^5C_1.$$

$\therefore$ the total number of ways in which two students of the same class and third from the others may be removed

$$= {}^2C_2 \times {}^7C_1 + {}^3C_2 \times {}^6C_1 + {}^4C_2 \times {}^5C_1 = 7 + 18 + 30 = 55.$$

Hence, the required chance $= \frac{55}{84}$.

(*iii*) Three students can be removed from 2nd year group in ${}^3C_3$, *i.e.* 1 way and from 3rd year group in ${}^4C_3$, *i.e.*, 4 ways.

$\therefore$ the total number of ways in which three students belong to the same class = 1 + 4 = 5.

Hence the required chance $= \frac{5}{84}$.

**Example 26.10.** *A has one share in a lottery in which there is 1 prize and 2 blanks ; B has three shares in a lottery in which there are 3 prizes and 6 blanks ; compare the probability of A's success to that of B's success.*

**Solution.** $A$ can draw a ticket in ${}^3C_1 = 3$ ways.

The number of cases in which $A$ can get a prize is clearly 1.

$\therefore$ the probability of $A$'s success $= \frac{1}{3}$.

Again $B$ can draw a ticket in ${}^9C_3 = \frac{9 \cdot 8 \cdot 7}{3 \cdot 2 \cdot 1} = 84$ ways.

The number of ways in which $B$ gets all blanks $= {}^6C_3 = \frac{6 \cdot 5 \cdot 4}{3 \cdot 2 \cdot 1} = 20$

$\therefore$ the number of ways of getting a prize $= 84 - 20 = 64$.

Thus the probability of $B$'s success $= 64/84 = 16/21$.

Hence $A$'s probability of success : $B$'s probability of success $= \frac{1}{3} : \frac{16}{21} = 7 : 16$.

## 26.3 PROBABILITY AND SET NOTATIONS

**(1) Random experiment.** Experiments which are performed essentially under the same conditions and whose results cannot be predicted are known as *random experiments. e.g.,* Tossing a coin or rolling a die are random experiments.

**(2) Sample space.** The set of all possible outcomes of a random experiment is called *sample space* for that experiment and is denoted by $S$.

The elements of the sample space $S$ are called the *sample points.*

*e.g.*, On tossing a coin, the possible outcomes are the head ($H$) and the tail ($T$). Thus $S = \{H, T\}$.

**(3) Event.** The outcome of a random experiment is called an *event*. Thus every subset of a sample space $S$ is an *event*.

*The null set* $\phi$ is also an event and is called an *impossible event. Probability of an impossible event* is zero *i.e.*, $P(\phi) = 0$.

**(4) Axioms**

(*i*) *The numerical value of probability lies between 0 and 1.*

*i.e.*, for any event $A$ of $S$, $0 \leq P(A) \leq 1$.

(*ii*) *The sum of probabilities of all sample events is unity i.e.,* $P(S) = 1$.

(*iii*) *Probability of an event made of two or more sample events is the sum of their probabilities.*

**(5) Notations**

(*i*) *Probability of happening of events* **A** *or* **B** is written as $P(A + B)$ or $P(A \cup B)$.

(*ii*) *Probability of happening of* **both the events A and B** is written as $P(AB)$ or $P(A \cap B)$.

(*iii*) *'Event A implies* ($\Rightarrow$) *event B'* is expressed as $A \subset B$.

(*iv*) *'Events A and B are mutually exclusive'* is expressed as $A \cap B = \phi$.

**(6)** *For any two events A and B,*

$$\mathbf{P(A \cap B') = P(A) - P(A \cap B)}$$

*Proof.* From Fig. 26.1,

$$(A \cap B') \cup (A \cap B) = A$$

$\therefore$ $P[(A \cap B') \cup (A \cap B)] = P(A)$

or $P(A \cap B') + P(A \cap B) = P(A)$

or $P(A \cap B') = P(A) - P(A \cap B)$

Similarly, $P(A' \cap B) = P(B) - P(A \cap B)$

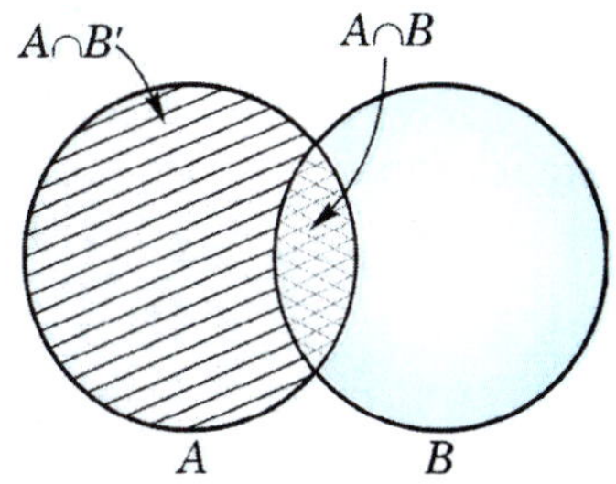

Fig. 26.1

## 26.4 ADDITION LAW OF PROBABILITY *or* THEOREM OF TOTAL PROBABILITY

**(1)** *If the probability of an event A happening as a result of a trial is P(A) and the probability of a* **mutually exclusive** *event B happening is P(B), then the probability of* **either** *of the events happening as a result of the trial is* **P(A + B) or P(A ∪ B) = P(A) + P(B).**

*Proof.* Let $n$ be the total number of equally likely cases and let $m_1$ be favourable to the event $A$ and $m_2$ be favourable to the event $B$. Then the number of cases favourable to $A$ or $B$ is $m_1 + m_2$. Hence the probability of $A$ or $B$ happening as a result of the trial

$$= \frac{m_1 + m_2}{n} = \frac{m_1}{n} + \frac{m_2}{n} = P(A) + P(B).$$

**(2)** *If A, B, are any two events* **(not mutually exclusive),** *then*

$$P(A + B) = P(A) + P(B) - P(AB)$$

or

$$P(A \cup B) = P(A) + P(B) - P(A \cap B)$$

If the events $A$ and $B$ are any two events then, there are some outcomes which favour both $A$ and $B$. If $m_3$ be their number, then these are included in both $m_1$ and $m_2$. Hence the total number of outcomes favouring either $A$ or $B$ or both is

$$m_1 + m_2 - m_3.$$

Thus the probability of occurrence of $A$ or $B$ or both

$$= \frac{m_1 + m_2 - m_3}{n} = \frac{m_1}{n} + \frac{m_2}{n} - \frac{m_3}{n}$$

Hence $$P(A + B) = P(A) + P(B) - P(AB)$$

or $$P(A \cup B) = P(A) + P(B) - P(A \cap B)$$ *(Andhra, 2019; V.T.U., 2016)*

**Obs.** *When A and B are mutually exclusive* $P(AB)$ or $P(A \cap B) = 0$ and we get

$$P(A + B) \text{ or } P(A \cup B) = P(A) + P(B).$$

In general, for a number of **mutually exclusive** events $A_1, A_2, \ldots A_n$, we have

$$P(A_1 + A_2 + \ldots + A_n) \text{ or } P(A_1 \cup A_2 \cup \ldots \cup A_n) = P(A_1) + P(A_2) + \ldots + P(A_n).$$

**(3)** *If A, B, C are any three events, then*

$$\mathbf{P(A + B + C) = P(A) + P(B) + P(C) - P(AB) - P(BC) - P(CA) + P(ABC).}$$

or

$$\mathbf{P(A \cup B \cup C) = P(A) + P(B) + P(C) - P(A \cap B) - P(B \cap C) - P(C \cap A) + P(A \cap B \cap C)}$$

*Proof.* Using the above result for any two events, we have

$$P(A \cup B \cup C) = P[(A \cup B) \cup C]$$
$$= P(A \cup B) + P(C) - P[(A \cup B) \cap C]$$
$$= [P(A) + P(B) - P(A \cap B)] + P(C) - P[(A \cap C) \cup (B \cap C)] \quad \text{(Distributive Law)}$$
$$= P(A) + P(B) + P(C) - P(A \cap B) - \{P(A \cap C) + P(B \cap C) - P(A \cap B \cap C)\}$$
$$[\because (A \cap C) \cap (B \cap C) = A \cap B \cap C]$$
$$= P(A) + P(B) + P(C) - P(A \cap B) - P(B \cap C) - P(C \cap A) + P(A \cap B \cap C) \quad [\because A \cap C = C \cap A.]$$

**Example 26.11.** *In a race, the odds in favour of the four horses $H_1, H_2, H_3, H_4$ are 1 : 4, 1 : 5, 1 : 6, 1 : 7 respectively. Assuming that a dead heat is not possible, find the chance that one of them wins the race.*

**Solution.** Since it is not possible for all the horses to cover the same distance in the same time (a dead heat), the events are mutually exclusive.

If $p_1, p_2, p_3, p_4$ be the probabilities of winning of the horses $H_1, H_2, H_3, H_4$ respectively, then

$$p_1 = \frac{1}{1+4} = \frac{1}{5} \quad [\because \text{ Odds in favour of } H_1 \text{ are } 1 : 4]$$

and $$p_2 = \frac{1}{6}, p_3 = \frac{1}{7}, p_4 = \frac{1}{8}.$$

Hence the chance that one of them wins $= p_1 + p_2 + p_3 + p_4$

$$= \frac{1}{5} + \frac{1}{6} + \frac{1}{7} + \frac{1}{8} = \frac{533}{840}.$$

**Example 26.12.** *A bag contains 8 white and 6 red balls. Find the probability of drawing two balls of the same colour.*

**Solution.** Two balls out of 14 can be drawn in $^{14}C_2$ ways which is the total number of outcomes.

Two white balls out of 8 can be drawn in $^8C_2$ ways. Thus the probability of drawing 2 white balls

$$= \frac{^8C_2}{^{14}C_2} = \frac{28}{91}$$

Similarly 2 red balls out of 6 can be drawn in $^6C_2$ ways. Thus the probability of drawing 2 red balls

$$= \frac{^6C_2}{^{14}C_2} = \frac{15}{91}.$$

Hence the probability of drawing 2 balls of the same colour (either both white or both red)

$$= \frac{28}{91} + \frac{15}{91} = \frac{43}{91}.$$

**Example 26.13.** *Find the probability of drawing an ace or a spade or both from a deck of cards* ?*

**Solution.** The probability of drawing an ace from a deck of 52 cards = 4/52.

Similarly the probability of drawing a card of spades = 13/52, and the probability of drawing an ace of spades = 1/52.

Since the two events (*i.e.*, a card being an ace and a card being of spades) are not mutually exclusive, therefore, the probability of drawing an ace or a spade

$$= \frac{4}{52} + \frac{13}{52} - \frac{1}{52} = \frac{4}{13}.$$

## 26.5 (1) INDEPENDENT EVENTS

Two events are said to be *independent,* if happening or failure of one does not affect the happening or failure of the other. Otherwise the events are said to be *dependent.*

*For two dependent events A and B, the symbol P(B / A) denotes the probability of occurrence of B, when A has already occurred.* It is known as the **conditional probability** and is read as a '*probability of B given A*'.

**(2) Multiplication law of probability** *or* **Theorem of compound probability.** *If the probability of an event A happening as a result of trial is P(A) and after A has happened the probability of an event B happening as a result of another trial* (*i.e.*, **conditional probability of B given A**) *is P(B / A), then the probability of* **both** *the events A and B happening as a result of two trials is* $\boldsymbol{P(AB)}$ **or** $\boldsymbol{P(A \cap B) = P(A) \,.\, P(B/A)}$.

*Proof.* Let $n$ be the total number of outcomes in the first trial and $m$ be favourable to the event $A$ so that $P(A) = m/n$.

Let $n_1$ be the total number of outcomes in the second trial of which $m_1$ are favourable to the event $B$ so that $P(B/A) = m_1/n_1$.

Now each of the $n$ outcomes can be associated with each of the $n_1$ outcomes. So the total number of outcomes in the combined trial is $nn_1$. Of these $mm_1$ are favourable to both the events $A$ and $B$. Hence

$$P(AB) \text{ or } P(A \cap B) = \frac{mm_1}{nn_1} = P(A) \,.\, P(B/A).$$

Similarly, the *conditional probability of A given B is* $P(A/B)$.

$$\therefore \qquad P(AB) \text{ or } P(A \cap B) = P(B) \,.\, P(A/B)$$

Thus $$P(A \cap B) = P(A) \,.\, P(B/A) = P(B) \,.\, P(A/B).$$

**(3)** *If the events A and B are* **independent,** *i.e.*, if the happening of $B$ does not depend on whether $A$ has happened or not, then $P(B / A) = P(B)$ and $P(A / B) = P(A)$.

$$\therefore \qquad \boldsymbol{P(AB) \text{ or } P(A \cap B) = P(A) \,.\, P(B).}$$

In general, $P(A_1A_2...A_n)$ or $P(A_1 \cap A_2 \cap ... \cap A_n) = P(A_1) \,.\, P(A_2) .... \,P(A_n)$.

* *Cards* : A pack of cards consists of four suits *i.e.*, Hearts, Diamonds, Spades and Clubs. Each suit has 13 cards : an Ace, a King, a Queen, a Jack and nine cards numbered 2, 3, 4, ..., 10. Hearts and Diamonds are *red* while Spades and Clubs are *black*.

**Cor.** If $p_1, p_2$ be the probabilities of happening of two independent events, then

(*i*) the probability that the first event happens and the second fails is $p_1(1 - p_2)$.

(*ii*) the probability that both events fail to happen is $(1 - p_1)(1 - p_2)$.

(*iii*) the probability that at least one of the events happens is

$1 - (1 - p_1)(1 - p_2)$. This is commonly known as their **cumulative probability.**

*In general, if $p_1, p_2, p_3, \ldots p_n$ be the chances of happening of n independent events, then their cumulative probability (i.e., the chance that at least one of the events will happen) is*

$$1 - (1 - p_1)(1 - p_2)(1 - p_3) \ldots (1 - p_n).$$

**Example 26.14.** *Two cards are drawn in succession from a pack of 52 cards. Find the chance that the first is a king and the second a queen if the first card is (i) replaced, (ii) not replaced.*

**Solution.** (*i*) The probability of drawing a king $= \frac{4}{52} = \frac{1}{13}$.

If the card is replaced, the pack will again have 52 cards so that the probability of drawing a queen is 1/13.

The two events being independent, the probability of drawing both cards in succession $= \frac{1}{13} \times \frac{1}{13} = \frac{1}{169}$.

(*ii*) The probability of drawing a king $= \frac{1}{13}$.

If the card is not replaced, the pack will have 51 cards only so that the chance of drawing a queen is 4/51.

Hence the probability of drawing both cards $= \frac{1}{13} \times \frac{4}{51} = \frac{4}{663}$.

**Example 26.15.** *A pair of dice is tossed twice. Find the probability of scoring 7 points (a) once, (b) at least once (c) twice.*

**Solution.** In a single toss of two dice, the sum 7 can be obtained as (1, 6), (2, 5), (3, 4), (4, 3), (5, 2), (6, 1) *i.e.*, in 6 ways, so that the probability of getting 7 = 6/36 = 1/6.

Also the probability of not getting 7 = 1 – 1/6 = 5/6.

(*a*) The probability of getting 7 in the first toss and not getting 7 in the second toss = 1/6 × 5/6 = 5/36.

Similarly, the probability of not getting 7 in the first toss and getting 7 in the second toss = 5/6 × 1/6 = 5/36.

Since these are mutually exclusive events, addition law of probability applies.

$\therefore$ required probability $= \frac{5}{36} + \frac{5}{36} = \frac{5}{18}$.

(*b*) The probability of not getting 7 in either toss $= \frac{5}{6} \times \frac{5}{6} = \frac{25}{36}$

$\therefore$ the probability of getting 7 at least once $= 1 - \frac{25}{36} = \frac{11}{36}$.

(*c*) The probability of getting 7 twice $= \frac{1}{6} \times \frac{1}{6} = \frac{1}{36}$.

**Example 26.16.** *There are two groups of objects : one of which consists of 5 science and 3 engineering subjects, and the other consists of 3 science and 5 engineering subjects. An unbiased die is cast. If the number 3 or number 5 turns up, a subject is selected at random from the first group, otherwise the subject is selected at random from the second group. Find the probability that an engineering subject is selected ultimately*

**Solution.** Prob. of turning up 3 or 5 $= \frac{2}{6} = \frac{1}{3}$.

Prob. of selecting an engg. subject from first group $= \frac{3}{8}$

$\therefore$ Prob of selecting an engg. subject from first group on turning up 3 or 5

$$= \frac{1}{3} \times \frac{3}{8} = \frac{1}{8} \qquad \ldots(i)$$

Now prob. of not turning 3 or 5 $= 1 - \frac{1}{3} = \frac{2}{3}\cdot$

Prob. of selecting an engg. subject from second group $= \frac{5}{8}$

$\therefore$ prob. of selecting an engg. subject from second group on turning up 3 or 5

$$= \frac{2}{3} \times \frac{5}{8} = \frac{5}{12} \qquad \text{...}(ii)$$

Thus the prob. of selecting an engg. subject

$$= \frac{1}{8} + \frac{5}{12} = \frac{13}{24}. \qquad \text{[From } (i) \text{ and } (ii)]$$

**Example 26.17.** *A box A contains 2 white and 4 black balls. Another box B contains 5 white and 7 black balls. A ball is transferred from the box A to the box B. Then a ball is drawn from the box B. Find the probability that it is white.* *(Andhra, 2020; V.T.U., 2015 ; Rohtak, 2011 S)*

**Solution.** The probability of drawing a white ball from box $B$ will depend on whether the transferred ball is black or white.

If a black ball is transferred, its probability is 4/6. There are now 5 white and 8 black balls in the box $B$. Then the probability of drawing white ball from box $B$ is $\frac{5}{13}\cdot$

Thus the probability of drawing a white ball from urn $B$, if the transferred ball is black

$$= \frac{4}{6} \times \frac{5}{13} = \frac{10}{39}.$$

Similarly the probability of drawing a white ball from urn $B$, if the transferred ball is white

$$= \frac{2}{6} \times \frac{6}{13} = \frac{2}{13}.$$

Hence required probability $= \frac{10}{39} + \frac{2}{13} = \frac{16}{39}.$

**Example 26.18.** *(a) A biased coin is tossed till a head appears for the first time. What is the probability that the number of required tosses is odd.*

*(b) Two persons A and B toss an unbiased coin alternately on the understanding that the first who gets the head wins. If A starts the game, find their respective chances of winning.*

**Solution.** (*a*) Let $p$ be the probability of getting a head and $q$ the probability of getting a tail in a single toss, so that $p + q = 1$.

Then probability of getting head on an odd toss

= Probability of getting head in the 1st toss
+ Probability of getting head in the 3rd toss
+ Probability of getting head in the 5th toss + ... ∞

$$= p + qqp + qqqqp + \ldots \infty$$

$$= p\,(1 + q^2 + q^4 + \ldots) = p\,.\,\frac{1}{1-q^2} \quad (q < 1)$$

$$= p\,.\,\frac{1}{(1-q)(1+q)} = p\cdot\frac{1}{p\,(1+q)} = \frac{1}{1+q}.$$

(*b*) Probability of getting a head = 1/2. Then $A$ can win in 1st, 3rd, 5th, ... throws.

$\therefore$ the chances of $A$'s winning $= \frac{1}{2} + \left(\frac{1}{2}\right)^2 \frac{1}{2} + \left(\frac{1}{2}\right)^4 \frac{1}{2} + \left(\frac{1}{2}\right)^6 \frac{1}{2} + \cdots$

$$= \frac{1/2}{1-(1/2)^2} = \frac{1}{2}\cdot\frac{4}{3} = \frac{2}{3}.$$

Hence the chance of $B$'s winning = 1 – 2/3 = 1/3.

**Example 26.19.** *Two cards are selected at random from 10 cards numbered 1 to 10. Find the probability p that the sum is odd, if*

*(i) the two cards are drawn together.*

*(ii) the two cards are drawn one after the other without replacement.*

*(iii) the two cards are drawn one after the other with replacement.*

**Solution.** (*i*) Two cards out of 10 can be selected in $^{10}C_2 = 45$ ways. The sum is odd if one number is odd and the other number is even. There being 5 odd numbers (1, 3, 5, 7, 9) and 5 even numbers (2, 4, 6, 8, 10), an odd and an even number is chosen in 5 × 5 = 25 ways.

Thus $$p = \frac{25}{45} = \frac{5}{9}.$$

(*ii*) Two cards out of 10 can be selected one after the other *without replacement* in 10 × 9 = 90 ways.

An odd number is selected in 5 × 5 = 25 ways and an even number in 5 × 5 = 25 ways

Thus $$p = \frac{25+25}{90} = \frac{5}{9}.$$

(*iii*) Two cards can be selected one after the other *with replacement* in 10 × 10 = 100 ways.

An odd number in selected in 5 × 5 = 25 ways and an even number in 5 × 5 = 25 ways.

Thus $$p = \frac{25+25}{100} = \frac{1}{2}.$$

**Example 26.20.** *Given P (A) = 1/4, P (B) = 1/3 and P (A ∪ B) = 1/2, evaluate P (A/B), P (B/A), P (A ∩ B′) and P (A/B′).* (*Andhra, 2022*)

**Solution.** (*i*) Since $P(A \cup B) = P(A) + P(B) - P(A \cap B)$

$$\therefore \quad \frac{1}{2} = \frac{1}{4} + \frac{1}{3} - P(A \cap B) \text{ or } P(A \cap B) = \frac{1}{12}.$$

Thus $$P(A/B) = \frac{P(A \cap B)}{P(B)} = \frac{1/12}{1/3} = \frac{1}{4}.$$

(*ii*) $$P(B/A) = \frac{P(A \cap B)}{P(A)} = \frac{1/12}{1/4} = \frac{1}{3}.$$

(*iii*) $$P(A \cap B') = P(A) - P(A \cap B) = \frac{1}{4} - \frac{1}{12} = \frac{1}{6}.$$

(*iv*) $$P(A/B') = \frac{P(A \cap B')}{P(B')} = \frac{1/6}{1 - P(B)} = \frac{1/6}{1 - 1/3} = \frac{1}{4}.$$

**Example 26.21.** *The odds that a book will be reviewed favourably by three independent critics are 5 to 2, 4 to 3 and 3 to 4. What is the probability that of the three reviews, a majority will be favourable.*

**Solution.** The probability that the book shall be reviewed favourably by first critic is 5/7, by second 4/7 and by third 3/7.

A majority of the three reviews will be favourable when two or three are favourable.

∴ prob. that the first two are favourable and the third unfavourable

$$= \frac{5}{7} \times \frac{4}{7} \times \left(1 - \frac{3}{7}\right) = \frac{80}{343}$$

Prob. that the first and third are favourable and second unfavourable

$$= \frac{5}{7} \times \frac{3}{7} \times \left(1 - \frac{4}{7}\right) = \frac{45}{343}$$

Prob. that the second and third are favourable and the first unfavourable

$$= \frac{4}{7} \times \frac{3}{7} \times \left(1 - \frac{5}{7}\right) = \frac{24}{343}$$

Finally, prob. that all the three are favourable $= \frac{5}{7} \times \frac{4}{7} \times \frac{3}{7} = \frac{60}{343}$

Since they are mutually exclusive events, the required prob.

$$= \frac{80}{343} + \frac{45}{343} + \frac{24}{343} + \frac{60}{343} = \frac{209}{343}.$$

**Example 26.22.** *In a competition, the odds that a Ph.D. thesis will be favourably reviewed by three independent examiners are 5 to 2, 4 to 3 and 3 to 4 respectively. What is the probability that, of the three examinars. (i) all reject the thesis, (ii) all approve the thesis and (iii) a majority approve the thesis.* (Pune. 2018)

**Solution.** Let the three examiners be $A$, $B$, $C$. Let $P_1$, $P_2$, $P_3$ be the probabilities that the thesis is approved and $q_1$, $q_2$, $q_3$ that the thesis is rejected by $A$, $B$, $C$ respectively, then

$$P_1 = \frac{5}{7}, P_2 = \frac{4}{7}, P_3 = \frac{3}{7} \text{ and } q_1 = \frac{2}{7}, q_2 = \frac{3}{7}\ q_3 = \frac{4}{7}$$

(*i*) the probability that all the examiner reject the thesis.

$$= q_1 q_2 q_3 = \frac{2}{7} \times \frac{3}{7} \times \frac{4}{7} = \frac{24}{343}$$

(*ii*) the probability that all of the examiners approve the thesis

$$= P_1 P_2 P_3 = \frac{5}{7} \times \frac{4}{7} \times \frac{3}{7} = \frac{60}{343}$$

(*iii*) A majority approve the thesis of atleast two examiners are favourable, that is

(*a*) If $A$ and $B$ approve and $C$ reject the probability $= P_1 P_2 . q_2 = \frac{5}{7} \times \frac{4}{7} \times \frac{4}{7} = \frac{80}{343}$

(*b*) If $A$ and $C$ approve and $B$ reject, the probability $= p_1 . q_2 . p_3 = \frac{5}{7} \times \frac{3}{7} \times \frac{3}{7} = \frac{45}{343}$.

(*c*) If $B$ and $C$ approve and $A$ reject, the probability $= q_1 . p_2 . p_3 = \frac{2}{7} \times \frac{4}{7} \times \frac{3}{7} = \frac{24}{343}$

(*d*) If $A$, $B$, $C$ all approve, the probability $= P_1 . P_2 . P_3 = \frac{60}{343}$

Hence the probability that a majority approve the thesis $= \frac{80}{343} + \frac{45}{343} + \frac{24}{343} + \frac{60}{343} = \frac{209}{343}$.

**Example 26.23.** *Three students A, B, C write an entrance examination. Their chances of passing are 1/2, 1/3, 1/4 respectively. Find the probability that atleast one of them passes.* (V.T.U., 2016)

**Solution.** The probability that $A$ can pass is 1/2.

The probability that $A$ cannot pass is $1 - \frac{1}{2}$.

Similarly the probabilities that $B$ and $C$ cannot pass are $1 - \frac{1}{3}$ and $1 - \frac{1}{4}$.

$\therefore$ the probability that $A$, $B$ and $C$ cannot pass is $\left(1-\frac{1}{2}\right)\left(1-\frac{1}{3}\right)\left(1-\frac{1}{4}\right)$.

Hence the probability that at least one student will pass

$$= 1 - \left(1-\frac{1}{2}\right)\left(1-\frac{1}{3}\right)\left(1-\frac{1}{4}\right) = \frac{3}{4}.$$

**Example 26.24.** *The students in a class are selected at random, one after the other, for an examination. Find the probability p that the boys and girls in the class alternate if*

*(i) the class consists of 4 boys and 3 girls,* *(ii) the class consists of 3 boys and 3 girls.*

**Solution.** (*i*) As there are 7 students in the class, the first examined must be a boy.

$\therefore$ prob. that first is a boy $= \dfrac{4}{7}$

Then the prob. that the second is a girl $= \dfrac{3}{6}\cdot$

$\therefore$ prob. of the next boy $= \dfrac{3}{5}$

Similarly the prob. that the fourth is a girl $= \dfrac{2}{4}$,

the prob. that the fifth is a boy $= \dfrac{2}{3}$,

the prob. that the sixth is a girl $= \dfrac{1}{2}$

and the last is a boy $= \dfrac{1}{1}\cdot$

Thus
$$p = \frac{4}{7}\cdot\frac{3}{6}\cdot\frac{3}{5}\cdot\frac{2}{4}\cdot\frac{2}{3}\cdot\frac{1}{2}\cdot\frac{1}{1} = \frac{1}{35}\cdot$$

(*ii*) The first student is a boy and the first student is a girl are two mutually exclusive cases. If the first student is a boy, then the probability $p_1$ that the students alternate is

$$p_1 = \frac{3}{6}\cdot\frac{3}{5}\cdot\frac{2}{4}\cdot\frac{2}{3}\cdot\frac{1}{2}\cdot\frac{1}{1} = \frac{1}{20}\cdot$$

If the first student is a girl, then the probability $p_2$ that the students alternate is

$$p_2 = \frac{3}{6}\cdot\frac{3}{5}\cdot\frac{2}{4}\cdot\frac{2}{3}\cdot\frac{1}{2}\cdot\frac{1}{1} = \frac{1}{20}\cdot$$

Thus the required prob. $p = p_1 + p_2 = \dfrac{1}{20} + \dfrac{1}{20} = \dfrac{1}{10}\cdot$

**Example 26.25.** (*Huyghen's problem*) *A and B throw alternately with a pair of dice. A wins if he throws 6 before B throws 7 and B wins if he throws 7 before A throws 6. If A begins, find his chance of winning.*

**Solution.** The sum 6 can be obtained as follows : (1, 5), (2, 4), (3, 3), (4, 2), (5, 1), *i.e.*, in 5 ways.

The probability of $A$'s throwing 6 with 2 dice is $\dfrac{5}{36}\cdot$

$\therefore$ the probability of $A$'s not throwing 6 is 31/36.

Similarly the probability of $B$'s throwing 7 is 6/36, *i.e.*, $\dfrac{1}{6}\cdot$

$\therefore$ the probability of $B$'s not throwing 7 is 5/6.

Now $A$ can win if he throws 6 in the first, third, fifth, seventh etc. throws.

$\therefore$ the chance of $A$'s wining

$$= \frac{5}{36} + \frac{31}{36}\times\frac{5}{6}\times\frac{5}{36} + \frac{31}{36}\times\frac{5}{6}\times\frac{31}{36}\times\frac{5}{6}\times\frac{5}{36} + \cdots$$

$$= \frac{5}{36}\left[1 + \left(\frac{31}{36}\times\frac{5}{6}\right) + \left(\frac{31}{36}\times\frac{5}{6}\right)^2 + \left(\frac{31}{36}\times\frac{5}{6}\right)^3 + \cdots\right]$$

$$= \frac{5}{36}\cdot\frac{1}{1-(31/36)\times(5/6)} = \frac{5}{36}\times\frac{36\times 6}{61} = \frac{30}{61}\cdot$$

**Example 26.26:** *If A, B and C toss a coin in succession on the understanding that the first one to throw a head wins. What are their respective chances of winning:* (*BHU, 2018*)

**Solution.** If $A$ is to win he should throw a head in the 1st or 4th or 7th.... throws. If $B$ is to win, he should throw is ½. If $A$ looses then $B$ will get a chance.

The probability that $C$ wins in the 3rd throw will be combined probability that $A$ and $B$ loose and then $C$ wins.

It is $$\left(\frac{1}{2}\times\frac{1}{2}\right)\times\frac{1}{2} = \frac{1}{8}$$

Similarly, the probability that $A$ wins in the 4th throw.

$$= \frac{1}{8}\times\frac{1}{2} = \frac{1}{2^4}$$

Hence the probability of $A$ success $= \frac{1}{2}+\frac{1}{2^4}+\frac{1}{2^7}+\dfrac{\frac{1}{2}}{1-\left(\frac{1}{2}\right)^3} = \frac{4}{7}$

the probability of $B$ success $= \frac{1}{2^2}+\frac{1}{2^5}+\frac{1}{2^8}+\dfrac{\left(\frac{1}{2}\right)^2}{1-\left(\frac{1}{2}\right)^3} - \frac{2}{7}$

The probability of $C$ success $= 1-\left(\frac{4}{7}+\frac{2}{7}\right) = \frac{1}{7}$.

**Example 26.27:** *A has 2 shares in a lottery in which there are 3 prizes and 5 blanks, and B has 3 shares in a lottery in which there are 4 prizes and 6 blanks. Show that A is chance of success is to B's as 27: 35.*

*(Marathwada, 2019)*

**Solution.** A can draw two tickets out of 8 in $^8C_2 = 28$ ways and he can get all blanks in $^5C_2 = 10$ ways.

$\therefore$ A can win a prize in $28 - 10 = 18$ ways

Hence chance of $A$ success $= \frac{18}{28} = \frac{9}{14}$

Next $B$ can draw 4 tickets in $^{10}C_3 = 120$ ways

and he can get all blanks in $^6C_3 = 20$ ways.

$\therefore$ $B$ can win a prize in $120 - 20 = 100$ ways.

Hence chance of $B$ success $= \frac{100}{120} = \frac{5}{6}$

$\therefore$ $A$'s chance of success : $B$'s change of success $= \frac{9}{14} : \frac{5}{6} : \frac{27}{35}$

## PROBLEMS 26.2

1. (*i*) Given $P(A) = 1/2$, $P(B) = 1/3$ and $P(AB) = 1/4$, find the value $P(A + B)$.
   (*ii*) Let $A$ and $B$ be two events with $P(A) = 1/2$, $P(B) = 1/3$ and $P(A \cap B) = 1/4$. Find $P(A/B)$, $P(A \cup B)$, $P(A'/B')$.
   *(Kurukshetra, 2013 ; V.T.U., 2012)*
2. In a single throw with two dice, what is the chance of throwing
   (*a*) two aces ? (*b*) 7 ? Is this probability the same as that for getting 7 in two throws of a single die ?
3. .Compare the chances of throwing 4 with one dice, 8 with two dice and 12 with three dice.
4. Find the probability that a non-leap year should have 53 Saturdays ?
5. When a coin is tossed four times, find the probability of getting (*i*) exactly one head, (*ii*) at most three heads and (*iii*) at least two heads ?
6. Ten coins are thrown simultaneously. Find the probability of getting at least seven heads.
7. If all the letters of word 'ENGINEER' be written at random, what is the probability that all the letters $E$ are found together.

8. A ten digit number is formed using the digits from zero to nine, every digit being used only once. Find the probability that the number is divisible by 4.
9. Four cards are drawn from a pack of 52 cards. What is the chance that
   (*i*) no two cards are of equal value ? (*ii*) each belongs to a different suit ?
10. Suppose 5 cards are drawn at random from a pack of 52 cards. If all cards are red what is the probability that all of them are hearts?
11. Out of 50 rare books, 3 of which are especially valuable, 5 are stolen at random by a thief. What is the probability that
    (*a*) none of the 3 is included ? (*b*) 2 of the 3 are included ?
12. Five men in a company of twenty are graduates. If 3 men are picked out of 20 at random, what is the probability that
    (*a*) they are all graduates ? (*b*) at least one is graduate ?
13. From 20 tickets marked from 1 to 20, one ticket is drawn at random. Find the probability that it is marked with a multiple of 3 or 5.
14. Five balls are drawn from a bag containing 6 white and 4 black balls. What is the chance that 3 white and 2 black balls are drawn ?
15. (*a*) The probability of $n$ independent events are $p_1, p_2, p_3, \ldots p_n$. Find the probability that at least one of the events will happen. Use this result to find the chance of getting at least one six in a throw of 4 dice.
    (*b*) The probability of four persons *A*, *B*, *C*, *D* hitting targets are respectively 1/2, 1/3, 1/4, 1/5. What is the probability that target is hit buy at least one person if all hit simultaneously. (*V.T.U., 2013*)
16. Find the probability of drawing 4 white balls and 2 black balls without replacement from a bag containing 1 red, 4 black and 6 white balls.
17. A bag contains 10 white and 15 black balls. Two balls are drawn in succession. What is the probability that one of them is black and the other white ?
18. A purse contains 2 silver and 4 copper coins and a second purse contains 4 silver and 4 copper coins. If a coin is selected at random from one of the two purses, what is the probability that it is a silver coin ?
19. A box I contains 5 white balls and 6 black balls. Another box II contains 6 white balls and 4 black balls. A box is selected at random and then a ball is drawn from it : (*i*) what is the probability that the ball drawn will be white ? (*ii*) Given that the ball drawn is white, what is the probability that it came from box I.
20. A party of $n$ persons take their seats at random at a round table ; find the probability that two specified persons do not sit together.
21. *A* speaks the truth in 75% cases, and *B* in 80% of the cases. In what percentage of cases, are they likely to contradict each other in stating the same fact ?
22. The probability that Sushil will solve a problem is 1/4 and the probability that Ram will solve it is 2/3. If Sushil and Ram work independently, what is the probability that the problem will be solved by (*a*) both of them, (*b*) at least one of them ?
23. A student takes his examination in four subjects, *P*, *Q*, *R*, *S*. He estimates his chances of passing in *P* as 4/5, in *Q* as 3/4, in *R* as 5/6 and in *S* as 2/3. To qualify, he must pass in *P* and at least two other subjects. What is the probability that he qualifies ?
24. The probability that a 50 year old man will be alive at 60 is 0.83 and the probability that a 45 year old women will be alive at 55 is 0.87. What is the probability that a man who is 50 and his wife who is 45 will both be alive 10 years hence ?
25. If on an average one birth in 80 is a case of twins, what is the probability that there will be at least one case of twins in a maternity hospital on a day when 20 births occur ?
26. Two persons *A* and *B* fire at a target independently and have a probability 0.6 and 0.7 respectively of hitting the target. Find the probability that the target is destroyed.
27. *A* and *B* throw alternately with a pair of dice. The one who throws 9 first wins. Show that the chances of their winning are 9 : 8.

## 26.6 BAYE'S THEOREM

*An event A corresponds to a number of exhaustive events $B_1, B_2, \ldots, B_n$. If $P(B_i)$ and $P(A/B_i)$ are given, then*

$$P(B_i/A) = \frac{P(B_i)\, P(A/B_i)}{\Sigma P(B_i) P(A/B_i)}.$$

*Proof.* By the multiplication law of probability,

$$P(AB_i) = P(A)\, P(B_i/A) = P(B_i)\, P(A/B_i) \qquad \ldots(1)$$

$$\therefore \qquad P(B_i/A) = \frac{P(B_i)\,P(A/B_i)}{P(A)} \qquad ...(2)$$

Since the event $A$ corresponds to $B_1, B_2, ..., B_n$, we have by the addition law of probability,

$$P(A) = P(AB_1) + P(AB_2) + ... + P(AB_n) = \Sigma P(AB_i) = \Sigma P(B_i)\,P(A/B_i) \qquad \text{[By Eqn. (1)]}$$

Hence from Eqn. (2), we have $\qquad P(B_i/A) = \dfrac{P(B_i)\,P(A/B_i)}{\Sigma P(B_i)\,P(A/B_i)}$

which is known as the *theorem of inverse probability*. *(Andhra., 2022, 2014 ; V.T.U., 2014)*

**Obs.** The probabilities $P(B_i)$, $i = 1, 2, ..., n$ are called *apriori probabilities* because these exist before we get any information from the experiment.

The probabilities $P(A/B_i)$, $i = 1, 2, ..., n$ are called *posteriori probabilities*, because these are found after the experiment results are known.

**Example 26.28.** *Three machines $M_1$, $M_2$ and $M_3$ produce identical items. Of their respective output 5%, 4% and 3% of items are faulty. On a certain day, $M_1$ has produced 25% of the total output, $M_2$ has produced 30% and $M_3$ the remainder. An item selected at random is found to be faulty. What are the chances that it was produced by the machine with the highest output?*

**Solution.** Let the event of drawing a faulty item from any of the machines be $A$, and the event that an item drawn at random was produced by $M_i$ be $B_i$. We have to find $P(B_i/A)$ for which we proceed as follows :

| | $M_1$ | $M_2$ | $M_3$ | *Remarks* |
|---|---|---|---|---|
| $P(B_i)$ | 0.25 | 0.30 | 0.45 | $\because$ sum = 1 |
| $P(A/B_i)$ | 0.05 | 0.04 | 0.03 | |
| $P(B_i)\,P(A/B_i)$ | 0.0125 | 0.012 | 0.0135 | sum = 0.38 |
| $P(B_i/A)$ | $\frac{0.0125}{0.038}$ | $\frac{0.012}{0.038}$ | $\frac{0.0135}{0.038}$ | by Baye's theorem |

The highest output being from $M_3$, the required probability = 0.0135/0.038 = 0.355.

**Example 26.29.** *There are three bags : first containing 1 white, 2 red, 3 green balls ; second 2 white, 3 red, 1 green balls and third 3 white, 1 red, 2 green balls. Two balls are drawn from a bag chosen at random. These are found to be one white and one red. Find the probability that the balls so drawn came from the second bag.*

**Solution.** Let $B_1, B_2, B_3$ pertain to the first, second, third bags chosen and $A$ : the two balls are white and red.

Now $\qquad P(B_1) = P(B_2) = P(B_3) = \dfrac{1}{3}$

$$P(A/B_1) = P(\text{a white and a red ball are drawn from first bag}) = ({}^1C_1 \times {}^2C_1)/{}^6C_2 = \frac{2}{15}$$

Similarly $\qquad P(A/B_2) = ({}^2C_1 \times {}^3C_1)/{}^6C_2 = \dfrac{2}{5}$, $P(A/B_3) = ({}^3C_1 \times {}^1C_1)/{}^6C_2 = \dfrac{1}{5}$

By Baye's theorem, $\qquad P(B_2/A) = \dfrac{P(B_2)\,P(A/B_2)}{P(B_1)\,P(A/B_1) + P(B_2)\,P(A/B_2) + P(B_3)\,P(A/B_3)}$

$$= \frac{\frac{1}{3} \times \frac{2}{5}}{\frac{1}{3} \times \frac{2}{15} + \frac{1}{3} \times \frac{2}{5} + \frac{1}{3} \times \frac{1}{5}} = \frac{6}{11}.$$

## PROBLEMS 26.3

1. In a certain college, 4% of the boys and 1% of girls are taller than 1.8 m. Further more 60% of the students are girls. If a student is selected at random and is found to be taller than 1.8 m., what is the probability that the student is a girl ? *(V.T.U., 2015 ; Kurukshetra, 2013)*
2. A student takes his examination in four subjects, $\alpha, \beta, \gamma, \delta$. He estimates his chance of passing in $\alpha$ as $\frac{4}{5}$, in $\beta$ as $\frac{3}{4}$, in $\gamma$ and $\frac{5}{6}$ and $\delta$ as $\frac{2}{3}$. To qualify he must pass in $\alpha$ and at least two other subjects. What is the probability that he qualifies ? *(GTU, 2019)*

3. A batch of 100 lamps is known to have 5% defectives. If 10 lamps are chosen at random, what is the probability that none of them will be defective? What is the probability that exactly two defectives will be found?

(*J.N.T.U.H, 2018*)

4. The contents of three urns are : 1 white, 2 red, 3 green balls ; 2 white, 1 red, 1 green balls and 4 white, 5 red, 3 green balls. Two balls are drawn from an urn chosen at random. These are found to be one white and one green. Find the probability that the balls so drawn came from the third urn.

5. A witness is known to speak truth in 60% cases and another witness in 70% cases. In what percentage of cases are they likely to contradict each other in stating the same fact? (*Kerala, 2018*)(**Ans.** 46%)

6. A can hit target 4 times in 5 shots, *B*, 3 times in 4 shots and *C* twice in 3 shots. They fire a volley. What is the probability that two shots at least hit? (**Ans.** 5/6)

7. A manufactures supplies cheap quarter horse power motors in lot of 25. A buyer, before taking a lot, tests at random a sample of 5 motors and accepts the lot if they are all good, otherwise he rejects the lot. Find the probability that

   (*a*) he will accepts, a lot containing 5 defective motor.

   (*b*) he will reject a lot containing only defective motor. (*Ranchi, 2019*)

## 26.7 RANDOM VARIABLE

If a real variable $X$ be associated with the outcome of a random experiment, then since the values which $X$ takes depend on chance, it is called a *random variable* or a *stochastic variable* or simply a *variate*. For instance, if a random experiment $E$ consists of tossing a pair of dice, the sum $X$ of the two numbers which turn up have the value 2, 3, 4, ..., 12 depending on chance. Then $X$ is the random variable. It is a function whose values are real numbers and depend on chance.

If in a random experiment, the event corresponding to a number $a$ occurs, then the corresponding random variable $X$ is said to assume the value $a$ and the probability of the event is denoted by $P(X = a)$. Similarly the probability of the event $X$ assuming any value in the interval $a < X < b$ is denoted by $P(a < X < b)$. The probability of the event $X \le c$ is written as $P(X \le c)$.

If a random variable takes a finite set of values, it is called a *discrete variate*. On the other hand, if it assumes an infinite number of uncountable values, it is called a *continuous variate*.

## 26.8 (1) DISCRETE PROBABILITY DISTRIBUTION

Suppose a discrete variate $X$ is the outcome of some experiment. If the probability that $X$ takes the values $x_i$, is $p_i$, then

$$P(X = x_i) = p_i \text{ or } p(x_i) \text{ for } i = 1, 2,...$$

where (*i*) $p(x_i) \ge 0$ for all values of $i$, (*ii*) $\Sigma p(x_i) = 1$

*The set of values* $x_i$ *with their* probabilities $p_i$ constitute a **discrete probability distribution** *of the discrete variate X.*

For example, the discrete probability distribution for $X$, the sum of the numbers which turn on tossing a pair of dice is given by the following table :

| $X = x_i$ | 2 | 3 | 4 | 5 | 6 | 7 | 8 | 9 | 10 | 11 | 12 |
|---|---|---|---|---|---|---|---|---|---|---|---|
| $p(x_i)$ | 1/36 | 2/36 | 3/36 | 4/36 | 5/36 | 6/36 | 5/36 | 4/36 | 3/36 | 2/36 | 1/36 |

[∵ There are 6 × 6 = 36 equally likely outcomes and therefore, each has the probability 1/36. We have $X = 2$ for one outcome, *i.e.* (1, 1) ; $X = 3$ for two outcomes (1, 2) and (2, 1); $X = 4$ for three outcomes (1, 3), (2, 2) and (3, 1) and so on.]

**(2) Distribution function.** *The distribution function F (x) of the discrete variate X is defined by*

$$F(x) = P(X \le x) = \sum_{i=1}^{x} p(x_i)$$ where $x$ is any integer. The graph of $F(x)$ will be stair step form (Fig. 26.2). The distribution function is also sometimes called *cumulative distribution function*.

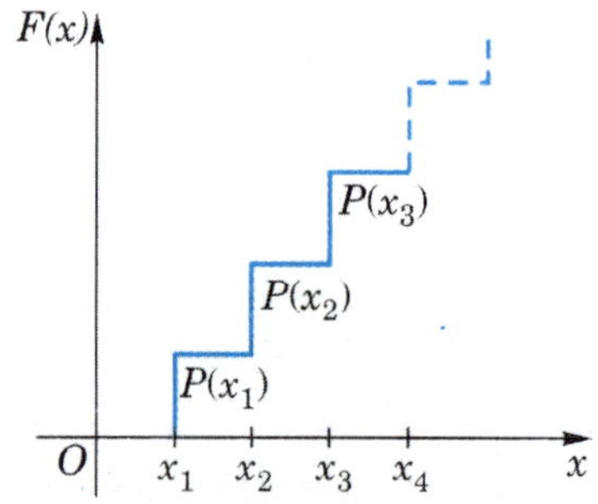

Fig. 26.2

**Example 26.30.** *A die is tossed thrice. A success is 'getting 1 or 6' on a toss. Find the mean and variance of the number of successes.* (V.T.U., 2011 S)

**Solution.** Probability of a success $= \frac{2}{6} = \frac{1}{3}$, Probability of failures $= 1 - \frac{1}{3} = \frac{2}{3}$.

$\therefore$ prob. of no success = Prob. of all 3 failures $= \frac{2}{3} \times \frac{2}{3} \times \frac{2}{3} = \frac{8}{27}$

Probability of one success and 2 failures $= 3c_1 \times \frac{1}{3} \times \frac{2}{3} \times \frac{2}{3} = \frac{4}{9}$

Probability of two successes and one failure $= 3c_2 \times \frac{1}{3} \times \frac{1}{3} \times \frac{2}{3} = \frac{2}{9}$

Probability of three successes $= \frac{1}{3} \times \frac{1}{3} \times \frac{1}{3} = \frac{1}{27}$

| Now | $x_i = 0$ | 1 | 2 | 3 |
|---|---|---|---|---|
| | $p_i = 8/27$ | 4/9 | 2/9 | 1/27 |

$\therefore$ mean $\quad \mu = \Sigma\, p_i x_i = 0 + \frac{4}{9} + \frac{4}{9} + \frac{1}{9} = 1.$

Also $\quad \Sigma\, p_i x_i^2 = 0 + \frac{4}{9} + \frac{8}{9} + \frac{9}{27} = \frac{5}{3}$

$\therefore$ variance $\quad \sigma^2 = \Sigma\, p_i x_i^2 - \mu^2 = \frac{5}{3} - 1 = \frac{2}{3}$.

**Example 26.31.** *The probability density function of a variate X is*

| $X$ : | 0 | 1 | 2 | 3 | 4 | 5 | 6 |
|---|---|---|---|---|---|---|---|
| $p(X)$ : | $k$ | $3k$ | $5k$ | $7k$ | $9k$ | $11k$ | $13k$ |

*(i) Find $P(X < 4)$, $P(X \geq 5)$, $P(3 < X \leq 6)$.* (V.T.U., 2013)

*(ii) What will be the minimum value of k so that $P(X \leq 2) > 3$.* (Andhra, 2021)

**Solution.** (*i*) If $X$ is a random variable, then

$$\sum_{i=0}^{6} p(x_i) = 1 \text{ i.e., } k + 3k + 5k + 7k + 9k + 11k + 13k = 1 \text{ or } k = 1/49.$$

$\therefore \quad P(X < 4) = k + 3k + 5k + 7k = 16k = 16/49.$

$P(X \geq 5) = 11k + 13k = 24k = 24/49.$

$P(3 < X \leq 6) = 9k + 11k + 13k = 33k = 33/49.$

(*ii*) $\quad P(X \leq 2) = k + 3k + 5k = 9k > 0.3$ or $k > 1/30$

Thus minimum value of $k = 1/30$.

**Example 26.32.** *A random variable X has the following probability function :*

| $x$ : | 0 | 1 | 2 | 3 | 4 | 5 | 6 | 7 |
|---|---|---|---|---|---|---|---|---|
| $p(x)$ : | 0 | $k$ | $2k$ | $2k$ | $3k$ | $k^2$ | $2k^2$ | $7k^2 + k$ |

*(i) Find the value of the k* *(ii) Evaluate $P(X < 6)$, $P(X \geq 6)$* (V.T.U., 2014)

*(iii) $P(0 < X < 5)$.* (Anna, 2012)

**Solution.** (*i*) If $X$ is a random variable, then

$$\sum_{i=0}^{7} p(x_i) = 1, \text{ i.e., } 0 + k + 2k + 2k + 3k + k^2 + 2k^2 + 7k^2 + k = 1$$

*i.e.,* $\quad 7k^2 + 9k - 1 = 0$ *i.e.* $(10 - k)(k + 1) = 0$ *i.e.,* $k = \frac{1}{10}$

(*ii*) $P(X < 6) = P(X = 0) + P(X = 1) + P(X = 2) + P(X = 3) + P(X = 4) + P(X = 5)$

$= 0 + k + 2k + 2k + 3k + k^2 = 8k + k^2 = \frac{8}{10} + \frac{1}{100} = \frac{81}{100}$

$$P(X \geq 6) = P(X = 6) + P(X = 7) = 2k^2 + 7k^2 + k = \frac{9}{100} + \frac{1}{10} = \frac{19}{100}$$

(*ii*) $P(0 < X < 5) = P(X = 1) + P(X = 2) + P(X = 3) + P(X = 4)$

$$= k + 2k + 2k + 3k = 8k = \frac{8}{10} = \frac{4}{5}.$$

## 26.9 (1) CONTINUOUS PROBABILITY DISTRIBUTION

When a variate $X$ takes every value in an interval, it gives rise to *continuous distribution* of $X$. The distributions defined by the variates like heights or weights are continuous distributions.

A major conceptual difference, however, exists between discrete and continuous probabilities. When thinking in discrete terms, the probability associated with an event is meaningful. With continuous events, however, where the number of events is infinitely large, the probability that a specific event will occur is practically zero. For this reason, continuous probability statements must be worded somewhat differently from discrete ones. Instead of finding the probability that $x$ equals some value, we find the probability of $x$ falling in a small interval.

Thus the probability distribution of a continuous variate $x$ is defined by a function $f(x)$ such that the probability of the variate $x$ falling in the small interval $x - \frac{1}{2}dx$ to $x + \frac{1}{2}dx$ is $f(x)\,dx$. Symbolically it can be expressed as $P\left(x - \frac{1}{2}dx \leq x \leq x + \frac{1}{2}dx\right) = f(x)\,dx$. Then $f(x)$ is called the *probability density function* and the continuous curve $y = f(x)$ is called the *probability curve.*

The range of the variable may be finite or infinite. But even when the range is finite, it is convenient to consider it as infinite by supposing the density function to be zero outside the given range. Thus if $f(x) = \phi(x)$ be the density function denoted for the variate $x$ in the interval $(a, b)$, then it can be written as

$$\begin{aligned} f(x) &= 0, & x &< a \\ &= \phi(x), & a &\leq x \leq b \\ &= 0, & x &> b. \end{aligned}$$

*The density function $f(x)$ is always positive and* $\int_{-\infty}^{\infty} f(x)\,dx = 1$ (*i.e.*, the total area under the probability curve and the $x$-axis is unity which corresponds to the requirements that the total probability of happening of an event is unity).

**(2) Distribution function**

*If* $F(x) = P(X \leq x) = \int_{-\infty}^{x} f(x)\,dx,$

*then $F(x)$ is defined as the* **cumulative distribution function** *or simply the* **distribution function** *of the continuous variate $X$*. It is the probability that the value of the variate $X$ will be $\leq x$. The graph of $F(x)$ in this case is as shown in Fig. 26.3(*b*).

The distribution function $F(x)$ has the following properties :

(*i*) $F'(x) = f(x) \geq 0$, so that $F(x)$ is a non-decreasing function.

(*i*) $F(-\infty) = 0$ ; (*iii*) $F(\infty) = 1$

(*iv*) $P(a \leq x \leq b) = \int_a^b f(x)\,dx = \int_{-\infty}^{b} f(x)\,dx - \int_{-\infty}^{a} f(x)\,dx = F(b) - F(a).$

**Example 26.33.** (*i*) *Is the function defined as follows a density function?*

$$\begin{aligned} f(x) &= e^{-x}, & x &\geq 0 \\ &= 0, & x &< 0, \end{aligned}$$

(*ii*) *If so, determine the probability that the variate having this density will fall in the interval (1, 2) ?*

(*iii*) *Also find the cumulative probability function F(2) ?*

**Solution.** (*i*) $f(x)$ is clearly $\geq 0$ for every $x$ in (1, 2) and

$$\int_{-\infty}^{\infty} f(x)\,dx = \int_{-\infty}^{0} 0 \cdot dx + \int_0^{\infty} e^{-x}\,dx = 1$$

Hence the function $f(x)$ satisfies the requirements for a density function.

(*ii*) Required probability $= P(1 \le x \le 2) = \int_1^2 e^{-x}\, dx = e^{-1} - e^{-2} = 0.368 - 0.135 = 0.233.$

This probability is equal to the shaded area in Fig. 26.3 (*a*).

(*iii*) Cumulative probability function $F(2)$

$$\int_{-\infty}^{2} f(x)\, dx = \int_{-\infty}^{0} 0 \cdot dx + \int_0^2 e^{-x}\, dx = 1 - e^{-2} = 1 - 0.135 = 0.865$$

which is shown in Fig. 26.3 (*b*).

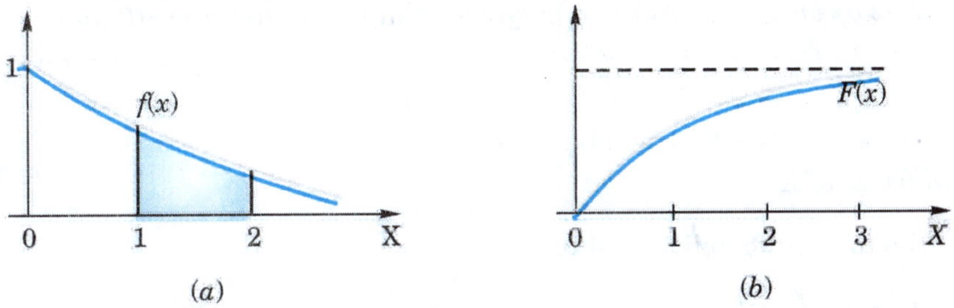


Fig. 26.3

## 26.10 (1) EXPECTATION

*The mean value* ($\mu$) *of the probability distribution of a variate X is commonly known as its* **expectation** *and is denoted by E(X).* If $f(x)$ is the probability density function of the variate $X$, then

$$\sum_i x_i f(x_i) \qquad \text{(discrete distribution)}$$

or $$E(X) = \int_{-\infty}^{\infty} x f(x)\, dx \qquad \text{(continuous distribution)}$$

In general, expectation of any function $\phi(x)$ is given by

$$E[\phi(x)] = \sum_i \phi(x_i) f(x_i) \qquad \text{(discrete distribution)}$$

or $$E[\phi(x)] = \int_{-\infty}^{\infty} \phi(x) f(x)\, dx \qquad \text{(continuous distribution)}$$

(2) **Variance** *of a distribution is given by*

$$\sigma^2 = \sum_i (x_i - \mu)^2 f(x_i) \qquad \text{(discrete distribution)}$$

or $$\sigma^2 = \int_{-\infty}^{\infty} (x - \mu)^2 f(x)\, dx \qquad \text{(continuous distribution)}$$

where $\sigma$ is the *standard deviation* of the distribution.

(3) *The* **rth moment** *about the mean* (denoted by $\mu_r$) *is defined by*

$$\mu_r = \Sigma (x_i - \mu)^r f(x_i) \qquad \text{(discrete distribution)}$$

or $$\mu_r = \int_{-\infty}^{\infty} (x - \mu)^r f(x)\, dx \qquad \text{(continuous distribution)}$$

(4) **Mean deviation from the mean** *is given by*

$$\Sigma\, |x_i - \mu|\, f(x_i) \qquad \text{(discrete distribution)}$$

or by $$\int_{-\infty}^{\infty} |x - \mu|\, f(x)\, dx \qquad \text{(continuous distribution)}$$

**Example 26.34.** *In a lottery, m tickets are drawn at a time out of n tickets numbered from 1 to n. Find the expected value of the sum of the numbers on the tickets drawn.* (*Rohtak, 2011*)

**Solution.** Let $x_1, x_2, ..., x_n$ be the variables representing the numbers on the first, second, ..., $n$th ticket. The probability of drawing a ticket out of $n$ tickets being in each case $1/n$, we have

$$E(x_i) = 1 \cdot \frac{1}{n} + 2 \cdot \frac{1}{n} + 3 \cdot \frac{1}{n} + ... + n \cdot \frac{1}{n} = \frac{1}{2}(n+1)$$

$\therefore$ expected value of the sum of the numbers on the tickets drawn

$$= E(x_1 + x_2 + ... + x_m) = E(x_1) + E(x_2) + ... + E(x_m)$$

$$= mE(x_i) = \frac{1}{2} m (n+1).$$

**Example 26.35.** *X is a continuous random variable with probability density function given by*

$$f(x) = kx \ (0 \le x < 2)$$
$$= 2k \ (2 \le x < 4)$$
$$= -kx + 6k \ (4 \le x < 6)$$

*Find k and mean value of X.*

**Solution.** Since the total probability is unity

$$\therefore \quad \int_0^6 f(x)\,dx = 1$$

*i.e.,*
$$\int_0^2 kx\,dx + \int_2^4 2k dx + \int_4^6 (-kx + 6k)\,dx = 1$$

or
$$k \left| x^2/2 \right|_0^2 + 2k \left| x \right|_2^4 + \left(-kx^2/2 + 6kx\right)_4^6 = 1$$

or
$$2k + 4k + (-10k + 12k) = 1 \text{ i.e., } k = 1/8.$$

$$\text{Mean of } X = \int_0^6 x\,f(x)\,dx$$

$$= \int_0^2 kx^2\,dx + \int_2^4 2kx\,dx + \int_4^6 x(-kx + 6k)dx$$

$$= k\left|x^3/3\right|_0^2 + 2k\left|x^2/2\right|_2^4 + \left(-k\left|x^3/3\right|_4^6 + 6k\left|x^2/2\right|_4^6\right)$$

$$= k(8/3) + k(12) - k(152/3) + 3k(20) = \frac{1}{8}(24) = 3.$$

**Example 26.36.** *A variate X has the probability distribution*

| $x$ : | −3 | 6 | 9 |
|---|---|---|---|
| $P(X = x)$ : | 1/6 | 1/2 | 1/3 |

*Find E (X) and E (X²). Hence evaluate E (2X + 1)².*

**Solution.**
$$E(X) = -3 \times \frac{1}{6} + 6 \times \frac{1}{2} + 9 \times \frac{1}{3} = 11/2.$$

$$E(X)^2 = 9 \times \frac{1}{6} + 36 \times \frac{1}{2} + 81 \times \frac{1}{3} = 93/2$$

$$\therefore \quad E(2X+1)^2 = E(4X^2 + 4X + 1) = 4E(X^2) + 4E(X) + 1$$
$$= 4(93/2) + 4(11/2) + 1 = 209.$$

**Example 26.37.** *The frequency distribution of a measurable characteristic varying between 0 and 2 is as under*

$$f(x) = x^3, \quad 0 \le x \le 1$$
$$= (2-x)^3, \quad 1 \le x \le 2.$$

*Calculate the standard deviation and also the mean deviation about the mean.*

**Solution.** Total frequency $N = \int_0^1 x^3\,dx + \int_1^2 (2-x)^3\,dx = \frac{1}{4} + \frac{1}{4} = \frac{1}{2}$

$\therefore$ $\mu_1'$ (about the origin) $= \frac{1}{N}\left[\int_0^1 x \cdot x^3\,dx + \int_1^2 x(2-x)^3\,dx\right]$

$$= 2\left\{\left|\frac{x^5}{5}\right|_0^1 + \left|-x\cdot\frac{(2-x)^4}{4}\right|_1^2 - \left|\frac{(2-x)^5}{20}\right|_1^2\right\} = 2\left(\frac{1}{5}+\frac{1}{4}+\frac{1}{20}\right) = 1$$

$\mu_2'$ (about the origin) $= \frac{1}{N}\left[\int_0^1 x^2 \cdot x^3 dx + \int_1^2 x^2\,(2-x)^3\,dx\right]$

$$= 2\left\{\left|\frac{x^6}{6}\right|_0^1 + \left|-x^2\,\frac{(2-x)^4}{4}\right|_1^2 + \frac{1}{2}\int_1^2 x\,(2-x)^4\,dx\right\}$$

$$= 2\left\{\frac{1}{6}+\frac{1}{4}+\frac{1}{2}\left[\frac{1}{5}+\frac{1}{30}\right]\right\} = \frac{16}{15}$$

Hence $\sigma^2 = \mu_2 = \mu_2' - (\mu_1')^2 = \frac{1}{15}$

*i.e.*, standard deviation $\sigma = \frac{1}{\sqrt{15}}$.

Mean deviation about the mean

$$= \frac{1}{N}\left\{\int_0^1 |x-1|\,x^3\,dx + \int_1^2 |x-1|\,(2-x)^3\,dx\right\}$$

$$= 2\left\{\int_0^1 (1-x)x^3\,dx + \int_1^2 (x-1)\,(2-x)^3\,dx\right\}$$

$$= 2\left\{\left(\frac{1}{4}-\frac{1}{5}\right)+\left(0+\frac{1}{20}\right)\right\} = \frac{1}{5}.$$

## 26.11 MOMENT GENERATING FUNCTION

(**1**) *The moment generating function (m.g.f.) of the discrete probability distribution of the variate X about the value x = a is defined as the expected value of* $e^{t(x-a)}$ and is denoted by $M_a(t)$. Thus

$$M_a(t) = \Sigma p_i\, e^{t(x_i - a)} \qquad ...(1)$$

which is a function of the parameter $t$ only.

Expanding the exponential in Eqn. (1), we get

$$M_a(t) = \Sigma p_i + t\Sigma p_i\,(x_i - a) + \frac{t^2}{2!}\,\Sigma p_i(x_i - a)^2 + ... + \frac{t^r}{r!}\,\Sigma x_i\,(x_i - a)^r + ...$$

$$= 1 + t\mu_1' + \frac{t^2}{2!}\,\mu'_2 + ... + \frac{t^r}{r!}\,\mu'_r + ... \qquad ...(2)$$

where $\mu'_r$ is the moment of order $r$ about $a$. Thus $M_a(t)$ generates moments and that is why it is called the moment generating function. From Eqn. (2), we find

$$\mu'_r = \text{coefficient of } t^r/r! \text{ in the expansion of } M_a(t).$$

Otherwise differentiating Eqn. (2) $r$ times with respect to $t$ and then putting $t = 0$, we get

$$\mu'_r = \left[\frac{d^r}{dt^r}\,M_a(t)\right]_{t=0} \qquad ...(3)$$

Thus the moment about any point $x = a$ can be found from Eqn. (2) or more conveniently from the formula (3).

Rewriting Eqn. (1) as

$$M_a(t) = e^{-at}\,\Sigma p_i\, e^{tx_i} \quad \text{or} \quad M_a(t) = e^{-at}\, M_0(t) \qquad ...(4)$$

*Thus the m.g.f. about the point a =* $e^{-at}$ *(m.g.f. about the origin).*

**Obs.** *The m.g.f. of the sum of two independent variables is the product of their m.g.fs.* ...(5)

**(2)** *If $f(x)$ is the density function of a continuous variate X, then the moment generating function of this continuous probability distribution about $x = a$ is given by*

$$M_a(t) = \int_{-\infty}^{\infty} e^{t(x-a)} f(x)\, dx.$$

**Example 26.38.** *Find the moment generating function of the exponential distribution*

$$f(x) = \frac{1}{c} e^{-x/c}, \; 0 \le x \le \infty, \; c > 0. \text{ Hence find its mean and S.D.}$$ *(Kurukshetra, 2009)*

**Solution.** The moment generating function about the origin is

$$M_0(t) = \int_0^{\infty} e^{tx} \cdot \frac{1}{c} e^{-x/c}\, dx = \frac{1}{c}\int_0^{\infty} e^{(t-1/c)x}\, dx \qquad \left[\because \; |t| < \frac{1}{c}\right]$$

$$= \frac{1}{c} \frac{\left| e^{(t-1/c)x} \right|_0^{\infty}}{|(t - 1/c)|} = (1 - ct)^{-1} = 1 + ct + c^2t^2 + c^3t^3 + \ldots$$

$$\therefore \quad \mu'_1 = \left[\frac{d}{dt} M_0(t)\right]_{t=0} = (c + 2c^2t + 3c^3t^2 + \ldots)_{t=0} = c$$

$$\mu'_2 = \left[\frac{d^2}{dt^2} M_0(t)\right]_{t=0} = 2c^2, \text{ and } \mu_2 = \mu'_2 - (\mu'_1)^2 = 2c^2 - c^2 = c^2.$$

Hence the mean is $c$ and S.D. is also $c$.

## 26.12 PROBABILITY GENERATING FUNCTION

*The probability generating function (p.g.f.) $P_x(t)$ for a random variable x which takes integral values 0, 1, 2, 3, ... only, is defined by*

$$P_x(t) = p_0 + p_1 t + p_2 t^2 + \ldots = \sum_{n=0}^{\infty} p_n t^n = E(t^x)$$

The coefficient of $t^n$ in the expansion of $P(t)$ in powers of $t$ gives $P(t)_{x=n}$.

$$\frac{\partial P}{\partial t} = \sum_{n=0}^{\infty} n p_n t^{n-1} \quad \text{or} \quad \left(\frac{\partial P}{\partial t}\right)_{t=1} = \Sigma\, n p_n = \mu_1'$$

$$\frac{\partial^2 P}{\partial t^2} = \sum_{n=0}^{\infty} n(n-1)\, p_n t^{n-2} \quad \text{or} \quad \left(\frac{\partial^2 P}{\partial t^2}\right)_{t=1} = \Sigma\, n(n-1)\, p_n = \mu_2' - \mu_1'$$

$$= \mu_2 + \mu_1'^2 - \mu_1' \text{ and so on}$$

Also $$\left(\frac{\partial^k P}{\partial t^k}\right)_{t=0} = n!\, p_n, \; k = 1, 2, \ldots n.$$

For integral valued variates, we have

$$P_x(e^t) = E(e^{tx}) = m.g.f. \text{ for } x.$$

**Obs.** The p.g.f. of the sum of two independent random variables is the product of their p.g.f.'s.

**Example 26.39.** *If x be a random variable with probability generating function $P_x(t)$, find the probability generating function of*

*(i)* $x + 2$ *(ii)* $2x$.

**Solution.** We have $P_x(t) = \sum_{k=0}^{\infty} p_k t^k$

(i) Probability generating function of $x + 2 = \sum_{k=0}^{\infty} p_k t^{k+2} = t^2 \sum_{k=0}^{\infty} p_k t^k = t^2 P_x(t)$.

(ii) Probability generating function of $2x = \sum_{k=0}^{\infty} p_k t^{2k} = \sum_{k=0}^{\infty} p_k (t^2)^k = P(t^2)$.

## PROBLEMS 26.4

1. A random variable $x$ has the following probability function :

| Values of $x$ : | –2 | –1 | 0 | 1 | 2 | 3 |
|---|---|---|---|---|---|---|
| $p(x)$ : | 0.1 | $k$ | 0.2 | $2k$ | 0.3 | $k$ |

Find the value of $k$ and calculate mean and variance. (*V.T.U., 2012 S*)

2. Find the standard deviation for the following discrete distribution :

| $x$ : | 8 | 12 | 16 | 20 | 24 |
|---|---|---|---|---|---|
| $p(x)$ : | 1/8 | 1/6 | 3/8 | 1/4 | 1/12 |

3. Obtain the distribution function of the total number of heads occurring in three tosses of an unbiased coin.
4. Show that for any discrete distribution $\beta_2 \geq 1$.
5. From an urn containing 3 red and 2 white balls, a man is to draw 2 balls at random without replacement, being promised ₹ 20 for each red ball he draws and ₹ 10 for each white one. Find his expectation.
6. Four coins are tossed. What is the expectation of the number of heads ?
7. The diameter of an electric cable is assumed to be a continuous variate with p.d.f. $f(x) = 6x(1-x), 0 \leq x \leq 1$. Verify that the above is a p.d.f. Also find the mean and variance.
8. A random variable gives measurements $X$ between 0 and 1 with a probability function

$$f(x) = 12x^3 - 21x^2 + 10x, \quad 0 \leq x \leq 1$$
$$= 0$$

(i) Find $P\left(X \leq \frac{1}{2}\right)$ and $P\left(X > \frac{1}{2}\right)$

(ii) Find a number $k$ such that $P(X \leq k) = \frac{1}{2}$

9. The power reflected by an aircraft that is received by a radar can be described by an exponential random variable $X$. The probability density of $X$ is given by $f(x) = \begin{cases} \frac{1}{x_0} e^{-x/x_0}, & x \geq 0 \\ 0, & x < 0 \end{cases}$

where $x_0$ is the average power received by the radar.

(i) What is the probability that the radar will receive power larger than the power received on the average ?
(ii) What is the probability that the radar will receive power less than the power received on the average ?

10. A function is defined as follows :

$$f(x) = 0, \quad x < 2$$
$$= \frac{1}{18}(2x+3), \quad 2 \leq x \leq 4$$
$$= 0, \quad x > 4.$$

Show that it is a density function. Find the probability that a variate having this density will fall in the interval $2 \leq x \leq 3$ ?

11. A continuous distribution of a variable $x$ in the range (–3, 3) is defined as

$$f(x) = \frac{1}{16}(3+x)^2, \quad -3 \leq x < -1$$
$$= \frac{1}{16}(2-6x^2), \quad -1 \leq x < 1$$
$$= \frac{1}{16}(3-x)^2, \quad 1 \leq x \leq 3.$$

Verify that the area under the curve is unity. Show that the mean is zero.

12. The frequency function of a continuous random variable is given by

$$f(x) = y_0 x\,(2 - x),\ 0 \le x \le 2.$$

Find the value of $y_0$, mean and variance of $x$.

13. The probability density $p(x)$ of a continuous random variable is given by

$$p(x) = y_0 e^{-|x|},\ -\infty < x < \infty.$$

Prove that $y_0 = 1/2$. Find the mean and variance of the distribution. (Calicut, 2012)

14. If $f(x) = \begin{cases} \dfrac{1}{2}(x+1), & -1 < x < 1 \\ 0, & \text{elsewhere} \end{cases}$

represents the density of a random variable $X$, find $E(X)$ and Var $(X)$.

15. A function is defined as under :

$$f(x) = 1/k,\ x_1 \le x \le x_2$$
$$= 0, \text{ elsewhere.}$$

Find the cumulative distribution of the variate $x$ when $k$ satisfies the requirements for $f(x)$ to be a density function.

## 26.13 REPEATED TRIALS

We know that the probability of getting a head or a tail on tossing a coin is $\frac{1}{2}$. If the coin is tossed thrice, the probability of getting one head and two tails can be combined as $H—T—T$, $T—H—T$, $T—T—H$. The probability of each one of these being $\frac{1}{2} \times \frac{1}{2} \times \frac{1}{2}$, *i.e.*, $\left(\frac{1}{2}\right)^3$, their total probability shall be $3(1/2)^3$.

Similarly if a trial is repeated $n$ times and if $p$ is the probability of a success and $q$ that of a failure, then the probability of $r$ successes and $n - r$ failures is given by $p^r q^{n-r}$.

But these $r$ successes and $n - r$ failures can occur in any of the ${}^nC_r$ ways in each of which the probability is same.

*Thus the probability of r successes is* ${}^nC_r\, p^r q^{n-r}$.

**Cor.** *The probabilities of at least r successes in n trials*

= the sum of the probabilities of $r, r + 1, \ldots, n$ successes

$= {}^nC_r\, p^r q^{n-r} + {}^nC_{r+1}\, p^{r+1} q^{n-r-1} + \ldots + {}^nC_n\, p^n.$

## 26.14 (1) BINOMIAL DISTRIBUTION*

*It is concerned with trials of a repetitive nature in which only the occurrence or non-occurrence, success or failure, acceptance or rejection, yes or no of a particular event is of interest.*

If we perform a series of independent trials such that for each trial $p$ is the probability of a success and $q$ that of a failure, then the probability of $r$ successes in a series of $n$ trials is given by ${}^nC_r\, p^r q^{n-r}$, where $r$ takes any integral value from 0 to $n$. The probabilities of 0, 1, 2, ... $r$, ..., $n$ successes are, therefore, given by

$$q^n,\ {}^nC_1 pq^{n-1},\ {}^nC_2 p^2 q^{n-2},\ \ldots,\ {}^nC_r\, p^r q^{n-r},\ \ldots,\ p^n.$$

*The probability of the number of successes so obtained* is called *the* **binomial distribution** for the simple reason that the probabilities are the successive terms in the expansion of the binomial $(q + p)^n$.

$\therefore$ the sum of the probabilities

$$= q^n + {}^nC_1 pq^{n-1} + {}^nC_2 p^2 q^{n-2} + \ldots + p^n = (q + p)^n = 1.$$

**(2) Constants of the binomial distribution.** The moment generating function about the origin is

$$M_0(t) = E(e^{tx}) = \Sigma\, {}^nC_x\, p^x q^{n-x} e^{tx} \qquad \text{[By (1) § 26.11]}$$
$$= \Sigma\, {}^nC_x\, (pe^t)^x q^{n-x} = (q + pe^t)^n$$

* It was discovered by a Swiss mathematician *Jacob Bernoulli* and was published posthumously in 1713.

Differentiating with respect to $t$ and putting $t = 0$ and using (3) § 26.11, we get the mean

$$\mu'_1 = np.$$

Since $M_a(t) = e^{-at} M_0(t)$, the m.g.f. of the binomial distribution about its mean $(m) = np$, is given by

$$M_m(t) = e^{-npt}(q + pe^t)^n = (qe^{-pt} + pe^{qt})^n$$

$$= \left(1 + pq\frac{t^2}{2!} + pq(q^2 - p^2)\frac{t^3}{3!} + pq(q^3 + p^3)\frac{t^4}{4!} + \cdots\right)^n \qquad \text{(Anna, 2012)}$$

or
$$1 + \mu_1 t + \mu_2 \frac{t^2}{2!} + \mu_3 \frac{t^3}{3!} + \mu_4 \frac{t^4}{4!} + \cdots$$

$$= 1 + npq\frac{t^2}{2!} + npq(q-p)\frac{t^3}{3!} + npq\,[1 + 3(n-2)pq]\frac{t^4}{4!} + \ldots$$

Equating the coefficients of like powers of $t$ on either side, we have

$$\mu_2 = npq,\ \mu_3 = npq(q-p),\ \mu_4 = npq\,[1 + 3(n-2)pq].$$

Also
$$\beta_1 = \frac{\mu_3^2}{\mu_2^3} = \frac{(q-p)^2}{npq} = \frac{(1-2p)^2}{npq} \quad \text{and} \quad \beta_2 = \frac{\mu_4}{\mu_2^2} = 3 + \frac{1-6pq}{npq}$$

Thus *mean* $= np$, *standard deviation* $= \sqrt{(npq)}$.

*skewness* $= (1-2p)/\sqrt{(npq)}$, *kurtosis* $= \beta_2$.

**Obs.** The skewness is positive for $p < \frac{1}{2}$ and negative for $p > \frac{1}{2}$. When $p = \frac{1}{2}$, the skewness is zero, *i.e.*, the probability curve of the binomial distribution will be symmetrical (bell-shaped).

As $n$ the number of trials increase indefinitely, $\beta_1 \to 0$, and $\beta_2 \to 3$.

**(3) Binomial frequency distribution.** If $n$ independent trials constitute one experiment and this experiment be repeated $N$ times, then the frequency of $r$ successes is $N\,{}^nC_r\,p^r\,q^{n-r}$. The possible number of successes together with these expected frequencies constitute the *binomial frequency distribution.*

**(4) Applications of Binomial distribution.** This distribution is applied to problems concerning : (*i*) Number of defectives in a sample from production line,

(*ii*) Estimation of reliability of systems,

(*iii*) Number of rounds fired from a gun hitting a target,

(*iii*) Radar detection.

**Example 26.40.** *The probability that a pen manufactured by a company will be defective is 1/10. If 12 such pens are manufactured, find the probability that*

(*a*) *exactly two will be defective.* (*b*) *at least two will be defective.*

(*c*) *none will be defective.* (*V.T.U., 2012 S*)

**Solution.** The probability of a defective pen is 1/10 = 0.1

∴ The probability of a non-defective pen is 1 – 0.1 = 0.9

(*a*) *The probability that exactly two will be defective*

$$= {}^{12}C_2\,(0.1)^2\,(0.9)^{10} = 0.2301$$

(*b*) *The probability that at least two will be defective*

= 1 – (prob. that either none or one is non-defective)

$$= 1 - [{}^{12}C_0(0.9)^{12} + {}^{12}C_1(0.1)\,(0.9)^{11}] = 0.3412$$

(*c*) *The probability that none will be defective*

$$= {}^{12}C_{12}(0.9)^{12} = 0.2833.$$

**Example 26.41.** *In 256 sets of 12 tosses of a coin, in how many cases one can expect 8 heads and 4 tails.*

**Solution.** $P(\text{head}) = \frac{1}{2}$ and $P(\text{tail}) = \frac{1}{2}$

By binomial distribution, probability of 8 heads and 4 tails in 12 trials is

$$P(X = 8) = {}^{12}C_8 \left(\frac{1}{2}\right)^8 \left(\frac{1}{2}\right)^4 = \frac{12!}{8!\,4!} \cdot \frac{1}{2^{12}} = \frac{495}{4096}$$

$\therefore$ the expected number of such cases in 256 sets

$$= 256 \times P(X = 8) = 256 \frac{495}{4096} = 30.9 = 31 \text{ (say)}.$$

**Example 26.42.** *In sampling a large number of parts manufactured by a machine, the mean number of defectives in a sample of 20 is 2. Out of 1000 such samples, how many would be expected to contain at least 3 defective parts.*

**Solution.** Mean number of defectives = 2 = $np$ = $20p$.

$\therefore$ The probability of a defective part is $p = 2/20 = 0.1$.

and the probability of a non-defective part = 0.9

$\therefore$ The probability of at least three defectives in a sample of 20.

$$= 1 - \text{(prob. that either none, or one, or two are non-defective parts)}$$
$$= 1 - [{}^{20}C_0(0.9)^{20} + {}^{20}C_1(0.1)(0.9)^{19} + {}^{20}C_2(0.1)^2(0.9)^{18}]$$
$$= 1 - (0.9)^{18} \times 4.51 = 0.323.$$

Thus *the number of samples having at least three defective parts* out of 1000 samples

$$= 1000 \times 0.323 = 323.$$

**Example 26.43.** *A set of 8 biased coins was tossed 256 times and the frequencies of throws were observed as follows.*

| No. of heads ($H$) | 0 | 1 | 2 | 3 | 4 | 5 | 6 | 7 | 8 |
|---|---|---|---|---|---|---|---|---|---|
| Frequency | 2 | 6 | 24 | 63 | 64 | 50 | 36 | 10 | 1 |

*what is the binomial distribution of the data?* [*Gujarat, 2018*]

**Solution.** Here $n = 8$, and $N = \Sigma f = 256$

$$\therefore \quad \text{Mean} = \frac{\Sigma fx}{\Sigma f} = \frac{1}{256}(0 + 6 + 48 + 189 + 256 + 250 + 216 + 70 + 8)$$
$$= \frac{1043}{256} = np = 8p \Rightarrow \frac{1043}{2048}$$
$$p = 0.5093$$
$$q = 1 - p = 0.4907$$

Hence the Binomial distribution to be filled to the data is

$$N(q + p)^n = 256(0.4907 + 0.5093)^8$$

This will give expected frequencies of the distribution.

**Example 26.44:** *The probability that a bomb dropped from a plane will strike the target is $\frac{1}{5}$. If fix bombs are dropped, find the probability that*

*(i) exactly two will strike the target*

*(ii) at least two will strike the target* (*Kolkata, 2018*)

**Solution.** The probability of 0, 1, 2.....

successes are given by respective terms in the expansion of

$$(q + p)^n = \left(\frac{4}{5} + \frac{1}{5}\right)^6, \text{ since } P = \frac{1}{5}, q = \frac{4}{5} \text{ and } n = 6$$

$\therefore P(2)$ = the probability that at least 2 bombs will strike the target.

$$= 1 - p(0) - p(1)$$

$= 1 - q^6 - 6C_1\, q^5 p = 1 - (0.8)^6 = 6(0.2)(0.8)^5$

$= 1 - 0.2621 - 0.3932 = 0.345.$

**Example 26.45:** *Find the probability that at most 5 defective components will be found in a lot of 200 if experience show that 2% of such components are defective. Also find probability of more than five defective components.*

*(VTU, 2018)*

**Solution.** Probability of a component being defective is $\frac{1}{50}$, which is small and hence the distribution can be taken as Poissonian.

$$P = \frac{1}{50},\ n = 200,\ m = np = 4$$

$$e^{-m} = e^{-4} = 0.0184$$

Hence the probability of 5 or less than 5 defective components

$$p(r \le 5) = P(0) + P(1) + \ldots.P(5)$$

$$= e^{-m}\left(1 + m + \frac{m^2}{2!} + \frac{m^3}{3!} + \frac{m^4}{4!} + \frac{m^5}{5!}\right)$$

$$= 0.0183\left(1 + 4 + 8 + \frac{32}{3} + \frac{32}{3} + \frac{128}{15}\right) = 0.785$$

The probability of more five defective components.

$$= P(r > 5) = P(6) + P(7) + \ldots.\ 1 = 1 - p(r \le 5),$$

$$= 1 - 0.785 = 0.215, \text{ since } P(r \le 5) + P(r > 5)$$

$$= 1.$$

## PROBLEMS 26.5

1. Determine the binomial distribution for which mean = 2 (variance) and mean + variance = 3. Also find $P(X \le 3)$.
2. (*a*) An ordinary six-faced die is thrown four times. What are the probabilities of obtaining 4, 3, 2, 1 and 0 faces ?

   (*b*) A die is thrown five times. If getting an odd number is a success, find the probability of getting at least four successes. *(U.P.T.U., 2012)*
3. If the chance that one of the ten telephone lines is busy at an instant is 0.2.

   (*a*) What is the chance that 5 of the lines are busy ?

   (*b*) What is the most probable number of busy lines and what is the probability of this number ?

   (*c*) What is the probability that all the lines are busy ?
4. If the probability that a new-born child is a male is 0.6, find the probability that in a family of 5 children there are exactly 3 boys.
5. If on an average 1 vessel in every 10 is wrecked, find the probability that out of 5 vessels expected to arrive, at least 4 will arrive safely.
6. The probability that a bomb dropped from a plane will strike the target is 1/5. If six bombs are dropped, find the probability that (*i*) exactly two will strike the target, (*ii*) at least two will strike the target.
7. A sortie of 20 aeroplanes is sent on an operational flight. The chances that an aeroplane fails to return is 5%. Find the probability that (*i*) one plane does not return (*ii*) at the most 5 planes do not return, and (*iii*) what is the most probable number of returns ?
8. The probability that an entering student will graduate is 0.4. Determine the probability that out of 5 students (*a*) none (*b*) one and (*c*) at least one will graduate.
9. Out of 800 families with 5 children each, how many would you expect to have (*a*) 3 boys, (*b*) 5 girls, (*c*) either 2 or 3 boys ? Assume equal probabilities for boys and girls.

10. If 10 per cent of the rivets produced by a machine are defective, find the probability that out of 5 rivets chosen at random (*i*) none will be defective, (*ii*) one will be defective, and (*iii*) at least two will be defective.

(*Kurukshetra, 2013*)

11. In a bombing action there is 50% chance that any bomb will strike the target. Two direct hits are needed to destroy the target completely. How many bombs are required to be dropped to give a 99% chance or better of completely destroying the target.

12. A product is 0.5% defective and is packed in cartons of 100. What percentage contains not more than 3 defectives ?

13. If in a lot of 500 solenoids 25 are defective, find the probability of 0, 1, 2, 3 defective solenoids in a random sample of 20 solenoids.

14. 500 articles were selected at random out of a batch containing 10,000 articles, and 30 were found to be defective. How many defectives articles would you reasonably expect to have in the whole batch ?

15. Fit a binomial distribution for the following data and compare the theoretical frequencies with the actual ones :

| $x$ : | 0 | 1 | 2 | 3 | 4 | 5 |
|---|---|---|---|---|---|---|
| $f$ : | 2 | 14 | 20 | 34 | 22 | 8 |

16. Fit a binomial distribution to the following frequency distribution :

(*a*)

| $x$ : | 0 | 1 | 3 | 4 |
|---|---|---|---|---|
| $y$ : | 28 | 62 | 10 | 4 |

(*U.P.T.U., 2011*)

(*b*)

| $x$ : | 0 | 1 | 2 | 3 | 4 | 5 | 6 |
|---|---|---|---|---|---|---|---|
| $f$ : | 13 | 25 | 52 | 58 | 32 | 16 | 4 |

## 26.15 (1) POISSON DISTRIBUTION*

*It is a distribution related to the probabilities of events which are extremely rare, but which have a large number of independent opportunities for occurrence.* The number of printing mistakes per page, number of accidents on a highway, number of dishonoured cheques at a bank, number of persons born blind per year in a large city and the number of deaths by horse kick in an army corps are some of the phenomena, in which this law is followed.

*This distribution can be derived as a limiting case of the binomial distribution by making n very large and p very small, keeping np fixed (= m, say).*

The probability of $r$ successes in a binomial-distribution is

$$P(r) = {}^nC_r\, p^r q^{n-r} = \frac{n(n-1)(n-2)\cdots(n-r+1)}{r!}\, p^r q^{n-r}$$

$$= \frac{np(np-p)(np-2p)\cdots(np-\overline{r-1}p)}{r!}\,(1-p)^{n-r}$$

As $n \to \infty$, $p \to 0$ $(np = m)$, we have

$$P(r) = \frac{m^r}{r!} \operatorname*{Lt}_{n\to\infty} \frac{(1-m/n)^n}{(1-m/n)^r} = \frac{m^r}{r!}\, e^{-m}$$

so that the probabilities of 0, 1, 2..., $r$,... successes in a Poisson distribution are given by

$$e^{-m},\ me^{-m},\ \frac{m^2e^{-m}}{2!},\ldots,\ \frac{m^re^{-m}}{r!},\ldots$$

The sum of these probabilities is unity as it should be.

* It was discovered in 1837 by a French mathematician Simon Denius Poisson (1781–1840).

**(2) Constants of the Poisson distribution.** These constants can easily be derived from the corresponding constants of the binomial distribution simply by making $n \to \infty$, $p \to 0$, $(q \to 1)$ and noting that $np = m$

$$Mean = \text{Lt}\,(np) = m$$

$$\mu_2 = \text{Lt}\,(npq) = m\ \text{Lt}\,(q) = m$$

$\therefore$ *Standard deviation* $= \sqrt{m}$ *(V.T.U., 2014)*

Also $\mu_3 = m$, $\mu_4 = m + 3m^2$

$\therefore$ *Skewness* $(= \sqrt{\beta_1}) = 1/m$, *Kurtosis* $(= \beta_2) = 3 + 1/m$.

Since $\mu_3$ is positive, Poisson distribution is positively skewed and since $\beta_2 > 3$, it is *Leptokurtic*.

**(3) Applications of Poisson distribution.** This distribution is applied to problems concerning : (*i*) Arrival pattern of 'defective vehicles in a workshop', 'patients in a hospital' or 'telephone calls'.

(*ii*) Demand pattern for certain spare parts.

(*iii*) Number of fragments from a shell hitting a target.

(*iv*) Spatial distribution of bomb hits.

**Example 26.46.** *If the probability of a bad reaction from a certain injection is 0.001, determine the chance that out of 2,000 individuals more than two will get a bad reaction.*

**Solution.** It follows a Poisson distribution as the probability of occurrence is very small.

Mean $m = np = 2000(0.001) = 2$

Probability that more than 2 will get a bad reaction

= 1 – [prob. that no one gets a bad reaction + prob. that one gets a bad reaction + prob. that two get bad reaction]

$$= 1 - \left[e^{-m} + \frac{m^1 e^{-m}}{1!} + \frac{m^2 e^{-m}}{2!}\right] = 1 - \left[\frac{1}{e^2} + \frac{2}{e^2} + \frac{2}{e^2}\right] \qquad [\because\ m = 2]$$

$$= 1 - \frac{5}{e^2} = 0.32. \qquad [\because\ e = 2.718]$$

**Example 26.47.** *In a certain factory turning out razor blades, there is a small chance of 0.002 for any blade to be defective. The blades are supplied in packets of 10, use Poisson distribution to calculate the approximate number of packets containing no defective, one defective and two defective blades respectively in a consignment of 10,000 packets.* *(V.T.U., 2016)*

**Solution.** We know that $m = np = 10 \times 0.002 = 0.02$

$$e^{-0.02} = 1 - 0.02 + \frac{(0.02)^2}{2!} - \ldots = 0.9802 \text{ approximately}$$

Probability of no defective blade is $e^{-m} = e^{-0.02} = 0.9802$

$\therefore$ no. of packets containing no defective blade is

$$10{,}000 \times 0.9802 = 9802$$

Similarly the number of packets containing one defective blade $= 10{,}000 \times me^{-m}$

$$= 10{,}000 \times (0.02) \times 0.9802 = 196$$

Finally the number of packets containing two defective blades

$$= 10{,}000 \times \frac{m^2 e^{-m}}{2!} = 10{,}000 \times \frac{(0.02)^2}{2!} \times 0.9802 = 2 \text{ approximately.}$$

**Example 26.48.** *Fit a Poisson distribution to the set of observations :*

| *x* : | 0 | 1 | 2 | 3 | 4 |
|---|---|---|---|---|---|
| *f* : | 122 | 60 | 15 | 2 | 1 |

*(U.T.U., 2010)*

**Solution.** Mean $= \dfrac{\Sigma f_i x_i}{\Sigma f_i} = \dfrac{60 + 36 + 6 + 4}{200} = 0.5.$

∴ mean of Poisson distribution *i.e.*, $m = 0.5$.

Hence the theoretical frequency for $r$ successes is

$$\frac{Ne^{-m}(m)^r}{r!} = \frac{200e^{-0.5}(0.5)^r}{r!} \quad \text{where } r = 0, 1, 2, 3, 4$$

∴ the theoretical frequencies are

| $x$ : | 0 | 1 | 2 | 3 | 4 |
|---|---|---|---|---|---|
| $f$ : | 121 | 61 | 15 | 2 | 0 |

($\because\ e^{-0.5} = 0.61$)

## PROBLEMS 26.6

1. If a random variable has a Poisson distribution such that $P(1) = P(2)$, find
(*i*) mean of the distribution. (*Rohtak, 2011 S*) (*ii*) $P(4)$.
2. $X$ is a Poisson variable and it is found that the probability that $X = 2$ is two-thirds of the probability that $X = 1$. Find the probability that $X = 0$ and the probability that $X = 3$. What is the probability that $X$ exceeds 3 ?
3. For Poisson distribution, prove that $m\,\mu_2\,\gamma_1\,\gamma_2 = 1$, where symbols have their usual meanings.
4. A certain screw making machine produces on average of 2 defective screws out of 100, and packs them in boxes of 500. Find the probability that a box contains 15 defective screws.
5. Using Poisson distribution, find the probability that all of the spades will be drawn from a pack of well-shuffled cards atleast once in 104 consecutive trials. (Given $e^{-2} = 0.136$). (*U.T.U., 2011*)
6. A manufacturer knows that the condensers he makes contain on the average 1 % defectives. He packs them in boxes of 100. What is the probability that a box picked at random will contain 3 or more faulty condensers ?
7. A car-hire firm has two cars which it hires out day by day. The number of demands for a car on each day is distributed as a Poisson distribution with mean 1.5. Calculate the proportion of days (*i*) on which there is no demand, (*ii*) on which demand is refused. ($e^{-1.5} = 0.2231$).
8. The incidence of occupational disease in an industry is such that the workmen have a 10% chance of suffering from it. What is probability that in a group of 7, five or more will suffer from it ?
9. The frequency of accidents per shift in a factory is as shown in the following table :

| Accidents per shift : | 0 | 1 | 2 | 3 | 4 |
|---|---|---|---|---|---|
| Frequency : | 180 | 92 | 24 | 3 | 1 |

Calculate the mean number of accidents per shift and the corresponding Poisson distribution and compare with actual observations.

10. A source of liquid is known to contain bacteria with the mean number of bacteria per cubic centimetre equal to 3. Ten 1 c.c., test-tubes are filled with the liquid. Assuming that Poisson distribution is applicable, calculate the probability that all the test-tubes will show growth *i.e.*, contain atleast 1 bacterium each.
11. Find the expectation of the function $\phi(x) = xe^{-x}$ in a Poisson distribution.
[**Hint :** If $m$ be the mean of the Poisson distribution, then expectation of

$$\phi(x) = \sum_{x=0}^{\infty} \frac{\phi(x) \cdot m^x e^{-m}}{x!} = m \exp. m\,(e^{-1} - m - 1)]$$

12. Fit a Poisson distribution to the following :

| $x$ : | 0 | 1 | 2 | 3 | 4 |
|---|---|---|---|---|---|
| $f$ : | 46 | 38 | 22 | 9 | 1 |

(*Rohtak, 2011*)

13. Fit a Poisson distribution to the following data given the number of yeast cells per square for 400 squares :

| No. of cells per sq. : | 0 | 1 | 2 | 3 | 4 | 5 | 6 | 7 | 8 | 9 | 10 |
|---|---|---|---|---|---|---|---|---|---|---|---|
| No. of squares : | 103 | 143 | 98 | 42 | 8 | 4 | 2 | 0 | 0 | 0 | 0 |

## 26.16 (1) NORMAL DISTRIBUTION*

Now we consider a continuous distribution of fundamental importance, namely the normal distribution. *Any quantity whose variation depends on random causes is distributed according to the normal law.* Its importance lies in the fact that a large number of distributions approximate to the normal distribution and as such it is a corner stone of modern statistics.

* In 1924, *Karl Pearson* found this distribution which *Abraham De Moivre* had discovered as early as 1733. It is also known as *Gaussian* distribution due to *Karl Friedrich Gauss*.

Let us define a variate $z = \dfrac{x - np}{\sqrt{(npq)}}$ ...(1)

where $x$ is a binomial variate with mean $np$ and S.D. $\sqrt{(npq)}$ so that $z$ is a variate with mean zero and variance unity. In the limit as $n$ tends to infinity, the distribution of $z$ becomes a continuous distribution extending from $-\infty$ to $\infty$.

It can be shown that *the limiting form of the binomial distribution Eqn. (1) for large values of n when neither p nor q is very small, is the normal distribution.* The normal curve is of the form

$$y = \frac{1}{\sigma\sqrt{(2\pi)}} e^{-(x-\mu)^2/2\sigma^2} \qquad \text{...(2)}$$

where $\mu$ and $\sigma$ are the mean and standard deviation respectively.

**(2) Properties of the normal distribution**

*I.* The normal curve Eqn. (2) is bell-shaped and is symmetrical about its mean. It is unimodal with ordinates decreasing rapidly on both sides of the mean (Fig. 26.3). The maximum ordinate is $1/\sigma\sqrt{(2\pi)}$, found by putting $x = \mu$ in Eqn. (2).

As it is symmetrical, its mean, median and mode are the same. Its points of inflexion (found by putting $d^2y/dx^2 = 0$ and verifying that at these points $d^3y/dx^3 \neq 0$) are given by $x = \mu \pm \sigma$, *i.e.*, these points are equidistant from the mean on either side.

*II. Mean deviation from the mean* $\mu$

$$= \int_{-\infty}^{\infty} |x - \mu| \cdot \frac{1}{\sigma\sqrt{(2\pi)}} e^{-(x-\mu)^2/2\sigma^2} dx \qquad [\text{Put } z = (x-\mu)/\sigma]$$

$$= \frac{\sigma}{\sqrt{(2\pi)}} \int_{-\infty}^{\infty} |z| e^{-z^2/2} dz$$

$$= \frac{\sigma}{\sqrt{(2\pi)}} \left[\int_{-\infty}^{0} -ze^{-z^2/2}dz + \int_{0}^{\infty} ze^{-z^2/2}dz\right] = \frac{2\sigma}{\sqrt{(2\pi)}} \int_0^{\infty} ze^{-z^2/2} dz$$

$$= \frac{2\sigma}{\sqrt{(2\pi)}} \left| -e^{-z^2/2} \right|_0^{\infty} = -\sqrt{\left(\frac{2}{\pi}\right)} \sigma(0-1) = 0.7979\,\sigma \approx (4/5)\,\sigma$$

*(Rohtak, 2011 S)*

*III. Moments about the mean*

$$\mu_{2n+1} = \int_{-\infty}^{\infty} (x-\mu)^{2n+1} \frac{1}{\sigma\sqrt{(2\pi)}} e^{-(x-\mu)^2/2\sigma^2} dx$$

$$= \frac{\sigma^{2n+1}}{\sqrt{(2\pi)}} \int_{-\infty}^{\infty} z^{2n+1} e^{-z^2/2} dz \text{ where } z = (x-\mu)/\sigma$$

= 0, since the integral is an odd function.

Thus all odd order moments about the mean vanish.

$$\mu_{2n} = \int_{-\infty}^{\infty} (x-\mu)^{2n} \frac{1}{\sigma\sqrt{(2\pi)}} e^{-(x-\mu)^2/2\sigma^2} dx$$

$$= \frac{\sigma^{2n}}{\sqrt{(2\pi)}} \int_{-\infty}^{\infty} z^{2n-1} e^{-z^2/2} \,.\, zdz \qquad [\text{Integrate by parts}]$$

$$= \frac{\sigma^{2n}}{\sqrt{(2\pi)}} \left[\left| -z^{2n-1} e^{-z^2/2} \right|_{-\infty}^{\infty} + \int_{-\infty}^{\infty} (2n-1) z^{2n-2} e^{-z^2/2} dz\right]$$

$$= \frac{\sigma^{2n}}{\sqrt{(2\pi)}} (0-0) + (2n-1)\sigma^2 \mu_{2n-2}$$

Repeated application of this reduction formula, gives

$$\mu_{2n} = (2n-1)(2n-3) \ldots 3 \,.\, 1\sigma^{2n}$$

In particular, $\mu_2 = \sigma^2$, $\mu_4 = 3\sigma^4$.

Hence $\beta_1 = \dfrac{\mu_3^2}{\mu_2^3} = 0$ and $\beta_2 = \dfrac{\mu_4}{\mu_2^2} = 3$ *(Andhra, 2014) i.e.*, the coefficient of skewness is zero (*i.e.* the curve is symmetrical) and the Kurtosis is 3. This is the basis for the choice of the value 3 in the definitions of platykurtic and leptokurtic.

*IV. The probability of x lying between* $x_1$ *and* $x_2$ is given by the area under the normal curve from $x_1$ to $x_2$, *i.e.*, $P(x_1 \le x \le x_2)$

$$= \frac{1}{\sigma\sqrt{(2\pi)}} \int_{x_1}^{x_2} e^{-(x-\mu)^2/2\sigma^2}\,dx$$

$$= \frac{1}{\sqrt{(2\pi)}} \int_{z_1}^{z_2} e^{-z^2/2}\,dz \text{ where } z = (x-\mu)/\sigma,\ dz = dx/\sigma \text{ and } z_1 = (x_1-\mu)/\sigma,\ z_2 = (x_2-\mu)/\sigma.$$

$$= \frac{1}{\sqrt{(2\pi)}} \left\{ \int_0^{z_2} e^{-z^2/2}\,dz - \int_0^{z_1} e^{-z^2/2}\,dz \right\} = P_2(z) - P_1(z)$$

The values of each of the above integrals can be found from the table IV–Appendix 2, which gives the values of

$$P(z) = \frac{1}{\sqrt{(2\pi)}} \int_0^{z} e^{-z^2/2}\,dz$$

for various values of $z$. This integral is called the *probability integral* or the *error function* due to its use in the theory of sampling and the theory of errors.

Using this table, we see that the area under the normal curve from $z = 0$ to $z = 1$, *i.e.* from $x = \mu$ to $\mu + \sigma$ is 0.3413.

∴ (*i*) The area under the normal curve between the ordinates $x = \mu - \sigma$ and $x = \mu + \sigma$ is 0.6826, ~ *68% nearly. Thus approximately 2/3 of the values lie within these limits.*

(*ii*) The area under the normal curve between $x = \mu - 2\sigma$ and $x = \mu + 2\sigma$ is 0.9544 ~ 95.5%, which implies that *about 4½% of the values lie outside these limits.*

(*ii*) *99.73% of the values lie between* $x = \mu - 3\sigma$ *and* $x = \mu + 3\sigma$ *i.e., only a quarter % of the whole lies outside these limits.*

(*iv*) *95% of the values lie between* $x = \mu - 1.96\sigma$ *and* $x = \mu + 1.96\,\sigma$ *i.e., only 5% of the values lie outside these limits.*

(*v*) *99% of the values lie between* $x = \mu - 2.58\sigma$ *and* $x = \mu + 2.58\sigma$ *i.e., only 1% of the values lie outside these limits.*

(*vi*) 99.9% of the values lie between $x = \mu - 3.29\sigma$ and $x = \mu + 3.29\sigma$.

In other words, a value that deviates more than σ from μ occurs about once in 3 trials. A value that deviates more than 2σ or 3σ from μ occurs about once in 20 or 400 trials. Almost all values lie within 3σ of the mean.

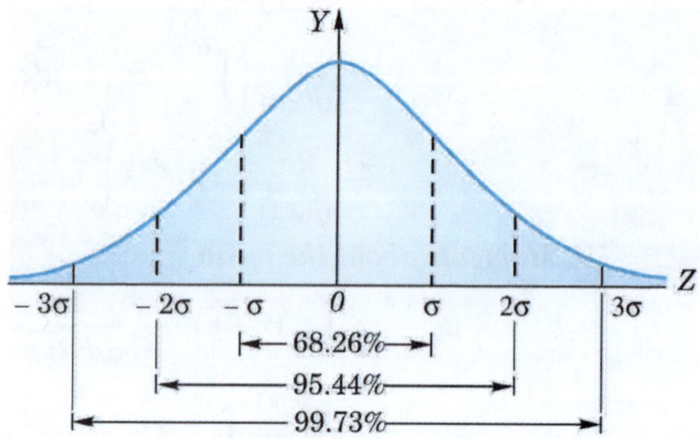

Fig. 26.4

The shape of the standardised normal curve is

$$y = \frac{1}{\sqrt{(2\pi)}} e^{-z^2/2}, \text{ where } z = (x-\mu)/\sigma \qquad ...(3)$$

and the respective areas are shown in Fig. 26.4. '*z*' is called a *normal variate*. Areas under the Normal curve are given in Table IV (page 1374).

**Obs.** 50% area lies in the $z$-interval (–0.745, 0.745) and 99% area lies in the $z$-interval (–2.58, 2.58).

**(3) Normal frequency distribution.** We can fit a normal curve to any distribution. If $N$ be the total frequency, μ the mean and σ the standard deviation of the given distribution then the curve

$$y = \frac{N}{\sigma\sqrt{(2\pi)}} e^{-(x-\mu)^2/2\sigma^2} \qquad ...(4)$$

will fit the given distribution as best as the data will permit. The frequency of the variate between $x_1$ and $x_2$ as given by the fitted curve, will be the area under Eqn. (1) from $x_1$ to $x_2$.

**(4) Applications of normal distribution.** This distribution is applied to problems concerning :

(*i*) Calculation of errors made by chance in experimental measurements.

(*ii*) Computation of hit probability of a shot.

(*iii*) Statistical inference in almost every branch of science.

## 26.17 PROBABLE ERROR

Any lot of articles manufactured to certain specifications is subject to small errors. In fact, measurement of any physical quantity shows slight error. In general, these errors of manufacture or experiment are of random nature and therefore, follow a normal distribution. While quoting a specification of an experimental result, we usually mention the *probable error* ($\lambda$). *It is such that the probability of an error falling within the limits* $\mu - \lambda$ *and* $\mu + \lambda$ *is exactly equal to the chance of an error falling outside these limits, i.e.* the chance of an error lying within $\mu - \lambda$ and $\mu + \lambda$ is $\frac{1}{2}$.

$$\therefore \quad \frac{1}{\sigma\sqrt{(2\pi)}} \int_{\mu-\lambda}^{\mu+\lambda} e^{-(x-\mu)^2/2\sigma^2}\, dx = \frac{1}{2}$$

or
$$\frac{1}{\sqrt{(2\pi)}} \int_{0}^{\lambda/\sigma} e^{-z^2/2} dz = \frac{1}{4} \qquad \left[ z = \frac{x-\mu}{\sigma} \right]$$

The table V, (Appendix 2) gives $\lambda/\sigma = 0.6745$

Hence the probable error $\lambda = 0.6745\sigma \sim \frac{2}{3}\sigma$.

**Obs.** Quartile deviation $= \frac{1}{2}(Q_3 - Q_1) \sim \frac{2}{3}\sigma$ ; Mean deviation $\approx \frac{4}{5}\sigma$ [p. 959]

$\therefore$ $Q.D. : M.D. : S.D. = 10 : 12 : 15.$

**Example 26.49.** *X is a normal variate with mean 30 and S.D. 5, find the probabilities that (i)* $26 \le X \le 40$, *(ii)* $X \ge 45$ *and (iii)* $|X - 30| > 5$.

**Solution.** We have $\mu = 30$ and $\sigma = 5$

$$\therefore \quad z = \frac{X-\mu}{\sigma} = \frac{X-30}{5}$$

(*i*) When $X = 26, z = -0.8$; when $X = 40$, $z = 2$

$$\begin{aligned} \therefore \quad P(26 \le X \le 40) &= P(-0.8 \le z \le 2) \\ &= P(-0.8 \le z \le 0) + P(0 \le z \le 2) \\ &= P(0 \le z \le 0.8) + 0.4772 \\ &= 0.2881 + 0.4772 = 0.7653 \end{aligned}$$

(*ii*) When $X = 45, z = 3$

$$\begin{aligned} \therefore \quad P(X \ge 45) &= P(z \ge 3) = 0.5 - P(0 \le z \le 3) \\ &= 0.5 - 0.4986 = 0.0014 \end{aligned}$$

(*iii*)
$$\begin{aligned} P[|X-30| \le 5] &= P[25 \le X \le 35] \\ &= P(-1 \le z \le 1) = 2P(0 \le z \le 1) \\ &= 2 \times 0.3413 = 0.6826 \end{aligned}$$

$$\begin{aligned} \therefore \quad P[|X-30| > 5] &= 1 - P[|X-30| \le 5] \\ &= 1 - 0.6826 = 0.3174. \end{aligned}$$

**Example 26.50.** *A certain number of articles manufactured in one batch were classified into three categories according to a particular characteristic, being less than 50, between 50 and 60 and greater than 60. If this characteristic is known to be normally distributed, determine the mean and standard deviation for this batch if 60%, 35% and 5% were found in these categories.*

**Solution.** Let $\mu$ be the mean (at $z = 0$) and $\sigma$ the standard deviation of the normal curve (Fig. 26.5).

Now 60% of the articles have the characteristic below 50, 35% between 50 and 60 and only 5% greater than 60.

Let the area to the left of the ordinate $PQ$ be 60% and that between the ordinates $PQ$ and $ST$ be 35% so that the areas to the left of $PQ$ ($z = z_1$) and $ST$ ($z = z_2$) are 0.6 and 0.95 respectively, *i.e.*, the area $OPQR = 0.6 - 0.5 = 0.1$ and the area $OSTR = 0.45$.

$\therefore$ area corresponding to $z_1 \left(= \dfrac{50 - \mu}{\sigma}\right) = 0.1$

and that corresponding to $z_2 \left(= \dfrac{60 - \mu}{\sigma}\right) = 0.45$

From the table IV, we have

$$(50 - \mu)/\sigma = 0.2533 \quad \text{and} \quad (60 - \mu)/\sigma = 1.645$$

whence $\sigma = 7.543$ and $\mu = 48.092.$

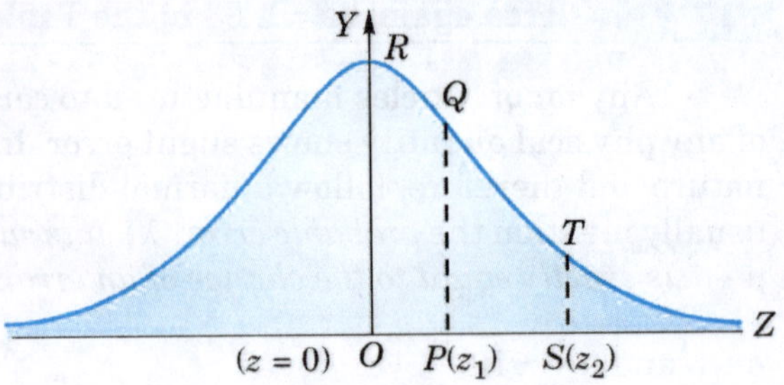

Fig. 26.5

**Example 26.51.** *In a normal distribution, 31% of the items are under 45 and 8% are over 64. Find the mean and S.D. of the distribution.* (V.T.U., 2016)

**Solution.** Let $\bar{x}$ be the mean and $\sigma$ the S.D. 31% of the items are under 45 means area to the left of the ordinate $x = 45$. (Fig. 26.6)

When $x = 45$, let $z = z_1$ so that $z_1 = \dfrac{45 - \bar{x}}{\sigma}$ ...(*i*)

$$\therefore \quad \int_{-\infty}^{z_1} \phi(z)\, dz = 0.31 \quad \text{or} \quad \int_{-\infty}^{0} \phi(z)\, dz - \int_{z_1}^{0} \phi(z)\, dz = 0.31$$

Hence $$\int_{z_1}^{0} \phi(z)\, dz = \int_{-\infty}^{0} \phi(z)\, dz - 0.31 = 0.5 - 0.31 = 0.19$$

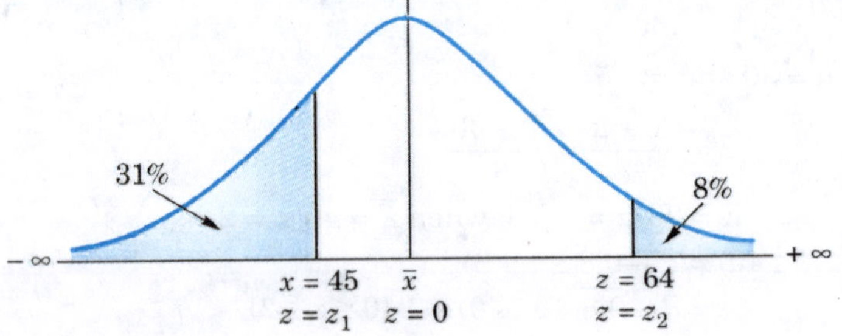

Fig. 26.6

From Table IV, $z_1 = -0.5$ ...(*ii*)

When $x = 64$, let $z = z_2$ so that $z_2 = (64 - \bar{x})/\sigma$ ...(*iii*)

$$\therefore \quad \int_{z_2}^{\infty} \phi(z)\, dz = 0.08 \quad \text{or} \quad \int_{0}^{\infty} \phi(z)\, dz - \int_{0}^{z_2} \phi(z)\, dz = 0.08$$

Hence $$\int_{0}^{z_2} \phi(z)\, dz = \int_{0}^{\infty} \phi(z)\, dz - 0.08 = 0.5 - 0.08 = 0.42$$

From Table IV, $z_2 = 1.4$ ...(*iv*)

From Eqns. (*i*) and (*ii*), $45 - \bar{x} = -0.5\sigma$

From Eqns. (*iii*) and (*iv*), $64 - \bar{x} = 1.4\sigma$

Solving these equations, we get $\bar{x} = 50$ and $\sigma = 10$.

**Example 26.52.** *In a test on 2000 electric bulbs, it was found that the life of a particular make, was normally distributed with an average life of 2040 hours and S.D. of 60 hours. Estimate the number of bulbs likely to burn for*

*(a) more than 2150 hours,* *(b) less than 1950 hours and*

*(c) more than 1920 hours and but less than 2160 hours.* (V.T.U., 2015)

**Solution.** Here $\mu = 2040$ hours and $\sigma = 60$ hours.

(*a*) For $x = 2150$, $z = \dfrac{x - \mu}{\sigma} = 1.833.$

$\therefore$ area against $z = 1.83$ in the Table III = 0.4664.

We, however, require the area to the right of the ordinate at $z = 1.83$. This area = 0.5 – 0.4664 = 0.0336. Thus the number of bulbs expected to burn for more than 2150 hours

$$= 0.0336 \times 2000 = 67 \text{ approximately.}$$

(*b*) For $x = 1950$, $z = \dfrac{x - \mu}{\sigma} = -1.5$

The area required in this case is to the left of $z = -1.33$

*i.e.*, $= 0.5 - 0.4082$ (Table value for $z = 1.33$)

$= 0.0918.$

$\therefore$ the number of bulbs expected to burn for less than 1950 hours

$$= 0.0918 \times 2000 = 184 \text{ approximately.}$$

(*c*) When $x = 1920$, $\quad z = \dfrac{1920 - 2040}{60} = -2$

When $x = 2160$, $\quad z = \dfrac{2160 - 2040}{60} = 2.$

The number of bulbs expected to burn for more than 1920 hours but less than 2160 hours will be represented by the area between $z = -2$ and $z = 2$. This is twice the area from the Table IV for $z = 2$, *i.e.*, $= 2 \times 0.4772 = 0.9544$.

Thus the required number of bulbs = 0.9544 × 2000 = 1909 nearly.

**Example 26.53.** *If the probability of committing an error of magnitude x is given by*

$$y = \frac{h}{\sqrt{\pi}} e^{-h^2x^2};$$

*compute the probable error from the following data :*

$m_1 = 1.305$ ; $m_2 = 1.301$ ; $m_3 = 1.295$ ; $m_4 = 1.286$ ;
$m_5 = 1.318$ ; $m_6 = 1.321$ ; $m_7 = 1.283$ ; $m_8 = 1.289$ ;
$m_9 = 1.300$ ; $m_{10} = 1.286$.

**Solution.** From the given data which is normally distributed, we have

$$\text{mean} = \frac{1}{10}\Sigma m_i = \frac{12.984}{10} = 1.2984$$

and

$$\sigma^2 = \frac{1}{10}\Sigma(m_i - \text{mean})^2$$

$$= \frac{1}{10}[(0.007)^2 + (0.003)^2 + (0.003)^2 + (0.012)^2 + (0.02)^2 + (0.023)^2 + (0.015)^2 + (0.009)^2 + (0.002)^2 + (0.012)^2]$$

$$= 0.0001594 \text{ whence } \sigma = 0.0126.$$

$\therefore$ *probable error* $= \dfrac{2}{3}\sigma = 0.0084$ approx.

**Example 26.54.** *Fit a normal curve to the following distribution.*

| *x* : | 2 | 4 | 6 | 8 | 10 |
|---|---|---|---|---|---|
| *f* : | 1 | 4 | 6 | 4 | 1 |

(*Rohtak, 2011*)

**Solution.**

$$\text{Mean} = \frac{\Sigma fx}{\Sigma f} = \frac{2 + 16 + 36 + 32 + 10}{16} = 6$$

$$\text{S.D.} = \sqrt{\left[\frac{\Sigma fx^2}{\Sigma f} - \left(\frac{\Sigma fx}{\Sigma f}\right)^2\right]} = \sqrt{(40 - 36)} = 2$$

Taking $\mu = 6$, $\sigma = 2$ and $N = 16$, the equation of the normal curve is

$$y = \frac{1}{\sigma\sqrt{2\pi}} e^{-(x-\mu)^2/2\sigma^2} \quad \text{or} \quad y = \frac{1}{2\sqrt{2\pi}} e^{-(x-6)^2/8} \qquad ...(i)$$

Area under Eqn. (*i*) in $(x_1, x_2)$ or $(z_1, z_2)$

$$= \frac{1}{\sqrt{2\pi}}\int_0^{z_2} e^{-z^2/2}dz - \frac{1}{\sqrt{2\pi}}\int_0^{z_1} e^{-z^2/2}\,dz \qquad \text{where } z = \frac{x-6}{2}$$

To evaluate these integrals, we refer to Table IV.

*Calculations :*

| *Mid x* | $(x_1, x_2)$ | $(z_1, z_2)$ | *Area under (i) in* $(z_1, z_2)$ | *Expected frequency* |
|---|---|---|---|---|
| 2 | (1, 3) | (–2.5, –1.5) | 0.4938 – 0.4332 | 16 × 0.606 = 0.97 |
| 4 | (3, 5) | (–1.5, –0.5) | 0.4332 – 0.1915 | 16 × 0.2417 = 3.9 |
| 6 | (5, 7) | (–0.5, 0.5) | 0.1915 + 0.1915 | 16 × 0.383 = 6.1 |
| 8 | (7, 9) | (0.5, 1.5) | 0.4332 – 0.1915 | 16 × 0.2417 = 3.9 |
| 10 | (9, 11) | (1.5, 2.5) | 0.4938 – 0.4332 | 16 × 0.606 = 0.97 |

Hence the expected (theoretical) frequencies corrected to nearest integer are 1, 4, 6, 4, 1 which agree with the observed frequencies. This shows that the normal curve Eqn. (*i*) is a proper fit to the given distribution.

## PROBLEMS 26.7

1. Show that the standard deviation for a normal distribution is approximately 25% more than the mean deviation.
2. For a normally distributed variate with mean 1 and S.D. 3, find the probabilities that
   (*i*) $3.43 \le x \le 6.19$ (*ii*) $-1.43 \le x \le 6.19$. (*Kurukshetra, 2013*)
3. If $z$ is normally distributed with mean 0 and variance 1, find
   (*i*) $P_r\{z \le -1.64\}$ ; (*ii*) $z_1$ if $P_r\{z \ge z_1\} = 0.84$.
4. In a certain examination, the percentage of candidates passing and getting distinctions were 45 and 9 respectively. Estimate the average marks obtained by the candidates, the minimum pass and distinction marks being 40 and 75 respectively. (Assume the distribution of marks to be normal).
5. A manufacturer of air-mail envelopes knows from experience that the weight of the envelopes is normally distributed with mean 1.95 gm and standard deviation 0.05 gm. About how many envelopes weighing (*i*) 2 gm or more ; (*ii*) 2.05 gm or more can be expected in a given packet of 100 envelopes.
6. The mean height of 500 students is 151 cm. and the standard deviation is 15 cm. Assuming that the heights are normally distributed, find how many students' heights lie between 120 and 155 cm.
7. The mean and standard deviation of the marks obtained by 1000 students in an examination are respectively 34.4 and 16.5. Assuming the normality of the distribution, find the approximate number of students expected to obtain marks between 30 and 60.
8. In an examination taken by 500 candidates, the average and the standard deviation of marks obtained (normally distributed) are 40% and 10%. Find approximately
   (*i*) how many will pass, if 50% is fixed as a minimum ?
   (*ii*) what should be the minimum if 350 candidates are to pass ?
   (*iii*) how many have scored marks above 60% ?
9. The mean inside diameter of a sample of 200 washers produced by a machine is 5.02 mm and the standard deviation is 0.05 mm. The purpose for which these washers are intended allows a maximum tolerance in the diameter of 4.96 to 5.08 mm, otherwise the washers are considered defective. Determine the percentage of defective washers produced by the machine, assuming the diameters are normally distributed.
   [**Hint.** 4.96 in standard units = (4.96 – 5.02)/0.05 = –1.2
   5.08 in standard units = (5.08 – 5.02)/0.05 = 1.2
   Proportion of non-defective washers = 2 (area between $z = 0$ and $z = 1.2$)
   = 0.7698 or 77% nearly.
   ∴ percentage of defective washers = 100 – 77 = 23%.]
10. Assuming that the diameters of 1000 brass plugs taken consecutively from a machine, form a normal distribution with mean 0.7515 cm. and standard deviation 0.0020 cm., how many of the plugs are likely to be rejected if the approved diameter is 0.752 ± 0.004 cm. ?

**11.** It is given that the age of thermostats of a particular make follow the normal law with mean 5 years and S.D. 2 years. 1000 units are sold out every month. How many of them will have to be replaced at the end of the second year.

**12.** The income of a group of 10,000 persons was found to be normally distributed with mean ₹ 750 p.m., and standard deviation of ₹ 50. Show that, of this group, about 95% had income exceeding ₹ 668 and only 5% had income exceeding ₹ 832. Also find the lowest income among the richest 100.

**13.** Find the equation of the best fitting normal curve to the following distribution :

| $x$ : | 0 | 1 | 2 | 3 | 4 | 5 |
|---|---|---|---|---|---|---|
| $y$ : | 13 | 23 | 34 | 15 | 11 | 4 |

**14.** Obtain the equation of the normal probability curve that may be fitted to the following data :

| Variable : | 4 | 6 | 8 | 10 | 12 | 14 | 16 | 18 | 20 | 22 | 24 |
|---|---|---|---|---|---|---|---|---|---|---|---|
| Frequency: | 1 | 7 | 15 | 22 | 35 | 43 | 38 | 20 | 13 | 5 | 1 |

**15.** A factory turns out an article by mass production and it is found that 10% of the product is rejected. Find the S.D. of the number of rejects and the equation to the normal curve to represent the number of rejects.

[**Hint.** $p = 0.1, q = 0.9, n = 100$.

$\therefore$ binomial distribution of rejects gives mean $= np = 10$, S.D. $= \sqrt{(npq)} = 3$

If this binomial distribution is approximated by a normal distribution, then the equation to the normal curve is

$$y = \frac{100}{\sigma\sqrt{(2\pi)}} e^{-(x-\mu)^2/2\sigma^2}, \quad \text{where } \mu = 10, \sigma = 3.]$$

**16.** Given that the probability of committing an error of magnitude $x$ is

$$y = \frac{h}{\sqrt{\pi}} e^{-h^2x^2}, \text{ show that the probable error is } 0.4769/h.$$

## 26.18 NORMAL APPROXIMATION TO BINOMIAL DISTRIBUTION

If the number of successes in a Binomial distribution range from $x_1$ to $x_2$, then the probability of getting these successes

$$= \sum_{r=x_1}^{x_2} {}^nC_r \, p^r \, q^{n-r}$$

As the number of trials increases, the Binomial distribution becomes approximated to the Normal distribution. The mean $np$ and the variance $npq$ of the binomial distribution will be quite close to the mean and standard deviation of thc approximated normal distribution. Thus *for n sufficiently large (≥ 30), the binomial distribution with probability of success p, is approximated by the normal distribution with* $\mu = np, \sigma = \sqrt{npq}$.

We must however, be careful to get the correct values of $z$. For any success $x$, real class interval is $(x - 1/2, x + 1/2)$. Hence

$$z_1 = \frac{x_1 - \frac{1}{2} - \mu}{\sigma} = \frac{x_1 - \frac{1}{2} - np}{\sqrt{npq}}; \; z_2 = \frac{x_2 + \frac{1}{2} - np}{\sqrt{npq}}$$

so that $P(x_1 < x < x_2) = P(z_1 < z < z_2) = \int_{z_1}^{z_2} \phi(z)\, dz$ which can be calculated by using table III–Appendix 2.

**Obs.** Normal approximation to the Binomial distribution will be fairly good even when
(*i*) $n$ is small and $p$ is near 1/2 (*ii*) $np$ and $nq$ are both ≥ 5.

**Example 26.55.** *In a referendum 60% of voters voted in favour. A random sample of 200 voters was selected. What is the probability that in the sample*

(*a*) *more than 130 voted in favour ?*

(*b*) *between 105 and 130 inclusive voted in favour ?*

(*c*) *120 voted in favour ?* (*Andhra, 2022*)

**Solution.** Here $n = 200, p = 0.6, q = 0.4$

$$\therefore \quad \mu = np = 200 \times 0.6 = 120; \; \sigma = \sqrt{npq} = \sqrt{48} = 6.928$$

(*a*) $P(x > 130) = P(x > 130.5) = P\left(x > \frac{130.5 - 120}{\sqrt{48}}\right) = P(z > 1.516) = 0.0648$

(*b*) $P(105 < x < 130) = P(105.5 < x < 129.5)$

$$= P\left(\frac{105.5 - 120}{\sqrt{48}} < z < \frac{129.5 - 120}{\sqrt{48}}\right) = P(-2.09 < z < 1.37) = 0.8964$$

(*c*) $P(x = 120) = P(119.5 < x < 120.5)$

$$= P(-0.072 < z < 0.072) = 0.0575.$$

## PROBLEMS 26.8

1. A pair of unbiased dice are rolled 180 times and their score recorded. Find
(*a*) $P(x \leq 20)$, (*b*) $P(20 \leq x \leq 40)$, (*c*) $P(20 < x \leq 30)$.
2. A marksman has a probability of 0.9 of hitting a target on a single shot. If the marksman has 40 shots, what is the probability that he hits the target (*a*) at least 35 times; (*b*) between 34 and 36 times; (*c*) 37 times.
3. A certain drug is effective in 72% of cases. Given 2000 people are treated with the drug, what is the probability that it will be effective for (*a*) at least 1400 patients, (*b*) less than 1390 patients, (*c*) 1420 patients.

## 26.19 SOME OTHER DISTRIBUTIONS

### Discrete distributions

**(1) Geometric distribution.** If $p$ be the probability of success and $k$ be the numbers of failures preceding the first success then this distribution is

$$P(k) = q^k p, \qquad k = 0, 1, 2, ..., q = 1 - p.$$

Obviously $\sum_{k=0}^{\infty} P(k) = p \sum_{k=0}^{\infty} q^k = p \cdot \frac{1}{1-q} = 1.$

It can easily be shown that mean $= q/p$, and variance $= q/p^2$.

**(2) Negative binomial distribution.** This distribution gives the probability that the event occurs for the $k$th time on the $r$th trial $(r \geq k)$. If $p$ be the probability of occurrence of an event then

$$P(k, r) = {}^{r-1}C_{k-1} p^k q^{r-k}.$$

It contains two parameters $p$ $(0 < p < 1)$ and $k$ (a positive integer). If $k = 1$, the Negative binomial distribution reduces to the geometric distribution.

**(3) Hypergeometric distribution.** Suppose a bag contains $m$ white and $n$ black balls. If $r$ balls are drawn one at a time (*with replacement*), then the probability that $k$ of them will be white is

$$P(k) = {}^mC_k \, {}^nC_{r-k} / {}^{m+n}C_r, \; k = 0, 1, ... r, r \leq m, r \leq n.$$

This distribution is known as *Hypergeometric distribution*.

For $\sum_{k=0}^{r} P(k) = 1$, since $\sum_{k=0}^{r} {}^mC_k \, {}^nC_{r-k} = {}^{m+n}C_r$

This can be proved by equating the coefficient of $t^r$ in

$$(1 + t)^m (t + 1)^n = (1 + t)^{m+n}$$

### Continuous distributions

**(4) Uniform (or Rectangular) distribution** (Fig. 26.7). A random variable $X$ is said to be uniformly distributed over the interval $-\infty < a < b < \infty$, if its density is given by

$$f(x) = \begin{cases} \frac{1}{b-a}, & a < x < b \\ 0, & \text{otherwise} \end{cases} \qquad ...(i)$$

The distribution given by Eqn. (*i*) is called a *uniform distribution*. In this distribution, $X$ takes the values with the same probability.

Its *mean* $\mu = \int_a^b x \cdot f(x)\, dx = \frac{1}{b-a} \left| \frac{x^2}{2} \right|_a^b = \frac{a+b}{2}.$

and *variance* $\sigma^2 = \mu_2' - (\mu)^2 = \int_a^b x^2 \cdot \frac{1}{b-a} dx - \left(\frac{a+b}{2}\right)^2 = \frac{1}{12}(b-a)^2.$

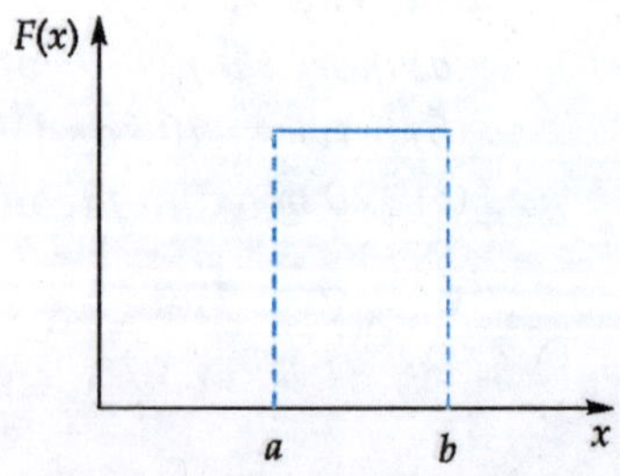

Fig. 26.7

**(5) Gamma distribution.** This continuous distribution is given by $f(x) = \dfrac{\lambda}{\Gamma(r)} (\lambda x)^{\gamma - 1} e^{-\lambda x}$ for all $x \geq, 0$,

where $\gamma$ and $\lambda$ (both > 0) are called the parameters of the *gamma distribution*. Its mean = $\gamma/\lambda$ and variance = $\gamma/\lambda^2$.

Gamma distribution tends to normal distribution as the parameter $\gamma$ tends to infinity.

**(6) Exponential distribution** (Fig. 26.8). This distribution is a special case of gamma distribution when

$\gamma = 1$ so that $f(x) = \begin{cases} \lambda e^{-\lambda x}, & \text{for } x \geq 0 \\ 0, & \text{for } x < 0 \end{cases}$, where $\lambda$ is a parameter. *(Andhra, 2019)*

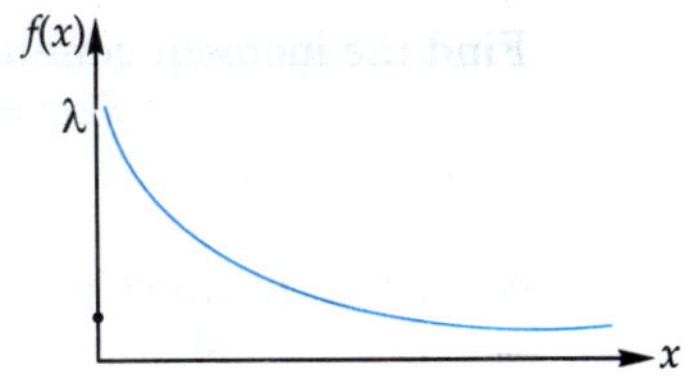

Fig. 26.8

It can be seen that mean = $1/\lambda$, standard deviation = $1/\lambda$.

This distribution plays an important role in the reliability and queuing theory.

**(7) Weibull distribution*.** This distribution is given by

$$f(x) = \frac{\alpha}{c} x^{\alpha - 1} e^{-x^{\alpha}/c}, x > 0, c > 0$$

where $c$ is a scale parameter and $\alpha$ a shape parameter.

Initially this distribution was used to describe experimentally observed variation in the fatigue resistance of steel and its elastic limits. But it has also been employed to study the variation of length of service of radio service equipment.

**Example 26.56.** *A die is cast until 6 appears. What is the probability that it must be cast more than 5 times ?*

**Solution.** Here probability of getting 6 is $p = \dfrac{1}{6}$. Then $q = \dfrac{5}{6}$.

If $X$ is the number of tosses required for the first success, then

$$P(X = x) = q^{x-1} p \text{ for } x = 1, 2, 3, \ldots$$

$\therefore$ required probability $= P(X > 5) = 1 - P(X \leq 5)$

$$= 1 - \sum_{x=1}^{5} \left(\frac{5}{6}\right)^{x-1} \cdot \left(\frac{1}{6}\right) = 1 - \frac{1}{6}\left\{1 + \frac{5}{6} + \left(\frac{5}{6}\right)^2 + \left(\frac{5}{6}\right)^3 + \left(\frac{5}{6}\right)^4\right\} = \left(\frac{5}{6}\right)^5.$$

**Example 26.57.** *A random variable X has a uniform distribution over (–3, 3), find k for which*

$$P(X > k) = \frac{1}{3}.$$

*Also evaluate* $P(X < 2)$ *and* $P[|X - 2| < 2]$.

**Solution.** (*i*) Density of $X = f(x) = \dfrac{1}{b-a} = \dfrac{1}{3-(-3)} = \dfrac{1}{6}$

$$\therefore \quad P(X > k) = 1 - P(X \leq k) = 1 - \int_{-3}^{k} f(x)\, dx$$

$$= 1 - \frac{1}{6}\int_{-3}^{k} dx = 1 - \frac{1}{6}(k + 3) = \frac{1}{3} \quad \text{(given)}$$

This gives $k = 1$.

(*ii*) $$P(X < 2) = \int_{-3}^{2} f(x)\, dx = \frac{1}{6}\int_{-3}^{2} dx = \frac{5}{6}.$$

(*iii*) $$P[|X - 2| < 2] = P[2 - 2 < X < 2 + 2] = P[0 < x < 4] = \int_0^3 f(x)\, dx = \frac{1}{6}\int_0^3 dx = \frac{1}{2}.$$

## PROBLEMS 26.9

1. Show that the mode of the *geometric distribution* $P(x) = \left(\dfrac{1}{2}\right)^x$, $x = 1, 2, 3, \ldots$, is unity.
2. Show that for the *rectangular distribution* $f(x) = 1, 0 \leq x \leq 1$,

$$\text{mean} = \frac{1}{2}, \text{ variance} = \frac{1}{12} \text{ and mean deviation} = \frac{1}{4}.$$

* It was first used by Swedish scientist *Weibull* in 1951.

3. Find the mean and variance of the *uniform distribution* given by $f(x) = 1/n$, $x = 1, 2, ..., n$.
4. Show that for the *exponential distribution*
$dP = y_0 e^{-x/\sigma}$, $0 \leq x \leq \infty$,
the mean and S.D. are both equal to $\sigma$.
5. Find the mean and variance of the *exponential distribution* $f(x) = \frac{1}{b} e^{-(x-a)/b}$ $x > a$.
6. Find the moment generating function for the *triangular distribution* given by
$$f(x) = x, \quad 0 \leq x \leq 1$$
$$= 2 - x, \quad 1 \leq x \leq 2.$$
7. Show that for the *Gamma distribution* $f(x) = \frac{e^{-x} x^{l-1}}{\Gamma(l)}$, $0 < x < \infty$, the mean and variance are both equal to $l$.
8. Find the moment generating function of the *Gamma distribution* $f(x) = \frac{1}{\Gamma(1/4)} e^{-x} x^{-3/4}$, $x \geq 0$, at the origin. *(Anna., 2016)*
9. Find the moment generating function of an exponential random variable and hence find its mean and variance. *(Anna, 2012)*

[**Chebyshev's inequality.** If $x$ is a continuous random variable with mean $\mu$ and variance $\sigma^2$, then for any positive real parameter $t$,
$$P(|x - 2| \geq t) \leq \sigma^2/t^2 \text{ or } P(|x - \mu| \leq t) \geq 1 - \sigma^2/t^2.$$
This result is known as *Chebyshev's inequality*. It gives limits to the probability that the value of the variate chosen at random will differ from mean by more than $t$.]

10. Obtain the moment generating function of negative binomial distribution and hence obtain its mean and variance. *(Andhra, 2016)*
11. For the points on a symmetrical die, prove that *Chebyshev's inequality* gives
$$P(|x - \bar{x}| > 2.5) < 0.478,$$
while the actual probability is zero.
12. For the *Geometrical distribution* $P(x) = 2^{-x}$, $x = 1, 2, 3, ...$, prove that *Chebyshev's inequality* gives
$$P(|x - 2| < 2) > \frac{1}{2},$$
while the actual probability is 15/16.

## 26.20 OBJECTIVE TYPE OF QUESTIONS

### PROBLEMS 26.10

*Select the correct answer or fill up the blanks in each of the following problems :*

1. The probability that $A$ happens is 1/3. The odds against happening of $A$ are
(*a*) 2 : 1 (*b*) 2 : 3 (*c*) 3 : 2 (*d*) 5 : 2.
2. The odds in favour of an event $A$ are 5 to 4. The probability of success of $A$ is
(*a*) 4/5 (*b*) 5/9 (*c*) 4/9.
3. The probability that $A$ passes a test is 2/3 and the probability that $B$ passes the same test is 3/5. The probability that only one of them passes is
(*a*) 2/5 (*b*) 4/15 (*c*) 2/15 (*d*) 7/15.
4. A buys a lottery ticket in which the chance of winning is 1/10 ; $B$ has a ticket in which his chance of winning is 1/20. The chance that atleast one of them wins is
(*a*) 1/200 (*b*) 29/200 (*c*) 30/200 (*d*) 170/200.
5. The probability that a non-leap year should have 53 Tuesdays is ...
6. The probability of getting 2 or 3 or 4 from a throw of single dice is ...
7. The mean of the Binomial distribution with $n$ observations and probability of success $p$, is
(*a*) $pq$ (*b*) $np$ (*c*) $\sqrt{np}$ (*d*) $\sqrt{pq}$.
8. If the mean of a Poisson distribution is $m$, then S.D. of this distribution is
(*a*) $m^2$ (*b*) $\sqrt{m}$ (*c*) $m$ (*d*) none of these.

9. The S.D. of the Binomial distribution is

(a) $\sqrt{npq}$ (b) $\sqrt{np}$ (c) $npq$ (d) $pq$.

10. In a Poisson distribution if $2P(x = 1) = P(x = 2)$, then the variance is

(a) 0 (b) –1 (c) 4 (d) 2.

11. If the probability of hitting a target by one shot be $p = 0.8$, then the probability that out of ten shots, seven will hit the target is ...
12. For a Poisson variate $x : P(x = 1) = P(x = 2)$, then the mean of $x$ is ...
13. If $P(A) = 0.35$, $P(B) = 0.73$ and $P(A \cap B) = 0.14$, then $P(A \cap B') = ...$
14. If $A$ and $B$ are independent, $P(B) = 0.14$ and $P(A/B) = 0.24$, then $P(A) = ...$
15. The probability distribution of the number of heads, when two coins are tossed, is ...
16. The parameters $p$ and $q$ of the Binomial distribution whose mean is 9 and variance is 9/4 are ... (*U.P.T.U., 2012*)
17. The area under the standard normal curve which lies between $z = 0.90$ and $z = -1.85$ is ...
[Given $P(0 < z < 1.85) = 0.4678$, $P(0 < z < 0.9) = 0.3159$]
18. The mean, median and mode of a normal distribution are ...
19. The mean and variance of a Poisson distribution are ...
20. If $A$ and $B$ are two mutually exclusive events, then $P(A \cup B) = ...$
21. For a normal distribution $\beta_1 = ...$ and $\beta_2 = ...$
22. The number of ways in which five people can be lined up to get on a bus are ...
23. A shipment of 10 television sets contains 3 defective sets. The number of ways in which one can purchase 4 of these sets and receive 2 defective sets are ...
24. The probability of getting a total of 5 when a pair of dice is tossed is ...
25. If $P(B) = 0.81$ and $P(A \cap B) = 0.18$, then $P(A/B) = ...$
26. If two unbiased dice are thrown simultaneously, the probability that the sum of the numbers on them is at least 10, is ...... .
27. If $X$ is a Poisson variate such that $P(X = 2) = P(X = 3)$, then $P(X = 0) = ...$
28. An unbiased die is tossed twice, then the probability of obtaining the sum 6, is ...
29. The variance of Poisson distribution with parameter $\lambda = 2$ is ...
30. The distribution in which mean, median, mode are equal is ...
31. For the Poisson variate, probability of getting at least one success is ...
32. Total number of events in rolling of an ideal die is ...
33. If $X$ be normal with mean 10 and variance 4, then $P(X < 11) = ...$
34. If $X$ is a binomial variate with parameters $n$ and $p$, then its m.g.f. about the origin is ...
35. In a normal distribution, mean deviation : standard deviation = ...
36. If $A$ and $B$ are independent and $P(A) = 1/2$, $P(B) = 1/3$ then $P(A \cap B) = ...$
37. If $X$ is the random variable representing the outcome of the roll of an ideal die, then $E(X) = ...$
38. If $X$ is a binomial variate with $p = 1/5$ for the experiment of 50 trials, then the standard deviation is ...
39. The area under the whole normal curve is ...
40. Given $X = B(n, p)$, then the conditions under which $X$ tends to a Poisson distribution, are ...
41. If $A$ and $B$ are mutually exclusive events then $P(A \cup B) = ...$
42. The probability of selecting $x$ white balls from a bag containing $y$ white and $z$ red balls is ...
43. If the mean and variance of the binomial distribution of 5 trials is 9/5, then $P(x \geq 1) = ...$
44. If $A$ and $B$ are mutually exclusive events, $P(A) = 0.29$, $P(B) = 0.43$, then $P(A \cup B) = ...$ and $P(A \cap B') = ...$
45. The mean of a Binomial distribution is 5, find the value of $E(2x - 9)$. (*W.B.T.U., 2015*)
46. If $x$ is a Poisson variable such that $P(x = 2) = 9\,P(x = 4) + 90\,P(x = 6)$, then the mean = ...
47. $\mu_r'$ the $r$th moment about the origin in terms of the m.g.f. is ...
48. The chance of throwing 7 in a single throw with two dice is ...
49. If $A$ and $B$ are any two events with $P(A) = 1/2$, $P(B) = 1/3$ and $P(A \cap B) = 1/4$, then $P(A/B) = ...$
50. In the roll of an ideal die, the probability of getting a prime number is ...
51. If $A$ and $B$ are mutually exclusive events, $P(A \cup B) = 0.6$, $P(B) = 0.4$, then $P(A) = ...$
52. The probability that a leap year should have 53 Sundays is ...
53. The probability density function of a binomial distribution is ...
54. The probable error is ... times S.D. approximately.
55. To fit a normal distribution, the parameters required are ...

56. A card is drawn from a well-shuffled pack of 52 cards, then the probability of this card being a red coloured ace is ...
57. If $P(1) = P(2)$, then the mean of the Poisson distribution is ...
58. Baye's theorem states that ...
59. If $x$ is a Poisson variate such that $P(x = 1) = 0.3$ and $P(x = 2) = 0.2$, then $P(x = 0) = ...$
60. If $A$ and $B$ are events such that $P(A \cup B) = 3/4$, $P(A \cap B) = 1/4$, and $P(A) = 2/3$, then $P(A) = ...$
61. The chance of throwing 7 in a single throw with two dice is the same as that of getting 7 in two throws of a single die. (True or False)
62. If the mean of a Poisson distribution is 5, then its variance is 10. (True or False)
63. If $X$ is normal with mean 3 and variance 1, then $X - 3$ is a variate with mean 0 and variance 1. (True or False)
64. If $X$ is a binomial variable with parameters $n = 10$, $p = 1/4$, then its standard deviation is 2.25. (True or False)
65. The mean of a binomial distribution mean is 5 and S.D. is 3. (True or False)
66. The mean and variance of Poisson distribution are equal. (True or False)
67. The graph of the normal distribution is symmetric with respect to the line $y = x$. (True or False)
68. The standard deviation of a binomial distribution is $np$. (True or False)
69. $f(x) = kx$ in $0 < x < 1$ is a valid probability density function, if $k = ...$
70. If $V(x) = 2$, then $V(2x + 3) = ...$
71. The p.d.f. of an exponential distribution is ...
72. If $X$ is uniformly distributed in $(-2, 3)$, then its variance is ...
73. The variance of Poisson distribution with parameter $\lambda = 2$ is ...
74. In Gamma distribution with parameter $l$, the variance is ...
75. If $f(x) = kx^3$, $0 < x < 1$ and 0 elsewhere, is a p.d.f., then $k = ...$
76. The m.g.f. of a random variable $X$ is $(1 - 2t)^{-4}$, then $E(x)$ is ...
77. A random variable $X$ has $F$-distribution with $(m, n)$ degrees of freedom, then $1/X$ has the same distribution with ... degrees of freedom.
78. If $X$ is a continuous random variable having the p.d.f. $f(x)$, then the m.g.f. about the origin is given by ...
79. If $f(X) = X + 2/k$, $X = 1, 2, 3, 4, 5$ is the probability distribution of a discrete random variable, then $k = ...$
80. The p.d.f. of Gamma variate is ...
81. If $X$ is uniformly distributed in $[a, b]$, then $E(X) = ...$
82. The marks obtained by students were found normally distributed with mean 75 and variance 100. The percentage of students who scored more than 75 marks is ...
83. When four unbiased coins are tossed, the probability of getting two heads is ...
84. If $f(x) = \begin{cases} kxe^{-x}, & x > 0 \\ 0, & \text{elsewhere} \end{cases}$, is the p.d.f. of $x$, then $k = ...$
85. If $x$ is a uniform distribution defined in the interval (4, 7), then its variance is ...
86. The p.d.f. of a continuous random variable is $f(x) = k/x^3$, $5 \le x \le 10$ ; 0 elsewhere, then the value of $k$ is
    (*a*) 1 (*b*) 50 (*c*) 200/3 (*d*) 200.
87. The relation between probability density function and cumulative density function of a random variable is ...
88. If $X$ has Poisson distribution with parameter $\lambda$, then $P(X \text{ is even}) = ...$
89. Range of $t$-distribution is ...
90. If the p.d.f. of $x$ is $f(x) = kx(1 - x)$, $0 < x < 1$, then $k = ...$
91. The function $f(x) = \begin{cases} kx^2, & 0 < x < 3 \\ 0, & \text{otherwise} \end{cases}$, is a probability function, then $k = ...$
92. If the random variable $x$ is uniformly distributed in $[0, 3]$, then its p.d.f. is $f(x) = 3$, $0 < x < 3$ ; 0, elsewhere. (True or False)
93. Exponential distribution $f(x)$ is defined by $f(x) = ae^{-2x}$, $0 < x < \infty$, then $a = ...$
94. The p.d.f. of Beta distribution with $\alpha = 1$, $\beta = 4$ is $f(x) = ...$
95. For a standard normal variate $z$, $P(-0.72 \le z \le 0) = ...$

CHAPTER 27

# Sampling and Inference



## 27.1 (1) INTRODUCTION

We know that a small section selected from the population is called a *sample* and the process of drawing a sample is called *sampling*. It is essential that a sample must be a *random* selection so that each member of the population has the same chance of being included in the sample. Thus the fundamental assumption underlying theory of sampling is *Random sampling*.

A special case of random sampling in which each event has the same probability $p$ of success and the chance of success of different events are independent whether previous trials have been made or not, is known as *simple sampling*.

The statistical constants of the population such as mean ($\mu$), standard deviation ($\sigma$) etc. are called the *parameters*. Similarly, constants for the *sample* drawn from the given population *i.e.*, mean ($\bar{x}$), standard deviation ($S$) etc. are called the *statistic*. The population parameters are in general, not known and their estimates given by the corresponding sample statistic are used. We use the Greek letters to denote the population parameters and Roman letters for sample statistic.

**(2) Objectives of sampling.** Sampling aims at gathering the maximum information about the population with the minimum effort, cost and time. The object of sampling studies is to obtain the best possible values of the parameters under specific conditions. Sampling determines the reliability of these estimates. The logic of the sampling theory is the logic of induction in which we pass from a particular (sample) to general (population). Such a generalisation from sample to population is called **Statistical Inference.**

## 27.2 SAMPLING DISTRIBUTION

Consider all possible samples of size $n$ which can be drawn from a given population at random. For each sample, we can compute the mean. The means of the samples will not be identical. If we group these different means according to their frequencies, the frequency distribution so formed is known as *sampling distribution of the mean*. Similarly we can have *sampling distribution of the standard deviation* etc.

While drawing each sample, we put back the previous sample so that the parent population remains the same. This is called *sampling with replacement* and all the subsequent formulae will pertain to sampling with replacement.

**(2) Standard error.** The standard deviation of the sampling distribution is called the *standard error* (*S.E.*). Thus the standard error of the sampling distribution of means is called standard error of means. The standard error is used to assess the difference between the expected and observed values. The reciprocal of the standard error is called *precision*.

If $n \geq 30$, a sample is called *large* otherwise *small*. The sampling distribution of large samples is assumed to be normal.

## 27.3 (1) TESTING A HYPOTHESIS*

To reach decisions about populations on the basis of sample information, we make certain assumptions about the populations involved. Such assumptions, which may or may not be true, are called *statistical hypothesis*. By testing a hypothesis is meant a process for deciding whether to accept or reject the hypothesis. The method consists in assuming the hypothesis as correct and then computing the probability of getting the observed sample. If this probability is less than a certain preassigned value the hypothesis is rejected.

**(2) Errors.** If a hypothesis is rejected while it should have been accepted, we say that a *Type I error* has been committed. On the other hand, if a hypothesis is accepted while it should have been rejected, we say that a *Type II error* has been made. The statistical testing of hypothesis aims at limiting the Type I error to a preassigned value (say : 5% or 1%) and to minimize the Type II error. The only way to reduce both types of errors is to increase the sample size, if possible.

**(3) Null hypothesis.** The hypothesis formulated for the sake of rejecting it, under the assumption that it is true, is called the *null hypothesis* and is denoted by $H_0$. To test whether one procedure is better than another, we assume that there is no difference between the procedures. Similarly to test whether there is a relationship between two variates, we take $H_0$ that there is no relationship. By accepting a null hypothesis, we mean that on the basis of the statistic calculated from the sample, we do not reject the hypothesis. It however, does not imply that the hypothesis is proved to be true. Nor its rejection implies that it is disproved.

## 27.4 (1) LEVEL OF SIGNIFICANCE

The probability level below which we reject the hypothesis is known as the *level of significance*. The region in which a sample value falling is rejected, is known as the *critical region*. We generally take two critical regions which cover 5% and 1% areas of the normal curve. The shaded portion in the figure corresponds to 5% level of significance. Thus the *probability of the value of the variate falling in the critical region is the level of significance*.

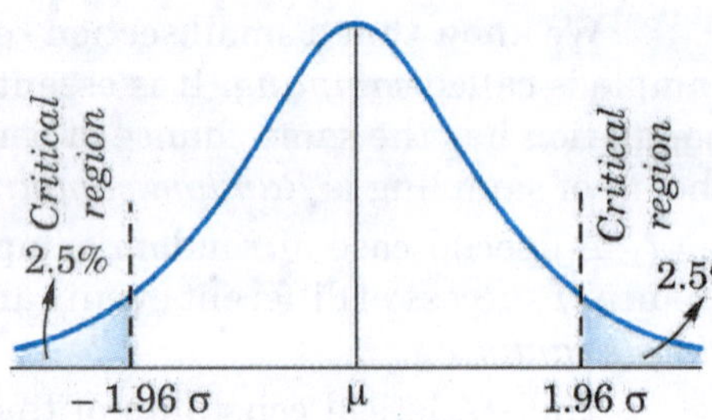

Fig. 27.1

Depending on the nature of the problem, we use a *single-tail test* or *double-tail test* to estimate the significance of a result. In a double-tail test, the areas of both the tails of the curve representing the sampling distribution are taken into account whereas in the single tail test, only the area on the right of an ordinate is taken into consideration. For instance, to test whether a coin is biased or not, double-tail test should be used, since a biased coin gives either more number of heads than tails (which corresponds to right tail), or more number of tails than heads (which corresponds to left tail only).

**(2) Tests of significance.** The procedure which enables us to decide whether to accept or reject the hypothesis is called the *test of significance*. Here we test whether the differences between the sample values and the population values (or the values given by two samples) are so large that they signify evidence against the hypothesis or these differences are so small as to account for fluctuations of sampling.

## 27.5 CONFIDENCE LIMITS**

Suppose that the sampling distribution of a statistic $S$ is normal with mean $\mu$ and standard deviation $\sigma$. As in the Fig. 27.1 the sample statistic $S$ can be expected to lie in the interval $(\mu - 1.96\sigma, \mu + 1.96\sigma)$ for 95% times *i.e.*, we can be confident of finding $\mu$ in the interval $(S - 1.96\sigma, S + 1.96\sigma)$ in 95% cases. Because of this, we call

*The American statistician *J. Neyman* (1894—1981) and the English statistician *E.S. Pearson* (1895—1980)–son of *Karl Pearson*, developed a systematic theory of tests around 1930.

***J. Neyman* developed the modern theory and terminology of confidence limits.

$(S - 1.96\sigma, S + 1.96\sigma)$ the 95% confidence interval for estimation of $\mu$. The ends of this interval (*i.e.* $S \pm 1.96\sigma$) are called 95% *confidence limits* (or *fiducial limits*) for $S$. Similarly $S \pm 2.58\sigma$ are 99% confidence limits. The numbers 1.96, 2.58 etc. are called *confidence coefficients*. The values of confidence coefficients corresponding to various levels of significance can be found from the normal curve area table IV – Appendix 2.

## 27.6 SIMPLE SAMPLING OF ATTRIBUTES

The sampling of attributes may be regarded as the selection of samples from a population whose members possess the attribute $K$ or not $K$. The presence of $K$ may be called a success and its absence a failure.

Suppose we draw a simple sample of $n$ items. Clearly it is same as a series of $n$ independent trials with the same probability $p$ of success. The probabilities of 0, 1, 2, ..., $n$ successes are the terms in the binomial expansion of $(q + p)^n$ where $q = 1 - p$.

We know that the mean of this distribution is $np$ and standard deviation is $\sqrt{(npq)}$ *i.e.*, the expected value of success in a sample of size $n$ is $np$ and the standard error is $\sqrt{(npq)}$.

If we consider the proportion of successes, then

(*i*) mean proportion of successes $= np/n = p$.

(*ii*) standard error of the proportion of successes

$$= \sqrt{\left(n \cdot \frac{p}{n} \cdot \frac{q}{n}\right)} = \sqrt{\left(\frac{pq}{n}\right)}$$

and (*iii*) precision of the proportion of successes $= \sqrt{(n/pq)}$, which varies as $\sqrt{n}$, since $p$ and $q$ are constants.

## 27.7 TEST OF SIGNIFICANCE FOR LARGE SAMPLES

We know that the binomial distribution tends to normal for large $n$. Suppose we wish to test the hypothesis that the probability of success in such trial is $p$. Assuming it to be true, the mean $\mu$ and the standard deviation $\sigma$ of the sampling distribution of number of successes are $np$ and $\sqrt{(npq)}$ respectively.

For a normal distribution, only 5% of the members lie outside $\mu \pm 1.96\sigma$ while only 1% of the members lie outside $\mu \pm 2.58\sigma$.

If $x$ be the observed number of successes in the sample and $z$ is the standard normal variate then $z = (x - \mu)/\sigma$.

*Thus we have the following test of significance :*

(*i*) *If* $|z| < 1.96$, *difference between the observed and expected number of successes is not significant.*

(*ii*) *If* $|z| > 1.96$, *difference is significant at 5% level of significance.*

(*iii*) *If* $|z| > 2.58$, *difference is significant at 1% level of significance.* (V.T.U., 2013)

**Example 27.1.** *A coin was tossed 400 times and the head turned up 216 times. Test the hypothesis that the coin is unbiased at 5% level of significance.* (Rohtak, 2011)

**Solution.** Suppose the coin is unbiased.

Then the probability of getting the head in a toss $= \frac{1}{2}$

$\therefore$ expected number of successes $= \frac{1}{2} \times 400 = 200$

and the observed value of successes = 216

Thus the excess of observed value over expected value = 216 – 200 = 16

Also S.D. of simple sampling $= \sqrt{npq} = \sqrt{\left(400 \times \frac{1}{2} \times \frac{1}{2}\right)} = 10$

Hence $z = \frac{x - np}{\sqrt{(npq)}} = \frac{16}{10} = 1.6$

As $z < 1.96$, the hypothesis is accepted at 5% level of significance *i.e.*, we conclude that the coin is unbiased at 5% level of significance.

**Example 27.2.** *A die was thrown 9000 times and a throw of 5 or 6 was obtained 3240 times. On the assumption of random throwing, do the data indicate an unbiased die ?* (V.T.U., 2010)

**Solution.** Suppose the die is unbiased.

Then the probability of throwing 5 or 6 with one die $= \frac{1}{3}$

The expected number of successes $= \frac{1}{3} \times 9000 = 3000$

and the observed value of successes = 3240

Thus the excess of observed value over expected value $3240 - 3000 = 240$

Also S.D. of simple sampling $= \sqrt{npq} = \sqrt{\left(9000 \times \frac{1}{3} \times \frac{2}{3}\right)} = 44.72$

Hence $z = \frac{x - np}{\sqrt{(npq)}} = \frac{240}{44.72} = 5.4$ nearly.

As $z > 2.58$, the hypothesis has to be rejected at 1% level of significance and we conclude that the die is biased.

**Example 27.3.** *In a locality containing 18000 families, a sample of 840 families was selected at random. Of these 840 families, 206 families were found to have a monthly income of ₹ 250 or less. It is desired to estimate how many out of 18,000 families have a monthly income of ₹ 250 or less. Within what limits would you place your estimate ?*

**Solution.** Here $p = \frac{206}{840} = \frac{103}{420}$ and $q = \frac{317}{420}$

$\therefore$ standard error of the population of families having a monthly income of ₹ 250 or less

$$= \sqrt{\left(\frac{pq}{n}\right)} = \sqrt{\left(\frac{103}{420} \times \frac{317}{420} \times \frac{1}{840}\right)} = 0.015 = 1.5\%$$

Hence taking $\frac{103}{420}$ (or 24.5%) to be the estimate of families having a monthly income of ₹ 250 or less in the locality, the limits are $(24.5 \pm 3 \times 1.5)\%$ *i.e.*, 20% and 29% approximately.

## 27.8 COMPARISON OF LARGE SAMPLES

Two large samples of sizes $n_1$, $n_2$ are taken from two populations giving proportions of attributes $A$'s as $p_1$, $p_2$ respectively.

(*a*) On the hypothesis that the populations are similar as regards the attribute $A$, we combine the two samples to find an estimate of the common value of proportion of $A$'s in the populations which is given by

$$p = \frac{n_1 p_1 + n_2 p_2}{n_1 + n_2}$$

If $e_1$, $e_2$ be the standard errors in the two samples then

$$e_1^2 = \frac{pq}{n_1} \text{ and } e_2^2 = \frac{pq}{n_2}$$

If $e$ be the standard error of the difference between $p_1$ and $p_2$, then

$$e^2 = e_1^2 + e_2^2 = pq\left(\frac{1}{n_1} + \frac{1}{n_2}\right) \quad \therefore \quad z = \frac{p_1 \sim p_2}{e}$$

If $z > 3$, the difference between $p_1$ and $p_2$ is real one.

If $z < 2$, the difference may be due to fluctuations of simple sampling.

But if $z$ lies between 2 and 3, then the difference is significant at 5% level of significance.

(*b*) If the proportions of $A$'s are not the same in the two populations from which the samples are drawn, but $p_1$ and $p_2$ are the true values of proportions then S.E. $e$ of the difference $p_1 \sim p_2$ is given by

$$e^2 = \frac{p_1q_1}{n_1} + \frac{p_2q_2}{n_2}$$

If $z = \frac{p_1 \sim p_2}{e} < 3$, the difference could have arisen due to fluctuations of simple sampling.

**Example 27.4.** *In a city A 20% of a random sample of 900 school boys had a certain slight physical defect. In another city B, 18.5% of a random sample of 1600 school boys had the same defect. Is the difference between the proportions significant ?* *(Andhra, 2015)*

**Solution.** We have $n_1 = 900, n_2 = 1600$

and $$p_1 = \frac{20}{100} = \frac{1}{5}, p_2 = \frac{18.5}{100}$$

$\therefore$ $$p = \frac{n_1p_1 + n_2p_2}{n_1 + n_2} = \frac{180 + 296}{900 + 1600} = 0.19$$

and $$q = 1 - 0.19 = 0.81$$

Thus $$e^2 = pq\left(\frac{1}{n_1} + \frac{1}{n_2}\right) = 0.19 \times 0.81\left(\frac{1}{900} + \frac{1}{1600}\right) = 0.0017$$

giving $$e = 0.04 \text{ nearly.}$$

Also $$p_1 - p_2 = \frac{1.5}{100} = 0.015 \quad \therefore \quad z = \frac{p_1 - p_2}{e} = \frac{.015}{.04} = 0.37$$

As $z < 1$, the difference between the proportions is not significant.

**Example 27.5.** *In two large populations there are 30% and 25% respectively of fair haired people. Is this difference likely to be hidden in samples of 1200 and 900 respectively from the two populations ?*

**Solution.** Here $p_1 = 0.3, p_2 = 0.25$ so that $p_1 - p_2 = 0.05$.

$\therefore$ $$e^2 = \frac{p_1q_1}{n_1} + \frac{p_2q_2}{n_2} = \frac{0.3 \times 0.7}{1200} + \frac{0.25 \times 0.75}{900}$$

so that $$e = 0.0195$$

$\therefore$ $$z = \frac{p_1 - p_2}{e} = \frac{0.05}{0.0195} = 2.5 \text{ nearly}$$

Hence it is unlikely that the real difference will be hidden.

## PROBLEMS 27.1

1. A die is tossed 960 times and it falls with 5 upwards 184 times. Is the die biased ?
2. 12 dice are thrown 3086 times and a throw of 2, 3, 4 is reckoned as a success. Suppose that 19142 throws of 2, 3, 4 have been made out. Do you think that this observed value deviates from the expected value ? If so, can the deviation from the expected value be due to fluctuations of simple sampling ?
3. Balls are drawn from a bag containing equal number of black and white balls, each ball being replaced before drawing another. In 2250 drawings 1018 black and 1232 white balls have been drawn. Do you suspect some bias on the part of the drawer ?
4. A sample of 1000 days is taken from meteorological records of a certain district and 120 of them are found to be foggy. What are the probable limits to the percentage of foggy days in the district ?
5. In a group of 50 first cousins there were found to be 27 males and 23 females. Ascertain if the observed proportions are inconsistent with the hypothesis that the sexes should be in equal proportion.
6. A random sample of 500 apples was taken from a large consignment and 65 were found to be bad. Estimate the proportion of the bad apples in the consignment as well as the standard error of the estimate. Deduce that the percentage of bad apples in the consignment almost certainly lies between 8.5 and 17.5.
7. 400 children are chosen in an industrial town and 150 are found to be under weight. Assuming the conditions of simple sampling, estimate the percentage of children who are under weight in the industrial town and assign limits within which the percentage probably lies ?

8. A machine produces 16 imperfect articles in a sample of 500. After machine is overhauled, it produces 3 imperfect articles in a batch of 100. Has the machine been improved ?
9. One type of aircraft is found to develop engine trouble in 5 flights out of a total of 100 and another type in 7 flights out of a total of 200 flights. Is there a significant difference in the two types of aircrafts so far as engine defects are concerned ?
10. In a sample of 600 men from a certain city, 450 are found smokers. In another sample of 900 men from another city, 450 are smokers. Do the data indicate that the cities are significantly different with respect to the habit of smoking among men ?
11. In a sample of 500 people from a state 280 take tea and rest take coffee. Can we assume that tea and coffee are equally popular in the state at 5% level of significance ?

## 27.9 (1) SAMPLING OF VARIABLES

We now consider sampling of a variable such as weight, height, etc. Each member of the population gives a value of the variable and the population is a frequency distribution of the variable. Thus a random sample of size $n$ from the population is same as selecting $n$ values of the variables from those of the distribution.

**(2) Sampling distribution of the mean.** *If a population is distributed normally with mean $\mu$ and standard deviation $\sigma$, then the means of all positive random samples of size n, are also distributed normally with mean $\mu$ and standard error $\sigma/\sqrt{n}$*. This result shows how the precision of a sample mean increases as the sample size increases.

## 27.10 CENTRAL LIMIT THEOREM

This is a very important theorem regarding the distribution of the mean of a sample *if the parent population is non-normal and the sample size is large.*

*If the variable X has a non-normal distribution with mean $\mu$ and standard deviation $\sigma$, then the limiting distribution of*

$$z = \frac{\bar{x} - \mu}{\sigma\sqrt{n}} \text{ as } n \to \infty,$$ *is the standard normal distribution* (*i.e.*, with mean 0 and unit S.D.)

There is no restriction upon the distribution of $X$ except that it has a finite mean and variance. This theorem holds well for a sample of 25 or more which is regarded as large.

Thus *if the population is normal, the sampling distribution of the mean is also normal with mean $\mu$ and S.E. $\sigma/\sqrt{n}$, while for large samples the same result holds even if the distribution of the population is non-normal.* This property is of universal use and throws light on the importance of normal distribution in statistical theory.

## 27.11 CONFIDENCE LIMITS FOR UNKNOWN MEAN

Let the population from which a random sample of size $n$ is drawn, have mean $\mu$ and S.D. $\sigma$. If $\mu$ is not known, there will be a range of values of $\mu$ for which observed mean $\bar{x}$ of the sample is not significant at any assigned level of probability. The relative deviation of $\bar{x}$ from $\mu$ is $(\bar{x} - \mu)/\sqrt{\sigma}$.

If $\bar{x}$ is not significant at 5% level of probability, then

$$|(\bar{x} - \mu)\sqrt{n}/\sigma| < 1.96 \quad i.e. \quad \bar{x} - 1.96\sigma/\sqrt{n} < \mu < \bar{x} + 1.96\sigma/\sqrt{n}$$

*Thus 95% confidence or fiducial limits for the mean of the population corresponding to given sample are $\bar{x} \pm 1.96\sigma/\sqrt{n}$.*

Similarly *99% confidence limits for $\mu$ are $\bar{x} \pm 2.58\sigma/\sqrt{n}$.*

**Example 27.6.** *A sample of 900 members is found to have a mean of 3.4 cm. Can it be reasonably regarded as a truly random sample from a large population with mean 3.25 cm and S.D. 1.61 cm.*

**Solution.** Here $\bar{x} = 3.4$ cm, $n = 900$, $\mu = 3.25$ and $\sigma = 1.61$ cm

$$\therefore \quad z = \frac{\bar{x} - \mu}{\sigma/\sqrt{n}} = \frac{3.4 - 3.25}{1.61/30} = 2.8$$

As $z > 1.96$, the deviation of the sample mean from the mean of the population is significant at 5% level of significance. Hence it cannot be regarded as a random sample.

**Example 27.7.** *The mean of a certain normal population is equal to the standard error of the mean of the samples of 100 from that distribution. Find the probability that the mean of the sample of 25 from the distribution will be negative ?*

**Solution.** If $\mu$ be the mean and $\sigma$ the S.D. of the distribution, then

$$\mu = \text{S.E. of the sample means} = \frac{\sigma}{\sqrt{100}} = \frac{\sigma}{10}$$

Also for a sample of size 25, we have $z = \dfrac{\bar{x} - \mu}{\sigma/\sqrt{25}} = \dfrac{\bar{x} - \sigma/10}{6/5} = \dfrac{5\bar{x}}{\sigma} - \dfrac{1}{2}$

Since $\bar{x}$ is negative, $z < -\frac{1}{2}$.

$\therefore$ the probability that a normal variate $z < -\frac{1}{2}$

$$= -\frac{1}{\sqrt{2\pi}}\int_{-\infty}^{-\frac{1}{2}} e^{-\frac{1}{2}z^2}\, dz = \frac{1}{\sqrt{2\pi}}\int_{1/2}^{\infty} e^{-\frac{1}{2}z^2}\, dz$$

$= 0.5 - 0.915 = 0.3085$, from the tables.

**Example 27.8.** *An unbiased coin is thrown n times. It is desired that the relative frequency of the appearance of heads should lie between 0.49 and 0.51. Find the smallest value of n that will ensure this result with 90% confidence.*

**Solution.** S.E. of the proportion of heads $= \sqrt{\dfrac{pq}{n}} = \sqrt{\dfrac{1}{2} \cdot \dfrac{1}{2} \cdot \dfrac{1}{n}} = \dfrac{1}{2\sqrt{n}}$

90% of confidence = 45% or 0.45 of the total area under the normal curve on each side of the mean.

$\therefore$ the corresponding value of $z = 1.645$, from the tables.

Thus $\qquad p \mp 1.645\sigma = 0.49 \quad \text{or} \quad 0.51.$

*i.e.,* $\qquad 0.5 - 1.645 \cdot \dfrac{1}{2\sqrt{n}} = 0.49 \quad \text{and} \quad 0.5 + 1.645 \cdot \dfrac{1}{2\sqrt{n}} = 0.51$

hence $\qquad \dfrac{1.645}{2\sqrt{n}} = 0.01 \quad \text{or} \quad \sqrt{n} = \dfrac{329}{4} \quad \text{or} \quad n = 6765$ approximately.

## 27.12 TEST OF SIGNIFICANCE FOR MEANS OF TWO LARGE SAMPLES

(*a*) Suppose two random samples of sizes $n_1$ and $n_2$ have been *drawn from the same population* with S.D. $\sigma$. We wish to test whether the difference between the sample means $\bar{x}_1$ and $\bar{x}_2$ is significant or is merely due to fluctuations of sampling.

If the samples are independent, then the standard error *e* of the difference of their means is given by

$$e^2 = e_1^2 + e_2^2$$

where $e_1 = \sigma/\sqrt{n_1}$, $e_2 = \sigma/\sqrt{n_2}$ are the S.E.s of the means of the two samples.

$\therefore \qquad e = \sigma\sqrt{\left(\dfrac{1}{n_1} + \dfrac{1}{n_2}\right)}$. Hence $\quad z = \dfrac{\bar{x}_1 - \bar{x}_2}{\sigma\sqrt{(1/n_1 + 1/n_2)}}$

is normally distributed with mean zero and S.D. 1.

*Test of significance* ($n_1$, $n_2$ being large) :

If $z > 1.96$, then the difference is significant at 5% level of significance.

If $z > 3$, it is highly probable that either the samples have not been drawn from the same population or the sampling is not simple.

(*b*) *If the samples are known to be drawn from different populations* with means $\mu_1$, $\mu_2$ and standard deviations $\sigma_1$ and $\sigma_2$. Then the standard error $e$ of their means is given by

$$e = \sqrt{\left(\frac{\sigma_1^2}{n_1} + \frac{\sigma_2^2}{n_2}\right)}$$

Assuming that the two populations have the same mean (*i.e.*, $\mu_1 = \mu_2$), the difference of the means of the samples will be normally distributed with mean zero and S.D. $e$. Now the same procedure of test of significance is applied.

**Example 27.9.** *The means of simple samples of sizes 1000 and 2000 are 67.5 and 68.0 cm respectively. Can the samples be regarded as drawn from the same population of S.D. 2.5 cm.* (*Rohtak, 2011*)

**Solution.** We have $\bar{x}_1 = 67.5$, $\bar{x}_2 = 68.0$

$n_1 = 1000$, $n_2 = 2000$.

On the hypothesis, that the samples are drawn from the same population of S.D. $\sigma = 2.5$, we get

$$z = \frac{\bar{x}_1 \sim \bar{x}_2}{\sigma\sqrt{\left(\frac{1}{n_1} + \frac{1}{n_2}\right)}} = \frac{67.5 \sim 68.0}{2.5\sqrt{\left(\frac{1}{1000} + \frac{1}{2000}\right)}} = \frac{0.5}{2.5 \times 0.0387} = \frac{0.5}{0.09675} = 5.1$$

Hence the difference between the sample means *i.e.*, 5.1 is very much greater than 1.96 and is therefore significant. Thus, the samples cannot be regarded as drawn from the same population.

**Example 27.10.** *A sample of height of 6400 soldiers has a mean of 67.85 inches and a standard deviation of 2.56 inches while a simple sample of heights of 1600 sailors has a mean of 68.55 inches and a standard deviation of 2.52 inches. Do the data indicate that the sailors are on the average taller than soldiers?* (*Andhra., 2015*)

**Solution.** Here $\bar{x}_1 = 67.85$, $\sigma_1 = 2.56$, $n_1 = 6400$

$\bar{x}_2 = 68.55$, $\sigma_2 = 2.52$, $n_2 = 1600$.

$\therefore$ S.E. of the difference of the mean heights is

$$e = \sqrt{\left[\frac{\sigma_1^2}{n_1} + \frac{\sigma_2^2}{n_2}\right]} = \sqrt{\left[\frac{(2.56)^2}{6400} + \frac{(2.52)^2}{1600}\right]}$$

$$= \sqrt{[0.001024 + 0.003969]} = 0.005 \text{ nearly.}$$

Also difference between the means $= \bar{x}_2 - \bar{x}_1 = 0.7$, which $> 10e$. This is highly significant. Hence the data indicates that the sailors are on the average taller than the soldiers.

## PROBLEMS 27.2

1. A sample of 400 items is taken from a normal population whose mean is 4 and variance 4. If the sample mean is 4.45, can the samples be regarded as a simple sample ?
2. To know the mean weights of all 10-year old boys in Delhi, a sample of 225 is taken. The mean weight of the sample is found to be 67 pounds with a S.D. of 12 pounds. Can you draw any inference from it about the mean weight of the population ?
3. A normal population has a mean 0.1 and a S.D. of 2.1. Find the probability that the mean of simple sample of 900 members will be negative.
4. If the mean breaking strength of copper wire is 575 lbs. with a standard deviation of 8.3 lbs., how large a sample must be used in order that there be one chance in 100 that the mean breaking strength of the sample is less than 572 lbs. ?

[**Hint.** $|z| = \left|\frac{\bar{x} - \mu}{\sigma}\sqrt{n}\right| = \frac{3}{8.3}\sqrt{n}$

Also from table IV, $z = 2.33$. Hence $n = 42$ nearly.]

5. A research worker wishes to estimate mean of a population by using sufficiently large sample. The probability is 95% that sample mean will not differ from the true mean by more than 25% of the S.D. How large a sample should be taken ?
6. The density function of a random variable $x$ is $f(x) = ke^{-2x^2+10x}$. Find the upper 5% point of the distribution of the means of the random sample of size 25 from the above population.
7. The means of two large samples of 1000 and 2000 members are 168.75 cms. and 170 cms. respectively. Can the samples be regarded as drawn from the same population of standard deviation 6.25 cms.
8. If 60 new entrants in a given university are found to have a mean height of 68.60 inches and 50 seniors a mean height of 69.51 inches ; is the evidence conclusive that the mean height of the seniors is greater than that of the new entrants ? Assume the standard deviation of height to be 2.48 inches.
9. A sample of 100 electric bulbs produced by manufacturer $A$ showed a mean life time of 1190 hours and a standard deviation of 90 hours. A sample of 75 bulbs produced by manufacturer $B$ showed a mean life time of 1230 hours, with a standard deviation of 120 hours. Is there a difference between the mean life time of two brands at a significance level of (*i*) 0.05 (*ii*) 0.01.
10. A random sample of 1000 men from North India shows that their mean wage is ₹ 5 per day with a S.D. of ₹ 1.50. A sample of 1500 men from South India gives a mean wage of ₹ 4.50 per day with a standard deviation of ₹ 2. Does the mean rate of wages varies as between the two regions ?

## 27.13 SAMPLING OF VARIABLES—SMALL SAMPLES

In case of large samples, sampling distribution approaches a normal distribution and values of sample statistic are considered best estimates of the parameters in a population. It will no longer be possible to assume that statistics computed from small samples are normally distributed. As such, *a new technique has been devised for small samples which involves the concept of 'degrees of freedom' which we explain below.*

**Number of degrees of freedom** *is the number of values in a set which may be assigned arbitrarily.* For instance, if $x_1 + x_2 + x_3 = 15$ and we assign any values of two of the variables (say : $x_1, x_2$), then the values of $x_3$ will be known. The two variables are therefore, free and independent choices for finding the third. Hence these are the degrees of freedom. *If there are n observations, the degrees of freedom (d.f.) are (n – 1).* In other words, while finding the mean of a small sample, one degree of freedom is used up and $(n - 1)$ *d.f.* are left to estimate the population variance.

## 27.14 (1) STUDENT'S t-DISTRIBUTION

Consider a small sample of size $n$, drawn from a normal population with mean $\mu$ and s.d. $\sigma$. If $\bar{x}$ and $\sigma_s$ be the sample mean and s.d., then the statistic, '$t$' is defined as

$$t = \frac{\bar{x} - \mu}{\sigma}\sqrt{n} \quad \text{or} \quad t = \frac{\bar{x} - \mu}{\sigma_s}\sqrt{(n-1)},$$

where $\nu = n - 1$ denotes the *df.* of $t$. If we calculate $t$ for each sample, we obtain the sampling distribution for $t$. This distribution known as *Student's t-distribution**, is given by

$$y = \frac{y_0}{(1 + t^2/\nu)^{(\nu+1)/2}} \quad \text{...(1)}$$

where $y_0$ is constant such that the area under the curve is unity.

**(2) Properties of t-distribution.**

*I.* This curve is symmetrical about the line $t = 0$, like the normal curve, since only even powers of $t$ appear in Eqn. (1). But it is more peaked than the normal curve with the same S.D. The $t$-curve approaches the horizontal axis less rapidly than the normal curve. Also $t$-curve attains its maximum value at $t = 0$ so that its mode coincides with the mean. (Fig. 27.2)

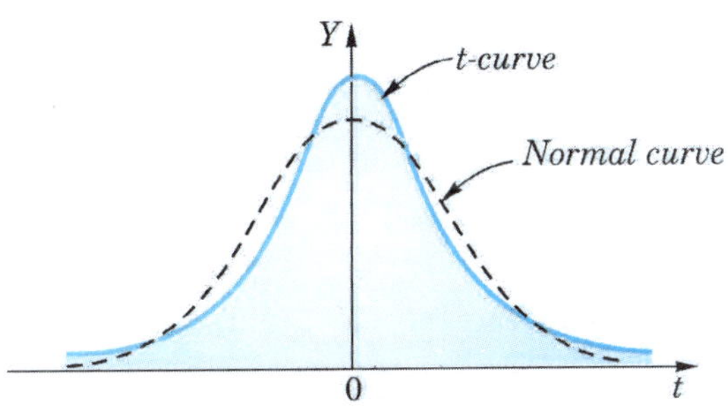

Fig. 27.2

*This distribution was first found by the English statistician *W.S. Gosset* in 1908 who wrote under the pen-name of '*Student*'. *R.A. Fisher* defined $t$ correctly and found its distribution in 1926.

*II.* The limiting form of $t$-distribution when $\nu \to \infty$ is given by $y = y_0 e^{-\frac{1}{2}t^2}$ which is a normal curve. This shows that *t is normally distributed for large samples.*

*III.* The probability $P$ that the value of $t$ will exceed $t_0$ is given by

$$P = \int_{t_0}^{\infty} y \, dx$$

The values of $t_0$ have been tabulated for various values of $P$ for various values of $\nu$ from 1 to 30 (Table V – Appendix 2).

*IV. Moments about the mean*

All the moments of odd order about the mean are zero, due to its symmetry about the line $t = 0$.

Even order moments about the mean are

$$\mu_2 = \frac{\nu}{\nu - 2}, \quad \mu_4 = \frac{3\nu^2}{(\nu-2)(\nu-4)}, \ldots$$

The $t$-distribution is often used in tests of hypothesis about the mean when the population standard deviation $\sigma$ is unknown.

## 27.15 SIGNIFICANCE TEST OF A SAMPLE MEAN

Given a random small sample $x_1, x_2, x_3 \ldots x_n$ from a normal population, we have to test the hypothesis that mean of the population is $\mu$. For this, we first calculate $t = (\bar{x} - \mu)\sqrt{n}/\sigma_s$

where
$$\bar{x} = \frac{1}{n}\sum_1^n x_i, \quad \sigma_s^2 = \frac{1}{n-1}\sum_1^n (x_i - \bar{x})^2$$

Then find the value of $P$ for the given $df$ from the table.

*If the calculated value of $t > t_{0.05}$, the difference between $\bar{x}$ and $\mu$ is said to be significant at 5% level of significance.*

If $t > t_{0.01}$, the difference is said to be significant at 1% level of significance.

*If $t < t_{0.05}$, the data is said to be consistent with the hypothesis that $\mu$ is the mean of the population.*

**Example 27.11.** *A certain stimulus administered to each of 12 patients resulted in the following increases of blood pressure : 5, 2, 8, –1, 3, 0, –2, 1, 5, 0, 4, 6. Can it be concluded that the stimulus will in general be accompanied by an increase in blood pressure.* *(V.T.U., 2012 S)*

**Solution.** Let us assume that the stimulus administered to all the 12 patients will increase the B.P. Taking the population to be normal with mean $\mu = 0$ and S.D. $\sigma$,

$$\bar{d} = \frac{5+2+8-1+3+0-2+1+5+0+4+6}{12} = 2.583$$

$$\sigma^2 = \frac{\Sigma d^2}{n} - \bar{d}^2 = \frac{1}{12}[5^2 + 2^2 + 8^2 + (-1^2) + 3^2 + 0^2 + (-2)^2 + 1^2 + 5^2 + 0^2 + 4^2 + 6^2] - (2.583)^2$$

$$= 8.744. \quad \therefore \quad \sigma = 2.9571$$

Now
$$t = \frac{\bar{d} - \mu}{\sigma_s}\sqrt{(n-1)} = \frac{2.583 - 0}{2.9571}\sqrt{(12-1)} = 2.897$$

Here $d.f.$ $\gamma = 12 - 1 = 11$.

For $\gamma = 11$, $t_{0.05} = 2.2$, from table V.

Since the $|t| > t_{0.05}$, our assumption is rejected *i.e.*, the stimulus does not increase the B.P.

**Example 27.12.** *The nine items of a sample have the following values : 45, 47, 50, 52, 48, 47, 49, 53, 51. Does the mean of these differ significantly from the assumed mean of 47.5 ?* *(V.T.U., 2013)*

**Solution.** We find the mean and standard deviation of the sample as follows :

| $x$ | $d = x - 48$ | $d^2$ |
|---|---|---|
| 45 | –3 | 9 |
| 47 | –1 | 1 |
| 50 | 2 | 4 |
| 52 | 2 | 4 |
| 48 | 0 | 0 |
| 47 | –1 | 1 |
| 49 | 1 | 1 |
| 53 | 5 | 25 |
| 51 | 3 | 9 |
| Total | 10 | 66 |

$$\therefore \qquad \bar{x} = \text{mean} = 48 + \frac{\Sigma d}{9} = 48 + \frac{10}{9} = 49.1$$

$$\sigma_s^2 = \frac{\Sigma d^2}{9} - \left(\frac{\Sigma d}{9}\right)^2 = \frac{66}{9} - \frac{100}{81} = \frac{494}{81}$$

$$\therefore \qquad \sigma_s = 2.47$$

Hence $$t = \frac{\bar{x} - \mu}{\sigma_s}\sqrt{(n-1)} = \frac{49.1 - 47.5}{2047}\sqrt{8} = 1.83$$

Here $\quad d.f.\ \nu = 9 - 1 = 8$

For $\nu = 8$, we get from table V, $t_{0.05} = 2.31$.

As calculated value of $t < t_{0.05}$, the value of $t$ is not significant at 5% level of significance which implies that there is no significant difference between $\bar{x}$ and $\mu$. Thus the test provides no evidence against the population mean being 47.5.

**Example 27.13.** *A machinist is making engine parts with axle diameter of 0.7 inch. A random sample of 10 parts shows mean diameter 0.742 inch with a standard deviation of 0.04 inch. On the basis of this sample, would you say that the work is inferior ?*

**Solution.** Here we have $\mu = 0.700$, $\bar{x} = 0.742$, $\sigma_x = 0.040$, $n = 10$.

Taking the hypothesis that the product is not inferior *i.e.*, there is no significant difference between $\bar{x}$ and $\mu$.

$$\therefore \qquad t = \frac{\bar{x} - \mu}{\sigma_x}\sqrt{(n-1)} = \frac{0.742 - 0.700}{0.040}\sqrt{(10-1)} = \frac{0.126}{0.040} = 3.16$$

Degrees of freedom $\rho = 10 - 1 = 9$.

For $\rho = 9$, we get from table V, $t_{0.05} = 2.262$.

As the calculated value of $t > t_{0.05}$, the value of $t$ is significant at 5% level of significance. This implies that $\bar{x}$ differs significantly from $\mu$ and the hypothesis is rejected. Hence the work is inferior. In fact, the work is inferior even at 2% level of significance.

**Example 27.14.** *Show that 95% confidence limits for the mean $\mu$ of the population are* $\bar{x} \pm \frac{\sigma_s}{\sqrt{n}} t_{0.05}$.

*Deduce that for a random sample of 16 values with mean 41.5 inches and the sum of the squares of the deviations from the mean 135 inches² and drawn from a normal population, 95% confidence limits for the mean of the population are 39.9 and 43.1 inches.*

**Solution.** (*a*) The probability $P$ that $t \le t_{0.05}$ is 0.95. Hence the 95% confidence limits for $\mu$ are given by

$$\left| \frac{\bar{x} - \mu}{\sigma_s}\sqrt{n} \right| \le t_{0.05}$$

or $$|\bar{x} - \mu| \le \frac{\sigma_s}{\sqrt{n}} t_{0.05} \quad \text{or} \quad \bar{x} - \frac{\sigma_s}{\sqrt{n}} t_{0.05} \le \mu \le \bar{x} + \frac{\sigma_s}{\sqrt{n}} t_{0.05}$$

We can, therefore, say with a confidence coefficient 0.95 that the confidence interval $\bar{x} \pm \frac{\sigma_s}{\sqrt{n}} t_{0.05}$ contains the population mean μ.

(*b*) Here, $n = 16$, $\nu = n - 1 = 15$, $\sigma_s = \sqrt{\frac{135}{15}} = 3.$

Also from table IV, $t_{0.05}$ (for ν = 15) = 2.13

$$\therefore \qquad \frac{\sigma_s}{\sqrt{n}} t_{0.05} = \frac{3}{4} \times 2.13 = 1.6 \text{ approx.}$$

Hence the required confidence limits are 41.5 ± 1.6 *i.e.*, 39.9 and 43.1 inches.

## 27.16 SIGNIFICANCE TEST OF DIFFERENCE BETWEEN SAMPLE MEANS

Given two independent samples $x_1, x_2, x_3, \ldots, x_{n_1}$ and $y_1, y_2, \ldots, y_{n_2}$ with means $\bar{x}$ and $\bar{y}$ and standard deviations $\sigma_x$ and $\sigma_y$ from a normal population with the same variance, we have to test the hypothesis that the population means $\mu_1$ and $\mu_2$ are the same.

For this, we calculate
$$t = \frac{\bar{x} \sim \bar{y}}{\sigma\sqrt{\left(\frac{1}{n_1} + \frac{1}{n_2}\right)}} \qquad \ldots(1)$$

where
$$\bar{x} = \frac{1}{n_1}\sum_1^{n_1} x_i,\ \bar{y} = \frac{1}{n_2}\sum_1^{n_2} y_i$$

and
$$\sigma_s^2 = \frac{1}{n_1 + n_2 - 2}\left[(n_1 - 1)\sigma_x^2 + (n_2 - 1)\sigma_y^2\right] = \frac{1}{n_1 + n_2 - 2}\left\{\sum_1^{n_1} (x_i - \bar{x})^2 + \sum_1^{n_2} (y_j - \bar{y})^2\right\}$$

It can be shown that the variate $t$ defined by Eqn. (1) follows the $t$-distribution with $n_1 + n_2 - 2$ degrees of freedom.

If the calculated value of $t > t_{0.05}$, the difference between the sample means is said to be significant at 5% level of significance.

If $t > t_{0.01}$, the difference is said to be significant at 1% level of significance.

If $t < t_{0.05}$, the data is said to be consistent with the hypothesis, that $\mu_1 = \mu_2$.

**Cor.** If the two samples are of the same size and the data are paired, then $t$ is defined by

$$t = \frac{\bar{d} - 0}{(\sigma/\sqrt{n})} \quad \text{where} \quad \sigma^2 = \frac{1}{n-1}\sum_1^n (d_i - \bar{d})^2$$

$d_i$ = difference of the $i$th members of the samples ;

$\bar{d}$ = mean of the differences = $\Sigma d_i/n$ ; and the number of *d.f.* = $n - 1$.

**Example 27.15.** *Eleven students were given a test in statistics. They were given a month's further tuition and a second test of equal difficulty was held at the end of it. Do the marks give evidence that the students have benefitted by extra coaching ?*

| *Boys* | 1 | 2 | 3 | 4 | 5 | 6 | 7 | 8 | 9 | 10 | 11 |
|---|---|---|---|---|---|---|---|---|---|---|---|
| *Marks I test* | 23 | 20 | 19 | 21 | 18 | 20 | 18 | 17 | 23 | 16 | 19 |
| *Marks II test* | 24 | 19 | 22 | 18 | 20 | 22 | 20 | 20 | 23 | 20 | 17 |

(*V.T.U., 2014*)

**Solution.** We compute the mean and the S.D. of the difference between the marks of the two tests as under :

$$\bar{d} = \text{mean of } d\text{'s} = \frac{11}{11} = 1\ ;\ \sigma_s^{\,2} = \frac{\Sigma(d - \bar{d})^2}{n - 1} = \frac{50}{10} = 5 \quad i.e., \quad \sigma_s = 2.24$$

Assuming that the students have not been benefited by extra coaching, it implies that the mean of the difference between the marks of the two tests is zero *i.e.*, μ = 0.

Then $t = \dfrac{\bar{d} - \mu}{\sigma_s}\sqrt{n} = \dfrac{1-0}{2.24}\sqrt{11} = 1.48$ nearly and $df\ \nu = 11 - 1 = 10$.

| *Students* | $x_1$ | $x_2$ | $d = x_2 - x_1$ | $d - \bar{d}$ | $(d - \bar{d})^2$ |
|---|---|---|---|---|---|
| 1 | 23 | 24 | 1 | 0 | 0 |
| 2 | 20 | 19 | –1 | –2 | 4 |
| 3 | 19 | 22 | 3 | 2 | 4 |
| 4 | 21 | 18 | –3 | –4 | 16 |
| 5 | 18 | 20 | 2 | 1 | 1 |
| 6 | 20 | 22 | 2 | 1 | 1 |
| 7 | 18 | 20 | 2 | 1 | 1 |
| 8 | 17 | 20 | 3 | 2 | 4 |
| 9 | 23 | 23 | — | –1 | 1 |
| 10 | 16 | 20 | 4 | 3 | 9 |
| 11 | 19 | 17 | –2 | –3 | 9 |
| | | | $\Sigma d = 11$ | | $\Sigma(d - \bar{d})^2 = 50$ |

From table V, we find that $t_{0.05}$ (for $\nu = 10$) = 2.228. As the calculated value of $t < t_{0.05}$, the value of $t$ is not significant at 5% level of significance *i.e.*, the test provides no evidence that the students have benefited by extra coaching.

**Example 27.16.** *From a random sample of 10 pigs fed on diet A, the increases in weight in a certain period were 10, 6, 16, 17, 13, 12, 8, 14, 15, 9 lbs. For another random sample of 12 pigs fed on diet B, the increases in the same period were 7, 13, 22, 15, 12, 14, 18, 8, 21, 23, 10, 17 lbs. Test whether diets A and B differ significantly as regards their effect on increases in weight ?*

**Solution.** We calculate the means and standard deviations of the samples as follows :

| | *Diet A* | | | *Diet B* | |
|---|---|---|---|---|---|
| $x_i$ | $x_i - \bar{x}$ | $(x_i - \bar{x})^2$ | $y_i$ | $y_i - \bar{y}$ | $(y_i - \bar{y})^2$ |
| 10 | –2 | 4 | 7 | –8 | 64 |
| 6 | –6 | 36 | 13 | –2 | 4 |
| 16 | 4 | 16 | 22 | 7 | 49 |
| 17 | 5 | 25 | 15 | 0 | 0 |
| 13 | 1 | 1 | 12 | –3 | 9 |
| 12 | 0 | 0 | 14 | –1 | 1 |
| 8 | –4 | 16 | 18 | 3 | 9 |
| 14 | 2 | 4 | 8 | –7 | 49 |
| 15 | 3 | 9 | 21 | 6 | 36 |
| 9 | –3 | 9 | 23 | 8 | 64 |
| | | | 10 | –5 | 25 |
| | | | 17 | 2 | 4 |
| 120 | 0 | 120 | 180 | 0 | 314 |

$$\bar{x} = \frac{120}{10} = 12 \text{ lbs.}, \ \bar{y} = \frac{180}{12} = 15 \text{ lbs.}$$

$$\sigma_s^2 = [\Sigma(x_i - \bar{x})^2 + \Sigma(y_i - \bar{y})^2]/(n_1 + n_2 - 2)$$

$$= (120 + 314)/(10 + 12 - 2) = (434/20) = 21.1$$

$$\therefore \quad \sigma_s = 4.65$$

Assuming that the samples do not differ in weight so far as the two diets are concerned *i.e.*, $\mu_1 - \mu_2 = 0$.

Hence $$t = \frac{(\bar{y} - \bar{x}) - 0}{\sigma_s\sqrt{\left(\dfrac{1}{n_1} + \dfrac{1}{n_2}\right)}} = \frac{15 - 12}{4.65\sqrt{\left(\dfrac{1}{10} + \dfrac{1}{12}\right)}} = \frac{3}{4.65}\sqrt{\frac{120}{22}} = 1.6 \text{ nearly}$$

Here $\quad d.f.\ \nu = n_1 + n_2 - 2 = 10 + 12 - 2 = 20.$

For $\nu = 20$, we find $t_{0.05} = 2.09$ [From Table V]

$\therefore$ the calculated value of $t < t_{0.05}$.

Hence the difference between the sample means is not significant *i.e.*, the two diets do not differ significantly as regards their effect on increase in weight.

## PROBLEMS 27.3

1. Find the student's $t$ for the following variable values in a sample of eight : –4, –2, –2, 0, 2, 2, 3, 3 ; taking the mean of the universe to be zero.
2. A random sample of 10 boys had the following I.Q. :

   70, 120, 110, 101, 88, 83, 95, 98, 107, 100.

   Do these data support the assumption of a population mean I.Q. of 100 (at 5% level of significance) ? *(V.T.U., 2006)*
3. A sample of 10 measurements of the diameter of a sphere gave a mean of 12 cm and a standard deviation 0.15 cm. Find 95% confidence limits for the actual diameter.
4. A random sample of size 25 from a normal population has the mean $\bar{x} = 47.5$ and s.d. $S = 8.4$. Does this information refute the claim that the mean of the population is $\mu = 42.1$.
5. A process for making certain bearings is under control if the diameter of the bearings have the mean 0.5 cm. What can we say about this process if a sample of 10 of these bearings has a mean diameter of 0.506 cm. and S.D. of 0.004 cm ?
6. A machine is supposed to produce washers of mean thickness 0.12 cm. A sample of 10 washers was found to have a mean thickness of 0.128 cm and standard deviation 0.008. Test whether the machine is working in proper order at 5% level of significance. *(Andhra, 2015)*
7. Find out the reliability of the sample mean of the following data : *Breaking strength of 10 specimens of 1.04 cms diameter hard-drawn copper wire :*

| Specimen | 1 | 2 | 3 | 4 | 5 | 6 | 7 | 8 | 9 | 10 |
|---|---|---|---|---|---|---|---|---|---|---|
| Breaking Strength (kgs) | 578 | 572 | 570 | 568 | 572 | 570 | 570 | 572 | 526 | 584 |

8. Test runs with 6 models of an experimental engine showed that they operated for 24, 28, 21, 23, 32 and 22 minutes with a gallon of fuel. If the probability of a Type I error is at the most 0.01, is this evidence against a hypothesis that on the average this kind of engine will operate for atleast 29 minutes per gallon of the same fuel. Assume normality.
9. Two horses $A$ and $B$ were tested according to the time (in seconds) to run a particular race with the following results :

| Horse $A$ : | 28 | 30 | 32 | 33 | 33 | 29 | and 34 |
|---|---|---|---|---|---|---|---|
| Horse $B$ : | 29 | 30 | 30 | 24 | 27 | and 29 | |

   Test whether you can discriminate between two horses ?

10. A group of 10 rats fed on a diet $A$ and another group of 8 rats fed on a different diet $B$, recorded the following increase in weights :

| Diet $A$ : | 5 | 6 | 8 | 1 | 12 | 4 | 3 | 9 | 6 | 10 | gm |
|---|---|---|---|---|---|---|---|---|---|---|---|
| Diet $B$ : | 2 | 3 | 6 | 8 | 10 | 1 | 2 | 8 | gm. | | |

   Does it show that superiority of diet $A$ over that of $B$ ? *(Andhra, 2014 ; Rohtak, 2011)*

11. A group of boys and girls were given an intelligence test. The mean score, S.D.s and numbers in each group are as follows :

| | *Boys* | *Girls* |
|---|---|---|
| Mean | 124 | 121 |
| S.D. | 12 | 10 |
| $n$ | 18 | 14 |

   Is the mean score of boys significantly different from that of girls ?

12. The means of two random samples of sizes 9 and 7 are 196.42 and 198.82 respectively. The sum of the squares of the deviations from the mean are 26.94 and 18.73 respectively. Can the sample be considered to have been drawn from the same normal population ?

## 27.17 (1) CHI-SQUARE ($\chi^2$) TEST

When a fair coin is tossed 80 times, we expect from theoretical considerations that heads will appear 40 times and tail 40 times. But this never happens in practice *i.e.,* the results obtained in an experiment do not agree exactly with the theoretical results. *The magnitude of discrepancy between observation and theory is given by the quantity* $\chi^2$ (pronounced as *chi-square*). If $\chi^2 = 0$, the observed and theoretical frequencies completely agree. As the value of $\chi^2$ increases, the discrepancy between the observed and theoretical frequencies increases.

**(2) Definition.** *If $O_1, O_2, ..., O_n$ be a set of observed (experimental) frequencies and $E_1, E_2, ..., E_n$ be the corresponding set of expected (theoretical) frequencies, then $\chi^2$ is defined by the relation*

$$\chi^2 = \frac{(O_1 - E_1)^2}{E_1} + \frac{(O_2 - E_2)^2}{E_2} + ... + \frac{(O_n - E_n)^2}{E_n}$$

$$= \sum \frac{(O_i - E_i)^2}{E_i} \qquad ...(1)$$

*with n – 1 degrees of freedom.* [$\Sigma O_i = \Sigma E_i = n$ the total frequency]

**(3) Chi-square distribution***

If $x_1, x_2, ..., x_n$ be $n$ independent normal variates with mean zero and s.d. unity, then it can be shown that $x_1^2 + x_2^2 + ... + x_n^2$, is a random variate having $\chi^2$-distribution with *ndf*.

The equation of the $\chi^2$-curve is

$$y = y_0 e^{-\chi^2/2} (\chi^2)^{(\nu - 1)/2} \qquad ...(2)$$

where $\nu = n - 1$ (Fig. 27.3).

**(4) Properties of $\chi^2$-distribution**

*I.* If $\nu = 1$, the $\chi^2$-curve (2) reduces to $y = y_0 e^{-\chi^2/2}$, which is the exponential distribution.

*II.* If $\nu > 1$, this curve is tangential to $x$-axis at the origin and is positively skewed as the mean is at $\nu$ and mode at $\nu - 2$.

*III.* The probability $P$ that the value of $\chi^2$ from a random sample will exceed $\chi_0^2$ is given by

$$P = \int_{\chi_0^2}^{\infty} y\, dx.$$

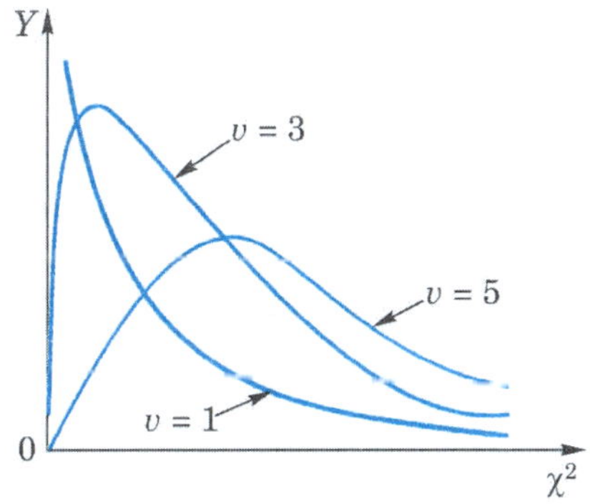

Fig. 27.3

The values of $\chi_0^2$ have been tabulated for various values of $P$ and for values of $\nu$ from 1 to 30. (Table-VI-Appendix 2)

For $\nu > 30$, the $\chi^2$-curve approximates to the normal curve and we should refer to normal distribution tables for significant values of $\chi^2$.

*IV.* Since the equation of $\chi^2$-curve does not involve any parameters of the population, *this distribution does not depend on the form of the population* and is therefore, very useful in a large number of problems.

*V. Mean* = $\gamma$ and *variance* = $2\gamma$.

## 27.18 GOODNESS OF FIT

The value of $\chi^2$ is used to test whether the deviations of the observed frequencies from the expected frequencies are significant or not. It is also used to test how will a set of observations fit a given distribution, $\chi^2$ therefore, provides a test of goodness of fit and may be used to examine the validity of some hypothesis about an observed frequency distribution. As a test of goodness of fit, it can be used to study the correspondence between theory and fact.

*This is a non-parametric distribution-free* test since in this we make no assumption about the distribution of the parent population.

---

**Hamlet* discovered this distribution in 1875. *Karl Pearson* rediscovered it independently in 1900 and applied it to test 'goodness of fit'.

**Procedure to test significance and goodness of fit.**

(*i*) Set up a 'null hypothesis' and calculate

$$\chi^2 = \sum \frac{(O_i - E_i)^2}{E_i}$$

(*ii*) Find the *df* and read the corresponding values of $\chi^2$ at a prescribed significance level from Table V.

(*iii*) From $\chi^2$-table, we can also find the probability $P$ corresponding to the calculated values of $\chi^2$ for the given *d.f.*

(*iv*) If $P < 0.05$, the observed value of $\chi^2$ is significant at 5% level of significance.

If $P < 0.01$, the value is significant at 1% level.

If $P > 0.05$, it is a good fit and the value is not significant.

**Example 27.17.** *In experiments on pea breeding, the following frequencies of seeds were obtained :*

| *Round and yellow* | *Wrinkled and yellow* | *Round and green* | *Wrinkled and green* | *Total* |
|---|---|---|---|---|
| *315* | *101* | *108* | *32* | *556* |

*Theory predicts that the frequencies should be in proportions 9:3:3:1. Examine the correspondence between theory and experiment.* (*V.T.U., 2013*)

**Solution.** The corresponding frequencies are

$$\frac{9}{16} \times 556, \frac{3}{16} \times 556, \frac{3}{16} \times 556, \frac{1}{16} \times 556 \text{ i.e., } 313, 104, 104, 35.$$

Hence $$\chi^2 = \frac{(315-313)^2}{313} + \frac{(101-104)^2}{104} + \frac{(108-104)^2}{104} + \frac{(32-35)^2}{35}$$

$$= \frac{4}{313} + \frac{9}{104} + \frac{16}{104} + \frac{9}{35} = 0.51 \quad \text{and} \quad df\ \nu = 4 - 1 = 3.$$

For $\nu = 3$, we have $\chi^2_{0.05} = 7.815$ [From Table VI]

Since the calculated value of $\chi^2$ is much less than $\chi^2_{0.05}$, there is a very high degree of agreement between theory and experiment.

**Example 27.18.** *A set of five similar coins is tossed 320 times and the result is*

| *No. of heads* : | *0* | *1* | *2* | *3* | *4* | *5* |
|---|---|---|---|---|---|---|
| *Frequency* : | *6* | *27* | *72* | *112* | *71* | *32* |

*Test the hypothesis that the data follow a binomial distribution.*

(*Andhra., 2015 ; V.T.U., 2015*)

**Solution.** For $\nu = 5$, we have $\chi^2_{0.05} = 11.07$.

$p$, probability of getting a head $= \frac{1}{2}$ ; $q$, probability of getting a tail $= \frac{1}{2}$.

Hence the theoretical frequencies of getting 0, 1, 2, 3, 4, 5 heads are the successive terms of the binomial expansion $320\,(p+q)^5$

$$= 320\,[p^5 + 5p^4q + 10p^3q^2 + 10p^2q^3 + 5pq^4 + q^5]$$

$$= 320\left[\frac{1}{32} + \frac{5}{32} + \frac{10}{32} + \frac{10}{32} + \frac{5}{32} + \frac{1}{32}\right] = 10 + 50 + 100 + 100 + 50 + 10$$

Thus the theoretical frequencies are 10, 50, 100, 100, 50, 10.

Hence

$$\chi^2 = \frac{(6-10)^2}{10} + \frac{(27-50)^2}{50} + \frac{(72-100)^2}{100} + \frac{(112-100)^2}{100} + \frac{(71-50)^2}{50} + \frac{(32-10)^2}{10}$$

$$= \frac{1}{100}(160 + 1058 + 784 + 144 + 882 + 4840) = \frac{7868}{100} = 78.68$$

and $$df\ \nu = 6 - 1 = 5.$$

Since the calculated value of $\chi^2$ is much greater than $\chi^2_{0.05}$, the hypothesis that the data follow the binomial law is rejected.

**Example 27.19.** *Fit a Poisson distribution to the following data and test for its goodness of fit at level of significance 0.05.*

| *x* : | 0 | 1 | 2 | 3 | 4 |
|---|---|---|---|---|---|
| *f* : | 419 | 352 | 154 | 56 | 19 |

**Solution.** Mean $m = \dfrac{\Sigma fx_i}{\Sigma f_i} = \dfrac{904}{1000} = 0.904. \quad \therefore \quad e^{-m} = e^{-0.904} = 0.4049.$

Hence the theoretical frequencies are $\dfrac{1000 \times e^{-m}(m)^r}{r!}$, $r = 0, 1, 2, 3, 4$

*i.e.*,

| *x* : | 0 | 1 | 2 | 3 | 4 | Total |
|---|---|---|---|---|---|---|
| *f* : | 404.9 | 366 | 165.4 | 49.8 | 11.3 | 997.4 |
| | (406.2) | | | | (12.6) | |

In order that the total observed and expected frequencies may agree, we take the first and last theoretical frequencies as 406.2 and 12.6 instead of 404.9 and 11.3 as shown in brackets. (In case, the expected frequencies are less than 10, we group together such classes. Here of course, none of the frequencies < 10).

Hence

$$\chi^2 = \frac{(419-406.2)^2}{406.2} + \frac{(352-366)^2}{366} + \frac{(154-165.4)^2}{165.4} + \frac{(56-49.8)^2}{49.8} + \frac{(19-12.6)^2}{12.6}$$
$$= 0.403 + 0.536 + 0.786 + 0.772 + 3.251 = 5.748$$

Since the mean of the theoretical distribution has been estimated from the given data and the totals have been made to agree, there are two constraints so that the number of degrees of freedom $\nu = 5 - 2 = 3$.

For $\nu = 3$, we have $\chi^2_{0.05} = 7.82$. [From Table VI]

Since the calculated value of $\chi^2 < \chi^2_{0.05}$, the agreement between the fact and theory is good and hence the Poisson distribution can be fitted to the data.

**Example 27.20.** *Fit a normal distribution to the following data of weights of 100 students of Delhi University and test the goodness of fit.*

| *Weights (kg)* : | 60-62 | 63-65 | 66-68 | 69-71 | 72-74 |
|---|---|---|---|---|---|
| *Frequency* : | 5 | 18 | 42 | 27 | 8 |

**Solution.** Here $N = 100$, mean $m = 67.45$ and S.D. $\sigma = 2.92$. The calculations are arranged as follows* :

| *Class boundary (x)* | $z = (x - m)/\sigma$ | *Area under normal curve from 0 to z* | *Area for each class (A)** | *Expected frequency ($f_e = N \times A$)* |
|---|---|---|---|---|
| 59.5 | –2.72 | 0.4967 | | |
| | | | 0.0413 | 4.13 |
| 62.5 | –1.70 | 0.4554 | | |
| | | | 0.2068 | 20.68 |
| 65.5 | –0.67 | 0.2486 | | |
| | | | 0.3892 | 38.92 |
| 68.5 | 0.36 | 0.1406 | | |
| | | | 0.2771 | 27.71 |
| 71.5 | 1.39 | 0.4177 | | |
| | | | 0.0743 | 7.43 |
| 74.5 | 2.41 | 0.4920 | | |

* *A* is obtained by subtracting the successive areas in the third column when the corresponding values of *z* have the same sign and adding them when the *z* values have opposite signs (which occurs only once).

$$\therefore \qquad \chi^2 = \frac{(5-4.13)^2}{4.13} + \frac{(18-20.68)^2}{20.68} + \frac{(42-38.92)^2}{38.92} + \frac{(27-27.71)^2}{27.71} + \frac{(8-7.43)^2}{7.43}$$
$$= 0.1833 + 0.3473 + 0.2437 + 0.0182 + 0.0437 = 0.8362$$

As regards the number of degrees of freedom ($\gamma$), there are three constraints ($i$) discrepancy between total observed and total estimated frequencies and ($ii$) mean ($m$) and standard deviation ($\sigma$) have been estimated from the sample data. $\therefore \quad r = 5 - 3 = 2$.

For $\gamma = 2$, $\chi^2_{0.05} = 0.103$ from Table VI.

Since $\chi^2 = 0.8362 > 0.103$. Hence the fit is not good.

## PROBLEMS 27.4

1. Five dice were thrown 96 times and the number of times 4, 5 or 6 were thrown were :

| No. of dice showing 4, 5 or 6 : | 5 | 4 | 3 | 2 | 1 | 0 |
|---|---|---|---|---|---|---|
| Frequency : | 8 | 18 | 35 | 24 | 10 | 1 |

Find the probability of getting this result by chance ?

2. Genetic theory states that children having one parent of blood type $M$ and other blood type $N$ will always be one of the three types $M$, $MN$, $N$ and that the proportions of these types will on average be 1 : 2 : 1. A report states that out of 300 children having one $M$ parent and one $N$ parent, 30% were found to be of type $M$, 45% of type $MN$ and remainder of type $N$. Test the hypothesis by $c^2$ test. *(V.T.U., 2012)*

3. A die was thrown 60 times and the following frequency distribution was observed :

| Faces : | 1 | 2 | 3 | 4 | 5 | 6 |
|---|---|---|---|---|---|---|
| $f_0$ : | 15 | 6 | 4 | 7 | 11 | 17 |

Test whether the die is unbiased ? *(V.T.U., 2012 S)*

4. The following table gives the number of aircraft accidents that occurred during the various days of the week. Find whether the accidents are uniformly distributed over the week ?

| Days : | Sun | Mon | Tue | Wed | Thu | Fri | Sat | Total |
|---|---|---|---|---|---|---|---|---|
| No. of accidents : | 14 | 16 | 8 | 12 | 11 | 9 | 14 | 84 |

5. Fit a binomial distribution to the data :

| $x$ : | 0 | 1 | 2 | 3 | 4 | 5 |
|---|---|---|---|---|---|---|
| $f$ : | 38 | 144 | 342 | 287 | 164 | 25 |

and test for goodness of fit, at the level of significance 0.05.

6. In 1000 extensive sets of trials for an event of small probability, the frequencies $f_0$ of the number $x$ of successes proved to be :

| $x$ : | 0 | 1 | 2 | 3 | 4 | 5 | 6 | 7 |
|---|---|---|---|---|---|---|---|---|
| $f_0$ : | 305 | 366 | 210 | 80 | 28 | 9 | 2 | 1 |

Fit a Poisson distribution to the data and test the goodness of fit.

7. The frequencies of localities according to the number of deaths due to cholera during eight years in 1000 localities is as follows :

| No. of deaths : | 0 | 1 | 2 | 3 | 4 | 5 | 6 | 7 |
|---|---|---|---|---|---|---|---|---|
| No. of localities : | 314 | 355 | 204 | 86 | 29 | 9 | 3 | 0 |

Fit a suitable distribution to the data and test the goodness of fit.

8. Obtain the equation of the normal curve that may be fitted to the data and test the goodness of fit.

| $x$ : | 4 | 6 | 8 | 10 | 12 | 14 | 16 | 18 | 20 | 22 | 24 | Total |
|---|---|---|---|---|---|---|---|---|---|---|---|---|
| $f(x)$ : | 1 | 7 | 15 | 22 | 35 | 43 | 38 | 20 | 13 | 5 | 1 | 200 |

## 27.19 (1) F-DISTRIBUTION*

Let $x_1, x_2, \ldots x_{n_1}$ and $y_1, y_2, \ldots y_{n_2}$ be the values of two independent random samples drawn from the normal populations $\sigma^2$ having equal variances.

* This distribution was introduced by the English statistician *Prof. R.A. Fisher* (1890–1962) who had greatly influenced the development of modern statistics.

Let $\bar{x}_1$ and $\bar{x}_2$ be the sample means and $s_1^2 = \frac{1}{n_1 - 1}\sum_1^{n_1}(x_i - \bar{x})^2$, $s_2^2 = \frac{1}{n_2 - 1}\sum_1^{n_2}(y_i - \bar{y})^2$ be the sample variances.

Then we define $F$ by the relation

$$F = \frac{s_1^2}{s_2^2} \qquad (s_1^2 > s_2^2)$$

This gives $F$-distribution (also known as variance ratio distribution) with $\gamma_1 = n_1 - 1$ and $\gamma_2 = n - 2$ degrees of freedom. *The larger of the variances is placed in the numerator.*

**(2) Properties.** I. The $F$-distribution curve lies entirely in the first quadrant and is *unimodal*.

II. The $F$-distribution is independent of the population variance $\sigma^2$ and depends on $\gamma_1$ and $\gamma_2$ only.

III. $F_\alpha(\gamma_1, \gamma_2)$ is the value of $F$ for $\gamma_1$ and $\gamma_2$ of such that the area to the right of $F_\alpha$ is $\alpha$.

IV. It can be shown that the mode of $F$-distribution is less than unity.

**(3) Significance test.** Snedecor's $F$-tables give 5% and 1% points of significance for $F$. (Table VII – Appendix 2). 5% points of $F$ mean that area under the $F$-curve to the right of the ordinate at a value of $F$, is 0.05. Clearly value of $F$ at 5% significance is lower than that at 1%. $F$-distribution is very useful for testing the equality of population means by comparing sample variances. As such it forms the basis of *analysis of variance*.

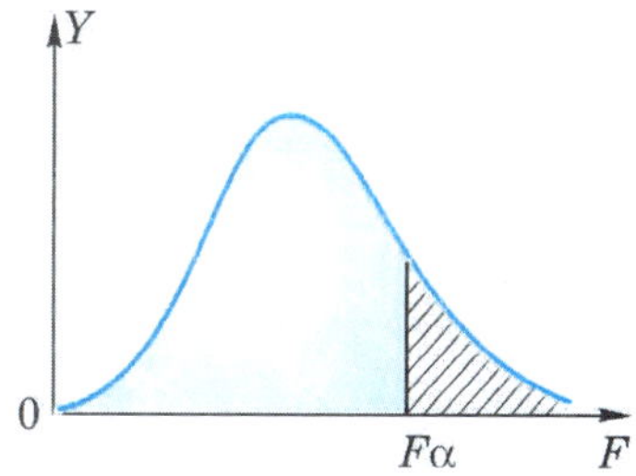

Fig. 27.4

**Example 27.21.** *Two samples of sizes 9 and 8 give the sum of squares of deviations from their respective means equal to 160 inches² and 91 inches² respectively. Can these be regarded as drawn from the same normal population ?*

**Solution.** We have $\Sigma(x - \bar{x})^2 = 160$ and $\Sigma(y - \bar{y})^2 = 91$

$$\therefore \qquad s_1^2 = \frac{160}{8} = 20$$

and

$$s_2^2 = \frac{91}{7} = 13.$$

Hence

$$F = \frac{s_1^2}{s_2^2} = \frac{20}{13} = 1.54 \text{ nearly.}$$

For $\gamma_1 = 8$, $\gamma_2 = 7$, we have $F_{0.05} = 3.73$. [From Table VII]

Since the calculated value of $F < F_{0.05}$, the population variances are not significantly different. Thus the two samples can be regarded as drawn from two normal populations with the same variance. If the two populations are to be same, their means should also be the same which can be verified by applying $t$-test provided the sample means are known.

**Example 27.22.** *Measurements on the length of a copper wire were taken in 2 experiments A and B as under :*

| | | | | | | | | | | |
|---|---|---|---|---|---|---|---|---|---|---|
| *A's measurements (mm)* | : | *12.29,* | *12.25,* | *11.86,* | *12.13,* | *12.44,* | *12.78,* | *12.77,* | *11.90,* | *12.47.* |
| *B's measurements (mm)* | : | *12.39,* | *12.46,* | *12.34,* | *12.22,* | *11.98,* | *12.46,* | *12.23,* | *12.06.* | |

*Test whether B's measurements are more accurate than A's. (The readings taken in both cases being unbiased)*

**Solution.** Readings in both cases being unbiased, $B$'s measurements will be taken more accurate if its population variance is less than that of $A$'s measurements.

Under the hypothesis that the two populations have the same variance (*i.e.* $\sigma_1^2 = \sigma_2^2$), we have

$$F = \frac{s_1^2}{s_2^2}$$

with $\gamma_1 = n_1 - 1 = 8$ and $\gamma_2 = n_2 - 1 = 7$.

We calculate the s.d.'s of the two series as follows :

| *A's measurements* | | | *B's measurements* | | |
|---|---|---|---|---|---|
| $x$ | $u = 100(x - 12)$ | $u^2$ | $y$ | $v = 100(y - 12)$ | $v^2$ |
| 12.29 | 29 | 841 | 12.39 | 39 | 1521 |
| 12.25 | 25 | 625 | 12.46 | 46 | 2116 |
| 11.86 | –14 | 196 | 12.34 | 34 | 1156 |
| 12.13 | 13 | 169 | 12.22 | 22 | 484 |
| 12.44 | 44 | 1936 | 11.98 | –2 | 4 |
| 12.78 | 78 | 6084 | 12.46 | 46 | 2116 |
| 12.77 | 77 | 5929 | 12.23 | 23 | 529 |
| 11.90 | –10 | 100 | 12.06 | 6 | 36 |
| 12.47 | 47 | 2209 | | | |
| | 289 | 18089 | | 214 | 7962 |

$$\therefore \quad s_1^2 = \frac{1}{n_1 - 1}\left[18089 - \frac{(289)^2}{n_1}\right] = \frac{1}{8}(18089 - 9280) = 1101.1$$

$$s_2^2 = \frac{1}{n_2 - 1}\left[7962 - \frac{(214)^2}{n_2}\right] = \frac{1}{7}(7962 - 5724) = 319.7$$

$$\therefore \quad F = \frac{s_1^2}{s_2^2} = \frac{1101.1}{319.7} = 3.44$$

For $\gamma_1 = 8$ and $\gamma_2 = 7$, from table VII, $F_{0.05} = 3.73$ and $F_{0.01} = 6.84$.

Since the calculated value of $F$ is less than both $F_{0.05}$ and $F_{0.01}$, the result is insignificant at both 5% and 1% level.

Hence there is no reason to say that $B$'s measurements are more accurate than those of $A$'s.

**Example 27.23:** *The record of weights of male populations follows normals distribution. Its mean and standard deviation are 70 kg and 15 kg respectively. If a researcher consider the records of 50 males, then what would be the mean and standard deviation of the chosen sample?* *(AKTU, 2019)*

**Solution:** Mean of population $\mu = 70$ kg

Standard deviation of the population = 15 kg

sample size $n = 50$

mean of the sample is given by $\mu_{\bar{x}} = 70$

$$\text{Standard deviation of sample } \sigma_{\bar{x}} = \frac{\sigma}{\sqrt{n}}$$

$$\sigma_{\bar{x}} = \frac{15}{\sqrt{50}}$$

$$\sigma_{\bar{x}} = 2.1 \text{ kg.}$$

## 27.20 (1) FISHER'S z-DISTRIBUTION

Changing the variable $F$ to $z$ by the substitution $z = \frac{1}{2}\log_e F$ or $F = e^{2z}$ in the $F$-distribution, we get the *Fisher's z-distribution.*

It is more nearly symmetrical than $F$-distribution. Table showing the values of $z$ that will be exceeded in simple sampling with probabilities 0.05 and 0.01 have been prepared for various values of $v_1$ and $v_2$.

**(2) Significance test.** As $z$-table give only critical values corresponding to right hand tail areas, therefore 5% (or 1%) points of $z$ imply that the area to the right of the ordinate at $z$ is 0.05 (or 0.01). In other words, *5% and 1% points of z correspond to 10% and 2% levels of significance respectively.*

**Example 27.24.** *Two gauge operations are tested for precision in making measurements. One operator completes a set of 26 readings with a standard deviations of 1.34 and the other does 34 readings with a standard deviations of 0.98. What is the level of significance of this difference.*

*(Given that for $\nu_1 = 25$ and $\nu_2 = 33$, $z_{0.05} = 0.305$, $z_{0.01} = 0.432$)*

**Solution.** We have $n_1 = 26, \sigma_x = 1.34$ ; $n_2 = 34, \sigma_y = 0.98$

$$\therefore \quad s_1^2 = \frac{n_1}{n_1 - 1} \cdot \sigma_x^2 = \frac{26}{25}(1.34)^2 \approx (1.34)^2 \quad \text{and} \quad s_2^2 = \frac{n_2}{n_2 - 1} \cdot \sigma_y^2 = \frac{34}{33}(0.98)^2 \approx (0.98)^2$$

$$\text{Hence} \quad F = \left(\frac{1.34}{0.98}\right)^2 = 1.8696 \quad \text{and} \quad z = \frac{1}{2}\log_e F = 1.1513 \log_{10} 1.8696 = 0.3129$$

Since the calculated value of $z$ is just greater than $z_{0.05}$ and less than $z_{0.01}$, the difference between the standard deviation is just significant at 5% level and insignificant at 1% level.

## PROBLEMS 27.5

1. Two samples of 9 and 7 individuals have variances 4.8 and 9.6 respectively. Is the variance 9.6 significantly greater than the variance 4.6 ?
2. Test for breaking strength were carried out on two lots of 5 and 9 steel wires. The variance of one lot was 230 and that of other was 492. Is there a significant difference in their variability ?
3. Show how you would use Fisher's $z$-test to decide whether the two sets of observations 17, 27, 18, 25, 27, 29, 27, 23, 17 and 16, 16, 20, 16, 20, 17, 15, 21, indicate samples from the same universe.
4. In two groups of ten children each, the increase in weight due to different diets during the same period, were in pounds

   3, 7, 5, 6, 5, 4, 4, 5, 3, 6

   8, 5, 7, 8, 3, 2, 7, 6, 5, 7.

   Is there a significant difference in their variability ?
5. The mean diameter of rivets produced by two firms $A$ and $B$ are practically the same but their standard deviations are different. For 16 rivets manufactured by firm $A$, the S.D. is 3.8 mm while for 22 rivets manufactured by firm $B$ is 2.9 mm. Do you think products of firm $B$ are of better quality than those of firm $A$.
6. The I.Q.'s of 25 students from one college showed a variance of 16 and those of an equal number from the other college had a variance of 8. Discuss whether there is any significant difference in variability of intelligence.
7. Two random samples from two normal populations are given below :

   | | | | | | | | |
   |---|---|---|---|---|---|---|---|
   | Sample I | : | 16 | 26 | 27 | 23 | 24 | 22 |
   | Sample II | : | 33 | 42 | 35 | 32 | 28 | 31 |

   Do the estimates of population variances differ significantly ?

   | | | | | |
   |---|---|---|---|---|
   | Degrees of freedom | : | (5, 5) | (5, 6) | (6, 5) |
   | 5% value of $F$ | : | 5.05 | 4.39 | 4.95. |

8. Two independent samples of sizes 7 and 6 have the following values :

   | | | | | | | | |
   |---|---|---|---|---|---|---|---|
   | Sample A : | 28 | 30 | 32 | 33 | 33 | 29 | 34 |
   | Sample B : | 29 | 30 | 30 | 24 | 27 | 29 | |

   Examine whether the samples have been drawn from normal populations having the same variance ?

   [Given that the values of $F$ at 5% level for (6, 5) d.f. is 4.95 and for (5, 6) d.f. is 4.39].

# 27.21 OBJECTIVE TYPE OF QUESTIONS

## PROBLEMS 27.6

*Select the correct answer or fill up the blanks in each of the following questions :*

1. The 'null hypothesis' implies that .........
2. The uses of $t$-distribution are .........
3. Type $I$ and type $II$ errors are such that .........
4. A single-tailed test is used when .........
5. Control limit theorem states that .........
6. A hypothesis is true, but is rejected. Then this is an error of type .........

7. If the standard deviation of a $\chi^2$ distribution is 10, then its degree of freedom is .........
8. Range of $F$-distribution is .........
9. A hypothesis is false but accepted, then there is an error of type .........
10. The mean and variance of a $\chi^2$ distribution with 8 degrees of freedom are ......... and ......... respectively.
11. In a $t$-distribution of sample size $n$, the degrees of freedom are .........
12. The test statistic $F = \dfrac{s_1^2}{s_2^2}$ is used when

    (i) $s_2^2 > s_1^2$ (ii) $s_2^2 < s_1^2$ (iii) $s_1^2 = s_2^2$ (iv) none of these.
13. The $t$-test is applicable to samples for which $n$ is .........
14. The two main uses of $\chi^2$-test are .........
15. Range of $t$-distribution is .........
16. If two samples are taken from two populations of unequal variances, we can apply $t$-test to test the difference of means. (True or False)
17. The Chi-square distribution is continuous. (True or False)

CHAPTER

# 28

# Numerical Solution of Equations

**1.** Introduction. **2.** Solution of algebraic and transcendental equations—Bisection method, Method of false position, Newton's method. **3.** Useful deductions from the Newton-Raphson Formula. **4.** Approximate solution of equations—Horner's method. **5.** Solution of linear simultaneous equations. **6.** Direct methods of solution—Gauss elimination method, Gauss-Jordan method, Factorization method. **7.** Iterative methods of solution—Jacobi's method, Gauss-Seidal method, Relaxation method. **8.** Solution of non-linear simultaneous equations—Newton-Raphson method. **9.** Determination of eigen values by iteration. **10.** Objective Type of Questions.

## 28.1 INTRODUCTION

The limitations of analytical methods have led the engineers and scientists to evolve graphical and numerical methods. As seen in § 1.8, the graphical methods, though simple, give results to a low degree of accuracy. Numerical methods can, however, be derived which are more accurate. With the advent of high speed digital computers and increasing demand for numerical answers to various problems, numerical techniques have become indispensible tool in the hands of engineers.

Numerical methods are often, of a repetitive nature. These consist in repeated execution of the same process where at each step the result of the preceding step is used. This is known as *iteration process* and is repeated till the result is obtained to a desired degree of accuracy.

In this chapter, we shall discuss some numerical methods for the solution of algebraic and transcendental equations and simultaneous linear and non-linear equations. We shall close the chapter by describing an iterative method for the solution of eigen-value problem. For a detailed study of these topics, the reader should refer to author's book '*Numerical Methods in Engineering & Science*'.

## 28.2 SOLUTION OF ALGEBRAIC AND TRANSCENDENTAL EQUATIONS

To find the roots of an equation $f(x) = 0$, we start with a known approximate solution and apply any of the following methods :

(1) **Bisection method.** This method consists in locating the root of the equation $f(x) = 0$ between $a$ and $b$. If $f(x)$ is continuous between $a$ and $b$, and $f(a)$ and $f(b)$ are of opposite signs then there is a root between $a$ and $b$. For definiteness, let $f(a)$ be negative and $f(b)$ be positive. Then the first approximation to the root is $x_1 = \frac{1}{2}(a + b)$.

If $f(x_1) = 0$, then $x_1$ is a root of $f(x) = 0$. Otherwise, the root lies between $a$ and $x_1$ or $x_1$ and $b$ according as $f(x_1)$ is positive or negative. Then we bisect the interval as before and continue the process until the root is found to desired accuracy.

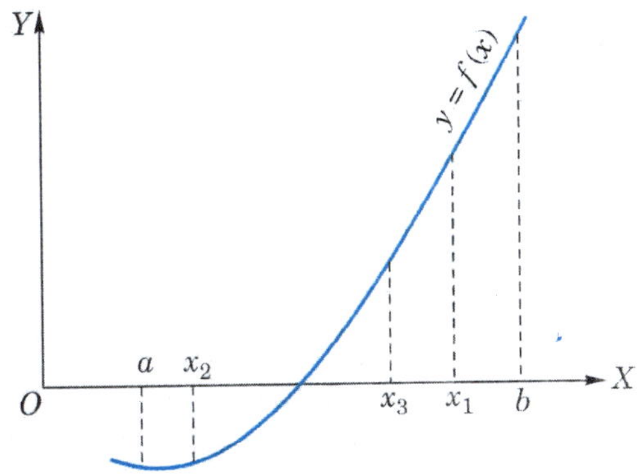


Fig. 28.1

In the Fig. 28.1, $f(x_1)$ is +ve, so that the root lies between $a$ and $x_1$. Then the second approximation to the root is $x_2 = \frac{1}{2}(a + x_1)$. If $f(x_2)$ is –ve, the root lies between $x_1$ and $x_2$. Then the third approximation to the root is $x_3 = \frac{1}{2}(x_1 + x_2)$ and so on.

**Example 28.1.** *(a) Find a root of the equation $x^3 - 4x - 9 = 0$, using the bisection method correct to three decimal places.*

*(b) Using bisection method, find the negative root of the equation $x^2 - 4x + 9 = 0$.* (J.N.T.U., 2009)

**Solution.** (*a*) Let $f(x) = x^3 - 4x - 9$

Since $f(2)$ is –ve and $f(3)$ is +ve, a root lies between 2 and 3

∴ first approximate to the root is

$$x_1 = \frac{1}{2}(2 + 3) = 2.5$$

Thus $\quad f(x_1) = (2.5)^3 - 4(2.5) - 9 = -3.375 \quad$ *i.e.,* –ve

∴ the root lies between $x_1$ and 3. Thus the second approximation to the root is

$$x_2 = \frac{1}{2}(x_1 + 3) = 2.75$$

Then $\quad f(x_2) = (2.75)^3 - 4(2.75) - 9 = 0.7969 \quad$ *i.e.,* +ve

∴ the root lies between $x_1$ and $x_2$. Thus the third approximation to the root is

$$x_3 = \frac{1}{2}(x_1 + x_2) = 2.625$$

Then $\quad f(x_3) = (2.625)^3 - 4(2.625) - 9 = -1.4121 \quad$ *i.e.,* –ve

∴ the root lies between $x_2$ and $x_3$. Thus the fourth approximation to the root is

$$x_4 = \frac{1}{2}(x_2 + x_3) = 2.6875$$

Repeating this process, the successive approximations are

$$x_5 = 2.71875, \quad x_6 = 2.70313, \quad x_7 = 2.71094$$
$$x_8 = 2.70703, \quad x_9 = 2.70508, \quad x_{10} = 2.70605$$
$$x_{11} = 2.70654, \quad x_{12} = 2.70642$$

Hence the root is 2.7064

(*b*) If α, β, γ are the roots of the given equation, then –α, –β, –γ are the roots of $(-x)^3 - 4(-x) + 9 = 0$

∴ the negative root of the given equation is the positive root of $x^3 - 4x - 9 = 0$ which we have found above to be 2.7064.

Hence the negative root for the given equation is –2.7064.

**Example 28.2.** *By using the bisection method, find an approximate root of the equation $\sin x = 1/x$, that lies between $x = 1$ and $x = 1.5$ (measured in radians). Carry out computations upto the 7th stage.* (JNTUK, 2022)

**Solution.** Let $f(x) = x \sin x - 1$. We know that $1^r = 57.3°$.

Since $\quad f(1) = 1 \times \sin(1) - 1 = \sin(57.3°) - 1 = -0.15849$

and $\quad f(1.5) = 1.5 \times \sin(1.5)^r - 1 = 1.5 \times \sin(85.95)° - 1 = 0.49625$ ;

a root lies between 1 and 1.5.

∴ first approximation to the root is $x_1 = \frac{1}{2}(1 + 1.5) = 1.25$.

Then $f(x_1) = (1.25)\sin(1.25) - 1 = 1.25 \sin(71.625°) - 1 = 0.18627$ and $f(1) < 0$.

∴ a root lies between 1 and $x_1 = 1.25$.

Thus the second approximation to the root is $x_2 = \frac{1}{2}(1 + 1.25) = 1.125$.

Then $f(x_2) = 1.125 \sin(1.125) - 1 = 1.125 \sin(64.46)° - 1 = 0.01509$ and $f(1) < 0$.

∴ a root lies between 1 and $x_2 = 1.125$.

Thus the third approximation to the root is $x_3 = \frac{1}{2}(1 + 1.125) = 1.0625$

Then $f(x_3) = 1.0625 \sin(1.0625) - 1 = 1.0625 \sin(60.88) - 1 = -0.0718 < 0$

and $f(x_2) > 0$, *i.e.* now the root lies between $x_3 = 1.0625$ and $x_2 = 1.125$.

$\therefore$ fourth approximation to the root is $x_4 = \frac{1}{2}(1.0625 + 1.125) = 1.09375$

Then $f(x_4) = -0.02836 < 0$ and $f(x_2) > 0$,

*i.e.*, the root lies between $x_4 = 1.09375$ and $x_2 = 1.125$.

$\therefore$ fifth approximation to the root is $x_5 = \frac{1}{2}(1.09375 + 1.125) = 1.10937$

Then $f(x_5) = -0.00664 < 0$ and $f(x_2) > 0$.

$\therefore$ the root lies between $x_5 = 1.10937$ and $x_2 = 1.125$.

Thus the sixth approximation to the root is

$$x_6 = \frac{1}{2}(1.10937 + 1.125) = 1.11719$$

Then $f(x_6) = 0.00421 > 0$. But $f(x_5) < 0$.

$\therefore$ the root lies between $x_5 = 1.10937$ and $x_6 = 1.11719$.

Thus the seventh approximation to the root is $x_7 = \frac{1}{2}(1.10937 + 1.11719) = 1.11328$

Hence the desired approximation to the root is 1.11328.

**(2) Method of false position** *or* **Regula-falsi method or Method of Interpolation.** This is the oldest method of finding the real root of an equation $f(x) = 0$ and closely resembles the bisection method. Here we choose two points $x_0$ and $x_1$ such that $f(x_0)$ and $f(x_1)$ are of opposite signs *i.e.*, the graph of $y = f(x)$ crosses the $x$-axis between these points (Fig. 28.2). This indicates that a root lies between $x_0$ and $x_1$ consequently $f(x_0) f(x_1) < 0$.

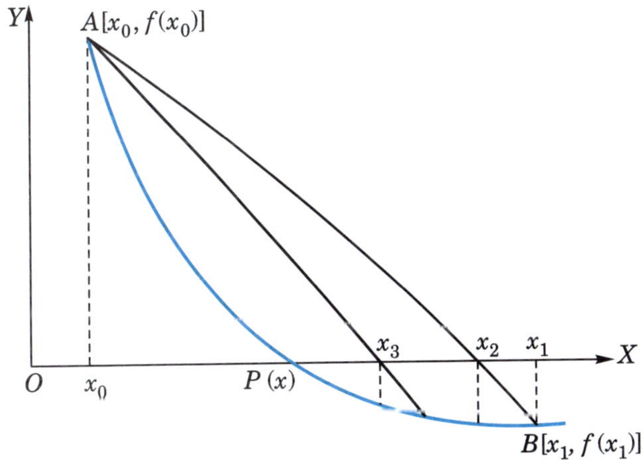

Fig. 28.2.

Equation of the chord joining the points $A[x_0, f(x_0)]$ and $B[x_1, f(x_1)]$ is

$$y - f(x_0) = \frac{f(x_1) - f(x_0)}{x_1 - x_0}(x - x_0) \quad \text{...(1)}$$

The method consists in replacing the curve $AB$ by means of the chord $AB$ and taking the point of intersection of the chord with the $x$-axis as an approximation to the root. So the abscissa of the point where the chord cuts the $x$-axis ($y = 0$) is given by

$$x_2 = x_0 - \frac{x_1 - x_0}{f(x_1) - f(x_0)} f(x_0) \quad \text{...(2)}$$

which is an approximation to the root.

If now $f(x_0)$ and $f(x_2)$ are of opposite signs, then the root lies between $x_0$ and $x_2$. So replacing $x_1$ by $x_2$ in Eqn. (2), we obtain the next approximation $x_3$. (The root could as well lie between $x_1$ and $x_2$ and we would obtain $x_3$ accordingly). This procedure is repeated till the root is found to desired accuracy. The iteration process based on Eqn. (1) is known as the *method of false position.*

**Obs.** The regula-falsi method (method of interpolation) has *linear rate of convergence.* *(Delhi, 2013)*

**Example 28.3.** *Find a real root of the equation $x^3 - 2x - 5 = 0$ by the method of false position correct to three decimal places.* *(JNTUK, 2022)*

**Solution.** Let $f(x) = x^3 - 2x - 5$

so that $f(2) = -1$ and $f(3) = 16$ *i.e.*, A root lies between 2 and 3.

$\therefore$ taking $x_0 = 2, x_1 = 3, f(x_0) = -1, f(x_1) = 16$, in the method of false position, we get

$$x_2 = x_0 - \frac{x_1 - x_0}{f(x_1) - f(x_0)} f(x_0) = 2 + \frac{1}{17} = 2.0588 \quad \text{...}(i)$$

Now $f(x_2) = f(2.0588) = -0.3908$ *i.e.*, the root lies between 2.0588 and 3.

$\therefore$ taking $x_0 = 2.0588$, $x_1 = 3$, $f(x_0) = -0.3908$, $f(x_1) = 16$, in Eqn. (*i*), we get

$$x_3 = 2.0588 - \frac{0.9412}{16.3908}(-0.3908) = 2.0813$$

Repeating this process, the successive approximations are

$$x_4 = 2.0862,\ x_5 = 2.0915,\ x_6 = 2.0934,\ x_7 = 2.0941,\ x_8 = 2.0943 \text{ etc.}$$

Hence the root is 2.094 correct to 3 decimal places.

**Example 28.4.** *Find the root of the equation cos x = xe^x using the regula-falsi method correct to four decimal places.* (*Delhi, 2015*)

**Solution.** Let $f(x) = \cos x - xe^x = 0$

So that $f(0) = 1, f(1) = \cos 1 - e = -2.17798$

*i.e.*, the root lies between 0 and 1.

$\therefore$ taking $x_0 = 0$, $x_1 = 1$, $f(x_0) = 1$ and $f(x_1) = -2.17798$ in the regula-falsi method, we get

$$x_2 = x_0 - \frac{x_1 - x_0}{f(x_1) - f(x_0)} f(x_0) = 0 + \frac{1}{3.17798} \times 1 = 0.31467 \quad ...(i)$$

Now $f(0.31467) = 0.51987$

*i.e.*, the root lies between 0.31467 and 1.

$\therefore$ taking $x_0 = 0.31467$, $x_1 = 1$, $f(x_0) = 0.51987$, $f(x_1) = -2.17798$ in Eqn. (*i*), we get

$$x_3 = 0.31467 + \frac{0.68533}{2.69785} \times 0.51987 = 0.44673$$

Now $f(0.44673) = 0.20356$

*i.e.*, the root lies between 0.44673 and 1.

$\therefore$ taking $x_0 = 0.44673$, $x_1 = 1$, $f(x_0) = 0.20356$, $f(x_1) = -2.17798$ in Eqn. (*i*), we get

$$x_4 = 0.44673 + \frac{0.55327}{2.38154} \times 0.20356 = 0.49402$$

Repeating this process, the successive approximations are

$$x_5 = 0.50995, \quad x_6 = 0.51520, \quad x_7 = 0.51692$$
$$x_8 = 0.51748, \quad x_9 = 0.51767, \quad x_{10} = 0.51775 \text{ etc.}$$

Hence the root is 0.5177 correct to 4 decimal places.

**Example 28.5.** *Find a real root of the equation x log$_{10}$ x = 1.2 by regula-falsi method correct to four decimal places.* (*JNTUK, 2022; V.T.U., 2010*)

**Solution.** Let $f(x) = x \log_{10} x - 1.2$

so that $f(1) = -\text{ve}, f(2) = -\text{ve}$ and $f(3) = +\text{ve}$.

$\therefore$ a root lies between 2 and 3.

Taking $x_0 = 2$ and $x_1 = 3$, $f(x_0) = -0.59794$ and $f(x_1) = 0.23136$, in the method of false position, we get

$$x_2 = x_0 - \frac{x_1 - x_0}{f(x_1) - f(x_0)} f(x_0) = 2.72102 \quad ...(i)$$

Now $f(x_2) = f(2.72102) = -0.01709$

*i.e.*, the root lies between 2.72102 and 3.

$\therefore$ taking $x_0 = 2.72102$, $x_1 = 3$, $f(x_0) = -0.01709$

and $f(x_1) = 0.23136$ in (*i*), we get

$$x_3 = 2.72102 + \frac{0.27898}{0.23136 + 0.01709} \times 0.01709 = 2.74021$$

Repeating this process, the successive approximations are

$$x_4 = 2.74024,\ x_5 = 2.74063 \text{ etc.}$$

Hence the root is 2.7406 correct to 4 decimal places.

**Example 28.6.** *Use the method of false position, to find the fourth root of 32 correct to three decimal places.*

**Solution.** Let $x = (32)^{1/4}$ so that $x^4 - 32 = 0$

Take $f(x) = x^4 - 32$. Then $f(2) = -16$ and $f(3) = 49$, *i.e.*, a root lies between 2 and 3.

$\therefore$ taking $x_0 = 2, x_1 = 3, f(x_0) = -16, f(x_1) = 49$ in the method of false position, we get

$$x_2 = x_0 - \frac{x_1 - x_0}{f(x_1) - f(x_0)} f(x_0) = 2 + \frac{16}{65} = 2.2462 \quad ...(i)$$

Now $f(x_2) = f(2.2462) = -6.5438$ *i.e.* the root lies between 2.2462 and 3.

$\therefore$ taking $x_0 = 2.2462, x_1 = 3, f(x_0) = -6.5438, f(x_1) = 49$

in Eqn. (*i*), we get $\quad x_3 = 2.2462 - \dfrac{3 - 2.2462}{49 + 6.5438}(-6.5438) = 2.335$

Now $f(x_3) = f(2.335) = -2.2732$ *i.e.* the root lies between 2.335 and 3.

$\therefore$ taking $x_0 = 2.335$ and $x_1 = 3, f(x_0) = -2.2732$ and $f(x_1) = 49$ in Eqn. (*i*), we obtain

$$x_4 = 2.335 - \frac{3 - 2.335}{49 + 2.2732}(-2.2732) = 2.3645$$

Repeating this process, the successive approximations are $x_5 = 2.3770, x_6 = 2.3779$ etc.

Since $x_5 = x_6$ upto 3 decimal places, we take $(32)^{1/4} = 2.378$.

**(3) Newton-Raphson method.** Let $x_0$ be an approximate root of the equation $f(x) = 0$. If $x_1 = x_0 + h$ be the exact root, then $f(x_1) = 0$.

$\therefore$ expanding $f(x_0 + h)$ by Taylor's series

$$f(x_0) + hf'(x_0) + \frac{h^2}{2!} f''(x_0) + \ldots\ldots = 0$$

Since $h$ is small, neglecting $h^2$ and higher powers of $h$, we get

$$f(x_0) + hf'(x_0) = 0 \quad \text{or} \quad h = -\frac{f(x_0)}{f'(x_0)} \quad ...(1)$$

$\therefore$ a closer approximation to the root is given by

$$x_1 = x_0 - \frac{f(x_0)}{f'(x_0)}.$$

Similarly, starting with $x_1$, a still better approximation $x_2$ is given by

$$x_2 = x_1 - \frac{f(x_1)}{f'(x_1)}.$$

In general, $\quad x_{n+1} = x_n - \dfrac{f(x_n)}{f'(x_n)} \quad ...(2)$

which is known as the *Newton-Raphson formula* or *Newton's iteration formula*.

**Obs. 1.** *Newton's method is useful in cases of large values of $f'(x)$ i.e.* when the graph of $f(x)$ while crossing the $x$-axis is nearly vertical.

**Obs. 2. Newton's method has a second order of quadratic convergence.** Suppose $x_n$ differs from the root $\alpha$ by a small quantity $\varepsilon_n$ so that $x_0 = \alpha + \varepsilon_n$ and $x_{n+1} = \alpha + \varepsilon_{n+1}$.

Then Eqn. (2) becomes $\alpha + \varepsilon_{n+1} = \alpha + \varepsilon_n - \dfrac{f(\alpha + \varepsilon_n)}{f'(\alpha + \varepsilon_n)}$

*i.e.,* $$\varepsilon_{n+1} = \varepsilon_n - \frac{f(\alpha + \varepsilon_n)}{f'(\alpha + \varepsilon_n)} = \varepsilon_n - \frac{f(\alpha) + \varepsilon_n f'(\alpha) + \frac{1}{2!}\varepsilon_n^2 f''(\alpha) + \ldots}{f'(\alpha) + \varepsilon_n f''(\alpha) + \ldots} \quad \text{[By Taylor's expansion.]}$$

$$= \varepsilon_n - \frac{\varepsilon_n f'(\alpha) + \frac{1}{2}\varepsilon_n^2 f''(\alpha) + \ldots}{f'(\alpha) + \varepsilon_n f''(\alpha) + \ldots} \quad [\because f(\alpha) = 0]$$

$$= \frac{\varepsilon_n^2 f''(\alpha)}{2[f'(\alpha) + \varepsilon_n f''(\alpha)]} \approx \frac{\varepsilon_n^2}{2} \cdot \frac{f''(\alpha)}{f'(\alpha)}. \quad \left[\begin{array}{l}\text{neglecting third and}\\ \text{higher powers of } \varepsilon_n.\end{array}\right]$$

This shows that the subsequent error at each step, is proportional to the square of the previous error and as such the convergence is quadratic. (*Anna, 2012*)

**Obs. 3. Geometrical interpretation.** Let $x_0$ be a point near the root $\alpha$ of the equation $f(x) = 0$ (Fig. 28.3). Then the equation of the tangent at $A_0\,[x_0, f(x_0)]$ is $y - f(x_0) = f'(x_0)(x - x_0)$.

It cuts the $x$-axis at $x_1 = x_0 - \frac{f(x_0)}{f'(x_0)}$ which is a first approximation to the root $\alpha$. If $A_1$ is the point corresponding to $x_1$ on the curve, then the tangent at $A_1$ will cut the $x$-axis of $x_2$ which is nearer to $\alpha$ and is, therefore, a second approximation to the root. Repeating this process, we approach to the root $\alpha$ quite rapidly. Hence *the method consists in replacing the part of the curve between the point $A_0$ and the $x$-axis by means of the tangent to the curve at $A_0$.* *(Delhi., 2015)*

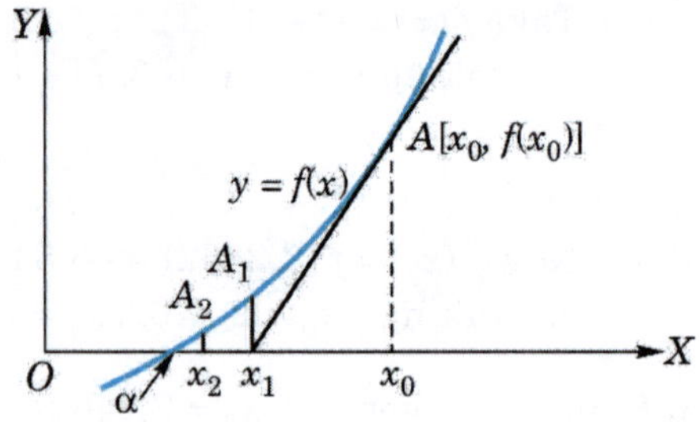

Fig. 28.3

**Example 28.7.** *Find the positive root of $x^4 - x = 10$ correct to three decimal places, using Newton-Raphson method.* *(J.N.T.U., 2008 ; Madras, 2006)*

**Solution.** Let $f(x) = x^4 - x - 10$

So that $f(1) = -10 = -\text{ve}, f(2) = 16 - 2 - 10 = 4 = +\text{ ve}$

$\therefore$ a root of $f(x) = 0$ lies between 1 and 2. Let us take $x_0 = 2$

Also $f'(x) = 4x^3 - 1$

Newton-Raphson's formula is

$$x_{n+1} = x_n - \frac{f(x_n)}{f'(x_n)} \quad \text{...}(i)$$

Putting $n = 0$, the first approximation $x_1$ is given by

$$x_1 = x_0 - \frac{f(x_0)}{f'(x_0)} = 2 - \frac{f(2)}{f'(2)} = 2 - \frac{4}{4 \times 2^3 - 1} = 2 - \frac{4}{31} = 1.871$$

Putting $n = 1$ in Eqn. $(i)$, the second approximation is

$$x_2 = x_1 - \frac{f(x_1)}{f'(x_1)} = 1.871 - \frac{f(1.871)}{f'(1.871)}$$

$$= 1.871 - \frac{(1.871)^4 - (1.871) - 10}{4(1.871)^3 - 1} = 1.871 - \frac{0.3835}{25.199} = 1.856$$

Putting $n = 2$ in Eqn. $(i)$, the third approximation is

$$x_3 = x_2 - \frac{f(x_2)}{f'(x_2)} = 1.856 - \frac{(1.856)^4 - (1.856) - 10}{4(1.856)^3 - 1}$$

$$= 1.856 - \frac{0.010}{24.574} - 1.856$$

Here $x_2 = x_3$. Hence the desired is 1.856 correct to three decimal places.

**Example 28.8.** *Find the Newton's method, the real root of the equation $3x = \cos x + 1$.* *(Anna, 2016 ; Delhi, 2015 ; V.T.U., 2013)*

**Solution.** Let $f(x) = 3x - \cos x - 1$

$f(0) = -2 = -\text{ve}, f(1) = 3 - 0.5403 - 1 = 1.4597 = +\text{ve}.$

So a root of $f(x) = 0$ lies between 0 and 1. It is nearer to 1. Let us take $x_0 = 0.6$.

Also $f'(x) = 3 + \sin x$

$\therefore$ Newton's iteration formula gives

$$x_{n+1} = x_n - \frac{f(x_n)}{f'(x_n)} = x_n - \frac{3x_n - \cos x_n - 1}{3 + \sin x_n}$$

$$= \frac{x_n \sin x_n + \cos x_n + 1}{3 + \sin x_n} \quad \text{...}(i)$$

Putting $n = 0$, the first approximation $x_1$ is given by

$$x_1 = \frac{x_0 \sin x_0 + \cos x_0 + 1}{3 + \sin x_0} = \frac{(0.6)\sin(0.6) + \cos(0.6) + 1}{3 \sin(0.6)}$$

$$= \frac{0.6 \times 0.5729 + 0.82533 + 1}{3 + 0.5729} = 0.6071$$

Putting $n = 1$ in Eqn. $(i)$, the second approximation is

$$x_2 = \frac{x_1 \sin x_1 + \cos x_1 + 1}{3 + \sin x_1} = \frac{0.6071 \sin(0.6071) + \cos(0.6071) + 1}{3 + \sin(0.6071)}$$

$$= \frac{0.6071 \times 0.57049 + 0.8213 + 1}{3 + 0.57049} = 0.6071 \quad \text{Clearly, } x_1 = x_2.$$

Hence the desired root is 0.6071 correct to four decimal places.

**Example 28.9.** *Find the double root of the equation $x^3 - x^2 - x + 1 = 0$.* *(VTU, 2017)*

**Solution.** Given double root of the equation

$$f(x) = x^3 - x^2 - x + 1 = 0$$
$$f'(x) = 3x^2 - 2x - 1$$
$$f''(x) = 6x - 2$$

If $x_{n+1}$ is double root of $f(x) = 0$, then it is also a root of $f'(x) = 0$

$$x_{n+1} = x_{n-2}, \frac{f(x_n)}{f'(x_n)} \quad \text{since } p = 2$$

$$\text{and also, } x_{n+1} = x_n - \frac{f'(x_n)}{f''(x_n)}$$

Now let us choose $x_0 = 0.8$

$$x_1 = x_0 - \frac{2f(x_0)}{f'(x_0)} = 0.8 - \frac{2 \times 0.072}{-0.68} = 1.012$$

$$x_1 = x_0 - \frac{f'(x_0)}{f''(x_0)} = 0.8 - \frac{(-0.68)}{2.8} = 1.043$$

The closeness of the two values of $x_1$

$x_1 = 1.01$ we obtain

$$x_2 = x_1 - \frac{f(x_1)}{f'(x_1)} = 1.01 - 0.0099 = 1.0001$$

$$x_2 = x_1 - \frac{f'(x)}{f''(x_1)} = 1.01 - 0.0099 = 1.0001$$

The exact double root is $x = 1$.

## 28.3 USEFUL DEDUCTIONS FROM THE NEWTON-RAPHSON FORMULA

(1) *Iterative formula to find 1/N is* $\quad x_{n+1} = x_n (2 - Nx_n)$

(2) *Iterative formula to find $\sqrt{N}$ is* $\quad x_{n+1} = \frac{1}{2}(x_n + N/x_n)$

(3) *Iterative formula to find $1/\sqrt{N}$ is* $\quad x_{n+1} = \frac{1}{2}(x_n + 1/Nx_n)$

(4) *Iterative formula to find $\sqrt[k]{N}$ is* $\quad x_{n+1} = \frac{1}{k}[(k-1)x_n + N/x_n^{k-1})]$

**Proofs.** (1) Let $x = 1/N$ or $1/x - N = 0$
Taking $f(x) = 1/x - N$, we have $f'(x) = -x^{-2}$
Then Newton's formula gives

$$x_{n+1} = x_n - \frac{f(x_n)}{f'(x_n)} = x_n - \frac{(1/x_n - N)}{-x_n^{-2}} = x_n + \left(\frac{1}{x^n} - N\right)x_n^2 = x_n + x_n - Nx_n^2 = x_n(2 - Nx_n)$$

*(J.N.T.U., 2015 ; Anna, 2013)*

**(2)** Let $x = \sqrt{N}$ or $x^2 - N = 0$

Taking $f(x) = x^2 - N$, we have $f'(x) = 2x$

Then Newton's formula gives

$$x_{n+1} = x_n - \frac{f(x_n)}{f'(x_n)} = x_n - \frac{x_n^2 - N}{2x_n} = \frac{1}{2}(x_n + N/x_n)$$ *(Anna., 2015 ; Madras, 2006)*

**(3)** Let $x = \frac{1}{\sqrt{N}}$ or $x^2 - \frac{1}{N} = 0$

Taking $f(x) = x^2 - 1/N$, we have $f'(x) = 2x$

Then Newton's formula gives

$$x_{n+1} = x_n - \frac{f(x_n)}{f'(x_n)} = x_n - \frac{x_n^2 - 1/N}{2x_n} = \frac{1}{2}\left(x_n + \frac{1}{Nx_n}\right)$$

**(4)** Let $x = \sqrt[k]{N}$ or $x^k - N = 0$

Taking $f(x) = x^k - N$, we have $f'(x) = kx^{k-1}$

Then Newton's formula gives

$$x_{n+1} = x_n - \frac{f(x_n)}{f'(x_n)} = x_n - \frac{x_n^k - N}{kx_n^{k-1}} = \frac{1}{k}\left[(k-1)x_n + \frac{N}{x_n^{k-1}}\right].$$ *(J.N.T.U., 2015)*

**Example 28.10.** *Evaluate the following (correct to four decimal places) by Newton's iteration method :*

*(i)* $1/31$ *(ii)* $\sqrt{5}$ *(Anna, 2012)*

*(iii)* $1/\sqrt{14}$ *(iv)* $\sqrt[3]{24}$ *(JNTUK, 2022)* *(v)* $(30)^{-1/5}$.

**Solution.** *(i)* Taking $N = 31$, the above formula (1) becomes

$$x_{n+1} = x_n(2 - 31x_n)$$

Since an approximate value of $1/31 = 0.03$, we take $x_0 = 0.03$

Then
$$x_1 = x_0(2 - 31x_0) = 0.03(2 - 31 \times 0.03) = 0.0321$$
$$x_2 = x_1(2 - 31x_1) = 0.0321(2 - 31 \times 0.0321) = 0.032257$$
$$x_3 = x_2(2 - 31x_2) = 0.032257(2 - 31 \times 0.032257) = 0.03226$$

Since $x_2 = x_3$ upto 4 decimal places, we have $1/31 = 0.0323$.

*(ii)* Taking $N = 5$, the above formula (2), becomes $x_{n+1} = \frac{1}{2}(x_n + 5/x_n)$

Since an approximate value of $\sqrt{5} = 2$, we take $x_0 = 2$

Then
$$x_1 = \frac{1}{2}(x_0 + 5/x_0) = \frac{1}{2}(2 + 5/2) = 2.25$$
$$x_2 = \frac{1}{2}(x_1 + 5/x_1) = 2.2361$$
$$x_3 = \frac{1}{2}(x_2 + 5/x_2) = 2.2361$$

Since $x_2 = x_3$ upto 4 decimal places, we have $\sqrt{5} = 2.2361$.

*(iii)* Taking $N = 14$, the above formula (3), becomes $x_{n+1} = \frac{1}{2}[x_n + 1/(14x_n)]$

Since an approximate value of $1/\sqrt{14} = 1/\sqrt{16} = \frac{1}{4} = 0.25$, we take $x_0 = 0.25$

Then
$$x_1 = \frac{1}{2}[x_0 + (14x_0)^{-1}] = \frac{1}{2}[0.25 + (14 \times 0.25)^{-1}] = 0.26785$$
$$x_2 = \frac{1}{2}[x_1 + (14x_1)^{-1}] = \frac{1}{2}[0.26785 + (14 \times 0.26785)^{-1}] = 0.2672618$$
$$x_3 = \frac{1}{2}[x_2 + (14x_2)^{-1}] = \frac{1}{2}[0.2672618 + (14 \times 0.2672618)^{-1}] = 0.2672612$$

Since $x_2 = x_3$ upto 4 decimal places, we take $1/\sqrt{14} = 0.2673$.

*(iv)* Taking $N = 24$ and $k = 3$, the above formula (4) becomes $x_{n+1} = \frac{1}{3}[2x_n + 24/x_n^2]$

Since an approximate value of $(24)^{1/3} = (27)^{1/3} = 3$, we take $x_0 = 3$.

Then $$x_1 = \frac{1}{3}(2x_0 + 24/x_0^2) = \frac{1}{3}(6 + 24/9) = 2.88889$$

$$x_2 = \frac{1}{3}(2x_1 + 24/x_1^2) = \frac{1}{3}[(2 \times 2.88889) + 24/(2.88889)^2] = 2.88451$$

$$x_3 = \frac{1}{3}(2x_2 + 24/x_2^2) = \frac{1}{3}[2 \times 2.88451 + 24/(2.88451)^2] = 2.8845$$

Since $x_2 = x_3$ upto 4 decimal places, we take $(24)^{1/3} = 2.8845$

(*v*) Taking $N = 30$ and $k = -5$, the above formula (4) becomes

$$x_{n+1} = \frac{1}{-5}(-6x_n + 30/x_n^{-6}) = \frac{x_n}{5}(6 - 30x_n^5)$$

Since an approximate value of $(30)^{-1/5} = (32)^{-1/5} = 1/2$, we take $x_0 = 1/2$

Then $$x_1 = \frac{x_0}{5}(6 - 30\,x_0^5) = \frac{1}{10}(6 - 30/2^5) = 0.50625$$

$$x_2 = \frac{x_1}{5}(6 - 30\,x_1^5) = \frac{0.50625}{5}[6 - 30(0.50625)^5] = 0.506495$$

$$x_3 = \frac{x_2}{5}(6 - 30\,x_2^5) = \frac{0.506495}{5}[6 - 30(0.506495)^5] = 0.506496.$$

Since $x_2 = x_3$ upto 4 decimal places, we take $(30)^{-1/5} = 0.5065$.

**Example. 28.11:** *Show that the Newton Raphson process has a quadratic convergence.* (*Marathwada, 2018*)

**Solution.** Let $\alpha$ be the root of equation $f(x) = 0$

$$f(\alpha) = 0 \quad \text{or} \quad f(x_n + \alpha - x_n) = 0$$

or $$f(x_n) + (\alpha - x_n)f'(x_n) + \frac{(\alpha - x_n)^2}{2!} f''x(n) + \ldots = 0$$

Expanding by a Taylor Series

$$\frac{-f(x_n)}{f'(x_n)} = (\alpha - x_n) + \frac{1}{2}(\alpha - x_n)^2 \frac{f''(x_n)}{f''(x_n)} \qquad \ldots(i)$$

Also $$x_{n+1} = x_n - \frac{f(x_n)}{f'(x_n)} \qquad \ldots(ii)$$

from Eqns. (*i*) and (*ii*), we get

$$x_{n+1} - \alpha = \frac{1}{2}(x_n - \alpha)^2 \frac{f''(x_n)}{f'(x_n)} \qquad \ldots(iii)$$

Now let, $\in_n = x_n - \alpha$

So that, $\in_{n+1} = x_{n+1} - \alpha$

Then Eqn. (*iii*) reduce to

$$\in_{n+1} \approx \frac{1}{2} \in_{n^2} \frac{f''(\alpha)}{f'(\alpha)}$$

This show that Newton–Raphson process has a second (quadratic) convergence.

## PROBLEMS 28.1

1. Find a root of the following equations, using the bisection method correct to three decimal places :
   (*i*) $x^3 - x - 11 = 0$ which lies between 2 and 3 (*ii*) $2x^3 + x^2 - 20x + 12 = 0$.
2. Using the bisection method, find a real root of the following equations correct to three decimal places :
   (*i*) $\cos x = xe^x$ (*ii*) $x \log_{10} x = 1.2$ lying between 2 and 3 (*JNTUK, 2022*)
   (*iii*) $e^x - x = 2$ lying between 1 and 1.4 (*iv*) $e^x = 4 \sin x$. (*J.N.T.U., 2015*)

3. Find a real root of the following equations correct to three decimal places by the method of false position (or interpolation) :
   (i) $x^3 - 5x + 1 = 0$ (*U.P.T.U., 2012*) (ii) $x^3 - 4x - 9 = 0$
   (iii) $x^3 - x^2 - 1 = 0$ (*Delhi., 2015*) (iv) $x^6 - x^4 - x^3 - 1 = 0$. (*Delhi, 2013*)
4. Using regula-falsi method, compute the real root of the following equations correct to three decimal places :
   (i) $xe^x = 2$ (*Andhra, 2021; J.N.T.U., 2015*) (ii) $\cos x = 3x - 1$
   (iii) $3x + \sin x - e^x = 0$ (*Andhra, 2021, 2022*) (iv) $2x - \log x = 7$ (*JNTUK, 2022; U.P.T.U., 2010*)
5. Find the fourth root of 12 correct to three decimal places using the method of false position. (*V.T.U., 2016*)
6. Find by Newton's method, a root of the following equations correct to 3 decimal places :
   (i) $x^3 - 2x - 5 = 0$ (ii) $3x^3 - 9x^2 + 8 = 0$ lying between 1 and 2.
7. Find a root of the following equations correct to three significant figures using Newton's iterative method :
   (i) $x^4 + x^3 - 7x^2 - x + 5 = 0$ lying between 2 and 3
   (ii) $x^5 - 5x^2 + 3 = 0$.
8. Find the negative root of the equation $x^3 - 21x + 3500 = 0$ correct to two decimal places by Newton's method.
9. Using Newton-Raphson method, find a root of the following equations correct to the three decimal places :
   (i) $xe^x - 2 = 0$ (ii) $x^2 + 4 \sin x = 0$ (*Delhi., 2015*)
   (iii) $x \sin x + \cos x = 0$ or $x \tan x + 1 = 0$ which is near $x = \pi$ (*V.T.U., 2015 ; Anna, 2012*)
10. Find by Newton's method, the root of the equations :
   (i) $\cos x = xe^x$ (*JNTUK, 2022*) (ii) $x \log_{10} x = 12.34$ (*Anna, 2012*)
   (iii) $10^x + x - 4 = 0$ (iv) $x + \log_{10} x = 3.375$ (*JNTUK, 2022*)
11. The equation $2e^{-x} = \dfrac{1}{x+2} + \dfrac{1}{x+1}$ has two roots greater than –1. Calculate these roots correct to five decimal places. (*U.P.T.U., 2012*)
12. Develop a recurrence formula for finding $\sqrt{N}$, using Newton-Raphson method and hence compute to three decimal places
   (i) $\sqrt{13}$ (ii) $\sqrt{10}$
13. Find the cube root of 17, using Newton-Raphson method.
14. Develop an algorithm using N-R method, to find the fourth root of a positive number $N$ and hence find.
   (i) $(11)^{1/4}$ (*J.N.T.U., 2015*) (ii) $(32)^{1/4}$
15. Evaluate the following (correct to 3 decimal places) by using the Newton-Raphson method :
   (i) 1/18 (ii) $1/\sqrt{15}$ (iii) $(28)^{-1/4}$.
16. Given, $5 \sin 4x = 1.8 e^{3x}$ solve the following equations by Regula-Falsi method. (*Burdwan, 2018*)
17. Given $x^3 - 3x + 4 = 0$, for the root which lies between –2 and –3, using the method three times in succession. (*AMIE, 2018*)
18. Given $2x - \log_{10} x = 7$, for the root which lies between 3.5 and 4, correct to 5 places of decimals. (*JNTUK, 2022; CES-2017*)
19. Prove that the equation $x^3 - 4x + 1 = 0$ has a root lying between 1 and 2 and apply Newton's method to find, it correct to decimal places. (*Ranchi-2018*)
20. Prove that iterative formula, for find the square root of $N$ is

$$x_{i+1} = \frac{1}{2}\left(x_i + \frac{N}{x_i}\right), \text{ Hence find the value of } \sqrt{135}.$$

## 28.4 APPROXIMATE SOLUTION OF EQUATIONS—HORNER'S METHOD

This is the best method of finding approximate values of both rational and irrational roots of a numerical equation. Horner's method consists in diminution of the root of an equation by successive digits occurring in the roots.

If the root of an equation lies between $a$ and $a + 1$, then the value of this root will be $a \,.\, bcd$ ..., where $b, c, d$ ... are digits in its decimal part. To obtain these, we proceed as follows :

(i) Diminish the roots of the given equation by $a$ so that the root of the new equation is $0 \,.\, bcd$ ...

(ii) Then multiply the roots of the transformed equation by 10 so that the root of the new equation is $b \,.\, cd$...

(iii) Now diminish the root by $b$ and multiply the roots of the resulting equation by 10 so that the root is $c. d$...

(*iv*) Next diminish the root by $c$ and so on. By continuing this process, the root may be evaluated to any desired degree of accuracy digit by digit. The method will be clear from the following example.

**Example 28.12.** *Find by Horner's method, the positive root of the equation $x^3 + x^2 + x - 100 = 0$ correct to three decimal places.*

**Solution.** *Step I.* Let $f(x) = x^3 + x^2 + x - 100$

By Descartes' rule of signs, there is only one positive root. Also $f(4) = -$ve and $f(5) = +$ve, therefore, the root lies between 4 and 5.

*Step II.* Diminish the roots of given equation by 4 so that the transformed equation is

$$x^3 + 13x^2 + 57x - 16 = 0 \qquad \ldots(i)$$

Its root lies between 0 and 1. (We draw a zig-zag line above the set of figures 13, 57, –16 which are the coefficients of the terms in Eqn. (*i*) as shown below. Now multiply the roots of Eqn. (*i*) by 10 for which multiply the second term by 10, the third term by 100 and the fourth term by 1000 (*i.e.* attach one zero to the second term, two zeros to the third term and three zeros to the fourth term). Then we get the equation

$$f_1(x) = x^3 + 130x^2 + 5700x - 16000 = 0 \qquad \ldots(ii)$$

| | | | | |
|---|---|---|---|---|
| 1 | 1 | 1 | – 100 | (4.264 |
| | 4 | 20 | 84 | |
| | 5 | 21 | – 16000 | |
| | 4 | 36 | 11928 | |
| | 9 | 5700 | – 4072000 | |
| | 4 | 264 | 3788376 | |
| | 130 | 5964 | – 283624000 | |
| | 2 | 268 | | |
| | 132 | 623200 | | |
| | 2 | 8196 | | |
| | 134 | 631396 | | |
| | 2 | 8232 | | |
| | 1360 | 63962800 | | |
| | 6 | | | |
| | 1366 | | | |
| | 6 | | | |
| | 1372 | | | |
| | 6 | | | |
| | 13780 | | | |

Its root lies between 0 and 10.

Clearly $f_1(2) = -$ve, $f_1(3) = +$ve

∴ the root of (*ii*) lies between 2 and 3 *i.e.*, first figure after decimal is 2.

*Step III.* Diminish the roots of $f_1(x) = 0$ by 2 so that the next transformed equation is

$$x^3 + 136x^2 + 6232x - 4072 = 0 \qquad \ldots(iii)$$

Its root lies between 0 and 1. (We draw the second zig-zag line above the set of figures 136, 6232, –4072). Multiply the roots of (*iii*), by 10, *i.e.* attach one zero to second term, two zeros to third term and three zeros to the fourth term. Then the new equation is

$$f_2(x) = x^3 + 1360x^2 + 623200x - 4072000 = 0$$

Its root lies between 0 and 10, which is nearly $= \dfrac{4072000}{623200} = 6$

Hence second figure after decimal place is 6.

*Step IV.* Diminish the roots of $f_2(x) = 0$ by 6, so that the transformed equation is

$$x^3 + 1378x^2 + 639628x - 283624 = 0.$$

Its root lies between 0 and 1. (We draw the third zig-zag line above the set of figures 1378, 639628, – 283624.) As before multiply its roots by 10, *i.e.* attach one zero to the second term, two zeros to the third term and three zeros to the fourth term. Then the equation becomes

$$f_3(x) = x^3 + 13780x^2 + 63962800x - 283624000 = 0$$

Its root lies between 0 and 10, which is nearly = $\dfrac{283624000}{63962800}$ = 4. Thus the roots of $f_3(x) = 0$ are to be diminished by 4 *i.e.* the third figure after decimal place is 4. But there is no need to proceed further as the root is required correct to three decimal places only. Hence the root is 4.264.

**Obs. 1.** *After two steps of diminishing, we apply the* **principle of trial divisor** *in which we divide the last coefficient by last but one coefficient to get the next integer by which the roots are to be diminished. These last two coefficients should have opposite signs.*

**Obs. 2.** *At any stage if the trial divisor suggests the next integer to be zero, then we should again multiply the roots by 10 and write zero in decimal place of the root.*

**Example 28.13.** *Find the cube root of 30 correct to 3 decimal places, using Horner's method.*

**Solution.** *Step I.* Let $x = \sqrt[3]{30}$ *i.e.* $f(x) = x^3 - 30 = 0$

Now $f(3) = -3$ (–ve), $f(4) = 34$ (+ve)

$\therefore$ the root lies between 3 and 4.

*Step II.* Diminish the roots of the given equation by 3 so that the transformed equation is

$$x^3 + 9x^2 + 27x - 3 = 0 \qquad \ldots(i)$$

Its roots lies between 0 and 1. (We draw a zig-zag line above the set of numbers 9, 27, –3 which are the coefficients of the terms in Eqn. (*i*). Now multiply the roots of Eqn. (*i*) by 10 for which attach one zero to the second term, two zeros to the third term and three zeros to the fourth term. Then we get the equation

$$f_1(x) = x^3 + 90x^2 + 2700x - 3000 = 0 \qquad \ldots(ii)$$

Its roots lies between 0 and 10.

Clearly $f_1(1) = -$ ve, $f_2(2) = +$ ve

$\therefore$ the root of Eqn. (*ii*) lies between 1 and 2 *i.e.*, first figure after decimal place is 1.

*Step III.* Diminish the roots of $f_1(x) = 0$ by 1, so that the next transformed equation is

$$x^3 + 93x^2 + 2883x - 209 = 0 \qquad \ldots(iii)$$

Its root lies between 0 and 1. (We draw a second zig-zag line above the set of figures 93, 2883, –209). Multiply the roots of Eqn. (*iii*) by 10 *i.e.*, attach one zero to second term, two zeros to third term and three zeros to the fourth term. Then the new equation is

$$f_2(x) = x^3 + 930x^2 + 288300x - 209000 = 0$$

Its root lies between 0 and 10, which is nearly

$$= 209000/288300 = 0.724 > 0 \text{ and } < 1.$$

Hence second figure after decimal place is 0.

| | | | | |
|---|---|---|---|---|
| 1 | 0 | 0 | –30 | (3.107 |
| | 3 | 9 | 27 | |
| | 3 | 9 | –30000 | |
| | 3 | 18 | 2791 | |
| | 6 | 2700 | –209000000 | |
| | 3 | 91 | | |
| | 90 | 2791 | | |
| | 1 | 92 | | |
| | 91 | 28830000 | | |
| | 1 | | | |
| | 92 | | | |
| | 1 | | | |
| | 9300 | | | |

*Step IV.* Diminish the root of $f_2(x) = 0$ by 0 and then multiply its roots by 10 so that

$$f_3(x) = x^3 + 9300x^2 + 28830000x - 209000000 = 0.$$

Its root lies between 0 and 10, which is nearly = 209000000/28830000 = 7.2 > 7 and < 8. Thus th
$f_3(x) = 0$ are to be diminished by 7, *i.e.*, the third figure after decimal is 7. Hence th

## PROBLEMS 28.2

1. Find by Horner's method, the root (correct to three decimal places) of the equation
   (i) $x^3 - 3x + 1 = 0$ which lies between 1 and 2
   (ii) $x^3 + x - 1 = 0$
   (iii) $x^3 - 6x - 13 = 0$
   (iv) $x^3 - 3x^2 + 2.5 = 0$ which lies between 1 and 2.
2. Using Horner's method, find the largest real root of $x^3 - 4x + 2 = 0$ correct to three decimal places.
3. Show that the root of the equation $x^4 + x^3 - 4x^2 - 16 = 0$ lies between 2 and 3. Find its value correct to two decimal places by Horner's method.
4. Find the negative root of the equation $x^3 - 9x^2 + 18 = 0$ correct to two decimal places by Horner's method.
5. Find the cube root of 25 by Horner's method correct to 3 decimal places.

## 28.5 SOLUTION OF LINEAR SIMULTANEOUS EQUATIONS

Simultaneous linear equations occur in various engineering problems. The student knows that a given system of linear equations can be solved by Cramer's rule or by Matrix method (§ 2.10). But these methods become tedious for large systems. However, there exist other numerical methods of solution which are well-suited for computing machines. We now explain some direct and iterative methods of solution.

## 28.6 DIRECT METHODS OF SOLUTION

(1) **Gauss elimination method.** In this method, the unknowns are eliminated successively and the system is reduced to an upper triangular system from which the unknowns are found by back substitution. The method is quite general and is well-adapted for computer operations. Here we shall explain it by considering a system of three equations for the sake of clarity.

Consider the equations

$$\left.\begin{aligned} a_1x + b_1y + c_1z &= d_1 \\ a_2x + b_2y + c_2z &= d_2 \\ a_3x + b_3y + c_3z &= d_3 \end{aligned}\right\} \quad \text{...(1)}$$

*Step I. To eliminate x from second and third rows.*

Assuming $a_1 \neq 0$, we eliminate $x$ from the second row by subtracting $(a_2/a_1)$ times the first row from the second row. Similarly we eliminate $x$ from the third row by eliminating $(a_3/a_1)$ times the first row from the third row. We thus, get the new system

$$\left.\begin{aligned} a_1x + b_1y + c_1z &= d_1 \\ b_2'y + c_2'z &= d_2' \\ b_3'y + c_3'z &= d_3' \end{aligned}\right\} \quad \text{...(2)}$$

Here the (1) equation is called the *pivotal equation* and $a_1$ is called the *first pivot.*

*Step II. To eliminate y from third row in (2).*

Assuming $b_2' \neq 0$, we eliminate $y$ from the third row of (2), by subtracting $(b_3'/b_2')$ times the second row from the third row. We thus, get the new system

$$\left.\begin{aligned} a_1x + b_1y + c_1z &= d_1 \\ b_2'y + c_2'z &= d_2' \\ c_3''z &= d_3'' \end{aligned}\right\} \quad \text{...(3)}$$

Here the (2) equation is the *pivotal equation* and $b_2'$ is the *new pivot.*

*Step III. To evaluate the unknowns.*

The values of $x, y, z$ are found from the reduced system (3) by back substitution.

**Obs. 1.** On writing the given equations as $\begin{bmatrix} a_1 & b_1 & c_1 \\ a_2 & b_2 & c_2 \\ a_3 & b_3 & c_3 \end{bmatrix}\begin{bmatrix} x \\ y \\ z \end{bmatrix} = \begin{bmatrix} d_1 \\ d_2 \\ d_3 \end{bmatrix}$ *i.e.*, $AX = D$, this method consists in ***transforming the coefficient matrix A to upper triangular matrix*** by elementary row transformations only.

**Obs. 2.** Clearly the method will fail if any one of the pivots $a_1$, $b_2'$ or $c_3''$ becomes zero. In such cases, we rewrite the equations in a different order so that the pivots are non-zero.

**Obs. 3.** *Partial and complete pivoting.* In the first step, the numerically largest coefficient of $x$ is chosen from all the equations and brought as the first pivot by interchanging the first equation with the equation having the largest coefficient of $x$. In the second step, the numerically largest coefficient of $y$ is chosen from the remaining equations (leaving the first equation) and brought as the *second pivot* by interchanging the second equation with the equation having the largest coefficient of $y'$. This process is continued till we arrive at the equation with the single variable. This modified procedure is called *partial pivoting.* (*U.P.T.U., 2012*)

If we are not keen about the elimination of $x, y, z$ in a specified order, then we choose at each stage the numerically largest coefficient of the entire matrix of coefficients. This requires not only an interchange of equations but also an interchange of the position of the variables. This method of elimination is called *complete pivoting.* It is more complicated and does not appreciably improve the accuracy.

**Example 28.14.** *Apply Gauss elimination method to solve the equations $x + 4y - z = -5$ ; $x + y - 6z = -12$ ; $3x - y - z = 4$.*

**Solution.** We have

| | Check sum | |
|---|---|---|
| $x + 4y - z = -5$ | $-1$ | ...(*i*) |
| $x + y - 6z = -12$ | $-16$ | ...(*ii*) |
| $3x - y - z = 4$ | $5$ | ...(*iii*) |

*Step I.* Operate Eqns. (*ii*) – (*i*) and Eqns. (*iii*) – 3(*i*) to eliminate $x$ :

| | Check sum | |
|---|---|---|
| $-3y - 5z = -7$ | $-15$ | ...(*iv*) |
| $-13y + 2z = 19$ | $8$ | ...(*v*) |

*Step II.* Operate Eqn. (*v*) $-\frac{13}{3}$ Eqn. (*iv*) to eliminate $y$ :

| | Check sum | |
|---|---|---|
| $\frac{71}{3} z = \frac{148}{3}$ | $73$ | ...(*vi*) |

*Step III.* By back-substitution, we get

From Eqn. (*vi*) : $z = \frac{148}{71} = 2.0845$

From Eqn. (*iv*) : $y = \frac{7}{3} - \frac{5}{3}\left(\frac{148}{71}\right) = -\frac{81}{71} = -1.1408$

From Eqn. (*i*) : $x = -5 - 4\left(-\frac{81}{71}\right) + \frac{148}{71} = \frac{117}{71} = 1.6479$

Hence $x = 1.6479, y = -1.1408, z = 2.0845$

**Note.** A useful check is provided by noting the sum of the coefficients and terms on the right, operating on those numbers as on the equations and checking that the derived equations have the correct sum.

**Otherwise :** We have $\begin{bmatrix} 1 & 4 & -1 \\ 1 & 1 & -6 \\ 3 & -1 & -1 \end{bmatrix}\begin{bmatrix} x \\ y \\ z \end{bmatrix} = \begin{bmatrix} -5 \\ -12 \\ 4 \end{bmatrix}$

Operate $R_2 - R_1$ and $R_3 - 3R_1$, $\begin{bmatrix} 1 & 4 & -1 \\ 0 & -3 & -5 \\ 0 & -13 & 2 \end{bmatrix}\begin{bmatrix} x \\ y \\ z \end{bmatrix} = \begin{bmatrix} -5 \\ -7 \\ 19 \end{bmatrix}$

Operate $R_3 - \frac{13}{3}R_2$, $\begin{bmatrix} 1 & 4 & -1 \\ 0 & -3 & -5 \\ 0 & 0 & 71/3 \end{bmatrix}\begin{bmatrix} x \\ y \\ z \end{bmatrix} = \begin{bmatrix} -5 \\ -7 \\ 148/3 \end{bmatrix}$

Thus, we have $z = 148/71 = 2.0845$,

$$3y = 7 - 5z = 7 - 10.4225 = -3.4225 \quad i.e., \quad y = -1.1408$$

and

$$x = -5 - 4y + z = -5 + 4(1.1408) + 2.0845 = 1.6479$$

Hence $x = 1.6479, y = -1.1408, z = 2.0845$.

**Example 28.15.** *Solve $10x - 7y + 3z + 5u = 6$, $-6x + 8y - z - 4u = 5$, $3x + y + 4z + 11u = 2$, $5x - 9y - 2z + 4u = 7$ by Gauss elimination method.*

**Solution.** We have

| | *Check sum* | |
|---|---|---|
| $10x - 7y + 3z + 5u = 6$ | 17 | ...(*i*) |
| $-6x + 8y - z - 4u = 5$ | 2 | ...(*ii*) |
| $3x + y + 4z + 11u = 2$ | 21 | ...(*iii*) |
| $5x - 9y - 2z + 4u = 7$ | 5 | ...(*iv*) |

*Step I.* To eliminate $x$, operate $\left[(ii) - \left(\frac{-6}{10}\right)(i)\right], \left[(iii) - \frac{3}{10}(i)\right], \left[(iv) - \frac{5}{10}(i)\right]$:

| | *Check sum* | |
|---|---|---|
| $3.8y + 0.8z - u = 8.6$ | 12.2 | ...(*v*) |
| $3.1y + 3.1z + 9.5u = 0.2$ | 15.9 | ...(*vi*) |
| $-5.5y - 3.5z + 1.5u = 4$ | $-3.5$ | ...(*vii*) |

*Step II.* To eliminate $y$, operate $\left[(vi) - \frac{3.1}{3.8}(v)\right], \left[(vii) - \left(\frac{-5.5}{3.8}\right)(v)\right]$:

$$2.4473684z + 10.315789u = -6.8157895 \qquad ...(viii)$$

$$-2.3421053z + 0.0526315u = 16.447368 \qquad ...(ix)$$

*Step III.* To eliminate $z$, operate $\left[(ix) - \left(\frac{-2.3421053}{2.4473684}\right)(viii)\right]$:

$$9.9249319u = 9.9245977$$

*Step IV.* By back-substitution, we get

$$u = 1, z = -7, y = 4 \text{ and } x = 5.$$

**(2) Gauss-Jordan method.** This is a modification of the Gauss elimination method. In this method, elimination of unknowns is performed not in the equations below but in the equations above also, ultimately **reducing the system to a diagonal matrix form *i.e.*,** *each equation involving only one unknown. From these equations the unknowns x, y, z can be obtained readily.*

Thus in this method, the labour of back-substitution for finding the unknowns is saved at the cost of additional calculations.

**Example 28.16.** *Apply Gauss-Jordan method to solve the equations*

$$x + y + z = 9 \; ; \; 2x - 3y + 4z = 13 \; ; \; 3x + 4y + 5z = 40.$$

**Solution.** We have

$$x + y + z = 9 \qquad ...(i)$$

$$2x - 3y + 4z = 13 \qquad ...(ii)$$

$$3x + 4y + 5z = 40 \qquad ...(iii)$$

*Step I.* Operate Eqns. (*ii*) – 2(*i*) and Eqns. (*iii*) – 3(*i*) to eliminate $x$ from Eqns. (*ii*) and (*iii*).

$$x + y + z = 9 \qquad ...(iv)$$

$$-5y + 2z = -5 \qquad ...(v)$$

$$y + 2z = 13 \qquad ...(vi)$$

*Step II.* Operate Eqns. (*iv*) + $\frac{1}{5}$ (*v*) and Eqns. (*vi*) + $\frac{1}{5}$ (*v*) to eliminate $y$ from Eqns. (*iv*) and (*vi*) :

$$x + \frac{7}{5}z = 8 \qquad ...(vii)$$

$$-5y + 2z = -5 \qquad ...(viii)$$

$$\frac{12}{5} z = 12 \qquad ...(ix)$$

*Step III.* Operate Eqns. $(vii) - \frac{7}{12}(ix)$ and Eqns. $(viii) - \frac{5}{6}(ix)$ to eliminate $z$ from Eqns. $(vii)$ and $(viii)$ :

$$x = 1$$
$$-5y = -15$$
$$\frac{12}{5} z = 12$$

Hence the solution is $x = 1, y = 3, z = 5$.

**Otherwise :** Rewriting the equations as $\begin{bmatrix} 1 & 1 & 1 \\ 2 & -3 & 4 \\ 3 & 4 & 5 \end{bmatrix} \begin{bmatrix} x \\ y \\ z \end{bmatrix} = \begin{bmatrix} 9 \\ 13 \\ 40 \end{bmatrix}$

Operate $R_2 - 2R_1, R_3 - 3R_1$, $\begin{bmatrix} 1 & 1 & 1 \\ 0 & -5 & 2 \\ 0 & 1 & 2 \end{bmatrix} \begin{bmatrix} x \\ y \\ z \end{bmatrix} = \begin{bmatrix} 9 \\ -5 \\ 13 \end{bmatrix}$

Operate $R_3 + \frac{1}{5} R_2$, $\begin{bmatrix} 1 & 1 & 1 \\ 0 & -5 & 2 \\ 0 & 0 & 12/5 \end{bmatrix} \begin{bmatrix} x \\ y \\ z \end{bmatrix} = \begin{bmatrix} 9 \\ -5 \\ 12 \end{bmatrix}$

Operate $-R_2, 5R_3$ $\begin{bmatrix} 1 & 1 & 1 \\ 0 & 5 & -2 \\ 0 & 0 & 12 \end{bmatrix} \begin{bmatrix} x \\ y \\ z \end{bmatrix} = \begin{bmatrix} 9 \\ 5 \\ 60 \end{bmatrix}$

Operate $R_2 + \frac{1}{6} R_3, \frac{1}{12} R_3$ $\begin{bmatrix} 1 & 1 & 1 \\ 0 & 5 & 0 \\ 0 & 0 & 1 \end{bmatrix} \begin{bmatrix} x \\ y \\ z \end{bmatrix} = \begin{bmatrix} 9 \\ 15 \\ 5 \end{bmatrix}$

Operate $\frac{1}{5} R_2$ $\begin{bmatrix} 1 & 1 & 1 \\ 0 & 1 & 0 \\ 0 & 0 & 1 \end{bmatrix} \begin{bmatrix} x \\ y \\ z \end{bmatrix} = \begin{bmatrix} 9 \\ 3 \\ 5 \end{bmatrix}$

Operate $R_1 - R_2 - R_3$ $\begin{bmatrix} 1 & 0 & 0 \\ 0 & 1 & 0 \\ 0 & 0 & 1 \end{bmatrix} \begin{bmatrix} x \\ y \\ z \end{bmatrix} = \begin{bmatrix} 1 \\ 3 \\ 5 \end{bmatrix}$

Hence, $x = 1, y = 3, z = 5$.

**Obs.** Here the process of elimination of variables amounts to *reducing the given coefficient metric to a* **diagonal matrix** *by elementary row transformations only.*

**Example 28.17.** *Solve the equations of example 28.15, by Gauss-Jordan method.*

**Solution.** We have

$$10x - 7y + 3z + 5u = 6 \qquad ...(i)$$
$$-6x + 8y - z - 4u = 5 \qquad ...(ii)$$
$$3x + y + 4z + 11u = 2 \qquad ...(iii)$$
$$5x - 9y - 2z + 4u = 7 \qquad ...(iv)$$

*Step I.* To eliminate $x$, operate $\left[(ii) - \left(\frac{-6}{10}\right)(i)\right], \left[(iii) - \left(\frac{3}{10}\right)(i)\right], \left[(iv) - \left(\frac{5}{10}\right)(i)\right]$ :

$$10x - 7y + 3z + 5u = 6 \qquad ...(v)$$
$$3.8y + 0.8z - u = 8.6 \qquad ...(vi)$$
$$3.1y + 3.1z + 9.5u = 0.2 \qquad ...(vii)$$
$$-5.5y - 3.5z + 1.5u = 4 \qquad ...(viii)$$

*Step II.* To eliminate $y$, operate $\left[(v) - \left(\frac{-7}{3.8}\right)(vi)\right], \left[(vii) - \left(\frac{3.1}{3.8}\right)(vi)\right], \left[(viii) - \left(\frac{-5.5}{3.8}\right)(vi)\right]$ :

$$10x + 4.4736842z + 3.1578947u = 21.842105 \quad ...(ix)$$

$$3.8y + 0.8z - u = 8.6 \quad ...(x)$$

$$2.4473684z + 10.315789u = -6.8157895 \quad ...(xi)$$

$$-2.3421053x + 0.0526315u = 16.447368 \quad ...(xii)$$

*Step III*. To eliminate $z$, operate $\left[(ix) - \left(\dfrac{4.473684}{2.4473684}\right)(xi)\right]$,

$$\left[(x) - \left(\frac{0.8}{2.4473684}\right)(xi)\right], \left[(xii) - \left(\frac{-2.3421053}{2.4473684}\right)(xi)\right]:$$

$$10x - 15.698923u = 34.301075$$

$$3.8y - 4.3720429u = 10.827957$$

$$2.4473684z + 10.315789u = -6.8157895$$

$$9.9247309u = 9.9245975$$

*Step IV*. From the last equation $u = 1$ nearly.

Substitution of $u = 1$ in the above three equations gives $x = 5, y = 4, z = -7$.

**(3) Factorization method*.** This method is based on the fact that every matrix $A$ can be expressed as the product of a lower triangular matrix and an upper triangular matrix, provided all the principal minors of $A$ are non-singular, *i.e.*, if $A = [a_{ij}]$, then

$$a_{11} \neq 0, \begin{vmatrix} a_{11} & a_{12} \\ a_{21} & a_{22} \end{vmatrix} \neq 0, \begin{vmatrix} a_{11} & a_{12} & a_{13} \\ a_{21} & a_{22} & a_{23} \\ a_{31} & a_{32} & a_{33} \end{vmatrix} \neq 0, \text{ etc.}$$

Also such a factorization if it exists, is unique.

Now consider the equations

$$a_{11}x_1 + a_{12}x_2 + a_{13}x_3 = b_1$$

$$a_{21}x_1 + a_{22}x_2 + a_{23}x_3 = b_2$$

$$a_{31}x_1 + a_{32}x_2 + a_{33}x_3 = b_3$$

which can be written as $\quad AX = B \quad ...(1)$

where $$A = \begin{bmatrix} a_{11} & a_{12} & a_{13} \\ a_{21} & a_{22} & a_{23} \\ a_{31} & a_{32} & a_{33} \end{bmatrix}, X = \begin{bmatrix} x_1 \\ x_2 \\ x_3 \end{bmatrix} \text{ and } B = \begin{bmatrix} b_1 \\ b_2 \\ b_3 \end{bmatrix}$$

Let $\quad A = LU, \quad ...(2)$

where $$L = \begin{bmatrix} 1 & 0 & 0 \\ l_{21} & 1 & 0 \\ l_{31} & l_{32} & 1 \end{bmatrix} \text{ and } U = \begin{bmatrix} u_{11} & u_{12} & u_{13} \\ 0 & u_{22} & u_{23} \\ 0 & 0 & u_{33} \end{bmatrix}$$

Then Eqn. (1) becomes $LUX = B \quad ...(3)$

Writing $\quad UX = V, \quad ...(4)$

Eqn. (3) becomes $\quad LV = B$ which is equivalent to the equations

$$v_1 = b_1 \,;\, l_{21}v_1 + v_2 = b_2 \,;\, l_{31}v_1 + l_{31}v_2 + v_3 = b_3$$

Solving these for $v_1, v_2, v_3$, we know $V$. Then, Eqn. (4) becomes

$$u_{11}x_1 + u_{12}x_2 + u_{13}x_3 = v_1 \,;\, u_{22}x_2 + u_{23}x_3 = v_2 \,;\, u_{33}x_3 = v_3,$$

from which $x_3, x_2$ and $x_1$ can be found by *back-substitution*.

*To compute the matrices L and U,* we write Eqn. (2) as

$$\begin{bmatrix} 1 & 0 & 0 \\ l_{21} & 1 & 0 \\ l_{31} & l_{32} & 0 \end{bmatrix} \begin{bmatrix} u_{11} & u_{12} & u_{13} \\ 0 & u_{22} & u_{23} \\ 0 & 0 & u_{33} \end{bmatrix} = \begin{bmatrix} a_{11} & a_{12} & a_{13} \\ a_{21} & a_{22} & a_{23} \\ a_{31} & a_{32} & a_{33} \end{bmatrix}$$

*Another name given to this discomposition is **Doolittle's method.**

Multiplying the matrices on the left and equating corresponding elements from both sides, we obtain

(i) $u_{11} = a_{11}$, $u_{12} = a_{12}$, $u_{13} = a_{13}$

(ii) $l_{21}u_{11} = a_{21}$ or $l_{21} = a_{21}/a_{11}$

$l_{31}u_{11} = a_{31}$ or $l_{31} = a_{31}/a_{11}$

(iii) $l_{21}u_{12} + u_{22} = a_{22}$ or $u_{22} = a_{22} - \dfrac{a_{21}}{a_{11}} a_{12}$

$l_{21}u_{13} + u_{23} = a_{23}$ or $u_{23} = a_{23} - \dfrac{a_{21}}{a_{11}} a_{13}$

(iv) $l_{31}u_{12} + l_{33}u_{22} = a_{32}$ or $l_{32} = \dfrac{1}{u_{22}}\left[a_{32} - \dfrac{a_{31}}{a_{11}}a_{12}\right]$

(v) $l_{31}u_{13} + l_{33}u_{23} + u_{33} = a_{33}$ which gives $u_{33}$.

*Thus we compute the elements of L and U in the following set order :*

(i) First row of $U$, (ii) First column of $L$,

(iii) Second row of $U$, (iv) Second column of $L$, (v) Third row of $U$.

This procedure can easily be generalised.

**Obs.** This method is superior to Gauss elimination method and is often used for the solution of linear systems and **for finding the inverse of a matrix.** Among the direct methods, Factorization method is also preferred as the software for computers.

**Example 28.18.** *Apply factorization method to solve the equations :*

$$3x + 2y + 7z = 4 \,;\, 2x + 3y + z = 5 \,;\, 3x + 4y + z = 7.$$ *(V.T.U., 2016)*

**Solution.** Let $\begin{bmatrix} 1 & 0 & 0 \\ l_{21} & 1 & 0 \\ l_{31} & l_{32} & 1 \end{bmatrix}\begin{bmatrix} u_{11} & u_{12} & u_{13} \\ 0 & u_{22} & u_{23} \\ 0 & 0 & u_{33} \end{bmatrix} = \begin{bmatrix} 3 & 2 & 7 \\ 2 & 3 & 1 \\ 3 & 4 & 1 \end{bmatrix}$ (*i.e.* $A$),

so that

(i) $u_{11} = 3$, $u_{12} = 2$, $u_{13} = 7$.

(ii) $l_{21}u_{11} = 2$, $\therefore\ l_{21} = 2/3$

$l_{31}u_{11} = 3$, $\therefore\ l_{31} = 1$.

(iii) $l_{21}u_{12} + u_{22} = 3$, $\therefore\ u_{22} = 5/3$,

$l_{21}u_{13} + u_{23} = 1$, $\therefore\ u_{23} = -11/3$.

(iv) $l_{31}u_{12} + l_{32}u_{22} = 4$, $\therefore\ l_{32} = 6/5$.

(v) $l_{31}u_{13} + l_{32}u_{23} + u_{33} = 1$

$\therefore$ $u_{33} = -8/5$

Thus $$A = \begin{bmatrix} 1 & 0 & 0 \\ 2/3 & 1 & 0 \\ 1 & 6/5 & 1 \end{bmatrix}\begin{bmatrix} 3 & 2 & 7 \\ 0 & 5/3 & -11/3 \\ 0 & 0 & -8/5 \end{bmatrix}$$

Writing $UX = V$, the given system becomes $\begin{bmatrix} 1 & 0 & 0 \\ 2/3 & 1 & 0 \\ 1 & 6/5 & 1 \end{bmatrix}\begin{bmatrix} v_1 \\ v_2 \\ v_3 \end{bmatrix} = \begin{bmatrix} 4 \\ 5 \\ 7 \end{bmatrix}$

Solving this system, we have $v_1 = 4$,

$\dfrac{2}{3}v_1 + v_2 = 5$ or $v_2 = \dfrac{7}{3}$

$v_1 + \dfrac{6}{5}v_2 + v_3 = 7$ or $v_3 = \dfrac{1}{5}$

Hence the original system becomes

$$\begin{bmatrix} 3 & 2 & 7 \\ 0 & 5/3 & -11/3 \\ 0 & 0 & -8/5 \end{bmatrix}\begin{bmatrix} x \\ y \\ z \end{bmatrix} = \begin{bmatrix} 4 \\ 7/3 \\ 1/5 \end{bmatrix}$$

*i.e.,* $$3x + 2y + 7z = 4\ ;\ \frac{5}{3}y - \frac{11}{3}z = \frac{7}{3}\ ;\ -\frac{8}{5}z = \frac{1}{5}$$

By back-substitution, we have $z = -1/8$, $y = 9/8$ and $x = 7/8$.

**Example 28.19.** *Solve the equations of Example 28.14 by factorization method.*

**Solution.** Let $\begin{bmatrix} 1 & 0 & 0 & 0 \\ l_{21} & 1 & 0 & 0 \\ l_{31} & l_{32} & 1 & 0 \\ l_{41} & l_{42} & l_{43} & 1 \end{bmatrix}\begin{bmatrix} u_{11} & u_{12} & u_{13} & u_{13} \\ 0 & u_{22} & u_{23} & u_{24} \\ 0 & 0 & u_{33} & u_{34} \\ 0 & 0 & 0 & u_{44} \end{bmatrix} = \begin{bmatrix} 10 & -7 & 3 & 5 \\ -6 & 8 & -1 & -4 \\ 3 & 1 & 4 & 11 \\ 5 & -9 & -2 & 4 \end{bmatrix}$ (*i.e.*, $A$)

so that

(*i*) $R_1$ of $U$ : $u_{11} = 10$, $u_{12} = -7$, $u_{13} = 3$, $u_{14} = 5$

(*ii*) $C_1$ of $L$ : $l_{21} = -0.6$, $l_{31} = 0.3$, $l_{41} = 0.5$

(*iii*) $R_2$ of $U$ : $u_{22} = 3.8$, $u_{23} = 0.8$, $u_{24} = -1$

(*iv*) $C_2$ of $L$ : $l_{32} = 0.81579$, $l_{42} = -1.44737$

(*v*) $R_3$ of $U$ : $u_{33} = 2.44737$, $u_{34} = 10.31579$

(*vi*) $C_3$ of $L$ : $l_{43} = -0.95699$

(*vii*) $R_4$ of $U$ : $u_{44} = 9.92474$

Thus

$$A = \begin{bmatrix} 1 & 0 & 0 & 0 \\ -0.6 & 1 & 0 & 0 \\ 0.3 & 0.81579 & 1 & 0 \\ 0.5 & -1.44737 & -0.95699 & 1 \end{bmatrix}\begin{bmatrix} 10 & -7 & 3 & 5 \\ 0 & 3.8 & 0.8 & -1 \\ 0 & 0 & 2.44737 & 10.31579 \\ 0 & 0 & 0 & 9.92474 \end{bmatrix}$$

Writing $UX = V$, the given system becomes

$$\begin{bmatrix} 1 & 0 & 0 & 0 \\ -0.6 & 1 & 0 & 0 \\ 0.3 & 0.81579 & 1 & 0 \\ 0.5 & -1.44737 & -0.95699 & 1 \end{bmatrix}\begin{bmatrix} v_1 \\ v_2 \\ v_3 \\ v_4 \end{bmatrix} = \begin{bmatrix} 6 \\ 5 \\ 2 \\ 7 \end{bmatrix}$$

Solving this system, we get

$$v_1 = 6,\ v_2 = 8.6,\ v_3 = -6.81579,\ v_4 = 9.92474.$$

Hence the original system becomes

$$\begin{bmatrix} 10 & -7 & 3 & 5 \\ 0 & 3.8 & 0.8 & -1 \\ 0 & 0 & 2.44737 & 10.31579 \\ 0 & 0 & 0 & 9.92474 \end{bmatrix}\begin{bmatrix} x \\ y \\ z \\ u \end{bmatrix} = \begin{bmatrix} 6 \\ 8.6 \\ -6.81579 \\ 9.92474 \end{bmatrix}$$

*i.e.,* $$10x - 7y + 3z + 5u = 6,\quad 3.8y + 0.8z - u = 8.6,$$

$$2.44737z + 10.31579u = -6.81579,\ u = 1.$$

By back-substitution, we get $u = 1$, $z = -7$, $y = 4$, $x = 5$.

## PROBLEMS 28.3

Solve the following equations by Gauss elimination method :

1. $x + y + z = 9, 2x - 3y + 4z = 13, 3x + 4y + 5z = 40.$ *(Anna, 2012)*
2. $3x - y + 2z = 12 ; x + 2y + 3z = 11, 2x - 2y - z = 2.$ *(V.T.U., 2015)*
3. $2x - y + 3z = 9 ; x + y + z = 6 ; x - y + z = 2.$
4. $2x_1 + 4x_2 + x_3 = 3 ; 3x_1 + 2x_2 - 2x_3 = -2 ; x_1 - x_2 + x_3 = 6.$
5. $5x_1 + x_2 + x_3 + x_4 = 4 ; x_1 + 7x_2 + x_3 + x_4 = 12 ;$
   $x_1 + x_2 + 6x_3 + x_4 = -5 ; x_1 + x_2 + x_3 + 4x_4 = -6.$ *(JNTUK, 2022)*

Solve the following equations by Gauss-Jordan method :

6. $2x + 5y + 7z = 52 ; 2x + y - z = 0 ; x + y + z = 9.$ *(V.T.U., 2016)*
7. $10x + y + z = 12 ; 2x + 10y + z = 13 ; x + y + 5z = 7.$ *(Anna., 2016)*
8. $x + 3y + 3z = 16 ; z + 4y + 3z = 18 ; x + 3y + 4z = 19.$ *(Andhra., 2015)*
9. $2x + y + z = 10 ; 3x + 2y + 3z = 18 ; x + 4y + 9z = 16.$ *(V.T.U., 2015)*
10. $2x_1 + x_2 + 5x_3 + x_4 = 5 ; x_1 + x_2 - 3x_3 + 4x_4 = -1 ;$
    $3x_1 + 6x_2 - 2x_3 + x_4 = 8 ; 2x_1 + 2x_2 + 2x_3 - 3x_4 = 2.$

Solve the following equations by factorization method :

11. $10x + y + z = 12 ; 2x + 10y + z = 13 ; 2x + 2y + 10z = 14.$
12. $x + 2y + 3z = 14 ; 2x + 3y + 4z = 20 ; 3x + 4y + z = 14.$
13. $2x + y + 4z = 12 ; 8x - 3y + 2z = 20 ; 4x + 11y - z = 33.$
    *(V.T.U., 2015)*
14. $2x_1 - x_2 + x_3 = -1 ; 2x_2 - x_3 + x_4 = 1 ; x_1 + 2x_3 - x_4 = -1 ; x_1 + x_2 + 2x_4 = 5.$
15. Find the inverse of the matrix $\begin{bmatrix} 1 & -1 & 1 \\ 1 & -2 & 4 \\ 1 & 2 & 2 \end{bmatrix}$ by Factorization method.

## 28.7 ITERATIVE METHODS OF SOLUTION

The preceding methods of solving simultaneous linear equations are known as *direct methods* as they yield exact solutions. On the other hand, an iterative method is that in which we start from an approximation to the true solution and obtain better and better approximations from a computation cycle repeated as often as may be necessary for achieving a desired accuracy.

Simple iteration methods can be devised for systems in which the coefficients of the leading diagonal are large compared to others. We now explain three such methods :

(1) **Jacobi's iteration method.** Consider the equations

$$\left.\begin{aligned} a_1x + b_1y + c_1z &= d_1 \\ a_2x + b_2y + c_2z &= d_2 \\ a_3x + b_3y + c_3z &= d_3 \end{aligned}\right\} \quad \text{...(1)}$$

If $a_1, b_2, c_3$ are large as compared to other coefficients, then solving these for $x, y, z$ respectively, the system can be written in the form

$$\left.\begin{aligned} x &= k_1 - l_1y - m_1z \\ y &= k_2 - l_2x - m_2z \\ z &= k_3 - l_3x - m_3y \end{aligned}\right\} \quad \text{...(2)}$$

Let us start with the initial approximations $x_0, y_0, z_0$ (each = 0) for the values of $x, y, z$. Substituting these on the right, we get the first approximations $x_1 = k_1, y_1 = k_2, z_1 = k_3$.

Substituting these on the right-hand sides of Eqn. (2), the second approximations are given by

$$x_2 = k_1 - l_1y_1 - m_1z_1$$
$$y_2 = k_2 - l_2x_1 - m_2z_1$$
$$z_2 = k_3 - l_3x_1 - m_3y_1$$

This process is repeated till the difference between two consecutive approximations is negligible.

**Example 28.20.** *Solve by Jacobi's iteration method, the equations $10x + y - z = 11.19$, $x + 10y + z = 28.08$, $-x + y + 10z = 35.61$, correct to two decimal places.*

**Solution.** Rewriting the given equations as

$$x = \frac{1}{10}(11.19 - y + z),\ y = \frac{1}{10}(28.08 - x - z),\ z = \frac{1}{10}(35.61 + x - y)$$

We start from an approximation, $x_0 = y_0 = z_0 = 0$.

*First iteration* $\quad x_1 = \dfrac{11.19}{10} = 1.119,\ y_1 = \dfrac{28.08}{10} = 2.808,\ z_1 = \dfrac{35.61}{10} = 3.561$

*Second iteration* $\quad x_2 = \dfrac{1}{10}(11.19 - y_1 + z_1) = 1.19$

$$y_2 = \frac{1}{10}(28.08 - x_1 - z_1) = 2.24$$

$$z_2 = \frac{1}{10}(35.61 + x_1 - y_1) = 3.39$$

*Third iteration* $\quad x_3 = \dfrac{1}{10}(11.19 - y_2 + z_2) = 1.22$

$$y_3 = \frac{1}{10}(28.03 - x_2 - z_2) = 2.35$$

$$z_3 = \frac{1}{10}(35.61 + x_2 - y_2) = 3.45$$

*Fourth iteration* $\quad x_4 = \dfrac{1}{10}(11.19 - y_3 + z_3) = 1.23$

$$y_4 = \frac{1}{10}(28.03 - x_3 - z_3) = 2.34$$

$$z_4 = \frac{1}{10}(35.61 + x_3 - y_3) = 3.45$$

*Fifth iteration* $\quad x_5 = \dfrac{1}{10}(11.19 - y_4 + z_4) = 1.23$

$$y_5 = \frac{1}{10}(28.08 - x_4 - z_4) = 2.34$$

$$z_5 = \frac{1}{10}(35.61 + x_4 - y_4) = 3.45$$

Hence $x = 1.23$, $y = 2.34$, $z = 3.45$.

**Example 28.21.** *Solve, by Jacobi's iteration method, the equations*

*$20x + y - 2z = 17$ ; $3x + 20y - z = -18$ ; $2x - 3y + 20z = 25$.* (*Anna, 2016*)

**Solution.** We write the given equations in the form

$$\left.\begin{aligned} x &= \frac{1}{20}(17 - y + 2z) \\ y &= \frac{1}{20}(-18 - 3x + z) \\ z &= \frac{1}{20}(25 - 2x + 3y) \end{aligned}\right\} \quad ...(i)$$

We start from an approximation $x_0 = y_0 = z_0 = 0$.

Substituting these on the right sides of the equations (*i*), we get

$$x_1 = \frac{17}{20} = 0.85\ ;\ y_1 = -\frac{18}{20} = -0.9\ ;\ z_1 = \frac{25}{20} = 1.25$$

Putting these values on the right of the Eqn. (*i*), we obtain

$$x_2 = \frac{1}{20}(17 - y_1 + 2z_1) = 1.02$$

$$y_2 = \frac{1}{20}(-18 - 3x_1 + z_1) = -0.965$$

$$z_2 = \frac{1}{20}(25 - 2x_1 + 3y_1) = 1.1515$$

Substituting these values in the right sides of the equations (*i*), we have

$$x_3 = \frac{1}{20}(17 - y_2 + 2z_2) = 1.0134$$

$$y_3 = \frac{1}{20}(-18 - 3x_2 + z_2) = -0.9954$$

$$z_3 = \frac{1}{20}(25 - 2x_2 + 3y_2) = 1.0032$$

Substituting these values, we get

$$x_4 = \frac{1}{20}(17 - y_3 + 2z_3) = 1.0009$$

$$y_4 = \frac{1}{20}(-18 - 3x_3 + z_3) = -1.0018$$

$$z_4 = \frac{1}{20}(25 - 2x_3 + 3y_3) = 0.9993$$

Putting these values, we have

$$x_5 = \frac{1}{20}(17 - y_4 + 2z_4) = 1.0000$$

$$y_5 = \frac{1}{20}(-18 - 3x_4 + z_4) = -1.0002$$

$$z_5 = \frac{1}{20}(25 - 2x_4 + 3y_4) = 0.9996$$

Again substituting these values, we get

$$x_6 = \frac{1}{20}(17 - y_5 + 2z_5) = 1.0000$$

$$y_6 = \frac{1}{20}(-18 - 3x_5 + z_5) = -1.0000$$

$$z_6 = \frac{1}{20}(25 - 2x_5 + 3y_5) = 1.0000$$

The values in the 5th and 6th iterations being practically the same, we can stop.

Hence the solution is $x = 1, y = -1, z = 1$.

(2) **Gauss-Seidel iteration method*.** This is a modification of the Jacobi's iteration method. As before, we start with initial approximations $x_0, y_0, z_0$ (each = 0) for $x, y, z$ respectively. Substituting $y = y_0, z = z_0$ in the first of the Eqn. (2) on page 956, we get

$$x_1 = k_1$$

Then putting $x = x_1, z = z_0$ in the second of the Eqn. (2) on page 956, we have

$$y_1 = k_2 - l_2 x_1 - m_2 z_0$$

Next substituting $x = x_1, y = y_1$ in the third of the Eqn. (2) on page 956, we obtain

$$z_1 = k_3 - l_3 x_1 - m_3 y_1$$

and so on, *i.e.*, as soon as new approximation for an unknown is found, it is immediately used in the next step.

This process of iteration is continued till convergency to the desired degree of accuracy is obtained.

**Obs 1.** Since the most recent approximation of the unknowns are used while proceeding to the next step, the convergence in the Gauss-Seidel method is faster than in Jacobi's method.

**Obs 2.** Gauss-Sedial method converges if in each equation, the absolute value of the largest coefficient is greater than the sum of the absolute values of the remaining coefficients.

**After Philipp Ludwig Von Seidel* (1821–1896) who also suggested a similar method.

**Example 28.22.** *Apply Gauss-Seidel iteration method to solve the equations of Ex. 28.21.*
*(Anna., 2016 ; V.T.U., 2016)*

**Solution.** We write the given equation in the form

$$x = \frac{1}{20}(17 - y + 2z)\ ;\ y = \frac{1}{20}(-18 - 3x + z)\ ;\ z = \frac{1}{20}(25 - 2x + 3y) \qquad ...(i)$$

We start from the approximation $x_0 = y_0 = z_0 = 0$. Substituting $y = y_0$, $z = z_0$ in the right side of the first of equations (*i*), we get

$$x_1 = \frac{1}{20}(17 - y_0 + 2z_0) = 0.8500$$

Putting $x = x_1$, $z = z_0$ in the second of the Eqn. (*i*), we have

$$y_1 = \frac{1}{20}(-18 - 3x_1 + z_0) = -1.0275$$

Putting $x = x_1$, $y = y_1$ in the last of the Eqn. (*i*), we obtain

$$z_1 = \frac{1}{20}(25 - 2x_1 + 3y_1) = 1.0109$$

For the second iteration, we have

$$x_2 = \frac{1}{20}(17 - y_1 + 2z_1) = 1.0025$$

$$y_2 = \frac{1}{20}(-18 - 3x_2 + z_1) = -0.9998$$

$$z_2 = \frac{1}{20}(25 - 2x_2 + 3y_2) = 0.9998$$

For the third iteration, we get

$$x_3 = \frac{1}{20}(17 - y_2 + 2z_2) = 1.0000$$

$$y_3 = \frac{1}{20}(-18 - 3x_3 + z_2) = -1.0000$$

$$z_3 = \frac{1}{20}(25 - 3x_3 + 2y_3) = 1.0000$$

The values in the 2nd and 3rd iterations being practically the same, we can stop.

Hence the solution is $x = 1, y = -1, z = 1$.

**Example 28.23.** *Solve the equations :*

$$10x_1 - 2x_2 - x_3 - x_4 = 3$$
$$-2x_1 + 10x_2 - x_3 - x_4 = 15$$
$$-x_1 - x_2 + 10x_3 - 2x_4 = 27$$
$$-x_1 - x_2 - 2x_3 + 10x_4 = -9$$

*by Gauss-Seidal iteration method.* *(JNTUK, 2022)*

**Solution.** Rewriting the given equations as

$$x_1 = 0.3 + 0.2x_2 + 0.1x_3 + 0.1x_4 \qquad ...(i)$$
$$x_2 = 1.5 + 0.2x_1 + 0.1x_3 + 0.1x_4 \qquad ...(ii)$$
$$x_3 = 2.7 + 0.1x_1 + 0.1x_2 + 0.2x_4 \qquad ...(iii)$$
$$x_4 = -0.9 + 0.1x_1 + 0.1x_2 + 0.2x_3 \qquad ...(iv)$$

*First iteration*

Putting $x_2 = 0, x_3 = 0, x_4 = 0$ in Eqn.(*i*), we get $x_1 = 0.3$

Putting $x_1 = 0.3, x_3 = 0, x_4 = 0$ in Eqn.(*ii*), we obtain $x_2 = 1.56$

Putting $x_1 = 0.3, x_2 = 1.56, x_4 = 0$ in Eqn.(*iii*), we obtain $x_3 = 2.886$

Putting $x_1 = 0.3, x_2 = 1.56, x_3 = 2.886$ in Eqn.(*iv*), we get $x_4 = -0.1368$

*Second iteration*

Putting $x_2 = 1.56$, $x_3 = 2.886$, $x_4 = -0.1368$ in Eqn.(*i*), we obtain

$x_1 = 0.8869$

Putting $x_1 = 0.8869$, $x_3 = 2.886$, $x_4 = -0.1368$ in Eqn.(*ii*), we obtain

$x_2 = 1.9523$

Putting $x_1 = 0.8869$, $x_2 = 1.9523$, $x_4 = -0.1368$ in Eqn.(*iii*), we have

$x_3 = 2.9566$

Putting $x_1 = 0.8869$, $x_2 = 1.9523$, $x_3 = 2.9566$ in Eqn.(*iv*), we get

$x_4 = -0.0248.$

*Third iteration*

Putting $x_2 = 1.9523$, $x_3 = 2.9566$, $x_4 = -0.0248$ in Eqn.(*i*), we obtain

$x_1 = 0.9836$

Putting $x_1 = 0.9836$, $x_3 = 2.9566$, $x_4 = -0.0248$ in Eqn.(*ii*), we obtain

$x_2 = 1.9899$

Putting $x_1 = 0.9836$, $x_2 = 1.9899$, $x_4 = -0.0248$ in Eqn.(*iii*), we get

$x_3 = 2.9924$

Putting $x_1 = 0.9836$, $x_1 = 1.9899$, $x_3 = 2.9924$ in Eqn.(*iv*), we get

$x_4 = -0.0042.$

*Fourth iteration.* Proceeding as above

$$x_1 = 0.9968,\ x_2 = 1.9982,\ x_3 = 2.9987,\ x_4 = -0.0008.$$

*Fifth iteration is*

$$x_1 = 0.9994,\ x_2 = 1.9997,\ x_3 = 2.9997,\ x_4 = -0.0001.$$

*Sixth iteration is*

$$x_1 = 0.9999,\ x_2 = 1.9999,\ x_3 = 2.9999,\ x_4 = -0.0001.$$

Hence the solution is $x_1 = 1$, $x_2 = 2$, $x_3 = 3$, $x_4 = 0$.

**(3) Relaxation method*.** Consider the equations

$$a_1x + b_1y + c_1z = d_1\ ;\ a_2x + b_2y + c_2z = d_2\ ;\ a_3x + b_3y + c_3z = d_3$$

We define the residuals $R_x$, $R_y$, $R_z$ by the relations

$$R_x = d_1 - a_1x - b_1y - c_1z\ ;\ R_y = d_2 - a_2x - b_2y - c_2z\ ;\ R_z = d_3 - a_3x - b_3y - c_3z \qquad ...(1)$$

To start with we assume $x = y = z = 0$ and calculate the initial residuals. Then the residuals are reduced step by step by giving increments to the variables. For this purpose, we construct the following *operation table :*

| | $\delta R_x$ | $\delta R_y$ | $\delta R_z$ |
|---|---|---|---|
| $\delta x = 1$ | $-a_1$ | $-a_2$ | $-a_3$ |
| $\delta y = 1$ | $-b_1$ | $-b_2$ | $-b_3$ |
| $\delta z = 1$ | $-c_1$ | $-c_2$ | $-c_3$ |

We note from the Eqn. (1) that if $x$ is increased by 1 (keeping $y$ and $z$ constant), $R_x$, $R_y$ and $R_z$ decrease by $a_1$, $a_2$, $a_3$ respectively. This is shown in the above table alongwith the effects on the residuals when $y$ and $z$ are given unit increments. (The table is the transpose of the coefficient matrix).

At each step, the numerically largest residual is reduced to almost zero. To reduce a particular residual, the value of the corresponding variable is changed ; *e.g.*, to reduce $R_x$ by $p$, $x$ should be increased by $p/a_1$.

When all the residuals have been reduced to almost zero, the increments in $x, y, z$ are added separately to give the desired solution.

**Obs.** As a check, the computed values of $x, y, z$ are substituted in Eqn. (1) and the residuals are calculated. If these residuals are not all negligible, then there is some mistake and the entire process should be rechecked.

**Example 28.24.** *Solve, by Relaxation method, the equations :*

$$9x - 2y + z = 50,\ x + 5y - 3z = 18,\ -2x + 2y + 7z = 19.$$

*This method was originally developed by *R.V. Southwell* in 1935, for application to structural engineering problems.

**Solution.** The residuals are given by

$R_x = 50 - 9x + 2y - z$ ; $R_y = 18 - x - 5y + 3z$ ; $R_z = 19 + 2x - 2y - 7z$

The operations table is

| | $\delta R_x$ | $\delta R_y$ | $\delta R_z$ |
|---|---|---|---|
| $\delta x = 1$ | –9 | –1 | 2 |
| $\delta y = 1$ | 2 | –5 | –2 |
| $\delta z = 1$ | –1 | 3 | –7 |

The relaxation table is

| | $R_x$ | $R_y$ | $R_z$ | |
|---|---|---|---|---|
| $x = y = z = 0$ | 50 | 18 | 19 | ...(*i*) |
| $\delta x = 5$ | 5 | 13 | 29 | ...(*ii*) |
| $\delta z = 4$ | 1 | 25 | 1 | ...(*iii*) |
| $\delta y = 5$ | 11 | 0 | –9 | ...(*iv*) |
| $\delta x = 1$ | 2 | –1 | –7 | ...(*v*) |
| $\delta z = -1$ | 3 | –4 | 0 | ...(*vi*) |
| $\delta y = -0.8$ | 1.4 | 0 | 1.6 | ...(*vii*) |
| $\delta z = 0.23$ | 1.17 | 0.69 | –0.09 | ...(*viii*) |
| $\delta x = 0.13$ | 0 | 0.56 | 0.17 | ...(*ix*) |
| $\delta y = 0.112$ | 0.224 | 0 | –0.054 | ...(*x*) |

$\Sigma\delta x = 6.13$, $\Sigma\delta y = 4.31$, $\Sigma\delta z = 3.23$

Thus $x = 6.13, y = 4.31, z = 3.23.$

[**Explanation.** In Eqn. (*i*), the largest residual is 50. To reduce it, we give an increment $\delta_x = 5$ and the resulting residuals are shown in Eqn. (*ii*). Of these $R_z = 29$ is the largest and we given an increment $\delta z = 4$ to get the results in Eqn. (*iii*). In Eqn. (*vi*), $R_y = -4$ is the (numerically) largest and we give an increment $\delta y = -4/5 = -0.8$ to obtain the results in Eqn. (*vii*). Similarly the other steps have been carried out.]

**Example 28.25.** *Solve by Relaxation method, the equations :*

$10x - 2y - 3z = 205$ ; $-2x + 10y - 2z = 154$ ; $-2x - y + 10z = 120.$ (*V.T.U., 2011 S*)

**Solution.** The residuals are given by

$R_x = 205 - 10x + 2y + 3z$ ; $R_y = 154 + 2x - 10y + 2z$ ; $R_z = 120 + 2x + y - 10z.$

The operations table is

| | $\delta R_x$ | $\delta R_y$ | $\delta R_z$ |
|---|---|---|---|
| $\delta x = 1$ | –10 | 2 | 2 |
| $\delta y = 1$ | 2 | –10 | –1 |
| $\delta z = 1$ | 3 | 2 | –10 |

The relaxation table is :

| | $R_x$ | $R_y$ | $R_z$ |
|---|---|---|---|
| $x = y = z = 0$ | 205 | 154 | 120 |
| $\delta x = 20$ | 5 | 194 | 160 |
| $\delta y = 19$ | 43 | 4 | 179 |
| $\delta z = 18$ | 97 | 40 | –1 |
| $\delta x = 10$ | –3 | 60 | 19 |
| $\delta y = 6$ | 9 | 0 | 25 |
| $\delta z = 2$ | 15 | 4 | 5 |
| $\delta x = 2$ | –5 | 8 | 9 |
| $\delta z = 1$ | –2 | 10 | –1 |
| $\delta y = 1$ | 0 | 0 | 0 |

$\Sigma\delta x = 32$, $\Sigma y = 26$, $\Sigma z = 21.$

Hence $x = 32, y = 26, z = 21.$

**Example 28.26:** *Solve the equations.*

$$x + y + z = 9 \quad ...(i)$$
$$2x + 5y + 7z = 52 \quad ...(ii)$$
$$2x + y - z = 0 \quad ...(iii)$$

*(Karaikal, 2018)*

**Solution.** Elimination $x$ from Eqns. (*ii*) and (*iii*) with the help of Eqn. (*i*),

$$x + y + z = 9 \quad ...(iv)$$
$$3y + 5z = 34 \quad ...(v)$$
$$-y - 3z = 34 \quad ...(vi)$$

Now eliminating $y$ from Eqn. (*vi*) with the help of Eqn. (*v*)

$$x + y + z = 9 \quad ...(vii)$$
$$3y + 5z = 34 \quad ...(viii)$$
$$z = 5 \quad ...(ix)$$

From Eqn. (*ix*), we get $z = 5$, putting this values of $z$ in Eqn. (*viii*)

we get $$3y + 25 = 34$$
$$y = 3$$

putting the values of $y$ and $z$ in Eqn. (*vii*), we get $x = 1$

So, we get $x = 1, y = 3, z = 5$

**Example 28.27:** *Solve the equations.* *(Uni. of Hyderbad, 2018)*

$$10x_1 - 7x_2 + 3x_3 + 5x_4 = 6$$
$$-6x_1 + 8x_2 - x_3 - 4x_4 = 5$$
$$3x_1 + x_2 + 4x_3 + 11x_4 = 2$$
$$5x_1 - 9x_2 - 2x_3 + 4x_4 = 7$$

*by Gauss alogorithm.*

**Solution.** Dividing the first equation by 10, the above equations can be written as

as $$x_1 - 0.7x_2 + 0.3\,x_3 + 0.5x_4 = 0.6 \quad ...(i)$$
$$-6x_1 + 8x_2 - x_3 - 4x_4 = 5 \quad ...(ii)$$
$$3x_1 + x_2 + 4x_3 + 11x_4 = 2 \quad ...(iii)$$
$$5x_1 - 9x_2 - 2x_3 + 4x_4 = 7 \quad ...(iv)$$

Now subtracting (–6) times the first equation from the second, 3 times the first equation from the third and 5 times the first equation from the fourth we have

$$x_1 - 0.7x_2 + 0.3x_3 + 0.5\,x_4 = 0.6$$
$$3.8x_2 + 0.8\,x_3 - x_4 = 8.6$$
$$3.1\,x_2 + 3.1\,x_3 + 9.5x_4 = 0.2$$
$$-5.5x_2 + 3.5x_3 + 1.5x_4 = 4$$

Now, we shall eliminate $x_2$ and then $x_3$

Finally, $$x_1 - 0.7x_3 + 0.3x_3 + 0.5x_4 = 0.6$$
$$x_2 + 0.63636\,x_3 - 0.27273\,x_4 = -0.72727$$
$$x_3 - 0.02247\,x_4 = -7.02247$$
$$10.370794\,x_4 = 10.37079$$

The above equations by back subsituation $x_1 = 5, x_2 = 4, x_3 = -7, x_4 = 1$.

**Example 28.28:** *Solve the equation by Gauss-Jordon method.*

$$2x_1 - 7x_2 + 4x_3 = 9$$
$$x_1 + 9x_2 - 6x_3 = 1$$
$$-3x_1 + 8x_2 + 5x_3 = 6$$

*(UKTU, 2018)*

**Solution.** Consider the augmented matrix

$$\begin{bmatrix} 2 & -7 & 4 & : & 9 \\ 1 & 9 & -6 & : & 1 \\ -3 & 8 & 5 & : & 6 \end{bmatrix}$$

The first row by dividing the pivot elements. Then reduce the remaining elements of the first column to zero by subtracting the new row from the second and also by subtracting (–3)times the new first row from the third row.

$$\begin{bmatrix} 1 & -\frac{7}{2} & 2 & : & \frac{9}{2} \\ 0 & \frac{25}{2} & -8 & : & -\frac{7}{2} \\ 0 & \frac{-5}{2} & 11 & : & -39/2 \end{bmatrix}$$

Now normalize the second row by dividing by the pivot element $\frac{25}{2}$

$$\begin{bmatrix} 1 & 0 & -\frac{6}{25} & : & \frac{88}{25} \\ 0 & 1 & -\frac{16}{25} & : & -\frac{7}{25} \\ 0 & 0 & \frac{7}{5} & : & \frac{94}{5} \end{bmatrix}$$

Next normalize the third row by dividing by the pivot elements $\frac{45}{5}$.

$$\begin{bmatrix} 1 & 0 & 0 & : & 4 \\ 0 & 1 & 0 & : & 1 \\ 0 & 0 & 1 & : & 2 \end{bmatrix}$$

So, the solution is $x_1 = 4, x_2 = 1, x_3 = 2$.

**Example 28.29:** *Solve the equations*

$$27x_1 + 6x_2 - x_3 = 85$$
$$6x_1 + 15x_2 + 2x_3 = 72$$
$$x_1 + x_2 + 54x_3 = 110$$

*(Andhra, 2018)*

**Solution.** Since in each equation of the coefficient is larger than the other satisfying the condition

$$x_1 = \frac{1}{27}(85 - 6x_1 + x_2)$$

$$x_2 = \frac{1}{15}(72 - 6x_1 - 2x_2)$$

$$x_3 = \frac{1}{54}(110 - x_1 - x_2)$$

We can start with any values of $x_2$ and $x_3$.

Let us take $x_2 = 0$

$$x_1^{(1)} = \frac{1}{27}(85 - 0 - 0) = 3.15$$

$$x_2^{(1)} = \frac{1}{15}(72 - 6 \times 3.15 - 2 \times 0) = 3.54$$

$$x_3^{(1)} = \frac{1}{54}(110 - 3.15 - 3.54) = 1.91$$

Now to obtain the second approximation

$$x_1^{(2)} = \frac{1}{27}(85 - 6 \times 3.54 + 1.91) = 2.43$$

$$x_2^{(2)} = \frac{1}{15}(72 - 6 \times 2.43 - 2 \times 1.91) = 3.57$$

$$x_3^{(2)} = \frac{1}{54}(110 - 2.43 - 3.57) = 1.92$$

proceeding

$$x_1^{(3)} = 2.426,\ x_2^{(3)} = 3.572,$$
$$x_3^{(3)} = 1.926$$
$$x_1^{(3)} = 2.426,\ x_2^{(3)} = 3.572,\ x_3^{(3)} = 1.926.$$

## PROBLEMS 28.4

1. Solve by Jacobi's method, the equations : $5x - y + z = 10$ ; $2x + 4y = 12$ ; $x + y + 5z = -1$. Start with the solution (2, 3, 0). *(V.T.U., 2015 S)*
2. Solve the equations $27x + 6y - z = 85$ ; $x + y + 54z = 110$ ; $6x + 15y + 2z = 72$.
   by (*a*) Jacobi's method (*b*) Gauss-Seidel method. *(JNTUK, 2022; Anna, 2013)*

   Solve the following equations by Gauss-Seidel method :
3. $2x + y + 6z = 9$ ; $8x + 3y + 2z = 13$ ; $x + 5y + z = 7$.
4. $28x + 4y - z = 32$ ; $x + 3y + 10z = 24$ ; $2x + 17y + 4z = 35$. *(Anna, 2013)*
5. $5x + 2y + z = 12$, $x + 4y + 2z = 15$ ; $x + 2y + 5z = 20$. *(Delhi., 2015)*
6. $1.2x + 2.1y + 4.2z = 9.9$ ; $5.3x + 6.1y + 4.7z = 21.6$ ; $9.2x + 8.3y + z = 15.2$.
7. $$\begin{bmatrix} 2 & -1 & 0 & 0 \\ -1 & 2 & -1 & 0 \\ 0 & -1 & 2 & -1 \\ 0 & 0 & -1 & 2 \end{bmatrix}\begin{bmatrix} x \\ y \\ z \\ u \end{bmatrix} = \begin{bmatrix} 0 \\ 0 \\ 0 \\ 5 \end{bmatrix}$$

   Solve by Relaxation method, the following sets of equations :
8. $3x + 9y - 2z = 11$ ; $4x + 2y + 13z = 24$ ; $4x - 4y + 3z = -8$.
9. $10x - 2y - 2z = 6$ ; $-x + 10y - 2z = 7$ ; $-x - y + 10z = 8$.
10. $-9x + 3y + 4z + 100 = 0$ ; $x - 7y + 3z + 80 = 0$ ; $2x + 3y - 5z + 60 = 0$.
11. $54x + y + z = 110$ ; $2x + 15y + 6z = 72$ ; $-x + 6y + 27z = 85$.

Solve the following equations by Gauss elimination as well as Gauss-Jordon method.

12. $$x_1 + x_2 + x_3 = 9$$ $$2x_1 - 3x_2 + 4x_3 = 13$$ $$3x_1 + 4x_2 + 5x_3 = 40$$ *(Osmania, 2018)*
13. $$x_1 + 2x_2 + x_3 = 8$$ $$2x_1 + 3x_2 + 4x_3 = 20$$ $$4x_1 + x_2 + 2x_3 = 16$$ *(Rourkela, 2018)*
14. $$15x_1 + 3x_2 - 4x_3 + 2x_4 = 49.207$$ $$3x_1 + 27x_2 - 5x_3 - x_4 = 18.024$$ $$-4x_1 + 5x_2 + 16x_3 + 5x_4 = -23.871$$ $$2x_1 - x_2 + 5x_3 + 19x_4 = 54.907$$ *(Andhra, M.E. 2018)*

## 28.8 SOLUTION OF NON-LINEAR SIMULTANEOUS EQUATIONS—NEWTON-RAPHSON METHOD

Consider the equations

$$f(x, y) = 0,\ g(x, y) = 0 \qquad \text{...(1)}$$

If an initial approximation $(x_0, y_0)$ to a solution has been found by graphical method or otherwise, then a better approximation $(x_1, y_1)$ can be obtained as follows :

Let $x_1 = x_0 + h, y_1 = y_0 + k$, so that $f(x_0 + h, y_0 + k) = 0, g(x_0 + h, y_0 + k) = 0$ ...(2)

Expanding each of the functions in Eqn. (2) by Taylor's series to first degree terms, we get approximately

$$\left.\begin{aligned} f_0 + h\frac{\partial f}{\partial x_0} + k\frac{\partial f}{\partial y_0} = 0 \\ g_0 + h\frac{\partial g}{\partial x_0} + k\frac{\partial g}{\partial y_0} = 0 \end{aligned}\right\} \quad ...(3)$$

where $\quad f_0 = f(x_0, y_0), \frac{\partial f}{\partial x_0} = \left(\frac{\partial f}{\partial x}\right)_{x_0, y_0}$ etc.

Solving the Eqn. (3) for $h$ and $k$, we get a new approximation to the root as

$$x_1 = x_0 + h, y_1 = y_0 + k$$

This process is repeated till we get the values to the desired accuracy.

**Example 28.30.** *Solve the system of non-linear equations :*

$$x^2 + y = 11, y^2 + x = 7.$$

**Solution.** An initial approximation to the solution is obtained from a rough graph of the given equations, as $x_0 = 3.5$ and $y_0 = -1.8$.

We have $\quad f = x^2 + y - 11$ and $g = y^2 + x - 7$ so that

$$\frac{\partial f}{\partial x} = 2x, \frac{\partial f}{\partial y} = 1 \quad \text{and} \quad \frac{\partial g}{\partial x} = 1, \frac{\partial g}{\partial y} = 2y.$$

Then Newton-Raphson's Eqn. (3) above will be

$$7h + k = 0.55, h - 3.6k = 0.26$$

Solving these, we get $h = 0.0855, k = -0.0485$

$\therefore$ the better approximation to the root is

$$x_1 = x_0 + h = 3.5855, y_1 = y_0 + k = -1.8485$$

Repeating the above process, replacing $(x_0, y_0)$ by $(x_1, y_1)$, we obtain $x_2 = 3.5844, y_2 = -1.8482$.

## PROBLEMS 28.5

1. Solve the equations $x^2 + y = 5$ ; $y^2 + x = 3$.
2. Solve the non-linear equations $x = 2(y + 1), y^2 = 3xy - 7$ correct to three decimals.
3. Use Newton-Raphson method to solve the equations $x = x^2 + y^2, y = x^2 - y^2$ correct to two decimals, starting with the approximation (0.8, 0.4).
4. Solve the non-linear equations $x^2 - y^2 = 4, x^2 + y^2 = 16$ numerically with $x_0 = y_0 = 2.828$ using N.R. method. Carry out two iterations.
5. Solve the equations $2x^2 + 3xy + y^2 = 3$ ; $4x^2 + 2xy + y^2 = 30$. Correct to three decimal places, using Newton-Raphson method, given that $x_0 = -3$, and $y_0 = 2$.

## 28.9 DETERMINATION OF EIGEN VALUES BY ITERATION

In § 2.14, we came across equations of the type

$$\left.\begin{aligned} (a_{11} - \lambda)x_1 + a_{12}x_2 + a_{13}x_3 = 0 \\ a_{21}x_1 + (a_{22} - \lambda)x_2 + a_{23}x_3 = 0 \\ a_{31}x_1 + a_{32}x_2 + (a_{33} - \lambda)x_3 = 0 \end{aligned}\right\} \quad ...(1)$$

which in matrix form, may be written as $[A - \lambda I] X = 0$ or $AX = \lambda X$ ...(2)

where $\quad A = [a_{ij}]$ and $X$ is the column matrix $[x_i]$.

Equation (1) will have a non-trivial solution if the coefficient matrix vanishes *e.g.*,

$$\begin{vmatrix} a_{11} - \lambda & a_{12} & a_{13} \\ a_{21} & a_{22} - \lambda & a_{23} \\ a_{31} & a_{32} & a_{33} - \lambda \end{vmatrix} = 0$$

This gives a cubic in $\lambda$ whose roots are *eigen values* of Eqn. (2) and corresponding to each *eigen value*, we have a non-zero solution $X = [x_1, x_2, x_3]$ which is called an *eigen vector*. Such an equation can ordinarily be solved easily.

In some applications, it is required to compute the numerically largest *eigen value* and the corresponding *eigen vector*. In such cases, the following iterative method is more convenient which is also well-suited for computing machines.

If $X_1, X_2, X_3$ be the eigen vectors corresponding to the eigen values $\lambda_1, \lambda_2, \lambda_3$, then an arbitrary column vector can be written as $X = k_1X_1 + k_2X_2 + k_3X_3$

Then $$AX = k_1AX_1 + k_2AX_2 + k_3AX_3 = k_1\lambda_1X_1 + k_2\lambda_2X_2 + k_3\lambda_3X_3$$

Similarly $$A^2X = k_1\lambda_1^2X_1 + k_2\lambda_2^2X_2 + k_3\lambda_3^2X_3$$

and $$A^rX = k_1\lambda_1^r X_1 + k_2\lambda_2^rX_2 + k_3\lambda_3^rX_3$$

If $|\lambda_1| > |\lambda_2| > |\lambda_3|$, then the contribution of the term $k_1\lambda_1^rX_1$ to the sum on the right increases with $r$ and therefore, every time we multiply a column vector by $A$, it becomes nearer to the eigen vector $X_1$. Then we make the largest component of the resulting column vector unity to avoid the factor $k_1$.

*Thus we start with a column vector X which is as near the solution as possible and evaluate AX which is written as $\lambda^{(1)} X^{(1)}$ after normalisation. This gives the first approximation $\lambda^{(1)}$ to the eigen value and $X^{(1)}$ to eigen vector. Similarly we evaluate $AX^{(1)} = \lambda^{(2)} X^{(2)}$ which gives the second approximation. We repeat this process till $[X^{(r)} - X^{(r-1)}]$ becomes negligible. Then $\lambda^{(r)}$ will be the largest eigen value of Eqn. (1) and $X^{(r)}$, the corresponding eigen vector.*

*This iterative procedure for finding the dominant eigen value of a matrix is known as Rayleigh's power method.**

**Example 28.31.** *Determine the largest eigen value and the corresponding eigen vector of the matrices using the power method :*

(i) $A = \begin{bmatrix} 5 & 4 \\ 1 & 2 \end{bmatrix}$ (ii) $A = \begin{bmatrix} 2 & -1 & 0 \\ -1 & 2 & -1 \\ 0 & -1 & 2 \end{bmatrix}$ *(V.T.U., 2016 ; Andhra, 2015 ; Anna, 2012)*

**Solution.** (*i*) Let the initial approximation to the eigen vector corresponding to the largest eigen value of $A$ be $X = \begin{bmatrix} 1 \\ 0 \end{bmatrix}$.

Then $$AX = \begin{bmatrix} 5 & 4 \\ 1 & 2 \end{bmatrix}\begin{bmatrix} 1 \\ 0 \end{bmatrix} = \begin{bmatrix} 5 \\ 1 \end{bmatrix} = 5\begin{bmatrix} 1 \\ 0.2 \end{bmatrix} = \lambda^{(1)}X^{(1)}$$

So the first approximation to the eigen value is $\lambda^{(1)} = 5$ and the corresponding eigen vector is $X^{(1)} = \begin{bmatrix} 1 \\ 0.2 \end{bmatrix}$.

Now $$AX^{(1)} = \begin{bmatrix} 5 & 4 \\ 1 & 2 \end{bmatrix}\begin{bmatrix} 1 \\ 0.2 \end{bmatrix} = \begin{bmatrix} 5.8 \\ 1.4 \end{bmatrix} = 5.8\begin{bmatrix} 1 \\ 0.241 \end{bmatrix} = \lambda^{(2)} X^{(2)}$$

Thus the second approximation to the eigen-value is $\lambda^{(2)} = 5.8$ and the corresponding eigen-vector is $X^{(2)} = \begin{bmatrix} 1 \\ 0.241 \end{bmatrix}$, repeating the above process, we get

Now $$AX^{(2)} = \begin{bmatrix} 5 & 4 \\ 1 & 2 \end{bmatrix}\begin{bmatrix} 1 \\ 0.241 \end{bmatrix} = 5.966\begin{bmatrix} 1 \\ 0.248 \end{bmatrix} = \lambda^{(3)} X^{(3)}$$

$$AX^{(3)} = \begin{bmatrix} 5 & 4 \\ 1 & 2 \end{bmatrix}\begin{bmatrix} 1 \\ 0.249 \end{bmatrix} = 5.994\begin{bmatrix} 1 \\ 0.250 \end{bmatrix} = \lambda^{(4)} X^{(4)}$$

$$AX^{(4)} = \begin{bmatrix} 5 & 4 \\ 1 & 2 \end{bmatrix}\begin{bmatrix} 1 \\ 0.250 \end{bmatrix} = 5.999\begin{bmatrix} 1 \\ 0.25 \end{bmatrix} = \lambda^{(5)} X^{(5)}$$

*After the English mathematician and physicist John William Strut known as *Lord Rayleigh* (1842–1919) who made important contributions to the theory of waves, elasticity and hydrodynamics. He was professor at Cambridge and London.

$$AX^{(5)} = \begin{bmatrix} 5 & 4 \\ 1 & 2 \end{bmatrix} \begin{bmatrix} 1 \\ 0.25 \end{bmatrix} = 6 \begin{bmatrix} 1 \\ 0.25 \end{bmatrix} = \lambda^{(6)} X^{(6)}$$

Clearly $\lambda^{(5)} = \lambda^{(6)}$ and $X^{(5)} = X^{(6)}$ upto 3 decimal places. Hence the largest eigen-value is 6 and the corresponding eigen vector is $\begin{bmatrix} 1 \\ 0.25 \end{bmatrix}$.

(*ii*) Let the initial approximation to the required eigen vector be $X = [1, 0, 0]'$.

Then $$AX = \begin{bmatrix} 2 & -1 & 0 \\ -1 & 2 & -1 \\ 0 & -1 & 2 \end{bmatrix} \begin{bmatrix} 1 \\ 0 \\ 0 \end{bmatrix} = \begin{bmatrix} 2 \\ -1 \\ 0 \end{bmatrix} = 2 \begin{bmatrix} 1 \\ -0.5 \\ 0 \end{bmatrix} = \lambda^{(1)} X^{(1)}.$$

So the first approximation to the eigen value is $\lambda^{(1)} = 2$ and the corresponding eigen vector

$$X^{(1)} = [1, -0.5, 0]'.$$

Hence $$AX^{(1)} = \begin{bmatrix} 2 & -1 & 0 \\ -1 & 2 & -1 \\ 0 & -1 & 2 \end{bmatrix} \begin{bmatrix} 1 \\ -0.5 \\ 0 \end{bmatrix} = \begin{bmatrix} 2.5 \\ -2 \\ 0.5 \end{bmatrix} = 2.5 \begin{bmatrix} 1 \\ -0.8 \\ 0.2 \end{bmatrix} = \lambda^{(2)} X^{(2)}.$$

Repeating the above process, we get

$$AX^{(2)} = 2.8 \begin{bmatrix} 1 \\ -1 \\ 0.43 \end{bmatrix} = \lambda^{(3)} X^{(3)} \,;\, AX^{(3)} = 3.43 \begin{bmatrix} 0.87 \\ -1 \\ 0.54 \end{bmatrix} = \lambda^{(4)} X^{(4)}$$

$$AX^{(4)} = 3.41 \begin{bmatrix} 0.80 \\ -1 \\ 0.61 \end{bmatrix} = \lambda^{(5)} X^{(5)} \,;\, AX^{(5)} = 3.41 \begin{bmatrix} 0.76 \\ -1 \\ 0.65 \end{bmatrix} = \lambda^{(6)} X^{(6)} \,;\, AX^{(6)} = 3.41 \begin{bmatrix} 0.74 \\ -1 \\ 0.67 \end{bmatrix} = \lambda^{(7)} X^{(7)}$$

Clearly $\lambda^{(6)} = \lambda^{(7)}$ and $X^{(6)} = X^{(7)}$ approximately.

Hence the largest eigen value is 3.41 and the corresponding eigen vector is $[0.74, -1, 0.67]'$.

## PROBLEMS 28.6

1. Find by power method, the larger eigen-value of the matrices :

   (*a*) $\begin{bmatrix} 1 & 2 \\ 3 & 4 \end{bmatrix}$ (*b*) $\begin{bmatrix} 4 & 1 \\ 1 & 3 \end{bmatrix}$.

2. Obtain the largest eigen-value and the corresponding eigen-vector for the equations

   $(2 - \lambda)x_1 - x_2 = 0 \,;\, -x_1 + (2 - \lambda)x_2 - x_3 = 0 \,;\, -x_2 + (2 - \lambda)x_3 = 0$

   by Rayleigh Quotient method.

3. Find the dominant eigen value and the corresponding eigen vector of the following matrices using the power method :

   (*a*) $\begin{bmatrix} 4 & 1 & -1 \\ 2 & 3 & -1 \\ -2 & 1 & 5 \end{bmatrix}$ (*V.T.U., 2015 S*) (*b*) $\begin{bmatrix} 6 & -2 & 2 \\ -2 & 3 & -1 \\ 2 & -1 & 3 \end{bmatrix}$ (*V.T.U., 2013*)

4. Find the largest eigen-value and the corresponding eigen-vector of the matrices :

   (*a*) $\begin{bmatrix} 1 & -3 & 2 \\ 4 & 4 & -1 \\ 6 & 3 & 5 \end{bmatrix}$ (*V.T.U., 2013 ; Anna, 2012*) (*b*) $\begin{bmatrix} 25 & 1 & 2 \\ 1 & 3 & 0 \\ 2 & 0 & -4 \end{bmatrix}$ (*Anna., 2016*)

   (*c*) $\begin{bmatrix} 1 & 3 & -1 \\ 3 & 2 & 4 \\ -1 & 4 & 10 \end{bmatrix}$ with initial approximation $[1, 1, 0]^T$. (*V.T.U., 2015*)

## 28.10 OBJECTIVE TYPE OF QUESTIONS

### PROBLEMS 28.7

*Fill up the blanks or select the correct answer to each of the following problems :*

1. Out of Regula-Falsi method and Newton-Rephson method, the rate of convergence is faster for ......... .
2. Using Newton's method, the value of $\sqrt[4]{N}$ is ......... . *(Anna, 2012)*
3. In the Regula-Falsi method of finding the real root of an equation, the curve $AB$ is replaced by ......... .
4. Newton's iterative formula to find the value of $\sqrt{N}$ is ......... .
5. Newton-Raphson formula converges when ......... .
6. In solving simultaneous equations by Gauss-Jordan method, the coefficient matrix is reduced to ......... matrix.
7. In the case of bisection method, the convergence is

   (*a*) linear (*b*) quadratic (*c*) very slow.
8. The order of convergence in Newton-Raphson method is

   (*a*) 2 (*b*) 3 (*c*) 0 (*d*) none.
9. The Newton-Raphson algorithm for finding the cube root of $N$ is ......... .
10. The bisection method for finding the root of an equation $f(x) = 0$ is ......... .
11. In Regula-falsi method, the first approximation is given by ......... .
12. The order of convergence in Method of interpolation is ......... . *(Delhi, 2013)*
13. The iterative formula for finding the reciprocal of $N$ is $x_{n+1}$ = ......... . *(Anna, 2013)*
14. As soon as a new value of a variable is found by iteration, it is used immediately in the following equations, this method is called

    (*a*) Gauss-Jordan method (*b*) Gauss-Seidal method

    (*c*) Jacobi's method (*d*) Relaxation method.
15. Out of Regula-falsi method and Newton-Raphson method, the rate of convergence is faster for ......... .
16. The difference between direct and iterative methods of solving simultaneous linear equations is ......... .
17. To which form the coefficient matrix is transformed when $AX = B$ is solved by Gauss elimination method.
18. Jacobi's iteration method can be used to solve a system of non-linear equations. (True or False)
19. The convergence in the Gauss-Seidal method is thrice as fast as in Jacobi's method. (True or False)
20. By Gauss elimination method, solve $x + y = 2$ and $2x + 3y = 5$.

CHAPTER

29

# Finite Differences and Interpolation

1. Finite differences. 2. Differences of a polynomial. 3. Factorial notation. 4. Relations between the operators. 5. To find one or more missing terms. 6. Newton's interpolation formulae. 7. Central difference interpolation formulae—Gauss's interpolation formulae ; Stirling's formula ; Bessel's formula ; Everett's formula. 8. Choice of an interpolation formula. 9. Interpolation with unequal intervals. 10. Lagrange's formula. 11. Divided differences. 12. Newton's divided difference formula. 13. Inverse interpolation. 14. Objective Type of Questions.

## 29.1 FINITE DIFFERENCES

Suppose we are given the following values of $y = f(x)$ for a set of values of $x$ :

| $x$ : | $x_0$ | $x_1$ | $x_2 \ldots x_n$. |
|---|---|---|---|
| $y$ : | $y_0$ | $y_1$ | $y_2 \ldots y_n$. |

Then the process of finding the values of $y$ corresponding to any value of $x = x_i$ between $x_0$ and $x_n$ is called *interpolation*. Thus *interpolation is the technique of estimating the value of a function for any intermediate value of the independent variable* while the process of computing the value of the function outside the given range is called *extrapolation*. The study of the interpolation is based on the concept of differences of a function which we proceed to discuss. For a detailed study, the reader should refer to author's book *'Numerical Methods in Engineering and Science'*.

Suppose that the function $y = f(x)$ is tabulated for the equally spaced values $x = x_0, x_0 + h, x_0 + 2h, \ldots, x_0 + nh$ giving $y = y_0, y_1, y_2, \ldots, y_n$. To determine the values of $f(x)$ or $f'(x)$ for some intermediate values of $x$, the following three types of differences are found useful :

(1) **Forward differences.** The differences $y_1 - y_0, y_2 - y_1, \ldots, y_n - y_{n-1}$ when denoted by $\Delta y_0, \Delta y_1, \ldots, \Delta y_{n-1}$ respectively are called the *first forward differences* where $\Delta$ is the *forward difference operator.* Thus the first forward differences are $\Delta y_r = y_{r+1} - y_r$.

Similarly, the second forward differences are defined by

$$\Delta^2 y_r = \Delta y_{r+1} - \Delta y_r$$

In general, $$\Delta^p y_r = \Delta^{p-1} y_{r+1} - \Delta^{p-1} y_r$$

defines the *pth forward differences.*

These differences are systematically set out as follows in what is called a *Forward Difference Table.*

In a difference table, $x$ is called the *argument* and $y$ the *function* or the *entry*. $y_0$, the first entry is called the *leading term* and $\Delta y_0, \Delta^2 y_0, \Delta^3 y_0$ etc. are called the *leading differences.*

**Obs.** Any higher order forward difference can be expressed in terms of the entries.

We have $\Delta^2 y_0 = \Delta y_1 - \Delta y_0 = (y_2 - y_1) - (y_1 - y_0) = y_2 - 2y_1 + y_0$

$\Delta^3 y_0 = \Delta^2 y_1 - \Delta^2 y_0 = (y_3 - 2y_2 + y_1) - (y_2 - 2y_1 + y_0) = y_3 - 3y_2 + 3y_1 - y_0$

$\Delta^4 y_0 = \Delta^3 y_1 - \Delta^3 y_0 = (y_4 - 3y_3 + 3y_2 - y_1) - (y_3 - 3y_2 + 3y_1 - y_0) = y_4 - 4y_3 + 6y_2 - 4y_1 + y_0$

The coefficients occurring on the right hand side being the binomial coefficient, we have in general,

$$\Delta^n y_0 = y_n - {}^n c_1 y_{n-1} + {}^n c_2 y_{n-2} - \ldots + (-1)^n y_0$$

**Forward Difference Table**

| *Value of x* | *Value of y* | *1st. diff.* | *2nd diff.* | *3rd diff.* | *4th diff.* | *5th diff.* |
|---|---|---|---|---|---|---|
| $x_0$ | $y_0$ | | | | | |
| | | $\Delta y_0$ | | | | |
| $x_0 + h$ | $y_1$ | | $\Delta^2 y_0$ | | | |
| | | $\Delta y_1$ | | $\Delta^3 y_0$ | | |
| $x_0 + 2h$ | $y_2$ | | $\Delta^2 y_1$ | | $\Delta^4 y_0$ | |
| | | $\Delta y_2$ | | $\Delta^3 y_1$ | | $\Delta^5 y_0$ |
| $x_0 + 3h$ | $y_3$ | | $\Delta^2 y_2$ | | $\Delta^4 y_1$ | |
| | | $\Delta y_3$ | | $\Delta^3 y_2$ | | |
| $x_0 + 4h$ | $y_4$ | | $\Delta^2 y_3$ | | | |
| | | $\Delta y_4$ | | | | |
| $x_0 + 5h$ | $y_5$ | | | | | |

**(2) Backward differences.** The differences $y_1 - y_0, y_2 - y_1, \ldots, y_n - y_{n-1}$ when denoted by $\nabla y_1, \nabla y_2, \ldots, \nabla y_n$ respectively, are called the *first backward differences* where $\nabla$ is the *backward difference operator*. Similarly we define higher order backward differences. Thus we have

$$\nabla y_r = y_r - y_{r-1},\ \nabla^2 y_r = \nabla y_r - \nabla y_{r-1},$$
$$\nabla^3 y_r = \nabla^2 y_r - \nabla^2 y_{r-1} \text{ etc.}$$

The differences are exhibited in the following :

**Backward Difference Table**

| *Value of x* | *Value of y* | *1st. diff.* | *2nd diff.* | *3rd diff.* | *4th diff.* | *5th diff.* |
|---|---|---|---|---|---|---|
| $x_0$ | $y_0$ | | | | | |
| | | $\nabla y_1$ | | | | |
| $x_0 + h$ | $y_1$ | | $\nabla^2 y_2$ | | | |
| | | $\nabla y_2$ | | $\nabla^3 y_3$ | | |
| $x_0 + 2h$ | $y_2$ | | $\nabla^2 y_3$ | | $\nabla^4 y_4$ | |
| | | $\nabla y_3$ | | $\nabla^3 y_4$ | | $\nabla^5 y_5$ |
| $x_0 + 3h$ | $y_3$ | | $\nabla^2 y_4$ | | $\nabla^4 y_5$ | |
| | | $\nabla y_4$ | | $\nabla^3 y_5$ | | |
| $x_0 + 4h$ | $y_4$ | | $\nabla^2 y_5$ | | | |
| | | $\nabla y_5$ | | | | |
| $x_0 + 5h$ | $y_5$ | | | | | |

**(3) Central differences.** Sometimes it is convenient to employ another system of differences known as *central differences*. In this system, the *central difference operator* $\delta$ is defined by the relations :

$$y_1 - y_0 = \delta y_{1/2},\ y_2 - y_1 = \delta y_{3/2},\ \ldots,\ y_n - y_{n-1} = \delta y_{n-1/2}$$

Similarly, higher order central differences are defined as

$$\delta y_{3/2} - \delta y_{1/2} = \delta^2 y_1,\ \delta y_{5/2} - \delta y_{3/2} = \delta^2 y_2,\ \ldots,$$
$$\delta^2 y_2 - \delta^2 y_1 = \delta^3 y_{3/2} \text{ and so on.}$$

These differences are shown in the following :

**Central Difference Table**

| *Value of x* | *Value of y* | *1st. diff.* | *2nd diff.* | *3rd diff.* | *4th diff.* | *5th diff.* |
|---|---|---|---|---|---|---|
| $x_0$ | $y_0$ | | | | | |
| | | $\delta y_{1/2}$ | | | | |
| $x_0 + h$ | $y_1$ | | $\delta^2 y_1$ | | | |
| | | $\delta y_{3/2}$ | | $\delta^3 y_{3/2}$ | | |
| $x_0 + 2h$ | $y_2$ | | $\delta^2 y_2$ | | $\delta^4 y_2$ | |
| | | $\delta y_{5/2}$ | | $\delta^3 y_{5/2}$ | | $\delta^5 y_{5/2}$ |
| $x_0 + 3h$ | $y_3$ | | $\delta^2 y_3$ | | $\delta^4 y_3$ | |
| | | $\delta y_{7/2}$ | | $\delta^3 y_{7/2}$ | | |
| $x_0 + 4h$ | $y_4$ | | $\delta^2 y_4$ | | | |
| | | $\delta y_{9/2}$ | | | | |
| $x_0 + 5h$ | $y_5$ | | | | | |

We see from this table that the central differences on the same horizontal line have the same suffix. Also the differences of odd order are known only for half values of the suffix and those of even order for only integral values of the suffix.

It is often required to find the mean of adjacent values in the same column of differences. We denote this mean by $\mu$. Thus

$$\mu\delta y_1 = \tfrac{1}{2}(\delta y_{1/2} + \delta y_{3/2}),\ \mu\delta^2 y_{3/2} = \tfrac{1}{2}(\delta^2 y_1 + \delta^2 y_2) \text{ etc.}$$

**Obs.** The reader should note that it is only the notation which changes and not the differences.

$$y_1 - y_0 = \Delta y_0 = \nabla y_1 = \delta y_{1/2}.$$

Of all the interpolation formulae, those involving central differences are most useful in practice as the coefficients in such formulae decrease much more rapidly.

**Example 29.1.** *Evaluate (i) $\Delta \tan^{-1} x$ (ii) $\Delta(e^x \log 2x)$ (iii) $\Delta(x^2/\cos 2x)$ (iv) $\Delta^2 \cos 2x$.*

**Solution.** (*i*) $\Delta \tan^{-1} x = \tan^{-1}(x+h) - \tan^{-1} x$

$$= \tan^{-1}\left\{\frac{x+h-x}{1+(x+h)x}\right\} = \tan^{-1}\left\{\frac{h}{1+hx+x^2}\right\}$$

(*ii*)
$$\Delta(e^x \log 2x) = e^{x+h}\log 2(x+h) - e^x \log 2x$$
$$= e^{x+h}\log 2(x+h) - e^{x+h}\log 2x + e^{x+h}\log 2x - e^x \log 2x$$
$$= e^{x+h}\log\frac{x+h}{x} + (e^{x+h} - e^x)\log 2x$$
$$= e^x\left[e^h \log\left(1+\frac{h}{x}\right) + (e^h - 1)\log 2x\right]$$

(*iii*)
$$\Delta\left(\frac{x^2}{\cos 2x}\right) = \frac{(x+h)^2}{\cos 2(x+h)} - \frac{x^2}{\cos 2x} = \frac{(x+h)^2\cos 2x - x^2\cos 2(x+h)}{\cos 2(x+h)\cos 2x}$$
$$= \frac{[(x+h)^2 - x^2]\cos 2x + x^2[\cos 2x - \cos 2(x+h)]}{\cos 2(x+h)\cos 2x}$$
$$= \frac{(2hx+h^2)\cos 2x + 2x^2\sin(h)\sin(2x+h)}{\cos 2(x+h)\cos 2x}$$

(*iv*)
$$\Delta^2 \cos 2x = \Delta\{\cos 2(x+h) - \cos 2x\}$$
$$= \Delta\cos 2(x+h) - \Delta\cos 2x$$
$$= [\cos 2(x+2h) - \cos 2(x+h)] - [\cos 2(x+h) - \cos 2x]$$
$$= -2\sin(2x+3h)\sin h + 2\sin(2x+h)\sin h$$
$$= -2\sin h\,[\sin(2x+3h) - \sin(2x+h)]$$
$$= -2\sin h\,[2\cos(2x+2h)\sin h] = -4\sin^2 h\cos(2x+2h).$$

**Example 29.2.** *Evaluate (i) $\Delta^2\left(\dfrac{5x+12}{x^2+5x+16}\right)$ (ii) $\Delta^2(ab^{cx})$*

*(iii) $\Delta^n(e^x)$ interval of differencing being unity.*

**Solution.** (*i*)
$$\Delta^2\left(\frac{5x+12}{x^2+5x+6}\right) = \Delta^2\left\{\frac{5x+12}{(x+2)(x+3)}\right\} = \Delta^2\left\{\frac{2}{x+2} + \frac{3}{x+3}\right\}$$
$$= \Delta\left\{\Delta\left(\frac{2}{x+2}\right) + \Delta\left(\frac{3}{x+3}\right)\right\} = \Delta\left\{2\left(\frac{1}{x+3} - \frac{1}{x+2}\right) + 3\left(\frac{1}{x+4} - \frac{1}{x+3}\right)\right\}$$
$$= -2\Delta\left\{\frac{1}{(x+2)(x+3)}\right\} - 3\Delta\left\{\frac{1}{(x+3)(x+4)}\right\}$$
$$= -2\left\{\frac{1}{(x+3)(x+4)} - \frac{1}{(x+2)(x+3)}\right\} - 3\left\{\frac{1}{(x+4)(x+5)} - \frac{1}{(x+3)(x+4)}\right\}$$

$$= \frac{4}{(x+2)(x+3)(x+4)} + \frac{6}{(x+3)(x+4)(x+5)} = \frac{2(5x+16)}{(x+2)(x+3)(x+4)(x+5)}$$

(ii) $\Delta(ab^{cx}) = a\,\Delta(b^{cx}) = a(b^{cx+1} - b^{cx}) = ab^{cx}(b-1)$

$$\Delta^2(ab^{cx}) = \Delta[\Delta ab^{cx}] = a(b-1)\,\Delta(b^{cx})$$
$$= a(b-1)(b^{cx+1} - b^{cx}) = a(b-1)^2\, b^{cx}.$$

(iii) $\Delta e^x = e^{x+1} - e^x = (e-1)e^x$

$$\Delta^2 e^x = \Delta(\Delta e^x) = \Delta[(e-1)\,e^x]$$
$$= (e-1)\,\Delta e^x = (e-1)(e-1)\,e^x = (e-1)^2\, e^x$$

Similarly $\Delta^3 e^x = (e-1)^3 e^x,\ \Delta^4 e^x = (e-1)^4 e^x, \ldots$ and $\Delta^n e^x = (e-1)^n e^x$.

## 29.2 DIFFERENCES OF A POLYNOMIAL

*The nth differences of a polynomial of the nth degree are constant and all higher order differences are zero.*

Let the polynomial of the $n$th degree in $x$, be

$$f(x) = ax^n + bx^{n-1} + cx^{n-2} + \ldots + k(x+h) + l$$

$\therefore$
$$\Delta f(x) = f(x+h) - f(x)$$
$$= a[(x+h)^n - x^n] + b[(x+h)^{n-1} - x^{n-1}] + \ldots + kh$$
$$= anhx^{n-1} + b'x^{n-2} + c'x^{n-3} + \ldots + k'x + l' \qquad \ldots(1)$$

where $b', c', \ldots, l'$ are new constant coefficients.

Thus the first differences of a polynomial of the $n$th degree is a polynomial of degree $(n-1)$.

Similarly
$$\Delta^2 f(x) = \Delta[f(x+h) - f(x)] = \Delta f(x+h) - \Delta f(x)$$
$$= anh[(x+h)^{n-1} - x^{n-1}] + b'[(x+h)^{n-2} - x^{n-2}] + \ldots + k'h$$
$$= an(n-1)\,h^2x^{n-2} + b''x^{n-3} + c''x^{n-4} + \ldots + k'', \qquad \text{[by (1)]}$$

$\therefore$ the second differences represent a polynomial of degree $(n-2)$.

Continuing this process, for the $n$th differences we get a polynomial of degree zero *i.e.*

$$\Delta^n f(x) = an(n-1)(n-2)\ldots 1\,.\, h^n = an!\, h^n \qquad \ldots(2)$$

which is a constant. Hence the $(n+1)$th and higher differences of a polynomial of $n$th degree will be zero.

**Obs.** The converse of this theorem is also true *i.e. if the nth differences of a function tabulated at equally spaced intervels are constant, the function is a polynomial of degree n.* This fact is important in numerical analysis as it enables us to approximate a function by a polynomial of $n$th degree, if its $n$th order differences become nearly constant.

**Example 29.3.** *Evaluate* $\Delta^{10}[(1-ax)(1-bx^2)(1-cx^3)(1-dx^4)]$.

**Solution.** $\Delta^{10}[(1-ax)(1-bx^2)(1-cx^3)(1-dx^4)] = \Delta^{10}[abcd\,x^{10} + (\ )\,x^9 + (\ )\,x^8 + \ldots + 1]$

$$= abcd\,\Delta^{10}(x^{10}) \qquad [\because\ \Delta^{10}(x^n) = 0 \text{ for } n < 10]$$
$$= abcd\,(10\,!). \qquad \text{[by Eqn. (2) above]}$$

## 29.3 (1) FACTORIAL NOTATION

*A product of the form $x(x-1)(x-2)\ldots(x-r+1)$ is denoted by $[x]^r$ and is called a* **factorial.**

In particular $[x] = x,\ [x]^2 = x(x-1)$

$$[x]^3 = x(x-1)(x-2), \text{ etc.}$$

In general $[x]^n = x(x-1)(x-2)\ldots(x-n+1)$

In case, the interval of differencing is $h$, then

$$[x]^n = x(x-h)(x-2h)\ldots(x-\overline{n-1}h)$$

which is called a *Factorial polynomial or function.*

*The factorial notation is of special utility in the theory of finite differences. It helps in finding the successive differences of a polynomial directly by simple rule of differentiation.*

*The result of differencing $[x]^r$ is analogous to that of differentiating $x^r$.*

**(2) To express a polynomial in the factorial notation**

(*i*) *arrange the coefficients of the powers of x in descending order, replacing missing powers by zeros ;*

(*ii*) *using detached coefficients divide by x, x – 1, x – 2, etc. successively.*

**Obs.** Every polynomial of degree $n$ can be expressed as a factorial polynomial of the same degree and *vice versa*.

**Example 29.4.** *Express* $y = 2x^3 - 3x^2 + 3x - 10$ *in a factorial notation and hence show that* $\Delta^3 y = 12$.

**Solution. First method :** Let $y = A[x]^3 + B[x]^2 + C[x] + D$.

Then

|  | $x^3$ | $x^2$ | $x$ |  |
|---|---|---|---|---|
| 1 | 2 | –3 | 3 | $-10 = D$ |
|  | — | 2 | –1 |  |
| 2 | 2 | –1 | $2 = C$ |  |
|  | — | 4 |  |  |
| 3 | 2 | $3 = B$ |  |  |
|  | — |  |  |  |
|  | $2 = A$ |  |  |  |

Hence $\qquad y = 2[x]^3 + 3[x]^2 + 2[x] - 10$

$\therefore \qquad \Delta y = 2 \times 3[x]^2 + 3 \times 2[x] + 2$

$\Delta^2 y = 6 \times 2[x] + 6$

$\Delta^3 y = 12$, which shows that the third differences of $y$ are constant, as they should be.

**Obs.** *The coefficient of the highest power of x remains unchanged while transforming a polynomial to factorial notation.*

**Second method** (*Direct method*) :

Let
$$y = 2x^3 - 3x^2 + 3x - 10$$
$$= 2x(x - 1)(x - 2) + Bx(x - 1) + Cx + D$$

Putting $x = 0, -10 = D$

Putting $x = 1, 2 - 3 + 3 - 10 = C + D$

$\therefore \quad C = -8 - D = -8 + 10 = 2$

Putting $x = 2, 16 - 12 + 6 - 10 = 2B + 2C + D$

$\therefore$
$$B = \frac{1}{2}(-2C - D) = \frac{1}{2}(-4 + 10) = 3.$$

Hence
$$y = 2x(x - 1)(x - 2) + 3x(x - 1) + 2x - 10 = 2[x]^3 + 3[x]^2 + 2[x] - 10$$

$\therefore$
$$\Delta y = 2 \times 3[x]^2 + 3 \times 2[x] + 2, \Delta^2 y = 6 \times 2[x] + 6, \Delta^3 y = 12.$$

**Example 29.5.** *Assuming that the following values of y belong to a polynomial of degree 4, compute the next three values :*

| $x$ : | 0 | 1 | 2 | 3 | 4 | 5 | 6 | 7 |
|---|---|---|---|---|---|---|---|---|
| $y$ : | 1 | –1 | 1 | –1 | 1 | — | — | — |

**Solution.** We construct the following difference table from the given data :

| $x$ | $y$ | $\Delta y$ | $\Delta^2 y$ | $\Delta^3 y$ | $\Delta^4 y$ |
|---|---|---|---|---|---|
| 0 | $y_0 = 1$ |  |  |  |  |
|  |  | –2 |  |  |  |
| 1 | $y_1 = -1$ |  | 4 |  |  |
|  |  | 2 |  | –8 |  |
| 2 | $y_2 = 1$ |  | –4 |  | 16 |
|  |  | –2 |  | 8 |  |

| | | | | | |
|---|---|---|---|---|---|
| 3 | $y_3 = -1$ | | 4 | | 16 |
| | | 2 | | $\Delta^3 y_2,$ | |
| 4 | $y_4 = 1$ | | $\Delta^2 y_3$ | | 16 |
| | | $\Delta y_4$ | | $\Delta^2 y_3,$ | |
| 5 | $y_5$ | | $\Delta^2 y_4$ | | 16 |
| | | $\Delta y_5$ | | $\Delta^3 y_4,$ | |
| 6 | $y_6$ | | $\Delta^2 y_5$ | | |
| | | $\Delta y_6$ | | | |
| 7 | $y_7$ | | | | |

Since the values of $y$ belong to a polynomial of degree 4, the fourth differences must be constant. But $\Delta^4 y = 16$.

$\therefore$ The other fourth order differences must also be 16. Thus

$$\Delta^4 y_1 = 16 = \Delta^3 y_2 - \Delta^3 y_1$$

*i.e.,*
$$\Delta^3 y_2 = \Delta^3 y_1 + \Delta^4 y_1 = 8 + 16 = 24$$
$$\Delta^2 y_3 = \Delta^2 y_2 + \Delta^3 y_2 = 4 + 24 = 28$$
$$\Delta y_4 = \Delta y_3 + \Delta^2 y_3 = 2 + 28 = 30$$

and
$$y_5 = y_4 + \Delta y_4 = 1 + 30 = 31$$

Similarly starting with $\Delta^4 y_2 = 16$, we get

$$\Delta^3 y_3 = 40, \Delta^2 y_4 = 68, \Delta y_5 = 98, y_6 = 129.$$

Starting with $\Delta^4 y_3 = 16$, we obtain

$$\Delta^3 y_4 = 56, \Delta^2 y_5 = 124, \Delta y_6 = 222, y_7 = 351.$$

## PROBLEMS 29.1

1. Construct the table of differences for the data below :

| $x$ : | 0 | 1 | 2 | 3 | 4 |
|---|---|---|---|---|---|
| $f(x)$ : | 1.0 | 1.5 | 2.2 | 3.1 | 4.6 |

Evaluate $\Delta^3 f(2)$.

2. If $u_0 = 3, u_1 = 12, u_2 = 18, u_3 = 2000, u_4 = 100$, calculate $\Delta u_0$.
3. Show that $\Delta^3 y_i = y_{i+3} - 3y_{i+2} + 3y_{i+1} - y_i$.
4. Form the table of backward differences of the function
$$f(x) = x^3 - 3x^2 - 5x - 7 \text{ for } x = -1, 0, 1, 2, 3, 4, 5.$$
5. If $y = x^3 + x^2 - 2x + 1$, calculate values of $y$ for $x = 0, 1, 2, 3, 4, 5$. Form the difference table, find the value of $y$ at $x = 6$ by extending the table and verify that same value is obtained by substitution. ***(Delhi, 2013)***
6. Extend the following table to two more terms on either side by constructing the difference table :

| $x$ : | $-.2$ | 0.0 | 0.2 | 0.4 | 0.6 | 0.8 | 1.0 |
|---|---|---|---|---|---|---|---|
| $y$ : | 2.6 | 3.0 | 3.4 | 4.28 | 7.08 | 14.2 | 29.0 |

7. Show that

(*i*) $\Delta\left[\dfrac{1}{f(x)}\right] = \dfrac{-\Delta f(x)}{f(x)\,f(x+1)}$ ; *(J.N.T.U., 2009)* (*ii*) $\Delta \log f(x) = \log\left\{1 + \dfrac{\Delta f(x)}{f(x)}\right\}$.

(*iii*) $\Delta\,[x(x+1)(x+2)(x+3)] = 4(x+1)(x+2)(x+3)$.

8. Evaluate :

(*i*) $\Delta(x + \cos x)$ (*ii*) $\Delta \tan^{-1}\left(\dfrac{n-1}{n}\right)$ (*iii*) $\Delta\left\{\dfrac{1}{x(x+4)(x+6)}\right\}$

(*iv*) $\Delta^2\left(\dfrac{1}{x^2 + 5x + 6}\right)$ (*v*) $\Delta^2 \sin(px + q)$ ***(J.N.T.U., 2015)***

9. Evaluate :

(*i*) $\Delta(e^{3x} \log 2x)$ (*ii*) $\Delta(2^x/x\,!)$ (*iii*) $\Delta^n(a^x)$ (*iv*) $\Delta^n\left(\dfrac{1}{x}\right)$.

10. If $f(x) = e^{ax+b}$, show that its leading differences form a geometric progression.
11. Prove that

(*i*) $y_3 = y_2 + \Delta y_1 + \Delta^2 y_0 + \Delta^3 y_0$ (*ii*) $\nabla^2 y_8 = y_8 - 2y_7 + y_6$ ; $\delta^2 y_5 = y_6 - 2y_5 + y_4$.

12. Evaluate : (*Delhi., 2015*)
(*i*) $\Delta^3 [(1-x)(1-2x)(1-3x)]$
(*ii*) $\Delta^{10}[(1-x)(1-2x^2)(1-3x^3)(1-4x^4)]$, if the interval of differencing is 2.
13. Express $x^3 - 2x^2 + x - 1$ into factorial polynomial. Hence show that $\Delta^4 f(x) = 0$
14. Express $u = x^4 - 12x^3 + 24x^2 - 30x + 9$ and its successive differences in factorial notation. Hence show that $\Delta^5 u = 0$.
15. Find the first and second differences of $x^4 - 6x^3 + 11x^2 - 5x + 8$ with $h = 1$. Show that the fourth difference is constant.
16. Obtain the function whose first difference is $2x^3 + 3x^2 - 5x + 4$.
17. Find the first term of the series whose second and subsequent terms are 8, 3, 0, –1, 0.
18. If $u(x)$ and $v(x)$ be two functions of $x$, prove that
(*i*) $\Delta[u(x)\,v(x)] = u(x)\,\Delta v(x) + v(x+1)\,\Delta u(x)$. (*ii*) $\Delta\left\{\frac{u(x)}{v(x)}\right\} = \frac{v(x)\,\Delta u(x) - u(x)\,\Delta v(x)}{v(x)\,v(x+1)}$.

## 29.4 (1) OTHER DIFFERENCE OPERATORS

We have already introduced the operators $\Delta$, $\nabla$ and $\delta$. Besides these, there are the operators $E$ and $\mu$, which we define below :

(*i*) **Shift operator E** *is the operation of increasing the argument $x$ by $h$* so that

$$Ef(x) = f(x+h),\ E^2 f(x) = f(x+2h),\ E^3 f(x) = f(x+3h) \text{ etc.}$$

The inverse operator $E^{-1}$ is defined by $E^{-1} f(x) = f(x-h)$

If $y_x$ is the function $f(x)$, then $Ey_x = y_{x+h}$, $E^{-1}y_x = y_{x-h}$, $E^n y_x = y_{x+nh}$, where $n$ may be any real number.

(*ii*) **Averaging operator** $\mu$ is defined by the equation $\mu y_x = \frac{1}{2}(y_{x+h/2} + y_{x-h/2})$

**Obs.** *In the difference calculus, $\Delta$ and $E$ are regarded as the fundamental operators and $\nabla$, $\delta$, $\mu$ can be expressed in terms of these.*

**(2) Relations between the operators.** We shall now establish the following identities :

(*i*) $\Delta = \mathbf{E} - \mathbf{1}$ (*ii*) $\nabla = \mathbf{1} - \mathbf{E}^{-1}$

(*iii*) $\delta = \mathbf{E}^{1/2} - \mathbf{E}^{-1/2}$ (*iv*) $\mu = \frac{1}{2}(\mathbf{E}^{1/2} + \mathbf{E}^{-1/2})$

(*v*) $\Delta = \mathbf{E}\nabla = \nabla\mathbf{E} = \delta\mathbf{E}^{1/2}$ (*JNTUK, 2022*) (*vi*) $\mathbf{E} = \mathbf{e}^{\mathbf{hD}}$.

*Proofs.* (*i*) $\Delta y_x = y_{x+h} - y_x = Ey_x - y_x = (E-1)\,y_x$.

This shows that the operators $\Delta$ and $E$ are connected by the symbolic relation

$$\Delta = E - 1 \quad \text{or} \quad E = 1 + \Delta$$

(*ii*) $$\nabla y_x = y_x - y_{x-h} = y_x - E^{-1} y_x = (1 - E^{-1}) y_x$$

$\therefore$ $$\nabla = 1 - E^{-1} \quad \text{or} \quad E = (1-\nabla)^{-1}$$

(*iii*) $$\delta y_x = y_{x+h/2} - y_{x-h/2} = E^{1/2} y_x - E^{-1/2} y_x = (E^{1/2} - E^{-1/2}) y_x$$

$\therefore$ $$\delta = E^{1/2} - E^{-1/2}.$$

(*iv*) $$\mu y_x = \frac{1}{2}(y_{x+h/2} + y_{x-y/2}) = \frac{1}{2}(E^{1/2} y_x + E^{-1/2} y_x) = \frac{1}{2}(E^{1/2} + E^{-1/2})\,y_x$$

$\therefore$ $$\mu = \frac{1}{2}(E^{1/2} + E^{-1/2}).$$

(*v*) $$E\nabla y_x = E(y_x - y_{x-h}) = Ey_x - Ey_{x-h} = y_{x+h} - y_x = \Delta y_x \quad \therefore \quad E\nabla = \Delta$$

Also $$\nabla E y_x = \nabla y_{x+h} = y_{x+h} - y_x = \Delta y_x \quad \therefore \quad \nabla E = \Delta$$

$$\delta E^{1/2} y_x = \delta y_{x+h/2} = y_{x+h/2+h/2} - y_{x+h/2-h/2} = y_{x+h} - y_x = \Delta y_x$$

$\therefore$ $$\delta E^{1/2} = \Delta$$

Hence $$\Delta = E\nabla = \nabla E = \delta E^{1/2}.$$

(*vi*) $$Ef(x) = f(x+h)$$

$$= f(x) + hf'(x) + \frac{h^2}{2!} f''(x) + \ldots \qquad \text{[By Taylor's series]}$$

$$= f(x) + hD\,f(x) + \frac{h^2}{2!} D^2 f(x) + \ldots = \left(1 + hD + \frac{h^2 D^2}{2!} + \frac{h^3 D^3}{3!} + \ldots\ldots\right) f(x) = e^{hD} f(x)$$

$\therefore$ $$E = e^{hD}$$

**Cor. 1.** $E = 1 + \Delta = e^{hD}$.

**2.** $D = \frac{1}{h}\log(1+\Delta) = \frac{1}{h}\left(\Delta - \frac{1}{2}\Delta^2 + \frac{1}{3}\Delta^3 - \ldots\right)$

**Note.** *A table showing the symbolic relations between the various operators is given below for ready reference. To prove such relations between the operators, always express each operator in terms of the fundamental operator E.*

**(3) Relations between the various operators**

| *In terms of* | $E$ | $\Delta$ | $\nabla$ | $\delta$ | $hD$ |
|---|---|---|---|---|---|
| $E$ | | $\Delta + 1$ | $(1+\nabla)^{-1}$ | $1 + \frac{1}{2}\delta^2 + \delta\sqrt{(1+\delta^2/4)}$ | $e^{hD}$ |
| $\Delta$ | $E - 1$ | — | $(1-\nabla)^{-1} - 1$ | $\frac{1}{2}\delta^2 + \delta\sqrt{(1+\delta^2/4)}$ | $e^{hD} - 1$ |
| $\nabla$ | $1 - E^{-1}$ | $1 - (1+)^{-1} - 1$ | — | $-\frac{1}{2}\delta^2 + \delta\sqrt{(1+\delta^2/4)}$ | $1 - e^{-hD}$ |
| $\delta$ | $E^{1/2} - E^{-1/2}$ | $\Delta(1+\Delta)^{-1/2}$ | $\nabla(1-\nabla)^{-1/2}$ | — | $2\sinh(hD/2)$ |
| $\mu$ | $\frac{1}{2}(E^{1/2} + E^{-1/2})$ | $(1+\Delta/2)(1+\Delta)^{-1/2}$ | $(1+\nabla/2)(1+\nabla)^{-1/2}$ | $\sqrt{(1+\delta^2/4)}$ | $\cosh(hD/2)$ |
| $hD$ | $\log E$ | $\log(1+\Delta)$ | $\log(1-)^{-1}$ | $2\sinh^{-1}(\delta/2)$ | |

**Example 29.6.** *Prove that*

$$e^x = \left(\frac{\Delta^2}{E}\right)e^x \cdot \frac{Ee^x}{\Delta^2 e^x},$$ *the interval of differencing being h.*

**Solution.** Since $\left(\frac{\Delta^2}{E}\right)e^x = \Delta^2 . E^{-1}e^x = \Delta^2 e^{x-h} = \Delta^2 e^x . e^{-h} = e^{-h}\Delta^2 e^x$

$\therefore$ $$\text{R.H.S.} = e^{-h}\Delta^2 e^x . \frac{Ee^x}{\Delta^2 e^x} = e^{-h} Ee^x = e^{-h} . e^{x+h} = e^x.$$

**Example 29.7.** *Prove with the usual notations, that*

*(i)* $hD = \log(1+\Delta) = -\log(1-\nabla) = \sinh^{-1}(\mu\delta)$

*(ii)* $(E^{1/2} + E^{-1/2})(1+\Delta)^{1/2} = 2 + \Delta$

*(iii)* $\Delta = \frac{1}{2}\delta^2 + \delta\sqrt{(1+\delta^2/4)}$ *(iv)* $\Delta^3 y_2 = \nabla^3 y_5$. **(Delhi, 2015)**

**Solution.** *(i)* We know that $e^{hD} = E = 1 + \Delta$ $\therefore$ $hD = \log(1+\Delta)$

Also $hD = \log E = -\log(E^{-1}) = -\log(1-\nabla)$ $[\because E^{-1} = 1 - \nabla]$

We have proved that $\mu = \frac{1}{2}(E^{1/2} + E^{-1/2})$ and $\delta = E^{1/2} - E^{-1/2}$

$\therefore$ $$\mu\delta = \frac{1}{2}(E^{1/2} + E^{-1/2})(E^{1/2} - E^{-1/2}) = \frac{1}{2}(E - E^{-1}) = \frac{1}{2}(e^{hD} - e^{-hD}) = \sinh(hD)$$

*i.e.* $hD = \sinh^{-1}(\mu\delta)$.

Hence $hD = \log(1+\Delta) = -\log(1-\nabla) = \sinh^{-1}(\mu\delta)$

*(ii)* $(E^{1/2} + E^{-1/2})(1+\Delta)^{1/2} = (E^{1/2} + E^{-1/2})E^{1/2} = E + 1 = 1 + \Delta + 1 = 2 + \Delta$.

*(iii)* $\frac{1}{2}\delta^2 + \delta\sqrt{(1+\delta^2/4)}$

$$= \frac{1}{2}(E^{1/2} - E^{-1/2})^2 + (E^{1/2} - E^{-1/2})\sqrt{[1 + (E^{1/2} - E^{-1/2})^2/4]}$$

$$= \frac{1}{2}(E + E^{-1} - 2) + (E^{1/2} - E^{-1/2})\sqrt{[(E + E^{-1} + 2)/4]}$$

$$= \tfrac{1}{2}(E + E^{-1} - 2) + \tfrac{1}{2}(E^{1/2} - E^{-1/2})(E^{1/2} + E^{-1/2})$$

$$= \tfrac{1}{2}[(E + E^{-1} - 2) + (E - E^{-1})] = \tfrac{1}{2}(2E - 2) = E - 1 = \Delta.$$

(*iv*) $\Delta^3 y_2 = (E - 1)^3 y_2$ $\quad [\because \ \Delta = E - 1]$

$= (E^3 - 3E^2 + 3E - 1)\, y_2 = y_5 - 3y_4 + 3y_3 - y_2$ ...(1)

$\nabla^3 y_5 = (1 - E^{-1})^3 y_5$ $\quad [\because \ \Delta = 1 - E^{-1}]$

$= (1 - 3E^{-1} + 3E^{-2} - E^{-3})\, y_5 = y_5 - 3y_4 + 3y_3 - y_2$ ...(2)

From Eqns. (1) and (2), $\Delta^3 y_2 = \nabla^3 y_5$.

## 29.5 TO FIND ONE OR MORE MISSING TERMS

When one or more values of $y = f(x)$ corresponding to the equidistant values of $x$ are missing, we can find these using any of the following two methods :

*First method :* We assume the missing term or terms as $a$, $b$ etc. and form the difference table. Assuming the last difference as zero, we solve these equations for $a$, $b$. These give the missing term/terms.

*Second method :* If $n$ entries of $y$ are given, $f(x)$ can be represented by $a(n-1)$th degree polynomial *i.e.*, $\Delta^n y = 0$. Since $\Delta = E - 1$, therefore $(E-1)^n y = 0$. Now expanding $(E - 1)^n$ and substituting the given values, we obtain the missing term/terms.

**Example 29.8.** *Find the missing term in the table :*

| $x$ : | 2 | 3 | 4 | 5 | 6 |
|---|---|---|---|---|---|
| $y$ : | 45.0 | 49.2 | 54.1 | ... | 67.4 |

**Solution.** Let the missing term be $a$. Then the difference table is as follows :

| $x$ | $y$ | $\Delta y$ | $\Delta^2 y$ | $\Delta^3 y$ | $\Delta^4 y$ |
|---|---|---|---|---|---|
| 2 | 45.0 $(= y_0)$ | | | | |
| | | 4.2 | | | |
| 3 | 49.2 $(= y_1)$ | | 0.7 | | |
| | | 49.9 | | $a - 59.7$ | |
| 4 | 54.1 $(= y_2)$ | | $a - 59.0$ | | $240.2 - 4a$ |
| | | $a - 54.1$ | | $180.5 - 3a$ | |
| 5 | $a (= y_3)$ | | $121.5 - a$ | | |
| | | $67.4 - a$ | | | |
| 6 | 67.4 $(= y_4)$ | | | | |

We know that $\Delta^4 y = 0$ *i.e.*, $240.2 - 4a = 0$.

Hence $a = 60.05$.

**Otherwise:** As only four entries $y_0, y_1, y_2, y_3$ are given, therefore $y = f(x)$ can be represented by a third degree polynomial.

$\therefore$ $\Delta^3 y =$ constant or $\Delta^4 y = 0$ *i.e.*, $(E - 1)^4 = 0$

*i.e.*, $(E^4 - 4E^3 + 6E^2 - 4E + 1) = 0$ or $y_4 - 4y_3 + 6y_2 - 4y_1 + y_0 = 0$

Let the missing entry $y_3$ be $a$ so that

$$67.4 - 4a + 6(54.1) - 4(49.2) + 45 = 0 \text{ or } -4a = -240.2$$

Hence $a = 60.05$.

**Example 29.9.** *Find the missing values in the following data :*

| $x$ : | 45 | 50 | 55 | 60 | 65 |
|---|---|---|---|---|---|
| $y$ : | 3.0 | ... | 2.0 | ... | −2.4 |

**Solution.** Let the missing value be $a$, $b$. Then the difference table is as follows :

| $x$ | $y$ | $\Delta y$ | $\Delta^2 y$ | $\Delta^3 y$ |
|---|---|---|---|---|
| 45 | $3(= y_0)$ | | | |
| | | $a - 3$ | | |
| 50 | $a(= y_1)$ | | $5 - 2a$ | |
| | | $2 - a$ | | $3a + b - 9$ |
| 55 | $2(= y_2)$ | | $b + a - 4$ | |
| | | $b - 2$ | | $3.6 - a - 36$ |
| 60 | $b(= y_3)$ | | $-0.4 - 2b$ | |
| | | $-2.4 - b$ | | |
| 65 | $-2.4\ (= y_4)$ | | | |

As only three entries $y_0, y_2, y_4$ are given, $y$ can be represented by a second degree polynomial having third differences as zero.

$\therefore$ $$\Delta^3 y_0 = 0 \text{ and } \Delta^3 y_1 = 0$$

*i.e.,* $$3a + b = 9,\ a + 3b = 3.6$$

Solving these, we get $a = 2.925$, $b + 0.0225$.

**Otherwise.** As only three entries $y_0 = 3, y_2 = 2, y_4 = -2.4$ are given, $y$ can be represented by a second degree polynomial having third differences as zero.

$\therefore$ $$\Delta^3 y_0 = 0 \quad \text{and} \quad \Delta^3 y_1 = 0$$

*i.e.,* $$(E - 1)^3\, y_0 = 0 \text{ and } (E - 1)^3\, y_1 = 0$$

*i.e.,* $$(E^3 - 3E^3 + 3E - 1)\, y_0 = 0\ ;\ (E^3 - 3E^2 + 3E - 1)\, .\, y_1 = 0$$

or $$y_3 - 3y_2 + 3y_1 - y_0 = 0\ ;\ y_4 - 3y_3 + 3y_2 - y_1 = 0$$

or $$y_3 + 3y_1 = 9\ ;\ 3y_3 + y_1 = 3.6$$

Solving three, we get $y_1 = 2.925$, $y_2 = 0.225$.

**Example 29.10.** *If $y_{10} = 3$, $y_{11} = 6$, $y_{12} = 11$, $y_{13} = 18$, $y_{14} = 27$, find $y_4$.*

**Solution.** Taking $y_{14}$ as $u_0$, we are required to find $y_4$ *i.e.,* $u_{-10}$. Then the difference table is

| $x$ | $u$ | $\Delta u$ | $\Delta^2 u$ | |
|---|---|---|---|---|
| $x_{-4}$ | $y_{10} = u_{-4} = 3$ | | | |
| | | 3 | | |
| $x_{-3}$ | $y_{11} = u_{-3} = 6$ | | 2 | |
| | | 5 | | 0 |
| $x_{-2}$ | $y_{12} = u_{-2} = 11$ | | 2 | |
| | | 7 | | 0 |
| $x_{-1}$ | $y_{13} = u_{-1} = 18$ | | 2 | |
| | | 9 | | |
| $x_0$ | $y_{14} = u_0 = 27$ | | | |

Then $$y_4 = u_{-10} = (E^{-1})^{10}\, u_0 = (1 - \nabla)^{10}\, u_8$$

$$= \left(1 - 10\nabla + \frac{10.9}{2}\nabla^2 - \frac{10.9.8}{1.2.3}\nabla^3 + \ldots\right) u_0$$

$$= u_0 - 10\nabla u_0 + 45\nabla^2 u_0 - 120\nabla^3 u_0$$

$$= 27 - 10 \times 9 + 45 \times 2 - 120 \times 0 = 27.$$

**Example 29.11.** *If $y_x$ is a polynomial for which fifth difference is constant and $y_1 + y_7 = -7845$, $y_2 + y_6 = 686$, $y_3 + y_5 = 1088$, find $y_4$.*

**Solution.** Starting with $y_1$ instead of $y_0$, we note that $\Delta^6 y_1 = 0$ [$\because$ $\Delta^5 y_1$ is constant.

*i.e.,* $$(E - 1)^6\, y_1 = (E^6 - 6E^5 + 15E^4 - 20E^3 + 15E^2 - 6E + 1)\, y_1 = 0$$

$\therefore$ $$y_7 - 6y_6 + 15y_5 - 20y_4 + 15y_3 - 6y_2 + y_1 = 0$$

or $$(y_7 + y_1) - 6(y_6 + y_2) + 15(y_5 + y_3) - 20y_4 = 0$$

*i.e.* $$y_4 = \frac{1}{20}[(y_1 + y_7) - 6(y_2 + y_6) + 15(y_3 + y_5)]$$

$$= \frac{1}{20}[-784 - 6(686) + 15(1088)] = 571.$$

**Example 29.12.** *Prove the following identities :*

(*i*) $u_1x + u_2x^2 + u_3x^3 + \ldots = \frac{x}{1-x}u_1 + \left(\frac{x}{1-x}\right)^2 \Delta u_1 + \left(\frac{x}{1-x}\right)^3 \Delta^2 u_1 + \ldots$

(*ii*) $u_0 + \frac{u_1x}{1!} + \frac{u_2x^2}{2!} + \frac{u_3x^3}{3!} + \ldots = e^x\left(u_0 + x\Delta u_0 + \frac{x^2}{2!}\Delta^2 u_0 + \frac{x^3}{3!}\Delta^3 u_0 + \ldots\right).$

**Solution.** (*i*) L.H.S. $= xu_1 + x^2Eu_1 + x^3E^2u_1 + \ldots = x(1 + xE + x^2E^2 + \ldots)\, u_1$ $[\because \; u_{x+h} = E^h u_x]$

$$= x \cdot \frac{1}{1 - xE}\, u_1, \text{ taking sum of infinite G.P.}$$

$$= x\left[\frac{1}{1 - x(1+\Delta)}\right]u_1 \qquad [\because \; E = 1 + \Delta]$$

$$= x\left(\frac{1}{1 - x - x\Delta}\right)u_1 = \frac{x}{1-x}\left(1 - \frac{x\Delta}{1-x}\right)^{-1} u_1 = \frac{x}{1-x}\left(1 + \frac{x\Delta}{1-x} + \frac{x^2\Delta^2}{(1-x^2)} + \ldots\ldots\right)u_1$$

$$= \frac{x}{1-x}u_1 + \frac{x^2}{(1-x)^2}\Delta u_1 + \frac{x^2}{(1-x)^3}\Delta^2 u_1 + \ldots = \text{R.H.S.}$$

(*ii*) $$\text{L.H.S.} = u_0 + \frac{x}{1!}Eu_0 + \frac{x^2}{2!}E^2u_0 + \frac{x^3}{3!}E^3u_0 + \ldots$$

$$= \left(1 + \frac{xE}{1!} + \frac{x^2E^2}{2!} + \frac{x^3E^3}{3!} + \ldots\right)u_0 = e^{xE}\,u_0 - e^{x(1+\Delta)}\,u_0$$

$$= e^x \cdot e^{x\Delta}\,u_0 = e^x\left(1 + \frac{x\Delta}{1!} + \frac{x^2\Delta^2}{2!} + \frac{x^3\Delta^3}{3!} + \ldots\right)u_0$$

$$= e^x\left(u_0 + \frac{x}{1!}\Delta u_0 + \frac{x^2}{2!}\Delta^2 u_0 + \frac{x^3}{3!}\Delta^3 u_0 + \ldots\right) = \text{R.H.S.}$$

## PROBLEMS 29.2

1. Explain the difference between $\left(\frac{\Delta^2}{E}\right)u_x$ and $\frac{\Delta^2 u_x}{Eu_x}$.

2. Evaluate taking $h$ as the interval of differencing :

   (*i*) $\frac{\Delta^2}{E}\sin x$ (*ii*) $\left(\frac{\Delta^2}{E}\right)x^4, (h = 1)$

   (*iii*) $\left(\frac{\Delta^2}{E}\right)\sin(x+h) + \frac{\Delta^2 \sin(x+h)}{E\sin(x+h)}$ (*iv*) $(\Delta + \nabla)^2(x^2 + x), (h = 1).$

3. With the usual notations, show that

   (*i*) $\nabla = 1 - e^{-hD}$ (*ii*) $D = \frac{2}{h}\sinh^{-1}\left(\frac{\delta}{2}\right)$

   (*iii*) $(1 + \Delta)(1 - \nabla) = 1.$ (*J.N.T.U., 2015*) (*iv*) $\Delta - \nabla = \nabla\Delta = \delta^2.$ (*JNTUK, 2022, 2015; Delhi., 2015*)

4. Prove that

(i) $\delta = \Delta(1 + \Delta)^{-1/2} = \nabla(1 - \nabla)^{-1/2}$ (ii) $\mu^2 = 1 + \dfrac{\delta^2}{4}$

(iii) $\delta(E^{1/2} + E^{-1/2}) = \Delta E^{-1} + \Delta$ (iv) $\nabla = \Delta E^{-1} = E^{-1}\Delta = 1 - E^{-1}$

5. Show that (i) $\mu\delta = \frac{1}{2}(\Delta + \nabla)$ (*Delhi., 2015 ; U.P.T.U., 2011*) (ii) $1 + \delta^2/2 = \sqrt{(1 + \delta^2\mu^2)}$

(iii) $\Delta + \nabla = \dfrac{\Delta}{\nabla} - \dfrac{\nabla}{\Delta}$ (*JNTUK, 2022*) (iv) $\nabla^2 = h^2D^2 - h^3D^3 + \dfrac{7}{12}h^4D^4 - \ldots\ldots$

6. Prove that

(i) $\nabla^r f_k = \Delta^r f_{k-r}$ (ii) $\Delta f_k^{\,2} = (f_k + f_{k+1})\,\Delta f_k$

(iii) $\Delta = \dfrac{1}{2}\delta^2 + \delta\sqrt{\left(1 + \dfrac{\delta^2}{4}\right)}$ (iv) $E^{1/2} = (1 + \delta^2/4)^{1/2} + \delta/2.$

7. Prove that $\nabla y_{n+1} = h\left(1 + \dfrac{1}{2}\nabla + \dfrac{5}{12}\nabla^2 + \ldots\right)y'_n.$

8. The following table gives the values of $y$ which is a polynomial of degree five. It is known that $f(3)$ is in error. Correct the error.

| $x$ : | 0 | 1 | 2 | 3 | 4 | 5 | 6 |
|---|---|---|---|---|---|---|---|
| $y$ : | 1 | 2 | 33 | 254 | 1025 | 3126 | 7777 |

9. Estimate the missing term in the following table :

| $x$ : | 0 | 1 | 2 | 3 | 4 |
|---|---|---|---|---|---|
| $f(x)$ : | 1 | 3 | 9 | — | 81 |

10. Find the missing terms of the following data :

| $x$ : | 1 | 1.5 | 2 | 2.5 | 3 | 3.5 | 4 |
|---|---|---|---|---|---|---|---|
| $f(x)$ : | 6 | ? | 10 | 20 | ? | 1.5 | 5 |

(*U.P.T.U., 2010*)

11. Find the missing values in the following table :

| $x$ : | 0 | 1 | 2 | 3 | 4 | 5 | 6 |
|---|---|---|---|---|---|---|---|
| $y$ : | 5 | 11 | 22 | 40 | ... | 140 | ... |

12. If $u_{13} = 1, u_{14} = -3, u_{15} = -1, u_{16} = 13$ find $u_8$.

13. Evaluate $y_4$ from the following data (stating the assumptions you make) :

$$y_0 + y_8 = 1.9243,\ y_1 + y_7 = 1.9590,\ y_2 + y_6 = 1.9823,\ y_3 + y_5 = 1.9956.$$

14. Using the method of separation of symbols, prove that

(i) $u_0 + u_1 + u_2 + \ldots\ldots + u_n = {}^{n+1}C_1 u_0 + {}^{n+1}C_3\,\Delta^2 u_0 + \ldots\ldots + {}^{n+1}C_{n+1}\,\Delta^n u_0.$

(ii) $y_x = y_n - {}^{n-x}C_1 \Delta y_{n-1} + {}^{n-x}C_2 \Delta^2 y_{n-2} - \ldots\ldots + (-1)^{n-x}\Delta^{n-x} y_{n-(n-x)}.$

15. Using the method of finite differences, sum the following series :

(i) 2.5 + 5.8 + 8.11 + 11.14 + ... to $n$ terms.

(ii) 1.2.3 + 2.3.4 + 3.4.5 + ... to $n$ terms.

16. Prove that $u_0 + u_1 x + u_2 x^2 + \ldots\ldots \infty = \dfrac{u_0}{1-x} + \dfrac{x\Delta u_0}{(1-x)^2} + \dfrac{x^2\Delta^2 u_0}{(1-x)^3} + \ldots\ldots \infty$

Hence sum the series $1.2 + 2.3x + 3.4x^2 + \ldots\ldots \infty$.

## 29.6 NEWTON'S INTERPOLATION FORMULAE

We now derive two important interpolation formulae by means of the forward and backward differences of a function. These formulae are often employed in engineering and scientific problems.

**(1) Newton's forward interpolation formula.** Let the function $y = f(x)$ take the values $y_0, y_1, y_2, \ldots$ corresponding to the values $x_0, x_0 + h, x_0 + 2h, \ldots$ of $x$. Suppose it is required to evaluate $f(x)$ for $x = x_0 + ph$, where $p$ is any real number.

For any real number $p$, we have defined $E$ such that

$$E^p f(x) = f(x + ph)$$

$$\therefore \quad y_p = f(x_0 + ph) = E^p f(x_0) = (1 + \Delta)^{-p}\, y_0 \qquad [\because\ E = 1 + \Delta]$$

$$= \left\{1 + p\Delta + \frac{p(p-1)}{2!}\Delta^2 + \frac{p(p-1)(p-2)}{3!}\Delta^3 + \ldots\right\} y_0 \qquad \text{[Using Binomial theorem]}$$

*i.e.,*
$$y_p = y_0 + p\Delta y_0 + \frac{p(p-1)}{2!}\Delta^2 y_0 + \frac{p(p-1)(p-2)}{3!}\Delta^3 y_0 + \ldots \qquad \ldots(1)$$

It is called **Newton's forward interpolation formula** as (1) contains $y_0$ and the forward differences of $y_0$.

**Obs.** *This formula is used for interpolating the values of y near the beginning of a set of tabulated values and extrapolating values of y a little backward (i.e. to the left) of $y_0$.*

**(2) Newton's backward interpolation formula.** Let the function $y = f(x)$ take the values $y_0, y_1, y_2, \ldots$ corresponding to the values $x_0, x_0 + h, x_0 + 2h, \ldots$ of $x$. Suppose it is required to evaluate $f(x)$ for $x = x_0 + ph$, where $p$ is any real number. Then we have

$$y_p = f(x_n + ph) = E^p f(x_n) = (1 - \nabla)^{-p} y_n \qquad [\because \; E^{-1} = 1 - \nabla]$$

$$= \left[1 + p\nabla + \frac{p(p+1)}{2!}\nabla^2 + \frac{p(p+1)(p+2)}{3!}\nabla^2 + \ldots\right] y_n \qquad \text{[Using Binomial theorem]}$$

*i.e.,*
$$y_p = y_n + p\nabla y_n + \frac{p(p+1)}{2!}\nabla^2 y_n + \frac{p(p+1)(p+2)}{3!}\nabla^3 y_n + \ldots \qquad \ldots(2)$$

It is called *Newton's backward interpolation formula as* Eqn. (2) contains $y_n$ and backward differences of $y_n$.

**Obs.** *This formula is used for interpolating the values of y near the end of a set of tabulated values and also for extrapolating values of y a little ahead* (to the right) *of $y_n$.*

**Example 29.13.** *Let $\frac{dy}{dx} = x + y$ and $y = 1$ when $x = 0$. Find y when $x = 0.05, 0.10, 0.1{:}$ and $0.20$.*

(*Marathwada 2018*)

**Solution.** Using Euler's method

$$y_{n+1} = y_n + hf(x_n, y_n) = y_n + h(x_n + y_n)$$

(*i*) $\quad y_1 = y_0 + h\,(x_0 + h_0) = 1 + 0.05\,(0 + 1) = 1.05 \qquad \ldots(i)$

(*ii*) $\quad y_2 = y_1 + h(x_1 + h_1) = 1 + 0.05(0 + 1) = 1.05$

(*iii*) $\quad y_3 = y_2 + h(x_2 + y_2) = 1.05 + 0.05\,(0.10 + 1.105) = 1.16525$

(*iv*) $\quad y_4 = y_3 + h(x_3 + y_3) = 1.16525 + 0.05\,(0.15 + 1.16525) = 1.2310$

Next, we will solve

$$y_1^{(2)} = y_0 + \frac{h}{2}\left[f(x_0, y_0) + f(x_1, y_1^{(1)})\right]$$

$$= y_0 + \frac{h}{2}\left[(x_0 + y_0) + (x_1 + y_1^{(1)})\right]$$

$$= 1 + \frac{0.05}{2}[(0 + 1) + (0.05 + 1.05)]$$

$$= 1 + 0.025 \times 2.10 = 1.0525$$

The third approximation is

$$y_1^{(4)} \pm y_0 + \frac{h}{2}\left[f(x_0, y_0) + f(x_1, y_1)^{(3)}\right]$$

$$= 1 + \frac{0.05}{2}[(0 + 1) + (0.05 + 1.0525)] = 1.05256$$

Now obtain, $y_2^{(1)} = y_1 + hf(x_1, y_1)$

first approximation, $= y_1 + h(x_1 + y_1)$

$= 1.0526 + 0.05\ (0.05 + 1.0526) = 1.1077$

the second approximation is

$$y_2^{(4)} = y_1 + \frac{h}{2}\left[f(x_1 y_1) + f(x_1, y_2)^{(1)}\right]$$

$$= 1.0526 + \frac{0.05}{2}[(0.5 + 1.05256) + (0.1 + 1.1077)]$$

$$= 1.0526 + 0.025\ (1.1026 + 1.2077) = 1.1104$$

The third approximation is

$$y_2^{(3)} = 1.0526 + \frac{0.05}{2}(1.1026 + 1.2104)$$

$$= 1.0526 + 0.025\ (1.1026 + 1.2077) = 1.1104$$

which is same as $y_2(2)$

Similarly we can obtain

$$y_3(2) = y_3(3) = 1.1737$$

$$y_4(2) = y_3(3) = 1.2428$$

So, $y_1 = 1.0526 \quad y_2 = 1.1104$

$y_3 = 1.1737 \quad y_4 = 1.2428$

Connect up to 3 places of decimals

$y_1 = 1.05254, \quad y_2 = 1.11034$

$y_3 = 1.17366, \quad y_4 = 1.24280.$

**Exercise 29.14:** *If $f(x) = x^3 - 3x^2 + 5x + 7$ find $\Delta f(x)$, $\Delta^2 f(x)$, $\Delta^3 f(x)$, $\Delta^4 f(x)$ when $h = 1$.* (AKTU, 2018)

**Solution.** We have

$$\Delta f(x) = f(x+1) - f(x), \quad \text{since } h = 1$$

$$= [(x+1)^3 - 3(x+1)^2 + 5(x+1) + 7] - (x^3 - 3x^2 + 5x + 7)$$

$$= 3(x^2 - x + 1)$$

$$\Delta^2 f(x) = \Delta[\Delta f(x)] = \Delta[3(x^2 - x + 1)] = 3[\Delta(x^2 - x + 1)]$$

$$= 3[\{(x+1)^2 - (x+1)^2 + 1]\} - (x^2 - x + 1)] = 6x$$

$$\Delta^3 f(x) = \Delta[\Delta^2 f(x)] = \Delta[6x] = 6\Delta(x)$$

$$= 6[(x+1) - x] = 6$$

$$\Delta^4 f(x) = \Delta[\Delta^2 f(x)] = \Delta(6)$$

$$= 6 - 6 = 0$$

Since $f(x)$ is a polynomial of degree $n$, fourth and all higher differences.

**Example 29.15.** *Find the cubic polynomial which takes the following values :*

| $x$ : | 0 | 1 | 2 | 3 |
|---|---|---|---|---|
| $f(x)$ : | 1 | 2 | 1 | 10 |

*Hence or otherwise evaluate $f(4)$.* (Anna, 2013)

**Solution.** The difference table is

| $x$ | $f(x)$ | $\Delta f(x)$ | $\Delta^2 f(x)$ | $\Delta^3 f(x)$ |
|---|---|---|---|---|
| 0 | **1** | | | |
| | | **1** | | |
| 1 | 2 | | **–2** | |
| | | –1 | | **12** |
| 2 | 1 | | 10 | |
| | | 9 | | |
| 3 | 10 | | | |

We take $x_0 = 0$ and $p = \dfrac{x - 0}{h} = x$ [$\because$ $h = 1$]

$\therefore$ using Newton's forward interpolation formula, we get

$$f(x) = f(0) + \frac{x}{1}\Delta f(0) + \frac{x(x-1)}{1.2}\Delta^2 f(0) + \frac{x(x-1)(x-2)}{1.2.3}\Delta^3 f(0)$$

$$= 1 + x(1) + \frac{x(x-1)}{2}(-2) + \frac{x(x-1)(x-2)}{6}(12)$$

$= 2x^3 - 7x^2 + 6x + 1$, which is the required polynomial.

To compute $f(4)$, we take $x_n = 3$, $x = 4$ so that $p = \dfrac{x - x_n}{h} = 1$ [$\because$ $h = 1$]

Using Newton's backward interpolation formula, we get

$$f(4) = f(3) + p\nabla f(3) + \frac{p(p+1)}{1.2}\nabla^2 f(3) + \frac{p(p+1)(p+2)}{1.2.3}\nabla^3 f(3)$$

$= 10 + 9 + 10 + 12 + 41.$

which is the same value as that obtained by substituting $x - 4$ in the cubic polynomial above.

**Obs.** The above example shows that *if a tabulated function is a polynomial, then interpolation and extrapolation give the same values.*

**Example 29.16.** *The following table gives corresponding values of x and y. Prepare a forward difference, table and express y as a function of x. Also obtain y when x = 2.5.*

| $x$ : | 0 | 1 | 2 | 3 | 4 |
|---|---|---|---|---|---|
| $y$ : | 7 | 10 | 13 | 22 | 43 |

*(Raipur, 2018)*

**Solution.** The following is the difference table:

| $x$ | $y$ | $\Delta y$ | $\Delta^2 y$ | $\Delta^3 y$ | $\Delta^4 y$ |
|---|---|---|---|---|---|
| 0 | 7 | | | | |
| | | 3 | | | |
| 1 | 10 | | 0 | | |
| | | 3 | | 6 | |
| 2 | 13 | | 6 | | 0 |
| | | 9 | | 6 | |
| 3 | 22 | | 12 | | |
| | | 21 | | | |
| 4 | 43 | | | | |

from the difference table, we have

$$y_0 = 7, \Delta y_0 = 3, \Delta^2 y_0 = 0, \Delta^3 y_0 = 6, \Delta_4 y_0 = 0$$

putting these value,

$$y_n = 7 + x \cdot 3 + \frac{x(x-1)}{2!} \cdot 0 + \frac{x(x-1)(x-2)}{3!}.6$$

$$y = x^3 - 3x^2 + 5x + 7$$

which is required equation of the curve which passes through the given points put $x = 2.5$ and $y = 16.375$.

**Exercise 29.17:** *Express $f(u) = u^4 - 3u^2 + 2u + 6$ in terms factorial polynomial. Hence show that $\Delta^2 f(u) = 24$.* *(Rourkela, 2018)*

**Solution.** Let $u^4 - 3u^2 + 2u + 6 = Au(u-1)(u-2)(u-3) + Bu(u-1)(u-2) + Cu(u-1) + Du + E$

Now put $u = 1$

$E = 6$

Next put $u = 1$

then $1 - 3 + 2 + 6 = D + E$

$D = 0$

So, putting $u = 2$, we get

$$14 = 2C + 2D + E \Rightarrow C = 4$$

and on putting $u = 3$, we have

$$66 = 6B + 6C + 3D + E$$

$$B = 6.$$

finally comparing coefficient of $u^4$,

we get $A = 1$

Hence $f(u) = u^{(4)} + 6u^{(3)} + 4u^{(2)} + 6$

Now $\Delta^4 f(u) = 24.$

## PROBLEMS 29.3

1. Find the number of men getting wage of ₹10 from the following table :

| *Wages (in ₹)* : | 5 | 15 | 25 | 35 |
|---|---|---|---|---|
| *No. of men* : | 9 | 30 | 35 | 42 |

*(U.P.T.U., 2011)*

2. State Newton's interpolation formula and use it to calculate the value of exp (1.85), given the following table :

| $x$ : | 1.7 | 1.8 | 1.9 | 2.0 | 2.1 | 2.2 | 2.3 |
|---|---|---|---|---|---|---|---|
| $f(x)$ : | 5.474 | 6.050 | 6.686 | 7.389 | 8.166 | 9.025 | 9.974 |

3. Express the value of θ in terms of $x$ using the following data :

| $x$ : | 40 | 50 | 60 | 70 | 80 | 90 |
|---|---|---|---|---|---|---|
| θ : | 184 | 204 | 226 | 250 | 276 | 304 |

Also find θ at $x = 43$. *(U.P.T.U., 2012)*

4. Given sin 45° = 0.7071, sin 50° = 0.7660, sin 55° = 0.8192, sin 60° = 0.8660, find sin 52°, using Newton's forward formula.

5. From the following table of half-yearly premium for policies maturing at different ages, estimate the premium for policies maturing at age of 46 :

| Age : | 45 | 50 | 55 | 60 | 65 |
|---|---|---|---|---|---|
| Premium (in rupees) : | 114.84 | 96.16 | 83.32 | 74.48 | 68.48 |

*(U.P.T.U., 2010)*

6. The area $A$ of a circle of diameter $d$ is given for the following values :

| $d$ : | 80 | 85 | 90 | 95 | 100 |
|---|---|---|---|---|---|
| $A$ : | 5026 | 5674 | 6362 | 7088 | 7854 |

*(V.T.U., 2010)*

Calculate the area of a circle of diameter 105.

7. Estimate the value of $f(22)$ and $f(42)$ from the following available data :

| $x$ : | 20 | 25 | 30 | 35 | 40 | 45 |
|---|---|---|---|---|---|---|
| $f(x)$ : | 354 | 332 | 291 | 260 | 231 | 204 |

*(Anna, 2012)*

8. From the following table :

| $x°$ : | 10 | 20 | 30 | 40 | 50 | 60 | 70 | 80 |
|---|---|---|---|---|---|---|---|---|
| cos $x$ : | 0.9848 | 0.9397 | 0.8660 | 0.7660 | 0.6428 | 0.5000 | 0.3420 | 0.1737 |

Calculate cos 25° and cos 73° using Gregory Newton formulae.

9. Find the number of men getting wages between ₹ 100 and ₹ 150 from the following data :

| Wages in ₹ : | 0—100 | 100—200 | 200—300 | 300—400 |
|---|---|---|---|---|
| Frequency : | 9 | 30 | 35 | 42 |

*(Delhi., 2015)*

10. Find the polynomial interpolating the data :

| $x$ : | 0 | 1 | 2 |
|---|---|---|---|
| $f(x)$ : | 0 | 5 | 2 |

11. Construct Newton's forward interpolation polynomial for the following data :

| $x$ : | 4 | 6 | 8 | 10 |
|---|---|---|---|---|
| $y$ : | 1 | 3 | 8 | 16 |

Hence evaluate $y$ for $x = 5$. *(Anna, 2012)*

12. Construct the difference table for the following data :

| $x$ : | 0.1 | 0.3 | 0.5 | 0.7 | 0.9 | 1.1 | 1.3 |
|---|---|---|---|---|---|---|---|
| $f(x)$ : | 0.003 | 0.067 | 0.148 | 0.248 | 0.370 | 0.518 | 0.697 |

Evaluate $f(0.6)$

13. Estimate from following table $f(3.8)$ to three significant figures using Gregory Newton backward interpolation formula:

| $x$ : | 0 | 1 | 2 | 3 | 4 |
|---|---|---|---|---|---|
| $f(x)$ : | 1 | 1.5 | 2.2 | 3.1 | 4.6 |

14. The following table gives the population of a town during the last six censuses. Estimate the increase in the population during the period from 1976 to 1978 :

| Year : | 1941 | 1951 | 1961 | 1971 | 1981 | 1991 |
|---|---|---|---|---|---|---|
| Population (in thousands) : | 12 | 15 | 20 | 27 | 39 | 52 |

15. In the following table, the values of $y$ are consecutive terms of a series of which 12.5 is the 5th term. Find the first and tenth terms of the series.

| $x$ : | 3 | 4 | 5 | 6 | 7 | 8 | 9 |
|---|---|---|---|---|---|---|---|
| $y$ : | 2.7 | 6.4 | 12.5 | 21.6 | 34.3 | 51.2 | 72.9 |

16. Given $u_1 = 40$, $u_3 = 45$, $u_5 = 54$, find $u_2$ and $u_4$.
17. If $u_{-1} = 10$, $u_1 = 8$, $u_2 = 10$, $u_4 = 50$, find $u_0$ and $u_3$.
18. Given $y_0 = 3$, $y_1 = 12$, $y_2 = 81$, $y_3 = 200$, $y_4 = 100$, $y_5 = 8$, without forming the difference table, find $\Delta^5 y_0$.

## 29.7 CENTRAL DIFFERENCE INTERPOLATION FORMULAE

In the preceding section, we derived Newton's forward and backward interpolation formulae which are applicable for interpolation near the beginning and end of tabulated values. Now we shall develop central difference formulae which are best suited for interpolation near the middle of the table.

If $x$ takes the values $x_0 - 2h$, $x_0 - h$, $x_0$, $x_0 + h$, $x_0 + 2h$ and the corresponding values of $y = f(x)$ are $y_{-2}$, $y_{-1}$, $y_0$, $y_1$, $y_2$, then we can write the difference table in the two notations as follows :

| $x$ | $y$ | *1st diff.* | *2nd diff.* | *3rd diff.* | *4th diff.* |
|---|---|---|---|---|---|
| $x_0 - 2h$ | $y_{-2}$ | | | | |
| | | $\Delta y_{-2} (= \delta y_{-3/2})$ | | | |
| $x_0 - h$ | $y_{-1}$ | | $\Delta^2 y_{-2} (= \delta^2 y_{-1})$ | | |
| | | $\Delta y_{-1} (= \delta y_{-1/2})$ | | $\Delta^3 y_{-2} (= \delta^3 y_{-1/2})$ | |
| $x_0$ | $y_0$ | | $\Delta^2 y_{-1} (\delta^2 y_0)$ | | $\Delta^4 y_{-2} (= \delta^4 y_0)$ |
| | | $\Delta y_0 (= \delta y_{1/2})$ | | $\Delta^3 y_{-1} (= \delta^3 y_{1/2})$ | |
| $x_0 + h$ | $y_1$ | | $\Delta^2 y_0 (= \delta^2 y_1)$ | | |
| | | $\Delta y_1 (= \delta y_{3/2})$ | | | |
| $x_0 + 2h$ | $y_2$ | | | | |

(1) **Gauss's forward interpolation formula.** The Newton's forward interpolation formula is

$$y_p = y_0 + p\Delta y_0 + \frac{p(p-1)}{1.2} \Delta^2 y_0 + \frac{p(p-1)(p-2)}{1.2.3} \Delta^3 y_0 + \dots \quad \dots(1)$$

We have $\Delta^2 y_0 - \Delta^2 y_{-1} = \Delta^3 y_{-1}$

*i.e.,* $\Delta^2 y_0 = \Delta^2 y_{-1} + \Delta^3 y_{-1}$ ...(2)

Similarly $\Delta^3 y_0 = \Delta^3 y_{-1} + \Delta^4 y_{-1}$ ...(3)

$\Delta^4 y_0 = \Delta^4 y_{-1} + \Delta^5 y_{-1}$ etc. ...(4)

Also $\Delta^3 y_{-1} - \Delta^3 y_{-2} = \Delta^4 y_{-2}$

*i.e.,* $\Delta^3 y_{-1} = \Delta^3 y_{-2} + \Delta^4 y_{-2}$

Similarly $\Delta^4 y_{-1} = \Delta^4 y_{-2} + \Delta^5 y_{-2}$ etc. ...(5)

Substituting for $\Delta^2 y_0, \Delta^3 y_0, \Delta^4 y_0$ ... from Eqns. (2), (3), (4) ... in Eqn. (1), we get

$$y_p = y_0 + p\Delta y_0 + \frac{p(p-1)}{1.2}(\Delta^2 y_{-1} + \Delta^3 y_{-1}) + \frac{p(p-1)(p-2)}{1.2.3}(\Delta^3 y_{-1} + \Delta^4 y_{-1})$$
$$+ \frac{p(p-1)(p-2)(p-3)}{1.2.3.4}(\Delta^4 y_{-1} + \Delta^5 y_{-1}) + \ldots.$$

Hence $$y_p = y_0 + p\Delta y_0 + \frac{p(p-1)}{2!}\Delta^2 y_{-1} + \frac{(p+1)p(p-1)}{3!}\Delta^3 y_{-1}$$
$$+ \frac{(p+1)p(p-1)(p-2)}{4!}\Delta^4 y_{-2} + \ldots \text{ [Using (5)]}$$

which is called *Gauss's forward interpolation formula.*

**Cor.** *In the central differences notation, this formula will be*

$$y_p = y_0 + p\delta y_{1/2} + \frac{p(p-1)}{2!}\delta^2 y_0 + \frac{(p+1)p(p-1)}{3!}\delta^3 y_{1/2} + \frac{(p+1)p(p-1)(p-2)}{4!}\delta^4 y_0 + \ldots$$

**Obs. 1.** *It employs odd differences just below the central line and even difference on the central line as shown below:*

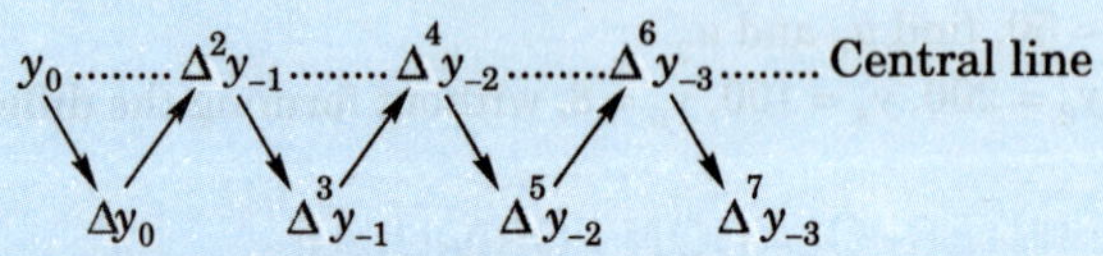


**Obs. 2.** This formula is used to interpolate the values of $y$ for $p$ $(0 < p < 1)$ measured forwardly from the origin.

**(2) Gauss's backward interpolation formula.** The Newton's forward interpolation formula is

$$y_p = y_0 + p\Delta y_0 + \frac{p(p-1)}{1.2}\Delta^2 y_0 + \frac{p(p-1)(p-2)}{1.2.3}\Delta^3 y_0 + \ldots \quad \ldots(1)$$

We have $\Delta y_0 - \Delta y_{-1} = \Delta^2 y_{-1}$

*i.e.,* $\Delta y_0 = \Delta y_{-1} + \Delta^2 y_{-1}$ ...(2)

Similarly $\Delta^2 y_0 = \Delta^2 y_{-1} + \Delta^3 y_{-1}$ ...(3)

$\Delta^3 y_0 = \Delta^3 y_{-1} + \Delta^4 y_{-1}$ etc. ...(4)

Also $\Delta^3 y_{-1} - \Delta^3 y_{-2} = \Delta^4 y_{-2}$

*i.e.,* $\Delta^3 y_{-1} = \Delta^3 y_{-2} + \Delta^4 y_{-2}$ ...(5)

Similarly $\Delta^4 y_{-1} = \Delta^4 y_{-2} + \Delta^5 y_{-2}$ etc. ...(6)

Substituting for $\Delta y_0, \Delta^2 y_0, \Delta^3 y_0$, ... from Eqns. (2), (3), (4) in Eqn. (1), we get

$$y_p = y_0 + p(\Delta y_{-1} + \Delta^2 y_{-1}) + \frac{p(p-1)}{1.2}(\Delta^2 y_{-1} + \Delta^3 y_{-1}) + \frac{p(p-1)(p-2)}{1.2.3}(\Delta^3 y_{-1} + \Delta^4 y_{-1})$$
$$+ \frac{p(p-1)(p-2)(p-3)}{1.2.3.4}(\Delta^4 y_{-1} + \Delta^5 y_{-1}) + \ldots$$

$$= y_0 + p\Delta y_{-1} + \frac{(p+1)p}{1.2}\Delta^2 y_{-1} + \frac{(p+1)p(p-1)}{1.2.3}\Delta^3 y_{-1} + \frac{(p+1)p(p-1)(p-2)}{1.2.3.4}\Delta^4 y_{-1}$$
$$+ \frac{p(p-1)(p-2)(p-3)}{1.2.3.4}\Delta^5 y_{-1} + \ldots$$

$$= y_0 + p\Delta y_{-1} + \frac{(p+1)p}{1.2}\Delta^2 y_{-1} + \frac{(p+1)\,p\,(p-1)}{1.2.3}(\Delta^3 y_{-2} + \Delta^4 y_{-2})$$
$$+ \frac{(p+1)\,p\,(p-1)\,(p-2)}{1.2.3.4}(\Delta^4 y_{-2} + \Delta^5 y_{-2}) + \ldots \quad \text{[Using (5) and (6)]}$$

Hence $$y_p + y_0 + p\Delta y_{-1} + \frac{(p+1)p}{2!}\Delta^2 y_{-1} + \frac{(p+1)\,p\,(p-1)}{3!}\Delta^3 y_{-2} + \frac{(p+2)\,(p+1)\,p\,(p-1)}{4!}\Delta^4 y_{-2} + \ldots$$

which is called *Gauss's backward interpolation formula.*

**Cor.** *In the central differences notation, this formula will be*

$$y_p = y_0 + p\delta y_{-1/2} + \frac{(p+1)p}{2!}\delta^2 y_0 + \frac{(p+1)\,p(p-1)}{3!}\delta^3 y_{-1/2} + \frac{(p+2)\,(p+1)\,p\,(p-1)}{4!}\delta^4 y_0 + \ldots$$

**Obs. 1.** This formula contains odd differences above the central line and even differences on the central line as shown below :

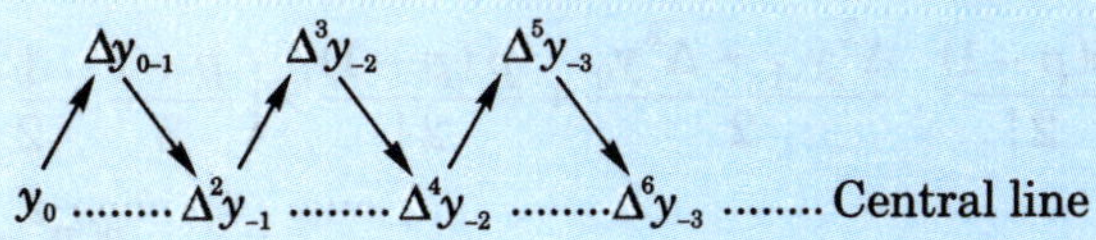


**Obs. 2.** It is used to interpolate the values of $y$ for a negative value of $p$ lying between –1 and 0.

**Obs. 3.** Gauss's forward and backward formulae are not of much practical use. However, these serve as intermediate steps for obtaining the important formulae of the following sections.

**(3) Stirling's formula.*** Gauss's forward interpolation formula is

$$y_p = y_0 + p\Delta y_0 + \frac{p\,(p-1)}{2!}\Delta^2 y_{-1} + \frac{(p+1)\,p(p-1)}{3!}\Delta^3 y_{-1} + \frac{(p+1)\,p(p-1)\,(p-2)}{4!}\Delta^4 y_{-2} + \ldots \quad \ldots(1)$$

Gauss's backward interpolation formula is

$$y_p = y_0 + p\Delta y_{-1} + \frac{(p+1)p}{2!}\Delta^2 y_{-1} + \frac{(p+1)\,p(p-1)}{3!}\Delta^3 y_{-2} + \frac{(p+2)\,(p+1)\,p\,(p-1)}{4!}\Delta^4 y_{-2} + \ldots \quad \ldots(2)$$

Taking the mean of Eqns. (1) and (2), we obtain

$$y_p = y_0 + p\left(\frac{\Delta y_0 + \Delta y_{-1}}{2}\right) + \frac{p^2}{2!}\Delta^2 y_{-1} + \frac{p\,(p^2-1)}{3!} \times \left(\frac{\Delta^3 y_{-1} + \Delta^3 y_{-2}}{2}\right) + \frac{p^2\,(p^2-1)}{4!}\Delta^4 y_{-2} + \ldots \quad \ldots(3)$$

which is called *Stirling's formula.*

**Cor.** *In the central differences notation, Eqn. (3) takes the form*

$$y_p = y_0 + p\,\mu\delta y_0 + \frac{p^2}{2!}\delta^2 y_0 + \frac{p\,(p^2+1^2)}{3!}\mu\delta^3 y_0 + \frac{p^2\,(p^2-1^2)}{4!}\delta^4 y_0 + \ldots \quad \ldots(4)$$

for $$\frac{1}{2}(\Delta y_0 + \Delta y_{-1}) = \frac{1}{2}(\delta y_{1/2} + \delta y_{-1/2}) = \mu\delta y_0$$

$$\frac{1}{2}(\Delta^3 y_{-1} + \Delta^3 y_{-2}) = \frac{1}{2}(\delta^3 y_{1/2} + \delta^3 y_{-1/2}) = \mu\delta^3 y_0 \text{ etc.}$$

**Obs.** *This formula involves means of the odd differences just above and below the central line and even differences on this line* as shown below :

$$\ldots y_0 \ldots \begin{pmatrix}\Delta y_{-1}\\ \Delta y_0\end{pmatrix} \ldots \Delta^2 y_{-1} \ldots \begin{pmatrix}\Delta^3 y_{-2}\\ \Delta^3 y_{-1}\end{pmatrix} \ldots \Delta^4 y_{-2} \ldots \begin{pmatrix}\Delta^5 y_{-3}\\ \Delta^5 y_{-2}\end{pmatrix} \ldots \Delta^6 y_{-3} \ldots \text{ Central line}$$

**(4) Bessel's formula.** Gauss's forward interpolation formula is

$$y_p = y_0 + p\Delta y_0 + \frac{p\,(p-1)}{2!}\Delta^2 y_{-1} + \frac{(p+1)\,p(p-1)}{3!}\Delta^3 y_{-1} + \frac{(p+1)\,p(p-1)\,(p-2)}{4!}\Delta^4 y_{-2} \ldots \quad \ldots(1)$$

*Named after the Scottish mathematicians *James Stirling* (1692-1770).

We have $\quad \Delta^2 y_0 - \Delta^2 y_{-1} = \Delta^3 y_{-1}$

*i.e.,* $\quad \Delta^2 y_{-1} = \Delta^2 y_0 - \Delta^3 y_{-1}$ ...(2)

Similarly $\quad \Delta^4 y_{-2} = \Delta^4 y_{-1} - \Delta^5 y_{-2}$ etc. ...(3)

Now Eqn. (1) can be written as

$$y_p = y_0 + p\Delta y_0 + \frac{p(p-1)}{2!}\left(\frac{1}{2}\Delta^2 y_{-1} + \frac{1}{2}\Delta^2 y_{-1}\right) + \frac{p(p^2-1)}{3!}\Delta^3 y_{-1} + \frac{p(p^2-1)(p-2)}{4!}\left(\frac{1}{2}\Delta^4 y_{-2} + \frac{1}{2}\Delta^4 y_{-2}\right) + \ldots$$

$$= y_0 + p\Delta y_0 + \frac{1}{2}\frac{p(p-1)}{2!}\Delta^2 y_{-1} + \frac{1}{2}\frac{p(p-1)}{2!}(\Delta^2 y_0 - \Delta^3 y_{-1}) + \frac{p(p^2-1)}{3!}\Delta^3 y_{-1}$$

$$+ \frac{1}{2}\frac{p(p^2-1)(p-2)}{4!}\Delta^4 y_{-2} + \frac{1}{2}\frac{p(p^2-1)(p-2)}{4!} \times (\Delta^4 y_{-1} - \Delta^5 y_{-2}) + \ldots \quad \text{[Using (2), (3) etc.]}$$

$$= y_0 + p\Delta y_0 + \frac{p(p-1)}{2!}\cdot\frac{\Delta^2 y_{-1} + \Delta^2 y_0}{2} + \frac{p(p-1)}{2!}\times\left(\frac{p+1}{3} - \frac{1}{2}\right)\Delta^3 y_{-1} + \frac{p(p^2-1)(p-2)}{4!}\cdot\frac{\Delta^4 y_{-2} + \Delta^4 y_{-1}}{2} + \ldots$$

Hence
$$y_p = y_0 + p\Delta y_0 + \frac{p(p-1)}{2!}\cdot\frac{\Delta^2 y_{-1} + \Delta^2 y_0}{2} + \frac{(p-1/2)\,p(p-1)}{3!}\Delta^3 y_{-1} + \frac{(p+1)\,p(p-1)(p-2)}{4!}\cdot\frac{\Delta^4 y_{-2} + \Delta^4 y_{-1}}{2} + \ldots \quad ...(4)$$

which is known as the *Bessel's formula.*

**Cor.** *In the central differences notation, Eqn. (4) becomes*

$$y_p = y_0 + p\delta y_{1/2} + \frac{p(p-1)}{2!}\mu\delta^2 y_{1/2} + \frac{(p-1/2)\,p(p-1)}{3!}\delta^3 y_{1/2} + \frac{(p+1)\,p(p-1)(p-2)}{4!}\mu\delta^4 y_{1/2} + \ldots \quad ...(5)$$

for $\quad \frac{1}{2}(\Delta^2 y_{-1} + \Delta^2 y_0) = y\delta^2 y_{1/2},\ \frac{1}{2}(\Delta^4 y_{-2} + \Delta^4 y_{-1}) = \mu\delta^4 y_{1/2}$ etc.

**Obs.** *This is a very useful formula for practical purposes. It involves odd differences below the central line and means of even differences of and below his line as shown below :*

$$\begin{array}{ccccccc} y_0 \cdots & \left\{\begin{array}{c}\Delta^2 y_{-1}\\ \\ \Delta^2 y_0\end{array}\right\} & \cdots & \left\{\begin{array}{c}\Delta^4 y_{-2}\\ \\ \Delta^4 y_{-1}\end{array}\right\} & \cdots & \left\{\begin{array}{c}\Delta^6 y_{-3}\\ \\ \Delta^6 y_{-2}\end{array}\right\} & \cdots \text{Central line} \\ \Delta y_0 & & \Delta^3 y_{-1} & & \Delta^5 y_{-2} & & \end{array}$$

**(5) Everett's formula.** Gauss's forward interpolation formula is

$$y_p = y_0 + p\Delta y_0 + \frac{p(p-1)}{2!}\Delta^2 y_{-1} + \frac{(p+1)\,p(p-1)}{3!}\Delta^3 y_{-1} + \frac{(p+1)\,p(p-1)(p-2)}{4!}\Delta^4 y_{-2} + \frac{(p+2)(p+1)\,p(p-1)(p-2)}{5!}\Delta^5 y_{-2} + \ldots \quad ...(1)$$

We eliminate the odd difference in Eqn. (1) by using the relations

$$\Delta y_0 = y_1 - y_0,\ \Delta^3 y_{-1} = \Delta^2 y_0 - \Delta^2 y_{-1},\ \Delta^5 y_{-2} = \Delta^4 y_{-1} - \Delta^4 y_{-2} \text{ etc.}$$

Then Eqn. (1) becomes

$$y_p = y_0 + p(y_1 - y_0) + \frac{p(p-1)}{2!}\Delta^2 y_{-1} + \frac{(p+1)\,p(p-1)}{3!}(\Delta^2 y_0 - \Delta^2 y_{-1})$$

$$+ \frac{(p+1)\,p(p-1)(p-2)}{4!}\Delta^4 y_{-2} + \frac{(p+2)(p+1)\,p(p-1)(p-2)}{5!} \times (\Delta^4 y_{-1} - \Delta^4 y_{-2}) + \ldots$$

$$= (1-p)\,y_0 + py_1 - \frac{p(p-1)(p-2)}{3!}\Delta^2 y_{-1} + \frac{(p+1)\,p(p-1)}{3!}\Delta^2 y_0$$

$$-\frac{(p+1)\,p\,(p-1)(p-2)(p-3)}{5!}\,\Delta^4 y_{-2} + \frac{(p+2)(p+1)\,p\,(p-1)(p-2)}{5!}\,\Delta^4 y_{-1} - \ldots$$

To change the terms with negative sign, putting $p = 1 - q$, we obtain

$$y_p = qy_0 + \frac{q(q^2-1^2)}{3!}\Delta^2 y_{-1} + \frac{q(q^2-1^2)(q^2-2^2)}{5!}\Delta^4 y_{-2} + \ldots\ldots + py_1 + \frac{p(p^2-1^2)}{3!}\Delta^2 y_0$$

$$+ \frac{p(p^2-1^2)(p^2-2^2)}{5!}\Delta^4 y_{-1} + \ldots$$

This is known as *Everett's formula.*

**Obs.** *This formula is extensively used and involves only even differences on and below the central line* as shown below :

$$y_0 \ldots\ldots \Delta^2 y_{-1} \ldots\ldots \Delta^4 y_{-2} \ldots\ldots \Delta^6 y_{-3} \ldots\ldots \text{Central line}$$

$$y_1 \qquad \Delta^2 y_0 \qquad \Delta^4 y_{-1} \qquad \Delta^6 y_{-2}$$

## 29.8 CHOICE OF AN INTERPOLATION FORMULA

The coefficients in the central difference formulae are smaller and converge faster than those in Newton's formulae. After a few terms, the coefficients in the Stirling's formula decrease more rapidly than those of the Bessel's formula and the coefficients of Bessel's formula decrease more rapidly than those of Newton's formula. As much, whenever possible, *central difference formulae should be used in preference to Newton's formulae.*

The right choice of an interpolation formula however, depends on the position of the interpolated value in the given data.

*The following rules will be found useful* :

1. *To find a tabulated value near the beginning of the table, use Newton's forward formula.*
2. *To find a value near the end of the table, use Newton's backward formula.*
3. *To find an interpolated value near the centre of the table, use either Stirling's or Bessel's or Everett's formula.*

If interpolation is required for $p$ lying between $-1/4$ and $1/4$, prefer Stirling's formula.

If interpolation is desired for $p$ lying between $1/4$ and $3/4$, use Bessel's or Everett's formula.

**Example 29.18.** *Find f(22) from the Gauss forward formula :*

| $x$ : | 20 | 25 | 30 | 35 | 40 | 45 |
|---|---|---|---|---|---|---|
| $f(x)$ : | 354 | 332 | 291 | 260 | 231 | 204 |

**Solution.** Taking $x_0 = 25$, $h = 5$, we have to find the value of $f(x)$ for $x = 22$.

*i.e.,* for $\quad p = \dfrac{x - x_0}{h} = \dfrac{22-25}{5} = -0.6$

The difference table is as follows :

| $x$ | $p$ | $y_p$ | $\Delta y_p$ | $\Delta^2 y_p$ | $\Delta^3 y_p$ | $\Delta^4 y_p$ | $\Delta^5 y_p$ |
|---|---|---|---|---|---|---|---|
| 20 | –1 | 354 $(= y_{-1})$ | | | | | |
| | | | –22 | | | | |
| 25 | 0 | <u>332</u> $(= y_0)$ | | <u>–19</u> | | | |
| | | | –41 | | 29 | | |
| 30 | 1 | 291 $(= y_1)$ | | 10 | | <u>–37</u> | |
| | | | –31 | | <u>–8</u> | | <u>45</u> |
| 35 | 2 | 260 $(= y_2)$ | | 2 | | 8 | |
| | | | –29 | | 0 | | |
| 40 | 3 | 231 $(= y_3)$ | | 2 | | | |
| | | | –27 | | | | |
| 45 | 4 | 204 $(= y_4)$ | | | | | |

Gauss forward formula is

$$y^p = y_0 + p\Delta y_0 + \frac{p(p-1)}{2!}\Delta y_{-1} + \frac{(p+1)\,p\,(p-1)}{3!}\Delta^3 y_{-1}$$

$$+ \frac{(p+1)\,p\,(p-1)(p-2)}{4!}\Delta^4 y_{-2} + (p+1)(p-1)(p-2)(p+2)\,\Delta^5 y_{-2}$$

$$\therefore \quad f(22) = 332 + (0.6)(-41) + \frac{(-0.6)(-0.6-1)}{2!}(-19) + \frac{(-0.6+1)(-0.6)(-0.6-1)}{3!}(-8)$$

$$+ \frac{(-0.6-1)(-0.6)(-0.6-1)(-0.6-2)}{4!}(-37)$$

$$+ \frac{(-0.6+1)(-0.6)(-0.6-1)(-0.6-2)(-0.6+2)}{5!}(45)$$

$$= 332 + 24.6 - 9.12 + 1.5392 - 0.5241$$

Hence $\quad f(22) = 347.983.$

**Example 29.19.** *Interpolate by means of Gauss's backward formula, the population of a town for the year 1974, given that :*

| *Year* : | *1939* | *1949* | *1959* | *1969* | *1979* | *1989* |
|---|---|---|---|---|---|---|
| *Population (in thousands)* : | *12* | *15* | *20* | *27* | *39* | *52* |

**Solution.** Taking $x_0 = 1969$, $h = 10$, the population of the town is to be found for $p = \frac{1974 - 1969}{10} = 0.5.$

The central difference table is

| $x$ | $p$ | $y_p$ | $\Delta y_p$ | $\Delta^2 y_p$ | $\Delta^3 y_p$ | $\Delta^4 y_p$ | $\Delta^5 y_p$ |
|---|---|---|---|---|---|---|---|
| 1939 | –3 | 12 | | | | | |
| | | | 3 | | | | |
| 1949 | –2 | 15 | | 2 | | | |
| | | | 5 | | 0 | | |
| 1959 | –1 | 20 | | 2 | | 3 | |
| | | | 7 | | 3 | | –10 |
| 1969 | 0 | 27 | | 5 | | –7 | |
| | | | 12 | | –4 | | |
| 1979 | 1 | 39 | | 1 | | | |
| | | | 13 | | | | |
| 1989 | 2 | 52 | | | | | |

Gauss's backward formula is

$$y_p = y_0 + p\Delta y_{-1} + \frac{(p+1)\,p}{2!}\Delta^2 y_{-1} + \frac{(p+1)\,p\,(p-1)}{3!}\Delta^3 y_{-2} + \frac{(p+1)(p+1)\,p\,(p-1)}{4!}\Delta^4 y_{-2}$$

$$+ \frac{(p+2)(p+1)\,p\,(p-1)(p-2)}{5!}\Delta^5 y_3 + \ldots$$

*i.e.,*

$$y_{.5} = 27 + (0.5)(7) + \frac{(1.5)(.5)}{2}(5) + \frac{(1.5)(.5)(-.5)}{6}(3) + \frac{(2.5)(1.5)(-.5)}{24}(-7)$$

$$+ \frac{(2.5)(1.5)(.5)(-.5)(-1.5)}{120}(-10)$$

$$= 27 + 3.5 + 1.875 - 0.1875 + 0.2743 - 0.1172 = 32.345 \text{ thousands approx.}$$

**Example 29.20.** *Given*

| $\theta°$ | : | 0 | 5 | 10 | 15 | 20 | 25 | 30 |
|---|---|---|---|---|---|---|---|---|
| $\tan\theta$ | : | 0 | 0.0875 | 0.1763 | 0.2679 | 0.3640 | 0.4663 | 0.5774 |

*Using Stirling's formula, estimate the value of tan 16°.*

**Solution.** Taking the origin at $\theta° = 15°$, $h = 5°$ and $p = \dfrac{\theta - 15}{5}$, we have the following central difference table :

| P | | | | | | |
|---|---|---|---|---|---|---|
| −3 | 0.0000 | | | | | |
| | | 0.08575 | | | | |
| −2 | 0.0875 | | 0.0013 | | | |
| | | 0.0888 | | | | |
| −1 | 0.1763 | | 0.0028 | | | |
| | | 0.0916 | | 0.0015 | | |
| 0 | 0.2679 | | 0.0045 | | 0.0002 | |
| | | 0.0961 | | 0.0017 | | −0.0002 |
| 1 | 0.3640 | | 0.0062 | | 0.0000 | |
| | | 0.1023 | | 0.0017 | | 0.0009 |
| 2 | 0.4663 | | 0.0088 | | 0.0009 | |
| | | 0.1111 | | 0.0026 | | |
| 3 | 0.5774 | | | | | |

At $\theta = 16°$, $\quad p = \dfrac{16 - 15}{5} = 0.2$

Stirling's formula is

$$y_p = y_0 + \frac{p}{1} \cdot \frac{\Delta y_{-1} + \Delta y_0}{2} + \frac{p^2}{2!}\Delta^2 y_{-1} + \frac{p(p^2-1)}{3!} \cdot \frac{\Delta^3 y_{-2} + \Delta^3 y_{-1}}{2} + \frac{p^2(p^2-1)}{4!}\Delta^4 y_{-2} + \dots$$

$$\therefore \quad y_{0.2} = 0.2679 + (0.2)\left(\frac{0.0916 + 0.0961}{2}\right) + \frac{(0.2)^2}{2!}(0.0045) + \dots$$

$$= 0.2679 + 0.01877 + 0.00009 + \dots = 0.28676$$

Hence $\quad \tan 16° = 0.28676.$

**Example 29.21.** *Employ Stirling's formula to compute $y_{12.2}$ from the following table ($y_x = 1 + \log_{10} \sin x$) :*

| $x°$ | : | 10 | 11 | 12 | 13 | 14 |
|---|---|---|---|---|---|---|
| $10^5 y_x$ | : | 23,967 | 28,060 | 31,788 | 35,209 | 38,368 |

**Solution.** Taking the origin at $x_0 = 12°$, $h = 1$ and $p = x - 12$, we have the following central table :

| $p$ | $y_x$ | $\Delta y_x$ | $\Delta^2 y_x$ | $\Delta^3 y_x$ | $\Delta^4 y_x$ |
|---|---|---|---|---|---|
| −2 | 0.23967 | | | | |
| | | 0.04093 | | | |
| −1 | 0.28060 | | −0.00365 | | |
| | | 0.03728 | | 0.00058 | |
| 0 | 0.31788 | | −0.00307 | | −0.00013 |
| | | 0.034121 | | −0.00045 | |
| 1 | 0.35209 | | −0.00062 | | |
| | | 0.03159 | | | |
| 2 | 0.38368 | | | | |

At $x = 12.2$, $p = 0.2$. (As $p$ lies between −1/4 and 1/4, the use of Stirling's formula will be quite suitable.)

Stirling's formula is

$$y_p = y_0 + \frac{p}{1}\cdot\frac{\Delta y_{-1}+\Delta y_{-0}}{2} + \frac{p^2}{2!}\Delta^2 y_{-1} + \frac{p(p^2-1)}{3!}\cdot\frac{\Delta^3 y_{-2}+\Delta^3 y_{-1}}{2} + \frac{p^2(p^2-1)}{4!}\Delta^4 y_{-2} + \ldots$$

When $p = 0.2$, we have

$$\therefore \quad y_{0.2} = 0.31788 + 0.2\left(\frac{0.03728+0.03421}{2}\right) + \frac{(0.2)^2}{2}(-0.00307)$$

$$+ \frac{(0.2)[(0.2)^2-1]}{6}\left(\frac{0.00058-0.00045}{2}\right) + \frac{(0.2)^2[(0.2)^2-1]}{24}(-0.00013)$$

$$= 0.31788 + 0.00715 - 0.00006 - 0.000002 + 0.0000002 = 0.32497.$$

**Example 29.22.** *Apply Bessel's formula to obtain $y_{25}$, given $y_{20} = 2854$, $y_{24} = 3162$, $y_{28} = 3544$, $y_{32} = 3992$.*

**Solution.** Taking the origin at $x_0 = 24$, $h = 4$, we have $p = \frac{1}{4}(x - 24)$.

$\therefore$ The central difference table is

| $p$ | $y$ | $\Delta y$ | $\Delta^2 y$ | $\Delta^3 y$ |
|---|---|---|---|---|
| –1 | 2854 | | | |
| | | 308 | | |
| 0 | <u>3162</u> | | <u>74</u> | |
| | | <u>382</u> | | <u>–8</u> |
| 1 | 3544 | | <u>66</u> | |
| | | 448 | | |
| 2 | 3992 | | | |

At $x = 25$, $p = (25 - 24)/4 = 1/4$. (As $p$ lies between 1/4 and 3/4, the use of Bessel's formula will yield accurate result.)

Bessel's formula is

$$y_p = y_0 + p\Delta y_0 + \frac{p(p-1)}{2!}\frac{\Delta^2 y_{-1}+\Delta^2 y_0}{2} + \frac{(p-1/2)\,p\,(p-1)}{3!}\Delta^3 y_{-1} + \ldots \qquad \ldots(1)$$

When $p = 0.25$, we have

$$y_p = 3162 + 0.25 \times 382 + \frac{0.25(-0.75)}{2}\left(\frac{74+66}{2}\right) + \frac{(-0.25)\,0.25\,(-0.75)}{6}(-8)$$

$$= 3162 + 95.5 - 6 - 5625 - 0.0625 = 3250.875 \text{ approx.}$$

**Example 29.23.** *Apply Bessel's formula to find the value of f(27.5) from the table :*

| $x$ : | 25 | 26 | 27 | 28 | 29 | 30 |
|---|---|---|---|---|---|---|
| $f(x)$ : | 4.000 | 3.846 | 3.704 | 3.571 | 3.448 | 3.333 |

**Solution.** Taking the origin at $x_0 = 27$, $h = 1$, we have $p = x - 27$

The central difference table is

| $x$ | $p$ | $y$ | $\Delta y$ | $\Delta^2 y$ | $\Delta^3 y$ | $\Delta^4 y$ |
|---|---|---|---|---|---|---|
| 25 | –2 | 4.000 | | | | |
| | | | –0.154 | | | |
| 26 | –1 | 3.846 | | 0.012 | | |
| | | | –0.142 | | –0.003 | |
| 27 | 0 | 3.704 | | 0.009 | | 0.004 |
| | | | –0.133 | | –0.001 | |
| 28 | 1 | 3.571 | | 0.010 | | –0.001 |
| | | | –0.123 | | –0.002 | |
| 29 | 2 | 3.448 | | 0.008 | | |
| | | | –0.115 | | | |
| 30 | 3 | 3.333 | | | | |

At $x = 27.5$, $p = 0.5$ (As $p$ lies between 1/4 and 3/4, the use of Bessel's formula will yield accurate result)
Bessel's formula is

$$y_p = y_0 + p\Delta y_0 + \frac{p(p-1)}{2!}\left(\frac{\Delta^2 y_{-1} + \Delta^2 y_0}{2}\right) + \frac{\left(p - \frac{1}{2}\right)p(p-1)}{3!}\Delta^3 y_{-1}$$

$$+ \frac{(p+1)\,p(p-1)\,(p-2)}{4!}\left(\frac{\Delta^4 y_{-2} + \Delta^4 y_{-1}}{2}\right) + \ldots$$

When $p = 0.5$, we have

$$y_p = 3.704 - \frac{(0.5)\,(0.5-1)}{2}\left(\frac{0.009 + 0.010}{2}\right) + 0$$

$$+ \frac{(0.5+1)\,(0.5)\,(0.5-1)\,(0.5-2)}{24}\left(\frac{-0.001 - 0.004}{2}\right)$$

$$= 3.704 - 0.11875 - 0.00006 = 3.585$$

Hence $f(27.5) = 3.585$.

**Example 29.24.** *Given the table*

| $x$ : | 310 | 320 | 330 | 340 | 350 | 360 |
|---|---|---|---|---|---|---|
| $\log x$ : | 2.49136 | 2.50515 | 2.51851 | 2.53148 | 2.54407 | 2.55630 |

*find the value of log 337.5 by Everett's formula.*

**Solution.** Taking the origin at $x_0 = 330$ and $h = 10$, we have $p = \dfrac{x - 330}{10}$

∴ The central difference table is

| $p$ | $y$ | $\Delta y$ | $\Delta^2 y$ | $\Delta^3 y$ | $\Delta^4 y$ | $\Delta^5 y$ |
|---|---|---|---|---|---|---|
| –2 | 2.49136 | | | | | |
| | | 0.01379 | | | | |
| –1 | 2.50515 | | –0.00043 | | | |
| | | 0.01336 | | 0.00004 | | |
| 0 | <u>2.51881</u> | | <u>–0.00039</u> | | <u>–0.00003</u> | |
| | | 0.01297 | | 0.00001 | | 0.00004 |
| 1 | <u>2.53148</u> | | <u>–0.00038</u> | | <u>0.00001</u> | |
| | | 0.01259 | | 0.00002 | | |
| 2 | 2.54407 | | –0.00036 | | | |
| | | 0.01223 | | | | |
| 3 | 2.55630 | | | | | |

To evaluate log 337.5 *i.e.* for $x = 337.5$, $p = \dfrac{337.5 - 330}{10} = 0.75$

(As $p > 0.5$ and $= 0.75$, Everett's formula will be quite suitable)
Everett's formula is

$$y_p = qy_0 + \frac{q(q^2-1^2)}{3!}\Delta^2 y_{-1} + \frac{q(q^2-1^2)\,(q^2-2^2)}{5!}\Delta^4 y_{-2} + \ldots + py_1 + \frac{p(p^2-1^2)}{3!}\Delta^2 y_0$$

$$+ \frac{p(p^2-1^2)\,(p^2-2^2)}{5!}\Delta^4 y_{-1} + \ldots$$

$$= 0.25 \times 2.51851 + \frac{0.25\,(0.0625-1)}{6} \times (-0.00039) + \frac{0.25\,(0.0625-1)\,(0.0625-4)}{120}$$

$$\times (-0.00003) + 0.75 \times 2.53148 + \frac{0.75\,(0.5625-1)}{6} \times (-0.00038)$$

$$+ \frac{0.75\,(0.5625-1)\,(0.5625-4)}{120} \times (0.00001)$$

$$= 0.62963 + 0.00002 - 0.0000002 + 1.89861 + 0.00002 + 0.0000001 = 2.52828 \text{ nearly.}$$

## PROBLEMS 29.4

1. Using Gauss's forward formula, evaluate $f(3.75)$ from the table :

| $x$ : | 2.5 | 3.0 | 3.5 | 4.0 | 4.5 | 5.0 |
|---|---|---|---|---|---|---|
| $y$ : | 24.145 | 22.043 | 20.225 | 18.644 | 17.262 | 16.047 |

2. Using Gauss's backward difference formula, find $y(8)$ from the following table :

| $x$ : | 0 | 5 | 10 | 15 | 20 | 25 |
|---|---|---|---|---|---|---|
| $y$ : | 7 | 11 | 14 | 18 | 24 | 32 |

3. Using Gauss's backward formula, estimate the number of persons earning wages between ₹ 60 and ₹ 70 from the following data :

| *Wages (₹)* : | Below 40 | 40–60 | 60–80 | 80–100 | 100–120 |
|---|---|---|---|---|---|
| *No. of persons : (in thousands)* | 250 | 120 | 100 | 70 | 50 |

4. From the following table :

| $x$ : | 1.00 | 1.05 | 1.10 | 1.15 | 1.20 | 1.25 | 1.30 |
|---|---|---|---|---|---|---|---|
| $e^x$ : | 2.7183 | 2.8577 | 3.0042 | 3.1582 | 3.3201 | 3.4903 | 3.6693 |

Find $e^{1.17}$, using Gauss forward formula.

5. The pressure $p$ of wind corresponding to velocity $v$ is given by the following data. Estimate $p$ when $y = 25$.

| $v$ : | 10 | 20 | 30 | 40 |
|---|---|---|---|---|
| $p$ : | 1.1 | 2 | 4.4 | 7.9 |

6. Using Stirling's formula find $y_{35}$, given $y_{20} = 512, y_{30} = 439, y_{40} = 346, y_{50} = 243$, where $y_x$ represents the number of persons at age $x$ years in a life table.

7. Employ Bessel's formula to find the value of $F$ at $x = 1.95$, given that

| $x$ : | 1.7 | 1.8 | 1.9 | 2.0 | 2.1 | 2.2 | 2.3 |
|---|---|---|---|---|---|---|---|
| $F$ : | 2.979 | 3.144 | 3.283 | 3.391 | 3.463 | 3.997 | 4.491 |

Which other interpolation formula can be used here ? Which is more appropriate ? Give reasons.

8. Calculate the value of $f(1.5)$ using Bessel's interpolation formula, from the following table :

| $x$ : | 0 | 1 | 2 | 3 |
|---|---|---|---|---|
| $f(x)$ : | 3 | 6 | 12 | 15 |

9. Apply Everett's formula to obtain $u_{25}$, given $u_{20} = 854, u_{24} = 3162, u_{28} = 3544, u_{32} = 3992$.

10. Using Everett's formula, evaluate $f(30)$, if $f(20) = 2854, f(28) = 3162, f(36) = 7088, f(44) = 7984$

11. Given the table :

| $x$ : | 310 | 320 | 330 | 340 | 350 | 360 |
|---|---|---|---|---|---|---|
| $\log x$ : | 2.4914 | 2.5052 | 2.5185 | 2.5315 | 2.5441 | 2.5563 |

Find the value of log 337.5 by Gauss's, Stirling's and Bessel's formulae.

## 29.9 INTERPOLATION WITH UNEQUAL INTERVALS

The various interpolation formulae derived so far possess the disadvantages of being applicable only to equally spaced values of the argument. It is, therefore, desirable to develop interpolation formulae for unequally spaced values of $x$. Now we shall study two such formulae :

(*i*) Lagrange's interpolation formula

(*ii*) Newton's general interpolation formula with divided differences.

## 29.10 LAGRANGE'S INTERPOLATION FORMULA

*If $y = f(x)$ takes the value $y_0, y_1, ..., y_n$ corresponding to $x = x_0, x_1, ..., x_n$, then*

$$f(x) = \frac{(x-x_1)(x-x_2)...(x-x_n)}{(x_0-x_1)(x_0-x_2)...(x_0-x_n)} y_0 + \frac{(x-x_0)(x-x_2)...(x-x_n)}{(x_1-x_0)(x_1-x_2)...(x_1-x_n)} y_1$$

$$+ ... + \frac{(x-x_0)(x-x_1)...(x-x_{n-1})}{(x_n-x_0)(x_n-x_1)...(x_n-x_{n-1})} y_n \qquad ...(1)$$

This is known as *Lagrange's interpolation formula for unequal intervals.*

**Proof.** Let $y = f(x)$ be a function which takes the values $(x_0, y_0), (x_1, y_1), ..., (x_n, y_n)$. Since there are $n + 1$ pairs of values of $x$ and $y$, we can represent $f(x)$ by a polynomial in $x$ of degree $n$. Let this polynomial be of the form

$$y = f(x) = a_0(x - x_1)(x - x_2) \dots (x - x_n) + a_1(x - x_0)(x - x_2) \dots (x - x_n) + a_2(x - x_0)(x - x_1)(x - x_3) \dots (x - x_n) + \dots + a_n(x - x_0)(x - x_1) \dots (x - x_{n-1}) \quad ...(2)$$

Putting $x = x_0, y = y_0$, in (2), we get

$$y_0 = a_0(x_0 - x_1)(x_0 - x_2)...(x_0 - x_n)$$

$$a_0 = y_0/[(x_0 - x_1)(x_0 - x_2)...(x_0 - x_n)]$$

Similarly putting $x = x_1, y = y_1$ in (2), we have $a_1 = y_1/[(x_1 - x_0)(x_1 - x_2)...(x_1 - x_n)]$

Proceeding the same way, we find $a_2, a_3 ... a_n$

Substituting the values of $a_0, a_1, ..., a_n$ in Eqn. (2), we get Eqn. (1).

**Obs.** Lagranges interpolation formula Eqn. (1) for $n$ points is a polynomial of degree $(n - 1)$ which is known as *Lagrangian polynomial* and is very simple to implement on a computer.

*This formula can also be used to split the given function into partial fractions.*

For on dividing both sides of Eqn. (1) by $(x - x_0)(x - x_1) \dots (x - x_n)$, we get

$$\frac{f(x)}{(x_0 - x_0)(x_0 - x_1)...(x_0 - x_n)} = \frac{y_0}{(x_0 - x_1)(x_0 - x_2)...(x_0 - x_n)} \cdot \frac{1}{x - x_0} + \frac{y_0}{(x_1 - x_0)(x_1 - x_2)...(x_1 - x_n)} \cdot \frac{1}{x - x_1} + \dots + \frac{y_n}{(x_n - x_0)(x_n - x_1)...(x_n - x_{n-1})} \cdot \frac{1}{x - x_n}.$$

**Example 29.25.** *Given the values*

| $x$ : | 5 | 7 | 11 | 13 | 17 |
|---|---|---|---|---|---|
| $f(x)$: | 150 | 392 | 1492 | 2366 | 5202 |

*evaluate f(9), using (i) Lagrange's formula.*

**Solution.** (*i*) Here $x_0 = 5, x_1 = 7, x_2 = 11, x_3 = 13, x_4 = 17$

and $y_0 = 150, y_1 = 392, y_2 = 1452, y_3 = 2366, y_4 = 5202.$

Putting $x = 9$ and substituting the above values in Lagrange's formula, we get

$$f(9) = \frac{(9-7)(9-11)(9-13)(9-17)}{(5-7)(5-11)(5-13)(5-17)} \times 150 + \frac{(9-5)(9-11)(9-13)(9-17)}{(7-5)(7-11)(7-13)(7-17)} \times 392$$

$$+ \frac{(9-5)(9-7)(9-13)(9-17)}{(11-5)(11-7)(11-13)(11-17)} \times 1452 + \frac{(9-5)(9-7)(9-11)(9-17)}{(13-5)(13-7)(13-11)(13-17)} \times 2366$$

$$+ \frac{(9-5)(9-7)(9-11)(9-13)}{(17-5)(17-7)(17-11)(17-13)} \times 5202 = -\frac{50}{3} + \frac{3136}{15} + \frac{3872}{3} - \frac{2366}{3} + \frac{578}{5} = 810.$$

**Example 29.26.** *Find the polynomial f(x) by using Lagrange's formula and hence find f (3) for*

| $x$ : | 0 | 1 | 2 | 5 |
|---|---|---|---|---|
| $f(x)$: | 2 | 3 | 12 | 147 |

(Anna, 2012)

**Solution.** Here $x_0 = 0, x_1, x_2 = 2, x_3 = 5$

and $y_0 = 2, y_1 = 3, y_2 = 12, y_3 = 147$

Lagrange's formula is

$$y = \frac{(x - x_1)(x - x_2)...(x - x_3)}{(x_0 - x_1)(x_0 - x_2)...(x_0 - x_3)} y_0 + \frac{(x - x_0)(x - x_2)...(x - x_3)}{(x_1 - x_0)(x_1 - x_2)...(x_1 - x_3)} y_1$$

$$+ \frac{(x - x_0)(x - x_1)...(x - x_3)}{(x_2 - x_0)(x_2 - x_1)...(x_2 - x_3)} y_2 + \frac{(x - x_0)(x - x_2)...(x - x_5)}{(x_3 - x_0)(x_3 - x_2)...(x_3 - x_2)} y_3$$

$$= \frac{(x-1)(x-2)(x-5)}{(0-1)(0-2)(0-5)}(2) + \frac{(x-0)(x-2)(x-5)}{(1-0)(1-2)(1-5)}(3)$$

$$+ \frac{(x-0)(x-1)(x-5)}{(2-0)(2-1)(2-5)}(12) + \frac{(x-0)(x-1)(x-2)}{(5-0)(5-1)(5-2)}(147)$$

Hence $\quad f(x) = x^3 + x^2 - x + 2$

$\therefore \quad f(3) = 27 + 9 - 3 + 2 = 35.$

**Example 29.27.** *A curve passes through the point (0, 18), (1, 10), (3, –18) and (6, 90). Find the slope of the curve at x = 2.*

**Solution.** Here $x_0 = 0, x_1 = 1, x_2 = 3, x_3 = 6$ and $y_0 = 18, y_1 = 10, y_2 = -18, y_3 = 90$

Since the values of $x$ are unequally spaced, we use the Lagrange's formula :

$$y = \frac{(x-x_1)(x-x_2)(x-x_3)}{(x_0-x_1)(x_0-x_2)(x_0-x_3)} y_0 + \frac{(x-x_0)(x-x_2)(x-x_3)}{(x_1-x_0)(x_1-x_2)(x_1-x_3)} y_1 + \frac{(x-x_0)(x-x_1)(x-x_3)}{(x_2-x_0)(x_2-x_1)(x_2-x_3)} y_2 + \frac{(x-x_0)(x-x_1)(x-x_2)}{(x_3-x_0)(x_3-x_1)(x_3-x_2)} y_3$$

$$= \frac{(x-1)(x-3)(x-6)}{(0-1)(0-2)(0-6)}(18) + \frac{(x-0)(x-3)(x-6)}{(1-0)(1-3)(1-6)}(10) + \frac{(x-0)(x-1)(x-6)}{(3-0)(3-1)(3-6)}(-18) + \frac{(x-0)(x-1)(x-3)}{(6-0)(6-1)(6-3)}(90)$$

$$= (-x^3 + 10x^2 - 27x + 18) + (x^3 - 9x^2 + 18x) + (x^3 - 7x^2 + 6x) + (x^3 - 4x^2 + 3x)$$

*i.e.,* $\quad y = 2x^3 - 10x^2 + 18$

Thus the slope of the curve at $(x = 2) = \left(\dfrac{dy}{dx}\right)_{x=2}$

$$= (6x^2 - 20x)_{x=2} = -16.$$

**Example 29.28.** *Using Lagrange's formula, express the function* $\dfrac{3x^2 + x + 1}{(x-1)(x-2)(x-3)}$ *as a sum of partial fractions.*

**Solution.** Let us evaluate $y = 3x^2 + x + 1$ for $x = 1, x = 2$ and $x = 3$

These values are

| $x$ : | $x_0 = 1$ | $x_1 = 2$ | $x_2 = 3$ |
|---|---|---|---|
| $y$ : | $y_0 = 5$ | $y_1 = 15$ | $y_2 = 31$ |

Lagrange's formula is

$$y = \frac{(x-x_1)(x-x_2)}{(x_0-x_1)(x_0-x_2)} y_0 + \frac{(x-x_0)(x-x_2)}{(x_1-x_0)(x_1-x_2)} y_1 + \frac{(x-x_0)(x-x_1)}{(x_2-x_0)(x_2-x_1)} y_2$$

$$= \frac{(x-1)(x-2)}{(0-1)(0-2)}(5) + \frac{(x-1)(x-3)}{(2-1)(2-3)}(15) + \frac{(x-1)(x-2)}{(3-1)(3-1)}(31)$$

Substituting the above values, we get

$$= \frac{(x-2)(x-3)}{(1-2)(1-3)}(5) + \frac{(x-1)(x-3)}{(2-1)(2-3)}(15) + \frac{(x-1)(x-2)}{(3-1)(3-2)}(31)$$

$$= 2.5(x-2)(x-3) - 15(x-1)(x-3) + 15.5(x-1)(x-2)$$

Thus $\quad \dfrac{3x^2 + x + 1}{(x-1)(x-2)(x-3)} = \dfrac{2.5(x-2)(x-3) - 15(x-1)(x-3) + 15.5(x-1)(x-2)}{(x-1)(x-2)(x-3)}$

$$= \frac{2.5}{x-1} - \frac{15}{x-2} + \frac{15.5}{x-3}.$$

**Example 29.29.** *Find the distance moved by a particle and its acceleration at the end of 4 seconds, if the time verses velocity data is as follows :*

| *t* : | *0* | *1* | *3* | *4* |
|---|---|---|---|---|
| *v* : | *21* | *15* | *12* | *10* |

**Solution.** Since the values of $t$ are not equispaced, we use Lagrange's formula :

$$v = \frac{(t-t_1)(t-t_2)(t-t_3)}{(t_0-t_1)(t_0-t_2)(t_0-t_3)} v_0 + \frac{(t-t_0)(t-t_2)(t-t_3)}{(t_1-t_0)(t_1-t_2)(t_1-t_3)} v_1$$

$$+ \frac{(t-t_0)(t-t_1)(t-t_3)}{(t_2-t_0)(t_2-t_1)(t_2-t_3)} v_2 + \frac{(t-t_0)(t-t_2)(t-t_5)}{(t_3-t_0)(t_3-t_2)(t_3-t_2)} v_3$$

$$v = \frac{(t-1)(t-3)(t-4)}{(-1)(-2)(-4)}(21) + \frac{t(t-3)(t-4)}{(1)(-2)(-3)}(15) + \frac{t(t-1)(t-4)}{(3)(2)(-1)}(12) + \frac{t(t-1)(t-3)}{(4)(3)(1)}(10)$$

*i.e.*, $$v = \frac{1}{12}(-5t^3 + 38t^2 - 105t + 252)$$

$\therefore$ Distance moved $s = \int_0^4 v\,dt = \frac{1}{12}\int_0^4 (-5t^3 + 38t^2 - 105t + 252)\,dt$ $\qquad \left[\because\ v = \frac{ds}{dt}\right]$

$$= \frac{1}{12}\left(-\frac{5t^4}{4} + \frac{38t^3}{3} - \frac{105t^2}{2} + 252t\right)_0^4$$

$$= \frac{1}{12}\left(-320 + \frac{2432}{3} - 840 + 1008\right) = 54.9$$

Also acceleration $= \frac{dv}{dt} = \frac{1}{2}(-15t^2 + 76t - 105 + 0)$

Hence acceleration at $(t = 4) = \frac{1}{12}(-15(16) + 76(4) - 105) = -3.4.$

## PROBLEMS 29.5

1. Use Lagrange's interpolation formula to find the value of $y$ when $x = 10$, if the following values of $x$ and $y$ are given :

| $x$: | 5 | 6 | 9 | 11 |
|---|---|---|---|---|
| $y$: | 12 | 13 | 14 | 16 |

(*JNTUK, 2022, 2015 ; Anna, 2012*)

2. Given $\log_{10} 654 = 2.8156$, $\log_{10} 658 = 2.8182$, $\log_{10} 659 = 2.8189$, $\log_{10} 661 = 2.8202$, find by using Lagrange's formula, the value of $\log_{10} 656$. (*Anna, 2012*)
3. The following are the measurements $T$ made on a curve recorded by oscilograph representing a change of current $I$ due to a change in the conditions of an electric current.

| $T$: | 1.2 | 2.0 | 2.5 | 3.0 |
|---|---|---|---|---|
| $I$: | 1.36 | 0.58 | 0.34 | 0.20 |

Using Lagrange's formula, find $I$ at $T = 16$.

4. Using Lagrange's interpolation, calculate the profit in the year 2000 from the following data :

| Year : | 1997 | 1999 | 2001 | 2002 |
|---|---|---|---|---|
| Profit in Lakhs of ₹ : | 43 | 65 | 159 | 248 |

(*Anna, 2012*)

5. Find the interpolating polynomial for (0, 2), (1, 3), (2, 12) and (5, 147) using Lagrange's interpolation formula. (*Delhi., 2015*)
6. If $y(1) = -3$, $y(3) = 9$, $y(4) = 30$, $y(6) = 132$, find the Lagrange's interpolation polynomial that takes the same values as $y$ at the given points.
7. Given $f(0) = -18$, $f(1) = 0$, $f(3) = 0$, $f(5) = -248$, $f(6) = 0$, $f(9) = 13104$, find $f(x)$.
8. Find the missing term in the following table using interpolation

| $x$: | 1 | 2 | 4 | 5 | 6 |
|---|---|---|---|---|---|
| $y$: | 14 | 15 | 5 | ... | 9 |

9. Using Lagrange's formula, express the following functions as sum of partial fractions:–

(*i*) $\frac{x^2 + x - 3}{x^3 - 2x^2 - x + 2}$ $\qquad$ (*ii*) $\frac{x^2 + 6x - 1}{(x^2 - 1)(x-4)(x-6)}$ $\qquad$ (*J.N.T.U., 2015*)

## 29.11 DIVIDED DIFFERENCES

The Lagrange's formula has the drawback that if another interpolation value were inserted, then the interpolation coefficients are required to be recalculated. The labour of recomputing the interpolation coefficients is saved by using Newton's general interpolation formula which employs what are called 'divided differences'. Before deriving this formula, we shall first define these differences.

If $(x_0, y_0)$, $(x_1, y_1)$, $(x_2, y_2)$, ... be given points, then the *first divided difference* for the arguments, $x_0, x_1$ is defined by the relation $[x_0, x_1] = \dfrac{y_1 - y_0}{x_1 - x_0}$.

Similarly $[x_1, x_2] = \dfrac{y_2 - y_1}{x_2 - x_1}$ and $[x_2, x_3] = \dfrac{y_3 - y_2}{x_3 - x_2}$ etc.

The *second divided difference* for $x_0, x_1, x_2$ is defined as $[x_0, x_1, x_2] = \dfrac{[x_1 - x_2] - [x_0, x_1]}{x_2 - x_0}$

The *third divided difference* for $x_0, x_1, x_2, x_3$ is defined as

$[x_0, x_1, x_2, x_3] = \dfrac{[x_1, x_2, x_3] - [x_0, x_1, x_2]}{x_3 - x_0}$ and so on.

**Obs. 1.** *The divided differences are symmetrical in their arguments i.e. independent of the order of the arguments.* For it is easy to write $[x_0, x_1] = \dfrac{y_0}{x_0 - x_1} + \dfrac{y_1}{x_1 - x_0} = [x_1, x_0]$ $[x_0, x_1, x_2]$

$$= \frac{y_0}{(x_0 - x_1)(x_0 - x_2)} + \frac{y_1}{(x_1 - x_0)(x_1 - x_2)} \cdot \frac{y_2}{(x_2 - x_0)(x_2 - x_1)}$$

$= [x_1, x_2, x_0]$ or $[x_2, x_0, x_1]$ and so on.

**Obs. 2.** *The nth divided differences of a polynomial of the nth degree are constant.*

Let the arguments be equally spaced so that, $x_1 - x_0 = x_2 - x_1 = ... = x_n - x_{n-1} = h$. Then

$$[x_0, x_1] = \frac{y_1 - y_0}{x_1 - x_0} = \frac{\Delta y_0}{h}$$

$$[x_0, x_1, x_2] = \frac{[x_1, x_2] - [x_0, x_1]}{x_2 - x_0} = \frac{1}{2h}\left\{\frac{\Delta y_1}{h} - \frac{\Delta y_0}{h}\right\}$$

$= \dfrac{1}{2!\,h^2}\Delta^2 y_0$ and in general, $[x_0, x_1, x_2, ... x_n] = \dfrac{1}{n!\,h^n}\Delta^n y_0$.

If the tabulated function is a $n$th degree polynomial, then $\Delta^n y_0$ will be constant. Hence the $n$th divided differences will also be constant.

## 29.12 NEWTON'S DIVIDED DIFFERENCE FORMULA

Let $y_0, y_1, ..., y_n$ be the values of $y = f(x)$ corresponding to the arguments $x_0, x_1, ..., x_n$. Then from the definition of divided differences, we have

$$[x, x_0] = \frac{y - y_0}{x - x_0}$$

so that $$y = y_0 + (x - x_0)[x, x_0] \quad ...(1)$$

Again $$[x, x_0, x_1] = \frac{[x, x_0] - [x_0, x_1]}{x - x_1}$$

which gives $$[x, x_0] = [x_0, x_1] + (x - x_1) + (x - x_1)[x, x_0, x_1]$$

Substituting this value of $[x, x_0]$ in Eqn. (1), we get

$$y = y_0 + (x - x_0)[x_0, x_1] + (x - x_0)(x - x_1)[x, x_0, x_1] \quad ...(2)$$

Also $$[x, x_0, x_1, x_2] = \frac{[x, x_0, x_1] - [x_0, x_1, x_2]}{x - x_2}$$

which gives $[x, x_0, x_1,] = [x_0, x_1, x_2] + (x, x_2)[x, x_0, x_1, x_2]$

Substituting this value of $[x, x_0, x_1]$ in Eqn. (2), we obtain

$$y = y_0 + (x - x_0)[x_0, x_1] + (x - x_0)(x - x_1)[x_0, x_1, x_2] + (x - x_0)(x - x_1)(x - x_2)[x, x_0, x_1, x_2]$$

Proceeding in this manner, we get

$$y = f(x) = y_0 + (x - x_0)[x_0, x_1] + (x - x_0)(x - x_1)[x_0, x_1, x_2] + (x - x_0)(x - x_1)(x - x_2)[x_0, x_1, x_2, x_3] + \dots + (x - x_0)(x - x_1)\dots(x - x_n)[x, x_0, x_1, \dots, x_n] \quad \dots(3)$$

which is called *Newton's general interpolation formula with divided differences.*

**Example 29.30.** *Given the values*

| $x$ : | 5 | 7 | 11 | 13 | 17 |
|---|---|---|---|---|---|
| $f(x)$ : | 150 | 392 | 1452 | 2366 | 5202, |

*evaluate f (9), using Newton's divided difference formula.* (*JNTUK, 2022; V.T.U., 2010*)

**Solution.** The divided difference table is

| $x$ | $y$ | *1st divided differences* | *2nd divided differences* | *3rd divided differences* |
|---|---|---|---|---|
| 5 | 150 | | | |
| | | $\frac{392-150}{7-5} = 121$ | | |
| 7 | 392 | | $\frac{265-121}{11-5} = 24$ | |
| | | $\frac{1452-392}{11-7} = 265$ | | $\frac{32-24}{13-5} = 1$ |
| 11 | 1452 | | $\frac{457-265}{13-7} = 32$ | |
| | | $\frac{2366-1452}{13-11} = 457$ | | $\frac{42-32}{17-7} = 1$ |
| 13 | 2366 | | $\frac{709-457}{17\quad 11} = 42$ | |
| | | $\frac{5202-2366}{17-13} = 709$ | | |
| 17 | 5202 | | | |

Taking $x = 9$ in the Newton's divided difference formula, we obtain

$$f(9) = 150 + (9 - 5) \times 121 + (9 - 5)(9 - 7) \times 24 + (9 - 5)(9 - 7)(9 - 11) \times 1$$
$$= 150 + 484 + 192 - 16 = 810.$$

**Example 29.31.** *Determine f (x) as a polynomial in x for the following data :*

| $x$ : | –4 | –1 | 0 | 2 | 5 |
|---|---|---|---|---|---|
| $f(x)$ : | 1245 | 33 | 5 | 9 | 1335 |

**Solution.** The divided differences table is

| $x$ | $f(x)$ | *1st divided differences* | *2nd divided differences* | *3rd divided differences* | *4th divided differences* |
|---|---|---|---|---|---|
| –4 | 1245 | | | | |
| | | –404 | | | |
| –1 | 33 | | 94 | | |
| | | –28 | | –14 | |
| 0 | 5 | | 10 | | 3 |
| | | 2 | | 13 | |
| 2 | 9 | | 88 | | |
| | | 442 | | | |
| 5 | 1335 | | | | |

Applying Newton's divided difference formula

$$f(x) = f(x_0) + (x - x_0)[x_0, x_1] + (x - x_0)(x - x_1)[x_0, x_1, x_2] + \ldots$$
$$= 1245 + (x + 4)(-404) + (x + 4)(x + 1)(94)$$
$$+ (x + 4)(x + 1)(x - 0)(-14) + (x + 4)(x + 1)x(x - 2)(3)$$
$$= 3x^4 - 5x^3 + 6x^2 - 14x + 5.$$

## PROBLEMS 29.6

1. Find the third divided difference with arguments 2, 4, 9, 10 of the function $f(x) = x^3 - 2x$.
2. Use Newton's divided difference method to compute $f(5.5)$ from the following data :

| $x$ : | 0 | 1 | 4 | 5 | 6 |
|---|---|---|---|---|---|
| $f(x)$ : | 1 | 14 | 15 | 6 | 3 |

(U.P.T.U., 2010)

3. Using Newton's divided difference formula, evaluate $f(8)$ and $f(15)$ given :

| $x$ : | 4 | 5 | 7 | 10 | 11 | 13 |
|---|---|---|---|---|---|---|
| $f(x)$ : | 48 | 100 | 294 | 900 | 1210 | 2028 |

(JNTUK, 2022; Anna, 2013)

4. Obtain the Newton's divided difference interpolation polynomial and hence find $f(6)$ :

| $x$ : | 3 | 7 | 9 | 10 |
|---|---|---|---|---|
| $f(x)$ : | 168 | 120 | 72 | 63 |

(U.P.T.U., 2012)

5. Using Newton's divided difference interpolation, find the polynomial of the given data :

| $x$ : | –1 | 0 | 1 | 3 |
|---|---|---|---|---|
| $f(x)$ : | 2 | 1 | 0 | –1 |

6. For the following table, find $f(x)$ as a polynomial in $x$ using Newton's divided difference formula:

| $x$ : | 5 | 6 | 9 | 11 |
|---|---|---|---|---|
| $f(x)$ : | 12 | 13 | 14 | 16 |

(JNTUK, 2022)

7. Using the following table, find $f(x)$ as a polynomial in $x$ :

| $x$ : | –1 | 0 | 3 | 6 | 7 |
|---|---|---|---|---|---|
| $f(x)$ : | 3 | –6 | 39 | 822 | 1611 |

(U.P.T.U., 2012)

8. Find the missing term in the following table using Newton's divided difference formula

| $x$ : | 0 | 1 | 2 | 3 | 4 |
|---|---|---|---|---|---|
| $y$ : | 1 | 3 | 9 | ... | 81 |

## 29.13 INVERSE INTERPOLATION

So far, given a set of values of $x$ and $y$, we have been finding the values of $y$ corresponding to a certain value of $x$. On the other hand, the process of estimating the value of $x$ for a value of $y$ (which is not in the table) is called the *inverse interpolation.*

Lagrange's formula is merely a relation between two variables either of which may be taken as the independent variable. Therefore, on inter-changing $x$ and $y$ in the Lagrange's formula, we obtain

$$x = \frac{(y - y_1)(y - y_2)\ldots(y - y_n)}{(y_0 - y_1)(y_0 - y_2)\ldots(y_0 - y_n)} x_0 + \frac{(y - y_0)(y - y_2)\ldots(y - y_n)}{(y_1 - y_0)(y_1 - y_2)\ldots(y_1 - y_n)} x_1$$
$$+ \ldots + \frac{(y - y_0)(y - y_1)\ldots(y - y_{n-1})}{(y_n - y_0)(y_n - y_1)\ldots(y_n - y_{n-1})} x_n \quad \ldots(1)$$

which is used for inverse interpolation.

**Example 29.32.** *The following table gives the values of x and y :*

| $x$ : | 1.2 | 2.1 | 2.8 | 4.1 | 4.9 | 6.2 |
|---|---|---|---|---|---|---|
| $y$ : | 4.2 | 6.8 | 9.8 | 13.4 | 15.5 | 19.6 |

*Find the value of x corresponding to y = 12, using Lagrange's technique.*

**Solution.** Here $x_0 = 1.2, x_1 = 2.1, x_2 = 2.8, x_3 = 4.1, x_4 = 4.9, x_5 = 6.2$

and $y_0 = 4.2, y_1 = 6.8, y_2 = 9.8, y_3 = 13.4, y_4 = 15.5, y_5 = 19.6$

Taking $y = 12$, the above formula (1) gives

$$x = \frac{(12-6.8)(12-9.8)(12-13.4)(12-15.5)(12-19.6)}{(4.2-6.8)(4.2-9.8)(4.2-13.4)(4.2-15.5)(4.2-19.6)} \times 1.2$$

$$+ \frac{(12-4.2)(12-9.8)(12-13.4)(12-15.5)(12-19.6)}{(6.8-4.2)(6.8-9.8)(6.8-13.4)(6.8-15.5)(6.8-19.6)} \times 2.1$$

$$+ \frac{(12-4.2)(12-6.8)(12-13.4)(12-15.5)(12-19.6)}{(9.8-4.2)(9.8-6.8)(9.8-13.4)(9.8-15.5)(9.8-19.6)} \times 2.8$$

$$+ \frac{(12-4.2)(12-6.8)(12-9.8)(12-15.5)(12-19.6)}{(13.4-4.2)(13.4-6.8)(13.4-9.8)(13.4-15.5)(13.4-19.6)} \times 4.1$$

$$+ \frac{(12-4.2)(12-6.8)(12-9.8)(12-13.4)(12-19.6)}{(15.5-4.2)(15.5-6.8)(15.5-9.8)(15.5-13.4)(15.5-19.6)} \times 4.9$$

$$+ \frac{(12-4.2)(12-6.8)(12-9.8)(12-13.4)(12-15.5)}{(19.6-4.2)(19.6-6.8)(19.6-9.8)(19.6-13.4)(19.6-15.5)} \times 6.2$$

$$= 0.022 - 0.234 + 1.252 + 3.419 - 0.964 + 0.055 = 3.55.$$

**Example 29.33.** *Apply Lagrange's formula inversely to obtain a root of the equation $f(x) = 0$, given that $f(30) = -30$, $f(34) = -13$, $f(38) = 3$, and $f(42) = 18$.*

**Solution.** Here $x_0 = 30, x_1 = 34, x_2 = 38, x_3 = 42$

and $y_0 = -30, y_1 = -13, y_2 = 3, y_3 = 18$

It is required to find $x$ corresponding to $y = f(x) = 0$.

Taking $y = 0$, the Lagrange's formula gives,

$$x = \frac{(y-y_1)(y-y_2)(y-y_3)}{(y_0-y_1)(y_0-y_2)(y_0-y_3)} x_0 + \frac{(y-y_0)(y-y_2)(y-y_3)}{(y_1-y_0)(y-y_2)(y_1-y_3)} x_1$$

$$+ \frac{(y-y_0)(y-y_1)(y-y_3)}{(y_2-y_0)(y_2-y_1)(y_2-y_3)} x_2 + \frac{(y-y_0)(y-y_1)(y-y_2)}{(y_3-y_0)(y_3-y_1)(y_3-y_2)} x_3$$

$$= \frac{13(-3)(-18)}{(-17)(-33)(-48)} \times 30 + \frac{30(-3)(-18)}{17(-16)(-31)} \times 34 + \frac{30(13)(-18)}{33(16)(-15)} \times 38 + \frac{30(13)(-3)}{48(31)(15)} \times 42$$

$$= -0.782 + 6.532 + 33.682 - 2.202 = 37.23$$

Hence the desired root of $f(x) = 0$ is 37.23.

**Example 29.34.** *Estimate f(42) from the following data using Newton backward interpolation.*

*(Karaikal University, 2021*

| *x :* | *20* | *25* | *30* | *35* | *40* | *45* |
|---|---|---|---|---|---|---|
| *f(x) :* | *354* | *332* | *291* | *260* | *231* | *204* |

**Solution.** The difference table is

| $x$ | $f$ | $\nabla f$ | $\nabla^2 f$ | $\nabla^3 f$ | $\nabla^4 f$ | $\nabla^5 f$ |
|---|---|---|---|---|---|---|
| 20 | 354 | | | | | |
| | | –22 | | | | |
| 25 | 332 | | –19 | | | |
| | | –41 | | 29 | | |
| 30 | 291 | | 10 | | –37 | |
| | | –31 | | –8 | | 45 |
| 35 | 260 | | 2 | | 8 | |
| | | –29 | | 0 | | |
| 40 | 231 | | 2 | | | |
| | | –27 | | | | |
| 45 | 204 | | | | | |

So, $X_n = 45$, $h = 5$, $x = 42$ and $p = -0.6$

Newton backward formula

$$P(x) = y_n + P\nabla Y_n + \frac{P(p+1)}{2!}\nabla^2 y_n + \frac{P(P+1)(P+2)}{3!}\nabla^3 y_n + \frac{P(P+1)(P+2)(P+3)}{4!}\nabla^4 y_n + \frac{P(P+1)(P+2)(P+3)(P+4)}{5!}\nabla^5 y_n$$

$$P(42) = 204 + (-0.6)(-27) + \frac{(-0.6)(0.4)}{2}\times 2\,\frac{(-0.6)(0.40)(1.4)}{6}\times 0 + \frac{(-0.6)(0.4)(1.4)(2.4)}{24}\times 8 + \frac{(-0.6)(0.4)(1.4)(2.4)(3.5)}{120}\times 45$$

$$= 219.1430$$

$$f(42) = 219.143$$

## PROBLEMS 29.7

1. Apply Lagrange's method to find the value of $x$ when $f(x) = 15$ from the given data :

| $x$ : | 5 | 6 | 9 | 11 |
|---|---|---|---|---|
| $f(x)$ : | 12 | 13 | 14 | 16 |

2. Obtain the value of $t$ when $A = 85$ from the following table, using Lagrange's method :

| $t$ : | 2 | 5 | 8 | 14 |
|---|---|---|---|---|
| $A$ : | 94.8 | 87.9 | 81.3 | 68.7 |

## 29.14 OBJECTIVE TYPE OF QUESTIONS

## PROBLEMS 29.8

*Select the correct answer or fill up the blanks in the following problems :*

1. Newton's backward interpolation formula is ......
2. Bessel's formula is most appropriate when $p$ lies between
   (*a*) – 0.25 and 0.25 (*b*) 0.25 and 0.75 (*c*) 0.75 and 1.00.
3. From the divided difference table for the following data :

| $x$: | 5 | 15 | 22 |
|---|---|---|---|
| $y$: | 7 | 36 | 160 |

4. Interpolation is the technique of estimating the value of a function for any .......
5. Bessel's formula for interpolation is .......
6. The 4th divided differences for $x_0, x_1, x_2, x_3, x_4$ = ....... .
7. Stirling's formula is best suited for $p$ lying between ....... .
8. Newton's divided differences formula is ....... .
9. Given $(x_0, y_0), (x_1, y_1), (x_2, y_2)$, Lagrange's interpolation formula is .......
10. If $f(0) = 1$, $f(2) = 5$, $f(3) = 10$ and $f(x) = 14$, then $x$ = ....... .
11. Gauss forward interpolation formula involves
    (*a*) even differences above the central line and odd differences on the central line
    (*b*) even differences below the central line and odd differences on the central line
    (*c*) odd differences below the central line and even differences on the central line
    (*d*) odd differences above the central line and even differences on the central line.
12. If $y(1) = 4$, $y(3) = 12$, $y(4) = 19$ and $y(x) = 7$ find $x$ using Lagrange's formula.
13. Extrapolation is defined as .......
14. The second divided difference of $f(x) = 1/x$, with arguments, $a, b, c$, is....... .
15. Gauss-forward interpolation formula is used to interpolate values of $y$ for
    (*a*) $0 < p < 1$ (*b*) $-1 < p < 0$
    (*c*) $0 < p < -\alpha$ (*d*) $-\alpha < p < 0$.
16. Given

| $x$: | 0 | 1 | 3 | 4 |
|---|---|---|---|---|
| $y$: | –12 | 0 | 6 | 12 |

    Using Lagrange's formula, a polynomial that can be fitted to the data is .......
17. The $n$th divided difference of a polynomial of degree $n$ is
    (*a*) zero (*b*) a constant
    (*c*) a variable (*d*) none of these.
18. If $h$ is the interval of differencing, $\Delta^2 x^3$ = ....... .
19. If $f(x) = 1/x$, show that divided differences $f[a, b] = -1/ab$ and $f[a, b, c] = 1/abc$. (*Delhi, 2015*)

CHAPTER 30

# Numerical Differentiation & Integration



## 30.1 NUMERICAL DIFFERENTIATION

It is the process of calculating the value of the derivative of a function at some assigned value of $x$ from the given set of values $(x_i, y_i)$. To compute $dy/dx$, we first replace the latter as many times as we desire. The choice of the interpolation formula to be used, will depend on the assigned value of $x$ at which $dy/dx$, is desired.

If the values of $x$ are equi-spaced and $dy/dx$, is required near the beginning of the table, we employ Newton's forward formula. If it is required near the end of the table, we use Newton's backward formula. For values near the middle of the table, $dy/dx$, is calculated by means of Stirling's or Bessel's formula. If the values of $x$ are not equi-spaced, we use Newton's divided difference formula to represent the function.

## 30.2 FORMULAE FOR DERIVATIVES

Consider the function $y = f(x)$ which is tabulated for the values $x_i\ (= x_0 + ih)$, $i = 0, 1, 2, \ldots n$.

**(1) Derivatives using forward difference formula.** Newton's forward interpolation formula is

$$y = y_0 + p\Delta y_0 + \frac{p(p-1)}{2!}\Delta^2 y_0 + \frac{p(p-1)(p-2)}{3!}\Delta^3 y_0 + \ldots$$

Differentiating both sides w.r.t. $p$, we have

$$\frac{dy}{dp} = \Delta y_0 + \frac{2p-1}{2!}\Delta^2 y_0 + \frac{3p^2-6p+2}{3!}\Delta^3 y_0 + \ldots$$

Since $p = \frac{(x - x_0)}{h}$, therefore $\frac{dp}{dx} = \frac{1}{h}\cdot$

Now
$$\frac{dy}{dx} = \frac{dy}{dp}\cdot\frac{dp}{dx} = \frac{1}{h}\left[\Delta y_0 + \frac{2p-1}{2!}\Delta^2 y_0 + \frac{3p^2-6p+2}{3!}\Delta^3 y_0 + \frac{4p^3-18p^2+22p-6}{4i}\Delta^4 y_0 + \cdots\right] \quad \ldots(1)$$

At $x = x_0, p = 0$. Hence putting $p = 0$,

$$\left(\frac{dy}{dx}\right)_{x_0} = \frac{1}{h}\left[\Delta y_0 - \frac{1}{2}\Delta^2 y_0 + \frac{1}{3}\Delta^3 y_0 - \frac{1}{4}\Delta^4 y_0 + \frac{1}{5}\Delta^5 y_0 - \frac{1}{6}\Delta^6 y_0 + \cdots\right] \quad \ldots(2)$$

Again differentiating Eqn. (1) w.r.t. $x$, we get

$$\frac{d^2y}{dx^2} = \frac{d}{dp}\left(\frac{dy}{dp}\right)\frac{dp}{dx}$$

$$= \frac{1}{h}\left[\frac{2}{2!}\Delta^2 y_0 + \frac{6p-6}{3!}\Delta^3 y_0 + \frac{12p^2 - 36p + 22}{4!}\Delta^4 y_0 + \cdots\right]\frac{1}{h}$$

Putting $p = 0$, we obtain

$$\left(\frac{d^2y}{dx^2}\right)_{x_0} = \frac{1}{h^2}\left[\Delta^2 y_0 - \Delta^3 y_0 + \frac{11}{12}\Delta^4 y_0 - \frac{5}{6}\Delta^5 y_0 + \frac{137}{180}\Delta^6 y_0 + \cdots\right] \quad ...(3)$$

Similarly $$\left(\frac{d^3y}{dx^3}\right)_{x_0} = \frac{1}{h^3}\left[\Delta^3 y_0 - \frac{3}{2}\Delta^4 y_0 + \cdots\right] \quad ...(4)$$

**Otherwise :** We know that $1 + \Delta = E = e^{hD}$

$$\therefore \quad hD = \log(1 + \Delta) = \Delta - \frac{1}{2}\Delta^2 + \frac{1}{3}\Delta^3 - \frac{1}{4}\Delta^4 + \ldots$$

or $$D = \frac{1}{h}\left[\Delta - \frac{1}{2}\Delta^2 + \frac{1}{3}\Delta^3 - \frac{1}{4}\Delta^4 + \cdots\right]$$

and $$D^2 = \frac{1}{h^2}\left[\Delta - \frac{1}{2}\Delta^2 + \frac{1}{3}\Delta^3 - \frac{1}{4}\Delta^4 + \cdots\right]^2 = \frac{1}{h^2}\left[\Delta^2 - \Delta^3 + \frac{11}{12}\Delta^4 + \cdots\right]$$

and $$D^3 = \frac{1}{h^2}\left[\Delta^3 - \frac{3}{2}\Delta^4 + \cdots\right]$$

Now applying the above identities to $y_0$, we get

$Dy_0$, *i.e.*, $$\left(\frac{dy}{dx}\right)_{x_0} = \frac{1}{h}\left[\Delta y_0 - \frac{1}{2}\Delta^2 y_0 + \frac{1}{3}\Delta^3 y_0 - \frac{1}{4}\Delta^4 y_0 + \frac{1}{5}\Delta^5 y_0 - \frac{1}{6}\Delta^6 y_0 + \cdots\right]$$

$$\left(\frac{d^2y}{dx^2}\right)_{x_0} = \frac{1}{h^2}\left[\Delta^2 y_0 - \Delta^3 y_0 + \frac{11}{12}\Delta^4 y_0 - \frac{5}{6}\Delta^5 y_0 + \frac{137}{180}\Delta^6 y_0 - \cdots\right]$$

and $$\left(\frac{d^3y}{dx^3}\right)_{x_0} = \frac{1}{h^3}\left[\Delta^2 y_0 - \frac{3}{2}\Delta^4 y_0 + \cdots\right]$$

which are the same as Eqns. (2), (3) and (4) respectively.

**(2) Derivatives using backward difference formula.** Newton's backward interpolation formula (p. 1089) is

$$y = y_n + p\nabla y_n + \frac{p(p+1)}{2!}\nabla^2 y_n + \frac{p(p+1)(p+2)}{3!}\nabla^3 y_n + \cdots$$

Differentiating both sides w.r.t. $p$, we get

$$\frac{dy}{dp} = \nabla y_n + \frac{2p+1}{2!}\nabla^2 y_n + \frac{3p^2 + 6p + 2}{3!}\nabla^3 y_n + \cdots$$

Since $p = \frac{x - x_n}{h}$, therefore $\frac{dp}{dx} = \frac{1}{h}$.

Now $$\frac{dy}{dx} = \frac{dy}{dp}\cdot\frac{dp}{dx} = \frac{1}{h}\left[\nabla y_n \; \frac{2p+1}{2!}\nabla^2 y_n + \frac{3p^3 + 6p + 2}{3!}\nabla^3 y_n + \cdots\right] \quad ...(5)$$

At $x = x_n$, $p = 0$. Hence putting $p = 0$, we get

$$\left(\frac{dy}{dx}\right)_{x_n} = \frac{1}{h}\left[\nabla y_n + \frac{1}{2}\nabla^2 y_n + \frac{1}{3}\nabla^3 y_n + \frac{1}{4}\nabla^4 y_n + \frac{1}{5}\nabla^5 y_n + \frac{1}{6}\nabla^6 y_n + \cdots\right] \quad ...(6)$$

Again differentiating (5) w.r.t. $x$, we have

$$\frac{d^2y}{dx^2} = \frac{d}{dp}\left(\frac{dy}{dx}\right)\frac{dp}{dx} = \frac{1}{h^2}\left[\nabla^2 y_n + \frac{6p+6}{3!}\nabla^3 y_n + \frac{6p^2 + 18p + 11}{12}\nabla^4 y_n + \cdots\right]$$

Putting $p = 0$, we obtain

$$\left(\frac{d^2y}{dx^2}\right)_{x_n} = \frac{1}{h^2}\left[\nabla^2 y_n + \nabla^3 y_n + \frac{11}{12}\nabla^4 y_n + \frac{5}{6}\nabla^5 y_n + \frac{137}{180}\nabla^6 y_n + \cdots\right] \quad ...(7)$$

Similarly, $$\left(\frac{d^3y}{dx^3}\right)_{x_n} = \frac{1}{h^3}\left[\nabla^3 y_n + \frac{3}{2}\nabla^4 y_n + \cdots\right] \quad ...(8)$$

**Otherwise :** We know that $1 - \nabla = E^{-1} = e^{-hD}$

$$\therefore \quad -hD = \log(1-\nabla) = -\left[\nabla + \frac{1}{2}\nabla^2 + \frac{1}{3}\nabla^3 + \frac{1}{4}\nabla^4 + \cdots\right]$$

or $$D = \frac{1}{h}\left[\nabla + \frac{1}{2}\nabla^2 + \frac{1}{3}\nabla^3 + \frac{1}{4}\nabla^4 + \cdots\right]$$

$$\therefore \quad D^2 = \frac{1}{h^2}\left[\nabla + \frac{1}{2}\nabla^2 + \frac{1}{3}\nabla^3 + \cdots\right]^2 = \frac{1}{h^2}\left[\nabla^2 + \nabla^3 + \frac{11}{12}\nabla^4 + \cdots\right]$$

Similarly, $$D^3 = \frac{1}{h^3}\left[\nabla^3 + \frac{3}{2}\nabla^4 + \cdots\right]$$

Applying these identities to $y_n$, we get

$Dy_n$, *i.e.*, $$\left(\frac{dy}{dx}\right)_{x_n} = \frac{1}{h}\left[\nabla y_n + \frac{1}{2}\nabla^2 y^n + \frac{1}{3}\nabla^3 y_n + \frac{1}{4}\nabla^4 y_n + \frac{1}{5}\nabla^5 y_n + \frac{1}{6}\nabla^6 y_n + \cdots\right]$$

$$\left(\frac{d^2y}{dx^2}\right)_{x_n} = \frac{1}{h^2}\left(\nabla^2 y_n + \nabla^3 y_n + \frac{11}{12}\nabla^4 y_n + \frac{5}{6}\nabla^5 y_n + \frac{137}{180}\nabla^6 y_n + \cdots\right)$$

and $$\left(\frac{d^3y}{dx^3}\right)_{x_n} = \frac{1}{h^3}\left[\nabla^3 y_n + \frac{3}{2}\nabla^4 y_n + \cdots\right]$$

which are the same as Eqns. (6), (7) and (8).

**(3) Derivatives using central difference formulae.** Stirling's formula is

$$y_p = y_0 + \frac{p}{1!}\left(\frac{\Delta y_0 + \Delta y_{-1}}{2}\right) + \frac{p^2}{2!}\Delta^2 y_{-1} + \frac{p(p^2-1^2)}{3!}\left(\frac{\Delta^3 y_{-1} + \Delta^3 y_{-2}}{2}\right) + \frac{p^2(p^2-1^2)}{4!}\Delta^4 y_{-2} + \ldots$$

Differentiating both sides w.r.t. $p$., we get

$$\frac{dy}{dp} = \left(\frac{\Delta y_0 + \Delta y_{-1}}{2}\right) + \frac{2p}{2!}\Delta^2 y_{-1} + \frac{3p^2-1}{3!}\left(\frac{\Delta^3 y_{-1} + \Delta^3 y_{-2}}{2}\right) + \frac{4p^3-2p}{4!}\Delta^4 y_{-2} + \ldots$$

Since $p = \dfrac{x - x_0}{h}$, $\therefore \quad \dfrac{dp}{dx} = \dfrac{1}{h}$.

Now $$\frac{dy}{dx} = \frac{dy}{dp}\cdot\frac{dp}{dx} = \frac{1}{h}\left[\left(\frac{\Delta y_0 + \Delta y_{-1}}{2}\right) + p\Delta^2 y_{-1} + \frac{3p^2-1}{6}\left(\frac{\Delta^3 y_{-1} + \Delta^3 y_{-2}}{2}\right) + \frac{2p^3-p}{12}\Delta^4 y_{-2} + \cdots\right]$$

At $x = x_0$, $p = 0$. Hence putting $p = 0$, we get

$$\left(\frac{dy}{dx}\right)_{x_0} = \frac{1}{h}\left[\frac{\Delta y_0 + \Delta y_{-1}}{2} - \frac{1}{6}\frac{\Delta^3 y_{-1} + \Delta^3 y_{-2}}{2} + \frac{1}{30}\frac{\Delta^5 y_{-2} + \Delta^5 y_{-3}}{2} + \cdots\right] \quad ...(9)$$

Similarly $$\left(\frac{d^2y}{dx^2}\right)_{x_0} = \frac{1}{h^2}\left[\Delta^2 y_{-1} - \frac{1}{12}\Delta^4 y_{-2} + \frac{1}{90}\Delta^6 y_{-3} - \cdots\right] \quad ...(10)$$

**Obs.** We can similarly use any other interpolation formula for computing the derivatives.

**Example 30.1.** *Given that*

| $x$ : | 1.0 | 1.1 | 1.2 | 1.3 | 1.4 | 1.5 | 1.6 |
|---|---|---|---|---|---|---|---|
| $y$ : | 7.989 | 8.403 | 8.781 | 9.129 | 9.451 | 9.750 | 10.031 |

*find* $\frac{dy}{dx}$ *and* $\frac{d^2y}{dx^2}$ *at* (*a*) $x = 1.1$ (U.P.T.U., 2012)

(*b*) $x = 1.6$. (U.P.T.U., 2012)

**Solution.** (*a*) The difference table is :

| $x$ | $y$ | $\Delta$ | $\Delta^2$ | $\Delta^3$ | $\Delta^4$ | $\Delta^5$ | $\Delta^6$ |
|---|---|---|---|---|---|---|---|
| 1.0 | 7.989 | | | | | | |
| | | 0.414 | | | | | |
| 1.1 | 8.403 | | –0.036 | | | | |
| | | 0.378 | | 0.006 | | | |
| 1.2 | 8.781 | | –0.030 | | –0.002 | | |
| | | 0.348 | | 0.004 | | 0.001 | |
| 1.3 | 9.129 | | –0.026 | | –0.001 | | 0.002 |
| | | 0.322 | | 0.003 | | 0.003 | |
| 1.4 | 9.451 | | –0.023 | | 0.002 | | |
| | | 0.299 | | 0.005 | | | |
| 1.5 | 9.750 | | –0.018 | | | | |
| | | 0.281 | | | | | |
| 1.6 | 10.031 | | | | | | |

We have

$$\left(\frac{dy}{dx}\right)_{x_0} = \frac{1}{h}\left[\Delta y_0 - \frac{1}{2}\Delta^2 y_0 + \frac{1}{3}\Delta^3 y_0 - \frac{1}{4}\Delta^4 y_0 + \frac{1}{5}\Delta^2 y_0 - \frac{1}{6}\Delta^6 y_0 + \cdots\right] \quad ...(i)$$

and

$$\left(\frac{d^2y}{dx^2}\right)_{x_0} = \frac{1}{h^2}\left[\Delta^2 y_0 - \Delta^3 y_0 + \frac{11}{12}\Delta^4 y_0 - \frac{5}{6}\Delta^5 y_0 - \cdots\right] \quad ...(ii)$$

Here $h = 0.1$, $x_0 = 1.1$, $\Delta y_0 = 0.378$, $\Delta^2 y_0 = -0.03$ etc.

Substituting these values in (*i*) and (*ii*), we get

$$\left(\frac{dy}{dx}\right)_{1.1} = \frac{1}{0.1}\left[0.378 - \frac{1}{2}(-0.03) + \frac{1}{3}(0.004) - \frac{1}{4}(-0.001) + \frac{1}{5}(0.003)\right] = 3.952$$

$$\left(\frac{d^2y}{dx^2}\right)_{1.1} = \frac{1}{(0.1)^2}\left[-0.03 - (0.004) + \frac{11}{12}(-0.001) - \frac{5}{6}(0.003)\right] = -3.74$$

(*b*) We use the above difference table and the backward difference operator $\nabla$ instead of $\Delta$.

$$\left(\frac{dy}{dx}\right)_{x_n} = \frac{1}{h}\left[\nabla y_n + \frac{1}{2}\nabla^2 y_n + \frac{1}{3}\nabla^3 y_n + \frac{1}{4}\nabla^4 y_n + \frac{1}{5}\nabla^2 y_n + \frac{1}{6}\nabla^6 y_n + \cdots\right] \quad ...(i)$$

and

$$\left(\frac{d^2y}{dx^2}\right)_{x_n} = \frac{1}{h^2}\left[\nabla^2 y_n + \nabla^3 y_n + \frac{11}{12}\nabla^4 y_n + \frac{5}{6}\nabla^5 y_n + \frac{137}{180}\nabla^6 y_n + \cdots\right] \quad ...(ii)$$

Here $h = 0.1$, $x_n = 1.6$, $\nabla y_n = 0.281$, $\nabla^2 y_n = -0.018$ etc.

Putting these values in (*i*) and (*ii*), we get

$$\left[\frac{dy}{dx}\right]_{1.6} = \frac{1}{0.1}\left[0.281 + \frac{1}{2}(-0.018) + \frac{1}{3}(0.005) + \frac{1}{4}(0.002) + \frac{1}{5}(0.003) + \frac{1}{6}(0.002)\right] = 2.75$$

$$\left(\frac{d^2y}{dx^2}\right)_{1.6} = \frac{1}{(0.1)^2}\left[-0.018 + 0.005 + \frac{11}{12}(0.002) + \frac{5}{6}(0.003) + \frac{137}{180}(0.002)\right]$$

$$= -0.715.$$

**Example 30.2.** *The following data gives the velocity of a particle for 20 seconds at an interval of 5 seconds. Find the initial acceleration using the entire data :*

| *Time t (sec)* : | 0 | 5 | 10 | 15 | 20 |
|---|---|---|---|---|---|
| *Velocity v (m/sec)* : | 0 | 3 | 14 | 69 | 228 |

**Solution.** The difference table is :

| $t$ | $v$ | $\Delta v$ | $\Delta^2 v$ | $\Delta^3 v$ | $\Delta^4 v$ |
|---|---|---|---|---|---|
| 0 | 0 | | | | |
| | | 3 | | | |
| 5 | 3 | | 8 | | |
| | | 11 | | 36 | |
| 10 | 14 | | 44 | | 24 |
| | | 55 | | 60 | |
| 15 | 69 | | 104 | | |
| | | 159 | | | |
| 20 | 228 | | | | |

An initial acceleration $\left(i.e. \dfrac{dv}{dt}\right)$ at $t = 0$ is required, we use Newton's forward formula :

$$\left(\frac{dv}{dt}\right)_{t=0} = \frac{1}{h}\left(\Delta v_0 - \frac{1}{2}\Delta^2 v_0 + \frac{1}{3}\Delta^3 v_0 - \frac{1}{4}\Delta^4 v_0 + \cdots\right)$$

$$\therefore \quad \left(\frac{dv}{dt}\right)_{t=0} = \frac{1}{5}\left[3 - \frac{1}{2}(8) + \frac{1}{3}(36) - \frac{1}{4}(24)\right] = \frac{1}{5}(3 - 4 + 12 - 6) = 1$$

Hence the initial acceleration is 1 m/sec$^2$.

**Example 30.3.** *A slider in a machine moves along a fixed straight rod. Its distance x cm. along the rod is given below for various values of the time t seconds. Find the velocity of the slider and its acceleration when t = 0.3 seconds.*

| *t* = | 0 | 0.1 | 0.2 | 0.3 | 0.4 | 0.5 | 0.6 |
|---|---|---|---|---|---|---|---|
| *x* = | 30.13 | 31.62 | 32.87 | 33.64 | 33.95 | 33.81 | 33.24 |

**Solution.** The difference table is :

| $t$ | $x$ | $\Delta$ | $\Delta^2$ | $\Delta^3$ | $\Delta^4$ | $\Delta^5$ | $\Delta^6$ |
|---|---|---|---|---|---|---|---|
| 0 | 30.13 | | | | | | |
| | | 1.49 | | | | | |
| 0.1 | 31.62 | | –0.24 | | | | |
| | | 1.25 | | –0.24 | | | |
| 0.2 | 32.87 | | –0.48 | | 0.26 | | |
| | | 0.77 | | 0.02 | | –0.27 | |
| 0.3 | 33.64 | | –0.46 | | –0.01 | | 0.29 |
| | | 0.31 | | 0.01 | | 0.02 | |
| 0.4 | 33.95 | | –0.45 | | 0.01 | | |
| | | –0.14 | | 0.02 | | | |
| 0.5 | 33.81 | | –0.43 | | | | |
| | | –0.57 | | | | | |
| 0.6 | 33.24 | | | | | | |

As the derivatives are required near the middle of the table, we use Stirling's formulae :

$$\left(\frac{dx}{dt}\right)_{t_0} = \frac{1}{h}\left(\frac{\Delta x_0 + \Delta x_{-1}}{2}\right) - \frac{1}{6}\left(\frac{\Delta^3 x_{-1} + \Delta^3 x_{-2}}{2}\right) + \frac{1}{30}\left(\frac{\Delta^5 x_{-2} + \Delta^5 x_{-3}}{2}\right) + \ldots \quad \ldots(i)$$

$$\left(\frac{d^2x}{dt^2}\right)_{t_0} = \frac{1}{h^2}\left[\Delta^2 x_{-1} - \frac{1}{12}\Delta^4 x_{-2} + \frac{1}{90}\Delta^6 x_{-3} \cdots\right] \quad \ldots(ii)$$

Here $h = 0.1$, $t_0 = 0.3$, $\Delta x_0 = 0.31$, $\Delta x_{-1} = 0.77$, $\Delta^2 x_{-1} = -0.46$ etc.

Putting these values in (*i*) and (*ii*), we get

$$\left(\frac{dx}{dt}\right)_{0.3} = \frac{1}{0.1}\left[\frac{0.31+0.77}{2} - \frac{1}{6}\left(\frac{0.01+0.02}{2}\right) + \frac{1}{30}\left(\frac{0.02-0.27}{2}\right) - \cdots\right] = 5.33$$

$$\left(\frac{d^2x}{dt^2}\right)_{0.3} = \frac{1}{(0.1)^2}\left[-0.46 - \frac{1}{12}(-0.01) + \frac{1}{90}(0.29) - \cdots\right] = -45.6$$

Hence the required velocity is 5.33 cm/sec and acceleration is – 45.6 cm/sec$^2$.

**Example 30.4.** *Using Bessel's formula, find f ' (7.5) from the following table :*

| *x* : | 7.47 | 7.48 | 7.49 | 7.50 | 7.51 | 7.52 | 7.53 |
|---|---|---|---|---|---|---|---|
| *f(x)* : | 0.193 | 0.195 | 0.198 | 0.201 | 0.203 | 0.206 | 0.208 |

**Solution.** Taking $x_0 = 7.50$, $h = 0.1$, we have $p = \frac{x - x_0}{h} = \frac{x - 7.50}{0.01}$

The difference table is :

| $x$ | $p$ | $y_p$ | $\Delta$ | $\Delta^2$ | $\Delta^3$ | $\Delta^4$ | $\Delta^5$ | $\Delta^6$ |
|---|---|---|---|---|---|---|---|---|
| 7.47 | –3 | 0.193 | | | | | | |
| | | | 0.002 | | | | | |
| 7.48 | –2 | 0.195 | | 0.001 | | | | |
| | | | 0.003 | | –0.001 | | | |
| 7.49 | –1 | 0.198 | | 0.000 | | 0.000 | | |
| | | | 0.003 | | –0.001 | | 0.003 | |
| 7.50 | 0 | 0.201 | | –0.001 | | 0.003 | | –0.01 |
| | | | 0.002 | | 0.002 | | –0.007 | |
| 7.51 | 1 | 0.203 | | 0.001 | | –0.004 | | |
| | | | 0.003 | | –0.002 | | | |
| 7.52 | 2 | 0.206 | | –0.001 | | | | |
| | | | 0.002 | | | | | |
| 7.53 | 3 | 0.208 | | | | | | |

According to Bessel's formula

$$y_p = y_0 + p\Delta y_0 + \frac{p(p-1)}{2!}\cdot\frac{\Delta^2 y_{-1} + \Delta^2 y_0}{2} + \frac{\left(p-\frac{1}{2}\right)p(p-1)}{3!}\cdot\Delta^3 y_{-1}$$
$$+ \frac{(p+1)p(p-1)(p-2)}{4!}\cdot\frac{\Delta^4 y_{-2} + \Delta^4 y_{-1}}{2} + \frac{\left(p-\frac{1}{2}\right)(p+1)p(p-1)(p-2)}{5!}\cdot\Delta^5 y_{-2}$$
$$+ \frac{(p+2)p(p+1)p(p-1)(p-2)(p-3)}{6!}\cdot\frac{\Delta^6 y_{-3} + \Delta^6 y_{-2}}{2} + \cdots \quad \ldots(i)$$

Since $p = \frac{x - x_0}{h}$, $\therefore$ $\frac{dp}{dx} = \frac{1}{h}$ and $\frac{dy}{dx} = \frac{dy}{dp}\cdot\frac{dp}{dx} = \frac{1}{h}\frac{dy}{dp}$

Differentiating Eqn. (*i*) w.r.t. *p* and putting $p = 0$, we get

$$\left(\frac{dy}{dx}\right)_{7.5} = \frac{1}{h}\left(\frac{dy}{dp}\right)_{p=0} = \frac{1}{h}\left[\Delta y_0 - \frac{1}{h}(\Delta^2 y_{-1} + \Delta^2 y_0) + \frac{1}{12}\Delta^3 y_{-1} + \frac{1}{24}(\Delta^4 y_{-2} + \Delta^4 y_{-1})\right.$$
$$\left. - \frac{1}{120}\Delta^5 y_{-2} - \frac{1}{240}(\Delta^6 y_{-3} + \Delta^6 y_{-2})\right]$$

$$\left(\frac{dy}{dx}\right)_{7.5} = \frac{1}{0.01}\left[0.002 - \frac{1}{4}(-0.001 + 0.001) + \frac{1}{12}(0.002)^1 + \frac{1}{24}(-0.004 + 0.003) - \frac{1}{120}(-0.007)\frac{-1}{240}(0.010 + 0)\right]$$

$[\because \ \Delta^6 y_{-2} = 0]$

$$= 0.2 + 0 + 0.01666 - 0.00583 + 0.00416 = 0.223.$$

**Example 30.5.** *Find f'(0) from the following data :*

| $x$ : | 3 | 5 | 11 | 27 | 34 |
|---|---|---|---|---|---|
| $f(x)$ : | –13 | 23 | 899 | 17315 | 35606 |

**Solution.** As the values of $x$ are not equi-spaced, we shall use Newton's divided difference formula. The divided difference table is

| $x$ | $f(x)$ | 1st div. diff. | 2nd div. diff. | 3rd div. diff. | 4th div. diff. |
|---|---|---|---|---|---|
| 3 | –13 | | | | |
| | | 18 | | | |
| 5 | 23 | | 16 | | |
| | | 146 | | 0.998 | |
| 11 | 899 | | 39.96 | | 0.0002 |
| | | 1025 | | 1.003 | |
| 27 | 17315 | | 69.04 | | |
| | | 2613 | | | |
| 34 | 35606 | | | | |

Fifth difference being zero, Newton's divided difference formula is

$$f(x) = f(x_0) + (x - x_0) f(x_0 - x_1) + (x - x_0)(x - x_1) f(x_0, x_1, x_2) + (x - x_0)(x - x_1)(x - x_2) f(x_0, x_1, x_2, x_3) + (x - x_0)(x - x_1) \times (x - x_2)(x - x_3) f(x_0, x_1, x_2 x_3, x_4)$$

Differentiating it w.r.t. $x$, we get

$$f'(x) = f(x_0, x_1) + (2x - x_0 - x_1) f(x_0, x_1, x_2) + [3x^2 - 2x(x_0 + x_1 + x_2) + (x_0x_1 + x_1x_2 + x_2x_0)] \times f(x_0, x_1, x_2, x_3) + [4x^3 - 3x^2(x_0 + x_1 + x_2 + x_3) + 2x(x_0x_1 + x_1x_2 + x_2x_3 + x_3x_0 + x_1x_3 + x_0x_2) - x_0x_1x_2 + x_1x_2x_3 + x_2x_3x_0 + x_0x_1x_3] f(x_0, x_1, x_2, x_3, x_4)$$

Putting $x_0 = 3, x_1 = 5, x_2 = 11, x_3 = 27$ and $x = 10$, we obtain

$$f'(x) = 18 + 12 \times 16 + 23 \times 0.998 - 426 \times 0.0002 = 232.869.$$

## 30.3 MAXIMA AND MINIMA OF A TABULATED FUNCTION

Newton's forward interpolation formula is

$$y = y_0 + p\Delta y_0 + \frac{p(p-1)}{2!}\Delta^2 y_0 + \frac{p(p-1)(p-2)}{3!}\Delta^3 y_0 + \dots$$

Differentiating it w.r.t. $p$, we get

$$\frac{dy}{dp} = \Delta y_0 + \frac{2p-1}{2}\Delta^2 y_0 + \frac{3p^2 - 6p + 2}{6}\Delta^3 y_0 + \dots \quad \dots(1)$$

For maxima or minima, $dy/dp = 0$. Hence equating the right hand side of Eqn. (1) to zero and retaining only upto third differences, we obtain

$$\Delta y_0 + \frac{2p-1}{2}\Delta^2 y_0 + \frac{3p^2 - 6p + 2}{6}\Delta^3 y_0 = 0$$

*i.e.,*

$$\left(\frac{1}{2}\Delta^3 y_0\right)p^2 + (\Delta^2 y_0 - \Delta^3 y_0)p + \left(\Delta y_0 - \frac{1}{2}\Delta^2 y_0 + \frac{1}{3}\Delta^3 y_0\right) = 0$$

Substituting the values of $\Delta y_0$, $\Delta^2 y_0$, $\Delta^3 y_0$ from the difference table, we solve this quadratic for $p$. Then the corresponding values of $x = x_0 + ph$ at which $y$ is maximum or minimum.

**Example 30.6.** *Find the maximum and minimum value of y from the following data :*

| $x$ : | $-2$ | $-1$ | $0$ | $1$ | $2$ | $3$ | $4$ |
|---|---|---|---|---|---|---|---|
| $y$ : | $2$ | $-0.25$ | $0$ | $-0.25$ | $2$ | $15.75$ | $56$ |

**Solution.** The difference table is :

| $x$ | $y$ | $\Delta y$ | $\Delta^2 y$ | $\Delta^3 y$ | $\Delta^4 y$ | $\Delta^5 y$ |
|---|---|---|---|---|---|---|
| –2 | 2 | | | | | |
| | | –2.25 | | | | |
| –1 | –0.25 | | 2.5 | | | |
| | | 0.25 | | –3 | | |
| 0 | 0 | | –0.5 | | 6 | |
| | | –0.25 | | 3 | | 0 |
| 1 | –0.25 | | 2.5 | | 6 | |
| | | 0.25 | | 9 | | 0 |
| 2 | 2 | | 11.5 | | 6 | |
| | | 13.75 | | 15 | | |
| 3 | 15.75 | | 26.5 | | | |
| | | 40.25 | | | | |
| 4 | 56 | | | | | |

Taking $x_0 = 0$, we have $y_0 = 0$, $\Delta y_0 = -0.25$, $\Delta^2 y_0 = 2.5$, $\Delta^3 y_0 = 9$, $\Delta^4 y_0 = 6$.
Newton's forward difference formula for the first derivative gives

$$\frac{dy}{dx} = \frac{1}{h}\left[\Delta y_0 - \frac{2p-1}{2!}\Delta^2 y_0 + \frac{3p^2 - 6p + 2}{3!}\Delta^3 y_0 - \frac{4p^3 - 18p^2 + 22p - 6}{4!}\Delta^4 y_0 - \cdots\right]$$

$$= \frac{1}{1}\left[-0.25 + \frac{2x-1}{2}(2.5) + \frac{1}{6}(3x^2 - 6x + 2)(9) + \frac{1}{24}(4x^2 - 18x^2 + 22x - 6)(6)\right]$$

$$= \frac{1}{1}[-0.25 + 2.5x - 1.25 + 4.5x^2 - 9x + 3 + x^3 - 4.5x^2 + 5.5x - 1.5] = x^3 - x$$

For $y$ to be maximum or minimum, $\dfrac{dy}{dx} = 0$ *i.e.*, $x^3 - x = 0$

*i.e.*, $x = 0, 1, -1$

Now $$\frac{d^2 y}{dx^2} = 3x^2 - 1 = -\text{ve for } x = 0$$
$$= +\text{ ve for } x = 1$$
$$= +\text{ ve for } x = -1$$

Since $$y = y_0 + x\Delta y_0 + \frac{x(x-1)}{2!}\Delta^2 y_0 + \ldots,\ y(0) = 0$$

Thus $y$ is maximum for $x = 0$, and maximum value $= y(0) = 0$.
Also $y$ is minimum for $x = 1$ and minimum value $= y(1) = -0.25$.

## PROBLEMS 30.1

1. Find $y'(0)$ and $y''$ from the following table :

| $x$ : | 0 | 1 | 2 | 3 | 4 | 5 |
|---|---|---|---|---|---|---|
| $y$ : | 4 | 8 | 15 | 7 | 6 | 2 |

*(Delhi., 2015)*

2. Find the first, second and third derivatives of $f(x)$ at $x = 1.5$ if

| $x$ : | 1.5 | 2.0 | 2.5 | 3.0 | 3.5 | 4.0 |
|---|---|---|---|---|---|---|
| $f(x)$ : | 3.375 | 7.000 | 13.625 | 24.000 | 38.875 | 59.000 |

*(Anna, 2016)*

3. Find the first and second derivatives of the function tabulated below, at the point $x = 1.1$ :

| $x$ : | 1.0 | 1.2 | 1.4 | 1.6 | 1.8 | 2.0 |
|---|---|---|---|---|---|---|
| $y$ : | 0 | 0.128 | 0.544 | 1.296 | 2.432 | 4.000 |

(*U.P.T.U., 2010*)

4. Given the following table of values of $x$ and $y$

| $x$ : | 1.00 | 1.05 | 1.10 | 1.15 | 1.20 | 1.25 | 1.30 |
|---|---|---|---|---|---|---|---|
| $y$ : | 1.000 | 1.025 | 1.049 | 1.072 | 1.095 | 1.118 | 1.140 |

find $\frac{dy}{dx}$ and $\frac{d^2y}{dx^2}$ at (*a*) $x = 1.05$ (*b*) $x = 1.25$ (*c*) $x = 1.15$.

5. For the following values of $x$ and $y$, find the first derivative at $x = 4$.

| $x$ : | 1 | 2 | 4 | 8 | 10 |
|---|---|---|---|---|---|
| $y$ : | 0 | 1 | 5 | 21 | 27 |

6. From the following table, find the values of $dy/dx$ and $d^2y/dx^2$ at $x = 2.03$.

| $x$ : | 1.96 | 1.98 | 2.00 | 2.02 | 2.04 |
|---|---|---|---|---|---|
| $y$ : | 0.7825 | 0.7739 | 0.7651 | 0.7563 | 0.7473 |

7. Find the value of cos 1.74 from the following table :

| $x$ : | 1.7 | 1.74 | 1.78 | 1.82 | 1.86 |
|---|---|---|---|---|---|
| $\sin x$ : | 0.9916 | 0.9857 | 0.9781 | 0.9691 | 0.9584 |

8. The distance covered by an athlete for the 50 metre is given in the following table :

| *Time (sec)* : | 0 | 1 | 2 | 3 | 4 | 5 | 6 |
|---|---|---|---|---|---|---|---|
| *Distance (metre)* : | 0 | 2.5 | 8.5 | 15.5 | 24.5 | 36.5 | 50 |

Determine the speed of the athlete at $t = 5$ sec. correct to two decimals.

9. The population of a certain town is given below. Find the rate of growth of the population in 1961 :

| *Year* : | 1931 | 1941 | 1951 | 1961 | 1971 |
|---|---|---|---|---|---|
| *Population (in lacs)* : | 40.62 | 60.80 | 71.95 | 103.56 | 132.65 |

(*Anna, 2012 ; U.P.T.U., 2012*)

10. The following data gives corresponding values of pressure and specific volume of a superheated stream.

| $v$ : | 2 | 4 | 6 | 8 | 10 |
|---|---|---|---|---|---|
| $p$ : | 105 | 42.7 | 25.3 | 16.7 | 13 |

Find the rate of change of

(*i*) pressure with respect to volume when $v = 2$,

(*ii*) volume with respect to pressure when $p = 105$.

11. The table below reveals the velocity $v$ of a body during the specific time $t$, find its acceleration at $t = 1.1$ ?

| $t$ : | 1.0 | 1.1 | 1.2 | 1.3 | 1.4 |
|---|---|---|---|---|---|
| $v$ : | 43.1 | 47.7 | 52.1 | 56.4 | 60.8 |

12. The elevation above a datum line of 7 points of a road is given below :

| $x$ : | 0 | 300 | 600 | 900 | 1200 | 1500 | 1800 |
|---|---|---|---|---|---|---|---|
| $y$ : | 135 | 149 | 157 | 183 | 201 | 205 | 193 |

Find the gradient of the road at the middle point.

13. A rod is rotating in a plane. The following table gives the angle θ (radians) through which the rod has turned for various values of the time $t$ second.

| $t$ : | 0 | 0.2 | 0.4 | 0.6 | 0.8 | 1.0 | 1.2 |
|---|---|---|---|---|---|---|---|
| $\theta$ : | 0 | 0.12 | 0.49 | 1.12 | 2.02 | 3.20 | 4.67 |

Calculate the angular velocity and the angular acceleration of the rod, when $t = 0.6$ second. (*U.P.T.U., 2011*)

14. Find the value of $f'(x)$ at $x = 0.4$ from the following table using Bessel's formula

| $x$ : | 0.01 | 0.02 | 0.03 | 0.04 | 0.05 | 0.06 |
|---|---|---|---|---|---|---|
| $f(x)$ : | 0.1023 | 0.1047 | 0.1071 | 0.1096 | 0.1122 | 0.1148 |

15. Derive the formula $y'_0 = (y_{-2} - 8y_{-1} + 8y_1 - y_2)/12h$, for numerical differentiation at mid-point of a table.

16. Find the value of $f'(8)$ from the table given below :

| $x$ : | 6 | 7 | 9 | 12 |
|---|---|---|---|---|
| $f(x)$ : | 1.556 | 1.690 | 1.908 | 2.158 |

17. Find the $f'$ (6) from the following data :

| $x$ : | 0 | 2 | 3 | 4 | 7 | 8 |
|---|---|---|---|---|---|---|
| $f(x)$ : | 4 | 26 | 58 | 112 | 466 | 922 |

18. Find the maximum and minimum values of $y$ from the following table :

| $x$ : | 0 | 1 | 2 | 3 | 4 | 5 |
|---|---|---|---|---|---|---|
| $f(x)$ : | 0 | 0.25 | 0 | 2.25 | 16 | 56.25 |

19. Find the value of $x$ for which $f(x)$ is minimum, using the table

| $x$ : | 9 | 10 | 11 | 12 | 13 | 14 |
|---|---|---|---|---|---|---|
| $f(x)$ : | 1330 | 1340 | 1320 | 1250 | 1120 | 930 |

Also find the maximum value of $f(x)$ ?

20. From the data $y(0.60) = 0.6221$, $y(0.65) = 0.6155$, $y(0.70) = 0.6138$ and $y(0.75) = 0.6170$, find the maximum or minimum value of $y$. *(Delhi., 2015)*

## 30.4 NUMERICAL INTEGRATION

The process of evaluating a definite integral from a set of tabulated values of the integrand $f(x)$ is called *numerical integration*. This process when applied to a function of a single variable, is known as *quadrature*.

The problem of numerical integration, like that of numerical differentiation, is solved by representing $f(x)$ by an interpolation formula and then integrating it between the given limits. In this way, we can derive quadrature formula for approximate integration of a function defined by a set of numerical values only.

## 30.5 NEWTON-COTES QUADRATURE FORMULA

Let $$I = \int_a^b f(x)\,dx$$

where $f(x)$ takes the values $y_0, y_1, y_2, \ldots y_n$ for $x = x_0, x_1, x_2, \ldots x_n$. (Fig. 30.1)

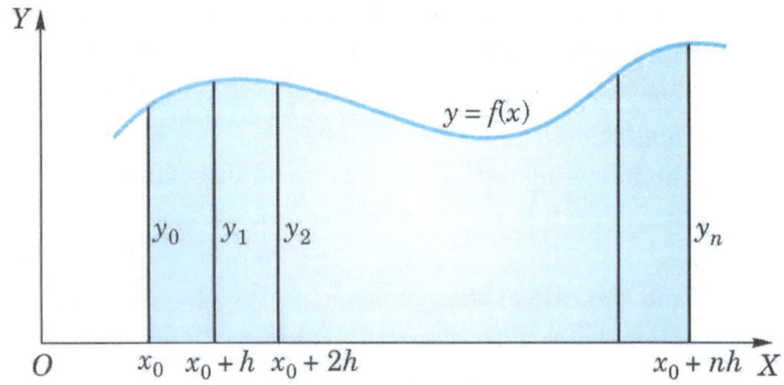

Fig. 30.1

Let us divide the interval $(a, b)$ into $n$ sub-intervals of width $h$ so that $x_0 = a, x_1 = x_0 + h, x_2 = x_0 + 2h, \ldots$ $x_n = x_0 + nh = b$. Then

$$I = \int_{x_0}^{x_0+nh} f(x)\,dx = h\int_0^n f(x_0 + rh)\,dr,$$

putting $x = x_0 + rh$, $dx = hdr$

$$= h\int_0^n \Bigg[ y_0 + r\,\Delta y_0 + \frac{r(r-1)}{2!}\Delta^2 y_0 + \frac{r(r-1)(r-2)}{3!}\Delta^3 y_0 + \frac{r(r-1)(r-2)(r-3)}{4!}\Delta^4 y_0 + \frac{r(r-1)(r-2)(r-3)(r-4)}{5!}\Delta^5 y_0 + \frac{r(r-1)(r-2)(r-3)(r-4)(r-5)}{6!}\Delta^6 y_0 + \cdots \Bigg] dr$$

[By Newton's forward interpolation formula]

Integrating term by, we obtain

$$\int_{x_0}^{x_0+nh} f(x)\,dx = nh\left[ y_0 + \frac{n}{2}\Delta y_0 + \frac{n(2n-3)}{12}\Delta^2 y_0 + \frac{n(n-2)^2}{24}\Delta^3 y_0 \right]$$
$$+ \left(\frac{n^4}{5} - \frac{3n^3}{2} + \frac{11n^2}{3} - 3n\right)\frac{\Delta^4 y_0}{4!} + \left(\frac{n^5}{6} - 2n^4 + \frac{35n^3}{4} - \frac{50n^2}{3} + 12n\right)\frac{\Delta^5 y_0}{5!}$$
$$+ \left(\frac{n^6}{7} - \frac{15n^5}{6} + 17n^4 - \frac{225n^3}{4} + \frac{274n^2}{3} - 60n\right)\frac{\Delta^6 y_0}{6!} + \cdots \quad \ldots(A)$$

This is known as *Newton-Cotes quadrature formula*. From this general formula, we deduce the following important quadrature rules by taking $n = 1, 2, 3 \ldots$

## 30.6 TRAPEZOIDAL RULE

Putting $n = 1$ in (A) § 30.5 and taking the curve through $(x_0, y_0)$ and $(x_1, y_1)$ as a straight line (Fig. 30.2) *i.e.* a polynomial of first order so that differences of order higher than first become zero, we get

$$\int_{x_0}^{x_0+h} f(x)\,dx = h\left(y_0 + \frac{1}{2}\Delta y_0\right) = \frac{h}{2}(y_0 + y_1)$$

Similarly
$$\int_{x_0}^{x_0+2h} f(x)\,dx = h\left(y_1 + \frac{1}{2}\Delta y_1\right) = \frac{h}{2}(y_1 + y_2)$$

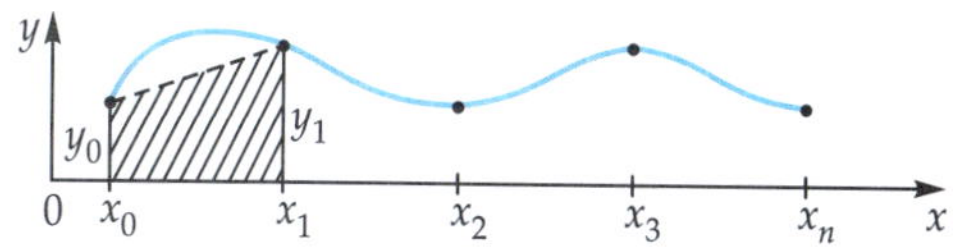

Fig. 30.2

.............................................

$$\int_{x_0+(n-1)h}^{x_0+nh} f(x)\,dx = \frac{h}{2}(y_{(n-1)} + y_n)$$

Adding these $n$ integrals, we obtain

$$\int_{x_0}^{x_0+nh} f(x)\,dx = \frac{h}{2}[(y_0 + y_n) + 2(y_1 + y_2 + \dots + y_{n-1})]$$

This is known as the ***trapezium rule.***

**Obs.** The area of each strip (trapezium) is found separately. Then the area under the curve and the *ordinates* at $x_0$ and $x_n$ is approximately equal to the areas of the trapeziums.

## 30.7 SIMPSON'S ONE-THIRD RULE

Putting $n = 2$ in (A) above and taking the curve through $(x_0, y_0)$, $(x_1, y_1)$ and $(x_2, y_2)$ as a parabola (Fig. 30.3) *i.e.*, a polynomial of second order so that differences of order higher than second vanish, we get

$$\int_{x_0}^{x_0+2h} f(x)\,dx = 2h\left(y_0 + \Delta y_0 + \frac{1}{6}\Delta^2 y_0\right)\frac{h}{3}(y_0 + 4y_1 + y_2)$$

Similarly,
$$\int_{x_0+2h}^{x_0+nh} f(x)\,dx = \frac{h}{3}(y_2 + 4y_3 + y_4)$$

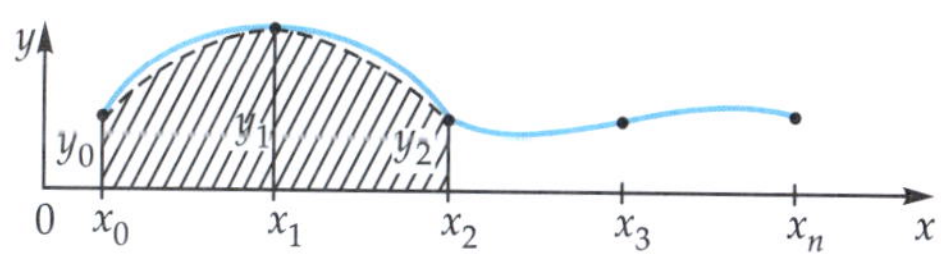

Fig. 30.3

.............................................

$$\int_{x_0+(n-2)h}^{x_0+nh} f(x)\,dx = \frac{h}{3}(y_{n-2} + 4y_{n-1} + y_n), \ n \text{ being even.}$$

Adding all these integrals, we have (when $n$ is even)

$$\int_{x_0}^{x_0+nh} f(x)\,dx = \frac{h}{3}[(y_0 + y_n) + 4(y_1 + y_3) + \dots + y_{n-1}) + 2(y_2 + y_4 + \dots + y_{n-2})]$$

This is known as the *Simpson's one-third rule* or simply *Simpson's rule* and is most commonly used.

**Obs.** While applying *Simpson's 1/3rd rule, the given interval must be divided into even number of equal subintervals*, since we find the area of two strips at a time.

## 30.8 SIMPSON'S THREE-EIGHTH RULE

Putting $n = 3$ in (A) above and taking the curve through $(x_i, y_i)$: $i = 0, 1, 2, 3$ as a polynomial of third order (Fig. 30.4) so that differences above the third order vanish, we get

$$\int_{x_0}^{x_0+3h} f(x)\,dx = 3h\left(y_0 + \frac{3}{2}\Delta y_0 + \frac{3}{4}\Delta^2 y_0 + \frac{1}{8}\Delta^3 y_0\right)$$

$$= \frac{3h}{8}(y_0 + 3y_1 + 3y_2 + y_3)$$

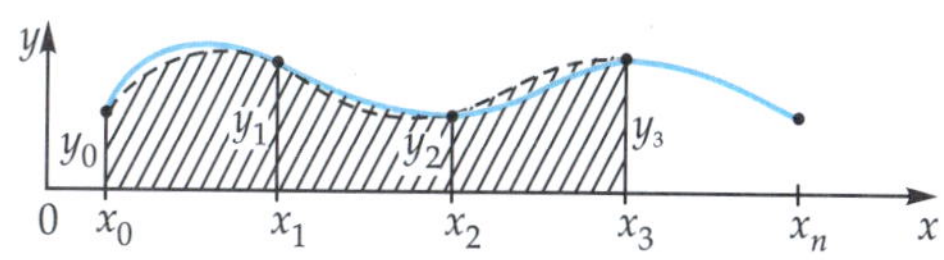

Fig. 30.4

Similarly,

$$\int_{x_0+3h}^{x_0+nh} f(x)\,dx = \frac{3h}{8}(y_3 + 3y_4 + 3y_5 + y_6) \text{ and so on.}$$

Adding all such expressions from $x_0$ to $x_0 + nh$, where $n$ is a multiple of 3, we obtain

$$\int_{x_0}^{x_0+nh} f(x)\, dx = \frac{3h}{8} [(y_0 + y_n) + 3(y_1 + y_2 + y_4 + y_5 + \dots + y_{n-1}) + 2(y_3 + y_6 + \dots + y_{n-3}]$$

which is known as *Simpson's three-eighth rule*.

**Obs.** While applying *Simpson's 3/8th rule, the number of sub-intervals should be taken as multiple of 3.*

## 30.9 BOOLE'S RULE

Putting $n = 4$ in (A) above and taking the curve through $(x_i, y_i)$, $i = 0, 1, 2, 3, 4$ as a polynomial of fourth order (Fig. 30.5) so that differences above the fourth order vanish, we obtain

$$\int_{x_0}^{x_0+4h} f(x)\, dx = 4h\left( y_0 + 2\Delta y_0 \frac{5}{3}\Delta^2 y_0 + \frac{2}{3}\Delta^3 y_0 + \frac{7}{90}\Delta^4 y_0 \right)$$

$$= \frac{2h}{45}(7y_0 + 32y_1 + 12y_2 + 32y_3 + 7y_8)$$

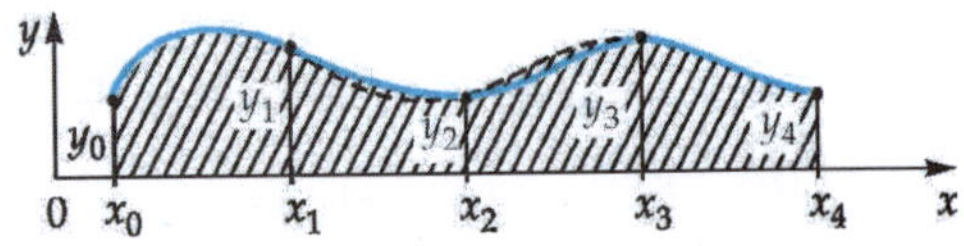


Fig. 30.5

Similarly

$$\int_{x_0+4h}^{x_0+8h} f(x)\, dx = \frac{2h}{45}(7y_4 + 32y_5 + 12y_6 + 32y_7 + 7y_8) \text{ and so on.}$$

Adding all these integrals from $x_0$ to $x_0 + nh$, where $n$ is a multiple of 4, we get

$$\int_{x_0}^{x_0+nh} f(x)\, dx = \frac{2h}{45}(7y_0 + 32y_1 + 12y_2 + 32y_3 + 14y_4 + 32y_5 + 12y_6 + 32y_7 + 14y_8 + \dots$$

This is known as *Boole's rule*.

**Obs.** While applying *Boole's rule*, the number of sub-intervals should be taken as a multiple of 4.

## 30.10 WEDDLE'S RULE

Putting $n = 6$ in (A) above and neglecting all differences above the sixth, we obtain

$$\int_{x_0}^{x_0+6h} f(x)\, dx = \left( y_0 + 3\Delta y_0 + \frac{9}{2}\Delta^2 y_0 + 4\Delta^3 y_0 + \frac{123}{60}\Delta^4 y_0 + \frac{11}{20}\Delta^5 x_0 + \frac{1}{6}\cdot\frac{41}{140}\Delta^6 y_0 \right)$$

If we replace $\frac{41}{140}\Delta^6 y_0$ by $\frac{3}{10}\Delta^6 y_0$, the error made will be negligible.

$$\therefore \quad \int_{x_0}^{x_0+6h} f(x)\, dx = \frac{3h}{10}(y_0 + 5y_1 + y_2 + 6y_3 + y_4 + 5y_5 + y_6)$$

Similarly

$$\int_{x_0+6h}^{x_0+12h} f(x)\, dx = \frac{3h}{10}(y_6 + 5y_7 + y_8 + 6y_9 + y_{10} + 5y_{11} + y_{12}) \text{ and so on.}$$

Adding all these integrals from $x_0$ to $x_0 + nh$, where $n$ is a multiple of 6, we get

$$\int_{x_0}^{x_0+nh} f(x)\, dx = \frac{3h}{10}(y_0 + 5y_1 + y_2 + 6y_3 + y_4 + 5y_5 + 2y_6 + 5y_7 + y_8 + \dots)$$

This is known as *Weddle's rule*.

**Obs.** While applying *Weddle's rule the number of sub-intervals should be taken as a multiple of 6*. Weddle's rule is generally more accurate than any of the others. Of the two Simpson rules, the 1/3 rule is better.

**Example 30.7.** *Evaluate* $\int_0^6 \frac{dx}{1+x^2}$ *by using (i) Trapezoidal rule,*

*(i) Simpson's 1/3 rule,* *(JNTUK, 2022)*

*(ii) Simpson's 3/8 rule,*

*(iii) Weddle's rule and compare the results with its actual value.* *(V.T.U., 2013)*

**Solution.** Divide the interval (0, 6) into six parts each of width $h = 1$. The values of $f(x) = \dfrac{1}{1+x^2}$ are given below :

| $x$ | 0 | 1 | 2 | 3 | 4 | 5 | 6 |
|---|---|---|---|---|---|---|---|
| $f(x)$ $= y$ | 1 $y_0$ | 0.5 $y_1$ | 0.2 $y_2$ | 0.1 $y_3$ | 0.05884 $y_4$ | 0.0385 $y_5$ | 0.027 $y_6$ |

(*i*) By Trapezoidal rule,

$$\int_0^6 \frac{1}{1+x^2} = \frac{h}{2}[(y_0 + y_6) + 2(y_1 + y_2 + y_3 + y_4 + y_5)]$$

$$= \frac{1}{2}[(1 + 0.027) + 2(0.5 + 0.2 + 0.1 + 0.0588 + 0.0385)] = 1.4108.$$

(*ii*) By Simpson's 1/3 rule,

$$\int_0^6 \frac{1}{1+x^2} = \frac{h}{3}[(y_0 + y_6) + 4(y_1 + y_3 + y_5) + 2(y_2 + y_4)]$$

$$= \frac{1}{3}[(1 + 0.027) + 4(0.5 + 0.1 + 0.0385) + 2(0.2 + 0.0588)] = 1.3662.$$

(*iii*) By Simpson's 3/8 rule,

$$\int_0^6 \frac{1}{1+x^2} = \frac{3h}{8}[(y_0 + y_6) + 3(y_1 + y_2 + y_4 + y_5) + 2y_3]$$

$$= \frac{3}{8}[(1 + 0.027) + 3(0.5 + 0.2 + 0.0588 + 0.0385) + 2(0.1)] = 1.3571.$$

(*iv*) By Weddle's rule,

$$\int_0^6 \frac{1}{1+x^2} = \frac{3h}{10}[y_0 + 5y_1 + y_2 + 6y_3 + y_4\ 5y_5 + y_6]$$

$$= 0.3[1 + 5(0.5) + 0.2 + 6(0.1) + 0.0588 + 5(0.0385) + 0.027] = 1.3735.$$

Also, $$\int_0^6 \frac{dx}{1+x^2} = \left|\tan^{-1} x\right|_0^6 = 1.4056$$

**Obs.** This shows that the value of the integral found by Weddle's rule is the nearest to the actual value followed by its value given by Simpson's 1/3rd.

**Example 30.8.** *Calculate the value of the integral* $I = \int_{0.2}^{1.4} (\sin x - \log_e x + e^x) \cdot dx$ *by (a) the Trapezoidal rule, (b) Simpson's one third rule (c) Simpson's three eighth rule (d) Eeddle's rule.* *(AKTU, 2019)*

**Solution.** Given $y = \sin x - \log_e x + e^x$ and take $h = 0.1$

Now tabulate the value of $y$ as follows.

| $x$ : 0.2 | 0.3 | 0.4 | 0.5 | 0.6 | 0.7 | 0.8 |
|---|---|---|---|---|---|---|
| $y$ : 3.02950 | 2.84935 | 2.79753 | 2.82130 | 2.89759 | 0.01464 | 0.16604 |

| $x$ : 0.9 | 1.0 | 1.1 | 1.2 | 1.3 | 1.4 |
|---|---|---|---|---|---|
| $y$ : 3.34830 | 3.55975 | 3.80008 | 4.06984 | 4.37050 | 4.70418 |

(*a*) By Trapezodial rule,

$$T = \frac{0.1}{2}[(3. - 02950 + 4.70418) + 2(2.84935 + ....4.37050)]$$

$$= \frac{1}{20}[(7.73368 + 2(36.69492)] = 4.05617$$

(*b*) By Simpson's one third rule,

$$I = \frac{(0.1)}{3}[7.73368 + 4(2.84935 + .....4.37050)] + 2(2.79753 + ..... + 4.06984)]$$

$$= \frac{1}{30}[7.73368 + 4(20.20417) + 2(16.49075)] = \frac{1}{30}(121.53186) = 4.05106$$

(*c*) By Simpson's three eighth rule

$$I = \frac{3(0.1)}{8}[7.73368 + 3(26.90750) + 2(9.78742)] = 4.05116.$$

By Weddl's rule, we have

$$I = \frac{3h}{10}\left[(y_0 + y_n) + 5(y_1 + y_5 + y_7 + y_{11}) + (y_2 + y_4 + y_8 + y_{10}) + 2y_6 + 6(y_3 + y_9)\right]$$

$$= \frac{3}{100}[7.73368 + 5(13.58879) + 13.32471 + 2(2.316604) + 6(6.62138)] = 4.05098.$$

**Example 30.9.** *Use Simpson's 1/3rd rule to find* $\int_0^{0.6} e^{-x^2}\, dx$ *by taking seven ordinates.*

(*V.T.U., 2011*)

**Solution.** Divide the interval (0, 0.6) into six parts each of width $h = 0.1$. The values of $y = f(x) = e^{-x^2}$ are given below :

| $x$ | 0 | 0.1 | 0.2 | 0.3 | 0.4 | 0.5 | 0.6 |
|---|---|---|---|---|---|---|---|
| $x^2$ | 0 | 0.01 | 0.04 | 0.09 | 0.16 | 0.25 | 0.36 |
| $y$ | 1 | 0.9900 | 0.9608 | 0.9139 | 0.8521 | 0.7788 | 0.6977 |
| | $y_0$ | $y_1$ | $y_2$ | $y_3$ | $y_4$ | $y_5$ | $y_6$ |

By Simpson's 1/3rd rule, we have

$$\int_0^{0.6} e^{-x^2} dx = \frac{h}{3}[(y_0 + y_6) + 4(y_1 + y_3 + y_5 + 2(y_2 + y_4)]$$

$$= \frac{0.1}{3}[(1 + 0.6977) + 4(0.99 + 0.9139 + 0.7788 + 2(0.9608 + 0.8521)]$$

$$= \frac{0.1}{3}[1.6977 + 10.7308 + 3.6258] = \frac{0.1}{3}(16.0543) = 0.5351.$$

**Example 30.10.** *Compute the value of* $\int_{0.2}^{1.4} (\sin x - \log x + e^x)\, dx$ *using Simpson's* $\frac{3}{8}$*th rule.*

**Solution.** Let $y = \sin x - \log_e x + e^x$ and $h = 0.2$, $n = 6$.

The values of $y$ are as given below :

| $x$ : | 0.2 | 0.4 | 0.6 | 0.8 | 1.0 | 1.2 | 1.4 |
|---|---|---|---|---|---|---|---|
| $y$ : | 3.0295 | 2.7975 | 2.8976 | 3.1660 | 3.5597 | 4.0698 | 4.4042 |
| | $y_0$ | $y_1$ | $y_2$ | $y_3$ | $y_4$ | $y_5$ | $y_6$ |

By Simpson's $\frac{3}{8}$th rule, we have

$$\int_{0.2}^{1.4} y\, dx = \frac{3h}{8}[(y_0 + y_6) + 2(y_3) + 3(y_1 + y_2 + y_4 + y_5)]$$

$$= \frac{3}{8}(0.2)[7.7336 + 2(3.1660) + 3(13.3247)] = 4.053$$

Hence $\int_{0.2}^{1.4} (\sin x - \log_e x + e^x)\, dx = 4.053.$

**Obs. Applications of Simpson's rule.** If the various ordinates in §30.5 represent equispaced cross-sectional areas, then Simpson's rule gives the volume of the solid. As such, Simpson's rule is very useful to civil engineers for calculating the amount of earth that must be moved to fill a depression or make a dam. Similar if the ordinates denote velocities at equal intervals of time, the Simpson's rule gives the distance travelled. The following examples illustrate these applications.

**Example 30.11.** *The velocity v (km / min) of a moped which starts from rests, is given at fixed intervals of time t (min) as follows :*

| *t :* | *2* | *4* | *6* | *8* | *10* | *12* | *14* | *16* | *18* | *20* |
|---|---|---|---|---|---|---|---|---|---|---|
| *y :* | *10* | *18* | *25* | *29* | *32* | *20* | *11* | *5* | *2* | *0* |

*Estimate approximately the distance covered in 20 minutes.*

**Solution.** If $s$ km be the distance covered in $t$ (min), then $\frac{ds}{dt} = v$

$$\therefore \qquad \left| s \right|_{t=0}^{20} = \int_0^{20} v\, dt = \frac{h}{3}[X + 4.O + 2E], \text{ by Simpson's rule}$$

Hence $h = 2$, $v_0 = 0$, $v_1 = 10$, $v_2 = 18$, $v_3 = 25$ etc.

$$\therefore \qquad X = v_0 + v_{10} = 0 + 0 = 0$$

$$O = v_1 + v_3 + v_5 + v_7 + v_9 = 10 + 25 + 32 + 11 + 2 = 80$$

$$E = v_2 + v_4 + v_6 + v_8 = 18 + 29 + 20 + 5 = 72$$

Hence the required distance $= \left| s \right|_{t-0}^{20} = \frac{2}{3}(0 + 4 \times 80 + 2 \times 72)$

$= 309.33$ km.

**Example 30.12.** *Calculate the values of* $\int_0^1 \frac{dx}{1+x^2}$ *using Weddle Rule (Take n = 6).*

*(Marathwada, 2019)*

**Solution.** $h = \frac{x_n - x_0}{n} = \frac{1-0}{6} = \frac{1}{6}$

$$f(x) = \frac{1}{1+x^2}$$

| $x$ | 0 | $\frac{1}{6}$ | $\frac{1}{3}$ | $\frac{1}{2}$ | $\frac{2}{3}$ | $\frac{5}{6}$ | 1 |
|---|---|---|---|---|---|---|---|
| $y = f(0)$ | 1 | 0.973 | 0.9 | 0.8 | 0.6923 | 0.5902 | 0.5 |

$$\int_0^1 \frac{dx}{1+x^2} = \frac{3h}{10}[y_0 + y_n + (y_2 + y_9) + 5(y_1 + y_5) + 6y_3 + 2y_6]$$

$$= \frac{3 \times \frac{1}{6}}{10}[1 + 0.5 + (0.9 + 0.6923) + 5(0.973 + 0.5902) + 6 \times 0.8] = 0.7854.$$

**Example 30.13.** *A solid of revolution is formed by rotating about the x-axis, the area between the x-axis, the lines x = 0 and x = 1 and a curve through the points with the following co-ordinates*

| *x :* | *0.00* | *0.25* | *0.50* | *0.75* | *1.00* |
|---|---|---|---|---|---|
| *y :* | *1.0000* | *0.9896* | *0.9589* | *0.9089* | *0.8415* |

*Estimate the volume of the solid formed using Simpson's rule.*

**Solution.** Here $h = 0.25, y_0 = 1, y_1 = 0.9896, y_2 = 0.9589$, etc.

$\therefore$ Required volume of the solid generated

$$= \int_0^1 \pi y^2\, dx = \pi \cdot \frac{h}{3}[(y_0^2 + y_4^2) + 4(y_1^2 + y_3^2) + 2y_2^2]$$

$$= 0.25\,\frac{\pi}{3}[\{1 + (0.8415)^2\} + 4\{(0.9896)^2 + (0.9089)^2\} + 2(0.0589)^2]$$

$$= \frac{0.25 \times 3.1416}{3}[1.7081 + 7.2216 + 1.839] = 0.2618\,(10.7687) = 2.8192.$$

## PROBLEMS 30.2

1. Using Simpson's Rule with $n = 4$. Subintervals approximate the area under the sine curve $f(x) = \sin x$ between $x = 0$, and $x = \pi$. (*Kerala. 2018*) (**Ans.** $\frac{\pi}{6}(2\sqrt{2} + 1)$)

2. Evaluate $\int_0^1 \frac{dx}{1+x^2}$ using (*i*) Trapezoidal rule taking $h = 1/4$ (*Delhi, 2015*)

   (*ii*) Simpson's 1/3rd rule taking $h = 1/4$. (*Delhi, 2015*)

   (*iii*) Simpson's 3/8th rule taking $h = 1/6$. (*Delhi, 2015 ; U.P.T.U, 2010*)

   (*iv*) Weddle's rule taking $h = 1/6$.

   Hence compute an approximate value of $\pi$ in each case. (*Delhi, 2013*)

3. Find the value of log $2^{1/3}$ from $\int_0^1 \frac{x^2}{1+x^3}dx$, using Simpson's 1/3 rule with $h = 0.25$. (*Anna., 2016*)

4. Evaluate $\int_0^6 x \sec x\, dx$ using eight intervals by Trapezoidal rule.

5. Using Simpon's Rule with $n = 4$, subintervals approximate the area under the inverse cosine curve $f(x)$ = arc cos $x$ between $x = -1$ and $x = 1$. (*Ujjain, 2018*)

6. Solve by Boole's Rule $\int_{0.2}^{0.6} \frac{1}{1+x^2}\,dx$, correct upto 3 decimal.

   Given $n = 4$. (*Karikal, 2018*)

7. Evaluate using Simpson's 1/3rd rule, (*i*) $\int_0^\pi \sin x\, dx$ using 11 ordinates.

   (*ii*) $\int_0^{\pi/2} \sqrt{\cos\theta}\, d\theta$ taking 9 ordinates.

   (*iii*) $\int_0^\pi \sqrt{(1 + 3\cos^2\theta)}\, d\theta$ using 6 sub-intervals. (*U.P.T.U., 2012*)

8. Evaluate correct to 4 decimal places, by Simpson's $\frac{3}{8}$th rule

   (*i*) $\int_0^9 \frac{dx}{1+x^3}$ (*U.P.T.U., M. Tech., 2010*) (*ii*) $\int_0^{\pi/2} \sin x\, dx$ (*Anna, 2016*) (*iii*) $\int_0^{\pi/2} e^{\sin x} dx$

9. A curve is drawn to pass through the points given by following table :

| $x$ : | 1 | 1.5 | 2 | 2.5 | 3 | 3.5 | 4 |
|---|---|---|---|---|---|---|---|
| $y$ : | 2 | 2.4 | 2.7 | 2.8 | 3 | 2.6 | 2.1 |

   Using Weddle's rule, estimate the area bounded by the curve, the $x$-axis and the lines $x = 1$, $x = 4$.(*V.T.U., 2011 S*)

10. A body is in the form of a solid of revolution. The diameter $D$ in cms of its sections at distances $x$ cm. from the one end are given below. Estimate the volume of the solid.

| $x$ : | 0 | 2.5 | 5.0 | 7.5 | 10.0 | 12.5 | 15.0 |
|---|---|---|---|---|---|---|---|
| $D$ : | 5 | 5.5 | 6.0 | 6.75 | 6.25 | 5.5 | 4.0 |

11. A rocket is launched from the ground. Its acceleration is registered during the first 80 seconds and is given in the table below. Using Simpson's $\frac{1}{3}$rd rule, find the velocity of the rocket at $t = 80$ seconds.

| $t$ sec : | 0 | 10 | 20 | 30 | 40 | 50 | 60 | 70 | 80 |
|---|---|---|---|---|---|---|---|---|---|
| $f$ (cm/sec$^2$) : | 30 | 31.63 | 33.34 | 35.47 | 37.75 | 40.33 | 43.25 | 46.69 | 50.67 |

12. A reservoir discharging water through sluices at a depth $h$ below the water surface has a surface area $A$ for various values of $h$ as given below :

| $h$ (ft.) : | 10 | 11 | 12 | 13 | 14 |
|---|---|---|---|---|---|
| $A$ (sq.ft.) : | 950 | 1070 | 1200 | 1350 | 1530 |

If $t$ denotes time in minutes, the rate of fall of the surface is given by $dh/dt = -48\sqrt{h}/A$.
Estimate the time taken for the water level to fall from 14 to 10 ft. above the sluices.

## 30.11 OBJECTIVE TYPE OF QUESTIONS

### PROBLEMS 30.3

*Select the correct answer or fill up the blanks in the following questions :*

1. The value of $\int_0^1 \frac{dx}{1+x}$ by Simpson's rule is

   (*a*) 0.96315 (*b*) 0.63915 (*c*) 0.69315 (*d*) 0.69351.

2. Using forward differences, the formula for $f'(a)$ = ...
3. In application of Simpson's 1/3 rule, the interval $h$ for closer approximation should be ...
4. $f(x)$ is given by

| $x$ : | 0 | 0.5 | 1 |
|---|---|---|---|
| $f(x)$ : | 1 | 0.8 | 0.5, |

   then using Trapezoidal rule, the value of $\int_0^1 f(x)\, dx$ is ...

5. If

| $x$ : | 0 | 0.5 | 1 | 1.5 | 2 |
|---|---|---|---|---|---|
| $f(x)$ : | 0 | 0.25 | 1 | 2.25 | 4 |

   then the value of $\int_0^2 f(x)\, dx$ by Simpson's 1/3rd rule is ...

6. Simpson's 3/8th rule states that ...
7. For the data :

| $t$ : | 3 | 6 | 9 | 12 |
|---|---|---|---|---|
| $y(t)$ : | –1 | 1 | 2 | 3, |

   the value of $\int_3^{12} y(t)\, dt$ when computed by Simpson's $\frac{1}{3}$rd rule is

   (*a*) 15 (*b*) 10 (*c*) 0 (*d*) 5.

8. While evaluating a definite integral by Trapezoidal rule, the accuracy can be increased by taking ...
9. The value of $\int_0^1 \frac{dx}{1+x^2}$ by Simpson's 1/3rd rule (taking $n$ = 1/4) is ...
10. For the data:

| $x$ : | 2 | 4 | 6 | 8 |
|---|---|---|---|---|
| $f(x)$ : | 3 | 5 | 6 | 7, |

   $\int_2^8 f(x)\, dx$ when found by the Trapezoidal rule is

   (*a*) 18 (*b*) 25 (*c*) 16 (*d*) 32.

11. The expression for $\left(\frac{dy}{dx}\right)_{x=x_0}$ using backward differences is ...
12. The number of strips required in Weddle's rule is ...
13. The number of strips required in Simpson's 3/8th rule is a multiple of

   (*a*) 1 (*b*) 2 (*c*) 3 (*d*) 6.

14. If $y_0 = 1, y_1 = \frac{16}{17}, y_2 = \frac{4}{5}, y_3 = \frac{16}{25}, y_4 = \frac{1}{2}$ and $h = \frac{1}{4}$, then using Trapezoidal rule, $\int_0^4 y dx$ = ...

15. Using Simpson's $\frac{1}{3}$rd rule, $\int_0^1 \frac{dx}{x}$ = ... (taking $n = 4$).

16. If $y_0 = 1, y_1 = 0.5, y_2 = 0.2, y_3 = 0.1, y_4 = 0.06, y_5 = 0.04$ and $y_6 = 0.03$, then $\int_0^4 y dx$ by Simpson's $\frac{3}{8}$th, rule is = ...

17. If $f(0) = 1, f(1) = 2.7, f(2) = 7.4, f(3) = 20.1, f(4) = 54.6$ and $h = 1$, then $\int_0^4 f(x)\, dx$ by Simpson's $\frac{1}{3}$rd rule = ...

18. Simpson's 1/3rd rule and direct integration give the same result if ...

19. To evaluate $\int_{x_0}^{x_n} y dx$ by Simpson's 1/3rd rule as well as Simpson's 3/8th rule, the number of intervals should be ....... and ....... respectively.

20. Whenever Trapezoidal rule is applicable, Simpson's 1/3rd rule can also be applied. (True or False)

CHAPTER 31

# Difference Equations

1. Introduction. 2. Definition. 3. Formation of difference equations. 4. Linear difference equations. 5. Rules for finding complementary function. 6. Rules for finding particular integral. 7. Simultaneous difference equations with constant coefficients. 8. Application to deflection of a loaded string. 9. Objective Type of Questions.

## 31.1 INTRODUCTION

Difference calculus also forms the basis of Difference equations. These equations arise in all situations in which sequential relation exists at various discrete values of the independent variable. The need to work with discrete functions arises because there are physical phenomena which are inherently of a discrete nature. In control engineering, it often happens that the input is in the form of discrete pulses of short duration. The radar tracking devices receive such discrete pulses from the target which is being tracked. As such differences equations arise in the study of electrical networks, in the theory of probability, in statistical problems and many other fields.

Just as the subject of Differential equations grew out Differential calculus to become one of the most powerful instruments in the hands of a practical mathematician when dealing with continuous processes in nature, so the subject of Difference equations is forcing its way to the fore for the treatment of discrete processes. Thus the difference equations may be thought of as the discrete counterparts of the differential equations.

## 31.2 DEFINITION

**(1) A difference equation** *is a relation between the differences of an unknown function at one or more general values of the argument.*

Thus $\Delta y_{(n+1)} + y_{(n)} = 2$ ...(1) and $\Delta y_{(n+1)} + \Delta^2 y_{(n-1)} = 1$ ...(2)

are difference equations.

An alternative way of writing a difference equation is as under :

Since $\Delta y_{(n+1)} = y_{(n+2)} - y_{(n+1)}$, therefore Eqn. (1) may be written as

$$y_{(n+2)} - y_{(n+1)} + y_{(n)} = 2 \qquad ...(3)$$

Also since, $\Delta^2 y_{(n-1)} = y_{(n+1)} - 2y_{(n)} + y_{(n-1)}$, therefore Eqn. (2) takes the form :

$$y_{(n+2)} - 2y_{(n)} + y_{(n-1)} = 1 \qquad ...(4)$$

Quite often, difference equations are met under the name of *recurrence relations.*

**(2) Order of a difference equation** *is the difference between the largest and the smallest arguments occuring in the difference equation divided by the unit of increment.*

Thus Eqn. (3) above is the *second order*, for

$$\frac{\text{largest argument} - \text{smallest argument}}{\text{unit of increment}} = \frac{(n+2)-n}{1} = 2,$$

and Eqn. (4) is of the *third order*, for $\dfrac{(n+2)-(n-1)}{1} = 3.$

**Obs.** *While finding the order of a difference equation, it must always be expressed in a form free of* $\Delta$*s,* for the highest power of $\Delta$ does not give order of the difference equation.

**(3) Solution** *of a difference equation is an expression for* $y_{(n)}$ *which satisfies the given difference equation.*

*The* general solution *of a difference equation is that in which the number of arbitrary constants is equal to the order of the difference equation.*

*A particular solution* or **particular integral** *is that solution which is obtained from the general solution by giving particular values to the constants.*

## 31.3 FORMATION OF DIFFERENCE EQUATIONS

The following examples illustrate the way in which difference equations arise and are formed.

**Example 31.1.** *Form the difference equation corresponding to the family of curves*

$$y = ax + bx^2 \qquad \text{(Delhi., 2015)} \ldots(i)$$

**Solution.** We have $\Delta y = a\Delta(x) + b\Delta(x^2) = a\,(x + 1 - x) + b[(x + 1)^2 - x^2]$

$$= a + b(2x + 1) \qquad \ldots(ii)$$

and $$\Delta^2 y = 2b[(x + 1) - x] = 2b \qquad \ldots(iii)$$

To eliminate $a$ and $b$, we have from Eqn. $(iii)$, $b = \frac{1}{2}\Delta^2 y$

and form Eqn. $(ii)$, $a = \Delta y - b\,(2x + 1) = \Delta y - \frac{1}{2}\Delta^2 y\,(2x + 1)$

Substituting these values of $a$ and $b$ in $(i)$, we get

$$y = \left[\Delta y - \frac{1}{2}\Delta^2 y\,(2x+1)\right]x + \frac{1}{2}\Delta^2 y \,.\, x^2$$

or $$(x^2 + x)\,\Delta^2 y - 2x\,\Delta y + 2y = 0$$

This is the desired difference equation which may equally well be written in terms of $E$ as

or $$(x^2 + x)\,y_{x+2} - (2x^2 + 4x)y_{x+1} + (x^2 + 3x + 2)y_x = 0.$$

**Example 31.2.** *From* $y_n = A2^n + B(-3)^n$*, derive a difference equation by eliminating the constants.*

**Solution.** We have $y_n = A.2^n + B(-3)^n,\ y_{n+1} = 2A.2^n - 3B(-3)^n$

and $$y_{n+2} = 4A.2^n + 9B(-3)^n.$$

Eliminating $A$ and $B$, we get

$$\begin{vmatrix} y_n & 1 & 1 \\ y_{n+1} & 2 & -3 \\ y_{n+2} & 4 & 9 \end{vmatrix} = 0 \quad \text{or} \quad y_{n+2} + y_{n+1} - 6y_n = 0$$

which is the desired difference equation.

### PROBLEMS 31.1

1. Write the difference equation $\Delta^3 y_x + \Delta^2 y_x + \Delta y_x + y_x = 0$ in the subscript notation. *(Delhi., 2015)*
2. Assuming $\frac{\log(1-z)}{1+z} = y_0 + y_1 z + y_2 z^2 + \ldots\ldots + y_n z^n, \ldots\ldots$, find the difference equations satisfied by $y_n$.
3. Form a difference equation by eliminating arbitrary constant from $u_n = a2^{n+1}$.
4. Find the difference equation satisfied by
   $(i)$ $y = a/x + b$ *(Delhi., 2015)* $(ii)$ $y = ax^2 - bx$.
5. Derive the difference equations in each of the following cases :
   $(i)$ $y_n = A.3^n + B.5^n$ *(Delhi., 2015)* $(ii)$ $y_n = (A + Bx)\,2^x$.
6. Form the difference equations generated by
   $(i)$ $y_n = ax + b2^x$ $(ii)$ $y_n = a2^n + b(-2)^n$ $(iii)$ $y_x = a2^x + b3^x + c$.

## 31.4 LINEAR DIFFERENCE EQUATIONS

**(1) Def.** *A **linear difference equation** is that in which $y_{n+1}, y_{n+2}$, etc. occur to the first degree only and are not multiplied together.*

*A **linear difference equation with constant coefficient** is of the form*

$$y_{n+r} + a_1 y_{n+r-1} + a_2 y_{n+r-2} + \dots + a_r y_n = f(n) \qquad ...(1)$$

where $a_1, a_2, \dots a_r$ are constants.

*Now we shall deal with linear difference equations with constant coefficients only.* Their properties are analogous to those of linear differential equations with constant co-efficients.

**(2) Elementary properties.** If $u_1(n), u_2(n), \dots, u_r(n)$ be $r$ independent solution of the equation

$$y_{n+r} + a_1 y_{n+r-1} + \dots + a_r y_n = 0 \qquad ...(2)$$

then its complete solution is $U_n = c_1 u_1(n) + \dots + c_r u_r(n)$

where $c_1, c_2, \dots, c_r$ are arbitrary constants.

If $V_n$ is a particular solution of Eqn. (1), then the complete solution of Eqn. (1) is $y_n = U_n + V_n$. *The part $U_n$ is called the **complementary function (C.F.)** and the part $V_n$ is called the **particular integral (P.I.)** of Eqn. (1).*

*Thus the **complete solution (C.S.)** of Eqn. (1) is* $\boldsymbol{y_n}$ **= *C.F.* + *P.I.***

## 31.5 RULES FOR FINDING THE COMPLEMENTARY FUNCTION

(*i.e., rules to solve a linear difference equation with constant coefficients having right hand side zero*).

**(1)** To begin with, consider the first order linear equation $y_{n+1} - \lambda y_n = 0$, where $\lambda$ is a constant.

Rewriting it as $\dfrac{y_{n+1}}{\lambda^{n+1}} - \dfrac{y_n}{\lambda^n} = 0$, we have $\Delta\left(\dfrac{y_n}{\lambda^n}\right) = 0$, which gives $y_n/\lambda^n = c$, a constant.

*Thus the solution of* $(E - \lambda)\, y_n = 0$ is $\boldsymbol{y_n = c.\lambda^n}$.

**(2)** Now consider the *second order linear equation* $y_{n+2} + ay_{n+1} + by_n = 0$ which in *symbolic form is*

$$(E^2 + aE + b)y_n = 0 \qquad ...(1)$$

Its symbolic co-efficient equated to zero *i.e.*, $E^2 + aE + b = 0$

is called the *auxiliary equation*. Let its roots be $\lambda_1, \lambda_2$.

**Case I.** *If these roots are real and distinct,* then Eqn. (1) is to equivalent to

$$(E - \lambda_1)(E - \lambda_2)\, y_n = 0 \qquad ...(2)$$

$$(E - \lambda_2)(E - \lambda_1)\, y_n = 0 \qquad ...(3)$$

If $y_n$ satisfies the subsidiary equation $(E - \lambda_1)y_n = 0$, then it will also satisfy Eqn. (3).

Similarly, if $y_n$ satisfies the subsidiary equation $(E - \lambda_2)y_n = 0$, then it will also satisfy Eqn. (2).

$\therefore$ it follows that we can derive two independent solutions of Eqn. (1), by solving the two subsidiary equations

$$(E - \lambda_1)\, y_n = 0 \quad \text{and} \quad (E - \lambda_2)\, y_n = 0$$

Their solutions are respectively, $y_n = c_1(\lambda_1)^n$ and $y_n = c_2(\lambda_2)^n$

where $c_1$ and $c_2$ are arbitrary constants.

Thus the general solution of Eqn. (1) is $y_n = c_1(\lambda_1)^n + c_2(\lambda_2)^n$

**Case II.** *If the roots are real and equal* (*i.e.*, $\lambda_1 = \lambda_2$), then Eqn. (2) becomes

$$(E - \lambda_1)^2 y_n = 0 \qquad ...(4)$$

Let $\qquad y_n = (\lambda_1)^n z_n$

where $z_n$ is a new dependent variable. Then Eqn. (4) takes the form

$$(\lambda_1)^{n+2} z_{n+2} - 2\lambda_1(\lambda_1)^{n+1} z_{n+1} + \lambda_1^2 \cdot (\lambda_1)^n z_n = 0$$

or $\qquad z_{n+2} - 2z_{n+1} + z_n = 0 \quad i.e., \quad \Delta^2 z_n = 0$

$\therefore \qquad z_n = c_1 + c_2 n$, where $c_1, c_2$ are arbitrary constants.

Thus the solution of Eqn. (1) becomes $y_n = (c_1 + c_2 n)(\lambda_1)^n$.

**Case III.** *If the roots are imaginary,* (*i.e.* $\lambda_1 = \alpha + i\beta, \lambda_2 = \alpha - i\beta$) then the solution of Eqn. (1) is

$$y_n = c_1(\alpha + i\beta)^n + c_2(\alpha - i\beta)^n \qquad [\text{Put } \alpha = r\cos\theta \text{ and } \beta = r\sin\theta]$$

$$= r^n\,[c_1(\cos n\theta + i\sin n\theta) + c_2(\cos n\theta - i\sin n\theta)]$$

$$= r^n [A_1 \cos n\theta + A_2 \sin n\theta]$$

where $A_1, A_2$ are arbitrary constants are $r = \sqrt{(\alpha^2 + \beta^2)}$,, $\theta = \tan^{-1}(\beta/\alpha)$.

**(3)** *In general, to solve the equation* $y_{n+r} + a_1 y_{n+r-2} + ... + a_r y_n = 0$ where $a$'s are constants :

(*i*) Write the equation in the *symbolic form* $(E^r + a_1 E^{r-1} + ... + a_r) y_n = 0$.

(*ii*) Write down the *auxiliary equation i.e.,* $E^r + a_1 E^{r-1} ... + a_r = 0$ and solve it for $E$.

(*iii*) Write the solution as follows :

| *Roots of A.E.* | *Solution, i.e. C.F.* |
|---|---|
| **1.** $\lambda_1, \lambda_2, \lambda_3, .......$ (real and distinct roots) | $c_1(\lambda_1)^n + c_2(\lambda_2)^n + c_3(\lambda_3)^n + .......$ |
| **2.** $\lambda_1, \lambda_1, \lambda_3, .......$ (2 real and equal roots) | $(c_1 + c_2 n)(\lambda_1)^n + c_3(\lambda_3)^n + .......$ |
| **3.** $\lambda_1, \lambda_1, \lambda_1, .......$ (3 real and equal roots) | $(c_1 + c_2 n + c_3 n^2)(\lambda_1)^n + .......$ |
| **4.** $\alpha + i\beta, \alpha - i\beta, .......$ (a pair of imaginary roots) | $r^n (c_1 \cos n\theta + c_2 \sin n\theta)$<br>where $r = \sqrt{(\alpha^2 + \beta^2)}$<br>and $\theta = \tan^{-1}(\beta/\alpha)$ |

**Example 31.3.** *Solve the difference equation* $u_{n+3} - 2u_{n+2} - 5u_{n+1} + 6u_n = 0$.

**Solution.** Given equation in *symbolic form* is $(E^3 - 2E^2 - 5E + 6)u_n = 0$

$\therefore$ its auxiliary equation is $E^3 - 2E^2 - 5E + 6 = 0$

or $(E - 1)(E + 2)(E - 3) = 0$. $\therefore E = 1, -2, 3$

Thus the complete solution is $u_n = c_1(1)^n + c_2(-2)^n + c_3(3)^n$.

**Example 31.4.** *Solve* $u_{n+2} - 2u_{n+1} + u_n = 0$.

**Solution.** Given difference equation in *symbolic form* is $(E^2 - 2E + 1)\, u_n = 0$.

$\therefore$ its auxiliary equation is $E^2 - 2E + 1 = 0$

or $(E - 1)^2 = 0$. $\therefore E = 1, 1$

Thus the required solution is $u_n = (c_1 + c_2 n)(1)^n$, *i.e.,* $u_n = c_1 + c_2 n$.

**Example 31.5.** *Solve* $y_{n+1} - 2y_n \cos \alpha + y_{n-1} = 0$.

**Solution.** This is a second order difference equation in $y_{n-1}$; which in *symbolic form* is

$$(E^2 - 2E \cos \alpha + 1)\, y_n = 0$$

The *auxiliary equation* is $E^2 - 2E \cos \alpha + 1 = 0$

$$E = \frac{2 \cos \alpha \pm \sqrt{(4 \cos^2 \alpha - 4)}}{4} = \cos \alpha \pm i \sin \alpha$$

Thus the solution is $y_{n-1} = (1)^{n-1} [c_1 \cos (n-1)\alpha + c_2 \sin (n-1)\alpha]$

or $y_n = c_1 \cos n\alpha + c_2 \sin n\alpha$.

**Example 31.6.** *The integers 0, 1, 1, 2, 3, 5, 8, 13, 21, .... are said to form a Fibonacci sequence. Form the Fibonacci difference equation and solve it.* (*Andhra., 2015*)

**Solution.** In this sequence, each number beyond the second, is the sum of its two previous number. If $y_n$ be the $n$th number then $y_n = y_{n-1} + y_{n-2}$ for $n > 2$.

or $y_{n+2} - y_{n+1} - y_n = 0$ (for $n > 0$)

or $(E^2 - E - 1)y_n = 0$ is the difference equation.

Its A.E. is $E^2 - E - 1 = 0$ which gives $E = \frac{1}{2}(1 \pm \sqrt{5})$.

Thus the solution is $y_n = c_1 \left(\frac{1 + \sqrt{5}}{2}\right)^2 + c_2 \left(\frac{1 - \sqrt{5}}{2}\right)^n$, for $n > 0$

When $n = 1, y = 0$

$$\therefore \qquad c_1\left(\frac{1+\sqrt{5}}{2}\right)+c_2\left(\frac{1-\sqrt{5}}{2}\right)=0 \qquad ...(i)$$

When $n = 2, y_2 = 0$

$$\therefore \qquad c_1\left(\frac{1+\sqrt{5}}{2}\right)^2+c_2\left(\frac{1-\sqrt{5}}{2}\right)^2=0 \qquad ...(ii)$$

Solving Eqns. (*i*) and (*ii*), we get

$$c_1=\frac{5-\sqrt{5}}{10} \text{ and } c_2=\frac{5+\sqrt{5}}{10}$$

Hence the complete solution is

$$y_n=\frac{5-\sqrt{5}}{10}\left(\frac{1+\sqrt{5}}{2}\right)^n+\frac{5+\sqrt{5}}{2}\left(\frac{1-\sqrt{5}}{2}\right)^n.$$

## PROBLEMS 31.2

Solve the following difference equations :

1. $u_{x+2}-4u_{x+1}+4u_x=0$ given $u_0=1, u_1=0$. (*Calicut, 2012*)
2. $y_{n+2}+2y_{n+1}+4y_n=0$. (*Delhi., 2015*)
3. $\Delta^2u_n+2\Delta u_n+u_n=0$.
4. $(\Delta^2-3\Delta+2)y_n=0$.
5. $4y_n-y_{n+2}=0$ given that $y_0=0, y_1=2$.
6. $y_{n+3}-3y_{n+2}+3y_{n+1}-y_n=0$. (*Anna., 2015*)
7. $f(x+3)-3f(x+1)-2f(x)=0$.
8. $u_{n+3}-3u_{n+1}+2u_n=0$, given $u_1=0, u_2=8$ and $u_3=-2$.
9. $(E^3-5E^2+8E-4)y_n=0$, given that $y_0=3, y_1=2, y_3=22$.
10. $u_{n+1}-2u_n+2u_{n-1}=0$.
11. $y_{m+3}+16y_{m-1}=0$.

    [**Hint.** $E^4=-16=16\,[\cos(2n+1)\pi+i\sin(2n+1)\pi]$ ; use De Moivre's theorem.]
12. Show that the difference equation $I_{m+1}-(2+r_o/r)\,I_m+I_{m-1}=0$ has the solution.

    $I_m=I_o\sinh(n-m)\,\alpha/\sinh(n-1)\,\alpha$, if $I=I_0$ and $I_n=0$, $\alpha$ being $=2\sinh^{-1}\frac{1}{2}(r_0/r)^{1/2}$.
13. A series of values of $y_n$ satisfy the relation, $y_{n+2}+ay_{n+1}+by_n$.

    Given that $y_0=0, y_1=1, y_2=y_3=2$. Show that $y_n=2^{n/2}\sin n\pi/4$.
14. A plant is such that each of its seeds when one year old produces 8-fold and produces 18-fold when two years old or more. A seed is planted and as soon as a new seed is produced it is planted. Taking $y_n$ to be the number of seeds produced at the end of the $n$th year, show that $y_{n+1}=8y_n+18(y_1+y_2+...+y_{n-1})$.

    Hence show that $y_{n+2}-9y_{n+1}-10y_n=0$ and find $y_n$.

## 31.6 RULES FOR FINDING THE PARTICULAR INTEGRAL

Consider the equation $y_{n+r}+a_1y_{n+r-1}+...+a_r\,y_n=f(n)$
which in *symbolic form* is $\phi(E)y_n=f(n)$ ...(1)

where $$\phi(E)=E^r+a_1\,E^{r-1}+...+a_r$$

Then the particular integral is given by P.I. $=\dfrac{1}{\phi(E)}f(n)$.

**Case I.** When $f(n)=a^n$ (*Power function*)

$$\text{P.I.}=\frac{1}{\phi(E)}\,a_n,\text{ put } E=a$$

$$=\frac{1}{\phi(a)}\,a^n,\text{ provided } \phi(a)\neq 0$$

If $\phi(a)=0$, *then for the equation*

(*i*) $(E-a)y_n=a^n$, $\qquad$ P.I. $=\dfrac{1}{E-a}a^n=na^{n-1}$

(*ii*) $(E-a)^2 y_n = a^n$, $\quad$ P.I. $= \dfrac{1}{(E-a)^2} a^n = \dfrac{n(n-1)}{2!} a^{n-2}$

(*iii*) $(E-a)^3 y^n = a^n$, $\quad$ P.I. $= \dfrac{1}{(E-a)^3} a^n = \dfrac{n(n-1)(n-2)}{3!} a^{n-3}$

and so on.

**Example 31.7.** *Solve* $y_{n+2} - 4y_{n+1} + 3y_n = 5^n$.

**Solution.** Given equation in *symbolic form* is $(E^2 - 4E + 3)y_n = 5^n$

$\therefore$ The auxiliary equation is $E^2 - 4E + 3 = 0$

or $\quad (E-1)(E-3) = 0. \quad \therefore \; E = 1, 3$

$\therefore \quad$ C.F. $= c_1(1)^n + c_2(3)^n = c_1 + c_2 \,.\, 3^n$

and $\quad$ P.I. $= \dfrac{1}{E^2 - 4E + 3} 5^n$ $\quad$ [Put $E = 5$]

$$= \frac{1}{25 - 4.5 + 3} 5^n = \frac{1}{8} \,.\, 5^n$$

Thus the complete solution is $\quad y_n = c_1 + c_2 \,.\, 3^n + 5^n/8$.

**Example 31.8.** *Solve* $u_{n+2} - 4u_{n+1} + 4u_n = 2^n$.

**Solution.** Given equation in *symbolic form* is $(E^2 - 4E + 4)u_n = 2^n$.

The auxiliary equation is $E^2 - 4E + 4 = 0. \quad \therefore \; E = 2, 2$.

C.F. $= (c_1 + c_2 n)2^n$

P.I. $= \dfrac{1}{(E-2)^2} \,.\, 2^n = \dfrac{n(n-1)}{2!} \,.\, 2^{n-2} = n(n-1)\, 2^{n-3}$

Hence the complete solution is $u_n = (c_1 + c_2 n)\, 2^n + n(n-1)\, 2^{n-3}$.

**Case II.** (1) *When* $f(n) = \sin kn$. $\quad$ (*Trigonometric function*)

$$\text{P.I.} = \frac{1}{\phi(E)} \sin kn = \frac{1}{\phi(E)} \left( \frac{e^{ikn} - e^{-ikn}}{2i} \right) = \frac{1}{2i} \left[ \frac{1}{\phi(E)} a^n - \frac{1}{\phi(E)} b^n \right]$$

where $a = e^{ik}$ and $b = e^{-ik}$.

Now proceed as in case I.

(2) *When* $f(n) = \cos kn$ $\quad$ P.I. $= \dfrac{1}{\phi(E)} \cos kn = \dfrac{1}{\phi(E)} \left( \dfrac{e^{ikn} + e^{-ikn}}{2} \right)$

$$= \frac{1}{2} \left[ \frac{1}{\phi(E)} a^n + \frac{1}{\phi(E)} b^n \right] \text{ as before}$$

Now proceed as in case I.

**Example 31.9.** *Solve* $y_{n+2} - 2\cos\alpha \,.\, y_{n+1} + y_n = \cos \alpha n$. $\quad$ (*Delhi, 2015*)

**Solution.** Given equation in *symbolic form* is $(E^2 - 2\cos\alpha \,.\, E + 1)y_n = \cos \alpha n$.

The *auxiliary equation is* $E^2 - 2\cos\alpha \,.\, E + 1 = 0$.

$$\therefore \quad E = \frac{2\cos\alpha \pm \sqrt{(4\cos^2\alpha - 4)}}{2} = \cos\alpha \pm i \sin\alpha$$

$\therefore \quad$ C.F. $= (1)^n [c_1 \cos \alpha n + c_2 \sin \alpha n]$ *i.e.*, $c_1 \cos \alpha n + c_2 \sin \alpha n$

$$\text{P.I.} = \frac{1}{E^2 - 2E\cos\alpha + 1} \cos \alpha n$$

$$= \frac{1}{E^2 - E(e^{i\alpha} + e^{-i\alpha}) + 1} \left( \frac{e^{ian} + e^{-ian}}{2} \right)$$

$$= \frac{1}{2}\left[\frac{1}{(E-e^{i\alpha})(E-e^{-i\alpha})}e^{i\alpha n} + \frac{1}{(E-e^{i\alpha})(E-e^{-i\alpha})}e^{-i\alpha n}\right]$$

[Put $E = e^{i\alpha}$] [Put $E = e^{-i\alpha}$]

$$= \frac{1}{2}\left[\frac{1}{(E-e^{i\alpha})}\cdot\frac{1}{e^{i\alpha}-e^{-i\alpha}}e^{i\alpha n} + \frac{1}{E-e^{-i\alpha}}\cdot\frac{1}{e^{-i\alpha}-e^{i\alpha}}e^{-i\alpha n}\right]$$

$$= \frac{1}{4i\sin\alpha}\left[\frac{1}{E-e^{i\alpha}}e^{i\alpha n} - \frac{1}{E-e^{-i\alpha}}e^{-i\alpha n}\right] = \frac{1}{4i\sin\alpha}[n\,.\,e^{i\alpha(n-1)} - n\,e^{-i\alpha(n-1)}]$$

$$= \frac{n}{2\sin\alpha}\left[\frac{e^{i\alpha(n-1)} - e^{-i\alpha(n-1)}}{2i}\right] = \frac{n\sin(n-1)\alpha}{2\sin\alpha}$$

Hence the complete solution is

$$y_n = c_1\cos\alpha n + c_2\sin\alpha n + \frac{n\sin(n-1)\alpha}{2\sin\alpha}.$$

**Case III.** *When* $f(n) = n^p$. P.I. $= \frac{1}{\phi(E)}n^p = \frac{1}{\phi(1+\Delta)}n^p$ (*Polynomial function*)

(1) Expand $[\phi(1+\Delta)]^{-1}$ in ascending powers of $\Delta$ by the Binomial theorem as far as the term in $\Delta^p$.

(2) Expand $n^p$ in the factorial form and operate on it with each term of the expansion.

**Example 31.10.** *Solve* $y_{n+2} - 4y_n = n^2 + n - 1$.

**Solution.** Given equation is $(E^2 - 4)y_n = n^2 + n - 1$.

The *auxiliary equation is* $E^2 - 4 = 0$, $\therefore$ $E = \pm 2$.

$\therefore$ C.F. $= c_1(2)^n + c_2(-2)^n$.

$\therefore$ P.I. $= \frac{1}{E^2-4}(n^2+n-1) = \frac{1}{(1+\Delta)^2-4}[n(n-1)+2n-1]$.

$$= \frac{1}{\Delta^2+2\Delta-3}([n]^2+2[n]-1) = -\frac{1}{3}\left[1-\left(\frac{2}{3}\Delta+\frac{\Delta^2}{3}\right)\right]^{-1}\{[n]^2+2[n]-1\}$$

$$= -\frac{1}{3}\left[1+\left(\frac{2}{3}\Delta+\frac{\Delta^2}{3}\right)+\left(\frac{2}{3}\Delta+\frac{\Delta^2}{3}\right)+\ldots\right]\{[n]^2+2[n]-1\}$$

$$= -\frac{1}{3}\left\{1+\frac{2}{3}\Delta+\frac{7}{9}\Delta^2+\ldots\right\}\{[n]^2+2[n]-1\} = -\frac{1}{3}\left\{[n]^2+2[n]-1+\frac{2}{3}(2[n]+2)+\frac{7}{9}\times 2\right\}$$

$$= -\frac{1}{3}\left\{[n]^2+\frac{10}{3}[n]+\frac{17}{9}\right\} = -\frac{n^2}{3}-\frac{7}{9}n-\frac{17}{27}.$$

Hence the *complete solution* is $y_n = c_1 2^n + c_2(-2)^n - \frac{n^2}{3} - \frac{7}{9}n - \frac{17}{27}$.

**Case IV.** *When* $f(n) = a^n F(n)$, $F(n)$, being a polynomial of finite degree in $n$.

$$\text{P.I.} = \frac{1}{\phi(E)}a^n F(n) = a^n\frac{1}{\phi(aE)}F(a)$$

Now $F(n)$ being a polynomial in $n$, proceed as in case III.

**Example 31.11.** *Solve* $y_{n+2} - 2y_{n+1} + y_n = n^2 . 2^n$.

**Solution.** Given equation is $(E^2 - 2E + 1)y_n = n^2 . 2^n$.

Its C.F. $= c_1 + c_2 n$

and P.I. $= \frac{1}{(E-1)^2}2^n . n^2 = 2^n\frac{1}{(2E-1)^2}n^2 = 2^n\frac{1}{(1+2\Delta)^2}n^2$

$$= 2^n (1 + 2\Delta)^{-2}\, n(n-1) + n = 2^n (1 - 4\Delta + 12\Delta^2 - \ldots)([n]^2 + [n])$$
$$= 2^n \{[n]^2 + [n] - 4(2[n] + 1) + 12 \times 2\}$$
$$= 2^n ([n]^2 - 7[n] + 20) = 2^n (n^2 - 8n + 20)$$

Hence the *complete solution* is $y_n = c_1 + c_2 n + 2^n (n^2 - 8n + 20)$.

## PROBLEMS 31.3

Solve the following difference equations :

1. $y_{n+2} - 5y_{n+1} - 6y_n = 4^n, y_0 = 0, y_1 = 1$.
2. $y_{p+3} - 3y_{p+2}\, 3y_{p+1} - y_p = 1$.
3. $y_{n+2} - 2y_{n+1} + 4y_n = 6$, given that $y_0 = 0$ and $y_1 = 2$.
4. $(E^2 - 4E + 3)y = 3^x$.
5. $y_{x+2} - 4y_{x+1} + 4y_x = 3.2^x + 5.4^x$.
6. $u_{n+2} - u_n = \cos n/2$.
7. $y_{p+2} - \left(2 \cos \dfrac{1}{2}\right) y_{p+1} + y_p = \sin p/2$.
8. $(E^2 - 4)y_x = x^2 - 1$.
9. $y_{n+3} + y_n = n^2 + 1, y_0 = y_1 = y_2 = 0$.
10. $y_{n+3} - 5y_{n+2} + 3y_{n+1} + 9y_n = 2^n + 3n$.
11. $(4E^2 - 4E + 1)y = 2^n + 2^{-n}$.
12. $u_{x+2} + 6u_{x+1} + 9u_x = x2^x + 3^x + 7$.
13. $y_{n+2} - 6y_{n+1} + 2y_n = 2^n + 6n$. (*Delhi., 2015*)
14. $u_{n+2} - 4u_{n+1}\, 4u_n = n^2 2^n$.
15. $(E^2 - 5E + 6)y_k = 4^k (k^2 - k + 5)$.
16. $(E^2 - 2E + 4)y_n = -2^n \left\{6 \cos \dfrac{n\pi}{3} + 2\sqrt{3} \sin \dfrac{n\pi}{3}\right\}$.
17. If $y = \left(x + \sqrt{1 + x^2}\right)^m$, find $y_n(10)$ (*Kumaun, 2018*)
18. If $y = \left[\log\left(x + \sqrt{1 + x^2}\right)\right]^2$, find all the derivatives of $y$ w.r.t. $x$ when $x = 0$ (*Bilaspur, 2018*)
19. If $y = \sin(m(\sin^{-1} x)$, find $y_n$ at $x = 0$ (*Jiwaji, 2018*)
20. If $y = x \log\left(\dfrac{x-1}{x+1}\right)$, Prove that $y_m (-1)^{m-1}(m-2)!\left[\dfrac{x-m}{(x-1)^m} - \dfrac{x+m}{(x+1)^m}\right]$
21. If $u = \dfrac{x^5}{Ax^2 + 2Bx + C}$, show that $(Ax^2 + 2Bx + C)\, U_n + 2n(Ax + B)\, U_{n-1} + b(b-1)\, A\, U_{n-2} = 0$, if $n > 5$
22. If $f(x) = \tan x$, show that $f^n(0) - {}^nC_2 f^{n-2}(0) + {}^nC_4 f^{n-4}(0) \ldots = \sin \dfrac{n\pi}{4}$.

## 31.7 SIMULTANEOUS DIFFERENCE EQUATIONS WITH CONSTANT COEFFICIENTS

The method used for solving simultaneous differential equations with constant coefficients also applies to simultaneous difference equations with constants coefficients. The following example illustrates the technique.

**Example 31.12.** *Solve the simultaneous difference equations*

$$u_{x+1} + v_x - 3u_x = x,\; 3u_x + v_{x+1} - 5v_x = 4^x$$

*subject to the conditions* $u_1 = 2, v_1 = 0$. (*Andhra, 2015 ; Delhi, 2015*)

**Solution.** Given equation in symbolic form, are

$$(E - 3)\, u_x + v_x = x \quad \ldots(i)$$
$$3u_x + (E - 5)v_x = 4^x \quad \ldots(ii)$$

Operating the first equation with $E - 5$ and subtracting the second from it, we get

$$[(E-5)(E-3)-3]u_x = (E-5)x - 4^x$$

or

$$(E^2 - 8E + 12)u_x = 1 - 4x - 4^x$$

Its solution is $$u_x = c_1 2^x + c_2 6^x - \frac{4}{5}x - \frac{19}{25} + \frac{4^x}{4} \quad ...(iii)$$

Substituting the value of $u_x$ from (*iii*) in (*i*), we get

$$v_x = c_1 2^x - 3c_2 6^x - \frac{3x}{5} - \frac{34}{25} - \frac{4^x}{4} \quad ...(iv)$$

Taking $u_1 = 2$, $v_1 = 0$, in (*iii*) and (*iv*), we obtain

$$2c_1 + 6c_2 = \frac{64}{25},\ 2c_1 - 18c_2 = \frac{74}{25}$$

when $$c_1 = 1.33,\ c_2 = -0.0167$$

Hence $$u_x = 1.33.2^x - 0.0167.6^x - 0.8x - 0.76 + 4^{x-1}$$

$$u_x = 1.33.2^x - 0.05.6^x - 0.6x - 1.36 - 4^{x-1}.$$

## PROBLEMS 31.4

Solve the following simultaneous difference equations :

1. $y_{x+1} - z_x = 2(x+1)$, $z_{x+1} - y_x = -2(x+1)$.
2. $y_{n+1} - y_n + 2z_{n+1} = 0$, $z_{n+1} - z_n - 2y_n = 2^n$.
3. $u_{n+1} + n = 3u_n + 2v_n$, $v_{n+1} - n = u_n + 2v_n$, given $u_0 = 0$, $v_0 = 3$.
4. $u_{x+1} + v_x + w_x = 1$, $u_x + v_{x+1} + w_x = x$, $u_x + v_x + w_{x+1} = 2x$.

## 31.8 APPLICATION TO DEFLECTION OF A LOADED STRING

Consider a light string of length $l$ stretched tightly between $A$ and $B$. Let the forces $P_i$ be acting at its equispaced points $x_i$ ($i = 1, 2, ..., n-1$) and perpendicular to $AB$ resulting in small transverse displacements $y_i$ at these points (Fig. 31.1). Assuming the angle $\theta_i$ made by the portion between $x_i$ and $x_{i+1}$ with the horizontal, to be small, we have

$$\sin\theta_i = \tan\theta_i = \theta_i \text{ and } \cos\theta_i = 1$$

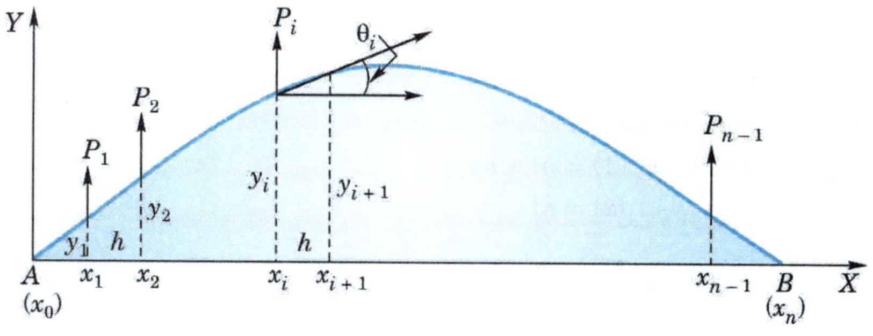


Fig. 31.1

If $T$ be the tension of the string at $x_i$, then $T\cos\theta_i = T$
*i.e.*, the tension may be taken as uniform.

Taking $x_{i+1} - x_i = h$, we have

$$y_{i+1} - y_i = h\tan\theta_i = h\theta_i \quad ...(1)$$

$$y_i - y_{i-1} = h\tan\theta_{i-1} = h\theta_{i-1} \quad ...(2)$$

Also resolving the forces in equilibrium at $(x_i, y_i) \perp$ to $AB$, we get

$$T\sin\theta_i - T\sin\theta_{i-1} + P_i = 0 \text{ i.e. } T(\theta_i - \theta_{i-1}) + P_i = 0 \quad ...(3)$$

Eliminating $\theta_i$ and $\theta_{i-1}$ from Eqns. (1), (2) and (3), we obtain

$$y_{i+1} - 2y_i + y_{i-1} = -\frac{hP_i}{T} \quad ...(4)$$

which is a difference equation and its solution gives the displacements $y_i$. To obtain the arbitrary constants in the solution, we take $y_0 = y_n = 0$ as the boundary conditions, since the ends $A$ and $B$ of the string are fixed.

**Example 31.13.** *A light string stretched between two fixed nails 120 cm apart, carries 11 loads of weight 5 gm each at equal intervals and the resulting tension is 500 gm weight. Show that the sag at the mid-point is 1.8 cm.*

**Solution.** Taking $h = 10$ cm, $P_i = 5$ gm and $T = 500$ gm wt.,
the above Eqns. (4) becomes $y_{i+1} - 2y_i + y_{i-1} = -1/10$

i.e.,
$$y_{i+2} - 2y_{i+1} + y_i = -\frac{1}{10}$$

Its A.E. is $(E-1)^2 = 0$ i.e. $E = 1, 1$. $\therefore$ C.F. $= c_1 + c_2 i$

and
$$\text{P.I.} = \frac{1}{(E-1)^2}\left(-\frac{1}{10}\right) = -\frac{1}{10}\frac{1}{(E-1)^2}(1)^i = -\frac{1}{10}\frac{i(i-1)}{2} = \frac{1}{20}(i - i^2)$$

Thus the C.S. is $y_i = c_1 + c_2 i + \frac{1}{20}(i - i^2)$

Since $y_0 = 0, \quad \therefore \quad c_1 = 0$

and $y_{12} = 0, \quad \therefore \quad c_2 = \frac{11}{20}$.

Hence
$$y_i = \frac{11}{20}i + \frac{1}{20}(i - i^2)$$

At the mid-point $i = 6$, we get $y_6 = 1.8$ cm.

**Example 31.14:** *If $y = x^n \log x$, Prove that*

*(a)* $y_{n+1} = \frac{n!}{x}$

*(b)* $x^2 y_{p+2} + (2p - 2n + 1)xy_{p+1} + (P - n)^2 yp = 0$ (UPTU, 2018)

**Solution:** (a) Given,
$$y = x^n \log x \quad \text{...(i)}$$
$$y_1 = x^n (1/n) + nx^{n-1} \log x$$
$$xy_1 = x^n + nx^n \log x$$
$$= x^n + ny \quad \text{...(ii)}$$

Differentiating Eqn. (ii) $n$ times we get
$$y_{n+1}(x) + {}^nC_1 y_n (1) = n! + ny_n$$
$$y_{n+1}(x) = n!$$
$$y_{n+1} = \frac{n!}{x}$$

(b)
$$y_1 = x^n \frac{1}{x} + nx^{n-1} \log x$$

Diff, $y$ w.r.t. $x$

$\Rightarrow$
$$xy_1 = x^n + nx^n \log x$$

Multiplying by $x$
$$xy_1 = x^n + ny \quad \text{...(iii)}$$

Differentiating Eqn. (iii) w.r.t. $x$, we get
$$xy_2 + y_1 = nx^{n-1} + ny_1$$
$$xy_2 + (1-n)y_1 = nx^{n-1}$$
$$x^2y^2 + (1-n)xy_1 = nx^n = n(xy_1 - xy)$$
$$\Rightarrow \quad x^2y^2 + (1-2n)xy_1 + n^2 y = 0 \quad \text{...(iv)}$$

Differentiating Eqn. $(iv)$ $p$ times by Leibnitzs theorem.

$$[y_{p+2}(x^2) + pc_1 y_{p+1}(2x) + pc_2 y_p(2)] + (1-2n)[y_{p+1}(x) + pc_1 y_p(1)] + n^{2y}p = 0$$

$$x^2 y_{p+2} + y_{p+1}[2p_x + (1-2n)x] + y_p\{p(p-1) + (1-2n)p + n^2\} = 0$$

$$x^2 y_{p+2} + 2(p-2n+1)x_{y+1} + (p-n)^{2y} p = 0$$

**Example 31.15:** *If $y = e^{m\cos^{-1}x}$. show that*

$$(1-x^2)y_{n+2} - (2n+1)x\,y_{n+1} - (n^2+m^2)\,y_n = 0$$

*and calculate $y_n(0)$* **(M.D.U. 2108)**

**Solution:** Let $$y = e^{m\cos^{-1}x} \quad ...(i)$$

$$y_1 = e^{m\cos^{-1}x}.\frac{-1}{\sqrt{1-x^2}} = \frac{-my}{\sqrt{1-x^2}} \quad ...(ii)$$

Squaring and cross multiplying

$$(1-x^2)y^2{}_1 = m^2y^2$$

Differentiating again w.r.t. $x$, we have

$$(1-x^2).2y_1y_2 - 2xy^2{}_1 = m^2 2yy_1$$

Dividing both sides by $2y_1$, we get

$$(1-x^2)y_2 - xy_1 = m^2y \quad ...(iii)$$

Differentiating $n$ times by Leibnitz's theorem

$$[y_{n+2}(1-x^2) + ny_{n+1}(-2x) + \frac{n(n-1)}{2!}y_n(-2)] - [y_{n+1}.x + {}^{ny}n.1] = m^2y_n$$

$$(1-x^2)y_{n+2} - (2n+1)xy_{n+1} = (m^2+n^2)\,y_n \quad ...(iv)$$

putting $x = 0$ in Eqns.$(i)$, $(ii)$, $(iii)$ and $(iv)$

$$y(0) = e^{m\cos^{-1}0} = e^{m\pi/2}$$

$$y_1(0) = -my(0) = -me^{m\pi/2}$$

$$y_2(0) = m^2y(0) = m^2e^{m\pi/2}$$

$$y_{n+2}(0) = (m^2+n^2)\,y_n(0) \quad ...(v)$$

putting $n = 1, 2, 3, 4....$ in Eqn. $(v)$

$$y_3(0) = (m^2+1^2)\,y_1(0) = -m(m^2+1^2)\,e^{m\pi/2}$$

$$y_4(0) = (m^2+2^2)\,y_2(0) = -m\,(m^2+2^2)\,e^{m\pi/2}$$

$$y_5(0) = (m^2+3^2)\,y_3(0) = -m(m^2+1^2)(m^2+3^2)\,e^{m\pi/2}$$

$$y_6(0) = (m^2+4^2)\,y_4(0) = m^2(m^2+2^2)(m^2+4^2)\,e^{m\pi/2}$$

- - - - - - - - - - - - - - - - - - - - - - - - - - - - - - - - - - - - - - -

- - - - - - - - - - - - - - - - - - - - - - - - - - - - - - - - - - - - - - -

- - - - - - - - - - - - - - - - - - - - - - - - - - - - - - - - - - - - - - -

Hence, when $n$ is odd.

$$y_n(0) = -m(m^2+1^2)(m^2+3^2)...[m^2+(n-2)]^2e^{m\pi/2}$$

and when $n$ is even.

$$y_n(0) = m^2(m^2+2^2)(m^2+4^2)...[m^2+(n-2)^2]\,e^{m\pi/2}$$

**Example 31.16:** *Solve the equation* $x_{t+2} - 5x_{t+1} + 6x_t = 4^t + t^2 + 3$ *(Nagpur, 2018)*

**Solution:** The homogeneous equation has characteristic equation $r^2 - 5r + 6 = 0$. With two different real roots (2, 3)

$$x_t = C4^t + Dt^2 + Et + F$$

$$C4^{t+2} + D(t+2)^2 + E(t+2) + F - 5(C4^{t+1} + D(t+1)^2 + E(t+1) + F) + 6(C4^t + Dt^2 + Et + F)$$
$$= 4t + t^2 + 3$$

Expanding and rearranging

$$2C4^t + 2Dt^2 + (-6D + 2E)t + (-D - 3E + 2F) = 4t + t^2 + 3$$

$$t = 0, 1, 2 \ldots$$
$$2C = 4$$
$$2D = 1$$
$$-6D + 2E = 0$$
$$-D - 3E + 2F = 3$$

$$C = \frac{1}{2}, \quad D = \frac{1}{2}, \quad E = \frac{3}{3} \quad \text{and} \quad F = 4$$

$$x_t = A2^t + B3^t + \frac{1}{2}4^t + \frac{1}{2}t^2 + \frac{3}{2}t + 4$$

**Example 31.17:** *Solve* $U_n = 2U_{n-1} - 3,\ n \geq 2$, *given* $n = 4$

**Solution:**
$$U_n = 2^{n-1} \times 4 - \frac{3(2^{n-1} - 1)}{2 - 1}$$
$$= 4 \times 2^{n-1} - 3 \times 2^{n-1} + 3 = 2^{n-1} + 3$$

Using the difference equation
$$U_2 = 2U_1 - 3 = 2 \times 4 - 3 = 8 - 5 = 5$$

Using the formula, $U_2 = 2^1 + 3 = 5$

**Example 31.18:** *Find the monthly repayment on a ₹500 loan at an interest rate 24% P.A. over 18 months.* *(Marathwada, 2018)*

**Solution:** Let $u_n$ be the amount owing after $n$ months

So, $u_0$ = ₹500 and $C$ is the repayment each month. Given that interest rate *P.A.* is 24% so rate per month is assumed to be 2%.

$$u_n = u_{n-1} + (2\% \text{ of } u_{n-1}) - C$$
$$\Rightarrow \quad u_n = u_{n-1} + 0.02\,u_{n-1} + C$$
$$u_n = 1.02\,u_{n-1} - C \qquad \ldots(i)$$
$$u_n = k^n u_0 + \frac{C(k^n - 1)}{(k - 1)}$$

where in the case $k = 1.02$
$$u_n = (1.02)^n u_0 - \frac{C(1.02^n - 1)}{1.02 - 1}$$
$$= 1.02^n\,u_0 - 50C\,(1.02^n - 1)$$
$$u_n = 1.02^n\,u_0 - 50C\,(1.02^n - 1)$$

At the end of the term $n = 18$
$$u_{18} = 1.02^{18}\,u_0 - 50C\,(1.02^{18} - 1)$$

But
$$u_0 = 500 \quad \text{and} \quad u_{18} = 0$$
$$\Rightarrow \quad 0 = 500 \times 1.02^{18} - 50C(1.02^{18} - 1)$$
$$\Rightarrow \quad C = \frac{500 \times 1.02^{18}}{50(1.02^{18} - 1)} = \frac{714.123}{21.4123}$$
$$C = \text{Rs. } 33.35$$

Repayment are made at the rate of ₹ 33.35 per month.

**Example 31.19:** *If* $y = \frac{\sin h^{-1}x}{\sqrt{1-x^2}}$ *prove that* $(1+x^2)\, y_{n+2} + (2n+3)xy_{n+1} + (n+1)^2 y_n = 0$ *(Kottayam, 2018)*

**Solution:** Consider

$$y = \frac{\sin h^{-1}x}{\sqrt{1-x^2}} \quad ...(i)$$

$$y^2(1+x^2) = (\sin^{-1} x)^2$$

Differentiate $y$ twice and find the relation between $y_2$, $y_1$ and $y$

Differentiating with respect to $x$

$$y^2\,(2x) + (1+x^2)2yy_1 = 2\sin^{-1} x \frac{1}{\sqrt{1+x^2}}$$

$$y^2\,(2x) + (1+x^2)\,2yy_1 = 2y$$

$$(1+x^2)y_1 + xy = 1 \quad ...(ii)$$

Differentiating Eqn. (*ii*) with respect

$$(1+x^2)y_2 + 2xy_1 + xy_1 + y = 0$$

$$(1+x^2)y_2 + 3xy_1 + y = 0$$

Now, apply Leibnitzs theorem

$$\therefore \quad \left[(1+x^2)y_{n+2} + n\,2x\,y_{n+1} + \frac{2n(n-1)}{2}y_n\right] + \left[3xy_{n+1} + 3ny_n\right] + y_n = 0$$

Now, simply

$$(1+x^2)y_{n+2} + (2n+3)\,xy_{n+1} + (n^2 - n + 3n + 1)y_n = 0$$

$$(1+x^2)y_{n+2} + (2n+3)xy_{n+1} + (n+1)^2\, y_n = 0$$

**Example 31.20:** *If* $y = (x^2-1)^n$, *Prove that*

$$(x^2-1)y_{n+2} + 2xy_{n+1} - n(n+1)y_n = 0$$ *(Lucknow, 2019)*

Hence if $P_n = \frac{d^n}{dx^n}(x^2-1)^n$. Show that $\frac{d}{dx}\left\{(1-x^2)\frac{dP_n}{dx}\right\} + n(n+1)P_n = 0.$

**Solution:** Consider

$$y = (x^2-1)^n \quad ...(1)$$

Differentiate $y$ twice and find the relation between $y_2$, $y_1$ and $y$

Differentiate with respect to $x$

$$y_n = n(x^2-1)^{n-1}\,(2x) = 2n\frac{(x^2-1)^n}{x^2-1} = \frac{2nxy}{x^2-1}$$

$$y_1\,(x^2-1) - 2nxy = 0 \quad ...(2)$$

Differentiate

$$y^2\,(x^2-1) + 2xy_1 - 2nxy_1 - 2xy = 0 \quad ...(3)$$

$$y_2(x^2-1) + 2(1-n)xy_1 - 2xy = 0$$

Apply Leibnitz's theorem

$$\therefore \quad y_{n+2}\,(x^2-1) + ({}^nc_1)y_{n+1}\,(2x) + {}^nc_2\,y_n\,(2) + 2(1-n)\,[y_{n+1}(x) + ({}^nc_1)(1)] - 2n^yn = 0$$

Now, solving

$$(x^2-1)y_{n+2} + 2xy_{n+1} - n(n+1)y_n = 0 \quad ...(4)$$

$$(x^2-1)y_{n+2} + 2x\,y_{n+1} - n(n+1)y_n = 0$$

$$\therefore \quad \frac{d}{dx}\left[(x^2-1)\frac{d}{dx}y_n\right] - n(n+1)y_n = 0$$

$$\therefore \quad -\frac{d}{dx}\left[(1-x^2)\frac{d}{dx}y_n\right] - n(n+1)y_n = 0$$

Given, $y_n = P_n$

$$\therefore \quad \frac{d}{dx}\left[(1-x^2)\frac{d}{dx}P_n\right] + n(n+1)P_n = 0$$

## PROBLEMS 31.5

1. A light string of length $(n + 1)l$ is stretched between two fixed points with a force $P$. It is loaded with $n$ equal masses $m$ at distance $l$. If the system starts rotating with angular velocity $\omega$, find the displacement $y_i$ of the $i$th mass.
2. If $y = \dfrac{\log x}{x}$, Prove that $y_n = \dfrac{(-1)^n n!}{x^{n+1}}\left\{\log x - \left\{1 - \dfrac{1}{1+2} + \dfrac{1}{3} + \dots \dfrac{1}{n}\right\}\right\}$
3. If $y = x \log (x + 1)$, Prove that $y_n = \dfrac{(-1)^{n-2}(n-2)!(x+n)}{(x+1)^n}$

## 31.9 OBJECTIVE TYPE OF QUESTIONS

## PROBLEMS 31.6

*Select the correct answer or fill up the blanks in the following questions :*

1. $y_n = A\,2^n + B\,3^n$, is the solution of the difference equation ...... .
2. The solution of $(E - 1)^3 u_n = 0$ is ...... .
3. The solution of the difference equation $u_{n+3} - 2u_{n+2} - 5u_{n+1} + 6u_n = 0$ is ...... .
4. The solution of $y_{n+1} - y_n = 2^n$ is ...... .
5. The difference equation $y_{n+1} - 2y_n = 0$ given that $y_0 = 2$ has $y_n =$ ...... as its solution. *(Anna, 2012)*
6. The difference equation corresponding to the family of curves $y = ax^2 + bx$ is ...... .
7. The particular integral of the equation $(E - 2)\, y_n = 1$.
8. The solution of $4y_n = y_{n+2}$ such that $y_0 = 0, y_1 = 2$, is ...... .
9. The equation $\Delta^2 u_{n+1} + \frac{1}{2}\Delta^2 u_n = 0$ is of order ...... .
10. The difference equation satisfied by $y = a + b/x$ is ...... .
11. The order of the difference equation $y_{n+2} - 2y_{n+1} + y_n = 0$ is ...... .
12. The solution of $y_{n+2} - 4y_{n+1} + 4y_n = 0$ is ...... .
13. The particular integral of $u_{x+2} - 6u_{x+1} + 9u_x = 3$ is ...... .
14. The difference equation generated by $u_n = (a + bn)\, 3^n$ is ...... .
15. Solution of $6y_{n+2} + 5y_{n+1} - 6y_n = 2^n$ is $y_n = A(2/3)^n + B(-3/2)^n + 2^n/28$. (True or False)

CHAPTER 32

# Numerical Solution of Ordinary Differential Equations



## 32.1 INTRODUCTION

The methods of solution so far presented are applicable to a limited class of differential equations. Frequently differential equations appearing in physical problems do not belong to any of these familiar types and one is obliged to resort to numerical methods. These methods are of even greater importance when we realise that computing machines are now available which reduce numerical work considerably.

A number of numerical methods are available for the solution of first order differential equations of the form :

$$\frac{dy}{dx} = f(x, y), \text{ given } y(x_0) = y_0 \qquad ...(1)$$

These methods yield solutions either as a power series in $x$ from which the values of $y$ can be found by direct substitution, or as a set of values of $x$ and $y$. The methods of Picard and Taylor series belong to the former class of solutions whereas those of Euler, Runge-Kutta, Milne, Adams-Bashforth etc. belong to the latter class. In these later methods, the values of $y$ are calculated in short steps for equal intervals of $x$ and are therefore, termed as *step-by-step methods.*

Euler and Runge-Kutta methods are used for computing $y$ over a limited range of $x$-values whereas Milne and Adams-Bashforth methods may be applied for finding $y$ over a wider range of $x$-values. These later methods require starting values which are found by Picard's or Taylor series or Runge-Kutta methods.

The initial condition in Eqn. (1) is specified at the point $x_0$. *Such problems in which all the initial conditions are given at the initial point only are called* ***initial value problems.*** *But there are problems involving second and higher order differential equations in which the conditions may be given at two or more points. These are known as* ***boundary value problems.*** In this chapter, we shall first explain methods for solving initial value problems and then give a method of solving boundary value problems.

## 32.2 PICARD'S METHOD*

Consider the first order equation $dy/dx = f(x, y)$ ...(1)

* Called after the French mathematician *Emile Picard* (1856—1941) who was professor in Paris since 1881 and is famous for his researches in the theory of functions.

It is required to find that particular solution of Eqn. (1) which assumes the value $y_0$ when $x = x_0$. Integrating Eqn. (1) between limits, we get

$$\int_{y_0}^{y} dy = \int_{x_0}^{x} f(x, y)\,dx \quad \text{or} \quad y = y_0 + \int_{x_0}^{x} f(x, y)\,dx \qquad ...(2)$$

This is an integral equation equivalent to Eqn. (1), for it contains the unknown $y$ under the integral sign. *As a first approximation* $y_1$ *to the solution*, we put $y = y_0$ in $f(x, y)$ and integrate Eqn. (2), giving

$$y_1 = y_0 + \int_{x_0}^{x} f(x, y_0)\,dx$$

*For a second approximation* $y_2$, we put $y = y_1$ in $f(x, y)$ and integrate Eqn. (2), giving

$$y_2 = y_0 + \int_{x_0}^{x} f(x, y_1)\,dx.$$

Similarly, *a third approximation is* $y_3 = y_0 + \int_{x_0}^{x} f(x, y_2)\,dx.$

Continuing this process, a sequence of functions of $x$, *i.e.*, $y_1, y_2, y_3$ ... is obtained each giving a better approximation of the desired solution than the preceding one.

**Obs.** Picard's method is of considerable theoretical value, but can be applied only to a limited class of equations in which the successive integrations can be performed easily. The method can be extended to simultaneous equations and equations of higher order (See § 32.11 and 32.12).

**Example 32.1.** *Using Picard's process of successive approximation, obtain a solution upto the fifth approximation of the equation dy/dx = y + x, such that y = 1 when x = 0. Check your answer by finding the exact particular solution.* *(JNTUK, 2022, 2015)*

**Solution.** (*a*) We have $y = 1 + \int_0^x (y + x)\,dx.$

*First approximation.* Put $y = 1$, in $y + x$, giving

$$y_1 = 1 + \int_0^x (1 + x)\,dx = 1 + x + x^2/2.$$

*Second approximation.* Put $y = 1 + x + x^2/2$ in $y + x$, giving

$$y_2 = 1 + \int_0^x (1 + 2x + x^2/2)\,dx = 1 + x + x^2 + x^3/6.$$

*Third approximation.* Put $y = 1 + x + x^2 + x^3/6$ in $y + x$, giving

$$y_3 = 1 + \int_0^x (1 + 2x + x^2 + x^3/6)\,dx = 1 + x + x^2 + \frac{x^3}{3} + \frac{x^4}{24}.$$

*Fourth approximation.* Put $y = y_3$ in $y + x$, giving

$$y_4 = 1 + \int_0^x \left(1 + 2x + x^2 + \frac{x^3}{3} + \frac{x^4}{24}\right) dx = 1 + x + x^2 + \frac{x^3}{3} + \frac{x^4}{12} + \frac{x^5}{120}.$$

*Fifth approximation.* Put $y = y_4$ in $y + x$, giving

$$y_5 = 1 + \int_0^x \left(1 + 2x + x^2 + \frac{x^3}{3} + \frac{x^4}{12} + \frac{x^5}{120}\right) dx = 1 + x + x^2 + \frac{x^3}{3} + \frac{x^4}{12} + \frac{x^5}{60} + \frac{x^6}{720} \qquad ...(i)$$

(*b*) Given equation :

$$\frac{dy}{dx} - y = x \text{ is a Leibnitz's linear in } x.$$

Its I.F. being $e^{-x}$, the solution is

$$ye^{-x} = \int xe^{-x}\,dx + c = -xe^{-x} - \int (-e^{-x})\,dx + c = -xe^{-x} - e^{-x} + c \qquad \text{[Integrate by parts]}$$

$\therefore \qquad y = ce^x - x - 1.$

Since $y = 1$, when $x = 0$, $\therefore \quad c = 2.$

Thus the desired particular solution is $y = 2e^x - x - 1$ ...(*ii*)

Or using the series : $e^x = 1 + x + \frac{x^2}{2!} + \frac{x^3}{3!} + \frac{x^4}{4!} + \ldots \infty,$

we get $$y = 1 + x + x^2 + \frac{x^3}{3} + \frac{x^4}{12} + \frac{x^5}{60} + \frac{x^6}{360} + \ldots \infty \qquad \ldots(iii)$$

Comparing Eqns. (*i*) and (*iii*), it is clear that Eqn. (*i*) approximates to the exact particular solution Eqn. (*ii*) upto the term in $x^5$.

**Obs.** At $x = 1$, the fourth approximation $y_4 = 3.433$ and the fifth approximation $y_5 = 3.434$ whereas exact value is 3.44.

**Example 32.2.** *What is the approximate value of x and y by using Picard's method for the particular solution of*

$$\frac{dy}{dx} = x + z, \quad \frac{dz}{dx} = x - y^2$$

*Given that* $y = 2, z = 1,$ *when* $x = 0$ *(M.Sc (KU), 2018)*

**Solution:** Here $x_0 = 0, y_0 = 2, z_0 = 1$

We have $$\frac{dy}{dx} = x + z \text{ or } dy = (x + z)\,dx$$

$$\int_0^y dy = \int_0^x (x + z)dx$$

$$y = 2 + \int_0^x (x + z)dx \qquad \ldots(i)$$

$$\frac{dz}{dx} = x - y^2$$

$$dz = (x - y^2)\,dx$$

$$\int_0^z dz = \int_0^x (x - y^2)dx$$

$$z = 1 + \int_0^x (x - y^2)dx$$

Now first $$y' = 2 + \int_0^x (x + z_0)dx = 2 + \int_0^x (x + 1)dx = 2 + x + \frac{x^2}{2}$$

$$z' = 1 + \int_0^x (x - y_0^2)dx = 1 + \int_0^x (x - y)dx = 1 - 4x + \frac{x^2}{2}$$

The second approximation of $y$ and $z$ are

$$y^{(2)} = 2 + \int_0^x (x + z_1)dx = 2 + \int_0^x \left(x + 1 - 4x + \frac{x^2}{2}\right)dx$$

$$= 2 + x - \frac{3}{2}x^2 + \frac{1}{6}x^3$$

$$z^{(2)} = 1 + \int_0^x (x - y_1^2)dx = 1 + \int_0^x \left[x - \left(2 + x + \frac{x^2}{2}\right)^2\right]dx$$

$$= 1 - 4x - \frac{3}{2}x^2 - x^3 - \frac{1}{4}x^4 - \frac{1}{20}x^5.$$

**Example 32.3:** *Solve the equation*

$$\frac{d^2y}{dx^2} - x^2\frac{dy}{dx} - x^4y = 0, \text{ given that}$$

$$y = 5 \text{ and } dy/dx = 1, \text{ when } x = 0$$ *(MSc Patna Uni., 2018)*

**Solution:** Let $\frac{dy}{dx} = z$

Let $\frac{d^2y}{dx^2} = \frac{dz}{dx}$

Thus the given equation reduces to

$$\frac{dz}{dx} - x^2z - x^4y = 0$$

$$\frac{dz}{dx} = x^2 z + x^4 y$$

along with the conditions that $x_0 = 0, \quad y = 5, z_0 = 1$

$$\frac{dy}{dz} = z \quad \text{or} \quad dy = z\,dx$$

$$\int_{y_0}^{y} dy = \int_{x_0}^{x} zdx$$

$$y = 5 + \int_0^x z\,dx$$

$$dz = (x^2z + x^4y)\,dx$$

$$\int_{z_o}^{z} dz = \int_{x_o}^{x} (x^2z + x^4y)dx$$

$$z = 1 + \int_o^x (x^2z + x^4z)dx \qquad \text{...(ii)}$$

First approximations are given by

$$y^{(1)} = 5 + \int_o^x z_o\,dx = 5 + \int_0^x dx = 5 + x$$

$$z^{(1)} = 1 + \int_o^x (x^2z_0 + x^4y_0)\,dx = 1 + \int_0^x (x^2 + 5x^4)dx = 1 + \frac{1}{3}x^3 + x^5$$

Second approximation are given by

$$y^{(2)} = 5 + \int_0^x z^{(1)}dx = 5 + \int_0^x \left(1 + \frac{x^3}{3} + x^5\right)dx = 5 + x + \frac{x^4}{12} + \frac{x^6}{6}$$

$$z^{(2)} = 1 + \int_0^x (x^2z^{(1)} + x^4\,y^{(1)}) \cdot dx$$

$$= 1 + \int_0^x \left[x^2\left(1 + \frac{x^3}{3} + x^5\right) + x^4(5 + x)\right]dx$$

$$= 1 + \frac{1}{3}x^3 + x^5 + \frac{2}{9}x^6 + \frac{1}{8}x^8$$

Find approxination; $y^{(3)} = 5 + x + \frac{1}{12}x^4 + \frac{1}{6}x^6 + \frac{2}{63}x^7 + \frac{1}{72}x^9$

$$z^{(3)} = 1 + \frac{1}{3}x^3 + x^5 + \frac{2}{9}x^6 + \frac{1}{8}x^8 + \frac{11}{224}x^9 + \frac{7}{264}x^{11}.$$

## 32.3 TAYLOR'S SERIES METHOD

Consider the first order equation $dy/dx = f(x, y)$ ...(1)

Differentiating Eqn. (1), we have

$$\frac{d^2y}{dx^2} = \frac{\partial f}{\partial x} + \frac{\partial f}{\partial y}\frac{dy}{dx} \quad i.e., \quad y'' = f_x + f_y f' \qquad ...(2)$$

Differentiating this successively, we can get $y'''$, $y^{iv}$ etc. Putting $x = x_0$ and $y = 0$, the values of $(y')_0$, $(y'')_0$, $(y''')_0$ can be obtained. Hence the Taylor's series

$$y(x) = y_0 + (x - x_0)(y')_0 + \frac{(x - x_0)^2}{2!}(y'')_0 + \frac{(x - x_0)^3}{3!}(y''')_0 + \ldots \qquad ...(3)$$

gives the values of $y$ for every value of $x$ for which Eqn. (3) converges.

On finding the value $y_1$ for $x = x_1$ from Eqn. (3), $y'$, $y''$ can be evaluated at $x = x_1$ by means of Eqn. (1), Eqn. (2) etc. Then $y$ can be expanded about $x = x_1$. In this way, the solution can be extended beyond the range of convergence of series Eqn. (3).

**Example 32.4.** *Find by Taylor's series method the value of y at x = 0.1 and x = 0.2 to five places of decimals from $dy/dx = x^2y - 1$, $y(0) = 1$.* (V.T.U., 2016)

**Solution.** Here $(y)_0 = 1$, $y' = x^2y - 1$, $(y')_0 = -1$

$\therefore$ Differentiating successively and substituting, we get

$$y'' = 2xy + x^2y', \qquad (y'')_0 = 0$$
$$y''' = 2y + 4xy' + x^2y'', \qquad (y''')_0 = 2$$
$$y^{iv} = 6y' + 6xy'' + x^2y''', \qquad (y^{iv})_0 = -6 \text{ etc.}$$

Putting these values in the Taylor's series,

$$y(x) = y_0 + xy'(0) + \frac{x^2}{2!}y''(0) + \frac{x^3}{3!}y'''(0) + \frac{x^4}{4!}y^{iv}(0) + \ldots,$$

we have $$y(x) = 1 + x(-1) + \frac{x^2}{2!}(0) + \frac{x^3}{3!}(2) + \frac{x^4}{4!}(-6) + \ldots = 1 - x + \frac{x^3}{3} - \frac{x^4}{4} + \ldots$$

Hence $y(0.1) = 0.90033$ and $y(0.2) = 0.80227$.

**Example 32.5.** *Employ Taylor's method to obtain approximate value of y at x = 0.2 for the differential equation $dy/dx = 2y + 3e^x$, $y(0) = 0$. Compare the numerical solution obtained with the exact solution.* (V.T.U., 2014)

**Solution.** (*a*) We have $y' = 2y + 3e^x$ $\qquad y'(0) = 2y(0) + 3e^0 = 3.$

Differentiating successively and substituting $x = 0$, $y = 0$, we get

$$y'' = 2y' + 3e^x, \qquad y''(0) = 2y'(0) + 3 = 9$$
$$y''' = 2y'' + 3e^x, \qquad y'''(0) = 2y''(0) + 3 = 21$$
$$y^{iv} = 2y''' + 3e^x, \qquad y^{iv}(0) = 2y'''(0) + 3 = 45 \text{ etc.}$$

Putting these values in the Taylor's series, we have

$$y(x) = y(0) + xy'(0) + \frac{x^2}{2!}y''(0) + \frac{x^3}{3!}y'''(0) + \frac{x^4}{4!}y^{iv}(0) + \ldots$$

$$= 0 + 3x + \frac{9}{2}x^2 + \frac{21}{6}x^3 + \frac{45}{24}x^4 + \ldots = 3x + \frac{9}{2}x^2 + \frac{7}{2}x^3 + \frac{15}{8}x^4 + \ldots$$

Hence $\quad y(0.2) = 3(0.2) + 4.5(0.2)^2 + 3.5(0.2)^3 + 1.875(0.4)^4 + \ldots = 0.8110$ ...(*i*)

(*b*) Now $\dfrac{dy}{dx} - 2y = 3e^x$ is a Leibnitz's linear in $x$.

Its I.F. being $e^{-2x}$, the solution is

$$ye^{-2x} = \int 3e^x \,.\, e^{-2x}\,dx + c = -3e^{-x} + c \quad \text{or} \quad y = -3e^x + ce^{2x}$$

Since $y = 0$ when $x = 0$, $\therefore$ $c = 3$.

Thus the exact solution is $y = 3(e^{2x} - e^x)$

When $x = 0.2$, $y = 3(e^{0.4} - e^{0.2}) = 0.8112$ ...(ii)

Comparing Eqns. (i) and (ii), it is clear that Eqn. (i) approximates to the exact value upto 3 decimal places.

**Example 32.6.** *Solve by Taylor's series method the equation* $\frac{dy}{dx} = \log(xy)$ *for y(1.1) and y(1.2), given y(1) = 2.*

**Solution.** We have $y' = \log x + \log y$ ; $y'(1) = \log 2$

Differentiating w.r.t. $x$ and substituting $x = 1$, $y = 2$, we get

$$y'' = \frac{1}{x} + \frac{1}{y}y' \; ; \; y''(1) = 1 + \frac{1}{2}\log 2$$

$$y''' = -\frac{1}{x^2} + \frac{1}{y} + y'' + y'\left(-\frac{1}{y^2}\right)y' \; ; \; y'''(1) = -1 + \frac{1}{2}\left(1 + \frac{1}{2}\log 2\right) - \frac{1}{4}(\log 2)^2$$

Substituting these values in the Taylor's series about $x = 1$, we have

$$y(x) = y(1) + (x-1)\,y'(1) + \frac{(x-1)^2}{2!}\,y''(1) + \frac{(x-1)^3}{3!}\,y'''(1) + \ldots$$

$$= 2 + (x-1)\log 2 + \frac{1}{2}(x-1)^2\left(1 + \frac{1}{2}\log 2\right) + \frac{1}{6}(x-1)^3\left[-\frac{1}{2} + \frac{1}{4}\log 2 - \frac{1}{4}(\log 2)^2\right]$$

$$\therefore \quad y(1.1) = 2 + (0.1)\log 2 + \frac{(0.1)^2}{2}\left(1 + \frac{1}{2}\log 2\right) + \frac{(0.1)^3}{6}\left[-\frac{1}{2} + \frac{1}{4}\log 2 - \frac{1}{4}(\log 2)^2\right] = 2.036$$

$$y(1.2) = 2 + (0.2)\log 2 + \frac{(0.2)^2}{2}\left(1 + \frac{1}{2}\log 2\right) + \frac{(0.2)^3}{6}\left[-\frac{1}{2} + \frac{1}{4}\log 2 - \frac{1}{4}(\log 2)^2\right] = 2.081.$$

## PROBLEMS 32.1

1. Using Picard's method, solve $dy/dx = -xy$ with $x_0 = 0$, $y_0 = 1$ upto third approximation.
2. Employ Picard's method to obtain, correct to four places of decimal, solution of the differential equation $dy/dx = x^2 + y^2$ for $x = 0.4$, given that $y = 0$ when $x = 0$.
3. Obtain Picard's second approximate solution of the initial value problem : $y' = x^2/(y^2 + 1)$, $y(0) = 0$. *(J.N.T.U.,2015)*
4. Find an approximate value of $y$ when $x = 0.1$, if $dy/dx = x - y^2$ and $y = 1$ at $x = 0$, using

   (a) Picard's method (b) Taylor's series. *(JNTUK, 2022; V.T.U., 2015)*
5. Solve $y' = x + y$ given $y(1) = 0$. Find $y(1.1)$ and $y(1.2)$ by Taylor's method. Compare the result with its exact value. *(Anna, 2015)*
6. Using Taylor's series method, compute the value of $y(0.2)$ correct to 3 decimal places from $\frac{dy}{dx} = 1 - 2xy$ given that $y(0) = 0$. *(Anna., 2016)*
7. Evaluate $y(0.1)$ correct to six places of decimals by Taylor's series method if $y(x)$ satisfies

   $y' = xy + 1$, $y(0) = 1$.
8. Solve $y' = 3x + y^2$, $y(0) = 1$ using Taylor's series method and compute $y(0.1)$. *(JNTUK, 2022)*
9. Using Taylor series method, find $y(0.1)$ correct to 3-decimal places given that

   $dy/dx = x^2 - y^2$, $y(0) = 1$.

10. Apply Picard's method to obtain the third approximation of the solutions of the equations

$$\frac{dy}{dx} = 2y - 2x^2 - 3$$

given that $y = 2$ when $x = 0$ (*Punjab Uni. (MSc), 2018*)

11. Use Picard's method to compute $y$ when $x = 0.4$ for the equation.

$$\frac{dy}{dx} = x^2 + y^2, \; y(0) = 0$$

Approximate $y$ and $z$ by using picards method for the solution of the equations.

12. Given $\frac{dy}{dx} = 2x + 3$, $\frac{dz}{dx} = 3xy + x^2z$ given that $y = 2$ and $z = 0$ when $x = 0$. Approximate $y$ by Picard's method for the solution of the equations. (*Kottayam 2018*)

13. Use Picards method to approximate $y$ when $x = 0.10$ given that $\frac{d^2y}{dx^2} + 2x\frac{dy}{dx} + y = 0$ and $y = 0.5$, $\frac{dy}{dx} = 0.1$. When $x = 0$.

## 32.4 EULER'S METHOD

Consider the equation $\frac{dy}{dx} = f(x, y)$ ...(1)

given that $y(x_0) = y_0$. Its curve of solution through $P(x_0, y_0)$ is shown dotted in Fig. 32.1. Now we have to find the ordinate of any other point $Q$ on this curve.

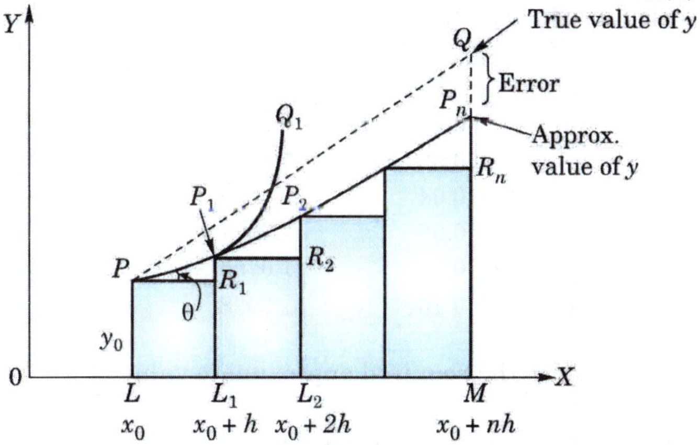

Fig. 32.1

Let us divide $LM$ into $n$ sub-intervals each of width $h$ at $L_1, L_2, \ldots$ so that $h$ is quite small. In the interval $LL_1$, we approximate the curve by the tangent at $P$. If the ordinate through $L_1$ meets this tangent in $P_1(x_0 + h, y_1)$, then

$$y_1 = L_1P_1 = LP + R_1P_1$$

$$= y_0 + PR_1 \tan\theta = y_0 + h\left(\frac{dy}{dx}\right)_P$$

$$= y_0 + h\,f(x_0, y_0)$$

Let $P_1Q_1$ be the curve of solution of (1) through $P_1$ and let its tangent at $P_1$ meet the ordinate through $L_2$ in $P_2(x_0 + 2h, y_2)$. Then

$$y_2 = y_1 + h\,f(x_0 + h, y_1) \quad \text{...(2)}$$

Repeating this process $n$ times, we finally reach an approximation $MP_n$ of $MQ$ given by

$$y_n = y_{n-1} + h\,f(x_0 + \overline{n-1}\,h, y_{n-1})$$

This is *Euler's method* of finding an approximate solution of (1).

**Obs.** In Euler's method, we approximate the curve of solution by the tangent in each interval, *i.e.* by a sequence of short lines. Unless $h$ is small, the error is bound to be quite significant. This sequence of lines may also deviate considerably from the curve of solution. Hence there is a modification of this method which is given in the next section.

**Example 32.7.** *Using Euler's method, find an approximate value of $y$ corresponding to $x = 1$, given that $dy/dx = x + y$ and $y = 1$ when $x = 0$.*

**Solution.** We take $n = 10$ and $h = 0.1$ which is sufficiently small. The various calculations are arranged as follows :

| $x$ | $y$ | $x + y = dy/dx$ | *Old* $y + 0.1\ (dy/dx) =$ *new* $y$ |
|---|---|---|---|
| 0.0 | 1.00 | 1.00 | 1.00 + 0.1(1.00) = 1.10 |
| 0.1 | 1.10 | 1.20 | 1.10 + 0.1(1.20) = 1.22 |
| 0.2 | 1.22 | 1.42 | 1.22 + 0.1(1.42) = 1.36 |
| 0.3 | 1.36 | 1.66 | 1.36 + 0.1(1.66) = 1.53 |
| 0.4 | 1.53 | 1.93 | 1.53 + 0.1(1.93) = 1.72 |
| 0.5 | 1.72 | 2.22 | 1.72 + 0.1(2.22) = 1.94 |
| 0.6 | 1.94 | 2.54 | 1.94 + 0.1(2.54) = 2.19 |
| 0.7 | 2.19 | 2.89 | 2.19 + 0.1(2.89) = 2.48 |
| 0.8 | 2.48 | 3.89 | 2.48 + 0.1(3.89) = 2.81 |
| 0.9 | 2.81 | 3.71 | 2.81 + 0.1(3.71) = 3.18 |
| 1.0 | 3.18 | | |

Thus the required approximate value of $y = 3.18$.

**Obs.** In example 32.1, the true value of $y$ from its exact solution at $x = 1$ is 3.44 whereas by Euler's method $y = 3.18$ and by Picard's method $y = 3.434$. In the above solution, had we chosen $n = 20$, the accuracy would have been considerably increased but at the expense of double the labour of computation. Euler's method is no doubt very simple but cannot be considered as one of the best.

**Example 32.8.** *Given* $\dfrac{dy}{dx} = \dfrac{y - x}{y + x}$ *with initial condition* $y = 1$ *at* $x = 0$ *; find* $y$ *for* $x = 0.1$ *by Euler's method.*

*(Delhi, 2015 ; Anna, 2013)*

**Solution.** We divide the interval (0, 0.1) into five steps *i.e.* we take $n = 5$ and $h = 0.02$. The various calculations are arranged as follows :

| $x$ | $y$ | $(y - x)/(y + x) = dy/dx$ | *Old* $y + 0.02\ (dy/dx) =$ *new* $y$ |
|---|---|---|---|
| 0.00 | 1.0000 | 1.0000 | 1.0000 + 0.02 (1.0000) = 1.0200 |
| 0.02 | 1.0200 | 0.9615 | 1.0200 + 0.02 (.9615) = 1.0392 |
| 0.04 | 1.0392 | 0.926 | 1.0392 + 0.02 (.926) = 1.0577 |
| 0.06 | 1.0577 | 0.893 | 1.0577 + 0.02 (.893) = 1.0756 |
| 0.08 | 1.0756 | 0.862 | 1.0756 + 0.02 (.862) = 1.0928 |
| 0.10 | 1.0928 | | |

Hence the required approximate value of $y = 1.0928$.

## 32.5 MODIFIED EULER'S METHOD

In the Euler's method, the curve of solution in the interval $LL_1$ is approximated by the tangent at $P$ (Fig. 32.1) such that at $P_1$, we have

$$y_1 = y_0 + h\, f(x_0, y_0) \qquad ...(1)$$

Then the slope of the curve of solution through $P_1$ [*i.e.* $(dy/dx)_{P_1} = f(x_0 + h, y_1)$] is computed and the tangent at $P_1$ to $P_1Q_1$ is drawn meeting the ordinate through $L_2$ in $P_2(x_0 + 2h, y_2)$.

Now we find a better approximation $y_1^{(1)}$ of $y(x_0 + h)$ by taking the slope of the curve as the mean of the slopes of the tangents at $P$ and $P_1$, *i.e.*

$$y_1^{(1)} = y_0 + \frac{h}{2}\,[f(x_0, y_0) + f(x_0 + h, y_1)] \qquad ...(2)$$

As the slope of the tangent at $P_1$ is not known, we take $y_1$ as found in Eqn. (1) by Euler's method and insert it on R.H.S. of (2) to obtain the first modified value $y_1^{(1)}$. The Eqn. (1) is therefore, called the *predictor* while Eqn. (2) serves as the *corrector of* $y_1$.

Again the corrector is applied and we find a still better value $y_1^{(2)}$ corresponding to $L_1$ as

$$y_1^{(2)} = y_0 + \frac{h}{2}[f(x_0, y_0) + f(x_0 + h, y_1^{(1)})]$$

We repeat this step, till two consecutive values of $y$ agree. This is then taken as the starting point for the next interval $L_1L_2$.

Once $y_1$ is obtained to desired degree of accuracy, $y$ corresponding to $L_2$ is found from the *predictor*

$$y_2 = y_1 + hf(x_0 + h, y_1)$$

and a better approximation $y_2^{(1)}$ is obtained from the *corrector*

$$y_2^{(1)} = y_1 + \frac{h}{2}[f(x_0 + h, y_1) + f(x_0 + 2h, y_2)].$$

We repeat this step until $y_2$ becomes stationary. Then we proceed to calculate $y_3$ as above and so on.

This is the *modified Euler's method* which is a predictor-corrector method.

**Example 32.9.** *Using modified Euler's method, find an approximate value of y when x = 0.3, given that dy/dx = x + y and y = 1 when x = 0.* (Delhi., 2015 ; V.T.U., 2013 ; Anna, 2012)

**Solution.** Taking $h = 0.1$, the various calculations are arranged as follows :

| $x$ | $x + y = y'$ | *Mean slope* | *Old y + 0.1 (mean slope) = new y* |
|---|---|---|---|
| 0.0 | 0 + 1 | — | 1.00 + 0.1 (1.00) = 1.10 |
| 0.1 | 0.1 + 1.1 | $\frac{1}{2}(1 + 1.2)$ | 1.00 + 0.1 (1.1) = 1.11 |
| 0.1 | 0.1 + 1.11 | $\frac{1}{2}(1 + 1.21)$ | 1.00 + 0.1 (1.105) = 1.1105 |
| 0.1 | 0.1 + 1.1105 | $\frac{1}{2}(1 + 1.2105)$ | 1.00 + 0.1 (1.1052) = 1.1105 |
| 0.1 | 1.2105 | — | 1.1105 + 0.1 (1.2105) = 1.2316 |
| 0.2 | 0.2 + 1.2316 | $\frac{1}{2}(1.2105 + 1.4316)$ | 1.1105 + 0.1 (1.3211) = 1.2426 |
| 0.2 | 0.2 + 1.2426 | $\frac{1}{2}(1.2105 + 1.4426)$ | 1.1105 + 0.1 (1.3266) = 1.2432 |
| 0.2 | 0.2 + 1.2432 | $\frac{1}{2}(1.2105 + 1.4432)$ | 1.1105 + 0.1 (1.3268) = 1.2432 |
| 0.2 | 1.4432 | — | 1.2432 + 0.1 (1.4432) = 1.3875 |
| 0.3 | 0.3 + 1.3875 | $\frac{1}{2}(1.4432 + 1.6875)$ | 1.2432 + 0.1 (1.5654) = 1.3997 |
| 0.3 | 0.3 + 1.3997 | $\frac{1}{2}(1.4432 + 1.6997)$ | 1.2432 + 0.1 (1.5715) = 1.4003 |
| 0.3 | 0.3 + 1.4003 | $\frac{1}{2}(1.4432 + 1.7003)$ | 1.2432 + 0.1 (1.5718) = 1.4004 |
| 0.3 | 0.3 + 1.4004 | $\frac{1}{2}(1.4432 + 1.7004)$ | 1.2432 + 0.1 (1.5718) = 1.4004 |

Hence $y(0.3) = 1.4004$ approximately.

**Obs.** In example 32.6, the approximate value of $y$ for $x = 0.3$ would be 1.53 whereas by modified Euler's method the corresponding value is 1.4004 which is nearer its true value 1.3997, obtained from its exact solution $y = 2e^x - x - 1$ by putting $x = 0.3$.

**Example 32.10.** *Using modified Euler's method, find y(0.2) and y(0.4) given*

$$y' = y + e^x,\ y(0) = 0.$$ (JNTUK, 2022)

**Solution.** We have $y' = y + e^x = f(x, y)$ ; $x = 0, y = 0$ and $h = 0.2$

The various calculations are arranged as under :

To calculate $y(0.2)$ :

| $x$ | $y + e^x = y'$ | *Mean slope* | *Old y + h (mean slope) = new y* |
|---|---|---|---|
| 0.0 | 1 | — | $0 + 0.2\,(1) \quad = 0.2$ |
| 0.2 | $0.2 + e^{0.2} = 1.4214$ | $\frac{1}{2}(1 + 1.4214) = 1.2107$ | $0 + 0.2\,(1.2107) = 0.2421$ |
| 0.2 | $0.2421 + e^{0.2} = 1.4635$ | $\frac{1}{2}(1 + 1.4635) = 1.2317$ | $0 + 0.2\,(1.2317) = 0.2463$ |
| 0.2 | $0.2463 + e^{0.2} = 1.4677$ | $\frac{1}{2}(1 + 1.4677) = 1.2338$ | $0 + 0.2\,(1.2338) = 0.2468$ |
| 0.2 | $0.2468 + e^{0.2} = 1.4682$ | $\frac{1}{2}(1 + 1.4682) = 1.2341$ | $0 + 0.2\,(1.2341) = 0.2468$ |

Since the last two values of $y$ are equal, we take $y(0.2) = 0.2468$.

To calculate $y(0.4)$.

| $x$ | $y + e^x = y'$ | *Mean slope* | *Old y + h (Mean slope) = new y* |
|---|---|---|---|
| 0.2 | $0.2468 + e^{0.2} = 1.4682$ | — | $0.2468 + 0.2\,(1.4682) = 0.5404$ |
| 0.4 | $0.5404 + e^{0.4} = 2.0322$ | $\frac{1}{2}(1.4682 + 2.0322) = 1.7502$ | $0.2468 + 0.2\,(1.7502) = 0.5968$ |
| 0.4 | $0.5968 + e^{0.4} = 2.0887$ | $\frac{1}{2}(1.4682 + 2.0887) = 1.7784$ | $0.2468 + 0.2\,(1.7784) = 0.6025$ |
| 0.4 | $0.6025 + e^{0.4} = 2.0943$ | $\frac{1}{2}(1.4682 + 2.0943) = 1.78125$ | $0.2468 + 0.2\,(1.78125) = 0.6030$ |
| 0.4 | $0.6030 + e^{0.4} = 2.0949$ | $\frac{1}{2}(1.4682 + 2.0949) = 1.7815$ | $0.2468 + 0.2\,(1.7815) = 0.6031$ |
| 0.4 | $0.6031 + e^{0.4} = 2.0949$ | $\frac{1}{2}(1.4682 + 2.0949) = 1.7816$ | $0.2468 + 0.2\,(1.7815) = 0.6031$ |

Since the last two value of $y$ are equal, we take $y(0.4) = 0.6031$.

Hence $y(0.2) = 0.2468$ and $y(0.4) = 0.6031$ approximately.

**Example 32.11.** *Solve the following by Euler's modified method :*

$$\frac{dy}{dx} = \log (x + y),\ y(0) = 2.$$

*at $x = 1.2$ and $1.4$ with $h = 0.2$.* *(V.T.U., 2015 ; Anna, 2012)*

**Solution.** The various calculations are arranged as follows :

| $x$ | $\log (x + y) = y'$ | *Mean slope* | *Old y + 0.2 (mean slope) = new y* |
|---|---|---|---|
| 0.0 | $\log (0 + 2)$ | — | $2 + 0.2\,(0.301) = 2.0602$ |
| 0.2 | $\log (0.2 + 2.0602)$ | $\frac{1}{2}(0.301 + 0.3541)$ | $2 + 0.2\,(0.3276) = 2.0655$ |
| 0.2 | $\log (0.2 + 2.0655)$ | $\frac{1}{2}(0.301 + 0.3552)$ | $2 + 0.2\,(0.3281) = 2.0656$ |
| 0.2 | 0.3552 | — | $2.0656 + 0.2\,(0.3552) = 2.1366$ |
| 0.4 | $\log (0.4 + 2.1366)$ | $\frac{1}{2}(0.3552 + 0.4042)$ | $2.0656 + 0.2\,(0.3797) = 2.1415$ |
| 0.4 | $\log (0.4 + 2.1415)$ | $\frac{1}{2}(0.3552 + 0.4051)$ | $2.0656 + 0.2\,(0.3801) = 2.1416$ |

| | | | |
|---|---|---|---|
| 0.4 | 0.4051 | — | 2.1416 + 0.2 (0.4051) = 2.2226 |
| 0.6 | log (0.6 + 2.2226) | $\frac{1}{2}$(0.4051 + 0.4506) | 2.1416 + 0.2 (0.4279) = 2.2272 |
| 0.6 | log (0.6 + 2.2272) | $\frac{1}{2}$(0.4051 + 0.4514) | 2.1416 + 0.2 (0.4282) = 2.2272 |
| 0.6 | 0.4514 | — | 2.2272 + 0.2 (0.4514) = 2.3175 |
| 0.8 | log (0.8 + 2.3175) | $\frac{1}{2}$(0.4514 + 0.4938) | 2.2272 + 0.2 (0.4726) = 2.3217 |
| 0.8 | log (0.8 + 2.3217) | $\frac{1}{2}$(0.4514 + 0.4943) | 2.2272 + 0.2 (0.4727) = 2.3217 |
| 0.8 | 0.4943 | — | 2.3217 + 0.2 (0.4943) = 2.4206 |
| 1.0 | log (1 + 2.4206) | $\frac{1}{2}$(0.4943 + 0.5341) | 2.3217 + 0.2 (0.5142) = 2.4245 |
| 1.0 | log (1 + 2.4245) | $\frac{1}{2}$(0.4943 + 0.5346) | 2.3217 + 0.2 (0.5144) = 2.4245 |
| 1.0 | 0.5346 | — | 2.4245 + 0.2 (0.5346) = 2.5314 |
| 1.2 | log (1.2 + 2.5314) | $\frac{1}{2}$(0.5346 + 0.5719) | 2.4245 + 0.2 (0.5532) = 2.5351 |
| 1.2 | log (1.2 + 2.5351) | $\frac{1}{2}$(0.5346 + 0.5723) | 2.4245 + 0.2 (0.5534) = 2.5351 |
| 1.2 | 0.5723 | — | 2.5351 + 0.2 (0.5723) = 2.6496 |
| 1.4 | log (1.4 + 2.6496) | $\frac{1}{2}$(0.5723 + 0.6074) | 2.5351 + 0.2 (0.5898) = 2.6531 |
| 1.4 | log (1.4 + 2.6531) | $\frac{1}{2}$(0.5723 + 0.6078) | 2.5351 + 0.2 (0.5900) = 2.6531 |

Hence $y(1.2) = 2.5351$ and $y(1.4) = 2.6531$ approximately.

**Example 32.12.** *Using Euler's modified method, obtain a solution of the equation $dy/dx = x + |\sqrt{y}|$, with initial conditions $y = 1$ at $x = 0$, for the range $0 \le x \le 0.6$ in steps of 0.2.*

**Solution.** The various calculations are arranged as follows :

| $x$ | $x + \|\sqrt{y}\| = y'$ | *Mean slope* | *Old y + .2 (mean slope) = new y* |
|---|---|---|---|
| 0.0 | 0 + 1 = 1 | — | 1 + 0.2 (1) = 1.2 |
| 0.2 | 0.2 + \|√(1.2)\| = 1.2954 | $\frac{1}{2}$(1 + 1.2954) = 1.1477 | 1 + 0.2 (1.1477) = 1.2295 |
| 0.2 | 0.2 + \|√(1.2295)\| = 1.3088 | $\frac{1}{2}$(1 + 1.3088) = 1.1544 | 1 + 0.2 (1.1544) = 1.2309 |
| 0.2 | 0.2 + \|√(1.2309)\| = 1.3094 | $\frac{1}{2}$(1 + 1.3094) = 1.1547 | 1 + 0.2 (1.1547) = 1.2309 |
| 0.2 | 1.3094 | — | 1.2309 + 0.2 (1.3094) = 1.4927 |
| 0.4 | 0.4 + \|√(1.4927)\| = 1.6218 | $\frac{1}{2}$(1.3094 + 1.6218) = 1.4654 | 1.2309 + 0.2 (1.4654) = 1.5240 |
| 0.4 | 0.2 + \|√(1.524)\| = 1.6345 | $\frac{1}{2}$(1.3094 + 1.6345) = 1.4718 | 1.2309 + 0.2 (1.4718) = 1.5253 |
| 0.4 | 0.4 + \|√(1.5253)\| = 1.6350 | $\frac{1}{2}$(1.3094 + 1.6350) = 1.4721 | 1.2309 + 0.2 (1.4721) = 1.5253 |
| 0.4 | 1.6350 | — | 1.5253 + 0.2 (1.635) = 1.8523 |

| | | | |
|---|---|---|---|
| 0.6 | $0.6 + \lvert\sqrt{(1.8523)}\rvert = 1.9610$ | $\frac{1}{2}(1.635 + 1.961) = 1.798$ | $1.5253 + 0.2\ (1.798) = 1.8849$ |
| 0.6 | $0.6 + \lvert\sqrt{(1.8849)}\rvert = 1.9729$ | $\frac{1}{2}(1.635 + 1.9729) = 1.8040$ | $1.5253 + 0.2\ (1.804) = 1.8861$ |
| 0.6 | $0.6 + \lvert\sqrt{(1.8861)}\rvert = 1.9734$ | $\frac{1}{2}(1.635 + 1.9734) = 1.8042$ | $1.5253 + 0.2\ (1.8042) = 1.8861$ |

Hence $y(0.6) = 1.8861$ approximately.

## PROBLEMS 32.2

1. Use a numerical method to obtain $y$ when $x = 0.1$ given that $y'' = xy' - 4y$, $y(0) = 3$, $y'(0) = 3$. Use Euler's modified method to find the value. *(AKTU, 2019)*
2. Using simple Euler's method solve for $y$ at $x = 0.1$ from $dy/dx = x + y + xy$, $y(0) = 1$, taking step size $h = 0.025$.
3. Using Euler's method, find the approximate value of $y$ when $dy/dx = x^2 + y^2$ and $y(0) = 1$ in five steps (*i.e.* $h = 0.2$). *(Anna, 2016)*
4. Use Euler's modified method to solve the equations given below
$$\frac{dy}{dx} = -xy^2, \text{ given } y = 2 \text{ at } x = 0$$
Obtain $y$ at $x = 0.2$ in two steps of 0.1 each *(UKTU, 2018)*
5. Using Euler's modified method to solve $\frac{dy}{dx} = 2 + \sqrt{xy}$, given that $y = 1$ when $x = 1$. Obtain $y$ at $x = 2.0$ in steps of 0.2. *(Annamalai, 2018)*
6. Given that $dy/dx = x^2 + y$ and $y(0) = 1$. Find an approximate value of $y(0.1)$ taking $h = 0.05$ by modified Euler's method. *(V.T.U., 2010)*
7. Using modified Euler's method, find $y(0.1)$ and $y(0.2)$ for the given equation $dy/dx = x^2 + y^2$, given that $y(0) = 1$. *(Anna., 2016)*
8. Given $\frac{dy}{dx} = \frac{y - x}{y + x}$ with boundary conditions $y = 1$ when $x = 0$, find approximately $y$ for $x = 0.1$, by Euler's modified method (5 steps). *(JNTUK, 2022, 2015)*
9. Given that $dy/dx = 2 + \sqrt{(xy)}$ and $y = 1$ when $x = 1$. Find approximate value of $y$ at $x = 2$ in steps of 0.2, using Euler's modified method. *(J.N.T.U., 2015)*
10. Given $y' = x + \sin y$, $y(0) = 1$. Compute $y(0.2)$ and $y(0.4)$ with $h = 0.2$ using Euler's Modfied Method *(JNTUK, 2022)*

## 32.6 RUNGE'S METHOD*

Consider the differential equation,

$$\frac{dy}{dx} = f(x, y),\ y(x_0) = y_0 \qquad \text{...(1)}$$

Clearly the slope of the curve through $P(x_0, y_0)$ is $f(x_0, y_0)$ (Fig. 32.2).

Integrate both sides of Eqn. (1) from $(x_0, y_0)$ to $(x_0 + h, y_0 + k)$, we have

$$\int_{y_0}^{y_0 + k} dy = \int_{x_0}^{x_0 + h} f(x, y)\ dx \qquad \text{...(2)}$$

To evaluate the integral on the right, we take $N$ as the mid-point of $LM$ and find the values of $f(x, y)$ (*i.e.* $dy/dx$) at the points $x_0, x_0 + h/2, x_0 + h$. For this purpose, we first determine the values of $y$ at these points.

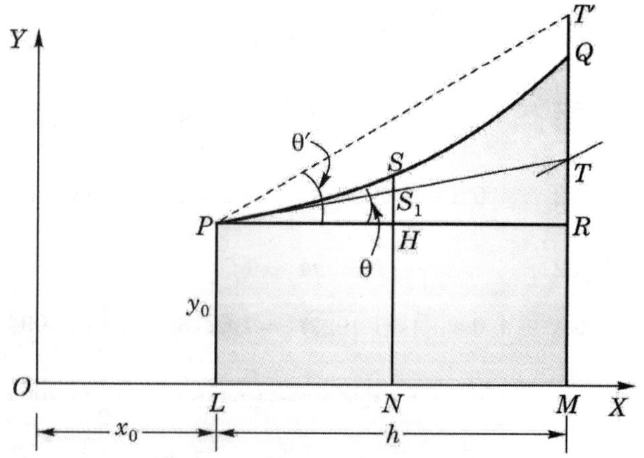


Fig. 32.2

* Called after the German mathematician *Carl Runge* (1856–1927) who was professor at Gottingen.

Let the ordinate through $N$ cut the curve $PQ$ in $S$ and the tangent $PT$ in $S_1$. The value of $y_s$ is given by the point $S_1$.

$\therefore \quad y_s = NS = LP + HS_1 = y_0 + PH \tan\theta$

$$= y_0 + \frac{h}{2}(dy/dx)_P = y_0 + \frac{h}{2} f(x_0, y_0) \quad ...(3)$$

Also $\quad y_T = MT = LP + RT = y_0 + PR \tan\theta = y_0 + hf(x_0, y_0)$.

Now the value of $y_Q$ at $x_0 + h$ is given by the point $T'$ where the line through $P$ drawn with slope at $T(x_0 + h, y_T)$ meets $MQ$.

$\because$ Slope at $T = \tan\theta' = f(x_0 + h, y_T) = f[x_0 + h, y_0 + hf(x_0, y_0)]$

$\therefore \quad y_Q = MR + RT' = y_0 + PT \tan\theta' = y_0 + hf[x_0 + h, y_0 + hf(x_0, y_0)] \quad ...(4)$

Thus the value of $f(x, y)$ at $P = f(x_0, y_0)$,

the value of $f(x, y)$ at $S = f(x_0 + h/2, y_s)$

and the value of $f(x, y)$ at $Q = f(x_0 + h, y_Q)$

where $y_S$ and $y_Q$ are given by Eqns. (3) and (4).

Hence from Eqn. (2), we obtain

$$k = \int_{x_0}^{x_0+h} f(x, y)\,dx = \frac{h}{6}[f_P + 4f_S + f_Q] \quad \text{[By Simpsons' rule]}$$

$$= \frac{h}{6}[f(x_0, y_0) + 4f(x_0 + h/2, y_S) + f(x_0 + h, y_Q)] \quad ...(5)$$

which gives a sufficiently accurate value of $k$ and also of $y = y_0 + k$.

The repeated application of Eqn. (5) gives the values of $y$ for equispaced points.

**Working rule** *to solve Eqn. (1) by Runge's method :*

*Calculate successively*

$$k_1 = hf(x_0, y_0)$$

$$k_2 = hf\left(x_0 + \tfrac{1}{2}h, y_0 + \tfrac{1}{2}k_1\right)$$

$$k' = hf(x_0 + h, y_0 + k_1)$$

and

$$k_3 = hf(x_0 + h, y_0 + k')$$

*Finally compute,* $\quad k = \frac{1}{6}(k_1 + 4k_2 + k_3)$.

*(Note that $k$ is the weighted mean of $k_1$, $k_2$ and $k_3$)*

**Example 32.13.** *Apply Runge's method to find an approximate value of $y$ when $x = 0.2$, given that $dy/dx = x + y$ and $y = 1$ when $x = 0$.*

**Solution.** Here we have $x_0 = 0, y_0 = 1, h = 0.2, f(x_0, y_0) = 1$

$\therefore \quad k_1 = hf(x_0, y_0) = 0.2\,(1) \quad = 0.200$

$k_2 = hf\left(x_0 + \tfrac{1}{2}h, y_0 + \tfrac{1}{2}k_1\right) = 0.2\,f(0.1, 1.1) \quad = 0.240$

$k' = hf(x_0 + h, y_0 + k_1) = 0.2\,f(0.2, 1.2) \quad = 0.280$

and $\quad k_3 = hf(x_0 + h, y_0 + k') = 0.2\,f(0.1, 1.28) \quad = 0.296$

$\therefore \quad k = \tfrac{1}{6}(k_1 + 4k_2 + k_3)$

$= \tfrac{1}{6}(0.200 + 0.960 + 0.296) \quad = 0.2426$

Hence the required approximate value of $y$ is 1.2426.

## 32.7 RUNGE-KUTTA METHOD*

The Taylor's series method of solving differential equations numerically is restricted by the labour involved in finding the higher order derivatives. However there is a class of methods known as Runge-Kutta methods which do not require the calculations of higher order derivatives. These methods agree with Taylor's series solution upto the terms in $h^r$, where $r$ differs from method to method and is called the *order of that method. Euler's method, Modified Euler's method and Runge's method are the Runge-Kutta methods of the first, second and third order respectively.*

*The fourth-order Runge-Kutta method is most commonly used and is often referred to as 'Runge-Kutta method' only.*

**Working rule** *for finding the increment k of y corresponding to an increment h of x by Runge-Kutta method from*

$$\frac{dy}{dx} = f(x, y),\ y(x_0) = y_0 \text{ is as follows :}$$

*Calculate successively*

$$k_1 = hf(x_0, y_0)$$

$$k_2 = hf\left(x_0 + \tfrac{1}{2}h,\ y_0 + \tfrac{1}{2}k_1\right)$$

$$k_3 = hf\left(x_0 + \tfrac{1}{2}h,\ y_0 + \tfrac{1}{2}k_2\right)$$

and

$$k_4 = hf(x_0 + h,\ y_0 + k_3)$$

*Finally compute*

$$k = \tfrac{1}{6}(k_1 + 2k_2 + 2k_3 + k_4)$$

which gives the required approximate value $y_1 = y_0 + k$.

*(Note that k is the weighted mean of $k_1$, $k_2$, $k_3$ and $k_4$)*

**Obs.** One of the advantages of these methods is that the operation is identical whether the differential equation is linear or non-linear.

**Example 32.14.** *Apply Runge-Kutta fourth order method, to find an approximate value of y when x = 0.2, given that dy / dx = x + y and y = 1 when x = 0.* *(Delhi, 2015 ; V.T.U., 2013)*

**Solution.** Here $x_0 = 0, y_0 = 1, h = 0.2, f(x_0, y_0) = 1$

$\therefore$ $k_1 = hf(x_0, y_0) = 0.2 \times 1 = 0.2000$

$k_2 = hf\left(x_0 + \tfrac{1}{2}h,\ y_0 + \tfrac{1}{2}k_1\right) = 0.2 \times f(0.1, 1.1) = 0.2400$

$k_3 = hf\left(x_0 + \tfrac{1}{2}h,\ y_0 + \tfrac{1}{2}k_2\right) = 0.2 \times f(0.1, 1.12) = 0.2440$

and $k_4 = hf(x_0 + h,\ y_0 + k_3) = 0.2 \times f(0.2, 1.244) = 0.2888$

$\therefore$ $k = \tfrac{1}{6}(k_1 + 2k_2 + 2k_3 + k_4)$

$= \tfrac{1}{6}(0.2000 + 0.4800 + 0.4880 + 0.2888) = \tfrac{1}{6} \times (1.4568) = 0.2468.$

Hence the required approximate value of $y$ is 1.2428.

**Example 32.15.** *Using Runge-Kutta method of fourth order, solve* $\dfrac{dy}{dx} = \dfrac{y^2 - x^2}{y^2 + x^2}$ *with y(0) = 1 at x = 0.2, 0.4.* *(Anna, 2015 ; V.T.U., 2012 S ; U.P.T.U., 2010)*

**Solution.** We have $f(x, y) = \dfrac{y^2 - x^2}{y^2 + x^2}$

*To find y(0.2) :*

Here $x_0 = 0, y_0 = 1, h = 0.2$

* See footnote p. 1154. Named after *Wilhelm Kutta* (1867—1944).

$$k_1 = h\,f(x_0, y_0) = 0.2\,f(0, 1) \qquad = 0.2000$$

$$k_2 = hf\left(x_0 + \frac{1}{2}h,\ y_0 + \frac{1}{2}k_1\right) = 0.2\,f(0.1, 1.1) \qquad = 0.19672$$

$$k_3 = hf\left(x_0 + \frac{1}{2}h,\ y_0 + \frac{1}{2}k_2\right) = 0.2\,f(0.1, 1.09836) \qquad = 0.1967$$

$$k_4 = hf(x_0 + h, y_0 + k_3) = 0.2\,f(0.2, 1.1967) \qquad = 0.1891$$

$$k = \frac{1}{6}(k_1 + 2k_2 + 2k_3 + k_4) = \frac{1}{6}[0.2 + 2(0.19672) + 2(0.1967) + 0.1891]$$

$$= 0.19599$$

Hence $\qquad y(0.2) = y_0 + k = 1.196.$

*To find y(0.4) :*

Here $\qquad x_1 = 0.2, y_1 = 1.196, h = 0.2$

$$k_1 = h\,f(x_1, y_1) \qquad = 0.1891$$

$$k_2 = hf\left(x_1 + \frac{1}{2}h,\ y_1 + \frac{1}{2}k_1\right) = 0.2\,f(0.3, 1.2906) \qquad = 0.1795$$

$$k_3 = hf\left(x_1 + \frac{1}{2}h,\ y_1 + \frac{1}{2}k_2\right) = 0.2\,f(0.3, 1.2858) \qquad = 0.1793$$

$$k_4 = hf(x_1 + h, y_1 + k_3) = 0.2\,f(0.4, 1.3753) \qquad = 0.1688$$

$$k = \frac{1}{6}(k_1 + 2k_2 + 2k_3 + k_4)$$

$$= \frac{1}{6}[0.1891 + 2(0.1795) + 2(0.1793) + 0.1688] \qquad = 0.1792$$

Hence $\qquad y(0.4) = y_1 + k = 1.196 + 0.1792 = 1.3752.$

**Example 32.16.** *Apply Runge-Kutta method to find an approximate value of y for x = 0.2 in steps of 0.1, if $dy/dx = x + y^2$, given that y = 1, where x = 0.*

**Solution.** Here we take $h = 0.1$ and carry out the calculations in two steps.

**Step I.** $x_0 = 0, y_0 = 1, h = 0.1$

$$\therefore \quad k_1 = hf(x_0, y_0) = 0.1\,f(0, 1) \qquad = 0.1000$$

$$k_2 = hf\left(x_0 + \tfrac{1}{2}h,\ y_0 + \tfrac{1}{2}k_1\right) = 0.1\,f(0.05, 1.1) \qquad = 0.1152$$

$$k_3 = hf\left(x_0 + \tfrac{1}{2}h,\ y_0 + \tfrac{1}{2}k_2\right) = 0.1\,f(0.05, 1.1152) \qquad = 0.1168$$

$$k_4 = hf(x_0 + h, y_0 + k_3) = 0.1\,f(0.1, 1.1168) \qquad = 0.1347$$

$$\therefore \quad k = \tfrac{1}{6}(k_1 + 2k_2 + 2k_3 + k_4)$$

$$= \tfrac{1}{6}(0.1000 + 0.2304 + 0.2336 + 0.1347) \qquad = 0.1165$$

giving $\qquad y(0.1) = y_0 + k = 1.1165.$

**Step II.** $x_1 = x_0 + h = 0.1, y_1 = 1.1165, h = 0.1$

$$\therefore \quad k_1 = hf(x_1, y_1) = 0.1\,f(0.1, 1.1165) \qquad = 0.1347$$

$$k_2 = hf\left(x_1 + \tfrac{1}{2}h,\ y_1 + \tfrac{1}{2}k_1\right) = 0.1\,f(0.15, 1.1838) \qquad = 0.1551$$

$$k_3 = hf\left(x_1 + \tfrac{1}{2}h,\ y_1 + \tfrac{1}{2}k_2\right) = 0.1\,f(0.15, 1.194) \qquad = 0.1576$$

$$k_4 = hf(x_1 + h, y_2 + k_3) = 0.1\,f(0.2, 1.1576) \qquad = 0.1823$$

$$\therefore \quad k = \tfrac{1}{6}(k_1 + 2k_2 + 2k_3 + k_4) \qquad = 0.1571$$

Hence $\qquad y(0.2) = y_1 + k = 1.2736.$

**Example 32.17.** *Use the following second order Runge-Kutta method described by*

$$y_{n+1} = y_n + \frac{1}{3}(2k_1 + k_2)$$

*where* $$k_1 = hf(x_n, y_n) \text{ and}$$

$$k_2 = hf\left(x_n + \frac{3}{2}h, y_n + \frac{3}{2}k_1\right)$$

*and find the numerical solution of the initial value problem described as*

$$\frac{dy}{dx} = \frac{y+x}{y+x}, y(0) = 1 \text{ at } x = 0.4 \text{ and taking } h = 0.2$$

*(JNTUK, 2022; Poona, 2018)*

**Solution:** Here, $f(x, y) = \frac{y+x}{y-x}$, $h = 0.2, x_0 = 0, y_0 = 1$

we $$k_1 = hf(x_0 y_0) = 0.2\,\frac{1+0}{1-0} = 0.2$$

$$k_2 = hf[x_0 + 0.3, y_0 + (1.5)(0.2)] = hf[0.3, 1.3] = 0.2\,\frac{1.3+0.3}{1.3-0.3} = 0.32$$

Now using the given *Runge – Kutta* method

$$y_{(0.2)} = y_1 = 1 + \frac{1}{3}(0.4 + 0.32) = 1.24$$

Now taking $$x_1 = 0.2, \quad y_1 = 1.24$$

$$K_1 = hf(x_1, y_1) = 0.2\,\frac{1.24+0.2}{1.24-0.2} = 0.2765$$

$$K_2 = hf\left(11 + \frac{3}{2}h, y_1 + \frac{3}{2}k_1\right) = h_f(0.5, 1.6554)$$

$$= 0.2\,\frac{1.6554+0.5}{1.6554-0.5} = 0.3731.$$

Again using the *Runge – Kutta* method $y(0.4) = y_2 = 1.24 + \frac{1}{3}[2(0.2769 + 0.3731] = 1.54897.$

## PROBLEMS 32.3

1. Use Runge's method to approximate $y$ when $x = 1.1$, given that $y = 1.2$, when $x = 1$ and $dy/dx = 3x + y^2$. *(J.N.T.U., 2015)*

2. Using Runge-Kutta method of order 4, find $y(0.2)$ given that $dy/dx = 3x + \frac{1}{2}y$, $y(0) = 1$, taking $h = 0.1$. *(V.T.U., 2016)*

3. Find the value of $y(1.1)$ using Runga-Kutta method of $4^{\text{th}}$ order for the given equation $\frac{dy}{dx} = y^2 + xy$ ; $y(1) = 1$. *(Anna., 2016)*

4. Use Runge-Kutta method to find $y$ when $x = 1.2$ in steps of 0.1, given that : $dy/dx = x^2 + y^2$ and $y(1) = 1.5$. *(JNTUK, 2022; Delhi., 2015)*

5. Using by Runge-Kutta method of fourth order so solve the following equation. $\frac{dy}{dx} = \frac{y}{x}$, $y(1)$, obtain $y$ when $x = 1.2$ in step of 0.1. *(KTU, 2018)*

6. Applying the Runge-Kutta method, solve the given equation $\frac{dy}{dx} = x + y^2$, $y(0) = 1$, find $y(0.\,2)$. *(Marathwada, 2018)*

7. Applying Runge-Kutta method solve the given equation

$$\frac{dy}{dx} = 3e^{-x} - 0.4y,$$

$$y(0) = 5, \text{ and } h1.5$$

Find the value of $y(3.0)$ (*Nagpur*, 2018)

8. Find by Runge-Kutta method an approximate value of $y$ for $x = 0.6$, given that $y = 0.41$ when $x = 0.4$ and $dy/dx = \sqrt{(x + y)}$. (*U.P.T.U., 2012*)

9. Using Runge-Kutta method of order 4, find $y(0.2)$ for the equation $\frac{dy}{dx} = \frac{y - x}{y + x}$, $y(0) = 1$. Take $h = 0.2$. (*V.T.U., 2011 S*)

10. Given that $dy/dx = (y^2 - 2x)/(y^2 + x)$ and $y = 1$ at $x = 0$ ; find $y$ for $x = 0.1, 0.2, 0.3, 0.4$ and $0.5$.

## 32.8 PREDICTOR-CORRECTOR METHODS

If $x_{i-1}$ and $x_i$ be two consecutive mesh points, we have $x_i = x_{i-1} + h$. In the Euler's method (§ 32.4), we have

$$y_i = y_{i-1} + hf(x_0 + \overline{i-1}\,h, y_{i-1}) \; ; i = 1, 2, 3, \ldots \qquad \ldots(1)$$

The modified Euler's method (§ 32.5), gives

$$y_i = y_{i-1} + \frac{h}{2}[f(x_{i-1}, y_{i-1}) + f(x_i, y_i)] \qquad \ldots(2)$$

The value of $y_i$ is first estimated by using Eqn. (1), then this value is inserted on the right side of Eqn. (2), giving a better approximation of $y_i$. This value of $y_i$ is again substituted in Eqn. (2) to find a still better approximation of $y_i$. This step is repeated till two consecutive values of $y_i$ agree. *This technique of refining an initially crude estimate of $y_i$ by means of a more accurate formula is known as* **predictor-corrector method.** The Eqn. (1) is therefore called the *predictor* while Eqn. (2) serves as a *corrector* of $y_i$.

In the methods so far explained, to solve a differential equation over an interval $(x_i, x_{i+1})$ only the value of $y$ at the beginning of the interval was required. In the *predictor-corrector* methods, four prior values are required for finding the value of $y$ at $x_{i+1}$. A predictor formula is used to predict the value of $y$ at $x_{i+1}$ and then a corrector formula is applied to improve this value.

We now describe two such methods, namely : Milne's method and Adams-Bashforth method.

## 32.9 MILNE'S METHOD

*Given $dy/dx = f(x, y)$ and $y = y_0, x = x_0$ ; to find an approximate value of $y$ for $x = x_0 + nh$ by Milne's method, we proceed as follows :*

The value $y_0 = y(x_0)$ being given, we compute

$$y_1 = y(x_0 + h), y_2 = y(x_0 + 2h), y_3 = y(x_0 + 3h),$$

by Picard's or Taylor's series method.

Next we calculate,

$$f_0 = f(x_0, y_0), f_1 = f(x_0 + h, y_1), f_2 = f(x_0 + 2h, y_2), f_3 = f(x_0 + 3h, y_3)$$

Then to find $y_4 = y(x_0 + 4h)$, we substitute Newton's forward interpolation formula

$$f(x, y) = f_0 + n\Delta f_0 + \frac{n(n-1)}{2}\Delta^2 f_0 + \frac{n(n-1)(n-2)}{6}\Delta^3 f_0 + \ldots$$

in the relation $y_4 = y_0 + \int_{x_0}^{x_0+4h} f(x, y)\, dx$

$$\therefore \quad y_4 = y_0 + \int_{x_0}^{x_0+4h}\left(f_0 + n\Delta f_0 + \frac{n(n-1)}{2}\Delta^2 f_0 + \ldots\right)dx \qquad [\text{Put } x = x_0 + nh, dx = hdn]$$

$$= y_0 + h\int_0^4\left(f_0 + n\Delta f_0 + \frac{n(n-1)}{2}\Delta^2 f_0 + \ldots\right)dn$$

$$= y_0 + h\left(4f_0 + 8\Delta f_0 + \frac{20}{3}\Delta^2 f_0 + \frac{8}{3}\Delta^3 f_0 + \ldots\right)$$

Neglecting fourth and higher order differences and expressing $\Delta f_0$, $\Delta^2 f_0$ and $\Delta^3 f_0$ in terms of the function values, we get

$$y_4^{(p)} = y_0 + \frac{4h}{3}(2f_1 - f_2 + 2f_3) \text{ which is called a } predictor.$$

Having found $y_4$, we obtain a first approximation to $f_4 = f(x_0 + 4h, y_4)$.

Then a better value of $y_4$ is found by Simpson's rule as

$$y_4^{(c)} = y_2 + \frac{h}{3}(f_2 + 4f_3 + f_4) \text{ which is called a } corrector.$$

Then an improved value of $f_4$ is computed and again the corrector is applied to find a still better value of $y_4$. We repeat this step until $y_4$ remains unchanged.

Once $y_4$ and $f_4$ are obtained to desired degree of accuracy, $y_5 = y(x_0 + 5h)$ is found from the *predictor* as

$$y_5^{(p)} = y_1 + \frac{4h}{3}(2f_2 - f_3 + 2f_4)$$

and $f_5 = f(x_0 + 5h, y_5)$ is calculated. Then a better approximation to the value of $y_5$ is obtained from the *corrector* as

$$y_5^{(c)} = y_3 + \frac{h}{3}(f_3 + 4f_4 + f_5).$$

We repeat this step till $y_5$ becomes stationary and we, then proceed to calculate $y_6$ as before.

This is *Milne's predictor-corrector method.* To ensure greater accuracy, we must first improve the accuracy of the starting values and then sub-divide the intervals.

**Example 32.18.** *Apply Milne's method, to find a solution of the differential equation $y' = x - y^2$ in the range $0 \le x \le 1$ for the boundary conditions $y = 0$ at $x = 0$.*

**Solution.** Using Picard's method, we have

$$y = y(0) + \int_0^x f(x, y)\, dx, \text{ where } f(x, y) = x - y^2.$$

To get the first approximation, we put $y = 0$ in $f(x, y)$,

giving $$y_1 = 0 + \int_0^x x\, dx = \frac{x^2}{2}$$

To find the second approximation, we put $y = x^2/2$ in $f(x, y)$,

giving $$y_2 = \int_0^x \left(x - \frac{x^4}{4}\right) dx = \frac{x^2}{2} - \frac{x^5}{20}$$

Similarly, the third approximation is

$$y_3 = \int_0^x \left[x - \left(\frac{x^2}{2} - \frac{x^5}{20}\right)^2\right] dx = \frac{x^2}{2} - \frac{x^5}{20} + \frac{x^8}{160} - \frac{x^{11}}{4400} \quad ...(i)$$

Now let us determine the starting values of the Milne's method from (*i*), by choosing $h = 0.2$.

$\therefore$ $x_0 = 0.0$, $y_0 = 0.0000$, $f_0 = 0.0000$

$x_1 = 0.2$, $y_1 = 0.020$, $f_1 = 0.1996$

$x_2 = 0.4$, $y_2 = 0.0795$, $f_2 = 0.3937$

$x_3 = 0.6$, $y_3 = 0.1762$, $f_3 = 0.5689$

Using the *predictor*, $$y_4^{(p)} = y_0 \frac{4h}{3}(2f_1 - f_2 + 2f_3)$$

$x = 0.8$, $y_4^{(p)} = 0.3049$, $f_4 = 0.7070$

and the *corrector*, $$y_4^{(c)} = y_2 + \frac{h}{3}(f_2 + 4f_3 + f_4), \text{ yields}$$

$$y_4^{(c)} = 0.3046, \quad f_4 = 0.7072 \quad ...(ii)$$

Again using the *corrector*, $y_4^{(c)} = 0.3046$, which is same as in Eqn. (*ii*)

Now using the *predictor*, $y_5^{(p)} = y_1 + \frac{4h}{3}(2f_2 - f_3 + 2f_4)$,

$x = 1.0, \qquad y_5^{(p)} = 0.4554, \qquad f_5 = 0.7926$

and the *corrector*, $y_5^{(c)} = y_3 + \frac{h}{3}(f_3 + 4f_4 + f_5)$, gives

$y_5^{(c)} = 0.4555, \qquad f_5 = 0.7925$

Again using the *corrector*,

$y_5^{(c)} = 0.4555$, a value which is the same as before.

Hence, $y(1) = 0.4555$.

**Example 32.19.** *Given* $y' = x(x^2 + y^2)\, e^{-x}$, $y(0) = 1$, *find y at x = 0.1, 0.2 and 0.3 by Taylor's series method and compute y(0.4) by Milne's method.*

**Solution.** Given $y(0) = 1$ and $h = 0.1$

We have $y'(x) = x\,(x^2 + y^2)\, e^{-x}$; $\qquad y'(0) = 0$

$$y''(x) = [(x^3 + xy^2)(-e^{-x}) + 3x^2 + y^2 + x(2y)\,y']\, e^{-x}$$

$$= e^{-x}\,[-x^3 - xy^2 + 3x^2 + y^2 + 2xyy'] \,; \qquad y''(0) = 1$$

$$y'''(x) = -e^{-x}\,[-x^3 - xy^2 + 3x^2 + y^2 + 2xyy' + 3x^2 + y^2 + 2xyy' - 6x - 2yy' - 2xy'^2 - 2xyy'] \qquad y'''(0) = -2$$

Substitute these values in the Taylor's series,

$$y(x) = y(0) + \frac{x}{1!}y'(0) + \frac{x^2}{2!}y'(0) + \frac{x^3}{3!}y'''(0) + \ldots$$

$$y(0.1) = 1 + (0.1)(0) + \frac{1}{2}(0.1)^2\,(1) + \frac{1}{6}(0.1)^3\,(-2) + \ldots$$

$$= 1 + 0.005 - 0.0003 = 1.0047 \quad \textit{i.e.,} \quad 1.005$$

Now taking $x = 0.1,\ y(0.1) = 1.005,\ h = 0.1$

$$y'(0.1) = 0.092,\ y''(0.1) = 0.849,\ y'''(0.1) = -1.247$$

Substituting these values in the Taylor's series about $x = 0.1$,

$$y(0.2) = y(0.1) + \frac{0.1}{1!}y'(0.1) + \frac{(0.1)^2}{2!}y''(0.1) + \frac{(0.1)^3}{3!}y'''(0.1) + \ldots$$

$$= 1.005 + (0.1)(0.092) + \frac{(0.1)^2}{2}(0.849) + \frac{(0.1)^3}{3}(-1.247) + \ldots$$

$$= 1.018$$

Now taking $x = 0.2,\ y(0.2) = 1.018,\ h = 0.1$

$$y'(0.2) = 0.176,\ y''(0.2) = 0.77,\ y'''(0.2) = 0.819$$

Substituting these values in the Taylor's series

$$y(0.3) = y(0.2) + \frac{0.1}{1!}y'(0.2) + \frac{(0.1)^2}{2!}y''(0.2) + \frac{(0.1)^3}{3!}y'''(0.2) + \ldots$$

$$= 1.018 + 0.0176 + 0.0039 + 0.0001 = 1.04$$

Thus the starting values of the Milne's method with $h = 0.1$ are

| | | |
|---|---|---|
| $x_0 = 0.0$ | $y_0 = 1$ | $f_0 = y'_0 = 0$ |
| $x_1 = 0.1$ | $y_1 = 1.005$ | $f_1 = 0.092$ |
| $x_2 = 0.2$ | $y_2 = 1.018$ | $f_2 = 0.176$ |
| $x_3 = 0.3$ | $y_3 = 1.04$ | $f_3 = 0.26$ |

Using the predictor, $y_4^{(p)} = y_0 + \frac{4h}{3}(2f_1 - f_2 + 2f_3)$

$$= 1 + \frac{4(0.1)}{3}[2(0.092) - (0.176) + 2(0.26)] = 1.09$$

$\therefore$ $x = 0.4$ $\quad y_4^{(p)} = 1.09 \quad f_4 = y'(0.4) = 0.362$

Using the corrector, $y_4^{(c)} = y_2 + \frac{h}{3}(f_2 + 4f_3 + f_4)$

$$\therefore \quad y_4^{(c)} = 0.018 + \frac{0.1}{3}(0.176 + 4(0.26) + 0.362) = 1.071$$

Hence $\quad y(0.4) = 1.071.$

**Example 32.20.** *Using Runge-Kutta method of order 4, find y for x = 0.1, 0.2, 0.3 given that $dy/dx = xy + y^2$, $y(0) = 1$. Continue the solution at x = 0.4 using Milne's method.* (V.T.U., 2012)

**Solution.** We have $\quad f(x, y) = xy + y^2.$

*To find y(0.1) :*

Here $\quad x_0 = 0, y_0 = 1, h = 0.1.$

$$\therefore \quad k_1 = h\,f(x_0, y_0) = (0.1)\,f(0.1) \qquad = 0.1000$$

$$k_2 = hf\left(x_0 + \frac{1}{2}h, y_0 + \frac{1}{2}k_1\right) = (0.1)\,f(0.05, 1.05) \qquad = 0.1155$$

$$k_3 = hf\left(x_0 + \frac{1}{2}h, y_0 + \frac{1}{2}k_2\right) = (0.1)\,f(0.05, 1.0577) \qquad = 0.1172$$

$$k_4 = hf(x_0 + h, y_0 + k_3) = (0.1)\,f(0.1, 1.1172) \qquad = 0.13598$$

$$k = \frac{1}{6}(k_1 + 2k_2 + 2k_3 + k_4)$$

$$= \frac{1}{6}(0.1 + 0.231 + 0.2348 + 0.13598) \qquad = 0.11687$$

Thus $y(0.1) = y_1 = y_0 + k = 1.1169.$

*To find y(0.2) :*

Here $x_1 = 0.1, y_1 = 1.1169, h = 0.1.$

$$k_1 = h\,f(x_1, y_1) = (0.1)\,f(0.1, 1.1169) \qquad = 0.1359$$

$$k_2 = hf\left(x_1 + \frac{1}{2}h, y_1 + \frac{1}{2}k_1\right) = (0.1)\,f(0.15, 1.1848) \qquad = 0.1581$$

$$k_3 = hf\left(x_1 + \frac{1}{2}h, y_1 + \frac{1}{2}k_2\right) = (0.1)\,f(0.15, 1.1959) \qquad = 0.1609$$

$$k_4 = hf(x_1 + h, y_1 + k_3) = (0.1)\,f(0.2, 1.2778) \qquad = 0.1888$$

$$k = \frac{1}{6}(k_1 + 2k_2 + 2k_3 + k_4) \qquad = 0.1605$$

Thus $y(0.2) = y_2 = y_1 + k = 1.2773.$

*To find y(0.3) :*

Here $x_2 = 0.2, y_2 = 1.2773, h = 0.1.$

$$k_1 = hf(x_2, y_2) = (0.1)\,f(0.2, 1.2773) \qquad = 0.1887$$

$$k_2 = hf\left(x_2 + \frac{1}{2}h, y_2 + \frac{1}{2}k_1\right) = (0.1)\,f(0.25, 1.3716) \qquad = 0.2224$$

$$k_3 = hf\left(x_2 + \frac{1}{2}h, y_2 + \frac{1}{2}k_2\right) = (0.1)\,f(0.25, 1.3885) \qquad = 0.2275$$

$$k_4 = hf(x_2 + h, y_2 + k_3) = (0.1)\,f(0.3, 1.5048) \qquad = 0.2716$$

$$k = \frac{1}{6}(k_1 + 2k_2 + 2k_3 + k_4) \qquad = 0.2267$$

Thus $y(0.3) = y_3 = y_2 + k = 1.504.$

Now the starting values of the Milne's method are :

| | | |
|---|---|---|
| $x_0 = 0.0$ | $y_0 = 1.0000$ | $f_0 = 1.0000$ |
| $x_1 = 0.1$ | $y_1 = 1.1169$ | $f_1 = 1.3591$ |
| $x_2 = 0.2$ | $y_2 = 1.2773$ | $f_2 = 1.8869$ |
| $x_3 = 0.3$ | $y_3 = 1.5049$ | $f_3 = 2.7132$ |

Using the *predictor*,

$$y_4^{(p)} = y_0 + \frac{4h}{3}(2f_1 - f_2 + 2f_3)$$

$$x_4 = 0.4 \qquad y_4^{(p)} = 1.8344 \qquad f_4 = 4.0988$$

and the *corrector*,

$$y_4^{(c)} = y_2 + \frac{h}{3}(f_2 + 4f_3 + f_4) \text{ yields}$$

$$y_4^{(c)} = 1.2773 + \frac{0.1}{3}[1.8869 + 4(2.7132) + 4.098]$$

$$= 1.8386 \qquad f_4 = 4.1159$$

Again using the *corrector,*

$$y_4^{(c)} = 1.2773 + \frac{0.1}{3}[1.8869 + 4(2.7132) + 4.1159]$$

$$= 1.8391 \qquad f_4 = 4.1182 \qquad ...(i)$$

Again using the *corrector*

$$y_4^{(c)} = 1.2773 + \frac{0.1}{3}[1.8869 + 4(2.7132) + 4.1182]$$

$= 1.8392$ which is same as (*i*).

Hence $y(0.4) = 1.8392$.

## PROBLEMS 32.4

1. Given $\frac{dy}{dx} = x^3 + y$, $y(0) = 2$. The value of $y(0.2) = 2.073$, $y(0.4) = 2.452$, and $y(0.6) = 3.023$ are got by R.K. Method of 4th order. Find $y(0.8)$ by Milne's predictor-corrector method taking $h = 0.2$. *(Anna, 2016)*
2. Given $2\, dy/dx = (1 + x^2)y^2$ and $y(0) = 1$, $y(0.1) = 1.06$, $y(0.2) = 1.12$, $y(0.3) = 1.21$, evaluate $y(0.4)$ by Milne's predictor-corrector method. *(V.T.U., 2011 S)*
3. From the data given below, find $y$ at $x = 1.4$, using Milne's predictor-corrector formula :

$$\frac{dy}{dx} = x^2 + \frac{y}{2}$$

| $x$ : | 1 | 1.1 | 1.2 | 1.3 |
|---|---|---|---|---|
| $y$ : | 2 | 2.2156 | 2.4549 | 2.7514 |

4. Using Milne's method, find $y(4.5)$ given $5xy' + y^2 - 2 = 0$ given $y(4) = 1$, $y(4.1) = 1.0049$, $y(4.2) = 1.0097$, $y(4.3) = 1.0143$, $y(4.4) = 1.0187$. *(Anna, 2015)*
5. If $\frac{dy}{dx} = 2e^x - y$, $y(0) = 2$, $y(0.1) = 2.010$, $y(0.2) = 2.04$ and $y(0.3) = 2.09$ ; find $y(0.4)$ using Milne's predictor-corrector method. *(V.T.U., 2010)*
6. Using Runge-Kutta method, calculate $y(0.1)$, $y(0.2)$, and $y(0.3)$ given that $\frac{dy}{dx} - \frac{2xy}{1+x^2} = 1$, $y(0) = 0$. Taking these values as starting values, find $y(0.4)$ by Milne's method.

## 32.10 ADAMS-BASHFORTH METHOD

Given $\frac{dy}{dx} = f(x, y)$ and $y_0 = y(x_0)$, we compute

$$y_{-1} = y(x_0 - h),\ y_{-2} = y(x_0 - 2h),\ y_{-3} = y(x_0 - 3h)$$

by Taylor's series of Euler's method or Runge-Kutta method.

Next we calculate $f_{-1} = f(x_0 - h, y_{-1})$, $f_{-2} = f(x_0 - 2h, y_{-2})$, $f_{-3} = f(x_0 - 3h, y_{-3})$.

Then to find $y_1$, we substitute Newton's backward interpolation formula

$$f(x, y) = f_0 + n\,\nabla f_0 + \frac{n(n+1)}{2}\nabla^2 f_0 + \frac{n(n+1)(n+2)}{6}\nabla^3 f_0 + \cdots$$

in $$y_1 = y_0 + \int_{x_0}^{x_0+h} f(x, y)\,dx \qquad \text{...(1)}$$

$$\therefore \quad y_1 = y_0 + \int_{x_0}^{x_1}\left(f_0 + n\nabla f_0 + \frac{n(n+1)}{2}\nabla^2 f_0 + \cdots\right)dx \qquad [\text{Put } x = x_0 + nh,\ dx = h\,dn]$$

$$= y_0 + h\int_0^1\left(f_0 + n\nabla f_0 + \frac{n(n+1)}{2}\nabla^2 f_0 + \cdots\right)dn$$

$$= y_0 + h\left(f_0 + \frac{1}{2}\nabla f_0 + \frac{5}{12}\nabla^2 f_0 + \frac{3}{8}\nabla^3 f_0 + \cdots\right)$$

Neglecting fourth and higher order differences and expressing $\nabla f_0, \nabla^2 f_0$ and $\nabla^3 f_0$ in terms of function values, we get

$$y_1^{(p)} = y_0 + \frac{h}{24}(55f_0 - 59f_{-1} + 37f_{-2} - 9f_{-3}) \qquad \text{...(2)}$$

This is called *Adams-Bashforth predictor formula.*

Having found $y_1$, we find $f_1 = f(x_0 + h_1, y_1)$.

Then to find a better value of $y_1$, we derive a *corrector formula* by substituting Newton's backward formula at $f_1$ *i.e.,*

$$f(x, y) = f_1 + n\nabla f_1 + \frac{n(n+1)}{2}\nabla^2 f_1 + \frac{n(n+1)(n+2)}{6}\nabla^3 f_1 + \cdots \text{ in Eqn. (1).}$$

$$\therefore \quad y_1 = y_0 + \int_{x_0}^{x_1}\left(f_1 + n\nabla f_1 + \frac{n(n+1)}{2}\nabla^2 f_1 + \cdots\right)dx \quad [\text{Put } x = x_1 + nh,\ dx = h\,dn]$$

$$= y_0 + \int_{-1}^{0}\left(f_1 + n\nabla f_1 + \frac{n(n+1)}{2}\nabla^2 f_1 + \cdots\right)dn$$

$$= y_0 + h\left(f_1 - \frac{1}{2}\nabla f_1 - \frac{1}{12}\nabla^2 f_1 - \frac{1}{24}\nabla^3 f_1 - \cdots\right)$$

Neglecting fourth and higher order differences and expressing $\nabla f_1, \nabla^2 f_1$ and $\nabla^3 f_1$ in terms of function values, we obtain

$$y_1^{(c)} = y_0 + \frac{h}{24}(9f_1 + 19f_0 - 5f_{-1} + f_{-2}) \qquad \text{...(3)}$$

which is called a *Adams-Moulton corrector formula.*

Then an improved value of $f_1$ is calculated and again the corrector Eqn. (3) is applied to find a still better value of $y_1$. This step is repeated till $y_1$ remains unchanged and then proceed to calculate $y_2$ as above.

**Obs.** To apply both Milne and Adams-Bashforth methods, we require four starting values of $y$ which are calculated by means of Picard's method or Taylor's series method or Euler's method or Runge-Kutta method. In practice, the Adams formulae (2) and (3) above together with fourth order Runge-Kutta formulae have been found to be most useful.

**Example 32.21.** *Given* $\frac{dy}{dx} = x^2(1 + y)$ *and* $y(1) = 1, y(1.1) = 1.233, y(1.2) = 1.548, y(1.3) = 1.979$, *evaluate* $y(1.4)$ *by Adams-Bashforth method.* (Anna., 2015 ; V.T.U., 2015)

**Solution.** Here $f(x, y) = x^2(1 + y)$.

Starting values of the Adams-Bashforth method with $h = 0.1$, are

$x = 1.0, y_{-3} = 1.000, f_{-3} = (1.0)^2(1 + 1.000) = 2.000$

$x = 1.1, y_{-2} = 1.233, f_{-2} = 2.702$

$x = 1.2, y_{-1} = 1.548, f_{-1} = 3.669$

$$x = 1.3, y_0 = 1.979, f_0 = 5.035$$

Using the predictor,

$$y_1^{(p)} = y_0 + \frac{h}{24}(55f_0 - 59f_{-1} + 37f_{-2} - 9f_{-3})$$

$$x = 1.4,\ y_1^{(p)} = 2.573, f_1 = 7.004$$

Using the corrector,

$$y_1^{(c)} = y_0 + \frac{h}{24}(9f_1 + 19f_0 - 5f_{-1} + f_{-2})$$

$$y_1^{(c)} = 1.979 + \frac{0.1}{24}(9 \times 7.004 + 19 \times 5.035 - 5 \times 3.669 + 2.702) = 2.575$$

Hence $y(1.4) = 2.575$.

**Example 32.22.** *If $\frac{dy}{dx} = 2e^x y$, $y(0) = 0$, find $y(4)$ using Adams predictor-corrector formula by calculating $y(1)$, $y(2)$ and $y(3)$ using Euler's modified formula.*

**Solution.** We have $f(x, y) = 2e^x y$.

*To find 0.1 :*

| $x$ | $2e^x y = y'$ | *Mean slope* | *Old y + h (Mean slope) = new y* |
|---|---|---|---|
| 0.0 | 4 | — | 2 + 0.1 (4) = 2.4 |
| 0.1 | $2e^{0.1}$ (2.4) = 5.305 | $\frac{1}{2}$(4 + 5.305) = 4.6524 | 2 + 0.1 (4.6524) = 2.465 |
| 0.1 | $2e^{0.1}$ (2.465) = 5.449 | $\frac{1}{2}$(4 + 5.449) = 4.7244 | 2 + 0.1 (4.7244) = 2.472 |
| 0.1 | $2e^{0.1}$ (2.4724) = 5.465 | $\frac{1}{2}$(4 + 5.465) = 4.7324 | 2 + 0.1 (4.7324) = 2.473 |
| 0.1 | $2e^{0.1}$ (2.473) = 5.467 | $\frac{1}{2}$(4 + 5.467) = 4.7333 | 2 + 0.1 (4.7333) = 2.473 |
| 0.1 | 5.467 | — | 2 + 0.1 (5.467) = 3.0199 |
| 0.2 | $2e^{0.2}$ (3.0199) = 7.377 | $\frac{1}{2}$(5.467 + 7.377) = 6.422 | 2.473 + 0.1 (6.422) = 3.1155 |
| 0.2 | 7.611 | $\frac{1}{2}$(5.467 + 7.611) = 6.539 | 2.473 + 0.1 (6.539) = 3.127 |
| 0.2 | 7.639 | $\frac{1}{2}$(5.467 + 7.639) = 6.553 | 2.473 + 0.1 (6.553) = 3.129 |
| 0.2 | 7.643 | $\frac{1}{2}$(5.467 + 7.643) = 6.555 | 2.473 + 0.1 (6.455) = 3.129 |
| 0.2 | 7.643 | — | 3.129 + 0.1 (7.643) = 3.893 |
| 0.3 | $2e^{0.3}$ (3.893) = 10.51 | $\frac{1}{2}$(7.643 + 10.51) = 9.076 | 3.129 + 0.1 (9.076) = 4.036 |
| 0.3 | 10.897 | $\frac{1}{2}$(7.643 + 10.897) = 9.266 | 3.129 + 0.1 (9.2696) = 4.056 |
| 0.3 | 10.949 | $\frac{1}{2}$(7.643 + 10.949) = 9.296 | 3.129 + 0.1 (9.296) = 4.058 |
| 0.3 | 10.956 | $\frac{1}{2}$(7.643 + 10.956) = 9.299 | 3.129 + 0.1 (9.299) = 4.0586 |

*To find y(0.4) by Adam's method,* the starting values with $h = 0.1$ are

$x = 0.0 \quad y_{-3} = 2.4 \quad f_{-3} = 4$

$x = 0.1 \quad y_{-2} = 2.473 \quad f_{-2} = 5.467$

$x = 0.2 \quad y_{-1} = 3.129 \quad f_{-1} = 7.643$

$x = 0.3 \quad y_0 = 4.059 \quad f_{-0} = 10.956$

Using the predictor formula

$$y_1^{(p)} = y_0 + \frac{h}{24}(55f_0 - 59f_{-1} + 37f_{-2} - 9f_{-3})$$

$$= 4.059 + \frac{0.1}{24}(55 \times 10.957 - 59 \times 7.643 + 37 \times 5.467 - 9 \times 4) = 5.383$$

Now $x = 0.4 \quad y_1 = 5.383 \quad f_1 = 2e^{0.4}(5.383) = 16.061$

Using the corrector formula,

$$y_1^{(c)} = y_0 + \frac{h}{24}(9f_1 + 19f_0 - 5f_{-1} + f_{-2})$$

$$= 4.0586 + \frac{0.1}{24}(9 \times 6.061 + 19 \times 10.956 - 5 \times 7.643 + 5.467) = 5.392$$

Hence $y(0.4) = 5.392$.

**Example 32.23.** *Solve the initial value problem $dy/dx = x - y^2$, $y(0) = 1$ to find $y(0.4)$ by Adam's method. Starting solutions required are to be obtained using Runge-Kutta method of order 4 using step value $h = 0.1$.*

**Solution.** We have $f(x, y) = x - y^2$.

*To find y (0.1) :*

Here $x_0 = 0, y_0 = 1, h = 0.1$

$$\therefore \quad k_1 = hf(x_0, y_0) = (0.1) f(0, 1) = -0.1000$$

$$k_2 = hf\left(x_0 + \frac{1}{2}h, y_0 + \frac{1}{2}k_1\right) = (0.1) f(0.05, 0.95) = -0.08525$$

$$k_3 = hf\left(x_0 + \frac{1}{2}h, y_0 + \frac{1}{2}k_2\right) = (0.1) f(0.05, 0.9574) = -0.0867$$

$$k_4 = hf(x_0 + h, y_0 + k_3) = (0.1) f(0.1, 0.9137) = -0.07341$$

$$k = \frac{1}{6}(k_1 + 2k_2 + 2k_3 + k_4) = -0.0883$$

Thus $\quad y(0.1) = y_1 = y_0 + k = 1 - 0.0883 = 0.9117$

*To find y(0.2) :*

Here $x_1 = 0.1, y_1 = 0.9117, h = 0.1$.

$$\therefore \quad k_1 = hf(x_1, y_1) = (0.1) f(0.1, 0.9117) = -0.0731$$

$$k_2 = hf\left(x_1 + \frac{1}{2}h, y_1 + \frac{1}{2}k_1\right) = (0.1) f(0.15, 0.8751) = -0.0616$$

$$k_3 = hf\left(x_1 + \frac{1}{2}h, y_1 + \frac{1}{2}k_2\right) = (0.1) f(0.15, 0.8809) = -0.0626$$

$$k_4 = hf(x_1 + h, y_1 + k_3) = (0.1) f(0.2, 0.8491) = -0.0521$$

$$k = \frac{1}{6}(k_1 + 2k_2 + 2k_3 + k_4) = -0.0623$$

Thus $\quad y(0.2) = y_2 = y_1 + k = 0.8494$.

*To find y (0.3) :*

Here $x_2 = 0.2, y_2 = 0.8494, y = 0.1$

$$k_1 = hf(x_2, y_2) = (0.1) f(0.2, 0.8494) = -0.0521$$

$$k_2 = hf\left(x_2 + \frac{1}{2}h, y_2 + \frac{1}{2}k_1\right) = (0.1) f(0.25, 0.8233) = -0.0428$$

$$k_3 = hf\left(x_2 + \frac{1}{2}h, y_2 + \frac{1}{2}k_2\right) = (0.1)\,f(0.25, 0.828) \qquad = -0.0436$$

$$k_4 = hf(x_2 + h, y_2 + k_3) = (0.1)\,f(0.3, 0.8058) \qquad = -0.0349$$

$$k = \frac{1}{6}(k_1 + 2k_2 + 2k_3 + k_4) \qquad = -0.0438$$

Thus $\quad y(0.3) = y_3 = y_2 + k = 0.8061$

Now the starting values of Adam's method with $h = 0.1$ are :

| | | | |
|---|---|---|---|
| $x = 0.0$ | $y_{-3} = 1.0000$ | $f_{-3} = 0.0 - (1.0)^2$ | $= -1.0000$ |
| $x = 0.1$ | $y_{-2} = 0.9117$ | $f_{-2} = 0.1 - (0.9117)^2$ | $= -1.7312$ |
| $x = 0.2$ | $y_{-1} = 0.8494$ | $f_{-1} = 0.2 - (0.8494)^2$ | $= -0.5215$ |
| $x = 0.3$ | $y_0 = 0.8061$ | $f_0 = 0.3 - (0.8061)^2$ | $= -0.3498$ |

Using the *predictor*,

$$y_1^{(p)} = y_0 + \frac{h}{24}(55f_0 - 59f_{-1} + 37f_{-2} - 9f_{-3})$$

$x = 0.4$
$$y_1^{(p)} = 0.8061 + \frac{0.1}{24}[55(-0.3498) - 59(-0.5215) + 37(-0.7312) - 9(-1)]$$

$$= 0.7789 \qquad f_1 = -0.2067$$

Using the *corrector*,

$$y_1^{(c)} = y_0 + \frac{h}{24}(9f_1 + 19f_0 - 5f_{-1} + f_{-2})$$

$$y_1^{(c)} = 0.8061 + \frac{0.1}{24}[9(-0.2067) + 19(-0.3498) - 5(-0.5215) - 0.7312] = 0.7785$$

Hence $\quad y(0.4) = 0.7785.$

## PROBLEMS 32.5

1. Using Adams-Bashforth method, obtain the solution of $dy/dx = x - y^2$ at $x = 0.8$, given the values

| $x$ : | 0 | 0.2 | 0.4 | 0.6 |
|---|---|---|---|---|
| $y$ : | 0 | 0.0200 | 0.0795 | 0.1762 |

*(V.T.U., 2016)*

2. Using Adams-Bashforth formulae, determine $y(0, 4)$ given the differential equation $dy/dx = \frac{1}{2}xy$ and the data

| $x$ : | 0 | 0.1 | 0.2 | 0.3 |
|---|---|---|---|---|
| $y$ : | 1 | 1.0025 | 1.0101 | 1.0228 |

*(Anna., 2016)*

3. Given $y' = x^2 - y$, $y(0) = 1$ and the starting values $y(0.1) = 0.90516$, $y(0.2) = 0.82127$, $y(0.3) = 0.74918$, evaluate $y(0, 4)$ using Adams-Bashforth method.
4. Using Adams-Bashforth method, find $y(4.4)$ given $5xy' + y^2 = 2$, $y(4) = 1$, $y(4, 1) = 1.0049$, $y(4, 2) = 1.0097$ and $y(4.3) = 1.0143$. *(Anna., 2016)*
5. Given the differential equation $dy/dx = x^2y + x^2$ and the data :

| $x$ : | 1 | 1.1 | 1.2 | 1.3 |
|---|---|---|---|---|
| $y$ : | 1 | 1.233 | 1.548488 | 1.978921 |

6. Using Adams-Bashforth method, evaluate $y(1.4)$, if $y$ satisfies $dy/dx + y/x = 1/x^2$ and $y(1) = 1$, $y(1.1) = 0.996$, $y(1.2) = 0.986$, $y(1.3) = 0.972$.

## 32.11 SIMULTANEOUS FIRST ORDER DIFFERENTIAL EQUATIONS

The simultaneous differential equations of the type

$$\frac{dy}{dx} = f(x, y, z) \qquad ...(1)$$

and
$$\frac{dz}{dx} = \phi(x, y, z) \qquad ...(2)$$

with initial conditions $y(x_0) = y_0$ and $z(x_0) = z_0$ can be solved by the methods discussed in the preceding sections, especially by Picard's or Runge-Kutta methods.

(i) *Picard's method gives*

$$y_1 = y_0 + \int f(x, y_0, z_0)\, dx,\ z_1 = z_0 + \int \phi(x, y_0, z_0)\, dx$$

$$y_2 = y_0 + \int f(x, y_1, z_1)\, dx,\ z_2 = z_0 + \int \phi(x, y_1, z_1)\, dx$$

$$y_3 = y_0 + \int f(x, y_2, z_2)\, dx,\ z_3 = z_0 + \int \phi(x, y_2, z_2)\, dx$$

and so on.

(ii) *Taylor's series* method is used as follows :

If $h$ be the step-size, $y_1 = y(x_0 + h)$ and $z_1 = z(x_0 + h)$. Then Taylor's algorithm for (1) and (2) gives

$$y_1 = y_0 + hy_0' + \frac{h^2}{2!} y_0'' + \frac{h^3}{3!} y_0''' + \ldots \quad \ldots(3)$$

$$z_1 = z_0 + hz_0' + \frac{h^2}{2!} z_0'' + \frac{h^3}{3!} z_0''' + \ldots \quad \ldots(4)$$

Differentiating Eqns. (1) and (2) successively, we get $y''$, $z''$, etc. So the values $y_0', y_0'', y_0'''\ldots$ and $z_0', z_0'', z_0'''\ldots$ are known. Substituting these in Eqns. (3) and (4), we obtain $y_1, z_1$ for the next step.

Similarly, we have the algorithms

$$y_2 = y_1 + hy_1' + \frac{h^2}{2!} y_1'' + \frac{h^3}{3!} y_1''' + \ldots \quad \ldots(5)$$

$$z_2 = z_1 + hz_1' + \frac{h^2}{2!} z_1'' + \frac{h^3}{3!} z_1''' + \ldots \quad \ldots(6)$$

Since $y_1$ and $z_1$ are known, we can calculate $y_1', y_1'', \ldots$ and $z_1', z_1'' \ldots$. Substituting these in (5) and (6), we get $y_2$ and $z_2$.

Proceeding further, we can calculate the other values of $y$ and $z$ step by step.

(iii) *Runge-Kutta method* is applied as follows :

Starting at $(x_0, y_0, z_0)$ and taking the step-sizes for $x, y, z$ to be $h, k, l$ respectively, the Runge-Kutta method gives,

$$k_1 = hf(x_0, y_0, z_0) \qquad l_1 = h\phi(x_0, y_0, z_0)$$

$$k_2 = hf\left(x_0 + \frac{1}{2}h, y_0 + \frac{1}{2}k_1, z_0 + \frac{1}{2}l_1\right) \qquad l_2 = h\phi\left(x_0 + \frac{1}{2}h, y_0 + \frac{1}{2}k_1, z_0 + \frac{1}{2}l_1\right)$$

$$k_3 = hf\left(x_0 + \frac{1}{2}h, y_0 + \frac{1}{2}k_2, z_0 + \frac{1}{2}l_2\right) \qquad l_3 = h\phi\left(x_0 + \frac{1}{2}h, y_0 + \frac{1}{2}k_2, z_0 + \frac{1}{2}l_2\right)$$

$$k_4 = hf(x_0 + h, y_0 + k_3, z_0 + l_3) \qquad l_4 = h\phi(x_0 + h, y_0 + k_3, z_0 + l_3)$$

Hence $\quad y_1 = y_0 + \frac{1}{6}(k_1 + 2k_2 + 2k_3 + k_4) \quad$ and $\quad z_1 = z_0 + \frac{1}{6}(l_1 + 2l_2 + 2l_3 + l_4)$

To compute $y_2$ and $z_2$, we simply replace $x_0, y_0, z_0$ by $x_1, y_1, z_1$ in the above formulae.

**Example 32.24.** *Using Picard's method find approximate values of y and z corresponding to x = 0.1, given that y(0) = 2, z(0) = 1 and dy/dx = x + z, dz/dx = x − y².*

**Solution.** Here $x_0 = 0, y_0 = 2, z_0 = 1$,

$$\frac{dy}{dx} = f(x, y, z) = x + z \text{ ;and } \quad \frac{dz}{dx} = \phi(x, y, z) = x - y^2$$

$$\therefore \quad y = y_0 + \int_{x_0}^{x} f(x, y\, z)\, dx \quad \text{and} \quad z = z_0 + \int_{x_0}^{x} \phi(x, y\, z)\, dx.$$

*First approximations* $\quad y_1 = y_0 + \int_{x_0}^{x} \phi(x, y_0, z_0)\, dx = 2 + \int_0^x (x + 1)\, dx = 2 + x + \frac{1}{2}x^2$

$$z_1 = z_0 + \int_{x_0}^{x} \phi(x, y_0, z_0)\, dx = 1 + \int_0^x (x - 4)\, dx = 1 - 4x + \frac{1}{2}x^2$$

*Second approximations* $y_2 = y_0 + \int_{x_0}^{x} f(x, y_1, z_1)\, dx = 2 + \int_0^x \left(x + 1 - 4x + \frac{1}{2}x^2\right) dx$

$$= 2 + x - \frac{3}{2}x^2 + \frac{x^3}{6}$$

$$z_2 = z_0 + \int_{x_0}^{x} \phi(x, y_1, z_1)\, dx$$

$$= 1 + \int_{x_0}^{x} \left[x - \left(2 + x + \frac{1}{2}x^2\right)^2\right] dx = 1 - 4x + \frac{3}{2}x^2 - x^3 - \frac{x^4}{4} - \frac{x^5}{20}.$$

*Third approximations* $y_3 = y_0 + \int_{x_0}^{x} f(x, y_2, z_2)\, dx$

$$= 2 + x - \frac{3}{2}x^2 - \frac{1}{2}x^3 - \frac{1}{4}x^4 - \frac{1}{20}x^5 - \frac{1}{120}x^6$$

$$z_3 = z_0 + \int_{x_0}^{x} \phi(x, y_2, z_2)\, dx$$

$$= 1 - 4x - \frac{3}{2}x^2 + \frac{5}{3}x^3 + \frac{7}{12}x^4 - \frac{31}{60}x^5 + \frac{1}{12}x^6 - \frac{1}{252}x^7$$

and so on.

When, $x = 0.1$,

$y_1 = 2.105, y_2 = 2.08517, y_3 = 2.08447$

$z_1 = 0.605, z_2 = 0.58397, z_3 = 0.58672.$

Hence, $y(0.1) = 2.0845, z(0.1) = 0.5867$

correct to four decimal places.

**Example 32.25.** *Solve the differential equations*

$$\frac{dy}{dx} = 1 + xz, \quad \frac{dz}{dx} = -xy \text{ for } x = 0.3,$$

*using fourth order Runge-Kutta method. Initial values are x = 0, y = 0, z = 1.* (V.T.U., 2015)

**Solution.** Here $f(x, y, z) = 1 + xz$, $\phi(x, y, z) = -xy$

$x_0 = 0, y_0 = 0, z_0 = 1$. Let us take $h = 0.3$.

$\therefore$ $k_1 = h f(x_0, y_0, z_0) = 0.3 f(0, 0, 1) = 0.3 (1 + 0) = 0.3$

$l_1 = h\, \phi(x_0, y_0, z_0) = 0.3(-0 \times 0) = 0$

$$k_2 = h f\left(x_0 + \frac{1}{2}h, y_0 + \frac{1}{2}k_1, z_0 + \frac{1}{2}l_1\right)$$

$$= (0.3) f(0.15, 0.15, 1) = 0.3(1 + 0.15) = 0.345$$

$$l_2 = h\, \phi\left(x_0 + \frac{1}{2}h, y_0 + \frac{1}{2}k_1, z_0 + \frac{1}{2}l_1\right)$$

$$= 0.3[-(0.15)(0.15)] = -0.00675.$$

$$k_3 = h f\left(x_0 + \frac{h}{2}, y_0 + \frac{k_2}{2}, z_0 + \frac{l_2}{2}\right)$$

$$= (0.3) f(0.15, 0.1725, 0.996625)$$

$$= 0.3[1 + 0.996625 \times 0.15] = 0.34485$$

$$l_3 = h\, \phi\left(x_0 + \frac{h}{2}, y_0 + \frac{k_2}{2}, z_0 + \frac{l_2}{2}\right)$$

$$= 0.3[-(0.15)(0.1725)] = -0.007762$$

$$k_4 = h\, f(x_0 + h, y_0 + k_3, z_0 + l_3) = (0.3)\, f(0.3, 0.34485, 0.99224) = 0.3893$$

$$l_4 = h\, \phi(x_0 + h, y_0 + k_3, z_0 + l_3) = 0.3\,[-(0.3)(0.34485)] = -0.03104$$

Hence,
$$y(x_0 + h) = y_0 + \frac{1}{6}(k_1 + 2k_2 + 2k_3 + k_4)$$

*i.e.*,
$$y(0.3) = 0 + \frac{1}{6}[0.3 + 2(0.345) + 2(0.34485) + 0.3893] = 0.34483$$

and
$$z(x_0 + h) = z_0 + \frac{1}{6}(l_1 + 2l_2 + 2l_3 + l_4)$$

*i.e.*,
$$z(0.3) = 1 + \frac{1}{6}[0 + 2 + (-0.00675) + 2(-0.0077625) + (-0.03104)] = 0.98999$$

## 32.12 SECOND ORDER DIFFERENTIAL EQUATIONS

Consider the second order differential equation $\dfrac{d^2y}{dx^2} = f\left(x, y, \dfrac{dy}{dx}\right)$

By writing $dy/dx = z$, it can be reduced to two first order simultaneous differential equations

$$\frac{dy}{dx} = z, \qquad \frac{dz}{dx} = f(x, y, z)$$

These equations can be solved as explained above.

**Example 32.26.** *Using Runge-Kutta method, solve $y'' = xy'^2 - y^2$ for $x = 0.2$ correct to 4 decimal places. Initial conditions are $x = 0$, $y = 1$, $y' = 0$.* *(Anna, 2013 ; V.T.U., 2013)*

**Solution.** Let $dy/dx = z = f(x, y, z)$. Then $dz/dx = xz^2 - y^2 = \phi(x, y, z)$

We have $x_0 = 0$, $y_0 = 1$, $z_0 = 0$, $h = 0.2$.

Using $k_1, k_2 \ldots$ for $f(x, y, z)$ and $l_1, l_2, \ldots$ for $\phi(x, y, z)$, Runge-Kutta formulae become

$$k_1 = hf(x_0, y_0, z_0) = 0.2\,(0) = 0 \qquad l_1 = h\phi(x_0, y_0, z_0) = 0.2(-1) = -0.2$$

$$k_2 = hf(x_0 + \tfrac{1}{2}h, y_0 + \tfrac{1}{2}k_1, z_0 + \tfrac{1}{2}l_1) = 0.2\,(-0.1) = -0.02 \qquad l_2 = h\phi(x_0 + \tfrac{1}{2}h, y_0 + \tfrac{1}{2}k_1, z_0 + \tfrac{1}{2}l_1) = 0.2(-0.999) = -0.1998$$

$$k_3 = hf(x_0 + \tfrac{1}{2}h, y_0 + \tfrac{1}{2}k_2, z_0 + \tfrac{1}{2}l_2) = 0.2\,(-0.0999) = -0.02 \qquad l_3 = h\phi(x_0 + \tfrac{1}{2}h, y_0 + \tfrac{1}{2}k_2, z_0 + \tfrac{1}{2}l_2) = 0.2(-0.9791) = -0.1958$$

$$k_4 = hf(x_0 + h, y_0 + k_3, z_0 + l_3) = 0.2\,(-0.1958) = -0.0392 \qquad l_4 = h\phi(x_0 + h, y_0 + k_3, z_0 + l_3) = 0.2(0.9527) = -0.1905$$

$$\therefore \quad k = \tfrac{1}{6}(k_1 + 2k_2 + 2k_3 + k_4) = -0.0199 \qquad l = \tfrac{1}{6}(l_1 + 2l_2 + 2l_3 + 2l_4) = -0.1970$$

Hence at $x = 0.2$,

$$y = y_0 + k = 1 - 0.0199 = 0.9801$$

and
$$y' = z = z_0 + l = 0 - 0.1970 = -0.1970.$$

**Example 32.27.** *Given $y'' + xy' + y = 0$, $y(0) = 1$, $y'(0) = 0$, obtain $y$ for $x = 0(0.1)0.3$ by any method. Further, continue the solution by Milne's method to calculate $y(0.4)$.* *(V.T.U., 2015 ; Anna, 2012)*

**Solution.** Putting $y' = z$, the given equation reduces to the simultaneous equations

$$z' + xz + y = 0, \; y' = z \qquad \ldots(i)$$

We employ Taylor's series method to find $y$.

Differentiating the given equation $n$ times, we get

$$y_{n+2} + xy_{n+1} + ny_n + y_n = 0$$

At
$$x = 0, \; (y_{n+2})_0 = -(n + 1)(y_n)_0$$

$$\therefore \quad y(0) = 1, \text{ gives } y_2(0) = -1, \; y_4(0) = 32, \; y_6(0) = -5 \times 3, \ldots$$

and $\qquad y_1(0) = 0$ yields $y_3(0) = y_5(0) = ... = 0.$

Expanding $y(x)$ by Taylor's series, we have

$$y(x) = y(0) + xy_1(0) + \frac{x^2}{2!} y_2(0) + \frac{x^3}{3!} y_3(0) + ...$$

$$\therefore \qquad y(x) = 1 - \frac{x^2}{2!} + \frac{3}{4!} x^4 - \frac{5 \times 3}{6!} x^6 + ... \qquad ...(ii)$$

and $$z(x) = y'(x) = -x + \frac{1}{2} x^3 - \frac{1}{8} x^5 = ... = -xy \qquad ...(iii)$$

From Eqn. (*ii*), we have

$$y(0.1) = 1 - \frac{(0.1)^2}{2} + \frac{1}{8}(0.1)^4 - ... = 0.995$$

$$y(0.2) = 1 - \frac{(0.2)^2}{2} + \frac{(0.2)^4}{8} - ... = 0.9802$$

$$y(0.3) = 1 - \frac{(0.3)^2}{2} + \frac{(0.3)^4}{8} - \frac{(0.3)^6}{48} + ... = 0.956$$

From Eqn. (*iii*), we have

$$z(0.1) = -0.0995, z(0.2) = -0.196, z(0.3) = -0.2863.$$

Also from Eqn. (*i*), $z'(x) = -(xz + y) \quad \therefore \quad z'(0.1) = 0.985, z'(.2) = -0.941, z'(0.3) = -0.87.$

Applying Milne's predictor formula, first to $z$ and then to $y$, we obtain

$$z(0.4) = z(0) + \frac{4}{3}(0.1)\{2z'(0.1) - z'(0.2) + 2z'(0.3)\}$$

$$= 0 + \left(\frac{0.4}{3}\right)\{-1.79 + 0.941 - 1.74\} = -0.3692$$

and $$y(0.4) = y(0) + \frac{4}{3}(0.1)\{2y'(0.1) - y'(0.2) + 2y'(0.3)\}$$

$$= 0 + \left(\frac{0.4}{3}\right)\{-0.199 + 0.196 - 0.5736\} = 0.9231 \qquad [\because \quad y' = z]$$

Also $z'(0.4) = -\{x(0.4)z(0.4) + y(0.4)\} = \{0.4(-0.3692) + 0.9231\} = -0.7754.$

Now applying Milne's corrector formula, we get

$$z(0.4) = z(0.2) + \frac{h}{3}\{z'(0.2) + 4z'(0.3) + z'(0.4)\}$$

$$= -0.196 + \left(\frac{0.1}{3}\right)\{-0.941 - 3.48 - 0.7754\} = -0.3692$$

and $$y(0.4) = y(0.2) + \frac{h}{3}\{y'(0.2) + 4y'(0.3) + y'(0.4)\}$$

$$= 0.9802 + \left(\frac{0.1}{3}\right)\{-0.196 - 1.1452 - 0.3692\} = 0.9232$$

Hence $\qquad y(0.4) = 0.9232$ and $z(0.4) = -0.3692.$

## PROBLEMS 32.6

1. Apply Picard's method to find the third approximation to the values of $y$ and $z$, given that $dy/dx = z$, $dz/dx = x^3(y + z)$, given $y = 1$, $z = \frac{1}{2}$ when $x = 0$. (*V.T.U., 2016*)
2. Solve the following differential equations using Taylor series method of the 4th order, for $x = 0.1$ and $0.2$, $\frac{dy}{dx} = xz + 1, \frac{dz}{dy} = -xy$; $y(0) = 0$ and $z(0) = 1$.
3. Find $y(0.1)$, $z(0.1)$, $y(0.2)$ and $z(0.2)$ from the system of equations $y' = x + z$, $z' = x - y^2$ given $y(0) = 0$, $z(0) = 1$ using 4th order of Runge-Kutta.

4. Using Picard's method, obtain the second approximation to the solution of

$$\frac{d^2y}{dx^2} = x^3\frac{dy}{dx} + x^3y \text{ so that } y(0) = 1, y'(0) = \frac{1}{2}.$$ (V.T.U., 2015)

5. Use Picard's method to approximate $y$ when $x = 0.1$, given that $\frac{d^2y}{dx^2} + 2x\frac{dy}{dx} + y = 0$ and $y = 0.5$, $\frac{dy}{dx} = 0.1$, when $x = 0$.
6. Using Runge-Kutta method of order four, solve $y'' = y + xy'$, $y(0) = 1$, $y'(0) = 0$ to find $y(0.2)$ & $y'(0.2)$. (V.T.U., 2016)
7. Consider the second order value problem $y'' - 2y' + 2y = e^{2t} \sin t$ with $y(0) = -0.4$ and $y'(0) = -0.6$. Using the fourth order Runge-Kutta method, find $y(0.2)$. (Anna, 2012)
8. The angular displacement $\theta$ of a simple pendulum is given by the equation

$$\frac{d^2\theta}{dt^2} + \frac{g}{l}\sin\theta = 0$$

where $l = 98$ cm and $g = 980$ cm/sec$^2$. If $\theta = 0$ and $d\theta/dt = 4.472$ at $t = 0$, use Runge-Kutta method to find $\theta$ and $d\theta/dt$ when $t = 0.2$ sec.

## 32.13 BOUNDARY VALUE PROBLEMS

Such a problem requires the solution of a differential equation in a region $R$ subject to the various conditions on the boundary of $R$. Practical applications give rise to many such problems. We shall discuss two-point linear boundary value problems of the following types :

(i) $\frac{d^2y}{dx^2} + \lambda(x)\frac{dy}{dx} + \mu(x)y = \gamma(x)$ with the conditions $y(x_0) = a$, $y(x_n) = b$.

(ii) $\frac{d^4y}{dx^4} + \lambda(x)y = \mu(x)$ with the conditions $y(x_0) = y'(x_0) = a$ and $y(x_n) = y'(x_n) = b$.

While there exist many numerical methods for solving such boundary value problems, the method of finite-differences is most commonly used. We shall explain this method in the next section.

## 32.14 FINITE-DIFFERENCE METHOD

In this method, the derivatives appearing in the differential equation and the boundary conditions are replaced by their finite-difference approximations and the resulting linear system of equations are solved by any standard procedure. These roots are the values of the required solution at the pivotal points.

*The finite-difference approximations to the various derivatives are derived as under :*

If $y(x)$ and its derivatives are single-valued continuous functions of $x$ then by Taylor's expansion, we have

$$y(x + h) = y(x) + hy'(x) + \frac{h^2}{2!}y''(x) + \frac{h^2}{3!}y'''(x) + \ldots \quad \ldots(1)$$

and

$$y(x - h) = y(x) - hy'(x) + \frac{h^2}{2!}y''(x) - \frac{h^3}{3!}y'''(x) + \ldots \quad \ldots(2)$$

Equation (1) gives $y'(x) = \frac{1}{h}[y(x + h) - y(x)] - \frac{h}{2}y''(x) - \ldots$

*i.e.,*

$$y'(x) = \frac{1}{h}[y(x + h) - y(x)] + O(h)$$

which is the *forward difference approximation of* $y'(x)$ with an error of the order $h$.

Similarly Eqn. (2) gives $y'(x) = \frac{1}{h}[y(x) - y(x - h)] + O(h)$

which is the *backward difference approximation of* $y'(x)$ with an error of the order $h$.

Subtracting Eqns. (2) from (1), we obtain

$$y'(x) = \frac{1}{2h}[y(x + h) - y(x - h)] + O(h^2)$$

which is the *central-difference approximation of* $y'(x)$ with an error of the order $h^2$. Clearly this central difference approximation to $y'(x)$ is better than the forward or backward difference approximations and hence should be preferred.

Adding Eqns. (1) and (2), we get

$$y''(x) = \frac{1}{h^2}[y(x+h) - 2y(x) + y(x-h)] + O(h^2)$$

which is the central difference approximation of $y''(x)$. Similarly we can derive central difference approximations to higher derivatives.

Hence *the working expressions for the central difference approximations to the first four derivatives of* $y_i$ *are as under :*

$$y'_i = \frac{1}{2h}(y_{i+1} - y_{i-1}) \quad ...(3)$$

$$y''_i = \frac{1}{h^2}(y_{i+1} - 2y_i + y_{i-1}) \quad ...(4)$$

$$y'''_i = \frac{1}{2h^3}(y_{i+2} - 2y_{i+1} + 2y_{i-1} - y_{i-2}) \quad ...(5)$$

$$y_i^{iv} = \frac{1}{h^4}(y_{i+2} - 4y_{i+1} + 6y_i - 4y_{i+1} + y_{i-2}) \quad ...(6)$$

**Obs.** The accuracy of this method depends on the size of the sub-interval $h$ and also on the order of approximation. As we reduce $h$, the accuracy improves but the number of equations to be solved also increases.

**Example 32.28.** *Solve the equation $y'' = x + y$ with the boundary conditions $y(0) = y(1) = 0$.*

**Solution.** We divide the interval (0, 1) into four sub-intervals so that $h = 1/4$ and the pivot points are $x_0 = 0, x_1 = 1/4, x_2 = 1/2, x_3 = 3/4$ and $x_4 = 1$.

The differential equation is approximated as

$$\frac{1}{h^2}[y_{i+1} - 2y_i + y_{i-1}] = x_i + y_i$$

or

$$16y_{i+1} - 33y_i + 16y_{i-1} = x_i,\ i = 1, 2, 3.$$

Using $y_0 = y_4 = 0$, we get the system of equations

$$16y_2 - 33y_1 = \frac{1}{4}$$

$$16y_3 - 33y_2 + 16y_1 = \frac{1}{2}$$

$$-33y_3 + 16y_2 = \frac{3}{4}$$

Their solution gives

$$y_1 = -0.03488,\ y_2 = -0.05632,\ y_3 = -0.05003.$$

**Obs.** The exact solution being $y(x) = \frac{\sinh x}{\sinh 1} - x$, the error at each nodal point is given in the table:

| $x$ | *Computed value y(x)* | *Exact value y(x)* | *Error* |
|---|---|---|---|
| 0.25 | –0.03488 | –0.03505 | 0.00017 |
| 0.5 | –0.05632 | –0.05659 | 0.00027 |
| 0.75 | –0.05003 | –0.05028 | 0.00025 |

**Example 32.29.** *Determine values of y at the pivotal points of the interval (0, 1), if y satisfies the boundary value problem $y^{iv} + 81y = 81x^2$, $y(0) = y(1) = y''(0) = y''(1) = 0$. (Take n = 3).*

**Solution.** Here $h = 1/3$ and the pivotal points are $x_0 = 0, x_1 = 1/3, x_2 = 2/3, x_3 = 1$. The corresponding $y$-values are $y_0 (= 0), y_1, y_2, y_3 (= 0)$.

Replacing $y^{iv}$ by its central difference approximation, the differential equation becomes

$$\frac{1}{h^4}(y_{i+2} - 4y_{i+1} + 6y_i - 4y_{i-1} + y_{i-2}) + 81y_i = 81x_i^2$$

or $$y_{i+2} - 4y_{i+1} + 7y_i - 4y_{i-1} + y_{i-2} = x_i^2,\ i = 1, 2$$

At $i = 1$, $$y_3 - 4y_2 + 7y_1 - 4y_0 + y_{-1} = 1/9$$

At $i = 2$, $$y_4 - 4y_3 + 7y_2 - 4y_1 + y_0 = 4/9$$

Using $y_0 = y_3 = 0$, we get $$-4y_2 + 7y_2 + y_{-1} = 1/9 \quad ...(i)$$

$$y_4 + 7y_1 - 4y_1 = 4/9 \quad ...(ii)$$

Regarding the conditions $y''_0 = y''_3 = 0$, we know that

$$x_i'' = \frac{1}{h^2}(y_{i+1} - 2y_i + y_{i-1})$$

At $i = 0$, $$y_0'' = 9(y_1 - 2y_0 + y_{-1})$$

or $$y_{-1} = -y_1 \quad [\because\ y_0 = y_0'' = 0] \quad ...(iii)$$

At $i = 3$, $$y_3'' = 9(y_4 - 2y_3 + y_2)$$

$$y_4 = -y_2 \quad [\because\ y_3 = y_3'' = 0] \quad ...(iv)$$

Using Eqn. (*iii*), the Eqn. (*i*) becomes

$$-4y_2 + 6y_1 = 1/9 \quad ...(v)$$

Using Eqn. (*iv*), the Eqn. (*ii*) reduces to

$$6y_2 - 4y_1 = 4/9 \quad ...(vi)$$

Solving Eqns. (*v*) and Eqn. (*vi*), we obtain

$$y_1 = 11/90 \text{ and } y_2 = 7/45.$$

Hence $y(1/3) = 0.1222$ and $y(2/3) = 0.1556$.

**Example 32.30.** *The deflection of a beam is governed by the equation*

$$\frac{d^4y}{dx^4} + 81y = \phi(x)$$

*where $\phi(x)$ is given by the table*

| $x$ : | 1/3 | 2/3 | 1, |
|---|---|---|---|
| $\phi(x)$ : | 81 | 162 | 243, |

*and boundary condition $y(0) = y'(0) = y''(1) = y'''(1) = 0$. Evaluate the deflection at the pivotal points of the beam using three sub-intervals.*

**Solution.** Here $h = 1/3$ and the pivotal points are $x_0 = 0$, $x_1 = 1/3$, $x_2 = 2/3$, $x_3 = 1$. The corresponding $y$-values are $y_0(= 0)$, $y_1$, $y_2$, $y_3$.

The given differential equation is approximated to

$$\frac{1}{h^4}(y_{i+2} - 4y_{i+1} + 6y_i - 4y_{i-1} + y_{i-2}) + 81y_i = \phi(x_i)$$

At $i = 1$, $$y_3 - 4y_2 + 7y_1 - 4y_0 + y_{-1} = 1 \quad ...(i)$$

At $i = 2$, $$y_4 - 4y_3 + 7y_2 - 4y_1 + y_0 = 2 \quad ...(ii)$$

At $i = 3$, $$y_5 - 4y_4 + 7y_3 - 4y_2 + y_1 = 3 \quad ...(iii)$$

We have $$y_0 = 0 \quad ...(iv)$$

Since $$y_i' = \frac{1}{2h}(y_{i+1} - y_{i-1})$$

$\therefore$ for $i = 0$, $$0 = y_0' = \frac{1}{2h}(y_1 - y_{-1}) \text{ i.e. } y_{-1} = y_1 \quad ...(v)$$

Since $$y_i'' = \frac{1}{h^2}(y_{i+1} - 2y_i + y_{i-1})$$

$\therefore$ for $i = 3$, $$0 = y_3'' = \frac{1}{h^2}(y_4 - 2y_3 + y_2), \text{ i.e. } y_4 = 2y_3 - y_2 \quad ...(vi)$$

Also $$y_i''' = \frac{1}{2h^3}(y_{i+2} - 2y_{i+1}\, 2y_{i-1} - y_{i-2})$$

$\therefore$ for $i = 3$, $$0 = y_3''' = \frac{1}{2h^3}(y_5 - 2y_4 + 2y_2 - y_1)$$

$$y_5 = 2y_4 - 2y + y_1 \qquad ...(vii)$$

Using Eqns. (*iv*) and (*v*), the Eqn. (*i*) reduces to

$$y_3 - 4y_2 + 8y_1 = 1 \qquad ...(viii)$$

Using Eqns. (*iv*) and (*vi*), the Eqn. (*ii*) becomes

$$-y_3 + 3y_2 - 2y_1 = 1 \qquad ...(ix)$$

Using Eqns. (*vi*) and (*vii*), the Eqn. (*iii*) reduces to

$$3y_3 - 4y_2 + 2y_1 = 3 \qquad ...(x)$$

Solving Eqns. (*viii*), (*ix*) and (*x*), we get

$$y_1 = 8/13,\ y_2 = 22/13,\ y_3 = 37/13.$$

Hence $$y(1/3) = 0.6154,\ y(2/3) = 1.6923,\ y(1) = 2.8462.$$

## PROBLEMS 32.7

1. Solve the boundary value problem for $x = 0.5$ :
$$\frac{d^2y}{dx^2} + y + 1 = 0,\ y(0) = y(1) = 0. \qquad \text{(Take } n = 4)$$
2. Find an approximate solution of the boundary value problem :
$$y'' + 8(\sin^2 \pi y)y = 0,\ 0 \le x \le 1,\ y(0) = y(1) = 1. \qquad \text{(Take } n = 4)$$
3. Solve the boundary value problem
$$xy'' + y = 0,\ y(1) = 1,\ y(2) = 2. \qquad \text{(Take } n = 4)$$
4. Solve the equation
$$y'' - 4y' + 4y = e^{3x}, \text{ with the conditions } y(0) = 0,\ y(1) = -2, \text{ taking } n = 4.$$
5. Solve the boundary value problem $y'' - 64y + 10 = 0$ with $y(0) = y(1) = 0$ by the finite difference method. Compute the value of $y(0.5)$ and compare with the true value.
6. Solve the boundary value problem
$$y'' + xy' + y = 3x^2 + 2,\ y(0) = 0,\ y(1) = 1.$$
7. The boundary value problem governing the deflection of a beam of length 3 metres is given by
$$\frac{d^4y}{dx^4} + 2y = \frac{1}{9}x^2 + \frac{2}{3}x + 4,\ y(0) = y'(0) = y(3) = y'(3) = 0.$$
The beam is built-in at the left end ($x = 0$) and simply supported at the right end ($x = 3$). Determine $y$ at the pivotal points $x = 1$ and $x = 2$.
8. Solve the boundary value problem,
$$\frac{d^4y}{dx^4} + 81y = 729x^2,\ y(0) = y'(0) = y''(1) = y'''(1) = 0. \qquad \text{(Use } n = 3)$$
9. Solve the equation $y^{iv} - y''' + y = x^2$ subject to the boundary conditions
$$y(0) = y'(0) = 0 \text{ and } y(1) = 2,\ y'(1) = 0. \qquad \text{(Take } n = 5)$$

## 32.15 OBJECTIVE TYPE OF QUESTIONS

## PROBLEMS 32.8

*Select the correct answer or fill up the blanks in the following questions :*

1. Which of the following is a step by step method :
(*a*) Taylor's (*b*) Adams-Bashforth (*c*) Picard's (*d*) None.
2. The finite difference scheme for the equation $2y'' + y = 5$ is ...... .
3. If $y'' = x + y$, $y(0) = 1$ and $y^{(1)}(x) = 1 + x + x^2/2$, then by Picard's method, the value of $y^{(2)}(x)$ is ...... .

4. The iterative formula of Euler's method for solving $y' = f(x, y)$ with $y(x_0) = y_0$, is ...... .
5. Taylor's series for solution of first order ordinary differential equations is ...... .
6. Using Runge-Kutta method of order four, the value of $y(0.1)$ for $y' = x - 2y$, $y(0) = 1$ taking $h = 0.1$ is
   (*a*) 0.813 (*b*) 0.825 (*c*) 0.0825 (*d*) none.
7. Given $y_0, y_1, y_2, y_3$, Milne's corrector formula to find $y_4$ for $dy/dx = f(x, y)$, is ...... .
8. The second order Runge-Kutta formula is ...... .
9. Adams-Bashforth predictor formula to solve $y' = f(x, y)$ given $y_0 = y(x_0)$ is ...... .
10. The multi-step methods available for solving ordinary differential equations are ...... .
11. To predict Adam's method atleast ...... values of $y$, prior to the desired value, are required.
12. Taylor's series solution of $y' = -xy$, $y(0) = 1$ upto $x^4$ is ...... .
13. Using modified Euler's method, the value of $y(0.1)$ for $\frac{dy}{dx} = x - y$, $y(0) = 1$ is ...... .
    (*a*) 0.809 (*b*) 0.909 (*c*) 0.0809 (*d*) none.
14. Milne's Predictor formula is ...... .
15. Adam's corrector formula is ...... .
16. Using Euler's method, $dy/dx = (y - 2x)/y$, $y(0) = 1$ ; gives $y(0.1) =$ ...... .
17. $\frac{d^2y}{dx^2} + y^2 \frac{dy}{dx} + y = 0$ is equivalent to a set of two first order differential equations ...... and ...... .
18. The formula for the 4th order Runge-Kutta method is ...... . (*Anna, 2012*)
19. Taylor's series method will be useful to give some ....... of Milne's method.
20. The name of two self-starting methods to solve $y' = f(x, y)$ given $y(x_0) = y_0$ are ......
21. In the derivation of fourth order Runge-Kutta formula, it is called fourth order because ......
22. If $y' = x$, $y(0) = 1$ then by Picard's method, the value of $y(1)$ is ......
    (*a*) 0.915 (*b*) 0.905 (*c*) 0.981 (*d*) none.
23. The finite difference scheme of the differential equation $y'' + 2y = 0$ is ......
24. If $y' = -y$, $y(0) = 1$, the Euler's method, the value of $y(1)$ is
    (*a*) 0.99 (*b*) 0.999 (*c*) 0.981 (*d*) none.
25. The finite difference formulae for $y'(x)$ and $y''(x)$ are ...... . (*Anna, 2012*)
26. In Euler's method if $h$ is small the method is too slow, if $h$ is large, it gives inaccurate value. (True or False)
27. Runge-Kutta method is a self-starting method. (True or False)
28. Predictor-corrector methods are self-starting methods. (True or False)

CHAPTER 33

# Numerical Solution of Partial Differential Equations



## 33.1 INTRODUCTION

There are many boundary value problems which involve partial differential equations. Only a few of these equations can be solved by analytical methods. In most cases, we depend on the numerical solution of such partial differential equations. Of the various numerical methods available for solving these equations, the method of finite differences is most commonly used.

In this method, the derivatives appearing in the equation and the boundary conditions are replaced by their finite-difference approximations. Then the given equation is changed to a difference equation which is solved by iterative procedures. This process is slow but gives good results of boundary value problems. An added advantage of this method is that the computation can be done by electronic computer. Here we shall apply this method to the solution of important applied partial differential equations. For a detailed study, the reader should refer to author's book *'Numerical Methods in Engineering and Science'*.

## 33.2 CLASSIFICATION OF SECOND ORDER EQUATIONS

The general linear partial differential equation of the second order in two independent variables is of the form.

$$A(x, y)\,\frac{\partial^2 u}{\partial x^2} + B(x, y)\,\frac{\partial^2 u}{\partial x \partial y} + C(x, y)\,\frac{\partial^2 u}{\partial y^2} + F\left(x, y, u, \frac{\partial u}{\partial x}, \frac{\partial u}{\partial y}\right) = 0$$

*Such a partial differential equation is said to be*

(*i*) **elliptic,** *if* $B^2 - 4AC < 0$,

(*ii*) **parabolic,** *if* $B^2 - 4AC = 0$,

and (*iii*) **hyperbolic,** *if* $B^2 - 4AC > 0$.

**Example 33.1.** *Classify the following equations :*

(*i*) $\dfrac{\partial^2 u}{\partial x^2} + 4\,\dfrac{\partial^2 u}{\partial x \partial y} + 4\,\dfrac{\partial^2 u}{\partial y^2} - \dfrac{\partial u}{\partial x} + 2\,\dfrac{\partial u}{\partial y} = 0$ (*ii*) $x^2\dfrac{\partial^2 u}{\partial x^2} + (1 - y^2)\dfrac{\partial^2 u}{\partial y^2} = 0, -\infty < x < \infty, -1 < y < 1$

(*iii*) $(1 + x^2)\dfrac{\partial^2 u}{\partial x^2} + (5 + 2x^2)\dfrac{\partial^2 u}{\partial x \partial t} + (4 + x^2)\dfrac{\partial^2 u}{\partial t^2} = 0.$

**Solution.** (*i*) Comparing this equation with (1) above, we find that

$$A = 1, B = 4, C = 4$$

$\therefore$ $$B^2 - 4AC = (4)^2 - 4 \times 1 \times 4 = 0$$

So the equation is parabolic.

(*ii*) Here $$A = x^2, B = 0, C = 1 - y^2$$

$\therefore$ $$B^2 - 4AC = 0 - 4x^2(1 - y^2) = 4x^2(y^2 - 1)$$

For all $x$ between $-\infty$ and $\infty$, $x^2$ is positive

For all $y$ between $-1$ and 1, $y^2 < 1$ $\therefore$ $B^2 - 4AC < 0$

Hence the equation is elliptic.

(*iii*) Here $A = 1 + x^2, B = 5 + 2x^2, C = 4 + x^2$

$\therefore$ $$B^2 - 4AC = (5 + 2x^2)^2 - 4(1 + x^2)(4 + x^2) = 9 \text{ i.e. } > 0$$

So the equation is hyperbolic.

## PROBLEMS 33.1

1. What is the classification of the equation $f_{xx} + 2f_{xy} + f_{yy} = 0$.
2. Determine whether the following equation is elliptic or hyperbolic ?
$$(x + 1)u_{xx} - 2(x + 2)u_{xy} + (x + 3)u_{yy} = 0.$$
3. Classify the equations (*i*) $y^2 u_{xx} - 2xy\, u_{xy} + x^2 u_{yy} + 2u_x - 3u = 0$

(*ii*) $x^2 \dfrac{\partial^2 u}{\partial x^2} + y^2 \dfrac{\partial^2 u}{\partial y^2} = x\dfrac{\partial u}{\partial x} - y\dfrac{\partial u}{\partial y}$

(*iii*) $\dfrac{3\partial^2 u}{\partial x^2} + 4\dfrac{\partial^2 u}{\partial x \partial y} + 6\dfrac{\partial^2 u}{\partial y^2} - 2\dfrac{\partial u}{\partial x} + \dfrac{\partial u}{\partial y} - u = 0.$

4. In which parts of the $(x, y)$ plane is the following equation elliptic ?
$$\partial^2 u/\partial x^2 + \partial^2 u/\partial x \partial y + (x^2 + 4y^2)\partial^2 u/\partial y^2 = 2\sin(xy).$$

## 33.3 FINITE-DIFFERENCE APPROXIMATIONS TO DERIVATIVES

Consider a rectangular region $R$ in the $x$-$y$ plane. Divide this region into a rectangular network of sides $\Delta x = h$ and $\Delta y = k$ as shown in Fig. 33.1. The points of intersection of the dividing lines are called *mesh points, nodal points* or *grid points.*

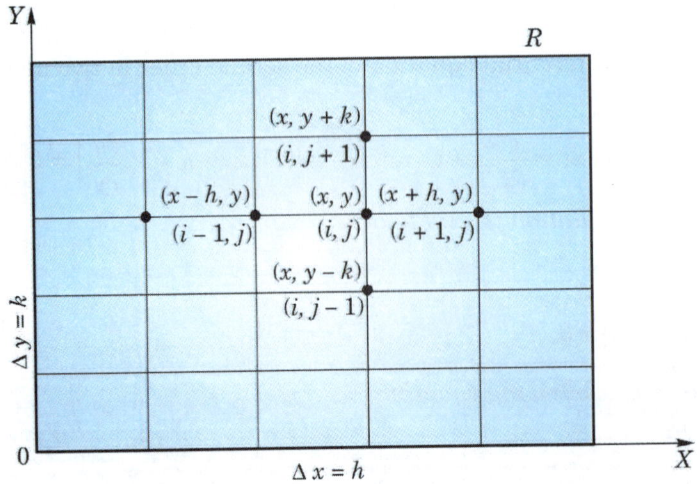


Fig. 33.1

Then we have the finite difference approximations for the partial derivatives in $x$-direction (§ 32.12) :

$$\frac{\partial u}{\partial x} = \frac{u(x + h, y) - u(x, y)}{h} + O(h)$$

$$= \frac{u(x, y) - u(x-h, y)}{h} + O(h) = \frac{u(x+h, y) - u(x-h, y)}{2h} + O(h^2)$$

and
$$\frac{\partial^2 u}{\partial x^2} = \frac{u(x-h, y) - 2u(x, y) + u(x+h, y)}{h^2} + O(h^2)$$

Writing $u(x, y) = u(ih, jk)$ as simply $u_{i, j}$, the above approximations become

$$u_x = \frac{u_{i+1, j} - u_{i, j}}{h} + O(h) \quad \text{...(1)}$$

$$= \frac{u_{i, j} - u_{i-1, j}}{h} + O(h) \quad \text{...(2)}$$

$$= \frac{u_{i+1, j} - u_{i-1, j}}{2h} + O(h^2) \quad \text{...(3)}$$

and
$$u_{xx} = \frac{u_{i-1, j} - 2u_{i, j} + u_{i+1, j}}{h^2} + O(h^2) \quad \text{...(4)}$$

Similarly we have the approximations for the derivatives w.r.t. $y$ :

$$u_y = \frac{u_{i, j+1} - u_{i, j}}{k} + O(k) \quad \text{...(5)}$$

$$= \frac{u_{i, j} - u_{i, j-1}}{k} + O(k) \quad \text{...(6)}$$

$$= \frac{u_{i, j+1} - u_{i, j-1}}{2k} + O(k^2) \quad \text{...(7)}$$

and
$$u_{yy} = \frac{u_{i, j-1} - 2u_{i, j} + u_{i, j+1}}{k^2} + O(k^2) \quad \text{...(8)}$$

Replacing the derivatives in any partial differential equation by their corresponding difference approximations Eqns. (1) to (8), we obtain the finite-difference analogues of the given equations.

## 33.4 ELLIPTIC EQUATIONS

The Laplace's equation

$$\nabla^2 u = \frac{\partial^2 u}{\partial x^2} + \frac{\partial^2 u}{\partial y^2} = 0$$

and the Poisson's equation

$$\frac{\partial^2 u}{\partial x^2} + \frac{\partial^2 u}{\partial y^2} = f(x, y)$$

are examples of elliptic partial differential equations. Laplace's equation arises in steady-state flow and potential problems. Poisson's equation arises in fluid mechanics, electricity and magnetism and torsion problems.

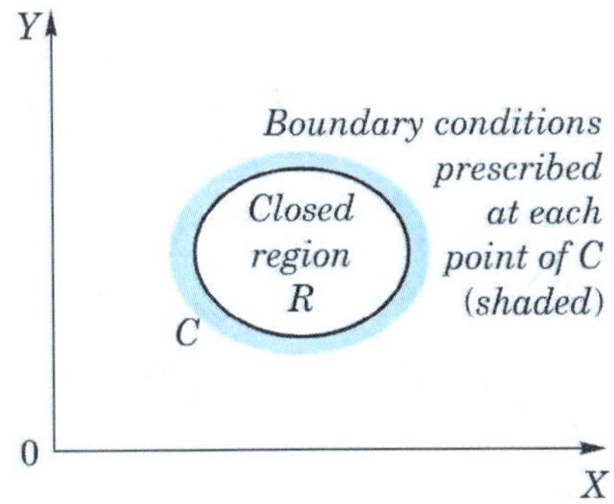


Fig. 33.2

The solution of these equations is a function $u(x, y)$ which is satisfied at every point of a region $R$ subject to certain boundary conditions specified on the closed curve $C$ (Fig. 33.2).

In general, problems concerning steady viscous flow, equilibrium stresses in elastic structures etc., lead to elliptic type of equations.

## 33.5 SOLUTION OF LAPLACE'S EQUATION

$$\frac{\partial^2 u}{\partial x^2} + \frac{\partial^2 u}{\partial y^2} = 0 \quad \text{...(1)}$$

Consider a rectangular region $R$ for which $u(x, y)$ is known at the boundary. Divide this region into a network of square mesh of side $h$, as shown in Fig. 33.3 (assuming that an exact sub-division of $R$ is possible). Replacing the derivatives in Eqn. (1) by their difference approximations, we have

$$\frac{1}{h^2}[u_{i-1,j} - 2u_{i,j} + u_{i+1,j}] + \frac{1}{h^2}[u_{i,j-1} - 2u_{i,j} + u_{i,j+1}] = 0$$

or

$$u_{i,j} = \frac{1}{4}[u_{i-1,j} + u_{i+1,j} + u_{i,j+1} + u_{i,j-1}] \quad ...(2)$$

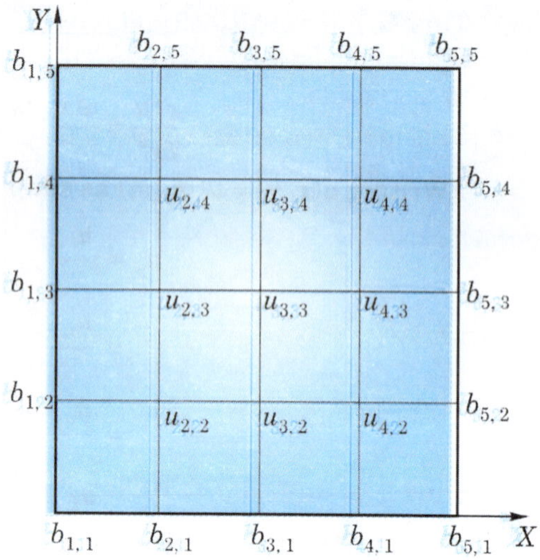

Fig. 33.3

This shows that the value of $u_{i,j}$ at any interior mesh point is the average of its values at four neighbouring points to the left, right, above and below. Equation (2) is called the **standard 5-point formula** which is exhibited in Fig. 33.4.

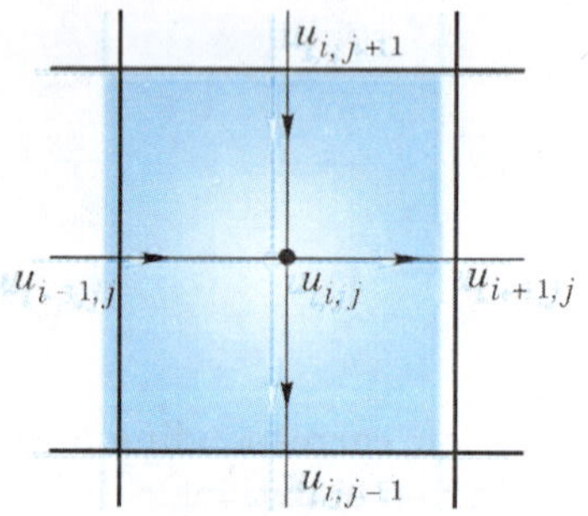

Fig. 33.4

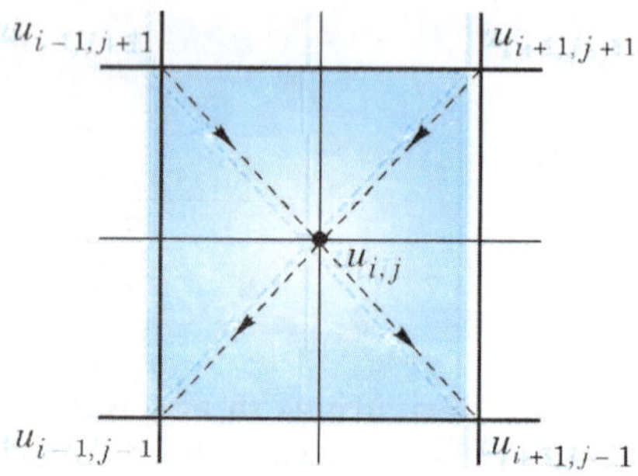

Fig. 33.5

Sometimes a formula similar to Eqn. (2) is used which is given by

$$u_{i,j} = \frac{1}{4}(u_{i-1,j+1} + u_{i+1,j-1} + u_{i+1,j+1} + u_{i-1,j-1}) \quad ...(3)$$

This shows that the value of $u_{i,j}$ is the average of its values at the four neighbouring diagonal mesh points. Equation (3) is called the **diagonal 5-point formula** which is represented in Fig. 33.5. Although Eqn. (3) is less accurate than Eqn. (2), yet it serves as a reasonably good approximation for obtaining the starting values at the mesh points.

Now to find the initial values of $u$ at the interior mesh points, we first use diagonal five point formula (3) and compute $u_{3,3}$, $u_{2,4}$, $u_{4,4}$, $u_{4,2}$ and $u_{2,2}$, in this order. Thus we get,

$$u_{3,3} = \frac{1}{4}(b_{1,5} + b_{5,1} + b_{5,5} + b_{1,1})\ ;\ u_{2,4} = \frac{1}{4}(b_{1,5} + u_{3,3} + b_{3,5} + b_{1,3})$$

$$u_{4,4} = \frac{1}{4}(b_{3,5} + b_{5,3} + b_{5,5} + u_{3,3})\ ;\ u_{4,2} = \frac{1}{4}(u_{3,3} + b_{5,1} + b_{3,1} + b_{5,3})$$

$$u_{2,2} = \frac{1}{4}(b_{1,3} + b_{3,1} + u_{3,3} + b_{1,1})$$

The values at the remaining interior points *i.e.* $u_{2,3}$, $u_{3,4}$, $u_{4,3}$ and $u_{3,2}$ are computed by the standard five-point formula (2). Thus, we obtain

$$u_{2,3} = \frac{1}{4}(b_{1,3} + u_{3,3} + u_{2,4} + u_{2,2}),\ u_{3,4} = \frac{1}{4}(u_{2,4} + u_{4,4} + b_{3,5} + u_{3,3})$$

$$u_{4,3} = \frac{1}{4}(u_{3,3} + b_{5,3} + u_{4,4} + u_{4,2}),\ u_{3,2} = \frac{1}{4}(u_{2,2} + u_{4,2} + u_{3,3} + u_{3,1})$$

Having found all the nine values of $u_{i,j}$ once, their accuracy is improved by repeated application of (2) in the form

$$u^{(n+1)}_{i,j} = \frac{1}{4}[u^{(n+1)}_{i-1,j} + u^{(n)}_{i+1,j} + u^{(n+1)}_{i,j+1} + u^{(n)}_{i,j-1}]$$

This formula utilises the latest iterative value available and scans the mesh points symmetrically from left to right along successive rows. This process is repeated till the difference of values in one round and the next becomes negligible.

This is known as *Liebmann's iteration process.*

**Example 33.2.** *Solve the elliptic equation $u_{xx} + u_{yy} = 0$ for the square mash of Fig. 33.6 with boundary values as shown.* (*V.T.U., 2016 ; Anna, 2012*)

**Solution.** Let $u_1, u_2, ..., u_9$ be the values of $u$ at the interior mesh-points. Since the boundary values of $u$ are symmetrical about $AB$.

$\therefore \quad u_7 = u_1, u_8 = u_2, u_9 = u_3.$

Also the values of $u$ being symmetrical about $CD$, $u_3 = u_1$, $u_6 = u_4, u_9 = u_7$.

Thus it is sufficient to find the values $u_1, u_2, u_4$ and $u_5$.

Now we find their initial value in the following order :

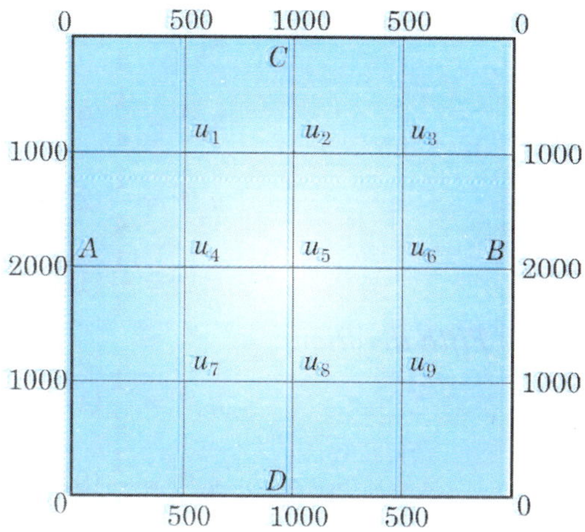

Fig. 33.6

$$u_5 = \frac{1}{4}(2000 + 2000 + 1000 + 1000) = 1500 \quad \text{(Std. formula)}$$

$$u_1 = \frac{1}{4}(0 + 1500 + 1000 + 2000) = 1125 \quad \text{(Diag. formula)}$$

$$u_2 = \frac{1}{4}(1125 + 1125 + 1000 + 1500) \approx 1188 \quad \text{(Std. formula)}$$

$$u_4 = \frac{1}{4}(2000 + 1500 + 1125 + 1125) \approx 1438 \quad \text{(Std. formula)}$$

We carry out the iteration process using the formulae :

$$u_1^{(n+1)} = \frac{1}{4}[1000 + u_2^{(n)} + 500 + u_4^{(n)}]$$

$$u_2^{(n+1)} = \frac{1}{4}[u_1^{(n+1)} + u_1^{(n)} + 1000 + u_5^{(n)}]$$

$$u_4^{(n+1)} = \frac{1}{4}[2000 + u_5^{(n)} + u_1^{(n+1)} + u_1^{(n)}]$$

$$u_5^{(n+1)} = \frac{1}{4}[u_4^{(n+1)} + u_4^{(n)} + u_2^{(n+1)} + u_2^{(n)}]$$

*First iteration* : (put $n = 0$)

$$u_1^{(1)} = \frac{1}{4}(1000 + 1188 + 500 + 1438) \approx 1032$$

$$u_2^{(1)} = \frac{1}{4}(1032 + 1032 + 1000 + 1500) = 1141$$

$$u_4^{(1)} = \frac{1}{4}(2000 + 1500 + 1032 + 1032) = 1391$$

$$u_5^{(1)} = \frac{1}{4}(1391 + 1391 + 1141 + 1141) = 1266$$

*Second iteration* : (put $n = 1$)

$$u_1^{(2)} = \frac{1}{4}(1000 + 1141 + 500 + 1391) = 1008$$

$$u_2^{(2)} = \frac{1}{4}(1008 + 1008 + 1000 + 1266) = 1069$$

$$u_4^{(2)} = \frac{1}{4}(2000 + 1266 + 1008 + 1008) = 1321$$

$$u_5^{(2)} = \frac{1}{4}(1321 + 1321 + 1069 + 1069) = 1195$$

*Third iteration* :

$$u_1^{(3)} = \frac{1}{4}(1000 + 1069 + 500 + 1321) = 973$$

$$u_2^{(3)} = \frac{1}{4}(973 + 973 + 1000 + 1195) = 1035$$

$$u_4^{(3)} = \frac{1}{4}(2000 + 1195 + 973 + 973) = 1288$$

$$u_5^{(3)} = \frac{1}{4}(1288 + 1288 + 1035 + 1035) = 1162$$

*Fourth iteration* :

$$u_1^{(4)} = \frac{1}{4}(1000 + 1135 + 500 + 1288) = 956$$

$$u_2^{(4)} = \frac{1}{4}(956 + 956 + 1000 + 1162) = 1019$$

$$u_4^{(4)} = \frac{1}{4}(2000 + 1162 + 956 + 956) = 1269$$

$$u_5^{(4)} = \frac{1}{4}(1269 + 1269 + 1019 + 1019) = 1144$$

*Fifth iteration* :

$$u_1^{(5)} = \frac{1}{4}(1000 + 1019 + 500 + 1269) = 947$$

$$u_2^{(5)} = \frac{1}{4}(947 + 947 + 1000 + 1144) \approx 1010$$

$$u_4^{(5)} = \frac{1}{4}(2000 + 1144 + 947 + 947) \approx 1260$$

$$u_5^{(5)} = \frac{1}{4}(1260 + 1260 + 1010 + 1010) = 1135$$

Similarly, $u_1^{(6)} = 942,\ u_2^{(6)} = 1005,\ u_4^{(6)} = 1255,\ u_5^{(6)} = 1130$

$u_1^{(7)} = 940,\ u_2^{(7)} = 1003,\ u_4^{(7)} = 1253,\ u_5^{(7)} = 1128$

$u_1^{(8)} = 939,\ u_2^{(8)} = 1002,\ u_4^{(8)} = 1252,\ u_5^{(8)} = 1127$

$u_1^{(9)} = 939,\ u_2^{(6)} = 1001,\ u_4^{(9)} = 1251,\ u_5^{(9)} = 1126$

Thus there is negligible difference between the values obtained in the eighth and ninth iterations.
Hence $u_1 = 939$, $u_2 = 1001$, $u_4 = 1251$ and $u_5 = 1126$.

**Example 33.3.** *Given the values of u(x, y) on the boundary of the square in the Fig. 33.7, evaluate the function u(x, y) satisfying the Laplace equation $\nabla^2 u = 0$ at the pivotal points of this figure.* *(V.T.U., 2013)*

**Solution.** To get the initial values of $u_1, u_2, u_3, u_4$, we assume $u_4 = 0$. Then

$$u_1 = \frac{1}{4}(1000 + 0 + 1000 + 2000) = 1000 \quad \text{(Diag. formula)}$$

$$u_2 = \frac{1}{4}(1000 + 500 + 1000 + 0) = 625 \quad \text{(Std. formula)}$$

$$u_3 = \frac{1}{4}(2000 + 0 + 1000 + 500) = 875 \quad \text{(Std. formula)}$$

$$u_4 = \frac{1}{4}(875 + 0 + 625 + 0) = 375 \quad \text{(Std. formula)}$$

We carry out the successive iterations, using the formulae

$$u_1^{(n+1)} = \frac{1}{4}[2000 + u_2^{(n)} + 1000 + u_3^{(n)}]$$

$$u_2^{(n+1)} = \frac{1}{4}[u_1^{(n+1)} + 500 + 1000 + u_4^{(n)}]$$

$$u_3^{(n+1)} = \frac{1}{4}[2000 + u_4^{(n)} + u_1^{(n+1)} + 500]$$

$$u_4^{(n+1)} = \frac{1}{4}[u_3^{(n+1)} + 0 + u_2^{(n+1)} + 0]$$

*First iteration* : (put $n = 0$)

$$u_1^{(1)} = \frac{1}{4}(2000 + 625 + 1000 + 875) = 1125$$

$$u_2^{(1)} = \frac{1}{4}(1125 + 500 + 1000 + 375) = 750$$

$$u_3^{(1)} = \frac{1}{4}(2000 + 375 + 1125 + 500) = 1000$$

$$u_4^{(1)} = \frac{1}{4}(1000 + 0 + 750 + 0) \approx 438$$

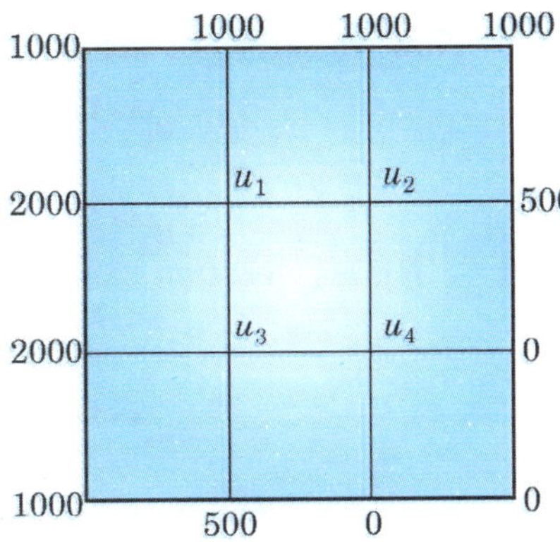

Fig. 33.7

*Second iteration* : (put $n = 1$)

$$u_1^{(2)} = \frac{1}{4}(2000 + 750 + 1000 + 1000) \approx 1188$$

$$u_2^{(2)} = \frac{1}{4}(1188 + 500 + 1000 + 438) \approx 782$$

$$u_3^{(2)} = \frac{1}{4}(2000 + 438 + 1188 + 500) \approx 1032$$

$$u_4^{(2)} = \frac{1}{4}(1032 + 0 + 782 + 0) \approx 454$$

*Third iteration* : (put $n = 2$)

$$u_1^{(3)} = \frac{1}{4}(2000 + 782 + 1000 + 1032) \approx 1204$$

$$u_2^{(3)} = \frac{1}{4}(1204 + 500 + 1000 + 454) \approx 789$$

$$u_3^{(3)} = \frac{1}{4}(2000 + 454 + 1204 + 500) \approx 1040$$

$$u_4^{(3)} = \frac{1}{4}(1040 + 0 + 789 + 0) \approx 458$$

Similarly, $u_1^{(4)} \approx 1207,\ u_2^{(4)} \approx 791,\ u_3^{(4)} \approx 1041,\ u_4^{(4)} = 458$

and $u_1^{(5)} = 1208,\ u_2^{(5)} = 791.5,\ u_3^{(5)} = 1041.5,\ u_4^{(5)} = 458.25$

Thus there is no significant difference between the fourth and fifth iteration values.

Hence $u_1 = 1208$, $u_2 = 792$, $u_3 = 1042$ and $u_4 = 458$.

**Example 33.4.** *Solve the Laplace equation $u_{xx} + u_{yy} = 0$ given that (Fig. 33.8).*

**Solution.** We first find the initial values in the following order :

$$u_5 = \frac{1}{4}(0 + 17 + 21 + 12.1) = 12.5 \qquad \text{(Std. formula)}$$

$$u_1 = \frac{1}{4}(0 + 12.5 + 0 + 17) = 7.4 \qquad \text{(Diag. formula)}$$

$$u_3 = \frac{1}{4}(12.5 + 18.6 + 17 + 21) = 17.28 \qquad \text{(Diag. formula)}$$

$$u_7 = \frac{1}{4}(12.5 + 0 + 0 + 12.1) = 6.15 \qquad \text{(Diag. formula)}$$

$$u_9 = \frac{1}{4}(12.5 + 9 + 21 + 12.1) = 13.65 \qquad \text{(Diag. formula)}$$

$$u_2 = \frac{1}{4}(17 + 12.5 + 7.4 + 17.3) = 13.55 \qquad \text{(Std. formula)}$$

$$u_4 = \frac{1}{4}(7.4 + 6.2 + 0 + 12.5) = 6.52 \qquad \text{(Std. formula)}$$

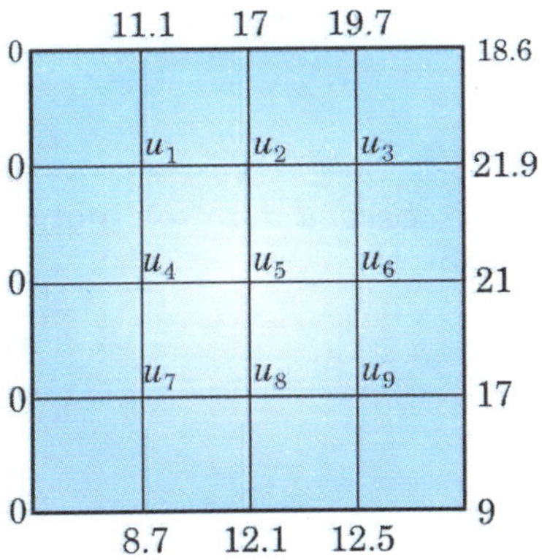

Fig. 33.8

$$u_6 = \frac{1}{4}(17.3 + 13.7 + 12.5 + 21) = 16.12 \quad \text{(Std. formula)}$$

$$u_8 = \frac{1}{4}(12.5 + 12.1 + 6.2 + 13.7) = 11.12 \quad \text{(Std. formula)}$$

Now we carry out the iteration process using the Standard formula :

$$u_1^{(n+1)} = \frac{1}{4}\left[(0 + 11.1 + u_4^{(n)} + u_2^{(n)})\right]$$

$$u_2^{(n+1)} = \frac{1}{4}\left[(u_1^{(n+1)} + 17 + u_5^{(n)} + u_3^{(n)})\right]$$

$$u_3^{(n+1)} = \frac{1}{4}\left[(u_2^{(n+1)} + 19.7 + u_6^{(n)} + 21.9)\right]$$

$$u_4^{(n+1)} = \frac{1}{4}\left[(u_1^{(n+1)} + 19.7 + u_7^{(n)} + u_5^{(n)})\right]$$

$$u_5^{(n+1)} = \frac{1}{4}\left[(u_4^{(n+1)} + u_2^{(n+1)} + u_8^{(n)} + u_6^{(n)})\right]$$

$$u_6^{(n+1)} = \frac{1}{4}\left[(u_5^{(n+1)} + u_3^{(n+1)} + u_9^{(n)} + 21)\right]$$

$$u_7^{(n+1)} = \frac{1}{4}\left[0 + (u_4^{(n+1)} + 8.7 + u_8^{(n)})\right]$$

$$u_8^{(n+1)} = \frac{1}{4}\left[(u_7^{(n+1)} + u_5^{(n+1)} + 12.1 + u_9^{(n)})\right]$$

$$u_9^{(n+1)} = \frac{1}{4}\left[(u_8^{(n+1)} + u_6^{(n)} + 12.8 + 17)\right]$$

*First iteration* (put $n = 0$, in the above results)

$$u_1^{(1)} = \frac{1}{4}\left(0 + 11.1 + u_4^{(o)} + u_2^{(o)}\right) = \frac{1}{4}(0 + 11.1 + 6.52 + 13.55) = 7.79$$

$$u_2^{(1)} = \frac{1}{4}(7.79 + 17 + 12.5 + 17.28) = 13.64$$

$$u_3^{(1)} = \frac{1}{4}(13.64 + 19.7 + 16.12 + 21.9) = 12.84$$

$$u_4^{(1)} = \frac{1}{4}(0 + 7.79 + 6.15 + 12.5) = 6.61$$

$$u_5^{(1)} = \frac{1}{4}(6.61 + 13.64 + 11.12 + 16.12) = 11.88$$

$$u_6^{(1)} = \frac{1}{4}(11.88 + 17.84 + 13.65 + 21) = 16.09$$

$$u_7^{(1)} = \frac{1}{4}(0 + 6.61 + 8.7 + 11.12) = 6.61$$

$$u_8^{(1)} = \frac{1}{4}(6.61 + 11.88 + 12.1 + 13.65) = 11.06$$

$$u_9^{(1)} = \frac{1}{4}(11.06 + 16.09 + 12.8 + 17) = 12.238$$

*Second iteration* (put $n = 1$)

$$u_1^{(2)} = \frac{1}{4}(0 + 11.1 + 6.61 + 13.64) = 7.84$$

$$u_2^{(2)} = \frac{1}{4}(7.84 + 17 + 11.88 + 17.84) = 16.64$$

$$u_3^{(2)} = \frac{1}{4}(13.64 + 19.7 + 16.09 + 21.9) = 17.83$$

$$u_4^{(2)} = \frac{1}{4}(0 + 7.84 + 6.61 + 11.88) = 6.58$$

$$u_5^{(2)} = \frac{1}{4}(6.58 + 13.64 + 11.06 + 16.09) = 11.84$$

$$u_6^{(2)} = \frac{1}{4}(11.84 + 17.83 + 14.24 + 21) = 16.23$$

$$u_7^{(2)} = \frac{1}{4}(0 + 6.58 + 8.7 + 11.06) = 6.58$$

$$u_8^{(2)} = \frac{1}{4}(6.58 + 11.84 + 12.1 + 14.24) = 11.19$$

$$u_9^{(2)} = \frac{1}{4}(11.19 + 16.23 + 12.8 + 17) = 14.30$$

*Third iteration* (put $n = 2$)

$$u_1^{(3)} = \frac{1}{4}(0 + 11.1 + 6.58 + 13.64) = 7.83$$

$$u_2^{(3)} = \frac{1}{4}(7.83 + 17 + 11.84 + 17.83) = 13.637$$

$$u_3^{(4)} = \frac{1}{4}(13.63 + 19.7 + 16.23 + 21.9) = 17.86$$

$$u_4^{(3)} = \frac{1}{4}(0 + 7.83 + 6.58 + 11.84) = 6.56$$

$$u_5^{(3)} = \frac{1}{4}(6.56 + 13.63 + 11.19 + 16.23) = 11.90$$

$$u_6^{(3)} = \frac{1}{4}(11.90 + 17.86 + 14.30 + 21) = 16.27$$

$$u_7^{(3)} = \frac{1}{4}(0 + 6.56 + 8.7 + 11.19) = 6.61$$

$$u_8^{(3)} = \frac{1}{4}(6.61 + 11.90 + 12.1 + 14.30) = 11.23$$

$$u_9^{(3)} = \frac{1}{4}(11.23 + 16.27 + 12.8 + 17) = 14.32$$

Similarly, $u_1^{(4)} = 7.82, u_2^{(4)} = 13.65, u_3^{(4)} = 17.88, u_4^{(4)} = 6.58, u_5^{(4)} = 11.94, u_6^{(4)} = 16.28,$

$u_7^{(4)} = 6.63, u_8^{(4)} = 11.25, u_9^{(4)} = 14.33$

$u_1^{(5)} = 7.83, u_2^{(5)} = 13.66, u_3^{(5)} = 17.89, u_4^{(5)} = 6.50, u_5^{(5)} = 11.95, u_6^{(5)} = 16.29,$

$u_7^{(5)} = 6.64, u_8^{(5)} = 11.25, u_9^{(5)} = 14.34$

## 33.6 SOLUTION OF POISSON'S EQUATION

$$\frac{\partial^2 \mathbf{u}}{\partial \mathbf{x}^2} + \frac{\partial^2 \mathbf{u}}{\partial \mathbf{y}^2} = \mathbf{f(x, y)} \qquad ...(1)$$

This is an *elliptic equation* which can be solved numerically at the interior mesh points of a square network when the boundary values are known. The standard 5-point formula for Eqn. (1) takes the form

$$u_{i-1,j} + u_{i+1,j} + u_{i,\,j+1} + u_{i,\,j-1} - 4u_{i,j} = h^2 f(ih, jh) \qquad ...(2)$$

By applying Eqn. (2) at each mesh-point, we arrive at linear equations in the pivotal values $i, j$. These equations can be solved by Gauss-Seidal iteration method.

**Example 33.5.** *Solve the Poisson equation $u_{xx} + u_{yy} = -81xy$, $0 < x < 1$, $0 < y < 1$ given that $u(0, y) = 0$, $u(x, 0) = 0$, $u(1, y) = 100$, $u(x, 1) = 100$ and $h = 1/3$.*

**Solution.** Here $h = 1/3$ $u_{i,\,j-1}$ $u_{i,\,j}$ (Fig. 33.9)

The standard 5-point formula for the given equation is

$$u_{i-1,j} + u_{i+1,j} + u_{i,j+1} + u_{i,j-1} - 4\,u_{i,j} = h^2 f(ih, jh) = h^2 [-81\,(ih\,.\,jh)] = h^4\,(-81)\,ij = -ij \quad ...(i)$$

For $u_1$ $(i = 1, j = 2)$, Eqn. $(i)$ gives $0 + u_2 + u_3 + 100 - 4u_1 = -2$

*i.e.*, $$-4u_1 + u_2 + u_3 = -102 \quad ...(ii)$$

For $u_2$ $(i = 2, j = 2)$, Eqn. $(i)$ gives $u_1 + 100 + u_4 + 100 - 4u_2 = -4$

*i.e.*, $$u_1 - 4u_2 + u_4 = -204 \quad ...(iii)$$

For $u_3$ $(i = 1, j = 1)$, Eqn. $(i)$ gives $0 + u_4 + 0 + u_1 - 4u_3 = -1$

*i.e.*, $$u_1 - 4u_3 + u_4 = -1 \quad ...(iv)$$

For $u_4$ $(i = 2, j = 1)$, gives $u_3 + 100 + u_2 - 4u_4 = -2$

$$u_2 + u_3 - 4u_4 = -102 \quad ...(v)$$

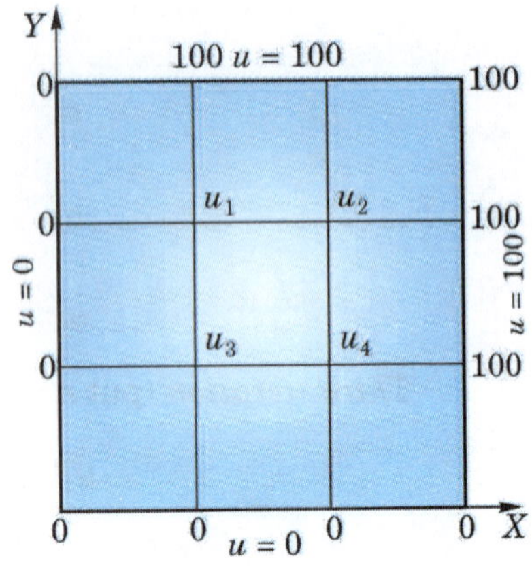

Fig. 33.9

Subtracting Eqns. $(v)$ from $(ii)$, $-4u_1 + 4u_4 = 0$ *i.e.* $u_1 = u_4$

Then Eqn. $(iii)$ becomes $2u_1 - 4u_2 = -204$ ...$(vi)$

and Eqn. $(iv)$ becomes $2u_1 - 4u_3 = -1$ ...$(vii)$

Now $(iv)$ × Eqns. $(ii) + (vi)$ gives $-14u_1 + 4u_3 = -612$ ...$(viii)$

Equation $(vii) + (viii)$ gives $-12u_1 = -613$

Thus $$u_1 = 613/12 = 51.0833 = u_4.$$

From Eqn. $(vi)$, $$u_2 = \frac{1}{2}\,(u_1 + 102) = 76.5477$$

From Eqn. $(vii)$, $$u_3 = \frac{1}{2}\left(u_1 + \frac{1}{2}\right) = 25.7916.$$

**Example 33.6.** *Solve the partial differential equation $\nabla^2 u = -10(x^2 + y^2 + 10)$ over the square with sides $x = 0 = y$, $x = 3 = y$ with $u = 0$ on the boundary and mesh length = 1.* *(Anna, 2012)*

**Solution.** Here $h = 1$ (Fig. 33.10).

∴ The standard 5-point formula for the given equation is

$$u_{i-1,j} + u_{i+1,j} + u_{i,j+1} + u_{i,j-1} - 4u_{i,j} = -10(i^2 + j^2 + 10) \quad ...(i)$$

*For* $u_1(i = 1, j = 2)$, Eqn. $(i)$ gives

$$0 + u_2 + 0 + u_3 - 4u_1 = -10(1 + 4 + 10)$$

*i.e.*, $$u_1 = \frac{1}{4}(u_2 + u_3 + 150) \quad ...(ii)$$

*For* $u_2(i = 2, j = 2)$, Eqn. $(i)$ gives $u_2 = \frac{1}{4}(u_1 + u_4 + 180)$ ...$(iii)$

*For* $u_3(i = 1, j = 1)$, we have $u_3 = \frac{1}{4}(u_1 + u_4 + 120)$ ...$(iv)$

*For* $u_4(i = 2, j = 1)$, we have $u_4 = \frac{1}{4}(u_2 + u_3 + 150)$ ...$(v)$

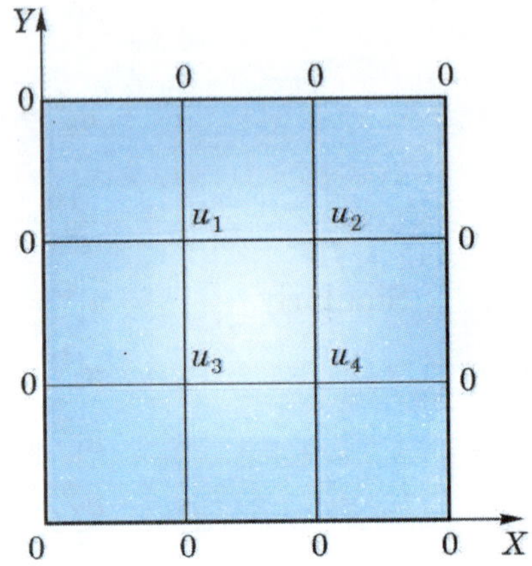

Fig. 33.10

Equations $(ii)$ and $(v)$ show that $u_4 = u_1$. Thus the above equations reduce to

$$u_1 = \frac{1}{4}(u_2 + u_3 + 150) \quad ...(vi)$$

$$u_2 = \frac{1}{2}(u_1 + 90) \quad ...(vii)$$

$$u_3 = \frac{1}{2}(u_1 + 60) \quad ...(viii)$$

Now let us solve these equations by Gauss-Seidal iteration method.

*First iteration* : Starting from the approximations $u_2 = 0$, $u_3 = 0$, we obtain

$$u_1^{(1)} = 37.5. \text{ Then } u_2^{(1)} = \frac{1}{2}(37.5 + 90) \approx 64\,;\ u_3^{(1)} = \frac{1}{2}(37.5 + 60) \approx 49$$

*Second iteration :*

$$u_1^{(2)} = \frac{1}{4}(64 + 49 + 150) \approx 66 \; ; u_2^{(2)} = \frac{1}{2}(66 + 90) = 78 \; ; u_3^{(2)} = \frac{1}{2}(66 + 60) = 63$$

*Third iteration :*

$$u_1^{(3)} = \frac{1}{4}(78 + 63 + 150) \approx 73 \; ; u_2^{(3)} = \frac{1}{2}(73 + 90) \approx 82 \; ; u_3^{(2)} = \frac{1}{2}(73 + 60) \approx 67$$

*Fourth iteration :*

$$u_1^{(4)} = \frac{1}{4}(82 + 67 + 150) \approx 75 \; ; u_2^{(4)} = \frac{1}{2}(75 + 90) = 82.5 \; ; u_3^{(4)} = \frac{1}{2}(75 + 60) = 67.5$$

*Fifth iteration :*

$$u_1^{(5)} = \frac{1}{4}(82.5 + 67.5 + 150) \approx 75 \; ; u_2^{(5)} = \frac{1}{2}(75 + 90) = 82.5 \; ; u_3^{(5)} = \frac{1}{2}(75 + 60) = 67.5$$

Since these values are the same as those of fourth iteration, we have

$$u_1 = 75, u_2 = 82.5, u_3 = 67.5 \text{ and } u_4 = 75.$$

## PROBLEMS 33.2

1. Solve the equation $u_{xx} + u_{yy} = 0$ for the following square mesh with boundary values as shown in Fig. 33.11. Iterate until the maximum difference between the successive values at any point is less than 0.001.
2. Solve $\nabla^2 u = 0$ under the conditions $(h = 1, k = 1)$, $u(0, y) = 0$, $u(4, y) = 12 + y$ for $0 \le y \le 4$ ; $u(x, 0) = 3x$, $u(x, 4) = x^2$ for $0 \le x \le 4$.
3. Solve the elliptic equation $u_{xx} + u_{yy} = 0$ for the square mesh with boundary values as shown in Fig. 33.12. Iterate until the maximum difference between successive values at any point is less than 0.005.

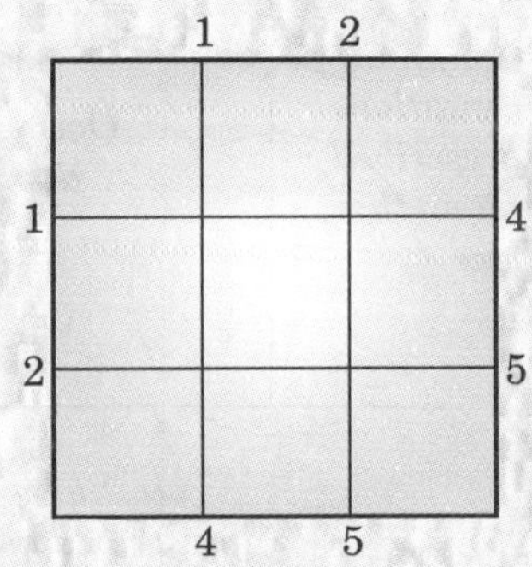

Fig. 33.11

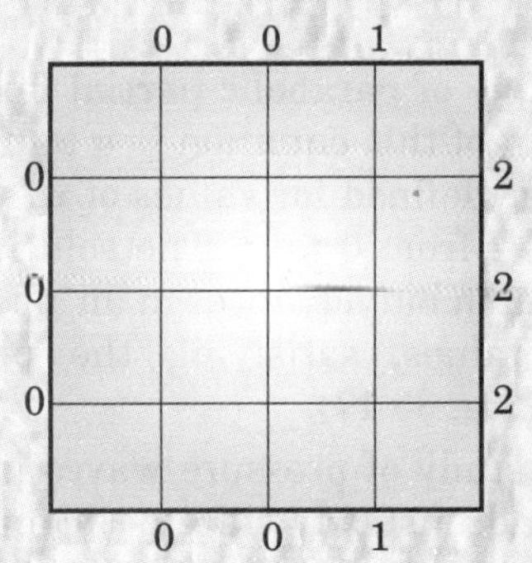

Fig. 33.12

4. Using central-difference approximation solve $\nabla^2 u = 0$ at the nodal points of the square grid of Fig. 33.13 using the boundary values indicated.

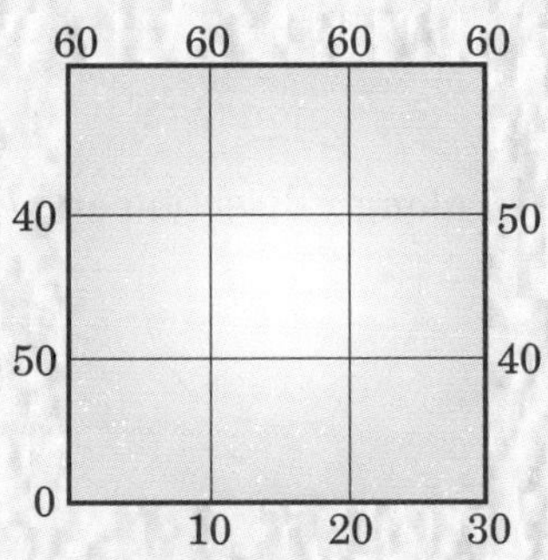

Fig. 33.13

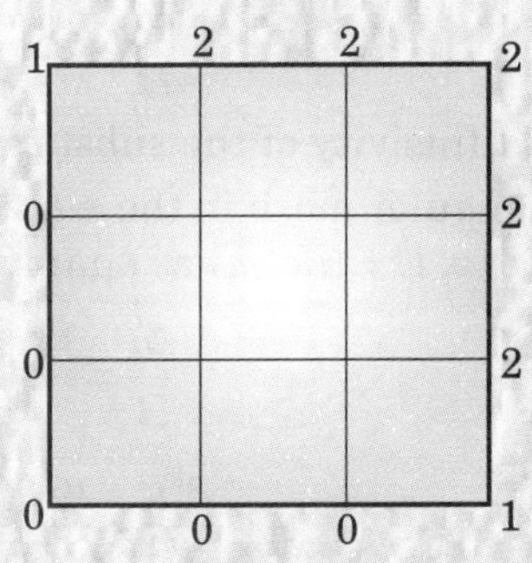

Fig. 33.14

5. Solve $u_{xx} + u_{yy} = 0$ for the square mesh with boundary values as shown in Fig. 33.14. Iterate till the mesh values are correct to two decimal places.

6. Solve the Laplace's equation $u_{xx} + u_{yy} = 0$ in the domain of Fig. 33.15.

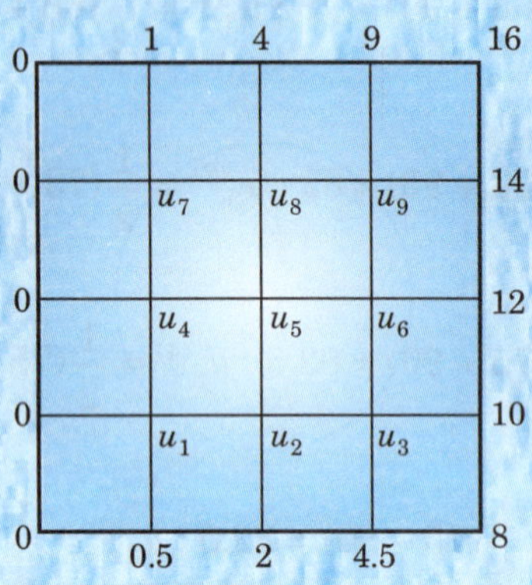


Fig. 33.15

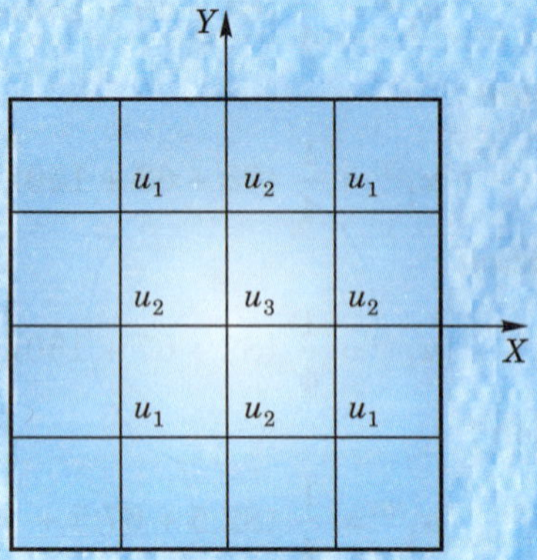


Fig. 33.16

7. Solve the Poisson's equation $\nabla^2 u = 8x^2y^2$ for the square mash of Fig. 33.16 with $u(x, y) = 0$ on the boundary and mesh length = 1. *(Anna, 2013)*

**Note.** Solution of elliptic equations by *Relaxation method* is given in author's book *'Numerical Methods in Engineering and Science'*.

## 33.7 PARABOLIC EQUATIONS

The one-dimensional heat conduction equation

$$\frac{\partial u}{\partial t} = c^2 \frac{\partial^2 u}{\partial x^2}$$

is a well-known example of parabolic partial differential equations. The solution of this equation is a temperature function $u(x, t)$ which is defined for values of $x$ from 0 to $l$ and for values of time $t$ from 0 to $\infty$. The solution is not defined in a closed domain but advances in an open-ended region from initial values, satisfying the prescribed boundary conditions. (Fig. 33.17).

In general, the study of pressure waves in a fluid, propagation of heat and unsteady state problems lead to parabolic type of equations.

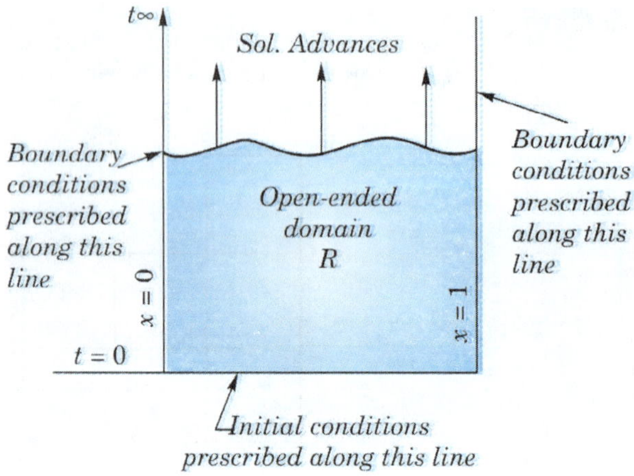


Fig. 33.17

## 33.8 SOLUTION OF HEAT EQUATION

$$\frac{\partial u}{\partial t} = c^2 \frac{\partial^2 u}{\partial x^2} \qquad ...(1)$$

where $c^2 = k/s\rho$ is the diffusivity of the substance (cm²/sec.)

Consider a rectangular mesh in the $x$-$t$ plane with spacing $h$ along $x$ direction and $k$ along time $t$ direction. Denoting a mesh point $(x, t) = (ih, jk)$ as simply $i, j$, we have

$$\frac{\partial u}{\partial t} = \frac{u_{i,\,j+1} - u_{i,\,j}}{k} \qquad \text{[By (5) § 33.3]}$$

and

$$\frac{\partial^2 u}{\partial x^2} = \frac{u_{i-1,\,j} - 2u_{i,\,j} + u_{i+1,\,j}}{h^2} \qquad \text{[By (4) § 33.3]}$$

Substituting these in Eqn. (1), we obtain

$$u_{i,\,j+1} - u_{i,\,j} = \frac{kc^2}{h^2}\,[u_{i-1,\,j} - 2u_{i,\,j} + u_{i+1,\,j}]$$

or $$u_{i,j+1} = \alpha u_{i-1,j} + (1-2\alpha)\,u_{i,j} + \alpha u_{i+1,j} \quad ...(2)$$
where $\alpha = kc^2/h^2$ is the mesh ratio parameter.

This formula enables us to determine the value of $u$ at the $(i, j+1)$th mesh point in terms of the known function values at the points $x_{i-1}$, $x_i$ and $x_{i+1}$ at the instant $t_j$. It is a relation between the function values at the two time levels $j+1$ and $j$ and is therefore, called a *2-level formula*. In schematic form Eqn. (2) is shown in Fig. 33.18 which is called the *Schmidt explicit formula* which is valid only for $0 < \alpha \le \frac{1}{2}$.

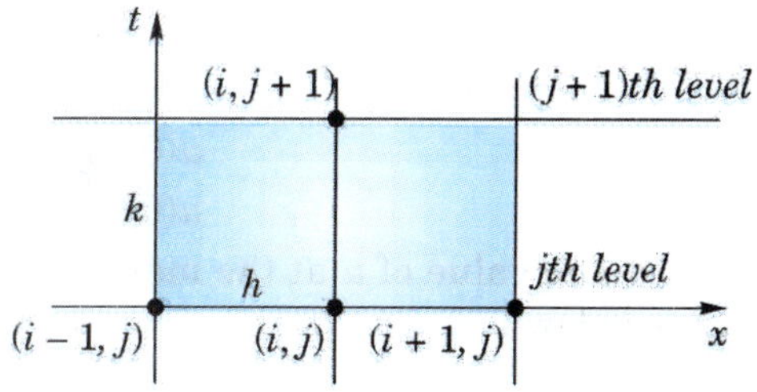


Fig. 33.18

**Obs.** In particular when $\alpha = \frac{1}{2}$, Eqn. (2) reduces to
$$u_{i,j+1} = \frac{1}{2}(u_{i-1,j} + u_{i+1,j}) \quad ...(3)$$
which shows that the *value of u at $x_i$ at time $t_{j+1}$ is the mean of the u-values at $x_{i-1}$ and $x_{i+1}$ at time $t_j$*. This relation, known as *Bendre-Schmidt recurrence relation,* gives the values of $u$ at the internal mesh points with the help of boundary conditions.

**Note.** The other formulae (*i.e., Crank-Nicolson formula* and *Du Fort-Frankel formula*) are given in author's book '*Numerical Methods in Engineering and Science*'.

**Example 33.7.** *Solve $\frac{\partial u}{\partial t} = \frac{\partial^2 u}{\partial^2 x^2}$ in $0 < x < 5, t \ge 0$ given that $u(x, 0) = 20, u(0, t) = 0, u(5, t) = 100$. Compute u for the time-step with h = 1 by Crank-Nicholson method.*

**Solution.** Here $c^2 = 1$ and $h = 1$.

Taking $\alpha$ (*i.e.*, $c^2k/h$) = 1, we get $k = 1$

Also we have

| j \ i | 0 | 1 | 2 | 3 | 4 | 5 |
|---|---|---|---|---|---|---|
| 0 | 0 | 20 | 20 | 20 | 20 | 100 |
| 1 | 0 | $u_1$ | $u_2$ | $u_3$ | $u_4$ | 100 |

Then Crank-Nicholson formula becomes
$$4u_{i,j+1} = u_{i-1,j+1} + u_{i+1,j+1} + u_{i-1,j} + u_{i+1,j}$$
$$4u_1 = 0 + 20 + 0 + u_2 \quad i.e. \quad 4u_1 - u_2 = 20 \quad ...(1)$$
$$4u_2 = 20 + 20 + u_1 + u_3 \quad i.e. \quad 4u_1 - 4u_2 + u_3 = -40 \quad ...(2)$$
$$4u_3 = 20 + 20 + u_2 + u_4 \quad i.e. \quad u_2 - 4u_3 + u_4 = -40 \quad ...(3)$$
$$4u_4 = 20 + 100 + u_3 + 100 \quad i.e. \quad 4u_3 - 4u_4 + = -220 \quad ...(4)$$
Now (1) – 4(2) gives $15u_2 - 4u_3 = 180$ ...(5)

4(3) + (4) gives $4u_2 - 15u_3 = -380$ ...(6)

Then 15(5) – 4(6) gives $209u_2 = 4220$ *i.e.* $u_2 = 20.2$

From (5), we get $4u_3 = 15 \times 20.2 - 180$ *i.e.* $u_3 = 30.75$

From (1), $4u_1 = 20 + 20.2$ *i.e.* $u_1 = 10.05$

From (4), $4u_4 = 220 + 30.75$ *i.e.* $u_4 = 62.69$

Thus the required values are 10.05, 20.2, 30.75 and 62.68.

**Example 33.8.** *Solve the boundary value problem $u_t = u_{xx}$ under the conditions $u(0, t) = u(1, t) = 0$ and $u(x, 0) = \sin \pi x$, $0 \le x \le 1$ using Schmidt method (Take $h = 0.2$ and $\alpha = 1/2$).* (V.T.U., 2013)

**Solution.** Since $h = 0.2$ and $\alpha = 1/2$

$\therefore$ $\alpha = \frac{k}{h^2}$ gives $k = 0.02$

Since $\alpha = 1/2$, we use Bendre-Schmidt relation

$$u_{i,j+1} = \frac{1}{2}(i_{i-1,j} + u_{i+1,j}) \qquad ...(i)$$

We have $u(0, 0) = 0$, $u(0.2, 0) = \sin \pi/5 = 0.5875$

$u(0.4, 0) = \sin 2\pi/5 = 0.95111$, $u(0.6, 0) = \sin 3\pi/5 = 0.9511$

$u(0.8, 0) = \sin 4\pi/5 = 0.5875$, $u(1, 0) = \sin \pi = 0$

The value of $u$ at the mesh points can be obtained by using the recurrence relation Eqn. $(i)$ as shown in table below :

| $t \downarrow$ | $x \rightarrow$ | 0 | 0.2 | 0.4 | 0.6 | 0.8 | 1.0 |
|---|---|---|---|---|---|---|---|
| | $j \backslash i$ | 0 | 1 | 2 | 3 | 4 | 5 |
| 0 | 0 | 0 | 0.5878 | 0.9511 | 0.9511 | 0.5878 | 0 |
| 0.02 | 1 | 0 | 0.4756 | 0.7695 | 0.7695 | 0.4756 | 0 |
| 0.04 | 2 | 0 | 0.3848 | 0.6225 | 0.6225 | 0.3848 | 0 |
| 0.06 | 3 | 0 | 0.3113 | 0.5036 | 0.5036 | 0.3113 | 0 |
| 0.08 | 4 | 0 | 0.2518 | 0.4074 | 0.4074 | 0.2518 | 0 |
| 0.1 | 5 | 0 | 0.2037 | 0.3296 | 0.3296 | 0.2037 | 0 |

**Example 33.9.** *Find the values of $u(x, t)$ satisfying the parabolic equation $\frac{\partial u}{\partial t} = 4\frac{\partial^2 u}{\partial x^2}$ and the boundary conditions $u(0, t) = 0 = u(8, t)$ and $u(x, 0) = 4x - \frac{1}{2}x^2$ at the points $x = i : i = 0, 1, 2, ..., 8$ and $t = \frac{1}{8}j : j = 0, 1, 2, ..., 5$.*

**Solution.** Here $c^2 = 4$, $h = 1$ and $k = 1/8$. Then $\alpha = c^2k/h^2 = 1/2$.

$\therefore$ we have Bendre-Schmidt's recurrence relation

$$u_{i,j+1} = \frac{1}{2}(u_{i-1,j} + u_{i+1,j}) \qquad ...(i)$$

Now since $u(0, t) = 0 = u(8, t)$

$\therefore$ $u_{0,i} = 0$ and $u_{8,j} = 0$ for all values of $j$, *i.e.*, the entries in the first and last column are zero.

Since $u(x, 0) = 4x - \frac{1}{2}x^2$

$$\therefore \quad u_{i,0} = 4i - \frac{1}{2}i^2 = 0, 3.5, 6, 7.5, 8, 7.5, 6, 3.5$$

for $i = 0, 1, 2, 3, 4, 5, 6, 7$ at $t = 0$

These are the entries of the first row.

| $j \backslash i$ | 0 | 1 | 2 | 3 | 4 | 5 | 6 | 7 | 8 |
|---|---|---|---|---|---|---|---|---|---|
| 0 | 0 | 3.5 | 6 | 7.5 | 8 | 7.5 | 6 | 3.5 | 0 |
| 1 | 0 | 3 | 5.5 | 7 | 7.5 | 7 | 5.5 | 3 | 0 |
| 2 | 0 | 2.75 | 5 | 6.5 | 7 | 6.5 | 5 | 2.75 | 0 |
| 3 | 0 | 2.5 | 4.625 | 6 | 6.5 | 6 | 4.625 | 2.5 | 0 |
| 4 | 0 | 2.3125 | 4.25 | 5.5625 | 6 | 5.5625 | 4.25 | 2.3125 | 0 |
| 5 | 0 | 2.125 | 3.9375 | 5.125 | 5.5625 | 5.125 | 3.9375 | 2.125 | 0 |

Putting $j = 0$ in Eqn. $(i)$, we have

$$u_{i,1} = \frac{1}{2}(u_{i-1,0} + u_{i+1,0})$$

Taking $i = 1, 2, ..., 7$ successively, we get

$$u_{1,1} = \frac{1}{2}(u_{0,0} + u_{2,0}) = \frac{1}{2}(0 + 6) = 3$$

$$u_{2,1} = \frac{1}{2}(u_{1,0} + u_{3,0}) = \frac{1}{2}(3.5 + 7.5) = 5.5$$

$$u_{3,1} = \frac{1}{2}(u_{2,0} + u_{4,0}) = \frac{1}{2}(6 + 8) = 7$$

$$u_{4,1} = 7.5, u_{5,1} = 7, u_{6,1} = 5.5, u_{7,1} = 3.$$

These are the entries in the second row.

Putting $j = 1$ in Eqn. $(i)$, the entries of the third row are given by

$$u_{i,2} = \frac{1}{2}(u_{i-1,1} + u_{i+1,1})$$

Similarly putting $j = 2, 3, 4$ successively in Eqn. $(i)$, the entries of the fourth, fifth and sixth rows are obtained.

Hence the values of $u_{i,j}$ are as given in the above table.

**Example 33.10.** *Solve the equation*

$$\frac{\partial u}{\partial t} = \frac{\partial^2 u}{\partial x^2}$$

*subject to the conditions* $u(x, 0) = \sin \pi x, 0 \le x \le 1$ ; $u(0, t) = u(1, t) = 0$. *Carry out computations for two levels, taking* $h = 1/3, k = 1/36$. *(V.T.U., 2016 ; Anna, 2012)*

**Solution.** Here $c^2 = 1, h = 1/3, k = 1/36$ so that

$$\alpha = kc^2/h^2 = 1/4.$$

Also $u_{1,0} = \sin \pi/3 = \sqrt{3}/2, u_{2,0} = \sin 2\pi/3 = \sqrt{3}/2$

and all other boundary values are zero as shown in Fig. 33.19.

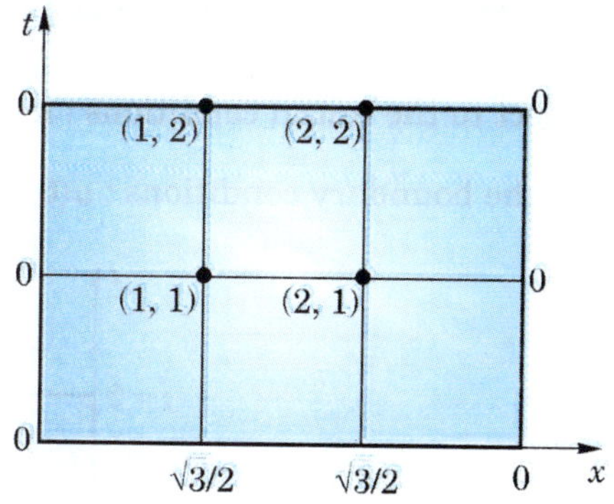


Fig. 33.19

*Schmidt's formula* [(2) of § 33.8]

$$u_{i,j+1} = \alpha u_{i-1,j} + (1 - 2\alpha) u_{i,j} + \alpha u_{i+1,j}$$

becomes

$$u_{i,j+1} = \frac{1}{4}[u_{i-1,j} + 2u_{i,j} + u_{i+1,j}]$$

For $i = 1, 2 ; j = 0$ :

$$u_{1,1} = \frac{1}{4}[u_{0,0} + 2u_{1,0} + u_{2,0}] = \frac{1}{4}(0 + 2 \times \sqrt{3/2} + \sqrt{3/2}) = 0.65$$

$$u_{2,1} = \frac{1}{4}[u_{1,0} + 2u_{2,0} + u_{3,0}] = \frac{1}{4}(\sqrt{3/2} + 2 \times \sqrt{3/2} + 0) = 0.65$$

For $i = 1, 2, j = 1$ :

$$u_{1,2} = \frac{1}{4}(u_{0,1} + 2u_{1,1} + u_{2,1}) = 0.49$$

$$u_{2,2} = \frac{1}{4}(u_{1,1} + 2u_{2,1} + u_{3,1}) = 0.49.$$

## PROBLEMS 33.3

1. Find the solution of the parabolic equation $u_{xx} = 2u_t$ when $u(0, t) = u(4, t) = 0$ and $u(x, 0) = x(4 - x)$, taking $h = 1$. Find the values upto $t = 5$.

2. Solve the equation $\frac{\partial^2 u}{\partial x^2} = \frac{\partial u}{\partial t}$ with the conditions $u(0, t) = 0$, $u(x, 0) = x(1 - x)$ and $u(1, t) = 0$. Assume $h = 0.1$. Tabulate $u$ for $t = k, 2k$ and $3k$ choosing an appropriate value of $k$.

**3.** Given $\frac{\partial^2 f}{\partial x^2} - \frac{\partial f}{\partial t} = 0$ ; $f(0, t) = f(5, t) = 0, f(x, 0) = x^2 (25 - x^2)$ ; find the values of $f$ for $x = ih(i = 0, 1, ..., 5)$ and $t = jk(j = 0, 1, ..., 6)$ with $h = 1$ and $k = \frac{1}{2}$, using the explicit method. *(Anna, 2015)*

**4.** Solve the heat equation $\partial u / \partial t = \partial^2 u / \partial x^2$ subject to the conditions $u(0, t) = u(1, t) = 0$ and

$$u(x, 0) = 2x \text{ for } 0 \le x \le \frac{1}{2}$$
$$= 2(1 - x) \text{ for } \frac{1}{2} \le x \le 1.$$

Take $h = 1/4$ and $k$ according to Bendre-Schmidt equation.

## 33.9 HYPERBOLIC EQUATIONS

The wave equation : $\frac{\partial^2 u}{dt^2} = c^2 \frac{\partial^2 u}{\partial x^2}$ is the simplest example of hyperbolic partial differential equations. Its solution is the displacement function $u(x, t)$ defined for values of $x$ from 0 to $l$ and for $t$ from 0 to $\infty$, satisfying the initial and boundary conditions. The solution as for parabolic equations, advances in an open-ended region (Fig. 33.17). In the case of hyperbolic equations however, we have two initial conditions and two boundary conditions.

Such equations arise from convective type of problems in vibrations, wave mechanics and gas dynamics.

## 33.10 SOLUTION OF WAVE EQUATION

$$\frac{\partial^2 u}{dt^2} = c^2 \frac{\partial^2 u}{\partial x^2} \quad ...(1)$$

subject to the initial conditions : $u = f(x), \frac{\partial u}{\partial t} = g(x), 0 \le x \le 1$ at $t = 0$ ...(2)

and the boundary conditions : $u(0, t) = \phi(t), u(1, t) = \psi(t)$. ...(3)

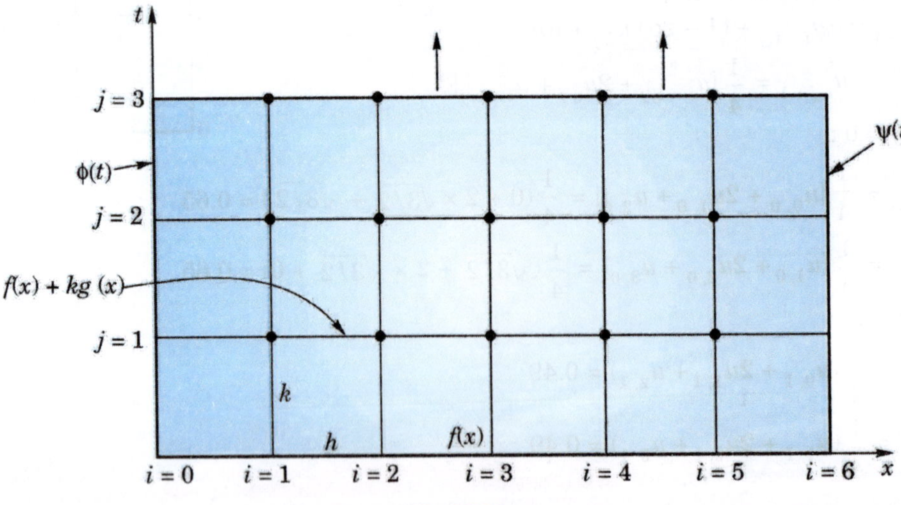


Fig. 33.20

Consider a rectangular mesh in the $x$-$t$ plane spacing $h$ along $x$ direction and $k$ along time $t$ direction. Denoting a mesh point $(x, t) = (ih, jk)$ as simply $i, j$, we have

$$\frac{\partial^2 u}{\partial x^2} = \frac{u_{i-1, j} - 2u_{i, j} + u_{i+1, j}}{h^2} \quad \text{and} \quad \frac{\partial^2 u}{\partial t^2} = \frac{u_{i,j-1} - 2u_{i, j} + u_{i, j+1}}{k^2}$$

Replacing the derivatives in Eqn. (1) by their above approximations, we obtain

$$u_{i,j-1} - 2u_{i,j} + u_{i,j+1} = \frac{c^2k^2}{h^2}(u_{i-1,j} - 2u_{i,j} + u_{i,j+1})$$

or $$u_{i,j+1} = 2(1 - \alpha^2c^2)\, u_{i,j} + \alpha^2c^2\, (u_{i-1,j} + u_{i+1,j} - u_{i,j-1}) \qquad ...(4)$$

where $\alpha = k/h$.

Now replacing the derivative in Eqn. (2) by its central difference approximation, we get

$$\frac{u_{i,j+1} - u_{i,j-1}}{2k} = \frac{\partial u}{\partial t} = g(x)$$ [See (7) p. 1179]

or $$u_{i,j+1} = u_{i,j-1} + 2k\, g(x) \quad \text{at } t = 0 \quad i.e. \ u_{i,1} = u_{i,-1} + 2kg(x) \text{ for } j = 0 \qquad ...(5)$$

Also initial condition $u = f(x)$ at $t = 0$ becomes $u_{i,0} = f(x)$ ...(6)

Combining Eqns. (5) and (6), we have $u_{i,1} = f(x) + 2kg(x)$ ...(7)

Also Eqn. (3) gives $u_{0,j} = \phi(t)$ and $u_{1,j} = \psi(t)$

Hence Eqn. (4) gives the values of $u_{i,j+1}$ at the $(j + 1)$th level when the nodal values at $(j - 1)$th and $j$th levels are known from Eqns. (6) and (7) as shown in Fig. 33.20. Thus Eqn. (4) gives an ***implicit scheme*** *for the solution of the wave equation.*

**A special case :** The coefficient of $u_{i,j}$ in Eqn. (4) will vanish if $\alpha = 1/c$ or $k = h/c$. Then Eqn. (4) reduces to the *simple form*

$$\mathbf{u_{i,j+1} = u_{i-1,j} + u_{i+1,j} - u_{i,j-1}} \qquad ...(8)$$

*This result provides an* ***explicit scheme*** *for the solution of the wave equation.*

**Obs. 1.** *For* $\alpha = 1/c$, the solution of Eqn. (4) is stable and coincides with the solution of Eqn. (1).
*For* $\alpha < 1/c$, the solution is stable but inaccurate.
*For* $\alpha > 1/c$, the solution is unstable.

**Obs. 2.** *The formula Eqn. (4) converges for $\alpha \le 1$ i.e. for $k \le h$.*

**Example 33.11.** *Evaluate the pivotal values of the equation $u_{tt} = 16u_{xx}$, taking $h = 1$ upto $t = 1.25$. The boundary conditions are $u(0, t) = u(5, t) = 0$, $u_t(x, 0) = 0$ and $u(x, 0) = x^2 (5 - x)$.* (V.T.U., 2015 S)

**Solution.** Here $c^2 = 16$.

$\therefore$ The difference equation for the given equation is

$$u_{i,j+1} = 2(1 - 16\alpha^2)\, u_{i,j} + 16\alpha^2\, (u_{i-1,j} + u_{i+1,j}) - u_{i,j-1} \quad \text{where } \alpha = k/h \qquad ...(i)$$

Taking $h = 1$ and choosing $k$ so that the coefficient of $u_{i,j}$ vanishes, we have $16\alpha^2 = 1$, *i.e.*, $k = h/4, = 1/4$.

$\therefore$ Eqn. (*i*) reduces to $$u_{i,j+1} = u_{i-1,j} + u_{i+1,j} - u_{i,j-1} \qquad ...(ii)$$

which gives a convergent solution (since $k/h < 1$). Its solution coincides with the solution of the given differential equation.

Now since $u(0, t) = u(5, t) = 0$.

$\therefore$ $u_{0,j} = 0$ and $u_{5,j} = 0$ for all values of $j$

*i.e.* the entries in the first and last columns are zero.

Since $u_{(x,0)} = x^2(5 - x)$

$\therefore$ $u_{i,0} = i^2(5 - i) = 4, 12, 18, 16$ for $i = 1, 2, 3, 4$ at $t = 0$.

These are the entries of the *first row*.

Finally the initial condition $u_t(x, 0) = 0$, becomes

$$\frac{u_{i,j+1} - u_{i,j-1}}{2k} = 0, \text{ when } j = 0, \text{ giving } u_{i,1} = u_{i,-1} \qquad ...(iii)$$

*Putting j = 0* in (*ii*),
$$u_{i,1} = u_{i-1,0} + u_{i+1,0} - u_{i,-1}$$
$$= u_{i-1,0} + u_{i+1,0} - u_{i,1} \quad \text{using } (iii)$$

or $$u_{i,1} = \frac{1}{2}(u_{i-1,0} + u_{i+1,0}) \qquad ...(iv)$$

Taking $i = 1, 2, 3, 4$ successively, we obtain

$$u_{1,1} = \frac{1}{2}(u_{0,0} + u_{2,0}) = \frac{1}{2}(0 + 12) = 6$$

$$u_{2,1} = \frac{1}{2}(u_{1,0} + u_{3,0}) = \frac{1}{2}(4 + 18) = 11$$

$$u_{3,1} = \frac{1}{2}(u_{2,0} + u_{4,0}) = \frac{1}{2}(12 + 16) = 14$$

$$u_{4,1} = \frac{1}{2}(u_{3,0} + u_{5,0}) = \frac{1}{2}(18 + 0) = 9$$

These are the entries of the *second row.*

*Putting* $j = 1$ in Eqn. (*ii*), we get

$$u_{i,2} = u_{i-1,1} + u_{i+1,1} - u_{i,0}$$

Taking $i = 1, 2, 3, 4$ successively, we obtain

$$u_{1,2} = u_{0,1} + u_{2,1} - u_{1,0} = 0 + 11 - 4 = 7$$
$$u_{2,2} = u_{1,1} + u_{3,1} - u_{2,0} = 6 + 14 - 12 = 8$$
$$u_{3,2} = u_{2,1} + u_{4,1} - u_{3,0} = 11 + 9 - 18 = 2$$
$$u_{4,2} = u_{3,1} + u_{5,1} - u_{4,0} = 14 + 0 - 16 = -2$$

These are the entries of the *third row.*

Similarly putting $j = 2, 3, 4$ successively in Eqn. (*ii*), the entries of the fourth, fifth and sixth rows are obtained. Hence the values of $u_{i,j}$ are as shown in the table below :

| $j$ \ $i$ | 0 | 1 | 2 | 3 | 4 | 5 |
|---|---|---|---|---|---|---|
| 0 | 0 | 4 | 12 | 18 | 16 | 0 |
| 1 | 0 | 6 | 11 | 14 | 9 | 0 |
| 2 | 0 | 7 | 8 | 2 | –2 | 0 |
| 3 | 0 | 2 | –2 | –8 | –7 | 0 |
| 4 | 0 | –9 | –14 | –11 | –6 | 0 |
| 5 | 0 | –16 | –18 | –12 | –4 | 0 |

**Example 33.12.** *The transverse displacement u of a point at a distance x from one end and at any time t of a vibrating string satisfies the equation* $\partial^2 u/\partial t^2 = 4\partial^2 u/\partial x^2$, *with boundary conditions* $u = 0$ *at* $x = 0, t > 0$ *and* $u = 0$ *at* $x = 4, t > 0$ *and initial conditions* $u = x(4 - x)$ *and* $\partial u/\partial t = 0$ *at* $t = 0, 0 \le x \le 4$. *Solve this equation numerically for one half period of vibration, taking* $h = 1$ *and* $k = 1/2$. (*V.T.U., 2016*)

**Solution.** Here, $h/k = 2 = c$.

∴ the difference equation for the given equation is

$$u_{i,j+1} = u_{i-1,j} + u_{i+1,j} - u_{i,j-1} \qquad ...(i)$$

which gives a convergent solution (since $k < h$).

Now since $u(0, t) = u(4, t) = 0$,

∴ $u_{0,j} = 0$ and $u_{4,j} = 0$ for all values of $j$.

*i.e.*, the entries in the first and last columns are zero.

Since $u_{(x,0)} = x(4 - x)$,

∴ $u_{i,0} = i(4 - i) = 3, 4, 3$ for $i = 1, 2, 3$ at $t = 0$.

These are the entries of the *first row.*

Also $u_t(x, 0) = 0$ becomes

$$\frac{u_{i,j+1} - u_{i,j-1}}{2k} = 0 \text{ when } j = 0, \text{ giving } u_{i,1} = u_{i,-1} \qquad ...(ii)$$

*Putting* $j = 0$ in Eqn. (*i*), $u_{i,1} = u_{i-1,0} + u_{i+1,0} - u_{i,-1}$

$= u_{i-1,0} + u_{i+1,0} - u_{i,1}$, using Eqn. (*ii*)

or

$$u_{i,1} = \frac{1}{2}(u_{i-1,0} + u_{i+1,0}) \qquad ...(iii)$$

Taking $i = 1, 2, 3$ successively, we obtain

$$u_{1,1} = \frac{1}{2}(u_{0,0} + u_{1,0}) = 2 \; ; \; u_{2,1} = \frac{1}{2}(u_{1,0} + u_{3,0}) = 3$$

$$u_{3,1} = \frac{1}{2}(u_{2,0} + u_{4,0}) = 2$$

These are the entries of the *2nd row*.

*Putting* $j = 1$ in Eqn. (*i*), $u_{i,2} = u_{i-1,1} + u_{i+1,1} - u_{i,0}$

Taking $i = 1, 2, 3$, successively, we get

$$u_{1,2} = u_{0,1} + u_{2,1} - u_{1,0} = 0 + 3 - 3 = 0$$
$$u_{2,2} = u_{1,1} + u_{3,1} - u_{2,0} = 2 + 2 - 4 = 0$$
$$u_{3,2} = u_{2,1} + u_{4,1} - u_{3,0} = 3 + 0 - 3 = 0$$

These are the entries of the *3rd row* and so on.

Now the equation of the vibrating string of length $l$ is $u_{tt} = c^2 u_{xx}$.

$\therefore$ Its period of vibration $= \frac{2l}{c} = \frac{2 \times 4}{2} = 4$ sec. [$\because$ $l = 4$ and $c = 2$]

This shows that we have to compute $u_{(x,t)}$ upto $t = 2$.

*i.e.*, similarly we obtain the values of $u_{i,2}$ (4th row) and $u_{i,3}$ (5th row).

Hence the values of $u_{i,j}$ are as shown in the table below :

| j \ i | 0 | 1 | 2 | 3 | 4 |
|---|---|---|---|---|---|
| 0 | 0 | 3 | 4 | 3 | 0 |
| 1 | 0 | 2 | 3 | 2 | 0 |
| 2 | 0 | 0 | 0 | 0 | 0 |
| 3 | 0 | –2 | –3 | –2 | 0 |
| 4 | 0 | –3 | –4 | –3 | 0 |

**Example 33.13.** *Find the solution of the initial boundary value problem ;* $\frac{\partial^2 u}{\partial t^2} = \frac{\partial^2 u}{\partial x^2}$, $0 \le x \le 1$ *; subject to the initial conditions* $u(x, 0) = \sin \pi x$, $0 \le x \le 1$ *and the boundary conditions* $u(0, t) = 0$, $u(1, t) = 0$, $t > 0$ *; by using in the (a) the explicit scheme.*

*(b) the implicit scheme.*

**Solution.** (*a*) *Explicit scheme*

Take $h = 0.2$, $k = \frac{h}{c} = 0.2$ [$\because$ $c = 1$]

$\therefore$ We use the formula, $u_{i,j+1} = u_{i-1,j} + u_{i+1,j} - u_{i,j-1}$ ...(*i*)

Since $u(0, t) = 0$, $u(1, t) = 0$, $u_{0,j} = 0$, $u_{1,j} = 0$ for all values of $j$

*i.e.*, the entries in the first and last columns are zero.

Since $u(x, 0) = \sin \pi x$, $u_{i,0} = \sin \pi x$

$\therefore$ $u_{1,0} = 0$, $u_{2,0} = \sin (0.2\pi) = 0.5878$, $u_{3,0} = \sin (0.4\pi) = 0.9511$, $u_{4,0} = \sin (0.6\pi) = 0.5878$.

These are the entries of the first row.

Since $u_t(x, 0) = 0$ we have $\frac{1}{2}(u_{i,j+1} - u_{i,j-1} = 0$, when $j = 0$

*i.e.*, $$u_{i,1} = u_{i,-1} \quad ...(ii)$$

Putting $j = 0$ in Eqn. (*i*), $u_{i,1} = u_{i-1,0} + u_{i+1,0} - u_{i,-1}$

Using Eqn. (*ii*) $$u_{i,1} = \frac{1}{2}(u_{i-1,0} + u_{i+1,} u_{i+1,0})$$

Taking $i = 1, 2, 3, 4$ successively, we obtain the entries of the second row.

Putting $j = 1$ in Eqn. (*i*), $u_{i,2} = u_{i-1,1} + u_{i+1,1} - u_{i,0}$

Now taking $i = 1, 2, 3, 4$ successively, we get the entries of the third row.

Similarly taking $j = 2, j = 3, j = 4$ successively, we obtain the entries of the fourth, fifth and sixth rows respectively.

| j \ i | 0 | 1 | 2 | 3 | 4 | 5 |
|---|---|---|---|---|---|---|
| 0 | 0 | 0.5878 | 0.9511 | 0.9511 | 0.5878 | 0 |
| 1 | 0 | 0.4756 | 0.7695 | 0.9511 | 0.7695 | 0 |
| 2 | 0 | 0.1817 | 0.4756 | 0.5878 | 0.3633 | 0 |
| 3 | 0 | 0 | 0.0001 | –0.1122 | –0.1816 | 0 |
| 4 | 0 | –0.1816 | –0.5878 | –0.7694 | 0.4755 | 0 |
| 5 | 0 | –0.5878 | –0.9511 | –0.9511 | –0.5878 | 0 |

(*b*) *Implicit scheme*

We have the formula :

$$u_{i,j+1} = 2(1-\alpha^2 c^2)\, u_{i,j} + \alpha^2 c^2 (u_{i-1,j} + u_{i+1,j}) - u_{i,j-1}, \text{ where } \alpha = k/h \quad ...(i)$$

Here $c^2 = 1$. Take $h = 0.25$ and $k = 0.5$ so that $\alpha = k/h = 2$.

$\therefore$ Eqn. (*i*) reduces to

$$u_{i,j+1} = -6u_{i,j} + 4(u_{i-1,j} + u_{i+1,j}) - u_{i,j-1} \quad ...(ii)$$

Since $\quad u_{i,0} = \sin \pi x$

$\therefore \quad u_{(1,0)} = 0.7071, u_{(2,0)} = 0.5, u_{3,0} = 0.7071$

There are the entries of the first row.

Since $u_t(x, 0) = 0$, we have $\frac{1}{2}(y_{i,i+1} - y_{i,i-1}) = 0$, where $j = 0$

$$\therefore \quad y_{i,1} = y_{i,-1} \quad ...(iii)$$

Put $j = 0$ and using Eqns. (*iii*), (*ii*) reduces to

$$u_{i,+1} = -3u_{i,0} + 2(u_{i-1,0} + 2u_{i+1,0})$$

Now taking

$$i = 1, u_{1,1} = -3u_{1,0} + 2(u_{0,0} + u_{2,0}) = 0.1213$$
$$i = 2, u_{2,1} = -3u_{2,0} + 2(u_{1,0} + u_{3,0}) = 0.1716$$
$$i = 3, u_{3,1} = -3u_{3,0} + 2(u_{2,0} + u_{4,0}) = 0.1213$$

These are the entries of the second row.

Putting $j = 1$, Eqn. (*ii*) reduces to

$$u_{i,2} = -6u_{i,1} + 4(u_{i-1,2} + u_{i+1,1})$$

Now taking

$$i = 1, u_{1,2} = -6u_{1,1} + 4(u_{0,1} + u_{2,1}) = 0.414$$
$$i = 2, u_{2,2} = -6u_{2,1} + 4(u_{1,1} + u_{3,1}) = 0.0592$$
$$i = 3, u_{3,2} = -6u_{3,1} + 4(u_{2,1} + u_{4,1}) = 0.414$$

These are the entries of the third row.

Putting $j = 2$, Eqn. (*ii*) reduces to

$$u_{i,3} = -6u_{i,2} + 4(u_{i-1,2} + u_{i+1,2}) - u_{i,1}$$

Now taking

$$i = 1, u_{1,3} = -6u_{1,2} + 4(u_{0,2} + u_{2,2}) - u_{1,1} = 0.1097$$
$$i = 2, u_{2,3} = -6u_{2,2} + 4(u_{1,2} + u_{3,2}) - u_{2,1} = 0.1476$$
$$i = 3, u_{3,3} = -6u_{3,2} + 4(u_{2,2} + u_{4,2}) - u_{4,2} = 0.1097$$

These are the entries of the third row.

Hence the value of $u_{i,j}$ are as tabulated below :

| j \ i | 0 | 1 | 2 | 3 | 4 |
|---|---|---|---|---|---|
| 0 | 0 | 0.7071 | 0.5 | 0.7071 | 0 |
| 1 | 0 | –0.1213 | –0.1716 | –0.1213 | 0 |
| 2 | 0 | 0.0414 | 0.0592 | 0.0414 | 0 |

**Example 33.14.** *Solve the equation $\nabla^2 u = -10(x^2 + y^2 + 10)$ over the square with sides $x = 0 = y$, $x = 3$, $= y$ with $u = 0$ on the boundary and mesh length $=1$.* (Kerala, 2018)

**Solution.** Given $h = 1$.

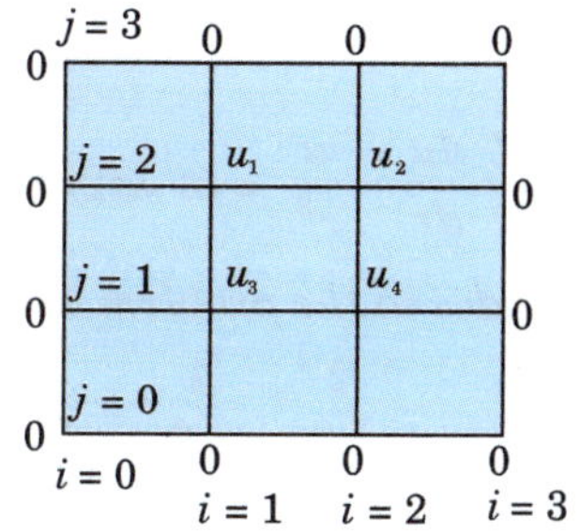


The standard 5-point formula 5-point formula for the given equation is

$$u_{i-1}, j + u_{i+1}, j + u_{i}, {}_{j+1} + u_i, j - 1 - 4u_{ij} = h^2 (ih, jh)$$
$$= -10(i^2 + j^2 + 10) \quad ...(i)$$

For $u_1 (i = 1, j = 2)$

from Eqn. (i) $\quad 0 + u_2 + 0 + u_3 - 4u_1 = -10(1^2 + D^2 + 10)$

$$u_1 = \frac{1}{4}(u_2 + u_3 + 150) \quad ...(ii)$$

For $u_2$ $(i = 2, j = 2)$

From Eqn. (i) $$u_2 = \frac{1}{4}(u_1 + u_4 + 180) \quad ...(iii)$$

For $u_3$ $(i = 1, j = 1)$, from Eqn. (i) $$u_3 = \frac{1}{4}(u_1 + u_4 + 120) \quad ...(iv)$$

For $u_4 (i = 2, j = 1)$ From Eqn. (i)

$$u_4 = \frac{1}{4}(u_2 + u_3 + 150) \quad ...(v)$$

from Eqns. (ii) and (iv), $u_1 = u_4$

$\therefore$ These equations are

$$u_1 = \frac{1}{4}(u_2 + u_3 + 150)$$

$$u_2 = \frac{1}{4}(u_1 + u_1 + 180) = \frac{1}{2}(u_1 + 90)$$

$$u_3 = \frac{1}{4}(u_1 + u_1 + 120) = \frac{1}{2}(u_1 + 60)$$

By Gauss Saidal Iteration method

**First iteration:** Put $\quad u_2 = 0, u_3 = 0$

$$u_1^{(1)} = 37.5,\ u_2^{(1)} = \frac{1}{2}(37.5 + 90) = 64,$$

$$u_3^{(1)} = \frac{1}{2}(37.5 + 60) = 49$$

**Second iteration** $$u_1^{(2)} = \frac{1}{4}(64 + 49 + 150) = 66$$

$$u_2^{(2)} = \frac{1}{2}(66 + 90) = 78,$$

$$u_3^{(2)} = \frac{1}{2}(66 + 60) = 63$$

3rd iteration : $u_1^{(3)} = 73$, $u_2^{(3)} = 82$, $u_3^{(3)} = 67$

4th iteration : $u_1^{(4)} = 75$, $u_2(4) = 82.5$, $u_3^{(4)} = 67.5$

5th iteration: $u_1^{(5)} = 75$, $u_2(5) = 82.5(5)$, $u_3^{(5)} = 67.5$

$$u_1 = 75$$
$$u_2 = 82.5$$
$$u_3 = 67.5$$
$$u_4 = u_1 = 75.$$

**Example 13.15 :** *The equation is* $\dfrac{d^2u}{dr^2} + \dfrac{1}{r}\dfrac{du}{dr} - \dfrac{u}{r^2} = 0,\ u(2) = 0.008,\ u(6.\ 5) = 0.003.$

*Find u as a partial equation of r, use u nodes to the problem equidisant:*

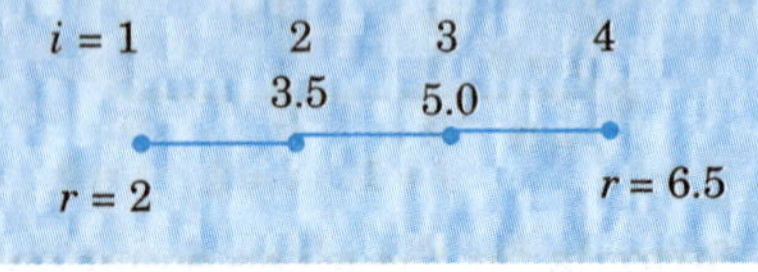


**Solution.**
$$\Delta r = \frac{6.5 - 2}{3} = 1.3$$

$$\left.\frac{d^2u}{dr^2}\right|_i \simeq \frac{u_{i+1} - 2u_i + u_{i-1}}{(\Delta r)^2}$$

$$\left.\frac{du}{dr}\right|_i \simeq \frac{u_{i+1} - 2u_i + u_i - 1}{(\Delta r)^2}$$

$$\left.\frac{du}{dr}\right|_i \simeq \frac{u_{i+1} - u_i}{\Delta r}$$

$$\frac{u_{i+1} - 2u_i + u_{i-1}}{(\Delta r)^2} + \frac{1}{r_i}\ \frac{u_{i+1} - u_i}{\Delta r} - \frac{4i}{r_i^2} = 0$$

**Note 1:** $u_1 = u(2) = 0.008$

**Note 2:** $i = 2,\ \dfrac{u_3 - 2u_2 + u_1}{(\Delta r)^2} + \dfrac{1}{r_2}\dfrac{u_3 - u_2}{\Delta r} - \dfrac{u_2}{r_2^2} = 0$

$$= \frac{u_3 - 2u_2 + u_1}{(1.5)^2} + \frac{1}{3.5}\,\frac{u_3 - u_2}{(1.5)^2} - \frac{u_2}{(3.5)^2} = 0$$

$$= 0\ 44444\, u_4 - 1.1610\, u_2 + 0.63492\, u_3 = 0 \qquad ...(i)$$

**Note 3:** $i = 3$

$$\frac{u_4 - 2u_3 + u_2}{(\Delta r)^2} + \frac{1}{r_3}\,\frac{u_4 - u_3}{\Delta r} - \frac{u_2}{r_3^{\,2}} = 0$$

$$\frac{u_4 - 2u_3 + u_2}{(1.5)^2} + \frac{1}{5}\,\frac{u_4 - u_3}{1.5} - \frac{u_2}{5^2} = 0$$

$$0.44444u_1 - 1.0622u_3 + 0.57778u_3 = 0 \qquad ...(ii)$$

**Note 4:** $i = 4$

$$u_4 = u(6.5) = 0.003 \qquad ...(iii)$$

$$\begin{bmatrix} 1 & 0 & 0 & 0 \\ 0.444444 & -1.1610 & 0.63492 & 0 \\ 0 & 0.44444 & -1.0622 & 0.57778 \\ 0 & 0 & 0 & 1 \end{bmatrix}\begin{bmatrix} u_1 \\ u_2 \\ u_3 \\ u_4 \end{bmatrix} = \begin{bmatrix} 0.008 \\ 0 \\ 0 \\ 0.003 \end{bmatrix}$$

This is tridiagonal Diagonally Dominant Thomas' algorithm

$$u_1 = 0.008$$
$$u_2 = 0.005128$$
$$u_3 = 0.003778$$
$$u_4 = 0.003$$

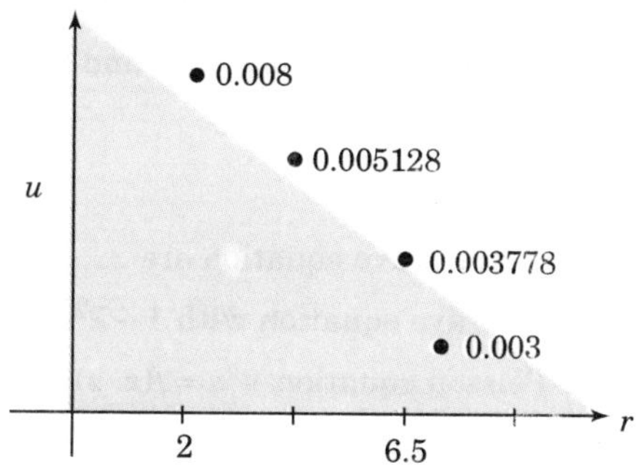


## PROBLEMS 33.4

1. Solve the boundary value problem $u_{tt} = u_{xx}$ with the conditions $u(0, t) = u(1, t) = 0$, $u(x, 0) = \frac{1}{2}x(1 - x)$ and $u_t(x, 0) = 0$, taking $h = k = 0.1$ for $0 \le t \le 0.4$. Compare your solution with the exact solution at $x = 0.5$ and $t = 0.3$.
2. The transverse displacement $u$ of a point at a distance $x$ from one end and at any time $t$ of a vibrating string satisfies the equation $\partial^2u/\partial t^2 = 25\ \partial^2u/\partial x^2$, with the boundary conditions $u(0, t) = u(5, t) = 0$ and the initial conditions $u(x, 0) = \begin{cases} 20x & \text{for } 0 \le x \le 1 \\ 5(5 - x) & \text{for } 1 \le x \le 5 \end{cases}$ and $u_t(x, 0) = 0$. Solve this equation numerically for one half period of vibration, taking $h = 1$, $k = 0.2$.
3. Solve $y_{tt} = y_{xx}$ upto $t = 0.5$ with a spacing of 0.1 subject to $y(0, t) = 0$, $y(1, t) = 0$, $y_t(x, 0) = 0$, and $y(x, 0) = 10 + x(1 - x)$.
4. The function $u$ satisfies the equation

$$\partial^2u / \partial t^2 = \partial^2u / \partial x^2$$

and the conditions : $u(x, 0) = \frac{1}{8}\sin \pi x$, $u_t(x, 0) = 0$ for $0 \le x \le 1$,

$$u(0, t) = u(1, t) = 0 \text{ for } t \ge 0.$$

Use the explicit scheme to calculate $u$ for $x = 0(0.1)\ 1$ and $t = 0(0.1)\ 0.5$.

## 33.11 OBJECTIVE TYPE OF QUESTIONS

## PROBLEMS 33.5

*Fill up the blanks or select the correct answer from each of the following questions :*

1. Which of the following equations is parabolic :
   (a) $f_{xy} - f_x = 0$ (b) $f_{xx} + 2f_{xy} + f_{yy} = 0$ (c) $f_{xx} + 2f_{xy} + 4f_{yy} = 0$ (d) none
2. $u_{ij} = \frac{1}{4}(u_{i+1,j} - u_{i-1,j} + u_{i,j+1} - u_{i,j-1})$ is Leibmann's five point formula. (True or False)
3. $u_{xx} + 3u_{xy} + u_{yy} = 0$ is classified as ...... (*U.P.T.U., 2012*)
4. $\nabla^2 u = f(x, y)$ is known as ......
5. The simplest formula to solve $u_{tt} = \alpha^2 u_{xx}$ is ......
6. The finite difference form of $\partial^2u / \partial x^2$ is ......
7. Schmidt's finite difference scheme to solve $u_t = c^2 u_{xx}$ is ......
8. The 5-point diagonal formula gives $u_{ij}$ = ......
9. The partial differential equation $(x + 1)u_{xx} - 2(x + 2)u_{xy} + (x + 3)u_{yy} = 0$ is classified as ......
10. $u_{i,j+1} = \frac{1}{2}(u_{i+1,j} + u_{i-1,j})$ is called ...... recurrence relation.
11. In terms of difference quotients $4u_{xx} = u_{tt}$ is ......
12. Bendre-Schmidt recurrence relation for one dimensional heat equation is ......

13. The diagonal 5-point formula to solve the Laplace equation $u_{xx} + u_{yy} = 0$ is ......
14. In the parabolic equation $u_t = \alpha^2 u_{xx}$ if $\lambda = k\alpha^2/h^2$, where $k = \Delta t$, and $h = \Delta x$, then explicit method is stable if $\lambda$ = ......
15. $2\frac{\partial^2 u}{\partial t^2} + 4\frac{\partial^2 u}{\partial x \partial y} + 3\frac{\partial^2 u}{\partial x^2} = 0$ is classified as ...... *(P.T.U., 2007)*
16. The boundary conditions of one-dimensional wave equation are ......
17. The explicit formula for one-dimensional wave equation with $1 - \lambda^2\alpha^2 = 0$ and $\lambda = k/h$ is ......
18. The difference scheme for solving the Poisson equation $\nabla^2 u = f(x, y)$ is ...... *(Anna, 2012)*
19. If $u$ satisfies Laplace equation and $u = 100$ on the boundary of a square, the value of $u$ at an interior grid point is ......
20. The Laplace equation $u_{xx} + u_{yy} = 0$ in difference quotients is ......
21. The equation $yu_{xx} + u_{yy} = 0$ is hyperbolic in the region ......
22. To solve $\frac{\partial u}{\partial t} = \frac{1}{2}\frac{\partial^2 u}{\partial x^2}$ by Bendre-Schmidt method with $h = 1$, the value of $k$ is ......

CHAPTER 34

# Linear Programming



## 34.1 INTRODUCTION

Linear programming deals with the optimization (maximization or minimization) of linear functions subject to linear constraints. This technique has found its applications to important areas of product mix, blending problems and diet problems. Oil refineries, chemical industries, steel industries and food processing industry are also using linear programming with considerable success.

In this chapter, our purpose is to present the principles of linear programming and the techniques of its application in a manner that will suit the engineering students. Beginning with the graphical method which provides a great deal of insight into the basic concepts, the simplex method of solving linear programming problems is developed. Then the reader is introduced to the Duality concept. Finally a special class of linear programming problems namely : Transportation and Assignment problems, is taken up. For a detailed study, the student should refer to author's book '*Numerical Methods in Engineering and Science*'.

## 34.2 FORMULATION OF THE PROBLEM

To begin with, a problem is to be presented in a linear programming form which requires defining the variables involved, establishing relationships between them and formulating the objective function and the constraints. We illustrate this through a few examples.

**Example 34.1.** *A manufacturer produces two types of models $M_1$ and $M_2$. Each $M_1$ model requires 4 hours of grinding and 2 hours of polishing ; whereas each $M_2$ model requires 2 hours of grinding and 5 hours of polishing. The manufacturer has 2 grinders and 3 polishers. Each grinder works for 40 hours a week and each polisher works for 60 hours a week. Profit on an $M_1$ model is ₹ 3 and on an $M_2$ model is ₹ 4. Whatever is produced in a week is sold in the market. How should the manufacturer allocate his production capacity to the two types of models so that he may make the maximum profit in a week.*

**Solution.** Let $x_1$ be the number of $M_1$ models and $x_2$ the number of $M_2$ models produced per week. Then the weekly profit (in ₹) is

$$Z = 3x_1 + 4x_2 \qquad ...(i)$$

To produce these number of models, the total number of grinding hours needed per week

$$= 4x_1 + 2x_2$$

and the total number of polishing hours required per week

$$= 2x_1 + 5x_2$$

Since the number of grinding hours available is not more than 80 and the number of polishing hours is not more than 180, therefore,

$$4x_1 + 2x_2 \leq 80 \quad ...(ii)$$

$$2x_1 + 5x_2 \leq 180 \quad ...(iii)$$

Also since the negative number of models are not produced, obviously we must have

$$x_1 \geq 0 \text{ and } x_2 \geq 0 \quad ...(iv)$$

Hence this allocation problem is, to find $x_1$, $x_2$ which

maximize $Z = 3x_1 + 4x_2$

subject to $4x_1 + 2x_2 \leq 80$, $2x_1 + 5x_2 \leq 180$, $x_1, x_2 \geq 0$.

**Obs.** *The variables that enter into the problem are called* **decision variables.**

*The expression Eqn. (i) showing the relationship between the manufacturer's goal and the decision variables, is called the* **objective function.**

*The inequalities Eqns. (ii), (iii) and (iv) are called the* **constraints.**

The objective function and the constraints being all linear, it is a *linear programming problem* (L.P.P). This is an example of a real situation from industry.

**Example 34.2.** *A firm making castings uses electric furnace to melt iron with the following specifications :*

| | *Minimum* | *Maximum* |
|---|---|---|
| *Carbon* | *3.20%* | *3.40%* |
| *Silicon* | *2.25%* | *2.35%* |

*Specifications and costs of various raw materials used for this purpose are given below :*

| *Material* | *Carbon %* | *Silicon %* | *Cost (₹)* |
|---|---|---|---|
| *Steel scrap* | *0.4* | *0.15* | *850/tonne* |
| *Cast iron scrap* | *3.80* | *2.40* | *900/tonne* |
| *Remelt from foundry* | *3.50* | *2.30* | *500/tonne* |

*If the total charge of iron metal required is 4 tonnes, find the weight in kg of each raw material that must be used in the optimal mix at minimum cost.*

**Solution.** Let $x_1, x_2, x_3$ be the amounts (in kg) of these raw materials. The objective is to minimize the cost *i.e.*,

minimize $$Z = \frac{850}{1000}x_1 + \frac{900}{1000}x_2 + \frac{500}{1000}x_3 \quad ...(i)$$

For iron melt to have a minimum of 3.2% carbon,

$$0.4x_1 + 3.8x_2 + 3.5x_3 \geq 3.2 \times 4{,}000 \quad ...(ii)$$

For iron melt to have a maximum of 3.4% carbon,

$$0.4x_1 + 3.8x_2 + 3.5x_3 \leq 3.4 \times 4{,}000 \quad ...(iii)$$

For iron melt to have a minimum of 2.25% silicon,

$$0.15x_1 + 2.41x_2 + 2.35x_3 \geq 2.25 \times 4{,}000 \quad ...(iv)$$

For iron melt to have a maximum of 2.35% silicon,

$$0.15x_1 + 2.41x_2 + 2.35x_3 \leq 2.35 \times 4{,}000 \quad ...(v)$$

Also, since the materials added up must be equal to the full charge weight of 4 tonnes.

$$\therefore \quad x_1 + x_2 + x_3 = 4{,}000 \quad ...(vi)$$

Finally since the amounts of raw material cannot be negative ...(vii)

$$x_1 \geq 0, x_2 \geq 0, x_3 \geq 0 \quad ...(vii)$$

Thus the linear programming problem is to find $x_1, x_2, x_3$ which

minimize $Z = 0.85x_1 + 0.9x_2 + 0.5x_3$

subject to $0.4x_1 + 3.8x_2 + 3.5x_3 \geq 12{,}800$

$0.4x_1 + 3.8x_2 + 3.5x_3 \leq 13{,}600$

$$0.15x_1 + 2.41x_2 + 2.35x_3 \geq 9{,}000$$
$$0.15x_1 + 2.41x_2 + 2.35x_3 \leq 9{,}400$$
$$x_1 + x_2 + x_3 = 4{,}000$$
$$x_1, x_2, x_3 \geq 0.$$

## PROBLEMS 34.1

1. A firm manufactures two items. It purchases castings which are then machined, bored and polished. Castings for items $A$ and $B$ cost ₹ 3 and ₹ 4 each and are sold at ₹ 6 and ₹ 7 each respectively. Running costs of these machines are ₹ 20, ₹ 14 and ₹ 17.50 per hour respectively. Formulate the problem so that the product mix maximizes the profit? Capacities of the machines are

| | *Part A* | *Part B* |
|---|---|---|
| *Machining capacity* | 25 per hr. | 40 per hr. |
| *Boring capacity* | 28 per hr. | 35 per hr. |
| *Polishing capacity* | 35 per hr. | 25 per hr. |

2. A firm manufactures 3 products $A$, $B$ and $C$. The profits are ₹ 3, ₹ 2 and ₹ 4 respectively. The firm has two machines $M_1$ and $M_2$ and below is the required processing time in minutes for each machine on each product.

| | | *Product* | | |
|---|---|---|---|---|
| | | A | B | C |
| *Machine* | $M_1$ | 4 | 3 | 5 |
| | $M_2$ | 2 | 2 | 4 |

Machines $M_1$ and $M_2$ have 2000 and 2500 machine-minutes respectively. The firm must manufacture 100 $A$'s, 200 $B$'s and 50 $C$'s but not more than 150 $A$'s. Set up an L.P.P. to maximize profit. (*Kurukshetra, 2009 S*)

3. Three products are processed through three different operations. The time (in minutes) required per unit of each product, the daily capacity of the operations (in minutes per day) and the profit per unit sold for each product (in rupees) are as follows:

| *Operation* | *Time per unit* | | | *Operation capacity* |
|---|---|---|---|---|
| | *Product I* | *Product II* | *Product III* | |
| 1 | 3 | 4 | 3 | 42 |
| 2 | 5 | 0 | 3 | 45 |
| 3 | 3 | 6 | 2 | 41 |
| Profit (₹) | 3 | 2 | 1 | |

The zero time indicates that the product does not require the given operation. The problem is to determine the optimum daily production for three products that maximize the profit. Formulate this production planning problem as a linear programming problem assuming that all units produced are sold.

4. An aeroplane can carry a maximum of 200 passengers. A profit of ₹ 400 is made on each first class ticket and a profit of ₹ 300 is made on each economy class ticket. The airline reserves at least 20 seats for first class. However, at least 4 times as many passengers prefer to travel by economy class than by the first class. How many tickets of each class must be sold in order to maximize profit for the airline? Formulate the problem as an L.P. model.

(*Rohtak, 2006*)

5. A firm manufactures headache pills in two sizes $A$ and $B$. Size $A$ contains 2 grains of asprin, 5 grains of bicarbonate and 1 grain of codeine. Size $B$ contains 1 grain of asprin, 8 grains of bicarbonate and 6 grains of codeine. It is found by users that it requires at least 12 grains of aspirin, 74 grains of bicarbonate and 24 grains of codeine for providing immediate effect. It is required to determine the least number of pills a patient should take to get immediate relief. Formulate the problem as a standard L.P.P.

6. Consider the following problem faced by a production planner in a soft-drink plant. He has two bottling machines $A$ and $B$. $A$ is designed for 8-ounce bottles and $B$ for 16-ounce bottles. However, each be can used on both types with some loss of efficiency. The following data is available:

| *Machine* | *8-ounce bottles* | *16-ounce bottles* |
|---|---|---|
| *A* | 100/minute | 40/minute |
| *B* | 60/minute | 75/minute |

The machines can be run 8 hour per day, 5 days per week. Profit on a 8-ounce bottle is 15 paise and on a 16-ounce bottle is 25 paise. Weekly production of the drink cannot exceed 300,000 ounces and the market can absorb 25,000

8-ounce bottles and 7,000 16-ounce bottles per week. The planner wishes to maximise his profit subject, of course, to all the production and marketing restrictions. Formulate this as a L.P.P.

7. A dairy feed company may purchase and mix one or more of three types of grains containing different amounts of nutritional elements. The data is given in the table below. The production manager specifies that any feed mix for his live stock must meet at least minimum nutritional requirements and seeks the least costly among all three mixes.

| | Item | *One unit weight of* | | | *Minimum requirement* |
|---|---|---|---|---|---|
| | | *Grain 1* | *Grain 2* | *Grain 3* | |
| | *A* | 2 | 3 | 7 | 1,250 |
| *Nutritional* | *B* | 1 | 1 | 0 | 250 |
| *ingredients* | *C* | 5 | 3 | 0 | 900 |
| | *D* | 6 | 25 | 1 | 232.5 |
| *Cost per weight of* | | 41 | 35 | 96 | |

Formulate the problem as a L.P. model.

8. A firm produces an alloy with the following specifications:

(*i*) specific gravity $\leq 0.97$; (*ii*) chromium content $\geq 15\%$; (*iii*) melting temperature $\geq 494°C$

The alloy requires three raw materials *A*, *B* and *C* whose properties are as follows:

| *Property* | *Properties of raw material* | | |
|---|---|---|---|
| | *A* | *B* | *C* |
| *Sp. gravity* | 0.94 | 1.00 | 1.05 |
| *Chromium* | 10% | 15% | 17% |
| *Melting pt.* | 470°C | 500°C | 520°C |

Find the values of *A*, *B*, *C* to be used to make 1 tonne of alloy of desired properties, keeping the raw material costs at the minimum when they are ₹ 105/tonne for *A*, ₹ 245/tonne for *B* and ₹ 165/tonne for *C*. Formulate an L.P. model for the problem.

## 34.3 GRAPHICAL METHOD

Linear programming problems involving only two variables can be effectively solved by a graphical technique which provides a pictorial representation of the solution and one gets insight into the basic concepts used in solving large L.P.P.

**Working procedure** *to solve a linear programming problem graphically:*

*Step 1.* Formulate the given problem as a linear programming problem.

*Step 2.* Plot the given constraints as equalities on $x_1 - x_2$ coordinate plane and determine the convex region* formed by them.

*Step 3.* Determine the vertices of the convex region and find the value of the objective function at each vertex. The vertex which gives the optimal (maximum or minimum) value of the objective function gives the desired optimal solution to the problem.

*Otherwise*. Draw the dotted line through the origin representing the objective function with $Z = 0$. As $Z$ is increased from zero, this line moves to the right remaining parallel to itself. We go on sliding this line (parallel to itself), till it is *farthest* away from the origin and passes through only one vertex of the convex region. This is the vertex where the maximum value of $Z$ is attained.

* *A* **region** *or a* **set** *of points is said to be* **convex** *if the line joining any two of its points lies completely in the region (or the set).* Figs. 34.1 and 34.2 represent convex regions while Figs. 34.3 and 34.4 do not form convex sets.

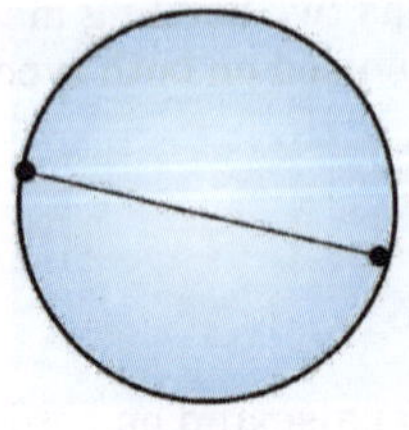

Fig. 34.1

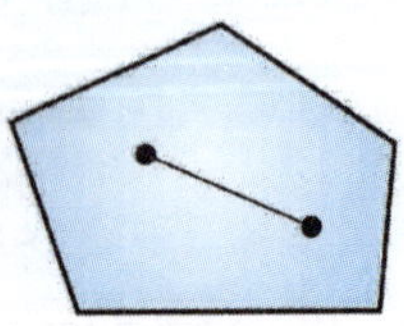

Fig. 34.2

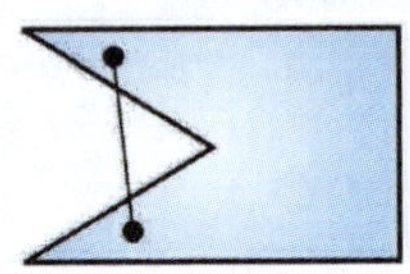

Fig. 34.3

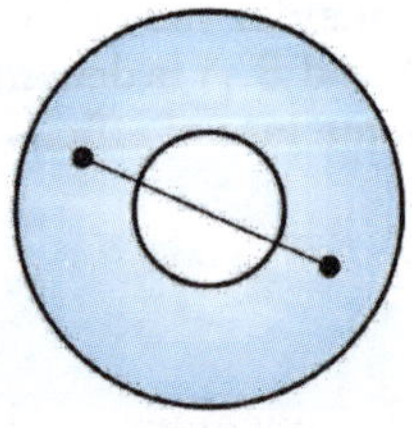

Fig. 34.4

When it is required to minimize $Z$, value of $Z$ is increased till the dotted line passes through the *nearest* vertex of the convex region.

**Example 34.3.** *Solve the L.P.P. of Ex. 34.1 graphically.*

**Solution.** The problem is:

Maximize $Z = 3x_1 + 4x_2$ ...(*i*)

subject to $4x_1 + 2x_2 \leq 80$ ...(*ii*)

$2x_1 + 5x_2 \leq 180$ ...(*iii*)

$x_1, x_2 \geq 0$ ...(*iv*)

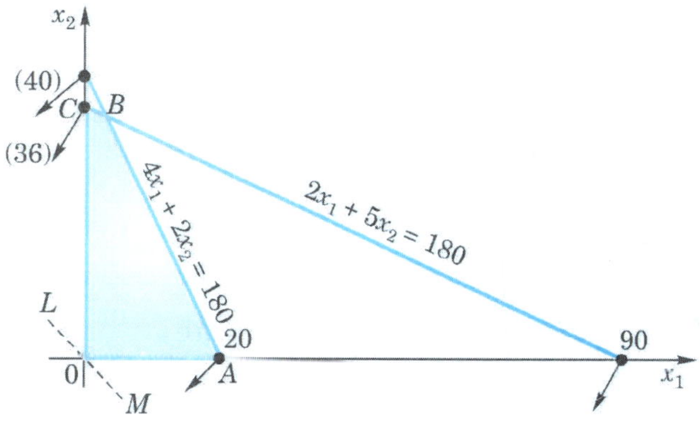

Fig. 34.5

Consider $x_1 - x_2$ coordinate system as shown in Fig. 34.5. The non-negativity restrictions Eqn. (*iv*) imply that the values of $x_1$, $x_2$ lie in the first quadrant only.

We plot the lines $4x_1 + 2x_2 = 80$ and $2x_1 + 5x_2 = 180$.

Then any point on or below $4x_1 + 2x_2 - 80$ satisfies Eqn. (*ii*) and any point on or below $2x_1 + 5x_2 = 180$ satisfies Eqn. (*iii*). This shows that the desired point $(x_1, x_2)$ must be somewhere in the shaded convex region $OABC$. This region is called the *solution space* or *region of feasible solutions* for the given problem. Its vertices are $O(0, 0)$, $A(20, 0)$ $B(2.5, 35)$ and $C(0, 36)$.

The values of the objective function Eqn. (*i*) at these points are

$Z(O) = 0$, $Z(A) = 60$, $Z(B) = 147.5$, $Z(C) = 144$.

Thus the maximum value of $Z$ is 147.5 and it occurs at $B$. Hence the optimal solution to the problem is

$$x_1 = 2.5,\ x_2 = 35 \text{ and } Z_{max} = 147.5.$$

**Otherwise.** Our aim is to find the point (or points) in the solution space which maximizes the profit function $Z$. To do this, we observe that on making $Z = 0$, Eqn. (*i*) becomes $3x_1 + 4x_2 = 0$ which is represented by the dotted line $LM$ through $O$. As the value of $Z$ is increased, the line $LM$ starts moving parallel to itself towards the right. Larger the value of $Z$, more will be the company's profit. In this way, we go on sliding $LM$ till it is farthest away from the origin and passes through one of the corners of the convex region. This is the point where the maximum value of $Z$ is attained. Just possible, such a line may be one of the edges of the solution space. In that case every point on that edge gives the same maximum value of $Z$.

Here $Z_{max}$ is attained at $B(2.5, 35)$. Hence the optimal solution is $x_1 = 2.5$, $x_2 = 35$ and $Z_{max} = 147.5$.

**Example 34.4.** *Find the maximum value of $Z = 2x + 3y$ subject to the constraints: $x + y \leq 30$, $y \geq 3, 0 \leq y \geq 12$, $x - y \geq 0$, and $0 \leq x \leq 20$.*

**Solution.** Any point $(x, y)$ satisfying the conditions $x \geq 0$, $y \geq 0$ lies in the first quadrant only. Also since $x + y \leq 30$, $y \geq 3$, $y \leq 12$, $x \geq y$ and $x \leq 20$, the desired point $(x, y)$ lies within the convex region $ABCDE$ (shown shaded in Fig. 34.6). Its vertices are $A(3, 3)$, $B(20, 3)$, $C(20, 10)$, $D(18, 12)$, and $E(12, 12)$.

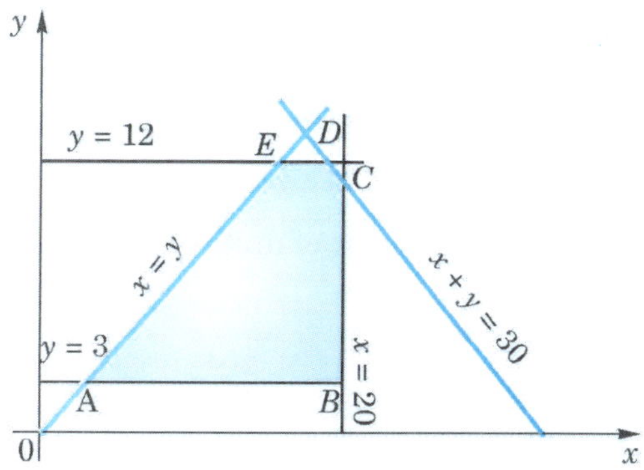

Fig. 34.6

The values of $Z$ at these five vertices are $Z(A) = 15$, $Z(B) = 49$, $Z(C) = 70$, $Z(D) = 72$, and $Z(E) = 60$.

Since the maximum value of $Z$ is 72 which occurs at the vertex $D$, the solution to the L.P.P. is $x = 18$, $y = 12$ and maximum $Z = 72$.

**Example 34.5.** *A company manufactures two types of cloth, using three different colours of wool. One yard length of type a cloth requires 4 oz of red wool, 5 oz of green wool and 3 oz of yellow wool. One yard length of type B cloth requires 5 oz of red wool, 2 oz of green wool and 8 oz of yellow wool. The wool available for manufacturer is 1000 oz of red wool, 1000 oz of green wool and 1200 oz of yellow wool. The manufacturer can make a profit of ₹ 5 on one yard of type A cloth and ₹ 3 on one yard of type B cloth. Find the best combination of the quantities of type A and type B cloth which gives him maximum profit by solving the L.P.P. graphically.*

**Solution.** Let the manufacturer decide to produce $x_1$ yards of type $A$ cloth and $x_2$ yards of type $B$ cloth. Then the total income in rupees, from these units of cloth is given by

$$Z = 5x_1 + 3x_2 \quad ...(i)$$

To produce these units of two types of cloth, he requires

red wool = $4x_1 + 5x_2$ oz,

green wool = $5x_1 + 2x_2$ oz,

and yellow wool = $3x_1 + 8x_2$ oz.

Since the manufacturer does not have more than 1000 oz of red wool, 1000 oz of green wool and 1200 oz of yellow wool, therefore

$$4x_1 + 5x_2 \leq 1000 \quad ...(ii)$$

$$5x_1 + 2x_2 \leq 1000 \quad ...(iii)$$

$$3x_1 + 8x_2 \leq 1200 \quad ...(iv)$$

Also $$x_1 \geq 0, x_2 \geq 0. \quad ...(v)$$

Thus the given problem is to maximize $Z$ subject to the constraints Eqns. (*ii*) to (*v*). (*V.T.U., 2004*)

Any point satisfying the condition Eqn. (*v*) lies in the first quadrant only. Also the desired point satisfying the constraints Eqns. (*ii*) to (*iv*) lies in the convex region $OABCD$ (Fig. 34.7). Its vertices are $O(0, 0)$ $A(200, 0)$, $B(3000/17, 1000/17)$, $C(2000/17, 1800/17)$ and $D(0, 150)$.

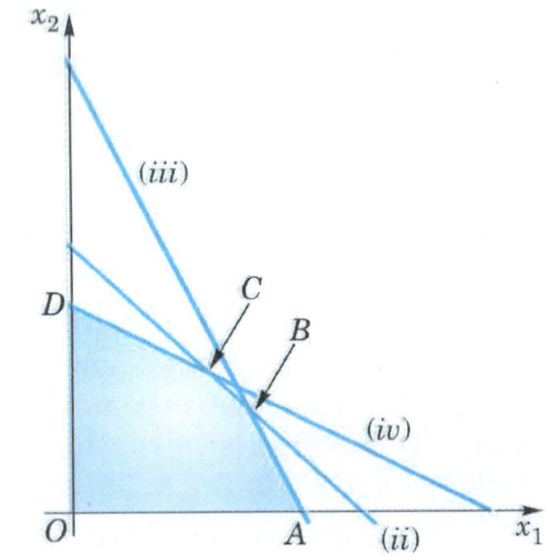

Fig. 34.7

The values of $Z$ at these vertices are given by $Z(O) = 0$, $Z(A) = 1000$, $Z(B) = 1057.6$, $Z(C) = 905.8$ and $Z(D) = 450$.

Since the maximum value of $Z$ is 1058.8 which occurs at the vertex $B$, the solution to the given problem is $x_1 = 3000/17$, $x_2 = 1000/17$ and max. $Z = 1058.8$.

Hence the manufacturer should produce 176.5 yards of type $A$ cloth, 58.8 yards of type $B$ cloth, so as to get the maximum profit of ₹ 1058.8.

**Example 34.6.** *A company makes two products (X and Y) using two machines (A and B). Each unit of X that is produced requires 50 minutes processing time on machine A and 30 minutes processing time on machine B. Each unit of Y that is produced requires 24 minutes processing time on machine A and 33 minutes processing time on machine B.*

*At the start of the current week there are 30 units of X and 90 units of Y in stock. Available processing time on machine A is forecast to be 40 hours and on machine B is forecast to be 35 hours.*

*The demand for X in the current week is forecast to be 75 units and for Y is forecast to be 95 units. Company policy is to maximise the combined sum of the units of X and the units of Y in stock at the end of the week.*

*(a) What is the problem formula that how much of each product to make in the current week as a linear program.*

*(b) Solve this linear program graphically.* (*UKTU, 2018*)

**Solution.** Let $x$ be the of number of units of $X$ produced in the current week.

$y$ be the number of units of $Y$ produced in the current week.

$$50x + 24y \leq 40\ (60) \text{ machine } A \text{ time}$$

$$30x + 33y \leq 35\ (60) \text{ machine } B \text{ time}$$

$$x \geq 75 - 30$$

$x \geq 45$ so, production of $x \geq$ demand (75) – initial stock (30)

$$y \geq 95 - 90$$

$y \geq 5$ so production of $y \geq$ demand (95) – initial stock (90),

The objective is maximise $(x + 30 - 75) + (y + 90 - 95) = (x + y - 50)$ maximise the number of units left in stock.

It is plain and maximum occurs at the intersection of $x = 45$ and $50x + 24y = 2400$.

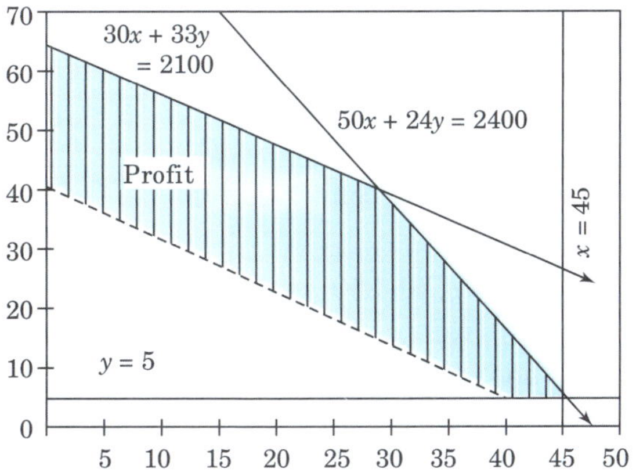


Solving,we get reading value of the graph,we have $x = 45$ and $y = 6.25$ with the value of the product being 1.25.

## 34.4 SOME EXCEPTIONAL CASES

The constraints generally, give region of feasible solution which may be bounded or unbounded. In problems involving two variables and having a finite solution, we observed that the optimal solution existed at a vertex of the feasible region. In fact, this is true for all *L.P.* problems for which solutions exist. Thus, it may be stated that *if there exists an optimal solution of an L.P.P., it will be at one of the vertices of the solution space*.

In each of the above examples, the optimal solution was unique. But it is not always so. In fact, *L.P.P. may have*

(*i*) *a unique optimal solution*, or (*ii*) *an infinite number of optimal solutions*, or (*iii*) *an unbounded solution*, or (*iv*) *no solution.*

We now give below a few examples to illustrate the exceptional cases (*ii*) to (*iv*).

**Example 34.7.** *A firm uses milling machines, grinding machines and lathes to produce two motor parts. The machining times required for each part, the machining times available on different machines and the profit on each motor part are given below:*

| *Type of machine* | *Machining time reqd. for the motor part (mts)* | | *Max. time available per week (minutes)* |
|---|---|---|---|
| | *I* | *II* | |
| *Milling machines* | *10* | *4* | *2,000* |
| *Grinding machines* | *3* | *2* | *900* |
| *Lathes* | *6* | *12* | *3,000* |
| *Profit/unit* (₹) | *100* | *40* | |

*Determine the number of parts I and II to be manufactured per week to maximize the profit.*

**Solution.** Let $x_1$, $x_2$ be the number of parts I and II manufactured per week. Then *objective* being to maximize the profit, we have maximize $Z = 100x_1 + 40x_2$ ...(*i*)

*Constraints* being on the time available on each machine, we obtain

for milling machines, $10x_1 + 4x_2 \le 2{,}000$ ...(*ii*)

for grinding machines, $3x_1 + 2x_2 \le 900$ ...(*iii*)

for lathes, $6x_1 + 12x_2 \le 3{,}000$ ...(*iv*)

Also $x_1, x_2 \ge 0$ ...(*v*)

Thus the problem is to determine $x_1$, $x_2$ which maximize Eqn. (*i*) subject to the constraints Eqns. (*ii*) to (*v*).

The solution space satisfying Eqns. (*ii*), (*iii*), (*iv*) and meeting the non-negativity restrictions Eqn. (*v*) is shown shaded in Fig. 34.9.

Note that Eqn. (*iii*) is a redundant constraint as it does not affect the solution space. The vertices of the convex region *OABC* are

$$O(0, 0), A(200, 0), B(125, 187.5), C(0, 250).$$

Values of the objective function Eqn. (*i*) at these vertices are $Z(O) = 0$, $Z(A) = 20{,}000$, $Z(B) = 20{,}000$ and $Z(C) = 10{,}000$.

Thus the maximum value of $Z$ occurs at two vertices $A$ and $B$.

∴ Any point on the line joining $A$ and $B$ will also give the same maximum value of $Z$ *i.e.*, there are infinite number of feasible solutions which yield the same maximum value of $Z$.

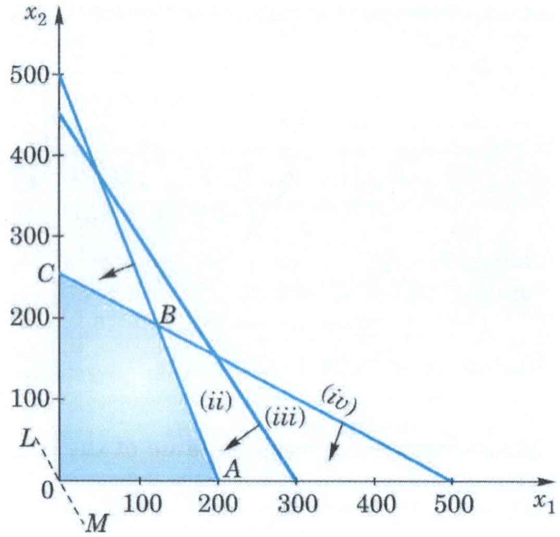

Fig. 34.9

Thus there is no unique optimal solution to the problem and any point on the line $AB$ can be taken to give the profit of ₹ 20,000.

**Obs.** An L.P.P. having more than one optimal solution, is said to have alternative or multiple optimal solutions. It implies that the resources can be combined in more than one way to maximize the profit.

**Example 34.8.** *Using graphical method, solve the following L.P.P.:*

*Maximize* $Z = 2x_1 + 3x_2$ ...(*i*)

*Subject to* $x_1 - x_2 \leq 2$ ...(*ii*)

$x_1 + x_2 \geq 4$ ...(*iii*)

$x_1, x_2 \geq 0.$ (*Rohtak, 2011 S*) ... (*iv*)

**Solution.** Consider $x_1 - x_2$ coordinate system. Any point $(x_1, x_2)$ satisfying the restrictions Eqn. (*iv*) lies in the first quadrant only. The solution space satisfying the constraints Eqns. (*ii*) and (*iii*) is the convex region shown shaded in Fig. 34.10.

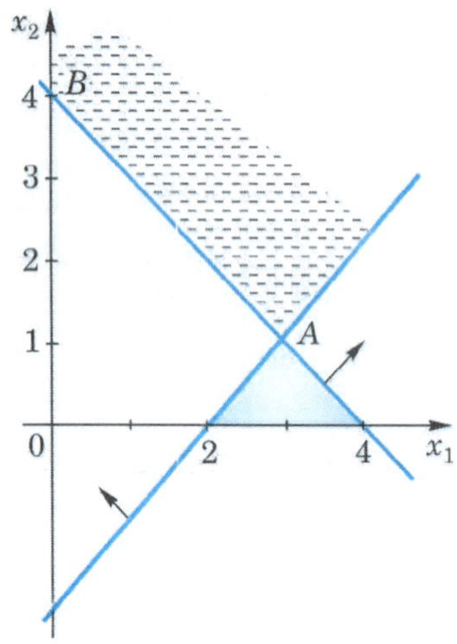

Fig. 34.10

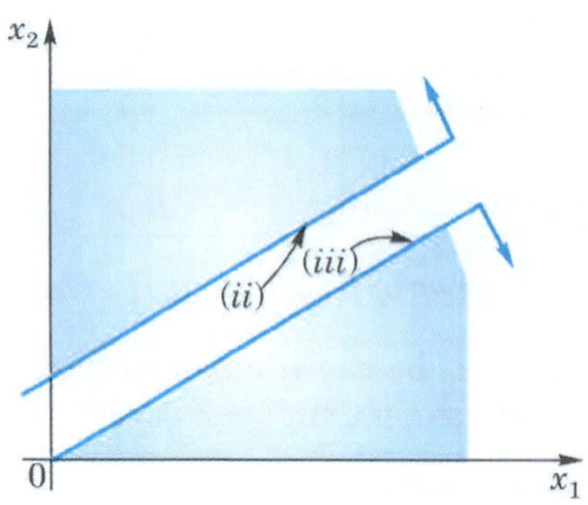

Fig. 34.11

Here the solution space is unbounded. The vertices of the feasible region (in the finite plane) are $A(3, 1)$ and $B(0, 4)$.

Values of the objective function Eqn. (*i*) at these vertices are $Z(A) = 9$ and $Z(B) = 12$.

But there are points in this convex region for which $Z$ will have much higher values. For instance, the point $(5, 5)$ lies in the shaded region and the value of $Z$ thereat is 25. In fact, the maximum value of $Z$ occurs at infinity. Thus the problem has an unbounded solution.

**Example 34.9.** *Solve graphically the following L.P.P.:*

*Maximize* $Z = 4x_1 + 3x_2$ ...(i)

*subject to* $x_1 - x_2 \le -1,$ ...(ii)

$-x_1 + x_2 \le 0$ ...(iii)

*and* $x_1, x_2 \ge 0$ ...(iv)

**Solution.** Consider $x_1 - x_2$ coordinate system. Any point $(x_1, x_2)$ satisfying Eqn. (*iv*) lies in the first quadrant only. The two solution spaces, one satisfying Eqn. (*ii*) and the other satisfying Eqn. (*iii*) are shown shaded in Fig. 34.11.

There being no point $(x_1, x_2)$ common to both the shaded regions, the problem cannot be solved. Hence the solution does not exist since the constraints are inconsistent.

**Obs.** The above problem had no solution because the constraints were incompatible. There may be cases in which the constraints are compatible but the problem may still have no feasible solution.

## PROBLEMS 34.2

Using graphical method, solve the following L.P. problems:

1. Max. $Z = 3x_1 + 5x_2$
   subject to $x_1 + 2x_2 \le 200, x_1 + x_2 \le 150, x_1 \le 60, x_1, x_2 \ge 0$
2. Max. $Z = 5x_1 + 7x_2$
   subject to $x_1 + x_2 \le 4, 5x_1 + 8x_2 \le 24, 10x_1 + 7x_2 \le 35$, and $x_1, x_2 \ge 0$.
3. Min. $Z = 20x_1 + 10x_2$
   subject to $x_1 + 2x_2 \le 40, 3x_1 + x_2 \ge 30, 4x_1 + 3x_2 \ge 60, x_1, x_2 \ge 0$. (*V.T.U., 2016*)
4. Max. $z = 3x + 5x_2$ subject to $x_1 + 2x_2 \le 2000, x_1 + x_2 \le 1500, x_2 \le 600$ and $x_1 \ge 0, x_2 \ge 0$.
5. A firm manufactures two products *A* and *B* on which the profits earned per unit are ₹ 3 and ₹ 4 respectively. Each product is processed on two machines $M_1$ and $M_2$. Product *A* requires one minute of processing time on $M_1$ and 2 minutes on $M_2$ while *B* requires one minute on $M_1$ and one minute on $M_2$. Machine $M_1$ is available for not more than 7 hours and 30 minutes while $M_2$ is available for 10 hours during any working day. Find the number of units of products *A* and *B* to be manufactured to get maximum profit.
6. Two spare parts *X* and *Y* are to be produced in a batch. Each one has to go through two processes *A* and *B*. The time required in hours per unit and total time available are given below:

|  | *X* | *Y* | *Total hours available* |
|---|---|---|---|
| *Process A* | 3 | 4 | 24 |
| *Process B* | 9 | 4 | 36 |

   Profits per unit of *X* and *Y* are ₹ 5 and ₹ 6 respectively. Find how many number of spare parts of *X* and *Y* are to be produced in this batch to maximize the profit. (Each batch is complete in all respects and one cannot produce fractional units and stop the batch).
7. A manufacturer has two products I and II both of which are made in steps by machines *A* and *B*. The process times per hundred for the two products on the two machines are:

| *Product* | *M/c. A* | *M/c. B* |
|---|---|---|
| I | 4 hrs. | 5 hrs. |
| II | 5 hrs. | 2 hrs. |

   Set-up times are negligible. For the coming period machine *A* has 100 hrs, and *B* has 80 hrs. The contribution for product I is ₹ 10 per 100 units and for product II is ₹ 5 per 100 units. The manufacturer is in a market which can absorb both products as much as he can produce for the immediate period ahead. Determine graphically, how much of products I and II, he should produce to maximize his contribution.
8. A production manager wants to determine the quantity to be produced per month of products *A* and *B* manufactured by his firm. The data on resources required and availability of resources are given below:

| *Resources* | *Requirements* | | *Available per month* |
|---|---|---|---|
| | *Product A* | *Product B* | |
| Raw material (kg.) | 60 | 120 | 12,000 |
| Machine hrs/piece | 8 | 5 | 600 |
| Assembly man hrs. | 3 | 4 | 500 |
| Sale price/piece | ₹ 30 | ₹ 40 | |

Formulate the problem as a standard L.P.P. Find product mix that would give maximum profit by graphical technique.

9. A pineapple firm produces two products: canned pineapple and canned juice. The specific amounts of material, labour and equipment required to produce each product and the availability of each of these resources are shown in the table given below:

| | *Canned Juice* | *Pineapple* | *Available Resources* |
|---|---|---|---|
| Labour (man hrs) | 3 | 2.0 | 12.0 |
| Equipment (m/c hrs) | 1 | 2.3 | 6.9 |
| Material (unit) | 1 | 1.4 | 4.9 |

Assuming one unit each of canned juice and canned pineapple has profit margins of ₹ 2 and ₹ 1 respectively. Formulate this as L.P. problem and solve it graphically.

Solve the following L.P. problems graphically:

10. Maximize $Z = 6x + 4y$ subject to $2x + y \geq 1$, $3x + 4y \geq 1.5$ and $x, y \geq 0$.

11. Minimize $Z = 8x_1 + 12x_2$ subject to $60x_1 + 30x_2 \geq 240$, $30x_1 + 60x_2 \geq 300$, $30x_1 + 180x_2 \geq 540$.
and $x_1, x_2 \geq 0$.

12. G.J. Breweries Ltd. have two bottling plants one located at '*G*' and other at '*J*'. Each plant produces three drinks: whiskey, beer and brandy. The number of bottles produced per day are as follows:

| *Drink* | *Plant at G* | *Plant at J* |
|---|---|---|
| Whiskey | 1500 | 1500 |
| Beer | 3000 | 1000 |
| Brandy | 2000 | 5000 |

A market survey indicates that during the month of July, there will be a demand of 20,000 bottles of whiskey, 40,000 bottles of beer and 44,000 bottles of brandy. The operating cost per day for plants at *G* and *J* are ₹ 600 and ₹ 400. For how many days each plant be run in July so as to minimize the production cost, while still meeting the market demand. Solve graphically.

## 34.5 GENERAL LINEAR PROGRAMMING PROBLEM

Any L.P.P. problem involving more than two variables may be expressed as follows:

Find the values of the variables $x_1, x_2, \ldots, x_n$ which maximize (or minimize) the objective function

$$Z = c_1x_1 + c_2x_2 + \ldots + c_nx_n \qquad \ldots(i)$$

subject to the constraints

$$\left.\begin{aligned} a_{11}x_1 + a_{12}x_2 + \ldots + a_{1n}x_n \leq b_1 \\ a_{21}x_1 + a_{22}x_2 + \ldots + a_{2n}x_n \leq b_2 \\ \ldots\ldots\ldots\ldots\ldots\ldots\ldots\ldots\ldots\ldots \\ a_{m1}x_1 + a_{m2}x_2 + \ldots + a_{mn}x_n \leq b_m \end{aligned}\right\} \qquad \ldots(ii)$$

and meet the non-negativity restrictions

$$x_1, x_2, \ldots, x_n \geq 0. \qquad \ldots(iii)$$

**Def. 1.** *A set of values $x_1, x_2, \ldots, x_n$ which satisfies the constraints of the L.P.P. is called its* **solution**.

**Def. 2.** *Any solution to a L.P.P. which satisfies the non-negativity restrictions of the problem is called its* **feasible solution**.

**Def. 3.** *Any feasible solution which maximizes (or minimizes) the objective function of the L.P.P. is called its* **optimal solution**.

Some of the constraints in Eqn. (*ii*) may be equalities, some others may be inequalities of (≤) type and remaining ones inequalities of (≥) type. The inequality constraints are changed to equalities by adding (or subtracting) non-negative variables to (from) the left hand side of such constraints.

**Def. 4.** *If the constraints of a general L.P.P. be*

$$\sum_{j=1}^{n} a_{ij}x_j \leq b_i$$ *(i = 1, 2, … k) then the non-negative variables $s_i$ which satisfy*

$$\sum_{j=1}^{n} a_{ij}x_j + s_i = b_i \ (i = 1, 2, \dots k),$$ *are called* **slack variables.**

**Def. 5.** *If the constraints of a general L.P.P. be*

$$\sum_{j=1}^{n} a_{ij}x_i \geq b_i, \ (i = k, k + 1, \dots)$$ *then the non-negative variables* $s_i$ *which satisfy*

$$\sum_{j=1}^{n} a_{ij}x_i - s_i = b_i, \ (i = k, k + 1, \dots),$$ *are called* **surplus variables.**

## 34.6 CANONICAL AND STANDARD FORMS OF L.P.P.

After the formulation of L.P.P., the next step is to obtain its solution. But before any method is used to find its solution, the problem must be presented in a suitable form. As such, we explain its following two forms:

**(1) Canonical form.** The general L.P.P. can always be expressed in the following form:

Maximize $Z = c_1x_1 + c_2x_2 + \dots + c_nx_n$

subject to the constraints $a_{i1}x_1 + a_{i2}x_2 + \dots + a_{in}x_n \leq b_i; i = 1, 2, \dots m$

$$x_1, x_2, \dots x_n \geq 0,$$

by making some elementary transformations. This form of the L.P.P. is called its **canonical form** and has the following characteristics:

(*i*) Objective function is of maximization type,

(*ii*) All constraints are of ($\leq$) type,

(*iii*) All variables $x_i$ are non-negative.

The canonical form is a format for a L.P.P. which finds its use in the Duality theory.

**(2) Standard form.** The general L.P.P. can also be put in the following form:

Maximize $Z = c_1x_1 + c_2x_2 + \dots c_nx_n$

subject to the constraints $a_{i1}x_1 + a_{i2}x_2 + \dots + a_{in}x_n - b_i; i = 1, 2, \dots m$

$$x_1, x_2, \dots x_n \geq 0,$$

This form of the L.P.P. is called its **standard form** and has the following characteristics:

(*i*) Objective function is of maximization type;

(*ii*) All constraints are expressed as equations;

(*iii*) Right hand side of each constraint is non-negative;

(*iv*) All variables are non-negative.

**Obs.** *Any L.P.P. can be expressed in the standard form.*

As minimize $Z = c_1x_1 + c_2x_2 + \dots + c_nx_n$

is equivalent to maximize $Z'(= -Z) = -c_1x_1 - c_2x_2 \dots - c_nx_n$,

the objective function can always be expressed in the maximization form.

The inequality constraints can always be converted to equalities by adding (or subtracting) the slack (or surplus) variables to the left hand sides of such constraints.

So far, the decision variables $x_1, x_2, \dots, x_n$ have been assumed to be all non-negative. In actual practice, these variables could also be zero or negative. If a variable is negative, it can always be expressed as the difference of two non-negative variables *e.g.* a variable $x_i$ can be written as

$$x_i = x_i' - x_i'' \qquad \text{where } x_i' \geq 0, x_i'' \geq 0.$$

**Example 34.10.** *Convert the following L.P.P. to the standard form:*

*Maximize* $Z = 3x_1 + 5x_2 + 7x_3$,

*subject to* $6x_1 - 4x_2 \leq 5, 3x_1 + 2x_2 + 5x_3 \geq 11, 4x_1 + 3x_3 \leq 2, x_1, x_2 \geq 0.$

**Solution.** As $x_3$ is unrestricted, let $x_3 = x_3' - x_3''$ where $x_3', x_3'' \geq 0$. Now the given constraints can be expressed as

$$6x_1 - 4x_2 \leq 5,$$

$$3x_1 + 2x_2 + 5x_3' - 5x_3'' \geq 11$$

$$4x_1 + 3x_3' - 3x_3'' \leq 2$$
$$x_1, x_2, x_3', x_3'' \geq 0.$$

Introducing the slack/surplus variables, the problem in standard form becomes:

Maximize $\quad Z = 3x_1 + 5x_2 + 7x_3' - 7x_3''$

subject to
$$6x_1 - 4x_2 + s_1 = 5,$$
$$3x_1 + 2x_2 + 5x_3' - 5x_3'' - s_2 = 11,$$
$$4x_1 + 3x_3' - 3x_3'' + s_3 = 2,$$
$$x_1, x_2, x_3', x_3'', s_1, s_2, s_3 \geq 0.$$

**Example 34.11.** *Express the following problem in the standard form:*

*Minimize* $\quad Z = 3x_1 + 4x_2$

*subject to* $\quad 2x_1 - x_2 - 3x_3 = -4,\ \ 3x_1 + 5x_2 + x_4 = 10,\ x_1 - 4x_2 = 12,\ x_1, x_3, x_4 \geq 0.$

**Solution.** Here $x_3$, $x_4$ are the slack/surplus variables and $x_1$, $x_2$ are the decision variables. As $x_2$ is unrestricted, let $x_2 = x_2' - x_2''$ where $x_2', x_2'' \geq 0$.

$\therefore$ The problem in standard form is

Maximize $\quad Z' (= -Z) = -3x_1 - 4x_2' + 4x_2''$

subject to
$$-2x_1 + x_2' - x_2'' + 3x_3 = 4$$
$$3x_1 + 5x_2' - 5x_2'' + x_4 = 10$$
$$x_1 - 4x_2' - 4x_2'' = 12$$
$$x_1, x_2', x_2'', x_3, x_4 \geq 0.$$

**Example 34.12:** *Machines X and Y, each unit of product 1 that is produced requires 15 minutes processing on machine X and 25 minutes processing on machine Y. Each unit of product 2 that is produced requires 7 minutes processing on machine X and 45 minutes processing on machine Y. The available time on machine X in week 5 is forecast to be 20 hours and on machine Y in week 5 is forecast to be 15 hours. Each unit of product 1 sold in week 5 gives a contribute to profit of \$ 10 and each unit of product 2 sold in week 5 gives a contribute to profit of \$ 4.*

*It may not be possible to produce enough to meet your forecast demand for these products in week 5 and each unit of unsatisfied demand for product 1 cost \$3, each unit of unsatisfied demand for product 2 cost \$1.*

*(a) Formulate the problem of deciding how much of each product to make in week 5 as linear program.*

*(b) Solve the linear program graphically.* *(UPTU, 2018)*

**Solution.** For product 1, applying exponential with a smoothing constant of 0.7.

$$M_1 = Y_1 = 23$$
$$M_2 = 0.7y_2 + 0.3\,M_1 = 0.7(27) + 0.3(23) = 25.80$$
$$M_3 = 0.7y_3 + 0.3\,M_2 = 0.7(34) + 0.3(25.80) = 31.54$$
$$M_4 = 0.7y_4 + 0.3\,M_3 = 0.7(40) + 0.3(31.54) = 37.46$$

The forecast for week five is just the average for week 4 $= M_4 = 37.46 = 31$.

For product 2,

$$M_1 = y_1 = 11$$
$$M_2 = 0.7y_2 + 0.3\,M_1 = 0.7(13) + 0.3(11) = 12.40$$
$$M_3 = 0.7y_3 + 0.3\,M_2 = 0.7(15) + 0.3(12.40) = 14.22$$
$$M_4 = 0.7y_4 + 0.3\,M_3 = 0.7(14) + 0.3(14.22) = 14.02$$

The forecast for week five is just the average for week 4 $= M_4 = 14.02 = 14$.

The constraints are

$$15x_1 + 7x_2 \leq 20(60) \text{ machine } X$$
$$25x_1 + 45x_2 \leq 15(60) \text{ machine } Y$$

**Maximise profit**

$10x_1 + 4x_2 - 3(37 - x_1) - 1(14 - x_2)$

Maximise, $13x_1 + 5x_2 - 125$

here $(x_2 = 0)$ at $x_1 = 36$,

at which point the maximum profit is $13(36) + 5(0) - 125 = \$\ 343$.

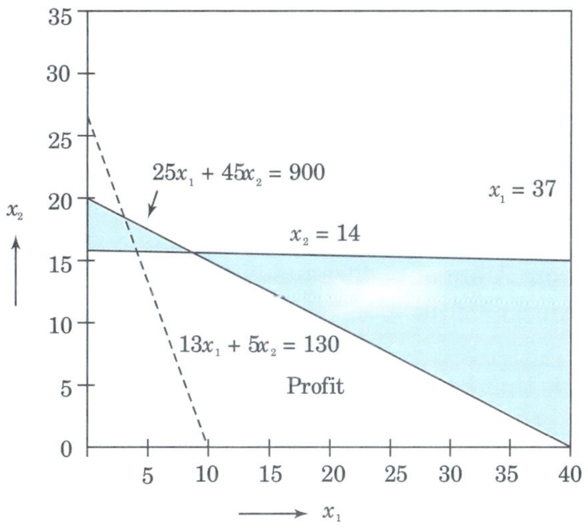


Fig. 34.12.

## 34.7 SIMPLEX METHOD

**(1)** While solving an L.P.P. graphically, the region of feasible solutions was found to be convex, bounded by vertices and edges joining them. The optimal solution occurred at some vertex. If the optimal solution was not unique, the optimal points were on an edge. These observations also hold true for the general L.P.P. Essentially the problem is that of finding the particular vertex of the convex region which corresponds to the optimal solution. The most commonly used method for locating the optimal vertex is the **simplex method**. This method consists in moving step by step from one vertex to the adjacent one. Of all the adjacent vertices, the one giving better value of the objective function over that of the preceding vertex, is chosen. This method of jumping from one vertex to the other is then repeated. Since the number of vertices is finite, the simplex method leads to an optimal vertex in a finite number of steps.

**(2)** *In simplex method, an infinite number of solutions is reduced to a finite number of promising solutions by using the following facts:*

(*i*) When there are $m$ constraints and $(m + n)$ variables ($m$ being $\leq n$), the starting solution is found by setting $n$ variables equal to zero and then solving the remaining $m$ equations, provided the solution exists and is unique. *The n zero variables are known as* **non-basic variables** *while the remaining m variables are called* **basic variables** *and they form a* **basic solution**.

(*ii*) In an L.P.P., the variables must always be non-negative. Some of the basic solutions may contain negative variables. Such solutions are called *basic infeasible solutions* and should not be considered. To achieve this, we start with a basic solution which is non-negative. The next basic solution must always be non-negative. This is ensured by feasibility condition. Such a solution is known as **basic feasible solution**.

*If all the variables in the basic feasible solution are positive, then it is called* **non-degenerate solution** *and if some of the variables are zero, it is called* **degenerate solution.**

(*iii*) A new basic feasible solution may be obtained from the previous one by equating one of the basic variables to zero and replacing it by a new non-basic variable. *The eliminated variable is called the* **outgoing variable** *while the new variable is known as the* **incoming variable.**

*The incoming variable must improve the value of the objective function* which is ensured by the optimality condition. *This process is repeated till no further improvement is possible. The resulting solution is called the* **optimal basic feasible solution** or simply **optimal solution.**

(**3**) *The simplex method is, therefore, based on the following two conditions:*

**I.** *Feasibility condition.* It ensures that if the starting solution is basic feasible, the subsequent will also be basic feasible.

**II.** *Optimality condition.* It ensures that only improved solutions will be obtained.

(**4**) *Now, we shall elaborate the above terms in relation to the general linear programming problem in standard form, i.e.,*

Maximize $$Z = c_1x_1 + c_2x_2 + \ldots + c_nx_n \qquad \ldots(1)$$

subject to $$\sum_{j=1}^{n} a_{ij}x_j + s_i = b_i\,,\ i = 1, 2, \ldots m \qquad \ldots(2)$$

and $$x_j \geq 0,\ s_i \geq 0,\ j = 1, 2, \ldots n \qquad \ldots(3)$$

(*i*) *Solution.* $x_1, x_2, \ldots x_n$ is a solution of the general L.P.P. if it satisfies the constraints Eqn. (2).

(*ii*) *Feasible solution,* $x_1, x_2, \ldots x_n$ is a feasible solution of the general L.P.P. if it satisfies both the constraints Eqn. (2) and the non-negativity restrictions Eqn. (3). The set $S$ of all feasible solutions is called the feasible region. A linear programme is said to be *infeasible* when the set $S$ is empty.

(*iii*) *Basic solution* is the solution of the $m$ basic variables when each of the $n$ non-basic variables is equated to zero.

(*iv*) *Basic feasible solution* is that *basic solution* which also satisfies the non-negativity restrictions Eqn. (3).

(*v*) *Optimal solution* is that basic feasible solution which also optimizes the objective function Eqn. (1) while satisfying the conditions Eqns. (2) and (3).

(*vi*) *Non-degenerate basic feasible solution* is that basic feasible solution which contains exactly $m$ non-zero basic variables. If any of the basic variables becomes zero, it is called a *degenerate basic feasible solution.*

**Example 34.13.** *Find all the basic solutions of the following system of equations identifying in each case the basic and non-basic variables:* $2x_1 + x_2 + 4x_3 = 11$, $3x_1 + x_2 + 5x_3 = 14$.

*Investigate whether the basic solutions are degenerate basic solutions or not. Hence find the basic-feasible solution of the system.*

**Solution.** Since there are $m + n = 3$ variables and there are $m = 2$ constraints in this problem, a basic solution can be obtained by setting any one variable equal to zero and then solving the resulting equations. Also the total number of basic solutions $= {}^{m+n}C_m = {}^3C_2 = 3$.

The characteristics of the various basic solutions are as given below:

| *No. of basic sol.* | *Basic variables* | *Nonbasic variables* | *Values of basic variables* | *Is the sol. feasible ? (Are all $x_j > 0$ ?)* | *Is the sol. degenerate ?* |
|---|---|---|---|---|---|
| 1. | $x_1, x_2$ | $x_3$ | $2x_1 + x_2 = 11$<br>$3x_1 + x_2 = 14$<br>$\therefore\ x_1 = 3, x_2 = 5$ | Yes | No |
| 2. | $x_2, x_3$ | $x_1$ | $x_2 + 4x_3 = 11$<br>$x_2 + 5x_3 = 14$<br>$\therefore\ x_2 = 3, x_3 = -1$ | No | Yes |
| 3. | $x_1, x_3$ | $x_2$ | $2x_1 + 4x_3 = 11$<br>$3x_1 + 5x_3 = 14$<br>$\therefore\ x_1 = 1/2, x_3 = 5/2$ | Yes | No |

The basic feasible solutions are:

(*i*) $x_1 = 3, x_2 = 5, x_3 = 0$; (*ii*) $x_1 = 1/2, x_2 = 0, x_3 = 5/2$

which are also non-degenerate basic solutions.

**Example 34.14.** *Find an optimal solution to the following L.P.P. by computing all basic solutions and then finding one that maximizes the objective function:*

$$2x_1 + 3x_2 - x_3 + 4x_4 = 8,\ x_1 - 2x_2 + 6x_3 - 7x_4 = -3,\ x_1, x_2, x_3, x_4 \geq 0,$$
$$\text{Max. } Z = 2x_1 + 3x_2 + 4x_3 + 7x_4.$$

**Solution.** Since there are four variables and two constraints, a basic solution can be obtained by setting any two variables equal to zero and then solving the resulting equations. Also the total number of basic solutions $= {}^4C_2 = 6$.

The characteristics of the various basic solutions are as given below:

| *No. of basic sol.* | *Basic variables* | *Non-basic variables* | *Values of basic variables* | *Is the sol. feasible ? (Are all $x_j \geq 0$ ?)* | *Value of Z* | *Is the sol. optimal ?* |
|---|---|---|---|---|---|---|
| 1. | $x_1, x_2$ | $x_3, x_4 = 0$ | $2x_1 + 3x_2 = 8$<br>$x_1 - 2x_2 = -3$<br>$\therefore\ x_1 = 1, x_2 = 2$ | Yes | 8 | No |
| 2. | $x_1, x_3$ | $x_2, x_4 = 0$ | $2x_1 - x_3 = 8$<br>$x_1 + 6x_3 = -3$<br>$\therefore\ x_1 = -14/13,$<br>$x_3 = -67/13$ | No | — | — |
| 3. | $x_1, x_4$ | $x_2, x_3 = 0$ | $2x_1 + 4x_4 = 8$<br>$x_1 - 7x_4 = -3$<br>$\therefore\ x_1 = 22/9,$<br>$x_4 = 7/9$ | Yes | 10.3 | No |
| 4. | $x_2, x_3$ | $x_1, x_4 = 0$ | $3x_2 - x_3 = 8$<br>$-2x_2 + 6x_3 = -3$<br>$\therefore\ x_2 = 45/16,$<br>$x_3 = 7/16$ | Yes | 10.2 | No |
| 5. | $x_2, x_4$ | $x_1, x_3 = 0$ | $3x_2 + 4x_4 = 8$<br>$-2x_2 - 7x_4 = -3$<br>$\therefore\ x_2 = 132/39,$<br>$x_4 = -7/13$ | No | — | — |
| 6. | $x_3, x_4$ | $x_1, x_2 = 0$ | $-x_3 + 4x_4 = 8$<br>$6x_3 - 7x_4 = -3$<br>$\therefore\ x_3 = 44/17,$<br>$x_4 = 45/17$ | Yes | 28.9 | Yes |

Hence the optimal basic feasible solution is

$$x_1 = 0, x_2 = 0, x_3 = 44/17, x_4 = 45/17 \text{ and the maximum value of } Z = 28.9.$$

## PROBLEMS 34.3

1. Reduce the following problem to the standard form:
   Determine $x_1 \geq 0, x_2 \geq 0, x_3 \geq 0$ so as to
   Maximize $Z = 3x_1 + 5x_2 + 8x_3$
   subject to the constraints $2x_1 - 5x_2 \leq 6$, $3x_1 + 2x_2 + x_3 \geq 5$, $3x_1 + 4x_3 \leq 3$.
2. Express the following L.P.P. in the standard form
   Maximize $Z = 3x_1 + 2x_2 + 5x_3$
   subject to $-5x_1 + 2x_2 \leq 5$, $2x_1 + 3x_2 + 4x_3 \geq 7$, $2x_1 + 5x_3 \leq 3$, $x_1, x_2, x_3 \geq 0$.
3. Convert the following L.P.P. to standard form:
   Maximize $Z = 3x_1 - 2x_2 + 4x_3$
   subject to $x_1 + 2x_2 + x_3 \leq 8$, $2x_1 - x_2 + x_3 \geq 2$, $4x_1 - 2x_2 - 3x_3 = -6$, $x_1, x_2 \geq 0$.
4. Obtain all the basic solutions to the following system of linear equations:
   $$x_1 + 2x_2 + x_3 = 4,\ 2x_1 + x_2 + 5x_3 = 5.$$
5. Show that the following system of linear equations has two degenerate feasible basic solutions and the non-degenerate basic solution is not feasible:
   $$2x_1 + x_2 - x_3 = 2,\ 3x_1 + 2x_2 + x_3 = 3.$$

6. Find all the basic solutions to the following problem:
Maximize $Z = x_1 + 3x_2 + 3x_3$,
subject to $x_1 + 2x_2 + 3x_3 = 4$, $2x_1 + 3x_2 + 5x_3 = 7$ and $x_1 \geq 0, x_2 \geq 0, x_3 \geq 0$.
Which of the basic solutions are (*a*) non-degenerate basic feasible, (*b*) optimal basic feasible ?

## 34.8 WORKING PROCEDURE OF THE SIMPLEX METHOD

*Assuming the existence of an initial basic feasible solution, an optimal solution to any L.P.P. by Simplex method is found as follows:*

*Step 1. (i) Check whether the objective function is to be maximized or minimized.*

If $$Z = c_1x_1 + c_2x_2 + c_3x_3 + \ldots + c_nx_n$$
is to be minimized, then convert it into a problem of maximization, by writing

Minimize Z = Maximize (–Z)

*(ii) Check whether all b's are positive.*

If any of the $b_i$'s is negative, multiply both sides of that constraint by –1 so as to make its right hand side positive.

*Step 2. Express the problem in the standard form.*

Convert all inequalities of constraints into equations by introducing slack/surplus variables in the constraints giving equations of the form

$$a_{11}x_1 + a_{12}x_2 + a_{13}x_3 + \ldots + s_1 + 0s_2 + 0s_3 + \ldots = b_1.$$

*Step 3. Find an initial basic feasible solution.*

If there are *m* equations involving *n* unknowns, then assign zero values to any $(n - m)$ of the variables for finding a solution. Starting with a basic solution for which $x_j : j = 1, 2, \ldots, (n - m)$ are each zero, find all $s_i$. If all $s_i$ are $\geq 0$, the basic solution is feasible and non-degenerate. If one or more of the $s_i$ values are zero, then the solution is degenerate.

The above information is conveniently expressed in the following Simplex table:

| | $c_j$ | $c_1$ | $c_2$ | $c_3 \ldots 0$ | 0 | $0 \ldots$ |
|---|---|---|---|---|---|---|
| $c_B$ | Basis | $x_1$ | $x_2$ | $x_3 \ldots s_1$ | $s_2$ | $s_3 \ldots b$ |
| 0 | $s_1$ | $a_{11}$ | $a_{12}$ | $a_{13} \ldots 1$ | 0 | $0 \ldots b_1$ |
| 0 | $s_2$ | $a_{21}$ | $a_{22}$ | $a_{23} \ldots 0$ | 1 | $0 \ldots b_2$ |
| 0 | $s_3$ | $a_{31}$ | $a_{32}$ | $a_{33} \ldots 0$ | 0 | $1 \ldots b_3$ |
| : | : | : | : | : : | : | : : |
| | | | Body matrix | | Unit matrix | |

[The variables $s_1, s_2, s_3$ etc. are called *basic variables* and variables $x_1, x_2, x_3$ etc. are called non-basic variables. *Basis* refers to the basic variables $s_1, s_2, s_3 \ldots c_j$ row denotes the coefficients of the variables in the objective function, while $c_B$-column denotes the coefficients of the basic variables only in the objective function. *b*-column denotes the values of the basic variables while remaining variables will always be zero. The coefficients of *x's* (decision variables) in the constraint equations constitute the *body matrix* while coefficients of slack variables constitute the *unit matrix*].

*Step 4. Apply optimality test.*

Compute $C_j = c_j - Z_j$ ; where $Z_j = \Sigma\, c_B a_{ij}$

[$C_j$-row is called *net evaluation row* and indicates the per unit increase in the objective function if the variable heading the column is brought into the solution.]

If all $C_j$ are negative, then the initial basic feasible solution is *optimal*.

If even one $C_j$ is positive, then the current feasible solution is not optimal (*i.e.*, can be improved) and proceed to the next step.

*Step 5. (i) Identify the incoming and outgoing variables.*

If there are more than one positive $C_j$, then the *incoming variable* is the one that heads the column containing maximum $C_j$. The column containing it is known as the *key column* which is shown marked with an arrow at the bottom. If more than one variable has the same maximum $C_j$, any of these variables may be selected arbitrarily as the incoming variable.

Now divide the elements under $b$-column by the corresponding elements of key column and choose the row containing the minimum positive ratio θ. Then replace the corresponding basic variable (by making its value zero). It is termed as the *outgoing variable*. The corresponding row is called the *key row* which is shown marked with an arrow on its right end. The element at the intersection of the key row and key column is called the *key element* which is shown bracketted. If all these ratios are ≤ 0, the incoming variable can be made as large as we please without violating the feasibility condition. Hence the problem has an *unbounded solution* and no further iteration is required.

(*ii*) *Iterate towards an optimal solution.*

Drop the outgoing variable and introduce the incoming variable alongwith its associated value under $c_B$ column. Convert the key element to unity by dividing the key row by the key element. Then make all other elements of the key column zero by subtracting proper multiples of key row from the other rows.

[This is nothing but the sweep-out process used to solve the linear equations. The operations performed are called *elementary row operations*.]

*Step 6. Go to step 4 and repeat the computational procedure* until either an optimal (or an unbounded) solution is obtained.

**Example 34.15.** *Using simplex method*

*Maximize* $Z = 5x_1 + 3x_2$

*subject to* $x_1 + x_2 \le 2,\ 5x_1 + 2x_2 \le 10,\ 3x_1 + 8x_2 \le 12,\ x_1, x_2 \ge 0.$ (*V.T.U., 2013*)

**Solution.** Consists of the following steps :

*Step 1. Check whether the objective function is to be maximized and all b's are positive.*

The problem being of maximization type and all $b$'s being ≥ 0, this step is not necessary.

*Step 2. Express the problem in the standard form.*

By introducing the slack variables $s_1, s_2, s_3$, the problem in standard form becomes

Max. $Z = 5x_1 + 3x_2 + 0s_1 + 0s_2 + 0s_3$

subject to
$$x_1 + x_2 + s_1 + 0s_2 + 0s_3 = 2 \quad ...(i)$$
$$5x_1 + 2x_2 + 0s_1 + s_2 + 0s_3 = 10 \quad ...(ii)$$
$$3x_1 + 8x_2 + 0s_1 + 0s_2 + s_3 = 12 \quad ...(iii)$$
$$x_1, x_2, s_1, s_2, s_3 \ge 0.$$

*Step 3. Find an initial basic feasible solution.*

There are three equations involving five unknowns and for obtaining a solution, we assign zero values to any two of the variables. We start with a basic solution for which we set $x_1 = 0$ and $x_2 = 0$. (This basic solution corresponds to the origin in the graphical method). Substituting $x_1 = x_2 = 0$ in (*i*), (*ii*) and (*iii*), we get the basic solution

$$s_1 = 2,\ s_2 = 10,\ s_3 = 12$$

Since all $s_1, s_2, s_3$ are positive, the basic solution is also feasible and non-degenerate.

∴ The basic feasible solution is

$$x_1 = x_2 = 0 \text{ (non-basic) and } s_1 = 2,\ s_2 = 10,\ s_3 = 12 \text{ (basic)}$$

∴ *Initial basic feasible solution* is given by the following table :

| | $c_j$ | 5 | 3 | 0 | 0 | 0 | | |
|---|---|---|---|---|---|---|---|---|
| $c_B$ | Basis | $x_1$ | $x_2$ | $s_1$ | $s_2$ | $s_3$ | $b$ | θ |
| 0 | $s_1$ | (1) | 1 | 1 | 0 | 0 | 2 | 2/1← |
| 0 | $s_2$ | 5 | 2 | 0 | 1 | 0 | 10 | 10/5 |
| 0 | $s_3$ | 3 | 8 | 0 | 0 | 1 | 12 | 12/3 |
| | $Z_j = \Sigma c_B a_{ij}$ | 0 | 0 | 0 | 0 | 0 | 0 | |
| | $C_j = c_j - Z_j$ | 5 | 3 | 0 | 0 | 0 | | |
| | | ↑ | | | | | | |

[For $x_1$-column ($j = 1$), $Z_j = \Sigma\, c_B\, a_{i1} = 0(1) + 0(5) + 0(3) = 0$

and for $x_2$-column ($j = 2$), $Z_j = \Sigma\, c_B\, a_{i2} = 0(1) + 0(2) + 0(8) = 0$

Similarly $Z_j(b) = 0(2) + 0(10) + 0(12) = 0$.]

*Step 4. Apply optimality test.*

As $C_j$ is positive under some columns, the initial basic feasible solution is not optimal (*i.e.,* can be improved) and we proceed to the next step.

*Step 5. (i) Identify the incoming and outgoing variables.*

The above table shows that $x_1$ is the *incoming variable* as its incremental contribution $C_j$ (= 5) is maximum and the column in which it appears is the *key column* (shown marked by an arrow at the bottom).

Dividing the elements under $b$-column by the corresponding elements of key-column, we find minimum positive ratio $\theta$ is 2 in two rows. We, therefore, arbitrarily choose the row containing $s_1$ as the *key row* (shown marked by an arrow on its right end). The element at the intersection of key row and the key column *i.e.*, (1), is the *key element*. $s_1$ is therefore, the *outgoing basic variable* which will now become non-basic.

Having decided that $x_1$ is to enter the solution, we have tried to find as to what maximum value $x_1$ could have without violating the constraints. So removing $s_1$, the new basis will contain $x_1$, $s_2$ and $s_3$ as the basic variables.

*(ii) Iterate towards the optimal solution.*

To transform the initial set of equations with a basic feasible solution into an equivalent set of equations with a different basic feasible solution, we make the key element unity. Here the key element being unity, we retain the key row as it is. Then to make all other elements in key column zero, we subtract proper multiples of key row from the other rows. Here we subtract 5 times the elements of key row from the second row and 3 times the elements of key row from the third row. These become the second and the third rows of the next table. We also change the corresponding value under $c_B$ column from 0 to 5, while replacing $s_1$ by $x_1$ under the basis. Thus the *second basic feasible solution* is given by the following table :

| | $c_j$ | 5 | 3 | 0 | 0 | 0 | | |
|---|---|---|---|---|---|---|---|---|
| $c_B$ | *Basis* | $x_1$ | $x_2$ | $s_1$ | $s_2$ | $s_3$ | $b$ | $\theta$ |
| 5 | $x_1$ | 1 | 1 | 1 | 0 | 0 | 2 | |
| 0 | $s_2$ | 0 | –3 | –5 | 1 | 0 | 0 | |
| 0 | $s_3$ | 0 | 5 | –3 | 0 | 1 | 6 | |
| | $Z_j = \Sigma\, c_B\, a_{ij}$ | 5 | 5 | 5 | 0 | 0 | 10 | |
| | $C_j = c_j - Z_j$ | 0 | –2 | –5 | 0 | 0 | | |

As $C_j$ is either zero or negative under all columns, the above table gives the optimal basic feasible solution. This optimal solution is $x_1 = 2$, $x_2 = 0$ and maximum $Z = 10$.

**Example 34.16.** *A firm produces three products which are processed on three machines. The relevant data is given below :*

| *Machine* | *Time per unit (minutes)* | | | *Machine capacity* |
|---|---|---|---|---|
| | *Product A* | *Product B* | *Product C* | *(minutes/day)* |
| $M_1$ | 2 | 3 | 2 | 440 |
| $M_2$ | 4 | — | 3 | 470 |
| $M_3$ | 2 | 5 | — | 430 |

*The profit per unit for products A, B, and C is ₹ 4, ₹ 3 and ₹ 6 respectively. Determine the daily number of units to be manufactured for each product. Assume that all the units produced are consumed in the market.*

**Solution.** Let the firm decide to produce $x_1$, $x_2$, $x_3$ units of products *A, B, C,* respectively. Then the L.P. model for this problem is :

Max. $Z = 4x_1 + 3x_2 + 6x_3$

subject to $2x_1 + 3x_2 + 2x_3 \leq 440$, $4x_1 + 3x_3 \leq 470$, $2x_1 + 5x_2 \leq 430$, $x_1, x_2, x_3 \geq 0$.

*Step 1. Check whether the objective function is to be maximized and all b's are non-negative.*

The problem being of maximization type and $b$'s being $\geq 0$, this step is not necessary.

*Step 2. Express the problem in the standard form.*

By introducing the slack variables $s_1$, $s_2$, $s_3$, the problem in standard form becomes :

Max. $Z = 4x_1 + 3x_2 + 6x_3 + 0s_1 + 0s_2 + 0s_3$

subject to $2x_1 + 3x_2 + 2x_3 + s_1 + 0s_2 + 0s_3 = 440$

$4x_1 + 0x_2 + 3x_3 + 0s_1 + s_2 + 0s_3 = 470$

$2x_1 + 5x_2 + 0x_3 + 0s_1 + 0s_2 + s_3 = 430$

$x_1, x_2, x_3, s_1, s_2, s_3 \geq 0.$

*Step 3. Find an initial basic feasible solution.*

The basic (non-degenerate) feasible solution is

$x_1 = x_2 = x_3 = 0$ (non-basic)

$s_1 = 440, s_2 = 470, s_3 = 430$ (basic)

$\therefore$ *Initial basic feasible solution* is given by the following table :

| | $c_j$ | 4 | 3 | 6 | 0 | 0 | 0 | | |
|---|---|---|---|---|---|---|---|---|---|
| $c_B$ | *Basis* | $x_1$ | $x_2$ | $x_3$ | $s_1$ | $s_2$ | $s_3$ | $b$ | $\theta$ |
| 0 | $s_1$ | 2 | 3 | 2 | 1 | 0 | 0 | 440 | 440/2 |
| 0 | $s_2$ | 4 | 0 | (3) | 0 | 1 | 0 | 470 | 470/3← |
| 0 | $s_3$ | 2 | 5 | 0 | 0 | 0 | 1 | 430 | 430/0 |
| $Z_j = \Sigma\, c_B\, a_{ij}$ | | 0 | 0 | 0 | 0 | 0 | 0 | | |
| $C_j = c_j - Z_j$ | | 4 | 3 | 6 | 0 | 0 | 0 | | |
| | | | | ↑ | | | | | |

*Step 4. Apply optimality test.*

As $C_j$ is positive under some columns, the initial basic feasible solution is not optimal and we proceed to the next step.

*Step 5. (i) Identify the incoming and outgoing variables.*

The above table shows that $x_3$ is the incoming variable while $s_2$ is the outgoing variable and (3) is the key element.

*(ii) Iterate towards the optimal solution.*

Drop $s_2$ and introduce $x_3$ with its associated value 6 under $c_B$ column. Convert the key element to unity and make all other elements of key column zero. Then the *second feasible solution* is given by the table below :

| | $c_j$ | 4 | 3 | 6 | 0 | 0 | 0 | | |
|---|---|---|---|---|---|---|---|---|---|
| $c_B$ | *Basis* | $x_1$ | $x_2$ | $x_3$ | $s_1$ | $s_2$ | $s_3$ | $b$ | $\theta$ |
| 0 | $s_1$ | – 2/3 | (3) | 0 | 1 | – 2/3 | 0 | 380/3 | 380/9← |
| 6 | $s_2$ | 4/3 | 0 | 1 | 0 | 1/3 | 0 | 470/3 | ∞ |
| 0 | $s_3$ | 2 | 5 | 0 | 0 | 0 | 1 | 430 | 86 |
| | $Z_j$ | 8 | 0 | 6 | 0 | 2 | 0 | 940 | |
| | $C_j$ | – 4 | 3 | 0 | 0 | – 2 | 0 | | |
| | | | ↑ | | | | | | |

*Step 6. As $C_j$ is positive under the second column, the solution is not optimal and we proceed further. Now $x_2$ is the incoming variable and $s_1$ is the outgoing variable and Eqn. (3) is the key element for the next iteration.*

Drop $s_1$ and introduce $x_2$ with its associated value 3 under $c_B$ column. Convert the key element to unity and make all other elements of the key column zero. Then the *third basic feasible solution* is given by the following table :

| | $c_j$ | 4 | 3 | 6 | 0 | 0 | 0 | | |
|---|---|---|---|---|---|---|---|---|---|
| $c_B$ | *Basis* | $x_1$ | $x_2$ | $x_3$ | $s_1$ | $s_2$ | $s_3$ | $b$ | $\theta$ |
| 3 | $x_2$ | –2/9 | 1 | 0 | 1/3 | –2/9 | 0 | 380/9 | |
| 6 | $x_3$ | 4/3 | 0 | 1 | 0 | 1/3 | 0 | 470/3 | |
| 0 | $s_3$ | 28/9 | 0 | 0 | –5/3 | 10/9 | 0 | 1970/9 | |
| | $Z_i$ | 22/3 | 3 | 6 | 1 | 4/3 | 0 | 3200/3 | |
| | $C_j$ | –10/3 | 0 | 0 | –1 | –4/3 | 0 | | |

Now since each $C_j \le 0$, therefore it gives the optimal solution

$$x_1 = 0, x_2 = 380/9, x_3 = 470/3 \text{ and } Z_{max} = 3200/3 \text{ i.e., } 1066.67 \text{ rupees.}$$

**Example 34.17.** *Solve the following L.P.P. by Simplex method :*

*Minimize* $Z = x_1 - 3x_2 + 3x_3$,

*subject to* $3x_1 - x_2 + 2x_3 \le 7, 2x_1 + 4x_2 \ge -12, -4x_1 + 3x_2 + 8x_3 \le 10, x_1, x_2, x_3 \ge 0.$

*(Kurukshetra, 2013 ; Rohtak, 2011 ; Bhopal, 2009)*

**Solution.** Consists of the following steps :

*Step 1. Check whether objective function is to be maximized and all b's are non-negative.*

As the problem is that of minimizing the objective function, converting it to the maximization type, we have Max. $Z' = -x_1 + 3x_2 - 3x_3$.

As the right hand side of the second constraint is negative, we write it as

$$-2x_1 - 4x_2 \le 12$$

*Step 2. Express the problem in the standard form.*

By introducing the slack variables $s_1, s_2, s_3$, the problem in the standard form becomes

Max. $Z' = -x_1 + 3x_2 - 3x_3 + 0s_1 + 0s_2 + 0s_3$

subject to

$$3x_1 - x_2 + 2x_3 + s_1 + 0s_2 + 0s_3 = 7$$
$$-2x_1 - 4x_2 + 0x_3 + 0s_1 + s_2 + 0s_3 = 12$$
$$-4x_1 + 3x_2 + 8x_3 + 0s_1 + 0s_2 + s_3 = 10$$
$$x_1, x_2, x_3, s_1, s_2, s_3 \ge 0.$$

*Step 3. Find initial basic feasible solution.*

The basic (non-degenerate) feasible solution is

$$x_1 = x_2 = x_3 = 0 \text{ (non-basic) ; } s_1 = 7, s_2 = 12, s_3 = 10 \text{ (basic)}$$

∴ *Initial basic feasible solution* is given by the table below :

| | $c_j$ | −1 | 3 | −3 | 0 | 0 | 0 | | |
|---|---|---|---|---|---|---|---|---|---|
| $c_B$ | *Basis* | $x_1$ | $x_2$ | $x_3$ | $s_1$ | $s_2$ | $s_3$ | $b$ | θ |
| 0 | $s_1$ | 3 | −1 | 2 | 1 | 0 | 0 | 7 | 7/(−1) |
| 0 | $s_2$ | −2 | −4 | 0 | 0 | 1 | 0 | 12 | 12/(−4) |
| 0 | $s_3$ | −4 | (3) | 8 | 0 | 0 | 1 | 10 | 10/3← |
| $Z_j = \Sigma c_B a_{ij}$ | | 0 | 0 | 0 | 0 | 0 | 0 | 0 | |
| $C_j = c_j - Z_j$ | | −1 | 3 | −3 | 0 | 0 | 0 | | |
| | | | ↑ | | | | | | |

*Step 4. Apply optimality test.*

As $C_j$ is positive under second column, the initial basic feasible solution is not optimal and we proceed further.

*Step 5. (i) Identify the incoming and outgoing variables.*

The above table shows that $x_2$ is the incoming variable, $s_3$ is the outgoing variable and (3) is the key element.

*(ii) Iterate towards the optimal solution.*

∴ Drop $s_3$ and introduce $x_2$ with its associated value 3 under $c_B$ column. Convert the key element to unity and make all other elements of the key column zero. Then the *second basic feasible solution* is given by the following table :

| | $c_j$ | −1 | 3 | −3 | 0 | 0 | 0 | | |
|---|---|---|---|---|---|---|---|---|---|
| $c_B$ | *Basis* | $x_1$ | $x_2$ | $x_3$ | $s_1$ | $s_2$ | $s_3$ | $b$ | θ |
| 0 | $s_1$ | (5/3) | 0 | 14/3 | 1 | 0 | 1/3 | 31/3 | 31/5← |
| 0 | $s_2$ | −22/3 | 0 | 32/3 | 0 | 1 | 4/3 | 76/3 | −38/11 |
| 3 | $x_2$ | −4/3 | 1 | 8.3 | 0 | 0 | 1/3 | 10/3 | −5/2 |
| | $Z_j$ | −4 | 3 | 8 | 0 | 0 | 1 | 10 | |
| | $C_j$ | 3 | 0 | −11 | 0 | 0 | −1 | | |
| | | ↑ | | | | | | | |

*Step 6.* As $C_j$ is positive under first column, the solution is not optimal and we proceed further $x_1$ is the incoming variable, $s_1$ is the outgoing variable and (5/3) is the key element.

$\therefore$ Drop $s_1$ and introduce $x_1$ with its associated value –1 under $c_B$ column. Convert the key element to unity and make all other elements of the key column zero. Then the *third basic feasible solution* is given by the table below :

| $c_B$ | $c_j$ Basis | –1 $x_1$ | 3 $x_2$ | –3 $x_3$ | 0 $s_1$ | 0 $s_2$ | 0 $s_3$ | $b$ |
|---|---|---|---|---|---|---|---|---|
| –1 | $x_1$ | 1 | 0 | 14/5 | 3/5 | 0 | 1/5 | 31/5 |
| 0 | $s_2$ | 0 | 0 | 156/5 | 22/5 | 1 | 14/5 | 354/5 |
| 3 | $x_2$ | 0 | 1 | 32/5 | 4/5 | 0 | 3/5 | 58/5 |
| | $Z_j$ | –1 | 3 | 82/5 | 9/5 | 0 | 8/5 | 143/5 |
| | $C_j$ | 0 | 0 | –97/5 | –9/5 | 0 | –8/5 | |

Now since each $C_j \le 0$, therefore it gives the optimal solution

$$x_1 = 31/5,\ x_2 = 58/5,\ x_3 = 0 \text{ (non-basic) and } Z'_{max} = 143/5$$

Hence $\quad Z_{min} = -\ 143/5.$

**Example 34.18.** *Maximize* $Z = 107x_1 + x_2 + 2x_3$, *subject to the constraints :* $14x_1 + x_2 - 6x_3 + 3x_4 = 7$,

$$16x_1 + \frac{1}{2}x_2 - 6x_3 \le 5,\ 3x_1 - x_2 - x_3 \le 0,\ x_1, x_2, x_3, x_4 \ge 0.$$

**Solution.** Consists of the following steps :

*Step 1. Check whether objective function is to be maximized and all b's are non-negative.*

This step is not necessary.

*Step 2. Express the problem in the standard form.*

Here $x_4$ is a slack variable. By introducing other slack variables $s_1$ and $s_2$ the problem in standard form becomes

Max. $\quad Z = 107x_1 + x_2 + 2x_3 + 0x_4 + 0s_1 + 0s_2$

subject to
$$\frac{14}{3}x_1 + \frac{1}{3}x_2 - 2x_3 + x_4 + 0s_1 + 0s_2 = \frac{7}{3}$$
$$16x_1 + \frac{1}{2}x_2 - 6x_3 + 0x_4 + s_1 + 0s_2 = 5$$
$$3x_1 - x_2 - x_3 + 0x_4 + 0s_1 + s_2 = 0$$
$$x_1, x_2, x_3, x_4, s_1, s_2 \ge 0.$$

*Step 3. Find initial basic feasible solution.*

The basic feasible solution is

$$x_1 = x_2 = x_3 = 0 \text{ (non-basic) ; } x_4 = 7/3,\ s_1 = 5,\ s_2 = 0 \text{ (basic)}$$

$\therefore$ *Initial basic feasible solution* is given in the table below :

| $c_B$ | $c_j$ Basis | 107 $x_1$ | 1 $x_2$ | 2 $x_3$ | 0 $x_4$ | 0 $s_1$ | 0 $s_2$ | $b$ | θ |
|---|---|---|---|---|---|---|---|---|---|
| 0 | $x_4$ | $\frac{14}{3}$ | $\frac{1}{3}$ | –2 | 1 | 0 | 0 | $\frac{7}{3}$ | $\frac{7}{3}/\frac{14}{3}$ |
| 0 | $s_1$ | 16 | $\frac{1}{2}$ | –6 | 0 | 1 | 0 | 5 | 5/16 |
| 0 | $s_2$ | (3) | –1 | –1 | 0 | 0 | 1 | 0 | 0/3← |
| $Z_j = \Sigma c_B a_{ij}$ | | 0 | 0 | 0 | 0 | 0 | 0 | 0 | |
| $C_j = c_j - Z_j$ | | 107 ↑ | 1 | 2 | 0 | 0 | 0 | | |

*Step 4. Apply optimality test.*

As $C_j$ is positive under some columns, the initial basic feasible solution is not optimal and we proceed further.

*Step 5. (i) Identify the incoming and outgoing variables.*

The above table shows that $x_1$ is the incoming variable, $s_2$ is the outgoing variable and (3) is the key element.

*(ii) Iterate towards the optimal solution.*

Drop $s_2$ and introduce $x_1$ with its associated value 107 under $c_B$ column. Convert key element to unity and make all other elements of the key column zeros. Then the *second basic feasible solution* is given by the following table :

| | $c_j$ | 107 | 1 | 2 | 0 | 0 | 0 | | |
|---|---|---|---|---|---|---|---|---|---|
| $c_B$ | Basis | $x_1$ | $x_2$ | $x_3$ | $x_4$ | $s_1$ | $s_2$ | $b$ | $\theta$ |
| 0 | $x_4$ | 0 | 17/9 | –4/9 | 1 | 0 | 14/9 | 7/3 | –21/4 |
| 0 | $s_1$ | 0 | 35/6 | –2/3 | 0 | 1 | –16/3 | 5 | –15/2 |
| 107 | $x_1$ | 1 | –1/3 | –1/3 | 0 | 0 | 1/3 | 0 | 0 |
| | $Z_j$ | 107 | –107/3 | –107/3 | 0 | 0 | 107/3 | | |
| | $C_j$ | 0 | 110/3 | 113/3 | 0 | 0 | –107/3 | | |
| | | | | ↑ | | | | | |

As $C_j$ is positive under some columns, the solution is not optimal. Here 113/3 being the largest positive value of $C_j$, $x_3$ is the incoming variable. But all the values of θ being ≤ 0, $x_3$ will not enter the basis. This indicates that the solution to the problem is unbounded.

[**Remember that** (*i*) *the incoming variable is the non-basic variable corresponding to the largest positive value of* $C_j$ *and*

(*ii*) *the outgoing variable is the basic-variable corresponding to the least positive ratio* θ, *obtained by dividing the b-column elements by the corresponding key-column elements.*]

## PROBLEMS 34.4

Using simplex method, solve the following *L.P.P.* (1–8) :

1. Maximize $Z = x_1 + 3x_2$,
   subject to $x_1 + 2x_2 \le 10,\ 0 \le x_1 \le 5,\ 0 \le x_2 \le 4$.
2. Maximize $Z = 4x_1 + 10x_2$
   subject to $2x_1 + x_2 \le 50,\ 2x_1 + 5x_2 \le 100,\ 2x_1 + 3x_2 \le 90,\ x_1, x_2 \ge 0$.
3. Maximize $Z = 4x_1 + 5x_2$,
   subject to $x_1 - 2x_2 \le 2,\ 2x_1 + x_2 \le 6,\ x_1 + 2x_2 \le 5,\ -x_1 + x_2 \le 2,\ x_1, x_2 \ge 0$.
4. Maximize $Z = 10x_1 + x_2 + 2x_3$
   subject to $x_1 + x_2 - 2x_3 \le 10,\ 4x_1 + x_2 + x_3 \le 20,\ x_1, x_2, x_3 \ge 0$. *(Rohtak, 2011 S)*
5. Maximize $Z = 3x_1 + 2x_2 + 5x_3$, subject to $x_1 + 2x_2 + x_3 \le 430,\ 3x_1 + 2x_3 \le 460,\ x_1 + 4x_2 \le 420,\ x_1, x_2, x_3 \ge 0$. *(V.T.U., 2015 S)*
6. Minimize $Z = 3x_1 + 5x_2 + 4x_3$
   subject to $2x_1 + 3x_2 \le 8,\ 2x_2 + 5x_3 \le 10,\ 3x_1 + 2x_2 + 4x_3 \le 15,\ x_1, x_2, x_3 \ge 0$.
7. Minimize $Z = x_1 - 3x_2 + 2x_3$,
   subject to $3x_1 - x_2 + 2x_3 \le 7,\ -2x_1 + 4x_2 \le 12,\ -4x_1 + 3x_2 + 8x_3 \le 10,\ x_1, x_2, x_3 \ge 0$.
8. Maximize $4x + 5y + 9z + 11t$,
   subject to $x + y + z + t \le 15,\ 7x + 5y + 3z + 2t \le 120,\ 3x + 5y + 10z + 15t \le 100,\ x, y, z, t \ge 0$. *(Delhi., 2015)*
9. A firm produces products *A* and *B* and sells them at a profit of ₹ 2 and ₹ 3 each respectively. Each product is processed on machines *G* and *H*. Product *A* requires 1 minute on *G* and 2 minutes on *H* whereas product *B* requires 1 minute on each of the machines. Machine *G* is not available for more than 6 hrs. 40 min/day whereas the time constraint for machine *H* is 10 hrs. Solve this problem *via* simplex method for maximizing the profit.
10. A company makes two types of products. Each product of the first type requires twice as much labour time as the second type. If all products are of second type only, the company can produce a total of 500 units a day. The market limits daily sales of the first and the second type to 150 and 250 units respectively. Assuming that the profits per

unit are ₹ 8 for type I and ₹ 5 for type II, determine the number of units of each type to be produced to maximize profit ?

11. The owner of a dairy is trying to determine the correct blend of two types of feed. Both contain various percentages of four essential ingredients. With the following data determine the least cost blend ?

| *Ingredient* | *% per kg of feed* | | *Min requirement in kg.* |
|---|---|---|---|
| | *Feed 1* | *Feed 2* | |
| 1 | 40 | 20 | 4 |
| 2 | 10 | 30 | 2 |
| 3 | 20 | 40 | 3 |
| 4 | 30 | 10 | 6 |
| Cost (₹/kg.) | 5 | 3 | |

12. A manufacturing firm has discontinued production of a certain unprofitable product line. This created considerable excess production capacity. Management is considering to devote their excess capacity to one or more of three product 1, 2, and 3. The available capacity on machines and the number of machine-hours required for each unit of the respective product, is given below :

| *Machine Type* | *Available Time (hrs/week)* | *Productivity (hrs/unit)* | | |
|---|---|---|---|---|
| | | *Product 1* | *Product 2* | *Product 3* |
| Milling machine | 250 | 8 | 2 | 3 |
| Lathe | 150 | 4 | 3 | 0 |
| Grinder | 50 | 2 | – | 1 |

The unit profit would be ₹ 20, ₹ 6 and ₹ 8 respectively for products 1, 2 and 3. Find how much of each product the firm should produce in order to maximize profit.

13. The following table gives the various vitamin contents of three types of food and daily requirements of vitamins alongwith cost per unit. Find the combination of food for minimum cost.

| *Vitamin (mg)* | *Food F* | *Food G* | *Food H* | *Minimum daily requirement (mg)* |
|---|---|---|---|---|
| *A* | 1 | 1 | 10 | 1 |
| *C* | 100 | 10 | 10 | 50 |
| *D* | 10 | 100 | 10 | 10 |
| Cost/unit (₹) | 10 | 15 | 5 | |

14. A farmer has 1,000 acres of land on which he can grow corn, wheat or soyabeans. Each acre of corn costs ₹ 100 for preparation, requires 7 man-days of work and yields a profit of ₹ 30. An acre of wheat costs ₹ 120 to prepare, requires 10 man-days of work and yields a profit of ₹ 40. An acre of soyabeans costs ₹70 to prepare, requires 8 man-days of work and yields a profit of ₹ 20. If the farmer has ₹ 100,000 for preparation and can count on 8,000 man-days of work, how many acres should be allocated to each crop to maximize profits ?

## 34.9 ARTIFICIAL VARIABLE TECHNIQUES

So far we have seen that the introduction of slack/surplus variables provided the initial basic feasible solution. But there are many problems wherein at least one of the constraints is of (≥) or (=) type and slack variables fail to give such a solution. There are two similar methods for solving such problems which we explain below :

**(1) M-method or Method of Penalties.** This method is due to *A. Charnes* and consists of the following steps :

*Step 1.* Express the problem in standard form.

*Step 2.* Add non-negative variables to the left hand side of all those constraints which are of (≥) or (=) type. Such new variables are called *artificial variables* and the purpose of introducing these is just to obtain an initial basic feasible solution. But their addition causes violation of the corresponding constraints. As such, we would like to get rid of these variables and would not allow them to appear in the final solution. For this purpose, we assign a very large penalty ($-M$) to these artificial variables in the objective function.

*Step* 3. Solve the modified L.P.P. by simplex method.

At any iteration of simplex method, one of the following three cases may arise :

(*i*) There remains no artificial variable in the basis and the optimality condition is satisfied. Then the solution is an optimal basic feasible solution to the problem.

(*ii*) There is at least one artificial variable in the basis at zero level (with zero value in $b$-column) and the optimality condition is satisfied. Then the solution is a degenerate optimal basic feasible solution.

(*iii*) There is at least one artificial variable in the basis at non-zero level (with positive value in $b$-column) and the optimality condition is satisfied. Then the problem has no feasible solution. The final solution is not optimal, since the objective function contains an unknown quantity $M$. Such a solution satisfies the constraints but does not optimize the objective function and is therefore, called *pseudo optimal solution*.

*Step 4.* Continue the simplex method until either an optimal basic feasible solution is obtained or an unbounded solution is indicated.

**Obs.** The artificial variables are only a computational device for getting a starting solution. Once an artificial variable leaves the basis, it has served its purpose and we forget about it *i.e.*, the column for this variable is omitted from the next simplex table.

**Example 34.19.** *Use Charne's penalty method to*

*Minimize* $Z = 2x_1 + x_2$

*subject to* $3x_1 + x_2 = 3,\ 4x_1 + 3x_2 \geq 6,\ x_1 + 2x_2 \leq 3,\ x_1, x_2 \geq 0.$

**Solution.** Consists of the following steps :

*Step 1. Express the problem in standard form.*

The second and third inequalities are converted into equations by introducing the surplus and slack variables $s_1$, $s_2$ respectively.

Also the first and second constraints being of (=) and (≥) type, we introduce two artificial variables $A_1, A_2$.

Converting the minimization problem to the maximization form, the L.P.P. can be rewritten as

$$\text{Max.} \quad Z' = -2x_1 - x_2 + 0s_1 + 0s_2 - MA_1 - MA_2$$

subject to

$$3x_1 + x_2 + 0s_1 + 0s_2 + A_1 + 0A_2 = 3$$
$$4x_1 + 3x_2 - s_1 + 0s_2 + 0A_1 + A_2 = 6$$
$$x_1 + 2x_2 + 0s_1 + s_2 + 0A_1 + 0A_2 = 3$$
$$x_1, x_2, s_1, s_2, A_1, A_2 \geq 0$$

*Step 2. Obtain an initial basic feasible solution.*

Surplus variable $s_1$ is not a basic variable since its value is $-6$. As negative quantities are not feasible, $s_1$ must be prevented from appearing in the initial solution. This is done by taking $s_1 = 0$. By setting the other non-basic variables $x_1$, $x_2$ each = 0, we obtain the initial basic feasible solution as

$$x_1 = x_2 = 0,\ s_1 = 0\ ;\ A_1 = 3,\ A_2 = 6,\ s_2 = 3$$

Thus the initial simplex table is

| | $c_j$ | $-2$ | $-1$ | $0$ | $0$ | $-M$ | $-M$ | | |
|---|---|---|---|---|---|---|---|---|---|
| $c_B$ | *Basis* | $x_1$ | $x_2$ | $s_1$ | $s_2$ | $A_1$ | $A_2$ | $b$ | $\theta$ |
| $-M$ | $A_1$ | (3) | 1 | 0 | 0 | 1 | 0 | 3 | 3/3← |
| $-M$ | $A_2$ | 4 | 3 | $-1$ | 0 | 0 | 1 | 6 | 6/4 |
| 0 | $s_2$ | 1 | 2 | 0 | 1 | 0 | 0 | 3 | 3/1 |
| $Z_j = \Sigma c_B a_{ij}$ | | $-7M$ | $-4M$ | $M$ | 0 | $-M$ | $-M$ | $-9M$ | |
| $C_j = c_j - Z_j$ | | $7M - 2$ | $4M - 1$ | $-M$ | 0 | 0 | 0 | | |
| | | ↑ | | | | | | | |

Since $C_j$ is positive under $x_1$ and $x_2$ columns, this is not an optimal solution.

*Step 3. Iterate towards optimal solution.*

Introduce $x_1$, and drop $A_1$ from basis.

∴ The new simplex table is

| $c_B$ | $c_j$ Basis | −2 $x_1$ | −1 $x_2$ | 0 $s_1$ | 0 $s_2$ | −M $A_2$ | $b$ | θ |
|---|---|---|---|---|---|---|---|---|
| −2 | $x_1$ | 1 | 1/3 | 0 | 0 | 0 | 1 | 3 |
| −M | $A_2$ | 0 | (5/3) | −1 | 0 | 1 | 2 | 6/5← |
| 0 | $s_2$ | 0 | 5/3 | 0 | 1 | 0 | 2 | 6/5 |
| | $Z_j$ | −2 | $-\frac{2}{3}-\frac{5M}{3}$ | $M$ | 0 | −M | −2 − 2M | |
| | $C_j$ | 0 | $-\frac{1}{3}+\frac{5M}{3}$ | −M | 0 | 0 | | |
| | | | ↑ | | | | | |

Since $C_j$ is positive under $x_2$ column, this is not an optimal solution.

∴ Introduce $x_2$ and drop $A_2$.

Then the revised simplex table is

| $c_B$ | $c_j$ Basis | −2 $x_1$ | −1 $x_2$ | 0 $s_1$ | 0 $s_2$ | $b$ |
|---|---|---|---|---|---|---|
| −2 | $x_1$ | 1 | 0 | 1/5 | 0 | 3/5 |
| −1 | $x_2$ | 0 | 1 | −3/5 | 0 | 6/5 |
| 0 | $s_2$ | 0 | 0 | 1 | 1 | 0 |
| | $Z_j$ | −2 | −1 | 1/5 | 0 | −12/5 |
| | $C_j$ | 0 | 0 | −1/5 | 0 | |

Since none of $C_j$ is positive, this an optimal solution. Thus, an optimal basic feasible solution to the problem is

$$x_1 = 3/5, x_2 = 6/5, \text{Max. } Z' = -12/5.$$

Hence the optimal value of the objective function is

$$\text{Min. } Z = -\text{Max. } Z' = -(-12/5) = 12/5$$

**Example 34.20.** *Maximize* $Z = 3x_1 + 2x_2$
*subject to the constraints* : $2x_1 + x_2 \le 2$, $3x_1 + 4x_2 \ge 12$, $x_1, x_2 \ge 0$.

**Solution.** Consists of the following steps :

*Step 1. Express the problem in standard form.*

The inequalities are converted into equations by introducing the slack and surplus variables $s_1$, $s_2$ respectively. Also the second constraint being of (≥) type, we introduce the artificial variable $A$. Thus the L.P.P. can be rewritten as

Max. $Z = 3x_1 + 2x_2 + 0s_1 + 0s_2 - MA$

subject to

$$2x_1 + x_2 + s_1 + 0s_2 + 0A = 2,$$
$$3x_1 + 4x_2 + 0s_1 - s_2 + A = 12,$$
$$x_1, x_2, s_1, A \ge 0$$

*Step 2. Find an initial basic feasible solution.*

Surplus variable $s_2$ is not a basic variable since its value is −12. Since a negative quantity is not feasible, $s_2$ must be prevented from appearing in the initial solution. This is done by letting $s_2 = 0$. By taking the other non-basic variables $x_1$ and $x_2$ each = 0, we obtain the initial basic feasible solution as

$$x_1 = x_2 = s_2 = 0, s_1 = 2, A = 12$$

∴ The initial simplex table is

| $c_B$ | $c_j$ Basis | 3 $x_1$ | 2 $x_2$ | 0 $s_1$ | 0 $s_2$ | −M $A$ | $b$ | θ |
|---|---|---|---|---|---|---|---|---|
| 0 | $s_1$ | 2 | (1) | 1 | 0 | 0 | 2 | 2 ← |
| −M | $A$ | 3 | 4 | 0 | −1 | 1 | 12 | 3 |
| $Z_j = \Sigma c_B a_{ij}$ | | −3M | −4 M | 0 | M | −M | −12M | |
| $C_j = c_j - Z_j$ | | 3 + 3M | 2 + 4M | 0 | −M | 0 | | |
| | | | ↑ | | | | | |

Since $C_j$ is positive under some columns, this is not an optimal solution.

*Step 3. Iterate towards optimal solution.*

Introduce $x_2$ and drop $s_1$.

$\therefore$ The new simplex table is

| | $c_j$ | 3 | 2 | 0 | 0 | $-M$ | |
|---|---|---|---|---|---|---|---|
| $c_B$ | Basis | $x_1$ | $x_2$ | $s_1$ | $s_2$ | $A$ | $b$ |
| 2 | $x_2$ | 2 | 1 | 1 | 0 | 0 | 2 |
| $-M$ | $A$ | $-5$ | 0 | $-4$ | $-1$ | 1 | 4 |
| $Z_j$ | | $4 + 5M$ | 2 | $2 + 4M$ | $M$ | $-M$ | $4 - 4M$ |
| $C_j$ | | $-(1 + 5M)$ | 0 | $-(2 + 4M)$ | $-M$ | 0 | |

Here each $C_j$ is negative and an artificial variable appears in the basis at non-zero level. Thus there exists a *pseudo optimal solution* to the problem.

**(2) Two-phase method.** This is another method to deal with the artificial variables wherein the L.P.P. is solved in two phases.

**Phase I.** *Step 1.* Express the given problem in the standard form by introducing slack, surplus and artificial variables.

*Step 2.* Formulate an artificial objective function

$$Z^* = -A_1 - A_2 \dots - A_m$$

by assigning (–1) cost to each of the artificial variables $A_i$ and zero cost to all other variables.

*Step 3.* Maximize $Z^*$ subject to the constraints of the original problem using the simplex method. Then three cases arise :

(*a*) *Max.* $Z^* < 0$ *and at least one artificial variable appears in the optimal basis at a positive level*

In this case, the original problem dosn't possess any feasible solution and the procedure comes to an end.

(*b*) *Max.* $Z^* = 0$ *and no artificial variable appears in the optimal basis.*

In this case, a basic feasible solution is obtained and we proceed to phase II for finding the optimal basic feasible solution to the original problem.

(*c*) *Max.* $Z^* = 0$ *and at least one artificial variable appears in the optimal basis at zero level.*

Here a feasible solution to the auxiliary L.P.P. is also a feasible solution to the original problem with all artificial variables set = 0.

To obtain a basic feasible solution, we prolong phase I for pushing all the artificial variables out of the basis (without proceeding on to phase II).

**Phase II.** The basic feasible solution found at the end of phase I is used as the starting solution for the original problem in this phase *i.e.*, the final simplex table of phase I is taken as the initial simplex table of phase II and the artificial objective function is replaced by the original objective function. Then we find the optimal solution.

**Example 34.21.** *Use two-phase method to*

*Minimize* $Z = 7.5x_1 - 3x_2$

*subject to the constraints* $3x_1 - x_2 - x_3 \geq 3,\ x_1 - x_2 + x_3 \geq 2,$

$$x_1, x_2, x_3 \geq 0.$$

**Phase I.** *Step 1. Express the problem in standard form.*

**Solution.** Introducing surplus variables $s_1$, $s_2$ and artificial variables $A_1$, $A_2$, the phase I problem in standard form becomes

Max. $\quad Z^* = 0x_1 + 0x_2 + 0x_3 + 0s_1 + 0s_2 - A_1 - A_2$

subject to
$$3x_1 - x_2 - x_3 - s_1 + 0s_2 + A_1 + 0A_2 = 3$$
$$x_1 - x_2 + x_3 + 0s_1 - s_2 + 0A_1 + A_2 = 2$$
$$x_1, x_2, x_3, s_1, s_2, A_1, A_2 \geq 0.$$

*Step 2. Find an initial basic feasible solution.*

Setting $x_1 = x_2 = x_3 = s_1 = s_2 = 0$,

we have $\quad A_1 = 3, A_2 = 2$ and $Z^* = -5$

∴ Initial simplex table is

| | $c_j$ | 0 | 0 | 0 | 0 | 0 | –1 | –1 | | |
|---|---|---|---|---|---|---|---|---|---|---|
| $c_B$ | *Basis* | $x_1$ | $x_2$ | $x_3$ | $s_1$ | $s_2$ | $A_1$ | $A_2$ | $b$ | θ |
| –1 | $A_1$ | (3) | –1 | –1 | –1 | 0 | 1 | 0 | 3 | 1← |
| –1 | $A_2$ | 1 | –1 | 1 | 0 | –1 | 0 | 1 | 2 | 2 |
| $Zj^* = \Sigma c_B a_{ij}$ | | –4 | 2 | 0 | 1 | 1 | –1 | –1 | –5 | |
| $C_j = c_j - Z_j^*$ | | 4<br>↑ | –2 | 0 | –1 | –1 | 0 | 0 | | |

As $C_j$ is positive under $x_1$ column, this solution is not optimal.

*Step 3. Iterate towards an optimal solution.*

Making key element (3) unity and replacing $A_1$ by $x_1$, we have the new simplex table :

| | $c_j$ | 0 | 0 | 0 | 0 | 0 | –1 | –1 | | |
|---|---|---|---|---|---|---|---|---|---|---|
| $c_B$ | Basis | $x_1$ | $x_2$ | $x_3$ | $s_1$ | $s_2$ | $A_1$ | $A_2$ | $b$ | θ |
| 0 | $x_1$ | 1 | $-\frac{1}{3}$ | $-\frac{1}{3}$ | $-\frac{1}{3}$ | 0 | $\frac{1}{3}$ | 0 | 1 | –3 |
| –1 | $A_2$ | 0 | $-\frac{2}{3}$ | $\left(\frac{4}{3}\right)$ | $\frac{1}{3}$ | –1 | $-\frac{1}{3}$ | 1 | 1 | $\frac{3}{4}$← |
| | $Z_j^*$ | 0 | $\frac{2}{3}$ | $-\frac{4}{3}$ | $-\frac{1}{3}$ | 1 | $\frac{1}{3}$ | –1 | –1 | |
| | $C_j$ | 0 | $-\frac{2}{3}$ | $\frac{4}{3}$<br>↑ | $\frac{1}{3}$ | –1 | $-\frac{1}{3}$ | 0 | | |

Since $C_j$ is positive under $x_3$ and $s_1$ columns, this solution is not optimal.

Making key element (4/3) unity and replacing $A_2$ by $x_3$, we obtain the revised simplex table :

| | $c_j$ | 0 | 0 | 0 | 0 | 0 | –1 | –1 | |
|---|---|---|---|---|---|---|---|---|---|
| $C_B$ | *Basis* | $x_1$ | $x_2$ | $x_3$ | $s_1$ | $s_2$ | $A_1$ | $A_2$ | $b$ |
| 0 | $x_1$ | 1 | –1/2 | 0 | –1/4 | –1/4 | 1/4 | 1/4 | 5/4 |
| 0 | $x_2$ | 0 | –1/2 | 1 | 1/4 | –3/4 | –1/4 | 3/4 | 3/4 |
| $Z_j^*$ | | 0 | 0 | 0 | 0 | 0 | 0 | 0 | 0 |
| $C_j$ | | 0 | 0 | 0 | 0 | 0 | –1 | –1 | |

Since all $C_j \le 0$, this table gives the optimal solution. Also $Z^*_{max} = 0$ and no artificial variable appears in the basis. Thus an optimal basic feasible solution to the auxiliary problem and therefore to the original problem, has been attained.

**Phase II.** Considering the actual costs associated with the original variables, the objective function is

Max. $Z' = -15/2x_1 + 3x_2 + 0x_3 + 0s_1 + 0s_2 - 0A_1 - 0A_2$

subject to $3x_1 - x_2 - x_3 - s_1 + 0s_2 + A_1 + 0A_2 = 3,$

$x_2 - x_2 + x_3 + 0s_1 - s_2 + 0A_1 + A_2 = 2,$

$x_2, x_2, x_3, s_1, s_2, A_1, A_2 \ge 0$

The optimal initial feasible solution thus obtained, will be an optimal basic feasible solution to the original L.P.P.

Using final table of phase I, the initial simplex table of phase II is as follows :

| | $c_j$ | –15/2 | 3 | 0 | 0 | 0 | |
|---|---|---|---|---|---|---|---|
| $c_B$ | *Basis* | $x_1$ | $x_2$ | $x_3$ | $s_1$ | $s_2$ | $b$ |
| –15/2 | $x_1$ | 1 | –1/2 | 0 | –1/4 | –1/4 | 5/4 |
| 0 | $x_3$ | 0 | –1/2 | 1 | 1/4 | –3/4 | 3/4 |
| $Z_j$ | | –15/2 | 15/4 | 0 | 15/8 | 15/8 | –75/8 |
| $C_j$ | | 0 | –3/4 | 0 | –15/8 | –15/8 | |

Since all $C_j \le 0$, this solution is optimal.

Hence an optimal basic feasible solution to the given problem is

$$x_1 = 5/4, x_2 = 0, x_3 = 3/4 \quad \text{and} \quad \min. Z = 75/8.$$

## 34.10 EXCEPTIONAL CASES

**(1) Tie for the incoming variable.** When more than one variable has the same largest positive value in $C_j$ row (in maximization problem), a tie for the choice of incoming variable occurs. As there is no method to break this tie, we choose any one of the prospective incoming variables arbitrarily. Such an arbitrary choice does not in any way affect the optimal solution.

**(2) Tie for the outgoing variable.** When more than one variable has the same least positive ratio under then θ-column, a tie for the choice of outgoing variable occurs. If the equal values of said ratio are > 1, choose any one of the prospective leaving variables arbitrarily. Such an arbitrary choice doesn't affect the optimal solution.

If the equal values of ratios are zero, the simplex method fails and we make use of the following degeneracy technique.

**(3) Degeneracy.** We know that a basic feasible solution is said to be degenerate if any of the basic variables vanishes. This phenomenon of getting a degenerate basic feasible solution is called *degeneracy* which may arise

(*i*) *at the initial state,* when atleast one basic variable is zero in the initial basic feasible solution

or (*ii*) *at any subsequent stage,* when the least positive ratios under θ-column are equal for two or more rows.

In this case, an arbitrary choice of one of these basic variables may result in one or more basic variables becoming zero in the next iteration. At times, the same sequence of simplex iterations is repeated endlessly without improving the solution. These are termed as *cycling* type of problems. Cycling occurs very rarely. In fact, cycling has seldom occurred in practical problems.

*To avoid cycling, we apply the following* **perturbation procedure :**

(*i*) Divide each element in the tied rows by the *positive coefficients* of the key column in that row.

(*ii*) Compare the resulting ratios (from left to right) first of unit matrix and then of the body matrix, column by column.

(*iii*) The outgoing variable lies in that row which first contains the smallest algebraic ratio.

**Example 34.22.** *Maximize* $Z = 5x_1 + 3x_2$
*subject to* $x_1 + x_2 \le 2$, $5x_1 + 2x_2 \le 10$, $3x_1 + 8x_2 \le 12$ ; $x_1, x_2 \ge 0$.

**Solution.** Consists of the following steps :

*Step 1. Express the problem in the standard form.*

Introducing the slack variables $s_1, s_2, s_3$, the problem in the standard form is

$$\text{Max. } Z = 5x_1 + 3x_2 + 0s_1 + 0s_2 + 0s_3$$
$$x_1 + x_2 + s_1 + 0s_2 + 0s_3 = 2$$
$$5x_1 + 2x_2 + 0s_1 + s_2 + 0s_3 = 10$$
$$3x_1 + 8x_2 + 0s_1 + 0s_2 + s_3 = 12$$
$$x_1, x_2, s_1, s_2, s_3 \ge 0.$$

*Step 2. Find the initial basic feasible solution.*

The initial basic feasible solution is

$$x_1 = x_2 = 0 \text{ (non-basic)}$$
$$s_1 = 2, s_2 = 10, s_3 = 12 \text{ (basic) and } Z = 0.$$

∴ Initial simplex table is

| | $c_j$ | 5 | 3 | 0 | 0 | 0 | | |
|---|---|---|---|---|---|---|---|---|
| $c_B$ | *Basis* | $x_1$ | $x_2$ | $s_1$ | $s_2$ | $s_3$ | $b$ | θ |
| 0 | $s_1$ | 1 | 1 | 1 | 0 | 0 | 2 | 2/1 |
| 0 | $s_2$ | (5) | 2 | 0 | 1 | 0 | 10 | 10/5← |
| 0 | $s_3$ | 3 | 8 | 0 | 0 | 1 | 12 | 12/3 |
| $Z_j = \Sigma\, c_B\, a_{ij}$ | | 0 | 0 | 0 | 0 | 0 | 0 | |
| $C_j = c_j - Z_j$ | | 5 | 3 | 0 | 0 | 0 | | |
| | | ↑ | | | | | | |

As $C_j$ is positive under $x_2$ columns, this solution is not optimal.

*Step 3. Iterate towards optimal solution.*

$x_1$ is the incoming variable. But the first two rows have the same ratio under θ-column. Therefore we apply *perturbation* method.

First column of the unit matrix has 1 and 0 in the tied rows. Dividing these by the corresponding elements of the key columns, we get 1/1 and 0/5, $s_2$-row gives the smaller ratio and therefore $s_2$ is the first outgoing variable and (5) is the key element.

Thus the new simplex table is

| | $c_j$ | 5 | 3 | 0 | 0 | 0 | | |
|---|---|---|---|---|---|---|---|---|
| $c_B$ | *Basis* | $x_1$ | $x_2$ | $s_1$ | $s_2$ | $s_3$ | $b$ | θ |
| 0 | $s_1$ | 0 | (3/5) | 1 | –1/5 | 0 | 0 | 0 |
| 5 | $x_1$ | 1 | 2/5 | 0 | 1/5 | 0 | 2 | 5← |
| 0 | $s_3$ | 0 | 34/5 | 0 | –3/5 | 1 | 6 | 15/17 |
| $Z_j$ | | 5 | 2 | 0 | 1 | 0 | 10 | |
| $C_j$ | | 0 | 1 | 0 | –1 | 0 | | |
| | | | ↑ | | | | | |

As $C_j$ is positive under $x_2$ column, this solution is not optimal.

Making key element (3/5) unity and replacing $s_1$ by $x_2$, we obtain the revised simplex table :

| | $c_j$ | 5 | 3 | 0 | 0 | 0 | |
|---|---|---|---|---|---|---|---|
| $c_B$ | *Basis* | $x_1$ | $x_2$ | $s_1$ | $s_2$ | $s_3$ | $b$ |
| 3 | $x_2$ | 0 | 1 | 5/3 | –1/3 | 0 | 0 |
| 5 | $x_1$ | 1 | 0 | –2/3 | 1/3 | 0 | 2 |
| 0 | $s_3$ | 0 | 0 | –34/3 | 5/3 | 1 | 6 |
| $Z_j$ | | 5 | 3 | 5/3 | 2/3 | 0 | 10 |
| $C_j$ | | 0 | 0 | –5/3 | –2/3 | 0 | |

As $C_j \le 0$ under all columns, this table gives the optimal solution. Hence an optimal basic feasible solution is $x_1 = 2$, $x_2 = 0$ and $Z_{max} = 10$.

## PROBLEMS 34.5

Solve the following L.P. problems using *M*-method :

1. Maximize $Z = 3x_1 + 2x_2 + 3x_3$
   subject to : $2x_1 + x_2 + x_3 \le 2$, $3x_1 + 4x_2 + 2x_3 \ge 8$, $x_1, x_2, x_3 \ge 0$.
2. Maximize $Z = 2x_1 + x_2 + 3x_3$
   subject to : $x_1 + x_2 + 2x_3 \le 5$, $2x_1 + 3x_2 + 4x_3 = 12$, $x_1, x_2, x_3 \ge 0$.
3. Maximize $Z = 8x_2$,
   subject to : $x_1 - x_2 \ge 0$, $2x_1 + 3x_2 \le -6$, $x_1, x_2$ unrestricted.
4. Maximize $Z = 5x_1 - 2x_2 + 3x_3$
   subject to : $2x_1 + 2x_2 - x_3 \ge 2$, $3x_1 - 4x_2 \le 3$, $x_2 + 3x_3 \le 5$, $x_1, x_2, x_3 \ge 0$.
5. Maximize $Z = x_1 + 2x_2 + 3x_3 - x_4$
   subject to : $x_1 + 2x_2 + 3x_3 = 15$, $2x_1 + x_2 + 5x_3 = 20$,
   $x_1 + 2x_2 + x_3 + x_4 = 10$, $x_1, x_2, x_3, x_4 \ge 0$.

Use two phase method to solve the following L.P. problems :

6. Minimize $Z = x_1 + x_2$
   subject to : $2x_1 + x_2 \ge 4$, $x_1 + 7x_2 \ge 7$,
   $x_1, x_2 \ge 0$.
7. Maximize $Z = 5x_1 + 3x_2$
   subject to : $2x_1 + x_2 \le 1$, $x_1 + 4x_2 \ge 6$,
   $x_1, x_2 \ge 0$.
8. Maximize $Z = 2x_1 + x_2 + x_3$,
   subject to $4x_1 + 6x_2 + 3x_3 \le 8$, $3x_1 - 6x_2 - 4x_3 \le 1$, $2x_1 + 3x_2 - 5x_3 \ge 4$, $x_1, x_2, x_3 \ge 0$. *(Anna., 2016)*

9. Maximize $Z = 5x_1 - 4x_2 + 3x_3$,
subject to : $2x_1 + 2x_2 - x_3 \geq 2$,
$3x_1 - 4x_2 \leq 3, x_2 + x_3 \leq 5$,
$x_1, x_2, x_3 \geq 0$. *(Mumbai, 2009)*

10. Maximize $Z = 5x_1 - 4x_2 + 3x_3$,
subject to : $2x_1 + x_2 - 6x_3 = 20$,
$6x_1 + 5x_2 + 10x_3 \leq 76$,
$8x_1 - 3x_2 + 6x_3 \leq 50$,
$x_1, x_2, x_3 \geq 0$.

Solve the following degenerate L.P. problems :

11. Maximize $Z = 9x_1 + 3x_2$
subject to : $4x_1 + x_2 \leq 8, 2x_1 + x_2 \leq 4$,
$x_1, x_2 \geq 0$.

12. Maximize $Z = 2x_1 + 3x_2 + 10x_3$
subject to : $x_1 + 2x_3 = 0, x_2 + x_3 = 1$,
$x_1, x_2, x_3 \geq 0$.

## 34.11 (1) DUALITY CONCEPT

One of the most interesting concepts in linear programming is the *duality* theory. Every linear programming problem has associated with it, another linear programming problem involving the same data and closely related optimal solutions. Such two problems are said to be *duals* of each other. While one of these is called the *primal,* the other the *dual.*

The importance of the duality concept is due to two main reasons. Firstly, if the primal contains a large number of constraints and a smaller number of variables, the labour of computation can be considerably reduced by converting it into the dual problem and then solving it. Secondly, the interpretation of the dual variables from the cost or economic point of view proves extremely useful in making future decisions in the activities being programmed.

**(2) Formulation of dual problem.** Consider the following L.P.P. :

Maximize $Z = c_1x_1 + c_2x_2 + \ldots + c_nx_n$,

subject to the constraints
$$a_{11}x_1 + a_{12}x_2 + \ldots + a_{1n}x_n \leq b_1,$$
$$a_{21}x_1 + a_{22}x_2 + \ldots + a_{2n}x_n \leq b_2,$$
$$\ldots\ldots\ldots\ldots\ldots\ldots\ldots\ldots$$
$$a_{m1}x_1 + a_{m2}x_2 + \ldots + a_{mn}x_n \leq b_m$$
$$x_1, x_2, \ldots, x_n \geq 0.$$

To construct the dual problem, we adopt the following guidelines :

(*i*) The maximization problem in the primal becomes the minimization problem in the dual and *vice versa*.

(*ii*) ($\leq$) type of constraints in the primal become ($\geq$) type of constraints in the dual and *vice versa*.

(*iii*) The coefficients $c_1, c_2, \ldots, c_n$ in the objective function of the primal become $b_1, b_2, \ldots, b_m$ in the objective function of the dual.

(*iv*) The constants $b_1, b_2, \ldots, b_m$ in the constraints of the primal become $c_1, c_2, \ldots, c_n$ in the constraints of the dual.

(*v*) If the primal has $n$ variables and $m$ constraints, the dual will have $m$ variables and $n$ constraints *i.e.* the transpose of the body matrix of the primal problem gives the body matrix of the dual.

(*vi*) The variables in both the primal and dual are non-negative.

Then the dual problem will be

Minimize $W = b_1y_1 + b_2y_2 + \ldots + b_my_m$

subject to the constraints
$$a_{11}y_1 + a_{21}y_2 + \ldots + a_{m1}y_m \geq c_1,$$
$$a_{12}y_1 + a_{22}y_2 + \ldots + a_{m2}y_m \geq c_2,$$
$$\ldots\ldots\ldots\ldots\ldots\ldots\ldots\ldots$$
$$a_{1n}y_1 + a_{2n}y_2 + \ldots + a_{mn}y_m \geq c_n,$$
$$y_1, y_2, \ldots, y_m \geq 0.$$

**Example 34.23.** *Write the dual of the following L.P.P.:*

*Minimize* $Z = 3x_1 - 2x_2 + 4x_3$,

*subject to* $3x_1 + 5x_2 + 4x_3 \geq 7, 6x_1 + x_2 + 3x_3 \geq 4, 7x_1 - 2x_2 - x_3 \leq 10,$
$x_1 - 2x_2 + 5x_3 \geq 3, 4x_1 + 7x_2 - 2x_3 \geq 2, x_1, x_2, x_3 \geq 0.$

**Solution.** Since the problem is of minimization, all constraints should be of $\geq$ type. We multiply the third constraint throughout by $-1$ so that $-7x_1 + 2x_2 + x_3 \geq -10$.

Let $y_1, y_2, y_3, y_4$ and $y_5$ be the dual variables associated with the above five constraints. Then the dual problem is given by

Maximize $W = 7y_1 + 4y_2 - 10y_3 + 3y_4 + 2y_5$

subject to $3y_1 + 6y_2 - 7y_3 + y_4 + 4y_5 \leq 3,\ 5y_1 + y_2 + 2y_3 - 2y_4 + 7y_5 \leq -2,$

$4y_1 + 3y_2 + y_3 + 5y_4 - 2y_5 \leq 4,\ y_1, y_2, y_3, y_4, y_5 \geq 0.$

**(3) Formulation of dual problem when the primal has equality constraints.** Consider the problem

Maximize $Z = c_1x_1 + c_2x_2$

subject to $a_{11}x_1 + a_{12}x_2 = b_1,\ a_{21}x_1 + a_{22}x_2 \leq b_2,\ x_1, x_2 \geq 0.$

The equality constraint can be written as

$$a_{11}x_1 + a_{12}x_2 \leq b_1 \text{ and } a_{11}x_1 + a_{12}x_2 \geq b_1$$

or
$$a_{11}x_1 + a_{12}x_2 \leq b_1 \text{ and } -a_{11}x_1 - a_{12}x_2 \leq -b_1,$$

Then the above problem can be restated as

Maximize $Z = c_1x_1 + c_2x_2$

subject to $a_{11}x_1 + a_{12}x_2 \leq b_1,\ -a_{11}x_1 - a_{12}x_2 \leq -b_1,$

$a_{21}x_1 + a_{22}x_2 \leq b_2,\ x_1, x_2 \geq 0.$

Now we form the dual using $y_1', y_1'', y_2$ as the dual variables. Then the dual problem is

Minimize $W = b_1(y_1' - y_1'') + b_2y_2,$

subject to $a_{11}(y_1' - y_1'') + a_{21}y_2 \geq c_1,\ a_{12}(y_1' - y_1'') + a_{22}y_2 \geq c_2,\ y_1' y_1'', y_2 \geq 0.$

The term $(y_1' - y_1'')$ appears in both the objective function and all the constraints of the dual. This will always happen whenever there is an equality constraint in the primal. Then the new variable $y_1' - y_1''$ $(= y_1)$ becomes unrestricted in sign being the difference of two non-negative variables and the above dual problem takes the form.

Minimize $W = b_1y_1 + b_2y_2,$

subject to $a_{11}y_1 + a_{21}y_2 \geq c_1,\ a_{12}y_1 + a_{22}y_2 > c_2,\ y_1$ unrestricted in sign, $y_2 \geq 0.$

In general, if the primal problem is

Maximize $Z = c_1x_1 + c_2x_2 + \ldots + c_nx_n,$

subject to

$$a_{11}x_1 + a_{12}x_2 + \ldots + a_{1n}x_n = b_1$$
$$a_{21}x_1 + a_{22}x_2 + \ldots + a_{2n}x_n = b_2$$
$$\ldots\ldots\ldots\ldots\ldots\ldots\ldots\ldots\ldots$$
$$a_{m1}x_1 + a_{m2}x_2 + \ldots + a_{mn}x_n = b_m$$
$$x_1, x_2, \ldots, x_n \geq 0,$$

then the dual problem is

Minimize $W = b_1y_1 + b_2y_2 + \ldots + b_my_m$

subject to

$$a_{11}y_1 + a_{21}y_2 + \ldots + a_{m1}y_m \geq c_1,$$
$$a_{12}y_1 + a_{22}y_2 + \ldots + a_{m2}y_m \geq c_2,$$
$$\ldots\ldots\ldots\ldots\ldots\ldots\ldots\ldots\ldots$$
$$a_{1n}y_1 + a_{2n}y_2 + \ldots + a_{mn}y_m \geq c_n$$
$$y_1, y_2, \ldots, y_m \text{ all unrestricted in sign.}$$

Thus *the dual variables corresponding to equality constraints are unrestricted in sign.* Conversely *when the primal variables are unrestricted in sign, corresponding dual constraints are equalities.*

**Example 34.24.** *Construct the dual of the L.P.P. :*

*Maximize* $Z = 4x_1 + 9x_2 + 2x_3,$

*subject to* $2x_1 + 3x_2 + 2x_3 \leq 7,\ 3x_1 - 2x_2 + 4x_3 = 5,\ x_1, x_2, x_3 \geq 0.$

**Solution.** Let $y_1$ and $y_2$ by the dual variables associated with the first and second constraints. Then the dual problem is

Minimize $W = 7y_1 + 5y_2,$

subject to $2y_1 + 3y_2 \leq 4,\ 3y_1 - 2y_2 \leq 9,\ 2y_1 + 4y_2 \leq 2,\ y_1 \geq 0,\ y_2$ is unrestricted in sign.

## PROBLEMS 34.6

Write the duals of the following problems (1 – 4) :

1. Maximize $Z = 10x_1 + 13x_2 + 19x_3$
   subject to $6x_1 + 5x_2 + 3x_3 \le 26$, $4x_1 + 2x_2 + 5x_3 \le 7$, $x_1, x_2, x_3 \ge 0$.
2. Minimize $Z = 2x_1 + 4x_2 + 3x_3$,
   subject to $3x_1 + 4x_2 + x_3 \ge 11$, $-2x_1 - 3x_2 + 2x_3 \le -7$, $x_1 - 2x_2 - 3x_3 \le -1$
   $3x_1 + 2x_2 + 2x_3 \ge 5$, $x_1, x_2, x_3 \ge 0$.
3. Maximize $Z = 3x_1 + 16x_2 + 7x_3$
   subject to $x_1 - x_2 + x_3 \ge 3$, $-3x_1 + 2x_3 \le 1$, $2x_1 + x_2 - x_3 = 4$, $x_1, x_2, x_3 \ge 0$.
4. Minimize $Z = 3x_1 - 3x_2 + x_3$
   subject to $2x_1 - 3x_2 + x_3 \le 5$, $4x_1 - 2x_2 \ge 9$, $-8x_1 + 4x_2 + 3x_3 = 8$,
   $x_1, x_2 \ge 0$ and $x_3$ is unrestricted.
5. Obtain the dual problem of the following L.P.P.
   Maximize $f(x) = 2x_1 + 5x_2 + 6x_3$
   subject to $5x_1 + 6x_2 - x_3 \le 3$, $-2x_1 + x_2 + 4x_3 \le 4$, $x_1 - 5x_2 + 3x_3 \le 1$,
   $-3x_1 - 3x_2 + 7x_3 \le 6$, $x_1, x_2, x_3 \ge 0$.
   Also verify that the dual of the dual problem is the primal problem.

## 34.12 (1) DUALITY PRINCIPLE

*If the primal and the dual problems have feasible solutions then both have optimal solutions and the optimal value of the primal objective function is equal to the optimal value of the dual objective function i.e.,*

$$\text{Max. } Z = \text{Min. } W$$

This is the fundamental theorem of duality. It suggests that an optimal solution to the primal problem can directly be obtained from that of the dual problem and *vice-versa*.

**(2) Working rules for obtaining an optimal solution to the primal (dual) problem from that of the dual (primal) :**

Suppose we have already found an optimal solution to the dual (primal) problem by simplex method.

*Rule I.* If the primal variable corresponds to a slack starting variable in the dual problem, then its optimal value is directly given by the coefficient of the slack variable with changed sign, in the $C_j$ row of the optimal dual simplex table and *vice-versa*.

*Rule II.* If the primal variable corresponds to an artificial variable in the dual problem, then its optimal value is directly given by the coefficient of the artificial variable, with changed sign, in the $C_j$ row of the optimal dual simplex table, after deleting the constant $M$ and *vice-versa*.

On the other hand, if the primal has an unbounded solution, then the dual problem will not have a feasible solution and *vice-versa*.

Now we shall workout two examples to demonstrate the primal dual relationships.

**Example 34.25.** *Construct the dual of the following problem and solve both the primal and the dual :*
*Maximize* $Z = 2x_1 + x_2$,
*subject to* $-x_1 + 2x_2 \le 2$ $x_1 + x_2 \le 4$, $x_1 \le 3$, $x_1, x_2 \ge 0$.

**Solution.** *Using the primal problem.* Since only two variables are involved, it is convenient to solve the problem graphically.

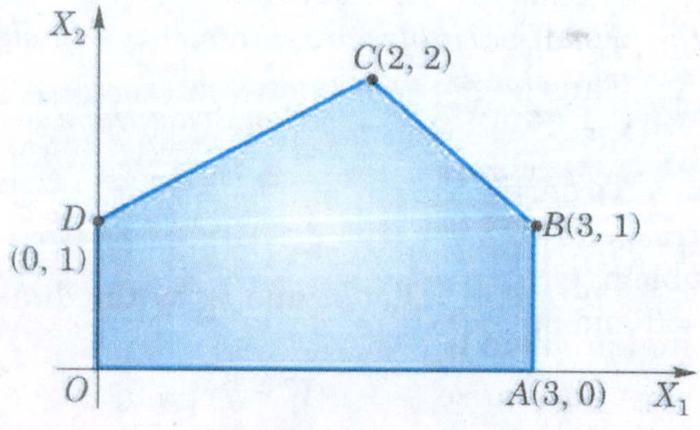

Fig. 34.12

In the $x_1, x_2$-plane, the five constraints show that the point $(x_1, x_2)$ lies within the shaded region $OABCD$ of Fig. 34.12. Values of the objective function $Z = 2x_1 + x_2$ at these corners are $Z(0) = 0$, $Z(A) = 6$, $Z(B) = 7$, $Z(C) = 6$ and $Z(D) = 1$. Hence the optimal solution is $x_1 = 3$, $x_2 = 1$ and max. $(Z) = 7$.

**Solution.** *Using the dual problem.* The dual problem of the given primal is :

Minimize $W = 2y_1 + 4y_2 + 3y_3$

subject to $-y_1 + y_2 + y_3 \geq 2,\ 2y_1 + y_2 \geq 1,\ y_1, y_2 \geq 0.$

*Step 1. Express the problem in the standard form.*

Introducing the slack and the artificial variables, the dual problem in the standard form is

Max. $W' = -2y_1 - 4y_2 - 3y_3 + 0s_1 + 0s_2 - MA_1 - MA_2$

subject to $-y_1 + y_2 + y_3 - s_1 + 0s_2 + A_1 + 0A_2 = 2,$

$2y_1 + y_2 + 0y_3 + 0s_1 - s_2 + 0A_1 + A_2 = 1$

*Step 2. Find an initial basic feasible solution.*

Setting the non-basic variables $y_1, y_2, y_3, s_1, s_2$, each equal to zero, we get the initial basic feasible solution as

$$y_1 = y_2 = y_3 = s_1 = s_2 = 0 \text{ (non-basic)};\ A_1 = 2,\ A_2 = 1. \quad \text{(basic)}$$

$\therefore$ Initial simplex table is

| | $c_j$ | $-2$ | $-4$ | $-3$ | $0$ | $0$ | $-M$ | $-M$ | | |
|---|---|---|---|---|---|---|---|---|---|---|
| $c_B$ | *Basis* | $y_1$ | $y_2$ | $y_3$ | $s_1$ | $s_2$ | $A_1$ | $A_2$ | $b$ | $\theta$ |
| $-M$ | $A_1$ | $-1$ | $1$ | $1$ | $-1$ | $0$ | $1$ | $0$ | $2$ | $2/1$ |
| $-M$ | $A_2$ | $2$ | $(1)$ | $0$ | $0$ | $-1$ | $0$ | $1$ | $1$ | $1/1\leftarrow$ |
| | $Z_j$ | $-M$ | $-2M$ | $-M$ | $M$ | $M$ | $-M$ | $-M$ | $-3M$ | |
| | $C_j$ | $M-2$ | $2M-4$ | $M-3$ | $-M$ | $-M$ | $0$ | $0$ | | |
| | | | $\uparrow$ | | | | | | | |

As $C_j$ is positive under some columns, the initial solution is not optimal.

*Step 3. Iterate towards an optimal solution.*

(*i*) Introduce $y_2$ and drop $A_2$. Then the new simplex table is

| | $c_j$ | $-2$ | $-4$ | $-3$ | $0$ | $0$ | $-M$ | $-M$ | | |
|---|---|---|---|---|---|---|---|---|---|---|
| $c_B$ | *Basis* | $y_1$ | $y_2$ | $y_3$ | $s_1$ | $s_2$ | $A_1$ | $A_2$ | $b$ | $\theta$ |
| $-M$ | $A_1$ | $-3$ | $0$ | $(1)$ | $-1$ | $1$ | $1$ | $-1$ | $1$ | $1/1\leftarrow$ |
| $-4$ | $y_2$ | $2$ | $1$ | $0$ | $0$ | $-1$ | $0$ | $1$ | $1$ | $1/0$ |
| | $Z_j$ | $3M-8$ | $-4$ | $-M$ | $M$ | $4-M$ | $-M$ | $M-4$ | $-M-4$ | |
| | $C_j$ | $6-3M$ | $0$ | $M-3$ | $-M$ | $M-4$ | $0$ | $4-2M$ | | |
| | | | | $\uparrow$ | | | | | | |

As $C_j$ is positive under some columns, this solution is not optimal.

(*ii*) Now introduce $y_3$ and drop $A_1$. Then the revised simplex table is

| | $c_j$ | $-2$ | $-4$ | $-3$ | $0$ | $0$ | $-M$ | $-M$ | |
|---|---|---|---|---|---|---|---|---|---|
| $c_B$ | *Basis* | $y_1$ | $y_2$ | $y_3$ | $s_1$ | $s_2$ | $A_1$ | $A_2$ | $b$ |
| $-3$ | $y_3$ | $-3$ | $0$ | $1$ | $-1$ | $1$ | $1$ | $-1$ | $1$ |
| $-4$ | $y_2$ | $2$ | $1$ | $0$ | $0$ | $-1$ | $0$ | $1$ | $1$ |
| | $Z_j$ | $1$ | $-4$ | $-3$ | $3$ | $1$ | $-3$ | $-1$ | $-7$ |
| | $C_j$ | $-3$ | $0$ | $0$ | $-3$ | $-1$ | $3-M$ | $1-M$ | |

As all $C_j \leq 0$, the optimal solution is attained.

Thus an optimal solution to the dual problem is

$$y_1 = 0,\ y_2 = 1,\ y_3 = 1,\ \text{Min. } W = -\text{Max. } (W') = 7.$$

To derive the optimal basic feasible solution to the primal problem, we note that the primal variables $x_1, x_2$ correspond to the artificial starting dual variables $A_1, A_2$ respectively. In the final simplex table of the dual problem, $C_j$ corresponding to $A_1$, and $A_2$ are 3 and 1 respectively after ignoring $M$. Thus by rule II, we get opt. $x_1 = 3$ and opt. $x_2 = 1$.

Hence an optimal basic feasible solution to the given primal is

$$x_1 = 3, x_2 = 1 \text{ ; max. } Z = 7.$$

**Obs.** The validity of the duality theorem is therefore, checked since max. $Z$ = min. $W$ = 7 from both the methods.

**Example 34.26.** *Using duality solve the following problem :*
*Minimize* $Z = 0.7x_1 + 0.5x_2$
*subject to* $x_1 \geq 4,\ x_2 \geq 6,\ x_1 + 2x_2 \geq 20,\ 2x_1 + x_2 \geq 18,\ x_1, x_2 \geq 0.$

**Solution.** The dual of the given problem is Max. $W = 4y_1 + 6y_2 + 20y_3 + 18y_4$,
subject to $y_1 + y_3 + 2y_4 \leq 0.7,\ y_2 + 2y_3 + y_4 \leq 0.5,\ y_1, y_2, y_3, y_4 \geq 0.$

*Step 1. Express the problem in the standard form.*

Introducing slack variables, the dual problem in the standard form becomes

Max. $W = 4y_1 + 6y_2 + 20y_3 + 18y_4 + 0s_1 + 0s_2$,
subject to $y_1 + 0y_2 + y_3 + 2y_4 + s_1 + 0s_2 = 0.7,$
$0y_1 + y_2 + 2y_3 + y_4 + 0s_1 + s_2 = 0.5,\ y_1, y_2, y_3, y_4 \geq 0.$

*Step 2. Find an initial basic feasible solution.*

Setting non-basic variables $y_1, y_2, y_3, y_4$ each equal to zero, the basic solution is

$$y_1 = y_2 = y_3 = y_4 = 0 \quad \text{(non-basic) ; } s_1 = 0.7,\ s_2 = 0.5 \quad \text{(basic)}$$

Since the basic variables $s_1, s_2 > 0$, the initial basic solution is feasible and non-degenerate.

Initial simplex table is

| | $c_j$ | 4 | 6 | 20 | 18 | 0 | 0 | | |
|---|---|---|---|---|---|---|---|---|---|
| $c_B$ | *Basis* | $y_1$ | $y_2$ | $y_3$ | $y_4$ | $s_1$ | $s_2$ | $b$ | θ |
| 0 | $s_1$ | 1 | 0 | 1 | 2 | 1 | 0 | 0.7 | 0.7/1 |
| 0 | $s_2$ | 0 | 1 | (2) | 1 | 0 | 1 | 0.5 | 0.5/2← |
| | $Z_j$ | 0 | 0 | 0 | 0 | 0 | 0 | 0 | |
| | $C_j$ | 4 | 6 | 20 | 18 | 0 | 0 | | |
| | | | | ↑ | | | | | |

As $C_j$ is positive in some columns, the initial basic solution is not optimal.

*Step 3. Iterate towards an optimal solution.*

(*i*) Introduce $y_3$ and drop $s_2$. Then the new simplex table is

| | $c_j$ | 4 | 6 | 20 | 18 | 0 | 0 | | |
|---|---|---|---|---|---|---|---|---|---|
| $c_B$ | *Basis* | $y_1$ | $y_2$ | $y_3$ | $y_4$ | $s_1$ | $s_2$ | $b$ | θ |
| 0 | $s_1$ | 1 | –1/2 | 0 | (3/2) | 1 | –1/2 | 9/20 | 3/10← |
| 20 | $y_3$ | 0 | 1/2 | 1 | $\frac{1}{2}$ | 0 | 1/2 | 1/4 | 1/2 |
| | $Z_j$ | 0 | 10 | 20 | 10 | 0 | 10 | 5 | |
| | $C_j$ | 4 | –4 | 0 | 8 | 0 | –10 | | |
| | | | | | ↑ | | | | |

As $C_j$, is positive under some of the columns, this solution is not optimal.

(*ii*) Introduce $y_4$ and drop $s_1$. Then the revised simplex table is

| | $c_j$ | 4 | 6 | 20 | 18 | 0 | 0 | |
|---|---|---|---|---|---|---|---|---|
| $c_B$ | *Basis* | $y_1$ | $y_2$ | $y_3$ | $y_4$ | $s_1$ | $s_2$ | $b$ |
| 18 | $y_4$ | 2/3 | –1/3 | 0 | 1 | 2/3 | –1/3 | 3/10 |
| 20 | $y_3$ | –1/3 | 2/3 | 1 | 0 | –1/3 | 2/3 | 1/10 |
| | $Z_j$ | 16/3 | 22/3 | 20 | 18 | 16/3 | 22/3 | 74/10 |
| | $C_j$ | –4/3 | –4/3 | 0 | 0 | –16/13 | –22/3 | |

As all $C_j \le 0$, this table gives the optimal solution.

Thus the optimal basic feasible solution is $y_1 = 0, y_2 = 0, y_3 = 20, y_4 = 18$, max. $W = 7.4$

*Step 4. Derive optimal solution to the primal.*

We note that the primal variables $x_1, x_2$ correspond to the slack starting dual variables $s_1, s_2$ respectively. In the final simplex table of the dual problem, $C_j$ values corresponding to $s_1$ and $s_2$ are $-16/3$ and $-22/3$ respectively.

Thus, by rule I, we conclude that opt. $x_1 = 16/3$ and opt. $x_2 = 22/3$.

Hence an optimal basic feasible solution to the given primal is

$$x_1 = 16/3, x_2 = 22/3; \text{ min. } Z = 7.4.$$

**Obs.** To check the validity of the duality theorem, the student is advised to solve the given L.P.P. directly by simplex method and see that min. $Z$ = max. $W = 7.4$.

**PROBLEMS 34.7**

Using duality solve the following problems (1 – 4) :

1. Minimize $Z = 2x_1 + 9x_2 + x_3$,
   subject to $x_1 + 4x_2 + 2x_3 \ge 5$, $3x_1 + x_2 + 2x_3 \ge 4$ and $x_1, x_2 \ge 0$.
2. Maximize $Z = 2x_1 + x_2$,
   subject to $x_1 + 2x_2 \le 10$, $x_1 + x_2 \le 6$, $x_1 - x_2 \le 2$, $x_1 - 2x_2 \le 1$, $x_1, x_2 \ge 0$.
3. Maximize $Z = 3x_1 + 2x_2$,
   subject to $x_1 + x_2 \ge 1$, $x_1 + x_2 \le 7$, $x_1 + 2x_2 \le 10$, $x_2 \le 3$, $x_1, x_2 \ge 0$.
4. Maximize $Z = 3x_1 + 2x_2 + 5x_3$
   subject to $x_1 + 2x_2 + x_3 \le 430$, $3x_1 + 2x_3 \le 460$, $x_1 + 4x_2 \le 420$, $x_1, x_2, x_3 \ge 0$.

## 34.13 (1) DUAL SIMPLEX METHOD

In § 34.9, we have seen that a set of basic variables giving a feasible solution can be found by introducing artificial variables and using $M$-method or Two phase method. Using the primal-dual relationships for a problem, we have another method (known as *Dual simplex method*) for finding an initial feasible solution. Whereas the regular simplex method starts with a basic feasible (but non-optimal) solution and works towards optimality, the dual simplex method starts with a basic unfeasible (but optimal) solution and works towards feasibility. The dual simplex method is quite similar to the regular simplex method, the only difference lies in the criterion used for selecting the incoming and outgoing variables. In the dual simplex method, we first determine the outgoing variable and then the incoming variable while in the case of regular simplex method reverse is done.

**(2) Working procedure for dual Simplex method :**

*Step 1. (i) Convert the problem to maximization form*, if it is not so.

*(ii) Convert* ($\ge$) *type constraints, if any to* ($\le$) *type* by multiplying such constraints by $-1$.

*(iii) Express the problem in standard form* by introducing slack variables.

*Step 2. Find the initial basic solution and express this information in the form of dual Simplex table.*

*Step 3. Test the nature of* $C_j = c_j - Z_j$ :

(*a*) If all $C_j \le 0$ and all $b_i \ge 0$, then optimal basic feasible solution has been attained.

(*b*) If all $C_j \le 0$ and at least one $b_i < 0$, then go to step 4.

(*c*) If any $C_j \ge 0$, the method fails.

*Step 4. Mark the outgoing variable.* Select the row that contains the most negative $b_i$. This will be the key row and the corresponding basic variable is the outgoing variable.

*Step 5. Test the nature of key row elements :*

(*a*) If all these elements are $\ge 0$, the problem does not have a feasible solution.

(*b*) If at least one element $< 0$, find the ratios of the corresponding elements of $C_j$-row to these elements. Choose the smallest of these ratios. The corresponding column is the key column and the associated variable is the *incoming variable*.

*Step 6. Iterate towards optimal feasible solution.* Make the key element unity. Perform row operations as in the regular simplex method and repeat iterations until either an optimal feasible solution is attained or there is an indication of non-existence of a feasible solution.

**Example 34.27.** *Using dual simplex method :*
*maximize* $-3x_1 - 2x_2$,
*subject to* $x_1 + x_2 \geq 1,\ x_1 + x_2 \leq 7,\ x_1 + 2x_2 \geq 10,\ x_2 \geq 3,\ x_1 \geq 0,\ x_2 \geq 0.$

*(Delhi., 2016 ; Kurukshetra, 2013)*

**Solution.** Consists of the following steps :

*Step 1. (i) Convert the first and third constraints into* (≤) *type. These constraints become*

$$-x_1 - x_2 \leq -1,\ -x_1 - 2x_2 \leq -10.$$

*(ii) Express the problem in standard form*

Introducing slack variables $s_1, s_2, s_3, s_4$ the given problem takes the form

Max. $Z = -3x_1 - 2x_2 + 0s_1 + 0s_2 + 0s_3 + 0s_4$

subject to $-x_1 - x_2 + s_1 = -1,\ x_1 + x_2 + s_2 = 7,\ -x_1 - 2x_2 + s_3 = -10,$
$x_2 + s_4 = 3,\ x_1, x_2, s_1, s_2, s_3, s_4 \geq 0.$

*Step 2. Find the initial basic solution*

Setting the decision variables $x_1, x_2$ each equal to zero, we get the basic solution

$$x_1 = x_2 = 0,\ s_1 = -1,\ s_2 = 7,\ s_3 = -10,\ s_4 = 3 \text{ and } Z = 0.$$

∴ Initial solution is given by the table below :

| | $c_j$ | −3 | −2 | 0 | 0 | 0 | 0 | |
|---|---|---|---|---|---|---|---|---|
| $c_B$ | *Basis* | $x_1$ | $x_2$ | $s_1$ | $s_2$ | $s_3$ | $s_4$ | $b$ |
| 0 | $s_1$ | −1 | −1 | 1 | 0 | 0 | 0 | −1 |
| 0 | $s_2$ | 1 | 1 | 0 | 1 | 0 | 0 | 7 |
| 0 | $s_3$ | −1 | (−2) | 0 | 0 | 1 | 0 | −10 ← |
| 0 | $s_4$ | 0 | 1 | 0 | 0 | 0 | 1 | 3 |
| | $Z_j = \Sigma c_B a_{ij}$ | 0 | 0 | 0 | 0 | 0 | 0 | 0 |
| | $C_j = c_j - Z_j$ | −3 | −2 ↑ | 0 | 0 | 0 | 0 | |

*Step 3. Test nature of* $C_j$.

Since all $C_j$ values are ≤ 0 and $b_1 = -1,\ b_3 = -10$, the initial solution is optimal but infeasible. We therefore, proceed further.

*Step 4. Mark the outgoing variable.*

Since $b_3$ is negative and numerically largest, the third row is the key row and $s_3$ is the outgoing variable.

*Step 5. Calculate ratios of elements in* $C_j$*-row to the corresponding negative elements of the key row.*

These ratios are −3/−1 = 3, −2/−2 = 1 (neglecting ratios corresponding to +ve or zero elements of key row). Since the smaller ratio is 1, therefore, $x_2$-column is the key column and (−2) is the key element.

*Step 6. Iterate towards optimal feasible solution.*

*(i)* Drop $s_3$ and introduce $x_2$ alongwith its associated value −2 under $c_B$ column. Convert the key element to unity and make all other elements of the key column zero. Then the second solution is given by the table below :

| | $c_j$ | −3 | −2 | 0 | 0 | 0 | 0 | |
|---|---|---|---|---|---|---|---|---|
| $c_B$ | *Basis* | $x_1$ | $x_2$ | $s_1$ | $s_2$ | $s_3$ | $s_4$ | $b$ |
| 0 | $s_1$ | $-\frac{1}{2}$ | 0 | 1 | 0 | $-\frac{1}{2}$ | 0 | 4 |
| 0 | $s_2$ | $\frac{1}{2}$ | 0 | 0 | 1 | $\frac{1}{2}$ | 0 | 2 |
| −2 | $x_2$ | $\frac{1}{2}$ | 1 | 0 | 0 | $-\frac{1}{2}$ | 0 | 5 |
| 0 | $s_4$ | $\left(-\frac{1}{2}\right)$ | 0 | 0 | 0 | $\frac{1}{2}$ | 1 | −2 ← |
| | $Z_j = \Sigma c_B a_{ij}$ | −1 | −2 | 0 | 0 | 1 | 0 | −10 |
| | $C_j = c_j - Z_j$ | −2 ↑ | 0 | 0 | 0 | −1 | 0 | |

Since all $C_j$ values are $\leq 0$ and $b_4 = -2$, this solution is optimal but infeasible. We therefore proceed further.

(ii) *Mark the outgoing variable.*

Since $b_4$ is negative, the fourth row is the key row and $s_4$ is the outgoing variable.

(iii) *Calculate ratios of elements in $C_j$-row to the corresponding negative elements of the key row.*

This ratios is $-2/-\frac{1}{2} = 4$ (neglecting other ratios corresponding to + ve or 0 elements of key row).

$\therefore$ $x_1$-column is the key column and $\left(-\frac{1}{2}\right)$ is the key element.

(iv) *Drop $s_4$ and introduce $x_1$* with its associated value $-3$ under the $c_B$ column. Convert the key element to unity and make all other elements of the key column zero. Then the third solution is given by the table below :

| | $c_j$ | $-3$ | $-2$ | 0 | 0 | 0 | 0 | |
|---|---|---|---|---|---|---|---|---|
| $c_B$ | *Basis* | $x_1$ | $x_2$ | $s_1$ | $s_2$ | $s_3$ | $s_4$ | $b$ |
| 0 | $s_1$ | 0 | 0 | 1 | 0 | $-1$ | $-1$ | 6 |
| 0 | $s_2$ | 0 | 0 | 0 | 1 | 1 | 1 | 0 |
| $-2$ | $x_2$ | 0 | 1 | 0 | 0 | 0 | 1 | 3 |
| $-3$ | $x_1$ | 1 | 0 | 0 | 0 | $-10$ | $-2$ | 4 |
| $Z_j$ | | $-3$ | $-2$ | 0 | 0 | 3 | 4 | $-18$ |
| $C_j$ | | 0 | 0 | 0 | 0 | $-3$ | $-4$ | |

Since all $C_j$ values are $\leq 0$ and all $b$'s are $\geq 0$, therefore this solution is optimal and feasible. Thus the *optimal solution* is $x_1 = 4$, $x_2 = 3$ and $Z_{max} = -18$.

**Example 34.28.** *Using dual simplex method, solve the following problem :*

*Minimize* $Z = 2x_1 + 2x_2 + 4x_3$

*subject to* $2x_1 + 3x_2 + 5x_3 \geq 2$, $3x_1 + x_2 + 7x_3 \leq 3$, $x_1 + 4x_2 + 6x_3 \leq 5$, $x_1, x_2, x_3 \geq 0$.

(*Rohtak, 2011*)

**Solution.** Consists of the following steps :

*Step 1.* (i) *Convert the given problem to maximization form* by writing

Maximize $Z' = -2x_1 - 2x_2 - 4x_3$.

(ii) *Convert the first constraint into* ($\leq$) *type*. Thus it is equivalent to

$$-2x_1 - 3x_2 - 4x_3 \leq -2$$

(iii) *Express the problem in standard form.*

Introducing slack variables, $s_1$, $s_2$, $s_3$, the given problem becomes

Max. $Z' = -2x_1 - 2x_2 - 4x_3 + 0s_1 + 0s_2 + 0s_3$

subject to
$$-2x_1 - 3x_2 - 5x_3 + s_1 + 0s_2 + 0s_3 = -2,$$
$$3x_1 + x_2 + 7x_3 + 0s_1 + s_2 + 0s_3 = 3,$$
$$x_1 + 4x_2 + 6x_3 + 0s_1 + 0s_2 + s_3 = 5,$$
$$x_1, x_2, x_3, s_1, s_2, s_3 \geq 0.$$

*Step 2. Find the initial basic solution.*

Setting the decision variables $x_1$, $x_2$, $x_3$ each equal to zero, we get the basic solution

$$x_1 = x_2 = x_3 = 0,\ s_1 = -2,\ s_2 = 3,\ s_3 = 5 \text{ and } Z' = 0.$$

$\therefore$ Initial solution is given by the table below :

| | $c_j$ | $-2$ | $-2$ | $-4$ | 0 | 0 | 0 | |
|---|---|---|---|---|---|---|---|---|
| $c_B$ | *Basis* | $x_1$ | $x_2$ | $x_3$ | $s_1$ | $s_2$ | $s_3$ | $b$ |
| 0 | $s_1$ | $-2$ | $(-3)$ | $-5$ | 1 | 0 | 0 | $-2 \leftarrow$ |
| 0 | $s_2$ | 3 | 1 | 7 | 0 | 1 | 0 | 3 |
| 0 | $s_3$ | 1 | 4 | 6 | 0 | 0 | 1 | 5 |
| $Z_j$ | | 0 | 0 | 0 | 0 | 0 | 0 | 0 |
| $C_j$ | | $-2$ | $-2$ $\uparrow$ | $-4$ | 0 | 0 | 0 | |

*Step 3. Test nature of $C_j$.*

Since all $C_j$ values are $\leq 0$ and $b_1 = -2$, the initial solution is optimal but infeasible.

*Step 4. Mark the outgoing variable.*

Since $b_1 < 0$, the first row is the key row and $s_1$ is the outgoing variable.

*Step 5. Calculate the ratio of elements of $C_j$-row to the corresponding negative elements of the key row.*

These ratios are $-2/-2 = 1$, $-2/-3 = 0.67$, $-4/-5 = 0.8$.

Since 0.67 is the smallest ratio, $x_2$-column is the key column and (–3) is the key element.

*Step 6. Iterate towards optimal feasible solution.*

Drop $s_1$ and introduce $x_2$ with its associated value –2 under $c_B$ column. Then the revised dual simplex table is

| | $c_j$ | –2 | –2 | –4 | 0 | 0 | 0 | |
|---|---|---|---|---|---|---|---|---|
| $c_B$ | *Basis* | $x_1$ | $x_2$ | $x_3$ | $s_1$ | $s_2$ | $s_3$ | $b$ |
| –2 | $x_2$ | 2/3 | 1 | 5/3 | –1/3 | 0 | 0 | 2/3 |
| 0 | $s_2$ | 7/3 | 0 | 16/3 | 1/3 | 1 | 0 | 7/3 |
| 0 | $s_3$ | –5/3 | 0 | –2/3 | 4/3 | 0 | 1 | 7/3 |
| | $Z_j$ | –4/3 | –2 | –10/3 | 2/3 | 0 | 0 | –4/3 |
| | $C_j$ | –2/3 | 0 | –2/3 | –2/3 | 0 | 0 | |

Since all $C_j \leq 0$ and all $b_i$'s are > 0, this solution is optimal and feasible. Thus the optimal solution is

$$x_1 = 0,\ x_2 = 2/3,\ x_3 = 0 \text{ and max. } Z' = -4/3 \text{ i.e., min. } Z = 4/3.$$

### PROBLEMS 34.8

Using dual simplex method, solve the following problems :

1. Maximize $Z = -3x_1 - x_2$
   subject to $x_1 + x_2 \geq 1,\ 2x_1 + 3x_2 \geq 2;\ x_1, x_2 \geq 0.$
2. Minimize $Z = 2x_1 + x_2$,
   subject to $3x_1 + x_2 \geq 3,\ 4x_1 + 3x_2 \geq 6,\ x_1 + 2x_2 \leq 3,\ x_1, x_2 \geq 0.$
3. Minimize $Z = x_1 + 2x_2 + 3x_3$,
   subject to $2x_1 - x_2 + x_3 \geq 4,\ x_1 + x_2 + 2x_3 \leq 8,\ x_2 - x_3 \geq 2;\ x_1, x_2, x_3 \geq 0.$
4. Minimize $Z = x_1 + 2x_2 + x_3 + 4x_4$
   subject to $2x_1 + 4x_2 + 5x_3 + x_4 \geq 10,\ 3x_1 - x_2 + 7x_3 - 2x_4 \geq 2$
   $5x_1 + 2x_2 + x_3 + 6x_4 \geq 15,\ x_1, x_2, x_3, x_4 \geq 0.$

## 34.14 (1) TRANSPORTATION PROBLEM

This is a special class of linear programming problems in which the objective is to transport a single commodity from various origins to different destinations at a minimum cost.

**(2) Formulation of a transportation problem.** *There are m plant locations (origins) and n distribution centres (destinations). The production capacity of the ith plant is $a_i$ and the number of units required at the jth destination is $b_j$. The transportation cost of one unit from the ith plant to the jth destination is $c_{ij}$. Our objective is to determine the number of units to be transported from the ith plant to jth destination so that the total transportation cost is minimum.*

Let $x_{ij}$ be the number of units shipped from $i$th plant to $j$th destination, then *the general transportation problem is :*

$$\text{Minimize } Z = \sum_{i=1}^{m}\sum_{j=1}^{n} c_{ij}\, x_{ij}$$

subject to the constraints

$x_{i1} + x_{i2} + \ldots + x_{in} = a_i$, for $i$th origin ($i = 1, 2, \ldots m$)

$x_{1j} + x_{2j} + \ldots + x_{mj} = b_j$, for $j$th destination ($j = 1, 2, \ldots n$)

$x_{ij} \geq 0$.

**Def. 1.** The two sets of constraints will be consistent if

$$\sum_{i=1}^{m} a_i = \sum_{j=1}^{n} b_j$$

which is the condition for a transportation problem to have a *feasible solution*. Problems satisfying this condition are called *balanced transportation problems.*

**2.** A feasible solution to a transportation problem is said to be a *basic feasible solution* if it contains at the most $(m + n - 1)$ strictly positive allocations, otherwise the solution will *degenerate.* If the total number of positive (non-zero) allocations is exactly $(m + n - 1)$, then the basic feasible solution is said to be *non-degenerate*.

**3.** A feasible solution which minimizes the transportation cost is called an *optimal solution.* This problem is explicitly represented in the following *transportation table :*

*Distribution centres (Destinations)*

| *Plants (origins)* | 1 | 2 | | $j$ | | $n$ | *Supply* |
|---|---|---|---|---|---|---|---|
| 1 | $c_{11}$ | $c_{12}$ | | $c_{1j}$ | | $c_{1n}$ | $a_1$ |
| 2 | $c_{21}$ | $c_{22}$ | | $c_{2j}$ | | $c_{2n}$ | $a_2$ |
| | | | | | | | $a_i$ |
| $i$ | $c_{i1}$ | $c_{i2}$ | | $c_{ij}$ | | $c_{in}$ | |
| | | | | | | | $a_m$ |
| $m$ | $c_{m1}$ | $c_{m2}$ | | $c_{mj}$ | | $c_{mn}$ | |
| *Demand* | $b_1$ | $b_2$ ……… | | $b_j$ ……… | | $b_n$ | $\Sigma a_i = \Sigma b_j$ |

The $mn$ squares are called *cells*. The per unit cost $c_{ij}$ of transporting from the $i$th origin to the $j$th destination is displayed in the *lower right side of the (i, j)th cell.* Any feasible solution is shown in the table by entering the value of $x_{ij}$ in the *small square at the upper left side of the (i, j)th cell.* The various $a$'s and $b$'s are called *rim requirements*. The feasibility of a solution can be verified by summing the values of $x_{ij}$ along the rows and down the columns.

**Obs. 1.** The special features of a transportation problem are that

(*i*) the coefficients of all $x_{ij}$ in the constraints are unity, and

(*ii*) the total supply $\Sigma a_i$ = total demand $\Sigma b_j$.

**Obs. 2.** The objective function and the constraints being all linear, the problem can be solved be simplex method. But the number of variables being large, there will be too many calculations. However, the coefficients of all $x_{ij}$ in the constraints being unity, we can look for some technique which would be simpler than the simplex method.

## 34.15 WORKING PROCEDURE FOR TRANSPORTATION PROBLEMS

*Step 1. Construct transportation table.* Express the supply from the origins $a_i$, demand at destinations $b_j$ and the unit shipping cost $c_{ij}$ in the form of a matrix, know as transportation table. If the supply and demand are equal, the problem is *balanced*.

*Step 2. Find the initial basic feasible solution.* We find an initial allocation which satisfies the demand at each project site without violating the capacities of the plants (origins) and also meeting the non-negativity restrictions. There are several methods for initial allocations *e.g.*, North-West corner rule, Row minima method, Least cost method, Vogel's approximation method. *The Vogel's approximation method (VAM) takes into account not only the least cost $c_{ij}$ but also the costs that just exceed the least cost $c_{ij}$ and therefore yields a better initial solution than obtained from other methods. As such we shall confine ourselves to VAM only which consists of the following steps :*

(*i*) Display the difference between the least and the next to least costs in each row, by enclosing them in brackets to the right of the row. Similarly display the differences for each column within brackets below that column.

(*ii*) Identify the row or column with the largest difference among all the rows and columns and allocate as much as possible under the rim requirements, to the lowest cost cell in that row or column. In case of a tie allocate to the cell associated with the lower cost.

If the greatest difference corresponds to *i*th row and $c_{ij}$ is the lowest cost in the *i*th row, allocate as much as possible *i.e.*, min $(a_i, b_j)$ in the $(i, j)$th cell and cross off the *i*th row or the *j*th column.

(*iii*) Recalculate the row and column differences for the reduced table and go to the previous step.

(*iv*) Repeat the procedure till all the rim requirements are satisfied. Note the solution in the upper left corner small squares of the basic cells.

*Step 3. Apply optimality check*

In the above solution, the number of allocations must be '$m + n - 1$', otherwise the basic solution degenerates.

*Now to test for optimality, we apply the modified distribution (MODI) method* and examine each unoccupied cell to determine whether making an allocation in it reduces the total transportation cost and then repeat this procedure until lowest possible transportation cost is obtained. This method consists of the following steps :

(*i*) Note the numbers $u_i$ along the left and $v_j$ along the top of the cost matrix such that their sums equal to the original costs of occupied cells *i.e.*, solve the equations $[u_i + v_j = c_{ij}]$ starting initially with some $u_i = 0$.

(*ii*) *Compute the net evaluations* $w_{ij} = u_i + v_j - c_{ij}$ for all the empty cells and enter them in upper right hand corners of the corresponding cells.

(*iii*) *Examine the sign of each* $w_{ij}$. If all $w_{ij} \le 0$, then the current basic feasible solution is optimal. If even one $w_{ij} > 0$, this solution is not optimal and we proceed further.

*Step 4. Iterate towards optimal solution*

(*i*) Choose the unoccupied cell with the largest $w_{ij}$ and mark $\theta$ in it.

(*ii*) Draw a closed path consisting of horizontal and vertical lines beginning and ending at $\theta$-cell and having its other corners at the allocated cells.

(*iii*) Add and subtract $\theta$ alternately to and from the transition cells of the loop subject to rim requirements. Assign a maximum value to $\theta$ so that one basic variable becomes zero and the other basic variables remain non-negative. Now the basic cell whose allocation has been reduced to zero leaves the basis.

*Step 5. Return to step 3 and repeat the process until an optimal basic feasible solution is obtained.*

**Example 34.29.** *Solve the following transportation problem :*

| | | *Destination* | | | | | |
|---|---|---|---|---|---|---|---|
| | | A | B | C | D | | |
| | I | 21 | 16 | 25 | 13 | 11 | |
| *Source* | II | 17 | 18 | 14 | 23 | 13 | *Availability* |
| | III | 33 | 27 | 18 | 41 | 19 | |
| | *Requirement* | 6 | 10 | 12 | 15 | 43 | |

(*Delhi., 2016*)

**Solution.** *Consists of the following steps :*

*Step 1. Transportation table.* Here the total availability and the total requirement being the same *i.e.* 43, the problem is balanced.

*Step 2. Find the initial basic feasible solution.* Following VAM, the differences between the smallest and next to the smallest costs in each row and each column are computed and displayed within brackets against the respective rows and columns (Table 1). The largest of these differences is (10) which is associated with the fourth column.

Table 1

| | | | | |
|---|---|---|---|---|
| 21 | 16 | 25 | [11] 13 | 11(3) |
| 17 | 18 | 14 | 23 | 13(3) |
| 32 | 27 | 18 | 41 | 19(9) |
| 6<br>(4) | 10<br>(2) | 12<br>(4) | 15<br>(10) | |

Table 2

| | | | | |
|---|---|---|---|---|
| 17 | 18 | 14 | [4] 23 | 13(3) |
| 32 | 27 | 18 | 41 | 19(9) |
| 6<br>(15) | 10<br>(9) | 12<br>(4) | 4 +<br>(18) | |

Since $c_{14}$ (= 13) is the minimum cost, we allocate $x_{14}$ = min (11, 15) = 11. This exhausts the availability of first row and therefore we cross it.

Table 3

| | | | |
|---|---|---|---|
| [6] 17 | 18 | 14 | 9(3) |
| 32 | 27 | 18 | 19(9) |
| 6<br>(15) | 10<br>(9) | 12<br>(4) | |

Table 4

| | | |
|---|---|---|
| [3] 18 | 14 | 3(4) |
| 27 | 18 | 19(9) |
| 10<br>(9) | 12<br>(4) | |

Table 5

| | | |
|---|---|---|
| [7] 27 | [12] 18 | 19 |
| 7 | 12 + | |

The row and column differences are now computed for reduced table 2 and displayed within brackets. The largest of these is (18) which is against the fourth column. Since $c_{14}$ (= 23) is the minimum cost, we allocate $x_{14}$ = min (13, 4) = 4.

This exhausts the availability of fourth column which we cross off. Proceeding in this way, the subsequent reduced transportation tables and differences for the remaining rows and columns are shown in Tables 3, 4 and 5.

Finally the initial basic feasible solution is as shown in Table 6.

Table 6

| | | | |
|---|---|---|---|
| 21 | 16 | 25 | [11] 13 |
| [6] 17 | [3] 18 | 14 | [4] 23 |
| 32 | [7] 27 | [12] 18 | 41 |

Table 7

| $u_i \backslash v_j$ | 17 | 18 | 9 | 23 |
|---|---|---|---|---|
| – 10 | (–) 21 | (–) 16 | (–) 25 | [11] 13 |
| 0 | [6] 17 | [3] 18 | (–) 14 | [4] 23 |
| 9 | (–) 32 | [7] 27 | [12] 18 | (–) 41 |

*Step 3. Apply optimality check*

As the number of allocations = $m + n - 1$ (*i.e.*, 6), we can apply MODI method.

(*i*) We have $u_2 + v_1 = 17$, $u_2 + v_2 = 18$, $u_3 + v_2 = 27$

$$u_3 + v_3 = 18,\ u_1 + v_4 = 13,\ u_2 + v_4 = 23$$

Let $u_2 = 0$, then $v_1 = 17$, $v_2 = 18$, $u_3 = 9$, $v_3 = 9$, $v_4 = 23$, $u_1 = -10$.

(*ii*) Net evaluations $w_{ij} = (u_i + v_j) - c_{ij}$ for all empty cells are

$$w_{11} = -14,\ w_{12} = -8,\ w_{13} = -26,\ w_{23} = -5,\ w_{31} = -6,\ w_{34} = -9.$$

(*iii*) Since all the net evaluations are negative, the current solution is optimal. Hence the optimal allocation is given by

$$x_{14} = 11,\ x_{21} = 6,\ x_{22} = 3,\ x_{24} = 4,\ x_{32} = 7 \text{ and } x_{33} = 12.$$

∴ The optimal (minimum) transportation cost

$$= 11 \times 13 + 6 \times 17 + 3 \times 18 + 4 \times 23 + 7 \times 27 + 12 \times 18 = ₹\ 796.$$

**Example 34.30.** *A company has three cement factories located in cities 1, 2, 3 which supply cement to four projects located in towns 1, 2, 3, 4. Each plant can supply 6, 1, 10 truck loads of cement daily respectively and the daily cement requirements of the projects are respectively 7, 5, 3, 2 truck loads. The transportation costs per truck load of cement (in hundreds of rupees) from each plant to each project site are as follows :*

| | | *Project sites* | | | |
|---|---|---|---|---|---|
| | | *1* | *2* | *3* | *4* |
| *Factories* | *1* | *2* | *3* | *11* | *7* |
| | *2* | *1* | *0* | *6* | *1* |
| | *3* | *5* | *8* | *15* | *9* |

*Determine the optimal distribution for the company so as to minimize the total transportation cost.*

**Solution.** Consists of the following steps :

*Step 1. Construct transportation table.* Express the supply from the factories, demands at sites and the unit shipping cost in the form of the following transportation table (Table 1). Here the supply being equal to the demand, the problem is balanced.

Table 1

| | | Project sites | | | | |
|---|---|---|---|---|---|---|
| | | 1 | 2 | 3 | 4 | Supply |
| Factories | 1 | 2 | 3 | 11 | 7 | 6 |
| | 2 | 1 | 0 | 6 | 1 | 1 |
| | 3 | 5 | 8 | 15 | 9 | 10 |
| Demand | | 7 | 5 | 3 | 2 | 17 |

*Step 2. Find the initial basic feasible solution.*

Using VAM, the initial basic feasible solution is as shown in Table 2. The transportation cost according to this route is given by

$$Z = ₹\,(1 \times 2 + 5 \times 3 + 1 \times 1 + 6 \times 5 + 3 \times 15 + 1 \times 9) \times 100 = ₹\,10{,}200.$$

*Step 3. Apply optimality check.*

As the numbers of allocations = $(m + n - 1)$ *i.e.*, 6, we can apply MODI method.

We now compute the net evaluations $w_{ij} = (u_i + v_j) - c_{ij}$ which are exhibited in Table 3. Since the net evaluations in two cells are positive, a better solution can be found.

Table 2

| | | | | |
|---|---|---|---|---|
| [1] 2 | [5] 3 | 11 | 7 | 6 |
| 1 | 0 | 6 | [1] 1 | 1 |
| [6] 5 | 8 | [3] 15 | [1] 9 | 10 |
| 7 | 5 | 3 | 2 | |

Table 3

| $u_i$ \ $v_j$ | 2 | 3 | 12 | 6 |
|---|---|---|---|---|
| 0 | [1] 2 | [5] 3 | (+) 11 | (–) 7 |
| – 5 | (–) 1 | (–) 0 | (+) 6 | [1] 1 |
| 3 | [6] 5 | (–) 8 | [3] 15 | [1] 9 |

*Step 4. Iterate towards optimal solution.*

*First iteration :*

(*a*) *Next basic feasible solution.*

(*i*) Choose the unoccupied cell with the maximum $w_{ij}$. In case of a tie, select the one with lower original cost. In Table 3, cells (1, 3) and (2, 3) each have $w_{ij}$ = 1 and out of these all (2, 3) has lower original cost 6, therefore we take this as the next basic cell and note θ in it.

(*ii*) Draw a closed path beginning and ending at θ-cell. Add and subtract θ, alternately to and from the transition cells of the loop subject the rim requirements. Assign a maximum value to θ so that one basic variable becomes zero and the other basic variables remain ≥ 0. Now the basic cell whose allocation has been reduced to zero leaves the basis. This gives the *second basic feasible solution* (Table 5).

Table 4

| | | | |
|---|---|---|---|
| [1] 2 | [5] 3 | 11 | 7 |
| 1 | 0 | [θ] 6 | [1] − θ 1 |
| [6] 5 | 8 | [3] − θ 15 | [1] + θ 9 |

Table 5

| | | | |
|---|---|---|---|
| [1] 2 | [5] 3 | 11 | 7 |
| 1 | 0 | [θ = 1] 6 | [1 − 1] 1 |
| [6] 5 | 8 | [3 − 1] 15 | [1 + 1] 9 |

∴ Total transportation cost of this revised solution.

$$= ₹\ (1 \times 2 + 5 \times 3 + 1 \times 6 + 6 \times 5 + 2 \times 15 + 2 \times 9) \times 100 = ₹\ 10{,}100.$$

(*b*) *Optimality check.* As the number of allocations in table 5 = $m + n - 1$ (*i.e.*, 6), we can apply MODI method. We compute the net evaluations which are shown in Table 6. Since the cell (1, 3) has a positive value, the second basic feasible solution is not optimal.

Table 6

| $u_i$ \ $v_j$ | 2 | 3 | 12 | 6 |
|---|---|---|---|---|
| 0 | [1] 2 | [5] 3 | (+) 11 | (–) 7 |
| – 6 | (–) 1 | (–) 0 | [1] 6 | (–) 1 |
| 3 | [6] 5 | (–) 8 | [2] 15 | [2] 9 |

Table 7

| | | | |
|---|---|---|---|
| [1 − 1] 2 | [5] 3 | [θ = 1] 11 | 7 |
| 1 | 0 | [1] 6 | 1 |
| [6 + 1] 5 | 8 | [2 − 1] 15 | [2] 9 |

*Second iteration :*

(*a*) *Next basic feasible solution.* In the second basic feasible solution introduce the cell (1, 3) taking θ = 1 and drop the cell (1, 1) giving Table 7. Thus we obtain the third basic feasible solution (Table 8).

Table 8

| | | | |
|---|---|---|---|
| 2 | [5] 3 | [1] 11 | 7 |
| 1 | 0 | [1] 6 | 1 |
| [7] 5 | 8 | [1] 15 | [2] 9 |

Table 9

| $u_i$ \ $v_j$ | 1 | 3 | 11 | 5 |
|---|---|---|---|---|
| 0 | (–) 2 | [5] 3 | [1] 11 | (–) 7 |
| – 5 | (–) 1 | (–) 0 | [1] 6 | (–) 1 |
| 4 | [7] 5 | (–) 8 | [1] 15 | [2] 9 |

(*b*) *Optimality Check.* As the number of allocations in Table 8 = $m + n - 1$ (*i.e.*, 6), we can apply MODI method.

We compute the net evaluations which are shown in Table 9. Since all the net evaluations are ≤ 0, this basic feasible solution is optimal.

Thus the optimal transportation policy is as shown in Table 9 and the optimal transportation cost

$$= ₹\ [5 \times 3 + 1 \times 11 + 1 \times 6 + 7 \times 5 + 1 \times 15 + 2 \times 9] \times 100 = ₹\ 10{,}000.$$

## 34.16 DEGENERACY IN TRANSPORTATION PROBLEMS

*When the number of basic cells in a non-transportation table, is less than '$m + n - 1$' the basic solution degenerates.* To remove the degeneracy, we assign a small positive value ε to as many zero-valued variables as may be necessary to complete '$m + n - 1$' basic variables. The cells containing ε are then treated like other basic cells and the problem is solved in the usual way. The ε's are kept till the optimum solution is attained. Then we let each ε → 0.

**Example 34.31.** *Solve the following transportation problem :*

| | | | *To* | | | | |
|---|---|---|---|---|---|---|---|
| *From* | 9 | 12 | 9 | 6 | 9 | 10 | 5 |
| | 7 | 3 | 7 | 7 | 5 | 5 | 6 |
| | 6 | 5 | 9 | 11 | 3 | 11 | 2 |
| | 6 | 8 | 11 | 2 | 2 | 10 | 9 |
| | 4 | 4 | 6 | 2 | 4 | 2 | 22 |

**Solution.** Consists of the following steps :

*Step 1. Transportation table.* The total supply and total demand being equal, the transportation problem is balanced.

*Step 2. Find the initial basic feasible solution.*

Using VAM, the initial basic feasible solution is as shown in Table 1.

*Step 3. Apply optimality check.* Since the number of basic cells is 8 which is less than $m + n - 1 = 9$, the basic solution degenerates. In order to complete the basis and thereby remove degeneracy, we require only one more positive basic variable. We select the variable $x_{23}$ and allocate a small positive quantity ε to the cell (2, 3).

Table 1

| | | | | | | Supply |
|---|---|---|---|---|---|---|
| 9 | 12 | [5] 9 | 6 | 9 | 10 | 5 |
| 7 | [4] 3 | [ε] 7 | 7 | 5 | [2] 5 | $6 + \varepsilon = 6$ |
| [1] 6 | 5 | [1] 9 | 11 | 3 | 11 | 2 |
| [3] 6 | 8 | 11 | [2] 2 | [4] 2 | 10 | 9 |
| 4 | 4 | $6 + \varepsilon = 6$ | 2 | 4 | 2 | + |

We now compute the net evaluations $w_{ij} = (u_i + v_j) - c_{ij}$ which are exhibited in Table 2. Since all the net evaluations are $\leq 0$, the current solutions is optimal. Hence the optimal allocation is

$$x_{13} = 5, x_{22} = 4, x_{26} = 2, x_{31} = 1, x_{33} = 1, x_{41} = 3, x_{44} = 2 \text{ and } x_{45} = 4.$$

∴ The minimum (optimal) transportation cost

$$= 5 \times 9 + 4 \times 3 + \varepsilon \times 7 + 2 \times 5 + 1 \times 6 + 1 \times 9 + 3 \times 6 + 2 \times 2 + 4 \times 2$$

$$= 112 + 7\varepsilon = ₹\ 112 \text{ as } \varepsilon \to 0.$$

Table 2

| $u_i$ \ $v_j$ | 4 | 3 | 7 | 0 | 0 | 5 |
|---|---|---|---|---|---|---|
| 2 | (–) 9 | (–) 12 | [5] 9 | (–) 6 | (–) 9 | (–) 10 |
| 0 | (–) 7 | [4] 3 | [ε] 7 | (–) 7 | (–) 5 | [2] 5 |
| 2 | [1] 6 | (0) 5 | [1] 9 | (–) 11 | (–) 3 | (–) 11 |
| 2 | [3] 6 | (–) 8 | (–) 11 | [2] 2 | [4] 2 | (–) 10 |

## PROBLEMS 34.9

1. Obtain an initial basic feasible solution to the following transportation problem :

| | | To | | | | |
|---|---|---|---|---|---|---|
| | | D | E | F | G | |
| From | A | 11 | 13 | 17 | 14 | 250 |
| | B | 16 | 18 | 14 | 10 | 300 |
| | C | 21 | 24 | 13 | 10 | 400 |
| | | 200 | 225 | 275 | 250 | |

2. Solve the following transportation problem :

| *Suppliers* / *Consumers* | *A* | *B* | *C* | *Available* |
|---|---|---|---|---|
| I | 6 | 8 | 4 | 14 |
| II | 4 | 9 | 8 | 12 |
| III | 1 | 2 | 6 | 5 |
| Required | 6 | 10 | 15 | 31 |

3. Consider four bases of operations $B_i$ and three targets $T_j$. The tons of bombs per aircraft from any base that can be delivered to any target are given in the following table :

| $B_i$ \ $T_j$ | 1 | 2 | 3 |
|---|---|---|---|
| 1 | 8 | 6 | 5 |
| 2 | 6 | 6 | 6 |
| 3 | 10 | 8 | 4 |
| 4 | 8 | 6 | 4 |

The daily sortie capability of each of the four bases is 150 sorties per day. The daily requirement in sorties over each target is 200. Find the allocation of sorties from each base to each target which maximizes the total tonnage over all the three targets.

4. A company has factories $F_1$, $F_2$, $F_3$ which supply warehouses at $W_1$, $W_2$ and $W_3$. Weekly factory capacities, weekly warehouse requirements and unit shipping costs (in rupees) are as follows :

| *Factories* | *Warehouses* | | | *Supply* |
|---|---|---|---|---|
| | $W_1$ | $W_2$ | $W_3$ | |
| $F_1$ | 16 | 20 | 12 | 200 |
| $F_2$ | 14 | 8 | 18 | 160 |
| $F_3$ | 26 | 24 | 16 | 90 |
| *Demand* | 180 | 120 | 150 | 450 |

Determine the optimal distribution for this company to minimize shipping costs.

5. A company is spending ₹ 1,000 on transportation of its units from plants to four distribution centres. The supply and demand of units, with unit cost of transportation are given below :

| *Plants* | *Distribution centres* | | | | *Availabilities* |
|---|---|---|---|---|---|
| | $D_1$ | $D_2$ | $D_3$ | $D_4$ | |
| $P_1$ | 19 | 30 | 50 | 12 | 7 |
| $P_2$ | 70 | 30 | 40 | 60 | 10 |
| $P_3$ | 40 | 10 | 60 | 20 | 18 |
| *Requirements* | 5 | 8 | 7 | 15 | |

What can be the maximum saving by optimal scheduling.

6. A departmental store wishes to stock the following quantities of a popular product in three types of containers :

| *Container type* | : | 1 | 2 | 3 |
|---|---|---|---|---|
| *Quantity* | : | 170 | 200 | 180 |

Tenders are submitted by four dealers who undertake to supply not more than the quantities shown below :

| *Dealer* | : | 1 | 2 | 3 | 4 |
|---|---|---|---|---|---|
| *Quantity* | : | 150 | 160 | 110 | 130 |

The store estimates that profit per unit will vary with the dealer as shown below :

| *Dealers* → *Container type* ↓ | 1 | 2 | 3 | 4 |
|---|---|---|---|---|
| 1 | 8 | 9 | 6 | 3 |
| 2 | 6 | 11 | 5 | 10 |
| 3 | 3 | 8 | 7 | 9 |

Find the maximum profit of the store.

7. Obtain an optimum basic feasible solution to the following transportation problem :

| | *To* | | | | |
|---|---|---|---|---|---|
| *From* | 7 | 3 | 4 | 2 | *Available* |
| | 2 | 1 | 3 | 3 | |
| | 3 | 4 | 6 | 5 | |
| *Demand* | 4 | 1 | 5 | 10 | |

8. A company has three plants at locations $A$, $B$ and $C$ which supply to warehouses located as $D$, $E$, $F$, $G$ and $H$. Monthly plant capacities are 800, 500 and 900 units respectively. Monthly warehouse requirements are 400, 400, 500, 400 and 800 units respectively. Unit transportation costs in rupees are given below :

| | | *To* | | | | |
|---|---|---|---|---|---|---|
| | | *D* | *E* | *F* | *G* | *H* |
| *From* | *A* | 5 | 8 | 6 | 6 | 3 |
| | *B* | 4 | 7 | 7 | 6 | 6 |
| | *C* | 8 | 4 | 6 | 6 | 3 |

Determine an optimum distribution for the company in order to minimize the total transportation cost.

## 34.17 (1) ASSIGNMENT PROBLEM

An assignment problem is a special type of transportation problem in which the objective is to assign a number of origins to an *equal* number of destinations at a minimum cost (or maximum profit).

**(2) Formulation of an assignment problem.** *There are n new machines $M_i$ ($i$ = 1, 2, ... n) which are to be installed in a machine shop. There are n vacant spaces $S_j$ ($j$ = 1, 2, ... n) available. The cost of installing the machine $M_i$ at space $S_j$ is $c_{ij}$ rupees. Let us formulate the problem of assigning machines to spaces so as to minimize the overall cost.*

Let $x_{ij}$ be the assignment of machine $M_i$ to space $S_j$ *i.e.*, let $x_{ij}$ be a variable such that

$$x_{ij} = \begin{cases} 1, \text{ if } i\text{th machine is installed at } j\text{th space} \\ 0, \text{ otherwise} \end{cases}$$

Since one machine can only be installed at each space, we have

$$x_{i1} + x_{i2} + ... + x_{in} = 1, \text{ for machine } M_i \; (i = 1, 2, ... n)$$
$$x_{1j} + x_{2j} + ... + x_{nj} = 1, \text{ for space } S_j \; (j = 1, 2, ... n)$$

Also the total installation cost is $\sum_{i=1}^{n} \sum_{j=1}^{n} c_{ij} x_{ij}$.

*Thus the assignment problem can be stated as follows :*

Determine $x_{ij} \geq 0$ ($j = 1, 2, ... n$) so as to

minimize $$Z = \sum_{i=1}^{n} \sum_{j=1}^{n} c_{ij}\, x_{ij}$$

subject to the constraints $\sum_{i=1}^{n} x_{ij} = 1, j = 1, 2, ... n$ and $\sum_{j=1}^{n} x_{ij} = 1, i = 1, 2, ... n.$

This problem is explicitly represented by the following $n \times n$ *cost matrix* :

| | | Spaces | | | | |
|---|---|---|---|---|---|---|
| | | $S_1$ | $S_2$ | $S_3$ | ... | $S_n$ |
| | $M_1$ | $c_{11}$ | $c_{12}$ | $c_{13}$ | ... | $c_{1n}$ |
| | $M_2$ | $c_{21}$ | $c_{22}$ | $c_{23}$ | ... | $c_{2n}$ |
| *Machines* | $M_3$ | $c_{31}$ | $c_{32}$ | $c_{33}$ | ... | $c_{3n}$ |
| | : | : | : | : | | : |
| | : | : | : | : | | : |
| | $M_n$ | $c_{n1}$ | $c_{n2}$ | $c_{n3}$ | | $c_{nn}$ |

Obs. This assignment problem constitutes $n$ ! possible ways of installing $n$ machines at $n$ spaces. If we enumerate all these $n$ ! alternatives and evaluate the cost of each one of them and select the one with the minimum cost, the problem would be solved. But this method would be very slow and time consuming, even for small value of $n$ and hence it is not at all suitable. However, a much more efficient method of solving such problems is available. This is the **Hungarian method** for solution of assignment problems which we describe below.

## 34.18 WORKING PROCEDURE TO SOLVE AN ASSIGNMENT PROBLEM

*Step 1. Reduce the matrix.* Subtract the smallest element of each row (of the given cost matrix) from all elements of that row. See if each row contains at least one zero. If not, subtract the smallest element of each column (not containing zero) from all the elements of that column. This gives the *reduced matrix.*

*Step 2. Assign the zeros*

(*a*) Examine rows (of the reduced matrix) successively until a row with exactly one unmarked zero is found. Make an assignment to this single zero by encircling it. Cross all other zeros in the column of this encircled zero, as these will not be considered for any future assignment. Continue in this way until all the rows have been examined.

(*b*) Now examine columns successively until a column with exactly one unmarked zero is found. Encircle this zero and make an assignment there. Then cross any other zero in its row. Continue in this way until all the columns have been examined.

In case, some rows or columns contain more than one unmarked zeros, encircle any unmarked zero arbitrarily and cross all other zeros in its row or column. Proceed in this way, till no zero is left unmarked.

*Step 3. Apply optimality check.*

Repeat step 2 (*a*) and (*b*) until one of the following occurs :

(*i*) If no row or no column is without assignment (encircled zero), then the current assignment is optimal.

(*ii*) If there is some row and/or column without an assignment, then the current assignment is not optimal and we go to next step.

*Step 4. Find minimum number of lines crossing all zeros.*

(*a*) Tick (✓) the rows which do not have assignments.

(*b*) Tick (✓) the columns (not already marked) which have zeros in the ticked row.

(*c*) Tick (✓) the rows (not already marked) which have assignments in ticked columns.

Repeat (*b*) and (*c*) until no more marking is required.

(*d*) Draw lines through all unticked rows and ticked columns. If the number of these lines is equal to the order of the matrix then it is an optimal solution otherwise not.

*Step 5. Iterate towards optimal solution.*

Select the smallest element and subtract it from all uncovered elements. Add this smallest element to every element lying at the intersection of two lines. The resulting matrix is the second basic feasible solution.

*Step 6.* Go to step 2 and repeat the procedure until the optimal solution is attained.

**Example 34.32.** *Four jobs are to be done on four different machines. The cost (in ₹) of producing ith job on the jth machine is given below :*

*Machines*

| | | $M_1$ | $M_2$ | $M_3$ | $M_4$ |
|---|---|---|---|---|---|
| | $J_1$ | 15 | 11 | 13 | 15 |
| | $J_2$ | 17 | 12 | 12 | 13 |
| *Jobs* | $J_3$ | 14 | 15 | 10 | 14 |
| | $J_4$ | 16 | 13 | 11 | 17 |

*Assign the jobs to different machines so as to minimize the total cost.*

**Solution.** Consists of the following steps :

*Step 1. Reduce the matrix.* Subtract the smallest element 11 of row 1 from all its elements. Similarly subtract 12, 10 and 11 from rows 2, 3 and 4 respectively. The resulting matrix is as shown in Table 1. Columns 1 and 4 do not have any zero element. Subtract the smallest element 4 of Col. 1 from all its elements and element 1 from all elements of Col. 4. The *reduced matrix* is as given in Table 1.

Table 1

| | $M_1$ | $M_2$ | $M_3$ | $M_4$ |
|---|---|---|---|---|
| $J_1$ | 4 | 0 | 2 | 4 |
| $J_2$ | 5 | 0 | 0 | 1 |
| $J_3$ | 4 | 5 | 0 | 4 |
| $J_4$ | 5 | 2 | 0 | 6 |

Table 2

| | $M_1$ | $M_2$ | $M_3$ | $M_4$ |
|---|---|---|---|---|
| $J_1$ | 0 | (0) | 2 | 3 |
| $J_2$ | 1 | ×0 | ×0 | (0) |
| $J_3$ | (0) | 5 | ×0 | 3 |
| $J_4$ | 1 | 2 | (0) | 5 |

*Step 2. Assign the zeros.* Row 4 has a single unmarked zero in Col. 3. Encircle it and cross all other zeros in Col. 3. Row 3 has a single unmarked zero in Col. 1. Encircle it and cross the other zero in col. 1. Row 1 has a single unmarked zero in Col. 2. Encircle it and cross the other zero in Col. 2. Finally row 2 has a single unmarked zero in Col. 4. Encircle it (Table 2).

*Step 3. Apply optimality check.* Since we have one encircled zero in each row and in each column, this gives the optimal solution.

∴ The optimal assignment policy is

Job 1 to machine 2, Job 2 to machine 4, Job 3 to machine 1, Job 4 to machine 3,

and the minimum assignment cost = ₹ (11 + 13 + 14 + 11) = ₹ 49.

**Example 34.33.** *A marketing manager has 5 salesmen and 5 sales districts. Considering the capabilities of the salesmen and the nature of districts, the marketing manager estimates that sales per month (in hundred rupees) for each salesman in each district would be as follows :*

*Sales districts*

| | | A | B | C | D | E |
|---|---|---|---|---|---|---|
| | 1 | 32 | 38 | 40 | 28 | 40 |
| | 2 | 40 | 24 | 28 | 21 | 36 |
| *Salesman* | 3 | 41 | 27 | 33 | 30 | 37 |
| | 4 | 22 | 38 | 41 | 36 | 36 |
| | 5 | 29 | 33 | 40 | 35 | 39 |

*Find the assignment of salesmen to districts that will result in maximum sales.* (Anna., 2016)

**Solution.** Consists of the following steps :

*Step 1. Reduce the matrix.* Convert the given maximization problem into a minimization problem, by making all the profits negative, since max. $Z = \min. (-Z)$. Then subtract the smallest element of each row from the elements of that row. Now subtract the smallest element of each col. (not containing zero) from the elements of that column. This gives the *reduced matrix* (Table 1).

Table 1

| | | | | |
|---|---|---|---|---|
| 8 | (0) | ~~0~~ | 7 | (0) |
| (0) | 14 | 12 | 14 | 4 |
| (0) | 12 | 8 | 6 | 4 |
| 19 | 1 | (0) | ~~0~~ | 5 |
| 11 | 5 | (0) | ~~0~~ | 1 |

Table 2

| | | | | |
|---|---|---|---|---|
| 12 | 0 | 0 | 7 | 0 |
| 0 | 10 | 8 | 10 | 0 |
| 0 | 8 | 4 | 2 | 0 |
| 23 | 1 | 0 | 0 | 5 |
| 15 | 5 | 0 | 0 | 1 |

*Step 2. Assign the zeros.* Rows 2 and 3 have each a single unmarked zero in Col. 1. Encircle these. Columns 2 and 5 have each a single unmarked zero in row 1. Encircle these and cross the zero in Row 1. Columns 3 and 4 have each unmarked zeros. Encircle the zeros in each of the rows 4 and 5 as shown in Table 1 and cross other zeros.

*Step 3. Apply optimality check.* As col. 4 is without assignment, this solution is not optimal. Therefore we go to next step.

*Step 4. Find minimum number of lines crossing all zeros.* Draw the least number of horizontal and vertical (dotted) lines which cover all the zeros. Since there are four dotted lines which are less than the order of the cost matrix (= 5), we got to step 5.

*Step 5. Iterate towards optimal solution.* Select the smallest element in the Table 1, not covered by the dotted lines. Such an element is 4 which lies at two different positions. Selecting the elements that lies at position (3, 5) arbitrarily, subtract it from all the uncovered elements of the cost matrix (Table 1) and add the same to the elements lying at the intersection of two dotted lines. Now draw more minimum number of dotted lines so as to cover the new zero. Here we draw such a line in Col. 5 (Table 2).

Table 3

| | A | B | C | D | E |
|---|---|---|---|---|---|
| 1 | | (0) | ~~0~~ | | |
| 2 | (0) | | | | ~~0~~ |
| 3 | ~~0~~ | | | | (0) |
| 4 | | | (0) | ~~0~~ | |
| 5 | | | ~~0~~ | (0) | |

Now, since the number of dotted lines is equal to the order for the cost matrix, the optimal solution is attained.

Finally, to determine this optimal assignment, we consider only the zero elements (Table 3) :

(*i*) Examine successively the Rows with exactly one zero. There is no such row.

(*ii*) Examine successively the columns with exactly one zero. Col. 2 has one zero, encircle it and cross all zeros of Row 1.

(*iii*) Encircle arbitrarily the zero in position (2, 1) and cross all zeros in row 2 and Col. 1. Then encircle the unmarked zero in row 3. Now encircle arbitrarily the zero in position (4, 3) and cross all zeros in Row 4 and Col. 3. Finally encircle the remaining unmarked zero in row 5.

Now each row and each column has one encircled zero, therefore the optimal assignment policy is :

Salesman 1 to district *B*, 2 to *A*, 3 to *E*, 4 to *C* and 5 to *D*.

Hence the maximum sales = ₹ (38 + 40 + 37 + 41 + 35) × 100 = ₹ 19,100.

## PROBLEMS 34.10

1. A firm plans to begin production of three new products on its three plants. The unit cost of producing $i$ at plant $j$ is as given below. Find the assignment that minimizes the total unit cost.

| | | Plant 1 | Plant 2 | Plant 3 |
|---|---|---|---|---|
| Product | 1 | 10 | 8 | 12 |
| | 2 | 18 | 6 | 14 |
| | 3 | 6 | 4 | 2 |

2. Solve the following assignment problem :

| | 1 | 2 | 3 | 4 |
|---|---|---|---|---|
| *A* | 10 | 12 | 19 | 11 |
| *B* | 5 | 10 | 7 | 8 |
| *C* | 12 | 14 | 13 | 11 |
| *D* | 8 | 15 | 11 | 9 |

3. A machine tool company decides to make four sub-assemblies through four contractors. Each contractor is to receive only one sub-assembly. The cost of each sub-assembly is determined by the bids submitted by each contractor and is shown in table below (in hundreds of rupees). Assign different assemblies to contractors so as to minimize the total cost.

| | | Contractor *A* | *B* | *C* | *D* |
|---|---|---|---|---|---|
| *Sub-assembly* | I | 15 | 13 | 14 | 17 |
| | II | 11 | 12 | 15 | 13 |
| | III | 18 | 12 | 10 | 11 |
| | IV | 15 | 17 | 14 | 16 |

4. Four professors are each capable of teaching any one of the four different courses. Class preparations time in hours for different topics varies from professor to professor and is given in the table below. Each professor is assigned only one course. Find the assignment policy schedule so as to minimize the total course preparation time for all courses.

| *Prof.* | *L.P.* | *Queuing Theory* | *Dynamic Programming* | *Regression analysis* |
|---|---|---|---|---|
| *A* | 2 | 10 | 9 | 7 |
| *B* | 15 | 4 | 14 | 8 |
| *C* | 13 | 14 | 16 | 11 |
| *D* | 3 | 15 | 13 | 8 |

5. Consider the problem of assigning four working labour units to four jobs. The assignment costs in thousands of rupees are given by the following matrix.

| *Labour unit* ↓ | *Job I* | *II* | *III* | *IV* |
|---|---|---|---|---|
| $L_1$ | 42 | 35 | 28 | 21 |
| $L_2$ | 30 | 25 | 20 | 15 |
| $L_3$ | 30 | 25 | 20 | 15 |
| $L_4$ | 24 | 20 | 16 | 12 |

Find the optimal assignment.

6. A company has six jobs to be processed by six mechanics. The following table gives the return in rupees when the $i$th job is assigned to the $j$th mechanic. How should the jobs be assigned to the mechanics so as to maximize the over all return ?

| *Mechanic* ↓ | *Job* | | | | | |
|---|---|---|---|---|---|---|
| | *I* | *II* | *III* | *IV* | *V* | *VI* |
| 1 | 9 | 22 | 58 | 11 | 19 | 27 |
| 2 | 43 | 78 | 72 | 50 | 63 | 48 |
| 3 | 41 | 28 | 91 | 37 | 45 | 33 |
| 4 | 74 | 42 | 27 | 49 | 39 | 32 |
| 5 | 36 | 11 | 57 | 22 | 25 | 18 |
| 6 | 13 | 56 | 53 | 31 | 17 | 28 |

CHAPTER 35

# Calculus of Variations



## 35.1 INTRODUCTION

The calculus of variations is a powerful technique for the solution of problems in dynamics of rigid bodies, optimization of orbits and vibration problems. The subject primarily concerns with finding maximum or minimum value of a definite integral involving a certain function. It is something beyond finding stationary values of a given function. Only an elementary exposition of the subject is given here with the sole aim of introducing the student to a topic whose importance is fast growing in science and engineering.

Before proceeding further, the student should revise § 5.12 concerning maxima and minima of functions of several variables.

## 35.2 FUNCTIONALS

Consider the problem of finding a curve through two points $(x_1, y_1)$ and $(x_2, y_2)$ whose length is a minimum (Fig. 35.1). It is same as determining the curve $y = y(x)$ for which $y(x_1) = y_1$, $y(x_2) = y_2$ such that $\int_{x_1}^{x_2} \sqrt{(1 + y'^2)}\, dx$ is a minimum.

In general terms, we wish to find the curve $y = y(x)$ where $y(x_1) = y_1$ and $y(x_2) = y_2$ such that for a given function $f(x, y, y')$,

$$\int_{x_1}^{x_2} f(x, y, y')\, dx \text{ is a stationery value or an extremum.} \qquad ...(1)$$

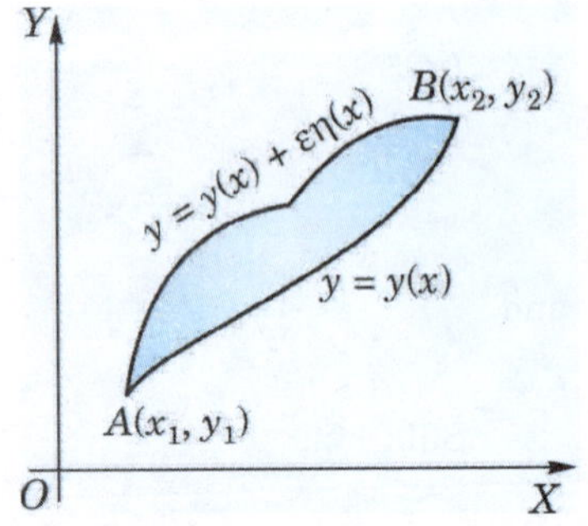


Fig. 35.1

*An integral such as Eqn. (1), which assumes a definite value for functions of the type $y = y(x)$ is called a* **functional.**

In differential calculus, we deal with the problems of maxima and minima of functions. The calculus of variations is however, concerned with maximizing or minimizing functionals.

## 35.3 EULER'S EQUATION

*A necessary condition for*

$$I = \int_{x_1}^{x_2} f(x, y, y')\, dx \text{ to be an extremum is that}$$

$$\frac{\partial \mathbf{f}}{\partial \mathbf{y}} - \frac{\mathbf{d}}{\mathbf{dx}}\left(\frac{\partial \mathbf{f}}{\partial \mathbf{y}'}\right) = \mathbf{0}$$

This is called *Euler's equation.*

*Proof.* Let $y = y(x)$ be the curve joining points $A(x_1, y_1)$, $B(x_2, y_2)$ which makes $I$ an extremum. Let

$$y = y(x) + \varepsilon\eta(x) \qquad ...(1)$$

be a neighbouring curve joining these points so that at $A$, $\eta(x_1) = 0$ and at $B$, $\eta(x_2) = 0$. ...(2)

The value of $I$ along Eqn. (1) is $I = \int_{x_1}^{x_2} f[x, y(x) + \varepsilon\eta(x), y'(x) + \varepsilon\eta'(x)]\, dx$

This being a function of $\varepsilon$, is a maximum or minimum for $\varepsilon = 0$, when

$$\frac{dI}{d\varepsilon} = 0 \text{ at } \varepsilon = 0 \qquad ...(3)$$

$\therefore$ Differentiating $I$ under the integral sign by Leibnitz's rule (p. 849), we have

$$\frac{dI}{d\varepsilon} = \int_{x_1}^{x_2}\left(\frac{\partial f}{\partial x}\frac{\partial x}{\partial \varepsilon} + \frac{\partial f}{\partial y}\frac{\partial y}{\partial \varepsilon} + \frac{\partial f}{\partial y'}\frac{\partial y'}{\partial \varepsilon}\right) dx \qquad ...(4)$$

But $\varepsilon$ being independent of $x$, $\frac{\partial x}{\partial \varepsilon} = 0$. Also from (1), $\frac{\partial y}{\partial \varepsilon} = \eta(x)$ and $\frac{\partial y'}{\partial \varepsilon} = \eta'(x)$.

Substituting these values in Eqn. (4), we get $\frac{dI}{d\varepsilon} = \int_{x_1}^{x_2}\left[\frac{\partial f}{\partial y}\eta(x) + \frac{\partial f}{\partial y'}\eta'(x)\right] dx$

Integrating the second term on the right by parts, we have

$$\frac{dI}{d\varepsilon} = \int_{x_1}^{x_2}\frac{\partial f}{\partial y}\eta(x)\, dx + \left[\left|\frac{\partial f}{\partial y'}\eta(x)\right|_{x_1}^{x_2} - \int_{x_1}^{x_2}\frac{d}{dx}\left(\frac{\partial f}{\partial y'}\right)\eta(x)\, dx\right]$$

$$= \int_{x_1}^{x_2}\left[\frac{\partial f}{\partial y} - \frac{d}{dx}\left(\frac{\partial f}{\partial y'}\right)\right]\eta(x)\, dx \qquad \text{[By (2)]}$$

Since this has to be zero by Eqn. (3),

$$\therefore \qquad \frac{\partial f}{\partial y} - \frac{d}{dx}\left(\frac{\partial f}{\partial y'}\right) = 0 \qquad ...(I)$$

which is the desired *Euler's equation.* (*V.T.U., 2013*)

**Obs. 1. Other forms of Euler's equation.**

(*a*) Since $f$ is a function of $x, y, y'$, we have $\frac{df}{dx} = \frac{\partial f}{\partial x} + \frac{\partial f}{\partial y}\frac{dy}{dx} + \frac{\partial f}{\partial y'}\frac{dy'}{dx}$

$$= \frac{\partial f}{\partial x} + \frac{\partial f}{\partial y}y' + \frac{\partial f}{\partial y'}y'' \qquad ...(5)$$

and

$$\frac{d}{dx}\left(y'\frac{\partial f}{\partial y'}\right) = y'\frac{d}{dx}\left(\frac{\partial f}{\partial y'}\right) + \frac{\partial f}{\partial y'}y'' \qquad ...(6)$$

Subtracting Eqn. (6) from Eqn. (5), we get $\frac{df}{dx} - \frac{d}{dx}\left(y'\frac{\partial f}{\partial y'}\right) = \frac{\partial f}{\partial x} + \frac{\partial f}{\partial y}y' - y'\frac{d}{dx}\left(\frac{\partial f}{\partial y'}\right)$

or

$$\frac{d}{dx}\left(f - y'\frac{\partial f}{\partial y'}\right) - \frac{\partial f}{\partial x} = y'\left\{\frac{\partial f}{\partial y} - \frac{d}{dx}\left(\frac{\partial f}{\partial y'}\right)\right\} = 0 \qquad \text{[By } (I)\text{]}$$

Hence

$$\frac{d}{dx}\left(f - y'\frac{\partial f}{\partial y'}\right) - \frac{\partial f}{\partial x} = 0 \qquad ...(II)$$

which is another form of Eqn. (*I*).

(*b*) Again since $\frac{\partial f}{\partial y'}$ is also a function of $x, y, y'$, say : $\psi(x, y, y')$.

$$\therefore \qquad \frac{d}{dx}\left(\frac{\partial f}{\partial y'}\right) = \frac{d\psi}{dx} = \frac{\partial \psi}{\partial y} + \frac{\partial \psi}{\partial y}\frac{dy}{dx} + \frac{\partial \psi}{\partial y'}\frac{dy'}{dx}$$

$$= \frac{\partial}{\partial x}\left(\frac{\partial f}{\partial y'}\right) + \frac{\partial}{\partial y}\left(\frac{\partial f}{\partial y'}\right) y' + \frac{\partial}{\partial y'}\left(\frac{\partial f}{\partial y'}\right) y'' = \frac{\partial^2 f}{\partial x \partial y'} + y' \frac{\partial^2 f}{\partial y \partial y'} + y'' \frac{\partial^2 f}{\partial y'^2}$$

Substituting this in Eqn. (*I*), we get $\frac{\partial f}{\partial y} - \frac{\partial^2 f}{\partial x \partial y'} - y' \frac{\partial^2 f}{\partial y \partial y'} - y'' \frac{\partial^2 f}{\partial y'^2} = 0$ ...(*III*)

which is an extended form of Eqn. (*I*).

**Obs. 2.** The above problem can easily be extended to the integral

$$= \int_{x_1}^{x_2} f(x, y_1, y_2 \dots y_n, y_1', y_2', \dots y_n')\, dx$$

involving $n$ functions $y_1, y_2, \dots y_n$ of $x$. Then the necessary condition for this integral to be stationary is

$$\frac{\partial f}{\partial y_i} - \frac{d}{dx}\left(\frac{\partial f}{\partial y_i'}\right) = 0, \quad i = 1, 2, \dots, n \qquad \dots(IV)$$

These are Euler's equations for the $n$ functions.

## 35.4 SOLUTIONS OF EULER'S EQUATION

Every solution of the Euler's equation which satisfies the boundary conditions, is called an *extremal* or a *stationary function* of the problem. The extremal can easily be obtained in the following cases :

**(1)** *When f is independent of x*

We have $\partial f/\partial x = 0$ and Euler's Eqn. (*II*) above becomes $\frac{d}{dx}\left(f - y' \frac{\partial f}{\partial y'}\right) = 0$

Integrating, we get $f - y' \frac{\partial f}{\partial y'}$ = constant. This directly gives a solution of Euler's equation.

**(2)** *When f is independent of y*

We have $\partial f/\partial y = 0$ and Euler's Eqn. (*I*) reduces to $\frac{d}{dx}\left(\frac{\partial f}{\partial y'}\right) = 0$

Integrating $\frac{\partial f}{\partial y}$ = constant which gives a solution directly.

**(3)** *When f is independent of y'.*

We have $\partial f/\partial y' = 0$ and the Eqn. (*I*) becomes $\frac{\partial f}{\partial y} = 0$ which gives the desired solution.

**(4)** *When f is independent of x and y*

We have $\quad \frac{\partial f}{\partial x} = 0, \frac{\partial f}{\partial y} = 0$ and $\frac{\partial^2 f}{\partial x \partial y'} = 0, \frac{\partial^2 f}{\partial y \partial y'} = 0.$

Then the Eqn. (*III*) above becomes $y'' \frac{\partial^2 f}{\partial y'^2} = 0.$

If $\frac{\partial^2 f}{\partial y'^2} \neq 0$, it reduces to $y'' = 0$ which gives a solution of the form $y = ax + b$.

**Example 35.1 :** *Using the Eular equation find the externals for the following function*

$\int_a^b y(x)^2 - 4y(x)\left(\frac{\delta}{\delta x} y(x)\right) + 4\left(\frac{\delta}{\delta x} y(x)^2\, dx\right)$ (*Kottayam, 2018*)

**Solution.** $\quad f(x, y(x), \frac{\delta}{\delta y} y(x)) = y(x)^2 - 4y(x)\left(\frac{\delta}{\delta y} y(x)\right) + \frac{\delta y}{\delta x} y(x)^2$ in the form

$$f(x, y, z) = y^2 - 4yz + 4z^2$$

We substituting $x$, $y(x)$ and $y'(x)$ for $x$, $y$ and $z$

$$\Rightarrow \qquad \frac{\delta}{\delta y} f(x, y(x), \frac{\delta}{\delta x} y(x)) = 2y - 4\left(\frac{\delta}{\delta x} y\right)$$

$$\Rightarrow \quad \frac{\delta f}{\delta z}(x, y(x), \frac{\delta y}{\delta x} y(x)) = 2y - 4\left(\frac{\delta}{\delta x} y\right)$$

$$\Rightarrow \quad \frac{\delta}{\delta x \delta z} f(x, y\,(x), \frac{\delta y}{\delta x} y(x)) = -4\left(\frac{\delta}{\delta x} y(x)\right) + 8\left(\frac{\delta}{\delta x} y\right)$$

$$\Rightarrow \quad \frac{\delta^2}{\delta x \delta z} f(x, y(x), \frac{\delta y}{\delta x}(x)) = -4\left(\frac{\delta y}{\delta x} y(x)\right) + 8\left(\frac{\delta^2}{\delta x^2} y(x)\right)$$

and finally obtain

$$2y(x) + 4\left(\frac{\delta}{\delta x} y(x)\right) - 8\left(\frac{\delta^2}{\delta x^2} y(x)\right) = 0$$

$$y(x) = -C_1 \cos h\left(\frac{1}{2} x\right) + C_2 \sin h\left(\frac{1}{2} x\right).$$

**Example 35.2.** *Find the extremals of the functional* $\int_{x_0}^{x_1} (y'^2/x^3)\,dx$.

**Solution.** We have $f = y'^2/x^3$ which is independent of $y$ *i.e.*, $\partial f/\partial y = 0$.

Also $$\frac{d}{dx}\left(\frac{\partial f}{\partial y'}\right) = \frac{d}{dx}\left(\frac{2y'}{x^3}\right) = 2\,\frac{x^3 y'' - y' \,.\, 3x^2}{x^6} = \frac{2}{x^4}(xy'' - 3y')$$

$\therefore$ Euler's equation reduces to $\frac{d}{dx}\left(\frac{\partial f}{\partial y'}\right) = 0$

*i.e.,* $$\frac{2}{x^4}(xy'' - 3y') = 0 \quad \text{or} \quad y''/y' = 3/x$$

Integrating both sides, $$\int \frac{y''}{y'}\,dy = 3\int \frac{dx}{x} + \log c$$

*i.e.,* $$\log y' = 3 \log x + \log c \quad \text{or} \quad y' = cx^3$$

Hence $$y = cx^4/4 + c' \quad \text{or} \quad y = c_1 x^4 + c_2$$

This is the required extremal.

**Example 35.3.** *Prove that the shortest distance between two points in a plane is a straight line.*

**Solution.** Let $A(x_1, y_1)$ and $B(x_2, y_2)$ be the given points and $s$ the arc length of a curve connecting them (Fig. 35.2). Then

$$s = \int_{x_1}^{x_2} ds = \int_{x_1}^{x_2} \sqrt{(1 + y'^2)}\,dx$$

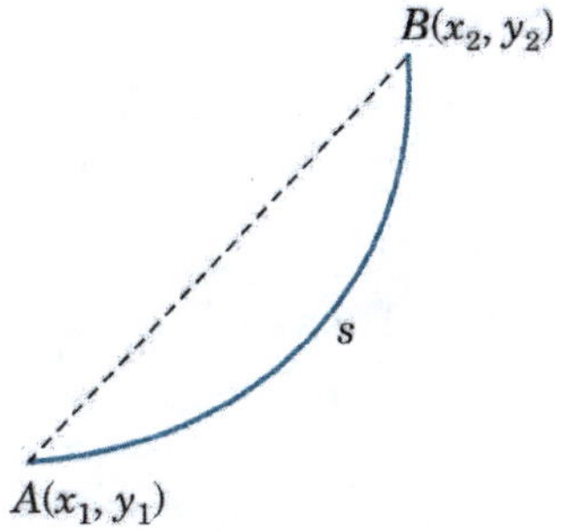

Fig. 35.2

Now $s$ will be minimum if it satisfies Euler's equation

$$\frac{\partial f}{\partial y} - \frac{d}{dx}\left(\frac{\partial f}{\partial y'}\right) = 0$$

Here $f = \sqrt{(1 + y'^2)}$ which is independent of $y$ *i.e.*, $\partial f/\partial y = 0$

$$\therefore \quad \frac{d}{dx}\left\{\frac{\partial}{\partial y'}\sqrt{(1 + y'^2)}\right\} = 0 \quad \text{or} \quad \frac{d}{dx}\left\{\frac{y'}{\sqrt{(1 + y'^2)}}\right\} = 0.$$

$\therefore$ On integration, we have $y'/\sqrt{(1 + y'^2)}$ = constant $y'$ = constant, $m$ say.

Integrating, we get $y = mx + c$, which is a straight line, the constants $m$ and $c$ are determined from the fact that the straight line passes through $A$ and $B$.

**Example 35.4.** *Find the curve passing through the points $(x_1, y_1)$ and $(x_2, y_2)$ which rotated about the x-axis gives a minimum surface area.*

**Solution.** In Fig. 35.3, the surface area $= \int_{x_1}^{x_2} 2\pi y \, ds$

$$= 2\pi \int_{x_1}^{x_2} y\sqrt{(1+y'^2)}\, dx. \text{ This has to be minimum.}$$

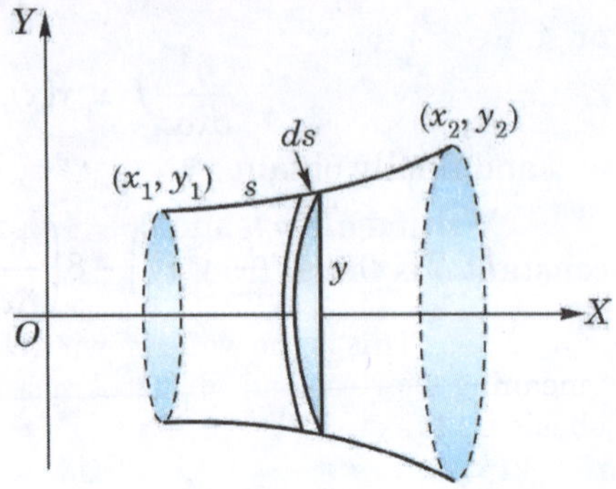

Fig. 35.3

Since $f = y\sqrt{(1+y'^2)}$ is independent of $x$, therefore, Euler's equation reduces to

$$f - y'\frac{\partial f}{\partial y'} = \text{constant, } c : \text{say} \qquad \text{[By § 35.4 (1)]}$$

$$\therefore \qquad y\sqrt{(1+y'^2)} - y'\frac{\partial}{\partial y'}\left\{y\sqrt{(1+y'^2)}\right\} = c$$

i.e.,
$$y\sqrt{(1+y'^2)} - y'\left\{\frac{y}{2}(1+y'^2)^{-1/2} \cdot 2y'\right\} = c$$

or
$$y/\sqrt{(1+y'^2)} = c \quad \text{or} \quad y' = \frac{dy}{dx} = \frac{\sqrt{(y^2-c^2)}}{c}$$

Separating the variables and integrating, we have

$$\int \frac{dy}{\sqrt{(y^2-c^2)}} = \int \frac{dx}{c} + c' \quad \text{or} \quad \cosh^{-1}\left(\frac{y}{c}\right) = \frac{x+a}{c}$$

i.e.,
$$y = c\cosh\left(\frac{x+a}{c}\right)$$

which is a *catenary*. The constants $a$ and $c$ are determined from the points $(x_1, y_1)$ and $(x_2, y_2)$.

**Obs.** This problem is also important in connection with *soap films* which are known to have shapes with minimum surface areas.

**Example 35.5.** *Find the path on which a particle in the absence of friction, will slide from one point to another in the shortest time under the action of gravity.*

**Solution.** Let the particle start sliding on the curve $OP_1$ from $O$ with zero velocity (Fig. 35.4). At time $t$, let the particle be $P(x, y)$ such that arc $OP = s$.

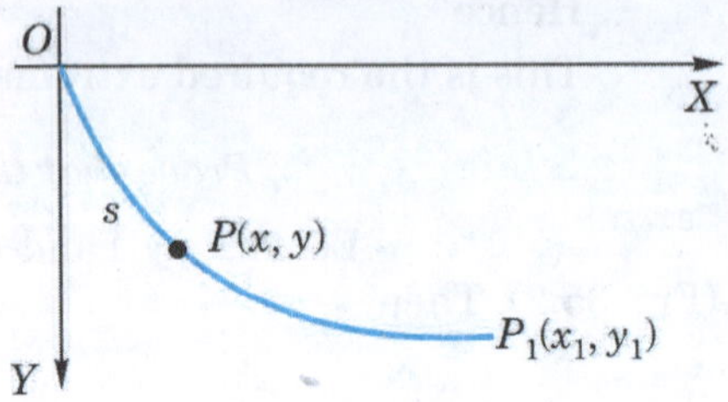

Fig. 35.4

By the principle of work and energy, we have

K.E. at $P$ – K.E. at $O$ = Work done in moving the particle from $O$ to $P$.

$$\frac{1}{2}m\left(\frac{ds}{dt}\right)^2 - 0 = mgy$$

or
$$ds/dt = \sqrt{(2gy)} \qquad \ldots(i)$$

Thus the time taken by the particle to move from $O$ to $P_1$ is

$$T = \int_0^T dt = \int_0^{x_1} \frac{ds}{\sqrt{(2gy)}} = \frac{1}{\sqrt{(2g)}}\int_0^{x_1} \frac{\sqrt{(1+y'^2)}}{\sqrt{y}}\, dx$$

Here $f = \sqrt{[(1+y'^2)/y]}$ is independent of $x$.

$\therefore$ Euler's equation reduces to $f - y'\,\partial f/\partial y'$ = constant, $c$ : say

i.e.,
$$\frac{\sqrt{(1+y'^2)}}{\sqrt{y}} - y'\frac{\partial}{\partial y'}\left\{\frac{\sqrt{(1+y'^2)}}{\sqrt{y}}\right\} = c \quad \text{or} \quad \frac{\sqrt{(1+y'^2)}}{\sqrt{y}} - y'\left\{\frac{y'}{\sqrt{(1+y'^2)}\sqrt{y}}\right\} = c$$

or
$$\sqrt{[y(1+y'^2)]} = 1/c = \sqrt{a}\text{, say.}$$

Solving for $y'$, we have
$$y' = \frac{dy}{dx} = \sqrt{\left(\frac{a-y}{y}\right)}$$

Separating the variables and integrating, we get

$$\int_0^x dx = \int_0^y \sqrt{\left(\frac{y}{a-y}\right)}\, dy \qquad \text{[Put } y = a \sin^2 \theta] \quad ...(i)$$

or
$$x = \int_0^\theta \sqrt{\left(\frac{a \sin^2 \theta}{a - a \sin^2 \theta}\right)}\, 2a \sin \theta \cos \theta \, d\theta$$

$$= a \int_0^\theta 2 \sin^2 \theta \, d\theta = a \int_0^\theta (1 - \cos 2\theta)\, d\theta = \frac{a}{2}(2\theta - \sin 2\theta) \quad ...(ii)$$

Writing $a/2 = b$ and $2\theta = \phi$, Eqns. (*ii*) and (*i*) become $x = b\,(\phi - \sin \phi)$, $y = b\,(1 - \cos \phi)$ which is a *cycloid*. The constant $b$ is found from the fact that the curve goes through $(x_1, y_1)$.

**Obs.** This is the well-known **brachistochrone problem** which derives its name from the Greek words 'brachistos' meaning shortest and 'chronos' meaning time. It was proposed by John Bernoulli in 1696 and its solution formed the basis for the study of the 'Calculus of Variations'. (*V.T.U., 2006*)

**Example 35.6.** *Find the curves on which the functional* $\int_0^1 [(y')^2 + 12xy]\, dx$ *with* $y(0) = 0$ *and* $y(1) = 1$ *can be extremised.* (*V.T.U., 2010*)

**Solution.** We have $\qquad f = y'^2 + 12xy$

$$\therefore \qquad \partial f/\partial y = 12x \,;\ \frac{\partial f}{\partial y'} = 2y' \,;\ \frac{d}{dx}\left(\frac{\partial f}{\partial y'}\right) = 2y''$$

Hence the Euler's equation $\dfrac{\partial f}{\partial y} - \dfrac{d}{dx}\left(\dfrac{\partial f}{\partial y'}\right) = 0$ becomes

$$12x - 2y'' = 0 \quad i.e., \quad y'' = 6x \qquad ...(i)$$

Integrating Eqn. (*i*), $\quad y' = 3x^2 + C \quad ...(ii) \quad$ and $\quad y = x^3 + Cx + C' \quad ...(iii)$

Using the boundary conditions, when $x = 0$, $y = 0$ Eqn. (*iii*) gives $C' = 0$.

When $x = 1$, $y = 1$, Eqn. (*iii*) again gives $C = 0$.

Hence Eqn. (*iii*) reduces to $y = x^3$ which is the only curve on which extremum can be attained.

**Example 35.7.** *On which curve the functional* $\int_0^{\pi/2} (y'^2 - y^2 + 2xy)\, dy$ *with* $y(0) = 0$ *and* $y(\pi/2) = 0$, *be extremized?* (*V.T.U., 2013*)

**Solution.** Let $\quad f = y'^2 - y^2 + 2xy$ so that $\dfrac{\partial f}{\partial y} = 0 - 2y + 2x$

and
$$\frac{d}{dx}\left(\frac{\partial f}{\partial y'}\right) = \frac{d}{dx}(2y') = 2y''$$

$\therefore\quad$ Euler's equation $\dfrac{\partial f}{\partial y} - \dfrac{d}{dx}\left(\dfrac{\partial f}{\partial y'}\right) = 0$ becomes

$$-2y + 2x - 2y'' = 0 \quad \text{or} \quad y'' + y = x \quad \text{or} \quad (D^2 + 1)\, y = x$$

Its A.E. $D^2 + 1 = 0\quad$ gives $\quad D = \pm\, i$.

$$\therefore \qquad \text{C.F.} = c_1 \cos x + c_2 \sin x$$

and
$$\text{P.I.} = \frac{1}{D^2 + 1} x = (1 + D^2)^{-1}\, x = (1 - D^2)\, x = x$$

Thus
$$y = c_1 \cos x + c_2 \sin x + x \qquad ...(i)$$

Using boundary conditions : when $x = 0$, $y = 0$, Eqn. (*i*) gives $c_1 = 0$ ;
when $x = \pi/2$, $y = 0$, Eqn. (*i*) gives $0 = c_2 + \pi/2$, *i.e.*, $c_2 = -\pi/2$.

Hence Eqn. (*i*) reduces to $y = x - \dfrac{\pi}{2} \sin x$, which is the only curve on which the given functional can be extremized.

**Example 35.8.** *Solve the variational problem*

$$\delta \int_1^2 [x^2 (y')^2 + 2y(x+y)]\, dx = 0, \text{ given } y(1) = y(2) = 0.$$

**Solution.** Let $f = x^2 (y')^2 + 2xy + 2y^2$ so that $\dfrac{\partial f}{\partial y} = 2x + 4y$, $\dfrac{\partial f}{\partial y} = 2x^2 y'$

$\therefore$ Euler's equation $\dfrac{\partial f}{\partial y} - \dfrac{d}{dx}\left(\dfrac{\partial f}{\partial y'}\right) = 0$ becomes

$$2x + 4y - \frac{d}{dx}(2x^2 y') = 0 \quad \text{or} \quad 2x + 4y - (2x^2 y'' + 4xy') = 0$$

or $\qquad x^2 y'' + 2xy' - 2y = x$. This is Cauchy's homogeneous linear (§ 13.9)

Putting $x = e^t$, it reduces to $(D^2 + D - 2)\, y = e^t$.

Its solution is $\qquad y = c_1 e^t + c_2 e^{-2t} + \dfrac{1}{3} te^t$ or $y = c_1 x + \dfrac{c_2}{x^2} + \dfrac{1}{3} x \log x$ $\qquad$ ...(*i*)

Since $y(1) = 0$, we have $c_1 + c_2 = 0$

and $\qquad y(2) = 0$ gives $0 = 2\,c_1 + \dfrac{1}{4} c_2 + \dfrac{2}{3} \log 2$

Solving these equations, we get $c_1 = -c_2 = \dfrac{-8}{21} \log 2.$

Putting the values of $c_1$ and $c_2$ in Eqn. (*i*), we get

$$y = \frac{1}{21}\{8 \log 2\, (x^{-2} - x) + 7x \log x\}$$

which is the required solution.

## 35.5 GEODESICS

*A geodesic on a surface is a curve along which the distance between any two points of the surface is a minimum.* To find the geodesics on a surface is a variational problem involving the conditional extremum. This problem was first studied by *Jacob Bernoulli* in 1698 and its general method of solution was given by *Euler*.

**Example 35.9.** *Show that the geodesics on a plane are straight lines.* (V.T.U., 2013)

**Solution.** Let $y = y(x)$ be a curve joining the points $A(x_1, y_1)$ and $B(x_2, y_2)$ in the $xy$-plane. Then the length of a curve joining $A$ and $B$ is given by

$$s = \int_A^B \frac{ds}{dx} dx = \int_{x_1}^{x_2} \sqrt{[1 + (dy/dx)^2]}\, dx \quad i.e., \quad s = \int_{x_1}^{x_2} \sqrt{(1 + y'^2)}\, dx$$

The geodesic on the $xy$-plane is the curve $y = y(x)$ for which $s$ is minimum.

We have $f(x, y, y') = \sqrt{(1 + y'^2)}$ which depends on $y'$ only. Hence the Euler's equation.

$$\frac{\partial f}{\partial y} - \frac{d}{dx}\left(\frac{\partial f}{\partial y'}\right) = 0 \text{ yields}$$

$$\frac{d}{dx}\left\{\frac{2y'}{2\sqrt{(1 + y'^2)}}\right\} = 0 \quad i.e., \quad y''\sqrt{(1 + y'^2)} - \frac{y'_2 2y'y''}{2\sqrt{(1 + y'^2)}} = 0$$

*i.e.* $\qquad y''(1 + y'^2) - y'^2 y'' = 0 \quad i.e., \quad \dfrac{d^2 y}{dx^2} = 0$

Integrating twice, we get $y = c_1 x + c_2$

which is a straight line.

Hence *the geodesics on a plane are straight lines.*

**Example 35.10.** *Find the geodesics on a right circular cylinder of radius a.*

**Solution.** In cylindrical coordinates $(\rho, \phi, z)$ we have

$$x = r\cos\phi,\ y = r\sin\phi,\ z = z.$$

$\therefore$ The element of arc on a right circular cylinder of radius $a$, is given by

$$ds^2 = (dx)^2 + (dy)^2 + (dz)^2 = (d\rho)^2 + (\rho d\phi)^2 + dz^2 = a^2\, d\phi^2 + dz^2 \qquad [\because\ \rho = a \text{ and } d\rho = 0]$$

or
$$ds = \sqrt{\{a^2 + (dz/d\phi)^2\}}\ .\ d\phi \quad \text{or} \quad s = \int_{\phi_1}^{\phi_2} \sqrt{(a^2 + z'^2)}\ d\phi \qquad \ldots(i)$$

Now the geodesic for the given cylinder is the curve for which $s$ is minimum. Here $f = \sqrt{(a^2 + z'^2)}$, which is a function of $\phi$ and $z'$ while $z$ is absent.

$\therefore$ Euler's equation for the functional Eqn. ($i$) reduces to

$$\frac{\partial f}{\partial z'} = \text{constant} \quad \text{or} \quad \frac{z'}{\sqrt{(a^2 + z'^2)}} = c, \text{ say.}$$

This simplifies to $(z')^2$ = constant or $dz/d\phi = c_1$, say.

Integrating, $z = c_1\phi + c_2$

*This is the desired geodesics on a circular cylinder which is a circular helix.* (Example 8.3, *p.* 368)

**Example 35.11.** *Show that the geodesics on a sphere of radius a are its great circles.*

**Solution.** In spherical coordinates $(r, \theta, \phi)$, we have

$$x = r\sin\theta\cos\phi,\ y = r\sin\theta\sin\phi,\ z = r\cos\theta. \qquad \ldots(i)$$

$\therefore$ The arc element on a sphere of radius $a$, is given by

$$ds^2 = dr^2 + (r\, d\theta)^2 + (r\sin\theta\, d\phi)^2 = a^2\, d\theta^2 + (a\sin\theta)^2\, d\phi^2 \qquad [\because\ r = a,\ dr = 0$$

or
$$ds = a\sqrt{[1 + \sin^2\theta\,(d\phi/d\theta)^2]}\ d\theta, \quad \text{or} \quad s = a\int_{\theta_1}^{\theta_2} \sqrt{[1 + \sin^2\theta\,.\,(\phi')^2]}\ d\theta$$

Now the geodesic on the sphere $r = a$ is the curve for which $s$ is minimum. Here $f = a\sqrt{(1 + \sin^2\theta\,.\,\phi'^2)}$ which is a function of $\theta$ and $\phi'$ while $\phi$ is absent.

$\therefore$ Euler's equation reduces to $\partial f/\partial\phi'$ = constant.

$$\frac{\partial f}{\partial \phi'} = \frac{a\sin^2\theta\,.\,\phi'}{\sqrt{(1 + \sin^2\theta\,.\,\phi'^2)}} = \text{constant.}$$

or
$$\frac{\sin^2\theta\,.\,\phi'}{\sqrt{(1 + \sin^2\theta\,.\,\phi'^2)}} = c \text{ (say)} \quad \text{or} \quad \sin^2\theta\,(\sin^2\theta - c^2)\,\phi'^2 = c^2$$

or
$$\frac{d\phi}{d\theta} = \frac{c}{\sin\theta\sqrt{(\sin^2\theta - c^2)}} = \frac{c\,\text{cosec}^2\,\theta}{\sqrt{(1 - c^2\,\text{cosec}^2\,\theta)}}$$

Integrating
$$\phi = \int \frac{c\,\text{cosec}^2\,\theta\,.\,d\theta}{\sqrt{[(1 - c^2) - (c\cot\theta)^2]}} + c' = -\sin^{-1}\left\{\frac{c\cot\theta}{\sqrt{(1 - c^2)}}\right\} + c'$$

or
$$\cot\theta = A\cos\phi + B\sin\phi \quad \text{or} \quad a\cos\theta = Aa\sin\theta\cos\phi + Ba\sin\theta\sin\phi$$

or
$$z = Ax + By \qquad \text{[By Eqn. } (i) \text{ when } r = a]$$

This is a plane through the centre (0, 0, 0) of the sphere which cuts the sphere along a great circle. Hence the required *geodesics are the arcs of the great circles.*

## PROBLEMS 35.1

1. Solve the Euler's equation for the following functionals :

($i$) $\int_{x_0}^{x_1} (x + y')\, y'\, dx$ ($ii$) $\int_{x_0}^{x_1} (1 + x^2 y')\, y'\, dx.$

2. Show that the general solution of the Euler's equation for the integral

$$\int_a^b \frac{1}{y}\sqrt{\left[1+\left(\frac{dy}{dx}\right)^2\right]}\, dx \text{ is } (x-h)^2 + y^2 = k^2.$$

Find the extremals of the following functionals :

3. $\int_{x_1}^{x_2} (y^2 + y'^2 + 2ye^x)\, dx.$

4. $\int_0^{\pi/2} (y^2 + y'^2 - 2y \sin x)\, dx,\ y(0) = y\,(\pi/2) = 0.$

5. $\int_0^{\pi} (y'^2 - y^2 + 4y \cos x)\, dx;\ y(0) = 0,\ y\,(\pi) = 0.$

6. $\int_{x_0}^{x_1} \frac{1+y^2}{y'^3}\, dx.$

7. $\int_1^2 \frac{x^3}{y'^2}\, dx\ ;\ y\,(1) = 0,\ y(2) = 3.$

8. $\int_1^2 \frac{\sqrt{(1+y'^2)}}{x}\, dx\ ;\ y\,(1) = 0,\ y(2) = 1.$

9. Solve the variational problem $\delta \int_0^{\pi/2} [y^2 - (y')^2]\, dx$ under the conditions $y(0) = 0,\ y(\pi/2) = 2.$ (*V.T.U., 2010*)

10. A heavy cable hangs freely under gravity between two fixed points. Show that the shape of the cable is a catenary.

11. A particle is moving with a force perpendicular to and proportional to its distance from the line of zero velocity. Show that the path of quickest descent (brachistochrone) is a circle.

12. Find the geodesics on a *right circular cone* of semi-vertical angle α.

## 35.6 ISOPERIMETRIC PROBLEMS

In certain problems, it is necessary to make a given integral.

$$I = \int_{x_1}^{x_2} f(x, y, y')\, dx \qquad ...(1)$$

maximum or minimum while keeping another integral

$$J = \int_{x_1}^{x_2} g(x, y, y')\, dx \qquad ...(2)$$

constant. Such problems involve one or more constraint conditions, just as $J = a$ constant. A typical example of this type is that of finding a closed curve of a given perimeter and maximum area. This being one of the earliest problems to engage attention, we often refer to problems of this type as *isoperimetric problems*.

Such problems are generally solved by the method of Lagrange multipliers. To extremize Eqn. (1), we multiply Eqn. (2) by λ and add to Eqn. (1) where λ is the Lagrange multiplier. Then the necessary condition for the integral $\int_{x_1}^{x_2} H\, dx$ to be an extremum is $\frac{\partial H}{\partial y} - \frac{d}{dx}\left(\frac{\partial H}{\partial y'}\right) = 0$ where $H = f + \lambda g$. The values of the two constants of integration and the parameter λ are determined from the three conditions namely : the two boundary conditions and the integral $J$ having given constant value.

**Example 35.12.** *Find the plane curve of fixed perimeter and maximum area.*

**Solution.** Let $l$ be the fixed perimeter of a plane curve between the points with abscissae $x_1$ and $x_2$ (Fig. 35.5). Then

$$l = \int_{x_1}^{x_2} \sqrt{(1+y'^2)}\, dx \qquad ...(i)$$

Also the area between the curve and the $x$-axis is

$$A = \int_{x_1}^{x_2} y\, dx \qquad ...(ii)$$

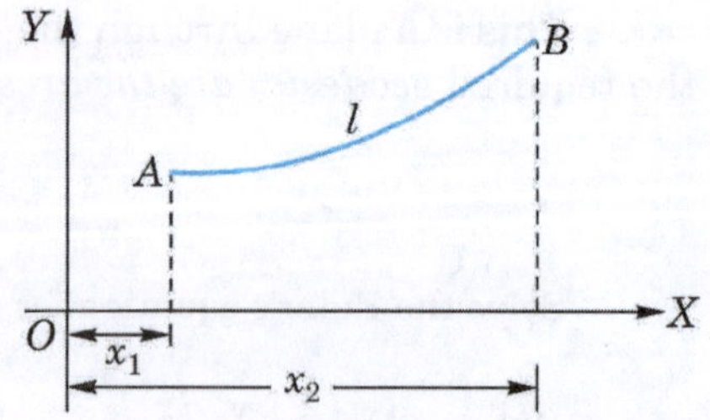

Fig. 35.5

We have to maximize Eqn. (*ii*) subject to constraint Eqn. (*i*).

$\therefore$ Taking $f = y$ and $g = \sqrt{(1 + y'^2)}$, we write $H = f + \lambda g = y + \lambda\sqrt{(1 + y'^2)}$

Now $H$ must satisfy the Euler's equation

$$\frac{\partial H}{\partial y} - \frac{d}{dx}\left(\frac{\partial H}{\partial y'}\right) = 0 \quad \therefore \quad 1 - \frac{d}{dx}\left[\frac{\lambda y'}{\sqrt{(1 + y'^2)}}\right] = 0$$

Integrating w.r.t. $x$, we have $x - \lambda y'/\sqrt{(1 + y'^2)} = 0$

Solving for $y'$, we get $\quad y' = \dfrac{x - a}{\sqrt{[\lambda^2 - (x - a)^2]}}$

Integrating again, $\quad y = \sqrt{[\lambda^2 - (x - a)^2]} + b$ *i.e.*, $(x - a)^2 + (y - b)^2 = \lambda^2$ which is a circle.

**Example 35.13.** *Prove that the sphere is the solid figure of revolution which, for a given surface area, has maximum volume.*

**Solution.** Consider the arc $OPA$ of the curve which rotates about the $x$-axis as shown in Fig. 35.6.

Then surface area $\quad S = \int_{x=0}^{a} 2\pi y\, ds = \int_0^a 2\pi y\sqrt{(1 + y'^2)}\, dx$

and volume of the solid so formed $\quad V = \int_0^a \pi y^2\, dx.$

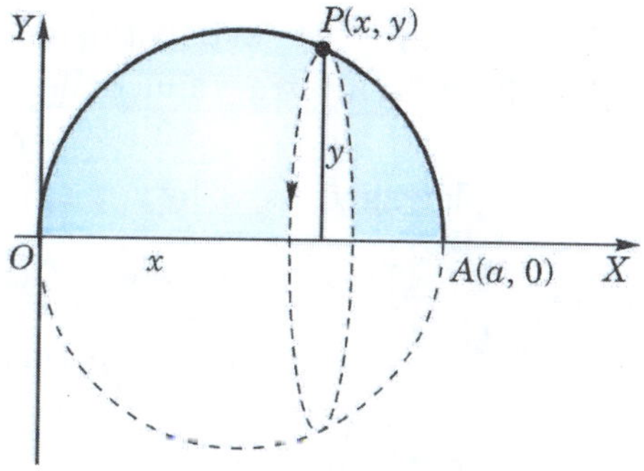

Fig. 35.6

Here we have to maximize $V$ subject to fixed $S$. Taking $f = \pi y^2$ and $g = 2\pi y\sqrt{(1 + y'^2)}$, we write $H = f + \lambda g = \pi y^2 + 2\pi\lambda y\sqrt{(1 + y'^2)}$.

Now $H$ has to satisfy Euler's equation. But it does not contain $x$ explicitly.

$$\therefore \quad H - y'\frac{\partial H}{\partial y'} = \text{constant, } c : \text{say}$$

*i.e.*,
$$\pi y^2 + 2\pi\lambda y\sqrt{(1 + y'^2)} - y' \,.\, 2\pi\lambda y\frac{y'}{\sqrt{(1 + y'^2)}} = c$$

or
$$\pi y^2 + \frac{2\pi\lambda y}{\sqrt{(1 + y'^2)}} = c \qquad \text{...}(i)$$

Since the curve passes through $O$ and $A$ for which $y = 0$, Eqn. (*i*) gives $c = 0$.

$$\therefore \quad y + 2\lambda/\sqrt{(1 + y'^2)} = 0$$

Solving for $y'$, $\quad y'\left(= \dfrac{dy}{dx}\right) = \dfrac{\sqrt{(4\lambda^2 - y^2)}}{y}$

Separating the variables and integrating, we get

$$\int dx = \int \frac{y\, dy}{\sqrt{(4\lambda^2 - y^2)}} + k \quad \text{or} \quad x = k - \sqrt{(4\lambda^2 - y^2)} \qquad \text{...}(ii)$$

When $x = 0, y = 0 \quad \therefore \quad k = 2\lambda$

$\therefore$ (*ii*) becomes $(x - 2\lambda)^2 + y^2 = (2\lambda)^2$ which is a circle with centre $(2\lambda, 0)$ and radius $2\lambda$.

Hence the figure formed by the revolution of given arc is a sphere.

## PROBLEMS 35.2

1. Find a function $y(x)$ for which $\int_0^1 (x^2 + y'^2)\, dx$ is stationary, given that $\int_0^1 y^2\, dx = 2$ ; $y(0) = 0, y(1) = 0$.
2. Find the extremals of the isoperimetric problem $v[y(x)] = \int_{x_0}^{x_1} y'^2\, dx$ given that $\int_{x_0}^{x_1} y\, dx = c$, a constant.

3. Show that the curve $c$ of given length $l$ which minimizes the curved surface area of the solid generated by the revolution of $c$ about the $x$-axis is a catenary.

4. Find the extremal of the functional $I = \int_0^{\pi} [(y')^2 - y^2]\, dx$ under the conditions $y(0) = 0, y(\pi) = 1$ and subject to the constraint $\int_0^{\pi} y\, dx = 1$.

5. Prove that the extremal of the isoperimetric problem $v[y(x)] = \int_1^4 y_2'\, dx, y(1) = 3, y(4) = 24$, subject to the condition $\int_1^4 y\, dx = 36$ is a parabola.

## 35.7 SEVERAL DEPENDENT VARIABLES

We now extend the variational problem of § 35.3 to a problem with several variables as functions of a single independent variable *i.e., A necessary condition for*

$$I = \int_{x_1}^{x_2} f(x, y_1, y_2, ..., y_n, y_1', y_2', ..., y_n')\, dx \qquad ...(1)$$

*to be an extremum is that*

$$\frac{\partial \mathbf{f}}{\partial \mathbf{y}_i} - \frac{\mathbf{d}}{\mathbf{dx}}\left(\frac{\partial \mathbf{f}}{\partial \mathbf{y}_i'}\right) = \mathbf{0},\ i = 1, 2, ..., n \qquad ...(2)$$

Let $y_1, y_2, ..., y_n$ satisfy the boundary conditions $y_i(x_1) = y_{i1}, y_i(x_2) = y_{i2}$

Consider arbitrary functions $\eta_1(x), \eta_2(x), ..., \eta_n(x)$ which are all zero on the boundary *i.e.,*

$$\eta_i(x_1) = 0 = \eta_i(x_2)$$

Replacing $y_1, y_2, ...$ by $y_1 + \varepsilon_1\eta_1, y_2 + \varepsilon_2\eta_2, ...$ in Eqn. (1), we get

$$I(\varepsilon) = \int_{x_1}^{x_2} f(x, y_1 + \varepsilon_1\eta_1, y_2 + \varepsilon_2\eta_2, ..., y_1' + \varepsilon_1\eta_1', y_2' + \varepsilon_2\eta_2', ...)\, dx \qquad ...(3)$$

This is a function of the parameters $\varepsilon_1, \varepsilon_2, ...$ and reduces to Eqn. (1) for $\varepsilon_1 = \varepsilon_2 = ... = 0$.

To find the stationary value of Eqn. (1), we find the stationary value of $I(\varepsilon)$ for $\varepsilon_1 = \varepsilon_2 = ... = 0$, $I(\varepsilon)$ will have a stationary value when

$$\frac{\partial I(\varepsilon)}{\partial \varepsilon_1} = 0,\ \frac{\partial I(\varepsilon)}{\partial \varepsilon_2} = 0, ...$$

Writing $\quad f = f(x, y_1, y_2, ..., y_1', y_2', ...)$

and $\quad F = f(x, y_1 + \varepsilon_1\eta_1, y_2 + \varepsilon_2\eta_2, ..., y_1' + \varepsilon_1\eta_1', y_2' + \varepsilon_2\eta_2', ...)$.

Equation (3) becomes $I(\varepsilon) = \int_{x_1}^{x_2} F\, dx$.

$x_1, x_2$ being constants independent of $\varepsilon_1$, differentiating under the integral sign, we get

$$\frac{\partial I(\varepsilon)}{\partial \varepsilon_1} = \int_{x_1}^{x_2} \frac{\partial F}{\partial \varepsilon_1}\, dx = \int_{x_1}^{x_2} \left(\frac{\partial F}{\partial y_1}\eta_1 + \frac{\partial F}{\partial y_1}\eta_1'\right) dx$$

$\therefore \quad \frac{\partial I(\varepsilon)}{\partial \varepsilon_1} = 0$, when $\varepsilon_1 = \varepsilon_2 = ... = 0$ gives

$$\int_{x_1}^{x_2} \left(\frac{\partial f}{\partial y_1}\eta_1 + \frac{\partial f}{\partial y_1'}\eta_1'\right) = 0$$

Integrating by parts the second term, we get

$$\int_{x_1}^{x_2} \frac{\partial f}{\partial y_1}\eta_1 + \left|\frac{\partial f}{\partial y_1'}\eta_1\right|_{x_1}^{x_2} - \int_{x_1}^{x_2} \frac{d}{dx}\left(\frac{\partial f}{\partial y_1'}\right)\eta_1(x)\, dx = 0$$

*i.e.,*

$$\int_{x_1}^{x_2} \left\{\frac{\partial f}{\partial y_1} - \frac{d}{dx}\left(\frac{\partial f}{\partial y_1'}\right)\right\}\eta_1(x)\, dx = 0 \qquad [\because \eta_1(x_1) = 0 = \eta_1(x_2)]$$

Since this equation must hold good for all values of $\eta_1(x)$, we get

$$\frac{\partial f}{\partial y_1} - \frac{d}{dx}\left(\frac{\partial f}{\partial y_1'}\right) = 0$$

Similarly $\dfrac{\partial I(\varepsilon)}{\partial \varepsilon_2} = 0$ when $\varepsilon_1 = \varepsilon_2 = \ldots = 0$, will give

$$\frac{\partial f}{\partial y_2} - \frac{d}{dx}\left(\frac{\partial f}{\partial y_2'}\right) = 0 \text{ and so on.}$$

All these conditions give a system of Euler's Eqn. (2). A solution of these equations leads to the desired curves.

**Example 35.14.** *Show that the functional* $\int_0^{\pi/2}\left\{2xy + \left(\frac{dx}{dt}\right)^2 + \left(\frac{dy}{dt}\right)^2\right\} dt$ *such that* $x(0) = 0$, $x(\pi/2) = -1$, $y(0) = 0$, $y(\pi/2) = 1$ *is stationary for* $x = -\sin t$, $y = \sin t$.

**Solution.** Euler's equations are $\dfrac{\partial f}{\partial x} - \dfrac{d}{dt}\left(\dfrac{\partial f}{\partial x'}\right) = 0$ ...(*i*)

$$\frac{\partial f}{\partial y} - \frac{d}{dt}\left(\frac{\partial f}{\partial y'}\right) = 0 \quad \text{...(ii)}$$

Here $f = 2xy + x'^2 + y'^2$. $\therefore$ $\dfrac{\partial f}{\partial x} = 2y$, $\dfrac{\partial f}{\partial x'} = 2x'$, $\dfrac{\partial f}{\partial y} = 2x$, $\dfrac{\partial f}{\partial y'} = 2y'$

(*i*) becomes $2y - \dfrac{d}{dt}(2x') = 0$ *i.e.*, $2y - 2\dfrac{d^2x}{dt^2} = 0$ or $\dfrac{d^2x}{dt^2} = y$ ...(*iii*)

(*ii*) becomes $2x - \dfrac{d}{dt}(2y') = 0$ *i.e.*, $2x - 2\dfrac{d^2y}{dt^2} = 0$ or $\dfrac{d^2y}{dt^2} = x$ ...(*iv*)

Now to solve these simultaneous differential equations, we differentiate Eqn. (*iii*) twice,

$$\frac{d^4x}{dt^4} = \frac{d^2y}{dt^2} = x \quad \text{[By (iv)]}$$

or $(D^4 - 1)\, x = 0$ which is a linear equation with constant coefficients.

Its solution is $x = c_1 e^x + c_2 e^{-x} + c_3 \cos x + c_4 \sin x$ ...(*v*)

From Eqn. (*iii*), $y = x'' = c_1 e^x + c_2 e^{-x} - c_3 \cos x - c_4 \sin x$ ...(*vi*)

Since $x = 0$ when $t = 0$ $\therefore$ $0 = c_1 + c_2 + c_3$ ...(*vii*)

$y = 0$ when $t = 0$ $\therefore$ $0 = c_1 + c_2 - c_3$ ...(*viii*)

Also $x = -1$ when $t = \pi/2$ $\therefore$ $-1 = c_1 e^{\pi/2} + c_2 e^{-\pi/2} + c_4$ ...(*ix*)

$y = 1$ when $t = \pi/2$ $\therefore$ $1 = c_1 e^{\pi/2} + c_2 e^{-\pi/2} - c_4$ ...(*x*)

Adding Eqns. (*vii*) and (*viii*), $c_1 + c_2 = 0$

Adding Eqns. (*ix*) and (*x*), $c_1 e^{\pi/2} + c_2 e^{-\pi/2} = 0$

Solving these equations, we get $c_1 = c_2 = 0$.

From Eqn. (*viii*), $c_3 = c_1 + c_2 = 0$. From Eqn. (*ix*), $c_4 = -1$.

Hence from Eqn. (*v*), $x = -\sin x$ and from Eqn. (*vi*), $y = \sin x$.

## 35.8 FUNCTIONALS INVOLVING HIGHER ORDER DERIVATIVES

*A necessary condition for*

$$I = \int_{x_1}^{x_2} f(x, y, y', y'')\, dx \quad \text{...(1)}$$

*to be extremum is* $\dfrac{\partial f}{\partial y} - \dfrac{d}{dx}\left(\dfrac{\partial f}{\partial y'}\right) + \dfrac{d^2}{dx^2}\left(\dfrac{\partial f}{\partial y''}\right) = 0.$

Let $y(x)$ be the function which makes Eqn. (1) stationary and satisfies the boundary conditions

$$y(x_1) = y_1, y(x_2) = y_2, y'(x_1) = y_1' \text{ and } y'(x_2) = y_2'.$$

Consider the differentiable function $\eta(x)$ such that

$$\eta(x_1) = 0 = \eta(x_2) \quad \text{and} \quad \eta'(x_1) = 0 = \eta'(x_2) \qquad ...(2)$$

Replacing $y$ by $y + \varepsilon\eta$ in Eqn. (1), we get

$$I(\varepsilon) = \int_{x_1}^{x_2} f(x, y + \varepsilon\eta, y' + \varepsilon\eta', y'' + \varepsilon\eta'')\,dx \qquad ...(3)$$

This is a function of the parameter $\varepsilon$ and reduces to Eqn. (1) for $\varepsilon = 0$.

To find the stationary value of Eqn. (1), we find the stationary value of $I(\varepsilon)$ for $\varepsilon = 0$. But $I(\varepsilon)$ will have a stationary value when $dI(\varepsilon)/d\varepsilon = 0$.

Writing $f = f(x, y, y', y'')$ and $F = f(x, y + \varepsilon\eta, y' + \varepsilon\eta', y'' + \varepsilon\eta'')$.

Equation (3) becomes $I(\varepsilon) = \int_{x_1}^{x_2} F\,dx$

$x_1, x_2$ being constants independent of $\varepsilon$, differentiating under the integral sign, we get

$$\frac{dI(\varepsilon)}{d\varepsilon} = \int_{x_1}^{x_2} \frac{dF}{d\varepsilon}\,dx = \int_{x_1}^{x_2} \left(\frac{\partial F}{\partial y}\eta + \frac{\partial F}{\partial y'}\eta' + \frac{\partial F}{\partial y''}\eta''\right) dx$$

$$\therefore \quad \frac{dI(\varepsilon)}{d\varepsilon} = 0 \text{ when } \varepsilon = 0 \text{ gives } \int_{x_1}^{x_2} \left(\frac{\partial f}{\partial y}\eta + \frac{\partial f}{\partial y'}\eta' + \frac{\partial f}{\partial y''}\eta''\right) dx = 0$$

Integrating by parts once the second term and twice the third term, we get

$$\int_{x_1}^{x_2} \frac{\partial f}{\partial y}\eta\,dx + \left|\frac{\partial f}{\partial y'}\eta\right|_{x_1}^{x_2} - \int_{x_1}^{x_2} \frac{d}{dx}\left(\frac{\partial f}{\partial y'}\right).\eta\,dx + \left|\frac{\partial f}{\partial y''}\eta' - \frac{d}{dx}\left(\frac{\partial f}{\partial y''}\right).\eta\right|_{x_1}^{x_2}$$

$$+ \int_{x_1}^{x_2} \frac{d^2}{dx^2}\left(\frac{\partial f}{\partial y''}\right).\eta\,dx = 0$$

or

$$\int_{x_1}^{x_2} \left[\frac{\partial f}{\partial y} - \frac{d}{dx}\left(\frac{\partial f}{\partial y'}\right) + \frac{d^2}{dx^2}\left(\frac{\partial f}{\partial y''}\right)\right]\eta(x)\,dx = 0 \qquad \text{[By (2)]}$$

Since this equation must hold good for all values of $\eta(x)$, we get

$$\frac{\partial f}{\partial y} - \frac{d}{dx}\left(\frac{\partial f}{\partial y'}\right) + \frac{d^2}{dx^2}\left(\frac{\partial f}{\partial y''}\right) = 0$$

In general, *a necessary condition for the functional* $I = \int_{x_1}^{x_2} f(x, y, y', y'', ... y^{(n)})\,dx$ *to be stationary will be*

$$\frac{\partial f}{\partial y} - \frac{d}{dx}\left(\frac{\partial f}{\partial y'}\right) + \frac{d^2}{dx^2}\left(\frac{\partial f}{\partial y''}\right) - ... + (-1)^n \frac{d^n}{dx^n}\left(\frac{\partial f}{\partial y^{(n)}}\right) = 0$$

which is called the *Euler-Poisson equation* and its solutions are called *extremals*.

**Example 35.15.** *Show that the curve which extremizes the functional* $I = \int_0^{\pi/4} (y''^2 - y^2 + x^2)\,dx$ *under the conditions* $y(0) = 0, y'(0) = 1, y(\pi/4) = y'(\pi/4) = 1/\sqrt{2}$ *is* $y = \sin x$.

**Solution.** The Euler-Poisson equation is

$$\frac{\partial f}{\partial y} - \frac{d}{dx}\left(\frac{\partial f}{\partial y'}\right) + \frac{d^2}{dx^2}\left(\frac{\partial f}{\partial y''}\right) = 0 \qquad ...(i)$$

Here $\quad f = (y'')^2 - y^2 + x^2, \quad \dfrac{\partial f}{\partial y} = -2y, \quad \dfrac{\partial f}{\partial y'} = 0, \quad \dfrac{\partial f}{\partial y''} = 2y''$

$\therefore$ Eqn. (*i*) becomes $-2y + \dfrac{d^2}{dx^2}(2y'') = 0$ or $y^{iv} - y = 0$ or $(D^4 - 1)\, y = 0$

Its solution is $\quad y(x) = c_1 e^x + c_2 e^{-x} + c_3 \cos x + c_4 \sin x$ ...(*ii*)

$\therefore \quad y'(x) = c_1 e^x - c_2 e^{-x} - c_3 \sin x + c_4 \cos x$ ...(*iii*)

Applying the given boundary conditions to Eqn. (*ii*) and (*iii*), we get

$$0 = y(0) = c_1 + c_2 + c_3,\ 1 = y'(0) = c_1 - c_2 + c_4$$

$$\frac{1}{\sqrt{2}} = y(\pi/4) = c_1 e^{\pi/4} + c_2 e^{-\pi/4} + \frac{1}{\sqrt{2}} c_3 + \frac{1}{\sqrt{2}} c_4 \qquad ...(iv)$$

$$\frac{1}{\sqrt{2}} = y'(\pi/4) = c_1 e^{\pi/4} - c_2 e^{-\pi/4} - \frac{1}{\sqrt{2}} c_3 + \frac{1}{\sqrt{2}} c_4$$

Solving the Eqn. (*iv*), we get $c_1 = c_2 = c_3 = 0$, $c_4 = 1$. Hence the required curve is $y = \sin x$.

## PROBLEMS 35.3

1. Show that the functional $\int_0^1 \left\{ 2x + \left(\dfrac{dx}{dt}\right)^2 + \left(\dfrac{dy}{dt}\right)^2 \right\} dt$, such that $x(0) = 1, y(0) = 1$ ; $x(1) = 1.5, y(1) = 1$ is stationary for $x = 1 + t^2/2$, $y = t$.
2. Find the extremals of the functional $v(y, z) = \int_{x_0}^{x_1} (2yz - 2y^2 + y'^2 - z'^2)\, dx$
3. Find a function $y(x)$ such that $\int_0^{\pi} y^2\, dx = 1$ which makes $\int_0^{\pi} y''^2\, dx$ a minimum if $y(0) = 0 = y(\pi)$, $y''(0) = 0 = y''(\pi)$.
4. Find the extremals of the functional $\int_0^{\pi/2} (y''^2 - y^2 + x^2)\, dx$ that satisfies the conditions $y(0) = 1$, $y'(0) = 0$, $y(\pi/2) = 0$, $y'(\pi/2) = -1$.
5. Find the extremals of the functional $\int_{-a}^{a} \left(\lambda y + \frac{1}{2}\mu y''^2\right) dx$ which satisfies the boundary conditions $y(-a) = 0$, $y'(-a) = 0$, $y(a) = 0$, $y'(a) = 0$.
6. Find the extremals of the following functionals :

   (*i*) $v[y(x)] = \int_{x_0}^{x_1} (16y^2 - y''^2 + x^2)\, dx$ (*Nagpur, 1997*) (*ii*) $v[y(x)] = \int_{x_0}^{x_1} (2xy + y'''^2)\, dx$.

## 35.9 APPROXIMATE SOLUTION OF BOUNDARY VALUE PROBLEMS — Rayleigh-Ritz Method

In § 35.3, we have seen that the solution of Euler's differential equation alongwith boundary conditions amounts to extremising a certain definite integral. This fact provides a technique of solving a boundary value problem approximately by assuming a trial solution satisfying the given boundary conditions and then extremising the integral whose integrand is found from the given differential equation.

To solve a boundary value problem of Rayleigh-Ritz method, we try to write the given differential equation as the Euler's equation of some variational problem. Then we reduce this variational problem to a minimizing problem assuming an approximate solution in the form

$$\bar{y}(x) = y_0(x) + \Sigma c_i\, \phi_i(x) \qquad ...(1)$$

where the *trial functions* $\phi_i(x)$ satisfy the boundary conditions and $\phi_i(x) = 0$ on the boundary $C$ of its region $R$.

Let the integral to be extremised be $I = \int_a^b f(y, y', x)\, dx$ ...(2)

such that $y(a) = A$ and $y(b) = B$.

Substituting Eqns. (1) in (2) by replacing $y$ in $\bar{y}$ in $I$, giving $\bar{I}$ as a function of the unknowns $c_i$. Then $c$'s become parameters which are so determined as to extremise $\bar{I}$. This requires

$$\frac{\partial \bar{I}}{\partial c_i} = 0, \qquad i = 1, 2, \ldots$$

Solving these equations, we get the values of $c_i$, which when substituted in Eqn. (1) give the desired solution.

**Example 35.16.** *Solve the boundary value problem $y'' - y + x = 0$ $(0 \le x \le 1)$, $y(0) = y(1) = 0$ by Rayleigh-Ritz method.*

**Solution.** Given differential equation is $y'' - y + x = 0$ ...(*i*)

Its solution is equivalent to extremising the integral $I = \int_0^1 F(x, y, y')\, dx$

where $$F(x, y, y') = 2xy - y^2 - y'^2, \quad \ldots(ii)$$

since the Euler's equation $\dfrac{\partial F}{\partial y} - \dfrac{d}{dx}\left(\dfrac{\partial F}{\partial y'}\right) = 0$ gives (*i*).

Assume that the *trial function* is $\bar{y}(x) = c_0 + c_1 x + c_2 x^2$ ...(*iii*)

To satisfy $y(0) = 0$, $y(1) = 0$, we require $c_0 = 0$, $c_2 = -c_1$.

$\therefore$ (*iii*) becomes $\bar{y}(x) = c_1 x(1 - x)$ ...(*iv*)

Substituting $\bar{y}$ and $\bar{y}'$ in $I$, we have

$$\bar{I} = \int_0^1 [2x\bar{y} - \bar{y}^2 - (\bar{y}')^2]\, dx = \int_0^1 [2c_1(x^2 - x^3) - c_1^2(x - x^2)^2 - c_1^2(1 - x)^2]\, dx$$

$$= \frac{1}{6}c_1 - \frac{11}{30}c_1^2$$

Its stationary values are given by $d\bar{I}/dc_1 = 0$. $\therefore \quad \dfrac{1}{6} - \dfrac{11}{15}c_1 = 0$ *i.e.*, $c_1 = \dfrac{5}{22}$.

Thus the approximate solution is $\bar{y}(x) = \dfrac{5}{22}x(1 - x)$ ...(*v*)

## 35.10 WEIGHTED RESIDUAL METHOD—Galerkin's Method

The starting point of this method is to guess a solution to the differential equation which satisfies the boundary conditions. This *trial solution* will contain certain parameters which can be adjusted to minimize the errors so that the trial solution is as close to the exact solution as possible.

Consider the boundary value problem

$$y'' = f(y, y', x) \text{ with } y(a) = A \text{ and } y(b) = B \quad \ldots(1)$$

We write the differential equation as $R = \bar{y}'' - f(\bar{y}, \bar{y}', x)$ ...(2)

where $R$ is the residual of the equation ($R = 0$ for the exact solution $y(x)$ only which will satisfy the boundary conditions).

Consider the trial solution as $\bar{y}(x) = c_1\, \phi_1(x) + c_2\, \phi_2(x) + \ldots$

where $\bar{y}(a) = A$ and $\bar{y}(b) = B$. The trial solution is differentiated twice and is substituted in (2).

To find $c_1, c_2, \ldots$, we weight the residual by trial functions $\phi_1(x), \phi_2(x), \ldots$ and set the integrals to zero. Thus we have $\int_c R\, \phi_1(x)\, dx = 0$, $\int_c R\, \phi_2(x)\, dx = 0, \ldots$

These lead to simultaneous equations in the unknowns.

Having found $c_1, c_2, \ldots$, the approximate solution $\bar{y}(x)$ is obtained.

**Example 35.17.** *Use Galerkin's method to solve the boundary value problem of Example 35.14. Compare your approximate solution with the exact solution.*

**Solution.** The residual is $R = \bar{y}'' - \bar{y} + x$ ...(*i*)

To find the trial solution which satisfies the boundary conditions, we derive a Lagrangian polynomial (§ 28.8) which passes through the points :

| $x$ | : | 0 | 1/2 | 1 |
|---|---|---|---|---|
| $y$ | : | 0 | $c$ | 0 |

The resulting polynomial is $\bar{y}(x) = 4cx(1-x)$, so that $\phi(x) = x(1-x)$.

Substituting $\bar{y}(x)$, $\bar{y}''(x)$ in (*i*), we get $R = 4cx^2 + (1-4c)x - 8c$

Thus $\int R\,\phi(x)\,dx = 0$ gives $\int_0^1 [4cx^2 + (1-4c)x - 8c]\,x(1-x)\,dx = 0$ whence $c = 5/88$.

Hence the approximate solution is $\bar{y}(x) = \dfrac{5}{22}x(1-x)$ which is same as found in Example 34.14.

*Exact solution.* Rewriting the given equation as $(D^2 - 1)y = -x$,

we find its solution as $y = c_1e^x + c_2e^{-x} + x$

Since $y(0) = 0$ and $y(1) = 0$, therefore $c_2 = -c_1 = 1/(e - e^{-1})$.

Hence the exact solution is $y = x - \dfrac{e^x - e^{-x}}{e - e^{-1}}$

The approximate and the exact solutions for some values of $x$ are given below for comparison :

| $x$ | *Approx. Sol.* | *Exact Sol.* |
|---|---|---|
| 0.25 | 0.043 | 0.035 |
| 0.50 | 0.057 | 0.057 |
| 0.75 | 0.043 | 0.05 |

**Obs.** To obtain a trial solution containing two unknown parameters, we derive a Lagrangian polynomial which passes through the points :

| $x$ | : | 0 | 1/3 | 2/3 | 1 |
|---|---|---|---|---|---|
| $y$ | : | 0 | $c_1$ | $c_2$ | 0 |

More the undetermined parameters, the more accurate is the solution, but it involves more labour to find their values.

**Example 35.18.** *Find the approximate deflection of a simply supported beam under a uniformly distributed load w Fig. 35.7, using Galerkin's method.*

**Solution.** The differential equation governing the deflection $y(x)$ of the beam is

$$EI\frac{d^4y}{dx^4} - w = 0,\ 0 < x < l \qquad (i)\ [\S\ 14.8]$$

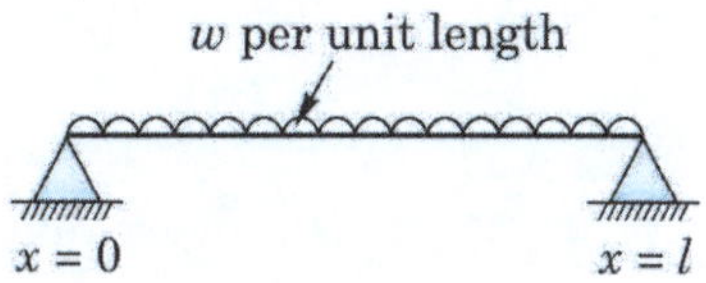


Fig. 35.7

The boundary conditions to be satisfied are

$$y(x=0) = y(x=l) = 0 \quad \text{(deflection is zero at ends)} \qquad ...(ii)$$

$$\frac{d^2y}{dx^2}(x=0) = \frac{d^2y}{dx^2}(x=l) = 0 \quad \text{(bending moment zero at ends)} \qquad ...(iii)$$

We assume the trial solution $\bar{y}(x) = c_1 \sin(\pi x/l) + c_2 \sin(3\pi x/l)$, which satisfies the boundary conditions Eqns. (*ii*) and (*iii*).

Substituting the trial solution in Eqn. (*i*), we obtain the residual

$$R = EIc_1(\pi/l)^4 \sin(\pi x/l) + EIc_2(3\pi/l)^4 \sin(3\pi x/l) - w$$

Thus $\int_0^l R\,.\,\sin(\pi x/l) = 0$ and $\int_0^l R\,.\,\sin(3\pi x/l) = 0$

Computing these integrals, we get

$$EIc_1(\pi/l)^4\, l/2 - w\,.\,2l/\pi = 0,\ EIc_2(3\pi/l)^4\, l/2 - w\,.\,2l/3\pi = 0$$

Solving these, we obtain $c_1 = \dfrac{4wl^4}{\pi^5 EI}$ and $c_2 = \dfrac{4wl^4}{243\pi^5 EI}$.

Hence the deflection of the beam is given by

$$\bar{y}(x) = \frac{4wl^4}{\pi^5 EI}\left\{\sin\left(\frac{\pi x}{l}\right) + \frac{1}{243}\sin\left(\frac{3\pi x}{l}\right)\right\}.$$

**Example 35.19 :** *Find the approximate deflection of a simply supported pendulum having mass m suspended in pendulum of length l. θ denote the angle from the vertical line and the arc length from the equilibrium point with S.* *(CSJM Kanpur Un., MSc. 2018)*

**Solution.** Given a mass suspended in pendulum of length $l$ and θ denote the angle from the vertical line and the arc length from the equilibrium point with $S$.

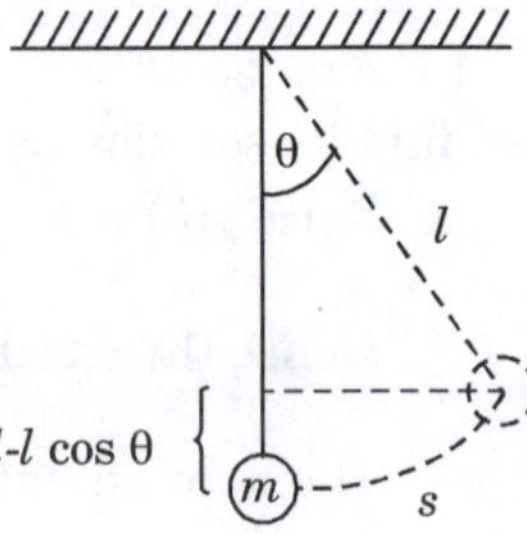


The arc length $S$ is connected to angle θ by

$$S = l\theta$$

*K.E.* $$T = \frac{1}{2} m \dot{S}^2 = \frac{1}{2} ml^2 \dot{\theta}^2$$

$$V = mg(l - l\cos\theta)$$

Now,

Hamilton's principal states that the motion will be such that the action integral

$$J(\theta) = \int_{t_0}^{t_1}\left(\frac{1}{2} ml\,\dot{\theta}^2 - mgl\,(1 - \cos\theta)\right)dt$$

has a stationary value. Lagrange' equation

$$L'\theta - \frac{d}{dt}L'\dot{\theta} = -mgl\sin\theta - ml\,\ddot{\theta} = 0$$

This implies that the equation of motion

$$\ddot{\theta} + \frac{g}{l}\sin\theta = 0$$

The Hamiltonian is in this case $$H = -L + \dot{\theta}\,L'\dot{\theta} = mgl(1 - \cos\theta) - \frac{ml^2\dot{\theta}^2}{2} + \dot{\theta}.\,ml^2\,\dot{\theta}$$

$$= mgl(1 - \cos\theta) + \frac{ml^2\dot{\theta}^2}{2} = V + T.$$

So, that is total energy of the system.

## PROBLEMS 35.4

1. Solve the boundary value problem :

$$y'' + y + x = 0\ (0 \le x \le 1),\ y(0) = y(1) = 0 \text{ by}$$

(*i*) Rayleigh-Ritz method, (*ii*) Galerkin's method. Compare your solution with the exact solution.

2. Using Galerkin's method, solve the boundary value problem $y'' = 3x + 4y$ ; $y(0) = 0$, $y(1) = 1$.
3. *Apply Galerkin's method to the boundary value problem* $y'' + y + x = 0\ (0 \le x \le 1)$ ; $y(0) = y(1) = 0$, *to find the coefficients of the approximate solution* $\bar{y}(x) = c_1 x(1 - x) + c_2 x^2(1 - x)$.

[***Hint.*** Substituting $\bar{y}(x)$, $\bar{y}''(x)$ in the given equation replacing $y$, $y''$ by $\bar{y}$, $y''$, we get the residual $R = -2c_1 + c_2(2 - 6x) + x(1 - x)(c_1 + c_2 x) + x$

Thus $$\int_0^1 R\,.\,x(1 - x)\,dx = 0 \text{ and } \int_0^1 R\,.\,x^2(1 - x)\,dx = 0.$$

Computing these integrals, we get

$$\frac{3}{10}c_1 + \frac{3}{20}c_2 = \frac{1}{12},\ \frac{3}{20}c_1 + \frac{13}{305}c_2 = \frac{1}{20}.$$

Solving these, we obtain $c_1 = 71/369$, $c_2 = 7/41$.]

4. Using Ritz method, find an approximate solution of the problem $y'' - y + 4xe^x = 0$, $y'(0) - y(0) = 1$, $y'(1) + y(1) = -e$.
5. Solve the boundary value problem : $y'' + (1 + x^2) y + 1 = 0$, $y(-1) = y(1) = 0$, by taking the approximate solution $\bar{y}(x) = c_1 (1 - x^2) + c_2 x^2 (1 - x^2)$ and using (*i*) Ritz method, (*ii*) Galerkin's method.
6. Given the boundary value problem : $y'' + \pi^2 y = x$, $y(0) = 1$, $y(1) = -0.9$.
   Use Galerkin's method to estimate $y(0.5)$, taking the trial solution :
   $$y = 1 - 1.9\,x + c_1 x(1 - x) + c_2 x^2 (1 - x).$$
7. Using Galerkin's method, obtain an approximate solution of the boundary value problem :
   $$\frac{d}{dx}\left(x\frac{dy}{dx}\right) + y = x,\ y(0) = 0,\ y(1) = 1,$$
   in the form $\bar{y}(x) = x + x(1 - x)(c_1 + c_2 x)$.
8. Of all the parabolas which pass through (0, 0) and (1, 1), determine the one, which when rotated about the $x$-axis, generates a solid of revolution with least possible volume between $x = 0$ and $x = 1$.
   [***Hint.*** Take the parabola as $y = x + cx(1 - x)$.]
9. *Using Rayleigh-Ritz method, find the potential at any point due to a charged sphere of radius a.*
   [***Hint.*** Potential at a distance $r$ from the centre of the sphere is $\phi = \phi_0 (r/a)^k$, where $\phi_0$ is the value of $\phi$ for $r = a$ and $k < 0$ so that $\phi \to 0$ as $r \to \infty$.
   Electrostatic field due to charged sphere being *conservative*, electrostatic intensity $E = \nabla\phi$.
   Also potential energy for unit volume $= \dfrac{1}{8\pi} E^2$
   $\therefore$ Total potential energy of the field in the entire region $R$ exterior to the given sphere is
   $$V = \frac{1}{8\pi}\iiint_R E^2\,dx\,dy\,dz = \frac{1}{8\pi}\iiint_R \left[\left(\frac{\partial\phi}{\partial x}\right)^2 + \left(\frac{\partial\phi}{\partial y}\right)^2 + \left(\frac{\partial\phi}{\partial z}\right)^2\right] dx\,dy\,dz$$
   $$= \frac{1}{8\pi}\int_a^\infty \int_0^\pi \int_0^{2\pi} \left(\frac{\partial\phi}{\partial r}\right)^2 r^2 \sin\theta\, d\phi\, d\theta\, dr.$$
   Electrostatic field will be in stable equilibrium if $V$ is minimum, *i.e.*, $dV/dp = 0$ and $d^2V/dp^2 > 0$.
   This gives $k = -1$. Hence $\phi = \phi_0\, a/r$.]

## 35.11 HAMILTON'S PRINCIPLE*

An important concept of mathematical physics is *Hamilton's principle* which states that *the time integral of the difference between the kinetic and potential energies of a dynamical system is stationary*

Consider a particle of mass $m$ moving from a fixed origin $O$ under the effect of a force $F$ (Fig. 35.8). At any time $t$, let its position vector be $\mathbf{R}$. Then by Newton's second law,

$$\frac{m d^2\mathbf{R}}{dt^2} = \mathbf{F} \qquad \text{...(1)}$$

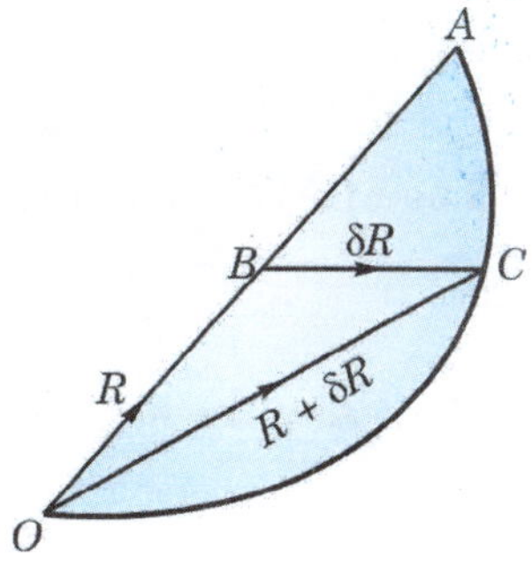

Fig. 35.8

Let the natural path $OBA$ of the particle be changed to another path $OCA$, end points remaining the same. Let this variation in path, often called *virtual displacement*, be $\delta\mathbf{R}$. Then the work done during this displacement is

$$\delta W = \mathbf{F} \,.\, \delta\mathbf{R} = \frac{m d^2\mathbf{R}}{dt^2} \,.\, \delta\mathbf{R} \qquad \text{[By (1)]}$$

Also the kinetic energy of the particle is

$$T = \frac{1}{2} m \left(\frac{d\mathbf{R}}{dt}\right)^2$$

* Named after the Irish mathematician *William Rowan Hamilton* (1805–1865) who is known for his contributions to dynamics.

$$\therefore \qquad \delta T = m\frac{d\mathbf{R}}{dt}\cdot\delta\left(\frac{d\mathbf{R}}{dt}\right) = m\frac{d\mathbf{R}}{dt}\cdot\frac{d}{dt}(\delta\mathbf{R})$$

Thus
$$\delta T + \delta W = m\frac{d\mathbf{R}}{dt}\cdot\frac{d}{dt}(\delta\mathbf{R}) + m\frac{d^2\mathbf{R}}{dt^2}\cdot\delta\mathbf{R} = m\frac{d}{dt}\left(\frac{d\mathbf{R}}{dt}\cdot\delta\mathbf{R}\right)$$

Integrating both sides w.r.t. $t$ from $t_0$ to $t_1$, we get

$$\int (\delta T + \delta W)\,dt = m\left|\frac{d\mathbf{R}}{dt}\cdot\delta\mathbf{R}\right|_{t_0}^{t_1} = 0 \qquad ...(2) \quad [\because \delta\mathbf{R} = 0 \text{ at } t_0 \text{ and } t_1]$$

If the force field is conservative, there exists a potential $V$ such that $W = -V$. Then Eqn. (2) takes the form

$$\int (\delta T - \delta V)\,dt = 0 \quad \text{or} \quad \delta\int (T - V)\,dt = 0 \quad i.e., \quad \int (T - V)\,dt \qquad ...(3)$$

is stationary. This proves the Hamilton's principle for a particle. Its derivation can be extended to a system of particles by summation and to a rigid body by integration. Hence the principle is true for any dynamical system.

**Obs.** It can be easily shown that *the integral Eqn. (3) is a minimum along the natural path* as compared to that along any other path joining $O$ to $A$.

**Def.** *The energy difference $T - V = L$ is called the* **kinetic potential** *or the* **Lagrangian function.**

## 35.12 LAGRANGE'S EQUATION

In a dynamical system, the position of a body can be specified by the quantities $q_1, q_2, \ldots q_n$ which are called the *generalised coordinates.*

The potential energy $V$, being a function of position only depends on these generalised coordinates $q_i$. The kinetic energy $T$, however, depends upon $q_i$ and the velocities $dq_i/dt$ (*i.e.*, $\dot{q}_i$) $i = 1, 2, \ldots n$. Therefore, Lagrangian function $L = T - V$ is also a function of $q_1$ and $q_i$, $i = 1, 2, \ldots n$.

Thus by Hamilton's principle, the system moves so that $\int_{t_0}^{t_1} L\,dt$ is stationary.

$\therefore$ Euler's equation must hold good, *i.e.*, $\dfrac{\partial L}{\partial q_i} - \dfrac{d}{dt}\left(\dfrac{\partial L}{\partial \dot{q}_i}\right) = 0$, $i = 1, 2, \ldots n$.

These are called *Lagrange's equations* which determine the motion of the system.

**Example 35.20.** *A mass, suspended at the end of a light spring having spring constant k, is set into vertical motion. Use Lagrange's equation, to find the equation of motion of the mass.*

**Solution.** At any time $t$, let the displacement of $m$ from the equilibrium position $O$ be $x$ (Fig. 35.9). Then the kinetic energy of $P$ is

$$T = \frac{1}{2}m\left(\frac{dx}{dt}\right)^2$$

Also the work done during its fall from $O$ to $P$ is

$$W = \int_0^x (mg - kx)\,dx = mgx - \tfrac{1}{2}kx^2.$$

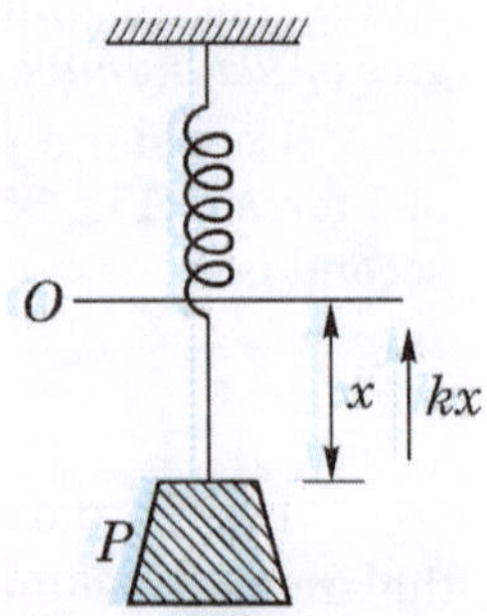

Fig. 35.9

If $V$ is the potential energy of the mass at $P$, then

$$V = -W = \tfrac{1}{2}kx^2 - mgx$$

$\therefore$ Lagrangian
$$L = T - V = \tfrac{1}{2}m\dot{x}^2 + mgx - \tfrac{1}{2}kx^2.$$

Thus the Lagrange's equation

$$\frac{\partial L}{\partial x} - \frac{d}{dt}\left(\frac{\partial L}{\partial \dot{x}}\right) = 0$$

becomes
$$(mg - kx) - \frac{d}{dt}(m\dot{x}) = 0 \quad \text{or} \quad m\frac{d^2x}{dt^2} = mg - kx$$

which is the required equation of motion.

**Example 35.21.** *Apply Lagrange's equations, to show that the equations of motion of the double pendulum of Fig. 35.10 are given by*

$$(m_1 + m_2)\, l_1 \ddot{\theta}_1 + m_2\, l_2 \ddot{\theta}_2 + (m_1 + m_2)\, g\theta_1 = 0$$

*and* $$l_1 \ddot{\theta}_1 + l_2 \ddot{\theta}_2 + g\theta_2 = 0$$

*for small angles* $\theta_1$, $\theta_2$.

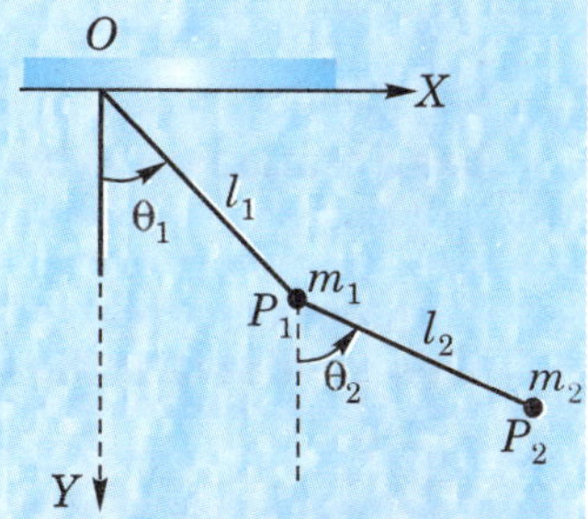

Fig. 35.10

**Solution.** At any time $t$, let the masses $m_1$, $m_2$ be at $P_1(x_1, y_1)$ and $P_2(x_2, y_2)$ (Fig. 35.10) so that

$$\left.\begin{aligned} x_1 &= l_1 \sin\theta_1,\ y_1 = l_1 \cos\theta_1 \\ x_2 &= l_1 \sin\theta_1 + l_2 \sin\theta_2,\ y_2 = l_1\cos\theta_1 + l_2 \cos\theta_2 \end{aligned}\right\} \quad ...(i)$$

Then total kinetic energy is

$$T = \tfrac{1}{2} m_1 (\dot{x}_1^{\,2} + \dot{y}_1^{\,2}) + \tfrac{1}{2} m_2 (\dot{x}_2^{\,2} + \dot{y}_2^{\,2})$$

$$= \tfrac{1}{2}(m_1 + m_2)\, l_1^2 \dot{\theta}_1^2 + \tfrac{1}{2} m_2 l_2^2 \dot{\theta}_2^2 + m_2 l_1 l_2 \dot{\theta}_1 \ddot{\theta}_2 \cos(\theta_1 - \theta_2) \qquad \text{[Using Eqn. } (i)]$$

Also total potential energy is

$$V = m_1 g l_1 (1 - \cos\theta_1) + m_2 g (l_1 + l_2 - l_1 \cos\theta_1 - l_2 \cos\theta_2)$$

$\therefore$ Lagrangian $$L = T - V = \tfrac{1}{2}(m_1 + m_2)\, l_1^2 \dot{\theta}_1^2 + \tfrac{1}{2} m_2 l_2^2 \dot{\theta}_2^2$$

$$+ m_2 l_1 l_2 \dot{\theta}_1 \dot{\theta}_2 \cos(\theta_1 - \theta_2) - (m_1 + m_2)\, g l_1 (1 - \cos\theta_1) - m_2 g l_2 (1 - \cos\theta_2) \quad ...(ii)$$

Thus the Lagrange's equation corresponding to $\theta_1$, is

$$\frac{\partial L}{\partial \theta_1} - \frac{d}{dt}\left(\frac{\partial L}{\partial \dot{\theta}_1}\right) = 0, \text{ becomes}$$

$$- m_2 l_1 l_2 \dot{\theta}_1 \dot{\theta}_2 \sin(\theta_1 - \theta_2) - (m_1 + m_2)\, g l_1 \sin\theta_1 - \frac{d}{dt}\left[(m_1 + m_2)\, l_1^2\, \dot{\theta}_1 + m_2 l_1 l_2 \dot{\theta}_2 \cos(\theta_1 + \theta_2)\right] = 0$$

or $$(m_1 + m_2)\, l_1 \ddot{\theta}_1 + m_2\, l_2 \ddot{\theta}_2 \cos(\theta_1 - \theta_2) + m_2 l_2 \dot{\theta}_2^{\,2} \sin(\theta_1 - \theta_2) + (m_1 + m_2)\, g \sin\theta_1 = 0 \quad ...(iii)$$

Similarly from Eqn. (*ii*), Lagrange's equation corresponding to $\theta_2$, *i.e.*,

$$\frac{\partial L}{\partial \theta_2} - \frac{d}{dt}\left(\frac{\partial L}{\partial \theta_2}\right) = 0$$

becomes $$m_2 l_1 l_2 \dot{\theta}_1 \dot{\theta}_2 \sin(\theta_1 - \theta_2) - m_2\, g l_2 \sin\theta_2 - \frac{d}{dt}\left[m_2 l_2^2\, \dot{\theta}_2 + m_2 l_1 l_2 \dot{\theta}_1 \cos(\theta_1 - \theta_2)\right] = 0$$

or $$l_2 \ddot{\theta}_2 + l_1 \ddot{\theta}_1 \cos(\theta_1 - \theta_2) - l_1 \dot{\theta}_1^2 \sin(\theta_1 - \theta_2) + g \sin\theta_2 = 0 \quad ...(iv)$$

Now $\dot{\theta}_1$ and $\dot{\theta}_2$ being small, retaining first order terms only, Eqns. (*iii*) and (*iv*) reduce to

$$(m_1 + m_2)\, l_1 \ddot{\theta}_1 + m_2\, l_2 \ddot{\theta}_2 + (m_1 + m_2)\, g\theta_1 = 0 \quad \text{and} \quad l_1 \ddot{\theta}_1 + l_2 \ddot{\theta}_2 + g\theta_2 = 0.$$

**Example 35.22:** *Find the approximate deflection of a simply supported pendulum having mass m suspended in pendulum of length l. θ denote the angle from the vertical line and the arc length from the equilibrium point with S.* (*CSJM Kanpur Un., MSc. 2018*)

**Solution.** Given a mass suspended in pendulum of length $l$ and $\theta$ denote the angle from the vertical line and the arc length from the equilibrium point with $S$.

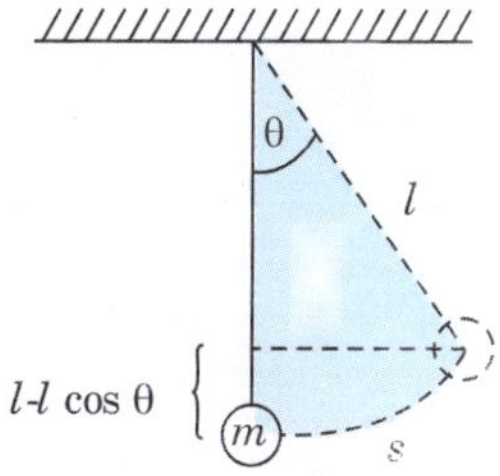

Fig. 35.11.

The arc length $S$ is connected to angle $\theta$ by

$$S = l\theta$$

*K.E.* $$T = \frac{1}{2} m\, S^2 = \frac{1}{2} m l^2\, \theta^2$$

$$V = mg(l - l\cos\theta)$$

Now,

Hamilton's principal states that the motion will be such that the action integral

$$J(\theta) = \int_{t_0}^{t_1} \left( \frac{1}{2} ml\, \theta^2 - mgl\,(1 - \cos\theta) \right) dt$$

has a stationary value. Lagrange' equation

$$L'\theta - \frac{d}{dt} L'\theta = -mgl \sin\theta - ml\,\ddot{\theta} = 0$$

This implies that the equation of motion

$$\ddot{\theta} + \frac{g}{l} \sin\theta = 0$$

The Hamiltonian is in this case

$$H = -L + \dot{\theta}\, L'\theta = mgl(1 - \cos\theta) - \frac{ml^2\dot{\theta}^2}{2} + \dot{\theta}.\, ml^2\, \dot{\theta}$$

$$= mgl(1 - \cos\theta) + \frac{ml^2\theta^2}{2} = V + T.$$

So, that is total energy of the system.

## PROBLEMS 35.5

1. In a single pendulum, a mass $m$ is suspended by a light rod of length $l$ and the system vibrates in a plane. Using Lagrange's equation, show that $\ddot{\theta} + (g/l) \sin\theta = 0$.

   Show that if $\theta$ is small, the period of oscillation is $2\pi\sqrt{(l/g)}$.
2. Two masses $m_1$ and $m_2$ are connected by an inextensible string which passes over a fixed pulley. Using Lagrange's equations, show that the acceleration of either mass is numerically
$$= (m_1 - m_2)\, g/(m_1 + m_2).$$
3. A perfectly flexible rope of uniform density per unit length is suspended with its end points fixed. Show that it assumes the shape of a catenary.
4. A bead of mass $m$ from rest slides without friction under gravity along a wire inclined at an angle $\alpha$ to the vertical and rotating with constant angular velocity $\omega$. Show that in times $t$, the bead has slided through a distance
$$\frac{g \cos\alpha}{\omega^2 \sin^2\alpha} \cosh\,(wt \sin\alpha - 1).$$

CHAPTER 36

# Integral Equations



## 36.1 INTRODUCTION

Integral equations play an effective role in the study of boundary value problems. Such equations also occur in many fields of mechanics and mathematical physics. Integral equations may be obtained directly from physical problems *e.g.*, radiation transfer problem and neutron diffusion problem etc. They also arise as representation formulae for the solutions of differential equations. A differential equation can be replaced by an integral equation with the help of initial and boundary conditions.

Integral equations were first encountered in the theory of Fourier integrals. In 1826, another integral equation was obtained by *Abel*. Actual development of the theory of integral equations began with the works of the Italian mathematician V. Volterra (1896) and the Swedish mathematician I. Fredholm (1900).

## 36.2 DEFINITION

*An integral equation is an equation in which an unknown function appears under the integral sign.* We shall take up integral equations in which only linear functions of the unknown function are involved. The general type of *linear integral equation* is of the form

$$y(x) = F(x) + \lambda \int_a^b K(x, t)\, y(t)\, dt$$

where $F(x)$ and $K(x, t)$ are known functions while $y(x)$ is to be determined. The function $K(x, t)$ is called the *Kernel of the integral equation.*

If $a$ and $b$ are constants, the equation is known as *Fredholm integral equation.*

If $a$ is a constant while $b$ is a variable, it is called a *Volterra integral equation.*

## 36.3 CONVERSION OF A LINEAR DIFFERENTIAL EQUATION TO AN INTEGRAL EQUATION AND VICE VERSA

*To make this transformation, the use of the following formula is necessary :*

$$\int_a^x \int_a^x f(x)\, dx\, dx = \int_a^x (x - t)\, f(t)\, dt \qquad ...(I)$$

In general, $$\int_a^x \int_a^x \ldots \int_a^x f(x)\, dx^n = \frac{1}{(n-1)!} \int_a^x (x - t)^{n-1}\, f(t)\, dt \qquad ...(II)$$

**Proof.** Let $$I_n = \int_a^x (x-t)^{n-1} f(t)\, dt \qquad ...(1)$$

where $n$ is a positive integer and $a$ is a constant.

Differentiating both sides w.r.t. $x$, using Leibnitz's rule (p. 1253)

$$\frac{dI_n}{dx} = \int_a^x \frac{\partial}{\partial x}(x-t)^{n-1} f(t)\, dt + [(x-t)^{n-1} f(t)]_{t=x} \cdot 1 - [(x-t)^{n-1} f(t)]_{t=a} \cdot 0$$

$$= \int_a^x (n-1)(x-t)^{n-2} f(t)\, dt = (n-1) I_{n-1}(x) \qquad ...(2)$$

Again differentiating (2) w.r.t. $x$

$$\frac{d^2 I_n}{dx^2} = (n-1)\frac{d}{dx}[I_{n-1}, (x)] = (n-1)(n-2) I_{n-2}, \text{ using (1)}$$

Proceeding in this way, we get

$$\frac{d^{n-1} I_n}{dx^{n-1}} = (n-1)(n-2) \ldots 1 \,.\, I_1(x) = (n-1)!\, I_1(x)$$

Now taking $n = 1$ in Eqn. (1), we get

$$I_1 = \int_a^x f(t)\, dt = \int_a^x f(x_1)\, dx_1 \qquad ...(3)$$

Putting $x = a$ in Eqn. (1), we obtain

$$I_n(a) = 0 \text{ for all } n$$

Taking $n = 2$ in Eqn. (2), we get $\frac{dI_2}{dx} = I_1(x)$

$$\therefore \quad I_2 = \int_a^x I_1(x_2)\, dx_2 \qquad [\because \ I_2(a) = 0]$$

$$= \int_a^x \int_a^{x_2} f(x_1)\, dx_1\, dx_2 \qquad \text{[Using (3)]} ...(4)$$

Putting $n = 3$ in Eqn. (2), we have $\frac{dI_3}{dx} = 2I_2(x)$

$$\therefore \quad I_3 = 2\int_a^x I_2(x)\, dx \qquad [\because \ I_3(a) = 0]$$

$$= 2\int_a^x \int_a^{x_3} \int_a^{x_2} f(x_1) dx_1\, dx_2\, dx_3 \qquad \text{[Using (4)]}$$

Proceeding in this way, we get

$$I_n = (n-1)! \int_a^x \int_a^{x_n} \cdots \int_a^{x_2} F(x_1)\, dx_1\, dx_2 \ldots dx_n$$

*i.e.,* $$\int_a^x \int_a^{x_n} \cdots \int_a^{x_2} f(x_1)\, dx_1\, dx_2 \ldots dx_n = \frac{1}{(n-1)!}\int_a^x (x-t)^{n-1} f(t)\, dt$$

If $x_2, x_3, \ldots x_n$ be the same as $x$, we get the formula (II) above.

**Example 36.1.** *Convert the differential equation $y''(x) - 3y'(x) + 2y(x) = 5 \sin x$, $y(0) = 1$, $y'(0) = -2$ into an integral equation.*

**Solution.** Integrating both sides of the given differential equation, we get

$$[y'(x) - y'(0)] - 3[y(x) - y(0)] + 2\int_0^x y(t)\, dt = 5(1 - \cos x)$$

Since $y'(0) = -2$ and $y(0) = 1$, it becomes

$$y'(x) + 2 - 3y(x) + 3 + 2\int_0^x y(t)\, dt = 5 - 5\cos x$$

*i.e.,* $$y'(x) - 3y(x) + 2\int_0^x y(t)\,dt = -5\cos x$$

Integrating again as before, we have

$$[y(x) - y(0)] - 3\int_0^x y(t)\,dt + 2\int_0^x\int_0^x y(t)\,dt = -5\sin x$$

or $$y(x) - 1 - 3\int_0^x y(t)\,dt + 2\int_0^x (x-t)\,y(t)\,dt = -5\sin x \qquad \text{[Using (1) p. 1274]}$$

or $$y(x) + \int_0^x [2(x-t) - 3]\,y(t)\,dt = 1 - 5\sin x$$

which is the desired integral equation.

**Example 36.2.** *Show that the integral equation*

$$y(x) = \int_0^x (x+t)\,y(t)\,dt + 1 \qquad \ldots(i)$$

*is equivalent to the differential equation*

$$y''(x) - 2x\,y'(x) - 3y(x) = 0,\ y(0) = 1,\ y'(0) = 0.$$

**Solution.** Differentiating (*i*) by Leibnitz's rule, we have

$$\frac{dy}{dx} = \int_0^x \frac{\partial}{\partial x}(x+t)\,y(t)\,dt + (x+x)\,y(x)\frac{d}{dx}(x) - (x+0)\,y(0)\frac{d}{dx}(0)$$

$$= \int_0^x y(t)\,dt + 2xy(x) = \int_0^x y(x)\,dx + 2xy(x) \qquad \ldots(ii)$$

Differentiating again w.r.t. $x$, we get

$$\frac{d^2y}{dx^2} = y(x) + 2[xy'(x) + 1\,.\,y(x)] = 2xy'(x) + 3y(x)$$

or $$\frac{d^2y}{dx^2} - 2x\frac{dy}{dx} - 3y(x) = 0 \qquad \ldots(iii)$$

Putting $x = 0$ in Eqn. (*i*), we obtain

$$y(0) = \int_0^0 (x+t)\,y(t)\,dt + 1 \quad i.e., \quad y(0) = 1$$

and putting $x = 0$ in Eqn. (*ii*), we get $y'(0) = 0$.

Hence Eqn. (*i*) is equivalent to Eqn. (*iii*) with initial conditions $y(0) = 1$, $y'(0) = 0$.

**Example 36.3.** *Show that the integral equation*

$$y(x) = \int_0^x t(t-x)\,y(t)\,dt + \frac{1}{2}x^2, \qquad \ldots(i)$$

*is equivalent to the differential equation*

$$\frac{d^2y}{dx^2} + xy = 1 \text{ and the conditions } y(0) = y'(0) = 0.$$

**Solution.** Differentiating Eqn. (*i*) w.r.t. $x$ by Leibnitz's rule, we have

$$\frac{dy}{dx} = \int_0^x \frac{\partial}{\partial x}[t(t-x)]\,y(t)\,dt + [t(t-x)\,y(t)]_{t=x}\cdot 1 + x - [t(t-x)\,y(t)]_{t=0}\cdot 0$$

$$= \int_0^x t(-1)\,y(t)\,dt + x = x - \int_0^x ty(t)\,dt \qquad \ldots(ii)$$

Differentiating Eqn. (*ii*) w.r.t. $x$, we get

$$\frac{d^2y}{dx^2} = 1 - \left\{\int_0^x \frac{\partial}{\partial x}[t\,y(t)]\,dt - [t\,y(t)]_{t=0}\cdot 0 + [ty(t)]_{t=x}\cdot 1\right\} = 1 - xy(x)$$

or $$\frac{d^2y}{dx^2} + xy = 1$$ which is the differential equation corresponding to Eqn. (*i*).

Also $y(0) = 0$ and $y'(0) = 0$.

**Example 36.4.** *Find the integral equation corresponding to the boundary value problem*

$$y''(x) + \lambda y(x) = 0, \qquad y(0) = y(1) = 0.$$

**Solution.** Integrating both sides of the given differential equation w.r.t. $x$ over $(0, x)$, we get

$$y'(x) - y'(0) + \lambda \int_0^x y(x)\, dx = 0$$

or $$y'(x) = c - \lambda \int_0^x y(x)\, dx, \text{ taking } y'(0) = c \qquad ...(i)$$

Again integrating Eqn. (*i*) w.r.t. $x$ in $(0, x)$, we obtain

$$y(x) - y(0) = cx - \lambda \int_0^x \int_0^x y(x)\, dx$$

$$y(x) = cx - \lambda \int_0^x (x - t)\, y(t)\, dt \qquad ...(ii)$$

[Using (I) of § 36.3 and noting that $y(0) = 0$]

Putting $x = 1$ in Eqn. (*ii*), we get

$$y(1) = c - \lambda \int_0^1 (1 - t)\, y(t)\, dt \qquad [\because \; y(1) = 0]$$

$$\therefore \qquad c = \lambda \int_0^1 (1 - t)\, y(t)\, dt \qquad ...(iii)$$

Substituting the value of $c$ from Eqns. (*iii*) in (*ii*), we get

$$y(x) = \lambda x \int_0^1 (1 - t)\, y(t)\, dt - \lambda \int_0^x (x - t)\, y(t)\, dt$$

$$= \lambda x \left\{ \int_0^x (1 - t)\, y(t)\, dt + \int_x^1 (1 - t)\, y(t)\, dt \right\} - \lambda \int_0^x (x - t)\, y(t)\, dt$$

$$= \lambda \int_0^x t(1 - x)\, y(t)\, dt + \lambda \int_x^1 (1 - t)\, y(t)\, dt$$

$$= \lambda \left\{ \int_0^x K(x, t)\, y(t)\, dt + \int_x^1 K(x, t)\, y(t)\, dt \right\}$$

where $$K(x, t) = \begin{cases} t(1 - x) & \text{when } t < x \\ x(1 - t) & \text{when } t > x \end{cases}$$

Hence $$y(x) = \lambda \int_0^1 K(x, t)\, y(t)\, dt$$

which is a *Fredholm integral equation* with a symmetric kernel $K(x, t)$.

## PROBLEMS 36.1

Transform each of the following boundary value problems into corresponding integral equations :

1. $y'' + xy' + y = 0$, given that $y(0) = 1, y'(0) = 0$.
2. $y'' - \sin xy' + e^y = x$, given that $y = 1$, $\frac{dy}{dx} = -1$ when $x = 0$.
3. $y'' + xy = 1$, given that $y(0) = y'(0) = 0$.
4. $y'' + (1 - x)\, y' + e^{-x} y = x^3 - 5x$, given that $y = -3$, $\frac{dy}{dx} = 4$ when $x = 0$.
5. $\frac{d^3y}{dx^3} + y = \cos x$ given that $y = 0, y' = 1, y'' = 2$ at $x = 0$.

6. $\frac{d^3y}{dx^3} + \frac{d^2y}{dx^2} - xy = \sin x$ given that $y = 1, y' = -1, y'' = \frac{1}{2}$ at $x = 0$.

7. $\frac{d^4y}{dx^4} - 4\frac{d^3y}{dx^3} + 6\frac{d^2y}{dx^2} - 4\frac{dy}{dx} + y = 4\cos 2x$ given that
$y = -1, y' = 4, y'' = 0, y''' = 2$ when $x = 0$.

Convert each of the following integral equations into differential equations alongwith initial conditions :

8. $y(x) = \int_0^x (x + t)\, y(t)\, dt - 1$.

9. $y(x) = \int_0^x (x - t)\, y(t)\, dt + 3\sin x$.

10. $y(x) + 3\int_0^x (x - t)^2\, y(t)\, dt = x^2 - 3x + 4$.

11. $y(x) + \int_0^x (x - t)^2 + 4(x - t) - 3\}\, y(t)\, dt = e^{-x}$.

12. If $y''(x) = f(x)$ ; $y(0) = y(l) = 0$, show that $y(x) = \int_0^l K(x, t)\, f(t)\, dt$,

where $K(x, t) = \begin{cases} \frac{t}{l}(x - l) & \text{when } t < x \\ \frac{x}{l}(t - l) & \text{when } t > x \end{cases}$

## 36.4 CONVERSION OF BOUNDARY VALUE PROBLEMS TO INTEGRAL EQUATIONS USING GREEN'S FUNCTION

Consider the linear homogeneous differential equation

$$L(y) + \phi(x) = 0 \quad ...(1)$$

where $$L(y) = \left[\frac{d}{dx}\left(p\frac{d}{dx}\right) + q\right] y = py'' + p'y' + qy \quad ...(2)$$

together with the homogeneous boundary conditions of the form

$$\alpha y + \beta \frac{dy}{dx} = 0 \quad ...(3)$$

Now let us find a function $G(x, t)$ which for a given number $t$, is given by $G_1(x)$ when $x < t$ and by $G_2(x)$ when $x > t$ and which has the following properties :

I. $G_1$ and $G_2$ satisfy the equation $L(G) = 0$ in their defined intervals *i.e.*, $L(G_1) = 0$ when $x < t$ ; $L(G_2) = 0$ when $x > t$.

II. $G_1$ and $G_2$ satisfy the boundary conditions at the end points $x = a$ and $x = b$ respectively.

III. $G(x, t)$ is continuous at $x = t$ *i.e.*, $G_1(t) = G_2(t)$.

IV. The derivative of $G$ is continuous at every point within the range of $x$ except at $x = t$ *i.e.*, $G_2'(t) - G_1'(t) = -1/p(t)$

**Def.** *$G(x, t)$ as defined above is called the* **Green's function.**

*If $G(x, t)$ exists, then the solution of the given problem can be transformed to the integral equation*

$$y(x) = \int_a^b G(x, t)\, \phi(t)\, dt \quad ...(4)$$

Let $y = y_1(x)$ and $y = y_2(x)$ be the non-trivial solutions of the equations $L(y) = 0$ which satisfy the homogeneous conditions at $x = a$ and $x = b$ respectively.

The above conditions *I* and *II* are satisfied if we write

$$G = \begin{cases} C_1 y_1(x), & \text{when } x < t \\ C_2 y_2(x), & \text{when } x > t \end{cases} \quad ...(5)$$

Imposing the condition III on (5), we get

$$C_2 y_2(t) - C_1 y_1(t) = 0 \quad ...(6)$$

Imposing the condition IV on (5), we have

$$C_2 y_2'(t) - C_1 y_1'(t) = -1/p(t) \quad ...(7)$$

Equations (6) and (7) give a unique solution, if

$$\begin{vmatrix} y_1(t) & y_2(t) \\ y_1'(t) & y_2'(t) \end{vmatrix} = y_1(t)\, y_2'(t) - y_2(t)\, y_1'(t) \neq 0$$

By Abel's formula*, we find that

$$y_1(t)\, y_2'(t) - y_2(t)\, y_1'(t) = C/p(t) \qquad \text{...(8)}$$

Now $[(7) \times y_2(t) - (6) \times y_1(t)]$ gives on using (8),

$$C_1 = -\frac{1}{C}\, y_2(t) \text{ and } C_2 = -\frac{1}{C} y_1(t)$$

Thus Eqn. (5) reduces to
$$G(x, t) = \begin{cases} -\dfrac{1}{C}\, y_1(x)\, y_2(t), & x < t \\ -\dfrac{1}{C}\, y_1(t)\, y_2(x), & x > t \end{cases} \qquad \text{...(9)}$$

*Conversely it can be shown that the integral equation.*

$$y(x) = \int_a^b G(x, t)\, \phi(t)\, dt$$

*where G(x, t) is as defined by Eqn. (9), satisfies the differential equation L(y) + φ(x) = 0 together with the prescribed boundary conditions.*

**Example 36.5.** *Transform the differential equation y″ + y = x, y (0) = 1, y′ (1) = 0 to a Fredholm integral equation, finding the corresponding Green's function.*

**Solution.** Given equation is $L(y) + \phi(x) = 0$ ...(i)

with the conditions $y(0) = 0$, $y'(1) = 0$, where $L(y) = y''$ and $\phi(x) = y - x$

The associated equation $L(y) = 0$ is $y''(y) = 0$ ...(ii)

Its solution is $y = C_1 x + C_2$ ...(iii)

Now $y_1(x)$ is a particular solution of (*ii*) satisfying the condition $y(0) = 0$.

∴ Taking $C_2 = 0$ and $C_1 = 1$, we get $y_1(x) = x$ ...(iv)

$y_2(x)$ is another solution of (*ii*) satisfying the condition $y'(1) = 0$.

∴ From (*iii*), $y'(1) = C_1 = 0$. Then $y(x) = C_2$

Taking $C_2 = 1$, a particular solution is $y_2(x) = 1$.

The constant $C$ is found from $y_1(x)\, y_2'(x) - y_2(x) y_1'(x) = \dfrac{C}{p(x)}$

Since $L(y) = py'' + p'y' + qy = y''$ (given), ∴ $p = 1$.

Thus $C = y_1(x)\, y_2'(x) - y_2(x)\, y_1'(x) = x \,.\, 0 - 1.1 = -1$.

∴ Green's function is given by

$$G(x, t) = \begin{cases} \dfrac{-y_1(x)\, y_2(t)}{C}, & x < t \\ \dfrac{-y_1(t)\, y_2(x)}{C}, & x > t \end{cases} = \begin{cases} x, & x < t \\ t, & x > t \end{cases} \qquad \text{...(v)}$$

Hence the equation (*i*) is equivalent to the integral equation

$$y(x) = \int_0^1 G(x, t)\, \phi(t)\, dt = \int_0^1 G(x, t) \,.\, (y - t)\, dt$$

$$= \int_0^1 G(x, t)\, y(t)\, dt - \left\{ \int_0^x x \,.\, t\, dt + \int_x^1 t \,.\, t\, dt \right\}$$

---

* The conditions that $y_1(x)$ and $y_2(x)$ satisfy the equation

$L(y) = 0$ are $(p\, y_1')' + qy_1 = 0$ ...(i), $(py_2')' + qy_2 = 0$ ...(ii)

$[(i) \times y_2 - (ii) \times y_1]$ gives $y_2(py_1')' - y_1(py'_2)' = 0$

or $[p(y_1 y_2' - y_2 y_1')]' = 0$ or $y_1 y_2' - y_2 y_1' = C/p$

$$= \int_0^1 G(x,t)\, y(t)\, dt - \left\{ x \left| \frac{t^2}{2} \right|_0^x + \left| \frac{t^3}{3} \right|_x^1 \right\}$$

$$= \int_0^1 G(x,t)\, y(t)\, dt - \frac{1}{6}(x^3 + 2)$$

where $G(x, t)$ is given by Eqn. ($v$).

**Example 36.6.** *Find the Green's function for the boundary value problem*

$$d^2y/dx^2 + \mu^2 x = 0, \qquad y(0) = 0 = y(1).$$

**Solution.** We observe that the solution $y_1 = \sin \mu x$ satisfies the boundary condition $y(0) = 0$ and the solution $y_2 = \sin \mu(x - 1)$ satisfies the second condition $y(1) = 0$. Also both these solutions are linearly independent.

The constant $C$ is found from $y_1 y_2' - y_2 y_1' = C/p\,(x)$.

Since $L(y) = py'' + p'y' + qy = y''$ (given), $\quad \therefore \quad p = 1$

$\therefore \quad C = y_1 y_2' - y_2 y_1' = \mu \sin \mu x \cos \mu(x - 1) - \mu \sin \mu(x - 1) \cos \mu x = \mu \sin \mu$

Hence the Green's function is given by

$$G(x,t) = \begin{cases} -\dfrac{y_1(x)\, y_2(t)}{C}, & x < t \\ -\dfrac{y_1(t)\, y_2(x)}{C}, & x > t \end{cases} = \begin{cases} -\dfrac{\sin \mu x \sin \mu(t-1)}{\mu \sin \mu}, & x < t \\ -\dfrac{\sin \mu t \sin \mu(x-1)}{\mu \sin \mu}, & x > t \end{cases}$$

## PROBLEMS 36.2

1. Transform the problem $d^2y/dx^2 + xy = 1$ ; $y(0) = 0 = y(1)$ to an integral equation, finding the corresponding Green's function.
2. Transform the problem $y'' + y = x$ ; $y(0) = 1$, $y'(1) = 0$ to an integral equation using Green's function.
3. Construct an integral equation corresponding to the boundary value problem

$$\frac{d^2u}{dx^2} + e^u\, u = x \;;\; u(0) = 0,\; u(1) = 1.$$

4. Find the Green's function for the boundary value problem $d^2y/dx^2 - y = 0$ with $y(0) = y(1) = 0$.
5. Transform the boundary value problem $x^2 \dfrac{d^2u}{dx^2} + x\dfrac{du}{dx} + (\lambda x^2 - 1)\, u = 0$ ; $u(0) = u(1) = 0$, to an integral equation.

[**Hint.** $u_1(x) = x$, $u_2(x) = \dfrac{1}{x} - x$ and $C = -2$]

## 36.5 SOLUTION OF AN INTEGRAL EQUATION

*The solution of the integral equation* $y(x) = F(x) + \lambda \int_a^b K(x, t)\, y(t)\, dt$ *is a function* $y(x)$ *which when substituted in the equation reduces it to an identity w.r.t.* $x$.

**Example 36.7.** *Show that* $y(x) = 2 - x$ *is a solution of the integral equation*

$$\int_0^x e^{x-t}\, y(t)\, dt = e^x + x - 1. \qquad \text{...}(i)$$

**Solution.** Since $\quad y(t) = 2 - t$

$$\therefore \quad \int_0^x e^{x-t}\, y(t)\, dt = \int_0^x e^{x-t}\, (2 - t)\, dt$$

$$= 2e^x \int_0^x e^{-t}\, dt - e^x \int_0^x t\, e^{-t}\, dt$$

$$= 2e^x \Big| -e^{-t} \Big|_0^x - e^x \left\{ \Big| t.(-e^{-t}) \Big|_0^x - \int_0^x 1.(-e^{-t})\, dt \right\}$$

$$= 2e^x(-e^{-x}+1) + e^x(xe^{-x}) - e^x \Big| -e^{-t} \Big|_0^x$$

$$= -2 + 2e^x + x + e^x(e^{-x} - 1) = e^x + x - 1.$$

Thus $(i)$ is identically satisfied by $y(x) = 2 - x$. Hence $y(x) = 2 - x$ is a solution of Eqn. $(i)$.

**Example 36.8.** *Show that the function $y(x) = (1 + x^2)^{-3/2}$ is a solution of the Volterra integral equation :*

$$y(x) = \frac{1}{1+x^2} - \int_0^x \frac{t}{1+x^2}\, y(t)\, dt. \qquad ...(i)$$

**Solution.** Substituting $y(x) = (1 + x^2)^{-3/2}$ in the R.H.S. of Eqn. $(i)$, we have

$$\frac{1}{1+x^2} - \int_0^x \frac{t}{1+x^2} \cdot \frac{1}{(1+t^2)^{3/2}}\, dt$$

$$= \frac{1}{1+x^2} + \frac{1}{1+x^2} \left\{ \frac{1}{(1+t^2)^{1/2}} \right\}_0^x$$

$$= \frac{1}{1+x^2} + \frac{1}{1+x^2} \cdot \frac{1}{(1+x^2)^{1/2}} - \frac{1}{1+x^2} = \frac{1}{(1+x^2)^{3/2}} = y(x)$$

Thus $y(x) = (1 + x^2)^{-3/2}$ is a solution of the integral equation $(i)$.

**Example 36.9.** *Show that the function $y(x) = xe^x$ is a solution of the integral equation*

$$y(x) = \sin x + 2\int_0^x \cos(x - t)\, y(t)\, dt. \qquad ...(i)$$

**Solution.** Substituting $y(x) = xe^x$ in the R.H.S. of Eqn. $(i)$, we have

$$\sin x + 2\int_0^x \cos(x-t)\,.\, te^t\, dt$$

$$= \sin x + 2\left\{ \cos x \int_0^x t\,.\,e^t \cos t\, dt + \sin x \int_0^x te^t \sin t\, dt \right\} \qquad \text{(Integrating by parts)}$$

$$= \sin x + 2\cos x \left\{ \frac{1}{2} \Big| t\,.\,e^t(\cos t + \sin t) \Big|_0^x - \frac{1}{2}\int_0^x e^t(\cos t + \sin t)\, dt \right\}$$

$$+ 2\sin x \left\{ \frac{1}{2} \Big| t\,.\,e^t(\sin t - \cos t) \Big|_0^x - \frac{1}{2}\int_0^x e^t(\sin t - \cos t)\, dt \right\}$$

$$= \sin x + xe^x(\cos^2 x + \cos x \sin x + \sin^2 x - \sin x \cos x)$$

$$- \cos x \int_0^x e^t(\cos t + \sin t)\, dt - \sin x \int_0^x t(\sin t - \cos t)\, dt$$

$$= \sin x + xe^x - \cos x \Big| e^t \sin t \Big|_0^x + \sin x \Big| e^t \cos t \Big|_0^x$$

$$= \sin x + xe^x - e^x \cos x \sin x + e^t \sin x \cos x - \sin x = xe^x$$

Thus $y(x) = xe^x$ is a solution of the integral equation $(i)$.

## 36.6 INTEGRAL EQUATIONS OF THE CONVOLUTION TYPE

$$y(x) = F(x) + \int_0^x K(x - t)\, y(t)\, dt$$

is an *integral equation of convolution type* and can be written as

$$y(x) = F(x) + K(x) * y(x)$$

It is a special integral equation of importance in applications.

Taking Laplace transform of both sides, assuming that $L\,F(x) = f(s)$ and $L[K(x)] = k(s)$ both exist, and using convolution theorem

$$\bar{y}(s) = f(s) + k(s) \,.\, y(s) \quad \text{or} \quad \bar{y}(s) = \frac{f(s)}{1 - k(s)}$$

Now taking the inverse transform of both sides, we get the required solution.

**Example 36.10.** *Solve the integral equation*

$$y(x) = 3x^2 + \int_0^x y(t) \sin(x - t)\, dt.$$

**Solution.** Given integral equation can be written as

$$y(x) = 3x^2 + y(x) * \sin(x - t) \qquad \text{...}(i)$$

Taking Laplace transform of both sides and using the convolution theorem, we get

$$\bar{y} = \frac{6}{s^3} + y \cdot \frac{s}{s^2 + 1} \quad \text{or} \quad \bar{y} = \frac{6(s^2 + 1)}{s^5} = 6\left(\frac{1}{s^3} + \frac{1}{s^5}\right)$$

On inversion, we get $\quad y = 6\left(\dfrac{x^2}{2!} + \dfrac{x^4}{4!}\right) = 3x^2 + x^4/4$

which is the required solution of Eqn. (*i*).

**Obs.** The above solution can also be verified by direct substitution in the given integral equation.

**Example 36.11.** *Solve* $y(t) = 2e^{-t} - 2\int_0^t y(u) \cos(t - u)\, du.$ *(Rohtak, 2011)*

**Solution.** The given equation can be written as

$$y(t) = 2e^{-t} - 2 \cos(t)\, y(u)$$

Taking Laplace transform of both sides and using convolution theorem, we get

$$\bar{y}(s) = \frac{2}{s + 1} - \frac{2s}{s^2 + 1}\bar{y}(s) \quad \text{or} \quad \bar{y}\left(1 + \frac{2s}{s^2 + 1}\right) = \frac{2}{s + 1}$$

or

$$\bar{y} = \frac{s^2 + 1}{(s + 1)^3} = \frac{1}{s + 1} - \frac{2}{(s + 1)^2} + \frac{2}{(s + 1)^3}$$

On inversion, we obtain $\quad y = e^{-t} - 2e^{-t} \,.\, t + \dfrac{2e^{-t} \,.\, t^2}{2!} = e^{-t}(t - 1)^2$

Hence $y = e^{-t}(t - 1)^2$ is the desired solution.

**Example 36.12.** *Solve the integral equation* $\int_0^x y(t)\, y(x - t)\, dt = 4 \sin 9x.$

**Solution.** The given integral equation can be written as

$$y(x) * y(x) = 4 \sin 9x \qquad \text{...}(i)$$

Taking Laplace transform of both sides and using the convolution theorem, we get

$$\{\bar{y}(s)\}^2 = \frac{36}{s^2 + 81} \quad \text{or} \quad \bar{y} = \pm \frac{6}{\sqrt{s^2 + 81}}$$

On inversion and noting that $L^{-1} \dfrac{1}{\sqrt{(s^2 + a^2)}} = J_0(ax)$, we get

$$y = \pm 6\, L^{-1}\left\{\frac{1}{\sqrt{(s^2 + 9^2)}}\right\} = \pm 6\, J_0(9x)$$

Thus $y = 6J_0(9x)$ and $y = -6\, J_0(9x)$ are both solutions of Eqn. (*i*).

## 36.7 ABEL'S INTEGRAL EQUATION

The integral equation $\int_0^x \frac{y(t)}{(x-t)^\alpha} dx = G(x)$

such that $G(x)$ is given and $\alpha$ is a constant $(0 < \alpha < 1)$, is known as *Abel's integral equation*. This is an important integral equation of the convolution type. An application of this equation is in finding the shape of a smooth wire lying in a vertical plane such that a bead placed anywhere on the wire slides to the lowest point in the same time. This is the well known *tautochrone problem* and the shape of the wire is a *cycloid*.

**Example 36.13.** *Solve the Abel's integral equation* $\int_0^x \frac{y(t)}{\sqrt{(x-t)}} dt = 1 + 2x - x^2$.

**Solution.** The given equation can be written as

$$y(x) * x^{-1/2} = 1 + 2x - x^2$$

Taking the Laplace transform of both sides and using convolution theorem, we get

$$\bar{y} \,.\, L(x^{-1/2}) = L(1 + 2x - x^2)$$

or $$\bar{y}\frac{\Gamma(1/2)}{s^{1/2}} = \frac{1}{s} + \frac{2}{s^2} - \frac{2}{s^3} \quad \text{or} \quad \bar{y} = \frac{1}{\Gamma(1/2)}\left(\frac{1}{s^{1/2}} + 2\cdot\frac{1}{s^{3/2}} - 2\frac{1}{s^{5/2}}\right)$$

On inversion, and noting that $L^{-1}\frac{1}{s^{n+1}} = \frac{x^n}{\Gamma(n+1)}$, we have

$$y = \frac{1}{\Gamma\left(\frac{1}{2}\right)}\frac{x^{-1/2}}{\Gamma\left(\frac{1}{2}\right)} + 2\,.\,\frac{x^{1/2}}{\Gamma\left(\frac{3}{2}\right)} - 2\frac{x^{3/2}}{\Gamma\left(\frac{5}{2}\right)}$$

$$= \frac{1}{\pi}\left(x^{-1/2} + 4x^{1/2} - \frac{8}{3}x^{3/2}\right)$$

Hence $$y = \frac{1}{3\pi\sqrt{x}}(3 + 12x - 8x^3) \text{ is the desired solution.}$$

## 36.8 INTEGRO-DIFFERENTIAL EQUATIONS

An integral equation in which various derivatives of the unknown function $y(x)$ are also present, is called an *integro-differential equation*. An example of such an equation is

$$y'(x) = y(x) - \cos x + \int_0^x \sin(x-t)\, y(t)\, dt$$

The solution of integro-differential equation subject to given initial conditions can also be found by Laplace transforms as illustrated below :

**Example 36.14.** *Solve* $\frac{dy}{dx} + 3y + 2\int_0^x y\, dx = x$, *given* $y(0) = 1$.

**Solution.** Given equation can be written as $y'(x) + 3y(x) + 2\int_0^x y\, dx = x$

Taking Laplace transform of both sides, we get

$$L[y'(x)] + 3\, L[y(x)] + 2L\left\{\int_0^x y(x)\, dx\right\} = L(x)$$

or $$\{s\,\bar{y}(s) - y(0)\} + 3\,\bar{y}(s) + 2\frac{1}{s}\,\bar{y}(s) = \frac{1}{s^2}$$ [Using § 21.6]

$$s\bar{y} + 3\bar{y} + 2\frac{1}{s}\,\bar{y} = 1 + \frac{1}{s^2}$$ [$\because$ $y(0) = 1$]

or $$\bar{y} = \frac{1+s^2}{s(s^2+3s+2)} = \frac{1+s^2}{s(s+1)(s+2)} = \frac{1}{2s} - \frac{2}{s+1} + \frac{5}{2(s+2)}$$

On inversion, we obtain $y = \frac{1}{2} L^{-1}\left(\frac{1}{s}\right) - 2L^{-1}\left(\frac{1}{s+1}\right) + \frac{5}{2} L^{-1}\left(\frac{1}{s+2}\right)$

Hence $y = \frac{1}{2} - 2e^{-x} + \frac{5}{2} e^{-2x}$ is the required solution.

**Example 36.15.** *Solve* $\frac{dy}{dx} = 3\int_0^x \cos 2(x-t)\, y(t)\, dt + 2$ *given* $y(0) = 1$.

**Solution.** Given equation can be written as

$$y'(x) = 3 \cos 2x * y(x) + 2$$

Taking Laplace transform of both sides, we get

$$L[(y'(x)] = 3L(\cos 2x) \,.\, \bar{y}(s) + \frac{2}{s}$$

$$s\,\bar{y}(s) - y(0) = \frac{3s\bar{y}(s)}{s^2+4} + \frac{2}{s} \quad \text{or} \quad \bar{y} = \frac{(s+2)(s^4+4)}{s^2(s^2+1)} \qquad [\because \; y(0) = 1]$$

$$= \frac{4}{s} + \frac{8}{s^2} - 3\frac{s}{s^2+1} - 6\frac{1}{s^2+1}$$

On inversion, we obtain $y = 4 + 8x - 3 \cos x - 6 \sin x$
which is the required solution.

**Obs.** The given integro-differential equation can be converted into the following integral equation by integrating it from 0 to $x$ and using $y(0) = 1$.

$$y(x) = 2x + 1 + 3\int_0^x (x-t) \cos 2(x-t)\, y(t)\, dt.$$

## PROBLEMS 36.3

1. Show that $y(x) = 1 - x$ is a solution of the integral equation $\int_0^x e^{x-t} y(t)\, dt = x$.
2. Show that $y(x) = 1$ is a solution of the Fredholm integral equation
$$y(x) + \int_0^1 x(e^{tx} - 1)\, y(t)\, dt = e^x - x.$$
3. Show that $y(x) = \frac{1}{\pi\sqrt{x}}$ is a solution of the integral equation $\int_0^x \frac{y(t)}{\sqrt{(x-t)}}\, dt = 1$.
4. Show that $y(x) = e^x(2x - 2/3)$ is a solution of the Fredholm integral equation
$$y(x) + \int_0^1 e^{x-t} y(t)\, dt = 2xe^x.$$

Solve each of the following integral equations :

5. $y(x) = x + \frac{1}{6}\int_0^x (x-t)^3 y(t)\, dt$.
6. $y(x) = x^2 + \int_0^x y(t) \sin(x-t)\, dt$.
7. $\int_0^x y(t)\, y(x-t)\, dt = 2y(x) + x - 2$.
8. $\int_0^x y(t)\, y(x-t)\, dt = 9 \sin 4x$.
9. Find a solution of the integral equation $y(x) = \frac{1}{2}\sin 2x + \int_0^x y(t)\; y(x-t)\, dt$.

Solve the following integral equations :

10. $\frac{dy}{dx} + 4y + 5\int_0^x y(t)\; dt = e^{-x}, y(0) = 0$.
11. $\frac{dy}{dx} + 2y + \int_0^x y(t)\, dt = \sin x, y(0) = 1$.

12. $y'(x) = x + \int_0^x y(x-t) \cos t \, dt, y(0) = 1.$

13. $y'(x) = \int_0^x y(t) \cos(x-t) \, dt, y(0) = 1.$

14. $\int_0^x \frac{y(t)}{\sqrt{(x-t)}} dt = \sqrt{x}.$

15. $\int_0^x \frac{y(t)}{(x-t)^{1/3}} dt = x(x+1).$

## 36.9 INTEGRAL EQUATIONS WITH SEPARABLE KERNELS

A kernel $K(x, t)$ of Fredholm integral equation is said to be *separable* (or *degenerate*) if it can be expressed as the sum of a finite number of terms, each of which is the product of a function of $x$ alone and a function of $t$ alone *i.e.*, if it is of the form $K(x, t) = \sum_{n=1}^{m} f_n(x) g_n(t)$.

Also since $\cos(x + t) = \cos x \cos t - \sin x \sin t$, $\cos(x + t)$ is a separable kernel.

## 36.10 SOLUTION OF FREDHOLM EQUATIONS WITH SEPARABLE KERNELS

Consider the integral equation

$$y(x) = \lambda \int_a^b K(x, t) \cdot y(t) \, dt + F(x) \qquad \text{...(1)}$$

where

$$K(x, t) = f_1(x) \cdot g_1(t) + f_2(x) \cdot g_2(t) + \dots + f_m(x) \cdot g_m(t) \qquad \text{...(2)}$$

Substituting Eqns. (2) in (1), we get

$$y(x) = \lambda \int_a^b \left[ \sum_{n=1}^{m} f_n(x) \cdot g_n(t) \right] y(t) \, dt + F(x)$$

$$= \lambda \sum_{n=1}^{m} f_n(x) \left\{ \int_a^b g_n(t) \, y(t) \, dt \right\} + F(x) \qquad \text{...(3)}$$

Evidently $\int_a^b g_n(t) y(t) \, dt = C_n$ (say), is a constant and will be different for different values of $n$. Then (3) takes the form

$$y(x) = \lambda \sum_{n=1}^{m} C_n f_n(x) + F(x) \qquad \text{...(4)}$$

This is a solution of Eqn. (1) in which $m$ constants $C_1, C_2, \dots, C_m$ are to be determined.

Now multiplying both sides of Eqn. (4) by $g_k(x)$ and integrating from $a$ to $b$, we obtain

$$\int_a^b y(x) \, g_k(x) \, dx = \lambda \int_a^b \sum_{n=1}^{m} C_n \, f_n(x) \, g_k(x) \, dx + \int_a^b F(x) \, g_k(x) \, dx$$

Writing $\int_a^b g_k(x) \, f_n(x) \, dx = \alpha_{kn}$ and $\int_a^b g_k(x) \, F(x) \, dx = \beta_k$, the above equation becomes

$$C_k = \lambda \sum_{n=1}^{m} C_n \alpha_{kn} + \beta_k$$

$$= \lambda(C_1 \alpha_{k1} + C_2 \alpha_{k2} + \dots + C_m \alpha_{km}) + \beta_k$$

Taking $k = 1, 2, \dots, m$, we get the following $m$ equations which determine $C_1, C_2, \dots C_m$ :

$$\left.\begin{array}{r} (1 - \lambda\alpha_{11}) C_1 - \lambda\alpha_{12} C_2 - \dots - \lambda \, \alpha_{1m} C_m = \beta_1 \\ - \lambda a_{21} C_1 + (1 - \lambda a_{22}) C_2 - \dots - \lambda \, a_{2m} \, C_m = \beta_2 \\ \dots\dots\dots\dots\dots\dots\dots\dots\dots\dots\dots\dots \\ - \lambda\alpha_{m1} C_1 - \lambda\alpha_{m2} C_2 - \dots + (1 - \lambda\alpha_{mm}) C_m = \beta_m \end{array}\right\} \qquad \text{...(5)}$$

Equations (5) will give a unique solution if the determinant $\Delta$ of the coefficients of $C_1, C_2, \dots C_m$ is not zero.

Now the following cases arise :

*I. When F(x) = 0,* Eqn. (1) is said to be a *homogeneous integral equation* and all β's are zero.

*(i) If* $\Delta \neq 0$, the only solution of Eqn. (4) is the trivial solution $C_1 = C_2 = ... = C_m = 0$. Then $y(x) = 0$ is the trivial solution of (1).

*(ii) If* $\Delta = 0$, at least one of the $C$'s can be assigned any value and the remaining $C$'s can be found accordingly. Then Eqn. (4) gives infinitely many solutions. The values of λ for which $\Delta = 0$ are known as the *eigen values*. Any non-trivial solution of the homogeneous integral equation for a certain value of λ is called the corresponding *eigen function*. The solutions corresponding to eigen values of λ can be expressed as arbitrary multiples of eigen functions.

*II. When* $F(x) \neq 0$. Let us assume that $\int_a^b g_m(x) F(x)\, dx = 0$ so that $\beta_m = 0$

*(i) If* $\Delta \neq 0$, the only solution of Eqn. (*i*) is the trivial solution $C_1 = C_2 = ... = C_m = 0$.

Then $y = F(x)$ is the desired solution of Eqn. (1).

*(ii) If* $\Delta = 0$, atleast one of the $C$'s can be given any value and the remaining $C$'s can be found.

Then (4) gives infinitely many solutions.

*III. When atleast one of the* β*'s* $\neq 0$

*(i) If* $\Delta \neq 0$, then Eqn. (5) give a unique solution of the constants $C$.

Hence there is a unique solution of Eqn. (1).

*(ii) If* $\Delta = 0$, then Eqn. (5) will be inconsistent.

∴ Either there is no solution or infinitely many solutions of (*i*) exist.

**Example 36.16.** *Find the eigen values and eigen functions of the following homogeneous integral equations with degenerate kernels :*

*(i)* $y(x) = \lambda \int_0^1 (2xt - 4x^2)\, y(t)\, dt$ *(ii)* $y(x) = \frac{1}{e^2 - 1} \int_0^1 e^{x+t}\, y(t)\, dt.$

**Solution.** (*i*) Given equation may be written as

$$y(x) = \lambda \left\{ 2x \int_0^1 ty(t)\, dt - 4x^2 \int_0^1 y(t)\, dt \right\}$$

or

$$y(x) = (2\lambda x)\, C_1 - (4\lambda x^2)\, C_2 \qquad ...(i)$$

where $C_1 = \int_0^1 ty(t)\, dt,\ C_2 = \int_0^1 y(t)\, dt$

Substituting $y(x)$ in $C_1, C_2$ we get

$$C_1 = \int_0^1 t\{(2\lambda t)\, C_1 - (4\lambda t^2)\, C_2\}\, dt$$

$$C_2 = \int_0^1 \{(2\lambda t)\, C_1 - (4\lambda t^2)\, C_2\}\, dt$$

or

$$C_1 \left\{ 1 - 2\lambda \int_0^1 t^2 dt \right\} + C_2 \left\{ 4\lambda \int_0^1 t^3\, dt \right\} = 0$$

$$C_1 \left\{ -2\lambda \int_0^1 t\, dt \right\} + C_2 \left\{ 1 + 4\lambda \int_0^1 t^2\, dt \right\} = 0$$

or

$$C_1 \left( 1 - \frac{2\lambda}{3} \right) + \lambda C_2 = 0\ ;\ -\lambda C_1 + C_2\, (1 + 4\lambda/3) = 0 \qquad ...(ii)$$

∴ The determinant of eigen values will be

$$\begin{vmatrix} 1 - 2\lambda/3 & \lambda \\ -\lambda & 1 + 4\lambda/3 \end{vmatrix} = 0 \text{ or } (\lambda + 3)^2 = 0$$

∴ The eigen values are $\lambda_1 = -3,\ \lambda_2 = -3$

For $\lambda_1 = \lambda_2 = -3$, the Eqn. (*ii*) reduce to

$$3C_1 - 3C_2 = 0\ ;\ 3C_1 - 3C_2 = 0 \ i.e.,\ C_1 = C_2$$

∴ From Eqn. (*i*) $y(x) = -6C_1(x - 2x^2) = x - 2x^2$ if $C_1 = -1/6$

Hence the eigen function corresponding to $\lambda_1 = \lambda_2 = -3$, is

$$y(x) = x - 2x^2$$

(*ii*) Given integral equation may be written as

$$y(x) = \frac{e^x}{e^2 - 1} \int_0^1 e^5 e^t \cdot y(t)\, dt = \frac{e^x}{e^2 - 1} C, \quad \text{...}(i)$$

where $C = \int_0^1 e^t \ y(t)\, dt$

Substituting the value of $y(t)$ from Eqn. (*i*) in $C$, we get

$$C = \int_0^1 e^t \left( \frac{e^t C}{e^2 - 1} \right) dt = \frac{C}{e^2 - 1} \int_0^1 e^{t^2}\, dt$$

or
$$C \left[ 1 - \frac{1}{e^2 - 1} \int_0^1 e^{t^2}\, dt \right] = 0 \quad i.e. \quad C = 0$$

Hence from Eqn. (*i*), $y(x) = 0$

which shows that the given integral equation has only trivial solution.

Thus it does not contain any eigen values or eigen functions.

**Example 36.17.** *Solve the integral equation*

$$y(x) = \cos x + \lambda \int_0^{\pi} \sin (x - t)\, y(t)\, dt.$$

**Solution.** Writing the given equation in the following form :

$$y(x) = \cos x + \lambda \left\{ \sin x \int_0^{\pi} \cos t\ y(t)\, dt - \cos x \int_0^{\pi} \sin t\ y(t)\, dt \right\}$$

or
$$y(x) = \cos x + (\lambda \sin x)\, C_1 - (\lambda \cos x)\, C_2 \quad \text{...}(i)$$

where $C_1 = \int_0^{\pi} \cos t \cdot y(t)\, dt$, $\quad C_2 = \int_0^{\pi} \sin t \cdot y(t)\, dt$

Substituting $y(x)$ in $C_1$ and $C_2$, we get

$$C_1 = \int_0^{\pi} \cos t\, \{\cos t + (\lambda \sin t)\, C_1 - (\lambda \cos t)\, C_2\}\, dt$$

$$C_2 = \int_0^{\pi} \sin t\, \{\cos t + (\lambda \sin t)\, C_1 - (\lambda \cos t)\, C_2\}\, dt$$

or
$$\left.\begin{aligned} C_1 \left\{ 1 - \lambda \int_0^{\pi} \cos t \sin t\, dt \right\} + C_2 \left( \lambda \int_0^{\pi} \cos^2 t\, dt \right) &= \int_0^{\pi} \cos^2 t\, dt \\ C_1 \left\{ -\lambda \int_0^{\pi} \sin^2 t\, dt \right\} + C_2 \left\{ 1 + \lambda \int_0^{\pi} \sin t \cos t\, dt \right\} &= \int_0^{\pi} \sin t \cos t\, dt \end{aligned} \right\} \quad \text{...}(ii)$$

By evaluating each of the integrals in Eqn. (*ii*), we get

$$C_1 + \frac{1}{2} C_2\, \lambda\pi = \frac{1}{2}\pi\, ; - \frac{1}{2} C_1\, \lambda\pi + C_2 = 0 \quad \text{...}(iii)$$

The determinant of the Eqn. (*iii*) is given by

$$\begin{vmatrix} 1 & \frac{1}{2}\lambda\pi \\ -\frac{1}{2}\lambda\pi & 1 \end{vmatrix} = 1 + \frac{1}{4}\lambda^2\pi^2 \neq 0$$

Thus the Eqn. (*iii*) have a unique solution

$$C_1 = \frac{2\pi}{4 + \lambda^2\pi^2}\, ; C_2 = \frac{\lambda\pi^2}{4 + \lambda^2\pi^2}$$

Substituting these values of $C_1$ and $C_2$ in Eqn. (*i*), we obtain the required solution

$$y(x) = \cos x + \frac{\lambda}{4 + \lambda^2\pi^2}(2\pi \sin x - \lambda\pi^2 \cos x)$$

or
$$y(x) = \frac{2}{4 + \lambda^2\pi^2}(2 \cos x + \pi\lambda \sin x).$$

## PROBLEMS 36.4

Determine the eigen values and eigen functions for the following homogeneous integral equations with degenerate kernels :

1. $y(x) = \lambda \int_0^1 (3x - 2)\ t \,.\, y(t)\, dt.$
2. $y(x) = \lambda \int_{-1}^1 (5x\,t^3 + 4x^2\,t + 3xt)\ y(t)\, dt.$
3. $y(x) = \lambda \int_0^{\pi/4} \sin^2 x \cdot y(t)\, dt.$
4. $y(x) - \lambda \int_0^{2\pi} \sin x\ \sin t\ y(t)\, dt = 0.$ (*Madras M.E., 2000 S*)
5. $y(x) = \lambda \int_0^{\pi} \sin x \cos t \,.\, y(t)\, dt.$
6. $y(x) = \lambda \int_0^{2\pi} \sin (x + t) \,\cdot\, y(t)\, dt.$ (*Madras M.E., 2000*)

Solve the following integral equations :

7. $y(x) = x + \lambda \int_0^1 (x - t)\ y(t)\, dt.$
8. $y(x) = x + \lambda \int_0^1 (1 + x + t)\ y(t)\, dt.$
9. $y(x) = x + \lambda \int_0^{\pi} (1 + \sin x \sin t)\ y(t)\, dt.$
10. $y(x) = (2x - \pi) + 4 \int_0^{\pi/2} \sin^2 x \cdot y(t)\, dt.$
11. $y(x) = \sin x + \lambda \int_0^{\pi/2} \sin x \cos t \cdot y(t)\, dt.$
12. $y(x) = x + \lambda \int_{-\pi}^{\pi} (x \cos t + t^2 \sin x + \cos x \sin t)\ y(t)\, dt.$
13. Obtain the solution of $y(x) = 1 + \lambda \int_0^1 xt \cdot y(t)\ dt$ in the form $y(x) = 1 + \frac{3\lambda x}{2(3 - \lambda)}$ $(\lambda \neq 3)$.

    What happens when $\lambda = 3$ ?
14. For the integral equation

    $$y(x) = F(x) + \lambda \int_0^1 (1 - 3xt)\ y(t)\, dt,$$

    find the eigen values of $\lambda$ and the corresponding eigen functions.
15. Obtain the most general solution of the equation

    $$y(x) = F(x) + \lambda \int_0^{2\pi} \sin (x + t)\ y(t)\, dt$$

    where (*i*) $F(x) = x$ (*ii*) $F(x) = 1$, under the assumption that $\lambda \neq \pm\, 1/\pi$.

## 36.11 SOLUTION OF FREDHOLM INTEGRAL EQUATION BY THE METHOD OF SUCCESSIVE APPROXIMATIONS

Consider the Fredholm equation $y(x) = F(x) + \lambda \int_a^b K(x, t)\ y(t)\, dt$ ...(1)

where $F(x)$ is continuous in $a \le x \le b$ and $K(x, t)$ is finite and continuous in the rectangle $a \le x \le b$ and $a \le t \le b$.

Replacing $y$ under the integral sign by an initial approximation $y(0)$, we get the first approximation

$$y^{(1)}(x) = F(x) + \lambda \int_a^b K(x, t)\ y^{(0)}(t)\, dt \qquad ...(2)$$

Replacing $y$ under the integral sign in Eqn. (1) by $y^{(1)}$, we get the next approximation

$$y^{(2)}(x) = F(x) + \lambda \int_a^b K(x, t)\ y^{(1)}(t)\, dt \qquad ...(3)$$

Proceeding in this manner, we get the general formula for successive approximations as

$$y^{(n)}(x) = F(x) + \lambda \int_a^b K(x, t)\, y^{(n-1)}(t)\, dt \qquad ...(4)$$

We now, obtain the condition for the convergency of the sequence $y^{(n)}(x)$.

Replacing $x$ by $t$ and $t$ by a dummy variable $t_1$, Eqn. (2) becomes

$$y^{(1)}(t) = F(t) + \lambda \int_a^b K(t, t_1)\, y^{(0)}(t_1)\, dt_1$$

Then Eqn. (3) takes the form

$$y^{(2)}(x) = F(x) + \lambda \int_a^b K(x, t) \left\{ F(t) + \lambda \int_a^b K(t, t_1)\, y^{(0)}(t_1)\, dt_1 \right\} dt$$

$$= F(x) + \lambda \int_a^b K(x, t)\, F(t)\, dt + \lambda^2 \int_a^b K(x, t) \,.\int_a^b K(t, t_1) y^{(0)}(t_1)\, dt_1\, dt \qquad ...(5)$$

Writing $K^* \phi(x) = \int_a^b K(x, t)\ \phi(t)\, dt$, the Eqns. (1), (2) and (5) become

$$y(x) = F(x) + \lambda K^* y(x)$$

$$y^{(1)}(x) = F(x) + \lambda K^* y^{(0)}(x)$$

$$y^{(2)}(x) = F(x) + \lambda K^* F(x) + \lambda^2 K^{*2} y^{(0)}(x)$$

Similarly $\quad y^{(3)}(x) = F(x) + \lambda K^* F(x) + \lambda^2 K^{*2} F(x) + \lambda^3 K^{*3} y^{(0)} x$

In general $\quad y^{(n)}(x) = F(x) + \lambda K^* F(x) + \lambda^2 K^{*2} F(x) + \lambda^3 K^{*3} F(x) + \dots + \lambda^n K^{*n} y^{(0)}(x)$

As $n \to \infty$, we get

$$y(x) = \underset{n \to \infty}{\text{Lt}}\ y^n(x) = F(x) + \underset{n \to \infty}{\text{Lt}}\ \{\lambda K^* F(x) + \lambda^2 K^{*2} F(x) + \dots \infty\} \qquad ...(6)$$

Now $F(x)$ and $K(x, t)$ being continuous for all values of $x$ and $t$ in $(a, b)$, $F(x) \le M$ and $|K(x, t)| \le m$ where $M, m$ are their respective maximum values in $(a, b)$.

$$\therefore \qquad |K^* F(x)| = \left| \int_a^b K(x, t)\ F(x)\, dt \right|$$

$$\le mM \left| \int_a^b dt \right| \le mM(b - a)$$

Similarly $\quad K^{*r} F(x) \le m^r M \,.(b - a)^r$

Then $\quad |\lambda^r K^{*r} F(x)| \le |\lambda^r| \,.\, m^r M(b - a)^r$

$$\le M\{|\lambda|\ m(b - a)\}^r$$

$\therefore$ In Eqn. (6), $\quad \sum_1^{\infty} \lambda^r K^{*r}\ F(x) \le M \sum_1^{\infty} [|\lambda| m(b - a)]^r$

Now the series on R.H.S. being a geometric series, converges for $|\lambda|\ m(b - a) < 1$.

Thus by comparison test, $\sum_1^{\infty} \lambda^r K^{*r}\ F(x)$ also converges for $|\lambda|\ m(b - a) < 1$

*i.e.*, for

$$|\lambda| < \frac{1}{m(b - a)} \qquad ...(7)$$

Hence the given integral equation Eqn. (1) will have a continuous solution when the condition Eqn. (7) is satisfied.

**Obs. 1.** To evaluate the successive terms in the series (6) conveniently, we choose $y^{(0)}(x) = F(x)$.

**Obs. 2.** Volterra integral equations can also be solved by following exactly similar procedure as above (See Example 36.19).

**Example 36.18.** *Solve, by using the method of successive approximations, the integral equation*

$$y(x) = 1 + \lambda \int_0^1 xt\ y(t)\, dt. \qquad ...(i)$$

**Solution.** Taking the initial approximation $y^{(0)}(x) = 1$ and substituting it in the R.H.S. of Eqn. (*i*), we get

$$y^{(1)}(x) = 1 + \lambda \int_0^1 xt \cdot 1\, dt = 1 + \lambda x \left| \frac{t^2}{2} \right|_0^1 = 1 + \frac{\lambda x}{2}$$

Substituting $y^{(1)}(x)$ in the R.H.S. of Eqn. (*i*), we have

$$y^{(2)}(x) = 1 + \lambda \int_0^1 xt\left(1 + \lambda \frac{t}{2}\right) dt = 1 + \lambda x \int_0^1 \left(t + \frac{\lambda t^2}{2}\right) dt$$

$$= 1 + \lambda x \left| \frac{t^2}{2} + \frac{\lambda}{2} \cdot \frac{t^3}{3} \right|_0^1 = 1 + \frac{\lambda x}{2} + \frac{\lambda^2 x}{6}$$

Substituting $y^{(2)}(x)$ in the R.H.S. of Eqn. (*i*), we get

$$y^{(3)}(x) = 1 + \lambda \int_0^1 xt\left(1 + \frac{\lambda t}{2} + \frac{\lambda^2 t}{6}\right) dt = 1 + \lambda x \left| \frac{t^2}{2} + \left(\frac{\lambda}{2} + \frac{\lambda^2}{6}\right) \frac{t^3}{3} \right|_0^1$$

$$= 1 + \frac{\lambda x}{2} + \frac{\lambda^2 x}{6} + \frac{\lambda^3 x}{18} = 1 + \frac{\lambda x}{2}\left(1 + \frac{\lambda}{3} + \frac{\lambda^2}{3^2} + \cdots\right)$$

Hence the solution of Eqn. (*i*) is

$$y(x) = 1 + \frac{\lambda x}{2}\left(1 + \frac{\lambda}{3} + \left(\frac{\lambda}{3}\right)^2 + \left(\frac{\lambda}{3}\right)^3 + \cdots\right)$$

As the number of terms tends to infinity, the exact solution is

$$y(x) = 1 + \frac{\lambda x}{2}\left(1 + \frac{\lambda}{3} + \left(\frac{\lambda}{3}\right)^2 + \left(\frac{\lambda}{3}\right)^3 + \cdots \infty\right)$$

$$= 1 + \frac{\lambda x}{2} \frac{1}{1 - \lambda/3}$$ [Summing up the G.P. which converges for $\lambda/3 < 1$]

or $$y(x) = 1 + \frac{3\lambda x}{2(3 - \lambda)}$$ only if $\lambda < 3$.

**Example 36.19.** *Using the method of successive approximations, solve the Volterra integral equation*

$$y(x) = 1 + x + \int_0^x (x - t)\, y(t)\, dt.$$

**Solution.** Taking the initial approximation $y^{(0)}(x) = 1 + x$ and substituting it in the R.H.S. of Eqn. (*i*), we get

$$y^{(1)}(x) = 1 + x + \int_0^x (x - t)(1 + t)\, dt$$

$$= 1 + x + x\left(x + \frac{x^2}{2}\right) - \left(\frac{x^2}{2} + \frac{x^3}{3}\right) = 1 + x + \frac{x^2}{2} + \frac{x^3}{6}$$

Substituting $y^{(1)}(x)$ in the R.H.S. of Eqn. (*i*), we obtain

$$y^{(2)}(x) = 1 + x + \int_0^x (x - t)\left(1 + t + \frac{t^2}{2} + \frac{t^3}{6}\right) dt$$

$$= 1 + x + x\left(x + \frac{x^2}{2} + \frac{x^3}{6} + \frac{x^4}{24}\right) - \left(\frac{x^2}{2} + \frac{x^3}{3} + \frac{x^4}{8} + \frac{x^5}{30}\right)$$

$$= 1 + x + \frac{x^2}{2} + \frac{x^3}{6} + \frac{x^4}{24} + \frac{x^5}{120}$$

Proceeding in this manner, we get

$$y^{(n)}(x) = 1 + \frac{x}{1!} + \frac{x^2}{2!} + \frac{x^3}{3!} + \frac{x^4}{4!} + \frac{x^5}{5!} + \cdots + \frac{x^n}{n!}$$

As $n \to \infty$, the exact solution of Eqn. (*i*) is

$$y(x) = \underset{n \to \infty}{\text{Lt}} \left(1 + \frac{x}{1!} + \frac{x^2}{2!} + \frac{x^3}{3!} + \cdots + \frac{x^n}{n!}\right) = 1 + \frac{x}{1!} + \frac{x^2}{2!} + \frac{x^3}{3!} + \ldots \infty = e^x.$$

**Example 36.20.** *Find the eigen value and the corresponding eigen functions of the homogenous Fredholm integral equation.* *(Kottaym, 2020)*

$$y(x) = \lambda \int_0^\pi (\cos^2 x \cos 2t + \cos 3x \cos^3 t)\, y(t)\, dt$$

**Solution.** The given integral equation can be written as

$$y(x) = \lambda \cos^2 x \int_0^\pi \cos 2t\; y(t) dt + \lambda \cos 3x \int_0^\pi \cos^3 t\; y(t)\, dt \qquad \ldots(i)$$

Now, let us define the following

$$C_1 = \int_0^\pi \cos 2t\; y(t) dt \qquad \ldots(i)$$

and

$$C_2 = \int_0^\pi \cos^3 t\; y(t) dt \qquad \ldots(ii)$$

Putting the value of $C_1$ and $C_2$ from Eqns. (*ii*) and (*iii*)
Putting the value of $C_1$ and $C_2$ from Eqns. (*ii*) and (*iii*) in Eqn. (*i*)

we get,
$$y(x) = \lambda C_1 \cos^2 x + \lambda\, C_2 \cos 3x \qquad \ldots(iv)$$

from which
$$y(t) = \lambda C_1 \cos^2 t + \lambda\, C_2 \cos 3x(t) \qquad \ldots(v)$$

Putting this value of (*t*) in Eqn. (*ii*), we get

$$C_1 = \int_0^\pi \cos 2t\, (\lambda C_1 \cos^2 t + \lambda\, C_2 \cos 3t\, dt$$

$$C_1 \left[1 - \lambda \int_0^\pi \cos t + \cos^2 t\; dt\right] - \lambda\, C_2 \int_0^\pi \cos 2t \cos 3t\; dt = 0$$

$$C_1 [1 - \lambda\, I_1] - \lambda\, C_2\, I_2 = 0 \qquad \ldots(vi)$$

where
$$I_1 = \frac{1}{2}\int_0^\pi \cos 2t\; dt + \frac{1}{2}\int_0^\pi \left[1 + \frac{\cos 4\pi}{2}\right] dt$$

Since
$$\cos^2 2t = 1 + \cos 4t$$

$$I_1 = 0 + \frac{1}{4}\left[t + \frac{\sin 4t}{4}\right]_0^\pi = \frac{\pi}{4} \qquad \ldots(vii)$$

$$I_2 = \int_0^\pi \cos 2t \cos 3t\; dt = \frac{1}{2}\int_0^\pi [\cos 5t + \cos t]\, dt$$

$$I_2 = \frac{1}{2}\left[\sin \frac{5t}{5} + \sin t\right]_0^\pi = 0 \qquad \ldots(viii)$$

Putting the value of $I_1$ and $I_2$ form (*ii*) and (*viii*) in (*vi*), we get

$$\left(1 - \frac{3\pi}{4}\right) C_1 - 0\,.\, C_2 = 0 \qquad \ldots(ix)$$

Now putting the value of $y(t)$ from Eqn. (*v*) in Eqn. (*iii*), we get,

$$C_2 = \int_0^\pi \cos^3 t (\lambda C_1 \cos^2 t + \lambda C_2 \cos 3t)\, dt$$

$$\lambda C_1 \int_0^\pi \cos^5 t\; dt + C_2\, [\lambda \int_0^\pi \cos^3 t + \cos 3t\; dt - 1] = 0$$

$$\lambda C_1 I_3 + C_2\,[\lambda]_4 - 1] = 0 \quad ...(x)$$

$$I_3 = \int_0^5 \cos^5 t\, dt = 0 \quad ...(xi)$$

where $I_3 - \int_0^5 \cos^5 t\, dt = 0$, by the property of definite integral Eqn. $(xi)$

$$I_4 = \int_0^\pi \frac{1+\cos 6t}{2}\,dt + \frac{3}{4}\int_0^\pi \frac{\cos 4t + \cos t}{2}\,dt$$

$$= \frac{1}{8}\left[t + \frac{\sin 6t}{3}\right]_0^\pi + \frac{3}{8}\left[\frac{\sin 4t}{4} + \sin t\right]_0^\pi = \frac{\pi}{8}$$

$$I_4 = \int_0^\pi \cos^3 t \cos 3t\, dt = \frac{\pi}{8} \quad ...(xii)$$

Using the value $I_3$ and $I_4$ from Eqns. $(xi)$ and $(xii)$

$$0.C_1 + \left(\frac{\lambda\pi}{8} - 1\right) C_2 = 0 \quad ...(xiii)$$

For non-zero solution, Eqn. $(ix)$ and $(xiii)$

$$\begin{vmatrix} 1 - \frac{\lambda\pi}{4} & 0 \\ 0 & \frac{\lambda\pi}{8} - 1 \end{vmatrix} = 0 \quad or \quad \left(1 - \frac{\lambda\pi}{t}\right)\left(\frac{\lambda\pi}{8} - 1\right) = 0$$

which $\lambda = \frac{4}{\pi}, \frac{8}{\pi}$

Hence the eigen value of the given integral equation are

$$\lambda_1 = \frac{4}{\pi} \quad \text{and} \quad \lambda_2 = \frac{8}{\pi}$$

## PROBLEMS 36.5

Apply the method of successive approximations, to solve the following Fredholm integral equations : (1 to 3)

1. $y(x) = 1 - \lambda \int_0^1 xt\, y(t)\, dt.$

2. $y(x) = \sin x + \lambda \int_0^{2\pi} \cos(x+t)\, y(t)\, dt.$

3. $y(x) = 1 + \lambda \int_0^1 (1 - 3xt)\, y(t)\, dt.$

Using the iterative method, solve the following Volterra integral equations : (4 to 6)

4. $y(x) = 1 + \int_0^x y(t)\, dt.$ *(Madras M.E., 2000 S)*

5. $y(x) = 1 + x - \int_0^x y(t)\, dt.$

6. $y(x) = x + \int_0^x (t - x)\, y(t)\, dt.$

7. $y(x) = 2(1 + x^2) - \int_0^x xy\,(t)\, dt.$

8. Choosing the initial approximation $y^{(0)}(x) = 0$, for the solution of the integral equation
$$y(x) = \int_0^x t(t-x)\, y(t)\, dt + \frac{x^2}{2}, \text{ show that } y^{(2)}(x) = \frac{x^2}{2} - \frac{x^5}{40}.$$

9. Starting with the initial approximation $y^{(0)}(x) = 1$, for the solution of the integral equation
$$y(x) = 1 + \int_0^x (x+t)\, y(t)\, dt,$$
show that $y^{(3)}(x) = 1 + \frac{3x^2}{2} + \frac{7x^4}{8} + \frac{77x^6}{240}.$ *(Madras M.E., 2000)*

CHAPTER 37

# Discrete Mathematics



## I. SET THEORY

### 37.1 SETS

**(1)** The concept and the language of sets play a very important role in expressing mathematical ideas concisely and precisely. It was Cantor* who first introduced and developed the notion of sets in mathematical investigations. It is, therefore, essential for a student of engineering to grasp the basic ideas of *Set Theory*.

**Def.** *A collection of objects defined by some property, is called a* **set**. *The objects belonging to a set are called its* **elements** *or* **members**.

*Examples* of a set are (*i*) the set of positive integers less than 25, (*ii*) set of pages in a book, and (*iii*) set of women students in a college.

A set is denoted by a single capital letter *e.g.* $A, B, ..., S, X, Y$ and the elements of a set are generally denoted by small letters $a, b, c, ..., x, y, z$.

*When e is an element of a set S, we write* $e \in S$ *and read as 'e belongs to S'. When e is not an element of S, we write* $e \notin S$.

If $S$ be a set of odd integers, $3, 7, 11 \in S$ but $4, 6 \notin S$.

**(2) Representation of a set**

(*i*) *Tabular form of a set.* In this, the elements are enclosed in curly brackets after separating them by commas, *e.g.*, the set of positive even integers less than 9 is written as $S = \{2, 4, 6, 8\}$ and the set of prime numbers between 4 and 14 is $T = \{5, 7, 11, 13\}$.

(*ii*) *Symbolic form of a set.* In this, the set is written as $\{x/P(x)\}$ where $x$ is a typical element of the set and $P(x)$ is the property satisfied by this element. In symbolic form, the above two sets are

$$S = \{x/x = a \text{ positive even number} < 9\}$$
$$T = \{x/x = a \text{ prime number between 4 and 14}\}.$$

*The great German mathematician George Cantor (1845–1918), the creator of Set theory.

**(3) Empty set or null set.** *A set which has no elements is called an* **empty set** *or the* **null set** *and is denoted by the symbol* $\phi$.

**(4) Finite** *and* **infinite sets**. *A set is said to be* **finite** *if it has a finite number of elements. Otherwise a set is said to be* **infinite**.

*The number of distinct elements in a finite set A is called its* **cardinality** *and is denoted by* $|A|$.

*For instance*, the set of days in a year is *finite*, the set of points in a line is an *infinite set*.

**(5) Subset.** *If every element of a set A is also an element of set B, then A is called a* **subset** *of B* and this relationship is denoted by $A \subset B$ or $B \supset A$ ; which is read as 'A is contained in B'.

*Another definition : If A and B are two sets such that*

$$x \in A \Rightarrow x \in B,$$

*then A is called a subset of B.*

The notation $\Rightarrow$ stands for the word '*implies*'.

*For instance*, the set $V$ of vowels is a subset of the set $A$ of the English alphabet and we write $V \subset A$.

**(6) Power set.** *For a set A, collection of all subsets of A is called the power set of A and is denoted by P(A).*

If $A = [1, 2, 3]$ then $P(A)$ consists of $2^3$ *i.e.* 8 elements $\phi$, [1], [2], [3], [1, 2], [2, 3], [3, 1] and [1, 2, 3].

In general, if $A$ has $n$ elements, then $P(A)$ has $2^n$ elements.

**(7) Equality of sets.** *Two sets A and B are said to be equal if the elements of both are the same i.e.*, if each element of $A$ is also an element of $B$ and *vice versa*, and we write $A = B$.

In other words, *if A and B are two sets such that*

$$A \subset B \quad \text{and} \quad B \subset A \Leftrightarrow A = B.$$

Here $\Leftrightarrow$ stands for '*implies and is implied by*' or '*if and only if*'.

*For instance*, $\{2, 3, 5\} = \{3, 2, 5, 3\} = \{2, 5, 3, 2\}$, since the change in the order of elements or the repetition of an element is immaterial and all these contain the same elements 2, 3, 5.

**(8) Proper and improper subsets.** *When the set B contains all the elements of A and some others, A is said to be a* **proper subset** *of B* and is denoted by $A \subseteq B$.

*i.e.*, if $A \subset B$ and $A \neq B$ then $A \subseteq B$.

*If* $A \subset B$ *and every element of B is also an element of A i.e.*, $B \subset A$, *then A is said to be an* **improper subset** *of B i.e.*, $A = B$.

*For instance*, the set of positive odd integers and the set of positive even integers are both proper subsets of the set of natural numbers.

**(9) Universal set** *is that which has all the sets under investigation as its subsets. It is generally denoted by 'U'.*

*For instance* the set of all letters of English alphabet is a universal set of the sets of the form $\{a, i, e, u\}$, $\{b, x, u, m\}$ etc.

**Example 37.1.** *If A, B, C are sets such that* $A \subseteq B$ *and* $B \subseteq C$, *then show that* $A \subseteq C$.

**Solution.** Let $x$ be any element of $A$.

Since $A \subseteq B$ *i.e.*, all the elements of $A$ belong to $B$,

so $$x \in A \Rightarrow x \in B \quad \text{...}(i)$$

Again as $B \subseteq C$ *i.e.*, all elements of $B$ belong to $C$,

so $$x \in B \Rightarrow x \in C \quad \text{...}(ii)$$

$\therefore$ It follows from (*i*) and (*ii*) that $x \in A \Rightarrow x \in C$

*i.e.*, $$A \subseteq C.$$

**Example 37.2.** *Which of the following sets are equal ?*

$$S_1 = \{1, 2, 2, 3\},\ S_2 = \{x : x^2 - 2x + 1 = 0\}$$
$$S_3 = \{3, 2, 1\} \text{ and } S_4 = \{x : x^3 - 6x^2 + 11x - 6 = 0\}.$$

**Solution.** Here $S_1 = \{1, 2, 2, 3\} = \{1, 2, 3\}$

$$S_2 = \{x : (x - 1)^2 = 0\} = \{1\},\ S_3 = \{1, 2, 3\}$$
$$S_4 = \{x : (x - 1)(x - 2)(x - 3) = 0\} = \{1, 2, 3\}$$

From these we find that $S_1$, $S_3$, $S_4$ are equal.

## 37.2 SET OPERATIONS

**(1) Union** *of two sets A and B is the set of all elements which belong to A or to B or to both. It is denoted by* $A \cup B$ read as '$A$ union $B$' and is represented by the shaded portion in Fig. 37.1.

Symbolically $A \cup B = \{x/x \in A \text{ or } x \in B\}$.

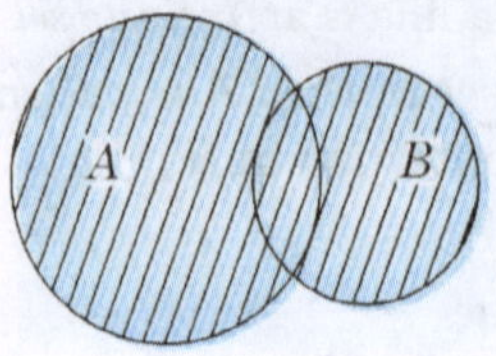


Fig. 37.1

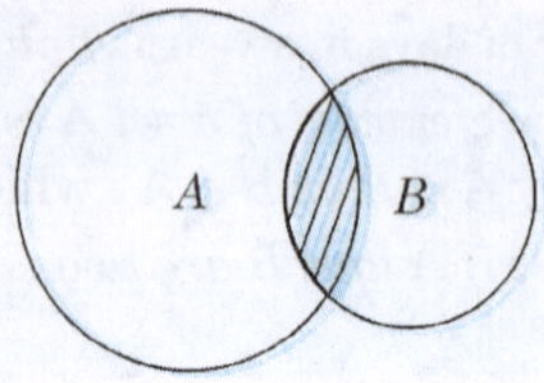


Fig. 37.2

**(2) Intersection** *of two sets A and B is the set of elements which are common to both A and B*. It is denoted by $A \cap B$ read as '$A$ intersection $B$' and is represented by the shaded portion in Fig. 37.2.

Symbolically $\quad A \cap B = \{x/x \in A \text{ and } x \in B\}$

*Such diagrams as Figs. 37.1 and 37.2 which exhibit the various relations between the sets are known as* **Venn diagrams**.

**(3) Disjoint sets.** *If the sets A and B have no common elements, they are called* **disjoint sets**. Their intersection is an empty set.

*For instance*, if $A$ be a set of boys in a college and $B$ the set of girls in the same college, then $A$ and $B$ are disjoint sets *i.e.* $A \cap B = \phi$.

**(4) Complement of a set.** *If* $B \subset A$, *the set of elements of A which are not in B is called the* **complement of B in A** *and is denoted by* $B^c$ *in A*. It is also known as the difference $A - B$ of sets $A$ and $B$. Thus

$$B^c \text{ in } A = \{x/x \in A \text{ and } x \notin B\}$$

which is shown shaded in Fig. 37.3(*i*).

*If U be a universal set, then the set '$U - A$' is called the complement of A and is denoted by* $A^c$, which is shown shaded in Fig. 37.3(*ii*).

*For instance*, if $U = \{1, 2, 3, 4, 5, \ldots\}$ and $A = (1, 3, 5, 7, \ldots)$, then $A^c = \{2, 4, 6, 8 \ldots\}$.

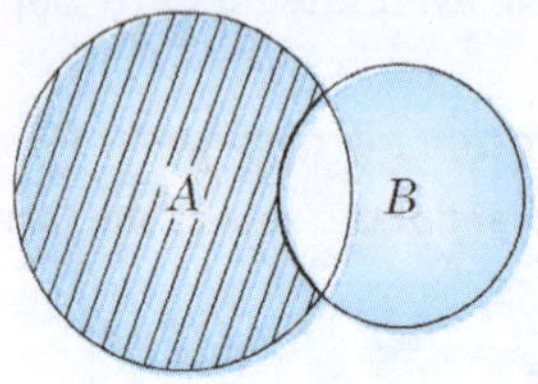


Fig. 37.3(i)

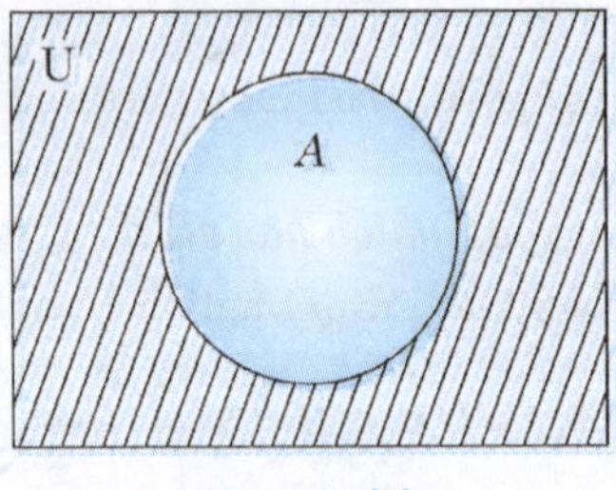


Fig. 37.3(ii)

**(5) Cartesian product** *of two sets A and B denoted by* $A \times B$ *is defined to be set of all ordered pairs (a, b), where* $a \in A$ *and* $b \in B$ *i.e.*,

$$A \times B = \{(a, b) : a \in A \quad \text{and} \quad b \in B\}$$

*For instance* if $A = \{1, 2\}$, $B = \{1, 2, 3\}$, then $A \times B = \{(1, 1), (1, 2), (1, 3), (2, 1), (2, 2), (2, 3)\}$, $B \times A = \{(1, 1), (1, 2), (2, 1), (2, 2), (3, 1), (3, 2)\}$ $\therefore \quad A \times B \neq B \times A$.

**Example 37.3.** *If* $A = \{2, 5, 6, 7\}$, $B = \{0, 2, 5, 7, 8\}$, $C = \{1, 2, 3, 5, 6\}$, *show that*

$$A \cup (B \cup C) = (A \cup B) \cap (A \cup C).$$

**Solution.** Here $\quad B \cap C = \{2, 5\}$

$\therefore \quad A \cup (B \cap C) = \{2, 5, 6, 7\} \quad \ldots(i)$

Again $\quad A \cup B = \{0, 2, 5, 6, 7, 8\}$,

$A \cup C = \{1, 2, 3, 5, 6, 7\}$

$\therefore \quad (A \cup B) \cap (A \cup C) = [2, 5, 6, 7] \quad \ldots(ii)$

Hence from Eqns. (*i*) and (*ii*), we get

$$A \cup (B \cap C) = (A \cup B) \cap (A \cup C).$$

**Example 37.4.** *With the help of Venn diagram, show that*

$$A \cap (B \cup C) = (A \cap B) \cup (A \cap C).$$

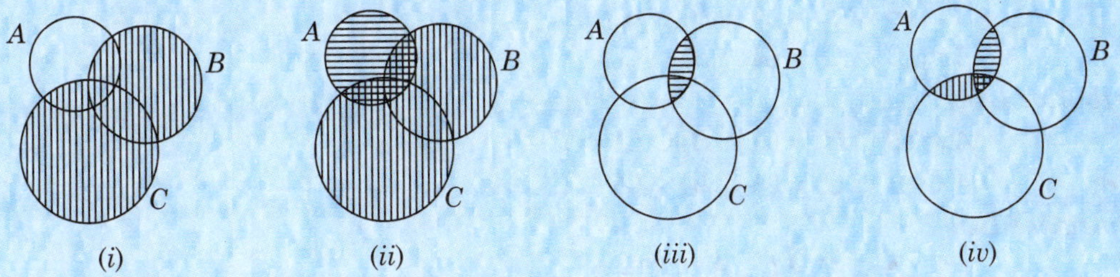


Fig. 37.4

**Solution.** First we draw vertical lines in the whole areas of $B$ and $C$ so as to represent $B \cup C$. [Fig. 37.4(*i*)].

Now draw horizontal lines in the whole area $A$. Thus the double hatched area in [Fig. 37.4(*ii*)] represents area common to $A$ and $B \cup C$ *i.e.*, $A \cap (B \cup C)$.

Again we draw horizontal lines in the area common to $A$ and $B$ so as to represent $A \cap B$ [Fig. 37.4(*iii*)].

Now draw vertical lines in the area common to $A$ and $C$, so as to represent $A \cap C$. Then the whole hatched area in [Fig. 37.4(*iv*)] represents $(A \cap B) \cup (A \cap C)$.

Hence we observe that the double hatched area in Fig. 37.4(*ii*) is equal to the total hatched area in Fig. 37.4(*iv*).

**Example 37.5.** *Prove that (i)* $A - (B \cap C) = (A - B) \cup (A - C)$.

*(ii)* $A \times (B \cap C) = (A \times B) \cap (A \times C)$. *(Tirupati, 2001)*

**Solution.** (*i*) Let $x$ be an arbitrary element of the set $A - (B \cap C)$, then

$$x \in A - (B \cap C) \Rightarrow x \in A \text{ and } x \notin (B \cap C) \qquad [\because\ x \notin (A - B) \Rightarrow x \in A \text{ and } x \notin B]$$
$$\Rightarrow x \in A \text{ and } (x \notin B \text{ or } x \notin C)$$
$$\Rightarrow (x \in A \text{ and } x \notin B) \text{ or } (x \in A \text{ and } x \notin C)$$
$$\Rightarrow x \in (A - B) \text{ or } x \in (A - C)$$
$$\Rightarrow x \in (A - B) \cup (A - C).$$
$$\therefore \quad A - (B \cap C) \subset (A - B) \cup (A - C) \qquad \ldots(i)$$

Again if $x$ be an arbitrary element of the set $(A - B) \cup (A - C)$, then

$$x \in (A - B) \cup (A - C) \Rightarrow x \in (A - B) \text{ or } x \in (A - C)$$
$$\Rightarrow (x \in A \text{ and } x \notin B) \text{ or } (x \in A \text{ and } x \notin C)$$
$$\Rightarrow x \in A \text{ and } (x \notin B \text{ or } x \notin C)$$
$$\Rightarrow x \in A \text{ and } x \notin B \cap C$$
$$\Rightarrow x \in A - (B \cap C)$$
$$\therefore \quad (A - B) \cup (A - C) \subset A - (B \cap C) \qquad \ldots(ii)$$

From Eqns. (*i*) and (*ii*), we get $A - (B \cap C) = (A - B) \cup (A - C)$.

(*ii*) $(x, y) \in A \times (B \cap C)$

$$\Rightarrow x \in A \text{ and } y \in (B \cap C)$$
$$\Rightarrow (x \in A \text{ and } y \in B) \text{ and } (x \in A \text{ and } y \in C)$$
$$\Rightarrow (x, y) \in (A \times B) \text{ and } (x, y) \in (A \times C)$$
$$\Rightarrow (x, y) \in (A \times B) \cap (A \times C)$$

Hence $A \times (B \cap C) = (A \times B) \cap (A \times C)$.

## 37.3 LAWS OF SET THEORY

**1.** *Commutative Law*

$$A \cup B = B \cup A \,;\, A \cap B = B \cap A.$$

**2.** *Associative Law*

$$A \cup (B \cup C) = (A \cup B) \cup C$$
$$A \cap (B \cap C) = (A \cap B) \cap C$$

**3.** *Distributive Law*

$$A \cap (B \cup C) = (A \cap B) \cup (A \cap C)$$
$$A \cup (B \cap C) = (A \cup B) \cap (A \cup C)$$

**4.** *Complement Law*

$$A \cup A^c = U \,;\, A \cap A^c = \phi.$$

**5.** *Identity Law*

$$A \cup \phi = A = \phi \cup A$$
$$A \cap U = A = U \cap A.$$

**6.** *Absorption Law*

$$A \cup (A \cap B) = A \,;\, A \cap (A \cap B) = A$$

**7.** *De Morgan's Law*

$$(A \cup B)^c = A^c \cap B^c \,;\, (A \cap B)^c = A^c \cup B^c$$

**8.** *Involution Law*

$$(A^c)^c = A.$$

## 37.4 PRINCIPLE OF INCLUSION

**(1)** *If A and B be sets with cardinalities* $|A|$ *and* $|B|$, *then*

$$|A \cup B| = |A| + |B| - |A \cup B|$$

*Proof.* The number of common elements in $A$ and $B$ is $|A \cap B|$. Each of these elements is counted twice in $|A| + |B|$, once in $|A|$ and once in $|B|$. This should be adjusted by subtracting the term $|A \cap B|$ from $|A| + |B|$.

Hence $$|A \cup B| = |A| + |B| - |A \cap B|.$$

**Obs.** Using the distributive law, we can extend the above result for three sets *A*, *B*, *C* so that

$$|A \cup B \cup C| = |A| + |B| + |C| - |A \cap B| - |B \cap C| - |C \cap A| + |A \cap B \cap C|$$

For
$$\begin{aligned} |A \cup B \cup C| &= |(A \cup B) \cup C| \\ &= |A \cup B| + |C| - |(A \cup B) \cap C| \\ &= |A| + |B| - |A \cap B| + |C| - |(A \cap C) \cup (B \cap C)| \\ &= |A| + |B| + |C| - |A \cap B| - [|A \cap C| + |B \cap C| - |A \cap B \cap C|] \end{aligned}$$

whence follows the result.

**(2) Duality.** *If S be any identity involving sets and operations* (*e.g.* complement, intersection $\cap$ and union $\cup$ etc.) *and a new set S* is obtained by replacing* $\cap$ *by* $\cup$, $\cup$ *by* $\cap$, $\phi$ *by U and U by* $\phi$ *in S, then the statement S* is true and is called the* **dual** *of the statement S.*

*For instance*, the dual of $A \cap (B \cup A) = A$ is $A \cup (B \cap A) = A$.

**Example 37.6.** *In a survey conducted on 250 persons, it was found that 180 drink tea and 70 drink coffee and 50 take both tea and coffee. How many drink atleast one beverage and how many drink neither ?*

**Solution.** Let $A$ be the set of tea drinkers and $B$ the set of coffee drinkers. Then

$$|A \cup B| = |A| + |B| - |A \cap B| = 180 + 70 - 50 = 200$$

Hence 200 persons drink at least one beverage and 250 – 200 = 50 persons drink neither tea nor coffee.

**Example 37.7.** *How many integers between 1 and 468 are divisible by 3 but not by 5.*

**Solution.** Number of integers between 1 and 468 which are divisible by 3 = $\left[\dfrac{468}{3}\right] = 156$

Number of integers between 1 and 468 which are divisible by 3 and 5 = $\left[\dfrac{468}{3\times 5}\right] = 31$

Hence the number of integers between 1 and 468 divisible by 3 but not by 5 = 156 – 31 = 125.

**Example 37.8.** *How many integers are between 1 and 200 which are divisible by any one of the integers 2, 3 and 5 ?*

**Solution.** Let $A_1$, $A_2$, $A_3$ denote the set of integers between 1 and 200 which are divisible by 2, 3, 5 respectively.

$$[A_1] = \left[\frac{200}{2}\right] = 100,\ [A_2] = \left[\frac{200}{3}\right] = 66,\ [A_3] = \left[\frac{200}{5}\right] = 40$$

$$[A_1 \cap A_2] = \left[\frac{200}{2\times 3}\right] = 33,\ [A_1 \cap A_3] = \left[\frac{200}{2\times 5}\right] = 20$$

$$[A_2 \cap A_3] = \left[\frac{200}{3\times 5}\right] = 13,\ [A_1 \cap A_2 \cap A_3] = \left[\frac{200}{2\times 3\times 5}\right] = 6$$

Hence
$$[A_1 \cup A_2 \cup A_3] = |A_1| + |A_2| + |A_3| - |A_1 \cap A_2| - |A_2 \cap A_3| - |A_3 \cap A_1| + [A_1 \cap A_2 \cap A_3]$$
$$= 100 + 66 + 40 - 33 - 13 - 20 + 6 = 146.$$

## PROBLEMS 37.1

1. Show that the following sets are equal :
$A = \{2, 1\}, B = \{1, 2, 1, 2, 1, 2\}, \quad C = \{x : x^2 - 3x + 2 - 0\}$.
2. Which of the following statements are true ? Give reason to support your answer.
(*i*) $\{a\} \subset \{a, b, c\}$ (*ii*) $a \subset \{a, b, c\}$ (*iii*) $a \subseteq \{a, b, c\}$
(*iv*) $\{a, b\} \subset \{a, b, c\}$ (*v*) $\{a, b\} \in \{a, b, c\}$ (*vi*) $\phi \subset \{a, b, c\}$.
3. Prove that
(*i*) $B - A$ is a subset of $A^c$. (*ii*) $B - A^c = B \cap A$
(*iii*) $\{A \subset B, B \subset C, C \subset A\} \Rightarrow A = C$.
4. If $A = \{a, b, c, d, e\}$, $B = \{a, c, e, g\}$ and $C = \{b, e, f, g\}$, prove that
(*i*) $(A \cup B) \cap C = (A \cap C) \cup (B \cap C)$ (*ii*) $(A \cap B) \cup (A \cap C) = A \cap (B \cup C)$.
5. If $A = A \cup B$ then prove that $B = A \cap B$.
6. Prove that $A \cup B = B \Leftrightarrow A \subset B$.
7. With the help of the Venn-diagram, prove that
(*i*) $(A \cup B)^c = A^c \cap B^c$ and $(A \cap B)^c = A^c \cup B^c$.
(*ii*) $A \cup (B \cap C) = (A \cup B) \cap (A \cup C)$.
8. If $B \subset A$, prove that
(*i*) $B \cup C \subset A \cup C$ (*ii*) $B \cap C \subset A \cap C$.
9. (*i*) If $A \cup B = A \cap B$, show that $A = B$.
(*ii*) If $A \cup B = A \cup C$ and $A \cap B = A \cap C$, show that $B = C$.
10. Prove that (*i*) $A - B = A - A \cap B$.
(*ii*) $A - (B \cup C) = (A - B) \cap (A - C)$.
11. Show that for any two sets $A$ and $B$
(*i*) $A - B = A \cap B^c$ (*ii*) $A \subseteq B \Leftrightarrow B^c \leq A^c$
(*iii*) $A \cup B = (A \cap \overline{B}) \cup (B \cap \overline{A}) \cup (A \cap B)$.
12. If $A, B, C$ be sets such that $A \subset B$, $B \cap C = \phi$, show that $A \cap C = \phi$.
13. Show that $A \cup (B \cup C)^c = (A \cup B^c) \cap (A \cup C^c)$.

14. Prove that $A \times (B \cup C) = (A \times B) \cup (A \times C)$.
15. If S is any set and $P(s)$ is its power set and $A$ and $B$ belong to $P(s)$, prove that $B \cap (A - B) = \phi$.
16. If $A$ and $B$ are finite sets then prove that $A \cup B$ and $A \cap B$ are finite sets and
$$n(A \cup B) = n(A) + n(B) - n(A \cap B).$$
17. In survey conducted on 200 people, it was found that 140 are smokers while 80 are alcoholic and 40 are both smokers and alcoholics. Find how many are neither smokers not alcoholics.
18. How many integers between 1 and 789 are divisible by 5 but not by 7.
19. How many integers are between 1 and 250 which are divisible by any of the integers 3, 5, and 7.
20. Out of a class of 153 students, 54 have taken History, 63 have taken Geography, 62 have taken Economics, and 43 have taken Geography and History, 45 have taken History and Economics, 46 have taken Geography and Economics and 37 have taken all the three subjects. How many of the students have not taken any of these three subjects ? Use a Venn diagram.

# II. ALGEBRA OF LOGIC

## 37.5 INTRODUCTION

**(1)** Logic is concerned with all types of reasoning such as valid statements, mathematical proofs, valid conclusions etc. Logical reasoning is used to prove theorems, to verify the correctness of computer programs and to draw conclusions from experiments. Later on, we shall observe that the algebra of sets and logic is analogous to the algebra of switching circuits which is similar to 'Boolean Algebra'.

**(2) Propositions and Statements.** A *proposition* is a declarative sentence which is either true (1) or false (0). Some authors use $T$ and $F$ respectively for 1 and 0. The truth or falsity of a proposition is defined as its *truth value*.

All the declarative sentences to which it is possible to assign one and only one of the two possible truth values are called *statements*.

**Example 37.9.** *Which of the following are statements : (a) Agra is in India (b) 3 + 4 = 5 (c) Where do you live ? (d) Do you speak Hindi ?*

**Solution.** (*a*) and (*b*) are statements that happen (*a*) is true and (*b*) is false.

(*c*) and (*d*) are questions so they are not statements.

**(3) Compound statements.** The statement which is composed of sub-statements and logical connectives is called a *compound statement*.

*e.g.*, 'It is raining and it is cold' is a compound statement as it is comprised of two sub-statements 'It is raining' and 'it is cold'.

**(4) Truth table.** The truth value of a compound statement is completely determined by the truth value of its substatements. *A convenient way to represent a compound statement is by means of the* **truth table** *wherein the values of a compound statement are specified for all possible choices of the values of the sub statements.*

We shall use the numbers 0 and 1 to denote the false and true statements. Also we use letters $p, q, r, \ldots$ to represent a proposition or logical variable.

## 37.6 LOGICAL OPERATORS

**(1) Conjunction.** *If p and q are two statements then their conjunction p and q written as $p \wedge q$, is defined by the truth table 1.*

Table 1

| $p$ | $q$ | $p \wedge q$ |
|---|---|---|
| 1 | 1 | 1 |
| 1 | 0 | 0 |
| 0 | 1 | 0 |
| 0 | 0 | 0 |

Table 2

| $p$ | $q$ | $p \vee q$ |
|---|---|---|
| 1 | 1 | 1 |
| 1 | 0 | 1 |
| 0 | 1 | 1 |
| 0 | 0 | 0 |

*For example,* the conjunction of $p$ : it is raining and $q$ : I am cold is $p \wedge q$ : It is raining and I am cold.

**(2) Disconjunction.** *If p and q are two statements, then their disconjunction p or q written as $p \vee q$ is defined by the truth table 2.*

*For example,* the disjunction of $p$ and $q$ for $p$ : it is raining today ; $q$ : 3 is an odd integer is

$p \vee q$ : it is raining today or 3 is an odd integer.

**(3) Negation.** *If p is a given statement and its negative 'not p', written as $\sim p$ (or Np or $\rceil p$) is defined by the following truth table :*

| $p$ | $\sim p$ |
|---|---|
| 0 | 1 |
| 1 | 0 |

*For example,* the negation of the following statement

(*a*) $p : 2 + 3 > 1$ is $\sim p : 2 + 3 \leq 1$

(*b*) $q$ : it is hot is $\sim q$ : it is cold.

**Example 37.10.** *If p be 'it is hot' and q be 'it is raining', describe each of the following statements by a sentence :*

(*a*) $q \vee \sim p$ (*b*) $\sim p \wedge \sim q$ (*c*) $\sim (\sim p \vee q)$.

**Solution.** (*a*) It is raining or it is not hot. (*b*) It is not hot and it is not raining.

(*c*) It is hot but not raining.

**(4) Conditional operator.** *The conditional statement 'if p then q' written as $p \to q$ is defined by the truth table 4* :

Table 4

| $p$ | $q$ | $p \to q$ |
|---|---|---|
| 1 | 1 | 1 |
| 1 | 0 | 0 |
| 0 | 1 | 1 |
| 0 | 0 | 1 |

Table 5

| $p$ | $q$ | $p \leftrightarrow q$ |
|---|---|---|
| 1 | 1 | 1 |
| 1 | 0 | 0 |
| 0 | 1 | 0 |
| 0 | 0 | 1 |

**Obs.** The contrapositive of conditional statement $p \to q$ is the statement $\sim p \to \sim q$.

**(5) Biconditional operator.** If $p$ and $q$ be two statements, then the statement '$p$ if and only if $q$' denoted by $p \leftrightarrow q$ and abbreviated as '$p$ if $q$' is called a *biconditional statement*. The truth table for biconditional statement is table 5.

**Example 37.11.** *Construct the truth tables for*

(*a*) $p \wedge \sim q$ (*b*) $(p \vee q) \vee \sim p$

(*c*) $(p \to q) \wedge (q \to p)$ (*d*) $(p \to q) \vee \sim (p \leftrightarrow \sim q)$.

**Solution.** (*a*) The truth table is

| $p$ | $q$ | $\sim q$ | $p \wedge \sim q$ |
|---|---|---|---|
| 1 | 1 | 0 | 0 |
| 1 | 0 | 1 | 1 |
| 0 | 1 | 0 | 0 |
| 0 | 0 | 1 | 0 |

(*b*) The truth table is

| $p$ | $q$ | $p \vee q$ | $\sim p$ | $(p \vee q) \vee \sim p$ |
|---|---|---|---|---|
| 1 | 1 | 1 | 0 | 1 |
| 1 | 0 | 1 | 0 | 1 |
| 0 | 1 | 1 | 1 | 1 |
| 0 | 0 | 0 | 1 | 1 |

(*c*) The truth table is

| $p$ | $q$ | $p \to q$ | $q \to p$ | $(p \to q) \wedge (q \to p)$ |
|---|---|---|---|---|
| 1 | 1 | 1 | 1 | 1 |
| 1 | 0 | 0 | 1 | 0 |
| 0 | 1 | 1 | 0 | 0 |
| 0 | 0 | 1 | 1 | 1 |

(*d*) The truth table in this case is

| $p$ | $q$ | $p \to q$ | $\sim q$ | $p \to \sim q$ | $(p \to \sim q)$ | $p \to q \vee \sim (p \to \sim q)$ |
|---|---|---|---|---|---|---|
| 1 | 1 | 1 | 0 | 0 | 1 | 1 |
| 1 | 0 | 0 | 1 | 1 | 0 | 0 |
| 0 | 1 | 1 | 0 | 1 | 0 | 1 |
| 0 | 0 | 1 | 1 | 0 | 1 | 1 |

## 37.7 STATEMENTS GENERATED BY A SET

**(1)** *If S be a set of statements, then any valid combination of statements in S with conjunction, disconjunction or negation is a statement generated by S.*

A statement generated by a set *S* need not include each element of *S* in its expression.

*For example, if p, q, r are statements in S then*

(*a*) $(p \wedge q) \wedge r$ (*b*) $\sim q \wedge r$ (*c*) $(p \wedge q) \vee (\sim q \wedge r)$

*are statements generated by S. Their truth tables are :*

| $p$ | $q$ | $r$ | $p \wedge q$ | $(p \wedge q) \wedge r$ (*a*) | $\sim q$ | $\sim q \wedge r$ (*b*) | (*c*) |
|---|---|---|---|---|---|---|---|
| 1 | 1 | 1 | 1 | 1 | 0 | 0 | 1 |
| 1 | 1 | 0 | 1 | 0 | 0 | 0 | 1 |
| 1 | 0 | 1 | 0 | 0 | 1 | 1 | 1 |
| 0 | 1 | 1 | 0 | 0 | 0 | 0 | 0 |
| 1 | 0 | 0 | 0 | 0 | 1 | 0 | 0 |
| 0 | 1 | 0 | 0 | 0 | 0 | 0 | 0 |
| 0 | 0 | 1 | 0 | 0 | 1 | 1 | 1 |
| 0 | 0 | 0 | 0 | 0 | 1 | 0 | 0 |

**(2) Tautology** *is an expression involving logical variables which is true for all cases in its truth table.* It is also called a *logical truth.*

**(3) Contradiction** is an expression involving logical variables which is false for all cases in its truth table. Obviously, the negation of a contradiction is a tautology.

In other words, a statement formula which is a tautology is identically true, while a formula which is a contradiction is identically false.

**Obs.** The conjunction of two tautologies is also a tautology.

**Example 37.12.** *Show that (a) $p \vee \sim p$ is a tautology (b) $p \to q \leftrightarrow (\sim p \vee q)$.*

**Solution.** (*a*) The truth table is

| $p$ | $\sim p$ | $p \vee \sim p$ |
|---|---|---|
| 1 | 0 | 1 |
| 0 | 1 | 1 |

Hence $p \vee \sim p$ is a tautology.

(b) The truth table is

| $p$ | $q$ | $p \to q$ | $\sim p$ | $\sim p \vee q$ | $(p \to q) \leftrightarrow (\sim p \vee q)$ |
|---|---|---|---|---|---|
| 1 | 1 | 1 | 0 | 1 | 1 |
| 1 | 0 | 0 | 0 | 0 | 1 |
| 0 | 1 | 1 | 1 | 1 | 1 |
| 0 | 0 | 1 | 1 | 1 | 1 |

Hence $(p \to q) \leftrightarrow (\sim p \vee q)$ is true.

## 37.8 EQUIVALENCE

**(1)** *If p and q be statements generated by the set of statements S, then p and q are equivalent if $p \leftrightarrow q$ is a tautology which is denoted by $p \Leftrightarrow q$.*

If $p \to q$ is a tautology, then we say that *p implies q* and write it as $p \Rightarrow q$.

**Obs.** All tautologies are equivalent to each other and all contradictions are equivalent to each other.

**(2) Equivalent formulae.** Some basic equivalent formulae are given below which can be proved by using truth tables :

1. $p \vee p \Leftrightarrow p$    $p \wedge p \Leftrightarrow p$    *Idempotent laws*
2. $p \vee q \Leftrightarrow q \wedge p$    $p \wedge q \Leftrightarrow q \wedge p$    *Commutative laws*
3. $(p \vee q) \vee r \Leftrightarrow p \vee (q \vee r)$    $(p \wedge q) \wedge r \Leftrightarrow p \wedge (q \wedge r)$    *Associative laws*
4. $p \vee (q \wedge r) \Leftrightarrow (p \vee q) \wedge (p \vee r)$    $p \wedge (q \vee r) \Leftrightarrow (p \wedge q) \vee (p \wedge r)$    *Distributive laws*
5. $p \vee \sim p \Leftrightarrow 1$    $p \wedge \sim p \Leftrightarrow 0$    *Negation law*
6. $p \vee 0 \Leftrightarrow p$    $p \wedge 1 \Leftrightarrow p$    *Identity laws*
7. $p \vee 1 \Leftrightarrow 1$    $p \wedge 0 \Leftrightarrow 0$    *Null laws*
8. $p \vee (p \wedge q) \Leftrightarrow p$    $p \wedge (p \vee q) \Leftrightarrow p$    *Absorption laws*
9. $\sim (p \vee q) \Leftrightarrow \sim p \wedge \sim q$    $\sim (p \vee q) \Leftrightarrow \sim p \vee \sim q$    *De Morgan's laws*
10. $p \Rightarrow p \vee q$    $q \Rightarrow p \vee q$    *Disjunctive addition*
11. $p \wedge q \Rightarrow q$    $p \wedge q \Rightarrow q$
12. $(p \vee q) \wedge \sim p \Rightarrow q$    $(p \vee q) \wedge \sim q \Rightarrow p$
13. $(p \to q) \wedge (q \to r) \Rightarrow (p \to r)$    *Chain rule*
14. $p \to q \Leftrightarrow \sim p \vee q$    *Conditional equivalence*
15. $p \leftrightarrow q \Leftrightarrow (p \to q) \wedge (q \to p)$    *Biconditional equivalence*

**Example 37.13.** *Show that*

*(a)* $p \to (q \to r) \quad \Leftrightarrow \quad p \to (\sim q \vee r) \quad \Leftrightarrow \quad (p \wedge q) \to r$

*(b)* $[\sim p \wedge (\sim q \wedge r)] \vee (q \wedge r) \vee (p \wedge r) \quad \Leftrightarrow \quad r$

**Solution.** (a) By conditional equivalence $q \to r \quad \Leftrightarrow \quad \sim q \vee r$

Replacing $q \to r$ by $\sim q \vee r$, we get $p \to (\sim q \vee r)$ which is equivalent to $\sim p \vee (\sim q \vee r)$ by the same rule.

Thus
$$\sim p \vee (\sim q \vee r) \Leftrightarrow (\sim p \vee \sim q) \vee r \qquad \text{[By (3)]}$$
$$\Leftrightarrow \sim (p \wedge q) \vee r \qquad \text{[By (9)]}$$
$$\Leftrightarrow (p \wedge q) \to r \qquad \text{[By (14)]}$$

(b) $[\sim p \wedge (\sim q \wedge r)] \vee (q \wedge r) \vee (p \wedge r)$
$$\Leftrightarrow [\sim p \wedge (\sim q \wedge r)] \vee (q \vee p) \wedge r \qquad \text{[By (4)]}$$
$$\Leftrightarrow [\sim p \wedge \sim q \wedge r] \vee (q \vee p) \wedge r \qquad \text{[By (3)]}$$
$$\Leftrightarrow [\sim (p \vee q) \wedge r] \vee (q \vee p) \wedge r \qquad \text{[By (9)]}$$
$$\Leftrightarrow [\sim (p \vee q) \vee (p \vee q)] \wedge r \qquad \text{[By (4)]}$$
$$\Leftrightarrow 1 \wedge r \qquad \text{[By (5)]}$$
$$\Leftrightarrow r \qquad \text{[By (6)]}$$

## 37.9 DUALITY LAW

**(1)** *Two formulae A and A* are said to be duals of each other if either one can be obtained from the other by replacing* $\wedge$ *by* $\vee$ *and* $\vee$ *by* $\wedge$.

If the formula A contains special variables 1 or 0, then its dual A* is obtained by replacing 1 by 0 and 0 by 1.

*e.g.,* (*i*) Dual of $(p \vee q) \wedge r$ is $(p \wedge q) \vee r$

(*ii*) Dual of $(p \wedge q) \vee 0$ is $(p \vee q) \wedge 1$.

**(2) Tautology implications.** *A statement A is said to tautologically imply a statement B if and only if* $A \to B$ *is a tautology which is read as* "*A implies B*".

*The implications listed below have important applications* which can be proved by truth tables :

| | |
|---|---|
| 1. $p \wedge q \Rightarrow p$ | $p \Rightarrow p \vee q$ |
| 2. $\sim p \Rightarrow p \to q$ | $q \Rightarrow p \to q$ |
| 3. $\sim (p \to q) \Rightarrow p$ | $\sim (p \to q) \Rightarrow \sim q$ |
| 4. $p \wedge (p \to q) \Rightarrow q$ | $\sim p \wedge (p \vee q) \Rightarrow q$ |
| 5. $(p \to q) \wedge (q \to r) \Rightarrow p \to r$ | |

## 37.10 ARGUMENTS

**(1)** *An argument is an assertion that a given set of propositions* $p_1, p_2, ..., p_n$ (called *premises*) *yields another proposition q* (called *conclusion*). The argument is symbolically written as "$p_1, p_2, ..., p_n \vdash q$".

An argument $p_1, p_2, ..., p_n \vdash q$ is true provided $q$ is true whenever all the premises $p_1, p_2, ..., p_n$ are true. An argument which is true is said to be '*valid argument*'. Otherwise it is called a *fallacy*.

**Example 37.14.** *Show that*

(*a*) *the argument* $p \leftrightarrow q, q \vdash p$ *is valid.*

(*b*) *the argument* $p \to q, q \vdash p$ *is a fallacy.*

**Solution.** (*a*) Let us first prepare the truth table as follows :

| $p$ | $q$ | $p \leftrightarrow q$ |
|---|---|---|
| 1 | 1 | 1 |
| 1 | 0 | 0 |
| 0 | 1 | 0 |
| 0 | 0 | 1 |

Since $p \leftrightarrow q$ is true in cases (rows) 1 and 4, and $q$ is true in cases 1 and 3, therefore $p \leftrightarrow q$ and $q$ both are true in case 1 only when $p$ is also true. This shows that the given argument is valid.

(*b*) Let us first prepare the truth table below :

| $p$ | $q$ | $p \to q$ |
|---|---|---|
| 1 | 1 | 1 |
| 1 | 0 | 0 |
| 0 | 1 | 1 |
| 0 | 0 | 1 |

This table shows that $p \to q$ and $q$ both are true in case 3 only while the conclusion $p$ is false. Hence the given argument is a fallacy.

**(2) Theorem.** *The argument* $p_1, p_2, ..., p_n \vdash q$ *is valid if and only if the proposition* $(p_1 \wedge p_2 \wedge p_3 \wedge ..., \wedge p_n) \to q$ *is a tautology.*

The proportions $p_1, p_2, ..., p_n$ are simultaneously true if and only if the proposition $p_1 \wedge p_2 \wedge ..., \wedge p_n$ is true *i.e.*, if the proposition $(p_1 \wedge p_2 \wedge ..., \wedge p_n) \to q$ is a tautology.

**Obs.** The validity of an argument depends upon the particular form of the argument, not on the truth values of the statement appearing in the argument.

**Example 37.15.** *Test the validity of the following argument :*

$S_1$ : *If 5 is less than 3, then 6 is not a prime number*

$S_2$ : *5 is not less than 6*

$S_3$ : *5 is a prime number.*

**Solution.** Let '5 is less than 3' be $p$ and '5 is a prime number' be $q$ . Then the given argument is of the form $p \to \sim q, \sim p \vdash q$.

Since in the last line of the truth table, the premises $p \to \sim q$ and $\sim q$ are true but the conclusion $q$ is false, therefore the argument is fallacy.

| $p$ | $q$ | $\sim q$ | $p \to \sim q$ | $\sim p$ |
|---|---|---|---|---|
| 1 | 1 | 0 | 0 | 0 |
| 1 | 0 | 1 | 1 | 0 |
| 0 | 1 | 0 | 1 | 1 |
| 0 | 0 | 1 | 1 | 1 |

## PROBLEMS 37.2

1. If $p \equiv$ Sam is a teacher, $q \equiv$ John is an honest boy, then translate the following into logical sentences :
   (a) $\sim (p \wedge q)$, (b) $p \vee \sim q$, (c) $\sim p \Leftrightarrow q$, (d) $p \Rightarrow \sim q$.
2. Change the following sentence into symbols :
   (a) 'If I do not have car or I do not wear good dress then I am not a millionaire'.
   (b) Everyone who is healthy can do all kinds of work.
3. Prepare truth tables for the following statements (a) $(p \Rightarrow q) \wedge \sim q$, (b) $(p \Leftrightarrow q) \wedge (r \vee q)$.
4. Write down the truth table of
   (a) $p \vee q$ *(Anna, 2013)* (b) $p \wedge (p \wedge q)$
5. Verify that the following are tautologies :
   (a) $p \to [q \to (p \wedge q)]$
   (b) $(p \wedge q \Rightarrow r) \Leftrightarrow (p \Rightarrow r) \vee (q \Rightarrow r)$ (c) $(p \Rightarrow q \wedge r) \Rightarrow (\sim r \Rightarrow \sim p)$.
6. Show that $Q \vee (P \wedge \sim Q) \vee (\sim P \wedge \sim Q)$ is a tautology.
7. Over the universe of positive integers
   $p(n)$ : $n$ is prime and $n < 32$.
   $q(n)$ : $n$ is a power of 3.
   $r(n)$ : $n$ is a divisor of 27.
   (i) What are the truth sets of these propositions ?
   (ii) Which of the three propositions implies one of the others ?
8. Given the propositions over the natural numbers
   $p : n < 4$, $q : 2n > 17$ and $r : n$ is a divisor of 18, what are the truth sets of
   (i) $q$, (ii) $p \wedge q$ ,
   (iii) $r$, (iv) $q \to r$.
9. Show that (a) $\sim Q, P \to Q \Rightarrow \sim P$.
   (b) $(P \to R) \wedge (Q \to R) \Leftrightarrow (P \vee Q) \to R$. *(Anna, 2013)*
10. Construct the truth table for (i) $(\sim p \to q) \wedge (q \rightleftarrows p)$.
    (ii) $\rceil[P \vee (Q \wedge R)] \rightleftarrows (P \vee Q) \wedge (P \vee R)$.
11. Prove that the following statement is a contradiction :
    $S = [(p \vee q) \wedge (p \vee \sim q) \wedge (\sim p \vee q) \wedge (\sim p \vee \sim q)]$.
12. If $p$, $q$, $r$ are three statements then prove that
    (a) $p \vee (q \wedge r) \equiv (p \vee q) \wedge (p \vee r)$ (b) $(p \Rightarrow q) \vee r \equiv (p \vee r) \Rightarrow (q \vee r)$
    (c) $\sim (p \vee q) \equiv \sim p \wedge \sim q$.
13. Define conjunction, conditional, biconditional and negation, with examples.

14. Show that (*i*) $p \wedge q$ logically implies $p \leftrightarrow q$.
    (*ii*) $p \leftrightarrow \sim q$ does not logically imply $p \to q$.
15. Write the duals of $(p \vee q) \wedge r$ and $(p \wedge q) \vee t$.
16. Show that $s \vee r$ is tautologically implied by $(p \vee q) \wedge (q \to r) \wedge (q \to s)$.
17. Let $P(n)$ be '$8^n - 3^n$ is a multiple of 5'. Prove that $P(n)$ is a tautology on $n$.
18. Prove that the following arguments are valid:
    (*a*) $P \to -Q, R \to Q, R \Rightarrow -P$. (*Anna, 2013*)
    (*b*) $P \to \rceil Q, R \to Q, R \vdash \rceil P$. (*Anna, 2012*)
19. Prove that $p \to (q \vee r) \Leftrightarrow (p \wedge \rceil q) \to r$.
20. Show that $R \vee S$ follows logically from the premises $C \vee D, C \vee D \to -H, H \to A \wedge \sim B, A \wedge \sim B \to R \vee S$.

## 37.11 PREDICATES

Statements involving variables such as '$x > 7$' and '$x = y + 7$' are neither true nor false so long as the values of the variable $x$, $y$ are not specified. We now, discuss the ways that propositions can be evolved from such statements. The statement '$x > 7$' has two parts : First part—'the variable $x$' is the *subject* of the statement ; Second part–'is greater than 7' is the *predicate* which refers to the property that the subject of statement can have. If $P(x)$ denotes the statement '$x > 7$' then $P$ is the *predicate* and $x$ is the *variable*. The statement $P(x)$ is also known as the value of the propositional function $P$ at $x$. The *predicates* are denoted by capital letters and the *objects* by the variables (denoted by small letters in brackets).

Thus *predicates are simple statements which turn out to be propositions involving variables whose values are not well specified.*

In other words, *predicate is a variable statement which becomes specific when particular values are assigned to the variables.*

There are statements which involves more than one variable consider the statement '$x = y + 7$' which is denoted by $Q(x, y)$ where $Q$ is the predicate and $x, y$ are the variables. When values are assigned to the variables $x, y$, the statement $Q(x, y)$ has the truth value.

Similarly $R(x, y, z)$ denotes the statement of the type '$x + y = z$'. When values are assigned to $x, y, z$, this statement has a truth value.

*For example, consider the statements (i) Ram is fair ; (ii) Sham is fair.*

Here in (*i*) and (*ii*) 'is fair' is the predicate while Ram and Sham are the objects. If we denote the predicate by $F$ and the objects by $r$ and $s$, then the above statements can be symbolically expressed as (*i*) $F(r)$ ; (*ii*) $F(s)$.

*Now consider the statement Ram is fair and the house is pink.*

Writing 'the house is pink' as $P(h)$, the given statement can be expressed as $F(r) \wedge P(h)$.

## 37.12 QUANTIFIERS

**(1)** In a propositional function, when all the variables are assigned values, the resulting statement has a truth value. However, there is another method to create a proposition from a propositional function which is called *quantification*. It is of two types : *Universal quantification* and *Existential quantification*.

**(2) Universal quantification.** Many statements assert that a property is true for all values of a variable in a certain domain. This domain is termed as the universe of discourse and such a statement is expressed using universal quantification.

Thus *the universal quantification of P(x) is the proposition 'P(x) is true for all values of x in the universe of discourse'*.

The universal quantification of $P(x)$ is denoted by $\forall\, xP(x)$. The symbol $\forall$ is called the *universal quantifier*.

**Obs.** When it is possible to list all the elements in the universe of discourse say : $x_1, x_2, ..., x_n$, then the universal quantification $\forall\, x\, P(x)$ is same as the conjunction $P(x_1) \wedge P(x_2) \wedge ..., P(x_n)$ for this conjunction is true if and only if $P(x_1)$, $P(x_2)$, ..., $P(x_n)$ are all true.

**Example 37.16.** *What is the truth value of the quantification $\forall\, x\, P(x)$ where*
*(a) P(x) is the statement '$x < 5$' and universe of discourse is the set of real numbers.*
*(b) P(x) is the statement '$x^2 < 18$' and the universe of discourse consists of positive integers not exceeding 5 ?*

**Solution.** (*a*) For instance, $P(6)$ is false ; therefore $P(x)$ is not true for all real numbers $x$.
Thus $\forall\, xP(x)$ is false.

(*b*) The statement $\forall\, xP(x)$ is same as the conjunction $P(1) \wedge P(2) \wedge P(3) \wedge P(4) \wedge P(5)$

Here the universe of discourse is 1, 2, 3, 4, 5 and $P(5)$ is the statement '$5^2 < 18$' which is false. Hence $\forall\, xP(x)$ is also false.

**(3) Existential quantification.** Many statements assert that there is an element with a certain property. To express such statements we use *existential quantification*. In such cases, we form a proposition which is true if and only if $P(x)$ is true for at least one value of $x$ in the universe of discourse.

Thus *the existential quantification of P(x) is the proposition 'these exists an element x in the universe of discourse such that P(x) is true'.*

The notation $\exists xP(x)$ is used for the *existential quantification* wherein $\exists$ is called the *existential quantifier*.

**Obs.** When it is possible to list all the elements in the universe say : $x_1, x_2, ..., x_n$, then the existential quantification $\exists x\, P(x)$ is same as the disjunction $P(x_1) \vee P(x_2) \vee ..., P(x_n)$.

**Example 37.17.** *What is the truth value of the quantification $\exists xP(x)$ where*

*(a) P(x) is the statement '$x > 5$' and universe of discourse is the set of real numbers.*

*(b) P(x) is the statement '$x^2 > 18$' and the universe of discourse consists of positive integers not exceeding 5 ?*

**Solution.** (*a*) Since '$x > 5$' is true, say : for $x = 6$, 8 etc.

$\therefore$ $\exists xP(x)$ is true.

(*b*) The statement $\exists xP(x)$ is same as the disjunction $P(1) \vee P(2) \vee P(3) \vee P(4) \vee P(5)$

The universe of discourse is 1, 2, 3, 4, 5 and $P(5)$ is the statement $5^2 > 18$, which is true.

Hence $\exists xP(x)$ is also true.

**Example 37.18.** *Find the truth value of each of the following statements :*

*(a)* $\exists x,\, x^2 = 1$ *(b)* $\forall x,\, |x| = x$ *(c)* $\exists x,\, x + 4 = x$.

**Solution.** (*a*) If $x_0 = 1$, then $x_0^2 = 1$, therefore the given statement is true.

(*b*) If $x_0 = -3$, then $|x_0| \neq x_0$, therefore the given statement is false.

(*c*) As there is no solution to $x + 4 = x$, the given statement is false.

**Example 37.19.** *Given P = {2, 3, 4, 5, 6}, state the truth values of each of the following statements :*

*(a)* $(\forall\, x \in P)\ (x + 3 < 10)$ *(b)* $(\forall\, x \in P)\ (x + 2 \leq 7)$

*(c)* $(\exists x \in P)\ (x + 3 = 10)$ *(d)* $(\exists x \in P)\ (x + 2 < 7)$.

**Solution.** (*a*) True, for each number in $P$ satisfies $x + 3 < 10$.

(*b*) False, for if $x_0 = 6$, then $x_0 + 2$ is not $\leq 7$.

(*c*) False, for no number in $P$ is a solution to $x + 3 = 10$.

(*d*) True, for if $x_0 = 2$ then $x_0 + 2 < 7$ is a solution.

**Example 37.20.** *Negate each of the following statements :*

*(a)* $\forall x,\, x^2 = x$, *(b)* $\forall\, x,\, x + 4 > x$ *(c)* $\forall x,\ |\ x\ | = x$.

**Solution.** (*a*) $\sim \forall\, x,\, x^2 = x \equiv \forall\, x \sim (x^2 = x) \equiv \forall\, x,\, x^2 \neq x$.

(*b*) $\sim \forall\, x,\, x + 4 > x \equiv \exists\, x \sim (x + 4 > x) \equiv \rightarrow x\,,\, x + 4 \leq x$.

(*c*) $\sim \forall\, x,\ |\ x\ | = x \equiv \exists\, x \sim (|\ x\ | = x) \equiv \exists\, x,\ |\ x\ | \neq x$.

**Example 37.21.** *Symbolise using quantifiers :*

*(i) Every even number is divisible by 2.*

*(ii) There is no prime number between 23 and 29.*

**Solution.** (*i*) $E(x)$ : $x$ is even number ; $D(x)$ : $x$ is divisible by 2.

$(\forall\, x)\ [E(x) \rightarrow D(x)]$

(*ii*) If $p$ denotes the set of prime numbers, then $(\nexists_n)_p\ (23 < n < 29)$.

**Example 37.22.** *Symbolise the expression 'All the world loves a lover.'*

**Solution.** Let $p(x) : x$ is a person ;

$L(x) : x$ is a lover.

and $Q(x, y) : x$ loves $y$

Then the required expression is

$$(\forall x)\ [p(x) \to (y)\ (p(y) \wedge L(y)] \to Q(x, y).$$

**Summary.** (*i*) $\forall\ Q(x)$ means that the predicate $Q(x)$ is true *for all values* in the universe of $x$.

(*ii*) $\exists Q(x)$ means that the predicate $Q(x)$ is satisfied if there is *at least one value* in the universe of $x$.

## 37.13 NORMAL FORMS

**(1)** For the given variables $p_1, p_2, ..., p_n$, we may form a statement $S(p_1, p_2, ..., p_n)$. The truth table for $S$ will contain $2^n$ rows for all possible truth values of the $n$ variables. The expression $S$ may have the truth value 1 in all cases or may have the truth value 0 in all cases or have the truth value 1 for at least one combination of truth values assigned to the $n$ variables. (Here $S$ is said to be satisfiable). The problem of finding in a finite number of steps whether a given expression is a tautology or a contradiction or at least satisfiable is known as a *decision problem*. As the formation of a truth table is quite cumbersome, we go for an alternate approach called *normal form*.

In this approach, we use the word 'sum' in place of disjunction and 'product' in place of conjunction.

A sum of the variables and their negations is called an *elementary sum*. Similarly a product of the variables and their negation is called an *elementary product*.

**(2) Disjunctive normal form** *of a given formula is the formula which is equivalent to the given formula and which contains the sum of the elementary products.* The disjunctive normal form of a given form is not unique. In fact, several disjunctive normal forms can be obtained for a given formula by applying the distributive laws in different ways.

A given formula is however, identically false if every elementary product appearing in its disjunctive normal form is identically false.

**Example 37.23.** *Obtain the disjunctive normal forms of*

(*i*) $p \wedge (p \to q)$ (*ii*) $\sim (p \vee q) \leftrightarrow (p \wedge q)$.

**Solution.** (*i*) $p \wedge (p \to q) \Leftrightarrow p \wedge (\sim p \vee q) \Leftrightarrow (p \wedge \sim p) \vee (p \wedge q)$.

which is the desired disjunctive normal form.

(*ii*) $\sim (p \vee q) \leftrightarrow (p \wedge q) \Leftrightarrow \sim (p \vee q) \wedge (p \wedge q) \vee (p \vee q) \wedge \sim (p \wedge q)$ $[\because\ E \leftrightarrow F \Leftrightarrow (E \wedge F) \vee (\sim E \wedge \sim F)$

$$\Leftrightarrow (\sim p \wedge \sim q \wedge p \wedge q) \vee [(p \vee q) \wedge (\sim p \vee \sim q)]$$

$$\Leftrightarrow (\sim p \wedge \sim q \wedge p \wedge q) \vee [(p \vee q) \wedge \sim p] \vee [(p \vee q) \wedge \sim q]$$

$$\Leftrightarrow (\sim p \wedge \sim q \wedge p \wedge q) \vee (p \wedge \sim p) \vee (q \wedge \sim p) \vee (p \wedge \sim q) \vee (q \wedge \sim q)$$

which is the desired disjunctive normal form.

**(3) Conjunctive normal form** *of a given formula is that formula which is equivalent to the given formula and contains the product of elementary sums.*

**Example 37.24.** *Find a conjunctive normal form of* $\sim (p \vee q) \leftrightarrow (p \wedge q)$.

**Solution.** $\sim (p \vee q) \leftrightarrow (p \wedge q) \Leftrightarrow [\sim (p \vee q) \to (p \wedge q)] \wedge [p \wedge q \to \sim (p \vee q)]$

$$\Leftrightarrow [(p \vee q) \vee (p \wedge q)] \vee [\sim (p \wedge q) \vee \sim (p \vee q)]$$ [By conditional equivalence]

$$\Leftrightarrow (p \vee q \vee p) \wedge (p \vee q \vee q) \wedge [(\sim p \vee \sim q) \vee (\sim p \wedge \sim q)]$$

$$\Leftrightarrow p \vee q \vee p \wedge (p \vee q \vee q) \wedge (\sim p \vee \sim q \vee \sim p) \wedge (\sim p \vee \sim q \vee \sim q)$$

which is the required conjunctive normal form.

**(4) Principal disjunctive normal form.** Consider a formula for the propositions $p$ and $q$ using conjunction as $p \wedge q, p \wedge \sim q, \sim p \wedge q, \sim p \wedge \sim q$. These terms are called *minterms* or *Boolean conjunction* of $p$ and $q$.

*An equivalent formula for a given formula, consisting of disjunctions of minterms only is called the* **principal disjunctive normal form** (*pdnf*) or **sum of products canonical form**.

**Procedure** *to obtain the principle disjunctive normal form* : (*i*) Replace the conditions and biconditions by their equivalent formulae containing $\wedge$, $\vee$, ~ only.

(*ii*) Using DeMorgans laws, apply negations to the variables.

(*iii*) Apply the distribution laws.

(*iv*) Introduce the missing factors to obtain minterms in the disjunctions.

(*v*) Delete identical minterms appearing in the disjunctions.

**Example 37.25.** *Obtain the pdnf for*

(*i*) $p \vee q$ (*ii*) $\sim(p \wedge q)$ (*iii*) $\sim p \vee q$ *i.e.*, $p \to q$.

**Solution.** (*i*)
$$p \vee q \Leftrightarrow [p \wedge (q \vee \sim q)] \vee [q \wedge (p \vee \sim p)]$$
$$\Leftrightarrow (p \wedge q) \vee (p \wedge \sim q) \vee (q \wedge p) \vee (q \wedge \sim p)$$
$$\Leftrightarrow (p \wedge q) \vee (p \wedge \sim q) \vee (q \wedge \sim p)$$

(*ii*)
$$\sim (p \wedge q) \Leftrightarrow (\sim p \vee \sim q) \Leftrightarrow [\sim p \wedge (\sim q \vee q)] \vee [\sim q \wedge (p \vee \sim p)]$$
$$\Leftrightarrow (\sim p \wedge \sim q) \vee (\sim p \wedge q) \vee (\sim q \wedge p) \vee (\sim q \wedge \sim p)$$
$$\Leftrightarrow (\sim p \wedge \sim q) \vee (\sim p \wedge q) \vee (\sim q \wedge p)$$

(*iii*)
$$\sim p \vee q \Leftrightarrow \sim p \wedge (q \vee \sim q) \vee [q \wedge (p \vee \sim p)]$$
$$\Leftrightarrow (\sim p \wedge q) \vee (\sim p \wedge \sim q) \vee (q \wedge p) \vee (q \wedge \sim p)$$
$$\Leftrightarrow (\sim p \wedge q) \vee (\sim p \wedge \sim q) \vee (q \wedge p).$$

**(5) Principal conjunctive normal form.** Consider a formula for the propositions $p$ and $q$ using disconjunction as $p \vee q$, $\sim p \vee q$, $p \vee \sim q$, $\sim p \vee \sim q$. These terms are called *max terms.*

*An equivalent formula for a given formula, consisting of conjunctions of the max terms only is called the* **principal conjunctive normal form** (*pcnf*) or **product of sum canonical form**.

**Procedure** *for obtaining pcnf* for a given formula is similar to the one for *pdnf* as all assertions made for *pdnf* can be made for *pcnf* using duality principle.

**Example 37.26.** *Obtain the principal disjunctive and conjunctive normal forms of*
$$p \to [(p \to q) \wedge \sim (\sim q \vee \sim p)].$$

**Solution.** (*i*)
$$p \to [(p \to q) \wedge \sim (\sim q \vee \sim p)]$$
$$\Leftrightarrow \sim p \wedge [(\sim p \vee q) \wedge (q \wedge p)]$$
[Using DeMorgan's law and equivalence $p \to q \Leftrightarrow \sim p \vee q$.]
$$\Leftrightarrow \sim p \vee [\sim p \wedge (q \wedge p)] \vee [q \wedge (q \wedge p)]$$
$$\Leftrightarrow \sim p \vee (q \wedge p)$$
$$\Leftrightarrow [\sim p \wedge (q \vee \sim q)] \vee (q \wedge p)$$
$$\Leftrightarrow (\sim p \wedge q) \vee (\sim p \wedge \sim q) \vee (q \wedge p)$$

This is the desired *pdnf.*

(*ii*)
$$p \to [(p \to q) \wedge \sim (\sim q \vee \sim p)]$$
$$\Leftrightarrow \sim p \vee [(\sim p \vee q) \wedge (q \wedge p)]$$
$$\Leftrightarrow [\sim p \vee (\sim p \vee q)] \wedge [\sim p \vee (q \wedge p)]$$
$$\Leftrightarrow (\sim p \vee q) \wedge (\sim p \vee q) \wedge (\sim p \vee p)$$
$$\Leftrightarrow \sim p \vee q$$

This is the desired *pcnf.*

**Example 37.27.** *Obtain the principal conjunctive normal form for* $(Q \to P) \wedge (\sim P \wedge Q)$.

**Solution.**
$$(Q \to P) \wedge (\sim P \wedge Q) \Leftrightarrow (\sim Q \vee P) \wedge (\sim P \wedge Q)$$
$$\Leftrightarrow (\sim Q \vee P) \wedge [\sim P \vee (Q \wedge \sim Q) \wedge Q \vee (P \wedge \sim P)]$$
$$\Leftrightarrow (\sim Q \vee P) \wedge (\sim P \vee Q) \wedge (\sim P \vee \sim Q) \wedge (Q \vee P) \wedge (Q \vee \sim P)$$
$$\Leftrightarrow (\sim Q \vee P) \wedge (\sim P \vee Q) \wedge (\sim P \vee \sim Q) \wedge (Q \vee P).$$

## 37.14 INFERENCE THEORY

Inferring the conclusions from certain premises is known as the *inference theory*. When conclusion is reached from a set of premises by using the accepted rules of reasoning, then this process is called a *deduction* or a *formal proof*. A proof of a theorem is a valid argument.

The criteria for finding 'whether an argument is valid' are called *rules* which are expressed in terms of premises and conclusions or in terms of statement formulae.

*The proofs are of two types : Direct or Indirect.*

(*i*) If in a proof the truth of the premises directly shows the truth of the conclusions, then it is called a *direct proof.*

(*ii*) *An indirect proof* proceeds by assuming that $p$ is true and also $C$ is false, and then deduce a contradiction using $p$ and $\sim C$, along with other premises.

The only difference between assumptions in a direct proof or an indirect proof is the negated conclusion.

*Rules for deriving direct and indirect proofs* :

(*i*) the proof should have finite steps only ;

(*ii*) each step must be either a premise or a proposition which is implied from previous steps using valid equivalence or implication ;

(*iii*) the last step for a direct proof must be conclusion while for an indirect proof it must be a contradiction.

**Example 37.28.** *State whether the conclusion C follows logically from the premises R and S*

| | | | |
|---|---|---|---|
| (*a*) | $R : p \to q,$ | $S : p,$ | $C : \sim q$ |
| (*b*) | $R : p,$ | $S : p \leftrightarrow q,$ | $C : \sim (p \wedge q)$ |
| (*c*) | $R : p \to q,$ | $S : p,$ | $C : q$ |

**Solution.** Let us first form the following truth table :

| $p$ | $q$ | $\sim p$ | $\sim q$ | $p \to q$ | $\sim (p \wedge q)$ | $p \leftrightarrow q$ |
|---|---|---|---|---|---|---|
| 1 | 1 | 0 | 0 | 1 | 0 | 1 |
| 1 | 0 | 0 | 1 | 0 | 1 | 0 |
| 0 | 1 | 1 | 0 | 1 | 0 | 0 |
| 0 | 0 | 1 | 1 | 1 | 1 | 1 |

(*a*) Here only the first row of premises $R$ and $S$ contains 1 but not the conclusion $C$.

Hence $C$ is not valid.

(*b*) Only the first row of both $R$ and $S$ contains 1 but not the conclusion $C$.

Hence $C$ is not valid.

(*c*) Here both $R$ and $S$ contain '1' only in the first row and the conclusion $C$ also has '1' in that row. Hence our conclusion is valid.

**Example 37.29.** *Find the direct and indirect proofs of* $p \to (q \to r),\ \sim r \vee p,\ q \Rightarrow r \to s.$

**Solution.** *Direct proof*

| | | | |
|---|---|---|---|
| (*i*) $\sim r \vee p$ | (premise) | (*ii*) $r$ | (another premise) |
| (*iii*) $p$ | [By (*i*) and (*ii*)] | (*iv*) $p \to (q \to r)$ | (premise) |
| (*v*) $q \to r$ | [By (*iii*) and (*iv*)] | (*vi*) $q$ | (premise) |
| (*vii*) $r$ | [By (*v*) and (*vi*)], which is a conclusion. | | |

*Indirect proof*

| | |
|---|---|
| (*i*) $\sim (s \to r)$ | (negative of conclusion) |
| (*ii*) $s \wedge \sim r$ | (By conditional equivalence) |
| (*iii*) $\sim r \vee p$ | (premise) |
| (*iv*) $\sim r$ and (*v*) $s$ | [By conjunctive simplification] |
| (*vi*) $p$ | [By disconjunction of (*iii*) and (*iv*)] |
| (*vii*) $p \to (q \to r)$ | (premise) |

(viii) $q \to r$ [By (vi) and (vii)]

(ix) $q$ (premise)

(x) $r$ [By (viii) and (ix)]

(xi) $r \wedge \sim r$ [By (x) and (iv)]

This is a contradiction.

## PROBLEMS 37.3

1. If $A = \{1, 2, 3, 4, 5\}$ be the universal set, determine the truth values of each of the following statements :
   (a) $(\forall x \in A)(x + 2 < 10)$ (b) $\exists x \in A (x + 2 = 10)$
2. Negate each of the following statements :
   (a) $\forall x, x^3 = x$ ; (b) $\forall x, x + 5 > x$
   (c) Some students are 26 or older. (d) All students live in the hostels.
3. What is the truth value of $\forall x P(x)$ where $P(x)$ is a statement '$x^2 < 10$' and the universe of discourse consists of positive integers not exceeding 4.
4. Use universal quantifier to state 'the sum of any two rational numbers is rational'.
5. Over the universe of real numbers, use quantifier to say that the equation $a + x = b$ has a solution for all values of $a$ and $b$.
6. Translate the following statements involving quantifiers, into formulae :
   (a) All rationals are reals. (b) No rationals are reals.
   (c) Some rationals are reals. (d) Some rationals are not reals.
7. Show that $Q \vee (P \wedge \sim Q) \vee (\sim P \wedge \sim Q)$ is a tautology.
8. Convert $\sim A \wedge (\sim B \to C) \Rightarrow 0$ into CNF. *(V.T.U., MCA, 2001)*
9. Without constructing truth tables, obtain the product of sums canonical form of the formula $(\neg P \to R) \wedge (Q \leftrightarrow P)$. Hence find the sum of products canonicals form.
10. Find the direct proof of $p \to r, q \to s, p \vee q \Rightarrow s \vee r$.
11. Prove that $P \to Q, Q \to R, P \vee R \Rightarrow R$ by using indirect method.
12. Using quantifier say $\sqrt{z}$ is not a real number ?
13. State whether the conclusion $C$ follows logically from the premises $R$ and $S$
    (a) $R : p \to q, \quad S : \sim q, \quad C : q$ (b) $R : p \to q, \quad S : q, \quad C : \sim p$.

# III. BOOLEAN ALGEBRA

## 37.15 INTRODUCTION

**(1)** The concept of Boolean algebra was first introduced by George Boole* in 1854 through his paper 'An investigation of the laws of thought'. It is basically two values *i.e.*, (0, 1) set. Earlier it had applications to statements and sets which are either true or false. In 1938 Claude Shannon showed that basic rules given by Boole could be used to design circuits. These days however, Boolean algebra has wide applications to switching circuits, electrical networks and electronic computers.

Basically there are three operations in the Boolean algebra (i) *AND*, (ii) *OR*, and (iii) *NOT*, which are symbolically represented by $\wedge$, $\vee$ and / respectively. Some authors, use the symbols (+), (.) and (/) for the same operations.

Here $'$ denotes the complement of an element and is defined by $0' = 1$ and $1' = 0$.

The operator $\wedge$ (*i.e.*, '*AND*') has the following values $1 \wedge 1 = 1, 1 \wedge 0 = 0, 0 \wedge 1 = 0, 0 \wedge 0 = 0$; while the operator $\vee$ (*i.e.*, *OR*) has the values $1 \vee 1 = 1, 1 \vee 0 = 1, 0 \vee 1 = 1, 0 \vee 0 = 0$.

**Def.** *Any non-empty set B with the binary operations '$\wedge$' and '$\vee$' and the uninary operation '/' is called the Boolean algebra [B, $\wedge$, $\vee$, /] if the following axioms hold where a, b, c are elements in B :*

1. *Commutative law* : $a \wedge b = b \wedge a$ ; $a \vee b = b \vee a$
2. *Associative law* : $a \wedge (b \wedge c) = (a \wedge b) \wedge c$
   $a \vee (b \vee c) = (a \vee b) \vee c$

*A British mathematician George Boole (1813–1864) who created Boolean algebra.

3. *Distributive law* : $a \vee (b \wedge c) = (a \vee b) \wedge (a \vee c)$

$a \wedge b \vee c = (a \wedge b) \vee (a \wedge c)$

4. *Complement law* : $a \wedge a' = 0, a \vee a' = 1$

The operations $\wedge$, $\vee$ and / are called sum, product, and complement respectively. We shall follow the usual practice that / has precedence over $\vee$, and $\vee$ has precedence over $\wedge$, unless guided by brackets. *e.g.*, $a \wedge b \vee c$ means $a \wedge (b \vee c)$ not $(a \wedge b) \vee c$ while $a \vee b'$ implies $a \vee b'$ not $(a \vee b)'$. *(Anna, 2012)*

**(2) Boolean function.** The variable $x$ is called a *Boolean variable* if it assumes values only from $B$ {0, 1}.

**Def.** *A function from the set* $\{(x_1, x_2, ..., x_n) : x_i \in B, 1 \le i \le n)$ *is called a Boolean function of degree n.* Boolean functions can be represented by expressions comprised of variables and Boolean operations.

*e.g.*, 0, 1, $x_1, x_2, ..., x_n$ are Boolean expressions in the variables $x_i$ $(1 \le i \le n)$. If $p$ and $q$ are Boolean expressions then $p \wedge q$, $p \vee q$ and $p'$ are also Boolean expressions and each represents a Boolean function.

By substituting 0 and 1 for the variables in the expression, the values of this function can be found.

**(3)** *If f and g be Boolean functions of degree n, then*

(*i*) *Complement of f* is the function $f'$ where

$$f'(x_1, x_2, ..., x_n) = [f(x_1, x_2, ..., x_n)]'$$

(*ii*) *f and g are equal* if $f(x_1, x_2, ..., x_n) = g(x_1, x_2, ..., x_n)$

(*iii*) *Boolean sum* $f \vee g$ *is*

$$(f \vee g)(x_1, x_2, ..., x_n) = f(x_1, x_2, ..., x_n) \vee g(x_1, x_2, ..., x_n)$$

(*iv*) *Boolean product* $f \wedge g$ *is*

$$(f \wedge g)(x_1, x_2, ..., x_n) = f(x_1, x_2, ..., x_n) \wedge g(x_1, x_2, ..., x_n).$$

**(4) Power in a Boolean function :** $x^2 = x \vee x = x$

$$x^3 = x^2 \vee x = x \vee x = x, ..., x^n = x$$

Similarly, $2x = x$, $3x = x$ etc.

**Example 37.30.** *Find the values of the Boolean function* $f = (x \wedge y') \vee z'$.

**Solution.** $f$ being third degree Boolean function has $2^3$ *i.e.* 8 values which are shown in the following table :

| $x$ | $y$ | $z$ | $y'$ | $z'$ | $x \wedge y'$ | $(x \wedge y') \vee z'$ |
|---|---|---|---|---|---|---|
| 1 | 1 | 1 | 0 | 0 | 0 | 0 |
| 1 | 1 | 0 | 0 | 1 | 0 | 1 |
| 0 | 1 | 1 | 0 | 0 | 0 | 0 |
| 1 | 0 | 1 | 1 | 0 | 1 | 1 |
| 0 | 0 | 1 | 1 | 0 | 0 | 0 |
| 1 | 0 | 0 | 1 | 1 | 1 | 1 |
| 0 | 1 | 0 | 0 | 1 | 0 | 1 |
| 0 | 0 | 0 | 1 | 1 | 0 | 1 |

## 37.16 DUALITY

**(1)** The dual of any Boolean function is obtained by interchanging Boolean sums and Boolean products along with the interchange of zeros and ones.

*For example* the dual of $x \vee (y \wedge 0)$ is $x \wedge (y \vee 1)$.

The dual of any theorem of a Boolean algebra is also its theorem. This implies that the dual of any theorem in Boolean algebra is always true.

**(2) Principle of duality.** *The dual of any theorem (or property) in Boolean algebra is also a theorem (or property).*

## 37.17 BOOLEAN IDENTITIES

There are many identities in Boolean algebra which are quite useful in simplifying electrical circuits. Some of the important ones are given below :

1. *Identity law* : $x \vee 0 = x$ ; $x \wedge 1 = x$
2. *Dominance laws* : $x \vee 1 = 1$ ; $x \wedge 0 = 0$
3. *Complement law* : $x \vee x' = 1$ ; $x \wedge x' = 0$
4. *Idempotent law* : $x \vee x = x$ ; $x \wedge x = x$
5. *Double complement law* : $(x')' = x$
6. *Commutative law* : $x \vee y = y \vee x$ ; $x \wedge y = y \wedge x$
7. *Associative law* : $x \vee (y \vee z) = (x \vee y) \vee z$
   $x \wedge (y \wedge z) = (x \wedge y) \wedge z$
8. *Distributive law* : $x \vee (y \wedge z) = (x \vee y) \wedge (x \vee z)$
   $x \wedge (y \vee z) = (x \wedge y) \vee (x \wedge z)$
9. *De-Morgan's law* : $x \wedge y = x' \vee y'$ ; $(x \vee y)' = x' \wedge y'$
10. *Absorption law* : $x \wedge (x \vee y) = x$

**Example 37.31.** *Let B be a Boolean algebra. Show that for all $a \in B$, there exists a unique complement $a'$.*

**Solution.** Let $b$ and $c$ be two complements of $\alpha$.

Then $b = b \wedge 1$ [∵ 0 is an additive identity]

$= b \wedge (a \vee c)$ [∵ $c$ is complement of $a$]

$= (b \wedge a) \vee (b \wedge c) = (a \wedge b) \vee (b \wedge c)$

$= 0 \vee (b \wedge c)$ [∵ $a \wedge b = a \wedge a' = 0$]

$= b \wedge c$ ...(*i*)

Similarly $c = c \wedge 1 = c \wedge (a \vee b)$ [∵ $a \vee b = a \vee a' = 1$]

$= (c \wedge a) \vee (c \wedge b)$

$= (a \wedge c) \vee (b \wedge c)$ [∵ $a \wedge c = a \wedge a' = 0$]

$= 0 \vee (b \wedge c)$

$= b \wedge c$ ...(*ii*)

From (*i*) and (*ii*), we find that $b = c$.

Thus the complement of $a$ is unique.

**Example 37.32.** *In a Boolean algebra, show that*

(*i*) $x + (x \,.\, y) = x$ (*ii*) $x \,.\, (x + y) = x$.

**Solution.** (*i*) $x + (x \,.\, y) = x \wedge (x \vee y) = (x \vee 0) \wedge (x \vee y)$ [∵ $x \vee 0 = x$]

$= x \vee (0 \wedge y)$ [By distributive law]

$= x \vee (y \wedge 0)$ [By commutative law]

$= x \vee 0$ [∵ $y \wedge 0 = 0$]

$= x$.

(*ii*) $x \,.\, (x + y) = x \vee (x \wedge y)$

$= (x \wedge 1) \vee (x \wedge y)$ [∵ $x \wedge 1 = x$]

$= x \wedge (1 \vee y)$ [By distributive law]

$= x \wedge (y \vee 1)$ [By commutative law]

$= x \wedge 1$ [∵ $y \vee 1 = 1$]

$= x$.

**Example 37.33.** *Simplify the following :*

(*i*) $(x + y) \,.\, x' \,.\, y'$ (*ii*) $x \vee y \wedge y \vee z \wedge y \vee z'$ (*iii*) $x \vee y \wedge [(x \wedge y') \vee y]'$.

**Solution.** (*i*) $(x \wedge y) \vee x' \vee y' = (x \wedge y) \vee (x' \vee y')$
$= (x \wedge y) \vee (x \wedge y)'$ [By De Morgan's law]
$= 1.$ [$\because \; p \vee p' = 1$]

(*ii*) $x \vee y \wedge y \vee z \wedge y \vee z' = (y \vee x) \wedge (y \vee z) \wedge (y \vee z')$ [By commutative law]
$= [y \vee (x \wedge z)] \wedge (y \vee z')$ [By distributive law]
$= y \vee [x \wedge z \wedge z']$
$= y \vee [x \wedge (z \wedge z')]$ [By associative law]
$= y \vee (x \wedge 0)$ [$\because \; z \wedge z' = 0$]
$= y \vee 0$ [$\because \; x \wedge 0 = 0$]
$= y.$

(*iii*) $x \vee y \wedge [(x \wedge y') \vee y]' = x \vee y \wedge [y \vee (x \wedge y')]'$ [By commutative law]
$= x \vee y \wedge [(y \vee x) \wedge (y \vee y')]'$ [By distributive law]
$= (x \vee y) \wedge [(x \vee y) \wedge 1]'$ [$\because \; y \vee y' = 1$]
$= (x \vee y) \wedge (x \vee y)' = 0.$

**Example 37.34.** *Show that*
(*i*) $x \vee y \wedge y \vee z \wedge z \vee x = (x \wedge y) \vee (y \wedge z) \vee (z \wedge x)$
(*ii*) $(x \wedge y) \vee (x' \wedge z) = (x' \vee y) \wedge (x \vee z).$

**Solution.** (*i*) R.H.S. $= (x \wedge y) \vee (y \wedge z) \vee (z \wedge x)$
$= (x \wedge y) \vee (z \wedge y) \vee (z \wedge x)$ [By commutative law]
$= (x \wedge y) \vee (z \wedge y \vee x)$ [By distributive law]
$= (x \vee z) \wedge (y \vee z) \wedge [x \vee (y \vee x)] \wedge [y \vee (y \wedge x)]$
$= (x \vee z) \wedge (y \vee z) \wedge [(x \vee y) \wedge (y \vee x)]$ [$\because \; x \vee x = x$ etc.]
$= (x \vee z) \wedge (y \vee z) \wedge [(x \vee y) \wedge (x \vee y)]$
$= (x \vee z) \wedge (y \vee z) \wedge (x \vee y)$ [$\because \; p \wedge p = p$
$= (x \vee y) \wedge (y \vee z) \wedge (z \vee x)$ (By commutative law)
= L.H.S.

(*ii*) L.H.S. $= (x \wedge y) \vee (x' \wedge z)$
$= [x \vee (x' \wedge z)] \wedge [y \vee (x' \wedge z)]$ [By distributive law]
$= [(x \vee x') \vee (x \vee z)] \wedge [(y \vee x') \wedge (y \vee z)]$
$= [1 \vee (x \vee z)] \wedge [(y \vee x') \wedge (y \vee z)]$
$= (x \vee z) \wedge (y \vee x') \wedge (y \vee z) \vee 1$
$= (x \vee z) \wedge (y \vee x') \wedge [(y \vee z) \vee (x \wedge x')]$
$= (x \vee z) \wedge (y \vee x') \wedge [(y \vee z) \vee x] \vee (y \vee z) \vee x']$
$= (x \vee z) \wedge [y \vee z \vee x] \wedge (x' \vee y) \vee [x' \wedge (y \vee z)]$
$= (x \vee z) \vee (1 \wedge y) \wedge (x' \vee y) \vee (1 \wedge z)$
$= [(x \vee z) \wedge 1] \wedge [(x' \vee y) \vee 1]$
$= (x \vee z) \wedge (x' \vee y) =$ R.H.S.

**Example 37.35.** *Show that*
(*i*) $x \vee y \wedge x' \vee y' = (x' \wedge y) \vee (x \wedge y')$
(*ii*) $[x \wedge (x' \vee y)] \vee [x' \wedge (x \vee y)] = y.$

**Solution.** (*i*) $(x \vee y) \wedge (x' \vee y') = [(x \vee y) \wedge x'] \vee [(x \vee y) \wedge y']$
$= [(x \wedge x') \vee (y \wedge x')] \vee \{(x \wedge y') \vee (y \wedge y')\}$ [By distributive law]
$= \{0 \vee (x' \wedge y)\} \vee \{(x \wedge y') \vee 0\}$ [$\because \; x \wedge x' = 0$
$= (x' \wedge y) \vee (x \wedge y').$

(*ii*) $[x \wedge (x' \vee y)] \vee [x' \wedge (x \vee y)]$
$= [(x \wedge x') \vee (x \wedge y)] \vee [(x' \wedge x) \vee (x' \wedge y)]$
$= [0 \vee (x \wedge y)] \vee [0 \vee (x' \wedge y)]$ [$\because \; x \wedge x' = 0$
$= (x \vee x') \wedge y = y \wedge 1 = y.$

**Example 37.36.** *If $a \vee x = b \vee x$ and $a \vee x' = b \vee x'$, then prove that $a = b$.*

**Solution.** Since $a \vee x = b \vee x$ and $a \vee x' = b \vee x'$

$\therefore$ $(a \vee x) \wedge (a \vee x') = (b \vee x) \wedge (b \vee x')$

*i.e.,* $a \vee (x \wedge x') = b \vee (x \wedge x')$ [By distributive law]

or $a \vee 0 = b \vee 0$ [$\because$ $x \wedge x' = 0$]

or $a = b.$

**Example 37.37.** *In Boolean algebra [B, +, ., /], show that*

$$(x \,.\, y' + y \,.\, z) \,.\, (x \,.\, z + y \,.\, z') = x \,.\, z$$

**Solution.** $\{(x \vee y') \wedge (y \vee z)\} \vee \{(x \vee z) \wedge (y \vee z')\}$

$= \{(x \vee y') \wedge y\} \vee \{(x \vee y') \wedge z\} \vee \{(x \wedge z) \wedge y\} \vee \{(x \vee z) \wedge z'\}$

$= \{(x \wedge y) \vee (y' \wedge y)\} \vee \{(x \wedge z) \vee (y' \wedge z)\} \vee \{(x \wedge y) \vee (z \wedge y)\} \vee \{(x \wedge z') \vee (z \wedge z')\}$

$= \{(x \wedge y) \vee 0\} \vee \{(x \wedge z) \vee (y' \wedge z)\} \vee \{(x \wedge y) \vee (z \wedge y)\} \vee \{(x \wedge z') \vee 0\}$

$= \{(x \wedge y) \vee (x \wedge y)\} \vee \{(x \wedge z) \vee (x \wedge z')\} \vee \{(z \wedge y) \vee (z \wedge y')\}$

$= (x \wedge y) \vee (x \wedge 1) \vee (z \wedge 1) = (x \wedge y) \vee x \vee z$

$= (x \vee x \vee z) \wedge (y \vee x \vee z) = (x \vee z) \wedge (y \vee x \vee z)$

$= (x \vee z) \wedge (1 \vee y) = (x \vee z) \wedge 1$

$= x \vee z.$

**PROBLEMS 37.4**

1. Find the truth table for the Boolean function $f(x, y, z) = (x \wedge y) \vee (y \wedge z')$.
2. Write the dual of the Boolean expression $x + x' \,.\, y = x + y$.
3. Simplify the following :
   (*i*) $(x \wedge y \wedge z)'$ (*ii*) $(x \vee y \vee z) \wedge (x' \wedge y' \wedge z')$.
4. In a Boolean algebra $[B, \wedge, \vee, /\,]$, prove that
   (*i*) $(x \wedge y) \vee (x \wedge y') = x$. (*ii*) $x' \wedge (x \vee y) = x' \wedge y$.
5. If $a \wedge x = b \wedge x$ and $a \wedge x' = b \wedge x'$, then show that $a = b$.
6. In a Boolean algebra $[B, \wedge, \vee, /]$, show that
   (*i*) $x \wedge (x \wedge y) = x \wedge y$ (*ii*) $x \vee (x \vee y) = x \vee y$.
7. In Boolean algebra, prove that
   (*i*) $x \wedge (x' \vee y) = x \wedge y$ (*ii*) $x' \wedge y = x' \wedge (x \vee y)$.
8. Show that $(x \wedge y') \vee (x' \wedge y) \vee (x' \wedge y') = x' \vee y'$.
9. If $B$ be a Boolean algebra and $x, y, z \in B$, prove that
   $$(x \vee y) \wedge (x \vee y') \wedge (x' \vee y) = x \wedge y$$
10. Prove that $(x \vee y) \wedge [z \vee (x' \wedge y')] = (x \vee y) \wedge z$.
11. In a Boolean algebra $B$, prove that $(a + b)' = a' \,.\, b' \; \forall \; a, b \in B$.
12. In any Boolean algebra, show that $a = b$ if and only if $a \,.\, b' + a' \,.\, b = 0$.
13. In Boolean algebra, show that
    (*i*) $(a + b) \,.\, (a' + c) = a \,.\, c + a' \,.\, b + b \,.\, c$.
    (*ii*) $(a + b') \,.\, (b + c') \,.\, (c + a') = (a' + b) \,.\, (b' + c) \,.\, (c' + a)$.
14. Give the truth table for the Boolean function
    $f : B_3 \to B$ determined by the polynomial
    $$P(x_1, x_2, x_3) = (x_1 \vee x_3) \wedge (x_1 \wedge (x_2 \vee x'_3)).$$

## 37.18 MINIMAL BOOLEAN FUNCTION

**Def.** *A minimal Boolean function in n variables is the product of $x_1, x_2, ..., x_n$.* It is also called *minterm*.

If $x, y$ are two variables and $x', y'$ are their complementary variables respectively, then $x \vee y, x' \vee y, x' \wedge y, x \wedge y', x' \vee y'$ are each a minimal Boolean function.

Similarly there are $2^3$ *i.e.* 8 minimal Boolean functions in the three variables $x, y, z$ *i.e.*, $x \vee y \vee z, x' \vee y \vee z, x \vee y' \vee z, x \vee y \vee z', x \vee y' \vee z, x' \vee y \vee z', x' \vee y' \vee z, x' \vee y' \vee z'$.

In general, *there are $2^n$ minimal Boolean functions (or minterms) in n variables.*

Similarly the join of the variables $x_1, x_2, .... x_n$ is called a *maxterm* and there will be $2^n$ *maxterms.*

## 37.19 DISJUNCTIVE NORMAL FORM

**(1) Def.** *A Boolean function which can be expressed as sum of minimal Boolean functions is called a* **Disjunctive normal form** *or* **minterm normal form** or **Canonical form.**

**(2)** *If the number of distinct terms in a disjunctive normal form of Boolean function in n variables are $2^n$, then it is called a* **complete disjunctive normal form.**

**(3) Complement function of a disjunctive normal form** *function f is the sum of all those terms of a complete disjunctive normal form which are not present in the disjunctive normal form of f.* The complement of $f$ is denoted by $f'$.

*For example,* if $f = (x \vee y) \wedge (x \vee y')$

then its complete disjunctive normal form in variables $x$ and $y$ is

$$(x \vee y) \wedge (x' \vee y) \wedge (x \wedge y') \wedge (x' \vee y')$$

$\therefore$ The complement function of this disjunctive normal form is $f'(x, y) = (x' \wedge y) \wedge (x' \vee y')$.

**Example 37.38.** *Find the value of the complete disjunctive normal form in three variables x, y, z.*

**Solution.** The complete disjunctive normal form in three variables $x, y, z$ is

$$f(x, y, z) = (x \vee y \vee z) \wedge (x \vee y \vee z') \wedge (x \vee y' \vee z) \wedge (x' \vee y \vee z)$$
$$\wedge (x \vee y' \vee z') \wedge (x' \vee y \vee z') \wedge (x' \vee y' \vee z) \wedge (x' \vee y' \vee z')$$
$$= [(x \vee y) \vee (z \wedge z')] \wedge [(x \wedge y') \vee (z \wedge z')] \wedge [(x' \wedge y) \vee (z \wedge z')] \wedge [(x' \vee y') \vee (z \wedge z')]$$
$$= [(x \vee y) \vee 0] \wedge [(x \wedge y') \vee 0] \wedge [(x' \wedge y) \vee 0] \wedge [(x' \vee y') \vee 0]$$
$$= [x \vee (y \wedge y')] \wedge [x' \vee (y \wedge y')]$$
$$= (x \vee 0) \wedge (x' \vee 0) = x \wedge x' = 0.$$

## 37.20 CONJUNCTIVE NORMAL FORM

**Def.** *If a Boolean function $f(x_1, x_2, ..., x_n)$ is expressed in the form of factors and each factor is the sum of all the n-variables, then such a function is called a* **conjunctive normal form** or **maxterm normal form** or **dual canonical form.**

**(2)** *If a conjective normal of a function of n variables contains all the $2^n$ distinct factors, then such a function is called a* **complete conjunctive normal form.**

**(3) Complement function of a conjunctive normal form** *function f is a Boolean function which is the product of all those terms of complete conjunctive normal form which are not present in conjunctive normal form of f.*

The complement of conjunctive normal form $f$ is denoted by $f'$.

*For example,* if $f(x, y) = (x \wedge y) \vee (x \wedge y')$

then its complete conjunctive normal form in $x$ and $y$ is

$$(x \wedge y) \vee (x \wedge y') \vee (x' \wedge y) \vee (x' \wedge y')$$

$\therefore$ The complement function of this conjunctive normal form is

$$f'(x, y) = (x' \wedge y) \vee (x' \wedge y')$$

**Example 37.39.** *Given Boolean expression f, where $f(x_1, x_2, x_3) = (x_3' \wedge x_2) \vee (x_1' \wedge x_3) \vee (x_2 \wedge x_3)$, simplify this expression stating the laws used and obtain the minterm normal form.*

**Solution.** Given $f(x_1, x_2, x_3) = (x_3' \wedge x_2) \vee (x_1' \wedge x_3) \vee (x_2 \wedge x_3)$

$= (x_3' \wedge x_2) \vee (x_2 \wedge x_3) \vee (x_1' \wedge x_3)$ [By commutative law]

$= (x_2 \wedge x_3') \vee (x_2 \wedge x_3) \vee (x_1' \wedge x_3)$

$= [x_2 \wedge (x_3' \vee x_3)] \vee (x_1' \wedge x_3)$ [By distributive law]

$= (x_2 \wedge 1) \vee (x_1' \wedge x_3)$ [$\because$ $p \vee p' = 1$]

$= x_2 \vee x_1' \wedge x_3$ [By identity law]

Minterm normal form of $f(x_1, x_2, x_3)$

$= [(x_3' \wedge x_2) \wedge (x_1 \vee x_1')] \vee [(x_1' \wedge x_3) \wedge (x_2 \vee x_2')] \vee [(x_2 \wedge x_3) \wedge (x_1 \vee x_1')]$

$= (x_3' \wedge x_2 \wedge x_1) \vee (x_3' \wedge x_2 \wedge x_1') \vee (x_1' \wedge x_3 \wedge x_2) \vee (x_1' \wedge x_3 \wedge x_2') \vee (x_2 \wedge x_3 \wedge x_1) \vee (x_2 \wedge x_3 \wedge x_1')$

$= (x_3' \wedge x_2 \wedge x_1) \vee (x_3' \wedge x_2 \wedge x_1') \vee (x_1' \wedge x_3 \wedge x_2) \vee (x_1' \wedge x_3 \wedge x_2') \vee (x_2 \wedge x_3 \wedge x_1).$

**Example 37.40.** *Express the following functions into conjunctive normal forms :*

*(i)* $x' \wedge y$ *(ii)* $(x \wedge y) \vee (x' \wedge y')$.

**Solution.** *(i)* $x' \wedge y = x' \wedge y \wedge (z \vee z') = (x' \wedge y \wedge z) \vee (x' \wedge y \wedge z')$.

*(ii)* $(x \wedge y) \vee (x' \wedge y') = [x \wedge y \wedge (z \vee z')] \vee [x' \wedge y' \wedge (z \vee z')]$

$= (x \wedge y \wedge z) \vee (x \wedge y \wedge z') \vee (x' \wedge y' \wedge z) \vee (x' \wedge y' \wedge z')$.

**Example 37.41.** *The function $f = (x \wedge y \wedge z) \vee (x \wedge y' \wedge z) \vee (x \wedge y \wedge z') \vee (x' \wedge y' \wedge z)$ is in conjunctive normal form. Write its complement ?*

**Solution.** The complete conjunctive normal form in three variables $x, y, z$ is $(x \wedge y \wedge z) \vee (x' \wedge y \wedge z) \vee (x \wedge y' \wedge z) \vee (x \vee y \vee z') \vee (x \wedge y' \wedge z') \vee (x' \wedge y \wedge z') \vee (x' \wedge y' \wedge z) \vee (x' \wedge y' \wedge z')$.

$\therefore$ The complement of the given function $F$ is

$$F' = (x' \wedge y \wedge z) \vee (x \wedge y' \wedge z') \vee (x' \wedge y \wedge z') \vee (x' \wedge y' \wedge z').$$

## PROBLEMS 37.5

1. Find the value of a complete disjunctive normal form in
   *(i)* two variables $x, y$. *(ii)* three variables $x, y, z$.
2. Express the following functions into disjunctive normal form :
   *(i)* $x \vee y$ *(ii)* $x \wedge (x' \vee y)$.
3. Express the Boolean function $F = A \vee (B' \wedge C)$ in a sum of minterms.
4. Convert the function $x \wedge y'$ to disjunctive normal form in three variables $x, y, z$.
5. Express the function $f = (x \vee y') \wedge (x \vee z) \wedge (x \vee y)$ into conjunctive normal form in which maximum number of variables are used.
6. Write the complement of the conjunctive normal form function $(x \wedge y \wedge z') \vee (x \wedge y' \wedge z') \vee (x \wedge y \wedge z')$.

## 37.21 SWITCHING CIRCUITS

**(1)** *A switching network is an arrangement of wires and switches* (or gates) *which connect two terminals.* A switch can be either closed or open. A closed switch permits and an open switch stops flow of current

**(2)** If $p$ denotes a switch, then $p'$ denotes that switch which is open when $p$ is closed and $p'$ is closed when $p$ is open.

If $x$ denotes the state of the switch $p$, then $x'$ represents the state of the switch $p'$. $x$ is called the Boolean variable which is a binary variable.

If $x = 1$ denotes the switch is closed or current flows, then $x = 0$ denotes that the switch is open or current stops.

**(3)** *Two switches $p_1$ and $p_2$ are either connected in series* (represented by $\wedge$) *or connected in parallel* (represented by $\vee$).

These are shown as follows :

*(i)*

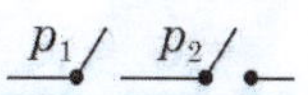


($p_1$ & $p_2$ in series : $p_1 \wedge p_2$)

*(ii)*

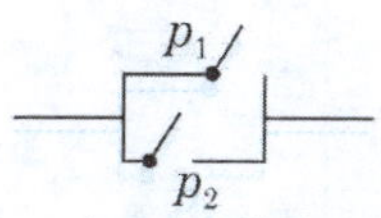


($p_1$ & $p_2$ in parallel : $p_1 \vee p_2$)

If $B$ [0, 1] is non-empty set and $\wedge$, $\vee$, $/$ are the operations on $B$, then the system [{0, 1}, $\wedge$, $\vee$, 1] is usually called *Boolean switching algebra.*

**(4) Simplification of circuits.** The simplification of a circuit means the least complicated circuit with minimum cost and best results. This depends on the cost of the equipment, number of switches and the type of the material used. Thus the simplification of circuits implies the use of lesser number of switches which can be achieved by using different properties of Boolean algebra. In other words, *the simplification of switching circuits is equivalent to simplification of the corresponding Boolean function.*

**Example 37.42.** *Draw the circuit which represents the Boolean function :*
(i) $(p_1 \wedge p_2) \wedge (p_3 \vee p_4)$ (ii) $(p_1 \wedge p_2) \vee (p_3 \vee p_4)$.

**Solution.** Here $p_1 \wedge p_2$ is a series circuit while $p_3 \vee p_4$ is a parallel circuit.

The required circuits are as follows :

(i) 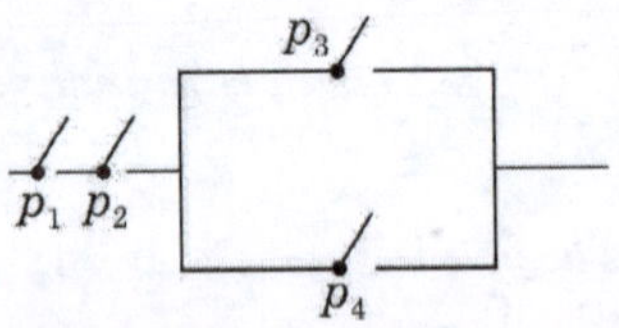


(ii) 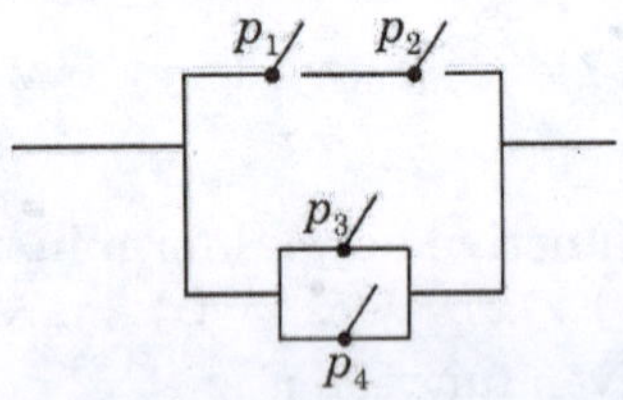


**Example 37.43.** *Write the Boolean functions representing the following circuits :*

(i) 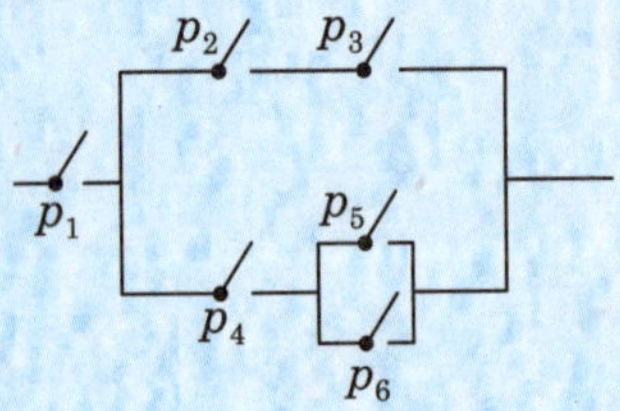


(ii) 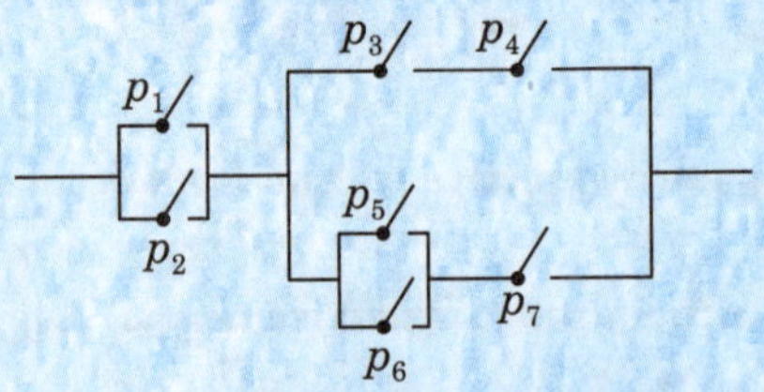


*Also draw the circuit diagram which would be the complement of the circuit in (ii).*

**Solution.** (i) The given circuit is represented by the Boolean function :

$$f = p_1 \wedge (p_2 \wedge p_3) \vee [p_4 \wedge (p_5 \vee p_6)]$$

(ii) The Boolean function for the given circuit is

$$f = (p_1 \vee p_2) \wedge [(p_3 \wedge p_4) \vee [(p_5 \vee p_6) \wedge p_7]$$

The complement of $f$ *i.e.,*

$$f' = (p_1 \vee p_2)' \vee [(p_3 \wedge p_4) \vee ((p_5 \vee p_6) \wedge p_7)]'$$
$$= (p_1' \wedge p_2') \vee [(p_3 \wedge p_4)' \wedge ((p_5 \vee p_6) \wedge p_7)']$$
$$= (p_1' \wedge p_2') \vee [(p_3' \vee p_4') \wedge ((p_5' \wedge p_6') \vee p_7')]$$

Its circuit diagram is as follows :

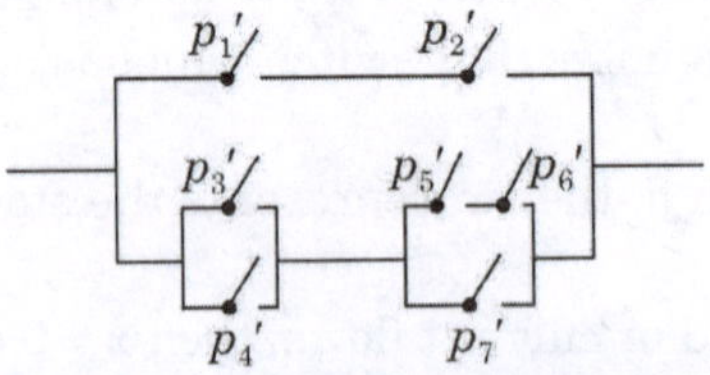


**Example 37.44.** *Simplify the following circuit and draw the diagram of the resulting circuit :*

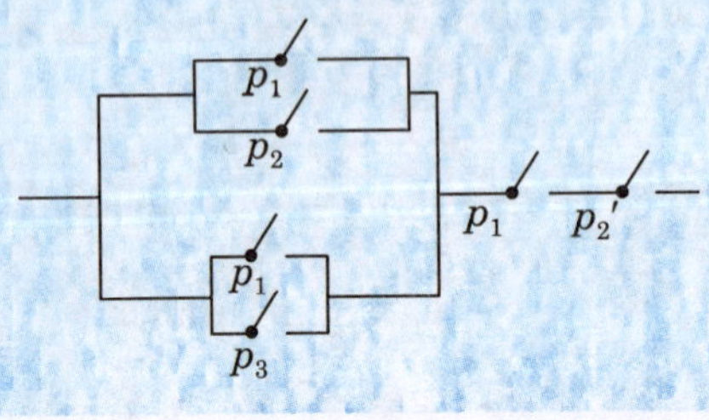

**Solution.** The given circuit is represented by the Boolean function $f$

$$= [(p_1 \vee p_2) \vee (p_1 \vee p_3)] \wedge (p_1 \wedge p_2') = (p_1 \vee p_2 \vee p_3) \wedge (p_1 \wedge p_2')$$

$$= (p_1 \wedge p_1 \wedge p_2') \vee (p_2 \wedge p_1 \wedge p_2') \vee (p_3 \wedge p_1 \wedge p_2') \quad \text{(By distributive law)}$$

$$= (p_1 \wedge p_2') \vee (p_3 \wedge p_1 \wedge p_2') = p_1 \wedge [p_2' \vee (p_3 \wedge p_2')]$$

$$= p_1 \wedge p_2' \quad \text{[By absorption law]}$$

Its circuit diagram is 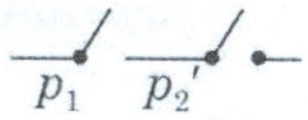

## PROBLEMS 37.6

1. Draw the circuit diagram represented by the Boolean functions :
   (i) $[p_1 \wedge (p_1 \vee p_2)] \vee [p_2 \wedge (p_1' \vee p_2)]$ (ii) $p_1 \wedge [(p_2 \vee p_4') \vee (p_3' \wedge (p_1 \vee p_4 \vee p_3'))] \wedge p_2$.
2. Write the Boolean functions representing the following circuits :

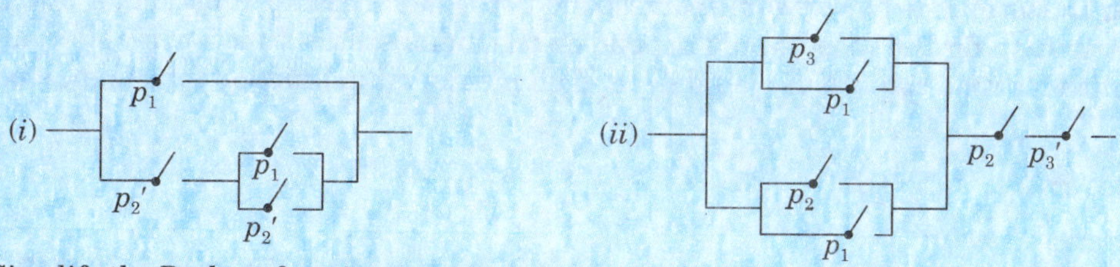

3. Simplify the Boolean functions, $p \vee (p' \wedge q) \vee (p \wedge q)$
4. Simplify the following circuit and draw the diagram of the resulting circuit :

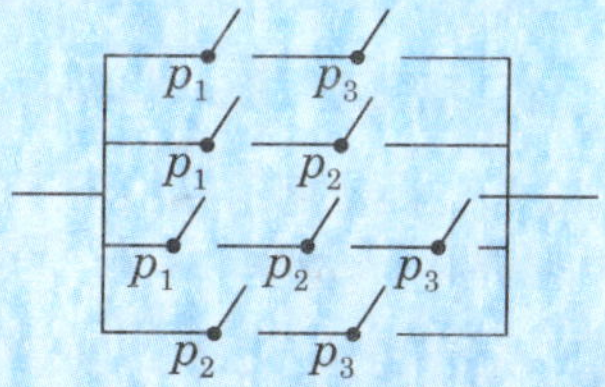

5. Draw the simplified network of $f(x, y, z) = (x \vee y \vee z) \wedge (x \vee y' \vee z) \wedge (x' \vee y' \vee z)$.
6. Consider the function $f(x_1, x_2, x_3) = [(x_1 \wedge x_2) \wedge (x_1 \wedge x_3)] \vee (x_1 \vee x_2')$
   (a) Simplify $f$ algebraically. (b) Draw the switching circuit of $f$.
   (c) Also find the minterm normal form of $f$.

# IV. FUZZY SETS

## 37.22 FUZZY LOGIC

We have so far dealt with the fundamentals of classical logic. Besides this, we have *crisp logic* which deals with propositions that are required to be either true or false. There is however another type of logic which includes not only the crisp values but all the values between true (1) and false (0). But there is some degree of vagueness about the exact value between [0, 1]. *The logic to infer a definite outcome from such vague inputs is called* **fuzzy logic**.

**(2) Fuzzy set.** To provide a mathematical modelling to fuzzy logic, L.A. Zudeh introduced the concept of 'Fuzzy sets' in 1965 on the basis of a *membership function*. The theory of 'fuzzy sets' is now fully developed.

**Def.** *A fuzzy set F of a non-zero set X(x) is defined as* $F = \{x, \mu_F(x)\} : x \in X$.

Here $\mu_F : X \to [0, 1]$ is a function called the *membership function* of $F$ and $\mu_F(x)$ is the degree of membership of $x \in X$ in $F$.

In particular $\mu(x) = 1$ implies full membership

$\mu(x) = 0$ implies non-membership

and $0 < \mu(x) < 1$ means intermediate membership.

A fuzzy set $F$ is, therefore, a set of pairs consisting of a particular element of the universe $X$ and its degree of membership *i.e.*, each $x$ is assigned a value in the range (0, 1) indicating the extent to which $x$ has the attribute $F$. It can also be represented as $F = \{[x_1, \mu_F(x_1)], [x_2, \mu_F(x_2)], ..., [x_n, \mu_F(x_n)]\}$.

*For example,* if $x$ is the number of cars in a lane, 'small' may be taken as a particular value of the fuzzy variable $x$ and to each $x$ is assigned a number in the range (0, 1) then $\mu_{small}(x) \in (0, 1)$ is the *membership function*.

**Example 37.45.** *In a car-race, all the cars complete the race in four time-groups : shortest time, moderate time, long time and longest time. If we note the time taken by each car in a group, it will give rise to a distribution of times. Now let us find the outcome of the race based on engine power, car speed and road conditions. Each of these variables may further be divided into :*

*(i) low, medium and high for the variable engine power,*

*(ii) slow, moderate and fast for the variable car speed,*

*(iii) rough, bumpy and smooth for the variable road conditions.*

Now we try to predict on some basis, in which of the four groups the car will finish, if it has low engine power, moderate speed and rough road.

Then the distribution for the engine power would correspond to the membership function for low, medium and high. Similarly the distribution for the speed would depend on the membership function for slow, moderate and fast, while the distribution for road conditions would depend on membership function for rough, bumpy and smooth.

## 37.23 FUZZY SET OPERATIONS

**(1)** A fuzzy set is said to be *normalised* when the largest element of the set (called *supremum*) is unity.

*For instance,* the set of members {5, 10 , 15, 20, 25} is normalised to {0.2, 0.4, 0.6, 0.8, 1} by dividing each member by 25, the supremum in the set.

*The normalization of a fuzzy set F is expressed as* $\underset{x \in X}{Sup}\, F(x) = 1$.

**(2) Complement.** *The complement of a fuzzy set F is the set $F^c$ with degree of membership of an element in $F^c$ equal to one minus degree of membership of this element in F. (Fig. 37.5)*

*For example,* if $F$ = [0.4 Ram, 0.6 Sham, 0.8 Jyoti, 0.9 Ritu]
be a set of intelligent students, then $F^c$ = [0.6 Ram, 0.4 Sham, 0.2 Jyoti, 0.1 Ritu]
is a set of non-intelligent students.

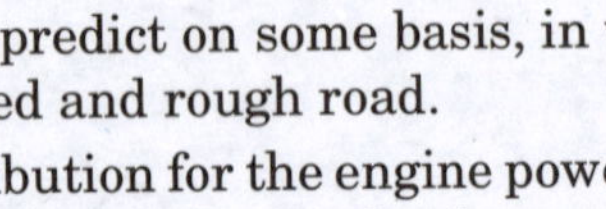

Fig. 37.5

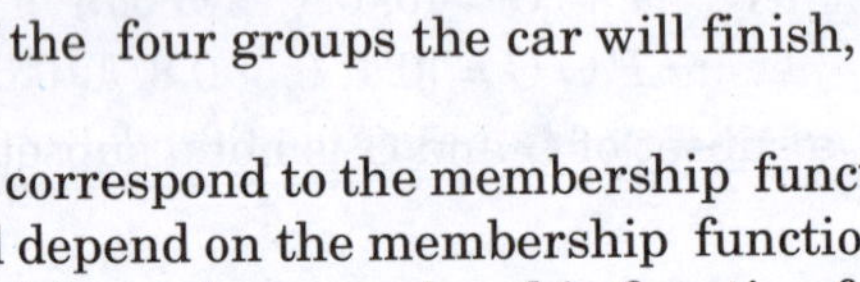

Fig. 37.6

**(3) Intersection.** *The intersection of two fuzzy sets F and G is the set $F \cap G$, where the degree of membership of an element in $F \cap G$ is the minimum of the degrees of membership of this element in F and G. (Fig. 37.6)*

**(4) Union.** *The union of two fuzzy sets F and G is the set $F \cup G$, where the degree of membership of an element in $F \cup G$, is the maximum of this element in F and G. (Fig. 37.7)*

*For example,* if

$F$ = [0.5 Rani, 0.2 Suman, 0.4 Anita, 0.8 Sunita]

be a set of fat girls, and

$G$ = [0.1 Rani, 0.6 Suman, 0.9 Anita, 0.5 Sunita]

be a set of tall girls, then

$F \cap G$ = [0.1 Rani, 0.2 Suman, 0.4 Anita, 0.5 Sunita]

and $F \cup G$ = [0.5 Rani, 0.6 Suman, 0.9 Anita, 0.8 Sunita]

**(5) Equality.** Two fuzzy sets $F$ and $G$ are said to be equal if and only if $F(x) = G(x)$ for all $x$ in $X$.

**(6) Subset.** The fuzzy set $F$ is said to be a subset of the fuzzy set $G$ (*i.e.*, $F \subseteq G$) if and only if $F(x) \le G(x)$ for all $x \in X$.

**(7) Double negation.** If $F$ is a fuzzy set, then $(F^c)^c = 1$.

**(8) De Morgan's laws.** If $F$ and $G$ are two fuzzy sets then

$$(F \cup G)^c = F^c \cap G^c \;;\; (F \cap G)^c = F^c \cup G^c.$$

**Example 37.46.** *Let the membership functions for the fuzzy sets F and G be as in the following table the find the value of $F \cap G$ and $F \cup G$. Are they subset of each other.*

| $X$ | 1 | 2 | 3 | 4 | 5 | 6 | 7 | 8 | 9 | 10 |
|---|---|---|---|---|---|---|---|---|---|---|
| $F$ | 0 | 0 | 0.1 | 0.5 | 0.8 | 1 | 0.3 | 0.5 | 0 | 0 |
| $G$ | 0 | 0 | 0 | 0 | 0.1 | 0.3 | 0.5 | 0.8 | 1 | 1 |
| $F^c$ | 1 | 1 | 0.9 | 0.5 | 0.2 | 0 | 0.7 | 0.5 | 1 | 1 |

**Solution :** Then the corresponding

$$F \cap G = [0, 0.1, 0.3, 0.5]$$

$$F \cup G = [0.1, 0.5, 0.8, 1, 0.5, 0.8, 1, 1]$$

Clearly $F$ is not a subset of $G$ and $G$ is not a subset of $F$.

## 37.24 TRUTH VALUE

**(1) Truth value** *of the negation of a proposition in fuzzy logic is one minus the truth value of the proposition.*

*For example,* if the truth value of the statement 'Ram is happy' is 0.8, then the truth value of 'Ram is not happy' is 0.2.

**(2) Truth value** *of the conjunction of two propositions in the fuzzy logic is the minimum of the truth values of the two propositions.*

**(3) Truth value of the disconjunction of two propositions** *in the fuzzy logic is the maximum of the truth values of two propositions.*

*For example,* if the truth value of 'Ram is fat' is 0.6, and the truth value of 'John is fat' is 0.9, then the truth value of the statement

(*a*) '*Ram and John are fat*' is 0.6.

(*b*) '*Ram or John is fat*' is 0.9.

(*c*) '*neither Ram nor John is fat*' is negation of minimum of negation of 'Ram is fat' (*i.e.*, 0.4) and negation of 'John is, fat' (*i.e.*, 0.1) = 0.1.

(*d*) '*Ram is not fat or John is not fat*' is maximum of 0.4 and 0.1 *i.e.*, 0.4.

## 37.25 ALGEBRAIC OPERATIONS ON FUZZY SETS

**(1) Algebraic sum** *of two fuzzy sets F and G is defined by the membership function*

$$\mu_{F+G}(x) = \mu_F(x) + \mu_G(x) - \mu_F(x)\,\mu_G(x)$$

and is written as $F + G$.

**Algebraic product** *of two fuzzy sets F and G is defined by the membership function.*

$$\mu_{F.G.}(x) = \mu_F(x).\mu_G(x)$$

and is written as $F.G.$

**(2) Properties of fuzzy set operations** which are common to crisp set operations, are as under

1. *Idempotent* : $F \cup F = F, \quad F \cap F = F$
2. *Identity* : $F \cup \phi = F, F \cap U = F$
3. *Commutative* : $F \cup G = G \cup F, \quad F \cap G = G \cap F$
4. *Distributive* : $F \cap (G \cap H) = (F \cap G) \cup (F \cap H), \quad F \cup (G \cap H) = (F \cup G) \cap (F \cup H)$
5. *Associative* : $(F \cup G) \cup H = F \cup (G \cup H), \quad (F \cap G) \cap H = F \cap (G \cap H)$
6. *Absorption* : $F \cup (F \cap G) = F, \quad F \cap (F \cup G) = F.$

## 37.26 GENERATION OF RULES FOR FUZZY PROBLEMS

We should know before hand all possible input-output relations while dealing with problems concerning fuzzy engines or fuzzy controls. These input-output rules are then expressed with 'if ... then' statements.

*For instance*, if $F$'s and $G$'s are inputs of fuzzy problems and $R$'s are the actions taken for each rule, then the set of 'if ... then' rules with two input variables $F_1$ and $G_1$ and the actions taken are shown in table 1.

*i.e.*, if $F_1$ and or $G_1$, then $R_{11}$, else
if $F_2$ and or $G_1$, then $R_{21}$, else
if $F_1$ and or $G_2$, then $R_{12}$, else
if $F_2$ and or $G_2$, then $R_{22}$.

Table 1

| | | |
|---|---|---|
| $F_1$ | $R_{11}$ | $R_{12}$ |
| $F_2$ | $R_{21}$ | $R_{22}$ |
| | $G_1$ | $G_2$ |

In case, the fuzzy statements have more variables, then 'if ... then' rules becomes more complicated to tabulate.

However such a tabulation can be simplified by following a *decomposition process* as follows :

The decomposition process of three fuzzy variables *F*, *G* and *H* with actions *R*'s taken is shown below :

Table 2

| | | |
|---|---|---|
| $F_1$ | $R_{11}$ | $R_{12}$ |
| $F_2$ | $R_{21}$ | $R_{22}$ |
| | $G_1$ | $G_2$ |

Table 3

| | | | | |
|---|---|---|---|---|
| $H_1$ | $R_{111}$ | $R_{121}$ | $R_{211}$ | $R_{221}$ |
| $H_2$ | $R_{112}$ | $R_{122}$ | $R_{212}$ | $R_{222}$ |
| | $R_{11}$ | $R_{12}$ | $R_{21}$ | $R_{22}$ |

Here the statements $F_2$ and $G_1$ and $H_1$ then $R_{211}$ is decomposed into 'if $F_2$ and $G_1$ then $R_{21}$' and 'if $R_{21}$ and $H_1$, then $R_{211}$.

Similarly $R_{21}$ and $H_2$ then $R_{212}$.

This decomposition process can easily be extended to any number of input variables.

## 37.27 FUZZY PROPOSITIONS

**(1)** *A fuzzy number is a fuzzy set* $R \to [0, 1]$. We can easily extend classical two-valued logic to three-valued logic. Fuzzy logic, however is an extension of multi-valued logic. It provides foundations for approximate reasoning with imprecise fuzzy propositions using fuzzy set theory.

The classical propositions are statements which are either true or false. In fuzzy logic, the truth or falsity of fuzzy propositions is assigned different degrees *i.e.*, the truth and falsity are expressed by numbers in [0, 1].

A variable whose values are 'words' or 'sentences' is called a *linguistic variable*. For example 'height' is a linguistic variable and its values are tall, very tall, quite tall, not tall, short, not very short, not quite tall etc.

**(2) Classification of fuzzy propositions.** The classification propositions are statements which are either true or false. In fuzzy logic, the truth or falsity of fuzzy propositions is assigned different degrees *i.e.*, the truth and falsity are expressed by numbers in [0, 1].

The fuzzy propositions of simple nature can be classified into the following four types. In each case, we introduce the relevant canonical form and then discuss its interpretation.

**Type I.** *Unconditional and unqualified propositions.*

The standard canonical form of this type of proposition is expressed as $p : u$ if $F$ ...(1)

Here $u$ is the variable that takes value $u$ from some universal set $U$ and $F$ is a fuzzy set on $U$ which represents a fuzzy predicate such as young, tall, low, high etc. Given a particular value $u$ (say $v$), this value belongs to $F$ with membership grade $F(v)$. This membership grade is then interpreted as the degree of truth $T(p)$ of proposition $p$

*i.e.*, $$T(p) = F(v) \qquad ...(2)$$

Here $T$ is a fuzzy set on [0, 1] which assigns the membership grade $F(v)$ to each value $v$ of $u$.

In some fuzzy propositions, values of variable $u$ in (1) are assigned to individuals in a given set $I$ *i.e.*, variable $u$ becomes a function $u : I \to u$ where $u(i)$ is the value of $v$ for individual $i$ in $U$. Accordingly the canonical form (2) is modified to the form

$$p : u\ (i) \text{ is } F \text{ where } i \in I \qquad ...(3)$$

**Example 37.47.** *Consider a set I of persons, each person is characterized by his 'age' and a fuzzy set expressing the predicate 'young' is given. Denoting our variable by 'age' and fuzzy set by 'young', the canonical form is*

$$p : \textit{age (i) is young.}$$

**Solution.** The degree of truth of this proposition $T(p)$ is then determined for each person $i$ in $I$ by means of the equation $T(p)$ = young [age $(i)$].

**Example 37.48.** *At a particular place on the earth, consider the air temperature u (in °C). Let Fig. 37.8 represent the membership function as predicate 'high'. Assuming that all relevant temperature readings are given, the corresponding fuzzy proposition is expressed as,*

$$p : \text{temp } (u) \text{ is high } (°C)$$

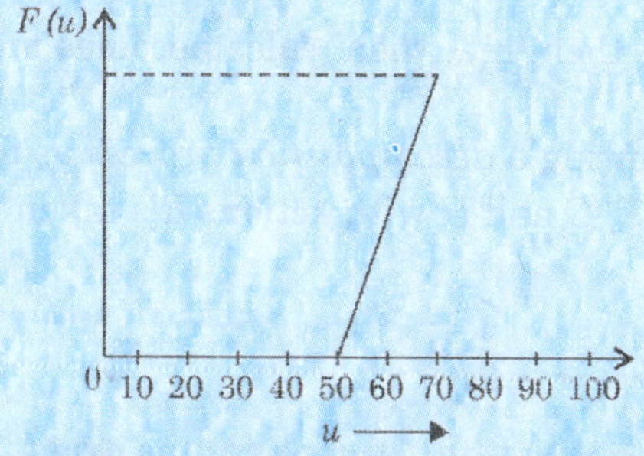

Fig. 37.8

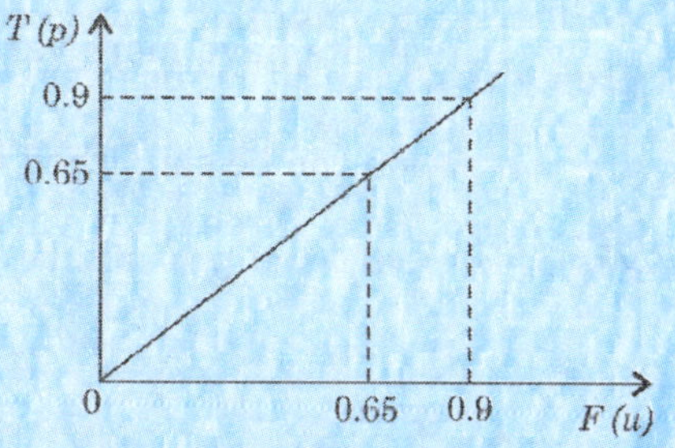

Fig. 37.9

**Solution.** The degree of truth $T(p)$ depends upon the actual value of the temperature and on the nature of predicate 'high' which is defined by the membership function $T$ in Fig. 37.9.

*e.g.*, if $u = 75$ then $F(75) = 0.65$ and $T(p) = 0.65$.

**Type II.** *Conditional and unqualified propositions.*

A proposition $p$ of this type is expressed by the canonical form

$$p : \text{If } x \text{ is } F \text{ then } y \text{ is } G \qquad ...(4)$$

where $x$, $y$ are variables whose values are in the sets $X$, $Y$ and $F$, $G$ are fuzzy sets on $X$ and $Y$ respectively. These propositions may also be viewed as propositions of the form

$$\{x, y\} \text{ is } R \qquad ...(5)$$

where $R$ is a fuzzy set on $X \times Y$ which is determined for each $x \in X$ and each $y \in Y$ by the formula

$$R(x, y) = B[F(x), G(y)], \qquad ...(6)$$

where $B$ is a binary operation as [0, 1] representing a suitable *fuzzy implication*.

**Type III.** *Unconditional and qualified propositions*

A proposition of this type is expressed by either of the following canonical forms :

$$p : u \text{ is } F \text{ is } S \qquad ...(7)$$

or

$$p : \text{Prob } (U \text{ is } F) \text{ is } P \qquad ...(8)$$

where $u$ is a variable that takes value $v$ from some universal set $U$ and $F$ is a fuzzy set on $U$ which represents a fuzzy predicate such as small, young, daughter etc.

Prob. ($U$ is $F$) is the probability of fuzzy set event '$u$ is $F$' ; $S$ is the fuzzy truth qualifier and $P$ is the fuzzy probability qualifier. $S$ and $P$ are both represented by fuzzy set on [0, 1].

**Example 37.49.** *Mary is 'young' is 'very true' where the predicate 'young' and the truth qualifier 'very true' are represented by the respective fuzzy sets shown in Fig. 37.10.*

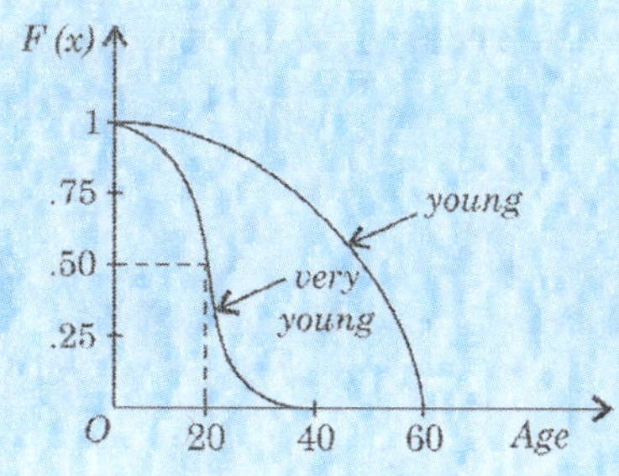

Fig. 37.10

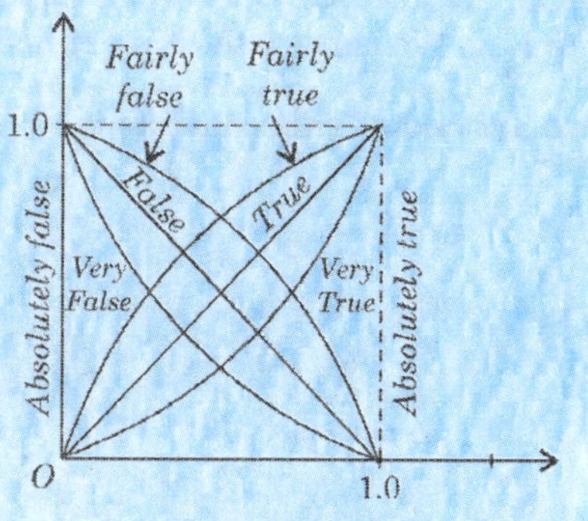

Fig. 37.11

**Solution.** The degree of truth $T(p)$ of any truth-qualified proposition $p$ is given for each $u \in U$ by $T(p) = S\{F(u)\}$.

Assuming that Mary's age is 20, she belongs to the set representing the predicate 'very young' with membership grade 0.50, our proposition belongs to the set of propositions which are 'very true' with membership grade 0.50 as shown in Fig. 37.11. This implies that the degree of truth of our truth qualified proposition is 0.50.

If the proposition be modified by changing the predicate to 'young' or the truth qualifier to 'fairly true', we would obtain the corresponding degree of truth of such propositions by the same method.

**Type IV.** *Conditional and qualified propositions*

This type of propositions can either be expressed in the canonical form

$$p : \text{If } x \text{ is } F, \text{ then } y \text{ is } G \text{ in } S$$

or

$$p : \text{Prob } \{x \text{ is } F/y \text{ is } G\} \text{ in } P,$$

where Prob $\{x$ is $F/y$ is $G\}$ is conditional probability.

## 37.28 APPLICATIONS OF FUZZY SETS

The concept of fuzzy sets has already influenced all engineering disciplines to various degrees.

*Electrical engineering* is the first such discipline where the utility of fuzzy logic and fuzzy sets has been recognized by developing controllers. Electronic circuits for fuzzy image processing have also been developed.

Some ideas regarding the application of fuzzy sets *in civil engineering* emerged around 1970. In the construction of bridges, dams, buildings etc. a designer has to take into account the safety factor for which the fuzzy theory has an effective role to play. Fuzzy set theory has also proved quite useful for assessing the life of existing constructions.

In *mechanical engineering* design problems, the utility of Fuzzy set theory was realised during mid 1980's. The membership function is expressed in terms of thermal expansion or corrosion or cost of different materials etc.

When the utility of fuzzy controllers was increasingly felt around mid 1980's, the need for computer hardware to implement the various operations involving fuzzy logic, had been recognized. In digital mode, fuzzy sets have been expressed as vectors of (0, 1) members.

Fuzzy control and fuzzy decision making are two well-developed areas of fuzzy set theory. These are directly relevant to *industrial engineering problems*. The utility of fuzzy sets has also been recognized for estimating the service life of given equipment under various conditions.

*Modern Reliability theory* has also been developed on the assumption of fuzzy sets. At any given time, an engineering product may be in functioning state to some degree or in failed state to another degree. The behaviour of an engineering product with respect to its functioning state and failed state has been characterized as based on fuzzy set theory.

The use of fuzzy set theory in *Robotics* includes approximate reasoning, fuzzy controllers, fuzzy pattern recognition and fuzzy data bases.

**Example 37.50.** *Solve the recurrence relation $a_n = a_{n-1} + h$ with initial term 90 = 4*

*(Karaikal University, 2020)*

**Solution.** Now first we get for recurrence relation, write out the first few terms of the sequence : 4, 5, 7, 10, 14, 10 ..... Look at the difference between terms $a_1 = a_0 = 1$ and $a_2 - a_1 = 2$ and so on. Here key things that the difference them is $n$. We can write this explicitly : $a_n = a_{n-1} + h$

We could have arrived at this conclusion directly from the recurrence relation by subtracting $a_{n-1}$ from both sides :

$$\begin{aligned} a_1 - a_0 &= 1 \\ a_2 - a_1 &= 2 \\ a_3 - a_2 &= 3 \\ &\vdots \\ &\vdots \\ a_n - a_{n-1} &= n \end{aligned}$$

Add all these equations. Wet get the sum 1 + 2 + .... $n$. We already know can be simplified to $\frac{n(n+1)}{2}$

$$(a_1 - a_0) + (a_2 - a_1) + .... (a_{n-1} - a_{n-2}) + (a_n - a_{n-1}) + ....$$

We are left with only the $- a_0$ from the first equation

We have $\quad -a_0 + a_n = \frac{n(n+1)}{2} \quad$ or $\quad a_n = \frac{n(n+1)}{2} + a_0$

But $\quad a_0 = 4$

So the solution to the recurrence relation

$$a_n = \frac{n(n+1)}{2} + 4$$

## PROBLEMS 37.7

1. Given fuzzy sets $F_1$ = [0.6 Sonu, 0.9 Renu, 0.7 Paul, 0.3 Sham]
   $F_2$ = [0.3 Sham, 0.8 Paul, 0.9 Renu, 0.5 Sonu]
   and $F_3$ = [0.8 Paul, 0.3 Sham, 0.5 Sonu, 0.9 Renu]
   Which of the above two sets are equal ?
2. Write the complement set of the fuzzy set $F$, if $F$ = [0.8 Ram, 0.3 Sham, 0.6 John, 0.7 Charu].
3. If $F = [0.3x_1, 0.7x_2, 0.5x_3, 0.8x_4]$ and $G = [0.4x_1, 0.6x_2, 0.1x_3, 0.9x_4]$
   been two fuzzy sets, then write down $F \cup G$ and $F \cap G$.
4. State the truth values of the negation of the following propositions :
   (*i*) Truth value of '$F$ is rich' is 0.8
   (*ii*) Truth value of '$G$ is fat' is 0.6
   (*iii*) Truth value of 'Mary is beautiful' is 0.7.
5. Let the membership functions of fuzzy sets $F$ and $G$ be as follows :
   $X$ : [1, 2, 3, 4, 5, 6, 7, 8, 9, 10]
   $F$ : [0, 0, 0, 0, 0.1, 0.3, 0.5, 0.9, 1, 1]
   $G$ : [0, 0, 0.1, 0.5, 0.9, 1, 0.9, 0.5, 0.0]
   State whether (*i*) $F = G$ (*ii*) $F$ is a subset of $G$.
   Also write down $F^c$, $F \cap G$ and $F \cup G$.
6. The truth values of the statements
   'Latif is a good player' is 0.7
   and 'John is a good player' is 0.6.
   What is the truth value of
   (*i*) the conjunction of the above two prepositions.
   (*ii*) the disjunction of the above prepositions.
7. Define a Fuzzy set and the standard operations on Fuzzy sets.
8. State the constituents of the pair in a fuzzy set.
9. Write a note on 'Fuzzy logic affects many disciplines' ?

CHAPTER

38

# Tensor Analysis



## 38.1 INTRODUCTION

Some physical quantities are specified by their magnitude only while others by their magnitude and direction. But certain quantities are associated with two or more directions. Such a quantity is called a *tensor*. The stress at a point of an elastic solid is an example of a tensor which depends on two directions—one normal to the area and other that of the force on it.

The properties of tensors are independent of the frames of reference used to describe them. That is why *Einstein* found tensors as a convenient tool for formulation of his Relativity theory. Since then, the subject of tensor analysis shot into prominence and is of great use in the study of Riemannian geometry, mechanics, elasticity, electro-magnet theory and numerous other fields of science and engineering. The emergence of tensor calculus as a symmetric subject is due to *Ricci* and his student *Levi-Cita*.

## 38.2 SUMMATION CONVENTION

Consider a sum of the type

$$a_1x_1 + a_2x_2 + \ldots + a_nx_n \quad i.e., \quad \sum_{i=1}^{n} a_i x_i \qquad \ldots(1)$$

In tensor analysis, the subscripts of the symbols $x_1, x_2, \ldots x_n$ are replaced by superscripts and we write these as $x^1, x^2, \ldots, x^n$. The superscripts do not stand for the various powers of $x$ but act as labels to distinguish different symbols. The power of a symbol (say : $x^i$) will be indicated as $(x^i)^2$, $(x^i)^3$ etc. Hence (1) is written as

$$\sum_{i=1}^{n} a_i x^i \,. \qquad \ldots(2)$$

A still simpler notation is to drop the $\Sigma$ sign and write (2) as $a_i x^i$ ...(3)

In this the repeated index $i$ successively takes up the values 1, 2, ..., $n$ and the expression (3) represents the sum of all such terms. The repeated index $i$ over which the summation is to be done, is called a *dummy* index since it doesn't appear in the final result. This notation, known as *summation convention*, is due to Einstein. We shall adopt this convention throughout this chapter and take the sum whenever a letter appears in a term once as a subscript and once as superscript.

**Example 38.1.** *Write the terms contained in $S = a_{ij}x^ix^j$ taking $n = 3$.*

**Solution.** Since the index $i$ occurs both as a subscript and as a superscript, we first sum on $i$ from 1 to 3.

$\therefore$ $$S = a_{1j}x^1x^j + a_{2j}x^2x^j + a_{3j}x^3x^j$$

Now each term in $S$ has to be summed up w.r.t. repeated index $j$ from 1 to 3.

$\therefore$ $$S = a_{11}x^1x^1 + a_{12}x^1x^2 + a_{13}x^1x^3 + a_{21}x^2x^1 + a_{22}x^2x^2 + a_{23}x^2x^3 + a_{31}x^3x^1 + a_{32}x^3x^2 + a_{33}x^3x^3$$

$$= a_{11}(x^1)^2 + a_{22}(x^2)^2 + a_{33}(x^3)^2 + (a_{12} + a_{21})x^1x^2 + (a_{13} + a_{31})x^1x^3 + (a_{23} + a_{32})x^2x^3.$$

**Example 38.2.** *If $f$ is a function of $n$ variables $x^i$, write the differential of $f$.*

**Solution.** Since $f = f(x^1, x^2, ..., x^n)$

$\therefore$ From Calculus, we have

$$df = \frac{\partial f}{\partial x^1}dx^1 + \frac{\partial f}{\partial x^2}dx^2 + ... + \frac{\partial f}{\partial x^n}dx^n = \frac{\partial f}{\partial x^i}dx^i.$$

## 38.3 (1) TRANSFORMATION OF COORDINATES

In a 3-dimensional space, the coordinates of a point are $(x^1, x^2, x^3)$ referred to a particular frame of reference. Similarly in an $n$-dimensional space, the coordinates of a point are $n$ independent variables $(x^1, x^2, ..., x^n)$ with respect to a certain frame of reference. Let $(\bar{x}^1, \bar{x}^2, ..., \bar{x}^n)$ be the coordinates of the same point referred to another frame of reference. Suppose, $\bar{x}^1, \bar{x}^2, ..., \bar{x}^n$ are independent single-valued functions of $x^1, x^2, ..., x^n$ so that

$$\bar{x}^1 = \phi^1(x^1, x^2, ..., x^n)$$
$$\bar{x}^2 = \phi^2(x^1, x^2, ..., x^n)$$
$$\cdots\cdots\cdots\cdots$$
$$\cdots\cdots\cdots\cdots$$
$$\bar{x}^n = \phi^n(x^1, x^2, ..., x^n)$$

or more briefly $$\bar{x}^i = \phi^i(x^1, x^2, ..., x^n) \quad ...(1)$$

We can solve the equations Eqn. (1) and express $x^i$ as functions of $\bar{x}^i$ so that

$$x^i = \psi^i(\bar{x}^1, \bar{x}^2, ..., \bar{x}^n) \quad ...(2)$$

*The Eqns. (1) and (2) are said to define a transformation of the coordinates from one frame of reference to another.*

**(2) Scalars or invariants.** *A function $\phi(x^1, x^2, x^3)$ is called a scalar or an invariant if its original value does not change upon transformation of coordinates from $x^1, x^2, x^3$ to $\bar{x}^1, \bar{x}^2, \bar{x}^3$.*

*i.e.,* $$\phi(x^1, x^2, x^3) = \psi(\bar{x}^1, \bar{x}^2, \bar{x}^3)$$

A scalar or invariant is also called a *tensor of order (or rank) zero.*

## 38.4 KRONECKER DELTA*

The quantity $\delta_i^j$ defined by the relations

$$\delta_i^j = 0, \quad \text{when } j \neq i$$

and $$\delta_i^j = 1, \quad \text{when } j = i,$$ is called *Kronecker delta.*

Evidently $$\delta_1^{\ 1} = \delta_2^{\ 2} = \delta_3^{\ 3} = ... = \delta_n^{\ n} = 1$$

while $$\delta_1^{\ 2} = \delta_2^{\ 3} = \delta_3^{\ 2} = ... = 0$$

*Called after the German mathematician *Leopold Kronecker* (1823–91) who made important contributions to number theory, algebra and group theory.

We note that by summing up w.r.t. the repeated index $j$,

$$\delta_{3j}\delta_2^j = a_{31}\delta_2^1 + a_{32}\delta_2^2 + a_{33}\delta_2^3 + a_{34}\delta_2^4 + \ldots$$
$$= 0 + a_{32} + 0 + 0 = a_{32}$$

In general,
$$a_{ij}\delta_k^j = a_{i1}\delta_k^1 + a_{i2}\delta_k^2 + \ldots + a_{ik}\delta_k^k + \ldots + a_{in}\delta_k^n$$
$$= 0 + 0 + \ldots + a_{ik} \cdot 1 + \ldots + 0 = a_{ik}.$$

**Example 38.3.** *Show that $a_{ij}A^{kj} = \Delta\delta_i^k$, where $\Delta$ is a determinant of order three and $A^{ij}$ are cofactors of $a^{ij}$.*

**Solution.** By expansion of determinants, we have

$$a_{11}A^{11} + a_{12}A^{12} + a_{13}A^{13} = \Delta$$
$$a_{11}A^{21} + a_{12}A^{22} + a_{13}A^{23} = 0$$
$$a_{11}A^{31} + a_{12}A^{32} + a_{13}A^{33} = 0$$

which can be compactly written as

$$a_{1j}A^{1j} = \Delta,\ a_{1j}A^{2j} = 0,\ a_{1j}A^{3j} = 0$$

Using Kronecker delta notation, these can be combined into a single equation

$$a_{1j}A^{kj} = \Delta\delta_1^k$$

Similarly $\qquad a_{2j}A^{kj} = \Delta\delta_2^k,\ a_{3j}A^{kj} = \Delta\delta_3^k$

All these nine equations are included in $a_{ij}A^{kj} = \Delta\delta_i^k$.

**Example 38.4.** *If $x^i$ and $\bar{x}^i$ are independent coordinates of a point, show that*

$$\frac{\partial x^j}{\partial \bar{x}^k}\frac{\partial \bar{x}^k}{\partial x^i} = \delta_i^j.$$

**Solution.** The partial derivatives of $\phi$ in the two coordinate systems are different and are connected by the following formula of Differential Calculus :

$$\frac{\partial \phi}{\partial x^i} = \frac{\partial \phi}{\partial \bar{x}^1}\cdot\frac{\partial \bar{x}^1}{\partial x^i} + \frac{\partial \phi}{\partial \bar{x}^2}\cdot\frac{\partial \bar{x}^2}{\partial x^i} + \frac{\partial \phi}{\partial \bar{x}^3}\frac{\partial \bar{x}^3}{\partial x^i} = \frac{\partial \phi}{\partial \bar{x}^k}\frac{\partial \bar{x}^k}{\partial x^i}$$

In particular, when $\phi = x^j$, we have $\dfrac{\partial x^j}{\partial x^i} = \dfrac{\partial x^j}{\partial \bar{x}^k}\dfrac{\partial \bar{x}^k}{\partial x^i}$ ...(*i*)

Since $x^j$ is independent of $x^i$, $\partial x^j/\partial x^i = 0$, when $j \neq i$
$= 1$, when $j = i$ ...(*ii*)

Hence the result follows from Eqns. (*i*) and (*ii*).

## 38.5 (1) CONTRAVARIANT VECTORS

Let $A^1, A^2, \ldots, A^n$ (*i.e.*, $A^i$) *be a set of n functions of the coordinate system* $x^1, x^2, \ldots, x^n$ (*i.e.*, $x^i$). *If these transform in another system of coordinates* $\bar{x}^1, \bar{x}^2, \ldots\ldots \bar{x}^n$ (*i.e.*, $\bar{x}^i$) *according to the law*

$$\bar{A}^i = \frac{\partial \bar{x}^i}{\partial x^j}A^j \qquad \text{...(1)}$$

*then $A^i$ are called components of a* **contravariant vector** *or* **contravariant tensor of order one.**

*An example of a contravariant vector.* Let us transform the coordinates of a point $x^i$ to $\bar{x}^i$ in a $n$-dimensional space.

Since $x$ is a function of $x^i$ (*i.e.*, $x_1, x_2, \ldots, x_n$), therefore,

$$d\bar{x}^i = \frac{\partial \bar{x}^i}{\partial x^1}dx^1 + \frac{\partial \bar{x}^i}{\partial x^2}dx^2 + \ldots + \frac{\partial \bar{x}^i}{\partial x^n}dx^n$$

$$= \frac{\partial \bar{x}^i}{\partial x^j}dx^j, \text{ using the summation convention.}$$

Comparing this with Eqn. (1), it follows that *the set of differentials $dx^1, dx^2, ..., dx^n$ is an example of a contravariant vector.* That is why the coordinates of a point are numbered by superscripts and not by subscripts.

**(2) Covariant vectors.** *Let $A_i$ be a set of n functions of the coordinate system $x^i$. If these transform in another system of coordinates $\bar{x}^i$ according to the law*

$$\bar{A}_i = \frac{\partial x^j}{\partial \bar{x}^i} A_j \quad ...(2)$$

*then $A_i$ are called the components of a* **covariant vector** *or* **covariant tensor of order one.**

*An example of a covariant vector.* Let $\phi$ be a function which has a fixed value at each point of space independent of the coordinate system employed. Therefore, $\phi$ is a function of the coordinates $x^i$ in the first system and a function of the coordinates $\bar{x}^i$ in the second system. By the chain rule

$$\frac{\partial \phi}{\partial \bar{x}^i} = \frac{\partial \phi}{\partial x^1}\frac{\partial x^1}{\partial x^i} + \frac{\partial \phi}{\partial x^2}\frac{\partial x^2}{\partial x^i} + ... + \frac{\partial \phi}{\partial x^n}\frac{\partial x^n}{\partial x^i} = \frac{\partial \phi}{\partial x^1}\frac{\partial x^1}{\partial \bar{x}^i} + \frac{\partial \phi}{\partial x^2}\frac{\partial x^2}{\partial \bar{x}^i} + ... + \frac{\partial \phi}{\partial x^n}\frac{\partial x^n}{\partial \bar{x}^i}$$

Comparing this equation with (2), it follows that *the set of derivatives,*

$$\partial\phi/\partial x^1, \partial\phi/\partial x^2, ..., \partial\phi/\partial x^n$$

*form a covariant vector.*

**Example 38.5.** *A covariant tensor has components $xy$, $2y - z^2$, $xz$ in rectangular coordinates. Find its covariant components in spherical coordinates.*

**Solution.** Here $x^1 = x, x^2 = y, x^3 = z$

and $\bar{x}^1 = r, \bar{x}^2 = \theta, \bar{x}^3 = \phi$ ...(*i*)

Let $A_1 = xy, A_2 = 2y - z^2, A_3 = xz$ ...(*ii*)

According to the law of transformation, we have $\bar{A}_i = \frac{\partial x^j}{\partial \bar{x}^i} A_j \quad (i = 1, 2, 3)$

and we wish to evaluate $\bar{A}_1, \bar{A}_2, \bar{A}_3$ where $A_1, A_2, A_3$ are known.

We know that $x = r \sin\theta \cos\phi, y = r \sin\theta \sin\phi, z = r = \cos\theta$ ...(*iii*)

Now $\bar{A}_1 = \frac{\partial x^1}{\partial \bar{x}^1} A_1 + \frac{\partial x^2}{\partial \bar{x}^1} A_2 + \frac{\partial x^3}{\partial \bar{x}^1} A_3 = \frac{\partial x}{\partial r} xy + \frac{\partial y}{\partial r}(2y - z^2) + \frac{\partial z}{\partial r} xz$ [From Eqns. (*i*) and (*ii*)]

$= \sin\theta\cos\phi \,.\, r\sin\theta\cos\phi \,.\, r\sin\theta\sin\phi + \sin\theta\sin\phi\,(2r\sin\theta\sin\phi - r^2\cos^2\theta)$
$+ \cos\theta \,.\, r\sin\theta\cos\phi \,.\, r\cos\theta$ [From Eqn. (*iii*)]

Similarly $\bar{A}_2 = \frac{\partial x^1}{\partial \bar{x}^2} A_1 + \frac{\partial x^2}{\partial \bar{x}^2} A_2 + \frac{\partial x^3}{\partial \bar{x}^2} A_3 = r\cos\theta\cos\phi \,.\, r\sin\theta\cos\phi \,.\, r\sin\theta\sin\phi$
$+ r\cos\theta\sin\phi\,(2r\sin\theta\sin\phi - r^2\cos^2\theta) + (-r\sin\theta)\, r\sin\theta\cos\phi \,.\, r\cos\theta$

and $\bar{A}_3 = \frac{\partial x^1}{\partial \bar{x}^3} A_1 + \frac{\partial x^2}{\partial \bar{x}^3} A_2 + \frac{\partial x^3}{\partial \bar{x}^3} A_3$

$= - r\sin\theta\sin\phi \,.\, r\sin\theta\cos\theta \,.\, r\sin\theta\sin\phi$
$+ r\sin\theta\cos\phi\,(2r\sin\theta\sin\phi - r^2\cos^2\theta) + 0$

## 38.6 TENSORS OF HIGHER ORDER

Let $i$ and $j$ be each given values 1 to $n$, then the symbol $A^{ij}$ will give rise to $n^2$ functions.

**(1)** *If $A^{ij}$ be a set of $n^2$ functions of the coordinates $x^1, x^2, ..., x^n$ which transforms to $\bar{A}^{ij}$ in another system of coordinates $\bar{x}^1, \bar{x}^2, ..., \bar{x}^n$, according to the law*

$$\bar{A}^{ij} = \frac{\partial \bar{x}^i}{\partial x^k}\frac{\partial \bar{x}^i}{\partial x^l} A^{kl} \quad ...(1)$$

*then the functions $\bar{A}^{ij}$ are said to be components of a* **contravariant tensor of the second order.**

**(2)** *If $A_{ij}$ be a set of $n^2$ functions of $x^1, x^2, ..., x^n$ which transform to $\bar{A}_{ij}$ in another system of coordinates, $\bar{x}^1, \bar{x}^2, ..., \bar{x}^n$ according to the law*

$$\bar{A}_{ij} = \frac{\partial x^k}{\partial \bar{x}^i}\frac{\partial x^l}{\partial \bar{x}^j}A_{kl}, \qquad ...(2)$$

*then the functions $\bar{A}_{ij}$ are said to be the components of a* **covariant tensor of the second order.**

**(3)** *If $A^i_{\ j}$ be a set of $n^2$ functions of $x^1, x^2, ..., x^n$ which transform to $\bar{A}^i_j$ in another system of coordinates, $\bar{x}^1, \bar{x}^2, ..., \bar{x}^n$, according to the law*

$$\bar{A}^i_j = \frac{\partial \bar{x}^i}{\partial x^k}\cdot\frac{\partial x^l}{\partial \bar{x}^j}A^k_{\ l}, \qquad ...(3)$$

*then $\bar{A}^i_j$ are said to be the components of a* **mixed tensor of the second order.** It transforms like a contravariant vector with respect to the index $i$ and like a covariant vector with regard to the index $j$. That is why $i$ is placed as a superscript and $j$ as subscript.

We can similarly define tensors of the orders higher than two.

**Obs.** Each of the above laws of transformation Eqns. (1) to (3), give rise to $n^2$ equations as $i$ and $j$ are each given the value 1 to $n$.

**Example 38.6.** *Show that the Kronecker delta is a mixed tensor of order two.*

**Solution.** If $\delta^j_i$ transforms to $\bar{\delta}^j_i$ in the coordinate system $\bar{x}^1, \bar{x}^2, ..., \bar{x}^n$ by the law for mixed tensors of order two, then

$$\bar{\delta}^j_i = \frac{\partial x^l}{\partial \bar{x}^i}\frac{\partial \bar{x}^j}{\partial x^m}\delta^m_l = \frac{\partial \bar{x}^j}{\partial x^m}\cdot\frac{\partial x^m}{\partial \bar{x}^i} = \frac{\partial \bar{x}^j}{\partial \bar{x}^i} = \delta_i^{\ j} \qquad [\because \ \delta^m_l = 0 \text{ for } l \neq m]$$

Hence $\delta^j_i$ is a mixed tensor of order two, having the same components in every coordinate system.

**Example 38.7.** *Show that the velocity of a fluid at any point is a contravariant tensor of rank one.*

**Solution.** Let $dx^1/dt, dx^2/dt, dx^3/dt$ be the components of fluid velocity of the point $(x^1, x^2, x^3)$, *i.e.*, $dx^i/dt$ be the components of velocity in the coordinate system $x^i$. Suppose the corresponding components of velocity in the coordinate system $\bar{x}^j$ are $d\bar{x}^j/dt$. Then $\bar{x}^1, \bar{x}^2, \bar{x}^3$ being the functions of $x^1, x^2, x^3$ which in turn are functions of $t$, we can write

$$\frac{d\bar{x}^j}{dt} = \frac{\partial \bar{x}^j}{\partial x^1}\frac{dx^1}{dt} + \frac{\partial \bar{x}^j}{\partial x^2}\frac{dx^2}{dt} + \frac{\partial \bar{x}^j}{\partial x^3}\frac{dx^3}{dt}$$

or
$$\frac{d\bar{x}^j}{dt} = \frac{\partial \bar{x}^j}{\partial x^i}\frac{dx^i}{dt} \qquad ...(i)$$

Now according to the law of tensor transformation, Eqn. ($i$) shows that the velocity of a fluid is a contravariant tensor of rank one.

**Example 38.8.** *Prove that there is no distribution between contravariant and covariant vectors if the transformation law is of the form $\bar{x}^i = a^i_m x^m + b^i$, where a's and b's are constants such that $a^i_{\ r} a^i_{\ m} = \delta^r_{\ m}$.*

**Solution.** Given transformation $\quad \bar{x}^i = a^i_m x^m + b^i \qquad ...(i)$

yields
$$\frac{\partial \bar{x}^i}{\partial x^m} = a^i_{\ m} \qquad ...(ii)$$

Also from ($i$),
$$a^i_{\ r}\bar{x}^i = a^i_{\ r}a^i_m x^m + a^i_{\ r}b^i$$
$$= \delta^r_m x^m + a^i_r b^i = x^r + a^i_{\ r}b^i.$$

$\therefore$
$$\frac{\partial x^r}{\partial \bar{x}^i} = a^i_r \quad i.e., \quad \frac{\partial x^m}{\partial \bar{x}^i} = a^i_m \qquad ...(iii)$$

From Eqn. (*ii*) and (*iii*), it is clear that any vector with components $a, b, c$ will on transformation give the same components whether transformed as a contravariant vector or as a covariant vector. Thus in this case, there is no distinction between the two.

## 38.7 SYMMETRIC AND SKEW-SYMMETRIC TENSORS

**(1)** *A tensor is said to be* **symmetric** *with respect to two contravariant (or two covariant) indices if its components remain unchanged on an interchange of the two indices.*

Thus the tensor $A^{ij}$ is symmetric if $A^{ij} = A^{ji}$, for every $i$ and $j$.

**(2)** *A tensor is said to be* **skew-symmetric** *with respect to two contravariant (or covariant) indices, if its components change sign on interchange of the two indices.*

Thus the tensor $A^{ij}$ is skew-symmetric if $A^{ji} = -A^{ij}$ for every $i$ and $j$.

In general, the tensor $A^{ijk}_{lm}$ is said to be symmetric or skew symmetric in $i$ and $j$ according as

$$A^{ijk}_{lm} = A^{jik}_{lm} \quad \text{or} \quad -A^{jik}_{lm}$$

**Example 38.9.** *Show that (i) a symmetric tensor of the second order has only $\frac{1}{2}n(n+1)$ different components.*

*(ii) A skew-symmetric tensor of the second order has only $\frac{1}{2}n(n-1)$ different non-zero components.*

**Solution.** (*i*) Let $A^{ij}$ be a symmetric tensor of order two so that $A^{ij} = A^{ji}$.

If each of the indices $i$ and $j$ take the values 1 to $n$, then $A^{ij}$ will have $n^2$ components. Out of these $n^2$ components, $n$ components $A_{11}, A_{22}, \ldots A_{nn}$ are independent.

Thus the remaining components are $(n^2 - n)$ which can be taken in pairs ($\because$ $A_{12} = A_{21}, A_{31} = A_{13}$ etc.)

Hence the total number of independent components

$$= n + \frac{1}{2}(n^2 - n) = \frac{1}{2}n(n+1)$$

(*ii*) Let $A^{ij}$ be a skew-symmetric tensor of order two so that $A^{ij} = -A^{ji}$. As above, $A^{ij}$ will have $n^2$ components. Out of these, $n$ components $A^{11}, A^{22}, \ldots A^{nn}$ are all zero. [$\because$ $A^{11} = -A^{11}$].

Omitting these, there are $(n^2 - n)$ components. Since $A^{12} = -A^{21}, A^{13} = -A^{31}$ etc., therefore ignoring the sign, $(n^2 - n)$ components can be taken in pairs.

Hence the total number of independent non-zero components

$$= \frac{1}{2}(n^2 - n) = \frac{1}{2}n(n-1).$$

### PROBLEMS 38.1

1. Write the following using the summation convention :

   (*i*) $\dfrac{d\phi}{dt} = \dfrac{\partial\phi}{\partial x^1}\dfrac{dx^1}{dt} + \dfrac{\partial\phi}{\partial x^2}\dfrac{dx^2}{dt} + \ldots + \dfrac{\partial\phi}{\partial x^n}\dfrac{dx^n}{dt}$ (*ii*) $(x^1)^2 + (x^2)^2 + (x^3)^2 + \ldots + (x^n)^2$.

2. Write out in full the following :

   (*i*) $a_{ij}x^i x^j$ $(i, j = 1, 2, 3)$ (*ii*) $g_{ij}dx^i dx^j (i, j = 1, 2, 3)$

   (*iii*) $g_{lm}g^{mp}$.

3. (*a*) Shows that $\delta^i_j$ is an invariant.

   (*b*) Evaluate (*i*) $\delta^i_j\, \delta^j_k$ (*ii*) $\delta^p_q\, \delta^q_r\, \delta^r_s$.

4. Show that (*i*) $\dfrac{\partial x^p}{\partial x^q} = \delta_q^{\ p}$ (*ii*) $\dfrac{\partial x^p}{\partial \bar{x}^q}\dfrac{\partial \bar{x}^p}{\partial x^r} = \delta_r^{\ p}$.

5. If the $\bar{x}$'s are $n$ independent functions of $x$'s and $i, j, k, l$ each take values from 1 to $n$, show that

$$\frac{\partial \bar{x}^i}{\partial x^k}\frac{\partial x^l}{\partial x^j} \cdot \delta^k_{\ l} = \delta^i_{\ j}.$$

6. Write down the law of transformation for the tensors

(i) $A_l^{jk}$ (ii) $C_{mn}$.

7. A quantity $A(i, j, k, l, m)$ which is a function of the coordinates $x^p$ transforms to another coordinate system $\bar{x}^p$ according to the law :

$$\bar{A}(r, s, t, u, v) = \frac{\partial \bar{x}^r}{\partial x^i} \frac{\partial \bar{x}^s}{\partial x^j} \frac{\partial x^k}{\partial \bar{x}^t} \frac{\partial x^l}{\partial \bar{x}^u} \frac{\partial \bar{x}^v}{\partial x^m} A(i, j, k, l, m)$$

Is this quantity a tensor ? If so express it suitably and state its nature and rank ?

8. If the components of two tensors are equal in one coordinate system, show that they are equal in all coordinate systems.

9. A covariant tensor has components $2x - z$, $x^2y$, $yz$ in cartesian coordinate system. Find its components in

(a) cylindrical coordinates (b) spherical coordinates.

10. If $g_{(i)}$ denotes the components of a covariant tensor of rank two, show that the product $g_{ij}\, dx^i\, dx^i$ is an invariant scalar.

11. A contravariant tensor has components $a$, $b$, $c$ in rectangular coordinates ; find the components in spherical coordinates.

12. Prove that $A_{ij}B^iC^j$ is an invariant, if $B^i$ and $C^j$ are contravariant vectors and $A_{ij}$ is a covariant tensor.

13. Show that $\partial A_p/\partial x^q$ is not a tensor even though $A_p$ is a covariant tensor of rank one.

14. If a tensor $A^{pqr}{}_{st}$ is a skew-symmetric with respect to the indices $p$ and $q$ in one coordinate system, show that it remains skew-symmetric with respect to $p$ and $q$ in any coordinate system.

## 38.8 ADDITION OF TENSORS

*The sum (or difference) of two tensors of the same order and type is another tensor of the same order and type.*

Let $A_{ij}$ and $B_{ij}$ be two tensors of the same order and same type. Their components in the coordinates system $\bar{x}^1, \bar{x}^2, ..., \bar{x}^n$ are $A_{ij}$ and $B_{ij}$, such that

$$\bar{A}_{ij} = \frac{\partial x^k}{\partial \bar{x}^i} \frac{\partial x^l}{\partial \bar{x}^j} A_{kl} \quad \text{and} \quad \bar{B}_{ij} = \frac{\partial x^k}{\partial \bar{x}^i} \frac{\partial x^l}{\partial \bar{x}^j} B_{kl}$$

$$\therefore \quad \bar{A}_{ij} \pm \bar{B}_{ij} = \frac{\partial x^k}{\partial \bar{x}^i} \frac{\partial x^l}{\partial \bar{x}^j} (A_{kl} \pm B_{kl}) \quad i.e., \quad \bar{C}_{ij} = \frac{\partial x^k}{\partial \bar{x}^i} \frac{\partial x^l}{\partial \bar{x}^j} C_{kl}$$

Thus $C_{ij}$ transforms in exactly the same manner as $A_{ij}$ and $B_{ij}$ and is, therefore, a tensor of the same order and same type.

## 38.9 OUTER PRODUCT OF TWO TENSORS

If $A^{ij}$ is a contravariant tensor of order two and $B_{kl}$ is a covariant tensor of order two then their product is a mixed tensor $C^{ij}{}_{kl}$ of order four such that

$$C_{kl}^{ij} = \bar{A}^{ij}\,\bar{B}_{kl} = \left(\frac{\partial \bar{x}^i}{\partial x^p} \frac{\partial \bar{x}^j}{\partial x^q} A^{pq}\right)\left(\frac{\partial x^r}{\partial \bar{x}^k} \frac{\partial x^s}{\partial \bar{x}^l} B_{rs}\right) = \frac{\partial \bar{x}^i}{\partial x^p} \frac{\partial \bar{x}^j}{\partial x^q} \frac{\partial x^r}{\partial \bar{x}^k} \frac{\partial x^s}{\partial \bar{x}^l} A^{pq}\, B_{rs} = \frac{\partial \bar{x}^i}{\partial x^p} \frac{\partial \bar{x}^j}{\partial x^q} \frac{\partial x^r}{\partial \bar{x}^k} \frac{\partial x^s}{\partial \bar{x}^l} C_{rs}^{pq}$$

But this is the law of transformation of a mixed tensor of order four. Therefore, $C^{ij}{}_{kl}$ is a mixed tensor of order four. Such products are called *outer products of two tensors.*

## 38.10 CONTRACTION OF A TENSOR

Consider a mixed tensor $A_l^{ijk}$ of order four. By the law of transformation, we have

$$\bar{A}_l^{ijk} = \frac{\partial \bar{x}^i}{\partial x^p} \frac{\partial \bar{x}^j}{\partial x^q} \frac{\partial \bar{x}^k}{\partial x^r} \frac{\partial x^s}{\partial \bar{x}^l} A_s^{pqr}$$

In this, put the covariant index $l = a$ contravariant index $i$, so that

$$\bar{A}_l^{ijk} = \frac{\partial \bar{x}^i}{\partial x^p}\frac{\partial \bar{x}^j}{\partial x^q}\frac{\partial \bar{x}^k}{\partial x^r}\frac{\partial x^s}{\partial \bar{x}^i}A_s^{pqr} = \frac{\partial \bar{x}^j}{\partial x^q}\frac{\partial \bar{x}^k}{\partial x^r}\frac{\partial x^s}{\partial x^p}A_s^{pqr}$$

$$= \frac{\partial \bar{x}^j}{\partial x^q}\frac{\partial \bar{x}^k}{\partial x^r}\delta_p^s A_s^{pqr} = \frac{\partial \bar{x}^j}{\partial x^q}\frac{\partial \bar{x}^k}{\partial x^r}A_p^{pqr}$$

This shows that $A_l^{ijk}$ is a contravariant tensor of order two.

*The process of getting a tensor of lower order (reduced by 2) by putting a covariant index equal to a contravariant index and performing the summation indicated is known as* **contraction.**

The tensors $A_i^{ijk}$ and $A_j^{ijk}$ obtained from contraction of the same tensor $A_l^{ijk}$ are generally different from each other unless the tensor $A_l^{ijk}$ is symmetric with respect to $i$ and $j$ (*i.e.*, $A_j^{ijk} = A_i^{ijk}$).

## 38.11 INNER PRODUCT OF TWO TENSORS

Given the tensors $A_k^{ij}$ and $B_{qr}^p$, if we first form their outer product $A_k^{ij}\, B_{qr}^p$ and contract this by putting $p = k$, then the result is $A^{ij}{}_k B_{qr}^k$ which is also a tensor, called the *inner product of the given tensors.*

*Hence the inner product of two tensors is obtained by first taking their outer product and then by contracting it.* We can get several inner products for the same two tensors by contracting in different ways.

**Example 38.10.** *Show that any inner product of the tensors $A_r^p$ and $B_t^{qs}$ is a tensor of rank three.*

**Solution.** The transformation laws for $A_r^p$ and $B_t^{qs}$ are

$$\bar{A}_r^p = \frac{\partial \bar{x}^p}{\partial x^i}\frac{\partial x^k}{\partial \bar{x}^r}A_k^i \quad ...(i) \qquad \text{and} \qquad \bar{B}^{qs}{}_t = \frac{\partial \bar{x}^q}{\partial x^j}\frac{\partial \bar{x}^s}{\partial x^l}\frac{\partial x^m}{\partial \bar{x}^t}B^{jl}{}_m \quad ...(ii)$$

$\therefore$ Inner product of $\bar{A}_q^p$ and $\bar{B}_t^{qs}$ is

$$\bar{A}_q^p\,\bar{B}_t^{qs} = \left(\frac{\partial \bar{x}^p}{\partial x^i}\frac{\partial x^k}{\partial \bar{x}^q}\right)\left(\frac{\partial \bar{x}^q}{\partial x^j}\frac{\partial \bar{x}^s}{\partial x^l}\frac{\partial x^m}{\partial \bar{x}^t}\right)A_k^i\,B_m^{jl}$$

$$= \frac{\partial \bar{x}^p}{\partial x^i}\frac{\partial \bar{x}^s}{\partial x^l}\frac{\partial x^m}{\partial \bar{x}^t}\delta_j^k\,A_k^l\,B_m^{jl} \qquad \left[\because \frac{\partial x^k}{\partial \bar{x}^q}\frac{\partial \bar{x}^q}{\partial x^j} = \frac{\partial x^k}{\partial x^j} = \delta_j^k\right.$$

$$= \frac{\partial \bar{x}^p}{\partial x^i}\frac{\partial \bar{x}^s}{\partial x^l}\frac{\partial x^m}{\partial \bar{x}^t}A_j^i\,B_m^{jl}$$

Hence the inner product of $\bar{A}_q^p$ and $\bar{B}_t^{qs}$ is a tensor of rank 3.

Similarly putting $p = t$ in the product of Eqn. (*i*) and (*ii*) and noting that

$$\frac{\partial \bar{x}^p}{\partial x^i}\frac{\partial x^m}{\partial \bar{x}^p} = \frac{\partial x^m}{\partial x^i} = \delta^m{}_i,$$

$A_r^p B_p^{qs}$ is found to be a tensor of rank 3.

Similarly, $A_r^p B_t^{qr}$ can also be shown to be a tensor of rank 3.

## 38.12 QUOTIENT LAW

To ascertain that a set of given functions forms the components of a tensor, we have to verify if the functions obey the tensor transformation laws. But this is a very tedious job. A simple test is provided by the quotient law which states that *if the inner product of a set of functions with an arbitrary tensor is a tensor, then these set of functions are the components of a tensor.*

The proof of this law is given below for a particular case.

**Example 38.11.** *Show that the expression $A(i, j, k)$ is a tensor if its inner product with an arbitrary tensor $B_k^{jl}$ is a tensor.*

**Solution.** Let $$A(i, j, k)\, B_k^{jl} = C_i^l \qquad ...(i)$$

where $C_i^l$ is a tensor. In the coordinate system $\bar{x}^i$, let Eqn. (*i*) transform to

$$\bar{A}(p, q, r)\bar{B}_r^{qs} = \bar{C}_p^s \qquad ...(ii)$$

where $\bar{B}_r^{qs}$ and $\bar{C}_p^s$ are the components of the tensors $B_k^{jl}$ and $C_i^l$. Expressing $B_r^{qs}$ in terms of $\bar{B}_k^{jl}$ and $\bar{C}_p^s$ in terms of $\bar{C}_i^l$, Eqn. (*ii*) takes the form

$$\bar{A}(p, q, r)\frac{\partial \bar{x}^q}{\partial x^i}\frac{\partial \bar{x}^s}{\partial \bar{x}^l}\frac{\partial x^k}{\partial \bar{x}^r}B_k^{jl} = \frac{\partial \bar{x}^s}{\partial x^l}\frac{\partial x^l}{\partial \bar{x}^p}C_i^l \qquad ...(iii)$$

Multiplying Eqn. (*i*) by $\dfrac{\partial \bar{x}^s}{\partial x^l}\dfrac{\partial x^i}{\partial \bar{x}^p}$ and subtracting from (*iii*), we get

$$\left\{\bar{A}(p, q, r)\frac{\partial \bar{x}^q}{\partial x^i}\frac{\partial \bar{x}^s}{\partial x^l}\frac{\partial x^k}{\partial \bar{x}^r} - A(i, j, k)\frac{\partial \bar{x}^s}{\partial x^l}\frac{\partial x^i}{\partial \bar{x}^p}\right\}B_k^{jl} = 0$$

Now $B_k^{jl}$ being an arbitrary tensor, the quantity within the brackets must be identically zero, *i.e.*,

$$\bar{A}(p, q, r)\frac{\partial \bar{x}^q}{\partial x^j}\frac{\partial \bar{x}^s}{\partial x^l}\frac{\partial x^k}{\partial \bar{x}^r} = A(i, j, k)\frac{\partial \bar{x}^s}{\partial x^l}\frac{\partial x^j}{\partial \bar{x}^p}$$

or $$\bar{A}(p, q, r) = \frac{\partial x^i}{\partial \bar{x}^p}\frac{\partial x^j}{\partial \bar{x}^q}\frac{\partial \bar{x}^r}{\partial x^k} = A(i, j, k)$$

But this is the law of tensor transformation. Hence $A(i, j, k)$ is a tensor of order three, with $i, j$ as covariant indices and $k$ as contravariant index.

## PROBLEMS 38.2

1. Prove that if a tensor equation is true for one coordinate system, it is true for all coordinate systems.
2. Show that every tensor can be expressed as the sum of two tensors, one of which is symmetric and the other skew-symmetric.
3. If $A^{pq}{}_r$ and $B^{pq}{}_r$ are tensors, prove that their sum and differences are also tensors.
4. Show that $A_{ij}$ is a tensor if its inner product with an arbitrary mixed tensor $B^j{}_k$ is a tensor.
5. Prove that (*a*) the contraction of the tensor $A^p{}_q$ is an invariant.
   (*b*) the contraction of the outer product of the tensors $A^p$ and $B_q$ is also an invariant.
6. Let $A^{pq}{}_{rst}$ be a tensor ; choose $p = t$ and $q = s$ and show that $A^{pq}{}_{rqp}$ is also a tensor. What is its rank ?

## 38.13 (1) RIEMANNIAN SPACE

The distance $ds$ between two adjacent points whose rectangular Cartesian coordinates are $(x, y, z)$ and $(x + dx, y + dy, z + dz)$ is given by $ds^2 = dx^2 + dy^2 + dz^2$.

Riemann extended the concept of distance to a space of $n$ dimensions and defined the distance $ds$ between two adjacent points $x^i$ and $x^i + dx^i$ $(i = 1, 2, ..., n)$ by the relation

$$ds^2 = a_{11}(dx^1)^2 + a_{22}(dx^2)^2 + ... + a_{nn}(dx^n)^2 + a_{12}dx^1dx^2 + ... + a_{lm}dx^l dx^m + ...$$
$$= a_{ij}dx^i dx^j, \text{ using summation convention.} \qquad ...(1)$$

The coefficients $a_{ij}$ are the functions of the coordinates $x^i$. The quadratic form Eqn. (1) is called a Riemannian metric and any space in which the distance is given by a such a metric is called a *Riemannian space*.

If in a particular coordinate system $X^i$, the quadratic form Eqn. (1) reduces to the form

$$ds^2 = (dX^1)^2 + (dX^2)^2 + ... (dX^n)^2,$$

then it is called a Euclidean metric and the corresponding space is called the *Euclidean space*.

**Obs.** The geometry based on the Riemannian metric is called the *Riemannian geometry* and that based on the Euclidean metric is called the *Euclidean geometry.*

**(2) Metric tensor.** As in the physical space, the distance $ds$ in the $n$-dimensional space is assumed to be independent of the coordinate system, *i.e.* a scalar invariant or a tensor of order zero. In the relation (1), $dx^i$ and $dx^j$ being displacements are components of a contravariant vector or a tensor of order one. Therefore, their outer product $dx^i dx^j$ is a contravariant tensor of order two. By the quotient law, the functions $a_{ij}$ must be components of a covariant tensor of order two.

Let us write $\quad a_{ij} = g_{ij} + h_{ij}$ where $g_{ij} = \frac{1}{2}(a_{ij} + a_{ji})$ and $h_{ij} = \frac{1}{2}(a_{ij} - a_{ji})$.

Interchanging $i$ and $j$, we have $g_{ji} = \frac{1}{2}(a_{ji} + a_{ij}) = g_{ij}$ and $h_{ji} = \frac{1}{2}(a_{ji} - a_{ij}) = -h_{ij}$

$\therefore$ $g_{ij}$ is symmetric and $h_{ij}$ is skew-symmetric. Thus (1) take the form

$$ds^2 = a_{ij} dx^i dx^j = (g_{ij} + h_{ij})\, dx^i dx^j$$

Now $h_{ij} dx^i dx^j$ is zero, since $h_{ij}$ is skew-symmetric. Hence $ds^2 = g_{ij} dx^i dx^j$ where $g_{ij}$ is a covariant symmetric tensor of order two. It is called the *metric tensor* or the *first fundamental tensor.*

## 38.14 CONJUGATE TENSOR

Let $g$ be the determinant $|g_{ij}|$ and $G_{ij}$ denote the cofactors of $g_{ij}$ in $g$. Define the function of $g^{ij}$ by the relation $g^{ij} = G_{ij}/g$ ...(1)

Since the functions $g_{ij}$ and $G_{ij}$ are symmetric in the subscripts, the functions $g^{ij}$ will be symmetric in the superscripts. Now

$$g_{ij} g^{ij} = g_{ij} \frac{G_{ij}}{g} = \frac{g}{g} = 1 \quad \text{and} \quad g_{ij} g^{lj} = g_{ij} \frac{G_{lj}}{g}$$

$$= 1, \text{ if } l = i, \; = 0, \text{ if } l \neq i$$

Thus

$$g_{ij}\, g^{lj} = \delta_i^l \qquad ...(2)$$

If $u^j$ be an arbitrary contravariant tensor, then its inner product with the tensor $g_{ij}$ will be an arbitrary covariant tensor due to contraction, *i.e.*,

$$g_{ij} u^j = v_i \qquad ...(3)$$

$$\therefore \quad g^{lj} v_l = g^{lj} g_{lj} u^j = u^j,$$

which is a contravariant tensor of order one. Therefore by quotient law, $g^{lj}$ are the components of a contravariant tensor of order two. Hence $g^{ij}$ is a symmetric contravariant tensor which is called the *conjugate tensor* or the *second fundamental tensor.*

**Obs.** In view of Eqn. (2), the relation between $g_{ij}$ and $g^{ij}$ is reciprocal. As such the *first and second fundamental tensors are also called* **reciprocal tensors.**

**Example 38.12.** *Find the components of the first and second fundamental tensors in spherical coordinates.*

**Solution.** Let $(x^1, x^2, x^3)$ be the rectangular cartesian coordinates and $\bar{x}^1, \bar{x}^2, \bar{x}^3$ be the spherical coordinates of a point so that

$$x^1 = x,\; x^2 = y,\; x^3 = z, \text{ and } \bar{x}^1 = r,\; \bar{x}^2 = \theta,\; \bar{x}^3 = \phi \qquad ...(i)$$

and

We know that $\quad x = r \sin\theta \cos\phi,\; y = r \sin\theta \sin\phi,\; z = r\cos\theta$ ...(*ii*)

Let $g_{pq}$ and $\bar{g}_{ij}$ be the metric tensors in cartesian and spherical coordinates respectively.

Then

$$ds^2 = (dx^1)^2 + (dx^2)^2 + (dx^3)^2 = g_{pq} dx^p dx^q$$

$$\therefore \quad g_{11} = 1 = g_{22} = g_{33}, \text{ and } g_{12} = 0 = g_{13} = g_{23} \text{ etc.} \qquad ...(iii)$$

On transformation

$$\bar{g}_{ij} = \frac{\partial x^p}{\partial \bar{x}^i} \frac{\partial x^q}{\partial \bar{x}^j} g_{pq} = \frac{\partial x^1}{\partial \bar{x}^i} \frac{\partial x^1}{\partial \bar{x}^j} g_{11} + \frac{\partial x^2}{\partial \bar{x}^i} \frac{\partial x^2}{\partial \bar{x}^j} g_{22} + \frac{\partial x^3}{\partial \bar{x}^i} \frac{\partial x^3}{\partial \bar{x}^j} g_{33} \qquad ...(iv)$$

Putting $i = j = 1$ in Eqn. (*iv*), we have

$$\bar{g}_{11} = \left(\frac{\partial x^1}{\partial \bar{x}^1}\right)^2 g_{11} + \left(\frac{\partial x^2}{\partial \bar{x}^1}\right)^2 g_{22} + \left(\frac{\partial x^3}{\partial \bar{x}^1}\right)^2 g_{33}$$

$$= \left(\frac{\partial x}{\partial r}\right)^2 + \left(\frac{\partial y}{\partial r}\right)^2 + \left(\frac{\partial z}{\partial r}\right)^2 \quad \text{[By Eqns. (i) and (iii)]}$$

$$= (\sin\theta\cos\phi)^2 + (\sin\theta\sin\phi)^2 + (\cos\theta)^2 = \sin^2\theta + \cos^2\theta = 1 \text{[By Eqn. (ii)]}$$

Putting $i = j = 2$ in Eqn. (*iv*), we have

$$\bar{g}_{22} = \left(\frac{\partial x^1}{\partial \bar{x}^2}\right)^2 g_{11} + \left(\frac{\partial x^2}{\partial \bar{x}^2}\right)^2 g_{22} + \left(\frac{\partial x^3}{\partial \bar{x}^2}\right)^2 g_{33}$$

$$= \left(\frac{\partial x}{\partial \theta}\right)^2 + \left(\frac{\partial y}{\partial \theta}\right)^2 + \left(\frac{\partial z}{\partial \theta}\right)^2 \quad \text{[By Eqns. (i) and (iii)]}$$

$$= (r\cos\theta\cos\phi)^2 + (r\cos\theta\sin\phi)^2 + (-r\sin\theta)^2 = r^2\cos^2\theta + r^2\sin^2\theta = r^2$$

Similarly $$\bar{g}_{33} = \left(\frac{\partial x^1}{\partial \bar{x}^3}\right)^2 g_{11} + \left(\frac{\partial x^2}{\partial \bar{x}^3}\right)^2 g_{22} + \left(\frac{\partial x^3}{\partial \bar{x}^3}\right)^2 g_{33} = r^2\sin^2\theta$$

and $$\bar{g}_{12} = 0 = \bar{g}_{13} = \bar{g}_{21} = \bar{g}_{23} = \bar{g}_{31} = \bar{g}_{32}$$

Hence the first fundamental tensor, written in matrix form, is

$$\begin{bmatrix} 1 & 0 & 0 \\ 0 & r^2 & 0 \\ 0 & 0 & r^2\sin^2\theta \end{bmatrix}$$

$$\therefore \quad g = \begin{vmatrix} 1 & 0 & 0 \\ 0 & r^2 & 0 \\ 0 & 0 & r^2\sin^2\theta \end{vmatrix} = r^4\sin^2\theta$$

and the cofactors in $g$ are given by

$$G_{11} = r^4\sin^2\theta,\ G_{22} = r^2\sin^2\theta,\ G_{33} = r^2;\ G_{12} = 0 = G_{13} = G_{21} = G_{23} = G_{31} = G_{32}$$

The components of the second fundamental tensor are given by $g^{ij} = G_{ij}/g$. Hence the second fundamental tensor in matrix form, is $\begin{bmatrix} 1 & 0 & 0 \\ 0 & 1/r^2 & 0 \\ 0 & 0 & 1/r^2\sin^2\theta \end{bmatrix}$.

**Example 38.13.** *Find the components of the metric tensor and the conjugate tensor in cylindrical coordinates.*

**Solution.** Let $(x^1, x^2, x^3)$ be the cartesian coordinates and $(\bar{x}^1, \bar{x}^2, \bar{x}^3)$ be the cylindrical coordinates of a point so that

$$x^1 = x,\ x^2 = y,\ x^3 = z \quad \text{and} \quad \bar{x}^1 = \rho,\ \bar{x}^2 = \phi,\ \bar{x}^3 = z \quad \text{...(i)}$$

We know that $$x = \rho\cos\phi,\ y = \rho\sin\phi,\ z = z \quad \text{...(ii)}$$

Let $g_{pq}$ and $\bar{g}_{ij}$ be the metric tensors in cartesian and cylindrical coordinates respectively.

Then $$ds^2 = (dx^1)^2 + (dx^2)^2 + (dx^3)^2 = g_{pq}dx^p$$

$$\therefore \quad g_{11} = 1 = g_{22} = g_{33} \quad \text{and} \quad g_{12} = 0 = g_{13} = g_{23} \text{ etc.} \quad \text{...(iii)}$$

On transformation,

$$\bar{g}_{ij} = \frac{\partial x^p}{\partial \bar{x}^i}\frac{\partial x^q}{\partial \bar{x}^j} g_{pq} = \frac{\partial x^1}{\partial \bar{x}^i}\frac{\partial x^1}{\partial \bar{x}^j} g_{11} + \frac{\partial x^2}{\partial \bar{x}^i}\frac{\partial x^2}{\partial \bar{x}^j} g_{22} + \frac{\partial x^3}{\partial \bar{x}^i}\frac{\partial x^3}{\partial \bar{x}^j} g_{33} \quad \text{...(iv)}$$

Putting $i = j = 1$ in Eqn. (*iv*), we have

$$\bar{g}_{11} = \left(\frac{\partial x^1}{\partial \bar{x}^1}\right)^2 g_{11} + \left(\frac{\partial x^2}{\partial \bar{x}^1}\right)^2 g_{22} + \left(\frac{\partial x^3}{\partial \bar{x}^1}\right)^2 g_{33} = \left(\frac{\partial x}{\partial \rho}\right)^2 + \left(\frac{\partial y}{\partial \rho}\right)^2 + \left(\frac{\partial z}{\partial \rho}\right)^2$$

$$= \cos^2 \phi + \sin^2 \phi + 0 = 1 \qquad \text{[By (i) and (ii)]}$$

Putting $i = j = 2$ in Eqn. (*iv*), we have

$$\bar{g}_{22} = \left(\frac{\partial x^1}{\partial \bar{x}^2}\right)^2 g_{11} + \left(\frac{\partial x^2}{\partial \bar{x}^2}\right)^2 g_{22} + \left(\frac{\partial x^3}{\partial \bar{x}^2}\right)^2 g_{33} = \left(\frac{\partial x}{\partial \phi}\right)^2 + \left(\frac{\partial y}{\partial \phi}\right)^2 + \left(\frac{\partial z}{\partial \phi}\right)^2$$

$$= (-\rho \sin \phi)^2 + (\rho \cos \phi)^2 + 0 = \rho^2 \qquad \text{[By (i) and (ii)]}$$

Similarly $$\bar{g}_{33} = \left(\frac{\partial x^1}{\partial \bar{x}^3}\right)^2 g_{11} + \left(\frac{\partial x^2}{\partial \bar{x}^3}\right)^2 g_{22} + \left(\frac{\partial x^3}{\partial \bar{x}^3}\right)^2 g_{33} = 0 + 0 + 1 = 1$$

and $$\bar{g}_{12} = 0 = \bar{g}_{13} = \bar{g}_{21} = \bar{g}_{23} = \bar{g}_{31} = \bar{g}_{32}$$

Hence the metric tensor, written in matrix form, is

$$\begin{bmatrix} 1 & 0 & 0 \\ 0 & \rho^2 & 0 \\ 0 & 0 & 1 \end{bmatrix} \qquad \therefore \qquad g = \begin{vmatrix} 1 & 0 & 0 \\ 0 & \rho^2 & 0 \\ 0 & 0 & 1 \end{vmatrix} = \rho^2$$

Also cofactors in $g$ are given by

$$G_{11} = \rho^2, G_{22} = 1, G_{33} = \rho^2 ; G_{12} = 0 = G_{13} = G_{21} = G_{23} = G_{31} = G_{32}$$

The components of the conjugate tensor are given by $g^{ij} = G_{ij}/g$.

Hence the conjugate tensor in matrix form is

$$\begin{bmatrix} 1 & 0 & 0 \\ 0 & 1/\rho^2 & 0 \\ 0 & 0 & 1 \end{bmatrix}$$

## 38.15 ASSOCIATED TENSORS

From Eqn. (3) of § 38.14, we have $$u^j . g_{ij} = v_i \qquad ...(1)$$

*i.e.*, the inner product of the tensor $u^j$ with the fundamental tensor $g_{ij}$ is another tensor $v_i$ which is called the *associated tensor* of $u^j$.

Similarly, we have $$v_l . g^{lj} = u^j \qquad ...(2)$$

Hence $u^j$ is the associated tensor of $v_l$.

Thus the indices of any tensor can be lowered or raised by forming its inner product with either of the fundamental tensors $g_{ij}$ or $g^{ij}$ as in Eqns. (1) or (2) above.

## 38.16 (1) LENGTH OF A VECTOR

The vector $\mathbf{A}$ is given by

$$\mathbf{A} = A^i g_i \quad \text{or} \quad \mathbf{A} = A_i g^i \qquad ...(1)$$

Also we have the associated vectors $A_i = g_{ij} A^j$ ...(2)

or $$A^i = g^{ij} A_j \qquad ...(3)$$

$\therefore$ Length of vector $$\mathbf{A} = (\mathbf{A} . \mathbf{A})^{1/2} = (A^i g_i . A^j g_j)^{1/2} \qquad \text{[By Eqn. (1)]}$$

$$= (g_{ij} A^i A^j)^{1/2} \qquad [\because g_i . g_k = g_{ij}]$$

$$= (A_i A^i)^{1/2} \qquad \text{[By Eqn. (2)]}$$

Also length of vector $\mathbf{A} = (\mathbf{A} . \mathbf{A})^{1/2} = (A_i g^i . A_j g^j)^{1/2} = (g^{ij} A_i A_j)^{1/2} = (A_i A^i)^{1/2}$ [By Eqn. (3)]

Hence the magnitude or length of the vector $A = \sqrt{(g_{ij} A^i A^j)} = \sqrt{(g^{ij} A_i A_j)} = \sqrt{(A_i A^i)}$ ...(4)

which is an invariant.

**Obs.** The length of a vector $A^1, 0, 0$ (in 3-dimensions) is $\sqrt{(g_{11}A^1A^1)}$, *i.e.* $\sqrt{g_{11}}A^1$. Similarly the length of the vector $0, A^2, 0$ is $\sqrt{g_{22}}A^2$ and the length of the vector $0, 0, A^3$ is $\sqrt{g_{33}}A^3$. Hence the physical components of a vector $A^i$ are $\sqrt{g_{11}}A^1, \sqrt{g_{22}}A^2, \sqrt{g_{33}}A^3$.

**(2) Angle between two vectors.** Let **A** and **B** be the given vectors such that

$$\mathbf{A} = A^i g_i \text{ and } B = B^j g_j \qquad \text{...(5)}$$

$\therefore$
$$\mathbf{A}\cdot\mathbf{B} = |\mathbf{A}|\,|\mathbf{B}|\cos\theta$$

or
$$\cos\theta = \frac{\mathbf{A}\cdot\mathbf{B}}{|\mathbf{A}||\mathbf{B}|} = \frac{A^iB^jg_{ij}}{\sqrt{(g_{ij}A^iA^j)}\sqrt{(g_{ij}B^iB^j)}} \qquad \text{[Using Eqns. (4) and (5)]}$$

In terms of associated vectors, we have

$$\cos\theta = \frac{A^iB_i}{\sqrt{(A^iA_i)}\sqrt{(B^iB_i)}}.$$

## PROBLEMS 38.3

1. If $ds^2 = 5(dx^1)^2 + 3(dx^2)^2 + 4(x^3)^2 - 6dx^1dx^2 + 4dx^2dx^3$, find the values of $g_{ij}$ and $g^{ij}$.
2. Find $g$ and $g^{ij}$ corresponding to the metric $ds^2 = \dfrac{dr^2}{1 - r^2/a^2} + r^2(d\theta^2 + \sin^2\theta\, d\phi^2)$.
3. The contravariant components of a vector **A** in plane cartesian coordinates $x = x^1$, $y = x^2$ are (33, 56). Using the tensor law of transformation, obtain the new components in polar coordinates $r = \bar{x}^1$ and $\theta = \bar{x}^2$.
4. Prove that the angles $\theta_{12}, \theta_{23}$ and $\theta_{31}$ between the coordinate curves in a 3-dimensional coordinate system are given by $\cos\theta_{12} = \dfrac{g_{12}}{\sqrt{(g_{11}g_{22})}}, \cos\theta_{23} = \dfrac{g_{23}}{\sqrt{(g_{22}g_{23})}}, \cos\theta_{31} = \dfrac{g_{31}}{\sqrt{(g_{33}g_{11})}}$.
5. Prove that for an orthogonal coordinate system
   (*i*) $g_{12} = g_{23} = g_{31} = 0$ (*ii*) $g^{11} = 1/g_{11}, g_{22} = 1/g^{22}, g_{33} = 1/g^{33}$.

## 38.17 CHRISTOFFEL SYMBOLS

*Christoffel symbol of the first kind* is denoted by $[ij, k]$ and is defined by

$$[ij, k] = \frac{1}{2}\left(\frac{\partial g_{jk}}{\partial x^i} + \frac{\partial g_{ik}}{\partial x^j} - \frac{\partial g_{ij}}{\partial x^k}\right) \qquad \text{...(1)}$$

where $g_{ij}$ are the components of the metric tensor.

*Christoffel symbol of the second kind* is denoted by $\begin{Bmatrix} k \\ ij \end{Bmatrix}$ and is defined by

$$\begin{Bmatrix} k \\ ij \end{Bmatrix} = g^{kl}\,[ij, l] \qquad \text{...(2)}$$

Some authors write Christoffel symbol of the second kind as $\{ij, k\}$ or $\Gamma_{ij}^k$.

**Obs. 1.** No summation is indicated in the Christoffel symbol of the first kind, but summation is to be made over $l$ in the Christoffel symbol of the second kind.

**Obs. 2.** It is evident from Eqn. (1) and (2) that the Christoffel symbols of both kinds are symmetric in the indices $i$ and $j$.

$$[ij, k] = [ji, k] \text{ and } \begin{Bmatrix} k \\ ij \end{Bmatrix} = \begin{Bmatrix} k \\ ji \end{Bmatrix}.$$

**Example 38.14.** *If* $(ds)^2 = (dr)^2 + r^2(d\theta)^2 + r^2\sin^2\theta\,(d\phi)^2$, *find the values of*

(*a*) $[22, 1]$ *and* $[13, 3]$ (*b*) $\begin{Bmatrix} 1 \\ 22 \end{Bmatrix}$ *and* $\begin{Bmatrix} 3 \\ 13 \end{Bmatrix}$.

**Solution.** It is a 3-dimensional space in spherical coordinates such that

$$x^1 = r, x^2 = \theta \text{ and } x^3 = \phi$$

Clearly $\qquad g_{11} = 1, g_{22} = r^2, g_{33} = r^2 \sin^2 \theta$ and $g_{ij} = 0$ for $i \neq j$. ...(*i*)

Also $g^{11} = 1, g^{22} = 1/r^2, g^{33} = 1/r^2 \sin^2 \theta$ (See Ex. 38.12) ...(*ii*)

(*a*) Christoffel symbols of the first kind are given by

$$[ij, k] = \frac{1}{2}\left[\frac{\partial g_{jk}}{\partial x^i} + \frac{\partial g_{ik}}{\partial x^j} - \frac{\partial g_{ij}}{\partial x^k}\right]_{i,\, j,\, k = 1, 2, 3} \qquad ...(iii)$$

Taking $i = 2, j = 2$, and $k = 1$ in Eqn. (*iii*), we get

$$[22, 1] = \frac{1}{2}\left[\frac{\partial g_{21}}{\partial x^2} + \frac{\partial g_{21}}{\partial x^2} - \frac{\partial g_{22}}{\partial x^1}\right] = \frac{1}{2}\left[\frac{\partial(0)}{\partial \theta} + \frac{\partial(0)}{\partial \theta} - \frac{\partial(r^2)}{\partial r}\right] = -r \qquad ...(iv)$$

Putting $i = 1, j = 3$ and $k = 3$ in Eqn. (*iii*), we obtain

$$[13, 3] = \frac{1}{2}\left[\frac{\partial g_{33}}{\partial x^1} + \frac{\partial g_{13}}{\partial x^3} - \frac{\partial g_{13}}{\partial x^3}\right] = \frac{1}{2}\left[\frac{\partial(r^2 \sin^2 \theta)}{\partial r} + \frac{\partial(0)}{\partial \phi} - \frac{\partial(0)}{\partial \phi}\right] = r \sin^2 \theta \qquad ...(v)$$

(*b*) Christoffel symbols of the second kind are defined by

$$\begin{Bmatrix} k \\ ij \end{Bmatrix} = g^{kl}\,[ij, l] = g^{k1}\,[ij, 1] + g^{k2}\,[ij, 2] + g^{k3}\,[ij, 3] \qquad ...(vi)$$

$$\therefore \quad \begin{Bmatrix} 1 \\ 22 \end{Bmatrix} = g^{11}[22, 1] + g^{12}[22, 2] + g^{13}[22, 3] = [22, 1] + 0[22, 2] + 0[22, 3]$$

$$= -r \qquad \text{[By Eqn. (iv)]}$$

$$\begin{Bmatrix} 3 \\ 13 \end{Bmatrix} = g^{31}[13, 1] + g^{32}[13, 2] + g^{33}[13, 3]$$

$$= 0[13, 1] + 0[13, 2] + \frac{1}{r^2 \sin^2 \theta}[13, 3] \qquad \text{[By Eqn. (ii)]}$$

$$= \frac{1}{r^2 \sin^2 \theta} \cdot r \sin^2 \theta = \frac{1}{r}. \qquad \text{[By Eqn. (v)]}$$

**Example 38.15.** *Prove that*

(*a*) $\dfrac{\partial g_{ij}}{\partial x^k} = [ik, j] + [jk, i]$ $\qquad$ (*b*) $\dfrac{\partial g^{ij}}{\partial x^k} = -\, g^{jl} \begin{Bmatrix} i \\ lk \end{Bmatrix} - g^{im} \begin{Bmatrix} j \\ mk \end{Bmatrix}$.

**Solution.** (*a*) By definition of Christoffel symbol of the first kind, we have

$$[ik, j] = \frac{1}{2}\left[\frac{\partial g_{kj}}{\partial x^i} + \frac{\partial g_{ij}}{\partial x^k} - \frac{\partial g_{ik}}{\partial x^j}\right] \qquad ...(i)$$

and

$$[jk, i] = \frac{1}{2}\left[\frac{\partial g_{ki}}{\partial x^j} + \frac{\partial g_{ji}}{\partial x^k} - \frac{\partial g_{jk}}{\partial x^i}\right] \qquad ...(ii)$$

Since $g_{ij}$ is a symmetric tensor, $\therefore \quad g_{ij} = g_{ji}, g_{jk} = g_{kj}, g_{ki} = g_{ik}$

Adding Eqns. (*i*) and (*ii*), we get the required result.

(*b*) We know that $g^{ij} g_{lj} = \delta^i_l$. [Refer to § 38.14 (2)]

Differentiating w.r.t. $x^k$, we get

$$g^{ij} \frac{\partial g_{lj}}{\partial x^k} + g_{lj} \frac{\partial g^{ij}}{\partial x^k} = 0 \qquad [\because \quad \delta^i_l = 1 \text{ or } 0]$$

Multiplying by $g^{lm}$ and transposing, we have

$$g^{lm} g_{lj} \frac{\partial g^{ij}}{\partial x^k} = -g^{lm} g^{ij} \frac{\partial g_{lj}}{\partial x^k} \quad \text{or} \quad \delta^m_j \frac{\partial g^{ij}}{\partial x^k} = -g^{lm} g^{ij} \{[lk, j] + [jk, l]\} \qquad \text{[From (a)]}$$

or $$\frac{\partial g^{im}}{\partial x^k} = -g^{lm}\left(g^{ij}[lk, j]\right) - g^{ij}\left(g^{lm}[jk, l]\right) = -g^{lm}\begin{Bmatrix} i \\ lk \end{Bmatrix} - g^{ij}\begin{Bmatrix} m \\ jk \end{Bmatrix}$$

Interchanging $m$ and $j$, we obtain the desired result.

## 38.18 TRANSFORMATION OF CHRISTOFFEL SYMBOLS

The fundamental tensors $g_{ij}$, $g^{ij}$ and also $[ij, k]$ are functions of the coordinates $x^i$. Let these transform to $\bar{g}_{ij}$, $\bar{g}^{ij}$ and $\overline{[ij, k]}$ in another coordinate system $\bar{x}^i$.

(1) *Law of transformation of Christoffel symbol of first kind.*

Let $[ij, k]$ which is a function of $x^i$, transform to $\overline{[ij, k]}$ in another coordinate system $\bar{x}^i$. Then

$$\overline{[ij, k]} = \frac{1}{2}\left(\frac{\partial \bar{g}_{jk}}{\partial \bar{x}^i} + \frac{\partial \bar{g}_{ik}}{\partial \bar{x}^j} - \frac{\partial \bar{g}_{ij}}{\partial \bar{x}^k}\right) \qquad ...(1)$$

Since $\bar{g}_{ij}$ is a covariant tensor of order two, we have

$$\bar{g}_{ij} = \frac{\partial x^p}{\partial \bar{x}^i}\frac{\partial x^q}{\partial \bar{x}^j} g_{pq} \qquad ...(2)$$

Differentiating both sides w.r.t. $\bar{x}^k$, we get

$$\frac{\partial \bar{g}_{ij}}{\partial \bar{x}^k} = \left(\frac{\partial^2 x^p}{\partial \bar{x}^k \partial \bar{x}^i}\frac{\partial x^q}{\partial \bar{x}^j} + \frac{\partial x^p}{\partial \bar{x}^i}\frac{\partial^2 x^q}{\partial \bar{x}^k \partial \bar{x}^j}\right) g_{pq} + \frac{\partial x^p}{\partial \bar{x}^i}\frac{\partial x^q}{\partial \bar{x}^j}\frac{\partial g_{pq}}{\partial x^r}\frac{\partial x^r}{\partial \bar{x}^k} \qquad ...(3)$$

[Note that $g_{pq}$ is in terms of original coordinates $x$ and to differentiate it w.r.t. $\bar{x}^k$, first we differentiate it w.r.t. $x^r$ and then differentiate $x^r$ w.r.t. $\bar{x}^k$.]

Interchanging $i$, $k$ and also $p$, $r$ in the last term of (3), we have

$$\frac{\partial \bar{g}_{ik}}{\partial \bar{x}^i} = \left(\frac{\partial^2 x^p}{\partial \bar{x}^i \partial \bar{x}^k}\frac{\partial x^q}{\partial \bar{x}^j} + \frac{\partial x^p}{\partial \bar{x}^k}\frac{\partial^2 x^q}{\partial \bar{x}^i \partial \bar{x}^j}\right) g_{pq} + \frac{\partial x^p}{\partial \bar{x}^k}\frac{\partial x^q}{\partial \bar{x}^j}\frac{\partial x^p}{\partial \bar{x}^i}\frac{\partial g_{qr}}{\partial x^p} \qquad ...(4)$$

Similarly interchanging $j$, $k$ and also $q$, $r$ in the last term of (3), we get

$$\frac{\partial \bar{g}_{ik}}{\partial \bar{x}^j} = \left(\frac{\partial^2 x^p}{\partial \bar{x}^j \partial \bar{x}^i}\frac{\partial x^q}{\partial \bar{x}^k} + \frac{\partial x^p}{\partial \bar{x}^i}\frac{\partial^2 x^q}{\partial \bar{x}^k \partial \bar{x}^j}\right) g_{pq} + \frac{\partial x^p}{\partial \bar{x}^i}\frac{\partial x^r}{\partial \bar{x}^k}\frac{\partial x^p}{\partial \bar{x}^j}\frac{\partial g_{pr}}{\partial x^q} \qquad ...(5)$$

Substituting the values from Eqns. (3), (4) and (5) in Eqn. (1), we obtain

$$\overline{[ij, k]} = \frac{\partial^2 x^p}{\partial \bar{x}^i \partial \bar{x}^j}\frac{\partial x^q}{\partial \bar{x}^k} g_{pq} + \frac{\partial x^p}{\partial \bar{x}^i}\frac{\partial x^q}{\partial \bar{x}^j}\frac{\partial x^r}{\partial \bar{x}^k}[pq, r] \qquad ...(6)$$

This is the desired law of transformation of Christoffel symbol of the first kind.

(2) *Law of transformation of Christoffel symbol of the second kind.*

Let $g^{kl}[ij, l]$ transform to $\bar{g}^{kl}\,\overline{[ij, l]}$.

Since $\bar{g}^{kl}$ is a contravariant tensor of order two.

$$\therefore \qquad \bar{g}^{kl} = \frac{\partial \bar{x}^k}{\partial x^s}\frac{\partial \bar{x}^l}{\partial x^t} g^{st} \qquad ...(7)$$

From Eqn. (6), we have $$\overline{[ij, l]} = \frac{\partial^2 x^p}{\partial \bar{x}^i \partial \bar{x}^j}\frac{\partial x^q}{\partial \bar{x}^l} g_{pq} + \frac{\partial x^p}{\partial \bar{x}^i}\frac{\partial x^q}{\partial \bar{x}^j}\frac{\partial x^r}{\partial \bar{x}^l}[pq, r] \qquad ...(8)$$

Multiplying the respective sides of Eqns. (7) and (8), we get

$$\overline{\begin{Bmatrix} l \\ ij \end{Bmatrix}} = \frac{\partial \bar{x}^t}{\partial x^t}\delta^q{}_s \frac{\partial^2 x^p}{\partial \bar{x}^i \partial \bar{x}^j} g^{st} g_{pq} + \frac{\partial \bar{x}^l}{\partial x^t}\delta^r{}_s \frac{\partial x^p}{\partial \bar{x}^i}\frac{\partial x^q}{\partial \bar{x}^j} g^{st}\,[pq, r] \qquad ...(9)$$

Since $\qquad \delta^q{}_s g^{st} g_{pq} = g^{qt} g_{pq} = \delta_p^{\ t}$

and $$\delta_s^{\ r} g^{st}[pq, r] = g^{rt}[pq, r] = \begin{Bmatrix} t \\ pq \end{Bmatrix}$$

$\therefore$ $$\begin{Bmatrix} \bar{l} \\ ij \end{Bmatrix} = \frac{\partial^2 x^p}{\partial \bar{x}^i \partial \bar{x}^j} \delta^t_{\ p} \frac{\partial \bar{x}^l}{\partial x^t} + \frac{\partial x^p}{\partial \bar{x}^i} \frac{\partial x^q}{\partial \bar{x}^j} \frac{\partial \bar{x}^l}{\partial x^t} \begin{Bmatrix} t \\ pq \end{Bmatrix}$$

or $$\begin{Bmatrix} \bar{l} \\ ij \end{Bmatrix} = \frac{\partial^2 x^t}{\partial \bar{x}^i \partial \bar{x}^j} \frac{\partial \bar{x}^l}{\partial x^t} + \frac{\partial x^p}{\partial \bar{x}^i} \frac{\partial x^q}{\partial \bar{x}^j} \frac{\partial \bar{x}^l}{\partial x^t} \begin{Bmatrix} t \\ pq \end{Bmatrix} \qquad ...(10)$$

This is the law of transformation of Christoffel symbol of the second kind.

**Obs. 1.** From Eqn. (10), we obtain the following important relation :

$$\frac{\partial^2 x^t}{\partial \bar{x}^i \partial \bar{x}^j} = \frac{\partial x^t}{\partial \bar{x}^l} \begin{Bmatrix} \bar{l} \\ ij \end{Bmatrix} - \frac{\partial x^p}{\partial \bar{x}^i} \frac{\partial x^q}{\partial \bar{x}^j} \begin{Bmatrix} t \\ pq \end{Bmatrix} \qquad ...(11)$$

**Obs. 2.** It is evident from Eqns. (6) and (10) that the Christoffel 3-index symbols are not tensors. These symbols transform like tensors only for linear transformation of coordinates.

**Example 38.16.** *Prove that* $\begin{Bmatrix} i \\ ij \end{Bmatrix} = \frac{\partial}{\partial x^i}(\log \sqrt{g})$.

**Solution.** Let $G_{jk}$ be the co-factor of $g_{ik}$ in $g$ so that $g = g_{ik} G_{ik}$ (summation over $k$ only)

$\therefore$ $$\frac{\partial g}{\partial g_{ik}} = G_{ik} \qquad [\because \ G_{ik} \text{ does not contain } g_{ik} \text{ explicitly}]$$

Also $$\frac{\partial g}{\partial g^j} = \frac{\partial g}{\partial g_{ik}} \cdot \frac{\partial g_{ik}}{\partial x^j} = G_{ik} \frac{\partial g_{ik}}{\partial x^j} \qquad \text{(summation over } i \text{ and } k) \quad ...(i)$$

We know that $$g^{ik} = \frac{G_{ik}}{g} \qquad ...(ii)$$

Substituting the value of $G_{ik}$ from Eqn. (*ii*) in Eqn. (*i*), we get

$$\frac{\partial g}{\partial x^j} = g g^{ik} \frac{\partial g_{ik}}{\partial x^j} \quad \text{or} \quad \frac{1}{g} \frac{\partial g}{\partial x^j} = g^{ik} \frac{\partial g_{ik}}{\partial x^j}$$

or $$\frac{\partial}{\partial x^j}(\log g) = g^{ik}([jk, i] + [ij, k]) \qquad \text{[By Ex. 38.15 (a)]}$$

$$= g^{ik}[jk, i] + g^{ik}[ji, k] = \begin{Bmatrix} k \\ jk \end{Bmatrix} + \begin{Bmatrix} i \\ ji \end{Bmatrix} = 2\begin{Bmatrix} i \\ ji \end{Bmatrix}$$

Hence $$\begin{Bmatrix} i \\ ij \end{Bmatrix} = \frac{1}{2} \frac{\partial}{\partial x^j}(\log g) = \frac{\partial}{\partial x^j} \log \sqrt{g}. \qquad ...(iii)$$

## 38.19 (1) COVARIANT DIFFERENTIATION OF A COVARIANT VECTOR

Let $A_i$ and $\bar{A}_i$ be the components of a covariant vector (*i.e.*, a tensor of first order) in the coordinate system $x^i$ and $\bar{x}^i$ respectively. Let us investigate the tensor character of the partial derivatives of $A_i$ w.r.t. the variables $x^j$. From the law of transformation. $\bar{A}_i = \frac{\partial x^p}{\partial \bar{x}^i} A_p$

Differentiating w.r.t. $\bar{x}^j$, we have $$\frac{\partial \bar{A}_i}{\partial \bar{x}^j} = \frac{\partial^2 x^p}{\partial \bar{x}^j \partial \bar{x}^i} A_p + \frac{\partial x^p}{\partial \bar{x}^i} \frac{\partial A_p}{\partial x^q} \frac{\partial x^q}{\partial \bar{x}^j} \qquad ...(1)$$

(Note that $A_p$ is not directly a function of $\bar{x}^j$).

…ce of the first term on the R.H.S. of Eqn. (1), it is evident that $\frac{\partial A_p}{\partial x^q}$ is not a tensor.

On replacing this term by $\dfrac{\partial^2 x^s}{\partial \bar{x}^i \partial \bar{x}^j} A_s$ and substituting for the second derivative from (11) of § 35.18, we get

$$\frac{\partial \bar{A}_i}{\partial \bar{x}^j} = \frac{\partial x^s}{\partial \bar{x}^l} A_s \left\{ \begin{matrix} \bar{l} \\ i\ j \end{matrix} \right\} - \frac{\partial x^p}{\partial \bar{x}^i} \frac{\partial x^q}{\partial \bar{x}^j} A_s \left\{ \begin{matrix} s \\ pq \end{matrix} \right\} + \frac{\partial x^p}{\partial \bar{x}^i} \frac{\partial x^q}{\partial \bar{x}^j} \frac{\partial A_p}{\partial \bar{x}^q}$$

or
$$\frac{\partial \bar{A}_i}{\partial \bar{x}^j} - \bar{A}_i \left\{ \begin{matrix} \bar{l} \\ i\ j \end{matrix} \right\} = \frac{\partial x^p}{\partial \bar{x}^i} \frac{\partial x^q}{\partial \bar{x}^j} \left[ \frac{\partial A_p}{\partial \bar{x}^q} - A_s \left\{ \begin{matrix} s \\ pq \end{matrix} \right\} \right]$$

This shows that the expression $\dfrac{\partial \bar{A}_i}{\partial \bar{x}^j} - \bar{A}_l \left\{ \begin{matrix} \bar{l} \\ i\ j \end{matrix} \right\}$

is a covariant tensor of the second order. This is called the covariant derivative of $\bar{A}_i$ w.r.t $\bar{x}^j$ and is denoted by $\bar{A}_{i,\,j}$.

**(2) Covariant differentiation of a contravariant vector.** Let $A^i$ and $\bar{A}^i$ be the components of a contravariant vector in the coordinate systems $x^i$ and $\bar{x}^i$. From the law of transformation $A^s = \dfrac{\partial x^s}{\partial \bar{x}^i} \bar{A}^i$.

Differentiating w.r.t. $\bar{x}^j$, we have $\dfrac{\partial A^s}{\partial \bar{x}^j} = \dfrac{\partial^2 x^s}{\partial \bar{x}^j \partial \bar{x}^i} \bar{A}^i + \dfrac{\partial x^s}{\partial \bar{x}^i} \dfrac{\partial \bar{A}^i}{\partial \bar{x}^j}$

Substituting for the second derivative from (11) of § 35.18, we get

$$\frac{\partial A^s}{\partial x^q} \frac{\partial x^q}{\partial \bar{x}^j} = \frac{\partial x^s}{\partial \bar{x}^l} \left\{ \begin{matrix} l \\ i\ j \end{matrix} \right\} - \frac{\partial x^p}{\partial \bar{x}^i} \frac{\partial x^q}{\partial \bar{x}^j} \left\{ \begin{matrix} s \\ p\ q \end{matrix} \right\} \bar{A}^i + \frac{\partial x^s}{\partial \bar{x}^i} \frac{\partial \bar{A}^i}{\partial \bar{x}^j}$$

Interchanging the dummy indices $i$, $l$ in the first term on the R.H.S. and putting

$\dfrac{\partial x^p}{\partial \bar{x}^i} \bar{A}^i = A^p$ in the second term, we obtain

$$\frac{\partial x^q}{\partial \bar{x}^j} \frac{\partial A^s}{\partial x^q} = \frac{\partial x^s}{\partial \bar{x}^i} \left\{ \begin{matrix} \bar{i} \\ l\ j \end{matrix} \right\} \bar{A}^l - \frac{\partial x^q}{\partial \bar{x}^j} A^p \left\{ \begin{matrix} s \\ p\ q \end{matrix} \right\} + \frac{\partial x^s}{\partial \bar{x}^i} \frac{\partial \bar{A}^i}{\partial \bar{x}^j}$$

Transposing the second term on the R.H.S. to the L.H.S., we get

$$\frac{\partial x^q}{\partial \bar{x}^j} \left[ \frac{\partial A^s}{\partial x^q} + A^p \left\{ \begin{matrix} s \\ p\ q \end{matrix} \right\} \right] = \frac{\partial x^s}{\partial \bar{x}^i} \left[ \frac{\partial \bar{A}^i}{\partial \bar{x}^j} + \bar{A}^l \left\{ \begin{matrix} \bar{i} \\ l\ j \end{matrix} \right\} \right]$$

or
$$\frac{\partial \bar{A}^i}{\partial \bar{x}^j} + \bar{A}^l \left\{ \begin{matrix} \bar{i} \\ l\ j \end{matrix} \right\} = \frac{\partial \bar{x}^i}{\partial x^s} \frac{\partial x^q}{\partial \bar{x}^j} \left[ \frac{\partial A^s}{\partial x^q} + A^p \left\{ \begin{matrix} s \\ p\ q \end{matrix} \right\} \right]$$

This shows that $\dfrac{\partial A^i}{\partial x^j} + A^l \left\{ \begin{matrix} i \\ l\ j \end{matrix} \right\}$ is a mixed tensor of the second order. This is called the *covariant derivative of $A^i$ w.r.t. $x^j$ and is denoted by $A^{i,}\,j$.*

**Obs.** *The following laws hold good for covariant differentiation :*
*(i) Covariant derivative of the sum (or difference) of two tensors = sum (or difference) of their covariant derivatives.*
*(ii) Covariant derivative of the product of two tensors = covariant derivative of first tensor × second tensor + covariant derivative of second tensor × first tensor.*

**Example 38.17.** *Prove that the covariant derivative of $g^{ij}$ is zero.*

**Solution.** Let $A_i$ denote a covariant vector which moves parallel to itself so that

$$A_{i,\,k} \text{ or } \frac{\partial A_i}{\partial x^k} - A_l \left\{ \begin{matrix} l \\ i\ k \end{matrix} \right\} = 0$$

Let $\phi = g^{ij} A_i A_j$ so that $\phi$ is a scalar invariant. Differentiating it w.r.t. $x^k$, we have

$$\frac{\partial \phi}{\partial x^k} = \frac{\partial g^{ij}}{\partial x^k} A_i A_j + g^{ij} \frac{\partial A_i}{\partial x^k} A_j + g^{ij} A_i \frac{\partial A_j}{\partial x^k}$$

$$= \frac{\partial g^{ij}}{\partial x^k} A_i A_j + g^{ij} \begin{Bmatrix} l \\ i \ k \end{Bmatrix} A_l A_j + g^{ij} \begin{Bmatrix} l \\ j \ k \end{Bmatrix} A_i A_l \qquad \text{[By } (i)]$$

Interchanging $i$ and $l$ in the second term and $j$ and $l$ in the last term on the right, we get

$$\frac{\partial \phi}{\partial x^k} = \left[ \frac{\partial g^{ij}}{\partial x^k} + g^{lj} \begin{Bmatrix} i \\ l \ k \end{Bmatrix} + g^{il} \begin{Bmatrix} j \\ l \ k \end{Bmatrix} \right] A_i A_j$$

Since $\partial\phi/\partial x^k$ is a covariant vector, the expression

$$\frac{\partial g^{ij}}{\partial x^k} + g^{lj} \begin{Bmatrix} i \\ l \ k \end{Bmatrix} + g^{il} \begin{Bmatrix} j \\ l \ k \end{Bmatrix}$$

is a tensor of the third order by quotient law. Thus it is the covariant derivative $g^{ij}{}_{,k}$.

$$\therefore \qquad g^{ij}{}_{,k} = \frac{\partial g^{ij}}{\partial x^k} + g^{lj} \begin{Bmatrix} i \\ l \ k \end{Bmatrix} + g^{il} \begin{Bmatrix} j \\ l \ k \end{Bmatrix}$$

$$= -g^{jl} \begin{Bmatrix} i \\ l \ k \end{Bmatrix} - g^{im} \begin{Bmatrix} j \\ m \ k \end{Bmatrix} + g^{lj} \begin{Bmatrix} i \\ l \ k \end{Bmatrix} + g^{il} \begin{Bmatrix} j \\ l \ k \end{Bmatrix} \qquad \text{[By Ex. 38.15 } (b)]$$

$$= -g^{im} \begin{Bmatrix} j \\ m \ k \end{Bmatrix} + g^{il} \begin{Bmatrix} j \\ l \ k \end{Bmatrix} = -g^{il} \begin{Bmatrix} j \\ l \ k \end{Bmatrix} + g^{il} \begin{Bmatrix} j \\ l \ k \end{Bmatrix} = 0$$

[Changing the dummy index $m$ to $l$]

## 38.20 (1) GRADIENT

If $\phi$ be a scalar function of the coordinates, then the gradient of $\phi$ is denoted by grad $\phi = \dfrac{\partial \phi}{\partial x^i}$ which is a *covariant vector.*

(2) **Divergence.** *The divergence of the contravariant vector $A^i$ is defined by*

$$div\ A^i = \frac{\partial A^i}{\partial x^i} + A^k \begin{Bmatrix} l \\ k \ i \end{Bmatrix} \text{ which is sometimes written as } A^i{}_{,i}.$$

*The divergence of the covariant vector $A_i$ is defined by div $A_i = g^{ik} A_{ik}$.*

(3) **Curl.** Let $A_i$ be a covariant vector, then

$$A_{i,j} = \frac{\partial A_i}{\partial x^j} - A_k \begin{Bmatrix} k \\ i \ j \end{Bmatrix} \text{ and } A_{j,i} = \frac{\partial A_j}{\partial x^i} - A_k \begin{Bmatrix} k \\ j \ i \end{Bmatrix} \text{ are covariant tensors.}$$

$$\therefore \qquad A_{i,j} - A_{j,i} = \frac{\partial A_i}{\partial x^j} - \frac{\partial A_j}{\partial x^i}$$

*is a covariant tensor of second order, which is called curl of $A_i$.*

Thus $\qquad curl\ A_i = A_{i,j} - A_{j,i}$.

**Obs 1.** *Curl $A_i$ is a skew-symmetric tensor.*

Since $\qquad A_{j,i} - A_{i,j} = -(A_{i,j} - A_{j,i})$.

**Obs. 2.** *Curl is a tensor and not a vector.* In a 3-dimensional space, however, curl has only three independent non-zero components and it can, therefore, be taken as a vector.

**Example 38.18.** *Prove that*

(a) $div\ A_i = \dfrac{1}{\sqrt{g}} \cdot \dfrac{\partial}{\partial x^k}(\sqrt{g}\, A^k)$ $\qquad$ (b) $\nabla^2 \phi = \dfrac{1}{\sqrt{g}} \dfrac{\partial}{\partial x^k}\left(\sqrt{g}\, g^{kr} \dfrac{\partial \phi}{\partial x^r}\right).$

**Solution.** (*a*) Using $\begin{Bmatrix} i \\ k\ i \end{Bmatrix} = \frac{\partial}{\partial x^k} \log \sqrt{g} = \frac{1}{\sqrt{g}} \frac{\partial \sqrt{g}}{\partial x^k}$ [By Ex. 38.16]

$$\text{div } A^i = \frac{\partial A^i}{\partial x^i} + A^k \begin{Bmatrix} i \\ k\ i \end{Bmatrix} = \frac{\partial A^i}{\partial x^i} + \frac{A^k}{\sqrt{g}} \frac{\partial \sqrt{g}}{\partial x^k} = \frac{\partial A^k}{\partial x^k} + \frac{A^k}{\sqrt{g}} \frac{\partial \sqrt{g}}{\partial x^k}$$

$$= \frac{1}{\sqrt{g}} \frac{\partial}{\partial x^k} (\sqrt{g}\, A^k) \quad ...(i)$$

(*ii*) We have $\nabla^2 \phi = \text{div grad } \phi$ ...(*ii*)

and $\text{grad } \phi = \frac{\partial \phi}{\partial x^r}$, which is a covariant vector.

The contravariant vector associated with $\partial\phi/\partial x^r$ is

$$A^k = g^{kr}\, \partial\phi/\partial x^r.$$

Then from Eqns. (*i*) and (*ii*), $\nabla^2 \phi = \text{div}\left(g^{kr} \frac{\partial \phi}{\partial x^r}\right) = \frac{1}{\sqrt{g}} \frac{\partial}{\partial x^k}\left(\sqrt{g}\, g^{kr} \frac{\partial \phi}{\partial x^r}\right)$.

**Example. 38.19:** *If $A_k^{ij}$ and $B_n^{lm}$ are tensor their sum and difference are tensor of the same rank and type.* [*MSc(AU), 2018*]

**Solution.** Given $A_k^{ij}$ and $B_k^{ij}$ are tensors. Then according to law of transformation

$$\bar{A}_k^{ij} = \frac{\delta \bar{x}^i}{\delta x^p} \frac{\delta \bar{x}^j}{\delta x^q} \frac{\delta x^r}{\delta x^{-k}} A_r^{pq}$$

$$\bar{B}_k^{ij} = \frac{\delta \bar{x}^i}{\delta x^p} \frac{\delta \bar{x}^j}{\delta x^q} \frac{\delta x^r}{\delta \bar{x}^k} B_r^{pq}$$

then $$\bar{A}_k^{ij} \pm \bar{B}_k^{ij} = \frac{\delta \bar{x}^i}{\delta x^p} \frac{\delta \bar{x}^j}{\delta x^q} - \frac{\delta x^r}{\delta \bar{x}^k} (A_r^{pq} \pm B_r^{pq})$$

If $$\bar{A}_k^{ij} \pm \bar{B}_k^{ij} = \frac{\delta \bar{x}^i}{\delta x^p} \frac{\delta \bar{x}^j}{\delta x^q} \frac{\delta x^r}{\delta \bar{x}^k} (A_r^{pq} \pm B_r^{pq})$$

$$\bar{A}_k^{ij} + \bar{B}_k^{ij} = \bar{C}_k^{ij} \text{ and } A_r^{pq} \pm B_r^{pq} = C_r^{pq}$$

$$\bar{C}_k^{ij} = \frac{\delta \bar{x}^i}{\delta x^p} \frac{\delta \bar{x}^j}{\delta x^q} \frac{\delta x^r}{\delta x^k} C_r^{pq}$$

The show that $C_k^{ij}$ in a tensors of same rank and type as $A_k^{ij}$ and $B_k^{ij}$.

**Example 38.20:** *If $T_i$ be the component of a covariant vector then show that $\left(\frac{\delta T_i}{\delta x^j} - \frac{\delta T_j}{\delta x^i}\right)$ are component of skewsymmetric covarient-tensor of rank two.* (*Chennai. 2018*)

**Solution.** Given $T_i$ is covariant-vector, then by law of transformation.

$$\bar{T}_i = \frac{\delta x^k}{\delta \bar{x}^i} T_k$$

Differentiating it. *w.r.t.* to $\bar{x}^j$

$$\frac{\delta \bar{T}_i}{\delta \bar{x}^j} = \frac{\delta}{\delta \bar{x}^j}\left(\frac{\delta x^k}{\delta \bar{x}^j} T_k\right) = \frac{\delta^2 x^k}{\delta \bar{x}^j \delta \bar{x}^i} T_k + \frac{\delta x^k}{\delta \bar{x}^i} \frac{\delta T_k}{\delta \bar{x}^j}$$

$$\frac{\delta \bar{T}_j}{\delta \bar{x}^j} = \frac{\delta^2 x^k}{\delta \bar{x}^j \delta \bar{x}^i} T_k + \frac{\delta x^k}{\delta \bar{x}^i} \frac{\delta x^e}{\delta \bar{x}^j} \frac{\delta T_k}{\delta \bar{x}^i} \quad ...(i)$$

$$\frac{\delta\bar{T}_j}{\delta\bar{x}^j} = \frac{\delta^2 x^k}{\delta\bar{x}^j\,\delta\bar{x}^j} T_k + \frac{\delta x^k\,\delta x^l\,\delta T_k}{\delta\bar{x}^i\delta\bar{x}^j\,\delta\bar{x}^i}$$

Interchaning indicies $k$ and $l$

$$\frac{\delta\bar{T}_j}{\delta\bar{x}_i} = \frac{\delta^2 x^k}{\delta\bar{x}^i\delta\bar{x}^j} T_k + \frac{\delta x^k\delta x^l\delta T_l}{\delta\bar{x}^i\delta\bar{x}^j\delta x^k} \quad ...(ii)$$

Substituting Eqns. (*i*) and (*ii*), we get

$$\frac{\delta\bar{T}_i}{\delta\bar{x}^j} - \frac{\delta\bar{T}_j}{\delta\bar{x}^i} = \frac{\delta x^k}{\delta\bar{x}^i}\frac{\delta x^l}{\delta\bar{x}^j}\left(\frac{\delta T_K}{\delta x^l} - \frac{\delta T_l}{\delta x^k}\right)$$

This is law of transformation of covariant tensor of rank two.

$\frac{\delta T_i}{\delta x^j} - \frac{\delta T_j}{\delta x^i}$ is skew symmetric tensor

$$A_{ij} = \frac{\delta T_i}{\delta x^j} - \frac{\delta T_j}{\delta x^i}$$

$$A_{ji} = \frac{\delta T_j}{\delta x^i} - \frac{\delta T_i}{\delta x^i}$$

$$A_{ji} = -A_{ij}$$

$$A_{ij} = -A_{ji}$$

$$A_{ij} = \frac{\delta T_i}{\delta x^j} - \frac{\delta T_j}{\delta x^i} \text{ is skew-symmetric}$$

So, $\frac{\delta T_i}{\delta\bar{x}^j} - \frac{\delta T_j}{\delta x^i}$ are component of skewsymmetric covariant tensor of rank two.

**Example 38.21:** *If the tensor $a_{ij}$ and $g_{ij}$ are symmetric and $u^i$, $v^j$ are components of contra variant vector satisfying the equations.*

$$(a_{ij} - kg_{ij})\,u^i = 0, \quad i, j = 1, 2, ..... n$$

$$(a_{ij} - k'g_{ij})\,v^j = 0, \; k \neq k'$$

*Prove that $g_{ij}\,u^i\,v^j = 0$, $a_{ij} = u^i\,v^j = 0$.* (Calicut, 2018)

**Solution.** Given,

$$(a_{ij} - kg_{ij})\,u^i = 0 \quad ...(i)$$

$$(a_{ij} - k'\,g_{ij})\,v^j = 0 \quad ...(ii)$$

Multiplying Eqns. (*i*) and (*ii*) by $u^j$ and $v^j$, we get $a_{ij}\,u^i\,v^j - a_{ij}\;v^i\,w^j - kg_{ij}\,u^i\,v^j + k'g_{ij}\,w^j\,v^i = 0$

Interchanging $i$ and $j$ in the second and fourth terms,

$$a_{ij}\,u^i\,v^j - a_{ji}\,v^j\,u^i - kg_{ij}\,u^i\,v^j + k'g_{ij}\,u^i\,v^j = 0$$

As $a_{ij}$ and $g_{ij}$ is symmetric *i.e.*, $a_{ij} = a_{ji}$ and $g_{ij} = g_{ji}$

$$-kg_{ij}\,v^ju^i + k'g_{ij}u^iv^j = 0$$

$$(k' - k)\,g_{ij}\,u^i\,v^j = 0$$

$$g_{ij}\,u^iv^j = 0 \quad \text{since } k \neq k' \Rightarrow k - k' \neq 0$$

Multipling Eqn. (*i*) by $v^j$, we get

$$a_{ij}\,u^j\,v^i - kg_{ij}\,u^i\,v^j = 0$$

$$a_{ij}\,u^i\,v^j = 0 \quad \text{and} \quad g_{ij}\,u^i\,v^j = 0$$

**Example 38.22:** *Given that $a_{mn}\,x^m\,x^n = b_{mn}\,x^m\,x^n$ for arbitrary values of $x^r$, show that $a_{(mn)} = b_{(mn)}$. i.e., $a_{mn} + a_{nm} = b_{mn} + b_{nm}$ also, $a_{mn}$ and $b_{mn}$ are symmetric tensors then further show the $a_{mn} = b_{mn}$.* (Maharastra, 2018)

**Solution.** We know that

$$a_{mn}\, x^m\, x^n = b_{mn}\, x^m\, x^n \qquad ...(i)$$

$$(a_{mn} - b_{mn})\, x^m\, x^n = 0$$

Differentiating *w.r.t.* $x^i$ partically

$$(a_{in} - b_{in})\, x^n + (a_{mi} - b_{mi}) x^m = 0$$

Differentiating agin *w.r.t.* $x^j$

$$(a_{ij} - b_{ij}) + (a_{ji} - b_{ji}) = 0$$

$$a_{ij} + a_{ji} = b_{ij} + b_{ji}$$

$$a_{mn} + a_{nm} = b_{mn} + b_{nm} \quad \text{or} \quad a_{(mn)} = b_{(mn)}$$

Also, $a_{mn}$ and $b_{mn}$ are symmetric

So, $$a_{mn} = a_{nm},\; b_{mn} = b_{nm}$$

So, $$2a_{mn} = 2b_{mn}$$

$$a_{mn} = b_{mn}.$$

## PROBLEMS 38.4

1. Determine Christoffel symbols of the first and second kind for an orthogonal curvilinear coordinate system.
2. Determine the Christoffel symbols of the first kind in (*a*) rectangular, (*b*) cylindrical and (*c*) spherical coordinates.
3. Evaluate the Christoffel symbols of the second kind in (*a*) rectangular, (*b*) cylindrical and (*c*) spherical coordinates.
4. Prove that (*a*) $[pq, r] = [qp, r]$, (*b*) $[pq, r] = g_{rs} \begin{Bmatrix} s \\ pq \end{Bmatrix}$
5. If $(ds)^2 = r^2 (d\theta)^2 + r^2 \sin^2 \theta\, (d\phi)^2$, find the values of

   (*a*) $[22, 1]$ and $[12, 2]$ (*b*) $\begin{bmatrix} 1 \\ 22 \end{bmatrix}$ and $\begin{bmatrix} 2 \\ 12 \end{bmatrix}$.
6. If $(ds)^2 = (dr)^2 + r^2 (d\theta)^2 + r^2 \sin^2 \theta\, (d\phi)^2$, find the values of

   (*a*) $[33, 1]$ and $[23, 3]$ (*b*) $\begin{bmatrix} 1 \\ 33 \end{bmatrix}$ and $\begin{bmatrix} 3 \\ 23 \end{bmatrix}$.
7. Show that the tensors $g_{ij}$, $g^{ij}$ and $\delta^i_j$ are constants with respect to covariant differentiation.
8. Write the covariant derivative w.r.t. $x^k$ of the tensors $u^{ij}$ and $A^k_{ij}$.
9. Show that the covariant derivative of $g_{ij}$ is zero.
10. Find the covariant derivative $A^j_k\, B^{lm}_n$ with respect to $x^q$.
11. Evaluate div $A^i$ in (*a*) cylindrical, (*b*) spherical coordinates.
12. Obtain the Laplace's equation in (*a*) cylindrical, (*b*) spherical coordinates.
13. If $A_{ij,\,k}$ is the curl of a covariant vector, prove that

    $$A_{ij,\,k} + A_{jk,\,i} + A_{ki,\,j} = 0.$$
14. Using tensor notation, show that

    (*a*) div curl $A^r = 0$ (*b*) curl grad $\phi = 0$.
15. Prove that in $n$-dimensional space, symmetric and skew-symmetric tensor have $\frac{n}{2}(n+1)$ and $\frac{n}{2}(n-1)$ independent components respectively. (*MDU. 2018*)
16. Given that $A\,(r, \Delta t)$ be a function of the co-ordinates in $n$-dimensional space such that for an arbitrary vector $B^r$ of the type indicated by the index $aA\,(r, s, t)\, B^r$ is equal to component $C^{st}$ of a contravariant tensor of order two. Prove that $A(r, s, t)$ are the components of a tensor of the form $A^{st}_r$. (*Agra, 2018*)
17. If $\phi = a_{jk}\, A^j A^k$, then show that $\phi = b_{jk}\, A^j A^k$, where $b_{jk}$ is symmetric.
18. If $|g_{ij}| \neq 0$, show that (*Annamalai, 2020*)

    $$g_{\alpha\beta} \frac{\delta}{\delta x^j} \begin{Bmatrix} \beta \\ i\,k \end{Bmatrix} = \frac{\delta}{\delta x^j} [ik, \alpha] - \begin{Bmatrix} \beta \\ i\,k \end{Bmatrix} ([B_j, \alpha] + [\alpha\, j, \beta])$$

APPENDIX 1

# Useful Information

## I. BASIC DATA

### 1. Basic Constants

| | | | |
|---|---|---|---|
| $e = 2.7183$ | $1/e = 0.3679$ | $\log_e 2 = 0.6931$ | $\log_e 3 = 1.0986$ |
| $\pi = 3.1416$ | $1/\pi = 0.3183$ | $\log_e 10 = 2.3026$ | $\log_{10} e = 0.4343$ |
| $\sqrt{2} = 1.4142$ | $\sqrt{3} = 1.732$ | 1 rad. = 57° 17′ 45″ | 1° = 0.0174 rad. |

### 2. Conversion Factors

| | |
|---|---|
| 1 ft. = 30.48 cm = 0.3048 m | 1 m = 100 cm = 3.2804 ft. |
| 1 $ft^2$ = 0.0929 $m^2$ | 1 acre = 4840 $yd^2$ = 4046.77 $m^2$ |
| 1 $ft^3$ = 0.0283 $m^3$ | 1 $m^3$ = 35.32 $ft^3$ |
| 1 m/sec = 3.2804 ft/sec. | 1 mile/h = 1.609 km/h. |

### 3. Systems of Units

| *Quantity* | *F.P.S. system* | *C.G.S. system* | *M.K.S. system* |
|---|---|---|---|
| Length | foot (ft) | centimetre (cm) | metre (m) |
| Mass | pound (lb) | gram (gm) | kilogram (kg) |
| Time | second (sec) | second (sec) | second (sec) |
| Force | lb. wt. | dyne | newton (nt) |

Note. The M.K.S. system is also known as the *International system of units (SI system)*.

### 4. Greek Letters Used

| | | | | | | | |
|---|---|---|---|---|---|---|---|
| $\alpha$ | alpha | $\theta$ | theta | $\kappa$ | kappa | $\tau$ | tau |
| $\beta$ | beta | $\phi$ | phi | $\mu$ | mu | $\chi$ | chi |
| $\gamma$ | gamma | $\psi$ | psi | $\nu$ | nu | $\omega$ | omega |
| $\delta$ | delta | $\xi$ | xi | $\pi$ | pi | $\Gamma$ | cap. gamma |
| $\varepsilon$ | epsilon | $\eta$ | eta | $\rho$ | rho | $\Delta$ | cap. delta |
| $\iota$ | iota | $\zeta$ | zeta | $\sigma$ | sigma | $\Sigma$ | cap. sigma |
| | | $\lambda$ | lambda | | | | |

### 5. Some Notations

| | | | |
|---|---|---|---|
| $\in$ | belongs to | $\cup$ | union |
| $\notin$ | does not belong to | $\cap$ | intersection |
| $\Rightarrow$ | implies | ∍/ | such that |
| $\Leftrightarrow$ | implies & implied by | | |

*Factorial n i.e.,* $n\,! = n(n-1)(n-2) \ldots\ldots 3\,.\,2\,.\,1.$

*Double factorials :* $(2n)!! = 2n\,(2n-2)\,(2n-4) \ldots\ldots 6\,.\,4\,.\,2.$

$$(2n-1)\,!! = (2n-1)\,(2n-3)\,(2n-5) \;\ldots\ldots\; 5\,.\,3\,.\,1.$$

*Stirling's approximation.* When $n$ is large $n\,! \sim \sqrt{2\pi n}\;.\,n^n\,e^{-n}$.

## II. ALGEBRA

**1. Quadratic equation :** $ax^2 + bx + c = 0$ has roots

$$\alpha = \frac{-b + \sqrt{b^2 - 4ac)}}{2a},\ \beta = \frac{-b - \sqrt{(b^2 - 4ac)}}{2a}$$

$$\alpha + \beta = -\frac{b}{a},\ \alpha\beta = \frac{c}{a}.$$

*Roots are equal if* $b^2 - 4ac = 0$

*Roots are real and distinct if* $b^2 - 4ac > 0$

*Roots are imaginary if* $b^2 - 4ac < 0$

**2. Cubic equation :** $x^3 + lx^2 + mx + n = 0$

*Cardan's method :*

(*i*) Remove $x^2$ term by putting $y = x - (-l/3)$

(*ii*) Equate coeffs. in the new equation and $y^3 - 3uvy - (u^3 + v^3) - 0$ $[\because\ y = u + v$

(*iii*) Find $u^3$ and $v^3$. Then find $u$ and $v$.

(*iv*) Get $y = u + v$ and $x = y - l/3$.

**3. Biquadratic equation :** $x^4 + kx^3 + lx^2 + mx + n = 0$

I. *Ferrari's method :*

(*i*) Combine $x^4$ and $x^3$ terms into a perfect square by adding a term in $\lambda$.

(*ii*) Make R.H.S. a perfect square to find $\lambda$.

(*iii*) Solve resulting quadratic equations.

II. *Descarte's method :*

(*i*) Remove $x^3$ term by putting $y = x - (-k/4)$

(*ii*) Equate transformed expression to $(y^2 + py + q)\,(y^2 - py + q')$

(*iii*) Equate coeffs. of like powers from both sides.

(*iv*) Find $p$, $q$ and $q'$ and solve resulting quadratics.

**4. Cross-multiplication :**

$$a_1x + b_1y + c_1z = 0$$
$$a_2x + b_2y + c_2z = 0$$

Then
$$\frac{x}{b_1c_2 - b_2c_1} = \frac{y}{c_1a_2 - c_2a_1} = \frac{z}{a_1b_2 - a_2b_1}$$

**5. Method of Least squares :**

(*i*) *Straight line of best fit* $y = a + bx$.

*Normal equations :* $\Sigma y = na + b\Sigma x,\ \Sigma xy = a\Sigma x + b\Sigma x^2$.

*To find a, b, solve these equations.*

(*ii*) *Parabola of best fit* $y = a + bx + cx^2$

*Normal equations :* $\Sigma y = na + b\Sigma x + c\Sigma x^2$,

$$\Sigma xy = a\Sigma x + b\Sigma x^2 + c\Sigma x^3,\ \Sigma x^2 y = a\Sigma x^2 + b\Sigma x^3 + c\Sigma x^4$$

To find $a$, $b$, $c$, solve these equations.

**6. Progressions :**

(*i*) Numbers $a$, $a + d$, $a + 2d$, ... are said to be in *Arithmetic progression (A.P.)*

Its $n$th term $T_n = a + \overline{n-1}\,d$ and *sum* $S_n = \frac{n}{2}\,(2a + \overline{n-1}\,d)$

(*ii*) Numbers $a$, $ar$, $ar^2$, ......, are said to be in *Geometric progression (G.P.)*

Its $n$th term $T_n = ar^{n-1}$ and *sum* $S_n = \frac{a(1-r^n)}{1-r},\ S_\infty = \frac{a}{1-r}\ (r < 1)$

(*iii*) Numbers $1/a$, $1/(a + d)$, $1/(a + 2d)$, ...... are said to be in *Harmonic progression (H.P.)* (*i.e.*, a sequence is said to be in H.P. if its reciprocals are in A.P.)

Its $n$th term $T_n = 1/(a + \overline{n-1}\, d)$

(*iv*) If $a$ and $b$ be two numbers then their

*Arithmetic mean* $= \frac{1}{2}(a + b)$, *Geometric mean* $= \sqrt{ab}$, *Harmonic mean* $= 2ab/(a + b)$

(*v*) *Natural numbers* are 1, 2, 3, ......, $n$.

$$\Sigma n = \frac{n(n+1)}{2},\ \Sigma n^2 = \frac{n(n+1)(2n+1)}{6},\ \Sigma n^3 = \left\{\frac{n(n+1)}{2}\right\}^2$$

## 7. Permutations and Combinations

$$^nP_r = \frac{n!}{(n-r)!};\ ^nC_r = \frac{n!}{r!(n-r)!} = \frac{^nP_r}{r!};\ ^nC_{n-r} = {^nC_r};\quad ^nC_0 = 1 = {^nC_n}.$$

## 8. Binomial theorem

(*i*) *When n is a positive integer*

$$(1 + x)^n = 1 + {^nC_1}\,x + {^nC_2}\,x^2 + {^nC_3}\,x^3 + \ldots\ldots + {^nC_n}\,x^n.$$

(*ii*) *When n is a negative integer or a fraction*

$$(1 + x)^n = 1 + nx + \frac{n(n-1)}{1.2}x^2 + \frac{n(n-1)(n-2)}{1.2.3}x^3 + \ldots\ldots \infty.$$

(*iii*) *Binomial coefficients* : $^nC_r = \dfrac{n!}{r!(n-r)!}$

## 9. Logarithms

(*i*) *Naturals logarithm* $\log x$ has base $e$ and is inverse of $e^x$.

*Common logarithm* $\log_{10} x = M \log x$  where $M = \log_{10} e = 0.4343$

(*ii*) $\log_a 1 = 0$; $\log_a 0 = -\infty\ (a > 1)$ ; $\log_a a = 1$.

(*iii*) $\log(mn) = \log m + \log n$ ; $\log(m/n) = \log m - \log n$ ; $\log(m^n) = n \log m$.

## 10. Partial Fractions

A fraction of the form $\dfrac{a_0x^m + a_1x^{m-1} + \ldots\ldots + a_m}{b_0x^n + b_1x^{n-1} + \ldots\ldots + b_n}$

in which $m$ and $n$ are positive integers, is called a *rational algebraic fraction.* When the numerator is of a lower degree than the denominator, it is called a *proper fraction.*

*To resolve a given fraction into partial fractions*, we first factorise the denominator into real factors. These will be either linear or quadratic, and some factors repeated. Then the proper fraction is resolved into a sum of partial fractions such that

(*i*) to a *non-repeated linear factor* $x - a$ in the denominator corresponds a partial fraction of the form $A/(x - a)$;

(*ii*) to a *repeated linear factor* $(x - a)^r$ in the denominator corresponds the sum of $r$ partial fractions of the form $\dfrac{A_1}{x-a} + \dfrac{A_2}{(x-a)^2} + \dfrac{A_3}{(x-a)^3} + \ldots + \dfrac{A_r}{(x-a)^r}$;

(*iii*) to a *non-repeated quadratic factor* $(x^2 + ax + b)$ in the denominator, corresponds a partial fraction of the form $\dfrac{Ax + B}{x^2 + ax + b}$;

(*iv*) to a *repeated quadratic factor* $(x^2 + ax + b)^r$ in the denominator, corresponds the sum of $r$ partial fractions of the form $\dfrac{A_1x + B_1}{x^2 + ax + b} + \dfrac{A_2x + B_2}{(x^2 + ax + b)^2} + \ldots\ldots + \dfrac{A_rx + B_r}{(x^2 + ax + b)^r}$.

Then we have to determine the unknown constants $A$, $A_1$, $B_1$ etc.

*To obtain the partial fraction corresponding to the non-repeated linear factor $x - a$ in the denominator, put $x = a$ everywhere in the given fraction except in the fac tor $x - a$ itself.*

In all other cases, equate the given fraction to a sum of suitable partial fractions in accordance with (*i*) to (*iv*) above, having found the partial fractions corresponding to the non-repeated linear factors by the above rule. Then multiply both sides by the denominator of the given fraction and equate the coefficients of like powers of $x$ or substitute convenient numerical values of $x$ on both sides. Finally solve the simplest of the resulting equations to find the unknown constants.

## 11. Matrices

(*i*) *A system of mn numbers arranged in a rectangular array of m rows and n columns is called a* **matrix** *of order $m \times n$.*

In particular if $m = n$, it is called a *square matrix of order n*.

(*ii*) *Two matrices of the same order can be added or subtracted* by adding or subtracting the corresponding elements.

(*iii*) *Product of a matrix A by a scalar k* is a matrix whose each element is $k$ times the corresponding elements of $A$.

(*iv*) *Two matrices can be multiplied* only when the number of columns in the first is equal to the number of rows in the second. If $A$ is of order $m \times n$ and $B$ is of order $n \times p$, then the product $AB$ is a matrix of order $m \times p$, obtained by multiplying and adding the row elements of $A$ with the corresponding column elements of $B$.

(*v*) *Transpose of a matrix A* is the matrix obtained by interchanging its rows and columns and is denoted by $A'$ or $A^T$.

A square matrix $A$ is said to be *symmetric* if $A = A'$ and *skew symmetric* if $A = -A'$.

(*vi*) If $A$ and $B$ are two square matrices such that $AB = I$ (*i.e.*, a unit matrix), then $B$ is called the *inverse* of $A$ and is denoted by $A^{-1}$. Then $AA^{-1} = A^{-1} A = I$.

**1.** $A^{-1} = \dfrac{1}{|A|} \, adj(A)$ **2.** $(A^T)^{-1} = (A^{-1})^T$ **3.** $(AB)^T = B^T A^T$ **4.** $(AB)^{-1} = B^{-1} A^{-1}$.

(*vii*) *Rank of a matrix* is the largest order of any non-vanishing minor of the matrix.

(*viii*) *Consistency of a system of equations in n unknowns.*

If the rank of the coefficient matrix $A$ be $r$ and that of the augmented matrix $K$ be $r'$, then

(*a*) the equations are inconsistent (*i.e.* there is no solution) when $r \neq r'$,

(*b*) the equations are consistent when $r = r'$.

(*c*) the equations are consistent and there are infinite number of solutions when $r = r' < n$.

(*ix*) *Eigen values:* If $A$ is any square matrix of order $n$, then the determinant of the matrix $A - \lambda I_n$ equated to zero is called the *Characterist equation* of $A$ and its roots are called the *eigen values of A*.

(*x*) *Cayley Hamilton theorem*: Every square matrix satisfies its own characteristic equation.

## 12. Determinants

(*i*) A **determinant** is defined for a square matrix $A$ and is denoted by $|A|$. Unlike a matrix it has a single value *e.g.*,

$$|A| = \begin{vmatrix} a_1 & b_1 & c_1 \\ a_2 & b_2 & c_2 \\ a_3 & b_3 & c_3 \end{vmatrix} = a_1 \begin{vmatrix} b_2 & c_2 \\ b_3 & c_3 \end{vmatrix} - b_1 \begin{vmatrix} a_2 & c_2 \\ a_3 & c_3 \end{vmatrix} + c_1 \begin{vmatrix} a_2 & b_2 \\ a_3 & b_3 \end{vmatrix}$$

$$= a_1 (b_2c_3 - b_3c_2) - b_1 (a_2c_3 - a_3c_2) + c_1 (a_2b_3 - a_3b_2)$$

In this way, determinant can be expanded in terms of any row or column.

(*ii*) **Properties :**

I. A determinant remains unaltered if its rows and columns are interchanged.

II. A determinant vanishes if two of its rows (or columns) are identical or proportional.

III. If each elements of a row (or column) consists of $m$ terms, the determinant can be expressed as the sum of $m$ determinants.

IV. If to each elements of a row (or column) be added equi-multiples of the corresponding elements of two or more rows (or columns), the determinant remains unaltered.

V. If the elements of a determinant $\Delta$ are functions of $x$ and two parallel lines become identical when $x = a$, then $x - a$ is a factor of $\Delta$.

## III. GEOMETRY

**1. Coordinates of a point :** *Cartesian* $(x, y)$ *and polar* $(r, \theta)$.

$$x = r\cos\theta, \qquad y = r\sin\theta$$

or $$r = \sqrt{(x^2 + y^2)}, \qquad \theta = \tan^{-1}(y/x). \text{ (Fig. AP-1.1).}$$

*Distance between two points* $(x_1, y_1)$ and $(x_2, y_2) = \sqrt{[(x_2 - x_1)^2 + (y_2 - y_1)^2]}$

*Point of division* of the line joining $(x_1, y_1)$ and $(x_2, y_2)$ in the ratio $m_1 : m_2$ is

$$\left(\frac{m_1x_2 + m_2x_1}{m_1 + m_2}, \frac{m_1y_2 + m_2y_1}{m_1 + m_2}\right)$$

*In a triangle* having vertices $(x_1, y_1)$, $(x_2, y_2)$ and $(x_3, y_3)$

(*i*) *Area* $= \dfrac{1}{2}\begin{vmatrix} x_1 & y_1 & 1 \\ x_2 & y_2 & 1 \\ x_3 & y_3 & 1 \end{vmatrix}$

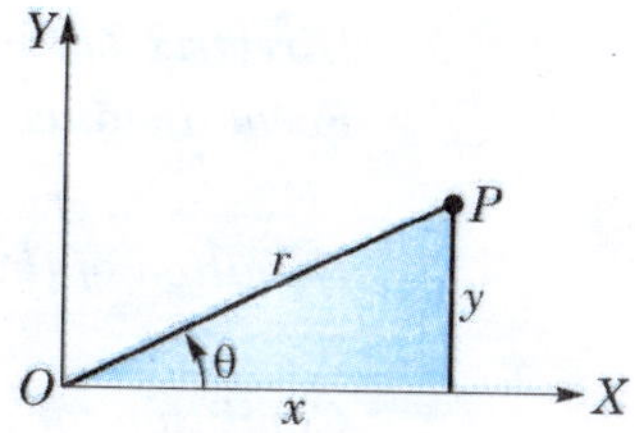

Fig. AP-1.1

(*ii*) *Centroid* (point of intersection of medians) is

$$\left(\frac{x_1 + x_2 + x_3}{3}, \frac{y_1 + y_2 + y_3}{3}\right)$$

(*iii*) *Incentre* (point of intersection of the internal bisectors of the angles) is

$$\left(\frac{ax_1 + bx_2 + cx_3}{a + b + c}, \frac{ay_1 + by_2 + cy_3}{a + b + c}\right)$$

where $a, b, c$ are the lengths of the sides of the triangle.

(*iv*) *Circumcentre* is the point of intersection of the right bisectors of the sides of the triangle.

(*v*) *Orthocentre* is the point of intersection of the perpendiculars drawn from the vertices to the opposite sides of the triangle.

**2. Straight Line**

(*i*) *Slope of the line* joining the points $(x_1, y_1)$ and $(x_2, y_2) = \dfrac{y_2 - y_1}{x_2 - x_1}$

Slope of the line $ax + by + c = 0$ is $-\dfrac{a}{b}$ *i.e.*, $-\dfrac{\text{coeff. of } x}{\text{coeff. of } y}$

(*ii*) *Equation of a line*

(*a*) having slope $m$ and cutting an intercept $c$ on $y$-axis is $y = mx + c$.

(*b*) cutting intercepts $a$ and $b$ from the axes is $\dfrac{x}{a} + \dfrac{y}{b} = 1$.

(*c*) passing through $(x_1, y_1)$ and having slope $m$ is $y - y_1 = m(x - x_1)$

(*d*) passing through $(x_1, y_1)$ and making an $\angle\theta$ with the $x$-axis is $\dfrac{x - x_1}{\cos\theta} = \dfrac{y - y_1}{\sin\theta} = r$

(*e*) through the point of intersection of the lines $a_1x + b_1y + c_1 = 0$ and $a_2x + b_2y + c_2 = 0$ is $a_1x + b_1y + c_1 + k(a_2x + b_2y + c_2) = 0$

(*iii*) *Angle between two lines* having slopes $m_1$ and $m_2$ is $\tan^{-1}\dfrac{m_1 - m_2}{1 + m_1m_2}$

Two lines are parallel if $m_1 = m_2$

Two lines are perpendicular if $m_1m_2 = -1$

Any line parallel to the line $ax + by + c = 0$ is $ax + by + k = 0$

Any line perpendicular to $ax + by + c = 0$ is $bx - ay + k = 0$

(*iv*) *Length of the perpendicular* from $(x_1, y_1)$ to the line $ax + by + c = 0$, is $\dfrac{ax_1 + by_1 + c}{\sqrt{(a^2 + b^2)}}$.

**3. Circle**

(*i*) *Equation of the circle* having centre $(h, k)$ and radius $r$ is $(x - h)^2 + (y - k)^2 = r^2$ (Fig. AP-1.2).

(*ii*) *Equation* $x^2 + y^2 + 2gx + 2fy + c = 0$ represents a circle having centre $(-g, -f)$ and radius $= \sqrt{(g^2 + f^2 - c)}$.

(*iii*) *Equation of the tangent* at the point $(x_1, y_1)$ to the circle $x^2 + y^2 = a^2$ is $xx_1 + yy_1 = a^2$.

(*iv*) *Condition for the line* $y = mx + c$ *to touch the circle* $x^2 + y^2 = a^2$ is $c = a\sqrt{(1 + m^2)}$.

(*v*) *Length of the tangent* from the point $(x_1, y_1)$ to the circle $x^2 + y^2 + 2gx + 2fy + c = 0$ is $\sqrt{(x_1^2 + y_1^2 + 2gx_1 + 2fy_1 + c)}$.

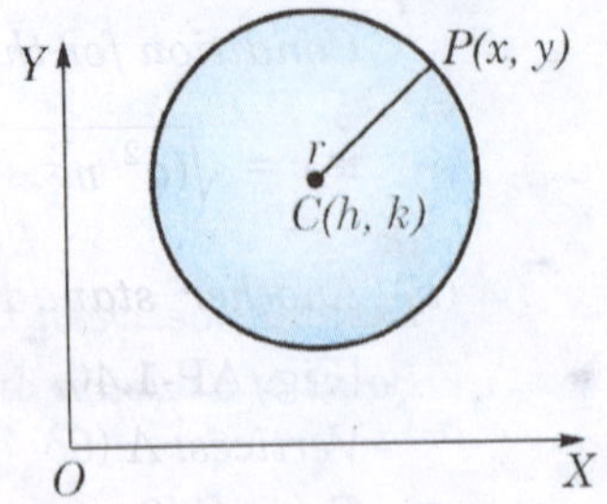

Fig. AP-1.2

### 4. Parabola

(*i*) *Standard equation of the parabola* $y^2 = 4ax$. (Fig. AP-1.3)

*Its parametric equations are* $x = at^2$, $y = 2at$.

*Latus-rectum* $LL' = 4a$, *Focus is* $S\ (a, 0)$

*Directrix ZM* is $x + a = 0$.

*Focal distance* of any point $P\ (x_1, y_1)$ on the parabola

$$y^2 = 4ax \text{ is } SP = x_1 + a$$

*Equation of the tangent* at $(x_1, y_1)$ to the parabola $y^2 = 4ax$ is

$$yy_1 = 2a\ (x + x_1)$$

*Condition for the line* $y = mx + c$ *to touch the parabola*

$$y^2 = 4ax \text{ is } c = a/m.$$

*Equation of the normal to the parabola* $y^2 = 4ax$ in terms of its slope *m* is

$$y = mx - 2am - am^3.$$

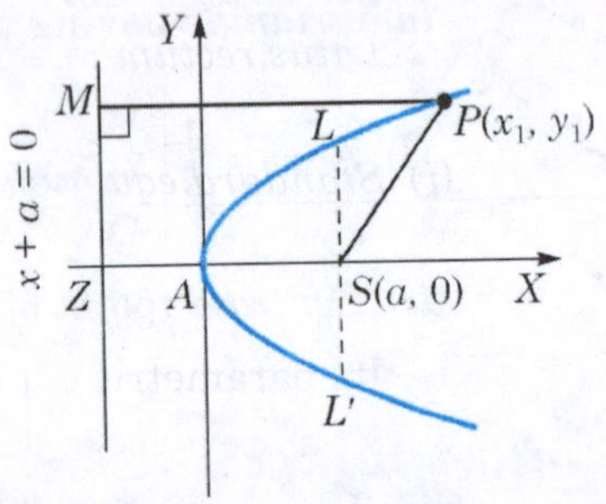

Fig. AP-1.3

(*ii*) *Other standard forms of parabola*

$y^2 = -4ax$

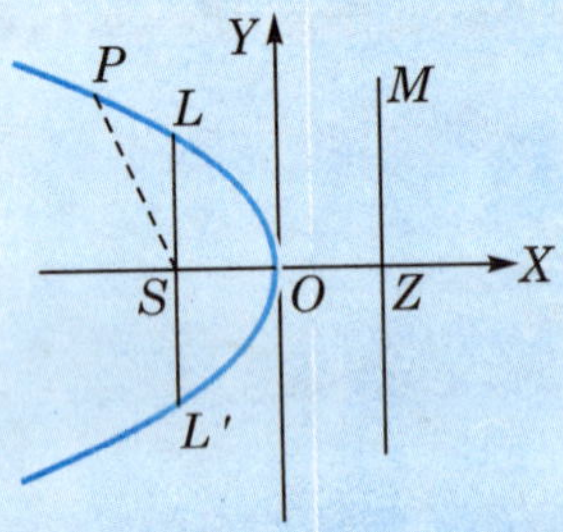


*Latus rectum* : $LL' = 4a$

*Focus* : $S\ (-a, 0)$

Directrix *ZM* : $x = a$

*Focal distance* : $SP = a - x$

$x^2 = 4ay$

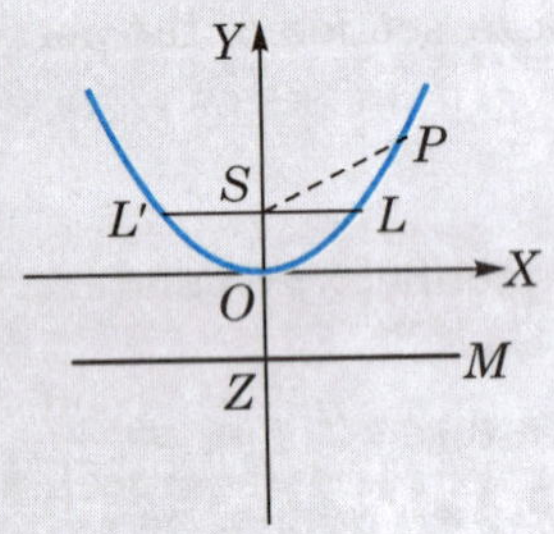


$LL' = 4a$

$S\ (0, a)$

$ZM : y = -a$

$SP = a + y$

$x^2 = -4ay$

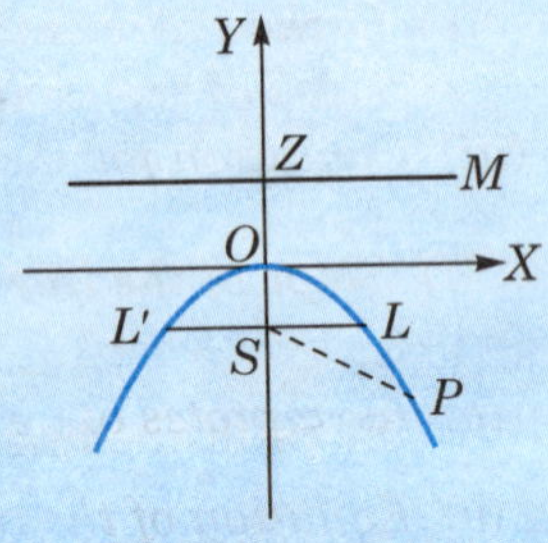


$LL' = 4a$

$S\ (0, -a)$

$ZM : y = a$

$SP = a - y$

### 5. Ellipse

(*i*) *Standard equation of the ellipse* $\dfrac{x^2}{a^2} + \dfrac{y^2}{b^2} = 1\ (a > b > 0)$. (Fig. AP-1.4)

*It parametric equations are* $x = a \cos\theta$, $y = b \sin\theta$.

*Eccentricity* $e = \sqrt{(1 - b^2/a^2)}$,

*Latus-rectum* $LSL' = 2b^2/a$.

Foci $S\ (-ae, 0)$ and $S'\ (ae, 0)$

*Directrices* $ZM\ (x = -a/e)$ and $Z'M'\ (x = a/e)$.

*Sum of the focal distances of any point on the ellipse is equal to the major axis i.e.,*

$$SP + S'P = 2a.$$

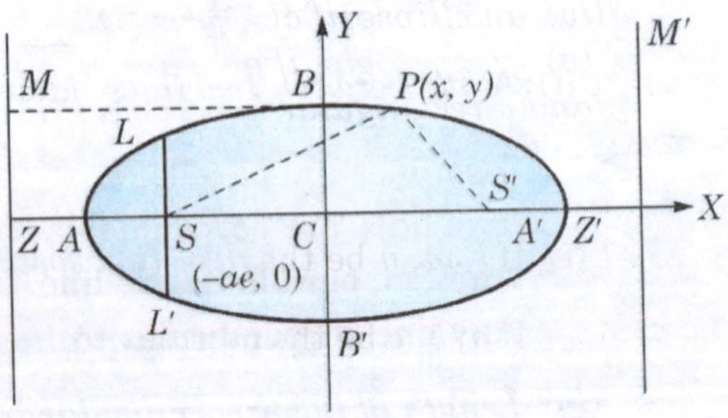

Fig. AP-1.4

*Equation of the tangent* at the point $(x_1, y_1)$ to the ellipse

$$\frac{x^2}{a^2} + \frac{y^2}{b^2} = 1 \text{ is } \frac{xx_1}{a^2} + \frac{yy_1}{b^2} = 1.$$

*Condition for the line* $y = mx + c$ *to touch the ellipse* $\frac{x^2}{a^2}+\frac{y^2}{b^2}=1$ is $c = \sqrt{(a^2 m^2 + b^2)}$

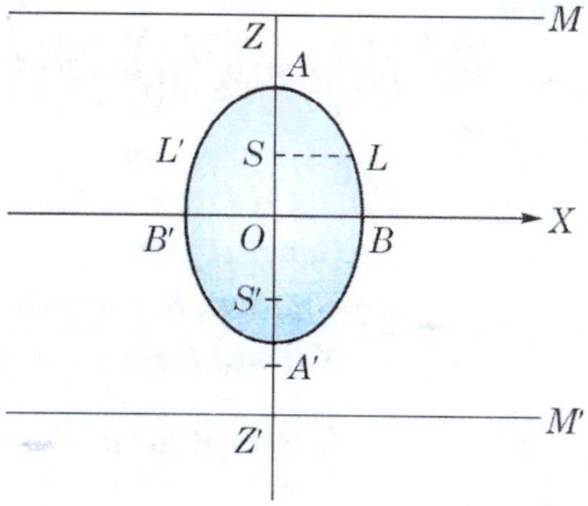


Fig. AP-1.4 (a)

(*ii*) *Another standard form of ellipse* $\frac{x^2}{a^2}+\frac{y^2}{b^2}=1$ $(a > b > 0)$ [Fig. AP-1.4(*a*)]

*Vertices:* $A\ (0o, a)$ ; $A'\ (0, -a)$

*Foci* : $S\ (0, ae)$ ; $S'\ (0, -ae)$

*Directrices* : $ZM : y = a/e$, $Z'M' : y = -a/e$

*Latus rectum* : $LSL' = 2b^2/a$

## 6. Hyperbola

(*i*) *Standard equations of the hyperbola is*

$$\frac{x^2}{a^2}-\frac{y^2}{b^2}=1. \text{ [Fig. AP-1.5]}$$

Its parametric equations are

$$x = a \sec \theta, \qquad y = b \tan \theta.$$

*Eccentricity* $e = \sqrt{(1 + b^2/a^2)}$,

*Latus-rectum* $LSL' = 2b^2/a$.

*Directrices*

$ZM\ (x = a/e)$ and $Z'M'\ (x = -a/e)$.

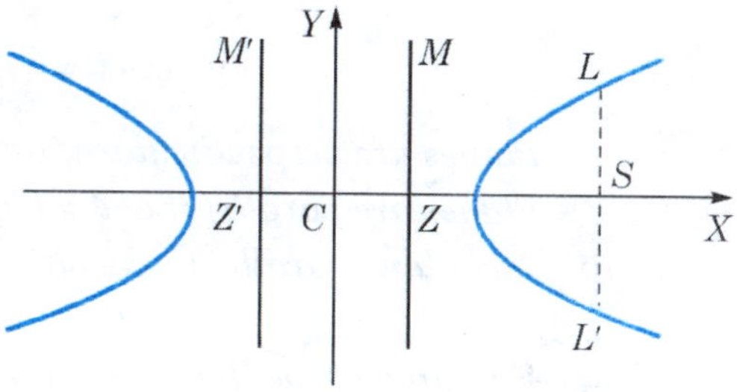


Fig. AP-1.5

(*ii*) *Equation of the tangent at the point* $(x_1, y_1)$ to the hyperbola

$$\frac{x^2}{a^2}-\frac{y^2}{b^2}=1 \text{ is } \frac{xx_1}{a^2}-\frac{yy_1}{b^2}=1.$$

(*iii*) *Condition for the line* $y = mx + c$ *to touch the hyperbola*

$$\frac{x^2}{a^2}-\frac{y^2}{b^2}=1 \text{ is } c = \sqrt{(a^2 m^2 - b^2)}$$

(*iv*) *Asymptotes of the hyperbola* $\frac{x^2}{a^2}-\frac{y^2}{b^2}=1$ are $\frac{x}{a}+\frac{y}{b}=0$ and $\frac{x}{a}-\frac{y}{b}=0$.

(*v*) *Equation of the rectangular hyperbola with asymptotes as axes is* $xy = c^2$.

Its parametric equations are $x = ct$, $y = c/t$.

## 7. Nature of a conic

*The equation* $ax^2 + 2hxy + by^2 + 2gx + 2fy + c = 0$ *represents*

(*i*) *a pair of lines, if* $\begin{vmatrix} a & h & g \\ h & b & f \\ g & f & c \end{vmatrix} (= \Delta) = 0$

(*ii*) *a circle, if* $a = b$, $h = 0$, $\Delta \neq 0$

(*iii*) *a parabola, if* $ab - h^2 = 0$, $\Delta \neq 0$.

(*iv*) *an ellipse, if* $ab - h^2 > 0$, $\Delta \neq 0$.

(*v*) *a hyperbola, if* $ab - h^2 < 0$, $\Delta \neq 0$,

*and a rectangular hyperbola if* in addition, $a + b = 0$.

## IV. SOLID GEOMETRY

**1.** (*i*) If $l, m, n$ be the *direction cosines* of a line then $l^2 + m^2 + n^2 = 1$.

If $a, b, c$ be the *direction ratios* of a line then $l = \frac{a}{\sqrt{\Sigma a^2}}$; $m = \frac{b}{\sqrt{\Sigma a^2}}$; $n = \frac{c}{\sqrt{\Sigma a^2}}$

(*ii*) If $\theta$ be the *angle between the lines having* d.c.'s $l, m, n$ and $l', m', n'$, then

$$\cos \theta = ll' + mm' + nn'$$

*Lines are perpendicular* if, $ll' + mm' + nn' = 0$

*Lines are parallel if* $l = l'$, $m = m'$, $n = n'$

*(iii) Projection of the line joining the points* $(x_1, y_1, z_1)$ *and* $(x_2, y_2, z_2)$ *on a line having d.c.'s* $l, m, n = l(x_1 - x_2) + m(y_1 - y_2) + n(z_1 - z_2)$.

**2. Plane**

*(i) Different forms of equation of a plane*

— *General form* : $ax + by + cz = d$

where $a, b, c$ are the d.r.s of a normal to the plane.

— *Normal form* : $lx + my + nz = p$

— *Intercept form* : $\frac{x}{a} + \frac{y}{b} + \frac{z}{c} = 1$.

— *Any plane passing through the point* $(x_1, y_1, z_1)$ is $a(x - x_1) + b(y - y_1) + c(z - z_1) = 0$

*(ii) Angle θ between the planes* $ax + by + cz = d$ and $a'x + b'y + c'z = d'$ is given by

$$\cos\theta = \frac{aa' + bb' + cc'}{\sqrt{(a^2 + b^2 + c^2)}\sqrt{(a'^2 + b'^2 + c'^2)}}$$

*Planes are perpendicular if* $aa' + bb' + cc' = 0$

*Planes are parallel if* $a/a' = b/b' = c/c'$

*(iii) Any plane parallel to the plane* $ax + by + cz = d$ is $ax + by + cz = k$.

**3. Straight line**

*(i) Equation of the line* through the point $(x_1, y_1, z_1)$ having d.r.s $a, b, c$ is

$$\frac{x - x_1}{a} = \frac{y - y_1}{b} = \frac{z - z_1}{c} \qquad \textit{(Symmetrical form)}$$

*(ii) Equation of the line* through the points $(x_1, y_1, z_1)$ and $(x_2, y_2, z_2)$ is

$$\frac{x - x_1}{x_2 - x_1} = \frac{y - y_1}{y_2 - y_1} = \frac{z - z_1}{z_2 - z_1} \qquad \textit{(Two points form)}$$

*(iii) Angle θ between the plane* $ax + by + cz = d$ *and the line*

$$\frac{x - x_1}{a'} = \frac{y - y_1}{b'} = \frac{z - z_1}{c'}$$

is

$$\sin\theta = \frac{aa' + bb' + cc'}{\sqrt{(a^2 + b^2 + c^2)}\sqrt{(a'^2 + b'^2 + c'^2)}}$$

*Line is parallel to the plane* if $aa' + bb' + cc' = 0$

*Line is perpendicular to the plane* if $a/a' = b/b' = c/c'$

*(iv) Coplanar lines*

Two lines $\frac{x - x_1}{l_1} = \frac{y - y_1}{m_1} = \frac{z - z_1}{n_1}$ and $\frac{x - x_2}{l_2} = \frac{y - y_2}{m_2} = \frac{z - z_2}{n_2}$

are coplanar if
$$\begin{vmatrix} x_2 - x_1 & y_2 - y_1 & z_2 - z_1 \\ l_1 & m_1 & n_1 \\ l_2 & m_2 & n_2 \end{vmatrix} = 0$$

and equation of the plane containing these lines is

$$\begin{vmatrix} x - x_1 & y - y_1 & z - z_1 \\ l_1 & m_1 & n_1 \\ l_2 & m_2 & n_2 \end{vmatrix} = 0$$

*(v) Shortest distance* between two skew lines

$$\frac{x - x_1}{l_1} = \frac{y - y_1}{m_1} = \frac{z - z_1}{n_1} \text{ and } \frac{x - x_2}{l_2} = \frac{y - y_2}{m_2} = \frac{z - z_2}{n_2}$$

is $\quad l(x_2 - x_1) + m(y_2 - y_1) + n(z_2 - z_1)$

where $l, m, n$ are given by $ll_1 + mm_1 + nn_1 = 0$ and $ll_2 + mm_2 + nn_2 = 0$

*Equation of the line of S.D.* is given by

$$\begin{vmatrix} x-x_1 & y-y_1 & z-z_1 \\ l_1 & m_1 & n_1 \\ l & m & n \end{vmatrix} = 0 \text{ and } \begin{vmatrix} x-x_2 & y-y_2 & z-z_2 \\ l_2 & m_2 & n_2 \\ l & m & n \end{vmatrix} = 0$$

**4. Sphere**

(*i*) *Equation of the sphere* having centre $(a, b, c)$ and radius $r$ is

$$(x - a)^2 + (y - b)^2 + (z - c)^2 = r^2$$

(*ii*) Equation $x^2 + y^2 + z^2 + 2ux + 2vy + 2wz + d = 0$ represents a sphere having *centre* $(-u, -v, -w)$ and *radius* $= \sqrt{(u^2 + v^2 + w^2 - d)}$

(*iii*) *Equation of the sphere* having the points $(x_1, y_1, z_1)$ and $(x_2, y_2, z_2)$ as the ends of a diameter is $(x - x_1)(x - x_2) + (y - y_1)(y - y_2) + (z - z_1)(z - z_2) = 0$

(*iv*) *Equation of a circle* (*i.e.*, section of a sphere $S = 0$ by the plane $U = 0$) is given by $S = 0$ and $U = 0$ taken together.

(*v*) *Equation of any sphere* through the circle of intersection of the sphere $S = 0$ and the plane $U = 0$ is $S + kU = 0$.

(*vi*) *Tangent plane* at any point $(x_1, y_1, z_1)$ of the sphere

$$x^2 + y^2 + z^2 + 2ux + 2vy + 2wz + d = 0 \text{ is}$$

$$xx_1 + yy_1 + zz_1 + u(x + x_1) + v(y + y_1) + w(z + z_1) + d = 0$$

(*vii*) Two spheres $x^2 + y^2 + z^2 + 2ux + 2vy + 2wz + d = 0$ and

$x^2 + y^2 + z^2 + 2u'x + 2v'y + 2w'z + d' = 0$ *cut orthogonally if* $2uu' + 2vv' + 2ww' = d + d'$.

**5. Cone**

(*i*) *Equation of a cone* with vertex at the origin is a homogeneous equation of the second degree in $x, y, z$.

(*ii*) *Enveloping cone* of the sphere $S = 0$ with vertex $(x_1, y_1, z_1)$ is $SS_1 = T^2$ where $S = x^2 + y^2 + z^2 - a^2$, $S_1 = x_1^2 + y_1^2 + z_1^2 - a^2$, $T = xx_1 + yy_1 + zz_1 - a^2$

**6. Quadric surfaces**

(*i*) *Ellipoid:* $\dfrac{x^2}{a^2} + \dfrac{y^2}{b^2} + \dfrac{z^2}{c^2} = 1$

(*ii*) *Hyperboloid* of one sheet: $\dfrac{x^2}{a^2} + \dfrac{y^2}{b^2} - \dfrac{z^2}{c^2} = 1$

*Hyperboloid* of two sheets: $\dfrac{x^2}{a^2} + \dfrac{y^2}{b^2} - \dfrac{z^2}{c^2} = -1$

(*iii*) *Cone:* $\dfrac{x^2}{a^2} + \dfrac{y^2}{b^2} - \dfrac{z^2}{c^2} = 0$

(*iv*) *Elliptic paraboloid:* $\dfrac{x^2}{a^2} + \dfrac{y^2}{b^2} = \dfrac{2z}{c}$

*Hyperbolic paraboloid:* $\dfrac{x^2}{a^2} - \dfrac{y^2}{b^2} = \dfrac{2z}{c}$

**7. Volumes and surface areas**

| *Solid* | *Volume* | *Curved surface area* | *Total surface area* |
|---|---|---|---|
| **Cube** (side $a$) | $a^3$ | $4a^2$ | $6a^2$ |
| **Cuboid** (length $l$, breadth $b$, height $h$) | $lbh$ | $2(l + b)h$ | $2(lb + bh + hl)$ |
| **Sphere** (radius $r$) | $\frac{4}{3}\pi r^3$ | — | $4\pi r^2$ |
| **Cylinder** (base radius $r$, height $h$) | $\pi r^2 h$ | $2\pi rh$ | $2\pi r(r + h)$ |
| **Cone** (base radius $r$, height $h$) | $\frac{1}{3}\pi r^2 h$ | $\pi rl$ | $\pi r(r + l)$ |

where slant height $l$ is given by $l = \sqrt{(r^2 + h^2)}$.

## V. TRIGONOMETRY

**1.**

| $\theta^\circ$ = | 0 | 30 | 45 | 60 | 90 | 180 | 270 | 360 |
|---|---|---|---|---|---|---|---|---|
| sin θ | 0 | 1/2 | $1/\sqrt{2}$ | $\sqrt{3}/2$ | 1 | 0 | –1 | 0 |
| cos θ | 1 | $\sqrt{3}/2$ | $1/\sqrt{2}$ | 1/2 | 0 | –1 | 0 | 1 |
| tan θ | 0 | $1/\sqrt{3}$ | 1 | $\sqrt{3}$ | ∞ | 0 | –∞ | 0 |

**2. Signs and variations of t-ratios**

| *Quadrant* | *sin* θ | *cos* θ | *tan* θ |
|---|---|---|---|
| I | + (0 to 1) | + (1 to 0) | + (0 to ∞) |
| II | + (1 to 0) | – (0 to – 1) | – (–∞ to 0) |
| III | – (0 to – 1) | – (–1 to 0) | + (0 to ∞) |
| IV | – (–1 to 0) | + (0 to 1) | – (–∞ to 0) |

**3.** *Any t-ratio of* $(n\,.\,90° \pm \theta) = \pm$ *same ratio of* θ, *when n is even.*
$= \pm$ *co-ratio of* θ, *when n is odd.*

*The sign + or – is to be decided from the quadrant in which* $n\,.\,90° \pm \theta$ *lies.*

*e.g.,* $\sin 570° = \sin(6 \times 90° + 30°) = -\sin 30° = -\frac{1}{2}$,

$\tan 315° = \tan(3 \times 90° + 45°) = -\cot 45° = -1$

**4.** $\sin(A \pm B) = \sin A \cos B \pm \cos A \sin B$

$\cos(A \pm B) = \cos A \cos B \mp \sin A \sin B$

$$\sin 2A = 2 \sin A \cos A = 2 \tan A/(1 + \tan^2 A)$$

$$\cos 2A = \cos^2 A - \sin^2 A = 1 - 2\sin^2 A = 2\cos^2 A - 1 = \frac{1-\tan^2 A}{1+\tan^2 A}$$

$$\tan(A \pm B) = \frac{\tan A \pm \tan B}{1 \mp \tan A \tan B};\ \tan 2A = \frac{2\tan A}{1-\tan^2 A}$$

**5.** $\sin A \cos B = \frac{1}{2}[\sin(A+B) + \sin(A-B)]$

$\cos A \sin B = \frac{1}{2}[\sin(A+B) - \sin(A-B)]$

$\cos A \cos B = \frac{1}{2}[\cos(A+B) + \cos(A-B)]$

$\sin A \sin B = \frac{1}{2}[\cos(A-B) - \cos(A+B)]$

**6.** $\sin C + \sin D = 2\sin\frac{C+D}{2}\cos\frac{C-D}{2}$

$\sin C - \sin D = 2\cos\frac{C+D}{2}\sin\frac{C-D}{2}$

$\cos C + \cos D = 2\cos\frac{C+D}{2}\cos\frac{C-D}{2}$

$\cos C - \cos D = -2\sin\frac{C+D}{2}\sin\frac{C-D}{2}$

**7.** $\sin 3A = 3\sin A - 4\sin^3 A$, $\cos 3A = 4\cos^3 A - 3\cos A$; $\tan 3A = \dfrac{3\tan A - \tan^3 A}{1 - 3\tan^2 A}$

**8.** $$a\sin x + b\cos x = r\sin(x+\theta)$$
$$a\cos x + b\sin x = r\cos(x-\theta)$$
where $a = r\cos\theta$, $b = r\sin\theta$ so that $r = \sqrt{(a^2+b^2)}$, $\theta = \tan^{-1}(b/a)$

**9.** *In any* $\Delta ABC$ :

(*i*) $a/\sin A = b/\sin B = c/\sin C$ (*Sine formula*)

(*ii*) $\cos A = \dfrac{b^2+c^2-a^2}{2bc}$ (*Cosine formula*)

(*iii*) $a = b\cos C + c\cos B$ (*Projection formula*)

(*iv*) *Area of* $\Delta ABC = \dfrac{1}{2}bc\sin A = \sqrt{s(s-a)(s-b)(s-c)}$ where $s = \dfrac{1}{2}(a+b+c)$.

**10. Series**

(*i*) *Exponential series* : $e^x = 1 + \dfrac{x}{1!} + \dfrac{x^2}{2!} + \dfrac{x^3}{3!} + \ldots\ldots \infty$

(*ii*) *Sin, cos, sinh, cosh series*

$$\sin x = x - \frac{x^3}{3!} + \frac{x^5}{5!} - \ldots\ldots \infty, \qquad \cos x = 1 - \frac{x^2}{2!} + \frac{x^4}{4!} - \ldots\ldots \infty$$

$$\sinh x = x + \frac{x^3}{3!} + \frac{x^5}{5!} + \ldots\ldots \infty, \qquad \cosh x = 1 + \frac{x^2}{2!} + \frac{x^4}{4!} + \ldots\ldots \infty$$

(*iii*) *Log series*

$$\log(1+x) = x - \frac{x^2}{2} + \frac{x^3}{3} - \ldots\ldots \infty, \qquad \log(1-x) = -\left(x + \frac{x^2}{2} + \frac{x^3}{3} + \ldots\ldots \infty\right)$$

(*iv*) *Gregory series*

$$\tan^{-1} x = x - \frac{x^3}{3} + \frac{x^5}{5} - \ldots\ldots \infty, \qquad \tanh^{-1} x = \frac{1}{2}\log\frac{1+x}{1-x} = x + \frac{x^3}{3} + \frac{x^5}{5} + \ldots\ldots \infty$$

**11.** (*i*) **Complex number :** $z = x + iy = r(\cos\theta + i\sin\theta) = re^{i\theta}$ [see Fig. AP-1.1]

(*ii*) **Euler's theorem :** $\cos\theta + i\sin\theta = e^{i\theta}$

(*iii*) **Demoivre's theorem :** $(\cos\theta + i\sin\theta)^n = \cos n\theta + i\sin n\theta$

**12.** (*i*) **Hyperbolic functions :** (*i*) $\sinh x = \dfrac{e^x - e^{-x}}{2}$; $\cosh x = \dfrac{e^x + e^{-x}}{2}$;

$$\tanh x = \frac{\sinh x}{\cosh x};\ \coth x = \frac{\cosh x}{\sinh x};\ \text{sech } x = \frac{1}{\cosh x};\ \text{cosech } x = \frac{1}{\sinh x}$$

(*ii*) *Relations between hyperbolic and circular functions :*

$$\sin ix = i\sinh x;\ \cos ix = \cosh x;\ \tan ix = i\tanh x.$$

(*iii*) *Inverse hyperbolic functions:*

$$\sinh^{-1} x = \log[x + \sqrt{x^2+1}];\ \cosh^{-1} x = \log[x + \sqrt{(x^2-1)}];\ \tanh^{-1} x = \frac{1}{2}\log\frac{1+x}{1-x}.$$

## VI. DIFFERENTIAL CALCULUS

**1. Standard limits :**

(*i*) $\underset{x\to a}{Lt}\ \dfrac{x^n - a^n}{x - a} = na^{n-1}$, $n$ any rational number

(*ii*) $\underset{x\to 0}{Lt}\ \dfrac{\sin x}{x} = 1$

(*iii*) $\underset{x\to 0}{Lt}\ (1+x)^{1/x} = e$

(*iv*) $\underset{x\to 0}{Lt}\ \dfrac{a^x - 1}{x} = \log_e a$

(*v*) $\underset{x\to\infty}{Lt}\ x^{1/x} = 1$

(*vi*) $\underset{n\to\infty}{Lt}\ (x)^{1/n} = 1$ ($x > 0$ and fixed)

(*vii*) $\underset{n\to\infty}{Lt}\left(\dfrac{x^n}{n!}\right) = 0$ ($x > 0$ and fixed)

(*viii*) $\underset{n\to\infty}{Lt}\left(\dfrac{\log n}{n}\right) = 0$

**2. Differentiation**

(i) $\frac{d}{dx}(uv) = u\frac{dv}{dx} + v\frac{du}{dx}$ $\qquad$ $\frac{d}{dx}\left(\frac{u}{v}\right) = \frac{v\,du/dx - u\,dv/dx}{v^2}$

$\frac{du}{dx} = \frac{du}{dy}\cdot\frac{dy}{dx}$ (*Chain Rule*) $\qquad$ $\frac{d}{dx}(ax+b)^n = n\,(ax+b)^{n-1}\,.\,a$

(ii) $\frac{d}{dx}(e^x) = e^x$ $\qquad$ $\frac{d}{dx}(a^x) = a^x \log_e a$

$\frac{d}{dx}(\log_e x) = 1/x$ $\qquad$ $\frac{d}{dx}(\log_a x) = \frac{1}{x\log a}$

(iii) $\frac{d}{dx}(\sin x) = \cos x$ $\qquad$ $\frac{d}{dx}(\cos x) = -\sin x$

$\frac{d}{dx}(\tan x) = \sec^2 x$ $\qquad$ $\frac{d}{dx}(\cot x) = -\text{cosec}^2 x$

$\frac{d}{dx}(\sec x) = \sec x\tan x$ $\qquad$ $\frac{d}{dx}(\text{cosec}\, x) = -\text{cosec}\, x\cot x$

(iv) $\frac{d}{dx}(\sin^{-1} x) = \frac{1}{\sqrt{(1-x^2)}}$ $\qquad$ $\frac{d}{dx}(\cos^{-1} x) = -\frac{1}{\sqrt{(1-x^2)}}$

$\frac{d}{dx}(\tan^{-1} x) = \frac{1}{1+x^2}$ $\qquad$ $\frac{d}{dx}(\cot^{-1} x) = \frac{-1}{1+x^2}$

$\frac{d}{dx}(\sec^{-1} x) = \frac{1}{x\sqrt{(x^2-1)}}$ $\qquad$ $\frac{d}{dx}(\text{cosec}^{-1}\, x) = \frac{-1}{x\sqrt{(x^2-1)}}$

(v) $\frac{d}{dx}(\sinh x) = \cosh x$ $\qquad$ $\frac{d}{dx}(\cosh x) = \sinh x$

$\frac{d}{dx}(\tanh x) = \text{sech}^2 x$ $\qquad$ $\frac{d}{dx}(\coth x) = -\text{cosech}^2 x$

(vi) $D^n(ax+b)^m = m(m-1)(m-2)\ldots\ldots(m-n+1)(ax+b)^{m-n}\,.\,a^n$

$D^n \log(ax+b) = (-1)^{n-1}(n-1)!\,a^n/(ax+b)^n$

$D^n(e^{mx}) = m^n e^{mx}$ $\qquad$ $D^n(a^{mx}) = m^n(\log a)^n\,.\,a^{mx}$

$$D^n\begin{bmatrix}\sin(ax+b)\\ \cos(ax+b)\end{bmatrix} = a^n\begin{bmatrix}\sin(ax+b+n\pi/2)\\ \cos(ax+b+n\pi/2)\end{bmatrix}$$

$$D^n e^{ax}\begin{bmatrix}\sin(bx+c)\\ \cos(bx+c)\end{bmatrix} = (a^2+b^2)^{n/2} e^{ax}\begin{bmatrix}\sin(bx+c+n\tan^{-1} b/a)\\ \cos(bx+c+n\tan^{-1} b/a)\end{bmatrix}$$

(vii) *Leibnitz theorem:* $(uv)_n = u_n + {}^nC_1 u_{n-1} v_1 + {}^nC_2 u_{n-2} v_2 + \ldots\ldots + {}^nC_r u_{n-r} v_r + \ldots\ldots + {}^nC_n v_n$

**3.** (i) *Maclaurin's series* : $f(x) = f(0) + \frac{x}{1!}f'(0) + \frac{x^2}{2!}f''(0) + \frac{x^3}{3!}f'''(0) + \ldots$

(ii) *Taylor's series* : $f(x+a) = f(a) + \frac{x}{1!}f'(a) + \frac{x^2}{2!}f''(a) + \frac{x^3}{3!}f'''(a) + \ldots$

**4. Curvature**

(i) *Radius of curvature* $\rho = \frac{(1+y_1^2)^{3/2}}{y_2}$, $\rho = \frac{(r^2+r_1^2)^{3/2}}{r^2+2r_1^2-rr_2}$; $\rho = r\frac{dr}{dp}$.

(ii) *Centre of curvature* : $\bar{x} = x - \frac{y_1(1+y_1^2)}{y_2}$, $\bar{y} = y + \frac{1}{y_2}(1+y_1^2)$.

(iii) **Evolute** *is the locus of the centre of curvature of a curve.* The curve is called the *involute* of the evolute.

(iv) **Envelope** *of a curve* $f(x, y, \alpha) = 0$ *is the 'α' eliminant from*

$$f(x, y, \alpha) = 0 \text{ and } \frac{\partial f}{\partial \alpha}(x, y, \alpha) = 0.$$

*The envelope of the normals to a curve is its* **evolute.**

**5. Asymptotes**

(*i*) *Asymptotes parallel to x-axis are obtained by equating to zero the coefficient of the highest power of x in the equation,* provided this is not merely a constant.

*Asymptotes parallel to y-axis are obtained by equating to zero the coefficient of highest power of y in the equation,* provided this is not merely a constant.

(*ii*) *Oblique asymptotes are obtained as follows:*

Put $x = 1, y = m$ in the highest degrees terms getting $\phi_n(m)$

Put $\phi_n(m) = 0$ and find the values of $m$.

Find $c$ from $c = -\phi_{n-1}(m)/\phi_n'(m)$

If two values of $m$ are equal, then find $c$ from

$$\frac{c^2}{2}\phi_n''(m) + c\phi'_{n-1}(m) + \phi_{n-2}(m) = 0$$

The asymptotes is $y = mx + c$.

(*iii*) *Asymptotes of polar curve* $1/r = f(\theta)$ is $r\sin(\theta - \alpha) = 1/f'(\alpha)$ where $\alpha$ is a root of $f(\theta) = 0$.

**6. Curve tracing**

(*i*) *A curve is symmetrical about x-axis,* if only even powers of $y$ occur in its equation.

(*ii*) *A curve is symmetrical about y-axis,* if only even powers of $x$ occur in its equation.

(*iii*) *A curve is symmetrical about the line* $y = x$, if on interchanging $x$ and $y$, its equation remains unchanged.

(*iv*) *A curve passes through the origin,* if there is no constant term in its equation.

(*v*) *Tangents to curve at the origin* are found by equating to zero the lowest degree terms.

**7. Partial Differentiation**

(*i*) *Euler's theorem.* If $u$ is a homogeneous function in $x$ and $y$ of degree $n$, then

$$x\frac{\partial u}{\partial x} + y\frac{\partial u}{\partial y} = nu\ ;\quad x^2\frac{\partial^2 u}{\partial x^2} + 2xy\frac{\partial^2 u}{\partial x\partial y} + y^2\frac{\partial^2 u}{\partial y^2} = n(n-1)\,u$$

(*ii*) *Chain rule* : $\dfrac{du}{dt} = \dfrac{\partial u}{\partial x}\dfrac{dx}{dt} + \dfrac{\partial u}{\partial y}\dfrac{dy}{dt}$, if $u = f(x, y), x = \phi(t), y = \psi(t)$.

(*iii*) $\dfrac{dy}{dx} = -\dfrac{\partial\phi}{\partial x}\Big/\dfrac{\partial\phi}{\partial y}$, if $\phi(x, y) = c$

(*iv*) *Jacobian* $J\left(\dfrac{u, v}{x, y}\right) = \dfrac{\partial(u, v)}{\partial(x, y)} = \begin{vmatrix} \partial u/\partial x & \partial u/\partial y \\ \partial v/\partial x & \partial v/\partial y \end{vmatrix}$

If $J = \partial(u, v)/\partial(x, y)$ and $J' = \partial(x, y)/\partial(u, v)$, then $JJ' = 1$

$$\frac{\partial(u, v)}{\partial(x, y)} = \frac{\partial(u, v)}{\partial(r, s)}\cdot\frac{\partial(r, s)}{\partial(x, y)}$$

(*v*) *Taylor's series* : $f(a + h, b + k) = f(a, b) + \left(h\dfrac{\partial}{\partial x} + k\dfrac{\partial}{\partial y}\right)f + \dfrac{1}{2!}\left(h\dfrac{\partial}{\partial x} + k\dfrac{\partial}{\partial y}\right)^2 f + \ldots\ldots$

(*vi*) *Maxima Minima* (*a*) $\dfrac{\partial f}{\partial x} = 0, \dfrac{\partial f}{\partial y} = 0$

(*b*) $\dfrac{\partial^2 f}{\partial x^2}\dfrac{\partial^2 f}{\partial y^2} > \dfrac{\partial^2 f}{\partial x\partial y}$ ; $\dfrac{\partial^2 f}{\partial x^2} < 0$ for maximum $\dfrac{\partial^2 f}{\partial x^2} > 0$ for minimum.

(*vii*) *Leibnitz's Rule* $\dfrac{d}{d\alpha}\left\{\int_a^b f(x, \alpha)\,d\alpha\right\} = \int_a^b \dfrac{\partial f(x, \alpha)}{\partial\alpha}\,d\alpha$

where $f(x, \alpha)$ and $\dfrac{\partial f(x, \alpha)}{\partial\alpha}$ are continuous functions of $x$ and $\alpha$ and $a, b$ are constants.

# VII. INTEGRAL CALCULUS

## 1. Integration

(i) $\int x^n \, dx = \frac{x^{n+1}}{n+1} \quad (n \neq -1)$ $\qquad \int \frac{1}{x} dx = \log_e x$

$\int e^x \, dx = e^x$ $\qquad \int a^x \, dx = a^x/\log_e a$

(ii) $\int \sin x \, dx = -\cos x$ $\qquad \int \cos x \, dx = \sin x$

$\int \tan x \, dx = -\log \cos x$ $\qquad \int \cot x \, dx = \log \sin x$

$\int \sec x \, dx = \log (\sec x + \tan x)$ $\qquad \int \text{cosec } x \, dx = \log (\text{cosec } x - \cot x)$

$\int \sec^2 x \, dx = \tan x$ $\qquad \int \text{cosec}^2 x \, dx = -\cot x$

(iii) $\int \frac{dx}{a^2 + x^2} = \frac{1}{a} \tan^{-1} \frac{x}{a}$ $\qquad \int \frac{dx}{\sqrt{(a^2 - x^2)}} = \sin^{-1} \frac{x}{a}$

$\int \frac{dx}{a^2 - x^2} = \frac{1}{2a} \log \frac{a+x}{a-x}$ $\qquad \int \frac{dx}{\sqrt{(a^2 + x^2)}} = \sinh^{-1} \frac{x}{a}$

$\int \frac{dx}{x^2 - a^2} = \frac{1}{2a} \log \frac{x-a}{x+a}$ $\qquad \int \frac{dx}{\sqrt{(x^2 - a^2)}} = \cosh^{-1} \frac{x}{a}$

(iv) $\int \sqrt{(a^2 - x^2)} \, dx = \frac{x\sqrt{(a^2 - x^2)}}{2} + \frac{a^2}{2} \sin^{-1} \frac{x}{a}$

$\int \sqrt{(a^2 + x^2)} \, dx = \frac{x\sqrt{(a^2 + x^2)}}{2} + \frac{a^2}{2} \sinh^{-1} \frac{x}{a}$

$\int \sqrt{(x^2 - a^2)} \, dx = \frac{x\sqrt{(x^2 - a^2)}}{2} - \frac{a^2}{2} \cosh^{-1} \frac{x}{a}$

(v) $\int e^{ax} \sin bx \, dx = \frac{e^{ax}}{a^2 + b^2} (a \sin bx - b \cos bx)$ $\qquad \int e^{ax} \cos bx \, dx = \frac{e^{ax}}{a^2 + b^2} (a \cos bx + b \sin bx)$

(vi) $\int_0^\infty e^{-ax} \sin bx \, dx = \frac{b}{a^2 + b^2}$ $\qquad \int_0^\infty e^{-ax} \cos bx \, dx = \frac{a}{a^2 + b^2}$

(vii) $\int \sinh x \, dx = \cosh x$ $\qquad \int \cosh x \, dx = \sinh x$

$\int \tanh x \, dx = \log \cosh x$ $\qquad \int \coth x \, dx = \log \sinh x$

$\int \text{sech}^2 x \, dx = \tanh x$ $\qquad \int \text{cosech}^2 x \, dx = -\coth x$

(viii) $\int_0^{\pi/2} \sin^n x \, dx = \int_0^{\pi/2} \cos^n x \, dx = \frac{(n-1)(n-3)(n-5) \ldots\ldots}{n(n-2)(n-4) \ldots\ldots} \times \left(\frac{\pi}{2}, \text{only if } n \text{ is even}\right)$

$\int_0^{\pi/2} \sin^m x \cos^n x \, dx = \frac{(m-1)(m-3) \ldots\ldots \times (n-1)(n-3) \ldots\ldots}{(m+n)(m+n-2)(m+n-4) \ldots\ldots} \times \left(\frac{\pi}{2}, \text{only if both } m \text{ and } n \text{ are even}\right)$

(ix) $\int_0^a f(x) \, dx = \int_0^a f(a-x) \, dx$

$\int_{-a}^a f(x) \, dx = 2\int_0^a f(x) \, dx,$ if $f(x)$ is an *even function*

$= 0,$ if $f(x)$ is an *odd function.*

$\int_0^{2a} f(x) \, dx = 2\int_0^a f(x) \, dx,$ if $f(2a - x) = f(x)$

$= 0,$ if $f(2a - x) = -f(x).$

(x) *Rule of Integration by parts :* $\int u(x) \, v(x) \, dx = uv_1 - \int u' v_1 \, dx$

(xi) *Leibnitz General Rule of Integration by parts :* $\int u(x) \, v(x) dx = uv_1 - u'v_2 + u''v_3 - u'''v_4 + \ldots\ldots$

where dashes denote differentiation of $u$ and number of subscripts denotes number of times $v$ is integrated.

## 2. Lengths of curves

(i) *Lenth of curve* $y = f(x)$ *between* $x = a, x = b$ *is* $\int_a^b \sqrt{\left\{1 + \left(\frac{dy}{dx}\right)^2\right\}}\, dx$

(ii) *Length of curve* $x = f(y)$ *between* $y = a, y = b$ *is* $\int_a^b \sqrt{\left\{1 + \left(\frac{dx}{dy}\right)^2\right\}}\, dy$

(iii) *Length of curve* $x = f(t), y = \phi(t)$ *between* $t = t_1, t = t_2$ *is* $\int_{t_1}^{t_2} \sqrt{\left\{\left(\frac{dx}{dt}\right)^2 + \left(\frac{dy}{dt}\right)^2\right\}}\, dt$

(iv) *Length of curve* $r = f(\theta)$ *between* $\theta = \alpha, \theta = \beta$ *is* $\int_\alpha^\beta \sqrt{\left\{r^2 + \left(\frac{dr}{d\theta}\right)^2\right\}}\, d\theta$

## 3. Areas of curves

(i) *Area bounded by* $y = f(x)$, *x-axis and* $x = a, x = b$ *is* $\int_a^b y\, dx$

(ii) *Area bounded by* $x = f(y)$, *y-axis* and $y = a, y = b$ *is* $\int_a^b x\, dy$

(iii) *Area bounded by* $r = f(\theta)$ and *lines* $\theta = \alpha, \theta = \beta$ *is* $\frac{1}{2}\int_\alpha^\beta r^2\, d\theta$

## 4. Volumes of revolution

(i) *Volume of revolution about x-axis of area bounded by* $y = f(x)$, *x-axis* and $x = a, x = b$ *is*

$$\int_a^b \pi y^2\, dx$$

(ii) *Volume of revolution about y-axis of area bounded by* $x = f(y)$, *y-axis and* $y = a, y = b$ *is*

$$\int_a^b \pi x^2\, dy$$

(iii) *Volume of revolution bounded by* $r = f(\theta)$ and $\theta = \alpha, \theta = \beta$

(a) *about OX* $= \int_\alpha^\beta \frac{2\pi}{3} r^3 \sin\theta\, d\theta$ (b) *about OY* $= \int_\alpha^\beta \frac{2\pi}{3} r^3 \cos\theta\, d\theta$

## 5. Surface areas of revolution

(i) *Surface area of revolution about x-axis of curve* $y = f(x)$ *from* $x = a$ *to* $x = b$ *is*

$$S = \int_{x=a}^{x=b} 2\pi y\, ds$$

*Cartesian form :* $S = \int_a^b 2\pi y \frac{ds}{dx}\, dx$ where $\frac{ds}{dx} = \sqrt{\left\{1 + \left(\frac{dy}{dx}\right)^2\right\}}$

*Parametric form :* $S = \int 2\pi y \frac{ds}{dt}\, dt$ where $\frac{ds}{dt} = \sqrt{\left\{\left(\frac{dx}{dt}\right)^2 + \left(\frac{dy}{dt}\right)^2\right\}}$

*Polar form :* $S = \int 2\pi y \frac{ds}{d\theta}\, d\theta$ where $\frac{ds}{d\theta} = \sqrt{\left\{r^2 + \left(\frac{dr}{d\theta}\right)^2\right\}}$

(ii) *Surface area of revolution about y-axis is* $\int 2\pi x\, ds$.

## 6. Multiple integrals

(i) *Area* $= \int_{x_1}^{x_2}\int_{y_1}^{y_2} dx\, dy$; *Volume* $= \int_{x_1}^{x_2}\int_{y_1}^{y_2} z\, dx\, dy$ *or* $\int_{x_1}^{x_2}\int_{y_1}^{y_2}\int_{z_1}^{z_2} dx\, dy\, dz$

(ii) *C.G. of a plane lamina:* $\bar{x} = \frac{\iint x\rho\, dx\, dy}{\iint \rho\, dx\, dy}$, $\bar{y} = \frac{\iint y\rho\, dx\, dy}{\iint \rho\, dx\, dy}$

*C.G. of a solid* $\bar{x} = \dfrac{\iiint x\rho\, dx\, dy\, dz}{\iiint \rho\, dx\, dy\, dz}$, $\bar{y} = \dfrac{\iiint y\rho\, dx\, dy\, dz}{\iiint \rho\, dx\, dy\, dz}$, $\bar{z} = \dfrac{\iiint z\rho\, dx\, dy\, dz}{\iiint \rho\, dx\, dy\, dz}$

(*iii*) *Centre of pressure* $\bar{x} = \dfrac{\iint_A px\, dx\, dy}{\iint_A p\, dx\, dy}$, $\bar{y} = \dfrac{\iint_A py\, dx\, dy}{\iint_A p\, dx\, dy}$

(*iv*) *M.I. about x-axis i.e.,* $I_x = \iiint_V \rho\,(y^2 + z^2)\, dx\, dy\, dz$

*M.I. about y-axis i.e.,* $I_y = \iiint_V \rho\,(z^2 + x^2)\, dx\, dy\, dz$

*M.I. about z-axis i.e.,* $I_z = \iiint_V \rho\,(x^2 + y^2)\, dx\, dy\, dz$

7. *Gamma function* $\Gamma(n) = \int_0^\infty e^{-x} x^{n-1}\, dx = (n-1)!,\ \Gamma(n+1) = n\,\Gamma(n) = n!,\ \Gamma(1/2) = \sqrt{\pi}$

*Beta function* $\beta(m, n) = \int_0^1 x^{m-1} (1-x)^{n-1}\, dx = \dfrac{\Gamma(m)\,\Gamma(n)}{\Gamma(m+n)}\ (m > 0, n > 0)$

## VIII. VECTORS

1. (*i*) If $\mathbf{R} = x\mathbf{I} + y\mathbf{J} + z\mathbf{K}$ then $r = |\mathbf{R}| = \sqrt{(x^2 + y^2 + z^2)}$

   (*ii*) $\overrightarrow{PQ}$ = Position vector of $Q$ – position vector of $P$.

2. If $\mathbf{A} = a_1\mathbf{I} + a_2\,\mathbf{J} + a_3\,\mathbf{K}$, $\mathbf{B} = b_1\,\mathbf{I} + b_2\,\mathbf{J} + b_3\,\mathbf{K}$, then

   (*i*) *Scalar product:* $\mathbf{A}\cdot\mathbf{B} = ab\cos\theta = a_1 b_1 + a_2 b_2 + a_3 b_3$

   (*ii*) *Vector product*: $\mathbf{A} \times \mathbf{B} = ab\sin\theta\,\hat{\mathbf{N}}$ = Area of the parallelogram having **A** and **B** as sides

$$= \begin{vmatrix} \mathbf{I} & \mathbf{J} & \mathbf{K} \\ a_1 & a_2 & a_3 \\ b_1 & b_2 & b_3 \end{vmatrix}$$

   (*iii*) $\mathbf{B} \perp \mathbf{A}$ if $\mathbf{A}\cdot\mathbf{B} = 0$ and **A** is parallel to **B** if $\mathbf{A} \times \mathbf{B} = \mathbf{0}$

3. (*i*) *Scalar triple product* $[\mathbf{A}\,\mathbf{B}\,\mathbf{C}] = \begin{vmatrix} a_1 & a_2 & a_3 \\ b_1 & b_2 & b_3 \\ c_1 & c_2 & c_3 \end{vmatrix}$ = Volume of parallelopiped

   (*ii*) If $[\mathbf{A}\,\mathbf{B}\,\mathbf{C}] = 0$, then **A, B, C** are coplanar.

   (*iii*) *Vector triple product* $\mathbf{A} \times (\mathbf{B} \times \mathbf{C}) = (\mathbf{A}\cdot\mathbf{C})\,\mathbf{B} - (\mathbf{A}\cdot\mathbf{B})\,\mathbf{C}$

   $(\mathbf{A} \times \mathbf{B}) \times \mathbf{C} = (\mathbf{C}\cdot\mathbf{A})\,\mathbf{B} - (\mathbf{C}\cdot\mathbf{B})\,\mathbf{A}$

4. (*i*) *grad* $f = \nabla f = \dfrac{\partial f}{\partial x}\mathbf{I} + \dfrac{\partial f}{\partial y}\mathbf{J} + \dfrac{\partial f}{\partial z}\mathbf{K}$

   *div* $\mathbf{F} = \nabla F = \dfrac{\partial f_1}{\partial x} + \dfrac{\partial f_2}{\partial y} + \dfrac{\partial f_3}{\partial z}$

   *curl* $\mathbf{F} = \nabla \times \mathbf{F} = \begin{vmatrix} \mathbf{I} & \mathbf{J} & \mathbf{K} \\ \dfrac{\partial}{\partial x} & \dfrac{\partial}{\partial y} & \dfrac{\partial}{\partial z} \\ f_1 & f_2 & f_3 \end{vmatrix}$ where $\mathbf{F} = f_1\,\mathbf{I} + f_2\,\mathbf{J} + f_3\,\mathbf{K}$

   (*ii*) If *div* $\mathbf{F} = 0$, then **F** is called a *solenoidal vector*.

   (*iii*) If *curl* $\mathbf{F} = \mathbf{0}$ then **F** is called an *irrotational vector*

5. *Velocity* $= d\mathbf{R}/dt$ ; *Acceleration* $= d^2\mathbf{R}/dt^2$ ; *Tangent vector* $= d\mathbf{R}/dt$ ; *Normal vector* $= \nabla\phi$

6. **Green's theorem:** $\int_C (\phi\, dx + \psi\, dy) = \iint_C \left(\dfrac{\partial\psi}{\partial x} - \dfrac{\partial\phi}{\partial y}\right) dxdy$

**Stoke's theorem:** $\int_C \mathbf{F} \cdot d\mathbf{R} = \int_S curl\ \mathbf{F} \cdot \mathbf{N}\ ds$

**Gauss divergence theorem:** $\int_S \mathbf{F} \cdot \mathbf{N}\ ds = \int_v div\ \mathbf{F}\ dv$

### 7. Coordinate systems

| | **Polar coordinates $(r, \theta)$** | **Cylindrical coordinates $(\rho, \phi, z)$** | **Spherical polar coordinates $(r, \theta, \phi)$** |
|---|---|---|---|
| *Coordinate transformations* | $X = r\cos\theta$<br>$y = r\sin\theta$ | $x = \rho\cos\phi$<br>$y = \rho\sin\phi$<br>$z = z$ | $x = r\sin\theta\cos\phi$<br>$y = r\sin\theta\sin\phi$<br>$z = r\cos\theta$ |
| *Jacobian* | $\frac{\partial(x, y)}{\partial(r, \theta)} = r$ | $\frac{\partial(x, y, z)}{\partial(\rho, \phi, z)} = \rho$ | $\frac{\partial(x, y, z)}{\partial(r, \theta, \phi)} = r^2 \sin\theta$ |
| *(Arc-length)*$^2$ | $(ds)^2 = (dr)^2 + r^2(d\theta)^2$<br>$dxdy = r\,d\theta\,dr$ | $(ds)^2 = (d\rho)^2 + \rho^2(d\phi)^2 + (dz)^2$ | $(ds)^2 = (dr)^2 + r^2(d\theta)^2 + (r\sin\theta)^2(d\phi)^2$ |
| *Volume-element* | | $dV = \rho\,d\rho\,d\phi\,dz$ | $dV = r^2 \sin\theta\,dr\,d\theta\,d\phi$ |

## IX. DIFFERENTIAL EQUATIONS

### 1. Equations of first order

(*i*) *Variables separable* : $f(y)\,dy/dx = \phi(x)$, $\int f(y)\,dy = \int \phi(x)\,dx + c$.

(*ii*) *Homogeneous equation* $dy/dx = f(x, y)/\phi(x, y)$ where $f(x, y)$ and $\phi(x, y)$ are of the same degree. Put $y = vx$ so that $dy/dx = v + x\,dv/dx$.

(*iii*) *Equations reducible to homogenous form* : $\frac{dy}{dx} = \frac{ax + by + c}{a'x + b'y + c'}$

When $a/a' \neq b/b'$, put $x = X + h$, $y = Y + k$

When $a/a' = b/b'$, put $ax + by = t$.

(*iv*) *Leibnitz's linear equation* : $\frac{dy}{dx} + Py = Q$ where $P$, $Q$ are functions of $x$.

$I.F. = e^{\int P\,dx}$, then *solution* is $y\,(I.F.) = \int Q\,(I.F.)\,dx + c$.

(*v*) *Bernoulli's equation* : $dy/dx + Py = Qy^n$, reducible to Leibnitz's equation by writing it as

$$y^{-n}\frac{dy}{dx} + Py^{1-n} = Q \text{ and putting } y^{1-n} = z.$$

(*vi*) *Exact equation* : $M(x, y)\,dx + N(x, y)\,dy = 0$

*Solution* is $\int_{(y\ cons.)} M\,dx + \int$ (terms of $N$ not containing $x$) $dy = c$, provided $\partial M/\partial y = \partial N/\partial x$.

(*vii*) *Clairaut's equation* : $y = px + f(p)$ where $p = dy/dx$.

Solution is obtained on replacing $p$ by $c$.

### 2. Linear equations with constant coefficients

$$\frac{d^n y}{dx^n} + k_1 \frac{d^{n-1}y}{dx^{n-1}} + \ldots\ldots + k_{n-1}\frac{dy}{dx} + k_n y = X$$

*Symbolic form* : $(D^n + k_1 D^{n-1} + \ldots. + k_{n-1} D + k_n)\,y = X$.

***I. To find C.F.***

| Roots of A.E. | C.F. |
|---|---|
| (i) $m_1, m_2, m_3, \ldots\ldots$ | $c_1e^{m_1x} + c_2e^{m_2x} + c_3e^{m_3x} + \ldots.$ |
| (ii) $m_1, m_1, m_3, \ldots\ldots$ | $(c_1 + c_2x)\, e^{m_1x} + c_3e^{m_3x} + \ldots.$ |
| (iii) $\alpha + i\beta, \alpha - i\beta, m_3, \ldots\ldots$ | $e^{\alpha x}(c_1 \cos \beta x + c_2 \sin \beta x) + c_3e^{m_3x} + \ldots.$ |
| (iv) $\alpha \pm i\beta, \alpha \pm i\beta, m_5, \ldots\ldots$ | $e^{\alpha x}\,[(c_1 + c_2x) \cos \beta x + (c_3 + c_4 x) \sin \beta x] + c_5\, e^{m_5x} + \ldots.$ |

***II. To find P.I.***

(i) $X = e^{ax}$, P.I. $= \dfrac{1}{f(D)} e^{ax}$, put $D = a$, [$f(a) \neq 0$

$= x \dfrac{1}{f'(D)} e^{ax}$, put $D = a$, [$f(a) = 0, f'(a) \neq 0$

$= x^2 \dfrac{1}{f''(D)} e^{ax}$, put $D = a$, [$f'(a) = 0, f''(a) \neq 0$

(ii) $X = \sin(ax + b)$ *or* $\cos(ax + b)$

P.I. $= \dfrac{1}{\phi(D^2)} \sin(ax + b)$ [or $\cos(ax + b)$], put $D^2 = -a^2$, [$\phi(-a)^2 \neq 0$]

$= x \dfrac{1}{\phi'(D^2)} \sin(ax + b)$ [or $(\cos ax + b)$], put $D^2 = -a^2$, [$\phi(-a^2) = 0, \phi'(-a^2) \neq 0$]

$= x^2 \dfrac{1}{\phi''(D^2)} \sin(ax + b)$ [or $\cos(ax + b)$], put $D^2 = -a^2$, [$\phi'(-a^2) = 0, \phi''(-a^2) \neq 0$

(iii) $X = x^m$, P.I. $= \dfrac{1}{f(D)} x^m = [f(D)]^{-1} x^m$. Expand $[f(D)]^{-1}$ in ascending powers of $D$ as for as $D^m$ and operate on $x^m$ term by term.

(iv) $X = e^{ax} V$, P.I. $= \dfrac{1}{f(D)} e^{ax} V = e^{ax} \dfrac{1}{f(D + a)} V.$

***III. Complete Solution :*** C.S. is $y$ = C.F. + P.I.

**3. Homogeneous linear equation :** $x^3 \dfrac{d^3y}{dx^3} + k_1x^2 \dfrac{d^2y}{dx^2} + k_2x \dfrac{dy}{dx} + k_3y = X$

reduces to linear equation with constant coefficients by putting

$$x = e^t,\ x\frac{dy}{dx} = Dy,\ x^2\frac{d^2y}{dx^2} = D(D-1)\,y,\ x^3\frac{d^3y}{dx^3} = D(D-1)(D-2)\,y$$

**4. Lagrange's linear partial differential equation**

$Pp + Qq = R$, $P, Q, R$ being functions of $x, y, z$.

*To solve it* (i) form the subsidiary equations $\dfrac{dx}{P} = \dfrac{dy}{Q} = \dfrac{dz}{R}$

(ii) solve these equations giving $u = a$, $v = b$.

(iii) Complete solution is $\phi(u, v) = 0$ or $u = f(v)$.

**5. Homogeneous linear partial differential equations with constant coefficients**

$$\frac{\partial^n z}{\partial x^n} + k_1 \frac{\partial^n z}{\partial x^{n-1} \partial y} + \ldots + k_n \frac{\partial^n z}{\partial y^n} = F(x, y)$$

*Symbolic form :* $(D^n + k_1D^{n-1} D' + \ldots + k_nD'^n)z = F(x, y)$

***To find C.F.***

| *Roots of A.E.* | C.F. |
|---|---|
| (i) $m_1, m_2, m_3, ......$ | $f_1(y + m_1x) + f_2(y + m_2x) + f_3(y + m_3x) + ......$ |
| (ii) $m_1, m_1, m_3, ......$ | $f_1(y + m_1x) + xf_2(y + m_1x) + f_3(y + m_3x) + ......$ |
| (iii) $m_1, m_1, m_1, ......$ | $f_1(y + m_1x) + xf_2(y + m_1x) + x^2f_3(y + m_3x) + ......$ |

***To find P.I.***

(i) $F(x, y) = e^{ax + by}$, $P.I. = \dfrac{1}{f(D, D')} e^{ax + by}$, put $D = a$, $D' = b$.

(ii) $F(x, y) = \sin(mx + ny)$ or $\cos(mx + ny)$

$$P.I. = \frac{1}{f(D^2, DD', D'^2)} \sin \text{ or } \cos(mx + ny), \text{ put } D^2 = -m^2, DD' = -mn, D^2 = -n^2$$

(iii) $F(x, y) = x^m y^n$, $P.I. = [f(D, D')]^{-1} x^m y^n$. Expand $[f(D, D')]^{-1}$ and operate on $x^m y^n$.]

(iv) $F(x, y)$ is any function of $x$ and $y$, $P.I. = \dfrac{1}{f(D, D')}, F(x, y)$.

Resolve $1/f(D, D')$ into partial fractions considering $f(D, D')$ as a function of $D$ alone and operate each partial fraction on $F(x, y)$ remembering that $\dfrac{1}{D - mD'} F(x, y) = \int F(x, c - mx)\, dx$.

***Complete solution:*** C.S. is $y = C.F. + P.I.$

## X. INFINITE SERIES

1. *Basic test:* If $\underset{n \to \infty}{Lt}\ u_n \neq 0$ then the series $\Sigma u_n$ diverges.
2. *G.P. Series:* $1 + r + r^2 + r^3 + ...... \infty$ converge if $|r| < 1$; diverges if $r \geq 1$ and oscillates if $r \leq -1$.
3. *p-series:* $\dfrac{1}{1^p} + \dfrac{1}{2^p} + \dfrac{1}{3^p} + ...... \infty$ converge for $p > 1$; diverges for $p \leq 1$.
4. *Comparison test:* If two positive term series $\Sigma u_n$ and $\Sigma v_n$ be such that $\underset{n \to \infty}{Lt} \dfrac{u_n}{v_n}$ = *finite quantity* ($\neq 0$), then $\Sigma u_n$ and $\Sigma v_n$ converge or diverge together.
5. *Ratio test:* In the positive term series $\Sigma u_n$, if $\underset{n \to \infty}{Lt} \dfrac{u_n}{u_{n+1}} = k$, then the series converges for $k > 1$, diverges for $k < 1$ and fails for $k = 1$.
6. *Raabes' test:* In the positive term series $\Sigma u_n$, if $\underset{n \to \infty}{Lt}\ n\left(\dfrac{u_n}{u_{n+1}} - 1\right) = k$, then the series converges for $k > 1$, diverges for $k < 1$ and fails for $k = 1$.
7. *Logarithmic test:* In the positive term series $\Sigma u_n$, if $\underset{n \to \infty}{Lt}\left(n \log \dfrac{u_n}{u_{n+1}}\right) = k$, then the series converges for $k > 1$, diverges for $k < 1$ and fails for $k = 1$.
8. If $u_n/u_{n+1}$ does not involve $n$ as an exponent or a logarithm, then the series $\Sigma u_n$ diverges.
9. *Cauchy's root test:* In a positive term series $\Sigma u_n$ if $\underset{n \to \infty}{Lt} (u_n)^{1/n} = \lambda$, then the series converges for $\lambda < 1$, diverges for $\lambda > 1$ and fails for $\lambda = 1$.
10. *Integral test:* A positive term series $\Sigma f(n)$ converges or diverges according as $\int_1^\infty f(x)\, dx$ is finite or infinite where $f(n)$ is continuous in $1 < x < \infty$ and decreases as $n$ increases.
11. *Leibnitz's test for alternating series:* An alternating series $u_1 - u_2 + u_3 - u_4 + ...... \infty$ converges if each term is numerically less than the previous term and $\underset{n \to \infty}{Lt}\ u_n = 0$.

    if $\underset{n \to \infty}{Lt} \neq 0$, then the given series is oscillatory.

**12.** *General Ratio test:* In an arbitrary term series $\Sigma u_n$ if $\underset{n\to\infty}{Lt} \frac{|u_{n+1}|}{|u_n|} = |k|$, then $\Sigma u_n$ is absolutely convergent if $|k| < 1$ and divergent if $|k| > 1$ and the test fails if $|k| = 1$.

## XI. FOURIER SERIES

**1.** $f(x) = \frac{1}{2} a_0 + a_1 \cos x + a_2 \cos 2x + ... + b_1 \sin x + b_2 \sin 2x + ......$ in $(0\ 2\pi)$.

where, $\alpha_0 = \frac{1}{\pi}\int_0^{2\pi} f(x)\,dx,\ a_n = \frac{1}{\pi}\int_0^{2\pi} f(x)\cos nx\,dx,\ b_n = \frac{1}{\pi}\int_0^{2\pi} f(x)\sin nx\,dx$

**2.** $f(x) = \frac{1}{2} a_0 + \sum_{n=1}^{\infty} a_n \cos nx + \sum_{n=1}^{\infty} b_n \cos nx$ in any interval $(0, 2c)$,

where, $a_0 = \frac{1}{c}\int_0^{2c} f(x)\,dx,\ a_n = \frac{1}{c}\int_0^{2c} f(x)\cos\frac{n\pi x}{c}\,dx,\ b_n = \frac{1}{c}\int_0^{c} f(x)\sin\frac{n\pi x}{c}\,dx$

**3.** *For even function f(x), Fourier expansion contains only cosine terms.*

*i.e.,* $a_0 = \frac{2}{c}\int_0^{c} f(x)\,dx,\ a_n = \frac{2}{c}\int_0^{c} f(x)\cos\frac{n\pi x}{c}\,dx,\ b_n = 0.$

*For odd function f (x), Fourier expansion contains only sine terms.*

*i.e.,* $a_0 = 0,\ a_n = 0,\ b_n = \frac{2}{c}\int_0^{c} f(x)\sin\frac{n\pi x}{c}\,dx.$

## XII. TRANSFORMS

**1. Laplace Transforms.** $L(f(t)) = \int_0^{\infty} e^{-st} f(t)\,dt$

(i) $L(1) = \frac{1}{s}$

(ii) $L(t^n) = \frac{n!}{s^{n+1}}$

(iii) $L(e^{at}) = \frac{1}{s-a}$

(iv) $L(\sin at) = \frac{a}{s^2+a^2}$

(v) $L(\cos at) = \frac{s}{s^2+a^2}$

(vi) $L(\sinh at) = \frac{a}{s^2-a^2}$

(vii) $L(\cosh at) = \frac{s}{s^2-a^2}$

(viii) $L\, e^{at} f(t) = F(s-a)$

(ix) $L f'(t) = sL f(t) - f(0)$

(x) $L[t^n f(t)] = (-1)^n \frac{d^n}{ds^n}[F(s)]$

(xi) $L\left[\frac{1}{t} f(t)\right] = \int_s^{\infty} F(s)\,ds$

(xii) $u(t-a) = \begin{cases} 0 \text{ when } t < a \\ 1 \text{ when } t > a \end{cases}$

(xiii) $L[u(t-a)] = \frac{e^{-as}}{s}$

(xiv) $L\delta(t-a) = e^{-as}$

(xv) $L f(t) = \frac{\int_0^T e^{-st} f(t)\,dt}{1-e^{-sT}}$ where $f(t)$ is a periodic function of period $T$.

**2. Inverse Laplace Transforms**

(i) $L^{-1}\left(\frac{1}{s}\right) = 1$

(ii) $L^{-1}\left(\frac{1}{s^n}\right) = \frac{t^{n-1}}{(n-1)!}$

(iii) $L^{-1}\left(\frac{1}{s-a}\right) = e^{at}$

(iv) $L^{-1}\left(\frac{1}{s^2+a^2}\right) = \frac{1}{a}\sin at$

(v) $L^{-1}\left(\frac{s}{s^2+a^2}\right) = \cos at$

(vi) $L^{-1}\left(\frac{s}{s^2-a^2}\right) = \frac{1}{a}\sinh at$

(vii) $L^{-1}\left(\frac{s}{s^2-a^2}\right) = \cosh at.$

(viii) $L^{-1}\left(\frac{s}{(s^2+a^2)^2}\right) = \frac{1}{2a}\, t \sin at$

(ix) $L^{-1}\left[\frac{s^2-a^2}{(s^2+a^2)^2}\right] = t \cos at.$

**3. Fourier Transforms :** $F(s) = \int_{-\infty}^{\infty} f(t)\, e^{ist}\, .dt$

*Fourier sine transform :* $F_s(s) = \int_0^{\infty} f(t) \sin st\, dt$

*Fourier cosine transform :* $F_c(s) = \int_0^{\infty} f(t) \cos st\, dt$

$$F\left(\frac{\partial^2 u}{\partial x^2}\right) = -s^2 F(u).$$

**4. Z-Transforms :** $Z(u_n) = \sum_{n=0}^{\infty} u_n z^{-n}$

(i) $Z(1) = \frac{z}{z-1}$

(ii) $Z(n) = \frac{z}{(z-1)^2}$

(iii) $Z(n^2) = \frac{z^2+z}{(z-1)^3}$

(iv) $Z(a^n) = \frac{z}{z-a}$

(v) $Z(na^n) = \frac{az}{(z-a)^2}$

(vi) $Z(\sin n\theta) = \frac{z \sin \theta}{z^2 - 2z\cos\theta + 1}$

(vii) $Z(\cos n\theta) = \frac{z(z-\cos\theta)}{z^2 - 2z\cos\theta + 1}$

(viii) $Z(\sinh n\theta) = \frac{z \sinh\theta}{z^2 - 2z\cosh\theta + 1}$

(ix) $Z(\cosh n\theta) = \frac{z(z-\cosh\theta)}{z^2 - 2z\cosh\theta + 1}$

## XIII. STATISTICS AND PROBABILITY

1. A.M. $\bar{x} = \frac{\Sigma f_i x_i}{\Sigma f_i}$

2. S.D. $\sigma = \sqrt{\frac{\Sigma f_i (x_i - \bar{x})^2}{\Sigma f_i}}$

3. *Moments about the mean :* $\mu_0 = 1$, $\mu_1 = 0$, $\mu_2 = \sigma^2$, $\mu_r = \frac{\Sigma f_i (x_i - \bar{x})^r}{\Sigma f_i}$

4. *Coeff. of skewness* = (mean – mode)/$\sigma$ which lies between – 1 and 1.

5. Kurtosis : $\beta_2 = \mu_4/\mu_2^2$.

6. *Coeff. of correlation* $r = \frac{n\,\Sigma d_x d_y - \Sigma d_x \Sigma d_y}{\sqrt{[\{n\,\Sigma d_x^2 - (\Sigma d_x)^2\}\{n\Sigma d_y^2 - (\Sigma d_y)^2\}]}}$; $-1 < r < 1$

7. *Line of regression of y on x :* $y - \bar{y} = r\,\frac{\sigma_y}{\sigma_x}(x - \bar{x})$

*Line of regression of x on y :* $x - \bar{x} = \frac{r\,\sigma_x}{\sigma_y}(y - \bar{y})$

8. **Probability** $p(A) = \frac{\text{No. of ways favourable to } A}{\text{Total no. of equally likely ways}}$, $p + q = 1$.

(i) $p(A \text{ or } B) = p(A) + p(B)$,

(ii) $p(A \text{ and } B) = p(A)\,.\,p(B)$

9. **Binomial distribution :** $p(r) = {}^nC_r\, p^r q^{n-r}$

*Mean* = $np$, *Variance* $(\sigma^2) = npq$

**10. Poisson distribution :** $p(r) = \frac{m^r}{r!} e^{-m}$

*Mean* $= m$, *Variance* $(\sigma^2) = m$.

**11. Normal distribution :** $f(x) = \frac{1}{\sigma\sqrt{2\pi}} e^{-\frac{(x-\mu)^2}{2\sigma^2}}$, *Standard variate* $= \frac{x-\mu}{\sigma}$

(*i*) *Probable error* $\lambda = 0.6745\ \sigma$.

(*ii*) 68% of values lie between $x = \mu - \sigma$ and $x = \mu + \sigma$.

95% of values lie between $x = \mu - 1.96\ \sigma$ and $x = \mu + 1.96\ \sigma$

99% of values lie between $x = \mu - 2.58\ \sigma$ and $x = \mu + 2.58\ \sigma$

## XIV. NUMERICAL TECHNIQUES

### 1. Solution of equations

(*i*) *Bisection method :* $x_n = \frac{1}{2}(x_{n-1} + x_{n-2})$

(*ii*) *Method of False position :* $x_2 = x_0 - \frac{x_1 - x_0}{f(x_1) - f(x_0)} f(x_0)$.

(*iii*) *Newton-Raphson method :* $x_{n+1} = x_n - \frac{f(x_n)}{f'(x_n)}$

(*iv*) *Iterative formula to find* $1/N$ is $x_{n+1} = x_n (2 - Nx_n)$

(*v*) *Iterative formula to find* $\sqrt{N}$ is $x_{n+1} = \frac{1}{2}(x_n + N/x_n)$

### 2. Solution of Linear Simultaneous equations

(*i*) *Matrix inversion method*. For the equations

$$a_1x + b_1y + c_1 z = d_1,\ a_2x + b_2y + c_2 z = d_2,\ a_3x + b_3y + c_3z = d_3$$

if $$A = \begin{vmatrix} a_1 & b_1 & c_1 \\ a_2 & b_2 & c_2 \\ a_3 & b_3 & c_3 \end{vmatrix}, X = \begin{bmatrix} x \\ y \\ z \end{bmatrix} \text{ and } D = \begin{bmatrix} d_1 \\ d_2 \\ d_3 \end{bmatrix}$$

then $$\begin{bmatrix} x \\ y \\ z \end{bmatrix} = \frac{1}{|A|} \begin{vmatrix} A_1 & A_2 & A_3 \\ B_1 & B_2 & B_3 \\ C_1 & C_2 & C_3 \end{vmatrix} \times \begin{bmatrix} d_1 \\ d_2 \\ d_3 \end{bmatrix}$$

where $A_1$, $B_1$, etc., are the co-factors of $a_1$, $b_1$, etc., in the determinant $|A|$.

(*ii*) *Gauss-elimination method*. In this method the coefficient matrix is transformed to *upper triangular matrix*.

(*iii*) *Gauss-Jordan method*. In this method the coefficient matrix is transformed to *diagonal matrix.*

(*iv*) *Gauss-Jordan method of finding the inverse of a matrix A*. The matrices $A$ and $I$ are written side by side and the same row transformations are performed on both till $A$ is reduced to $I$. Then the other matrix represents $A^{-1}$.

### 3. Finite differences and Interpolation

(*i*) *Forward differences:* $\Delta y_r = y_{r+1} - y_r$

*Backward differences:* $\nabla y_r = y_r - y_{r-1}$

*Central differences:* $\delta y_{n-1/2} = y_n - y_{n-1}$

(*ii*) *Relations between operations :*

$$\Delta = E - 1;\ \nabla = 1 - E^{-1};\ \delta = E^{1/2} - E^{-1/2}$$

$$\mu = \frac{1}{2}(E^{1/2} + E^{-1/2});\ \Delta = E\nabla = \nabla E = \delta E^{1/2};\ E = e^{hD}$$

(*iii*) *Factorial notation* $|\,x\,|^r = x(x-1)(x-2) \ldots (x-r+1)$.

*Factorial polynomial* $[x]^n = x(x-h)(x-2h) \ldots (x-h-1h)$

(iv) *Newton's forward interpolation formula*

$$y_p = y_0 + p\,\Delta y_0 + \frac{p(p-1)}{2!}\,\Delta^2 y_0 + \frac{p(p-1)(p-2)}{3!}\,\Delta^3 y_0 + \ldots \text{ where } p = (x - x_0)/h.$$

(v) *Newton's backward interpolation formula:*

$$y_p = y_n + p\,\nabla y_n + \frac{p(p+1)}{2!}\,\nabla^2 y_n + \frac{p(p+1)(p+2)}{3!}\,\nabla^3 y_n + \ldots\ldots \text{ where } p = (x - x_n)/h.$$

(vi) *Stirling's formula:*

$$y_p = y_0 + p\left(\frac{\Delta y_0 + \Delta y_{-1}}{2}\right) + \frac{p^2}{2!}\,\Delta^2 y_{-1} + \frac{p(p^2-1)}{3!}\left(\frac{\Delta^3 y_{-1} + \Delta^3 y_{-2}}{2}\right) + \frac{p^2(p^2-1)}{4!}\,\Delta^4 y_{-2} + \ldots\ldots$$

(vii) *Bessel's formula:*

$$y_p = y_0 + p\Delta y_0 + \frac{p(p-1)}{2!}\,\frac{\Delta^2 y_{-1} + \Delta^2 y_0}{2} + \frac{\left(p - \frac{1}{2}\right)p\,(p-1)}{3!}\,\Delta^3 y_{-1} + \frac{(p+1)\,p(p-1)(p-2)}{4!}\,\frac{\Delta^4 y_{-2} + \Delta^4 y_{-1}}{2} + \ldots$$

(viii) *Lagrange's interpolation formula:*

$$y = \frac{(x-x_1)(x-x_2)\ldots\ldots(x-x_n)}{(x_0-x_1)(x_0-x_2)\ldots\ldots(x_0-x_n)}\,y_0 + \frac{(x-x_0)(x-x_2)\ldots\ldots(x-x_n)}{(x_1-x_0)(x_1-x_2)\ldots\ldots(x_1-x_n)}\,y_1 + \ldots\ldots + \frac{(x-x_0)(x-x_1)\ldots\ldots(x-x_{n-1})}{(x_n-x_0)(x_n-x_1)\ldots\ldots(x_n-x_{n-1})}\,y_n$$

(ix) *Newton's divided difference formul*

$$y = f(x) = y_0 + (x - x_0)\,[x_0, x_1] + (x - x_0)(x - x_1)\,[x_0, x_1, x_2] + (x - x_0)(x - x_1)(x - x_2)\,[x_0, x_1, x_2, x_3] + \ldots$$

where $[x_0, x_1]\ \dfrac{y_1 - y_0}{x_1 - x_0}$, $[x_0, x_1, x_2] = \dfrac{[x_1, x_2] - [x_0, x_1]}{[x_2 - x_0]}$ and so on.

**4. Numerical differentiation**

(i) *Forward difference formulae:*

$$\left(\frac{dy}{dx}\right)_{x_o} = \frac{1}{h}\left[\Delta y_0 - \frac{1}{2}\Delta^2 y_0 + \frac{1}{3}\Delta^3 y_0 - \frac{1}{4}\Delta^4 y_0 + \ldots\ldots\right]$$

$$\left(\frac{d^2y}{dx^2}\right)_{x_o} = \frac{1}{h^2}\left[\Delta^2 y_0 - \Delta^3 y_0 + \frac{11}{12}\Delta^4 y_0 + \ldots\ldots\right] \text{ and so on.}$$

(ii) *Backward difference formulae:*

$$\left(\frac{dy}{dx}\right)_{x_o} = \frac{1}{h}\left[\nabla y_0 + \frac{1}{2}\nabla^2 y_0 + \frac{1}{3}\nabla^3 y_0 + \frac{1}{4}\nabla^4 y_0 + \ldots\ldots\right]$$

$$\left(\frac{d^2y}{dx^2}\right)_{x_o} = \frac{1}{h^2}\left[\nabla^2 y_0 + \nabla^3 y_0 + \frac{11}{12}\nabla^4 y_0 + \ldots\ldots\right] \text{ and so on.}$$

(iii) *Central difference formulae:*

$$\left(\frac{dy}{dx}\right)_{x_o} = \frac{1}{h}\left[\frac{\Delta y_0 + \Delta y_{-1}}{2} - \frac{1}{6}\,\frac{\Delta^3 y_{-1} + \Delta^3 y_{-2}}{2} + \frac{1}{30}\,\frac{\Delta^5 y_{-2} + \Delta^5 y_{-3}}{2} + \ldots\ldots\right]$$

$$\left(\frac{d^2y}{dx^2}\right)_{x_o} = \frac{1}{h^2}\left[\Delta^2 y_{-1} - \frac{1}{12}\Delta^4 y_{-2} + \frac{1}{90}\Delta^6 y_{-3} + \ldots\ldots\right]$$

**5. Numerical integration**

(i) *Trapezoidal rule:*

$$\int_{x_0}^{x_0+nh} f(x)\,dx = \frac{h}{2}\,[(y_0 + y_n) + 2\,(y_1 + y_2 + \ldots\ldots + y_{n-1})]$$

(*ii*) *Simpson's 1/3rd rule:*

$$\int_{x_0}^{x_0+nh} f(x)\,dx = \frac{h}{3}[(y_0 + y_n) + 4(y_1 + y_3 + \ldots\ldots + y_{n-1}) + 2(y_2 + y_4 + \ldots\ldots + y_{n-2})]$$

(Number of sub-intervals should be taken as even)

(*iii*) *Simpson's 3/8 th rule:*

$$\int_{x_0}^{x_0+nh} f(x)\,dx = \frac{3h}{8}[(y_0 + y_n) + 3(y_1 + y_2 + y_4 + y_5 + \ldots\ldots + y_{n-1}) + 2(y_3 + y_5 + \ldots\ldots + y_{n-2})]$$

(Number of sub-intervals should be taken as a multiple of 3)

(*iv*) *Weddle's rule:*

$$\int_{x_0}^{x_0+nh} f(x)\,dx = \frac{3h}{10}[y_0 + 5y_1 + y_2 + 6y_3 + y_4 + 5y_5 + 2y_6 + 5y_7 + \ldots\ldots]$$

(Number of sub-intervals should be taken as multiple of 6)

## 6. Numerical solution of ordinary differential equations

(*i*) *Picard's method:* $y_1 = y_0 + \int_{x_0}^{x} (x, y_0)\,dx$

$$y_2 = y_0 + \int_{x_0}^{x} f(x, y_1)\,dx \text{ etc.}$$

(*ii*) *Taylor's method:*

$$y = y_0 + (x - x_0)(y)_0 + \frac{(x - x_0)^2}{2!}(y'')_0 + \frac{(x - x_0)^3}{3!}(y''')_0 + \ldots$$

(*iii*) *Euler's method:* $y_2 = y_1 + h\,f(x_0 + h, y_1)$
Repeat this process till $y_2$ is stationary. Then calculate $y_3$ and so on.

(*iv*) *Modified Euler's method:* $y_2 = y_1 + \frac{1}{2}[f(x_0 + h, y_1) + f(x_0 + 2h, y_2)]$
Repeat this step till $y_2$ is stationary. Then calculate $y_3$ and so on.

(*v*) *Runge Kutta method:* $y_1 = y_0 + h$ where $h = \frac{1}{6}(k_1 + 2k_2, 2k_3 + k_4)$

such that $k_1 = h\,f(x_0, y_0)$; $k_2 = h\,f(x_0 + h/2, y_0 + k_1/2)$
$k_3 = h\,f(x_0 + h/2, y_0 + k_2/2)$; $k_4 = hf(x_0 + h, y_0 + k_3)$

(*vi*) *Milne's method:*

*Predictor formula:* $y_4 = y_0 + \frac{4h}{3}(2f_1 - f_2 + 2f_3)$

*Corrector formula:* $y_4 = y_2 + \frac{h}{3}(f_2 - 4f_3 + f_4)$

(Four prior values are required to find the next values)

(*vii*) *Adams-Bashforth method:*

*Predictor formula:* $y_1 = y_0 + \frac{h}{24}(55f_0 - 59f_{-1} + 37f_{-2} - 9f_{-3})$

*Corrector formula:* $y_1 = y_0 + \frac{h}{24}(9f_1 + 19f_0 - 5f_{-1} - f_{-2})$

(Four prior values are required to find the next values)

## 7. Numerical solution of partial differential equations

(*i*) *Classification of a second order equations:*

$$A(x, y)\frac{\partial^2 u}{\partial x^2} + B(x, y)\frac{\partial^2 u}{\partial x \partial y} + C(x, y)\frac{\partial^2 u}{\partial y^2} + \left(F(x, y, u,)\frac{\partial u}{\partial x}\frac{\partial u}{\partial y}\right) = 0$$

is said to be *elliptic* if $B^2 - 4AC < 0$
*parabolic* if $B^2 - 4AC = 0$
*hyperbolic* if $B^2 - 4AC > 0$

(*ii*) *Laplace equation:* $\dfrac{\partial^2 u}{\partial x^2} + \dfrac{\partial^2 u}{\partial y^2} = 0$

*Standard 5-point formula:* $u_{i,\,j} = \dfrac{1}{4}\,[u_{i-1,j} + u_{i+1,\,j} + u_{i,j+1} + u_{i,j-1}]$

*Diagonal 5-point formula:* $u_{i,\,j} = \dfrac{1}{4}\,[u_{i-1,j+1} + u_{i+1,\,j-1}\, u_{i+1,j+1} + u_{i-1,j-1}]$

(*iii*) *Poisson's equation:* $\dfrac{\partial^2 u}{\partial x^2} + \dfrac{\partial^2 u}{\partial y^2} = f\,(x, y).$

*Standard 5-point formula:*

$$u_{i-1,j} + u_{i+1,\,j} + u_{i,j+1} + u_{i,j-1} - 4u_{i,\,j} = h^2 f(ih, jh)$$

(*iv*) *One-dimensional heat equation:* $\dfrac{\partial u}{\partial t} = c^2\,\dfrac{\partial^2 u}{\partial x^2}$

*Schmidt formula:* $u_{i,j+1} = \alpha\, u_{i-1,\,j} + (1 - 2\alpha)\, u_{i,\,j} + \alpha\, u_{i+1,j}$ where $\alpha = kc^2/h^2$

When $\alpha = 1/2$, it reduces to

*Bendre-Schmidt relation:* $u_{i,j+1} = \dfrac{1}{2}(u_{i-1,\,j} + u_{i+1,j})$

(*v*) *Wave equation:* $\dfrac{\partial^2 u}{\partial t^2} = c^2\,\dfrac{\partial^2 u}{\partial x^2}$

*Explict formula for solution is*

$$u_{i,j+1} = 2(1 - \alpha^2 c^2)\, u_{i,j} + \alpha^2 c^2\,(u_{i-1,j} + u_{i+1,j} - u_{i,j-1}) \text{ where } \alpha = k/h$$

If $\alpha$ is so chosen that the coefficient of $u_{i,\,j}$ is zero *i.e.*, $k = h/c$ then the above explicit formula takes the *simplified form*

$$u_{i,j+1} = u_{i-1,j} + u_{i+1,j} - u_{i,j-1}$$

which provides an **explicit scheme** for the solution of the wave equation.

APPENDIX 2

# Tables

**Table I : Gamma Function, $\Gamma(\alpha) = \int_0^\infty e^{-t} t^{\alpha-1} dt$**

| α | Γ(α) | α | Γ(α) | α | Γ(α) | α | Γ(α) |
|---|---|---|---|---|---|---|---|
| 1.00 | 1.000000 | 1.26 | 0.904397 | 1.52 | 0.887039 | 1.78 | 0.926227 |
| 1.01 | 0.994326 | 1.27 | 0.902503 | 1.53 | 0.887568 | 1.79 | 0.928767 |
| 1.02 | 0.988844 | 1.28 | 0.900719 | 1.54 | 0.888178 | 1.80 | 0.931384 |
| 1.03 | 0.983550 | 1.29 | 0.899042 | 1.55 | 0.888869 | 1.81 | 0.934076 |
| 1.04 | 0.978438 | 1.30 | 0.897471 | 1.56 | 0.889639 | 1.82 | 0.936845 |
| 1.05 | 0.973504 | 1.31 | 0.896004 | 1.57 | 0.890490 | 1.83 | 0.939690 |
| 1.06 | 0.968744 | 1.32 | 0.894640 | 1.58 | 0.891420 | 1.84 | 0.942612 |
| 1.07 | 0.964152 | 1.33 | 0.893378 | 1.59 | 0.892428 | 1.85 | 0.945611 |
| 1.08 | 0.959725 | 1.34 | 0.892215 | 1.60 | 0.893516 | 1.86 | 0.948687 |
| 1.09 | 0.955459 | 1.35 | 0.891151 | 1.61 | 0.894681 | 1.87 | 0.951840 |
| 1.10 | 0.951351 | 1.36 | 0.890184 | 1.62 | 0.895924 | 1.88 | 0.955071 |
| 1.11 | 0.947395 | 1.37 | 0.889313 | 1.63 | 0.897244 | 1.89 | 0.958380 |
| 1.12 | 0.943590 | 1.38 | 0.888537 | 1.64 | 0.898642 | 1.90 | 0.961766 |
| 1.13 | 0.939931 | 1.39 | 0.887854 | 1.65 | 0.900117 | 1.91 | 0.965231 |
| 1.14 | 0.936416 | 1.40 | 0.887264 | 1.66 | 0.901668 | 1.92 | 0.968774 |
| 1.15 | 0.933041 | 1.41 | 0.886764 | 1.67 | 0.903296 | 1.93 | 0.972397 |
| 1.16 | 0.929803 | 1.42 | 0.886356 | 1.68 | 0.905001 | 1.94 | 0.976099 |
| 1.17 | 0.926700 | 1.43 | 0.886036 | 1.69 | 0.906782 | 1.95 | 0.979881 |
| 1.18 | 0.923728 | 1.44 | 0.885805 | 1.70 | 0.908639 | 1.96 | 0.983742 |
| 1.19 | 0.920885 | 1.45 | 0.885661 | 1.71 | 0.910572 | 1.97 | 0.987685 |
| 1.20 | 0.918169 | 1.46 | 0.885604 | 1.72 | 0.912580 | 1.98 | 0.991708 |
| 1.21 | 0.915577 | 1.47 | 0.885633 | 1.73 | 0.914665 | 1.99 | 0.995813 |
| 1.22 | 0.913106 | 1.48 | 0.885747 | 1.74 | 0.916826 | 2.00 | 1.000000 |
| 1.23 | 0.910755 | 1.49 | 0.885945 | 1.75 | 0.919062 | | |
| 1.24 | 0.908521 | 1.50 | 0.886227 | 1.76 | 0.921375 | | |
| 1.25 | 0.906403 | 1.51 | 0.886592 | 1.77 | 0.923763 | | |

**Table II : Bessel Functions**

| $x$ | $J_0(x)$ | $J_1(x)$ | $x$ | $J_0(x)$ | $J_1(x)$ |
|---|---|---|---|---|---|
| 0.0 | 1.0000 | 0.0000 | 3.0 | –0.2601 | 0.3991 |
| 0.1 | 0.9975 | 0.0499 | 3.1 | –0.2921 | 0.3009 |
| 0.2 | 0.9900 | 0.0995 | 3.2 | –0.3202 | 0.2613 |
| 0.3 | 0.9776 | 0.1483 | 3.3 | –0.3443 | 0.2207 |
| 0.4 | 0.9604 | 0.1960 | 3.4 | –0.3643 | 0.1792 |
| 0.5 | 0.9385 | 0.2423 | 3.5 | –0.3801 | 0.1374 |
| 0.6 | 0.9120 | 0.2867 | 3.6 | –0.3918 | 0.0955 |
| 0.7 | 0.8812 | 0.3290 | 3.7 | –0.3992 | 0.0538 |
| 0.8 | 0.8463 | 0.3688 | 3.8 | –0.4026 | 0.0128 |
| 0.9 | 0.8075 | 0.4059 | 3.9 | –0.4018 | –0.0272 |
| 1.0 | 0.7652 | 0.4401 | 4.0 | –0.3971 | –0.0660 |
| 1.1 | 0.7196 | 0.4709 | 4.1 | –0.3887 | – 0.1033 |
| 1.2 | 0.6711 | 0.4983 | 4.2 | –0.3766 | –0.1386 |
| 1.3 | 0.6201 | 0.5220 | 4.3 | –0.3610 | –0.1719 |
| 1.4 | 0.5669 | 0.5419 | 4.4 | –0.3423 | –0.2028 |
| 1.5 | 0.5118 | 0.5579 | 4.5 | –0.3205 | –0.2311 |
| 1.6 | 0.4554 | 0.5699 | 4.6 | –0.2961 | –0.2566 |
| 1.7 | 0.3980 | 0.5778 | 4.7 | –0.2693 | –0.2791 |
| 1.8 | 0.3400 | 0.5815 | 4.8 | –0.2404 | –0.2985 |
| 1.9 | 0.2818 | 0.5812 | 4.9 | –0.2097 | –0.3147 |
| 2.0 | 0.2239 | 0.5767 | 5.0 | –0.1776 | –0.3276 |
| 2.1 | 0.1666 | 0.5683 | 5.1 | –0.1443 | –0.3371 |
| 2.2 | 0.1104 | 0.5560 | 5.2 | –0.1103 | –0.3432 |
| 2.3 | 0.0555 | 0.5399 | 5.3 | –0.0758 | –0.3460 |
| 2.4 | 0.0025 | 0.5202 | 5.4 | –0.0412 | –0.3453 |
| 2.5 | –0.0484 | 0.4971 | 5.5 | –0.0068 | –0.3414 |
| 2.6 | –0.0968 | 0.4708 | 5.6 | 0.0270 | –0.3343 |
| 2.7 | –0.1424 | 0.4416 | 5.7 | 0.0599 | –0.3241 |
| 2.8 | –0.1850 | 0.4097 | 5.8 | 0.0917 | –0.3110 |
| 2.9 | –0.2243 | 0.3754 | 5.9 | 0.1220 | –0.2951 |

**Table III : Laplace Transforms**

| S. No. | Laplace Transform i.e. L | Inverse Laplace Transform i.e. $L^{-1}$ | Remarks |
|---|---|---|---|
| 1. | $L\,f(s) = \bar{f}(s) = \int_0^\infty e^{-st} f(t)\,dt$ | $L^{-1}[\,\bar{f}(s)\,] = f(t)$ | Definition |
| 2. | $a[f(t)] + b[g(t)] - ch(t)]$ | $a\,\bar{f}(s) + b\,\bar{g}(s) - c\bar{h}(s)$ | Linearity property |
| 3. | $e^{at} f(t)$ | $\bar{f}(s-a)$ | First shifting property |
| 4. | $f(at)$ | $\frac{1}{a}\bar{f}\left(\frac{s}{a}\right)$ | Change of scale |
| 5. | $f'(t)$ | $s\,\bar{f}(s) - f(o)$ | Derivative |
| 6. | $f''(t)$ | $s^n\,\bar{f}(s) - s^{n-1} f'(o)\ \ldots\ldots f^{n-1}(o)$ | $n$th derivative |
| 7. | $\int_0^t f(u)\,du$ | $\frac{1}{s}\bar{f}(s)$ | Integral division by $s$ |
| 8. | $t^n f(t)$ | $(-1)^n \frac{d^n}{ds^n}\bar{f}$, $n = 1, 2, 3, \ldots\ldots$ | Multiplication by $t^n$ |
| 9. | $\frac{1}{t} f(t)$ | $\int_0^\infty \bar{f}(s)\,ds$ | Division by $t$ |
| 10. | $\int_0^t f(u)\,g(t-u)\,du = F^*G$ | $L^{-1}[\bar{f}(s)\,\bar{g}(s)]$ | Convolution theorem |
| 11. | 1 | $1/s$ | $s > 0$ |
| 12. | $t^n$ | $n\,!/s^{n+1}$, $n = 1, 2, 3$ | Otherwise : $\frac{\Gamma(n+1)}{s^{n+1}}$ |
| 13. | $e^{at}$ | $1/(s-a)$ | $s > a$ |
| 14. | $e^{at}\,t^n$ | $n\,!/(s-a)^{n+1}$ | $s > a$ |
| 15. | $\sin at$ | $a/(s^2 + a^2)$ | $s > 0$ |
| 16. | $\cos at$ | $s/(s^2 + a^2)$ | $s > 0$ |
| 17. | $\sinh at$ | $a/(s^2 - a^2)$ | $s > \|a\|$ |
| 18. | $\cosh at$ | $s/(s^2 - a^2)$ | $s > \|a\|$ |
| 19. | $e^{at} \sin bt$ | $b/[(s-a)^2 + b^2]$ | $s > a$ |
| 20. | $e^{at} \cos bt$ | $(s-a)/[(s-a)^2 + b^2]$ | $s > a$ |
| 21. | $e^{at} \sinh bt$ | $b/[(s-a)^2 - b^2]$ | $s > a$ |
| 22. | $e^{at} \cosh bt$ | $(s-a)/[(s-a)^2 - b^2]$ | $s > a$ |
| 23. | $\frac{1}{2a}(t \sin at)$ | $s/(s^2 + a^2)^2$ | |
| 24. | $(t \cos at)$ | $(s^2 - a^2)/(s^2 + a^2)^2$ | |
| 25. | $\frac{1}{2a^3}(\sin at - at \cos at)$ | $1/(s^2 + a^2)^2$ | |
| 26. | $\frac{1}{2a}(\sin at + at \cos at)$ | $s^2/(s^2 + a^2)^2$ | |
| 27. | $\cos at - \frac{1}{2} at \sin at$ | $s^3/(s^2 + a^2)^2$ | |
| 28. | $\frac{1}{2a} t \sinh at$ | $s/(s^2 - a^2)^2$ | |
| 29. | $t \cosh at$ | $(s^2 + a^2)/(s^2 - a^2)^2$ | |
| 30. | $\frac{1}{2a}(\sinh at + at \cosh at)$ | $s^2/(s^2 - a^2)^2$ | |

| | | | |
|---|---|---|---|
| 31. | $\cosh at + \frac{1}{2}\, at \sinh at$ | $s^3/(s^2 - a^2)^2$ | |
| 32. | $\frac{1}{2a} t^2 \sin at$ | $(3s^2 - a^2)/(s^2 + a^2)^3$ | |
| 33. | $\frac{1}{2} t^2 \cos at$ | $(s^3 - 3a^2 s)/(s^2 + a^2)^3$ | |
| 34. | $\frac{e^{bt} - e^{at}}{b - a}$ | $\frac{1}{(s-a)(s-b)}$ | $a \neq b$ |
| 35. | $\frac{b\,e^{bt} - a\,e^{at}}{b - a}$ | $\frac{s}{(s-a)(s-b)}$ | $a \neq b$ |
| 36. | $\frac{e^{bt} - e^{at}}{t}$ | $\log\left(\frac{s-a}{s-b}\right)$ | |
| 37. | $\frac{\sin t}{t}$ | $\tan^{-1}\left(\frac{1}{s}\right)$ | |
| 38. | $u(t-a)$ | $e^{-as}/s$ | $u(t-a)$ is unit step function |
| 39. | $f(t-a)\,u(t-a)$ | $e^{-as}\,\bar{f}(s)$ | Second shifting property |
| 40. | $\delta(t-a)$ | $e^{-as}$ | $\delta(t-a)$ is unit impulse function |
| 41. | $f(t) = f(+T)$ | $\frac{1}{1 - e^{-st}} \int_0^T e^{-st} f(t)\,dt$ | Periodic function |
| 42. | $J_0(x)$ | $1/\sqrt{(s^2+1)}$ | Bessel function |
| 43. | $J_1(x)$ | $1 - s/\sqrt{(s+1)}$ | Bessel function |
| 44. | $erf(\sqrt{x})$ | $\frac{1}{s\sqrt{(s+1)}}$ | Error function |
| 45. | $L_n(x)$ | $\frac{n!\,(s-1)^n}{s^n + 1}$ | Laguerre's polynomial $(s > 1)$ |

**Table IV : Area under the Normal curve**

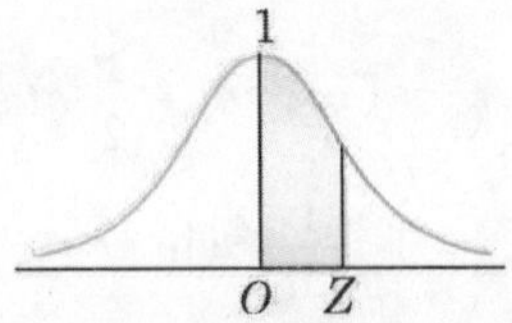


| x | 0 | 1 | 2 | 3 | 4 | 5 | 6 | 7 | 8 | 9 |
|---|---|---|---|---|---|---|---|---|---|---|
| 0.0 | 0.0000 | 0.0040 | 0.0080 | 0.0120 | 0.0160 | 0.0199 | 0.0239 | 0.0279 | 0.0319 | 0.0359 |
| 0.1 | 0.0398 | 0.0438 | 0.0478 | 0.0517 | 0.0557 | 0.0596 | 0.0636 | 0.0675 | 0.0714 | 0.0754 |
| 0.2 | 0.0793 | 0.0832 | 0.0871 | 0.0910 | 0.0948 | 0.0987 | 0.1026 | 0.1064 | 0.1103 | 0.1141 |
| 0.3 | 0.1179 | 0.1217 | 0.1255 | 0.1293 | 0.1331 | 0.1368 | 0.1406 | 0.1443 | 0.1480 | 0.1517 |
| 0.4 | 0.1554 | 0.1591 | 0.1628 | 0.1664 | 0.1700 | 0.1736 | 0.1772 | 0.1808 | 0.1844 | 0.1879 |
| 0.5 | 0.1915 | 0.1950 | 0.1985 | 0.2019 | 0.2054 | 0.2088 | 0.2123 | 0.2157 | 0.2190 | 0.2224 |
| 0.6 | 0.2258 | 0.2291 | 0.2324 | 0.2357 | 0.2389 | 0.2422 | 0.2454 | 0.2486 | 0.2518 | 0.2549 |
| 0.7 | 0.2580 | 0.2612 | 0.2642 | 0.2673 | 0.2704 | 0.2734 | 0.2764 | 0.2794 | 0.2823 | 0.2852 |
| 0.8 | 0.2881 | 0.2910 | 0.2939 | 0.2967 | 0.2996 | 0.3023 | 0.3051 | 0.3078 | 0.3106 | 0.3133 |
| 0.9 | 0.3159 | 0.3186 | 0.3212 | 0.3238 | 0.3264 | 0.3289 | 0.3315 | 0.3340 | 0.3365 | 0.3389 |
| 1.0 | 0.3413 | 0.3438 | 0.3461 | 0.3485 | 0.3508 | 0.3531 | 0.3554 | 0.3577 | 0.3599 | 0.3621 |
| 1.1 | 0.3643 | 0.3665 | 0.3686 | 0.3708 | 0.3729 | 0.3749 | 0.3770 | 0.3790 | 0.3810 | 0.3830 |
| 1.2 | 0.3849 | 0.3869 | 0.3888 | 0.3907 | 0.3925 | 0.3944 | 0.3962 | 0.3980 | 0.3997 | 0.4015 |
| 1.3 | 0.4032 | 0.4049 | 0.4066 | 0.4082 | 0.4099 | 0.4115 | 0.4131 | 0.4147 | 0.4162 | 0.4177 |
| 1.4 | 0.4192 | 0.4207 | 0.4222 | 0.4236 | 0.4251 | 0.4265 | 0.4279 | 0.4292 | 0.4306 | 0.4319 |
| 1.5 | 0.4332 | 0.4345 | 0.4357 | 0.4370 | 0.4382 | 0.4394 | 0.4406 | 0.4418 | 0.4429 | 0.4441 |
| 1.6 | 0.4452 | 0.4463 | 0.4474 | 0.4484 | 0.4495 | 0.4505 | 0.4515 | 0.4525 | 0.4535 | 0.4545 |
| 1.7 | 0.4554 | 0.4564 | 0.4573 | 0.4582 | 0.4591 | 0.4599 | 0.4608 | 0.4616 | 0.4625 | 0.4633 |
| 1.8 | 0.4641 | 0.4649 | 0.4656 | 0.4664 | 0.4671 | 0.4678 | 0.4686 | 0.4693 | 0.4699 | 0.4706 |
| 1.9 | 0.4713 | 0.4719 | 0.4726 | 0.4732 | 0.4738 | 0.4744 | 0.4750 | 0.4756 | 0.4761 | 0.4767 |
| 2.0 | 0.4772 | 0.4778 | 0.4783 | 0.4788 | 0.4793 | 0.4798 | 0.4803 | 0.4808 | 0.4812 | 0.4817 |
| 2.1 | 0.4821 | 0.4826 | 0.4830 | 0.4834 | 0.4838 | 0.4842 | 0.4846 | 0.4850 | 0.4854 | 0.4857 |
| 2.2 | 0.4861 | 0.4864 | 0.4868 | 0.4871 | 0.4875 | 0.4878 | 0.4881 | 0.4884 | 0.4887 | 0.4890 |
| 2.3 | 0.4893 | 0.4896 | 0.4898 | 0.4901 | 0.4904 | 0.4906 | 0.4909 | 0.4911 | 0.4913 | 0.4916 |
| 2.4 | 0.4918 | 0.4920 | 0.4922 | 0.4925 | 0.4927 | 0.4929 | 0.4931 | 0.4932 | 0.4934 | 0.4936 |
| 2.5 | 0.4938 | 0.4940 | 0.4941 | 0.4943 | 0.4945 | 0.4946 | 0.4948 | 0.4949 | 0.4951 | 0.4952 |
| 2.6 | 0.4953 | 0.4955 | 0.4956 | 0.4957 | 0.4959 | 0.4960 | 0.4961 | 0.4962 | 0.4963 | 0.4964 |
| 2.7 | 0.4965 | 0.4966 | 0.4967 | 0.4968 | 0.4969 | 0.4970 | 0.4971 | 0.4972 | 0.4973 | 0.4974 |
| 2.8 | 0.4974 | 0.4975 | 0.4976 | 0.4977 | 0.4977 | 0.4978 | 0.4979 | 0.4979 | 0.4980 | 0.4981 |
| 2.9 | 0.4981 | 0.4982 | 0.4982 | 0.4983 | 0.4984 | 0.4984 | 0.4985 | 0.4985 | 0.4986 | 0.4986 |
| 3.0 | 0.4987 | 0.4987 | 0.4987 | 0.4988 | 0.4988 | 0.4989 | 0.4989 | 0.4989 | 0.4990 | 0.4990 |
| 3.1 | 0.4990 | 0.4991 | 0.4991 | 0.4991 | 0.4992 | 0.4992 | 0.4992 | 0.4992 | 0.4993 | 0.4993 |

**Table V : Values of | $t$ | with probability $P$ and degrees of freedom ν**

| $P$ / ν | **0.50** | **0.10** | **0.05** | **0.02** | **0.01** |
|---|---|---|---|---|---|
| 1 | 1.000 | 6.34 | 12.71 | 31.82 | 63.66 |
| 2 | 0.816 | 2.92 | 4.30 | 6.96 | 9.92 |
| 3 | 0.765 | 2.35 | 3.18 | 4.54 | 5.84 |
| 4 | 0.741 | 2.13 | 2.78 | 3.75 | 4.60 |
| 5 | 0.727 | 2.02 | 2.57 | 3.36 | 4.03 |
| 6 | 0.718 | 1.94 | 2.45 | 3.14 | 3.71 |
| 7 | 0.711 | 1.90 | 2.36 | 3.00 | 3.50 |
| 8 | 0.706 | 1.86 | 2.31 | 2.90 | 3.36 |
| 9 | 0.703 | 1.83 | 2.26 | 2.82 | 3.25 |
| 10 | 0.700 | 1.81 | 2.23 | 2.76 | 3.17 |
| 11 | 0.697 | 1.80 | 2.20 | 2.72 | 3.11 |
| 12 | 0.695 | 1.78 | 2.18 | 2.68 | 3.06 |
| 13 | 0.694 | 1.77 | 2.16 | 2.65 | 3.01 |
| 14 | 0.692 | 1.76 | 2.14 | 2.62 | 2.98 |
| 15 | 0.691 | 1.75 | 2.13 | 2.60 | 2.95 |
| 16 | 0.690 | 1.75 | 2.12 | 2.58 | 2.92 |
| 17 | 0.689 | 1.74 | 2.11 | 2.57 | 2.90 |
| 18 | 0.688 | 1.73 | 2.10 | 2.55 | 2.88 |
| 19 | 0.688 | 1.73 | 2.09 | 2.54 | 2.86 |
| 20 | 0.687 | 1.72 | 2.09 | 2.53 | 2.84 |
| 21 | 0.686 | 1.72 | 2.08 | 2.52 | 2.83 |
| 22 | 0.686 | 1.72 | 2.07 | 2.51 | 2.82 |
| 23 | 0.685 | 1.71 | 2.07 | 2.50 | 2.81 |
| 24 | 0.685 | 1.71 | 2.06 | 2.49 | 2.80 |
| 25 | 0.684 | 1.71 | 2.06 | 2.48 | 2.79 |
| 26 | 0.684 | 1.71 | 2.06 | 2.48 | 2.78 |
| 27 | 0.684 | 1.70 | 2.05 | 2.47 | 2.77 |
| 28 | 0.683 | 1.70 | 2.05 | 2.47 | 2.76 |
| 29 | 0.683 | 1.70 | 2.04 | 2.46 | 2.76 |
| 30 | 0.683 | 1.70 | 2.04 | 2.46 | 2.75 |

**Table VI : Values of $\chi^2$ with probability $P$ and df $\nu$**

| $\nu$ \ $P$ | 0.99 | 0.95 | 0.50 | 0.30 | 0.20 | 0.10 | 0.05 | 0.01 |
|---|---|---|---|---|---|---|---|---|
| 1 | 0.0002 | 0.004 | 0.46 | 1.07 | 1.64 | 2.71 | 3.84 | 6.64 |
| 2 | 0.020 | 0.103 | 1.39 | 2.41 | 3.22 | 4.60 | 5.99 | 9.21 |
| 3 | 0.115 | 0.35 | 2.37 | 3.66 | 4.64 | 6.25 | 7.82 | 11.34 |
| 4 | 0.30 | 0.71 | 3.36 | 4.88 | 5.99 | 7.78 | 9.49 | 13.28 |
| 5 | 0.55 | 1.14 | 4.35 | 6.06 | 7.29 | 9.24 | 11.07 | 15.09 |
| 6 | 0.87 | 1.64 | 5.35 | 7.23 | 8.56 | 10.64 | 12.59 | 16.81 |
| 7 | 1.24 | 2.17 | 6.35 | 8.38 | 9.80 | 12.02 | 14.07 | 18.48 |
| 8 | 1.65 | 2.73 | 7.34 | 9.52 | 11.03 | 13.36 | 15.51 | 20.09 |
| 9 | 2.09 | 3.32 | 8.34 | 10.66 | 12.24 | 14.68 | 16.92 | 21.67 |
| 10 | 2.56 | 3.94 | 9.34 | 11.78 | 13.44 | 15.99 | 18.31 | 23.21 |
| 11 | 3.05 | 4.58 | 10.34 | 12.90 | 14.63 | 17.28 | 19.68 | 24.72 |
| 12 | 3.57 | 5.23 | 11.34 | 14.01 | 15.81 | 18.55 | 21.03 | 26.22 |
| 13 | 4.11 | 5.89 | 12.34 | 15.12 | 16.98 | 19.81 | 22.36 | 27.69 |
| 14 | 4.66 | 6.57 | 13.34 | 16.22 | 18.15 | 21.06 | 23.68 | 29.14 |
| 15 | 5.23 | 7.26 | 14.34 | 17.32 | 19.31 | 22.31 | 25.00 | 30.58 |
| 16 | 5.81 | 7.96 | 15.34 | 18.42 | 20.46 | 23.54 | 26.30 | 32.00 |
| 17 | 6.41 | 8.67 | 16.34 | 19.51 | 21.62 | 24.77 | 27.59 | 33.41 |
| 18 | 7.02 | 9.39 | 17.34 | 20.60 | 22.76 | 25.99 | 28.87 | 34.80 |
| 19 | 7.63 | 10.12 | 18.34 | 21.69 | 23.90 | 27.20 | 30.14 | 36.19 |
| 20 | 8.26 | 10.85 | 19.34 | 22.78 | 25.04 | 28.41 | 31.41 | 37.57 |
| 21 | 8.90 | 11.59 | 20.34 | 23.86 | 26.17 | 29.62 | 32.67 | 38.93 |
| 22 | 9.54 | 12.34 | 21.34 | 24.94 | 27.30 | 30.81 | 33.92 | 40.29 |
| 23 | 10.20 | 13.09 | 22.34 | 26.02 | 28.43 | 32.01 | 35.17 | 41.64 |
| 24 | 10.86 | 13.85 | 23.34 | 27.10 | 29.55 | 33.20 | 36.42 | 42.98 |
| 25 | 11.52 | 14.61 | 24.34 | 28.17 | 30.68 | 34.68 | 37.65 | 44.31 |
| 26 | 12.20 | 15.38 | 25.34 | 29.25 | 31.80 | 35.56 | 38.88 | 45.64 |
| 27 | 12.88 | 16.15 | 26.34 | 30.32 | 32.91 | 36.74 | 40.11 | 46.96 |
| 28 | 13.56 | 16.93 | 27.34 | 31.39 | 34.03 | 37.92 | 41.34 | 48.28 |
| 29 | 14.26 | 17.71 | 28.34 | 32.46 | 35.14 | 39.09 | 42.56 | 49.59 |
| 30 | 14.95 | 18.49 | 29.34 | 33.53 | 36.25 | 40.26 | 43.77 | 50.89 |

## Table VII : 5% and 1% points of F

| $v_2$ \ $v_1$ | 1 | 2 | 3 | 4 | 5 | 6 | 8 | 12 | 24 | ∞ |
|---|---|---|---|---|---|---|---|---|---|---|
| 2 | 18.51 | 19.00 | 19.16 | 19.25 | 19.30 | 19.32 | 19.37 | 19.41 | 19.45 | 19.50 |
| | 98.49 | 99.00 | 99.17 | 99.25 | 99.30 | 99.33 | 99.36 | 99.42 | 99.46 | 99.50 |
| 3 | 10.13 | 9.55 | 9.28 | 9.12 | 9.01 | 8.94 | 8.84 | 8.74 | 8.64 | 8.53 |
| | 34.12 | 30.82 | 29.46 | 28.71 | 28.24 | 27.91 | 27.49 | 27.05 | 26.60 | 26.12 |
| 4 | 7.71 | 6.94 | 6.59 | 6.39 | 6.26 | 6.16 | 6.04 | 5.91 | 5.77 | 5.63 |
| | 21.20 | 18.00 | 16.69 | 15.98 | 15.52 | 15.21 | 14.80 | 14.37 | 13.93 | 13.46 |
| 5 | 6.61 | 5.79 | 5.41 | 5.19 | 5.05 | 4.95 | 4.82 | 4.68 | 4.53 | 4.36 |
| | 16.26 | 13.27 | 12.06 | 11.39 | 10.97 | 10.67 | 10.27 | 9.89 | 9.47 | 9.02 |
| 6 | 5.99 | 5.14 | 4.76 | 4.53 | 4.39 | 4.28 | 4.15 | 4.00 | 3.84 | 3.67 |
| | 13.74 | 10.92 | 9.78 | 9.15 | 8.75 | 8.47 | 8.10 | 7.72 | 7.31 | 6.88 |
| 7 | 5.59 | 4.74 | 4.35 | 4.12 | 3.97 | 3.87 | 3.73 | 3.57 | 3.41 | 3.23 |
| | 12.25 | 9.55 | 8.45 | 7.85 | 7.46 | 7.19 | 6.84 | 6.47 | 6.07 | 5.65 |
| 8 | 5.32 | 4.46 | 4.07 | 3.84 | 3.69 | 3.58 | 3.44 | 3.28 | 3.12 | 2.93 |
| | 11.26 | 8.65 | 7.59 | 7.01 | 6.63 | 6.37 | 6.03 | 5.67 | 5.28 | 4.86 |
| 9 | 5.12 | 4.26 | 3.86 | 3.63 | 3.48 | 3.37 | 3.23 | 3.07 | 2.90 | 2.71 |
| | 10.56 | 8.02 | 6.99 | 6.42 | 6.06 | 5.80 | 5.47 | 5.11 | 4.73 | 4.31 |
| 10 | 4.96 | 4.10 | 3.71 | 3.48 | 3.33 | 3.22 | 3.07 | 2.91 | 2.74 | 2.54 |
| | 10.04 | 7.56 | 6.55 | 5.99 | 5.64 | 5.39 | 5.06 | 4.71 | 4.33 | 3.91 |
| 12 | 4.75 | 3.88 | 3.49 | 3.26 | 3.11 | 3.00 | 2.85 | 2.69 | 2.50 | 2.30 |
| | 9.33 | 6.93 | 5.95 | 5.41 | 5.06 | 4.82 | 4.50 | 4.16 | 3.78 | 3.36 |
| 14 | 4.60 | 3.74 | 3.34 | 3.11 | 2.96 | 2.85 | 2.70 | 2.53 | 2.35 | 2.13 |
| | 8.86 | 6.51 | 5.56 | 5.03 | 4.69 | 4.46 | 4.14 | 3.80 | 3.43 | 3.00 |
| 16 | 4.49 | 3.63 | 3.24 | 3.01 | 2.85 | 2.74 | 2.59 | 2.42 | 2.24 | 2.01 |
| | 8.53 | 6.23 | 5.29 | 4.77 | 4.44 | 4.20 | 3.89 | 3.55 | 3.18 | 2.75 |
| 18 | 4.41 | 3.55 | 3.16 | 2.93 | 2.77 | 2.66 | 2.51 | 2.34 | 2.15 | 1.92 |
| | 8.28 | 6.01 | 5.09 | 4.58 | 4.25 | 4.01 | 3.71 | 3.37 | 3.01 | 2.57 |
| 20 | 4.35 | 3.49 | 3.10 | 2.87 | 2.71 | 2.60 | 2.45 | 2.28 | 2.08 | 1.84 |
| | 8.10 | 5.85 | 4.94 | 4.43 | 4.10 | 3.87 | 3.56 | 3.23 | .2.86 | 2.42 |
| 25 | 4.24 | 3.38 | 2.99 | 2.76 | 2.60 | 2.49 | 2.34 | 2.16 | 1.96 | 1.71 |
| | 7.77 | 5.57 | 4.68 | 4.18 | 3.86 | 3.63 | 3.32 | 2.99 | 2.62 | 2.17 |
| 30 | 4.17 | 3.32 | 2.92 | 2.69 | 2.53 | 2.42 | 2.27 | 2.09 | 1.89 | 1.62 |
| | 7.56 | 5.39 | 4.51 | 4.02 | 3.70 | 3.47 | 3.17 | 2.84 | 2.47 | 2.01 |
| 40 | 4.08 | 3.23 | 2.84 | 2.61 | 2.45 | 2.34 | 2.18 | 2.00 | 1.79 | 1.51 |
| | 7.31 | 5.18 | 4.31 | 3.83 | 3.51 | 3.29 | 2.99 | 2.66 | 2.29 | 1.81 |
| 60 | 4.00 | 3.15 | 2.76 | 2.52 | 2.37 | 2.25 | 2.10 | 1.92 | 1.70 | 1.39 |
| | 7.08 | 4.98 | 4.13 | 3.65 | 3.34 | 3.12 | 2.82 | 2.50 | 2.12 | 1.60 |

APPENDIX 3

# Answers to Problems

## Problems 1.1, page 5

1. $x^4 - 6x^3 + 3x^2 + 42x - 70 = 0$
2. (*i*) $-2, 1 + 3i, 1 - 3i$ (*ii*) $2 \pm \sqrt{3}, 3, -5$
5. Two roots between (1, 2) and (–3, –4)
6. $2, 2, -\frac{1}{3}$
7. 3
8. $a = 2, b = 1$
9. 6, 4, –1
10. –4, 2, 6
11. 1, 1, 2, 2
12. $\frac{1}{2}(3 \pm \sqrt{5}; \frac{1}{2}(5 \pm \sqrt{5})$
13. 1, 4, 7
14. $1, \frac{1}{2}, \frac{1}{4}$
16. (*i*) –5, –2, 1, 4 (*ii*) 1, –2, 4, –8
17. (*i*) $m^2 - 2ln$, (*ii*) $lm - n$
18. 36
19. (*i*) 4/3, (*ii*) 16/9.

## Problems 1.2, page 8

1. $x^3 + 6x^2 - 36x + 27 = 0$
2. $6x^5 - 7x^4 - 13x^3 + 4x^2 - 2 = 0$
3. $10x^4 + 9x^3 + 8x^2 - 7x + 1 = 0$
4. –1, –2, –6, –7
5. (*i*) $x^3 - 9x^2 + 26x - 24 = 0$ ; (*ii*) $x^4 + 13x^3 + 60x^2 + 116x + 80 = 0$ ; (*iii*) $x^5 + 7 = 0$
6. $x^3 + 15x^2 + 52x - 36 = 0$
7. $y^3 + (p^3 + 3q) y^2 + 3q^2y + q^3 = 0$
8. $3x^3 - 11x^2 + 9x - 2 = 0$
9. $y^3 - qy^2 + pry - r^2 = 0$
10. (*a*) $y^3 + 4my - 8n = 0$ ; (*b*) $ny^3 + m^2y^2 - 2mny + n^2 = 0$ ; (*c*) $ny^3 - my^2 = 1$
11. $y^3 - 30y^2 + 225y - 68 = 0$
12. (*i*) $\frac{-5 \pm \sqrt{21}}{2}, \frac{5 \pm \sqrt{91}i}{12}$ (*ii*) 2, 2, 1/2, 1/2 (*iii*) 1, –2, 4, –1/2, 1/4 ;

    (*iv*) –1, – 2, 3, –1/2, 1/3 ; (*v*) $\pm 1, -3, -1/3, \frac{3 \pm \sqrt{5}}{2}$
13. $\frac{1}{2}(5 \pm \sqrt{21})$ ; $\frac{1}{2}(-3 \pm \sqrt{5})$

## Problems 1.3, page 11

1. –6, 3, 3
2. $5, \frac{1}{2}(-5 \pm i\sqrt{3})$
3. $6, -3 \pm 2\sqrt{(-3)}$
4. $-1, -2, \frac{1}{2}$
5. $\frac{1}{2}, -\frac{1}{6}(3 \pm i\sqrt{3})$
6. $5, \frac{1}{2}(1 + i\sqrt{3})$
7. $2\cos\frac{2\pi}{9}, 2\cos\frac{8\pi}{9}, 2\cos\frac{14\pi}{9}$
8. $\frac{1}{2}, \frac{-7 \pm 9i\sqrt{3}}{6}$.

## Problems 1.4, page 12

1. 1, 2, 3, 4
2. –3, 1, ± 2
3. 4, –2, $-1 \pm i$
4. –1, 3, $3 \pm \sqrt{30}$
5. $1 \pm \sqrt{7}$, $2 \pm \sqrt{3}$
6. $1 \pm 2i$, $-1 \pm \sqrt{2}$
7. $2 \pm \sqrt{3}$, $-2 \pm i\sqrt{3}$
8. 2, 4, $2 \pm 2i\sqrt{2}$.

## Problems 1.5, page 15

1. 1.32
2. 2.29
3. 0.45
4. (*i*) 0.71 rad
5. 1.81 rad
6. 0.26.

## Problems 1.6, page 15

1. (*d*)
2. (*c*)
3. (*c*)
4. (*c*)
5. (*c*)
6. (*a*)
7. (*d*)
8. (*b*)
9. (*c*)
10. (*a*)
11. (*c*)
12. minus
13. § 15.1 (*v*)
14. $p/q$
15. 21
16. –3 and –2
17. Conjugate pairs
18. $f(x)$ is continuous in $(a, b)$
19. $x^3 - 9x^2 + 29x - 24 = 0$
20. 3, 6, –2
21. $x^3 - 200x - 7000 = 0$
22. $p/r$
23. $x^4 + 2x^3 - x^2 - 6x - 6 = 0$
24. 6
25. minus
26. $pq = r$
27. 1, $\frac{1}{2}(-1 \pm \sqrt{3}i)$
28. (*iii*)
29. 1, 1, –2
30. $x^3 - 7x^2 + 12x - 10 = 0$
31. Zero and 2
32. 21
33. True
34. True.

## Problems 2.1 page 25

5. (*i*) 1 (*ii*) 0
13. $(a - b)(b - c)(c - a)$
14. $(a - b)(b - c)(c - a)(ab + bc + ca)$
15. $(b - c)(c - a)(a - b)(a - 1)(b - 1)(c - 1)$
16. $(a - b)(a - c)(a - d)(b - c)(b - d)(c - d)$
17. $x = 0, \pm\sqrt{(a^2 + b^2 + c^2 - ab - bc - ca)}$
18. $0, -\frac{1}{2}$.

## Problems 2.2 page 31

1. $x = 0, 3$
2. $x = -3, y = -2, z = -4, a = 3$
3. $x = 3, y = 8$
4. $-2\begin{bmatrix} 1 & 1 \\ 0 & 3 \end{bmatrix}$
5. $AB = \begin{bmatrix} 4 & 4 & 7 & 10 \\ 10 & 7 & 11 & 21 \\ 4 & 3 & 3 & 9 \end{bmatrix}$
7. (*i*) $[ax^2 + by^2 + cz^2 + 2fyz + 2gzx + 2hxy]$

(*ii*) $\begin{bmatrix} 8 & 7 \\ 122 & 104 \\ -365 & -131 \end{bmatrix}$ (*iii*) $\begin{bmatrix} 9 & 6 \\ -18 & -12 \\ 27 & 18 \end{bmatrix}$

10. $3I$
11. $\begin{bmatrix} -6 & 1 & 2 \\ 5 & 4 & 4 \\ 2 & 8 & -3 \end{bmatrix}$
15. $\begin{bmatrix} 1 & 0 & 0 \\ 7/5 & 1 & 0 \\ 3/5 & 41/19 & 1 \end{bmatrix}\begin{bmatrix} 5 & -2 & 1 \\ 0 & 19/5 & -32/5 \\ 0 & 0 & 327/19 \end{bmatrix}$

## Problems 2.3, page 36

**2.** (i) $\begin{bmatrix} 3 & 0 & 5.5 \\ 0 & 7 & 1.5 \\ 5.5 & 1.5 & 0 \end{bmatrix} + \begin{bmatrix} 0 & -2 & 5 \\ 2 & 0 & -2.5 \\ -0.5 & 2.5 & 0 \end{bmatrix}$ (ii) $\begin{vmatrix} a & \frac{1}{2}(a+c) & \frac{1}{2}(b+c) \\ \frac{1}{2}(a+c) & b & \frac{1}{2}(a+b) \\ \frac{1}{2}(b+c) & \frac{1}{2}(a+b) & c \end{vmatrix} + \begin{vmatrix} 0 & \frac{1}{2}(a-c) & \frac{1}{2}(b-c) \\ \frac{1}{2}(c-a) & 0 & \frac{1}{2}(b-a) \\ \frac{1}{2}(c-b) & \frac{1}{2}(a-b) & 0 \end{vmatrix}$

**5.** $B = \begin{bmatrix} 1 & 0 & 0 \\ 0 & 2 & 0 \\ 0 & 0 & 1 \end{bmatrix}$

**6.** (i) $\begin{bmatrix} 1 & -1 & 0 \\ -2 & 3 & -4 \\ -2 & 3 & -3 \end{bmatrix}$.

## Problems 2.4, page 44

**1.** 2

**2.** (i) 2 (ii) 2 (iii) 2

**3.** 2

**5.** No value of $p$ is possible.

**6.** (i) $\begin{bmatrix} -2 & 4/5 & 9/5 \\ 3 & -4/5 & -14/5 \\ -1 & 1/5 & 6/5 \end{bmatrix}$

(ii) $\begin{bmatrix} 3 & -5 & 6 \\ -2 & 4 & -5 \\ 1 & -2 & 3 \end{bmatrix}$

**7.** (i) 2, (ii) 2, (iii) 2, (iv) 2

**8.** $\begin{bmatrix} 1 & -1 & 0 \\ -2 & 3 & -4 \\ -2 & 3 & -3 \end{bmatrix}$

**9.** (i) 3 (ii) 2 (iii) 3.

## Problems 2.5, page 46

**1.** $\frac{1}{8}\begin{bmatrix} 2 & 2 & -2 \\ -9 & 11 & 5 \\ 5 & -7 & -1 \end{bmatrix}$

**2.** $\begin{bmatrix} 1 & -2 & 1 & 0 \\ 1 & -2 & 2 & -3 \\ 0 & 1 & -1 & 1 \\ -2 & 3 & -2 & 3 \end{bmatrix}$

## Problems 2.6, page 49

**1.** $x = 6, y = -7, z = 2$

**2.** $x = 2, y = -1, z = 1/2$

**3.** $x = 1.2, y = 2.2, z = 3.2$

**4.** $x = y = z = e^2$

**5.** $u = 1, v = 1/2, w = 1/3$

**6.** $x_1 = 1, x_2 = -5, x_3 = 5$

**7.** $x = 2, y = 1, z = 0$

**8.** $x = 1, y = 2, z = 3$

**9.** $x = y = z = 2$

**10.** $x_1 = 2, x_2 = 1, x_3 = -1, x_4 = 3$

**11.** $x = 1, y = -1, z = 1$

**12.** $i_1 = 1.5, i_3 = 2.5$

**13.** $i_1 = 369/175, i_2 = 24/25, i_3 = 72/175$

**14.** $x_1 = 2, x_2 = 1/5, x_3 = 0, x_4 = 4/5$.

## Problems 2.7, page 57

**1.** Consistent ; $x = 1, y = 3k - 2, z = k$ for all $k$

**2.** For $a = -1, b = 6$, the equations are consistant and have infinite number of solutions;
For $a \neq -1$, $b$ has any value, equations will have unique solutions.

**3.** (i) $\lambda = 3, \mu \neq 10$ ; (ii) $\lambda \neq 3$ ; (iii) $\lambda = 3, \mu = 10$

**4.** (i) Consistent ; $x = -1, y = 1, z = 2$ ;
(ii) Equations are inconsistent ; (iii) Consistent : $x = 2, y = 1, z = -4$

**5.** $a = 8, b \neq 6$; system possesses no solution, $a \neq 8, b$, may have any value, system possesses a unique system $a = 8, b = 6$ then system possesses an infinite number of solutions.

**6.** $\lambda \neq -5, x = 4/7, y = -9/7, z = 0$ ; $\lambda = -5, x = \frac{1}{7}(4 - 5k), y = \frac{1}{7}(13k - 9), z = k$ for all $k$

**9.** $\lambda = -1, 1, 12$ ; $x = -1/11, y = -15/11$ ; $x = -5, y = 1$ ; $x = \frac{1}{2}, y = 1$

**11.** $k = 3$ is the only real value for which $x = y = z$

**12.** $\lambda = 1, x_1 = 2t - s, x_2 = t, x_3 = s$ ; $\lambda = -3, x_1 = -t, x_2 = -2t, x_3 = t$

**13.** $(k = 1, 2)$

**14.** (*i*) Have infinite number of non-trivial solutions ; $x = \lambda - 5\mu/3, y = \lambda - 4\mu/3, z = \lambda, w = \mu$ for all values of $\lambda$ and $\mu$. (*ii*) $x = 11k_2 + 6k_1, y = -8k_2 - 3k_1, z = k_2, w = k_1$ where $k_1, k_2$ are arbitrary constants.

**16.** (*i*) $x = -1, y = -2, z = 4$, (*ii*) $x = 1, y = 2, z = 3$
(*iii*) $x_1 = 2, x_2 = 1, x_3 = -1, x_4 = 3$

## Problems 2.8, page 60

**1.** $\begin{bmatrix} x_1 \\ x_2 \end{bmatrix} = \begin{bmatrix} 9 & 6 \\ 11 & -2 \end{bmatrix} \begin{bmatrix} z_1 \\ z_2 \end{bmatrix}$

**2.** $A = \begin{bmatrix} \cos\alpha & -\sin\alpha \\ \sin\alpha & \cos\alpha \end{bmatrix}, A^{-1} = A' = \begin{bmatrix} \cos\alpha & \sin\alpha \\ -\sin\alpha & \cos\alpha \end{bmatrix}$

**3.** $Z = (BA)X$, where $BA = \begin{bmatrix} 1 & 4 & -1 \\ -1 & 9 & -1 \\ -3 & 14 & -1 \end{bmatrix}$

**4.** $x_1 = 19y_1 - 9y_2 + 2y_3$ ; $x_2 = -4y_1 + 2y_2 - y_3$ ; $x_3 = -2y_1 + y_2$

**5.** $x_1 = y_1 - y_3, x_2 = -y_1 + y_2 - 2y_3, x_3 = 3y_1 + 2y_2 + 2y_3$.

**7.** $a = \pm\frac{1}{\sqrt{2}}, b = \pm\frac{1}{\sqrt{6}}, c = \pm\frac{1}{\sqrt{3}}$

**10.** (*i*) No. (*ii*) No. (*iii*) Yes, $2\boldsymbol{x}_1 + \boldsymbol{x}_2 - \boldsymbol{x}_3 = 0$ (*iv*) Yes, $9\boldsymbol{x}_1 - 12\boldsymbol{x}_2 + 5\boldsymbol{x}_3 - 5\boldsymbol{x}_4 = 0$.

## Problems 2.9, page 72

**1.** 10 ; 30

**2.** (*a*) 10, 3 ; (1, 2), (3, –1) ; $(\lambda^3 - 16\lambda^2 + 60\lambda = 0$;

10, 6, 0; $\begin{bmatrix} 3 \\ 0 \\ 1 \end{bmatrix} \begin{bmatrix} 1 \\ 1 \\ 0 \end{bmatrix} \begin{bmatrix} 1 \\ -1 \\ 1 \end{bmatrix}$

**3.** (*a*) (1, 1 1) ; only one linear on independent eigen vector $[1, 0, 0]^T$.
(*b*) 0, 3, 5 ; (1, 2, 2), (2, 1, –2), (2, –2, 1) (*c*) 1, 2, 3 ; (1, 0, –1), (0, 1, 0), (1, 0, 1)
(*d*) 5, –3, –3 ; (1, 2, –1), (2, –1, 0), (3, 0, 1) (*e*) 8, 2, 2 ; (2, –1, 1), (1, 0, –2), (1, 2, 0)

**5.** (*i*) 8, 12, 6 (*ii*) 49, 121, 25

**6.** 1, 1, 1/5

**8.** $\begin{bmatrix} 2 & 1 & 1 \\ 1 & 2 & 1 \\ 0 & 0 & 1 \end{bmatrix}$

**9.** (*i*) $\begin{bmatrix} 2 & 3 \\ 3 & 5 \end{bmatrix}$ ; (*ii*) $\begin{bmatrix} 1 & 1/2 & -2/3 \\ 0 & -1/2 & 0 \\ 0 & 0 & 1/3 \end{bmatrix}$ (*iii*) $\begin{bmatrix} -1 & 3 & -2 \\ 3 & -3 & 1 \\ -2 & 1 & 0 \end{bmatrix}$ (*iv*) $\frac{1}{2}\begin{bmatrix} 1 & 1 & -1 \\ 1 & -1 & 1 \\ -1 & 1 & 1 \end{bmatrix}$

**11.** (*i*) $\lambda^3 - \lambda^2 - 2\lambda = 0$ ; $-1, 0, 2$ ; $\begin{bmatrix} 1 \\ 1 \\ -1 \end{bmatrix} \begin{bmatrix} 1 \\ 0 \\ -1 \end{bmatrix} \begin{bmatrix} 4 \\ 1 \\ -3 \end{bmatrix}$ (*ii*) $\lambda^3 - 9\lambda^2 - 24\lambda - 16 = 0$, 1, 4, 4; $\begin{bmatrix} -1 \\ 1 \\ 1 \end{bmatrix} \begin{bmatrix} 0 \\ 1 \\ -1 \end{bmatrix} \begin{bmatrix} 2 \\ 1 \\ 1 \end{bmatrix}$

**12.** (*a*) $\frac{1}{4}\begin{bmatrix} 3 & 1 & -1 \\ 1 & 3 & 1 \\ -1 & 1 & 3 \end{bmatrix}$ (*b*) $\begin{bmatrix} -3/40 & 1/8 & 11/40 \\ 7/20 & -1/4 & 1/20 \\ 1/8 & -1/8 & -1/8 \end{bmatrix}$ (*c*) $\begin{bmatrix} 4/15 & 2/15 & -1/30 \\ 4/3 & -5/3 & 1/3 \\ -11/15 & 13/15 & -1/30 \end{bmatrix}$ (*d*) $\frac{1}{3}\begin{bmatrix} -3 & -2 & 2 \\ 6 & 5 & -2 \\ -6 & -2 & 5 \end{bmatrix}$

**13.** $625I$

**14.** $\lambda^3 - 6\lambda^2 + 8\lambda - 3 = 0$, $\begin{bmatrix} 124 & -123 & 162 \\ -95 & 96 & -123 \\ 95 & -95 & 124 \end{bmatrix}$

**15.** $\begin{bmatrix} 1/5 & 0 & 0 \\ 0 & 1/5 & 0 \\ 0 & 0 & 1/5 \end{bmatrix}$

**16.** $A^{-1} = \frac{1}{4}\begin{bmatrix} 1 & 6 & 6 \\ -1 & 6 & 2 \\ 1 & -10 & -6 \end{bmatrix}, A^{-2} = \frac{1}{16}\begin{bmatrix} 1 & -18 & -18 \\ -5 & 10 & -6 \\ 5 & 6 & 22 \end{bmatrix}, A^{-3} = \frac{1}{64}\begin{bmatrix} 1 & 78 & 78 \\ -21 & 90 & 26 \\ 21 & -154 & -90 \end{bmatrix}$

**17.** $\begin{bmatrix} 1 & 0 & 0 \\ 25 & 1 & 0 \\ 25 & 0 & 1 \end{bmatrix}$.

**18.** $\lambda^3 - 6\lambda^2 + 11\lambda - 6 = 0;\ 1, 2, 3\ \begin{bmatrix} 1 \\ 0 \\ 1 \end{bmatrix}\begin{bmatrix} 20 \\ 1 \\ 15 \end{bmatrix}\begin{bmatrix} 10 \\ 1 \\ 6 \end{bmatrix}$

## Problems 2.10, page 80

**3.** (a) $\begin{bmatrix} -5 & 0 \\ 0 & 2 \end{bmatrix}$ (b) $\begin{bmatrix} 0 & 0 & 0 \\ 0 & 3 & 0 \\ 0 & 0 & 15 \end{bmatrix}$

**4.** $P = \begin{bmatrix} 0 & 1 & 1 \\ -3 & 0 & 1 \\ 2 & 0 & 0 \end{bmatrix}$

**5.** $A^n = \begin{bmatrix} 2^n + 3.6^n & -3.2^n + 3.6^n \\ -2^n + 6^n & 3.2^n + 6^n \end{bmatrix}; A^4 = \begin{bmatrix} 976 & 960 \\ 320 & 336 \end{bmatrix}$

**7.** (1, 1, –1), (1, 1, –1), (2, –1, 1) ; $4x^2 + y^2 + z^2$

**8.** $x^2 + y^2 - 2z^2$

**9.** (a) 1, 2, 4 ; (1, 0, 0), (0, 1, 1), (0, 1, –1) ; $\begin{bmatrix} 1 & 0 & 0 \\ 0 & 1 & 1 \\ 0 & 1 & -1 \end{bmatrix}$

(b) $x_1^2 + 2x_2^2 + 4x_3^2$

**10.** (i) $x_1^2 + 4x_2^2 + 4x_3^2$, $\begin{bmatrix} -1/\sqrt{3} & 0 & 2/\sqrt{6} \\ 1/\sqrt{3} & 1/\sqrt{2} & 1/\sqrt{6} \\ 1/\sqrt{3} & -1/\sqrt{2} & 1/\sqrt{6} \end{bmatrix}$

(ii) $-2x_1^2 + 3y_1^2 + 6z_1^2$ ; $\begin{bmatrix} -1/\sqrt{2} & -1/\sqrt{3} & 1/\sqrt{6} \\ 0 & 1/\sqrt{3} & 2/\sqrt{6} \\ 1/\sqrt{2} & -1/\sqrt{3} & 1/\sqrt{6} \end{bmatrix}$

**11.** 2, 1

**12.** Indefinite.

## Problems 2.11, page 84

**9.** $\begin{bmatrix} 0 & 2 & 1 \\ -2 & 0 & 0 \\ -1 & 0 & 0 \end{bmatrix} + i\begin{bmatrix} 2 & 1 & -1 \\ 1 & -1 & 3 \\ -1 & 3 & 0 \end{bmatrix}$

## Problems 2.12, page 85

**1.** (b) **2.** (a) **3.** (c) **4.** (c) **5.** (c) **6.** (a)
**7.** (a) **8.** (b) **9.** (d) **10.** (a) **11.** (d) **12.** (c)
**13.** (b) **14.** (d) **15.** (c) **16.** 2 **17.** sum **18.** 0, 8
**19.** 2 **20.** $\begin{bmatrix} 3 & 2 \\ 4 & 5 \end{bmatrix}$ **21.** 0

**22.** All the eigen values are ≥ 0 and at least one eigen value is zero.

**23.** (a) $n = p$, (b) $m = p, n = q$ **24.** (a) 8 (b) –36 **25.** (b)

**26.** $\begin{bmatrix} 2 & 3 & 1 \\ 4 & 6 & 2 \\ 6 & 9 & 3 \end{bmatrix}$ **27.** $x^2 + 4xy - 4y^2$ **28.** $x = y = z = 0$ **29.** $\frac{1}{3}, \frac{1}{6}, \frac{1}{4}$

**30.** $A^2 = A$ **31.** 2 **32.** 1, 4, 9 **33.** (iv)
**34.** 4 **35.** zero **36.** Indefinite **37.** 1 – 1

**38.** The elements of its leading diagonal **39.** 2 **40.** $\lambda_i$, $i = 1, 2, ..., n$

**41.** (*c*) **42.** $A$ or $A^T$ **43.** 1/3, 1/2, 1/5 **44.** $\lambda^3 - 7\lambda^2 + 16\lambda - 12 = 0$

**45.** 1, 1/3 **46.** $x = 3 - 1$ **47.** Symmetric ; skew-symmetric **48.** 7 ; 5

**49.** $\begin{bmatrix} \cos 3\theta & \sin 3\theta \\ -\sin 3\theta & \cos 3\theta \end{bmatrix}$ **50.** (*i*) *c*, (*ii*) (*b*) **51.** $\lambda^2 - 6\lambda + 3 = 0$ **52.** Augmented matrix

**53.** $\begin{bmatrix} 4 & -1 & 3 \\ -1 & -2 & 0 \\ 3 & 0 & 1 \end{bmatrix}$ **54.** $\lambda_1^3, \lambda_2^3, \lambda_3^3$ **55.** $1/\lambda$ **56.** 3

**57.** 38 **58.** 2 **59.** Index = 2, Signature = 1

**60.** False **61.** False **62.** False **63.** True

**64.** False **65.** True **66.** False **67.** True

**68.** True **69.** True **70.** True **71.** True

**72.** True **73.** $2x^2 + 2y^2 - 2z^2 + 2yz - 4zx$.

## Problems 3.1, page 93

**1.** (*i*) $\sqrt{159}$ ; $6/\sqrt{159}, 1/\sqrt{159}, 11/\sqrt{159}$ ; (*ii*) $\sqrt{131}$ ; $9/\sqrt{131}, -7/\sqrt{131}, 1/\sqrt{131}$

**3.** 90° **5.** $x = 1, y = -1$ **11.** 2 : 1.

## Problems 3.2, page 98

**1.** (*i*) 5 **2.** (*ii*) $A, B, C$ form a $\Delta$, rt. $\angle ed$ at $C$ **3.** $\dfrac{-2}{\sqrt{30}}, \dfrac{-5}{\sqrt{30}}, \dfrac{1}{\sqrt{30}}$

**4.** $\cos^{-1}(.62)$ **6.** 2.11 **7.** 13 ; 12/13, 4/13, 3/13

**13.** 60° **14.** $\cos^{-1}\left(-\dfrac{2}{3\sqrt{21}}\right)$.

## Problems 3.3, page 102

**1.** $-\mathbf{I} - 2\mathbf{J} + 2\mathbf{K}$ ; $\dfrac{1}{\sqrt{26}}$. **3.** $(2\mathbf{J} + \mathbf{K})/\sqrt{5}$

**5.** $\frac{1}{2}\sqrt{(94)}$ **6.** (*b*) $10\sqrt{3}$ **7.** $-2/\sqrt{26}$.

## Problems 3.4, page 106

**1.** 40 **2.** 17 ; $-24\mathbf{I} + 13\mathbf{J} + 4\mathbf{K}$

**4.** 70.5 **5.** $2\mathbf{I} - 7\mathbf{J} - 2\mathbf{K}$ ; $\sqrt{(57)}$ **6.** (1, 2, 2)

**7.** 6 **8.** 8.25 **9.** $\frac{5}{6}(-3\mathbf{I} + 2\mathbf{J} + 10\mathbf{K})$, $\frac{5}{6}\sqrt{113}$.

## Problems 3.5, page 110

**1.** 7 **2.** –4 **3.** (*ii*) Yes **4.** Not linearly dependent

**5.** 5/6 **6.** (*i*) 15 ; (*ii*) $1\frac{1}{3}$ **11.** (*i*) $-7\mathbf{I} - 11\mathbf{J} + 5\mathbf{K}$ ; (*ii*) $-30\mathbf{I} - 15\mathbf{J} + 15\mathbf{K}$

**15.** (*b*) $\frac{1}{6}abc \begin{vmatrix} 1 & \cos\psi & \cos\phi \\ \cos\psi & 1 & \cos\theta \\ \cos\phi & \cos\theta & 1 \end{vmatrix}$.

## Problems 3.6, page 115

**1.** $2x - y + 3z = 9$ **2.** (i) $\begin{vmatrix} x - x_1 & y - y_1 & z - z_1 \\ x_2 - x_1 & y_2 - y_1 & z_2 - z_1 \\ x_3 - x_1 & y_3 - y_1 & x_3 - z_1 \end{vmatrix}$ (ii) $2y + 3z - 1 = 0$

**3.** $\frac{1}{3}(2\mathbf{I} + 2\mathbf{J} + \mathbf{K})$ **4.** 3 **5.** $4x - 3y + 2z = 3$

**7.** $x - 5y - 2z + 7 = 0$ **8.** 2 ; $2x + 2y - 3z = 6$ **9.** $\Sigma(x_1 - x_2)\left\{x - \frac{1}{2}(x_1 + x_2)\right\} = 0$

**10.** $y = 2$ **11.** $7x + 6y = 8z - 11 = 0$ **12.** $k = 10.2$ ; $5x - 15y - 21z = 34$

**13.** 1/6 . $2x - 3y + 6z + 5 = 0$ **14.** $\cos^{-1}(\sqrt{2}/3)$

**15.** (i) $\frac{5}{\sqrt{83}}, \frac{-7}{\sqrt{83}}, \frac{-3}{\sqrt{83}}$ (ii) 83.7° (iii) $5x - 7y - 3z + 7 = 0$

**17.** $6x + 3y - 2z = 18$ ; $2x - 3y - 6z = 6$ **20.** $x^{-2} + y^{-2} + z^{-2} = 9p^{-2}$ **21.** $xyz = 6k^3$

**22.** (i) $25x + 17y + 62z - 78 = 0$ ; (ii) $x + 35y - 10z - 156 = 0$ ; (iii) bisects the acute angle.

## Problems 3.7, page 118

**1.** $\frac{x-3}{1} = \frac{y-2}{3} = \frac{z-4}{-2}$ **2.** 43°3′ **3.** 90° **4.** $x + 2 = \frac{y-3}{2} = \frac{z-4}{2}$

**5.** 3 **6.** $\frac{x-1}{1} = \frac{y-2}{-2} = \frac{z+1}{1}$ **7.** $\frac{x-1}{1} = \frac{y+2}{-19} = \frac{z-3}{3.5}$ ; $\frac{x-1}{11} = \frac{y+1}{13} = \frac{z-3}{-21.5}$

**8.** (3, 4, 5) **9.** 4.1 **10.** 8.57 **11.** (3, 4, 5) ; (ii) (26/7, –15/7, 17/7)

**12.** 40°15′ **13.** $29x - 27y - 22z = 85$ **14.** $2 - x = y + 1 = (z + 1)/3$.

## Problems 3.8, page 120

**1.** $7x - 2y - 3z = 0$ **2.** $2x + 3y + 6z = 38$ **3.** $11x + 12y - 8z = 5$

**4.** $3y - z = 2$ **5.** $\frac{x-4}{7} = \frac{y-6}{-13} = \frac{z+2/3}{9}$ **6.** $x + y + 2z = 1, x + y + (2/5)z = 1$

**7.** $\frac{x+4/15}{-11} = \frac{y-2/5}{9} = \frac{z}{15}$ **8.** $\frac{x-2}{3} = \frac{y-1}{-1} = \frac{z-1}{1}$.

## Problems 3.9, page 123

**1.** $x - 2y + z = 0$ **2.** (5, –7, 6)

**3.** $\frac{a\alpha + b\beta + c\gamma + d}{al + bm + cn} = \frac{a'\alpha + b'\beta + c'\gamma + d'}{a'l + b'm + c'n}$ **4.** (0, 1, 2) ; $4x + y - 2z + 3 = 0$

**7.** $-\frac{1}{6}(x - 5) = y - 3 = \frac{1}{2}(z - 13)$ **8.** $\frac{x-2}{7} = \frac{y-3}{4} = \frac{z-1}{-5}$

**9.** (2, 8, – 3) ; (0, 1, 2) ; 8.83.

## Problems 3.10, page 126

**1.** 14.

**2.** $1/\sqrt{6}$ ; $11x + 2y - 7z + 0 = 0$ ; $7x + y - 5z + 7 = 0$

**3.** 10.77 ; $\frac{1}{2}(x - 3) = \frac{1}{3}(y - 5) = \frac{1}{4}(z - 7)$ ; (3, 5, 7) ; (–1, –1, –1)

**4.** $\frac{1}{\sqrt{5}}$ ; $3x - 10y + 6z - 1 = 0 = x + 2z$.

## Problems 3.11, page 128

**4.** First and second planes cut along $x - 36 = -\frac{1}{2}(y + 22) = z$.

## Problems 3.12, page 131

**1.** $x^2 + y^2 + z^2 - 4x + 6y - 2z + 5 = 0$ ; $(2, -3, -1)$ ; 3

**2.** $x^2 + y^2 + z^2 - 2x + 2y - 4z = 0$ ; $(1, -1, 2)$ ; $\sqrt{6}$

**3.** (a) $x^2 + y^2 + z^2 - 4x - 4y - 4z + 3 = 0$ (b) $3(x^2 + y^2 + z^2) - 2(x + y + z) - 1 = 0$

**7.** (i) $x^2 + y^2 + z^2 - ax - by - cz = 0$,

(ii) $x^2 + y^2 + z^2 - ax - by - cz = 0$ and $\frac{x}{a} + \frac{y}{b} + \frac{z}{c} = 1$ ; $\left[\frac{a(b^{-2} + c^{-2})}{2\Sigma a^{-2}}, \frac{b(c^{-2} + a^{-2})}{2\Sigma a^{-2}}, \frac{c(a^{-2} + b^{-2})}{2\Sigma a^{-2}}\right]$

**8.** $(1, 3, 4)$ ; $\sqrt{7}$

**10.** $x^2 + y^2 + z^2 + 2(x + y + z + 1) = 0$

**11.** $13(x^2 + y^2 + z^2) - 35x - 21y + 43z + 176 = 0$

**12.** $3(x^2 + y^2 + z^2) - 7x - 8y + z + 10 = 0$

**13.** $x^2 + y^2 + z^2 + 7y - 8z + 24 = 0$.

## Problems 3.13, page 133

**1.** (i) $x + 4y + 8z - 3 = 0$, $x + 4y + 8z - 57 = 0$; $(5/3, -1/3, 1/3)$, $(7/3, 7/3, 17/3)$

(ii) $x + 2y + 2z = 9$, $2x + y - 2z = 9$

**2.** $x^2 + y^2 + z^2 + 2x + 4y + 6z - 11 = 0$ and $5(x^2 + y^2 + z^2) - 4x - 8y - 12z - 13 = 0$

**4.** $x^2 + y^2 + z^2 - 14(x + y + z) + 98 = 0$

**6.** $\frac{x - 0.6}{-2} = \frac{y - 2.4}{7} = \frac{z}{5}$

**8.** $3\sqrt{6}, \sqrt{6}$

**9.** $3x + y + z + 6 = 0$

**10.** $x^2 + y^2 + z^2 - x - y - z - 8 = 0$

**11.** $(12/5, 4, 9/5)$.

## Problems 3.14, page 137

**1.** $(\beta z - \gamma y)^2 = 4a(\alpha z - \gamma x)(z - \gamma)$

**2.** $528x^2 + 363y^2 + 76z^2 - 528xy - 264yz + 353zx + 704x + 1352z - 4436 = 0$

**3.** $5x^2 + 3y^2 + z^2 - 2xy - 6yz - 4zx + 6x + 8y + 10z - 26 = 0$

**4.** $x^2 + y^2 - 3z^2 - 2x - 2y + 6z - 1 = 0$

**5.** $x^2 + y^2 = z^2 \tan^2 \alpha$

**6.** $x^2 + 7y^2 + z^2 + 8xy + 8yz - 16zx = 0$

**7.** $4x^2 + 40y^2 + 19z^2 - 48xy - 72yz + 36zx = 0$

**8.** $4x^2 - 9y^2 + 4z^2 + 30y - 25 = 0$

**9.** $2(-x + y + z)^2 = 7(x^2 + y^2 + z^2)$

**10.** $yz \pm zx \pm xy = 0$, $\cos^{-1}(1/\sqrt{3})$ ; $x = y/\pm 1 = z/\pm 1$

**11.** $\cos^{-1} 4/\sqrt{(41)}$ ; $25x^2 - 16y^2 - 16z^2 = 0$

**12.** $4x^2 + 4y^2 - z^2 + 20z - 100 = 0$

**13.** $x = y/2 = z/-1$ ; $x/-2 = y = z$

**15.** $-2x^2 + y^2 - 2z^2 + 4xy - 8xz + 4yz + 8x - 10y + 8z - 3 = 0$.

## Problems 3.15, page 140

**1.** $5x^2 + 8y^2 + 5z^2 + 4yz + 8xz - 4xy - 144 = 0$

**2.** (i) $3x^2 + 6y^2 + 3z^2 + 8yz - 2zx + 6x - 24y - 18z + 24 = 0$

(ii) $18x^2 + 9y^2 + 12z^2 + 12yz + 24zx - 48x - 24y - 48z + 21 = 0$

**3.** $45x^2 + 40y^2 + 13z^2 + 12xy + 36yz - 24zx - 42x - 112y - 126z - 392 = 0$

**4.** $(a^{-2} + b^{-2} + c^{-2})(x^2 + y^2 + z^2 - ax - by - cz) = \left(\frac{x}{a} + \frac{y}{b} + \frac{z}{c} - 1\right)\left(\frac{x}{a} + \frac{y}{b} + \frac{z}{c} - 2\right)$

**5.** $9x^2 + 5y^2 + 9z^2 + 12xy + 6yz - 36x - 30y - 18z + 36 = 0$ ; $\pi$ units

**6.** $x^2 + y^2 - 2x - 4y - 11 = 0$

**7.** $10x^2 + 13y^2 + 5z^2 - 6yz - 12zx - 4xy - 36x - 18y + 30z = 135$.

**8.** $\left(\frac{lx}{a^2} + \frac{my}{b^2} + \frac{nz}{c^2}\right)^2 = \left(\frac{l^2}{a^2} + \frac{m^2}{b^2} + \frac{n^2}{c^2}\right)\left(\frac{x^2}{a^2} + \frac{y^2}{b^2} + \frac{z^2}{c^2} - 1\right)$

## Problems 3.16, page 145

1. Ellipsoid, 33.51
2. Hyperboloid of revolution of one sheet ; Hyperbola $5x^2 - y^2 = 6$. No area
3. Right circular cylinder with axis along $z$-axis
4. Hyperbolic paraboloid
5. Hyperboloid of two sheets
6. Parabolic cylinder
7. Right circular cylinder
8. Cone with vertex at the origin
9. Hyperbolic paraboloid
10. Cone

## Problems 3.17, page 145

1. (*b*) 2. (*a*) 3. (*c*) 4. (*b*) 5. (*b*) 6. (*a*)
7. (*c*) 8. (*d*) 9. (*c*) 10. (*c*) 11. (*d*) 12. (*c*)
13. (*b*) 14. (*c*) 15. (*c*) 16. (*c*) 17. (*c*) 18. (*b*)
19. (*c*) 20. (*c*) 21. (*b*) 22. (*b*) 23. (*c*) 24. (*c*)
25. (*b*) 26. (*a*)
27. $1/\sqrt{3}, 1/\sqrt{3}, 1/\sqrt{3}$
28. 0, 0, 1
29. $x = 0, y = 0$
30. (–3, 2, –1)
31. 8 or –10
32. $\begin{bmatrix} l_1 & m_1 & n_1 \\ l_2 & m_2 & n_2 \\ l_3 & m_3 & n_3 \end{bmatrix} = 0$
33. $y^2 + z^2 = (bx/a)^2$
34. $12x + 31y - 20z = 66$
35. 523.6
36. $\cos^{-1}(6/\sqrt{42})$
37. $\frac{x^2}{4} + \frac{y^2}{9} + z^2 = (x + y + z)^2$
38. (3, 5, 7), (5, 8, 11)
39. $-(x - 1) = y - 2 = \frac{1}{6}(z - 3)$
40. $x^2 + y^2 + z^2 - 2x - 4y + 2z + 16 = 0$
41. $al + bm + cn = 0,\ ax_1 + by_1 + cz_1 + d = 0$
42. $2/\sqrt{26}$
43. (3/2, –2, 2), $3\sqrt{5}/2$
44. $\sqrt{44}/3$
45. $\frac{x + 1/3}{1} = \frac{y + 2/3}{-2} = \frac{z}{1}$
46. Parabolic cylinder
47. Hyperboloid of two sheets
48. $\cos^{-1}\left(\frac{7}{\sqrt{84}}\right)$
49. 6, –4, 12
50. 6
51. True
52. True
53. True
54. Elliptic cylinder
55. $4(x^2 + y^2 + z^2) + 9(xy + yz + zx) = 0$
56. $(\mathbf{I} - 2\mathbf{J} - 8\mathbf{K})/\sqrt{69}$
57. $x + 3y + 5z + 5\sqrt{35} = 0$.

## Problems 4.1, page 150

6. $8t^3/(1 - t^2)^3$
7. $-\frac{3}{2}$
8. $\sin t/a \cos^4 t$.

## Problems 4.2, page 153

1. $(-1)^{n-1}(n-1)!\,2^n\,[(2x+1)^{-n} + (2x-1)^{-n}]$
2. $(-1)^n\left\{\frac{n!}{(x+1)^{n+1}} - \frac{(n-1)!}{(x+2)^n} + \frac{(n-1)!}{(x+1)^n}\right\}$
3. $\frac{1}{2}\left[5^n \cos(5x + n\pi/2) + \cos(x + n\pi/2)\right]$
4. $x^3 \cos\left(x + \frac{n\pi}{2}\right) + 3nx^2 \cos\left\{x + (n-2)\frac{\pi}{2}\right\} + 3\,n(n-1)x \cos\left\{x + (n-2)\frac{\pi}{2}\right\} + n(n-1)(n-2) \cos\left\{x + (n-3)\frac{\pi}{3}\right\}$,

**5.** $e^x[(2x+3)^3 + 6n(2x+3)^2 + 12n(n-1)(2x+3) - 8n(n-1)(n-2)]$,

**6.** $\frac{(20)^{n/2}}{2}[e^{2x}\sin(2x + n\tan^{-1}2) - e^{-2x}\sin(4x - n\tan^{-1}2)]$

**7.** $\frac{1}{2}e^{5x}S(41)^{n/2}\cos[4x + n\tan^{-1}(0.8)] + (29)^{n/2}\cos[2x + n\tan^{-1}(0.4)]$

**8.** $(-1)^n n!\left\{\frac{16}{(x-2)^{n+1}} - \frac{1}{(x-1)^{n+1}}\right\}$ **9.** $(-1)^n n!\left\{\frac{9(2)^{n-1}}{(2x+3)^{n+1}} - \frac{4}{(x+2)^{n+1}}\right\}$

**10.** $\frac{(-1)^n n!}{3}\left\{\frac{4}{(x+1)^{n+1}} + \frac{i-1}{4}\frac{1}{(x+i)^{n+1}} - \frac{i+1}{4}\frac{1}{(x-i)^{n+1}}\right\}$

**11.** $a^{n+2}e^{ax}x^2$,

**12.** $2(-1)^{n-1}(n-1)!\sin n\alpha \sin^n \alpha$, where $\alpha = \cot^{-1}x$.

## Problems 4.3, page 156

**1.** (i) $\frac{(-1)^{n-3}(n-3)!}{x^n}\{(n-1)(n-2) + n(3-n)x^2\}$ ;

(ii) $x^2\left\{5^n e^{3x}\sin\left(4x + n\tan^{-1}\frac{4}{5}\right)\right\} + n(2x)\left\{5^{n-1}e^{3x}\sin(4x + (n-1)\tan^{-1}\left(\frac{4}{5}\right)\right\} + n(n-1)$

$\left(\left\{5^{n-2}e^{3x}\sin 4x + (n-2)\tan^{-1}\frac{4}{5}\right\}\right)$

**3.** $y_{2m}(0) = 0, y_{2m+1}(0) = (-1)^m . (2m)!$

**5.** $(y_n)_o = 0$, if $n$ is even

$= m(1^2 - m^2)(3^2 - m^2) \ldots\ldots [(2n-1)^2 - m^2]$, if $n$ is odd

**6.** $(y_{2n})_0 = e^{m\pi/2}m^2(2^2 + m^2)(4^2 + m^2) \ldots\ldots [(2n-2)^2 + m^2]$

$(y_{2n+1})_o = -e^{m\pi/2}m(1^2 + m^2)(3^2 + m^2) \ldots\ldots [(2n-1)^2 + m^2]$

**11.** $\{m^2 - (n-2)^2\}\{m^2 - (n-4)^2\} \ldots\ldots (m^2 - 2^2)m^2$, $n$ even

$\{m^2 - (n-2)^2\}\{m^2 - (n-4)^2\} \ldots\ldots (m^2 - 1^2)m$, $n$ odd.

**14.** $3^3(3x^2 - 4)\sin 3x - 6^3 x\cos 2x$

**15.** $\frac{5!}{x^6}\left(1 + \frac{1}{2} + \frac{1}{3} + \frac{1}{4} + \frac{1}{5} - \log x\right)$

## Problems 4.4, page 162

**2.** $x = (2m-1)a/(2m+2n-1)$

**3.** (i) $c = 3.154, 0.846$ ; (ii) $c = \pi/2$ ; (iii) $c = e - 1$. (iv) $c = 0.5413$

**6.** 0.36 **12.** 0.25.

## Problems 4.5, page 167

**1.** $x \frac{x^2}{2} + \frac{x^3}{3} - \frac{x^4}{4} - + \ldots\ldots \infty$ **2.** $x + \frac{1}{2}\cdot\frac{x^3}{3} + \frac{1.3}{2.4}\cdot\frac{x^5}{5} + \frac{1.3.5}{2.4.6}\cdot\frac{x^7}{7} + \ldots\ldots \infty$

**3.** $x \frac{x^3}{3} + \frac{x^5}{5} - \frac{x^7}{7} - + \ldots\ldots \infty$ **4.** $1 + x \frac{x^2}{2} + \frac{x^3}{6} + \frac{x^4}{24} - + \ldots\ldots$

**5.** $\frac{x^2}{2} + \frac{x^4}{12} + \frac{x^6}{45} + \ldots\ldots$ **6.** $x + x^2 + \frac{x^3}{3} - \frac{x^5}{30} \ldots\ldots$

**24.** $\frac{m\sin\theta}{1!} - \frac{m(m^2 - 1^2)}{3!}\sin^3\theta + \frac{m(m^2-1^2)(m^2-3^2)}{5!}\sin^5\theta - \ldots\ldots$

**25.** $4 + 21(x-1) + 13(x-1)^2 + 2(x-1)^3$

**26.** (i) $e\left\{1+\frac{(x-1)}{1!}+\frac{(x-1)^2}{2!}+\frac{(x-1)^3}{3!}+\ldots\ldots\right\}$ (ii) $\frac{\pi}{4}+\frac{1}{2}(x-1)-\frac{1}{4}(x-1)^2+\frac{1}{12}(x-1)^3$

**27.** $1 \frac{(x-\pi/2)^2}{2!}+\frac{(x-\pi/2)^4}{4!}-\ldots\ldots$ ; 0.9998

**29.** (i) $\frac{1}{x}=1-(x-1)+(x-1)^2-(x-1)^3+(x-1)^4\ldots\ldots$

(ii) $\log(0.5)-\sqrt{3}(x-\pi/3)-2(x-\pi/3)^2-\frac{4\sqrt{3}}{3}(x-\pi/3)^3+\ldots\ldots$

**30.** 0.8482 **31.** (i) 2.6121. (ii) 1.12.

## Problems 4.6, page 170

**1.** $\log_e(a/b)$ **2.** –1/3 **3.** 1/3 **4.** $a \log a$ **5.** 1 **6.** 1/18
**7.** 1/2 **8.** 2 **9.** 3/2 **10.** 0 **11.** 1/30 **12.** 1
**13.** 1/3 **14.** 2 **15.** 1 **16.** 1 **17.** $11e/24$
**18.** $a = 2$ ; 1 **19.** $a = 5, b = -5$ **20.** $a = 1, b = 2, c = 1$.

## Problems 4.7, page 172

**1.** –1/3 **2.** 1/2 **3.** –2 **4.** –1/3 **5.** 2/3 **6.** $1/e$
**7.** $ae$ **8.** 1 **9.** $e$ **10.** $1/\sqrt{e}$ **11.** $1/e$ **12.** 0
**13.** 0 **14.** 1 **15.** $e^{-1/6}$ **16.** $e$ **17.** $e^{2/\pi}$ **18.** $e^{1/12}$
**19.** $-\frac{1}{2}$ **20.** $(6)^{1/3}$ **21.** $(abcd)^{1/4}$.

## Problems 4.8, page 176

**1.** (a) $x - 2y + 2 = 0$ ; (b) $x - 20y = 7$ ; $20x + y = 140$
**2.** $(a, b)$ **10.** $\pi/4$
**14.** $T = 2a \sin t/2$ ; $N = 2a \tan t/2. \sin t/2$ ; $S.T. = a \sin t$ ; $S.N. = 2a \sin^2 t/2. \tan t/2$
**15.** $a \sin^3\theta \tan\theta$.

## Problems 4.9, page 178

**7.** (i) $\pi/2$ ; (ii) $\pi/2$ (iii) $\pi/3$.

## Problems 4.10, page 182

**4.** $r^3 = 2ap^2$ **5.** $r^3 = a^2p$ **6.** $pa^m = r^{m+1}$ **7.** $r^{m+1} = \sqrt{2}\, a^m p$
**8.** $\frac{1}{p^2}=\frac{1}{r^2}\left\{2-\frac{r^{2n}}{a^{2n}}\right\}$ **9.** $(1 + m^2)p^2 = r^2$ **10.** (i) $\sqrt{(1 + 9x/4a)}$ ; (ii) $\cosh x/c$
**11.** $a\theta$ **12.** (i) $2a \sin\theta/2$ ; (ii) $a\sqrt{(\sec 2\theta)}$ ; (iii) $r\sqrt{(8r-3)}$.

## Problems 4.11, page 188

**1.** (i) $2a(1 + t^2)^{3/2}$ ; (ii) $y^2/c$ (iii) $(1 + a^3)^{3b}/6a^2$
**5.** (i) $(a^2\sin^2\theta + b^2\cos^2\theta)^{3/2}/ab$ ; (ii) $4a \sin\theta/2$ ; (iii) $at$
**12.** (i) 3/2 ; (ii) 1 ; (iii) $\sqrt{2a}$
**13.** (i) $\frac{4a}{3}\sin\frac{\theta}{2}$ ; (ii) $a^n/(n+1)r^{n-1}$ (iii) $\rho$ varies as $r^{1-n}$
**15.** $2\sqrt{(r^3/a)}$

## Problems 4.12, page 192

**1.** $a(2+3t)t^2, -4\sqrt{2}\,at^{3/2}$ **4.** (*i*) $x = a(t - \sin t), y - 2a = a(1 + \cos t)$, (*ii*) $x = a\cos\theta, y = a\sin\theta$

**5.** $(x+y)^{2/3} + (x-y)^{2/3} = 2a^{2/3}$

**7.** (*i*) $(x - 3a/4)^2 + (y + 3a/4)^2 = a^2/2$; (*ii*) $x^2 + y^2 \dfrac{21}{8} - (x+y) + \dfrac{432}{128} = 0$

**11.** $y^2 = 4ax$ **12.** $(x/a)^2 + (y/b)^2 = 1$ **13.** $27ay^2 = 4(x - 2a)^3$.

**14.** $(x/a)^2 + (y/b)^2 = 1$ **15.** $y = \dfrac{u^2}{2g} - \dfrac{gx^2}{2u^2}$

**16.** (*i*) $\sqrt{x} + \sqrt{y} = \sqrt{c}$ ; (*ii*) $4xy = c^2$ ; (*iii*) $x^{2/3} + y^{2/3} = c^{2/3}$ **17.** $x^{2/3} + y^{2/3} = c^{2/3}$

## Problems 4.13, page 197

**2.** $a = 1, b = 1/4$, Point of minima **4.** $x = 0.42l$ **5.** $v = (aw^2/3b)^{1/4}$

**8.** $\theta = \dfrac{\pi}{4} - \dfrac{\alpha}{2}$; $(1 - \sin\alpha)/(1 + \sin\alpha)$ **10.** Sq. with side $\sqrt{2}a$

**13.** $8 + 2\sqrt{7}, 2 + 2\sqrt{7}, 5 - \sqrt{7}$ **14.** Depth is half the width **15.** $(a^{2/3} + b^{2/3})^{3/2}$

**16.** $3\sqrt{3}a/4$ **25.** 2.5 km/hr.

## Problems 4.14, page 201

**1.** $x + y + a = 0$ **2.** $x = \pm a, y = \pm b$ **3.** $x = \pm a, y = \pm b$

**4.** $y = 0$ ; $x + 1 = 0$ ; $x + y = 0$ **5.** $y = x, y + 2x = 0, y + 2x + 1 = 0$

**6.** $x + a = 0$ ; $x - a = 0$ ; $x - y + \sqrt{2}a = 0$ ; $x - y\ \sqrt{2}a - = 0$

**7.** $x + 2y + 2 = 0, x + y = \pm\ 2\sqrt{2}$ **10.** $r\cos\theta = a$ ; $r\cos\theta = -a$ **11.** $r\cos\theta = 0$ ; $r\cos\theta = 2a$

**12.** $r\sin\theta = 2$ **13.** $r\sin(\theta - m\pi/n) = a/n\cos m\pi$.

## Problems 4.15, page 213

**1.** 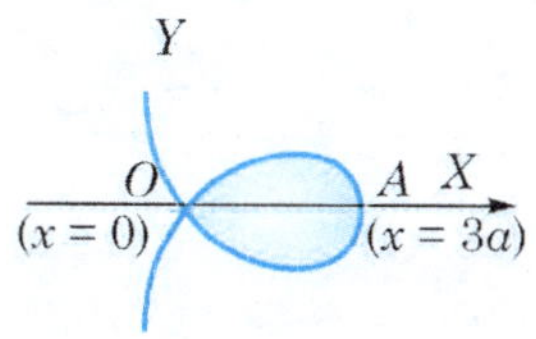


**2.** 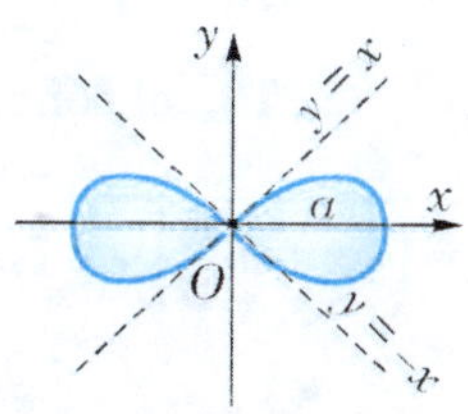


**3.** 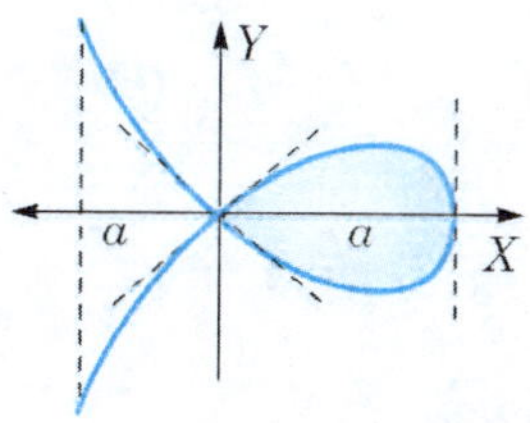


**4.** 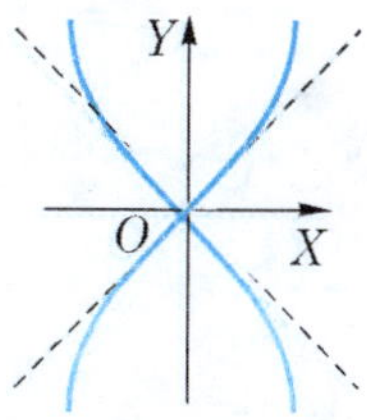


**5.** 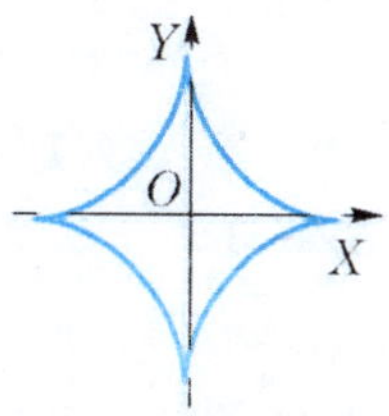


**6.** 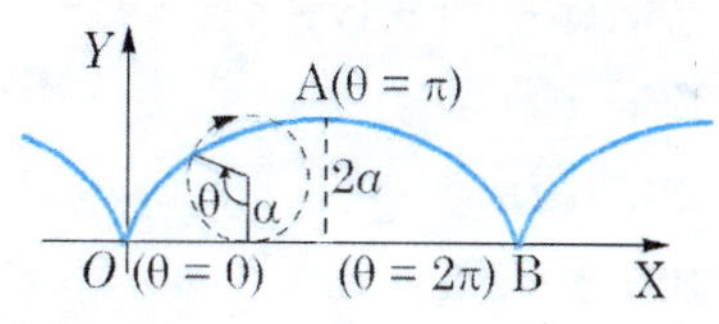


**7.** 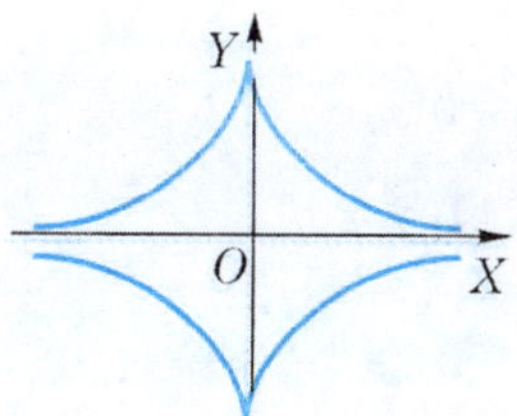


**8.** 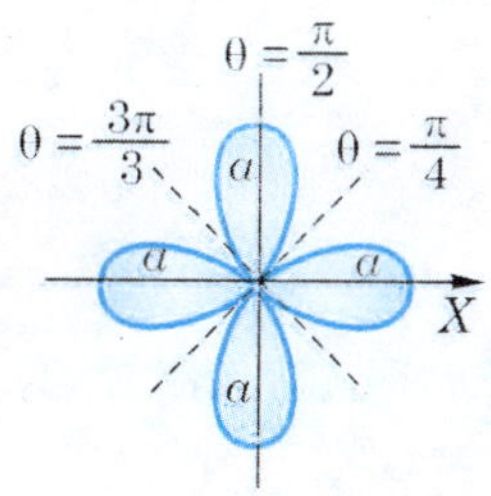


**9.** 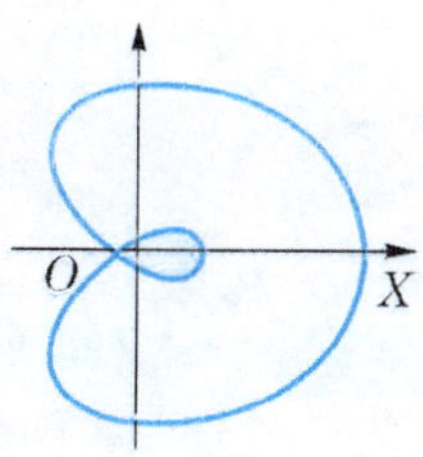

12. 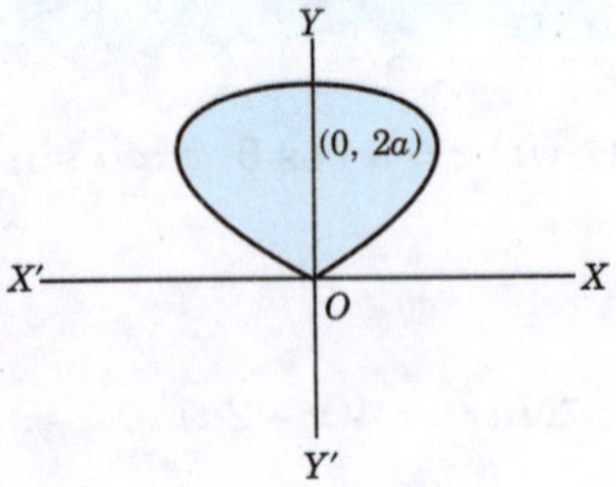


13. 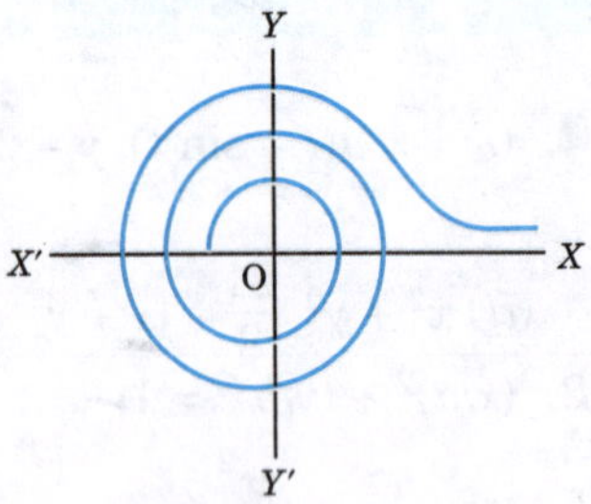


14. 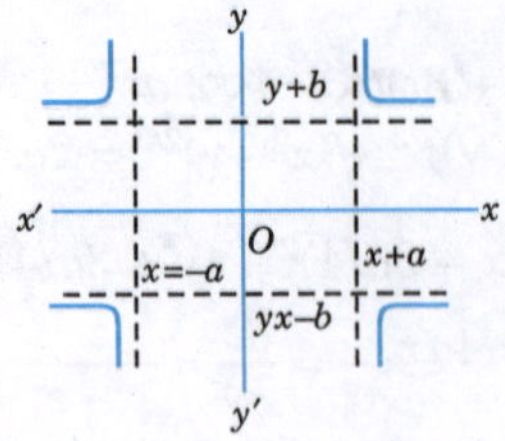


15. 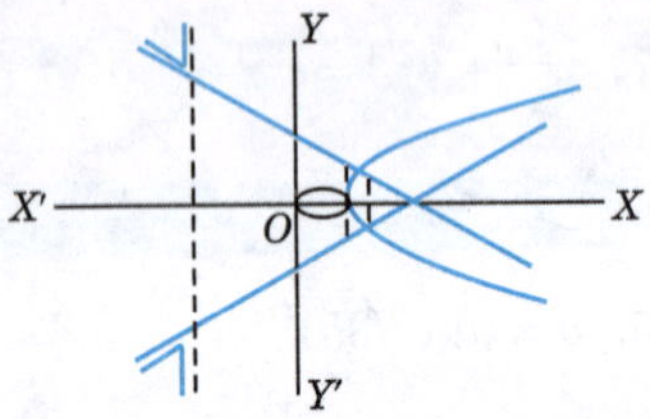


16. 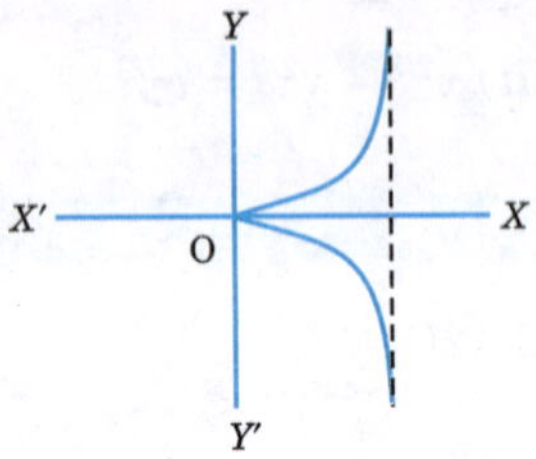


17. 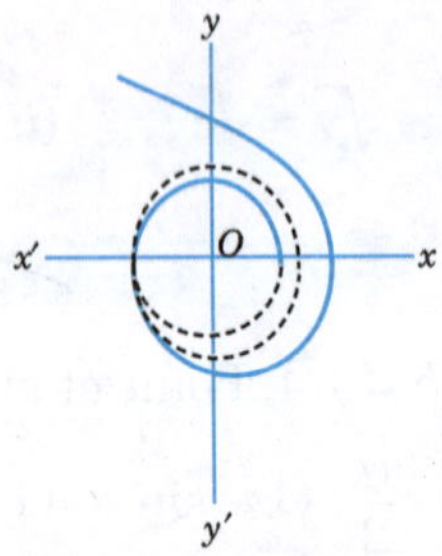


18. 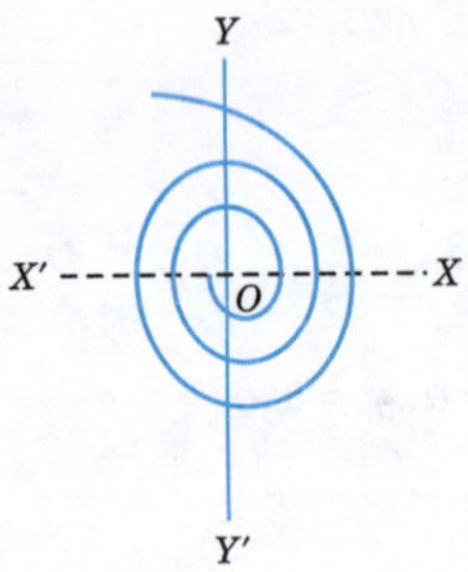


## Problems 4.16, page 214

**1.** $c$ **2.** $x^2 + 4ay = 0$ **3.** 1/5 **4.** $(c)$ **5.** $(a)$ **6.** $(b)$

**7.** $(c)$ **8.** $(b)$ **9.** $(b)$ **10.** $(c)$ **11.** $(b)$ **12.** $(c)$

**13.** $(c)$ **14.** $(b)$ **15.** $x^2 = 4y$ **16.** of constant length

**17.** $x + \frac{x^3}{3!} + \frac{x^5}{5!} + \ldots$ **18.** $\frac{1}{2}\left[8^n \cos\left(8x + n\frac{\pi}{2}\right) + 4^n \cos\left(4x + n\frac{\pi}{2}\right)\right]$

**19.** $-32/3a$ **20.** True **21.** $-a$ **22.** $\sqrt{13}$

**23.** $(x - a)^2 + (y - b)^2 = k^{-2}$ **24.** envelope **25.** $xy = c^2$

**26.** $\alpha$ **27.** $2a$ **28.** $e^x(x^3 - 12x^2 - 36x - 24)$

**29.** $(iii)$ **30.** $(x/a)^2 + (y/b)^2 = 1$ **31.** (B)

**32.** $c = 2.5$ **33.** $x = y$ **34.** node

**35.** Four loops of $r = a \sin 2\theta$ and three loops of $r = a \cos 3\theta$.

**36.** $y = \pm x$ **37.** $x = 4$ **38.** $4b$

**39.** (A) **40.** $r > a$ **41.** (D)

**42.** (D) **43.** (C) **44.** $x = \pm 1, y = \pm 1$.

## Problems 5.1, page 218

**1.** 2/3 **2.** Does not exist **3.** Zero **4.** Does not exist

**7.** Discontinuous.

## Problems 5.2, page 222

**1.** (*i*) $xy(2 - \cos xy) - \sin xy$ ; $x^2(1 - \cos xy)$ ;

(*ii*) $2x/(x^2 + y^2)$, $2y/(x^2 + y^2)$ ;

(*iii*) $(x^2 + 2xy - y^2)/\{(x^2 + y^2)^2 + (x + y)^2\}$ ; $(y^2 + 2xy - x^2)/\{(x^2 + y^2)^2 + (x + y)^2\}$ ;

(*iv*) $\dfrac{\partial z}{\partial x} = \dfrac{\partial z}{\partial y} = \dfrac{z}{1 - z}$

**11.** $n = 2, -3$

**18.** $e^{xyz}(x^2y^2z^2 + 3xyz + 1)$.

## Problems 5.4, page 230

**14.** $2u$.

## Problems 5.5, page 233

**2.** $-2/(e^{2t} + e^{-2t})$

**3.** $e^{3 \cos t + 2t^2}\,(4t - 3 \sin t)$

**4.** 6.5 sq. ft./sec

**6.** $2\rho(3 \sin^2 \phi + 1.)$

**7.** $8e^{4t}$.

**12.** $\dfrac{e^t}{\log t}\left(1 - \dfrac{1}{t \log t}\right)$

**13.** $3 \sin t \cos t\,(b^3 \sin t - a^3 \cos t)$

**14.** (*a*) $-\dfrac{b^2 x}{a^2 y}$, (*b*) $\dfrac{y}{x}\,\dfrac{2x + e^{xy} - \cos xy}{\cos xy - e^{xy} - x}$, (*c*) $\dfrac{e^x - 2^y}{2x - e^y}$ )

**15.** (*a*) $-\dfrac{(hx + by)^2 a - 2(ax + hy)(hx + by)h + (ax + hy)^2 b}{(hx + by)^3}$

(*b*) $-\dfrac{2a^2x^2}{y^5}$,

(*c*) $-\left[\dfrac{3x^2(y^3 - a^2x)^2 + 2a^2(x^3 - a^2y)(y^3 - a^2x) + 3y^2(x^3 - a^2y)^2}{(y^3 - a^2x)^3}\right]$

## Problems 5.6, page 239

**8.** 0 ; 0.

## Problems 5.7, page 244

**1.** $\dfrac{1}{1 + u + v}$

**4.** 20

**8.** Zero

**9.** $x(yv + 1 - w) + z - 2uv$

**12.** 0 ; $u = \tan v$

**14.** $w = v - 2u^2$

**15.** $-\dfrac{2(x - y)(y - z)(z - x)}{(u - v)(v - w)(w - u)}$.

**17.** $r^2 \sin \theta\, dr\, d\theta\, dz$

**19.** $1/2\,(x - y)(y - z)(z - x)$

## Problems 5.8, page 246

**1.** $4X + Y + Z = 6$ ; $\dfrac{X - 2}{4} = Y - 1 = Z + 3$

**2.** $3Y + 2Z - X - 3 = 0$, $1 - X = \dfrac{Y - 2}{3} = \dfrac{Z + 1}{2}$

**3.** $\dfrac{X}{x_1} + \dfrac{Y}{y_1} + \dfrac{Z}{z_1} = 3$ ; $x_1(X - x_1) = y_1(Y - y_1) = z_1(Z - z_1)$

**4.** $7X - 3Y + 8Z = 26$ ; $\dfrac{X - 1}{7} = \dfrac{Y + 1}{-3} = \dfrac{Z - 2}{8}$

**5.** $(-1, 2, 2/3)$

**7.** $\dfrac{X - x}{x} = \dfrac{Y - y}{y} = \dfrac{Z - z}{z}$.

## Problems 5.9, page 252

1. (*i*) $x - \frac{1}{6}(x^3 + 3xy^2)$

(*ii*) $\frac{1}{2\sqrt{2}}\left[1 + \{(x+1) + (y - \pi/4)\} + \frac{1}{2}\{(x+1)^2 - 2(x+1)(y-\pi/4) + (y-\pi/4)^2\}\right.$
$\left. + \frac{1}{6}\{(x+1)^3 + 3(x+1)^2(y-\pi/4) - 3(x+1)(y-\pi/4)^2 - (y-\pi/4)^3\} + \ldots\ldots\right]$

2. $1 + (x-1) + (x-1)(y-1) + \frac{1}{2}(x-1)^2(y-1) + \ldots$ 3. 2%

4. (*a*) $8\pi rx$ (*b*) 2% 7. $\frac{\alpha}{2}\cot\alpha + 2$

8. $-1\frac{1}{3}\%$ 10. 2.058% 15. 3% 16. $5r\%$

## Problems 5.10, page 265

1. (*i*) $(a, a)$ gives maximum if $a < 0$ and minimum if $a > 0$
(*ii*) Max. at $(\pm 1, 0)$ ; Min. at $(0, \pm 1)$ (*iii*) Max. at (4, 0), Min. at (6, 0)

2. 4, 2, 1 3. (*i*) $3a^2$ ; (*ii*) $p^2/(a^2 + b^2 + c^2)$ ; (*iii*) $3a^2$

4. $\frac{\sqrt{3}}{6}\lambda, \frac{\sqrt{4}}{6}\lambda, \frac{\sqrt{5}}{6}\lambda$ where $\lambda = (\sqrt{3} + \sqrt{4} + \sqrt{5})$ 5. $12 \times 12 \times 6$ cm

7. $(0, 0, \pm 1)$ 9. 4, 1 10. 50

11. 4, 8, 12 12. Two stationary values of $u$ are given by $\frac{l^2}{au-1} + \frac{m^2}{bu-1} + \frac{n^2}{cu-1} = 0.$

13. $\left(\frac{1}{2}, \frac{1}{4}\right); \left(\frac{11}{8}, \frac{-5}{8}\right); \frac{7\sqrt{2}}{8}$ 14. $a = \frac{\sqrt{5}}{2}b, h = \frac{1}{2}b$ 16. 50 units

17. Cube 18. $\left(I_1 : I_2 : I_3 = \frac{1}{R_1} : \frac{1}{R_2} : \frac{1}{R_3}\right)$

## Problems 5.11, page 268

1. $\frac{1}{2a^3}\tan^{-1}\frac{x}{a} + \frac{x}{2a^2(x^2 + a^2)}$ 2. $\frac{(-1)^n\, n!}{(m+1)^{n+1}}$

3. $\pi \log\left[\frac{1}{2} + \frac{1}{2}\sqrt{(1 - a^2)}\right]$ 4. $-\pi/(a^2 - 1)^{3/2}$ 15. $\frac{\sqrt{\pi}}{2}e^{-2a}$

## Problems 5.12, page 269

1. zero 2. (*a*) 3. 1 4. (*d*) 5. (*b*) 6. (*b*)
7. $y^x \log y$ ; $xy^{x-1}$ 8. (*c*) 9. (*d*) 10. (*d*) 11. (*b*) 12. (*d*)
13. (*a*) 14. (*d*) 15. (*c*) 16. (*b*) 17. zero 18. $2/(x + y)$
19. $rt - s^2 < 0$ 20. (*d*) 21. $4u$ 22. $\partial(u, v)/\partial(x, y)$
23. saddle point 24. (*c*) 25. $\frac{\partial u}{\partial x}\cdot\frac{dx}{dt} + \frac{\partial u}{\partial y}\cdot\frac{dy}{dt}$ 26. (*c*)
27. –1 28. (*c*) 29. equal 30. False.

## Problems 6.1, page 278

1. (*i*) 128/315 ; (*ii*) 8/45 2. (*i*) 128/315 ; (*ii*) $11\pi/192$

3. (*i*) $\frac{(2n-3)(2n-5)\ldots\ldots 3.1}{(2n-2)(2n-4)\ldots\ldots 4.2}\frac{\pi}{2}$ (*ii*) $\frac{3\pi}{1024}$ 4. (*i*) $35\pi/10240$ ; (*ii*) 16/495

**5.** (*i*) $3\pi/512$ (*ii*) $1/144$ **6.** (*i*) $5\pi/256$ ; (*ii*) $1/15$ **7.** (*i*) $35\pi a^4/8$ ; (*ii*) $5\pi a^3/2$

**8.** (*i*) $5\pi/8$ ; (*ii*) $28\pi$. (*iii*) $\dfrac{3\pi}{128}$ (*iv*) $12\pi$.

## Problems 6.2, page 280

**1.** (*i*) $\frac{1}{5}\tan^5 x\ \frac{1}{3} - \tan^3 x + \tan x - x;$ (*ii*) $\frac{1}{4} - \cot^4 x + \frac{1}{2}\cot x + \log \sin x$

**3.** $\frac{1}{2}\log 2 - \frac{1}{4}$ **4.** $\frac{\pi}{4} - \frac{2}{3}$ **5.** $I_n = \dfrac{(2)^{(n-2)/2}}{n-1} + \dfrac{n-2}{n-1} + I_{n-2}$

**6.** (*i*) $\frac{1}{5}\sec^4 x \,.\, \tan x + \frac{4}{15}(\sec^2 x + 2)\tan x;$ (*ii*) $\frac{11\sqrt{3}}{4} + \frac{3}{8}\log(2+\sqrt{3})$;

(*iii*) $\frac{1}{4} - \cot x \operatorname{cosec} x\ \frac{3}{8} - \cot x \operatorname{cosec} x + \frac{3}{8}\log(\operatorname{cosec} x - \cot x)$

**7.** $\left\{\frac{67\sqrt{2}}{48} + \frac{5}{16}\log(1+\sqrt{2})\right\} a^6$ **8.** $\frac{t^5}{5} - \frac{t^3}{3} + t - \tan^{-1} t$.

## Problems 6.3, page 283

**1.** $e^x(1 - x + x^2 - x^3 + x^4)$

**3.** $\int x^m (\log x)^n\, dx = \dfrac{x^{m+1}}{m+1}(\log x)^n - \dfrac{n}{m+1}\int x^m (\log x)^{n-1}\, dx\ ;\int_0^1 x^5 (\log x)^3\, dx = -1/216$

**5.** $149/225$ **6.** $3\pi^2/64 - 1/4$ **7.** $\frac{5}{16}\pi^4 - 15\pi^2 + 120$

**11.** $I_n = \dfrac{e^{ax}\cos^{n-1} x(a\cos x + n\sin x)}{a^2+n^2} + \dfrac{n(n-1)}{a^2+n^2} I_{n-2}\ ;\int_o^{\pi/2} e^{2x}\cos^3 x\, dx = \frac{2}{65}(3e^{\pi} - 11)$

**12.** $24/85$.

## Problems 6.4, page 288

**6.** (*i*) $3\pi/8$ ; (*ii*) $\dfrac{32}{35}$ (*iii*) $3\pi/256$ ; (*iv*) $15\pi/640$

**7.** (*i*) $\frac{1}{\sqrt{2}}\log\left(1+\sqrt{2}\right)$, (*ii*) $\frac{\pi}{2\sqrt{2}}\log\left(1+\sqrt{2}\right)$ **8.** (*i*) $16\pi/35$ ; (*ii*) $8\pi/315$.

**12.** (*i*) $\frac{\pi}{4}$, (*ii*) $\dfrac{\pi(\pi-2)}{2}$

## Problems 6.5, page 291

**1.** $\log 2$ **2.** $\frac{1}{3}\log 2$ **3.** $1/3$ **4.** $\pi/2$ **5.** $\frac{1}{4}\log 2$ **6.** $2e^{(\pi-4)/2}$.

## Problems 6.6, page 297

**2.** $21\frac{1}{12}$ **3.** $2a^2/5$ **4.** (*i*) $8a^2/15\sqrt{3}$ ; (*ii*) $(2 - \pi/2)a^2$

**5.** (*i*) $\pi a^2$ ; (*ii*) $\frac{\sqrt{2}}{3}a^2$ **6.** (*i*) $\pi a^2$ ; (*ii*) $4a^2$ **9.** $3\pi a^2$

**10.** $a^2/6$ **11.** $3\pi a^2$ ; $\pi a^2$ **12.** $\frac{11\pi}{3} - 2\sqrt{3}$

**13.** $(3\pi - 8)\, a/6$ **14.** $64a^2/3$

## Problems 6.7, page 300

**1.** (*i*) $3\pi a^2/2$ ; (*ii*) $a^2$ **2.** (*i*) $\pi a^2/8$ ; (*ii*) $\pi a^2/12$ **4.** $\frac{8a^3}{3}$

**5.** $\left(\frac{3}{2}a^2\right)$ **6.** $\pi b(2a + b)$ **8.** $\frac{\pi a^2}{2}$

## Problems 6.8, page 302

**1.** $12\frac{11}{27}a$ **2.** (*i*) $\log(2 + \sqrt{3})$, (*ii*) $\log_e(e + 1/e)$

**3.** (*i*) $a\left[\sqrt{2} + \log(1 + \sqrt{2})\right]$ ; (*ii*) $(15/16 + \log 2)\, a$ **4.** (*i*) $4a/\sqrt{3}$; (*ii*) $4\sqrt{3}$

**5.** 37.85 **8.** (*i*) $8a$ **9.** $6a$

**10.** $4\sqrt{3}$ **12.** $8a$

**13.** $\sqrt{2}\,\pi a\left\{1 + \left(\frac{1}{2}\right)^2 + \left(\frac{1.3}{2.4}\right)^2 \cdot \left(\frac{1}{2}\right)^2 + \left(\frac{1.3.5}{2.4.6}\right)^2 \cdot \left(\frac{1}{2}\right)^2 + \ldots\ldots\right\}$ **14.** $2a\left[\sqrt{2} + \log(\sqrt{2} + 1)\right]$.

## Problems 6.9, page 308

**1.** $\pi c^3(1 + \sinh 1 \cosh 1)$ **2.** $\pi h^2(a - h/3)$

**4.** $\pi a^3/12$ **5.** (*i*) $\frac{4}{3}\pi ab^2$ ; (*ii*) $\frac{4}{3}\pi a^2 b$ **6.** $\frac{\pi h}{3}(r^2 + rR + R^2)$

**7.** $48\pi a^3$ **8.** (*i*) $2\pi a^3(\log 2 - 2/3)$ ; (*ii*) $\pi a^3/24$ ; (*iii*) $\pi/48$

**9.** (*i*) $5\pi^2 a^3$ ; (*ii*) $5\pi^2 a^3$ **10.** $32\pi a^3/105$ **11.** $\frac{96\pi a^2}{5\sqrt{2}}$

**13.** (*i*) $\frac{4}{3}\pi a^3$ ; (*ii*) $\frac{8}{3}\pi a^3$ **14.** $\frac{4}{3}\pi a(a^2 + b^2)$ **15.** $\frac{\pi a^3}{4}\left\{\frac{1}{\sqrt{2}}\log(\sqrt{2} + 1) - \frac{1}{3}\right\}$.

**17.** $\frac{32\pi a^2}{5}$

## Problems 6.10, page 312

**1.** $\frac{\pi a^2}{2}(2 + \sinh 2)$ **2.** $\frac{8\pi a^2}{3}(2\sqrt{2} - 1)$ **3.** $2\pi ab\left\{\frac{b}{a} + \frac{a}{\sqrt{(a^2 - b^2)}}\sin^{-1}[\sqrt{(a^2 - b^2)}/a]\right\}$

**4.** $\frac{1}{2}\pi r^2 h$ ; $\pi r\sqrt{(r^2 + h^2)}$, where $r$ is the base-radius and $h$ the height of the cone

**5.** $4\pi a^2$ **8.** $\frac{64}{3}\pi a^2$ **9.** $\frac{64}{3}\pi a^2$ **12.** $4\pi a^2$

**13.** $\pi a^2\,[3\sqrt{2} - \log(\sqrt{2} + 1)]$.

## Problems 6.11, page 312

**1.** (*b*) **2.** (*c*) **3.** (*b*) **4.** (*c*) **5.** (*b*) **6.** (*c*)
**7.** (*c*) **8.** (*d*) **9.** (*b*) **10.** (*d*) **11.** (*c*) **12.** (*c*)
**13.** (*a*) **14.** $\frac{3\pi a^2}{2}$ **15.** (*iii*) **16.** $\pi a^2/12$ **17.** 1 **18.** $7\pi/8$

**19.** (*iii*) **20.** (*iii*) **21.** $\pi h(r_1^2 + r_1 r_2 + r_2^2)$ **22.** (*c*) **23.** (*a*)
**24.** (*b*) or (*c*) **25.** (*a*).

## Problems 7.1, page 322

**1.** $\frac{5}{6}$ **2.** $\left(\frac{b^3 - a^2}{3}\right)(\cos\alpha - \cos\beta)$ **3.** $\frac{\pi}{2b}$
**4.** $\frac{a^2(8 + \pi)}{8}$ **5.** $a^4/8$ **6.** $\frac{\pi}{4}ab(a^2 + b^2)$ **7.** $3/56$ **8.** $241/60$
**9.** $\pi a^2[\phi(a) - \phi(0)]$ **10.** $1 - 1/\sqrt{2}$ **11.** $\frac{\pi a^2}{4}(\log e - \frac{1}{2})$
**13.** $\pi a^2/b$ **14.** 1
**15.** (*i*) $\int_0^{2a}\int_{y\sqrt{2a}}^{2a} f(x, y)\,dx\,dy - \int_0^a \int_{a-\sqrt{(a^2-y^2}}^{a+\sqrt{(a^2-y^2)}} f(x, y)\,dx\,dy$ ; (*ii*) $\int_0^{\pi/2}\int_a^{ae^{\theta/2}} f(r, \theta)\,r dr\,d\theta$
**16.** $\frac{\pi}{4}$ **17.** $4a^2/3$ **18.** $45\pi/2$.

## Problems 7.2, page 324

**1.** 4.5 **2.** 7/6 **3.** $\pi a^2$ **4.** $\frac{3}{2}\log_e 3 - \frac{2}{3}$
**5.** $a^2$ **6.** $a^2(1 - \pi/4)$ **7.** 4/3 **8.** $a^2/4$.

## Problems 7.3, page 326

**1.** 18 **2.** $\frac{8}{3}abc(a^2 + b^2 + c^2)$ **3.** $\frac{1}{8}e^{4a} - \frac{3}{4}e^{2a} + e^a - \frac{3}{8}$
**4.** $\frac{1}{4}(13 - 8e + e^2)$ **5.** $\frac{e^{4a}}{8} - \frac{3}{4}e^{2a} + e^a - \frac{3}{8}$ **6.** $\frac{4}{35}$
**7.** $\frac{\pi^2}{8}$ **8.** $\frac{1}{8}$
**9.** $\frac{16}{3}$ **10.** $\frac{\pi}{3}$ **11.** $\pi$

## Problems 7.4, page 335

**1.** $\pi/8$ **2.** $\pi/2$ **3.** $8\left(\frac{\pi}{2} - \frac{5}{3}\right)a^2$ **4.** 0
**6.** $\pi$ **7.** $4\pi a$ **8.** $\frac{1}{2}\left(\log 2 - \frac{5}{8}\right)$ **10.** $3\pi$
**11.** $4\pi$ **12.** $16a^3/3$ **13.** $128a^3/15$ **14.** $3\pi a^3$
**15.** $4\sqrt{3}\,\pi$ **16.** $8a^4/3$ **18.** $\frac{1}{6}abc$. **19.** $2\pi^2 a^3$
**21.** $\frac{560}{140}$ **22.** $16\pi$ **23.** $\frac{4}{3}a^3\left(\frac{\pi}{2} - \frac{2}{3}\right)$ **25.** $\frac{10\pi}{3}$
**27.** $k\frac{a^2b^2}{20}$ **28.** $\frac{8\pi a^3}{3}$

## Problems 7.5, page 339

**2.** 64 **3.** $2(\pi - 2)a^2$ **4.** $2\pi a^2$ **5.** $\frac{3\pi a^2}{4}$.

## Problems 7.6, page 342

**1.** 2.84 units **2.** $\dfrac{\mu^2a^2b^2}{2}$ **4.** $\left(\dfrac{3a}{20}, \dfrac{3a}{16}\right)$

**5.** $\left[\dfrac{a(4a+3b)}{6(a+b)}, \dfrac{b(3a+b)}{6(a+b)}\right]$ **6.** $\left(\dfrac{\pi a\sqrt{2}}{8}, 0\right)$ **8.** $\left(\dfrac{128a}{429}, \dfrac{128b}{429}\right)$

**9.** (1/5, 1/5, 2/5). **10.** $\bar{x} = 3/4$ **11.** $\dfrac{k\pi a^4}{8}$ **12.** $\dfrac{\mu a^2b^2c^2}{720}$

**13.** $\dfrac{27}{26}$ metres **14.** $\left(\dfrac{21a}{128}, \dfrac{21b}{128}, \dfrac{21c}{128}\right)$ **15.** Divides the diagonal in the ratio 7 : 5

**16.** $\left(\dfrac{a}{2}, \dfrac{2}{3}h\right)$ where $a$ is the base, $h$ the depth

**17.** C.P. lies on the radius $\perp$ to the bounding diameter at a depth $32a/(15\pi)$ from the centre.

**18.** $\left(\dfrac{5a}{4}, 0, \dfrac{5a}{8}(m+n)\right)$ **19.** $\mu\pi a$ **20.** $\left(\dfrac{5a}{6}, 0\right)$ **21.** $\dfrac{1}{3}Ka^3h^2$

## Problems 7.7, page 349

**2.** $5Ma^2/4$ **3.** $2M/9$ **5.** $(21/32)\ \pi\rho a^4$ **6.** $\dfrac{2}{5}MR^2$ **7.** $\dfrac{1}{2}Mr^2$

**8.** (*i*) $\dfrac{3Mr^2}{10}$ ; (*ii*) $\dfrac{3M}{20}(r^2 + 4h^2)$ ; (*iii*) $\dfrac{M}{20}(3r^2 + 2h^2)$ **9.** 104803770$\rho$

**10.** $\dfrac{1}{30}$ **11.** $\dfrac{\pi\rho abc(a^2+b^2)}{30}$ **12.** $\dfrac{\rho a^2b^2}{8}$. **16.** $a^4$

**18.** $\left(\dfrac{-a^2}{3h}\sin\alpha, \dfrac{-b^2}{3h}\cos\alpha\right)$ **20.** (*i*) $\dfrac{\pi}{10}hr^4$ ; (*ii*) $\dfrac{\pi hr^2}{60}(2h^2+3r^2)$

## Problems 7.8, page 357

**1.** (*i*) 11.629 ; (*ii*) $\dfrac{4}{3}\sqrt{\pi}$ ; (*iii*) $\pi$; (*iv*) 0.1964 ; (*v*) $5\dfrac{\pi}{16}$

**2.** (*i*) $\sqrt{\pi}/2$ ; (*ii*) $\Gamma(5/4)$ ; (*iii*) $\sqrt{\pi}/3$ ; (*iv*) $2^{p+q-1}\beta(p, q)$

**4.** $\pi/4\sqrt{2}$ **7.** $-3/8$ **9.** $\dfrac{\Gamma\left(\dfrac{m+1}{n}\right)\Gamma(p+1)}{n\Gamma\left(\dfrac{m}{n}+p+1+\dfrac{1}{n}\right)}$, (*i*) $\dfrac{1}{396}$ (*ii*) $\dfrac{\sqrt{\pi}}{n}\cdot\dfrac{\Gamma\left(\dfrac{1}{n}\right)}{\Gamma\left(\dfrac{1}{n}+\dfrac{1}{2}\right)}$

**11.** 16/35 **14.** $\dfrac{2}{3}$ **17.** $\dfrac{ka^2b^2c^2}{48}$.

## Problems 7.9, page 362

**2.** $\dfrac{1}{2}K(\sqrt{3}/2)$ **3.** $\dfrac{2}{\sqrt{3}}\left\{F\left(\sqrt{\dfrac{2}{3}}, \dfrac{1}{2}\pi\right) - F\left(\sqrt{\dfrac{2}{3}}, \dfrac{1}{4}\pi\right)\right\}$ **4.** $2\sqrt{2}E(1/\sqrt{2}) - \sqrt{2}K(1/\sqrt{2})$

**5.** $erf(x) = \dfrac{2}{\pi}\left(x - \dfrac{x^3}{3} + \dfrac{x^5}{10} - \dfrac{x^7}{42} + \ldots\ldots\right)$; $erf(o) = 0$ **6.** (*i*) 0.3248 ; (*ii*) 0.5204.

**13.** $\dfrac{10}{3}$ **14.** $\dfrac{4}{3}$ **15.** $\int_0^5 \int_{2-x}^{2+x} f(x, y)\,dy\,dx$

**16.** $\int_1^2 \int_y^{y^2} f(x, y)\,dx\,dy$ **17.** $\int_0^1 \int_0^{\frac{1}{x}} f(x, y)\,dy\,dx$

## Problems 7.10, page 363

**1.** $\log a \log b$ **2.** Area of the triangle having vertices (0, 0), (0, 1), (1, 0)

**3.** $\sqrt{\pi}/2$ **4.** 3.1416 **5.** $15\sqrt{\pi}/8$ **6.** $92\pi$

**6.** 26 **8.** –1/3 **9.** $1/2\ \beta(4, 3/4)$ **10.** 1

**11.** 13 **12.** $\frac{\pi a^3}{6}$ **13.** (*d*) **14.** 27/4

**15.** $\pi a^2/12$ **16.** –1/3 **17.** $\infty$ **18.** 26/105

**19.** $r^2 \sin\theta\, dr\, d\theta\, d\phi$ **20.** $e^2 - 1$ **21.** $\frac{1}{4}\pi ab\,(a^2 + b^2)$. **22.** $\frac{1}{4}\pi \log(1 + \sqrt{2})$

**23.** 3/256 **24.** (*c*) **25.** $\frac{6}{25} + \frac{1}{2}\sin\frac{3}{5}$ **26.** 48/5

**27.** 1/6 **28.** 16/3 **29.** $\int_0^a \int_{-\sqrt{(a^2-x^2)}}^{\sqrt{(a^2-x^2)}} f(x, y)\, dy\, dx$

**30.** $\sqrt{2}\,\pi$ **31.** $3\pi/256$ **32.** 1/2 **33.** $\int_0^{\pi/2}\int_0^{2a\cos\theta} r\, d\theta\, dr$

**34.** $\frac{7h}{12}$ **35.** $\left(\frac{3}{20}, \frac{3}{16}\right)$ **36.** 16 **37.** $3Mr^2/10$

**38.** 1 **39.** log 2 **40.** $\frac{1}{2}\sqrt{\pi}$ **41.** (*c*) **42.** (*b*)

**43.** $\int_0^a \int_0^y f(x, y)\, dx\, dy$ .

## Problems 8.1, page 365

**3.** (*i*) $t^3 \sin t + 7t^2 \cos t + 20t \sin t - 10t$ ; (*ii*) $(20t^3 + t\sin t - \cos t)\,\mathbf{I}$
$-(2t\cos t + 2\sin t + 75t^2)\,\mathbf{J} - t(t\sin t + 2t^2\cos t + 10\cos t)\mathbf{K}$

**5.** $-4(\mathbf{I} + 2\mathbf{J})$ **6.** (*i*) $(ua^2 \sec\alpha)$; (*ii*) $a^3 \tan\alpha$ ; $(\cos t\,\mathbf{J} - \sin t\,\mathbf{I})\cos\alpha + \sin\alpha\,\mathbf{K}$

**7.** $\{t\mathbf{I} + 2\mathbf{J} + (2t - 3)\mathbf{K}\}/\sqrt{(5t^2 - 12t + 13)}$ ; $\frac{1}{3}(2\mathbf{I} + 3\mathbf{J} + \mathbf{K})$

**8.** $(x - a/\sqrt{2}) = y - a/\sqrt{2} = \left(z - \frac{a\pi}{4}\tan\alpha\right)/\sqrt{2}\tan\alpha$

**9.** (*i*) $ab/(a^2\sin^2 t + b^2\cos^2 t)^{3/2}$ ; (*ii*) $1/4\sqrt{2}$

**10.** (*i*) $\mathbf{R} = (p + q)\,\mathbf{I} + q\mathbf{J} + 2q\mathbf{K}$ ; $\frac{2\mathbf{J} - \mathbf{K}}{\sqrt{5}}$

(*ii*) $\mathbf{R} = p\mathbf{I} + (p + 2q)\,\mathbf{J} + (p + q)\mathbf{J} + (p + q)\mathbf{K}$ ; $(2\mathbf{K} - \mathbf{I} - \mathbf{J})/\sqrt{6}$ .

## Problems 8.2 page 372

**1.** In $1 \le t \le 5$, particle is displaced from $\mathbf{K} - 4\,\mathbf{I}$ to $5\mathbf{K} - 4\mathbf{I}$ around the circular helix.

**2.** $v(\text{at } t = 0) = \sqrt{37}$, $a(\text{at } t = 0) = \sqrt{325}$ **3.** $\alpha = \pm 1/\sqrt{6}$

**4.** $8\sqrt{14/7}$ ; $-\sqrt{14/7}$ **8.** 0 ; 3 **9.** $\cos^{-1}\left(\frac{9}{17}\right) = 58°2'$.

**12.** $\frac{ti + 2j + (2t - 3)k}{\sqrt{5t^2 - 12t + 13}}$ ; $\frac{1}{3}(2i + 2j + k)$

## Problems 8.3, page 378

**1.** $2(x\mathbf{I} + y\mathbf{J} + z\mathbf{K})/(x^2 + y^2 + z^2)$. **2.** $(-\mathbf{I} + 3\mathbf{J} + 2\mathbf{K})/\sqrt{14}$

**3.** (*i*) 12 (*ii*) $12\frac{1}{3}$ **4.** $15/\sqrt{17}$ **5.** $a = 6, b = 24, c = -8$

**7.** $a = \pm\frac{20}{9}, b = \pm\frac{55}{9}, c = \pm\frac{55}{9}$ **8.** $96(\mathbf{I} + 3\mathbf{J} - 3\mathbf{K})$ ; $96\sqrt{(19)}$

**10.** $\frac{1}{3}(2\mathbf{I}+2\mathbf{J}-\mathbf{K})$ **11.** $\cos^{-1}(1/\sqrt{22})$

## Problems 8.4, page 386

**1.** (i) $2(x+y+z);\ 0$; (ii) $2(xy-x+y);\ 2zi-(x^2+2z)k$; (iii) $4x;\ -xi-(2x-y)j+4yk$

**3.** $a=-2$; $4x(z-xy)\mathbf{I}+(y-2yz+4xy^2)\mathbf{J}+(2x^2+y^2-z^2-z)\mathbf{K}$

**9.** $\frac{1}{3}x^3-xy^2+\frac{1}{2}(x^2-y^2)$ **12.** $\frac{1}{2}x^2y^2+\frac{1}{3}(x^3+y^3)$

**13.** $\dfrac{2x(2x-1)}{(x^2+y^2+z^2)^{n+1}};\ n=\dfrac{1}{2}$ **14.** $a=4, b=2, c=-1$

**15.** (i) $2(y^3+3x^2y-6xy^2)z\mathbf{I}+2(3xy^2+x^3-6x^2y)z\mathbf{J}+2(xy^2+x^3-3x^2y)y\mathbf{K}$; (ii) Zero

**16.** $1724/\sqrt{21}$. **19.** $x^2yz^3$

## Problems 8.5, page 389

**1.** $75\frac{1}{3}\mathbf{I}+360\,\mathbf{J}-42\,\mathbf{K}$ **2.** $(t^4+2)\,\mathbf{I}-\left(\frac{t^3}{3}+4\right)\mathbf{J}$ **3.** $(t^3-t+2)\,\mathbf{I}+(1-t^4)\mathbf{J}+(4-4\cos t-3t)\,\mathbf{K}$

**4.** $\mathbf{V}=6\sin 2t\mathbf{I}+4(\cos 2t-1)\mathbf{J}+8t^2\mathbf{K}$; $\mathbf{R}=3(1-\cos 2t)\mathbf{I}+2\sin 2t\mathbf{J}+\frac{8t^3}{3}\mathbf{K}$.

## Problems 8.6, page 391

**1.** $3t+4.5t^2$ **2.** 0 **4.** $-2/3$ **5.** 5 **6.** $18\pi,\ \nabla\times\vec{F}\neq 0$

**7.** Zero **8.** 303 **9.** $8\frac{8}{35}$ **10.** $\left(4-\frac{\pi}{4}\right)i+\left(\pi-\frac{1}{2}\right)j-k$

**12.** $e^t i+\frac{1}{2}(e^{2t}+1)\,j+tk$ **14.** 0

## Problems 8.7, page 394

**2.** $3\frac{1}{3}$ **3.** 24. **5.** $\frac{-7}{24}$ **6.** 3

## Problems 8.8, page 397

**4.** $\pi ab$ **5.** $\frac{16}{3}$ **6.** Zero **7.** $\frac{8}{3}$ **8.** 128/5 **9.** 4/3

## Problems 8.9, page 403

**3.** $(\pi)$ **5.** $\frac{19}{2}\pi$ **10.** $-\frac{\pi a^2}{\sqrt{2}}$ **11.** 0 **12.** $\pi a^2$

**13.** $2\pi a^2$ **14.** 0

## Problems 8.10, page 408

**4.** $4\pi a^3$ **7.** (i) $\frac{12}{5}\pi a^5$ (ii) $12(e-e^{-1})$ **8.** $4\pi a^3$

**9.** $\frac{\pi a^6}{12}$ **10.** $\frac{5}{4}\pi a^4 b$ **11.** $-4\pi$ **12.** 8/3.

## Problems 8.11, page 412

3. $14\frac{2}{3}$
4. $x^3y - y^2z^2 + z^3$
5. (i) $\frac{1}{3}(x^3 + y^3 + z^3 - 3xyz)$ ; (ii) $x^2y + y^2z + z$ ; (iii) $xz^3 - yz + 3x^2y$. (iv) $\frac{x^3}{3} + y^3 + z^3$.
6. $\phi = x^2yz^3 + 20$
7. (i) Yes, $\frac{a}{2}(x^2 + y^2 - 2z^2)$ ; (ii) Yes
8. $xy \sin z + \cos x + y^2 z + c$
9. $x^2y + xz^3$ ; 202
10. $a = 4, b = 2, c = -1$
11. $a = 4$ ; $2x^2 y - xz^3$ ; 47.

## Problems 8.12, page 421

1. $x^2 + y^2 + z^2 - y = 0$
2. (i) $(\rho \sin 2\phi - z \sin \phi)\, \mathbf{T}_\rho - (2\rho \sin^2 \phi + z \cos \phi)\, \mathbf{T}_\phi + 3\rho \cos \phi\, \mathbf{T}_z$
   (ii) $(2p \cos^2 \phi - 3p^2 \sin^3 \phi)\, \mathbf{T}_p - (\rho \sin 2\phi + 3\rho^2 \sin^2 \phi \cos \phi)\, \mathbf{T}_\phi + \rho\, z \cos \phi\, \mathbf{T}_z$
3. (i) $r \sin\theta \Big[ \{\sin\theta\,(1 + \sin^2\phi) + r\cos^2\theta \sin\phi\}\, \mathbf{T}_r + \{\cos\theta\,(1 + \sin^2\phi) - r \sin\theta \cos\theta \sin\phi\}\, \mathbf{T}_\theta + \sin\phi \cos\phi\, \mathbf{T}_\phi \Big]$
   (ii) $r^2 \sin\theta\, \{(\sin^2\theta \cos^2\phi \sin\phi + \sin\theta\cos\theta \sin^2\phi + \cos^2\theta \cos\phi)\, \mathbf{T}_r + (\sin\theta\cos\theta\cos^2\phi \sin\phi + \cos^2\theta \sin^2\phi - \sin\theta\cos\theta\cos\phi)\, \mathbf{T}_\theta + (\cos\theta \sin\phi \cos\phi - \sin\theta \sin^2\phi \cos\phi)\, \mathbf{T}_\phi\}$
4. ***Cylinderical:*** $4\{(\cos\phi \sin\phi)\, T_e - (\sin\phi + \cos\phi)\, T_\phi + \sqrt{6}\, T_z\}$.
   ***Spherical:*** $4\{\sin\theta\cos\phi - \sin\theta\sin\phi + \sqrt{6}\cos\theta\}T_r + 4\{\cos\theta\cos\phi - \cos\theta\sin\phi - \sqrt{6}\sin\theta\}\, T_\theta - 4(\sin\phi + \cos\phi)\, T_\phi)$.
5. $\rho z \sin 2\phi\, \mathbf{T}_\rho + \rho z \cos 2\phi\, \mathbf{T}_\phi + \frac{1}{2}\rho^2 \sin 2\phi\, \mathbf{T}_z$.

## Problems 8.13, page 421

1. $1/\sqrt{14}$, $2/\sqrt{14}$, 3/14
2. $\frac{1}{4}(x - 2) = y - 1 = z + 3$
3. $dudv = \frac{1}{h_1 h_2}\, dxdy$
4. $4x - 3z + 2xz$
5. zero
6. $\frac{1}{2}\int_C (xdy - ydx)$
7. 3V
8. (i) $\mathbf{R}/r^3$ ; (ii) 3 ; (iii) 0.
9. Irrotational
10. $4\pi$
11. solenoidal
12. $-28/\sqrt{5}$
13. zero.
14. $-(y\mathbf{I} + z\mathbf{J} + x\mathbf{K})$.
15. $\frac{\partial f_1}{\partial x} + \frac{\partial f_2}{\partial y} + \frac{\partial f_3}{\partial z}$.
16. zero
17. $-(12\mathbf{I} + 5\mathbf{J} + 8\mathbf{K})$
18. zero
19. zero
20. zero
21. zero
22. $\frac{1}{r^2}$
23. §8.5(2)
24. $\frac{1}{\sqrt{21}}(2\mathbf{I} + 4\mathbf{J} - \mathbf{K})$.
25. 2, –2, 2
26. $6\frac{\sqrt{7}}{3}$
27. $2/r$
28. 7/3
29. zero
30. (c)
31. (c)
32. (b)
33. (c)
34. (a)
35. (a)
36. $5u$
37. zero
38. irrotational field
39. (a)
40. the rate at which fluid is originating at $P$ per unit volume.
41. (a)
42. it gives the maximum rate of change of $\phi$.
43. (iv)
44. (a)
45. (a)
46. (b)
47. (a)
48. (b)
49. (a)
50. Zero
51. True
52. True.
53. True.

## Problems 9.1, page 424

1. Convergent
2. Convergent
3. Convergent
4. Divergent
5. Convergent
6. Convergent
7. Convergent
8. Divergent.

## Problems 9.2, page 425

**1.** Convergent **2.** Convergent **3.** Oscillatory **4.** Convergent **5.** 15 m.

## Problems 9.3, page 432

**1.** Convergent **2.** Divergent **3.** Convergent **4.** Divergent
**5.** Convergent **6.** Divergent **7.** Divergent
**8.** Convergent **9.** Convergent **10.** Convergent **11.** Convergent
**12.** Divergent **13.** Divergent **14.** Convergent **15.** Convergent
**16.** Convergent **17.** Convergent **18.** Divergent **19.** Convergent
**20.** Convergent **21.** Converget for $x < 1$, Divergent for $x \geq 1$ **22.** Convergent
**23.** Convergent **24.** Convergent

## Problems 9.4, page 439

**1.** Conv. **2.** Conv. **3.** Conv. **4.** Conv. **5.** Conv. for $z \leq 1$ ; Div. for $z > 1$
**6.** Conv. for $x \geq 1$ ; Div. for $x < 1$ **7.** Conv. **8.** Conv. for $x < 1$ ; Div for $x \geq 1$.
**9.** Conv. for $x < 1$ ; Div. for $x \geq 1$ **10.** Conv. for $x \leq 1$, Div. for $x > 1$. **11.** Convergent for all values of $p$
**12.** Convergent **13.** Convergent **14.** Convergent **15.** Divergent
**16.** Conv. for $x < 1$, Div. for $x > 1$ ; Conv. for $p > 1$ and Div. for $p \leq 1$
**17.** Divergent **18.** Conv. if $\beta > \alpha > 0$ ; Div. if $\alpha \geq \beta > 0$.

## Problems 9.5, page 441

**1.** Conv. for $x \leq 1$ ; Div. for $x > 1$ **2.** Conv. for $x < 1$ ; Div. for $x \geq 1$
**3.** Conv. for $x < 2$ ; Div. for $x \geq 2$ **4.** Conv. for $x < e$ ; Div. for $x \geq e$
**5.** Conv. for $x \leq 1$ ; Div. for $x > 1$ **6.** Conv. for $x \leq 1$ ; Div. for $x > 1$
**7.** Divergent by Gauss test **8.** Conv. for $x < 1/e$ ; Div. for $x \geq 1/e$
**9.** Conv. for $x < 1$ ; Div. for $x \geq 1$ **10.** Divergent
**11.** Conv. for $x < 1$ ; Div. for $x > 1$. When $x = 1$, Conv. for $b - a > 1$, Div. for $b - a \leq 1$.
**12.** Conv. for $x < 2$ and Div. for $x \geq 2$ **13.** Conv. for $\frac{3}{2}$ and Divgt. for $x \geq \frac{3}{2}$
**14.** Conv. for $x < 1$ and Divgt.for $x \geq 1$ when $x = 1$ Conv. if $\gamma - \alpha - \beta > 0$ and Divgt. for $\gamma - \alpha - \beta \leq 0$
**15.** Conv. for $x \leq 1$ and Divgt. for $x > 1$ **16.** Conv. for $x \leq 1$ and Divgt. for $x > 1$
**17.** Convergent

## Problems 9.6, page 445

**1.** Convergent **2.** Convergent **3.** Convergent
**4.** Convergent **5.** Conv. for $x < 1$ ; Div. for $x \geq 1$
**6.** Conv. for $x < \frac{1}{2}$ ; Div. for $x \geq \frac{1}{2}$ **7.** Convergent.

## Problems 9.7, page 447

**1.** Oscillatory **2.** Convergent **3.** Convergent **4.** Oscillatory
**5.** Convergent **6.** Convergent **7.** Convergent **8.** Convergent
**9.** Convergent **10.** Convergent **11.** Oscillatory.

## Problems 9.8, page 457

1. (*i*) and (*ii*) conditionally convergent (*iii*) Convergent but not absolutely convergent.
3. (*i*) Conditionally convgt. for $0 < p \le 1$ ; (*ii*) Conditionally convgt ; (*iii*) Absolutely convergent if $p > 1$, conditionally convergent if $p \le 1$.
4. Absolutely convergent for (*i*) $0 < x < 1$ ; (*ii*) $-1 < x \le 1$ ; (*iii*) $|x| \le 1$.
5. Convergent for $x \le 1$ and not convergent for $x > 1$
6. (*i*) $-1 < x \le 1$ ; (*ii*) $-1 < x \le 1$ ;
7. $-e < x \le e$
8. (*i*) Absolutely convergent; (*ii*) convergent
9. Absolutely convergent.
10. Converges for $4 < x \le 6$. Absolutly convergent for $x = 4$.

## Problems 9.9, page 458

1. Convergent
2. Convergent
3. Divergent
4. Convergent
5. Divergent
6. Conv. for $x < 1$ ; Div. for $x \ge 1$
7. Conv. for $x < 1$ ; Div. for $x \ge 1$
8. Conv. for $x < 1$ ; Div. for $x \ge 1$
9. Conv. for $x < 1$ ; Div. for $x \ge 1$
10. Conv. for $x < 1/4$ ; Div. for $x \ge 1/4$
11. Conv. for $x < 2$ ; Div. for $x \ge 2$
12. Convergent for all $x$
13. Conv. for $x < 1$ ; Div. for $x \ge 1$
14. Convergent
15. Absolutely convergent
16. Convergent
17. Convergent for $p > 1$ ; divergent for $p \le 1$.
18. Conv. for $x \le 1$, div. for $x > 1$.

## Problems 9.10, page 461

1. Uniformly convergent for $0 \le x \le 1$.
2. to
5. Uniformly convergent for all real values of $x$
6. Uniformly convergent for $0 \le x \le 1/a$
10. (*i*) and (*ii*) Both converge uniformly for all real values of $x$.

## Problems 9.11, page 462

1. (*c*) 2. (*d*) 3. (*a*) 4. (*b*) 5. (*c*) 6. (*d*)
7. (*a*) 8. (*b*) 9. (*b*) 10. (*d*) 11. (*c*)
12. (*a*) (–1, 1) (*b*) (–1/2, 1/2) 13. $-1 < x \le 1$ 14. $k > 1$ 15. $a_n < k$ 16. Oscillatory
17. All values of $x$ 18. $k < 1$ 19. Convergent. 20. Divergent. 21. $q - p > 1$
22. Divergent 23. Convergent. 24. $0 < x < 4$ 25. yes 26. True 27. Convergent
28. Divergent 29. $x > 1$ 30. $0 \le x \le 1$ 31. (*b*) 32. (*c*) 33. (*d*)
34. (*b*) 35. True.

## Problems 10.1, page 472

1. $$\frac{2\sinh\pi}{\pi}\left\{\left(\frac{1}{2}-\frac{\cos x}{1^2+1}+\frac{\cos 2x}{2^2+1}-\ldots\right)-\left(\frac{\sin x}{1^2+1}-\frac{2\sin 2x}{2^2+1}+\frac{3\sin 3x}{3^2+1}\ldots\right)\right\}$$

$$\frac{\pi}{\sinh\pi}=2\left[\frac{1}{2^2+1}-\frac{1}{3^2+1}+\frac{1}{4^2+1}-\ldots\right].$$

## Problems 10.2, page 473

1. No
2. No
3. No, Function is not periodic and not defined at $t = 3$ in $(0, 2\pi)$

## Problems 10.3, page 476

**1.** $\frac{1}{2}\pi - \frac{4}{\pi}\left(\cos x + \frac{\cos 3x}{3^2} + \frac{\cos 5x}{5^2} + \ldots\ldots\right)$ **2.** $x = \pi - 2\left(\sin x + \frac{1}{2}\sin 2x + \frac{1}{3}\sin 3x + \ldots\ldots\right)$

**3.** $\frac{\pi^2}{6} - 2\left(\cos x - \frac{\cos 2x}{2^2} + \frac{\cos 3x}{3^2} - \ldots\ldots\right) - \frac{1}{\pi}\left\{\left(\frac{2}{1^3} - \frac{\pi^2}{1}\right)\sin x - \left(\frac{2}{2^3} - \frac{\pi^2}{2}\right)\sin 2x + \ldots\right\}$

**4.** $x - x^2 = \frac{-\pi^2}{3} + 4\left(\frac{\cos x}{1^2} - \frac{\cos 2x}{2^2} + \frac{\cos 3x}{3^2} - \frac{\cos 4x}{4^2} + \ldots\ldots\right) + 2\left(\frac{\sin x}{1} - \frac{\sin 2x}{2} + \frac{\sin 3x}{3} - \frac{\sin 4x}{4} + \ldots\ldots\right)$

**5.** $\frac{\pi}{2} + \frac{4}{\pi}\left(\frac{\cos x}{1^2} + \frac{\cos 3x}{3^2} + \ldots\ldots\right)$ **6.** $\frac{4}{\pi}\left(\sin x + \frac{\sin 3x}{3} + \frac{\sin 5x}{5} + \ldots\ldots \infty\right)$.

**7.** $e^{-x} = \left(\frac{1 - e^{-2\pi}}{\pi}\right)\left[\frac{1}{2} + \left(\frac{1}{2}\cos x + \frac{1}{5}\cos 2x + \frac{1}{10}\cos 3x + \ldots\right) + \left(\frac{1}{2}\sin x + \frac{2}{5}\sin 2x + \frac{3}{10}\sin 3x + \ldots\ldots\right)\right]$

**8.** $x \sin x = 1 - \frac{1}{2}\cos x - \frac{2}{1\cdot 3}\cos 2x + \frac{2}{2\cdot 4}\cos 3x - \frac{2}{3\cdot 5}\cos 4x + \ldots\ldots$

## Problems 10.4, page 483

**1.** $\frac{\pi - 1}{2} + \frac{1}{\pi}\sum_{n=1}^{\infty}\frac{1}{n}\sin n\pi x$

**2.** (*i*) $\frac{a^2}{3}\frac{a^2}{11^2}\left\{\frac{1}{12}\cos\frac{2\pi x}{a} + \frac{1}{2^2}\cos 4\pi\frac{x}{a} + \ldots\ldots\right\}\frac{a^2}{\pi}\left\{\frac{1}{1}\sin\frac{2\pi x}{a}\frac{1}{2}\sin\frac{4\pi x}{a} + \ldots\ldots\right\}$

(*ii*) $f(t) = \frac{2}{3} + \frac{4}{\pi^2}\left(\cos \pi t - \frac{\cos 2\pi t}{2^2} + \frac{\cos 3\pi t}{3^2} - \ldots\ldots\right)$

**4.** $f(x) = \frac{2}{\pi}\left(\sin \pi x + \frac{\sin 3\pi x}{3} + \ldots\right) - \frac{4}{\pi^2}\left(\cos \pi x + \frac{\cos 3\pi x}{3^2} + \frac{\cos 5\pi x}{5^3} + \ldots\right)$

**5.** $\frac{3}{2} - \frac{12}{\pi^2}\sum_{n=1}^{\infty}\frac{1}{(2n-1)^2}\left\{\frac{\cos(2n-1)\pi x}{3}\right\}$

**6.** $\frac{E}{\pi} + \frac{E}{2}\sin \omega t - \frac{2E}{\pi}\left[\frac{1}{1.3}\cos 2\omega t + \frac{1}{3.5}\cos 4\omega t + \frac{1}{5.7}\cos 6\omega t + \ldots\right]$

**7.** $f(x) = \frac{\pi}{4} + 2\sum_{n=1}^{\infty}(-1)^{n+1}\sin n\pi x$ ; put $x = 1/2$. **11.** $\frac{\pi^2}{6}$

## Problems 10.5, page 491

**1.** $f(x) = \frac{-\pi}{4} - \frac{2}{\pi}\left(\cos x + \frac{\cos 3x}{3^2} + \frac{\cos 5x}{5^2} + \ldots\right) + \left(3\sin x - \frac{\sin 2x}{2} + \frac{3\sin 3x}{3} - \frac{\sin 4x}{4} + \ldots\right)$

**3.** $f(x) = \frac{1}{\pi} + \frac{1}{2}\cos x + \frac{2}{\pi}\left(\frac{\cos 2x}{2^2 - 1} - \frac{\cos 4x}{4^2 - 1} + \frac{\cos 6x}{6^2 - 1} - \ldots\right)$

**4.** $f(x) = 2\left(\pi - \frac{4}{\pi}\right)\sin x - \pi \sin 2x + \frac{2}{3}\left(\pi - \frac{4}{9\pi}\right)\sin 3x - \frac{\pi}{2}\sin 4x + \ldots\ldots$

**5.** $1 + \sum_{n=1}^{\infty}\frac{4[(-1)^n - 1]}{n^2\pi^2}\cos\frac{n\pi x}{2}$

**6.** (*i*) $\frac{2}{\pi} - \frac{4}{\pi}\left(\frac{\cos 2x}{3} + \frac{\cos 4x}{15} + \ldots\ldots + \frac{\cos 2nx}{4n^2 - 1} + \ldots\ldots\right)$ (*ii*) $\frac{2}{\pi} + \sum_{n=1}^{\infty}\frac{4(-1)^{m+1}}{(4m^2 - 1)\pi}\cos\frac{2m\pi x}{l}$

**7.** $\frac{\pi}{2}+1-\frac{4}{\pi}\left(\cos x+\frac{1}{3^2}\cos 3x+\frac{1}{5^2}\cos 5x+...\right);\ \frac{\pi^2}{8}$ **8.** $\frac{4k}{\pi}\left(\sin x+\frac{\sin 3x}{3}+\frac{\sin 5x}{5}+......\infty\right)$.

## Problems 10.6, page 495

**1.** $\frac{\pi}{2}-\frac{4}{\pi}\left[\cos x+\frac{1}{3^2}.\cos 3x+\frac{1}{5^2}.\cos 5x+......\right]$ ; $2\left[\sin x-\frac{1}{2}\sin 2x+\frac{1}{3}\sin 3x-......\right]$

**2.** $f(x)=\frac{1}{2}+\frac{2}{\pi}\left(\sin\frac{\pi x}{2}+\frac{1}{3}\sin\frac{3\pi x}{2}+\frac{1}{5}\sin\frac{5\pi x}{2}+.....\right)$

**3.** $\sum_{n=2}^{\infty}\frac{1}{n}\sin 2nx$ **4.** $\frac{1}{3}+\frac{4}{\pi^2}\sum_{1}^{\infty}\frac{1}{n^2}\cos n\pi x$

**5.** $\frac{8}{\pi^3}\left(\frac{\sin\pi t}{1^3}+\frac{\sin 3\pi t}{3^3}+\frac{\sin 5\pi t}{5^3}+......\right)$

**6.** $-\frac{1}{2}\sin x+\sum_{n=2}^{\infty}\frac{(-1)^n.2n}{n^2-1}\sin nx$ **7.** $\sum_{n=1}^{\infty}\frac{2n\pi}{1+n^2\pi^2}(1-e\cos n\pi)\sin n\pi x$

**8.** $\frac{\pi^2}{6}-4\frac{\cos 2x}{2^2}+\frac{\cos 4x}{4^2}+\frac{\cos 6x}{6^2}+.....$ **9.** $\sum_{n=1}^{\infty}\frac{4}{\pi n^3}[1-(-1)^n]\sin nx$

**11.** $\frac{l}{4}+\sum_{n=1}^{\infty}\frac{2l}{(n\pi)^2}\left\{2\cos\frac{n\pi r}{2}-1-(-1)^n\right\}\cos\frac{n\pi x}{l}$ **12.** $\frac{8}{\pi}\cos\frac{\pi}{4}\left[\frac{\sin 2x}{1.3}-\frac{\sin 6x}{5.7}+\frac{\sin 10x}{9.11}+......\right]$

**13.** $\frac{2l^2h}{a(l-a)\pi^2}\left[\sin\frac{\pi a}{l}\sin\frac{\pi x}{l}+\frac{1}{2^2}\sin\frac{2\pi a}{l}\sin\frac{2\pi x}{l}+\frac{1}{3^2}\sin\frac{3\pi a}{l}\sin\frac{3\pi x}{l}+......\right]$.

**14.** $f(x)=2\left(\sin\pi x-\frac{1}{2}\sin^2\pi x+\frac{1}{3}\sin 3\pi x-\frac{1}{4}\sin 4\pi x+.....\right)$

**15.** $\pi x=2c\left(\frac{\sin\pi x}{c}-\frac{1}{2}\sin\frac{2\pi x}{c}+\frac{1}{3}\sin\frac{3\pi x}{c}-\frac{1}{4}\sin\frac{4\pi x}{c}+.....\right)$

**16.** $e^{-x}=\sin\text{h}\,a\left[\frac{1}{a}+\sum_{n=1}^{\infty}\frac{(-1)^n}{n^2\pi^2+a^2}\left(2a\cos\frac{n\pi x}{a}+2n\pi\sin\frac{\pi nx}{a}\right)\right]$

**17.** $x^2-2=\frac{-2}{3}-\frac{16}{\pi^2}\left(\cos\frac{\pi x}{2}-\frac{1}{2^2}\cos\pi x+\frac{1}{3^2}\cos\frac{3\pi x}{2}-\frac{1}{4^2}\cos 2\pi x+....\right)$

## Problems 10.7, page 498

**3.** $\pi^4/96$.

## Problems 10.8, page 502

**1.** $\sum_{-\infty}^{\infty}\frac{(\sinh al\cos n\pi l-i\cosh al\sin n\pi l)(a+in\pi)}{(a^2+n^2\pi^2)}e^{in\pi x/l}$

**2.** $\frac{2}{\pi}\left\{1-\frac{e^{2it}+e^{-2it}}{1.3}-\frac{e^{4it}+e^{-4it}}{3.5}-\frac{e^{6it}+e^{-6it}}{5.7}-......\right\}$

**3.** $\frac{a}{\pi} \sin a\pi \sum_{-\infty}^{\infty} \frac{(-1)^n e^{inx}}{a^2 - n^2}$

**4.** $\left(\frac{2}{\pi}\right)\left[(\pi^2 - 4)\sin x - \frac{1}{2}\pi^2 \sin 2x + \frac{1}{3}\left(\pi^2 - \frac{4}{3^2}\right)\sin 3x - \frac{1}{4}\pi^2 \sin 4x + \frac{1}{5}\left(\pi^2 - \frac{4}{5^2}\right)\sin 5x + .....\right]$

**5.** $\frac{a}{2} - \frac{a}{\pi}\left[(e^u - e^{-u}) + \frac{1}{3}(e^{3u} - e^{-3u}) + \frac{1}{5}(e^{5u} - e^{-5u}) + ......\right]$ where $u = i\pi x/l$.

**6.** $2\left(\sin x - \frac{1}{2}\sin 2x + \frac{1}{3}\sin 3x - \frac{1}{4}\sin 4x + .....\right), \frac{\pi}{2} - \frac{4}{\pi}\left(\cos x + \frac{\cos 3x}{3^2} + \frac{\cos 5x}{5^2} + .....\right)$

**7.** $\frac{8}{\pi^3}\left(\sin\frac{\pi t}{1^3} + \sin\frac{3\pi t}{3^3} + \sin\frac{5\pi t}{5^3} + .....\right); \frac{1}{6} - \frac{4}{\pi^2}\left(\cos\frac{2\pi t}{2^2} + \frac{\cos 4\pi t}{4^2} + \frac{\cos 6\pi t}{6^2} + .....\right)$

**8.** $\frac{-2}{\pi}\left(\sin 2\pi x + \frac{1}{2}\sin 4\pi x + \frac{1}{3}\sin 6\pi x + .....\right)\frac{-8}{\pi^2}\left(\cos \pi x + \frac{\cos 3\pi x}{3^2} + \frac{\cos 5\pi x}{5^2} + .....\right)$

**9.** $\frac{1}{\pi^2}\left[(\pi - 4)\sin \pi x + \frac{3\pi + 4}{3^2}\sin 3\pi x + \frac{5\pi - 4}{5^2}\sin 5\pi x + \frac{7\pi + 4}{7^2}\sin 7\pi x + .....\right]$

## Problems 10.9, page 507

**1.** $T = 7{\cdot}8 \sin\theta + 1{\cdot}5 \sin 2\theta - 9{\cdot}2 \sin 3\theta + 11{\cdot}6 \sin 4\theta .....$

**2.** $1.45 + (-0.37 \cos x + 0.17 \sin x) - (0.1 \cos 2x + 0.06 \sin 2x)$

**3.** $a_0 = 41.66, a_1 = -8.33, b_1 = -1.15$ **4.** $-0.0731$

**5.** $a_0 = 41{\cdot}66 ; a_1 = -8{\cdot}33 ; b_1 = -1{\cdot}16$.

**6.** $y = 48 \cdot 45 - 17 \cdot 70 \cos x - 3 \cdot 12 \cos 2x - 0 \cdot 35 \cos 3x + ..... -2 \cdot 23 \sin x - 1 \cdot 67 \sin 2x - 1 \cdot 75 \sin 3x + .....)$

**7.** $A = 1{\cdot}317, \alpha_1 = -0 \cdot 0083, B = -0{\cdot}1524, \alpha_2 = 0 \cdot 0315$.

8. $y = 1 \cdot 262 + 0 \cdot 041 \cos x + 0 \cdot 530 \cos 2x - 0 \cdot 103 \cos 3x +$ and $y = -0 \cdot 637 \sin x - 0 \cdot 231 \sin 2x + 0 \cdot 85 \sin 3x + .....$

## Problems 10.10, page 508

**1.** $2\pi/3$ **2.** $\frac{1}{2}\{f(c-0) + f(c+0)\}$ **3.** $(-1, 1)$ such that $f(x) = -f(-x)$

**4.** $f(x) = A$ when $0 < x < \pi$ and $f(x) = -A$ when $\pi < x < 2\pi$ **5.** Sine

**6.** § 10.11 (3) **7.** Zero **8.** not defined **9.** odd

**10.** Cosine **11.** even **12.** $x = k/n$ **13.** Zero

**14.** Cosine **15.** Zero **16.** $\int_0^2 x^2 \cos\frac{n\pi x}{2}\,dx$ **17.** $\frac{1}{T}\int_\alpha^{\alpha+2T} f(x)\sin\frac{n\pi x}{T}\,dx$

**18.** § 10.3 **19.** Zero **20.** $a_0 = \frac{2}{l}\int_0^l f(x)\,dx, a_n = \frac{2}{l}\int_0^l f(x)\cos\frac{n\pi x}{l}\,dx$

**21.** $\frac{4}{\pi}\left\{\sin x + \frac{\sin 3x}{3} + \frac{\sin 5x}{5} + ......\right\}$ **22.** $\pi$ **23.** Zero

**24.** $2l$ **25.** $\sum_{n=-\infty}^{\infty} (-1)^n \frac{(1 - in\pi)}{1 + n^2\pi^2} \sinh 1 e^{in\pi x}$ **26.** False

**27.** $-\pi/2$ **28.** odd **29.** Zero **30.** 3.5355

**31.** zero **32.** $-1/2$ **33.** $\frac{1}{2}a_n$ **34.** $\frac{\pi^2}{8}$

**35.** $\frac{\pi}{2} - \frac{4}{\pi}\left\{\cos x + \frac{1}{3^2}\cos 3x + \frac{1}{5^2}\cos 5x + ...\right\}$ **36.** $f(x) = \frac{3}{8} + \frac{1}{2}\cos 2x + \frac{1}{4}\cos 4x$

**37.** $x^2 - x$ **38.** $x(l + x)$ **39.** True **40.** False
**41.** False **42.** False.

## Problems 11.1, page 513

**1.** $x^2 \dfrac{d^2y}{dx^2} - 4x\dfrac{dy}{dx} + 6y = 0$ **2.** $\dfrac{d^2x}{dt^2} + a^2x = 0$ **3.** $\dfrac{d^2y}{dx^2} + 4y = 0$ **4.** $x\dfrac{d^2y}{dx^2} + 2\dfrac{dy}{dx} - xy = 0$

**5.** $\dfrac{d^2y}{dx^2} - 2\dfrac{dy}{dx} + 2y = 0$ **6.** $\dfrac{d^3y}{dx^3} - 7\dfrac{dy}{dx} + 6y = 0$ **7.** $2xy\dfrac{dy}{dx} + x^2 - y^2 = 0$

**8.** $(x^2 - 25)\left(\dfrac{dy}{dx}\right)^2 + x^2 = 0$ **9.** $y\dfrac{dy}{dx} = 2a.$ **11.** $y''' - 3y'' + 3y' - y = 0.$

## Problems 11.2, page 516

**1.** $\sqrt{(1-x^2)} + \sqrt{(1-y^2)} = c$ **2.** $\log\dfrac{x}{y} - \dfrac{1}{x} - \dfrac{1}{y} = c$ **3.** $\tan x \tan y = c$

**4.** $(1+x^2)^{3/2} - 3\sqrt{(1+y^2)} = c$ **5.** $\tan y = c(1 - e^x)$ **6.** $2e^{-y} = e^{-x^2} + 1$

**7.** $x = 2\cos y$ **8.** $x + \log(\cos y) = 0$ **9.** $(x^2+1)(y^2+1) = c$

**10.** $3e^{2x} - 2e^{3y} + 8x^3 = c$ **11.** $(1 - ay)(a + x) = cy$ **12.** $(x+1)(2 - e^y) = c$

**13.** $a\log\left(\dfrac{x-y-a}{x-y+a}\right) = 2y + c.$ **14.** $y = \tan^{-1}(x + y + 1) + c$ **15.** $\tan(x+y) = \sec(x+y) + x + c$

**16.** $x = \text{cosec}\,(x+y+1) - \cot(x+y+1) + c$

**17.** $\log\sin(y - x) = \dfrac{1}{2}x^2 + cz$ **18.** $\cos xy + \dfrac{1}{2x^2} = c.$

## Problems 11.3, page 518

**1.** $\log\dfrac{y}{x} = c + \dfrac{1}{4}\left(\dfrac{x}{y}\right)^2$ **2.** $(x/y)^3 = 3\log cy$ **3.** $cy^3 = x^2e^{-x/y}$

**4.** $xy^3 - x^3y = 0$ **5.** $(1 - ay)(a + x) = cy$ **6.** $2\cos x + \sec y = c$

**7.** $[\log(\sec x + \tan x)]^2 - [\log(\sec y + \tan y)]^2 = c)$ **8.** $y = 2x\tan^{-1}(cx)$

**9.** $e^{x/y} = y + c$

**10.** $\log y - \dfrac{x^2}{4y^2}\left(z\log\dfrac{y}{x} + 1\right) = c.$ **11.** $\log x = \dfrac{1}{2}\left[\dfrac{y}{x} - \dfrac{1}{2}\sin\left(\dfrac{2y}{x}\right)\right] + c$ **12.** $xy\cos(y/x) = c.$

**13.** $2x(\log x + c) = y\log[(y/a) - 1].$

**14.** $\tan y - c(1 - e^x)^3 = 0$ **15.** $y = a\tan\left(\sqrt{\dfrac{a}{x} - 1}\right)$ **16.** $y = \dfrac{a}{2}\log\dfrac{x-y-a}{x-y+a} + c$

**17.** $x^n = \left(n + \dfrac{bx^{1-n}}{1-n}\right)$

## Problems 11.4, page 521

**1.** $\log\dfrac{y^2 - x^2 + 2x - 1}{(x-1)^2} - \dfrac{b}{a}\log\dfrac{y - x + 1}{y + x + 1} = c$ **2.** $(y - x)^3 = c(y + x - 2)$

**3.** $(x + y)^7 = c(x - y - 2/3)^3$ **4.** $ax^2 + 2hxy + by^2 + 2gx + 2fy + c = 0$

**5.** $7x - 14y + 3\log(14x + 21y + 22) = c$ **6.** $x - y + \dfrac{3}{4}\log(8x - 12y - 5) = c$

**7.** $\log(x + y + \tfrac{1}{3}) + \tfrac{3}{2}(y - x) = c.$ **8.** $x^2 + 2yx - y^2 - 4x + 8y = c$

## Problems 11.5, page 525

**1.** $y = ce^{-\tan x} + \tan x - 1$
**2.** $y = \log x + c/\log x$
**3.** $y \sec^2 x = \sec x - 2$
**4.** $y \cosh x = c + \frac{2}{3} \cosh^3 x$
**5.** $y\sqrt{(1 - x^2)} = \sin^{-1} x + c$
**6.** $y = c(1 - x)^2 + (1 - x^2)$
**7.** $y(1 + \sin x) = c - x^2/2$
**8.** $2r \sin^2 \theta + \sin^4 \theta = c$
**9.** $ye^{x^2} = 2x + c$
**10.** $x = y^3 + cy$
**11.** $x = y^2 + 1 + ce^{y^2/2}$
**12.** $x = \sin^{-1} y - 1 + ce^{-\sin^{-1} y}$
**13.** $xe^{\tan^{-1} y} = \tan^{-1} y + c$

## Problems 11.6, page 528

**1.** $y^{-1} \sec x = \tan x + c$
**2.** $1/r = \sin \theta + c \cos \theta$
**3.** $x^2 + (4x^5 + c)y^4 = 0$
**4.** $1/y^2 = ce^{4x^2} - 1/4$
**5.** $\tan y = \frac{1}{2}(x^2 - 1) + Ce^{-x^2}$
**6.** $(x \log z)^{-1} = x^{-1} + c$
**7.** $e^{-y} = \frac{1}{2^x} + Cx$
**8.** $y/x = \log y + c$
**9.** $1/y = -x^2 + 2 + c \,.\, e^{-x^2/2}$
**10.** $e^{x+y} = \frac{1}{2}e^{2x} + c$
**11.** $\tan y = x^3 - 3x^2 + 6x - 6 + ce^{-x}$
**12.** $\cos y = \cos x\, (\sin x + c)$
**13.** $\sqrt{x} = \sqrt{y}(\log \sqrt{y} + c)$
**14.** $\frac{1}{xy} = -\frac{x^3}{3} + c$

## Problems 11.7, page 531

**1.** $x^3 + y^3 - 3axy = c$
**2.** $x^4 + 2x^2y^2 - y^4 - 2a^2x^2 - 2b^2y^2 = c$
**3.** $x^3 - 6x^2y - 6xy^2 + y^3 = c$
**4.** $\frac{x^5}{5} - x^2 y^2 + xy^4 + \cos y = c$
**5.** $x^5 + x^3y^2 - x^2y^3 - y^5 = c$
**6.** $x^3 + 3x^2y^2 + y^4 = c$
**7.** $x^2 - y^2 = cy^3$
**8.** $3y \cos 2x + 6y + 2y^3 = c$
**9.** $e^{xy} + y^2 = c.$
**10.** $x^2 + y^2 - 2xy - 2x - 2y = c$
**11.** $e^x = \sec x \tan y + c$
**12.** $x^2 y + xy - x \tan y + \tan y = c.$

## Problems 11.8, page 534

**1.** $ax + \tan^{-1} y/x = c$
**2.** $x^2 + y^2 - 2a^2 \tan^{-1} (y/x) = c$
**3.** $y + cx + \log x + 1 = 0$
**4.** $3 \log x - (y/x)^3 = c$
**5.** $\log (y/x) + \frac{1}{2}x^2y^2 = c$
**6.** $xy + \log (x/y) - (1/xy) = c$
**7.** $(y + 2/y^2)x + y^2 = c$
**8.** $x + ye^{x/y} = c$
**9.** $x^5 + 3x^4 + 3x^2y^2 = c$
**10.** $2 \cos (xy) + x^{-2} = c$
**11.** $\log\left(\frac{x}{y}\right) - xy = c$
**12.** $(x/y) + e^{x^3} = c$
**13.** $x^2 + y^2 - 2a^2 \tan^{-1}(y/x) = c$
**14.** $e^{xy^2} + x^4 - y^3 = c$
**15.** $y \sin x^2 - x^2 y + x = c$
**16.** $\sin x \tan y + \sin (x + y) = c$
**17.** $\frac{1}{3}e^{x^{-3}} - \frac{1}{2}\frac{y^2}{x^2} = c$

## Problems 11.9, page 535

**1.** $(x - y + c)(x^2 + y^2 + c) = 0$
**2.** $(y - 2x - c)(y - cx^3) = 0$
**3.** $(2y - x^2 + c)(y + x + ce^{-x} - 1) = 0$
**4.** $x^2 + y^2 = cx$
**5.** $(y - cx)(y^2 - x^2 - c) = 0$
**6.** $(y - c)(y + x^2 - c)(xy + cy + 1) = 0.$

## Problems 11.10, page 537

1. $x = \frac{2}{3}p + \frac{c}{p^2}$ & $y = \frac{p^2}{3} + \frac{c}{p}$ give the sol.

2. $y = \frac{2c}{p} + \frac{1-n}{1+n}p^n$

3. $x + c = \frac{a}{2}\left[\log \frac{p-1}{\sqrt{(1+p^2)}} - \tan^{-1} p\right]$, with the given relation

4. $y = 3x + \log(3 + ce^y)$

5. $2cy = c^2x^2 + 1$

6. $y = 2\sqrt{(xc)} + c^2$

7. $x = \sin p + c$, with the given relation.

8. $c^2 \log y = cx + 1$

## Problems 11.11, page 538

1. $y = c(x - c)^2$

2. $y^2 = 2cx + c^3$

3. $(y + ap)\sqrt{(p^2 - 1)} + a \cosh^{-1} p = c$, with the given relation

4. $y + (1 + p^2)^{-1} = c$, with the given relation.

## Problems 11.12, page 540

1. (*i*) Gen. sol. : $y = cx + a\sqrt{1 + c^2}$
   (*ii*) Gen. sol. : $c = \log(cx - y)$ ; Singular sol. : $y = x(\log x - 1)$
   (*iii*) Gen. sol. : $y = cx + 1/c^2$ ; Singular sol. : $4y^3 = 27x^2$
   (*iv*) Gen. sol. $y = cx - \sin^{-1} c$ ; Singular sol. $y = \sqrt{x^2 - 1} - \sin^{-1} \frac{\sqrt{x^2 - 1}}{x}$

2. $y = cx + (c - 2c^2)$

3. $(y - cx)(c - 1) = c$

4. $(y - cx)(c + 1) + ac^2 = 0$

5. $y^2 = cx^2 + c^2$ [***Hint***: Put $x^2 = u, y^2 = v$]

6. $xy = cy - c^2$ [***Hint*** : Put $u = y, v = xy$]

7. $y^2 = cx^2 - \frac{2c}{1+c}$.

## Problems 11.13, page 540

1. (*i*)
2. (*ii*)
3. (*iii*)
4. (*i*)
5. $\log y + c = x^2/2y^2$
6. $yx^2 = x^3 + c$
7. $e^x + x^2y + cy = 0$.
8. (*iii*)
9. $x^2 + y^2 + 2 \tan^{-1} y/x = c$
10. $\log x + c = y^3/3x^3$
11. (*i*)
12. $y^2 + 1/x + ce^{-y^2/2} = 2$
13. $y = cx + a/c^2$
14. (*d*)
15. $xy = c$ or $x^2 - y^2 = c$
16. 2
17. $xy = c$
18. (*b*)
19. (*b*)
20. $(1 + x^2)^{3/2} + (1 + y^2)^{3/2} = c$
21. $y = 5e^{-x}$
22. $x + y = u$
23. $x^{-5}$
24. § 11.11 (3)
25. $5x^4y^2 + 2(x^5 + y^5) = c$
26. $\sin(y/x) = cx$
27. (*a*)
28. (*c*)
29. $x + y\, dy/dx = 0$
30. $e^{-x^2} + 2\cos y = c$
31. (*c*)
32. False
33. False.
34. (*a*)
36. $\frac{d^3y}{dx^3} - 3\frac{d^2y}{dx^2} + 3\frac{dy}{dx} - y = 0$
37. $(e^x/y) + x^2 = c$.
38. $a = 1$, $b$ can have any value

## Problems 12.1, page 544

1. (*i*) $9y + 4x^2 = 0$ ; (*ii*) $3(x + 3y) = 2(1 - e^{3x})$

2. $y + 1 = 2e^{x^2/2}$

3. $x^2 + y^2 = cx$

4. $y = \sqrt{(a^2 - x^2)} + a \log\left(\frac{a - \sqrt{a^2 - x^2}}{x}\right) + c$

5. $y^2 = 4x$

6. $y = ae^{cx}$

7. $y = ax + b$

8. $x = 3y^2$

9. (*i*) $r(\theta - \alpha) = c$ ; (*ii*) $r = a + b\cos\theta$

10. $r^2 = a^2 \sin 2\theta$

11. $c^2x^2 = 2cy + 1$

12. $r = ae^{\theta \cot \alpha}$.

## Problems 12.2, page 547

1. $2x^2 + y^2 = c$
2. $x^2 + 2y^2 = c^2$
3. $3y^2 + 2x^2 = c^2$
4. $x^2 + y^2 + 2\mu y - c = 0$
5. This system is self-orthogonal
6. $x^{4/3} - y^{4/3} = b$
7. $r = c(1 - \cos\theta)$
8. $r = b\sin^2\dfrac{\theta}{2}$
9. $r = b(\cos\theta - \sin\theta)$
10. $r = 2b/(1 - \cos\theta)$
11. $r^2 = c^2 \sin 2\theta$
12. $r^n \sin\theta = b$
13. $r^n \cos n\theta = b$
16. $x^2 + y^2 + cx + 1 = 0$
17. $y = cx$.

## Problems 12.3, page 552

1. $V = \sqrt{\left(\dfrac{mg}{k}\right)} \tanh\left(\dfrac{9k}{m}t + c\right)$
3. $\dfrac{1}{k}\log_e 2$
5. $2\sqrt{v_o}/k$
6. $v^2 = 2gx - \dfrac{\lambda}{m}x^2$
10. $y = (\sqrt{150} - 0.001328t)^2$ ; $t_1 = 45$ min. 1 sec., $t_2 = 1$ hr. 16 min. 51 sec., $t_3 = 1$ hr. 38 min. 13 sec
11. 17 min. 4 sec.

## Problems 12.4, page 555

1. 0.0006931 sec
2. $\dfrac{10}{L^2 + R^2}(R\sin t - L\cos t + Le^{-Rt/L})$
4. $i = \dfrac{1}{5}(1 - e^{-100})$
5. $i = ke^{-t/RC} + \dfrac{\omega CE_m}{\sqrt{(1 + R^2C^2\omega^2)}}\sin(\omega t + \theta)$ where $\theta = \cot^{-1}(RC\omega)$.

## Problems 12.5, page 558

1. 52.5 mts
2. 48°C
3. 85 grams
4. 490,000 cal
5. 8.35 minutes.
6. $\dfrac{1}{1024}$
8. 50°C

## Problems 12.6, page 562

1. 0.000133 cal/hr
2. 2 log 3/log 2
3. $(1 - 1/p)^{21}$ times the original amount
4. 64.5 days
5. 1, 40,000 cal/hr.
6. $1.96 \times 10^7$ cal/hr, 112°C

## Problems 12.7, page 563

1. $6(1 - e^{-3})$
2. 54 m
3. 90.25%
4. $r(\theta - \alpha) = c$
5. $y = ae^{cx}$
6. rectangular hyperbola
7. $x^2 - y^2 = c$
8. The system is self-orthogonal
9. $2\sqrt{v_0}/k$
10. 2 log 3/log 2
11. Sunil
12. (*d*)
13. (*c*)
14. (*d*)
15. 2.21
16. (*c*)
17. (*c*)
18. (*a*)
19. False
20. True.

## Problems 13.1, page 567

1. $y = c_1e^{5x} + c_2e^{-x}$
2. $y = e^x(2\sin 2x - 3\cos 2x)$
3. $\dfrac{2}{3}e^{2t}\sin 3t$
4. $y = c_1e^{-x} + e^{x/2}\left(c_2\cos\dfrac{\sqrt{3}x}{2} + c_3\sin\dfrac{\sqrt{3}x}{2}\right)$
5. $y = c_1e^x + (c_2 + c_3x)e^{-x}$
6. $y = c_1e^x + c_2e^{2x} + c_3e^{3x}$

**7.** $y = (c_1 + c_2x + c_3x^2)\, e^x$

**8.** $y = c_1e^{\sqrt{2}x} + c_2e^{-\sqrt{2}x} + c_3 \cos\sqrt{2}x + c_4 \sin\sqrt{2}x$

**9.** $y = (c_1 + c_2x)\cos\sqrt{3}x + (c_3 + c_4x)\sin\sqrt{3}x$

**10.** $y = (c_1 + c_2x) \cos 2x + (c_3 + c_4 x) \sin 2x$

**11.** $y = c_1e^{-x} + c_2e^{2x} + e^{x/2}\,(c_3 \cos x/\sqrt{2} + c_4 \sin x/\sqrt{2}\,)$

**12.** $y = (c_1 + c_2x) \cos x + (c_3 + c_4x) \sin x + c_5e^x$.

**13.** $y = c_1e^{2x} + e^{-x/2} + \left\{(c_2 + c_3x) \cos \dfrac{\sqrt{3}}{2}x + (c_4 + c_5x) \sin \dfrac{\sqrt{3}}{2}x\right\}$

## Problems 13.2, page 579

**1.** $y = (c_1 + c_2x)\, e^{3x} + 3x^2\, e^{3x} + \dfrac{7}{25}\, e^{-2x} - \dfrac{1}{9} \log 2$

**2.** $y = \dfrac{3}{5} e^{-2x}\, (\cos x + 3 \sin x) - \dfrac{e^x}{10} - \dfrac{e^{-x}}{2}$

**3.** $y = c_1e^{2x} + c_2e^{-2x} + \dfrac{x}{4}\sinh(2x-1) + \dfrac{3^x}{(\log 3)^2 - 4}$

**4.** $y = c_1 + c_2e^{-x} - \dfrac{1}{10}(2 \cos 2x - \sin 2x)$

**5.** $y = c_1e^{-x} + c_2e^{-2x} + 1 + \dfrac{1}{10}(3 \sin 2x - \cos 2x)$

**6.** $y = c_1e^{-x} + c_2e^{2x} + \dfrac{1}{20}\,(\cos 2x - 3 \sin 2x)$

**7.** $y = c_1 + (c_2 + c_3x)\, e^{-x} - \dfrac{x^2}{2} e^{-x} + \dfrac{3}{50} \cos 2x - \dfrac{2}{25} \sin 2x$

**8.** $y = (c_1 + c_2x)e^{2x} - \dfrac{1}{8}\sin 2x + \dfrac{x^2}{2}e^{2x} + 1$

**9.** $y = (c_1 + c_2x)e^x + c_3\, e^{3x} + \dfrac{1}{8}\,(xe^{3x} - x^2e^x)$

**10.** $y = c_1e^x + c_2e^{-x} + \dfrac{e^x}{12}(2x^3 - 3x^2 + 9x)$

**11.** $y = c_1 + c_2e^x + c_3e^{-x} + xe^x - (x^2 + x) - 2 \sin x$

**12.** $y = \dfrac{2^x}{(\log 2)^3 - 6\,(\log 2)^2 + 11\,(\log 2) - 6} + \dfrac{1}{260}\,(7 \sin 2x + 9)$

**13.** $y = (c_1 + c_2x) \cos x + (c_3 + c_4 x) \sin x + x^4 - 24x^2 + 72 + \dfrac{1}{225} \sin 4x - \dfrac{1}{9} \sin 2x$

**14.** $y = c_1\, e^{-2x} + c_2e^{-3x} - \dfrac{e^{-2x}}{10}(\cos 2x + 2 \sin 2x)$

**15.** $y = c_1e^x + c_2e^{2x} - e^{2x} \sin (e^{-x})$

**16.** $y = e^{-x}(c_1 \cos \sqrt{2x} + c_2 \sin \sqrt{2x}\,) + \dfrac{e^x}{41}(4 \sin x + 5 \cos x)$.

**17.** $y = c_1\, e^{-x} + c_2e^{-3x} - \dfrac{e^{-x}}{5}(\sin x + 2 \cos x) + \dfrac{e^{3x}}{22}\left(x - \dfrac{5}{11}\right)$

**18.** $y = c_1 \cos \sqrt{2x} + c_2 \sin \sqrt{2x} + \dfrac{e^{3x}}{11}\left(x^2 - \dfrac{12}{11}x + \dfrac{50}{121}\right) + \dfrac{e^x}{17}(4 \sin 2x - \cos 2x)$

**19.** $y = c_1e^x + c_2e^{-x} + c_3 \cos x + c_4 \sin x - (1/5) \cos x \cosh x$

**20.** $y = c_1 \cos x + c_2 \sin x + e^x(1 - x + x^2/2) - \dfrac{x}{2}\cos x$

**21.** $y = c_1 \cos 4x + c_2 \sin 4x + \dfrac{1}{7}\left(x \sin 3x - \dfrac{6}{7} \cos 3x\right)$

**22.** $y = (c_1 + c_2x)\, e^{-x} + \dfrac{1}{2} \cos x + \dfrac{1}{2}\,(x - 1) \sin x$

**23.** $y = c_1e^x + c_2e^{-x} - \dfrac{1}{2}\,(x \sin x + \cos x) + (xe^x/12)\,(2x^2 - 3x + 9)$

**24.** $y = (c_1 + c_2x)e^{-x} - xe^{-x}\cos x$

**25.** $y = (c_1 + c_2x)e^{-x} + e^{-x}(6\cos x + 4x\sin x - x^2\cos x)$

**26.** $y = c_1e^{-x} + c_2e^{-2x} + c_3e^{-3x} + e^{-2x}\,.\,e^{e^x}$

**27.** $y = c_1 \cos ax + c_2 \sin ax - \frac{1}{a^2} \cos ax \log(\sec ax + \tan ax)$.

## Problems 13.3, page 583

**1.** $y = (c_1 - x/a)\cos ax + [c_2 + (1/a^2)\log \sin ax]\sin ax$

**2.** $y = c_1 \cos x + c_2 \sin x + \cos x \log(\cos x) + x \sin x$

**3.** $y = c_1 \cos x + c_2 \sin x - \cos x \log(\sec x + \tan x)$

**4.** $y = c_1 \cos x + c_2 \sin x + \frac{x}{2} \sin x - \frac{x^2}{4} \cos x$

**5.** $y = (c_1 + c_2 x)e^{2x} + \frac{1}{16} e^{-2x}$

**6.** $y = (c_1 + c_2 x)e^x + xe^x \log x$

**7.** $y = (c_1 + c_2 x + c_3 x^2)\cos x + (c_4 + c_5 x + c_6 x^2)\sin x + e^{-x/2}$
$\left[(c_7 + c_8 x)\cos\left(\frac{\sqrt{3}}{2}x\right) + (c_9 + c_{10}x)\sin\left(\frac{\sqrt{3}}{2}x\right)\right]$

**8.** $y = e^x(c_1 \cos x + c_2 \sin x) - e^x \cos x \log(\sec x + \tan x)$

**9.** $y = c_1 + c_2 e^{2x} - \frac{1}{2} e^x \sin x$

**10.** $y = c_1 \cos x + c_2 \sin x + \sin x \log(1 + \sin x) - x \cos x - 1$

**11.** $y = (c_1 + c_2 x)e^x + \frac{x^2 e^x}{4}(2 \log x - 3)$

**12.** $y = c_1 \cos x + c_2 \sin x - x \sin x$

**13.** $y = c_1 e^x + c_2 e^{-2x} - \frac{1}{4}(2x + 1) - \frac{1}{10}(\cos x + 3 \sin x)$

**14.** $y = e^x(c_1 \cos \sqrt{2}x + c_2 \sin \sqrt{2}x) + \frac{1}{27}(9x^3 + 18x^2 + 6x - 8) + \frac{1}{4}(\cos x - \sin x)$

**15.** $y = c_1 + c_2 e^{2x} - \frac{1}{2} e^x \sin x$.

**16.** $Q = Q_0 e^{-Rt/2L}\left(\cos \lambda t + \frac{12}{2\lambda L} \sin \lambda t\right)$; where $\lambda = \sqrt{\frac{1}{LC} - \frac{R^2}{4L^2}}$.

**18.** $y = e^{-1/4x}\left[c_1 \cos\left(\frac{\sqrt{2}}{4}x\right) + c_2 \sin\left(\frac{\sqrt{2}}{4}x\right)\right]$

## Problems 13.4, page 589

**1.** $y = c_1 \cos(\log x) + c_2 \sin(\log x) + \frac{x}{2}$

**2.** $y = c_1 x^4 + c_2 x^{-1} + \frac{x^4}{5} \log x$

**3.** $y = (c_1 + c_2 \log x)x^2 + \frac{1}{4} + 2x + \frac{1}{2}x^2(\log x)^2$

**4.** $y = c_1 x^2 + c_2 x^{-1} + \frac{1}{3}(x^2 - 1/x)\log x$

**5.** $u = \frac{kr}{8}(a^2 - r^2)$

**6.** $c_1 x^{-1} + c_2 x^{-2} + \frac{1}{2}\log x - \frac{3}{4}$

**7.** $y = c_1 x^{-1} + \sqrt{x}\{c_2 \cos[(\sqrt{3}/2)\log x] + c_3 \sin[(\sqrt{3}/2)\log x]\} + \frac{1}{2}x + \log x$

**8.** $y = c_1 x^{-2} + x[c_2 \cos(\sqrt{3}\log x) + c_3 \sin(\sqrt{3}\log x)] + 8\cos(\log x) - \sin(\log x)$

**9.** $y = (c_1 + e^x)\frac{1}{x} + (e^x - xe^x + c_2)\frac{1}{x^2}$

**10.** $y = c_1 x^{-1} + \{c_2 \cos(\log x) + c_3 \sin(\log x)\}x + 5x + 10(\log x)/x$

**11.** $y = \frac{1}{x}(c_1 + c_2 \log x) + \frac{1}{x}\log\frac{x}{1 - x}$

**12.** $y = x^{-2}(c_1 + c_2 \log x) + \frac{x}{9}\left(\log x - \frac{2}{3}\right)$

**13.** $y = c_1(x)^{1+\sqrt{3}} + c_2(x)^{1-\sqrt{3}} - \frac{1}{2}x^2(\log x + 1)$

**14.** $y = c_1 x^3 + c_2 x^{-4} + \frac{x^3}{98}\log x(7\log x - 2)$

**15.** $y = x\,[a\cos(\log x) + b\sin(\log x)] - \dfrac{1}{2}\log x\,(\cos(\log x))$

**16.** $y = c_1(2x+3)^a + c_2(2x+3)^b - \frac{3}{14}(2x+3) + \frac{3}{4}$ where $a, b = \dfrac{3 \pm \sqrt{57}}{4}$

**17.** $y = c_1(2x+7)^{1+\sqrt{2}} + c_2(2x+7)^{1-\sqrt{2}} - \dfrac{1}{2}$

**18.** $y = c_1\cos[\log(1+x)] + c_2\sin[\log(1+x)] + 2\log(1+x)\sin[\log(1+x)]$

**19.** $y = c_1\cos\log(1+x) + c_2\sin\log(1+x) - \frac{1}{3}\sin[2\log(1+x)]$

**20.** $y = c_1(3x+2)^{1/3} + c_2(3x+2)^{-1} + \dfrac{1}{27}\left[\dfrac{1}{15}(3x+2)^2 + \dfrac{1}{4}(3x+2) - 7\right]$

**21.** $y = c_1(x-1) + c_2(x-1)^2 + c_3(x-1)^{-2} + \log(x+1) + 1$

## Problems 13.6, page 594

**1.** $x = (c_1 + c_2x)\,e^{3x}$ ; $y = [(1-2x)(c_2 - 2c_1)]\,e^{3x}$

**2.** $x = \dfrac{e^{6t}}{2}\{(c_1+c_2)\cos t + (c_2 - c_1)\sin t\}$ ; $y = e^{6t}(c_1\cos t + c_2\sin t)$

**3.** $x = e^t + e^{-t}, y = e^{-t} - e^t + \sin t$

**4.** $x = c_1e^t + c_2e^{-5t} + \frac{6}{7}e^{2t}$ ; $y = c_2e^{-5t} - c_1e^t + \frac{8}{7}e^{2t}$

**5.** $x = 4e^{-2t} - 3e^{-3t}$ ; $y = 3e^{-3t} - 2e^{-2t}$

**6.** $x = \frac{1}{5}e^t + \frac{2}{5}e^{-t} - c_1\sin 2t + c_2\cos 2t,\ y = \frac{2}{5}e^t + \frac{1}{5}e^{-t} + c_1\cos 2t + c_2\sin 2t$

**7.** $x = c_1e^t + c_2e^{-5t} + \dfrac{3}{7}e^{2t} - \dfrac{2}{5}t - \dfrac{13}{25},\ y = c_1e^t - c_2e^{-5t} - \dfrac{4}{7}e^{2t} - \dfrac{3t}{5} - \dfrac{12}{25}$

**8.** $x = -t - \dfrac{2}{3},\ y = \dfrac{1}{2}t^2 + \dfrac{4}{3}t + c$

**9.** $y = c_1e^x + c_2e^{-2x} + 2e^{-x},\ z = 3c_1e^x + 2c_2e^{-2x} + 3e^{-x}$

**10.** $x = c_1e^{-t} + c_2e^{3t} - \frac{1}{5}(\cos t - 2\sin t)$ ; $y = 2c_1e^{-t} - 2c_2e^{3t} + \dfrac{1}{5}(\sin t + 2\cos t)$

**11.** $x = \dfrac{1}{2}\left(t + \dfrac{1}{t}\right),\ y = \dfrac{1}{2}\left(-t + \dfrac{1}{t}\right)$

**12.** $x = c_1 + c_2e^{-2t} + e^t$ ; $y = c_1 - c_2e^{-2t} + e^t$

**13.** $x = c_1e^{-t} + c_2e^{3t} - \dfrac{1}{5}(\cos t - 2\sin t),\ y = 2c_1e^{-t} - 2c_2e^{3t} + \dfrac{2}{5}\cos t + \dfrac{1}{5}\sin t$

**14.** $x = (c_1 + c_2t)\,e^{-t} + (c_3 + c_4t)\,e^t,\ y = -\dfrac{1}{2}[c_1 + c_2(1+t)]\,e^{-t} + \dfrac{1}{2}[c_4(1-t) - c_3]e^t$

**15.** $x = c_1e^x + c_2e^{-x} - c_3\cos x - c_4\sin x$ ; $y = c_1e^x + c_2e^{-x} + c_3\cos x + c_4\sin x$

**16.** $x = c_1e^t + c_2e^{-t} + c_3\cos t + c_4\sin t - \dfrac{t}{4}\cos t + \dfrac{t}{4}\sin t$

$y = -c_1e^t - c_2e^{-t} + c_3\cos t + c_4\sin t + \dfrac{1}{4}(2+t)(\sin t - \cos t)$

**17.** $x = \dfrac{8}{9}\left(1 - \cos\dfrac{3}{2}t\right),\ y = \dfrac{4}{3}t - \dfrac{8}{9}\sin\dfrac{3}{2}t.$

**18.** $y = \dfrac{c_1}{x} + c_2x^2 - \dfrac{1}{3}x^2\log x - \dfrac{2}{9}x^2$

**19.** $y = \dfrac{c_1}{x} + \dfrac{c_2}{x^2} + \dfrac{1}{6}x - \dfrac{1}{x^2}\sin x$

**20.** $y = c_1x + c_2x\log x + c_3x(\log x)^2 + c_4x(\log x)^4 + 4 + \log x$

**21.** $y = (c_1 + c_2\log x)\,x + \dfrac{C_3}{x} + \dfrac{\log x}{4x}$

**22.** $a = c_1x + c_2\cos\left(\log\sqrt{3}x + c_3\right) + \dfrac{x^2}{7}$

## Problems 13.7, page 595

**1.** $x = c_1 e^t + e^{-t_2}\left(c_2 \cos\frac{\sqrt{3}}{2}t + c_3 \sin\frac{\sqrt{3}}{2}t\right)$

**2.** $-\frac{1}{25}(3 \sin 2x + 4 \cos 2x)$ **3.** 1/6 **4.** $e^x(x - 1)$

**5.** (*b*) **6.** $y = c_1 + c_2 x + (c_3 + c_4 x) e^{-3x}$ **7.** (*a*)

**8.** $y = e^x (c_1 + c_2 x) \cos 2x + (c_3 + c_4 x) \sin 2x$ **9.** $y = \cos x + 2 \sin x$

**10.** (*ii*) **11.** $y = (c_1 + c_2 x) \cos x + (c_3 + c_4 x) \sin x$

**12.** $\frac{1}{10} \cosh 3x$ **13.** $y = a \log x + 6$

**14.** $y = (c_1 + c_2 x)\, e^{\sqrt{2}x} + (c_3 + c_4 x) e^{-\sqrt{2}x}$ **15.** $-e^x \sin x$

**16.** $\frac{1}{2} e^{2x}$ **17.** $\sin 2x$ **18.** $x\, 2^x/2\, [\log 2 - 2]$

**19.** $y = (c_1 + c_2 x) e^{-x/2} + c_3$ **20.** (*c*) **21.** (*a*)

**22.** $x e^{-t}$ **23.** $\frac{d^2y}{dt^2} + 7y = 2e^t$. **24.** (*c*)

**25.** (*a*) **26.** (*b*)

**27.** $y = (c_1 + c_2 \log x)\, x$ **28.** $x^2 y_2 + x y_1 - y = 0$ **29.** $\frac{1}{9} \log 2$

**30.** $e^t$ **31.** $\frac{1}{8} \cos 2x$ **32.** (*d*)

**33.** $x^3 e^x/6$ **34.** $\frac{d^3y}{dx^3} + 2\frac{d^2y}{dx^2} + \frac{dy}{dx} = 0$ **35.** (*a*)

**36.** False **37.** False.

## Problems 14.1, page 601

**1.** 38 sec **4.** $x = \frac{ue^{-\lambda nt}}{n\sqrt{1-\lambda^2}} \sin\, [nt \sqrt{(1 - \lambda^2)}\,]$

**6.** It must be shortened by 1/8640 of its length

**7.** It must be increased by 0.0074 ft./sec$^2$ **8.** 4321/4319

**9.** $k^2 > 4\mu$, $\theta = c_1 e^{\frac{-k+\lambda}{2}t} + c_2 e^{\frac{-k-\lambda}{2}t}$ where $\lambda = \sqrt{(k^2 - 4\mu)}$

$k^2 = 4\mu$, $\theta = (c_1 + c_2 t)\, e^{-kt/2}$

$k^2 < 4\mu$, $\theta = c_1\, e^{-kt/2} \cos\left(\frac{\sqrt{4\mu - k^2}}{2} t + c_2\right)$

**10.** $x = \frac{F_0}{2n^2}(\sin nt - n \cos nt)$.

## Problems 14.2, page 609

**2.** 1 ft. ; $\pi/2\sqrt{2}$ sec ; $4\sqrt{2}$ ft./sec **4.** $\pi/\sqrt{7}$

**5.** $x = e^{-5t}\left\{\cos\sqrt{220t} + (5/\sqrt{220})\sin\sqrt{220}\,t\right\}$. **6.** 0.45 sec. ; 1.15 sec

**8.** $x = \frac{10}{21}e^{-t} - \frac{17}{27}e^{-8t} - \frac{\sqrt{2}}{3}\sin\left(9t + \frac{\pi}{4}\right)\frac{\sqrt{2}}{3}, \frac{2\pi}{9}$ sec., $9/2\pi$ cycles/sec

**9.** $0.8\,(2\sin 4t - \cos 4t)$

**10.** (*i*) $x = Ae^{-kt}\cos\left\{t\sqrt{(b^2 - k^2)} + B\right\} + \{e^{-kt}/(b^2 + k^2 - n^2)\}\sin nt$

(*ii*) $x = Ae^{-kt}\cos\left\{t\sqrt{(b^2 - k^2)} + B\right\} - (te^{-kt}/2n)\cos nt$

## Problems 14.3, page 612

**2.** $i = 5(1 - e^{-0.1})$ **3.** $i = I\sin(T/\sqrt{LC})$

**4.** $i = 2Eke^{-RT/2L}\sin(kt/2L)$, where $k = \sqrt{\left(\frac{4L - CR^2}{C}\right)}$

**5.** $R^2 > 4L/C$ for over damping ; $R^2 = 4L/C$ for critical damping ; $R^2 < 4L/C$ for under damping ; critical resistance $= 2\sqrt{L/C}$

**6.** $q = e^{-500t}(0.002\cos 1323t + 0.0008\sin 1323t)$

**9.** (*i*) $i = Ae^{-\alpha t}\cosh(\beta t + \gamma)$ ; (*ii*) $i = Ae^{-\alpha t}\cos(\beta t + \gamma) + \frac{E}{R}\cos\phi\sin(pt + \phi)$

where $\alpha = -\frac{R}{2L}$, $\beta = \pm\left\{\left(\frac{R}{2L}\right)^2 - \frac{1}{CL}\right\}$ and $\phi = \tan^{-1}\{(1 - CLp^2)/CRp\}$.

**10.** $x = c_1\cos pt + c_2\sin pt + \frac{k}{p^2 - \omega^2}\sin\omega t$, $x = c_1\cos pt + c_2\sin pt - \frac{k}{2p}t\cos pt$

**11.** $S = l + (S_o - l)\cos t\sqrt{g/c}$

## Problems 14.4, page 620

**4.** $y = \frac{wl}{2Pn}\operatorname{cosec}\frac{nl}{2}\cos\left(nx - \frac{nl}{2}\right) - \frac{wl}{2nP}\cot\frac{nl}{2} + \frac{w}{2P}(x^2 - lx)$

**6.** $y = \frac{F}{P}[n\sin nx - l\cos nx + l - x]$ **7.** $\pi^2 EI/4l^2$ **8.** $\frac{W}{2a^2}\left(\operatorname{sech}\frac{al}{2} - \sec al\right)$

## Problems 14.5, page 624

**1.** $\frac{2u\sin\alpha}{g}; \frac{u^2\sin 2\alpha}{g}$ **2.** (*i*) $\frac{2u^2\sin(\alpha - \beta)\cos\alpha}{g\cos^2\beta}$ ; (*ii*) $\frac{u^2}{g(1 + \sin\beta)}$

3. $100\,e^{-0.2828t}$, 24.5 days **4.** $4x^2 + k^2y^2 = 4$

**5.** $i_1 = \frac{a}{p + \omega}\sin pt$, $i_2 = \frac{a}{p + \omega}\cos pt$. **7.** $x = a(nt - \sin nt)$, $y = a(1 - \cos nt)$

**8.** $x = \frac{E}{H\omega}(1 - \cos\omega t)$, $y = \frac{E}{H\omega}(\omega t - \sin\omega t)$, where $\omega = eH/m$.

**10.** 64.5 days **11.** 85 gms.

## Problems 14.6, page 624

**1.** (*b*) **2.** (*b*) **3.** (*c*) **4.** (*b*)
**5.** (*b*) **6.** (*b*) **7.** (*b*) **8.** 60 sec

**9.** $30/\pi\sqrt{LC}$ **10.** 0.0074 sec **11.** resonance **12.** $EI\frac{d^2y}{dx^2} - Py = \frac{w}{2}(x^2 - lx)$

**13.** $y = 0$ and $\frac{dy}{dx} = 0$

## Problems 15.1, page 626

**1.** $y = -x^2 \sin x - 4x \cos x + c_1x + c_2$ **2.** $y = \frac{x^4}{24} + \frac{x^3}{6}\log x - \frac{11}{35}x^3 + c_1x^2 + c_2x + c_3$

## Problems 15.2, page 628

**1.** $2(y^{1/4} - 1) = x$ **2.** $\sqrt{(y^2 - 8y)} + 4\cosh^{-1}(\frac{1}{4}y - 1) = 3x$

**3.** $r = \frac{\sqrt{(v^2 - a^2\omega^2)}}{\omega}\sinh\left[\omega t + \sinh^{-1}\frac{a\omega}{\sqrt{(v^2 - a^2\omega^2)}}\right]$

**4.** $t = \frac{h^{3/2}}{\sqrt{(2g)a}}\left[\cos^{-1}\sqrt{\frac{x}{h}} - \sqrt{\left(\frac{hx - x^2}{h}\right)}\right]$

## Problems 15.3 page 629

**1.** $y = c_1 - x^2 - c_2/x$ **2.** $y = c_1x + (c_1^2 + 1)\log(x - c_1) + c_2$
**3.** $15c_1^2y = 4(c_1x + a^2)^{5/2} + c_2x + c_3$ **4.** $x^2 + y^2 = a^2$
**5.** $\theta = \frac{m}{\mu a}\log\left(1 + \frac{\mu a\omega t}{m}\right)$ **6.** $v = \frac{1}{r_1 - r_2}\left[v_1r_1 - v_2r_2 - \frac{(v_2 - v_1)r_1r_2}{r}\right]$.

## Problems 15.4, page 631

**1.** $y = 2x - 2\log(1 - c_1e^{2x}) + c_2$ **2.** $y^2 = x^2 + c_1x + c_2$
**3.** $\log y = c_1e^x + c_2e^{-x}$ **4.** $(\log y - 1)(c_1x + c_2) = 1$
**5.** $(x - a)^2 + y^2 = c^2$, circles whose centres are on the *x*-axis.

## Problems 15.5, page 633

**1.** $y = e^{x^2}(c_1x + c_2)$ **2.** $y = (x^2 - x + c_1x)e^x + c_2x$
**3.** $y = e^x(c_1\log x + x + c_2)$ **4.** $cy = 1 + (k - x)\cot x$
**5.** $y = \left[c_1 - \frac{1}{2}\cos x - \frac{1}{5}c_2e^{-2x}(\cos x + 2\sin x)\right]e^x$.

## Problems 15.6, Page 634

**1.** $y = c_1\cos(\sin x) + c_2\sin(\sin x)$ **2.** $y = c_1\cos(1/x) + c_2\sin(1/x)$
**3.** $y = c_1e^t + c_2e^{-t} - t$ where $t = \cos x$ **4.** $y = c_1\cos(-n/x) + c_2\sin(-n/x)$

**5.** $y = c_1\cos(2\tan^{-1}x) + c_2\sin(2\tan^{-1}x)$ **6.** $y = c_1e^{\sqrt{2}\sin x} + c_2e^{-\sqrt{2}\sin x} + \sin^2 x$.

**7.** $y = e^{-\frac{x^2}{2}}\left[c_1e^{-3x} + c_2e^{3x} - \frac{1}{9}\left(x^2 + \frac{2}{9}\right)\right]$.

## Problems 15.7, page 636

1. $nx - lz = c(mz - ny)$
2. $x^2 + y^2 + z^2 = cx$
3. $xy^2 = cz^3$
4. $x^2 + y^2 - xz = cz$
5. $x^2 + y^2 = a \sin(2x + c)$
6. $x + y + z + \log(xyz) = c.$
7. $2xz = y^2(z^2 + c)$
8. $e^x y + e^y z + e^x.x = e^x$
9. $x + y + z = c(xy + yz + zx)$

## Problems 15.8, page 638

1. $x^3 - y^3 = c_1, x^2 - z^2 = c_2$
2. $lx + my + nz = c_1, x^2 + y^2 + z^2 = c_2$
3. $\frac{x-y}{y-z} = c_1, \frac{z-x}{y-z} = c_2$
4. $x^2 - y - 2xy = c_1, x^2 - y^2 - z^2 = c_2$
5. $xyz = c_1, x^2 + y^2 + z^2 = c_2$
6. $y = c_1 z, x^2 + y^2 + z^2 = c_2 z.$

## Problems 16.1, page 641

1. $y = a_1\left(x - \frac{x^3}{3!} + \frac{x^5}{5!} - \ldots\right)$
2. $y = C_1 x(1-x)^{-2} + C_2\{x(1-x)^{-2}\log x + (1-x)^{-1}\}$
3. $y = a_0\left(1 - \frac{x^2}{2} + \frac{x^4}{2\cdot 4} - \frac{x^6}{2\cdot 4\cdot 6} + \ldots\ldots\right) + a_1\left(x - \frac{x^3}{3} + \frac{x^5}{3\cdot 5} - \frac{x^7}{3\cdot 5\cdot 7} + \ldots\ldots\right)$
4. $y = 4 + 5x - 4x^2 - \frac{5}{3}x^3 - \frac{x^5}{3} - \frac{x^7}{7} - \ldots\ldots$
5. $y = a_0\left(1 + \frac{x^2}{2} - \frac{x^4}{8} + \frac{x^6}{16} - \frac{5x^8}{128} + \ldots\right) + a_1 x$
6. $y = a_0(1-2x^2) + a_1 x\left(1 - \frac{x^2}{2} - \frac{x^4}{8}\ldots\ldots\right)$

## Problems 16.2, page 648

1. $y = c_1 \cos\sqrt{x} + c_2 \sin\sqrt{x}$
2. $y = a_0\left(1 - x^2 + \frac{x^4}{4}\ldots\ldots\right) + a_1\left(x - \frac{x^3}{2} + \frac{3x^5}{10}\ldots\ldots\right)$
3. $y = (c_1 + c_2\log x)\left[1 + x + \frac{1}{(2!)^2}x^2 + \frac{1}{(3!)^3}x^3 + \ldots\ldots\right] - 2c_2\left[x + \frac{1}{(2!)^2}\left(1+\tfrac{1}{2}\right)x^2 + \frac{1}{(3!)^2}\left(1+\tfrac{1}{2}+\tfrac{1}{3}\right)x^3 + \ldots\ldots\right]$
4. $y = c_1(1 + x + x^2/4 + x^3/4\,.\,7 + \ldots) + c_2 x^{2/3}(1 + \tfrac{1}{3}x + x^2/3\,.\,6 + x^3/3\,.\,6\,.\,9 + \ldots)$
5. $y = a_0\left(1 - 2x + \frac{3}{2!}x^2 - \frac{4}{3!}x^3 + \ldots\right) + a_1\left[y_1 \log x + a_0\left(3x - \frac{13}{4}x^2 + \ldots\right)\right]$
6. $y = a_0 x\left(1 + \frac{x}{5} + \frac{x^2}{70} + \ldots\right) + a_1 x^{-1/2}\left(1 - x - \frac{x^2}{2} + \ldots\right)$
7. $y = c_1 x^{-\frac{1}{2}}\left(1 + \frac{x}{2} + \frac{x^2}{40} + \ldots\ldots\right) + c_2 x^{1/4}\left(1 + \frac{x}{14} + \frac{x^2}{616} + \ldots\ldots\right)$
   $+ c_2\sqrt{x}(x + x^2/2\,.\,3 + x^4/2\,.\,4\,.\,3\,.\,7 + x^6/2\,.\,4\,.\,6\,.\,3\,.\,7\,.\,11 + \ldots)$
8. $y = a_0\sqrt{x}(1-x) + a_1\left(1 - 3x + \frac{3x^2}{1\cdot 3} + \frac{3x^3}{3\cdot 5} + \frac{3x^4}{5\cdot 7} + \ldots\right)$
9. $y = a_0(1 - \tfrac{2}{3}x + \tfrac{1}{3}x^2 + \ldots) + a_1 x^4(1 - 2x + 3x^2 - 4x^3 + \ldots)$
10. $y = (C_1 + C_2\log x)\left[1 - \frac{x^3}{3^2} + \frac{x^6}{3^4(2!)^2} - \frac{x^9}{3^6(3!)^2} + \ldots\right]$
    $+2C_2\left[\frac{x^3}{3^3} + \frac{1}{3^5(2!)^2}\left(1+\frac{1}{2}\right)x^6 + \frac{1}{3^7(3!)^2} + \left(1+\frac{1}{2}+\frac{1}{3}\right)x^9 + \ldots\right]$
11. For $m = -3$, $y = a_0 x(1-2x)$.

**12.** $(C_1 + C_2 \log x)\left[1 + x + \frac{x^2}{(2!)^2} + \frac{x^3}{(3)^3} + \dots\right] - C_2\left[x + \frac{1}{(2!)^2}\left(1+\frac{1}{2}\right)x^2 + \frac{1}{(3!)^2}\left(1+\frac{1}{2}+\frac{1}{3}\right)x^3 + \dots\right]$

**13.** $C_1\left[1 - \frac{x^2}{2^2} + \frac{x^4}{2^2 4^2} - \frac{x^6}{2^2 4^2 6^2} + \dots\right] + C_2\left[y_1 \log x + a_0\left\{\frac{x^2}{2^2} - \frac{1}{2^2 4^2}\left(1+\frac{1}{2}\right)x^4 + \frac{1}{2^2 4^2 6^2}\left(1+\frac{1}{2}+\frac{1}{3}\right)x^6 \dots\right\}\right]$

## Problems 16.3, page 657

**1.** 0.224, 0.44.

## Problems 16.4, page 663

**1.** $y = c_1 J_{1/2}(x) + c_2 J_{-1/2}(x)$

**2.** $y = c_1 J_{2/5}(x) + c_2 J_{-2/5}(x)$

**3.** $y = x^n[c_1 J_n(kx) + c_2 y_n(kx)]$ where $n = \frac{1}{2}(1 - \alpha)$

**4.** $y = x[c_1 J_1(2x) + c_2 Y_1(2x)]$

**5.** $y = c_1\sqrt{x} J_1(2\sqrt{x}) + c_2\sqrt{x}\, Y_1(2\sqrt{x})$

**7.** $y = c_1\sqrt{x} J_n(x) + c_2\sqrt{x}\, J_{-n}(x)$

**11.** $x^2 = \sum_{n=1}^{\infty} \frac{2}{\alpha_n^2} \cdot \frac{1}{J_2^2(3\alpha_n)}(3\alpha_n J_1(3\alpha_n) - 2J_2(2\alpha_n))$

## Problems 16.5, page 673

**3.** (i) $\frac{2}{3}P_2 + 2P_1 + \frac{4}{3}P_0$ (ii) $\frac{2}{5}P_3 + \frac{4}{3}P_2 - \frac{2}{5}P_1 - \frac{7}{3}P_0$

(iii) $\frac{2}{5}P_3(x) + \frac{10}{3}P_2(x) + 6P_1(x) + \frac{8}{5}P_0(x)$

**8.** (i) $f(x) = -\frac{7}{3}P_0(x) - \frac{2}{5}P_1(x) + \frac{4}{3}P_2(x) + \frac{2}{5}P_3(x)$;

(ii) $f(x) = -\frac{32}{15}P_0(x) - \frac{4}{5}P_1(x) - \frac{40}{21}P_2(x) + \frac{2}{5}P_3(x) + \frac{8}{35}P_4(x)$.

## Problems 16.6, page 675

**1.** $x^3 = \frac{1}{4}(3T_1 + T_3)$.

## Problems 16.7, page 678

**1.** $y_n(x) = \sin nx,\ n = 1, 2, \dots$

**2.** $y_n(x) = \sin[(2n + 1)\pi x/2l],\ n = 0, 1, 2, \dots$

**3.** $y_n(x) = \cos nx,\ n = 0, 1, 2, \dots$

**4.** 1, $\sin x$, $\cos x$, $\sin 2x$, $\cos 2x$, ...

**5.** $y_n(x) = \sin\left[(2n + 1)\frac{\pi}{2}\log|x|\right],\ n = 0, 1, 2, \dots$

**6.** $[xe^{-x}y']' + ne^{-x^2}y = 0,\ p(x) = e^{-x}$

**7.** $[e^{-x^2}y']' + 2ne^{-x^2}y = 0,\ p(x) = e^{-x^2}$.

## Problems 16.8, page 678

**1.** $\frac{1}{3}(10 - 9P_1 + 8P_2)$

**2.** $\sqrt{(2/\pi x)}\cos x$

**3.** $\frac{2}{(2n+1)}$

**4.** zero

**5.** zero

**6.** $J_n(x) = \frac{x}{2n}(J_{n+1}(x) + J_{n-1}(x))$

**7.** $\int_0^1 xJ_n(\alpha x)J_n(\beta x)\,dx = 0$

**8.** $x\frac{d^2y}{dx^2} + \frac{dy}{dx} + xy = 0$

**9.** $\sqrt{(2/\pi x)}\sin x$

**14.** $z = f_1(y - ax) + xf_2(y - bx) + \frac{x^3y}{6} - \frac{(a+b)}{24}x^4$. **15.** $z = f_1(y - x) + xf_2(y - x) + \frac{1}{4}(x^4 - 2x^3y + 2x^2y^2)$

**16.** $z = f_1(y - x) + f_2(y + 2x) + ye^x$ **17.** $z = f_1(y - x) + xf_2(y - x) + f_3(y + x) + \frac{e^x}{25}(\cos 2y + 2\sin 2y)$

**18.** $z = f_1(y - x) + xf_2(y - x) + x\sin y$.

## Problems 17.7, page 703

**1.** $z = e^{-x}\phi_1(y) + e^x\phi_2(y - x) - \frac{xe^{-x}}{2}$ **2.** $z = e^x\phi_1(y + x) + e^{2x}\phi_2(y + x) + \frac{1}{2}e^{2x-y}$

**3.** $z = e^x\phi_1(y - x) + e^{3x}\phi_2(y - 2x) + x + 2y + 6$ **4.** $z = f_1(y+x) + e^{3x}f_2(y-x) - \frac{1}{3}\left(\frac{x^2y}{2} + \frac{xy}{3} + \frac{x^2}{3} + \frac{x^3}{6} + \frac{2x}{9}\right) - xe^{x+2y}$

**5.** $z = f_1(y) + e^{-x}f_2(y + x) + \frac{1}{3}x^3 - x^2 + xy^2 + 6x$ **6.** $w = f_1(y + ax) + x f_2(y + ax) = \frac{1}{2}x^2\psi(y + ax)]$

**7.** $z = f_1(x) + e^{3y}f_2(2y - x) + \frac{3}{50}[4\cos(3x - 2y) + 3\sin(3x - 2y)]$.

**8.** $z = f_1(y + x) + f_2(y - x) + \frac{1}{2}\tan x \tan y$ **9.** $\phi = f_1(y - x) + x f_2(y - x) + x\sin y$

**10.** $w = f_1(y + 3x) + x f_2(y + 3x) + x^2(3x + y)$ **11.** $u = f_1(y + x) + f_2(y - x) + f_3(y + 2x) - \frac{1}{2}xe^{x+y}$

## Problems 17.8, page 705

**1.** $z = \phi_1(x) + \phi_2(x + y + z)$ **2.** $z = \phi_1(y + \sin x) + \phi_2(y - \sin x)$

**3.** $y + xf(x + y + z) = F(x + y + z)$ **4.** $z = \phi_1(y/x) + \phi_2(x^2 + y^2) + xy$

**5.** $y = \phi_1(z) + e^x\phi_2(z)$ **6.** $y = \phi_1(x + y + z) + x\phi_2(x + y + z)$.

7. $z = f_1(x) + f_2(x + y + z)$ 8. $z = f_1(xy) + f_2(x + y)$

9. $y^3 - yf(y^2 + 2x) + F(y^2 + 2x)$, 10. $z = yf_1\left(\frac{y}{x}\right) + f_2\left(\frac{y}{x}\right)$

11. $y = xz - F_1(z) + F_2(x)$

## Problems 17.9, page 705

**1.** order two & degree two **2.** $z = f_1(y + 2x) + xf_2(y + 2x)$ **3.** $z = -x^2 \sin xy + yf(x) + \phi(x)$

**4.** $x^2 + y^2 + z^2 = f(x + y + z)$ **5.** $-\frac{1}{2}\sin(x + y)$ **6.** $xp + yq = z$

**7.** $z = ax + (1 - \sqrt{a})^2y + c$ **8.** $\sqrt{x} - \sqrt{y} = f(6/x - \sqrt{z})$ **9.** $x\log(x + y) = z + f(x + y)$

**10.** First **11.** $z = 2x + y\log x + f(xy)$ **12.** $\partial z/\partial x = \partial z/\partial y$

**13.** $z = f_1(y + x) + f_2(y + 6x)$ **14.** $4y^2p = q^2$ **15.** $u = \int f(y)\,dy + \phi(x)$

**16.** $c = 1$ **17.** $u = \frac{1}{6}x^3y + xf(y) + \phi(y)$ **18.** $f_1(y + x) + f_2(y + 6x)$

**19.** (iv) **20.** (iii) **21.** (iii) **22.** (iii) **23.** (ii) **24.** (iv)

**25.** (iv) **26.** (i) **27.** $y = e^x\phi_1(y) + e^{-x}\phi_2(y + x)$ **28.** $z = ax + by + 1/(a - b)$

**29.** False **30.** False **31.** True **32.** True **33.** False.

## Problems 18.1, page 708

1. $z = ce^{4ax^3} . e^{-3ay^4}$
2. $u = ce^{k(1/y - 1/x)}$
3. $u = 8e^{-12x-3y}$
4. $u = 3e^{x-y} - e^{2x-5y}$
5. $u = 3e^{-5x-3y} + 2e^{-3x-2y}$
6. $u = \frac{1}{\sqrt{2}} \sinh \sqrt{2}x + e^{-3y} \sin x.$

## Problems 18.2, page 716

1. $y(x, t) = \sum_{n=1}^{\infty} b_n \sin \frac{n\pi x}{l} \cos \frac{n\pi ct}{l}$, when $b_n = \frac{2}{l}\int_0^b f(x) . \sin \frac{n\pi x}{l} dx$

2. $y(x, t) = \sum_{n=1}^{\infty} \left( a_n \cos \frac{n\pi ct}{l} + b_n \sin \frac{n\pi ct}{l} \right) \sin \frac{n\pi x}{l}$ where

$$a_n = \frac{2}{l}\int_0^l f(x) \sin \frac{n\pi x}{l} dx,\ b_n = \frac{2}{n\pi c}\int_0^l g(x) \sin \frac{n\pi x}{l} dx$$

3. $u(x, t) = 2\left[ \cos at \sin x - \frac{1}{2} \cos 2\,at \sin 2x + \frac{1}{3} \cos 3\,at \sin 3x - \frac{1}{4} \cos 4\,at \cos 4x + ...... \right]$

4. $x = c_1 \cos pt + c_2 \sin pt + \frac{k}{p^2 - \omega^2} \sin \omega t\ (p \neq \omega)$ or $x = c_1 \cos pt + c_2 \sin pt - \frac{k}{2p} t \cos pt,\ (p = \omega)$

5. $y = \frac{8h}{\pi^2}\left( \sin \frac{\pi x}{l} \cos \frac{\pi ct}{l} - \frac{1}{3^2} \sin \frac{3\pi x}{l} \cos \frac{3\pi ct}{l} + ... \right)$

7. (i) $y(x, t) = a(x - x^2 - c^2t^2)$; (ii) $y(x, t) = \frac{a}{2}(1 - \cos 2\pi x \cos 2\pi ct).$

8. $S = l + (50 - l) \cos \left( t - \sqrt{g / c} \right)$

## Problems 18.3, page 724

1. $u(x, t) = \frac{4l}{\pi^2}\left[ \frac{1}{1^2} \sin \frac{\pi x}{l} \cdot e^{\frac{-a^2\pi^2 t}{l^2}} - \frac{1}{3^2} \sin \frac{3\pi x}{l} \cdot e^{\frac{-3a^2\pi^2 t}{l^2}} + \cdots \right]$

2. $u(x, t) = \frac{400}{\pi^2} \sum_{n=0}^{\infty} \frac{(-1)^n}{(2n+1)^2} e^{-[(2n+1)c\pi/100]^2 t} \sin \frac{(2n+1)\pi x}{100}$

3. $u(x, t) = -3x + 90 - \frac{80}{\pi} \sum_{1}^{\infty} \frac{1}{n} \sin \frac{n\pi x}{5} e^{-c^2n^2\pi^2 t/25}$

4. $u(x, t) = 50 - \frac{400}{\pi^2} \sum_{1}^{\infty} \frac{2}{(2n-1)^2} \cos \frac{(2n-1)\pi x}{50} . e^{-c^2\pi^2(2n-1)^2 t/2500}$

5. $V = V_0 e^{-\sqrt{(n/2k)x}} \sin [nt - \sqrt{(n/2k)}\, x].$

## Problems 18.4, page 729

1. $u = -\frac{8}{\pi} \sum_{n=1,3,5...}^{\infty} \frac{\sin nx \sinh n(\pi - y)}{n(n^2 - 4) \sinh nx}$

5. $u(x, y) = \frac{3200}{\pi^3} \sum_{1}^{\infty} \frac{\sin \frac{(2n-1)\pi x}{20} \sinh \frac{(n-1)\pi y}{20}}{(2n-1)^2 \sinh (2n-1)\pi}$

8. $u(x, y) = u_0 \cosh \frac{\pi x}{a} \cosh \frac{\pi}{a}(b - y) \operatorname{sech} \frac{\pi b}{a}.$

## Problems 18.5, page 734

**1.** $u(r, \theta) = \dfrac{8k}{\pi}\sum_{1}^{\infty}\left(\dfrac{r}{a}\right)^{2n-1}\dfrac{\sin(2n-1)\theta}{(2n-1)^3}$

**2.** $u(r, \theta) = \dfrac{3200}{\pi^2}\sum_{1}^{\infty}\left(\dfrac{r}{10}\right)^{2n-1}\dfrac{\sin(2n-1)\theta}{(2n-1)^3}$

**3.** $u(r, \pi) = \dfrac{2}{\pi}\sum_{n=1,3,5\ldots}^{\infty}\left(\dfrac{a}{r}\right)^{2n}\dfrac{r^{4n}-b^{4n}}{a^{4n}-b^{4n}}\cdot\dfrac{\sin 2n\theta}{n^3}$

**4.** $u(r, \theta) = \sum \dfrac{2k}{n\pi}\left(\dfrac{r}{a}\right)^{4n}(1-\cos n\pi)\sin 4n\theta$

**5.** $u(r, \theta) = 50 - \dfrac{200}{\pi}\sum_{1}^{\infty}\dfrac{1}{2n-1}\left(\dfrac{r}{a}\right)^{2n-1}\sin(2n-1)\theta$

**6.** $u(r, \theta) = \cos\theta\left(\dfrac{200}{r}-\dfrac{r}{2}\right)+\sin\theta\left(2r-\dfrac{200}{r}\right).$

**7.** $u(r, \theta) = 4\cos\theta\,(r - 1/r) + 4\sin\theta\,(r + 1/r).$

## Problems 18.6, page 738

**2.** $\dfrac{1}{\sqrt{2}}$ units

**3.** $x = c_1 e^{\sqrt{3}t} + c_2 e^{-\sqrt{3}t} + \left(c_3\cos\dfrac{t}{\sqrt{3}} + c_4\sin\dfrac{t}{\sqrt{3}}\right) - 12e^t$, and

$y = \dfrac{\sqrt{3}}{12}\left(c_2 e^{-\sqrt{3}t} - c_1 e^{\sqrt{3}t}\right) + \dfrac{1}{2\sqrt{3}}\left(c_3\sin\dfrac{t}{\sqrt{3}} - c_4\cos\dfrac{t}{\sqrt{3}}\right) + 2e^t$

## Problems 18.7, page 742

**1.** $e = e_0\sin\dfrac{\pi x}{l}\cos\dfrac{\pi t}{l\sqrt{(LC)}};\ i = i_0 - e_0\sqrt{\left(\dfrac{C}{L}\right)}\cos\dfrac{\pi x}{l}\sin\dfrac{\pi t}{l\sqrt{CL}}$

**3.** $v = \dfrac{20(l-x)}{l} + \dfrac{24}{\pi}\sum_{n-1}^{\infty}\dfrac{(-1)^{n+1}}{n}\sin\dfrac{n\pi x}{l}\exp(-n^2\pi^2 t/RCl^2)$

$i = \dfrac{20}{lR} + \dfrac{24}{lR}\sum_{n=1}^{\infty}(-1)^n\dfrac{n\pi x}{l}\exp(-n^2\pi^2 t/RCl^2)$

**4.** $v = V_0\cos\{pt - px\sqrt{(LC)}\}$.

## Problems 18.8, page 745

**4.** 34.315 kg

**5.** $x = \dfrac{m}{e}\dfrac{E}{H^2}\left[1-\cos\left(\dfrac{e}{m}Ht\right)\right], y = \dfrac{E}{H}\left[t - \dfrac{m}{(eH)}\sin\left(\dfrac{e}{m}HT\right)\right]$

**6.** $\dfrac{E}{2R}[2 - e^{-RT/L+M} - e^{RT/L-M}], \dfrac{E}{2R}[e^{-Rt/L+M} - e^{Rt/L-M}]$

## Problems 18.9, page 747

**1.** $\dfrac{\partial^2 v}{\partial x^2} = LC\dfrac{\partial^2 v}{\partial t^2}, \dfrac{\partial^2 i}{\partial t^2} = LC\dfrac{\partial^2 i}{\partial t^2}$

**2.** $\dfrac{\partial u}{\partial t} = c^2\left(\dfrac{\partial^2 u}{\partial x^2} + \dfrac{\partial^2 u}{\partial y^2} + \dfrac{\partial^2 u}{\partial z^2}\right)$

**3.** If $u(x, t)$ is the temperature, then temperature gradient at a point is $\partial u/\partial x$ for all $t$.

**4.** elliptic

**5.** $\partial^2 y/\partial t^2 = c^2\partial^2 y/\partial x^2$

**6.** $u = \dfrac{10}{l}x + 30$

**7.** parabolic partial differential equation

**8.** $r^2\dfrac{\partial^2 u}{\partial r^2} + r\dfrac{\partial u}{\partial r} + \dfrac{\partial^2 u}{\partial \theta^2} = 0$

**9.** $y = \dfrac{2h}{l}x, 0\ x < \dfrac{l}{2}; y = \dfrac{l}{2h}(2h - y), \dfrac{l}{2} < x < l$

**10.** $y(0, t) = 0, y(l, t) = 0, \left(\frac{\partial y}{\partial t}\right)_{t=0} = 0$

**11.** zero

**12.** $u(0, y) = 0, v(a, y) = 0, 0 < y < a; u_x(x, 0) = 0$, for all $t$ and $u(x, a) = u$ for $0 < x < a$

**13.** $\frac{\partial u(0,t)}{\partial x} = 0, \frac{\partial u(l,t)}{\partial x} = 0$ for all $t$

**14.** $y(x, t) = f(x + ct) + f(x - ct)$

**15.** $\frac{\partial u}{\partial t} = c^2\left(\frac{\partial^2 u}{\partial x^2} + \frac{\partial^2 u}{\partial y^2}\right)$ where $c^2$ is the diffusivity

**16.** $u(x) = x^2 + 20$

**17.** § 18.8-(6), (7), (8)

**18.** $u = 8e^{-12x - 3y}$

**19.** $z = 4e^{3x + t}$

**20.** $\alpha^2 (= k/s\rho)$ is called the diffusivity of the substance (cm²/sec)

**21.** $\frac{\partial^2 v}{\partial x^2} = RC\frac{\partial v}{\partial t}, \frac{\partial^2 i}{\partial x^2} = RC\frac{\partial c}{\partial t}$

**22.** $u(x, t) = \sum_{n=1}^{\infty} b_n \sin\frac{n\pi x}{l} e^{-c^2 n^2 \pi^2 t/l^2}$

**23.** § 18.7 – (3), (4), (5)

## Problems 19.1, page 756

**1.** $\frac{1}{5}(8 - 7i); \frac{1}{5}(8 + 7i)$.

**2.** $\sqrt{(26/5)}; \tan^{-1}(9/7) - \pi$

**3.** (*i*) $2\cos\frac{\theta}{2}\left(\cos\frac{\theta}{2} + i\sin\frac{\theta}{2}\right)$. (*ii*) $-8i/25$

**4.** $x = 1.5, y = \pm 2$

**5.** A circle: centre $(-1, 1)$ radius $\sqrt{2}$

**8.** $-1 + i\sqrt{3}, -1 - i\sqrt{3}, 1 - i\sqrt{3}; 4\sqrt{3}$

**9.** $-2 + 0.i, 1 - i\sqrt{3}$

**10.** $-1 - i, \sqrt{2}(\mp\sin 15° \pm i\cos 15°), \sqrt{2}(\mp\cos 15° \pm i\sin 15°)$

**11.** (*i*) Annular region between the circles of radii 2 and 4 with centre $(-3, 0)$ including boundary of inner circle; (*ii*) Region of complex plane above the line $y = 2$; (*iii*) Infinite region bounded by the lines $\theta = \pi/3$ and $\theta = \pi/2$; (*iv*) Real axis and region above it between $x = \pm 2$

**13.** (*i*) Ellipse with foci at $z = \pm 1$ and major for axis = 3
(*ii*) (*a*) Right bisector of the line joining $z = 3$ and $z = -1$;
(*b*) Circle through the points $z = 3$ and $z = -1$;

**14.** (*i*) Right bisector of the points 0 and 2; (*ii*) Circle through the points $\frac{1}{3}$ and 3

**15.** (*i*) A straight line; (*ii*) Circle with centre (1, 1/2) and radius $\sqrt{5/2}$.

**16.** (*i*) Interior and boundary of the circle $x^2 + y^2 + \frac{10}{3}y + 1 = 0$, (*ii*) circle $x^2 + y^2 = 1$

## Problems 19.2, page 760

**3.** $4m\pi/n(n + 1)$

**5.** $2^{n+1}\sin^n\frac{\alpha - \beta}{2}\cos n\left(\frac{\pi + \alpha + \beta}{2}\right)$.

## Problems 19.3, page 764

**1.** (*i*) $2^{1/4}$ cis $\pi/6$ ; $2^{1/4}$ cis $2\pi/3$ ; $2^{1/4}$ cis $7\pi/6$ ; $2^{1/4}$ cis $5\pi/3$.

(*ii*) $(2)^{1/5}\cos\frac{4n + 3}{10}\pi$, where $n = 0, 1, 2, 3, 4$; (*iii*) $\pm 2\sqrt{2}$

(*iv*) $2^3\sqrt{2}\cos r\pi/9$, where $r = 1, 7$ or $13$

**3.** $\pm i,\ \frac{1}{2}\left(\sqrt{3} \pm i\right), \frac{1}{2}\left(-\sqrt{3} \pm i\right)$

**5.** (*i*) $3;\ \sqrt{3}i, -\sqrt{3}i$.

(*ii*) $-1,\ \frac{1}{2}\left(1 \pm i\sqrt{3}\right), \pm(1+i)/\sqrt{2}, \pm(-1+i)/\sqrt{2}$

(*iii*) $\frac{1 \pm i}{\sqrt{2}}, \frac{-1 \pm i}{\sqrt{2}};\ 1, \cos\left(\pm\frac{2\pi}{5}\right), \cos\left(\pm\frac{4\pi}{5}\right)$.

(*iv*) $\cos(2n+1)\pi/6$, where $n = 0, 1, 2, 3, 4, 5$; $2(2m+1)\pi/3$, where $m = 0, 1, 2$

**6.** $(-1+i)/\sqrt{2}, (1-i)/\sqrt{2}$

**7.** $\pm 1, \pm i, \pm\left(\cos\frac{\pi}{6} \pm i\sin\frac{\pi}{6}\right), \pm\left(\cos\frac{\pi}{3} \pm i\sin\frac{\pi}{3}\right)$; Last four values

**9.** $x^3 + x^2 - 2x - 1 = 0$.

## Problems 19.4, page 766

**1.** $32\cos^5\theta - 24\cos^3\theta + 6\cos\theta$

**13.** $-(2)^{-11}(\sin 12\theta - 2\sin 10\theta - 4\sin 8\theta + 10\sin 6\theta + 5\sin 4\theta - 20\sin 2\theta)$

**14.** $\sin^5\theta = A\sin\theta - B\sin 3\theta + C\sin 5\theta$.

## Problems 19.5, page 772

**1.** (*i*) $e^{x^2-y^2}\cos 2xy,\ e^{x^2-y^2}\sin 2xy$ (*ii*) $ie^5$. (*iii*) $e^{16}\cos 30, e^{16}\sin 30$

**8.** $\frac{1}{64}(\cosh 7\theta + 7\cosh 5\theta + 21\cosh 3\theta + 35\cosh\theta)$

**14.** $z = \frac{1}{2}\log 2 + i\left(\frac{\pi}{4} + 2n\pi\right)$

**15.** (*i*) $z = (2n\pi + \pi/2) + i\cosh^{-1}2$. (*ii*) $-\log 3$.

**16.** $-13/12$.

## Problems 19.6, page 775

**10.** $\pm\frac{\pi}{4} + \frac{i}{4}\log\frac{1+\sin\theta}{1-\sin\theta}$ according as $\cos\theta$ is +ve or –ve

**11.** $\sin^{-1}\left(\sqrt{\sin\theta}\right) + i\log\left[\sqrt{(1+\sin\theta)} - \sqrt{\sin\theta}\right]$.

## Problems 19.7, page 778

**1.** (*i*) $\log_e 10 + i[\tan^{-1}(4/3) \pm 2n\pi]$; (*ii*) $\log_e 1 + i(\pi + 2n\pi)$ (*iii*) $\frac{1}{2}\log\frac{\pi^2}{2} + i\frac{3\pi}{4}$.

**4.** (*i*) $\sqrt{2}e^{-(2n-\frac{1}{4})\pi}, (2n - \frac{1}{4})\pi + \log\sqrt{2}$; (*ii*) $e^{-\pi^2/8}, (\pi/4)\log_e 2$

**9.** $\sqrt{[\frac{1}{2}(\cos 2x + \cosh 2y)]} - i\tan^{-1}(\tan x \tanh y)$

**10.** (*i*) $2n\pi \pm i\log(2 + \sqrt{3})$; (*ii*) $-\frac{1}{2}\log 3 + (n + \frac{1}{2})i\pi$.

## Problems 19.8, page 780

**1.** $e^{\sin\theta\cos\theta}\cos(\theta + \sin^2\theta)$

**2.** $\sin\alpha\cos(\cos\beta)\cosh(\sin\beta) - \cos\alpha\sin(\cos\beta)\sinh(\sin\beta)$

**3.** $1 + \left(2\cos\frac{\theta}{2}\right)^n . \sin\frac{n\theta}{2}$

**4.** $\log(2\cos\theta/2)$

5. $-\frac{1}{2}\tan^{-1}(\cos\beta \text{ cosech } \alpha)$

6. $\frac{1}{2}\tan^{-1}\dfrac{2c\sin\alpha}{1-c^2}$

7. $\left(2\sin\dfrac{\alpha}{2}\right)^{-n}\cos\left(\dfrac{n\pi}{2}-\dfrac{n\alpha}{2}\right)$

8. $\sin\dfrac{n(\pi-\alpha)}{2}/(2\sin\alpha/2)^n$

9. $\dfrac{x\sin\theta+(-1)^{n-1}x^{n+1}\{x\sin n\theta+\sin(n+1)\theta\}}{x^2+2x\cos\theta+1}$

10. $\dfrac{\cos\{\alpha+\frac{1}{2}(n-1)\beta\}\sin n\dfrac{\beta}{2}}{\sin\dfrac{1}{2}\beta}$

11. $\tan^{-1}\left(\dfrac{\sin\alpha\sin\beta}{1+\sin\alpha\cos\beta}\right)$

12. $\dfrac{1-x\cos\theta-x^n\cos n\theta+x^{n+1}\cos\overline{n-1}\,\theta}{1-2x\cos\theta+x^2}$.

## Problems 19.9, page 783

2. 0.053 radians
3. 1°59′
4. 39.7.

## Problems 19.10, page 783

1. (*b*)
2. (*c*)
3. (*b*)
4. (*c*)
5. (*b*)
6. (*d*)
7. odd
8. $32\cos^6\theta-48\cos^4\theta+18\cos^2\theta-1$
9. $6(1-2i)$
10. $2i\sin n\theta$
11. $\dfrac{1}{25}(-6+17i)$
12. $\cosh x\cos y$
13. $\dfrac{1}{19}$ radians
14. 1
15. $-\cos x\sinh y$
16. $e^{-\pi/4\sqrt{2}}$
17. $\alpha+\beta+\gamma$
18. $2i\,n\pi$
19. real
20. $\sinh\phi$
21. $\sinh 2\phi/(\cosh 2\theta+\cosh 2\phi)$
22. $16\cos^5\alpha-20\cos^3\alpha+5\cos\alpha$
23. an equilateral
24. $\pi/2$
25. $x=\pm 1, y=-4$.
26. a circle
27. True
28. True
29. True
30. True
31. False
32. True
33. False.

## Problems 20.1, page 795

4. $a=1, b=-6, c=1, d=2, e=4$

5. (*i*) $nz^{n-1}$ (*ii*) $\cosh z$.

6. (*i*) and (*ii*) Not analytic. (*iii*) Analytic

7. $p=-1$

11. & 12. $f(z)$ is not analytic at origin although C-R equations are satisfied there

14. (*i*) $z^3+3z^2+1+ic$ ; (*ii*) $\cos z$ ; (*iii*) $\log z$ ; (*iv*) $i/z$ ; (*v*) $e^z+i(c-z)$ ; (*vi*) $ze^{2z}+ic$ ; (*vii*) $z\sin z$ ; (*viii*) $x^2e^z+ic$

15. (*i*) $(1+i)/z+c$ ; (*ii*) $\cos z+c$ ; (*iii*) $e^z$ ; (*iv*) $\bar{z}e^{-\bar{z}}+c$ ; (*v*) $1+ize^{-z}$ ; (*vi*) $(2\cos x\cosh y)/(\cos 2x+\cosh 2y)$

16. (*i*) $f(z)=c-iz^3$ ; (*ii*) $f(z)=\left(\dfrac{1}{2}-\cot\dfrac{z}{2}\right)$ (*iii*) $\dfrac{\cot z}{1+c}+c\left(\dfrac{1+3i}{5}\right)e^z+c$

17. $\psi=3xy^2-x^3+c$

18. $2\tan^{-1}(y/x)$ ; $2\log z+c$

20. $v=C-e^{-2xy}\cos(x^2-y^2)$ ; $f(z)=C-i\,e^{iz^2}$

22. (*i*) $x^4-6x^2y^2+y^4=c$ (*ii*) $x^2-y^2+2e^x\sin y=c$ (*iii*) $r^2\sin 2\theta=c'$

23. (*i*) $x/(x^2+y^2)$ ; (*ii*) $\frac{1}{2}\log(x^2+y^2)$

24. (*i*) $a(1+\cos\theta+i\sin\theta\log r)$ ; (*ii*) $(r+1/r)\cos\theta+(r-1/r)\sin\theta+c$

27. $-2\tan^{-1}[(y-2)/(x-1)]$, $2i\log(z-1-2i)$.

## Problems 20.2, page 800

**1.** $z = \pm i$ **2.** $w = -1/z$

**3.** $w = (2i - 6z)/(iz - 3)$ ; fixed points $z = i, 2i$ ; no critical points

**4.** $w = \dfrac{(20 + 18i) - (32 + 12i)z}{(29 + 17i) - (11 - 3i)z}$

**5.** (*i*) $w = i(1 - z)/(1 + z)$ ; (*ii*) $w = \dfrac{(4i - 2) + (5 - 3i)z}{2i + (1 + i)z}$; (*iii*) $w = (1 - z)/(1 + z)$.

## Problems 20.3, page 805

**1.** (*i*) $I(w) > 0$ ; (*ii*) Region bounded by the parabolas $v^2 = 4(1 \pm u)$ ; and $u^2 = 1 - 2v$ ; (*iii*) Region bounded by parabolas $v^2 = \frac{1}{4} \pm u$, $v^2 = 4(1 \pm u)$ ; (*iv*) Region boundary $\rho = 2\sqrt{\rho}\cos\frac{\phi}{2} + 3$.

**2.** $w = z^6$

**3.** Lines parallel to $x$ and $y$ axes map into two families of rectangular hyperbolas in the $w$-plane which cut each other at right angles. Lines parallel to $u$ and $v$ axes map into two families of parabolas in the $z$ plane which cut each other orthogonally. It is conformal at $z = 0$

**4.** (*a*) Line $4v + 1 = 0$

**5.** (*b*) Every circle through the origin ($z = 0$) transforms into a st. line not passing through the origin ($w = 0$). If a line passes through $z = 0$, its image is a line through $w = 0$. (*b*) Circle with centre (1/2, 1/2) and radius $(1/\sqrt{2})$ (*c*) Lemniscate $p^2 = \cos 2\phi$

**10.** $z = \pm a$ **11.** See §20.10 (3) **14.** See § 20.10(4).

## Problems 20.4, page 807

**1.** $z = \cosh w$ **3.** $w = \sin z$ **4.** $w = \log z$.

## Problems 20.5, page 809

**1.** (*a*) $(5 - i)/6$ (*b*) $(5 + i)/6$ **2.** (*i*) $4 + (25/3)i$ ; (*ii*) $4 + 8i$ **3.** $\frac{1}{6}(64i - 103)$

**6.** (*i*) $i$ ; (*ii*) $2i$ ; (*iii*) 0 **7.** (*a*) $\frac{1}{3}(i - 1)$, (*b*) $\frac{1}{6}(5i - 3)$ **9.** (*i*) $\frac{2}{3}$; (*ii*) $-\frac{2}{3}$.

## Problems 20.6, page 815

**1.** (*i*) 0 ; (*ii*) $2\pi i$ **2.** 0 **3.** (*a*) zero ; (*b*) zero

**4.** (*i*) $2\pi i(e^4 - e^{-2})$ ; (*ii*) $\pi i/2$ (*iii*) $4\pi i$ **5.** (*i*) $-10\pi i$ (*ii*) $2\pi ie$ (*iii*) $\dfrac{8\pi i}{3e^2}$ **6.** (*i*) $4\pi i$ ; (*ii*) $2\pi ie^{-1}$ ; (*iii*) $-\pi i$

**7.** (*i*) $8\pi i$ ; (*ii*) 0 **8.** (*i*) 0 ; (*ii*) $2\pi(6 + 13i)$ ; (*iii*) $12\pi i$ **9.** zero.

## Problems 20.7, page 822

**1.** $\sum_{n=1}^{\infty} (-1)^{n+1} n(z - 1)^n$ ; Convgt. in $|z - 1| < 1$

**2.** $f(z) = \frac{1}{3} + \frac{u}{9}(z + i) - \frac{7}{27}(z + 1)^2 + \ldots\ldots$ Region of convergence is $|z + i| < 1$

**3.** (*i*) $\frac{1}{2}(z - 1) - \frac{1}{2^2}(z - 1)^2 + \frac{1}{2^3}(z - 1)^3 - \ldots\ldots$ (*ii*) $-(z - \pi/2) + (z - \pi/2)^3/3! - (z - \pi/2)^5/5! + \ldots$

(*iii*) (*a*) $f(z) = -\frac{1}{5}\sum_{n=0}^{\infty}(-1)^n (z+1)^n - \frac{1}{20}\sum_{n=0}^{\infty}\left(\frac{z+1}{4}\right)^n$ in the region $|z+1| < 1$. Also, $(z+1) < 4$

(*b*) $f(z) = -\frac{1}{5}\sum_{n=0}^{\infty}\left(\frac{z-1}{5}\right)^n \frac{1}{10}\sum_{n=0}^{\infty}\left(\frac{z-1}{2}\right)^n$ in the region $|z-1| < 2$. Also $|z-1| < 3$

**4.** (*i*) $\frac{1}{z+1} + \frac{1}{(z+1)^2} + \frac{1}{(z+1)^3} + \ldots + \frac{1}{2}\left[1 + \frac{z+1}{2} + \frac{(z+1)^2}{2^2} + \frac{(z+1)^3}{2^3}\right]$

(*ii*) $\frac{1}{4z}\left(1 + \frac{1}{z} + \frac{1}{z^2} + \ldots\right) - \frac{1}{12}\left(1 - \frac{z}{3} + \frac{z^2}{9} - \ldots\right)$. (*iii*) $-\frac{1}{2(z-1)} - 3\sum_{n=1}^{\infty}\frac{(z-1)^{n-1}}{2^{n+1}}$

**5.** (*i*) $e\left[(z-1)^{-2} + (z-1)^{-1} + \frac{1}{2!} + \frac{1}{3!}(z-1) + \frac{1}{4!}(z-1)^2 + \ldots\right.$

(*ii*) $e^2(z-1)^{-3} + 2e^2(z-1)^{-2} + 2e^2(z-1)^{-1} + \frac{4e^2}{3} + \frac{2e^2}{3}(z-1) + \ldots$

**6.** (*i*) $\sum_{n=1}^{\infty}(-1)^{n-1} n \,.\, (z-1)^{-n}$ for $|z-1| > 1$. (*ii*) $-\sum_{n=2}^{\infty}\frac{z^{2n-5}}{2(n-1)!}$

**7.** (*i*) $1 + \frac{3}{z}\left(1 - \frac{2}{z} + \frac{2^2}{z^2} - \frac{2^3}{z^3} + \ldots\ldots\right) - \frac{8}{3}\left(1 - \frac{z}{3} + \frac{z^2}{3^2} - \frac{z^3}{3^3} + \ldots\ldots\right)$

(*ii*) $\frac{2}{z+2} + \frac{3}{(z+2)^2} + \frac{3^2}{(z+2)^3} + \ldots + \frac{1}{5}\left(1 + \frac{z+2}{5} + \frac{(z+2)^2}{5^2} + \frac{(z+2)^3}{5^3} + \ldots\right)$

(*iii*) (*a*) $\frac{7}{z} - \frac{9}{z^2} - \frac{45}{z^3} - \frac{81}{z^4} + \ldots\ldots$ (*b*) $\frac{5}{2(z-3)} + \frac{7}{12} - \frac{z-3}{24} - \frac{5(z-3)^2}{432} + \frac{7(z-3)^2}{864} + \ldots\ldots$

**8.** (*a*) $\frac{z}{4} - \frac{5z^2}{16} + \frac{21}{64}z^5 - \ldots\ldots$; (*b*) $\frac{1}{3}\left(\frac{1}{z^5} - \frac{1}{z^3} - \frac{z}{4} + \frac{z^3}{16} - \frac{z^5}{64} + \ldots\ldots\right)$; (*c*) $\frac{1}{z^3} - \frac{3}{z^5} + \ldots\ldots$

**9.** $z = 0, z = 2$ are the isolated singularities **10.** $z = 0$ is an isolated essential singularity

**11.** $z = 0$, is a non-isolated essential singularity

**12.** $z = 1$ is a pole of order 2 **13.** $z = 1$ is a pole of order 4

**14.** $z = a$ is a double pole and $z = 0, \pm 1, \pm 2, \ldots\ldots$ are simple poles.

## Problems 20.8, page 828

**1.** $-\frac{1}{t} - 2i + 3t + 4i\,t^2 + \ldots$ where $t = z - i$ ; $-1$

**2.** (*i*) $3e/2$ ; (*ii*) Res $(z = -ai/2)$

**3.** (*i*) Res $f(0) = -1/2$, Res $f(2) = 2\frac{1}{2}$;

(*ii*) Res $f(-1) = 0$, Res $f(i) = \frac{1+2i}{2(1-i)}$, Res $f(-i) = \frac{2i-1}{2(i-1)}$

(*iii*) Res $f(-1) = 1$, Res $f(i) = \frac{2+i}{-1+i}$, Res $f(-i) = \frac{-2+i}{1+i}$

**4.** (*i*) Res $f(0) = -4/3$ ; (*ii*) Res $f(i) = \frac{1}{2}e^{-1}$ ; Res $f(-i) = \frac{1}{2}e$ ; (*iii*) $-i\,\pi/4$, Res $f(n\pi) = 1$, $n$ an integer.

**5.** (*i*) 0 ; (*ii*) $16\pi i/(2+3i)$

**6.** (*i*) $3\pi i$ (*ii*) $\frac{\pi i e^{-4}}{5}$ (*iii*) 0 ; (*iv*) $i\pi/4$ **7.** (*i*) $\pi i$ ; (*ii*) $\pi/2\,(3+2i)$, (*iii*) zero.

**8.** (*i*) $-2\pi i$ ; (*ii*) $\pi i$ ; (*iii*) $\pi/16$

**9.** (*i*) $-2\pi i$ ; (*ii*) $8\pi i/3e^2$ ; (*iii*) $\frac{\pi e^{2i}}{2}$ ; (*iv*) $\frac{8\pi}{3}ie^{-2}$

**10.** (*i*) $\frac{21\pi i}{16}$ ; (*ii*) $2\pi i$ sec 1(1 + tan 1) ; (*iii*) $-2\pi i$

**11.** $2\pi i$ **12.** $\frac{1}{z^3} - \frac{1}{6z} + \frac{7z}{360} - \frac{31z^6}{15120} + \ldots\ldots; -\frac{1}{3}\pi i$.

## Problems 20.10, page 836

**1.** (*i*) **2.** (*iii*) **3.** (*a*) **4.** (*iii*) **5.** (*ii*) **6.** (*ii*)
**7.** (*ii*) **8.** (*ii*) **9.** (*iii*) **10.** (*ii*) **11.** (*ii*) **12.** (*i*)
**13.** (*ii*) **14.** (*iii*) **15.** $v(x, y) = x^2 - y^2 + 2y + c$
**16.** $3x^2y - y^3 + c$ **17.** $u + iv$ is an analytic function **18.** $2i/3$
**19.** 1 **20.** $z = \frac{1}{2}(a + b)$ **21.** $2u + 1 = 0$
**22.** $z = 1, \frac{1}{2}(1 \pm \sqrt{3})$ **23.** $z = 0$ **24.** $-1$
**25.** $z - \frac{z^2}{2} + \frac{z^3}{3} - \frac{z^4}{4} + \ldots$ **26.** zero **27.** (*iii*)
**28.** $\frac{1}{2}\left[\frac{d^2}{dz^2}\{(z-a)^3 f(z)\}\right] z = a.$ **29.** zero **30.** $\frac{\partial u}{\partial x} = \frac{\partial v}{\partial y}, \frac{\partial u}{\partial y} = -\frac{\partial u}{\partial x}$
**31.** zero **32.** $e^x \sin y$ **33.** zero
**34.** $z = 2$ **35.** no point in the $z$-plane **36.** $\frac{\partial u}{\partial r} = \frac{1}{r}\frac{\partial v}{\partial \theta}, \frac{\partial v}{\partial r} = -\frac{1}{r}\frac{\partial u}{\partial \theta}$
**37.** zero
**38.** A simply connected region is one in which any closed curve, lying entirely within it, can be shrunk to a point without going out of the region
**39.** which is analytic or regular **40.** $z = 1, 2$
**41.** $(z_1 - z_2)(z_3 - z_4)/(z_1 - z_4)(z_3 - z_2)$ **42.** $i$
**43.** $\frac{1}{2} + \frac{3}{4}z + \frac{7}{8}z^2 + \frac{15}{16}z^3 + \ldots$ **44.** $\pm i$
**45.** 1 **46.** Magnification and rotation
**47.** The coefficient of $(z - a)^{-1}$ in the expansion of $f(z)$ around an isolated singularity $(z = a)$ is called the residue of $f(z)$ at that point.
**48.** $2\pi i$ **49.** zero **50.** $z = 1, 2$
**51.** $-\frac{1}{2}\left\{1 + \frac{z}{2} + \frac{z^2}{4} + \frac{z^3}{8} + \ldots\right\}$ **52.** $i/2$ **53.** $z = 0, 2$
**54.** § 20.14 **55.** $\pm i$ **56.** zero
**57.** Zeroes are at $z = \pm 1$, singularity is at $z = 0$ **58.** $-1$
**59.** essential singularity **60.** zero
**61.** $\sin z = \frac{1}{\sqrt{2}}\left\{1 + (z - \pi/4) - \frac{(z-\pi/4)^2}{2!} - \frac{(z-\pi/4)^3}{3!} + \frac{(z-\pi/4)^4}{4!} + \ldots\right\}$
**62.** a circle with centre (3, 2) and radius 2 in $w$-plane. **63.** $n\pi$, $n$ an integer
**64.** $\phi(a)/\psi'(a)$ **65.** circles **66.** True **67.** False **68.** True **69.** True
**70.** False **71.** True **72.** True **73.** True **74.** True **75.** True
**76.** True **77.** False **78.** False **79.** False **80.** True **81.** True
**82.** True.

## Problems 21.1, page 845

**1.** $\frac{1}{s-2}+\frac{24}{s^4}+\frac{3(s-2)}{s^4+9}$.

**2.** $\frac{1}{s}+\frac{\sqrt{\pi}}{s^{3/2}}+\sqrt[3]{\left(\frac{\pi}{s}\right)}$

**3.** $\frac{s\cos b-a\sin b}{s^2+a^2}$

**4.** $\frac{s^2-2s+4}{s(s^2+4)}$

**5.** $\frac{1}{4}\left[\frac{1}{s}+\frac{s}{s^2+4}+\frac{s}{s^2+16}+\frac{s}{s^2+36}\right]$

**6.** $\frac{1}{4}\left[\frac{2}{s^2+4}-\frac{6}{s^2+36}+\frac{4}{s^2+16}\right]$

**7.** $\frac{5}{4}\left\{\frac{1}{s^2+1}-\frac{3/2}{s^2+9}+\frac{1/2}{s^2+25}\right\}$

**8.** $\frac{s(s^2+28)}{(s^2+4)(s^2+36)}$

**9.** $\frac{b}{(s+a)^2-b^2}$

**10.** $\frac{1}{2s+2}+\frac{s+1}{2s^2+4s+10}$

**11.** $\frac{-s(s+2)}{2(s+1)(s^2+2s+37)}$

**12.** $\frac{a(s^2+2a^2)}{s^4+4a^4}$

**13.** $\frac{3}{2}\left[\frac{1}{s^2-9}+\frac{s^2-13}{s^4-10s^2+169}\right]$

**14.** $\frac{2}{(s+2)^3}$

**15.** $w\left[\frac{1}{s^2+w^2}+\frac{s^2-w^2}{(s^2+w^2)^2}\right]$

**16.** $4\frac{(4s^2+4s-1)}{(4s^2+1)^2}$

**17.** $\frac{4}{s}-\frac{e^{-s}}{s}$.

**18.** $\frac{1+e^{-\pi s}}{s^2+1}$

**19.** $e^{-2\pi s/3}\frac{s}{s^2+1}$

**20.** $\frac{4}{(s-1)(s^2-2s+5)}$.

## Problems 21.2, page 848

**1.** $(1/s^2T)-e^{-sT}/s(1-e^{sT})$

**2.** $\frac{\pi a}{(sa)^2+\pi^2}\coth\frac{sa}{2}$

**3.** $\frac{a}{2s}\tanh\frac{as}{4}$

**4.** $(1/s^2)\tanh\frac{1}{2}as$

**5.** $\frac{1}{\sqrt{(s^2+a^2)}}$

**6.** $(s^2-2as+a^2+b^2)^{-1/2}$

**7.** $\frac{2}{(s-2)\sqrt{(s+1)}}$.

## Problems 21.3, page 853

**2.** $\frac{2(3s^2+4)}{s^2(s^2+4)^2}$

**3.** $\frac{3s}{2}\left[\frac{1}{(s^2+1)^2}-\frac{1}{(s^2+9)^2}\right]+\frac{1}{4}\left[\frac{3(1-s^2)}{(s^2+1)^2}+\frac{9-s^2}{(s^2+9)^2}\right]$

**4.** $\frac{2s^3-6a^2s}{(s^2+a^2)^3}$

**5.** $60\left\{\frac{1}{(s-7)^6}+\frac{1}{(s-1)^6}\right\}$

**6.** $\frac{2a(s-a)}{(s^2-2as+2a^2)^2}$

**7.** $\frac{8(s+2)}{s^2+4s+20}$

**8.** $\frac{2(s+1)[2(s+1)^2-3]}{(s^2+2s+2)^3}$

**9.** $\log\{(s+b)/(s+a)\}$

**10.** $\cot^{-1}(s)$

**11.** $\frac{1}{2}\log\left(\frac{s^2+b^2}{(s-a)^2}\right)$

**12.** $\cot^{-1}(s+1)$

**13.** $\cot^{-1}s-\frac{1}{2}s\log(1+s^{-2})$

**14.** $\dfrac{1}{s-\log 2}+\dfrac{2s}{(s^2+1)^2}+\dfrac{1}{2}\log\left(\dfrac{s^2+9}{s^2+4}\right)$ **15.** (*i*) log 2/3 ; (*ii*) π/8 ; (*iii*) 12/169 ; (*iv*) 0.

**17.** (*i*) $\dfrac{1}{s(s^2-2s+2)}$ ; (*ii*) $\dfrac{1}{s}\cot^{-1}(s)$ ; (*iii*) $\dfrac{\cot^{-1}(s-1)}{s}$ ; (*iv*) $\dfrac{24(s+2)(s^2+4s+9)}{s[(s^2+4s+5)(s^2+4s+13)]^2}$.

## Problems 21.4, page 857

**1.** $\dfrac{1}{2}\left(\cos\dfrac{5t}{2}-\sin\dfrac{5t}{2}\right)-4\cosh 3t+6\sinh 3t$ **2.** $\cos at \sinh at$

**3.** $3e^{t/2}+2e^{t/3}$ **4.** $e^{2t}+2e^{-4t}$ **5.** $\dfrac{1}{3}(8e^{2t}-e^{-t})$

**6.** $4e^{3t}+16e^{-t}$ **7.** $(1+t^2-2t^3)c^{-2t}$

**9.** $e^{-t}\left(\dfrac{3}{2}t^2-\dfrac{1}{3}t^3\right)$ **10.** $\dfrac{1}{2}e^t-e^{2t}+\dfrac{5}{2}e^{3t}$

**11.** $\frac{1}{3}t(e^t-e^{-2t})$ **12.** $\frac{1}{13}(3e^{3t}-3\cos 2t+2\sin 2t)$ **13.** $\frac{1}{2}(\sin t-te^{-t})$

**14.** $(\frac{1}{3}a^2)\,[e^{at}-e^{-at/2}\{\cos(\sqrt{3}at/2)+\sqrt{3}\sin(\sqrt{3}at/2)\}]$

**15.** $\frac{1}{3}(5\sin t-\sin 2t)$ **16.** $\frac{1}{3}e^{-2t}(6\cos 3t-7\sin 3t)$

**17.** $\dfrac{1}{5}(1+e^{-t})\sin t+\dfrac{3}{5}(1-e^{-t})\cos t$ **18.** $(2/\sqrt{3})\sinh(\frac{1}{2}t)\sin(\frac{1}{2}\sqrt{3}t)$

**19.** $\cos at \sinh at$. **20.** $\dfrac{1}{2a^3}(\sinh at-\sin at)$ **21.** $\dfrac{1}{16}e^{-t}(\sin 2t-2t\cos 2t)$

## Problems 21.5, page 864

**1.** $\dfrac{1}{25}(e^{-5t}+5t-1)$ **2.** $\dfrac{1}{8}-\dfrac{1}{4}\left(t^2+t+\dfrac{1}{2}\right)e^{-2t}$ **3.** $\dfrac{1}{a^2}\cos\left(\dfrac{bt}{a}\right)$

**4.** $\dfrac{1}{a^2}\left(t-\dfrac{1}{a}\sin at\right)$ **5.** $\dfrac{1}{2}t^2+\cos t+1$ **6.** $\dfrac{1}{2}te^{-2t}\sin 2t$

**7.** $t\sin at$ **8.** $\dfrac{1}{2}(a^2t^2-4at+2)e^{-at}$ **9.** $\dfrac{1-e^t}{t}$

**10.** $\dfrac{1}{t}(e^{-bt}-e^{-at})$ **11.** $e^{-t}-e^{-2t}-e^{-3t}$ **12.** $\dfrac{1}{t}(\cos at-\cos bt)$

**13.** $\dfrac{2}{t}(1-\cos t)$ **14.** $\dfrac{2}{t}(e^t-\cos t)$ **15.** $\dfrac{\sin 2t}{t}$

**16.** $\dfrac{\sin t}{t}$ **17.** $\dfrac{2(\sinh t-t\cosh t)}{t^2}$ **18.** $\dfrac{e^{-bt}-e^{-at}}{a-b}$

**19.** $\dfrac{1}{2a^3}(\sin at-at\cos at)$ **20.** $\dfrac{1}{a^3}(at-\sin at)$ **21.** $t(e^{-t}+1)+2(e^{-t}-1)$

**22.** $\dfrac{1}{16}(e^{2t}-e^{-2t}-4te^{-2t})$ **23.** $\dfrac{1}{2}(e^t-\cos t-\sin t)$ **24.** $\dfrac{t^2}{2}+\cos t-1$

**25.** $\frac{t}{8a^3}(\sin at - at\cos at)$ **26.** $\frac{s}{(s^2+a^2)(s^2+b^2)}(a \neq b)$ **27.** $\frac{e^{-2t}}{54}(\sin 3t - 3t\cos 3t)$.

## Problems 21.6, page 869

**1.** $y = -\frac{1}{5}e^{-t} + \frac{1}{5}(2\sin 2t + \cos 2t)$ **2.** $y(t) = 2\cos t + \sin t$. **3.** $y = \frac{7}{4}e^{-t} - \frac{3}{4}e^{-3t} - \frac{1}{2}te^{-t}$.

**4.** $y = t - 3\sin t + \cos t$ **5.** $y = 2t + 3 + \frac{1}{2}(e^{3t} - e^t) - 2e^{2t}$ **6.** $y = 4e^{2t}(1 + t) - 7e^t$

**7.** $y = 2\cos 5t + t\sin 5t$ **8.** $y = \frac{1}{2\omega}\sin \omega t$

**9.** $y = \frac{1}{8}e^t - \frac{1}{40}e^{-3t} - \frac{1}{10}(2\sin t + \cos t)$ **10.** $y = \frac{1}{2}(\cos kt + \cosh kt)$

**11.** $y = \frac{1}{8}[(3 - t^2)\sin t - 3t\cos t]$ **12.** $x = \left[c_1\cos 2t + c_2\sin 2t + \frac{1}{3}\right]e^{-t}$ **13.** $y = 2 + 2e^{-t} + \frac{t^3}{3}$

**14.** $y = e^{2t}(x^2 - 6x + 12) - e^t(15x^2 + 7x + 11)$

**15.** $x = \frac{4}{9}\sin 2t - \frac{5}{9}\sin t - \frac{1}{3}t\cos 2t$ **16.** $y = \frac{1}{2}\left(\frac{3\sin t}{t} - \cos t\right)$

**17.** $y = e^{2t}$ **18.** $y = t$ **19.** $y = 3J_0(2t)$

**22.** $(n\sin at - a\sin nt)F_0/mn(n^2 - a^2)$, where $n^2 = k/m$.

## Problems 21.7, page 871

**1.** $x = \frac{1}{2}(e^t + \cos t + 2\sin t - t\cos t)$, $y = \frac{1}{2}(t\sin t - e^t + \cos t - \sin t)$

**2.** $x = e^t + e^{-t}$, $y = e^{-t} - e^t + \sin t$ **3.** $x = 2 + t^2/2$, $y = -1 - t^2/2$

**4.** $x = \frac{1}{10}(5 - 2e^{-t} - 3e^{-6t/11})$, $y = \frac{1}{5}(e^{-t} - e^{-6t/11})$ **5.** $x = e^6(1 + 2t) + 2e^{3t}$, $y = \sinh t + \cosh t - e^{-3t} - te^t$

**6.** $i_1 = \frac{a}{p+\omega}(\sin \omega t + \sin pt)$ ; $i_2 = \frac{a}{p-\omega}(\cos \omega t - \cos pt)$.

## Problems 21.8, page 879

**1.** $x = \frac{a}{2n^2}[\sin nt\cos\alpha - nt\cos(nt + \alpha)]$

**2.** (*i*) $(1 - 2t)\,u(t - \pi) + 2tu(t)$, $\frac{2}{s^2} + \left(\frac{1-2\pi}{s} - \frac{2}{s^2}\right)e^{-as}$

(*ii*) $t^2 + (4t - t^2)u(t-1)$, $\frac{2}{s^3} + e^{-s}\left(\frac{3}{s} + \frac{2}{s^2} - \frac{2}{s^3}\right)$

(*iii*) $\{u(t) - u(t - T)\}\cos(\omega t + \phi)$ ; $[(s\cos\phi - \omega\sin\phi) - e^{-sT} \times \{s\cos(\phi + \omega T) - \omega\sin(\phi + \omega T)\}]/(s^2 + \omega^2)$

**3.** (*i*) $\frac{s}{s^2+1} + \left(\frac{1}{s} + \frac{s}{s^2+1}\right)e^{-\pi s} - \left(\frac{1}{s} - \frac{1}{s^2+1}\right)e^{-2\pi s}$

(*ii*) $\frac{1}{s^2+1} + e^{-\pi s}\left(\frac{s}{s^2+4} - \frac{s}{s^2+1}\right) + e^{-2\pi s}\left(\frac{s}{s^2+9} - \frac{s}{s^2+4}\right)$

(*iii*) $\frac{2}{s^3}\{1 + e^{-2s}(2s^2 - 1) - 2e^{-4s}(1 + 4s)\}$ (*iv*) $\frac{2}{s^2+4}\left[e^{-2\pi s} - e^{-4\pi s}\right]$

**4.** (*i*) $e^{-s}/(s-1)$ ; (*ii*) $2e^{-s}/s^3$ ; (*iii*) $e^{-2s}\left\{\frac{24}{s^4}+\frac{42}{s^3}+\frac{28}{s^2}+\frac{25}{s}\right\}$ ; (*iv*) $e^{-s}(2+2s+s^2+s^3)/s^3$

**5.** $20e^{-2}$

**6.** (*i*) $-\sin t \,.\, u(t-\pi)$ ;

(*ii*) $\frac{1}{2}e^{-(t-1)}(t-1)^2 u(t-1)$ (*iii*) $\frac{e^{-(t-1)}}{\sqrt{\pi(t-1)}}u(t-1)$ (*iv*) $3-4(t-1)u(t-1)+4(t-3)u(t-3)$

**7.** $y=\frac{1}{2}\sin 2t+\frac{1}{4}(1-\cos 2t)-\left\{\frac{1}{4}1-\cos(t-1)u(t-1)\right\}$

**8.** $x=\frac{8}{9}\left(1-\cos\frac{3}{2}t\right); y=\frac{8}{9}\left(\frac{3}{2}t-\sin\frac{3}{2}t\right)$

**9.** $y(x)=\begin{cases}\dfrac{2Wx^2(3l-5x)}{81EI}, & 0<x<l/3\\[2mm] \dfrac{2Wx^2(3l-5x)}{81EI}+\dfrac{W}{6EI}\left(x-\dfrac{l}{3}\right)^3, & \dfrac{l}{3}<x<l\end{cases}$

**10.** $i_1=\frac{a}{p+\omega}(\sin\omega t+\sin pt); i_2=\frac{a}{p+\omega}(\cos\omega t-\cos pt)$

**11.** $x=\frac{I}{mn}e^{-\mu t/2m}\sin nt ; \frac{dx}{dt}=\frac{I}{m}e^{-\mu t/2m}\left(\cos nt-\frac{\mu}{2mn}\sin nt\right)$, where $n^2=\frac{k}{m}-\frac{\mu^2}{4m^2}$.

**12.** $x=2e^{2t}-4e^{-t}+\cos 2t-3\sin 2t$

**13.** $y=e^{-t}(4\cos 2t-\frac{1}{2}\sin 2t)-2\cos 2t+\frac{1}{2}\sin 2t$

**14.** $y=(1-t)e^{-t}+\frac{1}{2}\sin t$

**15.** $x=\frac{1}{4}(e^t-e^{3t}+2te^{3t}); y=\frac{1}{4}(5e^t-e^{3t}-2te^{3t})$

**16.** $x=2(1-e^{-t}-te^{-t}); y=-t(1+2e^{-t})+2(1-e^{-t})$

## Problems 21.9, page 880

**1.** $\frac{2s}{(s^2+1)^2}$

**2.** (*d*)

**3.** (*b*)

**4.** $\frac{s+3}{s^2+6s+34}$

**5.** $te^{-2t}$

**6.** $\frac{1}{3}e^{-2t}\sin 3t$

**7.** $s\bar{f}(s)-f(0)$

**8.** $\frac{1}{8}(2-3t)e^{-3t/2}$

**9.** $\frac{1}{2}\left[\frac{1}{s}+\frac{s}{s^2-16}\right]$

**10.** $\frac{1}{s-\log 2}$

**11.** $\frac{k!}{(s+1)^{k+1}}$

**12.** $\frac{2}{13}$

**13.** $e^{-at/s}$

**14.** $\Gamma(3/2)/s^{3/2}$

**15.** $s^2f(s)-sf(0)-f'(0)$

**16.** $\frac{1}{4}\left[\frac{3s}{s^2+16}+\frac{s}{s^2+144}\right]$

**17.** $\cot^{-1}(s/a)$

**18.** $\frac{e^{-3t}t^4}{24}$

**19.** $\frac{s\cos 3-2\sin 3}{s^2+4}$

**20.** (*c*)

**21.** $f(t-a)\,u(t-a)$

**22.** $e^{-as}$

**23.** $1-3t+2t^2$

**24.** $\int_0^t f(t)\,dt$

**25.** $\int_0^T \frac{e^{-st}f(t)\,dt}{(1-e^{-st})}$

**26.** $\frac{1}{(s+1)^2(s+2)}$

**27.** $\frac{2(s-3)}{s^2-6s+34}+\frac{12}{s^2-6s+25}$

**28.** $1/(s - \log 4)$ **29.** $\frac{e^{-3t}}{\sqrt{(\pi t)}}$. **30.** (*d*)

**31.** $\frac{t}{2}\sin\frac{t}{2}$ **32.** (*c*) **33.** (*c*) **34.** (*ii*)

**35.** (*iii*) **36.** (*ii*) **37.** (*iv*) **38.** (*i*)

**39.** $\int_a^b f(x)dx$ **40.** (*ii*) **41.** (*iii*) **42.** (*d*)

**43.** $\frac{1}{s^2+4}$ **44.** $2e^{-t} - e^{-2t} + e^{-3t}$ **45.** True **46.** False

**47.** False.

## Problems 22.1, page 894

**1.** $\frac{2}{\pi}\int_0^\infty \frac{\sin\lambda\cos\lambda x}{\lambda}d\lambda$ ; $\frac{\pi}{2}$ for $|x| < 1$, $\frac{\pi}{4}$ for $|x| = 1$, 0 for $|x| > 1$

**2.** (*i*) $\frac{4}{\pi}\int_0^\infty \frac{\sin\omega - \omega\cos\omega}{\omega^3}\cos\omega x\, d\omega$ (*ii*) $\frac{2}{\pi}\int_0^\infty \cos\omega x \frac{a}{a+\omega^2}d\omega$

**4.** (*i*) $\frac{2\sin as}{s}$, $\pi$ ; (*ii*) $2\{(a^2s^2 - 2)\sin as + 2as\cos as\}/s^3$

**5.** $\frac{4}{s^3}(\sin sa - sa\cos sa)$ **6.** (*i*) $\sqrt{(3\pi)}\,e^{-3s^2/4}$ (*ii*) $\frac{\sqrt{\pi}}{2}e^{(3is - s^2/4)}$ (*iii*) $\frac{-2}{s^2+1}$

**7.** $\frac{1}{1-is}$ **8.** $\frac{1-\cos 2s}{s}, \frac{\sin 2s}{s}$ **9.** $\sqrt{(\pi/2)}\,e^{-as}$

**10.** $\frac{s}{s^2+a^2}, \frac{\pi e^{-a}}{2a}$ **11.** $\frac{a}{a^2+s^2}$ ; $\frac{\pi}{2a}e^{-a\lambda}$ **12.** 1

**13.** $\frac{1}{a\sqrt{2}}e^{-s^2/4a^2}$ ; $\frac{1}{2a^3\sqrt{2}}e^{-s^2/4a^2}$ **15.** (*i*) $\frac{\pi}{2a^2}(1 - e^{-as})$ (*ii*) $\tan^{-1}(s/a)$

**16.** (*i*) $\frac{1}{2}\left\{\frac{\sin[a(1-s)]}{1-s} - \frac{\sin[a(1+s)]}{1+s}\right\}$ (*ii*) $(2\cos s - \cos 4s - 1)/s^2 - (2\sin s)/s$

**17.** $2/(\pi s^2)$ **18.** $F_s(p) = -32(-1)^p/p\pi$ ; $F_c(p) = 32\frac{(-1)^p - 1}{p^2\pi^2}$

**19.** $(\pi/s)\cos s/c$ **21.** $f(x) = (2 + 2\cos x - 4\cos 2x)/\pi x$ **22.** $2/\pi(1 + x^2)$.

## Problems 22.2, page 898

**2.** $\frac{1}{4}\int_{-\infty}^{\infty} e^{-[|x-t|+|t|]}dt$ **5.** $2\left(\frac{1-\cos s}{s^2}\right)$.

## Problems 22.3, page 902

**1.** $\frac{1}{9}[e^{-t} + e^{2t}(3t - 1)]$ **2.** $\frac{1}{5}(e^{-2t} - 2\sin t - \cos t)$ **3.** $(\sinh at - at)/a^2$

**4.** $\frac{1}{2}[e^x(x-1) + \cos x]$ **5.** $\frac{1}{2}(\sin t - t\cos t)$.

## Problems 22.4, page 911

**1.** $y = 30\, e^{-75t} \cos 5x$

**2.** $\sum_{n=1}^{\infty} \frac{V_o}{n\pi} (1 - \cos n\pi)\, e^{-n^2 t} \sin nx_0$

**3.** $u(x, t) = \frac{2}{\pi} \int_0^{\infty} e^{-s^2 t} \left\{ \frac{\sin(1 + x)\, s + \sin(1 - x)\, s}{s} \right\} ds$

**4.** $\frac{2}{\pi} \int_0^{\infty} \frac{s \sin s + \cos s - 1}{s^2} e^{-s^2 t} \cos sx\, ds$

**7.** $\theta(x, t) = \theta_0\, erf\left(\frac{x}{2\sqrt{(kt)}}\right) + \theta_0 \sum_{n=1}^{\infty} (-1)^n \left\{ erf \frac{nl - x}{2\sqrt{(kt)}} - erf \frac{nl + x}{2\sqrt{(kt)}} \right\}.$

## Problems 22.5, page 912

**1.** $F_c(s) = \int_0^{\infty} f(t) \cos st\, dt$

**2.** $s^2/2$

**3.** The Fourier transform of the convolution of $f(x)$ and $g(x)$ is the product of their Fourier transforms.

**4.** $f(x) = \frac{1}{2\pi} \int_{-\infty}^{\infty} F(s)\, e^{-isx}\, ds$

**5.** $\int_{-\infty}^{\infty} t^n f(t)\, e^{ist}\, dt$

**6.** $e^{isa} F(s)$

**7.** $\frac{2}{\pi} \int_0^{\infty} \sin \lambda x \int_0^{\infty} f(t) \sin \lambda t\, dt\, d\lambda$

**8.** $\frac{1}{a}$

**9.** $-s^2\, [F(u)]$

**10.** $\frac{2}{\pi} \int_0^{\infty} \sin(\lambda x)\, d\lambda \int_0^{\pi} \sin(\lambda t)\, dt$

**11.** $-\frac{n\pi}{l} \int_0^1 f(x) \cdot \cos \frac{n\pi x}{l} dx$

**12.** $\frac{1}{a} F\left(\frac{\lambda}{a}\right)$

**13.** $f(x) = \frac{2}{\pi^3} \sum_{p=1}^{\infty} \left( \frac{1 - \cos p\pi}{p^2} \right) \sin px$

**14.** $\frac{1}{2} F(s/2)$

**15.** $1/(s^2 + 1)$

**16.** True

**17.** False

**18.** True

**19.** True

**20.** False

**21.** True.

## Problems 23.1, page 920

**1.** (*i*) $e^{a/z}$ ; (*ii*) $z(e^{-z} - 1)$ ; (*iii*) $\frac{z}{z - e^{i\theta}}$

**2.** (*i*) $\frac{2z}{(z-1)^2} + \frac{z/\sqrt{2}}{z^2 - \sqrt{2z} + 1} + \frac{z}{z-1}$ ; (*ii*) $\frac{z^3 - 3z^2 + 4z}{(z-1)^3}$ ; (*iii*) $\frac{z(z+1)}{(z-1)^3} + \frac{3z}{(z-1)^2} + \frac{2z}{z-1}$

**6.** (*i*) $\frac{z^2 \sin \theta}{z^2 - 2z \cos \theta + 1}$ (*ii*) $\cos \alpha \frac{z(z - \cos \pi/8)}{z^2\ 2z \cos \pi/8 + 1} - \sin \alpha \frac{z \sin \pi/8}{z^2 - 22 \cos \pi/8 + 1}$

**7.** $\frac{z(z - \cos \theta)}{z^2 - 2z \cos \theta + 1}, |z| > 1; \frac{z(z^2 \cos \theta - 2z + \cos \theta)}{(z^2 - 2z \cos \theta + 1)^2}$

**8.** $\frac{z^2}{z^2 + 1}, |z| > 1; \frac{z^2}{z^2 + a^2}$

**9.** (*i*) $\frac{z}{z - e^{-a}}$ ; (*ii*) $\frac{z\, e^{-a}}{(z - e^{-a})^2}$ ; (*iii*) $\frac{(z + e^{-a})\, e^{-a}}{(z - e^{-a})^3}$

**12.** $\frac{z^2(1 + 3z^2)}{(1 - z)(1 + z^2)}$

**13.** $u_2 = 2, u_3 = 11.$

## Problems 23.2 page 924

**1.** $\frac{z}{z - 4}$ ; $|z| > 4$

**2.** $\frac{z}{2 - z}$ ; $|z| < 2$

**3.** $\frac{3z}{(4 - z)(z - 3)}$ ; $3 < |z| < 4.$

**4.** $\frac{5z}{(4 - 5)^2}$ ; $|z| > 5$

**5.** $-\log(1 - 3/z)$ ; $|z| > 3$

**6.** $e^{3/z}$, ROC is $z$-plane

**7.** $(1 - e^a/z)^{-1}$; $|z| > |e^a|$.

## Problems 23.3, page 927

**1.** $\frac{1}{2}(3^{n+1}-1)$ **2.** $(n+1)(-a)^n$ **3.** $\frac{1}{2}n(n-1)$

**5.** $4a^n$ **6.** $(1/3)^n - 2^n$ **7.** $n$

**8.** $\frac{3}{4}\left\{\frac{1}{(2)^{n-1}}+\frac{1}{(-4)^n}\right\}$ **9.** $1-\frac{(-i)^{n-1}}{i+1}+\frac{(-i)^{n-1}}{i-1}$ **10.** $(n^2+7n+4)(4)^{n-1}$

**11.** $u_n = (2)^{n-1}+(3)^{n-1}+(4)^{n-1}\ (n>0)$ **12.** $2+(2)^n+3(n-1)2^n,\ (n \ge 1)$

**13.** $(-1)^{n+1}-2n+\cos n\pi/2$ **14.** $1-e^{-at}$

**15.** (i) $\left(-\frac{1}{3}-\frac{z}{3^2}-\frac{z^2}{3^3}-\frac{z^3}{3^4}\cdots\right)+\left(\frac{1}{2}+\frac{z}{2^2}+\frac{z^2}{2^3}+\frac{z^3}{2^4}+\cdots\right)$; (ii) $(-2^{n-1})z^{-n},\ n>0$

(iii) $(3^{n-1}-2^{n-1})z^{-n},\ n\ge 1,\ 0,\ n\le 0$

**16.** $\frac{1}{2}(n-1)(n-2)5^{n-3},\ n\ge 3$ and $=0,\ n<3$ **17.** $2(-i)^{n-1}-(-2)^{n-1}$

**18.** $\frac{1}{3}+\frac{2}{3}\left(-\frac{1}{2}\right)^n$ **19.** $\frac{1}{3}-\frac{1}{3}\left(-\frac{1}{2}\right)^{n-1}$

**20.** $u_n = 1+\frac{1}{2}[(i)^{n-2}+(-i)^{n-2}]$ **21.** $(1+i)^{n-1}+(1-i)^{n-1}$ **22.** $2n\sin(n\pi/2),\ n=0,1,2,\ldots$

## Problems 23.4, page 931

**1.** $x_n = 2^n - 1$ **2.** $u_n = (1-n)(-1)^n + n^2 - 1$

**3.** $y_n = 2^{n-1}+(-2)^{n-1}$ **4.** $f(n) = 2+(-4)^n$

**5.** $\dot{y}_n = \frac{4}{3}\left[2(-1)^n+(2)^n\right]$ **6.** $36\left[\frac{1}{2}-(2)^n+\left(\frac{1}{2}\right)^n\right]$

**7.** $y_n = (c_1+c_2n)3^n+\frac{1}{2}n(n-1)3^{n-2}$ **8.** $y = c_1 2^n + c_2 3^n + \frac{1}{6}5^n$

**9.** $y = c_1\cos n\theta + c_2\sin n\theta + 1.$ **10.** $5(2^n)$

**11.** $y_n = \frac{1}{3}(-1)^n - \frac{2}{5}(-3)^n + \frac{1}{15}(2)^n$ **12.** $y_n = 1-2n+2^n$

**13.** $y_n = c_1 4^n + \left(c_2-\frac{n}{4}\right)2^n + 2n - \frac{8}{3}$ **14.** $y_k = \frac{1}{2}(k+2)\frac{1}{5^k}\cos\frac{k\pi}{2}$

**15.** $y_n = \left[\frac{1}{4}-\frac{9}{4}(-3)^n\right]\mu(n)$ **16.** $y_n = (-2)^{n-1},\ (n\ge 1).$

## Problems 23.5, page 931

**1.** $z/(z-1)$ **2.** $\sum_{n=0}^{\infty} u_n z^{-n}$ **3.** $z/(z-1)^2$ **4.** $az/(z-a)^2$

**5.** $\frac{z\sin\theta}{z^2-2z\cos\theta+1}$ **6.** $e^{1/z}$ **7.** $(z^2+z)/(z-1)^3$

**8.** $\frac{Z}{Z-e^{i\theta}}$ **9.** $2^{n-1}$ **10.** $(-1)^{n-1}n$ **11.** $u_0 = \underset{z\to\infty}{Lt}\{Z(u_n)\}$

**12.** False **13.** False **14.** True **15.** True

**16.** False **17.** False **18.** False.

## Problems 24.1, page 935

1. $a = 2.28, b = 6.1879, p = 30.46$
2. $a = 1120, b = 55.1$
3. $a = 0.2, b = 0.0044$
4. $a = 0.5012, n = 0.5$
5. $a = 0.115, b = 11.8$
6. $a = 4.1, b = 0.43$
7. $a = 0.0498, b = -0.02$.

## Problems 24.2, page 939

1. $y = 13.6x$
2. 15.2 thousand tons
3. $Y = 0.004P + 0.048$
4. $R = 70.052 + 0.292t$
5. $a = 0.545, b = 0.636$
6. (a) $y = 4.193 + 1.117x$ (b) $y = 8 - 0.5x$
7. $y = 1.243 - 0.004x + 0.22x^2$
8. $y = 0.34 - 0.78x + 0.99x^2$
9. $y = 18.866 + 66.158x - 4.333$
10. $R = 3.48 - 0.002V + 0.0029V^2$
11. $V = 2.593 - 0.326T + 0.023\,T^2$.

## Problems 24.3, page 943

1. 6.32, $b = 0.0095$
2. $y = (0.108)\,x^2 + (0.217)\,x$
3. $a = 3, b = 2$
4. $y = 7.187 - 5.16\,\dfrac{1}{x}$ ; 4.894
5. $a = 0.988, b = 3.275$
6. $y = 2.978\,x^{0.5143}$ ; 5.8769
7. $a = 0.1839, b = 0.0221$
8. $f(t) = 0.678\,e^{-3t} + 0.312\,e^{-2t}$
9. $a = 146.3, k = -0.412$
10. $a = 99.86, b = 1.2$.

## Problems 24.4, page 946

1. $a = 11.1, b = 0.71$
2. $y = 46.05 + 6.1\,x$
3. $a = 0.0028, b = 0.01, c = 4.18$
4. $a = 15.8, b = 2.1, c = -0.5$
5. $a = 1.459, b = 0.062$.

## Problems 24.5, page 948

1. $y = 0.12 + 0.47x$
2. $y = 1.184 + .523x$
3. $y = 1.53 + 0.063x + 0.074x^2$
4. $y = 0.485 + 0.397x + 0.124x^2$.

## Problems 24.6, page 949

1. $Y = aX + b$ where $X = x, Y = y/x$
2. $Y = A + BX$, where $X = \log_{10} p, Y = \log_{10} v, A = \dfrac{1}{r} \log b, B = -1/r$
3. § 24.4
4. $y = aX + c$, where $X = x^b$
5. (ii)
6. $\Sigma y = nA + B\Sigma x, \Sigma xy = A\Sigma x + B\Sigma x^2$ where $y = \log_{10} y, A = \log_{10} a, B = \log_{10} b$
7. § 24.12
8. Zero
9. $y = aX + b$ where $X = x^2/\log_{10} z, Y = y/\log_{10} x$
10. $a = 0.0167, b = 1.05$.
11. The moments of the observed values of $y$ are respectively equal to the moments of the calculated values of $y$
12. $a = 1.7, b = 1.26$
13. $y = a + bx$ where $x = 1/x, y = 1/y$
14. (r)
15. (b)

## Problems 25.1, page 957

1. 336.79
2. 64% get more than 50 marks ; median 54.7, $Q_1 = 46, Q_3 = 61.5$
3. Mean = 27.9 ; Median = 25.66 ; Mode = 21.85
4. Mean = 32.58 ; Median = 32.6 ; Mode = 35.1
5. 3.1%
6. 1.3%
7. 192 km/hr
8. 60 km/hr
9. 38.6 ; 36.2
10. Median = 12.2 days ; Mode = 11.4 days

## Problems 25.2, page 962

**1.** 4.45, 0.39 **3.** 4, 7 **4.** 10.04, 10.13, 11.69, 5.54, 2.35
**5.** 32, 32.6, 12.4 **6.** Q.D. = 10.9, S.D. = 15.26 **7.** 1.845 ; 1.8175
**8.** 0.55, 1.24 ; first, yes **9.** Height **10.** *A*
**11.** *B* is a better player and more consistent
**12.** *A* is more efficient, *B* is more consistent **13.** 161.3, 5.68.

## Problems 25.3, page 965

**1.** $\mu_1 = \mu_3 = 0$, $\mu_2 = 2$, $\mu_4 = 11$ ; $\beta_1 = 0$, $\beta_2 = 2.75$ **2.** 8.85 ; 5.25 ; 0.32 ; 1.09
**3.** –0.2064 **4.** 0.22 ; 1.157 **6.** 0 ; 2.9
**7.** $\beta_2 = 0.6541$ ; platykurtic
**8.** $\mu_1 = 0$, $\mu_2 = 1.72$, $\mu_3 = -1.32$, $\mu_4 = 9.4096$, kurtosis = 3.18, distributions is leptokurtic.

## Problems 25.4, page 978

**1.** (*a*) $r = 0.81$ ; $x = 0.5y + 0.5$, $y = 1.3x + 1.1$ ; (*b*) $r = 0.91$, $x = 1.3y - 4.4$, $y = 0.65x + 4.1$
**2.** $r = -0.92$ **3.** $r = 0.92$ **4.** $r = -0.055$ **5.** 0.7291
**6.** $r = 0.4517$ **7.** $y = 17.35 + 9.94\,x$
**9.** $m = (\beta - b)/(a - \alpha)$ **10.** $\bar{x} = 4$, $\bar{y} = 7$, $r = -0.5$ **11.** 134.5
**12.** 1.28 inch **13.** 0.8545 **15.** $r = 0.4517$
**16.** $\bar{x} = 67.6$, $\bar{y} = 68.61$, $r = 0.514$

## Problems 25.5, page 979

**1.** (*d*) **2.** (*d*) **3.** (*a*) **4.** (*a*)
**5.** (*b*) **6.** (*a*) **7.** (*b*) **8.** Coefficient of correlation
**9.** No **10.** 13.83 **11.** $\bar{x} = 2$, $\bar{y} = 3$, $r = \sqrt{3}$
**12.** Zero **13.** $\frac{1}{2}(Q_3 - Q_1)$ **14.** –1
**15.** § 25.9 **16.** $\frac{\Sigma XY}{n\,\sigma_x\,\sigma_y}$ **17.** $(\bar{x}, \bar{y})$
**18.** Reliability or consistency **19.** $\sqrt{\beta_1}$ **20.** degree of peakedness
**21.** $y - \bar{y} = r\,\frac{\sigma_y}{\sigma_x}(x - \bar{x})$ **22.** $100\,\sigma/\bar{x}$ **23.** $\tan^{-1}\left\{\frac{1 - r^2}{r}\cdot\frac{\sigma_x\,\sigma_y}{\sigma_x^2 + \sigma_y^2}\right\}$
**24.** 3 **25.** Two regression coefficients **26.** perpendicular
**27.** greater **28.** ±1 **29.** $1 - \frac{6\,\Sigma\,d_i^2}{n^3 - n}$
**30.** Coefficient of standard deviation **31.** zero
**32.** –1 and 1 **33.** –0.6 **34.** False
**35.** True **36.** False

## Problems 26.1, page 982

**1.** 4096 **2.** 360 ; 120 **3.** 120 ; 115 **4.** (*a*) 676000 (*b*) 468000

## Problems 26.2, page 993

**1.** (*i*) 7/12 ; (*ii*) $P(A/B) = 3/4$, $P(A \cup B) = 7/12$, $P(A'\,B') = 3/8$
**2.** (*a*) 1/36 ; (*b*) 1/6, Yes **3.** 36 : 30 : 25
**4.** 1/7 **5.** (*i*) 1/4, (*ii*) 7/8, (*iii*) 11/16 **6.** 15/1024

**7.** 3/28 **8.** 20/81 **9.** (*i*) 2816/4165 ; (*ii*) 2197/2025,
**10.** 0.11 **11.** (*a*) 6.739 ; (*b*) 0.024 **12.** (*a*) 1/114 ; (*b*) 685/1140
**13.** 2/801 **14.** 10/21 **15.** (*a*) $1-(1-p_1)(1-p_2)\ldots(1-p_n)$ ; 0.518 ; (*b*) 4/5
**16.** 15/17 **17.** 1/2 **18.** 5/12 **19.** (*i*) 83/110 (*ii*) 25/83
**20.** $1-2/(n-1)$ **21.** 7/20 **22.** (*a*) 1/6 ; (*b*) 3/4 **23.** 61/90
**24.** 0.72 **25.** 0.2223 **26.** 0.88

## Problems 26.3, page 995

**1.** 3/11 **2.** $\frac{61}{90}$ **3.** 0.584, 0.70 **4.** 15/59.
**5.** 46% **6.** $\frac{5}{6}$ **7.** (*a*) 0.292 (*b*) 0.20

## Problems 26.4, page 1003

**1.** $k = 1$ ; $\mu = 0.8$, $\sigma^2 = 2.232$ **2.** $2\sqrt{5}$
**3.** $F(x) = 0, -\infty < x < 0$
$= 1/8,\ 0 \le x \le 1$
$= 1/2,\ 1 \le x \le 2$
$= 7/8,\ 2 \le x \le 3$
$= 1,\ 3 \le x \le \infty$
**5.** ₹ 32
**6.** 2 **7.** $f(x)$ is a p.d.f. $\bar{x} = \frac{1}{2}, \sigma^2 = \frac{1}{20}$ **8.** (*i*) 9/16, 7/16 ; (*ii*) $k = 0.45$
**9.** (*i*) 0.37, (*ii*) 0.63 **10.** 4/9
**12.** $y_0 = 3/4$ ; Mean = 1 ; Variance = 1/5. **13.** 0.2
**14.** 1/3, 2/9 **15.** $F(x) = 0, x < x_1$ ; $(x - x_1)/(x_2 - x_1), x_1 \le x < x_2$ ; $1, x \ge x_2$.

## Problems 26.5, page 1007

**1.** $n = 4, p = q = \frac{1}{2}; \frac{15}{16}$ **2.** (*a*) ${}^4C_r(1/6)^{4-r}(5/6)^r$ ; $r = 0, 1, 2, 3, 4$ (*b*) 3/16
**3.** (*a*) 0.02579 ; (*b*) 0.04571 ; (*c*) $1.024 \times 10^{-7}$ **4.** 0.3456
**5.** 45927/50000 **6.** (*i*) 0.246 ; (*ii*) 0.345
**7.** (*i*) ${}^{20}C_1(1/20)(19/20)^{19}$ ; (*ii*) $\sum_{r=0}^{5} {}^{20}C_r (1/20)^r (19/20)^{20-r}$ (*iii*) 19
**8.** (*a*) 0.08 ; (*b*) 0.26 ; (*c*) 0.92 **9.** (*a*) 250 ; (*b*) 25 ; (*c*) 500
**10.** (*i*) 0.59049 ; (*ii*) 0.32805 ; (*iii*) 0.08146 **11.** 11
**12.** 99.83 **13.** 0.3585, 0.3773, 0.1887 ; 0.0596
**14.** 600 **15.** $100\,(.432 + .568)^5$
**16.** (*a*) $104\,(0.7404 + 0.2596)^4$ (*b*) $200\,(0.554 + 0.446)^6$.

## Problems 26.6, page 1010

**1.** (*i*) 2 ; (*ii*) $\frac{2}{3}e^{-2}$ **2.** $P(0) = 0.2636$, $P(3) = 0.1041$, $P(> 3) = 0.1506$
**4.** $(10)^{15}e^{-10}/15\,! = 0.035$ **5.** 0.865
**6.** 0.08 **7.** (*i*) 0.2231 (*ii*) 0.1913
**8.** 0.0008
**9.** $m = 0.51 = \sigma^2$ ; Poisson frequencies of 0, 1, 2, 3, 4 accidents are 180.1, 91.9, 23.4, 4, 0.6

**10.** 0.6
**12.** Theoretical frequencies are 44, 43, 21, 7, 1
**13.** Theoretical frequencies are 109, 142, 92, 40, 13, 3, 1, 0, 0, 0, 0.

## Problems 26.7, page 1016

**2.** (*i*) 0.1644 ; (*ii*) 0.7686
**3.** (*i*) 0.095 ; (*ii*) – 0.995
**4.** 36.4
**5.** (*i*) 16, (*ii*) 2
**6.** 294
**7.** 543
**8.** (*i*) 79 ; (*ii*) 35% ; (*iii*) 11
**10.** 52
**11.** 67
**12.** ₹ 866
**13.** $y = \dfrac{100}{\sqrt{(3.4\pi)}} e^{-\frac{(x-2)^2}{3.4}}$
**14.** $\mu = 13.64$. $\sigma = 3.98$.

## Problems 26.8, page 1018

**1.** (*a*) 0.0287, (*b*) 0.9672, (*c*) 0.5111
**2.** (*a*) 0.7854, (*b*) 0.1815, (*c*) 0.1815
**3.** (*a*) 0.97815, (*b*) 0.00595, (*c*) 0.01209.

## Problems 26.9, page 1019

**3.** $\frac{1}{2}(n+1), \frac{1}{12}(n^2-1)$
**5.** Mean = $a + b$, variance = $b^2$
**6.** $[(1-e^t)/t]^2$
**8.** $(1-t)^{3/4}$.
**9.** (See Example 26.38) Mean = $1/\lambda$, variance = $1/\lambda^2$.
**10.** Mean = k/p, variance = $\dfrac{k}{p}\left(\dfrac{1}{p}-1\right)$

## Problems 26.10, page 1020

**1.** (*a*)
**2.** (*b*)
**3.** (*d*)
**4.** (*b*)
**5.** 1/7
**6.** 1/2
**7.** (*b*)
**8.** (*b*)
**9.** (*a*)
**10.** (*c*)
**11.** 0.1288
**12.** 2
**13.** 0.21
**14.** 0.24
**15.**

| $X$: | 0 | 1 | 2 |
|---|---|---|---|
| $p(x)$: | 1/4 | 2/4 | 1/4 |

**16.** $p = \frac{3}{4}, q = \frac{1}{4}$
**17.** 0.7837
**18.** zero
**19.** equal
**20.** $P(A) + P(B)$
**21.** $\beta_1 = 0, \beta_2 = 3$
**22.** 120
**23.** 0.2646
**24.** 1/9
**25.** 0.2222
**26.** 1/6
**27.** $e^{-3}$
**28.** 5/36
**29.** 2
**30.** symmetrical
**31.** $1 - e^{-m}$
**32.** six
**33.** 0.6915
**34.** $(q + pe^t)^n$
**35.** 4 : 5
**36.** 1/6
**37.** 3.5
**38.** $\sqrt{2}$
**39.** unity
**40.** $n \to \infty, p \to 0$ such that $np$ is fixed
**41.** $P(A) + P(B)$
**42.** ${}^yC_x / {}^{y+z}C_x$
**43.** 0.6723
**44.** $P(A \cup B) = 0.72, P(A \cap B') = 0.1653$
**45.** 1
**46.** 1
**47.** $\mu'_r = \left[\dfrac{d^r}{dt^r}(\Sigma\, p_i\, e^{tx_i})\right]_{t=0}$
**48.** 1/6
**49.** 3/4
**50.** 1/2
**51.** 0.2
**52.** 2/7
**53.** $(q + p)^n$
**54.** 2/3
**55.** Mean and S.D.
**56.** 1/13
**57.** 2
**58.** § 26.6
**59.** $e^{-4/3}$
**60.** 1/3
**61.** True
**62.** False
**63.** True
**64.** False
**65.** False
**66.** True
**67.** False
**68.** False
**69.** $k = 2$
**70.** 8
**71.** $f(x) = \lambda e^{-\lambda x}$ for $x > 0$, $\lambda$ is a parameter
**72.** 25/12
**73.** 2
**74.** $l$
**75.** 4
**76.** 8
**77.** $n, m$ degrees of freedom
**78.** $\int_{-\infty}^{\infty} e^{t(x-a)} f(x)\, dx$

**79.** – 5/7 **80.** $\frac{e^{-x}x^{l-1}}{\Gamma(l)}$, $0 < x < \infty$ **81.** $\frac{1}{2}(b + a)$

**82.** 50% **83.** 3/8 **84.** 1 **85.** 3/4

**86.** (*iii*) **87.** $F(x) = \int_{-\infty}^{x} f(x)\,dx$ **88.** $\frac{\lambda^{2r}e^{-\lambda}}{(2r)!}$ **89.** $-\infty < t < \infty$

**90.** 6 **91.** 1/9 **92.** False **93.** 2

**94.** $4(1 - x)^3$ **95.** 0.264

## Problems 27.1, page 1027

**1.** Die is biased **2.** No **3.** Yes

**4.** 8.91% and 15.07% **5.** Consistent **6.** $p = 65/500$, S.E. = 0.015

**7.** 37.5% ; 30.3 and 44.7 respectively **8.** No

**9.** Difference is not significant

**10.** $Z \sim 6.56$ so that the difference is significant **11.** No.

## Problems 27.2, page 1030

**1.** No **2.** Mean weight lies between 64.6 and 69.4 lbs.

**3.** 0.0774 **5.** 62 **6.** 2.696 **7.** No

**8.** No **9.** (*i*) Yes, (*ii*) No **10.** Yes.

## Problems 27.3, page 1036

**1.** 0.25 **2.** $t = 0.62$, Yes **3.** 11.887 and 12.113 cm

**4.** Refute the claim **5.** Process is not under control **6.** No

**7.** Sample mean = 575.2 kg., S.E. = 2.75 kg **8.** Accept null hypothesis

**9.** Yes with 75% confidence. **10.** No **11.** No

**12.** No

## Problems 27.4, page 1040

**1.** 0.41 **2.** Hypothesis is correct

**3.** Significant at 5% level **4.** Yes

**5.** $f_e$ : 33.82 161.78 315.98 308.48 150.54 29.4

$\chi^2 = 7.97$. Binomial distribution gives a good fit at 5% level

**6.** $f_o$ : 305 365 210 80 18 12

$f_e$ : 301.2 361.4 216.8 86.7 26 7.9 ; $\chi^2 = 3.097$

Poisson distribution gives a good fit at 5% level

**7.** $f_o$ : 314 355 204 85 29 12

$f_e$ : 301 362.2 217.3 86.9 26.1 6.5

Poisson distribution can be fitted to the data

**8.** $\chi^2 = 1.2$. The fit is quite good at 5% level.

## Problems 27.5, page 1043

**1.** First variance cannot be regarded as significantly greater than the second

**2.** Not significant as $F = 2.1$ and $F_{0.05} = 4.15$ **4.** Not significant as $F = 2.4$ and $F_{0.01} = 3.2$

**5.** Product of firm *B* cannot be said to be of better quality than those of firm *A*.

**6.** Not significant at 1% level and just significant at 5% level as $F = 2$, $F_{0.01} = 2.62$ and $F_{0.05} = 1.98$

**7.** $F = 1.49$, Not-significant **8.** $F = 1.025$ ; Yes

## Problems 27.6, page 1043

**1.** § 27.3 (3) **2.** § 27.15 **3.** § 27.3 (2)

**4.** We are testing the hypothesis that one process is better than another

**5.** § 27.11 **6.** 1 **7.** 50 **8.** $(0, \infty)$

**9.** II **10.** 8 ; 16 **11.** $r = n - 1$ **12.** False

**13.** Less than 30 **14.** § 27.17 **15.** $-\infty < t < \infty$ **16.** (*ii*)

**17.** True

## Problems 28.1, page 1053

**1.** (*i*) 2.375, (*ii*) 2.875 **2.** (*i*) 0.519, (*ii*) 2.875, (*iii*) 1.146, (*iv*) 0.367

**3.** (*i*) 2.128, (*ii*) 2.7065, (*iii*) 1.375, (*iv*) 1.4036

**4.** (*i*) 0.853, (*ii*) 0.607, (*iii*) 2.3604 (*iv*) 3.789

**5.** 1.861 **6.** (*i*) 1.532, (*ii*) 1.226

**7.** (*i*) 1.855 (*ii*) 2.198 **8.** –16.56

**9.** (*i*) 0.853, (*ii*) –1.9338, (*iii*) 2.7985

**10.** (*i*) 0.518, (*ii*) 0.052, (*iii*) 0.695 (*iv*) 2.911

**11.** Root in interval (–0.8, 0.5) = 0.77009 ; Root in interval (0, 1) = 0.76839

**12.** $x_{n+1} = \frac{1}{2}(x_n + N/x_n)$ ; (*i*) 3.605 (*ii*) 3.162

**13.** 2.5713 **14.** (*i*) 1.8212 (*ii*) 2.3784 **15.** (*i*) 0.055 (*ii*) 0.258 (*iii*) 0.4347

**16.** 0.16; 0.32 **17.** –2.198

**18.** 3.78916 **19.** 2.93 **20.** 5.916

## Problems 28.2, page 1057

**1.** (*i*) 1.532, (*ii*) 0.684, (*iii*) 3.18, (*iv*) 1.168 **2.** 1.674

**3.** 2.231 **4.** –1.328 **5.** 2.924

## Problems 28.3, page 1064

**1.** $x = 1, y = 3, z = 5$ **2.** $x = 3, y = 1, z = 2$ **3.** $x = 1, y = 2, z = 3$

**4.** $x_1 = 2, x_2 = -1, x_3 = 3$ **5.** $x_1 = 1, x_2 = 2, x_3 = -1, x_4 = -2$ **6.** $x = 1, y = 3, z = 5$

**7.** $x = y = z = 1$ **8.** $x = 1, y = 2, z = 3$ **9.** $x = 7, y = -9, z = 5$

**10.** $x_1 = 2, x_2 = 1/5, x_3 = 0, x_4 = 4/5$ **11.** $x = y = z = 1$ **12.** $x = 1, y = 2, z = 3$

**13.** $x = 3, y = 2, z = 1$ **14.** $x_1 = -1, x_2 = 0, x_3 = 1, x_4 = 2$ **15.** $\begin{bmatrix} 1.2 & -0.4 & 0.2 \\ -0.2 & -0.1 & 0.3 \\ -0.4 & 0.3 & 0.1 \end{bmatrix}$.

## Problems 28.4, page 1072

**1.** $x = 2.556, y = 1.722, z = -1.055$

**2.** (*a*) $x = 2.426, y = 3.573, z = 1.926$ (*b*) $x = 2.426, y = 3.573, z = 1.926$

**3.** $x = 1, y = 1, z = 1$ **4.** $x = 0.998, y = 1.723, z = 2.024$

**5.** $x = 0.994, y = 2.000, z = 3.001$ **6.** $x = -13.223, y = 16.766, z = -2.306$

**7.** $x = 1, y = 2, z = 3, u = 4$ **8.** $x = 1.36, y = 2.103, z = 2.845$

**9.** $x = y = z = 1$
**10.** $x = 52.5, y = 44.5, z = 59.7$
**11.** $x = 1.93, y = 3.57, z = 2.43$
**12.** $x_1 = 1, x_2 = 3, x_3 = 5$
**13.** $x_1 = 1, x_2 = 3, x_3 = 3$
**14.** $x_1 = 2.031, x_2 = 2.683, x_3 = -1.118, x_4 = 3.111$

## Problems 28.5, page 1073

**1.** $x = 2, y = 1$
**2.** $x = -1.853, y = -1.927$
**3.** $x = 0.7974, y = 0.4006$
**4.** $x = 3.162, y = 6.45$
**5.** $x = -3.131, y = 2.362$

## Problems 28.6, page 1075

**1.** (*a*) 5.38, $\begin{bmatrix} 0.46 \\ 1 \end{bmatrix}$ ; (*b*) 4.418, $\begin{bmatrix} 1 \\ 0.618 \end{bmatrix}$
**2.** 3.41 ; [0.74, –1, 0.67]′
**3.** (*a*) 6, [1, 1, –1]′ (*b*) 8, [1, –0.5, 0.5]′
**4.** (*a*) 7 ; [2.099/7, 0.467/7, 1] (*b*) 25.182, [1, 0.045, 0.068]′ (*d*) 11.66 [0.025, 0.422, 1.000].

## Problems 28.7, page 1076

**1.** Newton-Raphson method
**2.** $x_{n+1} = \frac{1}{4}(3x_n + N/x_n{}^3)$
**3.** Chord AB
**4.** $x_{n+1} = \frac{1}{2}(x_n + N/x_n)$
**5.** initial approximation $x_0$ is chosen sufficiently close to the root
**6.** diagonal
**7.** (*c*)
**8.** (*a*)
**9.** $x_{n+1} = \frac{1}{3}(2x_n + N/x^2{}_n)$
**10.** $x_{n+1} = \frac{1}{2}(x_n + x_{n-1})$
**11.** $x_2 = x_0 - \dfrac{x_1 - x_0}{f(x_1) - f(x_0)} f(x_0)$
**12.** Linear
**13.** $x_{n+1} = x_n(2 - Nx_n)$
**14.** (*b*)
**15.** Newton-Raphson method
**16.** § 28.6
**17.** Upper triangular matrix
**18.** False
**19.** True
**20.** $x = 1, y = 1$.

## Problems 29.1, page 1082

**1.** 0.4
**2.** –7459
**5.** 241
**6.** 4.68, 2.68, 55.8, 99.88
**8.** (*i*) $1 - 2 \sin(x + 1/2) \sin 1/2$ ; (*ii*) $\tan^{-1}(1/2n^2)$ ; (*iii*) $192\,[x(x + 4)(x + 8)(x + 12)(x + 16)]$
(*iv*) $-2/[(x + 2)(x + 3)(x + 4)]$ (*v*) $-4 \sin(ph) \sin(2ph) \sin[4\{p(x + h) + q\}]$
**9.** (*i*) $e^{3x}[e^3 \log(1 + 1/x) + (e^3 - 1) \log 2x]$ (*ii*) $2^x(1 - x)/(1 + x)$
(*iii*) $(a - 1)^n a^x$ ; (*iv*) $(-1)^n n!/[x(x + 1)(x + 2) \ldots (x + n)]$.
**12.** (*i*) –36 ; (*ii*) $24 \times 2^{10} \times 10\,!$
**14.** $u = [x]^4 - 6[x]^3 + 13[x]^2 + x + 9$
**15.** $4x^3 - 12x^2 + 8x + 1$ ; $12x(x - 1)$
**16.** $\frac{1}{2}[x]^4 + 3[x]^3 + 4[x] + c$
**17.** $y(4) = 74, y(6) = 261$
**19.** 15.

## Problems 29.2, page 1087

**1.** $\left(\dfrac{\Delta^2}{E}\right) u_x = u_{x+h} - 2u_x + u_{x-h}$ ; $\dfrac{\Delta^2 u_x}{Eu_x} = \dfrac{u_{x+2h} - 2u_{x+h} + u_x}{u_{x+h}}$
**2.** (*i*) $2(\cos h - 1) \sin x$ ; (*ii*) $6x$ ; (*iii*) $2(\cos h - 1)[\sin(x + h) + 1]$ ; (*iv*) 8
**8.** *Error* = 10
**9.** 31
**10.** $f(1.5) = 0.222, f(5) = 22.022$
**11.** $y(4) = 74, y(6) = 261$
**12.** – 99
**13.** $y_4 = 1$ approx
**15.** (*i*) $n(3n^2 + 6n + 1)$ ; (*ii*) $\dfrac{n(n + 1)(n + 2)(n + 3)}{4}$
**16.** $2/(1 - x)^3$.

## Problems 29.3, page 1092

**1.** 23 **2.** 6.36 **3.** $\theta = 0.01x^2 + 1.01x + 130.1$; $\theta(\text{at } x = 43) = 192.02$
**4.** 0.788 **5.** ₹ 110.52 **6.** 8666 **7.** 352 ; 219
**8.** 0.9623, 0.2903 **9.** 15. **10.** $f(x) = 9x - 4x^2$ **11.** 1.625
**12.** 0.1955 **13.** 4.219 **14.** 2530 **15.** $y_1 = 0.1, y_{10} = 100$
**16.** $u_2 = 42, u_4 = 49$ **17.** 10, 22 **18.** 755.

## Problems 29.4, page 1102

**1.** 19.4 **2.** 12.826 **3.** 54000 **4.** 3.2219
**5.** 3.0375 **6.** 395 **7.** 3.347 **8.** 9
**9.** 3250.875 **11.** 2.5283 by all formulae.

## Problems 29.5, page 1105

**1.** 14.63 **2.** 2.8168 **3.** 0.89 **4.** 100
**5.** $x^3 + x^2 - x + 2$ **6.** $x^3 - 3x^2 + 5x - 6$ **7.** $x^5 - 9x^4 + 18x^3 - x^2 + 19x - 18$
**8.** 3 **9.** (*i*) $\dfrac{0.5}{x-1} - \dfrac{0.5}{x+1} + \dfrac{1}{x-2}$ (*ii*) $-\dfrac{1}{28(x+1)} - \dfrac{1}{30(x-1)} - \dfrac{17}{30(x-4)} + \dfrac{39}{70(x-6)}$

## Problems 29.6, page 1108

**1.** 1 **2.** 3.09 **3.** 448, 3150 **4.** 133.19
**5.** $f(x) = \dfrac{1}{24}x^3 - 25x + 24 - \dfrac{7}{6}x^2 + \dfrac{557}{60}x - 25$ **6.** $f(x) = \dfrac{1}{20}x^3$ **7.** $f(x) = x^4 - 3x^3 + 5x^2 - 6$
**8.** 31.

## Problems 29.7, page 1109

**1.** 11.5 **2.** 6.5928 **3.** 37.23

## Problems 29.8, page 1109

**1.** $ 7.3 **2.** (*b*)

**3.**

| $x$ | $f(x)$ | I.D.D. | II.D.D. |
|---|---|---|---|
| 5 | 7 | | |
| | | 2.9 | |
| 15 | 36 | | 0.87 |
| 22 | 160 | 17.7 | |

**4.** Intermediate value of the variables. **5.** $ 7.8

**6.** $\dfrac{[x_1, x_2, x_3, x_4] - [x_0, x_1, x_2, x_3]}{x_4 - x_0}$ **7.** $\dfrac{-1}{4}$ and $\dfrac{1}{4}$ **8.** $ 7.14

**9.** $f(x) = \dfrac{(x - x_1)(x - x_2)}{(x_0 - x_1)(x_0 - x_2)} + \dfrac{(x - x_0)(x - x_1)}{(x_1 - x_0)(x_1 - x_2)} + \dfrac{(x - x_0)(x - x_1)}{(x_2 - x_0)(x_2 - x_1)}$

**10.** $\dfrac{13}{5}$ **11.** (*c*) **12.** 1.857
**13.** *Extrapolation* is the process of estimating the value of a function outside the given range of values.
**14.** $1/(abc)$ **15.** (*a*) **16.** $x^3 - 7x^2 - 18x - 12$
**17.** (*b*) **18.** $6h^2(x + h)$.

## Problems 30.1, page 1118

**1.** –27.9, 117.67
**2.** 4.75, 9
**3.** 0.63, 6.6
**4.** (*a*) 0.493, –1.165 (*b*) 0.4473, –0.1583 ; (*c*) 0.4662, –0.2043
**5.** 2.8326
**6.** –0.06 ; 0.5
**7.** 0.175
**8.** 13.13 m/sec
**9.** 7.956
**10.** (*i*) –52.4, (*ii*) –0.0191
**11.** 44.92
**12.** 0.085
**13.** 3.82 rad/sec, 6.75 rad/sec$^2$
**14.** 0.2561
**16.** 0.0186
**17.** 135
**18.** $y_{max}(1) = 0.25, y_{min}(0) = 0$
**19.** *Max* $f(10.04) = 1340.03$
**20.** $y_{min} = 0.6137$

## Problems 30.2, page 1126

**1.** $\frac{\pi}{6}(2\sqrt{2}+1)$
**2.** (*i*) 0.7854 (*ii*) 0.7854, (*iii*) 0.78535, (*iv*) 0.7854
**3.** 0.2311
**4.** –6.436
**5.** $\pi$
**6.** 34.30
**7.** (*i*) 2.0009 ; (*ii*) 1.1873 ; (*iii*) 14.51086
**8.** (*i*) 1.1249 (*ii*) 0.911
**9.** 3.032
**10.** 408.8 cub. cm.
**11.** 30.87 m/sec.
**12.** 29 min nearly.

## Problems 30.3, page 1127

**1.** (*c*)
**2.** $\frac{1}{h}\left[\Delta f(a) - \frac{1}{2}\Delta^2 f(a) + \frac{1}{3}\Delta^3 f(a) - \cdots\right]$
**3.** *h* should be small
**4.** 0.775
**5.** $2\frac{2}{3}$
**6.** 30.8
**7.** (*b*)
**8.** larger number of sub-intervals
**9.** 0.7854
**10.** (*d*)
**11.** $y'_{x_n} \frac{1}{h} = \left[\nabla_{y_n} + \frac{1}{2}\nabla^2{}_{y_n} + \frac{1}{3}\nabla^3{}_{y_n} + \cdots\right]$
**12.** a multiple of 6
**13.** (*c*)
**14.** 0.783
**15.** 0.69
**16.** 1.36125
**17.** 1.36
**18.** if the entire curve is itself a parabola
**19.** even and multiples of 3
**20.** False

## Problems 31.1, page 1130

**1.** $y_{x+3} - 2y_{x+2} + 2y_{x+1} = 0$
**2.** $\Delta y_n = (-1)^{n+1}/(n+1)$
**3.** $u_{n+1} - 2u_n = 0$
**4.** (*i*) $(x+2)y_{x+2} - 2(2x+1)y_{x+1} + xy_x = 0$ ; (*ii*) $(x^2+x)y_{x+2} - (2x^2+4x)y_{x+1} + (x^2+3x+2)y_x = 0$
**5.** (*ii*) $y_{n+2} - 8y_{n+1} + 15y_n = 0$ ; (*ii*) $y_{n+2} - 6y_{n+1} + 4y_n = 0$
**6.** (*i*) $(x-1)y_{x+2} - (3x-2)y_{x+1} + 2xy_x = 0$ ; (*ii*) $y_{x+2} - 4y_x = 0$ ;
(*iii*) $y_{x+3} - 6y_{x+2} + 11y_{x+1} - 6y_x = 0$.

## Problems 31.2, page 1133

**1.** $u_n = (1-n)2^n$
**2.** $y_n = 2^n\left(c_1\cos\frac{2n\pi}{3} + c_2\sin\frac{2n\pi}{3}\right)$
**3.** $u_n = c_1\cos n\pi/2 + c_2\sin n\pi/2$
**4.** $y_n = c_1 . 2^n + c_2 . 3^n$.
**5.** $y_n = (2)^{n-1} + (-2)^{n-1}$
**6.** $y_n = c_1 + c_2 n + c_3 n^2$.
**7.** $f(x) = (c_1 + c_2 x)(-1)^x + c_3 . 2^x$
**8.** $u_n = 2n + (-2)^n$.
**9.** $y_n = 6 + (n-3)2^n$
**10.** $u_n = 2^{n/2}[c_1\cos n\pi/4 + c_2\sin n\pi/4]$
**11.** $y_m = 2^m\left\{c_1\cos\frac{m\pi}{4} + c_2 + \sin\frac{m\pi}{4} + c_3\cos\frac{3m\pi}{4} + c_4\sin\frac{3m\pi}{4}\right\}$
**14.** $y_n = c_1(-1)^n + c_2(10)^n$

## Problems 31.3, page 1136

**1.** $y_n = c_1(-1)^n + c_2(6)^n - 2^n/12$

**2.** $y_p = c_1 + c_2 p + c_3 p^2 + \frac{1}{6} p(p-1)(p-2)$

**3.** $y_n = 2^n \left( \frac{2}{\sqrt{3}} \sin \frac{n\pi}{3} - 2 \cos \frac{n\pi}{3} \right) + 2$

**4.** $y = c_1 + c_2 \,.\, 3^x + \frac{1}{2} x \,.\, 3^{x-1}$

**5.** $y_x = (c_1 + c_2 x) 2^x + 3x(x-1) 2^{x-3} + 5 \,.\, 4^{x-1}$

**6.** $u_n = c_1 + c_2(-1)^n + \frac{1}{2} \frac{\cos\left(\frac{n}{2} - 1\right) - \cos \frac{n}{2}}{1 - \cos 1}$

**7.** $y_p = c_1 \cos \frac{p}{2} + c_2 \sin \frac{p}{2} + \frac{p \cos\left(p - \frac{1}{2}\right)}{2 \sin \frac{1}{2}}$

**8.** $y_x = c_1 + 2^x + c_2(-2)^x - \frac{1}{27}(9x^2 + 12x + 11)$

**9.** $y_n = c_1(-1)^n + c_2 \cos \frac{n\pi}{3} + c_3 + \sin \frac{n\pi}{3} + \frac{1}{2} n(n-3)$

**10.** $y_n = (c_1 + c_2 n)(3)^n + c_3(-1)^n + \frac{1}{3}(2)^n - \frac{3n}{4}$

**11.** $y_n = (c_1 + c_2 n) 2^{-n} + \frac{2^n}{9} + n(n-1)\left(\frac{1}{2}\right)^{n-1}$

**12.** $u_x + (c_1 + c_2 x)(-3)^x + \frac{2^x}{25}(5x - 2) + \frac{2}{4} x^{x-2} + \frac{7}{16}$

**13.** $y_n = c_1(2)^n + c_2(4)^n + 2n(4^n - 2^n) - \frac{6}{5} n + \frac{24}{25}$

**14.** $u_n = \left\{ c_1 + c_2 n + \frac{1}{48} n(n-1)^2 (n-2) \right\} 2^n.$

**15.** $y_k = c_1 \,.\, 2^k + c_2 \,.\, 3^k + \frac{4^k}{2}(k^2 - 13k + 61)$

**16.** $y_n = 2^n \left\{ (c_1 + n) \cos \frac{n\pi}{3} + c_2 + \sin \frac{n\pi}{3} \right\}.$

**17.** $n$ is odd, $y_n(0) = m(m^2 - 1^2)(m^2 - 3^2) \ldots [m^2 - (n-2)^2]$, $n$ is even $y_n(0) = m^2(m^2 - 2^2)(m^2 - 4^2) \ldots [m^2 - (n-2)^2]$ )

**18.** $n$ is odd, $y_n(0) = 0$ and when $n$ is even $y_n(0) = (-1)^{n-2/2}\, 2.2^2.\, 4^2.\, 6^2 \ldots (n-2)^2$)

**19.** When $n$ is odd $y_n(0) = m(1^2 - m^2)(3^2 - m^2)(5^2 - m^2) \ldots [(n-2)^2 - m^2]$ and when $n$ is even $y_n(0) = 0$)

## Problems 31.4, page 1137

**1.** $y_x = a + b(-1)^x + x,\ z_x = a + b(-1)^{x+1} - (x+1)$

**2.** $y_x = (a + bx)(-1)^x - \frac{1}{9} \,.\, 2^{x+2},\ z_x = \frac{2^x}{9} - (-1)^x [a + b(x - \frac{1}{2})]$

**3.** $u_n = 2.4^n - 2\frac{1}{2} n(n-1),\ v_n = 4^n + 2 + \frac{1}{n} + n(n+1)$

**4.** $u_x = -2a + b(-2)^x - c + \frac{1}{2x}(3 - x),\ v_x = a + c + b(-2)^x + \frac{1}{2} x(x-1).$

## Problems 31.5, page 1142

**1.** $y_{i+1} - 2y_i + y_{i-1} = -\frac{l_m}{P} y_i$. Solve it for $y_i$.

## Problems 31.6, page 1142

**1.** $y_{n+2} - 5y_{n+1} + 6y_n = 0$

**2.** $u_n = c_1 + c_2 n + c_3 n^2$

**3.** $u_n = c_1 + c_2(-2)^n + c_3(3)^n$

**4.** $y_n = 2^{n+1}$

**5.** $y_n = c(2)^n - (n+1)$

**6.** $(x^2 + x) y_{x+2} - (2x^2 + 4x) y_{x+1} - (x^2 + 3x + 2) y_x = 0$

**7.** $y_n = c(2)^k + 1$

**8.** $y_n = (2)^{n+2} + (-2)^{n-1}$

**9.** Third

**10.** $(x+2) y_{n+2} - 2(n+1) y_{n+1} + n y_n = 0$

**11.** Second

**12.** $y_n = (C_1 + C_2 n) 2^n$

**13.** $\frac{1}{2} x(x-1)(3)^{x-2}$

**14.** $y_{n+2} - 6y_{n+1} + 9y_n = 0$

**15.** True.

## Problems 32.1, page 1148

1. $y = 1 - \frac{x^2}{2} + \frac{x^4}{8} - \frac{x^6}{48}$
2. 0.0214
3. $y = \frac{1}{3}x^3 - \frac{1}{81}x^9 + \ldots$
4. (*a*) 0.9138, (*b*) 0.1938
5. $y(1.1) = 0.1103, y(1.2) = 0.2428$. Exact $y(1.1) = 0.1103$ and $y(1.2) = 0.2428$
6. $y(x) = x - \frac{2}{3}x^3 + \frac{4}{15}x^5 \ldots$; 1.995
7. 1.1053425
8. 1.1272
9. 1.005.
10. $y^3 = 2 + x + x^2 - \frac{1}{3}x^4 - \frac{2}{25}x^5$
11. 0.214
12. $y^3 = 2 + x^2 + x^3 + \frac{3}{20}x^5 + \frac{1}{10}x^6, z^3 = 3x^2 + \frac{3}{4}x^4 + \frac{6}{5}x^5 + \frac{3}{28}x^7 + \frac{3}{40}x^8$
13. $y(0.1) = 0.5075$

## Problems 32.2, page 1154

1. 2.9399
2. 1.1448
3. 4.5559
4. 1.9227
5. 5.051
6. 2.2352
7. $y(0.1) = 1.222, y(0.2) = 1.375$
8. 1.0928
9. 5.051.
10. $y(0.1) = 0.9052, y(0.4) = 0.8213$

## Problems 32.3, page 1158

1. 1.7278
2. 1.1749
3. 1.2415
4. 2.5005
5. $y(0.1) = 0.9052, y(0.2) = 0.8213$
6. $y(0.1) = 2.9919, y(0.2) - 2.9627$
7. 0.3487
8. 0.57705
9. 1.1678
10. 1.0911, 1.1677, 1.2352, 1.2902, 1.338.

## Problems 32.4, page 1163

1. 3.795
2. 1.2797
3. 1.2415
4. $y(4.5) = 1.023$
5. $y(0.4) = 2.162$
6. $y(0.4) = 0.441$.

## Problems 32.5, page 1167

1. 0.2416
2. 1.0408
3. 0.6897
4. $y(4.4) = 1.019$
5. 2.5751
6. $y(1.4) = 0.949$.

## Problems 32.6, page 1171

1. $y_3 = 1 + \frac{x}{2} + \frac{3}{40}x^5 + \frac{1}{40}x^6 + \frac{1}{192}x^9 \cdot z_3 = \frac{1}{2} + \frac{3}{8}x^4 + \frac{1}{10}x^5 + \frac{3}{34}x^8 + \frac{7}{340}x^9 + \frac{1}{256}x^{12}$
2. $y(0.1) = 0.105, z(0.1) = 0.999$ ; $y(0.2) = 0.22, Z(0.2) = 0.997$
3. $y(0.1) = 2.0845, z(0.1) = 0.5867$
4. $y_2 = 1 + \frac{1}{2}x + \frac{3}{40}x^5$
5. $y(0.1) = 0.5075$
6. $y(0.2) = 0.9802, y(0.2) = -0.196$
7. $-0.5159$
8. $\theta(0.2) = 0.8367, (d\theta/dt)_{0.2} = 3.6545$.

## Problems 32.7, page 1175

**1.** 0.14031
**2.** $y(.25) = y(.75) = 2.4, y(.5) = 3.2$
**3.** $y(1.25) = 1.3513, y(1.5) = 1.635, y(1.75) = 1.8505$
**4.** $y(.25) = -0.3473, y(.5) = -0.9508, y(.75) = -1.7257$
**5.** $n = 2, y(0.5) = 0.1389$, true value $= 0.1505$ ; $n = 4, y(0.5) = 0.147$
**6.** $y(0.25) = 0.062, y(0.5) = 0.25, y(0.75) = 0.562$
**7.** $y(1) = 1.0171, y(2) = 1.094$
**8.** $y(1/3) = 1.1539, y(2/3) = 3.9231$ ; $y(1) = 7.4615$.

## Problems 32.8, page 1175

**1.** (*b*)
**2.** $y_{i+1} - 2y_i + y_{i-1} + h^2y_i = 5h^2$
**3.** $1 + x + x^2 + x^3/6$
**4.** § 31.4
**5.** § 31.3
**6.** (*b*)
**7.** $y_4 = y_2 + \frac{h}{3}(f_2 + 4f_3 + f_4)$
**8.** Modified Euler's method
**9.** $y_1 = y_0 + \frac{h}{24}(55f_0 - 59f_{-1} + 37f_{-2} - 9f_{-3})$
**10.** Milne's method and Adam-Bashforth method
**11.** four
**12.** $y = 1 + \frac{x^2}{2} + \frac{x^4}{8}$
**13.** (*b*)
**14.** $y_4 = y_0 + \frac{4h}{3}(2f_1 - f_2 + 2f_3)$
**15.** $y_1 = y_0 + \frac{h}{24}(9f_1 + 19f_0 - 5f_{-1} + f_{-2})$
**16.** 1.1818
**17.** $\frac{dy}{dx} = z, \frac{dz}{dx} + y(1 + yz) = 0$
**18.** § 31.7
**19.** starting values
**20.** Picards and Runge-Kutta methods
**21.** It agrees with Taylor's series solution upto the term in $h^4$
**22.** (*d*)
**23.** $y_{i+1} + 2(h^2 - 1)\, y_i + y_{i-1} = 0$
**24.** (*a*)
**25.** § 32.14 : (3) and (4)
**26.** True
**27.** False
**28.** False.

## Problems 33.1, page 1178

**1.** Parabolic
**2.** Hyperbolic
**3.** (*i*) Parabolic (*ii*) Elliptic (*iii*) Elliptic
**4.** Outside the ellipse $(x/0.5)^2 + (y/0.25)^2 = 1$.

## Problems 33.2, page 1187

**1.** $u_1 = 1.999, u_2 = 2.999, u_3 = 3.999$
**2.** 2.37, 5.6, 9.87, 2.89, 6.14, 9.89, 3.02, 6.17, 9.51
**3.** $u_1 = 10.188, u_2 = 0.5, u_3 = 1.188, u_4 = 0.25, u_5 = 0.625, u_6 = 1.25$
**4.** $u_1 = 26.66, u_2 = 33.33, u_3 = 43.33, u_4 = 46.66$
**5.** $u_1 = 0.99, u_2 = 1.49, u_3 = 0.49$
**6.** $u_1 = 1.57, u_2 = 3.71, u_3 = 6.57, u_4 = 2.06, u_5 = 4.69, u_6 = 8.06, u_7 = 2, u_8 = 4.92, u_9 = 9$
**7.** $u_1 = -3, u_2 = -2, u_3 = -2$.

## Problems 33.3, page 1191

**1.**

| j \ i | 0 | 1 | 2 | 3 | 4 |
|---|---|---|---|---|---|
| 0 | 0 | 3 | 4 | 3 | 0 |
| 1 | 0 | 2 | 3 | 2 | 0 |
| 2 | 0 | 1.5 | 2 | 1.5 | 0 |
| 3 | 0 | 1 | 1.5 | 1 | 0 |
| 4 | 0 | 0.75 | 1 | 0.75 | 0 |
| 5 | 0 | 0.5 | 0.75 | 0.5 | 0 |

2.

| j \ i | 0 | 1 | 2 | 3 | 4 | 5 | 6 | 7 | 8 | 9 | 10 |
|---|---|---|---|---|---|---|---|---|---|---|---|
| 0 | 0 | 0.09 | 0.16 | 0.21 | 0.24 | 0.25 | 0.24 | 0.21 | 0.16 | 0.09 | 0 |
| 1 | 0 | 0.08 | 0.15 | 0.20 | 0.23 | 0.24 | 0.23 | 0.20 | 0.15 | 0.08 | 0 |
| 2 | 0 | 0.075 | 0.14 | 0.19 | 0.22 | 0.23 | 0.22 | 0.19 | 0.14 | 0.075 | 0 |
| 3 | 0 | 0.07 | 0.133 | 0.18 | 0.21 | 0.22 | 0.21 | 0.18 | 0.133 | 0.07 | 0 |

3.

| j \ i | 0 | 1 | 2 | 3 | 4 | 5 |
|---|---|---|---|---|---|---|
| 0 | 0 | 24 | 84 | 144 | 144 | 0 |
| 1 | 0 | 42 | 84 | 114 | 72 | 0 |
| 2 | 0 | 42 | 78 | 78 | 57 | 0 |
| 3 | 0 | 39 | 60 | 67.5 | 39 | 0 |
| 4 | 0 | 30 | 53.25 | 49.5 | 33.75 | 0 |
| 5 | 0 | 26.6 | 39.75 | 43.5 | 24.75 | 0 |
| 6 | 0 | 19.88 | 35.06 | 32.25 | 21.75 | 0 |

4.

| j \ i | 0 | 1 | 2 | 3 | 4 |
|---|---|---|---|---|---|
| 0 | 0 | 0.5 | 1 | 0.5 | 0 |
| 1 | 0 | 0.5 | 0.5 | 0.5 | 0 |
| 2 | 0 | 0.25 | 0.5 | 0.25 | 0 |
| 3 | 0 | 0.25 | 0.25 | 0.25 | 0 |

## Problems 33.4, page 1199

1.

| $t = 0.3$ ; $x =$ | 0.1 | 0.2 | 0.3 | 0.4 | 0.5 |
|---|---|---|---|---|---|
| Numerical sol. $u =$ | 0.02 | 0.04 | 0.06 | 0.075 | 0.08 |
| Exact sol. $u =$ | 0.02 | 0.04 | 0.06 | 0.075 | 0.08 |

2.

| j \ i | 0 | 1 | 2 | 3 | 4 | 5 |
|---|---|---|---|---|---|---|
| 0 | 0 | 20 | 15 | 10 | 5 | 0 |
| 1 | 0 | 7.5 | 15 | 10 | 5 | 0 |
| 2 | 0 | – 5 | 2.5 | 10 | 5 | 0 |
| 3 | 0 | – 5 | – 10 | – 2.5 | 5 | 0 |
| 4 | 0 | – 5 | – 10 | – 15 | – 7.5 | 0 |
| 5 | 0 | – 5 | – 10 | – 15 | – 20 | 0 |

**3.**

| $j \backslash i$ | 0 | 1 | 2 | 3 | 4 | 5 | 6 | 7 | 8 | 9 | 10 |
|---|---|---|---|---|---|---|---|---|---|---|---|
| 0 | 0 | 10.19 | 10.16 | 10.21 | 10.24 | 10.25 | 10.24 | 10.21 | 10.16 | 10.09 | 0 |
| 1 | 0 | 5.08 | 10.15 | 10.20 | 10.23 | 10.24 | 10.23 | 10.20 | 10.15 | 5.08 | 0 |
| 2 | 0 | 0.06 | 5.12 | 10.17 | 10.20 | 10.21 | 10.20 | 10.17 | 10.12 | 0.06 | 0 |
| 3 | 0 | 0.04 | 0.08 | 5.12 | 10.15 | 10.16 | 10.15 | 10.12 | 10.08 | 0.04 | 0 |
| 4 | 0 | 0.02 | 0.04 | 0.06 | 5.08 | 10.09 | 10.08 | 10.06 | 10.04 | 0.02 | 0 |
| 5 | 0 | 0 | 0 | 0 | 0 | 0 | 0 | 0 | 0 | – 0.02 | 0 |

**4.**

| $j \backslash i$ | 0 | 0.1 | 0.2 | 0.3 | 0.4 | 0.5 |
|---|---|---|---|---|---|---|
| 0.1 | 0 | 0.037 | 0.07 | 0.096 | 0.113 | 0.119 |
| 0.2 | 0 | 0.031 | 0.059 | 0.082 | 0.096 | 0.101 |
| 0.3 | 0 | 0.023 | 0.043 | 0.059 | 0.07 | 0.074 |
| 0.4 | 0 | 0.012 | 0.023 | 0.031 | 0.037 | 0.039 |
| 0.5 | 0 | 0 | 0 | 0 | 0 | 0 |

## Problems 33.5, page 1199

**1.** (*b*)

**2.** False

**3.** A hyperbolic equation

**4.** Poisson's equation

**5.** $u_{i,j+1} = u_{i-1,j} + u_{i+1,j} - u_{i,j-1.}$

**6.** $(u_{i-1,j} - 2u_{i,j} + u_{i+1,j})/h^2$

**7.** § 32.8 (2)

**8.** $\frac{1}{4}(u_{i-1,j+1} + u_{i+1,j-1} + u_{i+1,j+1} + u_{i-1,j-1})$

**9.** Hyperbolic

**10.** Bendre-Schmidt

**11.** $u_{i,j+1} = 2(1 - 4\alpha^2)\, u_{i,j} + 4\alpha^2 (u_{i-1,j} + u_{i+1,j} - u_{i,j-1})$ where $\alpha = k/h$.

**12.** $u_{i,j+1} = \frac{1}{2}\,(u_{i-1,j} + u_{i+1,j})$

**13.** § 33.5(2)

**14.** $\lambda < \frac{1}{2}$

**15.** Elliptic

**16.** $u(0, t) = 0 = u\,(1, t), (t > 0)\,;\, u(x, 0) = f(x), 0 < x < l\,;\, \delta x/\delta t\,(x, 0) = 0, 0 < x < l$

**17.** $u_{i,j+1} = u_{i+1,j} - u_{i-1,j} - u_{i,j-1}$

**18.** § 33.6 (2)

**19.** 100

**20.** $\dfrac{u_{i-1,j} - 2u_{i,j} + u_{i+1,j}}{h^2} + \dfrac{u_{i,j-1} - 2u_{i,j} + u_{i,j+1}}{k^2} = 0$

**21.** $y < 0$

**22.** $k = 1/4$.

## Problems 34.1, page 1203

**1.** Max. $Z = 1.2x_1 + 1.4x_2$ ; subject to $40x_1 + 25x_2 \le 1000$,
$35x_1 + 28x_2 \le 980$, $25x_1 + 35x_2 \le 875$ and $x_1, x_2 \ge 0$

**2.** Max. $Z = 3x_1 + 2x_2 + 4x_3$ ; subject to $4x_1 + 3x_2 + 5x_3 \le 2000$,
$2x_1 + 2x_2 + 4x_3 \le 2500$, $100 \le x_1 \le 150$, $200 \le x_2$ and $50 \le x_3$

**3.** Max. $Z = 3x_1 + 2x_2 + x_3$ ; subject to $3x_1 + 4x_2 + 3x_3 \le 42$,
$5x_1 + 3x_3 \le 45$, $3x_1 + 6x_2 + 2x_3 \le 41$ and $x_1, x_2, x_3 \ge 0$.

**4.** Max. $Z = 400\,x + 300\,y$ ; subject to $x + y \le 200$, $x \ge 20$, $y \ge 4x$, $x \ge 0$, $y \ge 0$

**5.** Min. $Z = x_1 + x_2$ ; subject to $2x_1 + x_2 \ge 12$, $5x_1 + 8x_2 \ge 74$,
$x_1 + 6x_2 \ge 24$ and $x_1, x_2, x_3 \ge 0$.

**6.** Max. $Z = 0.15x_1 + 0.25x_2$ ; subject to
$2x_1 + 5x_2 \le 480{,}000$, $5x_1 + 4x_2 \le 720{,}000$,
$8x_1 + 16x_2 \le 300{,}000$, $0 \le x_1 \le 25{,}000$ and $0 \le x_2 \le 7{,}000$

**7.** Min. $Z = 41x_1 + 35x_2 + 96x_3$ ; subject to
$2x_1 + 3x_2 + 7x_3 \ge 1250$, $x_1 + x_2 \ge 250$, $5x_1 + 3x_2 \ge 900$,
$6x_1 + 25x_2 + x_3 \ge 232.5$ and $x_1, x_2, x_3 \ge 0$

**8.** Min. $Z = 100x_1 + 250x_2 + 160x_3$ ; subject to
$0.94x_1 + x_2 + 1.04x_3 \le 0.98$, $10x_1 + 15x_2 + 17x_3 \ge 14$,
$470x_1 + 500x_2 + 520x_3 \ge 495$, $x_1 + x_2 + x_3 = 1$ and $x_1, x_2, x_3 \ge 0$.

## Problems 34.2, page 1209

**1.** $x_1 = 100, x_2 = 50$ ; max. $Z = 550$

**2.** $x_1 = 8/15, x_2 = 12/5$, max. $Z = 24.8$

**3.** $x_1 = 6, x_2 = 12$ ; $Z_{min} = 240$

**4.** $x_1 = 1000, x_2 = 500$ ; $Z_{max} = 5500$

**5.** 450 units of product $B$ only ; max. profit = ₹ 1800

**6.** $X = 2, Y = 4.5$ ; max. profit = ₹ 37

**7.** $A = 1.18$ units, $B = 0.53$ units ; max. profit = ₹ 14.50 approx.

**8.** 2000/11 units of product $A$ and 1000/11 units of $B$ ; max. profit = ₹ 10,000

**9.** $x_1 = 4, x_2 = 0$ ; max. $z = 8$

**10.** Unbounded solution

**11.** $x_1 = 2, x_2 = 4$ ; min. $Z = 64$

**12.** Production cost will be min. if G and J run for 12 and 4 days respectively.

## Problems 34.3, page 1215

**1.** Max. $Z = 3x_1 + 5x_2 + 8x_3$ ; subject to $2x_1 - 5x_2 + s_1 = 6$,
$3x_1 + 2x_2 + x_3 - s_2 = 5$, $3x_1 + 4x_3 + s_3 = 3$ ;
$x_1, x_2, x_3, s_1, s_2, s_3 \ge 0$

**2.** Min. $Z = 3x_1 + 2x_2 + 5x_3$, subject to $-5x_1 + 2x_2 + s_1 = 5$,
$2x_1 + 3x_2 + 4x_3 - s_2 = 7$, $2x_1 + 5x_3 + s_3 \ge 3$
$x_1, x_2, x_3, s_1, s_2, s_3 \ge 0$.

**3.** Max. $Z = 3x_1 - 2x_2 + 4x_4 - 4x_5$ ; subject to
$x_1 + 2x_2 + x_4 - x_5 + s_1 = 8$, $2x_1 - x_2 + x_4 - x_5 - s_2 = 2$,
$-4x_1 + 2x_2 + 3x_4 - 3x_5 = 6$ ; $x_1, x_2, x_4, x_5, s_1, s_2 \ge 0$

**4.** (*i*) $x_1 = 2, x_3 = 1$ (Basic) ; $x_3 = 0$ (Non-basic). (*ii*) $x_1 = 5$, $x_3 = -1$ (Basic) ; $x_2 = 0$ (Non-basic) ; (*iii*) $x_2 = 5/3, x_3 = 2/3$ (Basic) ; $x_1 = 0$ (Non-basic). All the three basic solutions are non-degenerate

**6.** Basic solutions are (*i*) $x_1 = 2, x_2 = 1$ (Basic) and $x_3 = 0$ ;
(*ii*) $x_1 = x_3 = 1$ (Basic) and $x_2 = 0$ ; (*iii*) $x_2 = -1, x_3 = 2$ (Basic) and $x_1 = 0$
(*a*) First two solutions are non-degenerate basic feasible solutions
(*b*) First solution is optimal and $Z_{max} = 5$.

## Problems 34.4, page 1222

**1.** $x_1 = 2, x_2 = 4$, max. $Z = 14$

**2.** $x_1 = 0, x_2 = 20$ ; max. $Z = 200$

**3.** $x_1 = 7/3, x_2 = 4/3$ ; max. $Z = 16$

**4.** $x_1 = 5, x_2 = x_3 = 0$ ; max $Z = 50$

**5.** $x_1 = 0, x_2 = 100, x_3 = 230$ ; max. $Z = 1350$

**6.** $x_1 = 89/41, x_2 = 50/41, x_3 = 62/41$ ; max. $Z$ = ₹ 765/41

**7.** $x_1 = 4, x_2 = 5, x_3 = 0$ ; min $Z = -11$

**8.** $x_1 = 50, y = 0, z = 55/7, t = 0$ ; Max. = 695/7

**9.** $x_1 = 0, x_2 = 400$ units ; max. profit = ₹ 1200

**10.** $x_1 = 125, x_2 = 250$ units ; max profit = ₹ 2250

**11.** $x_1 = 400$ gms, $x_2 = 0$ ; min. cost = ₹ 2

**12.** $x_1 = 0, x_2 = x_3 = 50$ ; max. profit = ₹ 700

**13.** $x_1 = 0.5, x_2 = x_3 = 0.04$ units ; min. cost = ₹ 5.80

**14.** Averages for corn, wheat, soyabeans are 250, 625, zero respectively to achieve a max. profit of ₹ 32,000.

## Problems 34.5, page 1229

**1.** $x_1 = 0, x_2 = 2, x_3 = 0$ ; max. $Z = 4$

**2.** $x_1 = 3, x_2 = 2, x_3 = 0$ ; max. $Z = 8$

**3.** $x_1 = x_2 = -6/15$ ; max. $Z = -48/5$

**4.** $x_1 = 23/3, x_2 = 5, x_3 = 0$ ; max $Z = 85/3$

**5.** $x_1 = x_2 = x_3 = 5/2, x_4 = 0$ ; max. $Z = 15$

**6.** $x_1 = 21/13, x_2 = 10/13$ ; max. $Z = 31/13$

**7.** Infeasible

**9.** $x_1 = 23/3, x_2 = 5, x_3 = 0$ ; $Z_{max} = 85/3$

**10.** $x_1 = 55/7, x_2 = 30/7, x_3 = 0$ ; $Z_{max} = 155/7$

**11.** $x_1 = 2, x_2 = 0$ ; $Z_{max} = 18$

**12.** Degenerate solution : $x_1 = 0$ (non-basic) ; $x_2 = 1, x_3 = 0$ (basic) ; $Z_{max} = 3$.

## Problems 34.6, page 1232

**1.** Min. $W = 26y_1 + 7y_2$ ; subject to $6y_1 + 4y_2 \geq 10$
$5y_1 + 2y_2 \geq 13, 3y_1 + 5y_2 \geq 19$ ; $y_1, y_2, y_3 \geq 0$

**2.** Max. $W = 11y_1 + 7y_2 + y_3 + 5y_4$ ; subject to $3y_1 + 2y_2 - y_3 + 3y_4 \leq 2$,
$4y_1 + 3y_2 + 2y_3 + 2y_4 \leq 4, y_1 - 2y_2 + 3y_3 + 2y_4 \leq 3$ ;
$y_1, y_2, y_3, y_4 \geq 0$

**3.** Min. $W = -3y_1 + y_2 + 4y_3$ ; subject to $y_1 + 3y_2 - 2y_3 \leq -3$,
$y_1 + y_3 \geq 16, y_1 - 2y_2 + y_3 \leq -7$ ;
$y_1, y_2 \geq 0$, $y_3$, unrestricted in sign

**4.** Max. $W = -5y_1 + 9y_2 + 8y_3$ ; subject to $-2y_1 + 4y_2 - 8y_3 \leq 3$,
$3y_1 - 2y_2 + 4y_3 \leq -2, -y_1 + 3y_3 = 1$ ;
$y_1, y_2 \geq 0$, $y_3$ unrestricted

**5.** Min. $y = 3y_1 + 4y_2 + y_3 + 6y_4$ ; subject to $5y_1 - 2y_2 + y_3 - 3y_4 \geq 2$,
$6y_1 + y_2 - 5y_3 - 3y_4 \geq 5, -y_1 + 4y_2 + 3y_3 + 7y_4 \geq 6, y_1, y_2, y_3, y_4 \geq 0$.

## Problems 34.7, page 1235

**1.** $x_1 = x_2 = 0, x_3 = 5/2$ ; min. $Z = 2.5$

**2.** $x_1 = 4, x_2 = 2$ ; max. $Z = 10$.

**3.** $x_1 = 7, x_2 = 0$, max. $Z = 21$

**4.** $x_1 = 0, x_2 = 100, x_3 = 230$ ; max. $Z = 1350$.

## Problems 34.8, page 1238

**1.** $x_1 = 0, x_2 = 1$ ; max. $Z = -1$

**2.** $x_1 = 3/5, x_2 = 6/5$ ; min. $Z = 12/5$

**3.** $x_1 = 6, x_2 = 2, x_3 = 0$ ; min. $Z = 10$

**4.** $x_1 = 65/23, x_2 = 0, x_3 = 20/23, x_4 = 0$ ; min. $Z = 215/23$.

## Problems 34.9, page 1245

**1.** $x_{11} = 200, x_{12} = 50, x_{22} = 175, x_{24} = 125, x_{33} = 275, x_{34} = 125$; min. cost = ₹ 12075

**2.** $x_{13} = 14, x_{21} = 6, x_{22} = 5, x_{23} = 1, x_{32} = 5$ ; min. cost = 143

**3.** $x_{11} = 50, x_{12} = 100, x_{21} = 150, x_{33} = 150, x_{42} = 100, x_{43} = 50$ ; min. tonnage = 3300

**4.** $x_{11} = 140, x_{13} = 60, x_{21} = 40, x_{22} = 120, x_{33} = 90$ ; min. cost = ₹ 5920

**5.** $x_{11} = 5, x_{14} = 2, x_{22} = 3, x_{23} = 7, x_{32} = 5, x_{34} = 13$ ;
min. cost = ₹ 799 and maximum saving = ₹ 201

**6.** $x_{11} = 150, x_{13} = 20, x_{22} = 160, x_{24} = 40, x_{33} = 90, x_{34} = 90$ ; max. profit = 4920

**7.** $x_{13} = 2, x_{22} = 1, x_{23} = 2, x_{31} = 4, x_{33} = 1$ ; min. cost = 33

**8.** $x_{13} = 0, x_{15} = 800, x_{21} = 400, x_{24} = 100, x_{32} = 400, x_{33} = 200$,
$x_{34} = 300, x_{43} = 300$ ; min. cost = 9200.

## Problems 34.10, page 1250

**1.** $x_{11} = x_{22} = x_{33} = 1$ ; min. cost = ₹ 18
**2.** $A \to 2, B \to 3, C \to 4, D \to 1$ ; min. $Z = 38$
**3.** $I \to B, II \to A, III \to D, IV \to C$ ; min. cost = ₹ 49
**4.** $A \to$ Dyn. Prog., $B \to$ Queuing Th., $C \to$ Reg. Analysis, $D \to$ L.P. ; min. time = 28 hrs
**5.** (*i*) $A \to I, B \to II, C \to III, D \to IV$ ; (*ii*) $A \to I, B \to III, C \to II, D \to IV$
**6.** $1 \to IV, 2 \to II, 3 \to VI, 4 \to I, 5 \to III, 6 \to V$ ; max. profit = ₹ 270

## Problems 35.1, page 1259

**1.** (*i*) $y = \frac{1}{4}x^2 + c_1x + c_2$ ; (*ii*) $y = c_1x^{-1} + c_2$
**3.** $y = c_1 e^x + c_2 e^{-x} + \frac{xe^x}{2}$
**4.** $y = -x \cos x/2$
**5.** $y = (c + x) \sin x$
**6.** $y = \sinh (c_1x + c_2)$
**7.** $y = x^2 - 1$
**8.** $(x + c)^2 + y^2 = k^2$
**9.** $y = 2 \sin x$
**12.** The spirals of the family $r = a \sec (\phi \sin \alpha + b)$.

## Problems 35.2, page 1261

**1.** $y = \pm 2 \sin m\pi x$, where $m$ is an integer
**2.** $y = \lambda x^2 + ax + b$, where $\lambda, a, b$ are determined from the isoperimetric and boundary conditions.
**4.** $y(x) = \frac{1}{2}(1 - \cos x) + \frac{1}{4}(2 - \pi) \sin x$.

## Problems 35.3, page 1265

**2.** $y = (c_1 + c_2x) \cos x + (c_3 + c_4x) \sin x$,
$z = (c_1 + c_2x) \cos x + (c_3 + c_4x) \sin x - 2c_2 \sin x + 2c_4 \cos x$
**3.** $y = a_n \sin nx, n = 1, 2, 3 \ldots$
**4.** $y = \cos x$
**5.** $y = -\frac{\lambda}{24\mu}(x^2 - a^2)^2$
**6.** (*i*) $y = c_1e^{2x} + c_2e^{-2x} + c_3 \cos 2x + c_4 \sin 2x$ ;
(*ii*) $y = c_1 + c_2x + c_3x^2 + c_4x^3 + c_5x^4 + c_6x^5 + x^7/7\,!$.

## Problems 35.4, page 1268

**1.** (*i*) and (*ii*) $\bar{y} = \frac{5}{18}x(1 - x)$
**2.** $\bar{y} = \frac{x}{4}(5x - 1)$
**4.** $\bar{y} = 0.58 + 0.27x$
**5.** (*i*) and (*ii*) $c_1 = 0.93, c_2 = -0.05$
**6.** 0.05
**7.** $c_1 = 3.27, c_2 = -2.69$
**8.** $y = \frac{1}{2}(5x^2 - 3x)$.

## Problems 36.1, page 1276

**1.** $y(x) = 6x - 5 + \int_0^x (5 - 6x + 6t)\, y(t)\, dt$.

**2.** $y(x) = x - \sin x + e^x (x - 1) + \int_0^x [\sin x - e^x (x - t)]\, y(t)\, dt$

**3.** $y(x) = \int_0^x t(t - x)\, y(t)\, dt + \frac{1}{2}x^2$

**4.** $y(x) + \int_0^x [1 + x - 2t + (x - t) e^{-t}]\, y(t)\, dt = \frac{x^5}{20} - \frac{5x^3}{6} + x - 3$

**5.** $y(x) = \cos x - \frac{1}{2}x^2 - \frac{1}{3}x^3 - \frac{1}{2}\int_0^x (x-t)^2\, y(t)\, dt$

**6.** $y(x) = \sin x - \frac{1}{2} + \int_0^x \left\{\frac{1}{2}x(x-t)^2 - 1\right\} y(t)\, dt$

**7.** $y(x) - \int_0^x \{4 - 6(x-t) + 2(x-t)^2 - \frac{1}{6}(x-t)^3\}\, y(t)\, dt = \frac{1}{4}\cos 2x - \frac{19}{12} + \frac{32}{3}x - \frac{85}{6}x^2 + \frac{20}{3}x^3$

**8.** $y''(x) - 2xy'(x) - 3y(x) = 0$ ; $y(0) = 1, y'(0) = 0$

**9.** $y''(x) - y(x) + 3\sin x = 0$ ; $y(0) = 3, y'(0) = 0$

**10.** $y'''(x) + 6y(x) = 0$ ; $y(0) = 4, y'(0) = -3, y''(0) = 2$

**11.** $y'''(x) - 3y''(x) + 4y'(x) + 2y(x) + e^{-x} = 0$ ; $y(0) = 1, y'(0) = 2, y''(0) = 3$.

## Problems 36.2, page 1279

**1.** $y(x) = \int_0^1 G(x, t)\, ty(t)\, dt + \frac{1}{2}x(x-1)$, where $G(x, t) = x(1-t), (x < t)$ and $= t(1-x), x > t$

**2.** $y(x) = \int_0^1 G(x, t)\, y(t)\, dt + \frac{1}{6}(x^3 - 3x + 6)$, where $G(x, t) = x, x < t$ and $= t, x > t$

**3.** $u(x) = \int_0^1 G(x, t)\, e^t\, u(t)\, dt + \frac{1}{6}(x^3 + x)$, where $G(x, t) = x(1-2t), x < t$ and $= t(1-2x), x > t$

**4.** $G(x, t) = \frac{\sinh x \sinh (t-1)}{\sinh 1}, x < t$ and $= \frac{\sinh t \sinh (x-1)}{\sinh 1}, x > t$

**5.** $u(x) = \lambda \int_0^1 G(x, t)\, t \,.\, u(t)\, dt$, where $G(x, t) = \frac{1}{2}x\left(\frac{1}{t-t}\right), x < t$ and $= \frac{1}{2}\cdot t\left(\frac{1}{x-x}\right), x > t$

## Problems 36.3, page 1283

**5.** $y(x) = \frac{1}{2}(\sin x + \sinh x)$

**6.** $y(x) = x^2 + \frac{1}{12}x^4$

**7.** $y(x) = 1$

**8.** $y(x) = \pm 6J_0(4x)$

**9.** $y(x) = J_1(2x) \quad (x > 0)$

**10.** $y(x) = \frac{1}{2}e^{-2x}(\cos x + 3\sin x) - \frac{1}{2}e^{-x}$

**11.** $y(x) = (1-x)e^{-x} + 1/2 \sin x$.

**12.** $y(x) = 1 + x^2 + x^4/24$.

**13.** $y(x) = 1 + x^2/2$

**14.** $y(x) = 1/2$

**15.** $y(x) = \frac{3\sqrt{3}}{4\pi}x^{1/3}(3x + 2)$.

## Problems 36.4, page 1287

1. Has no eigen values and eigen functions
2. Eigen value $\lambda = 1/4$ ; eigen function is $y(x) = x^2 + 3x/2$
3. Eigen values $\lambda = 8/(\pi - 2)$ ; eigen function is $y(x) = \sin^2(x)$
4. Eigen value $\lambda = 1/\pi$, $y = \sin x$
5. Has no eigen values or eigen functions
6. Eigen values are $\lambda = \pm 1/\pi$ ; eigen functions are $y(x) = \cos x + \sin x$, $y(x) = \cos x - \sin x$

**7.** $y(x) = x + \frac{\lambda[(6-\lambda)x - 4]}{12 + \lambda^2}$

**8.** $y(x) = x + \frac{\lambda}{12(1-2\lambda) - \lambda^2}[10 + (6+\lambda)x]$

**9.** $y(x) = x + \frac{\lambda}{(1-\lambda\pi)\left(1 - \frac{1}{2}\lambda\pi\right) + 4\lambda^2}\left\{2\lambda\pi + \frac{1}{2}\pi^2\left(1 - \frac{1}{2}\lambda\pi\right) + \pi(1 - 2\lambda\pi)\sin x\right\}$

**10.** $y(x) = 2x - \pi + \frac{\pi^2 \sin^2 x}{\pi - 1}$

**11.** $y(x) = \frac{2}{2-\lambda}\sin x, \lambda \neq 2$

**12.** $y(x) = x + \frac{2\lambda\pi}{1 + 2\lambda^2\pi^2}(\lambda\pi x - 4\lambda\pi \sin x + \cos x)$

**13.** There is no solution to the integral equation when $\lambda = 3$

**14.** $\lambda_1 = 2, \lambda_2 = -2 ; y_1(x) = 1 - x, y_2(x) = 1 - 3x$

**15.** (*i*) When $F(x) = x$, solution is $y(x) = x + \lambda \left\{ \frac{2\lambda\pi^2}{\lambda^2\pi^2 - 1} \sin x + \frac{2\pi}{\lambda^2\pi^2 - 1} \cos x \right\}$

(*ii*) When $F(x) = 1$, solution is $y(x) = 1$.

## Problems 36.5, page 1291

**1.** $y(x) = 1 - \frac{3\lambda x}{2(3+\lambda)}$ $(\lambda \neq -3)$

**2.** $y(x) = \frac{\sin x}{1 + \lambda\pi}$ only if $| \lambda | < \frac{1}{\pi}$

**3.** $y(x) = \frac{4 + 2\lambda (2 - 3x)}{4 - \lambda^2}$ $(\lambda \neq 2)$

**4.** $y = e^x$

**5.** $y(x) = 1$

**6.** $y = \sin x$

**7.** $y(x) = 2$.

## Problems 37.1, page 1297

**2.** (*i*) *True*, since $\{a\}$ is a subset of the set $\{a, b, c\}$ ; (*ii*) and (*iii*) *False*, since the element $a$ cannot be a subset of the set $\{a, b, c\}$ ; (*iv*) *True*, since the set $\{a, b\}$ is a subset of the set $\{a, b, c\}$ ; (*v*) *False*, since the set $\{a, b\}$ is not an element of the set $\{a, b, c\}$ ; (*vi*) *True*, since the null set $\phi$ is a subset of every set.

**17.** 20

**18.** 105

**19.** 136

**20.** Number of students not taking any of these courses is 71.

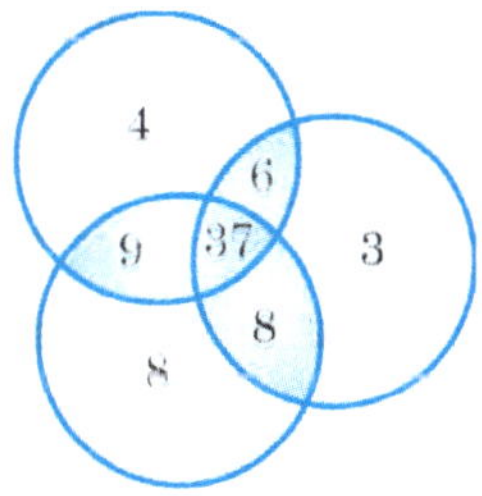


## Problems 37.2, page 1303

**1.** (*a*) It is not true that Sam is a teacher and John is an honest boy ; (*b*) Sam is a teacher and John is not an honest boy ; (*c*) Sam is not a teacher iff John is an honest boy ; (*d*) If Sam is a teacher then John is not an honest boy.

**2.** (*a*) $(p \vee q) \Rightarrow r$ where $p \equiv I$ have no car, $q = I$ do not wear good dress, $r = I$ am not, a millionaire.

**3.**

| $p$ | $q$ | $\sim q$ | $p \Rightarrow q$ | $p \Rightarrow q \wedge \sim q$ |
|---|---|---|---|---|
| 1 | 1 | 0 | 1 | 0 |
| 1 | 0 | 1 | 0 | 0 |
| 0 | 1 | 0 | 1 | 0 |
| 0 | 0 | 1 | 1 | 1 |

(*b*)

| $p$ | $q$ | $r$ | $p \Leftrightarrow q$ | $r \vee q$ | $(p \Leftrightarrow q) \wedge (r \vee q)$ |
|---|---|---|---|---|---|
| 1 | 1 | 1 | 1 | 1 | 1 |
| 1 | 1 | 0 | 1 | 1 | 0 |
| 1 | 0 | 1 | 0 | 1 | 0 |
| 1 | 0 | 0 | 0 | 0 | 0 |
| 0 | 1 | 1 | 0 | 1 | 0 |
| 0 | 1 | 0 | 0 | 1 | 0 |
| 0 | 0 | 1 | 1 | 1 | 1 |
| 0 | 0 | 0 | 1 | 0 | 0 |

**7.** (*i*) $T_p = [2, 3, 5, 7, 11, 13, 17, 19, 23, 29, 31]$

$T_q = [1, 3, 9, 27, ...]$, $T_r = [1, 3, 9, 7]$

(*ii*) $T_r \le T_q$

**10.** (*i*)

| $p$ | $q$ | $q \to p$ | $p \to (q \to p)$ |
|---|---|---|---|
| 0 | 0 | 1 | 1 |
| 0 | 1 | 1 | 1 |
| 1 | 0 | 0 | 0 |
| 1 | 1 | 1 | 1 |

**15.** (*i*) Dual of $(p \wedge q) \vee r$ is $(p \wedge q) \vee r$ (*ii*) Dual of $(p \wedge q) \vee t$ is $(p \vee q) \wedge t$

## Problems 37.3, page 1306

**1.** (*a*) $(\forall\, x \in A)\,(x + 2 < 10)$ (*b*) $(1\, x \in A)\,(x + 2 = 10)$

**2.** (*a*) $\forall\, x, (x^3 \neq x)$ (*b*) $\to x, (x + 5 \le x)$

(*c*) None of the students are 26 or older (*d*) Some students do not live in the hostels.

**3.** $\forall\, x\, P(x)$ is false **4.** $\forall\, (x_1, x_2)\; Q(x_1 + x_2)Q$

**5.** $\forall\, (a, b)\; R(a + x = b)$

**6.** (*a*) $\forall\, x[Q\,(x) \to R(x)]$ (*b*) $\sim \forall\, x[Q(x) - R(x)]$, (*c*) $\exists\, x[Q(x) \wedge R(x)]$, (*d*) $\exists x\; [Q\,(x) \wedge \sim R(x)]$

**8.** $(\sim A \vee \sim A) \wedge (B \vee \sim A) \wedge (\sim A \vee C) \wedge (B \vee C)$

**10.** 1. $p \vee q$ (Premise), 2. $\sim p \to q$ (conditional equivalence)

3. $q \to s$ (Premise) 4. $\sim p \to s$ (2, 3 chain rule)

5. $p \to r$ (Premise) 6. $\sim s \to p$ (4, conditional equivalence)

7. $\sim s \to r$(5, 6 chain rule) 8. $s \vee r$ (7, conditional equivalence)

**12.** $(b\; x)\; R\; (x = \sqrt{Z}\,)$ **13.** (*a*) Conclusion is not valid (*b*) Conclusion is not valid

## Problems 37.4, page 1313

**1.**

| $x$ | $y$ | $z$ | $x \wedge y$ | $z'$ | $y \wedge z'$ | $(x \wedge y) \vee (y \wedge z')$ |
|---|---|---|---|---|---|---|
| 0 | 0 | 0 | 0 | 1 | 0 | 0 |
| 0 | 0 | 1 | 0 | 0 | 0 | 0 |
| 0 | 1 | 0 | 0 | 1 | 1 | 1 |
| 1 | 0 | 0 | 0 | 1 | 0 | 0 |
| 1 | 1 | 0 | 1 | 1 | 1 | 1 |
| 1 | 0 | 1 | 0 | 0 | 0 | 0 |
| 0 | 1 | 1 | 0 | 0 | 0 | 0 |
| 1 | 1 | 1 | 1 | 0 | 0 | 1 |

**2.** $x \vee z' \wedge y = x \wedge y$ **3.** (*i*) $x' \vee y' \vee z'$ (*ii*) 0

**14.**

| $x_1$ | $x_2$ | $x_3$ | $x_1 \vee x_3$ | $x_3'$ | $x_2 \vee x_3'$ | $x_1 \wedge (x_2 \vee x_3')$ | $P$ |
|---|---|---|---|---|---|---|---|
| 0 | 0 | 1 | 1 | 0 | 0 | 0 | 0 |
| 0 | 1 | 0 | 0 | 1 | 1 | 0 | 0 |
| 0 | 1 | 0 | 0 | 1 | 1 | 0 | 0 |
| 1 | 0 | 0 | 1 | 1 | 1 | 1 | 1 |
| 1 | 1 | 0 | 1 | 1 | 1 | 1 | 1 |
| 0 | 1 | 1 | 1 | 0 | 1 | 0 | 0 |
| 1 | 0 | 1 | 1 | 0 | 0 | 0 | 0 |
| 1 | 1 | 1 | 1 | 0 | 1 | 1 | 1 |

## Problems 37.5, page 1315

**1.** (*i*) 0 (*ii*) 0

**2.** (*i*) $(x \vee y \vee z) \wedge (x \vee y \vee z')$ (*ii*) $x \vee y \wedge (x \vee y') \wedge (x' \vee y)$

**4.** $(x \vee y \vee z) \wedge (x \vee y \vee z') \wedge (x \vee y' \vee z) \wedge (x \vee y' \vee z') \wedge (x \vee y' \vee z') \wedge (x' \vee y' \vee z) \wedge (x' \vee y' \vee z')$

**5.** $(x \wedge y \wedge z) \vee (x \wedge y \wedge z') \vee (x \wedge y' \wedge z) \vee (x \wedge y' \wedge z')$

**6.** $F' = (x \wedge y \wedge z) \vee (x \wedge y' \wedge z) \vee (x' \wedge y \wedge z) \vee (x' \wedge y \wedge z') \vee (x' \wedge y' \wedge z)$

## Problems 37.6, page 1317

**1.** (*i*)

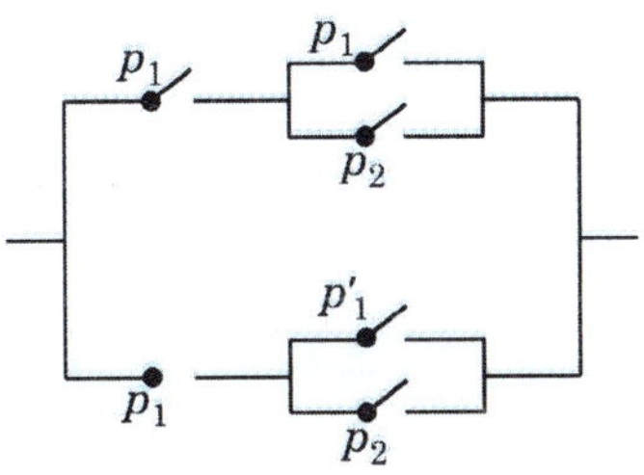


(*ii*)

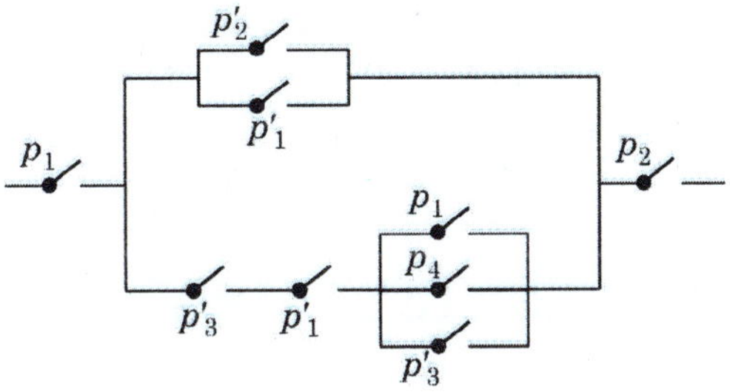


**2.** (*i*) $p_1 \vee [p_2' \wedge (p_1 \vee p_2')]$

(*ii*) $[(p_1 \vee p_2) \vee (p_1 \vee p_3)] \wedge (p_1 \wedge p_2')$

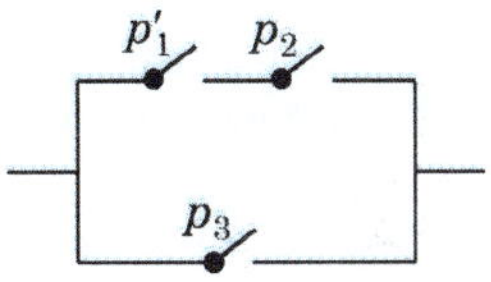


**3.** $x \wedge y$

**4.** $p_1' \wedge p_2 \vee p_3$ ;

**6.** (*a*) $x_1 \wedge x_2'$

(*b*)

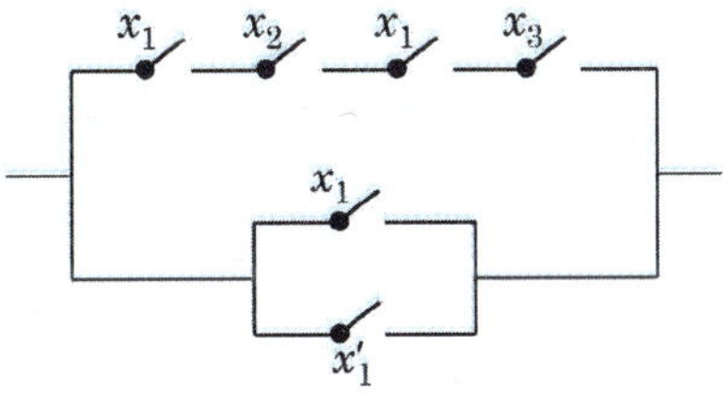


(*c*) $(x_1 \vee x_2' \vee x_3) \wedge (x_1 \vee x_2' \vee x_3') \wedge (x_3 \vee x_2' \vee x_1) \wedge (x_3 \vee x_2' \vee x_1')$

## Problems 37.7, page 1322

**1.** $F_2 = F_3$

**2.** $F^c$ = [0.2 Ram, 0.7 Sham, 0.4 John, 0.3 Charu]

**3.** $F \cup G = [0.4\, x_1, 0.7\, x_2, 0.5\, x_3, 0.9\, x_4]$

$F \cap G = [0.3\, x_1, 0.6\, x_2, 0.1\, x_3, 0.8\, x_4]$

**4.** (*i*) Truth value of '*F* is not rich' is 0.2

(*ii*) Truth value of '*G* is not fat' is 0.4

(*iii*) Truth value of 'Mary is not beautiful' is 0.3

**5.** (*i*) $F \neq G$

(*ii*) *F* is not a subset of *G* ; *G* is not a subset of *F*.

(*iii*) $F^c$ = [1, 1, 1, 1, 0.9, 0.7, 0.5, 0.1, 0, 0]

$F \cap G$ = [0, 0.1, 0.3, 0.5]

$F \cup G$ = [0.1, 0.5, 0.9, 1, 0.9, 0.9, 1, 1]

**6.** (*i*) Truth value of the cojunction of 'Latif and John are good players' is 0.6.

(*ii*) Truth value of the disjunction of 'Latif and John are good players' is 0.7.

**8.** Members and its degree of membership.

## Problems 38.1, page 1329

**1.** (*i*) $\dfrac{d\phi}{dt} = \dfrac{d\phi}{\partial x^i} \cdot \dfrac{dx^i}{dt}$ ; (*ii*) $x^i x^i$

**2.** (*i*) $a_{11}(x^1)^2 + a_{22}(x^2)^2 + a_{33}(x^3)^2 + (a_{12} + a_{21})\, x^1x^2 + (a_{13} + a_{31})x^1x^3 + (a_{23} + a_{32})x^2x^3$

(*ii*) $g_{11}(dx^1)^2 + g_{22}(dx^2)^2 + g_{33}(dx^3)^2 + 2g_{12}dx^1dx^2 + 2g_{23}dx^2dx^3 + 2g_{31}dx^3\, dx^1$.

(*iii*) $g_{11} = g_{12}\, g^{2p} + \ldots + g_{ln}\, g^{np}$

**3.** (*i*) $\delta^i_k$ ; (*ii*) $\delta^p_s$.

**6.** (*i*) $\bar{A}_p{}^{qr} = \dfrac{\partial \bar{x}^q}{\partial x^j} \dfrac{\partial \bar{x}^r}{\partial x^k} \dfrac{\partial x^i}{\partial \bar{x}^p} A_i^{jk}$ ; (*ii*) $\bar{C}_{pq} = \dfrac{\partial x^m}{\partial \bar{x}^p} \dfrac{\partial x^n}{\partial \bar{x}^q} C_{mn}$

**7.** Yes, $A^{ij}{}_{kl}{}^m$, contravariant order 3, covariant order 2, Rank 5

**9.** (*a*) $2\rho \cos^2 \phi - z \cos \phi + \rho^3 \sin^2 \phi \cos^2 \phi,\ - \rho^2 \sin 2\phi + \rho z \sin \phi + \rho^4 \sin \phi \cos^3 \phi,\ \rho z \sin \phi$ ;

(*b*) $2r \sin^2 \theta \cos^2 \phi - r \sin \theta \cos \theta \cos \phi + r^3 \sin^4 \theta \sin^2 \phi \cos^2 \phi + r^2 \sin \theta \cos^2 \theta \sin \phi$ ;

$2r^2 \sin \theta \cos \theta \cos^2 \phi - r^2 \cos^2 \theta \cos \phi + r^4 \sin^3 \theta \cos \theta \sin^2 \phi \cos^2 \phi - r^3 \sin^2 \theta \cos \theta \sin \phi$ ;

$-2r^2 \sin^2 \theta \sin \phi \cos \phi + r^2 \sin \theta \cos \theta \sin \phi + r^4 \sin^4 \theta \sin \phi \cos^3 \phi$

**11.** $(a \cos \phi + b \sin \phi) \sin \theta + c \cos \theta$ ; $\{(a \cos \phi + b \sin \phi) \cos \theta - c \sin \theta\}/r$ ; $(b \cos \phi - a \sin \phi)/r \sin \theta$.

## Problems 38.2, page 1332

**6.** Rank = 1

## Problems 38.3, page 1336

**1.** $g = 4, g^{11} = 2, g^{22} = 5, g^{33} = 1.5, g^{12} = 3, g^{23} = -2.5, g^{13} = -1.5,$

**2.** $g_{11} = 1, g_{22} = \rho^2, g_{33} = 1, g_{ij} = 0\ (i \neq j)$ ; $g^{11} = 1, g^{22} = \rho^{-2}, g^{33} = 1, g^{ij} = 0\ (i \neq j)$

**3.** $g = r^4 \sin^2 \theta/(1 - r^2/R^2)$ ; $g^{11} = 1 - r^2/R^2, g^{22} = 1/r^2, g^{33} = (r \sin \theta)^{-2}, g^{ij} = 0\ (i \neq j)$.

## Problems 38.4, page 1344

**1.** (*a*) $[ii, i] = \dfrac{1}{2} \partial g^{ij} / \partial x^i,\ [ii, k] = -\dfrac{1}{2} \partial g_{ii}/\partial x^k,$

$[ik, k] = [ki, k] = \dfrac{1}{2} \partial g_{kk}/\partial x^i,\ [ij, k] = 0$, when $i, j, k$ are all different

(*b*) $\begin{Bmatrix} i \\ ii \end{Bmatrix} = \dfrac{1}{2} g^{ij} \dfrac{\partial g_{ii}}{\partial x^i}, \begin{Bmatrix} k \\ ii \end{Bmatrix} = -\dfrac{1}{2} g^{kk} \dfrac{\partial g_{ii}}{\partial x^k}$

$\begin{Bmatrix} k \\ ik \end{Bmatrix} = \begin{Bmatrix} k \\ ki \end{Bmatrix} = \dfrac{1}{2} g^{kk} \dfrac{\partial g_{kk}}{\partial x^i}$ (no summation over $i$ or $k$)

$\begin{Bmatrix} k \\ ij \end{Bmatrix} = 0$, when $i, j, k$ are all different

**2.** (*a*) All are zero

(*b*) $[21, 2] = \rho = [12, 2]$ ; $[22, 1] = \rho$, all others are zero

(*c*) $[21, 2] = r = [12, 2]$ ; $[31, 3] = r \sin^2 \theta = [13, 3]$ ; $[32, 3]$ $= r^2 \sin \theta \cos \phi = [23, 3]$ ; $[22, 1] = -r$ ; $[33, 1] = -r \sin^2 \theta$ ; $[33, 2] = -r^2 \sin \theta \cos \phi$ ; all others are zero

**3.** (*a*) All are zero

(*b*) $\begin{Bmatrix} 1 \\ 22 \end{Bmatrix} = -\rho, \begin{Bmatrix} 2 \\ 21 \end{Bmatrix} = \begin{Bmatrix} 2 \\ 12 \end{Bmatrix} = \dfrac{1}{\rho}$, all others are zero

(*c*) $\begin{Bmatrix} 1 \\ 22 \end{Bmatrix} = -r, \begin{Bmatrix} 1 \\ 33 \end{Bmatrix} = -r \sin^2 \theta, \begin{Bmatrix} 2 \\ 21 \end{Bmatrix} = \begin{Bmatrix} 2 \\ 12 \end{Bmatrix} = \dfrac{1}{r},$

$$\begin{Bmatrix} 2 \\ 33 \end{Bmatrix} = -\sin\theta\cos\theta,\ \begin{Bmatrix} 3 \\ 31 \end{Bmatrix} = \begin{Bmatrix} 3 \\ 13 \end{Bmatrix} = \frac{1}{r},$$

$$\begin{Bmatrix} 3 \\ 32 \end{Bmatrix} = \begin{Bmatrix} 3 \\ 23 \end{Bmatrix} = \cot\theta, \text{ all others are zero}$$

**5.** (*a*) $r^2 \sin\theta \cos\theta$ ; $r^2 \sin\theta \cos\theta$ (*b*) $-\sin\theta\cos\theta$ ; $\cot\theta$

**6.** (*a*) $-r\sin^2\theta$ ; $r^2\sin\theta\cos\theta$ (*b*) $-r\sin^2\theta$ ; $\cot\theta$

**8.** (*a*) $u^{ij}{}_{,k} = \dfrac{\partial u^{ij}}{\partial x^k} + \begin{Bmatrix} i \\ ks \end{Bmatrix} u^{sj} + \begin{Bmatrix} i \\ ks \end{Bmatrix} u^{is}$ ; (*b*) $A^h{}_{ij,\,k} = \dfrac{\partial A^h{}_{ij}}{\partial x^k} - \begin{Bmatrix} s \\ ik \end{Bmatrix} A^h{}_{sj} + \begin{Bmatrix} h \\ ks \end{Bmatrix} A^s{}_{ij}$

**10.** $A^j{}_{k,\,q} B_n{}^{lm} + A^j{}_k B^{lm}{}_{n,\,q}$

**11.** (*a*) $\dfrac{1}{\rho}\left[\dfrac{\partial}{\partial p}(\rho A_\rho) + \dfrac{\partial}{\partial \phi}(A_\phi) + \dfrac{\partial}{\partial z}(\rho A_z)\right]$ ; (*b*) $\dfrac{1}{r^2}\dfrac{\partial}{\partial r}(r^2 A_r) + \dfrac{1}{r\sin\theta}\dfrac{\partial}{\partial r}(\sin\theta\, A_\theta) + \dfrac{1}{r\sin\theta}\dfrac{\partial A_\phi}{\partial \phi}$

**12.** (*a*) $\dfrac{\partial^2 v}{\partial \rho^2} + \dfrac{1}{\rho^2}\dfrac{\partial^2 v}{\partial \phi^2} + \dfrac{\partial^2 v}{\partial z^2} + \dfrac{1}{\rho}\dfrac{\partial v}{\partial \rho} = 0$ (*b*) $\dfrac{\partial^2 v}{\partial r^2} + \dfrac{1}{r^2}\dfrac{\partial^2 v}{\partial \theta^2} + \dfrac{1}{r^2\sin^2\theta}\dfrac{\partial^2 v}{\partial \phi^2} + \dfrac{2}{r}\dfrac{\partial v}{\partial r} + \dfrac{\cot\theta}{r^2}\dfrac{\partial v}{\partial \theta} = 0.$

# INDEX

## A



## B



## C

## D



## E



## F



## G



## H

## I



## J



## K



## L



## M



## N



## O

## P



## Q



## R



## S

## T



## U



## V



## W



## Z